D0215159

Principles of Developmental Genetics

Principles of Developmental Genetics

Second Edition

Edited by

Sally A. Moody
Department of Anatomy and Regenerative Biology
The George Washington University School
of Medicine and Health Sciences
Washington, D.C.
United States of America

AMSTERDAM • BOSTON • HEIDELBERG • LONDON • NEW YORK • OXFORD • PARIS
SAN DIEGO • SAN FRANCISCO • SINGAPORE • SYDNEY • TOKYO

Academic Press is an imprint of Elsevier

Academic Press is an imprint of Elsevier
32 Jamestown Road, London NW1 7BY, UK
525 B Street, Suite 1800, San Diego, CA 92101-4495, USA
225 Wyman Street, Waltham, MA 02451, USA
The Boulevard, Langford Lane, Kidlington, Oxford OX5 1GB, UK

First edition 2007

British Library Cataloguing-in-Publication Data
A catalogue record for this book is available from the British Library

Library of Congress Cataloging-in-Publication Data
A catalog record for this book is available from the Library of Congress

ISBN: 978-0-12-405945-0

For information on all Academic Press publications
visit our website at http://store.elsevier.com/

Typeset by TNQ Books and Journals
www.tnq.co.in

Printed and bound in the United States of America

Contents

4. Chemical Approaches to Controlling Cell Fate

Mingliang Zhang, Ke Li, Min Xie and Sheng Ding

5. BMP Signaling and Stem Cell Self-Renewal in the *Drosophila* Ovary

Darin Dolezal and Francesca Pignoni

6. Genomic Analyses of Neural Stem Cells

Nasir Malik, Soojung Shin and Mahendra S. Rao

7. Genomic and Evolutionary Insights into Chordate Origins

Shawn M. Luttrell and Billie J. Swalla

Section II

Early Embryology and Morphogenesis

8. Signaling Cascades, Gradients, and Gene Networks in Dorsal/Ventral Patterning

Girish S. Ratnaparkhi and Albert J. Courey

9. Building Dimorphic Forms: The Intersection of Sex Determination and Embryonic Patterning

Kristy L. Kenyon, Yanli Guo and Nathan Martin

Section III

Organogenesis

16. Neural Cell Fate Determination

Steven Moore and Frederick J. Livesey

17. Retinal Development

Andrea S. Viczian and Michael E. Zuber

18. Neural Crest Determination and Migration

Eric Theveneau and Roberto Mayor

Preface

Developmental Genetics: An Historical Perspective

Sally A. Moody

Department of Anatomy and Regenerative Biology, The George Washington University, School of Medicine and Health Sciences Washington DC, DC, USA

The ability of researchers to answer experimental questions greatly depends on the available technologies. New technologies lead to novel observations and field-changing discoveries, and influence the types of questions that can be asked. Today's recently available technologies include sequencing and analyzing the genomes of human and model organisms, genome-wide expression and epigenetic profiling, and high-throughput screening. The information provided by these approaches enables us to begin to understand the complexity of many biological processes through the elucidation of gene regulatory networks, signaling pathway networks, and epigenetic modifications. This book describes many lines of research that are being impacted by these new technologies, including developmental genetics and the related fields of clinical genetics, birth defects research, stem cell biology, regenerative medicine, and evolutionary biology.

The field of developmental genetics, or the study of how genes influence the developmental processes of an organism, has been influenced by new technologies and by interactions with other fields of study throughout its history. The concept of a genetic basis of development began in "modern" times at the intersection of descriptive embryology and cytology. Modern histological techniques were developed in the mid-19th century, largely by Wilhelm His so that he could study cell division in the neural tube, which enabled visualization of the cell nucleus, chromosomes, and the discrete steps of mitosis. Theodor Boveri cleverly applied these improved microscopic techniques to transparent marine embryos to demonstrate that each parent contributes equivalent groups of chromosomes to the zygote, and that each chromosome is an independently inherited unit. Importantly, he noted that if an embryo contains the incorrect number or improper combination of chromosomes it develops abnormally. However, many early embryologists rejected the idea that development is driven by prepackaged heritable particles because it seemed too similar to the idea of "preformation": the concept that development is driven by predetermined factors or "force" (sometimes described in rather mystical terms).

Wilhelm Roux, an advocate of studying the embryo from a mechanistic point of view, was a leader in manipulating the embryo with microsurgical techniques to elucidate causes and effects between component parts (experimental embryology). By using an animal model whose embryos were large, developed externally to the mother, could be surgically manipulated with sharpened forceps, and cultured in simple salt media (i.e., amphibians), he rejected the role of predetermined factors and demonstrated the importance of external (epigenetic) influences and cell-cell interactions in regulating developmental programs. Experimental embryologists further refined their skills in dissecting small bits of tissue from the embryo, recombining them with other tissues in culture or transplanting them to ectopic regions in the embryo. This work led to the invention of tissue culture by Ross Harrison and the discovery of tissue inductions by Hans Spemann and Hilde Mangold.

While experimental embryology was thriving, T. H. Morgan founded the field of *Drosophila* genetics. Also trained as an embryologist, Morgan was skeptical of Boveri's idea of heritable packets, and directed his studies towards understanding the principles of inheritance. For several decades, the two fields, experimental embryology and genetics, had little impact on one another. Interestingly, however, after a few decades of study of the fruit fly, Morgan's work supported the idea of discrete intracellular particles that directed heritable traits, which he named "genes." Nonetheless, experimental embryology and genetics remained fairly separate fields with distinct goals and points of view. Embryologists were elucidating the interactions that are important for the development of numerous tissues and organs, whereas geneticists were focused on the fundamentals of gene inheritance, regulation of expression, and discovering the genetic code.

Much of this work was carried out in non-mammalian animal models that develop external to the mother, and thus can be manipulated during during development and/or have very short life cycles. Elucidating the genetic basis of vertebrate development was delayed until new technologies in molecular biology and cloning were devised. The techniques for cloning eukaryotic genes and constructing vectors for controlling expression came from the field of bacterial and viral genetics. The rationale for and techniques of mutagenizing the entire genome and screening for developmental abnormalities came from the classical genetic studies in the fly and nematode. Important regulatory genes were discovered in these invertebrates, and their counterparts have been identified in many other animals, including mammals, by homology cloning approaches that were not common until the 1990s. Thus was born the modern field that we call developmental genetics.

An important advance has been the demonstration that homologues of the genes that regulate developmental processes in invertebrates also have important developmental functions in vertebrates. The wealth of information concerning the molecular genetic processes that regulate development in various animals demonstrates that developmental programs and biological processes are highly conserved, albeit not identical, from yeast to human. Indeed, the Human Genome Project has made it possible to identify the homologues in humans, and to demonstrate that many of these regulatory genes underlie human developmental disorders, birth defects, and aspects of adult diseases in which differentiation processes go awry. Currently, researchers are studying the fundamentals of developmental processes in the appropriate animal model and screening humans for mutations in the genes identified by the basic research to be likely causative candidates. Researchers are mutagenizing vertebrate animal models and screening for mutants that resemble known human syndromes. Stem cell biologists are using the molecular genetic information discovered by from developmental biology researchers to understand how to manipulate cellular differentiation *in vitro* and direct pluripotent cells to desired lineages for tissue replacement therapies. This cross-fertilization of fields is also impacting concepts in evolutionary biology, leading to a better understanding of "ancestral" species and tissue anlage via homologous gene expression profiles.

In recent years there have been significant technological advances in genetic, genomic, epigenetic, and protein expression analyses that are having a major impact on experimental approaches and analytical design. The intersection of developmental biology with these technologies offers a new view of developmental genetics that is only now beginning to be exploited in clinical approaches. It is this new intersection between modern developmental genetic approaches and potential clinical interventions that is the focus of this book. The book is organized into sections focused on different aspects of developmental genetics. Section I discusses the impact of new genetic and genomic technologies on development, stem cell biology, evolutionary biology, and understanding human birth defects. Section II discusses several major events in early embryogenesis, fate determination, and patterning, including cellular determinants (Boveri revisited?), gene cascades regulating embryonic axis formation, signaling molecules and transcription factors that regulate pattern formation, and the induction of the primary germ layers (ectoderm, mesoderm, and endoderm). Section III describes the morphogenetic and cellular movements that underlie the foundation of the organ systems, with a focus on the signaling cascades and transcriptional pathways that regulate organogenesis in representative systems derived from the embryonic ectoderm, mesoderm, and endoderm. These chapters illustrate how embryonic rudiments become organized into adult tissues, and how defects in these processes can result in congenital defects or disease. Each chapter demonstrates the usefulness of studying model organisms and discusses how this information applies to normal human development and clinical disorders. Several of the chapters in this section also discuss the utility of stem cells to repair damaged organs and the application of developmental genetics to the manipulation of stem cells for regenerative medicine. Section IV presents the developmental genetic underpinnings of a few selected clinical problems that commonly face pediatricians.

The goal of this book is to provide a resource for understanding the critical embryonic and prenatal developmental processes that are fundamental to the normal development of animals, including humans. It highlights new technologies to be used, new questions to be answered, and the important roles that invertebrate and vertebrate animal models have had in elucidating the genetic basis of human development and disease. Developmental genetics has re-emerged from its birth a century ago as a nexus of diverse fields that are using the common language of gene sequence and function. This is influencing what questions are posed and how the answers are used. New technologies are making it relatively easy to study gene expression and regulation at single cell, tissue, and embryonic levels. The conservation between the genomes of species that are separated by vast evolutionary time encourages us to more fully utilize animal models to gain important insights into the clinical relevance of the animal model data. It is our hope that this book will stimulate even more cross-fertilization and interactions between evolutionary biology, developmental biology, stem cell biology, basic scientists, and clinical scientists.

I wish to thank all of the authors for contributing such exciting and excellent chapters, and Catherine Van Der Laan for keeping all of us on schedule.

RECOMMENDED RESOURCES

Baltzer, F., 1967. Theodor Boveri, life and work of a great biologist 1862–1915. University of California Press, Los Angeles, CA.

Hamburger, V., 1988. The heritage of experimental embryology: Hans Spemann and the Organizer. Oxford University Press, New York.

Model Organisms for Biomedical Research (Web site): http://www.nih.gov/science/models (accessed 28/2/2014).

Morgan, T.H., 1926. The theory of the gene. Yale University Press, New Haven, CN.

Spemann, H., 1967. Embryonic development and induction. Hafner Publishing Co, New York.

Willier, B.H., Oppenheimer, J.M., 1964. Foundations of experimental embryology. Prentice-Hall, Englewood Cliffs, NJ.

Contributors

Kate G. Ackerman Department of Pediatrics and Biomedical Genetics, University of Rochester Medical Center, Rochester, NY, USA

Peter G. Alexander Center for Cellular and Molecular Engineering, Department of Orthopedic Surgery, University of Pittsburgh School of Medicine, Pittsburgh, PA, USA

Xiaoying Bai University of Texas Southwestern Medical Center, Dallas, TX, USA

Jennifer L. Baker Center for the Advanced Study of Hominid Paleobiology and Department of Anthropology, The George Washington University School of Medicine and Health Sciences, Washington DC, USA

David R. Beier Center for Developmental Biology and Regenerative Medicine, Seattle Children's Research Institute, Seattle, WA, USA

John W. Belmont Departments of Pathology and Immunology and Pediatrics, Baylor College of Medicine, Houston, TX, USA

David A. Billmire Division of Plastic Surgery, Department of Surgery, Cincinnati Childrens Hospital Medical Center, Cincinnati, OH, USA

Ira L. Blitz Department of Developmental and Cell Biology, University of California, Irvine, CA, USA

Samantha A. Brugmann Division of Plastic Surgery, Department of Surgery, Cincinnati Childrens Hospital Medical Center, Cincinnati, OH, USA; Division of Developmental Biology, Department of Pediatrics, Cincinnati Childrens Hospital Medical Center, Cincinnati, OH, USA

Matthew G. Butler Program in Genomics of Differentiation, National Institute of Child Health and Human Development, National Institutes of Health (NIH), Bethesda, MD, USA

Ching-Fang Chang Division of Plastic Surgery, Department of Surgery, Cincinnati Childrens Hospital Medical Center, Cincinnati, OH; Division of Developmental Biology, Department of Pediatrics, Cincinnati Childrens Hospital Medical Center, Cincinnati, OH, USA

Ajay B. Chitnis Section on Neural Developmental Dynamics, Program in Genomics of Differentiation, NICHD NIH, Bethesda, MD, USA

William T.Y. Chiu Department of Developmental and Cell Biology, University of California, Irvine, CA, USA

Ken W.Y. Cho Department of Developmental and Cell Biology, University of California, Irvine, CA, USA

Albert J. Courey Department of Chemistry and Biochemistry, University of California at Los Angeles, Los Angeles, CA, USA

Damian Dalle Nogare Section on Neural Developmental Dynamics, Program in Genomics of Differentiation, NICHD NIH, Bethesda, MD, USA

Jamie Davies University of Edinburgh, Edinburgh, UK

Sheng Ding Gladstone Institute of Cardiovascular Disease, Department of Pharmaceutical Chemistry, University of California, San Francisco, CA, USA

Darin Dolezal Dept. of Biochemistry and Molecular Biology, SUNY Upstate Medical University, Syracuse, NY, USA

Nataki C. Douglas Department of Genetics and Development, Columbia University Medical Center, New York, NY, USA

Hongling Du Division of Reproductive Endocrinology and Infertility, Department of Obstetrics, Gynecology and Reproductive Sciences, Yale University School of Medicine, New Haven, CT, USA

Benjamin Feldman Program on Genomics of Differentiation, Eunice Kennedy Shriver National Institute of Child Health and Human Development, Bethesda, MD, USA

Alejandra Fernandez GW Institute for Neuroscience; Department of Pharmacology and Physiology; GW Institute for Biomedical Sciences, The George Washington University School of Medicine and Health Sciences, Washington DC, USA

Maureen Gannon Department of Cell and Developmental Biology, Vanderbilt University Medical Center, Nashville, TN; VA Medical Center, Nashville, TN; Departments of Medicine and Molecular Physiology and Biophysics, Vanderbilt University Medical Center, Nashville, TN, USA

Mor Grinstein Department of Anatomy and Cell Biology, Rappaport Faculty of Medicine, Technion-Israel Institute of Technology, Haifa, Israel

Yanli Guo Department of Biology, Hobart and William Smith Colleges, Geneva, New York, NY

Renée V. Hoch Rubenstein Laboratory, Department of Psychiatry, University of California, San Francisco, CA, USA

Keiji Itoh Department of Developmental and Regenerative Biology, Mount Sinai School of Medicine, New York, NY, USA

Beverly A. Karpinski GW Institute for Neuroscience; Department of Anatomy and Regenerative Biology, The George Washington University School of Medicine and Health Sciences, Washington DC, USA

Matthew W. Kelley Laboratory of Cochlear Development, National Institute for Deafness and other Communication Disorders, NIH Porter Neuroscience Research Center, Bethesda, MD

Kristy L. Kenyon Department of Biology, Hobart and William Smith Colleges, Geneva, New York, NY, USA

Paul A. Krieg Department of Cellular and Molecular Medicine, University of Arizona College of Medicine, Tucson, AZ, USA

Anthony-Samuel LaMantia GW Institute for Neuroscience, Department of Pharmacology and Physiology, The George Washington University School of Medicine and Health Sciences, Washington DC, USA

Mark T. Langhans Center for Cellular and Molecular Engineering, Department of Orthopedic Surgery, University of Pittsburgh School of Medicine, Pittsburgh, PA, USA

Ke Li Gladstone Institute of Cardiovascular Disease, Department of Pharmaceutical Chemistry, University of California, San Francisco, CA, USA

Frederick J. Livesey Gurdon Institute, Department of Biochemistry and Cambridge Stem Cell Institute, University of Cambridge, Cambridge, UK

Mui Luong Department of Developmental and Cell Biology, University of California, Irvine, CA, USA

Shawn M. Luttrell Biology Department, University of Washington, Seattle, WA; Friday Harbor Laboratories, University of Washington, Friday Harbor, WA, USA

Nasir Malik National Institutes of Health, National Institute of Arthritis and Musculoskeletal and Skin Diseases, Bethesda, MD, USA

Zoë F. Mann Laboratory of Cochlear Development, National Institute for Deafness and other Communication Disorders, NIH Porter Neuroscience Research Center, Bethesda, MD, USA

Nathan Martin Department of Biology, Hobart and William Smith Colleges, Geneva, New York, NY, USA

Thomas M. Maynard GW Institute for Neuroscience; Department of Pharmacology and Physiology, The George Washington University School of Medicine and Health Sciences, Washington DC, USA

Roberto Mayor Department of Anatomy and Developmental Biology, University College London, London, UK

Daniel Meechan GW Institute for Neuroscience; Department of Pharmacology and Physiology, The George Washington University School of Medicine and Health Sciences, Washington DC, USA

Sigolène M. Meilhac Institut Pasteur, Department of Developmental and Stem Cell Biology, Paris, France; CNRS, URA2578, Paris, France

Anna M. Method Division of Developmental Biology, Cincinnati Children's Hospital Research Foundation and University of Cincinnati College of Medicine, Cincinnati, OH, USA

Sally A. Moody Department of Anatomy and Regenerative Biology, The George Washington University School of Medicine and Health Sciences, Washington DC, USA

Steven Moore Gurdon Institute, Department of Biochemistry and Cambridge Stem Cell Institute, University of Cambridge, Cambridge, UK

Timothy S. Mulligan Program in Genomics of Differentiation, National Institute of Child Health and Human Development, National Institutes of Health (NIH), Bethesda, MD, USA

L.A. Naiche National Cancer Institute, Cancer and Developmental Biology Lab, Frederick, MD, USA

Virginia E. Papaioannou Department of Genetics and Development, Columbia University Medical Center, New York, NY, USA

Aitana Perea-Gomez Institut Jacques Monod, Université Paris Diderot, Sorbonne Paris Cité, Paris, France; *CNRS, UMR7592, Paris, France*

Francesca Pignoni Dept. of Biochemistry and Molecular Biology; Dept. of Neuroscience and Physiology; Dept. of Ophthalmology, SUNY Upstate Medical University, Syracuse, NY, USA

Mahendra S. Rao National Institutes of Health, National Institute of Arthritis and Musculoskeletal and Skin Diseases, Bethesda, MD; National Institutes of Health Center for Regenerative Medicine, Bethesda, MD, USA

Girish S. Ratnaparkhi Biological Sciences, Indian Institute of Science Education & Research (IISER), Pune, India

Kimberly G. Riley Department of Cell and Developmental Biology, Vanderbilt University Medical Center, Nashville, TN, USA

Jean-Pierre Saint-Jeannet Department of Basic Science and Craniofacial Biology, New York University, College of Dentistry, New York City, NY, USA

Eric T. Sause Program in Genomics of Differentiation, National Institute of Child Health and Human Development, National Institutes of Health (NIH), Bethesda, MD, USA

Elizabeth N. Schock Division of Plastic Surgery, Department of Surgery, Cincinnati Childrens Hospital Medical Center, Cincinnati, OH; Division of Developmental Biology, Department of Pediatrics, Cincinnati Childrens Hospital Medical Center, Cincinnati, OH, USA

Thomas M. Schultheiss Department of Anatomy and Cell Biology, Rappaport Faculty of Medicine, Technion-Israel Institute of Technology, Haifa, Israel

Soojung Shin Invitrogen Corporation, Carlsbad, CA, USA

Sergei Y. Sokol Department of Developmental and Regenerative Biology, Mount Sinai School of Medicine, New York, NY, USA

Philippe Soriano Department of Developmental and Regenerative Biology, Mount Sinai School of Medicine, New York, NY, USA

Amber N. Stratman Program in Genomics of Differentiation, National Institute of Child Health and Human Development, National Institutes of Health (NIH), Bethesda, MD, USA

Billie J. Swalla Biology Department, University of Washington, Seattle, WA and Friday Harbor Laboratories, University of Washington, Friday Harbor, WA, USA

Hugh S. Taylor Department of Obstetrics, Gynecology and Reproductive Sciences Yale University School of Medicine, New Haven, CT, USA

Irma Thesleff Institute of Biotechnology, University of Helsinki, Finland

Eric Theveneau Department of Anatomy and Developmental Biology, University College London, London, UK

Rocky S. Tuan Center for Cellular and Molecular Engineering, Department of Orthopedic Surgery, University of Pittsburgh School of Medicine, Pittsburgh, PA, USA

Andrea S. Viczian Ophthalmology Department, Center for Vision Research, SUNY Eye Institute, Upstate Medical University, Syracuse, NY, USA

Andrew S. Warkman Department of Cellular and Molecular Medicine, University of Arizona College of Medicine, Tucson, AZ, USA

Andrew J. Washkowitz Department of Genetics and Development, Columbia University Medical Center, New York, NY, USA

Brant M. Weinstein Program in Genomics of Differentiation, National Institute of Child Health and Human Development, National Institutes of Health (NIH), Bethesda, MD, USA

James M. Wells Division of Developmental Biology, Cincinnati Children's Hospital Research Foundation and University of Cincinnati College of Medicine, Cincinnati, OH, USA

Marcin Wlizla Division of Developmental Biology, Cincinnati Children's Hospital Medical Center, Cincinnati, OH, USA

Min Xie Gladstone Institute of Cardiovascular Disease, Department of Pharmaceutical Chemistry, University of California, San Francisco, CA, USA

Yingzi Yang Genetic Disease Research Branch, National Human Genome Research Institute, NIH, Bethesda, MD, USA

Jianxin A. Yu Program in Genomics of Differentiation, National Institute of Child Health and Human Development, National Institutes of Health (NIH), Bethesda, MD, USA

Mingliang Zhang Gladstone Institute of Cardiovascular Disease, Department of Pharmaceutical Chemistry, University of California, San Francisco, CA, USA

Irene E. Zohn Center for Neuroscience Research, Children's Research Institute, Children's National Medical Center, Washington DC, USA

Leonard I. Zon Howard Hughes Medical Institute, Children's Hospital of Boston, Boston, MA

Aaron M. Zorn Division of Developmental Biology, Cincinnati Children's Hospital Medical Center, Cincinnati, OH, USA

Michael E. Zuber Ophthalmology Department, Center for Vision Research, SUNY Eye Institute, Upstate Medical University, Syracuse, NY, USA

Section I

Emerging Technologies and Systems Biology

Chapter 1

Generating Diversity and Specificity through Developmental Cell Signaling

Renée V. Hoch* **and Philippe Soriano**†

**Rubenstein Laboratory, Department of Psychiatry, University of California, San Francisco, CA;* †*Department of Developmental and Regenerative Biology, Mount Sinai School of Medicine, New York, NY*

Chapter Outline

GLOSSARY

ChIP (chromatin immunoprecipitation) A biochemical technique in which an antibody is bound to chromatin (DNA+protein), genomic DNA is cleaved into small fragments, and DNA bound by the antibody is pulled down and sequenced (ChIP-seq) or hybridized to a microarray (ChIP-chip). In this chapter, we discuss ChIP-seq using antibodies that bind transcription factors of interest to identify direct targets of a given signaling pathway, and antibodies that recognize transcriptional co-activators or chromatin modifications that can help to identify developmental enhancers.

Chromosome conformation capture (CCC) A method used to identify transcriptional targets of enhancer elements on the basis of physical interactions between enhancers and promoters.

Cytoneme Thin, actin-based cellular protrusion several cell diameters long, extending from the apical surface of a cell toward a source of signaling protein (ligand); first observed in *Drosophila* wing disc cells extending toward Dpp and Wg.

Feedback regulation A mechanism by which a signaling pathway regulates its own activity, e.g., by activating a regulatory factor that alters signal transduction, by altering sensitivity of the pathway to upstream signals, and/or by modifying the activity or interactions of proteins in the pathway. In some cases, multiple signaling pathways generate a feedback regulatory circuit, in which activation of one pathway results in regulation of another, and vice versa.

Filopodia Thin, short cellular protrusions containing both actin and microtubules that are not polarized toward a source of signaling protein.

Gene regulatory network (GRN) A network of transcription factors with cross-regulatory (e.g., repression, activation) relationships, e.g., that translate morphogen signals into discrete gene expression boundaries.

Genome-wide association study (GWAS) A genetic approach that uses high throughput sequencing to compare control and affected populations and identify genetic variants (e.g., SNPs, mutated alleles) that are associated with a given phenotype (disease).

Primary cilium Microtubule-based cytoplasmic extension comprised of a membrane ensheathed axoneme extending from the basal body, which is a modified centriole. In primary cilia, the axoneme is comprised of nine microtubule (MT) pairs in a 9+0 arrangement (in contrast to motile cilia which have 9 MT pairs surrounding two central fibrils, i.e., 9+2 arrangement). Proteins are moved into and out of the primary cilium via intraflagellar transport (IFT) along the axoneme. Plasma membranes around primary cilia are distinct/segregated from the cellular plasma membrane. Disruption of ciliary or IFT proteins results in a broad range of human ciliopathies.

Epistasis and genetic interactions Epistatic and genetic interactions are functional interactions between mutations in non-allelic genes suggestive of the gene products acting together in a given process or pathway. A *genetic interaction* is manifested through a compound mutant phenotype that is more pronounced than the sum of the two single mutant phenotypes, e.g., disrupted development in an animal heterozygous for two mutations for which either heterozygous mutation (in isolation) does not result in a developmental phenotype. An *epistatic relationship/interaction* is manifest through the ability of one allele to suppress the phenotypic consequences of a second mutation, typically indicating dominance of the epistatic mutation or the result of its being downstream in a common genetic pathway.

Endocytosis A process by which cell membranes envelop and internalize transmembrane and/or extracellular signaling proteins into cytoplasmic vesicles. Cytoplasmic proteins including dynamin mediate the inward budding of vesicles from the plasma membrane; dynamin is often targeted to block endocytosis in genetic studies. Once internalized, endocytic vesicles are sorted into different groups which undergo recycling to the plasma membrane or lysosomal degradation.

Enhancer element A region of DNA that impacts gene transcription in *cis* through the recruitment of transcription factors or other DNA binding/modifying proteins; also called cis-regulatory modules.

Heparin sulfate proteoglycan (HSPG) A macromolecule comprised of a core protein and glycosaminoglycan side chains of the heparin sulfate (polysaccharide) family, abundant in the extracellular matrix and sometimes associated with plasma membranes via lipid moieties. HSPGs are important for many signaling events as revealed by the effects of mutations in HSPG core proteins and synthesis enzymes (e.g., involved in appending the HS side groups).

Morphogen A protein that acts on target cells at a distance from its cell of origin, that forms an expression or activity gradient over a field of responsive cells, and that drives different cellular responses at different distances from the signal source. Morphogens can have non-synonymous or synonymous activity gradients, based on whether the signaling downstream of ligand-receptor activation exhibits a gradient that correlates with the extracellular ligand concentration.

RNA-seq An alternative to microarray analysis for expression profiling, in which RNA from a tissue of interest is isolated and subjected to high throughput sequencing, and the number of transcripts for each gene is quantified.

Transcriptional profiling Analysis of mRNA expression, e.g., using microarrays or RNA-seq; in cell signaling studies, comparative transcriptional profiling is often used to assess the transcriptional targets of a signaling pathway.

Yeast two-hybrid (Y2H) An experimental method used to assay for direct protein-protein interactions; in this method, a "bait" protein is fused to the DNA binding domain of a transcription factor (TF), and a series of "fish" proteins are fused to the activation domain of the TF. When the bait and fish proteins physically interact, the proximity of the two TF domains render the complex capable of driving expression of a reporter construct.

SUMMARY

- Highly conserved cell-cell signaling pathways are used reiteratively during development to instruct cellular proliferation, survival, migration, patterning, and differentiation.
- Responses to cell-cell signaling are dependent on spatial and temporal context.
- Related signaling proteins have become functionally specialized during evolution. For instance, members of the receptor tyrosine kinase (RTK) superfamily transduce distinct signals, and different transforming growth factor (TGF)β family ligands act as activators or inhibitors with distinct diffusion characteristics.
- Cytoplasmic extensions facilitate some cell-cell signaling interactions during development. In vertebrates, some signals are received and transduced via microtubule-based primary cilia, and in *Drosophila*, filopodia-like cytonemes bearing receptors extend toward some signaling sources.
- Extracellular signaling gradients guide cell migration and instruct positional identity or patterning in many developmental contexts. Multiple strategies are employed to establish and/or maintain these gradients, including diffusion of ligands and regulatory proteins, use of accessory proteins for transport and local activation of signaling proteins, regulated secretion, degradation, and recycling of ligands and receptors, and localized stabilization/retention of signals via heparin sulfate proteoglycans (HSPGs) or membrane localization motifs.
- Extracellular signaling gradients can generate different cellular responses through specialized signal transduction and transcriptional regulation mechanisms. These include protein modifications and feedback regulation sensitive to different signal thresholds, activation of gene regulatory networks, and the regulation of enhancers with different affinities for signal-activated transcription factors (TFs).
- Different types of signals coordinately instruct cellular responses *in vivo* through regulation of signal transduction proteins, feedback regulators, and genes/proteins involved in general cellular processes. In addition, crosstalk between pathways commonly occurs via co-regulation of common target genes. Co-regulation of target genes can be mediated by changes in chromatin, interactions between TFs, coordinate regulation of TF activity (e.g., by phosphorylation or sequestration), or coordinate regulation of enhancer elements.

1.1 INTRODUCTION

During development, cells communicate through a variety of extracellular signals to coordinate cellular behaviors with exquisite spatial and temporal precision. The developmental cell signaling field emerged from convergent work in developmental genetics and biochemistry. Landmark studies in the 1980s and 1990s used genetic screens to identify mutants that enhanced or suppressed receptor tyrosine kinase (RTK) loss-of-function phenotypes in *Drosophila* (Sevenless, Torso) and *C. elegans* (EGFR) embryos. Such mutants were arranged into hierarchies based on epistatic relationships and cell-autonomous versus -nonautonomous effects on RTK functions (reviewed in Furriols and Casanova, 2003; Moghal and Sternberg, 2003; Shilo, 2003; Nagaraj and Banerjee, 2004). In parallel, biochemical experiments validated the results of these screens and explored the molecular mechanisms underlying the observed genetic interactions. Thus genetically defined hierarchies were translated into a molecular signal transduction cascade connecting RTKs to activation of mitogen-activated protein kinase (MAPK; Figure 1.1A; reviewed in Porter and Vaillancourt, 1998; Schlessinger, 2000). These efforts and others collectively demonstrated that RTKs signal through an evolutionarily conserved biochemical pathway that is required during development and includes several proteins implicated in growth and oncogenesis.

RTK studies set the stage both conceptually and experimentally for subsequent studies that identified and characterized components of other cell signaling pathways (Figure 1.1, Table 1.1). These pathways are highly conserved and are used reiteratively during development. Cell-cell signals include secreted signals with varying ranges of action within tissues or

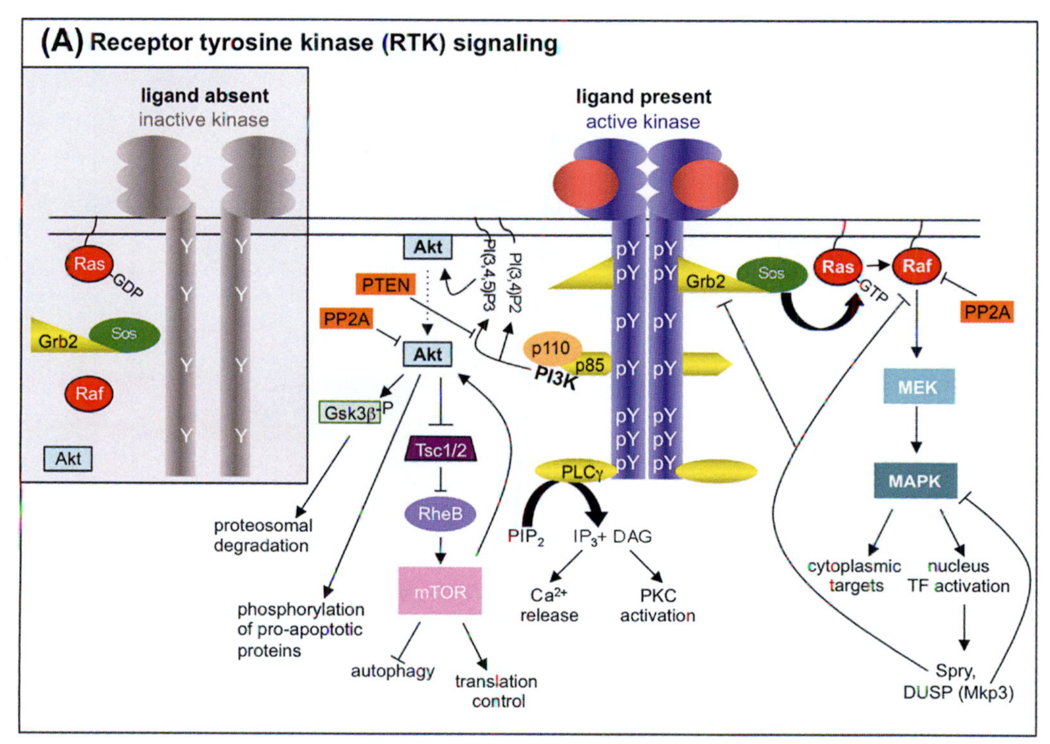

FIGURE 1.1 Basic Overview of Major Cell-Cell Signaling Pathways Discussed in this Chapter. (A) Receptor tyrosine kinases (RTK): extracellular ligand (red) binds the receptor and activates receptor cytoplasmic kinase domains. The receptors then autophosphorylate several tyrosine residues, generating docking sites for effector proteins (yellow). RTK family members differentially utilize effector proteins and signal transduction pathways, as discussed in the text. Effector proteins then initiate various signal transduction pathways such as the Ras-MAPK (mitogen-activated protein kinase), PI3K-Akt, and PLCγ pathways. Ras-MAPK signaling is often activated by recruitment of a Grb2-Sos complex to the receptor, enabling Sos to activate membrane-bound Ras through exchange of guanosine diphosphate (GDP) for guanosine triphosphate (GTP). Ras-GTP recruits Raf to the plasma membrane, then Ras hydrolyzes GTP and activates Raf, initiating a phosphorylation cascade: Raf phosphorylates MEK [MAPK/extracellular signal-regulated kinase (ERK)] which phosphorylates MAPK. Active phospho-MAPK can act on numerous cytoplasmic and nuclear targets. In the nucleus, MAPK activates transcription factors and induces expression of target genes, including feedback regulators such as Sprouty (Spry) and dual-specificity phosphatases (DUSP, also known as Mkp proteins). The PI3K pathway is activated by recruitment of the PI3K subunit p85 to phosphotyrosine-containing docking sites on RTKs. Once near the membrane, PI3K phosphorylates phospholipid targets to form PI(3,4)P2 and PI(3,4,5)P3. These molecules recruit PI3K effectors including Akt to the membrane for activation. Akt phosphorylates many targets including transcription factors (NFκB, CREB, FOXO, E2F), GSK3β (which it targets for proteosomal degradation), pro-apoptotic proteins BAD and Bim, and TSC2. Phosphorylation of TSC2 activates mTOR by relieving inhibition of RheB. mTOR inhibits autophagy, regulates translation, and feeds back to regulate Akt activity. PP2A negatively regulates signaling through both PI3K-Akt and Ras-MAPK pathways.

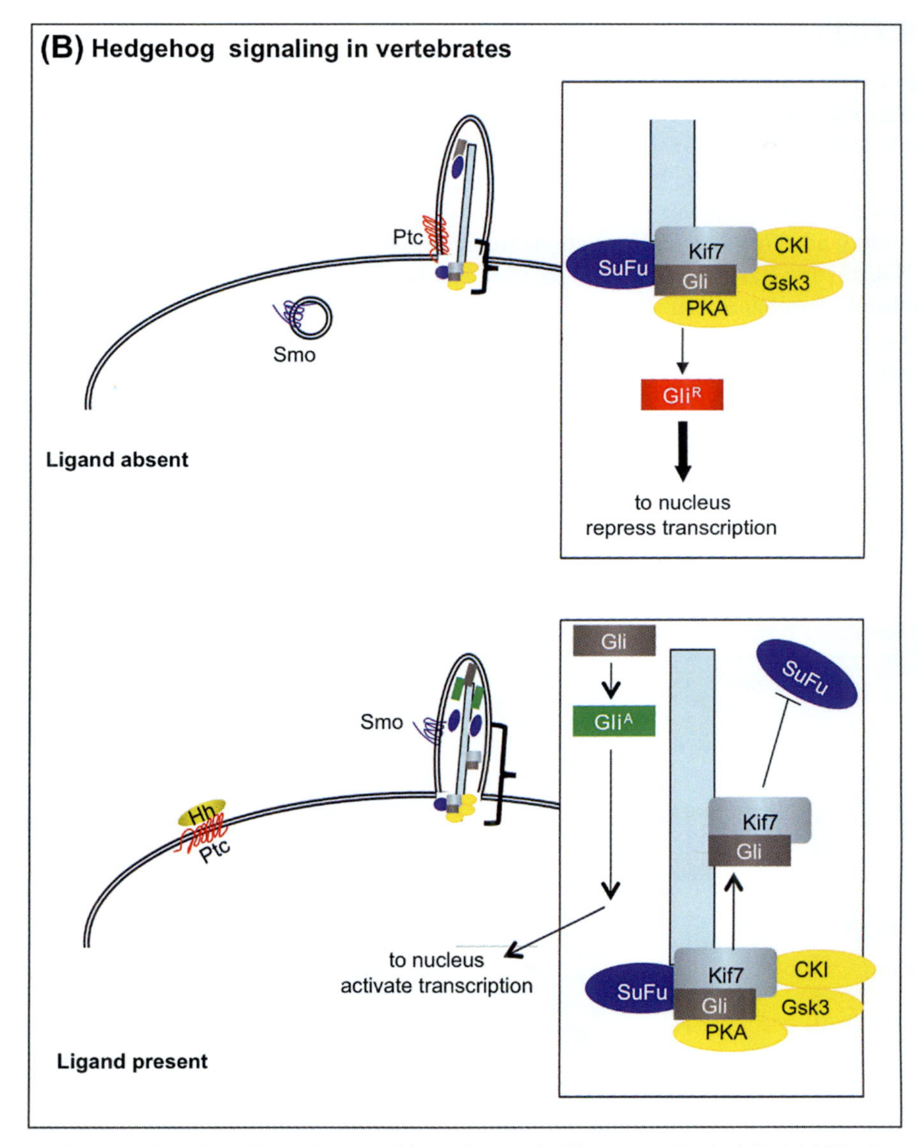

FIGURE 1.1 **(B, C) Hedgehog signaling**: in both vertebrates and invertebrates, the Hh receptor Patched (Ptc) inhibits Smoothened (Smo), a 7 pass transmembrane protein that activates the cytoplasmic signal transduction pathway. In the absence of Hh, the transcriptional effector Gli is processed to a repressive form (Gli^R). Hh-activated signaling culminates in preferential processing to the activator form, Gli^A. In vertebrates, different Gli isoforms are preferentially used for Hh target genes that require repression versus activation during development. (B) Hh-Gli signaling in vertebrates typically takes place at primary cilia. In the absence of ligand, Ptc localizes to the membrane at the base of the primary cilium. PKA also localizes to the ciliary base, where it phosphorylates Gli as a prerequisite for processing to GliR. In the presence of Hh, Ptc exits and Smo enters the ciliary membrane, there is an increase in tip-localized Gli, and Kif7 undergoes anterograde transport to cilia tips. Smo regulates Gli-SuFu localization in cilia tips and helps to suppress SuFu function. Kif7 likely antagonizes SuFu and/or promotes processing of Gli to GliA. The pathway culminates in tip-localized processing of Gli to Gli^A, which then undergoes active retrograde transport to the cytoplasm and then to the nucleus, where it activates target gene transcription. Boxes on the right highlight changes in the primary cilium regions bracketed on the left side of the diagrams.

between neighboring tissues, as well as other signals mediated by cell-cell contact. These signals bind to membrane-bound receptors on responding cells and thereby activate signal transduction pathways that result in context-specific responses, such as cell migration, differentiation, proliferation, and programmed cell death. In some cases, such as in Notch/Delta signaling, signal transduction involves protein cleavage and rapid nuclear translocation of a transcriptional regulator. In most other cases, signal transduction involves a cascade of post-translational modifications such as phosphorylation events which culminate in cytoskeletal changes and/or regulation of transcription factor activity. Recessive loss-of-function mutations in key signaling pathway components often result in severe developmental defects and embryonic lethality. However, studies of viable genetic lesions that alter protein function or expression levels have demonstrated that developmental signaling pathways contribute to the etiologies of numerous diseases and developmental disorders, including neurocristopathies, ciliopathies, and cancer.

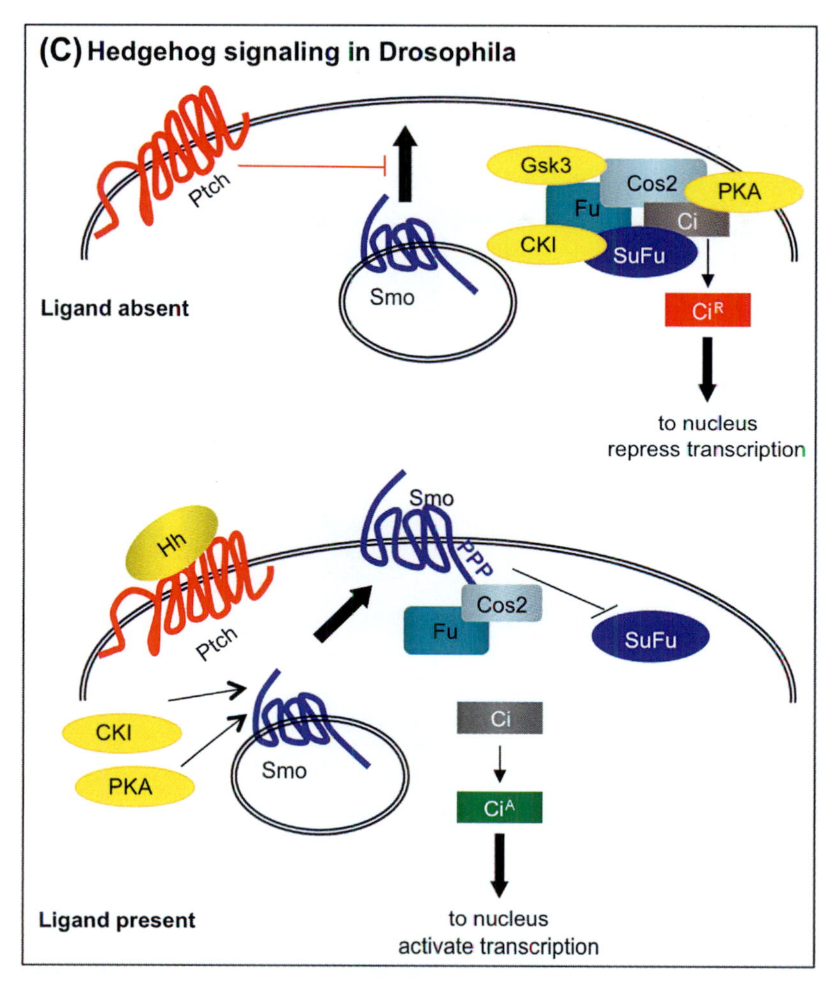

FIGURE 1.1 **(B, C) Hedgehog signaling**: Hedgehog signaling in *Drosophila*: in the absence of Hh, Ptc, but not Smo, localizes to the plasma membrane. A complex of proteins including Costal-2 (Cos2), Fused (Fu), Suppressor of Fused (SuFu), and Ci recruit kinases (PKA, CKI, Gsk3β) that phosphorylate Ci, which then is preferentially proteolyzed to a repressive form (Gli^R) that translocates to the nucleus and blocks transcription. Hh binding relieves Ptc inhibition of Smo. Smo then undergoes sequential phosphorylation by PKA and CKI, translocates to the plasma membrane, and recruits the Fu-Cos2 complex. This results in Ci release that then undergoes preferentially processing to Ci^A. Ci^A enters the nucleus and activates target gene transcription.

Increasingly sophisticated tools have enabled significant conceptual advances in our understanding of developmental cell signaling mechanisms. Genomic sequence data and deep sequencing technologies enable the rapid identification of genetic lesions in forward genetic screens and human genetic studies. Genomic sequence data have also enabled several types of comprehensive, unbiased, sequence-driven screens for modifiers and transcriptional targets of cell signaling pathways. Systems biology studies of gene expression and regulation (e.g., using microarrays, RNA-seq, ChIP-chip, ChIP-seq [defined in the Glossary]) facilitate identification of transcriptional targets and feedback loops induced by cell signaling. Proteomic technologies enable unbiased studies of protein-protein interactions relevant to signal transduction and regulation. In addition, imaging techniques have been developed that enable the visualization of signaling molecules and responses *in vivo* (Table 1.2). Complementing these modern techniques, classical genetics remain instrumental in addressing the ways that specific genes, proteins, and interactions contribute to cell signaling in distinct developmental contexts.

A full description of cell signaling roles in development is beyond the scope of this chapter. Instead, we will discuss examples from different model systems that illustrate diverse mechanisms of signal transmission, localization, transduction, and interpretation during development. Specifically, we will discuss the following:

- Identification of signaling pathway components,
- Functional diversification of related signaling proteins,
- Roles of cytoplasmic extensions in cell signaling,
- Formation and interpretation of signaling gradients, and
- Transcriptional regulation by developmental cell signaling pathways.

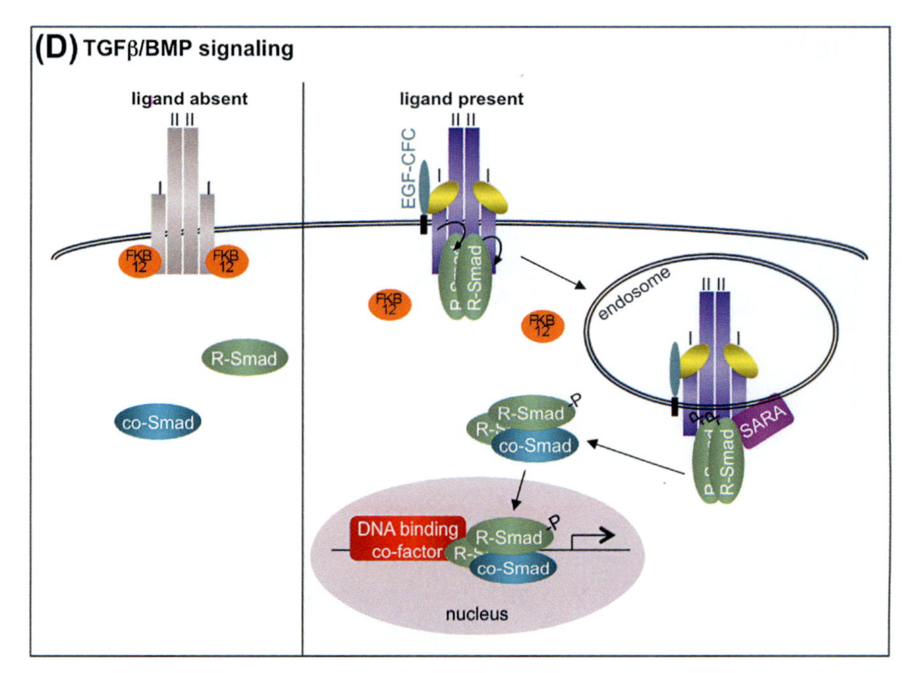

	TGFβ signaling	BMP signaling
Receptor type I	Alk4, 5, 7	Alk1, 2, 3, 6
Receptor type II	TGFβRII, ActRIIB	ActRIIA, ActRIIB, BMPRII, AMHRII
R-Smad	Smad2, 3	Smad1, 5, 8

FIGURE 1.1 **(D) TGFβ/BMP Signaling**: TGFβ and BMP ligands form complexes with heterotetrameric activin receptors to activate Smad signaling. Active BMP and TGFβ ligands bind type I (RI) and type II (RII) receptors, respectively, which are dual-specificity kinases with strong serine/threonine and weak tyrosine phosphorylation activity. These receptors then recruit their heterotetrameric RI/RII binding partners. For some ligands (e.g., Nodal, Gdf1), GPI-anchored EGF-CFC proteins are obligatory co-receptors that bind ligand and RI and can act in a cell-autonomous or -nonautonomous manner (dependent on GPI cleavage). Upon formation of the full ligand/receptor complex, RII serine/threonine phosphorylates RI, triggering a conformational change that activates its kinase domain, releases FKBP12 (which associates with inactive RI and helps maintain it in a silent state), and exposes the R-Smad binding site. R-Smads are recruited, the complex is internalized; the endosomal adaptor protein SARA stabilizes the receptor-R-Smad interaction. RI phosphorylates Smad on serine residues, allowing an intramolecular interaction in R-Smads that release them from the receptor. Phosphorylation of R-Smads also decreases their affinity for SARA and increases their affinity for co-Smads. R-Smads bind cytoplasmic co-Smad and trimeric Smad complexes translocate to the nucleus to regulate target genes, together with recruited DNA binding co-factors. Depending on the nature of the co-factors, Smads can positively or negatively regulate transcription of target genes. The table below the signaling diagram shows the different specificities of BMP versus TGFβ ligands for receptors and R-Smads. *Not shown:* in vertebrates, inhibitory Smads (Smad6, 7) can bind activated RI, block recruitment of R-Smads, and recruit E3 ubiquitin ligase to trigger RI degradation. In addition to this pathway, BMP/TGFβ receptors can signal via the MAPK (ERK) and p38MAPK pathways through recruitment of Shc and Grb2, respectively, to phospho-tyrosine residues on RI and RII. *Not shown:* both classes of ligands are regulated extracellularly. TGFβ ligands are secreted in an inactive (latent) state and require cleavage of a propeptide for activation; integrins and matrix metalloproteases together can mediate this cleavage. BMP ligands can form signaling-inactive extracellular complexes with Tsg and Chd for long range transport; localized Tolloid near target cells degrades Chd, releasing BMP from the complex and enabling it to bind with receptors. In addition, the secreted protein BMP-ER (Cv-2) binds cell surface associated HSPGs and complexes with BMP and type I receptors to facilitate short range signaling.

1.2 IDENTIFICATION OF SIGNALING PATHWAY COMPONENTS

Decades after the identification of core pathway components, we continue to expand our understanding of the breadth of molecules involved in cell signaling. Forward genetic screens employing chemical or insertional mutagenesis remain invaluable for identifying signaling pathway components and modifiers, and annotated genome sequence data make it possible to monitor screen saturation (i.e., the percentage of coding genes tested in the screen). These experiments can select for hypomorphic alleles that illuminate context-specific functions of broadly utilized genes, and, importantly, highlight *in vivo* consequences of genetic lesions. In parallel, modern sequencing and genomic resources enable genome-wide association studies (GWAS) that identify genotype-phenotype associations in human populations. Studies cross-referencing GWAS data with developmental genetics findings have uncovered novel roles of cell signaling genes in disease.

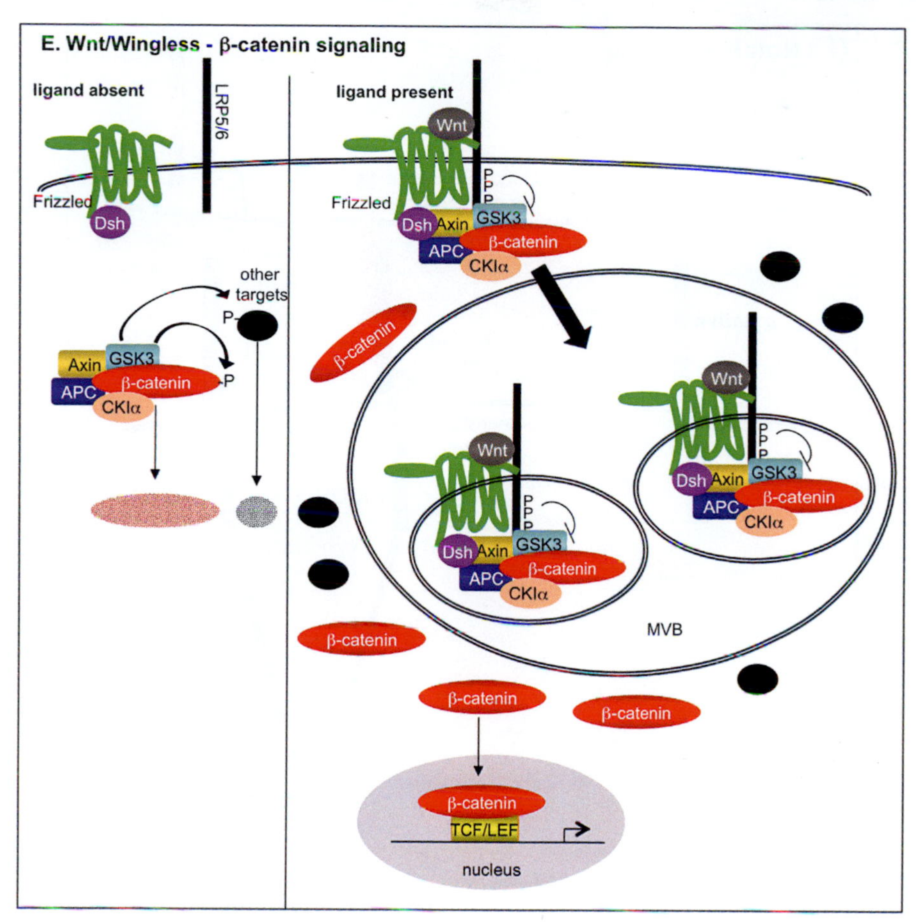

FIGURE 1.1 **(E) Wnt/Wingless – β-catenin pathway**: In the absence of ligand, a destruction complex comprised of GSK3, Axin, APC, CKIα, and other proteins (not shown) binds β-catenin. GSK3 phosphorylates β-catenin and other targets (including cytoskeletal and cell cycle proteins), targeting them for ubiquitin-mediated proteosomal degradation. When Wnt binds the Frizzled receptor, the ectodomain of an LRP co-receptor (vertebrate LRP5/6 or *Drosophila* Arrow) binds Wnt and Fz, stabilizing the trimeric complex. The intracellular domain of LRP is phosphorylated, enabling interactions with GSK3 and Axin. pLRP inhibits GSK3 activity, and Axin recruits the destruction complex. The adaptor protein Disheveled (Dsh) is thought to oligomerize and possibly facilitate clustering and/or endocytosis of signaling complexes. Active signaling complexes are endocytosed through the caveolin pathway and then sequestered in multivesicular bodies (MVBs). This blocks GSK3 phosphorylation of targets (including β-catenin), which then accumulate in the cytoplasm. β-catenin translocates into the nucleus where it activates target gene transcription together with TCF/LEF proteins (see also the Nusse lab's Wnt homepage: http://www.stanford.edu/group/nusselab/cgi-bin/wnt/ [accessed Nov 2013]). Importantly, Wnts can also signal via β-catenin-independent pathways not pictured here (Wnt-planar cell polarity, Wnt-calcium); these are described in reviews referenced in Table 1.1.

In the post-genomic era, screens for novel cell signaling pathway components and modifiers commonly employ sequence-driven, genome-wide approaches. These include RNAi or overexpression screens, interactome (large-scale protein-protein interaction) mapping, and developmental synexpression analysis. These approaches, discussed below, each provide a platform for comprehensively scanning the genome and generating new models of cell-cell signaling which can then be tested genetically and biochemically. Sequence-driven screens provide several advantages over classical techniques. They do not rely on chance to reach saturation of the genome. They selectively test effects of coding sequences on signaling outcome, whereas random mutagenesis hits can be intergenic. In addition, many sequence-driven screens do not rely on phenotypic assays to associate genes/proteins with a given pathway. Biological processes by nature tend to be characterized by a certain degree of robustness; i.e., developmental systems tend to compensate for subtle variations or deficiencies in genetic/biochemical pathways. Hence, factors that impact cell-cell signaling are not always essential for a normal developmental outcome and can be missed in phenotype-driven screens. Finally, sequence-driven screens are capable of identifying signaling functions of genes that contribute to multiple cellular processes or pathways. These genes would likely have pleiotropic mutant phenotypes and be discarded in screens for pathway-specific phenotypes, but can play important roles in orchestrating cellular responses to signals within complex *in vivo* microenvironments. For instance, such factors can mediate crosstalk with other active signaling pathways and/or modulate general cellular processes (e.g., transcription, translation, proliferation) to enable/mediate a cellular response.

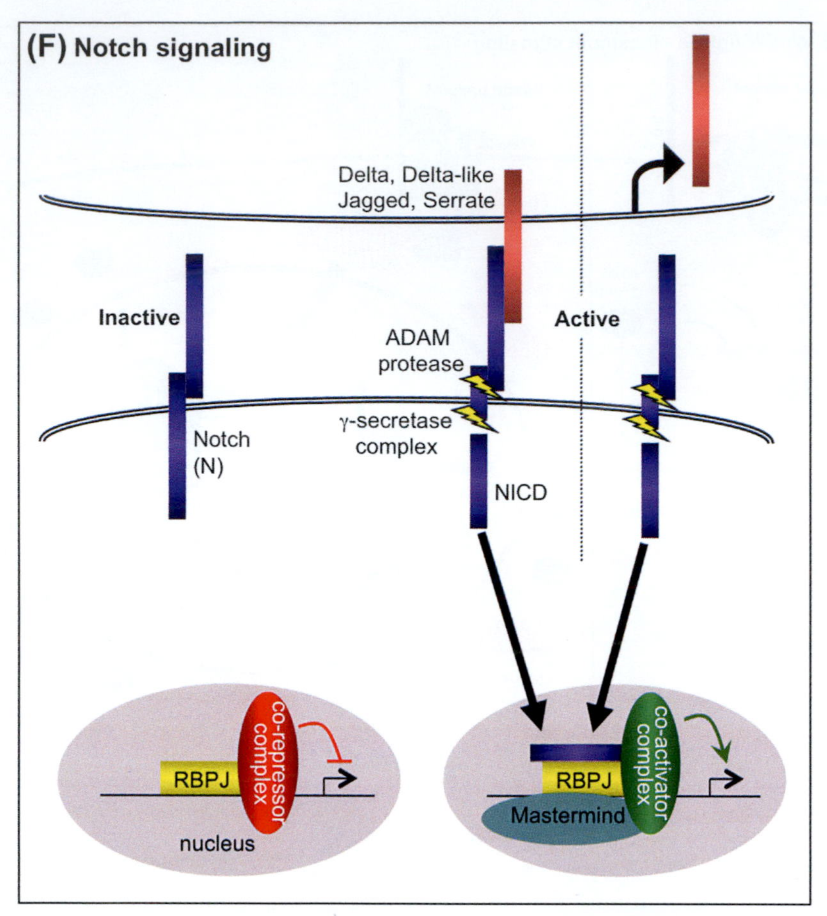

FIGURE 1.1 **(F) Notch signaling**: In the absence of ligand binding, RBPJ (also known as CSL transcription factors, Suppressor of Hairless [Su(H); *Drosophila*], and CBF1) interact with a co-repressor complex and inhibit transcription of Notch target genes. The Notch receptor can be activated either by interaction with ligands (Delta, Delta-like, Serrate, Jagged) or internalization of ligand into adjacent cells. Notch activation induces two cleavage events: ADAM (a disintegrin and metalloproteinase) proteases such as TNFα converting enzyme (TACE) shed the ectodomain and γ-secretase releases the Notch intracellular domain (NICD) into the cytoplasm. NICD translocates to the nucleus, and recruits a co-activator complex (including Mastermind/Mastermind-like) that displaces the RBPJ-co-repressor complex. The NICD complex then activates target gene transcription. Gene regulation by NICD occurs predominantly at enhancers, rather than promoters, and is affected by the chromatin environment. The composition of the co-activator complex is context-dependent. NICD has a very short-life in the nucleus: its C-terminal PEST domain (a peptide sequence rich in proline [P], glutamic acid [E], serine [S], threonine [T]) becomes phosphorylated, enabling ubiquitination and degradation. Association with Numb facilitates NICD ubiquitination/degradation, whereas acetylation (e.g., by p300 in a co-activator complex) can stabilize NICD to prolong signaling. In addition to being the effector of Notch signaling, NICD is regulated through phosphorylation by other kinases, including GSK3β; NICD phosphorylation can block its degradation. (Notch can also signal via "non-canonical" pathways described in reviews referenced in Table 1.1.)

RNAi and Overexpression Screens

RNA interference (RNAi) uses short double-stranded RNAs to trigger targeted degradation of specific mRNAs. This was developed in *C. elegans* for loss-of-function (LOF) studies and phenotype-driven screens (Fire et al., 1998; Wang and Barr, 2005). *In vivo* RNAi screens that assay for developmental phenotypes have since been conducted in several other model organisms. In addition, RNAi has been adapted for genome-wide LOF screens in cultured cells. In these experiments, cells are co-transfected with an RNAi construct and a pathway-responsive reporter, and then stimulated to activate a specific cell signaling pathway. The reporter provides a quantitative readout of the RNAi-targeted gene's effect on the signaling response. Parallel screens employing cDNA libraries have similarly been developed to study gain-of-function (GOF) effects due to overexpression. Numerous RNAi, siRNA (another targeted LOF technique), and cDNA libraries have been developed that include LOF/GOF constructs for every coding gene or for a functionally defined subset of genes (e.g., all kinases) in a genome.

The first genome-wide RNAi screen in *Drosophila* cells utilized Hh-responsive luciferase (transcriptional) reporters in cultured imaginal disc cells (Lum et al., 2003). RNAi of known Hh pathway genes altered Hh responses (luciferase activity) in this system, validating the approach. In addition, numerous genes previously unassociated with Hh signaling were found to modify Hh reporter activity and interact genetically with known Hh pathway members (Lum et al., 2003; Nybakken et al., 2005). Some of these genes belong to classes traditionally associated with cell-cell signaling and have since been validated by studies in other systems. These include the heparin sulfate proteoglycan Dally-like (Dlp, which had previously implicated in Wnt but not Hh

TABLE 1.1 Recent Reviews of Developmental Cell Signaling Pathways

Pathway and Review Topic	Year	Reference(s)
Receptor tyrosine kinase (**RTK**) signaling overview	2010	(Lemmon and Schlessinger, 2010)
Spatial regulation of **RTK** signaling (*localization within plasma membranes and endocytic trafficking*)	2012	(Casaletto and McClatchey, 2012)
Wnt/β-catenin signaling and disease	2012, 2013	(Clevers and Nusse, 2012; Niehrs, 2012; Kim et al., 2013)
Non-canonical **Wnt** signaling (*PCP, alternative ligands and receptors*)	2012	(Gao, 2012; van Amerongen, 2012)
Wnt secretion, modification, and trafficking	2011	(Buechling and Boutros, 2011)
Hedgehog signaling	2012, 2013	(Robbins et al., 2012; Chen and Jiang, 2013)
Hedgehog signaling and Gli proteins in human disease	2011	(Hui and Angers, 2011; Onishi and Katano, 2011)
Convergence of **Hedgehog** and **TGFβ** signaling in oncogenesis and metastasis	2012, 2013	(Javelaud et al., 2012; Perrot et al., 2013)
BMP signaling in development and disease	2010, 2011, 2012, 2013	(Walsh et al., 2010; Araujo et al., 2011; Ramel and Hill, 2012; Bandyopadhyay et al., 2013)
Notch signaling	2011, 2012	(Andersson et al., 2011; Borg-grefe and Liefke, 2012; Kan-dachar and Roegiers, 2012)
Non-canonical **Notch** signaling	2010	(Sanalkumar et al., 2010)
Notch signaling in development and disease	2012	(Louvi and Artavanis-Tsakonas, 2012; Penton et al., 2012)
Cilia in vertebrate development and disease	2012	(Oh and Katsanis, 2012)

TABLE 1.2 Live Imaging Techniques for Studying Cell Signaling *In Situ*

Technique	Use/Application	Reference(s)
General reviews		(Wang et al., 2008; Gell et al., 2012)
Photoactivatable/photo-convertible fusion proteins	study protein trafficking, kinetics and movement of molecules within cell	(Lukyanov et al., 2005; Gurskaya et al., 2006; Zhang et al., 2007)
FRET: Förster resonance energy transfer	study conformational change, post-translational modifications (e.g., phosphorylation), protein-protein interactions	(Aoki et al., 2013)
FRAP: Fluorescence recovery after photo-bleaching	study molecular movement (e.g., to measure diffusion constants)	(David et al., 2012)
FCS: Fluorescence correlation spectros-copy	visualize molecular dynamics, interactions, and diffusion	(Bacia and Schwille, 2007a; Bacia and Schwille, 2007b)
Two-focus FCS	visualize/model directional transport through cross-correlation of fluorescence in two nearby confocal planes	(Brinkmeier et al., 1999; Dertinger et al., 2007)
Modular scanning FCS	*in vivo* analysis of receptor-ligand interactions	(Ries et al., 2009)

signaling), three kinases (CK1α, Pitslre1, Cdk9) and a phosphatase (PP2A). Interestingly, the screens also indicated that Hh signaling is affected by factors involved in more general cellular processes including ribosome and proteosome function, RNA regulation and splicing, and vesicle trafficking. Such genes were also identified as Hh modifers, albeit at a low frequency, in *in vivo* screens (Eggenschwiler et al., 2001; Huangfu et al., 2003; Collins and Cohen, 2005; Huangfu and Anderson, 2005). Similar RNAi screens with different transcriptional reporters have been used to scan the genome for genes that impact JAK/STAT and Wnt signaling. Like the Hh studies, these screens also identified proteins utilized in other signaling pathways, as well as factors involved in general cellular processes (Baeg et al., 2005; DasGupta et al., 2005; Muller et al., 2005).

Parallel screens using the same cell type and RNAi library but different reporters can help identify points of crosstalk between pathways. Jacob et al. conducted a series of genome-wide RNAi screens in cultured mouse cells, and found that ~80% (280/346) of genes that modified Shh signaling also modified Wnt signaling (Jacob et al., 2011). By correlating RNAi results with data in the Online Mendelian Inheritance in Man (OMIM) database, this group identified ten Shh pathway modifying genes – that suppressed or had no effect on Wnt signaling – associated with human developmental syndromes, degenerative diseases, and cancer (Jacob et al., 2011). Among the disease-associated Shh-modifying genes were a glypican, a PKA subunit, and genes important for primary cilia and/or intraflagellar transport (IFT; Jacob et al., 2011).

Genome-wide screens in different cell types can also aid in identifying cell type- or species-specific mechanisms of signaling and signal regulation. This is exemplified by studies of Shh signaling, which utilizes ciliary IFT proteins, discussed below, and Sufu in vertebrates but not *Drosophila* (Figure 1.1B,C; Ohlmeyer and Kalderon, 1998; Ray et al., 1999; Han et al., 2003; Huangfu et al., 2003; Svard et al., 2006; Chen et al., 2009). Evangelista and colleagues conducted a mammalian kinome (all kinases and some kinase-regulatory factors) siRNA screen to identify genes that regulate Hh signaling. They identified nine candidates, including a regulatory subunit of PKA (required for Shh signaling in *Drosophila* and vertebrates) as well as three kinases that likely are specific to vertebrate Shh signaling: two kinases required for ciliogenesis (including Nek1, a human ciliopathy gene), and Cdc2l1, which binds Sufu and interferes with Sufu-Gli binding (Evangelista et al., 2008). Complementing this study, Varjosalo et al. used kinome and catalytically inactive kinome overexpression libraries as well as a kinome siRNA library to screen for Hh modifiers in mammalian cells. This screen identified PKA, MAP3K10, and DYRK2. *In vivo*, biochemical, and proteomics assays demonstrated that MAP3K10 positively regulates Shh-induced signaling, binds/phosphorylates DYRK2 and the IFT protein Kif3a, and affects GSK3β activity. DYRK2 negatively regulates Shh signaling, is required to repress signaling in the absence of Shh, and phosphorylates and induces proteosomal degradation of Gli2 (Varjosalo et al., 2008). This study used a proteomic approach to identify catalytic targets (assaying protein binding to and phosphorylation of microarrayed proteins) that could prove very useful in future studies of signaling transduction.

Collectively, these cell-based RNAi/cDNA screens are significantly broadening our understanding of cell signaling networks, although several modifiers identified in these screens still need to be evaluated *in vivo* for relevance to developmental signaling outcomes. Comparing results of LOF/GOF experiments in vertebrate versus invertebrate cell systems and in different cell types within a species will advance our knowledge of context-specific signaling mechanisms.

Interactome Mapping

Interactome mapping, in which signaling networks are modeled on the basis of physical protein-protein interactions, is not hindered by compensatory mechanisms that may mask roles of pathway members/modifiers in functional assays. Early interactome studies relied on genome-wide yeast two-hybrid (Y2H) assays using known signaling pathway proteins as baits. For example, Tewari and colleagues conducted a genome-wide Y2H screen for *C. elegans* proteins that interact physically with TGFβ pathway proteins. They generated an interactome map describing physical interactions among 59 proteins, only four of which had previously been assigned to the TGFβ signaling pathway. Novel components of this biochemically-defined interactome were validated *in vivo* based on expression in TGFβ-dependent contexts and genetic interaction with known TGFβ pathway genes. Thus several new proteins were modeled into the TGFβ signaling network including filamin, the TTX-1 homeobox protein, Swi/Snf chromatin remodeling factors, and Hsp90 (Tewari et al., 2004). Subsequent studies validated the interactome results in other systems, for instance demonstrating roles of Hsp90 and Swi/Snf factors in TGFβ signaling regulation (Wrighton et al., 2008; Xi et al., 2008). Genome-wide Y2H analyses have been reported for *C. elegans* and *Drosophila*, and protein-protein interaction data for multiple systems have been compiled into interactive public databases (Xenarios et al., 2002; Giot et al., 2003; Li et al., 2004; Formstecher et al., 2005). Comparative studies in different cell systems may illuminate context-specific mechanisms of signal transduction. Furthermore, Y2H is amenable to rapid, easy generation and analysis of mutant proteins, and so can be used to characterize protein-protein interaction domains.

Proteomics approaches employing tandem affinity purification-mass spectrometry (TAP-MS) have also been used for interactome analysis (Gavin et al., 2002). TAP-MS, like Y2H, entails generating protein fusion "baits": TAP tags are added to C- or N-termini of baits and used for sequential purification of protein complexes, which are then analyzed by MS. Compared to Y2H, TAP-MS is less sensitive for the detection of weak interactions or interaction partners expressed at low levels. However, it can be done using a greater range of physiological cell types and can also be used to identify protein-protein interactions dependent on cell signaling. In addition to the tradeoff in sensitivity, Y2H and TAP-MS differ in that Y2H identifies mostly direct protein-protein interactions, whereas TAP-MS characterizes protein complexes and therefore generates datasets including direct and indirect interactions. For these reasons, Y2H and TAP-MS can generate complementary datasets in interactome analyses.

Whereas cell stimulation/response assays have long been used to study responses to soluble ligands, biochemical responses to contact-mediated signals (e.g., Notch/Delta, Eph/ephrin) present a technical challenge in cultured cell systems. In some cases, soluble forms of membrane-bound ligands can drive signals normally mediated by cell contact. For instance, ephrins are transmembrane and glycophosphatidylinositol (GPI)-linked ligands for the Eph RTKs, and pre-clustered ephrin-Fc molecules have been used to stimulate Eph signaling in phosphoproteomic profiling experiments (Bush and Soriano, 2010). As an alternative approach, Jørgensen et al. used stable isotope labeling (SILAC; Ong et al., 2002) to differentially label Eph- versus ephrin-expressing cells, which they then co-cultured and used in phosphoproteomic analyses of bi-directional signaling (Jorgensen et al., 2009). Phosphotyrosine proteomics data were integrated with siRNA library screening data that identified which phosphoproteins impacted cell-cell sorting. Computational algorithms then generated phosphorylation network predictions (kinase>kinase substrate>phosphotyrosine binding partner) of signaling responses in Eph- versus ephrin- expressing cells (Jorgensen et al., 2009). These experiments identified many distinct and some common signaling modules activated in EphB2- versus ephrinB1-expressing cells. In some cases, commonly activated effectors differed in the two populations as to the amplitude/dynamics of tyrosine phosphorylation responses. In addition, several kinases/effectors were found to be differentially used in forward (Eph) versus reverse (ephrin) signaling (Jorgensen et al., 2009). For instance, IGF1R and PTK2 were identified as kinases that likely mediate signal transduction by ephrinB1, which does not have intrinsic kinase activity; EphB2 and ABL1 were identified as the major kinases driving EphB2 signal transduction (Jorgensen et al., 2009). Parallel experiments with signaling-mutant forms of Eph and ephrin identified cell-autonomous effects of the mutations on downstream signaling events, and also revealed that Eph mutations impact signaling in ephrin-expressing cells and vice versa (Jorgensen et al., 2009). Future studies using these techniques may reveal cell type-specific mechanisms of Eph-ephrin signaling and could uncover distinctions between signaling by different Eph-ephrin interaction pairs.

Developmental Synexpression Analysis

Signaling networks have increased in complexity during evolution due to gene duplication events and the incorporation of redundant or compensatory signaling events. Many proteins in these networks have modular and conserved protein interaction domains (e.g., phosphotyrosine binding domains, src homology domains) that are promiscuous in biochemical assays. Furthermore, *in vivo* analyses have indicated that many pathways use context-specific mechanisms of signal transduction during development. Therefore it has become a significant challenge to determine which proteins are functionally associated in distinct biological contexts. Conservation of expression patterns across species is highly suggestive of functional conservation, and so expression profiling in different model organisms can help to identify genes that likely function together in a given context. Hence, developmental synexpression analysis – identification of spatiotemporally co-expressed gene groups – has proven very useful in generating models of ligand-receptor relationships, signal transduction pathways, and regulatory events that comprise signaling modules *in vivo*. There is evidence that synexpressed genes are regulated by cis-regulatory modules (enhancers) that contain synexpression group (SG)-enriched sequence motifs hypothesized to recruit overlapping cohorts of transcription factors (TFs; Ramialison et al., 2012). In some cases SGs also include genes that are clustered in the genome, which likely also facilitates co-regulation (Ramialison et al., 2012).

An early genome-wide expression screen identified the evolutionarily conserved Fgf8 SG in zebrafish. This includes several regulators of the RTK-Ras-MAPK pathway, namely Sprouty proteins, the transmembrane protein Synexpressed with Fgf (Sef), and MAPK phosphatase 3 (Mkp3; Figure 1.1A; Kudoh et al., 2001; Furthauer et al., 2002; Lin et al., 2002; Tsang et al., 2002; Kawakami et al., 2003; Tsang et al., 2004). As might be expected for antagonists of a broadly utilized signal transduction cascade, these proteins do not exhibit strict RTK-specificity in biochemical assays (Camps et al., 1998; Reich et al., 1999; Tsang et al., 2002; Kovalenko et al., 2003; Preger et al., 2004; Torii et al., 2004). Yet, synexpression suggests that they are specifically involved in Fgf8-mediated responses, and functional studies have indicated that they antagonize Fgf signaling *in vivo* (Kramer et al., 1999; Furthauer et al., 2002; Tsang et al., 2002; Kawakami et al., 2003). This does not preclude the possibility that they inhibit signaling by other RTKs at sites of Fgf8 expression. Indeed, in *Drosophila*, Sprouty and Mkp3 also regulate EGFR signals, and Mkp3 is expressed in contexts dependent on multiple RTKs (Kramer et al., 1999; Kim et al., 2004; Gomez et al., 2005). Genetic interaction and/or proteomic studies are needed to determine the targets of Sef, Mkp3, and Sprouty regulation in vertebrates. Feedback inhibitors are commonly represented in developmental signaling SGs. Evidence from BMP SG studies suggests that synexpressed feedback regulators suppress phenotypic variability and confer robustness to morphogenetic developmental processes (Paulsen et al., 2011).

Large-scale *in situ* hybridization studies aimed at identifying SGs have also been conducted in *Drosophila*, *Xenopus*, mouse, and medaka; this developmental gene expression data is available online (Table 1.3). In addition, microarray analyses in mouse, zebrafish, and *Xenopus* have profiled gene expression at different embryonic and postnatal stages (Miki et al., 2001;

TABLE 1.3 Web-Based Resources for Developmental Biologists

Organism/Organ	URL	References
Gene Expression Databases		
(several of these websites also contain many other useful types of data/information)		
Cross-species mouse, zebrafish, medaka, *Drosophila*	4DXpress http://4dx.embl.de/4DXpress/guest/login.do;jsessionid=9E4F6641AA10CC9E54B529BE889A148F (accessed Spring 2013)	(Haudry et al., 2008)
Drosophila	Berkeley *Drosophila* Genome Project http://insitu.fruitfly.org/cgi-bin/ex/insitu.pl (accessed Spring 2013)	(Tomancak et al., 2002; Tomancak et al., 2007)
Xenopus	Xenbase http://www.xenbase.org/common/ (accessed Spring 2013)	(Bowes et al., 2008)
Xenopus microarray data	http://www.dkfz.de/en/mol_embryology/microarray.html (accessed Spring 2013)	(Baldessari et al., 2005)
Zebrafish	ZFIN http://zfin.org/ (accessed Spring 2013)	(Bradford et al., 2011)
Medaka	MEPD http://ani.embl.de:8080/mepd/ (accessed Spring 2013)	(Ramialison et al., 2012)
Mouse	See "Web-based digital gene expression atlases for the mouse" for a summary and comparison of mouse gene expression databases	(Geffers et al., 2012)
	Mouse gene expression database (GXD) http://www.informatics.jax.org/expression.shtml (accessed Spring 2013)	(Finger et al., 2011)
	EMAGE http://www.emouseatlas.org/emage/ (accessed Spring 2013)	(Richardson et al., 2010)
	Eurexpress http://www.eurexpress.org/ee/ (accessed Spring 2013)	(Diez-Roux et al., 2011)
	GenePaint http://www.genepaint.org/Frameset.html (accessed Spring 2013)	(Visel et al., 2004)
Brain	Allen Brain Atlas http://www.brain-map.org/ (accessed Spring 2013)	(Lein et al., 2007; Jones et al., 2009)
	GENSAT http://www.gensat.org/index.html (accessed Spring 2013)	(Gong et al., 2003)
Tooth	http://bite-it.helsinki.fi/ (accessed Spring 2013)	(Nieminen et al., 1998)
Craniofacial	Facebase https://www.facebase.org/ (accessed Spring 2013)	(Hochheiser et al., 2011)
Genitourinary tract	GUDMAP http://www.gudmap.org/index.html (accessed Spring 2013)	(Harding et al., 2011; Davies et al., 2012)

TABLE 1.3 Web-Based Resources for Developmental Biologists—cont'd

Organism/Organ	URL	References
Enhancer Element Data and Prediction Tools		
	Enhancer Element Locator http://www.cs.helsinki.fi/u/kpalin/EEL/ (accessed Spring 2013)	(Hallikas et al., 2006; Palin et al., 2006)
	VISTA enhancer browser http://enhancer.lbl.gov/ (accessed Spring 2013)	(Visel et al., 2007)
Genotype/Phenotype Association Data		
Man	Online Mendelian Inheritance in Man (OMIM) http://www.ncbi.nlm.nih.gov/omim (accessed Spring 2013)	(Amberger et al., 2009)
Man	Catalog of somatic mutations in cancer http://cancer.sanger.ac.uk/cancergenome/projects/cosmic/ (accessed Spring 2013)	(Forbes et al., 2006; Forbes et al., 2010)
Animals (excluding mouse, man)	Online Mendelian Inheritance in Animals (OMIA) http://omia.angis.org.au/ http://www.ncbi.nlm.nih.gov/entrez/query.fcgi?db=omia (accessed Spring 2013)	(Lenffer et al., 2006)

Lo et al., 2003; Baldessari et al., 2005). These large datasets can be mined to identify new SGs or expand on those previously identified. They can also provide clues as to context-specific, stage-specific, or species-specific signal transduction/modulation mechanisms, and to this end have been integrated into some computational algorithms to model signaling networks.

1.3 FUNCTIONAL DIVERSIFICATION OF RELATED SIGNALING PROTEINS

Over the course of evolution, members of gene families encoding key pathway components have adapted to have diverse, specialized functions while utilizing similar signal transduction pathways. Mechanisms of functional diversification have been investigated in studies of mammalian and *Drosophila* RTKs (mammalian: PDGFRα/β, Fgfr1, VEGFR1/2, EGFR; *Drosophila*: Torso, DER [EGFR], InR [Insulin receptor], PVR [PDGF/vascular endothelial growth factor (VEGF) receptor], Btl [Fgfr]). In some cases, functional diversification can be attributed to different gene expression patterns. However, chimeric receptor studies *in vivo* have demonstrated that even when placed in the same contexts, closely related signaling proteins often have distinct developmental potentials and transduce non-equivalent signals (although there are clearly many instances in which family members can compensate for one another). For example, the *Drosophila* Torso signaling domain incompletely rescues PDGFRα functions in knock-in mice and activates only a subset of PDGFRα-activated transduction pathways in primary cells (Hamilton et al., 2003). Likewise, the PDGFRβ signaling domain is unable to drive Fgfr1 responses during embryonic development (Hoch, 2005). In contrast, the Fgfr1 signaling domain activates more potent signaling responses than PDGFRα or Torso, and PDGFRα/Fgfr1 chimeric receptor-expressing embryos exhibit dominant gain-of-function phenotypes (Hamilton et al., 2003). Similarly, in *Drosophila*, the signaling domains of Torso and DER drive migration responses to Btl activation incompletely and to different degrees in chimeric receptor rescue experiments (Dossenbach et al., 2001).

RTK signals can be distinguished by the amplitude or duration of signaling through a common pathway, and/or by differential utilization of effector pathways (see Figure 1.1A for an overview of RTK signaling; [Tan and Kim, 1999]). During mouse development, PDGFRα transduces signals predominantly via a single effector (PI3K) recruitment site, whereas multiple effector pathways contribute additively to PDGFRβ functions (as has also been reported for Torso in *Drosophila*; Gayko et al., 1999; Klinghoffer et al., 2002; Tallquist et al., 2003). The selective use of one pathway may limit the amplitude and variability of PDGFRα responses, and may reflect the affinity or availability of effector proteins for this receptor. Notably, the PDGFRα signaling domain drives weaker MAPK responses than that of PDGFRβ in cultured embryonic cells, and the PDGFRβ signaling domain can fully rescue PDGFRα-dependent development *in vivo*, but the converse is not true (Klinghoffer et al., 2002). Differences in PDGFR accessibility to signal transduction proteins/pathways may in part be determined by their subcellular localization: in growth-arrested mouse fibroblasts, PDGFRrα and PDGFRβ are spatially segregated in ciliary membranes and the general plasma membrane, respectively (Schneider et al., 2005; Christensen et al., 2012).

Within the VEGFR subfamily of RTKs, VEGFR2 is the principal activator of signal transduction whose activity is regulated by VEGFR1. Interestingly, different VEGFR1 ligands specify inhibition versus potentiation of VEGFR2 signaling (Rahimi et al., 2000; Autiero et al., 2003; Meyer and Rahimi, 2003; Roberts et al., 2004). The functional specialization of VEGFRs has been attributed to an amino acid change in the activation loop of VEGFR1 at a residue that is highly conserved among other class III RTKs (Meyer et al., 2006). Within several RTK subfamilies, homo- and heterodimers can form *in vitro* but the significance of this observation *in vivo* is not known. Imaging techniques such as FRET and modular scanning FCS (Table 1.2) allow visualization of protein-protein interactions *in situ* and may shed light on this question. The VEGFR findings introduce the possibility that subunits within other multimeric receptor complexes have distinct functions and signaling potential.

Recently, systems biology approaches have shed additional light on biochemical mechanisms underlying functional specificity within the RTK superfamily. RNAi screens identified a core group of genes that commonly modulate MAPK signaling by multiple RTKs, as well as many that differentially modulate MAPK signaling by DER, InR, or PVR. Signal-specific MAPK modulators include the serotonin receptor, a zinc finger homeobox transcription factor, Akt-PI3K pathway components, and vesicular trafficking proteins (Friedman and Perrimon, 2006; Friedman et al., 2011). Interestingly, these experiments also identified cross-regulatory relationships among RTKs: PVR potentiated InR signaling and possibly antagonized DER signaling, whereas InR did not significantly impact DER signaling (Friedman and Perrimon, 2006). Friedman et al. (2011) also conducted TAP-MS experiments on unstimulated, EGF- and insulin-stimulated cells to generate RTK-MAPK pathway interactome maps in each condition (Friedman et al., 2011). The interactome maps were integrated with functional data to identify direct protein-protein interactions that impact MAPK signaling outcome. These data, which highlighted additional common and RTK-specific signal transduction mechanisms, are available to the community in multiple online formats (Friedman et al., 2011).

Phosphoproteomic studies employing quantitative mass spectrometry (MS), SILAC, and affinity purification with phosphorylated baits or phospho-specific antibodies have also provided new insight into mechanisms of RTK specialization. These experiments provide greater sensitivity than traditional co-immunoprecipitation methods and can be used to generate unbiased datasets specific to stimulated (signal-activated) cells. Schulze et al. used peptide pull-downs with quantitative MS to identify binding partners to all phosphorylated tyrosine (pTyr) residues on mammalian EGFRs (also known as the ErbB family; Schulze et al., 2005). They found some overlap in effector recruitment by the different family members: all four engaged the adaptor protein Shc at a juxtamembrane pTyr and recruited Grb2 and Shc to other pTyr residues. However, there were several notable differences among the family members. ErbB3 predominantly recruited PI3K, whereas ErbB2 (which had the fewest interaction sites) predominantly recruited Shc (Schulze et al., 2005). EGFR and ErbB4 had the most diverse arrays of interaction partners and also had more promiscuous pTyr residues (binding more than one effector) than ErbB2/B3. A time course experiment indicated that individual tyrosine residues on EGFR exhibit different phosphorylation dynamics (Schulze et al., 2005). ErbB2 lacks a functional ligand-binding domain and ErbB3 lacks an active kinase domain, suggesting these receptors are activated via heterodimers or clustering. Interestingly, microarray analysis indicated that ErbB2 and ErbB3 are commonly co-expressed with EGFR or ErbB4 (which are seldom co-expressed; Schulze et al., 2005).

Cell signaling proteins are implicated in many human diseases, and closely related proteins (e.g., different RTKs) tend to be associated with different types of disease. By understanding more fully the biochemical differences underlying functional specialization within signaling families, we may uncover genes/proteins that could be helpful drug targets to modulate signaling in specific cell types or downstream of specific receptors. This may help to avoid deleterious side effects of manipulating commonly utilized signaling effectors in therapeutic efforts.

1.4 ROLES OF CYTOPLASMIC EXTENSIONS IN CELL SIGNALING

The subcellular localization of signals, receptors, and signal transduction machinery is highly regulated during development and can polarize cells toward a signaling source, provide directionality to cell migration, and organize cells within epithelial tissues. Selective localization of cell signaling events can be mediated at the level of signal secretion (e.g., via directed exocytosis from apical versus basal membranes of an epithelium), extracellular modulation of signal stability/mobility, and signal reception/transduction (e.g., through subcellular localization of receptors and signal transduction effectors).

Genetic and live imaging studies have demonstrated that in some vertebrate and *Drosophila* contexts, dynamic actin- and microtubule-based cytoplasmic extensions are employed to localize receptors and signaling transduction machinery.

Cilia in Vertebrate Cell-Cell Signaling

In 2003, an ENU mutagenesis screen unexpectedly implicated primary cilia and IFT in Shh signaling: IFT mutations in mice resulted in neural tube and limb bud patterning phenotypes similar to those that result from Shh pathway disruption (Huangfu et al., 2003). Primary cilia are non-motile, microtubule-based organelles that protrude from most vertebrate cells. They extend from basal bodies, and consist of specialized membranes (structurally segregated from the plasma membrane) around a central axoneme along which IFT machinery facilitates retrograde and anterograde transport between the ciliary base and tip. Disruption of ciliogenesis or IFT results in a broad spectrum of human ciliopathy syndromes. Primary cilia disorder defects include polydactyly, craniofacial and brain malformations, situs inversus, polycystic kidney disease, obesity, diabetes, and subtle defects in the liver, spleen, and heart (Nigg and Raff, 2009; Goetz and Anderson, 2010; Hildebrandt et al., 2011).

Ciliogenesis and the disassembly of primary cilia are tightly coupled to the cell cycle machinery: primary cilia are present on quiescent cells but their disassembly is required for transition into the cell cycle. Fgf and Wnt signaling are required upstream of ciliogenesis in several vertebrate contexts, including the node/Kupffer's vesicle (KV), otic vesicles, and pronephric ducts (Basu and Brueckner, 2009; Liu et al., 2011; Caron et al., 2012; Qian et al., 2013). Transcriptional targets of these signaling pathways that mediate cilia formation, patterning, or function include Ift88, Enc1-like, and the ciliogenic TFs Foxj1 and Rfx2 (Neugebauer et al., 2009; Caron et al., 2012; Qian et al., 2013). Epistasis experiments studying KV have suggested that Wnt signaling functions downstream of Fgfs in ciliogenesis (Caron et al., 2012).

In vertebrates, the Shh pathway utilizes cilia as a specialized signaling organelle in which IFT mediates changes in protein localization essential for Gli processing and consequent transcriptional responses (Figure 1.1B). The Hh receptor Patched (Ptc) is at the base of the primary cilium in the absence of Hh. In the presence of Hh, Ptc exits the cilium and Smo enters, and there is an increase in tip-localized Gli. Gli-Sufu complexes are enriched in the tips of primary cilia in a Smo-dependent manner and need to be dissociated for Shh pathway activation (Humke et al., 2010; Tukachinsky et al., 2010). Kif7, whose ortholog promotes Ci/Gli processing and antagonizes Sufu in *Drosophila*, undergoes Hh-induced anterograde transport from the cilium base to the tip. PKA, which phosphorylates Gli as a prerequisite for proteolysis to GliR, localizes to the base of primary cilia and regulates levels of tip-localized Gli (Pan et al., 2006; 2009; Tuson et al., 2011).

Studies of Shh and cilia/IFT have indicated that some human ciliopathy traits are likely due to disrupted Shh signaling. In addition, ciliary defects can enhance or suppress oncogenesis due to Shh pathway mutations (Han et al., 2009). Cilia/IFT mutants often resemble LOF or GOF Shh mutants due to context-specific requirements for cilia-mediated processing of Gli to GliA or GliR, respectively. Hallmark phenotypes include ventralization of the neural tube (which utilizes GliR) and polydactyly (limb buds utilize GliA; discussed in Goetz and Anderson, 2010). Mutations in IFT and ciliopathy genes have helped to identify new contexts in which cilia are required for developmental cell signaling. In some cases, these experiments have shed light on the etiologies of poorly understood human ciliopathy traits. For example, studies of Rpgrip1l (a human ciliopathy gene) and Ttc21b mutant mice demonstrated the importance of primary cilia in mediating Shh signals in the ventral forebrain. Ttc21b mutants have abnormal primary cilia with partially defective retrograde transport but functional anterograde transport. This results in a GOF phenotype as evidenced by accumulation of Shh signaling machinery in cilia tips and expansion of Shh expression/activity in the ventral telencephalon and zona limitans intrathalamica (Tran et al., 2008; Stottmann et al., 2009). Rpgrip1l mutants have forebrain neuroepithelial cells that lack primary cilia and are deficient in production of processed Gli, as expected for Shh LOF (Besse et al., 2011). These mutations both result in expansion of the subpallium at the expense of the pallium, similar to Gli3 mutants, because the predominant output of Shh signaling in the ventral forebrain is processing of Gli3 to its repressor form (Stottmann et al., 2009; Besse et al., 2011). If active Gli3R protein is introduced genetically, then cilia are dispensable in this context (Besse et al., 2011). Defects in ventral forebrain patterning may underlie cognitive impairments in some ciliopathies.

Shh-related defects do not account for all phenotypes in human ciliopathies. It has become evident that IFT/cilia are used in some contexts to facilitate signaling by other pathways. During mammalian skin development, primary cilia are essential for terminal differentiation of suprabasal cells, a process initiated by Notch/Delta signaling (Figure 1F; Blanpain et al., 2006). In ~60–70% of suprabasal cells, Notch is concentrated in primary ciliary membranes, and presenilin, a protein in the γ-secretase complex which cleaves Notch, is localized at basal bodies (Ezratty et al., 2011). Knock down of IFT genes or elimination of cilia *in vivo* results in decreased Notch reporter activity and impaired suprabasal cell differentiation. Furthermore, NICD, the active cleaved form of Notch, was observed in the nuclei of ciliated but not unciliated cells in genetic chimeras (Ezratty et al., 2011).

Cilia have also been implicated in PDGF signaling. In cultured fibroblasts, PDGFRα has been observed in primary cilia membranes, and the RTK-activated signal transduction proteins MEK1/2 and Akt are localized along the cilium and at the ciliary base, respectively (Schneider et al., 2005). Studies of wild type and Ift88 mutant (no primary cilia) fibroblasts indicated that primary cilia of leading cells orient toward wound sites and are required for PDGF-AA-induced chemotaxis and phospho-Akt responses *in vitro* (Schneider et al., 2010). Notably, Ift88 mutant mice exhibit defects in wound repair (Schneider et al., 2010). PDGF signaling through PDGFRα has also been implicated in cilia disassembly and the G1-S phase transition (discussed in Christensen et al., 2012), though the *in vivo* significance of this observation is not yet known.

The role of cilia in canonical Wnt signaling has been controversial (reviewed in Wallingford and Mitchell, 2011). Studies of Ift88 and Ift72 mutant mice and zebrafish suggested that cilia are not essential for Wnt signaling during embryogenesis, and studies of Wnt signaling in Kif3a mutant mice yielded conflicting results (Huang and Schier, 2009; Ocbina et al., 2009; Lancaster et al., 2011b). In retrograde IFT mutants, one group reported no change in Wnt signaling, whereas another group demonstrated a decrease in Wnt activity in contexts that maintain cilia but an increase in Wnt signaling in contexts that lost cilia (Ocbina et al., 2009; Lancaster et al., 2011b). Studies of the ciliary/ciliopathy proteins Inversin, Chibby, and Jouberin support roles of cilia in canonical Wnt signaling in specific developmental contexts. Inversin, which is mutated in human patients with nephronophthisis type II, is expressed in vertebrate monocilia, interacts physically with Disheveled (Dsh), and disrupts canonical Wnt signaling downstream of Dsh *in vitro* (Watanabe et al., 2003; Simons et al., 2005). Inversin mutant mice exhibit renal and hair patterning phenotypes reminiscent of Wnt/Frz phenotypes, as well as left-right patterning phenotypes likely due to defects in node monocilia function (Watanabe et al., 2003; Simons et al., 2005). Chibby, a basal body protein required for ciliogenesis and basal body docking, physically interacts with β-catenin and negatively regulates canonical Wnt signaling in *Drosophila* embryos, mammalian lung epithelia, and cultured mammalian cells (Takemaru et al., 2003; Voronina et al., 2009; Love et al., 2010). Finally, the human ciliopathy protein Jouberin is required in mice for normal Wnt signaling and proliferation at the cerebellar midline. This likely contributes to the cerebellar hemisphere fusion defect observed in human Joubert syndrome patients and Jouberin deficient mice (Lancaster et al., 2011a). In ciliated cells exposed to Wnt, Jouberin is sequestered with β-catenin at basal bodies in an IFT-dependent manner. Jouberin facilitates nuclear translocation of β-catenin; its basal body sequestration hinders nuclear translocation of β-catenin and dampens canonical Wnt signaling responses (Lancaster et al., 2011b). Together, these findings suggest that cilia are not globally required for Wnt signaling during embryogenesis, but do impact Wnt signaling outcomes in select contexts.

It has been hypothesized that cilia provide a specialized sensory structure within which multiple types of signals are coordinated. An alternative model is that cilia exhibit pathway-specificity (as do cytonemes, discussed below) and thereby restrict cellular responsiveness and/or spatially segregate signaling responses. Pathway-specificity of ciliary signaling could be mediated by differential targeting of signaling proteins to cilia, and/or utilization of distinct subsets of IFT proteins or cargo-specific adaptors for axoneme transport (Boehlke et al., 2010; Mukhopadhyay et al., 2010; Christopher et al., 2012).

Cytonemes in *Drosophila* Signaling

Many, but not all, mechanisms of developmental cell signaling are evolutionarily conserved. Whereas primary cilia and IFT are essential for vertebrate Shh signaling, they are dispensable for Hh signaling in *Drosophila* (Ray et al., 1999; Han et al., 2003; Huangfu et al., 2003; Huangfu and Anderson, 2006). However, live imaging studies have revealed that long, narrow cytoplasmic extensions called cytonemes express signaling receptors and extend toward signaling centers in *Drosophila* (Ramirez-Weber and Kornberg, 1999; Hsiung et al., 2005; Roy et al., 2011). Interestingly, Hh- and Dpp [BMP]-oriented cytonemes in *Drosophila* wing discs are dependent on Fgf, similar to vertebrate cilia (Ramirez-Weber and Kornberg, 1999). In contrast to microtubule-based cilia, cytonemes are highly dynamic, actin-based structures, which generally extrude in a signal-dependent, polarized (e.g., from apical or basal face of epithelium) manner and orient toward signal sources in the wing disc, eye disc, and tracheal system. Interestingly, in the notum, where Dpp is required for cell survival but does not act as a morphogen, cytonemes are much shorter and are not polarized/directional (Hsiung et al., 2005). Vesicles containing fluorescent receptor fusion proteins have been observed in cytonemes, and Tkv (a *Drosophila* BMP/TGFβ type I receptor) has been observed in cytoneme-associated vesicles that move both anterogradely and retrogradely in an actin-dependent manner (Hsiung et al., 2005; Roy et al., 2011). It has been proposed that cytonemes sense and/or transport ligand across a field of cells and thus can facilitate long range cell-cell signaling across a morphogen gradient field or toward a chemoattractant. Importantly, signaling pathways appear to be spatially segregated into different cytoneme subtypes that differ in their ligand-specificities (Dpp, Bnl [Fgf], or Spitz [Egf]), physical properties (length, subcellular localization, polarity/orientation), and response dynamics (Hsiung et al., 2005; Roy et al., 2011). Cytonemes containing multiple types of receptor have not been observed, but it is possible that a single cell may extend different subtypes of cytonemes.

In addition to facilitating signals via secreted proteins, cytoneme-like extensions have been implicated in Notch/Delta (N/Dl) signaling. N/Dl signals (Figure 1F) mediate lateral inhibition in several contexts. Since Dl is a membrane-bound ligand, these signals are classically considered to be restricted to adjacent cells. However, adjacent cell-cell signaling does not lead to the observed N/Dl signaling outcome in mathematical models of notum bristle patterning (Cohen et al., 2010). In live imaging experiments, Cohen et al. observed dynamic, actin-based filopodia on the basal side of the notum that extend several cell diameters and sustain interactions with one another before N/Dl-induced cell state changes. N and Dl were also detected basally, and mathematical models incorporating filopodia/cytonemes to mediate N/Dl signaling between non-adjacent cells recapitulated *in vivo* observations (Cohen et al., 2010). Furthermore, genetic disruption of filopodia phenocopied N/Dl mutants with disrupted pattern refinement and excess bristle formation (Cohen et al., 2010). Together, these findings present a significant conceptual shift by suggesting that cytoneme/filopodial contacts enable membrane-bound ligands such as Dl to have a range of action that extends beyond immediately neighboring cells. Cohen et al. further suggested that filopodial contact-mediated signaling could introduce physical tension on the N/Dl interaction to enhance signaling output (Ahimou et al., 2004; Cohen et al., 2010).

Little is known about the genetic control of cytonemes and cytoneme dynamics, transport of signaling proteins within cytonemes, or the localization of signal transduction events downstream of cytoneme-mediated ligand-receptor contact. We await future studies to illuminate these areas, and to demonstrate definitively the functions of cytonemes in cell-cell signaling in *Drosophila*. Furthermore, high-resolution live imaging studies with membrane-expressed reporters are needed to determine whether cytonemes are employed in vertebrate cell-cell signaling.

1.5 FORMATION AND INTERPRETATION OF SIGNALING GRADIENTS

In numerous developmental contexts, secreted signaling proteins from a localized source are thought to form extracellular and/or intracellular signaling gradients that play integral roles in patterning and instructing cellular behaviors such as chemotaxis. We will focus our discussion of signaling gradients on morphogens, secreted proteins that induce different cellular responses at different distances from the source. However, the same principles and mechanisms may be used in other developmental contexts, e.g., to establish/maintain chemotactic signaling gradients and/or generate graded growth or patterning responses to secreted signals.

Morphogen Movement and Gradient Formation

In 1970, Francis Crick proposed a model of gradient formation employing diffusion and spatially uniform degradation (Crick, 1970). However, this model does not account for directionality of morphogen gradients, and *in vivo* studies have uncovered discrepancies between rates of morphogen diffusion and gradient formation. Some have suggested that modified diffusion models that incorporate feedback inhibitors and localized stabilization/degradation best explain experimental observations of gradient formation (Muller et al., 2013). Others argue in favor of a model in which active processes facilitate directed, graded spread of morphogens away from patterning centers (Erickson, 2011; Mii and Taira, 2011; Muller et al., 2012; 2013).

Different morphogens employ different strategies to traverse their active gradient fields *in vivo*. In the case of Dpp (*Drosophila* BMP), ligand diffusion and stability are notably affected by auxiliary factors including Short gastrulation (Sog), Twisted gastrulation (Tsg), and the protease Tolloid (Tld; reviewed in O'Connor et al., 2006). Tsg facilitates Sog/Dpp binding in a trimeric complex that enables Sog to keep Dpp inactive for extracellular transport across a tissue. At target sites, Tld cleaves Sog, releasing Dpp to act locally. Combining mathematical modeling with experimental genetics, Mizutani and colleagues demonstrated that a diffusion model incorporating the effects of these proteins can recapitulate the Dpp ligand gradient and non-synonymous Dpp activity (phospho-Smad, or pSmad) gradient observed in *Drosophila* embryos (Mizutani et al., 2005).

Feedback inhibitors enable responding cells to distinguish morphogen concentrations over a broad range, and can also expand morphogen distribution or range of action (Mii and Taira, 2011; Paulsen et al., 2011). Müller and colleagues combined zebrafish genetics, live imaging, and mathematical modeling to probe mechanisms underlying the effective range of action by TGFβ proteins in zebrafish embryos (Müller et al., 2012). Four TGFβ family members coordinately drive graded signaling responses during mesendoderm induction/patterning and left/right specification. The short range TGFβ proteins cyclops and squint activate signaling near the source, and are inhibited at more distal locations by long range feedback inhibitors Lefty1/2 (also TGFβ proteins). Photobleaching and live imaging experiments have demonstrated that TGFβ family members have pronounced differences in their diffusion constants ($D_{cyc} < D_{sqt} <<< D_{Ly}$) and more subtle differences in clearance rates (Müller et al., 2012). These physical properties of the closely related proteins likely help to determine their range of action: activators with lower diffusion constants accumulate near the signal source, while the inhibitors can disperse farther for long range action. Receptor binding affinities of the different TGFβ proteins and/or their interactions with other extracellular proteins may contribute to their effective diffusion rates and signaling ranges.

Internalization of extracellular signaling proteins can help to establish and maintain gradients by preventing morphogens from saturating the field of responsive tissue over time (Erickson, 2011). Following endocytosis, endosomes bearing receptors and/or ligands are sorted into different types of vesicles that are recycled to the plasma membrane or degraded via the lysosomal pathway. Cycles of endocytosis and re-secretion have been hypothesized to mediate morphogen movement in a "transcytosis" model of gradient formation. In support of this, Dpp colocalizes with endocytic vesicles and dynamin, a cytoplasmic protein that facilitates budding of endocytic vesicles, is required for Dpp movement in *Drosophila* imaginal discs (Teleman and Cohen, 2000; Kicheva et al., 2007).

Endocytosis and vesicular trafficking can also impact morphogen gradient interpretation. These processes can lead to internalization or lysosomal degradation of receptors, rendering the cell insensitive or less sensitive to extracellular signals. In some cases, ligand-receptor complexes are differentially competent to transduce signals at the plasma membrane or in endocytic vesicles, and so endocytosis and trafficking can affect responses to extracellular signals (discussed in Niehrs, 2012). Cbl is a cytoplasmic protein that can serve as an adaptor to activate dynamin-dependent endocytosis or as an E3 ligase for ubiquitylating proteins and/or sorting them to the lysosome for degradation. Cbl has been implicated in endocytosis and/or degradation of numerous cell signaling molecules, including ligands and receptors as well as cytoplasmic proteins such as the RTK-MAPK feedback inhibitor Sprouty. A comprehensive Cbl interactome map is presented in Schmidt and Dikic, 2005. In *Drosophila* egg chambers, Cbl and dynamin are essential for shaping the Gurken [EGF] morphogen gradient through endocytosis and lysosomal targeting of the EGFR (Chang et al., 2008). In zebrafish gastrulae, studies of dynamin and RAB5c, a GTPase required for endocytosis, revealed that endocytosis plays an important role in shaping the Fgf8 protein gradient emanating from the embryonic margin (Yu et al., 2009). In this context, Cbl mediates sorting of Fgf8–Fgfr complexes to degradative endosomal compartments. This does not affect the distribution of extracellular Fgf8 protein, but does restrict the range of Fgf8 action as assayed by target gene expression (Nowak et al., 2011). Intriguingly, this function of Cbl is dependent on distance from the Fgf8 source (Nowak et al., 2011).

There are seemingly contradictory reports on how endocytosis affects Wingless (Wg, or *Drosophila* Wnt) spread in *Drosophila* imaginal wing discs. Wg is secreted from a narrow band of cells at the dorsal/ventral compartment boundary of third instar wing discs (Figure 1.2), but it has been observed in vesicles up to ten cell diameters from producing cells and

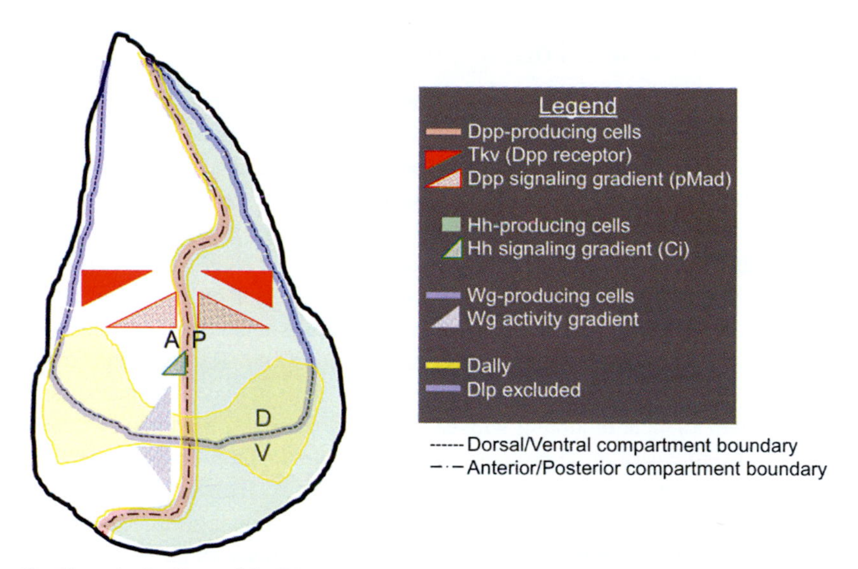

FIGURE 1.2 Morphogen Gradients in the *Drosophila* Wing Imaginal Disc (Third Instar). The *Drosophila* wing disc is subdivided by Anterior/Posterior (AP) and Dorsal/Ventral (DV) compartment boundaries, which serve as signaling centers for morphogenetic patterning of the wing disc epithelium. The AP boundary cells secrete Dpp, which acts at over a long range (~40 cells) to activate bi-directional pMad (Dpp activity) gradients emanating from the boundary. There is reduced pMad in cells at/near the boundary. The Dpp receptor type I, Tkv, is expressed in an opposing gradient to the pMad gradient, with highest expression near the margins. Wg (Wnt) is secreted by a band of cells 2–3 cells wide that straddle the DV boundary, and forms a bi-directional signaling gradient to regulate target gene expression. Wg has been observed in vesicles up to ten cell diameters from producing cells. Hh is produced by posterior compartment cells and in anterior cells very near to the AP boundary. However, Ptc is constitutively expressed in the anterior compartment, keeping the Hh signal "off" in these cells. Hh signals as a short range morphogen (range of ~10–15 cells wide) with a P>A gradient in anterior cells near the AP boundary. Expression patterns of the HSPGs Dally and Dlp are also shown. Dally is highly expressed in a narrow strip of cells at the AP boundary and in a broader zone around the DV boundary. Dlp is excluded (or expressed at low levels) in the Wg signaling zone at the DV boundary, but is expressed fairly uniformly elsewhere in the wing disc. Wg represses Dlp expression, and Dlp positively and negatively impacts low versus high [Wg] signaling, respectively. Dpp represses Dally expression, which is positively regulated by Wg and Hh.

induces expression of different genes at proximal, intermediate, and distal locations (Neumann and Cohen, 1997; Teleman and Cohen, 2000). In this context, Wg colocalizes with endocytic markers Rab5 and Rab7, but dynamin is not required for Wg spread (Marois et al., 2006; Kicheva et al., 2007). (In contrast, dynamin function does impact Wg signaling range in the *Drosophila* embryonic ectoderm; Bejsovec and Wieschaus, 1995.) These disparate findings can be explained by a model in which dynamin is not required for Wg movement across the epithelium but is required for Wg secretion, reuptake, or degradation (Strigini and Cohen, 2000). Wg can undergo receptor-dependent and -independent endocytosis (Marois et al., 2006). The receptors Frz/Frz2 negatively regulate levels of extracellular Wg, presumably by "capturing" Wg for endocytosis, and the Fz/Fz2 co-receptor Arrow targets Wg-Fz2 complexes for lysosomal degradation (Baeg et al., 2004; Han et al., 2005; Piddini et al., 2005; Marois et al., 2006). The lysosomal degradative pathway negatively regulates the range of intracellular (endosomal) Wg without affecting Wg phenotypic output or extracellular levels of Wg protein (Marois et al., 2006). An RNAi screen of protein localization, transport, and degradation genes identified UBPY (a ubiquitin-specific protease) as a positive modulator of canonical Wnt signaling. UBPY facilitates Fz de-ubiquitylation in a Wnt/Wg-independent manner and promotes receptor recycling of the receptor to the plasma membrane (Mukai et al., 2010). Surprisingly, endosome recycling to the plasma membrane is not required for intra- or extracellular Wg distribution (Marois et al., 2006). Together, these data support a model in which extracellular Wg distribution is regulated by Fz/Fz2 in an endocytosis-independent manner (e.g., by sequestration), but intracellular Wg distribution requires lysosomal degradation of internalized Wg. Importantly, Fz2 and Arrow transcription are negatively regulated by Wg signaling and are sensitive to disruption of endocytic trafficking (Marois et al., 2006). This feedback may confound results of trafficking experiments.

Drosophila studies have demonstrated that endocytic trafficking can regulate the subcellular localization of signaling responses. In *Drosophila* egg chambers, a cluster of border cells undergoes collective chemotactic migration toward the oocyte, which secretes PVF1 (a PDGF/VEGF ligand) and Gurken (EGF). Within the border cell cluster, PVR has a ligand-independent bias in localization toward the distal edges of front (nearest PVF1 source) and rear cells. Localized ligand secretion from the oocyte augments the front cell bias, so that there is more PVR in the leading edge of front cells than in distal edges of rear cells of the cluster (Janssens et al., 2010). Notably, there is also a very strong, dynamin-dependent bias of pTyr localization near the leading edge of front cells in the cluster, where phospho-PVR (pPVR, or active PVR) localizes with endosomes (Jekely et al., 2005; Janssens et al., 2010). Cbl negatively regulates PVR levels and is essential for polarized localization of PVR, pPVR, and pTyr in migrating border cells. This polarization is essential for chemotactic migration of the border cell cluster, is dependent on Cbl's E3 ubiquitin ligase activity, and based on Rab11 mutant studies is mediated in part by the endocytic recycling pathway (Jekely et al., 2005; Janssens et al., 2010). Based on these studies, Rørth and colleagues suggested a model in which Cbl-mediated endocytosis is required to sequester PVR (the *Drosophila* PDGF/VEGF receptor, an RTK) within select plasma membrane microdomains at the front of the border cell cluster for localized reception of the chemotactic signal.

Roles of Heparin Sulfate Proteoglycans (HSPGs) in Cell Signaling

The localization, stability, and mobility of extracellular signaling proteins can be notably affected by interactions with HSPGs. HSPGs are classified into transmembrane (TM), GPI-linked, and secreted ECM subtypes. HSPG production and modification are essential for Fgf-dependent development: global disruption of HSPG synthesis due to UDP-glucose dehydrogenase (Ugdh) mutations causes recessive phenotypes reminiscent of Fgf LOF mutants in mice and *Drosophila* (Lin et al., 1999; Garcia-Garcia and Anderson, 2003). In mice, disruption of the HS chain elongation factor Ext2 results in early extra-embryonic ectoderm defects, failure to gastrulate, and early embryonic lethality due to an inability of cells to respond to Fgf signals (Stickens et al., 2005; Shimokawa et al., 2011). Interestingly, mutations in Fgfr1 and Anosmin-1 (KAL1), a glycoprotein that modulates Fgf/HS complex formation, both cause Kallman's Syndrome in humans, characterized by impaired development of GnRH neurons and the olfactory bulb (Kim et al., 2008).

HSPGs have long been appreciated for their obligate role in forming signaling-competent HSPG-Fgf-Fgfr complexes (reviewed in Matsuo and Kimura-Yoshida, 2013). The cytoplasmic domain of syndecans (a subclass of TM HSPGs) can also promote internalization of Fgf/Fgfr complexes and enhance signaling (Matsuo and Kimura-Yoshida, 2013). In addition to these signaling functions, HSPGs are important for Fgf localization and/or movement. In zebrafish gastrula, a minor fraction (< 10%) of Fgf8 emanating from the embryonic margin is hindered in its diffusion by HSPGs upstream of receptor binding; this slow-moving fraction of Fgf8 significantly impacts the gradient's phenotypic output (Yu et al., 2009). In early stage chimeric mouse embryos, surface-tethered (TM, GPI-linked) HSPGs are essential for local retention of Fgfs but are also sensitive to proteinase/heparanase cleavage, which enables short range Fgf movement (Shimokawa et al., 2011). ECM HSPGs are also thought to restrict Fgf diffusion, although they too can be released from the ECM by heparanase. In *ex vivo* submandibular gland experiments, heparanase regulates Fgf-mediated branching morphogenesis, can free Fgf/Fgfr complexes from the ECM, and can cause dissolution of Fgf10/Fgfr complexes (Patel et al., 2007). Fgfs and Fgfrs have distinct affinities for different types of and differentially modified HSPGs (Ornitz, 2000; Mohammadi et al., 2005; Matsuo and Kimura-Yoshida, 2013). Interestingly, sulfation can reversibly switch HSPG activity/specificity: heavily sulfated HSPGs promote Fgf/Fgfr signaling, whereas desulfated HSPGs promote Wnt/Fz and down-regulate Fgf/Fgfr signaling (Matsuo and Kimura-Yoshida, 2013). Similarly, studies of HSPG modifications in *C. elegans* and mouse nervous system development have demonstrated that specific sulfations differentially affect and have context-specific effects on Fgf and Slit/Robo signaling (Bulow and Hobert, 2004; Conway et al., 2011).

Mechanisms of HSPG function have been studied extensively in *Drosophila* wing discs, where the glypicans Dally and Dally-like (Dlp) are required for proper localization of and responses to the Wg, Hh, and Dpp gradients (Figure 1.2; glypicans are a class of GPI-anchored HSPGs). Dally is expressed in the *Drosophila* wing disc in a zone at the dorsal/ventral (DV) boundary as well as in a stripe of Hh-responsive (Ci^+) cells at the anterior/posterior (AP) boundary, suggestive of its roles in supporting multiple morphogens (Han et al., 2005). Dlp, in contrast, is expressed uniformly in the wing disc except in the region near the Wg source, where it is expressed at low levels (Baeg et al., 2004; Kirkpatrick et al., 2004). Dally is positively regulated and Dlp is negatively regulated by Wg signaling, and the two factors differentially affect Wg protein distribution (Fujise et al., 2001; Han et al., 2005). Dlp is required for long range Wg signaling, and its transcription is sensitive to disruption of the endocytic pathway (Franch-Marro et al., 2005; Marois et al., 2006). Cell-based RNAi assays and clonal analyses *in vivo* demonstrated that Dlp positively regulates signaling by low [Wg] and negatively regulates signaling by high [Wg] (Baeg et al., 2004; Kirkpatrick et al., 2004). Models of Dlp function have been proposed in which Dlp stabilizes Wg in the extracellular matrix, restricts its mobility, facilitates Wg presentation among neighboring cells, and/or impacts Wg degradation/endocytosis (Franch-Marro et al., 2005).

Dally and Dlp have both been implicated in apical to basolateral trafficking and localized secretion of Hh in wing discs. To this end, the glypicans function together with Dispatched (Disp), a putative multipass transmembrane protein required in Hh-producing cells, and the Hh-binding transmembrane proteins Boi and Ihog [vertebrate homologs: CDO, BOC]. The Hh gradient forms on the basolateral face of the wing disc epithelium, but Hh protein is also observed apically. Apical Hh is recycled or transcytosed to the basolateral surface, where it is released for formation of the long range gradient (Callejo et al., 2011). Dally and Boi are essential for apical retention and internalization of Hh, and have been proposed to facilitate Hh recycling to the basolateral face of the epithelium: in endocytosis-defective mutants, Hh, Boi, and Dally accumulate apically (Bilioni et al., 2013). Dlp has been implicated in Hh release from the basolateral plane of the wing disc (Callejo et al., 2011). Here, Dlp may work in conjunction with Disp: Disp is concentrated at the basolateral membrane, colocalizes with Hh in vesicles proposed to traffic or "transcytose" from apical to basolateral membranes, and like Dlp has been implicated in basolateral Hh release (Callejo et al., 2011). Dally, Boi, and Ihog also impact expression of the secreted factor Wif, which is enriched basolaterally and is required for ECM stabilization and dispersion of Hh (Bilioni et al., 2013). Notably, Dally, Wif, Ihog, and Boi are all required for long range Hh signaling and are enriched in cytonemes. Ihog and Boi differentially localize to apical (Boi) versus basolateral (Ihog) cytoneme-like structures (Eugster et al., 2007; Ayers et al., 2010; Callejo et al., 2011; Bilioni et al., 2013). The mechanism of shuttling signaling proteins from the apical to basolateral surface of the wing disc epithelium is not restricted to Hh: Dlp has also been implicated in trafficking apically secreted and endocytosed Wg to the basolateral surface (Gallet et al., 2008).

In addition to these roles in Hh-secreting cells, Dally and Dlp facilitate Hh and Dpp signaling in responding cells. Dally and Dlp promote Hh-Ptc internalization, and there is evidence that this role of Dally is regulated cell-autonomously by Notum, an α,β-hydrolase first identified as a secreted feedback repressor of Wg (Giraldez et al., 2002; Gallet et al., 2008; Ayers et al., 2012). In the case of Dpp, Dally and Dlp are both required for ligand movement across the wing disc, and Dally is also required for Dpp signaling responses (Fujise et al., 2003; Belenkaya et al., 2004). Endocytosis is not required for Dpp movement but is required for Dpp signaling and impacts cell surface expression of its receptor, Tkv (Belenkaya et al., 2004). In contrast to its positive role in Hh-Ptc endocytosis, Dally has been proposed to stabilize Dpp on the cell surface by antagonizing Tkv (Akiyama et al., 2008). However, Dpp and Dally have also been observed in cytoplasmic vesicles, where they colocalize with Tkv and Pentagone (Pent), a secreted feedback inhibitor of Dpp required for long range Dpp movement (Vuilleumier et al., 2010). Vuilleumier et al. (2010) proposed a model in which Pent regulates the balance between two functions of Dally in mediating signaling (endocytosis) and facilitated transport (passing Dpp extracellularly between glypicans on neighboring cells). It is not yet known whether Pent regulates Dpp movement and/or signaling extracellularly or intracellularly in wing discs, but a vertebrate ortholog of Pent, Smoc-1, antagonizes BMP signaling downstream of the BMP receptor in *Xenopus* embryos (Thomas et al., 2009). Smoc-1 [Pent] has been implicated in mouse and human developmental syndromes characterized by eye and limb defects (Okada et al., 2011; Rainger et al., 2011).

Interpretation of Morphogen Gradients

The classical "French flag model" of morphogen gradient interpretation proposed that different threshold concentrations of morphogens induce different cell fate choices (Wolpert, 1969). This could be mediated by threshold-dependent signal transduction and/or transcriptional regulation downstream of the ligand-receptor interaction. In support of this model, Hh signaling in *Drosophila* wing discs induces sequential serine phosphorylation events on Smo that differentially affect low, intermediate, and high threshold target gene expression (Su et al., 2011). Phosphorylation by PKA is required to "stage" Smo for phosphorylation by CKI, which results in Smo membrane localization (Jia et al., 2004; Zhang et al., 2004; Su et al., 2011). PKA and CKI are both required for high threshold targets, PKA alone is required for intermediate threshold targets, and neither kinase is required for low threshold targets (Su et al., 2011). (Note that the cytoplasmic tail of Smo, PKA phosphorylation of Smo, and Hh-induced Smo membrane localization are not conserved in vertebrates; discussed in Tuson et al., 2011). Hh activation does not affect kinase levels or activity, but RNAi screens revealed that Hh signaling inhibits two phosphatases, PP1 and PP2A, which antagonize the first and second Smo phosphorylation events, respectively (Su et al., 2011). It is not yet clear how graded Hh levels are translated into differential regulation of these phosphatases and/or differential sensitivity of Smo phosphorylation sites to phosphatase activity. Su et al. speculated that differential phosphorylation of cell signaling proteins may be a mechanism used to generate other gradients of morphogen signaling activity (Su et al., 2011).

In many contexts, extracellular morphogen concentration does not directly dictate or correlate with the amplitude/dynamics of cell signaling responses and/or consequent cell fate choices, contrary to French flag model predictions. Mathematical modeling and phenotypic data have provided evidence that morphogen-activated gene regulatory networks (GRNs) play essential roles in translating graded morphogen signals into discrete domains of target gene expression. For example, a Hh-driven GRN in the developing *Drosophila* ocellar complex (simple eye) enables graded Hh signaling to give rise to two cell types. Initial Hh signals induce expression of engrailed (En) only in high [Hh] cells. These cells are only transiently responsive to Hh because En then represses Hh signaling and is maintained in a Hh-independent manner (Aguilar-Hidalgo et al., 2013). En is not induced in low [Hh] cells, which therefore remain responsive to Hh signaling over a longer duration (Aguilar-Hidalgo et al., 2013). Mathematical models of Hh in ocellar development suggest that a Hh-induced transcriptional repressor (En) which attenuates Hh signaling in one population of cells is important for the GRN's stability and robustness (Aguilar-Hidalgo et al., 2013). Interestingly, phenotypic data also implicate Notch in this GRN, through regulation of En in high [Hh] ocellar cells (Aguilar-Hidalgo et al., 2013).

In the vertebrate neural tube, Shh acts as a morphogen secreted by the notochord and floor plate that instructs distinct cell fates at ventral (high [Shh]), intermediate, and dorsal (low [Shh]) locations (Figure 1.3). Shh concentration does not predict the cellular signaling response (amplitude, duration) or cellular outcome during this process (Chamberlain et al., 2008; Balaskas et al., 2012). The Shh concentration gradient increases in amplitude even after the peak of Shh signaling dynamics, as measured by transcriptional responses (Balaskas et al., 2012). Furthermore, the absolute threshold level of

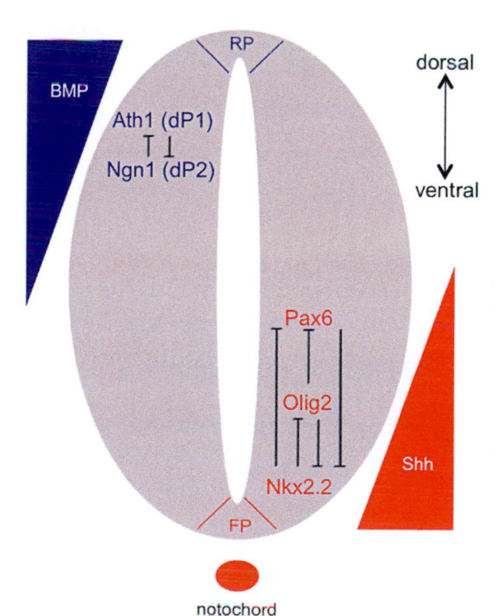

FIGURE 1.3 Gene Regulatory Networks (GRNs) in the Vertebrate Neural Tube. Shh from the notochord/floor plate and BMP from the roof plate drive opposing morphogenetic gradients that specify positional identity in the vertebrate neural tube. Positional identity transcription factors (TFs) form GRNs that integrate spatial and temporal signaling information. In the dorsal neural tube, Ngn1 is initially induced in a broad domain. Ath1 requires more sustained activation of BMP signaling and so is induced later. Mutual repression between Ath1 and Ngn1 as well as a late-onset negative feedback circuit help to refine positional identity TF expression domains: after Ath1 turns on dorsally, Ngn1 becomes restricted ventrally, so that the two TFs are appropriately localized to the dP1 and dP2 domains. Similarly, a GRN of Shh target genes including Nkx2.2, Olig2, and Pax6 orchestrates positional identity responses in the ventral neural tube. In this context, the Shh concentration and signaling gradients are non-synonymous: the [Shh] gradient continues to increase in amplitude even after the peak of Shh-activated transcription. In part, this can be explained by repression of Gli transcription by TFs that are induced by early Shh signals in the ventral neural tube. TFs in the Shh GRN differ in their requirements for signal amplitude and for transient versus sustained Shh signaling, and are activated through Gli binding sites with differential affinity for Gli.

Gli activity does not predict positional identity (Balaskas et al., 2012). In this context, Shh induces a GRN that includes positional identity transcription factors (TFs) Nkx2.2 (ventral), Olig2 (intermediate), and Pax6 (dorsal). Nkx2.2 represses Pax6 and Olig2, Olig2 and Pax6 repress Nkx2.2, and Olig2 represses Pax6 (Figure 1.3; Ericson et al., 1997; Briscoe et al., 1999; Briscoe et al., 2000; Novitch et al., 2001; Balaskas et al., 2012). Imaging studies revealed that a transient pulse of Hh response in the ventral neural tube initiates expression of ventral TFs, one of which represses Gli2 transcription (Peterson et al., 2012). Nkx2.2 requires transiently high Gli activity for the onset of expression, but has lower requirements for maintenance and requires more prolonged exposure to Shh than Olig2 (Balaskas et al., 2012). This GRN translates signal amplitude and temporal dynamics (i.e., duration of signal) into discrete domains of gene expression that are resistant to subtle fluctuations in overall Shh levels, and that are maintained even after Shh expression declines (Balaskas et al., 2012).

Peterson et al. conducted ChIP-seq and computational analyses of Gli binding sites (GBS) genome-wide, and determined that Shh differentially regulates target gene transcription through regulatory elements with distinct affinities for Gli (Peterson et al., 2012). The relative affinity of GBSs for Gli impacts the range of enhancer activation and facilitates ventral neural tube patterning *in vivo* (Peterson et al., 2012). This implies that the relative concentration of Shh required for target gene activation is in part hard-wired into genomic regulatory sequences. Differential affinity GBSs can affect temporal responses to Shh in addition to interpreting signal amplitude: a low affinity GBS is predicted to release the GliR more quickly upon Shh signal activation and favor rapid attenuation of GliA binding when Shh signals diminish (Peterson et al., 2012). In support of this, peak Shh signaling levels induce early and transient activation of Foxa2, a floor plate TF with low affinity GBS. Olig2, which is transcriptionally activated by Shh through de-repression of GliR, likewise has suboptimal GBS and is induced early. However, Nkx2.2 transcription is activated later through GliA and high affinity (optimal) GBS (Peterson et al., 2012). Interestingly, genome-wide analysis revealed that Gli target regions commonly have binding sites for other developmental transcription factors that may contribute to interpretation of Shh signals (Peterson et al., 2012).

In opposition to the ventral > dorsal Shh gradient, a dorsal > ventral gradient of morphogenetic BMP signaling emanates from the roof plate and patterns the dorsal neural tube (Figure 1.3). The BMP gradient activates a GRN to specify positional identity of interneuron subtypes including dP1 (dorsal) and dP2 (ventral; Figure 1.3). Ngn1, a marker of dP2 identity, is initially induced in a broad domain but becomes ventrally restricted as the dP1 marker Ath1 turns on dorsally (Tozer et al., 2013). Ath1 and Ngn1 repress one another (Gowan et al., 2001). Temporal dynamics – i.e., the duration of BMP exposure – play a critical role in determining BMP responses. Over time, a D > V gradient of BMP signaling response (pSmad) increases in amplitude and adopts a steeper slope, so that BMP signaling levels are higher and more sustained in dorsal dP1 populations than in ventral dP2 populations (Tozer et al., 2013). Positional identity specification requirements reflect these BMP activity gradient dynamics: dP2 progenitors respond rapidly to BMP and require only a transient signal for Ngn1 expression, whereas dP1 progenitors require sustained exposure to BMP for Ath1 induction. Reporter assays in explants revealed that during the first 12 h of BMP exposure, levels of BMP signaling levels progressively increase and do not require continuous exposure to ligand. Subsequently (18–24 h), BMP response in dP2 cells declines, dP1 progenitors begin to express Ath1, and high [BMP] triggers a late-onset negative feedback circuits that does not affect low [BMP] responses (Tozer et al., 2013). Wnt signals, which are also secreted from the roof plate, have been postulated to impact BMP signaling response duration via effects on stability of Smad proteins (Fuentealba et al., 2007; Tozer et al., 2013).

1.6 TRANSCRIPTIONAL REGULATION BY DEVELOPMENTAL CELL SIGNALING PATHWAYS

Transcriptional Responses to Cell Signaling

In most contexts, responses to cell-cell signaling are mediated at least in part by signal transduction pathways that culminate in transcriptional regulation of target genes. Often, target genes include feedback regulatory proteins that alter the responding cell's interactions with its environment by modulating signaling within and/or across pathways. For example, during *C. elegans* vulva induction, EGFR and Notch signaling induce transcription and/or activation of factors that establish reciprocal responsiveness to these pathways in neighboring cells. EGFR activation induces transcription-dependent internalization and degradation of Notch in the primary vulva cell. This enables Notch ligands to activate the receptor on neighboring cells, and so initiates lateral inhibition signaling (Shaye and Greenwald, 2002; 2005). Then, in secondary vulva cell precursors, Notch signaling induces the transcription of several MAPK pathway antagonists, thus inhibiting EGFR-Ras-MAPK signal transduction (Yoo et al., 2004). Transcription-mediated cross-regulation between pathways is commonly used in development, as evidenced by the finding that TGFβ/BMP and Wnt pathway genes were significantly enriched in results of a genome-wide analysis of Shh pathway (Gli) targets (Vokes et al., 2008).

Microarray and RNA-seq technologies allow genome-wide expression profiling to study signal-induced changes in transcription using cultured cells (+/– signal) or animal tissue (wild type versus signaling mutant). Transcriptional profiling results are dependent on the selection of informative tissue or cells. Cell culture systems enable assays of gene regulation at different time points in response to specific, acute stimuli, whereas *in vivo* tissues assay context-specific transcriptional changes in response to physiological signal levels. Fambrough and colleagues used microarrays to compare transcriptional responses downstream of different RTKs in cultured cells. PDGFRβ and Fgfr1 were found to induce the transcription of the same set of immediate early genes in this system (with some quantitative differences), whereas EGFR induced transcriptional responses that differed both qualitatively and quantitatively from these other RTKs (Fambrough et al., 1999). These findings, which may reflect the utilization of signal transduction pathways downstream of the different receptors, provide insight as to another mechanism underlying functional specificity within the RTK superfamily.

In a screen for Wnt target genes, comparative expression analyses were performed using gastrulation-stage wild type versus β-catenin mutant mouse embryos. In addition to known Wnt target genes, this study identified several novel targets, including components of other signaling pathways (e.g., Notch), and genes whose relevance to Wnt signaling was supported by their expression in domains of Wnt reporter activity during gastrulation. Some target genes identified in this study, including Grsf1 and Fragilis, were also functionally validated: embryos derived from RNAi knockdown embryonic stem cells recapitulated aspects of Wnt mutant phenotypes (Lickert et al., 2005). (Grsf1 and Fragilis encode RNA-binding and transmembrane proteins, respectively.) One caveat of this type of experiment is that it is difficult to discern direct targets of signaling pathways from transcriptional changes that are secondary to phenotypic changes.

Microarray/RNA-seq and ChIP-seq (described in the Glossary) complement one another in studies of transcriptional regulation by cell signaling. Microarray and RNA-seq identify genes whose expression is significantly affected by a cell signaling event. ChIP-seq (from cultured cells or live tissue) can then identify which of these genes are likely direct transcriptional targets of the signaling pathway on the basis of regulatory DNA (promoter/enhancer) association with signal-activated TFs. Transcriptional regulation by cell signaling can be evaluated in specific developmental contexts by using high fidelity amplification techniques to generate input material for microarray, RNA-seq, and ChIP-seq experiments from FACS-sorted cells or microdissected embryonic tissues. Results of these experiments can be used to develop new models of developmental and genetic pathways that explain how a given signal drives a specific developmental function.

Mechanisms of Transcriptional Regulation by Cell Signaling Pathways

Classical genetic screens and biochemical studies have elucidated basic mechanisms by which extracellular signals lead to transcriptional changes through "canonical" signal transduction pathways (Figure 1.1). For example, N/Dl signals induce cleavage of the N receptor, and the N intracellular domain (NICD) translocates into the nucleus where it functions with RPB-J proteins to recruit co-activator complexes and activate transcription. Wnt signaling leads to the stabilization of β-catenin, which translocates to the nucleus and activates TCF/LEF TFs. RTK signal transduction through Akt and MAPK pathways results in phosphorylation and activation of several TFs. These models, which provide a basic framework for understanding gene regulation by single pathways, have been significantly elaborated by studies investigating mechanisms of TF action and regulation, and by studies of coordinate transcriptional regulation by multiple signaling pathways. Insights from these studies help to explain how cell signaling differentially regulates target genes in distinct developmental contexts.

Proteomic approaches have provided novel insights into mechanisms of transcriptional regulation by Notch signaling. For example, an interactome study in T cell acute leukemia cells identified nuclear binding proteins of the transcriptionally active form of Notch (NICD). These experiments identified several "master regulator" type TFs, proteins suggestive of signaling crosstalk in the nucleus (e.g., Smad9, ERK2, ROCK1), as well as chromatin remodeling factors previously implicated in brain and palate development and left-right asymmetry (e.g., LSD1; Yatim et al., 2012). Follow-up studies indicated that NICD acts as a functional switch for the histone demethylase LSD1, changing it from a repressor (in the absence of Notch) to a co-activator (in the presence of Notch) and altering methylation site preferences (Yatim et al., 2012). Hence, Notch signaling can confer changes in chromatin accessibility to TF binding and can help recruit other TFs.

In many contexts, concurrent signaling pathways converge at the level of transcriptional regulation to coordinate cellular responses. For instance, expression of Prospero, a *Drosophila* TF, is coordinately regulated by TFs downstream of the EGFR and Sevenless RTKs, and Sox2 and Cdx3 are coordinately regulated by Wnt and Fgf signals in *Xenopus* (Xu et al., 2000; Haremaki et al., 2003; Takemoto et al., 2006). In *Drosophila* eyes, transcriptional co-regulation by EGFR and Notch signaling is mediated in part by opposing effects on the transcriptional co-repressor Groucho. EGFR-activated MAPK can phosphorylate the Notch (and Wnt) effector Groucho and thereby weaken its co-repressor activity. In this way, EGFR signaling de-represses transcription of Notch target genes (Hasson et al., 2005). Notch and EGFR also co-regulate Pax2 in cone cells of *Drosophila* eyes. In this case, TFs activated by the two pathways combinatorially regulate a Pax2 enhancer element, together with regional transcription factor, Lozenge (Flores et al., 2000). Notably, the organization of TF binding sites within this enhancer is important for determining its activity and cell type specificity (Swanson et al., 2010).

It is now well established that transcription-mediated crosstalk between pathways commonly occurs via combinatorial control of target gene enhancers. Indeed, a genome-wide screen *in silico* for transcription factor binding sites found significant overlap in putative targets of Tcf (Wnt) and Gli (Hh) transcriptional regulation (Hallikas et al., 2006). Curiously, genome-wide ChIP screens have demonstrated that several putative Gli/Hh targets – like the Engrailed2a (Eng2a) enhancer – are not associated with GBSs (Vokes et al., 2008). These targets may utilize non-canonical or low affinity GBSs not recognized in previous analyses, and/or they may recruit Gli to DNA indirectly through Gli-binding TFs. This latter mechanism was suggested by a recent study of Eng2a regulation in zebrafish myotome development. Eng2a is co-regulated by Hh and BMP signaling via coordinate enhancer regulation. Gli proteins (Hh effectors) induce Eng2a expression, but an Eng2a enhancer that recapitulates relevant myotome expression lacks canonical GBSs (Maurya et al., 2011). Maurya et al. employed a transregulation screen using a medaka cDNA library and identified Smad5 as a potent negative regulator of Eng2a expression. Gli2 and pSmad physically interact with the same region of the Eng2a enhancer, as demonstrated by ChIP-qPCR. Truncated Gli physically interacts with pSmad in the nucleus, and high levels of Hh signaling lead to depletion of nuclear pSmad (Maurya et al., 2011). The authors propose a model in which GliR is required for BMP-induced pSmad nuclear localization and Eng2a enhancer repression. In response to high Hh signaling (and Gli processing to GliA), pSmad is released from GliR and exits the nucleus, allowing GliA to activate the Eng2a enhancer (Maurya et al., 2011). This mechanism could be used in other developmental contexts in which BMP and Hh signals from nearby patterning centers/sources have opposing actions on target gene expression.

It has been postulated that deregulation of developmental enhancers contributes to human disease. This could explain why a significant proportion of human disease susceptibility loci identified in GWASs map to non-coding regions of the genome (Visel et al., 2009b). Putative enhancer elements can be identified computationally on the basis of evolutionary conservation, or biochemically through ChIP-chip/ChIP-seq experiments (e.g., that map epigenetic marks of active chromatin or DNA interactions with transcriptional co-activators). Visel and colleagues used both approaches in a genome-wide screen for developmental forebrain enhancers, which were then evaluated in transient transgenic reporter assays for their activity in mouse embryos (Visel et al., 2007; 2009a; 2013). This work generated a large resource for the scientific community: the enhancer sequences, genomic loci, and *in vivo* activity are publicly available on the VISTA browser (Table 1.3). They can be used as genetic tools with which to drive or study context-specific gene expression *in vivo*, and/or can be used in biochemical experiments to characterize transcriptional regulation downstream of cell-cell signaling events. Enhancers have been identified in intra- and intergenic regions at varying distances from transcriptional start sites, and so genetic (e.g., enhancer deletion *in vivo*) or biochemical (e.g., chromosome conformation capture, or CCC) experiments are needed to definitively identify transcriptional targets (Visel et al., 2009b).

1.7 TRANSCRIPTION-INDEPENDENT RESPONSES TO CELL SIGNALING

Although we have focused the majority of our discussion on transcription-mediated responses, cell signaling also induces many cytoplasmic and cytoskeletal responses that are essential for normal development. In many cases, transcription-independent responses utilize the same receptors but alternate cytoplasmic effectors, often classified as "non-canonical" pathways. Such pathways play integral roles in cell migration and in regulating cell polarity, adhesion, and ion flow. A prominent example of a non-canonical pathway is the Wnt-planar cell polarity (PCP) pathway, which is important for coordinating cell orientation and behavior within a tissue (reviewed in Gao, 2012). In vertebrates, a non-canonical Shh pathway that utilizes Smo but does not require cilia (or ciliary localization of Smo) facilitates chemoattraction toward Shh (Bijlsma et al., 2012). This Shh-migration pathway is genetically and pharmacologically separable from Shh-Gli signaling (Bijlsma et al., 2012). A third example of non-canonical cell signaling is the Notch/Abl pathway: during nervous system development, canonical Notch/Su(H) (transcription) signaling is essential for cell fate specification, whereas Notch regulation of Abl kinase in the cytoplasm is required for neuronal migration and axon patterning (reviewed in Giniger, 2012). The important question remains as to how signaling responses are diverted to transcription-mediated versus transcription-independent pathways downstream of receptor-ligand binding. Interestingly, in the case of Notch signaling, the two pathways can act simultaneously to coordinate nuclear and cytoplasmic responses to Notch activation (Giniger, 2012).

1.8 ROLES OF COMPUTATIONAL BIOLOGY IN DEVELOPMENTAL CELL SIGNALING STUDIES

Computational biology and mathematical modeling have become integral to developmental cell signaling studies. Bioinformatics and statistics are of paramount importance in gleaning significant data from the massive datasets generated by systems biology experiments. In addition, computational modeling enables researchers to predict the developmental consequences of multiple genetic changes with cross-regulatory effects. Hence, the integration of developmental signaling, genetics, and mathematical modeling has greatly facilitated studies of signal-activated GRNs. One challenge that remains in this arena is to expand the complexity of the models to incorporate increasing numbers of relevant factors including effects of genes, developmental time, and tissue growth on biological outcomes. As we deepen our understanding of biological processes, the algorithms used in mathematical modeling can be adapted to represent the *in vivo* signaling and regulatory networks more accurately.

Computational biology is also essential in integrating large datasets and thereby harnessing the power of available resources generated within the developmental biology and signaling communities. Vast amounts of data from the developmental biology community are now centralized in public databases and websites (Table 1.3). In addition, human genetics resources are available that can be used to cross-reference developmental findings with disease-association data. Computational integration of synexpression, biochemical, proteomic, and phenotypic data can notably expedite studies of developmental cell-cell signaling and significantly increase the discovery power of new datasets by helping to distinguish biologically relevant from artifactual data in systems biology experiments. Even without adding primary data from the laboratory, bioinformatic analyses of data from internet resources can generate testable models of signaling networks. For example, Zhong and Sternberg (2006) generated genome-wide predictions of functional interactions in *C. elegans* based on expression, phenotype, and physical and genetic interaction data from multiple model systems (Zhong and Sternberg, 2006). More recently, Lage et al. curated and integrated previously published phenotype and interactome data in the mouse to generate a testable model of spatio-temporal gene networks that underlie heart development (Lage et al., 2010). Several groups have developed algorithms that can be used to integrate expression data, protein interaction data, and loss-of-function data. There is an increasing need for an open-access resource that lists and annotates utilities of these algorithms so that researchers are aware of available tools.

Computational and experimental systems biology provide exciting complements to genetics and biochemistry that can greatly expand and accelerate studies of developmental cell signaling. The challenge now is to increase literacy in computational biology so that developmental biologists are empowered to take advantage of available resources and to interpret research articles that incorporate modeling and computational algorithms.

1.9 CLOSING REMARKS

In vivo, cells are commonly exposed to multiple concomitant cues. Thus, in order to fully understand cell-cell signaling, we need to consider how signaling pathways are integrated in comprehensive signaling networks within responding cells. We have highlighted some examples of mechanisms by which this integration occurs *in vivo*. Thanks to systems biology approaches and increasingly sophisticated and sensitive experimental methods, cell signaling models are becoming increasingly complex. This provides great power, e.g., to target very specific modules that can impact a particular pathway or process, but also poses the challenge of gleaning significant "big picture" results from large, complicated datasets. Furthermore, models generated in systems biology studies do not generally provide high level spatial information (e.g., localization of proteins within cells or tissues), which can be crucial in determining the significance of an interaction. Hence, additional genetic, biochemical, and imaging techniques are invaluable in identifying and interpreting the significance of data within network models.

We have highlighted only a small subset of developmental processes mediated by cell-cell signaling in order to showcase the diverse mechanisms used to localize, transduce, and interpret cell signals. We are constantly deepening our understanding of the intricate, interwoven molecular networks through which cell-cell signaling directs diverse, context-specific cellular responses during development. The increasing pace of research and new perspectives in cell signaling herald even more exciting times to come.

1.10 CLINICAL RELEVANCE: DEVELOPMENTAL CELL SIGNALING AND HUMAN DISEASE

Cell-cell signaling plays an integral role in the development, function, and homeostasis of many tissues and organ systems in multicellular organisms. Consequently, disruption of the major signaling pathways can lead to numerous diseases and developmental disorders. We have mentioned several human diseases associated with cell signaling mutations, and we refer readers to the reviews in Table 1.1 for in-depth discussions of disease associations for each pathway.

- Cell-cell signaling pathways are essential throughout development, and serve pleiotropic functions in different contexts. Deleterious mutations that completely abrogate the major developmental signaling pathways often result in embryonic lethality.
- Hypomorphic mutations (e.g., point mutations or single nucleotide polymorphisms (SNPs) altering expression level) in cell signaling genes disrupt the development of multiple organ systems and can thus result in pleiotropic phenotypes. In addition, some signaling mutations are associated with human diseases in which specific organ systems are affected. These include skeletal/bone disorders (Notch, RTK, BMP, Wnt), distal limb malformations (Shh), kidney disease (Notch, RTK), holoprosencephaly (Shh), inflammatory disease (RTK), cerebrovascular disease (Notch), diabetes (RTK, Wnt), and ciliopathies (Shh, Wnt).

- Many of the cell-cell signaling pathways used during development impact cell cycle, survival, and/or growth regulation, and consequently contribute to oncogenesis when de-regulated. Different pathways (Hh, Wnt, RTK, Notch, BMP/TGFβ) are associated with different types of human cancers, and some of these signaling pathways have been targeted for anti-cancer or anti-angiogenic therapies.
- By gaining a deeper understanding of mechanisms by which cell-cell signals are transduced and regulated in different *in vivo* contexts, we may gain insight into how cell signaling pathways can be selectively targeted for distinct therapeutic effects.

ACKNOWLEDGMENTS

We apologize to the many authors whose work we were unable to cite due to space limitations and the large scope of this chapter's subject area. We thank Bennett Penn, Todd Nystul, Grant Li, and Gabriel McKinsey for their comments on this manuscript. R.V.H. works in Dr. John L.R. Rubenstein's laboratory, supported by grant NS34661 (to J.L.R.R), and thanks J.L.R.R. for continued support during preparation of this chapter. Work in the laboratory of P.S. is supported by grants DE022363 and DE022778 from the National Institute of Dental and Craniofacial Research, and by NYSTEM grant N11G-131 from the New York State Health Department.

REFERENCES

Aguilar-Hidalgo, D., Dominguez-Cejudo, M.A., Amore, G., Brockmann, A., Lemos, M.C., Cordoba, A., Casares, F., 2013. A Hh-driven gene network controls specification, pattern and size of the *Drosophila* simple eyes. Development 140 (1), 82–92.

Ahimou, F., Mok, L.P., Bardot, B., Wesley, C., 2004. The adhesion force of Notch with Delta and the rate of Notch signaling. J. Cell Biol. 167 (6), 1217–1229.

Akiyama, T., Kamimura, K., Firkus, C., Takeo, S., Shimmi, O., Nakato, H., 2008. Dally regulates Dpp morphogen gradient formation by stabilizing Dpp on the cell surface. Dev. Biol. 313 (1), 408–419.

Amberger, J., Bocchini, C.A., Scott, A.F., Hamosh, A., 2009. McKusick's Online Mendelian Inheritance in Man (OMIM). Nucleic Acids Res. 37 (Database issue), D793–D796.

Andersson, E.R., Sandberg, R., Lendahl, U., 2011. Notch signaling: simplicity in design, versatility in function. Development 138 (17), 3593–3612.

Aoki, K., Kamioka, Y., Matsuda, M., 2013. Fluorescence resonance energy transfer imaging of cell signaling from *in vitro* to *in vivo*: Basis of biosensor construction, live imaging, and image processing. Dev. Growth Differ. 55 (4), 515–522.

Araujo, H., Fontenele, M.R., da Fonseca, R.N., 2011. Position matters: variability in the spatial pattern of BMP modulators generates functional diversity. Genesis 49 (9), 698–718.

Autiero, M., Waltenberger, J., Communi, D., Kranz, A., Moons, L., Lambrechts, D., Kroll, J., Plaisance, S., De Mol, M., Bono, F., et al., 2003. Role of PlGF in the intra- and intermolecular cross talk between the VEGF receptors Flt1 and Flk1. Nat. Med. 9 (7), 936–943.

Ayers, K.L., Gallet, A., Staccini-Lavenant, L., Therond, P.P., 2010. The long range activity of Hedgehog is regulated in the apical extracellular space by the glypican Dally and the hydrolase Notum. Dev. Cell 18 (4), 605–620.

Ayers, K.L., Mteirek, R., Cervantes, A., Lavenant-Staccini, L., Therond, P.P., Gallet, A., 2012. Dally and Notum regulate the switch between low and high level Hedgehog pathway signaling. Development 139 (17), 3168–3179.

Bacia, K., Schwille, P., 2007a. Fluorescence correlation spectroscopy. Methods. Mol. Biol. 398, 73–84.

Bacia, K., Schwille, P., 2007b. Practical guidelines for dual-color fluorescence cross-correlation spectroscopy. Nat. Protoc. 2 (11), 2842–2856.

Baeg, G.H., Selva, E.M., Goodman, R.M., Dasgupta, R., Perrimon, N., 2004. The Wingless morphogen gradient is established by the cooperative action of Frizzled and Heparan Sulfate Proteoglycan receptors. Dev. Biol. 276 (1), 89–100.

Baeg, G.H., Zhou, R., Perrimon, N., 2005. Genome-wide RNAi analysis of JAK/STAT signaling components in *Drosophila*. Genes Dev. 19 (16), 1861–1870.

Balaskas, N., Ribeiro, A., Panovska, J., Dessaud, E., Sasai, N., Page, K.M., Briscoe, J., Ribes, V., 2012. Gene regulatory logic for reading the Sonic Hedgehog signaling gradient in the vertebrate neural tube. Cell 148 (1–2), 273–284.

Baldessari, D., Shin, Y., Krebs, O., Konig, R., Koide, T., Vinayagam, A., Fenger, U., Mochii, M., Terasaka, C., Kitayama, A., et al., 2005. Global gene expression profiling and cluster analysis in *Xenopus laevis*. Mech. Dev. 122 (3), 441–475.

Bandyopadhyay, A., Yadav, P.S., Prashar, P., 2013. BMP signaling in development and diseases: a pharmacological perspective. Biochem. Pharmacol. 85 (7), 857–864.

Basu, B., Brueckner, M., 2009. Fibroblast "cilia growth" factor in the development of left-right asymmetry. Dev. Cell 16 (4), 489–490.

Bejsovec, A., Wieschaus, E., 1995. Signaling activities of the *Drosophila* wingless gene are separately mutable and appear to be transduced at the cell surface. Genetics 139 (1), 309–320.

Belenkaya, T.Y., Han, C., Yan, D., Opoka, R.J., Khodoun, M., Liu, H., Lin, X., 2004. *Drosophila* Dpp morphogen movement is independent of dynamin-mediated endocytosis but regulated by the glypican members of heparan sulfate proteoglycans. Cell 119 (2), 231–244.

Besse, L., Neti, M., Anselme, I., Gerhardt, C., Ruther, U., Laclef, C., Schneider-Maunoury, S., 2011. Primary cilia control telencephalic patterning and morphogenesis via Gli3 proteolytic processing. Development 138 (10), 2079–2088.

Bijlsma, M.F., Damhofer, H., Roelink, H., 2012. Hedgehog-stimulated chemotaxis is mediated by smoothened located outside the primary cilium. Sci. Signaling 5 (238), ra60.

Bilioni, A., Sanchez-Hernandez, D., Callejo, A., Gradilla, A.C., Ibanez, C., Mollica, E., Carmen Rodriguez-Navas, M., Simon, E., Guerrero, I., 2013. Balancing Hedgehog, a retention and release equilibrium given by Dally, Ihog, Boi and shifted/DmWif. Dev. Biol. 376 (2), 198–212.

Blanpain, C., Lowry, W.E., Pasolli, H.A., Fuchs, E., 2006. Canonical notch signaling functions as a commitment switch in the epidermal lineage. Genes Dev. 20 (21), 3022–3035.

Boehlke, C., Bashkurov, M., Buescher, A., Krick, T., John, A.K., Nitschke, R., Walz, G., Kuehn, E.W., 2010. Differential role of Rab proteins in ciliary trafficking: Rab23 regulates smoothened levels. J. Cell Sci. 123 (Pt 9), 1460–1467.

Borggrefe, T., Liefke, R., 2012. Fine-tuning of the intracellular canonical Notch signaling pathway. Cell Cycle 11 (2), 264–276.

Bowes, J.B., Snyder, K.A., Segerdell, E., Gibb, R., Jarabek, C., Noumen, E., Pollet, N., Vize, P.D., 2008. Xenbase: a *Xenopus* biology and genomics resource. Nucleic Acids Res. 36 (Database issue), D761–D767.

Bradford, Y., Conlin, T., Dunn, N., Fashena, D., Frazer, K., Howe, D.G., Knight, J., Mani, P., Martin, R., Moxon, S.A., et al., 2011. ZFIN: enhancements and updates to the Zebrafish Model Organism Database. Nucleic Acids Res. 39 (Database issue), D822–D829.

Brinkmeier, M., Dorre, K., Stephan, J., Eigen, M., 1999. Two-beam cross-correlation: a method to characterize transport phenomena in micrometer-sized structures. Anal. Chem. 71 (3), 609–616.

Briscoe, J., Pierani, A., Jessell, T.M., Ericson, J., 2000. A homeodomain protein code specifies progenitor cell identity and neuronal fate in the ventral neural tube. Cell 101 (4), 435–445.

Briscoe, J., Sussel, L., Serup, P., Hartigan-OConnor, D., Jessell, T.M., Rubenstein, J.L., Ericson, J., 1999. Homeobox gene Nkx2.2 and specification of neuronal identity by graded Sonic hedgehog signaling. Nature 398 (6728), 622–627.

Buechling, T., Boutros, M., 2011. Wnt signaling: signaling at and above the receptor level. Curr. Top. Dev. Biol. 97, 21–53.

Bulow, H.E., Hobert, O., 2004. Differential sulfations and epimerization define heparan sulfate specificity in nervous system development. Neuron 41 (5), 723–736.

Bush, J.O., Soriano, P., 2010. Ephrin-B1 forward signaling regulates craniofacial morphogenesis by controlling cell proliferation across Eph-ephrin boundaries. Genes Dev. 24 (18), 2068–2080.

Callejo, A., Bilioni, A., Mollica, E., Gorfinkiel, N., Andres, G., Ibanez, C., Torroja, C., Doglio, L., Sierra, J., Guerrero, I., 2011. Dispatched mediates Hedgehog basolateral release to form the long range morphogenetic gradient in the *Drosophila* wing disk epithelium. Proc. Natl. Acad. Sci. USA 108 (31), 12591–12598.

Camps, M., Chabert, C., Muda, M., Boschert, U., Gillieron, C., Arkinstall, S., 1998. Induction of the mitogen-activated protein kinase phosphatase MKP3 by nerve growth factor in differentiating PC12. FEBS Lett. 425 (2), 271–276.

Caron, A., Xu, X., Lin, X., 2012. Wnt/beta-catenin signaling directly regulates Foxj1 expression and ciliogenesis in zebrafish Kupffers vesicle. Development 139 (3), 514–524.

Casaletto, J.B., McClatchey, A.I., 2012. Spatial regulation of receptor tyrosine kinases in development and cancer. Nat. Rev. Cancer 12 (6), 387–400.

Chamberlain, C.E., Jeong, J., Guo, C., Allen, B.L., McMahon, A.P., 2008. Notochord-derived Shh concentrates in close association with the apically positioned basal body in neural target cells and forms a dynamic gradient during neural patterning. Development 135 (6), 1097–1106.

Chang, W.L., Liou, W., Pen, H.C., Chou, H.Y., Chang, Y.W., Li, W.H., Chiang, W., Pai, L.M., 2008. The gradient of Gurken, a long range morphogen, is directly regulated by Cbl-mediated endocytosis. Development 135 (11), 1923–1933.

Chen, M.H., Wilson, C.W., Li, Y.J., Law, K.K., Lu, C.S., Gacayan, R., Zhang, X., Hui, C.C., Chuang, P.T., 2009. Cilium-independent regulation of Gli protein function by Sufu in Hedgehog signaling is evolutionarily conserved. Genes Dev. 23 (16), 1910–1928.

Chen, Y., Jiang, J., 2013. Decoding the phosphorylation code in Hedgehog signal transduction. Cell Res. 23 (2), 186–200.

Christensen, S.T., Clement, C.A., Satir, P., Pedersen, L.B., 2012. Primary cilia and coordination of receptor tyrosine kinase (RTK) signaling. J. Pathol. 226 (2), 172–184.

Christopher, K.J., Wang, B., Kong, Y., Weatherbee, S.D., 2012. Forward genetics uncovers Transmembrane protein 107 as a novel factor required for ciliogenesis and Sonic hedgehog signaling. Dev. Biol. 368 (2), 382–392.

Clevers, H., Nusse, R., 2012. Wnt/beta-catenin signaling and disease. Cell 149 (6), 1192–1205.

Cohen, M., Georgiou, M., Stevenson, N.L., Miodownik, M., Baum, B., 2010. Dynamic filopodia transmit intermittent Delta-Notch signaling to drive pattern refinement during lateral inhibition. Dev. Cell 19 (1), 78–89.

Collins, R.T., Cohen, S.M., 2005. A genetic screen in *Drosophila* for identifying novel components of the hedgehog signaling pathway. Genetics 170 (1), 173–184.

Conway, C.D., Howe, K.M., Nettleton, N.K., Price, D.J., Mason, J.O., Pratt, T., 2011. Heparan sulfate sugar modifications mediate the functions of slits and other factors needed for mouse forebrain commissure development. J. Neurosci. 31 (6), 1955–1970.

Crick, F., 1970. Diffusion in embryogenesis. Nature 225 (5231), 420–422.

DasGupta, R., Kaykas, A., Moon, R.T., Perrimon, N., 2005. Functional genomic analysis of the Wnt-wingless signaling pathway. Science 308 (5723), 826–833.

David, D.J., McGill, M.A., McKinley, R.F., Harris, T.J., 2012. Live imaging of *Drosophila* embryos: quantifying protein numbers and dynamics at subcellular locations. Methods Mol. Biol. 839, 1–17.

Davies, J.A., Little, M.H., Aronow, B., Armstrong, J., Brennan, J., Lloyd-MacGilp, S., Armit, C., Harding, S., Piu, X., Roochun, Y., et al., 2012. Access and use of the GUDMAP database of genitourinary development. Methods Mol. Biol. 886, 185–201.

Dertinger, T., Pacheco, V., von der Hocht, I., Hartmann, R., Gregor, I., Enderlein, J., 2007. Two-focus fluorescence correlation spectroscopy: a new tool for accurate and absolute diffusion measurements. Chemphyschem 8 (3), 433–443.

Diez-Roux, G., Banfi, S., Sultan, M., Geffers, L., Anand, S., Rozado, D., Magen, A., Canidio, E., Pagani, M., Peluso, I., et al., 2011. A high-resolution anatomical atlas of the transcriptome in the mouse embryo. PLoS Biol. 9 (1), e1000582.

Dossenbach, C., Rock, S., Affolter, M., 2001. Specificity of FGF signaling in cell migration in *Drosophila*. Development 128 (22), 4563–4572.

Eggenschwiler, J.T., Espinoza, E., Anderson, K.V., 2001. Rab23 is an essential negative regulator of the mouse Sonic hedgehog signaling pathway. Nature 412 (6843), 194–198.

Erickson, J.L., 2011. Formation and maintenance of morphogen gradients: an essential role for the endomembrane system in *Drosophila melanogaster* wing development. Fly 5 (3), 266–271.

Ericson, J., Rashbass, P., Schedl, A., Brenner-Morton, S., Kawakami, A., van Heyningen, V., Jessell, T.M., Briscoe, J., 1997. Pax6 controls progenitor cell identity and neuronal fate in response to graded Shh signaling. Cell 90 (1), 169–180.

Eugster, C., Panakova, D., Mahmoud, A., Eaton, S., 2007. Lipoprotein-heparan sulfate interactions in the Hh pathway. Dev. Cell 13 (1), 57–71.

Evangelista, M., Lim, T.Y., Lee, J., Parker, L., Ashique, A., Peterson, A.S., Ye, W., Davis, D.P., de Sauvage, F.J., 2008. Kinome siRNA screen identifies regulators of ciliogenesis and hedgehog signal transduction. Sci. Signaling 1 (39), ra7.

Ezratty, E.J., Stokes, N., Chai, S., Shah, A.S., Williams, S.E., Fuchs, E., 2011. A role for the primary cilium in Notch signaling and epidermal differentiation during skin development. Cell 145 (7), 1129–1141.

Fambrough, D., McClure, K., Kazlauskas, A., Lander, E.S., 1999. Diverse signaling pathways activated by growth factor receptors induce broadly overlapping, rather than independent, sets of genes. Cell 97 (6), 727–741.

Finger, J.H., Smith, C.M., Hayamizu, T.F., McCright, I.J., Eppig, J.T., Kadin, J.A., Richardson, J.E., Ringwald, M., 2011. The mouse Gene Expression Database (GXD): 2011 update. Nucleic Acids Res. 39 (Database issue), D835–D841.

Fire, A., Xu, S., Montgomery, M.K., Kostas, S.A., Driver, S.E., Mello, C.C., 1998. Potent and specific genetic interference by double-stranded RNA in *Caenorhabditis elegans*. Nature 391 (6669), 806–811.

Flores, G.V., Duan, H., Yan, H., Nagaraj, R., Fu, W., Zou, Y., Noll, M., Banerjee, U., 2000. Combinatorial signaling in the specification of unique cell fates. Cell 103 (1), 75–85.

Forbes, S., Clements, J., Dawson, E., Bamford, S., Webb, T., Dogan, A., Flanagan, A., Teague, J., Wooster, R., Futreal, P.A., et al., 2006. Cosmic 2005. Br. J. Cancer 94 (2), 318–322.

Forbes, S.A., Tang, G., Bindal, N., Bamford, S., Dawson, E., Cole, C., Kok, C.Y., Jia, M., Ewing, R., Menzies, A., et al., 2010. COSMIC (the Catalogue of Somatic Mutations in Cancer): a resource to investigate acquired mutations in human cancer. Nucleic Acids Res. 38 (Database issue), D652–D657.

Formstecher, E., Aresta, S., Collura, V., Hamburger, A., Meil, A., Trehin, A., Reverdy, C., Betin, V., Maire, S., Brun, C., et al., 2005. Protein interaction mapping: a *Drosophila* case study. Genome Res. 15 (3), 376–384.

Franch-Marro, X., Marchand, O., Piddini, E., Ricardo, S., Alexandre, C., Vincent, J.P., 2005. Glypicans shunt the Wingless signal between local signaling and further transport. Development 132 (4), 659–666.

Friedman, A., Perrimon, N., 2006. A functional RNAi screen for regulators of receptor tyrosine kinase and ERK signaling. Nature 444 (7116), 230–234.

Friedman, A.A., Tucker, G., Singh, R., Yan, D., Vinayagam, A., Hu, Y., Binari, R., Hong, P., Sun, X., Porto, M., et al., 2011. Proteomic and functional genomic landscape of receptor tyrosine kinase and ras to extracellular signal-regulated kinase signaling. Sci. Signaling 4 (196), rs10.

Fuentealba, L.C., Eivers, E., Ikeda, A., Hurtado, C., Kuroda, H., Pera, E.M., De Robertis, E.M., 2007. Integrating patterning signals: Wnt/GSK3 regulates the duration of the BMP/Smad1 signal. Cell 131 (5), 980–993.

Fujise, M., Izumi, S., Selleck, S.B., Nakato, H., 2001. Regulation of dally, an integral membrane proteoglycan, and its function during adult sensory organ formation of *Drosophila*. Dev. Biol. 235 (2), 433–448.

Fujise, M., Takeo, S., Kamimura, K., Matsuo, T., Aigaki, T., Izumi, S., Nakato, H., 2003. Dally regulates Dpp morphogen gradient formation in the *Drosophila* wing. Development 130 (8), 1515–1522.

Furriols, M., Casanova, J., 2003. In and out of Torso RTK signaling. EMBO J. 22 (9), 1947–1952.

Furthauer, M., Lin, W., Ang, S.L., Thisse, B., Thisse, C., 2002. Sef is a feedback-induced antagonist of Ras/MAPK-mediated FGF signaling. Nat. Cell Biol. 4 (2), 170–174.

Gallet, A., Staccini-Lavenant, L., Therond, P.P., 2008. Cellular trafficking of the glypican Dally-like is required for full-strength Hedgehog signaling and wingless transcytosis. Dev. Cell 14 (5), 712–725.

Gao, B., 2012. Wnt regulation of planar cell polarity (PCP). Curr. Top. Dev. Biol. 101, 263–295.

Garcia-Garcia, M.J., Anderson, K.V., 2003. Essential role of glycosaminoglycans in Fgf signaling during mouse gastrulation. Cell 114 (6), 727–737.

Gavin, A.C., Bosche, M., Krause, R., Grandi, P., Marzioch, M., Bauer, A., Schultz, J., Rick, J.M., Michon, A.M., Cruciat, C.M., et al., 2002. Functional organization of the yeast proteome by systematic analysis of protein complexes. Nature 415 (6868), 141–147.

Gayko, U., Cleghon, V., Copeland, T., Morrison, D.K., Perrimon, N., 1999. Synergistic activities of multiple phosphotyrosine residues mediate full signaling from the *Drosophila* Torso receptor tyrosine kinase. Proc. Natl. Acad. Sci. USA 96 (2), 523–528.

Geffers, L., Herrmann, B., Eichele, G., 2012. Web-based digital gene expression atlases for the mouse. Mamm. Genome 23 (9–10), 525–538.

Gell, D.A., Grant, R.P., Mackay, J.P., 2012. The detection and quantitation of protein oligomerization. Adv. Exp. Med. Biol. 747, 19–41.

Giniger, E., 2012. Notch signaling and neural connectivity. Curr. Opin. Genet. Dev. 22 (4), 339–346.

Giot, L., Bader, J.S., Brouwer, C., Chaudhuri, A., Kuang, B., Li, Y., Hao, Y.L., Ooi, C.E., Godwin, B., Vitols, E., et al., 2003. A protein interaction map of *Drosophila melanogaster*. Science 302 (5651), 1727–1736.

Giraldez, A.J., Copley, R.R., Cohen, S.M., 2002. HSPG modification by the secreted enzyme Notum shapes the Wingless morphogen gradient. Dev. Cell 2 (5), 667–676.

Goetz, S.C., Anderson, K.V., 2010. The primary cilium: a signaling center during vertebrate development. Nat. Rev. Genetics 11 (5), 331–344.

Gomez, A.R., Lopez-Varea, A., Molnar, C., de la Calle-Mustienes, E., Ruiz-Gomez, M., Gomez-Skarmeta, J.L., de Celis, J.F., 2005. Conserved cross-interactions in *Drosophila* and *Xenopus* between Ras/MAPK signaling and the dual-specificity phosphatase MKP3. Dev. Dyn. 232 (3), 695–708.

Gong, S., Zheng, C., Doughty, M.L., Losos, K., Didkovsky, N., Schambra, U.B., Nowak, N.J., Joyner, A., Leblanc, G., Hatten, M.E., et al., 2003. A gene expression atlas of the central nervous system based on bacterial artificial chromosomes. Nature 425 (6961), 917–925.

Gowan, K., Helms, A.W., Hunsaker, T.L., Collisson, T., Ebert, P.J., Odom, R., Johnson, J.E., 2001. Crossinhibitory activities of Ngn1 and Math1 allow specification of distinct dorsal interneurons. Neuron 31 (2), 219–232.

Gurskaya, N.G., Verkhusha, V.V., Shcheglov, A.S., Staroverov, D.B., Chepurnykh, T.V., Fradkov, A.F., Lukyanov, S., Lukyanov, K.A., 2006. Engineering of a monomeric green-to-red photoactivatable fluorescent protein induced by blue light. Nat. Biotechnol. 24 (4), 461–465.

Hallikas, O., Palin, K., Sinjushina, N., Rautiainen, R., Partanen, J., Ukkonen, E., Taipale, J., 2006. Genome-wide prediction of mammalian enhancers based on analysis of transcription factor binding affinity. Cell 124 (1), 47–59.

Hamilton, T.G., Klinghoffer, R.A., Corrin, P.D., Soriano, P., 2003. Evolutionary divergence of platelet-derived growth factor alpha receptor signaling mechanisms. Mol. Cell Biol. 23 (11), 4013–4025.

Han, C., Yan, D., Belenkaya, T.Y., Lin, X., 2005. *Drosophila* glypicans Dally and Dally-like shape the extracellular Wingless morphogen gradient in the wing disc. Development 132 (4), 667–679.

Han, Y.G., Kim, H.J., Dlugosz, A.A., Ellison, D.W., Gilbertson, R.J., Alvarez-Buylla, A., 2009. Dual and opposing roles of primary cilia in medulloblastoma development. Nat. Med. 15 (9), 1062–1065.

Han, Y.G., Kwok, B.H., Kernan, M.J., 2003. Intraflagellar transport is required in *Drosophila* to differentiate sensory cilia but not sperm. Curr. Biol. 13 (19), 1679–1686.

Harding, S.D., Armit, C., Armstrong, J., Brennan, J., Cheng, Y., Haggarty, B., Houghton, D., Lloyd-MacGilp, S., Pi, X., Roochun, Y., et al., 2011. The GUDMAP database – an online resource for genitourinary research. Development 138 (13), 2845–2853.

Haremaki, T., Tanaka, Y., Hongo, I., Yuge, M., Okamoto, H., 2003. Integration of multiple signal transducing pathways on Fgf response elements of the *Xenopus* caudal homologue Xcad3. Development 130 (20), 4907–4917.

Hasson, P., Egoz, N., Winkler, C., Volohonsky, G., Jia, S., Dinur, T., Volk, T., Courey, A.J., Paroush, Z., 2005. EGFR signaling attenuates Groucho-dependent repression to antagonize Notch transcriptional output. Nat. Genet. 37 (1), 101–105.

Haudry, Y., Berube, H., Letunic, I., Weeber, P.D., Gagneur, J., Girardot, C., Kapushesky, M., Arendt, D., Bork, P., Brazma, A., et al., 2008. 4DXpress: a database for cross-species expression pattern comparisons. Nucleic Acids Res. 36 (Database issue), D847–D853.

Hildebrandt, F., Benzing, T., Katsanis, N., 2011. Ciliopathies. N. Engl. J. Med. 364 (16), 1533–1543.

Hoch, R.V., 2005. Distinctive mechanisms of receptor tyrosine kinase signal transduction required during mouse embryogenesis *Molecular and cellular biology*. vol. Ph.D. University of Washington, Seattle, WA.

Hochheiser, H., Aronow, B.J., Artinger, K., Beaty, T.H., Brinkley, J.F., Chai, Y., Clouthier, D., Cunningham, M.L., Dixon, M., Donahue, L.R., et al., 2011. The FaceBase Consortium: a comprehensive program to facilitate craniofacial research. Dev. Biol. 355 (2), 175–182.

Hsiung, F., Ramirez-Weber, F.A., Iwaki, D.D., Kornberg, T.B., 2005. Dependence of *Drosophila* wing imaginal disc cytonemes on Decapentaplegic. Nature 437 (7058), 560–563.

Huang, P., Schier, A.F., 2009. Dampened Hedgehog signaling but normal Wnt signaling in zebrafish without cilia. Development 136 (18), 3089–3098.

Huangfu, D., Anderson, K.V., 2005. Cilia and Hedgehog responsiveness in the mouse. Proc. Natl. Acad. Sci. USA 102 (32), 11325–11330.

Huangfu, D., Anderson, K.V., 2006. Signaling from Smo to Ci/Gli: conservation and divergence of Hedgehog pathways from Drosophila to vertebrates. Development 133 (1), 3–14.

Huangfu, D., Liu, A., Rakeman, A.S., Murcia, N.S., Niswander, L., Anderson, K.V., 2003. Hedgehog signaling in the mouse requires intraflagellar transport proteins. Nature 426 (6962), 83–87.

Hui, C.C., Angers, S., 2011. Gli proteins in development and disease. Annu. Rev. Cell Dev. Biol. 27, 513–537.

Humke, E.W., Dorn, K.V., Milenkovic, L., Scott, M.P., Rohatgi, R., 2010. The output of Hedgehog signaling is controlled by the dynamic association between Suppressor of Fused and the Gli proteins. Genes Dev. 24 (7), 670–682.

Jacob, L.S., Wu, X., Dodge, M.E., Fan, C.W., Kulak, O., Chen, B., Tang, W., Wang, B., Amatruda, J.F., Lum, L., 2011. Genome-wide RNAi screen reveals disease-associated genes that are common to Hedgehog and Wnt signaling. Sci. Signaling 4 (157), ra4.

Janssens, K., Sung, H.H., Rorth, P., 2010. Direct detection of guidance receptor activity during border cell migration. Proc. Natl. Acad. Sci. USA 107 (16), 7323–7328.

Javelaud, D., Pierrat, M.J., Mauviel, A., 2012. Crosstalk between TGF-beta and hedgehog signaling in cancer. FEBS Lett. 586 (14), 2016–2025.

Jekely, G., Sung, H.H., Luque, C.M., Rorth, P., 2005. Regulators of endocytosis maintain localized receptor tyrosine kinase signaling in guided migration. Dev. Cell 9 (2), 197–207.

Jia, J., Tong, C., Wang, B., Luo, L., Jiang, J., 2004. Hedgehog signaling activity of Smoothened requires phosphorylation by protein kinase A and casein kinase I. Nature 432 (7020), 1045–1050.

Jones, A.R., Overly, C.C., Sunkin, S.M., 2009. The Allen Brain Atlas: 5 years and beyond. Nat. Rev. Neurosci. 10 (11), 821–828.

Jorgensen, C., Sherman, A., Chen, G.I., Pasculescu, A., Poliakov, A., Hsiung, M., Larsen, B., Wilkinson, D.G., Linding, R., Pawson, T., 2009. Cell-specific information processing in segregating populations of Eph receptor ephrin-expressing cells. Science 326 (5959), 1502–1509.

Kandachar, V., Roegiers, F., 2012. Endocytosis and control of Notch signaling. Curr. Opin. Cell Biol. 24 (4), 534–540.

Kawakami, Y., Rodriguez-Leon, J., Koth, C.M., Buscher, D., Itoh, T., Raya, A., Ng, J.K., Esteban, C.R., Takahashi, S., Henrique, D., et al., 2003. MKP3 mediates the cellular response to FGF8 signaling in the vertebrate limb. Nat. Cell Biol. 5 (6), 513–519.

Kicheva, A., Pantazis, P., Bollenbach, T., Kalaidzidis, Y., Bittig, T., Julicher, F., Gonzalez-Gaitan, M., 2007. Kinetics of morphogen gradient formation. Science 315 (5811), 521–525.

Kim, M., Cha, G.H., Kim, S., Lee, J.H., Park, J., Koh, H., Choi, K.Y., Chung, J., 2004. MKP-3 has essential roles as a negative regulator of the Ras/mitogen-activated protein kinase pathway during *Drosophila* development. Mol. Cell Biol. 24 (2), 573–583.

Kim, S.H., Hu, Y., Cadman, S., Bouloux, P., 2008. Diversity in fibroblast growth factor receptor 1 regulation: learning from the investigation of Kallmann syndrome. J. Neuroendocrinol. 20 (2), 141–163.

Kim, W., Kim, M., Jho, E.H., 2013. Wnt/beta-catenin signaling: from plasma membrane to nucleus. Biochem. J. 450 (1), 9–21.

Kirkpatrick, C.A., Dimitroff, B.D., Rawson, J.M., Selleck, S.B., 2004. Spatial regulation of Wingless morphogen distribution and signaling by Dally-like protein. Dev. Cell 7 (4), 513–523.

Klinghoffer, R.A., Hamilton, T.G., Hoch, R., Soriano, P., 2002. An allelic series at the PDGFalphaR locus indicates unequal contributions of distinct signaling pathways during development. Dev. Cell 2 (1), 103–113.

Kovalenko, D., Yang, X., Nadeau, R.J., Harkins, L.K., Friesel, R., 2003. Sef inhibits fibroblast growth factor signaling by inhibiting FGFR1 tyrosine phosphorylation and subsequent ERK activation. J. Biol. Chem. 278 (16), 14087–14091.

Kramer, S., Okabe, M., Hacohen, N., Krasnow, M.A., Hiromi, Y., 1999. Sprouty: a common antagonist of FGF and EGF signaling pathways in *Drosophila*. Development 126 (11), 2515–2525.

Kudoh, T., Tsang, M., Hukriede, N.A., Chen, X., Dedekian, M., Clarke, C.J., Kiang, A., Schultz, S., Epstein, J.A., Toyama, R., et al., 2001. A gene expression screen in zebrafish embryogenesis. Genome Res. 11 (12), 1979–1987.

Lage, K., Mollgard, K., Greenway, S., Wakimoto, H., Gorham, J.M., Workman, C.T., Bendsen, E., Hansen, N.T., Rigina, O., Roque, F.S., et al., 2010. Dissecting spatio-temporal protein networks driving human heart development and related disorders. Mole. Syst. Biol. 6, 381.

Lancaster, M.A., Gopal, D.J., Kim, J., Saleem, S.N., Silhavy, J.L., Louie, C.M., Thacker, B.E., Williams, Y., Zaki, M.S., Gleeson, J.G., 2011a. Defective Wnt-dependent cerebellar midline fusion in a mouse model of Joubert syndrome. Nat. Med. 17 (6), 726–731.

Lancaster, M.A., Schroth, J., Gleeson, J.G., 2011b. Subcellular spatial regulation of canonical Wnt signaling at the primary cilium. Nat. Cell Biol. 13 (6), 700–707.

Lein, E.S., Hawrylycz, M.J., Ao, N., Ayres, M., Bensinger, A., Bernard, A., Boe, A.F., Boguski, M.S., Brockway, K.S., Byrnes, E.J., et al., 2007. Genome-wide atlas of gene expression in the adult mouse brain. Nature 445 (7124), 168–176.

Lemmon, M.A., Schlessinger, J., 2010. Cell signaling by receptor tyrosine kinases. Cell 141 (7), 1117–1134.

Lenffer, J., Nicholas, F.W., Castle, K., Rao, A., Gregory, S., Poidinger, M., Mailman, M.D., Ranganathan, S., 2006. OMIA (Online Mendelian Inheritance in Animals): an enhanced platform and integration into the Entrez search interface at NCBI. Nucleic Acids Res. 34 (Database issue), D599–D601.

Li, S., Armstrong, C.M., Bertin, N., Ge, H., Milstein, S., Boxem, M., Vidalain, P.O., Han, J.D., Chesneau, A., Hao, T., et al., 2004. A map of the interactome network of the metazoan *C. elegans*. Science 303 (5657), 540–543.

Lickert, H., Cox, B., Wehrle, C., Taketo, M.M., Kemler, R., Rossant, J., 2005. Dissecting Wnt/beta-catenin signaling during gastrulation using RNA interference in mouse embryos. Development 132 (11), 2599–2609.

Lin, W., Furthauer, M., Thisse, B., Thisse, C., Jing, N., Ang, S.L., 2002. Cloning of the mouse Sef gene and comparative analysis of its expression with Fgf8 and Spry2 during embryogenesis. Mech. Dev. 113 (2), 163–168.

Lin, X., Buff, E.M., Perrimon, N., Michelson, A.M., 1999. Heparan sulfate proteoglycans are essential for FGF receptor signaling during *Drosophila* embryonic development. Development 126 (17), 3715–3723.

Liu, D.W., Hsu, C.H., Tsai, S.M., Hsiao, C.D., Wang, W.P., 2011. A variant of fibroblast growth factor receptor 2 (Fgfr2) regulates left-right asymmetry in zebrafish. PLoS One 6 (7), e21793.

Lo, J., Lee, S., Xu, M., Liu, F., Ruan, H., Eun, A., He, Y., Ma, W., Wang, W., Wen, Z., et al., 2003. 15000 unique zebrafish EST clusters and their future use in microarray for profiling gene expression patterns during embryogenesis. Genome Res. 13 (3), 455–466.

Louvi, A., Artavanis-Tsakonas, S., 2012. Notch and disease: a growing field. Semin. Cell Dev. Biol. 23 (4), 473–480.

Love, D., Li, F.Q., Burke, M.C., Cyge, B., Ohmitsu, M., Cabello, J., Larson, J.E., Brody, S.L., Cohen, J.C., Takemaru, K., 2010. Altered lung morphogenesis, epithelial cell differentiation and mechanics in mice deficient in the Wnt/beta-catenin antagonist Chibby. PLoS One 5 (10), e13600.

Lukyanov, K.A., Chudakov, D.M., Lukyanov, S., Verkhusha, V.V., 2005. Innovation: Photoactivatable fluorescent proteins. Nat. Rev. Mol. Cell Biol. 6 (11), 885–891.

Lum, L., Yao, S., Mozer, B., Rovescalli, A., Von Kessler, D., Nirenberg, M., Beachy, P.A., 2003. Identification of Hedgehog pathway components by RNAi in *Drosophila* cultured cells. Science 299 (5615), 2039–2045.

Marois, E., Mahmoud, A., Eaton, S., 2006. The endocytic pathway and formation of the Wingless morphogen gradient. Development 133 (2), 307–317.

Matsuo, I., Kimura-Yoshida, C., 2013. Extracellular modulation of Fibroblast Growth Factor signaling through heparan sulfate proteoglycans in mammalian development. Curr. Opin. Genet. Dev.

Maurya, A.K., Tan, H., Souren, M., Wang, X., Wittbrodt, J., Ingham, P.W., 2011. Integration of Hedgehog and BMP signaling by the engrailed2a gene in the zebrafish myotome. Development 138 (4), 755–765.

Meyer, R.D., Mohammadi, M., Rahimi, N., 2006. A single amino acid substitution in the activation loop defines the decoy characteristic of VEGFR-1/FLT-1. J. Biol. Chem. 281 (2), 867–875.

Meyer, R.D., Rahimi, N., 2003. Comparative structure-function analysis of VEGFR-1 and VEGFR-2: What have we learned from chimeric systems? Ann. N. Y. Acad. Sci. 995, 200–207.

Mii, Y., Taira, M., 2011. Secreted Wnt "inhibitors" are not just inhibitors: regulation of extracellular Wnt by secreted Frizzled-related proteins. Dev. Growth Differ. 53 (8), 911–923.

Miki, R., Kadota, K., Bono, H., Mizuno, Y., Tomaru, Y., Carninci, P., Itoh, M., Shibata, K., Kawai, J., Konno, H., et al., 2001. Delineating developmental and metabolic pathways *in vivo* by expression profiling using the RIKEN set of 18,816 full-length enriched mouse cDNA arrays. Proc. Natl. Acad. Sci. USA 98 (5), 2199–2204.

Mizutani, C.M., Nie, Q., Wan, F.Y., Zhang, Y.T., Vilmos, P., Sousa-Neves, R., Bier, E., Marsh, J.L., Lander, A.D., 2005. Formation of the BMP activity gradient in the *Drosophila* embryo. Dev. Cell 8 (6), 915–924.

Moghal, N., Sternberg, P.W., 2003. The epidermal growth factor system in *Caenorhabditis elegans*. Exp. Cell Res. 284 (1), 150–159.

Mohammadi, M., Olsen, S.K., Goetz, R., 2005. A protein canyon in the FGF-FGF receptor dimer selects from an *a la carte* menu of heparan sulfate motifs. Curr. Opin. Struct. Biol. 15 (5), 506–516.

Mukai, A., Yamamoto-Hino, M., Awano, W., Watanabe, W., Komada, M., Goto, S., 2010. Balanced ubiquitylation and deubiquitylation of Frizzled regulate cellular responsiveness to Wg/Wnt. EMBO J. 29 (13), 2114–2125.

Mukhopadhyay, S., Wen, X., Chih, B., Nelson, C.D., Lane, W.S., Scales, S.J., Jackson, P.K., 2010. TULP3 bridges the IFT-A complex and membrane phosphoinositides to promote trafficking of G protein-coupled receptors into primary cilia. Genes Dev. 24 (19), 2180–2193.

Muller, P., Kuttenkeuler, D., Gesellchen, V., Zeidler, M.P., Boutros, M., 2005. Identification of JAK/STAT signaling components by genome-wide RNA interference. Nature 436 (7052), 871–875.

Muller, P., Rogers, K.W., Jordan, B.M., Lee, J.S., Robson, D., Ramanathan, S., Schier, A.F., 2012. Differential diffusivity of Nodal and Lefty underlies a reaction-diffusion patterning system. Science 336 (6082), 721–724.

Muller, P., Rogers, K.W., Yu, S.R., Brand, M., Schier, A.F., 2013. Morphogen transport. Development 140 (8), 1621–1638.

Nagaraj, R., Banerjee, U., 2004. The little R cell that could. Int. J. Dev. Biol. 48 (8–9), 755–760.

Neugebauer, J.M., Amack, J.D., Peterson, A.G., Bisgrove, B.W., Yost, H.J., 2009. FGF signaling during embryo development regulates cilia length in diverse epithelia. Nature 458 (7238), 651–654.

Neumann, C.J., Cohen, S.M., 1997. Long range action of Wingless organizes the dorsal-ventral axis of the *Drosophila* wing. Development 124 (4), 871–880.

Niehrs, C., 2012. The complex world of Wnt receptor signaling. Nat. Rev. Mol. Cell Biol. 13 (12), 767–779.

Nieminen, P., Pekkanen, M., Aberg, T., Thesleff, I., 1998. A graphical WWW-database on gene expression in tooth. Eur. J. Oral. Sci. 106 (Suppl. 1), 7–11.

Nigg, E.A., Raff, J.W., 2009. Centrioles, centrosomes, and cilia in health and disease. Cell 139 (4), 663–678.

Novitch, B.G., Chen, A.I., Jessell, T.M., 2001. Coordinate regulation of motor neuron subtype identity and pan-neuronal properties by the bHLH repressor Olig2. Neuron 31 (5), 773–789.

Nowak, M., Machate, A., Yu, S.R., Gupta, M., Brand, M., 2011. Interpretation of the FGF8 morphogen gradient is regulated by endocytic trafficking. Nat. Cell Biol. 13 (2), 153–158.

Nybakken, K., Vokes, S.A., Lin, T.Y., McMahon, A.P., Perrimon, N., 2005. A genome-wide RNA interference screen in *Drosophila melanogaster* cells for new components of the Hh signaling pathway. Nat. Genet. 37 (12), 1323–1332.

O'Connor, M.B., Umulis, D., Othmer, H.G., Blair, S.S., 2006. Shaping BMP morphogen gradients in the *Drosophila* embryo and pupal wing. Development 133 (2), 183–193.

Ocbina, P.J., Tuson, M., Anderson, K.V., 2009. Primary cilia are not required for normal canonical Wnt signaling in the mouse embryo. PLoS One 4 (8), e6839.

Oh, E.C., Katsanis, N., 2012. Cilia in vertebrate development and disease. Development 139 (3), 443–448.

Ohlmeyer, J.T., Kalderon, D., 1998. Hedgehog stimulates maturation of *Cubitus interruptus* into a labile transcriptional activator. Nature 396 (6713), 749–753.

Okada, I., Hamanoue, H., Terada, K., Tohma, T., Megarbane, A., Chouery, E., Abou-Ghoch, J., Jalkh, N., Cogulu, O., Ozkinay, F., et al., 2011. SMOC1 is essential for ocular and limb development in humans and mice. Am. J. Hum. Genet. 88 (1), 30–41.

Ong, S.E., Blagoev, B., Kratchmarova, I., Kristensen, D.B., Steen, H., Pandey, A., Mann, M., 2002. Stable isotope labeling by amino acids in cell culture, SILAC, as a simple and accurate approach to expression proteomics. Mol. Cell. Proteomics 1 (5), 376–386.

Onishi, H., Katano, M., 2011. Hedgehog signaling pathway as a therapeutic target in various types of cancer. Cancer Sci. 102 (10), 1756–1760.

Ornitz, D.M., 2000. FGFs, heparan sulfate and FGFRs: complex interactions essential for development, BioEssays: news and reviews in molecular. Cell. Dev. Biol. 22 (2), 108–112.

Palin, K., Taipale, J., Ukkonen, E., 2006. Locating potential enhancer elements by comparative genomics using the EEL software. Nat. Protoc. 1 (1), 368–374.

Pan, Y., Bai, C.B., Joyner, A.L., Wang, B., 2006. Sonic hedgehog signaling regulates Gli2 transcriptional activity by suppressing its processing and degradation. Mol. Cell. Biol. 26 (9), 3365–3377.

Pan, Y., Wang, C., Wang, B., 2009. Phosphorylation of Gli2 by protein kinase A is required for Gli2 processing and degradation and the Sonic Hedgehog-regulated mouse development. Dev. Biol. 326 (1), 177–189.

Patel, V.N., Knox, S.M., Likar, K.M., Lathrop, C.A., Hossain, R., Eftekhari, S., Whitelock, J.M., Elkin, M., Vlodavsky, I., Hoffman, M.P., 2007. Heparanase cleavage of perlecan heparan sulfate modulates FGF10 activity during *ex vivo* submandibular gland branching morphogenesis. Development 134 (23), 4177–4186.

Paulsen, M., Legewie, S., Eils, R., Karaulanov, E., Niehrs, C., 2011. Negative feedback in the bone morphogenetic protein 4 (BMP4) synexpression group governs its dynamic signaling range and canalizes development. Proc. Natl. Acad. Sci. USA 108 (25), 10202–10207.

Penton, A.L., Leonard, L.D., Spinner, N.B., 2012. Notch signaling in human development and disease. Semin. Cell Dev. Biol. 23 (4), 450–457.

Perrot, C.Y., Javelaud, D., Mauviel, A., 2013. Overlapping activities of TGF-beta and Hedgehog signaling in cancer: therapeutic targets for cancer treatment. Pharmacol. Ther. 137 (2), 183–199.

Peterson, K.A., Nishi, Y., Ma, W., Vedenko, A., Shokri, L., Zhang, X., McFarlane, M., Baizabal, J.M., Junker, J.P., van Oudenaarden, A., et al., 2012. Neural-specific Sox2 input and differential Gli binding affinity provide context and positional information in Shh-directed neural patterning. Genes Dev. 26 (24), 2802–2816.

Piddini, E., Marshall, F., Dubois, L., Hirst, E., Vincent, J.P., 2005. Arrow (LRP6) and Frizzled2 cooperate to degrade Wingless in *Drosophila* imaginal discs. Development 132 (24), 5479–5489.

Porter, A.C., Vaillancourt, R.R., 1998. Tyrosine kinase receptor-activated signal transduction pathways which lead to oncogenesis. Oncogene 17 (11 Reviews), 1343–1352.

Preger, E., Ziv, I., Shabtay, A., Sher, I., Tsang, M., Dawid, I.B., Altuvia, Y., Ron, D., 2004. Alternative splicing generates an isoform of the human Sef gene with altered subcellular localization and specificity. Proc. Natl. Acad. Sci. USA 101 (5), 1229–1234.

Qian, M., Yao, S., Jing, L., He, J., Xiao, C., Zhang, T., Meng, W., Zhu, H., Xu, H., Mo, X., 2013. ENC1-like integrates the retinoic acid/FGF signaling pathways to modulate ciliogenesis of Kupffers Vesicle during zebrafish embryonic development. Dev. Biol. 374 (1), 85–95.

Rahimi, N., Dayanir, V., Lashkari, K., 2000. Receptor chimeras indicate that the vascular endothelial growth factor receptor-1 (VEGFR-1) modulates mitogenic activity of VEGFR-2 in endothelial cells. J. Biol. Chem. 275 (22), 16986–16992.

Rainger, J., van Beusekom, E., Ramsay, J.K., McKie, L., Al-Gazali, L., Pallotta, R., Saponari, A., Branney, P., Fisher, M., Morrison, H., et al., 2011. Loss of the BMP antagonist, SMOC-1, causes Ophthalmo-acromelic (Waardenburg Anophthalmia) syndrome in humans and mice. PLoS Genetics 7 (7), e1002114.

Ramel, M.C., Hill, C.S., 2012. Spatial regulation of BMP activity. FEBS Lett. 586 (14), 1929–1941.

Ramialison, M., Reinhardt, R., Henrich, T., Wittbrodt, B., Kellner, T., Lowy, C.M., Wittbrodt, J., 2012. Cis-regulatory properties of medaka synexpression groups. Development 139 (5), 917–928.

Ramirez-Weber, F.A., Kornberg, T.B., 1999. Cytonemes: cellular processes that project to the principal signaling center in *Drosophila* imaginal discs. Cell 97 (5), 599–607.

Ray, K., Perez, S.E., Yang, Z., Xu, J., Ritchings, B.W., Steller, H., Goldstein, L.S., 1999. Kinesin-II is required for axonal transport of choline acetyltransferase in *Drosophila*. J. Cell Biol. 147 (3), 507–518.

Reich, A., Sapir, A., Shilo, B., 1999. Sprouty is a general inhibitor of receptor tyrosine kinase signaling. Development 126 (18), 4139–4147.

Richardson, L., Venkataraman, S., Stevenson, P., Yang, Y., Burton, N., Rao, J., Fisher, M., Baldock, R.A., Davidson, D.R., Christiansen, J.H., 2010. EMAGE mouse embryo spatial gene expression database: 2010 update. Nucleic Acids Res. 38 (Database issue), D703–D709.

Ries, J., Yu, S.R., Burkhardt, M., Brand, M., Schwille, P., 2009. Modular scanning FCS quantifies receptor-ligand interactions in living multicellular organisms. Nat. Methods 6 (9), 643–645.

Robbins, D.J., Fei, D.L., Riobo, N.A., 2012. The Hedgehog signal transduction network. Sci. Signaling 5 (246), re6.

Roberts, D.M., Kearney, J.B., Johnson, J.H., Rosenberg, M.P., Kumar, R., Bautch, V.L., 2004. The vascular endothelial growth factor (VEGF) receptor Flt-1 (VEGFR-1) modulates Flk-1 (VEGFR-2) signaling during blood vessel formation. Am. J. Pathol. 164 (5), 1531–1535.

Roy, S., Hsiung, F., Kornberg, T.B., 2011. Specificity of *Drosophila* cytonemes for distinct signaling pathways. Science 332 (6027), 354–358.

Sanalkumar, R., Dhanesh, S.B., James, J., 2010. Non-canonical activation of Notch signaling/target genes in vertebrates. Cell. Mol. Life. Sci. 67 (17), 2957–2968.

Schlessinger, J., 2000. Cell signaling by receptor tyrosine kinases. Cell 103 (2), 211–225.

Schmidt, M.H., Dikic, I., 2005. The Cbl interactome and its functions. Nat. Rev. Mol. Cell Biol. 6 (12), 907–918.

Schneider, L., Cammer, M., Lehman, J., Nielsen, S.K., Guerra, C.F., Veland, I.R., Stock, C., Hoffmann, E.K., Yoder, B.K., Schwab, A., et al., 2010. Directional cell migration and chemotaxis in wound healing response to PDGF-AA are coordinated by the primary cilium in fibroblasts. Cell. Physiol. Biochem. 25 (2–3), 279–292.

Schneider, L., Clement, C.A., Teilmann, S.C., Pazour, G.J., Hoffmann, E.K., Satir, P., Christensen, S.T., 2005. PDGFRalphaalpha signaling is regulated through the primary cilium in fibroblasts. Curr. Biol. 15 (20), 1861–1866.

Schulze, W.X., Deng, L., Mann, M., 2005. Phosphotyrosine interactome of the ErbB-receptor kinase family. Mol. Syst. Biol. 1. 2005 0008.

Shaye, D.D., Greenwald, I., 2002. Endocytosis-mediated down-regulation of LIN-12/Notch upon Ras activation in *Caenorhabditis elegans*. Nature 420 (6916), 686–690.

Shaye, D.D., Greenwald, I., 2005. LIN-12/Notch trafficking and regulation of DSL ligand activity during vulval induction in *Caenorhabditis elegans*. Development 132 (22), 5081–5092.

Shilo, B.Z., 2003. Signaling by the *Drosophila* epidermal growth factor receptor pathway during development. Exp. Cell Res. 284 (1), 140–149.

Shimokawa, K., Kimura-Yoshida, C., Nagai, N., Mukai, K., Matsubara, K., Watanabe, H., Matsuda, Y., Mochida, K., Matsuo, I., 2011. Cell surface heparan sulfate chains regulate local reception of FGF signaling in the mouse embryo. Dev. Cell 21 (2), 257–272.

Simons, M., Gloy, J., Ganner, A., Bullerkotte, A., Bashkurov, M., Kronig, C., Schermer, B., Benzing, T., Cabello, O.A., Jenny, A., et al., 2005. Inversin, the gene product mutated in nephronophthisis type II, functions as a molecular switch between Wnt signaling pathways. Nat. Genet. 37 (5), 537–543.

Stickens, D., Zak, B.M., Rougier, N., Esko, J.D., Werb, Z., 2005. Mice deficient in Ext2 lack heparan sulfate and develop exostoses. Development 132 (22), 5055–5068.

Stottmann, R.W., Tran, P.V., Turbe-Doan, A., Beier, D.R., 2009. Ttc21b is required to restrict sonic hedgehog activity in the developing mouse forebrain. Dev. Biol. 335 (1), 166–178.

Strigini, M., Cohen, S.M., 2000. Wingless gradient formation in the *Drosophila* wing. Curr. Biol. 10 (6), 293–300.

Su, Y., Ospina, J.K., Zhang, J., Michelson, A.P., Schoen, A.M., Zhu, A.J., 2011. Sequential phosphorylation of smoothened transduces graded hedgehog signaling. Sci. Signaling 4 (180), ra43.

Svard, J., Heby-Henricson, K., Persson-Lek, M., Rozell, B., Lauth, M., Bergstrom, A., Ericson, J., Toftgard, R., Teglund, S., 2006. Genetic elimination of Suppressor of fused reveals an essential repressor function in the mammalian Hedgehog signaling pathway. Dev. Cell 10 (2), 187–197.

Swanson, C.I., Evans, N.C., Barolo, S., 2010. Structural rules and complex regulatory circuitry constrain expression of a Notch- and EGFR-regulated eye enhancer. Dev. Cell 18 (3), 359–370.

Takemaru, K., Yamaguchi, S., Lee, Y.S., Zhang, Y., Carthew, R.W., Moon, R.T., 2003. Chibby, a nuclear beta-catenin-associated antagonist of the Wnt/Wingless pathway. Nature 422 (6934), 905–909.

Takemoto, T., Uchikawa, M., Kamachi, Y., Kondoh, H., 2006. Convergence of Wnt and FGF signals in the genesis of posterior neural plate through activation of the Sox2 enhancer N-1. Development 133 (2), 297–306.

Tallquist, M.D., French, W.J., Soriano, P., 2003. Additive effects of PDGF receptor beta signaling pathways in vascular smooth muscle cell development. PLoS Biol. 1 (2), E52.

Tan, P.B., Kim, S.K., 1999. Signaling specificity: the RTK/RAS/MAP kinase pathway in metazoans. Trends Genet. 15 (4), 145–149.

Teleman, A.A., Cohen, S.M., 2000. Dpp gradient formation in the *Drosophila* wing imaginal disc. Cell 103 (6), 971–980.

Tewari, M., Hu, P.J., Ahn, J.S., Ayivi-Guedehoussou, N., Vidalain, P.O., Li, S., Milstein, S., Armstrong, C.M., Boxem, M., Butler, M.D., et al., 2004. Systematic interactome mapping and genetic perturbation analysis of a *C. elegans* TGF-beta signaling network. Mol. Cell 13 (4), 469–482.

Thomas, J.T., Canelos, P., Luyten, F.P., Moos Jr., M., 2009. *Xenopus* SMOC-1 Inhibits bone morphogenetic protein signaling downstream of receptor binding and is essential for postgastrulation development in *Xenopus*. J. Biol. Chem. 284 (28), 18994–19005.

Tomancak, P., Beaton, A., Weiszmann, R., Kwan, E., Shu, S., Lewis, S.E., Richards, S., Ashburner, M., Hartenstein, V., Celniker, S.E., et al., 2002. Systematic determination of patterns of gene expression during *Drosophila* embryogenesis. Genome. Biol. 3 (12). RESEARCH0088.

Tomancak, P., Berman, B.P., Beaton, A., Weiszmann, R., Kwan, E., Hartenstein, V., Celniker, S.E., Rubin, G.M., 2007. Global analysis of patterns of gene expression during *Drosophila* embryogenesis. Genome Biol. 8 (7), R145.

Torii, S., Kusakabe, M., Yamamoto, T., Maekawa, M., Nishida, E., 2004. Sef is a spatial regulator for Ras/MAP kinase signaling. Dev. Cell 7 (1), 33–44.

Tozer, S., Le Dreau, G., Marti, E., Briscoe, J., 2013. Temporal control of BMP signaling determines neuronal subtype identity in the dorsal neural tube. Development 140 (7), 1467–1474.

Tran, P.V., Haycraft, C.J., Besschetnova, T.Y., Turbe-Doan, A., Stottmann, R.W., Herron, B.J., Chesebro, A.L., Qiu, H., Scherz, P.J., Shah, J.V., et al., 2008. THM1 negatively modulates mouse sonic hedgehog signal transduction and affects retrograde intraflagellar transport in cilia. Nat. Genet. 40 (4), 403–410.

Tsang, M., Friesel, R., Kudoh, T., Dawid, I.B., 2002. Identification of Sef, a novel modulator of FGF signaling. Nat. Cell. Biol. 4 (2), 165–169.

Tsang, M., Maegawa, S., Kiang, A., Habas, R., Weinberg, E., Dawid, I.B., 2004. A role for MKP3 in axial patterning of the zebrafish embryo. Development 131 (12), 2769–2779.

Tukachinsky, H., Lopez, L.V., Salic, A., 2010. A mechanism for vertebrate Hedgehog signaling: recruitment to cilia and dissociation of SuFu-Gli protein complexes. J. Cell Biol. 191 (2), 415–428.

Tuson, M., He, M., Anderson, K.V., 2011. Protein kinase A acts at the basal body of the primary cilium to prevent Gli2 activation and ventralization of the mouse neural tube. Development 138 (22), 4921–4930.

van Amerongen, R., 2012. Alternative Wnt pathways and receptors. Cold Spring Harb. Perspect. Biol. 4 (10).

Varjosalo, M., Bjorklund, M., Cheng, F., Syvanen, H., Kivioja, T., Kilpinen, S., Sun, Z., Kallioniemi, O., Stunnenberg, H.G., He, W.W., et al., 2008. Application of active and kinase-deficient kinome collection for identification of kinases regulating hedgehog signaling. Cell 133 (3), 537–548.

Visel, A., Blow, M.J., Li, Z., Zhang, T., Akiyama, J.A., Holt, A., Plajzer-Frick, I., Shoukry, M., Wright, C., Chen, F., et al., 2009a. ChIP-seq accurately predicts tissue-specific activity of enhancers. Nature 457 (7231), 854–858.

Visel, A., Minovitsky, S., Dubchak, I., Pennacchio, L.A., 2007. VISTA Enhancer Browser – a database of tissue-specific human enhancers. Nucleic Acids Res. 35 (Database issue), D88–D92.

Visel, A., Rubin, E.M., Pennacchio, L.A., 2009b. Genomic views of distant-acting enhancers. Nature 461 (7261), 199–205.

Visel, A., Taher, L., Girgis, H., May, D., Golonzhka, O., Hoch, R.V., McKinsey, G.L., Pattabiraman, K., Silberberg, S.N., Blow, M.J., et al., 2013. A high-resolution enhancer atlas of the developing telencephalon. Cell 152 (4), 895–908.

Visel, A., Thaller, C., Eichele, G., 2004. GenePaint.org: an atlas of gene expression patterns in the mouse embryo. Nucleic Acids Res. 32 (Database issue), D552–D556.

Vokes, S.A., Ji, H., Wong, W.H., McMahon, A.P., 2008. A genome-scale analysis of the cis-regulatory circuitry underlying sonic hedgehog-mediated patterning of the mammalian limb. Genes Dev. 22 (19), 2651–2663.

Voronina, V.A., Takemaru, K., Treuting, P., Love, D., Grubb, B.R., Hajjar, A.M., Adams, A., Li, F.Q., Moon, R.T., 2009. Inactivation of Chibby affects function of motile airway cilia. J. Cell Biol. 185 (2), 225–233.

Vuilleumier, R., Springhorn, A., Patterson, L., Koidl, S., Hammerschmidt, M., Affolter, M., Pyrowolakis, G., 2010. Control of Dpp morphogen signaling by a secreted feedback regulator. Nat. Cell Biol. 12 (6), 611–617.

Wallingford, J.B., Mitchell, B., 2011. Strange as it may seem: the many links between Wnt signaling, planar cell polarity, and cilia. Genes Dev. 25 (3), 201–213.

Walsh, D.W., Godson, C., Brazil, D.P., Martin, F., 2010. Extracellular BMP antagonist regulation in development and disease: tied up in knots. Trends Cell Biol. 20 (5), 244–256.

Wang, J., Barr, M.M., 2005. RNA interference in *Caenorhabditis elegans*. Meth. Enzymol. 392, 36–55.

Wang, Y., Shyy, J.Y., Chien, S., 2008. Fluorescence proteins, live-cell imaging, and mechanobiology: seeing is believing. Annu. Rev. Biomed. Eng. 10, 1–38.

Watanabe, D., Saijoh, Y., Nonaka, S., Sasaki, G., Ikawa, Y., Yokoyama, T., Hamada, H., 2003. The left-right determinant Inversin is a component of node monocilia and other 9+0 cilia. Development 130 (9), 1725–1734.

Wolpert, L., 1969. Positional information and the spatial pattern of cellular differentiation. J. Theor. Biol. 25 (1), 1–47.

Wrighton, K.H., Lin, X., Feng, X.H., 2008. Critical regulation of TGF-beta signaling by Hsp90. Proc. Natl. Acad. Sci. USA 105 (27), 9244–9249.

Xenarios, I., Salwinski, L., Duan, X.J., Higney, P., Kim, S.M., Eisenberg, D., 2002. DIP, the Database of Interacting Proteins: a research tool for studying cellular networks of protein interactions. Nucleic Acids Res. 30 (1), 303–305.

Xi, Q., He, W., Zhang, X.H., Le, H.V., Massague, J., 2008. Genome-wide impact of the BRG1 SWI/SNF chromatin remodeler on the transforming growth factor beta transcriptional program. J. Biol. Chem. 283 (2), 1146–1155.

Xu, C., Kauffmann, R.C., Zhang, J., Kladny, S., Carthew, R.W., 2000. Overlapping activators and repressors delimit transcriptional response to receptor tyrosine kinase signals in the *Drosophila* eye. Cell 103 (1), 87–97.

Yatim, A., Benne, C., Sobhian, B., Laurent-Chabalier, S., Deas, O., Judde, J.G., Lelievre, J.D., Levy, Y., Benkirane, M., 2012. NOTCH1 nuclear interactome reveals key regulators of its transcriptional activity and oncogenic function. Mol. Cell 48 (3), 445–458.

Yoo, A.S., Bais, C., Greenwald, I., 2004. Crosstalk between the EGFR and LIN-12/Notch pathways in *C. elegans* vulval development. Science 303 (5658), 663–666.

Yu, S.R., Burkhardt, M., Nowak, M., Ries, J., Petrasek, Z., Scholpp, S., Schwille, P., Brand, M., 2009. Fgf8 morphogen gradient forms by a source-sink mechanism with freely diffusing molecules. Nature 461 (7263), 533–536.

Zhang, C., Williams, E.H., Guo, Y., Lum, L., Beachy, P.A., 2004. Extensive phosphorylation of Smoothened in Hedgehog pathway activation. Proc. Natl. Acad. Sci. USA 101 (52), 17900–17907.

Zhang, L., Gurskaya, N.G., Merzlyak, E.M., Staroverov, D.B., Mudrik, N.N., Samarkina, O.N., Vinokurov, L.M., Lukyanov, S., Lukyanov, K.A., 2007. Method for real-time monitoring of protein degradation at the single cell level. BioTechniques 42 (4). 446, 448, 450.

Zhong, W., Sternberg, P.W., 2006. Genome-wide prediction of *C. elegans* genetic interactions. Science 311 (5766), 1481–1484.

Chapter 2

Applications of Deep Sequencing to Developmental Systems

L. Ira, Blitz, Mui Luong, T.Y. William, Chiu and Ken W.Y. Cho
Department of Developmental and Cell Biology, University of California, Irvine, CA

Chapter Outline

GLOSSARY

ChIA-PET An adaptation of 3C technologies that employs high throughput sequencing. This method utilizes ChIP to find regions of DNA looping associated with a specific transcription factor.

ChIP-seq A high throughput sequencing method employing immunoprecipitation of fragmented chromatin to identify sites bound by a protein, genome-wide.

Chromatin conformation capture (3C) An experimental method that employs quantitative polymerase chain reaction (PCR) to assay for suspected sites of DNA looping.

DNAse-seq A high throughput sequencing method to identify sites of open chromatin based on their relative accessibility to cleavage by DNAse I.

qPCR, or quantitative polymerase chain reaction (PCR) An assay based on detection of increasing fluorescence with cycle number as amplified DNA fragments increase in concentration in the reaction.

RNA-seq A high throughput sequencing method that sequences cDNA fragments derived from RNA populations. This method allows for determination of gene structure and quantitation of RNA abundances.

SUMMARY

- High throughput sequencing is a powerful new tool for obtaining information about nucleic acids.
- RNA-seq permits both new gene discovery and "digital" quantitation of all RNA levels.
- ChIP-seq assesses the genomic locations of bound proteins, protein modifications and DNA modifications.
- DNAse-seq identifies genomic regions that are in open chromatin states.
- Techniques like ChIA-PET can discover sites of short- and long-range DNA looping.
- These technologies provide critical insights into the mechanisms of gene regulation at the genome-wide level, revealing how cells make decisions both during normal development and in disease states.

2.1 INTRODUCTION

Fertilized eggs give rise to many millions of cells in the human embryo, comprising hundreds of different cell types. Since the genome is essentially identical between different cell types, biological differences between cells arise in large part due

Principles of Developmental Genetics.

to differences in gene expression. Each cell makes numerous decisions to express specific combinations of genes among the tens of thousands in its genome. If this tightly regulated process goes awry, developmental abnormalities may occur which, if not embryonically lethal, can lead to diseases in the adult. The regulatory program for turning on or off specific genes is embedded in the genome and currently, major efforts in biology are being devoted to understanding the regulatory mechanisms, which organize the expression of thousands of genes. Coordinated expression of groups of genes that function together is required to specify paths of cellular differentiation and also to maintain homeostasis. Recent advances in nucleic acid sequencing technologies have provided previously unforeseen opportunities to examine the regulatory landscape of the entire genome.

Thousands of transcription factors play a central role in controlling the expression of genes, and it is the interplay between sets of these factors in different cell types that controls differential gene activity. A given transcription factor can activate or suppress the expression of hundreds of target genes in a highly spatiotemporally coordinated manner. This is achieved via the binding of a specific transcription factor(s) to short sequences in the cis-regulatory regions of genes. In simple organisms such as the prokaryotes, these cis-acting motifs tend to be located relatively close to the basal promoter, a region where transcription is initiated. On the other hand, in eukaryotes, and especially those with larger genomes where genes are often separated by much longer intergenic distances, a class of cis-regulatory sequences called enhancers act to control gene expression and these may be dispersed over many tens of kilobases (and in some cases millions of bases) from the promoters of the genes that they regulate. Therefore, determining the physical location of active enhancers within the genome, as well as the function of each enhancer, is critical in uncovering the mechanisms of gene activation and cellular differentiation. In addition, gene expression can be regulated by epigenetic states of chromatin, the largely proteinaceous material in which the genome is packaged in the nucleus, which controls the access of transcripton factors to cis-regulatory elements in genic DNA.

Until recently, the tasks of identifying the thousands of genes in a genome, cataloging their expression, and mapping the enhancers responsible for gene activity changes, were unrealistic goals for biologists due to enormous labor requirements and prohibitive costs. However, recent advances in DNA sequencing and bioinformatics are bringing these formidable tasks into the realm of possibility for small research teams. Although it is still challenging, scientists can now examine the expression of all transcripts in a cell and obtain quantitative gene expression data. This genome level expression analysis combined with the mapping of active transcription factor binding sites on the genome is beginning to transform the way we view biological processes. In this chapter, we will discuss several fundamental genomic approaches that are currently used to study gene regulation in developing systems.

2.2 USING RNA-SEQ TO MAP AND QUANTIFY TRANSCRIPTS

Until relatively recently, DNA microarray approaches (e.g., Affymetrix GeneChip®) were the predominant choices for simultaneously examining gene expression changes in thousands of genes. Microarrays use DNA probes that are bound to platforms (slides or quartz wafers), and very large numbers of gene-specific probes are each arranged to create an array distributed over a very small area. Fluorescently tagged populations of cDNA fragments synthesized from the RNA under study are hybridized competitively to the platform and the hybridization efficiency of each probe is measured and compared between different experimental samples. The major drawback of the microarray approach has been the limitation of the analysis to genes that are already known. In contrast, high throughput sequencing technologies are independent of such limitations and therefore have rapidly supplanted microarrays as the method of choice. Illumina's sequencing technology has become the dominant method for gene expression profiling.

RNA-seq is a high throughput sequencing method for sequencing cDNA populations (Mortazavi et al., 2008; Nagal-akshmi et al., 2008; Wilhelm et al., 2008). Cellular RNAs are typically fragmented into short pieces and then these are converted to double-stranded cDNAs using random oligonucleotide primers to initiate first strand cDNA synthesis (Figure 2.1A). Once double-stranded cDNAs have been prepared, adaptor sequences are added to their ends, providing each fragment with a uniform end sequence. The entire population is then amplified by polymerase chain reaction (PCR) using oligonucleotide primers that are complementary to the adaptors. This amplification not only provides enough material for sequencing, but also ensures that all the cDNAs produced have the same sequence information on their ends, which is necessary for processing the cDNA fragments in the sequencer. The sequence reads generated from the amplified cDNA library are derived from the ends of the cDNA fragments and can be up to 150bp from one end (single-end sequencing) or both ends (paired-end sequencing). Currently one lane of single-end sequencing using an Illumina flow cell can yield up to ~400 million reads, which can then be compared to a reference genome sequence (or to known transcripts). A number of bioinformatics applications are available to perform these tasks, including Bowtie, BWA and ELAND (Garber et al., 2011). Mapping the reads to the genome allows us to identify the regions of the genome that are transcribed into RNA. An

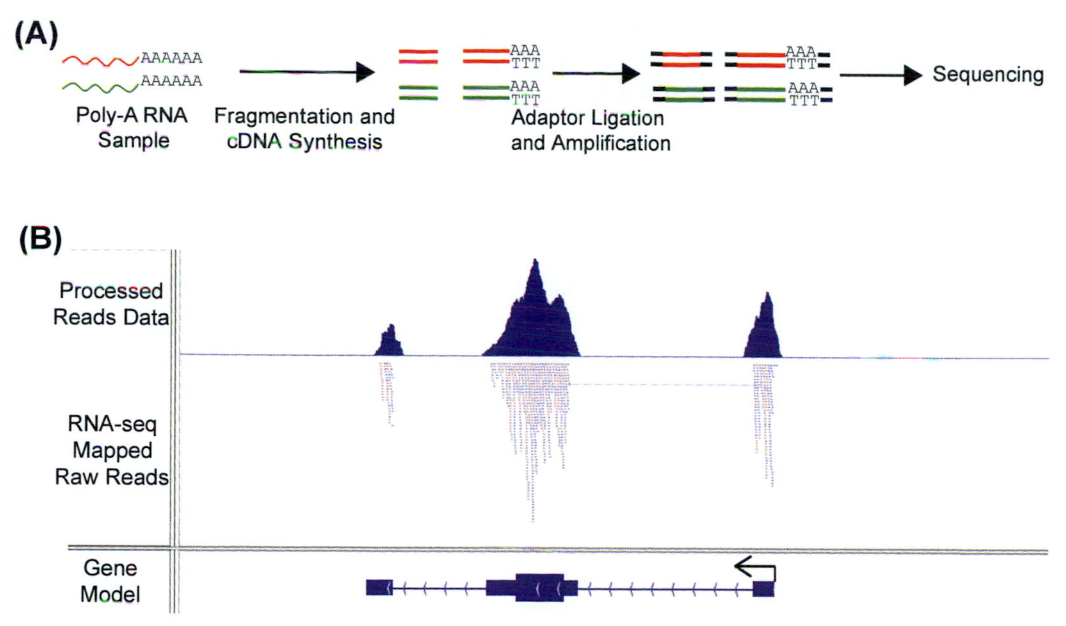

FIGURE 2.1 The RNA-seq Approach. Panel A is an overview of the workflow illustrating how single-stranded mRNAs are converted to double-stranded cDNA fragments with adaptors on their ends. These are the raw products that constitute a library for subsequent sequencing. Panel B contains a genome browser view of RNA-seq reads tiled across a gene. The bottom track depicts a gene with transcription oriented from right to left. Exons are shown as dark blue boxes with introns connecting them. The middle track shows the raw reads descending from the horizontal axis. Since sequences are randomly read from either end of the cDNAs in the library, and the reads from each strand are intermixed, red and blue highlighting is used to indicate which reads are derived from each strand. Note that the reads pile up over the exons. The upper track contains processed data, in which a smoothened curve shows the read "density."

alternative approach is to compare the read sequences to one another and assemble these, *de novo* (without the reference genomic sequence), into longer segments to generate a hypothetical set of transcripts using packages such as Trinity, Oases or Velvet (Zerbino and Birney, 2008; Grabherr et al., 2011; Schulz et al., 2012). Thus, RNA-seq can provide transcript information without genome information, and this is especially useful when working with organisms in which a complete genome sequence is not available. Either way, a genome-scale transcript set is then obtained that can be compared to the genome sequence to generate a transcriptome map.

A transcriptome map contains precise gene and exon boundaries (Figure 2.1B). A subset of RNA-seq reads span exon-exon junctions, thereby linking exons, but these are a small fraction of the total reads. Another approach to linking exons uses paired-end reads. cDNA fragments in which reads from one end map within one exon, and reads from the other end map within another exon, can provide the information necessary to link these exons together. RNA-seq can also identify transcript start sites, alternative splice sites, and polyA signals. Recently, RNA-seq has also contributed to the identification of microRNAs (a class of small non-coding RNAs) and lncRNAs (long non-coding RNAs), which seem to function in the transcriptional and post-transcriptional regulation of gene expression (Cabili et al., 2011; Morin et al., 2008).

RNA-seq is most frequently used to compare gene expression between different samples. In general, the reads are a representation of the original RNA population and the frequency occurrence of sequence tags is used to obtain relative expression levels of individual genes. The number of reads that map to any given gene or transcript can be tallied and reported directly (for example, as reads per gene); however, when comparing across different RNA-seq libraries one needs to consider the total number of reads generated for each library. Any two sequencing runs will yield different total numbers of reads and therefore the gene expression data need to take this into consideration. Data are typically reported as reads per given gene as a proportion of the number of millions of total reads. To illustrate this using an experimental dataset, in wild type *Xenopus tropicalis* early gastrulae, the gene *goosecoid* (*gsc*) was detected at a level of 1,985 mapped reads, where 79.6 million total reads were obtained from the sequencing run. Therefore the expression value for *gsc* is 1,985 reads divided by 79.6 million, or 25 reads per million. This measure does not readily permit comparisons between expression levels of different genes and therefore the abundance of sequence tags per gene or transcript is often described in the units Reads Per Kilobase of transcript per Million total reads (RPKM; Mortazavi et al., 2008). This measure takes into consideration the lengths of the transcripts, which can differ dramatically between genes. Thus, 1,985 reads mapped to *gsc*, which has a transcript length of 1.12kb, in a dataset of 79.6 million total reads, would yield an expression value in RPKM of 22.3. An advantage of RNA-seq is its large dynamic range (~5 orders of magnitude) of expression levels over which transcripts can be detected, thus this approach can robustly capture expression dynamics across different tissues or conditions, allowing comparisons between control and experimental samples. If *gsc* is expressed in control samples at an RPKM value of 22.3 but in experimental samples its expression level is 3.3, we can conclude that the level of RNA from *gsc* drops 6.8 fold in response to the experimental perturbation.

2.3 CHROMATIN IMMUNOPRECIPITATION FOR IDENTIFYING PROTEIN-DNA INTERACTIONS

Determining how proteins interact with genes to regulate their expression is essential for a full understanding of many biological processes and disease states. To understand how genes are regulated it is useful to obtain information about the genome-wide patterns of physical binding of transcription factors, interactions of RNA polymerase with genes, and the epigenetic states of chromatin. Chromatin immunoprecipitation (ChIP) is one of the most direct ways to identify the sites of interaction of these factors across the genome. In ChIP assays (Figure 2.2), cells, tissues or whole embryos are usually treated with formaldehyde to covalently crosslink proteins directly to the sites on DNA to which they are bound. The resulting crosslinked chromatin is sheared by sonication to obtain chromatin fragments containing DNA several hundred base pairs in length. Specific DNA-protein complexes are then immunoprecipitated with an antibody that recognizes the protein of interest. After removal of unbound chromatin, bound DNA is released from the protein-DNA complexes by reversing the crosslinking. This population of DNA fragments can then be analyzed by either quantitative polymerase chain reaction (ChIP-qPCR) or direct high throughput sequencing (ChIP-seq).

The ChIP-qPCR approach can be used to determine relative abundance of any DNA region of interest in the chromatin immunoprecipitated material (Figure 2.3). This method is based on the kinetics of the PCR reaction rather than on measuring the absolute levels of target DNA fragments. Oligonucleotide primer sequences are designed which flank a genomic region of interest to permit detection within the population of precipitated DNA fragments. During the PCR amplification, fluorescent dyes such as SYBR green are used, which only fluoresce when bound to double-stranded DNA (dsDNA). During PCR cycling the amount of SYBR green fluorescence is proportional to the amount of amplified fragment present, and its inclusion permits the continuous measurement of the accumulation of PCR product at every amplification cycle. The fluorescence accumulation is monitored and, once a threshold level is reached, the number of cycles required to reach this threshold is recorded as the Ct value, or threshold cycle value, for the reaction. Similarly, the region of interest is PCR amplified from sonicated genomic DNA that has not undergone immunoprecipitation. By using a dilution series of sonicated genomic ("input") DNA, one can obtain a series of Ct values for different amounts of starting material to generate a standard curve. Comparison between Ct values of immunoprecipitated chromatin and the standard curve permits determination of the percentage of input DNA for the genomic region that was recovered in the immunoprecipitate. As a control, different DNA regions, which are not suspected to bind the protein of interest, are similarly PCR amplified. Because the abundance of these nonspecific DNA fragments is significantly lower in the immunoprecipitate, the number of cycles required to reach the threshold is higher, and therefore the Ct value will be higher than for the region specifically immunoprecipitated. The Ct values of the immunoprecipitated region and the nonspecific control region first need to be converted into a percent input to control for differences in PCR efficiency inherent in PCR products with different sequence compositions. Comparison between of the percent inputs of the specifically immunoprecipitated region and nonspecific regions allows the enrichment of the region of interest to be assessed. Limitations of the ChIP-qPCR approach are the low throughput of the assay and the need to have *a priori* knowledge of the sites where a protein of interest might be bound.

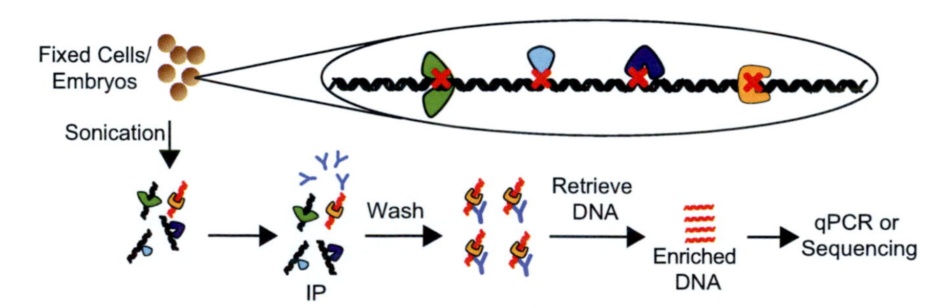

FIGURE 2.2 **An Overview of ChIP-qPCR and ChIP-seq Methods.** In the first step cells are fixed with formaldehyde to crosslink (red Xs) the bound proteins to DNA. Chromatin is fragmented by sonication and subjected to immunoprecipitation (IP) using antibodies (shown in blue) specific to the protein of interest. After washing the immunoprecipitates to remove non-specifically bound chromatin, crosslinks in the complexes are reversed and the DNA is isolated. This DNA is then subjected to either quantitative PCR or used to create a sequencing library for ChIP-seq analysis.

However, the assay and analysis are relatively straightforward, sensitivity is high, and relatively little material (in the range of a few hundred picograms of immunoprecipitated total genomic DNA) is required.

Chromatin immunoprecipitation (ChIP) can be combined with high throughput DNA sequencing (ChIP-seq) to identify locations of binding sites of a transcription factor to DNA (Johnson et al., 2007; Lin et al., 2007) or to map histone modifications along the entire genome. This technology maps reads generated from DNA fragments that are immunoprecipitated by specific antibodies against transcription factors, RNA polymerase, and histone modifications (Figure 2.4A). These reads are then tiled along the reference genome and their "piling up" along the genome can be represented graphically as a curve where the depth of coverage at each base is quantitated (Figure 2.4B). The positions of the peak summits are the approximate physical locations of protein binding to the genome. This approach thus provides a global view of transcription factor binding across the genome. Analyses of these data can also provide information such as the DNA sequence motifs that are bound *in vivo* by the transcription factor. Algorithms such as MEME search for enrichment of sequence motifs within bound regions that can be matched to databases of known binding preferences for well-characterized transcription factors (Bailey and Elkan, 1994). Therefore, this approach may also find other enriched sequence motifs nearby, implicating the identity of other transcription factor partners that may not have been suspected previously.

RNA polymerase II ChIP-seq provides the distribution of RNA polymerase along the length of genes (Figure 2.4B; Barski et al., 2007; Schones et al., 2008). While RNA-seq measures transcript levels, these are a function of both synthesis and degradation and therefore may not reflect a gene's current activity. On the other hand, ChIP-seq for RNA polymerase reveals the actual engagement of the gene in transcription. Similarly, antibodies recognizing post-translational modifications on specific amino acid residues of histones (e.g., methylation, acetylation) can be used in ChIP-seq experiments to determine their locations across the genome (Barski et al., 2007; Heintzman et al., 2009). Many of these histone "marks" are associated with states of chromatin activity and this information is useful for understanding changes in gene expression

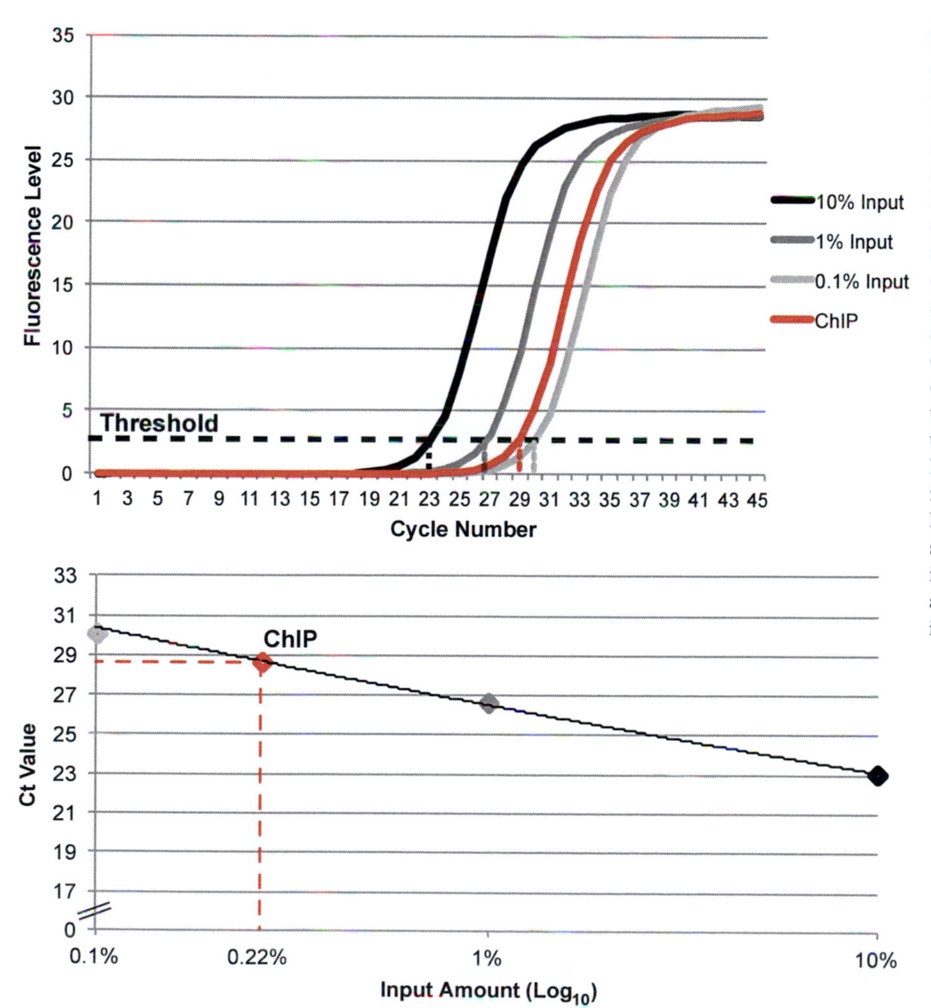

FIGURE 2.3 The Basics of ChIP-qPCR Analyses. The top panel shows the output from a series of four ChIP-qPCR reactions. Three reactions contain ten-fold dilutions of sonicated genomic DNA that has not been subjected to immunoprecipitation ("input" DNA). The fourth qPCR reaction (maroon) contained DNA from a chromatin immunoprecipitation experiment. The sigmoidal curves for each sample show the kinetics of accumulation of SYBR green fluorescence, and hence PCR products, over 45 cycles. The cycle number is recorded as the "threshold cycle" (Ct) value for a particular PCR reaction once its fluorescence curve crosses the threshold level (shown as a horizontal dashed line). In the bottom panel, the Ct values for the input DNA reactions are plotted as a straight line that serves as a standard curve. The ChIP sample's Ct value is plotted on this line and its value along the horizontal axis is recorded as its recovery expressed as a percentage of input DNA.

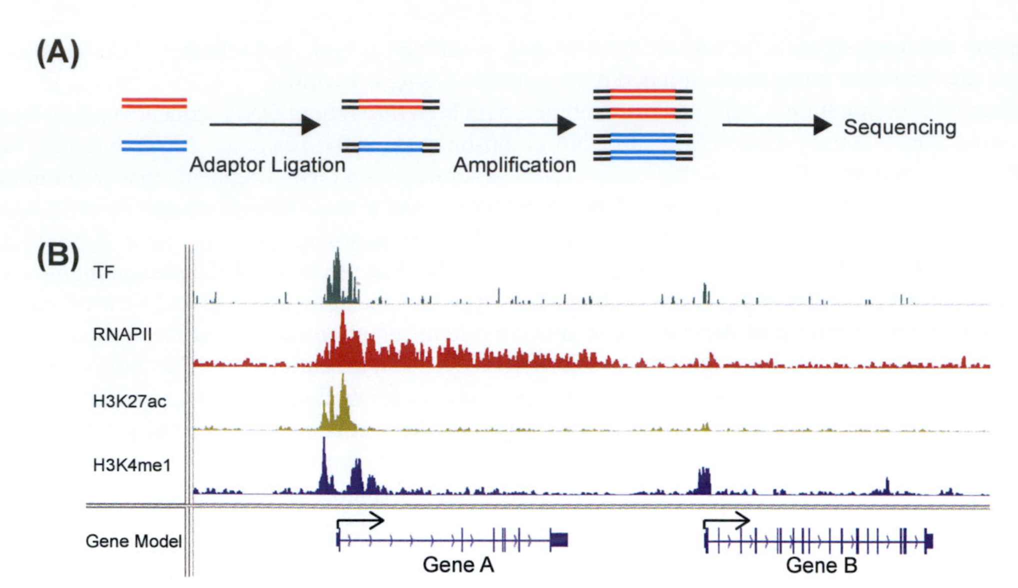

FIGURE 2.4 The ChIP-seq Approach. Panel A shows an overview of the making of a library from DNA that was derived from a ChIP experiment. Short DNA fragments are adaptor ligated and then amplified by PCR using oligonucleotide primers specific to the adaptor sequences. This produces the library that is subjected to high throughput sequencing. Panel B contains a genome browser view of multiple ChIP-seq experiments. The genomic interval shown contains two genes (A and B), both with transcription oriented from left to right. The top track (green) shows the binding pattern for a transcription factor that regulates Gene A. This factor is strongly bound near the promoter region of this gene, but not elsewhere along the genomic interval shown. The second track (red) shows the distribution of RNA polymerase II across this region. The polymerase appears to be bound along Gene A, but only background levels are detectable along Gene B. This strongly suggests that Gene A is being transcribed in the cells under analysis, while Gene B is silent. The bottom two tracks show ChIP-seq patterns for two different post-translational modifications to histone H3. The third track (light green) shows the pattern for acetylated lysine at position 27 of histone H3, whereas the fourth track (blue) shows mono-methylated lysine at position 4. The presence of both of these modifications is correlated with active enhancer and promoter regions. Gene A appears to be marked by both modifications in the region upstream of the transcription start site and in the first intron. Gene B shows weaker H3K4me1 but little if any H3K27ac, consistent with this gene being transcriptionally inactive.

and cellular differentiation. Genome-wide studies comparing patterns of histone marks between different cell lines have revealed that the location of enhancers is often correlated with specific histone H3 and H4 modifications (Heintzman et al., 2009; Rada-Iglesias et al., 2011). The human ENCODE (Encyclopedia of DNA Elements) project attempts to incorporate large numbers of transcription factor, polymerase and epigenetic mark ChIP-seq and other datasets to identify signatures of important cis-acting sequences functioning in regulating gene expression.

Two major drawbacks of ChIP-seq are the large amount of material (typically requiring 1–10 million cells per experiment) required and the limited availability of high-quality ChIP-grade antibodies that are capable of performing ChIP assays. The amount of starting material is particularly challenging when carrying out ChIP-seq for developing embryos or small tissue samples. Thus, ChIP-seq is so far limited to the model organisms where large numbers of embryos can be attained. *Xenopus* embryos provide a good example of a system where large numbers of embryos can be collected for ChIP-seq experiments to study genome-wide distributions of transcription factors, modified histones, RNA polymerase and other epigenetic marks. For example, Akkers et al. (2009) correlated the genomic locations of histone marks and RNA polymerase II with gene activity in *X. tropicalis* embryos to show, among other things, that certain histone modifications are associated with spatially regulated genes.

2.4 DNASE I HYPERSENSITIVE SITE MAPPING TO IDENTIFY CIS-REGULATORY REGIONS

The genome is packaged in chromosomes in the nucleus with different regions in open and closed chromatin states, linked to levels of transcriptional activity. DNA in these chromatin domains may be bound by nucleosomes and/or other non-histone proteins. Areas of active chromatin show hypersensitivity to DNAse I digestion, whereas regions more tightly wrapped in nucleosomes are both inaccessible to components of the transcriptional apparatus, and relatively resistant to DNAse I. Therefore, mapping regions of relative DNAse I hypersensitivity is valuable for identifying numerous types of regulatory elements, including promoters and enhancers.

Recently, a method for coupling DNAse I digestion to high throughput sequencing, DNAse-seq (Boyle et al., 2008), was developed to identify the location of DNAse I hypersensitive sites at the whole genome level (Figure 2.5). Nuclei are

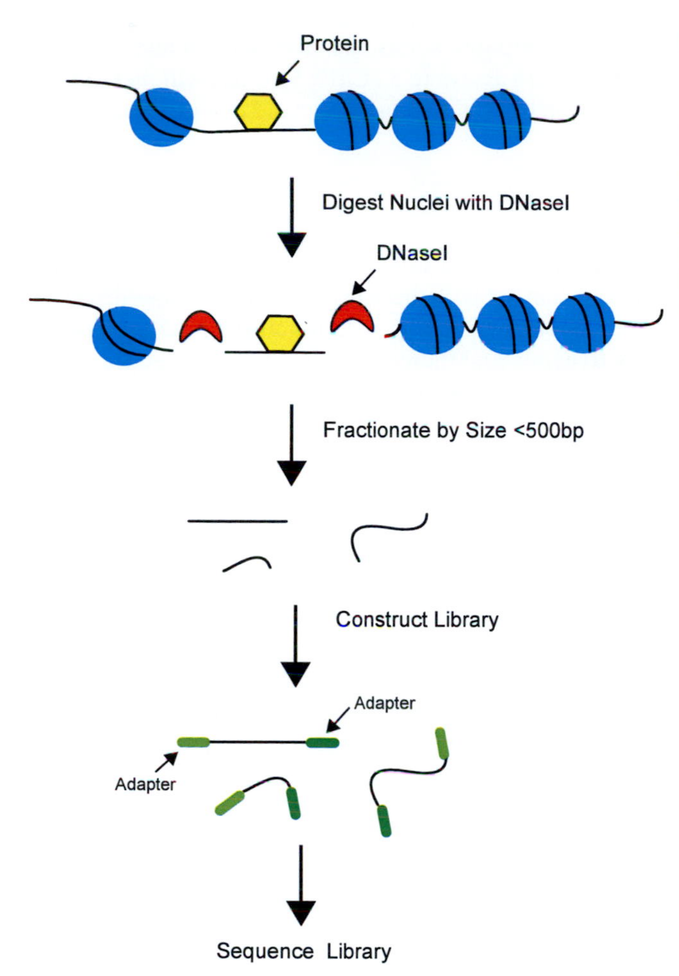

FIGURE 2.5 An Overview of the Workflow for DNAse-seq. The top of this illustration shows a double-stranded DNA molecule wrapped around four nucleosomes (blue) and a relatively open segment of chromatin containing a cis-regulatory region with a bound protein (yellow). DNAse (red) cleavage occurs preferentially in the open chromatin region releasing it from the bulk DNA. After size fractionation, the DNA is adaptor ligated and amplified to create a library that is ready for high throughput sequencing. Since DNA from open chromatin regions is released in greater amounts than from closed chromatin, open chromatin regions are over-represented in the library. Therefore, sequence reads tend to pile up in genomic intervals containing open chromatin.

subjected to a limited digestion by DNAse I to create breaks in open chromatin without completely digesting all unbound DNA. Since open chromatin is more sensitive to cleavage, DNA fragments from these regions are preferentially released from the bulk genomic DNA. These are isolated, tagged at both ends with adaptors, and subjected to high throughput DNA sequencing. The resulting sequence reads are then mapped against the reference genome, and accumulated sequence tags are analyzed. Reads accumulate in the regions where DNAse I cleaves the DNA and thus mark the locations of relatively open chromatin. This approach has the advantage of discovering new important enhancers without having *a priori* knowledge about the location and identity of bound transcription factors. Use of this technique in early *Drosophila* embryos has demonstrated the dynamic nature of chromatin architecture on developmentally regulated genes (Li et al., 2011; Thomas et al., 2011).

While DNAse-seq is gaining popularity, an alternative approach to identifying regulatory regions in chromosomes is Formaldehyde-Assisted Identification of Regulatory Elements (FAIRE-seq; Song et al., 2011), which is based on formaldehyde crosslinking in a similar way to ChIP. Instead of using a specific antibody to immunoprecipitate regions of bound DNA, fragmented chromatin is directly subjected to phenol-chloroform extraction. In this case, nucleosome-depleted fractions of DNA are preferentially segregated into the aqueous phase and subsequently identified by sequencing. Thus, FAIRE-enriched DNA fragments correspond to active cis-regulatory regions, similar to DNAse-seq.

DNAse-seq has been used for high-resolution footprint analysis. By mapping very large numbers of sequence reads (in the range of several hundred million per analysis instead of tens of millions of reads for ChIP-seq) over the genome, individual transcription factor footprints can be resolved at single base resolution (Boyle et al., 2011; Neph et al., 2012). The sequence read distribution within each DNAse-seq peak is non-uniformly distributed. Proteins bound to cis-regulatory elements within open chromatin act to protect the bound sequence (footprints) from DNAse I cleavage, and therefore reads mapping within DNAse-seq peaks flank protein binding sites. This high-resolution footprint analysis is remarkable as it gives a view of where transcription factors physically contact DNA and is likely to identify hundreds of thousands of *cis*-regulatory elements across the genome. A daunting task is to then determine the precise identities of the transcription factors bound within these cis-regulatory sequences from the ~1500 transcription factors encoded in vertebrate genomes. It is anticipated that continued motif discovery using both *in vivo* and *in vitro* assays (e.g., ChIP, SELEX) will increase our knowledge of binding preferences for individual transcription factors, enabling a more accurate and complete annotation of DNAse I footprints.

2.5 INTERACTIONS AT A DISTANCE

Enhancers act at a distance from the promoters of genes that they regulate, and in some cases enhancers can be located millions of base pairs away. The dominant theory for how enhancers can achieve these effects involves the formation of large DNA loops with the protein-bound enhancers and promoter elements being held together at the base of the loops. These looping interactions are presumed to be temporally dynamic and therefore only occur at specific stages of development or cell differentiation. Neither ChIP-seq nor DNAse-seq are designed to capture information on DNA looping.

Chromatin Conformation Capture (3C) was developed as an assay to examine DNA looping interactions (Dekker et al., 2002; Figure 2.6A). The 3C approach crosslinks proteins and DNA in complexes at the bases of DNA loops. These loops are then digested with restriction enzymes, and the multiple free ends created are subjected to intracomplex ligation. This results in covalent linkage of DNA ends from different genomic regions to one another within the complex that can subsequently be identified. After reversing protein-DNA crosslinks, the pool of resulting chimeric DNA fragments containing re-ligated DNA are subjected to PCR amplification using oligonucleotide primers specific to regions flanking the suspected looping. The 3C assay can be used to examine DNA looping over a time course of differentiation to determine when changes in long range enhancer interactions correlate with gene expression changes. Because this is a PCR-based approach, the throughput of the assay is low, and the approach is limited to regions of suspected interaction. There are several versions of the 3C assay that have increased throughput (de Wit and de Laat, 2012). Chromosome Conformation Capture "Carbon-Copy" (5C) experiments simultaneously use thousands of oligonucleotides in one experiment to detect millions of interactions, instead of using a few selected primers to analyze a handful of interactions (Dostie et al., 2006). 5C was successfully used in mammalian cell lines to analyze long-range interactions covering ~400kb between enhancers of the human β-globin locus that likely function during erythrocyte differentiation.

The most recent advance to this technology, Chromatin Interaction Analysis with Paired-End Tags (ChIA-PET; Fullwood et al., 2009) combines ChIP for a specific transcription factor with Chromosome Conformation Capture methods (Figure 2.6B). Immunoprecipitated complexes containing DNA loops are processed with modifications to Chromosome Conformation Capture methods and then the intramolecular ligation products are paired-end sequenced. The reads from the two ends of these ligation products are mapped back to the genome to identify pile ups of read pairs with an orientation consistent with regions of DNA looping (Figure 2.6B,C). ChIA-PET provides genome-wide information for DNA looping involving specific transcription factors of interest and reveals complicated enhancer interactions. While ChIA-PET assays do not require *a priori* knowledge of suspected looping, these experiments are very challenging, requiring large numbers of cells for each analysis (in the range of 10^9 cells), and sophisticated bioinformatics. Using such data, we may be able to address questions such as how cis-regulatory elements interact dynamically in the 3D architecture of chromatin, and how different enhancers decide when to interact with specific promoters and enhancers. The answers to these questions are likely to be complex, and will involve both the presence of other interacting transcription factors, availability of DNA target sites, and the state of local chromatin modifications. These analyses provide information on the function of transcription factors bound great distances away from gene promoters and add to the complex temporal regulation of gene transcription during cellular differentiation.

2.6 PROSPECTS

In the last five years the cost of DNA sequencing has plummeted, largely due to a substantial increase in throughput. The ability to sequence massive numbers of DNA fragments in parallel makes whole genome sequencing both cost and time effective. In this chapter we have presented some of the new sequencing technologies that have been developed to gain deeper insights into gene function at the genome-wide level. RNA-seq facilitates the identification of regions of the genome that are transcribed, can reveal the presence of novel genes, and allows for quantitation of RNA expression levels. ChIP-seq identifies regions of the genome that are protein bound, identifying sites used by transcription factors, where RNA polymerases are actively engaged in transcription, and where certain chromatin modifications are present. DNAse-seq reveals regions of relatively open chromatin, which are often cis-regulatory elements controlling gene expression. This method has the promise of also revealing discrete transcription factor binding information.

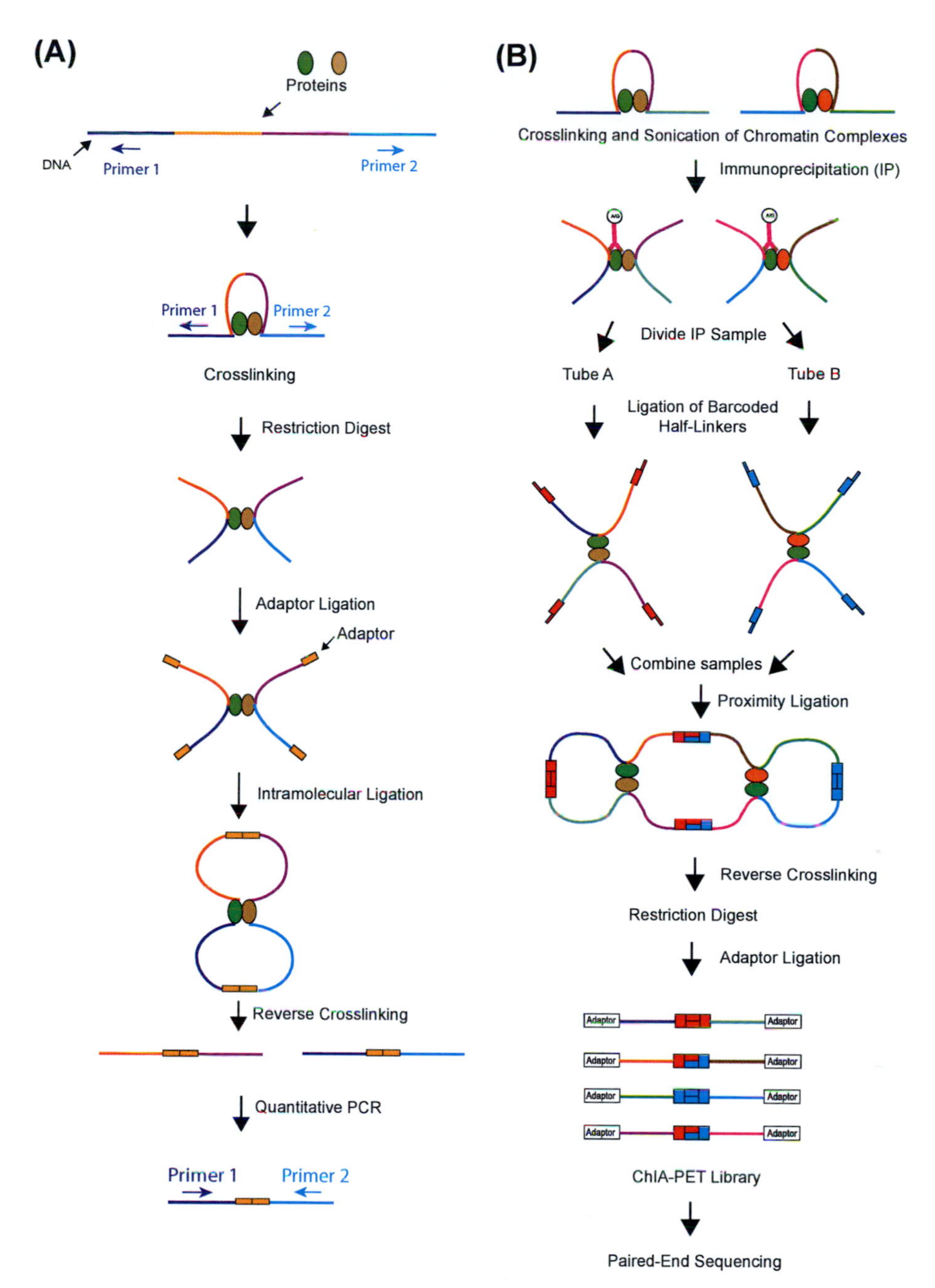

FIGURE 2.6 An Overview of Chromatin Conformation Capture (3C) and ChIA-PET Methods. Panel A shows a workflow for the Chromatin Conformation Capture (3C) method. A linear double-stranded DNA molecule forms a protein-DNA complex with a DNA loop. Segments of the loop are illustrated with different colors to assist in following the topology during subsequent steps. Following formaldehyde crosslinking of the protein-DNA complexes, a restriction enzyme is used to digest the loops and adaptors are ligated to the free ends. After adaptor addition, an intramolecular ligation step allows all combinations (only the informative combination is shown) of free ends within the protein-DNA complex to ligate. Following reverse crosslinking, specific oligonucleotide combinations can be used in qPCR reactions to demonstrate DNA fusions, hence confirming the formation of a suspected loop. Note that the inwardly oriented oligo pair at the bottom of the panel corresponds to an outwardly pointed orientation in the native DNA at the top of the figure. Panel B shows the approach for ChIA-PET. Two different DNA loops containing the green transcription factor are shown with different color-coded DNA segments. After crosslinking and restriction digestion of the loops, protein-DNA complexes are subjected to ChIP using antibodies specific to the green transcription factor, which is suspected to be involved in DNA looping. Following ChIP enrichment of green protein-DNA complexes, the sample is subdivided into two tubes. Each tube is then subjected to a ligation step using half-linkers (adaptors containing sticky ends) with different sequences (indicated with red and blue "bar codes"). The two samples are then mixed together and a ligation is performed at low DNA concentration to favor intramolecular "proximity" ligation. Many combinations of ligated DNA ends will be generated but informative ligation products are shown here. Infrequent intermolecular ligation products are marked by the hybrid red/blue linker ligations, whereas intramolecular ligations – the desired ligation products – are marked by red/red and blue/blue linker ligations. Following ligation, crosslinks in the protein-DNA complexes are broken and the DNA is digested with a restriction enzyme. Linearized DNA fragments are released containing internal ligated linkers marking fusion junctions. Adaptors are then ligated to these linear products, which are amplified to produce the final ChIA-PET library. Paired-end sequencing is performed and the reads are analyzed.

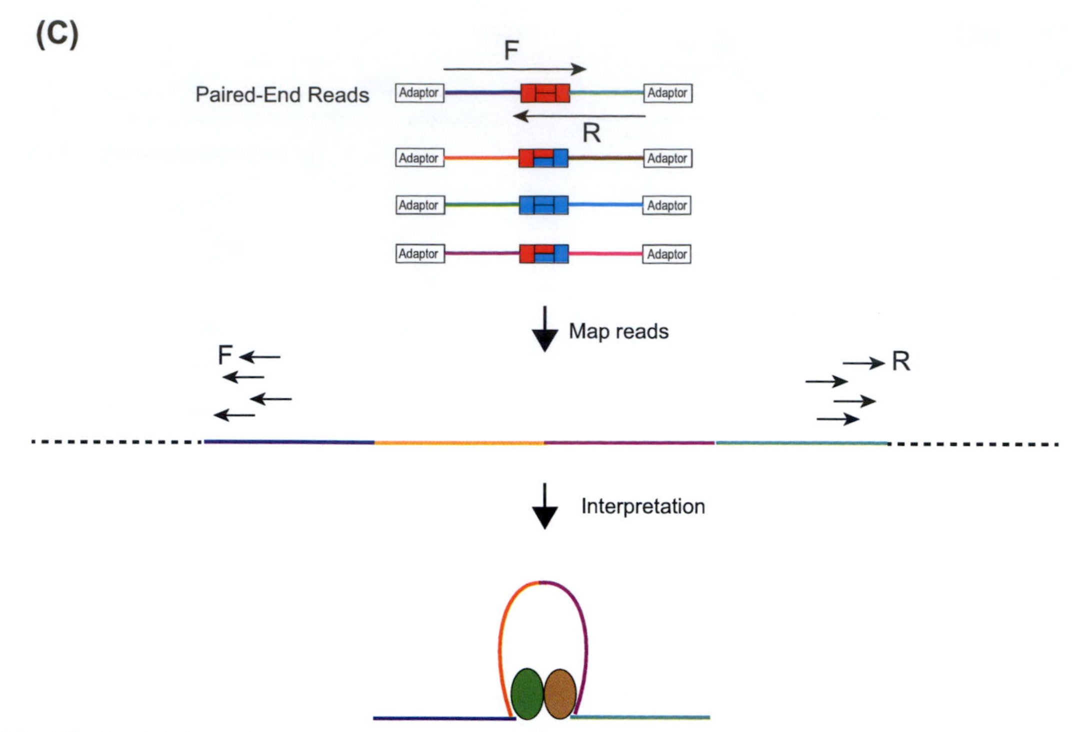

FIGURE 2.6 Cont'd Panel C shows the outcome of ChIA-PET sequencing. The library is sequenced from both ends of each fragment ("paired end" reads) and linker sequences inside the reads mark the ligation junctions. Any reads that contain red/blue hybrid adaptors will be identified and as products of intermolecular ligation and discarded. Quantitation of the frequency of red/blue hybrid ligations provides an estimate of the total percentage of reads that likely to contain false junctions. Read pairs (F and R, for forward and reverse) from intramolecular ligations are then mapped to the genome. The genomic interval from Panel B is shown in linear form. Two regions containing read pairs, with read orientations pointing outward, indicate that these were in close proximity in the intramolecular ligation as a result of DNA looping.

ChIA-PET, finds regions of DNA looping that involve a specific transcription factor. While technically challenging, this method shows great promise for understanding the dynamics of long-range cis-regulatory element functions.

Locating protein-DNA binding sites is important for identifying the inputs into gene regulation. Combined with loss-of-function studies to determine how bound proteins impact gene output (RNA), we can make genome-wide connections between transcription factors and target genes. Once these connections between numerous transcription factors and their targets are mapped, gene regulatory networks can be built to predict how cell fate and behaviors are controlled. With the availability of such networks, we can understand how mutations and disease states arise and determine the most efficient ways to reverse these processes.

There are many other applications for high throughput sequencing that are relevant to developmental genetics but have not been discussed in detail here. Great efforts are being made to connect phenotypic variations to mutations in the genome to provide genetic insights into human disease. Large-scale mapping of single nucleotide polymorphisms (SNPs) is providing the raw materials for genome-wide association studies (GWAS). These studies correlate the co-segregation of SNPs with human genetic diseases to permit identification of the genetic changes that cause the abnormalities. Cancer genomics is another area in which high throughput sequencing approaches are being applied to achieve better diagnostic tools, to create personalized therapies dependent on tumor gene expression profiles and mutations, and to understand the impact of complex genomic rearrangements common to many tumors.

2.7 CLINICAL RELEVANCE

- High throughput sequencing can be used to link gene activity to phenotype.
- Transcription factor binding sites are proposed to be a major site of variation between individuals. Finding these sites provides useful information relevant to phenotypic variation and disease.

RECOMMENDED RESOURCES

http://seqanswers.com/.

http://www.ebi.ac.uk/training/online/course/ebi-next-generation-sequencing-practical-course.

http://www.encodeproject.org/ENCODE/(all accessed 26 Feb 2014).

Zeng, W., Mortazavi, A., 2012. Technical considerations for functional sequencing assays. Nat. Immunol. 13, 802–807.

REFERENCES

Akkers, R.C., van Heeringen, S.J., Jacobi, U.G., Janssen-Megens, E.M., Françoijs, K.J., Stunnenberg, H.G., Veenstra, G.J., 2009. A hierarchy of H3K4me3 and H3K27me3 acquisition in spatial gene regulation in *Xenopus* embryos. Dev. Cell 17, 425–434.

Bailey, T.L., Elkan, C., 1994. Fitting a mixture model by expectation maximization to discover motifs in biopolymers. Proceedings of the Second International Conference on Intelligent Systems for Molecular Biology. 28–36. AAAI Press, Menlo Park, California.

Barski, A., Cuddapah, S., Cui, K., Roh, T.-Y., Schones, D.E., Wang, Z., Wei, G., Chepelev, I., Zhao, K., 2007. High-resolution profiling of histone methylations in the human genome. Cell 129, 823–837.

Boyle, A.P., Davis, S., Shulha, H.P., Meltzer, P., Margulies, E.H., Weng, Z., Furey, T.S., Crawford, G.E., 2008. High-resolution mapping and characterization of open chromatin across the genome. Cell 132, 311–322.

Boyle, A.P., Song, L., Lee, B.K., London, D., Keefe, D., Birney, E., Iyer, V.R., Crawford, G.E., Furey, T.S., 2011. High-resolution genome-wide *in vivo* footprinting of diverse transcription factors in human cells. Genome Res. 21, 456–464.

Cabili, M.N., Trapnell, C., Goff, L., Koziol, M., Tazon-Vega, B., Regev, A., Rinn, J.L., 2011. Integrative annotation of human large intergenic noncoding RNAs reveals global properties and specific subclasses. Genes Dev. 25, 1915–1927.

de Wit, E., de Laat, W., 2012. A decade of 3C technologies: insights into nuclear organization. Genes Dev. 26, 11–24.

Dekker, J., Rippe, K., Dekker, M., Kleckner, N., 2002. Capturing chromosome conformation. Science 295, 1306–1311.

Dostie, J., Richmond, T.A., Arnaout, R.A., Selzer, R.R., Lee, W.L., Honan, T.A., Rubio, E.D., Krumm, A., Lamb, J., Nusbaum, C., Green, R.D., Dekker, J., 2006. Chromosome Conformation Capture Carbon Copy (5C): a massively parallel solution for mapping interactions between genomic elements. Genome Res. 16, 1299–1309.

Fullwood, M.J., Liu, M.H., Pan, Y.F., Liu, J., Xu, H., Mohamed, Y.B., Orlov, Y.L., Velkov, S., Ho, A., Mei, P.H., Chew, E.G., Huang, P.Y., Welboren, W.J., Han, Y., Ooi, H.S., Ariyaratne, P.N., Vega, V.B., Luo, Y., Tan, P.Y., Choy, P.Y., Wansa, K.D., Zhao, B., Lim, K.S., Leow, S.C., Yow, J.S., Joseph, R., Li, H., Desai, K.V., Thomsen, J.S., Lee, Y.K., Karuturi, R.K., Herve, T., Bourque, G., Stunnenberg, H.G., Ruan, X., Cacheux-Rataboul, V., Sung, W.K., Liu, E.T., Wei, C.L., Cheung, E., Ruan, Y., 2009. An oestrogen-receptor-alpha-bound human chromatin interactome. Nature 462, 58–64.

Garber, M., Grabherr, M.G., Guttman, M., Trapnell, C., 2011. Computational methods for transcriptome annotation and quantification using RNA-seq. Nat. Methods 8, 469–477.

Grabherr, M.G., Haas, B.J., Yassour, M., Levin, J.Z., Thompson, D.A., Amit, I., Adiconis, X., Fan, L., Raychowdhury, R., Zeng, Q., Chen, Z., Mauceli, E., Hacohen, N., Gnirke, A., Rhind, N., di Palma, F., Birren, B.W., Nusbaum, C., Lindblad-Toh, K., Friedman, N., Regev, A., 2011. Full-length transcriptome assembly from RNA-Seq data without a reference genome. Nat. Biotechnol. 29, 644–652.

Heintzman, N.D., Hon, G.C., Hawkins, R.D., Kheradpour, P., Stark, A., Harp, L.F., Ye, Z., Lee, L.K., Stuart, R.K., Ching, C.W., Ching, K.A., Antosiewicz-Bourget, J.E., Liu, H., Zhang, X., Green, R.D., Lobanenkov, V.V., Stewart, R., Thomson, J.A., Crawford, G.E., Kellis, M., Ren, B., 2009. Histone modifications at human enhancers reflect global cell-type-specific gene expression. Nature 459, 108–112.

Johnson, D.S., Mortazavi, A., Myers, R.M., Wold, B., 2007. Genome-wide mapping of *in vivo* protein-DNA interactions. Science 316, 1497–1502.

Li, X.Y., Thomas, S., Sabo, P.J., Eisen, M.B., Stamatoyannopoulos, J.A., Biggin, M.D., 2011. The role of chromatin accessibility in directing the wide-spread, overlapping patterns of *Drosophila* transcription factor binding. Genome Biol. 12, R34.

Lin, C.Y., Vega, V.B., Thomsen, J.S., Zhang, T., Kong, S.L., Xie, M., Chiu, K.P., Lipovich, L., Barnett, D.H., Stossi, F., Yeo, A., George, J., Kuznetsov, V.A., Lee, Y.K., Charn, T.H., Palanisamy, N., Miller, L.D., Cheung, E., Katzenellenbogen, B.S., Ruan, Y., Bourque, G., Wei, C.L., Liu, E.T., 2007. Whole-genome cartography of estrogen receptor alpha binding sites. PLoS Genet. 3, e87.

Morin, R.D., O'Connor, M.D., Griffith, M., Kuchenbauer, F., Delaney, A., Prabhu, A.L., Zhao, Y., McDonald, H., Zeng, T., Hirst, M., Eaves, C.J., Marra, M.A., 2008. Application of massively parallel sequencing to microRNA profiling and discovery in human embryonic stem cells. Genome Res. 18, 610–621.

Mortazavi, A., Williams, B.A., McCue, K., Schaeffer, L., Wold, B., 2008. Mapping and quantifying mammalian transcriptomes by RNA-Seq. Nat. Methods 5, 621–628.

Nagalakshmi, U., Wang, Z., Waern, K., Shou, C., Raha, D., Gerstein, M., Snyder, M., 2008. The transcriptional landscape of the yeast genome defined by RNA sequencing. Science 320, 1344–1349.

Neph, S., Vierstra, J., Stergachis, A.B., Reynolds, A.P., Haugen, E., Vernot, B., Thurman, R.E., John, S., Sandstrom, R., Johnson, A.K., Maurano, M.T., Humbert, R., Rynes, E., Wang, H., Vong, S., Lee, K., Bates, D., Diegel, M., Roach, V., Dunn, D., Neri, J., Schafer, A., Hansen, R.S., Kutyavin, T., Giste, E., Weaver, M., Canfield, T., Sabo, P., Zhang, M., Balasundaram, G., Byron, R., MacCoss, M.J., Akey, J.M., Bender, M.A., Groudine, M., Kaul, R., Stamatoyannopoulos, J.A., 2012. An expansive human regulatory lexicon encoded in transcription factor footprints. Nature 489, 83–90.

Rada-Iglesias, A., Bajpai, R., Swigut, T., Brugmann, S.A., Flynn, R.A., Wysocka, J., 2011. A unique chromatin signature uncovers early developmental enhancers in humans. Nature 470, 279–283.

Schones, D.E., Cui, K., Cuddapah, S., Roh, T.-Y., Barski, A., Wang, Z., Wei, G., Zhao, K., 2008. Dynamic regulation of nucleosome positioning in the human genome. Cell 132, 887–898.

Schulz, M.H., Zerbino, D.R., Vingron, M., Birney, E., 2012. Oases: robust *de novo* RNA-seq assembly across the dynamic range of expression levels. Bioinformatics 28, 1086–1092.

Song, L., Zhang, Z., Grasfeder, L.L., Boyle, A.P., Giresi, P.G., Lee, B.K., Sheffield, N.C., Gräf, S., Huss, M., Keefe, D., Liu, Z., London, D., McDaniell, R.M., Shibata, Y., Showers, K.A., Simon, J.M., Vales, T., Wang, T., Winter, D., Zhang, Z., Clarke, N.D., Birney, E., Iyer, V.R., Crawford, G.E., Lieb, J.D., Furey, T.S., 2011. Open chromatin defined by DNaseI and FAIRE identifies regulatory elements that shape cell-type identity. Genome Res. 21, 1757–1767.

Thomas, S., Li, X.Y., Sabo, P.J., Sandstrom, R., Thurman, R.E., Canfield, T.K., Giste, E., Fisher, W., Hammonds, A., Celniker, S.E., Biggin, M.D., Stamatoyannopoulos, J.A., 2011. Dynamic reprogramming of chromatin accessibility during *Drosophila* embryo development. Genome Biol. 12, R43.

Wilhelm, B.T., Marguerat, S., Watt, S., Schubert, F., Wood, V., Goodhead, I., Penkett, C.J., Rogers, J., Bähler, J., 2008. Dynamic repertoire of a eukaryotic transcriptome surveyed at single-nucleotide resolution. Nature 453, 1239–1243.

Zerbino, D.R., Birney, E., 2008. Velvet: algorithms for *de novo* short read assembly using de Bruijn graphs. Genome Res. 18, 821–829.

Chapter 3

Using Mutagenesis in Mice for Developmental Gene Discovery

David R. Beier

Center for Developmental Biology and Regenerative Medicine, Seattle Children's Research Institute, Seattle, WA

Chapter Outline

GLOSSARY

Haploinsufficiency The condition in which having only single functional allele in a diploid organism results in pathology.

Metastable epiallele Genes that are variably expressed in genetically identical individuals due to epigenetic modifications established during early development.

SNP Single nucleotide polymorphism.

Whole exome sequencing Selective sequencing of the entire known coding sequence complement of a genome, usually after targeted capture by hybridization.

SUMMARY

- The use of *N*-ethyl-*N*-nitrosourea (ENU) mutagenesis for phenotype-driven genetic analysis has been an extremely productive means to investigate mammalian developmental biology and human congenital disorders.
- By use of sensitive assays for developmental processes, such as mice carrying transgenic reporter lines, very specific and subtle developmental processes can be queried.
- Rapid developments in tools for genomic analysis, in particular next-generation DNA sequencing methods, have dramatically decreased the effort required for the positional cloning of ENU-induced pathogenic mutations.
- A particular area of opportunity for ENU mutagenesis is the characterization of modifier loci affecting specific developmental defects.

3.1 USE OF ENU AS A MUTAGEN

The first characterizations of *N*-ethyl-*N*-nitrosourea (ENU) as an effective mutagen were undertaken by Russell and colleagues as part of his studies of DNA damaging agents at the Oak Ridge National Laboratory. Russell had developed the "specific locus test" as a method to quantify DNA mutation rates. In this analysis, wild-type mice subjected to various presumptive mutagenic regimens were crossed with a line of mice that were homozygous for seven different morphological or coat color mutations, which were readily scored by simple inspection. In an era when breeding of thousands of mice was not cost-prohibitive, this enabled the accurate assessment of mutation rates on a "per locus" basis, as the number of homozygous mutant progeny revealed the carrier frequency of new mutations in the treated parental mice.

These investigations showed that intraperitoneal injection of ENU could efficiently generate germ cell mutations without causing systemic morbidity. Additional studies determined that the efficiency of mutagenesis could be optimized by use of a fractionated dose (Russell et al., 1982). In this fashion, frequencies of 6–15 x 10^{-4} mutations per locus per gamete can be obtained, 1000-fold higher than the spontaneous rate of mutation of 0.5 x 10^{-6}. Characterization of ENU mutations in *Drosophila* revealed that the molecular basis of these mutants were single-base changes, and many recent studies have confirmed that this is the case in mammalian cells as well. The practical utility of ENU as a mutagen was first demonstrated by Bode and colleagues, who generated mouse models of hyperphenylalaninemia using a breeding strategy for uncovering recessive mutations that remains a paradigm for the field (see Figure 3.1) (McDonald et al., 1990; McDonald and Bode 1988).

However, while the utility of ENU mutagenesis for generating mutations was clear, it was not readily adopted as a means to study mammalian biology, primarily due to the difficulty at the time of identifying single nucleotide changes. Fortunately, the method was maintained and applied by investigators in a few laboratories; notably that of Guenet, for the generation of models of metabolic diseases (Montagutelli et al., 1994; Tutois et al., 1991), Favor, to genes causing cataracts (Favor et al., 1991), and Dove, who used it to generate the widely used *Min* mouse model of colorectal cancer that harbors a mutation in the APC gene (Moser et al., 1990). With the development of the tools for genomic analysis, which included the identification of strain-specific genetic markers and efficient methods to score them, ready access to DNA sequencing, and characterization of the mouse genome, it became feasible to pursue the positional cloning of ENU-induced mutant loci. For example, the development of panels of single nucleotide polymorphism (SNP) markers that were informative for many different strain combinations allowed genome-wide genotype analysis of multiple mice in a single experiment, which often enabled moderate-to-high resolution definition of recombinant intervals carrying the mutant locus (Moran et al., 2006). The generation of tools that simplified the substantial task of positional cloning led to the rapid implementation of this approach for phenotype-driven analysis of mouse biology.

As discussed below, this strategy has been particularly useful for analysis of mutations affecting development, likely due to the qualitative nature of these phenotypes and the relative simplicity of their ascertainment. However, given the development of potent new technologies for generating targeted mutations (Yang et al., 2013), it is necessary to examine the utility of an approach that requires a substantial mapping and cloning effort for mutant characterization; i.e., a phenotype-driven (vs. genotype-driven) strategy. In this regard, the expedience of the latter must be balanced against the utility of obtaining unbiased insight into the genetic basis of a trait. Many of the studies employing forward genetic analysis have yielded unexpected insights and have identified mutations in novel genes whose function were otherwise unknown. The importance of this cannot be overstated, as this approach is virtually guaranteed to identify genes that have fundamental biological roles, and further has the potential for identifying novel pathways for therapeutic intervention in human disease. While efficient knock-out methodology provides a powerful means for investigating mammalian biology, a phenotype-driven strategy is a complementary approach that will allow one to more directly query the genes that are required for specific developmental processes.

Another aspect of this approach is that the single-base changes induced by ENU often have functional consequences that are more varied than the null mutations most frequently generated by genomic targeting methods. For example, our discovery of the first gene that is causal for the human disorder of congenital diaphragmatic hernia was facilitated by our

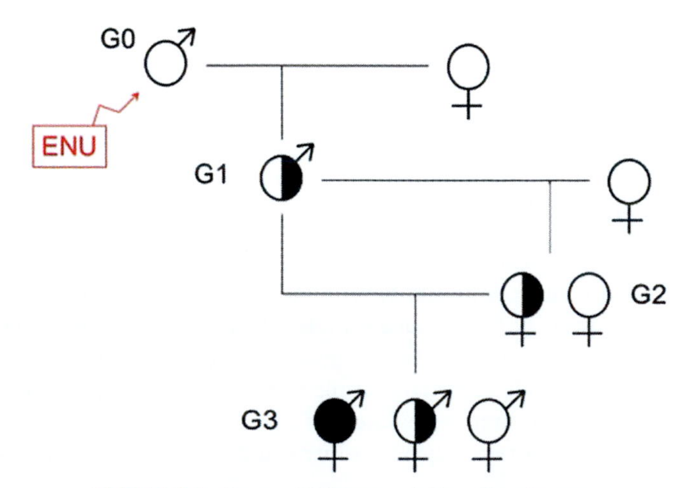

FIGURE 3.1 Forward Mutagenesis Breeding Strategy.

characterization of an ENU-induced mouse mutant carrying a splice-site mutation in the gene *Fog2* (*Zfpm2*) (Ackerman et al., 2005). *Fog2* null mutants had been generated and characterized by two different groups in detail; they were found to die in mid-gestation due to cardiac defects. While the ENU-induced allele also frequently resulted in heart abnormalities, a subset of mice survived to later stages of embryonic development; these were found to have poorly formed diaphragms. This subset likely survived because the mice expressed a small amount of normally spliced transcript, sufficient to mitigate the early consequences of *Fog2* deficiency. Hypomorphic alleles induced by ENU are quite common; due either to "leaky" splice-site mutations as described above, or due to missense mutations that do not completely abrogate gene function. Missense mutations induced by ENU can also result in dominant effects due to "gain-of-function", which can provide insight into protein domain function and roles in specific developmental pathways not readily obtained by characterization of null mutants. For example, transcription factor AP-2alpha (*Tcfap2a*) null homozygote mice show a large spectrum of developmental defects, among them missing middle ear bones and tympanic ring. Steel and colleagues identified a dominantly inherited mutant that has a missense mutation in the PY motif of the transactivation domain of *Tcfap2a*, which results in misshapen malleus, incus, and stapes without any other observable phenotype (Ahituv et al., 2004). *In vitro* assays suggested that this mutation causes a "gain-of-function" in the transcriptional activation of AP-2alpha.

3.2 ENU-INDUCED MUTATIONS IN MICE

A standard strategy for an ENU mutagenesis screen is shown in Figure 3.1 and is described as follows: Male mice are treated with a fractionated dose of ENU, which usually results in a period of infertility. Recovery from this occurs after 8–12 weeks, and the treated mice (designated G0 for Generation 0) are then bred. Dominantly or semi-dominantly inherited phenotypes can be assayed in their first generation (G1) progeny, to the extent that it is logistically feasible to assay large numbers of mice (Hrabe de Angelis et al., 2000; Nolan et al., 2000a). The productivity of this approach can be further enhanced by analyzing each mouse using a variety of tests (Gailus-Durner et al., 2005; Nolan et al., 2000b).

However, most mutations in the mouse have no or only subtle effects in heterozygotes. As such, a multi-generation breeding scheme is required to uncover recessive mutations. For reasons as yet unclear, this is particularly the case for developmental mutations; for example, haploinsufficiency for Sonic hedgehog (*Shh*) causes holoprosencephaly in humans (Roessler et al., 1996) but heterozygous mice appear essentially normal, even though homozygous embryos have profound patterning defects (Chiang et al., 1996). (My entirely untested speculation is that this is a function of size: most murine and human organs have essentially identical structure, but the latter are comprised of many more cells. Thus subtle defects in developmentally-relevant cellular processes caused by haploinsufficiency may be amplified and have phenotypic consequences in humans that are not detectable in mice.) Additionally, in contrast to dominant screens, recessive screens can identify mutations that result in lethality. While this additional breeding makes it less feasible to test larger numbers of G1 mice than in a screen for dominant mutations, this is offset by the fact that each G1 male carries many mutations. Russell's studies demonstrate that the per locus mutation frequency of optimized fractionated dose ENU treatment is about 1/1000. The nature of the specific locus test that Russell employed would primarily uncover mutations that were effectively null. Assuming 25,000 genes in the mouse genome, one can anticipate that a G1 male carries 25 null mutant alleles, and potentially many more with hypomorphic effects. The general strategy of husbandry described by Bode is a commonly used means of recovering recessive mutations, although other strategies utilizing G1 intercrosses are feasible.

The wealth of recent studies using ENU mutagenesis for a wide variety of phenotypic analyses demonstrates unequivocally that this approach is efficient (with respect to generating single nucleotide changes), that these mutations can be readily mapped (assuming reasonable penetrance), and that they can ultimately be identified using straightforward approaches of positional cloning. Thus the only unknown in determining the success of a mutagenesis experiment is the likelihood that a heritable phenotypic variant will be identified. At one level, this is a function of the number of genes that are required for the generation or maintenance of "normal," with respect to the phenotype in question. If, for example, it is a single gene, the probability of ascertaining this in a standard mutagenesis analysis is low. If, however, tens to hundreds of genes are required to generate or maintain the wild-type state, the likelihood of their discovery in a modest-size screen is quite reasonable. Given the latter scenario, the key determinant for success is whether the phenotype to be screened is amenable to reasonably high-throughput analysis (as even a modest-size screen still requires the analysis of hundreds to thousands of progeny) and whether the assay will unambiguously discriminate between normal variation and a truly mutant phenotype (as the latter will still be rare relative to the number of normal mice). With this general principle in mind, ENU mutagenesis can be applied for investigation of a wide variety of biological processes, as discussed in the following paragraphs.

As noted, the development of tools for genomic analysis has encouraged the utilization of this approach for a wide variety of phenotypes. The Mouse Genome Informatics database (MGI; http://www.informatics.jax.org) lists 2952 ENU-induced mutant alleles (i.e., ones in which the mutation has been characterized) corresponding to 2322 mutant genes

(Table 3.1). ENU mutagenesis has arguably been most successfully applied for discovering genes required for normal development. This is likely due to a number of factors: the biological complexity of the process (such that there are many target loci), the relative simplicity and qualitative nature of the assay (which may often require only inspection), and the fact that compensatory changes in gene expression are unlikely to "correct" a developmental defect and thus mask its occurrence (as compared to physiological/metabolic systems, which frequently have abundant buffering capacities). Furthermore, the present wealth of knowledge regarding the genetic basis of development often allows investigators to rapidly integrate novel findings into existing paradigms. This is highly advantageous, as otherwise it can be difficult to obtain a mechanistic understanding of the role of a gene with limited functional annotation.

3.3 ENU-INDUCED MUTATIONS AFFECTING DEVELOPMENT

An example of the importance of forward mutagenesis analysis for developmental gene discovery is its role in uncovering the requirement for primary cilia in the mediation of embryonic patterning. Much of this insight was obtained from studies by Kathryn Anderson and colleagues of mutants ascertained at mid-gestation with morphological defects (reviewed in Goetz et al., 2009). These were examined for abnormalities in neural tube patterning, and then more specifically by genetic and biochemical analysis for perturbation of the Hedgehog pathway. Remarkably, many of these proved to be genes previously implicated in flagellar function in Chlamydomonas, and, by extension, cilial function in vertebrates. Genes uncovered in mutagenesis studies by other investigators have contributed to our understanding of this pathway as well (Endoh-Yamagami et al., 2009; Ermakov et al., 2009; May et al., 2005; Tran et al., 2008). Complementing this are studies by Lo and colleagues in which many cilial mutations causing abnormalities of LR asymmetry were ascertained using ultrasound imaging of embryos *in utero* (Yu et al., 2004). The contribution of mutagenesis studies to the analysis of cilial function can be appreciated by considering the ENU-induced alleles exist for numerous genes that are relevant to its biology and pathology, including dyneins (*Dnah11, Dnah5, Dnaic1, Drc1, Dync1h1, Dync1i2, Dync1li1, Dync2h1*), intraflagellar transport proteins (*Ift88, Ift122, Ift 140, Ift 172, Ttc21b (Ift139), Wdr19 (Ift 144*), nephronophthisis proteins (*Nphp3, Nphp4, Cep290, Zfp423 Tmem67, Wdr19*), and other ciliopathy-associated genes (*Arl13b, Inpp5e, Mks1, Pkd1*). Importantly, while these studies underscore the requirement for primary cilia in developmental processes, the mechanism by which cilia mediate these signaling pathways remains largely unknown. This highlights how gene discovery in phenotype-driven analysis is usually just the first step in understanding biological function.

While the large number of ENU-induced mutations affecting development listed in Table 3.1 speaks to its utility, it is important to recognize that this approach, when applied broadly to developmental phenotypes, will yield causative

TABLE 3.1 ENU-induced Mutants in Mouse *Genome* Database

Phenotype	Alleles	Genes
Cardiovascular	403	354
Craniofacial	301	270
Digestive/alimentary	144	128
Embryogenesis	210	179
Endocrine/exocrine	105	95
Growth/size	1125	908
Hematopoietic	403	300
Immune	340	253
Integument	471	322
Limbs/digits/tail	564	486
Liver/biliary	96	83
Muscle	118	100
Nervous system	431	342
Total	2952	2322

mutations in all aspects of the relevant biology. For example, a query of MGI for ENU-induced mutations that have a skeletal phenotype generates a list of 136 mutants for which the affected gene has been identified by positional cloning. These include many genes that affect developmental patterning, such as *Shh, T, Wnt3a, Wnt7a, Pax6,* and *Hmx1*. Given that cilial genes mediate developmental signaling by *Shh*, many of the genes discussed in the paragraph above are also represented. Also included are genes that affect basic processes of chondrogenesis and osteogenesis, such as *Twist1, Trip11 (Gmap210),* and *Alx4*. Mutant alleles of genes than encode structural genes such as *Col1a1, Col2a1,* and *Frem1* are present, as are genes that mediate biochemical process relevant to bone, such as *Arsb, Phex, Galnt3* and *Tcirg1*. While this abundant yield illustrates the potential power of a phenotype-driven strategy, it also suggests that one may wish to constrain the phenotype so as to query developmental processes more specifically. The use of transgenic reporters for this purpose is discussed below. Alternatively, highly focused assays can be employed. For example, Hrabe de Angelis and colleagues utilized a screening panel of three clinical parameters that are commonly used as biochemical markers in patients with metabolic bone diseases (Sabrautzki et al., 2012). Screening of 9540 mice led to the identification of 71 new dominant mutant lines. Fifteen mutations among three genes (*Phex, Casr,* and *Alpl*) were identified by positional-candidate gene approaches and one mutation of the *Asgr1* gene was identified by next-generation sequencing, as discussed below.

As experience with ENU mutagenesis and confidence in its utility has grown, it has been applied for more targeted developmental analyses. For example, by introducing mice carrying transgenic reporter genes into mutagenesis screens, it is possible to query specific patterns of gene expression. This enables very sensitive ascertainment of abnormalities not apparent by simple morphological inspection. One area where this has proven highly productive is in the analysis of neural development, facilitated by the existence of a wide variety of transgenic lines with expression patterns corresponding to specific neuroanatomical tracts. One of the first studies in this arena was performed by Peterson and colleagues (Sabrautzki et al., 2012) using a transgenic reporter line that labeled the ganglionic eminences of the ventral forebrain and migrating cortical interneurons. The use of this reporter gene allowed the isolation of mutations that altered growth and morphogenesis of the cerebral cortex and, more specifically, allowed the identification of mutations disrupting interneuron migration from the ganglionic eminences into the cortex. This approach worked well, and the authors reported the identification of 13 mutants with defects in various aspects of cortical development. In another study that focused specifically on neural development, Cordes and colleagues used immunohistochemical visualization of cranial nerve patterning to screen for neurodevelopmental mutations, and identified seven mutant lines (Mar et al., 2005). The first neural developmental study we pursued employed a fortuitous transgenic insertion that expressed *Lacz* in the thalamocortical tract (Figure 3.2) (Dwyer et al., 2011). The expression of this reporter was highly reproducible

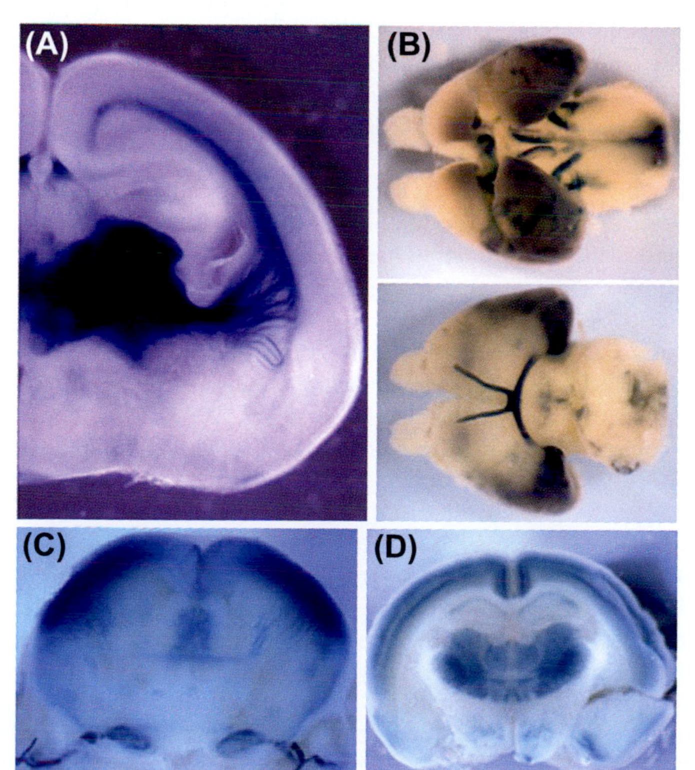

FIGURE 3.2 Transgenic Reporters of Neuroanatomy. (A) A transgenic insertion expresses *Lacz* in the thalamocortical tract (Dwyer et al., 2011). (B) *RARE-lacZ* transgenic mouse line expresses in the cerebral cortex, the hippocampus, and the optic tract (Rossant et al., 1991). (C) The *Tau-LacZ* transgenic line expresses *LacZ* in corticofugal axons (Ha et al., 2013). (D) The *Rgs4* insertional mutant expresses LacZ in layers II/III and V of the cortex (Ha et al., 2013).

and readily assayed by simply making a coronal slice across the exposed brain and staining with Xgal. In this study, six mutant lines were obtained that showed aberrant thalamo-cortical axon patterning phenotypes, including overfasciculation, misrouting to the ventral forebrain, and stalling at the cortical-striatal boundary. In our second study, we chose to use the RARE-*lacZ* transgenic mouse line (Rossant et al., 1991) to highlight several structures in the brain that we would otherwise not be able to readily score, including the optic nerves and hippocampus (Figure 3.2) (Stottmann et al., 2011). Unfortunately, this effort illustrated a potential hazard of mouse genetic analysis; namely, genetic background effects. While multiple mice with variant RARE-lacZ expression were initially ascertained, these patterns were found to have significant variability as we introduced different strains during the course of genetic mapping, and none of the abnormal phenotypes identified by the reporter proved consistent and heritable. Of note, advances in DNA sequencing may make outcrosses for genetic mapping unnecessary, as discussed in more detail below. In our most recent study, ENU mutagenesis was performed using mice expressing *LacZ* reporter genes in corticofugal axons or in layers II/III and V of the cortex (Figure 3.2) (Ha et al., 2013). This latter reporter is quite different in that it is expressed in cell bodies and not in axons, allowing a specific examination of cortical lamination. Multiple mutants were identified, including one carrying a splice-site mutation in reelin (*Reln*) that results in a premature stop codon and the truncation of its C-terminal region domain. Interestingly, this novel allele of *Reln* did not display cerebellar malformation or ataxia, and this is the first report of a *Reln* mutant without a cerebellar defect. As such, this mutant would not likely have been uncovered without the use of a cortical lamination reporter such as we used.

3.4 IDENTIFICATION OF MODIFIER LOCI

One application of mutagenesis that is likely to be increasingly employed is for the identification of modifier loci; that is, loci that either suppress or enhance the phenotypic consequences of a mutant locus. This is true for a number of reasons. Firstly, the substantial efforts in developmental analysis over the last two decades, combined with productive application of exome sequencing for gene discovery of human congenital defects, has resulted in the characterization of many of the loci that are crucial to mammalian development. While this annotation is by no means complete, screens for primary defects in developmental processes are increasingly identifying re-mutations of genes that have already been well characterized. There is certainly a component of ascertainment bias in this, but it is does suggest the yield of novel gene discovery will continue to decrease for the same amount of screening effort. The characterization of modifier loci is less likely to "compete" with analysis in human cohorts because of the difficulty of obtaining large numbers of individuals with specific genetic diseases that have been rigorously phenotyped, as well as the complexity of genetic analysis in a non-inbred population. Technically, the generation of modifiers using ENU mutagenesis is straightforward, as mice carrying specific mutations can either be mutagenized directly or readily introduced into the screen by breeding. Finally, as discussed below, the possibility of identifying modifiers by exome or genome sequencing should enable screens to be performed entirely in an inbred strain, removing as a confounder the strain variation previously required for genetic mapping.

An example of the successful utilization of this strategy is the work done by Pavan and colleagues analyzing genes that mediate neural crest development. Haploinsufficiency for the transcription factor SOX10 is associated with the pigmentary deficiencies of Waardenburg syndrome and is modeled in *Sox10* haploinsufficient mice (*Sox10(LacZ/+)*). Recognizing that genetic background affects WS severity in both humans and mice, they performed a mutagenesis screen to identify modifiers that increase the phenotypic severity of *Sox10(LacZ/+)* mice (Matera et al., 2008). This led to the identification of loci with quite different functions. For example, a truncation mutation in a hedgehog-signaling mediator, GLI-Kruppel family member 3 (*Gli3*) was identified, and additional genetic studies implicated the repressor activity of GLI3 as required for specification of neural crest to the melanocyte lineage (Matera et al., 2008). This screen also uncovered *Magoh*, a component of the exon junction complex, as required for normal melanoblast development (Silver et al., 2013). Separately from its interaction with *Sox10, Magoh* was found to control brain size by regulating neural stem cell division (Silver et al., 2010).

In another elegantly designed study, Whitelaw and colleagues developed a sensitized screen to identify genes involved in epigenetic gene silencing (Blewitt et al., 2008). Observing that multicopy transgene arrays are particularly sensitive to gene silencing, as evidenced by the fact that they often display variegation, they used a murine transgenic line that expresses GFP in a variegated manner to screen for genes involved in the establishment and maintenance of epigenetic state. These investigators were able to validate their presumptive modifying mutations by testing them in combination with *agouti viable yellow* (*Avy*), an endogenous "metastable epiallele," which is sensitive to epigenetic state. Analysis of approximately 5000 G1 offspring yielded 42 mutant lines and positional cloning of 29 of these resulted in the identification of mutations in 18 unique genes (Daxinger et al., 2013). This included mutations in DNA methyltransferases (*Dnmt1* and *Dnmt3b*), chromatin remodelers (*Smarca5* and *Baz1b*), a histone deacetylase (*Hdac1*), a transcriptional co-repressor (*Trim28*), a eukaryotic translation initiation factor (*eIF3h*), and *Smchd1*, a gene required for X-inactivation. The characterized lines were equally divided between those found to suppress the variegation phenotype, and those found to enhance it.

A recent modifier screen is of note because it provides not only insight into the fundamental biology of a disease process, but, exactly as one hopes for this kind of approach, a potential target for therapeutic intervention. Mutations in MECP2, encoding methyl CpG-binding protein 2, cause Rett syndrome, one of the most severe autism spectrum disorders. Justice and colleagues carried out a dominant ENU mutagenesis suppressor screen in *Mecp2*-null mice and isolated five suppressors that ameliorate the symptoms of *Mecp2* loss (Buchovecky et al., 2013). One of these proved to be a stop codon mutation in *Sqle*, encoding squalene epoxidase, a rate-limiting enzyme in cholesterol biosynthesis. The authors then showed that lipid metabolism is perturbed in the brains and livers of *Mecp2*-null male mice, and that statin drugs improved systemic perturbations of lipid metabolism, alleviated motor symptoms and increased longevity in *Mecp2* mutant mice. This genetic screen therefore points to cholesterol homeostasis as a potential target for the treatment of patients with Rett syndrome.

Any review of the present art and science of mutagenesis requires discussion of the profound change that advances in DNA sequencing technology have made in this arena. Until recently, a limiting factor in the task of positional cloning of ENU-induced mutations was the necessity for high resolution genetic mapping to narrow the recombinant interval as much as possible. Candidate genes would then be identified, based on presumptive function, expression patterns, or simple guesswork, and these would then be sequenced, using either genomic templates and analysis of exons and their flanking splice junctions, and/or cDNA templates and analysis of transcripts. This would typically require the analysis of large numbers of intercross or backcross progeny, which was both costly and slow. With the development of means for efficient sequencing of exons, generally by hybridization capture, as well as improved gene annotation, high resolution mapping is less compelling. That is, it is feasible to sequence the entire predicted exon complement from any specified region, or, with the development of comprehensive arrays, the entire genome. While it is certainly possible that a causal mutation will occur in non-transcribed sequences, there is presently only one report documenting an ENU-induced mutation occurring in a regulatory region (Sagai et al., 2004). There is also a single report describing an ENU-induced mutation in a microRNA seed region (Lewis et al., 2009). There is clearly bias in the conclusion that exons and their flanking regions are likely to contain the mutated nucleotide, as non-transcribed regions are rarely examined in detail. However, in the characterization of ENU mutants mapped to loci where known candidates reside, presumptively causal sequence variants can usually be found in coding regions or splice sites (e.g., Hart et al., 2005). Empirical experience supports the productivity of this approach for characterization of ENU-induced mutations (Fairfield et al., 2011).

Furthermore, there are sufficient SNPs in coding regions that this analysis can be used to delineate the map position directly from the sequence characterization of a pooled sample (in which all mice will be homozygous for the mutagenized parental background in the region of the mutant locus); that is, a separate mapping analysis may be unnecessary (but perhaps still advisable). In fact, whole exome sequencing may make it unnecessary to map at all. That is, one can contemplate sequencing a pooled sample of affected mice that were never out-crossed; while there will be many mutations segregating in the population, only one (or a few tightly linked variants) will be homozygous in all affected mice, and will appear as a difference from a reference genome (Andrews et al., 2012; Sun et al., 2012).

The feasibility of using whole-genome sequencing for simultaneous mapping and mutation discovery is potentially transformative for the modifier analyses discussed above. This is because it is now possible to pursue positional cloning on fully inbred backgrounds (i.e., without a cross to a mapping strain) using the mutagen-induced sequence variants as informative markers (Bull et al., 2013; our unpublished results). Given that moderate resolution mapping can be obtained using only hundreds of markers (Moran et al., 2006), the approximately 3000 variants introduced by standard mutagenesis protocols (1 mutation/Mb) should be ample for localization, even if they are incompletely ascertained (Figure 3.3). The utilization of mutagenesis on fully inbred backgrounds eliminates both the time and cost required for an outcross (if this was not built into the screen) and, more importantly, potentially confounding phenotypic variation due to strain-specific modifying loci.

In summary, phenotype-driven analysis utilizing screens of ENU-mutagenized mice has been and will likely continue to be a highly productive means of studying mammalian development. The ongoing development of technology for genomic analysis lowers a major barrier for mutation discovery, and a highly varied toolbox of transgenic reporter mice enables very sensitive analysis of specific developmental processes. The prospect of screening hundreds to thousands of progeny for mutants may seem daunting to the uninitiated; however, it can be unequivocally stated that the more challenging task will be the functional analysis of mutations after they are identified. Given this, it is worth investing effort into creating mutants that will be informative for the developmental biology of interest.

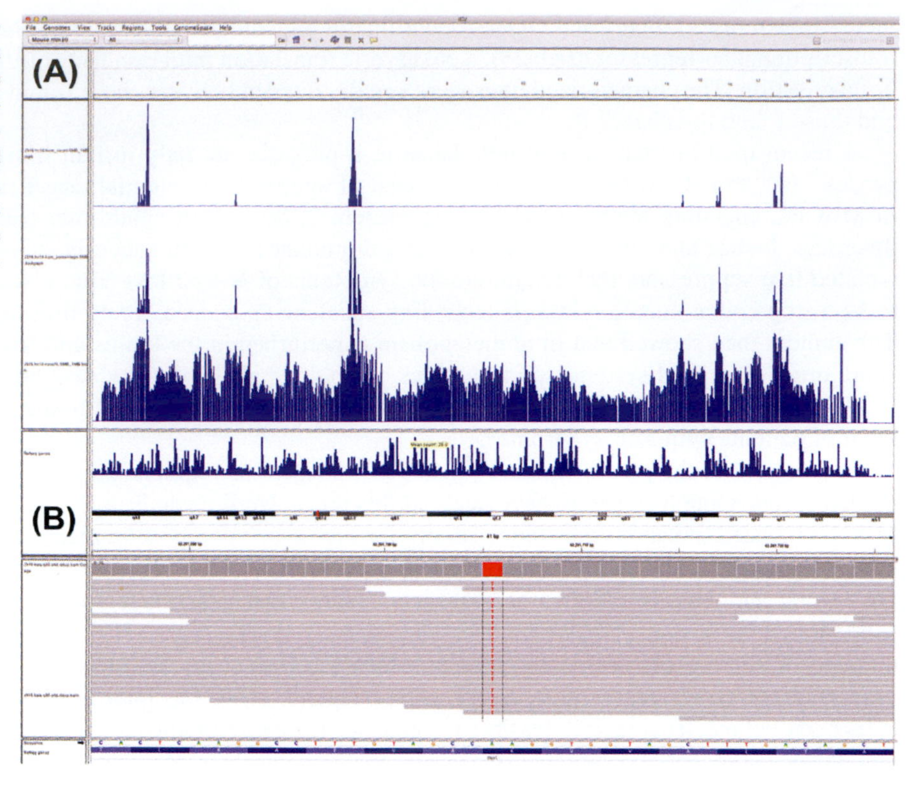

FIGURE 3.3 The genome was split into sliding windows of 5Mb with a 1Mb overlap. Within each window a number of calculations were made; total number of homozygous SNPs, percentage of SNPs that are homozygous, and average percentage of non-reference allele of called SNPs (tracks 1, 2 and 4 in panel A). In addition, regions in which only homozygous mutations occurred were mapped (track 3 of panel A). The putative phenotype causing homozygous mutation is seen in panel B.

3.5 CLINICAL RELEVANCE

- Phenotype-driven developmental genetics has the potential to enable gene and pathway discovery relevant to human congenital disorders.
- Useful animal models for study can be generated when mutations in known human disease genes are found.
- Characterization of modifier genes can both inform fundamental biology and potentially identify opportunities for therapeutic intervention.

RECOMMENDED RESOURCES

Mouse Genome Database, http://www.informatics.jax.org (accessed 21.03.14).

Protocol for Mouse Mutagenesis Using N-Ethyl-N-Nitrosourea (ENU), http://cshprotocols.cshlp.org/content/2008/4/pdb.prot4985.full (accessed 21.03.14).

REFERENCES

Ackerman, K.G., Herron, B.J., Vargas, S.O., Huang, H., Tevosian, S.G., et al., 2005. Fog2 is required for normal diaphragm and lung development in mice and humans. PLoS Genet. 1, 58–65.

Ahituv, N., Erven, A., Fuchs, H., Guy, K., Ashery-Padan, R., et al., 2004. An ENU-induced mutation in AP-2alpha leads to middle ear and ocular defects in Doarad mice. Mamm. Genome 15, 424–432.

Andrews, T.D., Whittle, B., Field, M.A., Balakishnan, B., Zhang, Y., et al., 2012. Massively parallel sequencing of the mouse exome to accurately identify rare, induced mutations: an immediate source for thousands of new mouse models. Open Biol. 2, 120061.

Blewitt, M.E., Gendrel, A.V., Pang, Z., Sparrow, D.B., Whitelaw, N., et al., 2008. SmcHD1, containing a structural-maintenance-of-chromosomes hinge domain, has a critical role in X inactivation. Nat. Genet. 40, 663–669.

Buchovecky, C.M., Turley, S.D., Brown, H.M., Kyle, S.M., McDonald, J.G., et al., 2013. A suppressor screen in Mecp2 mutant mice implicates cholesterol metabolism in Rett syndrome. Nat. Genet. 45, 1013–1020.

Bull, K.R., Rimmer, A.J., Siggs, O.M., Miosge, L.A., Roots, C.M., et al., 2013. Unlocking the bottleneck in forward genetics using whole-genome sequencing and identity by descent to isolate causative mutations. PLoS Genet. 9, e1003219.

Chiang, C., Litingtung, Y., Lee, E., Young, K.E., Corden, J.L., et al., 1996. Cyclopia and defective axial patterning in mice lacking Sonic hedgehog gene function. Nature 383, 407–413.

Daxinger, L., Harten, S.K., Oey, H., Epp, T., Isbel, L., et al., 2013. An ENU mutagenesis screen identifies novel and known genes involved in epigenetic processes in the mouse. Genome Biol. 14, R96.

Dwyer, N.D., Manning, D.K., Moran, J.L., Mudbhary, R., Fleming, M.S., et al., 2011. A forward genetic screen with a thalamocortical axon reporter mouse yields novel neurodevelopment mutants and a distinct emx2 mutant phenotype. Neural Dev. 6, 3.

Endoh-Yamagami, S., Evangelista, M., Wilson, D., Wen, X., Theunissen, J.W., et al., 2009. The mammalian Cos2 homolog Kif7 plays an essential role in modulating Hh signal transduction during development. Curr. Biol. 19, 1320–1326.

Ermakov, A., Stevens, J.L., Whitehill, E., Robson, J.E., Pieles, G., et al., 2009. Mouse mutagenesis identifies novel roles for left-right patterning genes in pulmonary, craniofacial, ocular, and limb development. Dev. Dyn. 238, 581–594.

Fairfield, H., Gilbert, G.J., Barter, M., Corrigan, R.R., Curtain, M., et al., 2011. Mutation discovery in mice by whole exome sequencing. Genome Biol. 12, R86.

Favor, J., Neuhauser-Klaus, A., Ehling, U.H., 1991. The induction of forward and reverse specific-locus mutations and dominant cataract mutations in spermatogonia of treated strain DBA/2 mice by ethylnitrosourea. Mutat. Res. 249, 293–300.

Gailus-Durner, V., Fuchs, H., Becker, L., Bolle, I., Brielmeier, M., et al., 2005. Introducing the German Mouse Clinic: open access platform for standardized phenotyping. Nat. Methods 2, 403–404.

Goetz, S.C., Ocbina, P.J., Anderson, K.V., 2009. The primary cilium as a Hedgehog signal transduction machine. Methods Cell Biol. 94, 199–222.

Ha, S., Stottmann, R.W., Furley, A.J., Beier, D.R., 2013. A forward genetic screen in mice identifies mutants with abnormal cortical patterning. Cereb. Cortex. epub Aug. 22.

Hart, A.W., McKie, L., Morgan, J.E., Gautier, P., West, K., et al., 2005. Genotype-phenotype correlation of mouse pde6b mutations. Invest. Ophthalmol. Vis. Sci. 46, 3443–3450.

Hrabe de Angelis, M.H., Flaswinkel, H., Fuchs, H., Rathkolb, B., Soewarto, D., et al., 2000. Genome-wide, large-scale production of mutant mice by ENU mutagenesis. Nat. Genet. 25, 444–447.

Lewis, M.A., Quint, E., Glazier, A.M., Fuchs, H., De Angelis, M.H., et al., 2009. An ENU-induced mutation of miR-96 associated with progressive hearing loss in mice. Nat. Genet. 41, 614–618.

Mar, L., Rivkin, E., Kim, D.Y., Yu, J.Y., Cordes, S.P., 2005. A genetic screen for mutations that affect cranial nerve development in the mouse. J. Neurosci. 25, 11787–11795.

Matera, I., Watkins-Chow, D.E., Loftus, S.K., Hou, L., Incao, A., et al., 2008. A sensitized mutagenesis screen identifies Gli3 as a modifier of Sox10 neurocristopathy. Hum. Mol. Genet. 17, 2118–2131.

May, S.R., Ashique, A.M., Karlen, M., Wang, B., Shen, Y., et al., 2005. Loss of the retrograde motor for IFT disrupts localization of Smo to cilia and prevents the expression of both activator and repressor functions of Gli. Dev. Biol. 287, 378–389.

McDonald, J., Shedlovsky, A., Dove, W., 1990. Investigating inborn errors of phenylalanine metabolism by efficient mutagenesis of the mouse germline, pp. In: Banbury Report 34: Biology of Mammalian Germ Cell Mutagenesis. Cold Spring Harbor Laboratory Press.

McDonald, J.D., Bode, V.C., 1988. Hyperphenylalaninemia in the hph-1 mouse mutant. Pediatr. Res. 23, 63–67.

Montagutelli, X., Lalouette, A., Coude, M., Kamoun, P., Forest, M., et al., 1994. aku, a mutation of the mouse homologous to human alkaptonuria, maps to chromosome 16. Genomics 19, 9–11.

Moran, J.L., Bolton, A.D., Tran, P.V., Brown, A., Dwyer, N.D., et al., 2006. Utilization of a whole genome SNP panel for efficient genetic mapping in the mouse. Genome Res. 16, 436–440.

Moser, A.R., Pitot, H.C., Dove, W.F., 1990. A dominant mutation that predisposes to multiple intestinal neoplasia in the mouse. Science 247, 322–324.

Nolan, P.M., Peters, J., Strivens, M., Rogers, D., Hagan, J., et al., 2000a. A systematic, genome-wide, phenotype-driven mutagenesis programme for gene function studies in the mouse. Nat. Genet. 25, 440–443.

Nolan, P.M., Peters, J., Vizor, L., Strivens, M., Washbourne, R., et al., 2000b. Implementation of a large-scale ENU mutagenesis program: towards increasing the mouse mutant resource. Mamm. Genome 11, 500–506.

Roessler, E., Belloni, E., Gaudenz, K., Jay, P., Berta, P., et al., 1996. Mutations in the human Sonic Hedgehog gene cause holoprosencephaly. Nat. Genet. 14, 357–360.

Rossant, J., Zirngibl, R., Cado, D., Shago, M., Giguere, V., 1991. Expression of a retinoic acid response element-hsplacZ transgene defines specific domains of transcriptional activity during mouse embryogenesis. Genes and Dev. 5, 1333–1344.

Russell, W., Hunsicker, P., Carpenter, D., Cornett, C., Guinn, G., 1982. Effect of dose-fractionation on the ethylnitrosurea induction of specific-locus mutations in mouse spermatogonia. Proc. Natl. Acad. Sci. USA 79, 3592–3593.

Sabrautzki, S., Rubio-Aliaga, I., Hans, W., Fuchs, H., Rathkolb, B., et al., 2012. New mouse models for metabolic bone diseases generated by genome-wide ENU mutagenesis. Mamm. Genome 23, 416–430.

Sagai, T., Masuya, H., Tamura, M., Shimizu, K., Yada, Y., et al., 2004. Phylogenetic conservation of a limb-specific, cis-acting regulator of Sonic hedgehog (Shh). Mamm. Genome 15, 23–34.

Silver, D.L., Leeds, K.E., Hwang, H.W., Miller, E.E., Pavan, W.J., 2013. The EJC component Magoh regulates proliferation and expansion of neural crest-derived melanocytes. Dev. Biol. 375, 172–181.

Silver, D.L., Watkins-Chow, D.E., Schreck, K.C., Pierfelice, T.J., Larson, D.M., et al., 2010. The exon junction complex component Magoh controls brain size by regulating neural stem cell division. Nat. Neurosci. 13, 551–558.

Stottmann, R.W., Moran, J.L., Turbe-Doan, A., Driver, E., Kelley, M., et al., 2011. Focusing forward genetics: a tripartite ENU screen for neurodevelopmental mutations in the mouse. Genetics 188, 615–624.

Sun, M., Mondal, K., Patel, V., Horner, V.L., Long, A.B., et al., 2012. Multiplex chromosomal exome sequencing accelerates identification of ENU-induced mutations in the mouse. G3 (Bethesda) 2, 143–150.

Tran, P.V., Haycraft, C.J., Besschetnova, T.Y., Turbe-Doan, A., Stottmann, R.W., et al., 2008. THM1 negatively modulates mouse sonic hedgehog signal transduction and affects retrograde intraflagellar transport in cilia. Nat. Genet. 40, 403–410.

Tutois, S., Montagutelli, X., Da Silva, V., Jouault, H., Rouyer-Fessard, P., et al., 1991. Erythropoietic protoporphyria in the house mouse. A recessive inherited ferrochelatase deficiency with anemia, photosensitivity, and liver disease. J. Clin. Invest. 88, 1730–1736.

Yang, H., Wang, H., Shivalila, C.S., Cheng, A.W., Shi, L., et al., 2013. One-step generation of mice carrying reporter and conditional alleles by CRISPR/Cas-mediated genome engineering. Cell 154, 1370–1379.

Yu, Q., Shen, Y., Chatterjee, B., Siegfried, B.H., Leatherbury, L., et al., 2004. ENU induced mutations causing congenital cardiovascular anomalies. Development 131, 6211–6223.

Chapter 4

Chemical Approaches to Controlling Cell Fate

Mingliang Zhang[1], Ke Li[1], Min Xie and Sheng Ding

Gladstone Institute of Cardiovascular Disease, Department of Pharmaceutical Chemistry, University of California, San Francisco, CA

Chapter Outline

GLOSSARY

Differentiation In developmental biology, differentiation is the process through which a stem cell or progenitor cell becomes more specialized.

EMT Epithelial-mesenchymal transition (EMT) describes the process through which epithelial cells lose their cell-cell and cell-extracellular matrix interactions and gain increased cell mobility together with the other properties of mesenchymal cells.

MoA The mechanism of action (MoA) of a small molecule refers to the mechanism by which the small molecule produces its effect.

Pluripotent stem cells Pluripotent stem cells, including the embryonic stem cells, epiblast stem cells, and induced pluripotent stem cells, possess the abilities of infinite self-renewal and differentiation into all the cell types of three germ layers.

SAR A structure-and-activity-relationship (SAR) usually refers to the relationship between the structure of a series of small molecules and their biological activity.

Self-renewal Self-renewal of stem cells describes the process of a stem cell undergoing mitotic division and giving rise to daughter cells that are identical to the mother cell.

SILAC Stable isotope labeling by amino acids in cell culture (SILAC) is a mass spectrometry based quantitative proteomic approach that is widely used for whole cellular proteome comparison. It relies on metabolic incorporation of normal amino acids and amino acids labeled with stable heavy isotopes into proteins.

Tissue-specific stem cells Tissue-specific stem cells are multipotent stem cells. They reside in specific tissues and possess limited abilities of self-renewal and differentiation into specialized cell types of the lineage they originate. Tissue-specific stem cells maintain tissue homeostasis and allow tissue regeneration in response to tissue injuries.

SUMMARY

- Efficient modulation of cell fate, state, or function, such as self-renewal, differentiation, mobility, polarization, and reprogramming, is critical to translate stem cell biology from bench to clinical therapies.
- The effect of small molecules is rapid, tunable, and reversible with exquisite temporal control. These advantages make chemical approaches desirable for modulating cell fate and probing underlying mechanisms, and highly suitable for development into conventional pharmaceuticals.

1 These authors contributed equally to this work

Principles of Developmental Genetics.

4.1 INTRODUCTION

Physiologically relevant stem cells, including embryonic stem cells (ESCs) and tissue-specific stem cells (TSCs), possess the abilities of self-renewal and differentiation. ESCs are isolated from the inner cell mass (ICM) of blastocysts of mammalian embryos and represent an unlimited resource of cells that can differentiate into all the cell types of primary germ layers and primordial germ cells, form teratomas when implanted into an immune-deficient mouse, and contribute to chimeras when injected into mouse blastocysts. This property is called pluripotency (Smith, 2001). To avoid the risk of tumorigenesis and enable the efficient engraftment after transplantation, the pluripotent ESCs are usually differentiated into desired populations, including TSCs for cell replacement therapy *in vivo*. TSCs reside in adult tissues, and their differentiation capacity is limited to the cell types within the particular lineage from which they originate, to maintain tissue renewal. Considering the unique abilities of self-renewal and differentiation, stem cells are regarded as the reservoir for tissue homeostasis and regeneration. Thus, stem cells are an excellent model for studying the fundamental biology of development, and they offer great promise for treating human diseases and injuries. However, despite the significant advances in stem cell biology in the last decades, the maintenance of stem cells and their directed differentiation to desired cell types are still challenging.

The breakthrough discovery of induced pluripotent stem cell (iPSC) technology that allows reprogramming of differentiated somatic cells into pluripotent cells (Takahashi and Yamanaka, 2006) firmly established the paradigm of ectopically expressing a defined set of target cell-specific transcription factors (TFs) to achieve cellular reprogramming. In addition to iPSCs, this strategy has been used to successfully generate a variety of cell types, including neuronal subtypes and neural progenitor cells, cardiomyocytes, hepatocytes, and endothelial cells. However, such an approach typically carries the significant risks associated with genetic manipulation.

A chemical approach may be particularly advantageous for addressing the challenges of cell fate control. Small molecules have proved to be useful in modulating cell fate and probing the underlying mechanisms, and offer distinct advantages over genetic manipulation. They typically provide a high degree of temporal control that is rapid and reversible. Their effects can be fine-tuned by varying their concentrations and combinations. In addition, synthetic chemistry could provide almost unlimited structural and functional features to small molecules for their interaction and regulation with biological targets. These advantages may further allow translation of small molecules from *in vitro* cell modulation into conventional pharmaceuticals that modify disease cells *in vivo*. Furthermore, such chemical approaches to stem cell biology and regeneration would improve our understanding of the mechanisms that control the developmental and regenerative potentials of various cell types, and, ultimately, facilitate their clinical application. In this chapter, we will focus on recent progress in the development of chemical approaches to controlling cell fate.

4.2 CHEMICAL APPROACHES TO CONTROLLING CELL FATE

Behaviors of Developmentally Relevant Stem Cells

Stem cells have the abilities to self-renew and differentiate into specialized cell types in response to instructive signals. Self-renewal defines the ability of stem cells to give rise to daughter cells identical to the parental cells during division, and differentiation refers generation of specialized cell types with limited developmental potential. Traditionally, physiologically relevant stem cells are classified as ESCs and TSCs, according to their origins and properties.

ESCs were the first pluripotent stem cells (PSCs) to be identified, and are derived from the ICM of mammalian preimplantation blastocysts (Evans and Kaufman, 1981). Although they both come from the ICM, ESCs from the mouse (mESCs) and human (hESCs) (Thomson et al., 1998) exhibit many striking differences, including distinct self-renewal requirements, gene expression, and cell behaviors such as colony morphology, responses to external signals, and sensitivity to single-cell dissociation. Studies suggest that hESCs may more closely resemble mouse epiblast stem cells (EpiSCs), which are derived from post-implantation egg-cylinder-stage epiblasts (Brons et al., 2007; Tesar et al., 2007). Both hESCs and EpiSCs grow as large, flat colonies, and require basic fibroblast growth factor (bFGF) and TGFβ-like signals to maintain their pluripotent state (James et al., 2005; Vallier et al., 2005). In contrast, mESC colonies are domed and much smaller, and their pluripotent state typically depends on leukemia inhibitory factor (LIF) and bone morphogenic protein 4 (BMP4) signals (Chambers and Smith, 2004; Niwa et al., 1998; Ying et al., 2003). Furthermore, hESCs/EpiSCs survive poorly after single-cell dissociation, and respond similarly to culture conditions that drive the differentiation, which also contrast with mESCs (Vallier et al., 2009; Nichols and Smith, 2009). However, there are also differences between hESCs and mouse EpiSCs in terms of some typical markers' expression. For example, mouse EpiSCs express stage-specific embryonic antigen 1 (SSEA1) on their surfaces but lack alkaline phosphatase (ALK) activity (Brons et al., 2007), whereas hESCs express SSEA3/4 on the cell surface and are positive for ALK activity (Thomson et al., 1995; Thomson and Marshall, 1998). Like mESCs, hESCs express Klf4, Dppa3, and Rex1, but EpiSCs do not (Chan et al., 2009; Clark et al., 2004). Of note, all three types of PSCs, mESCs, hESCs, and mouse EpiSCs, are able to form teratomas when implanted into immune-deficient mice. Mouse ESCs are able to contribute to chimeras when injected into pre-implantation embryos, the environment of which, however, is not permissive for EpiSCs to incorporate and contribute to cellular descendants. This is probably due to the stage mismatch between EpiSCs and the recipient embryo. Recent studies reveal that EpiSCs readily generate chimeras when grafted to post-implantation embryos, indicating the EpiSCs are functional equivalents of the early post-implantation epiblasts, as previously hypothesized (Huang et al., 2012). Nevertheless, these studies suggest that PSCs derived from distinct developmental stages or different species, may display different properties.

TSCs are multipotent cells with the ability to differentiate into the entire repertoire of cell types within the lineage from which they originate, and they are important for maintaining tissue homeostasis and mediating tissue/organ repair and regeneration throughout their lifetimes. Traditionally, TSCs are isolated from the differentiated tissue/organ in which they reside. Several types of TSCs have been successfully isolated, such as neural stem cells (NSCs), hematopoietic stem cells (HSCs), and mesenchymal stem cells (MSCs). TSCs have many advantageous qualities that make them attractive alternatives to PSC-derived cells for cell-based therapy, including a reduced tumorigenic potential and the ability to efficiently engraft and differentiate into desirable cell populations after transplantation. Encouraging therapeutic examples include the transplantation of HSCs or MSCs for treatment of hematologic or immune diseases, respectively. Although TSCs have considerable capacity of self-renewal, they typically exhibit limited lifespan and declined responsiveness to external signals over time. Consequently, developing selective methods for improving long-term expansion *in vitro*, engraftment *in vivo*, and specific differentiation of TSCs is highly desirable.

Clinical Relevance of Chemical Approaches to Stem Cell Fate Modulation

Stem cell fate is determined by intrinsic regulators and extrinsic signals. The physiological environment provides a perfect but complex combination of signaling, including the appropriate identity, abundance, location, and dynamics of stimuli that function in synergy with the intrinsic regulatory network to orchestrate the temporal-spatial control of self-renewal and differentiation. This developmental principle is highly ordered, and guides studies and therapeutic development of cell-based therapies and regenerative pharmaceuticals. The identification and characterization of small molecules that precisely control cell fate and behaviors with high specificity, and ultimately their tissue-specific delivery, would facilitate the development of therapies for treating human diseases and injuries.

Chemical Platform for Stem Cell Biology

The Principles of Chemical Approaches for Stem Cell Fate Control

Chemical Library

One approach to generating functional small molecules that control stem cell fate involves the use of phenotypic or pathway-specific screens of synthetic chemical or natural product libraries (Figure 4.1). The size and diversity of a given purified chemical library together with the selection method determine the chance of finding a desired "hit" compound. With recent advances in automation and detection technologies, millions of discrete compounds can be screened rapidly and cost-effectively. However, although combinatorial technologies allow the synthesis of a large number of molecules with immense structural diversity, it is impossible to saturate the chemical space. Because the diversity of chemical libraries is largely constrained by synthetic tractability, new synthetic technologies are the driving force to expand current chemical diversities for filling the chemical space. In addition, introducing a high level of structural variability to increase the molecular diversity of a chemical library drastically reduces the average fitness of the library to a given biological selection or screen, thereby resulting in most molecules being inactive (in analogy to population genetics). Consequently, the careful design of a chemical library becomes a critical aspect of combinatorial synthesis (diversity vs. fitness).

The concept of "privileged structures" describes selected structural motifs that can provide potent and selective ligands for multiple biologic targets by introducing different substitutions onto the same scaffold (Figure 4.1). Privileged structures typically exhibit good "drug-like" properties, such as good solubility, membrane permeability, oral bioavailability, and metabolic stability, which make the further development of "hits" into "leads" less problematic. Given the success of privileged structures, the diversification of these scaffolds using combinatorial techniques provides not only large numbers of compounds but also highly enriched "functional" molecules. Using key biological recognition motifs as the core scaffolds may represent one of the most straightforward and productive ways to generate "privileged" chemical libraries. Previously, we developed the concept of using privileged molecular scaffolds themselves as a diversity element for combinatorial synthesis to maximize the diversity while retaining a minimal threshold of fitness to biological screens. With this approach,

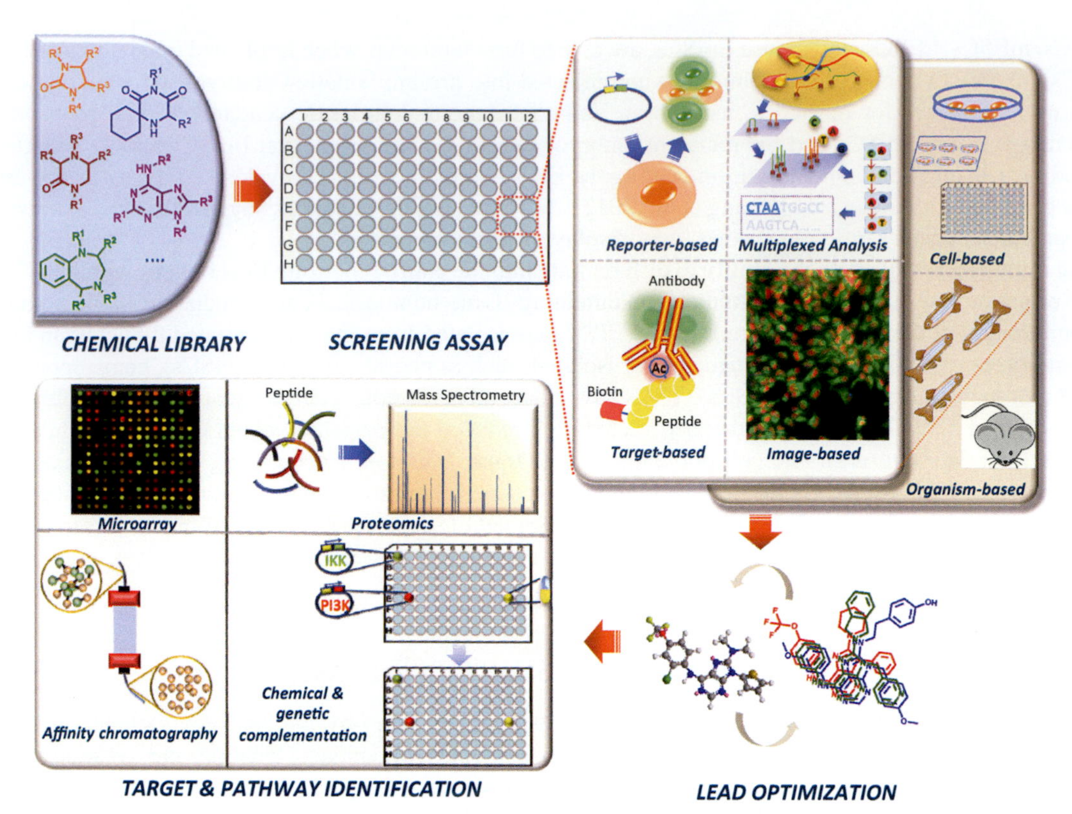

FIGURE 4.1 Chemical Library and Chemical Screening Approach. Chemical libraries for biological screens can be assembled from natural products and synthetic compounds through chemical synthesis and/or commercial sources. Chemical screening conventionally employs target-based assays (e.g., enzymatic or binding-based assays using purified proteins), and recently has increasingly utilized phenotypic readouts that more functionally examine particular cell behaviors, such as marker expression (by reporter or staining), biological pathway or process activity, or cell behaviors (e.g., cell proliferation, death, differentiation etc) in cell culture (e.g., primary cells or engineered cell lines) or whole-organism contexts (e.g., zebrafish or mice). The lead candidate compounds identified from a primary assay are usually confirmed by secondary assays, and optimized through structure-activity-relationship studies. To characterize the mechanism of action of novel small molecules identified from phenotypic screening, various strategies have been developed, including affinity-based target pull-down, transcriptome profiling, and genome-scale analyses. The functional interactions between the small molecule and proteins are typically validated by gain- or loss-of-function studies in cellular assays, as well as other relevant biochemical and cellular assays (shown here as chemical and genetic complementation). R denotes a chemical substitution moiety; IKK, inhibitor of nuclear factor-κB (IκB) kinase; PI3K, phosphatidylinositol-3-OH kinase.

a variety of naturally occurring and synthetic heterocycles that are known to interact with proteins involved in cell signaling (e.g., kinases, cell surface and nuclear receptors, enzymes) were used as the core molecular scaffolds. These included substituted purines, pyrimidines, indoles, quinazolines, pyrazines, pyrrolopyrimidines, pyrazolopyrimidines, phthalazines, pyridazines, pyridines, triazines, and quinoxalines (the first diversity elements). A general synthetic scheme was then developed that could be used in parallel reactions to introduce a variety of substituents into each of these scaffolds to create a diverse chemical library. The library synthesis involved introducing a second diversity element into these heterocyclic scaffolds using solution-phase alkylation or acylation reactions. This was followed by the capture of the modified heterocycles onto a solid support using different immobilized amines (introduced as a third diversity element). The resin-bound heterocycles could then be further modified (introduced as a fourth diversity element) through a variety of chemistries, including acylation, amination, and palladium-mediated cross-coupling reactions with amines, anilines, phenols, and boronic acids. Using these chemistries in conjunction with the "directed-sorting" method, we have generated diverse heterocycle libraries consisting of more than 100,000 discrete small molecules (representing more than 30 distinct structural classes), with an average purity of greater than 90%. These libraries have been proven to be a rich source of biologically active small molecules for targeting various proteins involved in a variety of signaling pathways.

Chemical Screening

Chemical screening is a central component of modern drug discovery. It has also been used in stem cell research where identification of potent and selective small molecules for stem cell fate modulation is desirable. High throughput chemical screening conventionally employs target-based assays (e.g., enzymatic or binding-based assays using purified proteins), and has been seen increasingly with phenotypic assays that examine particular marker expression (by reporter or staining), activity of proteins, or cell behaviors in cell culture (e.g., primary cells or engineered cell lines) or whole-organism contexts (e.g., zebrafish or mice) (Figure 4.1). Advances in automated microscopy and image analysis in particular have allowed the development and wide adoption of high throughput and high content imaging methods, which enable high-resolution multi-parameter acquisition at the single-cell level in multi-well plates and more functional screening.

Complementary to imaging-based functional assays is multiplexed molecular analysis for high throughput screening, as it addresses limitations in the lack of molecular understanding directly from phenotypic screening, and reduces biases resulting from using single or limited marker readout (Figure 4.1). In particular, the recently developed RASL-seq technology (RNA-mediated oligonucleotide Annealing, Selection, and Ligation with Next-Generation sequencing), which has evolved from a ligation-mediated assay by including an annealing step to offset random ligation, and is coupled with next-generation sequencing, enables highly multiplexed (e.g., examining hundreds to thousands of endogenous gene expression), quantitative, flexible (e.g., in terms of numbers of selected genes or screened samples in conventional screening format), fast, and cost-effective large-scale gene expression analysis for high throughput screening (Li et al., 2012). By examining the changes in cohorts of genes (e.g., gene signatures) that define specific phenotypes (e.g., perturbation of signaling pathway, loss-of-function or gain-of- function of a specific gene, cell state, or even disease state), RASL-seq provides a systems-biology approach to drug discovery.

There are generally two strategies guiding the chemical screening. Small molecules with known mechanisms are usually used to test specific hypotheseses in linking mechanism to phenotype (hypothesis-driven chemical screening). Alternatively, discovery-based chemical screening is particularly useful to identify small molecules that produce a desired phenotype without prior knowledge, and subsequently serve as tools to probe the underlying mechanisms.

Mechanism of Action of Small Molecules

Various strategies have been developed to characterize the mechanism of action (MoA) of novel small molecules identified from phenotypic screening, including affinity-based target pull-down, transcriptome profiling, and genome-scale analyses to examine functional interactions between the small molecule and proteins by gain- or loss-of-function in cellular assays (Figure 4.1). For target identification, the most widely practiced and straightforward approach is affinity pull-down using the small molecule probe that is attached to a solid matrix through a linker via permissive position on the small molecule without affecting its biological activity. In order to synthesize such a suitable affinity probe, the initially identified small molecules from the screening would first need to be optimized through structure-activity-relationship (SAR) studies for improved potency (means enhanced affinity with the target) and specificity (means less non-relevant/off-target binding partners). Most importantly, such a SAR study would also indicate where the small molecule can be attached via a linker to an affinity tag without affecting its biological activity (i.e., without affecting its interaction/binding to its target). With affinity probe, quantitative proteomics, methods such as SILAC (i.e., stable isotope labeling by amino acids in cell culture) can be used to examine the affinity enrichment of the target protein(s) and their associated complexes bound to the small molecule. The affinity-pull-downed target(s) will be further validated by genetic gain- or loss-of-function studies as well as alternative pharmacological treatments targeting the same pathway/mechanism to examine whether the putative target manipulation could recapitulate or modify the small molecule's effect in the cellular assays. Ultimately, biochemical studies (e.g., binding) would further establish the target and mechanism for the small molecule.

Small Molecules Modulating Stem Cell Fate

Chemical Control of Pluripotent Stem Cells

Small Molecules Modulating Self-Renewal of Pluripotent Stem Cells

In vitro established PSCs exist in at least two distinct stages: the pre-implantation ICM-like naïve pluripotent state, represented by the mESCs, and the post-implantation EpiSC-like primed pluripotent state, represented by the mEpiSCs and hESCs.

ESCs are conventionally maintained on feeder cells in the presence of serum fractions and cytokines, such as LIF for mESCs and bFGF for hESCs (Niwa et al., 1998; Thomson et al., 1998). However, feeder cells and serum products contain undefined factors, and their usage introduces significant variability in the culture conditions. Development of chemically defined medium conditions would be highly desirable. To this end, LIF/BMP combination and bFGF/TGFβ combination have been developed for culturing mESCs and hESCs in chemically defined conditions, respectively. To enable more robust mESCs expansion and uncover novel mechanisms in self-renewal, Chen et al. carried out a high throughput chemical screening and identified a synthetic compound, named as Pluripotin (and also known as SC1), which could maintain the long-term self-renewal of mESCs in the absence of feeder cells, serum and LIF (Chen et al., 2006). Pluripotin-expanded mESCs are able to differentiate into the cell types of the three germ layers *in vitro* and contribute to chimeras with germline transmission *in vivo*, indicative of naïve pluripotency. Further studies revealed that Pluripotin functions via a novel mechanism independent of the activation of essential self-renewal pathways, including LIF-STAT, BMP-SMAD, and Wnt-β-catenin signaling. Instead, Pluripotin inhibits two endogenous differentiation-inducing proteins RasGAP and extracellular-signal-regulated kinase 1(ERK1), without activating self-renewal pathways. Ying et al. also found that a combination of three specific chemical inhibitors of the protein kinases FGF receptor (FGFR), mitogen-activated protein kinase kinase (MEK), and glycogen synthase kinase 3 (GSK3) supports the derivation and long-term self-renewal of mESCs in the absence of exogenous cytokines (Ying et al., 2008). These findings indicate that ESCs have an innate program for self-renewal that does not require exogenous instruction, but instead is enabled by the elimination of differentiation-inducing signals.

As mESCs representing the naïve pluripotent state have been a powerful tool in biology and exhibit different and perhaps more useful properties than hESCs/EpiSCs representing the primed pluripotent state, there has been significant interest in reprogramming PSCs in the primed pluripotent state into the naïve pluripotent state, especially for non-rodent species.

Epithelial-mesenchymal transition (EMT) describes the process of epithelial cells losing their cell-cell and cell-extracellular matrix interactions and gaining increased cell mobility and other properties of mesenchymal cells. Differences in E-cadherin expression (i.e., higher E-cadherin expression in mESCs than in mEpiSCs) and cell-cell interaction (e.g., mESC's domed vs. EpiSC's flat colony morphology) indicate that transition from naïve pluripotency to primed pluripotency may exhibit features of EMT. In part through modulating EMT mechanisms, several small molecule conditions were developed to promote the transition of primed pluripotency toward naïve pluripotency. Xu et al. generated mESC-like hESCs by culturing them in mESC medium with addition of LIF, MEK inhibitor PD0325901 (Table 4.1), and p38 inhibitor SB203580 (Table 4.1). The converted hESCs are more similar to mESCs in morphology (i.e., they exhibit more compact, domed, and smaller colonies) and display naïve pluripotency-specific responses to signaling (e.g., require LIF for self-renewal, but not bFGF). Further study showed that the different cell culture signaling environments partly control E-cadherin expression and stability in the two pluripotent states (Xu et al., 2010). In another study, Li et al. generated mESC-like human iPSCs by combining genetic reprogramming and a chemical approach. Human iPSC colonies generated from human fibroblasts transduced with Oct4, Sox2, Nanog, and Lin28 were picked up and stably maintained in culture in the presence of human LIF, PD0325901, ALK5 inhibitor A-83-01 (Table 4.1), and GSK3 inhibitor CHIR99021 (Table 4.1). Such hiPSCs also exhibit similar cell behaviors and self-renewal requirements to those of mESCs (Li et al., 2009a).

Some approaches for converting EpiSCs to the naïve pluripotency state have been reported, such as ectopic expression of Klf4 or continuous culture of EpiSCs in the presence of LIF. However, the former approach requires genetic manipulation and the latter one is slow and stochastic. To solve this problem with a chemical approach, Zhou et al. developed a cocktail of small molecules called mAMFGi that includes inhibitors targeting the TGFβ pathway, mitogen-activated protein kinase kinase (MEK) pathway, fibroblast growth factor receptor (FGFR), and glycogen synthase kinase 3-β (GSK3β). It has been well characterized that the TGFβ pathway promotes EMT. Therefore, it was hypothesized that inhibition of TGFβ favors mesenchymal-to-epithelial transition (MET) and consequently promotes EpiSC-to-mESC reprogramming. Indeed, the TGFβ pathway inhibitor A83-01 greatly enhanced the conversion (Zhou et al., 2010). However, treatment of EpiSCs with mAMFGi produced mESC-like cells that could not efficiently contribute to chimera formation until the addition of Parnate (Table 4.1), a histone demethylase LSD1 inhibitor. This finding indicates that direct epigenetic modulation could synergize with signaling modulation to overcome the barriers towards the naïve pluripotency state.

Small Molecules Directing Differentiation of Pluripotent Stem Cells

Significant progress has been made in our ability to differentiate PSCs into a wide variety of cell types. Established differentiation conditions are essentially based on principles of embryonic development. We focus here on some newly developed strategies/applications for directing the differentiation of PSCs through the use of small molecules.

Differentiation Recapitulating Embryonic Development Developmental signaling pathways, including FGF, Wnt, Hedgehog, Notch, and TGFβ /BMP, play essential roles in embryonic development and PSC differentiation *in vivo* and *in vitro*. By recapitulating embryonic development, many stepwise differentiation protocols have been devised to differentiate PSCs into various lineages. Small molecules that can regulate these key developmental pathways have been shown to be particularly useful for the precise manipulation of PSC differentiation towards a desired cell type. For example, D'Amour et al. developed a stepwise protocol to differentiate hESCs into pancreatic endocrine-like cells by mimicking *in vivo* pancreatic

TABLE 4.1 Small Molecule Compounds Modulating Cell Fate

Epigenetic Modifiers			
Structure	**Name**	**Identity**	**Mechanism and Function**
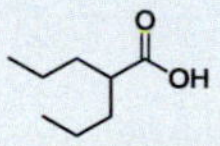	Valproic acid	Histone deacetylase inhibitor	Enable reprogramming of human fibroblasts transduced by Oct4 and Sox2 (Huangfu et al., 2008).
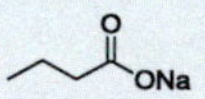	Sodium butyrate	Histone deacetylase inhibitor	Greatly enhance reprogramming of human fibroblasts transduced by 3 or 4 Yamanaka factors (Liskovykh et al., 2011).
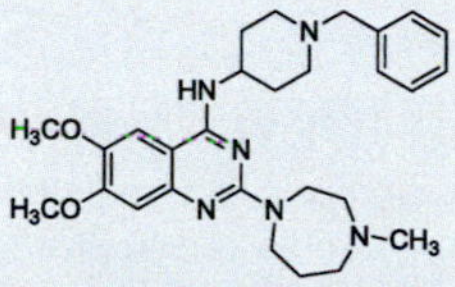	BIX-01294	G9a-like protein and G9a histone lysine methyltransferase inhibitor	Promote NPC and MEF reprogramming (Shi Y and Do JT et al., 2008)
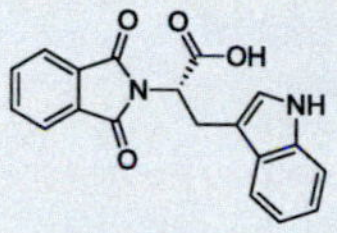	RG108	DNA methyltransferase inhibitor	Enable reprogramming of MEF transduced by Oct4 and Klf4 (Shi Y et al., 2008)
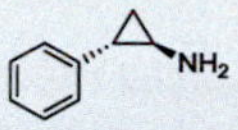	Parnate	Lysine-specific demethylase 1 (LSD1) inhibitor	Enable reprogramming of human keratinocytes transduced by Oct4 and Klf4 (Li et al., 2009)
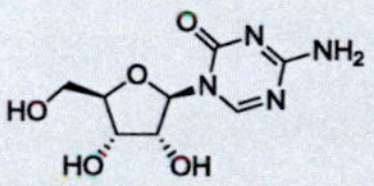	5-Azacytidine	DNA methyltransferase inhibitor	Enable reprogramming of MEF transduced by Oct4 only. (Yuan et al., 2011)
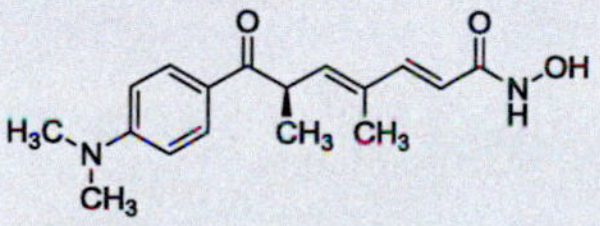	Trichostatin A	Histone deacetylase inhibitor	Enhance reprogramming of mouse and human somatic cells transduced by Yamanaka factors (Huangfu et al., 2008)
Metabolism Regulators			
Structure	**Name**	**Identity**	**Mechanism and Function**
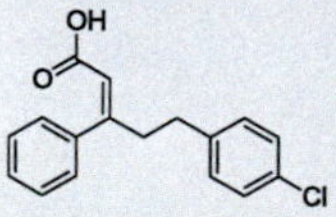	PS48	Phosphoinositide-dependent protein kinase 1 (PDK1) activator	Enhance reprogramming of human primary somatic cells (Zhu et al., 2010)
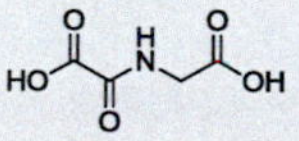	N-Oxaloylglycine	Inhibitor of α-ketoglutarate-dependent enzymes	Promotes glycolytic metabolism, promote human fibroblast reprogramming to hiPSCs. (Zhu et al., 2010)
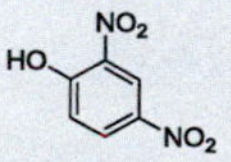	2,4-Dinitrophenol	Uncouples oxidative phosphorylation by carrying protons across the mitochondrial membrane	Promotes glycolytic metabolism, promote human fibroblast reprogramming to hiPSCs. (Zhu et al., 2010)
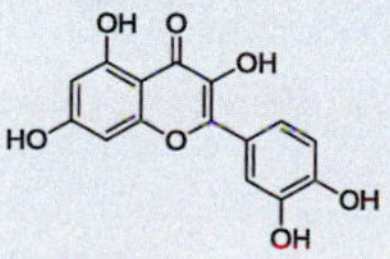	Quercetin	Inhibits many enzyme systems including mitochondrial ATPase, PI3-kinase and protein kinase C	Promotes glycolytic metabolism, promote human fibroblast reprogramming to hiPSCs. (Zhu et al., 2010)

Continued

TABLE 4.1 Small Molecule Compounds Modulating Cell Fate—cont'd

Genome Stability and Senescence			
Structure	**Name**	**Identity**	**Mechanism and Function**
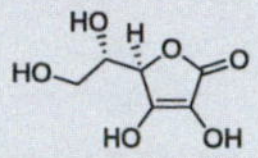	Vitamin C	Antioxidant/enzyme cofactor	Enhance reprogramming of mouse and human somatic cells in part by alleviating cell senescence (Esteban et al., 2010)
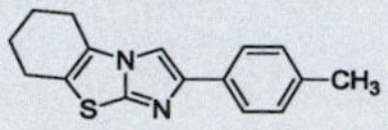	Cyclic Pifithrin-α	P53 inhibitor	enhance reprogramming efficiency of human and mouse somatic cells (Kawamura et al., 2009)
Cell Signaling Modulators			
Structure	**Name**	**Identity**	**Mechanism and Function**
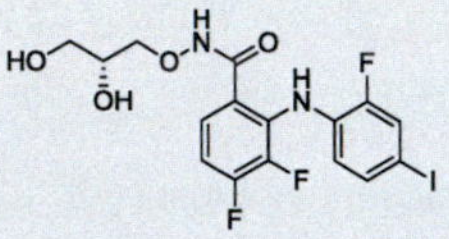	PD0325901	MEK1/2 inhibitor	Enhance reprogramming of human fibroblasts, stabilize and select true iPSC clones (Shi Y et al., 2008a; Shi Y et al., 2008b)
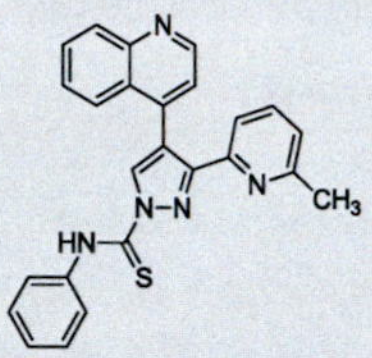	A83–01	ALK4/5/7 inhibitor	Enhance reprogramming of human fibroblasts (Lin et al., 2009)
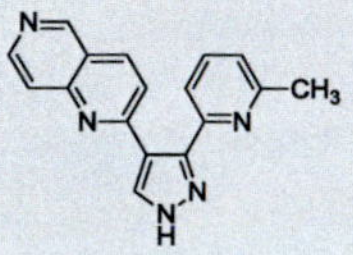	E-616452	ALK4/5/7 inhibitor	Replace Sox2 and enhance reprogramming (Ichida et al., 2009)
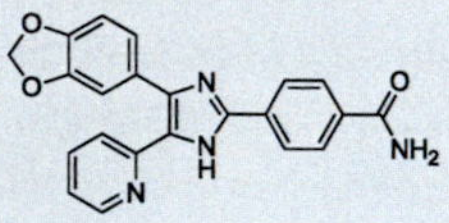	SB431542	ALK4/5/7 inhibitor	Replace Sox2 and enhance reprogramming (Ichida et al., 2009)
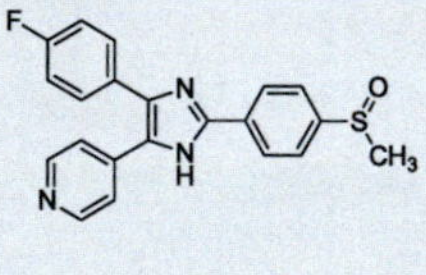	SB203580	p38 inhibitor	Promote the transition of primed pluripotency toward naïve pluripotency when combined with LIF and PD0325901(Xu et al., 2010)
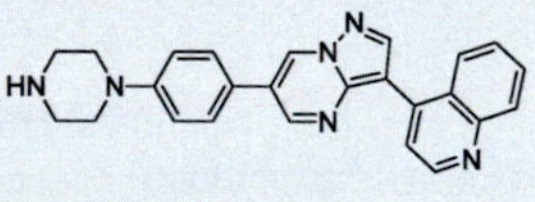	LDN193189	Inhibitor of BMP type I receptor ALK2/3	Promotes rapid neural induction of hESCs when combined with SB 431542 (Chambers et al., 2009)
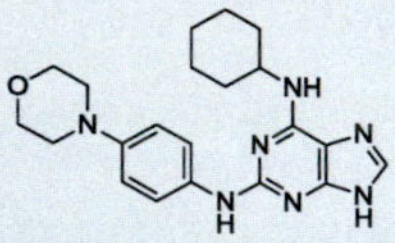	Reversine	A dual inhibitor of non-muscle myosin-II heavy chain and MEK1	Increases the plasticity of lineage-committed mammalian cells (Chen et al.,2004; Chen et al.,2007)
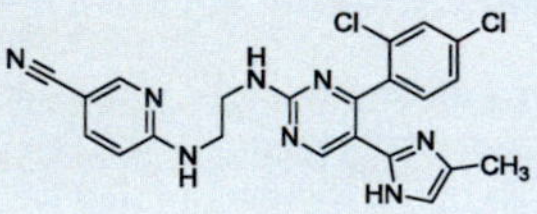	CHIR99021	GSK3β inhibitor	Enhance reprogramming of keratinocytes transduced by Oct4 and Klf4 (Li et al., 2009)

Continued

TABLE 4.1 Small Molecule Compounds Modulating Cell Fate—cont'd

	Cell Signaling Modulators		
	Kenpaullone	GSK3β and CDK1/cyclin B inhibitor	Replace Klf4 (Lyssiotis et al., 2009)
	(±)BayK 8644	L-type calcium channel agonist	Enable reprogramming of MEF transduced by Oct4 and Klf4 (Shi Y et al., 2008)
	Thiazovivin	ROCK inhibitor	Enhance reprogramming of human fibroblasts (Lin et al., 2009)
	Y-27632	Inhibitor of Rho-associated protein kinase p160ROCK	Enhance reprogramming of human somatic cells (Park et al., 2008)
	EI-275	Src kinase inhibitor	Replace Sox2 and enhance reprogramming (Ichida et al., 2009)
	Forskolin	Cell-permeable adenylyl cyclase activator	Promote hESCs transition toward mESC-like state (Hanna et al., 2010)
	AMD 3100	CXCR4 chemokine receptor antagonist	Induces rapid mobilization of both mouse and human hematopoietic progenitor cells. And switches Th2 type inflammatory responses to Th1 type (Broxmeyer et al., 2005; Larochelle et al., 2006)
	DMH-1	BMP ALK2 receptor inhibitor	Enhance the generation of neural progenitor cells from human urinal epithelial cells. Promotes neurogenesis in human induced pluripotent stem cells (iPSCs) (Wang et al., 2012)
	StemRegenin 1 (SR1)	Antagonist of the aryl hydrocarbon receptor	Promotes the self-renewal of human hematopoietic stem cells (Boitano et al., 2010)
	BIO5192	Integrin α4β1 (VLA-4) Antagonists	Induces rapid mobilization of murine hematopoietic stem and progenitors (Ramirez et al., 2009)

Continued

TABLE 4.1 Small Molecule Compounds Modulating Cell Fate—cont'd

	Cell Signaling Modulators		
	Ursolic acid	Antagonist of RORγt	Block Th17 cell differentiation (Xu et al., 2011)
	IDE1	Activator of Nodal pathway	Induces definitive endoderm formation in mouse and human embryonic stem cells (Borowiak et al., 2009). Induces Smad2 phosphorylation and increase Nodal expression (Filby et al., 2011)
	(-)-Indolactam V	Protein kinase C agonist	Directs differentiation of human embryonic stem cells into pancreatic progenitors (Chen et al., 2009)

organogenesis (D'Amour et al., 2006). This protocol entails five steps to differentiate hESCs to stages resembling definitive endoderm, primitive gut tube, posterior foregut, pancreatic endoderm, and pancreatic endocrine cells. In each step, a cocktail of small molecules and cytokines that could functionally recapitulate the developmental signaling was used to direct the lineage specification. The endocrine-like cells generated by this protocol express endocrine hormones and could release C-peptide in response to some secretory stimuli. In addition, pancreatic endoderm derived from hESCs by this protocol could efficiently generate glucose-responsive endocrine cells after implantation into mice. Human insulin and C-peptide were detected in the serum of implanted mice upon glucose stimulation (Kroon et al., 2008).

In addition, phenotypic screening was also conducted to identify novel small molecules that could functionally replace the cytokines involved in directing PSC differentiation. Using mESCs harboring a dTomato reporter gene driven by the Sox17 promoter, Borowiak et al. screened for small molecules that could induce definitive endoderm (DE)-like cells (Borowiak et al., 2009). They identified two structurally similar small molecules, IDE1 and IDE2, which could induce DE differentiation in up to 80% of mESCs and 50% of hESCs in the absence of Activin A (which is commonly used to induce hESC-to-DE differentiation *in vitro*). Both IDE1 and IDE2 induce Smad2 phosphorylation in mESCs, recapitulating the functions of Activin A, although their targets remain unknown. IDE1- and IDE2-induced endoderm-like cells could further differentiate into pancreatic progenitors *in vitro* and contribute to gut tube formation *in vivo* when injected into the developing gut tube of mouse embryos. Furthermore, these IDE1/IDE2-induced endoderm-like cells could be differentiated into a pancreatic lineage when subsequently treated with indolactam V, another small molecule identified from a phenotypic screening in hESC-derived DE cells that induces Pdx1 expression (Chen et al., 2009). Further mechanistic study revealed that indolactam V is an activator of protein kinase C (PKC), indicating a potential role for PKC during pancreatic development. Importantly, indolactam-V-induced cells could form mature pancreatic cells *in vitro* and *in vivo*.

Significant advances, particularly enabled by small molecules, have also been made in differentiating PSCs toward neural and subtype-specific neuronal cells. Chambers et al. developed a rapid and efficient monolayer neural induction method for hESCs that bypasses the conventional embryoid body approach (Chambers et al., 2009). They demonstrated that a combination of Noggin (a protein binding and inactivating BMP4) or LDN193189 (a synthetic small molecule inhibitor of BMP receptor, Table 4.1) with SB431542 (TGFβ receptor inhibitor, Table 4.1) promotes rapid neural induction of hESCs in a monolayer fashion with more than 80% purity. The two signaling pathway inhibitors appear to function synergistically to destabilize the self-renewal of hESCs, promote neural induction, and prevent cells from differentiating into trophectoderm, mesoderm, and endoderm lineages (for which BMP and TGFβ signaling have inductive effects). Based on this dual-Smad inhibition protocol, they further demonstrated that floor plate and midbrain dopamine neurons could be efficiently generated with additional instructive cues (e.g., SHH or SHH signaling pathway agonists). These studies suggest that ESC differentiation directed towards a specific lineage can be achieved by deliberately combining inductive signals for the desired cell lineage with inhibitory signals blocking ESC self-renewal and differentiation toward undesired lineages.

Further discovery of potent, highly selective, small molecule modulators of these key signaling players could be used to precisely manipulate PSC differentiation toward desired lineages, and reveal new mechanisms underlying some developmental processes.

Differentiation of Pluripotent Stem Cells to Self-Renewable Populations TSCs have the ability to maintain tissue homeostasis, and thus hold great promise for applications in regenerative medicine. Notably, TSCs derived from distinct developmental stages may display distinct potentials. The conventional approach, as mentioned above, to obtain TSCs through ESC differentiation is a "non-stop" process, the strategy/condition of which does not allow the capture of the intermediate progenitor cell population. Thus, to capture and stably expand ESC-derived TSCs remains a formidable challenge for translating ESCs toward various *in vitro* and therapeutic applications. An excellent study demonstrated the successful capture and stable expansion of primitive neuroepithelial cells (pNSCs) from hESCs by small molecules. The self-renewal of the definitive bFGF-responsive NSCs depends on the presence of bFGF and epidermal growth factor (EGF). However, these cells gradually transit into glial-restricted precursors during expansion, which are much less neurogenic. The ability to capture and expand NSCs long-term without losing neurogenic potential is of great clinical interest. To this end, Li et al. identified novel combinations of small molecules for either inducing or expanding pNSCs from hESCs in culture. They found that synergistic inhibition of the GSK3β, TGFβ and Notch signaling pathways by small molecule inhibitors could efficiently convert hESCs into pNSCs within one week. These pNSCs can stably self-renew in the presence of hLIF, GSK3 inhibitor (CHIR99021), and TGFβ receptor inhibitor (SB431542). These pNSCs differ from previously reported NSCs in their identity of primitive pre-rosette neuroepithelium, which may serve as an *in vitro* model to study stage-specific property of stem cells. Practically, this study provided an alternative way to obtain TSCs by differentiating PSCs into a renewable population, and also addressed critical issues of low yield and high heterogeneity of cells typically after sequential steps of non-stopping differentiation by providing stoppable intermediate stages. Importantly, the pNSCs exhibited highly neurogenic propensity and responsiveness to morphogenic signals to be patterned into region-specific neuronal subtypes, which could be of great clinical potential for neurodegenerative disease therapy, such as Parkinson's disease.

Chemical Control of Tissue-Specific Stem/Progenitor Cell Potential

Chemical approaches can also be used to regulate behaviors of TSCs *in vitro* and *in vivo*. TSCs are normally non-tumorigenic compared to PSCs, and they can efficiently engraft and differentiate into desirable cell types after transplantation. Therefore, they hold great promise for stem cell-based clinical applications. Due to the limited source of adult donors for TSCs and invasiveness involved in isolating primary cells from adults, *in vitro* differentiation of PSCs into desirable tissue-specific stem/progenitor cells represents an alternative approach for obtaining TSCs. Besides using small molecules for the *in vitro* derivation and maintenance of TSCs that can be used in cell-based therapy, directly delivering small molecules to diseased or damaged tissues/organs to control the behaviors of the patients' own cells within their *in vivo* niche may represent another attractive approach for regenerative medicine.

Modulation of Hematopoietic Stem Cell Expansion and Trafficking

HSCs are multipotent stem cells that reside in the bone marrow, and are able to differentiate into all the blood lineages. To date, HSC transplantation is perhaps the most successful application for stem cells in medicine, and yet for many disease conditions success remains low. This is largely due to imprecise molecular matching of donor and recipient, insufficient number of transplantable cells, and inadequate engraftability of donor cells. For example, in allogeneic HSC transplants, only 50% of candidates can find a human-leukocyte-antigen-matched adult donor. The *ex vivo* expansion of HSCs driven by cytokines is typically accompanied by the induction of differentiation, leading to loss of long-term self-renewal ability. To identify small molecules that promote the expansion of HSCs, Boitano et al. carried out an image-based screening of 100,000 heterocycle-containing compounds using human primary CD34+ HSCs (Boitano et al., 2010). They identified StemRegenin 1 (SR1, Table 4.1), a purine derivative that promotes the *ex vivo* expansion of CD34+ cells that retain the ability to sustain multi-lineage reconstitution and engraft secondary recipients. Mechanistic studies indicated that SR1 controls HSC expansion by antagonizing the aryl hydrocarbon receptor, which was not previously known to function in hematopoietic stem cell biology. This work not only enabled a more robust expansion of blood stem cells, but also identified the aryl hydrocarbon receptor as an important modulator of HSC self-renewal.

HSCs dynamically traffic between the bone marrow, peripheral blood, and secondary organs. As there are low numbers of HSCs in the blood under normal conditions, HSC mobilization into the peripheral circulation is important for the efficient collection of HSCs for autologous and allogeneic transplantation. This mobilization, and its reverse process – homing – were shown to be regulated by two receptor-ligand pathways: VLA4 and its ligand VCAM-1, and chemokine receptor CXCR4 and its ligand SDF-1. The studies that the SDF-1-CXCR4 signaling axis promotes the retention of HSCs within the marrow are crucial for the development of AMD3100 (Table 4.1) as a potent HSC mobilizing agent. AMD3100, a selective antagonist of CXCR4, specifically blocks the binding of SDF-1 to the receptor CXCR4, and consequently induces rapid mobilization of both mouse and human hematopoietic progenitor cells and synergistically augments G-CSF-induced mobilization of HSCs (Broxmeyer et al., 2005; Larochelle et al., 2006). Similarly, a small molecule that inhibits VLA-4 receptor, BIO5192 (Table 4.1), was also reported to mobilize HSCs (Ramirez et al., 2009). An additive effect on HSC mobilization was observed when BIO5192 was combined with AMD3100. It was recently found that the interaction between prostagliandin E2 (PGE2) and its G-protein-coupled receptor EP4 is also involved in mobilization of HSC/HPC (hematopoietic progenitor cell) from bone marrow to the circulation (Hoggatt et al., 2013). Treatment of mice with non-steroidal anti-inflammatory drugs (NSAID), which inhibit the endogenous level of PGE2, results in a significant increase in HPC number in circulation with modest HSC egress, indicating a different mechanism of HSC and HPC retention within bone marrow. These processes are independent of the SDF-1/CXCR4 axis, and could be enhanced when co-administered with G-CSF or CXCR4 antagonist. Notably, the function of PGE2 to HSC/HPC trafficking seems bidirectional. Brief stimulation with PGE2 increases HSC/HPC migration and engraftment by accelerating hematopoiesis, while long-term treatment, however, inhibits HPC expansion and prevents HSC/HPC trafficking. Therefore, it would be of clinical purpose for modulating the HSC/HPC mobilization protocol as part of therapies by optimizing the usage of NSAID, such as aspirin, indomethacin and Meloxicam.

Small Molecule Selectively Regulating CD4+ T Cell Differentiation

CD4+ T cells represent a unique branch of the adaptive immune system. Naïve CD4+ T cells are activated upon interaction with antigen-MHC complexes and differentiate into specific and functionally distinct subtypes, including T-helper 1(Th1), T-helper 2 (Th2), T-helper 17 (Th17), regulatory T, follicular helper T, and T-helper-9 cells, depending mainly on the cytokine milieu within the microenvironment and transcription factor profiles. Th17 cells are responsible for mounting immune responses against extracellular bacteria and fungi; however, they also participate in the development of autoimmunity. Furthermore, there is mounting evidence that Th17 cells influence inflammatory processes in humans. Therefore, identification of the pathways and small molecules that regulate Th17 cell differentiation has attracted significant interest. To this end, Xu et al. screened more than 2,000 known bioactive compounds using human naïve CD4+ T cells, and identified ursolic acid (UA, Table 4.1), a natural small molecule ubiquitously present in plants, as a selective inhibitor for blocking Th17 cell differentiation (Xu et al., 2011). Further characterization revealed that UA is a specific antagonist of RORγt, which is the Th17-specific master transcriptional regulator for IL-17. In addition, UA inhibited IL-17 production in both developing and mature Th17 cells. Importantly, UA treatment ameliorated the disease phenotype in mice induced with experimental autoimmune encephalomyelitis (EAE), a mouse model for multiple sclerosis, indicating the clinical potential of UA in Th17-mediated autoimmune diseases. This study is an excellent example of how it is possible to regulate the differentiation of endogenously existing tissue-specific precursor cells *in vivo* using a highly specific chemical identified from relevant *in vitro* assays. In conjunction with the progress in small molecule delivery technology, the ability to regulate cell fate *in vivo* could turn TSC-targeted clinical therapies into reality.

Chemical Control of Cellular Reprogramming

The breakthrough in reprogramming somatic cells into iPSCs opened up unprecedented opportunities in stem cell biology and regenerative medicine. In 2006 and 2007, Yamanaka and colleges reprogrammed mouse and human somatic cells into iPSCs through the ectopic expression of four defined TFs (i.e., Oct4, Sox2, Klf4, and cMyc) (Takahashi et al., 2007; Takahashi and Yamanaka, 2006). However, this genetic approach raised some safety concerns, such as the viral delivery system and permanent integration of exogenous oncogenes, which would hinder its potential therapeutic applications. To address these problems, researchers turned to non-integrating genetic manipulation and small molecules. Utilization of small molecules that can induce and enhance the reprogramming of cells into the pluripotent state would offer several significant advantages over genetic approaches, including reducing the risks of genetic manipulation with the oncogenic reprogramming factors, enhancing reprogramming efficiency, improving the quality of generated iPSCs, and ultimately changing the process to be directed/deterministic. A prevailing logic and practical strategy to develop small molecule-based reprogramming has been first to identify small molecules that can replace individual and combinations of reprogramming TFs in different cellular reprogramming contexts (e.g., using different starting somatic cell types that endogenously express some of the reprogramming TFs, with different ectopically expressed TFs, and/or with added small molecules and growth factors), and then to optimize sequential and combinatorial treatments with those small molecules to enable reprogramming without any other genetic factors. A series of small molecules that can functionally replace one or more of the exogenous reprogramming TFs, enhancing reprogramming efficiency, and accelerating reprogramming speed have been identified and characterized. According to their functional mechanisms, most of these compounds fall into one of the following categories: epigenetic modifiers, metabolism regulators, and cell signaling modulators. Representative examples will be discussed below.

Small Molecules Modulating Reprogramming Towards Pluripotent Stem Cells

Epigenetic Modifiers Somatic cells and pluripotent cells have drastic epigenetic differences across the epigenome, which must be changed during reprogramming. Some small molecules that could modulate epigenetic enzymes or mechanisms were found to modulate reprogramming

Shi et al. initially found that BIX-01294 (Table 4.1), an inhibitor of G9a histone methyltransferase, could significantly improve efficiency of reprogramming neural progenitor cells (NPCs) to iPSCs. With the addition of BIX-01294, mouse NPCs transduced with two TFs, Oct4, and Klf4 could be converted into iPSCs. Interestingly, BIX-01294 in conjunction with three TFs (i.e., Sox2/Klf4/c-Myc) could reprogram NPCs to iPSCs, bypassing the core pluripotency TF Oct4. It is possible that G9a mediated the epigenetic repression of Oct4, and its inhibition facilitates reactivation of Oct4. This example highlights the functional importance of epigenetic modifiers in cellular reprogramming (Shi et al., 2008b)

Some other epigenetic modifiers were also found to facilitate reprogramming. For example, valproic acid (Table 4.1), a histone deacetylase inhibitor, could enhance reprogramming and reactivate the silenced Dlk1-Dio3 to improve the quality of iPSCs (Stadtfeld et al., 2010). Vitamin C (Table 4.1), acting in part by promoting the activities of H3K36 histone demethylase and the TET family of methylcytosine dioxygenases, was found to promote reprogramming and prevent the loss of Dlk1-Dio3 imprinting and therefore help to generate all-iPSC mice (Stadtfeld et al., 2012). The DNA methyltransferase inhibitors RG108 (Table 4.1) and 5-azacytidine (Table 4.1), the histone demethylase inhibitor Parnate, and the histone deacetylase inhibitors trichostatin A (Table 4.1) and sodium butyrate (Table 4.1) were also reported to enhance reprogramming. Of note, the epigenetic modifiers are not specific for a particular cellular process, and are normally applied in combination with other TFs and/or small molecules of other categories, such as those that control signaling pathways. Combining different epigenetic compounds with small molecules of other regulatory functions could have desirable synergistic effects on reprogramming.

Metabolism Regulators Somatic cells typically use mitochondrial oxidation as their primary energy metabolism pathway, while PSCs mainly use glycolysis. Glycolysis could be more effective than oxidative phosphorylation at generating macromolecular precursors to meet the demand of rapid cell proliferation. For example, glycolytic intermediates could be used for the synthesis of nonessential amino acids and fatty acids. At the same time, glycolysis would generate less reactive oxygen species than mitochondrial oxidation and thereby avoid excessive oxidative damage to stem cells.

Initial evidence that metabolic reprogramming from mitochondrial oxidation to glycolysis is involved in reprogramming somatic cells into iPSCs came from the identification of PS48 (Table 4.1), a small molecule that significantly promotes the reprogramming of human primary somatic cells into iPSCs using only Oct4 (Zhu et al., 2010). PS48 is a small molecule allosteric activator of PDK1 that can lead to downstream Akt activation. In this study, it was found that PS48 treatment led to the up-regulation of several key glycolytic genes and consequently enhanced glycolysis. Several known small molecules that have been widely used to modulate mitochondrial oxidation and glycolytic metabolism were also tested. Consistently, it was found that compounds promoting glycolytic metabolism, such as 2,4-dinitrophenol (Table 4.1), Quercetin (Table 4.1) and *N*-oxaloylglycine (Table 4.1), are able to enhance reprogramming. In contrast, compounds that blocked glycolytic metabolism, such as oxalate and 2-DG, inhibited reprogramming. Further studies showed that human PSCs have functional oxidative phosphorylation machinery and an oxygen consumption rate similar to somatic cells, but oxidative phosphorylation is less coupled to ATP synthesis in these cells (Zhang et al., 2011). Indeed, Zhu et al. found that 2,4-dinitrophenol, an uncoupler of the mitochondrial respiratory chain, could greatly promote iPSC reprogramming. These studies indicate that a metabolic switch to glycolysis is an important early step in the reprogramming of somatic cells to iPSCs. Small molecules that can promote this metabolic shift would facilitate reprogramming.

Signaling Pathway Modulators Modulating reprogramming has also been achieved by small molecules that regulate key signaling pathways. Given the effect of MAPK pathway inhibition on naïve PSC self-renewal, it was found that PD0325901, a MEK inhibitor, could stabilize and select for authentic iPSCs when applied at a late stage of reprogramming (Shi et al., 2008a). Based on the notion that reprogramming fibroblast cells (i.e., mesenchymal type) into iPSCs (i.e., epithelial type) represents a MET process, small molecules promoting MET (or repressing EMT) were tested in iPSC reprogramming, including inhibitors of TGFβ receptor, MEK (which is downstream of TGFβ in EMT induction), and ROCK (which destabilizes E-cadherin). Consistent with this notion, it was found that TGFβ receptor inhibitor as well as its combination with MEK inhibitor and ROCK inhibitor could significantly enhance reprogramming efficiency and accelerate reprogramming speed (Lin et al., 2009).

Remarkably, the GSK3 inhibitor CHIR99021, which can activate Wnt/β-catenin signaling and alleviate TCF3's repressive function on pluripotency genes, was also found to significantly improve reprogramming efficiency. The reprogramming of MEFs transduced with Oct4 and Klf4 could be greatly enhanced by treatment with CHIR99021, and in combination with Parnate, CHIR99021 could also facilitate reprogramming of primary human keratinocytes transduced with Oct4 and Klf4 into iPSCs (Li et al., 2009b).In combination with BIX-01294, BayK 8644 (Table 4.1), a L-type calcium channel agonist, could enable the reprogramming of mouse embryonic fibroblasts transduced by Oct4 and Klf4 into iPSCs (Shi et al., 2008a). BayK 8644 plays its role upstream in cell signaling pathways and can compensate for viral transduction of Sox2.

Small Molecules that Control Trans-Differentiation

Besides the conventional paradigm of using target cell type-specific factors (e.g., transcription factors/TFs and miRNAs) for trans-differentiation, a novel paradigm of Cell Activation and Signaling-Directed (CASD) lineage conversion was established, which employs transient overexpression of iPSC-TFs (cell activation, CA) in conjunction with lineage-specific soluble signals (signal-directed, SD) to reprogram somatic cells into diverse lineage-specific cell types without entering the pluripotent state (Figure 4.2). This is based on the early events of induced pluripotency, which is established in a stepwise and stochastic fashion. Under extended expression of reprogramming factors and favorable culture conditions, only a rare subset among various intermediate cells finally become pluripotent. It was reasoned that those initial epigenetically unstable cells (induced by the iPSC reprogramming factors) could be re-directed into lineage-specific cell types under favorable condition without traversing pluripotency. Specifically it was found that by temporally restricting the ectopic overexpression of iPSC factors in fibroblasts, epigenetically "activated" cells could be generated rapidly. These cells could then be coaxed to "relax" back into certain differentiated states by specific culture conditions (that favor lineage-specific cell types and simultaneously inhibit the establishment of pluripotency using growth factors and small molecules), ultimately giving rise to somatic cells entirely distinct from the starting population. For example, it was found that with as little as four days of iPSC-TF expression (far shorter than what is required for induction of pluripotency), mouse fibroblasts can be directly reprogrammed into spontaneously contracting cardiomyocytes over 11–12 days under treatment with a small molecule JAK inhibitor for the first nine days (this blocks the establishment of pluripotency by inhibiting the LIF/STAT3 pathway), and BMP4 from day 9 onwards (this mediates cardiac mesoderm induction) (Efe et al., 2011). Applying the same concept and approach, neural (Kim et al., 2011; Lu et al., 2013; Thier et al., 2012; Wang et al., 2012), endothelial (Margariti et al., 2012), angioblast-like mesodermal progenitor cells (Kurian et al., 2012), and definitive endodermal cells were directly reprogrammed from fibroblasts rapidly and efficiently using transient expression of iPSC factors and treatments with FGFs/EGF (toward neural cells), BMP/VGEF (toward endothelial cells), or Activin A (toward definitive endodermal cells), respectively. Compared to trans-differentiation by overexpression of tissue-specific transcription factors, this CASD trans-differentiation paradigm has several advantages. A single combination of transcription factors is applicable to induce reprogramming toward various lineage-specific cell types. Its transient expression could be easily replaced by non-integrating or no-genetic methods. Most significantly, progenitor populations belonging to these lineages are generated in the process and can be isolated and expanded for various applications.

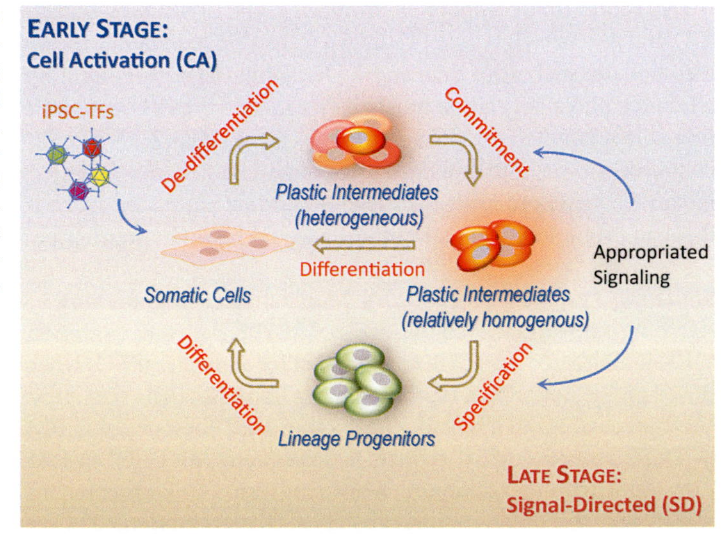

FIGURE 4.2 Cell Activation and Signaling-Directed (CASD) Lineage Conversion Model. Transient overexpression of iPSC-TFs (cell activation, CA) gradually de-differentiates the starting cells by erasing the epigenetic features and de-stabilizing of original transcription network, which is responsible for the cellular identity. These early initial events modify chromatin stochastically, and generate the "plastic intermediate" cells which are amendable to external lineage-specific soluble signals (signal-directed, SD) that could further commit and specify these "plastic intermediate" cells into diverse lineage-specific cell types, including lineage-specific stem/progenitor cells and terminally differentiated cell types, without entering the pluripotent state.

To improve the efficiency of CASD-based trans-differentiation, researchers have also turned to small molecules. A recent study by Wang et al. demonstrated that a small molecule cocktail containing CHIR99021, PD0325901, A83-01, thiazovivin, and DMH1 (Table 4.1) can enhance the generation of NPCs from human urinal epithelial cells that were delivered with episomal iPSC-TFs under the CASD paradigm (Wang et al., 2012). Interestingly, only one of the iPSC-TFs alone, Oct4, proved to be sufficient to direct the human fibroblasts to NSCs with treatment of small molecule cocktail, indicating the generality of small molecules' function in the CASD-based paradigm. This is consistent with previous work, indicating that Oct4 in conjunction with a small molecule cocktail enabled reprogramming of mouse and human somatic cells into iPSCs (Yuan et al., 2011; Zhu et al., 2010). These studies support the concept and strategy that combinations of small molecules could be developed for inducing lineage-specific reprogramming without any genetic factor under the CASD-based trans-differentiation paradigm.

Chemical Control of De-Differentiation of Lineage-Restricted Cells

For applications where lineage-specific stem or progenitor cells or other types of functional somatic cells are required, de-differentiation of lineage-restricted cells (just like what happens during regeneration in simpler organisms) may represent another attractive approach. Because they are lineage restricted, these de-differentiated cells may still exhibit restricted tissue identity, robust regenerative abilities, and lower tumorigenic potential *in vivo*.

Reversine (Table 4.1) is a small molecule that can facilitate the de-differentiation of lineage-restricted cells (Chen et al., 2004). To identify small molecules that induce mouse myogenic lineage-committed cells to de-differentiate *in vitro*, Chen et al. carried out a cell-based functional screen in which lineage-committed myoblasts were initially treated with compounds to induce de-differentiation, and then assayed for their differentiation ability to non-permissive cell types. This two-step protocol was based on the concept that de-differentiated cells could regain multipotency. The researchers discovered that reversine could indeed induce lineage reversal of C2C12 myogenic cells to become multipotent progenitor cells that could re-differentiate into osteoblasts and adipocytes under lineage-specific differentiation conditions. In addition, reversine treatment reprogrammed primary murine and human dermal fibroblasts into myogenic-competent cells that could further differentiate into skeletal muscle *in vitro* and *in vivo*. Mechanistic studies suggested that reversine acts as a dual inhibitor of non-muscle myosin-II heavy chain and MEK1, and that both activities were required for reversine's effects (Chen et al., 2007). Inhibition of MEK1 and non-muscle myosin-II heavy chain resulted in cell cycle alterations and changes in histone acetylation status. These data show how identification and characterization of reversine could provide new insights into the fundamental molecular mechanisms that control cellular de-differentiation. In so doing, the researchers also discovered a novel small molecule that may ultimately be useful for stem cell therapy.

4.3 CLINICAL RELEVANCE

- Improved understanding of stem cell biology guides the development of highly selective/potent approaches to stem cell fate control, and ultimately novel strategies in regenerative therapy.
- Chemical approaches avoid the need for genome modification by genetic approaches, and have the potential to be more safely, efficaciously, and quickly translated into clinical uses for the small molecule modulators of cell fate.
- The CASD-based cell trans-differentiation paradigm will allow the generation of different lineage-specific stem/progenitor cells under general and chemically defined conditions for various applications.

ACKNOWLEDGMENT/GRANT SUPPORT

Sheng Ding is supported by funding from National Institute of Child Health and Human Development, National Heart, Lung, and Blood Institute, and National Eye Institute/National Institute of Health [grant numbers HD064610, HL107436, EY021374], California Institute for Regenerative Medicine, and the J David Gladstone Institutes. We thank Anna Lisa Lucido and Gary Howard in J. David Gladstone Institutes for critical reading and editing of this manuscript. The authors apologize to all scientists whose research could not be properly discussed and cited in this review owing to space limitations.

RECOMMENDED RESOURCES

Web sites

Tocris Bioscience: http://www.tocris.com/ (accessed 20.03.14).
Cell Signaling Technology: http://www.cellsignal.com/ (accessed 20.03.14).

Reviews

Nichols, J., Smith, A., 2009. Naive and primed pluripotent states. Cell Stem Cell 4 (6), 487–492.

Song, M., Paul, S., Lim, H., Dayem, A.A., Cho, S.G., 2012 Feb. Induced pluripotent stem cell research: a revolutionary approach to face the challenges in drug screening. Arch. Pharm. Res. 35 (2), 245–260.

Li, W., Jiang, K., Ding, S., 2012 Jan. Concise review: A chemical approach to control cell fate and function. Stem Cells 30 (1), 61–68.

Xu, T., Zhang, M., Laurent, T., Xie, M., Ding, S., 2013. Concise review: Chemical approaches for modulating lineage-specific stem cells and progenitors. Stem Cells Trans. Med. 2, 355–361.

REFERENCES

Boitano, A.E., Wang, J., Romeo, R., Bouchez, L.C., Parker, A.E., Sutton, S.E., Walker, J.R., Flaveny, C.A., Perdew, G.H., Denison, M.S., et al., 2010. Aryl hydrocarbon receptor antagonists promote the expansion of human hematopoietic stem cells. Science 329, 1345–1348.

Borowiak, M., Maehr, R., Chen, S., Chen, A.E., Tang, W., Fox, J.L., Schreiber, S.L., Melton, D.A., 2009. Small molecules efficiently direct endodermal differentiation of mouse and human embryonic stem cells. Cell Stem Cell 4, 348–358.

Brons, I.G., Smithers, L.E., Trotter, M.W., Rugg-Gunn, P., Sun, B., Chuva de Sousa Lopes, S.M., Howlett, S.K., Clarkson, A., Ahrlund-Richter, L., Pedersen, R.A., et al., 2007. Derivation of pluripotent epiblast stem cells from mammalian embryos. Nature 448, 191–195.

Broxmeyer, H.E., Orschell, C.M., Clapp, D.W., Hangoc, G., Cooper, S., Plett, P.A., Liles, W.C., Li, X., Graham-Evans, B., Campbell, T.B., et al., 2005. Rapid mobilization of murine and human hematopoietic stem and progenitor cells with AMD3100, a CXCR4 antagonist. J. Exp. Med. 201, 1307–1318.

Chambers, I., Smith, A., 2004. Self-renewal of teratocarcinoma and embryonic stem cells. Oncogene 23, 7150–7160.

Chambers, S.M., Fasano, C.A., Papapetrou, E.P., Tomishima, M., Sadelain, M., Studer, L., 2009. Highly efficient neural conversion of human ES and iPS cells by dual inhibition of SMAD signaling. Nat. Biotechnol. 27, 275–280.

Chan, K.K., Zhang, J., Chia, N.Y., Chan, Y.S., Sim, H.S., Tan, K.S., Oh, S.K., Ng, H.H., Choo, A.B., 2009. KLF4 and PBX1 directly regulate NANOG expression in human embryonic stem cells. Stem Cells 27, 2114–2125.

Chen, S., Borowiak, M., Fox, J.L., Maehr, R., Osafune, K., Davidow, L., Lam, K., Peng, L.F., Schreiber, S.L., Rubin, L.L., et al., 2009. A small molecule that directs differentiation of human ESCs into the pancreatic lineage. Nat. Chem. Biol. 5, 258–265.

Chen, S., Do, J.T., Zhang, Q., Yao, S., Yan, F., Peters, E.C., Scholer, H.R., Schultz, P.G., Ding, S., 2006. Self-renewal of embryonic stem cells by a small molecule. Proc. Natl. Acad. Sci. USA 103, 17266–17271.

Chen, S., Takanashi, S., Zhang, Q., Xiong, W., Zhu, S., Peters, E.C., Ding, S., Schultz, P.G., 2007. Reversine increases the plasticity of lineage-committed mammalian cells. Proc. Natl. Acad. Sci. USA 104, 10482–10487.

Chen, S., Zhang, Q., Wu, X., Schultz, P.G., Ding, S., 2004. Dedifferentiation of lineage-committed cells by a small molecule. J. Am. Chem. Soc. 126, 410–411.

Clark, A.T., Rodriguez, R.T., Bodnar, M.S., Abeyta, M.J., Cedars, M.I., Turek, P.J., Firpo, M.T., Reijo Pera, R.A., 2004. Human STELLAR, NANOG, and GDF3 genes are expressed in pluripotent cells and map to chromosome 12p13, a hotspot for teratocarcinoma. Stem Cells 22, 169–179.

D'Amour, K.A., Bang, A.G., Eliazer, S., Kelly, O.G., Agulnick, A.D., Smart, N.G., Moorman, M.A., Kroon, E., Carpenter, M.K., Baetge, E.E., 2006. Production of pancreatic hormone-expressing endocrine cells from human embryonic stem cells. Nat. Biotechnol. 24, 1392–1401.

Efe, J.A., Hilcove, S., Kim, J., Zhou, H., Ouyang, K., Wang, G., Chen, J., Ding, S., 2011. Conversion of mouse fibroblasts into cardiomyocytes using a direct reprogramming strategy. Nat. Cell Biol. 13, 215–222.

Esteban, M.A., Wang, T., Qin, B., Yang, J., Qin, D., Cai, J., Li, W., Weng, Z., Chen, J., Ni, S., et al., 2010. Vitamin C enhances the generation of mouse and human induced pluripotent stem cells. Cell Stem Cell 6, 71–79.

Evans, M.J., Kaufman, M.H., 1981. Establishment in culture of pluripotential cells from mouse embryos. Nature 292, 154–156.

Filby, C.E., Williamson, R., van Kooy, P., Pebay, A., Dottori, M., Elwood, N.J., Zaibak, F., 2011. Stimulation of Activin A/Nodal signaling is insufficient to induce definitive endoderm formation of cord blood-derived unrestricted somatic stem cells. Stem Cell Res. Ther. 2, 16.

Hanna, J., Cheng, A.W., Saha, K., Kim, J., Lengner, C.J., Soldner, F., Cassady, J.P., Muffat, J., Carey, B.W., Jaenisch, R., 2010. Human embryonic stem cells with biological and epigenetic characteristics similar to those of mouse ESCs. Proc. Natl. Acad. Sci. USA 107, 9222–9227.

Hoggatt, J., Mohammad, K.S., Singh, P., Hoggatt, A.F., Chitteti, B.R., Speth, J.M., Hu, P., Poteat, B.A., Stilger, K.N., Ferraro, F., et al., 2013. Differential stem- and progenitor-cell trafficking by prostaglandin E2. Nature 495, 365–369.

Huang, Y., Osorno, R., Tsakiridis, A., Wilson, V., 2012. *In vivo* differentiation potential of epiblast stem cells revealed by chimeric embryo formation. Cell Rep. 2, 1571–1578.

Huangfu, D., Osafune, K., Maehr, R., Guo, W., Eijkelenboom, A., Chen, S., Muhlestein, W., Melton, D.A., 2008. Induction of pluripotent stem cells from primary human fibroblasts with only Oct4 and Sox2. Nat. Biotechnol. 26, 1269–1275.

Ichida, J.K., Blanchard, J., Lam, K., Son, E.Y., Chung, J.E., Egli, D., Loh, K.M., Carter, A.C., Di Giorgio, F.P., Koszka, K., et al., 2009. A small-molecule inhibitor of tgf-Beta signaling replaces sox2 in reprogramming by inducing nanog. Cell Stem Cell 5, 491–503.

James, D., Levine, A.J., Besser, D., Hemmati-Brivanlou, A., 2005. TGFbeta/activin/nodal signaling is necessary for the maintenance of pluripotency in human embryonic stem cells. Development 132, 1273–1282.

Kawamura, T., Suzuki, J., Wang, Y.V., Menendez, S., Morera, L.B., Raya, A., Wahl, G.M., Izpisua Belmonte, J.C., 2009. Linking the p53 tumour suppressor pathway to somatic cell reprogramming. Nature 460, 1140–1144.

Kim, J., Efe, J.A., Zhu, S., Talantova, M., Yuan, X., Wang, S., Lipton, S.A., Zhang, K., Ding, S., 2011. Direct reprogramming of mouse fibroblasts to neural progenitors. Proc. Natl. Acad. Sci. USA 108, 7838–7843.

Kroon, E., Martinson, L.A., Kadoya, K., Bang, A.G., Kelly, O.G., Eliazer, S., Young, H., Richardson, M., Smart, N.G., Cunningham, J., et al., 2008. Pancreatic endoderm derived from human embryonic stem cells generates glucose-responsive insulin-secreting cells *in vivo*. Nat. Biotechnol. 26, 443–452.

Kurian, L., Sancho-Martinez, I., Nivet, E., Aguirre, A., Moon, K., Pendaries, C., Volle-Challier, C., Bono, F., Herbert, J.M., Pulecio, J., et al., 2012. Conversion of human fibroblasts to angioblast-like progenitor cells. Nat. Methods 10, 77–83.

Larochelle, A., Krouse, A., Metzger, M., Orlic, D., Donahue, R.E., Fricker, S., Bridger, G., Dunbar, C.E., Hematti, P., 2006. AMD3100 mobilizes hematopoietic stem cells with long-term repopulating capacity in nonhuman primates. Blood 107, 3772–3778.

Li, H., Zhou, H., Wang, D., Qiu, J., Zhou, Y., Li, X., Rosenfeld, M.G., Ding, S., Fu, X.D., 2012. Versatile pathway-centric approach based on high-throughput sequencing to anticancer drug discovery. Proc. Natl. Acad. Sci. USA 109, 4609–4614.

Li, W., Wei, W., Zhu, S., Zhu, J., Shi, Y., Lin, T., Hao, E., Hayek, A., Deng, H., Ding, S., 2009a. Generation of rat and human induced pluripotent stem cells by combining genetic reprogramming and chemical inhibitors. Cell Stem Cell 4, 16–19.

Li, W., Zhou, H., Abujarour, R., Zhu, S., Young Joo, J., Lin, T., Hao, E., Scholer, H.R., Hayek, A., Ding, S., 2009b. Generation of human-induced pluripotent stem cells in the absence of exogenous Sox2. Stem Cells 27, 2992–3000.

Lin, T., Ambasudhan, R., Yuan, X., Li, W., Hilcove, S., Abujarour, R., Lin, X., Hahm, H.S., Hao, E., Hayek, A., et al., 2009. A chemical platform for improved induction of human iPSCs. Nat. Methods 6, 805–808.

Liskovykh, M.A., Chuikin, I.A., Ranjan, A., Popova, E., Tolkunova, E.N., Chechik, L.L., Malinin, A., Morozova, A.V., Mosienko, V., Bader, M., et al., 2011. Genetic manipulation and studying of differentiation properties of rat induced pluripotent stem cells. Tsitologiia 53, 946–951.

Lu, J., Liu, H., Huang, C.T., Chen, H., Du, Z., Liu, Y., Sherafat, M.A., Zhang, S.C., 2013. Generation of integration-free and region-specific neural progenitors from primate fibroblasts. Cell Rep. 3, 1580–1591.

Lyssiotis, C.A., Foreman, R.K., Staerk, J., Garcia, M., Mathur, D., Markoulaki, S., Hanna, J., Lairson, L.L., Charette, B.D., Bouchez, L.C., et al., 2009. Reprogramming of murine fibroblasts to induced pluripotent stem cells with chemical complementation of Klf4. Proc. Natl. Acad. Sci. USA 106, 8912–8917.

Margariti, A., Winkler, B., Karamariti, E., Zampetaki, A., Tsai, T.N., Baban, D., Ragoussis, J., Huang, Y., Han, J.D., Zeng, L., et al., 2012. Direct reprogramming of fibroblasts into endothelial cells capable of angiogenesis and reendothelialization in tissue-engineered vessels. Proc. Natl. Acad. Sci. USA 109, 13793–13798.

Nichols, J., Smith, A., 2009. Naive and primed pluripotent states. Cell Stem Cell 4, 487–492.

Niwa, H., Burdon, T., Chambers, I., Smith, A., 1998. Self-renewal of pluripotent embryonic stem cells is mediated via activation of STAT3. Genes Dev. 12, 2048–2060.

Park, I.H., Zhao, R., West, J.A., Yabuuchi, A., Huo, H., Ince, T.A., Lerou, P.H., Lensch, M.W., Daley, G.Q., 2008. Reprogramming of human somatic cells to pluripotency with defined factors. Nature 451, 141–146.

Ramirez, P., Rettig, M.P., Uy, G.L., Deych, E., Holt, M.S., Ritchey, J.K., DiPersio, J.F., 2009. BIO5192, a small molecule inhibitor of VLA-4, mobilizes hematopoietic stem and progenitor cells. Blood 114, 1340–1343.

Shi, Y., Desponts, C., Do, J.T., Hahm, H.S., Scholer, H.R., Ding, S., 2008a. Induction of pluripotent stem cells from mouse embryonic fibroblasts by Oct4 and Klf4 with small-molecule compounds. Cell Stem Cell 3, 568–574.

Shi, Y., Do, J.T., Desponts, C., Hahm, H.S., Scholer, H.R., Ding, S., 2008b. A combined chemical and genetic approach for the generation of induced pluripotent stem cells. Cell Stem Cell 2, 525–528.

Smith, A.G., 2001. Embryo-derived stem cells: of mice and men. Annu. Rev. Cell Dev. Biol. 17, 435–462.

Stadtfeld, M., Apostolou, E., Ferrari, F., Choi, J., Walsh, R.M., Chen, T., Ooi, S.S., Kim, S.Y., Bestor, T.H., Shioda, T., et al., 2012. Ascorbic acid prevents loss of Dlk1-Dio3 imprinting and facilitates generation of all-iPS cell mice from terminally differentiated B cells. Nat. Genet. 44 (398–405), S391–S392.

Stadtfeld, M., Apostolou, E., Akutsu, H., Fukuda, A., Follett, P., Natesan, S., Kono, T., Shioda, T., Hochedlinger, K., 2010. Aberrant silencing of imprinted genes on chromosome 12qF1 in mouse induced pluripotent stem cells. Nature. 465(7295),175–181.

Takahashi, K., Tanabe, K., Ohnuki, M., Narita, M., Ichisaka, T., Tomoda, K., Yamanaka, S., 2007. Induction of pluripotent stem cells from adult human fibroblasts by defined factors. Cell 131, 861–872.

Takahashi, K., Yamanaka, S., 2006. Induction of pluripotent stem cells from mouse embryonic and adult fibroblast cultures by defined factors. Cell 126, 663–676.

Tesar, P.J., Chenoweth, J.G., Brook, F.A., Davies, T.J., Evans, E.P., Mack, D.L., Gardner, R.L., McKay, R.D., 2007. New cell lines from mouse epiblast share defining features with human embryonic stem cells. Nature 448, 196–199.

Thier, M., Worsdorfer, P., Lakes, Y.B., Gorris, R., Herms, S., Opitz, T., Seiferling, D., Quandel, T., Hoffmann, P., Nothen, M.M., et al., 2012. Direct conversion of fibroblasts into stably expandable neural stem cells. Cell Stem Cell 10 (4), 473–479.

Thomson, J.A., Itskovitz-Eldor, J., Shapiro, S.S., Waknitz, M.A., Swiergiel, J.J., Marshall, V.S., Jones, J.M., 1998. Embryonic stem cell lines derived from human blastocysts. Science 282, 1145–1147.

Thomson, J.A., Kalishman, J., Golos, T.G., Durning, M., Harris, C.P., Becker, R.A., Hearn, J.P., 1995. Isolation of a primate embryonic stem cell line. Proc. Natl. Acad. Sci. USA 92, 7844–7848.

Thomson, J.A., Marshall, V.S., 1998. Primate embryonic stem cells. Curr. Top. Dev. Biol. 38, 133–165.

Vallier, L., Alexander, M., Pedersen, R.A., 2005. Activin/Nodal and FGF pathways cooperate to maintain pluripotency of human embryonic stem cells. J. Cell Sci. 118, 4495–4509.

Vallier, L., Touboul, T., Chng, Z., Brimpari, M., Hannan, N., Millan, E., Smithers, L.E., Trotter, M., Rugg-Gunn, P., Weber, A., et al., 2009. Early cell fate decisions of human embryonic stem cells and mouse epiblast stem cells are controlled by the same signalling pathways. PLoS One 4, e6082.

Wang, L., Huang, W., Su, H., Xue, Y., Su, Z., Liao, B., Wang, H., Bao, X., Qin, D., He, J., et al., 2012. Generation of integration-free neural progenitor cells from cells in human urine. Nat. Methods 10, 84–89.

Xu, T., Wang, X., Zhong, B., Nurieva, R.I., Ding, S., Dong, C., 2011. Ursolic acid suppresses interleukin-17 (IL-17) production by selectively antagonizing the function of RORγt protein. J. Biol. Chem. 286, 22707–22710.

Xu, Y., Zhu, X., Hahm, H.S., Wei, W., Hao, E., Hayek, A., Ding, S., 2010. Revealing a core signaling regulatory mechanism for pluripotent stem cell survival and self-renewal by small molecules. Proc. Natl. Acad. Sci. USA 107, 8129–8134.

Ying, Q.L., Nichols, J., Chambers, I., Smith, A., 2003. BMP induction of Id proteins suppresses differentiation and sustains embryonic stem cell self-renewal in collaboration with STAT3. Cell 115, 281–292.

Ying, Q.L., Wray, J., Nichols, J., Batlle-Morera, L., Doble, B., Woodgett, J., Cohen, P., Smith, A., 2008. The ground state of embryonic stem cell self-renewal. Nature 453, 519–523.

Yuan, X., Wan, H., Zhao, X., Zhu, S., Zhou, Q., Ding, S., 2011. Combined chemical treatment enables oct4-induced reprogramming from mouse embryonic fibroblasts. Stem Cells 29 (3), 549–553.

Zhang, J., Khvorostov, I., Hong, J.S., Oktay, Y., Vergnes, L., Nuebel, E., Wahjudi, P.N., Setoguchi, K., Wang, G., Do, A., et al., 2011. UCP2 regulates energy metabolism and differentiation potential of human pluripotent stem cells. EMBO J. 30, 4860–4873.

Zhou, H., Li, W., Zhu, S., Joo, J.Y., Do, J.T., Xiong, W., Kim, J.B., Zhang, K., Scholer, H.R., Ding, S., 2010. Conversion of mouse epiblast stem cells to an earlier pluripotency state by small molecules. J. Biol. Chem. 285, 29676–29680.

Zhu, S., Li, W., Zhou, H., Wei, W., Ambasudhan, R., Lin, T., Kim, J., Zhang, K., Ding, S., 2010. Reprogramming of human primary somatic cells by OCT4 and chemical compounds. Cell Stem Cell 7, 651–655.

Chapter 5

BMP Signaling and Stem Cell Self-Renewal in the *Drosophila* Ovary

Darin Dolezal* and Francesca Pignoni*,†,**

**Dept. of Biochemistry and Molecular Biology; †Dept. of Neuroscience and Physiology; **Dept. of Ophthalmology, SUNY Upstate Medical University, Syracuse, NY*

Chapter Outline

SUMMARY

- Bone morphogenetic protein (BMP) ligands from stromal niche cells suppress differentiation and promote germline stem cell self-renewal in the *Drosophila* ovary.
- Regulatory factors in the niche and the local extracellular matrix enhance BMP signaling specifically in the stem cell.
- Regulatory factors in the developing germline and surrounding stromal cells antagonize BMP signaling thereby permitting and promoting differentiation.
- Soma⇨soma and germline⇨soma intercellular communication contributes to the regulation of germline homeostasis.
- This complex regulation generates a steep BMP signaling gradient that ensures the tight control of self-renewal versus differentiation.

Principles of Developmental Genetics.

5.1 INTRODUCTION

Stem cells (SCs) play a crucial role in the maintenance and regeneration of adult tissues. The balance between self-renewal and differentiation is tightly controlled, and is critical for preventing tumorigenesis or tissue degeneration and precocious aging. Therapeutic applications require knowledge of the mechanisms regulating SC maintenance and differentiation at cellular, genetic and molecular levels. To this end, it is critical to study stem cells in their natural environment; the SC niche. However, the rarity of stem cells, the paucity of markers for their "undifferentiated" state, and the diversity of cell types constituting the niche complicates the study of stem cells in their natural setting in many organisms.

Invertebrate genetic models offer some of the most powerful platforms for the *in vivo* dissection of stem cell homeostasis, and one of the best understood stem cell systems is the self-renewing germline of the *Drosophila* female. The ovary contains easily identifiable germline stem cells (GSCs) within a well-characterized anatomical structure that is amenable to histological and biochemical analysis. Thanks to the advantages of the *Drosophila* model (availability of complex genetic techniques, less redundant genome compared to mammals, and wealth of reagents from decades of genetic screening), the fly ovary has emerged as a powerful system for studying fundamental aspects of stem cell biology.

In this chapter, we first describe the anatomy of the *Drosophila* ovary, including the specialized tissue microenvironment, or niche, in which the GSCs are located. Next, we discuss how bone morphogenetic proteins (BMPs) secreted by the niche function as self-renewal signals to ensure the maintenance of GSCs in round after round of cell division. We then focus on the critical role of multiple extrinsic and intrinsic mechanisms in promoting BMP signaling in GSCs and suppressing it in their differentiating progeny. Finally, we highlight how soma⇨soma and germline⇨soma communication also contributes to germline homeostasis, and conclude by touching upon the role of BMP signaling in vertebrate stem cells.

5.2 THE *DROSOPHILA* OVARY

An Interactive Stem Cell Niche

Each fly ovary is composed of approximately 20 individual egg-producing functional units, called ovarioles (Figure 5.1). The fully formed egg is found at the posterior end of the ovariole, closest to the oviduct, and it is preceded by maturing egg chambers arranged linearly as youngest to oldest from anterior to posterior. Each egg chamber consists of one developing oocyte and 15 nurse cells, surrounded by a layer of follicular epithelial cells (Figure 5.1). The nurse cells and oocyte constitute the germline and are derived from the GSCs (Wieschaus and Szabad, 1979), and although the follicle cells are somatic in origin, they too are derived from a stem cell population called the follicle stem cells (FSCs) (Margolis and Spradling, 1995).

Both GSCs and FSCs are found at the anterior tip of each ovariole in a structure called the germarium (Figure 5.1) (Lin and Spradling, 1993). Altogether the germarium comprises some 300 cells, including the GSCs and their differentiating progeny (called cystoblasts [CBs] and cysts), FSCs and their derived follicular epithelium (FCs), terminal filament cells (TFCs), cap cells (CCs), and escort cells (ECs) (Figure 5.1) (reviewed in Chen et al., 2011; Harris and Ashe, 2011; Losick et al., 2011). The TFCs, CCs, and ECs are important components of the GSC niche and, like the follicular epithelium, are somatic in origin (Xie and Spradling, 2000). A summary description of these cells and their roles in germline homeostasis is provided in Table 5.1.

A GSC Division Yields a Renewed GSC and a Differentiating CB

The GSC niche tightly regulates the balance between the self-renewing SCs and their differentiating progeny, thus its cellular composition remains fairly constant through most of the fly's lifespan. Two to three GSCs are positioned at the tip of each germarium where they are anchored to CCs via adherens junctions (Song et al., 2002). In a specialized asymmetric division, a GSC gives rise to two daughter cells that have different fates: a renewed SC and a differentiating CB (Figure 5.1). This division occurs along an axis that is more or less perpendicular to the CC:GSC interface (Morris and Spradling, 2011), resulting in the retention of one daughter cell in direct contact with the CCs and the placement of the other daughter cell outside the niche, one cell diameter away. Because CCs are a source of self-renewal factors, the former daughter cell continues to receive self-renewal signals and persists as a stem cell, whereas the latter ceases to respond and begins instead to undergo differentiation (Song et al., 2004; King et al., 2001; Cox et al., 2000; Xie and Spradling, 1998). In this manner, a single GSC division event produces both a replacement GSC and a differentiating CB.

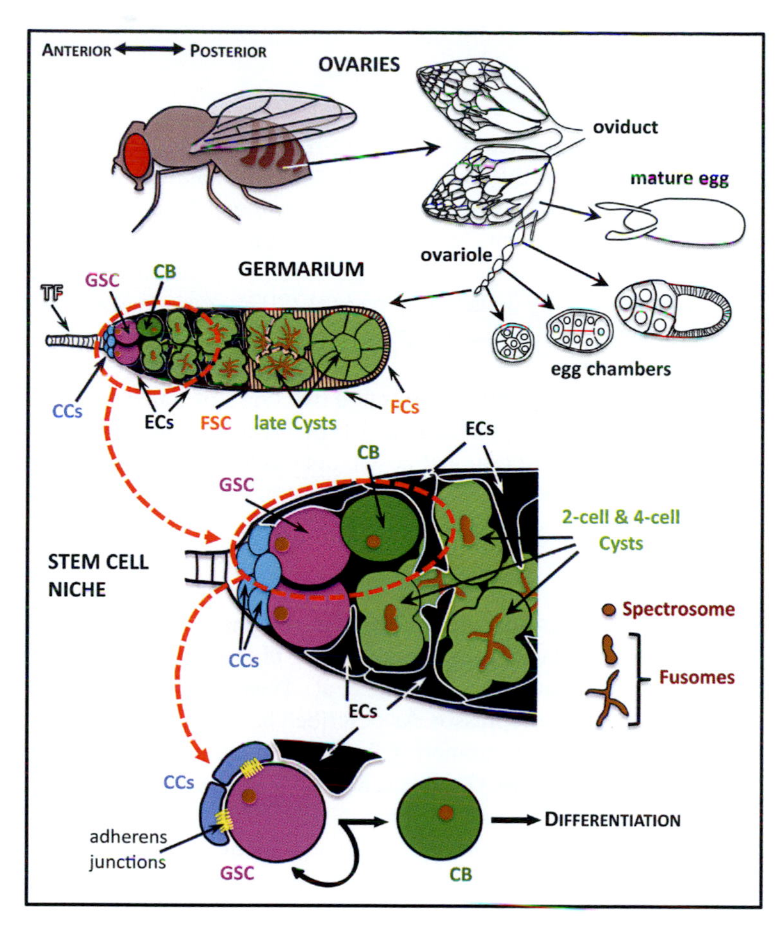

FIGURE 5.1 Anatomy of the *Drosophila* Ovary. The *Drosophila* ovaries occupy much of the female abdomen and each of the two **OVARIES** are composed of egg-producing units called **ovarioles**. The ovariole, in turn, is composed of the stem cell hosting **germarium** at the anterior tip, **egg chambers** at progressively more advances stages of maturation, and the **mature egg** at the posterior end near the oviduct. The cellular components of the **GERMARIUM** include cells of the germline lineage (**GSCs**, **CBs**, and **cysts**) and cells of the somatic lineage (**TFCs**, **CCs**, **ECs**, **FSCs** and **FCs**) (refer to Table 5.1 for full names and descriptions). In the germarium, GSCs divide into a renewed stem cell and a CB, which in turns undergoes four incomplete divisions to generate cysts of 16 cells connected by cytoplasmic bridges. During this process, germline stages can be distinguished by the evolving shape of the **spectrosome/fusome**, a cytoplasmic organelle important for orienting germ cell mitoses. In the GSC and CB, the structure appears as a single round sphere (**spectrosome**) located toward the anterior in the GSC. This structure then elongates and develops branches (**fusome**) in the **2-cell, 4-cell, 8-cell** and **16-cell cyst,** as it becomes distributed amongst all the interconnected cells through the cytoplasmic bridges. The **STEM CELL NICHE** is found at the anterior tip of the germarium where ~2–3 GSCs contact CCs anteriorly and ECs laterally. These interacting cells are the major somatic components of the **niche**, which ensures the production of one stem cell and one differentiating CB at each asymmetric division. CBs and cysts also form connections with other ECs that are critical to their continued differentiation.

TABLE 5.1 Cell Types of the Ovarian Stem Cell Niche

Acronym	Full Name	Origin	Description
CB	cystoblast	germline	differentiating daughter cell of the GSC
CC	cap cell	soma	niche cell that secretes self-renewal factors, including Dpp, Gbb and Dally
	cyst	germline	groups of differentiating germline cells connected through cytoplasmic bridges each cyst derives from a single CB and produces one egg chamber
EC	escort cell	soma	niche cell that promotes self-renewal of the GSC and differentiation of the CB
GSC	germline stem cell	germline	stem cell that divides to give rise to one GSC and one CB
TFC	terminal filament cell	soma	niche cell that secretes Upd ligand to induce *dpp* expression in CCs

Cyst Formation and Egg Production

Following GSC division, the CB undergoes four rounds of mitotic division with incomplete cytokinesis to produce cysts composed of two, four, eight, and finally 16 interconnected cells. Cysts migrate through the germarium away from the niche, making connections with the long cellular processes of the ECs (Figure 5.1) (Decotto and Spradling, 2005). Midway through the germarium, a developing cyst loses contact with the ECs and becomes surrounded by follicle cells produced by nearby FSCs (Figure 5.1) (Nystul and Spradling, 2007). The engulfed cyst then buds off from the germarium to become an egg chamber, which continues to grow larger as it moves toward the oviduct during oocyte maturation. The rate at which cysts are produced is dependent on environmental conditions, such as nutrients and sperm availability; in the absence of limiting factors, new egg chambers are produced every 12 hours, and the entire maturation process from GSC to egg takes about eight days (Drummond-Barbosa and Spradling, 2001).

5.3 THE BMP SIGNALING PATHWAY

Overview

Extensive studies have demonstrated that BMP signaling is critically important for GSC self-renewal (reviewed in Harris and Ashe, 2011). At a basic level, the role of BMP signaling is to prevent differentiation of stem cells naturally poised to divide and form cysts. Two BMP-type ligands, Dpp (BMP2/4) and Gbb (BMP6/7), are produced by the niche cells that surround the GSCs (Song et al., 2004; Xie and Spradling, 1998). These ligands are secreted into the extracellular space around the stem cells where they activate BMP receptors present on their plasma membranes (Figure 5.2A). In absence of these ligands, GSCs differentiate into cysts and the stem cell population is rapidly depleted (Song et al., 2004; Xie and Spradling, 1998). Thus, Dpp and Gbb function as "extrinsic" or "non-cell-autonomous" factors, produced by niche cells yet required by GSCs for self-renewal.

In *Drosophila*, as in vertebrates, BMP ligands trigger a phosphorylation cascade through two types of transmembrane serine/threonine kinase receptors. Following ligand binding to a Type II receptor, activated Type I receptors phosphorylate cytosolic signal transducers (R-SMADs) which in turn complex with transcriptional co-factors (co-SMADs) and translocate to the nucleus to regulate gene expression (Figure 5.2A) (Heldin et al., 1997; Wieser et al., 1995; Wrana et al., 1994) (also see Chapter 1).

In the *Drosophila* germline, the BMP ligands Dpp and Gbb bind to the Type II receptor Put, and transduction of the signal requires two Type I receptors, Sax and Tkv, all present on the GSC plasma membrane (Xie and Spradling, 1998; Haerry, 2010; Letsou et al., 1995). The activated form of the Tkv receptor mediates downstream signaling by phosphorylating Mad, a cytosolic R-SMAD (Haerry, 2010; Inoue et al., 1998). Phosphorylated Mad (pMad) then binds to the co-SMAD Med, and the pMad/Med complex in turn translocates into the nucleus (Figure 5.2A) (Inoue et al., 1998). Here, this complex interacts with additional factors to function as a transcriptional activator or repressor. As described below, a major target of BMP signaling in stem cells is the gene *bam*, which encodes a differentiation-promoting factor that must be silenced in the GSC (Figures 5.2A and 5.3) (Chen and McKearin, 2003a,b). Receptors, R-SMAD and co-SMAD function as "intrinsic" or "cell-autonomous" factors produced by, and acting in, the GSC to repress differentiation.

All core pathway components are evolutionarily conserved (see Table 5.2 for genes presented in this chapter and their vertebrate homologs). Somatic loss-of-function of the extrinsic ones (e.g., hypomorphic mutants or RNAi for Dpp) or germline loss-of-function of the intrinsic ones (e.g., germline clones for *tkv, Mad,* or *Med*; Figure 5.2B–C) results in differentiation instead of self-renewal leading to stem cell loss (Figure 5.2C) (Xie and Spradling, 1998; Liu et al., 2010). Conversely, excessive Dpp signaling, caused either by the overexpression of Dpp ligand or of a ligand-independent, constitutively activated, mutant form of the Tkv receptor (Tkv^{CA}), results in GSC tumors (Figure 5.2D) (Xie and Spradling, 1998; Casanueva and Ferguson, 2004).

BMP Signaling Blocks Differentiation in the GSC

The transcriptional repression of the differentiation-promoting gene *bam* is central to maintenance of the stem cell state, and is a major contribution of BMP signaling to GSC self-renewal (reviewed in Perinthottathil and Kim, 2011). In GSCs, the pMad/Med complex binds directly to a transcriptional silencer element in the *bam* promoter region and thus keeps the gene inactive (Figure 5.2A) (Chen and McKearin, 2003a,b). In CBs, decreased responsiveness to BMPs is accompanied by upregulation of *bam* expression (McKearin and Ohlstein, 1995). Loss of *bam* completely blocks cyst formation and results in the "tumorous" accumulation of single, undifferentiated germ cells within the germarium (McKearin and Spradling, 1990). Similar to the wild-type, *bam* mutant cells that maintain contact with the CCs continue to respond to BMP; whereas the other germline cells cease to respond, but do not differentiate into cysts (Casanueva and Ferguson, 2004). Conversely, targeted expression of Bam in GSCs causes the stem cells to differentiate into cysts while still in the niche, resulting in GSC depletion (Ohlstein and McKearin, 1997).

The Bam protein bears no characteristic motifs, but functions in a complex with the RNA-binding protein Bgcn to antagonize self-renewal factors (Figure 5.3) (Ohlstein et al., 2000; Lavoie et al., 1999; Rogers et al., 1986). Two such factors, Nos and Pum, function together in the GSC as a translational repressor complex (Sonoda and Wharton, 1999; Wharton et al., 1998). Nos/Pum bind Nos response elements in the 3' UTR of target mRNAs (Sonoda and Wharton, 1999; Wharton et al., 1998) and although the targets of Nos/Pum have not been identified in the GSC, it is likely that the complex silences the expression of differentiation factors (Wang and Lin, 2004; Forbes and Lehmann, 1998; Lin and Spradling, 1997). Interestingly, *bam* mutant germ cells undergo differentiation in the absence of *pum*, suggesting a model where *pum* is suppressed by *bam* (Chen and McKearin, 2005; Szakmary et al., 2005). Consistent with this model, Bam/Bgcn directly regulates Nos/Pum in two ways: it functions via the 3' UTR of the *nos* mRNA to

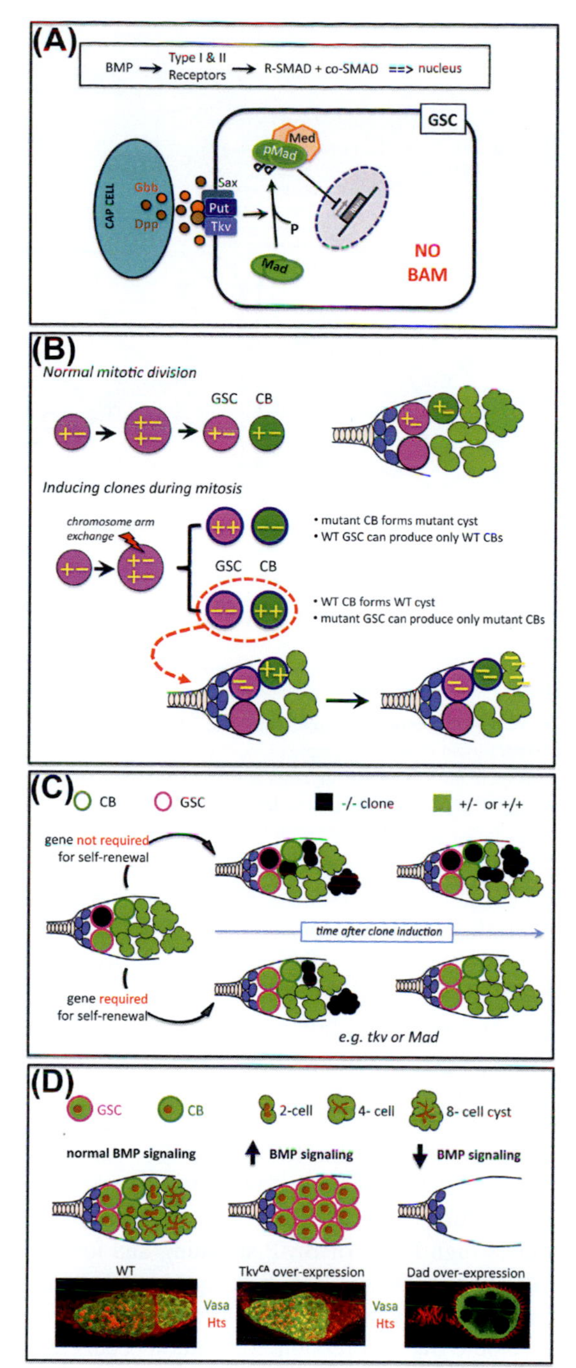

FIGURE 5.2 (A) The BMP Signaling Pathway and Gene Function Analysis in the *Drosophila* Female Germline. Core cellular and molecular players in the BMP signaling between CC and GSC. CCs express and secrete BMP ligands (Gbb and Dpp), which bind serine/threonine kinase receptors (Put, Sax, Tkv) on the surface of GSCs. Activated (phosphorylated) Tkv, in turn, catalyzes the phosphorylation of Mad into pMad. Upon binding of pMad to Med, the Mad/Med complex translocates to the nucleus where it suppresses transcription of the differentiation gene *bam*. (B) Genetic Analysis in the Developing Germarium. In *Drosophila*, genetic mosaics can be generated by inducing a recombinase-mediated exchange of chromosomal arms during mitosis (Fox et al. 2008). When chromosomal exchange occurs in the germline during the first asymmetric mitotic division (germline cells undergo meiosis only later, at the 16-cell cyst stage), either daughter cell (GSC or CB) may emerge as homozygous mutant. If that is the GSC, it will establish a mutant clonal GSC that continues to produce mutant CBs. Loss of a molecular markers (β-galactosidase or GFP) is usually used to mark the mutant cell and its progeny. (C) Gene Function can be Studied in the Germline Using the Mosaic Clonal Techniques Shown in (B). (Left) A single homozygous mutant 'marked clone' GSC is shown in black in an otherwise green wild-type (or heterozygous) background. (Top) A wild-type marked clonal GSC is maintained over time and continually produces clonal progeny. (Bottom) When a gene required for self-renewal is disrupted in a marked GSC, the mutant GSC exits the niche and differentiates. Eventually no marked clone germline cells are observed in the germarium, although mutant cysts and egg chambers may still be present. An example of genes that behave in this fashion are the core components of the BMP signaling pathway (**A**). (D) Dpp Signaling is Required and Sufficient for GSC Self-Renewal. (Left) A wild-type germarium contains both GSCs and early germ cell progeny (CBs and cysts). Vasa expression identifies all germline cells. Hts antibody identifies spectrosomes, fusomes, and the cell membranes of somatic cells. The spherical spectrosomes identify GSCs, whereas the branched fusomes identify differentiating cysts. (Middle) Enhanced or constitutive Dpp signaling (as when Tkv^{CA} is overexpressed in the GSCs) results in GSC tumors. No cysts are observed in the germarium and follicle cells alone occupy the posterior region (red). (Right) Loss or block of Dpp signaling in GSCs (as when Dad is overexpressed in the germline) results in complete GSC loss. In the Dad overexpression panel, an empty germarium is observed attached to a mid-late egg chamber.

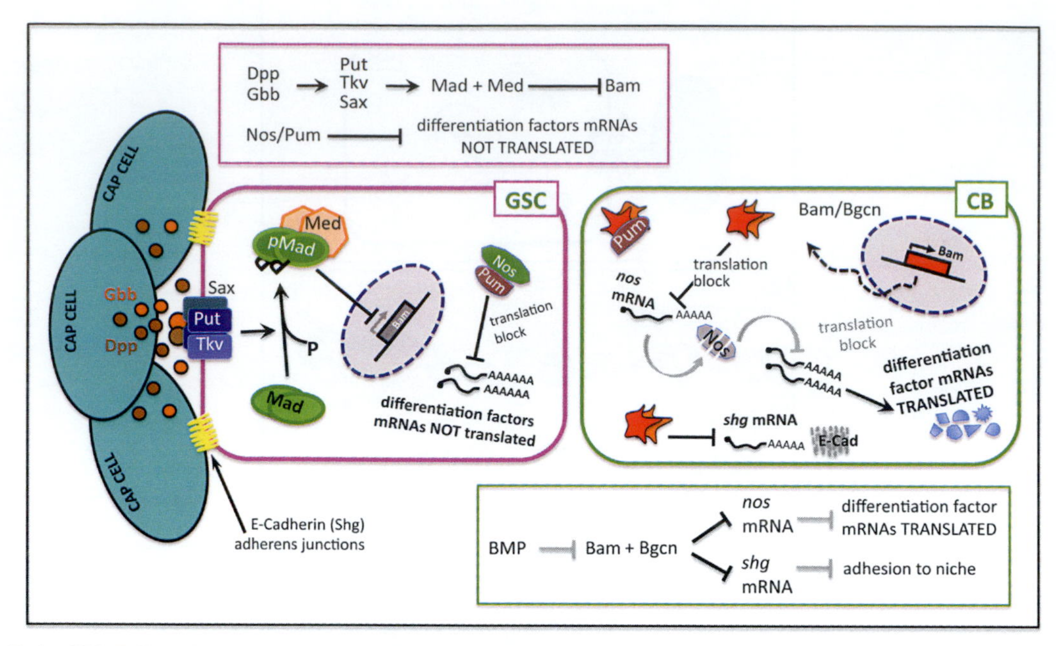

FIGURE 5.3 Role of BMP Signaling in GSC-CB Fate Choice. Phosphorylation of Mad by activated Tkv leads to repression of *bam* transcription by pMad/Med. In absence of Bam, the Nos/Pum translational repressor complex blocks the expression of differentiation-promoting mRNAs in the GSC. In the CB, loss of BMP signaling results in transcription of *bam*. Bam acts together with Bgcn to block Nos/Pum: Bam/Bgcn translationally represses *nos* and inhibits the activity of Pum, thereby de-repressing the expression of differentiation factors. In addition, Bam/Bgcn translationally represses *E-cadherin/shg*, permitting exit from the niche.

downregulate Nos protein expression (Li et al., 2009), and it physically interacts with Pum to inhibit its translational repressor activity (Figure 5.3) (Kim et al., 2010). Another self-renewal factor repressed ed by Bam is *shg*, which is the fly E-cadherin (Shen et al., 2009; Jin et al., 2008). E-cadherin is a critical component of the adherens junctions that anchor the GSCs to the CCs within the niche, and loss of E-cadherin causes the GSCs to exit the niche and differentiate (Song et al., 2002). Because Armadillo (the homolog of β-catenin) functions together with E-cadherin both in cell-cell adhesion and in Wingless (Wg/Wnt signal) signaling, Xie and colleagues studied several *arm* mutants that were either defective only in Wg signaling, or defective in both Wg signaling and E-cadherin-mediated cell adhesion, to show that anchoring by E-cadherin/Arm, but not Wg signaling, is required for GSC maintenance (Song et al., 2002). Subsequently, Tie Xie's group demonstrated that Bam/Bgcn acts via the 3' UTR of *shg* to down-regulate the translation of E-cadherin during differentiation (Shen et al., 2009). In summary, Bam functions in translational repression of target genes to promote differentiation.

Ultimately, GSC self-renewal requires high levels of BMP signaling and low levels of Bam expression. Conversely, differentiation proceeds when BMP signaling is low and Bam is derepressed. Over the past decade, genetic and biochemical experiments have revealed a complex web of regulators that function either positively, to reinforce BMP signaling, or negatively, to decrease signaling and promote Bam expression. Regulation occurs at every level of the BMP pathway from the deployment of ligands to the regulation of gene targets in the nucleus.

5.4 REGULATION OF BMP SIGNALING BY EXTRINSIC FACTORS

GSC self-renewal depends on the secretion of active BMP ligands from the surrounding niche cells. Both *dpp* and *gbb* are expressed in the somatic cells at the anterior region of the germarium but not in the germline (Song et al., 2004; Liu et al., 2010; Wang et al., 2008). Genetic experiments suggest that the CCs and to a lesser extent the neighboring ECs, at the apex of the germarium, produce BMP ligand (Xie and Spradling, 2000; Liu et al., 2010; Rojas-Rios et al., 2012; Song et al., 2007). For example, RNAi-mediated knockdown of *dpp* in CCs results in rapid GSC differentiation and loss from the niche, whereas the knockdown of *dpp* in ECs results in a milder reduction in the number of GSCs maintained over time (Rojas-Rios et al., 2012). These and other experiments suggest that the CCs are the primary source of Dpp and Gbb, and function as a localized source of BMP ligand.

Because Dpp is a secreted factor and is sufficient to block differentiation, the activity range of the Dpp signal must be tightly controlled to prevent the accumulation of stem cell progeny as GSCs or the dedifferentiation of recently formed

TABLE 5.2 **BMP Signaling and GSC Self-Renewal Factors Discussed in this Chapter**

Symbol	Full Name	Promotes	Functions in (Made in)	Molecular Function	Human Orthologs
bam	bag of marbles	differentiation	CB	translation repressor activity; mRNA 3′-UTR binding	GM114
brat	brain tumor	differentiation	CB	translation repressor activity; protein binding	No orthologs identified
Dad	Daughters against dpp	differentiation	GSC	BMP receptor inhibitor; I-SMAD	No orthologs identified
dally	division abnormally delayed	self-renewal	ECM (CC)	heparan sulfate proteoglycan	No orthologs identified
dpp	decapentaplegic	self-renewal	GSC (CC)	BMP2/4 type secreted ligand	BMP2; BMP4
fu	fused	differentiation	CB	serine/threonine kinase	STK36 - serine/threonine kinase 36
gbb	glass bottom boat	self-renewal	GSC (CC)	BMP6/7 type secreted ligand	BMP8A; BMP7; BMP6; BMP8B; BMP5
Lis-1	Lissencephaly-1	self-renewal	GSC	ATPase coupled dynein regulator	PAFAH1B1 - PAFAH1b, regulatory subunit 1
lsd1/Su(var)3-3	Suppressor of variegation 3-3	differentiation	EC	gene silencing; histone demethylase	KDM1A - lysine (K)-specific demethylase 1A
Mad	Mothers against dpp	self-renewal	GSC	sequence-specific DNA binding transcription factor; R-SMAD	SMAD2; SMAD9
Med	Medea	self-renewal	GSC	sequence-specific transcription cofactor; co-SMAD	SMAD4
nos	nanos	self-renewal	GSC	translation repressor activity; mRNA 3′-UTR binding	NANOS1; NANOS2; NANOS3
ote	otefin	self-renewal	GSC	nuclear lamin-binding protein; transcription corepressor activity	No orthologs identified
pum	pumilio	self-renewal / differentiation	GSC / CB	translation repressor activity; mRNA 3′-UTR binding	PUM1; PUM2
put	punt	self-renewal	GSC	transmembrane serine/threonine kinase	TGFBR2; ACVR2A; BMPR2; ACVR2B
sax	saxophone	self-renewal	GSC	transmembrane serine/threonine kinase	ACVR1; BMPR1A; BMPR1B; ACVR1C; ACVR1B
shn	schnurri	self-renewal	GSC	sequence-specific DNA binding transcription factor	No orthologs identified
smurf/lack	lethal with a checkpoint kinase	differentiation	CB	ubiquitin-protein ligase	SMURF1 & 2 - SMAD specific E3 ubiquitin protein ligases

Continued

TABLE 5.2 BMP Signaling and GSC Self-Renewal Factors Discussed in this Chapter—cont'd

Symbol	Full Name	Promotes	Functions in (Made in)	Molecular Function	Human Orthologs
tkv	thickveins	self-renewal	GSC	transmembrane serine/threonine kinase	ACVR1; BMPR1A; BMPR1B; ACVR1C; ACVR1B
vkg	viking	self-renewal	ECM (TFC/CC/EC)	type IV collagen; extracellular matrix structural constituent	No orthologs identified

CBs and cysts. Recently identified mechanisms that spatially restrict signaling between niche and stem cells fall into two broad categories:

1) Mechanisms that restrict production of the Dpp ligand to the somatic cells at the anterior end of the germarium, and
2) Mechanisms that restrict the distribution of Dpp to a narrow range of the extracellular space surrounding the GSCs.

Restricting Ligand Production in the Niche

The control of ligand production (including expression, processing, and secretion) is critical in the regulation of BMP pathway activity. Although little is known about the processing and secretion of Dpp from CCs, recent studies have uncovered genetic pathways that promote *dpp* transcription in CCs or largely suppress it in ECs.

dpp transcription in CCs is induced by the JAK/STAT signaling pathway (Figure 5.4A) (Decotto and Spradling, 2005; Wang et al., 2008; Lopez-Onieva et al., 2008). JAK/STAT signaling in *Drosophila* is activated at the cell surface by binding of Upd ligand to the transmembrane receptor Dome. Dome, in turn, transduces the signal to the intracellular Janus tyrosine kinase Hop, which phosphorylates Stat92E, a transcription factor that directly regulates gene expression (reviewed in Arbouzova and Zeidler, 2006). Within the germarium, expression of Upd is restricted to the TF and CCs, whereas JAK/STAT signaling activity is observed only in the CCs and nearby ECs (Wang et al., 2008; Lopez-Onieva et al., 2008). Interestingly, CCs that are mutant for *hop* or *Stat92E* fail to support GSCs maintenance (Figure 5.4A) (Wang et al., 2008; Lopez-Onieva et al., 2008). *dpp* transcripts are reduced or absent in cap cells with compromised JAK/STAT signaling, whereas *dpp* expression is increased following the overexpression of Upd (Wang et al., 2008; Lopez-Onieva et al., 2008). Furthermore, Upd was shown to induce transcription of a reporter containing *dpp* regulatory DNA *in vitro* (Wang et al., 2008). Together with additional evidence, these data suggest that the JAK/STAT pathway functions upstream of Dpp signaling by controlling *dpp* expression (Decotto and Spradling, 2005; Wang et al., 2008; Lopez-Onieva et al., 2008).

More importantly, negative regulation of JAK/STAT signaling outside of the niche is critical to avoid excessive production of Dpp ligand. In fact, the targeted expression of Upd in ECs results in ectopic *dpp* expression in the germarium and GSC tumors (Decotto and Spradling, 2005; Wang et al., 2008; Lopez-Onieva et al., 2008). Currently, it is not known how Upd ligand production and JAK/STAT activity are restricted within the niche. However, an epigenetic mechanism repressing *dpp* transcription in ECs has recently been uncovered. The highly conserved histone demethylase Lsd1 promotes heterochromatin formation and gene silencing (Eliazer et al., 2011). Lsd1 is required specifically in the ECs to prevent the formation of GSC tumors (Figure 5.2D). In brief, the loss of *Lsd1* function in ECs results in the upregulation of *dpp* expression in these cells, and the accumulation of Dpp-responsive GSC-like cells outside the natural niche (Figure 5.4A) (Eliazer et al., 2011). Interestingly, vertebrate homologs of Lsd1 also regulate the formation of heterochromatin, and it was shown recently that Lsd1 directly represses the transcription of TGFB1 in mammals (Wang et al., 2009). Whether Lsd1 silences *dpp* expression through the JAK/STAT or other pathways remains to be determined. However, the *Lsd1* mutant phenotype highlights the critical importance of restricting Dpp production to a single source at the anterior tip of the germarium, a process that likely involves additional mechanisms.

Restricting Ligand Distribution and Activity in the Extracellular Space

In the germarium, the range of ligand activity is restricted to a single cell, the GSC. Just one cell diameter away from the source, the CB must efficiently silence the pathway in order to initiate differentiation into a cyst. Since Dpp can

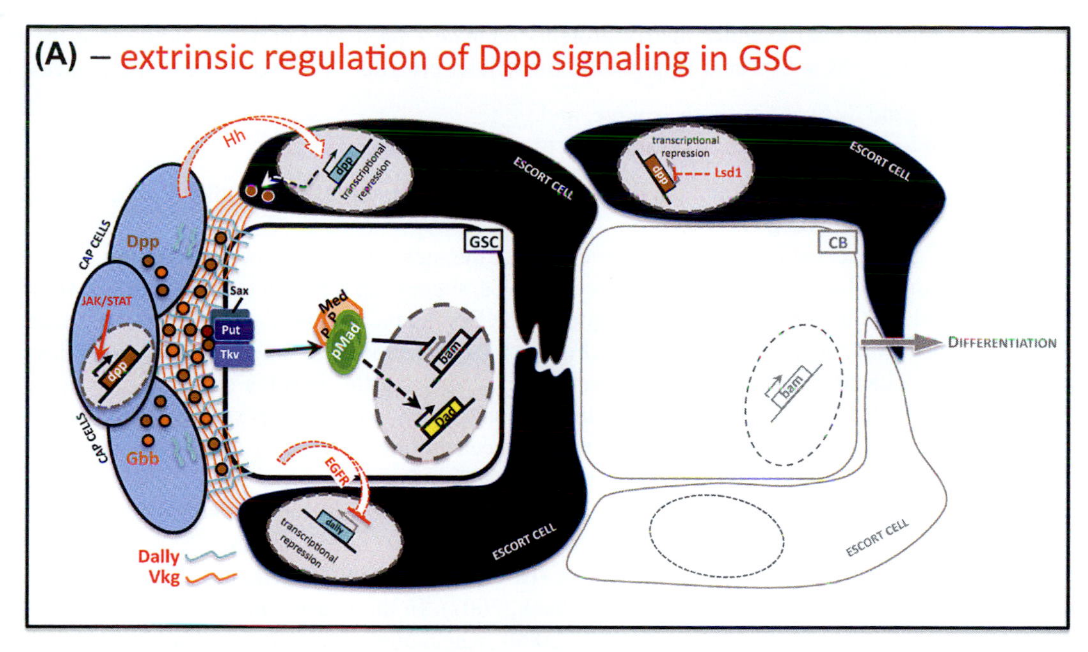

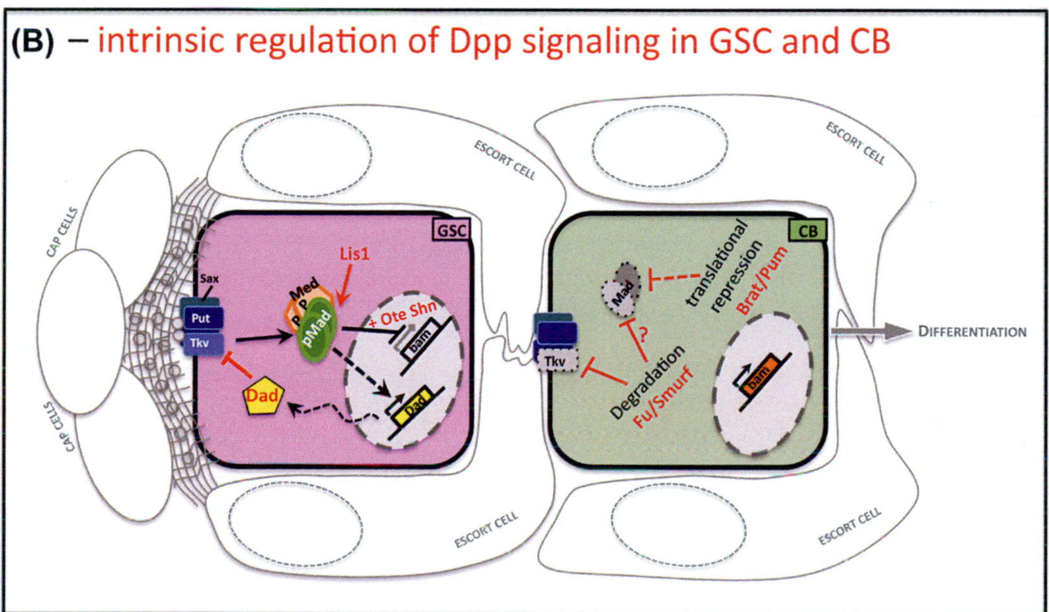

FIGURE 5.4 Extrinsic and Intrinsic Mechanisms Regulate the BMP Pathway at Multiple Levels. (A) Extrinsic Regulation of GSC Dpp Signaling at the Level of Dpp Production and Distribution. In CCs, the JAK/STAT pathway positively regulates *dpp* expression. The Upd ligand is produced by both TF cells and CCs (not shown). Expression of *dpp* in ECs that contact the CCs is induced through the Hh pathway by Hh ligand produced by the CCs. In ECs away from the niche, *dpp* transcription in negatively controlled by the chromatin-level regulator Lsd1. Dpp ligand distribution and activity is regulated by Vkg and Dally in the extracellular matrix between CCs and GSCs. Expression of *dally* is restricted to the CCs by EGFR-mediated silencing of the *dally* gene in ECs but not CCs. (B) Intrinsic Regulation of BMP Signal Transduction Occurs in Both GSCs and CBs. In GSCs, positive regulation occurs at the level of the Mad signal transducer by Lis-1 (promoting Mad/pMad stability and/or phosphorylation), and a negative feedback loop occurs at the level of the Tkv receptor by the I-SMAD Dad (inhibiting the phosphorylation of pMad). In CBs, negative regulation occurs at the level of the activated Tkv receptor by Fu/Smurf (through protein ubiquitination and degradation), and at the level of Mad by Brat (through translational repression of *Mad* mRNA) and possibly Fu/Smurf (through protein ubiquitination and degradation).

function as a long range signaling molecule (for instance, Dpp activates receptors up to 30 cell diameters away from the source in the developing wing epithelium (Lecuit et al., 1996; Nellen et al., 1996; Entchev et al., 2000; Teleman and Cohen, 2000) and reviewed in Affolter and Basler, 2007, and Umulis et al. 2009), tight control of not only ligand production but also delivery is necessary to produce the steep signaling gradient reflected in the GSC-ON/CB-OFF readout.

At the post-production level, extracellular matrix proteins regulate Dpp distribution and activity in the intercellular space between niche cells and GSCs. Secreted Dpp binds extracellular proteins with two major consequences:

1) Spatial restriction of ligand diffusion from the source, and
2) Enhancement of ligand-receptor interactions at the GSC plasma membrane.

Two factors, the Type IV collagen Vkg, and the heparan sulfate proteoglycan Dally, mediate these processes (Guo and Wang, 2009; Hayashi et al., 2009; Wang et al., 2008).

In the germarium, Vkg is detected around the somatic cells of the niche, including in between the niche cells and the GSCs. Loss of *vkg* function results in an increased number of GSCs, suggesting a broader reach of the BMP ligands across the germarium (Wang et al., 2008). Vkg is thought to sequester the Dpp ligand around the GSC through direct protein-protein interaction, thus limiting the range of Dpp signaling (Figure 5.4A). The ability of Vkg to bind Dpp resides in evolutionary conserved amino acids located near the C-terminus that are key to such binding in other organisms as well, e.g., between human type IV collagen and BMP4 (Wang et al., 2008). In addition, Ashe and colleagues showed that the presence of Vkg increases the ability of Dpp to bind the mouse BMP receptor 1A in cell culture, suggesting that Vkg may also promote BMP receptor interactions (Wang et al., 2008). In short, Vkg restricts the Dpp signaling range by binding and sequestering Dpp around the GSCs, and by possibly promoting Dpp-receptor interactions at the GSC membrane.

Another niche factor, the glypican Dally, also promotes Dpp signaling (Guo and Wang, 2009; Hayashi et al., 2009). Glypicans are members of the heparan sulfate proteoglycan (HSPG) family that are tethered to the cell surface by a GPI anchor (glycosylphosphatidylinositol) (Kirkpatrick and Selleck, 2007). In the ovary, Dally is expressed primarily in cap cells (Figure 5.4A) (Guo and Wang, 2009; Hayashi et al., 2009). Knockdown of Dally specifically in the somatic niche disrupts BMP signaling and causes GSC loss (Guo and Wang, 2009); and conversely, ectopic expression of Dally (in ECs or germline cells) enhances signaling thereby inducing GSC tumors (Guo and Wang, 2009; Hayashi et al., 2009). Interestingly, membrane tethering appears to be critical to Dally function, since overexpression of a secreted form of Dally does not noticeably affect GSC maintenance (Guo and Wang, 2009).

Experiments performed in cell culture demonstrated that Dally can physically interact with the Dpp protein and protect it from degradation (Dejima et al., 2011). In an elegant experiment by Nakato's group, S2 cells with Dpp bound on the cell surface were placed onto glass slides coated with either Dally or BSA, and the stability of Dpp protein was measured over time (Dejima et al., 2011). While Dpp degraded rapidly in the control sample, the presence of Dally slowed this process, demonstrating that Dally can enhance the stability of the extracellular ligand (Dejima et al., 2011). Interestingly, HSPGs have also been shown to regulate other signaling pathways (Hayashi et al., 2012, and reviewed in Kirkpatrick and Selleck, 2007), some of which do influence Dpp in the germarium (e.g., JAK/STAT, see "*Restricting Ligand Production in the Niche*" of Section 5.4 Regulation of BMP Signaling by Extrinsic Factors, and Hh, see "*ECs-CCs Communication in GSC Self-Renewal*" of Section 5.7 Selected Topics). Thus, Dally and/or other HSPGs may also modulate BMP signaling in additional, possibly indirect ways.

5.5 REGULATION OF BMP SIGNALING BY INTRINSIC FACTORS

In addition to the mechanisms controlling BMP ligand expression, distribution, and activity in the niche, most components of the intracellular signal transduction cascade are also regulated in both the GSC and the CB.

Regulation of BMP Signaling in GSCs

Mad/pMad is Stabilized in the Cytoplasm

Lissencephaly-1 (Lis-1) has a well-known neural developmental function in vertebrates (Wynshaw-Boris, 2007). In mice and humans, Lis-1 loss disrupts neuronal migration in the central nervous system and results in lissencephaly, a "smooth brain" phenotype (Hirotsune et al., 1998; Reiner et al., 1993). In the fly, Lis-1 is highly expressed in the nervous system where it is required in the proliferation of neuroblasts (Lei and Warrior, 2000). At a molecular level, Lis-1 binds to the dynactin-dynein motor protein complex and is important for regulating the microtubule cytoskeleton (Smith et al., 2000; Faulkner et al., 2000). Consistent with this molecular role, Lis-1 is required during *Drosophila* oogenesis to synchronize cell divisions during formation of mitotic cysts (Liu et al., 1999; Swan et al., 1999). Thus, a disorganized germline is a hallmark of the *Lis-1* mutant phenotype.

In the *Drosophila* GSC, however, Lis-1 appears to function in an additional capacity. It is clearly required for Dpp signaling and self-renewal, since *Lis-1* mutant GSCs show decreased pMad and increased *bam* expression (Chen et al., 2010).

This effect on BMP signal transduction can also be modeled in S2 cells in presence of exogenous Dpp. Under these conditions, the knockdown of *Lis-1* results in decreased Mad and pMad levels, suggesting a role in stabilizing Mad (Chen et al., 2010). However, when high Mad protein expression levels were restored through exogenous expression, pMad levels still remained low, indicating that Lis-1 is also required for pMad (Chen et al., 2010). Biochemical studies show that Lis-1 binds directly to Mad as well as pMad, and that these interactions are strengthened by the presence of the BMP receptor Tkv (Chen et al., 2010). These findings suggest that in addition to its other known roles in microtubule-dependent processes, Lis-1 plays a direct role in regulating the stability and/or phosphorylation of Mad/pMad in the GSC (Figure 5.4B).

pMad/Med Signaling is Promoted by Additional Factors in the Nucleus

Additional molecules have been shown to promote the activities of the pMad/Med complex in the GSC nucleus, such as the nuclear lamin Ote (Jiang et al., 2008) and the zinc finger transcription factor Shn (Figure 5.4B) (Xie and Spradling, 2000).

Loss of *ote* function results in GSC loss and, conversely, Ote overexpression can expand the size of the GSC population in the germarium (Jiang et al., 2008). Ote protein associates with the nuclear membrane of the GSC, and can directly bind to the co-SMAD Med, though not to the R-SMAD Mad. Ote/Med complexes can directly bind the *bam* transcriptional silencer element upstream of the promoter, strongly suggesting that an Ote/Med/pMad complex is required to repress *bam* expression in the GSC (Figure 5.4B) (Jiang et al., 2008). Consistent with Ote's function in modulating the Dpp-induced silencing of *bam*, *bam* expression is upregulated in *ote* mutant GSCs and *ote* alleles are dominant suppressors of *dpp* hyperactivation, i.e., loss of one wild-type *ote* allele (+/–) results in suppression of the tumorous phenotype induced by Dpp overexpression (Jiang et al., 2008). In summary, the silencing of *bam* by pMad/Med in the GSC requires the function of Ote, possibly for sequestering *bam* genomic DNA at an intranuclear location that promotes gene silencing (Jiang et al., 2008).

Another regulator of Dpp signaling that acts at the nuclear level is Shn (Xie, Spradling, 2000). Consistent with a role in silencing *bam*, *shn* mutant GSCs precociously differentiate and are lost from the niche (Xie and Spradling, 2000). Interestingly, a ternary protein complex containing Mad, Med, and Shn can bind to silencer elements in Dpp target genes *in vitro* (Figure 5.4B) (Dai et al., 2000; Udagawa et al., 2000). There is also evidence that a pMad/Med/Shn complex functions as a general mediator of Dpp-induced gene silencing in other tissues (Torres-Vazquez et al., 2000; 2001).

The interactions between pMad, Med, Ote and Shn underscore the complexity of target gene regulation in the stem cell nucleus and additional, unidentified factors may also contribute to the repression/expression of other genes targeted by Dpp.

Autoregulation of Receptor Activity at the Cell Membrane

Interestingly, one gene activated specifically in the GSC in response to Dpp encodes an inhibitory I-SMAD named Dad (Figure 5.4B) (Tsuneizumi et al., 1997). The Dad protein is structurally similar to the vertebrate inhibitory SMADs 6, 7, and 8, which have been shown to bind Type I BMP receptors and to block the phosphorylation of signaling R-SMADs (Hayashi et al., 1997; Imamura et al., 1997; Nakao et al., 1997). Similarly, Dad has been shown to physically interact with the Tkv receptor (Inoue et al., 1998). In cell culture experiments, the expression of Dad leads to decreases in the phosphorylation of Mad, the binding of pMad to Med, and the nuclear translocation of the pMad/Med complex, showing that Dad inhibits Dpp signaling (Inoue et al., 1998). Additional *in vitro* experiments have revealed that the physical interaction between Tkv and Mad is disrupted in the presence of Dad, suggesting that Dad may competitively inhibit the association between the BMP receptor and its signal transducer (Figure 5.4B) (Inoue et al., 1998). Consistent with this model, overexpression of Dad within the germline results in the rapid loss of GSCs, demonstrating that Dad can indeed inhibit self-renewal (Figure 5.2D) (Jiang et al., 2008). Yet, the loss of *Dad* does not induce the converse phenotype of GSC tumors (Xie and Spradling, 1998), though it does prolong GSC half-life. Because Dad is expressed specifically in GSCs (not in CBs), it is unclear how stem cells can benefit from expressing a factor that is antagonistic to self-renewal. One possibility is that this negative feedback loop tempers oscillations in signaling of an intrinsic nature or triggered by external factors not assessed under routine laboratory conditions. Alternatively, the absence of obvious tumors in *Dad* loss-of-function may simply reflect redundancy with other factors that downregulate the BMP pathway.

Regulation of BMP Signaling in CBs

Tkv is Targeted for Degradation During Germline Differentiation

The most striking evidence that a strong, active downregulation of BMP signal transduction occurs in the earliest differentiating germline cell, the CB, comes from Chen and colleagues (2010), who compared the targeted expression of the constitutively activated Tkv receptor, Tkv^{CA}, in GSCs versus CBs (Xia et al. 2010). As mentioned in the "*Overview*" of Section 5.3

The BMP Signaling Pathway, Tkv^{CA} is able to generate large tumors when expressed in the GSC (Casanueva and Ferguson, 2004). If Tkv^{CA} is expressed here, entire germarium becomes filled with GSC-like cells and the ovarioles show a complete lack of maturing egg chambers. This is not the case, however, if Tkv^{CA} is expressed specifically in the CB. In this latter case the germarium maintains an unperturbed, normal morphology (Xia et al., 2010). The inability of Tkv^{CA} to impose self-renewal if expressed in the CB illustrates that factors present in this cell can antagonize the signaling activity of the activated receptor.

One such factor is the serine/threonine kinase Fu, which together with the E3 ubiquitin ligase Smurf ubiquitinates the activated Tkv receptor, marking it for degradation (Figure 5.4B) (Xia et al., 2010). Loss of either *fu* or *smurf* gene function in the germline leads to the formation of GSC tumors (Casanueva and Ferguson, 2004; Preat et al., 1990; Smith and King, 1966). Extensive genetic analysis *in vivo* and biochemical studies in cell culture have revealed that Fu and Smurf bind directly to the activated form of the Tkv receptor to promote its ubiquitination and degradation (Xia et al., 2010). In one such *in vitro* experiment, a second point mutation was introduced into the Tkv^{CA} receptor that conferred resistance to Fu/Smurf mediated degradation. When this Fu/Smurf-resistant form of activated Tkv, $Tkv^{CA(S238A)}$, was expressed *in vivo*, it was sufficient to revert CBs and possibly early cysts into self-renewing GSC-like cells, generating tumorous masses of undifferentiated cells in the germaria (Xia et al., 2010). Recent evidence suggests that Fu/Smurf begins to function during GSC division in the daughter cell that is fated to differentiate even before the progeny cells have fully separated, and before Bam is upregulated (discussed in Section 5.7 – "From self-renewing SC to differentiating CB: the 'pre-CB'"). Thus, by promoting Tkv degradation, Fu/Smurf may play a major role in triggering the conversion to CB.

Interestingly, Fu and Smurf proteins have conserved roles in the regulation of vertebrate BMP signaling. In human cells, it has been shown that FU physically interacts with the Type I BMP receptor ALK3 to mediate ubiquitination (Xia et al., 2010). To determine whether vertebrate FU also facilitated the degradation of BMP receptors *in vivo*, an ALK3a-GFP fusion protein was expressed in zebrafish embryos with or without exogenous FU. While expression of the ALK3a transgene was observed in wild-type embryos, the signal was markedly diminished following injection of FU mRNA (Xia et al., 2010). This and other findings strongly suggest that FU, similar to its fly homolog, plays a role in downregulating BMP signaling at the level of the receptor. In *Xenopus*, Smurf1 has been shown to block the BMP transduction cascade and disrupt patterning during embryogenesis (Podos et al., 2001; Zhu et al., 1999). In mammalian cell culture, Smurf1 was also shown to physically interact with constitutively activated forms of the BMP Type I receptors ALK2, ALK3, or ALK6, and to target these receptors for ubiquitin-dependent degradation (Murakami et al., 2003). Interestingly, Smurf1 was found to bind directly to Smad1/5 and promote its degradation as well (Zhu et al., 1999). *Drosophila* Smurf can also physically interact with Mad and promote its ubiquitination in cell culture (Podos et al., 2001); recent genetic evidence suggests that Smurf may similarly regulate GSC maintenance *in vivo* (Figure 5.4B) (Lu et al., 2012).

Mad is Translationally Repressed in the CBs and Cysts

While Fu/Smurf mediates the downregulation of BMP signaling prior to the upregulation of Bam, another factor, the translational repressor Brat, plays an important role downstream of Bam in CBs and cysts to maintain their progress toward differentiation (Harris et al., 2011; Sonoda and Wharton, 2001).

Brat expression is not observed in the GSCs, but it is upregulated in CBs and cysts (Harris et al., 2011). In these cells, Brat forms a complex with Pum, and together Brat/Pum repress the expression of GSC self-renewal factors, including Mad (Figure 5.4B) (Harris et al., 2011). The ability of Brat to promote differentiation has been shown to depend on its ability to interact with Pum. While the ectopic overexpression of Brat in the GSCs resulted in their rapid loss through differentiation, a mutant form of Brat that failed to interact with Pum did not induce such a phenotype (Harris et al., 2011). Interestingly, and as discussed in "*BMP Signaling Blocks Differentiation in the GSC*" of Section 5.3 The BMP Signaling Pathway, Pum functions with Nos in the GSC where the Nos/Pum complex is thought to repress the expression of differentiation factors (Figure 5.3) (Wang and Lin, 2004; Forbes and Lehmann, 1998; Lin and Spradling, 1997). Thus, Pum has dual functions; it works together with Nos in the GSC to promote stem cell self-renewal, and together with Brat in the CB to promote differentiation. Consistent with this model, the loss of Brat results in small GSC tumors owing to the persistence of Mad and other self-renewal factors (Harris et al., 2011).

The functions of Nos, Pum, and Brat in the germline illustrate the important role of translational repression in regulating the switch between self-renewal and differentiation.

5.6 ADDITIONAL REGULATORS

Other potential regulators are less well understood and present more complex relationships with the BMP pathway. The loss of GSCs in mutants for the GTPase-activating factor *tuberous sclerosis 1* (TSC1) or the translational regulator *pelota* is accompanied by decreased levels of pMad but not *bam* induction (Sun et al., 2010; Xi et al., 2005). On the contrary, the chromatin remodeling factor Imitation SWI (ISWI) promotes GSC self-renewal by repressing *bam*, but does not affect pMad levels (Xi and Xie, 2005). Still others, such as the miRNA pathway components *Ago1* and *loqs*, appear to regulate GSC maintenance downstream of, or in parallel to, *bam* (Jin and Xie, 2007; Park et al., 2007; Yang et al., 2007a,b). Yet, one miRNA, miR-184, appears to promote differentiation through the Dpp receptor protein Sax (Iovino et al., 2009). Further studies are needed to clarify the roles of these factors in self-renewal and their contribution to BMP signaling.

5.7 SELECTED TOPICS

From Self-Renewing SC to Differentiating CB: the "Pre-CB"

How does a "BMP-silent" CB daughter cell emerge from a GSC mother with active BMP signaling? What is the mechanism that begins to shut down the pathway in the course of a single cell division?

Recent work has looked at daughter cells from single GSC divisions at a time when the two progeny cells have yet to complete cytokinesis and acquire clearly distinct identities. At this early stage, the daughter cell forming away from the niche does not show *bam* expression. However, the expression level of pMad is already decreased compared to the forming GSC sister (Xia et al., 2012). This intermediate stage, in which BMP signaling is downregulated but differentiation has not begun, is referred to as 'pre-cystoblast' (pre-CB) (Figure 5.5).

Interestingly, another change is detected at this stage: the upregulation of Fu expression specifically in the pre-CB (Xia et al., 2012). As discussed in "*Regulation of BMP Signaling in CBs*" of Section 5.5 Regulation of BMP Signaling by Intrinsic Factors, Fu is a kinase that functions together with the E3 ubiquitin ligase Smurf to target the activated form of the Tkv receptor for ubiquitination/degradation in the CB (Xia et al., 2010). The presence of Fu in the pre-CB is, therefore, not entirely surprising. In fact, degradation of activated Tkv would effectively initiate the shutdown of BMP signaling that is necessary for CB formation. However, it does raise the question of how Fu protein can be upregulated so rapidly in the pre-CB when it is undetectable in the mother GSC. To address this question, Chen and colleagues (2012) ectopically expressed Fu in GSCs under the control of a heat-shock promoter. Soon after a single heat-shock pulse, ectopic Fu protein could be observed in stem cells, however, the protein rapidly degraded in less than two hours after the heat shock (Xia et al., 2012). Since Fu mRNA is present in the GSC, the authors proposed that Fu protein is produced in the stem cell but that it is also continuously degraded. In this model, the presence of *fu* mRNA and protein (albeit unstable) underlines the "differentiation-ready" state of the GSC and contributes to the rapid transition from GSC to CB in the course of a single cell division (Figure 5.5). How Fu is degraded in the GSC is unknown. However, additional work by Chen and colleagues (2012) suggests that the degradation process is promoted by BMP signaling in the GSC, and a transient dip in pathway activity during GSC division may be sufficient to tip the scale in favor of Fu accumulation and toward the CB fate (Xia et al., 2012).

The mechanism that leads to the downregulation of the BMP pathway in the pre-CB would have to be *bam*-independent, since *bam* is not yet expressed at this stage. There is indeed genetic evidence for such a mechanism. In the *bam* mutant, GSC differentiation is blocked and intermediate GSC/CB-like cells accumulate in large numbers within the germarium

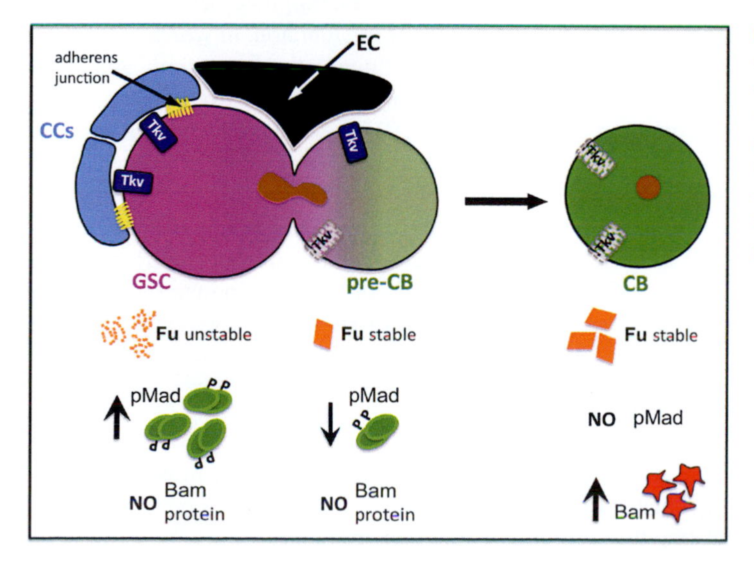

FIGURE 5.5 The Pre-CB Defines an Intermediate Stage that can be Observed Prior to Complete Cytokinesis. During GSC division, the daughter cell forming away from the niche downregulates BMP signaling (lower pMad levels compared to the GSC) prior to the expression of *bam*. Therefore, this cell represents the transition from GSC to CB, referred to as the pre-CB. Fu protein, which promotes the ubiquitination/degradation of Tkv, is observed in the pre-CB suggesting that Fu downregulates BMP signaling prior to the completion of GSC division.

(Casanueva and Ferguson, 2004; McKearin and Spradling, 1990). BMP signaling activity remains high in those two to three GSCs that remain in contact with the niche, but not in the undifferentiated germ cells that accumulate outside of the niche (Casanueva and Ferguson, 2004). This finding demonstrates the presence of a *bam*-independent mechanism of downregulation that mediates the shutdown of BMP signal transduction in cells outside the niche even in the absence of Bam. While the involvement of *fu* in this phenomenon has not been directly documented, the Fu partner Smurf is indeed required. In the absence of both Smurf and Bam, BMP signaling is upregulated in the GSC-like cells located outside the niche, and ectopic GSCs with high pMad accumulate throughout the germarium (Casanueva and Ferguson, 2004).

These findings strongly suggest that the Fu/Smurf complex provides the first intrinsic mechanism (prior to the expression of *bam)* for the downregulation of BMP signaling as daughter GSCs leave the niche. This very exciting work investigates the moment of transition from GSC to CB and future studies will better define this earliest stage of germ cell differentiation.

Escort Cells Dually Support Germline Self-Renewal and Differentiation

Whereas CCs promote self-renewal by anchoring the GSCs within the niche and secreting Dpp ligand, the ECs develop a more complex set of interactions with the germline. Also referred to as "inner germarium sheath cells," the ECs contact not only the GSCs, but also CCs, CBs and cysts (Decotto and Spradling, 2005; Kirilly et al., 2011).

ECs-CCs Communication in GSC Self-Renewal

As described in Section 5.4 Regulation of BMP Signaling by Extrinsic Factors, the most anterior EC cells differ from those located more posteriorly because they express *dpp* mRNA and are in contact with both GSCs and CCs (Song et al., 2004; Liu et al., 2010, Wang et al., 2008). Moreover, EC-targeted RNAi-mediated *dpp* silencing results in a partial loss of GSCs over time (Rojas-Rios et al., 2012). Together these findings suggest that Dpp production in the anterior-most ECs contributes to stem cell self-renewal. Interestingly, expression of *dpp* in these cells appears to depend on Hedgehog (Hh) signaling from neighboring CCs (Figure 5.4A). The secreted ligand Hh is specifically expressed in CCs (King et al., 2001), whereas the Hh receptor (the transmembrane protein Patched) is expressed in ECs (Rojas-Rios et al., 2012). Furthermore, loss of the Hh signaling component Smoothened in ECs results in reduced *dpp* mRNA levels in the germarium, lower BMP cascade activity in GSCs (i.e., decreased pMad), and GSC loss owing to precocious differentiation (Rojas-Rios et al., 2012).

This example illustrates the importance of cell-cell communication among somatic components of the niche and shows that multiple genetic pathways, JAK/STAT in CCs (see "*Restricting Ligand Production in the Niche*" of Section 5.4 Regulation of BMP Signaling by Extrinsic Factors) and Hh in ECs, ensure production of the self-renewal signal.

ECs Support Germline Differentiation

Another important aspect of ECs is that they also contact CBs and cysts. The association of the ECs with the CBs and cysts occurs mainly through thin cytoplasmic processes that extend from the EC cell body and interdigitate between the germ cells (Figure 5.1) (Kirilly et al., 2011). The name "escort" derives from an early model of EC function, in which these cells were thought to associate with each newly formed CB, migrate with the developing cyst through the germarium, and lastly die by apoptosis as the fully formed cyst became engulfed by FCs (Decotto and Spradling, 2005). However, extensive lineage tracing studies and live imaging of the germarium by Spradling and colleagues (2011) revealed instead that ECs are relatively stationary, and that cysts traverse the germarium by passing from one EC to another (Morris and Spradling, 2011). The long cellular extensions of the ECs are required to support differentiation possibly by restricting diffusion of the Dpp ligand away from the niche. Disruption of these cytoplasmic processes, by blocking either the GTPase Rho or the Formin-like actin regulator Capuccino, results in an expanded range of BMP signaling and the accumulation of undifferentiated GSC-like cells outside the niche (Kirilly et al., 2011). Since CBs and cysts can undergo de-differentiation in the presence of Dpp (Kai and Spradling, 2004), this function of the ECs may be critical to ensure oocyte production by preventing reversal of the differentiation program.

Germline to Soma Communication

Over the past few years, it has become clear that homeostasis of the germline involves not only niche-germline and niche-niche communication, but also feedback from the germline lineage to the somatic cells. The stem cells therefore play an active role in orchestrating their own maintenance by regulating the surrounding microenvironment.

Germline-ECs Signaling Controls Niche Size and GSC Numbers

In Section 5.4 Regulation of BMP Signaling by Extrinsic Factors, we described how Dally is an extracellular matrix protein that restricts and promotes ligand-receptor interactions by acting locally in the intracellular space between CCs and GSCs. Although the expression of *dally* in the CCs is required for GSC self-renewal, the downregulation of *dally* in ECs is just as important to prevent Dpp-receptor interactions outside the niche (Guo and Wang, 2009; Hayashi et al., 2009). This is achieved through germline to EC signaling.

Germline cells in the anterior region of the germarium secrete epithelial growth factor (EGF) ligands that activate receptors (EGFR) in the surrounding ECs (Liu et al., 2010). Activated EGFR triggers downstream signaling via the MAPK pathway in the EC, ultimately leading to the transcriptional silencing of *dally* in these cells (Figure 5.4A) (Liu et al., 2010). Germaria with GSCs triply mutant for the EGFR ligands *spitz*, *keren*, and *gurken* or singly mutant for *stet* (a protease required for the maturation of EGF ligands) show *dally* expression not only in the CCs, but also in ECs, and develop GSC tumors (Liu et al., 2010). Similarly, germaria with ECs mutant for transduction components of the EGFR-MAPK pathway (i.e., *egfr*, *ras*, or *mek*), also display an expanded *dally* expression and GSC tumors (Liu et al., 2010). This tumorous phenotype can be largely reverted by silencing *dally* in the EC or by restoring EGF ligand production in the germline (Liu et al., 2010).

These findings demonstrate that the germline exercises control on the size of the niche and consequently GSC numbers by restricting the expression of an extrinsic self-renewal factor.

Germline-CCs Signaling Promotes GSC Maintenance

Interestingly, evidence is also emerging that GSCs signal back to CCs to promote BMP signaling from the niche. GSCs mutant for the Notch ligands, Delta or Serrate, show reduced BMP signaling activity and are lost from the niche due to precocious differentiation (Ward et al., 2006). Interestingly, although the receiving components of Notch signaling, such as the Notch receptor (*N*), are dispensable in the GSC, CCs that are mutant for Notch intracellular signaling components fail to support GSC self-renewal (Ward et al., 2006). Furthermore, Notch activity is required in the CCs to maintain BMP pathway activity in the GSCs (Ward et al., 2006). These results suggest that Notch signaling from the GSC to the CC positively regulates the niche, such that the niche, in turn, promotes GSC maintenance. However, what lies downstream of activated Notch in the CC is not known. Potential target genes are *dpp*, *gbb*, and *dally*, which are all required in the niche to sustain BMP signaling in the GSC.

5.8 BMP SIGNALING AND STEM CELL HOMEOSTASIS IN VERTEBRATES

The BMP pathway is remarkably conserved from flies to vertebrates, as are its regulators, including Type IV collagens, HSPGs, Lis1, Fu, and Smurf. Although BMP signaling contributes to stem cell homeostasis in vertebrates, the complexity of many systems makes it difficult to discern its exact role. For example, BMPs are required for mouse embryonic stem cell self-renewal, but in combination with other cytokines; in absence of these factors, BMPs promote non-neural differentiation rather than stem cell self-renewal (Ying et al., 2003). In the mammalian intestinal epithelium, instead the pathway supports stem cell homeostasis by promoting differentiation rather than self-renewal (He et al., 2004). Nonetheless, in two mammalian SC systems, the hematopoietic stem cells (HSCs) and the spermatogonial stem cells (SSCs), evidence suggests that BMP signaling may regulate stem cell maintenance through self-renewal.

HSCs have been shown to reside in specialized microenvironments within the bone marrow. The BMP4 ligand is expressed in the various cells known to support HSC maintenance, including osteoblasts, perivascular cells, and megakaryocytes, and functions as an extrinsic factor in HSC self-renewal (Goldman et al., 2009). In BMP4 hypomorphic mice, HSC numbers are reduced, and abnormally low pSmad1/5/8 activity is observed specifically in HSCs and early hematopoietic progenitors (Goldman et al., 2009). Goldman and colleagues (2009) performed serial transplantation studies to show that transplanted wild-type HSCs have a decreased ability to repopulate BMP4-deficient niches and to function as stem cells when placed in a BMP4-deficient recipient, demonstrating that BMP4 is required specifically in the bone marrow niche to support HSC maintenance (Goldman et al., 2009). However, the diversity of BMPs, receptors, and other signaling components present in mouse may result in more complex contributions of the BMPs to HSC function (Zhang et al., 2003).

In the mammalian testes, spermatogenesis occurs at the epithelial layer of seminiferous tubules and BMP4 is required for the maintenance of spermatogenesis in adult mice (Goldman et al., 2006; Hu et al., 2004). Degenerating and "empty" seminiferous tubules are observed in testes from hypomorphic BMP4 mice (Goldman et al., 2006). The SSC niche is believed to include the Sertoli cells present in the seminiferous epithelium and BMP4 is preferentially expressed in these somatic cells, but not in the SSC or spermatogonial cells (Hu et al., 2004; Pellegrini et al., 2003). Moreover, phosphorylated

Smad1/5/8 is observed in spermatogonia, suggesting that the SSCs may respond to BMP ligands produced in the niche (Pellegrini et al., 2003). However, it remains to be established whether BMP4 signals through SMADs in the SSC. Recently, expression of a mammalian ortholog of Bam, GM114, was shown in early differentiating spermatogonia, but not SSCs, suggesting that this protein may play a similar role in mouse and fly (Tang et al., 2008). However, no phenotype was observed following loss of GM114 (Tang et al., 2008). One possibility is that GM114 is functionally redundant with other differentiation factors, a common difficulty of dissecting gene function in mammals.

These examples highlight some of the challenges associated with the study of stem cell biology in mammals as compared to studies in the fly ovary and other simpler systems.

5.9 CONCLUSIONS

Niches are complex microenvironments that must continually respond to environmental cues to control the divisions and maintenance of their stem cells. Thus, it stands to reason that maintaining a balance between self-renewal and differentiation requires a complex network of regulatory pathways that act at the level of the niche cells, the stem cells and their differentiating progeny. Over the past decade, research on the ovarian *Drosophila* germline has identified a complex web of extrinsic and intrinsic factors that regulates the production of, and response to, the primary self-renewal signal. As our understanding of mammalian stem cells continues to grow, invertebrate models will still play an important role in deciphering the fundamental strategies utilized in stem cell-driven tissue regeneration and reproduction.

5.10 CLINICAL RELEVANCE

- The germline stem cells in the *Drosophila* ovary serve as a platform for understanding the maintenance and behavior of stem cells in more complex systems.
- Stem cell maintenance requires communications with the surrounding microenvironment, and perturbations in these communications can lead to tumorigenesis and degenerative phenotypes.
- Understanding the cellular and molecular mechanisms regulating self-renewal will have important implications for stem cell-based therapies and cancer therapies.
- The BMP signaling pathway and its regulators are highly conserved and associated with multiple human disorders and cancer.

RECOMMENDED RESOURCES

Publications

Losick, V.P., Morris, L.X., Fox, D.T., Spradling, A., 2011. *Drosophila* stem cell niches: a decade of discovery suggests a unified view of stem cell regulation. Dev. Cell 21 (1), 159–171.

Xia, L., Jia, S., Huang, S., Wang, H., Zhu, Y., Mu, Y., Kan, L., Zheng, W., Wu, D., Li, X., Sun, Q., Meng, A., Chen, D., 2010. The Fused/Smurf complex controls the fate of *Drosophila* germline stem cells by generating a gradient BMP response. Cell 143 (6), 978–990.

Xia, L., Zheng, X., Zheng, W., Zhang, G., Wang, H., Tao, Y., Chen, D., 2012. The niche-dependent feedback loop generates a BMP activity gradient to determine the germline stem cell fate. Curr. Biol. 22 (6), 515–521.

Morris, L.X., Spradling, A.C., 2011. Long-term live imaging provides new insight into stem cell regulation and germline-soma coordination in the *Drosophila* ovary. Development (Cambridge, England) 138 (11), 2207–2215.

Fox, D.T., Morris, L.X., Nystul, T., Spradling, A.C., 2008. Lineage analysis of stem cells in *StemBook* D.T. Fox, L.X. Morris, T. Nystul, and A.C., Spradling, Cambridge (MA).

Websites

Flybase – A Database of Drosophila Genes and Genomes @ Flybase.org (accessed 08.10.13).

NIH Stem Cell Information @ http://stemcells.nih.gov/Pages/Default.aspx (accessed 08.10.13).

International Society for Stem Cell Research @ http://www.isscr.org (accessed 08.10.13).

REFERENCES

Affolter, M., Basler, K., 2007. The Decapentaplegic morphogen gradient: from pattern formation to growth regulation. Nat. Rev. Genet. 8 (9), 663–674.

Arbouzova, N.I., Zeidler, M.P., 2006. JAK/STAT signaling in *Drosophila*: insights into conserved regulatory and cellular functions. Development (Cambridge, England) 133 (14), 2605–2616.

Casanueva, M.O., Ferguson, E.L., 2004. Germline stem cell number in the *Drosophila* ovary is regulated by redundant mechanisms that control Dpp signaling. Development (Cambridge, England) 131 (9), 1881–1890.

Chen, D., McKearin, D., 2005. Gene circuitry controlling a stem cell niche. Curr. Biol. 15 (2), 179–184.

Chen, D., McKearin, D., 2003a. Dpp signaling silences bam transcription directly to establish asymmetric divisions of germline stem cells. Curr. Biol. 13 (20), 1786–1791.

Chen, D., McKearin, D.M., 2003b. A discrete transcriptional silencer in the bam gene determines asymmetric division of the *Drosophila* germline stem cell. Development (Cambridge, England) 130 (6), 1159–1170.

Chen, S., Kaneko, S., Ma, X., Chen, X., Ip, Y.T., Xu, L., Xie, T., 2010. Lissencephaly-1 controls germline stem cell self-renewal through modulating bone morphogenetic protein signaling and niche adhesion. Proc. Natl. Acad. Sci. USA 107 (46), 19939–19944.

Chen, S., Wang, S., Xie, T., 2011. Restricting self-renewal signals within the stem cell niche: multiple levels of control. Curr. Opin. Genet. Dev. 21 (6), 684–689.

Cox, D.N., Chao, A., Lin, H., 2000. Piwi encodes a nucleoplasmic factor whose activity modulates the number and division rate of germline stem cells. Development (Cambridge, England) 127 (3), 503–514.

Dai, H., Hogan, C., Gopalakrishnan, B., Torres-Vazquez, J., Nguyen, M., Park, S., Raftery, L.A., Warrior, R., Arora, K., 2000. The zinc finger protein schnurri acts as a Smad partner in mediating the transcriptional response to decapentaplegic. Dev. Biol. 227 (2), 373–387.

Decotto, E., Spradling, A.C., 2005. The *Drosophila* ovarian and testis stem cell niches: similar somatic stem cells and signals. Dev. Cell 9 (4), 501–510.

Dejima, K., Kanai, M.I., Akiyama, T., Levings, D.C., Nakato, H., 2011. Novel contact-dependent bone morphogenetic protein (BMP) signaling mediated by heparan sulfate proteoglycans. J. Biol. Chem. 286 (19), 17103–17111.

Drummond-Barbosa, D., Spradling, A.C., 2001. Stem cells and their progeny respond to nutritional changes during *Drosophila* oogenesis. Dev. Biol. 231 (1), 265–278.

Eliazer, S., Shalaby, N.A., Buszczak, M., 2011. Loss of lysine-specific demethylase 1 nonautonomously causes stem cell tumors in the *Drosophila* ovary. Proc. Natl. Acad. Sci. USA 108 (17), 7064–7069.

Entchev, E.V., Schwabedissen, A., Gonzalez-Gaitan, M., 2000. Gradient formation of the TGF-beta homolog Dpp. Cell 103 (6), 981–991.

Faulkner, N.E., Dujardin, D.L., Tai, C.Y., Vaughan, K.T., O'Connell, C.B., Wang, Y., Vallee, R.B., 2000. A role for the lissencephaly gene LIS1 in mitosis and cytoplasmic dynein function. Nat. Cell Biol. 2 (11), 784–791.

Forbes, A., Lehmann, R., 1998. Nanos and Pumilio have critical roles in the development and function of *Drosophila* germline stem cells. Development (Cambridge, England) 125 (4), 679–690.

Fox, D.T., Morris, L.X., Nystul, T., Spradling, A.C., 2008. Lineage analysis of stem cells in *StemBook* D.T. Fox, L.X. Morris, T. Nystul, and A.C, Spradling, Cambridge (MA).

Goldman, D.C., Bailey, A.S., Pfaffle, D.L., Al Masri, A., Christian, J.L., Fleming, W.H., 2009. BMP4 regulates the hematopoietic stem cell niche. Blood 114 (20), 4393–4401.

Goldman, D.C., Hackenmiller, R., Nakayama, T., Sopory, S., Wong, C., Kulessa, H., Christian, J.L., 2006. Mutation of an upstream cleavage site in the BMP4 prodomain leads to tissue-specific loss of activity. Development (Cambridge, England) 133 (10), 1933–1942.

Guo, Z., Wang, Z., 2009. The glypican Dally is required in the niche for the maintenance of germline stem cells and short-range BMP signaling in the *Drosophila* ovary. Development (Cambridge, England) 136 (21), 3627–3635.

Haerry, T.E., 2010. The interaction between two TGF-beta type I receptors plays important roles in ligand binding, SMAD activation, and gradient formation. Mech. Dev. 127 (7–8), 358–370.

Harris, R.E., Ashe, H.L., 2011. Cease and desist: modulating short-range Dpp signaling in the stem-cell niche. EMBO Rep. 12 (6), 519–526.

Harris, R.E., Pargett, M., Sutcliffe, C., Umulis, D., Ashe, H.L., 2011. Brat promotes stem cell differentiation via control of a bistable switch that restricts BMP signaling. Dev. Cell 20 (1), 72–83.

Hayashi, H., Abdollah, S., Qiu, Y., Cai, J., Xu, Y.Y., Grinnell, B.W., Richardson, M.A., Topper, J.N., Gimbrone Jr., M.A., Wrana, J.L., Falb, D., 1997. The MAD-related protein Smad7 associates with the TGFbeta receptor and functions as an antagonist of TGFbeta signaling. Cell 89 (7), 1165–1173.

Hayashi, Y., Kobayashi, S., Nakato, H., 2009. *Drosophila* glypicans regulate the germline stem cell niche. J. Cell Biol. 187 (4), 473–480.

Hayashi, Y., Sexton, T.R., Dejima, K., Perry, D.W., Takemura, M., Kobayashi, S., Nakato, H., Harrison, D.A., 2012. Glypicans regulate JAK/STAT signaling and distribution of the Unpaired morphogen. Development (Cambridge, England) 139 (22), 4162–4171.

He, X.C., Zhang, J., Tong, W.G., Tawfik, O., Ross, J., Scoville, D.H., Tian, Q., Zeng, X., He, X., Wiedemann, L.M., Mishina, Y., Li, L., 2004. BMP signaling inhibits intestinal stem cell self-renewal through suppression of Wnt-beta-catenin signaling. Nat. Genet. 36 (10), 1117–1121.

Heldin, C.H., Miyazono, K., ten Dijke, P., 1997. TGF-beta signaling from cell membrane to nucleus through SMAD proteins. Nature 390 (6659), 465–471.

Hirotsune, S., Fleck, M.W., Gambello, M.J., Bix, G.J., Chen, A., Clark, G.D., Ledbetter, D.H., McBain, C.J., Wynshaw-Boris, A., 1998. Graded reduction of Pafah1b1 (Lis1) activity results in neuronal migration defects and early embryonic lethality. Nat. Genet. 19 (4), 333–339.

Hu, J., Chen, Y.X., Wang, D., Qi, X., Li, T.G., Hao, J., Mishina, Y., Garbers, D.L., Zhao, G.Q., 2004. Developmental expression and function of Bmp4 in spermatogenesis and in maintaining epididymal integrity. Devel. Biol. 276 (1), 158–171.

Imamura, T., Takase, M., Nishihara, A., Oeda, E., Hanai, J., Kawabata, M., Miyazono, K., 1997. Smad6 inhibits signaling by the TGF-beta superfamily. Nature 389 (6651), 622–626.

Inoue, H., Imamura, T., Ishidou, Y., Takase, M., Udagawa, Y., Oka, Y., Tsuneizumi, K., Tabata, T., Miyazono, K., Kawabata, M., 1998. Interplay of signal mediators of decapentaplegic (Dpp): molecular characterization of mothers against dpp, Medea, and daughters against dpp. Mol. Biol. Cell 9 (8), 2145–2156.

Iovino, N., Pane, A., Gaul, U., 2009. miR-184 has multiple roles in *Drosophila* female germline development. Dev. Cell 17 (1), 123–133.

Jiang, X., Xia, L., Chen, D., Yang, Y., Huang, H., Yang, L., Zhao, Q., Shen, L., Wang, J., Chen, D., 2008. Otefin, a nuclear membrane protein, determines the fate of germline stem cells in *Drosophila* via interaction with Smad complexes. Dev. Cell 14 (4), 494–506.

Jin, Z., Kirilly, D., Weng, C., Kawase, E., Song, X., Smith, S., Schwartz, J., Xie, T., 2008. Differentiation-defective stem cells outcompete normal stem cells for niche occupancy in the *Drosophila* ovary. Cell Stem Cell 2 (1), 39–49.

Jin, Z., Xie, T., 2007. Dcr-1 maintains *Drosophila* ovarian stem cells. Curr. Biol. 17 (6), 539–544.

Kai, T., Spradling, A., 2004. Differentiating germ cells can revert into functional stem cells in *Drosophila melanogaster* ovaries. Nature 428 (6982), 564–569.

Kim, J.Y., Lee, Y.C., Kim, C., 2010. Direct inhibition of Pumilo activity by Bam and Bgcn in *Drosophila* germ line stem cell differentiation. J. Biol. Chem. 285 (7), 4741–4746.

King, F.J., Szakmary, A., Cox, D.N., Lin, H., 2001. Yb modulates the divisions of both germline and somatic stem cells through piwi- and hh-mediated mechanisms in the *Drosophila* ovary. Mol. Cell 7 (3), 497–508.

Kirilly, D., Wang, S., Xie, T., 2011. Self-maintained escort cells form a germline stem cell differentiation niche. Development (Cambridge, England) 138 (23), 5087–5097.

Kirkpatrick, C.A., Selleck, S.B., 2007. Heparan sulfate proteoglycans at a glance. J. Cell Sci. 120 (Pt 11), 1829–1832.

Lavoie, C.A., Ohlstein, B., McKearin, D.M., 1999. Localization and function of Bam protein require the benign gonial cell neoplasm gene product. Devel. Biol. 212 (2), 405–413.

Lecuit, T., Brook, W.J., Ng, M., Calleja, M., Sun, H., Cohen, S.M., 1996. Two distinct mechanisms for long-range patterning by Decapentaplegic in the *Drosophila* wing. Nature 381 (6581), 387–393.

Lei, Y., Warrior, R., 2000. The *Drosophila* Lissencephaly1 (DLis1) gene is required for nuclear migration. Devel. Biol. 226 (1), 57–72.

Letsou, A., Arora, K., Wrana, J.L., Simin, K., Twombly, V., Jamal, J., Staehling-Hampton, K., Hoffmann, F.M., Gelbart, W.M., Massague, J., 1995. *Drosophila* Dpp signaling is mediated by the punt gene product: a dual ligand-binding type II receptor of the TGF beta receptor family. Cell 80 (6), 899–908.

Li, Y., Minor, N.T., Park, J.K., McKearin, D.M., Maines, J.Z., 2009. Bam and Bgcn antagonize Nanos-dependent germ-line stem cell maintenance. Proc. Natl. Acad. Sci. USA 106 (23), 9304–9309.

Lin, H., Spradling, A.C., 1997. A novel group of pumilio mutations affects the asymmetric division of germline stem cells in the *Drosophila* ovary. Development (Cambridge, England) 124 (12), 2463–2476.

Lin, H., Spradling, A.C., 1993. Germline stem cell division and egg chamber development in transplanted *Drosophila* germaria. Dev. Biol. 159 (1), 140–152.

Liu, M., Lim, T.M., Cai, Y., 2010. The *Drosophila* female germline stem cell lineage acts to spatially restrict DPP function within the niche. Sci. Signal. 3 (132). pp. ra57.

Liu, Z., Xie, T., Steward, R., 1999. Lis1, the *Drosophila* homolog of a human lissencephaly disease gene, is required for germline cell division and oocyte differentiation. Development (Cambridge, England) 126 (20), 4477–4488.

Lopez-Onieva, L., Fernandez-Minan, A., Gonzalez-Reyes, A., 2008. Jak/Stat signaling in niche support cells regulates dpp transcription to control germline stem cell maintenance in the *Drosophila* ovary. Development (Cambridge, England) 135 (3), 533–540.

Losick, V.P., Morris, L.X., Fox, D.T., Spradling, A., 2011. *Drosophila* stem cell niches: a decade of discovery suggests a unified view of stem cell regulation. Dev. Cell 21 (1), 159–171.

Lu, W., Casanueva, M.O., Mahowald, A.P., Kato, M., Lauterbach, D., Ferguson, E.L., 2012. Niche-associated activation of rac promotes the asymmetric division of *Drosophila* female germline stem cells. PLoS Biol. 10 (7). e1001357.

Margolis, J., Spradling, A., 1995. Identification and behavior of epithelial stem cells in the *Drosophila* ovary. Development (Cambridge, England) 121 (11), 3797–3807.

McKearin, D., Ohlstein, B., 1995. A role for the *Drosophila* bag-of-marbles protein in the differentiation of cystoblasts from germline stem cells. Development (Cambridge, England) 121 (9), 2937–2947.

McKearin, D.M., Spradling, A.C., 1990. bag-of-marbles: a *Drosophila* gene required to initiate both male and female gametogenesis. Genes Dev. 4 (12B), 2242–2251.

Morris, L.X., Spradling, A.C., 2011. Long-term live imaging provides new insight into stem cell regulation and germline-soma coordination in the *Drosophila* ovary. Development (Cambridge, England) 138 (11), 2207–2215.

Murakami, G., Watabe, T., Takaoka, K., Miyazono, K., Imamura, T., 2003. Cooperative inhibition of bone morphogenetic protein signaling by Smurf1 and inhibitory Smads. Mol. Biol. Cell 14 (7), 2809–2817.

Nakao, A., Afrakhte, M., Moren, A., Nakayama, T., Christian, J.L., Heuchel, R., Itoh, S., Kawabata, M., Heldin, N.E., Heldin, C.H., ten Dijke, P., 1997. Identification of Smad7, a TGFbeta-inducible antagonist of TGF-beta signaling. Nature 389 (6651), 631–635.

Nellen, D., Burke, R., Struhl, G., Basler, K., 1996. Direct and long-range action of a DPP morphogen gradient. Cell 85 (3), 357–368.

Nystul, T., Spradling, A., 2007. An epithelial niche in the *Drosophila* ovary undergoes long-range stem cell replacement. Cell Stem Cell 1 (3), 277–285.

Ohlstein, B., Lavoie, C.A., Vef, O., Gateff, E., McKearin, D.M., 2000. The *Drosophila* cystoblast differentiation factor, benign gonial cell neoplasm, is related to DExH-box proteins and interacts genetically with bag-of-marbles. Genetics 155 (4), 1809–1819.

Ohlstein, B., McKearin, D., 1997. Ectopic expression of the *Drosophila* Bam protein eliminates oogenic germline stem cells. Development (Cambridge, England) 124 (18), 3651–3662.

Park, J.K., Liu, X., Strauss, T.J., McKearin, D.M., Liu, Q., 2007. The miRNA pathway intrinsically controls self-renewal of *Drosophila* germline stem cells. Curr. Biol. 17 (6), 533–538.

Pellegrini, M., Grimaldi, P., Rossi, P., Geremia, R., Dolci, S., 2003. Developmental expression of BMP4/ALK3/SMAD5 signaling pathway in the mouse testis: a potential role of BMP4 in spermatogonia differentiation. J. Cell Sci. 116 (Pt 16), 3363–3372.

Perinthottathil, S., Kim, C., 2011. Bam and Bgcn in *Drosophila* germline stem cell differentiation. Vitam. Horm 87, 399–416.

Podos, S.D., Hanson, K.K., Wang, Y.C., Ferguson, E.L., 2001. The DSmurf ubiquitin-protein ligase restricts BMP signaling spatially and temporally during *Drosophila* embryogenesis. Dev. Cell 1 (4), 567–578.

Preat, T., Therond, P., Lamour-Isnard, C., Limbourg-Bouchon, B., Tricoire, H., Erk, I., Mariol, M.C., Busson, D., 1990. A putative serine/threonine protein kinase encoded by the segment-polarity fused gene of *Drosophila*. Nature 347 (6288), 87–89.

Reiner, O., Carrozzo, R., Shen, Y., Wehnert, M., Faustinella, F., Dobyns, W.B., Caskey, C.T., Ledbetter, D.H., 1993. Isolation of a Miller-Dieker lissencephaly gene containing G protein beta-subunit-like repeats. Nature 364 (6439), 717–721.

Rogers, S., Wells, R., Rechsteiner, M., 1986. Amino acid sequences common to rapidly degraded proteins: the PEST hypothesis. Science (New York, N.Y.) 234 (4774), 364–368.

Rojas-Rios, P., Guerrero, I., Gonzalez-Reyes, A., 2012. Cytoneme-mediated delivery of hedgehog regulates the expression of bone morphogenetic proteins to maintain germline stem cells in *Drosophila*. PLoS Biol. 10 (4). e1001298.

Shen, R., Weng, C., Yu, J., Xie, T., 2009. eIF4A controls germline stem cell self-renewal by directly inhibiting BAM function in the *Drosophila* ovary. Proc. Natl. Acad. Sci. USA 106 (28), 11623–11628.

Smith, D.S., Niethammer, M., Ayala, R., Zhou, Y., Gambello, M.J., Wynshaw-Boris, A., Tsai, L.H., 2000. Regulation of cytoplasmic dynein behavior and microtubule organization by mammalian Lis1. Nat. Cell Biol. 2 (11), 767–775.

Smith, P.A., King, R.C., 1966. Studies on fused, a mutant gene producing ovarian tumors in *Drosophila melanogaster*. J. Natl. Cancer Inst 36 (3), 445–463.

Song, X., Call, G.B., Kirilly, D., Xie, T., 2007. Notch signaling controls germline stem cell niche formation in the *Drosophila* ovary. Development (Cambridge, England) 134 (6), 1071–1080.

Song, X., Wong, M.D., Kawase, E., Xi, R., Ding, B.C., McCarthy, J.J., Xie, T., 2004. Bmp signals from niche cells directly repress transcription of a differentiation-promoting gene, bag of marbles, in germline stem cells in the *Drosophila* ovary. Development (Cambridge, England) 131 (6), 1353–1364.

Song, X., Zhu, C.H., Doan, C., Xie, T., 2002. Germline stem cells anchored by adherens junctions in the *Drosophila* ovary niches. Science (New York, N.Y.) 296 (5574), 1855–1857.

Sonoda, J., Wharton, R.P., 2001. *Drosophila* Brain Tumor is a translational repressor. Genes Dev. 15 (6), 762–773.

Sonoda, J., Wharton, R.P., 1999. Recruitment of Nanos to hunchback mRNA by Pumilio. Genes Dev. 13 (20), 2704–2712.

Sun, P., Quan, Z., Zhang, B., Wu, T., Xi, R., 2010. TSC1/2 tumor suppressor complex maintains *Drosophila* germline stem cells by preventing differentiation. Development (Cambridge, England) 137 (15), 2461–2469.

Swan, A., Nguyen, T., Suter, B., 1999. *Drosophila* Lissencephaly-1 functions with Bic-D and dynein in oocyte determination and nuclear positioning. Nat. Cell Biol. 1 (7), 444–449.

Szakmary, A., Cox, D.N., Wang, Z., Lin, H., 2005. Regulatory relationship among piwi, pumilio, and bag-of-marbles in *Drosophila* germline stem cell self-renewal and differentiation. Curr. Biol. 15 (2), 171–178.

Tang, H., Ross, A., Capel, B., 2008. Expression and functional analysis of Gm114, a putative mammalian ortholog of *Drosophila* bam. Dev. Biol. 318 (1), 73–81.

Teleman, A.A., Cohen, S.M., 2000. Dpp gradient formation in the *Drosophila* wing imaginal disc. Cell 103 (6), 971–980.

Torres-Vazquez, J., Park, S., Warrior, R., Arora, K., 2001. The transcription factor Schnurri plays a dual role in mediating Dpp signaling during embryogenesis. Development (Cambridge, England) 128 (9), 1657–1670.

Torres-Vazquez, J., Warrior, R., Arora, K., 2000. schnurri is required for dpp-dependent patterning of the *Drosophila* wing. Dev. Biol. 227 (2), 388–402.

Tsuneizumi, K., Nakayama, T., Kamoshida, Y., Kornberg, T.B., Christian, J.L., Tabata, T., 1997. Daughters against dpp modulates dpp organizing activity in *Drosophila* wing development. Nature 389 (6651), 627–631.

Udagawa, Y., Hanai, J., Tada, K., Grieder, N.C., Momoeda, M., Taketani, Y., Affolter, M., Kawabata, M., Miyazono, K., 2000. Schnurri interacts with Mad in a Dpp-dependent manner. Genes. Cells 5 (5), 359–369.

Umulis, D., O'Connor, M.B., Blair, S.S., 2009. The extracellular regulation of bone morphogenetic protein signaling. Development (Cambridge, England) 136 (22), 3715–3728.

Wang, L., Li, Z., Cai, Y., 2008. The JAK/STAT pathway positively regulates DPP signaling in the *Drosophila* germline stem cell niche. J. Cell Biol. 180 (4), 721–728.

Wang, X., Harris, R.E., Bayston, L.J., Ashe, H.L., 2008. Type IV collagens regulate BMP signaling in *Drosophila*. Nature 455 (7209), 72–77.

Wang, Y., Zhang, H., Chen, Y., Sun, Y., Yang, F., Yu, W., Liang, J., Sun, L., Yang, X., Shi, L., Li, R., Li, Y., Zhang, Y., Li, Q., Yi, X., Shang, Y., 2009. LSD1 is a subunit of the NuRD complex and targets the metastasis programs in breast cancer. Cell 138 (4), 660–672.

Wang, Z., Lin, H., 2004. Nanos maintains germline stem cell self-renewal by preventing differentiation. Science (New York, N.Y.) 303 (5666), 2016–2019.

Ward, E.J., Shcherbata, H.R., Reynolds, S.H., Fischer, K.A., Hatfield, S.D., Ruohola-Baker, H., 2006. Stem cells signal to the niche through the Notch pathway in the *Drosophila* ovary. Curr. Biol. 16 (23), 2352–2358.

Wharton, R.P., Sonoda, J., Lee, T., Patterson, M., Murata, Y., 1998. The Pumilio RNA-binding domain is also a translational regulator. Mol. Cell 1 (6), 863–872.

Wieschaus, E., Szabad, J., 1979. The development and function of the female germ line in *Drosophila melanogaster*: a cell lineage study. Dev. Biol. 68 (1), 29–46.

Wieser, R., Wrana, J.L., Massague, J., 1995. GS domain mutations that constitutively activate T beta R-I, the downstream signaling component in the TGF-beta receptor complex. EMBO J. 14 (10), 2199–2208.

Wrana, J.L., Attisano, L., Wieser, R., Ventura, F., Massague, J., 1994. Mechanism of activation of the TGF-beta receptor. Nature 370 (6488), 341–347.

Wynshaw-Boris, A., 2007. Lissencephaly and LIS1: insights into the molecular mechanisms of neuronal migration and development. Clin. Genet. 72 (4), 296–304.

Xi, R., Doan, C., Liu, D., Xie, T., 2005. Pelota controls self-renewal of germline stem cells by repressing a Bam-independent differentiation pathway. Development (Cambridge, England) 132 (24), 5365–5374.

Xi, R., Xie, T., 2005. Stem cell self-renewal controlled by chromatin remodeling factors. Science (New York, N.Y.) 310 (5753), 1487–1489.

Xia, L., Jia, S., Huang, S., Wang, H., Zhu, Y., Mu, Y., Kan, L., Zheng, W., Wu, D., Li, X., Sun, Q., Meng, A., Chen, D., 2010. The Fused/Smurf complex controls the fate of *Drosophila* germline stem cells by generating a gradient BMP response. Cell 143 (6), 978–990.

Xia, L., Zheng, X., Zheng, W., Zhang, G., Wang, H., Tao, Y., Chen, D., 2012. The niche-dependent feedback loop generates a BMP activity gradient to determine the germline stem cell fate. Curr. Biol. 22 (6), 515–521.

Xie, T., Spradling, A.C., 2000. A niche maintaining germ line stem cells in the *Drosophila* ovary. Science (New York, N.Y.) 290 (5490), 328–330.

Xie, T., Spradling, A.C., 1998. Decapentaplegic is essential for the maintenance and division of germline stem cells in the *Drosophila* ovary. Cell 94 (2), 251–260.

Yang, L., Chen, D., Duan, R., Xia, L., Wang, J., Qurashi, A., Jin, P., Chen, D., 2007. Argonaute 1 regulates the fate of germline stem cells in *Drosophila*. Development (Cambridge, England) 134 (23), 4265–4272.

Yang, L., Duan, R., Chen, D., Wang, J., Chen, D., Jin, P., 2007. Fragile X mental retardation protein modulates the fate of germline stem cells in *Drosophila*. Hum. Mol. Genet. 16 (15), 1814–1820.

Ying, Q.L., Nichols, J., Chambers, I., Smith, A., 2003. BMP induction of Id proteins suppresses differentiation and sustains embryonic stem cell self-renewal in collaboration with STAT3. Cell 115 (3), 281–292.

Zhang, J., Niu, C., Ye, L., Huang, H., He, X., Tong, W.G., Ross, J., Haug, J., Johnson, T., Feng, J.Q., Harris, S., Wiedemann, L.M., Mishina, Y., Li, L., 2003. Identification of the haematopoietic stem cell niche and control of the niche size. Nature 425 (6960), 836–841.

Zhu, H., Kavsak, P., Abdollah, S., Wrana, J.L., Thomsen, G.H., 1999. A SMAD ubiquitin ligase targets the BMP pathway and affects embryonic pattern formation. Nature 400 (6745), 687–693.

Chapter 6

Genomic Analyses of Neural Stem Cells

Nasir Malik*, Soojung Shin† and Mahendra S. Rao*,**

**National Institutes of Health, National Institute of Arthritis and Musculoskeletal and Skin Diseases, Bethesda, MD; †Invitrogen Corporation, Carlsbad, CA; **National Institutes of Health Center for Regenerative Medicine, Bethesda, MD*

Chapter Outline

GLOSSARY

Neural stem cells Self-renewing, multipotent cells that can differentiate into neurons, oligodendrocytes and astrocytes.

Lineage restricted precursor cells Cells that are still dividing but have a more restricted developmental potential than stem cells.

Large scale genomic analysis Methods or techniques to analyze a significant fraction of the information coded in RNA or DNA present in cells.

Data mining The use of statistical and graphical tools to compare expression patterns between data sets.

Reference standard A reference sample that allows comparison of data across laboratories and different platforms.

SUMMARY

- Neural stem cells (NSC) can be derived from fetal neuroepithelial cells or pluripotent stem cells and differentiated into all central nervous system neuronal and glial lineages.
- Technologies that can be used to understand NSCs at the genomic and transcriptomic level include microarrays, RNA-seq, DNA-seq, and Chip-seq.
- Because each of these technologies has its own advantages and limitations, researchers must carefully determine which one best suit their needs.
- The amount of data generated is vast, and planning must be performed in advance to ensure that proper analyses are performed in a timely fashion.
- Information gathered from large scale genomic and transcriptomic analyses can provide valuable insights into the biology underlying neural function and how it can be altered in neurological disorders.

Principles of Developmental Genetics.
2015 Published by Elsevier Inc.

6.1 INTRODUCTION

The nervous system is one of the earliest organ systems to differentiate from the blastula stage embryo. Neural stem cells (NSCs) are derived from neuroepithelial cells in the neural tube and give rise to all neuronal and glial cells of the central nervous system (CNS) by symmetric and asymmetric divisions. This process can be mimicked in culture and NSCs can be derived from human pluripotent stem cell (PSC) cultures over a period of 2–3 weeks (Reubinoff et al., 2001; Zhang et al., 2001; Shin et al., 2006; Elkabetz et al 2008; Koch et al., 2009; Reinhardt et al., 2013). PSCs are the *in vitro* counterpart of the inner cell mass of the blastula stage embryo, and have the potential to generate any cell in the body. *In vivo*, the primitive neural tube forms by approximately the fourth week of gestation, and neurogenesis has commenced by the fifth week of development in humans (Kennea et al., 2002). At the time of neurulation, at about the fourth week of development, the neuroectoderm segregates from the ectoderm by a process called neural induction (see Chapter 11). The early neural plate then undergoes a stereotypical set of morphogenetic movements to form a hollow tube, which is comprised primarily of stem cells, by a process called primary neurulation (Rao, 1999). The neural crest, which will form the peripheral nervous system, segregates from the CNS at this stage (see Chapter 18). Stem cells that will generate the CNS reside in the ventricular zone (VZ) throughout the rostrocaudal axis, they appear to be regionally specified by the expression of different positional markers, and proliferate at different rates. The anterior neural tube undergoes a dramatic expansion and can be delineated into three primary vesicles: the forebrain (prosencephalon), midbrain (mesencephalon), and hindbrain (rhombencephalon). Differential growth and further segregation lead to the additional delineation of the prosencephalon into the telencephalon and the diencephalon, and of the rhombencephalon into the metencephalon and the myelencephalon. The caudal neural tube does not undergo a similar expansion, but it does increase in size in a similar proportion to the embryo, and it undergoes further differentiation to form the spinal cord. The properties of VZ stem cells have been characterized and they appear to be homogeneous, despite the acquisition of rostrocaudal and dorsoventral identity.

As development proceeds, the VZ becomes much reduced in size, and additional zones of mitotically active precursors can be identified. Mitotically active cells that accumulate adjacent to the VZ have been called subventricular zone (SVZ) cells. The VZ is later called the subependymal zone, as the VZ is reduced to a single layer of ependymal cells. The SVZ is prominent in the forebrain, and it can be identified as far back as the fourth ventricle. No SVZ is detectable in the more caudal regions of the brain, and, if it exists, it is likely a very small population of cells. An additional germinal matrix that is derived from the rhombic lip of the fourth ventricle, called the external granule layer, generates the granule cells of the cerebellum. Like the VZ, the SVZ can be divided into subdomains that express different rostrocaudal markers and that generate phenotypically distinct progeny. Discrete SVZ domains identified include the cortical SVZ, the medial ganglionic eminence, the lateral ganglionic eminence, and the caudal ganglionic eminence. The proportion of SVZ stem cells declines with development, and, in the adult, multipotent stem cells are likely present only in regions of ongoing neurogenesis (e.g., the anterior SVZ underlying the hippocampus). At this stage, marker expression is relatively heterogeneous (Bernier et al., 2000; Doetsch et al., 1996; Pevny and Rao, 2003).

Stem cells do not generate differentiated progeny directly, but rather generate the dividing populations of more restricted precursors that are analogous to the blast cells or restricted progenitors described in the hematopoietic lineages (Bedi et al., 1995; Katsura et al., 2001; Mujitaba et al., 1999). These precursors can divide and self-renew, but they are located in regions that are distinct from stem cell populations and they can be distinguished from the stem cell population by the expression of cell surface and cytoplasmic markers (Cai et al., 2004b; Kalyani et al., 1997; Liu et al., 2004). Investigators have begun to analyze NSC populations (Table 6.1) using a variety of techniques, with the idea that by understanding these populations and identifying the factors that regulate self-renewal and direct differentiation, one will be able to modulate the development and response of stem cells to environmental signals. Many of these approaches depend on large scale analytic tools that rely on comparisons of purified populations of cells that differ with regard to their stage of development or their exposure to factors or carry specific genetic abnormalities. In this chapter, we focus on general principles that should guide such an analysis, and how data mining efforts have provided important insights into the properties of NSCs.

6.2 THE IMPORTANCE OF GLOBAL ANALYSIS AND CAVEATS WHEN COMPARING CELL SAMPLES

The overall disposition of a cell depends on the steady state of a complex of interacting factors. The integration of these instructions occurs in the nucleus through combinations of signal-activated and tissue-restricted transcription factors binding to and controlling related enhancers or *cis*-regulatory modules of co-expressed genes. Additional regulation is provided by previously unappreciated epigenetic mechanisms, such as histone modulation, CpG island methylation, and by

TABLE 6.1 Methods that have been Used to Characterize Neural Stem Cell Populations

Cells Characterized	Methods Used	Reference
Neural stem cells and progenitor cells	Microarray	Luo et al., 2002
Neuroepithelial cells	Subtractive suppression hybridization	Cai et al., 2004b
Astrocyte-restricted precursors	Immunohistochemistry	Liu et al., 2004
Neural stem cells	Microarray	Wright et al, 2003
SSEA1-positive cells	Microarray	Abramova et al., 2005
Central nervous system progenitors	Microarray	Geschwind et al., 2001
Neurosphere-forming cells	Massively parallel sequencing	Cai et al., 2006
Fetal and ES-derived NSCs	Microarray	Shin et al., 2007
Spinal cord and brain derived neural stem cells	Microarray	Kelly et al., 2009
Embryonic stem cell, neural stem cells	RNA-seq	Sun et al., 2011
Adult neural stem cells	RNA-seq, Chip-Seq	Ramos et al., 2013

microRNAs. Thus, the response of the cell to any one perturbation is dependent on context, and this in part explains conflicting results reported in the literature. For example, the effect of Sonic hedgehog on NSCs is dependent on the presence or absence of fibroblast growth factor (Wechsler-Reya et al., 1999). Likewise, the response to bone morphogenetic protein depends on the density of the culture and the presence or absence of various regulatory genes (Rajan et al., 2003; Wilson et al., 2001). This context-dependent response suggests that the overall state of a cell needs to be understood before perturbation experiments are initiated so that consistent and meaningful analyses of the results can be obtained.

Several variables remain poorly understood for NSCs. For example, no distinction has been made between long-term and short-term self-renewing populations. Although there is evidence that stem cell age has an effect (Shen et al., 1998; Svendsen, 2000), no analysis so far has taken into account the effects of aging, acquisition of karyotypic abnormalities, the differences as a result of the acquisition of positional identity, or the differences between the types of NSCs that are present during development. For example, radial glia type stem cells, transdifferentiatied stem cell populations, VZ stem cells, SVZ-derived stem cells, and neurosphere-forming stem cells fulfill the criteria of NSC, such as self-renewal and the ability to differentiate into neurons and glia, but the comparisons between them have not been rigorous.

Two other observations have suggested that caution needs to be exercised when stem cell populations are analyzed. Stem cells propagated in culture stochastically differentiate, and, as such, they are invariably contaminated by various amounts of differentiated cells. For example, the proportion of stem cells in a neurosphere culture can vary greatly. Notably, the largest contaminating populations are astrocytes and astrocyte precursors which like NSCs divide and express Nestin. A confounding point is that these cells are difficult to distinguish from stem cells using the standard battery of tests. A second important observation that has been made is that there are species differences between stem cells, and, thus, extrapolating from mouse to human can lead to errors (Barker et al., 2003; Ginis et al., 2004). Each of these differences will add variability to the results and make cross-laboratory comparisons difficult unless attention is paid to the quality of the sample and detailed information is available regarding time of isolation, passage number in culture, and number of contaminating cells present. These differences need to be documented and accounted for when comparing data sets, because noise from such variability can skew the final analysis. In addition, when analyzing cells it is desirable to use a reliable and reproducible method that is cost-effective and sensitive enough to detect with high fidelity the global differences among the populations of cells as well as the subtle differences introduced as the cells are propagated in the culture or as they mature. Although several different methodologies have been proposed, none of the cross platform comparisons is very useful unless normalizing algorithms and considerations of technical variables inherent in large scale analyses are carefully considered (Table 6.2).

Nevertheless, as improvements have been made in our ability to obtain pure populations of cells, harvest RNA from single cells or small amounts of tissues, construct libraries, sort cells, and obtain high-quality genomic information, such large scale analyses are now used more commonly. However, it is recommended that analyses and comparisons be limited to one stage of development, in one species, with a single platform. The samples should be carefully examined for the presence of contaminating populations and the degree of contamination should be assessed (Figure 6.1). This initial quality control will be critical in yielding useful results.

TABLE 6.2 Comparison of Methods for Large Scale Genomic Analysis

	RNA-Seq	Chip-Seq	Microarray
Sample required	150 ng – 2 µg	2–10 ng	500 ng –1 µg
Data presentation	Reads per kb of exon model per million mapped reads	Peaks aligned to reference genome	Fluorescence hybridization intensity
Sensitivity	Very high (can increase or decrease by varying read number)	Very high (can increase or decrease by varying read number)	Moderate
Technical biases	Sequencing errors Correctly mapping reads	Antibody quality Sequencing errors Correctly mapping reads	Noise at low signal intensity Saturation at high signal intensity
Cost	High	High	Low
Turn-around time	2–4 weeks	2–4 weeks	1–2 weeks

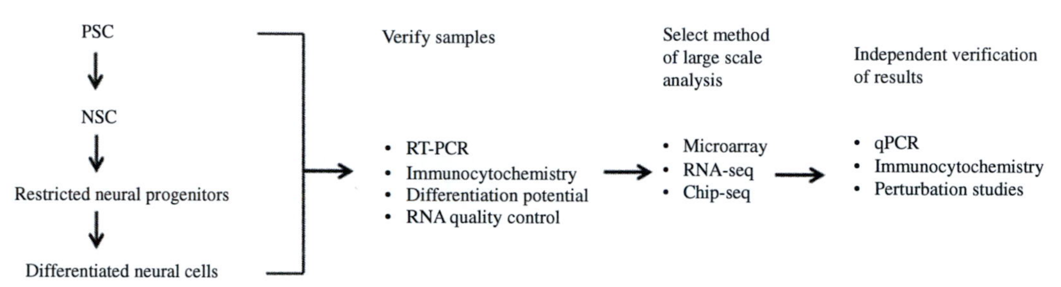

FIGURE 6.1 Flow Chart of Techniques Used to Characterize Cell Populations Using Large Scale Analysis. Samples for global analysis require verification in advance to assure dependable data production. Once the quality of the sample is controlled, one needs to decide upon a method of large scale analysis that is most appropriate for one's purpose. Once the generated dataset is processed and analyzed, then an independent method is selected to confirm the acquired results.

6.3 THE USE OF A REFERENCE STANDARD

When analyzing large batches of data generated in different laboratories using different techniques or slightly different cell culture protocols, one must determine the best way to make valid comparisons. Several strategies have been proposed including the idea of a reference standard (Dybkaer et al., 2004; Novoradovskaya et al., 2004; Loven et al., 2012). This idea, while not new, appears to be underappreciated in the stem cell field. Most researchers have however found it all but impossible to mine across data sets as there are too many variables that need to be normalized and too many assumptions that need to be made. Furthermore, there are often circumstances in which one simply lacks the data to make any appropriate assumption. In the ESC field, several strategies have been proposed (Loring et al., 2006). These include establishing a publicly available, well curated, data set that can be used as a ready reference, a set of standards that are readily available from a commercial or a not-for-profit provider or a control sample that all investigators can use as a standard. In principle each of these could be applied to the NSC field but to our knowledge no such common database exists as yet.

Immortalized or cancer stem cell lines such as C17.2, RT-4 or more recently identified cancer stem cell lines harvested from the appropriate species of interest have been proposed as possible standards (Imada et al., 1978; Snyder et al., 1992; Steindler, 2002). It is important however, when using such lines as a reference, to carefully assess the subclone that is being used. C17 subclones, for example, have shown remarkable variability and diametrically opposite results have been reported depending on the subclone used. The karyotype of this line is unstable, which may account for some of the differences seen. Nevertheless, since it has been so widely used, it could serve as a reference provided sufficient care was taken to use the same passage sample; banked at ATCC or some other responsible cell banking facility.

Fetal tissue samples from which pure populations of stem cells can be harvested at a defined stage in development in rodents may be an alternative choice for a reference standard. Many commercial entities provide such samples and these could therefore become a *de facto* standard. The equivalent stages of development are not readily accessible in humans however, and as such, an alternative control will need to be considered. Sorting or negative selection strategies have been

Ensuring that multiple data sets can be accurately compared

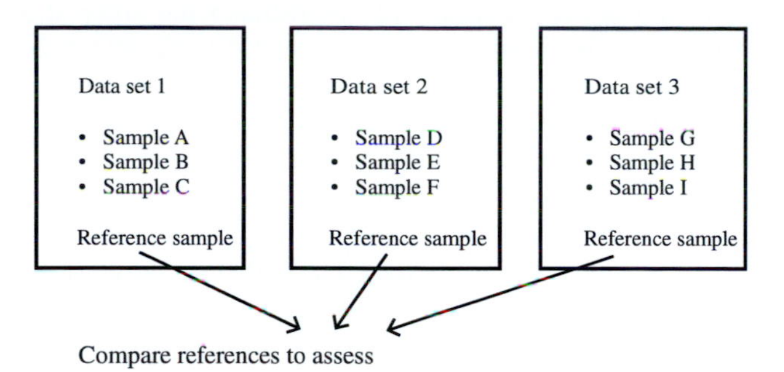

- Robustness of method
- Non-biological variation due to
  - Use of different methods
  - Lab-to-lab variation
  - Variation across chips/platforms

The results of comparison will determine whether data sets can be compared and the extent to which normalization will be required.

Reference characteristics

- Stable
- Readily accessible
- Widely available

Candidate references:

PSC or NSC lines which have shown high degree of reproducibility (r^2>0.95) across many runs in multiple labs.

FIGURE 6.2 The Importance of Using a Reference Standard. The reference standard makes it possible to compare data sets from different arrays, different time points, different laboratories and even different methods.

shown to enrich for stem cell populations and markers that define the stem cell stage are available, thus a reference standard for human stem cell analysis could be considered. As the ability to derive NSCs from PSCs has improved, stable NSC lines that have long-term renewal potential are now available. Lines which have generated high-quality data that is reproducible at both the global expression scale and in regards to specific markers known to be present in NSCs can serve as excellent reference standards that can be used every time an array is hybridized. Although this takes away one lane from the array that could have been used for an uncharacterized sample, the ability of the reference to make it possible to compare across different arrays makes this a desirable option. Alternatively, a publicly available, well curated data set could be generated from expression data gathered from NSCs from various sources as a digital reference standard.

In the absence of any of the above, we recommend obtaining RNA, genomic DNA, and miRNA from NTERA2. NTERA2 is a pluripotent line derived from a human embryonal testicular carcinoma that readily differentiates into neurons and glia. It has been carefully analyzed by several groups and labeled and unlabeled clones are commercially available as well as being available from ATCC. The line is often used as a comparator for ESC work and thus significant data on several different platforms are already available. These cells however are not optimal when detailed high resolution comparisons are required (Schwartz et al., 2005).

In our laboratory, having an internal reference standard has proven invaluable in allowing us to compare NSCs to each other, and to samples run at different times and to compare our results with those of several other colleagues without the necessity of repeating all the experiments. These standards have also allowed us to check the quality of markers, our Fluorescence Activated Cell Sorting (FACS) sorting efficiencies, the quality of our antibodies and have provided a basis for comparisons across different laboratories. By running a reference sample in one laboratory and sending results to another laboratory, our experiments can be easily compared and over time cross platform comparisons also become possible (Figure 6.2).

Once the concept of a standard is accepted widely, commercial providers can provide RNA, DNA and genomic material from such a reference that can be used as a comparator for all types of large scale studies. In the ESC field and in the microarray field, groups have been established to determine such standards; such uniformity has yielded useful results (Brazma et al., 2001; Husser et al., 2006; Wei et al., 2005) and we fully expect that a similar effort in the NSC field will be equally useful.

Methods of Analysis

The past several years have seen dramatic advances in technology and equally dramatic reductions in costs. Both genomic and proteomic methods are now available at costs that allow an average small laboratory to begin performing such experiments. The amount of material required for such an analysis has also become much smaller than was previously necessary, making these experimental approaches feasible even when the number of stem cells available is limited. For example, a gene expression profile using an Illumina chip costs about $150 per sample when calculating only the cost of the chip, and most institutions have core facilities that will perform all of the work associated with hybridizing the array at a reasonable cost. Even more importantly, the requirement for material has been dramatically reduced. We estimate that one can perform the entire battery of tests excluding proteomic analysis on about two million stem cells in any species, and newer technologies are bringing this figure down by another order of magnitude and also further reducing costs. Approximately ten million cells are enough for a mass spectrometry based analysis with equivalent amounts of material being required for Stable Isotopic Labeling using Amino acids in cell Culture (SILAC) and other similar methods. Indeed, it is expected that whole genome sequencing will soon cost less than US $1,000 and that profiling services will require even less material than is currently needed. These and other technical advances have facilitated the large scale gene expression analysis.

6.4 EPIGENETIC MODULATION

Over the past few years the importance of heritable epigenetic remodeling has been highlighted in regulating stem cell proliferation, cell fate determination, and carcinogenesis (Beaujean et al., 2004; Huntriss et al., 2004; Meehan, 2003; Ohgane et al., 2004; Vignon et al., 2002). Until, recently it has been difficult to study these events on a global scale; however, the ability to grow large numbers of cells and to differentiate them along specific pathways, coupled with the ability to perform such studies in a high throughput fashion has now made this possible. The concept of a bivalent chromatin state in PSCs, which facilitates expression of pluripotent genes but maintains lineage specific genes in a poised state ready to be activated when necessary, is now well established (Bernstein, 2006). Studies have begun to describe how this bivalent state is resolved during neurogenesis by derepression of poised genes required for neural commitment (Mikkelsen et al., 2007). Global methylation studies can be performed using a microarray or by whole genome bisulfite sequencing (Maitra et al., 2005). Illumina has described a bead array strategy to look at methylation patterns at 1500 loci encompassing regulatory elements in almost 400 genes. These include most genes known to be regulated during early embryonic development and those altered in tumorogenesis. These arrays have been used to examine methylation profiles in cancer stem cells and ESCs (Bibikova et al., 2006). Whole genome bisulfite sequencing of human ESCs differentiated to the three germ layers has been combined with Chip-seq and RNA-seq to understand the dynamic changes that take place at both the epigenetic and transcriptional levels (Gifford et al., 2013). The ENCODE consortium has also generated a vast dataset of genomic protein binding sites and epigenetic modifications in a host of cell types including human embryonic stem cells and neurons derived from them (ENCODE Project Consortium, 2012). These data are publicly available and can be viewed on the UCSC Genome Browser. Assessing the epigenetic profile of NSCs will be important prior to using them for transplant therapy, as maintenance of a particular epigenetic profile is probably critical for the appropriate function of cells.

6.5 MICRORNA

Micro (mi)RNAs are small non-coding RNA genes found in most eukaryotic genomes that are involved in the post-transcriptional regulation of gene expression. miRNAs are transcribed in the cell nucleus where they are processed into pre-miRNAs. Further processing occurs in the cytoplasm, where the pre-miRNAs are cleaved into their final ~22-nucleotide-long form which appears to regulate gene expression via transcriptional, translational or protein degradation regulation (Bartel, 2004; Szymanski et al., 2003). Recent reports have identified global strategies for identifying miRNAs, and over 450 such untranslated RNAs have been identified in humans, mice, and other species (Houbaviy et al., 2003; Lewis et al., 2003; Rajewsky et al., 2004). These approaches include computational analysis using sophisticated algorithms that recognize potential miRNA coding sequences and potential binding sites. Current estimates predict that there are ~2,000 miRNAs in the human genome. Other strategies have included making miRNA chips (Krichevsky et al., 2003). In addition, sequencing protocols analogous to massively parallel signature sequencing (MPSS) have been developed by Lynx Therapeutics that can be used to obtain quantitative data on miRNA made by a particular cell, which in turn predicts the overall state of the stem cell. Initial studies of the role of miRNAs in NSCs indicate that miR-9* and miR-124 repress the BAF53b and nBAF chromatin remodeling complexes in mice, resulting in the differentiation of NSCs to post-mitotic neurons in mice (Yoo et al., 2009). A recent paper reported that miR-17–92 and its paralogs are required for the proliferation of NSCs in mice, with knockout of these miRNAs resulting in transition of NSCs to more mature intermediate progenitors (Bian et al., 2013). Next generation sequencing (NGS) technologies such as RNA-seq, which can be used to identify all miRNAs in a sample, will be very useful for understanding the global role of miRNAs in NSCs.

6.6 MITOCHONDRIAL SEQUENCING

Structural and functional abnormalities in the mitochondria lead to functional defects in the nervous system, muscles and other organ systems. Somatic mitochondrial mutations are common in human cancers, aging cells and cells maintained in culture for prolonged periods. Mitochondrial DNA is also relevant to nuclear transfer, and estimating its stability is important in assessing the response of cells to stress and their ability to propagate in culture. Techniques to examine mitochondrial DNA mutations have been under development for some time. A recent description of a Polymerase Chain Reaction (PCR)-based approach for sequencing vertebrate mitochondrial genomes has attracted much attention, since it is faster and more economical than the traditional methods, which use cloned mtDNA and primer walking. Maitra and colleagues have developed a mitochondrial Custom RefSeq microarray as an array-based sequencing platform for the rapid and high-throughput analysis of mitochondrial DNA (Maitra et al., 2004). The MitoChip contains oligonucleotide probes synthesized using standard photolithography and solid-phase synthesis, and can sequence >29 kb of double-stranded DNA in a single assay.

It is useful to note that many mutations arise in the D-loop regions, and a simple PCR amplification and sequencing process can capture a large amount of information. No published data on baseline mitochondrial sequence and changes in it after culture of NSCs are currently available, but this information is likely to yield valuable insights about the role of the mitochondrial genome in NSC aging.

6.7 TRANSCRIPTOME MAPPING

Efforts have begun to identify the complete DNA binding sites and corresponding genes targeted by the transcriptional factors. One approach is to use an *in silico* computational strategy to retrieve putative genes with such binding sites. A direct and physiological approach is to perform chromatin immunoprecipitation (ChIP) of factors crosslinked *in vivo* to DNA targets, followed by identification of the specific DNA binding sites. Promoter chips, ranging from a focused selection of genes to complete gene sets, are now being made. NGS technologies have now matured such that it is relatively straightforward to determine all the genomic binding of a DNA binding protein with Chip-Seq. The findings from such a project will be of immense biological value in providing a description and understanding of the hierarchal relationships between groups of transcription factors and their target genes as they perform their tasks in embryonic development and specification of lineage fates and terminal differentiation.

Nuclear Run-On Assays

One analogous approach to identifying regulatory elements is to perform labeling of newly processed RNA to examine the genes induced only after a specific stimulus. Such hybridizations, while requiring larger amounts of material, are feasible with cell lines and with ESCs and can provide a global overview of the network of the transcriptional responses to a specific stimulus. More importantly, they provide an element of temporal control allowing one to better place individual genes in a transcriptional network. While such arrays have not been run with NSCs, experiments in other systems have yielded exciting results (Li et al., 2006) and we expect similar results from NSCs in the near future.

Proteomic Analysis, Glycosylation Maps and Other Protein Mapping Strategies

Most of our discussion has focused on methods for assessing genomic differences between cells. However, post-translational modifications play a crucial role in modifying genomic information and increasing the complexity of information that can be processed by a cell. The very complexity of the proteome has made it difficult to study on a large scale. Recently, however, multiple technical breakthroughs have begun to allow large scale analyses. These include advances in sensitivity of mass spectrometry, SILAC, developing variations on 2D gels, and labeling techniques to identify key proteins that are altered under different conditions, as well as the development of methods for isolating and sequencing small quantities of proteins (Elliott et al., 2004; Freeze, 2003; Ong et al., 2003). Proteomic analyses of NSCs have not been reported as yet, despite the medium-term availability of these cell lines.

6.8 DATA MINING: CHROMOSOME MAPPING, PATHWAY ANALYSIS, DATA REPRESENTATION

A major problem with large scale analysis has been knowing how to interpret the data, how to compare it, and how to extract meaningful biologically relevant information. Biologists, as a rule, cannot simply look at long lists of genes to identify critical information and examining the most abundant gene may not be of biological significance. For instance, changes in

Notch or ß-catenin signals of two-fold difference or less may be biologically significant, whereas ten or even one hundred fold differences in the expression of some genes may be irrelevant to the biological state of the cell. A particularly telling example of the relevancy of some differential gene expression data comes from the ESC literature, in which digital differential display identified TEX15 and DPPA5 as genes that are highly expressed in ESC but low or absent in all other populations examined (Adjaye et al., 2005; Kim et al., 2005; Lagarkova et al., 2006). These results were verified by PCR and immunocytochemistry, and yet data from knockout mice showed that these genes are dispensable for all assessed functions (Amano et al., 2006). On the other hand, a similar strategy identified Nanog, a previously unknown key regulator of ESC differentiation (Chambers et al., 2003; Mitsui et al., 2003).

In the stem cell field this has led to multiple attempts to consider how one should analyze data sets that are generated. The entire process is outlined in Figure 6.3. In our laboratory we have made the following assumptions: before pooling data or subjecting it to analysis, the quality of the sample used was tested. For example in NSCs, the harvested sample was assessed for its expression of known NSC markers and the absence of markers of differentiation. The presence or absence of these markers on the array is determined and the ability of the array to detect such differences is assessed by running the same sample on an array. If an array hybridization or MPSS analysis fails to detect an expected result, then that data are not used as all subsequent predictions are too uncertain. This analysis allows one to determine the expected sensitivity of the result and provides a rough idea of the sampling space (how much one will miss). We then examine carefully the intensity distribution of the expression levels of the genes present on the array. We have noted that in most cell types the distribution is quite similar and therefore any alteration suggests technical errors. While normalization algorithms can be employed in an attempt to use a particular anomalous data set, we generally red flag it, as most normalization algorithms tend to skew results.

Empirically we have determined that comparison across platforms is fraught with peril. Only positive results can be considered and negative results are generally not interpretable. For example, ESC-derived NSC samples were taken and examined by MPSS and Illumina bead arrays and a concordance of around 50% was shown, whereas the concordance rate for the sample run on a second Illumina array was close to 95% (unpublished observations). Even in such a comparison, only the presence or absence of an expression pattern can be considered and no attempts to compare expression levels

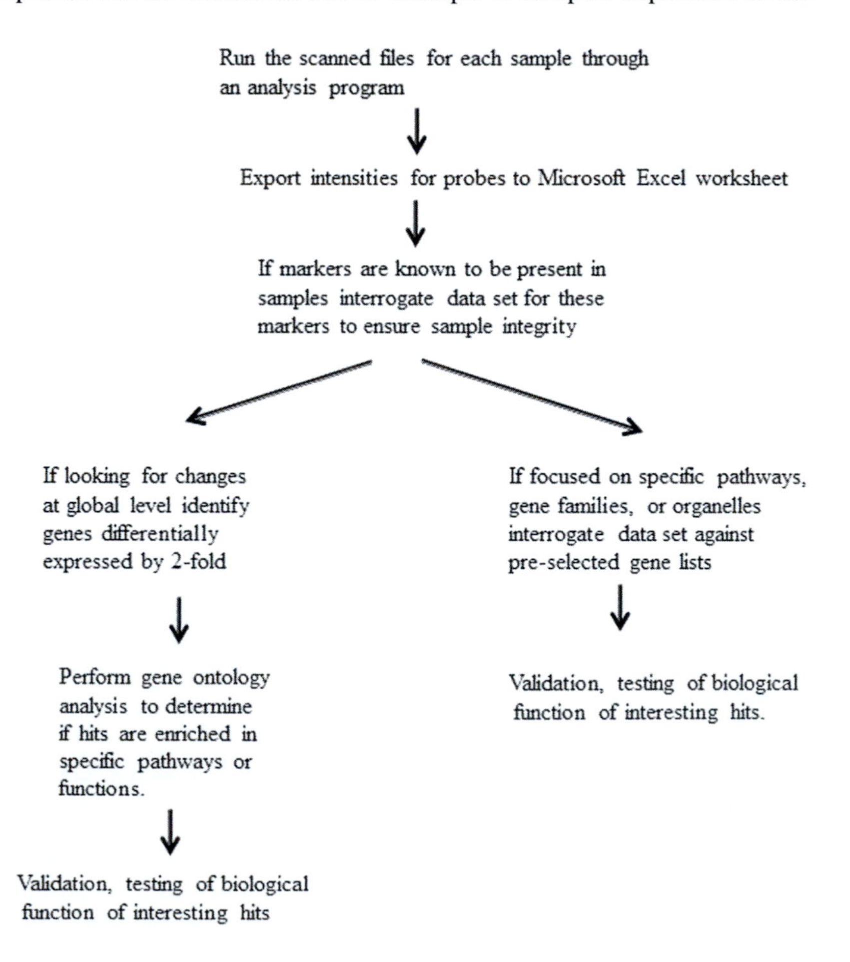

FIGURE 6.3 Analysis Pipeline for a Microarray Experiment. Once the hybridized array has been scanned the image files are run through an analysis file and intensities exported to Microsoft Excel worksheets. In these worksheets the data is interrogated for expression of known markers in the samples to ensure sample integrity. The data can then be analyzed at the global level for differentially expressed genes and also in a more focused manner to find important pathways that may be active. Once the analysis is complete interesting hits must be functionally validated.

between different methods should be made. We have also determined that amplification tends to provide a different pattern than using unamplified RNA even if the same sample is used. This appears not to be operator error, as amplification of different biological replicates performed at different times were closer than comparisons between amplified and unamplified samples. Therefore only samples that have been processed identically were compared as far as it is practical.

Once we are comfortable with the quality of each sample processed, we examine differential gene expression by examining pairwise comparisons rather than pooling the data or compare to a reference standard of baseline data that have been generated. This allows one to generate larger data sets and to compare across laboratories. For example, with NSCs we suggest using a well-characterized NSC sample as a potential reference standard. It is widely available, standardized, and has been run across multiple platforms and such data sets are publicly available. This method allows one to readily determine whether hybridization results are within the normal range, and whether the data are usable and comparable to results from other laboratories.

Once we have determined that we have a reasonable set of data, we then determine an appropriate cut-off for sensitivity that we are comfortable with. This ranges from array phenotype to phenotype and is an important criterion in any assessment. In MPSS for example, a theoretical sensitivity when three million tags are sequenced is three transcripts per million (tpm) (Miura et al., 2004a). However, we have empirically determined that testing or validating expression at such low levels is difficult and as such may or may not be useful to focus on. In our hands, an expression level of 50 tpm is readily verifiable and is a cut-off we use routinely (Cai et al., 2006); this does mean that we are potentially discarding useful information (which can be substantial as a majority of genes are expressed at low levels). Likewise, with Illumina bead arrays we use a cut-off of 50 to 100 arbitrary intensity units, even though the theoretical sensitivity of the arrays is much lower. It is however important that this be made clear in any publication, or alternatively the raw data be made available for independent analysis. This is particularly important as the genomic data bases are constantly being curated and EST tags assigned to a particular locus are being reassigned as better data becomes available. This curation sometimes means that a gene tag on an array may not represent the gene that it was originally thought to identify. Various groups have estimated this frequency and have recognized it is an important source of error in these types of analysis. We therefore strongly recommend that such curation and updating be a regular part of the data analysis.

Once we have determined a cut-off, established a set of phenotypes for analysis, and collected the curated data we then can begin to consider how best to analyze the data. This analysis is dependent on the biological questions one wishes to pose and while each strategy will be different, we can perhaps highlight some of the simple strategies that can be used. Currently we feel that the most complete genomic data available are human and mouse. We therefore have focused on large scale analysis in these two species. We find that mapping the expressed genes onto a genome browser (chromosome mapping) provides a very useful overview of global patterns of gene expression. It allows one to study regulation of genes on a global scale, identify genomic hotspots or correlate with known chromosomal breakpoints or SNP data sets or data developed by groups working in different disciplines. We would recommend the UCSF (University of California San Francisco) golden path browser for this purpose.

A second important strategy that we use routinely is to perform what we term a pathway analysis. Here, rather than looking at the expression of an individual gene, we examine an entire signaling pathway by mapping the expression of all genes detected for that pathway. A visual representation of the "on" and "off" state of the pathway can be easily obtained, and observation of the pathway as cells differentiate can provide a much clearer idea of whether the genes in that pathway are important in the process of differentiation. Such a pathway analysis, for example, clearly identified the LIF/GP130 pathway as being important in NSCs in humans but not in mice. It also helps identify key genes that must be tested in verification studies. Multiple commercial programs to perform such a pathway analysis exist and we recommend using any one of these.

Overall, our experience has been that if effort is made to develop well curated data sets, verification of observations is generally greater than 50% and one can glean important and reliable information. While a hit rate of 50% may sound low, it is useful to consider that this is much better than the one in 30,000 chance of finding a functionally useful gene that one started with prior to analysis. Using NSC biomarkers as an example, currently there are about ten markers that have been identified. By a large scale analysis one could readily identify perhaps two hundred of which perhaps 25% will be novel or unexpected genes (50); with a hit rate of 50% one could identify 25 new markers in a three month experiment, more than doubling the number of known NSC markers. In the next section we will discuss some general observations made about NSCs.

6.9 GENERAL OBSERVATIONS ABOUT THE PROPERTIES OF NEURAL STEM CELLS

We find that NSCs appear similar to other cells in synthesizing about 10,000–12,000 genes of an estimated 35,000 or 40,000 genes annotated in the RefSeq database. Average total RNA per cell tends to be higher in stem cells than in other cells (average of 5–10 pg) but is similar to levels seen in metabolically active cells. The distribution of transcript frequency suggests that most genes are transcribed at relatively low levels (less than 50 transcripts per cell) as assessed by MPSS. Mitochondrial, ribosomal and housekeeping genes tend to be more abundant, whereas transcription factors, growth factors and other cell type specific molecules are expressed at much lower levels. This pattern of gene expression is similar to that seen in most other cell types. Comparison of data with that for other cell types suggests that on average, the great majority of genes are shared or in common, while approximately 20% are different (by greater than ten-fold) in any two samples. Mapping of all expressed genes does not show a chromosomal bias as has been suggested in other stem cell populations. An example from an array experiment performed in our laboratory compared NSCs, astrocyte precursor cells (APCs), and oligodendrocyte precursor cells (OPCs) (Campanelli et al., 2008). The average correlation rate among NSC populations grown under different conditions was 0.90 (0.79–0.99) while the score dropped to 0.65 (0.57–0.71) and 0.77 (0.76–0.80) when NSCs were compared to APCs and OPCs, respectively. The correlation is much closer when identical samples are compared in two independent sequencing runs to 0.99 (0.96–0.99). In contrast, when identical samples are compared between two methods (Serial Analysis of Gene Expression (SAGE) and MPSS or Expressed Sequence Tag and MPSS) then the concordance rates are much lower (0.7), suggesting that those genes that are common between two methods are likely to be important. However, the lack of a high concordance when identical samples are analyzed by different methodologies suggests caution should be exercised in assuming that the failure to detect expression by any one method means that result is valid.

We have used the Illumina microarray platform to investigate the gene expression of normal and disease NSCs, neurons derived from these NSCs, astrocytes derived from a normal NSC line, and an immortalized astrocyte line. All the samples displayed similar intensity distributions, indicating that valid comparisons about gene expression could be made across this data set (Figure 6.4). As shown in Figure 6.5, at the global level, the normal NSCs and disease NSCs cluster together (r^2 = 0.95–0.96) and the neurons from these samples also group together (r^2 = 0.94–0.95). The NSC-derived astrocytes show greatest similarity to the NSC-derived neurons (r^2 = 0.91–0.92) while being more divergent from immortalized astrocytes (r^2 = 0.89), and with even less similarity to the parental NSC line from which they were derived (r^2 = 0.83). The data for the normal NSC lines are consistent with what we have seen in other array experiments as we generally see a correlation of around 0.95 for the various NSC lines in our laboratory, indicating that the lines differ very little in their gene expression profiles. It is not unexpected that the NSC-derived astrocytes and neurons are more similar to each other than the NSCs, as they primarily represent a population of differentiated cell types. It is perhaps surprising that the NSC-derived astrocytes are more similar to neurons than the immortalized NSC line, but this may be due to the possibility that the immortalization process has affected gene expression in the astrocytes. We also were able to identify a subset of ~20 genes that appeared to be unique to either NSCs, astrocytes, and neurons and via a pathway analysis find that Notch, TGF-β,and calcium signaling play important roles in these cell types (data not shown).

Examining NSC-enriched genes from previous studies suggests that several major pathways are active in NSCs (Abramova et al., 2005; Cai et al., 2006). These include the LIF/gp130 (in humans only), the FGF signaling pathway, the cell cycle regulatory pathways including myc and DNA repair and anti-apoptotic pathways. Intriguingly, the genes regulating the timing of differentiation, antisense RNA, siRNA, specific methylases and chromatin remodeling enzymes appear to be present at high levels, suggesting that epigenetic remodeling is an important aspect of NSC cell biology; as has also been shown for ESC cells.

6.10 SPECIES DIFFERENCES

An important finding that has been made clear from the availability of data sets from both mouse and human ESCs is the variability between species. While many key pathways are conserved, many differences have been highlighted as well. For example:

1) Oct3/4 homologs likely do not exist in chicken embryos (Soodeen-Karamath et al., 2001);
2) LIF signaling, which is critical for ESC self-renewal in rodents, does not appear to be critical or even required for human ESCs (Daheron et al., 2004; Ginis et al., 2004; Niwa, 2001);
3) No paralogs of E-Hox have been identified in humans and ES cell expressed Ras (Eras) appears to be a pseudogene in humans (Bhattacharya et al., 2004).

The low overall concordance rate between human and rodent ESCs (in one comparison around 40%) relative to that seen in human-to-human cell comparisons (90% between human ESC samples) provides additional support for these findings (Wei et al., personal communication). Although no analogous studies have been performed comparing mouse and human NSCs we expect similar species differences for this cell type.

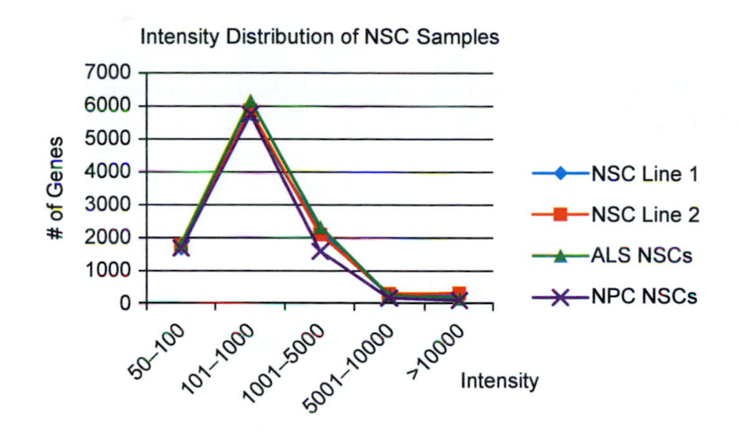

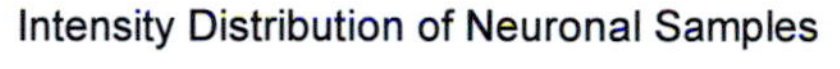

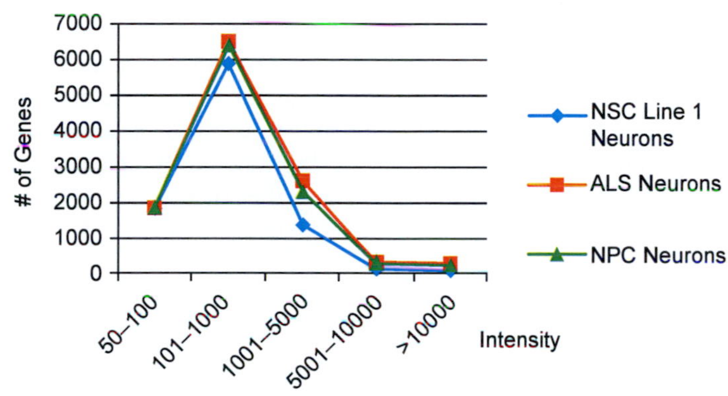

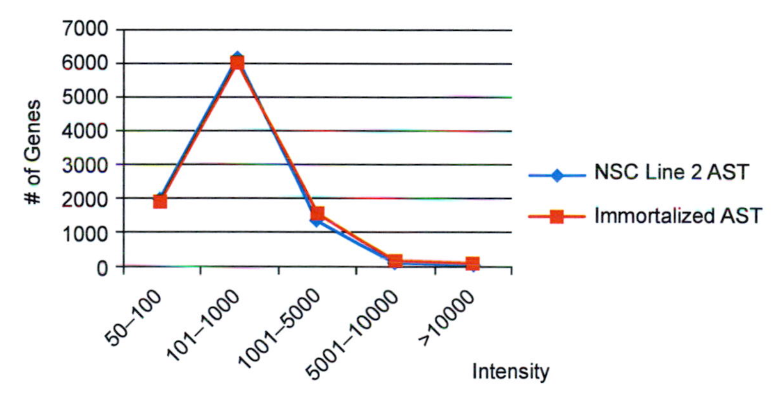

FIGURE 6.4 Intensity Distributions for NSC, Neuronal, and Astrocyte Samples. NSCs, NSC-derived neurons and astrocytes and an immortalized astrocyte line were assessed for intensity distribution across the entire range of values at an intensity above 50 and the intensity distributions were graphed. All samples had similar distributions both within and across subsets.

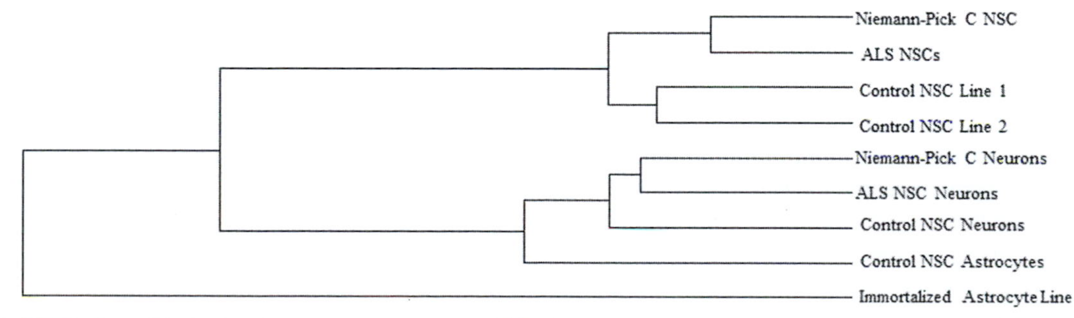

FIGURE 6.5 Dendrogram Displaying the Relationship Between NSCs, Neurons, and Astrocytes at the Global Level. The Gene Expression module of Illumina's Genome Studio software was used to generate a dendrogram to show the similarity of gene expression at the global level between several NSC lines, neurons and astrocytes derived from these lines and an immortalized astrocyte line.

6.11 LACK OF A "STEMNESS" PHENOTYPE

There is an operating bias in the literature that most stem cells will be similar to each other and indeed several manuscripts simply use the term "stem cell" rather than specifying which tissue they arise from and the stage of development used. This bias, however, is not supported by the large scale analytical data. Comparison of NSCs with other populations of stem cells has not identified any common stemness pathway (Cai et al., 2004a). Another important finding that has become obvious from such data set examination is the lack of any common stemness genes between ESCs and other somatic stem cells. Comparing data sets with expression in NSCs does not identify a common subset of genes and suggests that NSCs are quite different from other somatic stem cells (Fortunel et al., 2003; Ivanova et al., 2002; Ramalho-Santos et al., 2002). Our recent findings from a comparison of ESC-derived NSCs and fetal-derived NSCs (Shin et al., 2007) and more limited comparisons (Pevny and Rao, 2003) have shown clearly that stem cells can be readily distinguished from each other, and that stem cells or progenitor cells harvested at different stages of development behave differently.

6.12 ALLELIC VARIABILITY

It is clear that while individual stem cells share properties such as the ability to self-renew in culture and to differentiate into various phenotypes, there will nevertheless be differences between individual cell lines based on the genetic profile of the individual from whom they were derived. These allelic differences are not unexpected and merely reflect the diversity of phenotypes seen in the human population. Perhaps what has not been appreciated as much is how much variability this might impose on stem cell behavior, even if care is taken to isolate from the same region at the same developmental time using methods as similar as possible. The consequences of such allelic variability have been described in allergic responses, the ability to smell different odors, to metabolize alcohol, to digest milk, or to respond to toxins (Sultatos et al., 2004; Usuku et al., 1992).

When global pairwise comparisons are made between multiple stem cell lines using the same platform, one sees a high reproducibility when the same sample is run repeatedly (correlation coefficients in the range of 0.98–0.99). However, when two samples isolated by the same group at around the same time are run, the differences are much larger (correlation coefficient 0.92–0.94) indicating that a significant number of genes are differentially expressed between two stem cell lines. This sort of difference is consistently seen in all stem cell populations and is generally less than the difference between a cell line and its differentiated progeny and is less than the difference between a cell line from one species compared to another (Ginis et al., 2004). Nevertheless, at the individual gene level such changes can be quite dramatic (Bhattacharya et al., 2005). Such allelic variability likely underlies the differences in propagation, self-renewal and ability to differentiate into specific phenotypes that have been reported.

Large scale analysis of the kind we have reported allows one to map the underlying basis of such variability and make predictions about the behavior of cell lines (Lo et al., 2003; Ross et al., 2000; Yan et al., 2002a; Yan et al., 2002b). For example Lo et al. showed reliable measurement of allele specific gene expression based on large scale analysis. Selected genes were not only in the imprinted regions but also were distributed throughout the genome (Lo et al., 2003), which may be responsible for phenotypic variation.

Thus, as one begins considering stem cells for therapy one will need to carefully assess whether a lot of NSC prepared from one individual will behave in similar way to another lot of cells prepared from a different donor, even if identical protocols were used and the experiments were performed in a similar fashion. This inherent variability is of major concern when evaluating cells as therapy.

6.13 AGE DEPENDENT CHANGES IN NSCS

Examination of human cells in culture has shed some light on the cellular basis of aging. When grown in culture, normal human cells will undergo a limited number of divisions before entering a state of replicative senescence, in which they remain viable but are unable to divide further (Miura et al., 2004b). This change has been attributed to changes in mitochondria, protein expression, loss of genomic integrity, oxidative damage and progressive loss of DNA repair ability or erosion of telomere ends. Results from our laboratory and others have shown that ESCs appear to be immortal and can be propagated in continuous culture for a period of at least two years. Examination of the expression of immortality-associated genes has suggested that the expression of key regulators is important (Miura et al., 2004b). These include the expression of telomere-associated proteins, expression of some immortality-associated genes, high levels of DNA repair enzymes, the inhibition of p53, and the absence of Rb; thus altering cell cycle regulation. Comparing the phenotype of such genes between NSCs and ESCs shows significant differences. NSCs have shown active expression of Rb and their pattern of expression of telomere related genes and immortality-associated genes differs from that of ESCs. Differences are also apparent between adult stem cell populations and ESCs, which suggests that NSCs are not easily propagated indefinitely in culture. Indeed, Whittemore and colleagues have shown that long-term propagation of NSCs in culture decreased their ability to differentiate (Quinn et al., 1999). Pruit and colleagues have shown that NSCs undergo karyotypic changes as the animal ages, providing further confirmation of this hypothesis (Bailey et al., 2004). Our recent results examining the karyotypic stability of NSCs in culture have indicated that NSCs are less stable in culture than ESCs and can acquire karyotypic changes in as few as ten passages. Overall, large scale analysis suggests that there are fundamental differences in cell cycle and immortality-associated genes and these differences predict that NSCs will age in culture and senesce as do other adult stem cell populations.

6.14 CANCER STEM CELLS

The discovery that many cancers may be propagated by a small number of stem cells present in the tumor mass is an exciting finding (Al-Hajj et al., 2003; Huntly et al., 2005; Reya et al., 2001; Singh et al., 2004). This was first described in breast cancers and subsequently in a variety of solid tumors. With respect to the nervous system, several reports have suggested that cancer stem cells can be identified and that these cells bear a remarkable similarity to NSCs present in early development (Hemmati et al., 2003; Singh et al., 2004). Likewise, cells resembling glial progenitors have been isolated from some glial tumors (Kondo et al., 2004; Noble et al., 1995). In the case of glioblastoma, a subset of tumor cells that resemble neural stem cells were found to be responsible for the long-term growth of the tumor in mice (Chen et al., 2012). We and others have suggested that tumors can perhaps be phenotypically classified, and this classification correlates well with a cell of origin identified based on findings from lineage relationships that exist between stem and progenitor cells. The advantage of having a detailed database of gene expression from multiple normal cell lines and progenitor cells and the ability to compare their gene expression profiles with their transformed counterparts is obvious. In recent years, numerous gene expression studies have found that although brain cancers can be grouped into molecular subclasses, the signaling networks driving these cancers are remarkably similar, and normally function during early development (Brennan et al., 2009). Results such as these validate our belief that expression profiling at the global level can provide valuable insights regarding pathways that may be important for the development of a particular cell type or progression to a disease state.

6.15 CONCLUSIONS

Overall, the initial efforts at profiling NSCs from rodents and human have yielded useful insights, and have allowed the identification of key regulatory pathways and novel genes that are likely to play a role in regulating NSC self-renewal and differentiation. The data sets developed to date are an important resource, and can be mined with readily available tools. Better accessibility and curation of these data for the research community would greatly facilitate more advances in our knowledge of NSC biology. While profiling information has already yielded fruit in terms of understanding the basic biology of the cells, we believe that the availability of this information will have a utility far beyond current applications, including important future roles in the clinical setting in assessing the value of therapeutic interventions relating to advances in stem cell transplantation and treatment of tumors affecting the brain.

6.16 CLINICAL RELEVANCE

- Neural stem cells (NSCs) may have therapeutic potential as cellular replacement therapies in neurodegenerative disorders.
- Many tumors of the nervous system are derived from a small number of NSC-like cells that undergo unregulated proliferation.
- As the cost of large scale next generation sequencing technologies decreases, it will become a common practice to sequence tumors from patients to identify the best therapeutic options.
- Likewise, if NSCs become part of a treatment strategy, the ability to sequence the genomes of these cells before transplantation will make it possible to determine whether they carry any mutations that could be dangerous if the cells are transplanted into a patient.
- The ability to generate NSCs and differentiate them on a large scale will facilitate drug screens and the pathways modulated by positive hits from these screen can be identified by the use of large scale genomic technologies to further drug development for neurological disorders.

ACKNOWLEDGMENT

This work was supported by the NIH Common Fund.

RECOMMENDED RESOURCES

Reviews

Huse, J.T., Holland, E.C., 2010. Targeting brain cancer; advances in the molecular pathology of malignant glioma and medulloblastoma. Nat. Rev. 10, 319–331.

Hirabayashi, Y., Gotoh, Y., 2010. Epigenetic control of neural precursor fate during development. Nat. Rev. 11, 377–387.

Websites

Illumina's Next Generation Sequencing Tecnology:
http://www.illumina.com/technology/sequencing_technology.ilmn (Accessed June 2013).
Invitrogen's Next Generation Sequencing Technology:
http://www.invitrogen.com/site/us/en/home/Products-and-Services/Applications/Sequencing/Next-Generation-Sequencing.html (Accessed June 2013).
NHGRI DNA Microarray Technology Overview:
http://www.genome.gov/10000533 (Accessed June 2013).

REFERENCES

Abramova, N., Charniga, C., Goderie, S.K., Temple, S., 2005. Stage-specific changes in gene expression in acutely isolated mouse CNS progenitor cells. Dev. Biol. 283, 269–281.

Adjaye, J., Huntriss, J., Herwig, R., BenKahla, A., Brink, T.C., Wierling, C., Hultschig, C., Groth, D., Yaspo, M.L., Picton, H.M., et al., 2005. Primary differentiation in the human blastocyst: comparative molecular portraits of inner cell mass and trophectoderm cells. Stem. Cells. 23, 1514–1525.

Al-Hajj, M., Wicha, M.S., Benito-Hernandez, A., Morrison, S.J., Clarke, M.F., 2003. Prospective identification of tumorigenic breast cancer cells. Proc. Natl. Acad. Sci. USA 100, 3983–3988.

Amano, H., Itakura, K., Maruyama, M., Ichisaka, T., Nakagawa, M., Yamanaka, S., 2006. Identification and targeted disruption of the mouse gene encoding ESG1 (PH34/ECAT2/DPPA5). BMC Dev. Biol. 6, 11.

Bailey, K.J., Maslov, A.Y., Pruitt, S.C., 2004. Accumulation of mutations and somatic selection in aging neural stem/progenitor cells. Aging. Cell. 3, 391–397.

Barker, R.A., Jain, M., Armstrong, R.J., Caldwell, M.A., 2003. Stem cells and neurological disease. J. Neurol. Neurosurg. Psychiatry. 74, 553–557.

Bartel, D.P., 2004. MicroRNAs: genomics, biogenesis, mechanism, and function. Cell 116, 281–297.

Beaujean, N., Taylor, J., Gardner, J., Wilmut, I., Meehan, R., Young, L., 2004. Effect of limited DNA methylation reprogramming in the normal sheep embryo on somatic cell nuclear transfer. Biol. Reprod. 71, 185–193.

Bedi, A., Sharkis, S.J., 1995. Mechanisms of cell commitment in myeloid cell differentiation. Curr. Opin. Hematol. 2, 12–21.

Bernier, P.J., Vinet, J., Cossette, M., Parent, A., 2000. Characterization of the subventricular zone of the adult human brain: evidence for the involvement of Bcl-2. Neurosci. Res. 37, 67–78.

Bernstein, B.E., Mikkelsen, T.S., Xie, X., Kamal, M., Huebert, D.J., Cuff, J., Fry, B., Meissner, A.L., Wernig, M., Plath, K., et al., 2006. A bivalent chromatin structure marks key developmental genes in embryonic stem cells. Cell. 125, 315–326.

Bhattacharya, B., Cai, J., Luo, Y., Miura, T., Mejido, J., Brimble, S.N., Zeng, X., Schulz, T.C., Rao, M.S., Puri, R.K., 2005. Comparison of the gene expression profile of undifferentiated human embryonic stem cell lines and differentiating embryoid bodies. BMC. Dev. Biol. 5, 22.

Bhattacharya, B., Miura, T., Brandenberger, R., Mejido, J., Luo, Y., Yang, A.X., Joshi, B.H., Ginis, I., Thies, R.S., Amit, M., et al., 2004. Gene expression in human embryonic stem cell lines: unique molecular signature. Blood 103, 2956–2964.

Bian, S., Hong, J., Li, Q., Schebelle, L., Pollock, A., Knauss, J.L., Garg, V., Sun, T., 2013. MicroRNA cluster miR-17–92 regulates neural stem cell expansion and transition to intermediate progenitors in the developing mouse neocortex. Cell. Rep. 3, 1–9.

Bibikova, M., Chudin, E., Wu, B., Zhou, L., Wickham Garcia, E., Liu, Y., Shin, S., Plaia, T.W., Auerbach, J.M., Arking, D.E., et al., 2006. Human embryonic stem cells have a unique epigenetic signature. Genome. Res. 16, 1075–1083.

Brennan, C., Momota, H., Hambardzumyan, D., Ozawa, T., Tandon, A., Pedraza, A., Holland, E., 2009. Glioblastoma subclasses can be defined by activity among signal transduction pathways and associated genomic alterations. PLoS. One. 13, e7752.

Brazma, A., Hingamp, P., Quackenbush, J., Sherlock, G., Spellman, P., Stoeckert, C., Aach, J., Ansorge, W., Ball, C.A., Causton, H.C., et al., 2001. Minimum information about a microarray experiment (MIAME)-toward standards for microarray data. Nat. Genet. 29, 365–371.

Cai, J., Shin, S., Wright, L., Liu, Y., Zhou, D., Xue, H., Khrebtukova, I., Mattson, M.P., Svendsen, C., Rao, M.S., 2006. Massively parallel signature sequencing profiling of fetal human neural precursor cells. Stem. Cells. Dev. 15, 232–244.

Cai, J., Weiss, M.L., Rao, M.S., 2004. In search of "stemness". Exp. Hematol. 32, 585–598.

Cai, J., Xue, H., Zhan, M., Rao, M.S., 2004. Characterization of progenitor-cell-specific genes identified by subtractive suppression hybridization. Dev. Neurosci. 26, 131–147.

Campanelli, J.T., Sandrock, R.W., Wheatley, W., Xue, H., Zheng, J., Liang, F., Chesnut, J.D., Zhan, M., Rao, M.S., Liu, Y., 2008. Expression profiling of human glial precursors. BMC Dev. Biol. 8, 102.

Chambers, I., Colby, D., Robertson, M., Nichols, J., Lee, S., Tweedie, S., Smith, A., 2003. Functional expression cloning of Nanog, a pluripotency sustaining factor in embryonic stem cells. Cell 113, 643–655.

Chen, J., Li, Y., Yu, T.-S., McKay, R.M., Burns, D.K., Kernie, S.G., Parada, L.F., 2012. A restricted cell population propagates glioblastoma growth after chemotherapy. Nature 488, 522–526.

Daheron, L., Opitz, S.L., Zaehres, H., Lensch, W.M., Andrews, P.W., Itskovitz-Eldor, J., Daley, G.Q., 2004. LIF/STAT3 signaling fails to maintain self-renewal of human embryonic stem cells. Stem. Cells 22, 770–778.

Doetsch, F., Alvarez-Buylla, A., 1996. Network of tangential pathways for neuronal migration in adult mammalian brain. Proc. Natl. Acad. Sci. USA 93, 14895–14900.

Dybkaer, K., Zhou, G., Iqbal, J., Kelly, D., Xiao, L., Sherman, S., d'Amore, F., Chan, W.C., 2004. Suitability of stratagene reference RNA for analysis of lymphoid tissues. Biotechniques 37, 470–472. 474.

Elkabetz, Y., Panagiotakos, G., Al Shamy, G., Socci, N.D., Tabar, V., Studer, L., 2008. Human ES cell-derived neural rosettes reveal a functionally distinct early stem cell stage. Genes. Dev. 22, 152–165.

Elliott, S.T., Crider, D.G., Garnham, C.P., Boheler, K.R., Van Eyk, J.E., 2004. Two-dimensional gel electrophoresis database of murine R1 embryonic stem cells. Proteomics 4, 3813–3832.

ENCODE Project Consortium, 2012. An integrated encyclopedia of DNA elements in the human genome. Nature 489, 57–74.

Fortunel, N.O., Otu, H.H., Ng, H.H., Chen, J., Mu, X., Chevassut, T., Li, X., Joseph, M., Bailey, C., Hatzfeld, J.A., et al., 2003. Comment on "'Stemness': transcriptional profiling of embryonic and adult stem cells" and "a stem cell molecular signature". Science 302, 393. author reply 393.

Freeze, H., 2003. Mass spectrometry provides sweet inspiration. Nat. Biotechnol. 21, 627–629.

Geschwind, D.H., Ou, J., Easterday, M.C., Dougherty, J.D., Jackson, R.L., Chen, Z., Antoine, H., Terskikh, A., Weissman, I.L., Nelson, S.F., et al., 2001. A genetic analysis of neural progenitor differentiation. Neuron 29, 325–339.

Gifford, C.A., Ziller, M.J., Gu, H., Trapnell, C., Donaghey, J., Tsankov, A., Shalek, A.K., Kelley, R.R., Shishkin, A.A., Issner, R., et al., 2013. Transcriptional and epigenetic dynamics during specification of human embryonic stem cells. Cell 153, 1–15.

Ginis, I., Luo, Y., Miura, T., Thies, S., Brandenberger, R., Gerecht-Nir, S., Amit, M., Hoke, A., Carpenter, M.K., Itskovitz-Eldor, J., et al., 2004. Differences between human and mouse embryonic stem cells. Dev. Biol. 269, 360–380.

Hemmati, H.D., Nakano, I., Lazareff, J.A., Masterman-Smith, M., Geschwind, D.H., Bronner-Fraser, M., Kornblum, H.I., 2003. Cancerous stem cells can arise from pediatric brain tumors. Proc. Natl. Acad. Sci. USA 100, 15178–15183.

Houbaviy, H.B., Murray, M.F., Sharp, P.A., 2003. Embryonic stem cell-specific MicroRNAs. Dev. Cell 5, 351–358.

Huntly, B.J., Gilliland, D.G., 2005. Leukaemia stem cells and the evolution of cancer-stem-cell research. Nat. Rev. Cancer 5, 311–321.

Huntriss, J., Hinkins, M., Oliver, B., Harris, S.E., Beazley, J.C., Rutherford, A.J., Gosden, R.G., Lanzendorf, S.E., Picton, H.M., 2004. Expression of mRNAs for DNA methyltransferases and methyl-CpG-binding proteins in the human female germ line, preimplantation embryos, and embryonic stem cells. Mol. Reprod. Dev. 67, 323–336.

Husser, C.S., Buchhalter, J.R., Raffo, O.S., Shabo, A., Brown, S.H., Lee, K.E., Elkin, P.L., 2006. Standardization of microarray and pharmacogenomics data. Methods. Mol. Biol. 316, 111–157.

Imada, M., Sueoka, N., 1978. Clonal sublines of rat neurotumor RT4 and cell differentiation. I. Isolation and characterization of cell lines and cell type conversion. Dev. Biol. 66, 97–108.

Ivanova, N.B., Dimos, J.T., Schaniel, C., Hackney, J.A., Moore, K.A., Lemischka, I.R., 2002. A stem cell molecular signature. Science 298, 601–604.

Kalyani, A., Hobson, K., Rao, M.S., 1997. Neuroepithelial stem cells from the embryonic spinal cord: isolation, characterization, and clonal analysis. Dev. Biol. 186, 202–223.

Katsura, Y., Kawamoto, H., 2001. Stepwise lineage restriction of progenitors in lympho-myelopoiesis. Int. Rev. Immunol. 20, 1–20.

Kelly, T.K., Karsten, S.L., Geschwind, D.H., Kornblum, H.I., 2009. Cell lineage and regional identity of cultured spinal cord neural stem cells and comparison to brain-derived neural stem cells. PLoS One 4, e4213.

Kennea, N.L., Mehmet, H., 2002. Neural stem cells. J. Pathol. 197, 536–550.

Kim, S.K., Suh, M.R., Yoon, H.S., Lee, J.B., Oh, S.K., Moon, S.Y., Moon, S.H., Lee, J.Y., Hwang, J.H., Cho, W.J., et al., 2005. Identification of developmental pluripotency associated 5 expression in human pluripotent stem cells. Stem. Cells 23, 458–462.

Koch, P., Opitz, T., Steinbeck, J.A., Ladewig, J., Brustle, O., 2009. A rosette-type, self-renewing human ES cell-derived neural stem cell with potential for in vitro insturction and synaptic integration. Proc. Natl. Aca. Sci. USA 106, 3225–3230.

Kondo, T., Setoguchi, T., Taga, T., 2004. Persistence of a small subpopulation of cancer stem-like cells in the C6 glioma cell line. Proc. Natl. Acad. Sci. USA 101, 781–786.

Krichevsky, A.M., King, K.S., Donahue, C.P., Khrapko, K., Kosik, K.S., 2003. A microRNA array reveals extensive regulation of microRNAs during brain development. Rna 9, 1274–1281.

Lagarkova, M.A., Volchkov, P.Y., Lyakisheva, A.V., Philonenko, E.S., Kiselev, S.L., 2006. Diverse epigenetic profile of novel human embryonic stem cell lines. Cell Cycle 5, 416–420.

Lewis, B.P., Shih, I.H., Jones-Rhoades, M.W., Bartel, D.P., Burge, C.B., 2003. Prediction of mammalian microRNA targets. Cell 115, 787–798.

Li, H., Liu, Y., Shin, S., Sun, Y., Loring, J.F., Mattson, M.P., Rao, M.S., Zhan, M., 2006. Transcriptome coexpression map of human embryonic stem cells. BMC Genomics 7, 103.

Liu, Y., Han, S.S., Wu, Y., Tuohy, T.M., Xue, H., Cai, J., Back, S.A., Sherman, L.S., Fischer, I., Rao, M.S., 2004. CD44 expression identifies astrocyte-restricted precursor cells. Dev. Biol. 276, 31–46.

Lo, H.S., Wang, Z., Hu, Y., Yang, H.H., Gere, S., Buetow, K.H., Lee, M.P., 2003. Allelic variation in gene expression is common in the human genome. Genome. Res. 13, 1855–1862.

Loring, J.F., Rao, M.S., 2006. Establishing standards for the characterization of human embryonic stem cell lines. Stem. Cells 24, 145–150.

Loven, J., Orlando, D.A., Sigova, A.A., Lin, C.Y., Rahl, P.B., Burge, C.B., Levens, D.L., Lee, T.I., Young, R.A., 2012. Revisiting global gene expression analysis. Cell 151, 476–482.

Luo, Y., Cai, J., Liu, Y., Xue, H., Chrest, F.J., Wersto, R.P., Rao, M., 2002. Microarray analysis of selected genes in neural stem and progenitor cells. J. Neurochem. 83, 1481–1497.

Maitra, A., Arking, D.E., Shivapurkar, N., Ikeda, M., Stastny, V., Kassauei, K., Sui, G., Cutler, D.J., Liu, Y., Brimble, S.N., et al., 2005. Genomic alterations in cultured human embryonic stem cells. Nat. Genet. 37, 1099–1103.

Maitra, A., Cohen, Y., Gillespie, S.E., Mambo, E., Fukushima, N., Hoque, M.O., Shah, N., Goggins, M., Califano, J., Sidransky, D., et al., 2004. The Human MitoChip: a high-throughput sequencing microarray for mitochondrial mutation detection. Genome. Res. 14, 812–819.

Meehan, R.R., 2003. DNA methylation in animal development. Semin. Cell. Dev. Biol. 14, 53–65.

Mikkelsen, T.S., Ku, M., Jaffe, D.B., Issac, B., Lieberman, E., Giannoukos, G., Alvarez, P., Brockman, W., Kim, T.K., Roche, R.P., et al., 2007. Genome-wide maps of chromatin state in pluripotent and lineage-committed cells. Nature 448, 553–560.

Mitsui, K., Tokuzawa, Y., Itoh, H., Segawa, K., Murakami, M., Takahashi, K., Maruyama, M., Maeda, M., Yamanaka, S., 2003. The homeoprotein Nanog is required for maintenance of pluripotency in mouse epiblast and ES cells. Cell 113, 631–642.

Miura, T., Luo, Y., Khrebtukova, I., Brandenberger, R., Zhou, D., Thies, R.S., Vasicek, T., Young, H., Lebkowski, J., Carpenter, M.K., et al., 2004a. Monitoring early differentiation events in human embryonic stem cells by massively parallel signature sequencing and expressed sequence tag scan. Stem. Cells. Dev. 13, 694–715.

Miura, T., Mattson, M.P., Rao, M.S., 2004b. Cellular lifespan and senescence signaling in embryonic stem cells. Aging. Cell 3, 333–343.

Mujtaba, T., Piper, D.R., Kalyani, A., Groves, A.K., Lucero, M.T., Rao, M.S., 1999. Lineage-restricted neural precursors can be isolated from both the mouse neural tube and cultured ES cells. Dev. Biol. 214, 113–127.

Niwa, H., 2001. Molecular mechanism to maintain stem cell renewal of ES cells. Cell. Struct. Funct. 26, 137–148.

Noble, M., Gutowski, N., Bevan, K., Engel, U., Linskey, M., Urenjak, J., Bhakoo, K., Williams, S., 1995. From rodent glial precursor cell to human glial neoplasia in the oligodendrocyte-type-2 astrocyte lineage. Glia 15, 222–230.

Novoradovskaya, N., Whitfield, M.L., Basehore, L.S., Novoradovsky, A., Pesich, R., Usary, J., Karaca, M., Wong, W.K., Aprelikova, O., Fero, M., et al., 2004. Universal Reference RNA as a standard for microarray experiments. BMC Genomics. 5, 20.

Ohgane, J., Wakayama, T., Senda, S., Yamazaki, Y., Inoue, K., Ogura, A., Marh, J., Tanaka, S., Yanagimachi, R., Shiota, K., 2004. The Sall3 locus is an epigenetic hotspot of aberrant DNA methylation associated with placentomegaly of cloned mice. Genes. Cells 9, 253–260.

Ong, S.E., Foster, L.J., Mann, M., 2003. Mass spectrometric-based approaches in quantitative proteomics. Methods 29, 124–130.

Pevny, L., Rao, M.S., 2003. The stem-cell menagerie. Trends. Neurosci. 26, 351–359.

Quinn, S.M., Walters, W.M., Vescovi, A.L., Whittemore, S.R., 1999. Lineage restriction of neuroepithelial precursor cells from fetal human spinal cord. J. Neurosci. Res. 57, 590–602.

Rajan, P., Panchision, D.M., Newell, L.F., McKay, R.D., 2003. BMPs signal alternately through a SMAD or FRAP-STAT pathway to regulate fate choice in CNS stem cells. J. Cell. Biol. 161, 911–921.

Rajewsky, N., Socci, N.D., 2004. Computational identification of microRNA targets. Dev. Biol. 267, 529–535.

Ramalho-Santos, M., Yoon, S., Matsuzaki, Y., Mulligan, R.C., Melton, D.A., 2002. "Stemness": transcriptional profiling of embryonic and adult stem cells. Science 298, 597–600.

Ramos, A.D.., Diaz, A., Nellore, A., Delgado, R.N., Park, K.Y., Gonzales-Roybal, G., Oldham, M.C., Song, J.S., Lim, D.A., 2013. Integration of genome-wide approaches identifies lncRNAs of adult neural stem cells and their progeny in vivo. Cell Stem. Cell 12, 616–628.

Rao, M.S., 1999. Multipotent and restricted precursors in the central nervous system. Anat. Rec. 257, 137–148.

Reinhardt, P., Glatza, M., Hemmer, K., Tsytsyura, Y., Thiel, C.S., Hoing, S., Moritz, S., Parga, J.A., Wagner, L., Bruder, J.M., et al., 2013. Derivation and expansion using only small molecules of human neural progenitors for neurodegenerative disease modeling. PLoS One 8, e59252.

Reubinoff, B.E., Itsykson, P., Turetsky, T., Pera, M.F., Reinhartz, E., Itzik, A., Ben-Hur, T., 2001. Neural progenitors from human embryonic stem cells. Nat. Biotechnol. 19, 1134–1140.

Reya, T., Morrison, S.J., Clarke, M.F., Weissman, I.L., 2001. Stem cells, cancer, and cancer stem cells. Nature 414, 105–111.

Ross, D.T., Scherf, U., Eisen, M.B., Perou, C.M., Rees, C., Spellman, P., Iyer, V., Jeffrey, S.S., Van de Rijn, M., Waltham, M., et al., 2000. Systematic variation in gene expression patterns in human cancer cell lines. Nat. Genet. 24, 227–235.

Schwartz, C.M., Spivak, C.E., Baker, S.C., McDaniel, T.K., Loring, J.F., Nguyen, C., Chrest, F.J., Wersto, R., Arenas, E., Zeng, X., et al., 2005. NTera2: a model system to study dopaminergic differentiation of human embryonic stem cells. Stem. Cells Dev. 14, 517–534.

Shen, Q., Qian, X., Capela, A., Temple, S., 1998. Stem cells in the embryonic cerebral cortex: their role in histogenesis and patterning. J. Neurobiol. 36, 162–174.

Shin, S., Sun, Y., Liu, Y., Khaner, H., Svant, S., Cai, J., Xu, X.Q., Davidson, B.P., Stice, S.L., et al., 2007. Whole genome analysis of human neural stem cells derived from embryonic stem cells and stem and progenitor cells isolated from fetal tissue. Stem. Cells 25, 1298–1306.

Shin, S., Mitalipova, M., Noggle, S., Tibbitts, D., Venable, A., Rao, R., Stice, S.L., 2006. Long-term proliferation of human embryonic stem cell-derived neuroepithelial cells using defined adherent culture conditions. Stem. Cells 24, 125–138.

Shin, S., Sun, Y., Liu, Y., Khaner, H., Svant, S., Cai, J., Xu, Q.X., Davidson, B.P., Stice, S.L., Smith, A.K., 2007. Whole genome analysis of human neural stem cells derived from embryonic stem cells and stem and progenitor cells isolated from fetal tissue. Stem. Cells 25, 1298–1306.

Singh, S.K., Clarke, I.D., Hide, T., Dirks, P.B., 2004. Cancer stem cells in nervous system tumors. Oncogene 23, 7267–7273.

Snyder, E.Y., Deitcher, D.L., Walsh, C., Arnold-Aldea, S., Hartwieg, E.A., Cepko, C.L., 1992. Multipotent neural cell lines can engraft and participate in development of mouse cerebellum. Cell 68, 33–51.

Soodeen-Karamath, S., Gibbins, A.M., 2001. Apparent absence of oct 3/4 from the chicken genome. Mol. Reprod. Dev. 58, 137–148.

Steindler, D.A., 2002. Neural stem cells, scaffolds, and chaperones. Nat. Biotechnol. 20, 1091–1093.

Sultatos, L.G., Pastino, G.M., Rosenfeld, C.A., Flynn, E.J., 2004. Incorporation of the genetic control of alcohol dehydrogenase into a physiologically based pharmacokinetic model for ethanol in humans. Toxicol. Sci. 78, 20–31.

Sun, Y., Wang, Y., Hu, yY., Chen, G., Ma, H., 2011. Comparative analysis of neural transcriptomes and functional implication of unannotated intronic expression. BMC Genomics. 12, 494.

Svendsen, C., 2000. Adult versus embryonic stem cells: which is the way forward? Trends. Neurosci. 23, 450.

Szymanski, M., Barciszewski, J., 2003. Regulation by RNA. Int. Rev. Cytol. 231, 197–258.

Usuku, K., Joshi, N., Hauser, S.L., 1992. T-cell receptors: germline polymorphism and patterns of usage in demyelinating diseases. Crit. Rev. Immunol. 11, 381–393.

Vignon, X., Zhou, Q., Renard, J.P., 2002. Chromatin as a regulative architecture of the early developmental functions of mammalian embryos after fertilization or nuclear transfer. Cloning. Stem. Cells 4, 363–377.

Wechsler-Reya, R.J., Scott, M.P., 1999. Control of neuronal precursor proliferation in the cerebellum by Sonic Hedgehog. Neuron 22, 103–114.

Wei, C.L., Miura, T., Robson, P., Lim, S.K., Xu, X.Q., Lee, M.Y., Gupta, S., Stanton, L., Luo, Y., Schmitt, J., et al., 2005. Transcriptome profiling of human and murine ESCs identifies divergent paths required to maintain the stem cell state. Stem. Cells 23, 166–185.

Wilson, S.I., Edlund, T., 2001. Neural induction: toward a unifying mechanism. Nat. Neurosci. (Suppl. 4), 1161–1168.

Wright, L.S., Li, J., Caldwell, M.A., Wallace, K., Johnson, J.A., Svendsen, C.N., 2003. Gene expression in human neural stem cells: effects of leukemia inhibitory factor. J. Neurochem. 86, 179–195.

Yan, H., Dobbie, Z., Gruber, S.B., Markowitz, S., Romans, K., Giardiello, F.M., Kinzler, K.W., Vogelstein, B., 2002a. Small changes in expression affect predisposition to tumorigenesis. Nat. Genet. 30, 25–26.

Yan, H., Yuan, W., Velculescu, V.E., Vogelstein, B., Kinzler, K.W., 2002b. Allelic variation in human gene expression. Science 297, 1143.

Yoo, A.S., Staahl, B.T., Chen, L., Crabtree, G.R., 2009. MicroRNA-mediated switching of chromatin-remodeling complexes in neural development. Nature 469, 642046.

Zhang, S.C., Wernig, M., Duncan, I.D., Brustle, O., Thomson, J.A., 2001. In vitro differentiation of transplantable neural precursors from human embryonic stem cells. Nat. Biotechnol. 19, 1129–1133.

Chapter 7

Genomic and Evolutionary Insights into Chordate Origins

Shawn M. Luttrell and Billie J. Swalla

Biology Department, University of Washington, Seattle, WA and; Friday Harbor Laboratories, University of Washington, Friday Harbor, WA

Chapter Outline

GLOSSARY

Deuterostome Deuterostome literally means "second mouth" (deutero – two; stome – mouth). The blastopore is formed first during gastrulation and the mouth is formed secondarily. This mode of development applies to all deuterostomes except Tunicata. Echinodermata, Hemichordata, Xenoturbellida and Chordata are considered deuterostome phyla.

Endostyle This is an endodermal structure found in invertebrate chordates in the pharyngeal area. The endostyle secretes mucus to capture small particles and so increase the efficiency of filter feeding. In lancelets and tunicates, the endostyle also accumulates iodine and is considered homologous to the vertebrate thyroid gland.

Graptolites These abundant Cambrian fossils have been shown to be colonial hemichordates, or members of the hemichordate class Pterobranchia.

Hemichordates This phylum includes enteropneust worms and colonial pterobranchs. Hemichordates are tripartite as adults, having three body regions. The most anterior is the proboscis (protostome), then the collar (mesosome) and the posterior trunk (metasome). These three regions are reduced and modified in colonial Pterobranchia.

Lancelets The common name for cephalochordates. These animals are frequently referred to by the incorrect term amphioxus.

Notochord The key morphological character of chordates is the notochord. The notochord forms a stiff rod running from anterior to posterior in chordates beneath the dorsal neural tube, usually surrounded by a sheath of extracellular matrix. The gut is found just under the notochord in vertebrates and lancelets, but there is only an endodermal strand in the nonfeeding ascidian tadpole larvae. In lancelets and appendicularians, the notochord persists in the adult, while in ascidians the notochord undergoes apoptosis before metamorphosis and tail resorption. In vertebrates, the notochord persists as the intervertebral discs (nucleus pulposa) as the vertebrae develop from somites (Dahia et al., 2011; 2012).

Pharyngeal The area of the digestive system that serves as a respiratory and feeding organ in hemichordates, tunicates and lancelets. The vertebrate homolog is the pharynx, which develops into gills in early vertebrates, but is the area of the throat, including the thyroid gland and thymus in amniotes (birds and mammals).

Pharyngeal gill bars Cartilaginous elements made of extracellular matrix and located between the pharyngeal endoderm, giving structure to the pharynx of hemichordates, lancelets and vertebrates. Pharyngeal gill bars are secreted from endoderm in hemichordates and lancelets, but develop from neural crest cells in vertebrates.

Principles of Developmental Genetics.

Placodes An area of an ectodermal thickening from which cells delaminate and eventually achieve a cell fate that is not epidermal. There are both neurogenic and non-neurogenic cranial placodes, which are associated with the peripheral nervous system in vertebrates. Placodes were thought to be found only in vertebrates, but have recently been described in tunicates, using both molecular markers and careful morphological analyses.

Pterobranch Class Pterobranchia refers to a group of colonial hemichordates, or pterobranchs. Colonial hemichordates reproduce both sexually and asexually and have feeding tentacles to capture small particles for feeding. There are many fossil pterobranchs, called graptolites, but only two extant families, Rhabdopleuridae and Cephalodiscidae.

Stomochord The stomochord is a projection of the endoderm that juts forward into the hemichordate proboscis, against which the hemichordate heart beats. The hemichordate stomochord cells are vacuolated and make an extracellular sheath, cited as morphological evidence for the stomochord being a notochord homolog.

Tunicates A monophyletic group of animals that includes ascidians, appendicularians, and thaliaceans. This group of animals is also sometimes called "Urochordata," but Tunicata is the preferred term. There are over 2,800 described species of tunicates.

SUMMARY

- Hemichordates are the sister group to echinoderms, not chordates. Phylogenomics suggest that the tunicates are the sister group to vertebrates, while lancelets are more distantly related to vertebrates.
- *Hox* genes are expressed in an anterior to posterior manner in hemichordates and chordates. Tunicates have lost the middle *Hox* genes and show rather different tissue expression patterns. Echinoderms have a rearranged *Hox* cluster and show limited co-linearity of expression in the somatapleura.
- Pharyngeal gill slits in hemichordates and chordates are homologous. Pharyngeal gill bars develop similarly in hemichordates and lancelets, but differ from vertebrates in that they are acellular and endodermal in origin.
- Post-anal tails and endostyles in hemichordates and chordates are likely to be homologous.
- Chordates specify neural and non-neural ectoderm, while all ectoderm is neural in hemichordates. Hemichordates make a dorsal neural tube in their neck region.
- Recent gene expression data suggest the stomochord may be the notochord homolog in hemichordates.
- Tunicates contain migratory cells that develop into pigment cells, so may contain neural crest cells. Tunicates also have sensory placodes, and form a secondary anus after metamorphosis, so the excurrent siphon develops from the otic placode.

7.1 INTRODUCTION

History of Hypotheses of Chordate Origins

Vertebrates share several distinct morphological characters with three invertebrate groups: hemichordates (Figure 7.1A), tunicates (Figure 7.1B), and lancelets (Figure 7.1C). Tunicates, lancelets and vertebrates are a monophyletic group, the chordates (Swalla and Smith, 2008), which share five morphological features: a notochord, a dorsal neural tube, an endostyle, a muscular post-anal tail, and pharyngeal gill slits (Swalla, 2007). Hemichordates share some of these chordate features; the pharyngeal gill slits (Figure 7.1), an endostyle, dorsal neural tube, and a post-anal tail (Swalla, 2007; Brown et al., 2008). Developmental genetics and genomics have allowed a re-examination of the question of chordate origins by

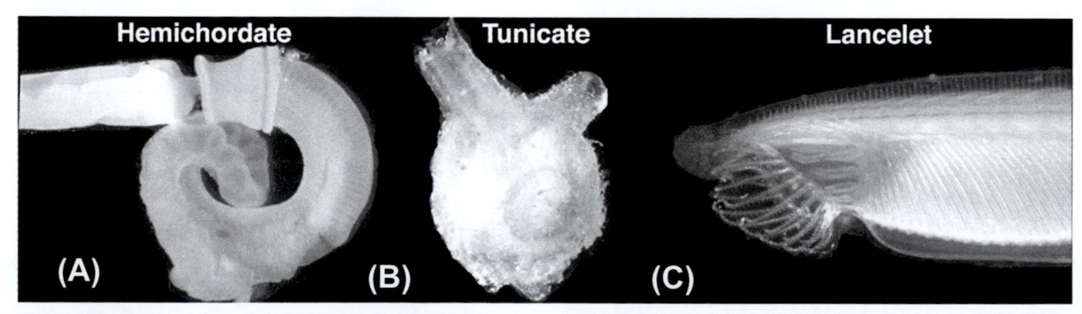

FIGURE 7.1 **Three Distinct Invertebrate Body Plans.** (A) An adult enteropneust hemichordate, *Saccoglossus kowalevskii*, (B) a tunicate molgulid ascidian, *Molgula provisionalis*, and (C) a lancelet, *Branchistoma viriginae*, showing the dramatic differences in their adult body plans. (A) The mouth of the enteropneust hemichordate is hidden in the collar region, directly behind the anterior proboscis, on the ventral side. Central nervous system and gill slits are dorsal, but the mouth is ventral. (B) The mouth of the tunicate is moved upwards at metamorphosis, shown here as the siphon to the top left, while the anus empties into the buccal siphon, shown to the top right. (C) The lancelet mouth has been modified for filter feeding, as shown by the cirri at the anterior, ventral side to the left. Central nervous system is dorsal, gill slits and mouth are ventral. All animals were collected and photographed at SMS at Fort Pierce, FL.

comparing developmental gene expression in embryos of different phyla. This powerful approach has allowed new insights into the molecular mechanisms underlying morphological changes. However, as we illuminate here, a comparison of the correct developmental stages is critical to interpreting the data. Genomics has allowed investigations into the phylogenetic relationships of the chordates and their invertebrate relatives, and comparisons of the shared genetic pathways in related embryos. We review current research and show that our view of the chordate ancestor has changed in the past 15 years. For many years, the chordate ancestor had been considered to be a filter feeding, tunicate-like animal with a tiny chordate tadpole larva (Figure 7.2). However, recent evidence from our lab and others has shown that the chordate ancestor is more likely a benthic worm, with a mouth and pharyngeal gill slits, supported by cartilaginous gill bars (Cameron et al., 2000; Gerhart et al., 2005; Rychel et al., 2006; Gillis et al., 2012). Further research in developmental genetics and genomics will be fruitful in resolving some of the remaining homologies between hemichordates and chordates.

Three disparate hypotheses of chordate origins are shown in Figure 7.2 (Garstang, 1928; Romer, 1967; Jeffries, 1986; Jollie, 1973; Gee, 1996; Gerhart et al., 2005; Brown et al., 2008). One early scenario of chordate origins, which was quite popular, is the Garstang view of chordate origins, first hypothesized near the turn of the 20th century (Figure 7.2; Garstang, 1928). This theory espouses the notion that the echinoderm and hemichordate larvae "evolved" into the ascidian tadpole larvae, and the adults of echinoderms, hemichordates and tunicates developed independently. However, developmental gene expression data show that hemichordates and chordates share adult features, not larval ones (Swalla, 2006; Brown et al., 2008). These results would favor the evolution of chordates from a direct developing hemichordate (Gerhart et al., 2005; Rychel et al., 2006), not from a hemichordate tornaria larva. Further comments on Garstang's theory, and the genetic evidence against it, are nicely summarized in Lacalli (2005).

Some textbooks still teach the scenario that all deuterostomes evolved from a colonial hemichordate, a pterobranch, as first published by Romer in 1967 (Figure 7.2). This theory was popularized because the fossil record has an abundance of colonial hemichordates, called graptolites, and because lophophorates were considered related to deuterostomes (Romer, 1967; Gee, 1996). Molecular phylogenetics has shown that the lophophorates are part of a large group of animals called the lophotrochozoa (Halanych, 2004), and that the colonial pterobranchs are derived hemichordates (Cameron et al., 2000). Collectively, these new data brought into question the widely held view of deuterostome evolution popularized by Romer.

We published a new hypothesis based on our molecular phylogenies and developmental gene expression patterns in 2000 (Figure 7.3; Cameron et al., 2000). In this scenario, the deuterostome ancestor is worm-like with gill slits, and the larvae of hemichordates and echinoderms developed independent of ascidian tadpole larvae. In the ensuing years, developmental gene expression data have continued to favor a worm-like deuterostome ancestor (Lowe et al., 2003; Gerhart et al., 2005; Rychel et al., 2006; Delsuc et al., 2006; Bourlat et al., 2006). Developmental genomics and genetics can provide key pieces of evidence for understanding chordate origins. Genomic information is available for at least a single member of each of the deuterostome monophyletic groups – echinoderms (the purple sea urchin *Stronglyocentrotus purpuratus*; Sodergren et al., 2006), hemichordates (an acorn worm, *Saccoglossus kowalevskii*; Freeman et al., 2008), tunicates (solitary

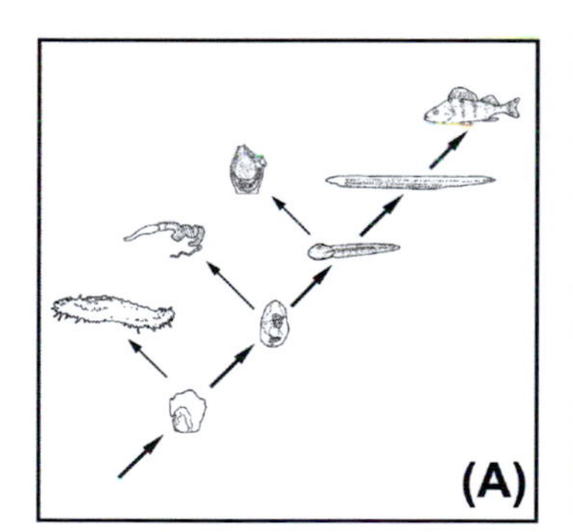

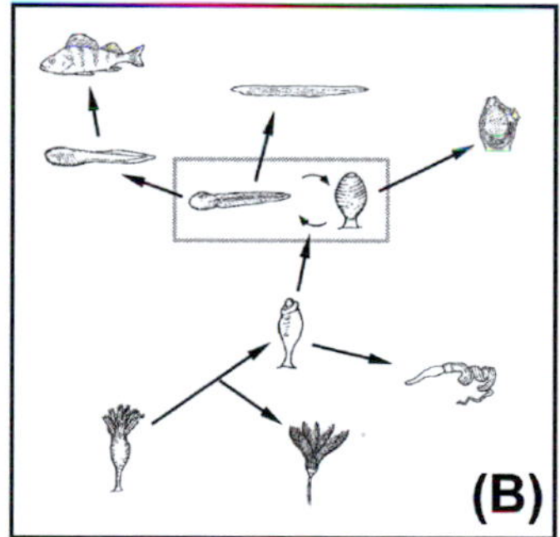

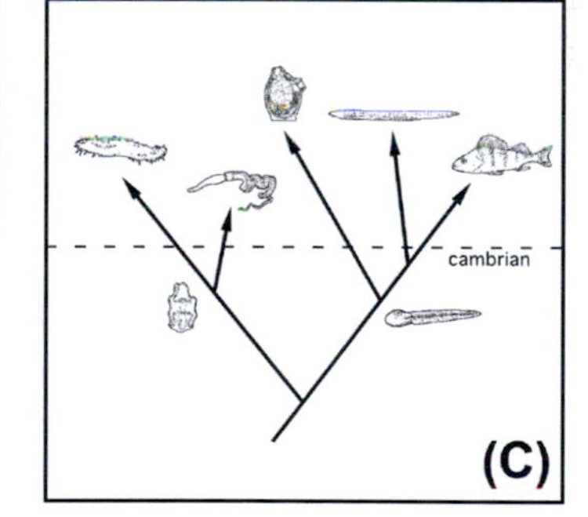

FIGURE 7.2 **Theories of Chordate Origins: Which Ones Fit the Available Data?** Several possible theories of chordate origins are depicted here. (A) Theory 1 was first proposed by Garstang (1928), and espouses the notion that the nonfeeding ascidian tadpole larva evolved from echinoderm-like and hemichordate-like larvae. However, developmental gene expression patterns in the different larvae show echinoderm larvae and hemichordate tornaria larvae are very similar, but both differ markedly from chordate expression patterns (Swalla, 2006). (B) Theory 2 was popularized by Romer (1967) and depicts the deuterostomes evolving from a pterobranch hemichordate. Phylogenetic and fossil evidence suggest that this is an unlikely scenario, since chordate, hemichordates and echinoderms all appear in the early Cambrian. (C) Theory 3 is a compilation of all available phylogenetic, fossil and gene expression data, first published in 2000 (Cameron et al., 2000). Parts of this theory were put first put forth by Jollie (1973) who considered Garstang's (1) and Romer's (2) theories extremely unlikely. The molecular evidence suggests that the chordate tadpole larva had an independent origin from an ancestor with a feeding dipleurulid larva. In this scenario, the chordate body plan would have evolved *de novo* in a direct-developing soft-bodied worm-like ancestor. The notochord would have evolved in the chordates from the co-option of genes used for other functions in the ancestor, or the notochord has been lost in hemichordates and echinoderms. We consider either scenario likely and are continuing to investigate.

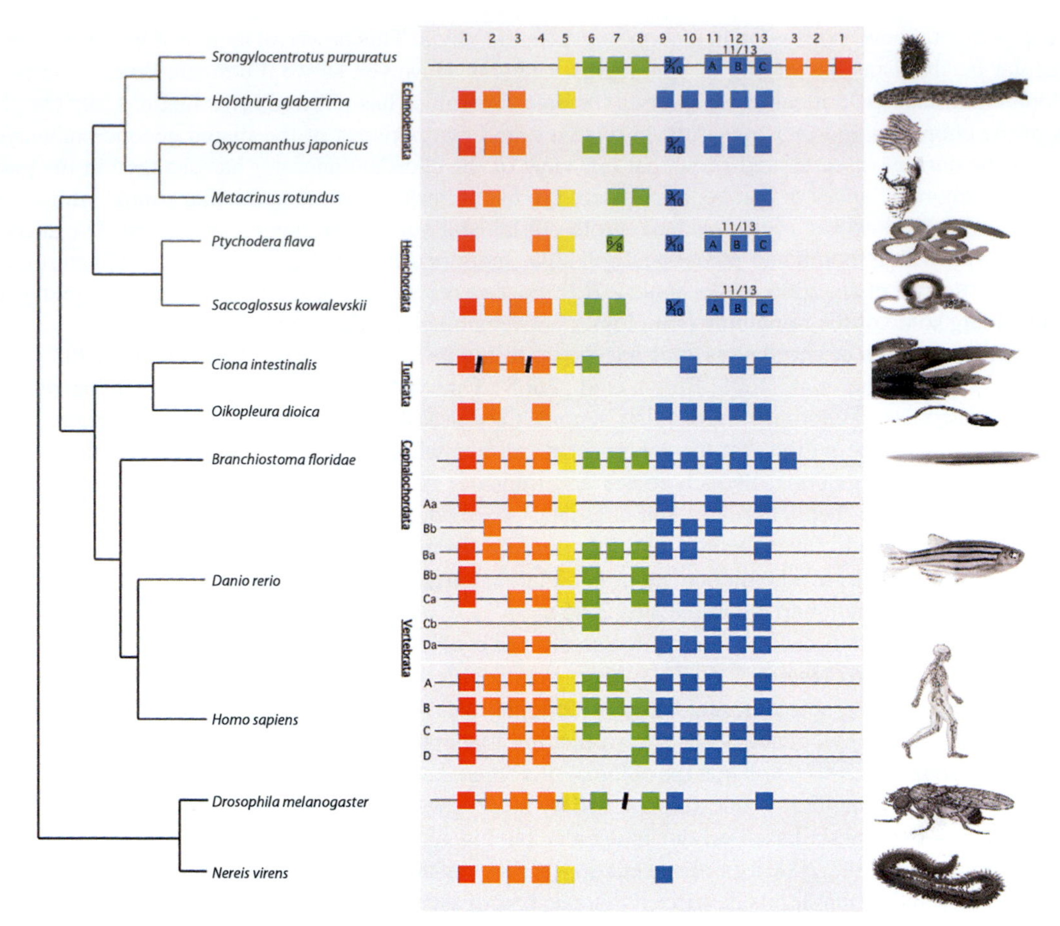

FIGURE 7.3 ***Hox* Gene Expression in the Deuterostomes.** An Ecdysozoa (fruit fly, *Drosophila melanogaster*) and a Lophotrochozoa (polychaete annelid, *Nereis virens*) are shown as outgroup protostome taxa. *Hox* clusters are shown here with all available hemichordate data and major chordate clades in 2008 (Brown et al., 2008). The two sequenced tunicates, *Oikopleura dioica* and *Ciona intestinalis* are shown with a cephalochordate, *Branchistoma floridae* and a couple of representative vertebrates. Humans have four duplicate clusters, as do all vertebrates, except teleost fishes, (*Danio rerio*), that had a second gene duplication event, resulting in eight clusters. Lines through the boxes depict that the genome has been completed, and the organization of the cluster is known. Slashes indicate genomic breaks in the clusters. Note the unusual inversion seen in sea urchins (top row). *References for figure updated from an earlier review (Brown et al., 2008; Aronowicz and Lowe, 2006; Cameron et al., 2006; Méndez et al., 2000; Mito and Endo, 2000; Passamaneck and Di Gregario, 2005; Peterson, 2004; Swalla, 2006).*

ascidians *Ciona intestinalis*, Dehal et al., 2002; *C. savignyi*, Vinson et al., 2005), the pelagic appendicularian *Oikopleura dioica* (Denoeud et al., 2010), cephalochordates (the lancelet *Branchistoma floridae*; Putnam et al., 2008), and many vertebrate species (Fong et al., 2013). Deuterostomes share many developmental genetic networks during early embryonic and larval development (Davidson and Erwin, 2006; Swalla, 2006). Developmental genetics can be highly informative by illuminating how these similar genetic pathways are expressed in different times and places in the process of elaborating the final morphology of the larvae and the adults (Swalla, 2006). We next review the latest findings of molecular phylogenetic and genomic analyses, examine developmental gene expression in different deuterostome phyla, and discuss the origin of the vertebrates in light of new data published in the past 15 years.

Molecular Phylogenetics of Deuterostomes

Phylogenetic relationships within the deuterostomes are critical to understanding the evolutionary changes that have occurred during chordate and vertebrate evolution (Figure 7.3; Swalla and Smith, 2008). Deuterostome phylogenetic relationships have been reviewed extensively elsewhere (Swalla and Smith, 2008), so they will be briefly summarized here. Schaeffer (1987) examined morphological and phylogenetic evidence and concluded that the deuterostomes, the group of animals that contains the vertebrates, were monophyletic. Later, in 1994, two papers examined deuterostome relationships using 18S rDNA for the first time. Wada and Satoh (1994) showed that deuterostomes were monophyletic, presented evidence that chaetognaths were not deuterostomes, and showed echinoderms and hemichordates to be sister groups, albeit with low bootstrap support. Turbeville et al., (1994) increased the deuterostome 18S rDNA data set and used the notochord as a morphological marker to place ascidians as chordates. Later, Cameron et al. (2000) greatly increased the number of tunicates and hemichordates in the deuterostome 18S rDNA database, and showed that echinoderms and hemichordates are sister groups with high bootstrap support. Morphological and molecular data since that time have continued to confirm the sister group relationship of echinoderms and hemichordates (Halanych, 2004; Smith et al., 2004; Bourlat et al., 2006; Swalla and Smith, 2008).

The tunicates, while monophyletic (Swalla et al., 2000; Tsagkogeorga et al., 2009), are difficult to place within the deuterostomes with 18S and 28S combined ribosomal sequence analysis (Swalla and Smith, 2008). Recent genome phylogenies, constructed using hundreds of genes, have suggested that tunicates are more closely related to vertebrates than lancelets are (Blair and Hedges, 2005; Bourlat et al., 2006; Delsuc et al., 2006). Many of the developmental programs and gene networks that are activated in ascidian embryos for specific tissues are quite similar to those activated during vertebrate development (Swalla, 2004; Passamaneck and Di Gregorio, 2005). Ascidians have a number of important transcription factors localized in the egg cytoplasm that are necessary for some tissue development, and thus have been described as having mosaic development (Nishida, 2005). In addition, ascidians have some unique features of tissue specification, such as cellulose production by the adult ectoderm, that are not found in vertebrates (Matthysse et al., 2004; Nakashima et al., 2008). These unique characteristics of ascidian development are thoroughly reviewed in Passamaneck and Di Gregorio (2005). Examination of the timing and spatial expression of homologous genes during development can be informative in understanding which morphological structures are homologous in animals with very different body plans (Gerhart et al., 2005; Rychel et al., 2006; Swalla, 2006). In the following sections, the expression of homologous genes in different deuterostome groups is discussed in the context of what these results tell us about the evolution of the vertebrates.

7.2 *HOX* GENE CLUSTER ORGANIZATION AND EXPRESSION IN DEUTEROSTOMES: ANTERIOR-POSTERIOR AXIS DEVELOPMENT

The *Hox* gene complex has shed light on both deuterostome relationships and the anterior-posterior homologies between body plans of the different phyla (Figure 7.3) (Aronowicz and Lowe, 2006; Swalla, 2006; Brown et al., 2008; Freeman et al., 2012; Pascual-Anaya et al., 2013). The *Hox* gene complex is a group of genes that are arranged from 3' to 5' co-linearly on the chromosome, and are also expressed from anterior to posterior during embryonic development. Invertebrate deuterostomes have a single *Hox* cluster, while vertebrates have multiple copies (Swalla, 2006). The sea urchin *Hox* cluster has been mapped and has undergone an inversion so that the most posterior gene, *Hox* 11/13c is next to *Hox* 3 (Cameron et al., 2006). Hemichordates and sea urchins share motifs in their three posterior *Hox* genes, called *Hox* 11/13a, 11/13b, and 11/13c, suggesting that these two groups had posterior gene duplications independent of the chordate lineage (Peterson, 2004; Cameron et al., 2006; Freeman et al., 2012; Pascual-Anaya et al., 2013).

Hox developmental gene expression provides evidence for anterior-posterior homologies between echinoderms, hemichordates, vertebrates, and lancelets (Lowe et al., 2003; Swalla, 2006; Pascual-Anaya et al., 2013). Developing hemichordates express their *Hox* genes in a co-linear fashion from anterior to posterior, but instead of expression only in the dorsal neural tube, expression is seen in the entire ectoderm of hemichordates (Lowe et al., 2003). These expression patterns reflect the fact that the hemichordate ectoderm has neural potential throughout the ectoderm, such as is seen in insects (Lowe et al., 2003). In echinoderms, co-linear expression has been reported in the developing adult somatapleura (Cameron et al., 2006), and in the adult nerve ring (Morris and Byrne, 2005). In contrast, tunicates show widely differing expression patterns of *Hox* genes, depending on whether the gene is expressed during the larval or adult stage (Spagnuolo et al., 2003; Passamaneck and Di Gregorio, 2005; Swalla, 2006).

7.3 PHARYNGEAL GILLS AND GILL BAR DEVELOPMENT

Pax 1/9 Expression and *Hox* Expression in Deuterostome Gill Slits

Pharyngeal gill slits in hemichordates were originally used as a morphological character that united the hemichordate enteropneust worms with chordates (Figure 7.1; Romer, 1967; Schaeffer, 1987; Rychel et al., 2006; Gillis et al., 2012). Structures are considered to be homologous if they have similar morphology and similar function. The clear homology of pharyngeal gill slit structures is what causes the hemichordates to fall between the echinoderms and chordates by morphological analysis (Figure 7.2 image 1; Schaeffer, 1987). The pharyngeal clefts and surrounding collagen skeleton of hemichordates, cephalochordates, and vertebrates are remarkably similar in form and function (Schaeffer, 1987; Rychel et al., 2006), making it most likely that this is the ancestral morphology. In contrast, tunicates lack any cartilage skeleton in their pharyngeal structures, suggesting that this has been lost evolutionarily (Figure 7.1). Developmental genetics allows the comparison of morphological structures at a new level – the level of genetic pathways expressed during the development of the structure. Recent work has shown that the pharyngeal slits in vertebrates, lancelets, and tunicates are elaborated after the expression of specific *Pax* genes. The single gene called *Pax 1/9* in invertebrate deuterostomes has been duplicated in vertebrates to form two genes; *Pax-1* and *Pax-9* (Neubüser et al., 1995; Holland and Holland, 1995; Ogasawara et al., 1999; Ogasawara et al., 2000a).

Expression of the paired box transcription factors *Pax-1* and *Pax-9* has been shown in endodermal pharyngeal pouches during vertebrate development (Neubüser et al., 1995; Wallin et al., 1996; Peters et al., 1998; Ogasawara et al., 2000a). Furthermore, these transcription factors are necessary for the proper development of the pharyngeal pouches and surrounding endodermal derivatives, such as the thymus, as seen by their absence in mice lacking either *Pax 1* or *Pax 9* (Wallin et al., 1996; Peters et al., 1998). In both chordates and hemichordates, *Pax 1/9* is expressed in endoderm of the pharynx and later in the pharyngeal slits. Notably, in ascidians, no expression was detected during embryogenesis. The first sign of *Pax 1/9* expression was in swimming tadpole larvae that were about to begin metamorphosis (Ogasawara et al., 1999). Likewise, expression in hemichordate adults was found to be highest in the gill endoderm (Ogasawara et al., 1999; Gillis et al., 2012). Not only was *Pax 1/9* expressed, but the downstream genes *eya* and *six 1* were also expressed (Gillis et al., 2012). These results suggest that the morphological and functional similarity between the pharyngeal gill slits in hemichordates (Ogasawara et al., 1999; Gillis et al., 2012), ascidians (Ogasawara et al., 1999), cephalochordates (Holland and Holland, 1995), and vertebrates (Neubüser et al., 1995; Wallin et al., 1996; Peters et al., 1998; Ogasawara et al., 2000a) is a reflection of similar genetic programs activated in pharyngeal endoderm at the time of differentiation by the *Pax 1/9* or *Pax 1* and *Pax 9* transcription factors. In the light of these results and the deuterostome phylogeny, the most parsimonious hypothesis is that the deuterostome ancestor had endodermally-derived gill slits and these were subsequently lost in the echinoderm lineage (Figure 7.2C). The mitrate carpoids, echinoderms from the Devonian era, do appear to have gill slits (Jeffries, 1986; Gee, 1996; Smith et al., 2004). Therefore, early echinoderms may have had pharyngeal gills and then lost them (Smith et al., 2004; Rychel et al., 2006; Swalla and Smith, 2008). Further examination of Cambrian echinoderms for evidence of pharyngeal gills will be informative, as will the cloning and characterization of the expression of *Pax 1/9* in echinoderms. No expression data for *Pax 1/9* have been reported in echinoderms to date, but it will be interesting to see whether this gene has expression reminiscent of gill slits or has been co-opted for other functions in echinoderms.

Not only do hemichordate pharyngeal gill slits share conserved transcription factors for their development as described above with vertebrates, they also share localization along the anterior-posterior axis. For example, in vertebrates, *Hox 1* is first expressed at the level between the first and second pharyngeal pouch (Lowe et al., 2003). When *Hox* gene expression was examined in hemichordates, *Hox 1* was expressed between the first and second pharyngeal pouch, suggesting that the location of the pharyngeal gills along the anterior-posterior axis is homologous between hemichordates, lancelets, and vertebrates (Lowe et al., 2003; Freeman et al., 2012).

Pharyngeal Gill Cartilage in Hemichordates and Lancelets is Acellular

The pharyngeal gill slits themselves are homologous between hemichordates and chordates, but what about the cartilaginous gill bars that lie between the gill openings? The morphology and development of the gill bars in hemichordates is similar to lancelets (Figure 7.1) (Schaeffer, 1987; Ruppert, 2005; Rychel et al., 2006). The bars develop as a thickening of the basal lamina between the pharyngeal endoderm, as first reported by Libbie Hyman (1959), but also recently shown by *in situ* hybridization of *Sox* and fibrillar collagen (Rychel and Swalla, 2007). The cartilaginous bars of hemichordates stain with alcian blue (Smith et al., 2003) and are acellular (Rychel et al., 2007), while the gill bars of lampreys are made by neural crest cells and are cellular (Zhang et al., 2006). The development of gill bars in hemichordates and lancelets needs to be examined in more detail, but it appears that the ancestral cartilage may have been acellular and was secreted by the endoderm (Rychel and Swalla, 2007). Later in evolution, neural crest cells in vertebrates may have migrated into those areas and replaced the acellular cartilage with cellular cartilage. More information on the development of gill bar cartilage is needed before any questions of homology between hemichordate and lancelet gill bars can be answered.

7.4 THE POST-ANAL TAIL AND THE ENDOSTYLE OF HEMICHORDATES: GENE EXPRESSION STUDIES

It is not clear how significant the post-anal tail is as a defining chordate feature. Ascidians do not have an open gut as larva, and thus do not have an anus, but both lancelets and vertebrates have a post-anal tail (Gerhart et al., 2005). The vertebrate and lancelet posterior *Hox* genes are expressed in the tissues of the post-anal tail. Phylogenetic analysis of hemichordate enteropneust worms show that they fall into two separate monophyletic groups; those that have feeding larvae similar to echinoderms and those that are direct developers (Cameron et al., 2000). The direct developing saccoglossids have post-anal tails that express the posterior *Hox* genes (Lowe et al., 2003; Freeman et al., 2012), whereas the ptychoderids lack a larval post-anal tail (Swalla, 2006). Instead, ptychoderid worms form an anus at the vegetal pole of the larvae which becomes the anus of the adult (Urata and Yamaguchi, 2004; Swalla, 2006; Röttinger and Lowe, 2012). These results could be interpreted as evidence that the vertebrates evolved from a direct developing hemichordate ancestor, as they are the only group of hemichordates that show a post-anal tail. However, since the hemichordates would have diverged from a chordate ancestor long before the Cambrian era (Blair and Hedges, 2005), there has been plenty of time for the independent evolution of a post-anal tail in both groups.

The endostyle present in lancelets and tunicates is thought to have homology with the vertebrate thyroid. For this reason, endostyle-specific genes have been isolated in an effort to examine this question using gene expression (Mazet, 2002; Ogasawara et al., 2000b; Sasaki et al., 2003). One of these genes is the homeobox gene *TTF-1* (thyroid transcription factor 1), which regulates thyroid peroxidase, the enzyme that iodinates thyroglobulin (Mazet, 2002; Ogasawara et al., 2000b; Sasaki et al., 2003). In lancelets, *TTF-1* is expressed throughout the six morphological zones of the endostyle (Mazet, 2002), whereas in tunicates expression is limited to particular zones (Sasaki et al., 2003). Both the tunicate and lancelet endostyles also bind iodine, so their endostyles are considered to be homologous to the vertebrate thyroid gland (Sasaki et al., 2003; Ruppert, 2005). When the hemichordate *TTF-1* was cloned and its gene expression was characterized, there was expression seen in the pharyngeal endoderm, stomochord and hindgut (Takacs et al., 2002). The pharyngeal endoderm of hemichordates also binds iodine throughout, even in the regions that do not morphologically resemble an endostyle (Ruppert, 2005). These results could indicate that the entire hemichordate pharynx fulfills an endostyle function (Rychel et al., 2006), or that the hemichordate endostyle is not homologous with the tunicate and lancelet endostyle (Ruppert, 2005). Further developmental and functional studies will be necessary to distinguish between these two hypotheses.

7.5 THE CENTRAL NERVOUS SYSTEM AND THE DORSAL-VENTRAL INVERSION HYPOTHESIS

Dorsal-ventral axis specification relies on some of the same signaling molecules in both vertebrates and invertebrates, however, there are differences in either the location or timing of expression of these molecules. Currently, there are two main hypotheses that explain these differences. In the first hypothesis, the signaling molecules directing axis specification are the same, but there is an inversion of expression localities; the Dorsal-Ventral Inversion hypothesis (Ruppert et al., 1999; Lowe et al., 2006). Bone morphogenetic proteins (BMPs) are expressed on the dorsal side of invertebrates, whereas chordate BMPs are expressed on the ventral side of the embryo (Lowe et al., 2006). However, later in vertebrate development, BMP expression is seen in the dorsal midline, following neural tube closure (Figure 7.4A). Our interpretation is that the dorsal midline BMP expression seen in hemichordates (Figure 7.4B) is identical to the vertebrate midline BMP expression, not the earlier vertebrate ventral expression (Figure 7.4A). Evidence supporting axis inversion in chordates is seen in *nodal* expression analysis. *Nodal* is a gene involved in left-right axis symmetry in chordates, and the gene is expressed on the left side of the chordate embryo. Conversely, *nodal* is expressed on the right side of sea urchins (Swalla and Smith, 2008). Collectively, this evidence supports the hypothesis that BMP expression was first expanded on the ventral side of a worm-like

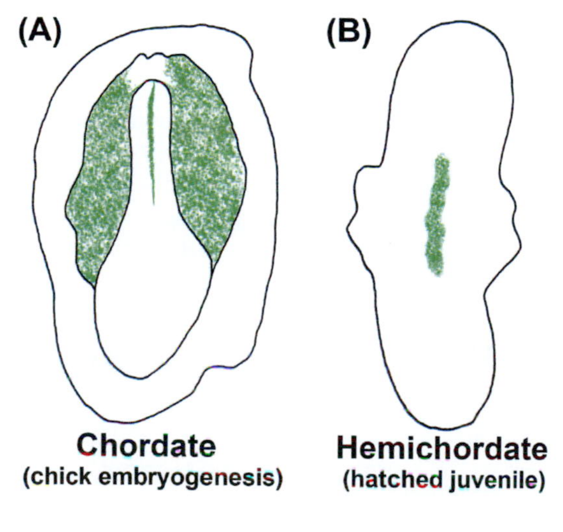

FIGURE 7.4 ***BMP*** **Expressed in Dorsal Midline During Early Development.** Illustrations showing the expression of *BMP* in green. (A) Dorsal view of chick embryogenesis after neurulation showing *BMP* expression in the putative ventral region and one dorsal strip in the midline of the neural tube. (B) Dorsal view of a juvenile hemichordate showing *BMP* expressed in the dorsal midline, just after neurulation. *Drawings made according to published photographs of* BMP *expression in the chordate chick (Martí, 2000) and hemichordate (Lowe et al., 2006).*

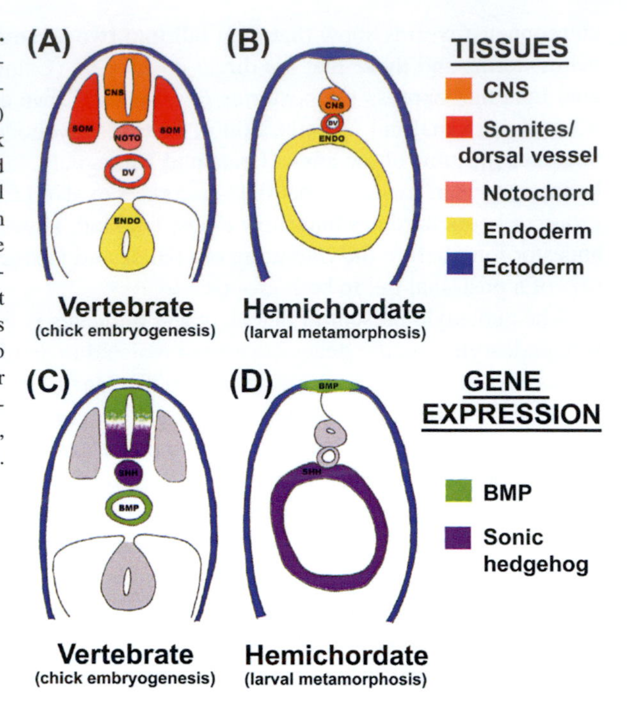

FIGURE 7.5 ***BMP* and *Sonic hedgehog* Expression During Neurulation.** Diagrams of cross-sections of a hemichordate and a vertebrate. In all pictures, dorsal is to the top and ventral is to the bottom. The upper two illustrations show specific tissues and structures, while the bottom two illustrations show *BMP* (green) and *Sonic hedgehog* (purple) expression. In the diagram of the developing chick embryo (A), the vertebrate neural tube (orange) develops above the notochord (pink). Somites (red) surround the neural tube and notochord. The dorsal vessel (red) is ventral to the notochord and dorsal to the endoderm (yellow). In diagram (B), the metamorphosing hemichordate neural tube (orange) develops above the dorsal vessel (red). The endoderm (yellow) is directly ventral to the dorsal vessel. Blue represents ectoderm in all images. The bottom two diagrams, show that *BMP* (green) is expressed in the dorsal ectoderm (blue) of both hemichordates (Lowe et al., 2006) and the chordate chick (Martí, 2000). *BMP* molecules are also expressed in the dorsal vessel (DV) of vertebrates (Rickman et al., 1985; Schneider et al., 1999; Young et al., 2004). *Sonic hedgehog* is expressed in the ventral central nervous system (CNS) and the notochord (NOTO) in vertebrates (Bardet et al., 2010) and expressed in the endoderm (ENDO) of hemichordates (Pani et al, 2012).

chordate ancestor. These changes in the chordate ancestor resulted in an animal with inverted dorsal-ventral and left-right axis symmetries compared to other invertebrate deuterostomes. This hypothesis assumes that the chordate ancestor did not possess a central nervous system, but merely a nerve net.

An alternative hypothesis to the inversion claim is supported by *BMP* expression studies and outlined in Figure 7.4. This hypothesis suggests that *BMP* expression is not inverted, but is exactly the same in hemichordates and chordates, in the dorsal midline, following neurulation. The timing of expression differs among developmental stages within and between chordates and hemichordates (Martí, 2000; Lowe et al., 2006). During chick embryogenesis, expression of *BMP* is found in a strip down the dorsal midline, just after neurulation (Figure 7.4; Martí, 2000). Juvenile hemichordates also express *BMP* in a strip down the dorsal midline (Figure 7.4; Lowe et al., 2006). The similar expression of *BMP* in a dorsal strip after neurulation appears to have a role in directing neural tube differentiation in both hemichordates and chordates. This hypothesis supports that the dorsal neural tube in hemichordates is homologous to the neural tube in chordates (Luttrell et al., 2012; Miyamoto and Wada, 2013).

The nervous system of larval ptychoderid hemichordates develops in a similar manner to chordates (Figure 7.5; Luttrell et al., 2012; Miyamoto and Wada, 2013). In both groups, the neural tube develops along the dorsal midline in an anterior to posterior fashion (Figure 7.4, dorsal views and Figure 7.5, cross-sections), but in hemichordates, the neural tube is found only in the collar region (Figure 7.4B; Kaul and Stach, 2010; Luttrell et al., 2012). The chordate central nervous system (CNS) forms by invaginating ectoderm that rolls up the neural tube dorsal to the notochord (Figure 7.5A). In hemichordates, collar ectoderm invaginates and rolls up to form the neural tube, after the development of the dorsal vessel (Figure 7.5B; Luttrell et al., 2012). Signaling molecules emanate from the notochord in chordates to induce neural tube formation (Figure 7.5C). The notochord also serves as structural support for the neural tube (Figure 7.5A). The notochord is dorsal to the dorsal vessel in chordates (Figure 7.5; Luttrell et al., 2012; Miyamoto and Wada, 2013). Hemichordates lack a notochord, so signaling molecules may emanate from the dorsal vessel and the endoderm to induce neural tube formation (Figure 7.5D; Miyamoto and Wada, 2013). These results have always been interpreted as the chordates gaining a notochord, but evolutionarily, it is equally likely that the hemichordate ancestor lost its notochord. In the latter scenario, the deuterostome ancestor may actually have been a chordate, with major losses in the echinoderm and tunicate lineages.

7.6 EVIDENCE FOR THE HEMICHORDATE STOMOCHORD HOMOLOGY TO CHORDATE NOTOCHORD

Ultrastructural studies of the hemichordate stomochord suggested that this structure could be the homolog of the chordate notochord (Balser and Ruppert, 1990). Gene expression studies of notochord-specific genes were expected to confirm this hypothesis. *Brachyury* is a T-box transcription factor, first isolated during mesoderm formation in vertebrates (Wilkinson et al., 1990; Holland et al., 1995) that is expressed exclusively in the ascidian notochord (Yasuo and Satoh, 1993; Swalla, 2006). When *Brachyury T* was cloned and described in echinoderms and hemichordates, it was expressed at the site of gastrulation at the vegetal pole, which later becomes the larval anus (Peterson et al., 1999; Swalla, 2006). These results suggest that the ancestral function of *Brachyury* as a transcription factor was in promoting gastrulation and formation of the three germ layers, and that the gene was later co-opted into notochord development (Swalla, 2006). However, recent results of Miyamoto and Wada (2013) show that the presumptive stomochord endoderm expresses *sonic hedgehog*, and may induce neurulation in the hemichordate neural tube. Therefore, these researchers present evidence that the stomochord is the notochord homolog in hemichordates. Candidate gene expression studies undertaken so far do not suggest any other hemichordate structure as a candidate for the notochord homolog (Gerhart et al., 2005), but more research will likely cast light on this intriguing possibility.

7.7 THE EVOLUTION OF PLACODES AND THE NEURAL CREST IN CHORDATES

The neural crest has been widely touted as a vertebrate innovation that allowed the development of complicated sensory structures in the anterior head, and the development of the skull and pharyngeal bars (Gans and Northcutt, 1983; see also Chapter 18). Therefore, it has long been assumed that tunicates and lancelets lack neural crest and cranial sensory placodes.

However, in recent years, using a combination of genetic and developmental techniques, there have been a number of reports of neural crest and placodes in tunicates, but not in lancelets (reviewed in Hall and Gillis, 2013). The sister group relationship of tunicates and vertebrates that has been reported for phylogenomic analyses supports the hypothesis that tunicates may contain neural crest and placodes (Bourlat et al., 2006; Delsuc et al., 2006). It is interesting that of 615 "neural crest genes," 91% of these genes are found in invertebrates that do not contain neural crest, suggesting that these genes were co-opted into neural crest cells in vertebrates (Hall and Gillis, 2013). Migratory and pigmented cells have been discovered in adult tunicates that develop from near the neural tube of the larvae and these have been suggested to be neural crest because they share parts of the neural crest gene network (Jeffery et al., 2008; Abitua et al., 2012). Recent work shows a cephalic melanocyte lineage in *Ciona intestinalis* larvae that can be reprogrammed into migrating cells by over-expression of the transcription factor *Twist* (Abitua et al., 2012). These results suggest that the gene network for neural crest is present in tunicates, and that the mesenchymal property of the neural crest was most recently evolved in vertebrates (Abitua et al., 2012). However, more work remains to be done to study the exact nature and potential of these "neural crest" cells in tunicates (Hall and Gillis, 2013).

There is excellent evidence from gene expression and morphological studies that tunicates have well-developed sensory placodes and lateral placodes (Manni et al., 2004; Bassham and Postlewait, 2005; Mazet et al., 2005). The buccal cavity and palps at the anterior of tunicates express *Six 1/2*, *Six 3/6*, *Eya* and *Pitx*, suggesting a homology with the hypophyseal and olfactory placodes of vertebrates (Manni et al., 2004; Bassham and Postlewait, 2005; Mazet et al., 2005). These results suggest that the common ancestor of vertebrates and tunicates had placodes, and that their anterior ends have homologous structures. A rather startling result is that the excurrent buccal opening in tunicates early on expresses *Six 1/2*, *Six 4/5*, *Eya*, and *Fox 1*, which are vertebrate markers for otic placodes, lateral line, and epibranchial placodes (Manni et al., 2004; Bassham and Postlewait, 2005; Mazet et al., 2005; see also Chapter 19). As mentioned before, tunicate larvae do not have an open gut, so do not have an anus during larval life. After metamorphosis, the gut is emptied out of the excurrent buccal siphon, after the tail has retracted, and the mouth forms at the anterior of the larvae. This would suggest that the adult tunicate is defecating out of its ear, an odd symmetry twist for a chordate.

7.8 STEM CELLS AND REGENERATION IN HEMICHORDATES

Regeneration is a fascinating, yet common occurrence among the metazoans. The majority of animal phyla have some species that reliably regenerate certain tissues and structures (Sánchez Alvarado, 2000). All groups within the deuterostomes, with the possible exception of cephalochordates and xenoturbella, have at least some species with regenerative capabilities. Echinoderms have been shown to have a remarkable capacity for regeneration (Candia Carnevali, 2006). Every extant class within the echinoderm lineage has been reported to regenerate to some degree (Candia Carnevali at al., 2009; D'Ancona Lunetta, 2009). Certain species of hemichordates, which are a sister group to the echinoderms, are capable of regenerating all body structures (Rychel and Swalla, 2008; Humphreys et al., 2010; Urata et al., 2012). Some species of both solitary and colonial tunicates have been shown to regenerate, either by budding or by re-growing missing or damaged structures (Sánchez Alvarado, 2000; Brown et al., 2009; Jeffery, 2012). Chordates possess numerous species that have some regenerative powers, including humans, which are able to regenerate bone and liver tissue (Sánchez Alvarado, 2000). While our overall understanding of regeneration in some species has improved, many of the key cellular and genetic processes controlling regeneration are still unknown.

Thomas H. Morgan defined two basic types of regeneration in metazoans. In the first mode, regeneration occurs by remodeling existing tissue without any active cell proliferation, termed morphallaxis. The second mode does require active cell proliferation and is known as epimorphosis (Morgan, 1898; 1901; Sánchez Alvarado, 2000). Depending on the organism, regeneration may employ both epimorphosis and morphallaxis. Agata et al., (2007) coined the terms distalization and intercalation in order to clarify the steps in regenerating new tissue. Wound healing is often the first step in regeneration. This allows new tissue, whether in the presence of a blastema or not, to be patterned on the correct axis. The process of wound healing is termed distalization. Intercalation is the second step of regeneration and it is characterized by the replacement of missing or damaged tissue. This latter process can occur by morphallaxis or epimorphosis. Regeneration can be categorized further by examining the final condition of the animal and the amputated tissue. If both regenerate completely to form two new individuals, then regeneration is considered to be bi-directional and essentially asexual reproduction. If, on the other hand, only one half of the animal regenerates to form a complete individual, regeneration is said to be unidirectional.

Ptychodera flava, a solitary enteropneust hemichordate, regenerates in a bi-directional fashion and employs both morphallaxis and epimorphosis to regenerate missing structures when bisected (Rychel and Swalla, 2008; Humphreys et al., 2010). *Glandiceps hacksi* has also been shown to regenerate with a combination of morphallaxis and epimorphosis (Urata et al., 2012). Because hemichordates are basal deuterostomes and employ both forms of regeneration, they present an exciting model to study the cellular and molecular mechanisms of regeneration. This ability to regenerate all structures in the body includes the CNS, hepatic sacs, heart, renal gland, and gonads (Rychel and Swalla, 2008).

The first step in hemichordate regeneration, as is seen in most animals, is wound healing. Rychel and Swalla (2008) documented that endoderm and ectoderm grow together to close the wound within two days post amputation in *Ptychodera flava*. Regeneration times vary based on the temperature of the seawater and overall health of the animal, with regeneration optimized between 26°C and 30°C (Humphreys et al., 2010). Following wound healing, a blastema forms and rapidly expands to establish proboscis and collar rudiments by the fifth day of regeneration (Humphreys et al., 2010). The blastema appears to be composed of multipotent stem-like cells. Rychel and Swalla (2008) used proliferating cell nuclear antigen (PCNA) antibody staining to show that actively dividing cells were present and scattered throughout the tissue at two days post-amputation. By the fourth day of regeneration, PCNA-positive cells were localized to the anterior regeneration site, where the proboscis and collar were forming (Rychel and Swalla, 2008). Humphreys et al., (2010) used *in situ* hybridization to show that the proboscis and collar rudiments of the blastema highly express *SoxB1* on day five of regeneration, while the surrounding non-regenerating tissue does not label with this probe. *SoxB1* is a gene that has been shown to be vital for generating induced pluripotent stem (iPS) cells (Takahashi et al., 2007). Additionally, the newly formed tissue is not pigmented, suggesting the structures were formed from newly dividing cells and not remodeled from existing cells (Rychel and Swalla, 2008; Humphreys et al., 2010). Collectively, this evidence supports the idea that the regeneration blastema is created by proliferating stem cells. Although the cells appear to be mesenchymal, it is an open question whether they are de-differentiated cells or a population of undifferentiated cells occupying an unknown stem cell niche. Further studies will be necessary to confirm the origin and character of the blastemal cells.

By the sixth day of regeneration in *P. flava*, complete proboscis and collar rudiments were formed and the mouth was opened on the ventral, posterior end of the proboscis (Rychel and Swalla, 2008; Humphreys et al., 2010). Furthermore, internal structures were being elaborated at this stage. Dorsal to the mouth, the stomochord develops from invaginating ectoderm. This structure serves to support the heart/kidney complex, which was seen to be forming two days later, by the eighth day of regeneration (Rychel and Swalla, 2008). All structures continued to develop and increase in size until around 12 days, when the newly formed tissue approximately matched the size of the old tissue. At this stage, the regenerating animal was capable of burrowing under sand, as it does in nature (Humphreys et al., 2010). The branchial region forms next and appears to combine both epimorphosis and morphallaxis. Abnormally high levels of apoptotic cells were detected approximately one centimeter away from the cut site in both the ectoderm and endoderm on days two through eight of regeneration using a TUNEL assay (Rychel and Swalla, 2008). However, the amount and location of programmed cell death varied over the time points sampled. In general, apoptosis was elevated in the endoderm farthest away from the cut site on day two and then progressed anteriorly towards the cut site. Additional sampling will need to be done beyond eight days to confirm the presence of apoptotic cells throughout the course of regeneration to confirm the extent of tissue remodeling during the formation of new structures. A blastema also forms between the collar and trunk between days twelve and twenty-five. The blastema goes on to differentiate and form the new branchial section of the animal (Humphreys et al., 2010). Complete regeneration is achieved approximately five weeks post amputation and it may take up to another five weeks for the newly formed tissue to achieve the same level of pigmentation as the old tissue (Humphreys et al., 2010). In conclusion, regeneration is impressive in the basal deuterostomes. Understanding the molecular basis of this process may yield clues to unlocking extensive regeneration in other deuterostomes, including humans. Future studies will be directed at identifying the source of stem cells in hemichordates and identifying the gene networks regulated during regeneration.

7.9 SUMMARY AND CONCLUSIONS

In summary, developmental genomics and genetics have allowed new insights to be gained about the question of chordate origins (Gerhart et al., 2005; Rychel et al., 2006). Genomics and gene expression studies have been informative in improving our understanding of the homology of various structures in invertebrate deuterostomes with vertebrates. Developmental gene expression data allows one to analyze the genetic pathways that are deployed to make similar structures in genetically different organisms. Gene expression data suggest that the anterior-posterior axis of hemichordates, lancelets and vertebrates are very similar, except that the neural genes are expressed throughout the ectoderm of hemichordates (Lowe et al., 2003). Tunicates have lost some of the middle *Hox* genes and express some of their *Hox* genes as larvae and some as adults, but only a few are expressed co-linearly (Spagnuolo et al., 2003; Passamaneck and Di Gregorio, 2005; Pascual-Anaya et al., 2013). The gill slits of hemichordates, lancelets and vertebrates are homologous (Rychel et al., 2006; Gillis et al., 2012), but the gill bars of lancelets and hemichordates are both acellular and endodermal in origin. In contrast, tunicates completely lack gill bars in their pharyngeal region, probably an adaptation that followed their acquisition of the cellulose tunic.

There is good recent evidence that tunicates have cell lineages that develop next to the neural tube, migrate and develop into pigment cells. These have been interpreted as neural crest cells, since they contain genes also identified in the neural crest gene network in vertebrates. Tunicates have been shown to have both neural and non-neural placodes, thought for many years to exist only in the vertebrates. While tunicates are clearly chordates, they have evolved some amazing changes in body plan, and are likely to have lost some structures evolutionarily at the time that the tunic evolved. Hemichordates have an anterior-posterior axis similar to chordates, but their dorsal central nervous system is restricted to the collar region (Luttrell et al., 2012; Miyamoto and Wada, 2013). Our view of the chordate ancestor is a benthic worm with gill slits and a mouth, which was able to filter feed, but also could ingest large particles. This ancestor also had a central nervous system and may or may not have had a notochord. Further research on developmental gene expression in lancelets, tunicates and hemichordates is likely to produce continuing insights into the evolution of vertebrates.

7.10 CLINICAL RELEVANCE

- **Chordate ancestor** – The chordate ancestor was a worm-like animal, with a through gut and the mouth at the anterior, anus at the posterior.
- **Hemichordate homologies** – Hemichordates have homologies with all of the major chordate features, except possibly the notochord, and are an excellent model system for the chordate and deuterostome ancestor.
- **Tunicate homologies** – Tunicates have all of the chordate features and are an excellent model system for understanding chordate gene networks.
- **Stem cells** – Chordate ancestors had stem cell populations that could be co-opted for specific functions.
- **Regeneration** – Chordate ancestors had remarkable regeneration capacities.

ACKNOWLEDGMENTS

A special thank you is owed to Sally Moody for putting this book together, and also for her many helpful and insightful suggestions during the writing of the chapter. I would like to acknowledge the members of the Swalla lab for their many contributions to my life and my work over the past twenty years.

RECOMMENDED RESOURCES

Biology of the Protochordata, 2005. A collection of reviews published in the *Canadian Journal of Zoology*, vol 83. http://cjz.nrc.ca (accessed 26.02.14).

Chordate Origins, 2008. A collection of reviews and theories on chordate origins published in *Genesis*. 46(11).

Matranga, Valeria, Rinkevich, Baruch (Eds.), 2009. Marine Stem Cells. Springer Publishing.

Ettensohn, C.A., Wessel, G.M., Wray, G.A. (Eds.), 2004. The Invertebrate Deuterostomes: An Introduction to their Phylogeny, Reproduction, Development, and Genomics. Methods Cell Biol, vol 74.

REFERENCES

Abitua, P.B., Wagner, E., Navarrete, I.A., Levine, M., 2012. Identification of a rudimentary neural crest in a non-vertebrate chordate. Nature 492 (7427), 104–107.

Agata, K., Saito, Y., Nakajima, E., 2007. Unifying principles of regeneration I: Epimorphosis versus morphallaxis. Dev. Growth Diff 49, 73–78.

Aronowicz, J., Lowe, C.J., 2006. Hox gene expression in the hemichordate *Saccoglossus kowalevskii* and the evolution of deuterostome nervous systems. Integr. Comp. Biol. 46 (6), 890–901.

Balser, E.J., Ruppert, E.E., 1990. Structure, ultrastructure, and function of the preoral heart-kidney in *Saccoglossus kowalevskii* (Hemichordata, Enteropneusta) including new data on the stomochord. Acta Zool. 71, 235–249.

Bardet, S.M., Ferran, J.L., Sanchez-Arrones, L., Puelles, L., 2010. Ontogenetic expression of *sonic hedgehog* in the chicken subpallium. Front Neuroanat. 4 (28), 1–21.

Bassham, S., Postlethwait, J.H., 2005. The evolutionary history of placodes: a molecular genetic investigation of the larvacean urochordate *Oikopleura dioica*. Development 132, 4259–4572.

Blair, J.E., Hedges, S.B., 2005. Molecular phylogeny and divergence times of deuterostome animals. Mol. Biol. Evol. 22, 2275–2284.

Bourlat, S.j., Juliusdottir, T., Lowe, C.J., Freeman, R., Aronowicz, J., Kirschner, M., Lander, E.S., Thorndyke, M., Nakano, H., Kohn, A.B., Heyland, A., Moroz, L.l., Copley, R.R., Telford, M.J., 2006. Deuterostome phylogeny reveals monophyletic chordates and the new phylum Xenoturbellida. Nature 444 (7115), 85–88.

Brown, F.D., Keeling, E.L., Le, A.D., Swalla, B.J., 2009. Whole body regeneration in a colonial ascidian, *Botrylloides violaceus*. J. Exp. Zool. B Mol. Dev. Evol. 312B (8), 885–900.

Brown, F.D., Prendergast, A., Swalla, B.J., 2008. Man is but a worm: Chordate origins. Genesis 46 (11), 605–613.

Cameron, C.B., Garey, J.R., Swalla, B.J., 2000. Evolution of the chordate body plan: new insights from phylogenetic analyses of deuterostome phyla. Proc. Natl. Acad. Sci. USA 97, 4469–4474.

Cameron, R.A., Rowen, L., Nesbitt, R., Bloom, S., Rast, J., Berney, K., Arenas-Mena, C., Martinez, P., Lucas, S., Richardson, P.M., Davidson, E.H., Peterson, K.J., Hood, L., 2006. Unusual gene order and organization of the sea urchin *Hox* cluster. J. Exp. Zoolog. B Mol. Dev. Evol. 306, 45–58.

Candia Carnevali, M.D., 2006. Regeneration in echinoderms: Repair, regrowth, cloning. Invert. Surv. J. 3, 64–76.

Candia Carnevali, M.D., Thorndyke, M.R., Matranga, V., 2009. Regenerating in echinoderms: A promise to understand stem cells potential. In: Rinkevich, R., Matranga, V. (Eds.), Stem Cells in Marine Organisms. Springer, New York.

Dahia, C.L., Mahoney, E.J., Durrani, A.A., Wylie, C., 2011. Intercellular signaling pathways active during and after growth and differentiation of the lumbar vertebral growth plate. Spine (Phila Pa 1976) 36 (14), 1071–1080.

Dahia, C.L., Mahoney, E., Wylie, C., 2012. Shh signaling from the nucleus pulposus is required for the postnatal growth and differentiation of the mouse intervertebral disc. PLoS One 7 (4), e35944.

D'Ancona Lunetta, G., 2009. Stem cells in *Holothuria polii* and *Sipunculus nudus*. In: Rinkevich, R., Matranga, V. (Eds.), Stem Cells in Marine Organisms. Springer, New York.

Davidson, E.H., Erwin, D.H., 2006. Gene regulatory networks and the evolution of animal body plans. Science. 311, 796–800.

Dehal, P., et al., 2002. The draft genome of *Ciona intestinalis*: insights into chordate and vertebrate origins. Science 298 (5601), 2157–2167.

Delsuc, F., Brinkmann, H., Chourrout, D., Philippe, H., 2006. Tunicates and not cephalochordates are the closest living relatives of vertebrates. Nature 439, 965–968.

Denoeud, F., et al., 2010. Plasticity of animal genome architecture unmasked by rapid evolution of a pelagic tunicate. Science 330 (6009), 1381–1385.

Fong, J.H., Murphy, T.D., Pruitt, K.D., 2013. Comparison of RefSeq protein-coding regions in human and vertebrate genomes. BMC Genomics 14 (1), 654.

Freeman, R., Ikuta, T., Wu, M., Koyanagi, R., Kawashima, T., Tagawa, K., Humphreys, T., Fang, G.C., Fujiyama, A., Saiga, H., Lowe, C., Worley, K., Jenkins, J., Schmutz, J., Kirschner, M., Rokhsar, D., Satoh, N., Gerhart, J., 2012. Identical genomic organization of two hemichordate *Hox* clusters. Curr. Biol. 22 (21), 2053–2058.

Freeman Jr., R.M., Wu, M., Cordonnier-Pratt, M.M., Pratt, L.H., Gruber, C.E., Smith, M., Lander, E.S., Stange-Thomann, N., Lowe, C.J., Gerhart, J., Kirschner, M., 2008. cDNA sequences for transcription factors and signaling proteins of the hemichordate *Saccoglossus kowalevskii*: efficacy of the expressed sequence tag (EST) approach for evolutionary and developmental studies of a new organism. Biol. Bull. 214 (3), 284–302.

Gans, C., Northcutt, R.G., 1983. Neural crest and the origin of vertebrates: a new head. Science 220, 268–274.

Garstang, W., 1928. The morphology of the Tunicata and its bearing on the phylogeny of the Chordata. Q. J. Micro. Sci. 72, 51–187.

Gee, H., 1996. Before the Backbone. Views on the Origin of the Vertebrates. Chapman and Hall, London.

Gerhart, J., Lowe, C., Kirschner, M., 2005. Hemichordates and the origin of chordates. Curr. Opin. Genet. Dev. 15, 461–467.

Gillis, A.J., Fritzenwanker, J.H., Lowe, C.J., 2012. A stem-deuterostome origin of the vertebrate pharyngeal transcriptional network. Proc. R. Soc. B. 279, 237–246.

Halanych, K.M., 2004. The new view of animal phylogeny. Ann. Rev. Ecol. Evol. Syst. 35, 229–256.

Hall, B.K., Gillis, J.A., 2013. Incremental evolution of the neural crest, neural crest cells and neural crest-derived skeletal tissues. J. Anat. 222 (1), 19–31.

Holland, N.D., Holland, L.Z., 1995. An amphioxus Pax gene, *AmphiPax-1*, expressed in embryonic endoderm, but not in mesoderm: implications for the evolution of class I paired box genes. Mol. Mar. Biol. Biotech. 4, 206–214.

Holland, P.W., Koschorz, B., Holland, L.Z., Herrmann, B.G., 1995. Conservation of *Brachyury (T)* genes in amphioxus and vertebrates: developmental and evolutionary implications. Development 121, 4283–4291.

Humphreys, T., Sasaki, A., Uenishi, G., Taparra, K., Arimoto, A., Tagawa, K., 2010. Regeneration in the hemichordate *Ptychodera flava*. Zoolog. Sci. 27 (2), 91–95.

Hyman, L.H., 1959. Hemichordata. In: The Invertebrates: Smaller Coelomate Groups. McGraw-Hill, New York, pp. 72–207.

Jeffery, W.R., 2012. Siphon regeneration capacity is compromised during aging in the ascidian *Ciona intestinalis*. Mech. Ageing Dev. 133 (9–10), 629–636.

Jeffery, W.R., Chiba, T., Krajka, F.R., Deyts, C., Satoh, N., Joly, J.S., 2008. Trunk lateral cells are neural crest-like cells in the ascidian *Ciona intestinalis*: insights into the ancestry and evolution of the neural crest. Dev. Biol. 324 (1), 152–160.

Jeffries, R.P.S., 1986. The Ancestry of Vertebrates. British Museum of Natural History, London.

Jollie, M., 1973. The origin of the chordates. Acta. Zool. 54, 81–100.

Kaul, S., Stach, T., 2010. Ontogeny of the collar cord: neurulation in the hemichordate Saccoglossus kowalevskii. J. Morphol. 271 (10), 1240–1259.

Lacalli, T.C., 2005. Protochordate body plan and the evolutionary role of larvae: old controversies resolved? Can. J. Zool. 83, 216–224.

Lowe, C.J., Terasaki, M., Wu, M., Freeman Jr., R.M., Runft, L., Kwan, K., Haigo, S., Aronowicz, J., Lander, E., Gruber, C., Smith, M., Kirschner, M., Gerhart, J., 2006. Dorsoventral patterning in hemichordates: insights into early chordate evolution. PLoS Biol. 4 (9), e291.

Lowe, C.J., Wu, M., Salic, A., Evans, L., Lander, E., Stange-Thomann, N., Gruber, C.E., Gerhart, J., Kirschner, M., 2003. Anteroposterior patterning in hemichordates and the origins of the chordate nervous system. Cell. 113, 853–865.

Luttrell, S.M., Konikoff, C., Byrne, A., Bengtsson, B., Swalla, B.J., 2012. Ptychoderid hemichordate neurulation without a notochord. Integr. Comp. Biol. 52 (6), 829–834.

Manni, L., Lane, N.J., Joly, J.S., Gasparini, F., Tiozzo, S., Caicci, F., Zaniolo, G., Burighel, P., 2004. Neurogenic and non-neurogenic placodes in ascidians. J. Exp. Zool. B Mol. Dev. Evol. 302, 483–504.

Martí, E., 2000. Expression of chick *BMP-1/Tolloid* during patterning of the neural tube and somites. Mech. Dev. 91, 415–419.

Matthysse, A.G., Deschet, K., Williams, M., Marry, M., White, A.R., Smith, W.C., 2004. A functional cellulose synthase from ascidian epidermis. Proc. Natl. Acad. Sci. 101 (4), 986–991.

Mazet, F., 2002. The Fox and the thyroid: the amphioxus perspective. Bioessays 24, 696–699.

Mazet, F., Hutt, J.A., Milloz, J., Millard, J., Graham, A., Shimeld, S.M., 2005. Molecular evidence from *Ciona intestinalis* for the evolutionary origin of vertebrate sensory placodes. Develop. Biol. 282, 494–508.

Méndez, A.T., Roig-López, J.L., Santiago, P., Santiago, C., García-Arrarás, J.E., 2000. Identification of *Hox* Gene Sequences in the Sea Cucumber *Holothuria glaberrima* Selenka (Holothuroidea: Echinodermata). Mar. Biotechnol. 2 (3), 231–240.

Mito, T., Endo, K., 2000. PCR survey of *Hox* genes in the crinoid and ophiuroid: evidence for anterior conservation and posterior expansion in the echinoderm *Hox* gene cluster. Mol. Phylogenet. Evol. 14 (3), 375–388.

Miyamoto, N., Wada, H., 2013. Hemichordate neurulation and the origin of the neural tube. Nat. Commun. 4 (2713), 1–8.

Morgan, T.H., 1898. Experimental studies of the regeneration of *Planaria maculata*. Arch. Entw Mech. Org 7, 364–397.

Morgan, T.H., 1901. Regeneration. The Macmillan Company, New York.

Morris, V.B., Byrne, M., 2005. Involvement of two *Hox* genes and *Otx* in echinoderm body-plan morphogenesis in the sea urchin *Holopneustes purpurescens*. J. Exp. Zool. B Mol. Dev. Evol. 304, 456–467.

Nakashima, K., Sugiyama, J., Satoh, N., 2008. A spectroscopic assessment of cellulose and the molecular mechanisms of cellulose biosynthesis in the ascidian *Ciona intestinalis*. Mar. Genomics 1 (1), 9–14.

Neubüser, A., Koseki, H., Balling, R., 1995. Characterization and developmental expression of *Pax 9*, a paired-box-containing gene related to Pax1. Develop. Biol. 170, 701–716.

Nishida, H., 2005. Specification of embryonic axis and mosaic development in ascidians. Dev. Dyn. 233, 1177–1193.

Ogasawara, M., Wada, H., Peters, H., Satoh, N., 1999. Developmental expression of *Pax1/9* genes in urochordate and hemichordate gills: insight into function and evolution of the pharyngeal epithelium. Development. 126, 2539–2550.

Ogasawara, M., Shigetani, Y., Hirano, S., Satoh, N., Kuratani, S., 2000a. *Pax1/Pax9*-Related genes in an agnathan vertebrate, *Lampetra japonica*: expression pattern of *LjPax9* implies sequential evolutionary events toward the gnathostome body plan. Develop. Biol. 223, 399–410.

Ogasawara, M., Shigetani, Y., Suzuki, S., Kuratani, S., Satoh, N., 2000b. Expression of thyroid transcription factor (*TTF-1*) gene in the ventral forebrain and endostyle of the agnathan vertebrate, *Lampetra japonica*. Genesis 30, 51–58.

Pani, A.M., Mullarkey, E.E., Aronowicz, J., Assimacopoulos, S., Grove, E.A., Lowe, C.J., 2012. Ancient deuterostome origins of vertebrate brain signaling centers. Nature 483 (7389), 289–294.

Pascual-Anaya, J., Aniello, S.D., Kuratani, S., Garcia-Fernàndez, J., 2013. Evolution of *Hox* gene clusters in deuterostomes. BMC Dev. Biol. 13 (1), 26.

Passamaneck, Y.J., Di Gregorio, A., 2005. *Ciona intestinalis*: Chordate development made simple. Develop. Dyn. 233, 1–19.

Peters, H., Neubüser, A., Kratochwil, K., Balling, R., 1998. *Pax9*-deficient mice lack pharyngeal pouch derivatives and teeth and exhibit craniofacial and limb abnormalities. Genes Dev. 12, 2735–2747.

Peterson, K.J., 2004. Isolation of *Hox* and *Parahox* genes in the hemichordate *Ptychodera flava* and the evolution of deuterostome *Hox* genes. Mol. Phy. Evol. 31, 1208–1215.

Peterson, K.J., Cameron, R.A., Tagawa, K., Satoh, N., Davidson, E.H., 1999. A comparative molecular approach to mesodermal patterning in basal deuterostomes: the expression pattern of *Brachyury* in the enteropneust hemichordate *Ptychodera flava*. Development 126, 85–95.

Putnam, N.H., et al., 2008. The amphioxus genome and the evolution of the chordate karyotype. Nature 453, 1064–1071.

Rickmann, M., Fawcett, J., Keynes, R., 1985. The migration of neural crest cells and the growth of motor axons through the rostral half of the chick somite. J. Embryol. Exp. Morphol. 90, 437–455.

Romer, A.S., 1967. Major steps in vertebrate evolution. Science 158, 1629–1637.

Röttinger, E., Lowe, C.J., 2012. Evolutionary crossroads in developmental biology: hemichordates. Development 139 (14), 2463–2475.

Ruppert, E.E., Cameron, C.B., Frick, J.E., 1999. Endostyle-like features of the dorsal epibranchial ridge of an enteropneust and the hypothesis of dorsal-ventral axis inversion in chordates. Invert. Biol. 118, 202–212.

Ruppert, E.E., 2005. Key characters uniting hemichordates and chordates: homologies or homoplasies? Can. J. Zool. 83, 8–23.

Rychel, A.L., Smith, S.E., Shimamoto, H.T., Swalla, B.J., 2006. Evolution and development of the chordates: Collagen and pharyngeal cartilage. Mol. Biol. Evol. 23, 1–9.

Rychel, A.L., Swalla, B.J., 2007. Development and evolution of chordate cartilage. J. Exp. Zool. B. 308 (3), 325–335.

Rychel, A.L., Swalla, B.J., 2008. Anterior regeneration in the hemichordate *Ptychodera flava*. Dev. Dyn. 237 (11), 3222–3232.

Sánchez Alvarado, A., 2000. Regeneration in the metazoans: Why does it happen? Bioessays 22, 578–590.

Sasaki, A., Miyamoto, Y., Satou, Y., Satoh, N., Ogasawara, M., 2003. Novel endostyle-specific genes in the ascidian *Ciona intestinalis*. Zoolog. Sci. 20, 1025–1030.

Schaeffer, B., 1987. Deuterostome monophyly and phylogeny. Evol. Biol. 21, 179–235.

Schneider, C., Wicht, H., Enderich, J., Wegner, M., Rohrer, H., 1999. Bone morphogenetic proteins are required *in vivo* for the generation of sympathetic neurons. Neuron 24, 861–870.

Smith, A.B., Peterson, K.J., Wray, G., Littlewood, D.T.J., 2004. From bilateral symmetry to pentaradiality: The phylogeny of hemichordates and echinoderms. In: Cracraft, J., Donoghue, M.J. (Eds.), Assembling the Tree of Life. Oxford Press, New York, pp. 365–383.

Smith, S.E., Douglas, R., Da Silva, K.B., Swalla, B.J., 2003. Morphological and molecular identification of *Saccoglossus* species (Hemichordata: Harrimaniidae) in the Pacific Northwest. Can. J. Zool. 81, 133–141.

Sodergren, et al., 2006. The genome of the sea urchin *Strongylocentrotus purpuratus*. Science 314 (5801), 941–952.

Spagnuolo, A., Ristoratore, F., Di Gregorio, A., Aniello, F., Branno, M., Di Lauro, R., 2003. Unusual number and genomic organization of *Hox* genes in the tunicate Ciona intestinalis. Gene 309, 71–79.

Swalla, B.J., 2004. Protochordate gastrulation: lancelets and ascidians. In: Stern, C. (Ed.), Gastrulation. Cold Spring Harbor Press, NY, pp. 139–149.

Swalla, B.J., 2006. Building divergent body plans with similar genetic pathways. Heredity 97, 235–243.

Swalla, B.J., 2007. New insights into vertebrate origins. In: Moody, Sally (Ed.), Principles of Developmental Genetics. Elsevier Press (San Diego, Elsevier), pp. 114–128.

Swalla, B.J., Cameron, C.B., Corley, L.S., Garey, J.R., 2000. Urochordates are monophyletic within the deuterostomes. Syst. Biol. 49, 122–134.

Swalla, B.J., Smith, A.B., 2008. Deciphering deuterostomes phylogeny: molecular, morphological and palaeontological perspectives. Phil. Trans. R. Soc. B. 363 (1496), 1557–1568.

Takacs, C.M., Moy, V.N., Peterson, K.J., 2002. Testing putative hemichordate homologues of the chordate dorsal nervous system and endostyle: expression of *NK2.1 (TTF-1)* in the acorn worm *Ptychodera flava* (Hemichordata, Ptychoderidae). Evol. Dev. 4, 405–417.

Takahashi, K., Tanabe, K., Ohnuki, M., Narita, M., Ichisaka, T., Tomoda, K., Yamanaka, S., 2007. Induction of pluripotent stem cells from adult human fibroblasts by defined factors. Cell 131 (5), 861–872.

Tsagkogeorga, G., Turon, X., Hopcroft, R.R., Tilak, M.K., Feldstein, T., Shenkar, N., Loya, Y., Huchon, D., Douzery, E.J., Delsuc, F., 2009. An updated 18S rRNA phylogeny of tunicates based on mixture and secondary structure models. BMC Evol. Biol. 9, 187.

Turbeville, J.M., Schulz, J.R., Raff, R.A., 1994. Deuterostome phylogeny and the sister group of the chordates: evidence for molecules and morphology. Mol. Biol. Evol. 11, 648–655.

Urata, M., Iwasaki, S., Ohtsuka, S., 2012. Biology of the swimming acorn worm *Glandiceps hacks*i from the Seto Inland Sea of Japan. Zoolog. Sci. 29 (5), 305–310.

Urata, M., Yamaguchi, M., 2004. The development of the enteropneust hemichordate *Balanoglossus misakiensis* KUWANO. Zoolog. Sci. 21, 533–540.

Vinson, J.P., Jaffe, D.B., O'Neill, K., Karlsson, E.K., Stange-Thomann, N., Anderson, S., Mesirov, J.P., Satoh, N., Satou, Y., Nusbaum, C., Birren, B., Galagan, J.E., Lander, E.S., 2005. Assembly of polymorphic genomes: Algorithms and application to Ciona savignyi. Genome Res. 15, 1127–1135.

Wada, H., Satoh, N., 1994. Details of the evolutionary history from invertebrates to vertebrates, as deduced from the sequences of 18S rDNA. Proc. Natl. Acad. Sci. USA 91, 1801–1804.

Wallin, J., Eibel, H., Neubüser, A., Wilting, J., Koseki, H., Balling, R., 1996. *Pax1* is expressed during development of the thymus epithelium and is required for normal T-cell maturation. Development 122, 23–30.

Wilkinson, D.G., Bhatt, S., Herrmann, B.G., 1990. Expression pattern of the mouse *T* gene and its role in mesoderm formation. Nature 343, 657–659.

Yasuo, H., Satoh, N., 1993. Function of the vertebrate *T* gene. Nature 364, 582–583.

Young, H.M., Anderson, R.B., Anderson, C.R., 2004. Guidance cues involved in the development of the peripheral autonomic nervous system. Auton. Neurosci. 112 (1–2). 1-1.

Zhang, G., Miyamoto, M.M., Cohn, M.J., 2006. Lamprey type II collagen and *Sox 9* reveal an ancient origin of the vertebrate collagenous skeleton. Proc. Natl. Acad. Sci. USA. 103 (9), 3180–3185.

Section II

Early Embryology and Morphogenesis

Chapter 8

Signaling Cascades, Gradients, and Gene Networks in Dorsal/Ventral Patterning

Girish S. Ratnaparkhi* **and Albert J. Courey†**

**Biological Sciences, Indian Institute of Science Education & Research (IISER), Pune, India; †Department of Chemistry and Biochemistry, University of California at Los Angeles, Los Angeles, CA*

Chapter Outline

SUMMARY

- Dorsal/ventral (DV) patterning is the process whereby embryonic cells assume different developmental fates as a function of their position along an organism's DV axis.
- In the *Drosophila* embryo, DV patterning begins during oogenesis and is completed during the early stages of embryogenesis. It requires cross-talk between three signal transduction pathways; the epidermal growth factor (EGF) receptor pathway, the Toll (Tl) pathway, and the Decapentaplegic (Dpp)/Short gastrulation (Sog) pathway.
- The *Drosophila* oocyte arises from germline stem cells present at the anterior tip of each ovariole. Each ovary contains ~16 such ovarioles, which consist of a linear series of egg chambers that gradually mature as they move from the anterior to the posterior end of the ovariole. The developing oocyte in each egg chamber is nourished by 15 nurse cells and is surrounded by a monolayer follicle cell epithelium.
- Gurken (Grk), a transforming growth factor (TGF)-α family protein, is secreted from one side of the developing oocyte close to its anterior end and signals via the EGF receptor in the adjacent follicle cells. This signal specifies DV polarity in the follicle cell epithelium, which in turn deposits a latent cue in the pervitelline space (the space between the eggshell and the oocyte plasma membrane) on what will become the ventral side of the embryo.
- After fertilization, activation of the Tl receptor by the ventral cue leads to formation of a ventral to dorsal nuclear concentration gradient of the protein Dorsal (DL).

Principles of Developmental Genetics.

- DL is a transcription factor that serves as a morphogen to direct cell fate as a function of position along the DV axis. It does so by regulating ~50 genes in the blastoderm embryo. Different genes are activated or repressed at different threshold DL concentrations, resulting in multiple domains of gene expression. The action of this gene network specifies the three primary territories along the DV axis; the presumptive mesoderm, the neurogenic ectoderm, and the dorsal ectoderm.
- Two of the genes regulated by DL, *dpp* and *sog*, encode critical components of a second morphogen system required for DV pattern formation. Dpp and Sog form opposing activity gradients along the DV axis, and through these gradients they subdivide the embryo into multiple developmental domains.
- Loss- and gain-of-function genetic analysis carried out in flies, frogs, and fish demonstrate that this role of the Dpp/Sog morphogen system in patterning the DV axis is conserved in both invertebrates and vertebrates.

8.1 INTRODUCTION

The specification of embryonic polarity, which results from the asymmetric distribution of gene products during oogenesis and early embryogenesis, is the first step in metazoan development. Subsequent to the polarization of the embryo, coordinate axes are specified, thereby allowing each cell to sense its position within the organism and differentiate in a manner appropriate to this position. The number of axes is variable: the embryos of radially symmetric coelenterates possess just a single axis, whereas bilaterian embryos possess two axes: an anterior–posterior (AP) axis and a dorsal–ventral (DV) axis (Figure 8.1A). This chapter introduces DV patterning in animals using the embryo of the fruit fly *Drosophila melanogaster* as an example. We describe the molecular events that are responsible for the specification and establishment of the DV axis from the oocyte to the blastoderm.

In *Drosophila*, polarity specification occurs during oogenesis (Roth and Schupbach, 1994; St Johnston and Nusslein-Volhard, 1992; van Eeden and St Johnston, 1999) (Figure 8.1). After fertilization, activity gradients of transcription factors and signaling molecules then form within the embryo specifying AP and DV coordinate axes (Morisato and Anderson, 1995; Nusslein-Volhard, 1991). The gradients are interpreted in terms of specific threshold concentrations (Driever and Nusslein-Volhard, 1988; Huang et al., 1997; Stathopoulos and Levine, 2002) leading to multiple discrete domains of gene expression along each axis. In many cases, these genes code for proteins that form additional gradients (Ferguson and Anderson, 1992; Hulskamp et al., 1990; Srinivasan et al., 2002) leading to the finer subdivision of the embryo.

Four gene networks define the polarity of the *Drosophila* embryo and establish the developmental axes. Three of these, the anterior, posterior, and terminal systems, collaborate to define AP polarity and the AP axis, while specification of DV

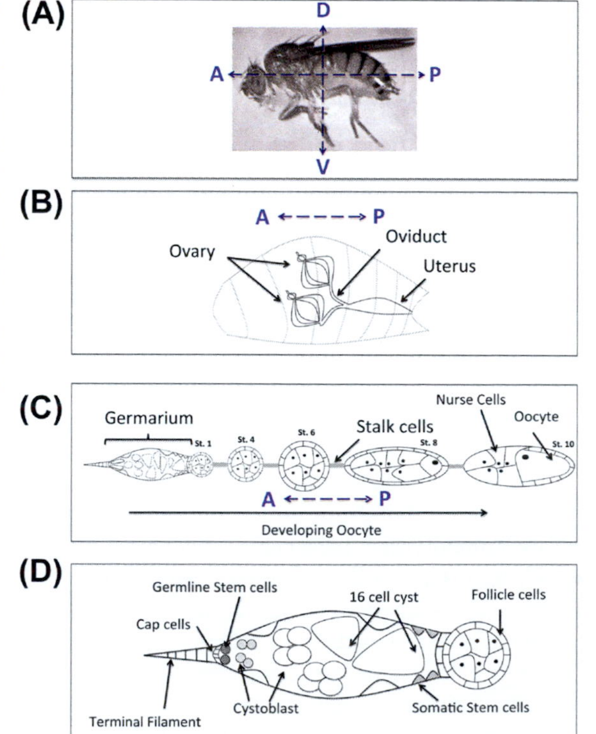

FIGURE 8.1 Oogenesis: from the Adult to the Mature Oocyte. (A) An adult female *Drosophila* with the AP and the DV axes indicated. In addition to these axes, the animal also has left/right asymmetry and a proximo-distal axis that patterns the appendages. (B) The female abdomen contains a pair of ovaries that connect via an oviduct to the uterus. Each ovary is a bundle of 16–18 ovarioles. (C) The ovariole consists of an anterior germarium, which contains the GSC's and a posterior series of egg chambers connected by stalk cells. Immature egg chambers are near the germarium and move towards the posterior as they mature. The specification of AP and DV polarity occurs during mid-oogenesis (stages 4–6). This simplified sketch is not to scale and many intermediate stages have been omitted. (D) The germarium. This structure, at the anterior end of each ovariole contains cap cells, which are part of the Germline Stem Cell (GSC) niche. 2–3 GSC's per germarium are found intimately associated with the niche. On asymmetric cell division, a cystoblast is formed that on further division forms the germline oocyte and nurse cells. The germline cells are surrounded by a follicle cell epithelium that arises from a separate somatic stem cell lineage.

polarity and the DV axis is the responsibility of the DV system (Anderson et al., 1985). Each system includes a set of maternally required genes, the mRNA products of which are synthesized in the nurse cells and transported into the oocyte late during oogenesis, and zygotically required genes, which are transcribed in the embryo a few hours after fertilization under the regulation of the maternal gene products. The maternal gene products are actively regulated during this maternal to zygotic transition (Benoit et al., 2009; Gouw et al., 2009; Tsurumi et al., 2011).

After fertilization, the embryo undergoes 13 rapid mitotic cycles. These nuclear divisions are not accompanied by cytokinesis, thus resulting in the production of a syncytium containing over 5,000 nuclei. During the eighth nuclear cycle, the nuclei migrate to the periphery of the embryo, resulting in the formation, by the start of the ninth cycle, of the syncytial blastoderm embryo in which the plasma membrane is lined with a single layer of nuclei. After completion of the 13th nuclear cycle, cell membranes cleave in from the surface of the embryo converting it into an ovoid-shaped monolayer epithelium called the cellular blastoderm embryo. The completion of cellularization is followed immediately by the cell movements that mark the onset of gastrulation, by which time DV axis formation and the initial subdivision of the embryo into broad developmental domains along its DV axis is complete.

During the syncytial blastoderm stage the embryo is divided into three primary domains along the DV axis. These include a ventral domain, which will give rise during gastrulation to the mesoderm, a ventrolateral domain (the neurogenic ectoderm), which will give rise to the ventral epidermis and the central nervous system, and a dorsal/dorsolateral domain (the dorsal ectoderm), which will give rise to the dorsal epidermis and amnioserosa (a cell sheet that covers the dorsal-most surface of the embryo during much of embryogenesis, but which is covered over by epidermal cells late during embryogenesis). The endoderm is derived from anterior and posterior regions after gastrulation is initiated by processes not described in this chapter.

The subdivision of the *Drosophila* embryo along its DV axis is directed by opposing activity gradients of two factors, these being the transcription factor Dorsal (DL), which is a member of the rel transcription factor family and a homolog of vertebrate NF-κB, along with the extracellular signaling protein Decapentaplegic (Dpp), a homolog of vertebrate BMP4/BMP2. The concentration of DL in the nucleus and therefore DL activity is maximal in the ventral regions of the syncytial blastoderm embryo, and decreases dorsally. In contrast, Dpp signaling activity is maximal at the dorsal midline and decreases ventrally. While the use of a rel family factor such as DL to regulate DV pattern formation may be unique to invertebrates, the use of Dpp/BMP2/4 in DV patterning is conserved in both invertebrates and vertebrates (Holley and Ferguson, 1997; Lall and Patel, 2001; Schmidt et al., 1995).

8.2 AP AND DV POLARITY IS SPECIFIED IN THE DEVELOPING OVARIOLE

The specification of the embryonic axes occurs during oogenesis prior to fertilization (Roth and Lynch, 2012; Roth and Schupbach, 1994; St Johnston and Nusslein-Volhard, 1992; van Eeden and St Johnston, 1999). The AP axis of the adult female and therefore of the ovary gives rise directly to the AP axis of the oocyte and therefore of the embryo. In contrast, the DV axis appears to be defined by a stochastic process, so that there is no correspondence between the DV axis of the mother and the DV axis of the oocytes developing inside her ovaries.

The female abdomen contains a pair of ovaries, each of which is a bundle of 16–18 ovarioles (Figure 8.1B). Each ovariole (Figure 8.1C) consists of an anterior germarium connected in a linear fashion to a series of differentiating egg chambers. The ~ten-day process of oogenesis leads to the production of the oocyte which is deposited in the oviduct on maturation.

Each germarium (Figure 8.1D) contains somatic terminal filament cells at its anterior end, followed by 4–6 non-mitotic somatic cap cells. Adjacent to the cap cells are the germline stem cells (GSCs) (Spradling et al., 2001; Xie, 2008) of the ovary. Each germarium has 2–3 GSCs. The cap cells and the local microenvironment are believed to form a niche (Xie, 2008) that supports their asymmetric cell division. At each division, each GSC renews itself and also generates a daughter cell, called a cystoblast. The renewed GSC is retained in the niche, while the cystoblast is displaced, thus allowing further symmetric division and differentiation. GSCs divide along the AP axis, which ensures that one cell maintains contact with the cap cells and remains in the niche, while the cystoblast is away from the niche. Dpp, a BMP2/4 homolog, has been shown to be the critical signal for maintaining stem cells (Xie, 2008; Xie and Spradling, 1998; Kirilly and Xie, 2007; Spradling et al., 2001), with *dpp* loss of function and concomitant loss of Dpp signal reception by the GSC leading to its differentiation, while overexpression of Dpp leads to an expansion of the number of stem cells. The source of Dpp is the cap cells, indicating the presence of a short-distance gradient of Dpp emanating from the cap cells and being critical for maintaining the niche microenvironment. In addition, genes such as *piwi, pumilio, nanos, bag of marbles* and *benign gonial cell neoplasm (bcgn)* are required to maintain GSCs (Xie, 2008).

Each cystoblast goes through four rounds of mitosis to form 16 cystocytes. Cytokinesis during these mitotic cycles is incomplete, and therefore the 16 germ line-derived cells are joined to one another via cytoplasmic connections termed ring canals. Of these 16, one cell becomes an oocyte whereas the remaining 15 serve as nurse cells. Each developing egg chamber thus contains one oocyte and 15 nurse cells, which are enclosed in a layer of somatic follicle cells. The follicle cells arise from a distinct population of somatic stem cells in the germarium. The niche for the follicle cells is also distinct from that of the germ cells. Renewal of the follicle stem cells requires a Wingless signal emanating from cap cells and Glass-bottom-boat, a Bone Morphogenetic Protein (BMP) homolog expressed by escort cells (Song et al., 2004). As the egg chambers (one oocyte, 15 nurse cells and follicle cells) mature, they are pushed to the posterior of the ovariole by more immature egg chambers. The egg chambers remain connected to each other by specialized cells named stalk cells. The ovariole with 6–10 egg chambers therefore consists of multiple snapshots of oogenesis with immature egg chambers towards the anterior germarium and the mature ones towards the posterior oviduct.

During oogenesis, the nurse cells manufacture the massive amounts of protein and mRNA that will later pattern and sustain the embryo during early embryogenesis. The contents of the nurse cells are transported into the growing oocyte via the ring canals. Some aspects of this system for generating a germline cyst and for nourishing the developing oocyte may be conserved in vertebrates, as recent studies of mouse oogenesis show that mouse ovaries also contain groups of interconnected germ line cells that result from synchronous mitotic division with incomplete cytokinesis (Lei and Spradling, 2013).

8.3 FROM THE OOCYTE TO THE FERTILIZED EGG: FORMATION OF THE DL NUCLEAR CONCENTRATION GRADIENT

The formation of the DL nuclear concentration gradient (Ip et al., 1991; Roth et al., 1989; Rushlow et al., 1989; Steward et al., 1988) occurs as a result of two maternally encoded signaling cascades. In the first, a signal is relayed from the oocyte to the layer of follicle cells surrounding the oocyte and then back to the eggshell (Gonzalez-Reyes et al., 1995; Gonzalez-Reyes and St Johnston, 1994). The net result is the deposition of a latent asymmetric signal (green arrows in Figure 8.2B) in the perivitelline space, i.e., the space between the inner eggshell membrane (the vitelline membrane) and the oocyte plasma membrane.

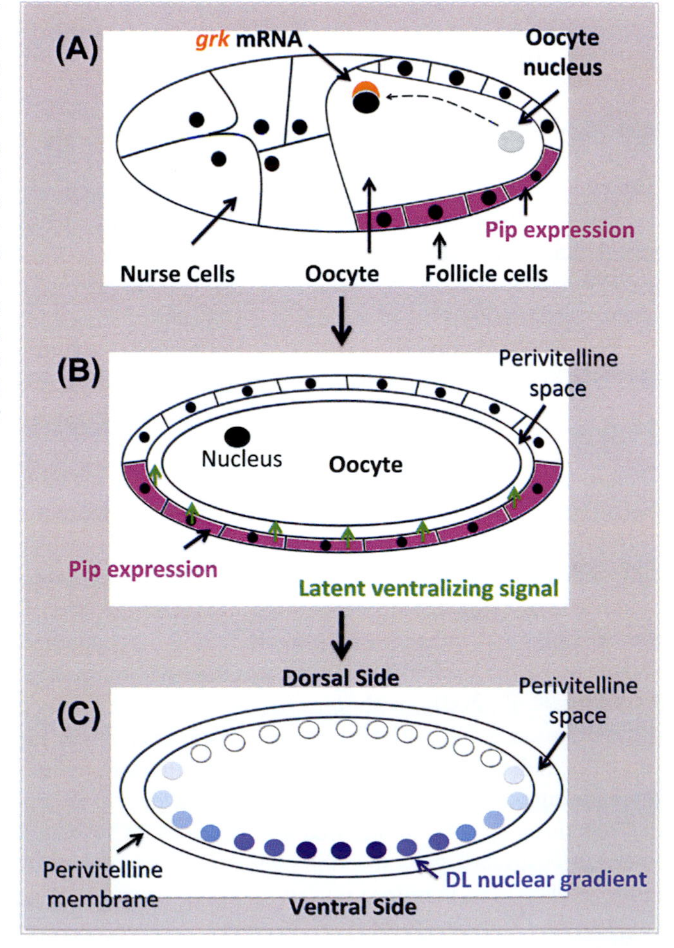

FIGURE 8.2 From Oocyte to Embryo. The Establishment of the DV Axis. (A) Within an egg chamber, DV polarity in the oocyte results from asymmetric localization of the oocyte nucleus. The oocyte nucleus moves (movement shown by a dashed arrow) from an initial posterior location (gray circle; dashed margin) to a final anterior location. The side to which the nucleus is closest upon its arrival at the anterior of the oocyte becomes the dorsal side of the future embryo. Specification of the dorsal side occurs when the oocyte nucleus produces *grk* mRNA which accumulates in the region between the nucleus and the oocyte plasma membrane. (B) Grk protein, which is secreted from the region which localizes *grk* mRNA, signals through the EGF receptor in the adjacent follicle cells leading to the repression of *pip* expression. *pip* expression is therefore restricted to the follicle cells on the ventral side of the future embryo (colored pink). Pip expression leads to the synthesis of a latent ventralizing signal (green arrows) that is deposited into the perivitelline space by the follicle cells before the egg shell is formed. (C) The dorsal gradient forms in the embryo a few hours after fertilization, within the *pip* expression domain.

After fertilization, this latent signal activates the second signaling cascade (Figure 8.3A), which transduces the signal to the interior of the zygote and leads to the ventral-specific nuclear uptake of DL.

Interactions Between Follicle Cells and the Germ Line Define DV Polarity

The 16 germ line cells in each *Drosophila* egg chamber are surrounded by a monolayer of follicle cells. These cells produce the eggshell, including the inner vitelline membrane and the outer chorion. In addition, as we will see, they receive signals from the oocyte and, in response, send a polarizing signal back to it.

The symmetry-breaking event leading to DV polarity is the movement of the oocyte nucleus from an initial posterior position (Figure 8.2A, dashed arrow) to an anterior location, a movement that is dependent on a polarized, intact microtubule (MT) network within the oocyte. Reciprocal signaling between the follicle cells and the oocyte regulates this movement. Gurken (Grk), a TGFα-like ligand, is secreted by the oocyte, and activates Torpedo (Top), the *Drosophila* Epidermal Growth Factor Receptor (EGFR), on the surface of follicle cells. An uncharacterized signal from the follicle cells back to the oocyte leads to the reorganization of the MT network placing the minus end of the microtubules at the anterior end of the oocyte (Gonzalez-Reyes et al., 1995; Gonzalez-Reyes and St Johnston, 1994). Two models have been put forward to explain the relationship between MT network reorganization and nuclear migration. In one model, the reorganization of the MT network occurs first, followed by the dynein motor-dependent transport of the passive nucleus toward the minus end of the microtubules (Roth and Lynch, 2012). In the second model, the nucleus migrates with the centrosomes, and this complex with its associated microtubule organizing center nucleates the formation of microtubules on the posterior side of the nucleus. The polymerizing microtubules then serve to push the nucleus/centrosomal complex toward the anterior of the oocyte (Roth and Lynch, 2012). Live imaging (Zhao et al., 2012) of nuclear movement in the oocyte, along with visualization of centrosomal and MT dynamics supports the second model.

When the nucleus reaches its correct anterior location it is in close proximity to the cell membrane, although its position is thought to be random relative to the maternal DV axis. However, by the mechanism described below, the side of the

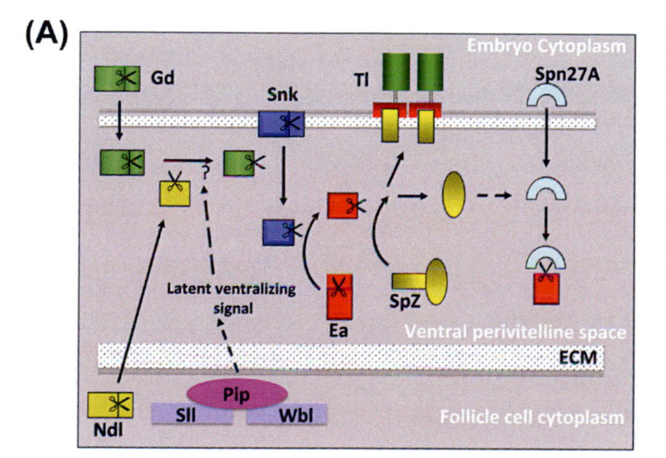

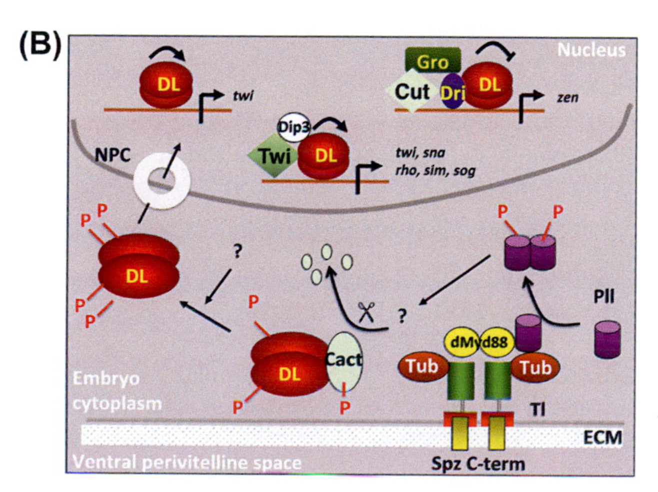

FIGURE 8.3 Transduction of the Latent Ventralizing Signal to the Interior of the Embryo Leads to Nuclear DL Import. (A) After fertilization, the latent ventralizing signal triggers a cascade of proteolytic cleavages by serine proteases, leading to the processing and activation of the Tl ligand Spz. The serine proteases are converted from an inactive state (shown as constrained scissors) to a proteolytically active state (scissors free to cleave). (B) Asymmetric activation of the Tl receptor initiates a signal transduction pathway that leads to the degradation of Cact and the graded nuclear import of DL. Once in the nucleus, DL activates and represses a number of genes that specify cell fate along the DV axis.

oocyte in closest proximity to the nucleus gives rise to the dorsal side of the oocyte and the embryo. Export of *grk* mRNA from the oocyte nucleus results in the accumulation of high levels of *grk* mRNA in a small arc between the nucleus and the plasma membrane (Figure 8.2A). *grk* ribonucleoprotein complexes are associated with the periphery of processing bodies (P bodies) along with translational activator Oo18 RNA binding protein (Orb) and anchoring factor Squid (Sqd) (Weil et al., 2012). Sqd associates with a number of other factors required for *grk* mRNA localization and translational control such as Hrb27C, Imp and Syncrip (Goodrich et al., 2004; McDermott et al., 2012). A second round of Grk signaling is then initiated from the oocyte to the overlying follicle cells resulting in a gradient of EGFR activation in the follicle cell epithelium. Highest levels of EGFR activity occur in the follicle cells closest to the nuclear source of grk mRNA. This high level of EGFR activity defines this side of the follicular epithelium as the dorsal side. EGFR signaling inside the follicle cells then represses *pipe* (*pip*) expression. As a consequence of the EGFR activation gradient, *pip* expression is restricted to the follicle cells adjacent to what will become the ventral side of the embryo (Figure 8.2A,B). *pip* is expressed uniformly in the ventral 40% of the follicle cells. The pip expression domain has a sharp border, suggesting that repression requires a certain threshold level of EGFR activity, which is only exceeded in the dorsal 60% of the egg chamber.

pip encodes a Golgi body-localized heparin sulfate 2-*O*-sulfotransferase that probably modifies glycosaminoglycans (GAG) of the extracellular matrix (ECM) during their transit through the Golgi apparatus (Sen et al., 2000). GAGs are long, unbranched sugar polymers that are attached to specific serine residues of proteins termed proteoglycans. Support for the idea that Pip catalyzes GAG sulfation comes from studies showing that the synthesis of 3'-phosphoadenosine 5'-phosphosulfate (PAPS), the sulfate group donor for sulfotransferase reactions, and the import of PAPS into the Golgi apparatus are required for DV patterning. It is assumed that Pip modifies unknown proteoglycan substrates and deposits them into the space outside the oocyte before the vitelline membrane and chorion are secreted. Because *pip* mRNA is only expressed ventrally, the modified substrate is presumably restricted to the perivitelline space on the ventral side of the embryo, where it serves as a latent ventral signal (Figure 8.2B, green arrows). Zhang et al. (2009) used mass spectrometry in combination with radioactive sulfate labeling to discover one possible target of Pip, a spatially localized protein called Vitelline membrane like (Vml). Vml mutants however do not show DV defects, indicating that other perhaps redundant Pip targets involved in DV patterning exist (Zhang et al., 2009a; Zhang et al., 2009b). It is possible that Pip sulfonates a number of proteins that are deposited in a restricted spatial distribution in the vitelline membrane, and that these proteins as a group represent the ventral latent signal. As we will see in a later section (Section "Asymmetric signaling by the Tl receptor leads to formation of a DL nuclear concentration gradient"), DL nuclear import after fertilization occurs in regions that abut the Pip expressing follicle cells during oogenesis (Figure 8.2B,C).

A Proteolytic Cascade in the Perivitelline Fluid Transduces the Ventral Signal from the Perivitelline Space to the Toll Ligand

The formation of a DL nuclear concentration gradient requires the activation of Toll (Tl), a receptor that is uniformly distributed throughout the plasma membrane of the syncytial embryo, by its ligand Spätzle (Spz). The production of activated Spz after fertilization requires a proteolytic cascade (Figure 8.3A) in the ventral perivitelline space (LeMosy et al., 1999; Morisato and Anderson, 1994; Schneider et al., 1994; Smith and DeLotto, 1994). This cascade is initiated by the Pip-dependent latent signals in the pervitelline space produced as described above. The proteolytic cascade involves the successive activation of three germ line encoded serine proteases: Gastrulation defective (Gd), Snake (Snk), and Easter (Ea). Gd, Snk and Ea form a protein complex (Cho et al., 2012). Activation of Gd requires unknown Pip modified substrates, including Vml and also Nudel (Nd), a serine protease that is secreted from follicle cells. Activated Gd then cleaves and activates Snk, which cleaves and activates Ea. Processed activated Ea proteolytically processes and thereby activates Spz. The resulting C-terminal fragment of processed Spz (Schneider et al., 1994) binds and activates the Tl receptor (Figure 8.3A,B). The restriction of Spz activation to the ventral perivitelline space appears to require the Pip-dependent localization of Gd to this region. In addition to catalyzing the cleavage of Snk, Gd as a component of a complex that includes both Snk and Ea may induce conformational changes in Snk that stimulate cleavage of Ea (Cho et al., 2012).

As we will see below, spatially restricted activation of the Tl receptor, which is uniformly distributed throughout the egg plasma membrane, results in the graded nuclear uptake of Dorsal (DL). It is not known how the uniform expression of *pip* in the ventral 40% of the egg chamber translates into the gradient of DL nuclear localization (Figures 8.2C, 8.4A). It is possible that the Ea inhibitor Serpin27A (Spn27a) plays a role in this process (Hashimoto et al., 2003). This inhibitor is secreted from the oocyte and may be preferentially secreted from the dorsal side of the embryo due to inhibition of secretion by Tl signaling. One possibility is that the ventral to dorsal diffusion of Ea together with the dorsal to ventral diffusion of Spn27A lead to a gradient of Ea activity (Chang and Morisato, 2002), which results in a gradient of processed Spz and therefore a gradient of Tl activity (Moussian and Roth, 2005; Stein et al., 1991).

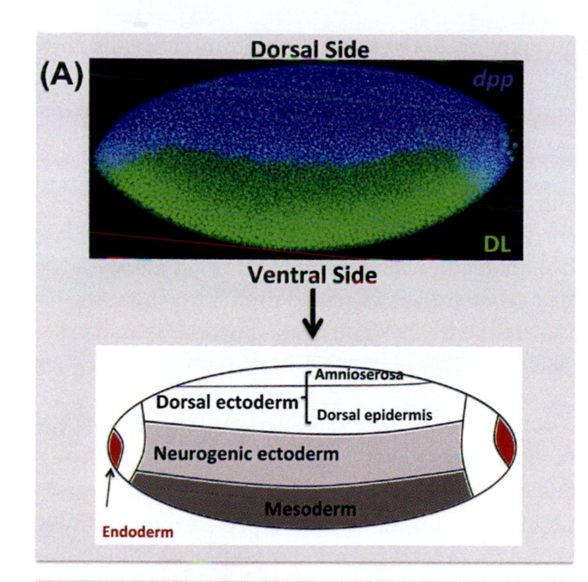

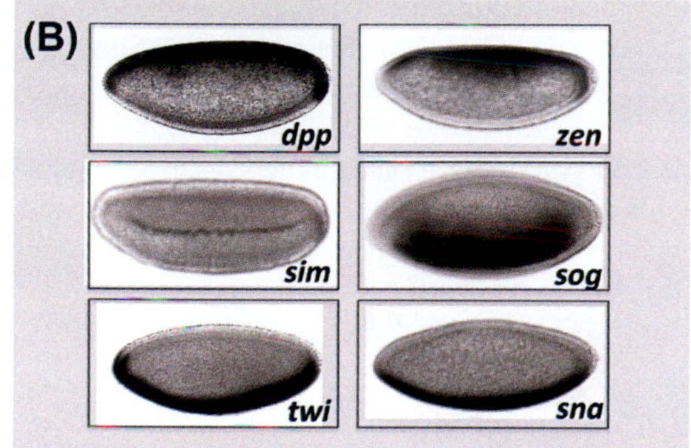

FIGURE 8.4 Activation and Repression by DL Leads to Specification of the Presumptive Germ Layers. (A) The relationship between the DL nuclear gradient and the presumptive germ layers at the cellular blastoderm stage. The DL gradient (green) is visualized using an antibody that recognizes the DL Rel Homology Domain. DL represses *dpp* (blue) transcription, restricting its expression to the dorsal and lateral regions of the embryo. (B) A subset of genes that are activated (*twi, sim sog, sna*) and repressed (*dpp, zen*) by DL. DL and DL target genes specify cell fate on the ventral and lateral sides of the blastoderm embryo. The mRNA is visualized by hybridization to labeled RNA probes.

Asymmetric Signaling by the Tl Receptor Leads to Formation of a DL Nuclear Concentration Gradient

Tl and its homologs in vertebrates (the Toll-like receptors) play conserved roles in the insect and vertebrate innate immune response (Anderson, 2000; Lemaitre et al., 1995). Thus, while the role of Tl in DV pattern formation may not be conserved in vertebrates (see section "The DV regulatory network"), studies of DV patterning in *Drosophila* and innate immunity have illuminated one another (Brennan and Anderson, 2004; Imler and Hoffmann, 2001). Tl signaling in the syncytial embryo is required for the nuclear import of DL. DL, a transcription factor, is encoded by a uniformly distributed maternally supplied mRNA, and therefore the translation of the mRNA in the syncytial embryo results in the accumulation of uniformly distributed DL protein. Tl regulates DL nuclear uptake in two ways. First, the Tl signal leads to degradation of Cactus (Cact), an inhibitory factor and a homolog of vertebrate I-κB that binds DL and sequesters it in the cytoplasm (Belvin et al., 1995; Bergmann et al., 1996; Govind et al., 1993; Roth et al., 1991). Second, the Tl signal may act directly on DL to enhance its nuclear import. The asymmetric distribution of the Tl ligand results in a DL nuclear concentration gradient, with highest levels of DL in ventral nuclei, low levels of DL in lateral nuclei, and little or no DL in dorsal nuclei (Roth et al., 1989; Rushlow et al., 1989; Steward et al., 1988) (Figure 8.4A).

Binding of activated Spz to the Tl receptor leads to dimerization of Tl (Figure 8.3B). Live imaging of Tl signaling using Tl-GFP fusions indicates that upon dimerization, Tl is endocytosed and signaling may occur from the endosomal compartment rather than the plasma membrane (Lund et al., 2010). The Tl signaling pathway is thought to have the following features: Endocytosis of Tl is required for Tl signaling, as blocking endocytosis using inhibitors reduces nuclear DL (Lund et al., 2010) while augmenting endocytosis increases Tl signaling (DeLotto, 2011). Also, dominant, constitutively active forms of Tl, such as Tl^{10B} preferentially partition into endosomes.

Dimerized Tl recruits dMyD88 via a shared TIR (Toll, IL-1 receptor, Resistance) motif. dMy88 complexes with Tube (Tub) via DEATH domain motifs present in both proteins (Feinstein et al., 1995). The Tl/dMyD88/Tub complex can then recruit the Pelle (Pll) kinase, which also contains a DEATH domain. The increase in the local concentration of these proteins near the Tl receptor leads to Pll autophosphorylation and also to phosphorylation of Tl and Tub by Pll (Norris and Manley, 1996; Shen and Manley, 2002; Sun et al., 2002). This has two consequences. First, phosphorylated Pll is released from the complex and thus signaling is rendered self-limiting. Second, the released phosphorylated Pll can now act on downstream components, leading to the phosphorylation of Cact.

The question of whether Pll acts on Cact directly or indirectly is controversial. Support for direct action comes from *in vitro* assays, which indicate that Pll can directly phosphorylate Cact. In vertebrates, however, the DL homolog NF-κB is sometimes phosphorylated by the I-κB kinase (IKK) complex, which includes the catalytic subunits IKKα and IKKβ. Support for the idea that similar kinases might regulate DL function comes from studies of innate immunity showing that IKKβ family kinase Ird5 is downstream of Pll and responsible for phosphorylation of Cact (Lu et al., 2001). Ird5 mutations do not, however, result in major defects in DV pattern formation, suggesting that there may be multiple partially redundant kinases required for Cact phosphorylation.

Cact blocks DL nuclear uptake perhaps by physically tethering DL in the cytoplasm (Figure 8.3B). Phosphorylation of Cact leads to its dissociation from DL and to its degradation (Govind et al., 1993; Reach et al., 1996; Roth et al., 1991) thereby allowing DL to interact with the nuclear importins, which escort DL into the nucleus via the nuclear pore complexes. In the case of I-κB, the vertebrate homolog of Cact, degradation depends on the ubiquitin/proteasome pathway. Phosphorylation of I-κB by IKK at two specific serine residues leads to recognition and polyubiquitylation by a ubiquitin ligase containing the F-box protein β-TrCP. The polyubiquitylated protein is then recognized and degraded by the 26S proteasome (Spencer et al., 1999). The phosphorylation sites in I-κB are not conserved in Cact. Nonetheless, embryos deficient in Slimb, the Drosophila homolog of β-TrCP, exhibit DV patterning defects, suggesting that Cactus degradation occurs by a pathway that is similar to the I-κB degradation pathway.

DL itself is phosphorylated (Drier et al., 1999) in a manner that is dependent upon activated Tl and this appears to be required for nuclear import (Figure 8.3B). Further evidence for the physiological significance of DL phosphorylation comes from studies of a mutant form of DL that is unable to bind Cact. Nuclear entry of this form of DL is not completely unregulated as one would expect if all Tll signaling went through Cact. Instead, although the DL gradient is extended, more nuclear uptake occurs ventrally than dorsally and this nuclear uptake is dependent upon the genes responsible for Spz processing (Drier et al., 2000).

The DL Nuclear Concentration Gradient Subdivides the Embryo into Multiple Developmental Domains

The DL nuclear concentration gradient (Figures 8.2C, 8.4A), detected by antibody staining, is first seen 90 minutes post fertilization, at nuclear cycle 10 in the syncytial embryo. The DL gradient persists for approximately 90 minutes, through nuclear cycle 14 and cellularization but disappears with the onset of gastrulation. Although the nuclear concentration gradient results from the ventral and ventrolateral specific nuclear uptake of DL, DL shuttles in a dynamic manner between the nucleus and the cytoplasm along the entire DV axis, implying the existence of mechanisms to maintain the gradient for this 90 minute period. With the development of fluorescent protein reporters such as GFP and the ability to image animal development live, it has been possible to follow the dynamics of the formation of the DL gradient in live embryos (DeLotto et al., 2007; Kanodia et al., 2009). Quantitative imaging, in conjunction with mathematical methods, allows the gradients to be modeled spatiotemporally. Analysis of the dynamics of the DL gradient reveals that it increases in amplitude from nuclear cycle 10–14, but maintains its shape. The DL nuclear protein concentration fluctuates during the cell cycles, but from interphase to interphase the levels increase in a monotonic fashion (Kanodia et al., 2009). This is in contrast to the AP Bicoid gradient, in which neither the amplitude nor the shape changes with time.

As discussed in the section "AP and DV polarity is specified in the developing ovariole", the blastoderm embryo contains three DV developmental domains: the mesoderm, the neurogenic ectoderm, and the dorsal ectoderm (Figure 8.4A). DL establishes these domains by functioning as both a transcriptional activator and a repressor to direct the spatially restricted expression of zygotically active DV patterning genes (reviewed in Stathopoulos and Levine, 2002). In general, DL activates the genes required for the mesodermal and neurogenic ectodermal fates (Figure 8.4A).

Dorsal Activates a Number of Genes in the Blastoderm Embryo

Based on a genomic promoter/enhancer sequence analysis, DL is believed to modulate the expression of 50–60 genes in the blastoderm embryo (Markstein et al., 2002; Stathopoulos et al., 2002). Of these, about 30 have been experimentally verified as DL targets. DL regulates these target genes via enhancer or silencer elements containing critical DL binding sites. DL-dependent enhancers are termed ventral activation regions (VARs), whereas DL-dependent silencers are termed ventral repression regions (VRRs) (Courey and Jia, 2001). Whether or not DL activates or represses any given target gene and the threshold concentration of DL required for this activation or repression depends on the quality and context of the binding sites (Ip et al., 1991; Jiang et al., 1991; Pan and Courey, 1992).

Genes such as *twist* (*twi*) and *snail* (*sna*), which encode mesodermal determinants, are only activated by the high concentrations of DL present in the ventral mesodermal anlage, while genes such as *short gastrulation* (*sog*) and *rhomboid* (*rho*), which are required for neurogenic ectodermal development, can be activated at lower DL concentrations (Figure 8.4B). *sog*, for example, is initially activated throughout the mesodermal and neurogenic ectodermal anlagen (the ventral 60% of the blastoderm embryo) (Stathopoulos and Levine, 2002). These neurogenic ectodermal genes are, however, very rapidly repressed by Sna in the presumptive mesoderm. *Single minded* (*sim*; Figure 8.4B) is expressed in a single layer of cells on each side of the embryo (Linne et al., 2012; Nambu et al., 1990). The restricted expression is a consequence of repression by Sna on the ventral side of the embryo and also by Notch mediated lateral inhibition. Sim expressing cells move to the ventral midline of the embryo by the end of gastrulation and Sim expression is required for patterning of ventral neurons and epidermis. DL functions as a repressor of genes such as *dpp*, *zerknüllt* (*zen*), and *tolloid* (*tld*), which are required for the dorsal ectodermal fate, in this way restricting their expression to the dorsal-most ~40% of the embryo (i.e., the region lacking nuclear DL) (Figure 8.4B) (Ip et al., 1992; St Johnston and Gelbart, 1987; Thisse and Thisse, 1992).

Dorsal Binding Affinity and Synergistic Interactions Determine the Borders of the Expression Domains

Binding sites that closely match the DL consensus recognition element are of high affinity and therefore interact with DL at lower concentrations. Thus, high affinity binding sites direct activation or repression in a broader ventral domain than do low affinity DL binding sites (Jiang and Levine, 1993). For example, the *twi* gene encodes a mesodermal determinant that is only expressed at high levels in the ventral 20% of the blastoderm embryo; where DL concentration is highest. Accordingly the two VARs in Twist contain only low affinity DL binding sites that match the consensus very poorly. In contrast, the *zen* gene encodes a dorsal ectodermal determinant that must be repressed by DL throughout the mesodermal and neurogenic ectodermal anlagen. Accordingly, the *zen* VRR contains several consensus and therefore high affinity DL binding sites (Stathopoulos and Levine, 2002).

The size of a domain of DL-mediated activation can be influenced by synergistic interactions between DL and other activators bound to nearby sites (Jiang and Levine, 1993). Expanded domains of activation of certain neurogenic ectodermal genes such as *rho* depend on binding sites for basic-helix-loop helix (bHLH) transcription factors close to the DL binding sites. bHLH factors such as Twi and Daughterless (Da) bound to these sites synergize with DL allowing activation in regions of lower DL concentration (Jiang and Levine, 1993). In general, the molecular mechanisms behind this synergy have not been well-defined.

Dorsal can also Function as a Groucho-Dependent Repressor

While DL is, by default, a transcriptional activator, it also actively represses the expression of some of the genes required for formation and patterning of the dorsal ectoderm including *dpp*, *zen* and *tld* (St Johnston and Gelbart, 1987). Repression of these targets requires DL-mediated recruitment of the co-repressor protein Groucho (Gro) to the VRRs in these genes (Dubnicoff et al., 1997) (Figure 8.3B). The affinity of DL for Gro appears to be quite low, thus explaining why DL is normally an activator and not a repressor. As expected, a DL variant with a strong Gro recruiting site does not activate DL targets (Ratnaparkhi et al., 2006). Repression of targets such as *zen* and *dpp* requires additional repressor proteins that binding to sites in the VRRs close to the DL binding sites and apparently assist DL in the recruitment Gro. This has been best studied in the case of the *zen* gene, which contains a VRR in the 5' flanking region. The *zen* VRR contains three high affinity DL binding sites as well as three evolutionarily conserved AT rich elements. These AT rich elements serve as binding sites for the sequence specific transcription factor Dead ringer (Dri); biochemical experiments show that DL and Dri bound to adjacent sites in DNA can cooperatively recruit Gro to the DNA. This has lead to the conclusion that DL and Dri serve to nucleate the formation of a Gro-containing multiprotein DNA-bound complex (a "repressosome") required for DL-mediated silencing (Valentine et al., 1998). Support for this conclusion comes from experiments showing that mutagenesis of the AT rich elements or elimination of Dri from the early embryo converts the zen VRR to a VAR (Jiang et al., 1993; Valentine et al., 1998). Furthermore, the activity of the VRR is critically dependent upon the spacing between the DL and Dri binding sites (Cai et al., 1996). Thus, when the spacing between one of the DL and one of the AT rich sites was increased by a non-integral multiple of the helical repeat, repression was lost – which is consistent with the idea that DL and Dri must be aligned on the same face of the helix to allow for efficient Gro recruitment. Consistent with the idea that DL-mediated repression requires cooperative recruitment of Gro by DL and additional repressors such as Dri, replacement of the DL C-terminal region (which contains both activation and repression domains) with the VP16 or p65 activation domain converts DL into a factor that is only capable of activation (Ratnaparkhi et al., 2006).

8.4 DPP/SOG ACTIVITY GRADIENTS ARE RESPONSIBLE FOR FURTHER PATTERNING OF THE DV AXIS

Repression and Activation by DL Establishes the Polarity of the Dpp/Sog Pattern Forming System

As discussed in the previous section, patterning of the *Drosophila* DV axis requires a DL nuclear concentration gradient (Figure 8.4A). This DL gradient serves two broad purposes. First, different threshold concentrations of DL activate genes required for the establishment and development of the ventral mesodermal domain and the ventrolateral neuroectodermal domain, including *sna*, *twi*, *rho*, etc. Second, the DL gradient activates and represses genes encoding some of the components of the Dpp/Sog morphogen system, which is required for subdividing the dorsal ectodermal domain. While the DL system may not be required for DV patterning in vertebrate embryos, we will see that the Dpp/Sog system plays a role in patterning the vertebrate DV axis (section "The DV regulatory network") that is homologous to its role in *Drosophila* embryogenesis.

As mentioned previously, Dpp is an extracellular signaling molecule with a high degree of similarity to vertebrate BMP 2 and 4 and is a member of the TGFβ superfamily (Ferguson and Anderson, 1992), while Sog (Francois et al., 1994; Holley et al., 1996; Schmidt et al., 1995) is also a secreted protein and is related to vertebrate Chordin. *dpp* transcription is activated by uniformly distributed activators, but repressed by DL, which restricts *dpp* transcription (Ray et al., 1991) to the dorsal ectodermal anlage (roughly the dorsal-most 40% of the embryo). *sog* transcription is activated by DL throughout the mesodermal and neurogenic ectodermal anlagen (roughly the ventral-most 60% of the embryo). Thus, because DL activates *sog*, but represses *dpp*, the transcripts of these two genes accumulate in non-overlapping, but abutting domains – *dpp* is transcribed where DL is absent, while *sog* is transcribed where DL is present (Figures 8.4B, 8.5). As we will see, despite their non-overlapping domains of expression, Sog works with Dpp to subdivide the dorsal ectoderm into amnioserosa and dorsal epidermis (reviewed in Ashe, 2005).

The BMP Signal Transduction System

In addition to Dpp, patterning of the dorsal ectoderm requires a second BMP family ligand, Screw (Scw) (Arora et al., 1994). Like all members of the TGFβ superfamily, Dpp and Scw function as dimers (either homodimers or heterodimers). The Dpp/Scw heterodimer is thought to be a much more potent signaling agent than either homodimer. Each BMP receptor contains type I and type II subunits (probably two molecules of each), and each subunit possesses serine/threonine kinase activity. The *Drosophila* genome encodes a single type II kinase called Punt (Pnt) that is common to all the *Drosophila* BMP receptors, and two type I kinases called Thickveins (Thv) and Saxophone (Sax), which are required for signaling by Dpp and Scw respectively (Nguyen et al., 1998; Ruberte et al., 1995).

Ligands are thought to recruit the constitutively active type II kinases to the type I kinases allowing phosphorylation and activation of the type I kinases. This, in turn, leads to the phosphorylation of the Smad family transcription factor Mad, which binds to another Smad family factor, Medea (Med), enters the nucleus, and activates and represses downstream genes that regulate developmental fate.

Using tagged forms of Dpp (Shimmi et al., 2005; Wang and Ferguson, 2005), it is possible to visualize the distribution of Dpp protein. During early cellular blastoderm formation, Dpp protein is distributed in a broad dorsal domain that reflects the broad distribution of the dpp transcript. However, as cellularization proceeds, a dramatic redistribution occurs – Dpp comes to be localized in a step gradient, with high concentrations in a stripe along the dorsal midline (in the presumptive amnioserosa) and lower concentrations in dorsolateral regions (the presumptive dorsal epidermis) (Figures 8.5A, 8.6A) (Ashe et al., 2000). This asymmetric distribution of Dpp is what drives the subdivision of the dorsal ectoderm into amnioserosa and dorsal epidermis (Figures 8.5A, 8.6A).

Diffusion of Sog Towards the Dorsal Midline Leads to the Formation of a Dpp Activity Gradient

The redistribution of Dpp is thought to depend on Sog (Ashe and Levine, 1999; Francois et al., 1994; Marques et al., 1997) (Figure 8.5A). Sog forms a gradient with high concentrations in the lateral neurogenic ectoderm (where the transcript is

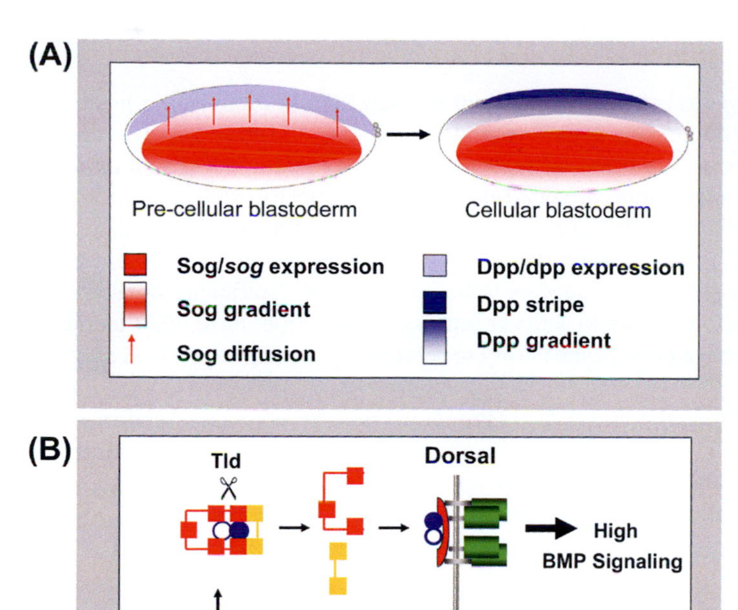

FIGURE 8.5 Formation of a DV Dpp Gradient. Interaction of the Sog gradient with Dpp leads to formation of an extracellular Dpp activity gradient. Dpp signaling via the Dpp receptors leads to activation and repression of genes that specify cell fate in dorsal and dorsolateral region of the blastoderm embryo. (A) *sog* is a DL target gene that is expressed in a broad lateral stripe in the blastoderm embryo. Sog is secreted into the extracellular space and dif- fuses away from the domain of *sog* expression, thus forming Sog gradients both ventral and dorsal to the zone of sog expression. In the dorsal region, the interaction of Sog with Dpp leads to the conversion of a homogenous Dpp expression domain in the dorsal 40% of the precellular embryo into an extracellular Dpp activity gradient in the cellular blastoderm embryo. Ultimately, extracellular Dpp comes to be distributed in a step gradient, with the highest concentrations at the dorsal midline and lower concentrations laterally. (B) Sog binds to Dpp/Scw heterodimers in a complex that also includes Tsg and that facilitates the diffusion of Dpp by blocking binding to the transmembrane receptor composed of *Sax, Tkv, and Pnt* gene products. At the dorsal midline, the degradation of Sog by Tld liberates the Dpp/Scw heterodimer. As a result of the low concentrations of intact Sog at the dorsal midline, the Dpp/Scw heterodimer does not rebind Sog but instead binds the receptor. This leads to the accumulation of Dpp/Scw at the dorsal midline, which results in the formation of the Dpp step gradient.

produced) and lower concentrations towards the dorsal and ventral midlines (Figure 8.5A) (Srinivasan et al., 2002). Genetic analysis suggests that Sog plays both negative and positive roles in Dpp regulation (Araujo and Bier, 2000; Biehs et al., 1996; Decotto and Ferguson, 2001; Francois et al., 1994). Sog binds to both Dpp homodimers and Dpp/Scw heterodimers in complexes that also contain the product of the *twisted gastrulation* (*tsg*) gene. When present in this complex, Dpp and Scw are unable to bind their receptors (Figure 8.5B). This prevents them from interfering with the normal developmental program in the neuroectodermal and mesodermal domains. At the same time, Sog facilitates diffusion of these ligands through the perivitelline space (red arrow, Figure 8.5A,B) to the dorsal midline resulting in the zone of high Dpp concentration at the dorsal midline (Ashe, 2005; Shimmi et al., 2005; Wang and Ferguson, 2005). Collagen IV, a component of the extracellular matrix that binds both Dpp and Sog, also modulates Dpp spreading (Sawala et al., 2012). Mutations in the two genes encoding Collagen IV variants (*viking* and *DCg1*) lead to spreading of the Dpp activity gradient indicating roles for Collagen IV in restraining or immobilizing Dpp.

How can diffusion of Sog generate a zone of concentrated Dpp around the dorsal midline? Key to this process is the protease encoded by the *tolloid* (*tld*) gene and expressed on the dorsal side of the embryo (scissors, Figure 8.5B). This protease cleaves Sog. Importantly, however, the protease does not cleave free Sog, but only Sog in association with the BMP ligands, and it is most active on complexes containing the Dpp/Scw heterodimer. This Tld cleavage of Sog liberates the Dpp/Scw heterodimer from the complex potentially allowing it to bind its receptor (Shimmi et al., 2005; Wang and Ferguson, 2005) and signal on the dorsal side or to bind to free Sog. Around the dorsal midline, the concentration of Sog is at a minimum and is too low to allow efficient rebinding of Sog to Dpp/Scw after Tld-mediated Sog degradation. Therefore, near the dorsal midline the consequence of Sog degradation is the immobilization of Dpp and Scw by binding to their receptors. The net result is the accumulation of high concentrations of the Dpp/Scw heterodimer around the dorsal midline (Figure 8.5B). While it is possible that this sharp step gradient of Dpp localization and signaling is due exclusively to the Sog gradient, it seems more likely that sharpening of the Dpp stripe requires positive autoregulation by the Dpp signaling pathway (Wang and Ferguson, 2005).

Redundancy in the BMP Signaling System

As we saw above, the activation of Sog by DL on the ventral side of the embryo is a key step in establishing positional information by the BMP morphogen system, since diffusion of Sog from a ventral source leads to formation of a gradient of signaling activity. At the same time, DL also represses both *dpp* and *tld* ventrally, producing additional spatial information that is at least partially redundant with the spatial information provided by the ventral-specific activation of *sog*. A DL variant that can activate but not repress transcription patterns the DV axis almost normally (Ratnaparkhi et al., 2006). This indicates redundancy in repression of Dpp signaling by indicating that DL-mediated repression is not essential for specification of the DV axis. DL activation (of *sog* and *brk*) by itself appears to be necessary and sufficient to pattern the DV axis. DL is one of the few rel family transcription factors known to actively repress transcription; most other members of this family appear to be dedicated activators. DL may have evolved the ability to repress *dpp* and *tld* expression after the vertebrate/invertebrate evolutionary split.

Additional redundancy in the BMP patterning system is also found downstream of receptor activation, and involves the action of the transcriptional repressor Brinker (Brk). The *brk* gene is activated by DL, and its expression is therefore limited to ventral regions where Dpp signaling is low or absent. Brk represses some of the same targets that are activated by pMad/Med. As a result, BMP targets are both activated dorsally by pMad/Med and repressed ventrally by Brk, redundantly ensuring that these targets will only be expressed in dorsal regions (Affolter et al., 2001; Jazwinska et al., 1999; Minami et al., 1999).

The redundancy of this spatial information is illustrated by experiments in which tagged forms of Dpp are ectopically expressed along the entire DV axis of the embryo. In these experiments, most of the tagged Dpp still manages to find its way to the dorsal midline of the embryo. Thus, asymmetric expression of the *dpp* gene appears to be dispensable for the ultimate asymmetric distribution of the Dpp protein, as long as the Dpp inhibitor Sog is asymmetrically distributed (Shimmi and O'Connor, 2003; Shimmi et al., 2005; Srinivasan et al., 2002).

Such redundant mechanisms are being found increasingly in other developmental pathways (Barolo and Posakony, 2002). While the exact reason for the multiple forms of redundancy in the BMP system is not clear, it is likely that it helps to render the pattern forming system robust in the face of variations in environmental conditions that might lead to minor perturbations of the Sog gradient.

8.5 THE DV REGULATORY NETWORK

The preceding sections of this chapter describe a complex gene regulatory network (GRN) involving the DL and Dpp/Sog gradients, the genes that direct the formation of these gradients, and the genes that are targeted by these gradients (Figure 8.6; reviewed in Levine and Davidson, 2005). The regulatory interactions that have been discussed have all been illuminated by traditional genetic and biochemical approaches. However, the use of bioinformatic methods to analyze the *Drosophila* genome, microarray methods, and the systematic examination of gene expression patterns are leading to the discovery of additional members of the DV GRN, which is now known to include 60 genes (Levine and Davidson, 2005; Stathopoulos et al., 2002) (Figure 8.6), about 40 of which are expressed at the cellular blastoderm stage.

Features of gene regulation and pattern formation that were not obvious from analysis of single genes can be appreciated at the level of the network. For example, by identifying genes with novel expression patterns, it may be possible to illuminate further the combinatorial interactions responsible for subdivision of the DV axis. In addition, the elucidation of gene networks that regulate the same process (e.g., germ layer specification) in diverse species will provide insights into mechanisms of molecular evolution.

8.6 COMPARISON OF DV PATTERNING IN *DROSOPHILA* AND VERTEBRATES

Tl Signaling

The Tl signaling pathway has at least two functions in the life of the fruit fly. In addition to its role in embryonic DV patterning, this pathway is also required for the *Drosophila* immune response (Anderson, 2000; Imler and Hoffmann, 2001). In insects, immunity means innate immunity, since insects lack the systems required for adaptive immunity (e.g., B and T lymphocytes). The innate immune response is triggered by the recognition of features common to many pathogens such as microbial cell wall derived lipopolysaccharides, peptidoglycans, and lipoproteins. These molecules trigger the production of antimicrobial peptides, which lyse the invading microbes (the humoral response), as well as the activation of macrophages, which engulf the microbes (the cellular response). Both of these responses are partially dependent on the Tl receptor, and

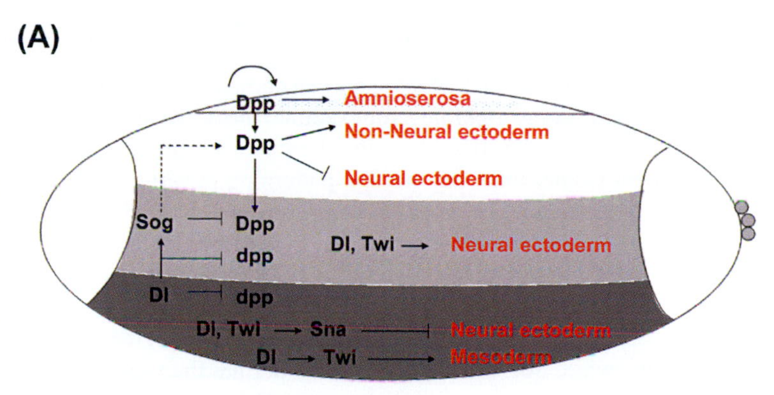

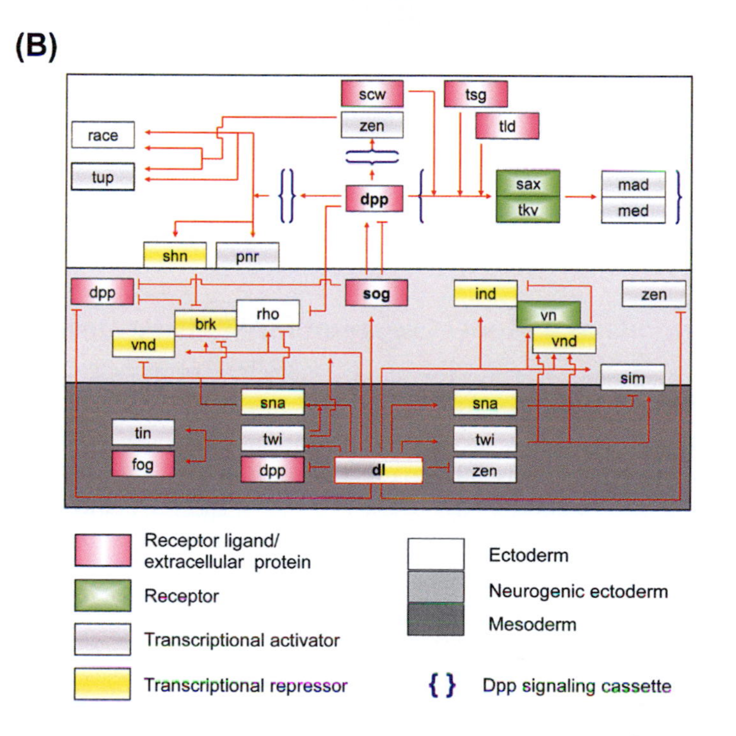

FIGURE 8.6 A Network of Gene Regulation in *Drosophila* DV Patterning. (A) The DL and Dpp activity gradients (Figure 4) lead to the spatially restricted activation and repression of a number of zygotically active DV patterning genes. DL represses *dpp* in ventral and lateral regions and also blocks Dpp signaling laterally by activating *sog* transcription. Sog binds to Dpp, blocks its binding to the Dpp receptor in laternal regions and generates an activity gradient of Dpp, with high Dpp signaling dorsally low signaling laterally. (B) Most of the genes regulated by DL or the Dpp signaling pathway encode transcription factors or signaling molecules that are all linked in a complex gene regulatory network (Levine and Davidson, 2005). Each developmental domain (dorsal ectoderm, neurogenic ectoderm, or mesoderm) expresses a set of DL or Dpp target genes that initially mark and later specify the domain. Arrows (→) indicate the activation of a gene; bars (—|) indicate the repression of a gene. DL is bifunctional and activates some genes (e.g., *twi, sna, sog*) while repressing others (e.g., *dpp, zen, tld*) when it is imported to the nucleus. DL blocks Dpp (BMP) signaling in the mesoderm and the neurogenic ectoderm by the transcriptional repression of dpp (Figure 8.4). This repression is supplemented by at least two additional pathways that inhibit Dpp signaling. These include the inhibition of the interaction between Dpp and its receptor by Sog and the repression of Dpp target genes by Brk. Dorsal ectoderm forms in the domain where Dpp signaling occurs, whereas neurogenic ectoderm forms in lateral regions that lack Dpp signaling and that also do not express Sna. Mesoderm forms in the ventral domain, where DL, Twi, and Sna are expressed together.

many other components of the Tl pathway. While DV pattern formation is directed by a single rel family protein, namely DL, the innate immune response in *Drosophila* is directed by DL and two additional rel family proteins termed Dif and Relish.

Vertebrate genomes encode a signaling pathway that is highly homologous to the *Drosophila* Tl pathway. Like the insect Tl pathway, the vertebrate pathway has critical roles in innate immunity. For example, vertebrate genomes encode multiple Tl-like receptors that recognize conserved microbial features and then trigger cellular and humoral innate immune responses. These signals are, in many cases, transduced by phosphorylation cascades that lead to the destruction of the Cact homolog I-κB and the consequent nuclear import of rel family transcription factors such as NF-κB.

While the role of the Tl pathway in innate immunity is clearly conserved in both vertebrates and invertebrates, the evidence to show a role for Tl signaling in vertebrate DV axis formation is not conclusive. *Drosophila* Spz and Tl can rescue DV pattern formation in *Xenopus* embryos after UV irradiation (a treatment that abolishes the DV pattern) (Armstrong et al., 1998). Also, expression of a dominant negative form of *Xenopus* MyD88 blocks Tl receptor activity, inhibits axis formation, and reduces the expression of pivotal organizer genes (Prothmann et al., 2000). Thus, the Tl pathway could have an ancient role in DV patterning that predates the evolutionary split between vertebrates and invertebrates. However, genetic loss-of-function studies in vertebrates to back up this conclusion have not been reported. Furthermore, studies of developmental evolution strongly suggest that the Tl pathway has only been co-opted for DV patterning during relatively recent arthropod evolution. While Tl is required for DV patterning in most or all homo-metabolous insects and probably also in hemi-metabolous insects, it does not seem to be used for this purpose in the arachnids or the crustaceans, this latter group being the arthropod clade most recently diverged from the insects (Lynch and Roth, 2011).

Dpp/Sog Orthologs Pattern the both the Invertebrate and Vertebrate DV Axis

The DV axis is reversed in vertebrates when compared to invertebrates (Arendt and Nubler-Jung, 1994; Gerhart, 2000; Holley et al., 1995). In invertebrates, for example, the nerve chord is a ventral structure, whereas the heart is a dorsal structure. In vertebrates, however, the nerve chord equivalent, the neural tube, is dorsal whereas the heart is ventral. As described below, recent studies showing that DV patterning in vertebrate embryos depends on the BMP morphogen system suggest that this difference may be superficial in nature (Gerhart, 2000).

Classical studies show that a piece of tissue (the Spemann organizer) from the dorsal lip of the amphibian blastopore can induce dorsal structures in ventral mesoderm and in embryos that have been ventralized by UV irradiation. Recent experimentation has shown that Chordin, the vertebrate ortholog of Sog, is partially responsible for the dorsalizing activity of the Spemann organizer, and functions by antagonizing BMP4 (a Dpp homolog) (Holley et al., 1995; Schmidt et al., 1995). Thus, while Sog/Chordin expression defines the ventral pole of the *Drosophila* embryo, it defines the dorsal pole of the frog embryo. Conversely, while Dpp/BMP4 expression defines the dorsal pole in *Drosophila*, it defines the ventral pole in frogs. Therefore, the apparent reversal of the body plan between vertebrates and invertebrates is likely an artifact of the way that early anatomists decided to define DV polarity in these two groups of animals.

Definitive proof that a BMP morphogen system organizes the vertebrate DV axis comes from loss-of-function genetic analysis carried out in zebrafish. Mutations in the genes encoding the zebrafish orthologs of BMP ligands and receptors result in dorsalization of the embryo, while a mutation in the gene encoding the zebrafish ortholog of Chordin leads to ventralization of the embryo. Furthermore, analysis of mutations in the gene encoding the zebrafish counterpart of Tld suggest that, like *Drosophila* Tld, zebrafish Tld potentiates BMP signaling presumably by degrading Chordin (Yamamoto and Oelgeschlager, 2004).

Drosophila as a Model Organism for Studying Development and Disease

Genome sequences of diverse animals, including *Drosophila melanogaster* (fruit fly), *Caenorhabditis elegans* (round worm), *Danio rerio* (zebrafish), *Mus musculus* (mouse), and *Homo sapiens* (man), confirm that invertebrates and mammals share many common genes. Genetic and biochemical analyses also show that GRNs are highly conserved across species. Frequently, homologous GRNs have homologous functions in different species, although sometimes a GRN evolves a new function that is unique to one species or group of species. For example, the DL/NF-κB network has homologous functions in innate immunity in vertebrates and invertebrates, whereas the function of this network in DV pattern formation may be unique to a subset of invertebrates. Nonetheless, the use of genetic approaches to characterize the genes that pattern the *Drosophila* DV axis has greatly informed our understanding of innate immunity in all organisms. Thus, studies of *Drosophila* DV patterning are relevant to an understanding of human immunodeficiency disorders (Table 8.1).

Similarly, the Dpp/BMP network has conserved functions in both vertebrate and insect DV patterning, but it also has many roles that are unique to insects, including roles in wing and leg development during metamorphosis. The powerful genetic approaches now available to study insect metamorphosis can therefore be used to great advantage to decipher the functional interactions that comprise this network. The BMP pathway also has roles that are unique to vertebrates, such as roles in bone morphogenesis. For example, a gain-of-function mutation in a gene encoding the human homolog of Thickveins (the type I Dpp receptor) is responsible for fibrodysplasia ossificans progressiva, a devastating disease that results from massive bone overgrowth (Shore et al., 2006). Studies of *Drosophila* embryogenesis and metamorphosis have greatly increased our understanding of the molecular basis for this disease.

8.7 CLINICAL RELEVANCE

- **Gene/Protein nomenclature:** *Drosophila* biologists usually name genes according to the mutant phenotype (the visible effect of the mutation). For example the *dorsal* (*dl*) gene is so named because embryos produced by mothers homozygous for a null allele of *dl* develop only dorsal structures (i.e., they are completely dorsalized). This finding suggests that *dl* is required for the formation of ventral structures and therefore, if *dl* had been named according to its function, it might have been called *ventral.* In general, the name of a *Drosophila* gene is italicized, while the name of the protein product is not italicized. The name of the protein is capitalized regardless of whether or not the gene name is capitalized. Hence the *dpp* gene encodes the Dpp protein, and the *Tl* gene encodes the Tl protein. In the case of *dl*, the protein product is spelled in all capital letters (DL), to avoid confusion with the product of the *Delta* (*Dl*) gene.
- **Of flies and men:** The sequences of the fly and human genomes reveal a strikingly high degree of similarity between these organisms (Adams et al., 2000; Venter et al., 2001). The fly genome contains ~13,000 protein coding genes as compared to

TABLE 8.1 ***Drosophila* Genes Discussed in this Chapter. The Genes are Arranged in Alphabetical Order**

Drosophila Gene (Shorthand)	Structure and Function	Vertebrate Counterpart
Bag of Marbles (bam)	Translational regulator in GSC maintenance.	
Benign gonial cell neoplasm (bgcn)	Ankyrin repeat containing protein. Translational regulator involved in GSC maintenance.	
Brinker (brk)	Transcriptional repressor. Expressed as a neuroectodermal stripe in the early embryo.	
Cactus (cact)	Binds to and sequesters DL in the cytoplasm.	Iκ-B
cut(ct)	Transcriptional activator. Homeodomain protein.	CUT, CUTL1
Cg25C (DCg1)	ECM component involved in immobilizing Dpp	Collagen IV
Decapentaplegic (dpp)	BMP ligand. Forms DV activity gradient. Cells receiving Dpp signals choose an ectodermal (non-neural) fate.	BMP-2/4. TGFβ superfamily.
dorsal (dl)	Rel homology domain (RHD) Transcription factor. Forms a DV nuclear gradient in *Drosophila* embryos.	Rel family proteins including NF-κB
Dri(dead ringer)/retained(retn)	Transcription factor binding AT rich domain. Part of a Gro recruiting platform.	DRIL1
easter (ea)	Serine protease.	Blood clotting proteases.
folded gastrulation (fog)	Secreted Ligand. Receptor not characterized.	
gastrulation defective (gd)	Serine protease.	Blood clotting proteases.
groucho (gro)	Co-repressor for DL.	TLE family proteins
gurken (grk)	TGFα like protein. Epidermal growth factor (EGF) ligand.	TGFα family
Heterogenous nuclear ribonucleoprotein at 27C (Hrb27C)	mRNA, 3′ UTR binding. Regulates grk mRNA.	
IGF-II mRNA binding protein (Imp)	mRNA binding. Regulates grk mRNA.	
intermediate neuroblasts defective (ind)	Homeobox gene.	Gsh1, Gsh2
Krapfen (Kra)/myd88 (myd88)	Adaptor protein, binds Tl, Tub.	Myd88
medea (med)	Transcription factor downstream of Dpp. Forms a complex with pMad.	Smad4, Smad family.
Mothers against dpp (Mad)	Transcription factor downstream of Dpp. Phosphorylated in response to Dpp signaling.	MADR1, Smad family.
nanos(nos)	A zinc finger containing translational repressor. Required for GSC formation and also to specify posterior identity	Nanos homolog 1
nudel (ndl)	Serine Protease.	
Oo18 RNA binding protein (orb)	RNA binding protein found in P bodies.	
Pelle (pll)	Serine/Threonine kinase.	IRAK

Continued

TABLE 8.1 *Drosophila* **Genes Discussed in this Chapter. The Genes are Arranged in Alphabetical Order—cont'd**

Drosophila Gene (Shorthand)	Structure and Function	Vertebrate Counterpart
Piwi (piwi)	Argonaute family protein regulating asymmetric cell division of GSCs.	Hiwi
pipe (pip)	Heparan-sulfate-2-*O*-sulfotransferase. Localized in Golgi complex.	
Pumilio (pum)	RNA binding protein. Regulates translation by binding to mRNA.	PUM1
pannier(pnr)	GATA Transcription factor.	
punt (pnt)	Transmembrane receptor kinase. Receptor for Dpp/Scw.	TGFβ receptor, Type II
related to angiotensin converting enzyme (race)	Zinc and Chloride dependent Peptidyl-dipeptidase activity.	Angiotensin-I-converting enzyme (ACE)
rhomboid (rho)	Intramembrane serine protease. Targets Spitz.	Rhomboid 1
saxophone (sax)	Transmembrane receptor kinase. Receptor for Dpp/Scw.	TGFβ receptor, Type I
schnurri (shn)	Transcription factor.	PRDII/MBPI/HIV-EP1
screw (scw)	BMP ligand, Unrestricted expression.	BMP-2/4. TGFβ superfamily
Serpin27A(Spn27a)	Serine protease inhibitor. Inhibits Ea.	
short Gastrulation (sog)	Bone Morphogenetic Protein (BMP) Signaling modulator. Interferes with Dpp signaling and thus confers neuro-ectodermal fate.	Chordin
single minded (sim)	bHLH-PAS protein.	SIM
slalom (sll)	3′-phosphoadenosine 5′-phosphosulfate transporter.	
snail (sn)	Transcriptionally Represses genes which would confer neuroectodermal fate.	SNAI1P.
snake (snk)	Serine protease.	Blood clotting proteases
spätzle (spz)	Tl ligand. Proteolytic cleavage of Spz activates the ligand.	
Squid (sqd)	RNA binding protein.	
Syncrip (Syp)	mRNA translation regulator.	
tailup (tup)	Transcription factor.	LIM-homeodomain family
thickveins (tkv)	Transmembrane receptor kinase. Receptor for Dpp/Scw.	TGFβ receptor, Type I
tinman (tin)	Transcription factor.	Nkx-2.5/CSX1. NK2 family.
Toll (Tl)	Transmembrane Receptor. Distributed evenly all over the embryonic membrane. Binds Spz.	Toll-interleukin like (TIL) superfamily
tolloid (tld)	Metalloendopeptidase. Cleaves Sog.	BMP-1
Torpedo (top)	EGF Receptor tyrosine kinase. Transmembrane receptor.	EGF family
tube (tub)	Functions as an adapter to recruit Pll.	

TABLE 8.1 ***Drosophila*** **Genes Discussed in this Chapter. The Genes are Arranged in Alphabetical Order—cont'd**

Drosophila Gene (Shorthand)	Structure and Function	Vertebrate Counterpart
tulip (tup)	3′-phosphoadenosine 5′-phosphosulfate transporter.	
twist (twi)	bHLH protein. Transcription factor.	TWIST
twisted gastrulation (tsg)	Heparin binding. Forms a complex with Dpp/Scw.	Human connective tissue growth factor
vein (vn)	EGF ligand	Neuregulins
ventral nervous system defective (vnd)	Homeobox gene. Gro dependent Repressor.	NK-2 family
Viking (vkg)	Collagen IV homolog, ECM component.	Collagen IV
Vitelline membrane like (vml)	Predicted target for Pip modification.	
windbeutel (wbl)	Endoplasmic reticulum protein.	
zerknüllt (zen)	Homeobox transcription factor.	Hox Class 3

~25,000 protein coding genes in humans. Current estimates indicate that at least 70% of the known 2,300 human disease genes have well-conserved counterparts in flies (Bier, 2005). The human disease genes include genes involved in cancer as well as neurological, cardiovascular, endocrine, and metabolic diseases. Many of the genes that play important roles in mammalian homeostasis were in fact initially discovered and characterized in flies. Because of the ease with which genetic screens can be carried out in *Drosophila*, flies continue to be one of the premier systems for discovering new gene functions. Such studies greatly aid our understanding of human development. *Drosophila* is also being increasingly used as a transgenic model for studying genes that cause human disease.

- ***Drosophila* in the study of early embryonic development:** *Drosophila* has a number of advantages that make it ideal for studying development by genetic approaches. These include: a short generation time (10 days); a simple genome (one-twentieth the size of the human genome); techniques for generating and analyzing genetic chimeras, and the availability of vectors for introducing modified genes and expressing them in a temporally and spatially regulated manner.

  Studies of *Drosophila* development were pioneered by Edward B. Lewis, Christiane Nusslein-Volhard and Eric Wieschaus, While Lewis' research was primarily concerned with characterization of the homeotic gene complex, Nusslein-Holhard and Weischaus carried out extensive screens for zygotic and maternal mutations that disrupt embryogenesis. Many of these mutations resulted in axial (DV, AP) patterning defects (Nusslein-Volhard, 1991; Nusslein-Volhard and Wieschaus, 1980). Genetic techniques were used to show that one set of genes defined in these screens, the "dorsal group" (which includes *dl*), encoded the maternally active components of a pathway that directed patterning of the DV axis (Ray et al., 1991). These studies led to prediction of the DL protein gradient in 1979, ten years before the gradient was finally visualized by three independent groups (Roth et al., 1989; Rushlow et al., 1989; Steward et al., 1988). Lewis, Nusslein-Volhard, and Wieschaus were honored "*for their discoveries concerning the genetic control of early embryonic development*" with the 1995 Nobel Prize for Physiology and Medicine.
- ***Drosophila* and Adult Stem cells:** The fertilized zygote is the original/master stem cell, as all cell types descend from this progenitor cell. In the adult animal some cells continue to retain their stemness; i.e., these cells undergo asymmetric self-renewing cell divisions and are responsible for maintaining a subpopulation of cells with high turnover. For example, in *Drosophila*, specific cells in the testis, midgut and ovary are stem cells that continuously produce sperm, gut cells, and oocytes (Losick et al., 2011). Each stem cell resides in a microenvironment, called the *niche* (Lander et al., 2012; Morrison and Spradling, 2008), which allows the cell to maintain its abilities. As the stem cell divides asymmetrically into two, one cell remains in the niche while the other sets off on the pathways to differentiation. In humans, for example, stem cell populations have been found in blood cell formation (Wilson and Trumpp, 2006), hair follicles (Fuchs, 2009), and the olfactory epithelium (Moore and Lemischka, 2006). Stem cells have great potential for human therapeutics with possibilities for use in replacement of damaged cells and tissues. A new development in cancer biology is the idea of cancer stem cells and a niche supporting their growth (Cabarcas et al., 2011), and therefore studies of stem cells have relevance to cancer therapy.

RECOMMENDED RESOURCES

1. From Egg to Embryo: Regional Specification in Early Development (1991). Edited by J. M. W. Slack, Jonathan B. L. Bard, Peter W. Barlow, David L. Kirk. Cambridge University Press.
2. The Making of a Fly. The Genetics of Animal Design (1992). Peter A. Lawrence. Blackwell Scientific Publications.
3. The Development of *Drosophila melanogaster* (1993). Volumes 1 and 2. Edited by Michael Bate and Volker Hartenstein. Cold Spring Harbor Laboratory Press, New York.
4. *Drosophila melanogaster*: From Oocyte to Blastoderm (1993). Edited by AS Wilkins. Wiley-Liss New York.
5. The Coiled Spring. How Life Begins (2000). Ethan Bier. Cold Spring Harbor Laboratory Press, New York.
6. Principles of Development (2001). Lewis Wolpert, Rosa Beddington, Thomas Jessell, Peter Lawrence, Elliot Meyerowitz, Jim Smith. Oxford University Press

Internet Resources for Exploring *Drosophila* Development

1. Flybase. A database of the *Drosophila* genome. www.flybase.org (accessed Sept. 2013).
2. Flymove. Images, movies and interactive Shockwaves of *Drosophila* development. http://flymove.uni-muenster.de/ (accessed Sept. 2013).
3. WWW-virtual library, *Drosophila* internet resources for research on the fruit fly. http://www.ceolas.org/fly/ (accessed Sept. 2013).
4. Berkeley *Drosophila* genome project (BDGP). http://www.fruitfly.org/EST/index.shtml (accessed Sept. 2013).
5. BDGP-Patterns of gene expression in *Drosophila* embryogenesis. http://www.fruitfly.org/cgi-bin/ex/insitu.pl (accessed Sept. 2013).
6. Flybase. Anatomy and images. http://flybase.org/anatomy/ (accessed Sept. 2013).
7. Flyview. A *Drosophila* Image Database. http://flyview.uni-muenster.de/ (accessed Sept. 2013).
8. Curagen – *Drosophila* interaction database. http://portal.curagen.com/cgi-bin/interaction/flyHome.pl (accessed Sept. 2013).
9. Flymine. An integrated database for *Drosophila* and *Anopheles* genomics. http://www.flymine.org/ (accessed Sept. 2013).
10. Homophila. Human disease to *Drosophila* gene database. http://superfly.ucsd.edu/homophila/ (accessed Sept. 2013).
11. Pubmed. NCBI's Digital online archive for journal literature using the *Entrez* integrated, text-based search and retrieval system. http://www.ncbi.nlm.nih.gov/entrez/query.fcgi?db=PubMed&itool=toolbar (accessed Sept. 2013).
12. Christiane Nüsslein-Volhard – Autobiography. http://nobelprize.org/medicine/laureates/1995/nusslein-volhard-autobio.html (accessed Sept. 2013).
13. Flybase. The interactive fly. http://flybase.bio.indiana.edu/allied-data/lk/interactive-fly/aimain/1aahome.htm (accessed Sept. 2013).
14. Drosphila Species Genomes (DroSpeGe). http://insects.eugenes.org/species/about/ (accessed Sept. 2013).

REFERENCES

Adams, M.D., Celniker, S.E., Holt, R.A., Evans, C.A., Gocayne, J.D., Amanatides, P.G., Scherer, S.E., Li, P.W., Hoskins, R.A., Galle, R.F., et al., 2000. The genome sequence of *Drosophila melanogaster*. Science 287, 2185–2195.

Affolter, M., Marty, T., Vigano, M.A., Jazwinska, A., 2001. Nuclear interpretation of Dpp signaling in *Drosophila*. Embo. J. 20, 3298–3305.

Anderson, K.V., 2000. Toll signaling pathways in the innate immune response. Curr. Opin. Immunol. 12, 13–19.

Anderson, K.V., Bokla, L., Nusslein-Volhard, C., 1985. Establishment of dorsal-ventral polarity in the *Drosophila* embryo: the induction of polarity by the Toll gene product. Cell 42, 791–798.

Araujo, H., Bier, E., 2000. sog and dpp exert opposing maternal functions to modify toll signaling and pattern the dorsoventral axis of the *Drosophila* embryo. Development 127, 3631–3644.

Arendt, D., Nubler-Jung, K., 1994. Inversion of dorsoventral axis? Nature 371, 26.

Armstrong, N.J., Steinbeisser, H., Prothmann, C., DeLotto, R., Rupp, R.A., 1998. Conserved Spatzle/Toll signaling in dorsoventral patterning of *Xenopus* embryos. Mech. Dev. 71, 99–105.

Arora, K., Levine, M.S., O'Connor, M.B., 1994. The screw gene encodes a ubiquitously expressed member of the TGF-beta family required for specification of dorsal cell fates in the *Drosophila* embryo. Genes. Dev. 8, 2588–2601.

Ashe, H.L., 2005. BMP signaling: synergy and feedback create a step gradient. Curr. Biol. 15, R375–377.

Ashe, H.L., Levine, M., 1999. Local inhibition and long-range enhancement of Dpp signal transduction by Sog. Nature 398, 427–431.

Ashe, H.L., Mannervik, M., Levine, M., 2000. Dpp signaling thresholds in the dorsal ectoderm of the *Drosophila* embryo. Development 127, 3305–3312.

Barolo, S., Posakony, J.W., 2002. Three habits of highly effective signaling pathways: principles of transcriptional control by developmental cell signaling. Genes. Dev. 16, 1167–1181.

Belvin, M.P., Jin, Y., Anderson, K.V., 1995. Cactus protein degradation mediates *Drosophila* dorsal-ventral signaling. Genes. Dev. 9, 783–793.

Benoit, B., He, C.H., Zhang, F., Votruba, S.M., Tadros, W., Westwood, J.T., Smibert, C.A., Lipshitz, H.D., Theurkauf, W.E., 2009. An essential role for the RNA-binding protein Smaug during the *Drosophila* maternal-to-zygotic transition. Development 136, 923–932.

Bergmann, A., Stein, D., Geisler, R., Hagenmaier, S., Schmid, B., Fernandez, N., Schnell, B., Nusslein-Volhard, C., 1996. A gradient of cytoplasmic Cactus degradation establishes the nuclear localization gradient of the dorsal morphogen in *Drosophila*. Mech. Dev. 60, 109–123.

Biehs, B., Francois, V., Bier, E., 1996. The *Drosophila* short gastrulation gene prevents Dpp from autoactivating and suppressing neurogenesis in the neuroectoderm. Genes. Dev. 10, 2922–2934.

Bier, E., 2005. *Drosophila*, the golden bug, emerges as a tool for human genetics. Nat. Rev. Genet. 6, 9–23.

Brennan, C.A., Anderson, K.V., 2004. *Drosophila*: the genetics of innate immune recognition and response. Annu. Rev. Immunol. 22, 457–483.

Cabarcas, S.M., Mathews, L.A., Farrar, W.L., 2011. The cancer stem cell niche – there goes the neighborhood? Int. J. Cancer. 129, 2315–2327.

Cai, H.N., Arnosti, D.N., Levine, M., 1996. Long-range repression in the *Drosophila* embryo. Proc. Natl. Acad. Sci. USA 93, 9309–9314.

Chang, A.J., Morisato, D., 2002. Regulation of Easter activity is required for shaping the Dorsal gradient in the *Drosophila* embryo. Development 129, 5635–5645.

Cho, Y.S., Stevens, L.M., Sieverman, K.J., Nguyen, J., Stein, D., 2012. A ventrally localized protease in the *Drosophila* egg controls embryo dorsoventral polarity. Curr. Biol. 22, 1013–1018.

Courey, A.J., Jia, S., 2001. Transcriptional repression: the long and the short of it. Genes. Dev. 15, 2786–2796.

Decotto, E., Ferguson, E.L., 2001. A positive role for Short gastrulation in modulating BMP signaling during dorsoventral patterning in the *Drosophila* embryo. Development 128, 3831–3841.

DeLotto, R., 2011. Shedding light on Toll signaling through live imaging. Fly. (Austin) 5, 141–146.

DeLotto, R., DeLotto, Y., Steward, R., Lippincott-Schwartz, J., 2007. Nucleocytoplasmic shuttling mediates the dynamic maintenance of nuclear Dorsal levels during *Drosophila* embryogenesis. Development 134, 4233–4241.

Drier, E.A., Govind, S., Steward, R., 2000. Cactus-independent regulation of Dorsal nuclear import by the ventral signal. Curr. Biol. 10, 23–26.

Drier, E.A., Huang, L.H., Steward, R., 1999. Nuclear import of the *Drosophila* Rel protein Dorsal is regulated by phosphorylation. Genes. Dev. 13, 556–568.

Driever, W., Nusslein-Volhard, C., 1988. A gradient of bicoid protein in *Drosophila* embryos. Cell 54, 83–93.

Dubnicoff, T., Valentine, S.A., Chen, G., Shi, T., Lengyel, J.A., Paroush, Z., Courey, A.J., 1997. Conversion of dorsal from an activator to a repressor by the global corepressor Groucho. Genes Dev. 11, 2952–2957.

Feinstein, E., Kimchi, A., Wallach, D., Boldin, M., Varfolomeev, E., 1995. The death domain: a module shared by proteins with diverse cellular functions. Trends. Biochem. Sci. 20, 342–344.

Ferguson, E.L., Anderson, K.V., 1992. Decapentaplegic acts as a morphogen to organize dorsal-ventral pattern in the *Drosophila* embryo. Cell 71, 451–461.

Francois, V., Solloway, M., O'Neill, J.W., Emery, J., Bier, E., 1994. Dorsal-ventral patterning of the *Drosophila* embryo depends on a putative negative growth factor encoded by the short gastrulation gene. Genes Dev. 8, 2602–2616.

Fuchs, E., 2009. The tortoise and the hair: slow-cycling cells in the stem cell race. Cell 137, 811–819.

Gerhart, J., 2000. Inversion of the chordate body axis: are there alternatives? Proc. Natl. Acad. Sci. USA 97, 4445–4448.

Gonzalez-Reyes, A., Elliott, H., St Johnston, D., 1995. Polarization of both major body axes in *Drosophila* by gurken-torpedo signaling. Nature 375, 654–658.

Gonzalez-Reyes, A., St Johnston, D., 1994. Role of oocyte position in establishment of anterior-posterior polarity in *Drosophila*. Science 266, 639–642.

Goodrich, J.S., Clouse, K.N., Schupbach, T., 2004. Hrb27C, Sqd and Otu cooperatively regulate gurken RNA localization and mediate nurse cell chromosome dispersion in *Drosophila* oogenesis. Development 131, 1949–1958.

Gouw, J.W., Pinkse, M.W., Vos, H.R., Moshkin, Y., Verrijzer, C.P., Heck, A.J., Krijgsveld, J., 2009. *In vivo* stable isotope labeling of fruit flies reveals post-transcriptional regulation in the maternal-to-zygotic transition. Mol. Cell. Proteomics 8, 1566–1578.

Govind, S., Brennan, L., Steward, R., 1993. Homeostatic balance between dorsal and cactus proteins in the *Drosophila* embryo. Development 117, 135–148.

Hashimoto, C., Kim, D.R., Weiss, L.A., Miller, J.W., Morisato, D., 2003. Spatial regulation of developmental signaling by a serpin. Dev. Cell 5, 945–950.

Holley, S.A., Ferguson, E.L., 1997. Fish are like flies are like frogs: conservation of dorsal-ventral patterning mechanisms. Bioessays 19, 281–284.

Holley, S.A., Jackson, P.D., Sasai, Y., Lu, B., De Robertis, E.M., Hoffmann, F.M., Ferguson, E.L., 1995. A conserved system for dorsal-ventral patterning in insects and vertebrates involving sog and chordin. Nature 376, 249–253.

Holley, S.A., Neul, J.L., Attisano, L., Wrana, J.L., Sasai, Y., O'Connor, M.B., De Robertis, E.M., Ferguson, E.L., 1996. The *Xenopus* dorsalizing factor noggin ventralizes *Drosophila* embryos by preventing DPP from activating its receptor. Cell 86, 607–617.

Huang, A.M., Rusch, J., Levine, M., 1997. An anteroposterior Dorsal gradient in the *Drosophila* embryo. Genes. Dev. 11, 1963–1973.

Hulskamp, M., Pfeifle, C., Tautz, D., 1990. A morphogenetic gradient of hunchback protein organizes the expression of the gap genes Kruppel and knirps in the early *Drosophila* embryo. Nature 346, 577–580.

Imler, J.L., Hoffmann, J.A., 2001. Toll receptors in innate immunity. Trends Cell. Biol. 11, 304–311.

Ip, Y.T., Kraut, R., Levine, M., Rushlow, C.A., 1991. The dorsal morphogen is a sequence-specific DNA-binding protein that interacts with a long-range repression element in *Drosophila*. Cell 64, 439–446.

Ip, Y.T., Park, R.E., Kosman, D., Yazdanbakhsh, K., Levine, M., 1992. dorsal-twist interactions establish snail expression in the presumptive mesoderm of the *Drosophila* embryo. Genes Dev. 6, 1518–1530.

Jazwinska, A., Rushlow, C., Roth, S., 1999. The role of brinker in mediating the graded response to Dpp in early *Drosophila* embryos. Development 126, 3323–3334.

Jiang, J., Cai, H., Zhou, Q., Levine, M., 1993. Conversion of a dorsal-dependent silencer into an enhancer: evidence for dorsal corepressors. Embo. J. 12, 3201–3209.

Jiang, J., Kosman, D., Ip, Y.T., Levine, M., 1991. The dorsal morphogen gradient regulates the mesoderm determinant twist in early *Drosophila* embryos. Genes Dev. 5, 1881–1891.

Jiang, J., Levine, M., 1993. Binding affinities and cooperative interactions with bHLH activators delimit threshold responses to the dorsal gradient morphogen. Cell 72, 741–752.

Kanodia, J.S., Rikhy, R., Kim, Y., Lund, V.K., DeLotto, R., Lippincott-Schwartz, J., Shvartsman, S.Y., 2009. Dynamics of the Dorsal morphogen gradient. Proc. Natl. Acad. Sci. USA 106, 21707–21712.

Kirilly, D., Xie, T., 2007. The *Drosophila* ovary: an active stem cell community. Cell. Res. 17, 15–25.

Lall, S., Patel, N.H., 2001. Conservation and divergence in molecular mechanisms of axis formation. Annu. Rev. Genet. 35, 407–437.

Lander, A.D., Kimble, J., Clevers, H., Fuchs, E., Montarras, D., Buckingham, M., Calof, A.L., Trumpp, A., Oskarsson, T., 2012. What does the concept of the stem cell niche really mean today? BMC. Biol. 10, 19.

Lei, L., Spradling, A.C., 2013. Mouse primordial germ cells produce cysts that partially fragment prior to meiosis. Development 140, 2075–2081.

Lemaitre, B., Meister, M., Govind, S., Georgel, P., Steward, R., Reichhart, J.M., Hoffmann, J.A., 1995. Functional analysis and regulation of nuclear import of dorsal during the immune response in *Drosophila*. Embo. J. 14, 536–545.

LeMosy, E.K., Hong, C.C., Hashimoto, C., 1999. Signal transduction by a protease cascade. Trends Cell. Biol. 9, 102–107.

Levine, M., Davidson, E.H., 2005. Gene regulatory networks for development. Proc. Natl. Acad. Sci. USA 102, 4936–4942.

Linne, V., Eriksson, B.J., Stollewerk, A., 2012. Single-minded and the evolution of the ventral midline in arthropods. Dev. Biol. 364, 66–76.

Losick, V.P., Morris, L.X., Fox, D.T., Spradling, A., 2011. *Drosophila* stem cell niches: a decade of discovery suggests a unified view of stem cell regulation. Dev. Cell. 21, 159–171.

Lu, Y., Wu, L.P., Anderson, K.V., 2001. The antibacterial arm of the *Drosophila* innate immune response requires an IkappaB kinase. Genes Dev. 15, 104–110.

Lund, V.K., DeLotto, Y., DeLotto, R., 2010. Endocytosis is required for Toll signaling and shaping of the Dorsal/NF-kappaB morphogen gradient during *Drosophila* embryogenesis. Proc. Natl. Acad. Sci. USA 107, 18028–18033.

Lynch, J.A., Roth, S., 2011. The evolution of dorsal-ventral patterning mechanisms in insects. Genes Dev. 25, 107–118.

Markstein, M., Markstein, P., Markstein, V., Levine, M.S., 2002. Genome-wide analysis of clustered Dorsal binding sites identifies putative target genes in the *Drosophila* embryo. Proc. Natl. Acad. Sci. USA 99, 763–768.

Marques, G., Musacchio, M., Shimell, M.J., Wunnenberg-Stapleton, K., Cho, K.W., O'Connor, M.B., 1997. Production of a DPP activity gradient in the early *Drosophila* embryo through the opposing actions of the SOG and TLD proteins. Cell 91, 417–426.

McDermott, S.M., Meignin, C., Rappsilber, J., Davis, I., 2012. *Drosophila* Syncrip binds the gurken mRNA localisation signal and regulates localised transcripts during axis specification. Biol. Open. 1, 488–497.

Minami, M., Kinoshita, N., Kamoshida, Y., Tanimoto, H., Tabata, T., 1999. Brinker is a target of Dpp in *Drosophila* that negatively regulates Dpp-dependent genes. Nature 398, 242–246.

Moore, K.A., Lemischka, I.R., 2006. Stem cells and their niches. Science 311, 1880–1885.

Morisato, D., Anderson, K.V., 1994. The spatzle gene encodes a component of the extracellular signaling pathway establishing the dorsal-ventral pattern of the *Drosophila* embryo. Cell 76, 677–688.

Morisato, D., Anderson, K.V., 1995. Signaling pathways that establish the dorsal-ventral pattern of the *Drosophila* embryo. Annu. Rev. Genet. 29, 371–399.

Morrison, S.J., Spradling, A.C., 2008. Stem cells and niches: mechanisms that promote stem cell maintenance throughout life. Cell 132, 598–611.

Moussian, B., Roth, S., 2005. Dorsoventral axis formation in the *Drosophila* embryo – shaping and transducing a morphogen gradient. Curr. Biol. 15, R887–899.

Nambu, J.R., Franks, R.G., Hu, S., Crews, S.T., 1990. The single-minded gene of *Drosophila* is required for the expression of genes important for the development of CNS midline cells. Cell 63, 63–75.

Nguyen, M., Park, S., Marques, G., Arora, K., 1998. Interpretation of a BMP activity gradient in *Drosophila* embryos depends on synergistic signaling by two type I receptors, SAX and TKV. Cell 95, 495–506.

Norris, J.L., Manley, J.L., 1996. Functional interactions between the pelle kinase, Toll receptor, and tube suggest a mechanism for activation of dorsal. Genes Dev. 10, 862–872.

Nusslein-Volhard, C., 1991. Determination of the embryonic axes of *Drosophila*. Dev. (Suppl. 1), 1–10.

Nusslein-Volhard, C., Wieschaus, E., 1980. Mutations affecting segment number and polarity in *Drosophila*. Nature 287, 795–801.

Pan, D., Courey, A.J., 1992. The same dorsal binding site mediates both activation and repression in a context-dependent manner. Embo. J. 11, 1837–1842.

Prothmann, C., Armstrong, N.J., Rupp, R.A., 2000. The Toll/IL-1 receptor binding protein MyD88 is required for Xenopus axis formation. Mech. Dev. 97, 85–92.

Ratnaparkhi, G.S., Jia, S., Courey, A.J., 2006. Uncoupling dorsal-mediated activation from dorsal-mediated repression in the *Drosophila* embryo. Development 133, 4409–4414.

Ray, R.P., Arora, K., Nusslein-Volhard, C., Gelbart, W.M., 1991. The control of cell fate along the dorsal-ventral axis of the *Drosophila* embryo. Development 113, 35–54.

Reach, M., Galindo, R.L., Towb, P., Allen, J.L., Karin, M., Wasserman, S.A., 1996. A gradient of cactus protein degradation establishes dorsoventral polarity in the *Drosophila* embryo. Dev. Biol. 180, 353–364.

Roth, S., Hiromi, Y., Godt, D., Nusslein-Volhard, C., 1991. Cactus, a maternal gene required for proper formation of the dorsoventral morphogen gradient in *Drosophila* embryos. Development 112, 371–388.

Roth, S., Lynch, J., 2012. Axis formation: microtubules push in the right direction. Curr. Biol. 22, R537–539.

Roth, S., Schupbach, T., 1994. The relationship between ovarian and embryonic dorsoventral patterning in *Drosophila*. Development 120, 2245–2257.

Roth, S., Stein, D., Nusslein-Volhard, C., 1989. A gradient of nuclear localization of the dorsal protein determines dorsoventral pattern in the *Drosophila* embryo. Cell 59, 1189–1202.

Ruberte, E., Marty, T., Nellen, D., Affolter, M., Basler, K., 1995. An absolute requirement for both the type II and type I receptors, punt and thick veins, for dpp signaling *in vivo*. Cell 80, 889–897.

Rushlow, C.A., Han, K., Manley, J.L., Levine, M., 1989. The graded distribution of the dorsal morphogen is initiated by selective nuclear transport in *Drosophila*. Cell 59, 1165–1177.

Sawala, A., Sutcliffe, C., Ashe, H.L., 2012. Multistep molecular mechanism for bone morphogenetic protein extracellular transport in the *Drosophila* embryo. Proc. Natl. Acad. Sci. USA 109, 11222–11227.

Schmidt, J., Francois, V., Bier, E., Kimelman, D., 1995. *Drosophila* short gastrulation induces an ectopic axis in *Xenopus*: evidence for conserved mechanisms of dorsal-ventral patterning. Development 121, 4319–4328.

Schneider, D.S., Jin, Y., Morisato, D., Anderson, K.V., 1994. A processed form of the Spatzle protein defines dorsal-ventral polarity in the *Drosophila* embryo. Development 120, 1243–1250.

Sen, J., Goltz, J.S., Konsolaki, M., Schupbach, T., Stein, D., 2000. Windbeutel is required for function and correct subcellular localization of the *Drosophila* patterning protein Pipe. Development 127, 5541–5550.

Shen, B., Manley, J.L., 2002. Pelle kinase is activated by autophosphorylation during Toll signaling in *Drosophila*. Development 129, 1925–1933.

Shimmi, O., O'Connor, M.B., 2003. Physical properties of Tld, Sog, Tsg and Dpp protein interactions are predicted to help create a sharp boundary in Bmp signals during dorsoventral patterning of the *Drosophila* embryo. Development 130, 4673–4682.

Shimmi, O., Umulis, D., Othmer, H., O'Connor, M.B., 2005. Facilitated transport of a Dpp/Scw heterodimer by Sog/Tsg leads to robust patterning of the *Drosophila* blastoderm embryo. Cell 120, 873–886.

Shore, E.M., Xu, M., Feldman, G.J., Fenstermacher, D.A., Cho, T.J., Choi, I.H., Connor, J.M., Delai, P., Glaser, D.L., LeMerrer, M., et al., 2006. A recurrent mutation in the BMP type I receptor ACVR1 causes inherited and sporadic fibrodysplasia ossificans progressiva. Nat. Genet. 38, 525–527.

Smith, C.L., DeLotto, R., 1994. Ventralizing signal determined by protease activation in *Drosophila* embryogenesis. Nature 368, 548–551.

Song, X., Wong, M.D., Kawase, E., Xi, R., Ding, B.C., McCarthy, J.J., Xie, T., 2004. Bmp signals from niche cells directly repress transcription of a differentiation-promoting gene, bag of marbles, in germline stem cells in the *Drosophila* ovary. Development 131, 1353–1364.

Spencer, E., Jiang, J., Chen, Z.J., 1999. Signal-induced ubiquitination of IkappaBalpha by the F-box protein Slimb/beta-TrCP. Genes Dev. 13, 284–294.

Spradling, A., Drummond-Barbosa, D., Kai, T., 2001. Stem cells find their niche. Nature 414, 98–104.

Srinivasan, S., Rashka, K.E., Bier, E., 2002. Creation of a Sog morphogen gradient in the *Drosophila* embryo. Dev. Cell 2, 91–101.

St Johnston, R.D., Gelbart, W.M., 1987. Decapentaplegic transcripts are localized along the dorsal-ventral axis of the *Drosophila* embryo. Embo. J. 6, 2785–2791.

St Johnston, D., Nusslein-Volhard, C., 1992. The origin of pattern and polarity in the *Drosophila* embryo. Cell 68, 201–219.

Stathopoulos, A., Levine, M., 2002. Dorsal gradient networks in the *Drosophila* embryo. Dev. Biol. 246, 57–67.

Stathopoulos, A., Van Drenth, M., Erives, A., Markstein, M., Levine, M., 2002. Whole-genome analysis of dorsal-ventral patterning in the *Drosophila* embryo. Cell 111, 687–701.

Stein, D., Roth, S., Vogelsang, E., Nusslein-Volhard, C., 1991. The polarity of the dorsoventral axis in the *Drosophila* embryo is defined by an extracellular signal. Cell 65, 725–735.

Steward, R., Zusman, S.B., Huang, L.H., Schedl, P., 1988. The dorsal protein is distributed in a gradient in early *Drosophila* embryos. Cell 55, 487–495.

Sun, H., Bristow, B.N., Qu, G., Wasserman, S.A., 2002. A heterotrimeric death domain complex in Toll signaling. Proc. Natl. Acad. Sci. USA 99, 12871–12876.

Thisse, C., Thisse, B., 1992. Dorsoventral development of the *Drosophila* embryo is controlled by a cascade of transcriptional regulators. Dev. Suppl, 173–181.

Tsurumi, A., Xia, F., Li, J., Larson, K., LaFrance, R., Li, W.X., 2011. STAT is an essential activator of the zygotic genome in the early *Drosophila* embryo. PLoS Genet. 7, e1002086.

Valentine, S.A., Chen, G., Shandala, T., Fernandez, J., Mische, S., Saint, R., Courey, A.J., 1998. Dorsal-mediated repression requires the formation of a multiprotein repression complex at the ventral silencer. Mol. Cell. Biol. 18, 6584–6594.

van Eeden, F., St Johnston, D., 1999. The polarisation of the anterior-posterior and dorsal-ventral axes during *Drosophila* oogenesis. Curr. Opin. Genet. Dev. 9, 396–404.

Venter, J.C., Adams, M.D., Myers, E.W., Li, P.W., Mural, R.J., Sutton, G.G., Smith, H.O., Yandell, M., Evans, C.A., Holt, R.A., et al., 2001. The sequence of the human genome. Science 291, 1304–1351.

Wang, Y.C., Ferguson, E.L., 2005. Spatial bistability of Dpp-receptor interactions during *Drosophila* dorsal-ventral patterning. Nature 434, 229–234.

Weil, T.T., Parton, R.M., Herpers, B., Soetaert, J., Veenendaal, T., Xanthakis, D., Dobbie, I.M., Halstead, J.M., Hayashi, R., Rabouille, C., et al., 2012. *Drosophila* patterning is established by differential association of mRNAs with P bodies. Nat. Cell. Biol. 14, 1305–1313.

Wilson, A., Trumpp, A., 2006. Bone-marrow haematopoietic-stem-cell niches. Nat. Rev. Immunol. 6, 93–106.

Xie, T., Song, X., Jin, Z., Pan, L., Weng, C., Chen, S., Zhang, N., 2008. Interactions between stem cells and their niche in the Drosophila ovary. Cold Spring Harb. Symp. Quant. Biol. 73, 39–47.

Xie, T., Spradling, A.C., 1998. Decapentaplegic is essential for the maintenance and division of germline stem cells in the *Drosophila* ovary. Cell 94, 251–260.

Yamamoto, Y., Oelgeschlager, M., 2004. Regulation of bone morphogenetic proteins in early embryonic development. Naturwissenschaften 91, 519–534.

Zhang, Z., Stevens, L.M., Stein, D., 2009a. Sulfation of eggshell components by Pipe defines dorsal-ventral polarity in the *Drosophila* embryo. Curr. Biol. 19, 1200–1205.

Zhang, Z., Zhu, X., Stevens, L.M., Stein, D., 2009b. Distinct functional specificities are associated with protein isoforms encoded by the *Drosophila* dorsal-ventral patterning gene pipe. Development 136, 2779–2789.

Zhao, T., Graham, O.S., Raposo, A., St Johnston, D., 2012. Growing microtubules push the oocyte nucleus to polarize the *Drosophila* dorsal-ventral axis. Science 336, 999–1003.

Chapter 9

Building Dimorphic Forms: The Intersection of Sex Determination and Embryonic Patterning

Kristy L. Kenyon, Yanli Guo and Nathan Martin
Department of Biology, Hobart and William Smith Colleges, Geneva, New York, NY

Chapter Outline

GLOSSARY

Alternative splicing The differential processing of pre-messenger (nuclear) RNA transcripts involving exon and intron sequences.

Cis-acting regulatory elements-regulatory elements (CREs) A distinct region of DNA of a gene that influences its transcription. CREs can be located upstream of the transcription start site or may be located downstream. These sequences can be modular in nature.

Dimorphism The expression of two different forms within a population or species; in contrast to monomorphism (one form).

Dosage compensation: The control of X chromosome gene expression dependent upon genotype (XX vs. XY).

Pleoitropy The phenomenon observed when one gene influence multiple traits and phenotypes.

Primary sex differentiation Refers to the specification and determination of the gonads (testes, ovaries) and associated production of sexually distinct gametes (spermatogenesis vs. oogenesis).

Secondary sex differentiation The development of external (genitalia) and internal characteristics that distinguish male and female forms.

SUMMARY

- Although sexual reproduction is fundamental to most animals, the mechanisms that determine sexual forms are highly variable.
- In *Drosophila*, alternative RNA splicing and translational control are two primary mechanisms involved in sex determination and dimorphic development.
- In mammals, transcription regulators and extracellular signaling direct the developmental trajectory of the bi-potential rudimentary gonads. Primary determination of gonadal sex requires both the activation and suppression of sex-specific genetic programs.
- At the cellular level, most animals are likely mosaics with regard to sexual identity; some cells "know" their sex and respond to sex-specific gene activity whereas other cells may be considered sex "neutral", demonstrating monomorphic development in both sexes.

Principles of Developmental Genetics.

- In the adult gonads, differentiated sexual fates can be labile at the cellular level. Maintenance of determination likely requires the continual activities of sex-determining mechanisms.
- The olfactory system of *Drosophila* offers a useful model system for studying how the sex determination pathway influences the dimorphic development of sex-specific neural circuitry.
- Dimorphic development requires the integration of sex-determining gene activities with embryonic programs that determine cell fate identity and function. Such integration can regulate processes such as cell proliferation, apoptosis and cell differentiation.
- Competency to respond to sex-inducing mechanisms can influence the developmental potential of embryonic and stem cell populations.

9.1 INTRODUCTION

A common and ancient feature of animal development is the expression of sexually dimorphic characteristics and traits. Males and females often demonstrate morphological structures, appearances and behaviors that contribute to reproductive processes. Such differences are generated during embryogenesis; sex determination involves the employment of genetic programs that lead to the differentiation of sexual characteristics. In animals, the mechanisms that determine sex are diverse and variable (see also Chapter 27). Sex-specific chromosomes (e.g., XX, XY) operate in some species, whereas environmental conditions (temperature, social cues) drive determination in others. Sexual differentiation is often distinguished at two levels, namely primary and secondary. Primary sexual differentiation refers to whether an organism possesses either male or female gonads, i.e., either testes or ovaries. Spermatogenesis and male hormone production are the primary activities of the testes, thus controlling many aspects of male development. Similarly, the ovaries produce the hormones necessary for oogenesis, further differentiation, and maintenance of the female reproductive system and associated structures.

Many sexually dimorphic, or sex-specific, structures are considered to be a part of secondary sexual differentiation. In mammals, the gonad controlled hormonal state of the organism determines most aspects of secondary differentiation. For insects such as *Drosophila*, somatic sex differentiation mostly depends upon cell autonomous expression of sex-specific genes and gene products. Examples of dimorphic physical features range from external genitalia to cellular specificities within the central nervous system. Animal behavior unique to each sex can be attributed to these biological differences. For example, specific gene expression patterns within neurons of the male CNS have been linked to proper courtship behaviors, including song production or copulation; misexpression of such genes in the female CNS causes them to display male courtship behaviors (Rideout at al., 2010). These differences can also impact how each sex responds to environmental conditions or external stimuli. It has been demonstrated that female *Drosophila* can survive better under stressful conditions, such as high exposure to ethanol, than males (Devineni and Heberlein, 2012).

The study of sex determination, especially in humans, has a complicated history given the societal implications of sexuality and gender issues. Yet, it is a fundamental aspect of animal development, with critical implications for understanding human health. Recent work in the field of stem cell biology indicates that the chromosomal sex of stem cells can influence their developmental potential (see Lecanu et al., 2011). Thus, understanding the intricacies of sex determination and its role in embryogenesis will be of great importance, especially considering the movement toward personalized medicine.

This chapter addresses the complexities of sex determination in both *Drosophila* and mammals, emphasizing the molecular mechanisms that act upon the genome to differentially control expression. The key points to be addressed include:

- The genetic mechanisms that drive sex determination in *Drosophila* and mammals;
- The cellular mechanisms that shape dimorphic development in the fly olfactory system;
- The intersection of sexual determination and embryonic patterning in *Drosophila*;
- The relevance of sex determination and dimorphism in human medicine and stem cell biology.

9.2 SEX DETERMINATION IN *DROSOPHILA MELANOGASTER*

In *Drosophila*, the sexual identity of all somatic cells is determined by the cellular response to X chromosome number as compared to the diploid state of the autosomes. If there is only one X chromosome, a diploid fly embryo will develop into a male (X:autosome set = 0.5 ratio). In contrast, the presence of two X chromosomes in a diploid fly yields a ratio of 1, resulting in female development. Chromosome information is interpreted through the activation of *sex-lethal (sxl)*, the master gene switch for sex identity development and dosage compensation (Burtis, 1993; Baker et al., 2001).

In females, X-linked genes such as *runt*, *sisterless-A* and *sisterless-B* are expressed by both X homologs, leading to higher protein levels than those observed in males (Kimura, 2011). These regulatory proteins bind the *sex-lethal* promoter and induce the activation of transcription. In males, whose cells have only one X chromosome, the X-linked activator proteins are not produced at a sufficient level to activate *sxl* transcription (Kimura, 2011). The expression of *sxl* affects downstream factors in the sex determination hierarchy, such *as transformer, fruitless, and doublesex*. The roles and the molecular activities of these genes in the *Drosophila* sex determination hierarchy will be discussed in detail below (see Recommended resources for details of the *Drosophila* genome).

Sex-Lethal

The ratio of X chromosomes to autosomes (X:A ratio) plays a pivotal role in the determination of the sex of the organism (Schüpbach, 1985). Numerator proteins are encoded by genes located on the X chromosome and are the X portion of the X:A ratio. These numerator proteins are believed to bind to the promoter of the *sex-lethal* gene immediately following fertilization (Kimura, 2011). The *myc* gene is one example of a numerator protein. Since they have two X chromosomes, females produce twice as much Myc protein as males, which aids in the activation of Sxl by its binding to the promoter Sxl-Pe (Kappes et. al., 2011). The activation of Sxl initiates the female development pathway. Conversely, without such activation of *sxl* gene expression, the male development pathway is initiated (reviewed in Salz and Erickson, 2010).

The differential expression of *sex-lethal* is an important example of control, at both the transcriptional and translational level. The transcriptional control of *sxl* involves two promoters, namely the "establishment" promoter (Pe) and the "maintenance" promoter (Pm). Pe is only activated in females, in response to X:A ratio. Transcripts from the Pe promoter are translated, thus resulting in an early source of functional Sxl. After transcription from the Pe promoter is completed, it becomes inactivated and transcription control switches to the Pm promoter. In males, which lack sufficient X chromosome linked proteins, no Pe-controlled transcripts are produced, resulting in the lack of early Sxl protein product (reviewed Camara et al., 2008, Figure 9.1).

The lack of early Sex-lethal protein in males and its presence in females has a significant downstream effect. Sex-lethal is an RNA splicing enzyme (reviewed in Salz and Erickson, 2010) and its function is required for further *sex-lethal* expression in females. Late expression of *sex-lethal* is achieved through the Pm promoter; this regulatory site is active and transcription from this promoter occurs in both sexes. However, Pm-produced transcripts require alternative RNA splicing to remove an early stop codon; without this splicing, a truncated and non-functional product is produced. In females, the early Sxl protein removes the early stop codon from Pm-produced transcripts. The stop codon removal allows these transcripts to encode a late functional Sxl protein in females. In males, the lack of an early Sxl product results in the inability to remove the stop codon from the Pm transcripts. Thus, males are translationally blocked from expressing Sxl during later stages (Figure 9.1, reviewed in Camara et al., 2008).

This auto-regulation of Sxl is significant for female development. Once established, this loop is maintained and allows females to continually express Sxl throughout their development, while in males such expression is quickly eliminated (Pomiankowski et al., 2004). Sxl has also been linked to the initiation of female development, for example initiating oogenesis (Hashiyama et al., 2011; Chau et al., 2012). Several functions of the Sxl protein have been described; one major target is the regulation of the *transformer* gene (Bell et al., 1988; reviewed in Camara et al., 2008).

Transformer

The alternatively spliced forms of Sxl expressed in each sex lead to the deployment of different downstream targets. Sxl directly affects the splicing of *transformer* post-transcription. According to Boggs and colleagues (1987), *transformer* RNA is spliced into a female-specific form and a non-functional/non-sex-specific form. Sxl is necessary for the proper splicing of *tra* into a functional form (Figure 9.1, Inoue et al., 1990), which occurs in females, but not males. In females, the functional Tra protein acts to regulate the splicing of the *doublesex* (*dsx*) gene. It has been determined that Tra interacts with another transformer protein (Tra-2), and these proteins function in concert as mRNA splicing factors to control the female-specific splicing of the *dsx* pre-mRNA (Mattox and Baker 1991; Waterbury et al., 1999; Kato et al., 2010; Verhulst et al., 2010). Tra and Tra-2 proteins are co-expressed only in females; if Tra/Tra-2 complexes are absent, then male-specific *doublesex* mRNA is produced, thus resulting in male fates (Sarno et al., 2009). Interestingly, the *tra-2* gene has also been linked to spermatogenesis in male *Drosophila*, but it has also been demonstrated that Tra is necessary to prevent the continued presence of male-specific gonadal precursors in the female somatic gonad (DeFalco et al., 2003). The specific splicing properties of the Tra and Tra-2 complex yield two isoforms of the *dsx* gene, each found uniquely in one sex or the other. However, unlike the upstream genes, both the male and female forms of the Dsx protein are functional and play important roles in the divergent pathways leading to sex determination and differentiation.

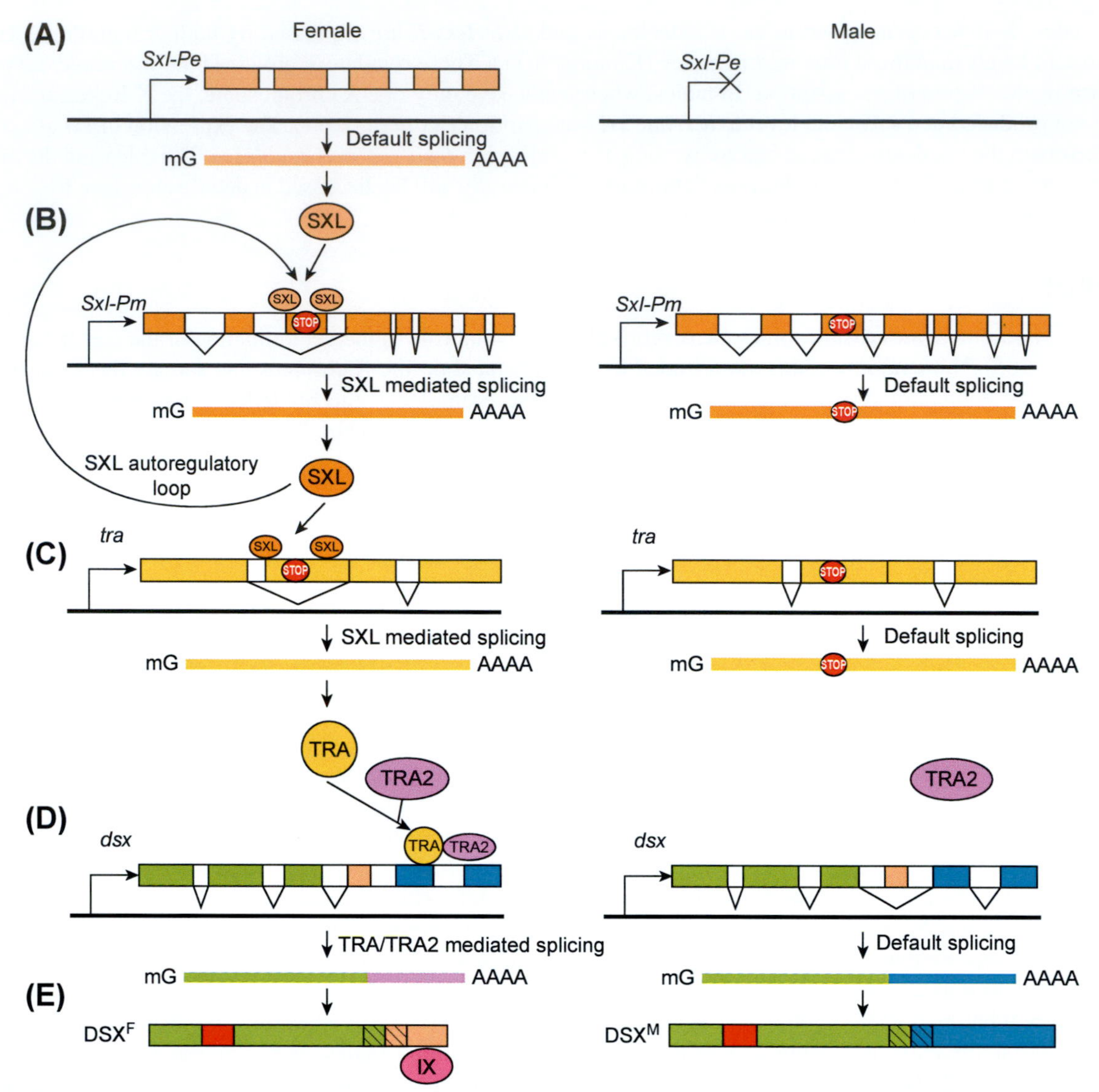

FIGURE 9.1 **Alternative Splicing of *Sex-lethal, Tra* and *Doublesex* Leads to Dimorphic Genetic Programs in *Drosophila Melanogaster.*** Early activation of the Pe promoter in females (A) leads to expression of a functional Sxl protein. Males do not achieve this early expression. In females, early Sxl activity acts on transcripts produced from the Pm promoter (B), removing a stop codon by alternative splicing that is necessary for the translation of a functional Sxl protein. The production of late Sxl protein production activates an auto-regulatory loop (B) that maintains Sxl expression and (C) regulates the alternative splicing of *tra* transcript in females. No Tra product is found in males. (D) Tra acts with Tra2 to regulate splicing necessary for the production of the female-specific isoform of Dsx^F (D, E). In males, absence of Tra results in default splicing and the expression of a male-specific isoform (Dsx^M). *Adapted from Camara et al., (2008), with permission from Elsevier.*

Doublesex

Doublesex is a founding member of the DM domain family of transcriptional regulators; many members of this gene family function at some level of sex determination or sex differentiation (reviewed in Matson and Zarkower, 2012). The Doublesex protein contains a highly conserved DM domain, consisting of two intertwined zinc fingers that include a cysteine-rich region (reviewed in Bellefroid et al., 2013). Unlike other transcriptional regulators, members of this family are believed to interact with its consensus sequence in the minor groove of DNA; this has led to speculation that the DM family works in part by influencing other proteins that bind in sites exposed within the major groove (Zhu et al., 2000; Zhang et al., 2006).

In females, Tra and Tra-2 regulate the splicing of *dsx* transcripts to produce a female-specific mRNA that leads to the translation of a sex-specific isoform, Dsx^F. In males, Tra/Tra2 complex is absent, thus the *dsx* transcript is translated into a male-specific isoform Dsx^M (Figure 9.1). Both Dsx^F and Dsx^M are functional transcription factors (Erdman et al., 1996; Waterbury et al., 1999; Shukla and Nagaraju, 2010; Siwicki and Kravitz, 2009). Both versions contain the DM type DNA binding domain located in the N-terminus; their structural differences occur in the C-terminal region (Narendra et al., 2002; Zhang et al., 2006). Dsx^F, shorter than Dsx^M by approximately 122 amino acids, requires a binding interaction with the cofactor Intersex for its sex-specific function (Garrett-Engele et al., 2002). Ix does not appear to be required for Dsx^M activity (Garrett-Engele et al., 2002), thus supporting that the differences in the C terminus mediates its specificity of function.

These two isoforms are critical for phenotypic differences in the gonads, in dimorphic characteristics including sex combs, courtship behavior, and in the development of the nervous system (Taylor and Truman, 1992; Villella and Hall, 1996; reviewed in Robinett et al., 2010). Recently, Luo and colleagues (2011) showed that there are approximately 23 target genes of the Dsx transcription factor. One major target is the gene known as *fruitless* (*fru*), whose function is most critical in the formation of neural circuitry required for courtship behaviors.

Fruitless

Similar to other members of the sex determination hierarchy (SDH), *fruitless* expression is complex and can be dimorphic. The *fruitless* gene codes for a transcription factor protein belonging to the BTB-zinc finger family (Demir and Dickson, 2005; Siwicki and Kravitz, 2009). BTB proteins are known to participate in numerous cellular processes, due to the high variability observed in the binding regions of the domain (Perez-Torrado et al., 2006).

Along with *doublesex, fruitless* is also under control of Tra/Tra2 mediated splicing in females. The *fruitless (fru)* gene is large, spanning at least 140 kilobases (kb) and it can be transcribed into at least seven classes of transcripts, involving four different promoters (P1, P2, P3 and P4) (Ryner et al., 1996). Pre-mRNAs transcribed from three (P2, P3, P4) promoters do not undergo sex-specific alternative splicing and are translated into several monomorphic Fru isoforms. They do not contribute to the sexual dimorphism in *Drosophila*. In contrast, transcripts generated by the activation of the P1 promoter are sex-specifically spliced by Tra and Tra2. In males, which lack the Tra protein, the default splicing creates a sex-specific mRNA, which in turn produces Fru^M, an isoform of the protein that includes 101 additional amino acids. In females, there is no expression of the Fru^M isoforms since Tra binds to P1 transcripts, thus preventing translation (Ryner et al., 1996).

Fru^M plays a significant role in specifying many male-specific behaviors, such as courtship song production (Rideout et al., 2007). Expression of Fru^M has been detected in the male brain, optic lobes and ventral nerve cord (Lee et al., 2000). Although the downstream target genes of fru remains unknown, it has been suggested that Fru^M may prevent cell death by inhibiting the genes (e.g., *head involution defective, reaper* and *grim*) that are responsible for programd cell death (Kimura et al., 2008).

Intersex

Intersex has been shown to play a later role in the sex determination pathway of *Drosophila*. Previous studies have shown that *intersex* is needed for proper female development but not for male development; however this topic is under some debate (Waterbury et al., 1999; Acharyya et al., 2002; Garrett-Engele et al., 2002). The expression of *intersex* is independent of Dsx; however, their interaction is crucial for female sex determination. Intersex acts as a transcriptional cofactor for the female isoform of Dsx and aids in the later activation of the female-specific genes, such as the yolk proteins, which are considered to be terminal in the SDH (Li and Baker, 1998; Garrett-Engele et al., 2002). This interaction is also associated with another gene known as *hermaphrodite* (*her*). Her proteins interact with Ix and Dsx to control female sex differentiation, but are not a part of the male sexual differentiation pathway (Li and Baker, 1998).

9.3 SEX DETERMINATION IN MAMMALS

The mechanisms of sex determination vary greatly across phyla, with evolutionary divergence observed at many levels. The primary determination of sex in mammals depends upon the chromosomal composition (XX or XY) of the embryonic precursors of the gonads (see also Chapter 27). During mammalian development, the mesodermal cells of the genital ridge exist in a bi-potential state, with the competence to develop into either testes or ovaries (Capal and Tanaka, 2013). The mammalian embryo demonstrates its duality by the presence of rudimentary precursor populations necessary for both reproductive systems prior to gonad determination (Chapter 27, this book). Sex-specific expression of a SRY, a Y-linked factor, acts as a primary switch that drives the cells of the genital ridge to form the male Serotoli cells of the testes. The Sertoli cell population and Leydig cell population of the developing testes release several hormones including testosterone (Leydig) and anti-Mullerian hormone (Sertoli) (Viger et al., 2005). Fetal exposure to these hormones results in the formation of male-specific structures, such as the vas deferens ducts that carry sperm, but also causes the development of female structures, including the Mullerian ducts, to cease (Viger et al., 2005). If the cells of the genital ridge lack SRY, then these embryonic cells generate the granulosa and thecal cells, which together form the follicles, the hormone-secreting cells that support and maintain the female gametes. Male development ceases concomitant with the differentiation of the female reproductive track. Determination of the gonads directs the primordial germ cells to enter either spermatogenesis or oogenesis. Maintenance of the sexual state and associated characteristics in the adult appears to be an important, yet poorly understood process (Viger et al., 2005).

Historically, the determination of female characteristics was often considered to be a "default" state in mammals, based on the observation that removal of the genital ridge prior to gonadal differentiation in rabbits resulted in female development regardless of the sex chromosome genotype (Jost, 1953). Yet, the complexity of the sex determination process at the molecular level has been revealed through the use of improving technology (see Chapter 27 for details). A more accurate view of sex determination would be that activation and suppression of sex-specific gene expression is essential for both male and female development (Figure 9.2; Nelson and Nusse, 2004; Kim et al., 2006; Sekido and Lovell-Badge, 2008).

The observed plasticity of adult gonadal cells has revealed insights into the mechanisms that sustain differentiated states. Two key genes that have been identified as being necessary for the maintenance of mammalian gonads are *Foxl2* and *Dmrt1*. *Foxl2* is an autosomal gene that is strongly upregulated in the ovaries during their formation (Verdin and De Baere, 2012). It has also been shown that *Foxl2* is continually expressed even after the ovaries have been fully formed, supporting the idea that the gene must play a role in the maintenance of ovarian cell fates (Bodega et al., 2004). The *Foxl2* gene belongs to the FOX gene family; the defining feature of this class of transcription factors is the conserved winged helix domain (reviewed in Hannenhalli and Kaestner, 2009). FOX proteins are involved in the development of many diverse structures including eyes, lungs, brain, cardiovascular system, and digestion system (reviewed in Jackson et al., 2010). Genotypic female (XX) mice lacking *Foxl2* demonstrate a female-to-male sex reversal phenotype and misexpression of *Foxl2* in male mice (XY) leads to abnormal formation of testicular tubules (Ottolenghi et. al. 2007). Loss of *Foxl2* in the ovaries of adult female mice caused the female-specific granulosa and thecal cell populations to transdifferentiate into Serotoli and Leydig-like cells, an outcome likely driven by the abnormal upregulation of Sox9 in the ovary (Uhlenhaut et al., 2009). These results suggest that *Foxl2* activity may be essential for inhibiting male-specific gene expression in the adult ovary, either by directly suppressing *Sox9* or by regulating other female-promoting regulators (e.g., *Wnt4*). Interestingly, lack of appropriate expression of *Foxl2* has been demonstrated in granulosa cell tumors, supporting its potential role in cell cycle control (Verdin and De Baere, 2012).

Unlike other embryonic processes, the upstream regulators of primary sex differentiation in mammals are not homologous to those of insects such as *Drosophila*. Yet, the conservation of downstream effectors that control sexually dimorphic development reveals deeper evolutionary connections. Dmrt1 (*Dsx-and mab-3-related transcription factor* 1) is a member of the conserved gene gamily that includes the Doublesex gene of *Drosophila* (reviewed in Matson and Zarkower, 2012). Across a broad range of vertebrates, including mammals, orthologs of Dmrt1 tend to be expressed in the developing gonads, indicating an early role in sex determination. Studies in humans and mice, however, have given a conflicting view of Dmrt1 function in primary sex determination. In humans, the *DMRT1* gene is located on chromosome 9, and it has been shown

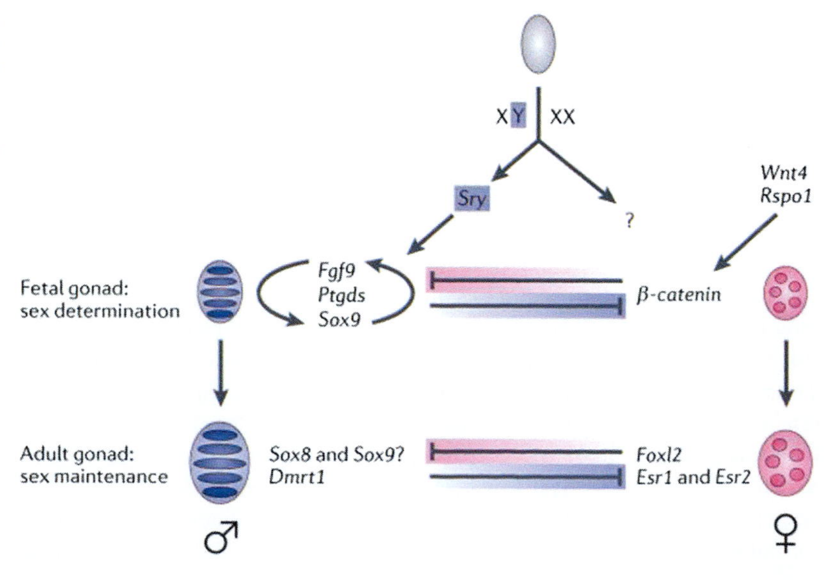

FIGURE 9.2 Determinaton of Sex in Mammals Requires the Activation and Supression of Transcriptional Regulators. Note that maintenance of sexual states requires the actions of Dmrt1 and Foxl2, though they are not required for the initial establishment of primary determination. *Adapted from Matson and Zarkower (2012), with permission from Nature.*

that the loss of a single copy (hemizygosity) leads to a sex reversal phenotype in XY individuals (Raymond et. al., 1998; Tannour-Louet, M. et al., 2010). Yet, loss of *Dmrt1* in XY mice does not cause such reversal, and testes formation occurs, although subsequent steps of differentiation are affected (Raymond et al., 2000). *Dmrt1* has been categorized as a tumor suppressor and its function is required for the proper regulation of mitosis and meiosis in germ cell populations of males and females (reviewed in Matson and Zarkower, 2012).

Recent studies have indicated that Dmrt1 function is essential for the maintenance of differentiated fates within the male testes. Post-natally, Dmrt1 null males lose *Sox9* expression within newly differentiating Sertoli cells, an outcome that is concomitant with an activation of female-specific gene expression, including *Foxl2* (Kim et al., 2007). The testes of such mice become populated with granulosa and theca-like cell populations indicative of transdifferentiation (Matson et al., 2011). It has been shown that *Dmrt1* can repress the expression of *Foxl2*, and other regulators of female sex determination. Thus, the opposing actions of Foxl2 and Dmrt1 seem to be essential for maintaining the sexual state of the adult gonad, despite the fact that neither seems to have a primary role in the sex determination in mice (Figure 9.2).

Overall, the studies involving *Foxl2* and *Dmrt1* indicate that the potency of the mammalian gonads is not fixed post-embryonically. The apparent plasticity of the adult gonads represents a rich area for investigation of the mechanisms that control pluripotency, in ways that may be sex-specific.

9.4 DIMORPHISM IN THE FLY OLFACTORY SYSTEM

Behaviors such as courtship, complicated dance, singing, display of beauty, and fighting are often critical for successful reproduction in different animal species. These reproductive behaviors are thought to originate from dimorphic processing dependent upon sex-specific neural populations and circuitry in the central nervous system (Simerly, 2002; Datta et al., 2008). For example, in *D. melanogaster*, the male sex pheromone cis-vaccenyl acetate (cVA) acts through the same set of olfactory receptor neurons in both sexes, but can elicit different reproductive behaviors in male and female flies. Dimorphism in the neural circuitry of the olfactory system in the fly brain likely mediates these opposing behaviors. In the formation of the nervous system of the fly, the somatic cells undergoing neurogenesis must be impacted by the genetic mechanisms that determine sex to achieve such differences in neural structure and function. The olfactory system of *Drosophila melanogaster* is an ideal model for studying the synergy between organogenesis and sex determination.

The Olfactory System of Flies

Sex pheromones and fruit odorants bind to olfactory receptor neurons (ORNs) within the sensilla of the *Drosophila* antennae. Binding with the odorants changes the basal firing rate of the ORNs (Pellegrino et al., 2010). These peripheral neurons then send information to the olfactory system in the central nervous system. The olfactory system within the *Drosophila* brain consists of three parts: the antennal lobe, the mushroom body, and the lateral horn. In the central nervous system, the antennal lobe receives the input from the ORNs of the antenna and then sends information to higher olfactory centers, including the mushroom body (MB) and the lateral horn (LH).

Approximately 1,200 olfactory receptor neurons from each antenna have been identified and they are categorized into 50 classes according to the odorants they detect (Laissue et al., 1999; reviewed in Vosshall and Stocker, 2007). Peripheral inputs from the antennae converge into approximately 43 glomeruli, sub-regions of the antennal lobe in the fly brain, and arborize on the central neurons that comprise the glomerular neuropil (Jefferis et al., 2007). Each glomerulus receives signals exclusively from each class of olfactory receptor neurons (Laissue et al., 1999). Two types of neurons exist in the antennal lobe: projection neurons and local neurons. Local neurons only form inter-glomerular connections within the antennal lobe. Projection neurons send axons to higher brain centers, primarily the lateral horn. A few neurons are connected to the mushroom body first, but the majority synapse directly with cells in the lateral horn (Jefferis et al., 2009)

Olfactory information is sent to two higher centers, namely the mushroom body and the lateral horn in the protocerebrum, via the projection neurons. Research indicates that the mushroom body is involved in associative olfactory learning, whereas the lateral horn is required for olfactory behaviors (Connolly et al., 1996; Stocker et al., 2001). One study has shown that there are more synapses in the lateral horn than in the mushroom body (Jefferis et al., 2007), which suggests that the majority of information flows to the lateral horn. When the mushroom body is damaged or inactivated, olfactory information can still access the lateral horn and the flies respond with appropriate behaviors to odorants (Connolly et al., 1996).

The lateral horn has been mapped through image registration, a technique that can combine multiple photographs to form a complete and coordinated image system (Rein et al., 2002). It has been discovered that the fruit odorants and pheromones activate different regions of the lateral horn (Jefferis et al., 2007). Fruit odorants from apple, banana and pineapple activate the dorsal and posterior lateral horn, whereas pheromones activate the anterior-ventromedial region.

Processing of Sex Pheromone Information

Sex pheromones play an important role in the sex-specific behaviors involved in reproductive behaviors (Jallon, 1984). Female flies produce three major sex pheromones including (7Z,11Z)-7,11-pentacosadiene, (7Z,11Z)-7,11-heptacosadiene and (7Z,11Z)-7,11-nonacosadiene (Mori et al., 2010). If a male detects the female attractants, he starts a series of courtship actions. Males orient to the female, follow her, tap her abdomen with their foreleg, sing a specific courtship song, lick her genitals, and finally attempt copulation (reviewed Siwicki and Kravitz, 2009). In contrast, males use cis-vaccenyl acetate (cVA) and CH503 to attract females (Bartelt et al., 1985; Yew et al., 2009). Male pheromone cVA is known to mediate male courtship behavior. Interestingly, even though cVA is detected by both female and male olfactory receptor neurons, it can have opposite effects on the two sexes. cVA inhibits male courtship behavior and stimulates aggression in other males, but it promotes mating behavior in females by increasing the female receptivity to males (Kurtovic et al., 2007). This reality presents a biological conundrum as to how a single chemical can have such dimorphic behavioral effects that are critical for sexual reproduction in flies.

Sexual Dimorphism in the Fly Olfactory System

Researchers have identified olfactory receptors (OR) that process male sex pheromones (cVA), namely the class known as Or67d (Ha and Smith, 2006). The *Drosophila* olfactory receptors are heteromeric ligand-gated ion channels, located within the fly antennae (Vosshall, 2000; Vosshall and Stocker, 2007). The ORNs that express Or67d terminate in three sexually dimorphic glomeruli: DA1, VA1lm, and VL2a (reviewed in Vosshall and Stocker, 2007). These glomeruli are enlarged in males. For example, in Hawaiian *Drosophila*, DA1 is approximately 5.9 times larger in males than in females (Kondoh et al., 2003). Despite the obvious dimorphism in the glomeruli, experiments have shown that the differential size does not account for the sex-specific behavioral response to cVA signaling (Datta et al., 2008).

If not based on size differences, what physical features of the glomeruli do contribute to these behaviors? The projection neurons of DA1 are cVA responsive and send axons directly to the anterior-ventromedial region of the lateral horn, which is devoted to processing pheromone information. The PNs of DA1 have sexually dimorphic neural connectivity with the lateral horn neurons (LHNs) (Datta et al., 2008). In addition to the dimorphic neural connectivity in the LH, significant difference in the relative size of the LH is found between the two sexes. It is 1% larger in males than females. Males and females also have their own enlarged area in the LH. The male and female enlarged areas correspond to 3% and 1.6% of the total LH volume respectively (Jefferis et al., 2007). Despite the obvious dimorphic size of the LH between the two sexes, no study has determined that size impacts the dimorphic cVA-related behaviors.

Transcriptional Control of Dimorphism in Olfactory Circuitry

Fruitless is a prominent candidate for mediating embryonic outcomes that underlie the sex-specific nature of pheromone detection and processing. Males lacking Fru^{M} do not perform most aspects of courtship behaviors, whereas females expressing Fru^{M} demonstrate male-specific behaviors, with the exception of copulation (reviewed in Siwicki and Kravitz, 2009). With a greater ability to probe gene expression and function at the single cell level, it has become clear that *fruitless* has a complex role in defining differentiated, sex-specific cell phenotypes within the central nervous system of the fly. Previous studies showed that Fru^{M} is expressed in roughly 1,700 neurons throughout the fly's central nervous system, including distinct sensory and processing populations within the olfactory system (Lee et al., 2000; reviewed in Cachero et al., 2010). Among the sensory neurons of the fly's antennae, the ORNs that express the cVA-receptor Or67d project to glomeruli that are *fru*+ (Stockinger et al., 2005). Recent studies have employed genetic tools to map the overall anatomy and architecture of the olfactory circuitry in the context of *fru* expression. Datta and colleagues defined a sex-specific circuit connecting pheromone-sensing neurons in the periphery with *fru*+ projection neurons in the antennal lobe (Datta et al., 2008; Figure 9.3). Specifically, they showed that *fru*+ projection neurons from the DA1 glomerulus demonstrate distinct, sex-specific terminal branching patterns in the lateral horn of the protocerebrum. In males, DA1 PN axons demonstrated a denser, male-specific axonal branching within the ventromedial region of the lateral horn that was absent in females. Moreover, they demonstrated that this dimorphic branching was dependent upon the expression of Fru^{M}. In males, loss of Fru^{M} significantly reduced the male-specific PN branching in the LH, whereas forcing females to express Fru^{M} resulted in the DA1 PNs exhibiting a branching pattern that was strikingly consistent with the male phenotype.

Cachero and colleagues utilized morphometric analyses to examine higher (second and third order) connections among neural cells with *fru*+ expression (Cachero et al., 2010). They described eight *Fru*+ LHN lineages: aIP-b, aIP-e, aSP-a, aSP-f, aSP-g, aSP-h, pMP-e and pSP-f. Seven out of eight LH neuronal populations have sexually dimorphic projections (aIP-b is the exception); the seven represent the principal neurons of the lateral horn (LHNs). By examining the overlap

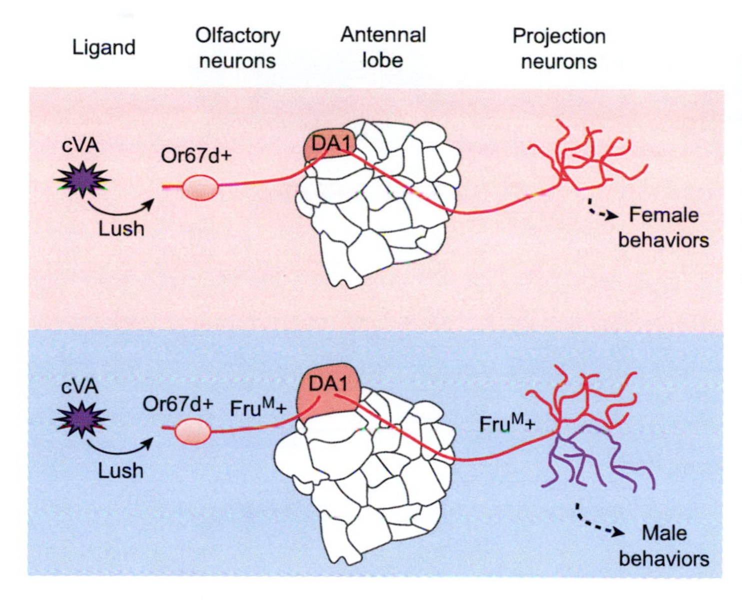

FIGURE 9.3 Sex-specific Circuitry Involving Higher Order Neurons may Mediate Dimorphic Behavior in Response to the Male Pheromone cVA. Both males and females detect cVA through Orb67-expressing olfactory neurons, transmitting to the DA1 glomerulus of the antennal lobe. The axons of the DA1 projection neurons demonstrate a denser and larger male-specific axonal branching pattern within the lateral horn that is absent in females. Development of the male-specific branching is dependent upon the expression of Fru^M (Datta et al., 2008). *Adapted from Stowers and Logan (2010), with permission from Elsevier.*

between the DA1 PN and the seven dimorphic LHN lineages, these researchers uncovered sex-specific connectivity within the *fru*+ inputs into the lateral horn. Two lineages, namely aSP-f and aSP-h, appear to only overlap with DA1 PNs axons in males; at least one lineage (aSP-g) demonstrates a female-specific intersection. Such sex-specific differences were found to extend at the output level of the LHNs, consistent with idea that these differences might mediate the behavioral differences associated with the cVA pheromone. Specifically, the axonal arbors of aSP-f and aSP-h were numerous within the male-enhanced region (MER) of the protocerebrum, but such connections were lacking in females. In contrast, the outputs from the LHN demonstrated female-specific arborization patterns in the dorsal, female-enhanced region (FER) of the protocerebrum.

In summary, the male sex pheromone cVA suppresses courtship behaviors and elicits aggressive behavior in males, but it increases female receptivity to males. No distinctive cVA signaling has been detected at the periphery level (sensory ORNs) nor at the initial processing level (antennal lobe/DA1 glomerulus). Taken together, the studies by Datta et al. (2008), and Cachero et al. (2010) indicate that sex-specific circuitry at the higher processing levels of the lateral horn are likely required for the opposing behavioral responses of male and female flies to cVA (Figure 9.3). What remains unknown is how such sex-specific circuitry is configured during development. Clearly, *fruitless* expression, specifically Fru^M, mediates key aspects, though there is discernible variability in phenotypes involving both gain-of-function (females) and loss-of-function (males) of *fruitless*. Penetrance and expressivity differences indicate that other factors must also contribute to sex-specific circuitry within the fly's brain. In addition, the targets of Fru^M are not well known, thus important questions remain unanswered.

9.5 INTEGRATION OF SEX DETERMINATION AND EMBRYONIC PATTERN FORMATION

In *Drosophila*, males and females share the same overall body plan, with specialization of sex-specific primary and secondary characteristics involving a subset of embryonic cell lineages and populations. Such sex-specific sculpting of the embryonic form requires cells to be able to respond and integrate information from those genetic programs that define sex (e.g., sxl ⇒ tra ⇒ dsx/fru), and those that control the determination of differentiated cell fates (e.g., Hox genes; reviewed in Mallo and Alonso, 2013). As discussed in the previous section, the dimorphic connectivity within the olfactory system leads to questions of how such differences in circuitry are generated, especially in relation to Fru^M activity.

Studies examining how sex-specific gene pathways integrate with embryonic patterning have begun to uncover the mechanisms employed in generating such differences in *D. melanogaster*. The most obvious examples are those that influence segment formation along the anterior-posterior axis. Segmentation of the fly body is controlled through the actions of the homeotic selector (Hox) genes of the *Antennapedia* and *Bithorax* gene clusters (reviewed in Malla and Alonso, 2013). Mutations in these genes cause dramatic phenotypes in which one body part/segment (e.g., antennae) becomes converted into another (e.g., legs), a phenomenon known as transformation (Lewis, 1978). As transcription factors, Hox proteins regulate each other's expression to define overlapping, but distinct domains along the AP axis (Figure 9.4). In general, transcriptional repression is critical for this patterning whereby posterior Hox genes inhibit the expression of those involved

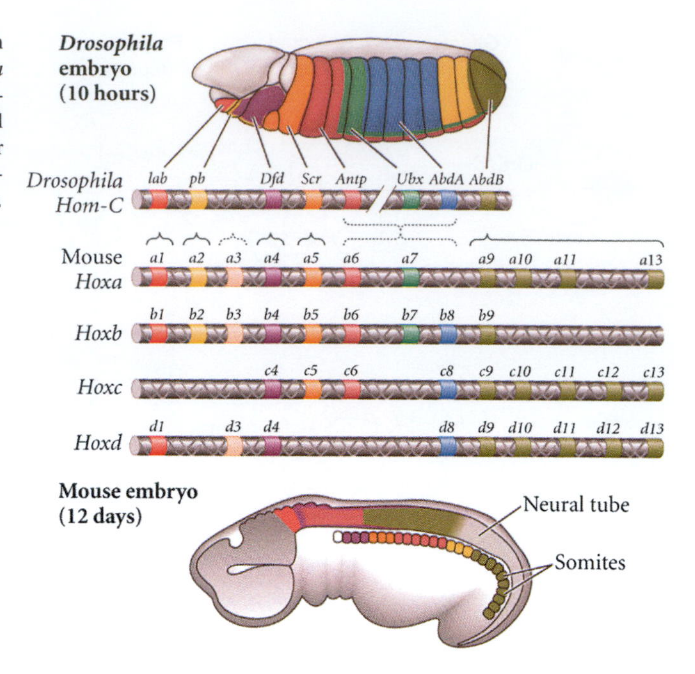

FIGURE 9.4 The Hox gene family is evolutionarily conserved, as shown by the distinctive similarity between the Hom-C complex in *Drosophila* and the four Hox gene complexes in mouse. In both species, the expression of these genes along the AP axis corresponds with their chromosomal orientation, a phenomenon known as colinearity. In the mouse, the four Hox clusters contain paralogous genes that are expressed in similar patterns. *Adapted with permission from Gilbert, S: Developmental Biology, 8th Edition.*

in specifying anterior fates. The chromosomal arrangement (colinearity), expression and activity of the homeotic selector genes and their vertebrate counterparts are remarkably conserved within the animal kingdom (Figure 9.4; reviewed in Malla and Alonso, 2013).

During embryogenesis, males and females demonstrate monomorphic development in the segmented body plan created during the embryonic and larval stages (specifically, three head, three thoracic, and eleven abdominal segments). In the pupal stages, the divergence of body formation occurs most acutely in the abdominal segments. The most posterior segments (A8 – A10) generate the male and female genitalia. In females, Dsx^F represses *branchless* (*bnl*), the fly version of fibroblast growth factor (FGF) in abdominal segment 9 (Ahmad and Baker, 2002); such repression does not occur in males, which express the male isoform of Dsx. The transcription factor *dachshund* (*dac*) is also a target of Dsx. In the A9 primordium of males, *dac* is expressed in the lateral domains, whereas in females *dac* is found in a medial domain corresponding to those giving rise to A8 (Keisman and Baker, 2001). Dsx modulates the sex-specific differences in *dac* expression within this tissue by influencing the signaling molecules (Dpp, Wg) that establish these domains (Keisman and Baker, 2001). More recently, Chatterjee and colleagues (2011) used a whole genome screening approach to identify downstream Dsx targets in the developing genital imaginal disc. Among the 23 candidate genes identified, three known transcription factors, namely *lozenge* (lz), *Drop* (Dr) and AP-2, demonstrated sex-specific patterns dependent upon Dsx activity. In females, Dsx activates *lozenge* in genitalia precursors while repressing *Dr* and *AP-2*. Both Dr and AP-2 are normally expressed only in the male genital disc; their repression in the female imaginal disc is dependent on Dsx^F function. It is currently unknown whether these transcriptional relationships are direct, though the identification of putative Dsx binding sites within each gene supports the possibility that they are (Chatterjee et al., 2011).

In addition to the external genitalia, other secondary dimorphic features of the abdomen require integration of sex determination and pattern formation. Pigmentation pattern and segment number are two such examples. Sex-specific pigmentation of the abdominal cuticle is controlled in part by Dsx isoforms in conjunction with the Hox gene member, Abdominal-B (Abd-B; see Figure 9.4). In females, Dsx^F and Abd-B act synergistically to activate the expression of *bric-a-brac* (*bab*) in abdominal segments A5–A7, whereas in males Dsx^M works with Adb-B to repress *bab* expression in this region (Kopp et al., 2000; Williams et al., 2008). This genetic regulation leads to the distinct differences in pigmentation, since Bab acts to inhibit genes necessary for pigment production. In males, the repression of Bab allows for pigmentation of the abdominal cuticle to occur, with the Hox genetic code (Adb-B) defining the limits along the anterior-posterior axis. In females, Dsx^F activates Bab, which then represses downstream genes, leading to an absence of pigment in the abdomen (Figure 9.5).

Dsx and Abd-B have also been shown to work collaboratively in shaping segment identity and number. Males generate fewer adult segments than females, thus resulting in a smaller abdomen; this dimorphism arises from changes in cell proliferation, apoptosis and the acquisition of segment identity (Wang et al., 2011). Some aspects of abdominal formation from the posterior segments appear to be monomorphic (e.g., apoptosis), but sex-specific differences in other processes (e.g., cell proliferation) can then drive divergence in development. Both Dsx and Abdominal-B appear to function in

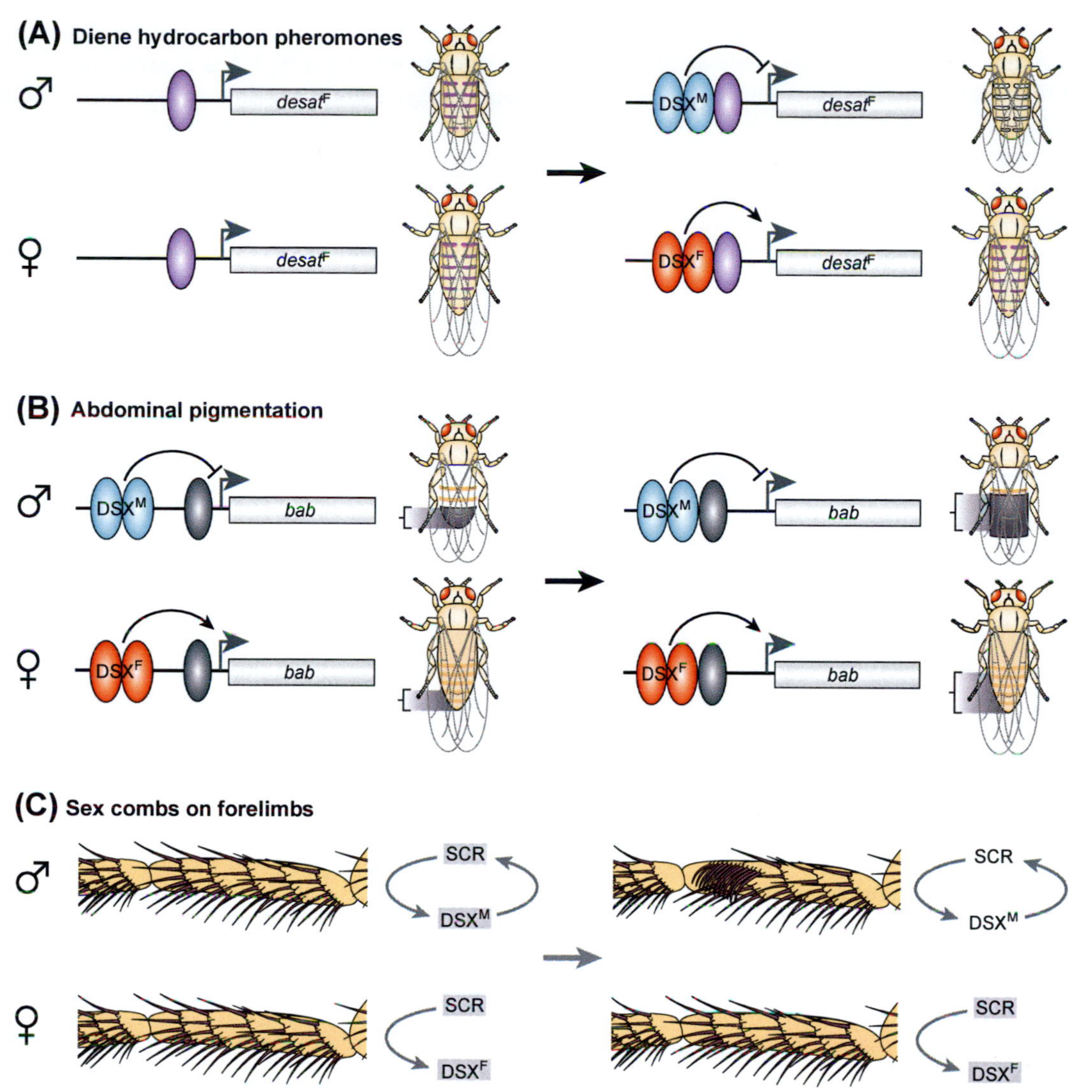

FIGURE 9.5 **Evolutionary Change in Form and Function can be Imparted by Changes in the Sex Determination Gene *Doublesex*.** (A) Enzymes involved in pheromone production become sex-specific by acquiring sex-specific transcriptional regulation in response to Double sex isoforms. In this example, comparative studies in *Drosophila* have revealed that the transcriptional regulation of the female-specific desaturase enzyme (desatF) has been modified by the acquisition of Dsx binding sites, placing it under control of the sex determination pathway (Shirangi et al., 2009). This enzyme plays an important role in female pheromone production. (B) Male and female differences in the abdominal pigmentation patterns are evolutionarily based in the acquisition and modification of binding sites for both Dsx and the Hox patterning protein Abd-D (black oval). Changes such as binding site number and relative proximity can influence the outcome of transcriptional control (Williams et al., 2008). (C) Temporal and spatial changes in the expression of double sex can lead to cells gaining competency for sex-specific development. In this example, the male-specific sex combs (brown patches) of the forelegs form in part because the embryonic cells of the segment giving rise to this section gained *doublesex* expression. Sex comb formation in males is dependent upon a mutual loop involving Dsx^M and the Hox gene Sex combs (Tanaka et al., 2011). *Adapted from Matson and Zarkower (2012), with permission from Nature.*

a sex-specific way that influences the fate of the A7 segment in males. In wild type males, the embryonic cells that comprise this segment do not express the signaling molecule Wingless, and it is these cells that are lost in the formation of the male abdomen (Wang et al. 2011). In this study, it was shown that either a reduction of Abd-B expression or the loss of Dsx in A7 resulted in restoration of Wingless expression in the A7 segment in males; conversely such changes in females had no effect (Wang et al., 2011). Lack of Wingless expression is associated with a suppression of cell proliferation, suggesting the Dsx and Abd-B may be affecting sex-specific segment growth by controlling Wg. A regulator of the anterior-posterior axis, Wg also acts in maintaining the polarity of abdominal segments as they form. Thus, lack of Wingless expression in males has multiple effects, which includes allowing for a partial transformation of A7 toward an A6 identity (Wang et al., 2011). In the precursors of A7, Dsx and Abd-B appear to provide a sex-specific context within which the other processes that are not sex-specific operate. It was observed that increased cell death in the developing A7 segment is a common attribute for both sexes, but the synergism with sex-specific differences in Wingless in males leads to a different developmental outcome for the A7 segment. It remains unknown whether Dsx and Abd-B directly influence Wingless expression.

Like the abdomen, the central nervous system of *Drosophila* displays significant dimorphism, which influences the reproductive behaviors exhibited by each sex. Differences between the sexes have been detected in specific populations with regard to neuronal numbers, axonal projections and synaptic density (Rideout et al., 2010; Robinett et al., 2010). Fru^M, for example, is expressed in roughly 2% of all neurons within the fly CNS (Usui-Aoki et al., 2000; Lee et al., 2000). Loss of Fru^M expression in these neuronal populations results in mutant male flies that exhibit aberrant courtship behaviors (Ito et al., 1996; Ryner et al., 1996; Goodwin et al., 2000). Loss of Dsx expression affects male courtship behavior, though Dsx^M is not sufficient to induce male courtship behaviors (Taylor et al., 1994; reviewed in Kimura, 2011). Previous studies have shown that *dsx* and *fru* are both expressed in the mesothoracic ganglion and loss-of-mutations in either gene causes song defects (reviewed in Siwicki and Kravitz, 2009; Kimura, 2011).

Cell proliferation and cell death are two main targets of sex-specific regulation in the CNS. Dsx^M, the isoform only active in males, promotes extra cell division in the neuroblasts of abdominal ganglion (Taylor et al., 1992) and two clusters (PC1, PC2) found the posterior brain with CNS (Sanders and Arbeitman, 2008) leading to dimorphic differences in cell number in these populations. One dimorphism shaped by cell death is the small neural cluster known as mAL, which functions in gustatory pheromone processing. In males, the mAL consists of 30 cells, whereas females have only five in this region (Kimura et al., 2005); this divergence has been shown to result from female-specific apoptosis. Fru^M is a key regulator of this difference, evidenced by the fact that removal of Fru^M in males leads to a reduction of mAL neurons, with corresponding changes in circuitry, causing these mutant males to display mAL architecture that is similar to that of wild type females (Kimura et al., 2005). Females expressing Fru^M splice variants demonstrate a male mAL pattern, both in cell number and in axonal projections. Kimura and colleagues have shown that Dsx^F functions in females to eliminate a male-specific population of neurons (P1) in the dorsal-posterior region of the brain (Kimura et al., 2008). One final example of dimorphism can be found in the mesothoraic ganglion (mSG) of the ventral nerve cord. Within this region, males contain a specific neuronal cluster, TN1, that is absent in females. This cluster plays a fundamental role in courtship behavior, controlling aspects of wing extension and movement (Lee et al., 2000; Rideout et al., 2007 Sanders and Arbeitman, 2008). *Fruitless* and *doublesex* are both expressed in the TN1 region (Rideout et al., 2007; Robinett et al., 2010). Dsx^F is necessary, though not fully sufficient, for the female-specific cell death that removes this population from the female Msg (Sanders and Arbeitman, 2008; Robinett et al., 2010). In males, expression of both Fru^M and Dsx^M are required for the behavioral outputs (e.g., courtship song) control by TN1, but their role in shaping the dimorphism is not fully understood (reviewed in Kimura, 2011).

Defining sexual differences, be they at the primary or secondary level, requires modification or refinement of developmental programs that may often be considered "sex-neutral". As the two terminal genes of the sex determination pathway, it is important to note that *fruitless* and *doublesex* are not expressed uniformly throughout all cells and tissues, at all stages. In fact, the expression of both genes is quite dynamic, both temporally and spatially. As argued by Robinett and colleagues (2010), fruit flies can be viewed as "mosaics", with some cells expressing terminal gene products that define sexual states (dsx^M, dsx^F, or fru^M) and other cells not doing so (presumptively being sex-neutral). From an evolutionary perspective, there is a benefit afforded to a state in which some cells are responsive to sex-specific programming, and others are not. Variation in gene expression does not have to affect development globally or systemically for the emergence of advantageous sexual traits during embryogenesis. Mosaicism of sexual states at the molecular level is likely to be a recurring theme in all sexually reproducing organisms, regardless of upstream genetic programs that determine sex.

Microevolutionary changes involving sexual differentiation can occur readily and by varied routes involving those genes that are involved in sex determination (Williams and Carroll, 2009). The acquisition of cis-acting regulatory elements (CREs) that place non-sex regulators of pattern under sexual control is one such mechanism (Figure 9.5). As discussed previously, the dimorphic nature of abdomen pigmentation has evolved in part due to the ability of a sex-specific protein to bind novel combinations of CREs, in concert with the actions of a Hox transcription factor. An alternative route to evolutionary change can occur when the CREs that control the expression of *doublesex* and *fruitless* lead to novel domains of expression, spatially and/or temporally. One such example is the gain of *dsx* expression that drives sex comb formation in males (Figure 9.5; Tanaka et al., 2011). One final point to be made is that the downstream effectors of sex determination demonstrate deeper levels of evolutionary conservation than those mechanisms that specify primary sexual states. For example, the DM gene family that includes *doublesex* and *Dmrt1* is quite large and growing as orthologs are identified from diverse species across metazoans. Strikingly, DM domain genes demonstrate a deeper, more ancient level of conservation, demonstrated by the fact that many orthologs function in sex determination. For example, invertebrates such as flat worms, coral, and the crustacean *Daphnia magna* use DM domain genes in sexual differentiation (reviewed in Matson and Zarkower, 2012). Members of this gene family may reveal the evolutionary changes that affect their deployment in sex-specific functions, such as those that influence CRE configurations. Likewise, dimorphisms can arise by the acquisition of CREs that allow DM domain proteins to bind and influence the transcription of those proteins that direct general pattern formation.

9.6 CLINICAL IMPLICATIONS OF SEXUAL DETERMINATION AND DIMORPHISM

Disorders of sexual development (DSD) have been well documented in human medicine, along with an associated history of controversial practices involved in the treatment of such conditions (see Chapter 27; also see Recommended Resources for website links involving human genetic disorders). Klinefelter syndrome (47, XXY) and Turner syndrome (46, X0) are two classic examples resulting often from meiotic errors and leading to abnormal sex chromosome number. These conditions affect the primary level of sex determination, as the establishment of sex hormones is not achieved, resulting from abnormal gonad formation and/or function. Gain-of-function and loss-of-function mutations affecting specific genes such as SRY result in DSDs, including sex reversal phenotypes (reviewed by Warr and Greenfield, 2012). As discussed previously, *Dmrt1* hemizygosity results in XY sex reversal (Raymond et. al., 1998; Tannour-Louet, M. et al., 2010). To date, at least one study has shown that mutation in *Foxl2* can contribute to a female-to-male sex reversal phenotype (Ottolenghi et al., 2007), but has been more consistently identified as being necessary for ovary differentiation in the adult (Uhlenhaut et al., 2009). Loss-of-function alleles for *Foxl2* have been identified in connection with a rare disease known as blepharophimosis-ptosis-epicanthus inversus syndrome (BPES), with and without ovarian abnormalities (Verdin and De Baere, 2012). In BPES, the eyelids do not form properly and some individuals demonstrate premature ovarian failure (POF) associated with mutations in *Foxl2*. Both Dmrt1 and Foxl2 have been associated with cancers stemming from the gonadal cell populations. Mutations in *Dmrt1* have been identified in testicular germ cell tumors (Kanetsky et al., 2011; Kratz et al., 2011) and *Foxl2* mutations have been identified in granulosa cell tumors of the ovary (Shah et al., 2009; reviewed in Verdin and De Baere, 2012).

In humans, chromosomal sex drives the establishment of primary sex (gonads) by directing gonadal precursors to form sex hormone-producing cell populations (testosterone, estrogen). Sex hormones in turn drive the dimorphic development of tissues and organs by binding to steroid receptors in responding cells; steroid receptor proteins function as transcriptional regulators based on their ability to bind DNA directly (Morris et al., 2004). Substantial work has been done to determine the effects that sex hormones have on the development of organs and tissue, especially the central nervous system. Sexual dimorphism and sex differences have been documented in a diverse array of human diseases and disorders (reviewed by Ray et al., 2008). Yet, despite the obvious connections, details about the ontological and mechanistic processes involved are poorly understood. Several recent studies highlight the importance of pushing questions about dimorphism into the forefront, especially when considering stem cell biology. Emerging data indicate that cell sex influences the properties and behaviors of stem cells, both *in vitro* and *in vivo* (reviewed by Ray et al., 2008). In one study (Deasy et al., 2007), researchers found that muscle-derived stem cells (MSDCs) derived from females (XX) regenerated at higher rate than those derived from males (XY) in a mouse model of muscular dystrophy. Host differences were also observed with females demonstrating more regenerative capacity than males (Deasy et al., 2007). Work by Leanu and colleagues (reviewed in 2011) showed that neural stem cells demonstrate sexual dimorphic properties that could potentially influence neurogenic potential. Other studies have addressed the roles of epigenetic phenomena such as genomic imprinting and X inactivation, both of which can influence gene expression in a sex-limited way. As with fruit flies, the mammalian cells need to be examined from the perspective that they may be mosaic in terms of sexual identity, perhaps determined by their competency to respond to sex hormones (through the expression of steroid receptors), their genetic variability in CREs that influence sex-specific transcriptional regulation and/or by the epigenomic profile imparted by those mechanisms that control chromatin structure.

9.7 CONCLUSIONS

The evolutionary importance of sexual reproduction is unquestionable, considering its ubiquity throughout the metazoan kingdom. Studies involving the model system *Drosophila* have provided significant insight into the mechanisms that drive dimorphic development at the primary and secondary levels. Comparative analyses involving diverse species have revealed that divergence may be the norm at the primary level of sex determination, but much conservation remains at the level of downstream effectors that regulate sexual differentiation. Dimorphic development requires the integration of sex-determining gene activities with embryonic programs that determine cell fate identity and function during pattern formation. Such integration can regulate processes such as cell proliferation, apoptosis, and cell differentiation. From a human perspective, understanding how sexual differentiation and dimorphism influence the physiological properties of diverse cell populations will be of critical importance as medicine moves toward cell-based strategies for drug discovery and disease treatments.

9.8 CLINICAL RELEVANCE

- Disorders of sexual development (DSDs), such as Klinefelter syndrome (47, XXY) and Turner syndrome (45, X0), often result from meiotic errors that disrupt the mechanisms required for primary and/or secondary sex determination.
- Sex reversal phenotypes that define some DSDs can be attributed to abnormal expression (loss or gain) of specific gene regulators required for male identify (*Sry, Sox9, Dmrt1*) or female identify (*Wnt4, Foxl2*).
- Maintenance of sexual fates requires the coordinated and sustained actions of sex-determining genes such as *Foxl2* and *Dmrt1*. Dysregulation of such factors in the adult gonads can cause differentiated cells to revert toward an undifferentiated state, indicative of their bi-potential lineage.
- Cancers stemming from gonadal somatic populations may be routed in abnormal expression of sex-determining genes.
- Emerging data indicate that sexual state of stem cells can influence the properties and behaviors exhibited, both *in vitro* and *in vivo.* Such findings support the need for better understanding of sexual dimorphic development at multiple levels.

ACKNOWLEDGMENTS

We thank S.A. Moody, C. Linn, and D. Droney for constructive feedback about this work. The authors have been supported by NICHD-NIH AREA grant (1R15HD060011–01), PI K.L. Kenyon, for research in this field.

RECOMMENDED RESOURCES

Flybase: http://www.bbc.co.uk/nature/adaptations/Sexual_dimorphism
Interactive Fly: http://www.sdbonline.org/fly/aimain/1aahome.htm
Berkeley Drosophila Genome Project: http://www.fruitfly.org/
Genetics Home Reference: http://ghr.nlm.nih.gov/
NCBI-MedGen: http://www.ncbi.nlm.nih.gov/medgen
World Health Organization: http://www.who.int/genomics/gender/en/index1.html
Online Mendelian Inheritance in Man (OMIM): http://omim.org/
Gendered Innovations: http://genderedinnovations.stanford.edu/what-is-gendered-innovations.html

REFERENCES

Acharyya, M., Chatterjee, R.N., 2002. Genetic analysis of an intersex allele (ix5) that regulates sexual phenotype of both female and male *Drosophila melanogaster*. Genet. Res. Aug; 80 (1), 7–14.

Ahmad, S.M., Baker, B.S., 2002. Sex-specific deployment of FGF signaling in *Drosophila* recruits mesodermal cells into the male genital imaginal disc. Cell May 31; 109 (5), 651–661.

Baker, B.S., Taylor, B.J., Hall, J.C., 2001. Are complex behaviors specified by dedicated regulatory genes? *Reasoning from Drosophila*. Cell 105, 13–24.

Bartelt, R.J., Schaner, A.M., Jackson, L.L., 1985. *Cis*-vaccenyl acetate as an aggregation pheromone in *Drosophila melanogaster*. J. Chem. Ecol. 11, 1747–1756.

Bell, L.R., Maine, E.M., Schedl, P., Cline, T.W., 1988. Sex-lethal, a *Drosophila* sex determination switch gene, exhibits sex-specific RNA splicing and sequence similarity to RNA binding proteins. Cell 55 (6), 1037.

Bellefroid, E.J., Leclère, L., Saulnier, A., Keruzore, M., Sirakov, M., Vervoort, M., De Clercq, S., 2013. Expanding roles for the evolutionarily conserved Dmrt sex transcriptional regulators during embryogenesis. Cell Mol. Life Sci. Oct; 70 (20), 3829–3845.

Bodega, B., Porta, C., Crosignani, P.G., Ginelli, E., Marozzi, A., 2004. Mutations in the coding region of the FOXL2 gene are not a major cause of idiopathic premature ovarian failure. Mol. Hum. Reprod. Aug; 10 (8), 555–557.

Boggs, R.T., Gregor, P., Idriss, S., Belote, J.M., McKeown, M., 1987. Regulation of sexual differentiation in *D. melanogaster* via alternative splicing of RNA from the transformer gene. Cell Aug 28; 50 (5), 739–747.

Burtis, K.C., 1993. The regulation of sex determination and sexually dimorphic differentiation in *Drosophila*. Curr. Opin. Cell Biol. 5, 1006–1014.

Cachero, S., Ostrovsky, A.D., Yu, J.Y., Dickson, B.J., Jefferis, G.S., 2010. Sexual dimorphism in the fly brain. Curr. Biol. Sep 28; 20 (18), 1589–1601.

Camara, N., Whiteworth, C., Doren, M., 2008. The creation of sexual dimorphism in the *Drosophila* soma. Cur. Top. Dev. Biol. (83), 65–107. 2008.

Capel, B., Tanaka, M., 2013. Forward to the special issue on sex determination. Dev. Dyn. 242 (4), 303–306.

Chatterjee, S.S., Uppendahl, L.D., Chowdhury, M.A., Ip, P.L., Siegal, M.L., 2011. The female-specific doublesex isoform regulates pleiotropic transcription factors to pattern genital development in *Drosophila*. Development Mar; 138 (6), 1099–1109. 2011.

Chau, J., Kulnane, L.S., Salz, H.K., 2012. Sex-lethal enables germline stem cell differentiation by down-regulating Nanos protein levels during *Drosophila* oogenesis. Pro. Natl. Acad. Sci. 109 (24), 9465–9470.

Connolly, J.B., Roberts, I.J., Armstrong, J.D., Kaiser, K., Forte, M., Tully, T., O'Kane, C.J., 1996. Associative learning disrupted by impaired Gs signaling in *Drosophila* mushroom bodies. Science Dec 20; 274 (5295), 2104–2107.

Datta, S.R., Vasconcelos, M.L., Ruta, V., Luo, S., Wong, A., Demir, E., Flores, J., Balonze, K., Dickson, B.J., Axel, R., 2008. The *Drosophila* pheromone cVA activates a sexually dimorphic neural circuit. Nature 452, 473–477.

DeFalco, T.J., Verney, G., Jenkins, A.B., McCaffery, J.M., Russell, S., Van Doren, M., 2003. Sex-specific apoptosis regulates sexual dimorphism in the *Drosophila* embryonic gonad. Dev. Cell Aug; 5 (2), 205–216.

Deasy, B., Lu, A., Rubin, R., Huard, J., Tebbets, J., Feduska, J., Schugar, R., Pollett, J., Sun, B., Urish, K., Gharaibeh, B., Coo, B., 2007. A role for cell sex in stem cell-mediated skeletal muscle regeneration: female cells have higher muscle regeneration efficiency. J. Cell Biol. 177 (1), 73–86.

Demir, E., Dickson, B.J., 2005. fruitless splicing specifies male courtship behavior in *Drosophila*. Cell Jun 3; 121 (5), 785–794.

Devineni, A.V., Heberlein, U., 2012. Acute ethanol responses in *Drosophila* are sexually dimorphic. Pro. Natl. Acad. Sci. 109 (51), 21087–21092.

Erdman, S.E., Chen, H.J., Burtis, K.C., 1996. Functional and genetic characterization of the oligomerization and DNA binding properties of the *Drosophila* doublesex proteins. Genetics 144 (4), 1639–1652.

Garrett-Engele, C.M., Siegal, M.L., Manoli, D.S., Williams, B.C., Li, H., Baker, B.S., 2002. intersex, a gene required for female sexual development in *Drosophila*, is expressed in both sexes and functions together with doublesex to regulate terminal differentiation. Development 129 (20), 4661–4675.

Goodwin, S.F., Taylor, B.J., Villella, A., Foss, M., Ryner, L.C., Baker, B.S., Hall, J.C., 2000. Aberrant splicing and altered spatial expression patterns in fruitless mutants of *Drosophila melanogaster*. Genetics Feb; 154 (2), 725–745.

Ha, T.S., Smith, D.P., 2006. A pheromone receptor mediates 11-cis-vaccenyl acetate-induced responses in *Drosophila*. J. Neurosci. Aug 23; 26 (34), 8727–8733.

Hannenhalli, S., Kaestner, K.H., 2009. The evolution of Fox genes and their role in development and disease. Nat. Rev. Genet. Apr; 10 (4), 233–240.

Hashiyama, K., Hayashi, Y., Kobayashi, S., 2011. *Drosophila* Sex lethal gene initiates female development in germline progenitors. Science 333 (6044), 885–888.

Inoue, K., Hoshijima, K., Sakamoto, H., Shimura, Y., 1990. Binding of the *Drosophila* sex-lethal gene product to the alternative splice site of transformer primary transcript. Nature Mar 29; 344 (6265), 461–463.

Ito, H., Fujitani, K., Usui, K., Shimizu-Nishikawa, K., Tanaka, S., Yamamoto, D., 1996. Sexual orientation in *Drosophila* is altered by the satori mutation in the sex-determination gene fruitless that encodes a zinc finger protein with a BTB domain. Proc. Natl. Acad. Sci. USA. Sep 3; 93 (18), 9687–9692.

Jackson, B.C., Carpenter, C., Nebert, D.W., Vasiliou, V., 2010. Update of human and mouse forkhead box (FOX) gene families. Hum. Genomics Jun; 4 (5), 345–352. Review.

Jallon, J.M., 1984. A few chemical words exchanged by *Drosophila during courtship and mating*. Behav. Genet. Sep; 14 (5), 441–478.

Jefferis, G., Turner, G., Masse, N., 2009. Olfactory Information processing in *Drosophila*. Curr. Biol. 19, 700–713.

Jefferis, G.S., Potter, C.J., Chan, A.M., Marin, E.C., Rohlfing, T., Maurer Jr, C.R., Luo, L., 2007. Comprehensive maps of *Drosophila* higher olfactory centers: spatially segregated fruit and pheromone representation. Cell Mar 23; 128 (6), 1187–1203.

Jost, A., 1953. Problems of fetal endocrinology: The gonadal and hypophyseal hormones. Recent Prog. Horm. Res. 8, 379–418.

Kanetsky, P.A., Mitra, N., Vardhanabhuti, S., Vaughn, D.J., Li, M., Ciosek, S.L., Letrero, R., D'Andrea, K., Vaddi, M., Doody, D.R., Weaver, J., Chen, C., Starr, J.R., Håkonarson, H., Rader, D.J., Godwin, A.K., Reilly, M.P., Schwartz, S.M., Nathanson, K.L., 2011. A second independent locus within DMRT1 is associated with testicular germ cell tumor susceptibility. Hum. Mol. Genet. Aug 1; 20 (15), 3109–3117.

Kappes, G., Deshpande, G., Mulvey, B.B., Horabin, J.I., Schedl, P., 2011. The *Drosophila* Myc gene, diminutive, is a positive regulator of the Sex-lethal establishment promoter, Sxl-Pe. Pro. Natl. Acad. Sci. 108 (4), 1543–1548.

Kato, Y., Kobayashi, K., Oda, S., Tatarazako, N., Watanabe, H., Iguchi, T., 2010. Sequence divergence and expression of a transformer gene in the branchiopod crustacean, *Daphnia magna*. Genomics Mar; 95 (3), 160–165.

Keisman, E.L., Baker, B.S., 2001. The *Drosophila* sex determination hierarchy modulates wingless and decapentaplegic signaling to deploy dachshund sex-specifically in the genital imaginal disc. Development May; 128 (9), 1643–1656.

Kim, S., Bardwell, V.J., Zarkower, D., 2007. Cell type-autonomous and non-autonomous requirements for Dmrt1 in postnatal testis differentiation. Dev. Biol. Jul 15; 307 (2), 314–327.

Kim, Y., Kobayashi, A., Sekido, R., DiNapoli, L., Brennan, J., Chaboissier, M.C., Poulat, F., Behringer, R.R., Lovell-Badge, R., Capel, B., 2006. Fgf9 and Wnt4 act as antagonistic signals to regulate mammalian sex determination. PLoS Biol. Jun;4 (6), e187.

Kimura, K., Ote, M., Tazawa, T., Yamamoto, D., 2005. Fruitless specifies sexually dimorphic neural circuitry in the Drosophila brain. Nature 438 (7065), 229–233.

Kimura, K., 2011. Role of cell death in the formation of sexual dimorphism in the *Drosophila* central nervous system. Dev. Growth Differ. Feb; 53 (2), 236–244.

Kimura, K., Ote, M., Tazawa, T., Hachiya, T., Yamamoto, T., 2008. Fruitless and doublesex coordinate to generate male-specific neurons that can initiate courtship. Neuron 59, 759–769.

Kondoh, Y., Kaneshiro, K.Y., Kimura, K., Yamamoto, D., 2003. Evolution of sexual dimorphism in the olfactory brain of Hawaiian *Drosophila*. Proc. Biol. Sci. May 22; 270 (1519), 1005–1013.

Kopp, A., Duncan, I., Godt, D., Carroll, S.B., 2000. Genetic control and evolution of sexually dimorphic characters in *Drosophila*. Nature Nov 30; 408 (6812), 553–559.

Kratz, C.P., Han, S.S., Rosenberg, P.S., Berndt, S.I., Burdett, L., Yeager, M., Korde, L.A., Mai, P.L., Pfeiffer, R., Greene, M.H., 2011. Variants in or near KITLG, BAK1, DMRT1, and TERT-CLPTM1L predispose to familial testicular germ cell tumor. J. Med. Genet. Jul; 48 (7), 473–476.

Kurtovic, A., Widmer, A., Dickson, B., 2007. A single class of olfactory neurons mediates behavioral responses to a *Drosophila* sex pheromone. Nature 446, 542–546.

Laissue, P.P., Reiter, C., Hiesinger, P.R., Halter, S., Fischbach, K.F., Stocker, R.F., 1999. Three-dimensional reconstruction of the antennal lobe in *Drosophila melanogaster*. J. Comp. Neurol. Mar 22; 405 (4), 543–552.

Lecanu, L., 2011. Sex, the underestimated potential determining factor in brain tissue repair strategy. Stem. Cells Dev. Dec; 20 (12), 2031–2035.

Lee, T., Yu, H., Chen, C., Shi, L., Huang, Y., 2009. Twin-spot MARCM to reveal the developmental origin and identity of neurons. Nat. Neurosci. 12, 947–953.

Lee, G., Foss, M., Goodwin, S.F., Carlo, T., Taylor, B.J., Hall, J.C., 2000 Jun 15. Spatial, temporal, and sexually dimorphic expression patterns of the fruitless gene in the *Drosophila* central nervous system. J. Neurobiol. 43 (4), 404–426.

Lewis, E.B., 1978. A gene complex controlling segmentation in *Drosophila*. Nature Dec 7; 276 (5688), 565–570.

Li, H., Baker, B.S., 1998. her, a gene required for sexual differentiation in *Drosophila*, encodes a zinc finger protein with characteristics of ZFY-like proteins and is expressed independently of the sex determination hierarchy. Development 125 (2), 225–235.

Luo, S.D., Shi, G.W., Baker, B.S., 2011. Direct targets of the *D. melanogaster* DSXF protein and the evolution of sexual development. Development 138 (13), 2761–2771.

Mallo, M., Alonso, C.R., 2013. The regulation of Hox gene expression during animal development. Development Oct; 140 (19), 3951–3963.

Matson, C.K., Murphy, M.W., Sarver, A.L., Griswold, M.D., Bardwell, V.J., Zarkower, D., 2011. DMRT1 prevents female reprogramming in the postnatal mammalian testis. Nature Jul 20; 476 (7358).

Matson, C.K., Zarkower, D., 2012. Sex and the singular DM domain: insights into sexual regulation, evolution and plasticity. Nat. Rev. Genet. Feb 7; 13 (3), 163–174.

Mattox, W., Baker, B.S., 1991. Autoregulation of the splicing of transcripts from the transformer-2 gene of *Drosophila*. Genes. Dev. 5 (5), 786–796.

Mori, K., Shikichi, Y., Shankar, S., Yew, J., 2010. Pheromone synthesis. Part 244: Synthesis of the racemate and enantiomers of (11Z,19Z)-CH503 (3-acetoxy-11,19-octacosadien-1-ol), a new sex pheromone of male *Drosophila melanogaster* to show its (S)-isomer and racemate as bioactive. Tetra 66, 7161–7168.

Morris, J.A., Jordan, C.L., Breedlove, S.M., 2004. Sexual differentiation of the vertebrate nervous system. Nat. Neurosci. Oct; 7 (10), 1034–1039.

Narendra, U., Zhu, L., Li, B., Wilken, J., Weiss, M.A., 2002. Sex-specific gene regulation. The Doublesex DM motif is a bipartite DNA-binding domain. J. Biol. Chem. 277 (45), 43463–43473.

Nelson, W.J., Nusse, R., 2004. Convergence of Wnt, ß-catenin, and cadherin pathways. Science 303 (5663), 1483–1487.

Ottolenghi, C., Pelosi, E., Tran, J., Colombino, M., Douglass, E., Nedorezov, T., Cao, A., Forabosco, A., Schlessinger, D., 2007 Dec 1. Loss of Wnt4 and Foxl2 leads to female-to-male sex reversal extending to germ cells. Hum. Mol. Genet. 16 (23), 2795–2804.

Pellegrino, M., Nakagawa, T., Vosshall, L.B., 2010. Single sensillum recordings in the insects *Drosophila melanogaster* and *Anopheles gambiae*. J. Vis. Exp. 36, 1–5.

Perez–Torrado, R., Yamada, D., Defossez, P.A., 2006. Born to bind: the BTB protein-protein interaction domain. Bioessays 28 (12), 1194–1202.

Pomiankowski, A., Nöthiger, R., Wilkins, A., 2004. The evolution of the *Drosophila* sex-determination pathway. Genetics 166 (4), 1761–1773.

Ray, R., Novotny, N., Crisostomo, P., Lahm, T., Abaranell, A., Meldrum, D., 2008. Sex steroids and stem cell function. Mol. Med. 14 (7), 493–501.

Raymond, C.S., Murphy, M.W., O'Sullivan, M.G., Bardwell, V.J., Zarkower, D., 2000. Dmrt1, a gene related to worm and fly sexual regulators, is required for mammalian testis differentiation. Genes. Dev. Oct 15; 14 (20), 2587–2595.

Raymond, C.S., Shamu, C.E., Shen, M.M., Seifert, K.J., Hirsch, B., Hodgkin, J., Zarkower, D., 1998. Evidence for evolutionary conservation of sex-determining genes. Nature Feb 12; 391 (6668), 691–695.

Rein, K., Zockler, M., Mader, M.T., Grubel, C., Heisenberg, M., 2002. The *Drosophila* Standard Brain. Curr. Biol. 12, 227–231.

Rideout, E., Dornan, A., Neville, M., Eadie, S., Goodwin, S., 2010. Control of sexual differentiation and behavior by the doublesex gene in *Drosophila melanogaster*. Natu. Neurosci. 13, 458–466.

Rideout, E.J., Billeter, J.C., Goodwin, S.F., 2007. The sex-determination genes fruitless and doublesex specify a neural substrate required for courtship song. Curr. Biol. 17 (17–3), 1473.

Robinett, C.C., Vaughan, A.G., Knapp, J.M., Baker, B.S., 2010. Sex and the single cell. II. There is a time and place for sex. PLoS Biol. May 4; 8 (5).

Ryner, L.C., Goodwin, S.F., Castrillon, D.H., Anand, A., Villella, A., Baker, B.S., Hall JC,Taylor, B.J., Wasserman, S.A., 1996. Control of male sexual behavior and sexual orientation in *Drosophila* by the fruitless gene. Cell Dec 13; 87 (6), 1079–1089.

Salz, H., Erickson, J.W., 2010. Sex determination in *Drosophila*: The view from the top. Fly 4 (1), 60–70.

Sanders, L.E., Arbeitman, M.N., 2008. Doublesex establishes sexual dimorphism in the *Drosophila* central nervous system in an isoform-dependent manner by directing cell number. Dev. Biol. Aug 15; 320 (2), 378–390.

Sarno, F., Ruiz, M.F., Sánchez, L., 2009. Effect of the transformer-2 gene of *Anastrepha* on the somatic sexual development of *Drosophila*. Int. J. Dev. Biol. 55 (10), 975–979.

Schüpbach, T., 1985. Normal female germ cell differentiation requires the female X chromosome to autosome ratio and expression of Sex-lethal in *Drosophila melanogaster*. Genetics 109 (3), 529–548.

Sekido, R., Lovell-Badge, R., 2008. Sex determination involves synergistic action of SRY and SF1 on a specific Sox9 enhancer. Nature 453 (7197), 930–934.

Shah, S.P., Köbel, M., Senz, J., Morin, R.D., Clarke, B.A., Wiegand, K.C., Leung, G., Zayed A,Mehl, E., Kalloger, S.E., Sun, M., Giuliany, R., Yorida, E., Jones, S., Varhol, R., Swenerton, K.D., Miller, D., Clement, P.B., Crane, C., Madore, J., Provencher, D., Leung, P., DeFazio, A., Khattra, J., Turashvili, G., Zhao, Y., Zeng, T., Glover, J.N., Vanderhyden, B., Zhao, C., Parkinson, C.A., Jimenez-Linan, M., Bowtell, D.D., Mes-Masson, A.M., Brenton, J.D., Aparicio, S.A., Boyd, N., Hirst, M., Gilks, C.B., Marra, M., Huntsman, D.G., 2009. Mutation of FOXL2 in granulosa-cell tumors of the ovary. N. Engl. J. Med. Jun 25; 360 (26), 2719–2729.

Shirangi, T.R., Dufour, H.D., Williams, T.M., Carroll, S.B., 2009. Rapid evolution of sex pheromone-producing enzyme expression in *Drosophila*. PLoS Biol. 7 (8).

Shukla, J.N., Nagaraju, J., 2010. Doublesex: a conserved downstream gene controlled by diverse upstream regulators. J. Genet. 89 (3), 341–356.

Simerly, R.B., 2002. Wired for reproduction: organization and development of sexually dimorphic circuits in the mammalian forebrain. Annu. Rev. Neurosci. 25, 507–536.

Siwicki, K.K., Kravitz, E.A., 2009. Fruitless, doublesex and the genetics of social behavior in *Drosophila melanogaster*. Curr. Opin. Neurobiol. 19 (2), 200–206.

Stocker, R.F., 2001. *Drosophila* as a focus in olfactory research: mapping of olfactory sensilla by fine structure, odor specificity, odorant receptor expression, and central connectivity. Microsc. Res. Tech. Dec 1; 55 (5), 284–296.

Stockinger, P., Kvitsiani, D., Rotkopf, S., Tirián, L., Dickson, B.J., 2005 Jun 3. Neural circuitry that governs *Drosophila* male courtship behavior. Cell 121 (5), 795–807.

Stowers, L., Logan, D.W., 2010. Sexual dimorphism in olfactory signaling. Curr. Opin. Neurobiol. 20, 770–775.

Tanaka, K., Barmina, O., Sanders, L.E., Arbeitman, M.N., Kopp, A., 2011. Evolution ofsex-specific traits through changes in HOX-dependent doublesex expression. PLoS Biol. 2011 Aug; 9(8).

Tannour-Louet, M., Han, S., Corbett, S.T., Louet, J.F., Yatsenko, S., Meyers, L., Shaw, C.A., Kang, S.H., Cheung, S.W., Lamb, D.J., 2010. Identification of *de novo* copy number variants associated with human disorders of sexual development. PLoS One Oct 26; 5 (10).

Taylor, B.J., Villella, A., Ryner, L.C., Baker, B.S., Hall, J.C., 1994. Behavioral and neurobiological implications of sex-determining factors in *Drosophila*. Dev. Genet. 15 (3), 275–296.

Taylor, B.J., 1992. Differentiation of a male-specific muscle in *Drosophila melanogaster* does not require the sex-determining genes doublesex or intersex. Genetics 132 (1), 179–191.

Taylor, B.J., Truman, J.W., 1992. Commitment of abdominal neuroblasts in *Drosophila* to a male or female fate is dependent on genes of the sex-determining hierarchy. Development 114 (3), 625–642.

Uhlenhaut, N.H., Jakob, S., Anlag, K., Eisenberger, T., Sekido, R., Kress, J., Treier, A.C., Klugmann, C., Klasen, C., Holter, N.I., Riethmacher, D., Schütz, G., Cooney, A.J., Lovell-Badge, R., Treier, M., 2009. Somatic sex reprogramming of adult ovaries to testes by FOXL2 ablation. Cell Dec 11; 139 (6), 1130–1142.

Usui-Aoki, K., Ito, H., Ui-Tei, K., Takahashi, K., Lukacsovich, T., Awano, W., Nakata, H., Piao, Z.F., Nilsson, E.E., Tomida, J., Yamamoto, D., 2000. Formation of the male-specific muscle in female *Drosophila* by ectopic fruitless expression. Nat. Cell Biol. Aug; 2 (8), 500–506.

Verdin, H., De Baere, E., 2012. Foxl2 impairment in human disease. Hormone Res. Pediatrics 77 (1), 2–11.

Verhulst, E.C., van de Zande, L., Beukeboom, L.W., 2010. Insect sex determination: it all evolves around transformer. Curr. Opin. Genet. Dev. Aug;20 (4), 376–383.

Viger, R.S., Silversides, D.W., Tremblay, J.J., 2005. New insights into the regulation of mammalian sex determination and male sex differentiation. Vitam. Horm. 70, 387–413.

Villella, A., Hall, J.C., 1996. Courtship anomalies caused by doublesex mutations in *Drosophila melanogaster*. Genetics 143 (1), 331–344.

Vosshall, L.B., Stocker, R.F., 2007. Molecular architecture of smell and taste in *Drosophila*. Annu. Rev. Neurosci. 30, 505–533.

Vosshall, L.B., 2000. Olfaction in *Drosophila*. Curr. Opin. Neurobiol. Aug; 10 (4), 498–503.

Wang, W., Kidd, B.J., Carroll, S.B., Yoder, J.H., 2011. Sexually dimorphic regulation of the Wingless morphogen controls sex-specific segment number in *Drosophila*. Proc. Natl. Acad. Sci. USA. Jul 5; 108 (27), 11139–11144.

Warr, N., Greenfield, A., 2012. The molecular and cellular basis of gonadal sex reversal in mice and humans. Wiley *Interdiscip*. Rev. Dev. Biol. Jul–Aug; 1 (4), 559–577.

Waterbury, J.A., Jackson, L.L., Schedl, P., 1999. Analysis of the doublesex female protein in *Drosophila melanogaster*: role in sexual differentiation and behavior and dependence on intersex. Genetics 152 (4), 1653–1667.

Williams, T.M., Carroll, S.B., 2009. Genetic and molecular insights into the development and evolution of sexual dimorphism. Nat. Rev. Genet. Nov; 10 (11), 797–804.

Williams, T.M., Selegue, J.E., Werner, T., Gompel, N., Kopp, A., Carroll, S.B., 2008. The regulation and evolution of a genetic switch controlling sexually dimorphic traits in *Drosophila*. Cell Aug 22; 134 (4), 610–623.

Yew, J.Y., Dreisewerd, K., Luftmann, H., Müthing, J., Pohlentz, G., Kravitz, E.A., 2009. A new male sex pheromone and novel cuticular cues for chemical communication in *Drosophila*. Curr. Biol. Aug 11; 19 (15), 1245–1254.

Zhang, W., Li, B., Singh, R., Narendra, U., Zhu, L., Weiss, M.A., 2006. Regulation of sexual dimorphism: mutational and chemogenetic analysis of the doublesex DM domain. Mol. Cell. Biol. Jan;26 (2), 535–547.

Zhu, L., Wilken, J., Phillips, N.B., Narendra, U., Chan, G., Stratton, S.M., Kent, S.B., Weiss, M.A., 2000. Sexual dimorphism in diverse metazoans is regulated by a novel class of intertwined zinc fingers. Genes. Dev. Jul 15; 14 (14), 1750–1764.

Chapter 10

Formation of the Anterior-Posterior Axis in Mammals

Aitana Perea-Gomez*,# and Sigolène M. Meilhac†,**

*Institut Jacques Monod, Université Paris Diderot, Sorbonne Paris Cité, Paris, France; †Institut Pasteur, Department of Developmental and Stem Cell Biology, Paris, France; **CNRS, URA2578, Paris, France; #CNRS, UMR7592, Paris, France

Chapter Outline

GLOSSARY

Anterior visceral endoderm (AVE) A population of visceral endoderm cells found in the anterior region of the conceptus, opposite the primitive streak at E6–E6.5, and characterized by the expression of a specific gene repertoire including *Hhex*, *Hesx1*, *Cer1* and *Lefty1*. The AVE is formed after cells of the distal visceral endoderm (DVE) have moved, thus breaking the radial symmetry of the egg cylinder. This population is heterogeneous as shown by distinct gene expression in medial and lateral domains, but also dynamic in nature, as its cellular composition changes during the general movement of the VE which occurs between E5.5 and E6.5.

Artiodactyls A group of herbivorous mammals with even-toed hooves, including pigs, hippopotamuses, and ruminants.

Conceptus Contains all the derivatives of the zygote, embryonic (epiblast) and extra-embryonic tissues.

Distal visceral endoderm (DVE) A group of visceral endoderm cells found at the distal tip of the egg cylinder at E5.5, characterized by its columnar morphology and the expression of a specific gene repertoire including *Hhex*, *Cer1* and *Lefty1*. The DVE is formed in two waves. The first wave consists of descendants of precursors already expressing *Lefty1* and *Cer1* in the primitive endoderm and migrate first anteriorly. The second wave is induced at the distal pole of the embryo after E5.5. As the DVE migrates towards one side, the anterior pole becomes defined, and the group of cells is referred to as the anterior visceral endoderm (AVE).

Egg cylinder The mouse conceptus after implantation, from E5 to gastrulation. The conceptus at these stages has the shape of a cylinder and comprises a proximal and a distal region.

Epiblast A tissue formed of pluripotent precursors that will generate all the fetal organs as well as the extra-embryonic mesoderm.

Extra-embryonic ectoderm (ExE) An extra-embryonic tissue derived from the (polar) trophectoderm, in contact with the inner cell mass in the blastocyst. Unlike cells of the ectoplacental cone (another trophectoderm derivative), which differentiate into giant cells, cells of the ExE retain stem cell potential. The ExE contributes to the placenta.

Primitive endoderm A polarized epithelium covering the epiblast precursors and facing the blastocoel cavity in the implanting blastocyst (E3.75 to E4.5). The primitive endoderm is an extra-embryonic tissue derived from inner cell mass precursors. It will give rise to the visceral endoderm.

Primitive streak The primitive streak forms opposite to the anterior visceral endoderm. It is a region of the epiblast along which precursor cells of the mesoderm and the definitive endoderm ingress during gastrulation when they undergo an epithelial to mesenchymal transition. Cells are progressively recruited into the primitive streak as it elongates distally, to give rise to regionally distinct precursors in an orderly fashion,

Principles of Developmental Genetics.

both along the medio-lateral and anterior–posterior axes. Epiblast cells located at the anterior pole, do not ingress through the primitive streak, remain epithelial and contribute to the anterior surface ectoderm and neurectoderm.

Visceral Endoderm (VE) A polarized absorptive epithelium of the egg cylinder derived from the primitive endoderm. It is specified in two regions covering the epiblast and the extra-embryonic ectoderm, referred to as EPI-VE and ExE-VE respectively. The ExE-VE and the proximal third of the EPI-VE contribute to the formation of the yolk sac. Cells in the EPI-VE are dispersed during gastrulation by definitive endoderm cells derived from the primitive streak and are incorporated into the embryonic gut.

SUMMARY

- Reciprocal interactions between extra-embryonic and embryonic tissues are required to establish the anterior-posterior axis of the mouse embryo.
- A clear morphological manifestation of anterior-posterior asymmetry in the embryo is the onset of gastrulation in the posterior region. Induction of the primitive streak depends on Nodal and Wnt signaling, and is modulated by the Anterior Visceral Endoderm.
- Before gastrulation, migration of the Distal Visceral Endoderm is the symmetry-breaking event, which defines the anterior pole of the mouse conceptus.
- Proximal-distal regionalization of the visceral endoderm precedes migration of the Distal Visceral Endoderm. The Distal Visceral Endoderm, which constitutes a heterogeneous population of cells, is induced by Nodal and restricted by Bmp signaling.
- The movement of the Distal Visceral Endoderm is driven by active cell migration and cell intercalation, accompanied by a global movement of visceral endoderm (VE) cells. The direction of the movement is controlled by Nodal and Wnt signaling.
- The observations in the mouse model may not fully apply to other mammalian species. The anterior visceral endoderm (AVE) has an equivalent in the rabbit and other vertebrate species. However, the equivalent of the extra-embryonic ectoderm is unclear.

10.1 INTRODUCTION

During development, before the formation of organs, cells acquire positional information along the three body axes: anterior-posterior (AP), dorsal-ventral (DV) and left-right. The AP axis, the main body axis of mammals, is defined along the head, trunk, and tail of the adult. The axis is formed when irreversible asymmetries are detected in the conceptus, and it is only later that embryonic cells are assigned a particular fate for a particular AP position, under the control of an organizer, the node (Beddington, 1994). This process is initiated at the time of gastrulation (see Chapter 12).

In many vertebrates, such as the fish or frog, the establishment of the body plan including the AP axis begins in the oocyte where maternal determinants are asymmetrically distributed. In contrast in the mouse, the AP axis becomes apparent five to six days after fertilization (Figure 10.1). Earlier in mouse development, extra-embryonic tissues are specified, which are necessary for the survival and patterning of the embryo. They segregate from the epiblast, which contains the precursors of all fetal organs maintained in a pluripotent state. Around the time of implantation into the uterus, between embryonic day (E)4 and E5, important morphological changes take place, leading to a particular arrangement of embryonic (epiblast) and extra-embryonic tissues, which conditions their interactions. The blastocyst is transformed into the egg cylinder, with a clear proximo-distal regionalization. At the distal pole, also referred to as the embryonic region, the epiblast cells are arranged as a pseudo-stratified epithelium around the amniotic cavity (Coucouvanis and Martin, 1999; Gardner and Cockroft, 1998). At the proximal pole, also referred to as the extra-embryonic region, the extra-embryonic ectoderm (ExE) forms and will later contribute to the fetal placenta (Copp, 1979; Perea-Gomez et al., 2007). The outer layer of the egg cylinder is an extra-embryonic tissue, the visceral endoderm (VE), which is a single layered epithelium with a basement membrane, covering the epiblast and the ExE. VE cells, which will contribute to the visceral yolk sac, have characteristics of absorbing epithelia, with extensive microvilli and large apical vacuoles. Their endocytic function is essential for nutrient supply before the formation of the embryonic vasculature and functional placenta (Belaoussoff et al., 1998; Wada et al., 2013).

At E5, no visible sign of anterior-posterior asymmetry can be distinguished. The egg cylinder has a radial symmetry around the proximo-distal axis, both in morphology and gene expression.

In this chapter we will describe how dynamic reciprocal signaling events between the epiblast and the surrounding extra-embryonic tissues regulate cell specification, cell movements and cell rearrangements. This results in the movement of cells of the distal visceral endoderm (DVE), which breaks the radial symmetry of the egg cylinder and defines the AP axis before the initiation of gastrulation. We will discuss current knowledge of the function, origin and directional movement of distal and anterior visceral endoderm cells, as well as the role of the anterior visceral endoderm (AVE) as a regulator of cell fate in the embryo. We shall focus mainly on the mouse, and discuss divergences among mammals and other vertebrates at the end.

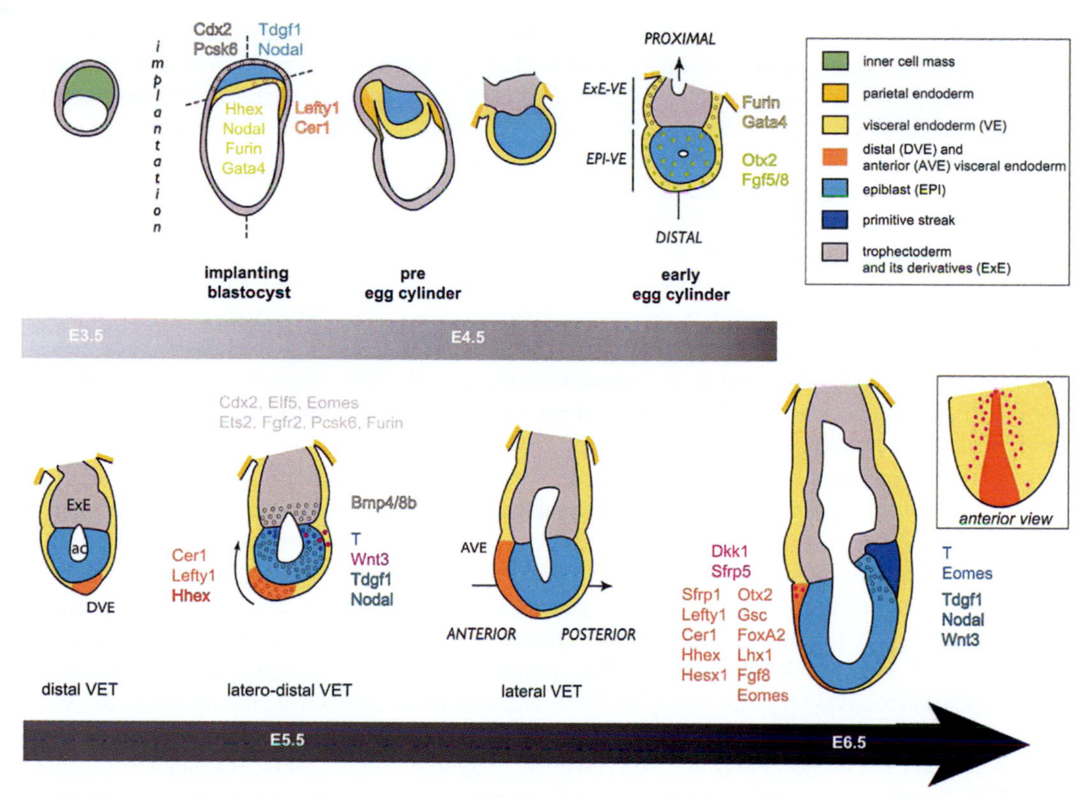

FIGURE 10.1 Stages of Development of the Mouse Embryo at the Time of the Formation of the AP Axis. Stages from the period of implantation into the uterus until gastrulation are expressed as number of days after fertilization. The tissues are drawn as distinct colors. After the pre-egg cylinder stage, the conceptus is shown without the parietal yolk sac (mural trophectoderm and parietal endoderm). The potential asymmetry of the implanting blastocyst is shown as gene expression and as a "tilt" of the embryonic (top)/abembryonic (bottom) axis (vertical dashed line) in relation to the plane of the primitive endoderm (oblique dashed line). The E3.5 blastocyst is transformed into an egg cylinder around embryonic day 4.5. It includes a distal region composed of epiblast (embryonic tissue) and visceral endoderm and a proximal region composed of extra-embryonic ectoderm (ExE) and visceral endoderm. The visceral endoderm expresses different markers in these regions and is therefore subdivided into EPI-VE and ExE-VE regions. The expression of some genes important for anterior–posterior patterning is schematized with dots (unless ubiquitous in a given tissue) and indicated on the sides. Upon migration of the distal visceral endoderm on embryonic day 5.5, the transformation of the proximal–distal axis into an anterior–posterior axis is shown by an arrow (substages according to the position of the visceral endoderm thickening, VET). At E6.5 the DVE cell population takes the name of anterior visceral endoderm and its heterogeneity is also shown from an anterior view (inset). Ac, amniotic cavity.

10.2 DISCOVERY AND IMPORTANCE OF THE AVE

An obvious sign of AP asymmetry in the mouse embryo is gastrulation, detectable from E6.5 by the appearance of the primitive streak, which defines the posterior pole. Until the mid-1990s, this was considered as the symmetry-breaking event that defined the AP axis of the embryo. However, studies of gene expression by the groups of Martin and Beddington, followed by others, challenged this view. They found that molecular markers expressed in the AVE, opposite the primitive streak, were already expressed on one side of the conceptus at least one day earlier (Rosenquist and Martin, 1995; Thomas and Beddington, 1996). These observations provided molecular evidence that the AP asymmetry is first established in the visceral endoderm before the initiation of gastrulation, and suggested for the first time that extra-embryonic tissues could be a source of patterning influences. Subsequent studies refined this view and revealed that the VE is first regionalized along the proximo-distal axis and that this axis is then transformed into the AP axis (Figure 10.1). Indeed soon after implantation, at E5.5, a population of distal VE cells is specified with distinct molecular and morphological properties. Unlike the rest of the VE, DVE cells are columnar and express a specific repertoire of genes encoding transcription factors such as Hhex or secreted proteins such as the Nodal antagonist Lefty1 and the Bmp, Nodal and Wnt antagonist Cer1 (Belo et al., 1997; Meno et al., 1996; Thomas et al., 1998; Yamamoto et al., 2004). Using DiI labeling, Beddington and colleagues found that between E5.5 and E6, DVE cells move asymmetrically towards one side of the embryo, which then becomes the anterior pole (Thomas et al., 1998). These findings indicate that the vectorial movement of DVE cells to the anterior region, where they take the name of AVE at E6.5, is the symmetry-breaking event that marks the establishment of the AP axis.

The movement of the DVE/AVE is not just a landmark of AP asymmetry: this group of cells exerts important functions for the early patterning of the mouse embryo along its AP axis. DVE/AVE cells play a major role in restricting the site of the formation of the primitive streak. In explant recombination experiments, AVE cells can repress the expression of posterior epiblast and mesoderm markers (Kimura et al., 2000). Conversely, when the induction of the DVE is impaired in mutants such as *Foxa2*$^{-/-}$*Lhx1*$^{-/-}$, or *Smad2*$^{-/-}$, ectopic and widespread mesoderm formation occurs in the entire epiblast (Perea-Gomez et al., 1999; Waldrip et al., 1998). Finally, in mutant embryos in which DVE cells fail to translocate to the anterior side, signs of AP asymmetry remain proximo-distal. In these mutants, affecting for example the transcription factors Otx2 or Foxh1, or the Nodal co-receptor Tdgf1 (also known as Cripto), the primitive streak forms as a ring involving the entire proximal epiblast, and the neurectoderm tissue is specified in the opposite distal region (Ding et al., 1998; Kimura et al., 2000; Perea-Gomez et al., 2001a; Yamamoto et al., 2001). Taken together, these results provide evidence that DVE/AVE cells can influence the development of the underlying epiblast and that one of their functions is to inhibit primitive streak formation in the anterior epiblast, protecting it from the signals involved in mesoderm formation (Kimura et al., 2000; Perea-Gomez et al., 2001a,b; 1999).

The effect of DVE and AVE cells is mainly mediated by the Nodal antagonists Cer1 and Lefty1. The growth factor Nodal is a major regulator of primitive streak and mesoderm formation (Brennan et al., 2001). At E5, Nodal is expressed in the entire epiblast (Figure 10.1), and this expression is progressively restricted to the proximal posterior epiblast where the primitive streak will form at E6.5 (Varlet et al., 1997). Nodal activity is regulated by extracellular factors such as the antagonists secreted by the DVE/AVE, or the proteases Pcsk6 (also known as Spc4) and Furin secreted by the ExE, and required to produce the mature and active form of Nodal (Beck et al., 2002; Mesnard et al., 2011). In embryos carrying mutations in both *Cer1* and *Lefty1*, *Nodal* expression is not downregulated in anterior epiblast cells and ectopic primitive streaks are formed (Perea-Gomez et al., 2002; Yamamoto et al., 2004). In addition to Nodal, Wnt signaling (in particular Wnt3) is also involved in primitive streak and mesoderm formation. This is exemplified by the absence of primitive streak and mesoderm formation in embryos with inactivation of β-*catenin* or *Wnt3,* and also in mutants for genes encoding the Wnt co-receptors Lrp5 and Lrp6, the Lrp chaperone Mesdc2, or the genes *Porcn* and *Gpr177* (*wntless* homolog) involved in the secretion of Wnt ligands (Biechele et al., 2013; Biechele et al., 2011; Fu et al., 2009; Hsieh et al., 2003; Huelsken et al., 2000; Kelly et al., 2004; Liu et al., 1999; Tortelote et al., 2012). Moreover, ectopic primitive streak structures form upon the ectopic activation of the Wnt pathway (Chazaud and Rossant, 2006; Kimura-Yoshida et al., 2005; Merrill et al., 2004; Popperl et al., 1997; Zeng et al., 1997). Wnt3 regulates primitive streak formation through feed-forward and autoregulatory loops involving reciprocal interactions between the epiblast, the ExE and the VE, and cross-regulation of the Nodal signaling pathway (Figure 10.2).

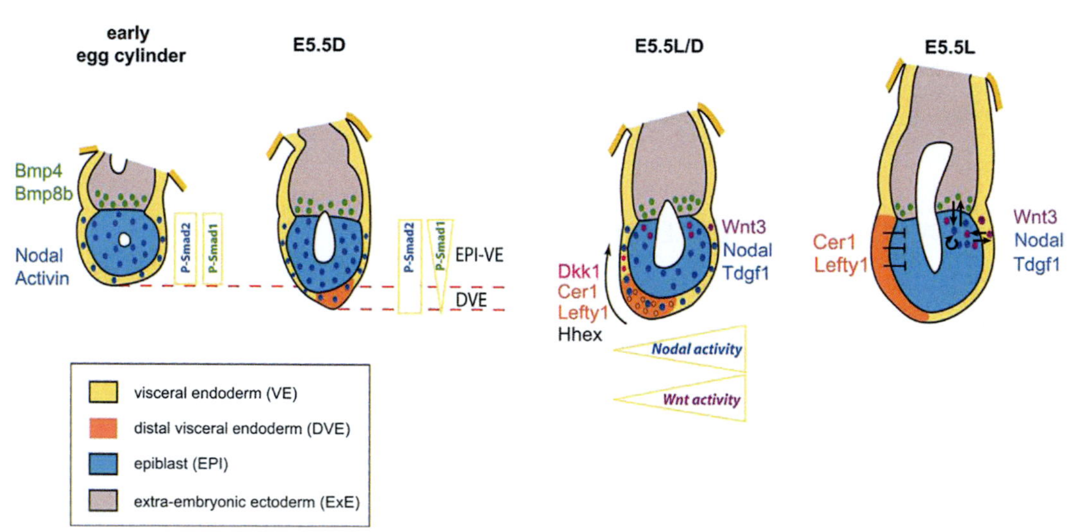

FIGURE 10.2 Reciprocal Interactions Between Embryonic and Extra-Embryonic Tissues Underlying the Induction, Movement and Function of the DVE/AVE. In the early egg cylinder, the combined actions of Nodal/Activin signaling derived from the epiblast, the visceral endoderm and the surrounding uterine environment, and Bmp signaling derived from the distal ExE, result in the specification of the EPI-VE region. These are transduced by nuclear phosphorylated Smad2 and phosphorylated Smad1 (P-Smad2 and P-Smad1) respectively in all VE cells covering the epiblast region. At E5.5D (Distal VET), presumably as a result of the growth of the egg cylinder, distal EPI-VE cells escape from the action of Bmp signals, so that P-Smad1 levels are reduced, and DVE specification occurs. The movement of the DVE is directed at E5.5L/D (latero-distal VET) by asymmetric Nodal and Wnt signaling, the former notably determined by Nodal, Tdgf1, Cer1, Lefty1 and the latter by Wnt3, Cer1 and Dkk1. After this movement, at E5.5L (Lateral VET), reciprocal interactions between the epiblast, the ExE and the VE mediated by Nodal, Wnt and Bmp signaling specify the future site of the formation of the primitive streak in the posterior epiblast. Arrows indicate activation and perpendicular strokes inhibition.

On the one hand, Wnt3 produced in the epiblast and in the posterior VE regulates its own expression (autoregulatory loop) as well as *Nodal* and *Tdgf1* expression in the posterior epiblast (cross-regulation). On the other hand, Nodal upregulates *Bmp4* in the ExE, which in turn activates *Wnt3* expression (feed-forward loop) (Barrow et al., 2007; Ben-Haim et al., 2006; Morkel et al., 2003; Tortelote et al., 2012; Vincent et al., 2003). Although genetic evidence is missing, it is possible that the action of the AVE in restricting primitive streak formation might also be mediated by the production of secreted Wnt antagonists like Dkk1, Cer1, Sfrp1, and Sfrp5 (Finley et al., 2003; Glinka et al., 1998; Kemp et al., 2005; Pearce et al., 1999).

In addition to its role in inhibiting Nodal and Wnt signaling, the AVE also acts as a morphogenetic barrier that maintains the integrity of the basement membrane. Thus, it prevents, in anterior epiblast cells, epithelial to mesenchymal transition, which is a characteristic of cell ingression in the primitive streak (Egea et al., 2008). The AVE is therefore a source of antagonistic and morphogenetic signals that prevent primitive streak and mesoderm formation in the anterior epiblast, preserving a pool of epithelial precursors for the ectodermal and neurectodermal embryonic tissues.

Since its discovery in the mid-1990s, the AVE has raised considerable interest owing to the possibility that it may display organizing properties essential for head formation in the mouse embryo. AVE ablation experiments and disruption of gene function specifically in the VE lead to forebrain truncations, indicating that the function of the AVE is essential for head development (Arkell and Tam, 2012; Miura and Mishina, 2007; Rhinn et al., 1998; Shawlot et al., 1999; Stern and Downs, 2012; Stuckey et al., 2011b; Thomas and Beddington, 1996; Varlet et al., 1997). However, explant recombination experiments and tissue grafts failed to uncover any anterior neural inducing activity of AVE cells. Therefore, the AVE cannot be considered as a head organizer (Kimura et al., 2000; Tam and Steiner, 1999). Is there an additional role for DVE and AVE cells in controlling head development? One example is the production of Wnt antagonists that may counteract the posteriorizing activity of Wnt, derived from the primitive streak, on the neurectoderm. This function is subsequently relayed by the anterior mesendoderm, which also expresses Wnt antagonists (reviewed by Arkell et al., 2013; Arkell and Tam, 2012; Stern and Downs, 2012). Ablation experiments have suggested that DVE cells may also regulate the translocation of anterior neural precursors, from distal to more proximal anterior regions. This could contribute to protecting these precursors from the influence of primitive streak derived signals (Miura and Mishina, 2007).

AVE cells exert antagonistic but also positive influences on other embryonic tissues. Ablation of a subset of AVE cells at E6.5, based on the expression of *Diphtheria Toxin A* under *Hhex* regulatory sequences, leads to abnormal development of the anterior primitive streak. However the cellular and molecular bases of this effect are not fully understood (Stuckey et al., 2011b). In addition, Bmp2 in the VE regulates the morphogenetic process of foregut invagination, as well as the rostrocaudal positioning of the head and heart. As yet, the cellular processes at play are unknown, and it is unclear whether this Bmp2 function is required specifically in the AVE or in the entire VE (Madabhushi and Lacy, 2011). The AVE not only provides signals for the morphogenesis of the embryonic foregut, but it also directly contributes cells, indicating that extra-embryonic cells can also be integrated in the embryo (Kwon et al., 2008). The AVE also participates in the induction of Primordial Germ Cells in the proximal posterior epiblast. This function may rely on the ability of AVE cells to produce secreted antagonists of the Bmp and Wnt signaling pathways (such as Cer1 and Dkk1) that antagonize the activity of Wnt3 and Bmp4 produced in the ExE and posterior epiblast respectively (Ohinata et al., 2009). Finally, signals derived from the visceral endoderm, and in particular from the AVE, can induce cardiomyocyte specification of mesoderm precursors (Arai et al., 1997; Nijmeijer et al., 2009).

Given the key role of DVE and AVE cells in the AP regionalization of the mouse embryo, it is important to understand how these cell populations are specified.

10.3 THE DVE IS A HETEROGENEOUS AND DYNAMIC CELL POPULATION, WHICH FORMS AFTER THE PROXIMO-DISTAL REGIONALIZATION OF THE VE

The DVE is a Heterogeneous Population of Cells

Since the initial discovery of asymmetric gene expression in the VE, numerous markers have been found to characterize both DVE cells at E5.5 and AVE cells at E6.5 (Figure 10.1). Recent results have highlighted the heterogeneity of these cell populations, also in terms of their origin and fate.

At E5.5, DVE cells can be identified by their specific shape. Unlike other cells in the VE, which are squamous, DVE cells are columnar. This morphological feature, referred to as Visceral Endoderm Thickening (VET), is instrumental to stage embryos according to the progression of the DVE movement (Rivera-Perez et al., 2003; Srinivas et al., 2004). A specific gene repertoire is expressed by DVE cells at E5.5 and AVE cells at E6.5 including genes coding for the homeodomain proteins Hhex and Hesx1, or for secreted proteins such as Lefty1 and Cer1. The expression of these genes largely overlaps. However, detailed analyses by double *in situ* hybridization, immunofluorescence, or based on the use of transgenic lines have uncovered heterogeneity in gene expression. *Cer1* and *Lefty1* are expressed more widely than *Hhex* at E5.5, in what appears as a slightly skewed distribution (Takaoka et al., 2011; Yamamoto et al., 2004). Still, after DVE migration has occurred, at E6.0, the expression domains of *Cer1*, *Lefty1* and *Hhex* do not overlap in the most anterior cells (Takaoka et al., 2011). Other AVE markers such as *Dkk1* and *Sfrp5*, are not expressed in a medial anterior stripe of the VE but rather in a horse-shoe shaped domain that seems to surround the medial domain of *Hhex*, *Cer1* and *Lefty1* expression (Finley et al., 2003; Kimura-Yoshida et al., 2005). The genetic ablation of cells expressing *Diphtheria Toxin subunit A* under *Hhex* regulatory sequences at E6.5 results in loss of *Cer1* and *Lefty1* expression, without affecting *Sfrp5* and *Dkk1* transcripts, further supporting the observation that these genes mark distinct cell populations (Stuckey et al., 2011b). Finally, genes such as those encoding the transcription factors Foxa2, Lhx1, Gsc, Otx2 and Eomesodermin or the secreted proteins Bmp2 and Fgf8, are first expressed in the entire VE covering the epiblast (referred to as EPI-VE) at E5.5 before being restricted to the AVE at E6.5 (Chazaud and Rossant, 2006; Ciruna and Rossant, 1999; Madabhushi and Lacy, 2011; Maruoka et al., 1998; Mesnard et al., 2006; Perea-Gomez et al., 1999). Although the DVE and AVE cell populations can be identified by the expression of an increasing number of specific transcripts (Goncalves et al., 2011), molecular heterogeneity thus exists between cells and also between stages.

Waves of DVE Specification

This heterogeneity in gene expression might suggest distinct origins of cells and/or dynamic gene regulation. To gain insight into the origin of DVE cells, clones of fluorescently labeled cells have been generated in the VE by mRNA microinjection of single progenitors into the E3.5 blastocyst. In all cases, the DVE, as defined by the expression of a *Cer1-GFP* transgene or by the position of the VET at E5.5, was composed of a mixture of labeled and unlabeled cells. This result indicates that the DVE is derived from more than one precursor between E3.5 and E5.5, and therefore has a polyclonal origin (Perea-Gomez et al., 2007).

Analysis of the expression of DVE markers before its formation at E5.5 has provided insight into the origin of this cell population. *Lefty1*, *Cer1* and *Hhex* are expressed in the primitive endoderm of E4.5 implanting blastocysts, with *Cer1* and *Lefty1* being restricted to a subset of precursors (Figure 10.3) (Chazaud and Rossant, 2006; Takaoka et al., 2011; Takaoka et al., 2006; Thomas et al., 1998; Torres-Padilla et al., 2007). Whether the spatial distribution of primitive endoderm cells expressing *Lefty1* or *Cer1* bears any relationship to morphological asymmetries (as for example the tilt, see Figure 10.1) in the implanting blastocyst that could relate to the AP axis in the post-implantation stages is the subject of some debate (Chazaud and Rossant, 2006; Gardner et al., 1992; Smith, 1985; Takaoka et al., 2006; Torres-Padilla et al., 2007). The observations of such an early expression raised the possibility that DVE precursors could be specified before the formation of the egg cylinder. Genetic tracing and time lapse studies have examined the fate of primitive endoderm cells expressing *Lefty1* and *Cer1* in relation to their contribution to the DVE at E5.5. Takaoka et al. (2011) generated a BAC transgenic line, *Lefty1-CreERT2*, which results in the activation of a *lacZ* reporter in *Lefty1* expressing cells upon tamoxifen induction. This experimental strategy has demonstrated that descendants of *Lefty1* expressing cells in the primitive endoderm of the implanting blastocyst (E4.5) are found in the DVE at E5.5 (Takaoka et al., 2011). Similarly, time lapse imaging revealed that cells expressing the transgene *Cer1-GFP* at the pre-egg cylinder stage (E4.75) contribute to the DVE domain at E5.5 (Torres-Padilla et al., 2007). *Lefty1* expressing cells at E4.5 and *Cer1-GFP* positive cells at E4.75 do not contribute exclusively to the DVE, as their descendants are also found in other regions of the proximal VE at E5.5, where *Lefty1* and *Cer1* have been downregulated (Takaoka et al., 2011; Torres-Padilla et al., 2007).

The DVE population at E5.5 also comprises cells that do not derive from the early precursors at the pre-egg cylinder stage and that start to express *Cer1* and *Lefty1* only around E5.5 (Takaoka et al., 2011; Torres-Padilla et al., 2007).

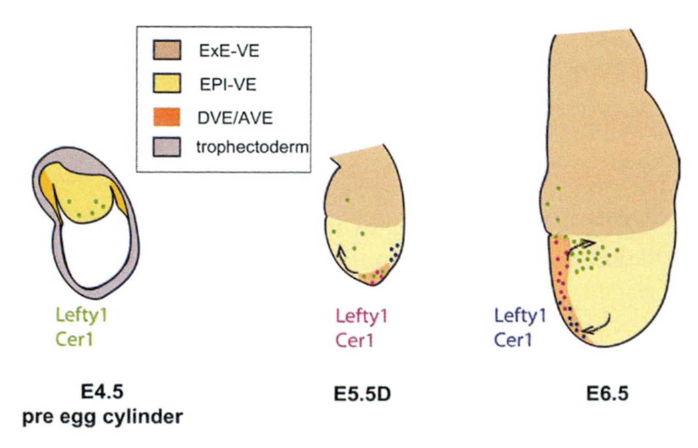

FIGURE 10.3 Dynamic Induction of DVE Cells. Waves of expression of *Lefty1* and *Cer1*, which mark the distal visceral endoderm (DVE), are shown as distinct colors. The distribution of their descendants at subsequent stages is represented by dots of corresponding colors. Arrows indicate the direction of cell movements. AVE, anterior visceral endoderm; EPI-VE, visceral endoderm over the epiblast; ExE-VE, visceral endoderm over the extra-embryonic ectoderm; E5.5D, E5.5 conceptus with a distal VET.

Genetic tracing has demonstrated that this second population gives rise to the midline AVE cells at E6.5 (Takaoka et al., 2011). In contrast DVE cells derived from the early E4.5 precursors do not contribute to the midline AVE but rather to two visceral endoderm bilateral stripes in the anterior lateral region that span the midline only in the most proximal region at E6.5. Taken together, these observations point to a model in which the DVE at E5.5 is a heterogeneous and dynamic population. A first wave of cells specified at the implanting blastocyst stage will move anteriorly ahead of a second wave of cells that is recruited after implantation and that will form the anterior region of the VE at E6.5 known as the AVE (Takaoka et al., 2007).

The recruitment of the second wave of *Cerl* and *Lefty1* expressing cells at the distal tip of E5.5 conceptuses is independent of the first wave. After ablation of *Lefty1* expressing cells between E5.0 and E5.5, based on the specific expression of *Diphtheria Toxin subunit A*, *Lefty1* expressing cells reappear in the distal visceral endoderm region (Takaoka et al., 2011). These results indicate that specific signaling events take place at the distal pole after E5.5, promoting *de novo* induction of DVE markers.

Induction of the DVE Depends on Balanced Nodal/Activin and Bmp Signaling

Genetic evidence points to a role of Nodal/Activin (also known as Inhibinβ) signaling in the induction of the DVE. The DVE is absent when Nodal signaling is impaired by inactivation of *Nodal* itself, or of genes encoding other components of the pathway such as the intracellular Nodal/Activin transducer Smad2, the proteases Furin and Pcsk6 required for Nodal maturation, or the Nodal co-receptors Tdgf1 and Cfc1 (also known as Cryptic) (Beck et al., 2002; Brennan et al., 2001; Camus et al., 2006; Chu and Shen, 2010; Waldrip et al., 1998). Conversely, excessive Nodal/Activin signaling results in an enlarged DVE, as observed for mutations in the extracellular antagonists Lefty1 and Cer1 or the intracellular inhibitor Trim33 (also known as ectodermin) (Morsut et al., 2010; Yamamoto et al., 2009; 2004). Conceptuses harboring mutations in the Nodal/Activin pathway have not been analyzed in detail at the implanting blastocyst stage and it is therefore unknown how the first wave of DVE specification is affected when Nodal/Activin signaling is impaired. Nevertheless, Foxh1, a transcription factor that mediates signaling, is required for *Lefty1* expression in the implanting blastocyst (Takaoka et al., 2006). Although the expression of components of the Nodal/Activin pathway is not restricted spatially at E4.5, the activity of a transgene reporter of Nodal/Activin signaling is confined to a subset of primitive endoderm cells (Granier et al., 2011; Takaoka et al., 2006). These observations indicate that DVE precursors may be specified by localized Nodal/Activin signaling in the implanting blastocyst at E4.5, which is an interesting perspective of research.

Current knowledge indicates that Nodal/Activin signaling influences DVE formation at two stages. Nodal signaling is first required in the early egg cylinder (E5.25) for the specification of the VE region covering the epiblast (EPI-VE) where the DVE will later arise. Mutations in Nodal, in its co-receptors Tdgf1 and Cfc1 or in Furin and Pcsk6 proteases produced by the conceptus, result in impaired expression of EPI-VE specific markers (Chu and Shen, 2010; Mesnard et al., 2011; 2006). Subsequently, Nodal/Activin signaling is required for the local DVE induction and maintenance at the distal tip of the conceptus. This is exemplified by the rapid downregulation of DVE markers in E5.5 cultured conceptuses treated with an inhibitor of Acvr1b/c and Tgfbr1 (also known as Alk4, 5, 7) that block Nodal/Activin signaling or with Follistatin, a specific inhibitor of Activin (Yamamoto et al., 2009). Importantly, Nodal/Activin signaling revealed by the presence of phosphorylated Smad2 is still active in the DVE of Nodal mutants, suggesting that other signals (Activin) or other sources of Nodal (uterus) are involved in DVE induction (Yamamoto et al., 2009). Indeed, Nodal and Activin are also produced by the maternal uterine environment (Albano et al., 1994; Jones et al., 2006; Park and Dufort, 2011). Taken together, these observations demonstrate that the formation of the DVE depends on Nodal and Activin. However, the relative contributions of the sources of signals (from the conceptus or the uterus) involved in this process have yet to be determined.

At the time of the formation of the DVE, Nodal is expressed in the entire epiblast and EPI-VE, as well as in the surrounding uterine epithelium. Similarly, Activin is ubiquitously distributed at this stage. How does widespread Nodal/Activin signaling result in local marker induction in just a small population of distal VE cells? Several studies have demonstrated that the entire VE is competent to express DVE markers, however DVE formation is inhibited proximally by signals derived from the extra-embryonic ectoderm. Ablation of the extra-embryonic region, and in particular of the posterior half at E5.5, or mutating genes involved in ExE maintenance, result in widespread expression of DVE markers in the entire EPI-VE (Donnison et al., 2005; Georgiades and Rossant, 2006; Mesnard et al., 2006; Richardson et al., 2006; Rodriguez et al., 2005). Conversely, grafting of extra-embryonic ectoderm cells into the distal region of the conceptus results in downregulation of DVE markers (Rodriguez et al., 2005). The effect of the extra-embryonic ectoderm is mediated by Bmp signaling. Bmp4 is able to restrict DVE formation both in explants and in cultured conceptuses, while treatment with the Bmp antagonist Noggin, or mutations in *Bmp4* and *Bmp8b* result in enlarged DVE domains (Ohinata et al., 2009; Soares et al., 2008; Yamamoto et al., 2009). The opposing influences of Nodal/Activin and Bmp on DVE formation might rely on competition for the type II Activin receptors Acvr2a and 2b, which are common signaling components of both pathways (Yamamoto et al., 2009). Additional antagonistic mechanisms, at the level of the ligands or the Smad4 effector, might also be at play, as shown for other developmental processes during which Nodal/Activin and Bmp signals have opposing effects (Furtado et al., 2008; Pereira et al., 2012). In addition, the action of Nodal/Activin signaling on DVE formation is further reinforced by an amplification loop involving p38 dependent activation of *Smad2* (Clements et al., 2011). DVE induction at E5.5 is therefore the result of the combined action of an inducing activity mediated by widespread Nodal/Activin signaling in the entire EPI-VE and an inhibiting activity of Bmp signals from the ExE on the proximal EPI-VE. Thus, the DVE can only be induced when the distal pole becomes released from the inhibitory influence of the ExE. This condition would potentially appear after E5.5 when the conceptus has grown and reached a threshold size (Mesnard et al., 2006; Stuckey et al., 2011a; Hiramatsu et al., 2013). In agreement with this scenario, the Nodal/Activin transducer phosphorylated Smad2 is present in all EPI-VE cells at E5.5 (Figure 10.2), whereas the Bmp activated phosphorylated Smad1 is excluded from DVE cells only after E5.5 (Yamamoto et al., 2009).

The gene regulatory network controlling *de novo* DVE marker induction downstream of Activin/Nodal signaling is starting to be unraveled. *Lefty1* is a direct target of Foxh1, an effector of the Nodal/Activin signaling pathway (Takaoka et al., 2006; Yamamoto et al., 2001), whereas *Cer1* is regulated by the transcription factors Eomesodermin and Foxa2 (Kimura-Yoshida et al., 2007; Nowotschin et al., 2013). In addition, mis-regulation of DVE marker expression is observed both in β-*catenin* mutants and in *Apc* mutants resulting in β-catenin gain of function (Chazaud and Rossant, 2006; Huelsken et al., 2000; Morkel et al., 2003). Interestingly, deletion of Porcn, which is required for the secretion of all Wnt ligands, does not lead to a similar phenotype, indicating that Wnt independent β–catenin signaling regulates DVE formation (Biechele et al., 2013).

The specification of DVE cells is a prerequisite for AP axis establishment in the mouse embryo. However, the symmetry-breaking event is the directional movement of DVE cells.

10.4 MECHANISMS OF DVE CELL MOVEMENT

The translocation of DVE cells from distal to anterior regions is the symmetry-breaking event for the establishment of the AP axis. Initially, DVE cells move away from the distal tip toward the proximal pole of the conceptus following direct trajectories. When DVE cells reach the embryonic/extra-embryonic junction, the movement, marked by convoluted trajectories, is reoriented laterally, as if it was transiently blocked by the junction (Thomas and Beddington, 1996; Srinivas et al., 2004). During migration, the VE remains a single layered epithelium, so that DVE cells do not migrate on top of other VE cells but rather between them (Srinivas et al., 2004). Accordingly, analysis of the distribution of tight and adherens junction proteins has demonstrated that the VE, including DVE, cells conserve their apico-basal polarity and epithelial integrity during DVE migration (Migeotte et al., 2010; Rivera-Perez et al., 2003; Trichas et al., 2011). These features are reminiscent of collective cell migration, in which cells pull themselves forward while retaining an epithelial organization exemplified by the persistence of junctions between adjacent cells and with the basal membrane.

Active Migration of the DVE Requires Actin Remodeling

Active migration of DVE cells was shown to be important for the distal to proximal movement. Time lapse imaging first revealed that DVE cells expressing the *Hhex-GFP* transgene move actively over a period of 4–5 h and adopt morphological features of migratory cells – with extensive protrusive activity in the direction of motion (Migeotte et al., 2010; Srinivas et al., 2004). These dynamic extensions suggest that DVE cells sample the environment for cues to direct their movement or to anchor their leading edge.

Several mutants have highlighted the role of actin dynamics regulation in DVE migration. The WAVE complex controls the formation of branched actin networks at the leading edges of migrating cells. Mutations in *Nap1,* encoding a component of the WAVE complex, disrupt DVE migration without affecting the tissue architecture (Rakeman and Anderson, 2006). The small GTPase Rac1 can act upstream of the WAVE complex, and in conceptuses specifically lacking *Rac1* in the VE, DVE cells do not form long protrusions and do not migrate (Migeotte et al., 2010). In addition, the WAVE complex is also regulated by PIP3. Depletion of the phosphatase Pten leads to abnormal VE migration associated with abnormal PIP3 levels and ectopic puncta of F-actin in VE cells (Bloomekatz et al., 2012). Therefore, translocation of the DVE involves active migration dependent on the production of projections based on actin remodeling.

The projections produced during the migration of the DVE cells expressing *Hhex-GFP* extend from the basal side of the cells, and show different characteristics in leading and trailing cells. The former produce longer lamellar extensions that can last for 1–2 h and that are ended by two small filopodia-like protrusions extending and collapsing dynamically (Migeotte et al., 2010). Cells located in the most anterior proximal domain of the DVE have been identified by their higher expression of the *Cer1-GFP* transgene and their elongated shape when compared to the more cuboidal follower cells (Morris et al., 2012). These results indicate that the most anterior DVE cells are potential leading cells that could respond to attracting or repulsive cues and orient the directional migration of groups of follower cells. In agreement with this hypothesis, laser ablation of the leading cells is sufficient to abolish DVE migration, whereas ablation of more posterior central cells does not affect the process (Morris et al., 2012). Similarly, genetic ablation of *Lefty1* expressing cells between E5.0 and E5.5 based on the specific expression of *Diphtheria Toxin subunit A*, prevents the migration of the second wave of DVE precursors appearing after E5.5 (Takaoka et al., 2011). Therefore, cell ablation experiments uncouple the movement of DVE/AVE cells from their specification, as movement is not recovered, whereas specification is. Uncovering the molecular events that control the specific properties of the leading cells will be crucial to understand the appearance of the AP axis in the mouse embryo. It will be important to determine whether the leading cells have a distinct gene repertoire, a distinct origin and/or whether they display their specific characteristics in response to extrinsic signals.

DVE Migration is Accompanied by a Re-Organization of the VE Epithelium with Distinct Mechanical Properties in the EPI-VE and ExE-VE

Once they reach the EPI/ExE border, *Hhex-GFP* expressing cells change shape and move laterally, while sending out fewer and shorter lateral projections (Migeotte et al., 2010; Srinivas et al., 2004). The analyses of cell shapes, cell contacts and neighbor relationships in fixed and living samples have shown how the VE epithelium is globally reorganized during the orderly migration of DVE cells. DVE migration involves neighbor exchange and cell intercalation, both between DVE cells and the rest of the VE, and among non-DVE cells (Migeotte et al., 2010; Trichas et al., 2011). Before DVE formation and migration, the EPI-VE and the VE covering the extra-embryonic ectoderm (ExE-VE) display similar cellular organizations, with the majority of VE cells of pentagonal or hexagonal shapes, which is a classical epithelial structure (Trichas et al., 2012). However, in conceptuses undergoing DVE migration, the organization in both regions of the VE becomes dramatically different. While the hexagonal arrangement is largely preserved in the ExE-VE, EPI-VE cells display a wide variety of shapes with a higher frequency of cells displaying four sides, in keeping with extensive cell intercalation (Migeotte et al., 2010; Perea-Gomez et al., 2007; Trichas et al., 2011, 2012). In mutant conceptuses in which DVE cells do not migrate, the reorganization of the EPI-VE epithelium does not occur, cell shapes are more homogeneous and cell intercalation is greatly reduced (Migeotte et al., 2010; Perea-Gomez et al., 2001a; Trichas et al., 2012). Taken together, these results indicate that the migration of the DVE is accompanied by major changes in the behavior of cells of the entire EPI-VE. Another manifestation of the coordinated cell intercalation in the EPI-VE corresponds to compensatory cellular rearrangements that may buffer the disequilibrium in cell packing accompanying the directional movement of the DVE. Rosettes have been described as arrangements of five or more VE cells meeting at a central point (Migeotte et al., 2010; Trichas et al., 2012). They are almost exclusively localized in the EPI-VE region, their density increases during DVE migration, and their occurrence is greatly reduced in mutants in which the DVE does not migrate. When rosette formation is specifically impaired in a computer simulation, the DVE migrates but does not maintain its structure as a coherent group, suggesting that the formation of the rosette arrangement is required for the orderly migration of DVE cells (Trichas et al., 2012).

The EPI-VE is permissive for DVE migration, whereas the ExE-VE is more rigid and acts as a barrier to migration. This is associated with a distinct behavior of cells in the EPI-VE and ExE-VE regions, the former displaying cell intercalation whereas in the latter, cells are relatively static (Figure 10.4). This suggests different molecular and mechanical properties of cells. Cell lineage analyses have demonstrated that EPI-VE and ExE-VE do not have distinct origins in the blastocyst primitive endoderm, as clones span the EPI/ExE border (Perea-Gomez et al., 2007). Thus, these regions are later induced. Differential gene expression in the regions of the VE can be observed from the early egg cylinder stage onwards, and is dependent on Nodal and Bmp signaling (Artus et al., 2012; Mesnard et al., 2006; Trichas et al., 2011; Yamamoto et al., 2009). The analysis of the distribution of components of the cell cortex involved in junctional remodeling in epithelia has shed novel insight into the cellular bases of the distinct behavior observed in EPI-VE and ExE-VE. F-actin and the molecular motor myosin IIA (Myh9) are abundant and arranged in an apical shroud in the ExE-VE cells, possibly locking the shape and location of these static cells. In contrast, EPI-VE cells show a cortical distribution of F-actin and Myh9 that might facilitate the intercalation required for DVE migration. Consistent with this, the F-actin shroud in the ExE-VE is reduced in mutants in which DVE cells fail to stop above the EPI/ExE border, suggesting that in this experimental case, the ExE-VE is a more permissive environment for migration (Trichas et al., 2011). Conversely, migration is abrogated in the EPI-VE upon deletion of the gene encoding the actin regulator Rac1, which impairs cell intercalation and cell shape changes (Migeotte et al., 2010). Further understanding of the role of actin dynamics in the differential behavior of VE cells requires the production of VE specific mutations or locally targeted manipulations of the different regulators of the process using techniques such as electroporation or lipofection.

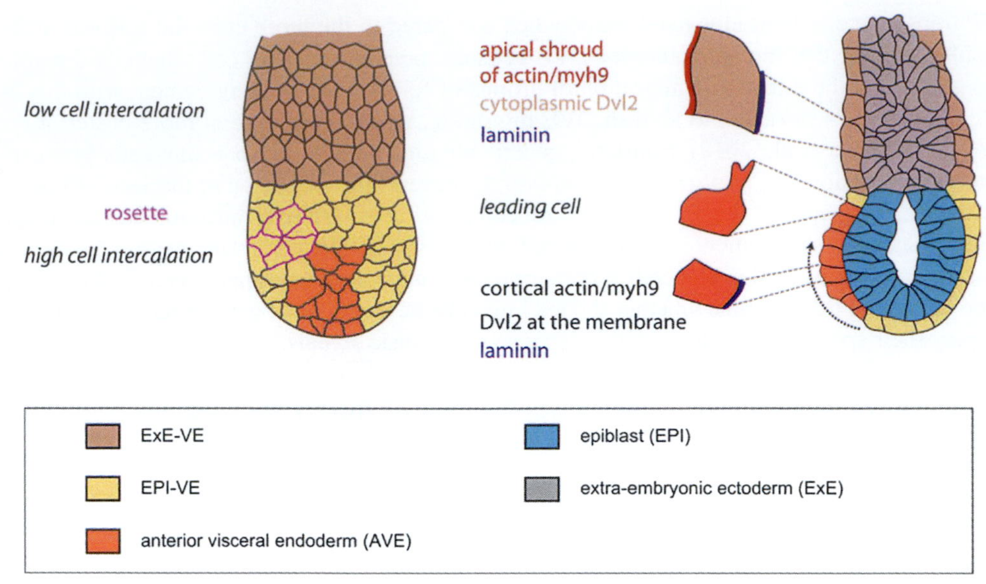

FIGURE 10.4 **Behavior of Visceral Endoderm Cells.** Visceral endoderm cells are epithelial with an apico-basal polarity. In the visceral endoderm covering the extra-embryonic ectoderm (ExE-VE), cells barely intercalate and are covered by a shroud of actin and myh9. In contrast in the visceral endoderm over the epiblast (EPI-VE), cells highly intercalate, as exemplified by the formation of rosettes, and have membrane localization of components of the PCP pathway such as Dvl2. The migration of AVE cells depends on leading cells with a protrusive activity. (Adapted from Figure 12.1 of Joyce and Srinivas, 2012.)

The planar cell polarity (PCP) pathway is a classical player in coordinating cellular rearrangements of epithelial tissues (Gray et al., 2011). In agreement with a role of this pathway during the movement of the DVE, the PCP effector Dvl2 is localized at the membrane specifically in the EPI-VE, whereas it is down-regulated and excluded from the membrane in ExE-VE cells (Trichas et al., 2011). This differential distribution is abrogated in *Nodal* mutants, thus indicating that Nodal controls both the identity and the mechanical properties of VE cells (Trichas et al., 2011). When PCP signaling is disrupted in conceptuses ubiquitously expressing a membrane tethered version of Celsr1, Dvl2 is no longer enriched at the membrane of EPI-VE cells and ExE-VE cells display abnormally persistent Dvl2. This is associated with an attenuated F-actin shroud. In these conceptuses, rosettes do not form, DVE cells abnormally disperse during migration and ignore the EPI/ExE border, suggesting that PCP signaling is involved in regulating the coordinated cell behavior in the VE epithelium (Trichas et al., 2011; 2012). However, PCP transgenic mice survive, which indicates that compensatory mechanisms exist, or that a disordered migration of the AVE can still be functional. Understanding the precise contribution of PCP proteins in the movement of the DVE and the reorganization of the VE will require the analysis of compound and tissue specific mutants at early post-implantation stages (Tao et al., 2012; 2009; Wallingford, 2012).

A Global Movement of the VE Takes Place at the Time of DVE Migration

The 5h process of DVE migration, described between E5.5D (distal VET) and E5.5L/D (latero-distal VET), is part of a global event involving movement of other EPI-VE cells at least until the onset of gastrulation. This was initially described as a "polonaise" movement (Weber et al., 1999). In a more recent study, when VE cells located in the distal region of the conceptuses at E5.5L/D were labeled after the DVE had migrated anteriorly, the descendants were found in the anterior visceral endoderm at E6.5. These results indicate that non-DVE cells also undergo an anterior movement after DVE migration is completed (Rivera-Perez et al., 2003).

Further understanding of the global movements in the VE came from cell tracking analyses in transgenic conceptuses harboring *Gata6-mTomato* and *Lefty1-mVenus* BACs marking respectively most VE cells and DVE cells between E5.5 and E6.5 (Takaoka et al., 2011). *Lefty1-mVenus* positive cells are initially located in the distal region at E5.5 and form the DVE population with a columnar morphology. As previously shown for *Hhex-GFP* positive cells, DVE cells expressing *Lefty1-mVenus* migrate proximally towards the future anterior region in 4 to 5h. When they reach the EPI/ExE border, *Lefty1-mVenus* cells become flat, stretch and move to more lateral positions so that at E6.5 DVE descendants are no longer found in the anterior visceral endoderm region, but rather in two bilateral VE stripes that span the midline only in the most proximal region (Figure 10.3). DVE descendants found in lateral positions at E6.5 do not express *Lefty1* and *Cer1*, however, they are still fluorescent due to the stability of the Venus protein. Strikingly, while *Lefty1-mVenus* positive DVE cells migrate, non-DVE cells, initially located in posterior and proximal regions, reach the distal pole where they begin to express *Lefty1-mVenus*, migrate behind DVE cells and occupy the anterior visceral endoderm region at E6.5. Taken together, these results point to a scenario in which a global movement affects the EPI-VE layer between E5.5 and E6.5. The process is initiated with the anterior migration of DVE cells, and pursued with the progressive recruitment into the distal pole of more posterior and proximal VE cells, which finally end anteriorly and form the AVE at E6.5. During this global movement, the expression of *Lefty1* and possibly of other DVE/AVE markers is dynamically regulated according to the position of VE cells. In mutant conceptuses in which DVE cells do not migrate (*Otx2*$^{-/-}$ mice) or the DVE is not specified at E5.5 (*Tdgf1*$^{-/-}$ mice) or when DVE cells are genetically ablated at E5.5, VE cells fail to undergo the global movement (Chu and Shen, 2010; Ding et al., 1998; Perea-Gomez et al., 2001a; Takaoka et al., 2011). These observations suggest that the specification and/or migration of DVE cells at E5.5 is required for the subsequent global movement of the VE to proceed (Takaoka and Hamada, 2012). It would be interesting to know whether VE cells other than DVE cells also produce similar projections during their migration, or whether the permissive mechanical properties of the EPI-VE are sufficient to account for all VE movements.

Cues Directing the Movement of the DVE

What are the molecular cues driving the direction of DVE migration? Wnt and Nodal have been identified as repulsive signals potentially involved in directing DVE movement (Figure 10.2). DVE cells fail to migrate and remain in distal positions in conceptuses lacking the canonical Wnt intracellular effector β-catenin or when *Wnt8a* is ectopically expressed in the entire epiblast (Huelsken et al., 2000; Kimura-Yoshida et al., 2005). In *Otx2* mutants, DVE cells are formed at E5.5 but fail to migrate, and this defect can be partially rescued by inhibition of Wnt signaling either with increased *Dkk1* expression or reduced *β-catenin* gene dosage (Kimura-Yoshida et al., 2005). These observations indicate that precise regulation of Wnt signaling is required for DVE migration or differentiation. Accordingly, embryological experiments involving the culture of E5.5 conceptuses in the presence of soaked beads have shown that DVE cells tend to migrate towards the source of the Wnt antagonist Dkk1 and away from a source of Wnt3a (Kimura-Yoshida et al., 2005). These results indicate that DVE cells can respond to Wnt signaling in their environment and migrate towards the region of lower Wnt signaling. The Wnt effector β-catenin and a Wnt transcriptional reporter are expressed in the VE, and enriched in the posterior region at E6.0 after DVE migration (Ferrer-Vaquer et al., 2010; Kimura-Yoshida et al., 2005). Yet, the precise localization of Wnt responsive cells at the onset of DVE migration remains to be elucidated. *Dkk1* is expressed in an anterior horseshoe-shaped domain in the foremost aspect of migrating DVE from E5.5L/D. However, it is unclear whether Dkk1 asymmetric distribution precedes the onset of DVE migration (Kimura-Yoshida et al., 2005; Miura et al., 2010). *Wnt3* has been found to be expressed in a restricted area of the prospective posterior EPI-VE at E5.5 presumably before DVE migration, suggesting that DVE could migrate away from this source of Wnt signaling (Rivera-Perez and Magnuson, 2005). However, mutations in *Wnt3* or *Dkk1* or in *Pcpn*, likely affecting the production of all Wnt ligands, do not impair the localization of the AVE to one side of the conceptus, indicating that other molecular cues can direct DVE migration (Biechele et al., 2013; Liu et al., 1999; Mukhopadhyay et al., 2001). Importantly, in addition to its role as an inhibitor of the LRP5/6 mediated canonical Wnt signaling, Dkk1 can also act as an agonist of Wnt-PCP signaling and as an inducer of cell extensions (Caneparo et al., 2007; Endo et al., 2008). Therefore, it will be important to determine whether PCP signaling mediates the effects of Dkk1 on DVE migration.

High levels of Nodal signaling were demonstrated to be required for DVE migration. Mutations affecting the level of *Nodal* transcription, the Nodal co-receptor Tdgf1, or the downstream transcription factor Foxh1, prevent DVE migration (Ding et al., 1998; Lowe et al., 2001; Norris et al., 2002; Yamamoto et al., 2001; 2004). Experiments involving localized ectopic expression by lipofection have demonstrated that DVE cells tend to migrate away from the side of higher *Nodal* expression and towards the side of higher expression of the genes encoding the Nodal antagonists Lefty1 and Cer1 (Yamamoto et al., 2004). This behavior was initially related to differential proliferation in the VE, regulated by Nodal signaling. According to this view, DVE cells migrate towards a region of lower proliferation created by the local activity of Cer1 and Lefty1, which are asymmetrically expressed at E5.5 (Yamamoto et al., 2004). However, this interpretation has been challenged by recent findings. Stuckey et al. (2011a) failed to observe regional proliferation differences in the VE at any time between E5.5 and E6.5. Moreover if *Nodal* overexpression is sufficient to increase VE proliferation in subpopulations of lipofected cells, inhibition of Nodal/Activin signaling in cultured E5.5 conceptuses significantly impairs the proliferation of just the epiblast layer (Stuckey et al., 2011a; Yamamoto et al., 2004). In *Foxh1*$^{-/-}$;*Nodal*$^{Lacz/+}$ conceptuses, DVE movement is rescued in a more efficient way when *Nodal* is overexpressed compared to *Cdk2* overexpression, which stimulates only cell proliferation (Yamamoto et al., 2004). In addition, increasing Nodal signaling by lowering the amount of Cer1 protein is sufficient to rescue DVE migration in *Tdgf1*$^{-/-}$; *Cer1*$^{+/-}$ mutants, indicating that Nodal can regulate DVE migration in a Tdgf1 independent fashion (Liguori et al., 2008). Taken together, these results highlight that Nodal might regulate DVE movement in other ways than the control of VE proliferation. One possibility is that Nodal controls epiblast growth and by doing so protects the DVE from inhibitory signals such as Wnt3 (Stuckey et al., 2011a). An open question is whether asymmetric Nodal signaling levels could influence the direction of DVE movement by locally regulating cellular mechanical properties in the VE (Trichas et al., 2011; Woo et al., 2012). The activity of the Nodal signaling pathway at the time of DVE migration has been inferred from the analysis of the distribution of phosphorylated Smad2. No evidence for differential levels of Nodal signaling in the EPI-VE has been described (Yamamoto et al., 2009). In addition, when *Cer1* and *Lefty1*, the two Nodal antagonists that are asymmetrically expressed at E5.5, are mutated, DVE migration still occurs (Perea-Gomez et al., 2002; Yamamoto et al., 2004). Understanding whether Nodal signaling is indeed asymmetrically activated in the early post-implantation conceptus, and the way that this might drive DVE migration will require novel tools to monitor and tune the activity of this pathway.

10.5 EVOLUTIONARY PERSPECTIVE

Although the mouse is widely used as a model organism, studies in other vertebrates are required to understand the evolutionary origin of the mechanism of AP patterning dependent on the anterior visceral endoderm. In other mammals, variations of the mouse model are observed (Eakin and Behringer, 2004; Selwood and Johnson, 2006). Blastocysts have a similar structure except in cases such as marsupials, in which the blastocyst is unilaminar without an inner cell mass. Implantation in the uterus does not always occur at this stage, because, in marsupials and artiodactyls (see glossary), gastrulation is complete before implantation. An extreme case is the horse conceptus, which does not implant until after the limb-buds are visible. This shows that in many cases AP patterning is intrinsic to the conceptus and independent of interactions with maternal tissues. Even in the mouse, in which AP patterning occurs at the time of implantation, it has been shown that the emergence of the AP axis does not correlate with the axes of the uterus but rather with the morphology of the embryo (Guo and Li, 2007; Mesnard et al., 2004; Perea-Gomez et al., 2004). The development of the blastocyst into an egg cylinder, including a cup-shaped epiblast, is peculiar to rodents. This permits a dramatic reduction in the volume of the conceptus (up to 50-fold) for embryos of similar size and thus an increase in the size of the litter. By contrast, the human gastrula (like that of most other mammals) remains unfolded, and the epiblast displays a planar morphology as a disc. This variation implies geometric changes such that the proximal–distal axis of the mouse egg cylinder becomes translated into a peripheral–central radius of the disc. Extra-embryonic tissues are present in all mammals, but their structures may differ, raising the question of how they interact with the epiblast. The equivalent of the ExE in non-rodent mammals is unclear. In species such as the guinea pig, the epiblast dissociates from trophectoderm derivatives, thereby preventing interactions between the tissues. In other cases, such as ungulates, the trophectoderm covering the inner cell mass (also referred to as Rauber's layer), from which the ExE derives in mouse, disappears; this leads to the exposure of the inner cell mass to the exterior. In this case, we can only speculate about the tissue equivalent to the ExE; it could be the remaining trophectoderm, which surrounds the epiblast. The other extra-embryonic structure essential for AP patterning in the mouse, the AVE, may be conserved in other mammals, although further characterization is required. An equivalent has been identified in the rabbit, and is known as the anterior marginal crescent. Similar to the mouse AVE, it is composed of columnar cells, which move. It is functional for the suppression of mesoderm formation and gradients of gene expression, including that of *Cer1* and *Dkk1*, are compatible with the mouse model (Idkowiak et al., 2004). In humans and pigs, a thickening of the prospective anterior endoderm has been observed, which is morphologically similar to the mouse AVE; however, its function is unknown (Hassoun et al., 2009).

Similarities to the mouse model can be identified not just in mammals but also in other amniotes. In the chick, the posterior marginal zone is able to induce the primitive streak. This is mediated by the TGFβ signal Vg1 and by cWnt8c, which in combination activate Nodal in the epiblast (Skromne and Stern, 2001; 2002). Cells of the posterior marginal zone do not contribute to adult tissues, and they are localized at the periphery of the area pellucida epiblast. This position of an extra-embryonic tissue, in addition to its early role as a Nodal inducer, suggests that the posterior marginal zone may be equivalent to the mouse ExE. An equivalent to the mouse AVE in the chick is the hypoblast, an extra-embryonic tissue that underlies the epiblast and expresses similar markers, including *Hesx1, Hhex, Otx2, Cerberus, Dkk1* and *Fgf8*. The hypoblast is also displaced anteriorly like the mouse AVE, leading to the removal of Nodal inhibition and the induction of the primitive streak in the posterior epiblast (Bertocchini et al., 2004). In addition, hypoblast cells can direct the movement of the adjacent epiblast layer through Fgf8 mediated induction of PCP gene expression resulting in mediolateral intercalation of epiblast cells (Foley et al., 2000; Voiculescu et al., 2007). Such an influence of the AVE in regulating epiblast cell movements has not been demonstrated in the mouse embryo. The anterior movement of the chick hypoblast is accompanied by the growth of another tissue, the endoblast. In light of the recent results concerning the global movements of the VE layer during DVE migration, it is tempting to speculate that the lateral and posterior EPI-VE regions constitute the murine equivalent of the avian endoblast (Takaoka et al., 2011). In anamniote vertebrates, AVE-like cells are also observed, such as the yolk cells of the vegetal region in the frog and the dorsal yolk syncytial cells in the zebrafish. These cells are mobile, express *Cerberus* or *Hhex*, and are important for AP patterning. Although the body plans of vertebrates are very similar and induced by similar pathways, differences are observed on how and when the AP axis is established. Human conceptuses, although genetically closer to the mouse model, are more reminiscent of the rabbit or chicken model in their flat morphology.

10.6 CONCLUSIONS

Establishment and patterning of the AP axis in the mouse requires reciprocal interactions between the embryo and the extra-embryonic tissues. The visceral endoderm emerges as a central player, with a fine and dynamic induction of regions of the tissue, both in terms of gene expression and cell behavior. Wnt and Nodal signaling mediate many of the morphogenetic processes in the visceral endoderm and their fine tuning, both in time and space, is important. This is achieved by the secretion of ligands, the activity of modifying enzymes (proteases), antagonists and co-receptors, all of which underlie the interactions between the embryonic and extra-embryonic tissues.

Many questions, however, remain unanswered. How is the symmetry-breaking event (the directional migration of DVE cells) controlled? Is the direction of migration pre-determined or is it a random process? In particular it will be important to address whether the direction of migration of the DVE bears any relationship to the asymmetric distribution of DVE precursors in the primitive endoderm before implantation. This will require a better understanding both at the morphological and molecular level of the transformation of the blastocyst into the egg cylinder, as well as further investigation of the spatial distribution of Wnt and Nodal signaling activities and other possible molecular cues before the onset of DVE migration. What is the significance of the molecular heterogeneity observed in DVE and AVE cell populations? Is this relevant to the initiation of directional DVE migration, the specification of leading versus follower cells? What are the mechanisms of the differential gene regulation? These challenges will be faced by dissecting the regulatory sequences of key markers, and by performing genetic perturbations at the clonal level, possibly using the Cas/Crispr technology (Yang et al., 2013).

10.7 CLINICAL RELEVANCE

- DVE migration is an accessible model for the *in vivo* study of collective epithelial cell migration in mammals, a process also at play in physiological situations, such as wound healing, and pathological conditions, such as cancer invasion and metastasis (Rorth, 2007).
- Nodal is an embryonic morphogen, which is absent from most adult tissues. However, it is re-expressed during tumorigenesis. The fine regulation of the Nodal pathway is central to the establishment of the AP axis in the mouse embryo. This is potentially relevant to tumor formation and progression in the adult organism (Massague, 2008; Strizzi et al., 2012). In addition, Nodal signaling is important in regenerative medicine, as it is a central player for the maintenance of mouse epiblast stem cells and human embryonic stem cells, as well as for the differentiation of cultured stem cells in endoderm (Kim and Ong, 2012; Pera and Tam, 2010).
- Extra-embryonic endoderm stem cell lines (XEN) can be directed towards VE and AVE differentiation. These cells are potentially useful for stem cell differentiation towards a cardiac cell fate (Artus et al., 2012; Brown et al., 2010; Kruithof-de Julio et al., 2011; Liu et al., 2012).
- Knowledge of mammalian development at the time of implantation has clinical implications for increasing the success of assisted reproductive strategies such as *in vitro* fertilization (IVF) and preimplantation genetic diagnosis (PGD), preventing the high rate of early pregnancy loss in humans.

ACKNOWLEDGMENTS

We thank S. Srinivas for kindly providing schemes for Figure 10.4, and I. Migeotte and C. Braendle for critical reading of the manuscript. APG and SMM are CNRS and INSERM research scientists respectively.

RECOMMENDED RESOURCES

Gastrulation, From cells to embryo, editor Stern CD, New York, CSHL Press, 2004. (Available at: http://www.gastrulation.org/, accessed 13.11.2013).

Mouse Genome Informatics: http://www.informatics.jax.org/ (accessed 13.11.2013) – Integrated access to data concerning the genetics, genomics, and biology of the laboratory mouse.

UNSW Embryology: http://embryology.med.unsw.edu.au/ (accessed 13.11.2013) – An educational resource for learning concepts in embryologic development.

EMAP: http://www.emouseatlas.org/emap/home.html (accessed 13.11.2013) – A website presenting 3D volumetric models of the embryo linked, in most cases, to comprehensive and detailed images of histological structure as well as a database of in situ gene expression data in the mouse embryo and an accompanying suite of tools to search and analyse the data.

Embryo Images Normal and Abnormal Mammalian Development: https://syllabus.med.unc.edu/courseware/embryo_images/unit-welcome/welcome_htms/contents.htm (accessed 13.11.2013) – A tutorial that uses scanning electron micrographs (SEMs) as the primary resource to teach mammalian embryology.

REFERENCES

Albano, R.M., Arkell, R., Beddington, R.S., Smith, J.C., 1994. Expression of inhibin subunits and follistatin during postimplantation mouse development: decidual expression of activin and expression of follistatin in primitive streak, somites and hindbrain. Development 120, 803–813.

Arai, A., Yamamoto, K., Toyama, J., 1997. Murine cardiac progenitor cells require visceral embryonic endoderm and primitive streak for terminal differentiation. Dev. Dyn. 210, 344–353.

Arkell, R.M., Fossat, N., Tam, P.P., 2013. Wnt signaling in mouse gastrulation and anterior development: new players in the pathway and signal output. Curr. Opin. Genet. Dev.

Arkell, R.M., Tam, P.P., 2012. Initiating head development in mouse embryos: integrating signaling and transcriptional activity. Open. Biol. 2, 120030.

Artus, J., Douvaras, P., Piliszek, A., Isern, J., Baron, M.H., Hadjantonakis, A.K., 2012. BMP4 signaling directs primitive endoderm-derived XEN cells to an extraembryonic visceral endoderm identity. Dev. Biol. 361, 245–262.

Barrow, J.R., Howell, W.D., Rule, M., Hayashi, S., Thomas, K.R., Capecchi, M.R., McMahon, A.P., 2007. Wnt3 signaling in the epiblast is required for proper orientation of the anteroposterior axis. Dev. Biol. 312, 312–320.

Beck, S., Le Good, J.A., Guzman, M., Ben Haim, N., Roy, K., Beermann, F., Constam, D.B., 2002. Extraembryonic proteases regulate Nodal signaling during gastrulation. Nat. Cell Biol. 4, 981–985.

Beddington, R.S., 1994. Induction of a second neural axis by the mouse node. Development 120, 613–620.

Belaoussoff, M., Farrington, S.M., Baron, M.H., 1998. Hematopoietic induction and respecification of A-P identity by visceral endoderm signaling in the mouse embryo. Development 125, 5009–5018.

Belo, J.A., Bouwmeester, T., Leyns, L., Kertesz, N., Gallo, M., Follettie, M., De Robertis, E.M., 1997. Cerberus-like is a secreted factor with neutralizing activity expressed in the anterior primitive endoderm of the mouse gastrula. Mech. Dev. 68, 45–57.

Ben-Haim, N., Lu, C., Guzman-Ayala, M., Pescatore, L., Mesnard, D., Bischofberger, M., Naef, F., Robertson, E.J., Constam, D.B., 2006. The nodal precursor acting via activin receptors induces mesoderm by maintaining a source of its convertases and BMP4. Dev. Cell 11, 313–323.

Bertocchini, F., Skromne, I., Wolpert, L., Stern, C.D., 2004. Determination of embryonic polarity in a regulative system: evidence for endogenous inhibitors acting sequentially during primitive streak formation in the chick embryo. Development 131, 3381–3390.

Biechele, S., Cockburn, K., Lanner, F., Cox, B.J., Rossant, J., 2013. Porcn-dependent Wnt signaling is not required prior to mouse gastrulation. Development 140, 2961–2971.

Biechele, S., Cox, B.J., Rossant, J., 2011. Porcupine homolog is required for canonical Wnt signaling and gastrulation in mouse embryos. Dev. Biol. 355, 275–285.

Bloomekatz, J., Grego-Bessa, J., Migeotte, I., Anderson, K.V., 2012. Pten regulates collective cell migration during specification of the anterior-posterior axis of the mouse embryo. Dev. Biol. 364, 192–201.

Brennan, J., Lu, C.C., Norris, D.P., Rodriguez, T.A., Beddington, R.S., Robertson, E.J., 2001. Nodal signaling in the epiblast patterns the early mouse embryo. Nature 411, 965–969.

Brown, K., Doss, M.X., Legros, S., Artus, J., Hadjantonakis, A.K., Foley, A.C., 2010. eXtraembryonic ENdoderm (XEN) stem cells produce factors that activate heart formation. PLoS One 5, e13446.

Camus, A., Perea-Gomez, A., Moreau, A., Collignon, J., 2006. Absence of Nodal signaling promotes precocious neural differentiation in the mouse embryo. Dev. Biol. 295, 743–755.

Caneparo, L., Huang, Y.L., Staudt, N., Tada, M., Ahrendt, R., Kazanskaya, O., Niehrs, C., Houart, C., 2007. Dickkopf-1 regulates gastrulation movements by coordinated modulation of Wnt/beta catenin and Wnt/PCP activities, through interaction with the Dally-like homolog Knypek. Genes Dev. 21, 465–480.

Chazaud, C., Rossant, J., 2006. Disruption of early proximodistal patterning and AVE formation in Apc mutants. Development 133, 3379–3387.

Chu, J., Shen, M.M., 2010. Functional redundancy of EGF-CFC genes in epiblast and extraembryonic patterning during early mouse embryogenesis. Dev. Biol. 342, 63–73.

Ciruna, B.G., Rossant, J., 1999. Expression of the T-box gene Eomesodermin during early mouse development. Mech. Dev. 81, 199–203.

Clements, M., Pernaute, B., Vella, F., Rodriguez, T.A., 2011. Crosstalk between Nodal/activin and MAPK p38 signaling is essential for anterior-posterior axis specification. Curr. Biol. 21, 1289–1295.

Copp, A.J., 1979. Interaction between inner cell mass and trophectoderm of the mouse blastocyst. II. The fate of the polar trophectoderm. J. Embryol. Exp. Morphol. 51, 109–120.

Coucouvanis, E., Martin, G.R., 1999. BMP signaling plays a role in visceral endoderm differentiation and cavitation in the early mouse embryo. Development 126, 535–546.

Ding, J., Yang, L., Yan, Y.T., Chen, A., Desai, N., Wynshaw-Boris, A., Shen, M.M., 1998. Cripto is required for correct orientation of the anterior-posterior axis in the mouse embryo. Nature 395, 702–707.

Donnison, M., Beaton, A., Davey, H.W., Broadhurst, R., L'Huillier, P., Pfeffer, P.L., 2005. Loss of the extraembryonic ectoderm in Elf5 mutants leads to defects in embryonic patterning. Development 132, 2299–2308.

Eakin, G.S., Behringer, R.R., 2004. Diversity of germ layer and axis formation among mammals. Semin. Cell Dev. Biol. 15, 619–629.

Egea, J., Erlacher, C., Montanez, E., Burtscher, I., Yamagishi, S., Hess, M., Hampel, F., Sanchez, R., Rodriguez-Manzaneque, M.T., Bosl, M.R., et al., 2008. Genetic ablation of FLRT3 reveals a novel morphogenetic function for the anterior visceral endoderm in suppressing mesoderm differentiation. Genes Dev. 22, 3349–3362.

Endo, Y., Beauchamp, E., Woods, D., Taylor, W.G., Toretsky, J.A., Uren, A., Rubin, J.S., 2008. Wnt-3a and Dickkopf-1 stimulate neurite outgrowth in Ewing tumor cells via a Frizzled3- and c-Jun N-terminal kinase-dependent mechanism. Mol. Cell Biol. 28, 2368–2379.

Ferrer-Vaquer, A., Piliszek, A., Tian, G., Aho, R.J., Dufort, D., Hadjantonakis, A.K., 2010. A sensitive and bright single-cell resolution live imaging reporter of Wnt/ss-catenin signaling in the mouse. BMC Dev. Biol. 10, 121.

Finley, K.R., Tennessen, J., Shawlot, W., 2003. The mouse secreted frizzled-related protein 5 gene is expressed in the anterior visceral endoderm and foregut endoderm during early post-implantation development. Gene Expr. Patterns 3, 681–684.

Foley, A.C., Skromne, I., Stern, C.D., 2000. Reconciling different models of forebrain induction and patterning: a dual role for the hypoblast. Development 127, 3839–3854.

Fu, J., Jiang, M., Mirando, A.J., Yu, H.M., Hsu, W., 2009. Reciprocal regulation of Wnt and Gpr177/mouse Wntless is required for embryonic axis formation. Proc. Natl. Acad. Sci. USA 106, 18598–18603.

Furtado, M.B., Solloway, M.J., Jones, V.J., Costa, M.W., Biben, C., Wolstein, O., Preis, J.I., Sparrow, D.B., Saga, Y., Dunwoodie, S.L., et al., 2008. BMP/SMAD1 signaling sets a threshold for the left/right pathway in lateral plate mesoderm and limits availability of SMAD4. Genes. Dev. 22, 3037–3049.

Gardner, R.L., Cockroft, D.L., 1998. Complete dissipation of coherent clonal growth occurs before gastrulation in mouse epiblast. Development 125, 2397–2402.

Gardner, R.L., Meredith, M.R., Altman, D.G., 1992. Is the anterior-posterior axis of the fetus specified before implantation in the mouse? J. Exp. Zool. 264, 437–443.

Georgiades, P., Rossant, J., 2006. Ets2 is necessary in trophoblast for normal embryonic anteroposterior axis development. Development 133, 1059–1068.

Glinka, A., Wu, W., Delius, H., Monaghan, A.P., Blumenstock, C., Niehrs, C., 1998. Dickkopf-1 is a member of a new family of secreted proteins and functions in head induction. Nature 391, 357–362.

Goncalves, L., Filipe, M., Marques, S., Salgueiro, A.M., Becker, J.D., Belo, J.A., 2011. Identification and functional analysis of novel genes expressed in the Anterior Visceral Endoderm. Int. J. Dev. Biol. 55, 281–295.

Granier, C., Gurchenkov, V., Perea-Gomez, A., Camus, A., Ott, S., Papanayotou, C., Iranzo, J., Moreau, A., Reid, J., Koentges, G., et al., 2011. Nodal cis-regulatory elements reveal epiblast and primitive endoderm heterogeneity in the peri-implantation mouse embryo. Dev. Biol. 349, 350–362.

Gray, R.S., Roszko, I., Solnica-Krezel, L., 2011. Planar cell polarity: coordinating morphogenetic cell behaviors with embryonic polarity. Dev. Cell 21, 120–133.

Guo, Q., Li, J.Y., 2007. Distinct functions of the major Fgf8 spliceform, Fgf8b, before and during mouse gastrulation. Development 134, 2251–2260.

Hiramatsu, R., Matsuoka, T., Kimura-Yoshida, C., Han, S.W, Mochida, K, Adachi, T, Takayama, S., Matsuo, I., 2013. External mechanical cues trigger the establishment of the anterior-posterior axis in early mouse embryos. Dev. Cell 27, 131–144.

Hassoun, R., Schwartz, P., Feistel, K., Blum, M., Viebahn, C., 2009. Axial differentiation and early gastrulation stages of the pig embryo. Differentiation 78, 301–311.

Hsieh, J.C., Lee, L., Zhang, L., Wefer, S., Brown, K., DeRossi, C., Wines, M.E., Rosenquist, T., Holdener, B.C., 2003. Mesd encodes an LRP5/6 chaperone essential for specification of mouse embryonic polarity. Cell 112, 355–367.

Huelsken, J., Vogel, R., Brinkmann, V., Erdmann, B., Birchmeier, C., Birchmeier, W., 2000. Requirement for beta-catenin in anterior-posterior axis formation in mice. J. Cell Biol. 148, 567–578.

Idkowiak, J., Weisheit, G., Plitzner, J., Viebahn, C., 2004. Hypoblast controls mesoderm generation and axial patterning in the gastrulating rabbit embryo. Dev. Genes Evol. 214, 591–605.

Jones, R.L., Kaitu'u-Lino, T.J., Nie, G., Sanchez-Partida, L.G., Findlay, J.K., Salamonsen, L.A., 2006. Complex expression patterns support potential roles for maternally derived activins in the establishment of pregnancy in mouse. Reproduction 132, 799–810.

Joyce, B., Srinivas, S., 2012. Cell movements in the egg cylinder stage mouse embryo. Results Probl. Cell Differ. 55, 219–229.

Kelly, O.G., Pinson, K.I., Skarnes, W.C., 2004. The Wnt co-receptors Lrp5 and Lrp6 are essential for gastrulation in mice. Development 131, 2803–2815.

Kemp, C., Willems, E., Abdo, S., Lambiv, L., Leyns, L., 2005. Expression of all Wnt genes and their secreted antagonists during mouse blastocyst and postimplantation development. Dev. Dyn. 233, 1064–1075.

Kim, P.T., Ong, C.J., 2012. Differentiation of definitive endoderm from mouse embryonic stem cells. Results. Probl. Cell Differ. 55, 303–319.

Kimura, C., Yoshinaga, K., Tian, E., Suzuki, M., Aizawa, S., Matsuo, I., 2000. Visceral endoderm mediates forebrain development by suppressing posteriorizing signals. Dev. Biol. 225, 304–321.

Kimura-Yoshida, C., Nakano, H., Okamura, D., Nakao, K., Yonemura, S., Belo, J.A., Aizawa, S., Matsui, Y., Matsuo, I., 2005. Canonical Wnt signaling and its antagonist regulate anterior-posterior axis polarization by guiding cell migration in mouse visceral endoderm. Dev. Cell 9, 639–650.

Kimura-Yoshida, C., Tian, E., Nakano, H., Amazaki, S., Shimokawa, K., Rossant, J., Aizawa, S., Matsuo, I., 2007. Crucial roles of Foxa2 in mouse anterior-posterior axis polarization via regulation of anterior visceral endoderm-specific genes. Proc. Natl. Acad. Sci. USA 104, 5919–5924.

Kruithof-de Julio, M., Alvarez, M.J., Galli, A., Chu, J., Price, S.M., Califano, A., Shen, M.M., 2011. Regulation of extra-embryonic endoderm stem cell differentiation by Nodal and Cripto signaling. Development 138, 3885–3895.

Kwon, G.S., Viotti, M., Hadjantonakis, A.K., 2008. The endoderm of the mouse embryo arises by dynamic widespread intercalation of embryonic and extraembryonic lineages. Dev. Cell 15, 509–520.

Liguori, G.L., Borges, A.C., D'Andrea, D., Liguoro, A., Goncalves, L., Salgueiro, A.M., Persico, M.G., Belo, J.A., 2008. Cripto-independent Nodal signaling promotes positioning of the A-P axis in the early mouse embryo. Dev. Biol. 315, 280–289.

Liu, P., Wakamiya, M., Shea, M.J., Albrecht, U., Behringer, R.R., Bradley, A., 1999. Requirement for Wnt3 in vertebrate axis formation. Nat. Genet. 22, 361–365.

Liu, W., Brown, K., Legros, S., Foley, A.C., 2012. Nodal mutant eXtraembryonic ENdoderm (XEN) stem cells upregulate markers for the anterior visceral endoderm and impact the timing of cardiac differentiation in mouse embryoid bodies. Biol. Open. 1, 208–219.

Lowe, L.A., Yamada, S., Kuehn, M.R., 2001. Genetic dissection of nodal function in patterning the mouse embryo. Development 128, 1831–1843.

Madabhushi, M., Lacy, E., 2011. Anterior visceral endoderm directs ventral morphogenesis and placement of head and heart via BMP2 expression. Dev. Cell 21, 907–919.

Maruoka, Y., Ohbayashi, N., Hoshikawa, M., Itoh, N., Hogan, B.L., Furuta, Y., 1998. Comparison of the expression of three highly related genes, Fgf8, Fgf17 and Fgf18, in the mouse embryo. Mech. Dev. 74, 175–177.

Massague, J., 2008. TGFbeta in Cancer. Cell 134, 215–230.

Meno, C., Saijoh, Y., Fujii, H., Ikeda, M., Yokoyama, T., Yokoyama, M., Toyoda, Y., Hamada, H., 1996. Left-right asymmetric expression of the TGF beta-family member lefty in mouse embryos. Nature 381, 151–155.

Merrill, B.J., Pasolli, H.A., Polak, L., Rendl, M., Garcia-Garcia, M.J., Anderson, K.V., Fuchs, E., 2004. Tcf3: a transcriptional regulator of axis induction in the early embryo. Development 131, 263–274.

Mesnard, D., Donnison, M., Fuerer, C., Pfeffer, P.L., Constam, D.B., 2011. The microenvironment patterns the pluripotent mouse epiblast through paracrine Furin and Pace4 proteolytic activities. Genes Dev. 25, 1871–1880.

Mesnard, D., Filipe, M., Belo, J.A., Zernicka-Goetz, M., 2004. The anterior-posterior axis emerges respecting the morphology of the mouse embryo that changes and aligns with the uterus before gastrulation. Curr. Biol. 14, 184–196.

Mesnard, D., Guzman-Ayala, M., Constam, D.B., 2006. Nodal specifies embryonic visceral endoderm and sustains pluripotent cells in the epiblast before overt axial patterning. Development 133, 2497–2505.

Migeotte, I., Omelchenko, T., Hall, A., Anderson, K.V., 2010. Rac1-dependent collective cell migration is required for specification of the anterior-posterior body axis of the mouse. PLoS Biol. 8, e1000442.

Miura, S., Mishina, Y., 2007. The DVE changes distal epiblast fate from definitive endoderm to neurectoderm by antagonizing nodal signaling. Dev. Dyn. 236, 1602–1610.

Miura, S., Singh, A.P., Mishina, Y., 2010. Bmpr1a is required for proper migration of the AVE through regulation of Dkk1 expression in the pre-streak mouse embryo. Dev. Biol. 341, 246–254.

Morkel, M., Huelsken, J., Wakamiya, M., Ding, J., van de Wetering, M., Clevers, H., Taketo, M.M., Behringer, R.R., Shen, M.M., Birchmeier, W., 2003. Beta-catenin regulates Cripto- and Wnt3-dependent gene expression programs in mouse axis and mesoderm formation. Development 130, 6283–6294.

Morris, S.A., Grewal, S., Barrios, F., Patankar, S.N., Strauss, B., Buttery, L., Alexander, M., Shakesheff, K.M., Zernicka-Goetz, M., 2012. Dynamics of anterior-posterior axis formation in the developing mouse embryo. Nat. Commun. 3, 673.

Morsut, L., Yan, K.P., Enzo, E., Aragona, M., Soligo, S.M., Wendling, O., Mark, M., Khetchoumian, K., Bressan, G., Chambon, P., et al., 2010. Negative control of Smad activity by ectodermin/Tif1gamma patterns the mammalian embryo. Development 137, 2571–2578.

Mukhopadhyay, M., Shtrom, S., Rodriguez-Esteban, C., Chen, L., Tsukui, T., Gomer, L., Dorward, D.W., Glinka, A., Grinberg, A., Huang, S.P., et al., 2001. Dickkopf1 is required for embryonic head induction and limb morphogenesis in the mouse. Dev. Cell 1, 423–434.

Nijmeijer, R.M., Leeuwis, J.W., DeLisio, A., Mummery, C.L., Chuva de Sousa Lopes, S.M., 2009. Visceral endoderm induces specification of cardiomyocytes in mice. Stem. Cell Res. 3, 170–178.

Norris, D.P., Brennan, J., Bikoff, E.K., Robertson, E.J., 2002. The Foxh1-dependent autoregulatory enhancer controls the level of Nodal signals in the mouse embryo. Development 129, 3455–3468.

Nowotschin, S., Costello, I., Piliszek, A., Kwon, G.S., Mao, C.A., Klein, W.H., Robertson, E.J., Hadjantonakis, A.K., 2013. The T-box transcription factor Eomesodermin is essential for AVE induction in the mouse embryo. Genes Dev. 27, 997–1002.

Ohinata, Y., Ohta, H., Shigeta, M., Yamanaka, K., Wakayama, T., Saitou, M., 2009. A signaling principle for the specification of the germ cell lineage in mice. Cell 137, 571–584.

Park, C.B., Dufort, D., 2011. Nodal expression in the uterus of the mouse is regulated by the embryo and correlates with implantation. Biol. Reprod. 84, 1103–1110.

Pearce, J.J., Penny, G., Rossant, J., 1999. A mouse cerberus/Dan-related gene family. Dev. Biol. 209, 98–110.

Pera, M.F., Tam, P.P., 2010. Extrinsic regulation of pluripotent stem cells. Nature 465, 713–720.

Perea-Gomez, A., Camus, A., Moreau, A., Grieve, K., Moneron, G., Dubois, A., Cibert, C., Collignon, J., 2004. Initiation of gastrulation in the mouse embryo is preceded by an apparent shift in the orientation of the anterior-posterior axis. Curr. Biol. 14, 197–207.

Perea-Gomez, A., Lawson, K.A., Rhinn, M., Zakin, L., Brulet, P., Mazan, S., Ang, S.L., 2001a. Otx2 is required for visceral endoderm movement and for the restriction of posterior signals in the epiblast of the mouse embryo. Development 128, 753–765.

Perea-Gomez, A., Meilhac, S.M., Piotrowska-Nitsche, K., Gray, D., Collignon, J., Zernicka-Goetz, M., 2007. Regionalization of the mouse visceral endoderm as the blastocyst transforms into the egg cylinder. BMC Dev. Biol. 7, 96.

Perea-Gomez, A., Rhinn, M., Ang, S.L., 2001b. Role of the anterior visceral endoderm in restricting posterior signals in the mouse embryo. Int. J. Dev. Biol. 45, 311–320.

Perea-Gomez, A., Shawlot, W., Sasaki, H., Behringer, R.R., Ang, S., 1999. HNF3beta and Lim1 interact in the visceral endoderm to regulate primitive streak formation and anterior-posterior polarity in the mouse embryo. Development 126, 4499–4511.

Perea-Gomez, A., Vella, F.D., Shawlot, W., Oulad-Abdelghani, M., Chazaud, C., Meno, C., Pfister, V., Chen, L., Robertson, E., Hamada, H., et al., 2002. Nodal antagonists in the anterior visceral endoderm prevent the formation of multiple primitive streaks. Dev. Cell 3, 745–756.

Pereira, P.N., Dobreva, M.P., Maas, E., Cornelis, F.M., Moya, I.M., Umans, L., Verfaillie, C.M., Camus, A., de Sousa Lopes, S.M., Huylebroeck, D., et al., 2012. Antagonism of Nodal signaling by BMP/Smad5 prevents ectopic primitive streak formation in the mouse amnion. Development 139, 3343–3354.

Popperl, H., Schmidt, C., Wilson, V., Hume, C.R., Dodd, J., Krumlauf, R., Beddington, R.S., 1997. Misexpression of Cwnt8C in the mouse induces an ectopic embryonic axis and causes a truncation of the anterior neuroectoderm. Development 124, 2997–3005.

Rakeman, A.S., Anderson, K.V., 2006. Axis specification and morphogenesis in the mouse embryo require Nap1, a regulator of WAVE-mediated actin branching. Development 133, 3075–3083.

Rhinn, M., Dierich, A., Shawlot, W., Behringer, R.R., Le Meur, M., Ang, S.L., 1998. Sequential roles for Otx2 in visceral endoderm and neuroectoderm for forebrain and midbrain induction and specification. Development 125, 845–856.

Richardson, L., Torres-Padilla, M.E., Zernicka-Goetz, M., 2006. Regionalised signaling within the extraembryonic ectoderm regulates anterior visceral endoderm positioning in the mouse embryo. Mech. Dev. 123, 288–296.

Rivera-Perez, J.A., Mager, J., Magnuson, T., 2003. Dynamic morphogenetic events characterize the mouse visceral endoderm. Dev. Biol. 261, 470–487.

Rivera-Perez, J.A., Magnuson, T., 2005. Primitive streak formation in mice is preceded by localized activation of Brachyury and Wnt3. Dev. Biol. 288, 363–371.

Rodriguez, T.A., Srinivas, S., Clements, M.P., Smith, J.C., Beddington, R.S., 2005. Induction and migration of the anterior visceral endoderm is regulated by the extra-embryonic ectoderm. Development 132, 2513–2520.

Rorth, P., 2007. Collective guidance of collective cell migration. Trends Cell Biol. 17, 575–579.

Rosenquist, T.A., Martin, G.R., 1995. Visceral endoderm-1 (VE-1): an antigen marker that distinguishes anterior from posterior embryonic visceral endoderm in the early post- implantation mouse embryo. Mech. Dev. 49, 117–121.

Selwood, L., Johnson, M.H., 2006. Trophoblast and hypoblast in the monotreme, marsupial and eutherian mammal: evolution and origins. Bioessays 28, 128–145.

Shawlot, W., Wakamiya, M., Kwan, K.M., Kania, A., Jessell, T.M., Behringer, R.R., 1999. Lim1 is required in both primitive streak-derived tissues and visceral endoderm for head formation in the mouse. Development 126, 4925–4932.

Skromne, I., Stern, C.D., 2001. Interactions between Wnt and Vg1 signaling pathways initiate primitive streak formation in the chick embryo. Development 128, 2915–2927.

Skromne, I., Stern, C.D., 2002. A hierarchy of gene expression accompanying induction of the primitive streak by Vg1 in the chick embryo. Mech. Dev. 114, 115–118.

Smith, L.J., 1985. Embryonic axis orientation in the mouse and its correlation with blastocyst relationships to the uterus. II. Relationships from 4 1/4 to 9 1/2 days. J. Embryol. Exp. Morphol. 89, 15–35.

Soares, M.L., Torres-Padilla, M.E., Zernicka-Goetz, M., 2008. Bone morphogenetic protein 4 signaling regulates development of the anterior visceral endoderm in the mouse embryo. Dev. Growth. Differ. 50, 615–621.

Srinivas, S., Rodriguez, T., Clements, M., Smith, J.C., Beddington, R.S., 2004. Active cell migration drives the unilateral movements of the anterior visceral endoderm. Development 131, 1157–1164.

Stern, C.D., Downs, K.M., 2012. The hypoblast (visceral endoderm): an evo-devo perspective. Development 139, 1059–1069.

Strizzi, L., Hardy, K.M., Kirschmann, D.A., Ahrlund-Richter, L., Hendrix, M.J., 2012. Nodal expression and detection in cancer: experience and challenges. Cancer Res. 72, 1915–1920.

Stuckey, D.W., Clements, M., Di-Gregorio, A., Senner, C.E., Le Tissier, P., Srinivas, S., Rodriguez, T.A., 2011a. Coordination of cell proliferation and anterior-posterior axis establishment in the mouse embryo. Development 138, 1521–1530.

Stuckey, D.W., Di Gregorio, A., Clements, M., Rodriguez, T.A., 2011b. Correct patterning of the primitive streak requires the anterior visceral endoderm. PLoS One 6, e17620.

Takaoka, K., Hamada, H., 2012. Cell fate decisions and axis determination in the early mouse embryo. Development 139, 3–14.

Takaoka, K., Yamamoto, M., Hamada, H., 2007. Origin of body axes in the mouse embryo. Curr. Opin. Genet. Dev. 17, 344–350.

Takaoka, K., Yamamoto, M., Hamada, H., 2011. Origin and role of distal visceral endoderm, a group of cells that determines anterior-posterior polarity of the mouse embryo. Nat. Cell Biol. 13, 743–752.

Takaoka, K., Yamamoto, M., Shiratori, H., Meno, C., Rossant, J., Saijoh, Y., Hamada, H., 2006. The mouse embryo autonomously acquires anterior-posterior polarity at implantation. Dev. Cell 10, 451–459.

Tam, P.P., Steiner, K.A., 1999. Anterior patterning by synergistic activity of the early gastrula organizer and the anterior germ layer tissues of the mouse embryo. Development 126, 5171–5179.

Tao, H., Inoue, K., Kiyonari, H., Bassuk, A.G., Axelrod, J.D., Sasaki, H., Aizawa, S., Ueno, N., 2012. Nuclear localization of Prickle2 is required to establish cell polarity during early mouse embryogenesis. Dev. Biol. 364, 138–148.

Tao, H., Suzuki, M., Kiyonari, H., Abe, T., Sasaoka, T., Ueno, N., 2009. Mouse prickle1, the homolog of a PCP gene, is essential for epiblast apical-basal polarity. Proc. Natl. Acad. Sci. USA 106, 14426–14431.

Thomas, P., Beddington, R., 1996. Anterior primitive endoderm may be responsible for patterning the anterior neural plate in the mouse embryo. Curr. Biol. 6, 1487–1496.

Thomas, P.Q., Brown, A., Beddington, R.S., 1998. Hex: a homeobox gene revealing peri-implantation asymmetry in the mouse embryo and an early transient marker of endothelial cell precursors. Development 125, 85–94.

Torres-Padilla, M.E., Richardson, L., Kolasinska, P., Meilhac, S.M., Luetke-Eversloh, M.V., Zernicka-Goetz, M., 2007. The anterior visceral endoderm of the mouse embryo is established from both preimplantation precursor cells and by de novo gene expression after implantation. Dev. Biol. 309, 97–112.

Tortelote, G.G., Hernandez-Hernandez, J.M., Quaresma, A.J., Nickerson, J.A., Imbalzano, A.N., Rivera-Perez, J.A., 2012. Wnt3 function in the epiblast is required for the maintenance but not the initiation of gastrulation in mice. Dev. Biol. 374, 164–173.

Trichas, G., Joyce, B., Crompton, L.A., Wilkins, V., Clements, M., Tada, M., Rodriguez, T.A., Srinivas, S., 2011. Nodal dependent differential localisation of dishevelled-2 demarcates regions of differing cell behavior in the visceral endoderm. PLoS Biol. 9, e1001019.

Trichas, G., Smith, A.M., White, N., Wilkins, V., Watanabe, T., Moore, A., Joyce, B., Sugnaseelan, J., Rodriguez, T.A., Kay, D., et al., 2012. Multi-cellular rosettes in the mouse visceral endoderm facilitate the ordered migration of anterior visceral endoderm cells. PLoS Biol. 10, e1001256.

Varlet, I., Collignon, J., Robertson, E.J., 1997. nodal expression in the primitive endoderm is required for specification of the anterior axis during mouse gastrulation. Development 124, 1033–1044.

Vincent, S.D., Dunn, N.R., Hayashi, S., Norris, D.P., Robertson, E.J., 2003. Cell fate decisions within the mouse organizer are governed by graded Nodal signals. Genes. Dev. 17, 1646–1662.

Voiculescu, O., Bertocchini, F., Wolpert, L., Keller, R.E., Stern, C.D., 2007. The amniote primitive streak is defined by epithelial cell intercalation before gastrulation. Nature 449, 1049–1052.

Wada, Y., Sun-Wada, G.H., Kawamura, N., 2013. Microautophagy in the visceral endoderm is essential for mouse early development. Autophagy 9, 252–254.

Waldrip, W.R., Bikoff, E.K., Hoodless, P.A., Wrana, J.L., Robertson, E.J., 1998. Smad2 signaling in extraembryonic tissues determines anterior-posterior polarity of the early mouse embryo. Cell 92, 797–808.

Wallingford, J.B., 2012. Planar cell polarity and the developmental control of cell behavior in vertebrate embryos. Annu. Rev. Cell Dev. Biol. 28, 627–653.

Weber, R.J., Pedersen, R.A., Wianny, F., Evans, M.J., Zernicka-Goetz, M., 1999. Polarity of the mouse embryo is anticipated before implantation. Development 126, 5591–5598.

Woo, S., Housley, M.P., Weiner, O.D., Stainier, D.Y., 2012. Nodal signaling regulates endodermal cell motility and actin dynamics via Rac1 and Prex1. J. Cell Biol. 198, 941–952.

Yang, H., Wang, H., Shivalila, C.S., Cheng, A.W., Shi, L., Jaenisch, R., 2013. One-step generation of mice carrying reporter and conditional alleles by CRISPR/Cas-mediated genome engineering. Cell 154, 1370–1379.

Yamamoto, M., Beppu, H., Takaoka, K., Meno, C., Li, E., Miyazono, K., Hamada, H., 2009. Antagonism between Smad1 and Smad2 signaling determines the site of distal visceral endoderm formation in the mouse embryo. J. Cell Biol. 184, 323–334.

Yamamoto, M., Meno, C., Sakai, Y., Shiratori, H., Mochida, K., Ikawa, Y., Saijoh, Y., Hamada, H., 2001. The transcription factor FoxH1 (FAST) mediates Nodal signaling during anterior-posterior patterning and node formation in the mouse. Genes Dev. 15, 1242–1256.

Yamamoto, M., Saijoh, Y., Perea-Gomez, A., Shawlot, W., Behringer, R.R., Ang, S.L., Hamada, H., Meno, C., 2004. Nodal antagonists regulate formation of the anteroposterior axis of the mouse embryo. Nature 428, 387–392.

Zeng, L., Fagotto, F., Zhang, T., Hsu, W., Vasicek, T.J., Perry, W.L., Lee, J.J., Tilghman, S.M., Gumbiner, B.M., Costantini, F., 1997. The mouse Fused locus encodes Axin, an inhibitor of the Wnt signaling pathway that regulates embryonic axis formation. Cell 90, 181–192.

Chapter 11

Early Development of Epidermis and Neural Tissue

Keiji Itoh and Sergei Y. Sokol

Department of Developmental and Regenerative Biology, Mount Sinai School of Medicine, New York, NY

Chapter Outline

GLOSSARY

Cytoplasmic determinant A cytoplasmic protein acting to promote a specific cell fate.
Epithelial polarity Regional differences within cells forming an epithelial tissue.
Neural induction Intercellular communication that converts ectodermal cells into neural tissue.
Neurogenesis Generation of neurons from uncommitted ectodermal cells.
Organizer A cell population that induces fate changes in neighboring cells and tissues.

SUMMARY

- The three germ layers are specified by multiple mechanisms.
- Dorsoventral polarity and inductive signals influence the differentiation of ectoderm into neural tissue and epidermis, and pattern the neural tube along the anteroposterior axis.
- Bone morphogenetic protein (BMP) signaling promotes epidermal development, whereas inhibition of BMP signaling leads to neural induction. Other patterning factors, including fibroblast growth factor (FGF) and Wnt pathways, also modulate this process.
- The differentiation of primary neurons and epidermal cell types is regulated by asymmetric divisions of polarized ectoderm progenitor cells resulting from the combined action of apical-basal polarity proteins and Notch signaling.

11.1 INTRODUCTION

Generation of the three germ layers, ectoderm, mesoderm and endoderm, is among the earliest and the most fundamental processes underlying animal development. The main derivatives of vertebrate ectoderm, the outer germ layer, are the central and the peripheral nervous system, the epidermis and the placodes. In the frog *Xenopus laevis*, which is commonly used as

Principles of Developmental Genetics.

a model vertebrate species, these ectodermal tissues develop from the animal pole region of the embryo. During gastrulation, ectoderm covers the whole surface of the embryo. Subsequently, ventral ectoderm develops into the epidermis, whereas dorsal ectoderm gives rise to the neural plate, which is later transformed into the neural tube. The neural tube is further subdivided into the forebrain, midbrain, hindbrain and the spinal cord along the anterior-posterior (AP) axis. Neural tissue produces neurons and glia, whereas non-neural ectoderm gives rise to epidermis, ectodermal glands, such as the cement gland and the hatching gland, and the placodes.

How does the ectodermal germ layer form in the vertebrate early embryo? Are there maternal cytoplasmic factors that define ectoderm as opposed to mesoderm and endoderm? Alternatively, ectoderm may develop as a default cell state, when other germ layers are not specified. This review will consider these alternative models of ectoderm specification and will describe studies revealing the existence of cytoplasmic determinants, which instruct embryonic cells to become ectoderm.

The diversification of ectoderm into epidermis and neural tissue is another issue addressed in this review. Neural induction, also known as primary embryonic induction, was first observed by Spemann and Mangold (Harland and Gerhart, 1997). In their experiments, a graft from the dorsal subequatorial region triggered neural plate formation at the ventral side of the host embryo. Lineage tracing experiments demonstrated that this ectopic neural plate originated from host ectoderm, indicating that it is a result of *induction,* a process by which one tissue (or group of cells) instructs another group of cells about their fate. Since the discovery of neural induction, the identification of its mediators has attracted the attention of many embryologists. Research during the last two decades has established that neural induction involves multiple signaling processes orchestrated by secreted factors which belong to the Wnt, fibroblast growth factor (FGF), and bone morphogenetic protein (BMP) families. We review the current understanding of signaling pathways that gives rise to neural tissue and epidermis in *Xenopus* embryos and other vertebrates. Given the multitude of transcription factors that serve as mediators of these signaling pathways in the regulation of early neural development, readers are referred to more comprehensive reviews on this topic (De Robertis and Kuroda, 2004; Hikasa and Sokol, 2013; Rogers et al., 2009).

Additionally, this review will focus on mechanisms underlying the differentiation of primary neurons from neuroectoderm as well as multiciliated cells and ionocytes from epidermal ectoderm. In vertebrate ectoderm, cell type differentiation is under the control of the Notch pathway, which helps to select individual cells from a group to undergo specific differentiation. Alternatively, individual cell types may form as a result of asymmetric cell divisions due to unequal distribution of cytoplasmic determinants. The process of cell fate specification is, therefore, distinct from the diversification of epidermal and neural tissue. Finally, we will discuss potential research directions needed to further understand ectoderm differentiation in normal development and disease.

11.2 SPECIFICATION OF ECTODERM AND MESENDODERM BY MUTUALLY ANTAGONISTIC FACTORS

Amphibian oocytes and fertilized eggs possess well-defined animal-vegetal polarity, which forms as a result of differential deposition of maternal proteins and mRNAs. The three germ layers form according to their position along the animal-vegetal axis (Figure 11.1). Therefore, ectoderm specification might be defined by maternally-derived factors, which are localized to the animal hemisphere and which differ from those distributed to the vegetal hemisphere (King et al., 2005). Cell progeny derived from the animal and the vegetal regions of the egg interact via secreted signaling factors to pattern the early embryo and generate the basic body plan. Molecular mechanisms for ectoderm specification appear to depend on both the inheritance of specific localized cytoplasmic determinants and on cell-cell interactions in the early embryo (Figure 11.1A–D).

When the ectodermal part of blastula (animal cap) is excised and cultured in isolation, it develops into atypical epidermis, whereas many secreted polypeptides – known as mesoderm inducing factors – are capable of re-specifying this tissue into mesoderm or endoderm (Harland and Gerhart, 1997). These observations suggest that mesoderm and endoderm (mesendoderm) originate from the embryological default state that corresponds to ectoderm. In agreement with this view, VegT, a T-box transcription factor, was shown to promote mesendoderm formation by activating transcription of Nodal-related mesoderm inducing factors of the transforming growth factor beta (TGFβ) superfamily (Zhang et al., 1998). Moreover, antisense oligonucleotide-mediated knockdown of maternal *VegT* RNA suppresses mesendoderm development and promotes the expansion of ectodermal fates, including epidermis and neural tissue (Zhang et al., 1998). Thus, in the absence of mesoderm and endoderm-inducing signals, ectoderm appears to develop as a default germ layer.

Alternatively, ectoderm may be specified by a localized cytoplasmic factor and this ectodermal determinant could be missing or suppressed in mesodermal and endodermal tissues. Indeed, Ectodermin, a RING-type ubiquitin ligase, was isolated as a maternal gene product with ectoderm-specific expression (Dupont et al., 2005). Ectodermin promotes ubiquitination of Smad4, a protein associating with Smad1 and Smad2, components of the BMP and Nodal signaling pathways, respectively. Ectodermin inhibits mesendoderm gene markers in presumptive ectodermal tissue and expands *Sox2*,

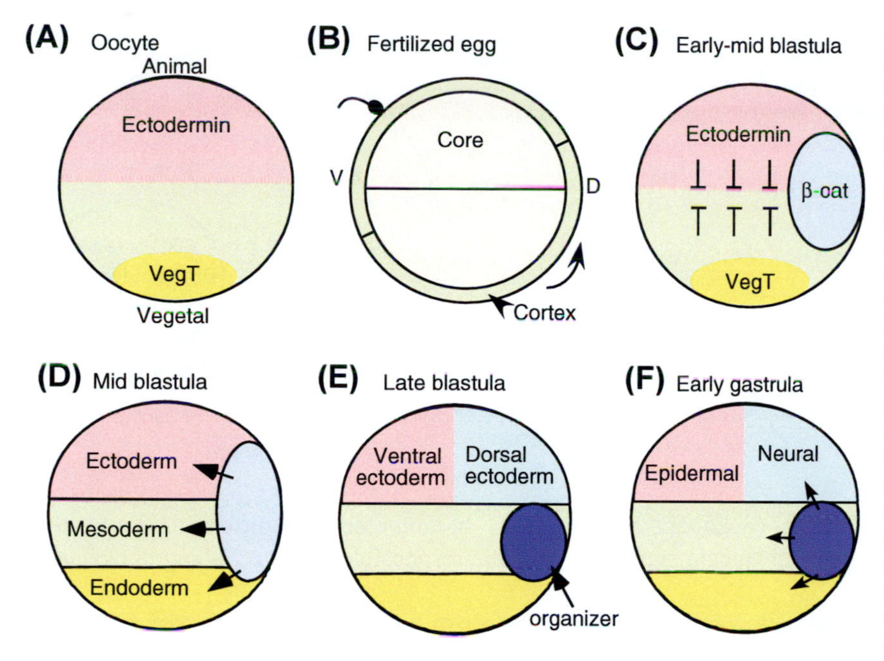

FIGURE 11.1 Early Ectoderm Development in *Xenopus* Embryos. (A) During oogenesis, the maternal determinants Ectodermin and VegT are deposited in the animal and the vegetal hemisphere, respectively. (B) After fertilization, the egg cortex rotates relative to the cytoplasm core in a microtubule-dependent process, known as the cortico-cytoplasmic rotation, which specifies future dorsal and ventral regions of the embryo. The side of sperm entry becomes ventral (V) and the opposite side becomes dorsal (D). (C) After cortical rotation, β-catenin accumulates on the dorsal side along the animal-vegetal axis. Ectodermin specifies ectoderm by counteracting mesendoderm induced by VegT. (D) All three germ layers are dorsalized by β-catenin signaling at blastula stages. (E) Dorsal ectoderm is influenced by β-catenin signaling and becomes predisposed to neural induction. The Spemann organizer forms in the dorsal subequatorial region as a result of combined action of VegT and β-catenin signaling. (F) At early gastrula stage, ventral ectoderm is specified as epidermis by BMP signaling, whereas dorsal ectoderm is specified as neural tissue due to BMP antagonists secreted by the organizer.

a neuroectodermal marker, presumably by suppressing Nodal and BMP signaling. Supporting this view, mesendodermal markers are expanded into the animal hemisphere at early gastrula stages in embryos depleted of Ectodermin with a specific antisense morpholino oligonucleotide (MO). In these embryos, epidermis is expanded at the expense of neural tissue due to upregulated BMP signaling, which is essential for epidermal development (see below and Dupont et al., 2005).

Xenopus ectodermally-expressed mesendoderm antagonist (Xema/Foxi1e) is another protein that is involved in ectoderm specification (Mir et al., 2007; Suri et al., 2005). Xema is Foxi-class transcription factor, which is first detectable in ectoderm at late blastula stages. Foxi1e suppresses mesendodermal gene markers, whereas Foxi1e MOs promote mesendoderm formation in the ectodermal territory. In contrast to Ectodermin, Foxi1e does not influence the conversion of epidermis into neural tissue, indicating that ectoderm development is independent from epidermal and neural specification. Since Foxi1e is a transcriptional activator (Mir et al., 2007; Suri et al., 2005), it is expected to induce additional mesendoderm inhibitors. Of note, Foxi transcription factors have been implicated in the specification of ectodermal cell types later in development (see Section 11.4). These studies suggest that ectoderm is specified via inhibition of mesendodermal fates by Ectodermin, Foxi1e and yet uncharacterized mesendoderm inhibitors. In conclusion, multiple molecular mechanisms operate to generate ectoderm and mesendoderm in the early vertebrate embryo (Figure 11.1A–D).

11.3 SPECIFICATION OF EPIDERMIS AND NEURAL TISSUE

The late blastula stages correspond to the first steps of diversification of the presumptive ectoderm into epidermis and neural tissue. This process is closely associated with dorsoventral patterning, since the epidermal tissue is derived from ventral ectoderm, whereas neural tissue forms dorsally. Dorsoventral polarity is generated by the cytoskeletal reorganization, also known as the cortico-cytoplasmic rotation, which occurs soon after fertilization (Harland and Gerhart, 1997). Dorsoventral polarization leads to the specification of future neural ectoderm in the dorsal animal region of the embryo and to the formation of the organizer, an early inducing center, in the dorsal vegetal region (Figure 11.1E,F). The organizer secretes a wide spectrum of specific signaling factors, including BMP antagonists, FGFs, Wnts and their antagonists, which induce and maintain the developing central nervous system in responding ectoderm (De Robertis and Kuroda, 2004). In contrast, BMP signaling on the ventral side suppresses neural tissue formation and promotes epidermal development (Figure 11.2).

Responding Tissue

Early Specification and Predisposition of Competent Ectoderm to Neural Induction

Early studies have shown that dorsal and ventral ectoderm cells are not equally responsive to organizer-derived signals. When isolated before gastrulation, dorsal ectoderm explants are more readily responsive to neural induction than ventral explants (Sharpe et al., 1987). Similarly, different mesodermal cell types are induced from dorsal and ventral blastula

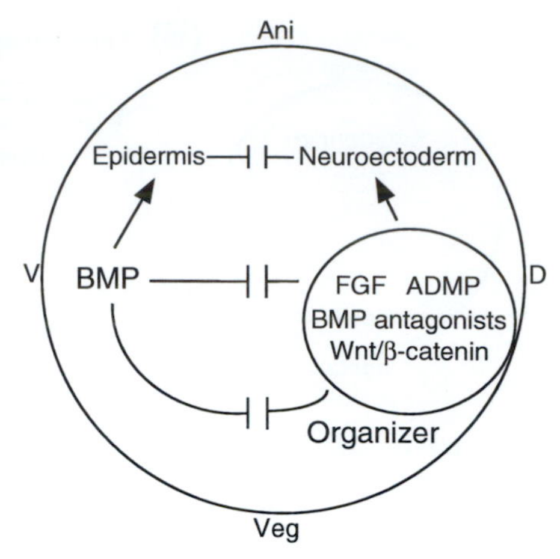

FIGURE 11.2 Regulatory Network of Pathways Involved in Neural Induction. BMP antagonists function together with Wnt/β-catenin and FGF in the organizer, a dorsal (D) signaling center to specify neuroectoderm. Ventral BMP signals specify epidermis. ADMP restricts organizer activity. The same pathways antagonize each other throughout the three germ layers. Ani – animal pole, Veg – vegetal pole.

ectoderm by Activin, a TGFβ-related growth factor (Sokol and Melton, 1991). In both cases, this differential responsiveness, or *competence*, was attributed to early Wnt/β-catenin signaling, which is responsible for establishing dorsoventral polarity (Sokol and Melton, 1992).

An ectodermal pre-pattern becomes detectable in dorsal ectoderm before gastrulation due to the selective expression of several genes, including *Sox2, Zic1, Xiro* and *FoxD5* (Bainter et al., 2001; Mizuseki et al., 1998a; Rogers et al., 2009; Sullivan et al., 2001). Other genes, such as *SoxD* and *Geminin*, are expressed throughout the ectoderm in the blastula stages, and become restricted to neuroectoderm during gastrulation (Bainter et al., 2001; Mizuseki et al., 1998b; Rogers et al., 2009). Both neuroectoderm-specific gene expression and the differential distribution of protein kinase C isoforms (Otte et al., 1991), reflect the labile specification of neuroectoderm, which requires continued organizer signaling for subsequent development of the vertebrate central nervous system.

The early expression of neuroectoderm-specific genes indicates that neural tissue development has already been initiated at blastula stages. Indeed, the Wnt/β-catenin pathway, which is thought to be involved in dorsoventral patterning, is activated soon after fertilization, as evidenced by the nuclear enrichment of β-catenin in the dorsal region of the embryo (Harland and Gerhart, 1997). By contrast, active FGF, TGFβ and BMP signaling pathways are not detectable until mid-blastula stages (Schohl and Fagotto, 2002). A promoter for the early dorsal mesendoderm-specific gene *Siamois* is directly induced by β-catenin, but fails to be activated by inhibitors of the BMP pathway that are also known to stimulate dorsal development (Fan et al., 1998). These findings suggest that the Wnt/β-catenin pathway functions in dorsoventral specification at an early stage, before other signaling pathways.

Although the interactions between the major signaling pathways have been studied in some detail with respect to mesoderm development, their role in neural specification is less well known. As many of these signaling factors influence dorsoventral polarity in all three germ layers (Harland and Gerhart, 1997), the mechanistic relationship between the pathways may be conserved during neural tissue specification. In support of this idea, dorsal expression of *geminin* is positively regulated by Wnt/β-catenin signaling and negatively controlled by Vent proteins, targets of the BMP pathway (Taylor et al., 2006). Additional studies of cis-regulatory elements of other early neuroectoderm-specific genes are necessary to provide more information on early neural tissue development.

Epidermal Development as a Result of BMP Signaling

Dissociation of blastula ectoderm leads to the suppression of epidermal markers, which can be overcome by exogenous BMP4. Similarly, blocking BMP signaling with a dominant negative BMP receptor inhibits epidermal differentiation and activated neural tissue markers in animal cap explants (Wilson and Hemmati-Brivanlou, 1997). Moreover, BMP2, 4 and 7 are expressed in the early *Xenopus* embryo and promote the epidermal fate (De Robertis and Kuroda, 2004; Harland, 2000; Wilson and Hemmati-Brivanlou, 1997). These findings support the notion that BMP signaling is responsible for epidermal development.

Mutagenesis screens in zebrafish implicated a large number of BMP signaling mediators in the specification of ventral cell fates (De Robertis and Kuroda, 2004). In *Xenopus* embryos, knockdown of BMP2, 4 and 7 resulted in the significant expansion of neural tissue, but epidermis still developed (Reversade and De Robertis, 2005). The same study also described anti-dorsalizing morphogenetic protein (ADMP), a member of the TGFβ superfamily with BMP4-like activity. Interestingly, a combined knockdown of ADMP, BMP2, BMP4 and BMP7 with specific MOs converts the entire epidermis into neural tissue (Reversade and De Robertis, 2005). Thus, multiple, redundant, BMP proteins direct epidermal development (Figure 11.2).

Inducing Tissue: The Organizer

Requirement for Organizer in Neural Development

The ability of the organizer, in presumptive dorsal mesoderm tissue, to induce the ectopic neural plate suggests that it may be essential for neural induction. In support of this notion, elimination of the organizer in frog embryos by ultraviolet irradiation leads to neural tissue defects (Harland and Gerhart, 1997). Similarly, interference with early Wnt/β-catenin signaling results in embryos without an organizer and deficient in neural development (Sokol, 1999). Animal pole explants, containing presumptive neuroectoderm and lacking dorsal mesoderm, fail to develop neural tissue when cultured *in vitro* (Harland and Gerhart, 1997). Moreover, in flat dorsal explants, which contain both the inducing and the responding tissue, neural tissue formation has been shown to entirely depend on organizer presence (Holowacz and Sokol, 1999). These studies document the requirement for organizer-derived signals in neural induction in *Xenopus*.

In contrast, organizer removal by microsurgical extirpation or the *one-eyed pinhead* mutation in the *EGF-CFC* gene, required for organizer formation, does not eliminate neural tissue in zebrafish (Davidson et al., 1999; Harland, 2000; Psychoyos and Stern, 1996; Shih and Fraser, 1996). Similarly, mechanical removal of the organizer equivalent in chick or mouse embryos did not eliminate neural development. Mouse embryos lacking *HNF-3b*, a gene essential for formation of the node, have neural tissue with essentially correct anteroposterior patterning (Ang and Rossant, 1994; Weinstein et al., 1994). These observations suggest that dorsal mesoderm, which is responsible for neural induction, may partially recover after the operation, or that more lateral tissues remaining after the operation retain the capacity to induce neural tissue. Alternatively, the organizer may play distinct roles in neural development of different vertebrate embryos.

Neural Inducing Factors of the Organizer

Since the discovery of the Spemann organizer, the identification of neural inducing factors was a major focus of research in vertebrate embryology (Harland and Gerhart, 1997). An important step toward this goal was the identification of molecular markers, specific for epidermal and neural cells. For example, Nrp1, Sox2 and NCAM are characteristic of neural tissue, whereas epidermal cytokeratin is a marker for epidermis (Bainter et al., 2001; Harland, 2000). Using these and other markers, several candidate neural inducers were identified. These include Noggin, Chordin and Follistatin, which are secreted proteins capable of inducing neural tissue in animal pole explants (Harland and Gerhart, 1997). These factors antagonize BMP signaling by directly binding BMP ligands. In addition, other secreted BMP antagonists, such as Cerberus and *Xenopus* nodal-related 3 (Xnr3), have been reported to trigger neural induction (Piccolo et al., 1999). Furthermore, neural inducing ability has been reported for proteins of the Wnt and FGF signaling pathways that are activated in the organizer, indicating that multiple signaling pathways besides BMP antagonists are involved (De Robertis and Kuroda, 2004; Harland, 2000; Schohl and Fagotto, 2002).

Signaling Pathways Regulating Neural Tissue Development

The BMP Pathway

Work by many different groups revealed that neural induction takes place upon inhibition of BMP signaling (De Robertis and Kuroda, 2004) (Figure 11.2). During BMP signal transduction, a ligand-activated BMP receptor, which encodes a serine/threonine protein kinase, recruits and phosphorylates Smad1. The phosphorylated Smad1 binds Smad4 and the complex translocates into the nucleus to regulate transcription (Massague and Wotton, 2000). Inhibition of BMP signaling during neural development is accomplished at different levels. Extracellularly, BMP signaling is blocked by the secreted factors Noggin, Chordin and Follistatin, which directly bind BMPs. Another level of regulation of the BMP pathway is via the inhibitory Smad6 and Smad7 proteins (Massague and Wotton, 2000). A third possible route for BMP signal inhibition is the phosphorylation of the Smad1 linker domain by MAP kinase leading to Smad1 degradation (Pera et al., 2003). Finally, BMP gene transcription is suppressed by Wnt and FGF signaling (Delaune et al., 2005; Harland, 2000). Thus, neural tissue development is accomplished via several molecular routes used to inhibit BMP signaling.

Accumulating evidence indicates that neural induction involves multiple cooperating inducing signals. Individual knockdowns of the BMP inhibitors Chordin, Follistatin or Noggin revealed rather mild defects in neural tissue formation, suggesting that these proteins have redundant functions in *Xenopus* (Khokha et al., 2005). Consistent with this notion, the triple knockdown of these BMP inhibitors resulted in dramatic defects in dorsal mesoderm and neuroectoderm, although neural tissue was not completely eliminated (Khokha et al., 2005). Similarly, mouse embryos with deleted *chordin* and *noggin* genes lack forebrain structures, but retain much of the remaining central nervous system (Bachiller et al., 2000;

De Robertis and Kuroda, 2004). Incomplete loss of neural tissue in embryos with depleted BMP antagonists indicates that BMP inhibition is only one of a number of molecular mechanisms involved in neural induction.

The FGF Pathway

Multiple lines of evidence indicate that the FGF pathway, in addition to BMP signaling, plays critical roles during neural tissue specification (Figure 11.2). The FGF pathway involves the tyrosine kinase receptors FGFR1–4, Ras GTPase, Raf and MAP kinase (Eswarakumar et al., 2005). In gain-of-function studies, FGFs have been shown to activate neural tissue markers in partially dissociated ectoderm cells (Harland and Gerhart, 1997), suggesting that FGF enhances the effect of BMP inhibition. Consistent with this view, neural tissue is readily induced by a combination of BMP antagonists and a low dose of FGF (Delaune et al., 2005). Activated MAPK has been detected in *Xenopus* neuroectoderm (Schohl and Fagotto, 2002), suggesting that FGF signaling is important for the response of the ectoderm to organizer-derived signals. Despite these predictions, blocking FGF signaling with a dominant negative FGFR-1 or Ras constructs failed to affect pan-neural markers (Holowacz and Sokol, 1999; Ribisi et al., 2000). It is possible that these reagents failed to block some FGFR-mediated responses. Indeed, a dominant negative FGFR-4 had a strong suppressive effect on neural development, suggesting that FGFR-4 is specifically involved (Delaune et al., 2005). Additionally, the pharmacological FGFR inhibitor SU5402 caused a complete loss of neural tissue when added to dorsal ectoderm at blastula stages, indicating an early requirement of FGF signaling before gastrulation (Delaune et al., 2005).

One may consider a number of mechanisms by which FGF signaling affects neural development. First, FGF can interfere with BMP transcription, because SU5402 upregulates dorsal BMP4 expression (Delaune et al., 2005). Second, FGF may suppress BMP signaling by MAPK-dependent phosphorylation of the linker domain of Smad1 (Pera et al., 2003). However, the role of FGF in neural development is unlikely to be exclusively due to inhibition of BMP signaling, because BMP antagonists failed to rescue neural deficiencies caused by SU5402 (Delaune et al., 2005). Thus, accumulating evidence supports the idea that FGF signaling cooperates with BMP antagonists in neural tissue formation (Figure 11.2).

This involvement of FGF in neural tissue development has been supported by studies in ascidian embryos, in which its role has been shown to be both necessary and sufficient for neural induction and transcriptional activation of the brain-specific Otx gene homolog (Bertrand et al., 2003). Similarly, FGFs, but not Chordin or Noggin, have been shown to induce neural tissue in the chick embryo (Stern, 2005). Consistent with this, both dominant interfering FGFR and SU5402 suppressed the node-mediated induction of neural tissue markers (Stern, 2005). Thus, the function of FGF signaling in vertebrate neural induction appears to be conserved.

Wnt Signaling

Wnt signaling operates in neural development in both positive and negative manners at different points during embryogenesis. The Wnt pathway involves signaling by Wnt ligands through Frizzled receptors, Disheveled (Dsh), glycogen synthase kinase 3 (GSK3), β-catenin and T cell factors (TCF) (Sokol, 1999). Over-expression of different pathway components in *Xenopus* embryos suggested that the activation of Wnt signaling leads to neural marker stimulation (Baker et al., 1999; Dominguez et al., 1995; Itoh and Sokol, 1997; Sokol et al., 1995). In a similar way to the FGF pathway, early Wnt/β-catenin signaling inhibits the transcription of BMP4 (Baker et al., 1999), and upregulates the BMP antagonists Chordin, Noggin and Follistatin (Domingos et al., 2001; Itoh and Sokol, 1997). Based on these results, it has been proposed that β-catenin predisposes dorsal ectoderm to neural induction at blastula stages (Figures 11.1, 11.2; Sokol and Melton, 1991; De Robertis and Kuroda, 2004; Stern, 2005). Surprisingly, Wnt signaling has been also reported to inhibit neural induction in both *Xenopus* and chick embryos (Stern, 2005; Itoh et al., 1995), arguably due to the different roles of the pathway at different developmental stages.

The role of Wnt signaling in neuralization has remained a subject of debate. Although Wnt8 and β-catenin can induce neural tissue in ectoderm explants (Baker et al., 1999), secreted Wnt inhibitors such as Dkks or FRPs do not inhibit neural induction *in vivo* (Baker et al., 1999; Glinka et al., 1998; Itoh and Sokol, 1999). Similarly, β-catenin MO does not inhibit general neural tissue markers (Heasman et al., 2000; Hikasa and Sokol, 2004). Perhaps β-catenin has to be depleted very early in development for the effect to be observed (Heasman et al., 2000). Since early β-catenin depletion results in ventralization, its effect on axial development is difficult to separate from its effect on neuroectoderm specification. Depletion of Frodo/DACT, a scaffold protein that associates with both Dsh and TCF3, leads to virtual elimination of the pan-neural markers Sox2 and Nrp1 (Hikasa and Sokol, 2004). Other studies show that inhibitors of the Wnt pathway and a dominant interfering TCF construct cause neural tissue expansion (Heeg-Truesdell and LaBonne, 2006; Itoh and Sokol, 1999; Itoh et al., 1995), arguing that Wnt signaling can serve to inhibit neural fate specification. Tissue- and stage-specific analysis is required to fully understand the involvement of Wnt signaling at different steps of neural development.

Anteroposterior Patterning of Neuroectoderm

The developing central nervous system becomes subdivided along the AP axis into forebrain, midbrain, hindbrain, and spinal cord. Pieter Nieuwkoop proposed that this regionalization is intimately coupled with the process of neuroectoderm specification and proposed the existence of a posteriorizing signal (or more than one) acting on newly formed neural tissue (Stern, 2005). This hypothesis has been supported by the demonstration that both FGF and Wnt signals function as posteriorizing signals.

Overexpression of FGF in the *Xenopus* embryo suppresses anterior development and induces posterior neural tissue markers in anterior neuroectoderm (Harland and Gerhart, 1997). In converse experiments, dominant negative forms of FGFR and Ras, and the FGFR inhibitor SU5402 downregulated posterior neural markers in whole embryos and dorsal explants (Delaune et al., 2005; Holowacz and Sokol, 1999; McGrew et al., 1997; Ribisi et al., 2000). These results indicate that FGFs are involved in patterning neural tissue along the AP axis.

Similar to FGF signaling, Wnt ligands cause anterior head defects (Harland and Gerhart, 1997; Itoh and Sokol, 1999). Anterior neural tissue is transformed into more posterior neural tissue upon overexpression of Wnt pathway components in Keller explants and neuralized animal caps in *Xenopus* (Domingos et al., 2001; Harland and Gerhart, 1997; Kiecker and Niehrs, 2001). Reflecting the posteriorizing activity of the Wnt pathway, graded amounts of Dsh specify distinct neural cell fates along the AP axis in animal caps (Itoh and Sokol, 1997). Moreover, negative regulators of Wnt signaling, including β-catenin MO, GSK-3, and secreted Wnt antagonists produce anteriorized embryos (Bang et al., 1999; Glinka et al., 1998; Heasman et al., 2000; Itoh and Sokol, 1999; Itoh et al., 1995; McGrew et al., 1997). Together, these studies support a role for Wnts in AP patterning.

The posteriorizing effect of Wnt signaling is traditionally considered to be due to posterior target gene activation by the β-catenin/TCF complex. Additionally, Wnt signaling derepresses several critical target genes through HIPK2-mediated phosphorylation of TCF3 (Hikasa et al., 2010) (Figure 11.3). The latter mechanism has been shown to modulate AP neural patterning in *Xenopus* embryos (Hikasa et al., 2010) and presumably functions in zebrafish (Kim et al., 2000).

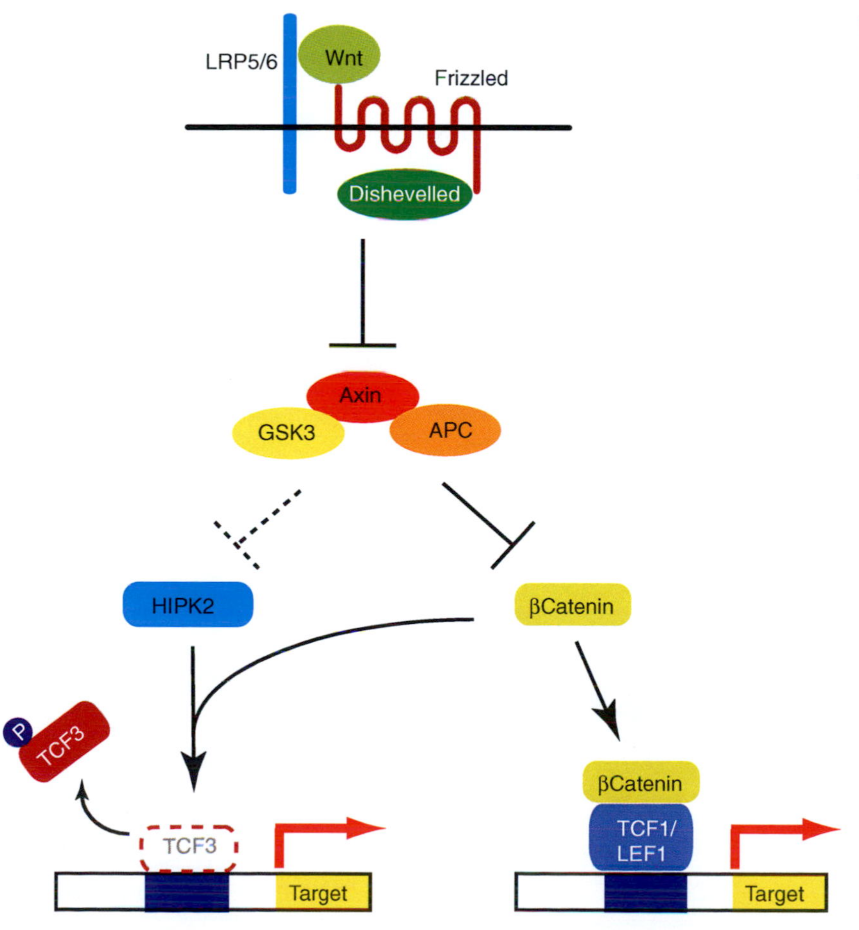

FIGURE 11.3 Wnt Signaling Mechanisms Operating during Anterior-Posterior Patterning. Wnt signals stabilize β-catenin and promote its ability to activate transcription. Alternatively, Wnt signals alleviate TCF3-dependent suppression of Wnt targets by stimulating TCF3 phosphorylation by HIPK2.

As the Wnt and FGF pathways have been implicated in both AP patterning and neural induction, one needs to consider a possibility that the two processes, neural induction and AP patterning, have a common molecular basis. According to Nieuwkoop, newly induced anterior neural tissue may simply represent the first step in AP patterning. However, the available evidence suggests that neural induction can be experimentally separated from AP patterning. First, inhibition of FGF signaling by a dominant negative FGFR1 suppresses posterior neural development, but not neural induction (Holowacz and Sokol, 1999), indicating that the posteriorization caused by FGF is different from neuralization. Similarly, secreted Wnt inhibitors inhibit posteriorization, but do not interfere with neural induction (McGrew et al., 1997). Supporting the view that neural induction is separable from AP patterning is the observation that BMP antagonists induce only anterior neural tissue, but are unable to stimulate posterior neural markers (Harland and Gerhart, 1997; Wilson and Hemmati-Brivanlou, 1997). The analysis of DNA regulatory sequences in promoters of regional neural target genes should prove useful to further understand the specific molecular mechanisms underlying AP patterning and neural induction.

In recent years, several studies have attempted to elucidate the mechanisms involved in AP patterning, which involve the cross-talk of BMP, FGF and Wnt signaling (Hikasa and Sokol, 2013). In addition to FGF/MAPK-dependent Smad1 phosphorylation, Smad-1 can be phosphorylated by GSK-3. The latter process is inhibited by the Wnt pathway, resulting in BMP target activation in cultured cells. However, the role of this phosphorylation *in vivo* still needs to be confirmed (Fuentealba et al., 2007).

11.4 ECTODERMAL CELL TYPE SPECIFICATION AND CELL POLARITY

In addition to AP patterning and neural induction, ectoderm undergoes cell differentiation to generate neuronal and glial cells in the central nervous system, a well as other specialized cells, such as ciliated cells and ionocytes in non-neural ectoderm of *Xenopus* embryos (Figure 11.4). In *Xenopus* gastrulae, the ectoderm is composed of superficial and deep cell layers. The superficial layer has a typical apical-basal polarity and is distinct from the deep layer. Different ectoderm layers express different molecular markers and generate distinct cell types. The molecular mechanisms underlying the

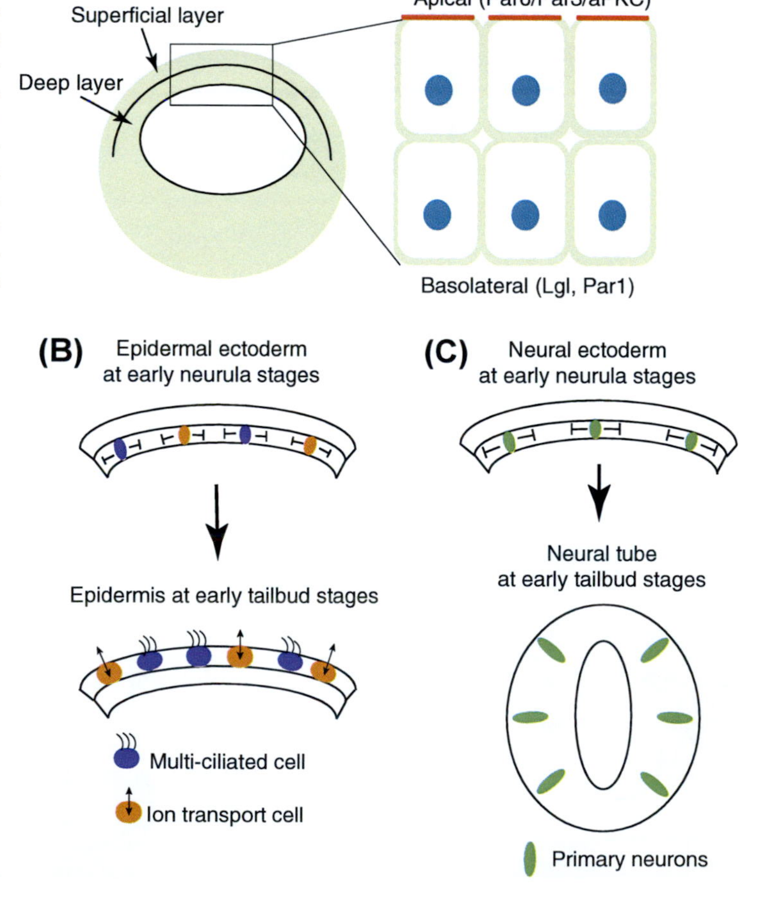

FIGURE 11.4 Specification of Ciliated Cells, Ionocytes and Primary Neurons in *Xenopus* Ectoderm. (A) Mid-sagittal section of a late blastula embryo is shown on the left. Animal pole ectoderm is composed of the superficial and the deep layers. The apical complex proteins Par6, Par3 and aPKC (orange) are localized primarily to the apical surface of the superficial layer, whereas the basolateral components Lethal giant larvae (Lgl) and Par1 (green) are present in both cell layers. (B) Ciliated cells and ionocytes are first specified in the deep layer of epidermal ectoderm in a scattered manner and move to the superficial layer at neurula stages. (C) Primary neurons are specified in the deep layer of the neural plate to occupy three distinct locations along the dorsoventral axis of the developing neural tube. The specification of these cell types involves inhibitory Notch signaling.

differentiation of specific cell types from distinct ectoderm layers are not fully understood, but might derive from the original apical-basal ectoderm polarity (Figure 11.4). Cell differentiation depends on polarized divisions of epithelial progenitor cells and may be controlled by specific signaling pathways, such as the Notch pathway.

The apical-basal polarity of epithelial cells is regulated by the *Par* genes, which were originally identified in a screen for *partitioning defective* mutations that interfere with asymmetric cell divisions in early *C. elegans* embryos (Bardin et al., 2004). An apical protein complex, including Par6, Par3 and atypical PKC (aPKC), antagonizes the basolateral domain of epithelial cells, which contains Lethal giant larvae (Lgl) and Par1 proteins. The two domains are separated by tight junctions (Bardin et al., 2004). In *Xenopus* ectoderm, aPKC counteracts Lgl and Par1 to determine the position of tight junctions in superficial cells and defines its apical-basal polarity (Chalmers et al., 2005; Dollar et al., 2005). The deep layer of *Xenopus* ectoderm lacks aPKC, indicating that it may have originated from asymmetric divisions of animal blastomeres (Chalmers et al., 2005). The Wnt pathway has been implicated in the establishment of the apical-basal polarity and the regulation of Lgl and Par1 localization (Choi and Sokol, 2009; Dollar et al., 2005; Ossipova et al., 2005). Nevertheless, the role of signaling processes in specifying ectodermal layers remains largely unknown. Future studies need to address the way that extracellular signaling controls intrinsic epithelial determinants and specifies cell fate in the vertebrate ectoderm.

Asymmetric Cell Division and Neurogenesis

In *Drosophila*, epithelial polarity proteins, including aPKC/Par3/Par6 and Lgl, are required for the asymmetric distribution of cell fate determinants in neural progenitors, known as neuroblasts (Bardin et al., 2004). Neuroblasts delaminate from epithelial cells and retain partial apical-basal polarity. When a neuroblast undergoes asymmetric cell division, the larger daughter cell, which inherits apical complex proteins, retains the neuroblast fate, whereas the smaller daughter cell differentiates into a neuron or a supporting cell. Thus, in neuroblasts, the apical-basal polarity may be associated with cell fate determination. Besides the modulation by apical-basal polarity proteins, the neuroblast fate is also regulated by the conserved Notch signaling pathway. The Notch receptor and the Delta ligand are transmembrane proteins that interact with each other in neighboring cells (Bardin et al., 2004). When Notch receives a signal from Delta, the intracellular domain of Notch is cleaved, translocates to the nucleus and associates with Suppressor of Hairless, a DNA binding protein, to control target gene expression. Only a single progenitor with inactive Notch signaling is selected to differentiate in a group of cells, whereas the differentiation of other cells is inhibited by Notch. After neuroblast specification, Notch signaling plays a role in progenitor differentiation by maintaining neuroblasts in an undifferentiated state.

The asymmetric cell division that leads to neuronal fates is causally connected to the Notch pathway. Fly neuronal progenitors, called ganglion mother cells, inherit Numb, an inhibitor of Notch, on the basal side. Numb is a cell fate determinant that allows neuronal differentiation by inhibiting Notch signaling in one of the daughter cells (Bardin et al., 2004). Thus, the establishment of the apical-basal polarity by aPKC/Par3/Par6 and Lgl, the asymmetric cell division and the localized distribution of Numb and Notch signaling may represent sequential steps of a conserved pathway leading to neurogenesis in fly embryos.

Mechanisms of vertebrate neurogenesis revealed a remarkable conservation to the mechanisms discovered in fly embryos. Similar to their role in *Drosophila*, vertebrate apical-basal polarity proteins regulate the asymmetric distribution of cell fate determinants, such as Numb. Mice lacking the *Lgl1* gene contain a large number of neural progenitor cells, but have a decreased number of neurons (Klezovitch et al., 2004). Moreover, a knockdown of AGS3, a mammalian protein implicated in asymmetric cell division and apical-basal polarity, decreased the number of neural progenitors, but stimulated neuronal differentiation (Sanada and Tsai, 2005). Thus, neurogenesis requires the activity of basolateral polarity determinants and the inhibition of Notch signaling, the two molecular systems that are not mutually exclusive, but are likely to function interdependently (Figures 11.4, 11.5).

During neurulation, the two layers of neuroectoderm intercalate to form a singled-layered neural tube (Davidson and Keller, 1999). Although both the superficial and deep ectoderm layers are competent to respond to neural induction, primary neurons originate from the deep layer only (Chalmers et al., 2002; Hartenstein, 1989). ESR-6e, a Notch target predominantly expressed in the superficial layer, has been shown to inhibit neurogenesis (Chalmers et al., 2002). This observation suggests that the lack of primary neurons in the superficial layer is due to active Notch signaling (Chalmers et al., 2002). Additionally, the constitutively active Notch receptor decreases, whereas a dominant negative form of Delta increases the number of primary neurons within the deep layer (Chitnis et al., 1995; Ma et al., 1996). More recent evidence shows that the basolateral polarity kinase Par1 (Ossipova et al., 2009) and the apical complex protein Par3 (Bultje et al., 2009) regulate neurogenesis by impinging on the Notch pathway. Par1 phosphorylates Mindbomb, a ubiquitin ligase that is required for

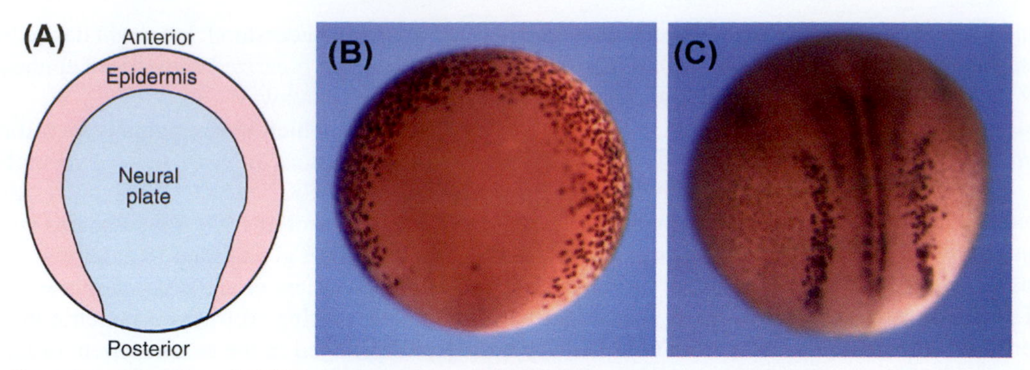

FIGURE 11.5 **Examples of Ectodermal Cell Types that are Specified by Notch Signaling** Whole mount *in situ* hybridization of early neurula embryos with *a-tubulin* (B) and *N-tubulin* (C) probes. (A) A schematic drawing of the equivalent stage. (B) Scattered *a-tubulin* staining marks multiciliated epidermal cells. (C) The three stripes of *N-tubulin*-positive cells in the neural plate correspond to primary neurons.

Notch pathway stimulation, whereas Par3 interacts with Numb, a Notch antagonist (Bultje et al., 2009; Ossipova et al., 2009). Both molecular interactions were proposed to be critical for the outcome of the asymmetric divisions of neural progenitors. Thus, the regulation of Notch signaling by apical-basal polarity determinants may be a major mechanism underlying neurogenesis in vertebrate ectoderm.

Specification of Ciliated Cells and Ionocytes in Epidermal Ectoderm

Xenopus skin consists of at least four cell types, including actively secreting goblet cells, multiciliated cells, and two different proton-secreting cells or ionocytes (Deblandre et al., 1999; Dubaissi and Papalopulu, 2011; Quigley et al., 2011). Multiciliated cells represent a specialized cell type that contains a large amount of α-tubulin and is distributed in an evenly spaced pattern within the superficial layer of non-neural ectoderm (Figure 11.4B, 11.5). Both multiciliated cells and ionocytes originate from the deep layer of late gastrula non-neural ectoderm and they move into the superficial layer at neurula stages (Deblandre et al., 1999; Dubaissi and Papalopulu, 2011; Quigley et al., 2011). The specification of multiciliated cells in the deep layer is controlled by Notch signaling, similarly to the specification of *Drosophila* neuroblasts and *Xenopus* primary neurons (Chalmers et al., 2002). Activated Notch pathway decreases, whereas Notch inhibitors increase the number of multiciliated cells. Overexpression of ESR-6e, a Notch target, also inhibits multiciliated cell differentiation, whereas a dominant negative form of ESR-6e has the opposite effect. Since ESR-6e is mainly expressed in the superficial layer, specification of multiciliated epidermal cells may be negatively controlled by Notch signaling (Deblandre et al., 1999). The same conclusions were reached after the analysis of Grainyhead-like 3 (Grhl-3), superficial layer-specific transcription factor that is regulated by Notch signaling and which itself is necessary for the control of keratin genes expressed during superficial layer differentiation (Chalmers et al., 2006; Quigley et al., 2011; Tao et al., 2005). Therefore, specification of multiciliated epidermal cells in the deep layer is closely associated with separation of superficial and deep layers of ectoderm as well as Notch signaling (Figures 11.4, 11.5).

Although the specification of ionocytes has not been studied in the same detail, it appears likely that the same mechanisms might operate during the differentiation of this cell type, for which Foxi transcription factors are also critical (Janicke et al., 2007; Quigley et al., 2011). Importantly, the different levels of Notch pathway activity, which regulate selective expression of *Esr-6e* and *Grhl-3* in the superficial cell layer, appear to be dictated by the antagonistic interactions of the apical-basal polarity proteins aPKC and Par1 (Ossipova et al., 2007). In summary, similarly to neuronal differentiation, initial apical-basal polarity in the epidermis is translated by the action of Par polarity proteins to differential activity of the Notch pathway and activation of specific target genes, which trigger the differentiation of multiciliated cells and ionocytes (Figure 11.4, 11.5).

Cell Fate Determination by Separate Patterning Processes

The available evidence indicates that vertebrate epidermal cell types, including ciliated cells and ionocytes, are specified by cell polarity determinants and by Notch signaling. Similarly, apicobasal polarization of the neuroectoderm appears to play a key role in neural cell fates, with basolateral determinants promoting neurogenesis and apical determinants suppressing it (Figure 11.4). An independent patterning system that involves BMP, Wnt and FGF signaling regulates the specification of epidermal versus neural ectoderm. We conclude that vertebrate ectoderm develops under the control of separate, conserved patterning systems, one involving BMP, Wnt and FGF signals, and the other based on unequal segregation of cytoplasmic determinants and Notch signaling between neighboring cells.

11.5 CLINICAL RELEVANCE

- The basic knowledge of factors involved in ectoderm development is essential for developing ways to generate specific populations of cells for regenerative medicine.
- Human neurodegenerative diseases are frequently linked to functional defects in specific cell populations.
- Basic mechanisms involved in cell polarization and asymmetric cell division may be employed to control the balance over proliferation and differentiation of progenitor/stem cell populations, and therefore, will be important for the design of future stem cell-based therapies.

ACKNOWLEDGMENTS

We regret that numerous original papers could not be cited due to space limitations. We thank Olga Ossipova and Maria Krivega for the images used in Figure 11.5. We also thank Jerome Ezan and Sylvie Janssens for comments on the manuscript. Work in the Sokol laboratory is supported by the National Institutes of Health.

RECOMMENDED RESOURCES

Hamburger, V., 1988. The Heritage of Experimental Embryology: Hans Spemann and the Organizer. Oxford Univ., Oxford, UK.

Keller, R., 1991. In: Kay, B.K., Peng, H.B. (Eds.), Early Embryonic Development of *Xenopus laevis*. Methods Cell Biol, 36. Acad. Press, pp. 61–113.

REFERENCES

Ang, S.L., Rossant, J., 1994. HNF-3 beta is essential for node and notochord formation in mouse development. Cell 78, 561–574.

Bachiller, D., Klingensmith, J., Kemp, C., Belo, J.A., Anderson, R.M., May, S.R., McMahon, J.A., McMahon, A.P., Harland, R.M., Rossant, J., et al., 2000. The organizer factors Chordin and Noggin are required for mouse forebrain development. Nature 403, 658–661.

Bainter, J.J., Boos, A., Kroll, K.L., 2001. Neural induction takes a transcriptional twist. Dev. Dyn. 222, 315–327.

Baker, J.C., Beddington, R.S., Harland, R.M., 1999. Wnt signaling in *Xenopus* embryos inhibits bmp4 expression and activates neural development. Genes. Dev. 13, 3149–3159.

Bang, A.G., Papalopulu, N., Goulding, M.D., Kintner, C., 1999. Expression of Pax-3 in the lateral neural plate is dependent on a Wnt-mediated signal from posterior nonaxial mesoderm. Dev. Biol. 212, 366–380.

Bardin, A.J., Le Borgne, R., Schweisguth, F., 2004. Asymmetric localization and function of cell-fate determinants: a fly's view. Curr. Opin. Neurobiol. 14, 6–14.

Bertrand, V., Hudson, C., Caillol, D., Popovici, C., Lemaire, P., 2003. Neural tissue in ascidian embryos is induced by FGF9/16/20, acting via a combination of maternal GATA and Ets transcription factors. Cell 115, 615–627.

Bultje, R.S., Castaneda-Castellanos, D.R., Jan, L.Y., Jan, Y.N., Kriegstein, A.R., Shi, S.H., 2009. Mammalian Par3 regulates progenitor cell asymmetric division via notch signaling in the developing neocortex. Neuron 63, 189–202.

Chalmers, A.D., Lachani, K., Shin, Y., Sherwood, V., Cho, K.W., Papalopulu, N., 2006. Grainyhead-like 3, a transcription factor identified in a microarray screen, promotes the specification of the superficial layer of the embryonic epidermis. Mech. Dev. 123, 702–718.

Chalmers, A.D., Pambos, M., Mason, J., Lang, S., Wylie, C., Papalopulu, N., 2005. aPKC, Crumbs3 and Lgl2 control apicobasal polarity in early vertebrate development. Development 132, 977–986.

Chalmers, A.D., Welchman, D., Papalopulu, N., 2002. Intrinsic differences between the superficial and deep layers of the *Xenopus* ectoderm control primary neuronal differentiation. Dev. Cell. 2, 171–182.

Chitnis, A., Henrique, D., Lewis, J., Ish-Horowicz, D., Kintner, C., 1995. Primary neurogenesis in *Xenopus* embryos regulated by a homologue of the Drosophila neurogenic gene Delta. Nature 375, 761–766.

Choi, S.C., Sokol, S.Y., 2009. The involvement of lethal giant larvae and Wnt signaling in bottle cell formation in *Xenopus* embryos. Dev. Biol. 336, 68–75.

Davidson, B.P., Kinder, S.J., Steiner, K., Schoenwolf, G.C., Tam, P.P., 1999. Impact of node ablation on the morphogenesis of the body axis and the lateral asymmetry of the mouse embryo during early organogenesis. Dev. Biol. 211, 11–26.

Davidson, L.A., Keller, R.E., 1999. Neural tube closure in *Xenopus laevis* involves medial migration, directed protrusive activity, cell intercalation and convergent extension. Development 126, 4547–4556.

De Robertis, E.M., Kuroda, H., 2004. Dorsal-ventral patterning and neural induction in *Xenopus* embryos. Annu. Rev. Cell. Dev. Biol. 20, 285–308.

Deblandre, G.A., Wettstein, D.A., Koyano-Nakagawa, N., Kintner, C., 1999. A two-step mechanism generates the spacing pattern of the ciliated cells in the skin of *Xenopus* embryos. Development 126, 4715–4728.

Delaune, E., Lemaire, P., Kodjabachian, L., 2005. Neural induction in *Xenopus* requires early FGF signaling in addition to BMP inhibition. Development 132, 299–310.

Dollar, G.L., Weber, U., Mlodzik, M., Sokol, S.Y., 2005. Regulation of Lethal giant larvae by Disheveled. Nature 437, 1376–1380.

Domingos, P.M., Itasaki, N., Jones, C.M., Mercurio, S., Sargent, M.G., Smith, J.C., Krumlauf, R., 2001. The Wnt/beta-catenin pathway posteriorizes neural tissue in *Xenopus* by an indirect mechanism requiring FGF signaling. Dev. Biol. 239, 148–160.

Dominguez, I., Itoh, K., Sokol, S.Y., 1995. Role of glycogen synthase kinase 3 beta as a negative regulator of dorsoventral axis formation in *Xenopus* embryos. Proc. Natl. Acad. Sci. USA 92, 8498–8502.

Dubaissi, E., Papalopulu, N., 2011. Embryonic frog epidermis: a model for the study of cell-cell interactions in the development of mucociliary disease. Dis. Model. Mech. 4, 179–192.

Dupont, S., Zacchigna, L., Cordenonsi, M., Soligo, S., Adorno, M., Rugge, M., Piccolo, S., 2005. Germ-layer specification and control of cell growth by Ectodermin, a Smad4 ubiquitin ligase. Cell 121, 87–99.

Eswarakumar, V.P., Lax, I., Schlessinger, J., 2005. Cellular signaling by fibroblast growth factor receptors. Cytokine Growth Factor Rev. 16, 139–149.

Fan, M.J., Gruning, W., Walz, G., Sokol, S.Y., 1998. Wnt signaling and transcriptional control of Siamois in *Xenopus* embryos. Proc. Natl. Acad. Sci. USA 95, 5626–5631.

Fuentealba, L.C., Eivers, E., Ikeda, A., Hurtado, C., Kuroda, H., Pera, E.M., De Robertis, E.M., 2007. Integrating patterning signals: Wnt/GSK3 regulates the duration of the BMP/Smad1 signal. Cell 131, 980–993.

Glinka, A., Wu, W., Delius, H., Monaghan, A.P., Blumenstock, C., Niehrs, C., 1998. Dickkopf-1 is a member of a new family of secreted proteins and functions in head induction. Nature 391, 357–362.

Harland, R., 2000. Neural induction. Curr. Opin. Genet. Dev. 10, 357–362.

Harland, R., Gerhart, J., 1997. Formation and function of Spemann's organizer. Annu. Rev. Cell. Dev. Biol. 13, 611–667.

Hartenstein, V., 1989. Early neurogenesis in *Xenopus*: the spatio-temporal pattern of proliferation and cell lineages in the embryonic spinal cord. Neuron 3, 399–411.

Heasman, J., Kofron, M., Wylie, C., 2000. Beta-catenin signaling activity dissected in the early *Xenopus* embryo: a novel antisense approach. Dev. Biol. 222, 124–134.

Heeg-Truesdell, E., LaBonne, C., 2006. Neural induction in *Xenopus* requires inhibition of Wnt-beta-catenin signaling. Dev. Biol. 298, 71–86.

Hikasa, H., Ezan, J., Itoh, K., Li, X., Klymkowsky, M.W., Sokol, S.Y., 2010. Regulation of TCF3 by Wnt-dependent phosphorylation during vertebrate axis specification. Dev. Cell. 19, 521–532.

Hikasa, H., Sokol, S.Y., 2004. The involvement of Frodo in TCF-dependent signaling and neural tissue development. Development 131, 4725–4734.

Hikasa, H., Sokol, S.Y., 2013. Wnt signaling in vertebrate axis specification. Cold Spring Harb. Perspect. Biol. 5. a007955.

Holowacz, T., Sokol, S., 1999. FGF is required for posterior neural patterning but not for neural induction. Dev. Biol. 205, 296–308.

Itoh, K., Sokol, S.Y., 1997. Graded amounts of *Xenopus* dishevelled specify discrete anteroposterior cell fates in prospective ectoderm. Mech. Dev. 61, 113–125.

Itoh, K., Sokol, S.Y., 1999. Axis determination by inhibition of Wnt signaling in *Xenopus*. Genes. Dev. 13, 2328–2336.

Itoh, K., Tang, T.L., Neel, B.G., Sokol, S.Y., 1995. Specific modulation of ectodermal cell fates in *Xenopus* embryos by glycogen synthase kinase. Development 121, 3979–3988.

Janicke, M., Carney, T.J., Hammerschmidt, M., 2007. Foxi3 transcription factors and Notch signaling control the formation of skin ionocytes from epidermal precursors of the zebrafish embryo. Dev. Biol. 307, 258–271.

Khokha, M.K., Yeh, J., Grammer, T.C., Harland, R.M., 2005. Depletion of three BMP antagonists from Spemann's organizer leads to a catastrophic loss of dorsal structures. Dev. Cell. 8, 401–411.

Kiecker, C., Niehrs, C., 2001. A morphogen gradient of Wnt/beta-catenin signaling regulates anteroposterior neural patterning in *Xenopus*. Development 128, 4189–4201.

Kim, C.H., Oda, T., Itoh, M., Jiang, D., Artinger, K.B., Chandrasekharappa, S.C., Driever, W., Chitnis, A.B., 2000. Repressor activity of Headless/Tcf3 is essential for vertebrate head formation. Nature 407, 913–916.

King, M.L., Messitt, T.J., Mowry, K.L., 2005. Putting RNAs in the right place at the right time: RNA localization in the frog oocyte. Biol. Cell. 97, 19–33.

Klezovitch, O., Fernandez, T.E., Tapscott, S.J., Vasioukhin, V., 2004. Loss of cell polarity causes severe brain dysplasia in Lgl1 knockout mice. Genes. Dev. 18, 559–571.

Ma, Q., Kintner, C., Anderson, D.J., 1996. Identification of neurogenin, a vertebrate neuronal determination gene. Cell 87, 43–52.

Massague, J., Wotton, D., 2000. Transcriptional control by the TGF-beta/Smad signaling system. EMBO. J. 19, 1745–1754.

McGrew, L.L., Hoppler, S., Moon, R.T., 1997. Wnt and FGF pathways cooperatively pattern anteroposterior neural ectoderm in *Xenopus*. Mech. Dev. 69, 105–114.

Mir, A., Kofron, M., Zorn, A.M., Bajzer, M., Haque, M., Heasman, J., Wylie, C.C., 2007. FoxI1e activates ectoderm formation and controls cell position in the *Xenopus* blastula. Development 134, 779–788.

Mizuseki, K., Kishi, M., Matsui, M., Nakanishi, S., Sasai, Y., 1998a. *Xenopus* Zic-related-1 and Sox-2, two factors induced by chordin, have distinct activities in the initiation of neural induction. Development 125, 579–587.

Mizuseki, K., Kishi, M., Shiota, K., Nakanishi, S., Sasai, Y., 1998b. SoxD: an essential mediator of induction of anterior neural tissues in *Xenopus* embryos. Neuron 21, 77–85.

Ossipova, O., Dhawan, S., Sokol, S., Green, J.B., 2005. Distinct PAR-1 proteins function in different branches of Wnt signaling during vertebrate development. Dev. Cell. 8, 829–841.

Ossipova, O., Ezan, J., Sokol, S.Y., 2009. PAR-1 phosphorylates Mind bomb to promote vertebrate neurogenesis. Dev. Cell. 17, 222–233.

Ossipova, O., Tabler, J., Green, J.B., Sokol, S.Y., 2007. PAR1 specifies ciliated cells in vertebrate ectoderm downstream of aPKC. Development 134, 4297–4306.

Otte, A.P., Kramer, I.M., Durston, A.J., 1991. Protein kinase C and regulation of the local competence of *Xenopus* ectoderm. Science 251, 570–573.

Pera, E.M., Ikeda, A., Eivers, E., De Robertis, E.M., 2003. Integration of IGF, FGF, and anti-BMP signals via Smad1 phosphorylation in neural induction. Genes Dev. 17, 3023–3028.

Piccolo, S., Agius, E., Leyns, L., Bhattacharyya, S., Grunz, H., Bouwmeester, T., De Robertis, E.M., 1999. The head inducer Cerberus is a multifunctional antagonist of Nodal, BMP and Wnt signals. Nature 397, 707–710.

Psychoyos, D., Stern, C.D., 1996. Restoration of the organizer after radical ablation of Hensen's node and the anterior primitive streak in the chick embryo. Development 122, 3263–3273.

Quigley, I.K., Stubbs, J.L., Kintner, C., 2011. Specification of ion transport cells in the *Xenopus* larval skin. Development 138, 705–714.

Reversade, B., De Robertis, E.M., 2005. Regulation of ADMP and BMP2/4/7 at opposite embryonic poles generates a self-regulating morphogenetic field. Cell 123, 1147–1160.

Ribisi Jr., S., Mariani, F.V., Aamar, E., Lamb, T.M., Frank, D., Harland, R.M., 2000. Ras-mediated FGF signaling is required for the formation of posterior but not anterior neural tissue in *Xenopus laevis*. Dev. Biol. 227, 183–196.

Rogers, C.D., Moody, S.A., Casey, E.S., 2009. Neural induction and factors that stabilize a neural fate. Birth Defects Res. C Embryo Today 87, 249–262.

Sanada, K., Tsai, L.H., 2005. G protein betagamma subunits and AGS3 control spindle orientation and asymmetric cell fate of cerebral cortical progenitors. Cell 122, 119–131.

Schohl, A., Fagotto, F., 2002. Beta-catenin, MAPK and Smad signaling during early *Xenopus* development. Development 129, 37–52.

Sharpe, C.R., Fritz, A., De Robertis, E.M., Gurdon, J.B., 1987. A homeobox-containing marker of posterior neural differentiation shows the importance of predetermination in neural induction. Cell 50, 749–758.

Shih, J., Fraser, S.E., 1996. Characterizing the zebrafish organizer: microsurgical analysis at the early-shield stage. Development 122, 1313–1322.

Sokol, S., Melton, D.A., 1991. Pre-existent pattern in *Xenopus* animal pole cells revealed by induction with activin. Nature 351, 409–411.

Sokol, S.Y., 1999. Wnt signaling and dorso-ventral axis specification in vertebrates. Curr. Opin. Genet. Dev. 9, 405–410.

Sokol, S.Y., Klingensmith, J., Perrimon, N., Itoh, K., 1995. Dorsalizing and neuralizing properties of Xdsh, a maternally expressed *Xenopus* homolog of dishevelled. Development 121, 1637–1647.

Sokol, S.Y., Melton, D.A., 1992. Interaction of Wnt and activin in dorsal mesoderm induction in *Xenopus*. Dev. Biol. 154, 348–355.

Stern, C.D., 2005. Neural induction: old problem, new findings, yet more questions. Development 132, 2007–2021.

Sullivan, S.A., Akers, L., Moody, S.A., 2001. foxD5a, a *Xenopus* winged helix gene, maintains an immature neural ectoderm via transcriptional repression that is dependent on the C-terminal domain. Dev. Biol. 232, 439–457.

Suri, C., Haremaki, T., Weinstein, D.C., 2005. Xema, a foxi-class gene expressed in the gastrula stage *Xenopus* ectoderm, is required for the suppression of mesendoderm. Development 132, 2733–2742.

Tao, J., Kuliyev, E., Wang, X., Li, X., Wilanowski, T., Jane, S.M., Mead, P.E., Cunningham, J.M., 2005. BMP4-dependent expression of *Xenopus* Grainyhead-like 1 is essential for epidermal differentiation. Development 132, 1021–1034.

Taylor, J.J., Wang, T., Kroll, K.L., 2006. Tcf- and Vent-binding sites regulate neural-specific geminin expression in the gastrula embryo. Dev. Biol. 289, 494–506.

Weinstein, D.C., Ruiz i Altaba, A., Chen, W.S., Hoodless, P., Prezioso, V.R., Jessell, T.M., Darnell Jr., J.E., 1994. The winged-helix transcription factor HNF-3 beta is required for notochord development in the mouse embryo. Cell 78, 575–588.

Wilson, P.A., Hemmati-Brivanlou, A., 1997. Vertebrate neural induction: inducers, inhibitors, and a new synthesis. Neuron 18, 699–710.

Zhang, J., Houston, D.W., King, M.L., Payne, C., Wylie, C., Heasman, J., 1998. The role of maternal VegT in establishing the primary germ layers in *Xenopus* embryos. Cell 94, 515–524.

Chapter 12

Taking the Middle Road: Vertebrate Mesoderm Formation and the Blastula-Gastrula Transition

Benjamin Feldman

Program on Genomics of Differentiation, Eunice Kennedy Shriver National Institute of Child Health and Human Development, Bethesda, MD

Chapter Outline

GLOSSARY

Blastoporal-ablastoporal axis An idealized pan-vertebrate axis that runs orthogonally to the zone of internalizing mesendodermal cells. The zone of internalizing mesendodermal cells defines the blastoporal end of this axis.

Commitment The degree to which a cell or tissue has undergone a differentiation step that is not readily reversed. Commitment is always discussed in the context of a particular differentiation step.

Fate domain An embryonic region whose cellular constituents undertake reproducible pathways of differentiation, *under normal circumstances*. These cells may or may not be committed and changes in their environment can reveal this.

GRN = Gene Regulatory Network A collection of genes and possibly other factors, each of which can regulate and/or be regulated by at least one other member. GRNs are often represented in circuit diagram-like formats.

Induction The phenomenon of an extrinsic signal triggering a cell or tissue to differentiate.

Pluripotent, multipotent Refers to cells or tissues that retain the potential to differentiate into more than one cell or tissue type.

Regulatory state The status of a cell's nucleus as defined by the sum total of trans-acting factors engaging it and the epigenetic marks it carries.

Principles of Developmental Genetics.
2015 Published by Elsevier Inc.

SMO proximodistal axis An idealized pan-vertebrate axis that runs parallel to the zone of internalizing mesendodermal cells. The Spemann-Mangold Organizer (SMO) defines the proximal end of this axis.

Triploblasts A classification encompassing animals that form all three embryonic germ layers (mesoderm, ectoderm, and endoderm) during embryogenesis. Most, if not all, triploblasts display bilateral symmetry during embryogenesis. As such, *Bilaterians* refers to the same subset of animals as *Triploblasts*.

If you come to a fork in the road, take it.

– Yogi Berra

I like a film to have a beginning, a middle and an end, but not necessarily in that order

– Jean-Luc Godard

SUMMARY

- Mesoderm is the middle of the three embryonic germ layers discovered and defined in the 19th century.
- Major phylogenetic classifications are based on the theory and evidence that the mesoderm is the most recently evolved germ layer.
- A reconsideration of conventional axes reveals a common vertebrate fate specification map that is coordinated by gradients of ligand signaling.
- Mesoderm differentiation mechanisms validate theoretical models of embryonic patterning from the pre-molecular era.
- Formation of vertebrate mesoderm for the entire rostrocaudal axis requires independent Nodal and T/Brachyury inputs.
- A caudal stem cell population gives rise to caudal neural tube neurons and to a mesodermal progenitor cell niche that gives rise to caudal paraxial mesoderm.

12.1 THE DISCOVERY OF MESODERM AND GERM LAYERS

Mesoderm as we understand it today refers to the middle of three embryonic layers that are visible in many animal embryos during the gastrula stage of development (Figure 12.1). Christian Pander described these three layers in chick embryos in 1817 (Figure 12.1), naming them "germ layers" (German: Keimblätter) based on his observation of their seed-like ability to give rise to the embryonic body through a remarkable combination of growth and morphogenesis (Pander, 1817). Over the next several decades, embryology emerged as a scientific discipline and researchers, notably Karl Ernst von Baer, went on to report the presence of germ layers in a number of other vertebrate species (von Baer, 1828). Over the next decades, germ layers were also reported in invertebrates and it was determined that the germ layers are comprised of cells, as opposed to being simple skins or membranes (Kovalevsky, 1868; Rathke, 1829; Remak, 1855). Thus by the middle of the 19th century it was firmly established that a number of animal bodies arise through the expansion and differentiation of cells arranged in deep, middle, and outer germ layers. It was further recognized that the outermost layer forms outer structures, e.g., skin, shell, or cuticle; the innermost layer forms deep structures, e.g., gut and lungs; and the middle layer forms the rest, e.g., muscle and bone.

Pander's original terms for the germ layers (Figure 12.1, legend) were ultimately replaced by *endoderm*, *ectoderm*, and *mesoderm*. These derive from George. J. Allman's nomenclature for the inner and outer layers of developing jellyfish (Allman, 1853). Allman combined the Greek words for inner (*endon*) and outer (*ektos*) with the Greek word for skin (*derma*). Ectoderm and endoderm became popular terms for a variety of adult and embryonic layered tissues in plants and animals alike. Mesoderm (Greek for middle: *mesos*) soon joined the lexicon, initially as a botanical term. Given the overwhelming evidence that the origin of plant multicellularity was separate from the origin of animal multicellularity, plant mesoderm will not be further discussed (Knoll, 2011).

12.2 GERM LAYER PHYLOGENY

In the 1870s, Ernst Haeckel and Ray Lankester independently formulated critical syntheses of embryology, taxonomy, and evolution (Haeckel, 1874; Lankester, 1873; 1877). Their ideas built upon the concept that all animals share a common ancestor, as had been recently and persuasively proposed by Charles Darwin, but also by previous natural scientists before him such as his grandfather Erasmus and Jean-Baptiste Lamarck (Darwin Charles, 1859; Darwin Erasmus, 1794; Lamarck,

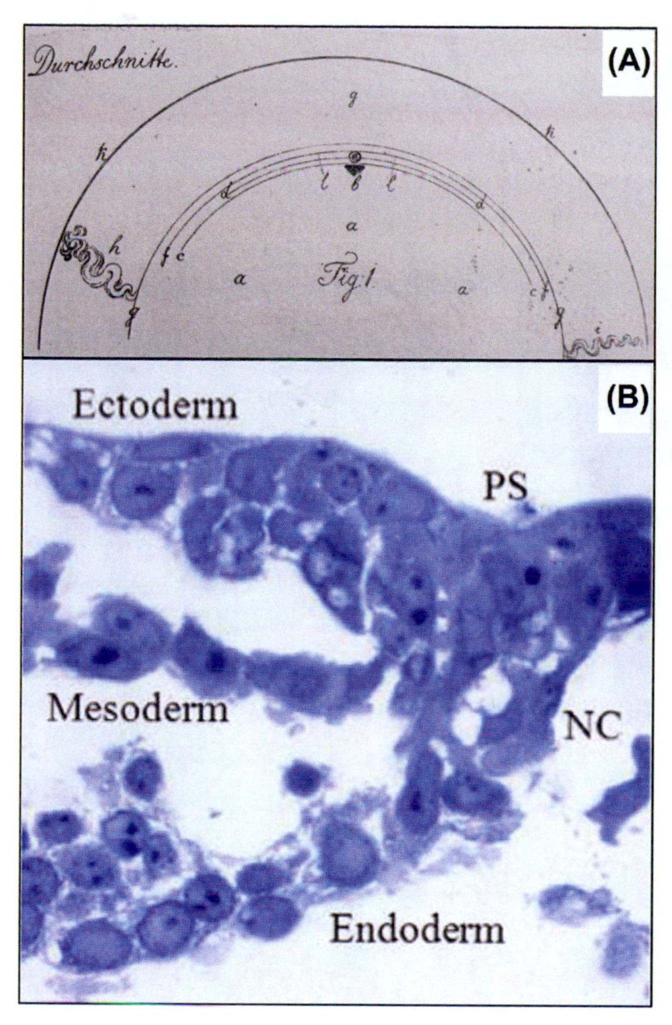

FIGURE 12.1 The Primary Germ Layers. A. Christian Pander's schematic sketch of a transverse section through a chick egg and embryo. The stated age "at the end of the second day" is ambiguous, but accompanying drawings place it within the intermediate to definitive primitive streak stages (Hamburger Hamilton Stage 3 to 4). ***a,a,a,*** yolk; ***b,*** center of germinal disc; ***c,c,*** endoderm (original German: *Schleimschichte* = mucous layer); ***d,d,*** mesoderm (original German: *Gefässhaut* = vascular skin); ***e,*** embryo (actually the *primitive streak* according to current nomenclature; the embryo spans the entire germinal disc, *i.e.,* up to the lateral limits of ***f,f***); ***f, f*** ectoderm (original German: *serose Schichte* = serous leaf); ***g,g,*** vitelline membrane; ***h,*** large chalaza; ***i,*** small chalaza; ***k,*** eggshell; ***l,l,*** not explained in the original, but likely indicate lateral boundaries of subsequent sketches. Photograph of original at the National Library of Medicine, Bethesda, MD (Tafel der Durchschnitte [cross-sections], Figure 1; Pander, 1817). **B.** Histological section through a human trilaminar blastodisc of the primitive streak stage (equivalent of day 19 postfertilization). Specimen was obtained from an ectopic oviductal pregnancy. Mesoderm and endoderm are indicated, having previously emerged from the primitive streak (**PS**). Overlying ectoderm and the mesoderm-derived notochord (**NC**) are also indicated. Reproduced with permission (Figure 9; Sathananthan et al., 2011).

1809). Haeckel and Lankester laid out detailed proposals for animal phylogenetic trees, with critical branches representing the acquisition of multicellularity and the acquisition of mesoderm. Their texts also stand among the earliest to explicitly use ectoderm, endoderm, and mesoderm to describe the primary animal germ layers. Except for the addition of an additional phylogenetic branch separating mesoderm-bearing animals that form their embryonic alimentary canals on the anterior side (Protostomes) from those that form it on the posterior side (Deuterostomes), Lankester's proposed phylogenetic tree has survived with remarkably few changes (Figure 12.2; Grobben, 1908; Huxley, 1875).

Sponges now represent a branch lacking any germ layers. Animals with two germ layers (ectoderm and endoderm) – Lankester's *diploblasts* – comprise Cnideria and Ctenophora: jellyfish and hydra being two examples. The remaining metazoan species display three germ layers (Lankester's *triploblasts*, also known as *bilaterians*) and are further sub-categorized into Protostomes and Deuterostomes. Thus, the proposal that mesoderm is the latest-evolved germ layer has withstood more than 135 years of scrutiny.

12.3 MESODERM'S FOSSIL RECORD

What do we know about the first mesoderm-forming animals? A huge diversity of bilaterian morphologies is seen in fossils from throughout the Cambrian explosion, spanning from ~540 to 485 million years ago (Mya). Clear bilaterian fossils have been found from at least 10 million years before the Cambrian, during the preceding Edicarian biota, and concurrent "trace fossils" have also been found that are believed to have arisen from the movements of animals with bilaterian body plans (Erwin et al., 1997; Droser et al., 2002; Martin M. W. et al., 2000). Fossils of earlier multicellular animals dating possibly as far back as 760 Mya have features consistent with diploblasts and sponges, both of which lack mesoderm (Brain et al., 2012; Droser and Gehling, 2008). Fossils of unicellular eukaryotes have been found dating as early as 1200 Mya (Butterfield, 2000). Thus, mesoderm emerged no later than 550 Mya and the overall fossil record supports Haeckel and

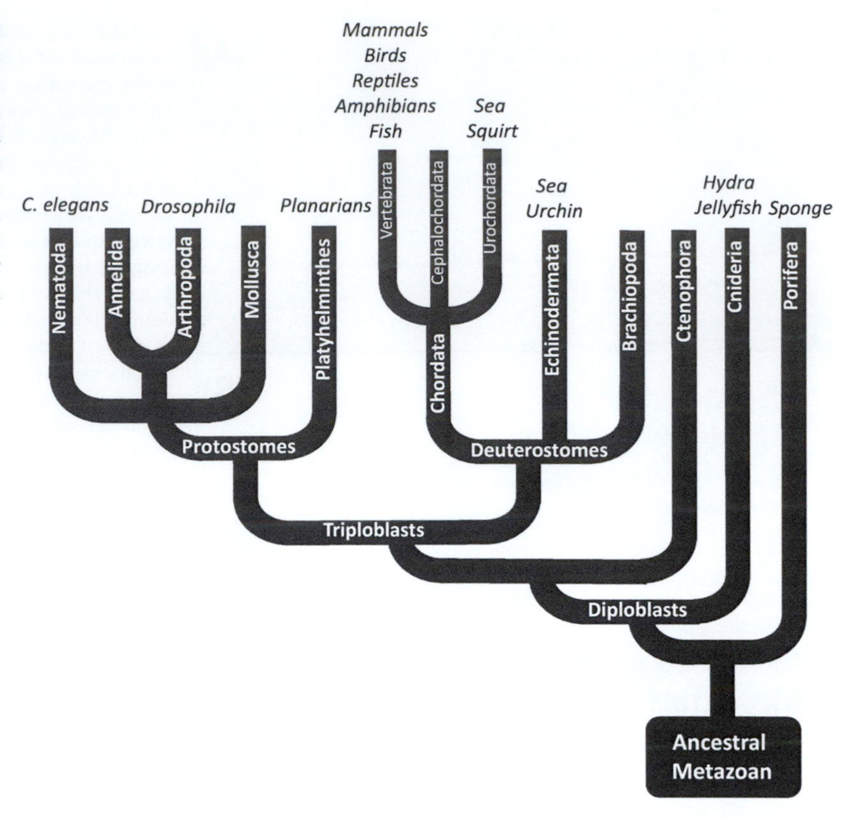

FIGURE 12.2 The Embryological Basis of Animal Phylogeny. All non-protist animals are presumed to have evolved from a single ancestral metazoan species. Eleven of the thirty-six animal phyla, including the nine largest, and the three subphyla of the Chordata phylum are indicated (vertical text in white). Branch points grouping diploblasts, triploblasts, protostomes, and deuterostomes are also indicated (horizontal text in white). The five vertebrate classes, as well as selected model invertebrates and others discussed in this chapter are listed as the top of their respective phyla or subphyla (horizontal text in black).

Lankester's proposed sequence of animal evolution. A very strong case has been made that the ~2500 million years it took for life to evolve from the first prokaryote to the first triploblast was not idle time; each functional component of a basic *metazoan toolkit* had to be built from scratch (Gerhart and Kirschner, 2007; Kirschner and Gerhart, 2005). But once that toolkit was assembled, animal diversification exploded.

12.4 EMBRYONIC ORGANIZERS AND INDUCTION

After more than fifty years of breakthroughs in descriptive embryology, the latter half of the 19th century saw the advent of experimental embryology, in which various interventions were used to label, dissect, and challenge embryos. Early experimental embryologists determined that certain isolated embryonic fragments could compensate for missing parts and go on to form nearly complete embryos, whereas other fragments could only form partial embryos. This suggested that some parts of embryos contained higher-level patterning functions than others.

In 1908, Ethel Browne published a series of experiments that identified and characterized the parts of hydra polyps that harbor a higher-level patterning function (Browne, 1908). When she grafted fragments of peristome tissue from the tentacle base of one polyp to a different location in a host polyp, a procedure generically known as heterotypic transplantation, a secondary polyp would form close to the site of transplantation. By performing heterotypic transplantations of peristomes between differently colored polyps, she determined that most of the secondary polyp was formed from host tissue, demonstrating that the peristome could somehow instruct host tissue to form a new polyp. This phenomena of one tissue recruiting another to a different fate is called *induction*.

The year prior to Browne's publication, Warren Lewis published what he interpreted to be the remarkable growth of a specific fragment from a frog embryo – the dorsal blastopore lip – when it was transplanted to an ectopic site (Lewis, 1907). We now know that Lewis was actually observing vertebrate induction, but this did not become clear until 1924 when the *organizer* experiments of Hans Spemann and Hilde Mangold were performed in newts (Spemann, 1935; Spemann and Mangold, 1924). They found that the dorsal blastopore lip – which contains all three germ layers – could cause the growth of a secondary conjoined embryo when heterotypically transplanted to a variety of sites. Transplanting between differently pigmented newts, as Browne had done with hydra, they further showed that most of the secondary embryo is derived from host cells. In recognition of the dorsal blastopore lips' remarkable induction abilities, they named it the *Organizer* – although it is now more commonly called the *Spemann-Mangold Organizer* (SMO; Figure 12.3F).

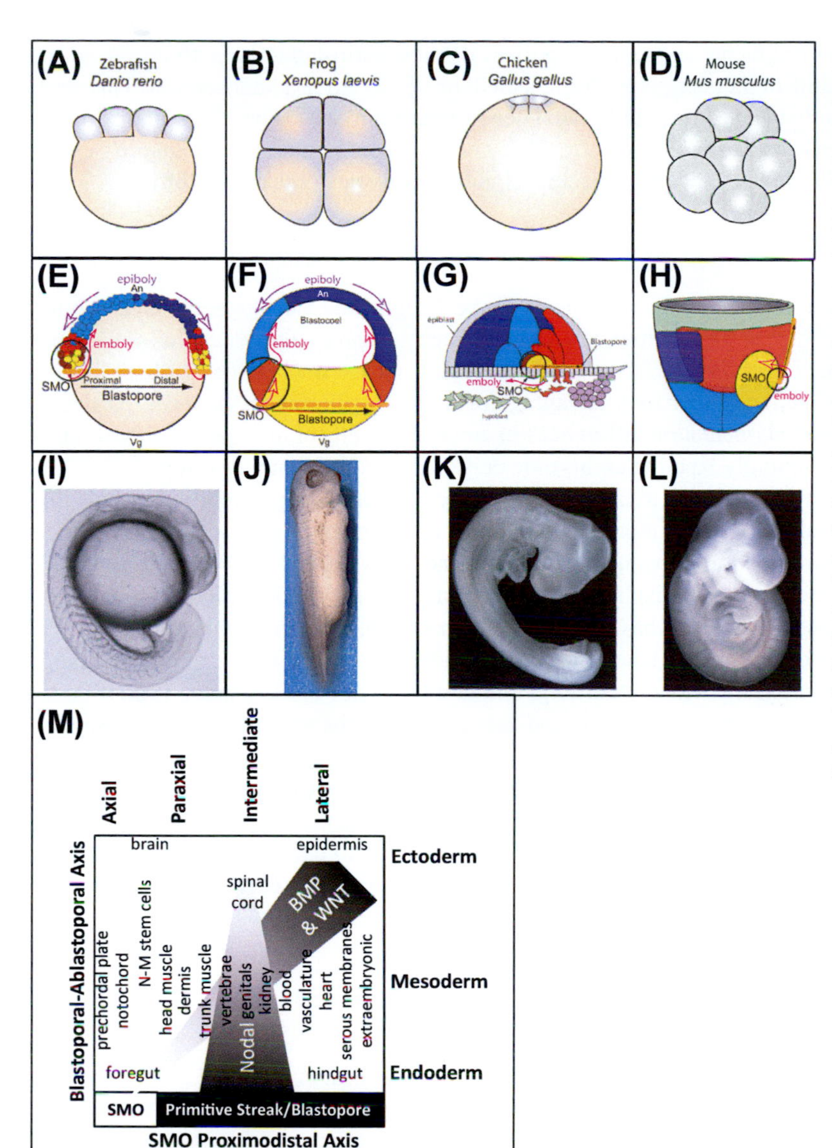

FIGURE 12.3 Early Development of Four Model Vertebrate Organisms. (A–D), illustrations of embryos at the 8-cell stage. Note the incomplete cleavage of zebrafish (A) and chick (C) and the complete cleavage with differently sized blastomeres in *X. laevis* (B), and uniformly sized blastomeres in mouse (D). Light gray indicates cytoplasm and beige indicates yolk. (E–H), illustrations of late blastula-early gastrula stage embryos. **An**, animal pole; **Vg**, vegetal pole; **SMO**, Spemann-Mangold Organizer or its equivalent. Red arrows indicate the pathway of emboly (internalization) that mesendoderm precursor cells take during gastrulation. Purple arrows in (E) and (F) indicate the pathway of epiboly, an independent gastrulation movement that animal pole cells take in zebrafish and *X. laevis*. Cellular fates are color coded as follows. **Red**, mesoderm and its precursors; **dark red**, prechordal mesoderm; **yellow**, definitive endoderm and its precursors; **dark blue**, epidermis, **lighter blue**, neural ectoderm; **green** and **violet**, extra-embryonic tissues. (I–L), photographs of phylotypic-stage embryos. Images and parts of text legend reprinted with permission (Figure 1; Solnica-Krezel, 2005). (M), schematic of a pan-vertebrate scheme for fate specification along SMO proximodistal and blastoporal-ablastoporal axes. An expanded list of mesodermal fates and literature sources is given in Table 12.1. While the Nodal and BMP/Wnt gradients have been clearly shown to be central to *X. laevis* and zebrafish fate specification, their role of in chick and mouse remain more speculative.

Embryonic SMO activity was subsequently located in other vertebrates: in *Hensen's node* of the chick (Figure 12.3G) and other avian embryos by Conrad Waddington, in the *shield* of killfish embryos by Jane Oppenheimer, and more recently in the zebrafish shield (Figure 12.3E), and in the *node* of mouse embryos (Figure 12.3H) by Rosa Beddington (Beddington, 1994; Oppenheimer, 1959; Shih J. and Fraser, 1996; Waddington, 1932; 1933). These studies indicate that the SMO is a fundamental component of vertebrate embryonic patterning. However the SMO contains all three germ layers and therefore cannot be responsible for the initiation of mesoderm formation.

Recombining a different set of newt embryonic fragments, it was Pieter Nieuwkoop who localized the inducers of mesoderm formation (Nieuwkoop, 1969). Nieuwkoop observed that mesoderm never forms from isolated animal pole fragments, but when fragments from the animal and vegetal poles are juxtaposed and grown together, some animal pole cells are induced to form mesoderm, seen by the formation of muscle. Therefore vegetal pole cells harbor mesoderm-inducing signals and animal pole cells are competent to respond to those signals. An analogous function has been found for the yolk syncytial layer (YSL), a teleost-specific extra-embryonic layer comprised of nuclei in a shared cytoplasm along the epiblast's vegetal face; the zebrafish YSL is necessary and sufficient for the induction of mesoderm and endoderm (Chen S. and Kimelman, 2000; Mizuno et al., 1996). Nieuwkoop went on to determine that dorsal vegetal pole newt embryo fragments induce dorsal mesoderm, including the SMO, irrespective of the original position of the ectoderm fragment, and that ventral vegetal pole fragments induced ventral fates in any ectoderm fragment (Boterenbrood and Nieuwkoop, 1973). The region of dorsal endoderm that induces the SMO has been subsequently termed the *Nieuwkoop Center*.

The strategy of using presumptive ectoderm fragments as passive reporters of mesoderm-inducing activity paved the way for the discovery of specific molecules capable of inducing mesoderm, as will be discussed in a later section. This strategy also led to the observation by John Gurdon that exposure of individual cells to the same mesoderm-inducing stimulus as groups of cells leads to mesoderm formation with far less efficiency. This phenomenon, named the *community effect,* indicates that signaling between responding cells can be a factor in successful inductive events (Gurdon, 1988).

12.5 CELL LINEAGE TRACING

The observations of the early 19th century had established key differences in the fates of the various germ layers, but a definitive accounting of each fate was lacking. Similar ambiguities remained concerning precisely which cells of earlier stage embryos would go on to contribute to which germ layer. Clarity came from cell lineage tracing. The earliest lineage tracing experiments were not interventional *per se*, but rather took advantage of tractable differences between particular living cells. For instance, Edwin Conklin exploited pigmentation differences in embryonic cells of one species of sea squirt (Conklin, 1905). Various artificial means of differentially labeling embryonic cells were subsequently developed. In 1929, Walther Vogt reported a method of labeling and tracing of individual frog embryo cells by applying small dye crystals that stained their membranes (Vogt, 1929). The contemporary and improved version of this approach involves direct injection of traceable, membrane permeable chemical markers into individual cells using fine glass needles (Kimmel et al., 1990; Moody, 1987a,b). Alternatively, groups of donor cells were transplanted into the same embryonic domain (i.e., homotypic transplants), but in hosts whose cells could be distinguished, as has been done, for instance, with chick-quail chimeras (Le Douarin, 1980). More recently, genetic strategies of cell lineage labeling have been added to the research toolbox (Bonnerot and Nicolas, 1993). Finally, improvements in microscopy, video recording, and computer power have made it possible to trace unlabeled cell lineages in transparent embryos such as *C. elegans* and zebrafish (Concha and Adams, 1998; Sulston et al., 1983).

More than a century of such studies made it clear that in vertebrates mesoderm forms not only muscles, heart, and bone, but also the blood, blood and lymph vessels, kidney, genitalia, mesenchyme, and serous membranes: the outer layers of organs that are otherwise endoderm derived (Table 12.1; Figure 12.3M). Such studies also enabled the creation of fate maps in which prospective tissue fates are mapped onto embryos at earlier stages. By locating prospective mesoderm cells before the formation of the mesoderm layer, fate maps helped researchers characterize the movements and changes these cells undergo during mesoderm differentiation.

12.6 THE BLASTULA-GASTRULA TRANSITION IN VERTEBRATES

The onset of overt mesoderm differentiation is characterized by changes in cell shape or adhesion followed by *internalization* movements, also called *emboly*. This broad pattern of behavior is shared by differentiating endoderm, except that the endoderm cells form a deeper layer beneath the mesoderm. Internalization is a hallmark movement of gastrulation and its onset marks the transition from the *blastula* to *gastrula* stages of embryogenesis. The full set of gastrulation movements, namely internalization, epiboly, and convergence-extension are described elsewhere (Solnica-Krezel and Sepich, 2012).

The cellular level details of internalization varies among the model vertebrates (Solnica-Krezel and Sepich, 2012). A common theme, however, is that mesoderm and endoderm precursor cells become intermingled during internalization, analogous to merging lanes of automobile traffic, and segregate after leaving the zone of internalization. As a consequence, mesoderm, and endoderm are sometimes collectively termed *mesendoderm* or *endomesoderm.*

In birds and mammals, embryonic cells of the blastula-stage epiblast are arranged in an epithelium above a layer of extra-embryonic cells known as the hypoblast or visceral endoderm (Figure 12.3G–H). Mesendoderm cells undergo epithelial-to-mesenchymal transitions (EMT), delaminate, and migrate as individuals beneath the epiblast (Harrisson et al., 1991; Tam et al., 1993); Figure 12.3G–H). The remaining epiblast cells will become ectoderm. This mode of internalization by individual mesendoderm precursor cells is called *ingression.* Ingression in the chick and mouse occurs along the posterior portion of the midline, forming a transitory ridge called the *primitive streak.* After passing through the primitive streak, mesendoderm cells disperse and segregate into definitive germ layers.

In zebrafish, blastula cells are arranged in a hemisphere of tightly packed mesenchyme, atop a spherical yolk (Figure 12.3E). Mesendoderm precursors located along the hemisphere's equator, known as the *blastula margin*, internalize as a group, then segregate into layers as they spread towards the top of the hemisphere, called the *animal pole*, while sandwiched between non-internalized ectoderm and the yolk. This movement is highly concerted, but the cells nonetheless exhibit individual migratory behaviors, a form of internalization that has been called *synchronous ingression* (Kane and Adams, 2002; Solnica-Krezel and Sepich, 2012).

TABLE 12.1 Mammalian Embryonic Mesoderm Derivatives

Primary Identity	Subsequent Identity (1)	Subsequent Identity (2)	Subsequent Identity (3)	*Nodal* null	*T* null
Axial	>>	Chorda-mesoderm	prechordal plate	absent	+
			notochord	absent	absent
Neuromesodermal (N-M) Stem Cells	>>	>>	caudal neural tube	+	++
	Mesodermal Progenitor Niche	Caudal paraxial mesoderm	caudal somite derivatives (see rostral somite derivatives below)	+/–	absent
		>>	caudal vascular endothelium	?	?
Rostral Paraxial Mesoderm	Neuromere		head mesenchyme	absent	+
	Rostral somites	Dorsolateral	body wall & limb muscles	absent	+
		Dermatome	dermis & subcutaneous skin	absent	+
		Myotome	epaxial (back) muscles	absent	+
		Sclerotome	tendons & vertebrae	absent	+
Intermediate	>>	>>	urogenital system	absent	caudal absent?
Lateral Plate	>>	>>	blood, blood & lymph vessels; vascular endothelium, heart; limb mesoderm (except muscle); serous membranes of body cavities, lungs, heart & abdominal organs	absent	caudal absent?

Table of mesoderm fates, based on multiple sources (Gilbert, 2006; Martin B. L. and Kimelman, 2012; Sadler and Langman, 2006; Tzouanacou et al., 2009). Placement of fates on the schematic pan-vertebrate fate map in Figure 12.3M is based upon Solnica-Krezel's original analysis as well as zebrafish fate maps and an overview of chick gastrulation by Claudio Stern (Kimmel et al., 1990; Solnica-Krezel, 2005; Stern, 2004; Warga and Nusslein-Volhard, 1999). The suggestion in this table that all caudal paraxial mesoderm and caudal neural tube derive from N-M cells is speculative. In the last two columns observed and unresolved outcomes are considered for individuals devoid of Nodal or T signals. Question marks (?) and mixed outcomes (+/–) are discussed in the main text.

The *X. laevis* blastula-stage epiblast is comprised of a spherical epithelium surrounding a fluid-filled cavity (Figure 12.3F). A small number of dorsal mesendoderm cells, the *bottle cells*, undergo apical constriction, causing an inward pocketing that forms the *blastopore*. The bottle cells internalize and travel towards the animal pole along the underside of non-internalized cells. They also remain in a cohesive epithelium, with lagging cells following the bottle cells in conveyer-belt fashion. This is called *involution*. The blastopore enlarges as involution progresses, leading to the internalization of mesendoderm precursors from the entire vegetal pole and equatorial zone. At the leading front of the involuted epithelium, cells undergo EMT and migrate as individuals to form the distinct mesoderm and endoderm layers.

12.7 THE CRYPTIC HOMOLOGY OF VERTEBRATE FATE MAPS

The time when cellular position and fates can first be linked, known as fate map emergence, varies considerably between *X. laevis,* for which the fate map emerges at cleavage stages, and zebrafish, for which the fate map emerges closer to the onset of gastrulation (Kimmel et al., 1990; Moody, 1987a,b). These differences are reconciled, however, when cellular fate *commitment* is considered. Commitment is queried by juxtaposing donor cells from one embryonic region with host cells from another embryonic region and observing whether the donor cells remain true to the fate associated with their origin or, alternatively, take on the fate of their host environment. Such experiments reveal that in both *X. laevis* and zebrafish, commitment to specific germ layers is not complete until the blastula-gastrula transition (David and Rosa, 2001; Heasman et al., 1985; Turner et al., 1989). There were some reports of cells in *X. laevis* as well as in chick and mouse that resisted commitment, but they appeared to be a minority and so limited attention was given to them by much of the developmental biology community, as will be discussed in a later section (Davis and Kirschner, 2000; Lawson et al., 1991; Selleck and Stern, 1991).

The differences between the *X. laevis* and zebrafish fate map emergence times are a function not only of when cells become committed but also of when their positions become fixed, with differing anatomies and cleavage modes (Figure 12.3A–D) allowing more (mouse, zebrafish) or less (*X. laevis,* chick) early randomization of cellular position (Wetts and Fraser, 1989). Embryonic cells that are not yet committed to a germ layer are called pluripotent. The blastula is thus a rich source of pluripotent cells, explaining the use of blastula-stage blastocysts as a source of murine and human embryonic stem cells (ESCs).

Due to genuine differences in the anatomy and cellular movements of different vertebrate embryo species, compounded by more arbitrary differences in how embryologists have named these species' embryonic structures and coordinate systems, the comparison of different vertebrate fate maps has been challenging (Figure 12.3E–H). However, a substantial homology is seen when a universal coordinate system is applied, for instance as proposed by Lila Solnica-Krezel (Solnica-Krezel, 2005). Solnica-Krezel's points of reference are the location of the future SMO (or node or shield) and the region where the future zone of internalization (blastopore, blastula margin, or primitive streak) will form. Two coordinate axes are proposed: the *SMO proximodistal axis* and the *blastoporal-ablastoporal axis* (Figure 12.3M). The SMO proximodistal axis runs parallel to the zone of internalization and extends from the most SMO-proximal to the most SMO-distal point. In zebrafish and *X. laevis* the SMO proximodistal axis is the same as the dorsoventral axis. In the chick it corresponds to the portion of the anteroposterior axis that is posterior to the SMO/node. Finally, the SMO proximodistal axis corresponds to the similarly named *proximodistal* axis in the mouse, but with the reverse polarity, since the *distal tip*, where the mouse node forms, must now be considered as SMO proximal. Within the prospective mesendoderm, cells proximal to the SMO along this axis have more axial and anterior fates and more distal cells have progressively more lateral and posterior fates (but see later sections for a discussion of posterior trunk and tail fates). Within the prospective ectoderm, cells proximal to the SMO will become neural and cells more distal will become epidermal. The blastoporal-ablastoporal axis extends from cells closest to the future internalization domain to cells furthest away, and the respective order in tissue fate is endoderm, mesoderm, and ectoderm. In *X. laevis* and zebrafish, the blastoporal-ablastoporal axis corresponds to the animal-vegetal axis. In the chick it corresponds to the mediolateral axes at the flanks of the primitive streak and also to the portion of the anteroposterior axis that is anterior to the SMO/node. In the mouse it corresponds roughly to the anteroposterior axis, wherein the primitive streak is on the posterior side.

The above comparison treats circular mesendoderm-forming domains (blastopores and blastula margins) and linear mesendoderm-forming domains (primitive streaks) as equivalent. In support of this line of thinking, two recent papers from Guojun Sheng and colleagues suggest that the switch between circular and linear internalization architecture is more facile than might have been imagined. The first paper documents reptilian species in which mesendoderm internalization occurs through a blastopore-like region and involves a combination of involution and ingression (Bertocchini et al., 2013). Since birds are evolutionarily closer to reptiles than mammals, this observation suggests an instance of convergent evolution whereby avian and mammalian primitive streaks independently evolved from an ancestral blastopore-like configuration. This possibility is rendered more plausible by the second paper in which the reverse process is demonstrated: ectopic activation of a single signaling pathway (FGF) converts avian primitive streaks from a linear to a circular arrangement (Alev et al., 2013).

12.8 EXCEPTIONS TO PRESUMED CELL LINEAGE RESTRICTIONS

Precise cell lineage studies have revealed exceptions to several presumed lineage barriers. One surprise was that the adult body is not entirely derived from the three germ layers. The gametes, also called germ cells, rather arise from the distinct primordial germ cell lineage (Lawson and Hage, 1994). A second "barrier-breaking" lineage of note is the neural crest, which is unique to vertebrates and discussed elsewhere in this book (see Chapter 18). The neural crest arises from ectoderm, but unlike other ectoderm it is highly migratory and contributes to a diverse set of tissues, including some, e.g., craniofacial bones, that might have been expected to be mesodermal. It has been proposed that the neural crest be viewed as a fourth primary germ layer, and vertebrates, accordingly, be viewed as quadroblasts rather than triploblasts (Hall, 2000).

Yet another presumed barrier, the idea that extra-embryonic tissues never contribute to adult tissues, has been disproved through several observations in the 21st century. With respect to mesoderm, adult blood and lymphoid cells were believed to derive exclusively from embryonic mesoderm, but it has been found that mature red blood cells and B cells can arise from extra-embryonic mesoderm of the yolk sac and that hematopoetic stem cells can arise from other extra-embryonic locales, namely the placenta, allantois, and chorion (Corbel et al., 2007; Gekas et al., 2005; Isern et al., 2008; Kingsley et al., 2004; Ottersbach and Dzierzak, 2005; Sugiyama et al., 2007; Zeigler et al., 2006). Finally, it has been demonstrated that a population of cells retains the plasticity to adopt mesodermal or neurectodermal fates well past stages of global gastrulation movements though the primitive streak or blastopore, as discussed in the next section.

12.9 BEYOND THE BLASTULA-GASTRULA TRANSITION: CAUDAL MESODERM AND THE LEFT-RIGHT AXIS

The rapidly evolving fields of left-right patterning and somitogenesis and their respective *Nodal Flow* and *Clock and Wavefront* mechanisms have been reviewed elsewhere (Dequeant and Pourquie, 2008; Hirokawa et al., 2012). This section offers an overview of the relevant embryonic structures, cellular movements, and timing of commitment. By the time gastrulation has commenced and the SMO has formed, most cells have already been specified to a germ layer in accordance to their blastoporal-ablastoporal position and further specification is ongoing based upon their SMO proximodistal position (Figure 12.3M). Next, global gastrulation movements transport precursors of the rostral axis close to their final rostrocaudal positions, but some caudal axis precursors are allocated to a multipotential pool of homogenous cells that ultimately congress at a caudal location known as the *tail bud* (or *caudal eminence* in humans). During the next stage of development, segmentation, SMO-proximal components of neurectoderm and mesoderm (the paraxial mesoderm) contribute to the body's segments in an orderly rostral-to-caudal progression. This begins with the pre-positioned paraxial mesoderm. The neuromeres form first; their non-neurectodermal paraxial mesoderm component will go on to produce head mesenchyme (Table 12.1). The rostral trunk somites form next, which, like all somites, are largely derived from paraxial mesoderm. These will form the muscles, non-ectodermal skin parts, and bones associated with more rostral levels of the future spine (Table 12.1; Figure 12.5E).

Many of the caudalmost embryonic structures are elaborated from the tail bud, which forms at a relatively earlier stage in anamniotes (*X. laevis* and zebrafish) than amniotes (mouse and chick), and accordingly harbors a larger fraction of the rostrocaudal axis (Wilson et al., 2009). After global gastrulation movements are completed, the tail bud begins a caudal progression, leaving a swathe of cells in its wake: raw material for the future caudal axis. Intriguingly, a superficial structure associated with the tail bud, the ventral ectodermal ridge, has been found to be essential for caudal progression (Goldman et al., 2000). Cellular internalization has been observed in the migrating tail bud of *X. laevis*, chicks, and zebrafish, which has led some to propose that the tail bud represents a zone of continuing gastrulation, termed *secondary gastrulation*, with others disagreeing. This long-standing debate was deeper than a quibble about how to name tail bud internalization movements (Handrigan, 2003). The question boiled down to whether the tail bud is more akin to an assemblage of pre-specified cells that will once again sort themselves out, or to a totipotent blastema from which any cell can be specified to a wide range of fates. The blastema model has gained increasing credibility over recent years, but the debate is not resolved. One of the key points of support for the blastema model came from a definitive fate mapping study in the mouse that involved genetic labeling of single cells and their progenies at somewhat randomized times and places. The patterns of labeled cells demonstrated presence of self-renewing bi-potent stem cells able to contribute to both paraxial mesoderm and the neural tube long after global gastrulation is completed (Tzouanacou et al., 2009). This validated the earlier evidence in the chick and *X. laevis,* and taken together it has become clear that these cells originate from a non-internalized population close to the SMO (the caudal lateral epiblast, or CLE, in amniotes) and that they then coalesce into a self-renewing stem cell niche in the tail bud from whence they later exit to form paraxial mesoderm and neural tube (Figure 12.3M; Cambray and Wilson, 2007; Davis and Kirschner, 2000; Gont et al., 1993; Lawson et al., 1991; Martin B. L. and Kimelman, 2012; Selleck and Stern, 1991; Stern et al., 2006; Wilson et al., 2009). These cells have been termed *axial stem cells* by some and by others *neuromesodermal stem cells*, or more succinctly N-M stem cells, as shall be used here. Although the tail bud forms relatively later in amniotes than anamniotes, CLE cells also persist at more rostral positions for relatively longer. Taking the entire population of N-M stem cells in the tail bud and/or the CLE together, the amniote and anamniote model organisms arguably contain progenitors for a similar (caudal) fraction of the rostrocaudal axis.

The transition from N-M cell to caudal paraxial mesoderm cell is not direct. After departing the N-M pool, but before going on to become paraxial mesoderm, N-M cells enter a new (and adjacent) niche of the tail bud: the *mesodermal progenitor cell* (MPC, also called paraxial mesoderm stem cells) niche, from whence cells go on to form caudal paraxial mesoderm, as well as, surprisingly, caudal vascular endothelium (Table 12.1; Dunty et al., 2008; Martin B. L. and Kimelman, 2010). While N-M stem cells clearly exist, whether or not they can be said to account for the entirety of caudal neural tube and caudal paraxial mesoderm remains unclear. However a "pan-N-M" origin of caudal tissues would bring a tantalizing maximum parsimony to the etiology of posterior defects in animal models and humans. As such, and following the principle of Occam's razor, this chapter examines some of the implications of a pan-N-M caudal axis (see Table 12.1, Figure 12.5E and the last section of this chapter), with caveats and alternate hypotheses hereby noted.

There has been less confusion regarding the timing of left-right axis commitment. Although symmetry-breaking mechanisms are still being elucidated and may begin prior to the blastula-gastrula transition, full commitment of left-right identity depends upon signals from the SMO and its midline derivatives. By this time, the relevant precursors have already been apportioned to their respective flanks through global gastrulation movements. In conclusion, it should be clear from the above that the shaping of a functional tail bud, SMO and midline from nascent mesoderm is critical to the later success of caudal development and the positioning and organization of asymmetric organs.

12.10 MOLECULAR-GENETIC DISSECTION OF MESODERM FORMATION*

Much of our current understanding of which gene products control mesoderm formation derives from *gain-* and *loss-of-function* experiments. Examples of gain-of-function experiments are: (1) incubating embryos or embryonic fragments in solutions supplemented with particular purified proteins and (2) exploiting embryos' natural protein synthesis apparatus to translate exogenous DNA or RNA that had been engineered at the molecular biology bench. Examples of loss-of-function experiments are (1) genetic screens that introduce random DNA mutations into a population followed by the identification and characterization of those mutations that produce recessive or haplo-insufficient phenotypes of interest and (2) targeted mutation or interference of a particular gene using DNA knockout or RNA-interfering technologies.

Such experiments have generated an increasingly large set of data points that must be integrated for a complete understanding of mesoderm formation. Recognizing that 50% of the approximately 20,000 protein-coding genes in humans (or a given model vertebrate organism) are expressed during embryogenesis, it is clear that this complexity is genuine and unavoidable (Yi et al., 2010).

One strategy for simplifying the quest for genes with roles in mesoderm formation has been to primarily consider genes for which there is (1) *opportunity*, by virtue of being expressed at the right place, time, and level, for instance among mesendoderm precursors, and (2) *sufficiency*, meaning that ectopic introduction of the gene product can induce or expand mesoderm formation, and (3) *necessity*, as evidenced, for instance, by mesoderm deficiencies in embryos in which the gene is disrupted.

As an example of sufficiency, the first molecules found to induce vertebrate mesoderm came from gain-of-function studies in *X. laevis*. It was found that FGF2, a ligand belonging to the Fibroblast Growth Factor (FGF) family, and Activin, a ligand belonging the Transforming Growth Factor beta (TGFβ) superfamily, are each sufficient to trigger mesoderm induction (Slack et al., 1988; Smith J. C. et al., 1990). However, it turns out that neither of these is robustly expressed at the right place and time, and neither has been found to be necessary for mesoderm formation.

Mouse Nodal, an Activin-related TGFβ ligand, more completely fulfills the triple criteria of opportunity, sufficiency, and necessity. *Nodal* RNA is transcribed in presumptive mesendoderm, its misexpression in frogs and zebrafish causes ectopic mesoderm formation, and homozygosity of a disrupted *Nodal* allele disrupts mouse primitive streak and anterior mesoderm formation, although some posterior mesoderm persists (Conlon et al., 1994; Jones et al., 1995; Toyama et al., 1995).

But lack of a mesoderm-related loss-of-function phenotype cannot rule out the possibility that a gene normally provides a redundant input. Conversely, a lack of sufficiency cannot rule out the possibility of a subtle synergistic or additive role in mesoderm differentiation for a given factor. As an example of redundancy, consider the two Nodal orthologs expressed by zebrafish mesendoderm precursors, Nodal-related 1 (Ndr1, also formerly called Znr2 and Squint), Ndr2 (also formerly called Znr1 and Cyclops) (Rebagliati et al., 1998b). In embryos homozygous for loss-of-function mutations in both genes, the SMO fails to form, cells along the blastula margin fail to internalize and head mesoderm, trunk mesoderm and endoderm derivatives never arise (Feldman et al., 1998; 2000). By contrast, embryos homozygous for just one of the *Ndr* mutations produce most classes of mesoderm and endoderm (Heisenberg and Nusslein-Volhard, 1997; Rebagliati et al., 1998a, Sampath et al., 1998).

In *X. laevis*, there are six Nodal-related genes, *Xnr1* through *Xnr6*, raising a high bar for combinatorial genetic loss-of-function experiments (Takahashi S. et al., 2000). A necessity for Nodal signaling in *X. laevis* mesendoderm formation has nonetheless been demonstrated through alternate strategies involving respective overexpression or treatment with an engineered dominant-negative construct, a native Nodal antagonist (the Lefty protein) or a Nodal receptor-blocking drug, each condition blocking the actions of multiple Xnrs in parallel (Agius et al., 2000; Cheng et al., 2000; Osada and Wright, 1999; Tanegashima et al., 2000).

---

* Note to readers: rather than follow each of the distinct conventions for gene and protein nomenclature that have been established for each organism, this chapter uses a "compromise" nomenclature in which the first letter of gene and protein names are capitalized, acronyms (but not other abbreviations) are capitalized and gene names are italicized.

Genetic redundancy is more common in zebrafish and *X. laevis* than in the mouse, chick, or human, due to an ancestral whole-genome duplication in teleost fish and to pseudotetraploidy in *X. laevis* (Kobel and Du Pasquier, 1979; Postlethwait et al., 2000). But genetic redundancy has also confounded the genetic study of mammalian mesoderm formation. For instance simultaneous mutation of two mouse Nodal receptor genes, Activin receptors *Acr2a* and *Acr2b,* is required to recapitulate the phenotype from loss of the single Nodal ligand (Song et al., 1999). Furthermore, although mice and chicks each have only one gene with high sequence identity to Nodal, their genomes encode other TGFβ ligands that signal through Nodal receptors, as do the zebrafish and *X. laevis* genomes. These include Vg1 orhtologs in chick, zebrafish, and *X. laevis* and GDF1 and GDF3 in mouse.

12.11 A GENE REGULATORY NETWORK VIEW OF DEVELOPMENT

The preceding genetic scenarios demonstrate the risk of simplifying one's focus to genes causing single mutant phenotypes. But simplification is nonetheless required. The remainder of this discussion of vertebrate mesoderm formation will be through the simplifying lens of gene regulatory networks (GRNs; Figure 12.4). GRNs have emerged as a valuable way of thinking about and representing development, and a common circuit diagram-like representation of GRNs has emerged through software tools such as BioTapestry (Davidson, 2006; Longabaugh et al., 2005). Reasonably complete GRNs have been published for the mesoderm formation of sea urchin and Drosophila (Levine and Davidson, 2005).

Figures 12.4A and B show GRNs for zebrafish and *X. laevis* mesoderm development that borrow heavily from four published mesendoderm GRNs for these organisms, but with modifications and updates (Chan et al., 2009; Koide et al., 2005; Loose and Patient, 2004; Morley et al., 2009). Any GRN component for which a reference is lacking in this text can be found in the original. Figures 12.4C and D show GRNs for aspects of mesoderm formation in the mouse and chick, adapted and modified from models that have been presented in different formats (Ben-Haim et al., 2006; Bertocchini et al., 2004; Perea-Gomez et al., 2002). Before discussing these GRNs, a general overview is in order.

Eric Davidson and Roy Britten introduced the GRN concept in 1969 as a thought experiment in which regulatory principles from Francois Jacob and Jacques Monod's *lac operon* were projected upon the complexity of multicellular embryogenesis (Britten and Davidson, 1969; Jacob, 1965). Core GRN components were hypothetical, but predictive enough to be updated over the years with the discovery of real-world transcription factors and eukaryotic enhancers. To fully appreciate a GRN, one must take the perspective of the embryo's nuclear genomes. A kind of tunnel vision is adopted in which development is equated to the temporal progression of *nuclear regulatory states*. The nuclear regulatory state is defined by the set of trans-acting factors engaging the genome at any given time. Because trans-acting factors are inherited by daughter cells, the nuclear regulatory state is also heritable.

A second level of GRN tunnel vision limits one's view to the fraction of trans-acting factors that are agents of differentiation and their targets that are also trans-acting agents of differentiation. What kinds of trans-acting nuclear molecules are agents of differentiation? A variety of molecules engage the nuclear genome, and many new insights into the epigenetic mechanisms of developmental regulatory control have recently emerged, some of which will be discussed in a later section. Enhancer-binding *transcription factors* (TFs) engage and activate discrete sets of target genes, thereby effecting discrete developmental changes. As such, TFs have been and will likely continue to be the principal actors of GRNs. Other chromatin or DNA modifying/binding enzymes engage broader swaths of the genome.

At the first approximation GRNs involve two classes of TF. First are those TFs that can re-engage and reprogram a cell's nucleus. Synthesis of such TFs drives cell autonomous cellular differentiation (e.g., Siamois – Figure 12.4A). The second class is comprised of pre-synthesized TFs that are poised to engage the nucleus as a result of incoming extracellular signals, for instance as part of a ligand-activated signaling relay (e.g., β-Catenin – Figure 12.4A–D). Such relays occur in *immediate-early* fashion, meaning that the nucleus is directly reprogrammed without any intermediary synthesis of additional gene products. This second class of TF and other molecules that relay incoming extracellular signals, e.g., receptors, are rarely shown in GRNs, but their actions are implicit. One reason that they are not displayed is that doing so would render the GRN illegibly dense, but the mechanistic links between known signal sources and their known downstream nuclear targets are also sometimes incompletely understood. Differentiation-promoting *outgoing* intercellular signals, frequently ligands (e.g., FGF4 – Figure 12.4A) are often represented in GRNs. These signals are central to the logic of multicellular development because they enable the differentiation of one nucleus to affect the differentiation of another.

Although by no means a general feature of GRNs, under special circumstances a TF can directly engage in inter-nuclear communication. This happens in nuclear syncytia in which multiple nuclei inhabit a common cytoplasm. Examples relevant to mesoderm formation are the zebrafish YSL (Figure 12.4B) and the *Drosophila* syncytial blastoderm. Several homeodomain-containing TFs have also been reported to translocate between cells through non-canonical mechanisms of secretion and uptake (Balayssac et al., 2006; Derossi et al., 1996; Wu X. and Gehring, 2014). The implications of this to cell autonomy during mesoderm formation have not been fully explored.

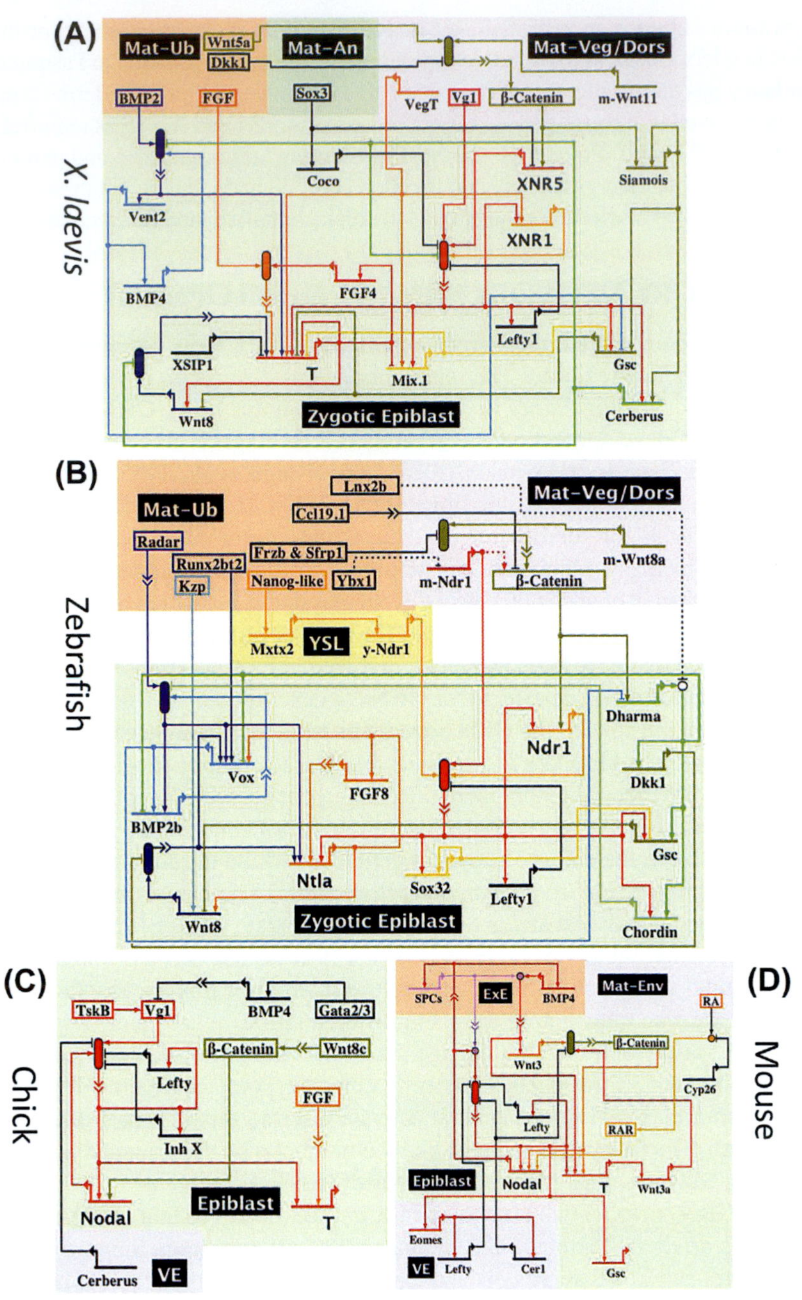

FIGURE 12.4 **Provisional GRNs of Vertebrate Early Mesoderm Formation and Patterning.** GRNs are presented for *X. laevis* (A), zebrafish (B), chick (C) and mouse (D). Colored background squares and their associated labels (black rectangles with white text) represent different classes of gene expression defined by either (1) spatial localization (**Epiblast**, **An** = animal, **Veg-Dors** = vegetal and dorsal, **YSL** = yolk syncytial layer, **VE** = visceral endoderm, **ExE** = extra-embryonic ectoderm and **VE** = visceral endoderm), (2) a lack of spatial localization (**Ub** = ubiquitous) or (3) the time of expression (**Zygotic** and **Mat** = maternal). Boxes with black text indicate proteins (or retinoic acid [RA]). Horizontal bars with black text represent genes, and the vertical arrows exiting the left or right side of each gene represent their gene products. The single gene product leaving each gene can have multiple targets, which are found by tracing all contiguous lines and forks that maintain the source gene's color. Regulatory links that involve non-cell autonomous (intercellular) signaling cascades contain double broken arrows (>>). Regulatory targets are either activated (arrowheads) or repressed (flat bars). Dotted links in **B** represent m-ndr1 RNA regulation of β-Catenin, Ybx1 regulation of m-ndr1 RNA, and Lnx2b ubiquitination of Dharma. Other links that impinge on genes represent the binding of a TF to a *cis*-regulatory enhancer. Although each gene's enhancers are schematized as lying 5' to the transcriptional start site, this is merely symbolic. The true position, order, and number of enhancers on each gene are not represented, and in a number of instances are incompletely determined. Indeed, the *provisional* status of these GRNs explicitly accounts for the possibility that intervening genetic links requiring *de novo* protein synthesis have been overlooked. Vertical ovals are nodes representing cumulative pathway signaling activity: blue for BMP, red for Nodal, orange for FGF and green for WNT. Small circles represent modifying interactions, with white in **B** representing ubiquitination and lilac and tangerine in **D** respectively indicating pro-domain removal and retinoic acid (RA) degradation. Gene names and their outgoing links are color coded with shades of red and orange marking mesendoderm-inducing processes, shades of blue marking factors contributing to SMO-distal patterning (i.e., ventral patterning in *X. laevis* and zebrafish), shades of green marking factors contributing to SMO-proximal patterning (i.e., dorsal patterning in *X. laevis* and zebrafish), black marking Nodal- and Wnt-antagonizing factors and lilac marking SPC actions.

12.12 GRNS OF VERTEBRATE MESODERM FORMATION

The GRNs of Figure 12.4 focus on the zygotic activation of two classes of mesodermal genes: those encoding Nodal-related ligands and those encoding orthologs of the T-box family TF Brachyury/T. A Nodal signaling gradient lies at the heart of blastoporal-ablastoporal patterning, with higher levels driving endoderm and mesoderm formation (Figure 12.3M), and some of the factors that help establish this gradient are represented. T is independently required for caudal axis development, including portions that would persist in the absence of Nodal alone (Table 12.1; Figure 12.5E). Several molecules acting in the initiation of *Nodal* and *T* activation have parallel, Nodal-independent roles in SMO formation. Gradients of Bone Morphogenetic Protein (BMP) and Wnt signaling lie at the heart of SMO proximodistal patterning, at least for *X. laevis* and zebrafish, with higher levels driving SMO-distal fates (Figure 12.3M), and some of the factors that help establish this gradient are also represented in Figure 12.4 as well as retinoic acid (RA) signaling in Figure 12.5. Additional species-specific roles of BMPs, Wnts and FGFs are represented. The molecular-cellular bases of Nodal, Wnt, BMP, FGF

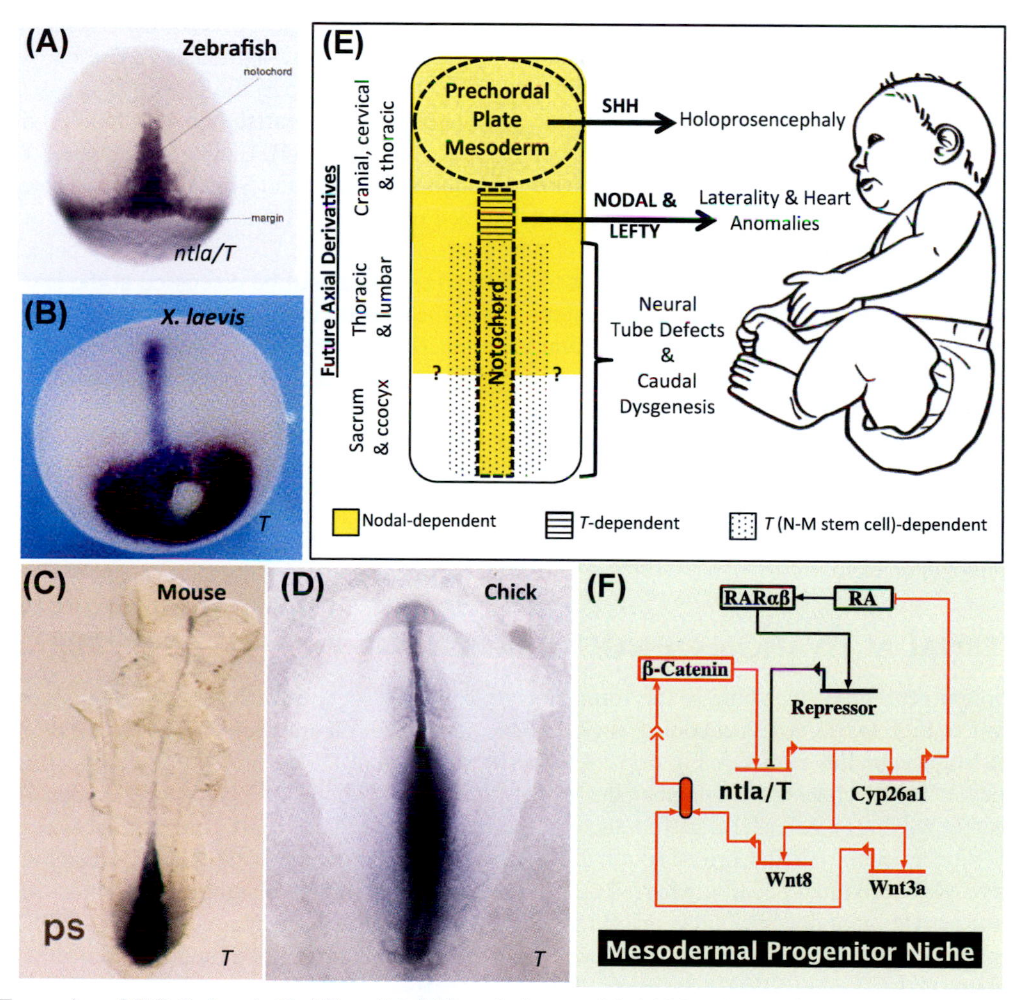

FIGURE 12.5 **Expression of *T* Orthologs in Undifferentiated Mesendoderm and Axial Mesoderm.** Whole-mount *in situ* hybridizations for *ntla* in zebrafish and *T* in *X. laevis*, mouse, and chick. Each is a dorsal view and developmental stages range from mid-gastrula (A–B) to early (D) and mid-somitogenesis (C). Images were reprinted, with permission, from the chick database GEISHA, the zebrafish database ZFIN and two publications (Bell G.W et al., 2004; Bowes et al., 2010; Greene et al., 1998; Thisse et al., 2001; Vignal et al., 2007). (E). Speculative schematic of a human embryo (left side) showing rostrocaudal positions of Nodal-dependent and T-dependent mesoderm and the birth defects likely to arise from deficiencies in said mesoderm (right side). SHH, NODAL and LEFTY on the horizontal arrows indicates these factors' known or proposed requirements to protect against the respective disease entity. The idea that N-M stem cells account for most of the caudal axis remains speculative. The indicated future axial identities are educated guesses based upon the positions of extant anterior somites in mouse brachyury/T mutants and extant posterior somites in zebrafish *MZoep* and *ndr1;ndr2* mutants (Chesley, 1935; Feldman et al., 1998; Gritsman et al., 1999). The correlation of these somite positions to vertebral identity is based on a correlative lineage tracing study in zebrafish (Morin-Kensicki et al., 2002). Question marks (?) refer to the unresolved question of whether or not caudal intermediate or lateral mesoderm progenitors are derived from N-M stem cells, as discussed in the text. The infant drawing is public domain. (F) GRN of the zebrafish mesodermal progenitor niche, adapted from Martin and Kimelman (Martin B.L. and Kimelman, 2010). Niche-promoting interactions are in red and niche-inhibitory interactions are in black. The red oval represents cumulative Wnt signaling.

and RA signaling are described elsewhere in this book (see also Chapter 1). The focus here being on GRN logic, only a few molecular-cellular details are provided. To minimize diagram complexity, a number of genes are mentioned in the text but not represented, particularly genes with similar actions to one that is shown. Other genes are not mentioned at all due to space limits or authorial ignorance (with apologies).

The essential role of Nodal signaling in mouse, *X. laevis* and zebrafish mesoderm was described above. A lack of primitive streak formation in chick embryos overexpressing Nodal antagonists extends the case to birds and the last of the four core vertebrate models, suggesting a pan-vertebral function (Yanagawa et al., 2011).

T was discovered through the mouse *Brachyury* mutation. First described in 1927, homozygosity of *Brachyury* causes a loss of the notochord and lethal posterior truncations, eliminating the axis caudal to the rostral-most 8–12 somites (Chesley, 1935; Dobrovolskaïa-Zavadskaïa, 1927). The cloning and molecular characterization of the affected gene, renamed as *T*, introduced a novel class of TFs: the T-box family (Kispert and Hermann, 1993). Several T-box proteins in addition to T/Brachyury have roles in mesoderm formation and posterior development, particularly VegT, Tbx6, Tbx16, and Eomesodermin in *X. laevis,* zebrafish, and the mouse (Showell et al., 2004; Wardle and Papaioannou, 2008). There has been a historical diversity of names for vertebrate T orthologs including Xbra (*X. laevis*), Ch-T (chick), and No Tail A and B (Ntla and Ntlb, zebrafish), however all but the zebrafish genes have been re-named as *T*. These model vertebrate *T* orthologs are each expressed by mesendoderm precursors and nascent mesoderm in the primitive streak and tail bud and in the axial chordamesoderm (Figure 12.5A–D). Simultaneous disruption of zebrafish Ntla and Ntlb produces notochord deletions and posterior truncations analogous to those seen in mouse and (Martin B. L. and Kimelman, 2008). Misexpression of *X. laevis T* induces formation of ectopic posterior mesoderm (Cunliffe and Smith, 1992). Chromatin imunoprecipitation in murine ES cells and zebrafish embryos shows that their respective T orthologs bind dozens of genes involved in muscle formation, gastrulation movements, posterior development (including *Tbx6* and *Tbx16*), notochord formation and left-right patterning (Evans et al., 2012; Morley et al., 2009). Finally, as will be discussed at the end of this chapter, T orthologs are central to the GRN that establishes and maintains the mesoderm progenitor cell niche from which caudal mesoderm is proposed to derive (Martin B. L. and Kimelman, 2010; Takada et al., 1994; Yamaguchi et al., 1999). For such reasons, *T* expression is commonly considered to be a pan-vertebrate proxy for mesoderm formation, in animal and embryonic stem cell studies alike (Fehling et al., 2003).

Thus, while other representatives might have been selected, it can be argued that Nodal and T orthologs are collectively involved in and required for all aspects of early mesoderm formation (Table 12.1). Consistent with this, simultaneous disruption of the Nodal receptor Oep and Ntla leads to mesoderm reduction or elimination along the entire zebrafish axis (Harvey et al., 2010; Schier et al., 1997). Figure 12.5E presents a summary of this Nodal/T-centric perspective in the context of an idealized human embryo.

12.13 MATERNAL ACTIVATION OF NODAL SIGNALING: *X. LAEVIS* AND ZEBRAFISH

Maternally supplied gene products, some in the form of asymmetrically localized RNAs, have fundamental roles for *X. laevis* and zebrafish. In *X. laevis*, RNAs encoding VegT (a T-box family TF; Figure 12.4A), Vg1 (a TGFβ family ligand that acts through the Nodal signaling pathway; Figure 12.4A) and Wnt11 (Figure 12.4A) are all initially localized to the vegetal cortex of the oocyte. The mechanisms are unclear, but localization of *Wnt11* and *Vg1* RNAs depend upon *VegT* RNA localization (Heasman et al., 2001). After fertilization, the egg's cortex rotates relative to its core, a process essential for localizing the SMO (Vincent et al., 1986). Cortical rotation is believed to deliver the maternal RNA source of the Wnt11 ligand to the prospective SMO-proximal position where it acts in concert with maternal Wnt5a to induce nuclear localization of β-Catenin and the activation of downstream genes (Figure 12.4A; Cha et al., 2008; Schroeder et al., 1999; Tao et al., 2005; Vonica and Gumbiner, 2007). The extent of Wnt-induced β-Catenin signaling is limited by ubiquitous maternal expression of the Wnt inhibitor Dkk1 (Figure 12.4A; Cha et al., 2008).

Among the *X. laevis* proteins whose synthesis are activated by β-Catenin figure the TFs Siamois (Figure 12.4A) and Twin and the Nodal pathway signaling ligands Xnr5 (Figure 12.4A) and Xnr6 (Carnac et al., 1996; Laurent et al., 1997; Yang J. et al., 2002). These factors trigger SMO formation, for instance via Siamois activation of the TF-encoding *Gsc* gene and the BMP/Nodal inhibitor-encoding *Cerberus* gene (Figure 12.4A; Bae et al., 2011; Carnac et al., 1996; Yamamoto et al., 2003). Transcriptional targets of the Nodal signaling pathway activated by Xnr5 include Nodal ligand genes themselves such as *Xnr1* (Figure 12.4A) and genes encoding Nodal antagonists such as *Lefty1* and *Cerberus* (Figure 12.4A; Cheng et al., 2000; Inui et al., 2012). This phenomenon of Nodal signaling simultaneously triggering auto-stimulatory and auto-inhibitory feedback loops is seen in each of the model vertebrates (Figure 12.4A–D; Shen, 2007). Additional Nodal pathway triggering in *X. laevis* comes from maternal Vg1 signaling through the Nodal pathway and from VegT, which directly activates Nodal ligand synthesis, including Xnr1 (Figure 12.4A, Kofron et al., 1999).

Some of the above are paralleled in zebrafish. RNA encoding Wnt8a is translocated from the vegetal pole of the oocyte to a future SMO-proximal site where its translation triggers localized nuclear accumulation of β-Catenin (Figure 12.4B; Lu et al., 2011). This occurs via an incompletely characterized microtubule-dependent process that appears to involve the cargo-linker protein Syntabulin (Nojima et al., 2010; Solnica-Krezel and Driever, 1994; Tran et al., 2012). As with *X. laevis*, zebrafish β-Catenin signaling triggers both Nodal signaling and Nodal-independent promotion of the SMO (shield). This occurs via synthesis of the TF Dharma and Ndr1, which in turn activate synthesis of Gsc and Chordin (another BMP inhibitor) (Kelly et al., 2000; Ryu et al., 2001; Shimizu et al., 2000). Also similar to *X. laevis*, zebrafish Wnt/β-Catenin signaling is delimited by maternal antagonists of Wnt signaling, in this case Frzb and Sfrp1 (Figure 12.4B; Lu et al., 2011).

The establishment of zebrafish Nodal signaling also involves non-Wnt inputs, but their anatomical sources and molecular nature differ substantially from *X. laevis*. In addition to zygotic *Ndr1* (Figure 12.4B) and *Ndr2* (not shown) transcripts in the embryo, there are maternal *Ndr1* transcripts (*m-Ndr1*; Figure 12.4B) and the extra-embryonic yolk syncytial layer (YSL) harbors additional zygotic *Ndr1* (*y-ndr1*; Figure 12.4B) and *ndr2* (*y-ndr2*; not shown) transcripts (Erter et al., 1998; Fan et al., 2007; Feldman et al., 1998; Gore and Sampath, 2002). Reminiscent of the translation-independent function of maternal *vegT* RNA in anchoring *Vg1* and *Wnt11*, *m-Ndr1* has a non-coding patterning function: activating β-Catenin signaling in prospective SMO-proximal cells (Figure 12.4B; Lim et al., 2012). Localization of *m-Ndr1* requires maternal Ybx1, a protein that also represses *m-ndr1* translation (Figure 12.4B; Kumari et al., 2013). In addition to positive regulation of β-Catenin by *m-Ndr1* RNA and maternal Wnt8a, the maternal chemokine Ccl19.1 (Figure 12.4B) acting through the G-protein coupled receptor Ccr7 (not shown) represses β-Catenin signaling (Wu S. Y. et al., 2012).

Synthesis of y-Ndr1 and y-Ndr2 depends upon the TF Mxtx2 whose own synthesis depends upon the TF Nanog-like (Hong et al., 2011; Xu et al., 2012). The YSL lacks the ability to respond to Nodal signals itself, yet simultaneous blocking of y-Ndr1 and y-Ndr2 translation recapitulates partial Nodal loss-of-function phenotypes, suggesting that these extra-embryonic Ndrs normally help trigger the Nodal auto-regulatory loop (Fan et al., 2007; Harvey and Smith, 2009). When the y-Ndrs are disrupted and WT *m-Ndr1* is replaced with a coding-null allele that retains its non-coding ability to activate β-Catenin, a phenocopy of complete *Nodal* loss-of-function is attained (Hong et al., 2011; Lim et al., 2012). This indicates that *m-Ndr1* normally also has a direct role in Nodal pathway activation and that Ybx1 repression of *m-Ndr1* translation is not absolute (Figure 12.4B).

12.14 ZYGOTIC REGULATION OF NODAL SIGNALING: CHICK AND MOUSE

The mechanisms that initiate zygotic Nodal gene transcription in the chick and mouse are less elucidated. But it is clear that chick and mouse embryos sculpt their zones of *Nodal* expression through the interplay of zygotic activators and repressors, with no obvious specific maternal triggers. As with most zebrafish and *X. laevis* Nodal genes, the unitary Nodal genes of these two organisms are expressed in prospective mesendoderm.

The chick epiblast harbors the potential to form multiple primitive streaks – a potential reined in by inhibitory signals from adjacent extra-embryonic tissue: the hypoblast, also called the visceral endoderm (VE, Figure 12.4C; Eyal-Giladi, 1970). Based on the above, on the timing of zygotic gene activation, and on a variety of embryological experiments, a model of chick mesendoderm formation has been proposed, re-represented here with several modifications (Figure 12.4C; Bertocchini et al., 2004).

The initial triggers come in the form of Vg1, which is expressed in the posterior marginal zone of the epiblast, and Wnt8c, which is more broadly expressed. More recent data from Bertocchini and Stern suggests that the Vg1 zone of expression might be positioned by Bmp4 antagonism driven by distal Gata2 and Gata3 expression (Figure 12.4C, Bertocchini and Stern, 2012). Returning to the 2004 model, Vg1 is proposed to induce an unidentified fast-diffusing inhibitor (Inh X) that outpaces Vg1 and Wnt8c and blocks their mesendoderm-inducing effects. But Vg1 and Wnt8c predominate locally, triggering *Nodal* synthesis. It seems likely that Vg1 and Inh X impinge directly on the Nodal signaling pathway, as indicated in Figure 12.4C, but the original model was agnostic on this point (Bertocchini et al., 2004). In support of a Nodal signaling role for chick Vg1, it is potentiated by binding to Tsukushi B (TskB) – a secreted factor whose *Xenopus* ortholog binds and potentiates Xnr2 signaling (Figure 12.4C; Morris et al., 2007; Ohta et al., 2006). Two subsequently expressed inhibitors are more clearly dedicated to the restriction of Nodal signaling. VE-derived Cerberus, the diffusible Nodal/BMP inhibitor, comes into play first and functions until its VE source is displaced by incoming extra-embryonic endoblast tissue. This allows Nodal-induced differentiation to proceed in the epiblast, which leads to synthesis of the diffusible Nodal inhibitor *lefty*. Lefty from the epiblast is also proposed to prevent the formation of supernumerary primitive streaks.

In the mouse, Nodal inhibitors have an analogous role in limiting the zone of effective Nodal signaling, and thereby mesendoderm and primitive streak formation (Figure 12.4D). Just as the Cerberus-secreting chick VE delimits the primitive streak in chick, Cerberus and Lefty from the mouse VE restrict Nodal signaling in the epiblast that would otherwise disrupt anterior patterning via anterior expansion of the primitive streak and/or formation of supernumerary primitive streaks (Perea-Gomez et al., 2002). This explains earlier observations that signals from the anterior VE are essential for the differentiation of anterior structures, namely the forebrain (Thomas and Beddington, 1996).

The VE expression of *Lefty* and *Cerberus* is triggered by early Nodal signals from the epiblast via activation of *Eomesodermin* (Figure 12.4D; Nowotschin et al., 2013) and maintained by VE *Nodal* expression triggered in parallel by epiblast-derived Nodal (not shown), but the triggers upstream of epiblast Nodal synthesis remain unknown. A possible clue may lie in the ability of retinoic acid (RA) from the maternal environment to induce Nodal expression unless opposed by the embryo's concerted expression of three *Cyp26* para-logs, which encode an RA-degrading enzyme (Figure 12.4D; Uehara et al., 2009). In contrast, the initial maturation of Nodal ligands during embryogenesis has been substantially elucidated in the mouse model. A key post-translational modification of Nodal and other TGFβ-family ligands is removal of their N-terminal pro-domain via cleavage by extracellular Subtilisin-like Proprotein Convertases (SPCs). Other TGFβ ligands undergo analogous processing by SPCs, including BMPs. Exploring the genetics of a cleavage-resistant form of Nodal, an unexpected signaling function for unprocessed Nodal protein was revealed and the authors proposed a model of Nodal signal amplification that has been incorporated into the mouse GRN of Figure 12.4D (Ben-Haim et al., 2006). By this model, unprocessed Nodal ligand translated in the epiblast diffuses to the extra-embryonic ectoderm (ExE) where it triggers synthesis of two SPCs (PACE4 and Furin) and the synthesis of BMP4. This establishes two parallel loops of Nodal signaling amplification. In one loop, BMP4's pro-domain is removed by Furin (SPCs; Figure 12.4D) and mature BMP4 diffuses to the epiblast where it triggers Wnt3 synthesis and downstream β-Catenin signaling, driving more Nodal synthesis. Through the other loop, the Nodal pro-domain is removed by secreted SPCs that diffuse to the epiblast, releasing mature Nodal that can signal along its conventional pathway (Mesnard et al., 2011). The roles of immature Nodal ligands and SPCs remain relatively unexplored in other model vertebrates.

12.15 ESTABLISHING T EXPRESSION

X. laevis T and zebrafish Ntla synthesis are driven by Nodal, BMP, FGF and Wnt signaling (Figure 12.4A–B). VegT is an additional driver in *X. laevis* (Figure 12.4A). In theory these drivers could operate upstream, as maternal or very early zygotic triggers or downstream, as later zygotic factors, to maintain T expression. There is evidence that maternal VegT acts upstream in *X. laevis* and that in both organisms Nodal signaling acts upstream, Wnt signaling downstream and BMP signaling up and downstream of T/Ntla synthesis (Chan et al., 2009; Koide et al., 2005). The role of FGF signaling appears to differ between the two organisms. Whereas in *X. laevis*, disruption of FGF signals causes a loss of *T* transcripts, FGF signaling in zebrafish is dispensable for the initiation of *Ntla* transcription (Figure 12.4A–B; Amaya et al., 1991; Fletcher and Harland, 2008; Griffin et al., 1995).

The above differences aside, a positive feedback maintenance loop has been delineated in both *X. laevis* and zebrafish in which T/Ntla triggers FGF4/FGF8 synthesis, triggering additional T/Ntla synthesis (Figure 12.4A–B; Loose and Patient, 2004; Morley et al., 2009). Analogously, T/Ntla triggers Wnt8 (and Wnt3a) synthesis, triggering additional T/Ntla synthesis; this loop is essential for maintenance of the mesodermal progenitor stem cell niche in the tail bud (Figure 12.4A–B; Loose and Patient, 2004; Martin B. L. and Kimelman, 2008; 2010; Morley et al., 2009). In zebrafish a positive feedback loop is also seen for BMP, whereby Ntla triggers Vox (Figure 12.4B and Vent (not shown) synthesis, triggering BMP2b synthesis, triggering more Ntla synthesis (Chan et al., 2009; Morley et al., 2009). The zebrafish BMP2b/Ntla and Wnt8/Ntla loops are also interlinked by virtue of Wnt8 activation of Vox and Vent synthesis (Figure 12.4B; Ramel and Lekven, 2004).

Similar links between the core signaling pathways and the initiation/maintenance of *T* transcription have been elucidated in the mouse, although BMP and Wnt signaling disruptions produce more severe phenotypes than what is seen in zebrafish and *X. laevis*. Disruption of Wnt pathway signaling in the mouse by genetic knockouts of *Wnt3* or a palmitoylating enzyme essential for maturation of all Wnt ligands eliminates *T* and *Nodal* transcripts, indicating a total loss of mesoderm (Biechele et al., 2013; Liu et al., 1999). By contrast, mesoderm formation persists in the face of Wnt/β-Catenin disruption in *X. laevis* and zebrafish. This is despite an astonishing array of (1) ventralization + anterior truncations, (2) dorsalization + posterior truncation and (3) ectopic or expanded head phenotypes that can arise, depending on whether the disruption affects (1) dorsal Wnt/β-Catenin-dependent SMO formation (Figure 12.4A–B), (2) ventral Wnt/β-Catenin-dependent SMO proximodistal patterning and feedback maintenance of T (Figure 12.4A–B), or (3) anterior Wnt/β-Catenin-dependent repression of forebrain development (not shown; Bellipanni et al., 2006; Fekany et al., 1999; Heasman et al., 1994; Houart et al., 2002; Kazanskaya et al., 2000; Martin B. L. and Kimelman, 2008; Shimizu et al., 2005).

Disruption of the mouse BMP Receptor 1 gene also causes a loss of *T* expression and decreased growth of the epiblast (Mishina et al., 1995). In *X. laevis* and zebrafish, BMP disruptions lead to losses of ventral mesoderm, but dorsal mesoderm still forms (Hild et al., 1999; Zimmerman et al., 1996). The requirement for murine Wnt and BMP signaling upstream of mesoderm formation might reflect dual roles for each pathway in *T* activation and in propagating the early unprocessed Nodal signal, as schematized in Figure 12.4D. The extent to which later functions for murine Wnt and BMP might be analogous to those in zebrafish and *X. laevis* remains an open question. One similarity with regard to *T* regulation, is the existence of a murine Wnt3a/T feedback loop, which is also required for maintenance of MPCs (Figure 12.4D; Dunty et al., 2008; Evans et al., 2012; Yamaguchi et al., 1999).

As with *X. laevis* and zebrafish, disruption of Nodal signaling in mouse leads to reduction, but not elimination, of *T* transcript levels (Conlon et al., 1994; Feldman et al., 1998; Osada and Wright, 1999). The role of murine FGF signaling in the initiation of *T* expression has been difficult to assess, as *FGF4* and *FGFR2* mutant embryos die prior to the initiation of *T* transcription (Arman et al., 1998; Feldman et al., 1995). The next earliest murine requirement for FGF signaling is seen in *FGFR1* and *FGF8* mutant embryos, which robustly express *T* and form primitive streaks, although subsequent mesodermal cell differentiation is lethally compromised (Sun et al., 1999).

As with other vertebrate models, chick *T* is a standard marker of primitive streak induction and Nodal signaling appears to induce primitive streak formation and *T* expression directly, whereas Wnt signaling likely contributes indirectly, via Nodal (Figure 12.4C). There is also evidence from pharmacological inhibition and gain-of-function experiments that chick primitive streak formation and *T* expression require independent and parallel inputs from early FGF8 signals (Figure 12.4C; Bertocchini et al., 2004). Unique to the chick among the four vertebrate models, BMP antagonism, effected by Chordin and Tsukushi A, is also critical for primitive streak formation and *T* expression (Bertocchini et al., 2004; Ohta et al., 2006; Streit et al., 1998).

12.16 ESTABLISHING HOMOGENOUS CELLULAR FIELDS

A fundamental and fascinating aspect of embryogenesis is that cellular fields arise in which all cells have essentially the same properties and, accordingly, essentially the same patterns of gene expression. One way such fields can form is through positive feedback loops that involve at least one non-cell-autonomous component. The non-cell-autonomous factors can "spread the word," thereby inducing the nuclei of naïve cells to adopt the same regulatory state. This type of network architecture also provides an explanation for the community effect (Bolouri and Davidson, 2010). Assume that a certain extracellular concentration of diffusible inducer is required for maintenance of certain regulatory state. A cell in isolation might not be able to produce the required ligand concentration. But a densely packed community of cells can achieve the necessary ligand concentration by sharing ligands and restricting their escape from the microenvironment.

The mesendoderm GRNs presented here include several examples of such community effect architecture. The following were already discussed: Nodal ligands inducing *nodal* gene expression in an immediate-early fashion (Figure 12.4A–D), T orthologs and Wnt ligands reinforcing each other's expression (Figure 12.4A–B and D) and T orthologs and FGF ligands reinforcing each other's expression (Figure 12.4A–B). Another example is BMP ligand autoinduction, either in an immediate-early fashion (Figure 12.4B) or *via* induced synthesis of an intermediary transcription factor (Figure 12.4A, Vent2; (Koide et al., 2005).

12.17 THE SMO PROXIMODISTAL AXIS

In zebrafish and *X. laevis*, the SMO proximodistal (D-V) axis is mirrored by parallel BMP and Wnt gradients, which are central to the establishment of D-V identities (Figure 12.3M; De Robertis and Kuroda, 2004; Langdon and Mullins, 2011). The GRNs presented here include a subset of the genes that act to establish these gradients. Ubiquitous maternal activators create a default scenario in which the regulatory state would become SMO-distal (ventral) unless disrupted by SMO-proximalizing (dorsalizing) signals. This default scenario is effected through maternal BMP ligand signaling (*X. laevis* Bmp2 and zebrafish Radar), activation of the Vox>Bmp2b cascade by the zebrafish TF Runx2bt2 and activation of Wnt8 by the zebrafish zinc-finger TF Kzp (Figure 12.4A–B; Clement et al., 1995; Flores et al., 2008; Sidi et al., 2003; Yao S. et al., 2010).

In *X. laevis*, disruption of the SMO-distal default state comes from β-Catenin's induction of the TFs Siamois (Figure 12.4A) and Twin (not shown), which induce the BMP and Wnt antagonist Cerberus (Figure 12.4A) and the BMP antagonist Chordin (not shown), which act non-cell autonomously via inactivating binding to their target ligands (Figure 12.4A). In zebrafish, β-Catenin induces Dharma, which induces Chordin and the Wnt antagonist Dkk1 (Figure 12.4B; Hashimoto et al., 2000). Like agonists, these non-cell autonomous antagonists can mediate the community effect. This is seen in the zebrafish (Figure 12.4B), in which Chordin antagonism of BMP signaling prevents activation of Vox, allowing Dharma to continue driving its own synthesis and the synthesis of Chordin (Chan et al., 2009). One constraint on Dharma that is not connected to BMP signaling is its targeting for degradation by the maternal ubiquitin kinase Lnx2b (Ro and Dawid, 2009).

Disruption of SMO-distalizing factors only goes so far, however. As the distance from the source of diffusible SMO-proximalizing signals increases, a self-perpetuating auto-regulatory BMP loop predominates to maintain the SMO-distal mesoderm field. This loop is direct in the case of zebrafish BMP2b (Figure 12.4B) and via Xvent2 synthesis (Figure 12.4A) in the case of *X. laevis* BMP4 (Chan et al., 2009; Koide et al., 2005). Cell autonomous loops can help reinforce the outcomes of this SMO-proximal-distal competition, "locking down" cells simultaneously exposed to SMO-proximalizing and SMO-distalizing signals into one regulatory state or the other. Thus in *X. laevis*, Xvent2 not only drives BMP4 synthesis, it represses the synthesis of the SMO-proximal TF Gsc (Koide et al., 2005). In the zebrafish, Vox and Dharma are mutually repressive and are each auto-regulatory, helping to lock cells into one fate or another (Chan et al., 2009). Such "either/or" circuits help establish sharp borders between cellular fields (Levine and Davidson, 2005).

Although the order of tissue fates along the SMO proximodistal axis is analogous for each of the model vertebrates (Figure 12.3E–H, M), the extent to which the amniotes might analogously employ BMP and Wnt gradients to effect these fates remains unclear. In support of a universal gradient scheme, BMP antagonists are robustly expressed by the SMO/node in the mouse (*Noggin* and *Chordin*) and chick (*Chordin*), and RNAs encoding both Wnt and BMP ligands are globally up-regulated in SMO-distal regions of the chick primitive streak (Alev et al., 2010; Bachiller et al., 2000; Streit et al., 1998). Emerging insights from regenerative medicine lend further support as Activin (mimicking Nodal) can induce ESCs to form mesendoderm with SMO-proximal-like high Gsc expression and co-treatment with BMP4 causes a shift to SMO-distal-like mesoderm with low Gsc and high T expression (Murry and Keller, 2008). An analogous choice between becoming a neural cell (SMO-proximal fate) or an epidermal cell can be biased to the epidermal option by BMP4 addition.

Against a universal gradient scheme, robust Wnt antagonist expression in chick and mouse SMOs has not been reported. Confounding the comparison further, pre-gastrula mice form a Wnt gradient (via Cer1 [Figure 12.4D] and Dkk1 antagonism from the anterior portion of the VE) as well as a Nodal gradient, both of which are orthogonal to the proposed scheme for anamniotes (Arnold and Robertson, 2009). However the Nodal gradient in mouse realigns with the onset of Nodal expression in the primitive streak and Wnt antagonism by the anterior VE may be a unique feature of mammalian head induction (de Souza and Niehrs, 2000). Thus, the notion of a pan-vertebrate fate specification framework involving at least BMP and Nodal gradients oriented to the SMO and the internalization domain merits further exploration (Figure 12.3M).

12.18 SEPARATING MESENDODERM FROM ECTODERM

Similar to SMO-distal default in the absence of SMO-proximalizing signals, there is an ectodermal default in the absence of mesendoderm-inducing signals. But once mesendoderm is induced, what prevents it from overrunning the presumptive ectodermal field of cells? Two basic mechanisms have emerged:

1) independent expression in presumptive ectoderm of molecules that counteract mesendoderm induction; and
2) a reaction-diffusion system in which the spread of Nodal antagonists outpaces the spread of Nodal agonists.

The first mechanism occurs through several scenarios in *X. laevis*. Maternal Sox3, which is enriched in the animal pole, represses mesendoderm by both repressing synthesis of Xnr5 (Figure 12.4A) and Xnr6 (not shown) and activating synthesis of anti-Nodal factors, namely the soluble Cerberus-related antagonist Coco (Figure 12.4A), as well as the TFs Ectodermin and Xema (Zhang and Klymkowsky, 2007; Zhang et al., 2004). Coco is also maternally expressed (Bell E. et al., 2003). Such "immunization" of cells against mesendoderm recruitment is not absolute and can be overcome, for instance, by the forced expression of VegT and Vg1 in animal pole cells (Yan and Moody, 2007).

Additional zygotically expressed anti-mesendoderm factors in the presumptive ectoderm of *X. laevis* include the TF Xsip (Figure 12.4A), which represses *T* transcription while activating neural genes and the TF XFDL156 (not shown) that represses transcription of T and other mesendoderm genes (Papin et al., 2002; Sasai et al., 2008). Zygotic activation of anti-mesendodermal factors in presumptive ectoderm, for example Sox3, also occurs in zebrafish (not shown; Shih Y. H. et al., 2010).

The second mechanism operates on the same principle as the one proposed for Vg1 and InhX in the chick: induction of a faster-diffusing or longer-range antagonist by a slower-diffusing or shorter-range agonist (Bertocchini et al., 2004). Such a scenario, known as a reaction-diffusion system, has the potential to generate a surprising diversity of patterns. Reaction-diffusion patterning was proposed by Alan Turing in 1952 and was subsequently refined and applied to simulations of hydra regeneration patterns by Alfred Gierer and Hans Meinhardt (Gierer and Meinhardt, 1972; Turing, 1952). It was later demonstrated that reaction-diffusion simulations, particularly scenarios of *local self-enhancement and long-range inhibition*, can replicate elaborate biological patterns such as the stripes and spots of seashells (Meinhardt, 2003).

Evidence has been found for a diversity of endogenous local self-enhancement long range inhibition systems, such as in the determination of where roof palette ruggae, feathers, and hair follicles will form (reviewed in Marcon and Sharpe,

2012). This system is also seen for Nodals and Lefties in left-right patterning and mesendoderm-ectoderm separation (Chen Y. and Schier, 2002; Nakamura et al., 2006). Mesendoderm-ectoderm separation is the simplest case, whereby the spread of the Lefty signal simply outpaces the spread of the Nodal signal, creating a gradient of Nodal/Lefty ratios that culminates in a zone that is highly buffered against mesendoderm induction, establishing the ablastoporal-blastoporal axis. An important feature of this system, which has been quantitatively modeled based on measured parameters of diffusion and clearance, is that it requires only one localized input, in contrast to the spatially distinct sources of BMP and Wnt ligands and agonists (Muller et al., 2012). A single cell or small group of cells can thus trigger the formation of a cellular field via the spread of Nodal reinforced by positive feedback, but the positive/negative feedback architecture also accounts for the Lefty signal that will ultimately limit the spread of that field. Such a system does not in theory require pre-localization of anti-mesendoderm factors in other domains, potentially explaining why animal pole pre-enrichment of such factors has not been documented in zebrafish.

12.19 SEPARATING MESODERM FROM ENDODERM

The separation of mesendoderm into mesoderm and endoderm and the separation of mesoderm into anterior and posterior subtypes appears to be effected through a combination of *morphogen* signaling and GRN architecture. The term morphogen was coined by Turing to mean any diffusible chemical able to trigger embryonic differentiation (Turing, 1952). The quest to identify such substances had been active at least since the discovery of SMO induction (Waddington et al., 1933). Albert Dalcq and Jean Pasteels had further proposed that complex patterns might arise from chemical inducers able to (1) act directly at a distance and (2) instruct naïve cells to adopt alternate states in a dose-dependent manner, and the contemporary definition of a morphogen includes these criteria (Dalcq and Pasteels, 1937; Neumann and Cohen, 1997; Wolpert, 1969).

Nodal ligands have been found to fulfill the above criteria for morphogens. First, Nodal-related proteins induce expression of different genes in a dose-dependent manner (Chen Y. and Schier, 2001; Jones et al., 1995). Second, zebrafish Ndr1 has the ability to diffuse and directly signal across many cell diameters (Chen Y. and Schier, 2001). Effective doses of Nodal stimulation can be attained, and are attained *in vivo*, as a function of time of exposure in addition to concentration (Hagos and Dougan, 2007). A likely explanation for this time-concentration equivalence comes from studies in which cellular memories of pulse exposures to Activin or BMP ligands could be measured as a function of proportional and longer-lasting changes in intracellular levels of downstream signal-transduction products, such as internalized receptor-ligand complexes and their phosphorylated Smad targets (Jullien and Gurdon, 2005; Simeoni and Gurdon, 2007). It should be noted that a morphogen gradient represents just one possible evolutionary solution to the problem of establishing a gradient of tissue fates; a distinct GRN-based solution has been described for endomesodermal specification in sea urchin (Smith J. et al., 2007).

What are the consequences of different Nodal exposure levels in terms of gene expression? A lack of Nodal signals allows expression of ectodermal genes, and increasingly higher levels elicit first markers of caudal mesoderm (e.g., *T* orthologs next markers of rostral axial mesendoderm (e.g., Gsc) and finally markers of endoderm (e.g., Sox17 and Mezzo; Chen Y. and Schier, 2001; Hagos and Dougan, 2007; Jones et al., 1995). This is precisely the order of germ layer fates seen along the ablastoporal-to-blastoporal axis of vertebrate fate maps.

Much remains to be learned about how differences in signaling strength produce initial differences in gene expression, but as with SMO proximodistal patterning, GRN circuits can help stabilize initial differences. For instance in *X. laevis*, T expression represses transcription of the endodermal gene *Mix.1* that can otherwise induce the rostral axial mesendoderm gene *Gsc* (Figure 12.4A; Latinkic and Smith, 1999; Lemaire et al., 1998). Thus, if an *X. laevis* cell is biased towards T expression, it will be inhibited from becoming endoderm or rostral mesendoderm. Reciprocally, if an *X. laevis* cell is biased towards Mix.1 or Gsc expression, these TFs' ability to repress T expression either directly (or *via Wnt8* repression in the case of Gsc, Figure 12.4A) will help the cell maintain its endodermal or rostral mesendodermal identity (Middleton et al., 2009; Yao J. and Kessler, 2001). Antagonism of Ntla expression *via* Gsc repression of Wnt gene synthesis has also been delineated in zebrafish (Figure 12.4B; Seiliez et al., 2006). An additional zebrafish GRN link shows how the TF Sox32 helps maintain the endodermal state by repressing synthesis of *Gsc* (Figure 12.4B; Chan et al., 2009; Sakaguchi et al., 2001).

12.20 EARLY ZYGOTIC GENES ARE POISED FOR ACTION

The GRNs above represent some of the earliest cell lineage restrictions that occur during embryogenesis. But they can only function if the naïve cells in which they act are responsive. This section briefly reviews some clues that have recently emerged regarding how the responsiveness of pluripotent embryonic cells is established.

The accessibility of genes to specific TFs and to the RNA transcription machinery depends upon a permissive combination of covalent (but reversible) modifications to the DNA and to the histone proteins around which the DNA is coiled. These are called *epigenetic marks*. Consistent with the specialized functions of male and female gametes, the egg and sperm nuclei that meet at fertilization have distinct epigenetic marks (Cantone and Fisher, 2013). But within a few short cell divisions, cells of the embryo have become pluripotent. This transition from a more committed regulatory state to a less committed regulatory state is unique and opposite to the process of progressive specialization discussed in this chapter thus far. It involves *epigenetic reprogramming*, which at the molecular level is seen as a global-scale reorganization of epigenetic marks (Cantone and Fisher, 2013).

It has been known since 1958 that the egg's cytoplasm harbors factors that can direct such reprogramming, based on Gurdon and colleagues' demonstration that nuclei from somatic, differentiated *X. laevis* cells can be "cloned" and raised as adult frogs by introducing them into unfertilized oocytes (Gurdon et al., 1958). Substantial insights into this process have emerged in recent years, including the discovery by Shinya Yamanaka and colleagues that forced expression of a set of four TFs that are normally expressed in ESCs can convert differentiated somatic cells into pluripotent cells, known as induced pluripotent stem cells (iPSCs) (Takahashi K. and Yamanaka, 2006).

Zygotic genome activation (ZGA) is a second transition that must occur in early embryonic cells before cell lineage determination can commence. The transition from zygotic quiescence to ZGA is accompanied by the degradation of maternal transcripts (Tadros and Lipshitz, 2005). The control of ZGA timing is incompletely understood, but it has been linked to increases to the nuclear/cytoplasm ratio and to the cell-cycle length of cleaving embryonic cells.

Specific mechanisms of transcriptional control of ZGA have also emerged (Lagha et al., 2012). First, the maternal *Drosophila* TF Zelda has been found to be essential for ZGA coordination and an analogous function was found for the maternal zebrafish TF Pou5f1 (an ortholog to one of the core iPSC TFs) (Leichsenring et al., 2013; Liang et al., 2008; Reim and Brand, 2006). Second, it was discovered in pre-ZGA *Drosophila* embryos and confirmed in human ESCs that RNA polymerase II is pre-loaded but paused on the promoters of a substantial number of early zygotic genes (Core et al., 2008; Zeitlinger et al., 2007). Third, a significant number of early zygotic genes have a pattern of two epigenetic histone marks on their promoters (Azuara et al., 2006; Bernstein et al., 2006; Lindeman et al., 2011; Vastenhouw and Schier, 2012). One of these histone marks (H3K4me3) has been associated with transcriptional activation and the other (H3K27me3) with repression, leading to the suggestion that such genes are set in a "poised" but inactive state in advance of ZGA. Finally, a significant number of early zygotic zebrafish genes bound by Pou5f1 were found to also have pre-loaded RNA polymerase II and H3K4me3/H3K27me3 marks, suggesting a functional link between these various phenomena (Leichsenring et al., 2013). Thus, while a great deal clearly remains to be elucidated, it can already be stated with some certainty that pluripotency is an actively acquired regulatory state in which key zygotic genes are poised for rapid deployment in response to inductive signals.

12.21 THE PHYLOTYPIC EGG TIMER

Despite disparate anatomies of eggs and early zygotes between vertebrate species (Figure 12.3A–D) they all converge towards the conserved anatomies of the phylotypic stage (Figure 12.3I–L). This convergence is counterintuitive: how and why do different forms become more similar only to become different again? In 1994, Denis Duboule answered this question with his "phylotypic egg timer" hypothesis; egg timer being a metaphor for the hourglass-shaped suite of interspecific divergence > convergence > divergence during the pre-phylotypic > phylotypic > post-phylotypic stages of development (Duboule, 1994). Duboule argues that the phylotypic stage is most conserved because it depends upon a high level of GRN interdependence, specifically "meta-*cis*" regulatory systems controlling expression of collinear Hox gene clusters across much of the body plan. By this scenario, viably reformulating the phylotypic GRN would require too many simultaneous evolutionary mutations.

Recent trancriptome profiling studies support Duboule's hypothesis. They show that the overall depth of evolutionary conservation of stage-specific zebrafish and *Drosophila* transcriptomes reach a maximum at their respective phylotypic stages, with earlier and later stages showing less conservation (Domazet-Loso and Tautz, 2010; Kalinka et al., 2010). Thus, at the molecular as well as the morphological level, the most highly conserved stage of life is the phylotypic one.

The converse to evolutionary phylotypic constraint is pre- and post-phylotypic evolutionary deconstraint. Post-phylotypic de-constraint has been attributed to *compartmentation* of the body plan, which can also be understood as a developmental shift away from global meta-cis control systems to locally coordinated networks, allowing mutational changes to act upon subsets of body compartments (Duboule, 1994; Gerhart and Kirschner, 2007; Schlosser and Wagner, 2004). With regard to pre-phylotypic constraints, although early embryos are certainly more global than compartmented by nature, the degree of interdependence is apparently still low enough for successful evolutionary tinkering, such as the deployment of maternal Nodal antagonists in *X. laevis* but not zebrafish (Figure 12.4A–B) or the mammalian use of the extra-embryonic AVE as a head inducer. Evolutionary deconstraint on both sides of the phylotypic bottleneck has likely been key to species diversification. After all, the successful occupation of new niches, e.g., from aqueous to terrestrial, must have sometimes required co-evolutionary changes in the anatomies of adults and their eggs.

12.22 MESODERM SPECIFICATION DEFECTS IN HUMANS

Nascent mesoderm comprises the precursors that will shape and act in every system of the post-embryonic body. Accordingly, genetic mutations and other teratogenic factors causing profound mesoderm deficiencies in animal models are generally embryonic lethal. For instance in mice, *Nodal* mutants show signs of death and maternal resorption by 9.5 days post-coitus (dpc) and *Brachyury* mutants die by 10.5 dpc (Chesley, 1935; Conlon et al., 1991). The corresponding human embryonic stages are attained between 25 and 35 days after conception. As such, human mesodermal anomalies as severe as those arising from complete *Nodal* or *Brachyury* loss-of-function could easily go undetected. This seems particularly true when one considers the large scale of lost conceptions that normally go unnoticed. An estimated 30% of fertilized human zygotes fail to implant and 30% are lost after implantation but prior to a pregnancy diagnosis (the stage corresponding to *Nodal* and *Brachyury*-like losses) (Macklon et al., 2002). An additional 10% of fertilized zygotes are estimated to be lost due to clinical miscarriage and 30% result in live birth. Of these neonates, about 3% are diagnosed with birth defects; i.e., about 0.9% of initial pool of fertilized zygotes (Rynn et al., 2008).

Given that radical loss of mesoderm tissue will cause undetected pregnancy loss, what developmental anomalies would arise from lesser mesoderm deficits? Figure 12.5E provides an overview of four classes of birth defects that are known or hypothesized to result from the following:

1) deficiencies of rostral axial (prechordal plate) mesendoderm leading to *holoprosencephaly* (HPE) (Roessler and Muenke, 2010);
2) deficiencies of axial mesoderm, leading to *laterality defects* including *congenital heart disease* (Peeters and Devriendt, 2006; Ramsdell, 2005);
3) deficiency of caudal axis extension or differentiation, leading to
   a) caudal dysgenesis, or
   b) neural tube defects (NTDs) (Copp and Greene, 2010; Garrido-Allepuz et al., 2011);
4) uncontrolled growth and differentiation of the vestigial notochord or the caudal eminence, respectively leading to standard *chordomas* or *sacrococcygeal chordomas* (Walcott et al., 2012).

The upstream insult(s) that cause a given mesodermal aberration could be diverse: environmental, chemical, genetic, or a combination thereof, with the downstream effects playing out through common cell- and tissue-level pathways (Ming and Muenke, 2002). Below is a summary of those pathways. In considering them, it should be noted that steps less proximal to mesoderm formation are nonetheless susceptible to insults that can cause the same class of clinical defect; this is discussed later in the context of Nodal signaling in laterality defects and Shh signaling in HPE. That said, the time window for any non-genetic or "later" genetic insult to contribute to the classes of birth defects considered here (with the exception of chordoma) is still quite specific: from the beginning of the third to the end of the eight week of pregnancy (Sadler and Langman, 2006).

(1) HPE (~1 in 15,000 newborns). Midline prechordal plate (PCP) mesendoderm deficits can lead to an inadequate dose of PCP-secreted Shh ligand, leading to insufficient ventralizing morphogenesis that normally bifurcates the prosencephalon (forebrain) and associated eye field (England et al., 2006). The consequent phenotypes can vary from a subtle narrowing of the eyes to non-viable fusions of the brain and eye that almost certainly inspired the ancient "myth" of the Cyclops (Stahl and Tourame, 2010).

(2) The establishment of laterality requires a healthy axial midline, which acts as a source of the left-sided Lefty1 signal, and a healthy SMO, which establishes asymmetric expression of Nodal (Nakamura et al., 2006; Raya and Izpisúa Belmonte, 2006). There are certainly other classes of molecules involved in laterality, but they do not have central roles in mesoderm formation. Errors in left-right patterning, known as *heterotaxy*, can be manifested as any combination of organ placement on the wrong side of the body, reversal of organ orientation or inversions, or duplications or deletions of organ parts. Scenarios of full mirror image inversions (~1 in 7,000 newborns) and more randomized forms (~1 in 10,000 newborns) are seen, with only the latter expected to arise from midline or SMO deficiencies. Randomization is also associated with a substantially elevated incidence of *congenital heart diseases* with partial inversions, such as *double outlet left ventricle* and *transposition* of the great arteries.

(3) Caudal dysgenesis incorporates a variety of posterior defects ranging from *caudal regression syndromes* that can comprise relatively subtle or at least viable deficits in the lower extremities to *sironomelia* (~1 in 50,000 newborns) – a severe and usually fatal syndrome characterized by truncations and or fusions of the lower limbs and posterior viscera, and a possible inspiration for myths of sirens and mermaids (Stahl and Tourame, 2010). It is easy to imagine a connection between posterior truncations and caudal axis extension; this line of reasoning led to the *defective blastogenesis hypothesis* as early as the 19th century and the possibility that N-M stem cells are indeed the major precursors of caudal tissues has reinvigorated this idea (Ferrer-Vaquer and Hadjantonakis, 2013; Stocker and Heifetz, 1987).
(4) NTDs (~1 in 800 newborns) are much more frequent at posterior locations, once again suggesting a link with caudal axis extension. But given the neural etiology of NTDs, it was not obvious how subtle deficits in caudal mesoderm would cause defects in caudal neurectoderm. The possibility that caudal mesoderm and caudal neural tissue derive from the *same proximal precursors*, the N-M stem cells, brings a fresh perspective to the etiology of NTDs and their caudal bias (see also Chapter 37).
(5) *Sacrococcygeal chordomas* are caudal embryonic tumors that have long been hypothesized to arise from continued growth of the caudal eminence. Derivatives of all three germ layers are typically present. Standard *chordomas* are posteriorly biased outgrowths from the spine with notochord-like histology, and therefore are believed to derive from vestigial fragments of notochord. The existence of an N-M stem cell population is also germane to Sacrococcygeal chordomas, as stem cells have a high potential for dysregulated growth.

12.23 HUMAN GENE VARIANTS ASSOCIATED WITH MESODERM SPECIFICATION DEFECTS

What kinds of genetic mutations cause these defects? Theoretical considerations lead to the following two predictions.

1) Due to evolutionary purifying selection of essential genes, the frequency and structural severity of heritable core mesoderm gene (CMG) variants should be diminished relative to other viable genes.
2) Given the roles of such CMGs as "master switches" in early development, large effects from hypomorphic variants are also expected.

These two predictions receive statistical support from a bioinformatic comparison of the mutational spectrum and human disease association of human genes orthologous to essential mouse genes vs. human genes orthologous to non-essential mouse genes (Park et al., 2008). Continuing with our Nodal/T-centric perspective on mesoderm formation, some disease associations will now be considered for these and related genes. In the course of this, some general issues relevant to the study of CMG variants in humans will be discussed.

Nodal Signaling Pathway Genes in Heterotaxy and HPE

Consistent with the prediction that CMG variants are rare and mild at the structural level, finding disease-associated mutations in human CMGs known to drive the underlying developmental processes can be a *needle in haystack*-like endeavor. Whereas a number of Nodal pathway mutations are associated with laterality defects, including congenital heart defects, as well as HPE, an extremely small number of these are in the *Nodal* gene itself (Peeters and Devriendt, 2006; Roessler et al., 2008; 2009). This might reflect a lack of a redundant/parallel human Nodal ligand, whereas more frequently mutated Nodal pathway genes share the task of Nodal signal propagation with other factors. As an example, heterotaxy and, less frequently, HPE are associated with mutations in the *Foxh1* gene whose protein product is a co-factor of the Smad2/Smad4 transcriptional complex that transmits the Nodal signal from receptor to nucleus. Animal studies indicate that the task of transmitting the totality of the Nodal signal across all cell types is shared by other TFs, namely Mixer and Eomesodermin, suggesting *Foxh1* mutations should be more viable than Nodal mutations (Germain et al., 2000; Kunwar et al., 2003; Picozzi et al., 2009; Slagle et al., 2011). A general complication in assessing whether *Nodal* pathway mutations cause heterotaxy or HPE is the dual role of Nodal signaling in establishing the PCP, other midline tissue and the SMO on the one hand, and on the other hand in establishing laterality downstream of these events (Nakamura et al., 2006).

In contrast to the scarcity of HPE-associated *Nodal* mutations, *SHH* mutations are frequently found (Roessler and Muenke, 2010). Despite this, Nodal pathway mutations often produce cyclopia in animal models (Constam and Robertson, 2000; Sampath et al., 1998; Song et al., 1999). Similar to Nodal pathway mutations causing heterotaxy only, one does not need to invoke a structural midline (PCP) defect to explain HPE arising from SHH pathway mutations, since SHH is the PCP signal itself (Figure 12.5E). It therefore seems that Nodal pathway mutations severe enough to cause midline defects have been thoroughly purged from neonatal genomes. The above suggests strong negative selection against strong Nodal pathway mutations in the gene pool, but why are more *de novo* variants not detected? One possibility is that a *Goldilocks effect* is driving the numbers down because the only complex allele combinations that would be detected must be weak enough to come to term but strong enough to cause HPE rather than merely heterotaxy (Roessler et al., 2008; 2009). In support of this idea, examination of the Kyoto collection of electively aborted first-trimester human embryos and fetuses revealed an holoprosencphaly incidence of ~1 in 250, which is about sixty times the incidence in newborns (Nishimura et al., 1968). This indicates that only ~2% of holoprosencephalic embryos and fetuses come to term. One has to wonder: if the Kyoto collection embryos were sequenced, how many more Nodal pathway vs. Shh pathway mutations would be found among the other 98% of more severely-affected individuals?

T/Brachyury in Posterior Defects

The possibility that the tail bud/caudal eminence harbors a niche of stem cells from whence the entire caudal paraxial mesoderm and caudal neural tube derives breathes new life into the *defective blastogenesis hypothesis*. Consistent with this, a small population of zebrafish tail bud cells that simultaneously express the "pan-mesodermal" marker *ntla/T* and the "pan-neural" marker *sox2* has been documented, as was previously observed for CLE cells in mouse and chick (Martin B. L. and Kimelman, 2012; Cambray and Wilson, 2007; Delfino-Machin et al., 2005). Furthermore, it is already clear in the mouse that Tbx6 is a major repressor of neural differentiation from N-M stem cells and that Sox2 is essential for N-M to N differentiation. Indeed, loss of murine Tbx6 causes each flank of paraxial mesoderm to become an ectopic neural tube that is presaged by ectopic expression of Sox2, but this does not occur if the only available Sox2 alleles lack a specific enhancer (Chapman and Papaioannou, 1998; Takemoto et al., 2011). Much remains to be learned about how N-M stem cells maintain their bi-potential state, for instance how is co-expression of T and Sox2 maintained? The role of Wnt signaling in N-M stem cell differentiation also needs to be resolved, with mouse studies implicating it as promoting N differentiation, but zebrafish studies implicating it as promoting M differentiation (Martin B. L. and Kimelman, 2012; Nowotschin et al., 2012; Takemoto et al., 2011).

A thorough model of how the mesodermal progenitor niche is maintained has been proposed for zebrafish (Figure 12.5F; Martin B. L. and Kimelman, 2010). Briefly, the community effect Wnt/T loop maintains the niche, but is threatened by retinoic acid (RA) emanating from more rostral cells with the potential to enter cells of the niche, bind the nuclear RA receptor RARαβ and repress *T* via a yet-to-be identified intermediary. T forestalls this from happening, however, by triggering synthesis of the RA-degrading enzyme Cyp26a.

From the above overview of N-M stem cells and the mesodermal progenitor niche, it should be clear that T is important to both types of cells. It is also tempting to speculate that disruptions of T, disruptions of Wnts or exposure to RA produce analogous posterior truncations precisely because they act through a common pathway: maintenance of the mesodermal progenitor niche (Chesley, 1935; Martin B. L. and Kimelman, 2008; Sive et al., 1990; Takada et al., 1994). The fact that one protocol of RA exposure to an animal model can replicate posterior truncations, whereas other RA-exposure protocols as well as a mouse knockout of Cyp26a1 can produce a range of NTDs, and even phenocopy sirenomelia, lends further support to the defective blastogenesis hypothesis as a unifying etiological pathway for a range of posterior defects (Abu-Abed et al., 2001; Kohga and Obata, 1992; Padmanabhan, 1998; Quemelo et al., 2007).

So, are *T* mutations found in humans with the range of posterior defects that might result from defects in an N-M stem cell or MPC pool? In fact, there is a scarcity of associated structural variants that is reminiscent of Nodal genes and HPE, and the same proposed negative selection factors could be at play. Three groups found an association of NTDs with a common *T* allele whose only variant feature is a single nucleotide polymorphism (SNP) in an intron (Jensen et al., 2004; Morrison et al., 1996; Shields et al., 2000). It will be important to learn whether the *T* locus associated with this SNP has altered regulation.

Much stronger associations are seen with chordomas. In one study, a non-synonymous SNP in the *T* gene was found to be associated, with biochemical studies showing that the protein forms T dimers of reduced stability and in another study entire duplication of the *T* gene was found to confer strong familial susceptibility (Papapetrou et al., 1997; Pillay et al., 2012; Yang X. R. et al., 2009). The idea that the duplication is pathogenic due to an excess of T activity seems straightforward and is consistent with the fact that T is expressed and required in chordamesoderm (Table 12.1; Figure 12.5E). But chordoma arising from the decreased T activity of the non-synonymous SNP allele is less easily explained.

12.24 IMPLICATIONS OF A PAN-N-M CAUDAL AXIS

In addition to supporting the defective blastogenesis hypothesis for several human caudal defects, if N-M stem cells account for all caudal neural and caudal paraxial mesoderm, a substantial re-ordering of the standard table of mesodermal (and caudal neural) tissue origins is required (Table 12.1, see also Figure 12.5E). Examining this re-ordered set of fates, the *T/*

Brachyury phenotype can be explained on the basis of three contributing factors: its requirement in chordamesoderm, its requirement in maintaining the MPC niche and its role in ensuring that a certain number of N-M stem cells adopt the M rather than N fate, the latter factor suggested by an excess of neural tube tissue in mouse *T/Brachyury* mutants, similar to Wnt3a mutants (Yamaguchi et al., 1999).

If the primary *Nodal* loss-of-function defect is equated to an inability to form internalizing primitive streak, the general persistence of T expression and associated tail formation in fish is also more clearly understood in the context of a pan-N-M caudal axis (Table 12.1; Figure 12.5E), because the CLE-derived N-M stem cells do not need to internalize (at least not initially) (Conlon et al., 1994; Feldman et al., 1998).

If a pan-N-M caudal axis hypothesis is correct, several questions emerge. First, do MPCs and/or N-M stem cells produce caudal intermediate or lateral mesoderm derivatives (question marks in Table 12.1 and Figure 12.5E)? Most fate mapping studies lack evidence for this, with one notable exception (see below). If N-M stem cells give rise to only caudal neural and caudal paraxial tissue, one would expect to see intermediate and lateral mesoderm derivatives in animals or humans with failures of paraxial mesoderm. This is in fact a proposed explanation for how sirenomelia arises. Given that limb buds derive from lateral mesoderm, it is proposed that they form even when paraxial derivatives such as vertebrae are lacking and in the absence of paraxial elements separating the limb buds, they merge as they grow (Garrido-Allepuz et al., 2011). Single limbs reminiscent of sirenomelia as well as disordered uritogenital elements can be observed in *T* and *Wnt3*a mutant mice (T. Yamaguchi, personal communication). The preceding arguments are consistent with the idea that intermediate and lateral mesoderm form in the absence of N-M stem cells, but differentiate in disordered fashion due to the absence of a paraxial mesoderm-derived "scaffold." On the other hand, it was recently reported that in zebrafish, MPCs, themselves derivatives of N-M stem cells, can form caudal vascular endothelium, which is normally a lateral mesoderm derivative (Martin B. L. and Kimelman, 2012).

The extent to which a pan-N-M caudal axis might or might not be Nodal dependent also needs to be resolved. Mouse and zebrafish *T* mutants lack all but the 8–12 rostral-most somites, yet zebrafish Nodal mutants produce only a small quantity of tail mesoderm (Chesley, 1935; Feldman et al., 1998; Martin B. L. and Kimelman, 2008). If the origins of Nodal-dependent (i.e., primitive streak) mesoderm and T-dependent (i.e., CLE) mesoderm were uncoupled, one would expect a substantially larger quantity of caudal mesoderm to form in Nodal mutants. Interestingly, there is substantial variability in the tail size of zebrafish Nodal mutants (B. F., unpublished observation), but not enough to balance the "missing" caudal mesoderm in Ntla/Ntlb-disrupted embryos. These observations suggest that in the context of a pan-N-M caudal axis, a substantial fraction of CLE cells must be Nodal- and T- dependent (Table 12.1, +/– and Figure 12.5E). Further supporting the idea that Nodal signals are required to establish a rostral fraction of CLE cells, in zebrafish trunk neural tube identities are not seen in Nodal mutants, the *ntla* gene has Nodal-dependent and Nodal-independent cis-regulatory enhancers, and MPCs of Nodal mutant origin fail to contribute to trunk mesoderm (Feldman et al., 2000; Harvey et al., 2010; Szeto and Kimelman, 2006).

12.25 CONCLUDING REMARKS

As I have endeavored to communicate, comparing developmental mechanisms between animal models can elucidate shared and species-specific mechanisms. Shared mechanisms tend to suggest etiologies of human disease, whereas species-specific mechanisms might suggest outside-the box cures, such as co-opting gene regulatory strategies of heart regeneration from zebrafish to address cardiovascular disease, or *planarian* strategies for controlled regenerative growth to address the runaway growth and differentiation of metastasized tumors and/or the risks of stem cell therapy (Kikuchi and Poss, 2012; Pellettieri and Sanchez Alvarado, 2007). I built BioTapestry GRN representations as guides to this chapter, because I am interested in their utility as tools for teaching, communicating, and comparing busy data sets. I make no claim of "systems level" insights, they are just a convenient way to track the directness/indirectness of epistatic relationships and the molecular classes of the players. The files are easily modified, and so I look forward to (polite) suggestions for improvements. The software is also free and the files are easily shared, so I also look forward to seeing them leave their niche of origin and begin a process of growth and differentiation.

12.26 CLINICAL RELEVANCE

- Severe mesoderm deficiencies are expected to be a significant cause of human pregnancy loss.
- Viable aberrations in rostral mesoderm formation are expected to cause clinical holoprosencephaly, heterotaxy, and congenital heart disorders.
- Viable aberrations in caudal mesoderm formation are expected to cause clinical caudal dysgenesis, caudal regression, caudal neural tube defects and chordomas.

- The possibility that most of the caudal axis derives from multipotent stem cells in the caudal lateral epiblast and caudal eminence suggests a unifying etiology to diverse caudal defects.
- The relative rarity among neonates with clinical birth defects of strong core mesoderm gene structural variants is likely due to the interrelated phenomenon of evolutionary negative selection and spontaneous abortion of the most adverse combinations of available alleles.

ACKNOWLEDGMENTS

I would like to thank to Erich Roessler and Terry Yamaguchi for helpful conversations.

RECOMMEND RESOURCES

Animal Model Databases

Mouse, http://www.informatics.jax.org (accessed 26.03.14).
Chick, http://geisha.arizona.edu (accessed 26.03.14).
laevis, X, http://www.xenbase.org (accessed 26.03.14).
Zebrafish, http://zfin.org.

BioTapestry Software

http://www.biotapestry.org/.

Reading

Developmental Biology (Gilbert, 2010)
The Plausibility of Life: Resolving Darwin's Dillemma (Kirschner and Gerhart, 2005)

Rare Biomedical Texts

The National Library of Medicine, Bethesda MD http://www.nlm.nih.gov/ (accessed 26.03.14).

REFERENCES

Abu-Abed, S., Dolle, P., Metzger, D., Beckett, B., Chambon, P., Petkovich, M., 2001. The retinoic acid-metabolizing enzyme, CYP26A1, is essential for normal hindbrain patterning, vertebral identity, and development of posterior structures. Genes Dev. 15, 226–240.

Agius, E., Oelgeschlager, M., Wessely, O., Kemp, C., De Robertis, E.M., 2000. Endodermal Nodal-related signals and mesoderm induction in Xenopus. Development 127, 1173–1183.

Alev, C., Wu, Y., Nakaya, Y., Sheng, G., 2013. Decoupling of amniote gastrulation and streak formation reveals a morphogenetic unity in vertebrate mesoderm induction. Development 140, 2691–2696.

Alev, C., Wu, Y., Kasukawa, T., Jakt, L.M., Ueda, H.R., Sheng, G., 2010. Transcriptomic landscape of the primitive streak. Development 137, 2863–2874.

Allman, G.J., 1853. On the Anatomy and Physiology of Cordylophora, a contribution to our knowledge of the Tubularian Zoophytes. Philos. Trans. R. Soc. London. 143, 367–384.

Amaya, E., Musci, T.J., Kirschner, M.W., 1991. Expression of a dominant negative mutant of the FGF receptor disrupts mesoderm formation in *Xenopus* embryos. Cell 66, 257–270.

Arman, E., Haffner-Krausz, R., Chen, Y., Heath, J.K., Lonai, P., 1998. Targeted disruption of fibroblast growth factor (FGF) receptor 2 suggests a role for FGF signaling in pregastrulation mammalian development. Proc. Natl. Acad. Sci. USA 95, 5082–5087.

Arnold, S.J., Robertson, E.J., 2009. Making a commitment: cell lineage allocation and axis patterning in the early mouse embryo. Nat. Rev. Mol. Cell Biol. 10, 91–103.

Azuara, V., Perry, P., Sauer, S., Spivakov, M., Jørgensen, H.F., John, R.M., Gouti, M., Casanova, M., Warnes, G., Merkenschlager, M., Fisher, A.G., 2006. Chromatin signatures of pluripotent cell lines. Nat. Cell Biol. 8, 532–538.

Bachiller, D., Klingensmith, J., Kemp, C., Belo, J.A., Anderson, R.M., May, S.R., McMahon, J.A., McMahon, A.P., Harland, R.M., Rossant, J., De Robertis, E.M., 2000. The organizer factors Chordin and Noggin are required for mouse forebrain development. Nature 403, 658–661.

Bae, S., Reid, C.D., Kessler, D.S., 2011. Siamois and Twin are redundant and essential in formation of the Spemann organizer. Dev. Biol. 352, 367–381.

Balayssac, S., Burlina, F., Convert, O., Bolbach, G., Chassaing, G., Lequin, O., 2006. Comparison of penetratin and other homeodomain-derived cell-penetrating peptides: interaction in a membrane-mimicking environment and cellular uptake efficiency. Biochemistry 45, 1408–1420.

Beddington, R.S., 1994. Induction of a second neural axis by the mouse node. Development 120, 613–620.

Bell, E., Munoz-Sanjuan, I., Altmann, C.R., Vonica, A., Brivanlou, A.H., 2003. Cell fate specification and competence by Coco, a maternal BMP, TGFbeta and Wnt inhibitor. Development 130, 1381–1389.

Bell, G.W., Yatskievych, T.A., Antin, P.B., 2004. GEISHA, a high throughput whole mount *in situ* hybridization screen in chick embryos. Dev. Dyn. 229, 677–687.

Bellipanni, G., Varga, M., Maegawa, S., Imai, Y., Kelly, C., Myers, A.P., Chu, F., Talbot, W.S., Weinberg, E.S., 2006. Essential and opposing roles of zebrafish beta-catenins in the formation of dorsal axial structures and neurectoderm. Development 133, 1299–1309.

Ben-Haim, N., Lu, C., Guzman-Ayala, M., Pescatore, L., Mesnard, D., Bischofberger, M., Naef, F., Robertson, E.J., Constam, D.B., 2006. The nodal precursor acting via activin receptors induces mesoderm by maintaining a source of its convertases and BMP4. Dev. Cell 11, 313–323.

Bernstein, B.E., Mikkelsen, T.S., Xie, X., Kamal, M., Huebert, D.J., Cuff, J., Fry, B., Meissner, A., Wernig, M., Plath, K., Jaenisch, R., Wagschal, A., Feil, R., Schreiber, S.L., Lander, E.S., 2006. A bivalent chromatin structure marks key developmental genes in embryonic stem cells. Cell 125, 315–326.

Bertocchini, F., Stern, C.D., 2012. Gata2 provides an early anterior bias and uncovers a global positioning system for polarity in the amniote embryo. Development 139, 4232–4238.

Bertocchini, F., Skromne, I., Wolpert, L., Stern, C.D., 2004. Determination of embryonic polarity in a regulative system: evidence for endogenous inhibitors acting sequentially during primitive streak formation in the chick embryo. Development 131, 3381–3390.

Bertocchini, F., Alev, C., Nakaya, Y., Sheng, G., 2013. A little winning streak: the reptilian-eye view of gastrulation in birds. Dev Growth Differ 55, 52–59.

Biechele, S., Cockburn, K., Lanner, F., Cox, B.J., Rossant, J., 2013. Porcn-dependent Wnt signaling is not required prior to mouse gastrulation. Development 140, 2961–2971.

Bolouri, H., Davidson, E.H., 2010. The gene regulatory network basis of the "community effect," and analysis of a sea urchin embryo example. Dev. Biol. 340, 170–178.

Bonnerot, C., Nicolas, J.F., 1993. Clonal analysis in the intact mouse embryo by intragenic homologous recombination. C R. Acad. Sci. III 316, 1207–1217.

Boterenbrood, E.C., Nieuwkoop, P.D., 1973. The formation of the mesoderm in urodelean amphibians. Wilhelm Roux' Archiv für Entwicklungsmechanik der Organismen 173, 319–332.

Bowes, J., Snyder, K., Segerdell, E., Jarabek, C., Azam, K., Zorn, A., Vize, P., 2010. Xenbase: gene expression and improved integration. Nucleic Acids Res. 38, D607–612.

Brain, C.K., Prave, A.R., Hoffmann, K.H., Fallick, A.E., Botha, A., Herd, D.A., Sturrock, C., Young, I., Condon, D.J., Allison, S.G., 2012. The first animals: ca. 760 million-year-old sponge-like fossils from Namibia. S. Afr. J. Sci. 108, 83–90.

Britten, R.J., Davidson, E.H., 1969. Gene regulation for higher cells: a theory. Science 165, 349–357.

Browne, E.N., 1908. The Production of New Hydranths in Hydra by the Insertion of Small Grafts. J. Exp. Zool. VIII, 1–35.

Butterfield, N.J., 2000. *Bangiomorpha pubescens* n. gen., n. sp.: implications for the evolution of sex, multicellularity, and the Mesoproterozoic/Neoproterozoic radiation of eukaryotes. Paleobiology 26, 386–404.

Cambray, N., Wilson, V., 2007. Two distinct sources for a population of maturing axial progenitors. Development 134, 2829–2840.

Cantone, I., Fisher, A.G., 2013. Epigenetic programming and reprogramming during development. Nat. Struct. Mol. Biol. 20, 282–289.

Carnac, G., Kodjabachian, L., Gurdon, J.B., Lemaire, P., 1996. The homeobox gene Siamois is a target of the Wnt dorsalisation pathway and triggers organiser activity in the absence of mesoderm. Development 122, 3055–3065.

Cha, S.W., Tadjuidje, E., Tao, Q., Wylie, C., Heasman, J., 2008. Wnt5a and Wnt11 interact in a maternal Dkk1-regulated fashion to activate both canonical and non-canonical signaling in *Xenopus* axis formation. Development 135, 3719–3729.

Chan, T.M., Longabaugh, W., Bolouri, H., Chen, H.L., Tseng, W.F., Chao, C.H., Jang, T.H., Lin, Y.I., Hung, S.C., Wang, H.D., Yuh, C.H., 2009. Developmental gene regulatory networks in the zebrafish embryo. Biochim. Biophys. Acta 1789, 279–298.

Chapman, D.L., Papaioannou, V.E., 1998. Three neural tubes in mouse embryos with mutations in the T-box gene Tbx6. Nature 391, 695–697.

Chen, S., Kimelman, D., 2000. The role of the yolk syncytial layer in germ layer patterning in zebrafish. Development 127, 4681–4689.

Chen, Y., Schier, A.F., 2001. The zebrafish Nodal signal Squint functions as a morphogen. Nature 411, 607–610.

Chen, Y., Schier, A.F., 2002. Lefty proteins are long-range inhibitors of squint-mediated nodal signaling. Curr. Biol. 12, 2124–2128.

Cheng, A.M., Thisse, B., Thisse, C., Wright, C.V., 2000. The lefty-related factor Xatv acts as a feedback inhibitor of nodal signaling in mesoderm induction and L-R axis development in *Xenopus*. Development 127, 1049–1061.

Chesley, P., 1935. *Development* of the short-tailed mutant in the house mouse. J. Exp. Zool. 70, 429–459.

Clement, J.H., Fettes, P., Knochel, S., Lef, J., Knochel, W., 1995. Bone morphogenetic protein 2 in the early development of *Xenopus laevis*. Mech. Dev. 52, 357–370.

Concha, M.L., Adams, R.J., 1998. Oriented cell divisions and cellular morphogenesis in the zebrafish gastrula and neurula: a time-lapse analysis. Development 125, 983–994.

Conklin, E.G., 1905. The Organizaton and *Cell* Lineage of the Ascidian Egg. J. Acad. Nat. Sci. Phila. 3, 1–119.

Conlon, F.L., Barth, K.S., Robertson, E.J., 1991. A novel retrovirally induced embryonic lethal mutation in the mouse: assessment of the developmental fate of embryonic stem cells homozygous for the 413.d proviral integration. Development 111, 969–981.

Conlon, F.L., Lyons, K.M., Takaesu, N., Barth, K.S., Kispert, A., Herrmann, B., Robertson, E.J., 1994. A primary requirement for nodal in the formation and maintenance of the primitive streak in the mouse. Development 120, 1919–1928.

Constam, D.B., Robertson, E.J., 2000. SPC4/PACE4 regulates a TGFbeta signaling network during axis formation. Genes Dev. 14, 1146–1155.

Copp, A.J., Greene, N.D., 2010. Genetics and development of neural tube defects. J. Pathol. 220, 217–230.

Corbel, C., Salaun, J., Belo-Diabangouaya, P., Dieterlen-Lievre, F., 2007. Hematopoietic potential of the pre-fusion allantois. Dev. Biol. 301, 478–488.

Core, L.J., Waterfall, J.J., Lis, J.T., 2008. Nascent RNA sequencing reveals widespread pausing and divergent initiation at human promoters. Science 322, 1845–1848.

Cunliffe, V., Smith, J.C., 1992. Ectopic mesoderm formation in *Xenopus* embryos caused by widespread expression of a Brachyury homologue. Nature 358, 427–430.

Dalcq, A., Pasteels, J., 1937. Une conception nouvelle des bases physiologiques de la morphogenese. Arch. Biol. 48, 669–710.

Darwin, C., 1859. On the origin of species by means of natural selection. J. Murray, London.

Darwin, E., 1794. Zoonomia; or the Laws of Organic Life. J. Johnson, London.

David, N.B., Rosa, F.M., 2001. Cell autonomous commitment to an endodermal fate and behaviour by activation of Nodal signalling. Development 128, 3937–3947.

Davidson, E.H., 2006. The regulatory genome: gene regulatory networks in development and evolution. Academic, Burlington, MA; San Diego.

Davis, R.L., Kirschner, M.W., 2000. The fate of cells in the tailbud of *Xenopus laevis*. Development 127, 255–267.

De Robertis, E.M., Kuroda, H., 2004. Dorsal-ventral patterning and neural induction in *Xenopus* embryos. Annu. Rev. Cell Dev. Biol. 20, 285–308.

de Souza, F.S., Niehrs, C., 2000. Anterior endoderm and head induction in early vertebrate embryos. Cell Tissue Res. 300, 207–217.

Delfino-Machin, M., Lunn, J.S., Breitkreuz, D.N., Akai, J., Storey, K.G., 2005. Specification and maintenance of the spinal cord stem zone. Development 132, 4273–4283.

Dequeant, M.L., Pourquie, O., 2008. Segmental patterning of the vertebrate embryonic axis. Nat. Rev. Genet. 9, 370–382.

Derossi, D., Calvet, S., Trembleau, A., Brunissen, A., Chassaing, G., Prochiantz, A., 1996. *Cell* internalization of the third helix of the Antennapedia homeodomain is receptor-independent. J. Biol. Chem. 271, 18188–18193.

Dobrovolskaïa-Zavadskaïa N, 1927. Sur la mortification spontanée de la queue chez la souris noveau-née et sur l'existence d'un caractère (facteur) héréditaire "non-viable". C. R. Seanc. Soc. Biol. 97, 114–116.

Domazet-Loso, T., Tautz, D., 2010. A phylogenetically based transcriptome age index mirrors ontogenetic divergence patterns. Nature 468, 815–818.

Droser, M.L., Gehling, J.G., 2008. Synchronous aggregate growth in an abundant new Ediacaran tubular organism. Science 319, 1660–1662.

Droser, M.L., Jensen, S., Gehling, J.G., 2002. Trace fossils and substrates of the terminal Proterozoic-Cambrian transition: implications for the record of early bilaterians and sediment mixing. Proc. Natl. Acad. Sci. USA 99, 12572–12576.

Duboule, D., 1994. Temporal colinearity and the phylotypic progression: a basis for the stability of a vertebrate Bauplan and the evolution of morphologies through heterochrony. Dev. Suppl., 135–142.

Dunty Jr., W.C., Biris, K.K., Chalamalasetty, R.B., Taketo, M.M., Lewandoski, M., Yamaguchi, T.P., 2008. Wnt3a/beta-catenin signaling controls posterior body development by coordinating mesoderm formation and segmentation. Development 135, 85–94.

England, S.J., Blanchard, G.B., Mahadevan, L., Adams, R.J., 2006. A dynamic fate map of the forebrain shows how vertebrate eyes form and explains two causes of cyclopia. Development 133, 4613–4617.

Erter, C.E., Solnica-Krezel, L., Wright, C.V., 1998. Zebrafish nodal-related 2 encodes an early mesendodermal inducer signaling from the extra-embryonic yolk syncytial layer. Dev. Biol. 204, 361–372.

Erwin, D., Valentine, J., Jablonski, D., 1997. The Origin of Animal Body Plans. Am. Sci. 85, 126–137.

Evans, A.L., Faial, T., Gilchrist, M.J., Down, T., Vallier, L., Pedersen, R.A., Wardle, F.C., Smith, J.C., 2012. Genomic targets of Brachyury (T) in differentiating mouse embryonic stem cells. PLoS One 7, e33346.

Eyal-Giladi, H., 1970. Differentiation potencies of the young chick blastoderm as revealed by different manipulations. II. Localized damage and hypoblast removal experiments. J. Embryol. Exp. Morphol. 23, 739–749.

Fan, X., Hagos, E.G., Xu, B., Sias, C., Kawakami, K., Burdine, R.D., Dougan, S.T., 2007. Nodal signals mediate interactions between the extra-embryonic and embryonic tissues in zebrafish. Dev. Biol. 310, 363–378.

Fehling, H.J., Lacaud, G., Kubo, A., Kennedy, M., Robertson, S., Keller, G., Kouskoff, V., 2003. Tracking mesoderm induction and its specification to the hemangioblast during embryonic stem cell differentiation. Development 130, 4217–4227.

Fekany, K., Yamanaka, Y., Leung, T., Sirotkin, H.I., Topczewski, J., Gates, M.A., Hibi, M., Renucci, A., Stemple, D., Radbill, A., Schier, A.F., Driever, W., Hirano, T., Talbot, W.S., Solnica-Krezel, L., 1999. The zebrafish bozozok locus encodes Dharma, a homeodomain protein essential for induction of gastrula organizer and dorsoanterior embryonic structures. Development 126, 1427–1438.

Feldman, B., Dougan, S.T., Schier, A.F., Talbot, W.S., 2000. Nodal-related signals establish mesendodermal fate and trunk neural identity in zebrafish. Curr. Biol. 10, 531–534.

Feldman, B., Poueymirou, W., Papaioannou, V.E., DeChiara, T.M., Goldfarb, M., 1995. Requirement of FGF-4 for postimplantation mouse development. Science 267, 246–249.

Feldman, B., Gates, M.A., Egan, E.S., Dougan, S.T., Rennebeck, G., Sirotkin, H.I., Schier, A.F., Talbot, W.S., 1998. Zebrafish organizer development and germ-layer formation require nodal-related signals. Nature 395, 181–185.

Ferrer-Vaquer, A., Hadjantonakis, A.K., 2013. Birth defects associated with perturbations in preimplantation, gastrulation, and axis extension: from conjoined twinning to caudal dysgenesis. Wiley Interdiscip. Rev. Dev. Biol. 2, 427–442.

Fletcher, R.B., Harland, R.M., 2008. The role of FGF signaling in the establishment and maintenance of mesodermal gene expression in *Xenopus*. Dev. Dyn. 237, 1243–1254.

Flores, M.V., Lam, E.Y., Crosier, K.E., Crosier, P.S., 2008. Osteogenic transcription factor Runx2 is a maternal determinant of dorsoventral patterning in zebrafish. Nat. Cell Biol. 10, 346–352.

Garrido-Allepuz, C., Haro, E., Gonzalez-Lamuno, D., Martinez-Frias, M.L., Bertocchini, F., Ros, M.A., 2011. A clinical and experimental overview of sirenomelia: insight into the mechanisms of congenital limb malformations. Dis. Model. Mech. 4, 289–299.

Gekas, C., Dieterlen-Lievre, F., Orkin, S.H., Mikkola, H.K., 2005. The placenta is a niche for hematopoietic stem cells. Dev. Cell 8, 365–375.

Gerhart, J., Kirschner, M., 2007. The theory of facilitated variation. Proc. Natl. Acad. Sci. USA 104 (Suppl. 1), 8582–8589.

Germain, S., Howell, M., Esslemont, G.M., Hill, C.S., 2000. Homeodomain and winged-helix transcription factors recruit activated Smads to distinct promoter elements via a common Smad interaction motif. Genes Dev. 14, 435–451.

Gierer, A., Meinhardt, H., 1972. A theory of biological pattern formation. Kybernetik 12, 30–39.

Gilbert, S.F., 2006. Developmental biology. Sinauer Associates, Inc. Publishers, Sunderland, Mass.

Gilbert, S.F., 2010. Developmental biology. Sinauer Associates, Sunderland, Mass.

Goldman, D.C., Martin, G.R., Tam, P.P., 2000. Fate and function of the ventral ectodermal ridge during mouse tail development. Development 127, 2113–2123.

Gont, L.K., Steinbeisser, H., Blumberg, B., de Robertis, E.M., 1993. Tail formation as a continuation of gastrulation: the multiple cell populations of the *Xenopus* tailbud derive from the late blastopore lip. Development 119, 991–1004.

Gore, A.V., Sampath, K., 2002. Localization of transcripts of the zebrafish morphogen Squint is dependent on egg activation and the microtubule cytoskeleton. Mech. Dev. 112, 153–156.

Greene, N.D., Gerrelli, D., Van Straaten, H.W., Copp, A.J., 1998. Abnormalities of floor plate, notochord and somite differentiation in the loop-tail (Lp) mouse: a model of severe neural tube defects. Mech. Dev. 73, 59–72.

Griffin, K., Patient, R., Holder, N., 1995. Analysis of FGF function in normal and no tail zebrafish embryos reveals separate mechanisms for formation of the trunk and the tail. Development 121, 2983–2994.

Gritsman, K., Zhang, J., Cheng, S., Heckscher, E., Talbot, W.S., Schier, A.F., 1999. The EGF-CFC protein one-eyed pinhead is essential for nodal signaling. Cell 97, 121–132.

Grobben, K., 1908. Die systematische Einteilung des Tierreiches. Verh. kaiserlich-koniglichen zoologisch-botanischen Ges. Wien 58, 491–511.

Gurdon, J.B., 1988. A community effect in animal development. Nature 336, 772–774.

Gurdon, J.B., Elsdale, T.R., Fischberg, M., 1958. Sexually mature individuals of *Xenopus laevis* from the transplantation of single somatic nuclei. Nature 182, 64–65.

Haeckel, E., 1874. Die Gastraea-Theorie, die phylogenetische Classification des Thierreichs und die Homologie der Keimblätter. Jenaische Zeitschr. Naturwiss. 8, 1–55.

Hagos, E.G., Dougan, S.T., 2007. Time-dependent patterning of the mesoderm and endoderm by Nodal signals in zebrafish. BMC Dev. Biol. 7, 22.

Hall, B.K., 2000. The neural crest as a fourth germ layer and vertebrates as quadroblastic not triploblastic. Evol. Dev. 2, 3–5.

Handrigan, G.R., 2003. Concordia discors: duality in the origin of the vertebrate tail. J. Anat. 202, 255–267.

Harrisson, F., Callebaut, M., Vakaet, L., 1991. Features of polyingression and primitive streak ingression through the basal lamina in the chicken blastoderm. Anat. Rec. 229, 369–383.

Harvey, S.A., Smith, J.C., 2009. Visualisation and quantification of morphogen gradient formation in the zebrafish. PLoS Biol. 7, e1000101.

Harvey, S.A., Tumpel, S., Dubrulle, J., Schier, A.F., Smith, J.C., 2010. no tail integrates two modes of mesoderm induction. Development 137, 1127–1135.

Hashimoto, H., Itoh, M., Yamanaka, Y., Yamashita, S., Shimizu, T., Solnica-Krezel, L., Hibi, M., Hirano, T., 2000. Zebrafish Dkk1 functions in forebrain specification and axial mesendoderm formation. Dev. Biol. 217, 138–152.

Heasman, J., Snape, A., Smith, J., Wylie, C.C., 1985. Single cell analysis of commitment in early embryogenesis. J. Embryol. Exp. Morphol. 89 (Suppl.), 297–316.

Heasman, J., Wessely, O., Langland, R., Craig, E.J., Kessler, D.S., 2001. Vegetal localization of maternal mRNAs is disrupted by VegT depletion. Dev. Biol. 240, 377–386.

Heasman, J., Crawford, A., Goldstone, K., Garner-Hamrick, P., Gumbiner, B., McCrea, P., Kintner, C., Noro, C.Y., Wylie, C., 1994. Overexpression of cadherins and underexpression of beta-catenin inhibit dorsal mesoderm induction in early *Xenopus* embryos. Cell 79, 791–803.

Heisenberg, C.P., Nusslein-Volhard, C., 1997. The function of silberblick in the positioning of the eye anlage in the zebrafish embryo. Dev. Biol. 184, 85–94.

Hild, M., Dick, A., Rauch, G.J., Meier, A., Bouwmeester, T., Haffter, P., Hammerschmidt, M., 1999. The smad5 mutation somitabun blocks Bmp2b signaling during early dorsoventral patterning of the zebrafish embryo. Development 126, 2149–2159.

Hirokawa, N., Tanaka, Y., Okada, Y., 2012. Cilia, KIF3 molecular motor and nodal flow. Curr. Opin. Cell Biol. 24, 31–39.

Hong, S.K., Jang, M.K., Brown, J.L., McBride, A.A., Feldman, B., 2011. Embryonic mesoderm and endoderm induction requires the actions of non-embryonic Nodal-related ligands and Mxtx2. Development 138, 787–795.

Houart, C., Caneparo, L., Heisenberg, C., Barth, K., Take-Uchi, M., Wilson, S., 2002. Establishment of the telencephalon during gastrulation by local antagonism of Wnt signaling. Neuron 35, 255–265.

Huxley, T.H., 1875. On the Classification of the Animal Kingdom. Am. Nat. 9, 65–70.

Inui, M., Montagner, M., Ben-Zvi, D., Martello, G., Soligo, S., Manfrin, A., Aragona, M., Enzo, E., Zacchigna, L., Zanconato, F., Azzolin, L., Dupont, S., Cordenonsi, M., Piccolo, S., 2012. Self-regulation of the head-inducing properties of the Spemann organizer. Proc. Natl. Acad. Sci. USA 109, 15354–15359.

Isern, J., Fraser, S.T., He, Z., Baron, M.H., 2008. The fetal liver is a niche for maturation of primitive erythroid cells. Proc. Natl. Acad. Sci. USA 105, 6662–6667.

Jacob, F., 1965. Genetics of the Bacterial Cell. Nobel Lect. http://www.nobelprize.org.

Jensen, L.E., Barbaux, S., Hoess, K., Fraterman, S., Whitehead, A.S., Mitchell, L.E., 2004. The human T locus and spina bifida risk. Hum. Genet. 115, 475–482.

Jones, C.M., Kuehn, M.R., Hogan, B.L., Smith, J.C., Wright, C.V., 1995. Nodal-related signals induce axial mesoderm and dorsalize mesoderm during gastrulation. Development 121, 3651–3662.

Jullien, J., Gurdon, J., 2005. Morphogen gradient interpretation by a regulated trafficking step during ligand-receptor transduction. Genes Dev. 19, 2682–2694.

Kalinka, A.T., Varga, K.M., Gerrard, D.T., Preibisch, S., Corcoran, D.L., Jarrells, J., Ohler, U., Bergman, C.M., Tomancak, P., 2010. Gene expression divergence recapitulates the developmental hourglass model. Nature 468, 811–814.

Kane, D., Adams, R., 2002. Life at the edge: epiboly and involution in the zebrafish. Results Probl. Cell Differ. 40, 117–135.

Kazanskaya, O., Glinka, A., Niehrs, C., 2000. The role of *Xenopus* dickkopf1 in prechordal plate specification and neural patterning. Development 127, 4981–4992.

Kelly, C., Chin, A.J., Leatherman, J.L., Kozlowski, D.J., Weinberg, E.S., 2000. Maternally controlled (beta)-catenin-mediated signaling is required for organizer formation in the zebrafish. Development 127, 3899–3911.

Kikuchi, K., Poss, K.D., 2012. Cardiac regenerative capacity and mechanisms. Annu. Rev. Cell Dev. Biol. 28, 719–741.

Kimmel, C.B., Warga, R.M., Schilling, T.F., 1990. Origin and organization of the zebrafish fate map. Development 108, 581–594.

Kingsley, P.D., Malik, J., Fantauzzo, K.A., Palis, J., 2004. Yolk sac-derived primitive erythroblasts enucleate during mammalian embryogenesis. Blood 104, 19–25.

Kirschner, M., Gerhart, J., 2005. The plausibility of life : resolving Darwin's dilemma. Yale University Press, New Haven.

Kispert, A., Hermann, B.G., 1993. The Brachyury gene encodes a novel DNA binding protein. EMBO J. 12, 4898–4899.

Knoll, A.H., 2011. The multiple origins of complex multicellularity. Annual Review of Earth and Planetary Sciences 39, 217–239.

Kobel, H.R., Du Pasquier, L., 1979. Hyperdiploid species hybrids for gene mapping in *Xenopus*. Nature 279, 157–158.

Kofron, M., Demel, T., Xanthos, J., Lohr, J., Sun, B., Sive, H., Osada, S., Wright, C., Wylie, C., Heasman, J., 1999. Mesoderm induction in *Xenopus* is a zygotic event regulated by maternal VegT via TGFbeta growth factors. Development 126, 5759–5770.

Kohga, H., Obata, K., 1992. Retinoic acid-induced neural tube defects with multiple canals in the chick: immunohistochemistry with monoclonal antibodies. Neurosci. Res. 13, 175–187.

Koide, T., Hayata, T., Cho, K.W., 2005. *Xenopus* as a model system to study transcriptional regulatory networks. Proc. Natl. Acad. Sci. USA 102, 4943–4948.

Kovalevsky, A., 1868. Beitrag zur Entwicklungsgeschichte der Tunicaten. Nachrichten Ges. Wiss. Göttingen 19, 401–415.

Kumari, P., Gilligan, P.C., Lim, S., Tran, L.D., Winkler, S., Philp, R., Sampath, K., 2013. An essential role for maternal control of Nodal signaling. Elife 2, e00683.

Kunwar, P.S., Zimmerman, S., Bennett, J.T., Chen, Y., Whitman, M., Schier, A.F., 2003. Mixer/Bon and FoxH1/Sur have overlapping and divergent roles in Nodal signaling and mesendoderm induction. Development 130, 5589–5599.

Lagha, M., Bothma, J.P., Levine, M., 2012. Mechanisms of transcriptional precision in animal development. Trends Genet. 28, 409–416.

Lamarck, J.B., 1809. Philosophie zoologique ou exposition des considérations relatives à l'histoire naturelle des animaux. Paris.

Langdon, Y.G., Mullins, M.C., 2011. Maternal and zygotic control of zebrafish dorsoventral axial patterning. Annu. Rev. Genet. 45, 357–377.

Lankester, E.R., 1873. On the primitive cell-layers of the embryo as the basis of genealogical classification of animals, and on the origin of vascular and lymph systems. Ann. Mag. Nat. Hist. 11, 321–338.

Lankester, E.R., 1877. Notes on the Embryology and classification of the Animal kingdom: comprising a revision of speculations relative to the origin and significance of the germ-layers. Q. J. Microscopical Sci. 68, 399–454.

Latinkic, B.V., Smith, J.C., 1999. Goosecoid and mix.1 repress Brachyury expression and are required for head formation in *Xenopus*. Development 126, 1769–1779.

Laurent, M.N., Blitz, I.L., Hashimoto, C., Rothbacher, U., Cho, K.W., 1997. The *Xenopus* homeobox gene twin mediates Wnt induction of goosecoid in establishment of Spemann's organizer. Development 124, 4905–4916.

Lawson, K.A., Hage, W.J., 1994. Clonal analysis of the origin of primordial germ cells in the mouse. Ciba Found. Symp. 182, 68–84. discussion 84–91.

Lawson, K.A., Meneses, J.J., Pedersen, R.A., 1991. Clonal analysis of epiblast fate during germ layer formation in the mouse embryo. Development 113, 891–911.

Le Douarin, N.M., 1980. The ontogeny of the neural crest in avian embryo chimaeras. Nature 286, 663–669.

Leichsenring, M., Maes, J., Mossner, R., Driever, W., Onichtchouk, D., 2013. Pou5f1 transcription factor controls zygotic gene activation in vertebrates. Science 341, 1005–1009.

Lemaire, P., Darras, S., Caillol, D., Kodjabachian, L., 1998. A role for the vegetally expressed *Xenopus* gene Mix.1 in endoderm formation and in the restriction of mesoderm to the marginal zone. Development 125, 2371–2380.

Levine, M., Davidson, E.H., 2005. Gene regulatory networks for development. Proc. Natl. Acad. Sci. USA 102, 4936–4942.

Lewis, W., 1907. Transplantation of the lips of the blastopore in Rana palustris. Am. J. Anat. 7, 137–143.

Liang, H.L., Nien, C.Y., Liu, H.Y., Metzstein, M.M., Kirov, N., Rushlow, C., 2008. The zinc-finger protein Zelda is a key activator of the early zygotic genome in Drosophila. Nature 456, 400–403.

Lim, S., Kumari, P., Gilligan, P., Quach, H.N., Mathavan, S., Sampath, K., 2012. Dorsal activity of maternal squint is mediated by a non-coding function of the RNA. Development 139, 2903–2915.

Lindeman, L.C., Andersen, I.S., Reiner, A.H., Li, N., Aanes, H., Østrup, O., Winata, C., Mathavan, S., Müller, F., Aleström, P., Collas, P., 2011. Prepatterning of developmental gene expression by modified histones before zygotic genome activation. Dev. Cell 21, 993–1004.

Liu, P., Wakamiya, M., Shea, M.J., Albrecht, U., Behringer, R.R., Bradley, A., 1999. Requirement for Wnt3 in vertebrate axis formation. Nat. Genet. 22, 361–365.

Longabaugh, W.J., Davidson, E.H., Bolouri, H., 2005. Computational representation of developmental genetic regulatory networks. Dev. Biol. 283, 1–16.

Loose, M., Patient, R., 2004. A genetic regulatory network for *Xenopus* mesendoderm formation. Dev. Biol. 271, 467–478.

Lu, F.I., Thisse, C., Thisse, B., 2011. Identification and mechanism of regulation of the zebrafish dorsal determinant. Proc. Natl. Acad. Sci. USA 108, 15876–15880.

Macklon, N.S., Geraedts, J.P., Fauser, B.C., 2002. Conception to ongoing pregnancy: the 'black box' of early pregnancy loss. Hum. Reprod. Update 8, 333–343.

Marcon, L., Sharpe, J., 2012. Turing patterns in development: what about the horse part? Curr. Opin. Genet. Dev. 22, 578–584.

Martin, B.L., Kimelman, D., 2008. Regulation of canonical Wnt signaling by Brachyury is essential for posterior mesoderm formation. Dev. Cell 15, 121–133.

Martin, B.L., Kimelman, D., 2010. Brachyury establishes the embryonic mesodermal progenitor niche. Genes Dev. 24, 2778–2783.

Martin, B.L., Kimelman, D., 2012. Canonical Wnt signaling dynamically controls multiple stem cell fate decisions during vertebrate body formation. Dev. Cell 22, 223–232.

Martin, M.W., Grazhdankin, D.V., Bowring, S.A., Evans, D.A., Fedonkin, M.A., Kirschvink, J.L., 2000. Age of Neoproterozoic bilatarian body and trace fossils, White Sea, Russia: implications for metazoan evolution. Science 288, 841–845.

Meinhardt, H., 2003. The algorithmic beauty of sea shells. Heidelberg, New York.

Mesnard, D., Donnison, M., Fuerer, C., Pfeffer, P.L., Constam, D.B., 2011. The microenvironment patterns the pluripotent mouse epiblast through paracrine Furin and Pace4 proteolytic activities. Genes Dev. 25, 1871–1880.

Middleton, A.M., King, J.R., Loose, M., 2009. Bistability in a model of mesoderm and anterior mesendoderm specification in *Xenopus* laevis. J. Theor. Biol. 260, 41–55.

Ming, J.E., Muenke, M., 2002. Multiple hits during early embryonic development: digenic diseases and holoprosencephaly. Am. J. Hum. Genet. 71, 1017–1032.

Mishina, Y., Suzuki, A., Ueno, N., Behringer, R.R., 1995. Bmpr encodes a type I bone morphogenetic protein receptor that is essential for gastrulation during mouse embryogenesis. Genes Dev. 9, 3027–3037.

Mizuno, T., Yamaha, E., Wakahara, M., Kuroiwa, M., Takeda, H., 1996. Mesoderm induction in zebrafish (commentary). Nature 383, 131–132.

Moody, S.A., 1987a. Fates of the blastomeres of the 16-cell stage *Xenopus* embryo. Dev. Biol. 119, 560–578.

Moody, S.A., 1987b. Fates of the blastomeres of the 32-cell-stage *Xenopus* embryo. Dev. Biol. 122, 300–319.

Morin-Kensicki, E.M., Melancon, E., Eisen, J.S., 2002. Segmental relationship between somites and vertebral column in zebrafish. Development 129, 3851–3860.

Morley, R.H., Lachani, K., Keefe, D., Gilchrist, M.J., Flicek, P., Smith, J.C., Wardle, F.C., 2009. A gene regulatory network directed by zebrafish No tail accounts for its roles in mesoderm formation. Proc. Natl. Acad. Sci. USA 106, 3829–3834.

Morris, S.A., Almeida, A.D., Tanaka, H., Ohta, K., Ohnuma, S., 2007. Tsukushi modulates Xnr2, FGF and BMP signaling: regulation of *Xenopus* germ layer formation. PLoS One 2, e1004.

Morrison, K., Papapetrou, C., Attwood, J., Hol, F., Lynch, S.A., Sampath, A., Hamel, B., Burn, J., Sowden, J., Stott, D., Mariman, E., Edwards, Y.H., 1996. Genetic mapping of the human homologue (T) of mouse T(Brachyury) and a search for allele association between human T and spina bifida. Hum. Mol. Genet. 5, 669–674.

Muller, P., Rogers, K.W., Jordan, B.M., Lee, J.S., Robson, D., Ramanathan, S., Schier, A.F., 2012. Differential diffusivity of Nodal and Lefty underlies a reaction-diffusion patterning system. Science 336, 721–724.

Murry, C.E., Keller, G., 2008. Differentiation of embryonic stem cells to clinically relevant populations: lessons from embryonic development. Cell 132, 661–680.

Nakamura, T., Mine, N., Nakaguchi, E., Mochizuki, A., Yamamoto, M., Yashiro, K., Meno, C., Hamada, H., 2006. Generation of robust left-right asymmetry in the mouse embryo requires a self-enhancement and lateral-inhibition system. Dev. Cell 11, 495–504.

Neumann, C., Cohen, S., 1997. Morphogens and pattern formation. Bioessays 19, 721–729.

Nieuwkoop, P.D., 1969. The formation of the mesoderm in the urodelean amphibians. I. Induction by the endoderm. Wilhelm Roux' Arch. Entwicklungsmechanik der Organismen 162, 341–373.

Nishimura, H., Takano, K., Tanimura, T., Yasuda, M., 1968. Normal and abnormal development of human embryos: first report of the analysis of 1,213 intact embryos. Teratology 1, 281–290.

Nojima, H., Rothhamel, S., Shimizu, T., Kim, C.H., Yonemura, S., Marlow, F.L., Hibi, M., 2010. Syntabulin, a motor protein linker, controls dorsal determination. Development 137, 923–933.

Nowotschin, S., Ferrer-Vaquer, A., Concepcion, D., Papaioannou, V.E., Hadjantonakis, A.K., 2012. Interaction of Wnt3a, Msgn1 and Tbx6 in neural versus paraxial mesoderm lineage commitment and paraxial mesoderm differentiation in the mouse embryo. Dev. Biol. 367, 1–14.

Nowotschin, S., Costello, I., Piliszek, A., Kwon, G.S., Mao, C.A., Klein, W.H., Robertson, E.J., Hadjantonakis, A.K., 2013. The T-box transcription factor Eomesodermin is essential for AVE induction in the mouse embryo. Genes Dev. 27, 997–1002.

Ohta, K., Kuriyama, S., Okafuji, T., Gejima, R., Ohnuma, S., Tanaka, H., 2006. Tsukushi cooperates with VG1 to induce primitive streak and Hensen's node formation in the chick embryo. Development 133, 3777–3786.

Oppenheimer, J.M., 1959. Extra-embryonic transplantation of sections of the fundulus embryonic shield. J. Exp. Zool. 140, 247–267.

Osada, S.I., Wright, C.V., 1999. *Xenopus* nodal-related signaling is essential for mesendodermal patterning during early embryogenesis. Development 126, 3229–3240.

Ottersbach, K., Dzierzak, E., 2005. The murine placenta contains hematopoietic stem cells within the vascular labyrinth region. Dev. Cell 8, 377–387.

Padmanabhan, R., 1998. Retinoic acid-induced caudal regression syndrome in the mouse fetus. Reprod. Toxicol. 12, 139–151.

Pander, C.H., 1817. Beiträge zur Entwicklungsgeschichte des Hünchens im Eye. Heinrich Ludwig Brönner, Würzburg.

Papapetrou, C., Edwards, Y.H., Sowden, J.C., 1997. The T transcription factor functions as a dimer and exhibits a common human polymorphism Gly-177-Asp in the conserved DNA-binding domain. FEBS Lett. 409, 201–206.

Papin, C., van Grunsven, L.A., Verschueren, K., Huylebroeck, D., Smith, J.C., 2002. Dynamic regulation of Brachyury expression in the amphibian embryo by XSIP1. Mech. Dev. 111, 37–46.

Park, D., Park, J., Park, S.G., Park, T., Choi, S.S., 2008. Analysis of human disease genes in the context of gene essentiality. Genomics 92, 414–418.

Peeters, H., Devriendt, K., 2006. Human laterality disorders. Eur. J. Med. Genet. 49, 349–362.

Pellettieri, J., Sanchez Alvarado, A., 2007. Cell turnover and adult tissue homeostasis: from humans to planarians. Annu. Rev. Genet. 41, 83–105.

Perea-Gomez, A., Vella, F.D., Shawlot, W., Oulad-Abdelghani, M., Chazaud, C., Meno, C., Pfister, V., Chen, L., Robertson, E., Hamada, H., Behringer, R.R., Ang, S.L., 2002. Nodal antagonists in the anterior visceral endoderm prevent the formation of multiple primitive streaks. Dev. Cell 3, 745–756.

Picozzi, P., Wang, F., Cronk, K., Ryan, K., 2009. Eomesodermin requires transforming growth factor-beta/activin signaling and binds Smad2 to activate mesodermal genes. J. Biol. Chem. 284, 2397–2408.

Pillay, N., Plagnol, V., Tarpey, P.S., Lobo, S.B., Presneau, N., Szuhai, K., Halai, D., Berisha, F., Cannon, S.R., Mead, S., Kasperaviciute, D., Palmen, J., Talmud, P.J., Kindblom, L.G., Amary, M.F., Tirabosco, R., Flanagan, A.M., 2012. A common single-nucleotide variant in T is strongly associated with chordoma. Nat. Genet. 44, 1185–1187.

Postlethwait, J.H., Woods, I.G., Ngo-Hazelett, P., Yan, Y.L., Kelly, P.D., Chu, F., Huang, H., Hill-Force, A., Talbot, W.S., 2000. Zebrafish comparative genomics and the origins of vertebrate chromosomes. Genome Res. 10, 1890–1902.

Quemelo, P.R., Lourenco, C.M., Peres, L.C., 2007. Teratogenic effect of retinoic acid in swiss mice. Acta Cir. Bras. 22, 451–456.

Ramel, M.C., Lekven, A.C., 2004. Repression of the vertebrate organizer by Wnt8 is mediated by Vent and Vox. Development 131, 3991–4000.

Ramsdell, A.F., 2005. Left-right asymmetry and congenital cardiac defects: getting to the heart of the matter in vertebrate left-right axis determination. Dev. Biol. 288, 1–20.

Rathke, H., 1829. Untersuchungen über die Bildung und Entwicklung des Flusskrebses. Voss Hirschfeld, Leipzig.

Raya, A., Izpisúa Belmonte, J.C., 2006. Left-right asymmetry in the vertebrate embryo: from early information to higher-level integration. Nat. Rev. Genet. 7, 283–293.

Rebagliati, M.R., Toyama, R., Haffter, P., Dawid, I.B., 1998a. Cyclops encodes a nodal-related factor involved in midline signaling. Proc. Natl. Acad. Sci. USA 95, 9932–9937.

Rebagliati, M.R., Toyama, R., Fricke, C., Haffter, P., Dawid, I.B., 1998b. Zebrafish nodal-related genes are implicated in axial patterning and establishing left-right asymmetry. Dev. Biol. 199, 261–272.

Reim, G., Brand, M., 2006. Maternal control of vertebrate dorsoventral axis formation and epiboly by the POU domain protein Spg/Pou2/Oct4. Development 133, 2757–2770.

Remak, R., 1855. Untersuchungen über die Entwicklung der Wirbeltiere. G. Reimer, Berlin.

Ro, H., Dawid, I.B., 2009. Organizer restriction through modulation of Bozozok stability by the E3 ubiquitin ligase Lnx-like. Nat. Cell Biol. 11, 1121–1127.

Roessler, E., Muenke, M., 2010. The molecular genetics of holoprosencephaly. Am. J. Med. Genet. C. Semin. Med. Genet. 154C, 52–61.

Roessler, E., Pei, W., Ouspenskaia, M.V., Karkera, J.D., Veléz, J.I., Banerjee-Basu, S., Gibney, G., Lupo, P.J., Mitchell, L.E., Towbin, J.A., Bowers, P., Belmont, J.W., Goldmuntz, E., Baxevanis, A.D., Feldman, B., Muenke, M., 2008. Reduced NODAL signaling strength via mutation of several pathway members including FOXH1 is linked to human heart defects and holoprosencephaly. Am. J. Hum. Genet. 83, 18–29.

Roessler, E., Pei, W., Ouspenskaia, M.V., Karkera, J.D., Veléz, J.I., Banerjee-Basu, S., Gibney, G., Lupo, P.J., Mitchell, L.E., Towbin, J.A., Bowers, P., Belmont, J.W., Goldmuntz, E., Baxevanis, A.D., Feldman, B., Muenke, M., 2009. Cumulative ligand activity of NODAL mutations and modifiers are linked to human heart defects and holoprosencephaly. Mol. Genet. Metab. 98, 225–234.

Rynn, L., Cragan, J., Correa, A., 2008. Update on overall prevalence of major birth defects – Atlanta, Georgia, 1978–2005. Morb. Mortal. Wkly. Rep. Centers for Dis. Control 57, 1–5.

Ryu, S.L., Fujii, R., Yamanaka, Y., Shimizu, T., Yabe, T., Hirata, T., Hibi, M., Hirano, T., 2001. Regulation of dharma/bozozok by the Wnt pathway. Dev. Biol. 231, 397–409.

Sadler, T.W., Langman, J., 2006. Langman's medical embryology. Lippincott Williams & Wilkins, Philadelphia.

Sadler, T.W., Langman, J., 2012. Langman's medical embryology. Wolters Kluwer Health/Lippincott Williams & Wilkins, Philadelphia.

Sakaguchi, T., Kuroiwa, A., Takeda, H., 2001. A novel sox gene, 226D7, acts downstream of Nodal signaling to specify endoderm precursors in zebrafish. Mech. Dev. 107, 25–38.

Sampath, K., Rubinstein, A.L., Cheng, A.M., Liang, J.O., Fekany, K., Solnica-Krezel, L., Korzh, V., Halpern, M.E., Wright, C.V., 1998. Induction of the zebrafish ventral brain and floorplate requires cyclops/nodal signalling. Nature 395, 185–189.

Sasai, N., Yakura, R., Kamiya, D., Nakazawa, Y., Sasai, Y., 2008. Ectodermal factor restricts mesoderm differentiation by inhibiting p53. Cell 133, 878–890.

Sathananthan, H., Selvaraj, K., Clark, J., 2011. The fine structure of human germ layers *in vivo*: clues to the early differentiation of embryonic stem cells *in vitro*. Reprod. Biomed. Online 23, 227–233.

Schier, A.F., Neuhauss, S.C., Helde, K.A., Talbot, W.S., Driever, W., 1997. The one-eyed pinhead gene functions in mesoderm and endoderm formation in zebrafish and interacts with no tail. Development 124, 327–342.

Schlosser, G., Wagner, GnP., 2004. Modularity in development and evolution. University of Chicago Press, Chicago.

Schroeder, K.E., Condic, M.L., Eisenberg, L.M., Yost, H.J., 1999. Spatially regulated translation in embryos: asymmetric expression of maternal Wnt-11 along the dorsal-ventral axis in *Xenopus*. Dev. Biol. 214, 288–297.

Seiliez, I., Thisse, B., Thisse, C., 2006. FoxA3 and goosecoid promote anterior neural fate through inhibition of Wnt8a activity before the onset of gastrulation. Dev. Biol. 290, 152–163.

Selleck, M.A., Stern, C.D., 1991. Fate mapping and cell lineage analysis of Hensen's node in the chick embryo. Development 112, 615–626.

Shen, M.M., 2007. Nodal signaling: developmental roles and regulation. Development 134, 1023–1034.

Shields, D.C., Ramsbottom, D., Donoghue, C., Pinjon, E., Kirke, P.N., Molloy, A.M., Edwards, Y.H., Mills, J.L., Mynett-Johnson, L., Weir, D.G., Scott, J.M., Whitehead, A.S., 2000. Association between historically high frequencies of neural tube defects and the human T homologue of mouse T (Brachyury). Am. J. Med. Genet. 92, 206–211.

Shih, J., Fraser, S.E., 1996. Characterizing the zebrafish organizer: microsurgical analysis at the early-shield stage. Development 122, 1313–1322.

Shih, Y.H., Kuo, C.L., Hirst, C.S., Dee, C.T., Liu, Y.R., Laghari, Z.A., Scotting, P.J., 2010. SoxB1 transcription factors restrict organizer gene expression by repressing multiple events downstream of Wnt signalling. Development 137, 2671–2681.

Shimizu, T., Bae, Y.K., Muraoka, O., Hibi, M., 2005. Interaction of Wnt and caudal-related genes in zebrafish posterior body formation. Dev. Biol. 279, 125–141.

Shimizu, T., Yamanaka, Y., Ryu, S.L., Hashimoto, H., Yabe, T., Hirata, T., Bae, Y.K., Hibi, M., Hirano, T., 2000. Cooperative roles of Bozozok/Dharma and Nodal-related proteins in the formation of the dorsal organizer in zebrafish. Mech. Dev. 91, 293–303.

Showell, C., Binder, O., Conlon, F.L., 2004. T-box genes in early embryogenesis. Dev. Dyn. 229, 201–218.

Sidi, S., Goutel, C., Peyrieras, N., Rosa, F.M., 2003. Maternal induction of ventral fate by zebrafish radar. Proc. Natl. Acad. Sci. USA 100, 3315–3320.

Simeoni, I., Gurdon, J.B., 2007. Interpretation of BMP signaling in early *Xenopus* development. Dev. Biol. 308, 82–92.

Sive, H.L., Draper, B.W., Harland, R.M., Weintraub, H., 1990. Identification of a retinoic acid-sensitive period during primary axis formation in *Xenopus laevis*. Genes Dev. 4, 932–942.

Slack, J.M., Isaacs, H.V., Darlington, B.G., 1988. Inductive effects of fibroblast growth factor and lithium ion on *Xenopus* blastula ectoderm. Development 103, 581–590.

Slagle, C.E., Aoki, T., Burdine, R.D., 2011. Nodal-dependent mesendoderm specification requires the combinatorial activities of FoxH1 and Eomesodermin. PLoS Genet. 7, e1002072.

Smith, J., Theodoris, C., Davidson, E.H., 2007. A gene regulatory network subcircuit drives a dynamic pattern of gene expression. Science 318, 794–797.

Smith, J.C., Price, B.M., Van Nimmen, K., Huylebroeck, D., 1990. Identification of a potent *Xenopus* mesoderm-inducing factor as a homologue of activin A. Nature 345, 729–731.

Solnica-Krezel, L., 2005. Conserved patterns of cell movements during vertebrate gastrulation. Curr. Biol. 15, R213–228.

Solnica-Krezel, L., Driever, W., 1994. Microtubule arrays of the zebrafish yolk cell: organization and function during epiboly. Development 120, 2443–2455.

Solnica-Krezel, L., Sepich, D.S., 2012. Gastrulation: making and shaping germ layers. Annu. Rev. Cell Dev. Biol. 28, 687–717.

Song, J., Oh, S.P., Schrewe, H., Nomura, M., Lei, H., Okano, M., Gridley, T., Li, E., 1999. The type II activin receptors are essential for egg cylinder growth, gastrulation, and rostral head development in mice. Dev. Biol. 213, 157–169.

Spemann, H., 1935. The Organizer-Effect in Embryonic Development. Nobel Lect. . http://www.nobelprize.org.

Spemann, H., Mangold, H., 1924. Über Induktion von Embryonalanlagen durch Implantation artfremder Organisatoren. Arch. Mikroskopische Anatomie und Entwicklungsmechanik 100, 599–638.

Stahl, A., Tourame, P., 2010. From teratology to mythology: ancient legends. Arch. Pediatr. 17, 1716–1724.

Stern, C.D., 2004. Gastrulation: from cells to embryo. Cold Spring Harbor Laboratory Press, Cold Spring Harbor, N.Y.

Stern, C.D., Charite, J., Deschamps, J., Duboule, D., Durston, A.J., Kmita, M., Nicolas, J.F., Palmeirim, I., Smith, J.C., Wolpert, L., 2006. Head-tail patterning of the vertebrate embryo: one, two or many unresolved problems? Int. J. Dev. Biol. 50, 3–15.

Stocker, J.T., Heifetz, S.A., 1987. Sirenomelia. A morphological study of 33 cases and review of the literature. Perspect. Pediatr. Pathol. 10, 7–50.

Streit, A., Lee, K.J., Woo, I., Roberts, C., Jessell, T.M., Stern, C.D., 1998. Chordin regulates primitive streak development and the stability of induced neural cells, but is not sufficient for neural induction in the chick embryo. Development 125, 507–519.

Sugiyama, D., Ogawa, M., Nakao, K., Osumi, N., Nishikawa, S., Nishikawa, S., Arai, K., Nakahata, T., Tsuji, K., 2007. B cell potential can be obtained from pre-circulatory yolk sac, but with low frequency. Dev. Biol. 301, 53–61.

Sulston, J.E., Schierenberg, E., White, J.G., Thomson, J.N., 1983. The embryonic cell lineage of the nematode Caenorhabditis elegans. Dev. Biol. 100, 64–119.

Sun, X., Meyers, E.N., Lewandoski, M., Martin, G.R., 1999. Targeted disruption of Fgf8 causes failure of cell migration in the gastrulating mouse embryo. Genes Dev. 13, 1834–1846.

Szeto, D.P., Kimelman, D., 2006. The regulation of mesodermal progenitor cell commitment to somitogenesis subdivides the zebrafish body musculature into distinct domains. Genes Dev. 20, 1923–1932.

Tadros, W., Lipshitz, H.D., 2005. Setting the stage for development: mRNA translation and stability during oocyte maturation and egg activation in Drosophila. Dev. Dyn. 232, 593–608.

Takada, S., Stark, K.L., Shea, M.J., Vassileva, G., McMahon, J.A., McMahon, A.P., 1994. Wnt-3a regulates somite and tailbud formation in the mouse embryo. Genes Dev. 8, 174–189.

Takahashi, K., Yamanaka, S., 2006. Induction of pluripotent stem cells from mouse embryonic and adult fibroblast cultures by defined factors. Cell 126, 663–676.

Takahashi, S., Yokota, C., Takano, K., Tanegashima, K., Onuma, Y., Goto, J., Asashima, M., 2000. Two novel nodal-related genes initiate early inductive events in *Xenopus* Nieuwkoop center. Development 127, 5319–5329.

Takemoto, T., Uchikawa, M., Yoshida, M., Bell, D.M., Lovell-Badge, R., Papaioannou, V.E., Kondoh, H., 2011. Tbx6-dependent Sox2 regulation determines neural or mesodermal fate in axial stem cells. Nature 470, 394–398.

Tam, P.P., Williams, E.A., Chan, W.Y., 1993. Gastrulation in the mouse embryo: ultrastructural and molecular aspects of germ layer morphogenesis. Microsc. Res. Tech. 26, 301–328.

Tanegashima, K., Yokota, C., Takahashi, S., Asashima, M., 2000. Expression cloning of Xantivin, a *Xenopus* lefty/antivin-related gene, involved in the regulation of activin signaling during mesoderm induction. Mech. Dev. 99, 3–14.

Tao, Q., Yokota, C., Puck, H., Kofron, M., Birsoy, B., Yan, D., Asashima, M., Wylie, C.C., Lin, X., Heasman, J., 2005. Maternal wnt11 activates the canonical wnt signaling pathway required for axis formation in *Xenopus* embryos. Cell 120, 857–871.

Thisse, B., Pflumio, S., Fürthauer, M., Loppin, B., Heyer, V., Degrave, A., Woehl, R., Lux, A., Steffan, T., Charbonnier, X.Q., Thisse, C., 2001. Expression of the zebrafish genome during embryogenesis (NIH R01 RR15402). ZFIN Direct Data Submission. http://zfin.org.

Thomas, P., Beddington, R., 1996. Anterior primitive endoderm may be responsible for patterning the anterior neural plate in the mouse embryo. Curr. Biol. 6, 1487–1496.

Toyama, R., O'Connell, M.L., Wright, C.V., Kuehn, M.R., Dawid, I.B., 1995. Nodal induces ectopic goosecoid and lim1 expression and axis duplication in zebrafish. Development 121, 383–391.

Tran, L.D., Hino, H., Quach, H., Lim, S., Shindo, A., Mimori-Kiyosue, Y., Mione, M., Ueno, N., Winkler, C., Hibi, M., Sampath, K., 2012. Dynamic microtubules at the vegetal cortex predict the embryonic axis in zebrafish. Development 139, 3644–3652.

Turing, A.M., 1952. The chemical basis of morphogenesis. Philos. Trans. R. Soc. London. 237, 37–72.

Turner, A., Snape, A.M., Wylie, C.C., Heasman, J., 1989. Regional identity is established before gastrulation in the *Xenopus* embryo. J. Exp. Zool. 251, 245–252.

Tzouanacou, E., Wegener, A., Wymeersch, F.J., Wilson, V., Nicolas, J.F., 2009. Redefining the progression of lineage segregations during mammalian embryogenesis by clonal analysis. Dev. Cell 17, 365–376.

Uehara, M., Yashiro, K., Takaoka, K., Yamamoto, M., Hamada, H., 2009. Removal of maternal retinoic acid by embryonic CYP26 is required for correct Nodal expression during early embryonic patterning. Genes Dev. 23, 1689–1698.

Vastenhouw, N.L., Schier, A.F., 2012. Bivalent histone modifications in early embryogenesis. Curr. Opin. Cell Biol. 24, 374–386.

Vignal, E., de Santa Barbara, P., Guemar, L., Donnay, J.M., Fort, P., Faure, S., 2007. Expression of RhoB in the developing *Xenopus laevis* embryo. Gene Expr. Patterns 7, 282–288.

Vincent, J.P., Oster, G.F., Gerhart, J.C., 1986. Kinematics of gray crescent formation in *Xenopus* eggs: the displacement of subcortical cytoplasm relative to the egg surface. Dev. Biol. 113, 484–500.

Vogt, W., 1929. Gestaltungsanalyse am Amphibienkeim mit Örtlicher Vitalfärbung 120, 384–706.

von Baer, K.E., 1828. Über Entwickelungsgeschichte der Thiere Beobachtung und Reflexion. Königsberg & Hannover: Bornträger Tierärztl. Hochsch.

Vonica, A., Gumbiner, B.M., 2007. The *Xenopus* Nieuwkoop center and Spemann-Mangold organizer share molecular components and a requirement for maternal Wnt activity. Dev. Biol. 312, 90–102.

Waddington, C.H., 1932. Experiments on the development of chick and duck embryos, cultivated *in vitro*. Phil. Trans. R. Soc. London. B. 221, 179–230.

Waddington, C.H., 1933. Induction by the primitive streak and its derivatives in the chick. J. Exp. Biol. 10, 38–46.

Waddington, C.H., Needham, J., Needham, D.M., 1933. Physico-chemical experiments on the amphibian organiser. Nature 132, 239.

Walcott, B.P., Nahed, B.V., Mohyeldin, A., Coumans, J.V., Kahle, K.T., Ferreira, M.J., 2012. Chordoma: current concepts, management, and future directions. Lancet Oncol. 13, e69–76.

Wardle, F.C., Papaioannou, V.E., 2008. Teasing out T-box targets in early mesoderm. Curr. Opin. Genet. Dev. 18, 418–425.

Warga, R.M., Nusslein-Volhard, C., 1999. Origin and development of the zebrafish endoderm. Development 126, 827–838.

Wetts, R., Fraser, S.E., 1989. Slow intermixing of cells during *Xenopus* embryogenesis contributes to the consistency of the blastomere fate map. Development 105, 9–15.

Wilson, V., Olivera-Martinez, I., Storey, K.G., 2009. Stem cells, signals and vertebrate body axis extension. Development 136, 1591–1604.

Wolpert, L., 1969. Positional information and the spatial pattern of cellular differentiation. J. Theor. Biol. 25, 1–47.

Wu, S.Y., Shin, J., Sepich, D.S., Solnica-Krezel, L., 2012. Chemokine GPCR signaling inhibits beta-catenin during zebrafish axis formation. PLoS Biol. 10, e1001403.

Wu, X., Gehring, W., 2014. Cellular uptake of the Antennapedia homeodomain polypeptide by macropinocytosis. Biochem. Biophys. Res. Commun. 443, 1136–1140.

Xu, C., Fan, Z.P., Müller, P., Fogley, R., DiBiase, A., Trompouki, E., Unternaehrer, J., Xiong, F., Torregroza, I., Evans, T., Megason, S.G., Daley, G.Q., Schier, A.F., Young, R.A., Zon, L.I., 2012. Nanog-like regulates endoderm formation through the Mxtx2-Nodal pathway. Dev. Cell 22, 625–638.

Yamaguchi, T.P., Takada, S., Yoshikawa, Y., Wu, N., McMahon, A.P., 1999. T (Brachyury) is a direct target of Wnt3a during paraxial mesoderm specification. Genes Dev. 13, 3185–3190.

Yamamoto, S., Hikasa, H., Ono, H., Taira, M., 2003. Molecular link in the sequential induction of the Spemann organizer: direct activation of the cerberus gene by Xlim-1, Xotx2, Mix.1, and Siamois, immediately downstream from Nodal and Wnt signaling. Dev. Biol. 257, 190–204.

Yan, B., Moody, S.A., 2007. The competence of *Xenopus* blastomeres to produce neural and retinal progeny is repressed by two endo-mesoderm promoting pathways. Dev. Biol. 305, 103–119.

Yanagawa, N., Sakabe, M., Sakata, H., Yamagishi, T., Nakajima, Y., 2011. Nodal signal is required for morphogenetic movements of epiblast layer in the pre-streak chick blastoderm. Dev. Growth. Differ. 53, 366–377.

Yang, J., Tan, C., Darken, R.S., Wilson, P.A., Klein, P.S., 2002. Beta-catenin/Tcf-regulated transcription prior to the midblastula transition. Development 129, 5743–5752.

Yang, X.R., Ng, D., Alcorta, D.A., Liebsch, N.J., Sheridan, E., Li, S., Goldstein, A.M., Parry, D.M., Kelley, M.J., 2009. T (brachyury) gene duplication confers major susceptibility to familial chordoma. Nat. Genet. 41, 1176–1178.

Yao, J., Kessler, D.S., 2001. Goosecoid promotes head organizer activity by direct repression of Xwnt8 in Spemann's organizer. Development 128, 2975–2987.

Yao, S., Qian, M., Deng, S., Xie, L., Yang, H., Xiao, C., Zhang, T., Xu, H., Zhao, X., Wei, Y.Q., Mo, X., 2010. Kzp controls canonical Wnt8 signaling to modulate dorsoventral patterning during zebrafish gastrulation. J. Biol. Chem. 285, 42086–42096.

Yi, H., Xue, L., Guo, M.X., Ma, J., Zeng, Y., Wang, W., Cai, J.Y., Hu, H.M., Shu, H.B., Shi, Y.B., Li, W.X., 2010. Gene expression atlas for human embryogenesis. FASEB J. 24, 3341–3350.

Zeigler, B.M., Sugiyama, D., Chen, M., Guo, Y., Downs, K.M., Speck, N.A., 2006. The allantois and chorion, when isolated before circulation or chorio-allantoic fusion, have hematopoietic potential. Development 133, 4183–4192.

Zeitlinger, J., Stark, A., Kellis, M., Hong, J.W., Nechaev, S., Adelman, K., Levine, M., Young, R.A., 2007. RNA polymerase stalling at developmental control genes in the *Drosophila* melanogaster embryo. Nat. Genet. 39, 1512–1516.

Zhang, C., Klymkowsky, M.W., 2007. The Sox axis, Nodal signaling, and germ layer specification. Differentiation 75, 536–545.

Zhang, C., Basta, T., Hernandez-Lagunas, L., Simpson, P., Stemple, D.L., Artinger, K.B., Klymkowsky, M.W., 2004. Repression of nodal expression by maternal B1-type SOXs regulates germ layer formation in *Xenopus* and zebrafish. Dev. Biol. 273, 23–37.

Zimmerman, L.B., De Jesus-Escobar, J.M., Harland, R.M., 1996. The Spemann organizer signal noggin binds and inactivates bone morphogenetic protein 4. Cell 86, 599–606.

Chapter 13

Vertebrate Endoderm Formation

Marcin Wlizla and Aaron M. Zorn

Division of Developmental Biology, Cincinnati Children's Hospital Medical Center, Cincinnati, OH

Chapter Outline

GLOSSARY

Endoderm The innermost germ layer of vertebrate embryos that gives rise to the epithelial cells lining the digestive and respiratory system as well as visceral organs including the liver, thyroid and pancreas.

Mesoderm One of the three vertebrate germ layers formed during gastrulation that gives rise to connective tissue, muscle, and the skeletal and circulatory systems.

Mesendoderm A transient vertebrate tissue that forms during gastrulation and is equally capable of differentiating into mesoderm and endoderm.

Definitive endoderm Endoderm tissue that contributes to the embryo proper, in contrast to the extra embryonic endoderm tissues (visceral and primitive endoderm) which may be required for correct development or embryo survival but do not contribute to the overall body plan.

Nodal A TGFβ signaling ligand that is a key regulator of the vertebrate endoderm GRN controlling first the induction of mesendoderm and then its segregation into distinct endoderm and mesoderm germ layers.

SUMMARY

- The embryonic endoderm contributes to the epithelial lining of the digestive and respiratory system and associated organs such as the liver and pancreas.
- Data from a number of animal models demonstrate that a conserved gene regulatory network (GRN) controls endoderm germ layer specification during gastrulation in vertebrates.
- The transforming growth factor (TGF)-β signaling ligand Nodal induces this endoderm GRN by activating the expression of a key group of transcription factors including Sox17, FoxA, and Mix-like proteins.
- These transcription factors interact in a complex auto-regulatory loop; with both distinct and overlapping downstream target genes they regulate the segregation of the endoderm and mesoderm lineages from a transient mesendoderm progenitor.
- Other signaling pathways including Wnt, Notch, and fibroblast growth factor (FGF) modulate the activity of the Nodal endoderm GRN but their relative importance appears to vary in different species.

- Our understanding of endoderm development in animal models has been translated into robust protocols to differentiate human endoderm *in vitro* from pluripotent stem cells.
- In addition to providing the possibility of regenerative medicine and a means of modeling human disease of endoderm tissues, the *in vitro* differentiation of endoderm from stem cells now provides a powerful model to further interrogate the molecular basis of human endoderm formation.

13.1 INTRODUCTION

Endoderm is the innermost germ layer of eumetazoan embryos, surrounded by mesoderm and ectoderm (Figure 13.1A). The endoderm contributes to the epithelial lining and associated organs of both the vertebrate digestive tract and respiratory system (Figure 13.1B) (Zorn and Wells, 2009). The study of endoderm as a distinct embryonic tissue dates back to investigations of early chick development by the German biologist and embryologist, Christian Heinrich Pander, in 1817 (Schmitt, 2005). Historically, the internal location of the endoderm within the embryo made studying it difficult relative to the other germ layers, yet in recent decades advanced molecular genetic and imaging techniques, as well as growing genomic data from diverse animals, has resulted in considerable progress being made in our understanding of the molecular basis of endoderm formation.

Studies of three model systems: mouse (*Mus musculus*), frog (*Xenopus laevis*), and zebrafish (*Danio rerio*), have been particularly instructive for understanding vertebrate endoderm development. Embryonic stem (ES) cell technology, gene targeting, and genetic lineage labeling have made the mouse a powerful model for understanding mammalian endoderm development. However, the internal gestation of mammalian embryos presents some limitations for embryological manipulations. Here, the experimental advantages of the externally developing frog and zebrafish embryos have provided a valuable complement to genetic studies in mouse, enabling an analysis of the earliest steps in vertebrate endoderm specification. In *Xenopus*, robust microsurgery, explant culture and microinjection have been particularly useful for dissecting the cell signaling pathways regulating endoderm induction during gastrulation. In zebrafish, the optically clear embryos and forward genetic screens have established it as a powerful system for analyzing the genetic pathways and cell behaviors controlling endoderm progenitor cells. Overall, a comparative approach utilizing the combined advantages of each of these model systems has greatly accelerated our understanding of the genetic programs controlling germ layer formation.

Data from these vertebrate animal models, in combination with observations made in non-vertebrate species, have revealed that a conserved gene regulatory network (GRN) controls specification of the endoderm germ layer during gastrulation. The core of this endoderm GRN is the secreted transforming growth factor (TGF)-β growth factor Nodal, which is both necessary and sufficient to induce endoderm in vertebrates. Nodal signaling promotes the expression of a group of key transcription factors including the Sox17, FoxA, Mix-like, and Gata families. These transcription factors interact in a complex auto-regulatory loop to regulate the expression of endoderm genes and to segregate the endoderm and mesoderm lineages, both of which are induced by Nodal signaling. The activity of this core Nodal GRN is modified by crosstalk with other signaling pathways including Wnt, Notch, and fibroblast growth factor (FGF), which influence endoderm differentiation, segregation of mesoderm and endoderm, and initial patterning of the nascent endoderm. Although each animal model has clear differences in endoderm morphogenesis and some aspects of the molecular pathways differ between species, overall a comparative approach has been key in piecing together the core elements of the GRN.

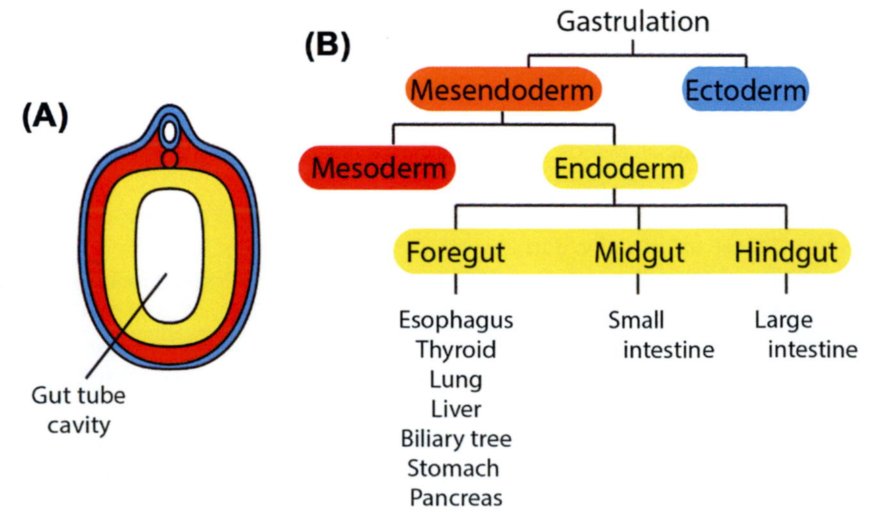

FIGURE 13.1 Germlayers and Vertebrate Endoderm Cell Lineages. Transverse section through a generalized post-gastrula stage vertebrate embryo (A) and diagrammatic representation of the progressive formation of endoderm derived tissues from gastrulation, through formation of distinct germ layers, followed by subdivision of endoderm along the anteroposterior axis into foregut, midgut, and hindgut domains eventually leading to formation of distinct organ domains. Germ layers: ectoderm (blue), mesoderm (red) and endoderm (yellow), as well as mesendoderm (orange) the bipotential pool that will generate both endoderm and mesoderm are indicated.

In the last decade, translation of the information on this conserved endoderm GRN from animal models has led to robust protocols to direct the differentiation of endoderm tissue from human pluripotent stem cells *in vitro*. With the revolution of induced pluripotent stem cells from human tissue (Takahashi and Yamanaka, 2006; Zorn and Wells, 2009), this has led to the exciting possibility of regenerative medicine and modeling of human disease in endoderm-derived organ systems. Finally, the experimental advantages of cell cultures have made *in vitro* differentiation of endoderm from stem cells an excellent system to further interrogate the molecular pathways controlling human endoderm development on a genome-wide scale (Cai et al., 2008; Nostro and Keller, 2012; Spence et al., 2011).

In this chapter we will synthesize the current understanding of endoderm germ layer formation in vertebrate model systems and describe how comparative studies have led our growing knowledge of the conserved GRN controlling this process. Finally, we will discuss how this understanding has led the directed differentiation of human endoderm tissue *in vitro,* with the ultimate goal of cell replacement therapy and modeling of human diseases of endodermal tissues.

13.2 OVERVIEW OF ENDODERM MORPHOGENESIS IN VERTEBRATES

In order to explain the molecular mechanisms regulating vertebrate endoderm development, it is important to first describe the morphogenesis of definitive endodermal tissue during gastrulation of the different model organisms.

Mouse

By the time of implantation at the blastocyst stage (embryonic day 4.0, E4.0) (Paria et al., 1993), three distinct cell lineages are present in the developing mouse conceptus: trophectoderm, primitive endoderm (PE), and the epiblast. The epiblast gives rise to the germ cells and the somatic tissues of the embryo proper, including the definitive endoderm (DE), mesoderm and ectoderm. PE and its derivative the visceral endoderm (VE) give rise to extra-embryonic tissue of the fetus including the yolk sac. This distinction between definitive (embryonic) and extra-embryonic endoderm is a hallmark of amniote embryos, but a recent genetic labeling and live imaging in mice indicated that this separation may not be so absolute, as some VE cells appear to contribute to the embryonic gut tube (Hogan and Zaret, 2002; Kwon et al., 2008).

The study of DE development has been somewhat complicated by a considerable overlap in the expression and function of genes in both the DE and VE (Hou et al., 2007; Sherwood et al., 2007). VE is necessary for correct implantation and for fetal gas, nutrient and waste exchange, and therefore VE defects often lead to early embryonic lethality before gastrulation, precluding the development of the primary germ layers (Yamamoto et al., 2004).

By E5.5 the epiblast is a single cell layered cup surrounded by a thin layer of VE cells. At E6.5 gastrulation begins as epiblast cells converge at the posterior proximal pole, then undergo a epithelial to mesenchymal transition and ingress through a transient embryonic structure, known as the primitive streak (PS) (Lawson et al., 1991). It is thought that as cells migrate through the PS they adopt a transient mesendoderm progenitor fate, which is rapidly segregated into distinct mesoderm and endoderm. Upon emergence from the PS, cells migrate into two distinct cell layers around the epiblast (Tam et al., 2007), with the endoderm forming the outermost layer and the mesoderm forming in between the DE and the ectoderm cells that do not ingress through the PS. Fate maps of embryos immediately prior to PS formation (E6.0–6.5) indicate that DE tissues originate from the posterior epiblast, the region where the PS will initially form, emerging mostly from the anterior tip of the PS (Tam et al., 2007). Furthermore, the order in which cells exit the PS also correlates to their eventual fate; *Hhex* positive ventral foregut progenitors move through by mid-streak stage and are then progressively followed by *Foxa2* expressing dorsal foregut, then midgut, and finally hindgut precursors (Tam et al., 2007; Zorn and Wells, 2009). Nascent DE cells emerging from the streak expand over the epiblast, displacing and intercalating with the overlaying VE (Figure 13.2A) (Kwon et al., 2008). How long these VE cells survive and whether they are actually capable of contributing to functional endodermal organs remains unclear. Cell lineage experiments indicate that germ layer fates, including DE, are rapidly determined during gastrulation (Lawson and Pedersen, 1987; Tam et al., 2003). As a result, in mice and other amniotes, the genetic programs regulating gastrulation are inseparably linked to the pathways regulating germ layer specification.

At the end of gastrulation, the mouse embryo is a three-layered cup of cells with a sheet of endoderm epithelium on the outside. At ~E8 in development the embryo "turns" inside-out, inverting the edges of the cup such that the endoderm becomes the inside of the embryo and begins to form a gut tube. This process starts at the anterior and posterior ends of the embryo, forming the foregut and hindgut pockets respectively (Franklin et al., 2008; Tremblay and Zaret, 2005). Once the pockets have formed, each rapidly expands towards the mid-section of the embryo as the lateral aspects of the intermediate regions of the epithelial sheet progressively close up on the ventral surface (Franklin et al., 2008; Tremblay and Zaret, 2005). By E9.5, the mouse embryonic endoderm is an epithelial gut tube with distinct regional identity along the A-P axis set to begin organ bud formation.

Xenopus

Fate mapping studies in *Xenopus* have shown that presumptive germ layer fates are distributed along the animal-vegetal (top-bottom) axis of the pre-gastrula embryo. As early as the 32-cell stage, when the embryo is organized into four stacked tiers of cells, the endoderm is primarily derived from the two vegetal tiers, the two animal tiers will primarily contribute to the ectoderm, and mesoderm is formed on the intersection of the endoderm and ectoderm and is derived from the two middle tiers (Dale and Slack, 1987; Moody, 1987). This general organization is maintained through the blastula and early gastrula stages, with the endoderm precursors originating from the yolky cells in the vegetal bottom half of the embryo, the presumptive mesoderm is located in an equatorial ring around the embryo, and the ectoderm is on the top of the embryo above the fluid-filled blastocoel cavity (Figure 13.2B) (Keller, 1976). For the most part, the zygotic genome becomes expressed around the mid-blastula stage (~4000 cells), and prior to this development is largely controlled by maternal factors deposited in the egg. Embryo microdissection, single cell isolations, and explant culture experiments have shown that germ layer fate is rapidly determined during gastrulation and that the mesoderm is induced in the equatorial tissue by factors secreted from the presumptive endoderm, which we now know are Nodals (Heasman et al., 1984; 1985; Nieuwkoop, 1969; Wylie et al., 1987; Zorn and Wells, 2009).

Gastrulation begins when cells invaginate on the future dorsal-anterior side of the embryo at the mes-endoderm boundary, in a region known as the dorsal blastopore lip or as the Spemann organizer; the source of many secreted factors that influence embryonic patterning including that of the nascent endoderm. Eventually, gastrulation movements spread around the entire circumference of the embryo. As in mouse, the cells that invaginate first contribute to the foregut, whereas cells located more laterally invaginate later and contribute to more posterior tissues (Keller, 1975; 1976; Nieuwkoop, 1997). Gene expression studies suggest that the equatorial region of the blastula and early gastrula has a transient bipotential mesendoderm fate, capable of differentiating into mesoderm and endoderm (Rodaway and Patient, 2001). Cells in this region express both mesoderm and endoderm markers in the blastula but by the end of gastrulation endoderm and mesoderm gene expression is largely mutually exclusive with a distinct boundary between the endoderm and mesoderm (Wardle and Smith, 2004). As we will see, one of the key functions of the endoderm GRN is to separate the endoderm and mesoderm lineages.

After gastrulation, most of the endoderm in the *Xenopus* embryo is a mass of tissue some 20 cells thick on the ventral side of the embryo, unlike in amniotes in which the endoderm is a single cell layer sheet. During somite stages of development this endodermal mass is patterned and a cavity forms in the middle of the tissue to form the gut lumen, resulting in a primitive gut tube that is strikingly similar to that in other vertebrate embryos (Chalmers and Slack, 2000).

Zebrafish

Following fertilization, the zebrafish embryo develops on top of a large yolk cell, forming a dome-shaped cap of cells termed the blastoderm. At the 512-cell stage, the edge, or the margin, of the blastoderm sinks into the yolk to form the yolk syncytial layer (YSL), an extra-embryonic multi-nucleated syncytium at the interface between the yolk and the blastoderm, while the rest of the blastoderm will form the epiblast, which contributes to the embryo proper (Kimmel and Law, 1985). The YSL itself is an important source of Nodal signaling ligands that induce both the endoderm and mesoderm in the adjacent epiblast (Rodaway et al., 1999).

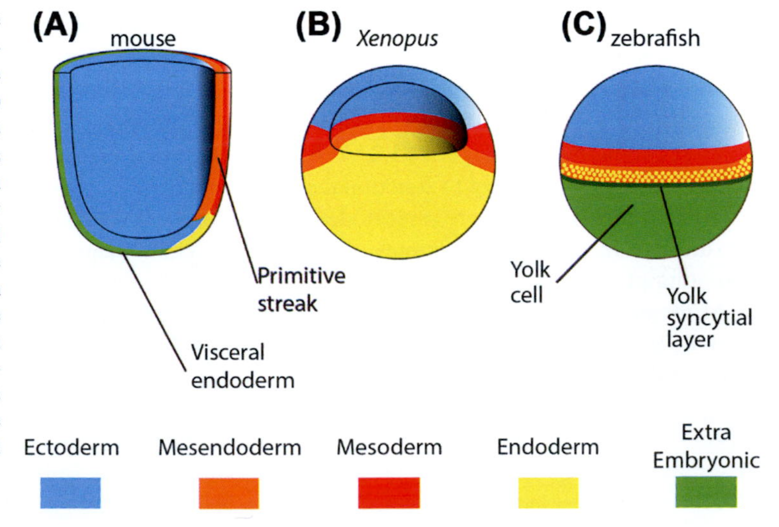

FIGURE 13.2 Fate Maps. Depictions of an early streak stage mouse embryo (A), a *Xenopus* blastula (B) and a zebrafish blastula (C). (A) The section through the early streak stage mouse gastrula shows the extra-embryonic visceral endoderm (green) being displaced by definitive endoderm (yellow) as it emerges from the anterior part of the primitive streak, which contains mesendodermal precursors (orange). (B) The section through the *Xenopus* blastula shows the blastocoel enclosed by the yolky endoderm on the vegetal side (yellow), the ectodermal animal cap (blue), and a ring of equatorial mesoderm (red), while the overlapping intersection between the mesoderm and endoderm contains the bipotential mesendoderm (orange). (C) A surface view of a zebrafish blastula immediately prior to gastrulation showing the embryonic epiblast sitting on top of the extra-embryonic yolk cell (green), the definitive endoderm precursors (yellow) are located within the bipotential mesendoderm tissue (orange) up to four cell diameters away from the yolk syncytial layer.

Fate mapping studies have shown that endoderm and mesoderm progenitor cells are intermingled in the marginal tissue next to the YSL (Figure 13.2C), with the endoderm precursors found closest to the YSL within four cell diameters of the embryonic margin (Kimmel et al., 1990; Warga and Nusslein-Volhard, 1999), whereas the mesoderm precursors can be found as far as eight cell diameters from the margin (Kimmel et al., 1990; Warga and Nusslein-Volhard, 1999).

During gastrulation, starting first on the future dorsal-anterior side, epiblast cells at the margin invaginate and migrate under the epiblast on top of the yolk surface. Eventually these morphogenetic movements spread around the entire circumference of the embryo. Similar to cells in the amniote PS, some of the invaginating cells form a hypoblast layer closest to the yolk, which will give rise to the endoderm, whereas some cells migrate between the hypoblast and epiblast and form the mesoderm. As in other vertebrates, the dorsoventral position of the mesendodermal cells at the onset of gastrulation corresponds to their eventual position in the gut, with the cells involuting first contributing to the foregut followed by more lateral cells which form the stomach and intestine (Kimmel et al., 1990; Warga and Nusslein-Volhard, 1999).

At the end of gastrulation in the zebrafish, the endoderm is a dispersed sheet of cells covering the yolk. As a result of non-canonical Wnt signaling through the Planar Cell Polarity (PCP) pathway, this sheet of endodermal cells then converges towards the midline to form a solid rod of endoderm cells (Horne-Badovinac et al., 2001; Matsui et al., 2005). Interestingly, the non-canonical Wnt/PCP pathway also appears to regulate early gut tube morphogenesis in *Xenopus* and mice, although in each case it appears to control distinct cell behaviors (Li et al., 2008; Matsuyama et al., 2009). This endodermal rod in zebrafish then undergoes coordinated epithelial remodeling to form a gut tube lumen (Bagnat et al., 2007; Horne-Badovinac et al., 2001), from which stereotypical organ buds emerge which resemble those in other vertebrate species (Ober et al., 2003).

13.3 MOLECULAR MECHANISMS OF ENDODERM DEVELOPMENT

Despite the diverse morphogenetic movements associated with endoderm formation among different vertebrate animals, many of the molecular mechanisms regulating them appear to be broadly conserved. The Nodal signaling pathway forms a core GRN that is necessary and sufficient for endoderm germ layer specification (Figure 13.3). Although the mechanisms regulating expression of *nodal* genes are divergent in different species, the downstream transcription factors induced by Nodal are very similar. These include the Sox17, FoxA, Mix-like, and Gata4–6 family of transcription factors, which regulate the transcription of endoderm differentiation genes and control the segregation of the endoderm and mesoderm lineages from the transient mesendoderm progenitor. They do this in part by repressing a mesoderm GRN that is induced by low levels of Nodal signaling. The activity of Nodal-induced endoderm GRN is modified by other signaling pathways including Wnt, FGF and Notch, which coordinate endoderm specification with morphogenesis and early patterning of the nascent endoderm (see Chapter 30). The robustness of this endoderm GRN is maintained by both positive and negative auto-regulatory loops between Nodal and the key endoderm transcription factors (Zorn and Wells, 2007; 2009).

Nodal Signal Transduction Pathway

The *Nodal* gene encoding a TGFβ family growth factor was first discovered in a genetic screen in mice (Conlon et al., 1991), and is critical for gastrulation and induction of both endoderm and mesoderm germ layers. High Nodal concentrations promote endoderm over mesoderm in all the vertebrate animals which have been examined. Nodal proteins are translated and secreted as pro-protein dimers, which are processed by proteolytic cleavage into mature secreted ligands by Furin/PACE, members of the subtilisin/kexin pro-protein convertase family (Beck et al., 2002). Proteolytic maturation, as well as

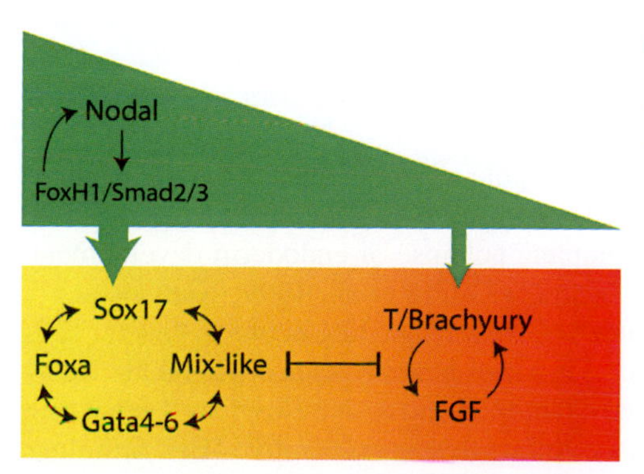

FIGURE 13.3 **Core Vertebrate Endoderm GRN.** Nodal signaling is at the apex of the endoderm GRN acting as a morphogen (green) with high levels close to source of the ligand promoting function of transcription factors involved in specification, maintenance, and patterning of endoderm (yellow) and lower levels promoting specification of mesoderm (red).

the activity of secreted Nodal-binding antagonists such as Lefty, regulate the bioavailability, stability, and range of Nodal diffusion in tissue (Ben-Haim et al., 2006; Eimon and Harland, 2002), all of which have been shown to influence the concentration-dependent activity of inducing endoderm versus mesoderm. Nodal dimers bind a receptor complex located at the cell surface and consisting of heteromeric complex of type I (Alk4 or Alk7) and type II (ActRIIA or ActRIIB) transmembrane receptor kinases and EGF-CFC co-receptors (uniquely necessary for Nodal signaling) (Reissmann et al., 2001; Song et al., 1999). Upon ligand-receptor activation, the receptor complex phosphorylates the key intracellular effector proteins, Smad2 or Smad3, which then form a multimeric complex with the related Smad4 (Nakao et al., 1997). This Smad2/4 complex then translocates into the nucleus to interact with transcription factors, most notably FoxH1 (FAST), some Mix-like factors, and p53 to directly activate the transcription of early mesendoderm genes, including the Sox17, FoxA, Mix-like, and Gata factors (Chen et al., 1997; Germain et al., 2000). Moreover, in all vertebrates, Smad/FoxH1 complexes directly regulate the transcription of both the *Nodal* gene and its antagonist *Lefty*, thus forming positive and negative auto-regulatory loops that precisely control the level of Nodal activity in the early embryo (Schier, 2009; Solnica-Krezel, 2003).

Nodal Signaling in Endoderm Induction

Explant cultures in *Xenopus,* as well as genetic manipulations to modulate Nodal signaling levels in mice and zebrafish, have demonstrated that Nodal signaling is necessary and sufficient to induce mesoderm and endoderm fate in a dose-dependent manner; with higher Nodal activity promoting endoderm and lower activity promoting mesoderm. In frog and fish there are multiple, largely redundant, *nodal-related (nr)* genes which perform the job of the single mammalian *Nodal* (Schier, 2009). In all vertebrates, *Nodal* genes are expressed in the presumptive endoderm and mesendoderm progenitors prior to gastrulation, with peak expression levels localized to the organizer and the site of gastrulation (dorsal blastopore, primitive streak and shield) (Conlon et al., 1994; Fan et al., 2007; Jones et al., 1995; Joseph and Melton, 1997; Takahashi et al., 2000). In zebrafish, maternally provided *nr* mRNA is also present in the non-embryonic YSL.

Experimental perturbations that abolish Nodal signaling result in failure of gastrulation, loss of all endoderm, and mesoderm deficiencies in mouse, *Xenopus*, and zebrafish (Feldman et al., 1998; Iannaccone et al., 1992; Osada and Wright, 1999). Endoderm development is much more sensitive to disruption of Nodal activity and even small reductions in signaling affect its specification, whereas much greater reduction in Nodal is required to abolish mesoderm formation (Agius et al., 2000; Dougan et al., 2003; Dunn et al., 2004). Studies in zebrafish have been instrumental in demonstrating that Nodal acts as a morphogen, with high Nodal concentrations inducing endoderm in the tissues within three cell diameters of the embryonic margin, whereas lower Nodal concentrations further from the ligand source induce mesoderm up to eight cells away from the margin. Nodal signaling promotes both its own expression as well as that of its antagonist Lefty, which diffuses more readily than Nodal ligands, and this sets up a Nodal activity gradient in the zebrafish margin to control the relative position of endoderm and mesoderm cells (Chen and Schier, 2001). A similar mechanisms appears to act in *Xenopus*. More recent analysis indicates that the eventual fate decision between mesoderm and endoderm depends on the integration of both the concentration of Nodal ligand and the duration of exposure to Nodal during the dynamic cell movements of gastrulation (Ben-Haim et al., 2006; Hagos and Dougan, 2007). Throughout vertebrates, loss of endoderm following Nodal disruption is likely not a simple secondary defect of failed gastrulation, since Nodal is sufficient for endoderm and mesoderm induction in experimental conditions where gastrulation does not occur, including mouse stem cell cultures and *Xenopus* explant experiments (Pfendler et al., 2005; Takahashi et al., 2000; Vallier et al., 2004).

Regulation of *Nodal* Gene Expression

Although Nodal is the key endoderm-inducing factor in all vertebrates, the mechanism by which *Nodal* gene expression is first activated in the early embryo varies considerably between species. In frogs and fish, the expression of *nodal-related (nr)* ligands is regulated by maternal factors, whereas in mice the genes that initiate gastrulation are critical for activating *Nodal* expression in the PS.

In *Xenopus*, the T-box transcription factor VegT, which is localized to the vegetal pole of the *Xenopus* egg, directly activates the transcription of *nodal-related* genes in the presumptive endoderm cells of the early blastula (Wardle and Smith, 2006). Depletion of maternal VegT from the *Xenopus* embryo results in a complete collapse of endoderm development, which can be largely rescued by adding back exogenous RNA encoding *nr* ligands (Kofron et al., 1999; Xanthos et al., 2001). Once VegT activates *nr* transcription in the presumptive endoderm, autocrine Nodal signaling between vegetal cells enhances and maintains *nr* expression through Smad2/FoxH1 DNA-binding sites in the *nr* genes (Osada et al., 2000). Many of the transcription factors expressed in the endoderm, including *Mix-like* genes, *Gata4*, *5* and *6*, and two *Sox17* genes, are also directly activated by VegT (Wardle and Smith, 2006; Xanthos et al., 2001). Although several putative *VegT* orthologs have been identified in mouse and zebrafish, including *tbx6* (Hug et al., 1997), *tbx16* (Ruvinsky et al., 1998), and *spadetail* (Griffin et al., 1998), none of them have been implicated in endoderm specification in those species. In contrast, it appears that the related T-box factor Eomesodermin (Eomes) is involved in endoderm development in these species.

In the mouse, an intricate auto-regulatory network is responsible for restricting *Nodal* expression to those cells in close proximity to and within the PS in the posterior epiblast. Prior to gastrulation, *Nodal* is initially expressed throughout the entire epiblast, where it activates the expression of the T-box gene, *Eomes,* in the Distal Visceral Endoderm (DVE) (Nowotschin et al., 2013). *Eomes* promotes Anterior Visceral Endoderm (AVE) fates in part by inducing expression of secreted Nodal inhibitors, *Lefty* and *Cerberus*, and by promoting the movement of the DVE towards the anterior side of the embryo (Nowotschin et al., 2013). As the source of Nodal inhibitors moves, it effectively restricts Nodal signaling to the posterior epiblast (Zorn and Wells, 2009), then the positive Nodal signaling feedback loop reinforces *Nodal* expression in the PS, where it promotes gastrulation and mesendoderm fate.

The mechanisms regulating the expression of *nodal-related* genes, *cyclops* (*ndr2/cyc*) and *squint* (*ndr1/sqt*) in the zebrafish are only partially understood. Recent studies have shown that the maternally-provided, pluripotency factor Nanog-like is important for the activation of Mxtx2, a member of the Mix-like family of transcription factors. The latter is both necessary and sufficient to induce the formation of the YSL, and zygotic expression of YSL genes including *nodal-related* genes, which then, through auto-regulatory interactions, induce their own expression in the nearby marginal cells (Hong et al., 2011; Xu et al., 2012). Although, *cyc* mRNA is also deposited in the egg maternally (Bennett et al., 2007b; Gore et al., 2005), this maternal contribution is not necessary for normal development, instead it is the zygotic expression of the *nodal-related* genes in the YSL and the margin that is necessary for induction of endoderm and mesoderm in the embryonic margin (Bennett et al., 2007b; Chen and Kimelman, 2000).

Despite the apparent lack of conservation in how *nodal* gene expression is first activated among different vertebrate taxa, the gene regulatory network controlling endoderm development downstream of Nodal is highly conserved.

13.4 ENDODERM GRN TRANSCRIPTION FACTORS

Nodal signaling activates the expression of a number of key transcription factors (TFs) including Sox17, Mix-like, FoxA, and Gata factors, which interact in a GRN to regulate endoderm specification and segregate the endoderm and mesoderm lineages from a transient mesendoderm state. Global transcriptional profiling, functional epistasis experiments, and genome-wide chromatin analysis in multiple model organisms indicate that these TFs regulate the transcription of both distinct and overlapping target genes, and that they interact with Nodal and other signaling pathways in complex feedback loops to control differentiation of the endoderm (Bennett et al., 2007a; Kim et al., 2011; Sinner et al., 2006; Tamplin et al., 2008). Although the overall role of these endoderm TFs is largely conserved in vertebrates, there are specie specific differences. For example Gata4–6 regulate endoderm development in fish and frogs but not in mammals, whereas other TFs such as Otx and the pluripotency stem cell factors, Oct and Nanog, appear to be involved in some species but not in others. In the following sections we will summarize the role of these TFs proving examples of their conserved and distinct functions.

Sox17

The SRY-related HMG-box 17 (Sox17) is one of the first transcription factors exclusively expressed in the endoderm during early development, and its identification was a key milestone in being able to unambiguously assay endoderm fate enabling studies of the endoderm GRN. In *Xenopus,* two closely related genes. *sox17α* and *sox17β,* are both expressed in the vegetal hemisphere of the blastula and then throughout the endoderm during gastrulation (Hudson et al., 1997). This expression is regulated directly by maternal VegT and zygotic Nodal/Smad2 signaling (Howard et al., 2007). In zebrafish, *sox17* is expressed in endoderm progenitors prior to gastrulation and then in early endoderm, but its function in endoderm specification has not been fully investigated (Dickmeis et al., 2001; Sakaguchi et al., 2001). Instead, studies have focused on an upstream regulator of *sox17*, a related Sox transcription factor called *casanova* (*cas/sox32*), the expression of which in endoderm progenitors is regulated by Nodal (Aoki et al., 2002; Chan et al., 2009). In the mouse, *Sox17* is expressed through gastrulation in both the VE and the nascent DE of the epiblast (Kanai-Azuma et al., 2002). In both *Xenopus* and the mouse, *Sox17* is critical for endoderm development. Blocking Sox17 function in frog embryos prevents endoderm formation, whereas *sox17* overexpression in animal caps results in induction of ectopic endoderm tissue (Clements et al., 2003; Hudson et al., 1997; Sinner et al., 2006). In *Sox17* null mice, endoderm formation is disrupted, particularly in the hindgut and midgut with increased level of apoptosis; suggesting that Sox17 has an important role in endoderm cell survival (Kanai-Azuma et al., 2002). In *cas* mutants zebrafish, mesoderm forms at the expense of endoderm and no *sox17* expression is detected (Alexander et al., 1999; Alexander and Stainier, 1999; Dickmeis et al., 2001; Kikuchi et al., 2001). Misexpression of *cas* is sufficient to both induce *sox17* expression and to transform mesoderm into endoderm (Kikuchi et al., 2001). Further study is required to determine whether the function of *cas* in endoderm is direct or whether it works via *sox17*.

Mix-Like

Mix-like proteins are a family of paired-like homeobox transcription factors. The mammalian genomes contain a single member of the Mix-like family, in *Xenopus*, where they were originally discovered, there are seven partially redundant Mix-like genes (*bix1/mix4, bix2/milk, bix3, bix4, mix1, mix2,* and *mix3/mixer*), and the zebrafish genome contains four Mix-like family members (*bonnie and clyde* (*bon*), *sebox* (*mezzo*), *mxtx1*, and *mxtx2*) (Pereira et al., 2012). In addition to being Nodal target genes, some Mix-like factors physically interact with Smad2/4 complexes and therefore are also effectors of Nodal signaling (Germain et al., 2000). *Mix-like* genes are expressed in both the transient mesendoderm progenitors as well as the DE, and they appear to be particularly important for segregating the endoderm and mesoderm lineages. In mice, *Mixl1* transcripts are present in the PS, but not in the node. In *Mixl1* mutant embryos gastrulation is disrupted, the amount of DE formed is reduced, and there appears to be excessive mesodermal tissue (Hart et al., 2002; Tam et al., 2007). In *Xenopus*, all seven mix-like genes are transiently expressed in the vegetal cells of the gastrula with high levels being found in the future endoderm-mesoderm boundary (Ecochard et al., 1998; Lemaire et al., 1998). Experimental manipulation of several of these factors has shown that they are sufficient to induce endoderm and repress mesoderm at high levels (Henry and Melton, 1998; Latinkic and Smith, 1999; Lemaire et al., 1998; Tada et al., 1998). For example, loss of *mixer* results in disrupted endoderm fate and expands expression of mesodermal markers (Henry and Melton, 1998; Kofron et al., 2004; Sinner et al., 2006). In zebrafish, zygotic *mxtx2* regulates expression of the endoderm inducing signals emanating from the YSL, which include the *nodal-related* genes (Hong et al., 2011; Xu et al., 2012). On the other hand, zygotic *bon* and *mezzo* are critical for endoderm development downstream of Nodal. *bon* and *mezzo* are both initially expressed in the mesendodermal precursors within six cell diameters of the embryonic margin, and then in the involuting axial mesendoderm during gastrulation (Alexander et al., 1999; Kikuchi et al., 2000; Poulain and Lepage, 2002). Mutations in *bon* or *mezzo* result in deficient endoderm development and reduced levels of *cas* and *sox17*, whereas morpholino-induced knockdown of *mezzo* in a *bon* mutant background shows complete loss of *sox17* expression (Poulain and Lepage, 2002).

FoxA

FoxA family genes are expressed in the endoderm in a broad range of animals including mammals, *Xenopus* and zebrafish (Monaghan et al., 1993; Odenthal and Nusslein-Volhard, 1998; Sinner et al., 2006), as well as in many non-vertebrates such as *Ciona* (Di Gregorio et al., 2001), sea urchin (Harada et al., 1996), *C. elegans* (Horner et al., 1998; Kalb et al., 1998), and *Drosophila* (Weigel et al., 1989) suggesting an evolutionarily ancient role in endoderm development. In all vertebrates, *foxa2* is expressed throughout the endoderm, with higher levels of expression being observed in the presumptive foregut. In the mouse there are three paralogs, *Foxa1*, *Foxa2* and *Foxa3*. Only *Foxa2* is necessary for early endoderm development, whereas *Foxa1* and *Foxa3* are involved in later endoderm organogenesis (Kaestner et al., 1998; 1999; Shih et al., 1999). In *Foxa2* mutant mice, gastrulation is disrupted and the formation of foregut and midgut is compromised, furthermore in chimeric embryos *Foxa2* null cells are only able to contribute to foregut and to a lesser extent to midgut endoderm (Ang et al., 1993; Dufort et al., 1998; McKnight et al., 2010; Sasaki and Hogan, 1993; Weinstein et al., 1994). Similarly, in *foxa2* mutant zebrafish only hindgut endoderm precursors remain at the late gastrula stage (Alexander and Stainier, 1999). In *Xenopus*, *foxa1*, *foxa2* and *foxa4* are all expressed in the gastrula endoderm but their function has not been characterized (D'Souza et al., 2003).

Gata4–6

Six members of the Gata family, class IV zinc finger domain containing GATA-binding transcription factors have been identified in vertebrates. Of these, Gata4, 5, and 6 promote endoderm development, downstream of Nodal in *Xenopus* and zebrafish (Alexander and Stainier, 1999; Reiter et al., 2001; Rodaway et al., 1999; Weber et al., 2000). In *Xenopus, gata4–6* genes are expressed in the endoderm of the gastrula, and each is sufficient to induce ectopic endoderm tissue in pluripotent cells of the animal caps (Afouda et al., 2005; Gao et al., 1998; Jiang and Evans, 1996; Shoichet et al., 2000; Weber et al., 2000). They appear to be largely redundant in *Xenopus*, promoting each other's expression. Depletion of all three is required to disrupt, though not completely abolish, endoderm formation (Afouda et al., 2005; Shoichet et al., 2000; Weber et al., 2000). The zebrafish *gata5* gene seems to have a similar role to that of the three *Xenopus gata* genes. Zebrafish *gata5* is initially expressed in the YSL and the margin, and then following gastrulation in the lateral plate and cardiac mesoderm, and nascent endoderm (Reiter et al., 2001) similarly to *Xenopus gata4–6*. The zebrafish gata5 mutant known as *faust* has only 60% of the normal number of endoderm cells (Reiter et al., 1999). Gata5 forms a transcriptional complex with Bon and Eomes that is involved in direct activation of *cas* (Bjornson et al., 2005).

In mice, Gata4–6 do not appear to regulate DE formation, rather they act in concert with Sox17 and Sox7 to control extra-embryonic endoderm development, and then later in embryogenesis they are required for endoderm organ differentiation (Jacobsen et al., 2002; Kuo et al., 1997; Morrisey et al., 1998; Narita et al., 1997; Watt et al., 2007; Zhao et al., 2005). In mouse ES cells, Gata4 and Gata6 are each sufficient to promote primitive endoderm differentiation (Fujikura et al., 2002), suggesting that the endoderm-promoting activity of Gatas that is observed in *Xenopus* and zebrafish, has been coopted to regulate extra embryonic endoderm in mammals. The role of Gata factors in endoderm development appears evolutionarily ancient, conserved in vertebrate species, as well as in sea urchins (Davidson et al., 2002; Yuh et al., 1998), *C. elegans* (Maduro et al., 2005; Zhu et al., 1997), and *Drosophila* (Lin et al., 1995; Okumura et al., 2005).

13.5 MODULATION OF THE ENDODERM GRN BY OTHER SIGNALING PATHWAYS

In vertebrates, the Wnt, FGF, and Notch signaling pathways act as modulators of endoderm formation by interacting with the Nodal pathway and by modifying the activity of the endoderm GRN transcription factors. In general, canonical Wnt signaling appears to promote high levels of Nodal enhancing mesendoderm formation and participates in initial endoderm patterning, which occurs around the same time as germ layer formation. In contrast, FGF and Notch appear to function in the endoderm versus mesoderm fate choice of the mesendodermal cells.

In all vertebrates, the cytoplasmic transducer of canonical Wnt signaling, β-catenin, promotes higher levels of *nodal* gene transcription in the pre-gastrula stage embryo (Granier et al., 2011). The activation of the canonical Wnt signaling pathway results in stabilization and accumulation of β-catenin in the cytoplasm and then its translocation into the nucleus. Here, through physical interactions with the TCF/LEF family of DNA-bound transcription factors it activates expression of Wnt target genes (Saito-Diaz et al., 2013). In *Xenopus* and zebrafish, β-catenin cooperates with maternal VegT and the YSL signal respectively to promote higher expression levels of mesendodermal genes including *nodal-related* ligands in the dorsoanterior region of the blastula and gastrula stage embryos, in some cases directly interacting with the *nodal-related* cis-regulatory elements (Bellipanni et al., 2006; Hyde and Old, 2000; Schohl and Fagotto, 2002; Xanthos et al., 2002). Recent data from *Xenopus* have also shown that following endoderm induction, foregut development requires repression, but not complete abolishment, of Wnt signaling activity (Li et al., 2008; McLin et al., 2007; Zhang et al., 2013).

In mice, Wnt signaling is necessary for gastrulation, and lack of β-catenin results in the failure of both the formation of the PS and the emergence of the definitive endoderm (Conlon et al., 1994; Huelsken et al., 2000). Furthermore, loss of *Adenomatous Polyposis Coli* (*APC*), a negative regulator of canonical Wnt signaling, results in production of excess mesendoderm (Chazaud and Rossant, 2006). It remains unclear whether β-catenin is required for mesendoderm induction or maintenance of endoderm in the mouse, as conditional loss of β-catenin in the mesendoderm cells results in the production of ectopic cardiac mesoderm at the expense of definitive endoderm (Lickert et al., 2002). However, consistent with a role in endoderm induction, the conditional deletion of *β-catenin* in both definitive and visceral endoderm results in the loss of *Sox17* expression (Engert et al., 2013). The spatiotemporal expression pattern of *Wnt3*, together with functional data, suggest it as a likely candidate regulating this process (Liu et al., 1999), possibly via a feedback loop with Nodal signaling. In *Wnt3* mutant embryos, the expression of *Nodal* is induced but is not maintained (Ben-Haim et al., 2006; Liu et al., 1999), whereas *Wnt3* expression is abolished in the absence of Nodal (Brennan et al., 2001). This interaction between Wnt and Nodal signaling pathways has also been observed in *Xenopus* and zebrafish, where high expression levels of *nodal-related* genes are maintained by Nodal autoregulation in cooperation with Wnt/β-catenin, and where expression of zygotic mesodermal *wnt8* expression during gastrulation is induced by Nodal signaling (Agathon et al., 2003; Cha et al., 2006).

In all vertebrates, FGF signaling is necessary and sufficient to induce mesoderm (Ciruna and Rossant, 2001; Griffin et al., 1995; Isaacs, 1997), and evidence from *Xenopus* and zebrafish demonstrated that FGF is a negative regulator of endodermal fates (Cha et al., 2004; Mizoguchi et al., 2006; Rodaway et al., 1999). The available data suggest a model in which Nodal signaling induces both endoderm and mesoderm fate. In the mesoderm, Nodal induces the expression of FGF and the T-box factor T-brachyury. Together FGF and T form the core of a mesoderm GRN and promote each other's expression, induce mesodermal gene expression and repress endodermal fate (Figure 13.3). A key role of the endoderm GRN transcription factors, particularly the Mix-like genes, is to repress the expression and activity of FGF and T (Latinkic et al., 1997). Thus, downstream of Nodal, mutual antagonism between the endoderm GRN and the mesoderm GRN appears to regulate the segregation of endoderm and mesoderm lineages from the mesendoderm progenitor induced by Nodal. However, this model remains to be rigorously validated and the detailed molecular mechanisms are yet to be determined.

In zebrafish and *Xenopus*, Notch signaling seems to regulate the segregation of endoderm and mesoderm fates and to control the number of cells allocated to each lineage. Experimental activation of Notch before gastrulation results in a reduced number of endoderm cells; however, an opposite effect is observed in *Xenopus* if Notch/Delta is disrupted during gastrulation (Contakos et al., 2005; Kikuchi et al., 2004). More studies are necessary to fully understand these temporally distinct roles. It is unclear whether this activity of Notch is conserved in mammals, since deletion of a transcription factor downstream of Notch, RBP-Jκ, or the Notch modifying enzyme required for signaling, Pofut1, does not affect early cell fate specification in mice (Shi et al., 2005; Souilhol et al., 2006).

13.6 HUMAN ENDODERM DIFFERENTIATION IN PLURIPOTENT STEM CELLS

Early human embryonic development is morphologically similar to that of the mouse, and defects in human embryonic endoderm specification will almost certainly result in lethality in the first trimester and thus will not reach the clinic. However, reduced or compromised endoderm formation, which allows fetal development to proceed further, might be the underlying cause of some severe birth defects such as foregut malformations, holoprosencephaly or cardiac defects, since the endoderm sends important signals to the developing head, nervous system and heart (Bouwmeester et al., 1996; Perea-Gomez et al., 2001; Piccolo et al., 1999; Vincent et al., 2003). Moreover, growing evidence suggests that disruptions in early development can impact adult physiology, including diseases in endoderm derived digestive organs, such as diabetes, which are a major burden on human health.

The most significant clinical contribution from studying endoderm development in animal models has been the translation of this information to the directed differentiation of human pluripotent stem cells. The recent revolution in ES cell technology has led to the promise of tissue replacement therapy and *in vitro* modeling of human disease. In particular the ability to make induced pluripotent stem (iPS) cells from virtually any cell type by induced ectopic expression of the four pluripotency factors, Oct3/4, Sox2, c-Myc and Klf4, (Takahashi and Yamanaka, 2006; Zorn and Wells, 2009) has opened the door to patient specific stem cells. The hope is that in the near future this will enable clinical researchers to study the physiology and pharmacological response of the patients own diseased cells *in vitro*, eventually allowing correction of genetic lesions in culture followed by reintroduction of corrected cells into the patients (Yusa et al., 2011). All of these stem cell based approaches require the ability to direct the differentiation of the appropriate tissues *in vitro*. Although this is still a work in progress for most cell types, in the last decade there has been tremendous success in the first step of this process; that is, inducing embryonic endoderm from ES/iPS cells in culture by recapitulating the signals that regulate tissue differentiation in animal models.

There are now robust protocols to direct the differentiation of human (and mouse) endoderm tissue from ES/iPS cells by treating cultures with high levels of the TGFβ family ligand Activin for 3–5 days, which mimics Nodal signaling (D'Amour et al., 2005; Kubo et al., 2004). Some protocols also include Wnt, which *in vivo* appears to promote the endoderm inducing activity of the Nodal/Activin signaling. Another critical step has been the identification of lineage-specific markers of germ layer specification in animal models, which can be used to benchmark the *in vitro* differentiation of the cells in culture. Successful differentiation of endoderm development in culture is characterized by more than 80% of the cells co-expressing Sox17 and Foxa2 (both DE markers in the vertebrate gastrula). Importantly, the culture should not co-express Sox17 with Gata4–6 or Sox7, which would indicate extra-embryonic endoderm (Yasunaga et al., 2005), nor express T-brachyury indicating mesoderm differentiation. Interestingly, in ES and iPS cells, Activin has concentration-dependent effects, promoting endoderm at high concentrations and mesoderm at low levels, just like Nodal does in animal models. Moreover, *in vitro* endoderm differentiation appears to have similar developmental kinetics, first transiting through a bipotential mesendoderm progenitor which transiently co-expressed Mix-like, brachyury, and Sox17 in the first few days followed by downregulation of the mesodermal genes (Figure 13.4) (Tada et al., 2005). This endoderm induction protocol is the standard first step in elaborate step-wise protocols to differentiate more mature endodermal tissue, such as hepatocytes, pancreatic beta-cells or intestine (Kroon et al., 2008; Nagaoka and Duncan, 2010; Spence et al., 2011).

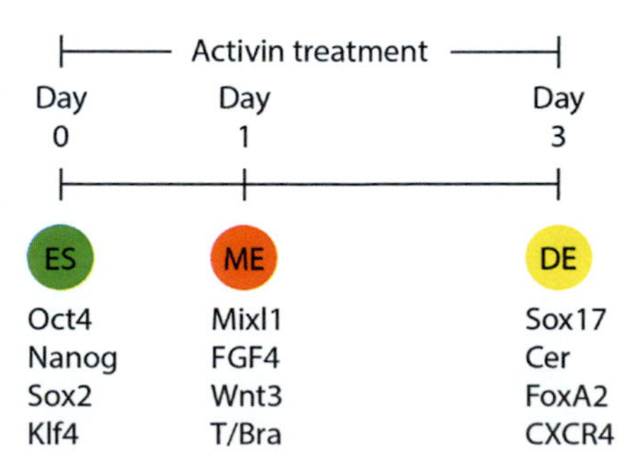

FIGURE 13.4 Activin-Induced Differentiation of Endoderm from ES Cells. Progressive differentiation of ES cells (green) towards definitive endoderm (DE; yellow) when treated with Activin. After one day of Activin treatment, the cells downregulate pluripotency genes and transition through a transient mesendoderm state (ME; orange) expressing markers typical of primitive streak as well as mesoderm and endoderm. DE forms after three days of treatment as the cells downregulate expression of mesoderm markers and upregulate markers specific to endoderm.

In addition to providing potential therapeutic tissue, endodermal differentiation in human stem cells has provided a powerful system to interrogate the molecular pathways regulating human endoderm development. In particular, stem cell culture offers biochemical approaches that are not possible even in animal embryos. For example, recent studies examining genome-wide epigenetic regulation of the endoderm GRN and dynamic chromatin association of Smad2 and Eomes during Activin-induced DE formation in ES cells has revealed an unexpected interaction with the pluripotency factors Nanog and Oct4 in regulating the transition from a pluripotent state to an endodermal cell state (Kartikasari et al., 2013; Kim et al., 2011; Teo et al., 2011). These results from human ES cells can then be validated *in vivo* using animal models (Yoon et al., 2011). Thus, human endoderm development from pluripotent stem cells has become another important tool for researchers investigating the complex GRN controlling development of definitive endoderm.

13.5 CLINICAL RELEVANCE

- Data from animal models have informed protocols for the differentiation of DE tissue from human pluripotent ES/iPS cells.
- Generation of DE from ES cells is the first critical step in growing human endoderm-derived tissues such as liver, pancreas and intestine for modeling human disease and ultimately for regenerative medicine.
- Experiments using ES and iPS cells to induce directed differentiation into distinct endodermal tissues have provided both a further understanding of genetic mechanisms regulating endoderm development as well as additional testable hypotheses for animal model experiments.

ACKNOWLEDGMENTS

We are grateful to the funding sources supporting the research in the Zorn lab: the NIH DK070858 and DK080823 grants to AMZ, and NIH T32 HL 7752–19 grant to Jeff Whitsett in part supporting MW.

RECOMMENDED RESOURCES

www.xenbase.org; www.zfin.org; www.emouseatlas.org; respectivew websites for *Xenopus*, zebrafish, and mouse designed to provide online access to comprehensive resources dealing with embryonic development and gene expression data for the major vertebrate model systems.

REFERENCES

Afouda, B.A., Ciau-Uitz, A., Patient, R., 2005. GATA4, 5 and 6 mediate TGFbeta maintenance of endodermal gene expression in *Xenopus* embryos. Development 132, 763–774.

Agathon, A., Thisse, C., Thisse, B., 2003. The molecular nature of the zebrafish tail organizer. Nature 424, 448–452.

Agius, E., Oelgeschlager, M., Wessely, O., Kemp, C., De Robertis, E.M., 2000. Endodermal Nodal-related signals and mesoderm induction in *Xenopus*. Development 127, 1173–1183.

Alexander, J., Rothenberg, M., Henry, G.L., Stainier, D.Y., 1999. casanova plays an early and essential role in endoderm formation in zebrafish. Dev. Biol. 215, 343–357.

Alexander, J., Stainier, D.Y., 1999. A molecular pathway leading to endoderm formation in zebrafish. Curr. Biol. 9, 1147–1157.

Ang, S.L., Wierda, A., Wong, D., Stevens, K.A., Cascio, S., Rossant, J., Zaret, K.S., 1993. The formation and maintenance of the definitive endoderm lineage in the mouse: involvement of HNF3/forkhead proteins. Development 119, 1301–1315.

Aoki, T.O., David, N.B., Minchiotti, G., Saint-Etienne, L., Dickmeis, T., Persico, G.M., Strahle, U., Mourrain, P., Rosa, F.M., 2002. Molecular integration of casanova in the Nodal signaling pathway controlling endoderm formation. Development 129, 275–286.

Bagnat, M., Cheung, I.D., Mostov, K.E., Stainier, D.Y., 2007. Genetic control of single lumen formation in the zebrafish gut. Nat. Cell. Biol. 9, 954–960.

Beck, S., Le Good, J.A., Guzman, M., Ben Haim, N., Roy, K., Beermann, F., Constam, D.B., 2002. Extra-embryonic proteases regulate Nodal signaling during gastrulation. Nat. Cell. Biol. 4, 981–985.

Bellipanni, G., Varga, M., Maegawa, S., Imai, Y., Kelly, C., Myers, A.P., Chu, F., Talbot, W.S., Weinberg, E.S., 2006. Essential and opposing roles of zebrafish beta-catenins in the formation of dorsal axial structures and neurectoderm. Development 133, 1299–1309.

Ben-Haim, N., Lu, C., Guzman-Ayala, M., Pescatore, L., Mesnard, D., Bischofberger, M., Naef, F., Robertson, E.J., Constam, D.B., 2006. The nodal precursor acting via activin receptors induces mesoderm by maintaining a source of its convertases and BMP4. Dev. Cell. 11, 313–323.

Bennett, J.T., Joubin, K., Cheng, S., Aanstad, P., Herwig, R., Clark, M., Lehrach, H., Schier, A.F., 2007a. Nodal signaling activates differentiation genes during zebrafish gastrulation. Dev. Biol. 304, 525–540.

Bennett, J.T., Stickney, H.L., Choi, W.Y., Ciruna, B., Talbot, W.S., Schier, A.F., 2007b. Maternal nodal and zebrafish embryogenesis. Nature 450, E1–E2. discussion E2–4.

Bjornson, C.R., Griffin, K.J., Farr 3rd, G.H., Terashima, A., Himeda, C., Kikuchi, Y., Kimelman, D., 2005. Eomesodermin is a localized maternal determinant required for endoderm induction in zebrafish. Dev. Cell. 9, 523–533.

Bouwmeester, T., Kim, S., Sasai, Y., Lu, B., De Robertis, E.M., 1996. Cerberus is a head-inducing secreted factor expressed in the anterior endoderm of Spemann's organizer. Nature 382, 595–601.

Brennan, J., Lu, C.C., Norris, D.P., Rodriguez, T.A., Beddington, R.S., Robertson, E.J., 2001. Nodal signaling in the epiblast patterns the early mouse embryo. Nature 411, 965–969.

Cai, J., DeLaForest, A., Fisher, J., Urick, A., Wagner, T., Twaroski, K., Cayo, M., Nagaoka, M., Duncan, S.A., 2008. Protocol for directed differentiation of human pluripotent stem cells toward a hepatocyte fate. In: StemBook(Cambridge (MA)).

Cha, S.W., Hwang, Y.S., Chae, J.P., Lee, S.Y., Lee, H.S., Daar, I., Park, M.J., Kim, J., 2004. Inhibition of FGF signaling causes expansion of the endoderm in *Xenopus*. Biochem. Biophys. Res. Commun. 315, 100–106.

Cha, Y.R., Takahashi, S., Wright, C.V., 2006. Cooperative non-cell and cell autonomous regulation of Nodal gene expression and signaling by Lefty/Antivin and Brachyury in *Xenopus*. Dev. Biol. 290, 246–264.

Chalmers, A.D., Slack, J.M., 2000. The *Xenopus* tadpole gut: fate maps and morphogenetic movements. Development 127, 381–392.

Chan, T.M., Chao, C.H., Wang, H.D., Yu, Y.J., Yuh, C.H., 2009. Functional analysis of the evolutionarily conserved cis-regulatory elements on the sox17 gene in zebrafish. Dev. Biol. 326, 456–470.

Chazaud, C., Rossant, J., 2006. Disruption of early proximodistal patterning and AVE formation in Apc mutants. Development 133, 3379–3387.

Chen, S., Kimelman, D., 2000. The role of the yolk syncytial layer in germ layer patterning in zebrafish. Development 127, 4681–4689.

Chen, X., Weisberg, E., Fridmacher, V., Watanabe, M., Naco, G., Whitman, M., 1997. Smad4 and FAST-1 in the assembly of activin-responsive factor. Nature 389, 85–89.

Chen, Y., Schier, A.F., 2001. The zebrafish Nodal signal Squint functions as a morphogen. Nature 411, 607–610.

Ciruna, B., Rossant, J., 2001. FGF signaling regulates mesoderm cell fate specification and morphogenetic movement at the primitive streak. Dev. Cell. 1, 37–49.

Clements, D., Cameleyre, I., Woodland, H.R., 2003. Redundant early and overlapping larval roles of Xsox17 subgroup genes in *Xenopus* endoderm development. Mech. Dev. 120, 337–348.

Conlon, F.L., Barth, K.S., Robertson, E.J., 1991. A novel retrovirally induced embryonic lethal mutation in the mouse: assessment of the developmental fate of embryonic stem cells homozygous for the 413.d proviral integration. Development 111, 969–981.

Conlon, F.L., Lyons, K.M., Takaesu, N., Barth, K.S., Kispert, A., Herrmann, B., Robertson, E.J., 1994. A primary requirement for nodal in the formation and maintenance of the primitive streak in the mouse. Development 120, 1919–1928.

Contakos, S.P., Gaydos, C.M., Pfeil, E.C., McLaughlin, K.A., 2005. Subdividing the embryo: a role for Notch signaling during germ layer patterning in *Xenopus laevis*. Dev. Biol. 288, 294–307.

D'Amour, K.A., Agulnick, A.D., Eliazer, S., Kelly, O.G., Kroon, E., Baetge, E.E., 2005. Efficient differentiation of human embryonic stem cells to definitive endoderm. Nat. Biotechnol. 23, 1534–1541.

D'Souza, A., Lee, M., Taverner, N., Mason, J., Carruthers, S., Smith, J.C., Amaya, E., Papalopulu, N., Zorn, A.M., 2003. Molecular components of the endoderm specification pathway in *Xenopus* tropicalis. Dev. Dyn. 226, 118–127.

Dale, L., Slack, J.M., 1987. Fate map for the 32-cell stage of *Xenopus laevis*. Development 99, 527–551.

Davidson, E.H., Rast, J.P., Oliveri, P., Ransick, A., Calestani, C., Yuh, C.H., Minokawa, T., Amore, G., Hinman, V., Arenas-Mena, C., et al., 2002. A provisional regulatory gene network for specification of endomesoderm in the sea urchin embryo. Dev. Biol. 246, 162–190.

Di Gregorio, A., Corbo, J.C., Levine, M., 2001. The regulation of forkhead/HNF-3beta expression in the Ciona embryo. Dev. Biol. 229, 31–43.

Dickmeis, T., Mourrain, P., Saint-Etienne, L., Fischer, N., Aanstad, P., Clark, M., Strahle, U., Rosa, F., 2001. A crucial component of the endoderm formation pathway, CASANOVA, is encoded by a novel sox-related gene. Genes. Dev. 15, 1487–1492.

Dougan, S.T., Warga, R.M., Kane, D.A., Schier, A.F., Talbot, W.S., 2003. The role of the zebrafish nodal-related genes squint and cyclops in patterning of mesendoderm. Development 130, 1837–1851.

Dufort, D., Schwartz, L., Harpal, K., Rossant, J., 1998. The transcription factor HNF3beta is required in visceral endoderm for normal primitive streak morphogenesis. Development 125, 3015–3025.

Dunn, N.R., Vincent, S.D., Oxburgh, L., Robertson, E.J., Bikoff, E.K., 2004. Combinatorial activities of Smad2 and Smad3 regulate mesoderm formation and patterning in the mouse embryo. Development 131, 1717–1728.

Ecochard, V., Cayrol, C., Rey, S., Foulquier, F., Caillol, D., Lemaire, P., Duprat, A.M., 1998. A novel *Xenopus* mix-like gene milk involved in the control of the endomesodermal fates. Development 125, 2577–2585.

Eimon, P.M., Harland, R.M., 2002. Effects of heterodimerization and proteolytic processing on Derriere and Nodal activity: implications for mesoderm induction in *Xenopus*. Development 129, 3089–3103.

Engert, S., Burtscher, I., Liao, W.P., Dulev, S., Schotta, G., Lickert, H., 2013. Wnt/beta-catenin signaling regulates Sox17 expression and is essential for organizer and endoderm formation in the mouse. Development 140, 3128–3138.

Fan, X., Hagos, E.G., Xu, B., Sias, C., Kawakami, K., Burdine, R.D., Dougan, S.T., 2007. Nodal signals mediate interactions between the extra-embryonic and embryonic tissues in zebrafish. Dev. Biol. 310, 363–378.

Feldman, B., Gates, M.A., Egan, E.S., Dougan, S.T., Rennebeck, G., Sirotkin, H.I., Schier, A.F., Talbot, W.S., 1998. Zebrafish organizer development and germ-layer formation require nodal-related signals. Nature 395, 181–185.

Franklin, V., Khoo, P.L., Bildsoe, H., Wong, N., Lewis, S., Tam, P.P., 2008. Regionalisation of the endoderm progenitors and morphogenesis of the gut portals of the mouse embryo. Mech. Dev. 125, 587–600.

Fujikura, J., Yamato, E., Yonemura, S., Hosoda, K., Masui, S., Nakao, K., Miyazaki Ji, J., Niwa, H., 2002. Differentiation of embryonic stem cells is induced by GATA factors. Genes. Dev. 16, 784–789.

Gao, X., Sedgwick, T., Shi, Y.B., Evans, T., 1998. Distinct functions are implicated for the GATA-4, -5, and -6 transcription factors in the regulation of intestine epithelial cell differentiation. Mol. Cell. Biol. 18, 2901–2911.

Germain, S., Howell, M., Esslemont, G.M., Hill, C.S., 2000. Homeodomain and winged-helix transcription factors recruit activated Smads to distinct promoter elements via a common Smad interaction motif. Genes. Dev. 14, 435–451.

Gore, A.V., Maegawa, S., Cheong, A., Gilligan, P.C., Weinberg, E.S., Sampath, K., 2005. The zebrafish dorsal axis is apparent at the four-cell stage. Nature 438, 1030–1035.

Granier, C., Gurchenkov, V., Perea-Gomez, A., Camus, A., Ott, S., Papanayotou, C., Iranzo, J., Moreau, A., Reid, J., Koentges, G., et al., 2011. Nodal cis-regulatory elements reveal epiblast and primitive endoderm heterogeneity in the peri-implantation mouse embryo. Dev. Biol. 349, 350–362.

Griffin, K., Patient, R., Holder, N., 1995. Analysis of FGF function in normal and no tail zebrafish embryos reveals separate mechanisms for formation of the trunk and the tail. Development 121, 2983–2994.

Griffin, K.J., Amacher, S.L., Kimmel, C.B., Kimelman, D., 1998. Molecular identification of spadetail: regulation of zebrafish trunk and tail mesoderm formation by T-box genes. Development 125, 3379–3388.

Hagos, E.G., Dougan, S.T., 2007. Time-dependent patterning of the mesoderm and endoderm by Nodal signals in zebrafish. BMC. Dev. Biol. 7, 22.

Harada, Y., Akasaka, K., Shimada, H., Peterson, K.J., Davidson, E.H., Satoh, N., 1996. Spatial expression of a forkhead homologue in the sea urchin embryo. Mech. Dev. 60, 163–173.

Hart, A.H., Hartley, L., Sourris, K., Stadler, E.S., Li, R., Stanley, E.G., Tam, P.P., Elefanty, A.G., Robb, L., 2002. Mixl1 is required for axial mesendoderm morphogenesis and patterning in the murine embryo. Development 129, 3597–3608.

Heasman, J., Snape, A., Smith, J., Wylie, C.C., 1985. Single cell analysis of commitment in early embryogenesis. J. Embryol. Exp. Morphol. 89 (Suppl.), 297–316.

Heasman, J., Wylie, C.C., Hausen, P., Smith, J.C., 1984. Fates and states of determination of single vegetal pole blastomeres of *X. laevis*. Cell 37, 185–194.

Henry, G.L., Melton, D.A., 1998. Mixer, a homeobox gene required for endoderm development. Science 281, 91–96.

Hogan, B.L.M., Zaret, K.S., 2002. Development of the endoderm and its tissue derivatives. In: Rossant, J., Tam, P.L. (Eds.), Mouse Development. Academic Press, San Diego, CA, pp. 301–330.

Hong, S.K., Jang, M.K., Brown, J.L., McBride, A.A., Feldman, B., 2011. Embryonic mesoderm and endoderm induction requires the actions of non-embryonic Nodal-related ligands and Mxtx2. Development 138, 787–795.

Horne-Badovinac, S., Lin, D., Waldron, S., Schwarz, M., Mbamalu, G., Pawson, T., Jan, Y., Stainier, D.Y., Abdelilah-Seyfried, S., 2001. Positional cloning of heart and soul reveals multiple roles for PKC lambda in zebrafish organogenesis. Curr. Biol. 11, 1492–1502.

Horner, M.A., Quintin, S., Domeier, M.E., Kimble, J., Labouesse, M., Mango, S.E., 1998. pha-4, an HNF-3 homolog, specifies pharyngeal organ identity in Caenorhabditis elegans. Genes. Dev. 12, 1947–1952.

Hou, J., Charters, A.M., Lee, S.C., Zhao, Y., Wu, M.K., Jones, S.J., Marra, M.A., Hoodless, P.A., 2007. A systematic screen for genes expressed in definitive endoderm by Serial Analysis of Gene Expression (SAGE). BMC. Dev. Biol. 7, 92.

Howard, L., Rex, M., Clements, D., Woodland, H.R., 2007. Regulation of the *Xenopus* Xsox17alpha(1) promoter by co-operating VegT and Sox17 sites. Dev. Biol. 310, 402–415.

Hudson, C., Clements, D., Friday, R.V., Stott, D., Woodland, H.R., 1997. Xsox17alpha and -beta mediate endoderm formation in *Xenopus*. Cell 91, 397–405.

Huelsken, J., Vogel, R., Brinkmann, V., Erdmann, B., Birchmeier, C., Birchmeier, W., 2000. Requirement for beta-catenin in anterior-posterior axis formation in mice. J. Cell. Biol. 148, 567–578.

Hug, B., Walter, V., Grunwald, D.J., 1997. tbx6, a Brachyury-related gene expressed by ventral mesendodermal precursors in the zebrafish embryo. Dev. Biol. 183, 61–73.

Hyde, C.E., Old, R.W., 2000. Regulation of the early expression of the *Xenopus* nodal-related 1 gene, Xnr1. Development 127, 1221–1229.

Iannaccone, P.M., Zhou, X., Khokha, M., Boucher, D., Kuehn, M.R., 1992. Insertional mutation of a gene involved in growth regulation of the early mouse embryo. Dev. Dyn. 194, 198–208.

Isaacs, H.V., 1997. New perspectives on the role of the fibroblast growth factor family in amphibian development. Cell. Mol. Life. Sci. 53, 350–361.

Jacobsen, C.M., Narita, N., Bielinska, M., Syder, A.J., Gordon, J.I., Wilson, D.B., 2002. Genetic mosaic analysis reveals that GATA-4 is required for proper differentiation of mouse gastric epithelium. Dev. Biol. 241, 34–46.

Jiang, Y., Evans, T., 1996. The *Xenopus* GATA-4/5/6 genes are associated with cardiac specification and can regulate cardiac-specific transcription during embryogenesis. Dev. Biol. 174, 258–270.

Jones, C.M., Kuehn, M.R., Hogan, B.L., Smith, J.C., Wright, C.V., 1995. Nodal-related signals induce axial mesoderm and dorsalize mesoderm during gastrulation. Development 121, 3651–3662.

Joseph, E.M., Melton, D.A., 1997. Xnr4: a *Xenopus* nodal-related gene expressed in the Spemann organizer. Dev. Biol. 184, 367–372.

Kaestner, K.H., Hiemisch, H., Schutz, G., 1998. Targeted disruption of the gene encoding hepatocyte nuclear factor 3gamma results in reduced transcription of hepatocyte-specific genes. Mol. Cell. Biol. 18, 4245–4251.

Kaestner, K.H., Katz, J., Liu, Y., Drucker, D.J., Schutz, G., 1999. Inactivation of the winged helix transcription factor HNF3alpha affects glucose homeostasis and islet glucagon gene expression in vivo. Genes. Dev. 13, 495–504.

Kalb, J.M., Lau, K.K., Goszczynski, B., Fukushige, T., Moons, D., Okkema, P.G., McGhee, J.D., 1998. pha-4 is Ce-fkh-1, a fork head/HNF-3alpha,beta,gamma homolog that functions in organogenesis of the C. elegans pharynx. Development 125, 2171–2180.

Kanai-Azuma, M., Kanai, Y., Gad, J.M., Tajima, Y., Taya, C., Kurohmaru, M., Sanai, Y., Yonekawa, H., Yazaki, K., Tam, P.P., et al., 2002. Depletion of definitive gut endoderm in Sox17-null mutant mice. Development 129, 2367–2379.

Kartikasari, A.E., Zhou, J.X., Kanji, M.S., Chan, D.N., Sinha, A., Grapin-Botton, A., Magnuson, M.A., Lowry, W.E., Bhushan, A., 2013. The histone demethylase Jmjd3 sequentially associates with the transcription factors Tbx3 and Eomes to drive endoderm differentiation. EMBO. J. 32, 1393–1408.

Keller, R.E., 1975. Vital dye mapping of the gastrula and neurula of *Xenopus laevis*. I. Prospective areas and morphogenetic movements of the superficial layer. Dev. Biol. 42, 222–241.

Keller, R.E., 1976. Vital dye mapping of the gastrula and neurula of *Xenopus laevis*. II. Prospective areas and morphogenetic movements of the deep layer. Dev. Biol. 51, 118–137.

Kikuchi, Y., Agathon, A., Alexander, J., Thisse, C., Waldron, S., Yelon, D., Thisse, B., Stainier, D.Y., 2001. casanova encodes a novel Sox-related protein necessary and sufficient for early endoderm formation in zebrafish. Genes. Dev. 15, 1493–1505.

Kikuchi, Y., Trinh, L.A., Reiter, J.F., Alexander, J., Yelon, D., Stainier, D.Y., 2000. The zebrafish bonnie and clyde gene encodes a Mix family homeodomain protein that regulates the generation of endodermal precursors. Genes. Dev. 14, 1279–1289.

Kikuchi, Y., Verkade, H., Reiter, J.F., Kim, C.H., Chitnis, A.B., Kuroiwa, A., Stainier, D.Y., 2004. Notch signaling can regulate endoderm formation in zebrafish. Dev. Dyn. 229, 756–762.

Kim, S.W., Yoon, S.J., Chuong, E., Oyolu, C., Wills, A.E., Gupta, R., Baker, J., 2011. Chromatin and transcriptional signatures for Nodal signaling during endoderm formation in hESCs. Dev. Biol. 357, 492–504.

Kimmel, C.B., Law, R.D., 1985. Cell lineage of zebrafish blastomeres. II. Formation of the yolk syncytial layer. Dev. Biol. 108, 86–93.

Kimmel, C.B., Warga, R.M., Schilling, T.F., 1990. Origin and organization of the zebrafish fate map. Development 108, 581–594.

Kofron, M., Demel, T., Xanthos, J., Lohr, J., Sun, B., Sive, H., Osada, S., Wright, C., Wylie, C., Heasman, J., 1999. Mesoderm induction in *Xenopus* is a zygotic event regulated by maternal VegT via TGFbeta growth factors. Development 126, 5759–5770.

Kofron, M., Wylie, C., Heasman, J., 2004. The role of Mixer in patterning the early *Xenopus* embryo. Development 131, 2431–2441.

Kroon, E., Martinson, L.A., Kadoya, K., Bang, A.G., Kelly, O.G., Eliazer, S., Young, H., Richardson, M., Smart, N.G., Cunningham, J., et al., 2008. Pancreatic endoderm derived from human embryonic stem cells generates glucose-responsive insulin-secreting cells in vivo. Nat. Biotechnol. 26, 443–452.

Kubo, A., Shinozaki, K., Shannon, J.M., Kouskoff, V., Kennedy, M., Woo, S., Fehling, H.J., Keller, G., 2004. Development of definitive endoderm from embryonic stem cells in culture. Development 131, 1651–1662.

Kuo, C.T., Morrisey, E.E., Anandappa, R., Sigrist, K., Lu, M.M., Parmacek, M.S., Soudais, C., Leiden, J.M., 1997. GATA4 transcription factor is required for ventral morphogenesis and heart tube formation. Genes. Dev. 11, 1048–1060.

Kwon, G.S., Viotti, M., Hadjantonakis, A.K., 2008. The endoderm of the mouse embryo arises by dynamic widespread intercalation of embryonic and extra-embryonic lineages. Dev. Cell. 15, 509–520.

Latinkic, B.V., Smith, J.C., 1999. Goosecoid and mix.1 repress Brachyury expression and are required for head formation in *Xenopus*. Development 126, 1769–1779.

Latinkic, B.V., Umbhauer, M., Neal, K.A., Lerchner, W., Smith, J.C., Cunliffe, V., 1997. The *Xenopus* Brachyury promoter is activated by FGF and low concentrations of activin and suppressed by high concentrations of activin and by paired-type homeodomain proteins. Genes. Dev. 11, 3265–3276.

Lawson, K.A., Meneses, J.J., Pedersen, R.A., 1991. Clonal analysis of epiblast fate during germ layer formation in the mouse embryo. Development 113, 891–911.

Lawson, K.A., Pedersen, R.A., 1987. Cell fate, morphogenetic movement and population kinetics of embryonic endoderm at the time of germ layer formation in the mouse. Development 101, 627–652.

Lemaire, P., Darras, S., Caillol, D., Kodjabachian, L., 1998. A role for the vegetally expressed *Xenopus* gene Mix.1 in endoderm formation and in the restriction of mesoderm to the marginal zone. Development 125, 2371–2380.

Li, Y., Rankin, S.A., Sinner, D., Kenny, A.P., Krieg, P.A., Zorn, A.M., 2008. Sfrp5 coordinates foregut specification and morphogenesis by antagonizing both canonical and noncanonical Wnt11 signaling. Genes. Dev. 22, 3050–3063.

Lickert, H., Kutsch, S., Kanzler, B., Tamai, Y., Taketo, M.M., Kemler, R., 2002. Formation of multiple hearts in mice following deletion of beta-catenin in the embryonic endoderm. Dev. Cell. 3, 171–181.

Lin, W.H., Huang, L.H., Yeh, J.Y., Hoheisel, J., Lehrach, H., Sun, Y.H., Tsai, S.F., 1995. Expression of a *Drosophila* GATA transcription factor in multiple tissues in the developing embryos. Identification of homozygous lethal mutants with P-element insertion at the promoter region. J. Biol. Chem. 270, 25150–25158.

Liu, P., Wakamiya, M., Shea, M.J., Albrecht, U., Behringer, R.R., Bradley, A., 1999. Requirement for Wnt3 in vertebrate axis formation. Nat. Genet. 22, 361–365.

Maduro, M.F., Hill, R.J., Heid, P.J., Newman-Smith, E.D., Zhu, J., Priess, J.R., Rothman, J.H., 2005. Genetic redundancy in endoderm specification within the genus Caenorhabditis. Dev. Biol. 284, 509–522.

Matsui, T., Raya, A., Kawakami, Y., Callol-Massot, C., Capdevila, J., Rodriguez-Esteban, C., Izpisua Belmonte, J.C., 2005. Noncanonical Wnt signaling regulates midline convergence of organ primordia during zebrafish development. Genes. Dev. 19, 164–175.

Matsuyama, M., Aizawa, S., Shimono, A., 2009. Sfrp controls apicobasal polarity and oriented cell division in developing gut epithelium. PLoS Genet. 5, e1000427.

McKnight, K.D., Hou, J., Hoodless, P.A., 2010. Foxh1 and Foxa2 are not required for formation of the midgut and hindgut definitive endoderm. Dev. Biol. 337, 471–481.

McLin, V.A., Rankin, S.A., Zorn, A.M., 2007. Repression of Wnt/beta-catenin signaling in the anterior endoderm is essential for liver and pancreas development. Development 134, 2207–2217.

Mizoguchi, T., Izawa, T., Kuroiwa, A., Kikuchi, Y., 2006. Fgf signaling negatively regulates Nodal-dependent endoderm induction in zebrafish. Dev. Biol. 300, 612–622.

Monaghan, A.P., Kaestner, K.H., Grau, E., Schutz, G., 1993. Postimplantation expression patterns indicate a role for the mouse forkhead/HNF-3 alpha, beta and gamma genes in determination of the definitive endoderm, chordamesoderm and neuroectoderm. Development 119, 567–578.

Moody, S.A., 1987. Fates of the blastomeres of the 32-cell-stage *Xenopus* embryo. Dev. Biol. 122, 300–319.

Morrisey, E.E., Tang, Z., Sigrist, K., Lu, M.M., Jiang, F., Ip, H.S., Parmacek, M.S., 1998. GATA6 regulates HNF4 and is required for differentiation of visceral endoderm in the mouse embryo. Genes. Dev. 12, 3579–3590.

Nagaoka, M., Duncan, S.A., 2010. Transcriptional control of hepatocyte differentiation. Prog. Mol. Biol. Transl. Sci. 97, 79–101.

Nakao, A., Imamura, T., Souchelnytskyi, S., Kawabata, M., Ishisaki, A., Oeda, E., Tamaki, K., Hanai, J., Heldin, C.H., Miyazono, K., et al., 1997. TGF-beta receptor-mediated signaling through Smad2, Smad3 and Smad4. EMBO. J. 16, 5353–5362.

Narita, N., Bielinska, M., Wilson, D.B., 1997. Wild-type endoderm abrogates the ventral developmental defects associated with GATA-4 deficiency in the mouse. Dev. Biol. 189, 270–274.

Nieuwkoop, P.D., 1969. The formation of the mesoderm in the urodelean amphibians. I. Induction by the endoderm. Wilhelm Roux Arch EntwMech Org. 162, 341–373.

Nieuwkoop, P.D., 1997. Short historical survey of pattern formation in the endo-mesoderm and the neural anlage in the vertebrates: the role of vertical and planar inductive actions. Cell. Mol. Life. Sci. 53, 305–318.

Nostro, M.C., Keller, G., 2012. Generation of beta cells from human pluripotent stem cells: Potential for regenerative medicine. Semin. Cell. Dev. Biol. 23, 701–710.

Nowotschin, S., Costello, I., Piliszek, A., Kwon, G.S., Mao, C.A., Klein, W.H., Robertson, E.J., Hadjantonakis, A.K., 2013. The T-box transcription factor Eomesodermin is essential for AVE induction in the mouse embryo. Genes. Dev. 27, 997–1002.

Ober, E.A., Field, H.A., Stainier, D.Y., 2003. From endoderm formation to liver and pancreas development in zebrafish. Mech. Dev. 120, 5–18.

Odenthal, J., Nusslein-Volhard, C., 1998. Fork head domain genes in zebrafish. Dev. Genes. Evol. 208, 245–258.

Okumura, T., Matsumoto, A., Tanimura, T., Murakami, R., 2005. An endoderm-specific GATA factor gene, dGATAe, is required for the terminal differentiation of the Drosophila endoderm. Dev. Biol. 278, 576–586.

Osada, S.I., Saijoh, Y., Frisch, A., Yeo, C.Y., Adachi, H., Watanabe, M., Whitman, M., Hamada, H., Wright, C.V., 2000. Activin/nodal responsiveness and asymmetric expression of a *Xenopus* nodal-related gene converge on a FAST-regulated module in intron 1. Development 127, 2503–2514.

Osada, S.I., Wright, C.V., 1999. *Xenopus* nodal-related signaling is essential for mesendodermal patterning during early embryogenesis. Development 126, 3229–3240.

Paria, B.C., Huet-Hudson, Y.M., Dey, S.K., 1993. Blastocyst's state of activity determines the "window" of implantation in the receptive mouse uterus. Proc. Natl. Acad. Sci. USA 90, 10159–10162.

Perea-Gomez, A., Rhinn, M., Ang, S.L., 2001. Role of the anterior visceral endoderm in restricting posterior signals in the mouse embryo. Int. J. Dev. Biol. 45, 311–320.

Pereira, L.A., Wong, M.S., Mei Lim, S., Stanley, E.G., Elefanty, A.G., 2012. The Mix family of homeobox genes – key regulators of mesendoderm formation during vertebrate development. Dev. Biol. 367, 163–177.

Pfendler, K.C., Catuar, C.S., Meneses, J.J., Pedersen, R.A., 2005. Overexpression of Nodal promotes differentiation of mouse embryonic stem cells into mesoderm and endoderm at the expense of neuroectoderm formation. Stem Cells Dev. 14, 162–172.

Piccolo, S., Agius, E., Leyns, L., Bhattacharyya, S., Grunz, H., Bouwmeester, T., De Robertis, E.M., 1999. The head inducer Cerberus is a multifunctional antagonist of Nodal, BMP and Wnt signals. Nature 397, 707–710.

Poulain, M., Lepage, T., 2002. Mezzo, a paired-like homeobox protein is an immediate target of Nodal signaling and regulates endoderm specification in zebrafish. Development 129, 4901–4914.

Reissmann, E., Jornvall, H., Blokzijl, A., Andersson, O., Chang, C., Minchiotti, G., Persico, M.G., Ibanez, C.F., Brivanlou, A.H., 2001. The orphan receptor ALK7 and the Activin receptor ALK4 mediate signaling by Nodal proteins during vertebrate fevelopment. Genes. Dev. 15, 2010–2022.

Reiter, J.F., Alexander, J., Rodaway, A., Yelon, D., Patient, R., Holder, N., Stainier, D.Y., 1999. Gata5 is required for the development of the heart and endoderm in zebrafish. Genes. Dev. 13, 2983–2995.

Reiter, J.F., Kikuchi, Y., Stainier, D.Y., 2001. Multiple roles for Gata5 in zebrafish endoderm formation. Development 128, 125–135.

Rodaway, A., Patient, R., 2001. Mesendoderm. an ancient germ layer? Cell 105, 169–172.

Rodaway, A., Takeda, H., Koshida, S., Broadbent, J., Price, B., Smith, J.C., Patient, R., Holder, N., 1999. Induction of the mesendoderm in the zebrafish germ ring by yolk cell-derived TGF-beta family signals and discrimination of mesoderm and endoderm by FGF. Development 126, 3067–3078.

Ruvinsky, I., Silver, L.M., Ho, R.K., 1998. Characterization of the zebrafish tbx16 gene and evolution of the vertebrate T-box family. Dev. Genes. Evol. 208, 94–99.

Saito-Diaz, K., Chen, T.W., Wang, X., Thorne, C.A., Wallace, H.A., Page-McCaw, A., Lee, E., 2013. The way Wnt works: components and mechanism. Growth. Factors. 31, 1–31.

Sakaguchi, T., Kuroiwa, A., Takeda, H., 2001. A novel sox gene, 226D7, acts downstream of Nodal signaling to specify endoderm precursors in zebrafish. Mech. Dev. 107, 25–38.

Sasaki, H., Hogan, B.L., 1993. Differential expression of multiple fork head related genes during gastrulation and axial pattern formation in the mouse embryo. Development 118, 47–59.

Schier, A.F., 2009. Nodal morphogens. Cold Spring Harb. Perspect. Biol. 1. a003459.

Schmitt, S., 2005. From eggs to fossils: epigenesis and transformation of species in Pander's biology. Int. J. Dev. Biol. 49, 1–8.

Schohl, A., Fagotto, F., 2002. Beta-catenin, MAPK and Smad signaling during early *Xenopus* development. Development 129, 37–52.

Sherwood, R.I., Jitianu, C., Cleaver, O., Shaywitz, D.A., Lamenzo, J.O., Chen, A.E., Golub, T.R., Melton, D.A., 2007. Prospective isolation and global gene expression analysis of definitive and visceral endoderm. Dev. Biol. 304, 541–555.

Shi, S., Stahl, M., Lu, L., Stanley, P., 2005. Canonical Notch signaling is dispensable for early cell fate specifications in mammals. Mol. Cell. Biol. 25, 9503–9508.

Shih, D.Q., Navas, M.A., Kuwajima, S., Duncan, S.A., Stoffel, M., 1999. Impaired glucose homeostasis and neonatal mortality in hepatocyte nuclear factor 3alpha-deficient mice. Proc. Natl. Acad. Sci. USA 96, 10152–10157.

Shoichet, S.A., Malik, T.H., Rothman, J.H., Shivdasani, R.A., 2000. Action of the Caenorhabditis elegans GATA factor END-1 in *Xenopus* suggests that similar mechanisms initiate endoderm *Development* in ecdysozoa and vertebrates. Proc. Natl. Acad. Sci. USA 97, 4076–4081.

Sinner, D., Kirilenko, P., Rankin, S., Wei, E., Howard, L., Kofron, M., Heasman, J., Woodland, H.R., Zorn, A.M., 2006. Global analysis of the transcriptional network controlling *Xenopus* endoderm formation. Development 133, 1955–1966.

Solnica-Krezel, L., 2003. Vertebrate development: taming the nodal waves. Curr. Biol. 13, R7–R9.

Song, J., Oh, S.P., Schrewe, H., Nomura, M., Lei, H., Okano, M., Gridley, T., Li, E., 1999. The type II activin receptors are essential for egg cylinder growth, gastrulation, and rostral head development in mice. Dev. Biol. 213, 157–169.

Souilhol, C., Cormier, S., Tanigaki, K., Babinet, C., Cohen-Tannoudji, M., 2006. RBP-Jkappa-dependent notch signaling is dispensable for mouse early embryonic development. Mol. Cell. Biol. 26, 4769–4774.

Spence, J.R., Mayhew, C.N., Rankin, S.A., Kuhar, M.F., Vallance, J.E., Tolle, K., Hoskins, E.E., Kalinichenko, V.V., Wells, S.I., Zorn, A.M., et al., 2011. Directed differentiation of human pluripotent stem cells into intestinal tissue in vitro. Nature 470, 105–109.

Tada, M., Casey, E.S., Fairclough, L., Smith, J.C., 1998. Bix1, a direct target of *Xenopus* T-box genes, causes formation of ventral mesoderm and endoderm. Development 125, 3997–4006.

Tada, S., Era, T., Furusawa, C., Sakurai, H., Nishikawa, S., Kinoshita, M., Nakao, K., Chiba, T., 2005. Characterization of mesendoderm: a diverging point of the definitive endoderm and mesoderm in embryonic stem cell differentiation culture. Development 132, 4363–4374.

Takahashi, K., Yamanaka, S., 2006. Induction of pluripotent stem cells from mouse embryonic and adult fibroblast cultures by defined factors. Cell 126, 663–676.

Takahashi, S., Yokota, C., Takano, K., Tanegashima, K., Onuma, Y., Goto, J., Asashima, M., 2000. Two novel nodal-related genes initiate early inductive events in *Xenopus* Nieuwkoop center. Development 127, 5319–5329.

Tam, P.P., Kanai-Azuma, M., Kanai, Y., 2003. Early endoderm *Development* in vertebrates: lineage differentiation and morphogenetic function. Curr. Opin. Genet. Dev. 13, 393–400.

Tam, P.P., Khoo, P.L., Lewis, S.L., Bildsoe, H., Wong, N., Tsang, T.E., Gad, J.M., Robb, L., 2007. Sequential allocation and global pattern of movement of the definitive endoderm in the mouse embryo during gastrulation. Development 134, 251–260.

Tamplin, O.J., Kinzel, D., Cox, B.J., Bell, C.E., Rossant, J., Lickert, H., 2008. Microarray analysis of Foxa2 mutant mouse embryos reveals novel gene expression and inductive roles for the gastrula organizer and its derivatives. BMC Genomics 9, 511.

Teo, A.K., Arnold, S.J., Trotter, M.W., Brown, S., Ang, L.T., Chng, Z., Robertson, E.J., Dunn, N.R., Vallier, L., 2011. Pluripotency factors regulate definitive endoderm specification through eomesodermin. Genes. Dev. 25, 238–250.

Tremblay, K.D., Zaret, K.S., 2005. Distinct populations of endoderm cells converge to generate the embryonic liver bud and ventral foregut tissues. Dev. Biol. 280, 87–99.

Vallier, L., Reynolds, D., Pedersen, R.A., 2004. Nodal inhibits differentiation of human embryonic stem cells along the neuroectodermal default pathway. Dev. Biol. 275, 403–421.

Vincent, S.D., Dunn, N.R., Hayashi, S., Norris, D.P., Robertson, E.J., 2003. Cell fate decisions within the mouse organizer are governed by graded Nodal signals. Genes. Dev. 17, 1646–1662.

Wardle, F.C., Smith, J.C., 2004. Refinement of gene expression patterns in the early *Xenopus* embryo. Development 131, 4687–4696.

Wardle, F.C., Smith, J.C., 2006. Transcriptional regulation of mesendoderm formation in *Xenopus*. Semin. Cell. Dev. Biol. 17, 99–109.

Warga, R.M., Nusslein-Volhard, C., 1999. Origin and development of the zebrafish endoderm. Development 126, 827–838.

Watt, A.J., Zhao, R., Li, J., Duncan, S.A., 2007. Development of the mammalian liver and ventral pancreas is dependent on GATA4. BMC. Dev. Biol. 7, 37.

Weber, H., Symes, C.E., Walmsley, M.E., Rodaway, A.R., Patient, R.K., 2000. A role for GATA5 in *Xenopus* endoderm specification. Development 127, 4345–4360.

Weigel, D., Jurgens, G., Kuttner, F., Seifert, E., Jackle, H., 1989. The homeotic gene fork head encodes a nuclear protein and is expressed in the terminal regions of the Drosophila embryo. Cell 57, 645–658.

Weinstein, D.C., Ruiz i Altaba, A., Chen, W.S., Hoodless, P., Prezioso, V.R., Jessell, T.M., Darnell Jr., J.E., 1994. The winged-helix transcription factor HNF-3 beta is required for notochord development in the mouse embryo. Cell 78, 575–588.

Wylie, C.C., Snape, A., Heasman, J., Smith, J.C., 1987. Vegetal pole cells and commitment to form endoderm in *Xenopus laevis*. Dev. Biol. 119, 496–502.

Xanthos, J.B., Kofron, M., Tao, Q., Schaible, K., Wylie, C., Heasman, J., 2002. The roles of three signaling pathways in the formation and function of the Spemann Organizer. Development 129, 4027–4043.

Xanthos, J.B., Kofron, M., Wylie, C., Heasman, J., 2001. Maternal VegT is the initiator of a molecular network specifying endoderm in *Xenopus laevis*. Development 128, 167–180.

Xu, C., Fan, Z.P., Muller, P., Fogley, R., DiBiase, A., Trompouki, E., Unternaehrer, J., Xiong, F., Torregroza, I., Evans, T., et al., 2012. Nanog-like regulates endoderm formation through the Mxtx2-Nodal pathway. Dev. Cell. 22, 625–638.

Yamamoto, M., Saijoh, Y., Perea-Gomez, A., Shawlot, W., Behringer, R.R., Ang, S.L., Hamada, H., Meno, C., 2004. Nodal antagonists regulate formation of the anteroposterior axis of the mouse embryo. Nature 428, 387–392.

Yasunaga, M., Tada, S., Torikai-Nishikawa, S., Nakano, Y., Okada, M., Jakt, L.M., Nishikawa, S., Chiba, T., Era, T., 2005. Induction and monitoring of definitive and visceral endoderm differentiation of mouse ES cells. Nat. Biotechnol. 23, 1542–1550.

Yoon, S.J., Wills, A.E., Chuong, E., Gupta, R., Baker, J.C., 2011. HEB and E2A function as SMAD/FOXH1 cofactors. Genes. Dev. 25, 1654–1661.

Yuh, C.H., Bolouri, H., Davidson, E.H., 1998. Genomic cis-regulatory logic: experimental and computational analysis of a sea urchin gene. Science 279, 1896–1902.

Yusa, K., Rashid, S.T., Strick-Marchand, H., Varela, I., Liu, P.Q., Paschon, D.E., Miranda, E., Ordonez, A., Hannan, N.R., Rouhani, F.J., et al., 2011. Targeted gene correction of alpha1-antitrypsin deficiency in induced pluripotent stem cells. Nature 478, 391–394.

Zhang, Z., Rankin, S.A., Zorn, A.M., 2013. Different thresholds of Wnt-Frizzled 7 signaling coordinate proliferation, morphogenesis and fate of endoderm progenitor cells. Dev. Biol. 378, 1–12.

Zhao, R., Watt, A.J., Li, J., Luebke-Wheeler, J., Morrisey, E.E., Duncan, S.A., 2005. GATA6 is essential for embryonic development of the liver but dispensable for early heart formation. Mol. Cell. Biol. 25, 2622–2631.

Zhu, J., Hill, R.J., Heid, P.J., Fukuyama, M., Sugimoto, A., Priess, J.R., Rothman, J.H., 1997. end-1 encodes an apparent GATA factor that specifies the endoderm precursor in *Caenorhabditis elegans* embryos. Genes. Dev. 11, 2883–2896.

Zorn, A.M., Wells, J.M., 2007. Molecular basis of vertebrate endoderm development. Int. Rev. Cytol. 259, 49–111.

Zorn, A.M., Wells, J.M., 2009. Vertebrate endoderm sevelopment and organ formation. Annu. Rev. Cell. Dev. Biol. 25, 221–251.

Chapter 14

Epithelial Branching: Mechanisms of Patterning and Self-Organization

Jamie Davies
University of Edinburgh, Edinburgh, UK

Chapter Outline

GLOSSARY

Ampulla The swollen end of a branch, that will give rise to new branches.
Bifurcation Division into two (as in one branch bifurcating to become two).
Clefting Division of an ampulla by ingrowth of a sharp cleft.
Dipodial Branching in which all branches are equal, with no central trunk.
Fractal Self-similar (looks the same at all scales).
Intussusceptive branching Branching by longitudinal division of a tube (by local ingrowth and meeting of the walls).
Monopodial Branching in which side-branches form from a main trunk.
Ramogen A signaling molecule that promotes branching.

SUMMARY

- Epithelial trees are common in glandular organs because they allow the efficient transport and exchange of substances.
- There are several ways of making trees: convergence, sprouting, cleavage and intussusception. Sprouting can result in either monopodial (side-branches from a central trunk) or dipodial (all branches equal) trees.
- Through symmetry-breaking, initially equal cells become branch leaders and followers, using positive feedback.
- Tip and stalk cells become different in terms of gene expression.
- The basic patterning of the tree appears to use self-avoidance based on autocrine secretion, to encourage spread and avoid tangling.
- Organ-specific patterning uses paracrine secretions by surrounding mesenchyme, with negative feedback keeping the epithelium and mesenchyme balanced.

14.1 INTRODUCTION: THE IMPORTANCE OF EPITHELIAL BRANCHING TO ORGANOGENESIS

The influences exerted by natural selection on the architecture of organisms vary greatly across different species and habitats, but some evolutionary pressures are almost universal. One is the tendency to maximize the controlled exchange of substances (oxygen, water, food, waste) with the environment, because the capacity for such exchange is a serious limit on

Principles of Developmental Genetics.

the rate of an organism's metabolism. Another is the tendency of maximize the efficiency of the internal transport of the same resources. Both problems can be solved by organizing the exchange and transport systems of the organism as approximately fractal (scale-free) structures (Mandelbrot, 1977). Indeed, by assuming that fractal structures dominate exchange and transport, one can derive the well-known "scaling laws" by which measures such as metabolic rate, heartbeat, etc., vary with body mass within a type of organism (West et al., 1997; 1999).

One of the most common fractal structures, effectively identified as such by Leonardo da Vinci (Long, 2004), is a branching network in which tube diameter and internode length decrease by a constant proportion as branching progresses. Such branching networks are a very common feature of multicellular animals, particularly large ones. In humans, for example, internal transport is achieved mainly through an approximately fractal branching system of blood vessels, while communication of gases and waste products proceeds via approximately fractally branched air tubes and urine collecting ducts. Exocrine organs such as the pancreas, salivary and mammary glands also use branched ductal systems. The absorptive epithelium of the alimentary canal is a rare exception to this rule, being an essentially unbranched structure, probably because it has to pass solids, and any thinning of lumen diameter would risk blockage. Its surface is still reminiscent of fractal geometry, however, consisting of microvilli on villi on ridges.

Branched structures, then, are essential to the body plan of all but the simplest animals, and their development is an important feature of metazoan organogenesis (Davies, 2005).

14.2 TYPES AND SCALES OF BRANCHING MORPHOGENESIS

Branching morphogenesis of animals takes place across a huge range of scales, varying from subcellular structures such as dendritic trees, to branched cells such as astrocytes, to small cell collectives such as *Drosophila* tracheae, to very large multicellular tissues such as the lungs of a blue whale.

The production of branched structures by single cells is outside the scope of this chapter, but a discussion of the relevant subcellular mechanisms can be found elsewhere (Davies, 2013). The production of multicellular branched structures takes place by four general mechanisms; confluence, intussusception, clefting and sprouting (Figure 14.1). Although it results in the formation of a branched structure, confluence (Figure 14.1A) is not strictly "branching morphogenesis" because nothing actually branches; rather, individual tubules meet and fuse to form a branched system. The development of the pronephric kidney in *Xenopus laevis* proceeds by a mechanism very much like that of Figure 14.1A (Vize et al., 2003) and confluence is also important in the production of capillary networks by vasculogenesis (Drake, 2003).

Intussusceptive branching (Figure 14.1B) is a process by which a large-diameter tube is split into multiple small-diameter tubes, with a similar total cross-sectional area, in much the same way that a large river might split into a delta. It is the only one of these mechanisms that can create a branched network from a tube in which fluid continues to flow, and is therefore very important in development and remodeling of the vascular system (Patan, 2004). Its mechanisms are so different from clefting and sprouting that it will not be considered further in this chapter.

Clefting (Figure 14.1C) and sprouting (Figure 14.1D) are responsible for the majority of branched epithelial systems, and the rest of this chapter will concentrate on these processes. There are two basic ways in which epithelial buds

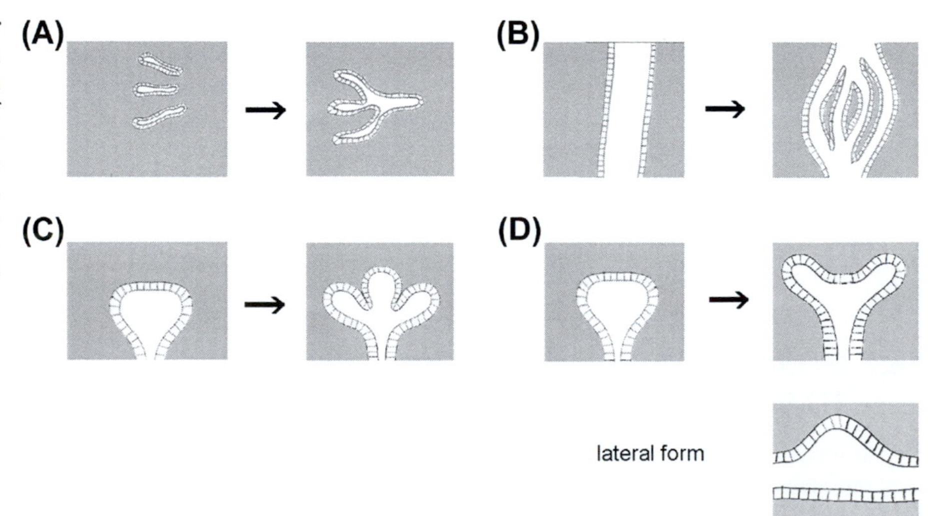

FIGURE 14.1 The Different forms of Branching. (A) in confluence, independent tubules flow together to create a branched system; (B) in intussussception, the walls of the tube fold inwards and fuse to form elongated pillars that divide a flow into multiple branches; (C) in clefting, deep clefts divide the end bud of an epithelial tubule into different tips; (D) in sprouting, new tips sprout outwards from either the end (top diagram) or the side (bottom diagram) of an existing tubule.

produce a branched system; monopodial and dipodial. In monopodial branching, a principal "trunk" extends forwards and lateral branches form from its sides (Figure 14.2); a classical Christmas tree grows by monopodial branching. In dipodial branching, the bud bifurcates to form two equal branches, which bifurcate again and so on (Figure 14.2); many seaweeds grow by dipodial branching. Sometimes, branching can occur by trifurcation rather than bifirucation, and organs such as the kidney show a mixture of both bifurcation and trifurcation (Watanabe and Costantini, 2004). The epithelial tubules of some organs can use both methods in sequence. Mammary glands, for example, use dipodial branching to create the tree of milk ducts and then use a form of monopodial branching to create groups of alveoli along them (Fata et al., 2004).

Monopodial branching always occurs by a process of sprouting, in which part of the wall of an existing duct bulges outwards and becomes the growing tip of a new duct (Figure 14.1D). Many examples of dipodial branching also appear to take place by sprouting. In the ureteric bud of the kidney, for example, movies made of branching morphogenesis using GFP-tagged tissues (Watnabe and Cosantini, 2004) clearly show outward bulging of new branches with little evidence of clefts. In other organs, such as the salivary gland, the broad ends of the branching tubules (sometimes called "ampullae", because of their shape) seem to be split by narrow clefts (Figure 14.1C) (Nakanishi and Ishii, 1989; Miyazaki, 1990; Nakanishi et al., 1987). It is possible that these clefts represent a stationary restriction around which the expanding epithelium has to divide (Nakanishi et al., 1988).

In the branching epithelia of mammals, ampullae and the sprouts that arise from them are usually composed of a population of cells rather than beginning as processes from one single cell (this is in marked contrast to the tracheal system of *Drosophila melanogaster*, which is the most closely-studied invertebrate branching epithelium).

14.3 THE ROLE OF GENETICS IN STUDYING MECHANISMS OF BRANCHING MORPHOGENESIS

Genetics is used in the study of branching morphogenesis in much the same way that it is used in other developing systems. Sporadic mutants that generate a phenotype with aberrant branching of some or all arborizing epithelia are used to identify genes that play an essential role in normal branching morphogenesis. Also, precisely induced mutations in transgenic animals can be used to test specific hypotheses about the involvement of particular molecules by a reverse-genetic approach. Genetics is only one approach to the problem, and a variety of pharmacological and culture techniques are also used, because these have the power to reach cellular processes that do not depend on changes in gene expression. Cellular signal transduction systems are examples of such processes.

Classical and reverse genetics have identified a vast range of mutations that have an effect on branching. Attempting to describe them all here would not be particularly informative. Instead, this chapter will consider the principal problems in branching morphogenesis and the genes that are involved in them, without attempting to be comprehensive.

14.4 THE PROBLEMS: WHAT MOST NEEDS TO BE EXPLAINED

Common as it is in animal development, branching morphogenesis raises a number of challenging questions. Why is it, for example, that some cells "decide" to lead the advance of a branch while their neighbors do not, remaining as an ordinary tubule side-wall or becoming the base of a valley between branches? How do the branches of a growing tree spread out properly and avoid tangles? How does the shape of a tree adapt itself to environmental constraints? How does the network of genes that drive branching allow rapid changes to the scale of, and number of branch generations in, epithelial trees over evolutionary time? How does the growth of an epithelial tree remain in pace with the growth of its surrounding mesenchymal tissue, never running too far ahead or behind? How do the branches and trunk of an epithelial tree know how much to grow in girth to allow easy flow of fluid? None of these questions, all of which will come up in the briefest of consideration of branching, have yet been answered completely: nevertheless, enough has now been discovered to provide at least an outline understanding. These questions provide the topics of the rest of this chapter.

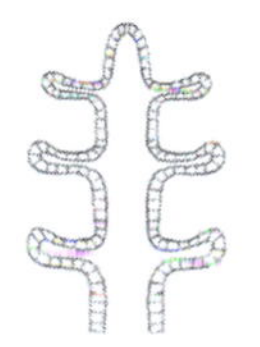
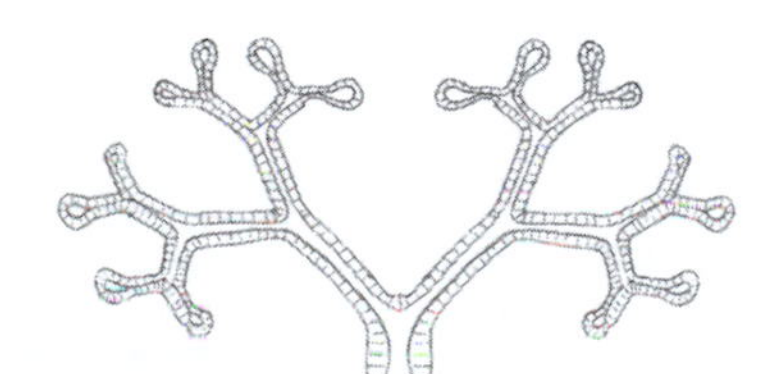

FIGURE 14.2 Monopodial branching (left), in which lateral branches are emitted from a principal "trunk", is contrasted with dipodial branching (right) in which a tree is formed by successive divisions of branch tips.

14.5 SYMMETRY-BREAKING: WHY DO EPITHELIA BRANCH RATHER THAN JUST BALLOON?

The problem of symmetry-breaking, inherent in all branching systems, is illustrated most clearly by a simple culture model. In this model, MDCK renal collecting duct cells are suspended in three-dimensional collagen-rich gels, and grow clonally to produce rounded, polarized cysts (McAteer et al., 1986). If a ramogen (=branch-promoting signaling molecule) such as Hepatocyte Growth Factor (HGF) is added, the cysts elongate to form branching tubes (Santos and Nigam, 1993). All the cells start alike: by the time tubulogenesis is visible, some cells are advancing quickly to make the tubes while others are not, being left behind in the valleys between these tubes (Figure 14.3). Gaining heterogeneity from apparent homogeneity is generally referred to as symmetry-breaking.

In many systems, symmetry-breaking involves positive feedback. For example, a cell departing stochastically from the original ground state to a new state may produce signals that reinforce its own decision while simultaneously inhibiting neighboring cells from following suit. A simple example is provided by the growth of sand dunes in initially flat sand: random chance determines where the wind drops aeolian sand grains. Wherever a sand grain has been dropped, it will create a small lee behind it, slowing the wind and increasing the chances of another grain being dropped there. As the pile gets higher, this effect is stronger and it keeps growing. Downwind of it, there is little air movement so little sand is picked up, and formation of a new dune is inhibited for some distance. The effect of positive feedback on the growth of existing dunes and of lateral inhibition blocking the growth of valleys turns an initially flat sandscape into dunes.

There is evidence for this type of feedback in branching systems. When mammary epithelial cells are cultured in shaped wells in invadable gels, they are least invasive on concave surfaces of wells (for example, the right hand side of a 'C'-shaped well) and most invasive on convex surfaces (Nelson et al., 2006). This seems to be due to the autocrine secretion of transforming growth factor beta (TGFβ), an inhibitor of invasion: the concentration of this molecule will be highest on concave surfaces, where many cells secrete their TGFβ toward the same space, while it will be lowest on convex surfaces where the contribution of the nearest cells can spread out in all directions (Figure 14.4). Any small (random) protrusion of a cell from the edge of a cyst will take it to lower concentrations of TGFβ, thus encouraging it to advance further, where it will perceive even lower concentrations of TGFβ and so on. Cells left behind will then be in a concave area, and this will inhibit their advance strongly so they remain.

A mechanism such as this, whether working through diffusion of an inhibitor or direct sensing of curvature, makes a cyst unstable. Balloon-like growth would be difficult: in the face of any instability, the cyst will break its symmetry spontaneously into out-pushings and valleys. Along the side of a sufficiently long out-pushing tube, the same thing will apply, and side-branches develop.

14.6 TIPS AND STALKS

In most branching systems, there is a difference between the cells at the tips of growing branches and those in the stalks behind them. They differ in the expression of a number of key genes, many of which are connected with signaling (Kuure et al., 2010). Examples are shown in Table 14.1; many more have been identified using microarrays (Lu et al., 2004; Schmidt-Ott et al., 2005). In some cases, such as tracheae of *Drosophila melanogaster*, the tip is a single cell

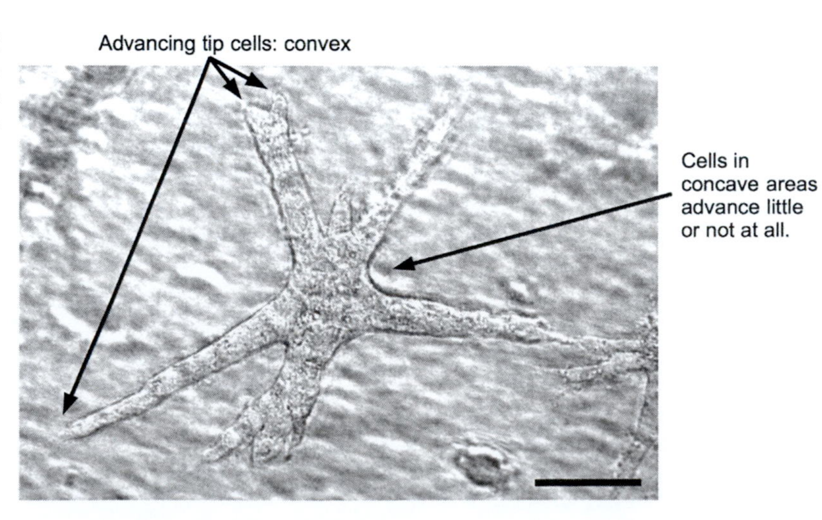

FIGURE 14.3 **A Multicellular Branching System of mIMCD3 Cells in a 3-Dimensional Gel.** The cells were a multicellular cyst before the addition of ramogens: the symmetry was then broken, some cells leading new tubes and others not. Scale bar – 100μm.

(Maruyama and Andrew, 2012): in mammalian systems, tips are collectives of tens or hundreds of cells. Tip cells might be different for several possible reasons, a fact that can make functional analysis difficult.

The tip-specific functions most relevant to this chapter are guidance and growth. In typical mammalian epithelial trees, the tip area is the location of the most mitosis, driving growth (Goldin and Opperman, 1980; Michael and Davies, 2004). In insect tracheae there is no mitosis, but the tip cells are important in guidance, exploring their environment with thin cell

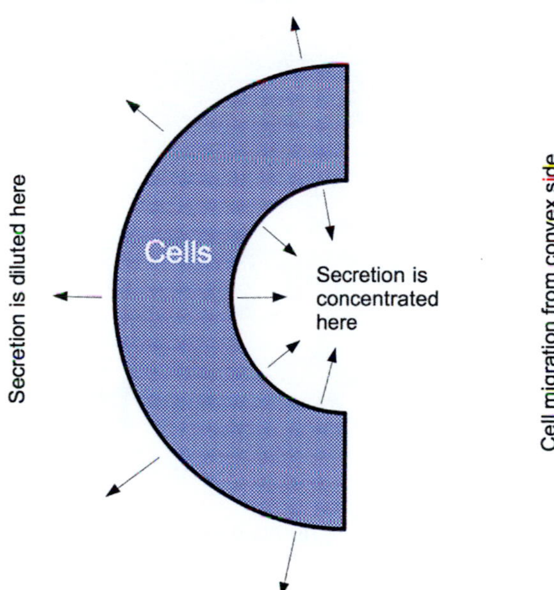

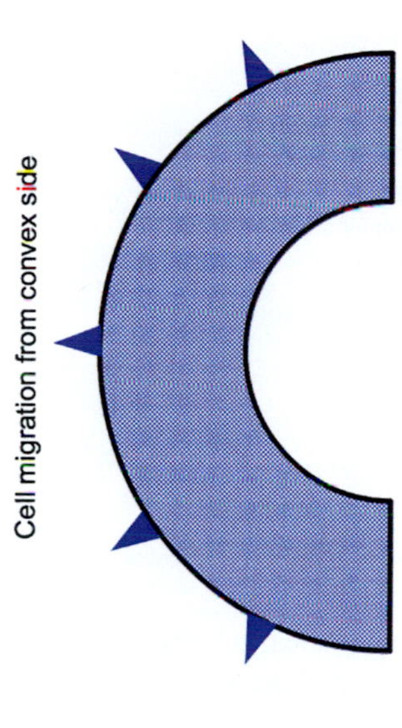

FIGURE 14.4 A Diagrammatic Representation of the Experiments of Nelson et al., 2006), in which Mammary Epithelial Cells were Cultured in Shaped Wells. Geometry determines how concentrated a secreted molecule will be in the matrix outside the well, and this regulates the probability of cell invasion of this matrix: the cells tend to move much more from the convex surfaces, at which there is a lower concentration of autocrine inhibitor.

TABLE 14.1

Organ	Gene	Tip/Stalk	Ref
Kidney	Wnt11	Tip	(Kispert et al., 1996)
	Sox9	Tip	(Kent et al., 1996)
	Collagen XVIII	Stalk	(Lin et al., 2001)
Lung	BMP4	Tip	(Li et al., 2005)
	Wnt7b	Tip	
	Spry2	Tip	(Hashimoto et al., 2002)
	Shh	Tip	(Li et al., 2005)
	Notch	Tip	(Post et al., 2000)
	Netrin1	Stalk	(Liu et al., 2004)
	SP-C	Tip	(Hashimoto et al., 2002)
Prostate	BMP7	Tip	(Grishina et al., 2005)
	Shh	Tip	(Puet al., 2004)
Pancreas	EphB2,B3,B4	Tip	(van Eyll et al., 2006)
	Ngn3	Stalk	(van Eyll et al., 2006)
Salivary gland	Smo	Tip	(Jaskoll et al., 2004b)
	Gli3	Stalk	(Jaskoll et al., 2004a)

Evidence from gene expression studies in a number of branching epithelia that tips and stalks are in distinct states of differentiation. This table lists representative examples of differentially-expressed genes, and is not intended to be a complete list.

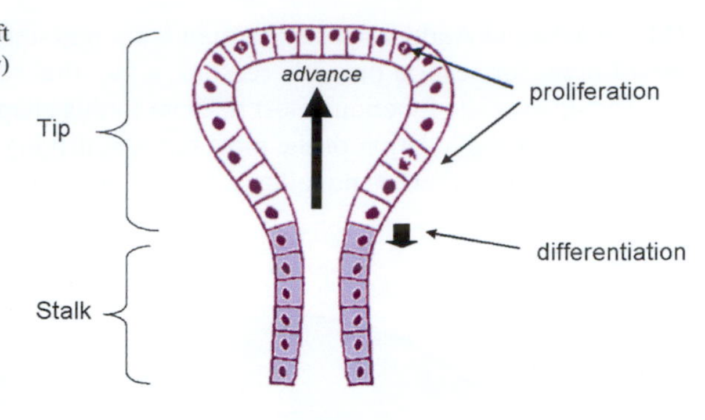

FIGURE 14.5 Proliferation Takes Place Mainly in the Tip Regions. Cells "left behind" by the advancing tip (perhaps passively, perhaps by their own activity) differentiate into stalks.

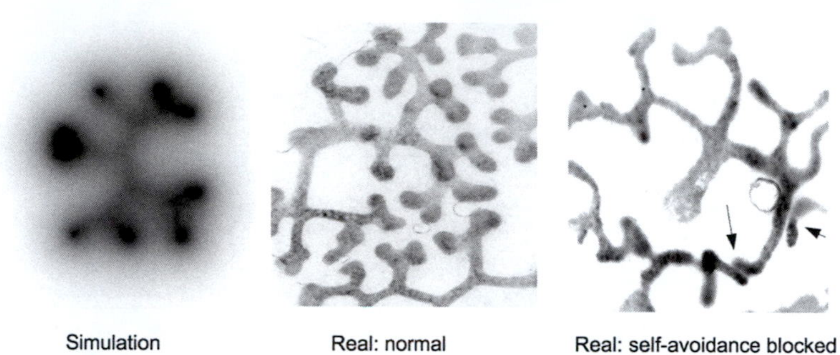

FIGURE 14.6 Self-Avoidance in Branching of the Renal Collecting Duct. The left-most panel shows a computer model of branching using rules based on avoidance of a molecule secreted by branch tips: the darker the color, the more concentrated the molecule. This simulation began with a single unbranched tube. The center panel shows mutual avoidance by the developing collecting duct trees of three normal kidneys in culture. The right-most panel shows the collisions (arrows) that result from pharmacological inhibition of the self-avoidance system.

processes (cytonemes: Roy et al., 2011). It is not clear whether mammalian tips also operate in this way, or whether guidance is mediated by alterations in the probability of mitosis.

Cells left behind by advancing tips become stalk cells (Figure 14.5), with a different pattern of gene expression. Some changes are simple loss of tip-specific functions. Others may reflect stalk-specific functions. In the kidney, for example, stalk cells express Wnt proteins distinct from those of the tip and these alter the behavior of the surrounding mesenchyme. The combination of self-renewal and differentiation implied by this model further suggests that the branch tips may harbor stem cells. Careful study of the mammary gland has indeed identified a population of slowly-cycling cells that have the properties of stem cells (Kenney et al., 2001) and that can form a complete mammary tree if transplanted into the empty fat pad of a host mouse (Smith, 1996; Shackleton et al., 2006). They are, however, distributed in both the tips and the stalks. This location might allow the formation of lateral branches later. In the kidney, the tips allow the mesenchyme that surrounds them to remain in a stem-cell state (the "cap mesenchyme"), while the stalks cause cells abutting them to enter the nephron-producing pathway (Park et al., 2012).

Yet other changes reflect the maturation of the stalk regions for physiological functions (for example, expression of aquaporin 2 in renal collecting ducts: Devuyst et al., 1996) or for transport (for example, acquisition of a smooth muscle layer). The division between tips and stalks seems to be maintained by systems of feedback and lateral inhibition. In the kidney, for example, removing the tip from a stalk causes stalk cells to produce a new tip and vice versa, suggestion that only the presence of a tip prevents all stalk cells being tip-like (Sweeney et al., 2008).

14.7 PATTERNING: HOW TO MAKE A TREE NOT A TANGLE

Branching epithelia make well-spaced trees in which branches ramify efficiently through space without colliding and tangling. What is more, they do so even when the shape of an organ is altered, for example by growing it in flat culture systems rather than in the embryo (Sebinger et al., 2010). Production of an atypical tree that is suited to an atypical environment suggests that its patterning is achieved by a mechanism of adaptive self-organization rather than simply reading a deterministic "genetic blueprint."

One simple way in which a tree may pattern itself automatically is self-avoidance: the secretion of an autocrine inhibitor of tip advance. Clearly, if the inhibitor were too powerful it would prevent the advance of the tip that secretes it, and all branching would cease. With a less powerful diffusible inhibitor, though, each tip would still be able to move, the probability of advance being greatest in directions of lowest inhibitor concentration. If such a situation is modeled on a computer (with the added "rule" that, when concentration is below a critical threshold, a tip bifurcation happens), the result is the production of a well-spaced tree (Figure 14.6). Significantly, such a model *automatically* creates first branch point angles that are different from those of subsequent branches; for example producing a 'T' shaped first branch and 'L'-shaped second ones, just as happens in the real renal collecting duct. In the model, no special rules are required for the "unique" first branching point, suggesting that in real life they may not be required either.

An implication of the self-avoidance model is that the tips of a growing epithelial tree will avoid mutual collisions even if they are cultured in a manner that aims several trees at one another. When this experiment is performed with real renal collecting duct trees, tips do indeed avoid colliding even when this avoidance makes each tree highly distorted (Figure 14.6). In this case, blocking the relevant cell-cell signaling pathways relaxes the self-avoidance, and collisions occur (Figure 14.6). Together, the model and the results of real culture experiments suggest that basic patterning of a tree can be accomplished by a very simple mechanism of repulsion. This mechanism will adjust automatically to unusual environments and it is scalable, being equally suited to making a small or large version of the tree. The ability of mammalian organs to make changes of scale that are rapid in developmental (e.g., pregnancy and lactation) and evolutionary time (Miniature Terrier or Great Dane) is striking, and almost certainly depends on an underlying scalable patterning mechanism such as self-avoidance.

14.8 IMPOSING SUBTLETY: MAKING ORGAN-SPECIFIC PATTERNS

It has been known for many decades – certainly for much longer than the existence of autocrine signals has been suspected – that branching epithelia depend on the presence of paracrine signals from the mesenchyme that surrounds them. Genetics has proved to be a powerful method for identifying these signals (indeed, the similarity of phenotype of two mutations has often been the first hint that a particular growth factor is the ligand for an "orphan" receptor).

Many of these signals promote branching: they are referred to as ramogens. The branching epithelia of many organs (lung, pancreas, salivary gland, mammary gland, prostate and kidney) express the FGFRIIIb receptor tyrosine kinase (RTK). The lungs, pancreas and salivary gland use FGF10 as the main ligand to act on FGFRIIIb, whereas the mammary gland and prostate use FGF7. The kidney uses FGF7 as well, but in that organ the whole fibroblast growth factor (FGF) system assumes a relatively minor importance and the Ret RTK and its ligand glial cell-derived neurotrophic factor (GDNF) are the major stimulators of branching. In addition, all of these organs express epidermal growth factor receptor (EGFR) and branching is stimulated by its ligands epidermal growth factor (EGF) and/or TGFβ. Most also express the Met RTK, and branching and/or elongation is stimulated by its ligand HGF. In addition to these ligand-receptor complexes, others are used by various organs. These include amphiregulin, neuregulin, EGF, TGFβ, colony stimulating factor, IGFs, etc. (Wang and Laurie, 2004).

As well as having receptors for stimulators of branching, the cells of branching epithelia also contain receptors for inhibitors of branching from the surrounding mesenchyme. Many of these inhibitors are members of the TGFβ superfamily. The lungs, mammary glands and kidneys bear receptors for TGFβ itself, which signals via R-Smads 2/3 (Bartram and Speer, 2004) and acts as a powerful inhibitor of branching (Bartram and Speer, 2004; Ritvos et al., 1995). Branching in the pancreas, salivary gland, mammary gland, salivary gland, and prostate is also inhibited by Activin, which also signals via R-Smads 2/3. Bone morphogenetic proteins (BMPs), which signal via R-Smad1, are inhibitory in some systems. In the kidney, for example, BMP2 acts as a powerful inhibitor of branching, while BMP7 acts as a stimulant at low concentrations and an inhibitor at high concentrations; the inhibitory action depends on R-Smad1 (Piscione et al., 2001), while the stimulatory effect proceeds via p38 MAP kinase (Hu et al., 2004).

Various Wnt proteins are expressed by epithelia and by their surrounding mesenchymes in different organs. Some organs express few Wnts. The developing prostate, for example, seems to express just Wnt4 (Zhang et al., 2006). The lungs express Wnt2 in the mesenchyme, Wnt7b in the epithelium and Wnt11 in both; some Wnt knockouts result in branching abnormalities (Pongracz and Stockley, 2006). In kidneys, the branching epithelium expresses Wnt9b and Wnt11, while the mesenchyme expresses Wnt4 and Wnt7b in different places (Carroll et al., 2005; Stark et al., 1994; Gavin et al., 1990). Mammary glands express Wnt2, Wnt4, Wnt5a, Wnt5b, Wnt6, Wnt7b and Wnt10b in complex patterns during their development (Lane and Leder. 1997; Weber-Hall et al, 1994; Bradbury et al., 1995). Genetic and culture experiments have shown that Wnt1 and Wnt4 both encourage branching morphogenesis in the mammary gland.

These are not the only ligands that control branching morphogenesis, and the roles of other families (such as Notch) are currently the subject of close investigation. It should be noted, however, that conventional genetics has not proved to be as useful at finding molecules that control branching as might be hoped. Perhaps because of redundancy in a living embryo, the effects of knocking out a gene *in vivo* out tend to be much less than that of inhibiting gene function in culture (Davies, 2002).

The dependency of branching epithelia on mesenchyme-derived signals creates the potential for elaborate feedback networks. This is demonstrated very well in the developing kidney, in which the developmental biology of the mesenchyme has been studied rather more than in other organs (because the mesenchyme is the ultimate source of most of the clinically-important parts of the kidney). Mesenchyme that has not yet been penetrated by the branching epithelium is a potent source of GDNF, the main stimulant of branching in this organ (see above). Once a particular area of mesenchyme is penetrated by branching epithelium, though, Wnt 9b signals from the epithelium (Carroll et al., 2005), and perhaps others, induce the mesenchyme to change its state of differentiation. The mesenchyme undergoes a complex series of developmental changes, beyond the scope of this chapter, that result in it producing epithelial nephrons and mature renal stroma. Critically, it ceases to produce branch-promoting GDNF, and begins to produce branch-inhibiting molecules such as BMP2 and, eventually, TGFβ. The system is therefore arranged so that the virgin mesenchyme ahead of branching epithelium encourages production of new branches (and probably also attracts them), but any areas that have already been penetrated inhibit any further branching activity. This ensures that branching is not excessive, and tends to maintain a constant ratio of branches per unit volume of mesenchyme.

In lungs, the main stimulator of branching is FGF10. The epithelium produces Shh, which does not diffuse far but which inhibits the mesenchymal production of FGF10 (Bellusci et al., 1997; Miller et al., 2001). Again, the effect is to limit expansion of the branching epithelium to virgin mesenchyme.

The use of feedback mechanisms such as these makes branching systems error-tolerant; if an area of mesenchyme is "missed" first time round, it will remain attractive and may therefore promote a lateral branch, so that it is eventually served by the branching system. Indeed, this is probably the reason for the lateral branching that is observed, infrequently, in developing kidneys (Srinivas et al., 1999).

Feedback can go beyond simple tolerance of errors, however; it makes branching systems effectively self-organizing, and therefore makes them qualitatively different from many of the other aspects of developmental biology studied by geneticists. In wild type animals, the number and exact location of such aspects of the body plan as fingers, vertebrae and teeth is constant and predictable. The number and exact location of epithelial branches in an organ is far less reproducible and predictable, however, even in left-right "duplicate" organs of the same animal. The same is true of micro-vasculature. This is because the precise arrangement of the branches is not specified in the genetic program; rather, that program specifies a system of feedback that can organize branching tissues into patterns that are statistically predictable but not predictable in precise detail. This method of embryonic development – adaptive self-organization – is discussed in detail elsewhere (Davies, 2013).

Branching morphogenesis, then, is a complex process that is regulated by genetic and other controls that operate at a variety of nested levels. This chapter, along with most of the research done over the last few decades, has emphasized common themes that can be found in most or all branching epithelia of mammalian organs. There are, however, differences and it is likely that future research will pay more attention to these differences, because they may explain why a mammary gland, for example, does not look like a lung.

14.9 CLINICAL RELEVANCE

- Malformations of epithelial trees underlie clinically significant diseases, especially cystic diseases of the kidney.
- Branched epithelial trees are a critical component of many internal organs: understanding how they develop is necessary for tissue engineering replacements.
- A large proportion of cancers, including those of the mammary, lung and prostate, arise from branched epithelia: there is probably a close relationship between the cellular systems used for invasive behavior by a normal growing branch and for invasion by a metastatic tumor.

RECOMMENDED RESOURCES

Book: Branching Morphogenesis, J. Davies, 2005, Springer.

Video of branching morphogenesis in culture; http://www.youtube.com/watch?v=dxsWnNxpRBU (accessed 18.11.13).

GUDMAP database (very large database of gene expression in renal development, including that of the branching collecting duct): www.gudmap.org (accessed 18.11.13).

GO Ontology entry for branching morphogenesis (a useful springboard for exploring molecules involved): https://www.ebi.ac.uk/QuickGO/GTerm?id=GO:0048754 (accessed 18.11.13).

REFERENCES

Bartram, U., Speer, C.P., 2004. The role of transforming growth factor beta in lung development and disease. Chest 125, 754–765.

Bellusci, S., Grindley, J., Emoto, H., Itoh, N., Hogan, B.L., 1997. Fibroblast growth factor 10 (FGF10) and branching morphogenesis in the embryonic mouse lung. Development 124, 4867–4878.

Bradbury, J.M., Edwards, P.A., Niemeyer, C.C., Dale, T.C., 1995. Wnt-4 expression induces a pregnancy-like growth pattern in reconstituted mammary glands in virgin mice. Dev. Biol. 170, 553–563.

Carroll, T.J., Park, J.S., Hayashi, S., Majumdar, A., McMahon, A.P., 2005. Wnt9b plays a central role in the regulation of mesenchymal to epithelial transitions underlying organogenesis of the mammalian urogenital system. Dev. Cell. 9, 283–292.

Davies, J.A., 2002. Do different branching epithelia use a conserved developmental mechanism? Bioessays 24, 937–948.

Davies, J., 2005. Branching morphogenesis. Landes Biomedical.

Davies, J., 2013. Mechanisms of morphogenesis, second ed. Academic Press.

Devuyst, O., Burrow, C.R., Smith, B.L., Agre, P., Knepper, M.A., Wilson, P.D., 1996. Expression of aquaporins-1 and -2 during nephrogenesis and in autosomal dominant polycystic kidney disease. Am. J. Phys. 271, F169–F183.

Drake, C.J., 2003. Embryonic and adult vasculogenesis. Birth Defects Res. C. Embryo. Today 69, 73–82.

Fata, J.E., Werb, Z., Bissell, M.J., 2004. Regulation of mammary gland branching morphogenesis by the extracellular matrix and its remodeling enzymes. Breast. Cancer. Res. 6, 1–11.

Gavin, B.J., McMahon, J.A., McMahon, A.P., 1990. Expression of multiple novel Wnt-1/int-1-related genes during fetal and adult mouse development. Genes Dev. 4, 2319–2332.

Goldin, G.V., Opperman, L.A., 1980. Induction of supernumerary tracheal buds and the stimulation of DNA synthesis in the embryonic chick lung and trachea by epidermal growth factor. J. Embryol. Exp. Morphol. 60, 235–243.

Grishina, I.B., Kim, S.Y., Ferrara, C., Makarenkova, H.P., Walden, P.D., 2005. BMP7 inhibits branching morphogenesis in the prostate gland and interferes with Notch signaling. Dev. Biol. 288, 334–347.

Hu, M.C., Wasserman, D., Hartwig, S., Rosenblum, N.D., 2004. p38MAPK acts in the BMP7-dependent stimulatory pathway during epithelial cell morphogenesis and is regulated by Smad1. J. Biol. Chem. 279, 12051–12059.

Hashimoto, S., Nakano, H., Singh, G., Katyal, S., 2002. Expression of Spred and Sprouty in developing rat lung. Mech. Dev. 119 (Suppl. 1), S303–S309.

Jaskoll, T., Witcher, D., Toreno, L., Bringas, P., Moon, A.M., Melnick, M., 2004a. FGF8 dose-dependent regulation of embryonic submandibular salivary gland morphogenesis. Dev. Biol. 268 (2), 457–469.

Jaskoll, T., Leo, T., Witcher, D., Ormestad, M., Astorga, J., Bringas Jr., P., Carlsson, P., Melnick, M., 2004b. Sonic hedgehog signaling plays an essential role during embryonic salivary gland epithelial branching morphogenesis. Dev. Dyn. 229, 722–732.

Kenney, N.J., Smith, G.H., Lawrence, E., Barrett, J.C., Salomon, D.S., 2001. Identification of Stem Cell Units in the Terminal End Bud and Duct of the Mouse Mammary Gland. J. Biomed.Biotechnol. 1, 133–143.

Kent, J., Wheatley, S.C., Andrews, J.E., Sinclair, A.H., Koopman, P., 1996. A male-specific role for SOX9 in vertebrate sex determination. Development 122, 2813–2822.

Kispert, A., Vainio, S., Shen, L., Rowitch, D.H., McMahon, A.P., 1996. Proteoglycans are required for maintenance of Wnt-11 expression in the ureter tips. Development 122, 3627–3637.

Kuure, S., Chi, X., Lu, B., Costantini, F., 2010. The transcription factors Etv4 and Etv5 mediate formation of the ureteric bud tip domain during kidney development. Development 137, 1975–1979.

Lane, T.F., Leder, P., 1997. Wnt-10b directs hypermorphic development and transformation in mammary glands of male and female mice. Oncogene 15, 2133–2144.

Li, C., Hu, L., Xiao, J., Chen, H., Li, J.T., Bellusci, S., Delanghe, S., Minoo, P., 2005. Wnt5a regulates Shh and Fgf10 signaling during lung development. Dev. Biol. 287, 86–97.

Lin, Y., Zhang, S., Rehn, M., Itaranta, P., Tuukkanen, J., Heljasvaara, R., Peltoketo, H., Pihlajaniemi, T., Vainio, S., 2001. Induced repatterning of type XVIII collagen expression in ureter bud from kidney to lung type: association with sonic hedgehog and ectopic surfactant protein C. Development 128, 1573–1585.

Liu, Y., Suzuki, Y.J., Day, R.M., Fanburg, B.L., 2004. Rho kinase-induced nuclear translocation of ERK1/ERK2 in smooth muscle cell mitogenesis caused by serotonin. Circ. Res. 95, 579–586.

Long, C., 2004. Leonardo da Vinci's rule and fractal complexity in dichotomous trees. J. Theor. Biol. 167, 107–113.

Lu, J., Qian, J., Izvolsky, K.I., Cardoso, W.V., 2004. Global analysis of genes differentially expressed in branching and non-branching regions of the mouse embryonic lung. Dev. Biol. 273, 418–435.

Mandelbrot, B., 1997. The fractal geometry of nature. Freeman, Cranbury, New Jersey.

Maruyama, R., Andrew, D.J., 2012. Drosophila as a model for epithelial tube formation. Dev. Dyn. 241, 119–135.

McAteer, J.A., Dougherty, G.S., Gardner Jr., K.D., Evan, A.P., 1986. Scanning electron microscopy of kidney cells in culture: surface features of polarized epithelia. Scan. Electron. Microsc. 1986, 1135–1150.

Michael, L., Davies, J.A., 2004. Pattern and regulation of cell proliferation during murine ureteric bud development. J. Anat. 204, 241–255.

Miller, L.A., Wert, S.E., Whitsett, J.A., 2001. Immunolocalization of sonic hedgehog (Shh) in developing mouse lung. J. Histochem. Cytochem. 49, 1593–1604.

Miyazaki, M., 1990. Branching morphogenesis in the embryonic mouse submandibular gland: a scanning electron microscopic study. Arch. Histol. Cytol. 53, 157–165.

Nakanishi, Y., Ishii, T., 1989. Epithelial shape change in mouse embryonic submandibular gland: modulation by extracellular matrix components. Bioessays 11, 163–167.

Nakanishi, Y., Morita, T., Nogawa, H., 1987. Cell proliferation is not required for the initiation of early cleft formation in mouse embryonic submandibular epithelium *in vitro*. Development 99, 429–437.

Nakanishi, Y., Nogawa, H., Hashimoto, Y., Kishi, J., Hayakawa, T., 1988. Accumulation of collagen III at the cleft points of developing mouse submandibular epithelium. Development 104, 51–59.

Nelson, C.M., Vanduijn, M.M., Inman, J.L., Fletcher, D.A., Bissell, M.J., 2006. Tissue geometry determines sites of mammary branching morphogenesis in organotypic cultures. Science 314, 298–300.

Park, J.S., Ma, W., O'Brien, L.L., Chung, E., Guo, J.J., Cheng, J.G., Valerius, M.T., McMahon, J.A., Wong, W.H., McMahon, A.P., 2012. Six2 and Wnt regulate self-renewal and commitment of nephron progenitors through shared gene regulatory networks. Dev. Cell. 23, 637–651.

Patan, S., 2004. How is branching of animal blood vessels implemented? In: Davies, J. (Ed.), Branching morphogenesis. Landes Bioscience. Texas, Georgetown.

Piscione, T.D., Phan, T., Rosenblum, N.D., 2001. BMP7 controls collecting tubule cell proliferation and apoptosis via Smad1-dependent and -independent pathways. Am. J. Physiol. Renal. Physiol. 280, F19–F33.

Pongracz, J.E., Stockley, R.A., 2006. Wnt signaling in lung development and diseases. Respir. Res. 7, 15.

Post, L.C., Ternet, M., Hogan, B.L., 2000. Notch/Delta expression in the developing mouse lung. Mech. Dev. 98, 95–98.

Pu, Y., Huang, L., Prins, G.S., 2004. Sonic hedgehog-patched Gli signaling in the developing rat prostate gland: lobe-specific suppression by neonatal estrogens reduces ductal growth and branching. Dev. Biol. 273, 257–275.

Ritvos, O., Tuuri, T., Eramaa, M., Sainio, K., Hilden, K., Saxen, L., Gilbert, S.F., 1995. Activin disrupts epithelial branching morphogenesis in developing glandular organs of the mouse. Mech. Dev. 50, 229–245.

Roy, S., Hsiung, F., Kornberg, T.B., 2011. Specificity of Drosophila cytonemes for distinct signaling pathways. Science 332, 354–358.

Santos, O.F., Nigam, S.K., 1993. HGF-induced tubulogenesis and branching of epithelial cells is modulated by extracellular matrix and TGF-beta. Dev. Biol. 160, 293–302.

Schmidt-Ott, K.M., Yang, J., Chen, X., Wang, H., Paragas, N., Mori, K., Li, J.Y., Lu, B., Costantini, F., Schiffer, M., Bottinger, E., Barasch, J., 2005. Novel regulators of kidney development from the tips of the ureteric bud. J. Am. Soc. Nephrol. 16, 1993–2002.

Sebinger, D.D., Unbekandt, M., Ganeva, V.V., Ofenbauer, A., Werner, C., Davies, J.A., 2010. A novel, low-volume method for organ culture of embryonic kidneys that allows development of cortico-medullary anatomical organization. PLoS One 2010 May 10;5(5):e10550.

Shackleton, M., Vaillant, F., Simpson, K.J., Stingl, J., Smyth, G.K., Asselin-Labat, M.L., Wu, L., Lindeman, G.J., Visvader, J.E., 2006. Generation of a functional mammary gland from a single stem cell. Nature 439, 84–88.

Smith, G.H., 1996. Experimental mammary epithelial morphogenesis in an *in vivo* model: evidence for distinct cellular progenitors of the ductal and lobular phenotype. Breast Cancer Res. Treat. 39, 21–31.

Srinivas, S., Goldberg, M.R., Watanabe, T., D'Agati, V., al Awqati, Q., Costantini, F., 1999. Expression of green fluorescent protein in the ureteric bud of transgenic mice: a new tool for the analysis of ureteric bud morphogenesis. Dev. Genet. 24, 241–251.

Stark, K., Vainio, S., Vassileva, G., McMahon, A.P., 1994. Epithelial transformation of metanephric mesenchyme in the developing kidney regulated by Wnt-4. Nature 372, 679–683.

Sweeney, D., Lindström, N., Davies, J.A., 2008. Developmental plasticity and regenerative capacity in the renal ureteric bud/collecting duct system. Development 135, 2505–2510.

van Eyll, J.M., Passante, L., Pierreux, C.E., Lemaigre, F.P., Vanderhaeghen, P., Rousseau, G.G., 2006. Eph receptors and their ephrin ligands are expressed in developing mouse pancreas. Gene. Expr. Patterns. 6, 353–359.

Vize, P., Carroll, T., Wallingford, J., 2003. Induction, development and physiololgy of the pronephric tubules. In: Vize, P., Woolf, A., Bard, J.B.L. (Eds.), The kidney. Elsevier, London, pp. 19–47.

Wang, J., Laurie, G.W., 2004. Organogenesis of the exocrine gland. Dev. Biol. 273, 1–22.

Watanabe, T., Costantini, F., 2004. Real-time analysis of ureteric bud branching morphogenesis *in vitro*. Dev. Biol. 271, 98–108.

Weber-Hall, S.J., Phippard, D.J., Niemeyer, C.C., Dale, T.C., 1994. Developmental and hormonal regulation of Wnt gene expression in the mouse mammary gland. Differentiation 57, 205–214.

West, G.B., Brown, J.H., Enquist, B.J., 1997. A general model for the origin of allometric scaling laws in biology. Science 276, 122–126.

West, G.B., Brown, J.H., Enquist, B.J., 1999. The fourth dimension of life: fractal geometry and allometric scaling of organisms. Science 284, 1677–1679.

Zhang, T.J., Hoffman, B.G., Ruiz,d, A., Helgason, C.D., 2006. SAGE reveals expression of Wnt signaling pathway members during mouse prostate development. Gene Expr. Patterns. 6, 310–324.

Chapter 15

Lessons from the Zebrafish Lateral Line System

Ajay B. Chitnis and Damian Dalle Nogare

Section on Neural Developmental Dynamics, Program in Genomics of Differentiation, NICHD NIH, Bethesda, MD

Chapter Outline

GLOSSARY

Atoh1 Atonal Homolog 1a (Atoh1a) and Atonal Homolog 1b (Atoh1b) are the zebrafish homologs of *Drosophila* Atonal, a bHLH, transcription factor that also determines the specification of mechanoreceptors in *Drosophila*. Atoh1a expression gives cells in protoneuromasts or neuromasts the potential to become sensory hair cell progenitors.

Chemokines A class of chemotactic cytokine signaling molecules secreted by cells that direct the chemotactic response of other cells. They are divided into CXC, CC, CX3C and XC subfamilies.

FGFs Fibroblast Growth Factors are a family of growth factors, whose members are involved in angiogenesis, wound healing, embryonic development and various endocrine signaling pathways.

Lef1 Lymphoid enhancer-binding factor 1, a transcription factor that binds Lef/TCF sites in the DNA to mediate Wnt/ß-catenin signaling.

Mantle cells (outer support cells) These are cells that surround the inner support cells and serve as a proliferative population that can replenish the population of inner support cells, which are progressively lost as they are specified as sensory hair cell progenitors.

Neuromasts Sensory organs deposited by the migrating PLLp to initiate establishment of the Posterior Lateral Line (PLL) system. They have paired sensory hair cells at the center. **Support cells** (also called inner support cells) surround sensory hair cells and **mantle cells** (also called outer support cells) surround support cells.

PLLp Posterior Lateral Line Primordium, a cluster of approximately 100 cells that migrates collectively from the ear to the tip of the tail periodically depositing neuromasts.

Protoneuromasts These are nascent neuromasts forming within the migrating PLLp. The migrating PLLp contains 2–3 protoneuromasts at progressive stages of maturation. It consists of a cluster of about 15–30 cells that form epithelial rosettes in which a sensory hair cell progenitor gets specified at the center.

Sensory hair cell progenitor A cell expressing relatively high levels of Atoh1a that has the potential to undergo a final division to produce a pair of sensory hair cells. The first sensory hair cell progenitor is specified at the center of a protoneuromast. However, additional sensory hair cell progenitors continue to be specified from support cells that surround the central sensory hair cells during growth and regeneration of the neuromasts.

Principles of Developmental Genetics.

Support cells (inner support cells) A population of cells that surround the central sensory hair cell progenitors or central sensory hair cells. While providing support to inner sensory hair cells, they also serve as a population of undifferentiated progenitors from which sensory hair cell progenitors are specified during the course of growth and regeneration of neuromasts. In contrast to sensory hair cell progenitors, these cells do not express high levels *atoh1a* and have relatively high levels of Notch activity.

Wnts Signaling molecules in an evolutionarily conserved signaling pathway that get their name from the homolog, Wingless, in *Drosophila*, and mammalian Int1, discovered in mice.

SUMMARY

- The Posterior Lateral Line system (PLL) in zebrafish is a remarkably effective model system for understanding how interactions between cells and integration of input from multiple signaling systems regulates pattern formation, morphogenesis, growth, and collective migration during the self-organization of tissues during development.
- The PLL is established by the PLL primordium (PLLp), a collection of about 100 cells, which migrates from the ear to the tip of the tail periodically depositing sensory organs called neuromasts.
- FGF ligands, expressed in response to Wnt signaling in a leading domain, establish center-biased FGF signaling centers in an adjacent trailing domain to coordinate periodic formation of nascent neuromasts or protoneuromasts within the migrating PLLp.
- FGF signaling initiates center-biased *atoh1a* expression in protoneuromasts, giving them the potential to become sensory hair cell progenitors. Lateral inhibition mediated by Notch signaling operates in this context to restrict *atoh1a* expression to the central cell, specifying it as a sensory hair cell progenitor, while neighboring cells, in which atoh1a has been suppressed, become support cells.
- Polarized expression of chemokine receptors, Cxcr4b and Cxcr7b, in leading and trailing cells, respectively, determines directed migration of the PLLp along a path defined by the relatively uniform expression of the chemokine ligand, Cxcl12a.
- Cxcl12a ligand binds the Cxcr4b receptor to engage the G protein coupled signaling pathway and determine migratory behavior of leading cells, while Cxcr7b receptors in trailing cells play a key role in reducing availability of the Cxcl12a to Cxcr4b in trailing cells.

15.1 INTRODUCTION

The lateral line is a sensory system that has evolved to allow fish and amphibians to detect the pattern of water flow over their body surface (Coombs and van Netten, 2006). Information about water flow is used for a variety of purposes, including the detection of objects without touch or "touch-at-a-distance" (Dijkgraaf, 1963) and to provide sensory feedback during swimming. The lateral line system detects water flow with sensory organs called neuromasts, which are small cell clusters with sensory hair cells at their center. Many recent studies have focused on the Posterior Lateral Line (PLL), a part of the zebrafish lateral line system, which has neuromasts distributed in a stereotyped pattern over the trunk and tail. Information from the sensory hair cells of the PLL system is conveyed to the brain by the sensory neurons of the Posterior Lateral Line ganglion located next to the ear. These sensory neurons have peripheral processes that innervate the neuromasts, while their central processes carry information into the hindbrain (Chitnis et al., 2012; Ghysen and Dambly-Chaudiere, 2007).

Formation of the PLL system in zebrafish is pioneered by the end of the first day of development by the PLL primordium (PLLp), a cluster of about a hundred cells (Figure 15.1A) that migrates under the skin from near the ear to the tip of the tail along the horizontal myoseptum (Gompel et al., 2001; Kimmel et al., 1995). During this migration the PLLp periodically generates and deposits 5–6 neuromasts (L1, L2 etc) to initiate formation of the PLL system (Figure 15.1B). At the end of its migration it resolves to form 2–3 terminal neuromasts (TN).

Each neuromast has sensory hair cells at its center (Figure 15.1C). The sensory hair cells are sequentially generated from central hair cell progenitors whose final division produces pairs of differentiated sensory hair cells with sensitivity to water movement in opposing directions. Support cells surround the central sensory hair cells. While providing structural support to sensory hair cells, the support cells also serve as a pool of undifferentiated cells from which sensory hair cell progenitors are sequentially specified during growth and regeneration of the neuromast (Wibowo et al., 2011). These inner support cells are themselves surrounded by mantle cells (also called outer support cells). Proliferating mantle cells replenish inner support cells that are lost as they are specified as hair cell progenitors (Harris et al., 2003; Williams and Holder, 2000).

There are striking similarities in the mechanisms that determine the development and function of sensory hair cells in neuromasts and sensory hair cells located in our inner ear. In addition, the accessibility of this system and the remarkable ability of neuromasts to replace their sensory hair cells following damage has made the lateral line system an attractive

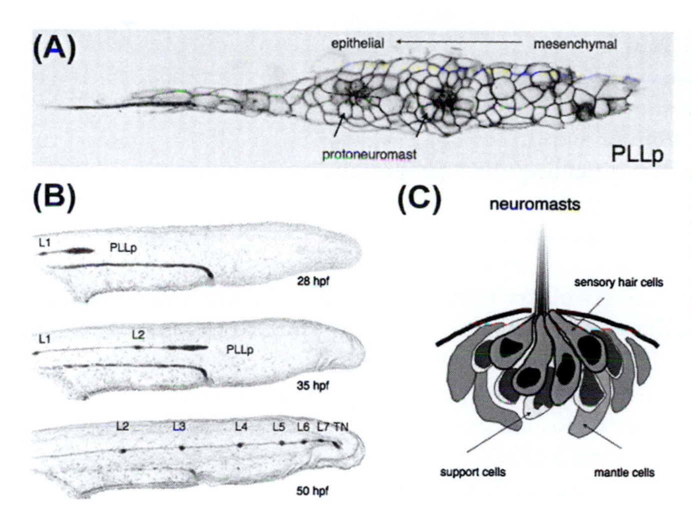

FIGURE 15.1 The PLLp Migrates From the Ear to the Tip of the Tail Periodically Depositing Neuromasts. (A) Confocal slice through the PLLp in a transgenic Cldn:lyn-GFP embryo shows distinct changes in morphology from the leading end (right) to the trailing end of the PLLp. (B) Images of the migrating PLLp between 28 and 50 hours post fertilization (hpf) showing periodic deposition of neuromasts (L1-L7, TN terminal neuromasts). The original fluorescent photomicrographs have been inverted to the create gray scale illustrations in A and B. (C) Schematic of a deposited neuromast showing central sensory hair cells and surrounding support and mantle cells (also referred to as inner and outer support cells). Adapted from Moon *et al* 2011.

model for studying the biology of sensory hair cells and to identify strategies for promoting regeneration of our own sensory hair cells (Behra et al., 2009; Coffin et al., 2010; Dufourcq et al., 2006; Harris et al., 2003; Hernandez et al., 2006; 2007; Moon et al., 2011). Recent studies, however, have shown that the PLL also serves as an extraordinary model system for exploring a broad range of fundamental mechanisms operating during the self-organization and development of organ systems (Chitnis et al., 2012; Friedl and Gilmour, 2009; Ghysen and Dambly-Chaudiere, 2007).

15.2 EMERGENCE OF THE PLL SYSTEM

The PLL system emerges as an ectodermal thickening, the PLL placode, posterior to the otic vesicle, and is one of a number of such placodal structures that form within an initial horseshoe-shaped pre-placodal domain to establish paired sensory organs (olfactory epithelium, lens, inner ear, lateral line), cranial ganglia and the adenohypophysis (see also Chapter 19). The potential mechanisms involved in induction of the PLL placode will not be discussed here because they are described in other reviews (Schlosser, 2010; Streit, 2007).

Once the PLL placode delaminates, its cells divide rapidly and it separates into a caudal compartment of about 100 cells that forms the PLLp, where sensory neuromasts form. The rostral compartment of about 20 cells forms the sensory neurons of the PLL sensory ganglion, which innervate the neuromasts. The mechanisms that lead to the segregation of these two fates in the placodal cells remain poorly understood. However, Notch signaling determines how many cells will become part of the sensory ganglion and how many will instead remain part of the PLLp (Mizoguchi et al., 2011). Inhibition of Notch signaling increases the number of sensory neurons but reduces the number of cells that are retained in the rostral part of the PLLp, where the first neuromast forms. On the other hand, activation of Notch signaling reduces the number of lateral line sensory neurons, while increasing the number of cells in the PLLp that contribute to the formation of the first neuromast. This role of Notch signaling is reminiscent of its early role in the mouse and chick ear, where Notch signaling determines how many cells are allowed to remain part of the sensory epithelium, within which sensory hair cells form, and how many are selected to become the sensory neurons that eventually innervate these mechanoreceptors (Brooker et al., 2006) (see also Chapter 21). As in the zebrafish PLL placode, early loss of Notch signaling in the mouse allows more cells to become part of the acoustico-vestibular sensory ganglion at the cost of cells that contribute to formation of sensory hair cells in the utricular and saccular macula. Interestingly, however, in the zebrafish ear itself, Notch does not have an obviously similar role in determining how many cells are allowed to form part of the sensory epithelium. In contrast to the chick and mouse, reduced Notch signaling results in both more sensory neuron cells in the acoustic ganglion and more sensory hair cells in the ear (Haddon et al., 1998).

15.3 MORPHOGENESIS AND SEQUENTIAL FORMATION OF PROTONEUROMASTS IN THE PLLp

The establishment of a number of transgenic zebrafish lines with expression of fluorescent proteins in the entire PLLp, in specific types of cells or reporting specific signaling activities has provided unique insight about the dynamic behavior of the PLLp as it makes its journey from near the ear to the tip of the tail (Dorsky et al., 2002; Faucherre et al., 2009; Gallardo et al., 2010; Germana et al., 2010; Gilmour et al., 2002; Hernandez et al., 2007; Kim et al., 2013;

McDermott et al., 2010; Molina et al., 2007; Shimizu et al., 2012; Xiao et al., 2005). One line in particular, Tg Cldn: Lyn-GFP, with membrane-tethered green fluorescent protein (Lyn-GFP) expression in the PLLp, has transformed our view of the PLLp (Haas and Gilmour, 2006). Examination of the PLLp at high magnification using this line reveals distinct changes in its morphology from its leading to its trailing end (Figure 15.1A). Leading cells are flatter with more of a mesenchymal morphology. Trailing cells are taller, reflecting their progressive epithelialization, as they acquire apical basal polarity and reorganize as epithelial rosettes to form nascent neuromasts or "protoneuromasts." The migrating PLLp has 2–3 protoneuromasts at progressive stages of maturation. Formation of protoneuromasts is initiated prior to migration when the PLLp is adjacent to the ear. The first protoneuromast forms at the trailing end of the PLLp and the next ones form sequentially, progressively closer to the leading end (Figure 15.2A). As the migrating PLLp begins depositing neuromasts from its trailing end, formation of new protoneuromasts is initiated progressively closer to the leading end (Matsuda et al., 2013; Nechiporuk and Raible, 2008). It is important to note that while neuromast deposition is periodic, the migrating PLLp continuously deposits cells in its wake. While cells incorporated into protoneuromasts within the migrating PLLp are deposited as neuromasts, cells not incorporated into protoneuromasts are deposited as interneuromast cells and they form a continuous chain connecting deposited neuromasts. These interneuromast cells serve as a pool of progenitors from which additional neuromasts are formed as the fish grows (Grant et al., 2005; Hernandez et al., 2007; Lopez-Schier and Hudspeth, 2005). The loss of cells from the PLLp as a consequence of cell deposition is compensated for in part by proliferation within the PLLp. However, the PLLp does become smaller as it migrates to the tip of the tail, where its migration terminates with the remaining PLLp cells resolving into 2–3 terminal neuromasts.

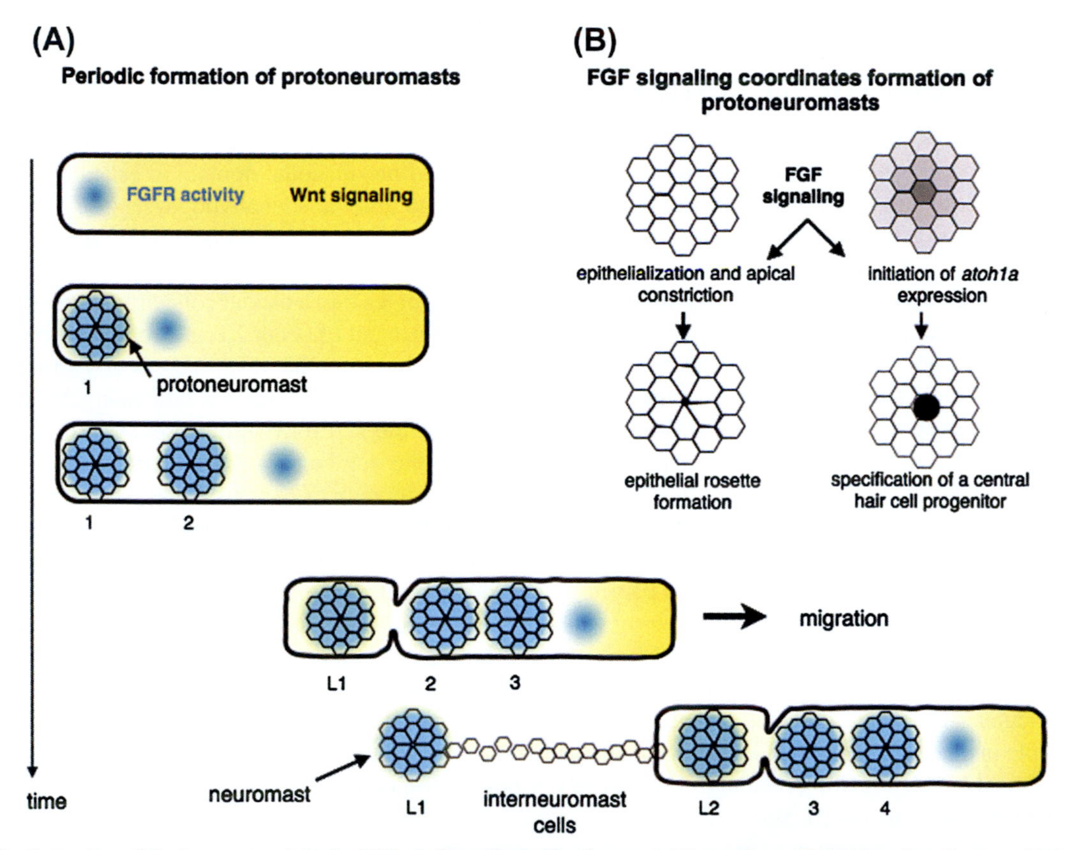

FIGURE 15.2 Formation of Protoneuromasts in the PLLp is Coordinated by Sequential Formation of FGF Signaling Centers. (A) Protoneuromast formation is initiated at the trailing end and progresses periodically toward the leading end (1,2, 3, etc.). The PLLp is initially characterized by broad Wnt signaling (yellow). Formation of a trailing FGF signaling center (cyan) initiates protoneuromast formation. Additional FGF signaling centers are then sequentially formed progressively closer to the leading end. This is accompanied by a progressive reduction in the size of the leading Wnt active domain. FGF signaling coordinates formation of protoneuromasts. Protoneuromasts progressively mature and are eventually deposited as neuromasts (L1, L2, etc.) from the trailing end of the migrating PLLp. Cells not incorporated into protoneuromasts by FGF signaling are deposited as interneuromast cells. (B) FGF signaling coordinates formation of protoneuromasts by promoting morphogenesis of epithelial rosettes and by initiating expression of *atoh1a*, which gives cells the potential to become sensory hair cell progenitors. It also determines expression of components of the Notch signaling pathway that ensure *atoh1a* expression, hence sensory hair cell progenitor fate, is eventually restricted to the central cell.

15.4 ESTABLISHMENT OF POLARIZED WNT AND FGF SIGNALING SYSTEMS IN THE PLLp

Morphological distinctions and systematic changes in the shape, fate and behavior of cells, from the leading to the trailing end, are coordinated, at least in part, by the establishment of polarized Wnt and FGF signaling systems in the PLLp (Figure 15.3) (Aman and Piotrowski, 2008; Lecaudey et al., 2008; Nechiporuk and Raible, 2008). Although it remains unclear which Wnt ligands are responsible, Wnt/ß-catenin signaling dominates the entire PLLp prior to migration and initiation of protoneuromast formation. This has been inferred from the initial broad expression of genes like *lef1* whose expression is dependent on Wnt/ß-catenin signaling (Aman and Piotrowski, 2008; Gamba et al., 2010). Activation of the Wnt signaling system has a number of important consequences in the PLLp (Figure 15.3). First, as noted above, it promotes expression of Lef1, a key mediator of Wnt signaling in the PLLp, and of additional factors like Rspo3 that facilitate Wnt signaling (Matsuda et al., 2013). At the same time, Wnt signaling promotes expression of two FGF ligands (Aman and Piotrowski, 2008). However, these FGF ligands are unable to effectively activate the FGF receptor in Wnt active cells where they are produced because Wnt signaling also drives expression of a number of factors that locally inhibit activation of the FGF signaling cascade. For example, Wnt signaling promotes expression of Sef (Aman and Piotrowski, 2008), a transmembrane protein that inhibits activation of FGF receptors (Tsang et al., 2002). In addition, Lef1-mediated Wnt signaling determines the expression of the FGF signaling inhibitor, Dusp6 (Tsang et al., 2004), in the Wnt active domain (Matsuda et al., 2013).

As activation of the FGF signaling cascade is locally inhibited by the activation of the Wnt/ß-catenin signaling pathway, FGF3 and FGF10 diffuse away to activate FGF receptors in an adjacent domain where there is less inhibition by the Wnt pathway. Once initiated, FGF signaling drives increased expression of the FGF receptor (FGFR), which facilitates FGF signaling. In addition, initiation of FGF signaling drives expression of Dkk1 (Aman and Piotrowski, 2008), a diffusible Wnt inhibitor, which limits the inhibitory effect of Wnt signaling. Together, these changes help establish a stable, trailing, FGF signaling system and restrict Wnt signaling to a smaller leading domain.

Although FGF signaling eventually becomes established in a progressively larger trailing part of the PLLp, *dkk1* expression is restricted to a zone just adjacent to the leading Wnt active zone. This suggests that its expression may not only be dependent on FGF signaling but also on a limited amount of Wnt signaling from the leading zone. This dual dependence on Wnt and FGF signaling may help mediate a switch from Wnt signaling in leading cells to FGF signaling in trailing cells. As *dkk1* is not expressed in the trailing zone, where Wnt signaling nevertheless appears to be suppressed, it is possible that there are additional mechanisms, independent of Dkk1, that limit Wnt signaling in trailing cells of the PLLp.

The establishment of a trailing FGF signaling system is key to initiation of protoneuromast formation (Figures 15.2B and 3B). It promotes changes that facilitate reorganization of cells to form epithelial rosettes and it initiates expression of

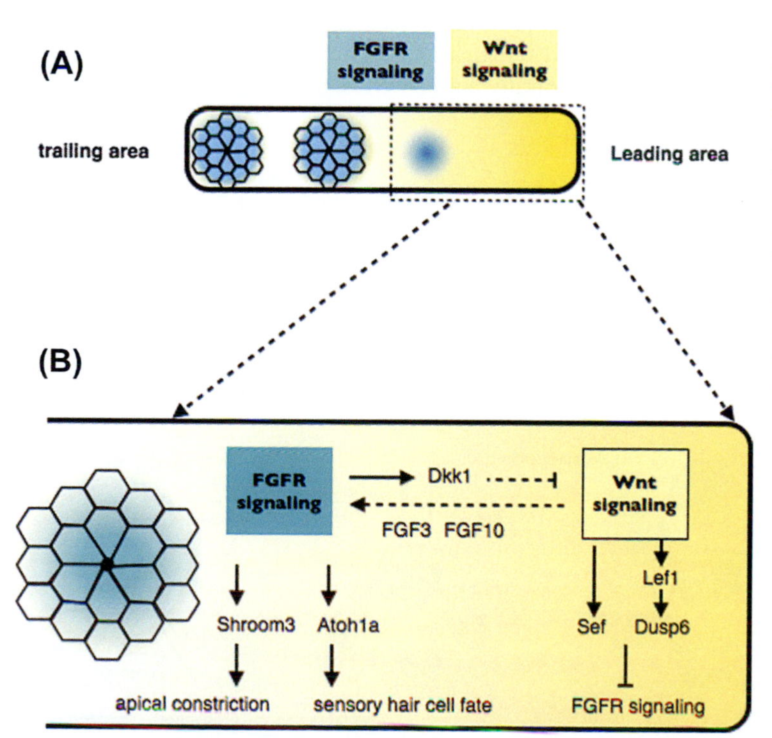

FIGURE 15.3 Coupled Yet Mutually Inhibitory Wnt FGF Signaling Systems Facilitate Periodic Formation of FGF Signaling Centers. (A) Schematic of PLLp showing a leading Wnt active zone (yellow) with trailing FGF signaling centers (cyan). (B) Detail of boxed area in A showing how Wnt signaling system initiates the formation of an adjacent FGF signaling system. Wnt signaling in a leading domain drives FGF3 and FGF10 expression but locally inhibits FGFR signaling by also determining expression of FGFR signaling inhibitors Sef and Dusp6. FGF signaling promotes apical constriction by initiating Shroom3 expression and gives cells the potential to become hair cell progenitors by promoting atoh1a expression. Once FGF signaling is initiated, it is stabilized by the FGF-signaling-dependent expression of the Wnt inhibitor Dkk1.

factors that lead to specification of a central hair cell progenitor (Figures 15.2B and 3B) (Lecaudey et al., 2008; Nechiporuk and Raible, 2008). Morphogenesis of epithelial rosettes is mediated by ERK 1/2 signaling downstream of the FGF receptor and Shroom3, whose FGF-dependent expression in the PLLp promotes apical constriction and rosette assembly (Ernst et al., 2012). FGF signaling also promotes expression of Atoh1a, a basic helix-loop-helix transcription factor that gives cells in the protoneuromasts the potential to become sensory hair cell progenitors. However, lateral inhibition mediated by Notch signaling restricts *atoh1a* expression to a central cell in the developing protoneuromast (Figures 15.2B and 4) and this cell becomes specified as a sensory hair cell progenitor (Itoh and Chitnis, 2001; Sarrazin et al., 2006).

15.5 SPECIFICATION OF A CENTRAL HAIR CELL PROGENITOR BY LATERAL INHIBITION MEDIATED BY NOTCH SIGNALING

FGF signaling initiates *atoh1a* expression in protoneuromast cells and Atoh1a drives expression of *deltaD* (Matsuda and Chitnis, 2010). Expression of the Notch ligand, DeltaD, on the surface of Atoh1a-expressing cells allows these cells to activate Notch in neighboring cells, which in turn inhibits them from expressing *atoh1a* (Figure 15.4B). In protoneuromasts this process of lateral inhibition mediated by Notch signaling ensures *atoh1a* expression remains restricted to a central cell, specifying it as a hair cell progenitor. On the other hand, surrounding cells, inhibited from expressing *atoh1a*, are maintained as support cells (Figure 15.4C).

But how does lateral inhibition restrict *atoh1a* expression to a central cell? It is sometimes assumed that if all cells in a protoneuromast cell cluster were to initially express similar levels of Atoh1a and Delta, then activation of Notch in neighboring cells would be adequate to explain how *atoh1a* expression becomes restricted to a central cell. However, in such a situation, the central cell would actually be the least likely to retain *atoh1a* expression following lateral inhibition, and a cell at the edge of the protoneuromast cluster would be more likely to retain *atoh1a* expression. This is because a central cell, surrounded by Delta-expressing neighbors, would receive more inhibition than the edge cells, which receive no inhibition from cells outside the protoneuromast cell cluster that do not express Delta (Chitnis, 1995). If on the other hand, expression of Atoh1a and Delta was initiated in a centrally biased pattern, with highest expression in the central cell of a protoneuromast cluster, then lateral inhibition could ensure that this cell retains *atoh1* expression while Notch activation ensures it remains off in surrounding neighbors (Figure 15.4). This is indeed what appears to happen in the PLLp, as FGF signaling is initiated in a centrally biased pattern in nascent protoneuromasts (Chitnis et al., 2012), where it initiates centrally biased expression of both *atoh1a* and *deltaA*. Atoh1a initiates expression of *deltaD* and, by activating Notch in the neighbors, the centrally biased expression of both DeltaA and DeltaD ensures *atoh1a* remains restricted to a central cell in the maturing protoneuromasts (Matsuda and Chitnis, 2010).

Although the centrally biased activation of FGF signaling in nascent protoneuromast accounts for *atoh1a* dependent specification of a sensory hair cell at the center of a protoneuromast, it remains unclear why FGFs, secreted in a Wnt-dependent manner in the leading zone of the PLLp, would first initiate centrally biased FGF signaling in an adjacent trailing

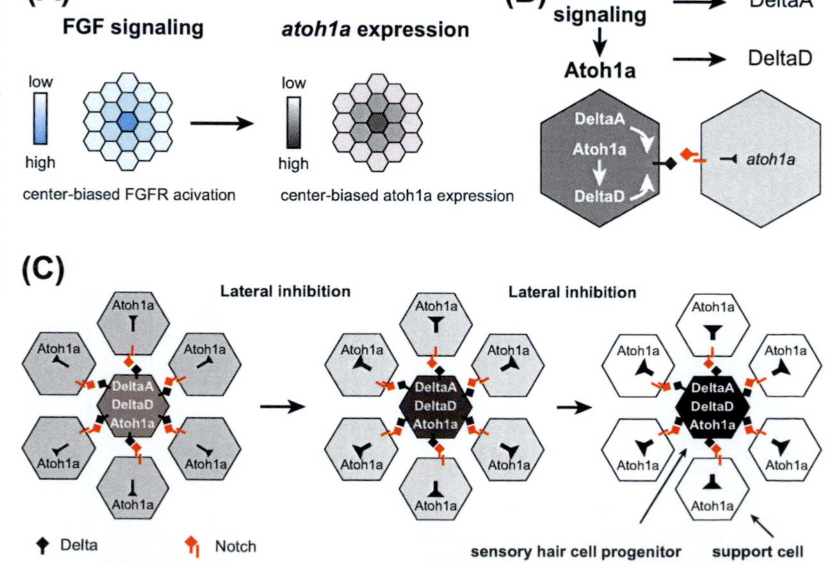

FIGURE 15.4 Specification of a Central Sensory Hair Cell Progenitor by Lateral Inhibition Mediated by Notch Signaling. (A) Center-biased FGFR activation initiates center-biased *atoh1a* expression. (B) FGF signaling determines the initial expression of both atoh1a and the Notch ligand DeltaA. Atoh1a, in turn, determines expression of an additional Delta homolog, DeltaD. DeltaA and DeltaD both contribute to lateral inhibition mediated by Notch signaling. (C) The initial center-biased pattern of DeltaA, DeltaD and Atoh1a expression is critical for lateral inhibition to eventually restrict Atoh1a expression to the central cell, which becomes specified as the sensory hair cell progenitor.

domain. One intriguing possibility is that the centrally biased pattern of FGF signaling is determined by the Wnt and FGF signaling systems interacting to form a reaction-diffusion system capable of self-organizing a focal center of FGF signaling. A reaction-diffusion system with local activation and long-range inhibition (see Resources section for more details) locally enhances its own production; for example, an activator system, A, with a relatively short range of diffusion. At the same time it drives an inhibitory signaling system, B, with a much longer range of diffusion. In such a system, local activation can stabilize the activator system "A" in one location, while the long-range inhibition mediated by system "B" prevents establishment of the activator system at a distance. Such a system has the potential to spontaneously generate focal signaling centers from an initially diffuse broad pattern of activity (Meinhardt, 2012; Meinhardt and Gierer, 2000). The possibility that the Wnt and FGF signaling pathways interact to function as part of a reaction-diffusion system remains speculative at this time and is under investigation in our laboratory.

15.6 THE HAIR CELL PROGENITOR BECOMES A NEW LOCALIZED SOURCE OF FGF

Although Wnt signaling initiates the expression of FGF ligands FGF3 and FGF10 at the leading end of the PLLp, an independent mechanism maintains expression of FGFs in the trailing PLLp, where protoneuromasts mature and where there is too little Wnt signaling to maintain FGF production. While Atoh1a expression specifies a cell as a hair cell progenitor, it also makes the cell express FGF10 (Lecaudey et al., 2008; Matsuda and Chitnis, 2010). Therefore, as the central protoneuromast cell becomes specified as a hair cell progenitor, it also becomes a new source of FGF at the center of the maturing protoneuromast (Figure 15.5B). As a source of FGF, it has the potential to activate the FGF receptor and induce *atoh1a* in neighboring cells. However, Atoh1a in the central cell also makes this cell express *deltaD*. DeltaD activates Notch in neighboring cells and ensures that they do not express *atoh1a*, which allows them to be maintained as support cells. Though Notch activation in these support cells prevents them from expressing *atoh1a*, activation of FGF signaling in these cells promotes expression of Notch3, ensuring the support cells remain competent to respond to DeltaD expressed by the central hair cell progenitor.

The interactions between the central hair cell progenitor, which is a source of FGF10 and DeltaD, and surrounding support cells, in which relatively high levels of FGF and Notch activation are maintained, helps establish a robust self-sustaining system that appears poised to produce new *atoh1a*-expressing hair cell progenitors when the system demands it. The central hair cell progenitor in the mature trailing protoneuromast eventually shuts off *atoh1a* when it divides to form two terminal sensory hair cells. When the central cells differentiate and stop expressing Atoh1a, the surrounding support cells are transiently relieved from lateral inhibition, as the central cells are no longer a source of Atoh1a-dependent DeltaD.

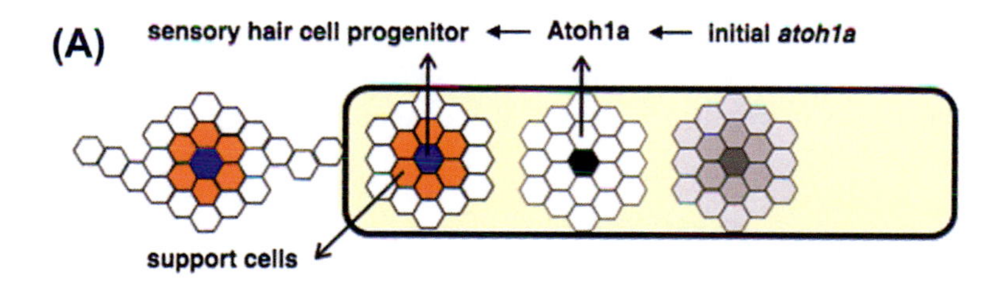

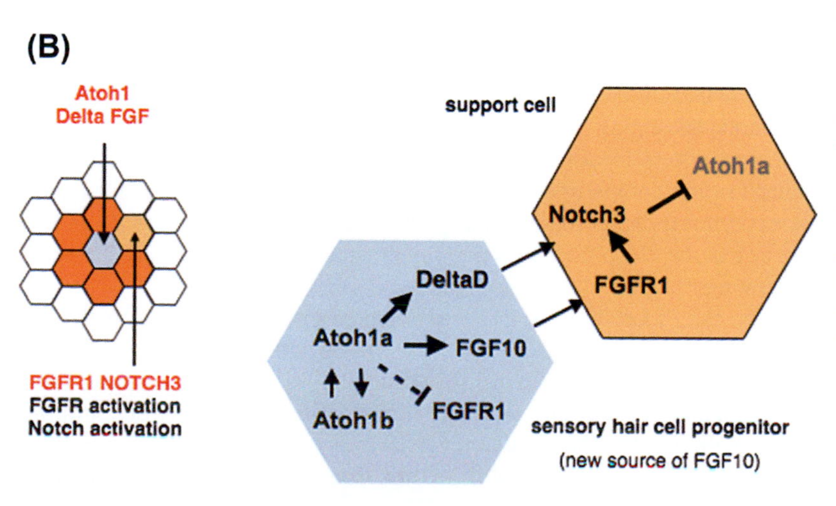

FIGURE 15.5 The Central Sensory Hair Cell Progenitor Eventually Becomes a New Source of FGF10, Which Maintains FGF Signaling in Maturing Protoneuromasts at the Trailing End of the PLLp. (A) The initial centrally biased pattern of Atoh1 becomes restricted to the central cell as a protoneuromast matures. The cell that maintains Atoh1a expression becomes specified as a sensory hair cell progenitor (blue), while the neighbors, in which *atoh1a* expression has been suppressed by Notch signaling, are maintained as support cells (orange). (B) Atoh1a drives expression of FGF10 and DeltaD in the central hair cell progenitor. This allows the central hair cell progenitor to maintain relatively high levels of FGF receptor (FGFR) and Notch activation in surrounding support cells. FGFR activation helps maintain Notch3 expression in surrounding support cells and Notch activation prevents support cells from initiating *atoh1a* expression. Atoh1a also inhibits expression of FGFR1 in the hair cell progenitor, making this cell less responsive to FGF ligands. As FGFR activation is lost in the central cell, Atoh1b is maintains Atoh1a expression via autoregulation

This is likely to allow some of the surrounding support cells to begin expressing *atoh1a*. Interactions between these cells could then ensure that only one of the support cells acquires high enough levels of *atoh1a* expression to allow it to be specified as the next sensory hair cell progenitor.

Though Atoh1a promotes FGF10 expression, it inhibits the expression of the FGFR1 receptor (Matsuda and Chitnis, 2010). Hence the central progenitor cell, which becomes a source of FGF10, is itself less responsive to FGF signaling. Hence, while centrally biased FGF signaling initiates *atoh1a* expression in nascent protoneuromasts, FGF signaling eventually becomes attenuated in the central hair cell progenitor and hence *atoh1a* expression cannot be effectively maintained in this cell by FGF signaling. Instead, Atoh1a initiates the expression of another Atoh1 homolog, *atoh1b*, and Atoh1b in turn promotes expression of *atoh1a* (Figure 15.5). In this manner, reminiscent of the role of these Atoh1 homologs in the ear (Millimaki et al., 2007), Atoh1a expression becomes less dependent on FGF signaling and more dependent on autoregulation (Matsuda and Chitnis, 2010).

While *atoh1a* is expressed in protoneuromasts as they begin to form at the leading end of the PLLp, Atoh1a only initiates *atoh1b* expression in relatively mature protoneuromasts at the trailing end of the PLLp. The maturation of the protoneuromasts appears to be determined by the progressive loss of Wnt signaling in trailing cells, as inhibition of Wnt signaling with chemical inhibitor IWR-1 or reduction of Wnt signaling in *lef1* deficient embryos is associated with earlier initiation of *atoh1b* in developing protoneuromasts in the PLLp (Matsuda et al., 2013).

15.7 NOTCH SIGNALING – AN ESSENTIAL NODE IN THE SELF-ORGANIZATION OF THE PLLp

Notch signaling plays a key role in determining the integrity of the self-sustaining system described above. When Notch signaling is lost the system progressively unravels (Matsuda and Chitnis, 2010). In the absence of Notch signaling, *atoh1a* expression is still initiated in a centrally biased pattern by FGF signaling and the central cell becomes a source of FGF10. It activates FGF signaling in neighboring cells, and in the absence of Notch signaling the neighbors cannot be inhibited from expressing *atoh1a*. These cells become a new source of FGF10 and the cycle repeats with these cells again inducing *atoh1a* expression in their neighbors. In the wake of this progressive induction of *atoh1a* expression, the *atoh1a* expressing cells reduce their own expression of the FGF receptor. So although the number of cells expressing FGF10 increases, their expression of the FGF receptor is lost and FGF signaling becomes progressively attenuated in the migrating PLLp. The progressive loss of FGF signaling ultimately leads to the loss of FGF-dependent expression of Dkk1 in the PLLp. Loss of this Wnt inhibitor, and possibly other mechanisms that limit Wnt activity, allows dramatic expansion of the Wnt active domain in the PLLp. In addition, as cells express *atoh1a* they are inappropriately specified as hair cell progenitors and the support cell population is lost. The progressive loss of FGF signaling, support cells and the eventual reduction of *cdh1* (E-cadherin) in the trailing part of the PLLp, all ultimately contribute to a morphological collapse of the PLLp system. Together these changes illustrate how Notch signaling is a key hub in the genetic regulatory network that determines self-organization of the PLLp system (Matsuda and Chitnis, 2010).

15.8 PERIODIC FORMATION OF PROTONEUROMASTS

While the PLLp starts out with Wnt signaling throughout the PLLp, an effective FGF signaling system eventually becomes established at the trailing end, where the first protoneuromast forms (Figure 15.2). Why FGF signaling first takes hold at the trailing end of the PLLp remains unclear, but it may be related to additional, as yet unidentified, patterning mechanisms operating within the PLLp that weaken Wnt signaling and facilitate the establishment of a trailing FGF signaling system. Once the FGF signaling system is established, it promotes the expression of factors like Dkk1 that help restrict Wnt signaling to a smaller, leading zone of the PLLp. As the Wnt system locally inhibits activation of the FGF signaling pathway, progressive restriction of Wnt signaling to a smaller leading compartment eventually facilitates the sequential formation of additional FGF signaling centers and new protoneuromasts closer to the leading end of the PLLp (Figure 15.2). In this manner the changing balance of Wnt and FGF signaling in the PLLp facilitates periodic formation of FGF signaling centers and their accompanying protoneuromasts in the migrating PLLp.

Manipulations that alter Wnt and FGF signaling in the PLLp illustrate how the balance of these signaling pathways influences the periodic formation of protoneuromasts in the PLLp. For example, loss of Lef1, a mediator of canonical Wnt/ß-catenin signaling, facilitates establishment of an FGF signaling system and subsequent formation of a nascent protoneuromast closer to the leading edge. This is at least in part because Lef1 normally limits FGF signaling by determining expression of DuSP6 (Matsuda et al., 2013), a dual specificity phosphatase that reverses the FGF-dependent phosphorylation mediated by MAP kinase ERK1/2, which is essential for protoneuromast formation. Hence, loss of *lef1* or inhibition of Dusp6 function results in loss of at least one mechanism that normally suppresses FGF-signaling-dependent protoneuromast formation. This allows the formation of a nascent protoneuromast closer to the leading edge.

While loss of Wnt signaling allows protoneuromasts to form closer to the leading end of the PLLp, attenuation of the FGF signaling system delays the formation of the first protoneuromast. This is illustrated by FGF10 morphants in which weakened FGF signaling delays establishment of the trailing FGF system, Wnt signaling dominates the entire PLLp for a longer period of time and formation of the first protoneuromast at the trailing end of the PLLp is delayed (Matsuda et al., 2013; Nechiporuk and Raible, 2008). In this context, weakening the Wnt system with IWR-1, a chemical inhibitor of Wnt signaling, restores earlier formation of the first protoneuromast (Matsuda et al., 2013). This illustrates how the balance of Wnt and FGF signaling influences when and where new protoneuromasts form.

15.9 TERMINATION OF THE PLLp SYSTEM

As protoneuromasts sequentially form in the migrating PLLp, the Wnt active domain progressively shrinks and is eventually lost (Matsuda et al., 2013; Valdivia et al., 2011). In the absence of Wnt-dependent inhibition of FGF signaling, the remaining cells in the PLLp establish multiple FGF signaling centers to form 2–3 terminal neuromasts. Though the precise pattern of neuromast deposition along its journey is somewhat variable, the PLLp reproducibly ends its migration at the tip of the tail. Its termination correlates with the PLLp reaching the end of the path that guides its migration, as discussed below. However, there appear to be additional mechanisms that ensure the demise of the Wnt system precisely when it reaches the tip of the tail.

Attenuation of Wnt signaling in the PLLp results in premature termination of the PLLp system. When the PLLp is exposed to the Wnt inhibitor IWR-1, for example, the *lef1* expression domain is smaller and disappears prematurely (Matsuda et al., 2013). Similarly, premature termination is seen when *lef1* function is lost (Valdivia et al., 2011). Premature termination is not as prominent, however, in *rspo3* morphants that have a reduction in proliferation similar to that in *lef1* morphants (Matsuda et al., 2013). This suggests premature termination is not simply due to the PLLp running out of cells due to reduced proliferation. Rather, Lef1 is critical for sustaining an active Wnt system in the migrating PLLp until it migrates to the tip of the tail. The Wnt activity mediated by Lef1 helps to maintain a population of cells that normally contributes to the formation of neuromast and inter-neuromasts cells in the later phase of its migration.

15.10 POLARIZED MIGRATION OF THE PLLp ALONG A PATH DEFINED BY CHEMOKINE SIGNALS

The PLLp migrates from the ear to the tip of the tail following a path defined by the chemokine (C-X-C motif) ligand 12a (Cxcl12a), also known as Stromal cell derived factor 1a (Sdf1a), expressed by muscle pioneer cells along the horizontal myoseptum (David et al., 2002; Li et al., 2004) (Figure 15.6). Although the expression of *cxcl12a* along the horizontal myoseptum determines the path followed by the migrating PLLp, it does not determine the direction of migration. In fact, in *fused somite* mutant embryos, in which the stripe of *cxcl12a* ends prematurely, the PLLp sometimes makes a "U"-turn at the end of the *cxcl12a* path and begins to migrate in the opposite direction, illustrating that there is no inherent directional information in the stripe of *cxcl12a* expression (Haas and Gilmour, 2006). The direction of PLLp migration is, instead, determined by the polarized expression of two chemokine receptors, Cxcr4b and Cxcr7b, in the leading zone and in the trailing zone of the PLLp, respectively (Figure 15.6) (Dambly-Chaudiere et al., 2007; Valentin et al., 2007). Although transcripts for *cxcr4b* are in a leading zone, they overlap a little with *cxcr7b* in the trailing zone, where Cxcr4b protein is also likely to be expressed. The domain of *cxcr4b* expression is relatively constant in the PLLp during a deposition cycle. However, the relative size of the *cxcr7b* expression domain waxes and wanes, as a large fraction of *cxcr7b*-expressing cells in the trailing part of the PLLp is lost every time a neuromast is deposited by the migrating PLLp (Aman et al., 2011).

Differences in the function of these two chemokine receptors are key to the polarized behavior of the PLLp. Cxcl12a is thought to interact with the Cxcr4b receptor and engage the G protein-coupled signaling pathway to stabilize protrusive activity in the direction in which the cells encounter sufficiently high levels of Cxcl12a. Cxcr7b, on the other hand, has been described as a decoy receptor: it binds Cxcl12a, but does not have intracellular components that allow it to engage the G protein-coupled signaling pathway to promote protrusive activity (Naumann et al., 2010; Rajagopal et al., 2010). However, the complex of Cxcl12a bound to Cxcr7b is internalized and Cxcl12a is degraded in the lysosome. This leads to a local depletion of Cxcl12a, which ensures that leading cells, expressing Cxcr4b and responsive to Cxcl12a, always encounter higher levels of Cxcl12a in the leading direction of the PLLp. At the same time, Cxcr7b-mediated depletion of Cxcl12a is likely to ensure that Cxcr4b receptors, also expressed by trailing cells, are not exposed to high enough levels of Cxcl12a to determine movement in the opposite direction.

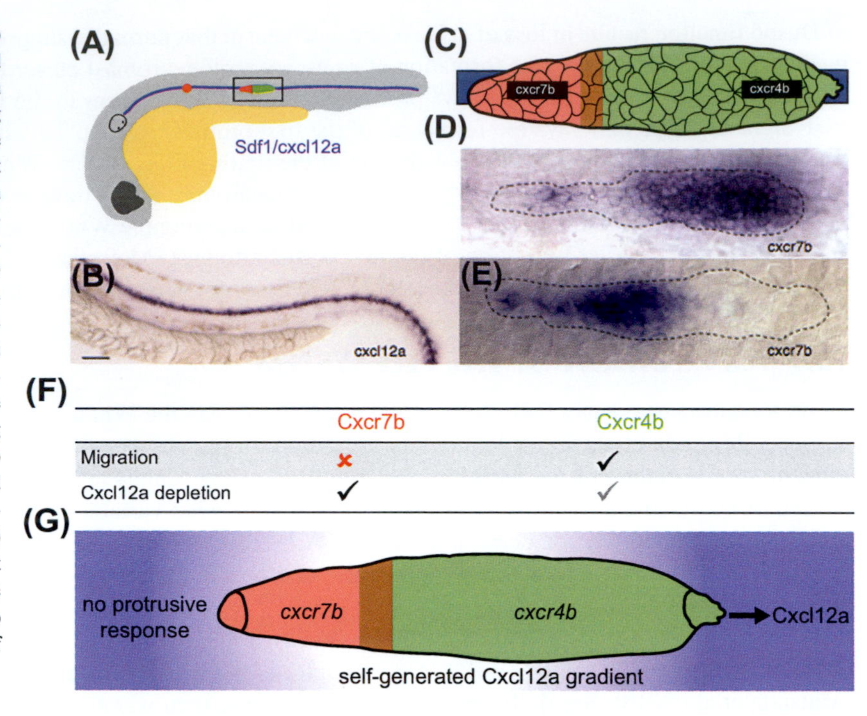

FIGURE 15.6 The Polarized Expression of Cxcr4b and Cxcr7b in the PLLp Determines Unidirectional Migration of the PLLp Along a Relatively Uniform Stripe of cxcl12a Expression. (A) Schematic showing migration of the PLLp (green and red) along a stripe of *cxcl12a* expression (blue). A deposited neuromast is shown in red. (B) Whole mount *in situ* hybridization (WISH) photomicrograph of *cxcl12a* along the horizontal myoseptum in the trunk and tail of an approximately 32 hpf embryo. (C) Schematic showing *cxcr4b* (green) and *cxcr7b* (red) expression in leading and trailing zones of the PLLp, respectively. Overlap of transcripts shown in brown. Cxcr4b receptor expression is expected to extend to the trailing end. (D, E) WISH photomicrographs of *cxcr4b* and *cxcr7b* transcripts in the PLLp (outlined with dotted line). (F) Table comparing function of Cxcr4b and Cxcr7b receptors. Only Cxcr4b can mediate migratory behavior. Both Cxcr4b and Cxcr7b are likely to deplete Cxcl12a from the environment, though depletion is primarily due to Cxcr7b. (G) Local depletion of Cxcl12a generates a local gradient of Cxcl12a. Leading Cxcr4b expressing cells mediate a migratory response to Cxcl12a, while trailing Cxcr7b cells prevent access to Cxcl12a, reducing the ability of trailing cells to respond with migratory behavior.

The self-generated gradient of Cxcl12a has recently been "visualized" using an elegant Cxcr4b lifetime reporter line (Dona et al., 2013)*. This reporter line takes advantage of the fact that the Cxcr4b receptor is internalized after it binds to Cxcl12a, and higher levels of Cxcl12a in the environment are expected to correlate with higher levels of Cxcr4b internalization. The reporter line uses a "tandem fluorescent protein timer (TFt)" in which the Cxcr4b receptor is tagged with two tandem fluorescent fusion proteins; a fast maturing green fluorescent protein and a slower maturing red fluorescent protein. The ratio of red to green fluorescence correlates with the age of the Cxcr4b-TFt receptor, and its observed ratio along the length of the primordium provides a measure of the average lifetime of the Cxcr4b-TFt receptor along the length of the system. Cxcr4b-TFt is internalized and degraded following its interaction with Cxcl12a. This leaves a larger fraction of recently synthesized, primarily green, Cxcr4b-TFt receptor in the cell, which in turn reduces the ratio of red to green fluorescence, called the "lifetime ratio." In domains where the Cxcr4b- TFt receptor is not turned over following its interaction with Cxcl12a, both green and red fluorescent proteins have time to mature and the ratio of red to green fluorescence, the "lifetime ratio," is relatively high. In this manner the lifetime ratio is inversely proportionate to the level of Cxcl12a encountered by Cxcr4b receptors in the PLLp. Measurement of this ratio confirms that the highest levels of Cxcl12a are encountered by Cxcr4b receptors at the leading end of PLLp, where there is most turnover of the Cxcr4b receptor, while Cxcr4b in trailing cells encounter the least amount of Cxcl12a. The study also demonstrates that Cxcr7b in trailing cells plays a key role in determining the amount of Cxcl12a available for interaction with Cxcr4b receptors (Dona et al., 2013). However, the study leaves open an important question: Is the primary role of Cxcr7b in trailing cells to (1) generate a local gradient of Cxcl12a that determines the direction of migration of leading Cxcr4b expressing cells or (2) to polarize the response of the PLLp to Cxcl12a in its path, by ensuring that Cxcr4b receptors in trailing cells are unable to bind Cxcl12a and determine protrusive behavior in the opposite direction?

While it has been emphasized that interaction of Cxcl12a with Cxcr7b in trailing cells results in local internalization and degradation of Cxcl12a, interaction of Cxcl12a with Cxcr4b in the leading zone also results in internalization of the bound ligand receptor complex, which results in a transient depletion of Cxcl12a from the path over which the PLLp has migrated. Hence it remains possible that Cxcr4b-mediated Cxcl12a internalization in the leading zone contributes to the shape and length of the self-generated gradient and ensures that cells at the leading end always see higher levels of Cxcl12a ahead in their direction of migration.

Though *cxcr4b* and *cxcr7b* are expressed broadly in leading and trailing domains, cell transplantation experiments have shown that their function is critical only at the edges of the PLLp. For example, in *cxcr4b* mutant embryos, polarized migratory behavior resumes only when a wild type cell transplanted into the *cxcr4b* mutant PLLp finds its way to the leading edge of the PLLp (Haas and Gilmour, 2006). Similarly, rescue of a *cxcr7b* mutant PLLp is achieved when a wild type cell finds its way to the trailing end (Valentin et al., 2007). This suggests that while *cxcr4b* cells have the potential to steer the PLLp, they cannot do so unless they are at the leading edge. Similarly, the *cxcr7b* expressing cells need to be at the trailing end before they can effectively polarize the migratory response of the PLLp.

Although Cxcr7b does not engage the G protein-coupled signaling pathway it promotes ß-arrestin dependent signaling, which may contribute to the effective recycling of the Cxcr7b receptor (Decaillot et al., 2011; Luker et al., 2010; Rajagopal et al., 2010). This is likely to facilitate persistent clearing of Cxcl12a without the receptor itself being depleted. It remains possible that Cxcr7b-ß arrestin dependent signaling contributes in additional ways to ensuring effective polarized collective migration of the PLLp.

While the polarized expression of chemokine receptors steers the unidirectional migration of the PLLp, the collective migration of the approximately 100 PLLp cells involves active migration of the trailing cells in a direction dictated by the leading cells. Although trailing cells are epithelialized and have apical basal polarity, they extend cryptic basal lamellipodia directed toward the leading cells (Farooqui and Fenteany, 2005; Lecaudey et al., 2008). The mechanism that directs this active migration remains poorly understood. When *cxcr4b* mutant cells are transplanted into a wild type PLLp they migrate normally with adjacent wild type cells. This suggests that chemokine signaling mediated by Cxcr4b is not required for migration of trailing cells. However, it is likely to involve FGF signaling since inhibition of FGF signaling using SU5402 results in less polarized extension of lamellipodia (Lecaudey et al., 2008).

15.11 REGULATION OF NEUROMAST SPACING

A number of studies have identified conditions in which the pattern of cell deposition is altered; however, many aspects of the mechanisms that regulate cell deposition remain obscure. As is to be expected, the pattern of neuromast deposition is related to the periodic generation of neuromasts and to the speed of PLLp migration. In *fgf10* morphants the establishment of the FGF signaling system at the trailing end is delayed, and this delays formation of stable protoneuromasts, which in turn is reflected by a delay in the deposition of the first stable neuromast (Matsuda et al., 2013; Nechiporuk and Raible, 2008). Similarly, in *lef1* mutants it has been shown that the closer spacing of especially the last few neuromasts correlates with slower migration of the PLLp, which reduces spacing between deposited neuromasts (Matsuda et al., 2013; Valdivia et al., 2011). However, in both of these situations it is not clear that the actual process of cell deposition is significantly altered. For example, in *fgf10* morphants although there is a delay in deposition of the first neuromast, the PLLp, whose migration does not appear to be significantly affected by loss of one of the two FGF ligands expressed in the PLLp, continuously deposits interneuromast cells.

One study suggests that proliferation within the PLLp is a key determinant of the pattern of neuromast deposition (Aman et al., 2011). This study suggests that the PLLp can be divided into two zones based on the pattern of *cxcr7b* expression:

a) a leading migratory zone characterized by absence of *cxcr7b* expression, whose size, it was suggested, remains relatively constant in the context of the cycle of neuromast deposition, and
b) a trailing zone with *cxcr7b* expression, associated with depositing or deposited cells.

As the PLLp lengthens following cell proliferation, cells that were originally in the leading *cxcr7b*-free domain eventually find themselves in the trailing compartment with *cxcr7b* expression, associated with deposition, as some as yet undefined homeostatic mechanism keeps the size of leading *cxcr7b*-free domain relatively constant in size. Hence this model suggests that the rate of proliferation determines how quickly cells in migratory *cxcr7b*-free zone find themselves in the trailing domain with *cxcr7b* expression, and are deposited.

The proliferation-dependent deposition model helps to explain how changes in cell number in the PLLp can influence the pattern of neuromast deposition in specific contexts. For example, it accounts for the wider-spaced deposition of neuromasts in conditions that increase cell death. Apoptosis reduces the rate of PLLp elongation and this slows the entry of cells into a trailing zone, where expression of Cxcr7b is associated with cell deposition. However, the proliferation patterns do not always correlate with deposition patterns and correlations are not consistent for all the deposited neuromasts. In *lef1* deficient embryos, in which proliferation is significantly reduced, neuromasts are deposited closer rather than further apart, which is the opposite of what the proliferation-dependent deposition model would predict. In these embryos, however, as noted above, it is the slower migration that appears to account for the closer spacing (Matsuda et al., 2013). On the other hand, in *rspo3* morphants, in which there is a similar reduction in proliferation, it is not clear that neuromast spacing clearly correlates with altered proliferation patterns either (Matsuda et al., 2013).

15.12 CONCLUSIONS

In this chapter we have reviewed mechanisms that determine the self-organization of patterning, morphogenesis and migration in the PLLp system. The relative simplicity of this system allows it to serve as a model for the self-organization of organ systems. By comparing the role of specific signaling pathways in the PLLp to their role in other developmental contexts we learn about fundamental, evolutionarily conserved, molecular mechanisms operating during metazoan development. The establishment of coupled yet mutually antagonistic Wnt-FGF signaling systems is reminiscent of their role in fin, bone, and other developmental contexts (Ambrosetti et al., 2008; Fraser et al., 2013; Stoick-Cooper et al., 2007). On the other hand, similarities in the rules that govern collective migration illustrate how characteristic behaviors emerge when cells follow similar rules, independent of the signaling mechanisms determining these interactions. In both the PLLp and neural crest systems, for example, cells migrate over long distances guided by signaling factors that define their pathway, Cxcl12a in the case of the PLLp system and VEGF in the case of neural crest cells (McLennan et al., 2012); in both contexts local degradation of these factors determines formation of a self-generated gradient which is thought to facilitates directional migration (see also Chapter 18). Similarly, in both contexts trailing cells play follow-the-leader as their active migration is facilitated by additional factors secreted by leading cells (Kasemeier-Kulesa et al., 2010; McLennan et al., 2012; Olesnicky Killian et al., 2009; Theveneau et al., 2010, 2013).

Agent-based computational models have been used to visualize how, when a large number of cells interact following a specific set of simple rules, predictable complex behaviors emerge. Such models have already been built for exploring the collective migration of neural crest cells and epithelial sheets (Bindschadler and McGrath, 2007; McLennan et al., 2012; Vitorino et al., 2011). They are also being developed in our laboratory to explore both the patterning and morphogenesis of the PLLp, where we expect they will provide a platform to validate existing models, define deficiencies in current models and lead to a better understanding of the cellular strategies that determine similar behaviors in various developmental contexts.

Together these studies illustrate how the PLLp system in zebrafish serves as an extraordinary system for exploring the mechanisms that determine the self-organization of fate and behavior in developing organ systems.

15.13 CLINICAL RELEVANCE

- Mechanisms that determine the formation of sensory hair cell progenitors in the lateral line system are remarkably similar to those that determine formation of the sensory hair cells in human ears.
- Sensory hair cells in neuromasts and in our ears are susceptible to damage from similar environmental insults. When sensory hair cells in our ear die, they cannot be regenerated; however, sensory hair cells in neuromasts are constantly being turned over and are remarkably good at regeneration. Understanding the mechanisms that determine this regeneration is likely to define strategies that will facilitate regeneration of sensory hair cells in our ears.
- Interactions between Wnt and FGF signaling regulate the pattern of growth and differentiation in a wide range of tissues. Dysfunction of these control mechanisms can lead to uncontrolled growth, as in cancer. The PLLp system provides an excellent platform for understanding how interactions between these signaling systems are regulated to determine pattern formation in developing tissues.
- The mechanisms that regulate collective migration of PLLp cells are very similar to those determining migration of metastatic cancer cells. The ability to observe and manipulate the reproducible migratory behavior of the PLLp makes it a particularly attractive model system for developing strategies for preventing or controlling metastatic behavior.

RECOMMENDED RESOURCES

The Wnt Pathway: http://en.wikipedia.org/wiki/Wnt_signaling_pathway (last modified 31st Jan 2014).

The FGF signaling pathway: http://en.wikipedia.org/wiki/Fibroblast_growth_factor (last modified 21st Jan 2014).

Reaction-diffusion systems and Activator inhibitor systems http://www.eb.tuebingen.mpg.de/research/emeriti/hans-meinhardt/home.html (last modified 21st Jan 2014).

Chemokines http://en.wikipedia.org/wiki/Chemokine (last updated March 24th 2014).

REFERENCES

Aman, A., Nguyen, M., Piotrowski, T., 2011. Wnt/β-catenin dependent cell proliferation underlies segmented lateral line morphogenesis. Dev. Biol. 349, 470–482.

Aman, A., Piotrowski, T., 2008. Wnt/β-catenin and Fgf signaling control collective cell migration by restricting chemokine receptor expression. Dev. cell 15, 749–761.

Ambrosetti, D., Holmes, G., Mansukhani, A., Basilico, C., 2008. Fibroblast growth factor signaling uses multiple mechanisms to inhibit Wnt-induced transcription in osteoblasts. Mol. Cell Biol. 28, 4759–4771.

Behra, M., Bradsher, J., Sougrat, R., Gallardo, V., Allende, M.L., Burgess, S.M., 2009. Phoenix is required for mechanosensory hair cell regeneration in the zebrafish lateral line. PLoS genetics 5, e1000455.

Bindschadler, M., McGrath, J.L., 2007. Sheet migration by wounded monolayers as an emergent property of single-cell dynamics. J.cell sci. 120, 876–884.

Brooker, R., Hozumi, K., Lewis, J., 2006. Notch ligands with contrasting functions: Jagged1 and Delta1 in the mouse inner ear. Development (Cambridge, England) 133, 1277–1286.

Chitnis, A.B., 1995. The role of Notch in lateral inhibition and cell fate specification. Mol. and Cell. Neurosci. 6, 311–321.

Chitnis, A.B., Nogare, D.D., Matsuda, M., 2012. Building the posterior lateral line system in zebrafish. Dev. Neurobiol. 72, 234–255.

Coffin, A.B., Ou, H., Owens, K.N., Santos, F., Simon, J.A., Rubel, E.W., Raible, D.W., 2010. Chemical screening for hair cell loss and protection in the zebrafish lateral line. Zebrafish 7, 3–11.

Coombs, S., van Netten, S.M., 2006. The hydrodynamics and structural mechanics of the Lateral line system. In: Shadwick, R.E., Lauder, G.V. (Eds.), Fish Biomechanics. Elsevier, New York, pp. 103–139.

Dambly-Chaudiere, C., Cubedo, N., Ghysen, A., 2007. Control of cell migration in the development of the posterior lateral line: antagonistic interactions between the chemokine receptors CXCR4 and CXCR7/RDC1. BMC Dev. Biol. 7, 23.

David, N.B., Sapede, D., Saint-Etienne, L., Thisse, C., Thisse, B., Dambly-Chaudiere, C., Rosa, F.M., Ghysen, A., 2002. Molecular basis of cell migration in the fish lateral line: role of the chemokine receptor CXCR4 and of its ligand, SDF1. Proc. Natl. Acad. Sci. U.S.A. 99, 16297–16302.

Decaillot, F.M., Kazmi, M.A., Lin, Y., Ray-Saha, S., Sakmar, T.P., Sachdev, P., 2011. CXCR7/CXCR4 heterodimer constitutively recruits β-arrestin to enhance cell migration. 2011 Sep 16 J. Biol. Chem. 286 (37), 32188–32197.

Dijkgraaf, S., 1963. The functioning and significance of the lateral-line organs. Biol. Rev. Camb. Philos. Soc. 38, 51–105.

Dona, E., Barry, J.D., Valentin, G., Quirin, C., Khmelinskii, A., Kunze, A., Durdu, S., Newton, L.R., Fernandez-Minan, A., Huber, W., et al., 2013. Directional tissue migration through a self-generated chemokine gradient. 2013 Nov 14 Nature 503 (7475), 285–289.

Dorsky, R.I., Sheldahl, L.C., Moon, R.T., 2002. A transgenic Lef1/β-catenin-dependent reporter is expressed in spatially restricted domains throughout zebrafish development. Dev. Biol. 241, 229–237.

Dufourcq, P., Roussigne, M., Blader, P., Rosa, F., Peyrieras, N., Vriz, S., 2006. Mechano-sensory organ regeneration in adults: the zebrafish lateral line as a model. Mol. Cell. Neurosci. 33, 180–187.

Ernst, S., Liu, K., Agarwala, S., Moratscheck, N., Avci, M.E., Dalle Nogare, D., Chitnis, A.B., Ronneberger, O., Lecaudey, V., 2012. Shroom3 is required downstream of FGF signaling to mediate proneuromast assembly in zebrafish. Development (Cambridge, England) 139, 4571–4581.

Farooqui, R., Fenteany, G., 2005. Multiple rows of cells behind an epithelial wound edge extend cryptic lamellipodia to collectively drive cell-sheet movement. J. cell Sci. 118, 51–63.

Faucherre, A., Pujol-Marti, J., Kawakami, K., Lopez-Schier, H., 2009. Afferent neurons of the zebrafish lateral line are strict selectors of hair-cell orientation. PloS one 4, e4477.

Fraser, G.J., Bloomquist, R.F., Streelman, J.T., 2013. Common developmental pathways link tooth shape to regeneration. Dev. Biol. 377, 399–414.

Friedl, P., Gilmour, D., 2009. Collective cell migration in morphogenesis, regeneration and cancer. Nat. Rev. 10, 445–457.

Gallardo, V.E., Liang, J., Behra, M., Elkahloun, A., Villablanca, E.J., Russo, V., Allende, M.L., Burgess, S.M., 2010. Molecular dissection of the migrating posterior lateral line primordium during early development in zebrafish. BMC Dev. Biol. 10, 120.

Gamba, L., Cubedo, N., Lutfalla, G., Ghysen, A., Dambly-Chaudiere, C., 2010. Lef1 controls patterning and proliferation in the posterior lateral line system of zebrafish. Dev. Dyn. 239, 3163–3171.

Germana, A., Laura, R., Montalbano, G., Guerrera, M.C., Amato, V., Zichichi, R., Campo, S., Ciriaco, E., Vega, J.A., 2010. Expression of brain-derived neurotrophic factor and TrkB in the lateral line system of zebrafish during development. Cell.Mol. Neurobiol. 30, 787–793.

Ghysen, A., Dambly-Chaudiere, C., 2007. The lateral line microcosmos. Genes & development 21, 2118–2130.

Gilmour, D.T., Maischein, H.M., Nusslein-Volhard, C., 2002. Migration and function of a glial subtype in the vertebrate peripheral nervous system. Neuron 34, 577–588.

Gompel, N., Cubedo, N., Thisse, C., Thisse, B., Dambly-Chaudiere, C., Ghysen, A., 2001. Pattern formation in the lateral line of zebrafish. Mech. Dev. 105, 69–77.

Grant, K.A., Raible, D.W., Piotrowski, T., 2005. Regulation of latent sensory hair cell precursors by glia in the zebrafish lateral line. Neuron 45, 69–80.

Haas, P., Gilmour, D., 2006. Chemokine signaling mediates self-organizing tissue migration in the zebrafish lateral line. Deve. cell 10, 673–680.

Haddon, C., Jiang, Y.J., Smithers, L., Lewis, J., 1998. Delta-Notch signaling and the patterning of sensory cell differentiation in the zebrafish ear: evidence from the mind bomb mutant. Development (Cambridge, England) 125, 4637–4644.

Harris, J.A., Cheng, A.G., Cunningham, L.L., MacDonald, G., Raible, D.W., Rubel, E.W., 2003. Neomycin-induced hair cell death and rapid regeneration in the lateral line of zebrafish (*Danio rerio*). J. Assoc. Res. Otolaryngol. 4, 219–234.

Hernandez, P.P., Moreno, V., Olivari, F.A., Allende, M.L., 2006. Sub-lethal concentrations of waterborne copper are toxic to lateral line neuromasts in zebrafish (*Danio rerio*). Hear. Res. 213, 1–10.

Hernandez, P.P., Olivari, F.A., Sarrazin, A.F., Sandoval, P.C., Allende, M.L., 2007. Regeneration in zebrafish lateral line neuromasts: expression of the neural progenitor cell marker sox2 and proliferation-dependent and-independent mechanisms of hair cell renewal. Dev. Neurobiol. 67, 637–654.

Itoh, M., Chitnis, A.B., 2001. Expression of proneural and neurogenic genes in the zebrafish lateral line primordium correlates with selection of hair cell fate in neuromasts. Mech. Dev. 102, 263–266.

Kasemeier-Kulesa, J.C., McLennan, R., Romine, M.H., Kulesa, P.M., Lefcort, F., 2010. CXCR4 controls ventral migration of sympathetic precursor cells. J. Neurosci. 30, 13078–13088.

Kim, K.H., Park, H.J., Kim, J.H., Kim, S., Williams, D.R., Kim, M.K., Jung, Y.D., Teraoka, H., Park, H.C., Choy, H.E., et al., 2013. Cyp1a reporter zebrafish reveals target tissues for dioxin. Aquat. Toxicol. (Amsterdam, Netherlands) 134–135, 57–65.

Kimmel, C.B., Ballard, W.W., Kimmel, S.R., Ullmann, B., Schilling, T.F., 1995. Stages of embryonic development of the zebrafish. Dev. Dyn. 203, 253–310.

Lecaudey, V., Cakan-Akdogan, G., Norton, W.H., Gilmour, D., 2008. Dynamic Fgf signaling couples morphogenesis and migration in the zebrafish lateral line primordium. Development (Cambridge, England) 135, 2695–2705.

Li, Q., Shirabe, K., Kuwada, J.Y., 2004. Chemokine signaling regulates sensory cell migration in zebrafish. Dev. Biol. 269, 123–136.

Lopez-Schier, H., Hudspeth, A.J., 2005. Supernumerary neuromasts in the posterior lateral line of zebrafish lacking peripheral glia. Proc. Natl. Acad. Sci. U. S. A. 102, 1496–1501.

Luker, K.E., Steele, J.M., Mihalko, L.A., Ray, P., Luker, G.D., 2010. Constitutive and chemokine-dependent internalization and recycling of CXCR7 in breast cancer cells to degrade chemokine ligands. Oncogene 29, 4599–4610.

Matsuda, M., Chitnis, A.B., 2010. Atoh1a expression must be restricted by Notch signaling for effective morphogenesis of the posterior lateral line primordium in zebrafish. Development (Cambridge, England) 137, 3477–3487.

Matsuda, M., Nogare, D.D., Somers, K., Martin, K., Wang, C., Chitnis, A.B., 2013. Lef1 regulates Dusp6 to influence neuromast formation and spacing in the zebrafish posterior lateral line primordium. Development (Cambridge, England) 140, 2387–2397.

McDermott Jr., B.M., Asai, Y., Baucom, J.M., Jani, S.D., Castellanos, Y., Gomez, G., McClintock, J.M., Starr, C.J., Hudspeth, A.J., 2010. Transgenic labeling of hair cells in the zebrafish acousticolateralis system. Gene. Expr. Patterns 10, 113–118.

McLennan, R., Dyson, L., Prather, K.W., Morrison, J.A., Baker, R.E., Maini, P.K., Kulesa, P.M., 2012. Multiscale mechanisms of cell migration during development: theory and experiment. Development (Cambridge, England) 139, 2935–2944.

Meinhardt, H., 2012. Turing's theory of morphogenesis of 1952 and the subsequent discovery of the crucial role of local self-enhancement and long-range inhibition. Interface Focus 2, 407–416.

Meinhardt, H., Gierer, A., 2000. Pattern formation by local self-activation and lateral inhibition. Bioessays 22, 753–760.

Millimaki, B.B., Sweet, E.M., Dhason, M.S., Riley, B.B., 2007. Zebrafish atoh1 genes: classic proneural activity in the inner ear and regulation by Fgf and Notch. Development (Cambridge, England) 134, 295–305.

Mizoguchi, T., Togawa, S., Kawakami, K., Itoh, M., 2011. Neuron and sensory epithelial cell fate is sequentially determined by Notch signaling in zebrafish lateral line development. J. Neurosci. 31, 15522–15530.

Molina, G.A., Watkins, S.C., Tsang, M., 2007. Generation of FGF reporter transgenic zebrafish and their utility in chemical screens. BMC Dev. Biol. 7, 62.

Moon, I.S., So, J.H., Jung, Y.M., Lee, W.S., Kim, E.Y., Choi, J.H., Kim, C.H., Choi, J.Y., 2011. Fucoidan promotes mechanosensory hair cell regeneration following amino glycoside-induced cell death. Hear. Res. 282, 236–242.

Naumann, U., Cameroni, E., Pruenster, M., Mahabaleshwar, H., Raz, E., Zerwes, H.G., Rot, A., Thelen, M., 2010. CXCR7 functions as a scavenger for CXCL12 and CXCL11. PloS one 5, e9175.

Nechiporuk, A., Raible, D.W., 2008. FGF-dependent mechanosensory organ patterning in zebrafish. Science (New York, NY 320, 1774–1777.

Olesnicky Killian, E.C., Birkholz, D.A., Artinger, K.B., 2009. A role for chemokine signaling in neural crest cell migration and craniofacial development. Dev. Biol. 333, 161–172.

Rajagopal, S., Kim, J., Ahn, S., Craig, S., Lam, C.M., Gerard, N.P., Gerard, C., Lefkowitz, R.J., 2010. β-arrestin- but not G protein-mediated signaling by the "decoy" receptor CXCR7. Proc. Natl. Acad. Sci. U.S.A. 107, 628–632.

Sarrazin, A.F., Villablanca, E.J., Nunez, V.A., Sandoval, P.C., Ghysen, A., Allende, M.L., 2006. Proneural gene requirement for hair cell differentiation in the zebrafish lateral line. Dev. Biol. 295, 534–545.

Schlosser, G., 2010. Making senses development of vertebrate cranial placodes. Int. Rev. Cell Mol. Biol. 283, 129–234.

Shimizu, N., Kawakami, K., Ishitani, T., 2012. Visualization and exploration of Tcf/Lef function using a highly responsive Wnt/β-catenin signaling-reporter transgenic zebrafish. Dev. Biol. 370, 71–85.

Stoick-Cooper, C.L., Weidinger, G., Riehle, K.J., Hubbert, C., Major, M.B., Fausto, N., Moon, R.T., 2007. Distinct Wnt signaling pathways have opposing roles in appendage regeneration. Development (Cambridge, England) 134, 479–489.

Streit, A., 2007. The preplacodal region: an ectodermal domain with multipotential progenitors that contribute to sense organs and cranial sensory ganglia. Int. J. Dev. Biol. 51, 447–461.

Theveneau, E., Marchant, L., Kuriyama, S., Gull, M., Moepps, B., Parsons, M., Mayor, R., 2010. Collective chemotaxis requires contact-dependent cell polarity. Dev. cell 19, 39–53.

Theveneau, E., Steventon, B., Scarpa, E., Garcia, S., Trepat, X., Streit, A., Mayor, R., 2013. Chase-and-run between adjacent cell populations promotes directional collective migration. Nature cell Biol 15, 763–772.

Tsang, M., Friesel, R., Kudoh, T., Dawid, I.B., 2002. Identification of Sef, a novel modulator of FGF signaling. Nature cell biology 4, 165–169.

Tsang, M., Maegawa, S., Kiang, A., Habas, R., Weinberg, E., Dawid, I.B., 2004. A role for MKP3 in axial patterning of the zebrafish embryo. Development (Cambridge, England) 131, 2769–2779.

Valdivia, L.E., Young, R.M., Hawkins, T.A., Stickney, H.L., Cavodeassi, F., Schwarz, Q., Pullin, L.M., Villegas, R., Moro, E., Argenton, F., et al., 2011. Lef1-dependent Wnt/β-catenin signaling drives the proliferative engine that maintains tissue homeostasis during lateral line development. Development (Cambridge, England) 138, 3931–3941.

Valentin, G., Haas, P., Gilmour, D., 2007. The chemokine SDF1a coordinates tissue migration through the spatially restricted activation of Cxcr7 and Cxcr4b. Curr. Biol. 17, 1026–1031.

Vitorino, P., Hammer, M., Kim, J., Meyer, T., 2011. A steering model of endothelial sheet migration recapitulates monolayer integrity and directed collective migration. Mol. Cell Biol. 31, 342–350.

Wibowo, I., Pinto-Teixeira, F., Satou, C., Higashijima, S., Lopez-Schier, H., 2011. Compartmentalized Notch signaling sustains epithelial mirror symmetry. Development (Cambridge, England) 138, 1143–1152.

Williams, J.A., Holder, N., 2000. Cell turnover in neuromasts of zebrafish larvae. Hear. Res. 143, 171–181.

Xiao, T., Roeser, T., Staub, W., Baier, H., 2005. A GFP-based genetic screen reveals mutations that disrupt the architecture of the zebrafish retinotectal projection. Development (Cambridge, England) 132, 2955–2967.

NOTE ADDED IN PROOF

Elegant evidence for a self-generated Cxcl12a signaling gradient has also been provided by Venkiteswaran et al., 2013. Venkiteswaran G, Lewellis SW, Wang J, Reynolds E, Nicholson C, Knaut H. Generation and dynamics of an endogenous, self-generated signaling gradient across a migrating tissue. Cell. 2013 Oct 24;155(3):674–687

Section III

Organogenesis

Chapter 16

Neural Cell Fate Determination

Steven Moore and Frederick J. Livesey
Gurdon Institute, Department of Biochemistry and Cambridge Stem Cell Institute, University of Cambridge, Cambridge, UK

Chapter Outline

GLOSSARY

Cell fate The neuronal subtype that a progenitor will become. This term does not imply commitment or differentiation, only that the cell will eventually become a certain type.

Commitment An irreversible decision to produce or become a particular cell type. This is defined operationally as the inability of a cell to change its fate when exposed to different signaling cues.

Competence The ability of a cell to respond to a cue or set of cues to produce a defined outcome.

Differentiation The elaboration of particular characteristics expressed by an end-stage cell type, or by a cell en route to becoming an end-stage cell. This term is not synonymous with commitment, but differentiation features are used to determine when a cell is committed.

Gene regulatory network The interactions between transcription factors and their targets that govern developmental processes.

Multipotent The ability of a progenitor to generate more than one class of terminally differentiated cells.

Progenitor cell A mitotic cell that is not capable of indefinite self-renewal and which will produce a limited repertoire of cell types.

Proneural gene A basic helix-loop-helix transcription factor that is both necessary and sufficient to initiate the development of neuronal lineages and to promote the generation of progenitor cells which are committed to neuronal differentiation.

Specification The cell intrinsic integration of both extra and intracellular patterning cues to determine a specific transcriptional identity.

SUMMARY

- Proneural genes encode transcription factors, which are responsible for initiating the development of neuronal lineages by promoting the generation of neural progenitor cells.
- Neuronal differentiation genes encode transcription factors which function downstream of proneural genes and are responsible for neural cell fate determination and differentiation.
- Neural progenitor cells acquire a distinct spatial identity first through the division of the nascent central nervous system (CNS) into discrete territories and regions, and then through the spatial patterning of individual regions.
- Within each region, neural progenitor cells are multipotent and give rise to different types of neurons over time, through the process of asymmetric division.
- The competence of neural progenitor cells to generate different neuronal progeny changes over time due to the action of both intrinsic and extrinsic mechanisms.

Principles of Developmental Genetics.

- The principles of embryonic developmental neurobiology can be applied to the *in vitro* study of neurogenesis and the cell biology underlying neurodegenerative disease.

16.1 INTRODUCTION

The mature nervous system in any organism contains many distinct types of neurons, as defined by their morphology, connectivity, neurotransmitter phenotype, and electrophysiological properties. Neurogenesis, the process by which postmitotic neurons are generated from pools of mitotic progenitor cells, is a highly regulated process (Edlund and Jessell, 1999; Livesey and Cepko, 2001). Different classes of neurons are produced in a temporal sequence that is conserved in different species, and each region of the nervous system is composed of specific neural subtypes (Cepko et al., 1996). Discrete phenotypes or identities are assigned to the postmitotic progeny of neural progenitor cells through the process of cell fate determination. To a significant degree, the fates of those progeny appear to be decided within mitotic progenitors before cell cycle exit. Thus, progenitor cells have an integrative function, whereby they combine extrinsic information in the form of extracellular signals with information intrinsic to the cell to decide the fates of their daughter cells, as will be discussed in more detail below. Terminal differentiation of those postmitotic, neuronal daughter cells is implemented by transcriptional networks, which have been shown to exhibit a degree of plasticity by recent studies (De la Rossa et al., 2013; Rouaux and Arlotta, 2013). This property of neuronal differentiation is one aspect of development that can be manipulated to generate specific cell types *in vitro* for research and therapeutic purposes.

The developmental biology of neural cell fate determination can be broadly divided into a series of processes: the induction or appearance of neurogenic tissue(s) – that is tissue containing neural stem and progenitor cells (see also Chapter 11); the division of this tissue into distinct territories or regions that go on to form different components of the adult nervous system; and the ordered production of region-specific neurons within each territory. Several striking recent studies have clearly shown that this process can be recapitulated *in vitro,* generating particular classes of neurons from embryonic stem (ES) cells through a series of discrete steps aimed at guiding cells through each stage in this process (Kim et al., 2002; Wichterle et al., 2002). Furthermore, understanding of the transcriptional control of neuronal differentiation and terminal neuronal identity has been used to directly transdifferentiate somatic cells into neurons, as well as to alter neuronal identities *in vivo*. We review here the principles by which neural cell fates are determined, using selected examples to highlight fundamental processes and concepts, and the application of those principles for *ex vivo* studies of human neuronal development and neurological disease modeling.

16.2 FUNDAMENTALS OF NEUROGENESIS

All of the neurons that comprise the nervous system are generated by multipotent neural stem or progenitor cells (Brand and Livesey, 2011). For example, after delaminating from the neuroectoderm, *Drosophila* neuroblasts undergo a series of apically/basally orientated asymmetric divisions. These asymmetric divisions give rise to a smaller daughter cell, the ganglion mother cell (GMC), which buds off from the dorsal/lateral cortex of the self renewing neuroblast. GMCs then divide terminally to give rise to two neurons or glia.

By comparison, in the mammalian cerebral cortex, neural progenitor cells of the ventricular zone (VZ) give rise to the outer radial layers, which are composed of differentiated neurons of distinct identities. This process of layer formation in the cerebral cortex is known to involve the asymmetric division of neural progenitor cells in the VZ followed by the outward migration of postmitotic neurons. It is thought that the formation of the different radial layers is due to changes in, and the restriction of, progenitor cell competence in the VZ over time (Desai and McConnell, 2000).

Some basic principles of nervous system development are common throughout the brain: neural progenitor cells are multipotent, generating multiple classes of neurons; there is a temporal order to the generation of different neuronal types; and progenitor cells give rise to neurons before generating glia. As discussed below, the production of specific cell types is encoded within the progenitor cells, in part by early patterning of the neuraxis that gives progenitor cells a spatial or regional identity. For example, retinal progenitor cells are only competent to produce the classes of neurons that are appropriate for the organization of the retina. However, neural progenitor cells can respond to extrinsic factors that regulate their cell fate choices. Cortical progenitors are competent to respond to extrinsic signals until late S/early G2 in the cell cycle (McConnell, 1988), in agreement with the finding that retinal progenitors make cell fate choices prior to M phase (Belliveau and Cepko, 1999).

As in many developing systems, several intercellular signaling systems regulate neural progenitor cell self-renewal and differentiation, including Delta/Notch, Fgfs, TGFβ and Wnt/wingless. Rather than imposing specific cell fates on differentiating neurons, all of these signaling pathways appear to regulate core decisions about self-renewal and neurogenesis. This is best illustrated in the case of Notch signaling, which acts to positively promote self-renewal, and whose inhibition promotes neurogenesis. However, the neurons made in the absence of Notch signaling are typically appropriate to the developmental stage and brain region being studied: in the case of the retina, for example, blocking Notch signaling early in development in the period during which retinal ganglion neurons are being produced results in an increased production of the same cell type, rather than inducing the production of later born neuronal types (Austin et al., 1995).

Now classic work has identified groups of transcription factors that regulate fundamental aspects of neurogenesis, conserved across species. Proneural genes are transcription factors which contain a basic helix-loop-helix (bHLH) domain which confers dimerization and DNA binding properties (Murre et al., 1989). The proneural genes were originally identified in *Drosophila* in the early 1970s as a complex of genes involved in the early stages of neural development (Garcia-Bellido, 1979; Ghysen and Dambly-Chaudiere, 1988). This genomic interval, termed the *achaete-scute complex*, shares homology with the vertebrate achaete-scute like (ASCL) gene family. Further studies identified the *Drosophila atonal* family, which is represented in vertebrate genomes by the atonal homologs (ATOH), neurogenin (Ngn), NeuroD and Olig genes (Bertrand et al., 2002; Jarman et al., 1993). Proneural genes are both necessary and sufficient to initiate the development of neuronal lineages and to promote the generation of committed neural progenitors.

Proneural genes function by binding to DNA as heterodimers with the ubiquitously expressed bHLH 'E' proteins: E2A, HEB and E2–2 in vertebrates, and *daughterless* (*da*) in *Drosophila* (Cabrera and Alonso, 1991; Johnson et al., 1992; Massari and Murre, 2000). The bHLH domain of the proneural genes contains a stretch of ten DNA binding residues, of which nine are conserved between all proneural genes (Bertrand et al., 2002; Chien et al., 1996). These conserved residues recognize the E-box (CANNT) DNA binding site in the regulatory regions of target genes. Most proneural genes function as activators of target gene transcription, with the exception of *Olig*2, which is a repressor (Cabrera and Alonso, 1991; Johnson et al., 1992; Mizuguchi et al., 2001; Novitch et al., 2001). Repression of proneural function can be achieved by disruption of their heterodimerization with the ubiquitous E proteins. The *Drosophila extra macrochaetae* (*emc*) and vertebrate *inhibitor of differentiation* (*Id*) genes possess bHLH domains but lack DNA binding motifs, and are thought to compete with proneural proteins for E proteins, thus inhibiting proneural gene function (Cabrera and Alonso, 1991; Campuzano, 2001; Yokota, 2001). The *Drosophila hairy* and *enhancer of split* (*Espl*), and the vertebrate *hairy and enhancer of split* homolog (*Hes*) (Andersson et al., 2006b), *hairy*, and *enhancer of split related (Her), and enhancer of split related* (*Esr*) genes are transcriptional repressors of proneural genes and are also thought to repress proneural function by the disruption of heterodimer formation (Davis and Turner, 2001; Kageyama and Nakanishi, 1997).

16.3 THE GENERATION OF SPECIFIC CELL TYPES WITHIN THE VERTEBRATE NERVOUS SYSTEM: SPINAL CORD DEVELOPMENT

The developing vertebrate nervous system is initially specified to an anterior identity following gastrulation and is subsequently posteriorized by the graded activities of morphogens expressed in the caudal regions of the neural plate. Anterior neural identities are maintained by the action of secreted antagonists, which limit the posteriorizing activity of morphogens (see Chapter 11). The patterning of neural progenitors across the anterior-posterior axis of the CNS gives rise to domains that are fated to become the forebrain, midbrain, hindbrain, and spinal cord of the adult (Figure 16.1). The highly stereotypical and complex array of neuronal subtypes within each tissue is specified by the coordinated actions of extrinsically secreted ligands, known as morphogens, and intrinsic transcriptional networks. To illustrate the mechanisms by which progenitor cells in the nervous system become regionally specified, and how this integrates with developmental time to produce defined cell types, we will focus on the spinal cord, where a substantial body of work has uncovered many fundamental principles.

Progenitors distributed across the dorsoventral axis of the developing neural tube are divided into 11 molecularly distinct domains by the actions of morphogens, secreted from organizing centers at the dorsal and ventral midlines. The dorsal progenitor domains, known as dP1–6, give rise to specific classes of sensory interneurons within the functional spinal cord (Helms and Johnson, 2003). Similarly, the ventral region of the neural tube is divided into five distinct domains, each of which is directly correlated with the neuronal output of its progenitors (Briscoe et al., 2000). The pMN domain gives rise to spinal motor neurons, whereas the p0, p1, p2, and p3 domains generate the interneurons that coordinate motor activity (Briscoe et al., 2000; Ericson et al., 1997; Pierani et al., 2001; Sharma et al., 1998).

The formation of progenitor domains within the dorsal neural tube is dependent on the activity of the bone morphogenetic protein (BMP) and Wnt signaling pathways, whose ligands are secreted from the roof plate at the dorsal midline (Helms and Johnson, 2003). The mechanisms acting downstream of morphogen signaling to define and maintain specific domains of neural progenitors in the dorsal neural tube are currently poorly understood. By contrast, the

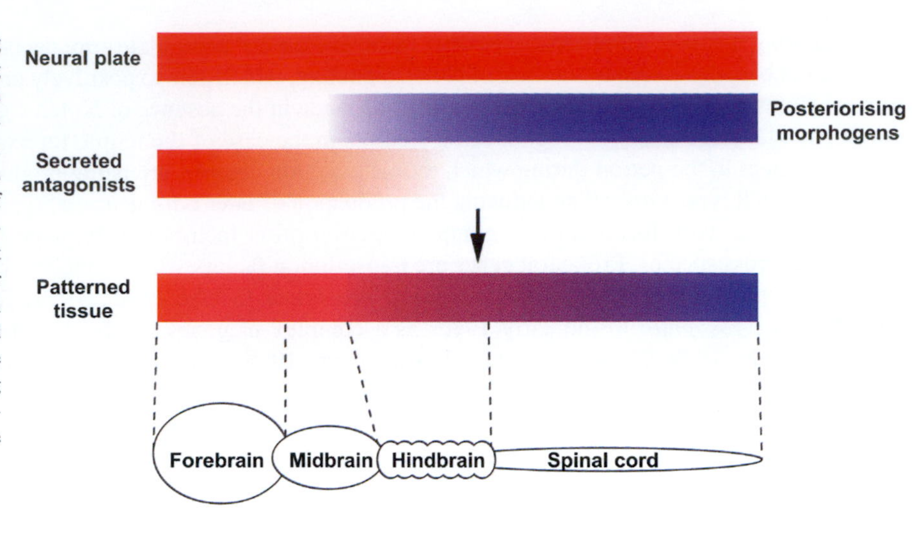

FIGURE 16.1 Anterior-Posterior Patterning of the Central Nervous System. Following gastrulation, the presumptive central nervous system resides dorsally in embryo as a neuroepithelial sheet, known as the neural plate. Neural progenitors initially express markers of an anterior identity, however morphogens of the Wnt, retinoic acid and FGF families posteriorizes the tissue in concentration dependent manner. Secreted antagonists act in the anterior of the tissue to limit the activity of morphogen induced signaling, resulting in a tissue pattered across the anterior-posterior axis. Following the folding and closing of the neural tube during neurulation, the initial patterning of neural progenitors plays a major role in establishing the subdivisions of the central nervous system.

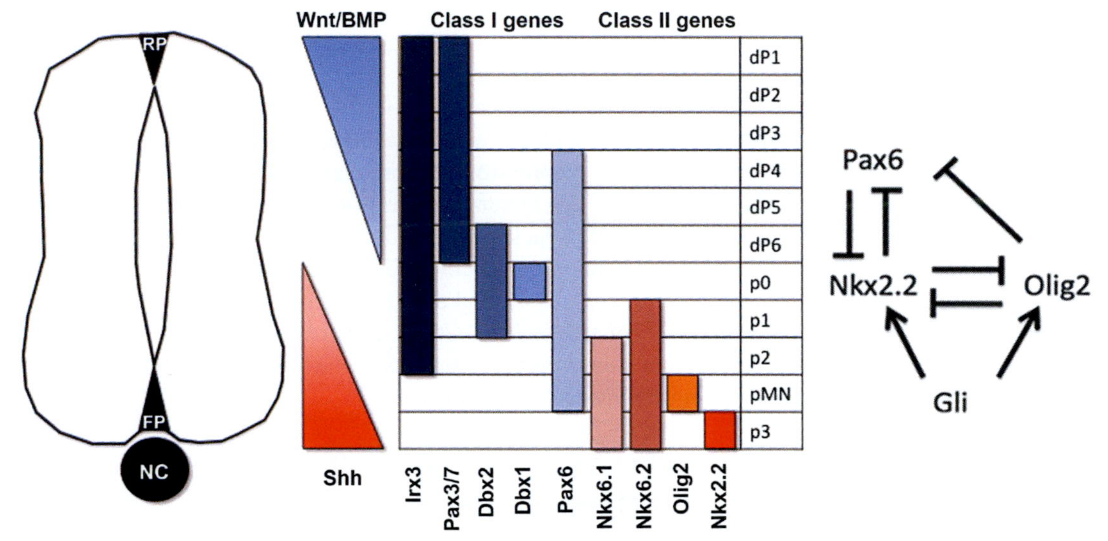

FIGURE 16.2 Patterning the Dorsoventral Axis of the Embryonic Spinal Cord. Neural tube progenitors are exposed to opposing gradients of Wnt/BMP and Shh signaling, whose ligands are respectively secreted from the roof plate (RP) dorsally and both the floor plate (FP) and notochord (NC) ventrally. Morphogen signaling acts to impart cells with a unique transcriptional identity based upon their localization with the dorsoventral axis of the tissue. Class I patterning genes, which are repressed by Shh signaling, are expressed in the dorsal progenitor (dP) domains. Conversely type II genes, induced by activated Shh signaling, define the ventral progenitor domains. The transcriptional profile of progenitors in a given domain is directly related to their neuronal output during neurogenesis. Selective repressive interactions between transcription factors expressed in adjacent domains form the basis of a gene regulatory network, which acts to define and maintain the boundaries between distinct progenitor subtypes. This network of selective repression is best exemplified by Pax6, Olig2, and Nkx2.2, whose interactions not only the segregate the pMN and p3 domains in the ventral neural tube but also enable the cellular interpretation of activated Shh signaling over time.

specification of progenitor identity in the ventral half of the tissue represents the most complete model of vertebrate neural patterning.

The five ventral progenitor domains are specified by a gradient of the signaling molecule sonic hedgehog (Shh), which is secreted first from the notochord and subsequently from the floor plate (Ericson et al., 1996; Roelink et al., 1995). Shh signaling acts to induce the expression of a set of transcription factors, commonly known as Class 2 genes (Nkx2.2/2.9, Olig2, Nkx6.1, and Nkx6.2), in a dose dependent manner within the ventral neural tube (Briscoe et al., 2000). Conversely, dorsally expressed Class 1 genes (Pax6, Irx3, Dbx2, Dbx1, and Pax3/7) are repressed by increasing concentrations of Shh ligand in accordance with their expression domains (Briscoe et al., 2000; Ericson et al., 1996) (Figure 16.2).

The intracellular interpretation of the Shh gradient is dependent on the binding of the ligand to its receptor, patched (ptc). This interaction relieves the repression that ptc exerts upon smoothened (smo), the transducer of the pathway. Smo functions to promote the conversion of Gli family transcription factors from their repressive form to activators. Gli factors subsequently activate the expression of ventral Shh target genes by directly binding to their associated regulatory elements, in combination with general neural transcription factors such as the SoxB1 family (Oosterveen et al., 2012; Peterson et al., 2012). A large subset of Shh regulated genes are de-repressed by a loss of repressive Gli binding in the presence of activated signaling, adding to the complexity of gradient interpretation (Oosterveen et al., 2012; Peterson et al., 2012). In addition to neural patterning genes, activated Gli factors increase the expression of the Shh receptor ptc, which acts to exert negative feedback on the pathway in a process known as temporal adaptation (Dessaud et al., 2007). Thus, the response of progenitors to Shh signaling is dependent on both the extracellular concentration of the ligand and the duration of intracellular pathway activation (Dessaud et al., 2007).

Shh signaling acts to establish a gene regulatory network in the ventral tube, in which transcription factors expressed in adjacent domains exhibit selective repressive interactions (Briscoe et al., 2000; Ericson et al., 1997; Mizuguchi et al., 2001; Novitch et al., 2001; Sander et al., 2000; Vallstedt et al., 2001). This repressive network functions to establish and maintain progenitor identity by enforcing sharp boundaries of gene expression, ensuring the correct distribution of neurons in the functional spinal cord. The best-described motif within the ventral neural tube gene regulatory network involves the genes Pax6, Olig2, and Nkx2.2, which repress each other to define the p2, pMN, and p3 domains respectively (Balaskas et al., 2012; Briscoe et al., 2000; Ericson et al., 1997; Novitch et al., 2001). In addition to establishing and maintaining these three progenitor domains, recent studies have shown that cellular interpretation of Shh signaling in the ventral neural tube is an emergent property of the regulatory network that connects these three transcription factors (Figure 16.2) (Balaskas et al., 2012).

How do Spinal Cord TFs Function in the Specification of Neuronal Identity?

Spinal cord transcription factors (TFs) guide the differentiation of progenitors by repressing the determinants of alternative fates. This is achieved, at least in part, by the presence of conserved domains within many patterning factors which share homology with the engrailed repressor (Muhr et al., 2001; Smith and Jaynes, 1996). This domain is sufficient to recruit Groucho-TLE (Gro/TLE) family co-repressors, which are broadly expressed in the neural tube, to the targets of patterning genes (Muhr et al., 2001). Gro/TLE repressors are thought to mediate gene regulation by promoting interaction with histone deacetylases to modulate chromatin structure, or possibly by more direct interaction with the transcription machinery (Chen et al., 1999; Edmondson and Roth, 1998; Edmondson et al., 1996; Lee and Pfaff, 2001). It is possible that additional co-repressors function along with spinal cord TFs in the developing neural tube.

The prevailing model for neuronal fate specification in the developing spinal cord is therefore based upon the presence of general activators and domain specific repression (Lee and Pfaff, 2001). In this model, the regulatory elements of genes expressed in postmitotic neurons are primed for activation by broadly expressed transcription factors. Permissive domains of expression are generated by the absence of repressor binding sites within these regulatory regions, leading to the acquisition of unique cell fates. This hypothesis is supported by the respecification of neuronal subtypes in a variety of patterning gene mutants but requires further investigation to be validated at the molecular level. These determinants of postmitotic neuronal identity in the ventral neural tube include *evx1* in V0 cells, *en1* in V1 cells, *Lhx3/4* and *Chx10* in V2 cells, MNR2/HB0 and *Isl1/2* in motor neurons, and *Sim1* in V3 interneurons (Lee and Pfaff, 2001). Studies of mice harboring mutations in these spinal cord TF target genes have revealed that while some dictate all aspects of cell identity, others only act to specify certain features of cell fate, such as axon guidance properties (Matise and Joyner, 1997; Moran-Rivard et al., 2001; Saueressig et al., 1999).

How is the Switch from Neuronal to Glial Production Encoded?

One of the fundamental properties of neurogenesis is that progenitor cells produce neurons before glial subtypes. Recent studies examining the post-translational modification of the key pMN determinant Olig2 have provided insight into the molecular mechanisms regulating this switch. These studies reveal that the phosphorylation status of Olig2 is developmentally regulated, and directs the formation of specific transcription factor complexes (Li et al., 2011; Sun et al., 2011). During neural tube patterning, phosphorylated Olig2 protein forms homodimers that are required for the repression of Pax6 from the pMN domain (Li et al., 2011). Following neurogenesis, dephosphorylated Olig2 forms complexes with the proneural gene Ngn2 which permits oligodendrocyte specification. Ectopic expression of mutant forms of Olig2 that cannot be phosphorylated results in a complete loss of motor neuron generation. This is due to the formation of Olig2/Ngn2 complexes during the critical period of neuronal specification and failure to induce key motor neuron fate determinants. Thus, the post-translational status of transcription factors employed at multiple developmental stages is a mechanism by which progenitors can give rise to both neurons and glia. Furthermore, the epigenetic status of promoters controlling the expression of proneural and glial genes is inversely correlated toward the end of neurogenesis, coordinating the decrease in neuronal production with the start of gliogenesis in multiple CNS tissues (Hirabayashi et al., 2009; Naka et al., 2008).

16.4 COMMON THEMES IN CNS DEVELOPMENT

The signaling pathways that specify progenitor identity in the neural tube are also responsible for the segregation of the developing forebrain into pallial and subpallial divisions. In dorsal regions, Wnt signaling induces the expression of Pax6 to define progenitors that will give rise to the neocortex of the pallium (Gunhaga et al., 2003). Shh secreted from the ventral midline of the forebrain specifies the progenitor domains of the subpallium in both a dose and time dependent manner, although this mechanism appears to be based primarily on the temporal competence of cells rather than pathway-mediated feedback (Sousa and Fishell, 2010). A subset of progenitors within the medial and caudal ganglionic eminences of the subpallium are fated to become gabaergic interneurons, which migrate into the neocortex to form local circuits with glutamatergic pyramidal cells (Wonders and Anderson, 2006).

Cortical progenitors proliferate in the VZ of the dorsal forebrain before producing the first postmitotic neurons by asymmetric cell division. At this stage, cortical neuroepithelial cells become fate restricted and acquire a radial glial morphology (Noctor et al., 2001). The coordinated mitoses of radial glia act to expand the pool of progenitors within the VZ and create a secondary source of intermediate progenitors within the subventricular zone (SVZ) (Noctor et al., 2004). The increase in cortical complexity observed during evolution is thought to be due to the expansion of secondary progenitors within the SVZ, both in terms of individual cellular subtypes and the dynamics of their division profiles (Lui et al., 2011). The VZ and SVZ produce neurons that migrate along the processes of radial glia to form the cortical plate. Over the course of cortical development, newly born neurons migrate beyond the existing populations to take up more superficial positions within the cortical plate. This process results in the production of six cortical layers in an "inside out" manner, with deeper layers representing neurons born during the early stages of neurogenesis and upper layers composed of later born cells (Figure 16.3).

Groundbreaking experiments utilizing *ex vivo* cultures of mouse cortical progenitor cells at clonal density revealed that the coordinated division profiles of progenitors enables them to first give rise to deep layer neurons before switching their output to upper layer neurons and then glia (Qian et al., 2000; Shen et al., 2006). This observation suggested that cortical progenitors transit through competence states, in which they form a given class of neuron or glial cell depending

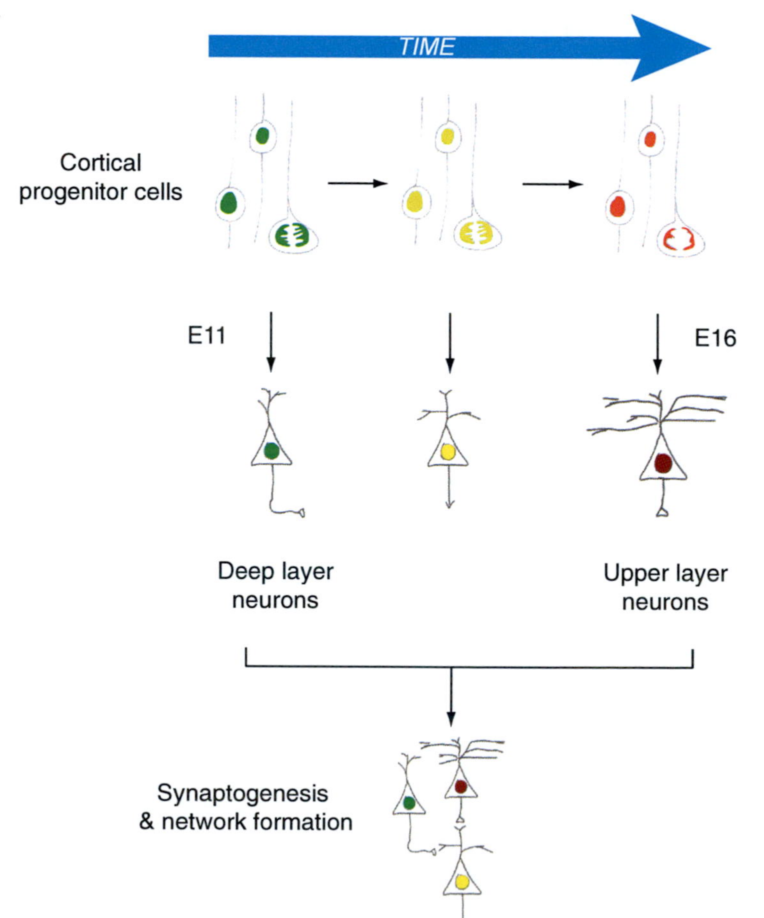

FIGURE 16.3 The Temporal Competence Model of Corticogenesis. During the early stages of corticogenesis, for example at E11 in mice, the asymmetric division of radial glia results in self-renewal and the production of deep layer neurons. These neurons migrate along the processes of radial glia towards the cortical plate. Over time, the competence of radial glia changes, such that they begin to switch their neuronal output to upper layer subtypes. These newly born neurons also migrate along radial glia to take up more superficial positions in the cortical plate than older, deep layer, neurons. This process results in the construction of a layered neocortex, which undergoes synaptogenesis to create a functional tissue.

on their developmental potential, as was proposed in the retina (Cepko et al., 1996). Molecular mechanisms underpinning the competence model were provided by the identification of a transcriptional cascade that regulates the temporal order of neurogenesis in the *Drosophil*a embryo (Isshiki et al., 2001; Pearson and Doe, 2003). Those experiments demonstrated that *Drosophila* neuroblasts sequentially express the transcription factors Hunchback, Krüppel, Pdm and Castor during development, and that the neurons formed during these expression windows maintain the transcriptional profile of their parent progenitor (Isshiki et al., 2001). Misexpression of these transcription factors is sufficient to alter the competence state of progenitors, resulting in predictable changes their neuronal output (Isshiki et al., 2001; Pearson and Doe, 2003).

Ikaros is one of the vertebrate homologs of *Drosophila* Hunchback, and it has been shown to have a similar role in regulating the temporal order of neurogenesis in the mammalian retina (Elliott et al., 2008). In the developing cerebral cortex, it is highly expressed at the start of corticogenesis and progressively decreases over time, suggesting that the core components of the temporal competence mechanism are conserved across evolution and employed in multiple tissues (Alsio et al., 2013). In support of this hypothesis, forced expression of Ikaros during the competence window of upper layer neuron generation is sufficient to alter the output of progenitors to deep layer subtypes (Alsio et al., 2013). The cellular interpretation of the temporal competence model appears to be executed at the level of chromatin reorganization, however further studies are required in order to fully elucidate this mechanism (Hirabayashi et al., 2009; Kohwi et al., 2013; Pereira et al., 2010).

Within the cortex, neurons are regionally specified along the rostrocaudal axis in the classical pattern of neocortical areas (Mountcastle, 1998). In this model, distinct neuronal domains are generated by an intrinsic spatial pattern within progenitors known as the neocortical protomap (Rakic, 1988). In contrast to the combinatorial expression of transcription factors that patterns the majority of the CNS, the neocortical protomap is manifest as gradients of transcription factor expression across the field of progenitors (Bishop et al., 2000; Mallamaci et al., 2000; Sansom et al., 2005). The core set of these factors, Emx2, Pax6, Coup-TF1, and Sp8, are expressed in a graded manner across the rostro-caudal and mediolateral axes of the neocortex in response to Fgf, BMP, and Wnt ligands secreted from multiple organizing centers within the tissue (O'Leary and Sahara, 2008; Sansom and Livesey, 2009). The morphogens and transcription factors that act to pattern the neocortex are proposed to form the basis of a gene regulatory network, a hypothesis supported by gain and loss of function studies in mice, however the molecular basis of these interactions remains largely unknown (Bishop et al., 2000; Hamasaki et al., 2004; Mallamaci et al., 2000).

16.5 POSTMITOTIC REFINEMENT OF SUBTYPE IDENTITY

Cortical function relies on the acquisition and maintenance of specific neuronal subtype identities. Recent work has begun to describe a regulatory network that segregates neuronal subtypes within the cortical plate based upon their postmitotic expression profiles. Layer 5 neurons, the major output of the cortex to the brainstem and spinal cord, express the transcription factors Fezf2 and Ctip2. Layer 2/3 neurons connect the two hemispheres of the cerebral cortex and express Satb2. Loss of Fezf2 results in the ectopic expression of Satb2 in layer 5 neurons and a switch in neuronal identity (Chen et al., 2008). Overexpression of Fezf2 specifically in postmitotic upper layer neurons results in the acquisition of a deep layer transcriptional identity, projection trajectory, and connectivity (De la Rossa et al., 2013; Rouaux and Arlotta, 2013). In Satb2 mutants, Ctip2 is ectopically expressed, and upper layer cortical neurons project towards the brainstem and spinal cord (Alcamo et al., 2008; Britanova et al., 2008). Ctip2 mutants display a respecification of layer 5 neurons to Tbr1 positive layer 6 neurons, and furthermore, Tbr1 directly represses the layer 5 marker Fezf2 (Han et al., 2011; Srinivasan et al., 2012). These studies highlight a complex network of selective repression operating in cortical neurons that acts to define and maintain specific subtypes. The postmitotic specification and maintenance of cell fate in the cortex contrasts with the network operating within spinal cord progenitors, however the regulatory logic of selective repression appears to be conserved.

16.6 APPLYING DEVELOPMENTAL PRINCIPLES TO STEM CELL RESEARCH: DIRECTED DIFFERENTIATION FROM PLURIPOTENT STEM CELLS

The studies described in this chapter have led to a working understanding of how an initially equivalent population of neural progenitors generates the highly specialized cell types of the central nervous system. These principles of embryonic neural development form the basis of groundbreaking experiments in which pluripotent stem cells are directed to specific neuronal fates *in vitro*. These models of neural development permit the study of progenitor specification, neurogenesis, and network formation at a resolution that is not possible in animal models. Furthermore, the ability to reprogram somatic cells from donor patients into an induced pluripotent state (Takahashi and Yamanaka, 2006) provides an opportunity to gain insight into the mechanisms underlying neurodevelopmental syndromes and degenerative diseases.

In Vitro Models of Cortex Development

Mouse ES cells grown as a single layer and cultured in defined media can perform neural induction to give rise to progenitors with a forebrain transcriptional profile, in agreement with the default identity of neuroepithelial cells *in vivo* (Ying et al., 2003). Modification of this process by signaling pathway manipulation biases forebrain progenitors to specific regional identities. Stimulation of the Shh pathway directs progenitors to ventral forebrain subtypes, including cortical interneurons, in a temporally restricted manner (Danjo et al., 2011; Watanabe et al., 2005). By contrast, neural induction of mouse ES cells in the presence of a Shh pathway inhibitor prevents the specification of subpallial progenitors and consequently results in the production of cortical progenitors (Gaspard et al., 2008). These progenitors produce excitatory pyramidal neurons before switching to gliogenesis, recapitulating a fundamental property of neuronal development. Furthermore, the sequential generation of neurons with specific layer identities is preserved *in vitro*, strengthening the previously discussed temporal competence model of corticogenesis (Gaspard et al., 2008). Cortical progenitors are also produced by aggregate cultures of mouse ES cells in the presence of Wnt and TGFβ pathway inhibitors (Eiraku et al., 2008). Remarkably, cells grown in this 3D culture model self-organize to form polarized structures that resemble the laminated developing neocortex (Eiraku et al., 2008). Over time, these tissues produce at least one class of secondary progenitor and both deep and upper layer neurons. Together, these studies established directed differentiation of stem cells to dorsal forebrain progenitors as a means to investigate corticogenesis *in vitro*.

The directed differentiation of human pluripotent cells to cortical tissue is highly desirable as a tool to investigate both developmental processes and the mechanisms by which diseases and syndromes affect the function of the cortex. The application of mouse stem cell culture protocols to human tissue is sufficient to specify cortical progenitors, however a major limitation of these methods is the observed bias towards the production of deep layer neurons compared to upper layers (Eiraku et al., 2008; Espuny-Camacho et al., 2013). This observation can be attributed to the inability to recapitulate the primate SVZ *in vitro*, which comprises multiple secondary progenitor subtypes that are not present in the rodent brain (Lui et al., 2011). A recent study has made progress in this area by demonstrating that a combination of TGFβ inhibition and stimulation of the retinoic acid signaling pathway is sufficient to direct human stem cells to cortical progenitors (Shi et al., 2012b). Importantly, this protocol results in the generation of the full complement of secondary progenitors, and consequently both deep and upper layer neurons are produced in the expected numbers over a timeframe that mirrors human development (Shi et al., 2012b).

Recapitulation of Anterior-Posterior Patterning

As previously discussed, neural progenitors are initially specified to an anterior identity and subsequently posteriorized by the graded activities of the Wnt, Fgf, and retinoic acid signaling pathways (see Chapter 11). In agreement with *in vivo* data, temporal manipulation of developmental signaling pathways can also posteriorize progenitors to defined neuronal fates *in vitro*. Parkinson's disease is primarily caused by the selective loss of dopaminergic neurons within the substantia nigra of the midbrain, making it an attractive model for stem cell based replacement therapies. During embryonic development, dopaminergic neuronal progenitors are specified within the ventral midline of the midbrain by the posteriorizing activity of Fgf8, secreted from the organizing region known as the isthmus, and ventralization by Shh signaling (Ye et al., 1998). Pioneering studies using mouse ES cells have demonstrated that following neural induction, recapitulation of these signaling events *in vitro* is sufficient for the production of dopaminergic neurons (Lee et al., 2000). Furthermore, transplantation of stem cell derived dopaminergic neuronal progenitors into mice in which dopaminergeric neurons had been chemically ablated results in the production of neurons that are sufficient to reduce the motor deficits associated with this animal disease model (Kim et al., 2002). Application of these protocols to human stem cells is sufficient for the production of dopaminergic neurons. However, activation of the Wnt pathway in the context of Fgf8 and Shh treatments greatly increases the specification of dopaminergic neuron progenitors by recapitulating the complex signaling environment of the midbrain floor plate (Joksimovic et al., 2009; Kriks et al., 2011). These human stem cell derived dopaminergic neurons reduce Parkinsonian symptoms in rodents and survive as grafts in primate models of the disease (Kriks et al., 2011).

Similarly, recapitulation of developmental signaling events *in vitro* is sufficient to direct pluripotent cells to a motor neuron fate. Following neural induction, mouse ES cells caudalized with a high dose of retinoic acid and subsequently ventralized by Shh signaling express the pMN progenitor markers Nkx6.1 and Olig2 (Wichterle et al., 2002). These progenitors give rise to neurons that express the full complement of motor neuron markers and survive as grafts when implanted into the developing chick spinal cord (Wichterle et al., 2002). The same pathway manipulations produce motor neuron progenitors from human pluripotent cells (Li et al., 2005; 2008). However, human cells require approximately five weeks to acquire a mature motor neuron identity compared to the five days observed in mice, highlighting the species-specific temporal regulation of neurogenesis that is evident throughout the CNS (Li et al., 2005; 2008; Wichterle et al., 2002).

Three-Dimensional Models of Neural Tissues

As mentioned above, in addition to generating specific neuronal cell types and replaying key aspects of development, stem cell systems are also potentially powerful for modeling morphogenesis and neural circuit formation. In this case, 3D models that reflect *in vivo* tissue architecture are a major goal of the stem cell field. Pioneering work from Yoshiki Sasai and colleagues, initially modeling mouse cortex and retina/optic cup development from ES cells, demonstrated that replaying development via embryoid body-like structures uncovered the capacity for self-organization within 3D *in vitro* developing structures (Eiraku et al., 2008; 2011). In the absence of signaling pathway manipulation, neural tissues of mixed anterior-posterior and dorsoventral identities are generated from mouse and human cells applied to these systems (Lancaster et al., 2013). However, inhibition of the posteriorizing activity of the Wnt signaling pathway has enabled the robust development of human 3D cerebral cortex *in vitro* (Kadoshima et al., 2013).

16.7 USING TRANSCRIPTION FACTOR CODES TO DIRECTLY SPECIFY CELL FATE

The temporal manipulation of developmental signaling pathways is a robust method of directing multipotent cells to specific neural lineages. However, several studies have shown that the ectopic expression of key fate determinants is sufficient to direct cells to defined fates in the absence of activated signaling, highlighting the fact that cell identity is largely determined by specific transcriptional profiles. The process of employing transcriptional codes to specify the fate of pluripotent cells is termed direct differentiation (Figure 16.4). The ectopic expression of the transcription factors Brn2, Ascl1 and Myt1l (BAM) is sufficient to produce functional human neurons from ES cells in approximately 15 days (Pang et al., 2011). In a similar study, overexpression of Ngn2 alone was shown to direct human stem cells to a neuronal identity in approximately 14 days (Zhang et al., 2013). Remarkably, neurons induced by this method exhibit a homogenous transcriptional profile that correlates with that observed within layer 2/3 cortical neurons and form functional synapses *in vitro* (Zhang et al., 2013). Direct differentiation of dopaminergic neurons is achieved by overexpressing Ascl1 with the fate determinants Nurr1 and Lmx1a, recapitulating the transcription factor cascade that specifies and then maintains the dopaminergic identity (Andersson et al., 2006a; Theka et al., 2013). These recent studies suggest that specific combinations of proneural genes and transcription factors are sufficient to specify neuronal lineages in the absence of developmental signaling pathways, and that they form the basis of an experimental paradigm that is likely to be widely utilized by the research community.

The logic of transcriptionally specifying neural identity can also be extended to differentiated, non-neural cell types. This process, termed transdifferentiation, was established in a seminal study in which the previously described BAM factors were overexpressed in mouse fibroblasts, giving rise to functional neurons (Figure 16.4) (Vierbuchen and Wernig, 2011). Strikingly, neuronal transdifferentition appears to represent a direct lineage conversion, as cycling progenitors are not observed during the respecification process (Caiazzo et al., 2011; Pang et al., 2011; Son et al., 2011; Vierbuchen and Wernig, 2011). The BAM factor combination is sufficient to generate neurons from human fibroblasts, however addition of the proneural gene NeuroD1 greatly enhances the conversion of cells to a neural fate (Pang et al., 2011). Cells induced by these methods exhibit a transcriptional profile broadly consistent with forebrain glutamatergic neurons, however a

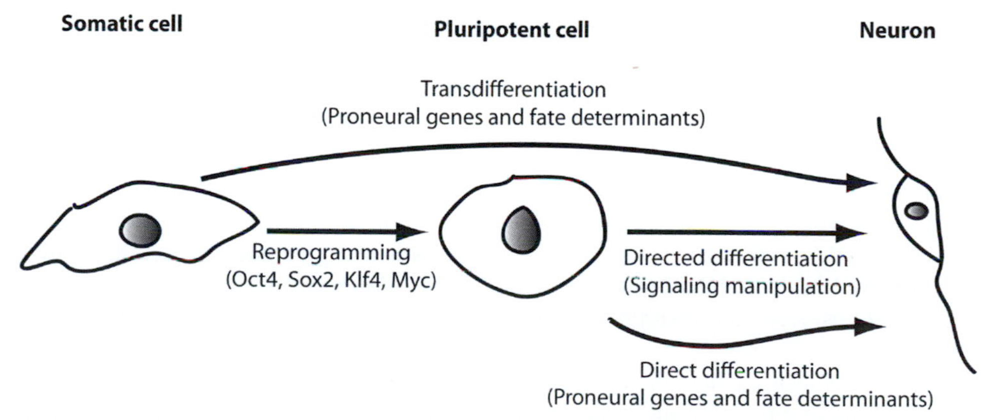

FIGURE 16.4 **Defining Neuronal Subtypes *In Vitro*.** Directed differentiation describes the process in which a pluripotent cell, derived either from an embryo or somatic cells induced to pluripotent state, is specified to a particular fate by recapitulating key developmental signaling events. By contrast, the use of transcription factor codes to specify the fate of pluripotent cells is termed direct differentiation. Transdifferentiation applies the logic of using transcription factor codes to specify cell identity to somatic cells, inducing a direct lineage conversion to the desired neuronal subtype.

homogenous subtype is not produced. By contrast, the ectopic expression of Ascl1, Nurr1, and Lmx1a is sufficient to directly transdifferentiate human fibroblasts to functional dopaminergic neurons, consistent with the results of direct differentiation experiments (Caiazzo et al., 2011; Theka et al., 2013). Similarly, the combinatorial overexpression of BAM factors, patterning genes expressed in the pMN, and terminal differentiation markers is sufficient to convert fibroblasts to spinal motor neurons (Son et al., 2011).

16.8 MODELING HUMAN NEUROLOGICAL DISORDERS AND DISEASES

Stem cell systems that replay neural development or enable the production of defined neuronal cell types are potentially powerful platforms for modeling human developmental disorders and diseases of the nervous system. Down's syndrome is a common genetic disorder, caused by the triplication of chromosome 21. Patients with Down's syndrome exhibit an array of developmental phenotypes, including varying degrees of intellectual disability.

Directed differentiation of Down's syndrome/Trisomy 21 induced pluripotent stem cells to cortical neurons suggested that impaired progenitor proliferation and differentiation could result in changes in cortical development resulting in intellectual disabilities (Hibaoui et al., 2014). Furthermore, reducing the expression of the DYRK1A protein kinase that is on Chr21 partially rescued these deficits in trisomy 21 neural progenitor cells, providing mechanistic insight into the etiology of neurodevelopmental aspects of this condition (Hibaoui et al., 2014). An independent human stem cell study of trisomy 21 neurons suggested that the synaptic connections between stem cell derived neurons were compromised in Down's syndrome, however the molecular basis of this phenotype requires further investigation (Weick et al., 2013).

Abnormal neural development and function are also observed in patients with autistic spectrum disorders. Rett syndrome, considered part of the autistic spectrum, is caused by mutations in the X-linked methyl-CpG binding protein, MeCP2, which has been shown to function as both an activator and repressor of transcription (Amir et al., 1999; Chahrour et al., 2008). Stem cell derived Rett syndrome neurons exhibit synaptic defects and alterations in calcium signaling, consistent with the manifestation of the mutation in patients (Marchetto et al., 2010). Moreover, these neuronal defects can be recapitulated in control neurons by reducing the expression of MeCP2 using RNA interference (Marchetto et al., 2010). Consistent with its proposed role in transcriptional regulation, MeCP2 mutant neurons were subsequently shown to exhibit a global transcriptional repression, providing insight into the mechanism of neuronal dysfunction (Li et al., 2013).

In contrast to the developmental disorders discussed above, Alzheimer's disease is characterized by the progressive loss of glutamatergic pyramidal neurons in the cortex as a consequence of the altered processing of the amyloid precursor protein (APP) and the hyperphosphorylation of the microtubule-associated protein tau (Selkoe, 2011). Alzheimer's pathology is therefore associated with the increased pathogenic processing of APP to the amyloid-β peptide, which forms extracellular plaques and intracellular neurofibrillary tangles composed of hyperphosphorylated tau (Selkoe, 2011). The ability to robustly differentiate human stem cells to cortical neurons represents an opportunity to study Alzheimer's pathology and progression that is not possible in patients or animal models. Consequently, several directed differentiation and transdifferentiation approaches have been used to study aspects of the cell biology that underpins the disease pathology.

Fibroblasts donated from patients with a duplication of the APP gene, a genetic cause of familial Alzheimers' disease, and reprogrammed to a pluripotent state, can be subsequently differentiated to forebrain neurons (Israel et al., 2012; Takahashi and Yamanaka, 2006). These neurons exhibit both increased production of the amyloid-β peptide and phosphorylation of tau, in comparison to non-diseased neurons and the patients' fibroblasts (Israel et al., 2012). Down's syndrome also predisposes patients to Alzheimer's disease, primarily due to the triplication of the APP locus on Chromosome 21. Differentiation of Down's syndrome pluripotent cells by TGFβ inhibition and activation of retinoic acid signaling generates glutamatergic pyramidal neurons of both lower and upper cortical layers (Shi et al., 2012a, b). Down's syndrome neurons increase the production of amyloid-β peptide, which aggregates extracellularly in structures analogous to the plaques observed in the human brain (Shi et al., 2012a). Furthermore, Down's cortical neurons exhibit tau hyperphosphorylation and mislocalization within dendrites, both classical hallmarks of Alzheimer's disease (Gotz et al., 1995; Shi et al., 2012a).

Together, these studies demonstrate how the recapitulation of developmental processes *in vitro* is able to model complex, poorly understood, diseases and syndromes affecting the nervous system and provide a novel system in which to assay potential therapeutics. These systems all share a reliance on our fundamental understanding of the cellular and molecular biology of neural development to enable the production of the cellular model, and, in the case of neurodevelopmental disorders, to interpret the differences in early development manifest in those models.

CLINICAL RELEVANCE

- An understanding of neurogenesis and neural cell fate determination is central to understanding the etiology and pathogenesis of neurodevelopmental disorders.
- Human stem cell models of neural development have been created, informed by our understanding of neural cell fate determination.
- Using patient-specific stem cells, these models can be used for mechanistic studies of the pathobiology of neurodevelopmental disorders, such as autism, Rett syndrome and Down's syndrome.
- These approaches also enable the production of functional human neural networks, and the development and application of human cellular models of adult-onset neurological diseases, including Alzheimer's disease, motor neuron disease/ALS and Parkinson's disease.

RECOMMENDED RESOURCES

Brand, A.H., Livesey, F.J., 2011. Neural stem cell biology in vertebrates and invertebrates: more alike than different? Neuron 70, 719–729.

Han, S.S., Williams, L.A., Eggan, K.C., 2011. Constructing and deconstructing stem cell models of neurological disease. Neuron 70, 626–644.

Martynoga, B., Drechsel, D., Guillemot, F., 2012. Molecular control of neurogenesis: a view from the mammalian cerebral cortex. Cold Spring Harb. perspect. biol. 4.

The Allen Brain Atlas. www.brain-map.org (accessed 01.04.14). A database of gene expression profiles in the rodent, primate, and human central nervous systems at various stages of development and adulthood.

REFERENCES

Alcamo, E.A., Chirivella, L., Dautzenberg, M., Dobreva, G., Farinas, I., Grosschedl, R., McConnell, S.K., 2008. Satb2 regulates callosal projection neuron identity in the developing cerebral cortex. Neuron 57, 364–377.

Alsio, J.M., Tarchini, B., Cayouette, M., Livesey, F.J., 2013. Ikaros promotes early-born neuronal fates in the cerebral cortex. Proc. Natl. Acad. Sci. USA 110, E716–725.

Amir, R.E., Van den Veyver, I.B., Wan, M., Tran, C.Q., Francke, U., Zoghbi, H.Y., 1999. Rett syndrome is caused by mutations in X-linked MECP2, encoding methyl-CpG-binding protein 2. Nat. genet. 23, 185–188.

Andersson, E., Tryggvason, U., Deng, Q., Friling, S., Alekseenko, Z., Robert, B., Perlmann, T., Ericson, J., 2006a. Identification of intrinsic determinants of midbrain dopamine neurons. Cell 124, 393–405.

Andersson, L., Bostrom, P., Ericson, J., Rutberg, M., Magnusson, B., Marchesan, D., Ruiz, M., Asp, L., Huang, P., Frohman, M.A., et al., 2006b. PLD1 and ERK2 regulate cytosolic lipid droplet formation. J. cell sci. 119, 2246–2257.

Austin, C.P., Feldman, D.E., Ida Jr., J.A., Cepko, C.L., 1995. Vertebrate retinal ganglion cells are selected from competent progenitors by the action of Notch. Development 121, 3637–3650.

Balaskas, N., Ribeiro, A., Panovska, J., Dessaud, E., Sasai, N., Page, K.M., Briscoe, J., Ribes, V., 2012. Gene regulatory logic for reading the Sonic Hedgehog signaling gradient in the vertebrate neural tube. Cell 148, 273–284.

Belliveau, M.J., Cepko, C.L., 1999. Extrinsic and intrinsic factors control the genesis of amacrine and cone cells in the rat retina. Development. In press.

Bertrand, N., Castro, D.S., Guillemot, F., 2002. Proneural genes and the specification of neural cell types. Nat. Rev. Neurosci. 3, 517–530.

Bishop, K.M., Goudreau, G., O'Leary, D.D., 2000. Regulation of area identity in the mammalian neocortex by Emx2 and Pax6. Science 288, 344–349.

Brand, A.H., Livesey, F.J., 2011. Neural stem cell biology in vertebrates and invertebrates: more alike than different? Neuron 70, 719–729.

Briscoe, J., Pierani, A., Jessell, T.M., Ericson, J., 2000. A homeodomain protein code specifies progenitor cell identity and neuronal fate in the ventral neural tube. Cell 101, 435–445.

Britanova, O., de Juan Romero, C., Cheung, A., Kwan, K.Y., Schwark, M., Gyorgy, A., Vogel, T., Akopov, S., Mitkovski, M., Agoston, D., et al., 2008. Satb2 is a postmitotic determinant for upper-layer neuron specification in the neocortex. Neuron 57, 378–392.

Cabrera, C.V., Alonso, M.C., 1991. Transcriptional activation by heterodimers of the achaete-scute and daughterless gene products of *Drosophila*. Embo J. 10, 2965–2973.

Caiazzo, M., Dell'Anno, M.T., Dvoretskova, E., Lazarevic, D., Taverna, S., Leo, D., Sotnikova, T.D., Menegon, A., Roncaglia, P., Colciago, G., et al., 2011. Direct generation of functional dopaminergic neurons from mouse and human fibroblasts. Nature 476, 224–227.

Campuzano, S., 2001. Emc, a negative HLH regulator with multiple functions in *Drosophila* development. Oncogene 20, 8299–8307.

Cepko, C.L., Austin, C.P., Yang, X., Alexiades, M., Ezzeddine, D., 1996. Cell fate determination in the vertebrate retina. Proc. Natl. Acad. Sci. USA 93, 589–595.

Chahrour, M., Jung, S.Y., Shaw, C., Zhou, X., Wong, S.T., Qin, J., Zoghbi, H.Y., 2008. MeCP2, a key contributor to neurological disease, activates and represses transcription. Science 320, 1224–1229.

Chen, B., Wang, S.S., Hattox, A.M., Rayburn, H., Nelson, S.B., McConnell, S.K., 2008. The Fezf2-Ctip2 genetic pathway regulates the fate choice of subcortical projection neurons in the developing cerebral cortex. Proc. Natl. Acad. Sci. USA 105, 11382–11387.

Chen, G., Fernandez, J., Mische, S., Courey, A.J., 1999. A functional interaction between the histone deacetylase Rpd3 and the corepressor groucho in *Drosophila* development. Genes Dev. 13, 2218–2230.

Chien, C.T., Hsiao, C.D., Jan, L.Y., Jan, Y.N., 1996. Neuronal type information encoded in the basic helix-loop-helix domain of proneural genes. Proc. Natl. Acad. Sci. USA 93, 13239–13244.

Danjo, T., Eiraku, M., Muguruma, K., Watanabe, K., Kawada, M., Yanagawa, Y., Rubenstein, J.L., Sasai, Y., 2011. Subregional specification of embryonic stem cell-derived ventral telencephalic tissues by timed and combinatory treatment with extrinsic signals. J. Neurosci. 31, 1919–1933.

Davis, R.L., Turner, D.L., 2001. Vertebrate hairy and Enhancer of split related proteins: transcriptional repressors regulating cellular differentiation and embryonic patterning. Oncogene 20, 8342–8357.

De la Rossa, A., Bellone, C., Golding, B., Vitali, I., Moss, J., Toni, N., Luscher, C., Jabaudon, D., 2013. *In vivo* reprogramming of circuit connectivity in postmitotic neocortical neurons. Nat. Neurosci. 16, 193–200.

Desai, A.R., McConnell, S.K., 2000. Progressive restriction in fate potential by neural progenitors during cerebral cortical development. Development 127, 2863–2872.

Dessaud, E., Yang, L.L., Hill, K., Cox, B., Ulloa, F., Ribeiro, A., Mynett, A., Novitch, B.G., Briscoe, J., 2007. Interpretation of the sonic hedgehog morphogen gradient by a temporal adaptation mechanism. Nature 450, 717–720.

Edlund, T., Jessell, T.M., 1999. Progression from extrinsic to intrinsic signaling in cell fate specification: a view from the nervous system. Cell 96, 211–224.

Edmondson, D.G., Roth, S.Y., 1998. Interactions of transcriptional regulators with histones. Methods 15, 355–364.

Edmondson, D.G., Smith, M.M., Roth, S.Y., 1996. Repression domain of the yeast global repressor Tup1 interacts directly with histones H3 and H4. Genes Dev. 10, 1247–1259.

Eiraku, M., Takata, N., Ishibashi, H., Kawada, M., Sakakura, E., Okuda, S., Sekiguchi, K., Adachi, T., Sasai, Y., 2011. Self-organizing optic-cup morphogenesis in three-dimensional culture. Nature 472, 51–56.

Eiraku, M., Watanabe, K., Matsuo-Takasaki, M., Kawada, M., Yonemura, S., Matsumura, M., Wataya, T., Nishiyama, A., Muguruma, K., Sasai, Y., 2008. Self-organized formation of polarized cortical tissues from ESCs and its active manipulation by extrinsic signals. Cell stem cell 3, 519–532.

Elliott, J., Jolicoeur, C., Ramamurthy, V., Cayouette, M., 2008. Ikaros confers early temporal competence to mouse retinal progenitor cells. Neuron 60, 26–39.

Ericson, J., Morton, S., Kawakami, A., Roelink, H., Jessell, T.M., 1996. Two critical periods of Sonic Hedgehog signaling required for the specification of motor neuron identity. Cell 87, 661–673.

Ericson, J., Rashbass, P., Schedl, A., Brenner-Morton, S., Kawakami, A., van Heyningen, V., Jessell, T.M., Briscoe, J., 1997. Pax6 controls progenitor cell identity and neuronal fate in response to graded Shh signaling. Cell 90, 169–180.

Espuny-Camacho, I., Michelsen, K.A., Gall, D., Linaro, D., Hasche, A., Bonnefont, J., Bali, C., Orduz, D., Bilheu, A., Herpoel, A., et al., 2013. Pyramidal neurons derived from human pluripotent stem cells integrate efficiently into mouse brain circuits *in vivo*. Neuron 77, 440–456.

Garcia-Bellido, A., 1979. Genetic analysis of the *achaete-scute* system of *Drosophila Melanogaster*. Genetics 91, 491–520.

Gaspard, N., Bouschet, T., Hourez, R., Dimidschstein, J., Naeije, G., van den Ameele, J., Espuny-Camacho, I., Herpoel, A., Passante, L., Schiffmann, S.N., et al., 2008. An intrinsic mechanism of corticogenesis from embryonic stem cells. Nature 455, 351–357.

Ghysen, A., Dambly-Chaudiere, C., 1988. From DNA to form: the achaete-scute complex. Genes Dev. 2, 495–501.

Gotz, J., Probst, A., Spillantini, M.G., Schafer, T., Jakes, R., Burki, K., Goedert, M., 1995. Somatodendritic localization and hyperphosphorylation of tau protein in transgenic mice expressing the longest human brain tau isoform. Embo J. 14, 1304–1313.

Gunhaga, L., Marklund, M., Sjodal, M., Hsieh, J.C., Jessell, T.M., Edlund, T., 2003. Specification of dorsal telencephalic character by sequential Wnt and FGF signaling. Nat. Neurosci. 6, 701–707.

Hamasaki, T., Leingartner, A., Ringstedt, T., O'Leary, D.D., 2004. EMX2 regulates sizes and positioning of the primary sensory and motor areas in neocortex by direct specification of cortical progenitors. Neuron 43, 359–372.

Han, W., Kwan, K.Y., Shim, S., Lam, M.M., Shin, Y., Xu, X., Zhu, Y., Li, M., Sestan, N., 2011. TBR1 directly represses Fezf2 to control the laminar origin and development of the corticospinal tract. Proc. Natl. Acad. Sci. USA 108, 3041–3046.

Helms, A.W., Johnson, J.E., 2003. Specification of dorsal spinal cord interneurons. Curr. opin. neurobiol. 13, 42–49.

Hibaoui, Y., Grad, I., Letourneau, A., Sailani, M.R., Dahoun, S., Santoni, F.A., Gimelli, S., Guipponi, M., Pelte, M.F., Bena, F., et al., 2014. Modelling and rescuing neurodevelopmental defect of Down syndrome using induced pluripotent stem cells from monozygotic twins discordant for trisomy 21. EMBO mol. med..

Hirabayashi, Y., Suzki, N., Tsuboi, M., Endo, T.A., Toyoda, T., Shinga, J., Koseki, H., Vidal, M., Gotoh, Y., 2009. Polycomb limits the neurogenic competence of neural precursor cells to promote astrogenic fate transition. Neuron 63, 600–613.

Israel, M.A., Yuan, S.H., Bardy, C., Reyna, S.M., Mu, Y., Herrera, C., Hefferan, M.P., Van Gorp, S., Nazor, K.L., Boscolo, F.S., et al., 2012. Probing sporadic and familial Alzheimer's disease using induced pluripotent stem cells. Nature 482, 216–220.

Isshiki, T., Pearson, B., Holbrook, S., Doe, C.Q., 2001. *Drosophila* neuroblasts sequentially express transcription factors which specify the temporal identity of their neuronal progeny. Cell 106, 511–521.

Jarman, A.P., Grau, Y., Jan, L.Y., Jan, Y.N., 1993. atonal is a proneural gene that directs chordotonal organ formation in the *Drosophila* peripheral nervous system. Cell 73, 1307–1321.

Johnson, J.E., Birren, S.J., Saito, T., Anderson, D.J., 1992. DNA binding and transcriptional regulatory activity of mammalian achaete-scute homologous (MASH) proteins revealed by interaction with a muscle-specific enhancer. Proc. Natl. Acad. Sci. USA 89, 3596–3600.

Joksimovic, M., Yun, B.A., Kittappa, R., Anderegg, A.M., Chang, W.W., Taketo, M.M., McKay, R.D., Awatramani, R.B., 2009. Wnt antagonism of Shh facilitates midbrain floor plate neurogenesis. Nat. Neurosci. 12, 125–131.

Kadoshima, T., Sakaguchi, H., Nakano, T., Soen, M., Ando, S., Eiraku, M., Sasai, Y., 2013. Self-organization of axial polarity, inside-out layer pattern, and species-specific progenitor dynamics in human ES cell-derived neocortex. Proc. Natl. Acad. Sci. USA 110, 20284–20289.

Kageyama, R., Nakanishi, S., 1997. Helix-loop-helix factors in growth and differentiation of the vertebrate nervous system. Curr. Opin. Genet. Dev. 7, 659–665.

Kim, J.H., Auerbach, J.M., Rodriguez-Gomez, J.A., Velasco, I., Gavin, D., Lumelsky, N., Lee, S.H., Nguyen, J., Sanchez-Pernaute, R., Bankiewicz, K., et al., 2002. Dopamine neurons derived from embryonic stem cells function in an animal model of Parkinson's disease. Nature 418, 50–56.

Kohwi, M., Lupton, J.R., Lai, S.L., Miller, M.R., Doe, C.Q., 2013. Developmentally regulated subnuclear genome reorganization restricts neural progenitor competence in *Drosophila*. Cell 152, 97–108.

Kriks, S., Shim, J.W., Piao, J., Ganat, Y.M., Wakeman, D.R., Xie, Z., Carrillo-Reid, L., Auyeung, G., Antonacci, C., Buch, A., et al., 2011. Dopamine neurons derived from human ES cells efficiently engraft in animal models of Parkinson's disease. Nature 480, 547–551.

Lancaster, M.A., Renner, M., Martin, C.A., Wenzel, D., Bicknell, L.S., Hurles, M.E., Homfray, T., Penninger, J.M., Jackson, A.P., Knoblich, J.A., 2013. Cerebral organoids model human brain development and microcephaly. Nature 501, 373–379.

Lee, S.H., Lumelsky, N., Studer, L., Auerbach, J.M., McKay, R.D., 2000. Efficient generation of midbrain and hindbrain neurons from mouse embryonic stem cells. Nat. biotechnol. 18, 675–679.

Lee, S.K., Pfaff, S.L., 2001. Transcriptional networks regulating neuronal identity in the developing spinal cord. Nat. Neurosci. 4 (Suppl.), 1183–1191.

Li, H., de Faria, J.P., Andrew, P., Nitarska, J., Richardson, W.D., 2011. Phosphorylation regulates OLIG2 cofactor choice and the motor neuron-oligodendrocyte fate switch. Neuron 69, 918–929.

Li, X.J., Du, Z.W., Zarnowska, E.D., Pankratz, M., Hansen, L.O., Pearce, R.A., Zhang, S.C., 2005. Specification of motoneurons from human embryonic stem cells. Nat. biotechnol. 23, 215–221.

Li, X.J., Hu, B.Y., Jones, S.A., Zhang, Y.S., Lavaute, T., Du, Z.W., Zhang, S.C., 2008. Directed differentiation of ventral spinal progenitors and motor neurons from human embryonic stem cells by small molecules. Stem Cells 26, 886–893.

Li, Y., Wang, H., Muffat, J., Cheng, A.W., Orlando, D.A., Loven, J., Kwok, S.M., Feldman, D.A., Bateup, H.S., Gao, Q., et al., 2013. Global transcriptional and translational repression in human-embryonic-stem-cell-derived Rett syndrome neurons. Cell stem cell 13, 446–458.

Livesey, F.J., Cepko, C.L., 2001. Vertebrate neural cell-fate determination: lessons from the retina. Nat. Rev. Neurosci. 2, 109–118.

Lui, J.H., Hansen, D.V., Kriegstein, A.R., 2011. *Development* and evolution of the human neocortex. Cell 146, 18–36.

Mallamaci, A., Muzio, L., Chan, C.H., Parnavelas, J., Boncinelli, E., 2000. Area identity shifts in the early cerebral cortex of Emx2-/- mutant mice [see comments]. Nat. Neurosci. 3, 679–686.

Marchetto, M.C., Carromeu, C., Acab, A., Yu, D., Yeo, G.W., Mu, Y., Chen, G., Gage, F.H., Muotri, A.R., 2010. A model for neural development and treatment of Rett syndrome using human induced pluripotent stem cells. Cell 143, 527–539.

Massari, M.E., Murre, C., 2000. Helix-loop-helix proteins: regulators of transcription in eucaryotic organisms. Mol. Cell Biol. 20, 429–440.

Matise, M.P., Joyner, A.L., 1997. Expression patterns of developmental control genes in normal and Engrailed-1 mutant mouse spinal cord reveal early diversity in developing interneurons. J. Neurosci. 17, 7805–7816.

McConnell, S.K., 1988. Fates of visual cortical neurons in the ferret after isochronic and heterochronic transplantation. J. Neurosci. 8, 945–974.

Mizuguchi, R., Sugimori, M., Takebayashi, H., Kosako, H., Nagao, M., Yoshida, S., Nabeshima, Y., Shimamura, K., Nakafuku, M., 2001. Combinatorial roles of olig2 and neurogenin2 in the coordinated induction of pan-neuronal and subtype-specific properties of motoneurons. Neuron 31, 757–771.

Moran-Rivard, L., Kagawa, T., Saueressig, H., Gross, M.K., Burrill, J., Goulding, M., 2001. Evx1 is a postmitotic determinant of v0 interneuron identity in the spinal cord. Neuron 29, 385–399.

Mountcastle, V.B., 1998. The Cerebral Cortex. Harvard University Press, Cambridge, Mass.

Muhr, J., Andersson, E., Persson, M., Jessell, T.M., Ericson, J., 2001. Groucho-mediated transcriptional repression establishes progenitor cell pattern and neuronal fate in the ventral neural tube. Cell 104, 861–873.

Murre, C., McCaw, P.S., Vaessin, H., Caudy, M., Jan, L.Y., Jan, Y.N., Cabrera, C.V., Buskin, J.N., Hauschka, S.D., Lassar, A.B., et al., 1989. Interactions between heterologous helix-loop-helix proteins generate complexes that bind specifically to a common DNA sequence. Cell 58, 537–544.

Naka, H., Nakamura, S., Shimazaki, T., Okano, H., 2008. Requirement for COUP-TFI and II in the temporal specification of neural stem cells in CNS development. Nat. Neurosci. 11, 1014–1023.

Noctor, S.C., Flint, A.C., Weissman, T.A., Dammerman, R.S., Kriegstein, A.R., 2001. Neurons derived from radial glial cells establish radial units in neocortex. Nature 409, 714–720.

Noctor, S.C., Martinez-Cerdeno, V., Ivic, L., Kriegstein, A.R., 2004. Cortical neurons arise in symmetric and asymmetric division zones and migrate through specific phases. Nat. Neurosci. 7, 136–144.

Novitch, B.G., Chen, A.I., Jessell, T.M., 2001. Coordinate regulation of motor neuron subtype identity and pan-neuronal properties by the bHLH repressor Olig2. Neuron 31, 773–789.

O'Leary, D.D., Sahara, S., 2008. Genetic regulation of arealization of the neocortex. Curr. opin. neurobiol. 18, 90–100.

Oosterveen, T., Kurdija, S., Alekseenko, Z., Uhde, C.W., Bergsland, M., Sandberg, M., Andersson, E., Dias, J.M., Muhr, J., Ericson, J., 2012. Mechanistic differences in the transcriptional interpretation of local and long-range Shh morphogen signaling. Dev. cell 23, 1006–1019.

Pang, Z.P., Yang, N., Vierbuchen, T., Ostermeier, A., Fuentes, D.R., Yang, T.Q., Citri, A., Sebastiano, V., Marro, S., Sudhof, T.C., et al., 2011. Induction of human neuronal cells by defined transcription factors. Nature 476, 220–223.

Pearson, B.J., Doe, C.Q., 2003. Regulation of neuroblast competence in *Drosophila*. Nature 425, 624–628.

Pereira, J.D., Sansom, S.N., Smith, J., Dobenecker, M.W., Tarakhovsky, A., Livesey, F.J., 2010. Ezh2, the histone methyltransferase of PRC2, regulates the balance between self-renewal and differentiation in the cerebral cortex. Proc. Natl. Acad. Sci. USA 107, 15957–15962.

Peterson, K.A., Nishi, Y., Ma, W., Vedenko, A., Shokri, L., Zhang, X., McFarlane, M., Baizabal, J.M., Junker, J.P., van Oudenaarden, A., et al., 2012. Neural-specific Sox2 input and differential Gli-binding affinity provide context and positional information in Shh-directed neural patterning. Genes Dev. 26, 2802–2816.

Pierani, A., Moran-Rivard, L., Sunshine, M.J., Littman, D.R., Goulding, M., Jessell, T.M., 2001. Control of interneuron fate in the developing spinal cord by the progenitor homeodomain protein Dbx1. Neuron 29, 367–384.

Qian, X., Shen, Q., Goderie, S.K., He, W., Capela, A., Davis, A.A., Temple, S., 2000. Timing of CNS cell generation: a programmed sequence of neuron and glial cell production from isolated murine cortical stem cells. Neuron 28, 69–80.

Rakic, P., 1988. Specification of cerebral cortical areas. Science 241, 170–176.

Roelink, H., Porter, J.A., Chiang, C., Tanabe, Y., Chang, D.T., Beachy, P.A., Jessell, T.M., 1995. Floor plate and motor neuron induction by different concentrations of the amino-terminal cleavage product of sonic hedgehog autoproteolysis. Cell 81, 445–455.

Rouaux, C., Arlotta, P., 2013. Direct lineage reprogramming of post-mitotic callosal neurons into corticofugal neurons *in vivo*. Nat. cell biol. 15, 214–221.

Sander, M., Paydar, S., Ericson, J., Briscoe, J., Berber, E., German, M., Jessell, T.M., Rubenstein, J.L., 2000. Ventral neural patterning by Nkx homeobox genes: Nkx6.1 controls somatic motor neuron and ventral interneuron fates. Genes Dev. 14, 2134–2139.

Sansom, S.N., Hebert, J.M., Thammongkol, U., Smith, J., Nisbet, G., Surani, M.A., McConnell, S.K., Livesey, F.J., 2005. Genomic characterisation of a Fgf-regulated gradient-based neocortical protomap. Development 132, 3947–3961.

Sansom, S.N., Livesey, F.J., 2009. Gradients in the brain: the control of the development of form and function in the cerebral cortex. Cold Spring Harb. perspect. biol. 1, a002519.

Saueressig, H., Burrill, J., Goulding, M., 1999. Engrailed-1 and netrin-1 regulate axon pathfinding by association interneurons that project to motor neurons. Development 126, 4201–4212.

Selkoe, D.J., 2011. Alzheimer's disease. Cold Spring Harb. perspect. biol. 3.

Sharma, K., Sheng, H.Z., Lettieri, K., Li, H., Karavanov, A., Potter, S., Westphal, H., Pfaff, S.L., 1998. LIM homeodomain factors Lhx3 and Lhx4 assign subtype identities for motor neurons. Cell 95, 817–828.

Shen, Q., Wang, Y., Dimos, J.T., Fasano, C.A., Phoenix, T.N., Lemischka, I.R., Ivanova, N.B., Stifani, S., Morrisey, E.E., Temple, S., 2006. The timing of cortical neurogenesis is encoded within lineages of individual progenitor cells. Nat. Neurosci. 9, 743–751.

Shi, Y., Kirwan, P., Smith, J., MacLean, G., Orkin, S.H., Livesey, F.J., 2012a. A human stem cell model of early Alzheimer's disease pathology in Down's syndrome. Sci. transl. med. 4. 124ra129.

Shi, Y., Kirwan, P., Smith, J., Robinson, H.P., Livesey, F.J., 2012b. Human cerebral cortex development from pluripotent stem cells to functional excitatory synapses. Nat. Neurosci. 15, 477–486. S471.

Smith, S.T., Jaynes, J.B., 1996. A conserved region of engrailed, shared among all en-, gsc-, Nk1-, Nk2- and msh-class homeoproteins, mediates active transcriptional repression *in vivo*. Development 122, 3141–3150.

Son, E.Y., Ichida, J.K., Wainger, B.J., Toma, J.S., Rafuse, V.F., Woolf, C.J., Eggan, K., 2011. Conversion of mouse and human fibroblasts into functional spinal motor neurons. Cell stem cell 9, 205–218.

Sousa, V.H., Fishell, G., 2010. Sonic hedgehog functions through dynamic changes in temporal competence in the developing forebrain. Curr. opin. genet. dev. 20, 391–399.

Srinivasan, K., Leone, D.P., Bateson, R.K., Dobreva, G., Kohwi, Y., Kohwi-Shigematsu, T., Grosschedl, R., McConnell, S.K., 2012. A network of genetic repression and derepression specifies projection fates in the developing neocortex. Proc. Natl. Acad. Sci. USA 109, 19071–19078.

Sun, Y., Meijer, D.H., Alberta, J.A., Mehta, S., Kane, M.F., Tien, A.C., Fu, H., Petryniak, M.A., Potter, G.B., Liu, Z., et al., 2011. Phosphorylation state of Olig2 regulates proliferation of neural progenitors. Neuron 69, 906–917.

Takahashi, K., Yamanaka, S., 2006. Induction of pluripotent stem cells from mouse embryonic and adult fibroblast cultures by defined factors. Cell 126, 663–676.

Theka, I., Caiazzo, M., Dvoretskova, E., Leo, D., Ungaro, F., Curreli, S., Manago, F., Dell'Anno, M.T., Pezzoli, G., Gainetdinov, R.R., et al., 2013. Rapid generation of functional dopaminergic neurons from human induced pluripotent stem cells through a single-step procedure using cell lineage transcription factors. Stem cells transl. med. 2, 473–479.

Vallstedt, A., Muhr, J., Pattyn, A., Pierani, A., Mendelsohn, M., Sander, M., Jessell, T.M., Ericson, J., 2001. Different levels of repressor activity assign redundant and specific roles to Nkx6 genes in motor neuron and interneuron specification. Neuron 31, 743–755.

Vierbuchen, T., Wernig, M., 2011. Direct lineage conversions: unnatural but useful? Nat. biotechnol. 29, 892–907.

Watanabe, K., Kamiya, D., Nishiyama, A., Katayama, T., Nozaki, S., Kawasaki, H., Watanabe, Y., Mizuseki, K., Sasai, Y., 2005. Directed differentiation of telencephalic precursors from embryonic stem cells. Nat. Neurosci. 8, 288–296.

Weick, J.P., Held, D.L., Bonadurer 3rd, G.F., Doers, M.E., Liu, Y., Maguire, C., Clark, A., Knackert, J.A., Molinarolo, K., Musser, M., et al., 2013. Deficits in human trisomy 21 iPSCs and neurons. Proc. Natl. Acad. Sci. USA 110, 9962–9967.

Wichterle, H., Lieberam, I., Porter, J.A., Jessell, T.M., 2002. Directed differentiation of embryonic stem cells into motor neurons. Cell 110, 385–397.

Wonders, C.P., Anderson, S.A., 2006. The origin and specification of cortical interneurons. Nat. Rev. Neurosci. 7, 687–696.

Ye, W., Shimamura, K., Rubenstein, J.L., Hynes, M.A., Rosenthal, A., 1998. FGF and Shh signals control dopaminergic and serotonergic cell fate in the anterior neural plate. Cell 93, 755–766.

Ying, Q.L., Stavridis, M., Griffiths, D., Li, M., Smith, A., 2003. Conversion of embryonic stem cells into neuroectodermal precursors in adherent monoculture. Nat. biotechnol. 21, 183–186.

Yokota, Y., 2001. Id and development. Oncogene 20, 8290–8298.

Zhang, Y., Pak, C., Han, Y., Ahlenius, H., Zhang, Z., Chanda, S., Marro, S., Patzke, C., Acuna, C., Covy, J., et al., 2013. Rapid single-step induction of functional neurons from human pluripotent stem cells. Neuron 78, 785–798.

Chapter 17

Retinal Development

Andrea S. Viczian and Michael E. Zuber

Ophthalmology Department, Center for Vision Research, SUNY Eye Institute, Upstate Medical University, Syracuse, NY

Chapter Outline

GLOSSARY

Ciliary margin The most peripheral region of the retina. In the frog and fish, this region contains dividing retinal progenitor cells that contribute to retinal growth and/or repair.

Extrinsic factors Extracellular molecules such as secreted signaling proteins and their transmembrane receptors that provide a cell with information about its environment.

Eye field or eye primordia A domain in the anterior neural plate that is specified to give rise to eye tissue. This domain expresses a combination of eye field transcription factors that are essential for eye formation.

Induced Pluripotent Stem (iPS) Cells A type of stem cell that is derived from somatic or non-pluripotent cell sources by misexpressing transcription factors present in pluripotent stem cells.

Misexpression The addition of RNA, DNA or protein into a developing embryo that naturally receives less or none of that same product.

Neural retina The sensory neuroepithelium of the eye that contains multiple neuronal cell types organized into three cellular layers. This tissue is part of the central nervous system, and it is specialized for transducing light information and transmitting it to the brain.

Optic vesicle An evagination of the eye domain from the lateral walls of the embryonic forebrain that will give rise to eye tissues, including the neural retina and retinal pigment epithelium.

Retinal pigment epithelium (RPE) A single layer of pigmented epithelial cells that lie behind the neural retina. These cells are necessary for the maintenance of visual function and form a part of the blood/brain barrier in the eye.

Retinal progenitor cell A dividing cell that will give rise to multiple cell types in the neural retina but that is limited in potential and not capable of prolonged self-renewal.

Transdifferentiation When differentiated cells transform directly from one cell type to another by the addition of extrinsic or intrinsic factors.

SUMMARY

- The retina contains a tri-layered structure, made of seven unique classes of cells, that is evolutionarily conserved in all vertebrates.

Principles of Developmental Genetics.

- The retina and pigmented epithelium form from the same region of the anterior neural plate.
- Signaling pathways and transcription factors necessary for retina formation are conserved across the vertebrate species studied.
- Transplantation of fetal retinal sheets or individual retinal progenitor cells, derived from pluripotent cells, can repair vision.
- Induced pluripotent (iPS) cells have been successfully used to model a human retinal degenerative disease.

17.1 INTRODUCTION

The retina is an outcrop of the central nervous system. Even though it is a relatively simple tissue, containing only three layers of cells with six classes of neurons and one class of glia, it has been a struggle to understand how this tissue forms. As early as 1832, the chick embryo was used to study early eye formation. At that time, controversy existed over whether eyes form from a single field of cells that splits into two, or start out separately. This was an important question as cylopia was a condition suffered by newborns. By carefully examining early chick eye formation, Emil Huschke concluded that cyclopia is the result of a single eye field failing to separate (Adelmann, 1936). Throughout the decades that followed, others have confirmed this fact and observed a single eye primordia in mouse, as well as in human fetuses. Over the years, model organisms, like the fly, frog, fish, mouse and chick, have provided a basic understanding of the mechanisms driving eye formation. This work has pioneered our understanding of human eye development.

Each model organism provides a different set of answers to questions about eye development. The fly is an excellent model for understanding signaling pathways and the molecules that drive them. The frog and chick have been used to understand the early events in neural patterning and eye formation, since their embryos are accessible during these early stages of development. The teleost fish, particularly zebrafish, is amenable to genetic alterations and can be visualized due to their transparency during embryonic development. Finally, the mouse has been genetically manipulated to better understand the function of various proteins of the mammalian eye. Altogether, studying how the eye grows in each organism has led to significant gains in our understanding of the molecular mechanisms that drive eye formation. This knowledge has recently been used to generate human retinal cells from pluripotent cell sources. These cells are now being tested for their ability to repair vision in blind patients. Therefore, the knowledge gained from studying model organisms has been extremely valuable.

In this chapter we will discuss the molecules and signaling pathways that drive eye formation in vertebrates. We will first discuss the early patterning that occurs in the neural plate, through the formation of the optic vesicle and cup, up to the formation of the neural retina and pigmented epithelium. We will also discuss the work done in cell culture studies to generate retinal cells from pluripotent cell sources and how this work is being used to model retinal disease and repair the damaged or degenerating retina.

17.2 EYE FIELD SPECIFICATION

The vertebrate eye begins as a small region of the anterior neural plate called the eye field (Figure 17.1). Eye field cells are multipotent, and during normal development they generate the cell types of the neural retina as well as retinal pigment epithelium. The process by which one region of the vertebrate neural plate becomes destined to form the retina is called

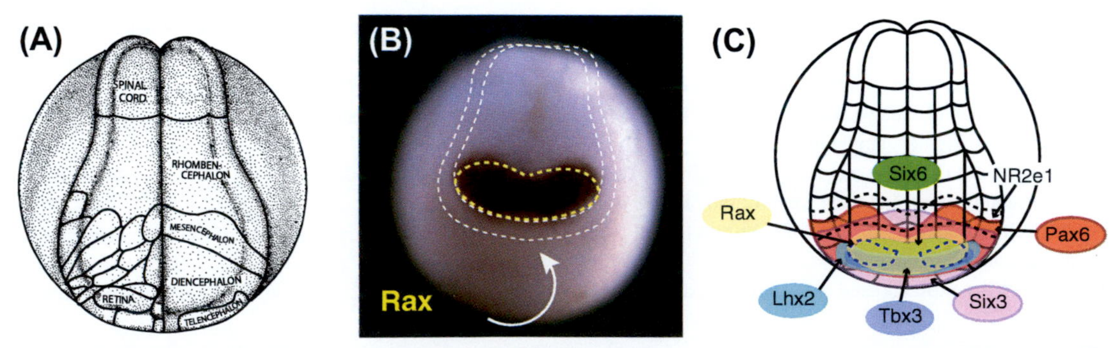

FIGURE 17.1 Eye Field Transcription Factors have an Overlapping Expression Pattern. (A) Fate map of the neural plate in *Xenopus laevis* embryo before gastrulation, when the eye is specified (modified from Eagleson and Harris, 1990). (B) Image of a *Xenopus* embryo stained with an anti-sense RNA probe for Rax. The embryo has been rotated (arrow) to visualize the anterior neural plate, where Rax is expressed (dark area outlined in yellow). Dashed white lines trace the border of the neural plate. (C) The expression pattern of eye field transcription factors, or EFTFs, *Pax6*, *Rax*, *Tbx3*, *Six3*, *Six6* and *Nr2e1,* with anterior neural patterning transcription factor, *Otx2*. Note that six of the seven transcription factors are expressed in the eye field.

eye field specification. Embryos from which the eye field has been removed do not form eyes. If cultured alone or transplanted to another location on the body, amphibian eye fields will form a morphologically normal retina (Adelmann, 1929; King, 1905; Spemann, 1901 and reviewed in Spemann, 1938). From these experiments, and similar studies in the chick (Kobayashi et al., 2009), it can be concluded that very early in development, the eye field is specified to generate cells of the eye, and that all the genes required for this process have been expressed prior to, or at, neural plate stages. This section outlines the genes and signaling systems that specify the eye field within the neural plate and separate it into the two distinct eye primordia that will eventually form the two eyes.

Early Neural Plate Patterning

The frog, *Xenopus laevis*, has been used as a tool to study early events in eye development for over a hundred years. It is an excellent model system since the developing embryos are relatively large (~1.5 mm) and grow in a simple salt solution at room temperature. Similar developmental stages in mammals are difficult, if not impossible, to study in the tiny, mammalian embryos that develop *in utero*. Remarkably, the molecular mechanisms discovered in *Xenopus* have largely been conserved during mammalian development.

The eye field is not established until late gastrulation in the frog. Nevertheless, shortly after fertilization, only a subset of the blastomeres normally generate progeny that go on to form retinal cells. After only the fifth cleavage (32 cell stage), a "retinogenic zone" is already established (Huang and Moody, 1993; Moody, 1987; Moore and Moody, 1999). Only 9 of the 32 blastomeres at this stage are competent (but are not yet committed) to form retina. Cell-cell interactions are important in restricting the location of the retinogenic zone to the animal half of the embryo. If a blastomere from the retinogenic zone is transplanted from the animal to the vegetal side of the embryo, it will no longer generate retinal-destined progeny (Moore and Moody, 1999). Since zygotic transcription is not active until later developmental stages, maternal determinants are thought to regulate retinal competence at these early stages. Consistent with this idea, the vegetally-enriched maternal transcription factor VegT and the growth factor Gdf1 (also known as Vg1) can inhibit retinogenic blastomeres from forming retina. Conversely reducing VegT and Gdf1 levels enlarges the retina, and can even induce a retinal fate in vegetal blastomeres if Bone Morphogenic (BMP) and Wingless-Int (Wnt) protein signaling are simultaneously repressed (Yan and Moody, 2007).

During gastrulation, cells that involute and migrate along the inside of the blastocoel toward the animal pole induce the overlying cells to form neural tissue. Neural induction is regulated by Wnts, Fibroblast Growth Factors (FGFs), and BMP inhibitors (see Chapter 11). Early Wnt signaling induces the formation of the gastrula organizer (known as the shield, blastopore and node in fish, frogs and mammals, respectively). Diffusible FGF ligands and BMP antagonists (including Chordin, Noggin, Activin and Norrin) from the endoderm and gastrula organizer initiate and maintain a neural fate in the neuroectoderm. In their absence, BMP signaling in the non-neural ectoderm would prevent neural induction. Soon after neural induction is initiated, the role of Wnt signaling switches from being inductive to repressive with respect to eye formation. In addition to neural plate formation, the surrounding non-neural ectoderm must also be maintained for the appropriate formation of tissues bordering the neural plate. After neural induction, Wnts and BMPs both function to maintain these non-neural ectodermal regions. As mentioned earlier, BMP inhibitors restrict BMP signaling. Wnt antagonists (including Dickkopf [Dkk] and Secreted Frizzled Related Proteins [SFRPs]) that are secreted by the gastrula organizer are involved in holding Wnt signaling at bay.

As a derivative of the neural plate, eye formation is tightly coordinated with neural induction. In fact, at early developmental stages, almost all of the neural plate has the capacity to form eyes, since in this neural induction or activation phase, the entire dorsal ectoderm has an anterior neuroectoderm bias (Saha and Grainger, 1992). The activation phase, however, is soon followed by a transformation phase, in which more caudal regions of the neural plate are patterned to form the midbrain and hindbrain, whereas more rostral regions are fated to generate the forebrain; including the eyes.

The signaling systems (Wnt, BMP and FGF) required for neural induction continue to function during neural patterning stages. In addition, Nodal and Retinoic Acid signaling also regulate the anterior-posterior patterning of the neural plate (see Chapter 11). These signaling systems also activate secondary signaling centers, and together they pattern the presumptive forebrain (prosencephalon) into the tissues it will eventually form; including the telencephalon, eyes, hypothalamus and diencephalon. Inactivation or modification of any one of these signaling systems can result in abnormal eye formation.

As three of the five signaling systems mentioned are also required for normal neural induction, distinguishing direct versus indirect consequences for eye formation has been difficult. Most of the information about the role of these signaling systems has come from animal model systems, primarily fish and frogs. For example, fish mutants have clearly demonstrated a requirement for very precise control of Wnt signaling (reviewed in Fuhrmann, 2008). Even small changes in the level of Wnt signaling (resulting in either an increase or a decrease) result in changes in neural patterning and eye formation.

Canonical Wnt signaling results in the accumulation of the transcriptional regulator β-catenin, which interacts with transcription factors to regulate the expression of Wnt target genes. If cells expressing excessive Wnt1 or Wnt8b are placed in the presumptive forebrain of a zebrafish, the anterior telencephalon and eyes are reduced, while the more posterior diencephalon is expanded in size. A phenotype similar to excessive Wnt signaling is also observed in the zebrafish mutant called *masterblind (mbl)*. This mutation affects the gene Axin, which is required to degrade β-catenin. As a result, *mbl*$^{-/-}$ fish have excessive Wnt signaling, again resulting in a smaller telencephalon and eyes and a larger diencephalon. These results alone could be interpreted to mean Wnt signaling is incompatible with eye field and eye formation. However, dire consequences are also observed when too little Wnt signaling takes place during neural patterning. Wnt antagonists are expressed by the most anterior cells of the neural plate. These include secreted proteins, such as Dkk1, that mimic Wnts and bind to (yet do not activate) Wnt receptors, as well as secreted proteins that mimic Wnt receptors, such as SFRPs, that bind to Wnts and prevent them from interacting with Wnt receptors. Excessive expression of Wnt signaling antagonists results in expansion of the telencephalon and eyes at the expense of diencephalon. In the complete absence of Wnt antagonists, the forebrain degenerates. In summary, disrupting Wnt activity by modifying the ligands or their signaling cascade has dramatic effects on eye formation. Either too much, or too little Wnt activity can lead to abnormal eye formation, and all evidence suggests that a precise Wnt activity gradient is required to appropriately position the eye field in the forebrain.

Retinal Specification

The extrinsic factors and signaling systems that first induce and then pattern the early neural plate also regulate the expression and activity of key intracellular molecules. These include a group of transcription factors known collectively as the eye field transcription factors (EFTFs), which are required together, and coordinate the specification of the most anterior neural plate to form the eye field. The EFTFs have distinct expression patterns during embryonic development; however, at the time of eye field specification, the expression patterns of these genes coincide with the location and timing of eye field specification (Figure 17.1B, C). This is true not only in fish and frogs, where their roles in vertebrate eye formation have been most extensively studied, but also in mammals. Functional inactivation of Pax6, Six3, Six6, Rax, Lhx2 and Nr2e1 results in abnormal or the absence of eyes in flies, frogs, fish, and/or rodents (reviewed in Zuber, 2010). Mutations in human EFTFs also affect development of the eyes. Although null mutations are rarely observed due to the severity of the phenotypes, point mutations, insertions and deletions that result in altered protein function result in a variety of genetic disorders including cornea and iris abnormalities, cataracts, aniridia, microphthalmia, anophthalmia and holoprosencephaly (reviewed in Gregory-Evans et al., 2013).

The evidence indicates that individually, the EFTFs play distinct roles during early eye formation. For example, the Sine oculis-related homeobox-3 (*Six3*) gene is critical not only for eye formation, but also for early patterning of the neural plate. As discussed previously, Wnt and BMP signaling must be held at bay in the neural plate in order for the eye field to form. Work in both mice and fish demonstrates that Six3 is required to repress Wnt and BMP signaling. *Six3*-null mice die at birth and have a severely reduced prosencephalon and a corresponding expansion of the Wnt1 expression domain. Experiments in both chick and fish embryos show that Six3 is a direct negative regulator of Wnt1 expression. In fish, Six3 also binds directly to the *Bmp4* promoter to repress its expression, thereby protecting the eye field from BMP signaling. If *Six3* expression is lost, the eye field and eyes are reduced in size. The zebrafish mutant *chordino* has a mutation in the BMP-inhibitor *chordin gene*, which results in an increase in BMP signaling and a reduction in eye size. Expression of additional Six3 in *chordino* fish can rescue the small eye phenotype. Six3 is also clearly required for normal human forebrain development, as a variety of *Six3* allelic variants have been linked to holoprosencephaly in which the forebrain fails to develop normally.

In addition to providing an environment in which the eye field can be specified, EFTFs also regulate the proliferation within the anterior neural plate that is necessary to generate the appropriate number of retinal cells required for normal eye formation. An excess of the EFTF Six6 in the frog stimulates the proliferation of eye field cells, resulting in a dramatic increase in the size of both the eye field and eventually the eyes to nearly twice their normal size. Mice lacking functional Six6 have eyes of reduced size. Six6 forms a functional complex with the transcription factor Dachshund (Dach) and together they directly co-repress the expression of the cyclin-dependent kinase inhibitor p27Kip1, which is required for cell cycle progression. Consistent with these results, *Six6* mutations have been linked to microphthalmia in a human patient.

In addition to being individually required for normal eye formation, the EFTFs are sufficient, when combined, to form eyes. Coordinated expression of the EFTFs Tbx3, Rax, Pax6, Six3, Nr2e1, and Six6 with the anterior neural patterning transcription factor Otx2 is sufficient to induce ectopic eye fields and eyes in frog tadpoles. When expressed in pluripotent frog stem cells, then transplanted to host tadpoles, these same transcription factors were shown to generate not only morphologically normal, but also functional eyes. Their requirement for normal eye formation in multiple species and their ability to convert pluripotent cells to an eye field/retinal stem cell fate, has made EFTF expression a key indicator for the successful conversion of other pluripotent cells to retinal progenitor/stem cell lineages (see the section "Embryonic stem cell-derived retinal progenitor").

It is still unclear how the neural patterning events direct the expression of the eye field transcription factors. The neural and head inducers Noggin, Wnts and FGFs can induce the expression of EFTFs and ectopic eyes in frog embryos, however, the mechanisms by which they accomplish this remain unclear. A purine-mediated signaling pathway and the modulation of transmembrane voltage potential have both been shown to be required and sufficient for the expression of EFTFs and eye formation. However, the genetic and molecular mechanisms by which these pathways drive EFTF expression and eye formation remain elusive, as do the downstream targets of the EFTFs that are required to form the eye. Nevertheless, these results demonstrate that EFTFs form a genetic network that has been partially conserved during evolution, which, in the anterior neural plate of vertebrate embryos, is required for eye primordia formation.

Morphogenetic Movements

In all the vertebrates studied, the eye field at neural plate stages is a single, bilaterally symmetric crescent of cells crossing the embryonic midline. Consequently, the retinal progenitor cells of the eye field must be separated into two distinct fields of cells that will eventually form the right and left eye. If this does not happen, a single large midline eye will form – a condition known as cyclopia. The precise mechanisms that regulate eye field separation are species specific, however, it is known that both the Nodal and Hedgehog signaling systems are required. The zebrafish *cyclops (cyc)* gene is a Nodal-related member of the Transforming Growth Factor-β (TGFβ) family of secreted signaling molecules. The morphogenic movements of fish forebrain cells required for eye field separation do not take place in *cyc* mutant fish. Mutations in another Nodal ligand (*squint*), Nodal cofactors (*one-eyed pinhead*) and transcription factors that modulate Nodal signaling (*schmalspur*) all result in cyclopic fish. Although mice null for Nodal signaling die early in development, genetic combinations that reduce, but do not ablate, Nodal signaling give similar results. If $Nodal^{+/-}$ heterozygous mice are crossed with mice heterozygous for *Nodal receptor*, the animals are cyclopic. Nodal signaling regulates eye field separation at least in part via the Hedgehog (Hh) family of secreted proteins. The transcriptional regulator Smad2, which is controlled by Nodal signaling, directly regulates Sonic hedgehog (Shh) signaling in fish and chickens. Hedgehog signaling represses the midline expression of *Pax6*, and other EFTFs, while activating the expression of the optic stalk marker Pax2, which can itself, repress *Pax6* expression. Fish, mice and humans lacking functional Hedgehog signaling are cyclopic. In addition to genetic mutations, cyclopia can also result from environmental toxins. Research in the 1950s identified the corn lily (*Veratrum californicum*) as the source of the toxin 11-deoxyjervine (also known as cyclopamine). Pregnant ewes, mares and other mammals feeding on corn lilies were found to give birth to one-eyed offspring. Cyclopamine regulates the activity of Smoothened, a receptor for the Hedgehog ligand. Consequently, the action of cyclopamine also appears to be through the Hedgehog signaling pathway.

Vignette: *Drosophila* Genetics: What the Fly told the Vertebrate Eye

As described briefly above, many of the vertebrate EFTFs were originally identified as *Drosophila* genes required for eye formation in the fly, or have fly homologs. In some instances, the vertebrate versions can functionally restore the corresponding fly mutant or induce ectopic eyes in the fly. For example, misexpression of mouse *eyes absent* (*Eya*) and *Pax6* in flies can induce ectopic eye development and rescue eye formation in loss-of-function mutants. Based on its remarkable evolutionary conservation, requirement for normal eye formation in multiple species, and the ability of mammalian *Pax6* homologs to induce ectopic fly eyes, *eyeless/Pax6* was originally hailed as a potential master regulator of eye formation. In initial models, *eyeless* was placed at the top of a hierarchy of genes that drove eye formation in the eye portion of the eye-antennal imaginal disk complex. However, a much more complex set of interactions has emerged. The genes *twin-of-eyeless* (*toy*), *sine oculis* (*so*), *optix*, *eyes absent* (*eya*), *dachshund* (*dac*), *eye gone* (*eyg*), and *twin-of-eyegone* (*toe*) are all required for fly eye formation. Individually or in combinations, these genes can also induce ectopic eyes, sometimes in the absence of *eyeless*. For instance, both *toy* and *optix* can induce ectopic eyes via an *eya*-independent mechanism. In the fly, a model has developed in which the retinal determination genes described above behave as a network with both hierarchical components and regulatory feedback loops

17.3 OPTIC VESICLE AND CUP FORMATION

The eye field undergoes morphological changes to eventually form the eye cup (Figure 17.2). As with eye field specification, they changes are driven by both extrinsic and intrinsic mechanisms. What are the gross morphological changes and the underlying molecular mechanisms that direct these changes?

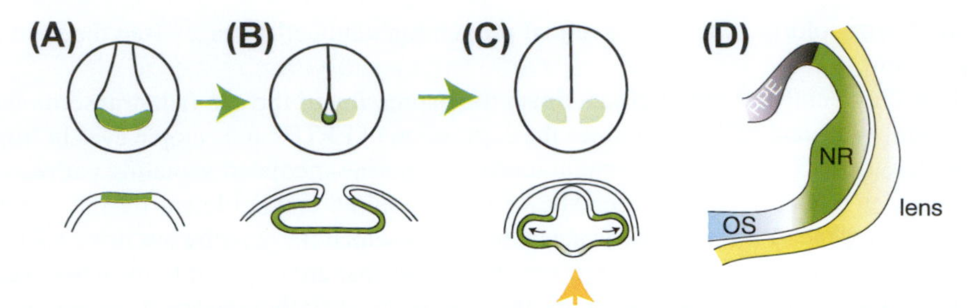

FIGURE 17.2 The Eye Field Develops from the Anterior Neural Plate and Eventually forms the Retina and Retinal Pigment Epithelium, or RPE, and the Optic Stalk. (A) The eye primordium, or eye field (green), develops from the anterior region of the neural plate. The neural plate is outlined in black. (B) As the neural tube closes, the eye field expands. (C) The eye field continues to grow toward the lateral ectoderm while sonic hedgehog signaling at the midline (yellow) separates the eye field into two regions on either side of the midline. (D) This cross-section shows how the eye field has been specified to form the retinal pigment epithelium (RPE), neural retina (NR) and optic stalk (OS). Thickening of both the neural retina and lens primordium (lens) occurs when they come into contact. This is an evolutionarily conserved process.

Optic Vesicle Evagination

The optic vesicle begins forming at around human embryonic day 22 (7–8 paired somites) or mouse embryonic day 8.25 (somite stage 8–14; [Muller and O'Rahilly, 1985; Yun et al., 2009]). At the beginning, the optic sulcus or optic grooves forms bilateral divots in the forebrain region of the neural plate, which expand laterally as the neural plate curls up to form the neural tube (Figure 17.2B,C). The optic vesicle evaginates from the diencephalon until it comes into contact with the overlying ectoderm (Figure 17.2D). When evagination or optic vesicle formation is arrested, it leads to anophthalmia or the absence of eyes (one or both). Anophthalmia is caused by mutations in many different genes. Careful examination of animal models carrying genetic mutations can reveal the developmental stage at which they are required. For example, *Otx2* mutants fail to form forebrain/midbrain, *Six3* mutants fail to form forebrain whereas *Rax* mutants fail to form optic grooves, suggesting these transcription factors are necessary in specifying early anterior neural plate, forebrain, and just eye field, respectively (Acampora et al., 1995; Carl et al., 2002; Lagutin et al., 2003; Mathers et al., 1997). In contrast, null mutants of *Lhx2* develop optic vesicles but not optic cups (Porter et al., 1997). This is because Lhx2 controls, among other things, *BMP7* expression. Whereas BMP signaling needs to be repressed in eye field formation, it is required for proper optic vesicle patterning. BMP signaling activates *Pax2* expression in the pre-optic stalk region of the optic vesicle. Lhx2 also regulates the genes necessary to specify retinal pigment epithelium (*Mitf*) and neural retina (*Vsx2*), as well as those essential for establishing the dorsal/ventral retinal patterning (Yun et al., 2009). When *Lhx2* expression is ablated during the late developmental stages of optic vesicle morphogenesis, the mutant expresses ectopic markers for the thalamic eminence and anterodorsal hypothalamus, suggesting that Lhx2 is necessary to maintain optic vesicle identity (Roy et al., 2013).

When the optic vesicle contacts the overlying ectoderm, the overlying ectoderm thickens, as does the optic vesicle. The overlying ectoderm forms the lens placode while the distal optic vesicle folds back upon itself (invaginates) to become the optic cup. Three dimensional eye cups cultured from mouse embryonic stem cells have provided a great advance toward creating retinal tissue in a petri dish (Eiraku et al., 2011). For the first time, optic cup evagination could be studied outside the embryo, away from the overlying ectoderm. Neuroepithelial cells, expressing GFP under the control of the *Rax* promoter, were observed to invaginate on their own, suggesting that these cells contain the intrinsic instructions for this type of morphogenetic movement (Eiraku et al., 2011).

Neural Retina Versus Retinal Pigment Epithelium Versus Optic Stalk Specification

Patterning of the optic vesicle begins before optic cup invagination (Figure 17.2D). From lineage analyses, we know that RPE forms from the proximal/dorsal region; neural retina forms from the distal/dorsal region and the optic stalk forms from the proximal/ventral region. Activin signaling from the extraocular mesenchyme promotes RPE formation, FGF from the lens placode promotes retina and Shh signaling from the ventral midline promotes optic stalk formation (Chiang et al., 1996; de Iongh and McAvoy, 1993; Fuhrmann et al., 2000). After invagination, the neural retina and RPE continue to grow circumferentially. Where the optic stalk and the ventral retina meet, the optic fissure forms and is progressively sealed, first proximally, then it zips closed until it is sealed distally. After joining together, the optic stalk and neural retina interact through the newly formed optic disc. Ganglion cell axons exit the retina through the optic disc, and the hyaloid artery enters the retina through the same structure. If the optic fissure fails to close in humans, this leads to ocular coloboma. In severe cases, this disease causes blindness in children (reviewed in Heavner and Pevny, 2012).

It was recently discovered that FGF signaling (FGF-Frs2α-Shp2-Ras-ERK) plays a crucial role in optic fissure closure and optic disc specification. Deletion of the FGF receptors 1 and 2 causes downregulation in *Pax2* expression in the optic fissure and upregulation of *Mitf* (Cai et al., 2013). This leads to ocular coloboma in mice. In the neural retina, Vsx2 (formerly Chx10) represses *Mitf*; if *Vsx2* is knocked out, then the neural retina becomes RPE (Horsford et al., 2005). If Mitf fails to function, then the RPE becomes neural retina (Scholtz and Chan, 1987). The antagonistic regulation between these two transcription factors is at the heart of the specification of neural retina and RPE.

Axial Patterning of the Neural Retina

The dorsal retina expresses BMP and the ventral retina receives a Shh signal. These two pathways antagonize each other to generate the correct dorsal/ventral pattern necessary for eye formation. BMP signaling induces dorsal *Tbx5* expression, whereas Shh signaling induces ventral *Pax2* and *Vsx2* expression. If Tbx5 is overexpressed ventrally, *Pax2* and *Vsx2* expression is repressed. Similarly, if dorsal Vsx2 expression is increased, *Tbx5* expression is repressed. Similar mechanisms are at play in the creation of the anterior (nasal) to posterior (temporal) gradient of the optic vesicle. Foxg1 regulates the expression of the anterior genes, whereas Foxd1 regulates posterior genes. All four of these transcription factors work together to establish their own quadrant of retinal tissue (reviewed in Gregory-Evans et al., 2013).

17.4 RETINAL PROGENITOR PROLIFERATION

Retinal progenitor cells proliferate and then exit the cell cycle to form differentiated retinal cells. Retinal malformation occurs if too many or too few retinal progenitor cells fail to exit the cell cycle. The cell cycle contains two gap phases, G_1 and G_2. These gap phases are separated by a DNA synthesis (S-phase) and cell division phase (M-phase). Cyclins, cyclin-dependent kinases (CDK) and their inhibitors control when proliferating cells undergo cell cycle exit, which leads to their differentiation. Transcription factors that play a role in differentiation have also been found to regulate or be regulated by these same cell cycle affectors.

Cell Cycle Regulation and Maintenance

The generation of differentiated neural cells is controlled by the length of the G_1 phase of the cell cycle. With each division, the G_1 phase increases. This is due to the accumulation of CDK inhibitors in the cell. If a chemical is added to a cultured mouse embryo that mimics the action of the CDK inhibitor, thus shortening the G_1 phase, the mouse will undergo premature neurogenesis. This regulation can be directly linked to the transcription factor, Neurogenin2 (Ngn2), which activates the genes required for neural differentiation. CDKs phosphorylate multiple sites on the Ngn2 protein. When Ngn2 is heavily phosphorylated, it fails to bind its target DNA, which leads to a loss of neural differentiation. During a short G_1 phase, CDKs accumulate in the cell and phosphorylate Ngn2, keeping the neural differentiation genes turned off. When the G_1 phase lengthens, CDK levels are low, and Ngn2 can reach its targets and promote neural differentiation (reviewed in Hindley and Philpott, 2012).

During M-phase, DNA replicates when the protein, Geminin, loads mini-chromosome maintenance proteins onto replication origin sites. Geminin also directly interacts with Six3, an eye field transcription factor that is necessary for normal eye development (see the section "Retinal specification"). These two proteins seem to have antagonist roles in retinal cell proliferation. When overexpressed, Geminin causes eye and forebrain defects that are rescued by misexpression of Six3 in the medaka fish. If Geminin protein levels are reduced, then the eye/forebrain is enlarged. This is augmented by overexpression of Six3 (Del Bene et al., 2004). Interestingly, the close family member of Six3, Six6, also causes enlarged eyes when misexpressed (Zuber et al., 1999). Future studies are needed to determine whether Six6 also regulates proliferation through an interaction with Geminin.

Regulation of Cell Death

As the optic cup is forming, retinal progenitor cells divide and occupy the neuroblastic layer. It is during this time and shortly after birth that significant apoptosis is observed in the retina. Consistent with this observation, the pro-apoptotic Bcl-2 family members, including Bax and Bak, are also highly expressed during this period. Removal of these two pro-apoptotic factors during retinal development (double Bax/Bak knockout mice) causes the retinas to be thickened and diminishes developmental apoptosis, suggesting that these proteins play a pivotal role in pruning the developing retina (reviewed in Doonan et al., 2012).

Cell death can also be caused by the expression of a protein at the wrong time or in the wrong amount. CyclinD1 is a cyclin that promotes retinal cell proliferation during development, but if overexpressed in differentiated retinal cells, causes cell death (Skapek et al., 2001). In mutant mice with no CyclinD1, retinal progenitor proliferation is decreased and photoreceptors undergo apoptosis postnatally (Ma et al., 1998). The decreased proliferation in *CyclinD1* knockout mice causes the cell cycle to lengthen in early retina progenitor cells. This leads to early differentiation and a reduction in the number of dividing cells (Das et al., 2009). There is clearly a delicate balance between differentiation, proliferation and apoptosis during retinal development.

17.5 RETINOGENESIS

The retina contains six classes of neurons (rod and cone photoreceptors, horizontal, bipolar, amacrine and ganglion cells) and one glial cell type, the Müller glia (Figure 17.3A). What are the molecular processes that direct the seemingly homogeneous neuroectoderm toward these diverse retinal cell types?

Lineage Restriction

Retinal cells are born in a sequential but overlapping order that is conserved across all vertebrates studied (Figure 17.3B; reviewed in Cepko et al., 1996). At the point where the optic stalk meets the optic cup, retinal differentiation begins in a small patch in the central retina and spreads to the periphery. Ganglion cells are born first but are closely followed by horizontal and cone photoreceptors. The late-born retinal cell types include the rod photoreceptors, amacrine cells, bipolar cells, and, as in the cerebral cortex, the glia cell (Müller glia) differentiates last (Wong and Rapaport, 2009). Early lineage tracing experiments have found that retinal progenitor cell progeny varied in clone size and the cell types made (Holt et al., 1988; Turner et al., 1990). As the retinal progenitor progresses through development, it gradually becomes more restricted until it can only produce late-born retinal cell types. This area of study led scientists to propose the "competence model" of retinal differentiation (reviewed in Gregory-Evans et al., 2013; Livesey and Cepko, 2001).

In the competence model, the retinal progenitor cell changes its competence state over time by changing intrinsic transcription factors. Transcription factor(s) turn on genes that alter the response of the retinal progenitor to extrinsic factors. This model predicts that the transcription factor(s) expressed by any class of retinal progenitor (ganglion progenitor, rod progenitor, etc.) would be the same. A careful examination of individual progenitor cells tells a different story; the heterogeneity of transcription factor levels found among these cells is striking (Trimarchi et al., 2008). When individual rat progenitor cells were cultured and tracked, the lineage order also failed to follow the strict progression seen during retinal development. That is, some progenitors were observed to generate rod photoreceptors *after* Müller glia (Gomes et al., 2011).

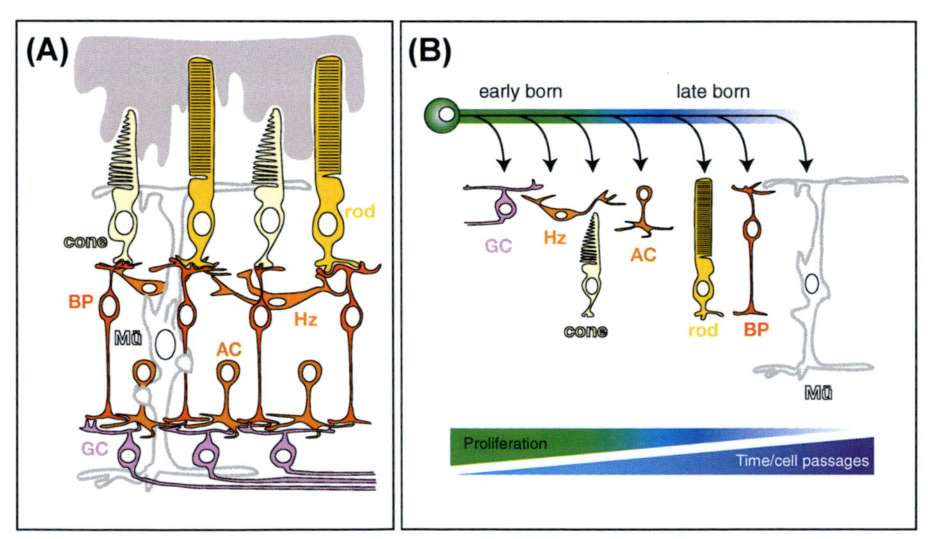

FIGURE 17.3 **The Vertebrate Retina is made of Cells that Generally follow the same Birth Order.** (A) The retina is a tri-layered structure made of six types of *Neurons*: ganglion (GC), amacrine (AC), bipolar (BP), horizontal (Hz) cells and rod (R) and cone (C) photoreceptors; and one type of glia, called Müller glia (Mü). The retinal pigment epithelium sits above the photoreceptor layer (gray area). (B) Retinal cells have a birth order that is conserved among all vertebrates studied. Ganglion cells are born first, closely followed by horizontal, cone and amacrine cells. These are the early-born cell types. The late-born retinal cells, rods, bipolar and Müller glia, emerge from retinal progenitors in an overlapping pattern with the early-born cells. The graded triangles at the bottom indicate how proliferation of the retinal progenitors decreases coincident with the increasing time and cell divisions.

To see if this also occurred *in vivo*, an elegant approach used the transparent, live zebrafish embryo for lineage analysis (He et al., 2012). Progenitors fall into three groups. Those that give rise to: 1) two progenitors (PP); 2) one progenitor and one differentiated cell (PD); or 3) two differentiated cells (DD). The authors observed that ganglion cells overwhelmingly arise from PD progenitors whereas other retinal cell types mostly come from DD progenitors (He et al., 2012). The bHLH transcription factor, Ath5, is required for ganglion cell specification. In *Ath5* mutants, besides losing ganglion cells, retinal differentiation is delayed and the size of the retina increases. Without the Ath5 instruction, the PD progenitor is thought to revert to a PP progenitor, allowing for the observed retinal size increase. This also explains why ganglion cells are observed to be the first retinal cell type born – they are the most frequent cell type to differentiate alongside mitotic progenitors. A stochastic mathematical model, in which retinal cell fate was randomly assigned, produced results that precisely fit the retinal cell fate choices observed in the living retina (He et al., 2012). However, Ath5 is expressed in a particular time/place, arguing that deterministic forces are at work. Most likely, future studies will reveal that both random and deterministic forces are controlling retinal cell fate.

Retinal Cell Class Specification

Many extrinsic factors with complex signaling pathways, affect retinal specification. Several growth factors perpetuate proliferation in retinal progenitors, namely FGF, TGFα and epidermal growth factor (EGF). Shh signaling affects retinal progenitors; however, it acts differently in different model systems. In fish, Shh induces a wave of ganglion cell differentiation. In contrast, in the mouse and chick, Shh is expressed by ganglion cells and promotes retinal progenitor proliferation. Shh acts by inducing the expression of Hes1, a bHLH transcription factor that negatively regulates differentiation. Shh also plays a role in the multipotency of the retinal progenitor cell. If Shh is prematurely activated in retinal progenitor cells, then the cells erroneously differentiate into amacrine and horizontal cells (reviewed in Gregory-Evans et al., 2013).

Other extrinsic factors also affect retinal cell specification. T3, an active form of the thyroid hormone, causes an enrichment of cone photoreceptors in retina progenitor cultures (Kelley et al., 1995). Rod photoreceptor fate is enriched in rodent retinal progenitor cells treated with Taurine, Retinoic Acid, Activin and/or FGF2 (reviewed in Levine et al., 2000). First identified by studying retinal progenitor cell cultures, these factors are now being used to bias embryonic stem cells toward photoreceptors (see Section 17.7 below).

Another chemical inhibitor, DAPT, blocks Notch signaling and promotes photoreceptor cell fate in stem cell cultures. The idea of blocking Notch in stem cells came from a tremendous amount of work in model organisms. First identified in the fly, Notch signaling occurs when the Notch ligand, Delta-like 1 (Dll1) on the membrane of one cell, binds to the Notch receptor of a neighboring cell. Notch signaling prevents retinal progenitor differentiation. Like Shh, Notch signaling acts through Hes1, which keeps the progenitor cell proliferating (reviewed in Kageyama et al., 2008). When Hes1 expression increases, Dll1 decreases, thereby decreasing Notch signaling in neighboring cells. Overexpression of Dll1 in *Xenopus* retinal progenitors causes an increase in photoreceptor cell fate (Dorsky et al., 1997). Similarly, in the mouse, knockdown of the Notch signal-promoting, DNA-binding protein, RBPj, causes an increase in cone photoreceptors, but here this is seen at the expense of rod photoreceptors (Jadhav et al., 2006; Riesenberg et al., 2009). The battle between these two cell types is also observed by altering internal transcription factors.

Transcription factors regulate genes that restrict retinal progenitors to a particular retinal cell fate (for a recent list, see Andreazzoli, 2009). For example, once a retinal progenitor exits the cell cycle and becomes fated toward photoreceptors, postmitotic cells express the cone-rod homeobox (*Crx*) gene; a subset of Crx-positive cells express the neural retina leucine zipper (NRL) protein. Knockdown of *Nrl* causes the formation of an all-cone retina, as NRL activates rod-specific genes in the putative cone precursor default state of the photoreceptor progenitor. An orphan nuclear receptor, NR2E3, suppresses cone genes while co-activating rod genes with NRL and Crx. Although these downstream factors have been well characterized, only a few proteins that specify photoreceptor progenitor fate have been identified, namely, orthodenticle homeobox 2 (Otx2), Blimp1 and, in *Xenopus*, Otx5b (Nishida et al., 2003; Viczian et al., 2003; Brzezinski et al., 2010; Katoh et al., 2010). Mouse Otx2 was found in dividing progenitor cells that give rise to bipolar, horizontal and photoreceptors, whereas Blimp1 was only expressed by postmitotic photoreceptor precursor cells. In this way, the mitotic retinal progenitor becomes gradually more restricted in fate choice. The next step is to more closely examine the molecular mechanism that distinguishes a mitotic from a postmitotic photoreceptor progenitor.

Transcription factors also regulate cell cycle exit. It has long been known that early cell cycle exit causes early-born retinal progenitors to form early-born retinal cell fates (Ohnuma et al., 2002). Also, heterochronic cultures (early-born incubated with late-born cells) fail to induce early cell fates in late-born cells, and vice versa. This suggests that an internal clock is ticking inside the retinal progenitor. In part, Shh signaling controls this clock by regulating the cell cycle length via CyclinD1 (see the section "Regulation of cell death"). The clock is also regulated by intrinsic factors. Ikaros, a zinc finger transcription factor, is normally expressed in early-born but not late-born retinal progenitors. Misexpressing Ikaros can change the competence of late-born cells to form early-born cells (Elliott et al., 2008). Interestingly, cells from the Ikaros-expressing lineage can form all retinal cell types, suggesting that this transcription factor is permissive for the formation of all retinal cells, not only those fated to become early-born cells (Tarchini et al., 2012).

Transcription factors precisely function in a well-regulated time window. RNA binding factors and microRNAs control this activity. In *Xenopus*, *Otx2*, *Otx5b* and *Vsx1* are all transcribed early in development but are not translated until much later (Decembrini et al., 2006). In *Xenopus*, Otx2 and Vsx1 misexpression causes retinal progenitors to form bipolar cells, whereas Otx5b specifies photoreceptors. If the 3' untranslated region (UTR) of these genes was overexpressed with the coding region, the overexpression phenotype was much more subtle. The 3' UTR of each gene was individually fused to GFP. Remarkably, just this region replicated the spatiotemporal expression pattern of the protein. This process is linked to the internal clock. Adding back these proteins to individual retinal cells, which were forced out of the cell cycle and failed to form rods and bipolar cells, overcame this restraint and they differentiated into these cell types. The authors identified four microRNAs (miR-129, miR-155, miR-214 and miR-222) that bind to the 3' UTR of Otx2 and Vsx1 mRNAs and block their translation, thereby inhibiting premature bipolar cell differentiation (Decembrini et al., 2009). The microRNA processing enzyme, Dicer, has been shown to be essential for retinal cell cycle progression through its regulation of the Notch pathway (Davis et al., 2011; Georgi and Reh, 2011). Shh was used to speed up the cell cycle, whereas cyclopamine, a Shh inhibitor, slowed it down. Therefore, it is likely that both Notch and Shh signaling are tightly coupled to microRNA regulation. This underlines the importance of both extrinsic signaling and intrinsic genetic regulation in generating the correct number and timing of differentiated retinal cells.

Retinal Subtype Specification

In a superficial characterization, the retina is a simple, neatly organized tissue, divided into three layers of cells with six types of neurons and one type of glia. A more careful examination shows that within these seven classes of retinal cells are subclasses or subtypes. While mouse rod photoreceptors have a single wavelength sensitivity, mouse cone cells can function as short-wavelength (S-cone) or middle-wavelength sensitive cones (M-cone), depending on the type of Opsin that they express. The nuclear receptor, Thyroid hormone receptor β2 (TRβ2), is one of the earliest indicators of cone formation. Removal of this receptor causes formation of only the S-cones, not the M-cones (reviewed in Swaroop et al., 2010).

There is also diversity among the other retinal cell classes. So far, it has been determined that there are at least three subtypes of horizontal cells, ten subtypes of bipolar cells, 30 subtypes of amacrine cells and 12 subtypes of ganglion cells (Chen et al., 2012). As with the cones, each subtype has its own molecular signature. The 30 identified amacrine cell subtypes can be categorized into three major classes based on the neurotransmitters they utilize. Transcription factors have been identified that gradually restrict cell fate; first retinal progenitor to amacrine cell, then amacrine cell to each major subclass (reviewed in Gregory-Evans et al., 2013).

In ganglion cell subtype specification, the cell adhesion molecule, Cadherin 6 (Cdh6) is involved. Cdh6 is maintained both in multipotent retinal progenitors and in the progeny that become retinal ganglion, bipolar and amacrine cells. This protein marks a subset of retinal ganglion cells that detects vertical motion, yet it is also expressed in multiple subsets of bipolar and amacrine cells (De la Huerta et al., 2012). This suggests that there are other adhesion proteins or transcription factors that specify these subtypes. Members of the family of POU-domain transcription factors (also known as Brn3) have been shown to be involved. Each Brn3 family member (Brn3a, Brn3b and Brn3c) controls the differentiation of subsets of retinal ganglion cells (Badea et al., 2009; 2012; Badea and Nathans, 2011). While this is a good start, many more genes will need to be identified to understand how all these subtypes are generated.

17.6 ADULT RETINAL STEM CELLS AND RETINAL REGENERATION

Adult retinal stem cells give hope to those affected with damaged or degenerating retinas. Instead of invasive surgical procedures, the idea is to awaken the endogenous repair mechanism stored in these quiescent cells. The study of model organisms with robust retinal regeneration has led to the identification of key signaling pathways that can be exploited in mammals, where these cells naturally fail to repair the blind eye. In order to begin discussing these important findings, we need to first define progenitor and stem cells, and indicate how they differ.

Defining Retinal Progenitors and Stem Cells

Retinal progenitors give rise to specific retinal cell types, as well as new multipotent retinal progenitor cells during the early stages of eye formation. As these cells progress through development, multipotent retinal progenitors become gradually restricted in the types of retinal cells that they will become. In this chapter, we will refer to cells in the embryo as retinal progenitor cells. In contrast, stem cells are defined as cells that not only give rise to restricted cell types but also continually renew themselves (Lanza, 2009). Adult retinal stem cells can clearly be identified in the periphery of the differentiated retina in simpler vertebrates (amphibians and teleost fish). This area, called the ciliary marginal zone in the frog (Figure 17.4A,B) or the circumferential germinal zone in fish, contributes to the growth of the retina while maintaining retinal stem cells in this region (reviewed by Amato et al., 2004). Therefore, we refer to these cells as adult retinal stem cells.

Adult Retinal Stem Cells

Adult retinal stem cells repair the damaged retina in teleost fish and amphibians, generating all retinal cell types. In the post-hatch chick, these peripheral dividing cells are more restricted (Figure 17.4D). Although they are clearly undergoing mitosis, their progeny are limited to two retinal cell types (Fischer and Reh, 2000). If these retinas are treated with growth factors, additional retinal cell types are observed (Fischer and Reh, 2003). The identification of these restricted, yet transformable, stem cells in the post-hatch chick led to the hypotheses that mammalian peripheral retinal cells could have similar characteristics.

At the outer edge of the mammalian retina, the ciliary body contains two epithelial layers; the pigmented and non-pigmented ciliary epithelium (CE; Figure 17.4E). During development, the retinal pigmented epithelial (RPE) progenitors give rise to the pigmented CE, whereas the retinal progenitors give rise to the non-pigmented cells. When isolated and grown in culture, a small population of mouse and human pigmented CE cells proliferate, form floating spheres and appear to express genes normally found in retinal cells. This discovery led to the idea that mouse and human RPE-derived adult retinal stem cells reside in a quiescent state in the periphery of the retina. This idea is in line with studies on embryonic chick retina (~E4). Researchers observed that RPE cells and peripheral retinal cells do not repair the retina on their own, but can do so if they are treated with Shh, BMP and FGF signaling pathway components (reviewed in Fischer and Bongini, 2010). Recent studies of isolated and cultured adult human RPE have also shown that these cells can dedifferentiate, undergo extensive proliferation and, when differentiated again, can become RPE or express markers for mesenchyme and neural cells (Salero et al., 2012). However, careful examination by electron microscopy of the murine RPE-derived CE cultured spheres revealed that all cells maintained their CE characteristics (pigmentation and morphology); that is, they failed to differentiate into *bona fide* retinal cells even after many passages and culture conditions that favor retinal progenitor differentiation (Cicero et al., 2009). This suggests that either mammalian adult retinal stem cells exist in another area of the retina, or that the culture conditions to coax retinal stem cells out of their CE state are yet to be discovered (reviewed by Locker et al., 2010).

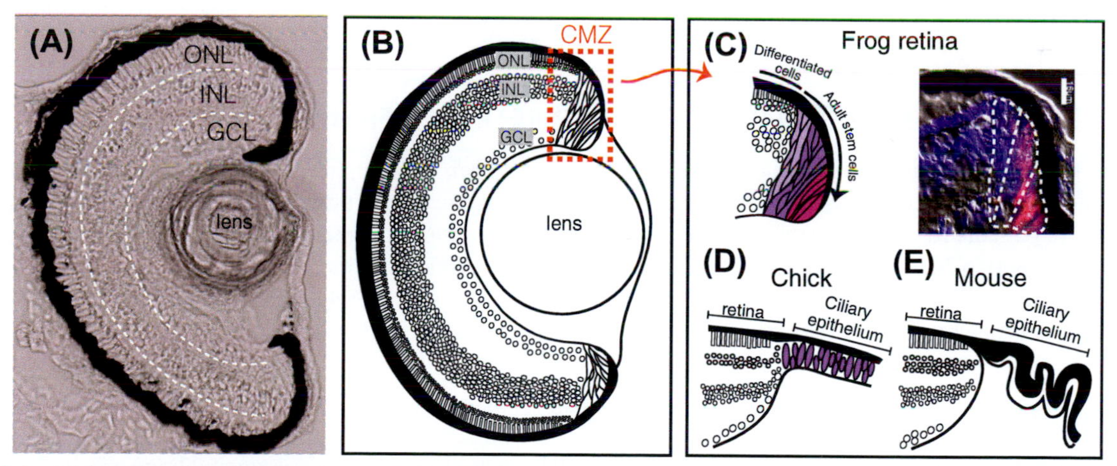

FIGURE 17.4 Adult Retinal Stem Cells are Present in Frog Retina. (A) Cross-section of a *Xenopus laevis* eye reveals the three layers of the retina: the outer nuclear layer (ONL), inner nuclear layer (INL) and ganglion cell layer (GCL). The retina is oriented with dorsal side up. (B) Illustration shows the ciliary marginal zones (CMZ) at the outer edge of the retina, which contain the adult retinal stem cells. These stem cells divide and their progeny add to the retina as the animal grows. (C) The boxed region in panel B is expanded to show how the CMZ is divided into regions. The most peripheral region (dark pink) contains the fastest dividing stem cells while the proximal region (light purple) has the slowest, nearly differentiated cells. A tadpole was injected with EdU, a marker that is incorporated into DNA during mitosis. Four hours later, the animals were anesthetized, fixed and the tissue processed to visualize the EdU-positive cells (magenta dye). The nuclei of the cells is stained with DAPI (dark blue). The peripheral CMZ cells have incorporated this chemical indicating that they underwent cell division. (D–E) Illustration of the peripheral region of the post-hatch chick (D) and mouse retina (E). *Drawings were based on images of retinal slices that were previously published (Cicero et al., 2009; Fischer and Reh, 2000).*

Retinal Regeneration

Retinal regeneration mostly relies on other sources of cells in the retina, besides the adult retinal stem cells. The retinal pigment epithelium and Müller glia are the primary cell source for retinal repair in those animals in which retinal regeneration naturally occurs. These cells undergo a transdifferentiation process in which they stop expressing cell-specific markers, proliferate, express retinal progenitor markers, and then differentiate into the lost retinal cell type(s). Different cells are the primary source of new retinal cells in different species. In fish, the Müller glia are the source of cells for retinal repair, whereas in amphibians, the RPE cells are the primary cells activated. It is interesting to note that both RPE and Müller glia are also required for retinal homeostasis and proper retinal functioning. The exact molecular mechanisms activating these cells during retinal injury is still under investigation. Unsurprisingly, signaling pathways and transcription factors that have been found in the developing retina also play a role in this type of transdifferentiation repair (reviewed in Barbosa-Sabanero et al., 2012).

Classically, retinal regeneration via the RPE has been studied using the newt retina. In these remarkable animals, removal of the entire retina causes the RPE to proliferate and form two new epithelial layers. One of the new layers (non-pigmented layer) continues to proliferate and forms an entirely new retina. If the embryonic chick retina is removed, the RPE undergoes a similar process. This is enhanced by the addition of FGF. Similarly to the effect of FGF, overexpression of the *Pax6* gene causes an increase in neurons from the RPE, suggesting that *Pax6* expression is downstream of FGF signaling. Another eye field transcription factor, Rax, is also required for proper retinal regeneration in *Xenopus* (Martinez-De Luna et al., 2011). Extracellular matrix also plays a role in RPE-derived retinal repair. Regeneration of the *Xenopus* retina in post-metamorphic eyes depends on whether the vascular membrane, rich with laminins, is intact. Activin signaling, which promotes RPE formation during development, inhibits RPE-mediated retinal regeneration (reviewed in Barbosa-Sabanero et al., 2012; Lamba et al., 2008). Future studies will be necessary to better understand the molecular mechanisms driving RPE-derived retinal repair, as they may lead to insights about reactivating these cells in the mammalian retina.

Retinal regeneration via the Müller glia is most evident in teleost fish but also is found in chick retina. In the chick, 80% of Müller cell progeny achieve the proliferation stage, express retinal progenitor markers (*Pax6* and *Vsx2*) but fail to differentiate. The few that go on to differentiate become bipolar cells, amacrine cells and Müller glia. As in RPE-derived regeneration, FGF signaling is key to Müller glia-mediated regeneration in chicks (reviewed in Fischer and Bongini, 2010). In stark contrast, the Müller glia of teleost fish (even in animals without damage) are dividing slowly. The dedifferentiated Müller glia can directly generate any differentiated retinal cell type except rod photoreceptors. To make new rods, Müller glia first form rod progenitor cells, which then give rise to this differentiated cell. A key factor in Müller cell mediated regeneration is the bHLH transcription factor, Achaete-scute homolog 1 (Ascl1). When the fish retina is damaged, Ascl1 is upregulated in proliferating Müller glia; proliferation is blocked when Ascl1 is knocked down (Fausett et al., 2008). Furthermore, when *Ascl1* overexpression is targeted to murine Müller cells in retinal explants, the Müller glia divide into neuron-like cells that express retinal markers and function like neurons (Pollak et al., 2013).

17.7 RETINAL STEM/PROGENITOR CELLS FOR RETINAL REPAIR

The idea of using embryonic retinal tissue to repair the retina has been around since the 1980s. Fetal RPE/retinal sheets isolated from late stage rat embryos (E19–20) integrate and form synaptic connections with the endogenous retina. Phase II clinical trials in patients receiving human fetal sheet transplants showed improvement in seven out of ten patients one year after surgery (20/800 to 20/160 in the most vision-improved patient). While the ethical dilemma of using human fetal tissue makes this type of treatment controversial, the improvement in patient vision clearly identifies the need to develop *in vitro* generated retinal progenitor sheets for optimal vision recovery (reviewed in Seiler and Aramant, 2012).

Embryonic Stem Cell-Derived Retinal Progenitor

Blindness due to Retinitis Pigmentosa, Age-related Macular Degeneration or Leber's congenital amaurosis (LCA) is primarily caused by the loss of photoreceptor cells. Therefore, rather than use sheets of retinal tissue, attempts have also been made to replace individual cells. Transplantation studies on mouse GFP-expressing rod progenitor cells indicate that cells isolated at early postnatal ages have the best chance of integrating and forming characteristic photoreceptor outer segments. As cells of comparable ages exist in a third trimester human fetus, it became obvious that the conditions to grow these cells *in vitro* were necessary. Work in *Xenopus* showed that misexpression of only seven transcription factors (*Pax6*, *Rax*, *Tbx3*, *Six3*, *Six6*, *Nr2e1*, *Otx2*) could transform pluripotent primitive ectoderm into a functional retina (Viczian et al., 2009). Activating these genes in human embryonic stem cells, by adding factors found in the developing eye field (IGF-1, Dkk-1, and Noggin) drove the ES cells to a retinal progenitor cell fate. When transplanted, these human cells were able to partially repair the vision of a mouse model of LCA (*crx*$^{-/-}$ mouse). Independently, a similar protocol was developed in which Wnt and Nodal/Activin signaling was reduced (via Dkk-1 and Lefty-A) and then cells were treated with photoreceptor-inducing extrinsic factor; this generated photoreceptors from human, monkey and mouse embryonic stem cells (reviewed in Lamba et al., 2009). However, transplanting these cells into the adult mouse retina produced tumors rather than differentiated photoreceptor cells (West et al., 2012). By dissociating ES cells and then forcing quick reaggregation in the presence of extracellular matrix proteins and Nodal, the Sasai group has generated beautiful self-organizing three-dimensional eye cups, complete with the retinal lamination pattern seen in the early developing optic cup (reviewed in Sasai et al., 2012). Similar results were observed when they used human ES cells (Nakano et al., 2012). These results suggest a new way forward: generating retinal sheets from artificially generated eye cups. Another possibility is to generate artificial scaffolds to grow retinal sheets of ES-derived retinal cell (McUsic et al., 2012). Recent studies show that ES-derived photoreceptors generated in three-dimensional eye cups can integrate into the degenerating retina (Gonzalez-Cordero et al., 2013).

Induced Pluripotent Stem Cell Derived Retinal Progenitor

A less ethically charged pluripotent cell source is the induced pluripotent stem cell or iPS cell. The somatic cell, obtained from a biopsy of a patient's skin or blood, can be reprogramd to form ES-like cells. This has the added advantage in cell replacement therapy that immune rejection will not be an issue. Each iPS cell line is slightly different from the next and needs to be tested prior to use. Retinal cells and RPE can be generated by culturing iPS cells in the same retinal cell-inducing extrinsic factor mix that was described for ES cells (Hirami et al., 2009; Lamba et al., 2010). Another retinal cell promoting protocol was developed to take advantage the fact that iPS cell lines express higher levels of Pax6 during differentiation. Higher levels of Pax6 corresponded with cells that naturally adopted an anterior neuroectoderm default state, which is also found for human ES cells. Both human ES cells and this iPS cell line were successfully transformed into retinal cells and RPE by culturing them through a "stepwise differentiation process" that does not require extrinsic factors (Buchholz et al., 2009; Meyer et al., 2009). During the differentiation procedure, the optic vesicle-like iPS cells could be identified by morphology (Meyer et al., 2011), and they formed laminated structures, reminiscent of an early eye cup (Phillips et al., 2012). RPE cells generated this way were also used in a cell replacement therapy to improve vision in rat and mouse models of Retinitis Pigmentosa (Carr et al., 2009; Li et al., 2012). The Japanese government has recently approved the use of these cells for cell replacement therapy in humans (Cyranoski, 2013).

Modeling Retinal Disease

Another advantage of iPS cells is that they can be used to model retinal degenerative diseases, as was nicely demonstrated in a recent proof-of-principle report (Meyer et al., 2011). In this study, the authors generated iPS cell lines from a patient with gyrate atrophy, a disease that primarily affects RPE but eventually leads to blindness. The patient's fibroblasts were used to generate iPS cells which were then transformed into RPE-like cells. These newly generated cells exhibited the expected disease-related problems. Using higher than normal pharmacological intervention to treat the disease, the dysfunctional iPS-derived RPE recovered. Genetically correcting the known mutation relieved the cells of the disease-related symptoms. These results show how iPS cells can be used to model and treat diseases of individual patients (Meyer et al., 2011). However, bringing this technology to individual patients would be very costly, since iPS cell line production and screening is a lengthy process. Future studies could focus on yielding faster and more reliable ways to generate these cells for individualized medicine.

17.8 CONCLUSIONS AND FUTURE DIRECTIONS

The study of model organisms has clearly led to a better understanding of neural retinal and RPE development in humans. We now have the tools to generate pluripotent-derived cells in culture from a patient's own skin or blood cells. This remarkable advance is firmly rooted in evidence gathered from studies on fish, fly, chick, frog and mouse eye development. To show that pluripotent stem cells could be generated from somatic tissue, Sir John Gurdon experimented on the frog, *Xenopus laevis*; Dr. Shinya Yamanaka identified the transcription factors that transform somatic cells into pluripotent cells in the mouse; and this work was collectively awarded the Nobel Prize in Physiology or Medicine in 2012. From this work and many other studies, scientists have discovered how to generate RPE and neural retina in the culture dish. These cells are now being tested in clinical trials of cell replacement therapy. If sight restoration is successful, the cure can be directly attributed to the hard work of basic scientists and clinicians alike.

17.9 CLINICAL RELEVANCE

- Fetal sheet transplantations have undergone Phase II clinical trials, in which they have been shown to improve vision in seven out of ten patients.
- Skin or blood can be used to generate induced pluripotent (iPS) cells, which can be further cultured to form retina and RPE.
- Induced pluripotent (iPS) cells naturally form RPE in culture, and these are being tested in cell replacement therapy protocols.
- Clinical trials using retinal progenitors derived from iPS cells for cell replacement therapy have been approved in Japan.

RECOMMENDED RESOURCES

Beccari, L., Marco-Ferreres, R., Bovolenta, P., 2013. The logic of gene regulatory networks in early vertebrate forebrain patterning. Mech. Dev. 130, 95–111.

Gregory-Evans, C.Y., Wallace, V.A., Gregory-Evans, K., 2013. Gene networks: Dissecting pathways in retinal development and disease. Prog. Retin. Eye Res. 33, 40–66.

Kiecker, C., Lumsden, A., 2012. The role of organizers in patterning the nervous system. Annu. Rev. Neurosci. 35, 347–367.

Sinn, R., Wittbrodt, J., 2013. An eye on eye development. Mech. Dev. 130, 347–358.

Movie showing morphogenetic movements in eye development [Trevor D. Lamb, Shaun P. Collin & Edward N. Pugh, Jr., 2007 Evolution of the vertebrate eye: opsins, photoreceptors, retina and eye cup. Nat. Rev. Neurosci. 8, 960–976]: http://www.nature.com/nrn/journal/v8/n12/extref/nrn2283-s1.swf (accessed 18.11.13).

REFERENCES

Acampora, D., Mazan, S., Lallemand, Y., Avantaggiato, V., Maury, M., Simeone, A., Brulet, P., 1995. Forebrain and midbrain regions are deleted in Otx2–/– mutants due to a defective anterior neuroectoderm specification during gastrulation. Development 121, 3279–3390.

Adelmann, H., 1936. The problem of cyclopia. Q. Rev. Biol. 11, 161–182.

Adelmann, H.B., 1929. Experimental studies on the development of the eye. I. The effect of the removal of the median and lateral areas of the anterior end of the urodelan neural plate on the development of the eyes (*Triton teniatus* and *Amblystoma punctatum*). J. Exp. Zool. 54, 249–290.

Amato, M.A., Arnault, E., Perron, M., 2004. Retinal stem cells in vertebrates: parallels and divergences. Int. J. Dev. Biol. 48, 993–1001.

Andreazzoli, M., 2009. Molecular regulation of vertebrate retina cell fate. Birth Defects Res. C. Embryo. Today 87, 284–295.

Badea, T.C., Cahill, H., Ecker, J., Hattar, S., Nathans, J., 2009. Distinct roles of transcription factors brn3a and brn3b in controlling the development, morphology, and function of retinal ganglion cells. Neuron 61, 852–864.

Badea, T.C., Nathans, J., 2011. Morphologies of mouse retinal ganglion cells expressing transcription factors Brn3a, Brn3b, and Brn3c: analysis of wild type and mutant cells using genetically-directed sparse labeling. Vision Res. 51, 269–279.

Badea, T.C., Williams, J., Smallwood, P., Shi, M., Motajo, O., Nathans, J., 2012. Combinatorial expression of Brn3 transcription factors in somatosensory Neurons: genetic and morphologic analysis. J. Neurosci. 32, 995–1007.

Barbosa-Sabanero, K., Hoffmann, A., Judge, C., Lightcap, N., Tsonis, P.A., Del Rio-Tsonis, K., 2012. Lens and retina regeneration: new perspectives from model organisms. Biochem. J. 447, 321–334.

Brzezinski, J.A., Lamba, D.A., Reh, T.A., 2010. Blimp1 controls photoreceptor versus bipolar cell fate choice during retinal development. Development 137, 619–629.

Buchholz, D.E., Hikita, S.T., Rowland, T.J., Friedrich, A.M., Hinman, C.R., Johnson, L.V., Clegg, D.O., 2009. Derivation of functional retinal pigmented epithelium from induced pluripotent stem cells. Stem Cells 27, 2427–2434.

Cai, Z., Tao, C., Li, H., Ladher, R., Gotoh, N., Feng, G.S., Wang, F., Zhang, X., 2013. Deficient FGF signaling causes optic nerve dysgenesis and ocular coloboma. Development.

Carl, M., Loosli, F., Wittbrodt, J., 2002. Six3 inactivation reveals its essential role for the formation and patterning of the vertebrate eye. Development 129, 4057–4063.

Carr, A.J., Vugler, A.A., Hikita, S.T., Lawrence, J.M., Gias, C., Chen, L.L., Buchholz, D.E., Ahmado, A., Semo, M., Smart, M.J., Hasan, S., da Cruz, L., Johnson, L.V., Clegg, D.O., Coffey, P.J., 2009. Protective effects of human iPS-derived retinal pigment epithelium cell transplantation in the retinal dystrophic rat. PLoS One 4, e8152.

Cepko, C.L., Austin, C.P., Yang, X., Alexiades, M., Ezzeddine, D., 1996. Cell fate determination in the vertebrate retina. Proc. Natl. Acad. Sci. USA 93, 589–595.

Chen, Z., Li, X., Desplan, C., 2012. Deterministic or stochastic choices in retinal neuron specification. Neuron 75, 739–742.

Chiang, C., Litingtung, Y., Lee, E., Young, K.E., Corden, J.L., Westphal, H., Beachy, P.A., 1996. Cyclopia and defective axial patterning in mice lacking Sonic hedgehog gene function. Nature 383, 407–413.

Cicero, S.A., Johnson, D., Reyntjens, S., Frase, S., Connell, S., Chow, L.M., Baker, S.J., Sorrentino, B.P., Dyer, M.A., 2009. Cells previously identified as retinal stem cells are pigmented ciliary epithelial cells. Proc. Natl. Acad. Sci. USA 106, 6685–6690.

Cyranoski, D., 2013. Stem cells cruise to clinic. Nature 494, 413.

Das, G., Choi, Y., Sicinski, P., Levine, E.M., 2009. Cyclin D1 fine-tunes the neurogenic output of embryonic retinal progenitor cells. Neural. Dev. 4, 15.

Davis, N., Mor, E., Ashery-Padan, R., 2011. Roles for Dicer1 in the patterning and differentiation of the optic cup neuroepithelium. Development 138, 127–138.

de Iongh, R., McAvoy, J.W., 1993. Spatio-temporal distribution of acidic and basic FGF indicates a role for FGF in rat lens morphogenesis. Dev. Dyn. 198, 190–202.

De la Huerta, I., Kim, I.J., Voinescu, P.E., Sanes, J.R., 2012. Direction-selective retinal ganglion cells arise from molecularly specified multipotential progenitors. Proc. Natl. Acad. Sci. USA 109, 17663–17668.

Decembrini, S., Andreazzoli, M., Vignali, R., Barsacchi, G., Cremisi, F., 2006. Timing the generation of distinct retinal cells by homeobox proteins. PLoS Biol 4, e272.

Decembrini, S., Bressan, D., Vignali, R., Pitto, L., Mariotti, S., Rainaldi, G., Wang, X., Evangelista, M., Barsacchi, G., Cremisi, F., 2009. MicroRNAs couple cell fate and developmental timing in retina. Proc. Natl. Acad. Sci. USA 106, 21179–21184.

Del Bene, F., Tessmar-Raible, K., Wittbrodt, J., 2004. Direct interaction of geminin and Six3 in eye development. Nature 427, 745–749.

Doonan, F., Groeger, G., Cotter, T.G., 2012. Preventing retinal apoptosis–is there a common therapeutic theme? Exp. Cell. Res. 318, 1278–1284.

Dorsky, R.I., Chang, W.S., Rapaport, D.H., Harris, W.A., 1997. Regulation of Neuronal diversity in the *Xenopus* retina by Delta signaling. Nature 385, 67–70.

Eagleson, G.W., Harris, W.A., 1990. Mapping of the presumptive brain regions in the neural plate of *Xenopus* laevis. J. Neurobiol. 21, 427–440.

Eiraku, M., Takata, N., Ishibashi, H., Kawada, M., Sakakura, E., Okuda, S., Sekiguchi, K., Adachi, T., Sasai, Y., 2011. Self-organizing optic-cup morphogenesis in three-dimensional culture. Nature 472, 51–56.

Elliott, J., Jolicoeur, C., Ramamurthy, V., Cayouette, M., 2008. Ikaros confers early temporal competence to mouse retinal progenitor cells. Neuron 60, 26–39.

Fausett, B.V., Gumerson, J.D., Goldman, D., 2008. The proneural basic helix-loop-helix gene ascl1a is required for retina regeneration. J. Neurosci. 28, 1109–1117.

Fischer, A.J., Bongini, R., 2010. Turning Muller glia into neural progenitors in the retina. Mol. Neurobiol.

Fischer, A.J., Reh, T.A., 2000. Identification of a proliferating marginal zone of retinal progenitors in postnatal chickens. Dev. Biol. 220, 197–210.

Fischer, A.J., Reh, T.A., 2003. Growth factors induce neurogenesis in the ciliary body. Dev. Biol. 259, 225–240.

Fuhrmann, S., 2008. Wnt signaling in eye organogenesis. Organ 4, 60–67.

Fuhrmann, S., Levine, E.M., Reh, T.A., 2000. Extraocular mesenchyme patterns the optic vesicle during early eye development in the embryonic chick. Development 127, 4599–4609.

Georgi, S.A., Reh, T.A., 2011. Dicer is required for the maintenance of notch signaling and gliogenic competence during mouse retinal development. Dev. Neurobiol. 71, 1153–1169.

Gomes, F.L., Zhang, G., Carbonell, F., Correa, J.A., Harris, W.A., Simons, B.D., Cayouette, M., 2011. Reconstruction of rat retinal progenitor cell lineages *in vitro* reveals a surprising degree of stochasticity in cell fate decisions. Development 138, 227–235.

Gonzalez-Cordero, A., West, E.L., Pearson, R.A., Duran, Y., Carvalho, L.S., Chu, C.J., Naeem, A., Blackford, S.J., Georgiadis, A., Lakowski, J., Hubank, M., Smith, A.J., Bainbridge, J.W., Sowden, J.C., Ali, R.R., 2013. Photoreceptor precursors derived from three-dimensional embryonic stem cell cultures integrate and mature within adult degenerate retina. Nat. Biotechnol. 31, 741–747.

Gregory-Evans, C.Y., Wallace, V.A., Gregory-Evans, K., 2013. Gene networks: Dissecting pathways in retinal development and disease. Prog. Retin. Eye Res. 33, 40–66.

He, J., Zhang, G., Almeida, A.D., Cayouette, M., Simons, B.D., Harris, W.A., 2012. How variable clones build an invariant retina. Neuron 75, 786–798.

Heavner, W., Pevny, L., 2012. Eye development and retinogenesis. Cold Spring Harb. Perspect. Biol. 1–17.

Hindley, C., Philpott, A., 2012. Co-ordination of cell cycle and differentiation in the developing nervous system. Biochem. J. 444, 375–382.

Hirami, Y., Osakada, F., Takahashi, K., Okita, K., Yamanaka, S., Ikeda, H., Yoshimura, N., Takahashi, M., 2009. Generation of retinal cells from mouse and human induced pluripotent stem cells. Neurosci. Lett. 458, 126–131.

Holt, C.E., Bertsch, T.W., Ellis, H.M., Harris, W.A., 1988. Cellular determination in the *Xenopus* retina is independent of lineage and birth date. Neuron 1, 15–26.

Horsford, D.J., Nguyen, M.T., Sellar, G.C., Kothary, R., Arnheiter, H., McInnes, R.R., 2005. Chx10 repression of Mitf is required for the maintenance of mammalian neuroretinal identity. Development 132, 177–187.

Huang, S., Moody, S.A., 1993. The retinal fate of *Xenopus* cleavage stage progenitors is dependent upon blastomere position and competence: studies of normal and regulated clones. J. Neurosci. 13, 3193–3210.

Jadhav, A.P., Mason, H.A., Cepko, C.L., 2006. Notch 1 inhibits photoreceptor production in the developing mammalian retina. Development 133, 913–923.

Kageyama, R., Ohtsuka, T., Shimojo, H., Imayoshi, I., 2008. Dynamic Notch signaling in neural progenitor cells and a revised view of lateral inhibition. Nat. Neurosci. 11, 1247–1251.

Katoh, K., Omori, Y., Onishi, A., Sato, S., Kondo, M., Furukawa, T., 2010. Blimp1 suppresses Chx10 expression in differentiating retinal photoreceptor precursors to ensure proper photoreceptor development. J. Neurosci. 30, 6515–6526.

Kelley, M.W., Turner, J.K., Reh, T.A., 1995. Ligands of steroid/thyroid receptors induce cone photoreceptors in vertebrate retina. Development 121, 3777–3785.

King, H.D., 1905. Experimental studies on the eye for the frog embryo. Archiv für Entwicklungsmechanik der Organismen 19, 85–107.

Kobayashi, T., Yasuda, K., Araki, M., 2009. Generation of a second eye by embryonic transplantation of the antero-ventral hemicephalon. Dev. Growth Differ. 51, 723–733.

Lagutin, O.V., Zhu, C.C., Kobayashi, D., Topczewski, J., Shimamura, K., Puelles, L., Russell, H.R., McKinnon, P.J., Solnica-Krezel, L., Oliver, G., 2003. Six3 repression of Wnt signaling in the anterior neuroectoderm is essential for vertebrate forebrain development. Genes. Dev. 17, 368–379.

Lamba, D., Karl, M., Reh, T., 2008. Neural regeneration and cell replacement: a view from the eye. Cell Stem Cell 2, 538–549.

Lamba, D.A., Karl, M.O., Reh, T.A., 2009. Strategies for retinal repair: cell replacement and regeneration. Prog. Brain Res. 175, 23–31.

Lamba, D.A., McUsic, A., Hirata, R.K., Wang, P.R., Russell, D., Reh, T.A., 2010. Generation, purification and transplantation of photoreceptors derived from human induced pluripotent stem cells. PLoS One 5, e8763.

Lanza, R.P., 2009. Essentials of stem cell biology. Elsevier/Academic Press, Amsterdam; Boston.

Levine, E.M., Fuhrmann, S., Reh, T.A., 2000. Soluble factors and the development of rod photoreceptors. Cell. Mol. Life Sci. 57, 224–234.

Li, Y., Tsai, Y.T., Hsu, C.W., Erol, D., Yang, J., Wu, W.H., Davis, R.J., Egli, D., Tsang, S.H., 2012. Long-term safety and efficacy of human-induced pluripotent stem cell (iPS) grafts in a preclinical model of retinitis pigmentosa. Mol. Med. 18, 1312–1319.

Livesey, F.J., Cepko, C.L., 2001. Vertebrate neural cell-fate determination: lessons from the retina. Nat. Rev. Neurosci. 2, 109–118.

Locker, M., El Yakoubi, W., Mazurier, N., Dullin, J.P., Perron, M., 2010. A decade of mammalian retinal stem cell research. Arch. Ital. Biol. 148, 59–72.

Ma, C., Papermaster, D., Cepko, C.L., 1998. A unique pattern of photoreceptor degeneration in cyclin D1 mutant mice. Proc. Natl. Acad. Sci. USA 95, 9938–9943.

Martinez-De Luna, R.I., Kelly, L.E., El-Hodiri, H.M., 2011. The Retinal Homeobox (Rx) gene is necessary for retinal regeneration. Dev. Biol. 353, 10–18.

Mathers, P.H., Grinberg, A., Mahon, K.A., Jamrich, M., 1997. The Rx homeobox gene is essential for vertebrate eye development. Nature 387, 603–607.

McUsic, A.C., Lamba, D.A., Reh, T.A., 2012. Guiding the morphogenesis of dissociated newborn mouse retinal cells and hES cell-derived retinal cells by soft lithography-patterned microchannel PLGA scaffolds. Biomaterials 33, 1396–1405.

Meyer, J.S., Howden, S.E., Wallace, K.A., Verhoeven, A.D., Wright, L.S., Capowski, E.E., Pinilla, I., Martin, J.M., Tian, S., Stewart, R., Pattnaik, B., Thomson, J.A., Gamm, D.M., 2011. Optic vesicle-like structures derived from human pluripotent stem cells facilitate a customized approach to retinal disease treatment. Stem Cells 29, 1206–1218.

Meyer, J.S., Shearer, R.L., Capowski, E.E., Wright, L.S., Wallace, K.A., McMillan, E.L., Zhang, S.C., Gamm, D.M., 2009. Modeling early retinal development with human embryonic and induced pluripotent stem cells. Proc. Natl. Acad. Sci. USA 106, 16698–16703.

Moody, S.A., 1987. Fates of the blastomeres of the 32-cell-stage *Xenopus* embryo. Dev. Biol. 122, 300–319.

Moore, K.B., Moody, S.A., 1999. Animal-vegetal asymmetries influence the earliest steps in retina fate commitment in *Xenopus*. Dev. Biol. 212, 25–41.

Muller, F., O'Rahilly, R., 1985. The first appearance of the neural tube and optic primordium in the human embryo at stage 10. Anat. Embryol. (Berl) 172, 157–169.

Nakano, T., Ando, S., Takata, N., Kawada, M., Muguruma, K., Sekiguchi, K., Saito, K., Yonemura, S., Eiraku, M., Sasai, Y., 2012. Self-formation of optic cups and storable stratified neural retina from human ESCs. Cell Stem Cell 10, 771–785.

Nishida, A., Furukawa, A., Koike, C., Tano, Y., Aizawa, S., Matsuo, I., Furukawa, T., 2003. Otx2 homeobox gene controls retinal photoreceptor cell fate and pineal gland development. Nat. Neurosci. 6, 1255–1263.

Ohnuma, S., Hopper, S., Wang, K.C., Philpott, A., Harris, W.A., 2002. Co-ordinating retinal histogenesis: early cell cycle exit enhances early cell fate determination in the *Xenopus* retina. Development 129. 2435–246.

Phillips, M.J., Wallace, K.A., Dickerson, S.J., Miller, M.J., Verhoeven, A.D., Martin, J.M., Wright, L.S., Shen, W., Capowski, E.E., Percin, E.F., Perez, E.T., Zhong, X., Canto-Soler, M.V., Gamm, D.M., 2012. Blood-derived human iPS cells generate optic vesicle-like structures with the capacity to form retinal laminae and develop synapses. Invest. Ophthalmol. Vis. Sci. 53, 2007–2019.

Pollak, J., Wilken, M.S., Ueki, Y., Cox, K.E., Sullivan, J.M., Taylor, R.J., Levine, E.M., Reh, T.A., 2013. Ascl1 reprograms mouse Muller glia into neurogenic retinal progenitors. Development.

Porter, F.D., Drago, J., Xu, Y., Cheema, S.S., Wassif, C., Huang, S.P., Lee, E., Grinberg, A., Massalas, J.S., Bodine, D., Alt, F., Westphal, H., 1997. Lhx2, a LIM homeobox gene, is required for eye, forebrain, and definitive erythrocyte development. Development 124. 2935–244.

Riesenberg, A.N., Liu, Z., Kopan, R., Brown, N.L., 2009. Rbpj cell autonomous regulation of retinal ganglion cell and cone photoreceptor fates in the mouse retina. J. Neurosci. 29, 12865–12877.

Roy, A., de Melo, J., Chaturvedi, D., Thein, T., Cabrera-Socorro, A., Houart, C., Meyer, G., Blackshaw, S., Tole, S., 2013. LHX2 is necessary for the maintenance of optic identity and for the progression of optic morphogenesis. J. Neurosci. 33, 6877–6884.

Saha, M.S., Grainger, R.M., 1992. A labile period in the determination of the anterior-posterior axis during early neural development in *Xenopus*. Neuron 8, 1003–1014.

Salero, E., Blenkinsop, T.A., Corneo, B., Harris, A., Rabin, D., Stern, J.H., Temple, S., 2012. Adult human RPE can be activated into a multipotent stem cell that produces mesenchymal derivatives. Cell Stem Cell 10, 88–95.

Sasai, Y., Eiraku, M., Suga, H., 2012. *In vitro* organogenesis in three dimensions: self-organising stem cells. Development 139, 4111–4121.

Scholtz, C.L., Chan, K.K., 1987. Complicated colobomatous microphthalmia in the microphthalmic (mi/mi) mouse. Development 99, 501–508.

Seiler, M.J., Aramant, R.B., 2012. Cell replacement and visual restoration by retinal sheet transplants. Prog. Retin. Eye Res. 31 (6), 661–687.

Skapek, S.X., Lin, S.C., Jablonski, M.M., McKeller, R.N., Tan, M., Hu, N., Lee, E.Y., 2001. Persistent expression of cyclin D1 disrupts normal photoreceptor differentiation and retina development. Oncogene 20, 6742–6751.

Spemann, H., 1901. Correlationen in der Entwickelung des Auges (On interactions during eye development (Translated from German)). Verhandl. d. anat. Gesellsch 15, 61–79.

Spemann, H., 1938. Embryonic development and induction. Yale University Press H. Milford, Oxford University Press, New Haven London.

Swaroop, A., Kim, D., Forrest, D., 2010. Transcriptional regulation of photoreceptor development and homeostasis in the mammalian retina. Nat. Rev. Neurosci. 11, 563–576.

Tarchini, B., Jolicoeur, C., Cayouette, M., 2012. *In vivo* evidence for unbiased Ikaros retinal lineages using an Ikaros-Cre mouse line driving clonal recombination. Dev. Dyn. 241, 1973–1985.

Trimarchi, J.M., Stadler, M.B., Cepko, C.L., 2008. Individual retinal progenitor cells display extensive heterogeneity of gene expression. PLoS One 3, e1588.

Turner, D.L., Snyder, E.Y., Cepko, C.L., 1990. Lineage-independent determination of cell type in the embryonic mouse retina. Neuron 4, 833–845.

Viczian, A.S., Solessio, E.C., Lyou, Y., Zuber, M.E., 2009. Generation of functional eyes from pluripotent cells. PLoS Biol 7, e1000174.

Viczian, A.S., Vignali, R., Zuber, M.E., Barsacchi, G., Harris, W.A., 2003. XOtx5b and XOtx2 regulate photoreceptor and bipolar fates in the *Xenopus* retina. Development 130, 1281–1294.

West, E.L., Gonzalez-Cordero, A., Hippert, C., Osakada, F., Martinez-Barbera, J.P., Pearson, R.A., Sowden, J.C., Takahashi, M., Ali, R.R., 2012. Defining the integration capacity of ES cell-derived photoreceptor precursors. Stem Cells 30 (7), 1424–1435.

Wong, L.L., Rapaport, D.H., 2009. Defining retinal progenitor cell competence in *Xenopus laevis* by clonal analysis. Development 136, 1707–1715.

Yan, B., Moody, S.A., 2007. The competence of *Xenopus* blastomeres to produce neural and retinal progeny is repressed by two endo-mesoderm promoting pathways. Dev. Biol. 305, 103–119.

Yun, S., Saijoh, Y., Hirokawa, K.E., Kopinke, D., Murtaugh, L.C., Monuki, E.S., Levine, E.M., 2009. Lhx2 links the intrinsic and extrinsic factors that control optic cup formation. Development 136, 3895–3906.

Zuber, M.E., 2010. Eye Field Specification in *Xenopus laevis*. Curr. Top. Dev. Biol. 93, 29–60.

Zuber, M.E., Perron, M., Philpott, A., Bang, A., Harris, W.A., 1999. Giant eyes in *Xenopus laevis* by overexpression of XOptx2. Cell 98, 341–352.

Chapter 18

Neural Crest Determination and Migration

Eric Theveneau and Roberto Mayor
Department of Anatomy and Developmental Biology, University College London, London, UK

Chapter Outline

GLOSSARY

Cell migration A process during which cells translocate their cell body to a new location according to various signals present in their local environment. Such signals include extracellular matrix, guidance cues and contacts with other cells.

Determination An irreversible decision made by an undifferentiated cell that restricts its differentiation.

Ectoderm One of the three germ layers of the embryo. It gives rise to the epidermis, the neural plate, and the neural crest.

Embryonic induction The interaction between an inducing tissue and a responding tissue. As a result, the responding tissue undergoes a change in fate.

Epithelial-mesenchymal transition A series of changes at the molecular level that converts an epithelial cell into a mesenchymal cell. Epithelial-mesenchymal transition (EMT) involves changes in cell-cell and cell-matrix adhesion properties as well as changes in cell polarity.

Neural crest The transient tissue found in vertebrate embryos that is formed at the border of the neural plate and that undergoes a transformation into migrating cells, which differentiate into a huge variety of cells. It gives rise to most of the cartilage and bones of the head, the peripheral nervous system, and the pigments of the skin, among other things.

Neurulation The developmental process in vertebrate embryos that results in the formation of the neural plate and that terminates with its closure into the neural tube.

SUMMARY

- The neural crest is induced at the border of the neural plate by signals that come from the epidermis, the neural plate, and the underlying mesoderm.
- Two kinds of signals are required to induce the neural crest: an intermediate level of bone morphogenetic protein (BMP) activity and a second signal involving Wnt, fibroblast growth factors (FGF) and retinoic acid (RA).
- After the neural crest has induced, its separation from the surrounding tissues involves an epithelial-mesenchymal transition which is controlled by a network of transcription factors.
- Neural crest cells migrate following stereotypical routes that are defined by the availability of permissive extracellular matrix, complementary distribution of negative and positive guidance cues, and cell-cell interactions.
- Neural crest cells undergo solitary and collective cell migration.
- Anterior and posterior neural crest gives rise to very different kinds of cells, tissues, and organs.
- Failure during neural crest development is associated with several human pathologies.

18.1 INTRODUCTION

Although the neural crest is a discrete and transient structure that comprises only a few cells, it embodies many of the crucial issues of developmental biology. Questions related to embryonic induction, morphogenesis, differentiation, and even evolution have to be addressed as part of the understanding of neural crest development. The neural crest is formed at the border of the neural plate (Figure 18.1). How is this group of cells induced at that particular position? During and after the closure of the neural tube, the neural crest cells start one of the most dramatic morphogenetic events during early embryogenesis: they segregate from the neuroepithelium and migrate throughout the embryo. How is this segregation from the epithelia and subsequent cell migration controlled? Is it similar to what happens during tumor progression and metastasis? The neural crest differentiates into a prodigious array of cell types (Table 18.1). How is this differentiation regulated? What determines the fate of each of the neural crest derivatives? Finally, the neural crest represents a key tissue in vertebrate evolution, because many of the typical features of vertebrates (e.g., the jaw) are neural crest derivatives. How did this small group of cells emerge during evolution, and how was it able to play such an important role in vertebrate evolution? These are questions that have fascinated developmental and evolutionary biologists for more than a century.

The neural crest has sometimes been referred to as the fourth germ layer (the first three layers being the ectoderm, the endoderm and the mesoderm). The crest gives rise to cells that are typical of ectoderm (e.g., neurons, epidermal cells) as well as those with mesodermal or endodermal characteristics (e.g., muscle, cartilage, endocrine cells). The cell types include the following (see Table 18.1):

1) The neurons and glial cells of the sensory, sympathetic, and parasympathetic nervous system;
2) The epinephrine-producing (medulla) cells of the adrenal gland;
3) The pigment-containing cells of the epidermis;
4) Many of the skeletal and connective tissue components of the head; and
5) Some of the connective tissue of the heart.

The neural crest was discovered in the chick in 1868 by Wilhelm His. His identified a band of cells lying between the neural tube and the epidermal ectoderm as the origin of the spinal and cranial ganglia; he called this band *Zwischentrang* or *intermediate cord.* The term *neural crest* was used for the first time by Arthur Milnes Marshall in 1879, according to Brian Hall (Hall, 1999). Since then, considerable progress has been made in the neural crest field. The cellular and molecular mechanisms that control neural crest induction and differentiation are now being unraveled. A number of genes that are

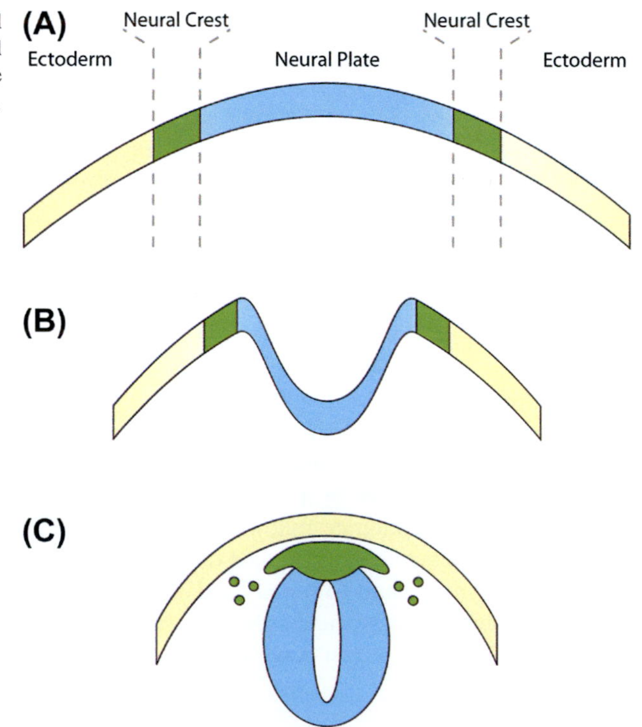

FIGURE 18.1 Neurulation and Neural Crest Formation. (A) Open neural plate. (B) Neural folds rise at the border of the neural plate. (C) The neural tube closes and the neural crest remains between the dorsal neural tube and the epidermis, from where they start their migration. *Blue,* Neural plate/tube; *green,* neural crest; *yellow,* epidermis.

TABLE 18.1 Neural Crest Derivatives

NC Population	Origin Along the Anterior-Posterior (AP) Axis	Organ/Region Invaded	Neural Crest Contribution (Non-Exhaustive)
Cephalic	From diencephalon to s3/s4 boundary	Face and neck	Cartilage, bone, connective tissue, dermis, tendons, neurons and glia of the cephalic parasympathetic ganglia.
		Ear	Ossicles of the middle ear, melanocytes of the inner ear.
		Eye	Corneal stroma and endothelium, iridial stroma and muscles, ciliary muscles, ossicles and cartilage, muscles of eyelids, pericytes of ocular blood vessels, connective tissue and mesenchyme.
		Skin	Pigment cells
Cardiac	From r4/r5 boundary to s3/s4 boundary	Heart	Connective tissue of aorta and pulmonary artery, condense to form the septum of the outflow track.
Enteric (Vagal)	From s1 to s7 and s28 caudalward	Gut	Neuron and glia of the parasympathetic ganglia
		Lung	Neurons
Trunk	From s4 caudalward	Somites and mesenchyme around the aorta	Neuron and glia of dorsal root ganglia and sympathetic ganglia
	From s4 caudalward	Skin	Pigment cells
	From s18 to s24	Adrenal gland	Medullary cells

S: somites; r: rhombomeres; AP: anterior-posterior, NC: neural crest.

expressed specifically in the neural crest have been identified and their function studied. Many of the molecules that control neural crest migration have been discovered. We will describe some of these findings in this chapter, but first we will describe some experimental approaches to studying the neural crest.

18.2 TECHNIQUES TO IDENTIFY NEURAL CREST DEVELOPMENT

One of the requisites for the study of neural crest cells is the ability to identify them; however, this was a serious problem for many years. Neural crest cells do not exist as a typical tissue, because, very soon after they are induced, they disperse into individual, highly migratory cells. After they reach their final target, they differentiate into a huge variety of different kinds of cells. Three general methods have been used to identify and study neural crest cells:

(1) *The destruction or ablation of neural crest cells.* With this technique, the neural fold or the prospective neural crest is removed or destroyed, and the resulting deficiencies in the embryos are analyzed. This kind of technique was widely used in amphibian and chick studies, and it contributed to the identification of neural crest derivatives. However, it has a major drawback: the extirpation of the neural fold can trigger a regulatory mechanism in the embryo and the regeneration of neural crest precursors from the surrounding tissues.

(2) *Labeling techniques.*

  2.1) *The quail-chick marker system.* This technique was developed by Nicole LeDouarin in 1969, and it is based on the ability to distinguish the nucleus of quail cells from chick cells. Quail neural fold cells are grafted in the neural fold of a chick embryo, and their migration and fate are analyzed later by identifying the quail cells in the chick host. A complete and accurate fate map of the neural crest was developed through the use of this technique, and most of the neural crest derivatives were traced back to the neural folds of chick embryos.

2.2) *Dye labeling.* Vital dye staining has long been used to study neural crest migration and differentiation. Several lipophilic dyes have been recently developed, such as DiI and DiO. In this technique, the dyes are applied *in vivo* to the premigratory neural crest cells, and migration and cell fate can then be studied. Other cell lineage markers in common use are lysinated fluorescein, rhodamine dextran, and, more recently, green and red fluorescent proteins.

(3) *Endogenous neural crest markers.* Some enzymes and antigens are expressed more or less specifically in neural crest cells, such as acetylcholinesterase and the antigen HNK-1/NC-1. However, a true revolution came with the identification of the gene products present specifically in neural crest cells, which can be analyzed by *in situ* hybridization. The first of these genes was identified almost 20 years ago, and it codes for members of the *Snail* gene family (Nieto et al., 1994; Mayor et al., 1995). Since then, several other genes (mostly transcription factors) have been identified as specific markers for the neural crest. Many of these factors are expressed during all of the steps of neural crest development, induction, migration, and differentiation, and they can be used as unequivocal markers for these different processes. More recently, some of the regulatory elements controlling the expression of these genes in the neural crest were identified and used to drive the expression of reporter molecules such as the Green Fluorescent Protein (GFP) or alkaline-phosphatase in fish and mice. These transgenic lines of animals have been used to repeat some of the classical studies that were originally performed using the techniques described in points 2.1 and 2.2 above. They have the main advantage of allowing neural crest lineage tracing without the need for extensive surgery. However, none of the currently available transgenic animals fully recapitulates the expression of the neural crest marker used to drive the reporter's expression. In addition, some of these animals have been shown to have developmental defects linked to the genetic manipulation performed (Lewis et al., 2013).

18.3 SPECIFICATION OF NEURAL CREST CELLS

Embryologic Studies of Neural Crest Specification

The neural crest is specified at the border of the neural plate, between the neural plate and the epidermis. Thus, many different tissues are adjacent to the neural crest, and they can be potential sources of inductive signals. The tissues that are in close apposition to the neural crest are the underlying mesoderm, the epidermis, and the neural plate (Figure 18.2A–C). Two kinds of experiments can be used to identify the tissue that induces the neural crest:

1) The tissue can be removed mechanically or genetically, and the development of the neural crest can be analyzed; or
2) The tissue can be co-cultured with a competent ectoderm, and the induction of neural crest can be assayed.

Pioneer experiments in amphibians showed that the mesoderm underlying the neural crest has the ability to induce neural crest cells (Raven and Kloos, 1945). Raven and Kloos dissected different pieces of mesoderm from the neurula of an amphibian embryo and grafted them into the cavity of a gastrula. After the gastrula developed, the neural crest derivatives were identified by their morphology. They found that medial pieces of mesoderm (dorsal mesoderm or notochord) induce neural plate and neural crest but that a specific induction of neural crest was obtained when more lateral (paraxial and intermediate) mesoderm was grafted. More recent experiments have confirmed this early observation. In amphibians, when the dorsolateral marginal zone (prospective dorsolateral mesoderm) is removed from the embryo, no neural crest develops (Bonstein et al., 1998; Marchant et al., 1998; Steventon et al., 2009). If the same dorsolateral marginal zone is conjugated with competent ectoderm, a strong induction of neural crest markers and some neural crest derivatives (e.g., melanocytes) are observed (Bonstein et al., 1998; Marchant et al., 1998; Monsoro-Burq et al., 2003; Steventon et al., 2009). Interestingly, the dorsolateral marginal zone is fated to underlie the neural crest, and this is the same tissue that Raven and Kloos described as the *neural crest inducer.* However, it has been shown recently in zebrafish that the involution of the mesoderm is not required for neural crest induction (Ragland and Raible, 2004). Several zebrafish mutants, such as the *One-eyed pinhead,* do not develop dorsal mesoderm, and the more lateral mesoderm does not involute; thus, because these mutants do not have the mesoderm underlying the neural crest, they are the equivalent of a genetic ablation of this mesoderm. It seems that neural crest is still induced in those mutants; therefore, the mesoderm does not have to underlie the neural crest to induce it. It should be pointed out that, although the dorsal mesoderm is greatly reduced in these mutants and there is no involution of the remaining mesoderm, they still have lateral mesoderm that is capable of sending its inductive signals. In conclusion, a specific dorsolateral mesoderm plays an important role in neural crest induction. The inductive activity of this mesoderm starts before this tissue is involuted, and it continues at least until it is underlying the neural crest. Recent work performed in chick and amphibian embryos has shown that neural crest induction is a multistep process, and that the dorsolateral mesoderm (fated to become intermediate mesoderm) is required for

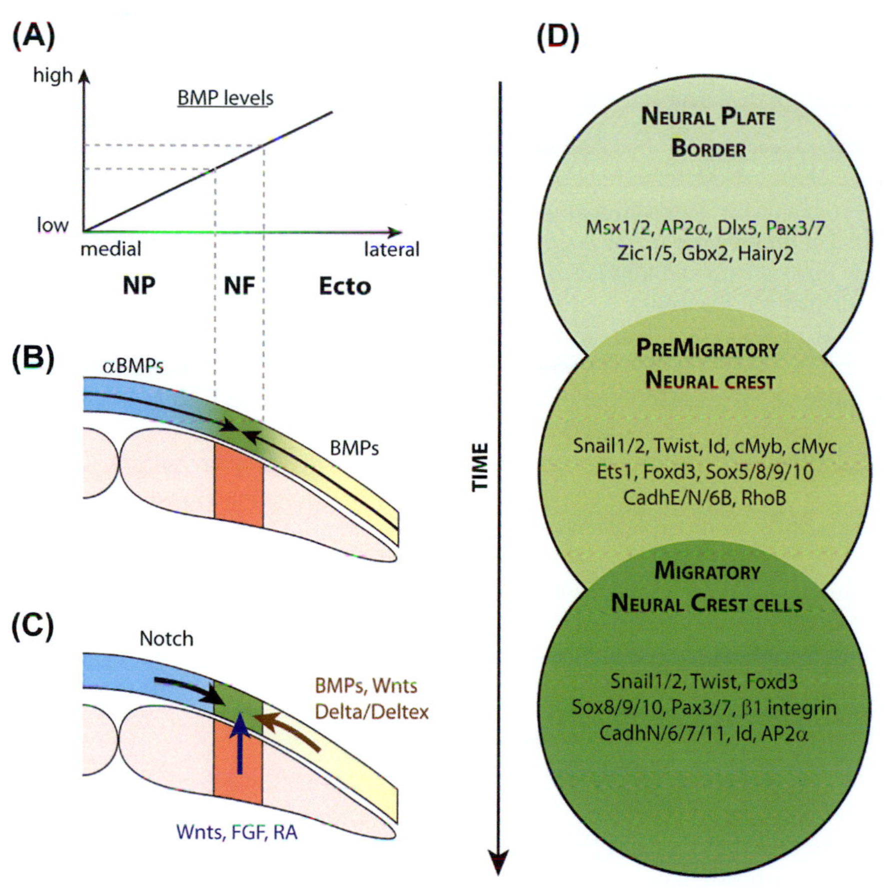

FIGURE 18.2 Neural Crest Induction. (A, B) Neural crest cells are induced within the neural fold (NF), which is exposed to intermediate levels of BMP. (C) Secondary signals maintain the neural crest. These signals include Notch, BMP, Wnt and RA coming from the ectoderm and the underlying mesoderm. Blue, neural plate; green, neural fold/neural crest; pink, mesoderm; yellow, ectoderm. (D) Neural crest genetic cascade. Each circle represent different levels in the neural crest gene regulatory network; however they are fused to emphasize the idea that the definition of these different categories is only artificial, and many genes can be found in more than one level.

maintenance of the neural crest fate (Patthey et al., 2009; Steventon et al., 2009). Furthermore, before neural crest induction, the neural fold (which contains the neural crest precursors) is induced by a combination of signals produced by the adjacent tissues (Steventon and Mayor, 2012).

The other tissues that are adjacent to the neural crest are the neural plate and the epidermis, and it has been shown than an interaction between these two tissues is able to induce neural crest cells. Grafts of neural plate cells into epidermis induce neural crest cells at the border of the graft in amphibian and chick embryos (Mancilla and Mayor, 1996; Moury and Jacobson, 1989; Selleck and Bronner-Fraser, 1995). By using lineage markers in the grafted neural plate, it was possible to determine that the neural crest was induced at the border originated from the neural plate and from the epidermis. Therefore, it seems that signals from the epidermis are able to transform neural plate cells into neural crest cells and that signals from the neural plate do the same in the epidermal tissue. However, recent experiments have shown that the neural plate has a higher potency than the epidermis to differentiate into neural crest (Pieper et al., 2012). *In vitro* cocultures of neural plate attached to epidermis also exhibit the expression of neural crest markers. This indicates that the interaction between neural plate and epidermis is sufficient to specify the neural crest cells (Mancilla and Mayor, 1996; Selleck and Bronner-Fraser, 1995).

In conclusion, the neural crest is induced by a combination of signals coming from three different sources (see Figure 18.2B,C). An interaction between the neural plate and the epidermis changes the fate of these two tissues, causing them to become neural crest cells. In addition, signals from the lateral mesoderm are also involved in neural crest induction and maintenance. The observation that it is possible to induce neural crest by the simple juxtaposition of neural plate and epidermis, without mesoderm, suggests that the inductive activity of the mesoderm may simply reinforce the initial induction set up by the interaction between the neural plate and the epidermis.

Molecular Studies on Neural Crest Specification

Inductive Signals

The search for the neural crest inductive signals has been long but very fruitful. Five different families of extracellular inductive molecules have been found to be involved in neural crest induction: the bone morphogenetic proteins (BMPs) and their antagonists, like noggin and chordin; different Wnts; several fibroblast growth factors (FGFs); retinoic acid (RA); and Notch/Delta.

For all of these molecules, two general kinds of experimental approaches have been used:

(1) *Loss-of-function experiments.* The putative inducer is inhibited, and the induction of the neural crest markers is assayed. The inducer can be inhibited by specific chemical inhibitors that are added to the medium in which the embryos are grown or in which the prospective isolated neural crest are cultured. Another common method of blocking the inductive signals is the expression of mutated forms of the receptor for each signal. Such truncated forms can bind to their ligands but do not activate the signaling cascade downstream and thus act as dominant negatives. The expression of the dominant negative receptors will block specific pathways activated by the putative inducer. Consequently, if the tested signal is required for neural crest induction, the expression of neural crest markers should be inhibited. There are also dominant negative constructs for molecules that work within the pathway. Finally, it is also possible to directly inhibit the expression of genes by using antisense mRNA or oligonucleotides that specifically block translation (i.e., Morpholinos), or short RNA sequences that target a specific mRNA and degrade it (i.e., siRNA, shRNA or miRNA).

(2) *Gain-of-function experiments.* The putative inducer is added to competent ectoderm, or the pathway is activated; the induction of neural crest markers is then analyzed. The inducer can be added directly to the medium in which the ectoderm is cultured, it can be added to the embryos, or its expression can be forced into a given tissue by direct transfection of mRNA or DNA. Alternatively, cells expressing the inducer or beads soaked with the inducer can be grafted in competent ectoderm to test whether or not they are able to induce neural crest. Finally, another important requirement is that an inducer needs to be expressed in the correct place and time. If the putative inducer is never expressed in the tissues that have been identified as the source of neural crest inducers, it is unlikely that it could be the real inducer of the neural crest.

The molecules that have thus far been identified as neural crest inducers are described below.

Bone Morphogenetic Protein

According to the default model of neural induction, the ectoderm produces BMPs, which inhibit the specification of the ectoderm as neural cells, whereas the organizer or dorsal mesoderm secretes anti-BMPs (noggin, chordin, nodal), which block the neural inhibition and allow the ectodermal cells to adopt a neural fate (De Robertis and Kuroda, 2004; Kuroda et al., 2004) (see also Chapter 11 in this book). According to this model (Aybar and Mayor, 2002), the neural crest is specified at the border of the neural plate, where an intermediate level of BMP is reached (Figure 18.2A,B). Thus, the BMP activity in that region is not as high as in the epidermis and not as low as in the neural plate. As predicted by this model, inhibition of BMP activity in the ectoderm will expand the neural crest domain, because the thresholds that specify these cells are now reached across a wider ectodermal domain. The expression of a dominant negative BMP receptor and expression of the anti-BMP molecules noggin and chordin in amphibian embryos and different BMP mutants in zebrafish exhibit a phenotype with an expanded neural crest domain (Marchant et al., 1998; Nguyen et al., 1998). In addition, treatments of competent ectoderm with increasing levels of BMP are able to induce the expression of neural plate, neural folds, and epidermal markers, sequentially (Marchant et al., 1998; Neave et al., 1997; Nguyen et al., 1998; Wilson and Hemmati-Brivanlou, 1997). The addition of BMPs to dissected neural plate induces the expression of neural crest markers (Liem et al., 1995). Taken together, all of these data support the notion that a specific level of BMP, within a BMP gradient, plays an important role in the specification of the neural plate border. We should point out here that this precise level of BMP is not sufficient to induce neural crest, and that additional signals are required to induce the neural crest within the neural plate border.

Fibroblast Growth Factor

One of these additional signals is FGF. The treatment of competent ectoderm with a combination of BMP inhibitors and FGF is able to trigger the expression of neural crest markers (Mayor et al., 1997; Monsoro-Burq et al., 2003). The activation of FGF signaling in whole embryos leads to an expansion of neural crest markers. The inhibition of FGF activity by the expression of dominant negative receptors or by treatment with FGF inhibitors completely blocks neural crest induction. Thus, gain- and loss-of-function experiments support the notion that FGF is involved in neural crest induction.

In situ hybridization of *fgf8* demonstrates that it is expressed in the mesoderm underlying the neural crest, and that the specific inhibition of FGF8 in the dorsolateral mesoderm abolishes neural crest induction (Monsoro-Burq et al., 2003). However, recent work in amphibians has shown that FGF is not a direct neural crest inducer, but instead it is able to induce the expression of Wnt8 (Hong et al., 2008), a proper neural crest inducer – as described below.

Wnts

The treatment of ectoderm with a precise level of BMP and Wnt also induces the expression of neural crest markers (Bastidas et al., 2004; Deardorff et al., 2001; Garcia-Castro et al., 2002; LaBonne and Bronner-Fraser, 1998; Lekven et al., 2001; Lewis et al., 2004; Saint-Jeannet et al., 1997; Tribulo et al., 2003; Villanueva et al., 2002). The inhibition of Wnt signaling by expressing different Wnt inhibitors or dominant negative versions of Wnt pathway proteins produces a strong inhibition in neural crest induction. Grafts of cells expressing Wnt1 lead to an up-regulation of neural crest markers in the chick (Garcia-Castro et al., 2002).

It is clear that at least two tissues that are adjacent to the prospective neural crest cells express Wnt ligands. The dorsal epidermis in chick expresses Wnt6, and it has been proposed as one of the inductive signals produced by the epidermis. The dorsolateral mesoderm, which has been identified as a source of inductive signals for the neural crest, expresses Wnt8 in *Xenopus*, zebrafish, and chick embryos. Thus, there is compelling evidence that several members of the Wnt family work as neural crest inducers.

Retinoic Acid

The treatment of amphibian embryos with RA produces the ectopic expression of neural crest markers, and so does the expression of an activated form of the RA receptor (Begemann et al., 2001; Villanueva et al., 2002). The *in vitro* culture of anterior neural folds or neuralized ectoderm, which does not differentiate into neural crest cells, exhibits an upregulation of neural crest markers after treatment with RA. The expression of dominant negative forms of RA receptors inhibits neural crest induction *in vitro*. Taken together, these experiments indicate that RA is also involved in neural crest specification.

Is RA present at the right time and place to be a neural crest inducer? Because RA is not a protein, it is not possible to analyze the expression of a single gene to know where RA is present in the embryo, and measurements of RA in whole embryos have been difficult. It is possible to obtain an idea of where RA could be active by analyzing the expression of the enzymes involved in its synthesis and degradation. Raldh2 is the key enzyme required for RA synthesis. This enzyme is expressed in posterior and paraxial mesoderm. Thus, this mesoderm, which has inductive activity, could be a source of RA. The anterior region of the embryo, where no neural crest is formed, expresses Cyp26, which is one of the enzymes responsible for the degradation of RA. In conclusion, RA has neural crest inductive activity, and it is present in some of the tissues in which the inducers are produced.

Notch/Delta

Notch is a large, single-transmembrane domain protein that acts as a receptor for the ligand Delta (or Serrate). The ligand is also a transmembrane protein, and, after ligand stimulation, the intracellular domain of Notch is cleaved and transported into the nucleus (see Chapter 1). In the nucleus, the intracellular domain of Notch interacts with other transcription factors and regulates the expression of target genes. It has been shown that Notch is involved in neural crest induction in zebrafish, chick, and *Xenopus* embryos (Cornell and Eisen, 2000; 2002; Endo et al., 2002; 2003; Glavic et al., 2004). The activation of Notch leads to an expansion of the neural crest domain, whereas its inhibition produces a reduced population of neural crest cells. Although there is consensus regarding the role of Notch during neural crest induction in these three animal models, the specific molecular mechanism as well as the expression of the different components of this pathway seems to be different among them (Cornell and Eisen, 2005).

A Model of Neural Crest Specification

The current model for neural crest induction proposes that it is specified at the border of the neural plate in two steps. During the first step, a gradient of BMP activity is established in the ectoderm (see Figure 18.2A,B). This gradient is formed as a consequence of the interaction between BMP that is produced in the ventral or lateral ectoderm and BMP-binding molecules produced in the dorsal or medial mesoderm. Within this gradient, high levels of BMP activity specify epidermis, low levels specify neural plate, and intermediate levels are required for early neural fold specification. These early neural folds are not completely induced to become neural crest, because additional signals are required. Additional signals, Wnts and RA, work together to transform the early neural fold into neural crest cells (Figure 18.2C). Wnt signals, produced by the mesoderm underlying the neural crest at slightly later stages, are also involved in neural crest maintenance (Steventon et al., 2009).

Others signaling molecules such as Notch/Delta and FGF are also important for neural crest specification. However, experiments in *Xenopus* and chick embryos suggest that Notch works by controlling the levels of BMP in the neural fold region, whereas FGF promotes Wnt signaling as described above. Thus, it is very likely that Notch/Delta works downstream of the BMP gradient, but that it forms a loop of activity that controls the maintenance of the BMP levels. It is interesting to note that these three transforming molecules (FGF, Wnts, and RA) are also involved in the posteriorization of the neural plate, so it seems that the posteriorization of the neural plate is very much linked with the specification of the neural crest. It is well known that the neural crest is not formed in the most anterior neural fold, and it has been proposed that these three molecules work by transforming the anterior neural fold, which is induced within the BMP gradient, into a more posterior neural fold (posteriorization), which corresponds to neural crest cells.

Genetic Network in the Neural Crest

After the neural crest has been specified by the combination of extracellular signals described previously, a large group of genes is expressed in its cells. Many of these genes code for transcription factors, and a cascade or network of transcriptional regulation is set up in the neural crest cells. It is not clear which are the first members and which are the last ones of this cascade; however, on the basis of the time of expression, the region of expression, and some functional experiments, a possible regulatory network has been proposed (Figure 18.2D). For simplicity, the network has been divided in three groups of genes (Mayor and Aybar, 2001; Meulemans and Bronner-Fraser, 2004; Sauka-Spengler and Bronner-Fraser, 2008; Steventon et al., 2005). The first group, called *neural plate border specifier genes,* contains the earliest genes that are expressed in the neural plate border. Their expression is not restricted to the prospective neural crest, because they are expressed in a wider domain that includes prospective epidermal and neural plate cells. They control the expression of the second group of genes, called *PreMigratory* or *neural crest specifier genes.* The expression of this second group of genes is restricted to the prospective neural crest cells, and most of them control the specification and survival of the neural crest cells. The third group of genes is called *Migratory neural crest genes*, and they are expressed when the neural crest starts to migrate. They are the target of the two previous groups of genes and control different cellular activities related to cell migration, such as adhesion, motility, polarity, etc. It should be pointed out that many of these genes are expressed from early neural induction to neural crest migration, making the distinction between these three categories rather diffuse. Thus, more than the expression of a given gene, it is the co-expression of a particular set of genes within a given region of the competent ectoderm at a specific time that defines these three broad steps of neural crest development. Finally, many of these genes are directly involved in neural crest differentiation or they control the expression of genes that are the direct regulators of cell differentiation; these have been called *neural crest effector genes* (Meulemans and Bronner-Fraser, 2004).

We will describe briefly some of the genes in each of the first three groups.

Neural Plate Border Specifier Genes

This group contains the following genes: *Msx, Ap2,* and *Dlx, Zic1/5, Pax3/7, Gbx2* and *Hairy2.* They are initially expressed in the non-neural ectoderm, and they are later restricted to the neural fold region in *Xenopus* and chick embryos. Their expression patterns correlate with the assumed ventral-dorsal gradient of BMP activity, and the identification of cis-regulatory elements in some of these genes indicates that they are very likely to be direct targets of BMP. Loss-of-function experiments in mouse, zebrafish, and *Xenopus* embryos show that these genes are required for the early specification of the neural crest.

Pre-Migratory Neural Crest Genes

These genes are expressed later than the neural plate border genes, and they are only expressed in the neural crest territory. They encode for transcription factors of the *Snail* and *SoxD/E* families of genes and for the genes *FoxD3, Id3, Twist, Myc, Cadherin E/N/6B* and *RhoB*. It is also important to note that *Snail1, Snail2 (Slug),* and *FoxD3* all function as transcriptional repressors; this means that there are additional transcription factors that mediate their activity. Gain- and loss-of-function studies for each of these genes show that they are essential for neural crest development and that they probably control different aspects of neural crest specification, migration, and/or differentiation. In addition, many of them are involved in cell survival.

Migratory Neural Crest Genes

These genes are expressed during neural crest migration and functional experiments have shown that they confer migratory abilities on the neural crest cells. Some of these genes overlap with the pre-migratory genes, such as *Snail1/2, Twist, FoxD3, Sox8/9/10* and *Id*; and some are even expressed in the neural plate border, such as *Pax3/7* and *AP2*; but others are specifically turned on at this stage, like *Cadherins 7/11* and *integrins.*

18.4 NEURAL CREST CELL MIGRATION

Epithelial-Mesenchymal Transition and Delamination

Induced neural crest cells start their journey throughout the embryo by breaking away from their surrounding tissues in a process called delamination (Duband, 2010; Theveneau and Mayor, 2012a). During this phase, the cells progressively detach from the neuroepithelium and invade the local extracellular matrix. The delamination requires changes in cell-cell and cell-matrix adhesion properties as part of an epithelial-mesenchymal transition (Figure 18.3A). In the trunk, these changes are regulated by a series of transcription factors such as *Snail2*, *Foxd3* and *Sox9/10,* which modulate expression of Cadherins and Integrins (Figure 18.3B). The acquisition of a new repertoire of cell-cell adhesion molecules plays two roles: it allows neural crest cells to separate from the neuroepithelium, as cells with differential adhesion properties barely mix; it also triggers a motile behavior by promoting a mesenchymal phenotype with more dynamic adhesion to the extracellular matrix. This cascade of transcriptional regulation is downstream of a Wnt/BMP-dependent signaling pathway and is doubled by a direct regulation at the protein level. Some Cadherins are directly cleaved off by enzymes that render them inactive, thus weakening the cell-cell adhesion. Furthermore, directly inhibitory crosstalk between Cadherins and Integrins favors motility over stable cell clustering (Alfandari et al., 2010; McKeown et al., 2012; Theveneau and Mayor, 2012a).

While in the trunk, neural crest cells exit a closed neural tube; in the head, neural crest delamination and neural tube closure are practically concomitant. Cells located in the neural folds are still connected with both the neural plate cells and the ectodermal cells of the prospective epidermis. In addition to N-Cadherin and several type II-Cadherins such as Cadherin 6/6B/7/11, early migratory cephalic neural crest cells retain a substantial amount of E-Cadherin which is repressed by the transcription factor Twist to allow epithelial-mesenchymal transition (Barriga et al., 2013; Dady et al., 2012). Another key difference between cephalic and trunk neural crest cells is that cephalic neural crest delaminate *en masse*. It is thought that additional factors such as Ets1 and LSox5 are involved in triggering this all-at-once onset of migration, but the downstream mechanisms remain elusive (Perez-Alcala et al., 2004; Theveneau et al., 2007).

Pathways of Neural Crest Migration

Migratory neural crest cells follow stereotypical routes (Figure 18.4). In the head, these pathways are very well conserved across species and the overall organization of the migratory neural crest cells can be summarized as follows. A continuous wave of neural crest cells delaminates from the diencephalon to the caudal part of the rhombencephalon, which corresponds to the region of the neural tube facing the first few somites. This wave is rapidly restricted into three main streams by physical and chemical constraints present in the local environment. The first stream colonizes the region around the eye, the ventral portion of the face and the first branchial arch. The second stream migrates anterior to the otic vesicle and invades the second branchial arch. The third stream migrates posterior to the otic vesicle and invades the third and fourth branchial arches. Some of these post-otic neural crest cells are called cardiac neural crest cells because they will continue further ventrally to colonize the heart. Another group of post-otic cells, called the enteric neural crest, delaminates from the neural tube facing somites 1 to 7 and migrates past the aorta to colonize the gut (Alfandari et al., 2010; Hall, 2008; Kulesa et al., 2010; Kuo and Erickson, 2010; Le Douarin and Kalcheim, 1999; Mayor and Theveneau, 2013; Theveneau and Mayor, 2012b).

In the trunk, neural crest cells have been shown to follow two main routes: a ventro-medial one that is dictated by the organization of the paraxial mesoderm into somites; and a dorso-lateral pathway in which cells migrate under the prospective epidermis in a non-segmented manner. In amniotes, the newly formed epithelial somites separate into an epithelial

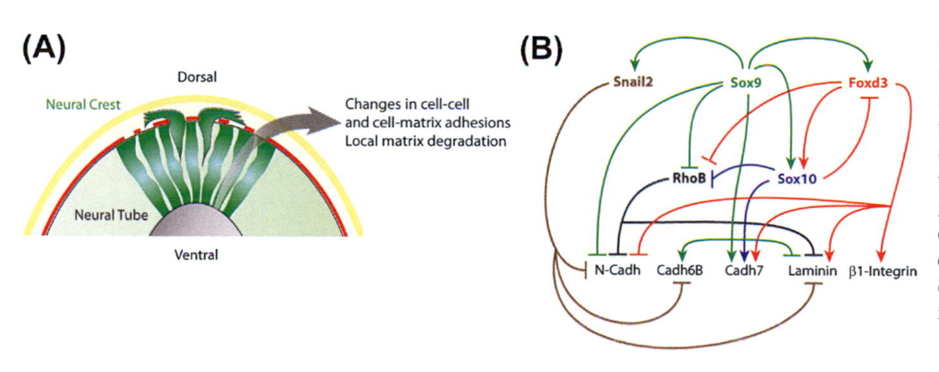

FIGURE 18.3 Neural Crest Delamination and Epithelial-Mesenchymal Transition. (A) Dorsal neural tube region at the time of trunk neural crest emigration. Neural crest cells detach from their neighbors and invade the local extracellular matrix. (B) Partial genetic cascade controlling the switch from epithelial to mesenchymal phenotype. *Yellow*, ectoderm/epidermis; *light green*, neural tube; dark *green*, neural crest; *red*, extracellular matrix.

dermomyotome and a mesenchymal sclerotome roughly around the time of neural crest emigration (Kuo and Erickson, 2010; Mayor and Theveneau, 2013; Theveneau and Mayor, 2012a). Neural crests cells first start by migrating ventrally along the neural tube and progressively invade the anterior part of the sclerotome as it undergoes epithelial-mesenchymal transition. In amphibians and fish, the somite is mostly composed of muscle precursors with very little sclerotome, and neural crest cells migrate along the somite rather than invading it. Despite these anatomical differences, a segmented pattern of migration imposed by the somites nonetheless emerges. Amphibian neural crest cells align along the posterior somites whereas in fish they migrate along the middle portion of each somite. The neural crest undertaking the ventral path will later aggregate to form the dorsal root and sympathetic ganglia of the peripheral nervous system (Figure 18.4B,C), whereas the neural crest migrating under the skin will later differentiate as pigment cells. Finally, there are two additional subpopulations emerging from the trunk neural crest. One emigrates from the mid-trunk and gives rise to endocrine cells within the adrenal gland. Another, which delaminates from the caudal neural tube, is a second group of enteric neural crest cells that colonizes the gut from its most caudal part migrating anterior-ward to mix with enteric crest from the post-otic region.

Molecular Control of Neural Crest Migration

Neural crest cells migrate using the local extracellular matrix available to them. Matrices of various compositions have been described within the migratory routes with a major contribution from Fibronectin, Laminins and Collagens (Alfandari et al., 2010; Perris and Perissinotto, 2000). The ability of neural crest cells to use these matrices for migration depends on the expression of specific transmembrane receptors for each of these matrix components (i.e., Integrins, Syndecans) and proteases involved in matrix remodeling (i.e., ADAMs, MMPs). The extracellular matrix is needed to provide support but can also locally act as a negative cue that prevents invasion of a given region or structure; such as Versicans and Collagen IX in the posterior half of the somites in chick and mouse embryos. Physical constraints such as the eye and the otic vesicle, which are dense epithelial structures that cannot be invaded, force cells to move around them, hence shaping the streams.

In addition, several molecules acting as inhibitors of cell migration are released in the local extracellular matrix and prevent cell motility in some areas. In the head, odd-numbered rhombomeres (i.e., r3 and r5, see Figure 18.4A) are known to secrete members of the semaphorin-3 family (sema3A/D/F). Migratory neural crest cells express the neuropilin-plexin complexes required to sense these molecules, and abolishing semaphorin signaling leads to ectopic migration into otherwise neural crest-free regions (Eickholt et al., 1999; Gammill et al., 2007; Gammill et al., 2006; Osborne et al., 2005; Schwarz et al., 2009; Yu and Moens, 2005). Cells are also gathered into discrete streams by affinity with one another and their local environment. Different repertoires of Eph-ephrin molecules are expressed by the different streams of cephalic neural crest, and altering Eph-ephrin expression patterns forces some cells to move into the wrong stream. Similarly, neural crest cells can only colonize tissues with compatible Eph-ephrin codes; modifying the Eph-ephrin patterns might prevent

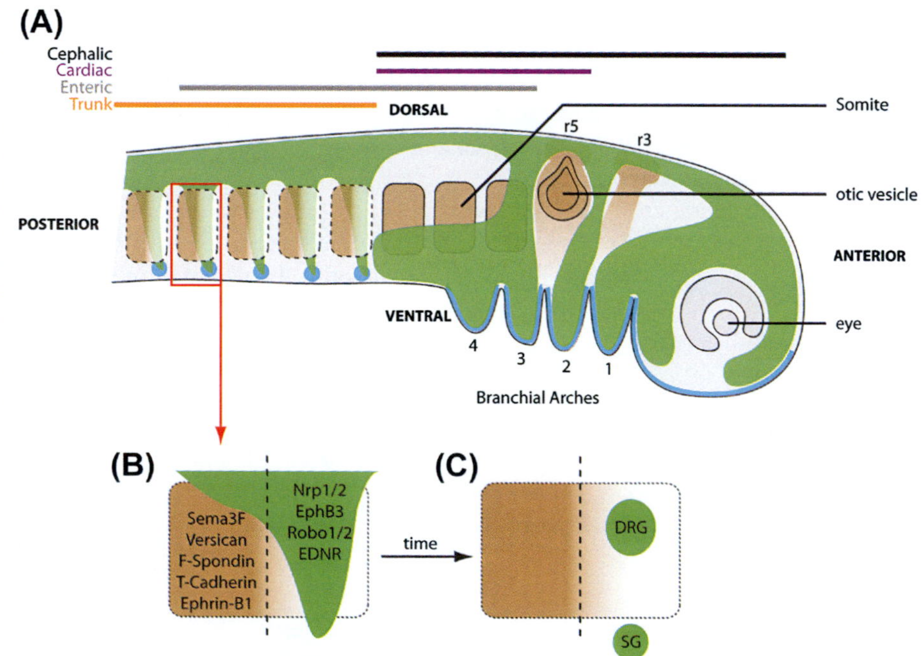

FIGURE 18.4 Neural Crest Migration. (A) Overview of the different neural crest subpopulations and their migration paths. (B) Neural crest migration through the anterior part of the somite as an example of how the differential distribution of guidance molecules in the environment and their receptors in migratory neural crest cells restricts neural crest migration. (C) Aggregation of migratory neural crest cells into peripheral ganglia at the end of migration. *Green,* neural crest cells; *brown,* inhibitory cues; *blue,* positive cues. DRG, dorsal root ganglion; EDNR, endothelin receptor; Nrp, neuropilin; r3/5; rhombomeres 3 and 5; Robo, roundabout; Sema, Semaphorin; SG, sympathetic ganglion.

invasion or induce ectopic migration of the neural crest (Davy et al., 2004; Krull et al., 1997; Mellott and Burke, 2008; Santiago and Erickson, 2002; Smith et al., 1997).

Control of trunk neural crest migration into the dorsolateral pathway or through the somites relies on similar mechanisms mediated by semaphorins, ephrins, endothelins and distribution of permissive versus non-permissive extracellular matrix components (Figure 18.4B). For instance, in the chick, entry into the dorsolateral path depends on which endothelin receptor is expressed. Endothelin receptor (EDNR)-positive cells will take the ventral route whereas EDNR2-positive cells will be able to enter the dorsolateral path (Kelsh et al., 2009; Pla et al., 2005; Shin et al., 1999).

Finally, more precise targeting into specific regions is controlled by positive cues that may act locally as attractants, although there is a lack of *in vivo* evidence for such action. Members of the FGF, and PDGF families as well as VEGFA, GDNF and Sdf1 are all positive regulators of neural crest migration and have been proposed as chemoattractants (Belmadani et al., 2005; 2009; Kasemeier-Kulesa et al., 2010; Kubota and Ito, 2000; McLennan et al., 2010; Olesnicky Killian et al., 2009; Saito et al., 2011; 2012; Smith and Tallquist, 2010; Svetic et al., 2007; Theveneau et al., 2010; Trokovic et al., 2005). Sdf1 for instance is required for targeting neural crest cells into the anlagen of peripheral ganglia or the hair follicles, and FGFs and VEGFA control entry into the branchial arches, whereas GDNF seems to control migration along the gut (Sasselli et al., 2012; Young et al., 2001). However, none of these molecules presents an expression pattern that fits with their putative role as attractants, and loss-of-function experiments lead to an early arrest of migration rather than a broad dispersion of cells due to a lack of attraction into a given region. Therefore, much remains to be understood about their actual role in promoting neural crest cell motility and guidance *in vivo*.

One key aspect of neural crest migration is that nearly all neural crest cells migrate as part of a large group in which cells make repeated cell-cell contacts during their journey (Theveneau and Mayor, 2011). Such contacts are sufficient to polarize the cells such that neural crest cells tend to move away from one another. This mechanism is known as contact-inhibition of locomotion and accounts for the overall dispersion of neural crest cells (Abercrombie and Heaysman, 1953; Carmona-Fontaine et al., 2008; Erickson, 1985; Mayor and Carmona-Fontaine, 2010; Teddy and Kulesa, 2004). It works as follows. A contact between two neural crest cells triggers a local inhibition of cell protrusions. This leads to a transient arrest of cell movement directly followed by a change of direction away from the contact. Studies in amphibians and fish have shown that contact-inhibition of locomotion requires a contact involving N-Cadherin and activation of the non-canonical Wnt-planar cell polarity pathway which triggers a local activation of RhoA and an inhibition of Rac1 accounting for the inhibition of cell protrusions (Carmona-Fontaine et al., 2008; Theveneau et al., 2010). RhoA and Rac1 are small GTPases that regulate acto-myosin contractility and actin polymerization; making them essential regulators of cell polarity and motility. Importantly, at least in the cephalic neural crest, the overall pro-dispersion effect of contact-inhibition is counterbalanced by a local paracrine chemotactic system relying on Complement factor C3a signaling (Carmona-Fontaine et al., 2011). Each neural crest cell secretes C3a while expressing its receptor C3aR. This means that regions of high neural crest cell density have also a higher concentration of C3a. Thus, neural crest cells tend to follow one another by short-range C3a-dependent chemotaxis. The combination of contact-inhibition and local mutual attraction means that the neural crest population is a loose mesenchymal group that nonetheless moves around as a collective structure whose cell density is kept relatively constant by means of mutual attraction.

Interestingly, contact-dependent polarity and guidance cues seem to be integrated rather than working in parallel. Cell-cell interactions mediated by N-Cadherin or gap junctions have been shown to be required for both Sdf1-dependent chemotaxis and semaphorin-mediated repulsion (Theveneau et al., 2010; Xu et al., 2006). These suggest that cell-cell cooperation through contact-inhibition and co-attraction is essential for cell guidance, thus favoring collective movement rather than individual migration. One has to keep in mind, however, that the neural crest cell population is extremely diverse and encounters many different environments while migrating. Neural crest cells are seen migrating as collective groups, strands, streams or isolated cells depending on the region of the embryo and the stage of migration that is monitored. Therefore, some of the mechanisms described for one subpopulation of neural crest may not fully apply to all the others. For instance, contact-inhibition-dependent migration has been shown for cephalic neural crest and enteric neural crest along the gut; but while the cephalic crest clearly migrates as a collective population of cells, the enteric neural crest seem to undergo a more disorganized directional spreading with little over time coordination among direct neighbors.

The overall conclusion is that neural crest migration is controlled in time and space by four main players:

i) Interaction with the local extracellular matrix;
ii) Negative cues (chemical and physical barriers);
iii) Positive cues;
iv) Cell-cell interactions.

The actual balance of these signals at any given time point dictates where cells go.

18.5 HUMAN PATHOLOGIES

Several human pathologies, collectively known as neurocristopathies, are produced by a failure of neural crest development. The defect can occur during the early specification of the neural crest as part of its migration or differentiation. One of the abnormalities associated with the cephalic neural crest is DiGeorge syndrome (see also Chapter 36). Patients with this syndrome have an incomplete migration of the neural crest in the pharyngeal arches. They have hypocalcemia that arises from defects in the parathyroid glands, thymic hypoplasia, and outflow tract defects of the heart. The facial anomalies include low-set ears, a small mouth and philtrum (the space between the upper lip and the nose), and cleft palates. DiGeorge syndrome is associated with deletions on the long arm of chromosome 22, and at least two genes present in that region are involved in this syndrome: *Tbx1* and *Tuple1* (Baldini, 2005).

A failure of vagal neural crest cell migration to the colon results in the absence of enteric ganglia and, thus, the absence of peristaltic movement in the bowel. The human pathology associated with this failure is called Hirschsprung disease (or aganglionic megacolon), and it has been demonstrated that the gene *Sox10* plays a key role in the generation of this pathology (Farlie et al., 2004).

Several animal models have been developed which reproduce some or all aspects of these and other neural crest pathologies. The most frequent organism used to study these pathologies is the mouse, because is a mammal; however, organisms such as *Xenopus* and zebrafish have also been very useful to study the role of specific genes in a particular phenotype that mimics the human syndrome. Understanding the normal cellular/molecular/genetics mechanisms of neural crest development in different animal models will be essential for the further understanding of human syndromes.

Finally, neural crest cells are a good model to study mechanisms of cancer progression (Lim and Thiery, 2012; Strobl-Mazzulla and Bronner, 2012). Indeed numerous factors involved in cancer metastasis and epithelial-mesenchymal transition (i.e., Snail/Snail2) were first identified as regulators of neural crest development. Some cancers directly stem from neural crest derivatives such as melanoma and neuroblastoma. Moreover, in many ways, cancer metastasis recapitulates neural crest development in a dysregulated manner (Figure 18.5) making the neural crest a cell population of choice to analyze the function of oncogenes in their normal physiological context (Mayor and Theveneau, 2013; Thiery et al., 2009). Studying the normal function of genes is essential to understand which abilities they confer onto cells when activated.

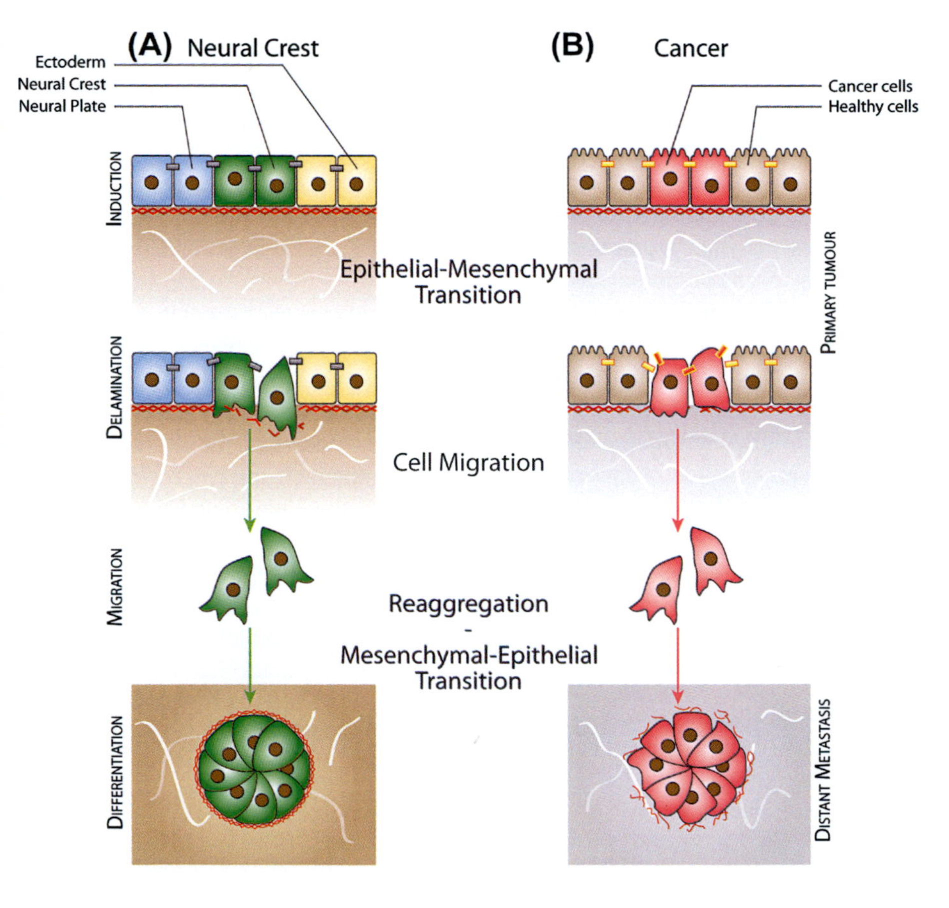

FIGURE 18.5 Similarities between Neural Crest Development and Cancer Progression. (A) Neural crest development. Neural crest induction provide neural crest cells with set of genes that control their survival and allow them to separate from their surrounding tissues, migrate throughout their local environment and finally settle into new locations. (B) Cancer progression. Similarly to neural crest cells, tumor cells acquire a specific set of genes that will favor their survival and growth and allow them to undergo long distance migration to establish secondary tumors. Many of the key players in metastatic cancer are involved in neural crest development (i.e., Snail, Twist, Ets transcription factors).

Embryo development is a great system to study cell behavior in complex environment. In addition, during cancer it is nearly impossible to predict where and when migration and epithelial-mesenchymal transition will occur. Therefore, embryos provide a reproducible and predictable system in which it is possible to first monitor epithelial-mesenchymal transition and cell migration *in vivo*, which then can be manipulated by altering gene expression and function. In addition, valuable insights into how cell migration and homing into specific tissues are regulated during cancer progression can be gained by analyzing this phenomenon during embryo development.

18.6 CLINICAL RELEVANCE

- Neural crest developmental defects are at the basis of many human syndromes and genetic disorders collectively known as neurocristopathies.
- Metastatic cancer recapitulates neural crest development in a dysregulated manner, hence making the neural crest a good model to study the function of oncogenes in their normal physiological context.
- Adult tissues (i.e., enteric nervous system and hair follicles) contain neural crest-derived stem cells making the neural crest an interesting population to study in the context of regenerative medicine.

ACKNOWLEDGMENTS

We apologize to our colleagues in the field whose work could not be cited owing to a lack of space. Investigations performed in our laboratory are supported by grants to R.M. from the UK Medical Research Council, the Biotechnology and Biological Sciences Research Council and Wellcome Trust. E.T. was supported by a Wellcome Trust Value in People award.

RECOMMENDED RESOURCES

Sauka-Spengler, Bronner, 2010. SnapShot: Neural Crest. Cell 143. 486–486.

Mayor, R., Theveneau, E., 2013. The neural crest. Development 140, 2247–2251.

Special issue of Developmental Biology on Neural Crest. Dev. Biol. 366, 2012, 1–100.

REFERENCES

Abercrombie, M., Heaysman, J.E., 1953. Observations on the social behavior of cells in tissue culture. I. Speed of movement of chick heart fibroblasts in relation to their mutual contacts. Exp. Cell Res. 5, 111–131.

Alfandari, D., Cousin, H., Marsden, M., 2010. Mechanism of *Xenopus* cranial neural crest cell migration. Cell Adh. Migr. 4.

Aybar, M.J., Mayor, R., 2002. Early induction of neural crest cells: lessons learned from frog, fish and chick. Curr. Opin. Genet. Dev. 12, 452–458.

Baldini, A., 2005. Dissecting contiguous gene defects: TBX1. Curr. Opin. Genet. Dev. 15, 279–284.

Barriga, E.H., Maxwell, P.H., Reyes, A.E., Mayor, R., 2013. The hypoxia factor Hif-1alpha controls neural crest chemotaxis and epithelial to mesenchymal transition. J. Cell Biol. in press.

Bastidas, F., De Calisto, J., Mayor, R., 2004. Identification of neural crest competence territory: role of Wnt signaling. Dev. Dyn. 229, 109–117.

Begemann, G., Schilling, T.F., Rauch, G.J., Geisler, R., Ingham, P.W., 2001. The zebrafish neckless mutation reveals a requirement for raldh2 in mesodermal signals that pattern the hindbrain. Development 128, 3081–3094.

Belmadani, A., Jung, H., Ren, D., Miller, R.J., 2009. The chemokine SDF-1/CXCL12 regulates the migration of melanocyte progenitors in mouse hair follicles. Differentiation 77, 395–411.

Belmadani, A., Tran, P.B., Ren, D., Assimacopoulos, S., Grove, E.A., Miller, R.J., 2005. The chemokine stromal cell-derived factor-1 regulates the migration of sensory neuron progenitors. J. Neurosci. 25, 3995–4003.

Bonstein, L., Elias, S., Frank, D., 1998. Paraxial-fated mesoderm is required for neural crest induction in *Xenopus* embryos. Dev. Biol. 193, 156–168.

Carmona-Fontaine, C., Matthews, H.K., Kuriyama, S., Moreno, M., Dunn, G.A., Parsons, M., Stern, C.D., Mayor, R., 2008. Contact inhibition of locomotion *in vivo* controls neural crest directional migration. Nature 456, 957–961.

Carmona-Fontaine, C., Theveneau, E., Tzekou, A., Tada, M., Woods, M., Page, K.M., Parsons, M., Lambris, J.D., Mayor, R., 2011. Complement fragment C3a controls mutual cell attraction during collective cell migration. Dev. Cell 21, 1026–1037.

Cornell, R.A., Eisen, J.S., 2000. Delta signaling mediates segregation of neural crest and spinal sensory neurons from zebrafish lateral neural plate. Development 127, 2873–2882.

Cornell, R.A., Eisen, J.S., 2002. Delta/Notch signaling promotes formation of zebrafish neural crest by repressing Neurogenin 1 function. Development 129, 2639–2648.

Cornell, R.A., Eisen, J.S., 2005. Notch in the pathway: the roles of Notch signaling in neural crest development. Semin. Cell Dev. Biol. 16, 663–672.

Dady, A., Blavet, C., Duband, J.L., 2012. Timing and kinetics of E- to N-cadherin switch during neurulation in the avian embryo. Dev. Dyn. 241, 1333–1349.

Davy, A., Aubin, J., Soriano, P., 2004. Ephrin-B1 forward and reverse signaling are required during mouse development. Genes Dev. 18, 572–583.

De Robertis, E.M., Kuroda, H., 2004. Dorsal-ventral patterning and neural induction in *Xenopus* embryos. Annu. Rev. Cell Dev. Biol. 20, 285–308.

Deardorff, M.A., Tan, C., Saint-Jeannet, J.P., Klein, P.S., 2001. A role for frizzled 3 in neural crest development. Development 128, 3655–3663.

Duband, J.L., 2010. Diversity in the molecular and cellular strategies of epithelium-to-mesenchyme transitions: Insights from the neural crest. Cell Adh. Migr. 4, 458–482.

Eickholt, B.J., Mackenzie, S.L., Graham, A., Walsh, F.S., Doherty, P., 1999. Evidence for collapsin-1 functioning in the control of neural crest migration in both trunk and hindbrain regions. Development 126, 2181–2189.

Endo, Y., Osumi, N., Wakamatsu, Y., 2002. Bimodal functions of Notch-mediated signaling are involved in neural crest formation during avian ectoderm development. Development 129, 863–873.

Endo, Y., Osumi, N., Wakamatsu, Y., 2003. Deltex/Dtx mediates NOTCH signaling in regulation of Bmp4 expression in cranial neural crest formation during avian development. Dev. Growth Differ. 45, 241–248.

Erickson, C.A., 1985. Control of neural crest cell dispersion in the trunk of the avian embryo. Dev. Biol. 111, 138–157.

Farlie, P.G., McKeown, S.J., Newgreen, D.F., 2004. The neural crest: basic biology and clinical relationships in the craniofacial and enteric nervous systems. Birth Defects. Res. C Embryo. Today 72, 173–189.

Gammill, L.S., Gonzalez, C., Bronner-Fraser, M., 2007. Neuropilin 2/semaphorin 3F signaling is essential for cranial neural crest migration and trigeminal ganglion condensation. Dev. Neurobiol. 67, 47–56.

Gammill, L.S., Gonzalez, C., Gu, C., Bronner-Fraser, M., 2006. Guidance of trunk neural crest migration requires neuropilin 2/semaphorin 3F signaling. Development 133, 99–106.

Garcia-Castro, M.I., Marcelle, C., Bronner-Fraser, M., 2002. Ectodermal Wnt function as a neural crest inducer. Science 297, 848–851.

Glavic, A., Silva, F., Aybar, M.J., Bastidas, F., Mayor, R., 2004. Interplay between Notch signaling and the homeoprotein Xiro1 is required for neural crest induction in *Xenopus* embryos. Development 131, 347–359.

Hall, B., 2008. The neural crest and neural crest cells in vertebrate development and evolution. Springer, New York.

Hall, B.K., 1999. The neural crest in development and evolution. Springer, New York.

Hong, C.S., Park, B.Y., Saint-Jeannet, J.P., 2008. Fgf8a induces neural crest indirectly through the activation of Wnt8 in the paraxial mesoderm. Development 135, 3903–3910.

Kasemeier-Kulesa, J.C., McLennan, R., Romine, M.H., Kulesa, P.M., Lefcort, F., 2010. CXCR4 controls ventral migration of sympathetic precursor cells. J. Neurosci. 30, 13078–13088.

Kelsh, R.N., Harris, M.L., Colanesi, S., Erickson, C.A., 2009. Stripes and belly-spots – a review of pigment cell morphogenesis in vertebrates. Semin. Cell Dev. Biol. 20, 90–104.

Krull, C.E., Lansford, R., Gale, N.W., Collazo, A., Marcelle, C., Yancopoulos, G.D., Fraser, S.E., Bronner-Fraser, M., 1997. Interactions of Eph-related receptors and ligands confer rostrocaudal pattern to trunk neural crest migration. Curr. Biol. 7, 571–580.

Kubota, Y., Ito, K., 2000. Chemotactic migration of mesencephalic neural crest cells in the mouse. Dev. Dyn. 217, 170–179.

Kulesa, P.M., Bailey, C.M., Kasemeier-Kulesa, J.C., McLennan, R., 2010. Cranial neural crest migration: new rules for an old road. Dev. Biol. 344, 543–554.

Kuo, B.R., Erickson, C.A., 2010. Regional differences in neural crest morphogenesis. Cell Adh. Migr. 4.

Kuroda, H., Wessely, O., De Robertis, E.M., 2004. Neural induction in *Xenopus*: requirement for ectodermal and endomesodermal signals via Chordin, Noggin, beta-Catenin, and Cerberus. PLoS Biol. 2, E92.

LaBonne, C., Bronner-Fraser, M., 1998. Neural crest induction in *Xenopus*: evidence for a two-signal model. Development 125, 2403–2414.

Le Douarin, N., Kalcheim, C., 1999. The neural crest. Cambridge University Press, Cambridge, UK; New York, NY, USA.

Lekven, A.C., Thorpe, C.J., Waxman, J.S., Moon, R.T., 2001. Zebrafish wnt8 encodes two wnt8 proteins on a bicistronic transcript and is required for mesoderm and neurectoderm patterning. Dev. Cell 1, 103–114.

Lewis, A.E., Vasudevan, H.N., O'Neill, A.K., Soriano, P., Bush, J.O., 2013. The widely used Wnt1-Cre transgene causes developmental phenotypes by ectopic activation of Wnt signaling. Dev. Biol.

Lewis, J.L., Bonner, J., Modrell, M., Ragland, J.W., Moon, R.T., Dorsky, R.I., Raible, D.W., 2004. Reiterated Wnt signaling during zebrafish neural crest development. Development 131, 1299–1308.

Liem Jr., K.F., Tremml, G., Roelink, H., Jessell, T.M., 1995. Dorsal differentiation of neural plate cells induced by BMP-mediated signals from epidermal ectoderm. Cell 82, 969–979.

Lim, J., Thiery, J.P., 2012. Epithelial-mesenchymal transitions: insights from development. Development 139, 3471–3486.

Mancilla, A., Mayor, R., 1996. Neural crest formation in *Xenopus laevis*: mechanisms of Xslug induction. Dev. Biol. 177, 580–589.

Marchant, L., Linker, C., Ruiz, P., Guerrero, N., Mayor, R., 1998. The inductive properties of mesoderm suggest that the neural crest cells are specified by a BMP gradient. Dev. Biol. 198, 319–329.

Mayor, R., Aybar, M.J., 2001. Induction and development of neural crest in *Xenopus laevis.* Cell Tissue Res. 305, 203–209.

Mayor, R., Carmona-Fontaine, C., 2010. Keeping in touch with contact inhibition of locomotion. Trends Cell Biol. 20, 319–328.

Mayor, R., Guerrero, N., Martinez, C., 1997. Role of FGF and noggin in neural crest induction. Dev. Biol. 189, 1–12.

Mayor, R., Theveneau, E., 2013. The neural crest. Development 140, 2247–2251.

Mayor, R., Morgan, R., Sargent, M.G., 1995. Induction of the prospective neural crest of *Xenopus*.. Dev. Mar. 121 (3), 767–777.

McKeown, S.J., Wallace, A.S., Anderson, R.B., 2012. Expression and function of cell adhesion molecules during neural crest migration. Dev. Biol. 373, 244–257.

McLennan, R., Teddy, J.M., Kasemeier-Kulesa, J.C., Romine, M.H., Kulesa, P.M., 2010. Vascular endothelial growth factor (VEGF) regulates cranial neural crest migration *in vivo*. Dev. Biol. 339, 114–125.

Mellott, D.O., Burke, R.D., 2008. Divergent roles for Eph and ephrin in avian cranial neural crest. BMC Dev. Biol. 8, 56.

Meulemans, D., Bronner-Fraser, M., 2004. Gene-regulatory interactions in neural crest evolution and development. Dev. Cell 7, 291–299.

Monsoro-Burq, A.H., Fletcher, R.B., Harland, R.M., 2003. Neural crest induction by paraxial mesoderm in *Xenopus* embryos requires FGF signals. Development 130, 3111–3124.

Moury, J.D., Jacobson, A.G., 1989. Neural fold formation at newly created boundaries between neural plate and epidermis in the axolotl. Dev. Biol. 133, 44–57.

Neave, B., Holder, N., Patient, R., 1997. A graded response to BMP-4 spatially coordinates patterning of the mesoderm and ectoderm in the zebrafish. Mech. Dev. 62, 183–195.

Nieto, M.A., Sargent, M.G., Wilkinson, D.G., Cooke, J., 1994. Control of cell behavior during vertebrate development by Slug, a zinc finger gene. Sci. May 6 264 (5160), 835–839.

Nguyen, V.H., Schmid, B., Trout, J., Connors, S.A., Ekker, M., Mullins, M.C., 1998. Ventral and lateral regions of the zebrafish gastrula, including the neural crest progenitors, are established by a bmp2b/swirl pathway of genes. Dev. Biol. 199, 93–110.

Olesnicky Killian, E.C., Birkholz, D.A., Artinger, K.B., 2009. A role for chemokine signaling in neural crest cell migration and craniofacial development. Dev. Biol. 333, 161–172.

Osborne, N.J., Begbie, J., Chilton, J.K., Schmidt, H., Eickholt, B.J., 2005. Semaphorin/neuropilin signaling influences the positioning of migratory neural crest cells within the hindbrain region of the chick. Dev. Dyn. 232, 939–949.

Patthey, C., Edlund, T., Gunhaga, L., 2009. Wnt-regulated temporal control of BMP exposure directs the choice between neural plate border and epidermal fate. Development 136, 73–83.

Perez-Alcala, S., Nieto, M.A., Barbas, J.A., 2004. LSox5 regulates RhoB expression in the neural tube and promotes generation of the neural crest. Development 131, 4455–4465.

Perris, R., Perissinotto, D., 2000. Role of the extracellular matrix during neural crest cell migration. Mech. Dev. 95, 3–21.

Pieper, M., Ahrens, K., Rink, E., Peter, A., Schlosser, G., 2012. Differential distribution of competence for panplacodal and neural crest induction to non-neural and neural ectoderm. Development 139, 1175–1187.

Pla, P., Alberti, C., Solov'eva, O., Pasdar, M., Kunisada, T., Larue, L., 2005. Ednrb2 orients cell migration towards the dorsolateral neural crest pathway and promotes melanocyte differentiation. Pigment Cell Res. 18, 181–187.

Ragland, J.W., Raible, D.W., 2004. Signals derived from the underlying mesoderm are dispensable for zebrafish neural crest induction. Dev. Biol. 276, 16–30.

Raven, C., Kloos, J., 1945. Induction by medial and lateral pieces of archenteron roof, with special reference to the determination of the neural crest. Acta Neerl. Norm Pathol. 55, 348–362.

Saint-Jeannet, J.P., He, X., Varmus, H.E., Dawid, I.B., 1997. Regulation of dorsal fate in the neuraxis by Wnt-1 and Wnt-3a. Proc. Natl. Acad. Sci. USA 94, 13713–13718.

Saito, D., Takase, Y., Murai, H., Takahashi, Y., 2012. The dorsal aorta initiates a molecular cascade that instructs sympatho-adrenal specification. Science 336, 1578–1581.

Santiago, A., Erickson, C.A., 2002. Ephrin-B ligands play a dual role in the control of neural crest cell migration. Development 129, 3621–3632.

Sasselli, V., Pachnis, V., Burns, A.J., 2012. The enteric nervous system. Dev. Biol. 366, 64–73.

Sato, A., Scholl, A.M., Kuhn, E.B., Stadt, H.A., Decker, J.R., Pegram, K., Hutson, M.R., Kirby, M.L., 2011. FGF8 signaling is chemotactic for cardiac neural crest cells. Dev. Biol.

Sauka-Spengler, T., Bronner-Fraser, M., 2008. A gene regulatory network orchestrates neural crest formation. Nat. Rev. Mol. Cell Biol. 9, 557–568.

Schwarz, Q., Maden, C.H., Davidson, K., Ruhrberg, C., 2009. Neuropilin-mediated neural crest cell guidance is essential to organize sensory neurons into segmented dorsal root ganglia. Development 136, 1785–1789.

Selleck, M.A., Bronner-Fraser, M., 1995. Origins of the avian neural crest: the role of neural plate-epidermal interactions. Development 121, 525–538.

Shin, M.K., Levorse, J.M., Ingram, R.S., Tilghman, S.M., 1999. The temporal requirement for endothelin receptor-B signaling during neural crest development. Nature 402, 496–501.

Smith, A., Robinson, V., Patel, K., Wilkinson, D.G., 1997. The EphA4 and EphB1 receptor tyrosine kinases and ephrin-B2 ligand regulate targeted migration of branchial neural crest cells. Curr. Biol. 7, 561–570.

Smith, C.L., Tallquist, M.D., 2010. PDGF function in diverse neural crest cell populations. Cell Adh. Migr. 4, 561–566.

Steventon, B., Araya, C., Linker, C., Kuriyama, S., Mayor, R., 2009. Differential requirements of BMP and Wnt signaling during gastrulation and neurulation define two steps in neural crest induction. Development 136, 771–779.

Steventon, B., Carmona-Fontaine, C., Mayor, R., 2005. Genetic network during neural crest induction: from cell specification to cell survival. Semin Cell Dev. Biol. 16, 647–654.

Steventon, B., Mayor, R., 2012. Early neural crest induction requires an initial inhibition of Wnt signals. Dev. Biol. 365, 196–207.

Strobl-Mazzulla, P.H., Bronner, M.E., 2012. Epithelial to mesenchymal transition: new and old insights from the classical neural crest model. Semin Cancer Biol. 22, 411–416.

Svetic, V., Hollway, G.E., Elworthy, S., Chipperfield, T.R., Davison, C., Adams, R.J., Eisen, J.S., Ingham, P.W., Currie, P.D., Kelsh, R.N., 2007. Sdf1a patterns zebrafish melanophores and links the somite and melanophore pattern defects in choker mutants. Development 134, 1011–1022.

Teddy, J.M., Kulesa, P.M., 2004. *In vivo* evidence for short- and long-range cell communication in cranial neural crest cells. Development 131, 6141–6151.

Theveneau, E., Duband, J.L., Altabef, M., 2007. Ets-1 confers cranial features on neural crest delamination. PLoS One 2, e1142.

Theveneau, E., Marchant, L., Kuriyama, S., Gull, M., Moepps, B., Parsons, M., Mayor, R., 2010. Collective chemotaxis requires contact-dependent cell polarity. Dev. Cell 19, 39–53.

Theveneau, E., Mayor, R., 2011. Can mesenchymal cells undergo collective cell migration? The case of the neural crest. Cell Adh. Migr. 5, 490–498.

Theveneau, E., Mayor, R., 2012a. Neural crest delamination and migration: from epithelium-to-mesenchyme transition to collective cell migration. Dev. Biol. 366, 34–54.

Theveneau, E., Mayor, R., 2012b. Neural crest migration: interplay between chemorepellents, chemoattractants, contact inhibition, epithelial-mesenchymal transition, and collective cell migration. Wiley Interdiscip. Rev. Dev. Biol. 1, 435–445.

Thiery, J.P., Acloque, H., Huang, R.Y., Nieto, M.A., 2009. Epithelial-mesenchymal transitions in development and disease. Cell 139, 871–890.

Tribulo, C., Aybar, M.J., Nguyen, V.H., Mullins, M.C., Mayor, R., 2003. Regulation of Msx genes by a Bmp gradient is essential for neural crest specification. Development 130, 6441–6452.

Trokovic, N., Trokovic, R., Partanen, J., 2005. Fibroblast growth factor signaling and regional specification of the pharyngeal ectoderm. Int. J. Dev. Biol. 49, 797–805.

Villanueva, S., Glavic, A., Ruiz, P., Mayor, R., 2002. Posteriorization by FGF, Wnt, and retinoic acid is required for neural crest induction. Dev. Biol. 241, 289–301.

Wilson, P.A., Hemmati-Brivanlou, A., 1997. Vertebrate neural induction: inducers, inhibitors, and a new synthesis. Neuron 18, 699–710.

Xu, X., Francis, R., Wei, C.J., Linask, K.L., Lo, C.W., 2006. Connexin 43-mediated modulation of polarized cell movement and the directional migration of cardiac neural crest cells. Development 133, 3629–3639.

Young, H.M., Hearn, C.J., Farlie, P.G., Canty, A.J., Thomas, P.Q., Newgreen, D.F., 2001. GDNF is a chemoattractant for enteric neural cells. Dev. Biol. 229, 503–516.

Yu, H.H., Moens, C.B., 2005. Semaphorin signaling guides cranial neural crest cell migration in zebrafish. Dev. Biol. 280, 373–385.

Chapter 19

Development of the Pre-Placodal Ectoderm and Cranial Sensory Placodes

Sally A. Moody* **and Jean-Pierre Saint-Jeannet**†

**Department of Anatomy and Regenerative Biology, The George Washington University School of Medicine and Health Sciences, Washington DC;*
†Department of Basic Science and Craniofacial Biology, New York University, College of Dentistry, New York City, NY

Chapter Outline

GLOSSARY

Adenohypophysis The anterior pituitary gland, derived from the adenohypophyseal placode, which contains cells that secrete a number of peptide hormones.

Branchio-otic syndrome An autosomal dominant syndrome in humans that presents with branchial cleft fistulas and hearing loss. Some cases are caused by mutations in the *Eya1* gene and some are caused by mutations in *Six* genes.

Branchio-otic renal syndrome An autosomal dominant syndrome in humans that presents with craniofacial defects, hearing loss, and kidney or urinary tract defects.

Neurulation The process by which the flat, disc-shaped neural plate ectoderm folds into an elongated tube, thereby becoming the precursor of the central nervous system.

Organizer The mesodermal region of the vertebrate embryo that is the source of signaling molecules that dorsalize and pattern both the mesoderm and the ectoderm.

Oto-facial-cervical syndrome Patients present with hearing loss, a long narrow face and various facial and cervical structural abnormalities. Some cases are caused by mutations in the *Eya1* gene.

Pre-placodal ectoderm The region of the embryonic ectoderm that surrounds the anterior neural plate and that will give rise to all of the cranial sensory placodes. It is characterized by the expression of *Six* and *Eya* genes.

SUMMARY

- Cranial sensory placodes arise from a common precursor field, the pre-placodal ectoderm (PPE), which surrounds the anterior neural plate just lateral to the neural crest. They give rise to secretory cells and numerous sensory structures in the vertebrate head.
- The PPE forms in response to interactions between the neural and non-neural ectoderm and to signaling factors (FGF, BMP, Wnt and retinoic acid) from adjacent tissues (underlying mesoderm, endoderm, anterior neural ridge).

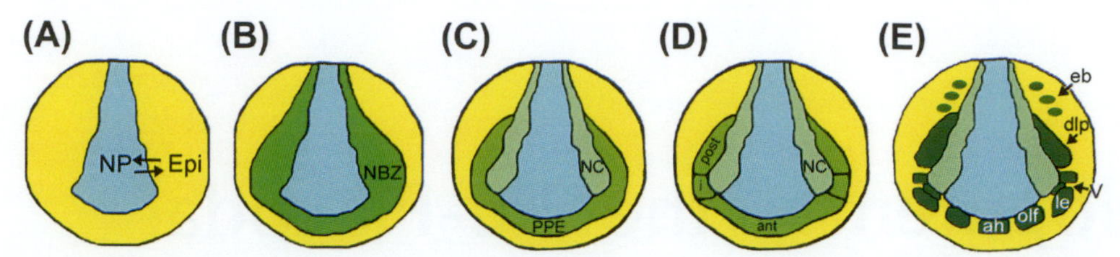

FIGURE 19.1 **The Ectodermal Domains of the *Xenopus* Embryo at Different Developmental Stages.** (A) At gastrulation, the early embryonic ectoderm is divided into domains that will give rise to the neural plate (NP, blue) and epidermis (Epi, yellow). Interactions between these two domains (arrows), as well as signaling from underlying tissues establish a neural border zone (NBZ; green) at early neural plate stages (B). (C) The NBZ subsequently gives rise to the medially located neural crest (NC; light green) and the laterally located pre-placodal ectoderm (PPE, dark green). (D) At mid-neural plate stages, the PPE acquires initial patterning, and can be divided into three separate fields: anterior (ant), intermediate (i) and posterior (post). (E) In response to local signaling and differential transcription factor expression at later neural plate stages, the pre-placodal ectoderm breaks up into individual placodes, including: those derived from the anterior field (the adenohypophyseal (ah), olfactory (olf), and lens (le)); those derived from the intermediate field (V; ophthalmic placode is dorsal to lens placode, and maxillomandibular placode is posterior to it); and those derived from the posterior field: (epibranchial (eb) and in *Xenopus*, the dorsolateral placode (dlp), which later breaks up into several lateral line placodes and the otic placode).

- Members of the Six family of transcription factors, particularly Six1 and Six4, are major regulators of PPE fate specification. Six1 upregulates other placode genes and downregulates epidermal and neural crest genes, thus maintaining PPE/placodal fate. Several co-factor proteins, including members of the Eya, Groucho and Dachshund families, modify Six protein activity.
- The PPE is initially competent to form all of the different placodes. Signaling from adjacent tissues along its anterior-posterior extent activates different sets of transcription factors that cause the PPE to separate into individual placodes with different developmental fates.
- Six1 and Eya1 maintain placode cells in an undifferentiated, proliferative state. They are downregulated when differentiation is initiated. For placode-derived neurons, this involves a dose-dependent regulation of SoxB1 transcription factors.
- Mutations in *SIX* and *EYA* genes underlie several human syndromes whose phenotypes often include hearing loss.

19.1 INTRODUCTION

During the evolution of the vertebrate head, a number of specialized sensory organs arose that are derived from thickenings in the embryonic ectoderm called cranial sensory placodes (von Kupffer, 1895; Platt, 1896; Knouff, 1935; reviewed by Webb and Noden, 1993; Baker and Bonner-Fraser, 2001; Streit, 2004; Schlosser 2005; Schlosser, 2007; Schlosser, 2008). As gastrulation is completed, the ectoderm surrounding the nascent neural ectoderm, called the neural border (NB) zone, is specified to give rise to peripheral sensory structures that derive from the neural crest and the more laterally located pre-placodal ectoderm (PPE) (Figure 19.1). The specific derivatives of the neural crest are described elsewhere (see Chapter 18); the induction of the PPE, its development into specific cranial sensory placodes, and the developmental genetic regulation of neurogenesis in the cranial sensory ganglia are the subjects of this chapter.

During neural tube closure, signals from adjacent tissues cause the PPE to separate into many discrete placodes, which are histologically recognized as patches of thickened ectoderm and have distinct developmental fates (Figure 19.1). These placodes will then produce secretory cells and both the structural and neural elements of numerous cranial sensory organs (Figure 19.2). These include the anterior pituitary gland, the olfactory epithelium (see Chapter 20), the lens, and the auditory (see Chapter 21) and vestibular organs, and in aquatic species the lateral line organs (see Chapter 15). In addition, cranial nerve sensory ganglia contain cells derived from both placodes and neural crest. Thus, the PPE gives rise to many important structures in the vertebrate head. However, although the cranial sensory placodes have been histologically recognized for more than a century (reviewed in Knouff, 1935; Streit, 2004; Schlosser, 2005), it is only in the past two decades that we have begun to understand the molecular mechanisms that specify the development of these important sensory precursors. Several laboratories have identified genes that are highly expressed in the PPE and in individual placodes, allowing researchers to begin to reveal the molecular pathways that induce and specify the fate of these important embryonic cells. In this chapter we will review the signaling pathways and transcription factors that specify the PPE, maintain its boundaries, cause individual placodes to form, and regulate neurogenesis in the cranial sensory ganglia.

19.2 CRANIAL SENSORY PLACODES GIVE RISE TO DIVERSE STRUCTURES

Histologically, much is known about the development of the various cranial sensory placodes (reviewed in Webb and Noden, 1993; LeDouarin et al., 1986; Baker and Bronner-Fraser, 2001). Each placode is first recognized as a tall, columnar

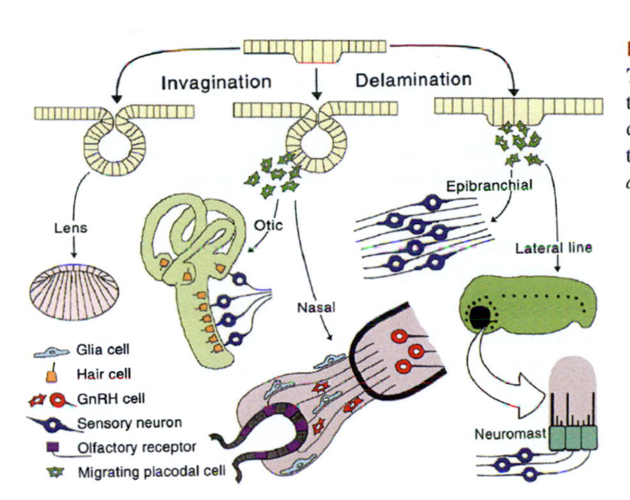

FIGURE 19.2 Placodes give Rise to Several Cranial Structures and Cell Types. The initial epithelial thickening (top of figure) can invaginate to give rise to a pit (olfactory/nasal) or a vesicle (adenohypophyseal, lens, otic), or cells can delaminate and migrate to a secondary position (cranial ganglia, lateral lines). Note the numerous cell types that derive from the cranial placodes. *(Modified from Webb and Noden, 1993, and from Brugmann and Moody, 2005).*

epithelium in the surface ectoderm that then undergoes extensive morphogenetic rearrangements. Some placodes (i.e., adenohypophyseal, olfactory, lens, otic) invaginate as cup-shaped structures (Figure 19.2). The olfactory cup, which is also called the olfactory or nasal pit, further folds to line the nasal cavity as a sensory epithelium. The adenohypophyseal, lens and otic placodes, however, continue to invaginate and eventually pinch off from the surface ectoderm to form epithelial-lined vesicles that then further differentiate into highly specialized structures. In other placodes, the underlying basement membrane of the thickened epithelium becomes fragmented, thereby allowing cells to delaminate. Some migrate within the surface ectoderm to form patches of sensory organs (e.g., the lateral lines; see Chapter 15), and some migrate away from the surface ectoderm to coalesce within the head mesenchyme as primary sensory neurons within a cranial nerve ganglion.

In each placode the cells can adopt a variety of cell fates, including secretory cells, sensory receptor cells, neurons, glia, or supporting cells, depending upon their placode of origin (Figure 19.2). The single adenohypophyseal placode invaginates when it occupies the dorsal midline of the oral ectoderm to form Rathke's pouch; the pouch then pinches off as a vesicle to form the adenohypophysis (anterior pituitary gland) whose cells secrete a number of peptide hormones (e.g., somatotropin, prolactin, gonadotropin, thyrotropin, corticotropin). The bilateral olfactory placodes give rise to primary olfactory receptor neurons that detect odorants, basal stem cells, glia, mucus-secreting cells, and structural cells (see Chapter 20); in some animals, they also give rise to the related vomeronasal epithelium, which detects pheromones. In addition, some cells migrate from the olfactory placode into the forebrain to become neuropeptide-secreting and gonadotropin releasing hormone-secreting neurons (Murakami and Arai, 1994; Northcutt and Muske, 1994; Hilal et al., 1996; Sabado et al., 2012). The lens placode gives rise to the lens vesicle in the anterior segment of the forming eye cup. The anterior layer of the lens vesicle gives rise to stem cells and supporting cells, and the posterior layer gives rise to the crystalline-producing cells that focus light on the neural retina. Two placodes contribute primary sensory neurons to the cranial ganglion associated with the trigeminal nerve: the ophthalmic placode (also called profundal), which is located dorsal to the eye, and the trigeminal placode (also called maxillomandibular), which is located just caudal to the eye (Figure 19.1E). These cells provide the sensory innervation of the face, the oral cavity, and the scalp. The large, distal neurons of the trigeminal ganglion are derived from the placodes, whereas the small, proximal neurons and the glia derive from the neural crest (LeDouarin et al., 1986). The otic placode gives rise to both the auditory and vestibular parts of the entire inner ear, including the mechanosensory hair cells, the supporting cells, the endolymph secreting cells, the biomineralized otoliths, and the vestibulocochlear sensory ganglia (see Chapter 21). A series of epibranchial placodes form just dorsal to the branchial (pharyngeal) ectodermal clefts. Cells from the epibranchial placodes migrate into the branchial mesenchyme to become the large neurons in the distal sensory ganglia of three cranial nerves. Those of the facial nerve (called the geniculate ganglion) innervate the taste buds; those of the glossopharyngeal nerve (called the petrosal ganglion) innervate the taste buds, heart, and visceral organs; and those of the vagus nerve (called the nodose ganglion) innervate the heart and other visceral organs. As described for the trigeminal ganglion, the small, proximal neurons and the glia of these ganglia are derived from the neural crest.

All vertebrates have these placodes in common, but there are numerous species variations (reviewed by Baker and Bronner-Fraser, 2001; Streit, 2004; Schlosser, 2005; 2007; 2010). For example, in some animals, the derivatives of the profundal and trigeminal placodes are maintained as separate sensory ganglia, whereas in most they coalescence into a single, crescent-shaped, trigeminal ganglion (Schlosser and Northcutt, 2000). In most vertebrates there are three epibranchial placodes, but in some there are up to six separate placodes with associated ganglia. In some animals a placode dorsal to the geniculate placode gives rise to the paratympanic organ (birds) and the spiracular organ (fish) (O'Neill et al., 2012). In amphibians and fish there is an additional lateral line sensory system that is specialized for aquatic life (see Chapter 15). This system consists of islands of sensory organs (receptors and supporting cells as well as the sensory neurons that innervate them) distributed across the head and trunk epidermis that have striking similarities to inner ear receptor organs. Mechanoreceptive neuromast organs detect water turbulence, and electroreceptive ampullary organs detect electrical fields. In some amphibians, there also are hypobranchial placodes located ventral to the second and third endodermal pouches; these give rise to the hypobranchial ganglia, whose function is presently unknown (Schlosser, 2003). We do not yet understand the evolutionary mechanisms that have given rise to the diversity of these structures across species (reviewed by Bassham and Postlthwait, 2005; Schlosser, 2005; 2007; 2008). However, many of the genes involved in placode development are highly conserved from invertebrates to vertebrates (Schlosser, 2005; 2008; Gasparini et al., 2013; Graham and Shimeld, 2013; see also Chapter 7). Therefore, the potential to form ectodermal sensory structures appears to be a basal characteristic, whereas the diversity of structures across species is likely to have arisen in response to unique environmental niches.

19.3 SPECIFICATION OF THE PRE-PLACODAL ECTODERM

Placodes are Derived from a Common Precursor Field in the Embryonic Ectoderm

The classic, histological descriptions of cranial sensory placode formation proposed that these diverse structures and cell types derive from a common precursor region called the pre-placodal ectoderm (PPE; also called the "pan-placodal region" or "common placodal field"), which forms as a U-shaped band of ectoderm surrounding the anterior margin of the neural plate and extending approximately to the posterior limit of the hindbrain (reviewed in Knouff, 1935; Schlosser, 2005) (Figure 19.1). In amphibians, this band is histologically distinct as a thickened epithelium, but it is not so easily discerned in amniote species. More recently, four sets of experimental evidence provide strong support for the idea that all of the cranial sensory placodes initially derive from a common field (reviewed by Streit, 2004; Schlosser and Ahrens, 2004; Schlosser, 2010; Saint-Jeannet and Moody, 2014): transplantation studies; fate mapping studies; explant studies; and a common molecular signature.

Prior to the availability of molecular markers, Anton Jacobson (Jacobson, 1963) transplanted pieces of the amphibian PPE into novel anterior-posterior (A-P) positions along the margin of the neural plate (the NB zone). When the manipulations were done at early stages, PPE cells gave rise to placodes that were appropriate for their new A-P position. However, when manipulations were done a few hours later, PPE cells gave rise to placodes that were appropriate for their original A-P position. He proposed that cells within the PPE are initially competent to form any placode, and only after interactions with adjacent tissues at specific A-P addresses do they become specified to particular placode fates (Jacobson, 1966). This idea has since been supported by several similar transplantation studies using molecular markers. For example, any region of the PPE transplanted to the posterior hindbrain region at early neural plate stages can produce structures expressing otic genes (Groves and Bronner-Fraser, 2000). Fate mapping studies in the chick and several amphibians also show that all placodes originate from a common PPE field (Couly and LeDouarin, 1987; Couly and LeDouarin, 1990; Streit, 2002; reviewed in Schlosser and Ahrens, 2004; Streit, 2004; Pieper et al., 2011). Further, when very small groups of cells are labeled in the NB zone of the chick, fish, and frog, those labeled at early PPE stages contribute to multiple placodes whereas those labeled at later stages contribute to distinct placodes (Streit, 2002; Bhattacharyya et al., 2004; Toro and Varga, 2007; Xu et al., 2008; Ladher et al., 2010; Pieper et al., 2011). Explant studies, which reveal the developmental potential of a tissue, also support a common PPE field. For example, Bailey et al. (2006) explanted different A-P regions of the chick PPE and later assayed them for which placode genes they autonomously expressed. They showed that without additional interactions the entire PPE shares an initial Pax6-positive molecular state that only has the potential to form lens. This "ground" state must then be acted upon by local signaling for other placode fates to be initiated. Similarly, Martin and Groves (2005) showed that NB zone explants must first express PPE genes before otic placode fate can be induced. Consistent with this interpretation, in whole embryos placodes can be ectopically induced by gene overexpression for the most part only in anterior ectoderm lateral to the neural plate, i.e., in the PPE territory (e.g., Oliver et al., 1996; Schlosser et al., 2008; reviewed in Streit, 2004). Thus, these studies indicate that cells must first attain a pre-placodal state before a specific placode identity can be achieved. Finally, PPE cells appear to have a common molecular signature: at early stages of chick, frog, fish and mouse development, the entire PPE field expresses *Six* and *Eya* genes (Esteve and Bovolenta, 1999; Kobayashi et al., 2000; Pandur and Moody, 2000; David et al., 2001; Ghanbari et al., 2001; McLarren et al., 2003; Schlosser and Ahrens, 2004; Streit, 2004; Litsiou et al., 2005; Christophorou et al., 2009; Sato et al., 2010), and in both frog and chick the domains of these genes overlap with the fate maps of the placodes (reviewed in Streit, 2004; Pieper et al., 2011). As described later (Section 19.4), in both model animals and humans there is evidence that these pan-PPE genes are required for proper development of several placodes and their derivatives.

Are the PPE and the Neural Crest Induced by Similar Mechanisms?

Because both the neural crest and cranial sensory placodes are derived from the ectoderm that surrounds the lateral edge of the neural plate (the NB zone), and because both tissues contribute to the peripheral nervous system, it was suggested that the PPE is induced by mechanisms similar to those that induce the neural crest (Baker and Bronner-Fraser, 2001). However, there are several reasons that there also should be differences in neural crest versus PPE induction:

1) the PPE forms lateral to the neural crest;
2) only the PPE extends around the most anterior tip of the neural plate; and
3) the placodes form only in the head whereas the neural crest extends through the entire trunk (Figure 19.1).

Furthermore, by late neural plate stages the PPE and the neural crest express distinctly different sets of genes (reviewed in Bhattacharyya and Bronner-Fraser, 2004; Meulemans and Bronner-Fraser, 2004; Schlosser and Ahrens, 2004; Grocott et al., 2012). Over the past several years, researchers have used PPE-specific gene expression to experimentally examine the similarities and differences in the mechanisms that induce the PPE and the neural crest. As we shall see, while the neural crest and placodes share many early developmental aspects, they are differentially responsive to local signals, may have independently arisen during evolution and by neural plate stages are derived from separate ectodermal lineages (reviewed in Grocott et al., 2012; Pieper et al., 2012; Groves and LaBonne, 2014).

One developmental step that is in common between the PPE and the neural crest is the formation of the NB zone (Figure 19.1). Several studies in chicks and *Xenopus* have shown that when small pieces of the neural plate are transplanted to ectopic regions of the embryonic non-neural ectoderm, neural crest and PPE markers are induced (Selleck and Bronner-Fraser, 1995; 2000; Mancilla and Mayor, 1996; Woda et al, 2003; Glavic et al., 2004; Litisou et al., 2005; Ahrens and Schlosser, 2005). For both cell types, members of the Dlx family of transcription factors are required for this neural plate-induced phenotype (McLarren et al., 2003; Woda et al. 2003; Pieper et al., 2012). Several fate mapping studies support the notion that the NB zone comprises a common field of mixed PPE and neural crest progenitors (Figure 19.1B). In chick, dye-labeling of small groups of cells in the NB zone near the hindbrain region of the neural plate showed that otic placode precursors are scattered over a wide region and are intermingled with cells that will give rise to neural plate, neural crest and epidermis (Streit, 2002). Even at neural fold stages, there is no evidence for spatial segregation of these precursor populations. In *Xenopus*, similar intermingling was observed at neural plate stages across the entire cranial NB zone (Pieper et al., 2011). Thus, both the ectopic transplantation studies and fate mapping studies have led to the conclusion that the PPE and neural crest share an early developmental path.

However, there also is evidence that the NB zone gives rise to two separate precursor populations (Figure 19.1C). First, neural crest and placodes may have separately arisen during evolution (Schlosser, 2005; 2008; Gasparini et al., 2013; Graham and Shimeld, 2013; see Chapter 7). Second, there is evidence that anterior NB zone explants preferentially make placode cells whereas posterior NB zone explants preferentially make neural crest cells (Basch et al., 2006; Sjodal et al., 2007; Patthey et al., 2008). Third, it is not common for both placode cells and neural crest cells to arise from the same small group of cells labeled at neural plate stages (~10–15%; Streit, 2002; Pieper et al., 2011); these "mixed fate" groups were not observed when labeling was performed at neural fold or neural tube stages. These observations suggest that although PPE and neural crest cells arise from a common field, it is unlikely that they derive directly from a common progenitor. To resolve this issue, single cell lineage analysis will be required, as has been done for zebrafish and chick placode progenitors. In these studies, single cell labeling has shown that even when precursors for lens, adenohypophyseal and olfactory placodes are highly intermingled, they only give rise to a single placode fate (Dutta et al., 2005; Toro and Varga, 2007; Bhattacharyya and Bronner, 2013). Fourth, performing neural plate transplantation with lineage labeled tissue suggests that the neural and non-neural ectoderms of *Xenopus* are differentially competent to give rise to neural crest versus PPE cells (Pieper et al., 2012). The neural plate tissue grafted into ventral epidermis expressed neural plate, neural border, and neural crest genes, whereas PPE genes were only expressed in the adjacent host epidermis. If epidermis was grafted into the NB zone, PPE genes were expressed; if neural plate was grafted into the NB zone, neural crest genes were expressed. These experiments provide convincing evidence that different ectodermal domains in *Xenopus* are differentially competent to give rise to PPE and neural crest by neural plate stages. However, previous work in both the chick and frog indicates that neural crest genes can be induced in both early neural and non-neural ectoderm (Selleck and Bronner-Fraser, 1995; 2000; Mancilla and Mayor, 1996). Similarly, Pieper et al. (2012) show that early ectoderm (from blastula and gastrula stages) can give rise to both neural crest and placodes. Thus, and consistent with the fate mapping studies, the segregation of PPE and neural crest fates seems to gradually occur over developmental time. However, it remains to be determined whether predetermined precursors physically segregate, or if common precursors become lineage restricted (see Groves and LaBonne, 2014).

Specification of the PPE by Signaling Pathways

The neural plate grafting experiments discussed above indicate that interactions between the neural and non-neural ectoderm are an important source of signals for the formation of both neural crest and PPE markers (Figure 19.1A). In the chick and *Xenopus*, in which transplantation experiments are readily feasible, it has been elegantly demonstrated that tissues in addition to the neural plate are required for PPE induction (Litsiou et al., 2005; Ahrens and Schlosser, 2005) (Figure 19.3). Ablation and transplantation experiments in chicks showed that the neural plate signal is not sufficient to induce PPE markers in extra-embryonic ectoderm, and that head mesoderm, including cardiac but not paraxial precursors, can provide an additional PPE-inducing signal (Litsiou et al., 2005). Similar manipulations in *Xenopus* have shown that the endomesoderm, including cardiac but not axial/paraxial precursors, can provide the required additional signal (Ahrens and Schlosser, 2005). Both studies identified fibroblast growth factor (FGF) as the likely additional signal, but also showed that FGF signaling alone is not sufficient to induce ectopic PPE markers in the epidermis. Consistent with these interpretations, studies in fish also show that FGF signaling promotes PPE gene expression (Esterberg and Fritz, 2009; Kwon et al., 2010). It is interesting that in several animals FGF8 is expressed in the anterior neural ridge and in an anterior U-shape of mesoderm approximating the PPE. Grocott et al. (2012) propose that FGF promotes PPE formation by both inhibiting PPE-repressing factors and upregulating PPE-specific gene expression.

There are some caveats to the studies showing a critical role for FGF signaling in PPE induction, however. First, while FGF signaling favors PPE formation in fish (Esterberg and Fritz, 2009; Kwon et al., 2010), FGF8 and FGF3 are not required for the expression of some PPE marker genes (Phillips et al., 2001; Maroon et al., 2002; Leger and Brand, 2002; Liu et al., 2003; Solomon et al., 2004). Second, activating the FGF pathway with a constitutively active receptor represses endogenous PPE gene expression (Brugmann et al., 2004). There also is evidence that while low levels of FGF signaling promotes

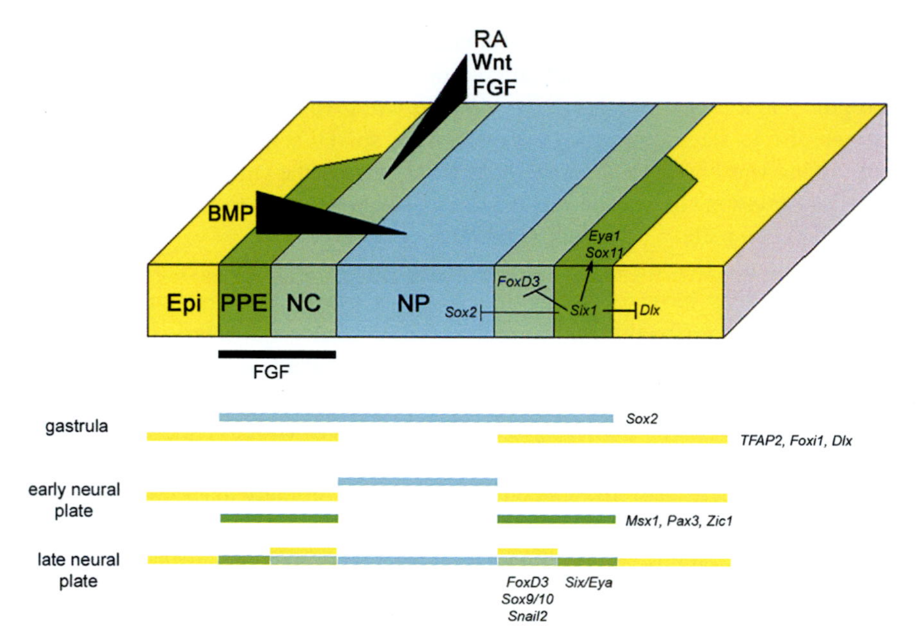

FIGURE 19.3 Several Steps are Involved in Forming the Pre-Placodal Ectoderm. Top: the embryonic ectoderm of a neurula stage has been flattened into a sheet and divided into its four major domains: epidermis (Epi), pre-placodal ectoderm (PPE), neural crest, (NC) and neural plate (NP). Left side: The PPE and NC are induced by interactions between the epidermis and neural plate, as well as FGF signals from underlying tissues. Both PPE and NC appear to require lowered levels of BMP signaling. NC cells also require Wnt and posterior FGF signals, whereas Wnt signaling and high levels of FGF signaling inhibit PPE formation. RA signaling also appears to set the posterior boundary of the PPE. Right side: Some of the genes that are expressed specifically in the different ectodermal domains are shown. Knockdown and overexpression studies indicate that Six1 promotes the expression of other placode genes (e.g., *Eya1*, *Sox11*) at the expense of neural crest (e.g., *FoxD3*), neural plate (*Sox2*), and epidermal (*Dlx*) genes. **Bottom**: As a consequence of the signaling interactions depicted above, different sets of transcription factors are regionally expressed over developmental time. At gastrula stages, the expression domains of neural plate genes (blue bar; e.g., *Sox2*) and epidermal genes (yellow bar; e.g., *TFAP2, Foxi, Dlx*) overlap in a region that will become the neural border zone. At early neural plate stages, *TFAP2, Foxi,* and *Dlx* genes continue to be expressed in the NB zone and epidermis (yellow bar), but neural plate gene expression withdraws from the NB zone, concomitant with the upregulation of "neural border specifying" genes (dark green bar; e.g., *Msx1, Pax3, Zic1*). At late neural plate stages differential gene expression and signaling upregulates *Six* and *Eya* genes in the PPE domain (medium green bars) and "neural crest specifying" genes (e.g., *FoxD3, Sox9/10, Snail2*) in the neural crest domain (light green bars); *TFAP2, Foxi,* and *Dlx genes* are expressed in the neural crest and epidermis (yellow bars), but are excluded from the PPE. Boundaries appear to be maintained by mutual repression between domain-specific transcription factors (see Grocott et al., 2012 and Groves and LaBonne, 2014 for details).

PPE fate, high levels inhibit it (Hong and Saint-Jeannet, 2007). Third, two recently characterized *Six1* gene PPE enhancers, which are conserved across humans, mice, chicks, and frogs, do not contain any obvious FGF pathway binding sites (Sato et al., 2010; 2012); however, because there are six other enhancers that promote PPE-specific expression, further study may reveal the anticipated FGF pathway input. These discrepancies should be sorted out by experiments that distinguish the relevant timing and levels of FGF signaling that specifically regulate PPE gene expression.

Several studies have shown that high levels of bone morphogenetic protein (BMP) signaling induce an epidermal fate, whereas intermediate levels are required for neural crest induction (see Chapter 18), and low levels are required for neural plate induction (Figure 19.3). There are several experiments in *Xenopus* that suggest that reduced levels of BMP signaling also are necessary for PPE formation. First, Ahrens and Schlosser (2005) showed that the effectiveness of FGF8 for inducing PPE genes in the non-neural ectoderm, either by FGF8-coated beads or by FGF8-expressing animal cap grafts, depended on the concomitant reduction of the level of BMP signaling in the epidermis. Second, in the chick, frog, and fish, *in vivo* expression of dominant-negative BMP receptors or of BMP antagonists (e.g., Noggin, Chordin, Cerberus) expanded the domains of PPE genes (Brugmann et al., 2004; Glavic et al., 2004; Ahrens and Schlosser, 2005; Litsiou et al., 2005; Kwon et al., 2010). However, several experiments using animal cap ectodermal explants indicate that some BMP signaling appears to be necessary for this effect; experimentally reducing BMP levels induces PPE gene expression, whereas eliminating BMP signaling leads to neural plate gene expression (Brugmann et al., 2004; Hong and Saint-Jeannet, 2007; Park and Saint-Jeannet, 2008). Interestingly, the levels of BMP reduction also may distinguish between induction of neural crest versus PPE fate in both fish and frog. When ectodermal explants were cultured in different concentrations of Noggin protein, *Six1* and *Eya1* were highly expressed at very low concentrations (1–5 ng/mL), and their expression was dramatically reduced as the Noggin concentration increased to intermediate levels that induced a neural crest gene (*FoxD3*) or to high levels that induced a neural plate gene (*Sox2*) (Brugmann et al., 2004). In fish, BMP levels that induce a neural crest fate activate *TFAP2*, whereas BMP levels that induce a PPE fate activate *TFAP2* as well as *Foxi1* and *Gata3* (Bhat et al., 2013). However, two observations are not concordant with a simple gradient model for BMP induction of PPE:

1) non-neural ectoderm transplanted into the neural plate, which is presumed to be a region of high BMP antagonist expression, strongly expresses *Six1* (Ahrens and Schlosser, 2005); and
2) BMP4 mRNA levels are relatively high along the neural plate border (Hemmati-Brivanlou and Thomsen, 1995; Streit and Stern, 1999).

In none of these studies has the level of BMP protein in the areas of PPE gene expression been measured, so the issue requires further testing. Recent studies demonstrate that the timing of the BMP signaling also appears to be critical. In fish, high levels of BMP at gastrula stages activate NB zone genes, which are necessary for both PPE and neural crest formation, whereas at neural plate stages lower levels of BMP are necessary to activate PPE genes (Kwon et al., 2010; Bhat et al., 2013). Similarly, human embryonic stem cells directed to a placode fate require an early pulse of BMP to activate NB zone genes, and a later attenuation of BMP signaling to activate PPE genes (Leung et al., 2013; Dincer et al., 2013).

Several studies also have shown that Wnt signaling is required for neural crest induction (see Chapter 18). In contrast, Wnt signaling appears to antagonize PPE induction (Figure 19.3). In both chick and *Xenopus* embryos, local elevation of Wnt signaling represses PPE gene expression, whereas local reduction of Wnt signaling expands it (Brugmann et al., 2004; Litsiou et al., 2005; Matsuo-Takasaki et al., 2005; Hong and Saint-Jeannet, 2007). It is interesting to note that secreted anti-Wnt factors are highly expressed in the anterior neural plate and the underlying chordomesoderm (Bradley et al., 2000; Pera and De Robertis, 2000; Carmona-Fontaine et al., 2007; Takai et al., 2010), which may account for the presence of PPE genes and the absence of neural crest genes in the ectoderm surrounding the anterior tip of the neural plate (Figure 19.1C). In support of this idea, if a secondary axis with an obvious A-P axis is induced by ectopically expressed Noggin or Chordin or by grafting the Organizer, *Six1* is expressed only at the anterior pole of the ectopic axis (Brugmann et al., 2004; Ahrens and Schlosser, 2005). In chick NB zone explants, if Wnt signaling is attenuated in the presence of low BMP, placode markers are expressed, whereas if Wnt signaling persists with low BMP, neural crest markers are expressed (Patthey et al., 2009). Thus, it is a combination of signaling that is critical for PPE gene induction. In fact, Litsiou et al. (2005) showed that the additive effect of reducing both BMP and Wnt signaling on PPE induction in the extra-embryonic ectoderm works best when there is a simultaneous early pulse of FGF. They proposed that the FGF pulse confers a "neural border" state on the ectoderm, and that those cells within the NB zone that are protected from Wnt signaling become PPE, whereas those that are exposed to Wnt signaling become neural crest. It is interesting to note that one of the conserved *Six1* PPE enhancers contains a Wnt/Lef1 site (Sato et al., 2012), suggesting that Wnt regulation may be direct. But, this site has not yet been functionally tested. There also is evidence that anterior and posterior domains of the PPE may have different requirements with regard to Wnt signaling. In the zebrafish *masterblind* mutants, which carry a point mutation in the GSK3-binding domain of axin1 leading to increased Wnt activity, the most anterior placodes (olfactory and lens) are lost, whereas posterior placodes are expanded (Heisenberg et al., 2001). In *Xenopus,* expression of FGF8a activates expression of the olfactory placode gene *Dmrt4* in animal caps injected with Noggin, while the simultaneous activation of canonical Wnt signaling in these explants promotes expression of the otic/epibranchial gene *Pax8* and reduced *Dmrt4* expression (Park and Saint-Jeannet, 2008).

Another important signal in PPE formation is retinoic acid (RA). Raldh2, the RA-synthesizing enzyme, is expressed, among other places, in a discrete U-shaped ectodermal domain around the anterior neural plate that resembles the PPE (Chen et al., 2001). Decreasing RA signaling during PPE formation expands the posterior limit of the U-shaped FGF8 expression in the cranial mesoderm (Shiotsugu et al., 2004), suggesting that endogenous RA contributes to limiting the PPE to the head (Figure 19.3). But RA signaling also induces two genes expressed in the PPE: *Tbx1*, a T-box transcription factor, and *Ripply3/Dscr6*, a Groucho-associated co-repressor (Arima et al., 2005). *Ripply3* and *Tbx1* expression domains in the PPE partially overlap; in regions where only *Tbx1* is expressed, PPE genes are induced, whereas in regions where *Tbx1* and *Ripply3* overlap, PPE genes are repressed (Janesick et al., 2012) (Figure 19.5). These studies indicate that RA signaling is both required for formation of the posterior, *Tbx1*-positive part of the PPE, and for restricting the posterior boundary of the PPE. Studies of the mouse pharyngeal arch and cardiac development indicate that Tbx1 acts upstream of Six1/Eya1, which in turn directly regulate *FGF8* expression (Guo et al., 2011); perhaps this pathway also regulates posterior PPE formation.

19.4 GENES THAT SPECIFY PRE-PLACODAL ECTODERM FATE

In response to neural inductive signaling, the embryonic ectoderm is divided into two transcriptionally distinct fields: the neural ectoderm and non-neural ectoderm (see Chapter 11). Interactions between these two fields and with the underlying endomesoderm initiate the formation of the NB zone, as described above. As a result, the cells in this zone begin to express a distinct set of genes, called "NB specifying" genes, that are required for the later expression of neural crest-specific and/or PPE-specific genes (reviewed in Meulemans and Bronner-Fraser, 2004; Sargent, 2006; Park and Saint-Jeannet, 2010; Grocott et al., 2012).

NB Zone Genes

Based on their expression domains surrounding the neural plate, and their ability to regulate neural crest gene expression, members of the *Dlx*, *Msx*, *Pax*, and *Zic* families were called "NB specifying" genes (Meulemans and Bronner-Fraser, 2004). In addition, *TFAP2α*, *GATA* and *Foxi* genes have also been implicated in regulating the formation of the NB zone and/or its derivatives (Sargent, 2006; Grocott et al., 2012). For example, in human embryonic stem cell cultures, the expression of TFAP2α, GATA and DLX genes necessarily precede the expression of PPE genes (Leung et al., 2013). In most vertebrates examined, *Dlx*, *GATA*, *Foxi*, *Msx1,* and *TFAP2α* genes are expressed broadly in the non-neural ectoderm at blastula and gastrula stages, but their expression also overlaps with the border of the neural ectodermal domain (reviewed in Grocott et al., 2012) (Figure 19.3, bottom). As gastrulation proceeds and the neural plate forms, definitive neural plate genes, like *Sox2*, are excluded from the overlap region, *Msx1* and *Pax3* are confined to the NB zone, and *Dlx*, *GATA*, *Foxi* and *TFAP2α* no longer overlap with neural plate genes. Eventually, *Msx1* and *Pax3* expression is confined to the neural crest domain, some *Dlx* genes and *TFAP2α* are expressed in the neural crest, PPE and epidermal domains, and other *Dlx*, *GATA* and *Foxi* factors are only expressed in the PPE and epidermal domains (Figure 19.4). Functionally, how do these various transcription factors lead to distinct neural plate, neural crest, PPE and epidermal domains?

Three *Dlx* genes (*Dlx3*, *Dlx5*, *Dlx6*), which are related to *Drosophila distal-less*, are upregulated in response to BMP (Luo et al., 2001a,b). Gain-of-function studies in *Xenopus* indicate that *Dlx* genes repress neural plate genes and are required for the expression of both neural crest and PPE genes (Feledy et al., 1999; Beanan and Sargent, 2000; Luo et al., 2001a; Woda et al., 2003). In the chick, *Dlx5* represses neural plate genes, but when expressed in the epidermis it promotes the PPE gene *Six4* (McLarren et al., 2003). Loss of Dlx function in fish leads to reduced PPE gene expression and smaller placodes (Solomon and Fritz, 2002; Kaji and Artinger, 2004; Esterberg and Fritz, 2009). Neural plate transplantation experiments have confirmed that *Dlx* genes are required for NB zone formation; when grafted into non-neural ectoderm that expressed *Dlx* genes, both neural crest and PPE markers were induced, whereas when grafted into non-neural ectoderm in which the activities of all *Dlx* genes were downregulated by a pan-*Dlx*-repressor construct, neither was induced (Woda et al., 2003). Dlx factors may promote PPE formation by regulating a BMP antagonist in the NB zone (Esterberg and Fritz, 2009), and/or by interacting with GATA factors (Solomon and Fritz 2002; McLarren et al., 2003; Phillips et al., 2006; Pieper et al., 2012). However, although studies show that loss of either Dlx3 or GATA2 significantly reduces PPE gene expression, only increased Dlx3 expands PPE gene expression (Pieper et al., 2012), suggesting that these two NB zone genes are not equivalent in activity. In accord, an anterior PPE enhancer in the *Six1* gene contains both GATA and homeodomain (HD) putative

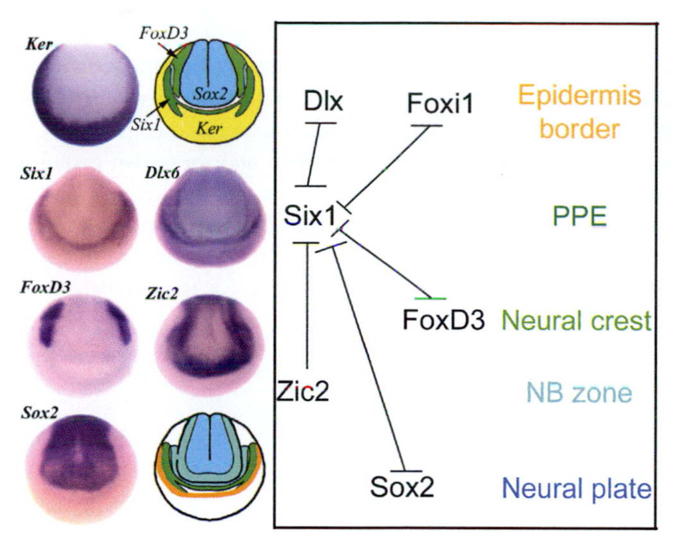

FIGURE 19.4 Boundaries Between the Four Ectodermal Domains may be Formed by Mutually Repressive Interactions. The left column shows *Xenopus* embryos that are stained for endogenous expression domains at neural plate stages to demonstrate the four ectodermal domains: *Ker*, epidermis; *Six1*, PPE; *FoxD3*, neural crest; *Sox2*, neural plate. A summary of these domains is shown on top of the next column of embryos. In the second column the expression domains of two neural border zone genes (*Dlx6, Zic2*) are shown, below which is a summary of the relative expression domains for neural plate genes (blue), neural border zone genes (aqua), PPE genes (green) and genes at the epidermal border (orange). Experiments, which are summarized in the box to the right, demonstrate that Six1 maintains PPE fate in part through the mutual repression with *genes* expressed in the other domains and in the border zones. See Brugmann et al. (2004) and Groves and LaBonne (2014) for details.

binding sites (Sato et al., 2010). Mutating the GATA sites reduced expression of a reporter, whereas mutating the HD sites eliminated it. GST assays indicated that both Dlx5 and Msx1 can bind to the HD sites, and reporter assays indicated that Dlx5 activates *Six1* whereas Msx1 represses it (Sato et al., 2010).

Msx1 is a direct target of BMP, and is required for neural crest formation in numerous animals (see Chapter 18). The expression domain of *Msx1* overlaps with those of the *Dlx* genes in the NB zone, although *Msx1* tends to be expressed more posteriorly (Arkell and Beddington, 1997; Feledy et al., 1999; McLarren et al., 2003; Schlosser and Ahrens, 2004; Monsoro-Burq et al., 2005). Interestingly, Msx and Dlx proteins can inhibit each other through the formation of heterodimers (Zhang et al., 1997; Givens et al., 2005). Experiments that knocked down multiple members of the *Msx* and *Dlx* families demonstrated that different levels of Msx and Dlx, regulated by their mutual antagonism, bias NB zone cells towards either a neural crest fate (Msx-high, Dlx-low) or a PPE fate (Msx-low, Dlx-high) (Phillips et al., 2006). Msx1 also activates other "NB specifying" genes: *Pax3* and *Zic1*.

Several studies have shown that interactions between Pax3 and Zic1 are required to initiate neural crest gene expression (Tribulo et al., 2003; Monsoro-Burq et al., 2005; Sato et al., 2005). Another study reported that they also affect PPE development. Hong and Saint-Jeannet (2007) regulated the timing of *Pax3* expression with inducible constructs to demonstrate that:

1) Pax3 has an early role in NB zone formation and a second later role in neural crest specification;
2) while Zic1 synergizes with Pax3 to produce neural crest, in the absence of Pax3 it promotes PPE gene expression;
3) Zic1 is required for *Six1* and *Sox2* expression in the PPE.

They also showed that the Zic1 effect on the PPE occurs after the BMP/Wnt signaling that establishes the NB zone.

Foxi1 and TFAP2α are also required for both neural crest and PPE formation. Foxi1 is a member of the *Drosophila* Forkhead family. Knockdown of *Foxi1* in *Xenopus* embryos expands the *Sox2*-expressing neural plate domain into the NB zone, and reduces "NB specifying" genes and epidermis genes (Matsuo-Takasaki et al., 2005). Foxi1 gain-of-function reduces both neural plate and "NB specifying" genes and expands epidermal genes. By utilizing a hormone-inducible version of Foxi1, these authors showed that the early expression of *Foxi1* at the border between the neural and non-neural ectoderm is required for NB zone formation, but its subsequent expression represses NB zone genes. It should be noted that another Fox gene, *Foxg1* (aka. *BF-1*) also is expressed in all placodes in the mouse (Hatini et al., 1999), and conditional knockouts in the mouse can have olfactory and auditory defects (Pauley et al., 2006; Duggan et al., 2008; Hwang et al., 2009; Kawauchi et al., 2009), but to our knowledge its potential role in NB zone formation has not yet been studied. *TFAP2α* is a key regulator of epidermis-specific genes (Nguyen et al., 1998; Luo et al., 2002; Zhang et al., 2006). Loss-of-function of TFAP2α causes a downregulation of epidermis-specific *Keratin* and "NB specifying" genes and upregulation of neural plate genes (Luo et al., 2002; de Croze et al., 2011). In contrast, TFAP2α gain-of-function expands "NB specifying" genes and neural crest genes and reduces neural plate genes (Luo et al., 2003; Hoffman et al., 2007; de Croze et al., 2011). Epistasis analyses show that TFAP2α acts upstream of other "NB specifying" genes, with *Pax3* being a direct transcriptional target (Luo et al., 2002; de Croze et al., 2011). Interestingly, it is later expressed at high levels in pre-migratory neural crest cells but not in PPE cells. Genome-wide profiling of chromatin markings associated with active enhancers and analyses of predicted transcription factors binding sites indicate that human neural crest enhancers are preferentially occupied by TFAP2A (Rada-Iglesias et al., 2012), which further highlights its critical role in the regulation of neural crest fate.

Taken together, the preponderance of data indicate that "NB specifying" genes produce a NB state from which either neural crest or PPE cells can arise, depending upon the timing and combinations of genes that are expressed (Figures 19.3, 19.4). Specifically, *Msx* is required for neural crest formation, *Dlx* is required for PPE formation, and these two proteins regulate each other's activity. In addition, *Pax3* and *Zic1* are co-required for neural crest formation, whereas *Zic1* alone is required for PPE formation. A recent study showed that a maintained expression of the NB genes *TFAP2α, Foxi1* and *GATA* in the NB zone is required for PPE formation, and this is accomplished by crossregulation between these factors after BMP signaling is attenuated (Bhat et al., 2013). Thus, this NB transcriptional state appears to set the stage for the expression of PPE-specific gene expression.

PPE Genes

Just as several transcription factors are required for neural crest specification (Meulemans and Bronner-Fraser, 2004), the PPE also expresses distinct sets of genes (reviewed in Bhattacharyya and Bronner-Fraser, 2004; Schlosser and Ahrens, 2004; Grocott et al., 2012; Saint-Jeannet and Moody, 2014). In particular, members of two gene families (*Six, Eya*) are candidates for specifying the early pre-placode state because several are expressed in the characteristic U-shaped domain that surrounds the anterior neural plate corresponding to the morphologic description of the PPE in the early embryo (Esteve and Bovolenta, 1999; Kobayashi et al., 2000; Pandur and Moody, 2000; David et al., 2001; Ghanbari et al., 2001; McLarren et al., 2003; Bessarab et al., 2004; Schlosser and Ahrens, 2004; Sato et al., 2010). Recent functional studies indicate that these two gene families are necessary for the initial specification of the PPE and formation of some of its derivatives (Figures 19.3, 19.4).

Six *Genes*

Vertebrate Six proteins are highly related to *Drosophila* Sine oculis (SO). Although SO is essential for fly visual system formation (Cheyette et al., 1994; Serikaku and O'Tousa, 1994), vertebrate *Six* genes play major roles in several organs, including eye, muscle, kidney, and craniofacial development (Kawakami et al., 1996; Brodbeck and Englert, 2004). All Six/SO proteins contain a highly conserved Six-type homeodomain, which binds DNA, and an adjacent N-terminal Six domain (SD), which appears to increase DNA binding specificity by interacting with co-factors (Pignoni et al., 1997; Kawakami et al., 2000; Kobayashi et al., 2001). Vertebrate *Six* genes have been grouped into three subfamilies (*Six1/Six2; Six3/Six4; Six5/Six6*) on the basis of sequence variations in both the homeodomain and the SD regions (Kawakami et al., 2000). These genes are clustered in vertebrate genomes, but with some species variations. Using the Genomicus program (http://www.genomicus.biologie.ens.fr), we find that in humans, *SIX1* is flanked by *SIX4*, oriented in the same direction, and *SIX6*, oriented in the opposite direction (Chromosome [Ch] 14); the same arrangement is conserved in the mouse (Ch12) and frog (*Xenopus tropicalis*) genomes. This arrangement also occurs in the zebrafish genome, but has been duplicated (Ch13, Ch20). In the chicken genome, a predicted gene separates *Six4* and *Six1* (Ch5). In humans, *SIX2* is flanked by *SRBD1*, oriented in the same direction, and *SIX3*, oriented in the opposite direction (Ch2); the same arrangement is conserved in the mouse (Ch17) and frog (*Xenopus tropicalis*) genomes. In zebrafish, there is a different gene in the position of *Srbd1*, and in the chicken the *Six3* homolog is absent. In humans, *SIX5* is flanked by *DMPK* and a predicted gene (Ch19); the same arrangement is found in mouse (Ch7). In zebrafish, a gene called *Six4.3* is the closest homolog and it is adjacent to the *Six9* gene (Ch18). In the chicken and frog, no clear homolog of *SIX5* has yet been identified. Thus, although the *Six* genes are highly conserved across vertebrates, there are some genomic differences that might contribute to functional differences during development.

Three *Six* genes (*Six1*, *Six2*, *Six4*) are expressed in vertebrate PPE, placodes, and/or placode derivatives. However, their expression patterns across vertebrates are not identical (reviewed by Brugmann and Moody, 2005). We do not yet know whether the differences result from the different gene arrangements described above that could alter tissue specific enhancers, or from different experimental techniques and/or developmental stages that make the patterns appear disparate. In general, *Six1* and *Six2* are expressed in the PPE, placodes (except the lens, which expresses *Six3*), lateral line organs, muscle precursors, kidneys, genitalia, and limb buds. *Six4* is typically expressed in the PPE, placodes, muscle precursors, kidneys, brain, and eye. It will be very important to fully describe the developmental expression patterns of these *Six* genes across all of the animal models and humans to fully understand their roles in placode development and related congenital syndromes.

Several experiments indicate that Six1 has a central role in PPE/placode development. Six1 loss-of-function in *Xenopus* results in the loss of other early PPE markers (*Eya1*, *Sox11*) and several placode-specific markers (Brugmann et al., 2004; Schlosser et al., 2008). Similarly, the in chick, repression of Six1 targets leads to a loss of placode-specific genes and otic placode defects (Christophorou et al., 2009), and in fish, Six1 knockdown results in loss of inner ear hair cells (Bricaud and Collazo, 2006; 2011). Although *Six1* is expressed in the mouse PPE (Sato et al., 2010), its loss seems to affect later stages of placode development. For example, the initial formation of the olfactory placode appears normal, but morphogenesis, progenitor production, and neurogenesis are disrupted (Ikeda et al., 2007; 2010; Chen et al., 2009). The lack of earlier defects may be due to gene redundancy because in *Six1/Six4* compound null mice the olfactory placode does not form (Chen et al., 2009); however, *Eya2* expression is unaffected suggesting the PPE still forms. There also are defects in the inner ear and in numerous cranial ganglia, potentially as a result of increased cell death and reduced proliferation (Oliver et al., 1995; Laclef et al., 2003; Zheng et al., 2003; Ozaki et al., 2004; Zou et al., 2004; Konishi et al., 2006). Even *Six1*-heterozygous mice have hearing loss due to cochlear defects (Zheng et al., 2003).

Increased expression of wild type Six1 by mRNA injection into the precursors of the NB zone in *Xenopus* embryos expands the expression domains of other early PPE genes, and represses the adjacent epidermal and neural crest domains (Brugmann et al., 2004). Similarly, in the chick, Six1 activation in the NB zone is required for PPE and specific placode gene expression and it represses neural plate and neural crest genes (Christophorou et al., 2009). These results demonstrate that elevated Six1 expression in the NB zone promotes PPE gene expression at the expense of the other ectodermal domains (Figures 19.3 and 19.4). However, it is important to note that ectopic expression of Six1 outside the NB zone is not sufficient to cause ectopic expression of other PPE/placode markers (Brugmann et al., 2004; Schlosser et al., 2008; Christophorou et al., 2009), supporting the idea that the appropriate signaling environment of the NB zone is prerequisite.

In humans, *SIX1* mutations lead to some cases of branchio-otic syndrome (BOS, specifically BOS3), whose phenotypes include mild craniofacial defects and significant hearing loss (Ruf et al., 2004). Nine mutations in BOS3 patients from 16 unrelated families have been reported to date; seven are missense mutations in the SD and two are missense or deletion mutations in the HD (Ruf et al., 2004; Ito et al., 2006; Sanggaard et al., 2007; Kochhar et al., 2008; Noguchi et al., 2011). The mutations either disrupt the Six1-Eya1 interaction or the ability of Six1 to bind to DNA (Patrick et al., 2009). In zebrafish, expression of mutant *Six1* mRNA that carries a BOS patient mutation (R110W) interferes with the Six1/Eya1 interaction that promotes cell proliferation and hair cell formation (Bricaud and Collazo, 2011). The *Catweasel* (*Cwe*) mouse mutant is thought to be a good model for the related brachio-otic-renal syndrome (BOR), which is similar to BOS but also includes kidney defects, because it harbors a missense mutation in the SD (Bosman et al., 2009) that is similar to at least one BOR family (Mosrati et al., 2011). Heterozygous-*Cwe* mice have an ectopic row of hair cells in the cochlea and homozygous-*Cwe* mice have a loss of hair cells in the cochlea, semicircular canals and utricle. Recently the structure of the interaction between human SIX1 and EYA2 has been solved; this work identified how EYA2 modifies the way in which the SIX1 homeodomain interacts with DNA (Patrick et al., 2013). These authors showed that this modification in DNA binding is essential for the SIX1-EYA driven metastasis that occurs in several adult cancers, but it has not yet been studied in a placode development context.

The functional roles of Six2 and Six4, which are also expressed in the PPE, have yet to be described in any detail, although they are frequently used as placode marker genes in a number of animal models. To our knowledge, no human syndromes have been assigned to mutations in *SIX2* or *SIX4*. Interestingly, BOR2 is ascribed to a mutation in SIX5 (Hoskins et al., 2007), but curiously this gene is not expressed in the PPE of animal models (Brugmann and Moody, 2005), and a recent study questioned whether there are mutations in SIX5 in these patients (Krug et al., 2011). Craniofacial phenotypes have not been reported in either *Six2*-null or in *Six4*-null mice (Ozaki et al., 2001; Self et al., 2006), but this lack of phenotype may be the result of redundant functions between the family members. A role for Six4 involvement in placode development is supported by reports that *Six1/Six4*-double null mutant mice have more severe defects than either of the single mutants (Grifone et al., 2005; Zou et al., 2006a; Chen et al., 2009). Because Six1, Six2, and Six4 have distinct roles in myogenic differentiation and in kidney development (Ohto et al., 1998; Spitz et al., 1998; Fougerousse et al., 2002; Xu et al., 2003; Brodbeck and Englert, 2004; Himeda et al., 2004), we anticipate that they will have distinct functions in PPE and sensory placode development as well. Therefore, it will be important to perform both gain- and loss-of-function studies with Six2 and Six4, alone and in combination with Six1, to better understand their separate and perhaps cooperative functions in PPE and placode development.

Eya *Genes*

There are four vertebrate *Eya* genes, located on four different chromosomes, that are homologous to *Drosophila Eyes absent* (*Eya*). The latter plays an essential role in fly eye development as a co-factor for SO (Bonini et al., 1993). Using the Genomicus program (http://www.genomicus.biologie.ens.fr), we find that in humans *EYA1* is flanked by *RP11* and *XKR9* (Ch8), whereas in mouse (Ch1), chicken (Ch2), frog and zebrafish (Ch24), *Eya1* is flanked by *Msc* and *Xkr9*. In humans *EYA2* is flanked by *SLC2A10* and *ZMYND8* (Ch20). The same arrangement is found in the mouse (Ch2), chicken (Ch26), and frog, but in zebrafish (Ch6), different genes flank *Eya2*. In humans *EYA3* is flanked by *PTAFR* and *XKR8* (Ch1). The same arrangement is found in the mouse (Ch1), chicken (Ch23), and frog, but in zebrafish (Ch19), *Ptafr* is missing and there are four copies of *Xkr8*. In humans *EYA4* is flanked by *RPS12* and *TCF21* (Ch6), and the same arrangement is found in the mouse (Ch10), chicken (Ch3), and frog. In zebrafish (Ch23), *Rps12* and *Tcf21* are found in opposite positions relative to *Eya4* and the orientation of *Rps12* is reversed. Thus, the orientations of the vertebrate *Eya* genes are mostly conserved; among the model systems the most variation is found in the partially duplicated genome of zebrafish.

Vertebrate *Eya* genes are expressed in numerous embryonic tissues, including the eyes, somites, kidneys, and hypaxial muscle precursors; *Eya1*, *Eya2*, and *Eya4* are also expressed in the PPE and the placodes (Duncan et al., 1997; Xu et al., 1997; Sahly et al., 1999; David et al., 2001; Ishihara et al., 2008; Neilson et al., 2010; Modrell and Baker, 2012). In *Xenopus*, chick and zebrafish PPE, the expression patterns of *Eya1* and/or *Eya2* are remarkably similar to that of *Six1* (Sahly et al., 1999; David et al., 2001; McLarren et al., 2003; Bane et al., 2005; Neilson et al., 2010), suggesting that they have important roles in PPE/placode development. The Eya proteins do not bind to DNA, but they contain a highly conserved protein/protein binding domain called the Eya domain (ED) located at the C-terminus. In *Drosophila*, the ED participates in protein/protein binding with the SD of the SO protein (Pignoni et al., 1997). In vertebrates, the interaction between the Six1 SD and the Eya1 ED regions of the proteins is essential for Eya1 nuclear translocation and for exerting the transcriptional function of the complex (Ohto et al., 1999; Ikeda et al., 2002; Patrick et al., 2013). However, Eya1 can bind to several proteins in addition to Six1. It can act as a co-factor for other Six proteins (Six2, Six4, and Six5; Heanue et al., 1999; Ohto et al., 1999), and protein interaction studies in *Drosophila* have identified several other potential Eya binding partners (Giot et al., 2003; Kenyon et al., 2005; Database of Interacting Proteins (DIP): http://dip.doe-mbi.ucla.edu/dip/Main.cgi). Eya also functions as a phosphatase; this activity is necessary for *Drosophila* eye development (Rayapureddi et al., 2003; Tootle et al., 2003), and it may regulate whether the Six1- Dachshund complex (described in the next section) acts as a transcriptional repressor or activator (Li et al., 2003). There also is evidence that Eya1 is a substrate for mitogen-activated protein kinase in the receptor tyrosine kinase signaling pathway (Hsiao et al., 2001). These potential multiple cellular functions will need to be kept in mind when evaluating the consequences of *Eya* mutations in animal models and human diseases.

Experimental data for the role of Eya proteins in PPE/placode development are most abundant for Eya1. *Eya1*-null mutant mice show defects in the inner ear, some of the cranial ganglia, the thymus, thyroid, parathyroid, kidney, and skeletal muscles (Johnson et al., 1999; Xu et al., 1999; 2002; Zou et al., 2004; 2006b). In *Xenopus*, expression of mutant versions of Eya1 that mimic human mutations lead to multiple defects in the developing ear (Li et al., 2010). In zebrafish, knockdown of Eya1 by MOs or by the *dogeared* (*dog*) mutation, which is caused by a point mutation in the ED, results in defects of the inner ear and lateral line, but the defects appear to prevent survival of the sensory cell precursors rather than perturb the establishment of the PPE (Kozlowski et al., 2005). Analysis of the effects of two zebrafish *Eya1* mutants (*aal* and *dog*) on the anterior pituitary, which is derived from the adenohypophyseal placode, showed that three of the four cell lineages are dependent on *Eya1* expression (Nica et al., 2006; Pogoda and Hammerschmidt, 2007). To date *Eya2*-null mouse phenotypes have not been reported. *Eya3*-null mice have behavioral abnormalities, consistent with its regulation of neural plate cell survival in *Xenopus* (Kriebel et al., 2007), but craniofacial deficits were not noted in these mice (Soker et al., 2008). The potential role of Eya2 and Eya3 in PPE/placode development is ripe for further experimentation. *Eya4*-null mice have inner ear defects (Depreux et al., 2008). Knockdown of *Eya4* in zebrafish causes loss of hair cells in the otic vesicle and lateral line and heart defects (Schonberger et al., 2005; Wang et al., 2008).

Patients diagnosed with BOS1 and BOR1 harbor *EYA1* mutations (Abdelhak et al., 1997; Kumar et al., 1997; Rodriguez-Soriano, 2003; Spruijt et al., 2006); cataracts and isolated defects in the anterior segment of the eye also have been associated with *EYA1* mutations (Azuma et al., 2000). Often the defects in the protein lie in the ED, where they act to inhibit the interaction between Eya1 and Six proteins (Buller et al., 2001; Ozaki et al., 2002). Partial deletions of *EYA1* that include other genes cause oto-facio-cervical syndrome (Rickard et al., 2001; Estefania et al., 2006). To date, no mutations in *EYA2* or *EYA3* have been reported in humans. However, mutations in *EYA4* are involved in Autosomal Dominant Non-Syndromic Sensorineural Deafness 10 (DFNA10; OMIM #601316) (Wayne et al., 2001; Makishima et al., 2007) and Dilated Cardiomyopathy with Sensorineural Hearing Loss, Autosomal Dominant (CMD1J; OMIM #605362) (Schonberger et al., 2005), which implicate developmental defects in the otic placode. The described mutations, which include small insertions, amino acid substitutions, and large deletions, are all predicted to truncate the EYA4 protein within the ED, which is required for interactions with SIX (and perhaps other) proteins.

Are there other Six/Eya Interacting Proteins Involved in PPE Specification?

It is well documented that Six proteins bind, via their SDs, to several proteins that lack the ability to bind to DNA, and there is evidence that some of these proteins modulate Six function as either co-activators or co-repressors (Zhu et al.,

2002; Tessmar et al., 2002; Giot et al., 2003; Brugmann et al., 2004; Bricaud and Collazo, 2011). In fact, a gene network that includes *Pax*, *Six*, *Eya*, and *Fox* genes has been described to be essential in eye, lens, muscle, and kidney development (reviewed by Bhattacharyya and Bronner-Fraser, 2004; Brodbeck and Englert, 2004; Streit, 2004). In *Drosophila*, yeast two-hybrid experiments identified 25 proteins that are likely to specifically bind to SO (Giot et al., 2003; Kenyon et al., 2005). Because SO belongs to the same Six subfamily as vertebrate Six1/Six2, several of these proteins may have important roles in PPE fate specification. The top two interactors were Eya and Groucho, whose functional relationships with Six genes have already been experimentally documented. Six1/Six2 can act as both transcriptional activators and repressors, depending on the presence of Eya or Groucho co-factors (Silver et al., 2003). To answer the question of how *Six1* expression in the NB zone can both up-regulate PPE genes and downregulate epidermal and neural crest genes, activating (Six1VP16) and repressing (Six1EnR) Six1 constructs were expressed in *Xenopus* embryos (Brugmann et al., 2004). This study concluded that Six1 transcriptionally activates PPE genes because Six1VP16 induced the same phenotypes as co-expression of wild type Six1 and Eya1. Similarly, co-expression of Six1 and Eya2 in chick upregulates PPE genes (Christophorou et al., 2009). Conversely, Six1 appears to transcriptionally repress epidermal and neural crest genes because Six1EnR induced the same phenotypes as co-expression of wild type Six1 with Groucho (Brugmann et al., 2004; Christophorou et al., 2009). In medaka fish, Groucho function is required for otocyst formation (Bajoghli et al., 2005), and in the zebrafish otocyst, combined Six1/Eya1 activation of target genes is required for hair cell fate, whereas combined Six1/Groucho repression of target genes is required for neural cell fate (Bricaud and Collazo, 2011). Because both *Eya* and *Groucho* genes are endogenously expressed in the NB zone, these data indicate that Six1 functions in PPE development as both a transcriptional activator and a repressor, depending on the co-factor with which it interacts.

Because the *Drosophila* interactome data indicate that there are several other potential SO co-factors, there may be additional modifiers of Six transcriptional activity that are developmentally relevant to PPE/placode development. We proposed that these putative co-factors may be involved in the nearly 50% of patients presenting with BOS and BOR phenotypes that do not harbor mutations in SIX1 or EYA1 (Neilson et al., 2010). Therefore, we embarked on a screen of vertebrate homologs of putative fly SO-interactors. We were especially interested in these proteins because the SD of fly SO differs by only four non-conserved amino acids from the SD of Six1 and Six2 in *Xenopus* and in humans; all other changes involve conserved or semi-conserved amino acids. Using database searches (prior to the publication of the *Xenopus* genome) we identified between one and four *Xenopus* genes (either *X. laevis* or *X. tropicalis*) with amino acid sequences homologs to 20/25 of the fly factors that can interact with SO. We analyzed the expression patterns of the homologs of 11/20 of these genes, and found that most overlap extensively with the expression patterns of Six1/Six2 (Neilson et al., 2010). Besides members of the Eya and Groucho families discussed above, functional data are only available for Sine oculis binding protein (Sobp). In the fly, this protein is co-expressed with SO in the eye imaginal disc, where its mis-expression interferes with normal eye neurogenesis (Kenyon et al., 2005). In the mouse, a recessive mutation in *Sobp* causes deafness due to abnormal cochlear development (Chen et al., 2008). In humans, a truncating mutation of SOBP caused intellectual disability and craniofacial abnormalities in a family, and one family member had hearing loss (Basel-Vanagaite et al., 2007; Birk et al., 2010). We now await functional data for the other putative Six1 co-factors to understand their potential roles in PPE/placode development.

There also are likely to be protein regulators that modify Six1 function by binding to or modifying the activity of Eya or other co-factor proteins. For example, Dachshund (Dac) has an important role in *Drosophila* eye development in cooperation with Eya and SO (Chen et al., 1997). Dach can bind to both Eya and DNA, but it does not have a direct interaction with SO. Vertebrate *Dach* is expressed widely in embryonic tissues, including placodes (McLarren et al., 2003; Schlosser, 2006; Grocott et al., 2012), and it can regulate the transcriptional effectiveness of Six/Eya complexes (Heanue et al., 1999; Ikeda et al., 2002; Li et al., 2003). However, a specific role for Dach or for other potential members of the Six/Eya transcriptional complex in PPE/placode development remains to be revealed. It is anticipated that the genome, proteome, and interactome databases of simpler organisms (in particular *Caenorhabditis elegans* and *Drosophila*) will provide important clues regarding which additional factors are important in PPE/placode development. These types of interrogations will help prioritize functional studies in vertebrate animal models, and thereby identify new causal genes in human congenital syndromes, such as BOS and BOR, that affect the cranial sensory organs.

19.5 MAINTAINING THE BOUNDARIES OF THE PRE-PLACODAL ECTODERM

It has been proposed that "NB specifying" genes interpret the neural inductive and anterior-posterior signals in the locale of the NB zone and subsequently activate "neural crest fate-specifying" genes (*FoxD3*, *Sox9/10*, *Snail2*) (Meulemans and Bronner-Fraser, 2004). There also is evidence that these genes subsequently function to maintain the boundaries between the various ectodermal domains. In the chick, for example, *Msx1* and Pax3/7 are first expressed in a broad stripe that defines the NB zone, and subsequently become restricted to the medially located neural crest, suggesting that later interactions between Msx1/Pax3/Pax7 and PPE genes lead to the segregation of the neural crest and PPE domains (reviewed by Streit, 2004). Likewise, neural plate and neural crest genes are downregulated by the co-expression of PPE genes (Six1, Eya2) or by the repression of Six1 targets (Christophorou et al., 2009). Refinements in the domains of the genes that are first expressed in overlapping zones and then expressed in discrete stripes around the neural plate also have been described in *Xenopus* (reviewed by Schlosser and Ahrens, 2004; Schlosser, 2006) (Figure 19.4). For example, as discussed above, early expression of *Dlx3/5/6* and *Zic* genes is necessary for the formation of the NB zone and for the expression of both neural crest and PPE marker genes. However, at neural plate stages the expression domains of the *Dlx* genes resolve into stripes that border the PPE and *Zic* genes are expressed in the cranial neural crest (Figure 19.4). Their lack of expression in the *Six1*-expressing PPE suggests that *Dlx* and *Zic* genes later repress PPE genes, leading to a boundary between the neural crest and placode domains. Experiments in the frog support this idea. Increasing the expression of all three *Dlx* genes (Woda et al., 2003) or of only *Dlx5* or only *Dlx6* (Brugmann et al., 2004) in the NB zone reduces the size of the PPE. Increased *Zic2* diminishes the PPE domain and expands the neural crest domain (Brugmann et al., 2004), whereas loss of *Zic1*, which reduces PPE and neural crest, expands neural plate (Hong and Saint-Jeannet, 2007). Six1 in turn represses the late expression domains of *Dlx* and *Zic* genes (Brugmann et al., 2004); in fact, epidermal, neural crest, and neural plate genes also have mutually repressive interactions with *Six1* (Brugmann et al., 2004; Matsuo-Takasaki et al., 2005). Thus, after the fates of the four major ectodermal sub-domains are specified and express unique transcription factors, these same factors appear to continue as maintenance factors to preserve the boundaries between the sub-domains (Figures 19.3 and 19.4). It is worth noting that recent studies in the fly show that in order for SO (Six1/2) and Optix (Six3/6) to trigger ectopic eye formation, they both must first function as repressors of non-retinal fates (Anderson et al., 2012). The same is likely to be true for establishing the borders between PPE and its neighboring ectodermal domains (Figure 19.4; see Groves and LaBonne, 2014 for details).

It should be kept in mind, however, that the types of experiments that have been performed to date do not sufficiently control the timing or spatial localization of overexpression and loss-of-function. Many of these genes likely have changing roles as the embryonic ectoderm becomes specified to different sub-domain fates. For example, *Foxi1* is initially expressed throughout the non-neural ectoderm, and it is required for the expression of later PPE and neural crest genes (Matsuo-Takasaki et al., 2005). During neural plate stages, however, when the PPE is fully established, the *Foxi1* expression domain is mostly lateral to *Six1/Eya1* domains. When Foxi1 is overexpressed at this time with a hormone-inducible construct it represses *Six1* and *Eya1* expression and reduces the size of the PPE. Thus, early *Foxi1* expression is required to specify the NB zone, whereas its later expression maintains the border between the PPE and the epidermis. Further manipulations involving the use of constructs that can be temporally and spatially controlled need to be performed to fully understand the molecular interactions that establish and maintain the boundaries between the different ectodermal sub-domains.

19.6 PLACODE IDENTITY

After the PPE is established as a separate sub-domain in the embryonic ectoderm, with distinct boundaries from the other sub-domains (epidermis, neural crest, neural plate), the tissue undergoes several steps of regionalization (reviewed in Bailey and Streit, 2006; Grocott et al., 2012; Saint-Jeannet and Moody, 2014). Although the early PPE is competent to form all types of placodes and has a unified molecular ground state, a few hours later it begins to acquire separate anterior (adenohypophyseal, olfactory, lens), intermediate (trigeminal) and posterior (otic, lateral line, epibranchial) domains (reviewed in Bailey et al., 2006; Grocott et al., 2012) (Figure 19.5). Fate mapping studies show that at late PPE stages, precursors of anterior and posterior regions begin to sort out (Streit, 2002; Bhattacharyya et al., 2004; Dutta et al., 2005; Toro and Varga, 2007; Xu et al., 2008; Ladher et al., 2010; Pieper et al., 2011), concomitant with the differential expression of several sets of transcription factors that are placode-specific (Figure 19.5). Specifically, studies in chick and fish embryos show that adjacent ectodermal cell populations within the PPE can contribute to distinct placodes, suggesting that PPE progenitor cells are undergoing extensive cell movements at this stage (Streit, 2002; Bhattacharyya et al., 2004; Dutta et al., 2005; Xu et al., 2008; reviewed in Toro and Varga, 2007; Ladher et al., 2010). However, most of these studies have been using labeling techniques that do not allow one to follow the movement of single cells. Therefore, it remains possible that the movements of these labeled groups of cells may represent random cell movements associated with differential growth of the ectoderm, rather than true cell sorting based on cell identity. A more recent work in *Xenopus* suggests that cell rearrangements are minimal in the PPE (Pieper et al., 2011); using time-lapse imaging to follow the movement of single cells in *Xenopus* head ectoderm, these authors reported very little directed, large-scale cell rearrangements within the PPE at early neurula stages, and during the initial segregation of placodal domains. Rather the cells appeared to move without changing their relative positions within the PPE and with adjacent ectodermal territories (Pieper et al., 2011). Another study recently performed at later developmental stages showed that cells of the epibranchial placode move actively to segregate

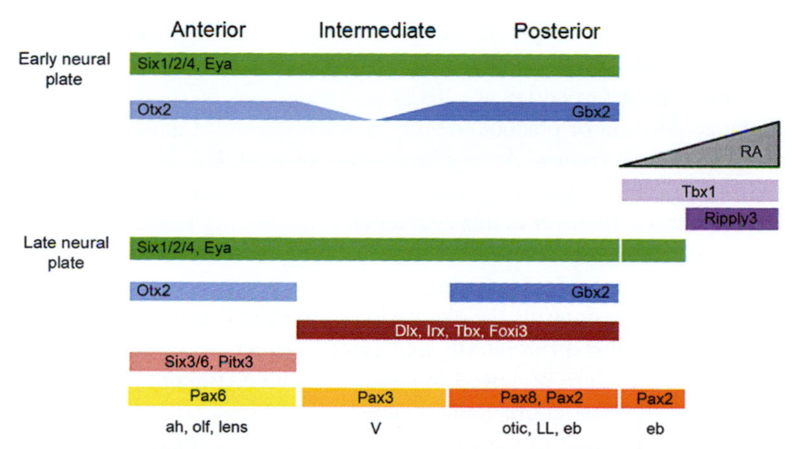

FIGURE 19.5 A Summary of the Anterior-Posterior (A-P) Patterning of the PPE into Anterior, Intermediate and Posterior Domains. At early neural plate stages the anterior-specific expression of *Otx2* and posterior-specific expression of *Gbx2*, and their mutual repression set the stage for A-P patterning of the PPE, which uniformly expresses *Six* and *Eya* genes. Subsequently, RA signaling that locally upregulates *Tbx1* and *Ripply3* expression likely causes new *Six/Eya* expression in a more posterior domain that will give rise to the most posterior epibranchial (eb) placodes. At late neural plate stages, other transcription factors become restricted to either the anterior domain or the intermediate/posterior domains. Presumably these transcriptional combinations, together with further local signaling, results in region-specific *Pax* gene expression, leading to the specification of distinct placode fates (ah, adenohypophyseal; olf, olfactory; lens; V, trigeminal; otic; LL, lateral line, eb, epibranchial). See Grocott et al. (2012) for details.

into distinct subpopulations. Here, the interaction between placode and neural crest cells is the driving force in directing the coordinated migration of epibranchial placode cells to their final position (Theveneau et al., 2013). In zebrafish, time-lapse recordings of transgenic embryos expressing *Pax2a:GFP* in the otic/epibranchial placode show that cells are continuously recruited from adjacent regions into the *Pax2a* expression domain, through directed migration (Bhat and Riley, 2011). The cell recruitment in the posterior placode is severely impaired in integrin-α5 depleted embryos, indicating that this process is guided by integrin-ECM (extracellular matrix) interactions. Interestingly, the most anterior placodes (pituitary, olfactory, and lens) were unaffected in these embryos, suggesting that this phenomenon may not regulate cell movements in other regions of the PPE (Bhat and Riley, 2011). Altogether these studies point to possible species-specific differences in the relative importance and nature of the cell arrangements associated with sensory placodes separation. These differences could also be strictly experimental, due to differences in the exact developmental time point at which these analyses are performed.

There is evidence that the anterior versus posterior expression of *Six1* is differentially regulated. First, in *Xenopus* the most posterior lateral line and epibranchial placodes form posterior to the initial Six1/Eya1 PPE expression domain; these placodes acquire Six1/Eya1 expression only after neural tube formation (Schlosser and Ahrens, 2004). We speculate that this later expression occurs in response to retinoic acid regulation of *Rippley3/Tbx1* expression, as discussed in Section IIC (Janesick et al., 2012) (Figure 19.5). Second, Kawakami's group has identified an enhancer in the *Six1* gene that is conserved across tetrapods and activates reporter expression only in the anterior part of the PPE (Sato et al., 2010). Other conserved enhancers activate reporter expression in subsets of placodes, for example, intermediate + posterior; adenohypophyseal + posterior, or anterior + posterior (Sato et al., 2012). It will be interesting to analyze the binding sites in these enhancers to determine the direct regulators of region-specific expression of *Six1*.

An important contribution to the early regionalization of the PPE, downstream of Six/Eya genes, is the differential expression of *Otx2* in the anterior domain and *Gbx2* in the posterior domain (Steventon et al., 2012); where their expression overlaps (intermediate domain) both genes are downregulated by their mutual inhibitory activity (Figure 19.5). It will be interesting to determine whether the regional expression of Otx2/Gbx2 leads to the later, mostly posterior expression of the *Dlx*, *Irx, Tbx* and *Foxi3* genes and the mostly anterior expression of *Six3, Six6* and *Pitx3* (Khatri and Groves, 2003; Schlosser and Ahrens, 2004; Schlosser, 2006; reviewed in Grocott et al., 2012).

It has long been appreciated that different members of the *Pax* gene family are expressed in different placodes (Baker and Bronner-Fraser, 2001; Schlosser and Ahrens, 2004): *Pax6* in anterior placodes, *Pax3* in the ophthalmic/profundal placode, and *Pax2* and *Pax8* in posterior placodes. Transplantation experiments in the chick indicate that the onset of *Pax* expression correlates with the acquisition of placode identity (Baker et al., 1999; Baker and Bronner-Fraser, 2000), thereby leading these authors to propose that the combination of *Pax* genes expressed by an individual placode (the "Pax code") provides identity to that placode. Recent work supports this idea: the appropriate Pax genes are expressed in the PPE domains that fate map to specific placodes prior to placode separation (Pieper et al., 2012; Grocott et al., 2012). Further, several studies show that the *Pax* genes are required for the development of these respective placodes (e.g., Hans et al., 2004; Mackereth et al., 2005; Dude et al., 2009; Shaham et al., 2012), and that placode domains segregate by mutual repression between the *Pax* genes (reviewed in Grocott et al., 2012).

An important aspect of the acquisition of placode identity is the response of naïve PPE cells to regional signals from underlying mesodermal and endodermal tissues. A very large body of work has defined the relative roles of BMP, FGF, Wnt, RA, Platelet Derived Growth Factor (trigeminal), and Sonic Hedgehog (adenohypophyseal) signaling in placode identity. While this literature is too extensive to cover in this chapter and has recently been reviewed (Saint-Jeannet and Moody, 2014), it is notable that a novel signaling pathway involved in olfactory and lens induction has recently been discovered (Llera-Forero et al., 2013). The neuropeptide Somatostatin, originating from the anterior mesendoderm, signals the anterior PPE to produce Nociceptin, which in turn controls the expression of Pax6 in the olfactory and lens progenitors.

Other PPE transcription factors are also differentially expressed during placode segregation, which suggests that they also contribute to placode identity; these are reviewed in detail elsewhere (Baker and Bronner-Fraser, 2001; Schlosser and Ahrens, 2004; Streit, 2004; Schlosser, 2006; Saint-Jeannet and Moody, 2014). But, it is interesting to note that some early genes that are broadly expressed in the PPE (e.g., *Six*, *Eya*, *Dlx*, *Foxi, Irx*) have maintained expression in some placodes and/or in their differentiating derivatives. Because these genes are expressed both broadly during early stages and in specific placodes during later stages, it is necessary to use temporally and spatially controlled constructs to determine their specific roles in initial placode identity versus later differentiation processes. The continued expression of Six1 in placode-derived structures has been most extensively studied in the otic vesicle, likely because of the hearing loss associated with human patients harboring *SIX1* mutations (reviewed in Wong et al., 2013; see also Chapter 21). In zebrafish, Six1 acts as a transcriptional activator to promote proliferation and hair cell formation and as a transcriptional repressor to inhibit cell death and neuron formation (Bricaud and Collazo, 2011). In the lens placode, however, *Six1* and *Eya1* genes are silenced before the lens begins to differentiate. The continued roles of PPE genes and placode-specific genes during differentiation importantly relates to identifying the subtle phenotypes observed (or missed) in human patients.

19.7 REGULATION OF PLACODE-DERIVED SENSORY NEURON DIFFERENTIATION

The sets of genes that are critical for setting up individual placode identity (Figure 19.5) also are involved in regulating the differentiation of placode precursor cells into the distinct cell types found in the mature placode-derived organs. For some structures this process is dealt with in detail elsewhere (pituitary, Pogoda and Hammerschmidt, 2007; olfactory, Chapter 20; lens, Charlton-Perkins et al., 2011; otocyst, Chapter 21). Here, we focus on the so-called "neurogenic" placodes (trigeminal, epibranchial) that give rise to the sensory neurons in the ganglia of cranial nerves V, VII, IX and X. These placodes are specialized niches that produce sensory neurons over a prolonged period (see Lassiter et al., 2013 for details). Unlike the neural crest cells that also contribute to most of the cranial nerve ganglia, the placode-derived neurons are specified within the placode, delaminate from it and migrate to the coalescing ganglion as post-mitotic neurons (Graham et al., 2007). The signaling pathways involved in neurogenesis in these placodes have recently been reviewed in detail (Lassiter et al., 2013). In general, the Notch pathway plays a critical role in selecting which cells in the neurogenic field will become neurons, the FGF pathway regulates delamination, the Wnt pathway contributes to delamination, *Pax* and *bHLH* gene expression, and the BMP pathway regulates *bHLH* and other neural differentiation genes.

Prior to delamination of neurons, three sets of genes are expressed in each of the neurogenic placodes: 1) *Six* and *Eya*; 2) *SoxB1* neural stem cell genes (*Sox2*, *Sox3*); and 3) *basic-Helix-Loop-Helix* (*bHLH*) neural progenitor genes. Here we review their potential roles upstream of the critical signaling pathways mentioned above that regulate neurogenesis. Three sets of observations suggest that these genes interact in a regulatory pathway that controls the onset of differentiation of placode-derived cranial sensory neurons (Figure 19.6): expression patterns, loss-of-function studies and gain-of-function studies.

First, the spatial and temporal expression patterns of these three sets of genes are overlapping but distinct. As the placodes separate, *Six* and *Eya* genes are expressed in each neurogenic placode (Pandur and Moody, 2000; Ghanbari et al., 2001; Bhattacharyya and Bronner-Fraser, 2004; Schlosser and Ahrens, 2004; Brugmann and Moody, 2005; Schlosser, 2006). *Six1* expression, for example, is maintained in the outer, proliferative layer of the placode and is downregulated as cells delaminate from the epithelium and coalesce into ganglia (Pandur and Moody, 2000; Schlosser et al., 2008). *SoxB1* genes, which are important in maintaining neural stem cells in the central nervous system (reviewed in Bergsland et al., 2011; Wegner, 2013; Moody et al., 2013; Thiel, 2013) also are expressed in neurogenic placodes. They are first detected in the PPE, after *Six* and *Eya* genes but before individual placodes segregate (Schlosser and Ahrens, 2004) (Figure 19.4). In neurogenic placodes, cells expressing high levels of *SoxB1* genes are located in the inner layer (Schlosser et al., 2008)

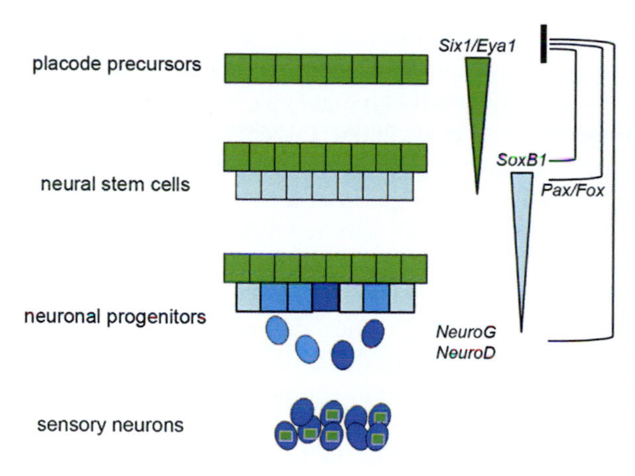

FIGURE 19.6 A Model of the Gene Regulatory Pathway Proposed to Regulate the Onset of Neurogenesis in Neurogenic Placodes that form Cranial Sensory Ganglia. **Six/Eya maintain proliferative, undifferentiated placode precursors (dark green). Six/Eya are required for the expression of *SoxB1* genes in neural stem cells (light blue), which are only detected in the deeper-positioned cells in the placode once Six/Eya protein levels decrease. *SoxB1* genes are required for the expression of *bHLH* neural progenitor genes (darker blue), which are only detected once the levels of SoxB1 protein decrease. NeuroG (medium blue) is detected in the deeper layer of the placode and in delaminating cells. Later, NeuroD (dark blue) is detected in these same cells and in the sensory neurons in the coalescing ganglion. These neurons re-express *Six/Eya* genes (green squares), which are required for cell survival. Several studies indicate that the genes expressed at each step of the process, including *Pax/Fox* genes that are involved in placode identity, feedback to negatively regulate *Six/Eya* genes, thus promoting differentiation.**

(Figure 19.6). bHLH transcription factors promote the generation of neural progenitors, cause them to exit the cell cycle, and promote neuronal differentiation in both the central nervous system and in the neurogenic placodes (reviewed in Schlosser, 2006; Castro and Guillemot, 2011). They can be grouped into two classes: those that are expressed early in the neural fate cascade (the neural determination factors such as Neurogenin [NeuroG]) and those that are expressed later in the cascade (the neural differentiation factors, such as NeuroD). In *Xenopus*, the expression of *NeuroG* and *NeuroD* genes in the neurogenic placodes is complimentary to that of *Six1/Eya1* (Schlosser and Northcutt, 2000; Schlosser and Ahrens, 2004; Schlosser et al., 2008), indicating opposing roles in neurogenesis. In several frog neurogenic placodes, *NeuroG* expression is detected first in the inner ectodermal layer and later in the neural progenitor cells that delaminate and migrate away from the placode; *NeuroG* expression is lost as the coalescing progenitor cells differentiate into neurons. The expression of *NeuroD* occurs later than that of *NeuroG*, also in scattered cells within the inner ectodermal layer, and it remains expressed in most or all of the placode-derived ganglion cells. Thus, the expression patterns suggest that Six1/Eya1 maintain placode cells in a proliferative, undifferentiated "precursor" state, SoxB1 factors transition these cells to a neural stem cell state, and bHLH factors specify neural progenitors (with Notch pathway input; Lassiter et al., 2013) and regulate their differentiation into the sensory neurons of the cranial ganglia (Figure 19.6).

What experimental observations place Six/Eya genes upstream in this pathway? First, knockdown experiments in *Xenopus* show that Six1 and Eya1 are required for the expression of *SoxB1* genes, *bHLH* genes and differentiated neuron markers (Schlosser et al., 2008). In mouse, bHLH factors also appear to be regulated by Six1 and Eya1 in neurogenic placodes. For example, the loss of Six1 in the olfactory epithelium expands bHLH factors (Ikeda et al., 2010). In *Eya1*-null neurogenic placodes, *NeuroG*- and *NeuroD*-positive cells are depleted, delamination is blocked, and sensory neuron differentiation markers (*Phox2a/b*, *SCG10*) are missing from ganglion cells (Zou et al., 2004). Similar but weaker phenotypes are seen in *Six1*-null embryos (Zou et al., 2004), likely due to the remaining activity of *Six4* in these embryos (Konishi et al., 2006). Second, gain-of-function assays in *Xenopus* indicate that the level of Six1/Eya1 expression is critical for their function in the neural differentiation pathway. Increasing Six1/Eya1 in the neurogenic placodes to high levels increases proliferation, resulting in a larger domain of *SoxB1*-expressing neural stem cells, but fewer *bHLH*-expressing neural progenitors (Schlosser et al., 2008). When Six1/Eya1 levels are lower, the transition of *SoxB1*-positive cells to *bHLH*-positive cells is promoted. Similar results were observed in the zebrafish otocyst (Bricaud and Collazo, 2006; Bricaud and Collazo, 2011); increased *Six1* expression in the otocyst increased the proliferating cell population and repressed sensory neuron differentiation. These observations indicate that Six/Eya genes maintain a proliferative precursor cell, and that they need to be downregulated in order for neural differentiation to proceed.

In other systems *Six* genes also keep precursor cells in a proliferative state prior to cell type differentiation. The loss of *Six1* in mice appears to decrease proliferation, which results in apoptosis (Li et al., 2003; Ozaki et al., 2004). In humans, SIX1 overexpression was identified in hyper-proliferating cell populations (e.g., primary breast cancers and metastatic lesions) (Ford et al., 1998). These authors showed that human SIX1 overexpression allows DNA damage to go unchecked by causing an attenuation of the DNA damage-induced G2 checkpoint. Six1 overexpression also influences cell proliferation by directly activating the transcription of CyclinA1 (Coletta et al., 2004), indicating that Six1 may maintain cells in an immature state by influencing cell cycle regulation. In fact, reactivating Six1 in adult tissues leads to misregulated cell proliferation and a number of cancers in humans (e.g., Li et al., 2013; Patrick et al., 2009; 2013).

Gene knockout studies in mice show that *Six* and *Eya* genes have an additional, later role in cranial ganglion neurogenesis. *Six1*-null, *Eya1*-null and *Six1/Six4*-null embryos have small trigeminal, vestibulocochlear and epibranchial sensory ganglia (Xu et al., 1999; Zheng et al., 2003; Zou et al., 2004; Konishi et al., 2006). Study of these mutants at embryonic stages showed that the ganglia begin to form, but the cells subsequently die, suggesting that the Six/Eya genes have a later role in cranial ganglion neurogenesis. This was borne out by antibody staining of wild type ganglia, which showed both Six1 and Six4 are expressed in subsets of differentiating sensory neurons (Konishi et al., 2006). Loss of *Six1* and *Six4* leads to cell autonomous apoptosis of newly formed cranial ganglion neurons. The authors present evidence that *Six* genes may contribute to sensory neuron survival by regulating the expression of the anti-apoptotic gene *Bcl-x*.

There are several questions that remain to be answered concerning the regulation of neuronal differentiation from the neurogenic placodes. First, what other neural stem cell genes are involved? Although SoxB1 factors are "neural stem cell" factors, they also play important roles in lens placode development, which has no neural derivatives (Kamachi et al., 1998), and in the differentiation of hair cells and supporting cells in the ear (Ahmed et al., 2012). Thus, they must have roles other than neurogenesis. In addition, neither *Sox2* nor *Sox3* is expressed in the profundal or trigeminal placodes, which give rise to sensory neurons (Schlosser and Ahrens, 2004). Thus, there must be other genes that act as neural stem cell genes to initiate neurogenesis. One potential candidate in the frog is *Sox11*, which is expressed in both the neural plate and PPE and which is upregulated by Six1 (Brugmann et al., 2004). Second, while it is well accepted that the levels of transcription factors can have varying transcriptional consequences due to binding kinetics with the DNA and with potential co-factors, this has not yet been explored rigorously in ganglion neuron differentiation. This has the potential of being clinically relevant, because the BOS and BOR mutations affect only one allele of either *Six1* or *Eya1*. Third, studies in the mouse indicate that the *Six1*-21 enhancer functions to integrate input from Sox, Pax, Fox, and bHLH factors, suggesting a transcriptional mechanism by which *Six1* expression can be downregulated to permit differentiation to proceed (Sato et al., 2012) (Figure 19.6). This now needs to be experimentally tested. Finally, not all cranial ganglia are equally affected by the loss of Six/Eya genes (Zou et al., 2004), suggesting positional input into the regulatory program. In the future it will be important to identify all the factors involved in this process and determine precisely how they interact to regulate placode neurogenesis. We expect that the late steps in this process will be very similar to neural plate and neural crest neurogenesis, but the regulatory inputs will obviously be placode-specific (e.g., Grocott et al., 2012).

19.8 FUTURE DIRECTIONS

Determining the molecular mechanisms of placode gene function is important for both understanding normal development and interpreting human congenital syndromes. First, the differential roles of all of the Six proteins in PPE/ placode development need to be determined. All three (Six1, Six2, Six4) are expressed during placode development, and one affected locus in BOS patients contains the *Six1*, *Six4*, and *Six6* genes (Ruf et al., 2004). Second, a comprehensive understanding of which proteins interact with Six proteins as co-factors is needed. The penetrance of BOS and BOR is variable, and studies in *Six1* heterozygous mice suggest that there are additional modifier genes that influence Six1 activity or function, thereby modulating the mutant phenotype (Xu et al., 2003; Ruf et al., 2004). The interactome data from *C. elegans* and *Drosophila* indicate that there are several proteins yet to be experimentally tested that could potentially influence Six and Eya functions. Third, understanding the function of all of the genes involved in PPE specification and placode differentiation pathways will have a major impact on craniofacial tissue repair efforts. Elucidating the basic molecular mechanisms by which PPE cells are induced and transformed from the embryonic ectoderm into numerous differentiated cell types and how the process differs from that described for the closely related neural crest will be critical for designing techniques for sensory organ replacement from various stem and progenitor cell sources. It is noteworthy that recent *in vitro* studies have successfully differentiated human embryonic stem cells into placode cells (Chen et al., 2012; Leung et al., 2013), in some instances generating trigeminal sensory neurons capable of *in vivo* engraftment in chick and mouse embryos, mature lens fibers, and anterior pituitary cells capable of producing human growth hormone (GH) and adrenocorticotropic hormone (ACTH) *in vivo* (Dincer et al., 2013). Understanding the placode gene regulatory network in the embryo will surely enhance *in vitro* differentiation steps for regenerating placode derivatives, making it possible to repair craniofacial defects that result from congenital abnormalities, trauma, and disease.

19.9 CLINICAL RELEVANCE

- Placodes contribute to important secretory cells and cranial sensory organs that are critical for the normal behavior of an animal.
- Mutation of genes involves in placode development lead to a variety of congenital syndromes.
- Branchio-otic syndrome (BOS) and branchio-otic renal syndrome (BOR), most notably characterized by hearing loss, are among the best-studied human syndromes involving placode development. The Six/Eya genes are causative in about 50% of patients.

- Future work identifying the regulatory network involved in PPE specification, placode development and differentiation should reveal further genetic causes of craniofacial syndromes.
- Human embryonic stem cells have been induced to express placode genes in culture (Chen et al., 2012; Leung et al., 2013; Dincer et al., 2013). Understanding the genetic pathways that regulate placode development will be critical for the use of stem cells to repair and regenerate cranial sensory organs.

ACKNOWLEDGMENTS

We apologize to those authors whose work was not cited due to space limitations. Work from the authors' laboratories is supported by NIH grant R01 DE022065 (S. A. M.) and R01 DE014212 (J-P. S-J.) We wish to thank members of the Moody and Saint-Jeannet labs for helpful comments on the manuscript.

RECOMMENDED RESOURCES

Database of Interacting Proteins (DIP). http://dip.doe-mbi.ucla.edu/dip/Main.cgi (accessed 28.02.14).

Online Mendelian Inheritance in Man (OMIM). www.omim.org (accessed 28.02.14).

Grocott, T., Tambalo, M., Streit, A., 2012. The peripheral sensory nervous system in the vertebrate head: A gene regulatory perspective. Dev. Biol. 370, 3–23.

Park, B.Y., Saint-Jeannet, J.-P., 2010. Induction and segregation of the vertebrate cranial placodes. Colloquium Series on Dev. Biol. 1 (3). Morgan & Claypool Publishers.

REFERENCES

Abdelhak, S., Kalatzis, V., Heilig, R., Compain, S., Samson, D., Vincent, C., Weil, D., Cruaud, C., Sahly, I., Leibovici, M., Bitner-Glindzicz, M., Francis, M., Lacombe, D., Vigneron, J., Charachon, R., Boven, K., Bedbeder, P., Van Regemorter, N., Weissenbach, J., Petit, C., 1997. A human homologue of the *Drosophila eyes absent* gene underlies branchio-oto-renal (BOR) syndrome and identifies a novel gene family. Nature Genet. 15, 157–164.

Ahmed, M., Wong, E.Y., Sun, J., Xu, J., Wang, F., Xu, P.X., 2012. Eya1-Six1 interaction is sufficient to induce hair cell fate in the cochlea by activating Atoh1 expression in cooperation with Sox2. Dev. Cell 22, 377–390.

Ahrens, K., Schlosser, G., 2005. Tissues and signals involved in the induction of placodal Six1 expression in. *Xenopus laevis*. Dev. Biol. 288, 40–59.

Anderson, A.M., Weasner, B.M., Weasner, B.P., Kumar, J.P., 2012. Dual transcriptional activities of SIX proteins define their roles in normal and ectopic eye development. Development 139, 991–1000.

Arima, K., Shiotsugu, J., Niu, R., Khandpur, R., Martinez, M., Shin, Y., Koide, T., Cho, K.W., Kitayama, A., Ueno, N., et al., 2005. Global analysis of RAR- responsive genes in the *Xenopus* neurula using cDNA microarrays. Dev. Dyn. 232, 414–431.

Arkell, R., Beddington, R.S., 1997. BMP-7 influences pattern and growth of the developing hindbrain of mouse embryos. Development 124, 1–12.

Azuma, N., Hirakiyama, A., Inoue, T., Asaka, A., Yamada, M., 2000. Mutations of a human homologue of the *Drosophila eyes absent* gene (EYA1) detected in patients with congenital cataracts and ocular anterior segment anomalies. Hum. Molec. Genet. 9, 363–366.

Bailey, A.P., Streit, a, 2006. Sensory organs: Making and breaking the pre-placodal region. Curr. Top. Dev. Biol. 72, 167–204.

Bailey, A.P., Bhattacharyya, S., Bronner-Fraser, M., Streit, A., 2006. Lens specification is the ground state of all sensory placodes, from which Fgf promotes olfactory identity. Dev. Cell 11, 505–517.

Bajoghli, B., Aghaallaei, N., Czerby, T., 2005. Groucho corepressor proteins regulate otic vesicle outgrowth. Dev. Dyn. 233, 760–771.

Baker, C.V.H., Bronner-Fraser, M., 2000. Establishing neuronal identity in vertebrate neurogenic placodes. Development 127, 3045–3056.

Baker, C.V., Bronner-Fraser, M., 2001. Vertebrate cranial placodes I. Embryonic induction. Dev. Biol. 232, 1–61.

Baker, C.V., Stark, M.R., Marcelle, C., Bronner-Fraser, M., 1999. Competence, specification and induction of Pax-3 in the trigeminal placode. Development 126, 147–156.

Bane, B.C., Van Rybroek, J.M., Kolker, S.J., Weeks, D.L., Manaligod, J.M., 2005. Eya1 expression in the developing inner ear. Ann. Otol. Rhinol. Laryngol. 114, 853–858.

Basch, M.L., Bronner-Fraser, M., Garcia-Castro, M.I., 2006. Specification of the neural crest occurs during gastrulation and required Pax7. Nature 441, 218–222.

Basel-Vanagaite, L., Rainschtein, L., Inbar, D., Gothelf, D., Hennekam, R., Straussberg, R., 2007. Autosomal recessive mental retardation syndrome with anterior maxillary protrusion and strabismus: Mrams syndrome. Am. J. Med. Genet. 143a, 1687–1691.

Bassham, S., Postlthwait, J.H., 2005. The evolutionary history of placodes: A molecular genetic investigation of the larvacean urochordate. Oikoleura dioica. Development 132, 4259–4272.

Beanan, M.J., Sargent, T.D., 2000. Regulation and function of Dlx3 in vertebrate development. Dev. Dyn. 218. 545-5.

Bergsland, M., Ramskold, D., Zaouter, C., Klum, S., Sandberg, R., Muhr, J., 2011. Sequentially acting Sox transcription factors in neural lineage development. Genes Dev. 25, 2453–2464.

Bessarab, D.A., Chong, S.W., Korzh, V., 2004. Expression of zebrafish *Six1* during sensory organ development and myogenesis. Dev. Dyn. 230, 781–786.

Bhat, N., Riley, B., 2011. Integrin-A5 coordinates assembly of posterior cranial placodes in zebrafish and enhances Fgf-dependent regulation of otic/epibranchial cells. Plos One 6, E27778.

Bhat, N., Kwon, H.J., Riley, B.B., 2013. A gene network that coordinates preplacodal competence and neural crest specification in zebrafish. Dev. Biol. 373, 107–117.

Bhattacharyya, S., Bronner, M.E., 2013. Clonal analysis in the anterior pre-placodal region: Implications for the early lineage bias of placodal precursors. Int. J. Dev. Biol. 57(9–10), 753–757.

Bhattacharyya, S., Bronner-Fraser, M., 2004. Hierarchy of regulatory events in sensory placode development. Curr. Opin. Genet. Dev. 14, 520–526.

Bhattacharyya, S., Bailey, A.P., Bronner-Fraser, M., Streit, A., 2004. Segregation of lens and olfactory precursors from a common territory: cell sorting and reciprocity of Dlx5 and Pax6 expression. Dev. Biol. 271, 403–414.

Birk, E., Har-Zahav, A., Manzini, C.M., Pasmanik-Chor, M., Kornreich, L., Walsh, C.A., Noben-Trauth, K., Albin, A., Simon, A.J., Colleaux, L., Morad, Y., Rainshtein, L., Tischfield, D.J., Wang, P., Magai, N., Shoshani, N., Rechavi, G., Gothelf, D., Maydan, G., Shohat, M., Basel-Vanagaite, L., 2010. SOBP is mutated in syndromic and non-syndromic intellectual disability and is highly expressed in the brain limbic system. Am. J. Hum. Genet. 87, 694–700.

Bonini, N.M., Leiserson, W.M., Benzer, S., 1993. The eyes absent gene: genetic control of cell survival and differentiation in the developing *Drosophila* eye. Cell 72, 379–395.

Bosman, E.A., Quint, E., Fuchs, E., Hrabe de Angelis, M., Steel, K.P., 2009. Catweasel mice: a novel role for Six1 in sensory patch development and a model for branchio-otic-renal syndrome. Dev. Biol. 328, 285–296.

Bradley, L., Sun, B., Collins-Racie, L., LaVallie, E., McCoy, J., Sive, H., 2000. Different activities of the frizzled-related proteins frzb2 and sizzled2 during *Xenopus* anteroposterior patterning. Dev. Biol. 227, 118–132.

Bricaud, O., Collazo, A., 2006. The transcription factor six1 inhibits neuronal and promotes hair cell fates in the developing zebrafish (*Danio rerio*) inner ear. J. Neurosci. 26, 10438–10451.

Bricaud, O., Collazo, A., 2011. Balancing cell numbers during organogenesis: Six1a differentially affects neurons and sensory hair cells in the inner ear. Dev. Biol. 357, 191–201.

Brodbeck, S., Englert, C., 2004. Genetic determination of nephrogenesis: the Pax/Eya/Six gene network. Pediatr. Nephrol 19, 249–255.

Brugmann, S.A., Moody, S.A., 2005. Induction and specification of the vertebrate ectodermal placodes: precursors of the cranial sensory organs. Biol. Cell. 97, 303–319.

Brugmann, S.A., Pandur, P.D., Kenyon, K.L., Pignoni, F., Moody, S.A., 2004. Six1 promotes a placodal fate within the lateral neurogenic ectoderm by functioning as both a transcriptional activator and repressor. Development 131, 5871–5881.

Buller, C., Xu, X., Marquis, V., Schwanke, R., Xu, P.X., 2001. Molecular effects of Eya1 domain mutations causing organ defects in BOR syndrome. Hum. Mol. Genet. 10, 2775–2781.

Carmona-Fontaine, C., Acuna, G., Ellwanger, K., Niehrs, C., Mayor, R., 2007. Neural crests are actively precluded from the anterior neural fold by a novel inhibitory mechanism dependent on Dickkopf1 secreted by the prechordal mesoderm. Dev. Biol. 309, 208–221.

Castro, D.S., Guillemot, F., 2011. Old and new functions of proneural factors revealed by the genome-wide characterization of their transcriptional targets. Cell Cycle 10, 4026–4031.

Charlton-Perkins, M., Brown, N.L., Cook, T.A., 2011. The lens in focus: a comparison of lens development in *Drosophila* and vertebrates. Mol. Genet. Genomics 286, 798–811.

Chen, R., Amoui, M., Zhang, Z., Mardon, G., 1997. Dachshund and eyes absent proteins form a complex and function synergistically to induce ectopic eye development in. *Drosophila*. Cell 91, 893–903.

Chen, Y., Pollet, N., Niehrs, C., Pieler, T., 2001. Increased XRALDH2 activity has a posteriorizing effect on the central nervous system of *Xenopus* embryos. Mech. Dev. 101, 91–103.

Chen, Z., Montcouquioi, M., Calderon, R., Jenkins, N.A., Copeland, N.G., Kelley, M.W., Noben-Trauth, K., 2008. Jacl/Sobp, encoding a nuclear zinc finger protein, is critical for cochlear growth, cell fate and patterning of the organ of Corti. J. Neurosci. 28, 6633–6641.

Chen, B., Kim, E.H., Xu, P.X., 2009. Initiation of olfactory placode development and neurogenesis is blocked in mice lacking both Six1 and Six4. Dev. Biol. 326, 75–85.

Chen, W., Jongkamonwiwat, N., Abbas, L., Eshtan, S.J., Johnson, S.L., Kuhn, S., Milo, M., Thurlow, J.K., Andrews, P.W., Marcotti, W., Moore, H.D., Rivolta, M.N., 2012. Restoration of auditory evoked response by human ES-derived otic progenitors. Nature 490, 278–282.

Cheyette, B.N., Green, P.J., Martin, K., Garren, H., Hartenstein, V., Zipursky, S.L., 1994. The *Drosophila sine oculis* locus encodes a homeodomain-containing protein required for the development of the entire visual system. Neuron 12, 977–996.

Christophorou, N.A., Bailey, A.P., Hanson, S., Streit, A., 2009. Activation of Six1 target genes is required for sensory placode formation. Dev. Biol. 336, 327–336.

Coletta, R.D., Christensen, K., Reichenberger, K.J., Lamb, J., Micomonaco, D., Huang, L., Wolf, D.M., Muller-Tidow, C., Golub, T.R., Kawakami, K., Ford, H.L., 2004. The Six1 homeoprotein stimulates tumorigenesis by reactivation of cyclin A1. Proc. Natl. Acad. Sci. USA 101, 6478–6483.

Couly, G.F., LeDouarin, N.M., 1987. Mapping of the early neural primordium in quail-chick chimeras. II. The prosencephalic neural plate and neural folds: implications for the genesis of cephalic human congenital abnormalities. Dev. Biol. 120, 198–214.

Couly, G.F., LeDouarin, N.M., 1990. Head morphogenesis in embryonic avian chimeras: evidence for a segmental pattern in the ectoderm corresponding to the neuromeres. Development 108, 543–558.

de Croze, N., Maczkowiak, F., Monsoro-Burq, A.H., 2011. Reiterative AP2a activity controls sequential steps in the neural crest gene regulatory network. Proc. Natl. Acad. Sci. USA 108, 155–160.

David, R., Ahrens, K., Wedlich, D., Schlosser, G., 2001. *Xenopus* Eya1 demarcates all neurogenic placodes as well as migrating hypaxial muscle precursors. Mech. Dev. 103, 189–192.

Depreux, F.F., Darrow, K., Conner, D.A., Eavey, R.D., Liberman, M.C., Seidman, C.E., Seidman, J.G., 2008. Eya4-deficient mice are a model for heritable otitis media. J. Clin. Invest. 118, 651–658.

Dincer, Z., Piao, J., Niu, L., Ganat, Y., Kriks, S., Zimmer, B., Shi, S.-H., Tabar, V., Studer, L., 2013. Specification of functional cranial placode derivatives from human pluripotent stem cells. Cell Reports 5, 1–16.

Dude, C.M., Kuan, C.Y., Bradshaw, J.R., Greene, N.D., Relaix, F., Stark, M.R., Baker, C.V., 2009. Activation of Pax3 target genes is necessary but not sufficient for neurogenesis in the ophthalmic trigeminal placode. Dev. Biol. 326, 314–326.

Duggan, C.D., DeMaria, S., Baudhuin, A., Stafford, D., Ngai, J., 2008. Foxg1 is required for development of the vertebrate olfactory system. J. Neurosci. 28, 5229–5239.

Duncan, M.K., Kos, L., Jenkins, N.A., Gilbert, D.J., Copeland, N.G., Tomarev, S.I., 1997. Eyes absent, a gene family found in several metazoan phyla. Mammal. Genome 8, 479–485.

Dutta, S., Dietrich, J.E., Aspock, G., Burdine, R.D., Schier, A., Westerfield, M., Varga, Z.M., 2005. Pitx3 defines an equivalence domain for lens and anterior pituitary placode. Development 132, 1579–1590.

Estefania, E., Ramirez-Camacho, R., Gomar, M., Trinidad, A., Arellano, B., Garcia-Berrocal, J.R., Verdaguer, J.M., Vilches, C., 2006. Point mutation of an EYA1-gene splice site in a patient with oto-facial-cervical syndrome. Ann. Hum. Genet. 70, 140–144.

Esterberg, R., Fritz, A., 2009. Dlx3b/4b are required for the formation of the preplacodal region and otic placode through local modulation of BMP activity. Dev. Biol. 325, 189–199.

Esteve, P., Bovolenta, P., 1999. cSix4, a member of the six gene family of transcription factors, is expressed during placode and somite development. Mech, Dev., 85,161–85,165.

Feledy, J.A., Beanan, M.J., Sandoval, J.J., Goodrich, J.S., Lim, J.H., Matsuo-Takasaki, M., Sato, S.M., Sargent, T.D., 1999. Inhibitory patterning of the anterior neural plate in *Xenopus* by homeodomain factors Dlx3 and Msx1. Dev. Biol. 212, 455–464.

Ford, H.L., Kabingu, E.N., Bump, E.A., Mutter, G.L., Pardee, A.B., 1998. Abrogation of the G2 *cell* cycle checkpoint associated with overexpression of HSIX1: a possible mechanism of breast carcinogenesis. Proc. Natl. Acad. Sci. USA 95, 12608–12613.

Fougerousse, F., Durand, M., Lopez, S., Suel, L., Demignon, J., Thornton, C., Ozaki, H., Kawakami, K., Barbet, P., Beckmann, J.S., Maire, P., 2002. Six and Eya expression during human somitogenesis and MyoD family activation. J. Muscle Res. Cell Motil 23, 255–264.

Gasparini, F., Degasperi, V., Shimeld, S.M., Burighel, P., Manni, L., 2013. Evolutionary conservation of the placodal transcriptional network during sexual and asexual development of chordates. Dev Dyn. 242, 752–766.

Ghanbari, H., Seo, H.C., Fjose, A., Brändli, A.W., 2001. Molecular cloning and embryonic expression of *Xenopus* Six homeobox genes. Mech. Dev. 101, 271–277.

Giot, L., Bader, J.S., Brouwer, C., Chaudhuri, A., Kuang, B., Li, Y., Hao, Y.L., Ooi, C.E., Godwin, B., et al., 2003. A protein interaction map of. Drosophila melanogaster 302, 1727–1736.

Givens, M.L., Rave-Harel, N., Goonewardena, V.D., Kurotani, R., Berdy, S.E., Swan, C.H., Rubenstein, J.L., Robert, B., Mellon, P.L., 2005. Developmental regulation of gonadotropin-releasing hormone gene expression by the Msx and Dlx homeodomain protein families. J. Biol. Chem. 280, 19156–19165.

Glavic, A., Maris Honoré, S., Gloria Feijóo, C., Bastidas, F., Allende, M.L., Mayor, R., 2004. Role of Bmp signaling and the homeoprotein iroquois in the specification of the cranial placodal field. Dev. Biol. 272, 89–103.

Graham, A., Blentic, A., Dusque, S., Begbie, J., 2007. Delamination of cells from neurogenic placodes does not involve an epithelial-to-mesenchymal transition. Development 134, 4141–4145.

Graham, A., Shimeld, S.M., 2013. The origin and evolution of the ectodermal placodes. J. Anat. 222, 32–40.

Grifone, R., Demignon, J., Houbron, C., Souil, E., Niro, C., Seller, M.J., Hamard, G., Maire, P., 2005. Six1 and Six4 homeoproteins are required for Pax3 and Mrf expression during myogenesis in the mouse embryo. Development 132, 2235–2249.

Grocott, T., Tambalo, M., Streit, A., 2012. The peripheral sensory nervous system in the vertebrate head: A gene regulatory perspective. Dev. Biol. 370, 3–23.

Groves, A.K., Bronner-Fraser, M., 2000. Competence, specification and commitment in otic placode induction. Development 127, 3489–3499.

Groves, A.K., Labonne, C., 2014. Setting appropriate boundaries: Fate, patterning and competence at the neural plate border. Dev. Biol. 389(1), 2–12. http://dx.doi.org/ 10.1016/j.ydbio.2013.11.027. [Epub ahead of print].

Guo, C., Sun, Y., Zhou, B., Adam, R.M., Li, X., Pu, W.T., Morrow, B.E., Moon, A., Li, X., 2011. A Tbx1-Six1/Eya1-Fgf8 genetic pathway controls mammalian cardiovascular and craniofacial morphogenesis. J. Clin. Invest. 121, 1585–1595.

Hans, S., Liu, D., Westerfield, M., 2004. Pax8 and Pax2a function synergistically in otic specification, downstream of the Foxi1 and Dlx3b transcription factors. Development 131, 5091–5102.

Hatini, V., Xin, Y., Balas, G., Lai, E., 1999. Dynamics of placodal lineage development revealed by targeted transgene expression. Dev. Dyn. 215, 332–343.

Heanue, T.A., Reshef, R., Davis, R.J., Mardon, G., Oliver, G., Tomarev, S., Lassar, A.B., Tabin, C.J., 1999. Synergistic regulation of vertebrate muscle development by Dach2, Eya2, and Six1, homologs of genes required for *Drosophila* eye formation. Genes Dev. 15, 3231–3243.

Heisenberg, C.P., Houart, C., Take-Uchi, M., Rauch, G.J., Young, N., Coutinho, P., Masai, I., Caneparo, L., Concha, M.L., Geisler, R., Dale, T.C., Wilson, S.W., Stemple, D.L., 2001. A mutation in the Gsk3-binding domain of zebrafish Masterblind/Axin1 leads to a fate transformation of telencephalon and eyes to diencephalon. Genes Dev. 15, 1427–1434.

Hemmati-Brivanlou, A., Thomsen, G.H., 1995. Ventral mesodermal patterning in *Xenopus* embryos: expression patterns and activities of BMP-2 and BMP-4. Dev. Genet. 17, 78–89.

Hilal, E.M., Chen, J.H., Silverman, A.J., 1996. Joint migration of gonadotropin-releasing hormone (GnRH) and neuropeptide Y (NPY) neurons from olfactory placode to central nervous system. J. Neurobiol. 31, 487–502.

Himeda, C.L., Ranish, J.A., Angello, J.C., Maire, P., Aebersold, R., Hauschka, S.D., 2004. Quantitative proteomic identification of six4 as the trex-binding factor in the muscle creatine kinase enhancer. Mol. Cell Biol. 24, 2132–2143.

Hoffman, T.L., Javier, A.L., Campeau, S.A., Knight, R.D., Schilling, T.F., 2007. Tfap2 transcription factors in zebrafish neural crest development and ectodermal evolution. J. Exp. Zool. B *Mol. Dev.* Evol. 308, 679–691.

Hong, C.-S., Saint-Jeannet, J.-P., 2007. The activity of Pax3 and Zic1 regulates three distinct cell fates at the neural plate border. Mol. Biol. Cell 18, 2192–2202.

Hoskins, B.E., Cramer, C.H., Silvius, D., Zou, D., Raymond, R.M., Orten, D.J., Kimberling, W.J., Smith, R.J., Weil, D., Petit, C., Otto, E.A., Xu, P.X., Hildebrandt, F., 2007. Transcription factor Six5 is mutated in patients with branchio-otic-renal syndrome. *Am.* J. Hum. Genet. 80, 800–804.

Hsiao, F.C., Williams, A., Davies, E.L., Rebay, I., 2001. Eyes absent mediates cross-talk between retinal determination genes and the receptor tyrosine kinase signaling pathway. Dev. Cell 1, 51–61.

Hwang, C.H., Simeone, A., Lai, E., Wu, D.K., 2009. Foxg1 is required for proper separation and formation of sensory cristae during inner ear development. Dev. Dyn. 238, 2725–2734.

Ikeda, K., Watanabe, Y., Ohto, H., Kawakami, K., 2002. Molecular interaction and synergistic activation of a promoter by Six, Eya, and Dach proteins mediated through CREB binding protein. Mol. Cell. Biol. 22, 6759–6766.

Ikeda, K., Ookawara, S., Sato, S., Ando, Z., Kageyama, R., Kawakami, K., 2007. Six1 is essential for early neurogenesis in the development of olfactory epithelium. Dev. Biol. 311, 53–68.

Ikeda, K., Kageyama, R., Suzuki, Y., Kawakami, K., 2010. Six1 is indispensable for production of functional progenitor cells during olfactory epithelial development. *Int.* J. Dev. Biol. 54, 1453–1464.

Ishihara, T., Ikeda, K., Sato, S., Yajia, H., Kawakami, K., 2008. Differential expression of Eya1 and Eya2 during chick early embryonic development. Gene Expr. Patterns 8, 357–367.

Ito, T., Noguchi, Y., Yashima, T., Kitamura, K., 2006. Six1 mutation associated with enlargement of the vestibular aqueduct in a patient with branchio-oto syndrome. Laryngoscope 116, 797–799.

Jacobson, A.G., 1963. The determination and positioning of the nose, lens and ear. III. Effects of reversing the antero-posterior axis of epidermis, neural plate and neural fold. J. Exp. Zool. 154, 293–303.

Jacobson, A.G., 1966. Inductive processes in embryonic development. Science 152, 25–34.

Janesick, A., Shiotsugu, J., Taketani, M., Blumberg, B., 2012. RIPPLY3 is a retinoic acid-inducible repressor required for setting the borders of the pre-placodal ectoderm. Development 139, 1213–1224.

Johnson, K.R., Cook, S.A., Erway, L.C., Matthews, A.N., Sanford, L.P., Paradies, N.E., Friedman, R.A., 1999. Inner ear and kidney anomalies caused by IAP insertion in an intron of the Eya1 gene in a mouse model of BOR syndrome. Hum. Mol. Genet. 8, 645–653.

Kaji, T., Artinger, K.B., 2004. dlx3b and dlx4b function in the development of Rohon-Beard sensory neurons and trigeminal placode in the zebrafish neurula. Dev. Biol. 276, 523–540.

Kamachi, Y., Uchikawa, M., Collignon, J., Lovell-Badge, R., Kondoh, H., 1998. Involvement of Sox1, 2 and 3 in the early and subsequent molecular events of lens induction. Development 125, 2521–2532.

Kawakami, K., Ohto, H., Ikeda, K., Roeder, R.G., 1996. Structure, function and expression of a murine homeobox protein AREC3, a homologue of *Drosophila sine oculis* gene product, and implication in development. Nucleic Acids Res. 24, 303–310.

Kawakami, K., Sato, S., Ozaki, H., Ikeda, K., 2000. Six family genes-structure and function as transcription factors and their roles in development. Bioessays 22, 616–626.

Kawauchi, S., Kim, J., Santos, R., Wu, H.H., Lander, A.D., Calof, A.L., 2009. Foxg1 promotes olfactory neurogenesis by antagonizing Gdf11. Development 136, 1453–1464.

Kenyon, K.L., Li, D.J., Clouser, C., Tran, S., Pignoni, F., 2005. Fly SIX-type homeodomain proteins Sine oculis and Optix partner with different cofactors during eye development. Dev. Dyn. 234, 497–504.

Khatri, S.B., Groves, A.K., 2013. Expression of the Foxi2 and Foxi3 transcription factors during development of chicken sensory placodes and pharyngeal arches. Gene Expr. Patterns 13, 38–42.

Knouff, R.A., 1935. The developmental pattern of ectodermal placodes in. *Rana pipiens*. J. Comp. Neurol. 62, 17–71.

Kobayashi, M., Osanai, H., Kawakami, K., Yamamoto, M., 2000. Expression of three zebrafish Six4 genes in the cranial sensory placodes and the developing somites. Mech. Dev. 98, 151–155.

Kobayashi, M., Nishikawa, K., Suzuki, T., Yamamoto, M., 2001. The homeobox protein Six3 interacts with the Groucho corepressor and acts as a transcriptional repressor in eye and forebrain formation. Dev. Biol. 232, 315–326.

Kochhar, A., Orten, D.J., Sorensen, J.L., Fischer, S.M., Cremers, C.W., Kimberling, W.J., Smith, R.J., 2008. SIX1 mutation screening in 247 branchio-otic-renal syndrome families: a recurrent missense mutation associated with BOR. Hum. Mutat. 29, 565.

Konishi, Y., Ikeda, K., Iwakura, Y., Kawakami, K., 2006. Six1 and Six4 promote survival of sensory neurons during early trigeminal gangliogenesis. Brain Res. 1116, 93–102.

Kozlowski, D.J., Whitfield, T.T., Hukriede, N.A., Lam, W.K., Weinberg, E.S., 2005. The zebrafish *dog-eared* mutation disrupts *eya1*, a gene required for cell survival and differentiation in the inner ear and lateral line. Dev. Biol. 277, 27–41.

Kriebel, M., Muller, F., Hollemann, T., 2007. Xeya3 regulates survival and proliferation of neural progenitor cells within the anterior neural plate of *Xenopus* embryos. Dev. Dyn. 236, 1526–1534.

Krug, P., Moriniere, V., Marlin, S., Koubi, V., Gabriel, H.D., Colin, E., Bonneau, D., Salomon, R., Antignac, C., Heidet, L., 2011. Mutation screening of the EYA1, SIX1, and SIX5 genes in a large cohort of patients harboring branchio-otic-renal syndrome call into question the pathogenic role of SIX5 mutations. Hum. Mutat. 32, 183–190.

Kumar, S., Deffenbacher, K., Cremers, C.W., Van Camp, G., Kimberling, W.J., 1997. Brachio-oto-renal syndrome: identification of novel mutations, molecular characterization, mutation distribution and prospects for genetic testing. Genet. Test 1, 243–251.

Kwon, H.J., Bhat, N., Sweet, E.M., Cornell, R.A., Riley, B.B., 2010. Identification of early requirements for preplacodal ectoderm and sensory organ development. PLoS Genet. 6:e1001133.

Laclef, C., Souil, E., Demignon, J., Maire, P., 2003. Thymus, kidney and craniofacial abnormalities in Six 1 deficient mice. Mech. Dev. 120, 669–679.

Ladher, R.K., O'Neill, P., Begbie, J., 2010. From shared lineage to distinct functions: development of the inner ear and epibranchial placodes. Development 137, 1777–1785.

Lassiter, R.N.T., Stark, M.R., Zhao, T., Zhou, C.J., 2013. Signaling mechanisms controlling cranial placode neurogenesis and delamination. Dev. Biol. (in press) Dec 3. pii: S0012–1606(13)00642–8. http://dx.doi.org/10.1016/j.ydbio.2013.11.025.

LeDouarin, N.M., Fontaine-Perus, J., Couly, G., 1986. Cephalic ectodermal placodes and neurogenesis. Trends Neurosci. 9, 175–180.

Leger, S., Brand, M., 2002. Fgf8 and Fgf3 are required for zebrafish ear placode induction, maintenance and inner ear patterning. Mech. Dev. 119, 91–108.

Leung, A.W., Morest, K.D., Li, J.Y., 2013. Differential BMP signaling controls formation and differentiation of multipotent preplacodal progenitors from human embryonic stem cells. Dev. Biol. 379, 208–220.

Li, X., Oghi, K.A., Zhang, J., Krones, A., Bush, K.T., Glass, C.K., Nigam, S.K., Aggarwal, A.K., Maas, R., Rose, D.W., Rosenfeld, M.G., 2003. Eya protein phosphatase activity regulates Six1-Dach-Eya transcriptional effects in mammalian organogenesis. Nature 426, 247–254.

Li, Y., Manaligod, J.M., Weeks, D.L., 2010. EYA1 mutations associated with the branchio-otic renal syndrome result in defective otic development in *Xenopus laevis*. Biol. Cell 102, 277–292.

Li, Z., Tian, T., Lv, F., Chang, Y., Wang, X., Zhang, L., Li, X., Li, L., Ma, W., Wu, J., Zhang, M., 2013. Six1 promotes proliferation of pancreatic cancer cells via up-regulation of Cyclin D1 expression. PLoS One 8. e59203.

Litsiou, A., Hanson, S., Streit, A., 2005. A balance of FGF, BMP and WNT signalling positions the future placode territory in the head. Development 132, 4051–4062.

Liu, D., Chu, H., Maves, L., Yan, Y.L., Morcos, P.A., Postlethwait, J.H., Westerfield, M., 2003. Fgf3 and Fgf8 dependent and independent transcription factors are required for otic placode specification. Development 130, 2213–2224.

Llera-Forero, L., Tambalo, M., Christophorou, M., Chambers, D., Houart, C., Streit, A., 2013. Neuropeptides: developmental signals in placode progenitor formation. Dev. Cell 26, 195–203.

Luo, T., Matsuo-Takasaki, M., Sargent, T.D., 2001a. Distinct roles for Distal-less genes Dlx3 and Dlx5 in regulating ectodermal development in. *Xenopus*. Mol. Reprod. Dev. 60, 331–337.

Luo, T., Matsuo-Takasaki, M., Lim, J.H., Sargent, T.D., 2001b. Differential regulation of Dlx gene expression by a BMP morphogenetic gradient. Int. J. Dev. Biol. 45, 681–684.

Luo, T., Matsuo-Takasaki, M., Thomas, M.L., Weeks, D.L., Sargent, T.D., 2002. Transcription factor AP-2 is an essential and direct regulator of epidermal development in *Xenopus*. Dev. Biol. 245, 136–144.

Luo, T., Lee, Y.-H., Saint-Jeannet, J.-P., Sargent, T.D., 2003. Induction of neural crest in *Xenopus* by transcription factor AP2. Proc. Natl. Acad. Sci. USA 100, 532–537.

Mackereth, M.D., Kwak, S.J., Fritz, A., Riley, B.B., 2005. Zebrafish pax8 is required for otic placode induction and plays a redundant role with Pax2 genes in the maintenance of the otic placode. Development 132, 371–382.

Makishima, T., Madeo, A.C., Brewer, C.C., Zalewski, C.K., Butman, J.A., Sachdev, V., Arai, A.E., Holbrook, B.M., Rosing, D.R., Griffith, A.J., 2007. Nonsyndromic hearing loss DFNA10 and a novel mutation of EYA4: evidence for correlation of normal cardiac phenotype with truncating mutations of the Eya domain. Am. J. Med. Genet. 143A, 1592–1598.

Mancilla, A., Mayor, R., 1996. Neural crest formation in *Xenopus laevis*: mechanism of Xslug induction. Dev. Biol. 177, 580–589.

Maroon, H., Walshe, J., Mahmood, R., Kiefer, P., Dickson, C., Mason, I., 2002. Fgf3 and Fgf8 are required together for formation of the otic placode and vesicle. Development 129, 2099–2108.

Martin, K., Groves, A.K., 2005. Competence of cranial ectoderm to respond to Fgf signaling suggests a two-step model of otic placode induction. Development 133, 877–887.

Matsuo-Takasaki, M., Matsumura, M., Sasai, Y., 2005. As essential role of Foxi1a for ventral specification of the cephalic ectoderm during gastrulation. Development 132, 3885–3894.

McLarren, K.W., Litsiou, A., Streit, A., 2003. DLX5 positions the neural crest and preplacode region at the border of the neural plate. Dev. Biol. 259, 34–47.

Meulemans, D., Bronner-Fraser, M., 2004. Gene-regulatory interactions in neural crest evolution and development. Dev. Cell 7, 291–299.

Modrell, M.S., Baker, C.V., 2012. Evolution of electrosensory ampullary organs: conservation of Eya4 expression during lateral line development in jawed vertebrates. Evol. Dev. 14, 277–285.

Moody, S.A., Klein, S.L., Karpinski, B.A., Maynard, T.M., LaMantia, A.S., 2013. On becoming neural: what the embryo can tell us about differentiating neural stem cells. Amer. J. Stem Cells 2, 74–94.

Monsoro-Burq, A.H., Wang, E., Harland, R., 2005. Msx1 and Pax3 cooperate to mediate FGF8 and Wnt signals during *Xenopus* neural crest induction. Dev. Cell 8, 167–178.

Mosrati, M.A., Hammami, B., Rebeh, I.B., Ayadi, L., Dhouib, L., Ben Mahfoudh, K., Hakim, B., Charfeddine, I., Mnif, J., Ghorbel, A., Masmoudi, S., 2011. A novel dominant mutation in SIX1, affecting a highly conserved residue, results in only auditory defects in humans. Eur. J. Med. Genet. 54, e484–e488.

Murakami, S., Arai, Y., 1994. Transient expression of somatostatin immunoreactivity in the olfactory-forebrain region in the chick embryo. Brain Res. Dev. Brain Res. 82, 277–285.

Neilson, K.M., Pignoni, F., Yan, B., Moody, S.A., 2010. Developmental expression patterns of candidate co-factors for vertebrate Six family transcription factors. Dev. Dyn. 239, 3446–3466.

Nguyen, V.H., Schmid, B., Trout, J., Connors, S.A., Ekker, M., Mullins, M.C., 1998. Ventral and lateral regions of the zebrafish gastrula, including the neural crest progenitors, are established by a bmp2b/swirl pathway of genes. Dev. Biol. 199, 93–110.

Nica, G., Herzog, W., Sonntag, C., Nowak, M., Schwarz, H., Zapata, A.G., Hammerschmidt, M., 2006. Eya1 is required for lineage-specific differentiation, but not for cell survival in the zebrafish adenohypophysis. Dev. Biol. 292, 189–204.

Noguchi, Y., Ito, T., Nishio, A., Honda, K., Kitamura, K., 2011. Audiovestibular findings in a branchio-oto syndrome patient with a SIX1 mutation. Acta Otolaryngol 131, 413–418.

Northcutt, R.G., Muske, L.E., 1994. Multiple embryonic origins of gonadotropin-releasing hormone (GnRH) immunoreactive neurons. Dev. Brain Res. 78, 279–290.

Ohto, H., Takizawa, T., Saito, T., Kobayashi, M., Ikeda, K., Kawakami, K., 1998. Tissue and developmental distribution of Six family gene products. Int. J. Dev. Biol. 42, 141–148.

Ohto, H., Kamada, S., Tago, K., Tominaga, S., Ozaki, H., Sato, S., Kawakami, K., 1999. Cooperation of Six and Eya in activation of their target genes through nuclear translocation of Eya. Mol. Cell. Biol. 19, 6815–6824.

Oliver, G., Loosli, F., Koster, R., Wittbrodt, J., Gruss, P., 1996. Ectopic lens induction in fish in response to the murine homeobox gene Six3. Mech. Dev. 60, 233–239.

Oliver, G., Mailhos, A., Wehr, R., Copeland, N.G., Jenkins, N.A., Gruss, P., 1995. *Six3*, a murine homologue of the *sine oculis* gene, demarcates the most anterior border of the developing neural plate and is expressed during eye development. Development 121, 4045–4055.

O'Neill, P., Mak, S.S., Fritzsch, B., Ladher, R.K., Baker, C.V., 2012. The amniote paratympanic organ develops from a previously undiscovered sensory placode. Nat. Commun 3, 1041.

Ozaki, H., Watanabe, Y., Takahashi, K., Kitamura, K., Tanaka, A., Urase, K., Momoi, T., Sudo, K., Sakagami, J., Asano, M., Iwakura, Y., Kawakami, K., 2001. Six4, a putative myogenin gene regulator, is not essential for mouse embryonic development. Mol. Cell Biol. 21, 3343–3350.

Ozaki, H., Watanabe, Y., Ikeda, K., Kawakami, K., 2002. Impaired interactions between mouse Eya1 harboring mutations found in patients with brachio-oto-renal syndrome and Six, Dach, and G proteins. J. Hum. Genet. 47, 107–116.

Ozaki, H., Nakamura, K., Funahashi, J., Ikeda, K., Yamada, G., Tokano, H., Okamura, H.O., Kitamura, K., Muto, S., Kotaki, H., Sudo, K., Horai, R., Iwakura, Y., Kawakami, K., 2004. Six1 controls patterning of the mouse otic vesicle. Development 131, 551–562.

Pandur, P.D., Moody, S.A., 2000. *Xenopus* Six1 gene is expressed in neurogenic cranial placodes and maintained in the differentiating lateral lines. Mech. Dev. 96, 253–257.

Park, B.-Y., Saint-Jeannet, J.-P., 2008. Hindbrain-derived Wnt and Fgf signals cooperate to specify the otic placode in. *Xenopus*. Dev. Biol. 324, 108–121.

Park, B.Y., Saint-Jeannet, J.-P., 2010. Induction and segregation of the vertebrate cranial placodes. Colloquium Series on Developmental Biology 1 (3). Morgan & Claypool Publishers.

Patthey, C., Gunhaga, L., Edlund, T., 2008. Early development of the central and peripheral nervous systems is coordinated by Wnt and BMP signals. PLoS One 3. e1625.

Patthey, C., Edlund, T., Gunhaga, L., 2009. Wnt-regulated temporal control of BMP exposure directs the choice between neural plate border and epidermal fate. Development 136, 73–83.

Patrick, A.N., Schiemann, B.J., Yang, K., Zhao, R., Ford, H.L., 2009. Biochemical and functional characterization for six SIX1 Branchio-otic-renal syndrome mutations. J. Biol. Chem. 284, 20781–20790.

Patrick, A.N., Cabrera, J.H., Smaith, A.L., Chen, X.S., Ford, H.L., Zhao, R., 2013. Structure function analyses of the human SIX1-EYA2 complex reveal insights into metastasis and BOR Syndrome. Nature Struct. Mol. Biol. 20, 447–453.

Pauley, S., Lai, E., Fritzsch, B., 2006. Foxg1 is required for morphogenesis and histogenesis of the mammalian inner ear. Dev. Dyn. 235, 2470–2482.

Pera, E.M., De Robertis, E.M., 2000. A direct screen for secreted proteins in *Xenopus* embryos identifies distinct activities for the Wnt antagonists Crescent and Frzb-1. Mech. Dev. 96, 183–195.

Phillips, B.T., Bolding, K., Riley, B.B., 2001. Zebrafish fgf3 and fgf8 encode redundant functions required for otic placode induction. Dev. Biol. 235, 351–365.

Phillips, B.T., Kwon, H.J., Melton, C., Houghtaling, P., Fritz, A., Riley, B.B., 2006. Zebrafish msxB, msxC and msxE function together to refine the neural-nonneural border and regulate cranial placodes and neural crest development. Dev. Biol. 294, 376–390.

Pieper, M., Eagleson, G.W., Wosniok, W., Schlosser, G., 2011. Origin and segregation of cranial placodes in. *Xenopus laevis*. Dev. Biol. 360, 257–275.

Pieper, M., Ahrens, K., Rink, E., Peter, A., Schlosser, G., 2012. Differential distribution of competence for panplacodal and neural crest induction to non-neural and neural ectoderm. Development 139, 1175–1187.

Pignoni, F., Hu, B., Zavitz, K.H., Xiao, J., Garrity, P.A., Zipursky, S.L., 1997. The eye-specification proteins So and Eya form a complex and regulate multiple steps in *Drosophila* eye development. Cell 91, 881–891.

Platt, J.B., 1896. Ontogenetic differentiation of the ectoderm in necturus. II. On the development of the peripheral nervous system. Quart. J. Mic. Sci. 38, 485–547.

Pogoda, H.M., Hammerschmidt, M., 2007. Molecular genetics of pituitary development in zebrafish. Semin. Cell Dev. Biol. 18, 543–558.

Rada-Iglesias, A., Bajpai, R., Prescott, S., Brugmann, S.A., Swigut, T., Wysocka, J., 2012. Epigenomic annotation of enhancers predicts transcriptional regulators of human neural crest. Cell Stem Cell 11, 633–648.

Rayapureddi, J.P., Kattamuri, C., Steinmetz, B.D., Frankfort, B.J., Ostrin, E.J., Mardon, G., Hegde, R.S., 2003. Eyes absent represents a class of protein tyrosine phosphatases. Nature 426, 295–298.

Rickard, S., Parker, M., van't Hoff, W., Barnicoat, A., Russell-Eggitt, I., Winter, R.M., Bitner-Glindzicz, M., 2001. Oto-facial-cervical (OFC) syndrome is a contiguous gene deletion syndrome involving EYA1: molecular analysis confirms allelism with BOR syndrome and further narrows the Duane syndrome critical regions to 1cM. Hum. Genet. 108, 398–403.

Rodriguez-Soriano, J., 2003. Brachio-oto-renal syndrome. J. Nephrol. 16, 603–605.

Ruf, R.G., Xu, P.X., Silvius, D., Otto, E.A., Beekmann, F., Muerb, U.T., Kumar, S., Neuhaus, T.J., Kemper, M.J., Raymond, R.M.J., et al., 2004. SIX1 mutations cause branchio-oto-renal syndrome by disruption of EYA1-SIX1-DNA complexes. Proc. Natl. Acad. Sci. USA 101, 8090–8095.

Sabado, V., Barraud, P., Baker, C.V., Streit, A., 2012. Specification of GnRH-1 *neurons* by antagonistic FGF and retinoic acid signaling. Dev. Biol. 362, 254–262.

Sahly, I., Andermann, P., Petit, C., 1999. The zebrafish *eya1* gene and its expression pattern during embryogenesis. Dev. Genes Evol. 209, 399–410.

Saint-Jeannet, J.-P., Moody, S.A., 2014. Establishing the pre-piacodal region and breaking it into placodes with distinct identities. Dev. Biol. (under revision).

Sanggaard, K.M., Rendtorff, N.D., Kjaer, K.W., Eiberg, H., Johnsen, T., Gimsing, S., Dyrmose, J., Nielsen, K.O., Lage, K., Tranebjaerg, L., 2007. Branchio-otic-renal syndrome: detection of EYA1 and SIX1 mutations in five out of six Danish families by combining linkage, MLPA and sequencing analyses. Eur. J. Hum. Genet. 15, 1121–1131.

Sargent, T.D., 2006. Transcriptional regulation at the neural plate border. Adv. Exp. Med. Biol. 589, 32–44.

Sato, T., Sasai, N., Sasai, Y., 2005. Neural crest determination by co-activation of Pax3 and Zic1 genes in *Xenopus* ectoderm. Development 132, 2355–2363.

Sato, S., Ikeda, K., Ochi, H., Ogino, H., Yajima, H., Kawakami, K., 2010. Conserved expression of mouse Six1 in the pre-placodal region (PPR) and identification of an enhancer for the rostral PPR. Dev. Biol. 344, 158–171.

Sato, S., Ikeda, K., Shioi, G., Nakao, K., Yajima, H., Kawakami, K., 2012. Regulation of Six1 expression by evolutionarily conserved enhancers in tetrapods. Dev. Biol. 368, 95–108.

Schlosser, G., 2003. Hypobranchial placodes in *Xenopus laevis* give rise to hypobranchial ganglia, a novel type of cranial ganglia. Cell Tiss. Res. 312, 21–29.

Schlosser, G., 2005. Evolutionary origins of vertebrate placodes: insights from developmental studies and from comparisons with other deuterostomes. J. Exp. Zool. B. Dev. Evol. 304, 347–399.

Schlosser, G., 2006. Induction and specification of cranial placodes. Dev. Biol. 294, 303–351.

Schlosser, G., 2007. How old genes make a new head: redeployment of Six and Eya genes during the evolution of vertebrate cranial placodes. Integr. Comp. Biol. 47, 343–359.

Schlosser, G., 2008. Do vertebrate neural crest and cranial placodes have a common evolutionary origin? Bioessays 30, 659–672.

Schlosser, G., 2010. Making senses development of vertebrate cranial placodes. Int. Rev. Cell Mol. Biol. 283, 129–234.

Schlosser, G., Ahrens, K., 2004. Molecular anatomy of placode development in. *Xenopus laevis*. Dev. Biol. 271, 439–466.

Schlosser, G., Awtry, T., Brugmann, S.A., Jensen, E.D., Neilson, K., Ruan, G., Stammler, A., Voelker, D., Yan, B., Zhang, C., Klymkowsky, M.W., Moody, S.A., 2008. Eya1 and Six1 promote neurogenesis in the cranial placodes in a SoxB1-dependent fashion. Dev. Biol. 320, 199–214.

Schlosser, G., Northcutt, R.G., 2000. Development of neurogenic placodes in *Xenopus laevis*. J. Comp. Neurol 418, 121–146.

Schonberger, J., Wang, L., Shin, J.T., Kim, S.D., Depreux, F.F.S., Zhu, H., Zon, L., Pizard, A., Kim, J.B., MacRae, C.A., Mungall, A.J., Seidman, J.G., Seidman, C.E., 2005. Mutation in the transcriptional coactivator EYA4 causes dilated cardiomyopathy and sensorineural hearing loss. Nature Genetics 37, 418–422.

Self, M., Lsagutin, O.V., Bowling, B., Hendrix, J., Cal, Y., Dressler, G.R., Oliver, G., 2006. Six2 is required for suppression of nephrogenesis and progenitor renewal in the developing kidney. EMBO J 25, 5214–5228.

Selleck, M.A., Bronner-Fraser, M., 1995. Origins of the avian neural crest: The role of neural plate-epidermal interactions. Development 121, 525–538.

Selleck, M.A., Bronner-Fraser, M., 2000. Avian neural crest cell fate decisions: a diffusible signal mediates induction of neural crest by the ectoderm. Int. J. Dev. Neurosci. 18, 621–627.

Serikaku, M.A., O'Tousa, J.E., 1994. *Sine oculis* is a homeobox gene required for *Drosophila* visual system development. Genetics 138, 1137–1150.

Shaham, O., Menuchin, Y., Farhy, C., Ashery-Padan, R., 2012. Pax6: a multilevel regulator of ocular development. Prog. Retin. Eye Res. 31, 351–376.

Shiotsugu, J., Katsuyama, Y., Arima, K., Baxter, A., Koide, T., Song, J., Chandraratna, R.A., Blumberg, B., 2004. Multiple points of interaction between retinoic acid and FGF signaling during embryonic axis formation. Development 131, 2653–2667.

Silver, S.J., Davies, E.L., Doyon, L., Rebay, I., 2003. Functional dissection of eyes absent reveals new modes of regulation within the retinal determination gene network. Mol. Cell Biol. 23, 5989–5999.

Sjodal, M., Edlund, T., Gunhaga, L., 2007. Time of exposure to BMP signals plays a key role in the specification of the olfactory and lens placodes. ex vivo. Dev. Cell 13, 141–149.

Solomon, K.S., Kwak, S.J., Fritz, A., 2004. Genetic interactions underlying otic placode induction and formation. Dev. Dyn. 230, 419–433.

Solomon, K.S., Fritz, A., 2002. Concerted action of two dlx paralogs in sensory placode formation. Development 129, 3127–3136.

Soker, T., Dalke, C., Puk, O., Floss, T., Becker, L., Bolle, I., Favor, J., Hans, W., Holter, S.M., Horsch, M., Kallnik, M., Kling, E., Moerth, C., Schrewe, A., Stigloher, C., Topp, S., Gailus-Durner, V., Naton, B., Beckers, J., Fuchs, H., Ivandic, B., Klopstock, T., Schulz, H., Wolf, E., Wurst, W., Bally-Cuif, L., de Angelis, M.H., Graw, J., 2008. Pleio-tropic effects in Eya3 knockout mice. BMC Dev. Biol. 8, 118.

Spitz, F., Demignon, J., Porteu, A., Kahn, A., Concordet, J.P., Daegelen, D., Maire, P., 1998. Expression of myogenin during embryogenesis is controlled by Six/sine oculis homeoproteins through a conserved MEF3 binding site. Proc. Natl. Acad. Sci. U S A 95, 14220–14225.

Spruijt, L., Hoefsloot, L.H., van Schaijk, G.H., van Waardenburg, D., Kremer, B., Brackel, H.J., de Die-Smulders, C.E., 2006. Identification of a novel EYA1 mutation presenting in a newborn with laryngomalacia, glossoptosis, retrognathia, and pectus excavatum. Am. J. *Med. Genet.* A 140, 1343–1345.

Streit, A., 2002. Extensive cell movements accompany formation of the otic placode. Dev. Biol. 249, 237–254.

Streit, A., 2004. Early development of the cranial sensory nervous system: from a common field to individual placodes. Dev. Biol. 276, 1–15.

Streit, A., Stern, C., 1999. Establishment and maintenance of the border of the neural plate in the chick: involvement of FGF and BMP activity. Mech. Dev. 82, 51–66.

Steventon, B., Mayor, R., Streit, A., 2012. Mutual repression between Gbx2 and Otx2 in sensory placodes reveals a general mechanism for ectodermal patterning. Dev. Biol. 367, 55–65.

Takai, A., Inomata, H., Arakawa, A., Yakura, R., Matsuo-Takasaki, M., Sasai, Y., 2010. Anterior neural development requires Del1, a matrix-associated protein that attenuates canonical Wnt signaling via the Ror2 pathway. Development 137, 3293–3302.

Tessmar, K., Loosli, F., Wittbrodt, J., 2002. A screen for co-factors of Six3. Mech. Dev. 117, 103–113.

Theveneau, E., Steventon, B., Scarpa, E., Garcia, S., Trepat, X., Streit, A., Mayor, R., 2013. Chase-and-run between adjacent cell populations promotes directional collective migration. Nature Cell Biol. 15, 763–772.

Thiel, G., 2013. How Sox2 maintains neural stem cell identity. Biochem. J. 450, e1–e2.

Tootle, T.L., Silver, S.J., Davies, E.L., Newman, V., Latek, R.R., Mills, I.A., Selengut, J.D., Parlikar, B.E., Rebay, I., 2003. The transcription factor Eyes absent is a protein tyrosine phosphatase. Nature 426, 299–302.

Toro, S., Varga, Z.M., 2007. Equivalent progenitor cells in the zebrafish anterior pre-placodal field give rise to adenohypophysis, lens, and olfactory placodes. Semin. Cell Dev. Biol. 18, 534–542.

Tribulo, C., Ayba, M.J., Nguyen, V.H., Mullins, M.C., Mayor, R., 2003. Regulation of Msx genes by Bmp gradient is essential for neural crest specification. Development 130, 6441–6452.

von Kuffler, C., 1895. Studien zur vergleichenden Entwicklungsgeschichte des Kpofes des Cranioten. 3. Heft: Die Entwicklung der Kopfnerven von Ammocoetes planeri. Munchen.

Wang, L., Sewell, W.F., Kim, S.D., Shin, J.T., MacRae, C.A., Zon, L.I., Seidman, J.G., Seidman, C.E., 2008. Eya4 regulation of Na+/K+-ATPase is required for sensory system development in zebrafish. Development 135, 3425–3434.

Wayne, S., Robertson, N.G., DeClau, F., Chen, N., Verhoeven, K., Prasad, S., Tranebjarg, L., Morton, C.C., Ryan, A.F., Van Camp, G., Smith, R.J., 2001. Mutations in the transcriptional activator Eya4 cause late-onset deafness at the DFN10 locus. Hum. Mol. Genet. 10, 195–200.

Webb, J.F., Noden, D.M., 1993. Ectodermal placodes: Contributions to the development of the vertebrate head. Am. Zool 33, 434–447.

Wegner, M., 2013. SOX after SOX: SOXession regulates neurogenesis. Genes Dev. 25, 2423–2428.

Woda, J.M., Pastagia, J., Mercola, M., Artinger, K.B., 2003. Dlx proteins position the neural plate border and determine adjacent cell fates. Development 130, 331–342.

Wong, E.Y.M., Ahmed, M., Xu, P.-X., 2013. EYA1-SIX1 complex in neurosensory cell fate induction in the mammalian inner ear. Hearing Research 297, 13–19.

Xu, P.X., Cheng, J., Epstein, J.A., Maas, R.L., 1997. Mouse Eya genes are expressed during limb tendon development and encode a transcriptional activation function. Proc. Natl. Acad. Sci. USA 94, 11974–11979.

Xu, P.X., Adams, J., Peters, H., Brown, M.C., Heaney, S., Maas, R., 1999. Eya1-deficient mice lack ears and kidneys and show abnormal apoptosis of organ primordia. Nat. Genet. 23, 113–117.

Xu, P.X., Zheng, W., Laclef, C., Maire, P., Maas, R.L., Peters, H., Xu, X., 2002. Eya1 is required for the morphogenesis of mammalian thymus, parathyroid and thyroid. Development 129, 3033–3044.

Xu, P.X., Zheng, W., Huang, L., Maire, P., Laclef, C., Silvius, D., 2003. Six1 is required for the early organogenesis of mammalian kidney. Development 130, 3085–3094.

Xu, H., Dude, C.M., Baker, C.V., 2008. Fine-grained fate maps for the ophthalmic and maxillomandibular trigeminal placodes in the chick embryo. Dev. Biol. 317, 174–186.

Zhang, H., Hu, G., Wang, H., Sciavolino, P., Iler, N., Shen, M.M., Abate-Shen, C., 1997. Heterodimerization of Msx and Dlx homeoproteins results in functional antagonism. Mol. Cell Biol. 17, 2920–2932.

Zhang, Y., Luo, T., Sargent, T.D., 2006. Expression of TFAP2beta and TFAP2gamma genes in *Xenopus laevis*. Gene Expr. Patterns 6, 589–595.

Zheng, W., Huang, L., Wei, Z.B., Silvius, D., Tang, B., Xu, P.X., 2003. The role of Six1 in mammalian auditory system development. Development 130, 3989–4000.

Zhu, C.C., Dyer, M.A., Uchikawa, M., Kondoh, H., Lagutin, O.V., Oliver, G., 2002. Six3-mediated auto repression and eye development requires its interaction with members of the Groucho-related family of co-repressors. Development 129, 2835–2849.

Zou, D., Silvius, D., Fritzsch, B., Xu, P.X., 2004. Eya1 and Six1 are essential for early steps of sensory neurogenesis in mammalian cranial placodes. Development 131, 5561–5572.

Zou, D., Silvius, D., Davenport, J., Grifone, R., Maire, P., Xu, P.X., 2006a. Patterning of the third pharyngeal pouch into thymus/parathyroid by Six and Eya1. Dev. Biol. 293, 499–512.

Zou, D., Silvius, D., Rodrigo-Blomqvist, S., Enerback, S., Xu, P.X., 2006b. Eya1 regulates the growth of otic epithelium and interacts with Pax2 during the development of all sensory areas of the inner ear. Dev. Biol. 298, 430–441.

Chapter 20

Building the Olfactory System

Morphogenesis and Stem Cell Specification in the Olfactory Epithelium and Olfactory Bulb

Anthony-Samuel LaMantia

GW Institute for Neuroscience, Department of Pharmacology and Physiology, The George Washington University School of Medicine and Health Sciences, Washington DC

Chapter Outline

GLOSSARY

Glomeruli A specialized "globe" of neuropil that consists of dendrites from olfactory bulb mitral cells, dendrites from olfactory bulb granule cells, axon endings and synaptic boutons from olfactory receptor neurons, and axonal and dendritic processes from periglomerular cells, whose cell bodies define the limits of each glomerulus.

Gonadotropin Releasing Hormone (GnRH) Also know as luteinizing releasing hormone. This peptide hormone is synthesized by a small number of cells in the hypothalamus that are generated initially in the olfactory epithelium. They migrate subsequently to the hypothalamus. GnRH is released in pulses from the hypothalamic neurons that produce it, in both males and females. It regulates the secretion of follicle stimulating hormone (FSH) and luteinizing hormone (LH), which are essential for generation of gametes in both females and males.

Neuropil A collection of axons, dendrites, synaptic specialization, and supporting glial processes. Cell bodies are not included in the neuropil. Most synapses in invertebrate as well as vertebrate brains are made in the neuropil.

Neurosphere An *in vitro*, usually clonally derived, accumulation of proliferative neuronal precursors grown from primary stem or progenitor cells taken from the central or peripheral nervous system. Primary neurosphere cultures rely upon selection by beta fibroblast growth factor (bFGF) and epidermal growth factor (EGF) to stabilize and propagate dividing cells from the primary neural tissue. Once established, primary neurospheres can be dissociated and these dissociated cells will go onto generate additional secondary neurospheres.

Retinoic acid Abbreviated as RA, it is the acidic form of vitamin A (retinol), usually synthesized from dietary intake of vitamin A. RA is a member of the steroid/thyroid family of lipid-soluble ligands that activate receptor transcription factors to directly control gene expression. RA is a major morphogenetic signal – one of the first to be discovered – as well as a regulator of adult regeneration and tissue repair.

Semaphorin3a A cell surface adhesion molecule, one of a family of semaphorins. The semaphorins are considered to be primarily anti-adhesion or repulsive molecules that cause growing axons to stop and turn in an opposite direction to the location of the semaphorin. Semaphorin receptors include two families of molecules, the plexins, and the neuropilins.

SUMMARY

- Olfactory system development is essential for vertebrates to recognize edible food, distinguish conspecifics or benign species from competitors or predators, and reproduce successfully.

- The mechanisms that underlie the development of this sensory system are remarkably similar to those that mediate differentiation of bilaterally symmetric appendages that emerge from the vertebrate midline during embryogenesis: the limbs, the face and jaws, and aortic arches of the heart. Thus, olfactory morphogenesis – for both peripheral and central nervous system components – engages signaling pathways and transcriptional regulators that influence the development of appendages that are key either for interaction with the environment (limbs, jaw, and face), or homeostasis and survival (aortic arches).
- Subsequent differentiation of the olfactory pathway as a neural rather than musculo-skeletal or vascular structure reflects local modifications of signaling that leads to specification of distinct olfactory epithelial (OE) stem and progenitor cell populations, as well as parallel specification of forebrain structures that receive and process peripheral olfactory sensory information.
- The functions of some molecular regulators, and consequences of mutations in key genes, are similar to those seen in the limb, the face, and the heart. Others may be unique to the developing olfactory pathway or deployed in a quantitatively distinctive fashion. The concerted action of these genes results in coordinated development of peripheral olfactory receptor neurons and target regions in the central nervous system that represent, relay, and process olfactory information.

20.1 INTRODUCTION

The initial specification and subsequent differentiation of the olfactory system – which variably includes the "main" olfactory pathway (Figure 20.1) and the vomeronasal component of the pathway – is a critical event for all vertebrates, on a par with gastrulation at an earlier point in development. The reason for this is simple: the olfactory pathway ensures each individual's access to food, friends, and sex. Without the appropriate initial differentiation of the olfactory placode in the periphery and the subsequent development of the central olfactory pathway to receive, represent, and process olfactory information, feeding, social, and reproductive behaviors are disabled or impossible (Stowers et al., 2002, 2013; Ferrero and Liberles, 2010; Kavaliers et al., 2005). Initial feeding and maternal/offspring interactions are disrupted if olfactory development is compromised (Levy and Keller, 2009). Recognition of conspecifics versus predators is disordered (Apfelbach et al., 2005; Dewan et al., 2013). Reproduction is impossible if Gonadotropin Releasing Hormone (GnRH) secretory neurons are not specified in the olfactory placode during the early stages of cranial morphogenesis, prior to their migration to the hypothalamus (Bouligand et al., 2010). The repercussions of failed olfactory development (or maintenance) are seen not only in

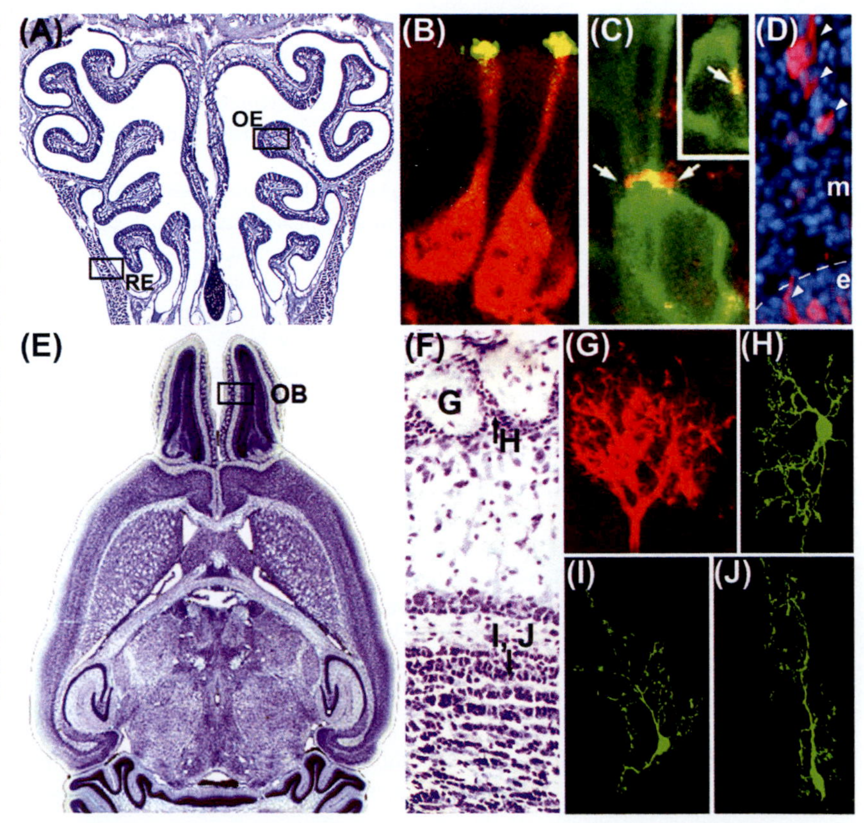

FIGURE 20.1 Principal Structures and Cell Classes in the Mature Vertebrate Olfactory Pathway. (A) A coronal section through the nasal cavity of the adult mouse. The box indicates the olfactory epithelium (OE), which is far thicker than the respiratory epithelium (RE) that is also found in the nasal cavity. (B) The olfactory receptor neuron (ORN), the primary peripheral chemosensory neuron, immunostained for the olfactory marker protein (red) and adenylcyclase III (green), which is limited to the cilia where odorant transduction occurs. (C) Vomeronasal receptor neurons (VRN) stained here with OMP (green) and Trpc3 (a TRP channel found selectively on VRNs). (D) GnRH neurons in the E11.5 OE (e) exit the epithelium (arrowhead) and enter the mesenchyme, where they migrate toward the forebrain. (E) The olfactory bulb (OB) is seen at the anterior end of a horizontal section through the forebrain. The box indicates the area shown in panel F. (F) Cellular organization of the OB show the glomeruli (G), accumulations of dendrites from mitral cells, the OB projection neurons. (G) A single mitral cell dendrite, found within an OB glomerulus. (H) A periglomerular interneuron, found in approximately the position indicated the arrow and (H). (I, J) Two examples of granule cells found in the granule cell layer of the OB.

animals which are thought to depend heavily on olfaction as a primary sensory interface with the environment, but also in humans as part of the spectrum of deficits that define disorders including autism (Schecklmann et al., 2013), schizophrenia (Rupp, 2010), Parkinson's disease (Doty, 2012), and Alzheimer's disease (Sun et al., 2012). Pathological changes in these disorders are seen both in the peripheral olfactory epithelium (OE; Figure 20.1) as well as the primary central relay region for the olfactory pathway, the olfactory bulb (OB; Figure 20.1). Thus, there can be little doubt of the critical nature of olfactory development for establishing the capacity of any vertebrate – including humans – to survive and thrive in its environment.

The distinctive developmental mechanisms that underlie the specification, cellular differentiation, assembly, and maintenance of the olfactory pathway reflect the unique nature of olfactory sensation. Unlike all other special sensory systems – vision, where rods and cones transduce, and ganglion cells relay, or audition, where hair cells transduce, and sensory neurons of the eighth nerve relay – transduction and central relay of olfactory stimuli are accomplished by a single cell class. The olfactory receptor neurons (ORN) in the OE are both transducers and relay neurons for primary stimuli (Figure 20.1). The axons of ORNs project directly from the OE to OB. The monoallelic expression of single odorant receptors from a family of approximately 500–2000, depending on the species, seven transmembrane G-protein coupled odorant receptors found in most vertebrate genomes (Sullivan et al., 1995; Reed, 2004) distinguishes ORNs as perhaps the most molecularly diverse class of vertebrate sensory receptors. This receptor diversity is reflected in the organization of the axonal projections from the OE to the OB (Sakano, 2010). In the OB, axons from ORNs that express single odorant receptor genes segregate into small subsets of glomeruli (Figure 20.1), the specialized accumulation of dendritic neuropil that receives direct ORN input. Thus, olfactory development must be regulated to ensure maximal diversity as well as remarkable specificity of the final projections and circuits that ensure appropriate olfactory information processing.

Finally, the unique exposure of the ORNs directly to the environment – these nerve cells are only protected from the outside world by a layer of mucus – is likely responsible for another remarkable distinction of vertebrate olfactory pathway. Development is not limited to embryogenesis, but is recapitulated throughout life. ORNs degenerate and die, and then are replaced by the division of stem or progenitor cells that retain the capacity to generate new ORNs in the adult sensory epithelium (Schwob, 2002). A parallel process is also seen in the OB, where variable numbers of periglomerular and granule cells are replaced throughout much of life. These OB interneurons are generated in a distal region of the adult brain – the forebrain anterior subventricular zone (SVZ) and adjacent ventricular domains (Lledo et al., 2008; Luskin, 1993). They then migrate to the OB via the rostral migratory stream, a glial-rich scaffold that spans the territory between the forebrain lateral ventricle and the OB (Ghashghaei et al., 2007; Lois and Alvarez-Buylla, 1994). Thus, the developmental imperatives for construction, and ongoing reconstruction of the olfactory pathway are significant. These mechanisms must be capable of specifying diverse initial cell lineages and final cell identity, and then recapitulating these events to replace neurons in the olfactory pathway over an individual's lifetime.

20.2 THE INITIAL DEVELOPMENT OF THE OLFACTORY PATHWAY

The key components of the vertebrate olfactory pathway, the OE and OB, develop from the most anterior portion of the head (Figure 20.2). In most vertebrates, the OE apparently differentiates from the surface ectoderm, just anterior and lateral to the margin of the anterior neural plate. In the mouse, the OE can be recognized on the ninth day of gestation (Embryonic day E9) as a local thickening in the lateral surface ectoderm of the head. This thickened epithelium, plus the underlying mesenchyme, constitutes the olfactory placode, a transient structure that prefigures the subsequent differentiation of the OE. The initial derivation of the surface ectoderm that becomes the OE remains uncertain. Early embryological work indicated that the OE is derived from a distinct anterior domain of the surface ectoderm of the head, albeit with potential interactions between adjacent germ layers (Jacobson AG 1963a,b). It is possible that this ectoderm is established by a planar migration of cells from the anterior-most margin of the neural plate (Couly and Le Douarin, 1985), the splitting of a common field shared with the lens placode (Bhattacharyya et al., 2004), or a secondary migration of neural crest that provides additional neural precursors to the presumptive OE (Forni et al., 2011). Nevertheless, the OE shares molecular characteristics with other cranial placodes, including expression and functional activity of *Pax*, *Dlx* and *Six* genes (Ikeda et al., 2007; Bhattacharyya and Bronner-Fraser, 2008; Nomura et al., 2007). Thus, the available evidence supports the conclusion that the presumptive OE is initially established as a placodal domain in the anterior-most portion of the vertebrate head quite early in development (also see Chapter 19).

Within another 24 hours in the mouse, the olfactory placode has undergone a dramatic morphogenetic change. The nascent OE is identifiable as a distinct invagination of the surface epithelium, and the entire placode has differentiated into a nasal prominence with lateral and medial nasal processes (Figure 20.2) that will eventually constitute the lateral part of the nares (lateral) and the fused nasal septum (medial). Thus, two tissues, the surface ectoderm and the underlying mesenchyme seem to be necessary for the initial morphogenesis of the OE. Subsequently, it will undergo further morphogenetic

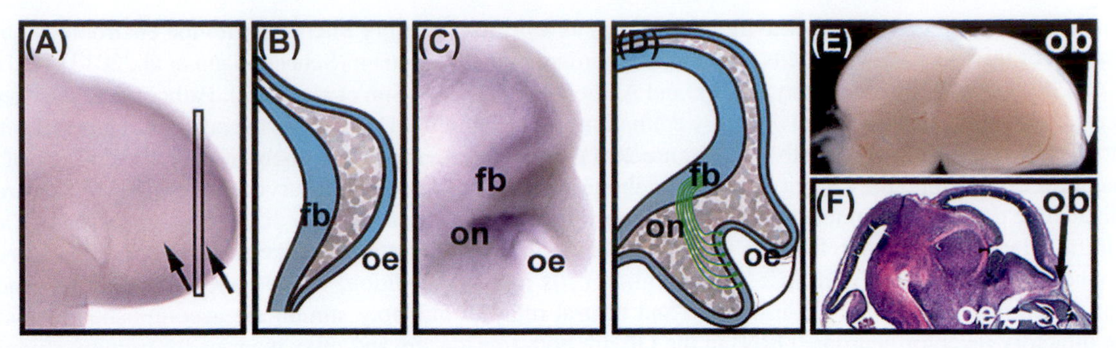

FIGURE 20.2 **The Key Components of the Vertebrate Olfactory Pathway: The OE and OB, Develop from the Most Anterior Portion of the Head.** (A) The head, viewed from the side, of an embryonic day (E) 9.0 mouse embryo. The arrows and box indicate structures seen in a cartoon coronal section through the anterior pole of the head, shown in panel B. (B) A cartoon coronal section though the E9.0 head showing the three tissue compartments that interact to establish the olfactory pathway. The forebrain neuroepithelium (fb), the presumptive olfactory epithelium on the surface of the head (oe), and the frontonasal mesenchyme (gray stipple) which is adjacent to both epithelia. (C) A front view of the E10.5 mouse head, labeled immunohistochemically for NCAM (dark purple label), which is expressed in the axons of ORNs as they grow toward the forebrain. (D) A cartoon showing the morphogenesis of the olfactory epithelium (oe), the olfactory nerve (on). (E) Lateral view of the mouse brain at E14.5 showing the olfactory bulb (ob) at the anterior pole of the forebrain, ventral to the cerebral cortex. (F) A sagittal section through the E14.5 head, stained with hematoxylin and eosin shows the location of the olfactory bulb (ob), and the olfactory epithelium (oe).

changes, including an initial medial evagination of the epithelium that becomes a distinct sensory domain in many, but not all, vertebrates – the vomeronasal organ (VNO; humans and most other primates do not have a recognizable VNO). Sensory neurons in the VNO express distinct receptors, and are thought to be essential for recognizing pheromonal and other "social" chemical signals, including predators and competitors in the environment (Isogai et al., 2011). The region of the embryonic OE from which the VNO differentiates is also the location where GnRH neurons are generated (Wray et al., 1989; Schwanzel-Fukuda and Pfaff, 1989). These cells then delaminate from the epithelium and migrate through the mesenchyme between the developing OE and the forebrain. They then enter the ventral forebrain, and eventually reach the hypothalamus. Thus, OE morphogenesis results in the differentiation of a broad surface of ORNs arrayed across the nasal cavity, the development of the VNO in many, but not all, vertebrates, and the genesis of GnRH neurons in the OE and their subsequent migration to the hypothalamus.

The OB emerges, apparently, from the antero-ventral/lateral region of the telencephalic vesicle. Nevertheless, its morphogenesis relies upon an additional mechanism. The mitral cells, the primary projection neurons of the OB (see Figure 20.1) are generated from the anterolateral telencephalon (Blanchart et al., 2006; Lopez-Mascaraque and de Castro, 2002; Jimenez et al., 2000). These cells are the earliest generated neurons of the OB (Hinds, 1968), and are among the first populations of post-mitotic neurons in the entire telencephalon. In contrast, most OB interneurons are generated at a distinct location in the ventro-medial telencephalon, the lateral ganglionic eminence (LGE; Figure 20.2). Interneuron progenitors are specified in the LGE, and they undergo terminal neurogenic cell divisions at this site (Wichterle et al., 2001; Stenman et al., 2003). The post-mitotic interneurons then become migratory, and travel via the embryological equivalent of the rostral migratory stream (RMS) (Tucker et al., 2006) into the OB. This event – the distal genesis and subsequent migration of interneurons into the OB – is essential for OB morphogenesis. The addition of large numbers of interneurons drives the evagination of the OB from the antero-lateral/ventral region of the telencephalon. Unlike mitral cell genesis, which begins early in gestation and is completed rapidly, OB interneuron genesis begins slightly later during embryogenesis and continues throughout much of early postnatal life (Batista-Brito et al., 2008; Lemasson et al., 2005). The initial genesis of OB interneurons eventually yields to an apparent ongoing replacement of subsets of interneurons generated in the anterior SVZ that migrate to the OB via the RMS. Thus, the morphogenesis and cytogenesis of the OB relies upon progenitor specification and subsequent neurogenesis in two distinct domains in the developing forebrain. One, in the antero-ventral/lateral forebrain, gives rise to mitral cells, OB projection neurons. The other, in the medial ventral forebrain, gives rise to most OB interneurons that subsequently migrate into the nascent OB to drive morphogenesis.

20.3 OE INDUCTION: NON-AXIAL MESENCHYMAL/EPITHELIAL INTERACTIONS DRIVE DIFFERENTIATION

Key structures in the olfactory pathway, like most other structures in the embryo, rely upon local cell-cell interactions – generally referred to as induction – in apparently undifferentiated anlagen to establish molecular patterning, and to drive subsequent morphogenesis. Until the early 1990s there was very little appreciation of whether inductive interactions were essential for olfactory development. Earlier embryological studies suggested that there was significant capacity for local morphogenetic signaling from olfactory structures. In the 1930s, Balinsky (referenced in Slack, 1995) showed that transplantation of the olfactory placode beneath the flank ectoderm resulted in a supernumerary limb. This observation indicated that the morphogenetic signaling capacity of the olfactory placode must be similar to that of tissues that constitute the limb bud; however, the cellular or molecular foundations of this observation remained obscure for an additional 60 years. In the intervening period, Graziadei (Stout and Graziadei, 1980; Graziadei and Monti-Graziadei, 1992; Dryer and Graziadei, 1994) did parallel experiments to assess the "organizing" capacity of the olfactory placode for its central neural targets: transplantation of the olfactory placode to sites adjacent to the neural tube. The transplanted placode differentiated into an OE and "induced" a local OB-like structure innervated by ORNs in the ectopic OE. Apparently, the rudimentary tissues of the olfactory placode (surface epithelium and underlying mesenchyme) had the capacity to maintain OE differentiation, as well as to organize local morphogenesis and subsequent differentiation of a target structure in the central nervous system.

These intriguing observations remained unexplained until the 1990s. The definition of molecular signals that underlie the tissue-tissue interactions that define morphogenetic induction generally (Slack, 1993; Ingham and Placzek, 2006; Kelly and Melton, 1995), as well as new tools to identify and characterize where these signals act in an embryo, reinvigorated the analysis of olfactory morphogenetic interactions. *In vitro* and *in vivo* observations implicated at least two families of inductive signaling molecules, the fibroblast growth factors (FGFs; Kawauchi et al., 2005; Lassiter et al., 2013; Sabado et al., 2012; Murdoch and Roskams, 2013; Hebert et al., 2003; Falardeau et al., 2008; DeHamer et al., 1994) and the bone morphogenetic proteins (BMPs; Shou et al., 2000; Maier et al., 2010; Peretto P et al., 2002) in ORN specification. Similarly, inactivation of the gene encoding Sonic hedgehog (Shh) in mice resulted in dramatically disrupted nasal morphogenesis (Chiang C et al., 1996) accompanied by altered differentiation of the olfactory nerve (Balmer and LaMantia, 2004). In parallel, observations made in teratogenized rodents (Shenefelt, 2010) suggested that retinoic acid (RA), another key inductive signaling molecule (Chambon, 1994; Evans 1994; Thaller and Eichele, 1987; Balkan et al., 1992), might play an essential role in guiding olfactory pathway early development. Shenefelt (1972) found that RA exposure prior to initial morphogenesis of olfactory structures in rodent embryos resulted in their absence at later stages of development. This observation of exogenous RA's teratogenic effects suggested that endogenous RA might influence the morphogenesis of both the OE and the OB. We used RA transcriptional indicator transgenic mice (Balkan et al., 1992) to determine whether there was endogenous RA signaling in the surface epithelium and forebrain neuroepithelium that accompanied the morphogenesis and cellular differentiation of the OE and the OB (LaMantia et al., 1993). We found two distinct domains of RA activated reporter gene expression: one in the ventrolateral forebrain, and a second in the surface ectoderm that defines the olfactory placode. This RA signaling apparently was mediated by a local source of the ligand in the mesenchyme interposed between the surface ectoderm that becomes the OE, and the ventral forebrain domain that gives rise to the OB (Figure 20.3, see also Figure 20.5; LaMantia et al., 1993).

The identification of RA and other signals in and around early olfactory structures raised an intriguing parallel – implied by the work of Balinsky in the 1930s, but then left unexplored for an additional 60 years. The organizing capacity of the tissues that constitute the olfactory placode was likely parallel to that of the tissues that constitute the limb bud. The molecular mechanisms that defined limb bud organizing interactions were understood in outline by the mid to late 1990s. FGFs, BMPs, Shh, and RA had been defined as "cardinal" signals for establishing inductive interactions and morphogenetic axes in the vertebrate limb bud, and parallel cellular and molecular mechanisms also seemed to operate in the craniofacial primordia (Figure 20.3; Thesleff et al., 1991; Neubuser et al., 1997; Capdevila and Izpisua Belmonte, 2001; see Chapter 29 and Chapter 35). Based upon Balinsky's classic observation, as well as the embryological experiments of Graziadei, it seemed possible that the initial morphogenesis of the olfactory pathway might rely upon cellular mechanisms and molecular signals similar to those used for induction and morphogenesis for a number of bilaterally symmetric, non-axial structures along the embryonic anterior-posterior (A-P) axis. A key aspect of morphogenesis and stem cell specification in the limb buds, aortic arches, and craniofacial primordial (branchial arches/maxillary processes) is the apposition of discrete tissue compartments and obligate interactions between them that drive morphogenesis – a process referred to as secondary induction. Thus, distinct accumulations of *mesenchymal* cells adjacent to specialized *epithelial* zones – the best known of these include the apical ectodermal ridge in the limb, and local "organizers" in face (Duboc and Logan, 2009; Fernandez-Teran et al., 1999; Richman and Tickle, 1989, 1992; Wedden, 1987) constitute local morphogenetic fields in the limbs and the craniofacial primordia. It therefore seemed possible that mesenchymal/epithelial (M/E) interactions mediated by the same molecular signals might guide initial olfactory pathway development.

We sought to determine whether M/E interactions mediated by cardinal inductive signals were essential for olfactory morphogenesis and cellular specification. We first asked whether interactions between mesenchymal and epithelial tissues of the olfactory placode were necessary for OE differentiation and the formation of an olfactory nerve (ON). We found that separation of the two compartments at E9.0 of mouse development, prior to initial morphogenesis and neurogenesis in the

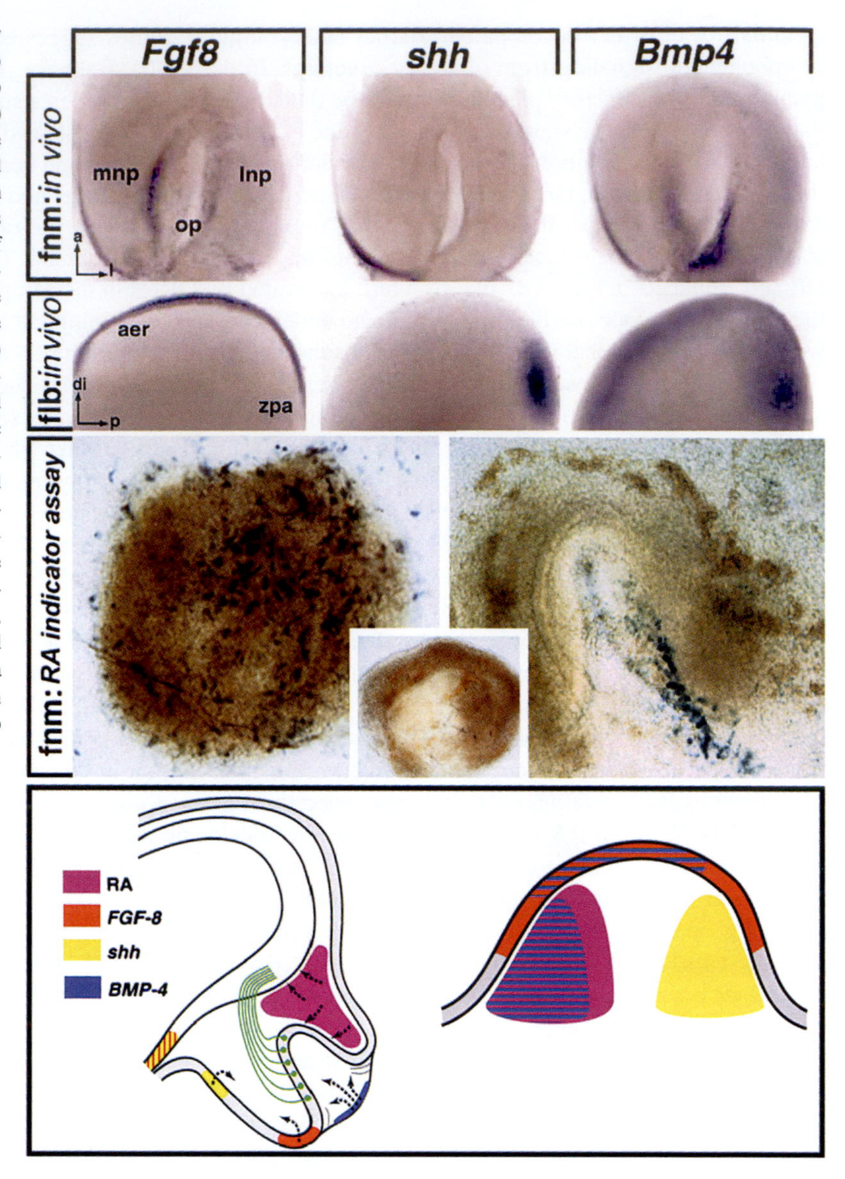

FIGURE 20.3 Inductive Signals in the Olfactory Primordia and Their Similarity to those in the Limb (from LaMantia 2000, Bhasin et al., 2003). The top row shows isolated nasal processes (see Figure 20.1C) after *in situ* hybridization for *Fgf8, shh, and Bmp4.* The medial nasal process (mnp) is found at the midline and will fuse with its counterpart on the opposite side to form the nasal septum. The lateral nasal process (lnp) becomes the nares. *Fgf8* is seen primarily in the medial aspect of the invaginating OE. *Shh* is seen in the antero-medial epithelium, and Bmp4 is seen in the posterior aspect. The second row shows similar limited expression of the same three signals in epithelial (apical ectodermal ridge:aer) or epithelial and mesenchymal (zone of polarizing activity: zpa) "organizer" domains in the limb bud. The third row shows an "indicator" assay for production of retinoic acid. A live frontonasal explant has been placed, mesenchyme side down, on a monolayer of clonally derived RA-responsive cells. At left, the E10.5 frontonasal mesenchyme provides RA that activates RA-responsive indicator cells (blue). The inset shows the result of the same explant experiment when the explant is placed epithelial side down. At right, the production of RA at E11.5, based upon activation of RA indicator cells, is localized to the lateral nasal process. The bottom panel shows a comparison of cardinal signals and their localization in the frontonasal process and forebrain (left) and the limb bud (right).

olfactory placode, resulted in a failure of OE differentiation *in vitro* (LaMantia et al., 2000). In contrast, when olfactory placodal epithelium and frontonasal mesenchyme are apposed to one another, a fully patterned OE with a coherent olfactory nerve emerges (Figure 20.4). Accordingly, we next asked whether the mesenchymal and epithelial tissues that constitute the nascent olfactory placode during early gestation in the mouse (E9) interact via cardinal morphogenetic signals (RA, Shh, BMP, FGF) to facilitate olfactory pathway morphogenesis and OE neurogenesis. Our results were quite clear: these four signals, available from distinct zones in the OE and frontonasal mesenchyme influence axial patterning and neuronal differentiation in the OE (LaMantia et al., 2000; Figure 20.4). Thus, there is a shared M/E inductive mechanism that guides the bilateral morphogenesis of appendages from the midline body axis in the trunk and the face. In the trunk this mechanism results in musculo-skeletal differentiation, in the face it results in odontogenesis, cartilage, and bone differentiation, and in the nose it results in the genesis and differentiation of ORNs and their capacity to grow into the OB.

The central role of M/E interactions raises a key question: how does each compartment acquire the specificity necessary for successful induction of olfactory structures? We recognized that the mesenchymal compartment in the OE primordium has a unique cellular identity: it is derived nearly entirely from the neural crest (Serbedzija et al., 1992; Osumi-Yamashita et al., 1994). Indeed, as in cardiac and craniofacial primordia (Keyte and Hutson, 2012; Cordero et al., 2011; Trainor, 2010; Johnston and Bronsky, 1995), when neural crest migration or specification is disrupted, olfactory morphogenesis fails (Figure 20.5). In the olfactory pathway, this failure reflects, in part, the normal localization of RA production to the neural crest-derived mesenchyme adjacent to the olfactory placodal ectoderm (LaMantia et al., 1993;

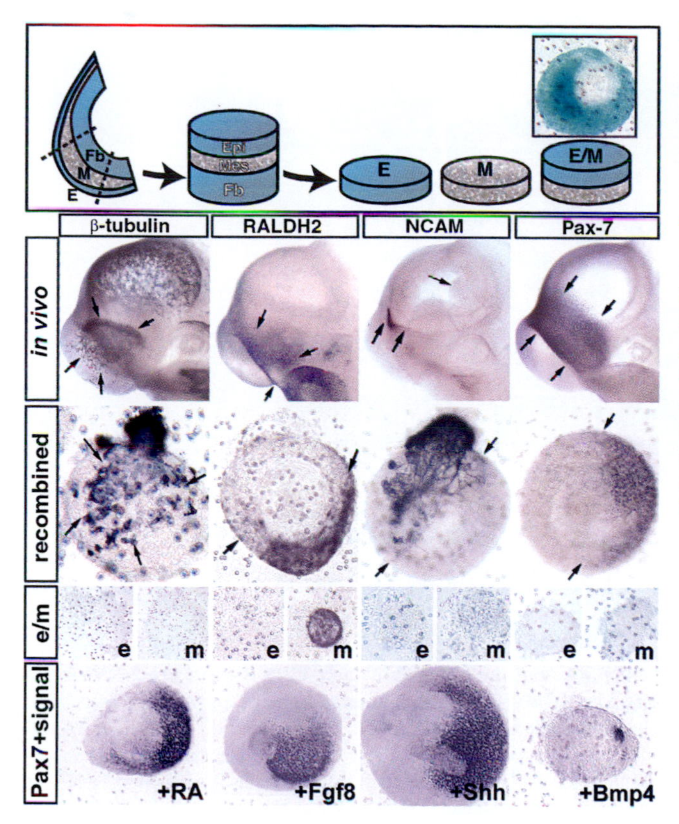

FIGURE 20.4 M/E Induction Drives OE and Olfactory Nerve (ON) Development. The top row shows the *in vitro* mesenchymal/epithelial explant culture system used to assess the role of M/E interactions in OE and ON development. E9.0 embryos are used, and the time in culture is approximately 24 hours. The inset shows that when M (turquoise; ROSA26-labeled) and E (gray; no label) tissues are recombined, there is no mixing of cells between the two compartments. The second row shows the expression of several markers of OE neurogenesis (Β-tubulin, NCAM) or mesenchymal patterning (RALDH2, Pax7) in the olfactory primordia *in vivo*. The third row shows expression of the same markers shown in the second row in the recombined M/E explants. All markers are expressed in patterns that approximate the *in vivo* distribution. None of these markers are expressed at E9.0, when the explant culture is established. The fourth row shows that with the exception of Raldh2, none of the other markers are expressed when epithelial (e) and mesenchymal (m) tissues are separated and cultured. The fifth row shows that when the cardinal signals that define lateral, medial and posterior axes are enhanced by bath application, the lateral mesenchymal marker Pax7 (see third row, far right for control) is shifted in expected ways based upon the normal axial position of the signaling molecule.

2000; Anchan et al., 1997; Bhasin et al., 2003; Figure 20.5). Indeed, RA production by neural crest is seen not only in the developing olfactory placode. RA is also produced by the neural crest (Niederreither et al., 2003) in the developing face, heart, and to a lesser extent, limbs (Schneider et al., 2001; Sandell et al., 2007). Thus, when the neural crest is disrupted or absent, RA signals are no longer available to drive initial induction and differentiation, and olfactory pathway morphogenesis is compromised (LaMantia et al., 2000; Anchan et al., 1997; LaMantia, 1999). Together, these observations indicate that signals from the neural crest derived frontonasal mesenchyme, as well as presumptive olfactory epithelium, are key for olfactory pathway morphogenesis, parallel to essential roles for mesenchyme and epithelia in limb, heart, and craniofacial development.

20.4 OB INDUCTION: PARALLEL M/E REGULATION OF INITIAL MORPHOGENESIS

Much of the initial work on M/E induction in the olfactory pathway focused on the consequences of this process for OE and olfactory nerve development. Nevertheless, the available evidence indicates that M/E inductive interactions between the frontonasal mesenchyme and the forebrain neuroepithelium are essential for OB differentiation. Early evidence for the role of M/E signaling in OB morphogenesis came from observations of the consequences of two well-studied morphogenetic mutations: *Pax6* (the *small eye* or *Sey* mouse; Grindley et al., 1995) and *Gli3* (the *extra toes* or Xt^j mouse; Hui and Joyner, 1993; Maynard et al., 2002). Initial characterization of both mutants showed that the OB was not recognizable in either mutant from initial morphogenesis through late gestation when homozygotes of both genotypes die (although an apparent ectopic remnant remains in the *Sey/Sey* mouse; Jimenez et al., 2000). We showed that in both *Sey/Sey* (Figure 20.5) and Xt^j/Xt^j mice, RA signaling from the mesenchyme to the forebrain is absent, prior to either disrupted OE or ON differentiation as well as OB morphogenetic failure (Maynard et al., 2002; Anchan et al., 1997; Balmer and LaMantia, 2004). In the *Sey/Sey* mouse, disruption of M/E mediated RA signaling, and presumably failure of OB differentiation, is due to the absence of RA producing neural crest derived mesenchyme. In contrast, the disruption in the *Xtj/Xtj* mouse is due to the lack of RA responsiveness of the forebrain, but not the olfactory periphery. Accordingly, in the *Xtj/Xtj* mouse, the OE develops normally, but the OB is absent. Several additional mutations, either in transcription factors (Long et al., 2003; Shaker et al., 2012; Ragancokova et al. 2014; Laub et al., 2005; Yoshihara et al., 2005; Bishop et al., 2003; Yoshida et al., 1997), some morphogenetic signals themselves (Shh; Balmer and LaMantia, 2004) or their receptors (FGFr; Hebert et al., 2003) compromise initial OB development. The relationship of these genes to M/E signaling-dependent OB morphogenesis versus

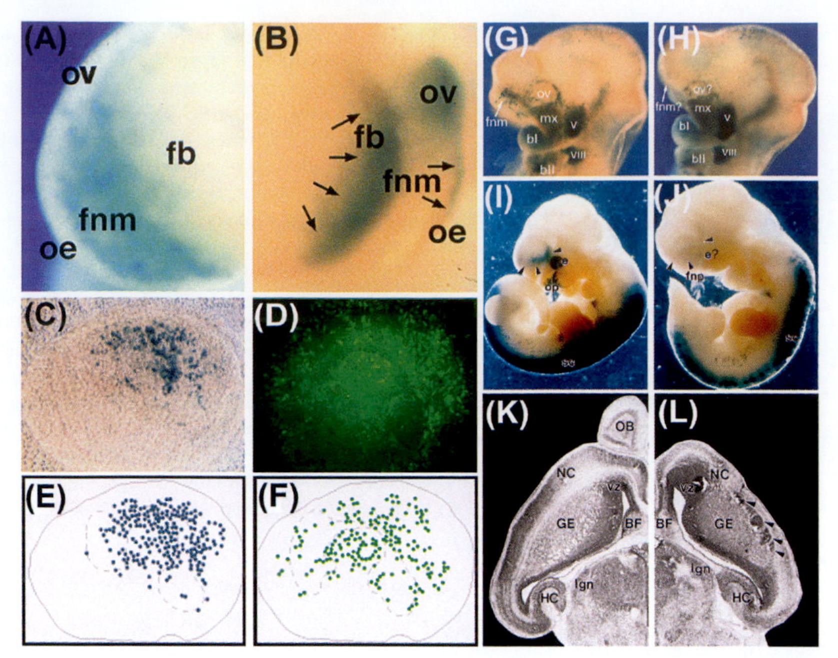

FIGURE 20.5 Neural Crest Cells in the Lateral Nasal Mesenchyme Produce RA, and the Absence of these Cells in the *Pax6*$^{Sey/Sey}$ Mutant Results in Failure of OE and OB Development. (A) Neural crest cells (turquoise, visualized with the β-geo6 transgene) in the frontonasal mesenchyme (fnm) are positioned between the developing olfactory epithelium (oe) and forebrain (fb). The optic vesicle (ov) is further posterior. (B) Sites of RA signaling (turquoise, detected with a DR5 RARE RA-responsive transgene) include the oe, fb, and ov; all are adjacent to neural crest cells in the fnm. (C) Explant of the nasal prominence placed mesenchyme side down on an RA indicator cell monolayer. Neural crest cells (turquoise, β-geo6 transgene) are concentrated in the lateral half of the structure. (D) Pattern of RA-activation in the RA indicator monolayer beneath the explant shown in C (reporter cells express GFP, green). (E, F) Plots of the neural crest cells (blue) and RA activated indicator cells (green) show the correspondence between neural crest cells and RA production in the lateral frontonasal mesenchyme. (G) The localization of neural crest cells in the lateral nasal process (fnm, arrow) of an E10.5 wild-type (WT) embryo. These cells are also seen in the cranial ganglia (V, VIII) the maxillary process (mx), and the branchial arches (bI, bII). (H) Neural crest cells are absent from the frontonasal process (fnm?) in the *Pax6*$^{Sey/Sey}$ mutant. (I) The pattern of RA activation in the forebrain (arrowheads) and olfactory placode (op) of a WT E10.5 mouse embryo. There is also RA signaling in the eye (e). (J) Altered pattern of RA signaling in the Pax6$^{Sey/Sey}$ embryo where RA producing neural crest are absent. Signaling is absent from the olfactory placode (fnp) as well as the forebrain (arrowheads) and the eye (e?). (K) Section through the forebrain of an E14.5 WT mouse shows the olfactory bulb (OB) at the anterior aspect of the forebrain. Additional forebrain subdivisions (NC: neocortex, GE: ganglionic eminences, HC: hippocampus, BF: basal forebrain) and the lateral geniculate nucleus (lgn) can be seen. (L) In the *Pax6*$^{Sey/Sey}$ E14.5 embryo, the OB is selectively absent – other forebrain and thalamic structures can be seen clearly.

additional mechanisms including anterior-posterior or dorso-ventral forebrain patterning, regulation of cell proliferation or survival, and migration of early generated OB interneurons remain to be determined (see below).

Our data indicate that M/E induction is particularly critical for the specification of the LGE-derived OB interneurons, whose migration into the OB anlagen likely drives much of the initial morphogenesis of the OB (see Figure 20.2). More than a decade of elegant developmental and molecular analysis has defined the LGE as a distinct compartment from the adjacent medial ganglionic eminence (the MGE which gives rise to interneurons for the cerebral cortex). The LGE is the source of most OB interneurons during embryonic development. Local organizing centers within the forebrain, particularly those mediated by Shh and FGFs in the antero-ventral and anterior prosencephalon/telencephalon clearly influence OB morphogenesis, as do a small set of transcription factors associated with the anterior forebrain during early embryogenesis. Nevertheless, there has been little insight gained into the mechanisms by which these patterning centers in the forebrain are established to support distinctions between the LGE and MGE, and therefore, OB morphogenesis. Our work showed that migratory specificity for LGE versus MGE cells for the nascent OB is established quite early (Figure 20.6; Tucker et al., 2006). Subsequently, we found that M/E interactions between the frontonasal mesenchyme and the ventro-anterior forebrain are essential for establishing Shh and FGF organizing centers in the forebrain, and thus for patterning LGE versus MGE gene expression (Tucker et al. 2008; Figure 20.6). Indeed, the frontonasal mesenchyme is apparently essential for maintaining robust, patterned expression of Shh, as well as inducing and patterning restricted FGF8 expression. M/E signaling regulates LGE gene expression, and thus OB morphogenesis and differentiation, using a "two step" mechanism. First, mesenchyme induces and patterns signaling centers, and second, these signaling centers then influence LGE/MGE gene expression. Thus, pharmacological blocking of Shh or FGF signaling prevents the M/E mediated patterning of LGE/MGE specific genes that also rely upon mesenchymal signals for their appropriate expression (Figure 20.6). In the absence

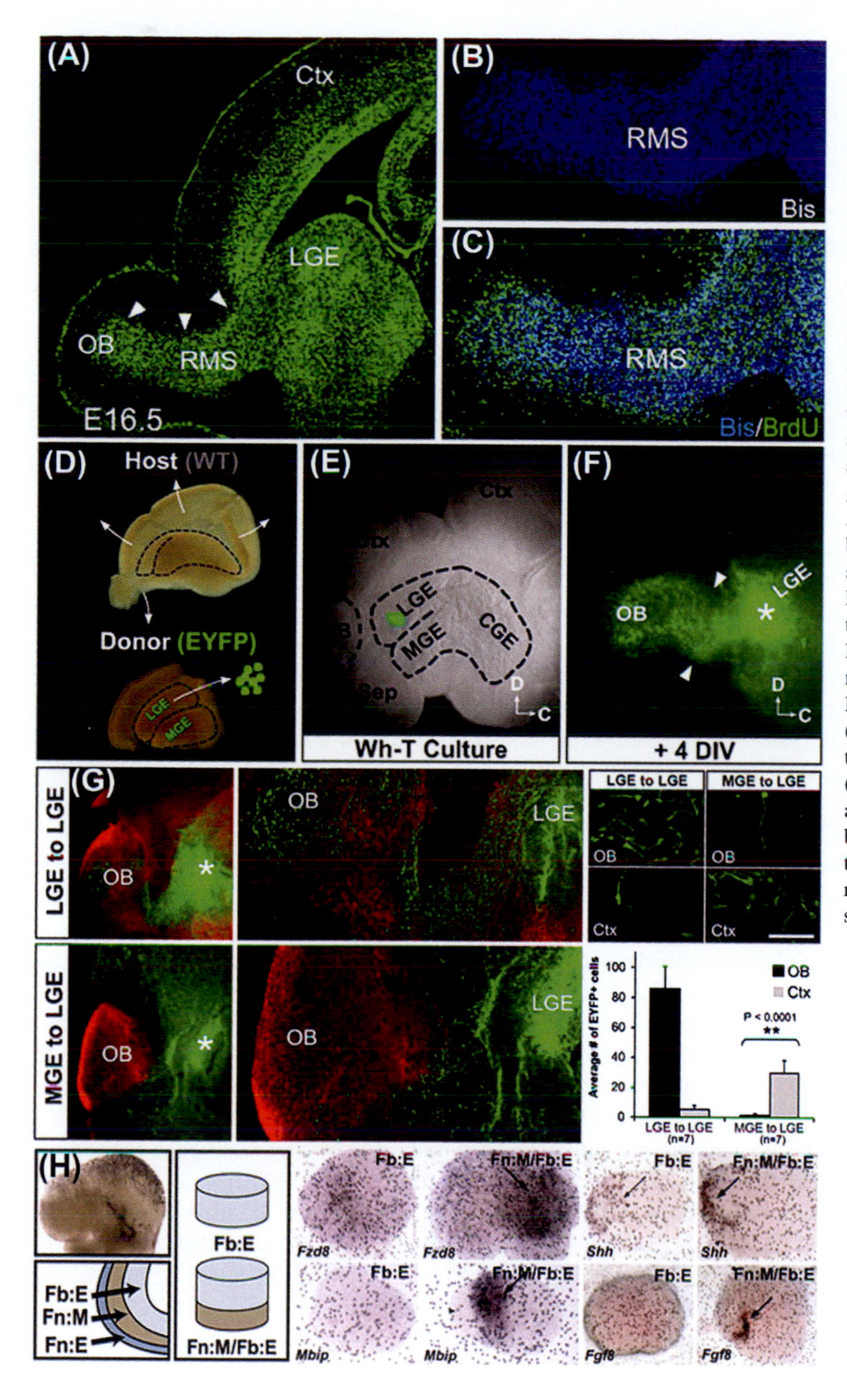

FIGURE 20.6 Migration, Morphogenesis and M/E Interactions in the Olfactory Bulb. (A) A coronal section through the E16.5 forebrain showing post-mitotic migratory cells (labeled with a BrdU pulse two days earlier) as they traverse the territory from the lateral ganglionic eminence (LGE) to the olfactory bulb (OB) via a distinct pathway, the rostral migratory stream (RMS, arrowheads). (B) The RMS is a cell dense zone (bisbezamide nuclear label, blue). (C) The E14.5 BrdU birthdated cells, presumably from the LGE, are intermingled in the cell dense environment of the RMS. (D) An *in vitro* assay for migratory specificity of LGE versus medial ganglionic eminence (MGE) cells. The entire forebrain vesicle from an E14.5 "host" embryo is opened and flattened (top). Donor LGE or MGE cells, as small explants, are collected from an b-actin EYFP embryo of identical age (the transgene labels all neurons). (E) Donor EYFP grafts from the LGE or MGE are placed in the LGE. (F) After four days *in vitro* (DIV) specific migration can be seen from the graft into the olfactory bulb. (G) The specificity of migration of LGE cells that start within the LGE for the OB. When LGE cells are placed in the LGE, they migrate with remarkable fidelity to the OB. When MGE cells are placed in the LGE, they migrate into their normal target the cortex (Ctx), but not to the OB. (H) M/E induction specifies the LGE. An *in vitro* M/E assay (left panels), begun at E9.0, prior to the emergence of the LGE and MGE shows that frontonasal mesenchyme (Fn:M) is crucial for patterned expression of LGE (Fzd8) and MGE (Mbip) markers (middle panels) in the forebrain epithelium (Fb:E). Shh and Fgf8 regulates the patterned expression of these genes, and Fn:M is apparently responsible for establishing local sources of both of these signals (right panels).

of mesenchyme, FGF provided by an extrinsic source like an agarose bead can elicit some of the patterning that normally depends upon M/E interactions (Figure 20.6). Finally, the presence of inductive signals from the frontonasal mesenchyme apparently influences migratory capacity for LGE generated interneurons that migrate selectively into the OB (Figure 20.6).

Together, these observations show that interactions between the frontonasal mesenchyme and forebrain neuroepithelium contribute to OB induction and morphogenesis, in parallel with those between frontonasal mesenchyme and surface ectoderm that underlie OE induction and morphogenesis. This shared inductive mechanism for establishing the peripheral source of olfactory transduction and relay – the OE – and central target for olfactory information – the OB – suggests that M/E interactions coordinate the initial development of the entire olfactory pathway. The consequences of M/E patterning in the peripheral and central components of the olfactory pathway may influence the parallel patterning and cellular specification mechanisms that constrain axon/target recognition and circuit formation. This mechanism would be similar to that seen for the limb and spinal cord (Dasen and Jessell, 2009), or the face and the hindbrain (Alexander et al, 2009); both are dependent on the concerted activity of Hox genes and other transcriptional regulators in peripheral and central nervous system. For the olfactory pathway, however, there is no known equivalent of these Hox-mediated mechanisms that operate both in the OE and OB. The parallel establishing of RA activated domains of gene expression in the OE and OB via M/E induction, however, suggests that such coordination may exist.

20.5 STEM CELLS, CELL LINEAGES AND NEURONAL SPECIFICATION IN THE OE

The identity of specific stem cells for peripheral and central olfactory structures is clearly an essential question. Given the developmental significance of olfactory pathway differentiation – no food, friends, or sex without it – as well as the potential for understanding how stem cells, or their proliferative progeny, can be maintained as a reserve for neuronal replacement in the adult nervous system, interest in these cells seems warranted. Finally, the OE provides an ideal model system to evaluate the relationship between early patterning mechanisms via M/E induction and subsequent specification of stem and progenitor cells. Recently, a new appreciation of stem and progenitor cell identities in the developing OE has emerged; however, the relationship between these cells and the progenitors that give rise to ORNs and other OE cell classes in the adult remain uncertain (Leung et al., 2007; Murdoch and Roskams, 2007; Beites et al., 2005; Jang et al., 2014).

The identity of true OE stem cells that generate the initial embryonic OE neuronal lineage remained uncertain until recently. For several decades, the location and cellular characteristics of embryonic OE stem cells – slowly, symmetrically dividing stem cells in a specific location or niche within the OE that have the capacity to generate all differentiated cell types of the mature tissue – were undefined. Early work indicated that progenitors modulated by a hierarchical cascade of neuronal bHLH transcription factors, including Ascl1, NeuroG1, and NeuroD1, were important for OE neurogenesis (Balmer and LaMantia, 2005). From the earliest observations (Guillemot et al., 1993) through several studies of loss-of-function mutants (Murray et al., 2003; Cau et al., 1997, 2000, 2002; Krolewski et al., 2012), it was clear that Ascl1 and the other bHLH genes were important for the expansion of the numbers of ORNs and other OE cell classes. Nevertheless, in the absence of Ascl1, ORNs are still produced. Indeed, every neuronal and non-neuronal cell type found in the OE can be identified in these mutants (Guillemot et al., 1993; Tucker et al., 2010), although their numbers appear to be dramatically diminished. Thus, it seemed unlikely, even from the early 1990s, that Ascl1-expressing cells were OE stem cells. Instead, Ascl1, and perhaps the other bHLH genes, are more likely important for expanding the numbers of ORNs and other OE cells generated from an initial pool of fate-specified stem cells.

The OE stem cells themselves are a distinct population of slowly, symmetrically dividing cells in a specific location in the nascent OE (Tucker et al., 2010). These cells, found in the lateral aspect of the OE as it invaginates, express the transcription factors Meis1 and Pax7, in combination with low levels of the transcription factor Sox2 and high levels of the transcriptional regulator Six1 (Figure 20.7). They divide slowly and symmetrically and apparently can generate all OE cell classes (Tucker et al., 2010; Murdoch et al., 2010). Meis1 and Pax7, at present, appear to be primarily markers for this stem cell population; mutations in either gene do not appear to significantly disrupt initial OE differentiation (Hisa et al., 2004; Mansouri et al., 1996). Nevertheless, Meis1 can apparently influence the maintenance of OE stem cell properties by suppressing the expression of Ascl1, which identifies rapidly dividing transit amplifying precursors (Figure 20.7; Tucker et al., 2010). In contrast, Sox2 and Six1 have been functionally associated with OE neurogenesis. Sox2 is the primary SoxB1 factor found in the OE (Donner et al., 2007; Wood and Episkopou, 1999), as is also the case in the retina (Taranova et al., 2006). When Sox2 expression is reduced by a hypomorphic allele (Figure 20.7) or enhanced by electroporation (Figure 20.7) the OE stem cell population is respectively expanded or diminished (Tucker et al., 2010). These observations, plus the graded expression of Sox2 in the normal embryonic OE – low in the apparent stem cells, enhanced in transit amplifying cells and ORNs (Figure 20.7), establish Sox2 as a key regulator of lineage progression in the OE. Genetic analyses of Six1 and other Six genes indicate that these genes are essential for initial OE morphogenesis and subsequent neurogenesis (Ikeda et al., 2007, 2010; Chen et al., 2009). Nevertheless, the specific role of the Six genes in initial OE lineage progression or ORN specification remains unknown.

This new appreciation of embryonic OE stem cells, subsequent progenitor classes, and mechanisms of lineage progression indicates that much of the basic identity of OE neurons is established at an early stage by the axial inductive signals provided by M/E interactions. Indeed, earlier evidence along with new data suggest that RA and Fgfs are key antagonists for establishing precursor populations including OE stem cells and Ascl1 transit amplifying populations (Kawauchi et al., 2005; Tucker et al., 2010; Song et al., 2004; Rawson et al., 2010). Thus, in the olfactory placode/OE, as at other sites of M/E induction, signaling interactions are key for establishing stem cell identity. A key question that remains to be answered is the relationship between embryonic OE stem and transit amplifying cells and the progenitors retained in the mature OE that give rise to new ORNs throughout life. Many of the transcription factors and signaling molecules that influence embryonic OE stem cell specification are retained in the adult OE (Whitman and Greer 2009; Rawson and LaMantia, 2007). Thus, it seems possible that adult OE progenitors retain or redeploy mechanisms that are first used in the embryo.

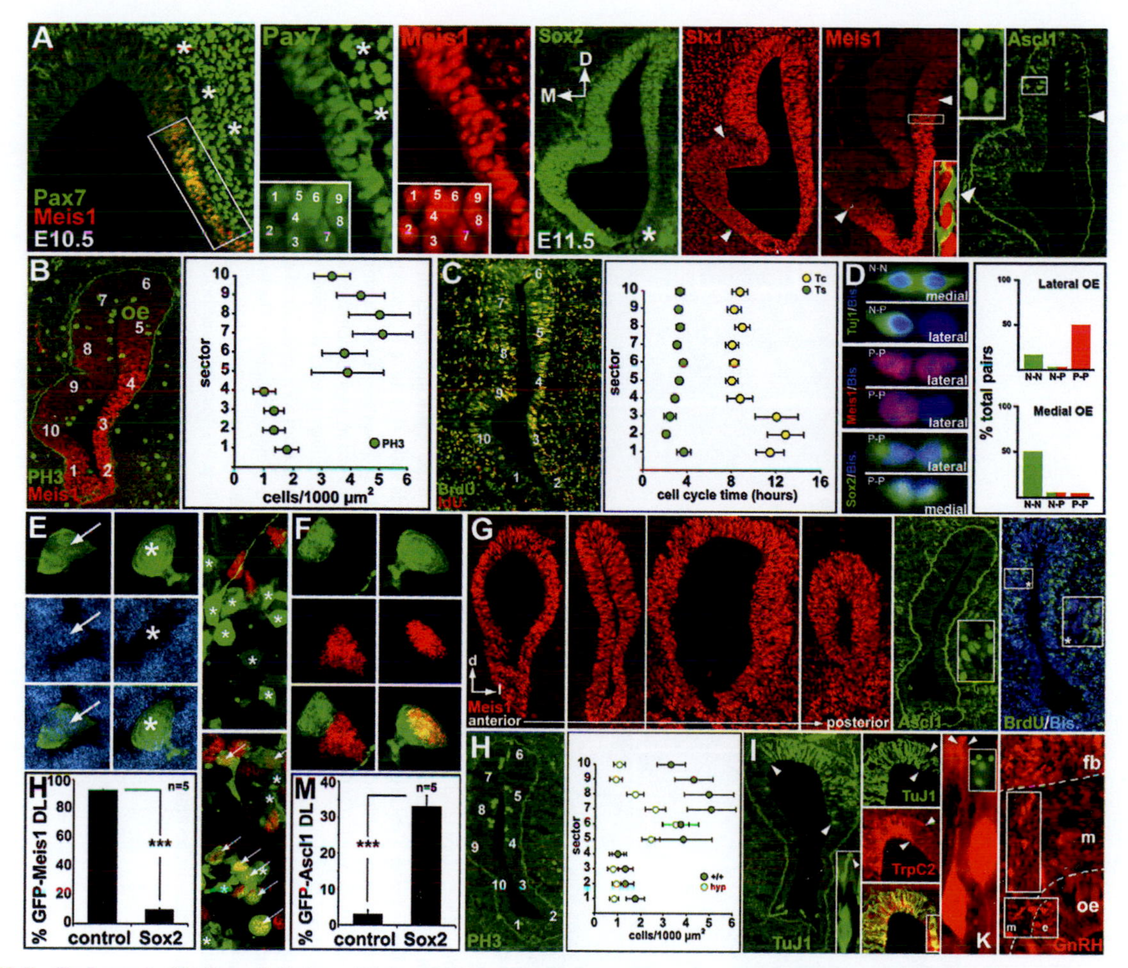

FIGURE 20.7 **Embryonic OE Stem Cells have Distinct Molecular and Proliferative Properties.** (A) There is a distinct population of Pax7/Meis1 labeled proliferative cells (based upon BrdU labeling, not shown) in the lateral/ventral aspect of the OE at E10.5. The Pax7/Meis1 population, clearly located in the lateral OE by E11.5, expresses low levels of Sox2 and high levels of Six1. The remaining cells in the medial OE express higher levels of Sox2, variably lower levels of Six1, and Ascl1 in a mosaic. (B) Proliferative activity is in register with molecular distinctions in the OE. Mitotic activity, quantified based upon frequency of PH3-labeled cells, is substantially lower in the lateral OE where Pax7/Meis1 cells are found. (C) Cell cycle times (Tc) but not S-phase (Ts), measured by staggered injections of distinct thymidine analogus, are substantially higher in the OE region where Pax7/Meis1 cells are found. (D) Dividing cells in the lateral OE, isolated and observed in a pair cell assay, divide symmetrically to yield additional precursor cells (labeled with Meis1), while dividing cells in the medial OE primarily divided terminally and symmetrically to yield neurons. (E) Sox2 dosage is a key regulator of molecular identity of OE precursors. When Sox2 dosage is elevated, Meis1 expression (blue label in left hand panels) is suppressed. (F) Sox2 dosage is a positive regulator of Ascl1. When Sox2 is elevated, Ascl1 expression is enhanced. (G) Loss of Sox2 dosage in a *Sox2*$^{hyp/-}$ mutant E11.5 embryo results in expansion of the Meis1 population throughout the entire OE, and a loss of the Ascl1 population. In addition, BrdU incorporation appears diminished. (H) Proliferation is diminished to low levels throughout the *Sox2*$^{hyp/-}$ OE, based upon frequency of PH3-labeled mitotic cells. (I) Despite changes in precursor distribution and kinetics, all of the major neuronal classes are generated in the *Sox2*$^{hyp/-}$ OE.

20.6 CELL LINEAGES, MIGRATION AND NEURONAL SPECIFICATION IN THE DEVELOPING AND ADULT OB

In contrast to the still incomplete description of embryonic or adult OE stem cells and their niches, a plethora of data defines the stem cells in the adult forebrain subventricular zone (Figure 20.8), which apparently generate granule cells for the adult OB (Whitman and Greer, 2009; Alvarez-Buylla et al., 2008; Doetsch et al., 1999), and their subsequent migration to the OB via the RMS. There is less specific information about the embryonic stem cells that generate the initial complement of OB granule cells (Tucker et al., 2006), and other OB cell classes including mitral cells and other OB interneurons.

Ongoing neurogenesis in the adult olfactory bulb was initially reported in the 1960s (Altman and Das, 1965), and the phenomenon was further characterized in the 1970s and 1980s (Kaplan and Hinds, 1977; Kaplan et al., 1985); however, the location of the stem cell niche, and the identity of the stem cells themselves remained unknown. From the mid-1990s onward, a fairly complete description of OB granule cell emerged based on *in vitro* assessments of stem cell proliferative potential (Gottlieb, 2002), viral tracing methods (Yoon et al., 1996), and the discovery of the general glial-like molecular

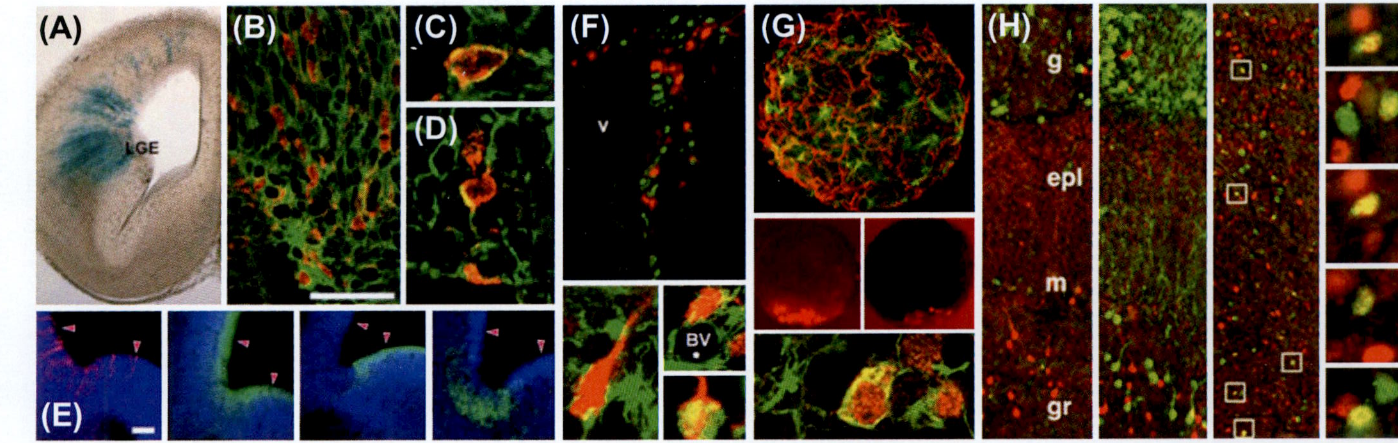

FIGURE 20.8 **Local Signaling Via Retinoic Acid Influences LGE Progenitors, and their Counterparts in the Adult Brain that Continue to Generate OB Interneurons.** (A) A large number of apparent radial glial precursors in the LGE are RA activated, based upon expression of an RA-reporter transgene (turquoise labeling). (B, C, D) These cells (shown in these and subsequent panels with a red immunofluorescent label) express the embryonic radial glial marker RC2. (E) These RA activated cells (red label, denoted by arrows at left) are coincident with precursors that express the LGE-selective makers Pax6, Gsh2, and ER81 (shown from left to right). (F) The RA activated cells (red) in the adult subventricular zone (SVZ, which arises from the ventricular surface of the LGE and is found at the anterior end of the mature forebrain) are not rapidly proliferative. None of them are double labeled after a pulse of BrdU (green). The lower panels show that these RA activated cells (red) are instead a subset of the slowly dividing stem cells in the SVZ. They co-express nestin (left), are often found adjacent to blood vessels (BV*; upper right panel), and they are "label retaining"; therefore, RA activated cells (red) can be heavily labeled by BrdU exposure provided up two months prior to sacrificing the animal. (G) RA activated cells can be found *in vitro* in neurospheres grown from the dissected lining of the SVZ. At top, a neurosphere cultured from a cell suspension of SVZ cells (and thought to be "clonal") has progenitor cells that express GFAP, another marker of slowly dividing stem cells, and the EGF receptor, which recognizes more rapidly dividing SVZ transit amplifying cells. The middle panels show a primary neurosphere (directly cultured from SVZ dissociated cells) as well as a secondary neurosphere (cultured from a suspension of primary neurosphere cells). In both instances, there is a subset of RA activated cells in a distinct location in the neurosphere. The bottom panel shows that these RA activated cells express GFAP. (H) Periglomerular neurons (in the glomerular layer of the OB, labeled "g") and granule neuroins (in the granular layer of the OB, labeled "gr"), are RA activated (red). Although they do not overlap extensively with two established marker for periglomerular and granule cells (calbindin and calretinin, first two panels in H), they are nonetheless functional neurons, based upon their expression of the immediate early gene c-fos (green) after olfactory stimulation. At far right are several examples of periglomerular and granule cells that are both RA activated and c-fos labeled.

identity of neural stem cells in the central nervous system (Doetsch, 2003). These multiple lines of evidence securely identified a population of astrocyte-like neural stem cells in the anterior subventricular zone as the progenitors for OB granule cells and related OB interneurons. These SVZ glia/neuronal stem cells divide slowly and symmetrically for self-renewal, and presumably via asymmetric division give rise to rapidly dividing transit amplifying precursors with a distinct molecular identity that generate terminally post-mitotic OB interneurons locally in the anterior forebrain SVZ. Many of these post-mitotic neurons die in the SVZ, but those that survive must migrate to the OB. In the adult brain, there is a distinct cellular scaffold that defines the migratory pathway from the SVZ to the OB – the glial conduits and interneurons in transit collectively define the RMS. The RMS is seen in many adult mammals; however, it is not present in mature humans (Sanai et al., 2011). Accordingly, it is thought that the human OB does not have extensive replacement of OB interneurons from SVZ sources during adult life. The functional significance of this distinction in humans versus other mammals remains unclear.

The clear definition of a cellular mechanism for ongoing replacement of OB granule cells raised a key question: does this process retain essential aspects of signaling that establish LGE OB precursors during initial embryonic specification and morphogenesis of the OB? We therefore asked whether RA signaling similar to that seen in embryonic domains that give rise to the OB – particularly the LGE – was retained in the mature OB or SVZ (Figure 20.8). There is robust RA signaling in the LGE at midgestation, the time when the earliest OB interneurons are generated (Figure 20.8; Haskell and LaMantia, 2005). The RA activated LGE cells, which have a radial glial-like morphology and molecular identity, are seen in the same regions as cells that express additional markers of LGE precursor identity, including Gsh2 and ER81 (Stenman et al., 2003; Corbin et al., 2000). The presence of an RA activated subset of apparent LGE precursors in the embryo raised the question of whether similar cells are retained in the adult SVZ, which is thought to be the adult remnant of the LGE ventricular/subventricular zone. We found a population of RA activated cells in the adult SVZ. These cells are slowly dividing, often found near blood vessels – a preferred location for SVZ neural stem cells – and they express glial fibrillary acidic protein, a diagnostic marker for neural stem cells (Figure 20.8; Haskell and LaMantia, 2005). This suggests that the RA activated cells are a sub-population of the slowly dividing radial glial neuronal stem cells of the SVZ. To further evaluate this possibility, we determined whether the RA activated cells could be seen in neurospheres – the clonally related progenitors that arise as aggregates from dissociated SVZ cultures. We found RA activated cells in both primary neurospheres and the secondary neurospheres that are formed from dispersed primary neurosphere cells. Thus, RA activated cells are a sub-class of SVZ neural stem cells. There may be a relationship between RA signaling in OB interneuron progenitors and the progeny of these cells, OB granule and periglomerular cells. We found that a subset of adult OB interneurons is RA activated (Haskell and LaMantia, 2005; Thompson Haskell et al., 2002). These interneurons are generated throughout postnatal life. They are functionally active in the OB based upon expression of c-fos, a marker of acute neuronal activity throughout the nervous system. Thus, it is possible that RA signaling establishes, and then continues to influence the maintenance and renewal of OB interneurons throughout life.

20.7 OLFACTORY PATHWAY DEVELOPMENT BEYOND MORPHOGENESIS, STEM CELLS AND NEURONAL SPECIFICATION

Once the OE and OB are established as distinct structures with clearly specified cell classes, neural circuits for chemosensation must be established. As a first step in this process, ORN axons must grow from the nascent OE through the frontonasal mesenchyme, enter the ventro-medial forebrain, and then turn laterally to reach the anlagen of the OB (Figure 20.9). This challenging program of axon guidance relies upon the concerted expression of cell adhesion molecules on the ORN axons, as well as a distinct pattern of extracellular matrix molecules that are provided by migratory cells from the OE that travel with the pioneer axons from the developing nose to the forebrain (Figure 20.9; Whitesides and LaMantia, 1996; Cloutier et al., 2002; 2004; Nakashima et al., 2013; Miller et al., 2010; Nguyen-Ba-Charvet et al., 2008; Gong and Shipley, 1996). Many of these studies focus on later stages of ORN axon guidance. Our work, however, examined the initial development of the olfactory nerve as the first axons grow from the OE to the OB.

The identity of the mesenchyme is essential for establishing the pathway from OE to OB. When the limb bud mesenchyme, which contains several of the same molecular constituents found in frontonasal mesenchyme, is apposed to the nascent OE, neurogenesis occurs. These neurons, however, lack the functional properties of ORNs, and their axons cannot penetrate the basal lamina between the anomalously induced OE and the underlying heterologous mesenchyme (Figure 20.9; Rawson NE et al., 2010). Apparently, appropriate ORN identity, a specific permissive environment at the OE/frontonasal mesenchyme interface, and the capacity to pattern a specific trajectory from nose to brain within the mesenchyme are crucial for initial establishment of the olfactory pathway. Similar cues also guide the migration of GnRH neurons to the ventral forebrain (Parkash et al., 2012; Messina et al., 2011; Cariboni et al., 2011; Gamble et al., 2005; Schwarting et al.,

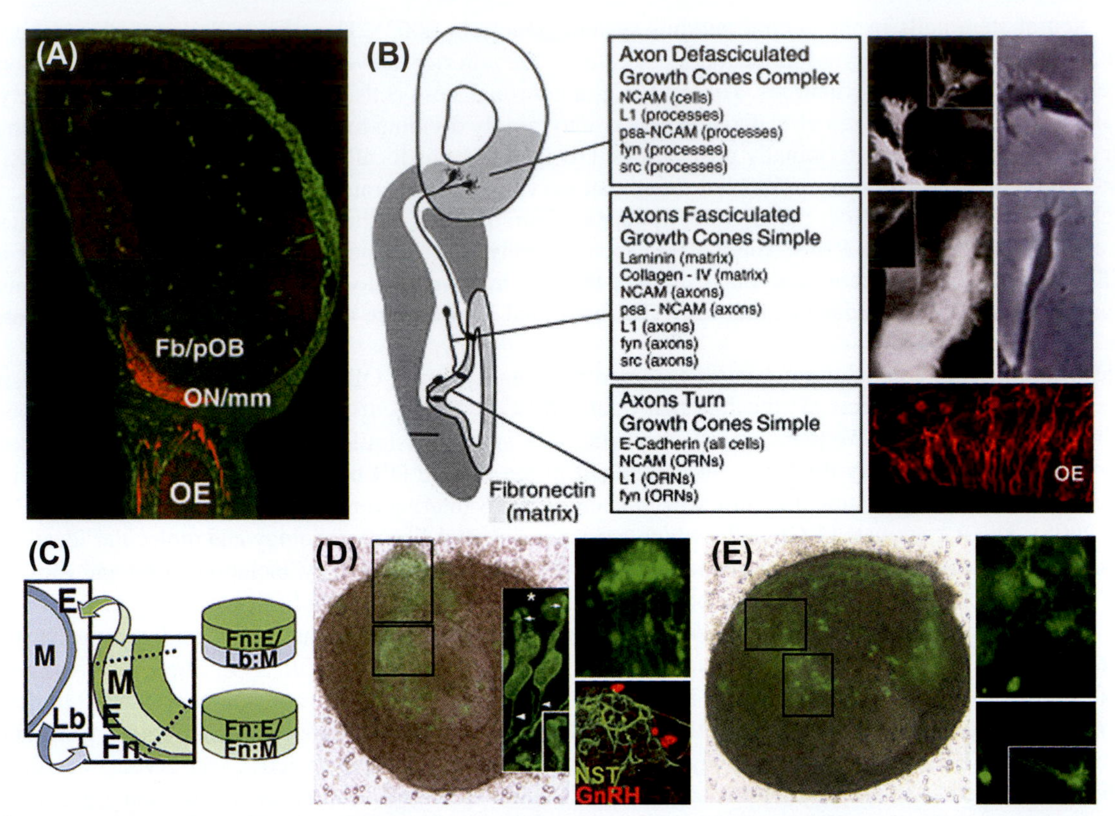

FIGURE 20.9 Mesenchymal Patterning, Axon Growth, and Specificity of Frontonasal M/E Interactions for Initial Growth and Targeting of the Olfactory Nerve. (A) The trajectory of olfactory axons (red, immunolabeled for NCAM) from the OE to the OB at E11.5. (B) A schematic of the distribution of major cell surface adhesion and extracellular matrix adhesion molecules, and some signaling intermediates (fyn, src), and their relationship to ORN growth cone behavior in the pathway from OE to OB. (C) A schematic of a "swapping" experiment: frontonasal mesenchyme is replaced with limb bud mesenchyme to induce the presumptive OE from an E9.0 embryo. (D) Fn:E/Fn:M explants are appropriately patterned (green labeling indicates a Ngn1:eGFP transgene marker that labels ORNs and their axons). Inset shows appropriate cytological differentiation of ORNs, including the differentiation of single dendrites that extend into the lumen (*), dendritic knobs (arrows and smaller inset), and single axons that extend into the olfactory nerve (arrowheads). At right, ORN axons and olfactory ensheathing cells form a coherent olfactory nerve (top), and that nerve serves as a migratory substrate for GnRH neurons generated in the OE as they migrate toward their absent forebrain target (bottom). (E) Fn:E/Lb:M explants do not pattern appropriately. Neurons are scattered over much of the placodal epithelium. At right, these neurons lack single dendrites, knobs, or other characteristics of ORNs (top). They have long single neurites that end in complex growth cones (inset), and these axons never enter the mesenchyme.

2001) however, once the GnRH cells enter the forebrain, they depart from ORN axons, and continue migrating medially and ventrally until they reach the hypothalamus. A number of mutations have been described that disrupt the migration or differentiation of GnRH neurons in mice or humans. These include transcription factors like Sox10 (Pingault et al., 2013), adhesion molecules like Semaphorin3a (Hanchate et al., 2012), and a unique adhesion molecule ANOSMIN-1 (Hu and Bouloux, 2011), which was the first mutant gene associated with Kallman's syndrome (and remarkably, is not found in the mouse genome).

In the OE, the onset of expression of the odorant receptor genes, and the capacity for odor transduction is a key event for further organization of axonal projections to the OB. Odorant receptor expression results in truly remarkable cellular diversity among ORNs – each mature receptor expresses one allele of a subset of between 600 and 2000 genes that encode the G-protein coupled odorant receptor family. This diversity arises via a feedback mechanism for allelic selection that as yet is not well understood (Nguyen et al., 2007; Serizawa et al., 2003). The consequence of this remarkable cellular diversification is the distribution of ORNs that express subsets of odorant receptors in at least four fairly broad zones (Ressler et al., 1993; Mori et al., 2000). Each zone has roughly topologically restricted projections to the surface of the OB. Moreover, axons from ORNs that express each odorant receptor sort themselves so that they uniquely innervate the distinctive spherical accumulation of dendrites that define the glomeruli of the OB (Mombaerts et al., 1996; Vassar et al., 1994). This molecular precision is thought to rely upon the odorant receptors themselves, as well as additional adhesion signals (Mori and Sakano, 2011). Nevertheless, the mechanisms that ensure this precise odorant receptor "map" in the OB remain unresolved.

20.8 PERSPECTIVE

The available evidence indicates that olfactory pathway development is coordinated by a shared M/E inductive mechanism that initially establishes peripheral ectodermal placodal and central forebrain domains to generate the OE and the OB. This mechanism then continues to operate to establish distinct stem cell niches for neural stem cells in the OE and the forebrain SVZ that will generate new ORNs and OB interneurons throughout life. This coordination likely extends to an inductive influence on migratory pathways for ORN axons as they form the ON, and OB interneurons as they migrate from the LGE (and then the SVZ in the mature brain) to the OB via the RMS. The molecular and genetic regulation of this coordinated process is remarkably similar to that in the limbs, aortic arches, and craniofacial primordia. Cardinal inductive signals, most essentially RA, Fgfs, Bmps, and to a lesser extent, Shh, mediate M/E induction. An array of transcriptional regulators including Pax6, Gli3, Emx1 and 2, Dlx 1 and 2, Sox2, Meis1, Ascl1, and Six1 all are regulated by M/E interactions and influence morphogenesis or neuronal specification in the OE and OB. Thus, the genome is deployed to build the olfactory pathway – and ensure its integrity for the ongoing appreciation of food, friends, and sex for each individual – in a unique way. An M/E inductive mechanism engages a common genetic program for secondary induction to guide the differentiation of the sensory periphery: the OE and ON, and the essential central processing center for olfactory information: the OB.

20.9 CLINICAL RELEVANCE

- The consequences of disrupted olfactory development include disorders like Kallman's syndrome, which results in anosmia and infertility. Olfactory pathway developmental deficits may contribute to altered sensory capacities – especially those that influence food intake and social interaction – in a broad range of developmental disorders including autism.
- Development in the olfactory epithelium and bulb, or a variant of developmental mechanisms, is recapitulated throughout life to facilitate the ongoing replacement of olfactory receptor neurons over the entire lifespan, and the more modest addition of olfactory bulb granule cells.
- Neural stem cells – or progenitors that are descended from olfactory epithelial or bulb stem cells – persist in distinct molecular environments and specific locations ("niches") in the adult nose and brain. It is important to note, however, that the in the adult human olfactory bulb, addition/replacement of olfactory bulb granule cells does not seem to occur.
- Olfaction is the first sense to be compromised in a number of neurodegenerative diseases including Parkinson's, Huntington's and Alzheimer's diseases. This may reflect the disruption of the integrity of the stem cell niche in the periphery, or the capacity of the olfactory bulb to accommodate new connections made by newly generated olfactory receptor neurons.

RECOMMENDED RESOURCES

For information about the functional organization of the olfactory system:

Sullivan, S.L., Ressler, K.J., Buck, L.B., 1995. Spatial patterning and information coding in the olfactory system. Curr. opin. Genet. Devel. 5 (4), 516–523.

For information about adult neurogenesis in the olfactory pathway

Whitman, M.C., Greer, C.A., 2009. Adult neurogenesis and the olfactory system. Prog. Neurobiol. 89 (2), 162–175.

For information about odorant receptor-selective mapping in the olfactory pathway

Sakano, H., 2010. Neural map formation in the mouse olfactory system. Neuron 67 (4), 530–542.

REFERENCES

Alexander, T., Nolte, C., Krumlauf, R., 2009. Hox genes and segmentation of the hindbrain and axial skeleton. Annu. Rev. Cell Dev. Biol. 25, 431–456.

Altman, J., Das, G.D., 1965. Post-natal origin of microneurones in the rat brain. Nature 207 (5000), 953–956.

Alvarez-Buylla, A., Kohwi, M., Nguyen, T.M., Merkle, F.T., 2008. The heterogeneity of adult neural stem cells and the emerging complexity of their niche. Cold Spring Harb. Symp. Quant. Biol. 73, 357–365.

Anchan, R.M., Drake, D.P., Haines, C.F., Gerwe, E.A., LaMantia, A.S., 1997. Disruption of local retinoid-mediated gene expression accompanies abnormal development in the mammalian olfactory pathway. J. Comp. Neurol. 379 (2), 171–184.

Apfelbach, R., Blanchard, C.D., Blanchard, R.J., Hayes, R.A., McGregor, I.S., 2005. The effects of predator odors in mammalian prey species: a review of field and laboratory studies. Neurosci. Biobehav. Rev. 29 (8), 1123–1144.

Balkan, W., Colbert, M., Bock, C., Linney, E., 1992. Transgenic indicator mice for studying activated retinoic acid receptors during development. Proc. Natl. Acad. Sci. U S A. 89 (8), 3347–3351.

Balmer, C.W., LaMantia, A.S., 2004. Loss of Gli3 and Shh function disrupts olfactory axon trajectories. J. Comp. Neurol. 472 (3), 292–307.

Balmer, C.W., LaMantia, A.S., 2005. Noses and neurons: induction, morphogenesis, and neuronal differentiation in the peripheral olfactory pathway. Dev. Dyn. 234 (3), 464–481.

Batista-Brito, R., Close, J., Machold, R., Fishell, G., 2008. The distinct temporal origins of olfactory bulb interneuron subtypes. J. Neurosci. 28 (15), 3966–3975.

Beites, C.L., Kawauchi, S., Crocker, C.E., Calof, A.L., 2005. Identification and molecular regulation of neural stem cells in the olfactory epithelium. Exp. Cell Res. 306 (2), 309–316.

Bhasin, N., Maynard, T.M., Gallagher, P.A., LaMantia, A.S., 2003. Mesenchymal/epithelial regulation of retinoic acid signaling in the olfactory placode. Dev. Biol. 261 (1), 82–98.

Bhattacharyya, S., Bronner-Fraser, M., 2008. Competence, specification and commitment to an olfactory placode fate. Development 135 (24), 4165–4177.

Bhattacharyya, S., Bailey, A.P., Bronner-Fraser, M., Streit, A., 2004. Segregation of lens and olfactory precursors from a common territory: cell sorting and reciprocity of Dlx5 and Pax6 expression. Dev. Biol. 271 (2), 403–414.

Bishop, K.M., Garel, S., Nakagawa, Y., Rubenstein, J.L., O'Leary, D.D., 2003. Emx1 and Emx2 cooperate to regulate cortical size, lamination, neuronal differentiation, development of cortical efferents, and thalamocortical pathfinding. J. Comp. Neurol. 457 (4), 345–360.

Blanchart, A., De Carlos, J.A., Lopez-Mascaraque, L., 2006. Time frame of mitral cell development in the mice olfactory bulb. J. Comp. Neurol. 496 (4), 529–543.

Bouligand, J., et al., 2010. Genetics defects in GNRH1: a paradigm of hypothalamic congenital gonadotropin deficiency. Brain Res. 1364, 3–9.

Capdevila, J., Izpisua Belmonte, J.C., 2001. Patterning mechanisms controlling vertebrate limb development. Annu. Rev. Cell Dev. Biol. 17, 87–132.

Cariboni, A., et al., 2011. Defective gonadotropin-releasing hormone neuron migration in mice lacking SEMA3A signalling through NRP1 and NRP2: implications for the aetiology of hypogonadotropic hypogonadism. Hum. Mol. Genet. 20 (2), 336–344.

Cau, E., Casarosa, S., Guillemot, F., 2002. Mash1 and Ngn1 control distinct steps of determination and differentiation in the olfactory sensory neuron lineage. Development 129 (8), 1871–1880.

Cau, E., Gradwohl, G., Casarosa, S., Kageyama, R., Guillemot, F., 2000. Hes genes regulate sequential stages of neurogenesis in the olfactory epithelium. Development 127 (11), 2323–2332.

Cau, E., Gradwohl, G., Fode, C., Guillemot, F., 1997. Mash1 activates a cascade of bHLH regulators in olfactory neuron progenitors. Development 124 (8), 1611–1621.

Chambon, P., 1994. The retinoid signaling pathway: molecular and genetic analyses. Semin. Cell Biol. 5 (2), 115–125.

Chen, B., Kim, E.H., Xu, P.X., 2009. Initiation of olfactory placode development and neurogenesis is blocked in mice lacking both Six1 and Six4. Dev. Biol. 326 (1), 75–85.

Chiang, C., et al., 1996. Cyclopia and defective axial patterning in mice lacking Sonic hedgehog gene function. Nature 383 (6599), 407–413.

Cloutier, J.F., et al., 2002. Neuropilin-2 mediates axonal fasciculation, zonal segregation, but not axonal convergence, of primary accessory olfactory neurons. Neuron 33 (6), 877–892.

Cloutier, J.F., et al., 2004. Differential requirements for semaphorin 3F and Slit-1 in axonal targeting, fasciculation, and segregation of olfactory sensory neuron projections. J. Neurosci. 24 (41), 9087–9096.

Corbin, J.G., Gaiano, N., Machold, R.P., Langston, A., Fishell, G., 2000. The Gsh2 homeodomain gene controls multiple aspects of telencephalic development. Development 127 (23), 5007–5020.

Cordero, D.R., et al., 2011. Cranial neural crest cells on the move: their roles in craniofacial development. Am. J. Med. Genet. Part A 155A (2), 270–279.

Couly, G.F., Le Douarin, N.M., 1985. Mapping of the early neural primordium in quail-chick chimeras. I. Developmental relationships between placodes, facial ectoderm, and prosencephalon. Dev. Biology 110 (2), 422–439.

Dasen, J.S., Jessell, T.M., 2009. Hox networks and the origins of motor neuron diversity. Curr. Top. Dev. Biol. 88, 169–200.

DeHamer, M.K., Guevara, J.L., Hannon, K., Olwin, B.B., Calof, A.L., 1994. Genesis of olfactory receptor neurons *in vitro*: regulation of progenitor cell divisions by fibroblast growth factors. Neuron 13 (5), 1083–1097.

Dewan, A., Pacifico, R., Zhan, R., Rinberg, D., Bozza, T., 2013. Non-redundant coding of aversive odours in the main olfactory pathway. Nature 497 (7450), 486–489.

Doetsch, F., 2003. The glial identity of neural stem cells. Nat. Neurosci. 6 (11), 1127–1134.

Doetsch, F., Caille, I., Lim, D.A., Garcia-Verdugo, J.M., Alvarez-Buylla, A., 1999. Subventricular zone astrocytes are neural stem cells in the adult mammalian brain. Cell 97 (6), 703–716.

Donner, A.L., Episkopou, V., Maas, R.L., 2007. Sox2 and Pou2f1 interact to control lens and olfactory placode development. Dev. Biol. 303 (2), 784–799.

Doty, R.L., 2012. Olfaction in Parkinson's disease and related disorders. Neurobiol. Dis. 46 (3), 527–552.

Dryer, L., Graziadei, P.P., 1994. Influence of the olfactory organ on brain development. Perspect. Dev. Neurobiol. 2 (2), 163–174.

Duboc, V., Logan, M.P., 2009. Building limb morphology through integration of signalling modules. Curr. Opin. Genet. Dev. 19 (5), 497–503.

Evans, R.M., 1994. The molecular basis of signaling by vitamin A and its metabolites. Harvey Lect. 90, 105–117.

Falardeau, J., et al., 2008. Decreased FGF8 signaling causes deficiency of gonadotropin-releasing hormone in humans and mice. J. Clin. Invest. 118 (8), 2822–2831.

Fernandez-Teran, M., Piedra, M.E., Ros, M.A., Fallon, J.F., 1999. The recombinant limb as a model for the study of limb patterning, and its application to muscle development. Cell Tissue Res. 296 (1), 121–129.

Ferrero, D.M., Liberles, S.D., 2010. The secret codes of mammalian scents. Wiley interdisciplinary Rev. Syst. Biol. Med. 2 (1), 23–33.

Forni, P.E., Taylor-Burds, C., Melvin, V.S., Williams, T., Wray, S., 2011. Neural crest and ectodermal cells intermix in the nasal placode to give rise to GnRH-1 neurons, sensory neurons, and olfactory ensheathing cells. J. Neurosci. 31 (18), 6915–6927.

Gamble, J.A., et al., 2005. Disruption of ephrin signaling associates with disordered axophilic migration of the gonadotropin-releasing hormone neurons. J. Neurosci. 25 (12), 3142–3150.

Ghashghaei, H.T., Lai, C., Anton, E.S., 2007. Neuronal migration in the adult brain: are we there yet? Nat. Rev. Neurosci. 8 (2), 141–151.

Gong, Q., Shipley, M.T., 1996. Expression of extracellular matrix molecules and cell surface molecules in the olfactory nerve pathway during early development. J. Comp. Neurol. 366 (1), 1–14.

Gottlieb, D.I., 2002. Large-scale sources of neural stem cells. Annu. Rev. Neurosci. 25, 381–407.

Graziadei, P.P., Monti-Graziadei, A.G., 1992. The influence of the olfactory placode on the development of the telencephalon in Xenopus laevis. Neurosci. 46 (3), 617–629.

Grindley, J.C., Davidson, D.R., Hill, R.E., 1995. The role of Pax-6 in eye and nasal development. Development 121 (5), 1433–1442.

Guillemot, F., Lo, L.C., Johnson, J.E., Auerbach, A., Anderson, D.J., Joyner, A., 1993. Mammalian achaete-scute homolog 1 is required for the early development of olfactory and autonomic neurons. Cell 75 (3), 463–476.

Hanchate, N.K., et al., 2012. SEMA3A, a gene involved in axonal pathfinding, is mutated in patients with Kallmann syndrome. PLoS genetics 8 (8), e1002896.

Haskell, G.T., LaMantia, A.S., 2005. Retinoic acid signaling identifies a distinct precursor population in the developing and adult forebrain. J. Neurosci. 25 (33), 7636–7647.

Hebert, J.M., Lin, M., Partanen, J., Rossant, J., McConnell, S.K., 2003. FGF signaling through FGFR1 is required for olfactory bulb morphogenesis. Development 130 (6), 1101–1111.

Hinds, J.W., 1968. Autoradiographic study of histogenesis in the mouse olfactory bulb. I. Time of origin of neurons and neuroglia. J. Comp. Neurol. 134 (3), 287–304.

Hisa, T., et al., 2004. Hematopoietic, angiogenic and eye defects in Meis1 mutant animals. The EMBO J. 23 (2), 450–459.

Hu, Y., Bouloux, P.M., 2011. X-linked GnRH deficiency: role of KAL-1 mutations in GnRH deficiency. Mol. Cell. Endocrinol. 346 (1–2), 13–20.

Hui, C.C., Joyner, A.L., 1993. A mouse model of greig cephalopolysyndactyly syndrome: the extra-toesJ mutation contains an intragenic deletion of the Gli3 gene. Nat. Genet. 3 (3), 241–246.

Ikeda, K., et al., 2007. Six1 is essential for early neurogenesis in the development of olfactory epithelium. Dev. Biol. 311 (1), 53–68.

Ikeda, K., Kageyama, R., Suzuki, Y., Kawakami, K., 2010. Six1 is indispensable for production of functional progenitor cells during olfactory epithelial development. Int. J. Dev. Biol. 54 (10), 1453–1464.

Ingham, P.W., Placzek, M., 2006. Orchestrating ontogenesis: variations on a theme by sonic hedgehog. Nat. Rev. Genet. 7 (11), 841–850.

Isogai, Y., et al., 2011. Molecular organization of vomeronasal chemoreception. Nature 478 (7368), 241–245.

Jacobson, A.G., 1963a. The determination and positioning of the nose, lens and ear. I. Interactions within the ectoderm, and between the ectoderm and underlying tissues. J. Exp. Zool. 154, 273–283.

Jacobson, A.G., 1963b. The determination and positioning of the nose, lens and ear. II. The role of the endoderm. J. Exp. Zool. 154, 285–291.

Jang, W., Chen, X., Flis, D., Harris, M., Schwob, J.E., 2014. Label-retaining, quiescent globose basal cells are found in the olfactory epithelium. J. Comp. Neurol. 522 (4), 731–749.

Jimenez, D., et al., 2000. Evidence for intrinsic development of olfactory structures in Pax-6 mutant mice. J. Comp. Neurol. 428 (3), 511–526.

Johnston, M.C., Bronsky, P.T., 1995. Prenatal craniofacial development: new insights on normal and abnormal mechanisms. Crit. Rev. Oral Biol. Med. 6 (4), 368–422.

Kaplan, M.S., Hinds, J.W., 1977. Neurogenesis in the adult rat: electron microscopic analysis of light radioautographs. Science 197 (4308), 1092–1094.

Kaplan, M.S., McNelly, N.A., Hinds, J.W., 1985. Population dynamics of adult-formed granule neurons of the rat olfactory bulb. J. Comp. Neurol. 239 (1), 117–125.

Kavaliers, M., Choleris, E., Pfaff, D.W., 2005. Genes, odours and the recognition of parasitized individuals by rodents. Trends Parasitol. 21 (9), 423–429.

Kawauchi, S., et al., 2005. Fgf8 expression defines a morphogenetic center required for olfactory neurogenesis and nasal cavity development in the mouse. Development 132 (23), 5211–5223.

Kelly, O.G., Melton, D.A., 1995. Induction and patterning of the vertebrate nervous system. Trends in genetics: TIG 11 (7), 273–278.

Keyte, A., Hutson, M.R., 2012. The neural crest in cardiac congenital anomalies. Differentiation 84 (1), 25–40.

Krolewski, R.C., Packard, A., Jang, W., Wildner, H., Schwob, J.E., 2012. Ascl1 (Mash1) knockout perturbs differentiation of nonneuronal cells in olfactory epithelium. PloS one 7 (12), e51737.

LaMantia, A.S., 1999. Forebrain induction, retinoic acid, and vulnerability to schizophrenia: insights from molecular and genetic analysis in developing mice. Biol. Psychiatry 46 (1), 19–30.

LaMantia, A.S., Bhasin, N., Rhodes, K., Heemskerk, J., 2000. Mesenchymal/epithelial induction mediates olfactory pathway formation. Neuron 28 (2), 411–425.

LaMantia, A.S., Colbert, M.C., Linney, E., 1993. Retinoic acid induction and regional differentiation prefigure olfactory pathway formation in the mammalian forebrain. Neuron 10 (6), 1035–1048.

Lassiter, R.N., Stark, M.R., Zhao, T., Zhou, C.J., 2013. Signaling mechanisms controlling cranial placode neurogenesis and delamination. Dev. Biol.

Laub, F., et al., 2005. Transcription factor KLF7 is important for neuronal morphogenesis in selected regions of the nervous system. Mol. Cell. Biol. 25 (13), 5699–5711.

Lemasson, M., Saghatelyan, A., Olivo-Marin, J.C., Lledo, P.M., 2005. Neonatal and adult neurogenesis provide two distinct populations of newborn neurons to the mouse olfactory bulb. J. Neurosci. 25 (29), 6816–6825.

Leung, C.T., Coulombe, P.A., Reed, R.R., 2007. Contribution of olfactory neural stem cells to tissue maintenance and regeneration. Nat. Neurosci. 10 (6), 720–726.

Levy, F., Keller, M., 2009. Olfactory mediation of maternal behavior in selected mammalian species. Behav. Brain Res. 200 (2), 336–345.

Lledo, P.M., Merkle, F.T., Alvarez-Buylla, A., 2008. Origin and function of olfactory bulb interneuron diversity. Trends Neurosci. 31 (8), 392–400.

Lois, C., Alvarez-Buylla, A., 1994. Long-distance neuronal migration in the adult mammalian brain. Science 264 (5162), 1145–1148.

Long, J.E., Garel, S., Depew, M.J., Tobet, S., Rubenstein, J.L., 2003. DLX5 regulates development of peripheral and central components of the olfactory system. J. Neurosci. 23 (2), 568–578.

Lopez-Mascaraque, L., de Castro, F., 2002. The olfactory bulb as an independent developmental domain. Cell Death Differ. 9 (12), 1279–1286.

Luskin, M.B., 1993. Restricted proliferation and migration of postnatally generated neurons derived from the forebrain subventricular zone. Neuron 11 (1), 173–189.

Maier, E., et al., 2010. Opposing Fgf and Bmp activities regulate the specification of olfactory sensory and respiratory epithelial cell fates. Development 137 (10), 1601–1611.

Mansouri, A., Stoykova, A., Torres, M., Gruss, P., 1996. Dysgenesis of cephalic neural crest derivatives in Pax7–/– mutant mice. Development 122 (3), 831–838.

Maynard, T.M., Jain, M.D., Balmer, C.W., LaMantia, A.S., 2002. High-resolution mapping of the Gli3 mutation extra-toes reveals a 51.5kb deletion. Mamm. Genome 13 (1), 58–61.

Messina, A., et al., 2011. Dysregulation of Semaphorin7A/beta1-integrin signaling leads to defective GnRH-1 cell migration, abnormal gonadal development and altered fertility. Hum. Mol. Genet. 20 (24), 4759–4774.

Miller, A.M., Treloar, H.B., Greer, C.A., 2010. Composition of the migratory mass during development of the olfactory nerve. J. Comp. Neurol. 518 (24), 4825–4841.

Mombaerts, P., et al., 1996. Visualizing an olfactory sensory map. Cell 87 (4), 675–686.

Mori, K., Sakano, H., 2011. How is the olfactory map formed and interpreted in the mammalian brain? Annu. Rev. Neurosci. 34, 467–499.

Mori, K., von Campenhause, H., Yoshihara, Y., 2000. Zonal organization of the mammalian main and accessory olfactory systems. Philos. Trans. R. Soc. Lond. B. Biol. Sci. 355 (1404), 1801–1812.

Murdoch, B., Roskams, A.J., 2007. Olfactory epithelium progenitors: insights from transgenic mice and *in vitro* biology. J. Mol. Histol. 38 (6), 581–599.

Murdoch, B., Roskams, A.J., 2013. Fibroblast growth factor signaling regulates neurogenesis at multiple stages in the embryonic olfactory epithelium. Stem cells and development 22 (4), 525–537.

Murdoch, B., DelConte, C., Garcia-Castro, M.I., 2010. Embryonic Pax7-expressing progenitors contribute multiple cell types to the postnatal olfactory epithelium. J. Neurosci. 30 (28), 9523–9532.

Murray, R.C., Navi, D., Fesenko, J., Lander, A.D., Calof, A.L., 2003. Widespread defects in the primary olfactory pathway caused by loss of Mash1 function. J. Neurosci. 23 (5), 1769–1780.

Nakashima, A., et al., 2013. Agonist-independent GPCR activity regulates anterior-posterior targeting of olfactory sensory neurons. Cell 154 (6), 1314–1325.

Neubuser, A., Peters, H., Balling, R., Martin, G.R., 1997. Antagonistic interactions between FGF and BMP signaling pathways: a mechanism for positioning the sites of tooth formation. Cell 90 (2), 247–255.

Nguyen, M.Q., Zhou, Z., Marks, C.A., Ryba, N.J., Belluscio, L., 2007. Prominent roles for odorant receptor coding sequences in allelic exclusion. Cell 131 (5), 1009–1017.

Nguyen-Ba-Charvet, K.T., Di Meglio, T., Fouquet, C., Chedotal, A., 2008. Robos and slits control the pathfinding and targeting of mouse olfactory sensory axons. J. Neurosci. 28 (16), 4244–4249.

Niederreither, K., et al., 2003. The regional pattern of retinoic acid synthesis by RALDH2 is essential for the development of posterior pharyngeal arches and the enteric nervous system. Development 130 (11), 2525–2534.

Nomura, T., Haba, H., Osumi, N., 2007. Role of a transcription factor Pax6 in the developing vertebrate olfactory system. Dev. growth Differ. 49 (9), 683–690.

Osumi-Yamashita, N., Ninomiya, Y., Doi, H., Eto, K., 1994. The contribution of both forebrain and midbrain crest cells to the mesenchyme in the frontonasal mass of mouse embryos. Dev. Biol. 164 (2), 409–419.

Parkash, J., et al., 2012. Suppression of beta1-integrin in gonadotropin-releasing hormone cells disrupts migration and axonal extension resulting in severe reproductive alterations. J. Neurosci. 32 (47), 16992–17002.

Peretto, P., et al., 2002. BMP mRNA and protein expression in the developing mouse olfactory system. J. Comp. Neurol. 451 (3), 267–278.

Pingault, V., et al., 2013. Loss-of-function mutations in SOX10 cause Kallmann syndrome with deafness. Am. J. Hum. Genet. 92 (5), 707–724.

Ragancokova, D., et al., 2014. TSHZ1-dependent gene regulation is essential for olfactory bulb development and olfaction. J. Clin. Invest.

Rawson, N.E., LaMantia, A.S., 2007. A speculative essay on retinoic acid regulation of neural stem cells in the developing and aging olfactory system. Exp. Gerontol. 42 (1–2), 46–53.

Rawson, N.E., et al., 2010. Specific mesenchymal/epithelial induction of olfactory receptor, vomeronasal, and gonadotropin-releasing hormone (GnRH) neurons. Dev. Dyn. 239 (6), 1723–1738.

Reed, R.R., 2004. After the holy grail: establishing a molecular basis for mammalian olfaction. Cell 116 (2), 329–336.

Ressler, K.J., Sullivan, S.L., Buck, L.B., 1993. A zonal organization of odorant receptor gene expression in the olfactory epithelium. Cell 73 (3), 597–609.

Richman, J.M., Tickle, C., 1989. Epithelia are interchangeable between facial primordia of chick embryos and morphogenesis is controlled by the mesenchyme. Dev. Biol. 136 (1), 201–210.

Richman, J.M., Tickle, C., 1992. Epithelial-mesenchymal interactions in the outgrowth of limb buds and facial primordia in chick embryos. Dev. Biol. 154 (2), 299–308.

Rupp, C.I., 2010. Olfactory function and schizophrenia: an update. Curr. Opin.Psychiatry 23 (2), 97–102.

Sabado, V., Barraud, P., Baker, C.V., Streit, A., 2012. Specification of GnRH-1 neurons by antagonistic FGF and retinoic acid signaling. Dev. Biol. 362 (2), 254–262.

Sakano, H., 2010. Neural map formation in the mouse olfactory system. Neuron 67 (4), 530–542.

Sanai, N., et al., 2011. Corridors of migrating neurons in the human brain and their decline during infancy. Nature 478 (7369), 382–386.

Sandell, L.L., et al., 2007. RDH10 is essential for synthesis of embryonic retinoic acid and is required for limb, craniofacial, and organ development. Genes Dev. 21 (9), 1113–1124.

Schecklmann, M., et al., 2013. A systematic review on olfaction in child and adolescent psychiatric disorders. J. Neural. Transm. 120 (1), 121–130.

Schneider, R.A., Hu, D., Rubenstein, J.L., Maden, M., Helms, J.A., 2001. Local retinoid signaling coordinates forebrain and facial morphogenesis by maintaining FGF8 and SHH. Development 128 (14), 2755–2767.

Schwanzel-Fukuda, M., Pfaff, D.W., 1989. Origin of luteinizing hormone-releasing hormone neurons. Nature 338 (6211), 161–164.

Schwarting, G.A., Kostek, C., Bless, E.P., Ahmad, N., Tobet, S.A., 2001. Deleted in colorectal cancer (DCC) regulates the migration of luteinizing hormone-releasing hormone neurons to the basal forebrain. J. Neurosci. 21 (3), 911–919.

Schwob, J.E., 2002. Neural regeneration and the peripheral olfactory system. Anat. Rec. 269 (1), 33–49.

Serbedzija, G.N., Bronner-Fraser, M., Fraser, S.E., 1992. Vital dye analysis of cranial neural crest cell migration in the mouse embryo. Development 116 (2), 297–307.

Serizawa, S., et al., 2003. Negative feedback regulation ensures the one receptor-one olfactory neuron rule in mouse. Science 302 (5653), 2088–2094.

Shaker, T., Dennis, D., Kurrasch, D.M., Schuurmans, C., 2012. Neurog1 and Neurog2 coordinately regulate development of the olfactory system. Neural. Dev. 7, 28.

Shenefelt, R.E., 1972. Morphogenesis of malformations in hamsters caused by retinoic acid: relation to dose and stage at treatment. Teratology 5 (1), 103–118.

Shenefelt, R.E., 2010. Morphogenesis of malformations in hamsters caused by retinoic acid: relation to dose and stage at treatment. Teratology 5:103–18. 1972. Birth defects Res. Part A, Clin. Mol. Teratol. 88 (10), 847–862.

Shou, J., Murray, R.C., Rim, P.C., Calof, A.L., 2000. Opposing effects of bone morphogenetic proteins on neuron production and survival in the olfactory receptor neuron lineage. Development 127 (24), 5403–5413.

Slack, J.M., 1993. Embryonic induction. Mech. Dev. 41 (2–3), 91–107.

Slack, J.M., 1995. Developmental biology. Growth factor lends a hand. Nature 374 (6519), 217–218.

Song, Y., Hui, J.N., Fu, K.K., Richman, J.M., 2004. Control of retinoic acid synthesis and FGF expression in the nasal pit is required to pattern the craniofacial skeleton. Dev. Biol. 276 (2), 313–329.

Stenman, J., Toresson, H., Campbell, K., 2003. Identification of two distinct progenitor populations in the lateral ganglionic eminence: implications for striatal and olfactory bulb neurogenesis. J. Neurosci. 23 (1), 167–174.

Stout, R.P., Graziadei, P.P., 1980. Influence of the olfactory placode on the development of the brain in *Xenopus laevis* (Daudin). I. Axonal growth and connections of the transplanted olfactory placode. Neuroscience 5 (12), 2175–2186.

Stowers, L., Cameron, P., Keller, J.A., 2013. Ominous odors: olfactory control of instinctive fear and aggression in mice. Curr. Opin. Neurobiol. 23 (3), 339–345.

Stowers, L., Holy, T.E., Meister, M., Dulac, C., Koentges, G., 2002. Loss of sex discrimination and male-male aggression in mice deficient for TRP2. Science 295 (5559), 1493–1500.

Sullivan, S.L., Ressler, K.J., Buck, L.B., 1995. Spatial patterning and information coding in the olfactory system. Curr. Opin. Genet. Dev. 5 (4), 516–523.

Sun, G.H., Raji, C.A., Maceachern, M.P., Burke, J.F., 2012. Olfactory identification testing as a predictor of the development of Alzheimer's dementia: a systematic review. Laryngoscope 122 (7), 1455–1462.

Taranova, O.V., et al., 2006. SOX2 is a dose-dependent regulator of retinal neural progenitor competence. Genes.Dev. 20 (9), 1187–1202.

Thaller, C., Eichele, G., 1987. Identification and spatial distribution of retinoids in the developing chick limb bud. Nature 327 (6123), 625–628.

Thesleff, I., Partanen, A.M., Vainio, S., 1991. Epithelial-mesenchymal interactions in tooth morphogenesis: the roles of extracellular matrix, growth factors, and cell surface receptors. J. Craniofac. Genet.Dev. Biol. 11 (4), 229–237.

Thompson Haskell, G., Maynard, T.M., Shatzmiller, R.A., Lamantia, A.S., 2002. Retinoic acid signaling at sites of plasticity in the mature central nervous system. J. Comp. Neurol. 452 (3), 228–241.

Trainor, P.A., 2010. Craniofacial birth defects: The role of neural crest cells in the etiology and pathogenesis of Treacher Collins syndrome and the potential for prevention. Am. J. Med. Gene. Part A 152A (12), 2984–2994.

Tucker, E.S., et al., 2008. Molecular specification and patterning of progenitor cells in the lateral and medial ganglionic eminences. J. Neurosci. 28 (38), 9504–9518.

Tucker, E.S., et al., 2010. Proliferative and transcriptional identity of distinct classes of neural precursors in the mammalian olfactory epithelium. Development 137 (15), 2471–2481.

Tucker, E.S., Polleux, F., LaMantia, A.S., 2006. Position and time specify the migration of a pioneering population of olfactory bulb interneurons. Dev. Biol. 297 (2), 387–401.

Vassar, R., et al., 1994. Topographic organization of sensory projections to the olfactory bulb. Cell 79 (6), 981–991.

Wedden, S.E., 1987. Epithelial-mesenchymal interactions in the development of chick facial primordia and the target of retinoid action. Development 99 (3), 341–351.

Whitesides 3rd, J.G., LaMantia, A.S., 1996. Differential adhesion and the initial assembly of the mammalian olfactory nerve. J. Comp. Neurol. 373 (2), 240–254.

Whitman, M.C., Greer, C.A., 2009. Adult neurogenesis and the olfactory system. Prog. Neurobiol. 89 (2), 162–175.

Wichterle, H., Turnbull, D.H., Nery, S., Fishell, G., Alvarez-Buylla, A., 2001. *In utero* fate mapping reveals distinct migratory pathways and fates of neurons born in the mammalian basal forebrain. Development 128 (19), 3759–3771.

Wood, H.B., Episkopou, V., 1999. Comparative expression of the mouse Sox1, Sox2 and Sox3 genes from pre-gastrulation to early somite stages. Mech. Dev. 86 (1–2), 197–201.

Wray, S., Grant, P., Gainer, H., 1989. Evidence that cells expressing luteinizing hormone-releasing hormone mRNA in the mouse are derived from progenitor cells in the olfactory placode. Proc. Natl. Acad. Sci. U. S. A. 86 (20), 8132–8136.

Yoon, S.O., et al., 1996. Adenovirus-mediated gene delivery into neuronal precursors of the adult mouse brain. Proc. Natl. Acad. Sci. U. S. A. 93 (21), 11974–11979.

Yoshida, M., et al., 1997. Emx1 and Emx2 functions in development of dorsal telencephalon. Development 124 (1), 101–111.

Yoshihara, S., Omichi, K., Yanazawa, M., Kitamura, K., Yoshihara, Y., 2005. Arx homeobox gene is essential for development of mouse olfactory system. Development 132 (4), 751–762.

Chapter 21

Development of the Inner Ear

Zoë F. Mann and Matthew W. Kelley

Laboratory of Cochlear Development, National Institute for Deafness and other Communication Disorders, NIH Porter Neuroscience Research Center, Bethesda, MD

Chapter Outline

GLOSSARY

Atoh1 A basic helix-loop-helix transcription factor that is necessary and sufficient for hair cell formation.

Lateral inhibition A conserved developmental process in which a population of equipotential cells become sorted into one of more unique phenotypes through a process in which some cells acquire a particular fate and then express membrane bound ligands that bind to membrane receptors on neighboring cells to prevent those cells from developing with the same fate.

Mechanosensory hair cell Specialized sensory cells found in the inner ear of birds, fish, amphibians, reptiles and mammals. Modified microvilli located at the lumenal surface of each cell detect small movements or pressure waves.

Neurosensory progenitor cells Multipotent progenitor cells found in the ventral anterior region of the otocyst. These cells have the ability to develop as both the neurons comprising the VIIIth ganglion or the hair cells and supporting cells that constitute the sensory epithelia of the inner ear.

Organ of Corti The specialized sensory epithelium of the mammalian cochlea. It contains a highly regular mosaic of hair cells and supporting cells. Hair cells are arranged in four distinct rows while supporting cells separate each hair cell from each of its neighbors.

Otic placode An ectodermal thickening that develops in the dorsal anterior region of the embryo adjacent to the developing hindbrain

Otocyst A hollow sphere derived from the otic placode. As development proceeds, it will undergo a series of morphological changes to give rise to all of the structures within the inner ear.

Spiral ganglion A cluster of afferent neurons that develop from neurosensory cell progenitors and act as the primary relay between hair cells and the central nervous system.

SUMMARY

- The inner ear acts as the primary mediator for the sensations of hearing, balance and acceleration.
- Most of the structures within the inner ear develop from the otic placodes, paired epithelial thickenings located adjacent to the developing hind brain.
- A key step in development of the inner ear is specification of identity along the main body axes at the otocyst stage. Recent advances have identified roles for the Hedgehog, Wnt and Retinoic acid signaling pathways in this process.

Principles of Developmental Genetics.
2015 Published by Elsevier Inc.

- The anterior ventral region of the otocyst contains neurosensory progenitor cells that will develop as both the neurons that form the VIIIth cranial nerve and the sensory structures within the ear itself.
- Each sensory structure within the inner ear contains a mosaic of mechanosensory hair cells, a unique type of sensory cell, and supporting cells, a non-sensory cell type that provides trophic and structural support to the hair cells.
- The development of this mosaic utilizes a highly conserved signaling cassette which includes bHLH transcription factors and the Notch signaling pathway. Through positive and inhibitory interactions, subsets of cells within the developing mosaics are specified as either hair cells or supporting cells.

21.1 INTRODUCTION

The vertebrate inner ear acts as the primary transducer for the perception of hearing, balance and acceleration. As a result of its diverse functions, the mature structure is an elaborate combination of uniquely shaped regions, including the semicircular canals and cochlear duct. However, embryonically all of these structures are derived from the otic placode, a thickening of the surface ectoderm located immediately adjacent to the developing hindbrain. As development proceeds, the placode invaginates and separates from the surface ectoderm to form a spherical otocyst. Soon after closure, a subset of cells delaminates from the otocyst to give rise to the cochleovestibular ganglion (VIIIth cranial nerve, also called the vestibulocochlear nerve), while the remaining cells undergo an elaborate morphogenetic process to form the rest of the inner ear. Although the morphological development of this structure has been described in detail, the understanding of the molecular and genetic pathways, both intrinsic and extrinsic, that regulate its development has remained limited until recently. However, the advent of advanced molecular techniques, in combination with mouse genetics, has resulted in a rapid increase in the understanding of these processes over the last few years. This chapter will review the specific developmental events required for inner ear formation, and discusses the genetic understanding of those events.

21.2 ANATOMY OF THE INNER EAR

The inner ear is a fluid-filled structure containing a complex series of epithelium-lined ducts and canals, all connected by a central lumen (Figure 21.1). This membranous labyrinth is surrounded by an osseous structure referred to as the bony

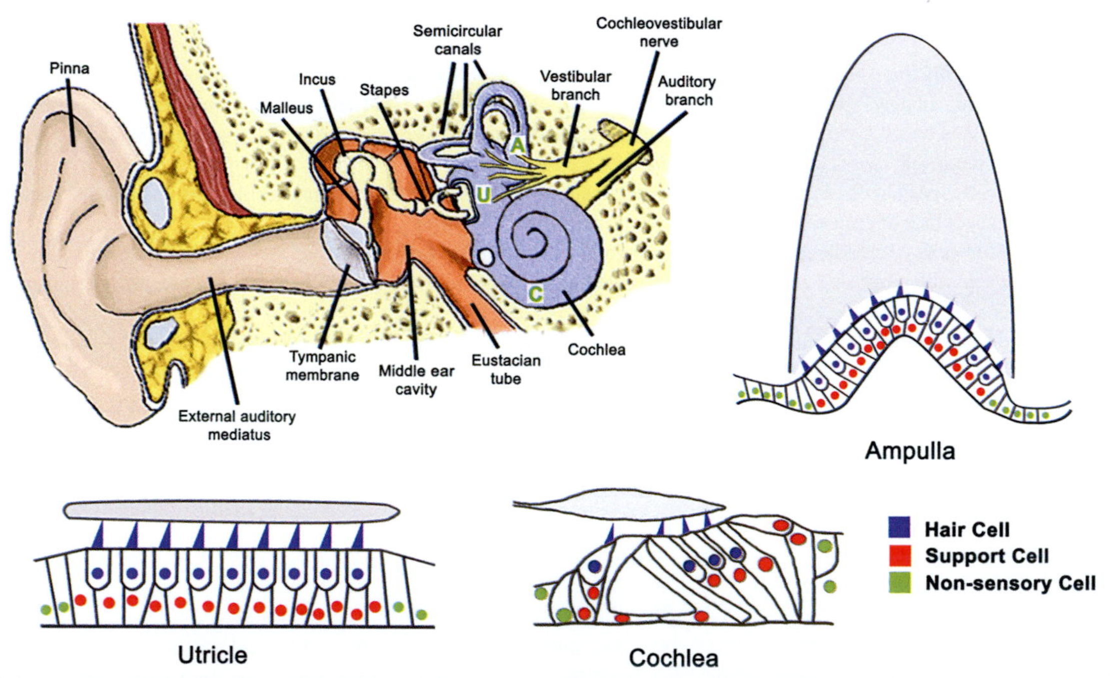

FIGURE 21.1 Anatomy of the Ear. Illustration of the inner ear in humans. The pinna and auditory meatus concentrate sound waves onto the tympanic membrane. Vibrations of the membrane are conducted through the ossicular chain (malleus, incus, stapes) to the oval window and into the inner ear and cochlea. A cross-section of the sensory epithelium of the cochlea, the organ of Corti, is illustrated with sensory hair cells, supporting cells and non-sensory cells indicated. In addition to the organ of Corti, the inner ear also contains the sacculus and utriculus, which detect gravitational acceleration and three ampullae, each of which is associated with a specific semicircular canal, and regulates sense of balance. The structure of each of these epithelia is illustrated as well.

labyrinth. In general, the ear can be divided into three distinct regions. Dorsally, three semicircular canals are oriented orthogonally to one another in order to mediate balance in all three planes of movement. The medial region of the structure contains two otolithic organs (containing calcium-carbonate based otoliths), the utriculus and sacculus, which are responsible for the detection of linear and gravitational acceleration. Structures related to auditory perception are located ventrally. In virtually all mammals, the auditory portion of the inner ear is represented by a coiled cochlear duct which extends between one and three quarter and three and three quarter turns, depending on the species. In non-mammalian vertebrates, such as birds and some classes of reptile, the auditory region of the inner ear exists as a curved, elongated tube referred to as the basilar papilla. Each of the structures described above includes a sensory epithelium (see below), which receives innervation from branches of the VIIIth cranial nerve (Figure 21.1). The nerve contains bipolar afferent neurons derived from the otocyst, and glia cells derived from the migrating neural crest.

Each of the inner ear structures described above includes a sensory epithelium comprised of mechanosensory hair cells and non-sensory supporting cells arranged in a regular mosaic. The epithelia are pseudostratified, with hair cells located at the lumenal surface, whereas supporting cells form one or more layers between the hair cells and a basement membrane. Each hair cell contains a cluster of modified microvilli, called stereocilia, that project from the lumenal surface. Movement of these stereocilia in response to movement or sound causes a depolarization of the cell. The basal surface of each hair cell forms multiple synapses with neurons from the VIIIth nerve. Depolarization of the hair cell increases the rate of tonic release of neurotransmitter, which leads to generation of action potentials in the post-synaptic neurons.

Formation of the Otic Placode and Otocyst

The majority of the cells that give rise to the membranous labyrinth of the inner ear are derived from the otic placode. Development of this placode begins as a transient thickening of the ectodermal epithelium at the edge of the neural plate adjacent to the developing hindbrain. Initial formation generally occurs during mid-somatogenesis. Shortly after formation, the otic placode invaginates to give rise to an otic cup that pinches off to form the otocyst or otic vesicle. Although a complete review of otic induction is beyond the scope of this chapter, it is important to discuss several aspects of this process. The otic placode forms within the neural border zone that surrounds the anterior neural plate and is referred to as the pre-placodal region because these cells are poised to develop as cranial placodes (see Chapter 19). In response to different inductive signals, different populations of pre-placodal cells give rise to unique placode types (Chen and Streit, 2013). The otic placode is characterized by the unique expression of a number of genes including, but not limited to, *Pax2*, *Pax8*, *Eya1* and *Bmp4* (Chen and Streit, 2013). Its location is specified as a result of inductive signals emanating from a number of structures located in the same region of the embryo, including the developing hindbrain, surrounding mesenchyme (referred to as periotic mesenchyme) and even nearby endoderm (Bok et al., 2005; Wu and Kelley, 2012). The specific molecular signaling pathways that induce otic formation apparently vary somewhat between species, but members of the fibroblast growth factor (Fgf) pathway derived from mesenchyme, endoderm and neural tube play prominent roles (Yang et al., 2013). Roles for Wnt signaling have also been shown in some species but not in others (Freter et al., 2008).

Determination of Axes

Once the otocyst has formed, the next key step in its formation is the specification of identity along its three axes; dorsal-ventral (DV), anterior-posterior (AP) and medial-lateral (ML). Transplant studies in which developing otocysts were rotated around one axis demonstrated that fixation of the axes occurs fairly early after otocyst closure (Wu et al., 1998; Bok et al., 2005). However as will be discussed below, early markers of asymmetry within the otocyst are present prior to axis fixation, suggesting the existence of an initial period of plasticity. The results of these studies also indicated differences in the timing of fixation, with the AP axis being established first, followed by the DV and finally the LM axes. The factors that are known to play a role in the specification of each of these axes will be discussed below.

Formation of the AP Axis

The first indication of the formation of the AP axis is the asymmetric expression of specific genes within the early otocyst. Strong expression of *Lunatic fringe (Lfng), Fibroblast growth factor 10 (Fgf10), Six Homeobox 1 (Six1),* and *SRY-box containing gene 2 (Sox2)* is evident in the anterior region of the otocyst, whereas expression of each of these genes is weak in the posterior region (Morsli et al., 1998; Kiernan et al., 2005b; Bok et al., 2007; Wu and Kelley, 2012). This anterior region contains the neuro-sensory precursors that will give rise to both neural and sensory cells within the mature ear. The posterior region, by comparison, primarily gives rise to non-sensory structures, such as the endolymphatic duct, but does also contribute to the posterior crista. Recent work with both the chicken and the mouse has implicated an important role for Retinoic acid (RA), a morphogen critical for multiple stages of embryogenesis in specification of AP identity (Shimozono et al., 2013). RA is a derivative of Vitamin A, and its biological levels are regulated by the expression of specific metabolic and catabolic enzymes in different cell types. At the otocyst stage, an RA gradient is present along the AP axis as a result of differential expression of the RA synthesizing enzyme retinalydehyde dehydrogenase 2 (*Raldh2*) posteriorly and the degrading enzyme cytochrome P450-associated RA catabolizing enzyme (*Cyp26c1*) anteriorly (Reijntjes et al., 2004; Sirbu et al., 2005; Bok et al., 2011b). As a result, cells located in the anterior otocyst have only a brief exposure to RA, leading to their development as neurons or sensory structures, whereas cells in the posterior otocyst experience higher levels of RA and develop as non-sensory structures (Bok et al., 2011a). Experimental manipulations of the levels of RA along the AP axis lead to changes in gene expression and morphology of the inner ear that are consistent with posteriorization of the otocyst in the presence of high levels of RA and anteriorization of the otocyst when RA synthesis is inhibited. While the specific targets of RA have not been determined, one hypothesis is that RA induces a posterior fate through activation of the T-box transcription factor *Tbx1* (Bok et al., 2011a), a known negative regulator of otic neurogenesis and thus of anterior fate (Raft et al., 2004). In addition to RA, Fgfs and Bone morphogenetic proteins (Bmps) also play roles in specifying the AP axis. Fgf8 is expressed in the anterior ectoderm in the chicken where it promotes neurogenic (anterior) cell fates through upregulation of *Sox3* (Abello et al., 2010). Developmental studies in other systems, such as the limb bud (Duester and Zhao, 2008) have demonstrated functional antagonism between Fgfs and RA (Cunningham and Duester, 2012), a phenomenon that could occur in the otocyst as well. Finally, Bmps play an inductive role in the specification of non-neuronal (posterior) cell fates through activation of the transcription factor *Lmx1b* (Abello et al., 2010).

Patterning of the DV Axis

Overall patterning of the DV axis relies largely on neuronally derived external signals in both mammalian and avian systems (Giraldez, 1998; Bok et al., 2005). In particular, Wnts secreted from the dorsal hindbrain and Sonic hedgehog secreted from the ventral floor plate and notochord act to pattern multiple structures along the DV axis including the spinal cord and inner ear (Riccomagno et al., 2005b). Within the inner ear, Shh secreted from the ventral midline specifies different cell fates long the DV axis by controlling the relative levels of Gli2/Gli3 activator and repressor activity (Bok et al., 2007). The ventral-most portion of the cochlea requires high levels of Shh to induce Gli2/Gli3 activator activity while the proximal region of the cochlea, along with the medially-located utricle and saccule require significantly lower levels of Shh and only limited Gli2/Gli3 activator activity for normal development. Finally, more dorsal structures are dependent upon higher levels of Gli3 repressor activity for correct formation (Bok et al., 2007; Brown and Epstein, 2011b). Opposing the ventral Shh signaling center is a dorsal center located in the hindbrain that generates Wnt1 and Wnt3a. These factors induce expression of dorsal specific genes such as Dlx5/6 and Gbx2 while also regulating expression of Gli3 (Ohyama et al., 2006; Alvarez-Medina et al., 2008; Riccomagno et al., 2005b).

Formation of the Medial-Lateral Axis

Considerably less is known about the factors that specify the ML axis. *Kreisler* mutants which have hindbrain defects resulting from a mutation in *MafB*, have inner ear defects that are similar to those observed following deletion of *Gastrulation Brain Homeobox 2 (Gbx2)*, a gene that is restricted to the medial side of the otocyst (Lin et al., 2005; Choo et al., 2006; Miyazaki et al., 2006). In both cases, development of lateral structures within the inner ear appeared relatively normal while medial structures were significantly disrupted. *Gbx2*, as well as *Paired Box Gene 2 (Pax2)* are expressed at early time points in otocyst development, suggesting that the medial portion of the otocyst is specified quite early, perhaps even before any other axes (Wu and Kelley, 2012). Finally, the medial-lateral axis may play an important part in development of the VIIIth nerve as neuroblasts that will develop as the vestibular portion of the nerve delaminate from the lateral side of the otocyst whereas auditory neuroblasts arise from the medial half (Koundakjian et al., 2007; Bell et al., 2008).

21.3 SPECIFICATION OF NEURAL AND SENSORY FATES: A COMMON ORIGIN FOR NEURONAL AND PROSENSORY CELLS

A key effect of the establishment of the AP axis is the specification of the neuro-sensory domain (NSD) in the anterior otocyst. Signaling along the ML axis then specifies the types of neurons formed within the NSD (i.e., vestibular or auditory) and the DV axis acts to restrict the NSD to the ventral portion of the otocyst (Riccomagno et al., 2005b; Ohyama et al., 2006; Brown and Epstein, 2011b). As discussed, overlapping expression domains of specific genes, including *Six1*, *Eya1*,

Sox2 and *Lfng,* can be used to roughly define the NSD in the developing otocyst. Fate-mapping studies have demonstrated that in both avian and mammalian systems, cells within the cochleovestibular ganglion (CVG) give rise to CVG neurons, sensory hair cells and associated supporting cells through a common lineage (Satoh and Fekete, 2005; Raft et al., 2007). However, individual clones show extensive variability with some clones comprising only neurons, some hair cells and supporting cells and others a mixture of all three. These results suggest that the NSD may contain regional heterogeneity, in which some areas contain cells that will develop with a uniform cell fate (neurons or hair cells/supporting cells) while other regions may be more variable.

Specification of Inner Ear Cell Fates

The earliest markers associated with specification of an individual cell fate within the otocyst are the neurogenic basic helix-loop-helix transcription factors *Ngn1* and *NeuroD* in a subset of cells within the NSD, expressed soon after otocyst closure. Deletion of either gene results in either a complete (*Ngn1*) (Ma et al., 2000; Matei et al., 2005) or nearly complete (*NeuroD1*) (Liu et al., 2000) loss of all neurons within the CVG. Similarly, deletion of *Sox2* also leads to an absence of CVG neurons (Dabdoub et al., 2008; Evsen et al., 2013). Following the onset of *Ngn1* expression, these cells delaminate from the otic epithelium, migrate a short distance, and coalesce to form the nascent CVG (Ma et al., 1998, 2000). Subsequent rounds of mitosis increase the size of the CVG prior to neuronal differentiation. In addition to *Ngn1* and *NeuroD*, neural precursor cells within the otocyst express the Notch ligand *Delta-like-1(Dll1)* (Adam et al., 1998; Brooker et al., 2006). While the specific role of *Dll1* has not been determined, Notch signaling plays an important role in sorting of common progenitors toward different fates. Based on this, it may be the case that binding of Dll1 to its receptor Notch in adjacent cells prevents some cells within the NSD from developing as neurons. A preliminary report indicated an increase in the size of the CVG in *Dll1* mutants, but this observation has not been confirmed.

Once the CVG precursors have delaminated from the otocyst, the majority of the remaining cells within the NSD are believed to give rise to the prosensory cells that will develop as hair cells and supporting cells. As is the case for the neuronal population, the specification of cell fate within this population is directly dependent on the expression of Sox2, as an otocyst-specific deletion of *Sox2* leads to a complete loss of prosensory cells and prosensory-derived structures (Kiernan et al., 2005b). Similarly, deletion of *Eya1* also leads to a loss of prosensory formation, either through loss of *Sox2* or because other, as yet unidentified, *eya1* target genes are downregulated (Ahmed et al., 2012). Prosensory formation is also altered in the absence of the Notch ligand, Jagged1 (Jag1). Jag1 is expressed in prosensory cells from relatively early time points and is maintained in the subset of prosensory cells that go on to develop as supporting cells. Deletion of *Jag1* or the Notch target gene *Rbpj* leads to a decrease in the size of the prosensory population and therefore to decreased numbers of hair cells and supporting cells (Kiernan et al., 2005a; Brooker et al., 2006; Basch et al., 2011; Yamamoto et al., 2011). However, examination of the expression of *Sox2* indicates that the initial formation of the prosensory population is normal, at least in *Rbpj* mutants, but as development continues, *Sox2* expression is lost suggesting that the cells fail to maintain a prosensory identity.

Overall, these results indicate that axial specification leads to the formation of the NSD in the anterior region of the otocyst. Once the NSD is formed, a subset of those cells becomes specified to develop as CVG precursors, with the majority of the remaining cells developing as prosensory cells. The demonstration of varied clonal lineages within this population suggests that extrinsic factors play a role in regulating the relative proportion of cells that are directed to the neuronal versus prosensory populations.

21.4 DEVELOPMENT OF THE SEMICIRCULAR CANALS AND CRISTAE

Following closure of the otocyst and delamination of the neural progenitors, the cells that will form the membranous labyrinth undergo extensive morphological growth and modification to give rise to the elaborate structures of the inner ear, including the three sensory cristae and associated semicircular canals that are required for detection of angular head movements and maintenance of balance, and the coiled cochlear duct, required for hearing (Figure 21.2). The development of these structures will be discussed in turn. The three cristae and their associated non-sensory canals are derived from flattened outgrowths originating from the dorsal surfaces of the otocyst, which are referred to as vertical and horizontal canal pouches (Wu and Kelley, 2012). The anterior and posterior canals arise from the vertical canal pouch, while the lateral canal develops from the horizontal pouch. Prosensory cells from the NSD give rise to the sensory cristae while cells contributing to the inner ear canals originate from a region near the presumptive cristae, known as the canal genesis zone (Chang et al., 2004). Overall, it is thought that the sensory cristae govern development of the semicircular canals and that in the absence of correct cristae formation the canals will no longer form correctly (Kiernan et al., 2005b; Chang et al., 2008). The precise

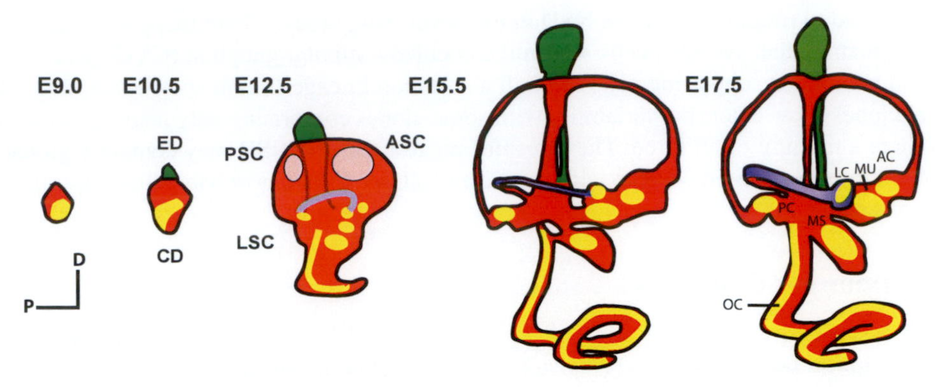

FIGURE 21.2 Development of the Inner Ear. Illustrations of the developmental changes in the membranous labyrinth of the inner ear as it progresses from otocyst to maturity. Ages are for mouse. Following closure of the otic cup to form an otocyst, the neurosensory domain (yellow) becomes specified in the anterior ventral region of the duct. By E10.5, two extensions, a dorsal protrusion that will give rise to the endolymphatic duct (ED) and a ventral protrusion that will give rise to the cochlear duct (CD), are present. Cells derived from the neurosensory domain will give rise to the sensory epithelia of the inner ear. By E12.5, the cochlear and endolymphatic ducts are more pronounced and the three semicircular canals have begun to form through remodeling of the tissue (pink areas). As development continues, the inner ear assumes its mature morphology with formation of the canals, continued extension of the cochlear duct and refinement of the sensory epithelia.

mechanism by which the cristae dictate canal formation is not presently known, however there are data implicating important roles for Bmp and Fgf signaling pathways (Gerlach et al., 2000; Chang et al., 2004; 2008). The development of the inner ear cristae and semicircular canals was recently reviewed in more detail (Wu and Kelley, 2012).

The development of the cristae themselves has been shown to be dependent on expression of Wnt1 and Wnt3a derived from the dorsal hindbrain (Riccomagno et al., 2005a; Wu and Kelley, 2012). Wnt signaling induces expression of *Dlx5* (Acampora et al., 1999) while Fgf3, also originating from the dorsal hindbrain, plays a role in the expression of *Hmx3* (Wang et al., 1998), both of which are required for development of the cristae and the semicircular canals. Another signaling molecule important for cristae and canal formation is Shh. As discussed, Shh is released from the ventral hindbrain, and in *Shh*$^{-/-}$ mutants, the lateral canal is missing and the anterior and posterior canals develop with abnormal morphologies (Brown and Epstein, 2011a). However, this severe phenotype was not observed in inner ear specific-conditional *Smoothened* (*Smo*) knockouts. Since Smo is a necessary receptor for Shh signaling, this result suggests that the requirement for Shh in canal development is most probably indirect (Brown and Epstein, 2011a). A likely intermediary is the periotic mesenchyme surrounding the developing inner ear. Surgical manipulations *in ovo* have demonstrated a role for these cells in both canal and cristae development (Liang et al., 2010). The precise molecular mechanisms underlying the developmental role of the mesenchyme in canal formation remain unclear, but defects in two mesenchymal genes *Pou3f4* and *Prx* have been shown to alter canal formation as well (ten Berge et al., 1998; Phippard et al., 1999; Sobol et al., 2005).

21.5 DEVELOPMENT OF THE COCHLEAR DUCT AND ORGAN OF CORTI

In mammals, the primary auditory sensory epithelium, the organ of Corti, is contained in the coiled cochlear duct. Overall, this structure is a remarkable example of cellular organization and patterning. The duct initiates as a ventral out-pocketing of the otocyst that extends and coils as it develops (Figure 21.3). Outgrowth and coiling are dependent on numerous extrinsic factors and genetic pathways, many of which arise from surrounding tissues. In particular, Shh emanating from the notochord and floor plate induces ventral (auditory) identity and regulates cochlear outgrowth (Riccomagno et al., 2002). In addition, mutations in two transcription factors expressed in periotic mesenchyme, *Tbx1* and *Pou3f4,* also result in abnormal coiling and significant shortening of the cochlear duct (Phippard et al., 1999; Braunstein et al., 2008; 2009), although the nature of the specific signals have not been determined.

Formation of the Organ of Corti

As the duct extends, the epithelial cells that line the lumen become regionalized. Cells on the ventral surface will develop as non-sensory structures, including the stria vascularis, spiral ligament and Reissner's membrane, while cells on the dorsal surface will give rise to the organ of Corti, and the inner and outer sulci. When outgrowth begins, the cells along the floor of the duct are morphologically homogenous (Figure 21.3). However, a subset of cells already expresses prosensory markers including *Jag1*, *Sox2* and *Lfng*. The exact mechanisms that regulate the boundaries of the prosensory domain are unknown,

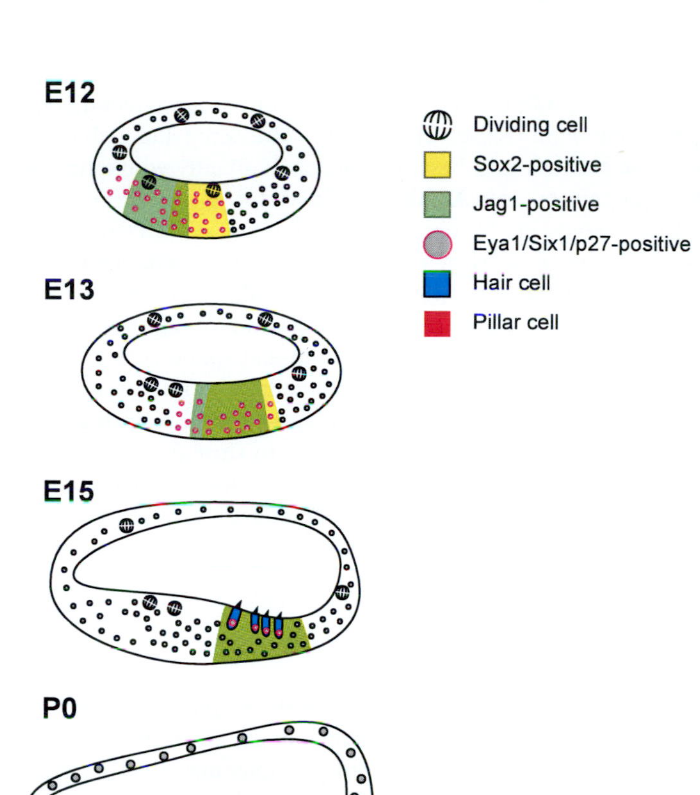

FIGURE 21.3 Development of the Cochlear Duct. Cross-sectional views of the developing mouse cochlear duct with developmental ages for mouse illustrated. The floor of the duct, which contains the organ of Corti, is illustrated at the bottom of each section. At E12 the duct appears homogenous with ongoing proliferation throughout. However, expression data indicate that *Jagged1* (green) and *Sox2* (yellow) are expressed in overlapping populations of cells within the floor of the duct. By E13, the prosensory domain (positive for *Jag1*, *Sox2*, *Eya1*, *Six1* and *p27*) has formed and the cells within this domain have become post-mitotic. By E15, developing hair cells (blue) are present within the prosensory domain and overall proliferation within the duct has decreased. Finally, by P0, the developing organ of Corti (OC), containing hair cells (blue), pillar cells (magenta) and supporting cells can be identified.

however roles for both Bmp4, originating on the lateral side of the duct, and Shh, originating near the medial side of the duct, have been proposed (Driver et al., 2008; Ohyama et al., 2010).

As development continues, prosensory cells undergo a wave a terminal mitosis that begins in the apex and extends progressively towards the base. This pattern of cell cycle exit is dependent on expression of the cell cycle inhibitor $p27^{klp1}$, beginning in the apex and proceeding towards the base (Chen and Segil, 1999). Deletion of $p27^{klp1}$ results in prolonged proliferation of prosensory cells, an overproduction of both hair cells and supporting cells, and deafness. Terminal mitosis still occurs in the absence of $p27^{klp1}$, but it is delayed and proceeds in a reverse gradient, beginning in the base of the cochlea and extending towards the apex. Cellular mitosis is a complex process that is controlled through expression of combinations of both cell cycle activators (CDKs) and cell cycle inhibitors (CKIs), many of which are expressed in the developing cochlea. Because of the importance of cell cycle regulation in the development of regenerative strategies for hair cell loss, numerous studies have generated a nearly comprehensive understanding of which CDKs and CKIs are most important for regulation of cell cycle in the cochlea (Schimmang and Pirvola, 2013). Perhaps two of the most intriguing aspects of these studies have been the observations that additional cell divisions lead to supernumerary cells within the cochlea, suggesting that cell death is not an important part of normal cochlear development (Schimmang and Pirvola, 2013), and that forced cell cycle re-entry in the adult cochlea often leads to cell death, demonstrating that simply driving HCs to divide has significant consequences.

Development of Cochlear Hair Cells

Following terminal mitosis, individual cells within the duct begin to differentiate. Surprisingly, differentiation occurs in a basal-to-apical gradient; the reverse of the gradient of terminal mitosis (Anniko, 1983). As a result, cells located in the apex of the cochlea remain in a post-mitotic, undifferentiated state for an extended period of time. The first cells to differentiate are the mechanosensory hair cells. Two types of hair cells are present in the organ of Corti; inner hair cells (IHCs), which function as the primary sensory receptors, and outer hair cells (OHCs), which modulate the sensitivity and selectivity of the cochlea to a given sound stimulus (Figure 21.3). During development, IHCs develop before OHCs at any given position along the basal-to-apical axis. The earliest marker for hair cell formation is the expression of the bHLH transcription factor Atoh1 (Anniko, 1983; Bermingham et al., 1999; Lanford et al., 2000; Chen et al., 2002). Atoh1 is a proneural gene that regulates the development of multiple neuronal subtypes in the brain (Klisch et al., 2011; Mulvaney and Dabdoub, 2012). Atoh1 expression is first observed in prosensory cells at the cochlear mid-base, nearly concomitantly with terminal mitosis, and then this rapidly extends towards the apex with developmental age (Anniko, 1983; Lanford et al., 2000; Chen et al., 2002; Driver et al., 2013). As development proceeds, Atoh1 expression becomes restricted to hair cells. Deletion of *Atoh1* leads to a complete absence of hair cells, while forced expression of Atoh1 in prosensory cells, or even in other cells within the cochlear duct, can induce hair cell formation at early developmental time points (Woods et al., 2004; Zhang et al., 2007). Although the precise downstream targets of Atoh1 in hair cells have not been determined, several recent studies have identified targets of Atoh1, referred to as a targetome, in other tissues including the cerebellum and spinal cord (Klisch et al., 2011). Several of those targets, including *Barhl1* and *Hes6*, are expressed in cochlear hair cells as well (Klisch et al., 2011). However, a similar "targetome" has not yet been determined for Atoh1 in hair cells. Nor have the factors the regulate expression of Atoh1 within prosensory cells been fully characterized. However, Atoh1 is known to strongly promote its own expression through a positive feedback loop (Helms et al., 2000). This mechanism probably accounts for much of the ongoing expression of Atoh1 but does not explain how initial onset occurs. Recent work has demonstrated a key role for β-catenin and the canonical Wnt signaling pathway (Shi et al., 2010; Jacques et al., 2012), however which specific Wnt molecules are involved and the nature of their signaling has not been determined.

Atoh1, Sox2, Six1 and Eya1

As discussed, Sox2, an SRY-related HMG-box transcription factor, is required for prosensory development. Moreover, Sox2 has also been shown to induce prosensory and, in some cases, HC development (Dabdoub et al., 2008; Neves et al., 2013). The inductive abilities of Sox2 in this context are related to its ability to bind the Atoh1 promoter directly. However, the interaction between Sox2 and Atoh1 is not that straightforward, in that Sox2 also antagonizes Atoh1 expression under some circumstances (Dabdoub et al., 2008; Neves et al., 2013). Early in cochlear development, Sox2 is instructive for prosensory formation and is required for Atoh1 expression. However, as development continues, Sox2 is downregulated in those prosensory cells that will develop as HCs (Dabdoub et al., 2008; Neves et al., 2011; 2013; Freeman and Daudet, 2012). Moreover, mouse models with reduced, but not absent, Sox2 expression, actually show premature hair cell differentiation along with an increase in the number of HCs (Dabdoub et al., 2008). These results suggest the hypothesis that Sox2 acts to antagonize expression of Atoh1 and that downregulation of Sox2 is a key step in HC formation (Dabdoub et al., 2008; Liu et al., 2012).

More recently, Six1, a homeodomain protein, and Eya1, a phosphatase and co-activator, both of which are required for normal development of the cochlea and inner ear, have been shown to induce HC formation when ectopically co-expressed in cells located in close proximity to the prosensory region. Interestingly, this effect was inhibited by co-expression of Sox2, even though interactions between the three factors, Sox2, Eya1 and Six1 were shown to facilitate expression of Atoh1. These results highlight the complexity of the role of Sox2 during HC development (Ahmed et al., 2012; Liu et al., 2012).

Notch Signaling in HC-Supporting Cell (SC) Formation

The Notch signaling pathway is a highly conserved mechanism of cell-cell communication that regulates cell fate and differentiation in multiple tissues (Fiuza and Arias, 2007; see also Chapter 1). Both Notch ligands and their receptors are transmembrane proteins meaning that cell-cell contact is necessary for signaling. The "classical" Notch signaling loop has been revealed through a series of elegant experiments in *Drosophila* (Martinez Arias et al., 2002; D'Souza et al., 2010; Heitzler, 2010; Guruharsha et al., 2012). Briefly, one or more of the Notch receptors are expressed within a group of cells with equivalent potential to develop as one of multiple cell types (an equivalence group). As development proceeds, a proneural bHLH protein, such as Atoh1, is upregulated within some or all of the cells within the group. Expression of the proneural molecule leads to expression of a Notch ligand, which then binds to Notch on a neighboring cell, inducing the expression of an inhibitory bHLH molecule and ultimately downregulation of expression of the original proneural bHLH protein (Fiuza and Arias, 2007). *Notch1* is expressed throughout the developing cochlear sensory epithelium and four Notch ligands, *Jagged2 (Jag2), Delta-like 1(Dll1), Delta/Notch-like EGF-related receptor (Dner)* and *Delta-like 3 (Dll3)* are expressed in the developing HCs (Lanford et al., 2000; Kiernan et al., 2005a; Hartman et al., 2007; 2010). Similarly, multiple inhibitory bHLH molecules including *Hes1, Hes5* and *Hey1* are expressed in developing SCs (Lanford et al., 2000; Zine et al., 2001; Hayashi et al., 2008). Deletion of many of these factors leads to an overproduction of HCs, as would be predicted if Notch signaling acts to limit HC production (Zhang et al., 2000; Zine et al., 2001; Kiernan et al., 2005b). Overall, these results demonstrate a canonical role for Notch signaling in the sorting of prosensory cells into HCs and SCs.

It is important to consider the multiple roles of Notch signaling during development of the inner ear (summarized in Figure 21.4). As discussed, at the early otocyst stage, Jagged1-Notch interactions maintain the expression of both *Sox2* and *Prox1* in prosensory patches, and forced ectopic expression Jagged1 or activated Notch can induce ectopic prosensory domains (Daudet and Lewis, 2005; Kiernan et al., 2006; Daudet et al., 2007; Basch et al., 2011; Yamamoto et al., 2011; Ahmed et al., 2012). However, at later time points, expression of multiple Notch ligands in developing HCs acts to establish the characteristic mosaic structure of the organ of Corti through lateral inhibition (Lanford et al., 1999; Daudet and Lewis, 2005; Kiernan et al., 2005b). The molecular basis accounting for the multiple roles of Notch signaling have not been determined. The inhibitory bHLH molecules that are expressed at the two different stages differ, but it remains unclear how those targets are determined and regulated. In addition, while Notch1 seems to be the most likely receptor for both stages, nuclear localization of the intracellular domain of Notch1 is only observed at time points that coincide with the second phase of Notch signaling (Murata et al., 2006). Hopefully additional, more focused experiments will reveal important aspects of the two different effects of Notch during inner ear development.

Development of Cochlear Supporting Cells

As discussed, all inner ear sensory epithelia are comprised of a mosaic of HCs and SCs. While HCs are similar in function to neurons, SC function is similar to glia in that they provide trophic support to the HCs. In the vestibular epithelia, all SCs have a morphology resembling undifferentiated epithelial cells, whereas in the cochlea, SCs develop into multiple differentiated cell types including Deiters' cells, Hensen's cells, border cells and pillar cells. In stark comparison with HCs, significantly less is known about SC development (Wu and Kelley, 2012). Previous work has demonstrated that ectopic HCs can induce neighboring cells to develop as SCs (Woods et al., 2004), leading to the hypothesis that HCs release a factor or factors that induce SC formation.

While the nature of SC-inducing factors is largely unknown, recent work has identified some of the signaling pathways that lead to the formation of pillar cells. *Fibroblast growth factor receptor 3* (*Fgfr3*), a tyrosine kinase, is expressed by

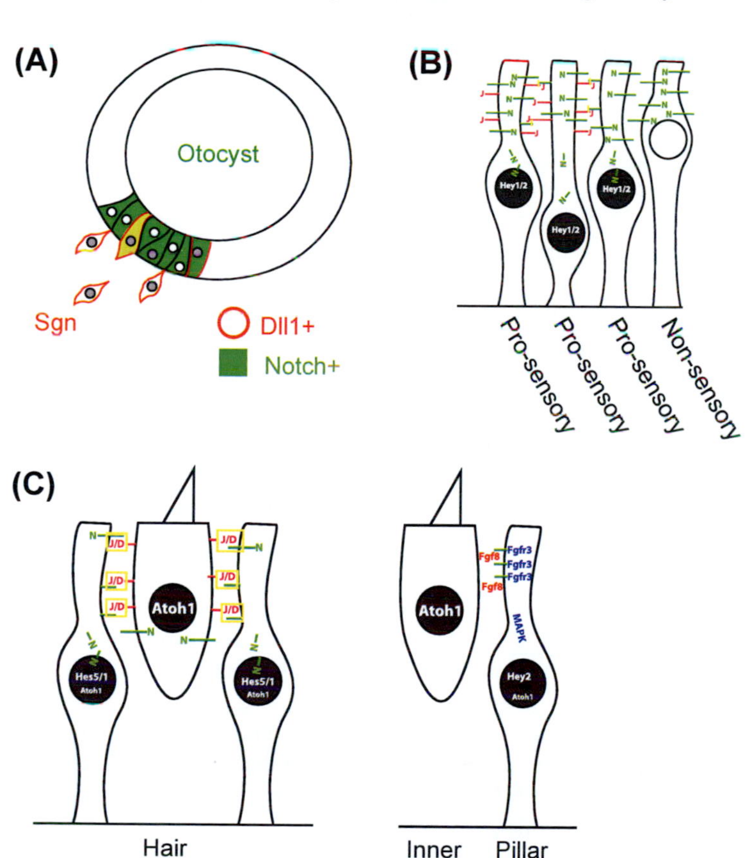

FIGURE 21.4 Roles of Notch Signaling During Inner Ear Development. (A) At the otocyst stage, Notch is expressed broadly within the otocyst while Delta-like 1 (Dll1) is expressed in a subset of cells within the neurosensory region. These cells delaminate from the otocyst to become the neural precursors that will give rise to the spiral ganglion (Sgn). (B) Following delamination, expression of Jagged1 in populations of cells induces those cells to develop as prosensory cells. While the nature of the interaction is not fully understood, the likely signaling pathway is illustrated. Cells within the Jagged1-positive (J) population reciprocally activate Notch (N) in neighboring cells to induce prosensory identity, most likely through expression of Hey1 and/or Hey2. In contrast, cells located one cell diameter from the edge of the Jagged1-positive population fail to activate Notch and develop as non-sensory cells. (C) Finally, as development continues, prosensory cells become partitioned into hair cells and supporting cells. The primary pathway for this sorting is expression of multiple Notch ligands, including Jagged2, Delta-like 1 (J/D) in hair cells. Binding to Notch1 leads to induction of Hes5 and Hes1 in neighboring cells. Hes expression inhibits expression of Atoh1, preventing these cells from developing as hair cells. Finally, in the case of inner hair cells and pillar cells, activation of a Notch target, Hey2, is controlled through binding of Fgf8, generated by inner hair cells, to Fgfr3, expressed by developing pillar cells.

pillar cells beginning at early stages of cochlear development and mutations in *Fgfr3* lead to deafness in both humans and mice (Colvin et al., 1996; Puligilla et al., 2007). Fibroblast growth factor 8 (Fgf8), a ligand for Fgfr3, is expressed exclusively within the cochlea in IHCs, which are located directly adjacent to developing pillar cells (Jacques et al., 2007). This spatial arrangement is consistent with an inductive interaction between the two cell types, a hypothesis that was confirmed by experiments showing that targeted deletion of either *Fgfr3* or *Fgf8* leads to defects in pillar cell formation (Colvin et al., 1996; Mueller et al., 2002; Jacques et al., 2007; Szarama et al., 2013). The Fgf8/Fgfr3 signaling interaction has two functions; firstly, it induces prosensory cells to develop as pillar cells (Jacques et al., 2007; Szarama et al., 2012) and secondly, it prevents the same cell from developing as HCs, as cochleae from *Fgfr3* mutants contain extra OHCs (Hayashi et al., 2007; Puligilla et al., 2007). Interestingly, the role of Fgf signaling in the regulation of pillar cell fate seems to continue beyond the embryonic period. *Sprouty2*, a member of a family of negative regulators of FGF is expressed in developing pillar cells and Deiters' cells, and seems to act to limit the effects of Fgfr3 activation to those cells located closest to the IHCs. Deletion of *Sprouty2* results in an increase in the number of pillar cells, which apparently occurs through transformation of Deiters' cells to pillar cells during the post-natal time period (Shim et al., 2005).

Finally, while the signaling pathways that are activated through Fgfr3 to initiate pillar cell formation have not been determined, the mechanism underlying inhibition of HC formation is understood. As discussed, activation of Notch is a key factor in preventing HC formation. However, pillar cells still form in the organ of Corti even following inhibition or disruption of Notch signaling. Interestingly, developing pillar cells uniquely express the Notch target gene, *Hey2*, but expression of this factor is controlled through FGF, rather than Notch, signaling (Doetzlhofer et al., 2009). This interaction demonstrates an interesting modification of signaling pathways, presumably in response to developmental and/or evolutionary influences (Doetzlhofer et al., 2009).

Innervation of the Inner Ear

Sensory hair cells in both the vestibular and auditory epithelia of the inner ear receive afferent innervation from the cochleovestibular (also called the vestibulocochlear) nerve. As discussed, the progenitors of these fibers originate in the otocyst and delaminate fairly soon after otocyst closure. The developing neuroblasts migrate away from the otocyst and then coalesce to form the primary neurons of the CVG. As development proceeds, the CVG can be subdivided into the spiral ganglion (SG), which innervates the auditory system, and the vestibular ganglion, which innervates HCs in the vestibular system.

Within the auditory system, developing SG neurons extend peripheral axons that reach the epithelium at about the same time that HCs begin to differentiate. While initially indistinguishable, these fibers eventually develop as either type I fibers which innervate primarily IHCs, or type II fibers which primarily innervate OHCs. Type I SGNs are large bipolar cells that comprise 90–95% of the total SGN population. Type II SGNs are smaller, bipolar or pseudobipolar, and make up only 5–10% of the total SGN population (Nayagam et al., 2011). In general, each type I neuron synapses with a single IHC, whereas each type II neuron forms synapses with roughly 5–10 OHCs (Rubel and Fritzsch, 2002). Moreover, type II fibers typically extend along a single row of OHCs, making *en passant* synapses as they pass towards the basal end of the cochlea.

By comparison with the organ of Corti, significantly less is known regarding the molecular mechanisms and signaling pathways that regulate development of type I and type II SG neurons or how their unique patterns of innervation develop. A number of outstanding reviews have been published recently (Appler and Goodrich, 2011; Nayagam et al., 2011; Coate and Kelley, 2013), therefore, only a brief discussion will be presented here. Developing neurites that are fated to develop as either type I or type II neurons pass through periotic mesenchyme before penetrating the basement membrane of the sensory epithelium near the developing IHCs. Individual neurites, regardless of whether they will develop as type I or type II neurons, remain in contact with developing IHCs until OHCs begin to develop. At that point, many fibers extend through the pillar cells to make contact with OHCs. Whether an individual cell that extends a fiber towards the OHCs at this time is already determined to develop as a type II fiber is not clear. Reports of both exuberant fiber extension followed by retraction, and fibers with clear type I or type II morphologies have been published, with the caveat that both the methods and species used were not consistent between all studies (Echteler, 1992; Huang et al., 2007; Appler and Goodrich, 2011). At the molecular level, the best characterized signaling interactions known to influence outgrowth and innervation, are related to two neurotrophins, Brain-derived neurotrophic factor (BDNF) and Neurotrophin-3 (NT3), and their receptors TrkB and TrkC (Fritzsch et al., 1997b; 1999; Farinas et al., 2001). Both BDNF and NT3 are expressed in dynamic and overlapping regions of the developing sensory epithelium beginning at time points which precede the arrival of most SG neurites and extending into adulthood. Similarly, developing SG neurons express TrkB and TrkC mRNA beginning at early time points and extending throughout development (Ernfors et al., 1992; Ylikoski et al., 1993). Deletion of any one of the four interacting genes, or of combinations, leads to significant and complicated defects in cochlear innervation patterns, including mis-routed fibers and extensive SG loss (Fritzsch et al., 1997a; Postigo et al., 2002; Tessarollo et al., 2004). More recently, an interaction between the axon guidance molecules EphA4, expressed on SG neurons, and Ephrin-A5, expressed on HCs, has been shown to prevent a subset of type I fibers from crossing the pillar cells and entering the OHC region (Defourny et al., 2013). Many other members of various axon guidance pathways have been shown to be expressed in the developing cochlea, suggesting potential signaling roles, but functional confirmation of the involvement of these pathways has been limited (Murakami et al., 2001; Wong et al., 2002; Fekete and Campero, 2007; Coate et al., 2012). With the development of more precise tools for both the temporal and spatial deletion of particular genes, it seems likely that a better understanding of cochlear innervation should soon be achieved.

CONCLUSIONS

Most of the cells that will give rise to the inner ear arise from the otocyst, a placode derived structure that forms adjacent to the developing hindbrain. As development proceeds, the otocyst progresses from an equipotential spherical structure into an elaborate labyrinth of unique cell types and morphologies. The first step in this process is the specification of identity within the otocyst along the three major axes of the body. Once this is completed, unique regions of the otocyst develop as specific inner ear structures including the semicircular canals, which mediate balance, and the coiled cochlea, which acts as the auditory organ. Within the developing cochlea, individual progenitor cells become specified to develop as one of the entire range of cell types required for auditory function, including mechanosensory hair cells and associated supporting cells. Finally, developing neurons within the cochleovestibular nerve, an early derivative of the otocyst, extend neurites that form contacts with HCs to conduct auditory and vestibular information to the brain. Significant progress has been made in the identification of the molecular signaling pathways that mediate many of these events, adding to a more comprehensive knowledge of the development of this system.

CLINICAL RELEVANCE

- Loss of mechanosensory hair cells is the primary cause of both age-related and congenital hearing impairment. Therefore, identification of the factors that induce hair cell formation, such as Atoh1, has potential clinical implications in terms of the development of potential regenerative therapies.
- In some cases of hearing impairment, both hair cells and supporting cells are lost. In this situation, an understanding of the factors that induce the formation of progenitors that can develop as either hair cells or supporting cells will be important.
- Cochlear implants represent the best clinical option for individuals with profound hearing loss or deafness. But outcomes of cochlear implantation are highly dependent on the number of existing spiral ganglion neurons and their connectivity to the implant. Therefore, the identification of factors that promote spiral ganglion neuron survival and guidance could be applied to improve interactions between implants and remaining ganglion neurons, hopefully leading to better outcomes for implant recipients.
- Similarly, decreased vestibular function is an increasing problem in the aging population, and like hearing loss, is thought to result primarily from the loss of sensory hair cells. In contrast with hearing, there are currently no prosthetic therapies available to treat vestibular conditions, highlighting the need for a better understanding of how hair cells develop and might be regenerated.

REFERENCES

Abello, G., Khatri, S., Radosevic, M., Scotting, P.J., Giraldez, F., Alsina, B., 2010. Independent regulation of Sox3 and Lmx1b by FGF and BMP signaling influences the neurogenic and non-neurogenic domains in the chick otic placode. Dev. Biol. 339, 166–178.

Acampora, D., Merlo, G.R., Paleari, L., Zerega, B., Postiglione, M.P., Mantero, S., Bober, E., Barbieri, O., Simeone, A., Levi, G., 1999. Craniofacial, vestibular and bone defects in mice lacking the Distal-less-related gene Dlx5. Development 126, 3795–3809.

Adam, J., Myat, A., Le Roux, I., Eddison, M., Henrique, D., Ish-Horowicz, D., Lewis, J., 1998. Cell fate choices and the expression of Notch, Delta and Serrate homologues in the chick inner ear: parallels with Drosophila sense-organ development. Development 125, 4645–4654.

Ahmed, M., Wong, E.Y., Sun, J., Xu, J., Wang, F., Xu, P.X., 2012. Eya1-Six1 interaction is sufficient to induce hair cell fate in the cochlea by activating Atoh1 expression in cooperation with Sox2. Dev. cell 22, 377–390.

Alvarez-Medina, R., Cayuso, J., Okubo, T., Takada, S., Marti, E., 2008. Wnt canonical pathway restricts graded Shh/Gli patterning activity through the regulation of Gli3 expression. Development 135, 237–247.

Anniko, M., 1983. Cytodifferentiation of cochlear hair cells. Am. J. Otolaryngol. 4, 375–388.

Appler, J.M., Goodrich, L.V., 2011. Connecting the ear to the brain: Molecular mechanisms of auditory circuit assembly. Prog. Neurobiol. 93, 488–508.

Basch, M.L., Ohyama, T., Segil, N., Groves, A.K., 2011. Canonical Notch signaling is not necessary for prosensory induction in the mouse cochlea: insights from a conditional mutant of RBPjkappa. J. Neurosci. 31, 8046–8058.

Bell, D., Streit, A., Gorospe, I., Varela-Nieto, I., Alsina, B., Giraldez, F., 2008. Spatial and temporal segregation of auditory and vestibular neurons in the otic placode. Dev. Biol. 322, 109–120.

Bermingham, N.A., Hassan, B.A., Price, S.D., Vollrath, M.A., Ben-Arie, N., Eatock, R.A., Bellen, H.J., Lysakowski, A., Zoghbi, H.Y., 1999. Math1: an essential gene for the generation of inner ear hair cells. Science 284, 1837–1841.

Bok, J., Chang, W., Wu, D.K., 2007. Patterning and morphogenesis of the vertebrate inner ear. Int. J. Dev. Biol. 51, 521–533.

Bok, J., Raft, S., Kong, K.A., Koo, S.K., Drager, U.C., Wu, D.K., 2011a. Transient retinoic acid signaling confers anterior-posterior polarity to the inner ear. Proc. Natl. Acad. Sci. USA 108, 161–166.

Bok, J., Raft, S., Kong, K.A., Koo, S.K., Drager, U.C., Wu, D.K., 2011b. Transient retinoic acid signaling confers anterior-posterior polarity to the inner ear. Proc. Natl. Acad. Sci. USA 108, 161–166.

Bok, J.W., Bronner-Fraser, M., Wu, D.K., 2005. Role of the hindbrain in dorsoventral but not anteroposterior axial specification of the inner ear. Development 132, 2115–2124.

Braunstein, E.M., Crenshaw 3rd, E.B., Morrow, B.E., Adams, J.C., 2008. Cooperative function of Tbx1 and Brn4 in the periotic mesenchyme is necessary for cochlea formation. J. Assoc. Res. Otolaryngol. 9, 33–43.

Braunstein, E.M., Monks, D.C., Aggarwal, V.S., Arnold, J.S., Morrow, B.E., 2009. Tbx1 and Brn4 regulate retinoic acid metabolic genes during cochlear morphogenesis. BMC Dev. Biol. 9, 31.

Brooker, R., Hozumi, K., Lewis, J., 2006. Notch ligands with contrasting functions: Jagged1 and Delta1 in the mouse inner ear. Development 133, 1277–1286.

Brown, A.S., Epstein, D.J., 2011a. Otic ablation of smoothened reveals direct and indirect requirements for Hedgehog signaling in inner ear development. Development 138, 3967–3976.

Brown, A.S., Epstein, D.J., 2011b. Otic ablation of smoothened reveals direct and indirect requirements for Hedgehog signaling in inner ear development. Development 138, 3967–3976.

Chang, W., Brigande, J.V., Fekete, D.M., Wu, D.K., 2004. The development of semicircular canals in the inner ear: role of FGFs in sensory cristae. Development 131, 4201–4211.

Chang, W., Lin, Z., Kulessa, H., Hebert, J., Hogan, B.L., Wu, D.K., 2008. Bmp4 is essential for the formation of the vestibular apparatus that detects angular head movements. PLoS Genet. 4, e1000050.

Chen, J., Streit, A., 2013. Induction of the inner ear: stepwise specification of otic fate from multipotent progenitors. Hear. Res. 297, 3–12.

Chen, P., Segil, N., 1999. p27(Kip1) links cell proliferation to morphogenesis in the developing organ of Corti. Development 126, 1581–1590.

Chen, P., Johnson, J.E., Zoghbi, H.Y., Segil, N., 2002. The role of Math1 in inner ear development: Uncoupling the establishment of the sensory primordium from hair cell fate determination. Development 129, 2495–2505.

Choo, D., Ward, J., Reece, A., Dou, H.W., Lin, Z.S., Greinwald, J., 2006. Molecular mechanisms underlying inner ear patterning defects in kreisler mutants. Dev. Biol. 289, 308–317.

Coate, T.M., Kelley, M.W., 2013. Making connections in the inner ear: Recent insights into the development of spiral ganglion neurons and their connectivity with sensory hair cells. Semin. Cell. Dev. Biol. 24, 460–469.

Coate, T.M., Raft, S., Zhao, X., Ryan, A.K., Crenshaw 3rd, E.B., Kelley, M.W., 2012. Otic mesenchyme cells regulate spiral ganglion axon fasciculation through a Pou3f4/EphA4 signaling pathway. Neuron 73, 49–63.

Colvin, J.S., Bohne, B.A., Harding, G.W., McEwen, D.G., Ornitz, D.M., 1996. Skeletal overgrowth and deafness in mice lacking fibroblast growth factor receptor 3. Nat. Genet. 12, 390–397.

Cunningham, T.J., Zhao, X., Sandell, L.L., Evans, S.M., Trainor, P.A., Duester, G., 2013. Antagonism between retinoic acid and fibroblast growth factor during limb bud development. Cell Rep. 3(5), 1503–1511.

D'Souza, B., Meloty-Kapella, L., Weinmaster, G., 2010. Canonical and non-canonical Notch ligands. Curr. Top. Dev. Biol. 92, 73–129.

Dabdoub, A., Puligilla, C., Jones, J.M., Fritzsch, B., Cheah, K.S., Pevny, L.H., Kelley, M.W., 2008. Sox2 signaling in prosensory domain specification and subsequent hair cell differentiation in the developing cochlea. Proc. Natl. Acad. Sci. USA 105, 18396–18401.

Daudet, N., Lewis, J., 2005. Two contrasting roles for Notch activity in chick inner ear development: specification of prosensory patches and lateral inhibition of hair-cell differentiation. Development 132, 541–551.

Daudet, N., Ariza-McNaughton, L., Lewis, J., 2007. Notch signaling is needed to maintain, but not to initiate, the formation of prosensory patches in the chick inner ear. Development 134, 2369–2378.

Defourny, J., Poirrier, A.L., Lallemend, F., Mateo Sanchez, S., Neef, J., Vanderhaeghen, P., Soriano, E., Peuckert, C., Kullander, K., Fritzsch, B., Nguyen, L., Moonen, G., Moser, T., Malgrange, B., 2013. Ephrin-A5/EphA4 signaling controls specific afferent targeting to cochlear hair cells. Nat Commun 4, 1438.

Doetzlhofer, A., Basch, M.L., Ohyama, T., Gessler, M., Groves, A.K., Segil, N., 2009. Hey2 regulation by FGF provides a Notch-independent mechanism for maintaining pillar cell fate in the organ of Corti. Dev. Cell. 16, 58–69.

Driver, E.C., Sillers, L., Coate, T.M., Rose, M.F., Kelley, M.W., 2013. The Atoh1-lineage gives rise to hair cells and supporting cells within the mammalian cochlea. Dev. Biol. 376, 86–98.

Driver, E.C., Pryor, S.P., Hill, P., Turner, J., Ruther, U., Biesecker, L.G., Griffith, A.J., Kelley, M.W., 2008. Hedgehog signaling regulates sensory cell formation and auditory function in mice and humans. J. Neurosci. 28, 7350–7358.

Duester, G., Zhao, X.L., 2008. Retinoic acid gives limb development a hand. Faseb J 22.

Echteler, S.M., 1992. Developmental segregation in the afferent projections to mammalian auditory hair cells. Proc. Natl. Acad. Sci. USA 89, 6324–6327.

Ernfors, P., Merlio, J.P., Persson, H., 1992. Cells expressing mRNA for neurotrophins and their receptors during embryonic rat development. Eur. J. Neurosci. 4, 1140–1158.

Evsen, L., Sugahara, S., Uchikawa, M., Kondoh, H., Wu, D.K., 2013. Progression of neurogenesis in the inner ear requires inhibition of Sox2 transcription by neurogenin1 and neurod1. J. Neurosci. 33, 3879–3890.

XXFarinas, I., Jones, K.R., Tessarollo, L., Vigers, A.J., Huang, E., Kirstein, M., de Caprona, D.C., Coppola, V., Backus, C., Reichardt, L.F., Fritzsch, B., 2001. Spatial shaping of cochlear innervation by temporally regulated neurotrophin expression. J. Neurosci. 21, 6170–6180.

Fekete, D.M., Campero, A.M., 2007. Axon guidance in the inner ear. Int. J. Dev. Biol. 51, 549–556.

Fiuza, U.M., Arias, A.M., 2007. Cell and molecular biology of Notch. J. Endocrinol. 194, 459–474.

Freeman, S.D., Daudet, N., 2012. Artificial induction of Sox21 regulates sensory cell formation in the embryonic chicken inner ear. PloS One 7, e46387.

Freter, S., Muta, Y., Mak, S.S., Rinkwitz, S., Ladher, R.K., 2008. Progressive restriction of otic fate: the role of FGF and Wnt in resolving inner ear potential. Development 135, 3415–3424.

Fritzsch, B., Pirvola, U., Ylikoski, J., 1999. Making and breaking the innervation of the ear: neurotrophic support during ear development and its clinical implications. Cell. Tissue Res. 295, 369–382.

Fritzsch, B., Silos-Santiago II, Bianchi, L.M., Farinas II, 1997a. Effects of neurotrophin and neurotrophin receptor disruption on the afferent inner ear innervation. Semin. Cell. Dev. Biol. 8, 277–284.

Fritzsch, B., Silos-Santiago, I., Bianchi, L.M., Farinas, I., 1997b. Effects of neurotrophin and neurotrophin receptor disruption on the afferent inner ear innervation. Semin. Cell. Dev. Biol. 8, 277–284.

Gerlach, L.M., Hutson, M.R., Germiller, J.A., Nguyen-Luu, D., Victor, J.C., Barald, K.F., 2000. Addition of the BMP4 antagonist, noggin, disrupts avian inner ear development. Development 127, 45–54.

Giraldez, F., 1998. Regionalized organizing activity of the neural tube revealed by the regulation of lmx1 in the otic vesicle. Dev. Biol. 203, 189–200.

Guruharsha, K.G., Kankel, M.W., Artavanis-Tsakonas, S., 2012. The Notch signaling system: recent insights into the complexity of a conserved pathway. Nat. Rev. Genet. 13, 654–666.

Hartman, B.H., Nelson, B.R., Reh, T.A., Bermingham-McDonogh, O., 2010. Delta/notch-like EGF-related receptor (DNER) is expressed in hair cells and neurons in the developing and adult mouse inner ear. J. Assoc. Res. Otolaryngol. 11, 187–201.

Hartman, B.H., Hayashi, T., Nelson, B.R., Bermingham-McDonogh, O., Reh, T.A., 2007. Dll3 is expressed in developing hair cells in the mammalian cochlea. Dev. Dyn. 236, 2875–2883.

Hayashi, T., Cunningham, D., Bermingham-McDonogh, O., 2007. Loss of Fgfr3 leads to excess hair cell development in the mouse organ of Corti. Dev. Dyn. 236, 525–533.

Hayashi, T., Kokubo, H., Hartman, B.H., Ray, C.A., Reh, T.A., Bermingham-McDonogh, O., 2008. Hesr1 and Hesr2 may act as early effectors of Notch signaling in the developing cochlea. Dev. Biol. 316, 87–99.

Heitzler, P., 2010. Biodiversity and noncanonical Notch signaling. Curr. Top. Dev. Biol. 92, 457–481.

Helms, A.W., Abney, A.L., Ben-Arie, N., Zoghbi, H.Y., Johnson, J.E., 2000. Autoregulation and multiple enhancers control Math1 expression in the developing nervous system. Development 127, 1185–1196.

Huang, L.C., Thorne, P.R., Housley, G.D., Montgomery, J.M., 2007. Spatiotemporal definition of neurite outgrowth, refinement and retraction in the developing mouse cochlea. Development 134, 2925–2933.

Jacques, B.E., Montcouquiol, M.E., Layman, E.M., Lewandoski, M., Kelley, M.W., 2007. Fgf8 induces pillar cell fate and regulates cellular patterning in the mammalian cochlea. Development 134, 3021–3029.

Jacques, B.E., Puligilla, C., Weichert, R.M., Ferrer-Vaquer, A., Hadjantonakis, A.K., Kelley, M.W., Dabdoub, A., 2012. A dual function for canonical Wnt/beta-catenin signaling in the developing mammalian cochlea. Development 139, 4395–4404.

Kiernan, A.E., Xu, J., Gridley, T., 2006. The Notch ligand JAG1 is required for sensory progenitor development in the mammalian inner ear. PLoS Genet. 2, e4.

Kiernan, A.E., Cordes, R., Kopan, R., Gossler, A., Gridley, T., 2005a. The Notch ligands DLL1 and JAG2 act synergistically to regulate hair cell development in the mammalian inner ear. Development 132, 4353–4362.

Kiernan, A.E., Pelling, A.L., Leung, K.K., Tang, A.S., Bell, D.M., Tease, C., Lovell-Badge, R., Steel, K.P., Cheah, K.S., 2005b. Sox2 is required for sensory organ development in the mammalian inner ear. Nature 434, 1031–1035.

Klisch, T.J., Xi, Y., Flora, A., Wang, L., Li, W., Zoghbi, H.Y., 2011. *In vivo* Atoh1 targetome reveals how a proneural transcription factor regulates cerebellar development. Proc. Natl. Acad. Sci. USA 108, 3288–3293.

Koundakjian, E.J., Appler, J.L., Goodrich, L.V., 2007. Auditory neurons make stereotyped wiring decisions before maturation of their targets. J. Neurosci. 27, 14078–14088.

Lanford, P.J., Shailam, R., Norton, C.R., Gridley, T., Kelley, M.W., 2000. Expression of Math1 and HES5 in the cochleae of wildtype and Jag2 mutant mice. J. Assoc. Res. Otolaryngol. 1, 161–171.

Lanford, P.J., Lan, Y., Jiang, R., Lindsell, C., Weinmaster, G., Gridley, T., Kelley, M.W., 1999. Notch signaling pathway mediates hair cell development in mammalian cochlea. Nat. Genet. 21, 289–292.

Liang, J.K., Bok, J., Wu, D.K., 2010. Distinct contributions from the hindbrain and mesenchyme to inner ear morphogenesis. Dev. Biol. 337, 324–334.

Lin, Z., Cantos, R., Patente, M., Wu, D.K., 2005. Gbx2 is required for the morphogenesis of the mouse inner ear: a downstream candidate of hindbrain signaling. Development 132, 2309–2318.

Liu, M., Pereira, F.A., Price, S.D., Chu, M.J., Shope, C., Himes, D., Eatock, R.A., Brownell, W.E., Lysakowski, A., Tsai, M.J., 2000. Essential role of BETA2/NeuroD1 in development of the vestibular and auditory systems. Genes Dev 14, 2839–2854.

Liu, Z., Owen, T., Fang, J., Srinivasan, R.S., Zuo, J., 2012. *In vivo* Notch reactivation in differentiating cochlear hair cells induces Sox2 and Prox1 expression but does not disrupt hair cell maturation. Dev. Dyn. 241, 684–696.

Ma, Q., Anderson, D.J., Fritzsch, B., 2000. Neurogenin 1 null mutant ears develop fewer, morphologically normal hair cells in smaller sensory epithelia devoid of innervation. J. Assoc. Res. Otolaryngol. 1, 129–143.

Ma, Q., Chen, Z., del Barco Barrantes, I., de la Pompa, J.L., Anderson, D.J., 1998. neurogenin1 is essential for the determination of neuronal precursors for proximal cranial sensory ganglia. Neuron 20, 469–482.

Martinez Arias, A., Zecchini, V., Brennan, K., 2002. CSL-independent Notch signaling: a checkpoint in cell fate decisions during development? Curr. Opin. Genet. Dev. 12, 524–533.

Matei, V., Pauley, S., Kaing, S., Rowitch, D., Beisel, K.W., Morris, K., Feng, F., Jones, K., Lee, J., Fritzsch, B., 2005. Smaller inner ear sensory epithelia in Neurog 1 null mice are related to earlier hair cell cycle exit. Dev. Dyn. 234, 633–650.

Miyazaki, H., Kobayashi, T., Nakamura, H., Funahashi, J., 2006. Role of Gbx2 and Otx2 in the formation of cochlear ganglion and endolymphatic duct. Dev. Growth Differ. 48, 429–438.

Morsli, H., Choo, D., Ryan, A., Johnson, R., Wu, D.K., 1998. Development of the mouse inner ear and origin of its sensory organs. J. Neurosci. 18, 3327–3335.

Mueller, K.L., Jacques, B.E., Kelley, M.W., 2002. Fibroblast growth factor signaling regulates pillar cell development in the organ of corti. J. Neurosci. 22, 9368–9377.

Mulvaney, J., Dabdoub, A., 2012. Atoh1, an essential transcription factor in neurogenesis and intestinal and inner ear development: function, regulation, and context dependency. J. Assoc. Res. Otolaryngol. 13, 281–293.

Murakami, Y., Suto, F., Shimizu, M., Shinoda, T., Kameyama, T., Fujisawa, H., 2001. Differential expression of plexin-A subfamily members in the mouse nervous system. Dev. Dyn. 220, 246–258.

Murata, J., Tokunaga, A., Okano, H., Kubo, T., 2006. Mapping of notch activation during cochlear development in mice: implications for determination of prosensory domain and cell fate diversification. J. Comp. Neurol. 497, 502–518.

Nayagam, B.A., Muniak, M.A., Ryugo, D.K., 2011. The spiral ganglion: connecting the peripheral and central auditory systems. Hear. Res. 278, 2–20.

Neves, J., Vachkov, I., Giraldez, F., 2013. Sox2 regulation of hair cell development: incoherence makes sense. Hear. Res. 297, 20–29.

Neves, J., Parada, C., Chamizo, M., Giraldez, F., 2011. Jagged 1 regulates the restriction of Sox2 expression in the developing chicken inner ear: a mechanism for sensory organ specification. Development 138, 735–744.

Ohyama, T., Mohamed, O.A., Taketo, M.M., Dufort, D., Groves, A.K., 2006. Wnt signals mediate a fate decision between otic placode and epidermis. Development 133, 865–875.

Ohyama, T., Basch, M.L., Mishina, Y., Lyons, K.M., Segil, N., Groves, A.K., 2010. BMP signaling is necessary for patterning the sensory and nonsensory regions of the developing mammalian cochlea. J. Neurosci. 30, 15044–15051.

Phippard, D., Lu, L., Lee, D., Saunders, J.C., Crenshaw 3rd, E.B., 1999. Targeted mutagenesis of the POU-domain gene Brn4/Pou3f4 causes developmental defects in the inner ear. J. Neurosci. 19, 5980–5989.

Postigo, A., Calella, A.M., Fritzsch, B., Knipper, M., Katz, D., Eilers, A., Schimmang, T., Lewin, G.R., Klein, R., Minichiello, L., 2002. Distinct requirements for TrkB and TrkC signaling in target innervation by sensory neurons. Genes Dev 16, 633–645.

Puligilla, C., Dabdoub, A., Brenowitz, S.D., Kelley, M.W., 2010. Sox2 induces neuronal formation in the developing mammalian cochlea. J. Neurosci. 30, 714–722.

Puligilla, C., Feng, F., Ishikawa, K., Bertuzzi, S., Dabdoub, A., Griffith, A.J., Fritzsch, B., Kelley, M.W., 2007. Disruption of fibroblast growth factor receptor 3 signaling results in defects in cellular differentiation, neuronal patterning, and hearing impairment. Dev. Dyn. 236, 1905–1917.

Raft, S., Nowotschin, S., Liao, J., Morrow, B.E., 2004. Suppression of neural fate and control of inner ear morphogenesis by Tbx1. Development 131, 1801–1812.

Raft, S., Koundakjian, E.J., Quinones, H., Jayasena, C.S., Goodrich, L.V., Johnson, J.E., Segil, N., Groves, A.K., 2007. Cross-regulation of Ngn1 and Math1 coordinates the production of neurons and sensory hair cells during inner ear development. Development 134, 4405–4415.

Reijntjes, S., Gale, E., Maden, M., 2004. Generating gradients of retinoic acid in the chick embryo: Cyp26C1 expression and a comparative analysis of the Cyp26 enzymes. Dev. Dyn. 230, 509–517.

Riccomagno, M.M., Takada, S., Epstein, D.J., 2005a. Wnt-dependent regulation of inner ear morphogenesis is balanced by the opposing and supporting roles of Shh. Gene. Dev. 19, 1612–1623.

Riccomagno, M.M., Takada, S., Epstein, D.J., 2005b. Wnt-dependent regulation of inner ear morphogenesis is balanced by the opposing and supporting roles of Shh. Gene. Dev. 19, 1612–1623.

Riccomagno, M.M., Martinu, L., Mulheisen, M., Wu, D.K., Epstein, D.J., 2002. Specification of the mammalian cochlea is dependent on Sonic hedgehog. Gene. Dev. 16, 2365–2378.

Rubel, E.W., Fritzsch, B., 2002. Auditory system development: primary auditory neurons and their targets. Annu. Rev. Neurosci. 25, 51–101.

Satoh, T., Fekete, D.M., 2005. Clonal analysis of the relationships between mechanosensory cells and the neurons that innervate them in the chicken ear. Development 132, 1687–1697.

Schimmang, T., Pirvola, U., 2013. Coupling the cell cycle to development and regeneration of the inner ear. Semin. Cell. Dev. Biol. 24, 507–513.

Shi, F., Cheng, Y.F., Wang, X.L., Edge, A.S., 2010. Beta-catenin up-regulates Atoh1 expression in neural progenitor cells by interaction with an Atoh1 3' enhancer. J. Biol. Chem. 285, 392–400.

Shim, K., Minowada, G., Coling, D.E., Martin, G.R., 2005. Sprouty2, a mouse deafness gene, regulates cell fate decisions in the auditory sensory epithelium by antagonizing FGF signaling. Dev. Cell. 8, 553–564.

Shimozono, S., Iimura, T., Kitaguchi, T., Higashijima, S., Miyawaki, A., 2013. Visualization of an endogenous retinoic acid gradient across embryonic development. Nature 496, 363–366.

Sirbu, I.O., Gresh, L., Barra, J., Duester, G., 2005. Shifting boundaries of retinoic acid activity control hindbrain segmental gene expression. Dev. Biol. 283. 623–623.

Sobol, S.E., Teng, X., Crenshaw 3rd, E.B., 2005. Abnormal mesenchymal differentiation in the superior semicircular canal of brn4/pou3f4 knockout mice. Arch. Otolaryngol. Head Neck Surg. 131, 41–45.

Szarama, K.B., Gavara, N., Petralia, R.S., Chadwick, R.S., Kelley, M.W., 2013. Thyroid hormone increases fibroblast growth factor receptor expression and disrupts cell mechanics in the developing organ of corti. BMC Dev. Biol. 13, 6.

ten Berge, D., Brouwer, A., Korving, J., Martin, J.F., Meijlink, F., 1998. Prx1 and Prx2 in skeletogenesis: roles in the craniofacial region, inner ear and limbs. Development 125, 3831–3842.

Tessarollo, L., Coppola, V., Fritzsch, B., 2004. NT-3 replacement with brain-derived neurotrophic factor redirects vestibular nerve fibers to the cochlea. J. Neurosci. 24, 2575–2584.

Wang, W., Van De Water, T., Lufkin, T., 1998. Inner ear and maternal reproductive defects in mice lacking the Hmx3 homeobox gene. Development 125, 621–634.

Wong, K., Park, H.T., Wu, J.Y., Rao, Y., 2002. Slit proteins: molecular guidance cues for cells ranging from neurons to leukocytes. Curr. Opin. Genet. Dev. 12, 583–591.

Woods, C., Montcouquiol, M., Kelley, M.W., 2004. Math1 regulates development of the sensory epithelium in the mammalian cochlea. Nat. Neurosci. 7, 1310–1318.

Wu, D.K., Kelley, M.W., 2012. Molecular mechanisms of inner ear development. Cold Spring Harbor. Perspect. Biol. 4, a008409.

Wu, D.K., Nunes, F.D., Choo, D., 1998. Axial specification for sensory organs versus non-sensory structures of the chicken inner ear. Development 125, 11–20.

Yamamoto, N., Chang, W., Kelley, M.W., 2011. Rbpj regulates development of prosensory cells in the mammalian inner ear. Dev. Biol. 353, 367–379.

Yang, L., O'Neill, P., Martin, K., Maass, J.C., Vassilev, V., Ladher, R., Groves, A.K., 2013. Analysis of FGF-dependent and FGF-independent pathways in otic placode induction. PloS One 8, e55011.

Ylikoski, J., Pirvola, U., Moshnyakov, M., Palgi, J., Arumae, U., Saarma, M., 1993. Expression patterns of neurotrophin and their receptor mRNAs in the rat inner ear. Hear. Res. 65, 69–78.

Zhang, N., Martin, G.V., Kelley, M.W., Gridley, T., 2000. A mutation in the Lunatic fringe gene suppresses the effects of a Jagged2 mutation on inner hair cell development in the cochlea. Curr. Biol. 10, 659–662.

Zhang, Y., Zhai, S.Q., Shou, J., Song, W., Sun, J.H., Guo, W., Zheng, G.L., Hu, Y.Y., Gao, W.Q., 2007. Isolation, growth and differentiation of hair cell progenitors from the newborn rat cochlear greater epithelial ridge. J. Neurosci. Methods 164, 271–279.

Zine, A., Aubert, A., Qiu, J., Therianos, S., Guillemot, F., Kageyama, R., de Ribaupierre, F., 2001. Hes1 and Hes5 activities are required for the normal development of the hair cells in the mammalian inner ear. J. Neurosci. 21, 4712–4720.

Chapter 22

Molecular Genetics of Tooth Development

Irma Thesleff
Institute of Biotechnology, University of Helsinki, Finland

Chapter Outline

GLOSSARY

Ameloblasts Epithelial cells producing dental enamel.
Ectodermal dysplasias Syndromes that affect two or more ectodermal organs; the most common syndromes involving dental defects.
Enamel knots Signaling centers in dental epithelium that regulate tooth morphogenesis.
Hypodontia Tooth agenesis (the congenital absence of one or more teeth).
Odontogenesis Tooth development.
Odontoblasts Mesenchymal cells producing dentin.
Oligodontia Severe tooth agenesis that affects more than six teeth (wisdom teeth not included).

SUMMARY

- Tooth morphogenesis and cell differentiation are regulated by a chain of reciprocal interactions between the oral epithelium and underlying mesenchyme.
- Most genes that regulate tooth development seem to be associated with the four major signaling pathways: bone morphogenetic protein (BMP)/TGFβ, Wnt, Shh, and fibroblast growth factor (FGF). They are integrated at many levels and affect the functions of each other.
- Placodes and enamel knots are signaling centers in the dental epithelium and they regulate the initiation of teeth and the morphogenesis of the tooth crown, respectively.
- Ectodysplasin (Eda) (a tumor necrosis factor family signal) is a specific regulator of teeth and other organs developing as ectodermal appendages, and it affects the activity of the other major signal pathways. The disruption of the Eda pathway causes the syndrome hypohidrotic ectodermal dysplasia (HED).
- Modulating the various signal pathways in genetically modified mice generates differences in tooth number, shape and size. The initiation of teeth, as well as tooth replacement, appears to depend on the activity of Wnt signaling.
- Mutations in dozens of genes have been shown to cause aberrations in tooth development both in mice and humans. The most common dental defects in humans are hypodontia (missing teeth) and structural aberrations in dentin and enamel.

Principles of Developmental Genetics.

22.1 INTRODUCTION

Teeth develop as appendages of the ectoderm, and their early development shares marked morphological as well as molecular similarities with other ectodermal organs, such as hair, feathers and many glands. Teeth are only found in vertebrates, but they have been lost during evolution in some vertebrate species, e.g., birds and turtles. The ways in which dentitions are organized in different vertebrate species vary greatly. Most fish, amphibians, and reptiles have a homodont dentition (all teeth have a similar shape), and their teeth are replaced throughout life. Mammalian teeth, on the other hand, are sequentially organized into distinct groups from front to back: incisors, canines, premolars, and molars, and they are mostly replaced once during the lifetime (deciduous or milk teeth are replaced by permanent or secondary teeth). The different tooth groups show characteristic differences in morphology (heterodonty). However, there are extensive modifications in the dental formulae within mammals, and, because the dentition is characteristic for each species, the variations in the patterning, numbers, and shapes of teeth have formed the basis for the analysis of fossil records and the understanding of mammalian evolution.

Most of our knowledge of the developmental anatomy of teeth, and almost all of our knowledge of their developmental genetics has come from studies of mice. The central features of the morphogenesis of individual teeth are basically similar in all vertebrates, and, therefore, the information derived from mice is readily applicable to other animals as well as to humans. However, mouse dentition differs from that seen in most other mammals in some aspects. First, they have only one incisor in each half of the jaw (humans have two incisors, and many other mammals have three), and the mouse incisors grow continuously. Second, although mice have three molars like humans and most other mammals, they completely lack canines and premolars. Third, most mammals have two dentitions, but mice have only one dentition, which is not replaced.

There is a long tradition of research into the mechanisms of tooth development, and classic tissue recombination studies have shown that interactions between the epithelium and the underlying neural-crest-derived mesenchyme are instrumental regulators of tooth morphogenesis (Kollar and Baird, 1970; Mina and Kollar, 1987; Lumsden, 1988). On the basis of such studies, together with some more recent work, it is now known that the epithelial-mesenchymal interactions are sequential and reciprocal, and that they regulate both tooth morphogenesis and the differentiation of the cells forming the dental hard tissues (Bei, 2009a; Jussila and Thesleff, 2012; Jussila et al., 2013, reviews). As a result of the advances in mouse genetics and research using informative mouse models we are beginning to understand in greater detail the molecular and genetic bases of tooth morphogenesis.

22.2 DEVELOPMENTAL ANATOMY

The initiation of individual teeth is preceded by the formation of an ectodermal ridge called the *dental lamina* (or *odontogenic band).* In all vertebrates, the dental lamina forms as an epithelial stripe in the mandibular, maxillary, and frontonasal prominences of the embryo, and it marks the future tooth rows and dental arches. In the mouse embryonic jaws, the dental lamina thickens in the incisor and molar regions and forms placodes, which are multilayered epithelial condensations that resemble both morphologically and functionally the placodes of other organs developing as ectodermal appendages. The placodes then proliferate and form buds that intrude into the condensed dental mesenchyme (Figure 22.1). The transition of the bud to the cap stage starts when the epithelial bud invaginates at its tip. The enamel knot, which is a signaling center, forms at this location as an aggregation of epithelial cells, and it regulates the folding and growth of the epithelium. The epithelium flanking the enamel knot grows down, forming the cervical loops. The mesenchymal cells, which become surrounded by the epithelial cervical loops, form the dental papilla. These events determine the extent of the tooth crown. The epithelium differentiates into distinct cell layers and forms the enamel organ. The peripheral part of the condensed dental mesenchyme generates the dental follicle that surrounds the enamel organ epithelium and gives rise to periodontal tissues (Nanci, 2008).

During the following bell stage, the tooth germ grows rapidly, and the shape of the tooth crown becomes evident. The locations of tooth cusps are determined by the secondary enamel knots. These form as epithelial thickenings (similarly to the primary enamel knot), and they specify the points of epithelial folding. During the bell stage, the mesenchymal cells of the dental papilla directly underlining the dental epithelium differentiate into odontoblasts laying down the organic matrix of dentin, and the juxtaposed epithelial cells differentiate into ameloblasts depositing the enamel matrix (see Figure 22.1). Cell differentiation and matrix deposition always start at the tips of the future cusps (i.e., at the sites of the enamel knots). During the entire morphogenesis of the tooth crown, a gradient of differentiation is seen, during which the stage of differentiation decreases in the cusp tip-to-cervical direction.

The root forms after the completion of crown development in mouse molars as well as in all human teeth. Root morphogenesis is guided by the growth of the cervical part of the dental epithelium, and this epithelium does not differentiate into

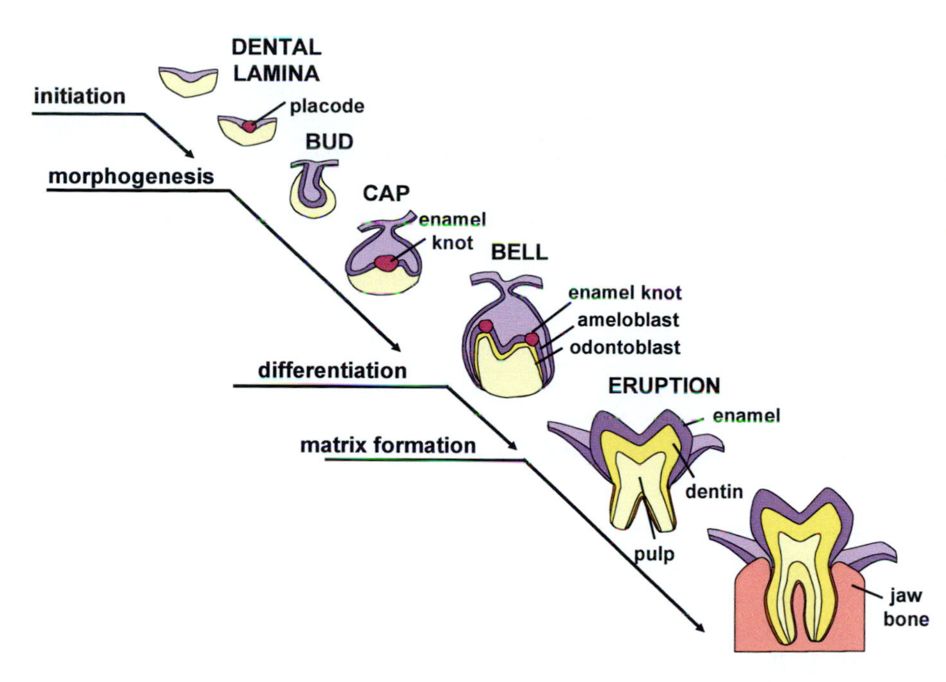

FIGURE 22.1 Development of a Molar Tooth Interactions between the epithelium *(violet)* and the underlying mesenchyme *(yellow)* regulate development. The shape of the tooth crown is determined by the folding and growth of the epithelium. The placodes and enamel knots are signaling centers and important regulators of morphogenesis.

ameloblasts. Instead, the epithelium fragments, and this allows for contacts between the dental follicle cells and the root surface, and for the differentiation of the follicle cells into cementoblasts secreting a thin layer of bone-like cementum. The dental follicle cells also form the periodontal ligament that links the tooth to alveolar bone. The tooth subsequently erupts into the oral cavity (see Figure 22.1).

Dentin resembles bone in its biochemical composition, although its histological appearance is different. Unlike the bone-forming osteoblasts, the odontoblasts do not get incorporated into the dentin matrix. Instead, each odontoblast leaves behind a cytoplasmic process, which becomes embedded in dentin and thereby contributes to the formation of a dentinal tubule. Odontoblast cell bodies remain as a confluent layer between the dentin and the cells of the dental pulp. The enamel matrix is composed of unique enamel proteins, including amelogenin, enamelin, and ameloblastin, which direct the formation and mineralization of enamel into the hardest tissue in the body. After the end of the secretory phase, the ameloblasts regulate the maturation of enamel, and they degenerate with the other layers of enamel epithelium during tooth eruption (Nanci, 2008).

22.3 GENE EXPRESSION PATTERN DATABASES

Dynamic expression patterns during tooth development have been reported for over 300 different genes, and this information has been collected into a graphic database (http://bite-it.helsinki.fi). Most of this information is derived from *in situ* hybridization analyses of embryonic mouse teeth, and, curiously, in the majority of cases, the tooth examined has been the lower (mandibular) first molar. It is noteworthy that, so far, no regulatory gene unique to the teeth has been identified, which apparently reflects the conservation of the genetic regulation of animal development and the multiple functions that developmental regulatory genes have in the embryo.

Most of the genes in the database encode molecules that are associated with various signaling pathways. Although this certainly reflects the interests of the researchers in intercellular communication, it conceivably is an indication that inductive cell interactions constitute the single most important mechanism regulating embryonic development (Gurdon, 1987; see also Chapter 1). The more recent ToothCODE database (http://compbio.med.harvard.edu/ToothCODE/) contains gene expression profile data of early mouse tooth development derived from microarray analyses. These are combined with literature-derived genetic regulatory evidence and signaling network annotations, as well as the Wnt/bone morphogenetic protein (BMP) gene regulatory network underlying epithelial-mesenchymal interactions in tooth development.

The expression of signaling pathway genes is typically reiterated during advancing tooth morphogenesis (Tummers and Thesleff, 2009, O'Connell et al., 2012). Most signaling molecules belong to the four widely used conserved families: transforming growth factor (TGF)-β (includingbone morphogenetic protein, BMP), fibroblast growth factor (FGF), Wnt, and hedgehog (Hh). With the exception of sonic hedgehog (Shh), which is expressed only in the epithelium, the signaling molecules are expressed in both epithelium and mesenchyme, and the receptors are often seen in the adjacent tissue, which indicates that they may have roles in the mediation of tissue interactions. The gene expression patterns have suggested functions for all signaling families in the regulation of the initiation, budding, and epithelial morphogenesis as well as in the differentiation of the dental cells, and many of the suggested roles have been confirmed in functional studies (Tummers and Thesleff, 2009; Bei, 2009a; Jussila et al., 2013, reviews; also described later in this chapter).

Although the gene expression patterns are not direct indicators of functional significance, the information in the gene expression databases is very useful in many ways. For example, the data provide tools for examining the co-expression of genes, for the discovery of possible interactions between genes and molecules, and for the study of signaling networks. Furthermore, as the mouse genome and the genomes of many other vertebrates have been sequenced, the gene expression databases have particular value as a tool for bioinformatics studies. For example, different sets of co-expressed genes can be selected and searched for common gene regulatory elements for the exploration of the principles of tissue-specific gene regulation.

22.4 THE DISRUPTION OF SIGNALING PATHWAYS ARRESTS MOUSE TOOTH DEVELOPMENT

Tooth morphogenesis is arrested at an early stage in a large number of knockout mice, and the reason for this is usually the blocking of one or more signaling pathways (Bei, 2009a). The first reported such mice were the *Msx1* null mutants, and the arrest in tooth development occurred at the transition from the bud to the cap stage (Figure 22.2; Satokata and Maas, 1994). *Msx1* is a transcription factor that functions both downstream and upstream of BMP-4, and its mutant phenotype was later rescued by BMP-4 (Bei et al., 2000). In *Pax9* null mutants, teeth are also arrested at the bud stage, and it was shown that *Pax9* is regulated by FGF and that it is also required for BMP-4 expression in the mesenchyme (Peters et al., 1998). The bud-to-cap stage transition is also blocked in *Lef1* and *Runx2* null mutant mice. *Lef1* is a transcription factor that mediates Wnt signaling, and its function is required in the epithelium, whereas *Runx2* functions in the mesenchyme. Both *Lef1* and *Runx2* mediate FGF signaling between the dental epithelium and the mesenchyme (Kratochwil et al., 2002; Åberg et al., 2004).

In many null mutant mice, tooth development is arrested during initiation and before the formation of the dental placodes (see Figure 22.2). Such arrest in *Gli2/Gli3* double-mutant mice indicated a requirement for Shh signaling (Hardcastle et al., 1998), and a similar phenotype in *Msx1/Msx2* double mutants suggested a role of BMP signaling in tooth initiation (Bei and Maas, 1998). The role of FGF signaling in early tooth development was indicated by conditional deletion of *Fgf8* in the oral epithelium, which resulted in tooth arrest during the initiation stage (Trumpp et al., 1999). The necessary role of Wnt signaling at this stage was demonstrated by the overexpression of the Wnt inhibitor Dkk1 in the ectoderm using the Keratin 14 (K14) promoter (Andl et al., 2002). Taken together, the phenotypes indicate that all four conserved signaling pathways are necessary during the very early stages of tooth development (see Figure 22.2).

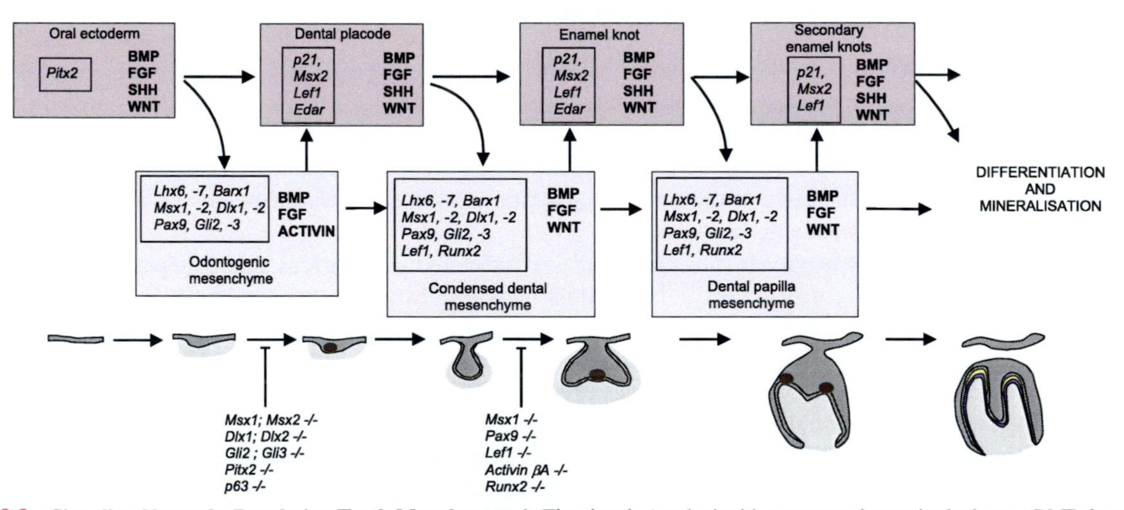

FIGURE 22.2 **Signaling Networks Regulating Tooth Morphogenesis** The signals (marked with uppercase letters in the boxes: BMP, bone morphogenetic protein; FGF, fibroblast growth factor; SHH, sonic hedgehog); WNT, and their target genes (marked in italics in the boxes) expressed in epithelial *(grey)* and mesenchymal *(white)* tissues, reiteratively mediate communication between or within the epithelium and the mesenchyme *(arrows)*. In the epithelium, the signals are mostly expressed in the signaling centers (the dental placode and the primary and secondary enamel knots, marked in black). Knockouts of several genes in mice (listed under the drawing) result in the arrest of tooth morphogenesis at the initiation or bud stage.

The development of all teeth is blocked in the previously described mutants, which indicates that the respective genes are required for the formation of all individual teeth. In the mouse, this means the single incisor and the three molars in each half of the two jaws. However, there are also examples of null mutant mice in which only some teeth fail to form, which suggests that there are differences in genetic regulation among different jaws and tooth types. When the function of the TGFβ protein Activin is knocked out, all teeth except the maxillary molars are arrested before the bud stage, despite the fact that Activin is expressed in all teeth (Ferguson et al., 1998). The *Dlx1/Dlx2* double mutants display the opposite phenotype, and they lack only the maxillary molars (Thomas et al., 1997). In the case of the *Dlx* null mutants, the likely reason is the redundancy of different *Dlx* genes, because *Dlx6* and *Dlx7* are additionally expressed in the maxilla, where they apparently rescue molar development.

22.5 THE GENETIC BASIS OF HUMAN TOOTH AGENESIS

The genes that were shown to be necessary for tooth development in knockout mouse studies have provided candidate genes in the search for gene mutations that cause human tooth agenesis. *Msx1* was the first gene that was shown to be required for mouse tooth development, and it was also the first gene identified behind human oligodontia (Vastardis et al., 1996). *Oligodontia* refers to severe tooth agenesis that affects more than six teeth, besides third molars (Thesleff and Pirinen, 2006; Nieminen, 2009). In addition to missing teeth, the *Msx1* null mice also have a cleft palate, which is also associated with oligodontia in some human patients. *Pax9* is another gene that is required for mouse tooth formation, and it was linked with human tooth agenesis by the candidate gene approach (Stockton et al., 2000). Interestingly, although the *Pax9* null mutant mice have severe malformations in a number of other organs as well, no additional defects have been reported in patients with human oligodontia caused by *Pax9* mutations.

The mutations that have so far been identified in the *Msx1* and *Pax9* genes are loss-of-function mutations, and these result in haplo-insufficiency. The function of the respective genes is reduced, and hypodontia never affects all teeth in the affected individuals. The missing teeth typically represent the last teeth to form in the different tooth classes; this feature characterizes all types of human tooth agenesis. However, there is a remarkable variation in the number of missing teeth between the patients with the same genotypes. The dental phenotypes of *Msx1* and *Pax9* mutations differ from each other in some aspects; the *Pax9* mutations particularly affect the molars. The human *Pax9* oligodontia phenotype was reproduced in mice by gradually reducing the gene dosage in an allelic series of *Pax9* mutant mice (Kist et al., 2005). It was shown that *Pax9* is required during multiple stages of tooth development and that the minimal *Pax9* gene dosage required for the formation of individual teeth increased from the anterior to the posterior molars. These mice provide a useful mouse model for human oligodontia.

Another gene that has been linked with human oligodontia with no associated congenital malformations is *Axin2*. Interestingly, the patients were predisposed to colorectal cancer (Lammi et al., 2004). However, no link to cancer was found in three oligodontia patients with Axin2 mutations (Bergendal et al., 2011). *Axin2* functions in the mediation of Wnt-βcatenin signal together with *APC,* which is the gene most commonly associated with colorectal cancer. The phenotype of oligodontia in these patients is different from those caused by *Pax9* and *Msx1* mutations, because *Axin2* mutations affect the secondary teeth and posterior molars almost exclusively. The normal development of deciduous teeth in these patients suggests that *Axin2* may be required for tooth renewal.

Recently, mutations in the *Wnt10A* gene have been discovered to be the most common cause of oligodontia. In a cohort of patients with non-syndromic oligodontia, more than half exhibited mutations in *Wnt10A* (van den Boogaard et al., 2012). *Wnt10A* mutations had been previously reported to cause severe hypodontia associated with various other clinical symptoms, in particular with features of ectodermal dysplasia syndromes (Adaimy et al. 2007; Bohring et al., 2009).

22.6 DENTAL PLACODES AND THE PATHOGENESIS OF ECTODERMAL DYSPLASIA SYNDROMES

As described previously, tooth development is arrested in many knockout mice before the formation of the dental placodes. Because similar epithelial placodes initiate the development of all organs that form as appendages of the ectoderm (e.g., hair and nails; mammary, salivary, sweat, and sebaceous glands), it is not surprising that tooth defects are often associated with congenital defects in other ectodermal organs (Pispa and Thesleff, 2003; Thesleff and Pirinen, 2006; Mikkola, 2009). It has become evident that the same genes regulate the formation and function of placodes in different ectodermal organs. Typically, signals from all of the four conserved families are required for placode development. Studies mainly on hair and feathers have identified Wnts and FGFs as activators of placode formation and BMPs as inhibitors (Mikkola, 2009; Stark et al., 2007). The available evidence indicates that most of these functions are similar in dental placodes.

Ectodermal dysplasia syndromes are defined as conditions in which two or more types of ectodermal organs are affected, and dental defects in these syndromes typically include multiple missing teeth (oligodontia) and small, misshapen teeth (Figure 22.3). Many genes have been identified in which mutations cause ectodermal dysplasias. Given the similarities in the early development of various ectodermal organs, it is not surprising that these genes encode molecules that regulate placode formation and function. The analysis of the functions of the human ectodermal dysplasia genes in mouse models has elucidated the molecular mechanisms underlying ectodermal placode formation and increased the understanding of the pathogenesis of the human conditions (Mikkola, 2009).

Mutations in the transcription factor p63 cause the EEC syndrome, which is characterized by ectodermal dysplasia, ectrodactyly, and cleft lip and palate (Celli et al., 1999). A typical patient has a severe dental phenotype with multiple missing and misshapen teeth. The *p63* knockout mice lack all ectodermal organs and die at birth (Mills et al., 1999). Detailed analysis of the tooth and hair phenotypes in these mice showed that development is arrested before placode development. The dental lamina forms normally as a normal multilayered epithelium, but the dental placodes fail to form (Laurikkala et al., 2006). Similarly, hair placodes are completely absent. It was shown that the ΔN isoform of *p63* specifically is required for the mediation of several signaling pathways regulating placode formation. ΔNp63 function was necessary for FGF, BMP, and Notch1 signaling, and *Fgfr2b, Bmp7,* and *Notch1* were identified as potential targets of *p63* (Laurikkala et al., 2006).

The positional cloning of genes behind hypohidrotic ectodermal dysplasia (HED) led to the discovery of the ectodysplasin (Eda) signaling pathway, a tumor necrosis factor (TNF) pathway regulating ectodermal organ development (see Figure 22.3A). The characteristic features of HED are oligodontia (see Figure 22.3B), thin and sparse hair, and a severe lack of sweat glands; additional ectodermal defects in the nails and the salivary glands are also common. Mutations in the gene encoding the ligand Eda cause the most common, X-chromosomal form of HED, whereas mutations in the genes encoding the Eda receptor Edar and the signal mediator Edaradd are responsible for two autosomal forms of HED with a similar phenotype (Kere et al., 1996; Headon and Overbeek, 1999). Like other TNF signals, Eda signaling is mediated in the nucleus by the NFκB transcription factor (see Figure 22.3A). HED-ID, a syndrome with all of the features of HED and associated immunodeficiency, is caused by impaired NFκB function as a result of mutations in IKKγ (NEMO), which is an intracellular key component in TNF signaling pathways.

The role of the Eda pathway in the development of teeth and other ectodermal organs has been studied in detail in mice (Mikkola, 2009). The mouse model for X-chromosomal HED is the spontaneous *Eda* null mutant *Tabby*. It has a tooth phenotype that is characterized by missing third molars and sometimes also missing incisors and by small and misshapen crowns of the first molars. The *Eda* null mouse has defects in many ectodermal glands, and lacks the first wave of hair follicles which generate the long guard hairs. Transgenic mice overexpressing Eda in the ectoderm (K14-Eda) have been informative with regard to the role of the Eda pathway and the pathogenesis of the ectodermal defects in HED (Mustonen et al, 2003). It is now known that Eda signaling promotes the formation and growth of ectodermal placodes. The Eda receptor is expressed in the placodes of all ectodermal organs, and, when it is overactivated, the placodes grow larger than normal. This results in the stimulation of ectodermal organ development that is seen as longer hairs, increased sweat excretion, extra mammary glands, and supernumerary teeth (Mustonen et al., 2003). The teeth form in front of the first molar and represent premolars, which were lost early during rodent evolution.

The mechanisms by which Eda regulates placode growth have been addressed in *in vitro* experiments, in which Eda recombinant protein is applied on skin dissected from embryonic mice. Low concentrations of Eda stimulate placode growth, and in high concentrations, it causes fusions of the enlarged placodes. Interestingly, Eda does not stimulate cell proliferation in the

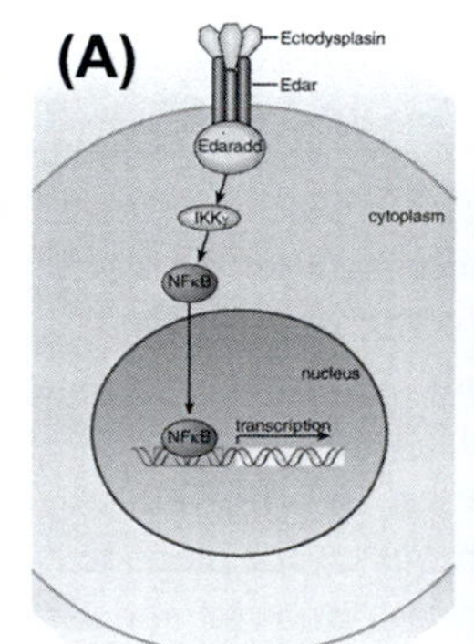

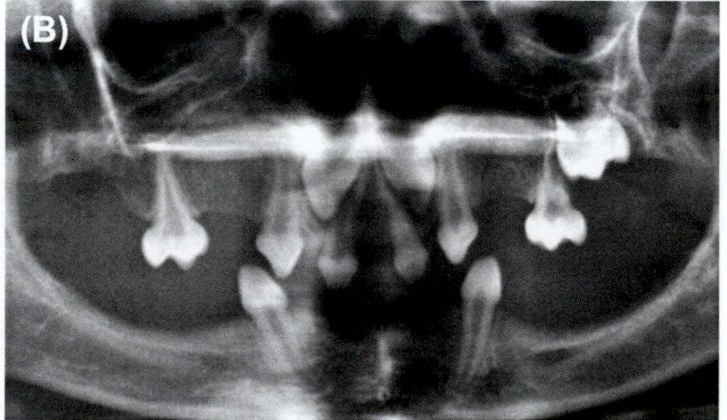

FIGURE 22.3 **Ectodysplasin Regulates the Development of All Ectodermal Organs** (A) The ectodysplasin–Edar signaling pathway. Ectodysplasin is a signal in the tumor necrosis factor family. The disruption of any one of the indicated genes in this pathway results in hypohidrotic ectodermal dysplasia. (B) Severe tooth agenesis resulting from a loss-of-function mutation of the ectodysplasin gene.

placode but rather promotes the incorporation of surrounding ectodermal cells into the placode and the change of their fate from epidermal to placodal cells (Mustonen et al., 2004; Ahtiainen et al., 2014). The molecular mechanisms behind the function of Eda signaling have started to emerge through the discovery of the target genes of Edar. Interestingly, the targets include the BMP inhibitors CCN2 and follistatin, the Wnt inhibitor Dkk4, as well as the signaling molecules Shh, Wnt10b and Fgf20 (Pummila et al., 2007; Fliniaux et al., 2008, Häärä et al., 2012). Hence, Eda signaling apparently regulates the formation of ectodermal placodes by modulating the actions of the four conserved signal pathways. In addition, Eda has functions during the morphogenesis of the ectodermal organs. In teeth, Eda regulates enamel knots and tooth crown pattering (described later).

Intriguingly, prenatal and neonatal injections of recombinant Eda protein rescued the tooth, hair and sweat gland phenotypes of the *Eda* mutant mice (Gaide and Schneider, 2003). Neonatal Eda protein injections had even more dramatic effects in dogs. The $Eda^{-/-}$ dogs have a very severe tooth phenotype, characterized by absence of most permanent premolars and incisors, and Eda protein apparently rescued the development of all these teeth completely (Casal et al., 2007). Ectodermal dysplasia is the first genetic malformation that has been permanently corrected by recombinant protein treatment in animals, and these findings have already led to the preparations for a clinical trial to test whether human X-linked HED can be prevented by neonatal injections of Eda protein.

22.7 ENAMEL KNOTS, TOOTH SHAPES, AND THE FINE-TUNING OF SIGNAL PATHWAYS

In addition to an arrest before placode formation, tooth development is blocked in many knockout mice at the late bud stage, before enamel knot formation and the transition to the cap stage (see Figure 22.2). Like the placodes, the enamel knots are signaling centers (Jernvall et al., 1994), and their function is apparently required for the bud-to-cap stage transition. In addition, the enamel knots have a key role in the regulation of the shape of the molar tooth crown by the induction of secondary enamel knots initiating cusp development (Jernvall et al., 2000). The enamel knots induce the folding and growth of the flanking enamel epithelium and thereby determine the shape of the tooth crown. At least a dozen different signaling molecules belonging to the four major signaling pathways are locally expressed in the primary enamel knot and most of them have also been localized in the secondary enamel knots (Jernvall and Thesleff, 2012).

The function of many enamel knot signals in tooth crown patterning has been demonstrated in genetically modified mouse lines. For example, knocking out *Shh* results in the formation of numerous cusps and in the fusion of molars, whereas in the *Fgf20* knockouts some cusps are missing and the teeth are smaller than normal (Dassule et al., 2000; Häärä et al., 2012). Another example is the knockout of *Sostdc1* (also called ectodin and wise), which is an inhibitor of Wnt and BMP signaling and is expressed around the enamel knot. The first and second molars are fused and there are extra cusps (Kassai et al., 2005; Ahn et al., 2010). The *Eda* null mutant mice (described previously) are characterized by missing teeth as well as small first molars with reduced cusp number. This phenotype is caused by the small primary enamel knot and the fusion of the secondary enamel knots (Pispa et al., 1999). By contrast, the K14-Eda mice overexpressing ectodysplasin in the ectoderm have large molar placodes and enamel knots that explain the extra teeth in front of the first molars and the abnormal cusp patterns (Kangas et al., 2004). Since Eda signaling modulates other signaling pathways (described previously) the effects of Eda overexpression on tooth patterning likely result from their fine-tuning.

The variation of the molar crown phenotypes in mammals supports the hypothesis that the patterning of the tooth cusps is regulated by enamel knot signals via the reaction diffusion mechanism, where the interplay of activators and inhibitors determine the patterns (Salazar-Ciudad and Jernvall, 2002). Curiously, the cusp pattern of the K14-Eda mouse resembles that of kangaroos, and the molar of the *Sostdc1* knockout has some similarity with rhinoceros molars (Figure 22.4,

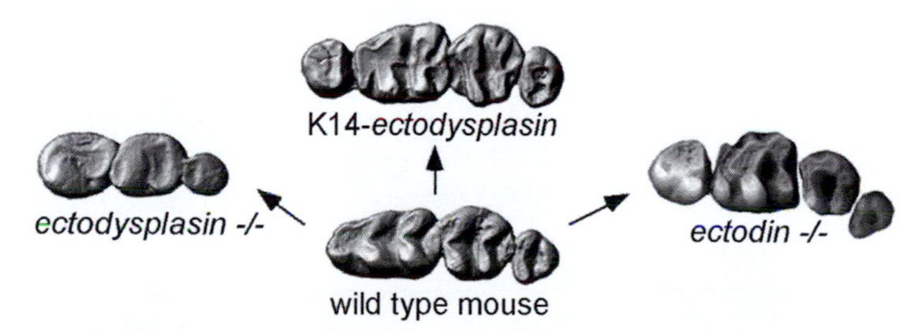

FIGURE 22.4 Fine-Tuning of Cell-Cell Signaling Alters Dental Patterning Molar teeth of transgenic mice show changes in tooth number, size, and cusp pattern. The ectodin *(Sostdc1)* null mutant (Wnt and BMP pathways inhibited) has an extra tooth (premolar) and fused first and second molars, and the cusp pattern resembles that of a rhinoceros. The ectodysplasin null mutant (several other signaling pathways affected) has small teeth with reduced cusps (in addition, the third molar is often missing), and the transgenic mouse overexpressing ectodysplasin (K14-ectodysplasin) has an extra tooth and an altered cusp pattern that resembles that of a kangaroo. *Courtesy of Jukka Jernvall.*

Kangas et al., 2004; Kassai et al., 2005). These phenotypes indicate that changing the dose of a signaling molecule can profoundly alter the species-specific cusp pattern, and thus the current thinking is that fine-tuning of enamel knot signaling may account for the variation of mammalian cusp patterns during evolution (Jernvall and Thesleff, 2012).

22.8 THE GENETIC BASIS OF TOOTH REPLACEMENT

The capacity to generate new teeth has been reduced during evolution. Humans, like most other mammals, can replace their teeth once, while reptiles have the ability to replace their teeth continuously. The genetic and molecular mechanisms of tooth replacement have remained poorly understood. This is mainly because the mouse, which is practically the only model animal used thus far, does not have a secondary dentition. The histological features of replacement tooth formation have been studied in some mammals, such as ferrets and humans (Ooë, 1981; Järvinen et al., 2009), as well as in reptiles (Richman and Handrigan, 2011), and these studies indicate that replacement teeth originate from the so-called successional dental lamina associated with the preceding tooth.

The successional dental lamina apparently contains stem cells and/or progenitor cells which have the capacity to form teeth. In reptiles, including the gecko and the alligator, label-retaining cells expressing some stem cell marker genes have been localized in specific locations in the successional dental lamina (Handrigan et al., 2010; Wu et al., 2013). On the other hand, Sox2 was discovered as a marker gene for all dental lamina cells in several mammalian and reptile species (Juuri et al., 2013); Sox2 is expressed in the primary dental lamina of mouse embryos as well as in the successional dental lamina in ferret and human replacement teeth. It is also present in the dental lamina in reptiles including alligator and gecko. It was suggested that Sox2 is a general marker for naïve dental progenitor cells, which have a capacity for new tooth generation (Juuri et al., 2013). Sox2 is also expressed in the epithelial stem cells which fuel the continuous growth of mouse incisors (Juuri et al., 2012). The mechanisms of the maintenance, proliferation and differentiation of the mouse incisor stem cells are under active investigation in many laboratories around the world (Michon et al., 2013).

Expression of some of the same genes that regulate (primary) tooth development in the mouse has been localized during early formation of replacement teeth in ferrets and reptiles (Järvinen et al., 2009; Handrigan and Richman, 2009; Jussila and Thesleff, 2013; Wu et al., 2013), and interestingly, the activation of the Wnt signaling pathway seems to coincide with the initiation of replacement teeth from the successional dental lamina in all these species. The theory of a key role for Wnt signaling in tooth renewal is supported by the observation that forced activation of the Wnt pathway in the oral epithelium of mouse embryos (*β-cat* $^{ex3K14/+}$) results in the generation of dozens of teeth: when one tooth germ of a mutant embryo was grown as a transplant under the kidney capsule of a nude mouse, it generated more than 40 new teeth during three weeks (Figure 22.5; Järvinen et al., 2006). Transplanted incisor tooth buds formed new incisors, and molars generated molars. When the mutant tooth germs were cultured *in vitro* they produced new tooth buds continuously over 19 days, thus indicating a process of successional tooth formation. It was concluded that the capacity for tooth replacement may depend on the

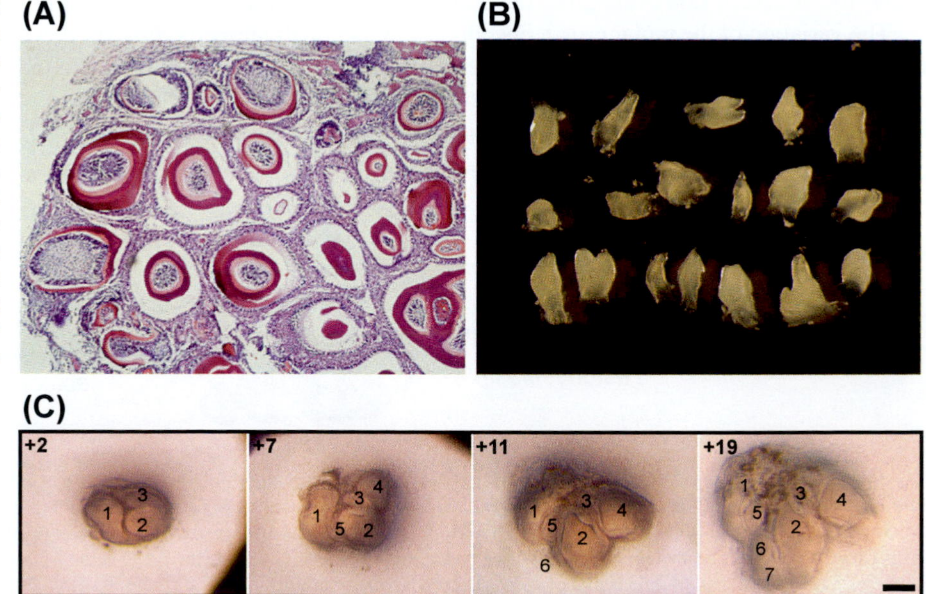

FIGURE 22.5 Activation of Wnt Signaling Stimulates Tooth Renewal (A) Histologic section through a tumor-like outgrowth that developed from one transplanted molar tooth bud of a mouse embryo expressing a stabilized form of β-catenin (*βcat*$^{ex3K14/+}$). Dozens of teeth have formed, and they represent different developmental stages. (B) Mineralized teeth that were dissected from a similar outgrowth. The teeth are small and have roots and mostly simple conical crowns. (C) When a (*βcat*$^{ex3K14/+}$) mutant tooth bud was cultured *in vitro* for 19 days, new teeth were initiated continuously.

activity of Wnt signaling, and that the reduced ability of tooth renewal during mammalian evolution may have involved changes in Wnt signaling (Järvinen et al., 2006).

Although human teeth are normally replaced only once, there is a rare syndrome called *cleidocranial dysplasia* (CCD) that is characterized by multiple supernumerary teeth representing a partial third dentition (Jensen and Kreiborg, 1990). CCD is caused by heterozygous loss-of-function mutations in the *Runx2* gene (Mundlos et al., 1997). The examination of the dental phenotype of the mouse model of CCD (i.e., *Runx2* heterozygote mice) showed that Shh-expressing extra buds form at the lingual aspect of the first molar (Wang et al., 2005). *Runx2* is required for the mediation of FGF signaling during early tooth formation (Åberg et al., 2004), but it is also associated with Wnt signaling in dental mesenchyme (James et al., 2006); hence, the role of *Runx2* in tooth renewal may well be associated with the Wnt pathway.

In addition to tooth replacement, successional formation of teeth takes place when new primary teeth are added within a tooth class; for example when the second and third molars develop from the posterior aspect of the previously developed molar. The two forms of successional tooth development apparently represent variations of the same developmental process. This suggestion is supported by histological similarity of replacement tooth formation and molar addition, and by the demonstration that the second and third molars of the mouse develop from Sox2 positive dental lamina associated with the first molar (Juuri et al., 2013). In addition, more than a third of CCD patients have one or more extra posterior molars in addition to the supernumerary replacement teeth, indicating a similar role of Runx2 in the two processes (Jensen and Kreiborg, 1990).

22.9 THE DEVELOPMENTAL GENETICS OF DENTIN AND ENAMEL FORMATION

Inherited defects in dentin and enamel structure are rare. Dentinogenesis imperfecta and dentin dysplasia are hereditary defects in dentin formation. These are mostly autosomal dominant genetic conditions, in which severe dentin defects affect both the crown and the roots of the teeth (Barron et al., 2008). Mutations in type I collagen, which is the main component of both bone and dentin matrix, cause dentinogenesis imperfecta type 1 which is inherited with osteogenesis imperfecta, and affects both bones and teeth. Most other dentin defects are caused by mutations in the dentin matrix component dentin sialophosphoprotein (DSPP) (Xiao et al., 2001; Barron et al., 2008).

Amelogenesis imperfecta refers to hereditary defects in enamel formation. Mutations have been identified in human patients with amelogenesis imperfecta in genes that encode enamel proteins, such as amelogenin and enamelin (Lagerstrom et al., 1991; Rajpar et al., 2001) and regulatory molecules including KLK4 and MMP20 (Hart et al., 2004; Kim et al., 2005). The mutations are inherited either in an autosomal dominant, autosomal recessive or X-linked manner, and the patients have no other congenital abnormalities (Bei, 2009b; Poulter et al., 2014). Enamel defects occur also as traits in several syndromes, mostly in association with skin diseases and metabolic diseases (Thesleff and Pirinen, 2006). As an example, mutations in the basement membrane component laminin-5 causes amelogenesis imperfecta which is associated usually with the severe skin disease junctional epidermolysis bullosa but may occur also without the skin defects (Poulter et al., 2014).

Many mouse models have been generated for the human dentinogenesis and amelogenesis imperfecta types described above, but there are additionally many transgenic and mutant mouse lines exhibiting abnormal or lacking enamel formation (Bei, 2009b), and a few with defects in dentin development. The mutations in these mice are mostly in genes regulating cell differentiation. Dentin and enamel are produced by columnar mesenchymal and epithelial cells which are unique for teeth: the odontoblasts and the ameloblasts, respectively (see Figure 22.1). These cells are initially juxtaposed in the bell stage tooth germ and separated by a basement membrane, and their differentiation is regulated by reciprocal epithelial-mesenchymal interactions. Most of the same signaling molecules that regulate tooth morphogenesis have also been shown to be important for dental cell differentiation. Signals in the BMP, TGFβ, FGF and Wnt families have been implicated in odontoblast differentiation (Unda et al., 2000; Yamashiro et al., 2007; Kim et al., 2011, 2013). BMPs were originally identified as bone inducers and they stimulate odontoblast differentiation and dentin matrix production as well. This capacity of BMPs has long been examined *in vivo* in attempts to promote dentin regeneration (Nakashima and Reddi, 2003). Wnt/β-catenin signaling is active in odontoblasts and is required to differentiate odontoblasts for root formation (Kim et al., 2013). On the other hand, persistent stabilization of β-catenin in the mouse dental mesenchyme leads to premature differentiation of odontoblasts and differentiation of cementoblasts, and induces excessive dentin and cementum formation *in vivo,* indicating that fine-tuning of Wnt signaling is important for normal dentinogenesis (Kim et al., 2011).

BMPs and Shh have been associated with ameloblast differentiation. Shh signaling is required for the proper polarization of ameloblasts (Gritli-Linde et al., 2002), and it was shown that *Msx2* is required upstream of Shh in ameloblast differentiation (Bei et al., 2004). Evidence from the incisors of some transgenic mice indicates that BMP-4 is the major signaling molecule that regulates ameloblast differentiation and enamel formation. Inhibition of BMP function by overexpression of either follistatin or noggin in the incisor epithelium prevents enamel formation (Wang et al., 2004b; Plikus et al., 2005).

The mouse incisor is traditionally used as a model to study the formation of dentin and enamel, because the incisors grow continuously, and the dental hard tissues are generated throughout the life of the animal.

Enamel formation depends on proper junctions between the columnar ameloblasts and between the ameloblast and the stratum intermedium cell layers. The deletion of the cell adhesion proteins nectin-1 and nectin-3, which are required for desmosome formation, cause mild enamel defects, whereas the deletion of the membrane protein PERP results in more dramatic enamel defects (Barron et al., 2008; Yoshida et al., 2010; Jheon et al., 2011). The nectins as well as PERP were shown to be necessary for the integrity between the ameloblast and stratum intermedium layers.

In many of the mouse mutants exhibiting enamel defects, the morphogenesis of teeth is also disturbed, and this results in missing or extra teeth and aberrant tooth shapes. Examples are mice lacking Shh or *Msx2* expression and mice overexpressing ectodysplasin, Edar, follistatin, or noggin (Bei et al., 2004; Mustonen et al., 2003; Pispa et al., 2004; Wang et al., 2004a,b; Plikus et al., 2005). Accordingly in humans, enamel defects are often associated with other dental anomalies, typically with the reduction of tooth number and size. These observations reflect the fact that the same genes regulate different aspects of tooth development (including initiation, morphogenesis, and the differentiation of the hard-tissue-forming cells) and that the genes are iteratively used during the advancing stages of tooth development.

22.10 CLINICAL RELEVANCE

- The genes and mechanisms that regulate tooth development are remarkably similar to those that regulate other tissues and organs, particularly other organs that develop as ectodermal appendages. In fact, no regulatory genes are known that regulate the morphogenesis of teeth exclusively. Therefore, severe dental aberrations can be expected to be associated with defects in other tissues and organs.
- Different genes may have similar functions, and they act in gene regulatory networks. Therefore it is mostly impossible to predict the causative gene from the phenotype of a dental defect. In addition, many genes (particularly those participating in the mediation of cell-cell signaling) are used reiteratively during tooth development and regulate morphogenesis as well as cell differentiation, and thus their mutations may affect the numbers, shapes, as well as hard-tissue structure of teeth.
- The new knowledge on the genetic mechanisms of tooth initiation and morphogenesis, for example the accumulating evidence indicating that Wnt signals are potent inducers of tooth formation, will hopefully lead to novel therapies aiming at the biological replacement of missing teeth.

RECOMMENDED RESOURCES

Jussila, M., Juuri, E., Thesleff, I., 2013. Tooth development and regeneration. In: Huang, G.T.-J., Thesleff, I. (Eds.), Stem cells in craniofacial development and regeneration. Wiley-Blackwell, John Wiley & Sons, pp. 125–153.

Tooth Gene Expression Database: http://bite-it.helsinki.fi

ToothCODE Database: http://compbio.med.harvard.edu/ToothCODE/

REFERENCES

Åberg, T., Wang, X., Kim, J., Yamashiro, T., Bei, M., Rice, R., Ryoo, H., Thesleff, I., 2004. Runx2 mediates FGF signaling from epithelium to mesenchyme during tooth morphogenesis. Dev. Biol. 270, 76–93.

Adaimy, L., Chouery, E., Megarbane, H., Mroueh, S., Delague, V., Nicolas, E., Belguith, H., de Mazancourt, P., Megarbane, A., 2007. Mutation in WNT10A is associated with an autosomal recessive ectodermal dysplasia: the odonto-onycho-dermal dysplasia. Am. J. Hum. Genet. 81, 821–828.

Ahn, Y., Sanderson, B.W., Klein, O.D., Krumlauf, R., 2010. Inhibition of Wnt signaling by Wise (Sostdc1) and negative feedback from Shh controls tooth number and patterning. Development 137, 3221–3231.

Ahtiainen, L., Lefebvre, S., Lindfors, P.H., Renvoisé, E., Shirokova, V., Vartiainen, M.K., Thesleff, I., Mikkola, M.L., 2014. Directional cell migration but not proliferation drives hair placode morphogenesis. Dev. Cell. 28, 588–602.

Andl, T., Reddy, S.T., Gaddapara, T., Millar, S.E., 2002. WNT signals are required for the initiation of hair follicle development. Dev. Cell. 2, 643–653.

Barron, M.J., Brookes, S.J., Draper, C.E., Garrod, D., Kirkham, J., Shore, R.C., Dixon, M.J., 2008. The cell adhesion molecule nectin-1 is critical for normal enamel formation in mice. Hum. Mol. Genet. 17, 3509–3520.

Bei, M., 2009a. Molecular genetics of tooth development. Curr. Opin. Genet. Dev. 19, 504–510.

Bei, M., 2009b. Molecular genetics of ameloblast cell lineage. J Exp Zool (Mol Dev Evol) 312B, 437–444.

Bei, M., Kratochwil, K., Maas, R.L., 2000. BMP4 rescues a non-cell-autonomous function of Msx1 in tooth development. Development 127, 4711–4718.

Bei, M., Maas, R.L., 1998. FGFs and BMP4 induce both Msx1-independent and Msx1-dependent signaling pathways in early tooth development. Development 125, 4325–4333.

Bei, M., Stowell, S., Maas, R., 2004. Msx2 controls ameloblast terminal differentiation. Dev. Dyn. 231, 758–765.

Bergendal, B., Klar, J., Stecksén-Blicks, C., et al., 2011. Isolated oligodontia associated with mutations in EDARADD, AXIN2, MSX1, and PAX9 genes. Am. J. Med. Genet. A. 155A, 1616–1622.

Bohring, A., Stamm, T., Spaich, C., Haase, C., Spree, K., Hehr, U., Hoffmann, M., Ledig, S., Se, l.S., Wieacker, P., Röpke, A., 2009. WNT10A mutations are a frequent cause of a broad spectrum of ectodermal dysplasias with sex-biased manifestation pattern in heterozygotes. Am. J. Hum. Genet. 85, 97–105.

van den Boogaard, M.J., Créton, M., Bronkhorst, Y., van der Hout, A., Hennekam, E., Lindhout, D., Cune, M., Ploos van Amstel, H.K., 2012. Mutations in Wnt10A are present in more than half of isolated hypodontia cases. J. Med. Genet. 49, 327–331.

Casal, M.L., Lewis, J.R., Mauldin, E.A., et al., 2007. Significant correction of disease after postnatal administration of recombinant ectodysplasin A in canine X-linked ectodermal dysplasia. Am. J. Hum. Genet. 81, 1050–1056.

Celli, J., Duijf, P., Hamel, B.C.J., Bamshad, M., Kramer, B., Smits, A.P.T., Newbury-Ecob, R., Hennekam, R.C.M., Van Buggenhout, G., van Haeringen, B., et al., 1999. Heterozygous germline mutations in the p53 homolog p63 are the cause of EEC syndrome. Cellule 99, 143–153.

Dassule, H.R., Lewis, P., Bei, M., Maas, R., McMahon, A.P., 2000. Sonic hedgehog regulates growth and morphogenesis of the tooth. Development 127, 4775–4785.

Ferguson, C.A., Tucker, A.S., Christensen, L., Lau, A.L., Matzuk, M.M., Sharpe, P.T., et al., 1998. Activin is an essential early mesenchymal signal in tooth development that is required for patterning of the murine dentition. Genes. Dev. 12, 2636–2649.

Fliniaux, I., Mikkola, M.L., Lefebvre, S., Thesleff, I., 2008. Identification of dkk4 as a target of Eda-A1/Edar pathway reveals an unexpected role of ectodysplasin as inhibitor of Wnt signaling in ectodermal placodes. Dev. Biol. 320, 60–71.

Gaide, O., Schneider, P., 2003. Permanent correction of an inherited ectodermal dysplasia with recombinant EDA. Nat. Med. 9, 614–618.

Gritli-Linde, A., Bei, M., Maas, R., Zhang, X.M., Linde, A., McMahon, A.P., 2002. Shh signaling within the dental epithelium is necessary for cell proliferation, growth and polarization. Development 129, 5323–5337.

Gurdon, J.B., 1987. Embryonic induction – molecular prospects. Development 99, 285–306.

Häärä, O., Harjunmaa, E., Lindfors, P., Huh, S.H., Fliniaux, I., Åberg, T., Jernvall, J., Ornitz, D.M., Mikkola, M.L., Thesleff, I., 2012. *Ectodysplasin* regulates activator-inhibitor balance in murine tooth development through *Fgf20* signaling. Development 139, 3189–3199.

Handrigan, G.R., Leung, K.J., Richman, J.M., et al., 2010. Identification of putative dental epithelial stem cells in a lizard with life-long tooth replacement. Development 137, 3545–3549.

Handrigan, G.R., Richman, J.M., 2009. A network of Wnt, hedgehog and BMP signaling pathways regulates tooth replacement in snakes. Dev. Biol. 348, 130–141.

Hardcastle, Z., Mo, R., Hui, C.C., Sharpe, P.T., 1998. The Shh signaling pathway in tooth development – defects in Gli2 and Gli3 mutants. Development 125, 2803–2811.

Hart, P.S., Hart, T.C., Michalec, M.D., Ryu, O.H., Simmons, D., Hong, S., Wright, J.T., 2004. Mutation in kallikrein 4 causes autosomal recessive hypomaturation amelogenesis imperfecta. J. Med. Genet. 41, 545–549.

Headon, D.J., Overbeek, P.A., 1999. Involvement of a novel TNF receptor homologue in hair follicle induction. Nat. Genet. 22, 370–374.

James, M.J., Järvinen, E., Wang, X.-P., Thesleff, I., 2006. The different roles of Runx2 during early neural crest-derived bone and tooth development. J. Bone. Miner. Res. 21, 1034–1044.

Järvinen, E., Närhi, K., Birchmeier, W., Taketo, M.M., Jernvall, J., Thesleff, I., 2006. Continuous tooth generation in mouse is induced by activated epithelial Wnt/β-catenin signaling. J. Bone. Miner. Res. 103, 18627–18632.

Järvinen, E., Tummers, M., Thesleff, I., 2009. The role of the dental lamina in mammalian tooth replacement. J Exp. Zool. (Mol. Dev. Evol.) 312B, 281–291.

Jensen, B.L., Kreiborg, S., 1990. Development of the dentition in cleidocranial dysplasia. J. Oral. Pathol. Med. 19, 89–93.

Jernvall, J., Keränen, S.V.E., Thesleff, I., 2000. Evolutionary modification of development in mammalian teeth: Quantifying gene expression patterns and topography. Proc. Natl. Acad. Sci. USA 97, 14444–14448.

Jernvall, J., Kettunen, P., Karavanova, I., Martin, L.B., Thesleff, I., 1994. Evidence for the role of the enamel knot as a control center in mammalian tooth cusp formation: non-dividing cells express growth stimulating Fgf-4 gene. Int. J. Dev. Biol. 38, 463–469.

Jernvall, J., Thesleff, I., 2012. Tooth shape formation and tooth renewal: evolving with the same signals. Development 139, 3487–3497.

Jheon, A.H., Mostowfi, P., Snead, M.L., Ihrie, R.A., Sone, E., Pramparo, T., Attardi, L.D., Klein, O.D., et al., 2011. PERP regulates enamel formation via effects on cell-cell adhesion and gene expression. J. Cell. Sci. 124, 745–754.

Jussila, M., Juuri, E., Thesleff, I., 2013. Tooth development and regeneration. In: Huang, G.T.-J., Thesleff, I. (Eds.), Stem cells in craniofacial development and regeneration. Wiley-Blackwell, John Wiley & Sons, pp. 125–153.

Jussila, M., Thesleff, I., 2012. Signaling networks regulating tooth organogenesis and regeneration, and the specification of dental mesenchymal and epithelial cell lineages. In: Rossant, J., Tam, P.P.L., Nelson, W.J. (Eds.), Signals, switches and networks in mammalian developmentCold Spring Harb Perspect Biol, 4. (4):a008425.

Juuri, E., Jussila, M., Seide, K., Holmes, S., Wu, P., Richman, J., Heikinheimo, K., Chuong, C.M., Arnold, K., Hochedlinge, r.K., Klein, O., Michon, F., Thesleff, I., 2013. Sox2 marks epithelial competence to generate teeth in mammals and reptiles. Development 140, 1424–1432.

Juuri, E., Saito, K., Ahtiainen, L., Seidel, K., Tummers, M., Hochedlinger, K., Klein, O.D., Thesleff, I., Michon, F., et al., 2012. Sox2+ Stem Cells Contribute to All Epithelial Lineages of the Tooth via Sfrp5+ Progenitors. Dev. Cell. 23, 317–328.

Kangas, A.T., Evans, A.R., Thesleff, I., Jernvall, J., 2004. Nonindependence of mammalian dental characters. Nature 432, 211–214.

Kassai, Y., Munne, P., Hotta, Y., Penttilä, E., Kavanagh, K., Ohbayashi, N., Takeda, S., Thesleff, I., Jernvall, J., Itoh, N., 2005. Regulation of mammalian tooth cusp patterning by ectodin. Science 309, 2067–2070.

Kere, J., Srivastava, A.K., Montonen, O., Zonana, J., Thomas, N., Ferguson, B., Munoz, F., Morgan, D., Clarke, A., Baybayan, P., et al., 1996. X-linked anhidrotic (hypohidrotic) ectodermal dysplasia is caused by mutation in a novel transmembrane protein. Nat. Genet. 13, 409–416.

Kim, J.W., Simmer, J.P., Hart, T.C., et al., 2005. MMP-20 mutation in autosomal recessive pigmented hypomaturation amelogenesis imperfecta. J. Med. Genet. 42, 271–275.

Kim, T.H., Lee, J.Y., Baek, J.A., Lee, J.C., Yang, X., Taketo, M.M., Jiang, R., Cho, E.S., et al., 2011. Consitutive stabilization of b-catenin in the dental mesenchyme leads to excessive dentin and cementum formation. Biochem. Biophys. Res. Commun. 412, 549–555.

Kim, T.H., Bae, C.H., Lee, J.C., Ko, S.O., Yang, X., Jiang, R., Cho, E.S., et al., 2013. β-catenin is required in odontoblasts for tooth root formation. J. Dent. Res. 92, 215–221.

Kist, R., Watson, M., Wang, X., Cairns, P., Miles, C., Reid, D.J., Peters, H., 2005. Reduction of Pax9 gene dosage in an allelic series of mouse mutants causes hypodontia and oligodontia. Hum. Mol. Genet. 14, 3605–3617.

Kollar, E.J., Baird, G.R., 1970. Tissue interactions in embryonic mouse tooth germs. II. The inductive role of the dental papilla. J. Embryol. Exp. Morphol. 24, 173–186.

Kratochwil, K., Galceran, J., Tontsch, S., Roth, W., Grosschedl, R., 2002. FGF4, a direct target of LEF1 and Wnt signaling, can rescue the arrest of tooth organogenesis in Lef1(-/-) mice. Genes. Dev. 16, 3173–3185.

Lagerström, M., Dahl, N., Nakahori, Y., Nakagome, Y., Bäckman, B., Landegren, U., Pettersson, U., et al., 1991. A deletion in the amelogenin gene (AMG) causes X-linked amelogenesis imperfecta (AIH1). Genomics 10, 971–975.

Lammi, L., Arte, S., Somer, M., Jarvinen, H., Lahermo, P., Thesleff, I., Pirinen, S., Nieminen, P., 2004. Mutations in AXIN2 cause familial tooth agenesis and predispose to colorectal cancer. Am. J. Hum. Genet. 74, 1043–1050.

Laurikkala, J., Mikkola, M.L., James, M., Tummers, M., Mills, A., Thesleff, I., 2006. P63 regulates multiple signaling pathways required for ectodermal organogenesis and differentiation. Development 133, 1553–1563.

Lumsden, A.G., 1988. Spatial organization of the epithelium and the role of neural crest cells in the initiation of the mammalian tooth germ. Development 103 (Suppl.), 155–169.

Michon, F., Jheon, H., Seidel, K., Klein, O.D., 2013. An incisive look at stem cells: The mouse incisor as an emerging model for tooth renewal. In: Huang, G.T.-J., Thesleff, I. (Eds.), Stem cells in craniofacial development and regeneration. Wiley Blackwell, Hoboken, New Jersey, pp. 315–328.

Mikkola, M.L., 2009. Molecular aspects of hypohidrotic ectodermal dysplasia. Am. J. Med. Genet. A. 149A, 2031–2036.

Mills, A.A., Zheng, B.H., Wang, X.J., Vogel, H., Roop, D.R., Bradley, A., et al., 1999. P63 is a p53 homologue required for limb and epidermal morphogenesis. Nature 398, 708–713.

Mina, M., Kollar, E.J., 1987. The induction of odontogenesis in non-dental mesenchyme combined with early murine mandibular arch epithelium. Arch. Oral. Biol. 32, 123–127.

Mundlos, S., Otto, F., Mundlos, C., Mulliken, J.B., Aylsworth, A.S., Albright, S., Lindhout, D., Cole, W.G., Henn, W., Knoll, J.H., et al., 1997. Mutations involving the transcription factor CBFA1 cause cleidocranial dysplasia. Cellule 89, 677–680.

Mustonen, T., Ilmonen, M., Pummila, M., Kangas, A.T., Laurikkala, J., Jaatinen, R., Pispa, J., Gaide, O., Schneider, P., Thesleff, I., Mikkola, M., 2004. Ectodysplasin A1 promotes placodal cell fate during early morphogenesis of ectodermal appendages. Development 131, 4907–4919.

Mustonen, T., Pispa, J., Mikkola, M.L., Pummila, M., Kangas, A.T., Jaatinen, R., Thesleff, I., 2003. Stimulation of ectodermal organ development by ectodysplasin-A1. Dev. Biol. 259, 123–136.

Nakashima, M., Reddi, A.H., 2003. The application of bone morphogenetic proteins to dental tissue engineering. Nat. Biotechnol. 21, 1025–1032.

Nieminen, P., 2009. Genetic basis of tooth agenesis. J. Exp. Zool. B. (Mol. Dev. Evol.) 312B, 320–342.

Nanci, A., 2008. Ten Cate's oral histology: development, structure, and function. Mosby Elsevier, St Louis.

O'Connell, D.J., Ho, J.W., Mammoto, T., Turbe-Doan, A., O'Connell, J.T., Haseley, P.S., Koo, S., Kamiya, N., Ingber, D.E., Park, P.J., Maas, R.L., et al., 2012. A Wnt-BMP feedback circuit controls inter-tissue signaling dynamics in tooth organogenesis. Sci. Signal 5, ra4.

Ooë, T., 1981. Human Tooth and Dental Arch Development. Ishiyaku Publishers Inc, Tokyo, Japan.

Peters, H., Neubüser, A., Kratochwil, K., Balling, R., 1998. Pax9-deficient mice lack pharyngeal pouch derivatives and teeth and exhibit craniofacial and limb abnormalities. Genes. Dev. 12, 2735–2747.

Pispa, J., Jung, H-S., Jernvall, J., Kettunen, P., Mustonen, T., Tabata, M.J., Kere, J., Thesleff, I., et al., 1999. Cusp patterning defect in Tabby mouse teeth and its partial rescue by FGF. Dev. Biol. 216, 521–534.

Pispa, J., Mustonen, T., Mikkola, M.L., Kangas, A.T., Koppinen, P., Lukinmaa, P.L., Jernvall, J., Thesleff, I., 2004. Tooth patterning and enamel formation can be manipulated by misexpression of TNF receptor Edar. Dev. Dyn. 231, 432–440.

Pispa, J., Thesleff, I., 2003. Mechanisms of ectodermal organogenesis. Dev. Biol. 262, 195–205.

Plikus, M.V., Zeichner-David, M., Mayer, J.A., Reyna, J., Bringas, P., Thewissen, J.G., Snead, M.L., Chai, Y., Chuong, C.M., 2005. Morphoregulation of teeth: modulating the number, size, shape and differentiation by tuning Bmp activity. Evol. Dev. 7, 440–457.

Poulter, J.A., El-Sayed, W., Shore, R.C., Kirkham, J., Inglehearn, C.F., Mighell, A.J., 2014. Whole-exome sequencing, without prior linkage, identifies a mutation in LAMB3 as a cause of dominant hypoplastic amelogenesis imperfect. Eur. J. Hum. Genet. 22, 132–135.

Pummila, M., Fliniaux, I., Jaatinen, R., James, M., Laurikkala, J., Schneider, P., Thesleff, I., Mikkola, M.L., 2007. Ectodysplasin has a dual role in ectodermal organogenesis: inhibition of BMP activity and induction of Shh expression. Development 134, 117–125.

Rajpar, M.H., Harley, K., Laing, C., Davies, R.M., Dixon, M.J., 2001. Mutation of the gene encoding the enamel-specific protein, enamelin, causes autosomal-dominant amelogenesis imperfecta. Hum. Mol. Genet. 10, 1673–1677.

Richman, J.M., Handrigan, G.R., 2011. Reptilian tooth development. Genesis 49, 247–260.

Salazar-Ciudad, I., Jernvall, J., 2002. A gene network model accounting for development and evolution of mammalian teeth. Proc. Natl. Acad. Sci. USA 99, 8116–8120.

Stockton, D.W., Das, P., Goldenberg, M., D'Souza, R.N., Patel, P.I., 2000. Msx1 deficient mice exhibit cleft palate and abnormalities of craniofacial and tooth development. Nat. Genet. 6, 348–356.

Stark, J., Andl, T., Millar, S.E., 2007. Hairy math: insights into hair-follicle spacing and orientation. Cellule 128, 17–20.

Stockton, D.W., Das, P., Goldenberg, M., et al., 2000. Mutation of PAX9 is associated with oligodontia. Nat. Genet. 24, 18–19.

Thesleff, I., Pirinen, S., 2006. Dental anomalies: Genetics. In: Encyclopedia of life sciences. John Wiley & Sons Ltd, Chichester.

Thomas, B.L., Tucker, A.S., Qui, M., Ferguson, C.A., Hardcastle, Z., Rubenstein, J.L., Sharpe, P.T., et al., 1997. Role of Dlx-1 and Dlx-2 genes in patterning of the murine dentition. Development 124, 4811–4818.

Tummers, M., Thesleff, I., 2009. The importance of signal pathway modulation in all aspects of tooth development. J. Exp. Zool. (Mol. Dev. Evol.) 310B, 309–319.

Trumpp, A., Depew, M.J., Rubenstein, J.L., Bishop, J.M., Martin, G.R., et al., 1999. Cre-mediated gene inactivation demonstrates that FGF8 is required for cell survival and patterning of the first branchial arch. Genes. Dev. 13, 3136–3148.

Unda, F.J., Martin, A., Hilario, E., Begue-Kirn, C., Ruch, J.V., Arechaga, J., 2000. Dissection of the odontoblast differentiation process *in vitro* by a combination of FGF1, FGF2, and TGFbeta1. Dev. Dyn. 218, 480–489.

Vastardis, H., Karimbux, N., Guthua, S.W., Seidman, J.G., Seidman, C.E., et al., 1996. A human MSX1 homeodomain missense mutation causes selective tooth agenesis. Nat. Genet. 13, 417–421.

Wang, X.P., Suomalainen, M., Jorgez, C.J., Matzuk, M.M., Wankell, M., Werner, S., Thesleff, I., 2004a. Modulation of activin/BMP signaling by follistatin is required for the morphogenesis of mouse molar teeth. Dev. Dyn. 231, 98–108.

Wang, X.P., Suomalainen, M., Jorgez, C.J., Matzuk, M.M., Werner, S., Thesleff, I., 2004b. Follistatin regulates enamel patterning in mouse incisors by asymmetrically inhibiting BMP signaling and ameloblast differentiation. Dev. Cell. 7, 719–730.

Wang, X.P., Åberg, T., James, M.J., Levanon, D., Groner, Y., Thesleff, I., 2005. Runx2 (Cbfa1) inhibits Shh signaling in the lower but not upper molars of mouse embryos and prevents the budding of putative successional teeth. J. Dent. Res. 84, 138–143.

Wu, P., Wu, ., Jiang, T.X., Elsey, R.M., Temple, B.L., Divers, S.J., Glenn, T.C., Yuan, K., Chen, M.H., Widelitz, R.B., Chuong, C.M., et al., 2013. Specialized stem cell niche enables repetitive renewal of alligator teeth. Proc. Natl. Acad. Sci. USA 110, E2009–2018.

Xiao, S., Yu, C., Chou, X., Yuan, W., Wang, Y., Bu, L., Fu, G., Qian, M., Yang, J., Shi, Y., et al., 2001. Dentinogenesis imperfecta 1 with or without progressive hearing loss is associated with distinct mutations in DSPP. Nat. Genet. 27, 201–204. Erratum in: Nat. Genet. 27:345, 2001.

Yamashiro, T., Zheng, L., Shitaku, Y., Saito, M., Tsubakimoto, T., Takada, K., Takano-Yamamoto, T., Thesleff, I., 2007. Wnt10a regulates dentin sialophosphoprotein mRNA expression and possibly links odontoblasts differentiation and tooth morphogenesis. Differentiation 75, 452–462.

Yoshida, T., Miyoshi, J., Takai, Y., Thesleff, I., 2010. Cooperation of nectin-1 and nectin-3 is required for normal ameloblast function but not for early tooth morphogenesis. Dev. Dyn. 239, 2558–2569.

Chapter 23

Early Heart Development

Paul A. Krieg and Andrew S. Warkman

Department of Cellular and Molecular Medicine, University of Arizona College of Medicine, Tucson, AZ

Chapter Outline

GLOSSARY

Cardia bifida The formation of two beating and looping heart tubes as a result of the improper fusion of the heart primordia.

Determined Cells are considered to be determined if they are able to assume their natural fate when placed in an antagonistic environment. Therefore, determination represents a higher level of commitment.

Induction The process by which one group of cells signals to a second group in a way that influences their development. This is achieved through signaling by diffusible growth factors.

Specified Cells are considered to be specified if they are able to assume their natural fate when placed in a neutral environment (e.g., tissue culture media).

First heart field (FHF) Early differentiating myocardial cells that contribute to the linear heart tube and ultimately to the left ventricle.

Second heart field (SHF) Later differentiating myocardial cells that contribute primarily to the right ventricle, the atria and outflow tract.

SUMMARY

- The process of heart induction appears to be highly conserved among vertebrates.
- The endoderm is important for heart induction in vertebrates. The details of the mechanism remain poorly understood.
- The localized inhibition of canonical Wnt signaling, which is regulated by endogenous Wnt antagonists, is required for heart induction.
- The size of the developing heart is regulated by inhibitory signals.
- The bone morphogenetic protein (BMP), Wnt, fibroblast growth factor (FGF) and Notch signaling pathways are important during early heart development.
- Different transcription factor profiles, in different regions of the original heart field, direct differentiation of the individual subdomains of the developing heart.
- Using a defined set of transcription factors, it is now possible to reprogram non-cardiac cells to the myocardial lineage.

23.1 INTRODUCTION

The efficient functioning of the heart and the associated circulatory system is essential for distributing oxygen and nutrients to the rapidly developing vertebrate embryo. In all vertebrates, the heart primordia arise in the anterior mesoderm as symmetrically paired regions located on either side of the embryonic midline. The combined movements of gastrulation and morphogenic folding cause the two patches of precardiac tissue to meet at the ventral midline, where they fuse to form a single cardiac primordium. The nascent heart is a simple linear tube containing a thick outer wall of muscle tissue, the myocardium, and a thin inner lining of endothelial tissue, the endocardium. In zebrafish, fusion of the cardiac primordia initially forms a conical structure, but this resolves to a linear tube quite soon after (Stainier et al., 1993). A significant proportion of the cells within the cardiac primordium have yet to undergo differentiation at this time of development, and remain in a state of active proliferation. These are the cells of the second heart field, which will be discussed below.

Although beating has yet to commence at the early heart tube stage, the precardiac tissue has already established a complex transcriptional program and has begun to express a large number of myocardial proteins that will assemble into the functional cardiac muscle. Transcripts for both the cardiac specific transcription factors and the muscle proteins function as sensitive markers for the developing cardiac tissue throughout early development. As cardiogenesis proceeds, the linear heart tube undergoes a complicated looping process that establishes the relative location of the atria and ventricles and the approximate location of the major vessels entering and leaving the heart. An enormous amount of information has been learned about the molecular genetic regulation of heart development over the last decade, but despite the degree of detail now available, numerous steps in the process that leads to a functioning heart are poorly understood. In recent years increasingly sophisticated techniques, mostly using mouse genetic approaches, have resulted in a dramatic increase in our knowledge of the later events in heart development. It is these later events that, while producing rather subtle defects from an embryological point of view, can lead to devastating health problems for a child born with the defect. While the mechanisms leading to later stage defects are of great interest from a medical point of view, they are not the focus of this review, which will rather address the very early events of heart development. Therefore we will not discuss the role of neural crest, formation of cardiac valves, positioning of the great vessels, or establishment of cardiac left/right asymmetry. Instead this review will provide a larger overview of the fundamental process of heart formation and will take a closer look at those areas where our understanding remains limited.

23.2 EMBRYOLOGY OF HEART DEVELOPMENT

Classic embryological studies of early heart development have generally used chick and amphibian embryos, both of which develop outside of their mothers and are extremely resilient to manipulations. More recently, embryological methods have been applied to zebrafish and ascidian embryos, particularly through the use of tissue specific lineage tracing using fluorescent markers in live embryos. For practical purposes however, these embryos are much less suited for experiments involving microsurgery or extended culture of tissue explants. Although the final structure of the heart is somewhat different in these different model organisms (particularly in the number of chambers) the hearts are very similar at the linear tube stage, suggesting that the initial events in heart formation are highly conserved. Through the study of these model systems, it has become apparent that commitment to the heart lineage in vertebrates involves a complex interplay of multiple tissue interactions that provide both inductive and inhibitory signals.

Fate-mapping studies in the chick embryo indicate that the cells destined to become the heart are among the first cells to involute during gastrulation. Before gastrulation, these cells are bilaterally distributed in the epiblast lateral to the mid-portion of the primitive streak (Figure 23.1). Incubation of explanted epiblast tissue in culture demonstrates that these cells are yet to be specified, because they are unable to differentiate into heart muscle (Holtzer et al., 1990; Yatskievych et al., 1997). At the onset of gastrulation (stage 3; Hamburger and Hamilton, 1951), cells fated to become the heart are located within the anterior third of the primitive streak (Garcia-Martinez and Schoenwolf, 1993). Transplant assays of the precardiac mesoderm after it has moved through the streak and started to migrate anteriorly and laterally (stage 4+ or 5) reveal that these cells are specified for the cardiac lineage but not yet determined. Experimentally, this can be demonstrated by showing that precardiac mesoderm is able to differentiate into heart tissue when it is cultured in isolation (i.e., *in vitro* culture) but not when it is transplanted into noncardiogenic regions of a similarly-staged embryo (Antin et al., 1994; Gannon and Bader, 1995; Gonzalez-Sanchez and Bader, 1990; Montgomery et al., 1994). Interestingly, similar explant studies at later stages of development have demonstrated that mesoderm immediately medial to the cardiac mesoderm, but which does not normally contribute to the heart, is able to give rise to cardiac tissue *in vitro* (Rawles, 1943; Rosenquist and De Haan, 1966). These were some of the earliest indications of a role for inhibitory signals in delimiting the size of the precardiac precursor population in the embryo. The heart precursors become committed to the cardiac lineage at around stage 6 (Montgomery

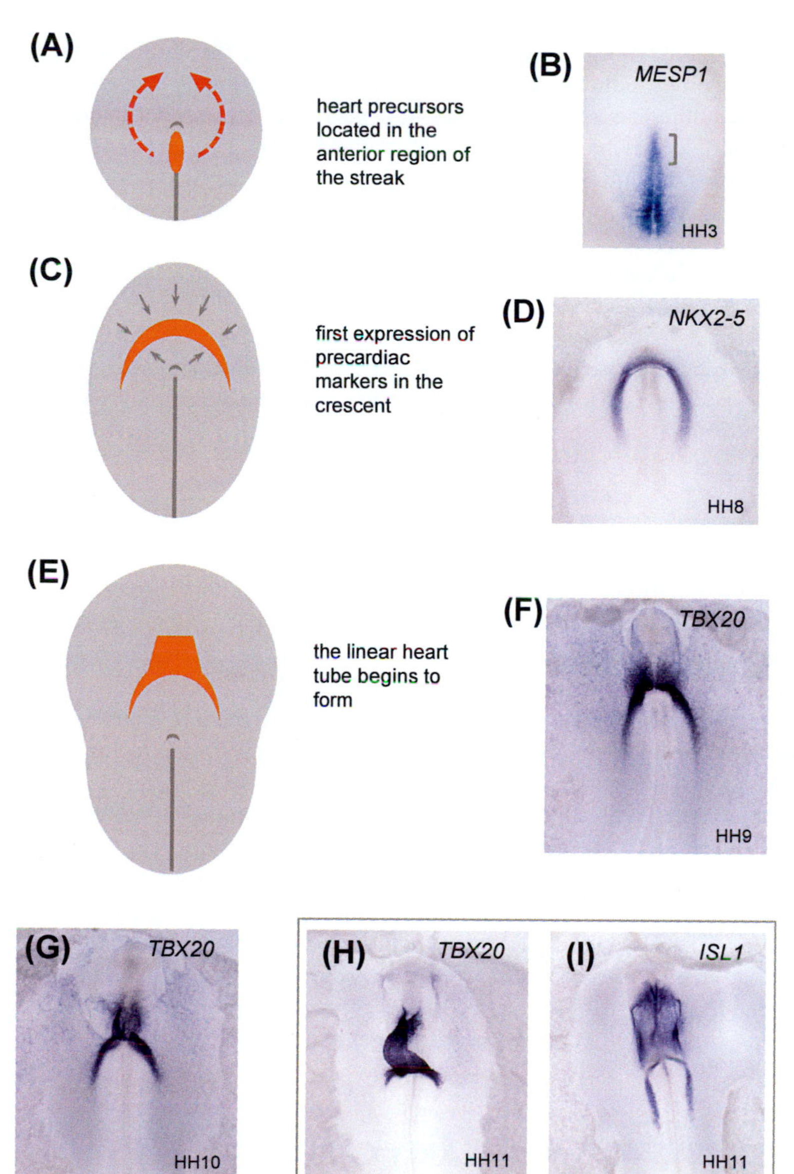

FIGURE 23.1 Early Development of the Heart. (A–F) Images at left show a ventral view of a stylized planar embryo, such as chick or human. The developmental stage is indicated at lower right of each panel. (A, B). Early gastrulation embryo. Cells fated to become heart are shown in orange at the cranial end of the primitive streak (gray line). The approximate path of migration of the precardiac mesoderm is indicated by the dashed lines. There are no specific markers of the precardiac tissues at this developmental stage, but the streak tissue is indicated by expression of MESP1. (C, D). Mid-gastrulation embryo. Cells with the potential to become heart are located in a crescent shape at the cranial end of the embryo. Arrows represent inhibitory signals acting to limit the size of the precardiac mesoderm. Precardiac tissue is specifically marked by expression of the transcription factor, NKX2–5. (E, F). Later stage mid-gastrulation embryo. Cells at the cranial end of embryo have migrated to the ventral midline and formed a primitive heart tube. Expression of the transcription factor TBX20 marks the developing myocardial tissue. G-I. Continuing development of the early heart tube. Expression of TBX20 shows continued growth of the myocardium tissue of the heart tube and beginnings cardiac looping (in H). I. Isl1 is a marker of the second heart field, which differentiates into myocardium later during development. Note that the ISL1 pattern, marking undifferentiated tissue, is quite distinct from the differentiated heart tissue expressing TBX20 at the same developmental stage (compare H and I). *All images are gene expression patterns of chick embryos assayed by in situ hybridization from the Geisha Chicken Embryo Gene Expression Database (see Recommended resources).*

et al., 1994), and the expression of terminal differentiation markers (e.g., myosin heavy chains) is first detected in the cardiac mesoderm at stage 7 (Han et al., 1992). Recently, beautiful time lapse movies have been prepared using quail embryos, showing the migration route of precardiac cells, all the way from the primitive streak to their final positions within the linear heart tube (Cui et al., 2009). Analysis of the trajectories of migratory cells showed that the precardiac mesoderm does not migrate as a cohesive sheet. Instead, individual cells show independent migratory paths. These observations demonstrate that the starting location of a cell within the primitive streak does not rigorously define the region of the heart that it will finally populate.

In *Xenopus*, the paired heart primordia are initially located on the dorsal side of the embryo on either side of the blastopore lip, and they are among the first mesodermal tissues to involute during gastrulation (Keller, 1976). Studies of tissue explanted from *Xenopus* early gastrula embryos show that precardiac cells already exhibit some degree of specification before involution, because they are capable of differentiating into heart and other cells types in explant culture (Nascone and Mercola, 1995). Similar studies using late gastrula embryos (stage 12.5) demonstrate a higher degree of specification of the heart primordia, because explants exhibit robust cardiac differentiation in cell culture (Sater and Jacobson, 1989).

Formation of precardiac tissue requires inductive interactions between the endoderm and the overlying mesodermal tissue (Orts-Llorca, 1963). Subsequent studies using mid-gastrula chick embryos showed that the anterior but not the posterior endoderm is able to induce cardiogenesis in posterior primitive streak explants that would not normally give rise to heart

(Schultheiss et al., 1995). Finer mapping of the inductive properties of the anterior endoderm reveal that both the lateral and medial portions of this tissue exhibit inductive properties, although cardiogenesis *in vivo* occurs only in mesoderm that is associated with the lateral anterior endoderm (Schultheiss et al., 1995). Thus, it appears that the anterior endoderm generates a rather broad cardiac field in the overlying mesoderm, which is subsequently refined and restricted by additional signals. Broadly similar explant studies in *Xenopus* have shown that endoderm lying adjacent to the cardiac primordia in gastrula stage *Xenopus* embryos functions as heart-inducing tissue when it is cultured with general mesodermal tissue (Nascone and Mercola, 1995).

In both chick and *Xenopus* embryos, the mesoderm adjacent to heart primordia is able to differentiate into heart in culture but not within the embryo. Studies using co-culture of embryonic tissue explants have shown the presence of a potent inhibitory signal to cardiogenesis within neural tissue (Climent et al., 1995; Garriock and Drysdale, 2003; Jacobson, 1961) and notochord (Schultheiss and Lassar, 1997). Overall, it is clear that both positive and negative signals resulting from various tissue interactions act in concert to induce cardiogenesis within a defined region of the developing embryo. In the sections below we will explore the cell biology of the inductive and repressive signals identified by these classical embryology studies.

The First and Second Heart Fields

The embryonic induction events described above result in the formation of the heart field, a region of embryonic tissue that has the potential to express cardiac differentiation genes and to develop into myocardium. Approximately ten years ago it was discovered that the initial heart field could be separated into two broad domains (marked by different precardiac gene expression patterns), which would ultimately contribute to different regions of the heart (Cai et al., 2003). The region that differentiated first and formed the heart tube, and which ultimately contributed most of the tissue to the left ventricle, was called the first heart field (FHF). A later differentiating region, marked by expression of the Islet1 transcription factor (Isl1), contributed primarily to the right ventricle, the atria and the outflow region of the heart. This region was called the second heart field (SHF). In mouse embryos, at the time when the FHF cells differentiate and form the linear heart tube, the SHF cells assume a position dorsal to the myocardial tissue. The SHF cells continue to divide and contribute myocardial tissue to the atria, right ventricle and outflow tract during subsequent development. In chick, *Xenopus* and zebrafish, the SHF contribution to the myocardium is much less than in mammals and is largely restricted to the outflow tract region (Gessert and Kühl, 2009; Waldo et al., 2005; Zhou et al., 2011). The FHF and SHF regions are not sharply demarcated and gene expression patterns show a significant degree of overlap, but lineage tracing studies have clearly demonstrated the broadly different developmental fates of these two major domains. Additional information on the cellular and molecular biology of the first and second heart fields can be found in several comprehensive reviews (Francou et al., 2013; Rochais et al., 2009; Vincent and Buckingham, 2010).

23.3 MULTIPLE SIGNALING PATHWAYS ARE INVOLVED IN EARLY CARDIOGENESIS

A combination of embryology and cellular biology approaches has lead to a reasonably clear picture of the signaling pathways essential for early heart development (Figure 23.2). While there is general consensus on the pathways involved, working out the mechanisms that regulate the precise timing and balance of the pathways is an area of intense research, which is likely to take many years to understand completely. The sections below provide a brief overview of the current understanding of the growth factor function during early heart formation.

Bone Morphogenetic Protein Signaling

Studies in *Drosophila* have demonstrated an important role for the bone morphogenetic protein (BMP)-related signaling protein, decapentaplegic, in the formation of the dorsal vessel, an insect organ broadly analogous to the heart (Frasch, 1995). Studies in vertebrates have shown that at least three members of the BMP family are expressed within the cardiogenic region of a primitive streak chick embryo (Schultheiss et al., 1997). BMP2 is expressed in the lateral endoderm, whereas BMP4 and BMP7 are expressed in the ectoderm immediately adjacent to the precardiac mesoderm. Functional studies have shown that BMP2 or BMP4 is able to induce cardiac differentiation and beating tissue when added to normally non-cardiac mesoderm from anterior regions of the embryo. Furthermore, inhibition of endogenous BMP signaling in cardiogenic mesoderm tissue results in a complete block to expression of cardiac differentiation markers. In gain-of-function studies, addition of BMP alone was not sufficient to induce cardiac differentiation in all mesoderm tissue (Schultheiss et al., 1997) indicating that additional signaling factors are required for normal cardiogenesis.

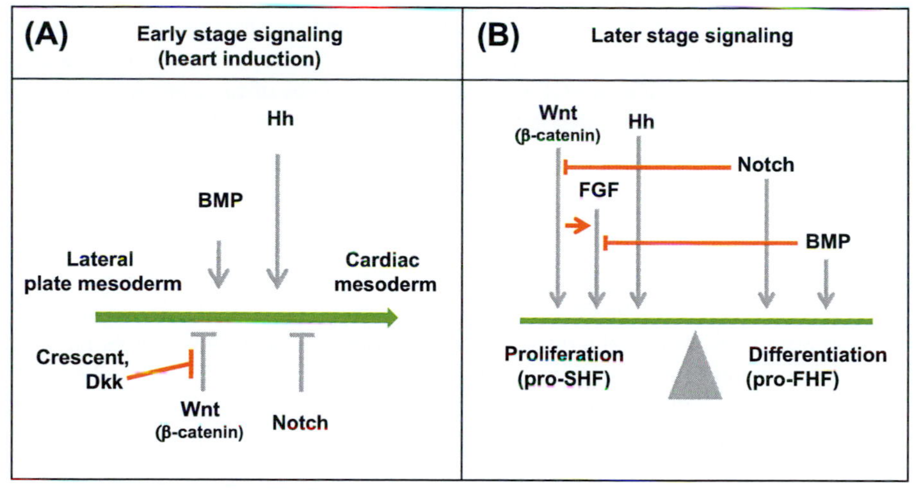

FIGURE 23.2 **Signaling Pathways during Early Heart Development.** (A) Signaling pathways active when cardiac mesoderm is initially specified. Signaling by BMP and Hh proteins, promote heart development, whereas canonical Wnt signaling (through β-catenin) and Notch signaling, inhibit heart development. Blocking of canonical Wnt signaling by Wnt inhibitors (Dkk and Crescent) is a critical step in allowing specification of the cardiac lineage to occur. (B) Signaling pathways during later heart development. Later stage signaling can be viewed as a balance between pro-differentiation signals, favoring formation of ventricular myocardium (first heart field; FHF), and pro-proliferation signals that favor later development of the atrial and outflow tract myocardium (second heart field; SHF). Horizontal orange lines show positive and negative interactions between signaling pathways.

Studies in other organisms have confirmed the importance of BMP signaling for cardiac development, although perhaps not at the level of cardiac specification. In *Xenopus*, BMP signals are essential for the migration and/or fusion of the heart primordia and for cardiomyocyte differentiation, as evidenced by the downregulation of the expression of *Nkx2–5* and the reduced expression of *myosin light chain-2* and *cardiac troponin I* (Shi et al., 2000; Walters et al., 2001). In *Xenopus* however, BMPs are not normally expressed in tissues with cardiac induction activity suggesting that BMP signaling does not play a pivotal role in the earliest steps of heart induction. These observations can be reconciled if BMP signaling is necessary for later, post-gastrulation aspects of heart development, after the heart primordia have migrated a more anterior position within the embryo. The expression of Noggin (a potent BMP inhibitor) within the developing heart patches of both frog (Fletcher et al., 2004) and mouse (Yuasa et al., 2005) embryos provides a further indication that the timing of BMP signaling is important and tightly regulated. Genetic analysis of BMP signaling during early heart development in the mouse is complicated by the multiple roles for BMP signaling during early embryogenesis and by functional redundancy between BMP isoforms. For example, ablation of BMP2 activity results in embryos with differentiated myocardial tissue but numerous structural malformations of the heart (Zhang and Bradley, 1996), and ablation of BMP4 activity results in the death of the embryo before heart development (Winnier et al., 1995). The redundancy problems have been overcome by tissue specific ablation of BMP receptor activity in the cardiac primordia (Klaus et al., 2007). Specifically, when a *Mesp1*-Cre driver was used to ablate expression of the BMP receptor, BMPR1a, heart development was severely reduced, with just a small amount of differentiated cardiac tissue remaining. Further analysis suggested that derivatives of the first heart field were missing and only second heart field derivatives remained (Klaus et al., 2007). Therefore, BMP signaling regulates normal heart development in all model systems, although its precise role may vary between species.

Fibroblast Growth Factor Signaling

Understanding the role of fibroblast growth factor (FGF) signaling in heart development has been hampered by the functional redundancy of the large number of FGF family members, and by the fact that genetic ablation of FGF genes can lead to very early embryonic lethality. Studies in chick embryos suggest that despite expression of several FGF ligands at the right place at the right time, FGF signaling does not play an essential role during the initial induction of cardiac tissue (Alsan and Schultheiss, 2002). On the other hand, a series of experiments in *Xenopus* (Samuel and Latinkic, 2009) indicate that FGF plus Nodal signaling is sufficient to induce cardiac precursors and that this induction is independent of Wnt signaling. This is an intriguing result, since a number of other studies suggest that inhibition of canonical Wnt signaling is a crucial first step in initiation of heart development (see the next section below).

Apart from the events surrounding cardiac induction, studies in several model systems make it clear that FGF signaling does play an essential role in later heart development, primarily in regulating the proliferation of cardiac precursors. In zebrafish, FGF signaling is required for overall correct size of the heart and also for regulating the proportional sizes of the atrial and ventricular chambers (Marques et al., 2008). In mammals, the primary role of FGF appears to be the regulation of cell proliferation within the SHF. Ablation of FGF8 results in deficiencies of the outflow tract regions of the developing heart (Frank et al., 2002). These are tissues that are contributed by the later differentiating cells of the SHF. Targeted deletion of both FGF8 and FGF10 in the heart forming region results in even more severe defects (Watanabe et al., 2010).

Wnt Signaling

At different times during heart development, signaling through the Wnt pathway plays both positive and negative roles. As described above, cardiac development is regulated at least in part by inhibitory signals emanating from the notochord and neural tissue. The inhibitory signal from neural structures probably involves Wnt family proteins. In the chick, Wnt1 and Wnt3a, both of which activate the canonical β-catenin pathway, are expressed in dorsal neural tissue at the time of heart induction (Tzahor and Lassar, 2001). Ectopic expression of either of these molecules mimics the repressive effects of neural tissue, while reduction of Wnt activity in neural tissues using Wnt antagonists results in an expansion of cardiac tissues into areas that are not normally fated to form heart (Tzahor and Lassar, 2001).

In the chick embryo, the Wnt inhibitor protein, Crescent, is expressed in the anterior endoderm, immediately adjacent to the precardiac mesoderm. Ectopic expression of Crescent in the chick embryo results in formation of cardiac tissue in regions of the embryo that do not normally form heart (Marvin et al., 2001). These studies suggest that Crescent functions to induce cardiac tissue by repressing the canonical activity of Wnt3a and Wnt8c within the anterior portion of the chick embryo. Studies using *Xenopus* support the Wnt inhibitor model for cardiac induction. For example, it appears that the endogenous Wnt antagonists, Crescent and Dickkopf-1, act to inhibit Wnt3a and Wnt8 activity originating from the organizer (Schneider and Mercola, 2001). This inhibition leads to the induction of precardiac tissue in the mesoderm on either side of the blastopore lip. By contrast, ectopic expression of either Wnt3a or Wnt8 was able to block cardiogenesis in tissues that would normally form heart (Schneider and Mercola, 2001). Although the inhibition of Wnt3a and Wnt8 activity appears to be required for heart induction in both *Xenopus* and chick embryos, it is interesting to note that the proposed source of the inhibitor signal differs in these organisms; Crescent and Dickkopf-1 are expressed in the organizer region of the frog embryo, but Crescent is expressed broadly within the anterior endoderm of the chick. Despite the differences in the domains of expression, both model systems suggest a single unifying mechanism: the blocking of canonical Wnt signaling leads to the induction of cardiac tissue in responsive mesodermal tissue. At present, the requirement for inhibition of canonical Wnt signaling for heart induction has not been demonstrated in mice, probably due to the redundancy of Wnt ligands, Wnt inhibitors and Wnt receptors. However, cell culture studies have demonstrated that inhibition of Wnt signaling strongly upregulates expression of myocardial differentiation markers in mouse ES cells (Wang et al., 2011). Perhaps the most dramatic demonstration of the importance of the inhibition of canonical Wnt signaling comes from mouse knockout research, in which β-catenin activity was abolished in different embryonic tissues. When β-catenin expression was ablated in the embryonic endoderm, the resulting embryos formed numerous heart-like structures at ectopic locations (Lickert et al., 2002). Note that the knockout was in endoderm rather than mesoderm, which implies that still additional factors are likely to be involved in transmitting the cardiac induction signal to the precardiac tissues.

Other studies have suggested that signaling through non-canonical (i.e., non β-catenin-dependent) Wnt pathways may be involved in cardiac induction. These studies have focused on Wnt11, which is expressed in the involuting mesoderm of the gastrula stage frog embryo, including the precardiac mesoderm adjacent to the organizer. Studies in frog embryos (Pandur et al., 2002) reveal that the inhibition of endogenous Wnt11 signaling results in the decreased expression of both precardiac markers and cardiac differentiation markers and aberrant heart tube formation. In over-expression experiments, Wnt11 ligand induced cardiac differentiation in regions of the *Xenopus* (Pandur et al., 2002) and chick embryo (Eisenberg and Eisenberg, 1999) that do not normally form heart tissue. The interpretation of these results is complicated by the fact that the expression of high levels of non-canonical Wnt ligands appears to inhibit the canonical Wnt signaling pathway (Maye et al., 2004; Topol et al., 2003; Weidinger and Moon, 2003). Therefore, it is possible that the heart induction observed after Wnt11 over-expression is in fact mimicking the effects of the Wnt inhibitors Dickkopf and Crescent in inducing cardiac tissue. This possible role of Wnt11 in cardiac induction is not shared in the mouse; Wnt11 knockout animals form largely normal hearts (Majumdar et al., 2003).

After initial induction of the precardiac lineage, Wnt signaling continues to function in regulation of heart development. In *Xenopus*, Wnt6 is expressed in the ectodermal tissue immediately overlying developing cardiac tissue and plays an inhibitory role in regulating the size of the heart (Lavery et al., 2008). Therefore, when Wnt6 signaling is inhibited, the heart becomes larger, perhaps through upregulation of expression of GATA4. In a negative feedback mechanism, the developing *Xenopus* cardiac tissue itself expresses a Wnt inhibitor protein, Sfrp1, which partially opposes the activity of Wnt6 signaling (Gibb et al., 2013). Ablation of Sfrp1 activity results in almost complete loss of differentiated myocardial tissue, whereas over-expression of Sfrp1 results in a larger heart.

In contrast to these inhibitory activities, Wnt signaling through the canonical pathway also plays an essential positive role in heart development. Work by a number of different laboratories provides an extremely consistent view of the role of Wnt signaling in proliferation of SHF progenitors in the mouse embryo. Using a number of different approaches, these studies demonstrated that blocking the Wnt/β-catenin pathway in SHF precursors results in a dramatic reduction of SHF derived tissue, especially the right ventricular myocardium and the outflow tract (Ai et al, 2007; Cohen et al., 2007; Klaus et al., 2007; Kwon et al., 2007; Lin et al., 2007; Qyang et al., 2007). In addition, Wnt signaling also plays a positive role in proliferation of FHF progenitors in the mouse embryo, and it appears that this Wnt-induced proliferation is normally kept in check by the Hippo mitosis inhibitory pathway (Heallen et al., 2011). When essential components of the Hippo pathway are conditionally ablated in cardiac progenitors using an *Nkx2–5*-Cre driver, a massively oversized heart results. Taken together, these studies indicate essential roles for Wnt signaling throughout the cardiac development program, with Wnt playing both positive and negative roles in promoting cardiac growth at different times during development.

Hedgehog Signaling

Hedgehog signaling plays an important role in regulating the proliferation of cardiac progenitor cells at multiple times during the heart development program. Studies in zebrafish indicate that Hh signaling, acting at about the time that the cardiac lineage is specified, regulates the number of cells in the heart field. When Hh signaling is inhibited, the size of the cardiac precursor population is diminished, whereas stimulation of Hh signaling results in a larger pool of cardiac precursors (Thomas et al., 2008). Gene ablation studies show that Hh signaling is also important for heart development in the mouse. Animals that are deficient for the hedgehog co-receptor, Smoothened, or that lack activity of Shh, Ihh, or both, exhibit multiple defects in cardiac development, including reduced heart size (Zhang et al., 2001). Tissue specific ablation of Hh signaling components in different cardiac progenitors indicates that Hh signaling is required within the SHF since ablation of Hh signaling in this cell population leads to severe morphological defects in the outflow tract (Goddeeris et al., 2007; Lin et al., 2006; Washington Smoak et al., 2005). In the chick, inhibition of Hh signaling leads to failure of outflow tract development due to reduced SHF cell proliferation in the undifferentiated precursor population (Dyer and Kirby, 2009).

Notch Signaling

Signaling via the Notch pathway may play a role in regulating the size of the heart field. In fish (Serbedzija et al., 1998) and amphibia (Sater and Jacobson, 1990), the original cardiac primordia contain cells that will not ultimately contribute to the heart but that are capable of forming cardiac tissue if the definitive heart progenitors are removed. These latent cardiac cells are gradually lost as development proceeds. Analysis of the mechanism behind this loss of plasticity in *Xenopus* has implicated Notch1 and its ligand Serrate (Rones et al., 2000). Increased Notch signaling results in the decreased expression of cardiac differentiation markers, whereas a reduction of Notch signaling using dominant-negative constructions results in increased myocardial gene expression. Subsequent studies have confirmed the role of Notch in regulating the number of cardiac precursor cells, and suggested that the effects may be mediated through regulation of the pro-cardiogenic transcription factor GATA4 (Miazga and McLaughlin, 2009). In cell culture experiments using mouse ES cells, the presence of Notch signaling results in inhibition of cardiomyocyte development, whereas reduction of Notch signaling allows more heart cells to develop (Nemir et al., 2006). However, numerous studies of the role of Notch signaling during zebrafish, chick and mouse embryonic development have failed to reveal any significant role for Notch in regulation of the size of the heart field, which suggests that this function is not shared by all vertebrates.

Later in heart development, Notch is one of the signaling pathways that regulate the size of the SHF. Specifically, Notch appears to function as a brake on the proliferation of SHF progenitors stimulated by the Wnt pathway (Kwon et al., 2009). When Notch 1 expression was conditionally knocked-out in the SHF, the right ventricle failed to develop. This was apparently because the normal precursors of the right ventricle continued to proliferate under the influence of Wnt signaling, rather than differentiate into right ventricular myocardium. Increased canonical Wnt signaling in the absence of Notch1 function was confirmed by the detection of higher levels of β-catenin in the nuclei of SHF cells.

23.4 TRANSCRIPTIONAL REGULATION OF EARLY HEART DEVELOPMENT

The previous sections of this chapter have outlined how various tissue interactions serve to activate or inhibit the signaling cascades required for cardiac induction. However, it is the transcriptional response within the responding tissues that

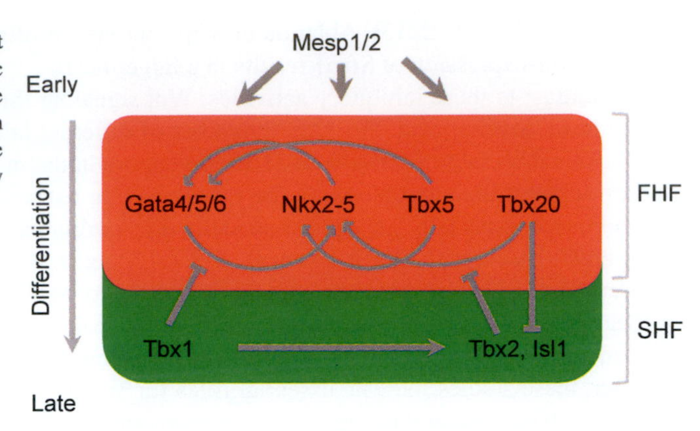

FIGURE 23.3 Transcriptional Regulation of Early Heart Development. Mespl/2 is required to specify formation of precardiac mesoderm. Expression of different transcription factors then delineates the original heart field into the early differentiating first heart field (FHF, in red) and the later differentiating second heart field (SHF, in green). Positive and negative interactions between transcription factors are indicated by gray lines.

ultimately leads to development of the different tissues of the heart. Knowledge of the transcription factors expressed within precardiac tissue has provided an outline of the gene regulatory pathways lying downstream of the heart-inducing signals and has allowed the partitioning of precardiac tissue into different domains with different developmental fates (Figure 23.3).

Mesp Transcription Factors

Mesp transcription factors are members of the helix-loop-helix family of DNA binding proteins. The ascidian, *Ciona*, contains a single *Mesp* gene which is required for development of the heart (Satou et al., 2004). Vertebrates have at least two highly related *Mesp* genes and double knockout of the *Mesp1* and *Mesp2* genes in mouse results in a complete absence of heart formation (Kitajima et al., 2000). However, interpretation of these results is complicated by the fact that Mesp1 and Mesp2 are rather broadly expressed in mesodermal tissues, and development of quite a number of other mesodermal derivatives is severely compromised in the double knockout animals. Subsequent experiments have indicated that Mesp1 is indeed capable of playing an instructive role in heart development. For example, expression of Mesp protein in *Xenopus* embryos leads to ectopic expression of cardiac marker genes (David et al., 2008; Kriegmair et al., 2013). Furthermore, cell culture studies show that Mesp1 (usually in cooperation with other transcription factors) is a crucial factor in programming ES cells (Bondue et al., 2008; David et al., 2008; Hartung et al., 2013) or reprogramming dermal fibroblasts to the cardiac lineage (Islas et al., 2012). In the embryo, the ability of Mesp to regulate the cardiogenic program may be at least partly mediated through transcriptional activation of the Wnt inhibitor Dkk1 (David et al., 2008), which functions as a cardiac inducer by blocking canonical Wnt signaling in precardiac tissue (see the section "Wnt signaling," above).

The Nkx2 Family of Homeodomain Transcription Factors

In *Drosophila*, the homeodomain transcription factor tinman plays a central role in the initial formation of the dorsal vessel, a contractile muscular tube that is analogous to the vertebrate heart (Bodmer, 1993). Vertebrate genomes contain a number of *tinman*-related genes (reviewed in Evans, 1999), but the most important is *Nkx2–5*. All vertebrates examined express *Nkx2–5* in precardiac tissue, and the onset of expression coincides with the timing of heart specification. However, the vertebrate Nkx2–5 does not function equivalently to tinman in *Drosophila*. Whereas *Drosophila tinman* mutants completely lack a dorsal vessel and no heart markers are expressed (Bodmer, 1993), mice lacking Nkx2–5 function make a normal linear heart tube that expresses most cardiac differentiation markers, but embryos die with multiple severe heart defects as the heart tube starts to loop (Lyons et al., 1995). This less dramatic phenotype does not appear to be the result of functional redundancy, because *Nkx2–5/Nkx2–6* double knockout mice exhibit a phenotype very similar to that of the *Nkx2–5* knockout (Tanaka et al., 2001). Furthermore, experiments in *Xenopus* show that *Nkx2–5* expression is not sufficient to direct cells toward the cardiac lineage. The ectopic expression of either *Nkx2–5* or *Nkx2–3* in frog embryos caused an increase in size of the heart at the normal location, but it was unable to induce precocious expression of cardiac differentiation markers in the cardiac region or ectopic expression of cardiac genes outside of the heart (Cleaver et al., 1996).

The regulation of *Nkx2–5* transcription in vertebrates is complex, and the gene is potentially downstream of numerous signaling pathways. For example, evidence indicates that BMP is important for the early expression of *Nkx2–5*. Experiments in chick embryos demonstrate that BMP4 can activate *Nkx2–5* expression, even in noncardiogenic mesoderm (Schultheiss et al., 1997). In addition, transcription-regulating sequences of the *Nkx2–5* gene contain an evolutionarily conserved SMAD binding site that activates *Nkx2–5* transcription in response to BMP signaling (Liberatore et al., 2002; Lien et al., 2002).

GATA Family

The zinc finger transcription factor pannier is essential for heart development in *Drosophila* (Gajewski et al., 1999). The vertebrate homologs of pannier are the GATA transcription factors, three of which (GATA4, GATA5 and GATA6) are expressed in the developing heart (reviewed in Peterkin et al., 2005). Analysis of the function of GATA factors in vertebrate heart development is complicated by their overlapping expression domains and the fact that they are expressed in both the precardiac mesoderm and adjacent endodermal tissue. Mice deficient in GATA4 exhibit cardia bifida, reduction in expression of other precardiac marker genes, including *Nkx2–5*, and the number of cardiac precursor cells is reduced (Kuo et al., 1999; Reiter et al., 1999). *GATA5* knockout mice generate a normal heart (Nemer et al., 1999), indicating that GATA5 function is not required for normal cardiac development. However, in zebrafish *GATA5* mutants, the number of myocardial precursor cells is reduced, and so is the expression of several other cardiac factors, including Nkx2–5 (Reiter et al., 1999). Depletion of GATA6 function in *Xenopus* embryos using antisense oligonucleotides results in decreased myocardial differentiation and an inability to maintain the expression of *Nkx2–5* and *Nkx2–3* in the developing heart (Peterkin et al., 2003). *GATA6* knockout mice die very early in embryogenesis due to defects in extra-embryonic endoderm development (Koutsourakis et al., 1999), but use of tetraploid complementation methods to circumvent these extra-embryonic defects demonstrates that early heart development is fairly normal in *GATA6* deficient animals (Zhao et al., 2005). However, the absolute requirement for GATA function for heart development can be demonstrated when both GATA4 and GATA6 are depleted in the mouse embryo, resulting in a complete absence of cardiac tissue (Zhao et al., 2008).

The requirement for GATA function for heart formation can also be demonstrated by gain-of-function studies. In *Xenopus* experiments, ectopic expression of *Xenopus* GATA-4 in isolated ectodermal explants (i.e., animal caps) is able to induce beating tissue (Latinkic et al., 2003), whereas the ectopic expression of GATA5 in zebrafish embryo results in the ectopic expression of *Nkx2–5* and ultimately in the presence of clusters of beating cardiac cells (Reiter et al., 1999). Although the precise role of different GATA orthologs appears to vary in different organisms, the weight of evidence suggests that the expression of GATA and Nkx2–5 family proteins is essential for the cardiogenic program in the embryo. Indeed, it appears that GATA factors directly regulate the expression of *Nkx2–5* family genes and vice versa within precardiac tissues (Davis et al., 2000; Lien et al., 1999; Molkentin et al., 2000; Searcy et al., 1998). This interplay of transcription factors represents a potential mechanism for the amplification and stabilization of the cardiac transcriptional program in cardiac progenitor cells.

Isl1

Islet-1 (Isl1) is a member of the LIM/homeodomain family of transcription factors and is expressed in a subdomain of the precardiac mesoderm (Karlsson et al., 1990). It derives its name from its expression in the islet cells of the pancreas, but it is also widely recognized as a marker for the cells of the SHF. *Isl1* knockout mice die on embryonic day 10.5 as a result of severe cardiac defects, including the complete absence of outflow tract, right ventricle, and much of the atria (Cai et al., 2003). The mechanism underlying this phenotype involves altered proliferation, survival, and migration of SHF cardiac progenitor cells. *Isl1* expression is integrated within the growth factor signaling network functioning during early cardiac development. Knockout studies in the mouse show that *Isl1* expression is activated by canonical Wnt signaling (Klaus et al., 2012) and that Isl1 functions as a direct transcriptional regulator of FGF10 expression in the SHF (Golzio et al., 2012). Overall, Isl1 is regarded as the primary regulator of later-developing SHF myocardium.

T-box Transcription Factors

The *T-box* (*Tbx*) genes encode a large family of transcription factors, many of which play important roles in embryonic patterning and organogenesis (see Chapter 34). The first indication that Tbx factors were important for heart development came from human genetic studies, in which it was discovered that mutations underlying the human congenital heart diseases DiGeorge syndrome and Holt-Oram syndrome were located in the *Tbx1* (Chieffo et al., 1997) and *Tbx5* genes (Li et al., 1997), respectively. Subsequent studies have revealed that six of the 17 *Tbx* family genes are expressed in the cardiac tissue of vertebrates before and during cardiac differentiation (Greulich et al., 2011; Plageman and Yutzey, 2005; Showell et al., 2006; Yamada et al., 2000). Interpretation of the specific role of Tbx factors during heart development is complicated by several factors. First, some members of the Tbx family act as transcriptional activators, whereas others are transcriptional repressors (reviewed in Plageman and Yutzey, 2005). Second, Tbx family proteins are capable of forming heterodimers, both with other Tbx proteins and other cardiac transcription factors, including GATA4 and GATA5 and Nkx2–5 (Brown et al., 2005; Stennard et al., 2003).

Tbx5 is amongst the first transcription factors to be expressed in the FHF region of the precardiac mesoderm. Knockout of *Tbx5* resulted in embryonic death at around E10 (Bruneau et al., 2001). Although many myocardial differentiation markers were expressed, the hearts were severely malformed. Expression levels of two other cardiac transcription factors, GATA4 and Nkx2–5, were downregulation in the *Tbx5* deficient hearts, thereby contributing to the general disruption of the cardiac development program (Bruneau et al., 2001). Similarly, mice lacking Tbx20 function formed a linear heart tube, but the subsequent looping and chamber formation did not occur normally, resulting in embryonic death (Stennard et al., 2005). In the *Tbx20*-null animals, expression of Tbx2, a Tbx family protein with repressor activity, was upregulated, demonstrating the degree of interconnectedness of the *Tbx* gene program. Additional knockout studies in mouse suggest a high degree of functional redundancy between other cardiac-expressed *Tbx* genes because ablation of individual *Tbx* genes did not result in early cardiac defects. Instead myocardial tissue was formed fairly normally, but alterations to cardiac morphology and to the relative size of heart structures lead to impaired cardiac function and embryonic/neonatal death. For example, mice lacking Tbx1 function survive almost until birth, but have a spectrum of cardiac defects consistent with a diminished population of SHF cells (Jerome and Papaioannou, 2001).

23.5 PROGRAMMING AND REPROGRAMMING OF CELLS TOWARDS A CARDIAC FATE

Despite the general success in elucidating the signaling pathways and transcription factors involved in embryonic heart development, it has proved to be extremely difficult to identify the correct complement of transcription factors to program differentiation of cardiac tissue. Possible exceptions to this general observation are reports regarding zebrafish (Reiter et al., 2001) and *Xenopus* (Latinkic et al., 2003), in which over-expression of the GATA transcription factors, GATA5 and GATA4 respectively, brought about the ectopic expression of cardiac markers and, if allowed to develop further, the presence of ectopic foci of beating cardiac tissue. Exciting recent studies indicate major advances in reprogramming towards cardiac differentiation, specifically the cardiomyocyte cell type, in mammalian cells in culture and mammalian embryos. One of the approaches included a cardiac specific chromatin remodeling factor, Baf60c, together with the sequence specific transcription activators, GATA4 and Tbx5 (Takeuchi and Bruneau, 2009). This cocktail of factors resulted in ectopic expression of cardiomyocyte differentiation markers and the presence of beating tissue at ectopic locations in the mouse embryo. Additional studies with human and mouse cells in culture have shown that other combinations of chromatin remodeling proteins and transcription factors are capable of reprogramming ES cells and fibroblasts into cardiomyocytes (Hartung et al., 2013; Ieda et al., 2010; Islas et al., 2012; Wada et al., 2013). The implications of these studies for regenerative medicine are extremely positive. With continued improvement of the methods, it seems likely that it will become possible to improve the function of damaged hearts by implanting cells engineered specifically towards myocardial differentiation.

23.6 CONCLUSIONS

The conservation of the underlying mechanisms (i.e., cell signaling and transcriptional regulation) involved in heart induction has enabled researchers to further our understanding of this complicated process using a variety of animal models ranging from fruit flies to rodents. Although no animal model system is ideal for all types of experiments, over recent years we have gained considerable insight into both human heart development and disease by capitalizing on the advantages that each model system provides. Although our understanding of this complicated process is far from complete, remarkable progress has been made in understanding the overall process and harnessing the knowledge of the underlying molecular mechanisms to reprogram the differentiation of human cells. These studies are a compelling demonstration of the way in which basic science investigation of developmental processes in model organisms has led us to the doorstep of effective treatment of human heart disease.

23.7 CLINICAL SIGNIFICANCE

- Cardiac development genes discovered and characterized in model organisms are causally related to human congenital heart defects.
- Different signaling pathways and different transcription factor profiles direct development of sub-regions of the original heart field, explaining the origin of human congenital heart defects.
- Defined transcription factors are sufficient to program embryonic stem cells, or reprogram differentiated non-cardiac cells, to the myocardial lineage.

- Knowledge of growth factor signaling pathways active in the embryo will facilitate growth and directed differentiation of reprogramd myocardial cells for treatment of human heart disease.

ACKNOWLEDGMENTS

PAK is the Allan C. Hudson and Helen Lovaas Endowed Professor of the Sarver Heart Center at the University of Arizona College of Medicine. The Krieg Laboratory is funded by the Sarver Heart Center and by the National Heart, Lung, and Blood Institute of the National Institutes of Health, grant #HL093694.

RECOMMENDED RESOURCES

GEISHA – A Chicken Embryo Gene Expression Database. (http://geisha.arizona.edu/geisha/ [Accessed Nov 2013]). Provides developmental expression patterns for numerous genes, including most genes involved in early heart development.

Heart Development and Regeneration, 2010. In: Rosenthal, N., Harvey, R. (Eds.), A comprehensive two volume work covering all aspects of early heart development. Academic Press.

REFERENCES

Ai, D., Fu, X., Wang, J., Lu, M.F., Chen, L., Baldini, A., Klein, W.H., Martin, J.F., 2007. Canonical Wnt signaling functions in second heart field to promote right ventricular growth. Proc. Natl. Acad. Sci. USA 104, 9319–9324.

Alsan, B.H., Schultheiss, T.M., 2002. Regulation of avian cardiogenesis by Fgf8 signaling. Development 129, 1935–1943.

Antin, P.B., Taylor, R.G., Yatskievych, T., 1994. Precardiac mesoderm is specified during gastrulation in quail. Dev. Dyn. 200, 144–154.

Bodmer, R., 1993. The gene tinman is required for specification of the heart and visceral muscles in *Drosophila*. Development 118, 719–729.

Bondue, A., Lapouge, G., Paulissen, C., Semeraro, C., Iacovino, M., Kyba, M., Blanpain, C., 2008. Mesp1 acts as a master regulator of multipotent cardiovascular progenitor specification. Cell. Stem Cell. 3, 69–84.

Brown, D.D., Martz, S.N., Binder, O., Goetz, S.C., Price, B.M., Smith, J.C., 2005. Conlon FL Tbx5 and Tbx20 act synergistically to control vertebrate heart morphogenesis. Development 132, 553–563.

Bruneau, B.G., Nemer, G., Schmitt, J.P., Charron, F., Robitaille, L., Caron, S., Conner, D.A., Gessler, M., Nemer, M., Seidman, C.E., Seidman, J.G., 2001. A murine model of Holt-Oram syndrome defines roles of the T-box transcription factor Tbx5 in cardiogenesis and disease. Cell 106, 709–721.

Cai, C.L., Liang, X., Shi, Y., Chu, P.H., Pfaff, S.L., Chen, J., Evans, S., 2003. Isl1 identifies a cardiac progenitor population that proliferates prior to differentiation and contributes a majority of cells to the heart. Dev. Cell. 5, 877–889.

Chieffo, C., Garvey, N., Gong, W., Roe, B., Zhang, G., Silver, L., Emanuel, B.S., Budarf, M.L., 1997. Isolation and characterization of a gene from the DiGeorge chromosomal region homologous to the mouse Tbx1 gene. Genomics 43, 267–277.

Cleaver, O.B., Patterson, K.D., Krieg, P.A., 1996. Overexpression of the tinman-related genes XNkx-2.5 and XNkx-2.3 in *Xenopus* embryos results in myocardial hyperplasia. Development 122, 3549–3556.

Climent, S., Sarasa, M., Villar, J.M., Murillo-Ferrol, N.L., 1995. Neurogenic cells inhibit the differentiation of cardiogenic cells. Dev. Biol. 171, 130–148.

Cohen, E.D., Wang, Z., Lepore, J.J., Lu, M.M., Taketo, M.M., Epstein, D.J., Morrisey, E.E., 2007. Wnt/b-catenin signaling promotes expansion of Isl-1-positive cardiac progenitor cells through regulation of FGF signaling. J. Clin. Invest. 117, 1794–1804.

Cui, C., Cheuvront, T.J., Lansford, R.D., Moreno-Rodriguez, R.A., Schultheiss, T.M., Rongish, B.J., 2009. Dynamic positional fate map of the primary heart-forming region. Dev. Biol. 332, 212–222.

David, R., Brenner, C., Stieber, J., Schwarz, F., Brunner, S., Vollmer, M., Mentele, E., Müller-Höcker, J., Kitajima, S., Lickert, H., Rupp, R., Franz, W.M., 2008. MesP1 drives vertebrate cardiovascular differentiation through Dkk-1-mediated blockade of Wnt-signaling. Nat. Cell. Biol. 10, 338–345.

Davis, D.L., Wessels, A., Burch, J.B., 2000. An Nkx-dependent enhancer regulates cGATA-6 gene expression during early stages of heart development. Dev. Biol. 217, 310–322.

Dyer, L.A., Kirby, M.L., 2009. Sonic hedgehog maintains proliferation in secondary heart field progenitors and is required for normal arterial pole formation. Dev. Biol. 330, 305–317.

Eisenberg, C.A., Eisenberg, L.M., 1999. WNT11 promotes cardiac tissue formation of early mesoderm. Dev. Dyn. 216, 45–58.

Evans, S.M., 1999. Vertebrate tinman homologues and cardiac differentiation. Semin. Cell. Dev. Biol. 10, 73–83.

Fletcher, R.B., Watson, A.L., Harland, R.M., 2004. Expression of *Xenopus tropicalis* noggin1 and noggin2 in early development: two noggin genes in a tetrapod. Gene Expr. Patterns 5, 225–230.

Foley, A.C., Mercola, M., 2005. Heart induction by Wnt antagonists depends on the homeodomain transcription factor Hex. Genes Dev. 19, 387–396.

Francou, A., Saint-Michel, E., Mesbah, K., Théveniau-Ruissy, M., Rana, M.S., Christoffels, V.M., Kelly, R.G., 2013. Second heart field cardiac progenitor cells in the early mouse embryo. Biochim. Biophys. Acta. 1833, 795–798.

Frank, D.U., Fotheringham, L.K., Brewer, J.A., Muglia, L.J., Tristani-Firouzi, M., Capecchi, M.R., Moon, A.M., 2002. An Fgf8 mouse mutant phenocopies human 22q11 deletion syndrome. Development 129, 4591–4603.

Frasch, M., 1995. Induction of visceral and cardiac mesoderm by ectodermal Dpp in the early *Drosophila* embryo. Nature 374, 464–467.

Gajewski, K., Fossett, N., Molkentin, J.D., Schulz, R.A., 1999. The zinc finger proteins Pannier and GATA4 function as cardiogenic factors in *Drosophila*. Development 126, 5679–5688.

Gannon, M., Bader, D., 1995. Initiation of cardiac differentiation occurs in the absence of anterior endoderm. Development 121, 2439–2450.

Garcia-Martinez, V., Schoenwolf, G.C., 1993. Primitive-streak origin of the cardiovascular system in avian embryos. Dev. Biol. 159, 706–719.

Garriock, R.J., Drysdale, T.A., 2003. Regulation of heart size in *Xenopus laevis*. Differentiation 71, 506–515.

Gessert, S., Kühl, M., 2009. Comparative gene expression analysis and fate mapping studies suggest an early segregation of cardiogenic lineages in *Xenopus laevis*. Dev. Biol. 334, 395–408.

Gibb, N., Lavery, D.L., Hoppler, S., 2013. sfrp1 promotes cardiomyocyte differentiation in *Xenopus* via negative-feedback regulation of Wnt signaling. Development 140, 1537–1549.

Goddeeris, M.M., Schwartz, R., Klingensmith, J., Meyers, E.N., 2007. Independent requirements for Hedgehog signaling by both the anterior heart field and neural crest cells for outflow tract development. Development 134, 1593–1604.

Golzio, C., Havis, E., Daubas, P., Nuel, G., Babarit, C., Munnich, A., Vekemans, M., Zaffran, S., Lyonnet, S., Etchevers, H.C., 2012. ISL1 directly regulates FGF10 transcription during human cardiac outflow formation. PLoS One 7, e30677.

Gonzalez-Sanchez, A., Bader, D., 1990. *In vitro* analysis of cardiac progenitor cell differentiation. Dev. Biol. 139, 197–209.

Greulich, F., Rudat, C., Kispert, A., 2011. Mechanisms of T-box gene function in the developing heart. Cardiovasc. Res. 91, 212–222.

Hamburger, V., Hamilton, H.L., 1951. A series of normal stages in the development of the chick embryo. J. Morphol. 88, 49–92.

Han, Y., Dennis, J.E., Cohen-Gould, L., Bader, D.M., Fischman, D.A., 1992. Expression of sarcomeric myosin in the presumptive myocardium of chicken embryos occurs within six hours of myocyte commitment. Dev. Dyn. 193, 257–265.

Hartung, S., Schwanke, K., Haase, A., David, R., Franz, W.M., Martin, U., Zweigerdt, R., 2013. Directing cardiomyogenic differentiation of human pluripotent stem cells by plasmid-based transient overexpression of cardiac transcription factors. Stem Cells Dev. 22, 1112–1125.

Heallen, T., Zhang, M., Wang, J., Bonilla-Claudio, M., Klysik, E., Johnson, R.L., Martin, J.F., 2011. Hippo pathway inhibits Wnt signaling to restrain cardiomyocyte proliferation and heart size. Science 332, 458–461.

Holtzer, H., Schultheiss, T., Dilullo, C., Choi, J., Costa, M., Lu, M., Holtzer, S., 1990. Autonomous expression of the differentiation programs of cells in the cardiac and skeletal myogenic lineages. Ann. N. Y. Acad. Sci. 599, 158–169.

Ieda, M., Fu, J.D., Delgado-Olguin, P., Vedantham, V., Hayashi, Y., Bruneau, B.G., Srivastava, D., 2010. Direct reprogramming of fibroblasts into functional cardiomyocytes by defined factors. Cell 142, 375–386.

Islas, J.F., Liu, Y., Weng, K.C., Robertson, M.J., Zhang, S., Prejusa, A., Harger, J., Tikhomirova, D., Chopra, M., Iyer, D., Mercola, M., Oshima, R.G., Willerson, J.T., Potaman, V.N., Schwartz, R.J., 2012. Transcription factors ETS2 and MESP1 transdifferentiate human dermal fibroblasts into cardiac progenitors. Proc. Natl. Acad. Sci. U.S.A. 109, 13016–13021.

Jacobson, A.G., 1961. Heart determination in the newt. J. Exp. Zool. 146, 139–151.

Jerome, L.A., Papaioannou, V.E., 2001. DiGeorge syndrome phenotype in mice mutant for the T-box gene, Tbx1. Nat. Genet. 27, 286–291.

Karlsson, O., Thor, S., Norberg, T., Ohlsson, H., Edlund, T., 1990. Insulin gene enhancer binding protein Isl-1 is a member of a novel class of proteins containing both a homeo- and a Cys-His domain. Nature 344, 879–882.

Keller, R.E., 1976. Vital dye mapping of the gastrula and neurula of *Xenopus laevis.* II. Prospective areas and morphogenetic movements of the deep layer. Dev. Biol. 51, 118–137.

Kitajima, S., Takagi, A., Inoue, T., Saga, Y., 2000. MesP1 and MesP2 are essential for the development of cardiac mesoderm. Development 127, 3215–3226.

Klaus, A., Saga, Y., Taketo, M.M., Tzahor, E., Birchmeier, W., 2007. Distinct roles of Wnt/b-catenin and Bmp signaling during early cardiogenesis. Proc. Natl. Acad. Sci. U.S.A. 104, 18531–18536.

Klaus, A., Müller, M., Schulz, H., Saga, Y., Martin, J.F., Birchmeier, W., 2012. Wnt/β-catenin and Bmp signals control distinct sets of transcription factors in cardiac progenitor cells. Proc. Natl. Acad. Sci. U.S.A. 109, 10921–10926.

Koutsourakis, M., Langeveld, A., Patient, R., et al., 1999. The transcription factor GATA6 is essential for early extraembryonic development. Development 126, 723–732.

Kriegmair, M.C., Frenz, S., Dusl, M., Franz, W.M., David, R., Rupp, R.A., 2013. Cardiac differentiation in *Xenopus* is initiated by mespa. Cardiovasc. Res. 97, 454–463.

Kuo, H., Chen, J., Ruiz-Lozano, P., Zou, Y., Nemer, M., Chien, K.R., 1999. Control of segmental expression of the cardiac-restricted ankyrin repeat protein gene by distinct regulatory pathways in murine cardiogenesis. Development 126, 4223–4234.

Kwon, C., Arnold, J., Hsiao, E.C., Taketo, M.M., Conklin, B.R., Srivastava, D., 2007. Canonical Wnt signaling is a positive regulator of mammalian cardiac progenitors. Proc. Natl. Acad. Sci. U.S.A. 104, 10894–10899.

Kwon, C., Qian, L., Cheng, P., Nigam, V., Arnold, J., Srivastava, D., 2009. A regulatory pathway involving Notch1/β-catenin/Isl1 determines cardiac progenitor cell fate. Nat. Cell. Biol. 11, 951–957.

Latinkic, B.V., Kotecha, S., Mohun, T.J., 2003. Induction of cardiomyocytes by GATA4 in *Xenopus* ectodermal explants. Development 130, 3865–3876.

Lavery, D.L., Martin, J., Turnbull, Y.D., Hoppler, S., 2008. Wnt6 signaling regulates heart muscle development during organogenesis. Dev. Biol. 323, 177–188.

Li, Q.Y., Newbury-Ecob, R.A., Terrett, J.A., Wilson, D.I., Curtis, A.R., Yi, C.H., Gebuhr, T., Bullen, P.J., Robson, S.C., Strachan, T., Bonnet, D., Lyonnet, S., Young, I.D., Raeburn, J.A., Buckler, A.J., Law, D.J., Brook, J.D., 1997. Holt-Oram syndrome is caused by mutations in TBX5, a member of the Brachyury (T) gene family. Nat. Genet. 15, 21–29.

Liberatore, C.M., Searcy-Schrick, R.D., Vincent, E.B., Yutzey, K.E., 2002. Nkx-2.5 gene induction in mice is mediated by a Smad consensus regulatory region. Dev. Biol. 244, 243–256.

Lickert, H., Kutsch, S., Kanzler, B., Tamai, Y., Taketo, M.M., Kemler, R., 2002. Formation of multiple hearts in mice following deletion of β-catenin in the embryonic endoderm. Dev. Cell. 3, 171–181.

Lien, C.L., McAnally, J., Richardson, J.A., Olson, E.N., 2002. Cardiac-specific activity of an Nkx2–5 enhancer requires an evolutionarily conserved Smad binding site. Dev. Biol. 244, 257–266.

Lien, C.L., Wu, C., Mercer, B., et al., 1999. Control of early cardiac-specific transcription of Nkx2–5 by a GATA-dependent enhancer. Development 126, 75–84.

Lin, L., Bu, L., Cai, C.L., Zhang, X., Evans, S., 2006. Isl1 is upstream of sonic hedgehog in a pathway required for cardiac morphogenesis. Dev. Biol. 295, 756–763.

Lin, L., Cui, L., Zhou, W., Dufort, D., Zhang, X., Cai, C.L., Bu, L., Yang, L., Martin, J., Kemler, R., Rosenfeld, M.G., Chen, J., Evans, S.M., 2007. B-catenin directly regulates Islet1 expression in cardiovascular progenitors and is required for multiple aspects of cardiogenesis. Proc. Natl. Acad. Sci. U.S.A. 104, 9313–9318.

Lyons, I., Parsons, L.M., Hartley, L., Li, R., Andrews, J.E., Robb, L., Harvey, R.P., 1995. Myogenic and morphogenetic defects in the heart tubes of murine embryos lacking the homeo box gene Nkx2–5. Genes Dev. 9, 1654–1666.

Majumdar, A., Vainio, S., Kispert, A., McMahon, J., McMahon, A.P., 2003. Wnt11 and Ret/Gdnf pathways cooperate in regulating ureteric branching during metanephric kidney development. Development 130, 3175–3185.

Marques, S.R., Lee, Y., Poss, K.D., Yelon, D., 2008. Reiterative roles for FGF signaling in the establishment of size and proportion of the zebrafish heart. Dev. Biol. 321, 397–406.

Marvin, M.J., Di Rocco, G., Gardiner, A., et al., 2001. Inhibition of Wnt activity induces heart formation from posterior mesoderm. Genes Dev. 15, 316–327.

Maye, P., Zheng, J., Li, L., Wu, D., 2004. Multiple mechanisms for Wnt11-mediated repression of the canonical Wnt signaling pathway. J. Biol. Chem. 279, 24659–24665.

Miazga, C.M., McLaughlin, K.A., 2009. Coordinating the timing of cardiac precursor development during gastrulation: a new role for Notch signaling. Dev. Biol. 333, 285–296.

Molkentin, J.D., Antos, C., Mercer, B., Taigen, T., Miano, J.M., Olson, E.N., 2000. Direct activation of a GATA6 cardiac enhancer by Nkx2.5: evidence for a reinforcing regulatory network of Nkx2.5 and GATA transcription factors in the developing heart. Dev. Biol. 217, 301–309.

Montgomery, M.O., Litvin, J., Gonzalez-Sanchez, A., Bader, D., 1994. Staging of commitment and differentiation of avian cardiac myocytes. Dev. Biol. 164, 63–71.

Nascone, N., Mercola, M., 1995. An inductive role for the endoderm in *Xenopus* cardiogenesis. Development 121, 515–523.

Nemer, G., Qureshi, S.T., Malo, D., Nemer, M., 1999. Functional analysis and chromosomal mapping of Gata5, a gene encoding a zinc finger DNA-binding protein. Mamm. Genome. 10, 993–999.

Nemir, M., Croquelois, A., Pedrazzini, T., Radtke, F., 2006. Induction of cardiogenesis in embryonic stem cells via downregulation of Notch1 signaling. Circ. Res. 98, 1471–1478.

Orts-Llorca, 1963. Influence of the endoderm on heart differentiation during the early stages of development of the chick embryo. Roux. Arch. Dev. Biol. 154, 533–551.

Pandur, P., Lasche, M., Eisenberg, L.M., Kuhl, M., 2002. Wnt-11 activation of a non-canonical Wnt signaling pathway is required for cardiogenesis. Nature 418, 636–641.

Peterkin, T., Gibson, A., Loose, M., Patient, R., 2005. The roles of GATA-4, -5 and -6 in vertebrate heart development. Semin. Cell. Dev. Biol. 16, 83–94.

Peterkin, T., Gibson, A., Patient, R., 2003. GATA-6 maintains BMP-4 and Nkx2 expression during cardiomyocyte precursor maturation. EMBO. J. 22, 4260–4273.

Plageman Jr., T.F., Yutzey, K.E., 2005. T-box genes and heart development: putting the "T" in heart. Dev. Dyn. 232, 11–20.

Qyang, Y., Martin-Puig, S., Chiravuri, M., Chen, S., Xu, H., Bu, L., Jiang, X., Lin, L., Granger, A., Moretti, A., Caron, L., Wu, X., Clarke, J., Taketo, M.M., Laugwitz, K.L., Moon, R.T., Gruber, P., Evans, S.M., Ding, S., Chien, K.R., 2007. The renewal and differentiation of Isl1+ cardiovascular progenitors are controlled by a Wnt/b-catenin pathway. Cell Stem Cell. 1, 165–179.

Rawles, M.E., 1943. The heart-forming areas of the early chick blastoderm. Physiol. Zool. 41, 22–42.

Reiter, J.F., Alexander, J., Rodaway, A., Yelon, D., Patient, R., Holder, N., Stainier, D.Y., 1999. Gata5 is required for the development of the heart and endoderm in zebrafish. Genes Dev. 13, 2983–2995.

Reiter, J.F., Verkade, H., Stainier, D.Y., 2001. Bmp2b and Oep promote early myocardial differentiation through their regulation of gata5. Dev. Biol. 234, 330–338.

Rochais, F., Mesbah, K., Kelly, R.G., 2009. Signaling pathways controlling second heart field development. Circ. Res. 104, 933–942.

Rones, M.S., McLaughlin, K.A., Raffin, M., Mercola, M., 2000. Serrate and Notch specify cell fates in the heart field by suppressing cardiomyogenesis. Development 127, 3865–3876.

Rosenquist, G.C., De Haan, R.L., 1966. Migration of the precardiac cells in the chick embryo. A radioautographic study, Carnegie Inst Washington. Contrib. Embryol, 71–110.

Samuel, L.J., Latinkić, B.V., 2009. Early activation of FGF and nodal pathways mediates cardiac specification independently of Wnt/b-catenin signaling. PLoS One 4, e7650.

Sater, A.K., Jacobson, A.G., 1989. The specification of heart mesoderm occurs during gastrulation in *Xenopus laevis*. Development 105, 821–830.

Sater, A.K., Jacobson, A.G., 1990. The restriction of the heart morphogenetic field in *Xenopus laevis*. Dev. Biol. 140, 328–336.

Satou, Y., Imai, K.S., Satoh, N., 2004. The ascidian MesP gene specifies heart precursor cells. Development 131, 2533–2541.

Schneider, V.A., Mercola, M., 2001. Wnt antagonism initiates cardiogenesis in *Xenopus laevis*. Genes Dev. 15, 304–315.

Schultheiss, T.M., Burch, J.B., Lassar, A.B., 1997. A role for bone morphogenetic proteins in the induction of cardiac myogenesis. Genes Dev. 11, 451–462.

Schultheiss, T.M., Lassar, A.B., 1997. Induction of chick cardiac myogenesis by bone morphogenetic proteins. Cold. Spring. Harb. Symp. Quant. Biol. 62, 413–419.

Schultheiss, T.M., Xydas, S., Lassar, A.B., 1995. Induction of avian cardiac myogenesis by anterior endoderm. Development 121, 4203–4214.

Searcy, R.D., Vincent, E.B., Liberatore, C.M., Yutzey, K.E., 1998. A GATA-dependent nkx-2.5 regulatory element activates early cardiac gene expression in transgenic mice. Development 125, 4461–4470.

Serbedzija, G.N., Chen, J.N., Fishman, M.C., 1998. Regulation in the heart field of zebrafish. Development 125, 1095–1101.

Shi, Y., Katsev, S., Cai, C., Evans, S., 2000. BMP signaling is required for heart formation in vertebrates. Dev. Biol. 224, 226–237.

Showell, C., Christine, K.S., Mandel, E.M., Conlon, F.L., 2006. Developmental expression patterns of Tbx1, Tbx2, Tbx5, and Tbx20 in *Xenopus tropicalis*. Dev. Dyn. 235, 1623–1630.

Stainier, D.Y.R., Lee, R.K., Fishman, M.C., 1993. Cardiovascular development in the zebrafish. I. Myocardial fate map and heart tube formation. Development 119, 31–40.

Stennard, F.A., Costa, M.W., Elliott, D.A., Rankin, S., Haast, S.J., Lai, D., McDonald, L.P., Niederreither, K., Dolle, P., Bruneau, B.G., Zorn, A.M., Harvey, R.P., 2003. Cardiac T-box factor Tbx20 directly interacts with Nkx2–5, GATA4, and GATA5 in regulation of gene expression in the developing heart. Dev. Biol. 262, 206–224.

Stennard, F.A., Costa, M.W., Lai, D., Biben, C., Furtado, M.B., Solloway, M.J., McCulley, D.J., Leimena, C., Preis, J.I., Dunwoodie, S.L., Elliott, D.E., Prall, O.W., Black, B.L., Fatkin, D., Harvey, R.P., 2005. Murine T-box transcription factor Tbx20 acts as a repressor during heart development, and is essential for adult heart integrity, function and adaptation. Development 132, 2451–2462.

Takeuchi, J.K., Bruneau, B.G., 2009. Directed transdifferentiation of mouse mesoderm to heart tissue by defined factors. Nature 459, 708–711.

Tanaka, M., Schinke, M., Liao, H.S., Yamasaki, N., Izumo, S., 2001. Nkx2.5 and Nkx2.6, homologs of *Drosophila* tinman, are required for development of the pharynx. Mol. Cell. Biol. 21, 4391–4398.

Thomas, N.A., Koudijs, M., van Eeden, F.J., Joyner, A.L., Yelon, D., 2008. Hedgehog signaling plays a cell-autonomous role in maximizing cardiac developmental potential. Development 135, 3789–3799.

Topol, L., Jiang, X., Choi, H., Garrett-Beal, L., Carolan, P.J., Yang, Y., 2003. Wnt-5a inhibits the canonical Wnt pathway by promoting GSK-3-independent β-catenin degradation. J. Cell. Biol. 162, 899–908.

Tzahor, E., Lassar, A.B., 2001. Wnt signals from the neural tube block ectopic cardiogenesis. Genes. Dev. 15, 255–260.

Vincent, S.D., Buckingham, M.E., 2010. How to make a heart: the origin and regulation of cardiac progenitor cells. Curr. Top. Dev. Biol. 90, 1–41.

Wada, R., Muraoka, N., Inagawa, K., Yamakawa, H., Miyamoto, K., Sadahiro, T., Umei, T., Kaneda, R., Suzuki, T., Kamiya, K., Tohyama, S., Yuasa, S., Kokaji, K., Aeba, R., Yozu, R., Yamagishi, H., Kitamura, T., Fukuda, K., Ieda, M., 2013. Induction of human cardiomyocyte-like cells from fibroblasts by defined factors. Proc. Natl. Acad. Sci. U.S.A. 110, 12667–12672.

Waldo, K.L., Hutson, M.R., Ward, C.C., Zdanowicz, M., Stadt, H.A., Kumiski, D., Abu-Issa, R., Kirby, M.L., 2005. Secondary heart field contributes myocardium and smooth muscle to the arterial pole of the developing heart. Dev. Biol. 281, 78–90.

Walters, M.J., Wayman, G.A., Christian, J.L., 2001. Bone morphogenetic protein function is required for terminal differentiation of the heart but not for early expression of cardiac marker genes. Mech. Dev. 100, 263–273.

Wang, H., Hao, J., Hong, C.C., 2011. Cardiac induction of embryonic stem cells by a small molecule inhibitor of Wnt/β-catenin signaling. ACS Chem. Biol. 6, 192–197.

Washington Smoak, I., Byrd, N.A., Abu-Issa, R., Goddeeris, M.M., Anderson, R., Morris, J., Yamamura, K., Klingensmith, J., Meyers, E.N., 2005. Sonic hedgehog is required for cardiac outflow tract and neural crest cell development. Dev. Biol. 283, 357–372.

Watanabe, Y., Miyagawa-Tomita, S., Vincent, S.D., Kelly, R.G., Moon, A.M., Buckingham, M.E., 2010. Role of mesodermal FGF8 and FGF10 overlaps in the development of the arterial pole of the heart and pharyngeal arch arteries. Circ. Res. 106, 495–503.

Weidinger, G., Moon, R.T., 2003. When Wnts antagonize Wnts. J. Cell. Biol. 162, 753–755.

Winnier, G., Blessing, M., Labosky, P.A., Hogan, B.L., 1995. Bone morphogenetic protein-4 is required for mesoderm formation and patterning in the mouse. Genes Dev. 9, 2105–2116.

Yamada, M., Revelli, J.P., Eichele, G., Yamada, M., Revelli, J.P., Eichele, G., Barron, M., Schwartz, R.J., 2000. Expression of chick Tbx-2, Tbx-3, and Tbx-5 genes during early heart development: evidence for BMP2 induction of Tbx2. Dev. Biol. 228, 95–105.

Yatskievych, T.A., Ladd, A.N., Antin, P.B., 1997. Induction of cardiac myogenesis in avian pregastrula epiblast: the role of the hypoblast and activin. Development 124, 2561–2570.

Yuasa, S., Itabashi, Y., Koshimizu, U., Tanaka, T., Sugimura, K., Kinoshita, M., Hattori, F., Fukami, S., Shimazaki, T., Ogawa, S., Okano, H., Fukuda, K., 2005. Transient inhibition of BMP signaling by Noggin induces cardiomyocyte differentiation of mouse embryonic stem cells. Nat. Biotechnol. 23, 607–611.

Zhang, H., Bradley, A., 1996. Mice deficient for BMP2 are nonviable and have defects in amnion/chorion and cardiac development. Development 122, 2977–2986.

Zhang, X.M., Ramalho-Santos, M., McMahon, A.P., 2001. Smoothened mutants reveal redundant roles for Shh and Ihh signaling including regulation of L/R asymmetry by the mouse node. Cell 105, 781–792.

Zhao, R., Watt, A.J., Li, J., Luebke-Wheeler, J., Morrisey, E.E., Duncan, S.A., 2005. GATA6 is essential for embryonic development of the liver but dispensable for early heart formation. Mol. Cell. Biol. 25, 2622–2631.

Zhao, R., Watt, A.J., Battle, M.A., Li, J., Bondow, B.J., Duncan, S.A., 2008. Loss of both GATA4 and GATA6 blocks cardiac myocyte differentiation and results in acardia in mice. Dev. Biol. 317, 614–619.

Zhou, Y., Cashman, T.J., Nevis, K.R., Obregon, P., Carney, S.A., Liu, Y., Gu, A., Mosimann, C., Sondalle, S., Peterson, R.E., Heideman, W., Burns, C.E., Burns, C.G., 2011. Latent TGF-β binding protein 3 identifies a second heart field in zebrafish. Nature 474, 645–648.

Chapter 24

Blood Vessel Formation

Amber N. Stratman, Jianxin A. Yu, Timothy S. Mulligan, Matthew G. Butler, Eric T. Sause and Brant M. Weinstein

Program in Genomics of Differentiation, National Institute of Child Health and Human Development, National Institutes of Health (NIH), Bethesda, MD

Chapter Outline

GLOSSARY

Angioblast Mesoderm-derived endothelial precursors that have certain characteristics of endothelial cells but that have not yet assembled into functional vessels.

Angiogenesis The formation of new blood vessels from pre-existing vessels by endothelial sprouting and splitting. This process of remodeling the primary capillary network leads to the formation of mature arteries and veins.

Lymphatic system A blind-ended system of vessels that protects and maintains the fluid environment of the body by filtering and draining away lymphatic fluid. Under normal conditions, the lymphatic vascular system is necessary for the return of extravasated interstitial fluid and macromolecules to the blood circulation, for immune defense, and for the uptake of dietary fats.

Vasculogenesis The process of the formation of primitive vessels in vertebrate embryos during which mesodermal cells differentiate into endothelial precursor cells called angioblasts. These angioblasts then differentiate *in situ* into endothelial cells and coalesce to form the earliest vessels.

Vascular endothelial growth factor The first growth factor described as being a specific mitogen for endothelial cells. It is important for the migration, proliferation, maintenance, and survival of endothelial cells, and it is critical during both vasculogenesis and angiogenesis. It also plays an important role in arterial differentiation.

SUMMARY

- Primitive blood vessels in vertebrate embryos form by a process called vasculogenesis, during which mesodermal cells differentiate into endothelial precursor cells called angioblasts. These angioblasts then differentiate into endothelial cells (ECs) and coalesce to form the earliest vessels. The subsequent growth and remodeling of the vasculature occurs by angiogenesis, during which new blood vessels sprout from pre-existing vessels.
- The vascular endothelial growth factor (VEGF) signaling pathway has been shown to be important for the migration, proliferation, maintenance, and survival of ECs and is critical during both vasculogenesis and angiogenesis. In addition to its important function during embryonic development, VEGF is thought to play a major role in tumor-induced neovascularization. Angiopoietin-Tie signaling is thought to be critical for the maturation and stabilization of the nascent vascular network.
- One of the most fundamental steps in the differentiation of the vasculature is the specification of the arterial and venous endothelium. Recent work has highlighted the role of Notch, VEGF and Hedgehog signaling in the specification of arterial identity. Classically, differences between arteries and veins were attributed to physiological factors such as the direction and pressure of blood flow. However, recent work has shown that molecular distinctions between arterial and venous endothelium can appear before the onset of blood flow.
- Vertebrates possess a second, blind-ended vascular system, the lymphatic system, which is responsible for clearing and draining fluids and macromolecules that leak from blood vessels into the interstitial spaces of tissues and organs. Evidence suggests that the first lymphatic endothelial cells (LECs) emerge by transdifferentiation from primitive veins. The transcription factor Prox1 is a critical regulator of LEC specification, whereas the VEGF family members VEGF-C and VEGF-D are important for LEC migration and lymphangiogenesis.
- During early development, newly formed vessels follow a highly stereotypic and evolutionarily well-conserved pattern of network assembly that is reminiscent of that followed by the nervous system and axon tracts. A variety of recent studies have shown that many well-known molecular pathways used for the guidance of migrating axons play an important role in the guidance and patterning of developing blood vessels.

24.1 INTRODUCTION

The development of the vascular system is one of the earliest and most essential events to long-term successful organogenesis. All organs depend on a vascular supply for the delivery of nutrients and oxygen, and the clearance of waste. Severe disruptions in vascular network formation during the early stages of embryonic development are lethal, while the maintenance of vessel integrity and vessel physiology is critically important, not only during embryonic development but also in postnatal adult life. Many of the events that regulate vascular development are subsequently reactivated during adult neo-angiogenesis, including during tissue regeneration, wound healing, and in disease states such as formation of the vascular beds supporting tumors. A full understanding of the mechanisms and signaling pathways controlling vascular development is essential for identifying therapeutic targets for these adult pathologies. The following chapter summarizes our current understanding of major areas of knowledge in developmental vascular biology.

24.2 EMERGENCE OF THE BLOOD VASCULAR SYSTEM

Basic Concepts

The cardiovascular system could be considered to be composed of three major components – the heart, the blood vascular system, and the lymphatic vascular system. Each part of the system will be discussed separately throughout this chapter. The cardiovascular system is the first functional organ system to develop during vertebrate embryogenesis. Embryonic growth and development rely on the transportation of oxygen, nutrients, waste products and other cellular and humoral factors by the cardiovascular system. The heart pumps these components through the blood vessels, forming a closed circulatory network with blood remaining contained within the vessels, except through leakage or hemorrhage.

Blood is pumped away from the heart through arteries, reaches the body tissues via the capillaries, and is returned to the heart through veins. Modifications and additions to this basic plan have evolved to meet the requirements of different vertebrate species. Fish, for instance, utilize a single circulation in which blood is oxygenated at the gill capillaries and is then sent directly to the body tissues before it returns to the heart (i.e., heart-gills-body-heart). Because of the simplicity of this system, the fish heart only has a single pump with two chambers: the atrium and the ventricle. The blood enters the heart through the atrium and exits through the ventricle. In air-breathing vertebrates, an additional circulatory loop to the lungs has been acquired for blood oxygenation. In this pulmonary circuit, blood is pumped from the heart to the lungs, and then back to the heart again before it is delivered to the body tissues. This is known as a double circulation (i.e., heart-lungs-heart-body-heart). Utilizing a double circulation does not necessarily mean that the heart is divided into two separate pumps, however. Amphibians and reptiles have blood enter a three-chambered heart via two separate atria, but these atria send the blood to a single common ventricle where the oxygenated and deoxygenated blood mix before being pumped to the body and lungs from the same source. In reptiles, the beginnings of a ventricular septum has begun to form, but it is not completed, therefore the blood sent to the body is slightly more oxygenated than the blood sent to the lungs (Koshiba-Takeuchi et al., 2009). In birds and mammals, the ventricular septum is complete, forming a four-chambered heart. Deoxygenated blood enters the right atrium of the heart from the vena cava. It is then emptied into the right ventricle before it is pumped through the pulmonary artery (one of the few arteries to carry deoxygenated blood) to the lungs, where it is oxygenated. The blood then returns to the heart through the pulmonary vein (one of the few veins to carry oxygenated blood) into the left atrium. From there, the blood travels to the left ventricle and is pumped through the aorta to the body tissues. The oxygenated blood leaves the aorta through the aortic arch, which branches out and disperses throughout the body into ever-smaller arteries and arterioles. Blood eventually reaches all living cells of the body through microscopic vessels called capillaries, which are so narrow that blood cells travel through them in single file. Here, the blood and tissues undergo nutrient and waste exchange. From the capillaries, the blood drains into small venules, which progressively fuse to form larger veins and eventually the vena cava, which returns the blood to the heart at the right atrium (in mammals).

Arteries and veins share a basic histological structure (Figure 24.1). The vessel walls are composed of two basic cell types, vascular endothelial cells (ECs) and vascular smooth muscle cells (VSMCs). The inner lining of the blood vessels

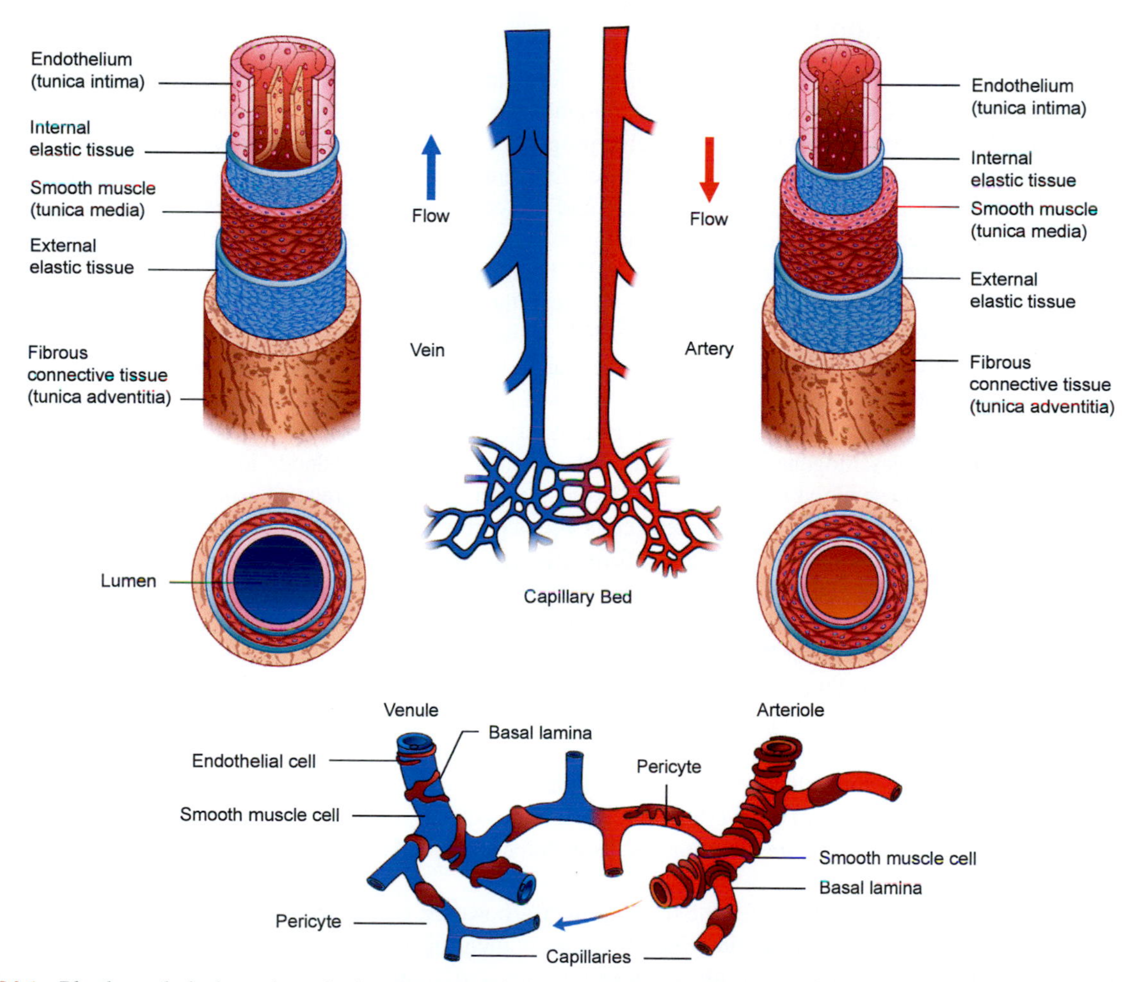

FIGURE 24.1 Blood vessels, both arteries and veins, are composed of an endothelial layer (tunica intima) that is surrounded by elastic tissue, a smooth muscle cell layer (tunica media), and external elastic tissue and finally a layer of fibrous connective tissue (tunica adventitia). Large caliber arteries tend to have thicker smooth muscle cell layers as compared to capillaries and veins to help regulate flow and vessel tone, while large caliber veins tend to possess specialized internal structures such as valves. The two types of vessels are connected through a system of capillaries, where most of the body's oxygen and waste exchange occurs. *Reproduced from Cleaver and Krieg (1999) with permission.*

adjacent to the lumen is a thin, single layer of ECs called the intimal layer, or tunica intima. A layer of sub-endothelial connective tissue encompasses the intima and is in turn surrounded by multiple layers of VSMCs and/or pericytes embedded in an extracellular matrix that is rich in elastin. This muscular layer is called the medial layer, and is highly developed in the arteries. The medial layer is itself surrounded by an outer layer of connective tissue called the adventitial layer, or tunica adventitia. Nerves run through the adventitia and innervate both the smooth muscle layer and the vessels themselves.

The earliest vessels that develop in vertebrates form though the process of vasculogenesis; i.e., the *de novo* generation of new vessels from mesodermal cells that differentiate into endothelial precursor cells called angioblasts. Angioblasts then differentiate further into ECs before coalescing to form the first tubular vessels. In mammalian and avian embryos, these first vessels appear and form into a primary vascular plexus. New vessels must continue to form during embryogenesis and do so though the process of angiogenesis. In non-sprouting angiogenesis or intussusception, a trans-vascular post or pillar forms to subdivide a vessel, whereas in sprouting angiogenesis a new vessel will sprout off the side of a pre-existing one.

Once the primitive endothelial tubes are formed, the endothelium secretes factors that contribute to the recruitment and/or differentiation of smooth muscle through a process called *vascular myogenesis*. Several reviews have described the differentiation of these VSMCs, as well as the effects of their differentiated state on proliferation, contractility, wound healing, vascular stability and vascular disease (Owens et al., 2004; Kawai-Kowase and Owens, 2007; Orlandi and Bennett, 2010; Stratman and Davis, 2012). Different models have been proposed to describe the way in which the VSMC phenotype is determined, including the effects of different growth factors, cytokines, miRNA, epigenetic markers, and other environmental cues (Owens et al., 2004; Davis-Dusenbery et al., 2011).

Emergence and Specification of Endothelial Cells

Before blood vessels begin to form, mesodermal cells first differentiate into endothelial precursor cells called angioblasts. Though some characteristics are shared between ECs and angioblasts, angioblasts have yet to assemble into functional vessels (Flamme et al., 1997). The source of these endothelial progenitors has been identified in various species. Transplantation experiments in quail and chick embryos have identified the somatic and splanchnopleuric subsets of mesoderm that give rise to angioblasts (Coffin and Poole, 1988; Pardanaud and Dieterlen-Lievre, 1999). In zebrafish, angioblasts segregate from the lateral plate mesoderm in a vascular endothelial growth factor receptor (VEGFR)-2/flk1/kdr dependent manner at the 7-somite stage (Fouquet et al., 1997; Liao et al., 1997; Thompson et al., 1998). Cell-lineage analysis suggests that these angioblasts contribute to the primordia of the dorsal aorta, the cardinal veins, and even the intersegmental vessels in the zebrafish trunk (Zhong et al., 2001; Childs et al., 2002). However new work suggests an alternative mechanism of vessel formation in the fish by which the first artery and vein arise from a single common precursor vessel. In this model, EphrinB2 is expressed in cells that have an arterial fate, whereas EphB4 is expressed in cells with a venous fate, and these species interact to drive vessel migration and separation from the single parental vessel (Herbert et al., 2009). In developing mouse embryos, the onset of vasculogenesis is marked by the formation of blood islands in the extraembryonic yolk sac (Risau and Flamme, 1995). These blood islands develop from mesodermal cell aggregates around embryonic day 7.5 to 8, and consist of an inner layer of primitive hematopoietic cells with a peripheral population of angioblasts surrounding them. These angioblasts then differentiate into ECs, migrate, form a lumen, and interconnect to form a primary vascular plexus (Risau and Flamme, 1995; Kamei et al., 2006). The close spatial relationship, as well as the simultaneous emergence of the hematopoietic cells and ECs within the blood islands has led to the hypothesis that they share a common precursor, the hemangioblast (His, 1900; Murray, 1932; Williams and Burgess, 1980).

Several lines of evidence support the existence of a hematopoietic/endothelial progenitor with dual potential. For example, angioblasts and hematopoietic progenitors express many of the same transcription factors and cell surface receptors, such as CD34 (Fina et al., 1990), Flk, Flt1, Tie1, Tie2 (Dumont et al., 1995), and SCL/tal1 (Kallianpur et al., 1994; Stratman et al., 2011). Moreover, the development of both endothelial and hematopoietic lineages is impaired in embryos bearing mutations or dominant-negative forms of one of these relevant genes (Shalaby et al., 1995; Shivdasani et al., 1995; Visvader et al., 1998). Similarly, the zebrafish *cloche* mutation results in defects in blood cells as well as blood vessels (Stainier et al., 1995; Liao et al., 1997).

Probably the most compelling evidence for the hemangioblast, however, comes from studies involving the *in vitro* differentiation of mouse embryonic stem cells, in which the development of the endothelial and hematopoietic lineages in embryoid bodies is thought to recapitulate events that take place *in vivo* in the yolk sac blood islands (Doetschman et al., 1985; Wiles and Keller 1991; Wang et al., 1992; Vittet et al., 1996). Using this model system, Choi and colleagues (Choi, 1998) and Chung and colleagues (Chung et al., 2002) isolated a transient cell population that expresses markers common to both cell populations (SCL/tal-1, CD34, and Flk1), and determined that single cells could give rise to clones containing both hematopoietic cells and ECs. More recently, an Flk1-positive population of cells was identified in the posterior primitive streak of embryonic day 7 to 7.5 mouse embryos, thus providing further support to the idea of a common progenitor for hematopoietic cells and ECs *in vivo* (Huber et al., 2004).

Although many studies looking at spatial development, gene expression, and mutant phenotypes support the concept of a common hematopoietic cell and EC progenitor in the yolk sac *in vivo*, other studies have argued against its existence. In avian embryos, a clonal differentiation assay of VEGFR-2-positive cells sorted out from the very early blastodisc failed to give rise to mixed colonies of endothelial and hematopoietic cells (Eichmann et al., 1997). Additionally, the lineage tracing of cells from the primitive streak to the yolk sac has failed to identify cells with dual endothelial and hematopoietic potential (Kinder et al., 2001). Finally, Ferkowicz and colleagues (Ferkowicz et al., 2003) showed that hematopoietic and endothelial progenitors can be distinguished by the differential expression of CD41 as soon as they exit from the primitive streak. Overall, this has proven to be a very controversial area of study and continues to yield conflicting results (reviewed by Bertrand and Traver, 2009; Hirschi, 2012).

Clearly, there is still much work to be done to reconcile the differences between these apparently contradictory findings. However, some of the discrepancies may lie in subtle differences in the timing and commitment of precursor cells to the different vascular lineages among organisms. Another explanation may be that there are distinct populations of precursor cells, in which some have multilineage "hemangioblastic" potential and others have only single lineage commitment capabilities. EC differentiation in the embryo proper does not occur in close association with hematopoiesis, except for on the floor of the aorta (reviewed by Jaffredo et al., 2005), where the endothelial and hematopoietic clusters are present in close proximity. Recent data have suggested that the endothelium of this region might be hemogenic (i.e., capable of giving rise to definitive hematopoietic cells through an endothelial intermediate). However, other studies suggest that hematopoietic precursor cells originate from the surrounding mesenchyme and migrate through the aorta wall before subsequently entering the circulation (reviewed by Jaffredo et al., 2005a,b; Zovein et al., 2008; Zape and Zovein, 2011; Hirschi, 2012). Continued work in this area is rapidly shedding light on the cellular origins and molecular pathways regulating hematopoietic cell and EC lineage differentiation.

De novo Formation of a Primary Vascular Plexus: Vasculogenesis

The first blood vessels are formed through a *de novo* process called vasculogenesis in which endothelial progenitors migrate and coalesce to form vascular tubes. These tubes then interconnect to form a vascular network capable of circulating blood. After extensive growth and remodeling, this relatively unstructured network becomes a mature vascular plexus (Risau, 1997). The process of vasculogenesis has been observed in multiple species. In mice, the cranial vessels and the aortic primordia begin forming at embryonic day 7.5 with the emergence of endocardial progenitor cells (Drake and Fleming, 2000). Chick embryos show differentiation of the dorsal aorta and several capillaries by the 12-somite stage, when the heart first begins to beat (reviewed by Eichmann et al., 2005). The first angioblasts of the zebrafish originate from the lateral plate mesoderm and migrate to the trunk midline between the 10- and 15-somite stages, before coalescing to form the primary axial vessels of the trunk (reviewed by Weinstein, 2002). As shown in a detailed atlas of developing zebrafish blood vessels (Isogai et al., 2001), the same initial circulatory circuits that develop in mammals and avian species are also present in the fish (Figure 24.2), making this organism, with its optically clear embryos and larvae, an important model for comparative studies of vascular development.

Remodeling and Maturation of the Vascular Plexus: Angiogenesis

Once vasculogenesis has established a primary vascular plexus, it must be extensively remodeled and grow into a more mature network in coordination with overall embryonic and larval development. The growth of new blood vessels from pre-existing ones is called angiogenesis, and is essential for both proper embryonic development and tissue/organ growth and repair. Improper angiogenic growth and remodeling contributes to numerous malignant, inflammatory, ischemic, infectious, and immune disorders. There are two forms of angiogenesis: sprouting and non-sprouting (reviewed by Scappaticci, 2002); see also Chapter 14). During sprouting angiogenesis, new blood vessels form by budding or "sprouting" from a pre-existing vessel, and then growing and extending to form a new branch. This process involves the proteolytic degradation of the extracellular matrix adjacent to an existing vessel, migration and proliferation of ECs from the wall of the vessel, and finally lumenization of the new vessel. Non-sprouting angiogenesis occurs by intussusception, in which previously established vessels are split by the formation of trans-vascular pillars. Intussusceptive and sprouting angiogenesis are both employed to remodel and expand on an initial vascular plexus. This occurs concomitantly with the acquisition of VSMCs or pericytes, which then stabilize the nascent vasculature and are essential for the maturation phase of vessel development.

Throughout development, blood vessels are constantly expanding, remodeling, and adapting to meet the increasing demands for nutrient and oxygen delivery, and waste removal. Organs and tissues signal to the vessels to promote their

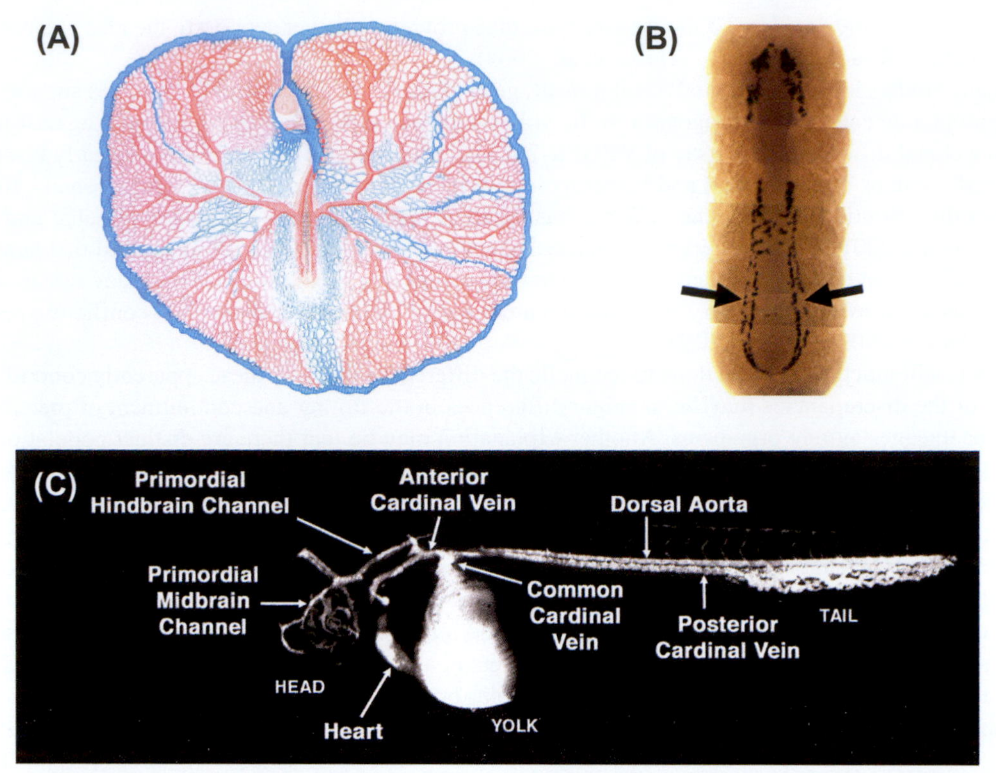

FIGURE 24.2 The earliest vessels formed during embryonic development form though the process of vasculogenesis. (A) Avian yolk sac vascular plexus, with arterial-venous vessel patterning already apparent (modified from Popoff, 1894). (B) *In situ* hybridization of the fli1 gene (arrows) on 10-somite stage zebrafish embryos. Fli1 marks emerging vascular and hematopoietic precursors in the early lateral mesoderm (composite image, dorsal view, anterior up). (C) Confocal microangiography of a 24 hour post-fertilization zebrafish embryo demonstrating the early vasculogenic derived vessels.

growth, repair, and if necessary, their regression. They can also provide cues that cause ECs to adopt specialized phenotypes and features that particular organs need to interact properly with the circulatory system (Nikolova and Lammert, 2003). In this way, ECs along the vascular tree become functionally and molecularly heterogeneous, with gene expression differing among ECs as they make arterial-venous distinctions and adapt to the needs of nearby tissues and organs. Increasing evidence suggests that ECs, in turn, also provide instructive cues to the surrounding organs that help determine their location, differentiation and morphology during development and into adulthood. Blood vessels and organ-specific cells interact with each other continuously throughout development and postnatal life to coordinate the formation of a functional organism with a vascular system to meet its ever-changing needs.

Molecular Regulation of Vascular Development

Vascular development is dictated by a number of different signals that are received and processed by ECs during the entire vascular differentiation and patterning process. A number of different signaling pathways involved in the development of the vasculature have been described over the years, (reviewed by Coultas et al., 2005), and although not the only important players, the best characterized of these signaling cascades involve receptor tyrosine kinases, (reviewed by Folkman and D'Amore, 1996; Ilan et al., 1998; Takakura et al., 1998; Yancopoulos et al., 1998). In the following sections we will review two pathways: the vascular endothelial growth factor (VEGF) and the angiopoietin/Tie receptor pathways. Both of these pathways have been shown to be critical for vascular development, and seem to dictate their mainly vascular-specific effects though restriction of the receptors for both classes of molecules primarily to endothelial cells. Some additional signaling factors and pathways important for the guidance and patterning of the developing vasculature will be discussed in greater detail in a later portion of this chapter.

Vascular Endothelial Growth Factor Signaling

VEGF ligands have been shown to be essential for the migration, proliferation, maintenance and survival of ECs, and are critical during both vasculogenesis and angiogenesis (reviewed by Carmeliet and Collen, 2000; Yancopoulos et al., 2000).

The best described VEGF isoform, VEGF-A, was the first growth factor identified to be a mitogen specific for ECs (Ferrara, 1999; Carmeliet and Conway, 2001; Ferrara et al., 2003), and was initially defined, characterized and purified based on its ability to induce vascular permeability and EC proliferation. The VEGF family of ligands includes five characterized isoforms in mammals (VEGF-A through VEGF-D and placental growth factor) that display differential interactions with three receptor tyrosine kinase receptors (VEGFR-1/Flt-1, VEGFR-2/Flk-1/KDR, and VEGFR3/Flt-4), as well as with a number of ancillary receptor components, such as the neuropilins (reviewed by Goishi and Klagsbrun, 2004).

VEGF-A signals by binding to two of these receptors: VEGFR-1 and VEGFR-2. Expression of both receptors is largely restricted to the vasculature, which accounts for the discrete nature of VEGF-A signaling to blood ECs. By contrast, the VEGFR-3 receptor has conventionally been thought to be more restricted to lymphatic ECs (Kukk et al., 1996), although recent data have shown that it plays an indispensible role in promoting angiogenesis within the developing vasculature, and it might be a strong permissive cue to initiate angiogenic sprouting (Gore et al., 2011; Benedito et al., 2012; Partanen et al., 2013). VEGF ligands, but VEGF-A in particular, are produced by a number of different cell types both developmentally and pathologically, including macrophages, T cells, smooth muscle cells and tumor cells (Klagsbrun and D'Amore, 1996). VEGF isoforms are thought to play a major role in tumor-induced neovascularization, and as such have been the target of a number of different clinical treatments, including a humanized monoclonal antibody directed against VEGF-A for the treatment of colorectal and renal tumors (Willett et al., 2004). Clinical trials of a wide variety of VEGF inhibiting compounds have been ongoing for a number of years, but as yet there is still a significant amount of data that needs to be obtained to fully determine the efficacy of these treatments.

What is clear, though, is that during embryonic development, the expression of both VEGF-A and its receptor VEGFR-2 correlate closely with sites of vessel formation (Jakeman et al., 1993; Shweiki et al., 1993; Liang et al., 1998). The most conclusive evidence for the critical role of VEGF-A as a key regulator of both vasculogenesis and angiogenesis has come from studies in knockout mice (Shalaby et al., 1995; Carmeliet et al., 1996; Ferrara, 1996). In murine embryos lacking either VEGF-A or VEGFR-2, blood islands, ECs and major vessels fail to develop in appreciable numbers, resulting in embryonic lethality between embryonic days 8.5 and 9.5. The disruption of just one of the two VEGF-A alleles results in embryonic lethality between embryonic days 11 and 12, demonstrating its indispensable role in vascular development and making this gene one of very few that shows haplo-insufficiency during murine development. Recent studies in the zebrafish have uncovered mutations in a number of signaling pathways that regulate VEGFR availability and/or expression, which also lead to disruptions of the developing vasculature and altered downstream signaling (Gore et al., 2011; Pan et al., 2012). One study identified zebrafish embryos with defects in angiogenic sprouting that had mutations in Cds2, an enzyme required for phosphoinositide recycling. One of the phosphoinositides regenerated through the activity of this enzyme is phosphoinositide 4,5 bisphosphate (PIP2), a critical substrate required for signaling downstream from VEGFR-2 (Pan et al., 2012). In conjunction, ongoing studies in both murine and cell culture systems also suggest the critical nature of VEGFR internalization and receptor recycling for proper activation of downstream signaling pathways (Pitulescu and Adams, 2010; Simons, 2012; Nakayama and Berger, 2013).

Knockdown of VEGFR-1 in mice also leads to embryonic lethality (Fong et al., 1995; Vajkoczy et al., 1999). Although ECs are found at embryonic and extra-embryonic sites, the vessels that are formed are mispatterned and abnormally organized, apparently as a result of an overproliferation of the contributing ECs (Fong et al., 1995; Kearney et al., 2002). Many reports also now provide data that suggest that the soluble form of this receptor, sVEGR-1, negatively regulates or restrains angiogenesis to localized sites by working as a VEGF-A ligand "trap" (Ambati et al., 2006; Kappas et al., 2008; Chappell et al., 2009; Avraham-Davidi et al., 2012). A third VEGFR, VEGFR3/Flt-4, is essential for lymphatic development and is activated by binding to VEGF-C, although VEGFR3/Flt-4-deficient mice or zebrafish embryos also show defects in blood vessel formation (Dumont et al., 1998; Gore et al., 2011; Benedito et al., 2012). VEGF-A has also been shown to play an important role during early postnatal life. The partial inhibition of VEGF-A achieved by Cre-loxP-mediated gene excision results in increased mortality, stunted body growth, and impaired organ development in targeted mice (Gerber et al., 1999), and more recently, a critical role for this factor has also been demonstrated during adult neovascularization (Grunewald et al., 2006).

Angiopoietin/Tie Signaling

Following the discovery of VEGF-A, a second family of growth factors important for EC survival and vascular remodeling was identified, with members of this family termed the angiopoietins (Davis et al., 1996; Gale and Yancopoulos, 1999). Like the VEGF receptors, the specificity of angiopoietins for vascular endothelium results from a restricted distribution of their tyrosine kinase receptors Tie1 and Tie2/Tek on endothelial cells (Dumont et al., 1995; Sato et al., 1995). Angiopoietin-1 (Ang-1) seems to be important in stabilizing vessel walls by promoting interactions between endothelial cells (EC) and surrounding pericytes and smooth muscle cells. Ang1 signaling opposes VEGF-mediated internalization of VE-cadherin to maintain endothelial barrier function, possibly accounting for some of the stabilizing effects noted (Gavard et al., 2008). Consistent with a constitutive stabilizing role, Ang1 is widely expressed in adult normal tissues (Suri et al., 1996). In addition to stabilization, exogenous postnatal Ang1 treatment in mice increased EC proliferation independent of sprouting in venular capillaries, suggesting the ability of Ang1 to induce EC proliferation (Thurston et al., 2005). In *Angpt1*$^{-/-}$ mouse embryos, the early stages of VEGF-dependent vascular development seem to occur normally. However, remodeling and stabilization of the primitive vascular plexus are severely perturbed, leading to embryonic lethality. A similar phenotype, though more evident in the brain capillary plexus, was reported for murine embryos lacking the Tie2 receptor (Sato et al., 1995). On the other hand, transgenic overexpression of Ang1 leads to striking hypervascularization by promoting vascular maturation and inhibiting normal vascular pruning (Suri et al., 1998). Altogether, these results suggest a critical role for the Ang1/Tie2 system in normal remodeling, maturation and stabilization of the developing vasculature.

In sharp contrast to Ang1, angiopoietin 2 (Ang2) also binds to the Tie2 receptor but is unable to activate it and acts as a natural antagonist for the Ang1/Tie2 interaction. Transgenic overexpression of Ang2 results in a lethal phenotype reminiscent of that seen in Ang1 or Tie2 knockout mice (Maisonpierre et al., 1997). Ang2 is highly expressed at sites of vascular remodeling and is hypothesized to destabilize mature vessels, thus rendering them more amenable to remodeling, regression or additional angiogenic growth depending on other signals the vessels are receiving, notably VEGF (Yancopoulos et al., 2000). Ang2 knockout mice have a complex phenotype (Gale et al., 2002) that includes some vascular defects, supporting a role for Ang2 in postnatal angiogenesis and/or vascular remodeling, but prominently includes malformations of the lymphatic vasculature. Large and small lymphatic vessels are generally formed, but have defects in their overall organization. The mice appear normal at birth, but soon after the start of feeding they develop severe defects due to lymphatic dysfunction and die around postnatal day 14 (P14). To help clarify the role of Ang2 in blood and lymphatic vessel development, Gale et al. (2002) generated mice in which *Ang1* was knocked into the *Ang2* locus. These mice showed an almost complete rescue of the lymphatic defects, but the postnatal remodeling defects of the retinal blood vasculature persisted. One interpretation of this phenotype is that Ang2 normally acts as a Tie2 agonist in the lymphatic vasculature (because Ang1 can rescue the defect), but acts as a Tie2 antagonist in the retinal vasculature (because Ang1 cannot rescue the defect). In addition to antagonizing Ang1 signaling, postnatal inhibition of Ang2 by antibody injection demonstrated a reduction of retinal angiogenesis characterized by fewer tip cells and filipodia (Felcht et al., 2012). The absence of Tie2 within actively angiogenic VECs at the leading edge showed that Ang2 signaled in a Tie2-independent manner (Felcht et al., 2012).

Of the various Ang/Tie signaling components, only *Tie2 (aka Tek)* mutations have been shown to cause human disease. A missense mutation was found in *Tie2* in individuals suffering from venous malformations (Vikkula et al., 1996). The mutant Tie2 had increased kinase activity relative to wild type Tie2, a finding seen in several additional Tie2 mutations (Vikkula et al., 1996; Wouters et al., 2010). Somatic, second-hit *Tie2* mutations were identified in lesions suggesting a complete loss of normal Tie2 may be required for the manifestation of venous malformations (Limaye et al., 2009).

Like Tie2, the closely related Tie1 receptor is primarily expressed in vascular endothelial cells (Sato et al., 1993), but until recently ligands for this receptor had not been identified and it remained an orphan receptor. Vascular development proceeds normally in Tie1-deficient mice up to approximately E13.0, but shortly thereafter they begin to show signs of edema, local hemorrhage, and rupturing of microvessels, and die between E13.5 and E18.5 (Puri et al., 1995; Sato et al., 1995). Although initial studies failed to demonstrate binding of angiopoietins to Tie1, more recent work has indicated that Ang1 and Ang4 may function as activating ligands for Tie1 (Saharinen et al., 2005), and that Tie1 activation is promoted by the formation of heterodimeric complexes with Tie2 (Marron et al., 2000; Tsiamis et al., 2002; Saharinen et al., 2005). The mechanisms underlying the effects of this binding are not yet known, and it still remains unclear whether Tie1 can function as an independent receptor. Generation of mice with conditional null alleles or the development of more effective and specific of inhibitors may reveal the precise roles of Tie signaling and the different ligands and receptors in later development and in mature vessels.

24.3 ARTERIAL-VENOUS DIFFERENTIATION

One of the very earliest steps in the differentiation of the endothelium is the specification of arterial versus venous identity (Lawson and Weinstein, 2002; Rossant and Hirashima, 2003; Torres-Vazquez et al., 2003; Eichmann et al., 2005). For many years it was assumed that arterial-venous specification was one of the last events in development of the vascular network and primarily dictated by anatomical location of the vessel and hemodynamic forces across the vessel wall. However, recent work suggests that the identity of ECs is likely genetically determined before the onset of circulation and vessel assembly. Beginning with studies in the mouse showing that ephrinB2 and its receptor EphB4 are specifically expressed in arterial and venous endothelium respectively prior to circulation (Wang et al., 1998), a large number of studies have now highlighted the signaling pathways involved in the differentiation of arteries and veins.

(Gridley, 2003), and non-cardiac vascular defects such as stenosis, aneurysm, and hemorrhage are also frequently observed in AGS patients, accounting for a third or more of their mortality (Kamath et al., 2004; Emerick et al., 2005).

Further evidence supporting the importance of the Notch signaling pathway in the formation and/or maintenance of the vasculature has come from *in vivo* studies carried out in animal models including mice, rats and zebrafish. Notch1 and Notch1/Notch4 mutant mouse embryos display severe defects in angiogenic vascular remodeling, which lead to vascular defects and hemorrhaging at around E10.5 (Swiatek et al., 1994; Krebs et al., 2000; Rossant and Howard, 2002). In contrast, the specific expression of an active form of Notch4 in the endothelium leads to a disorganized vascular network and a reduction in the number of small vessels, also resulting in embryonic lethality at E10.5 (Uyttendaele et al., 2001). In a similar fashion, mice lacking the Notch ligands Jagged-1 and Dll1 also die during early gestation due to severe defects in vascular plexus remodeling (Barrantes et al., 1999; Xue et al., 1999). Mutations in Dll4, a Notch ligand specifically expressed in developing arteries, leads to defective development of the dorsal aorta and cardinal vein, formation of arterial-venous shunts, downregulation of arterial markers and upregulation of venous markers in the dorsal aorta (Duarte et al., 2004; Gale et al., 2004). These results suggest a causative relationship between Notch signaling and abnormalities in vascular morphogenesis, and well as a potential role for Notch signaling in arterial-venous differentiation.

The Notch Signaling Pathway is Required for Arterial-Venous Cell Fate Determination

The arterial-restricted expression pattern of Notch signaling components suggests a specific role in arterial-venous cell fate determination, but it was not until relatively recently that this role was confirmed by functional studies in zebrafish and mice. Zebrafish embryos deficient in Notch signaling due to a mutation in ubiquitin ligase gene *mindbomb* (*mib*) (Jiang et al., 1996) or microinjection with a dominant-negative form of the transcription repressor Suppressor of Hairless (*Su(H),* the common downstream effector of Notch signaling) (Wettstein et al., 1997) display diminishing arterial markers such as *ephrinB2a* or *notch3* (Lawson et al., 2001; Zhong et al., 2001) accompanied by ectopic expression of venous markers in the artery locations (Figure 24.3A). Conversely, forced activation of the Notch pathway by either ubiquitous or endothelial-specific expression of Notch-ICD induces ectopic expression of artery markers in veins (Lawson et al., 2002). These observations strongly suggest that pro-arterial fate and anti-venous fate cues are in part specified and maintained by the Notch pathway.

The *gridlock* (*grl*) gene (Zhong et al., 2001) has also been reported to be a target of the Notch pathway in the zebrafish vasculature. The *grl* gene encodes a basic helix-loop-helix protein that belongs to the *hairy and enhancer of split (hes)* family of transcriptional repressors (Nakagawa et al., 1999). Vascular expression of *grl* is restricted to the aorta, and zebrafish embryos with mutations in the gene fail to establish trunk circulation as a result of incomplete aorta formation (Stainier et al., 1995; Zhong et al., 2000). A number of studies have reported different results regarding the potential role of *grl* in arterial differentiation downstream from Notch. Injection of *grl* mRNA into wild type zebrafish embryos represses expression of the venous markers *flt4* and *ephb4* and enhances expression of the arterial marker *ephrinb2* (Zhong et al., 2001). However, it has also been reported that *grl* is normally expressed in the dorsal aorta of embryos that lack Notch activity, despite ectopic expression of other artery-vein markers and clear effects on vascular morphology (Lawson et al., 2001). The controversy indicates that *grl* might not be the functional repressor of venous markers directly *in vivo* and suggests the existence of potentially other *hes* related transcription factors to repress venous cell fate downstream of the Notch pathway. The mammalian ortholog of *grl*, *hey2*, is expressed in the developing cardiovascular system and has been shown to be a direct target of Notch signaling *in vitro* (Nakagawa et al., 2000). Although *hey2* knockout does not seem to affect artery-vein fate in mice (Donovan et al., 2002; Fischer et al., 2002; Sakata et al., 2002), double mutation of *hey2* and *hes1* (another *hairy* related transcription factor) does reduce arterial markers as well as resulting in vascular shunts and defects of dorsal aorta formation (Fisher and Caudy, 1998). Together, these data suggest that activation of Notch signaling seems to be an essential requirement for the specification of arterial cell fate in vertebrates.

The Role of Hedgehog Signaling in Vascular Development

Hedgehogs are a class of morphogen proteins that interact with heparin on the cell surface through an N-terminal basic domain. They are tethered to the surface through cholesterol and fatty acyl modification, and their signaling is thought to be crucial throughout development. Sonic hedgehog (Shh) is important for the determination of cell differentiation in structures that lie adjacent to the notochord, including blood vessels (Lawson and Weinstein, 2002). Studies in zebrafish have shown that Shh acts upstream from VEGF to induce arterial endothelial cell differentiation (Lawson et al., 2002). The expression of VEGF is upregulated by Shh in both mouse (Pola et al., 2001) and zebrafish (Lawson et al., 2002) embryos, whereas its expression is lost in the somites and hypochord of zebrafish lacking Shh function, resulting in the formation of a single midline axial vessel expressing venous markers only (Lawson et al., 2002). In mice, Shh administered to aged animals induces new vessel growth in injured ischemic hind limbs. However, Shh seems to have no effect on endothelial cell migration or proliferation *in vitro*, although it does induce expression of pro-angiogenic VEGF and angiopoietins-1 and -2 from interstitial mesenchymal cells (Pola et al., 2001). These observations suggest that Shh plays an indirect role in angiogenesis, potentially promoting other downstream angiogenic signaling pathways (Pola et al., 2001; Lawson et al., 2002). However, other data suggest that hedgehog signals may also be received directly by endothelial cells to promote their proper morphogenesis into tubular vessels (Vokes et al., 2004).

Taken together, the studies described above suggest that a molecular pathway consisting of sequential activation of hedgehog, VEGF, and Notch signaling regulates arterial differentiation in the developing vasculature (Figure 24.3B).

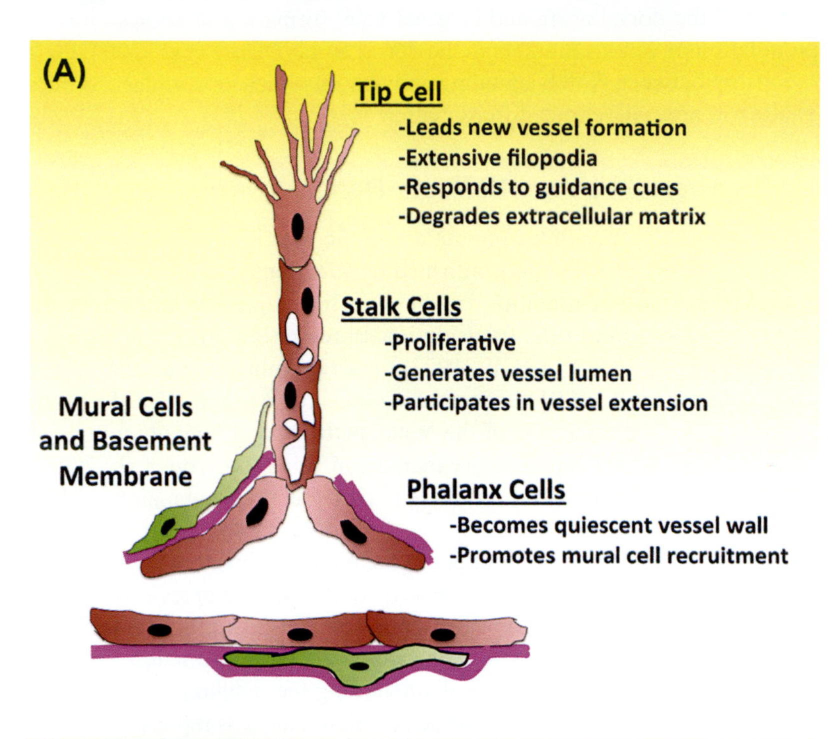

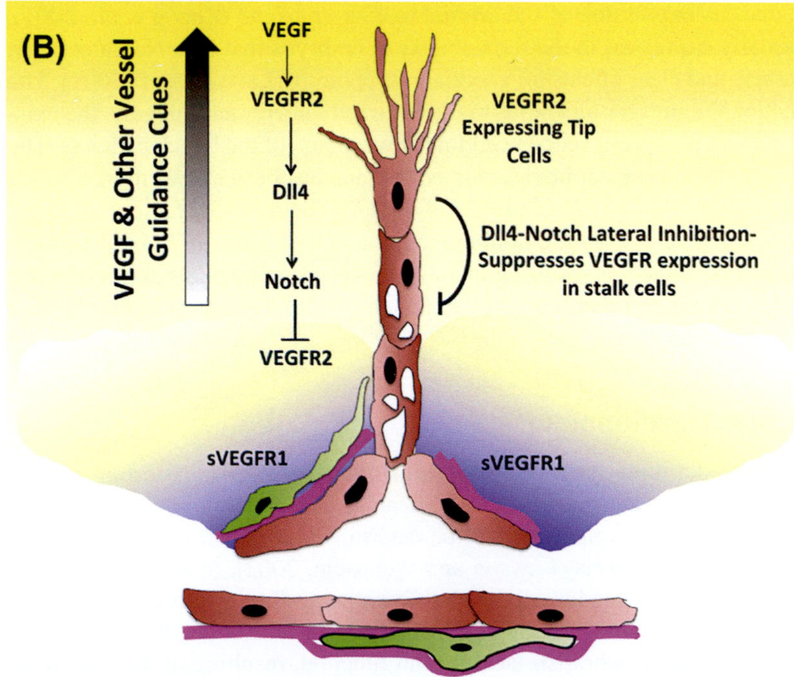

FIGURE 24.4 Molecular Regulation and Specification of Endothelial Cells During Angiogenesis. (A) During sprouting angiogenesis ECs adopt a number of different phenotypes, including the cells in the leading position, the tip cells, the cells following the tip cells, called stalk cells, and the cells that become quiescent to form the new vascular wall at the base of the sprout, called phalanx cells. Each cell type adopts some specialized functions, such as tip cells having excessive filopodia and enhanced proteolytic capabilities as compared to stalk or phalanx cells, to be able to "lead" new vessel formation. Mural cells, such as pericytes or smooth muscle cells, work in conjunction with ECs to stabilize the vascular wall following angiogenic sprouting events. (B) The molecular regulation of angiogenesis is initiated through increased VEGFR2 levels in tip cells. Increased VEGF signaling is thought to induce levels of Dll4 and trans-activate Notch signaling in stalk cells. Notch activation leads to downregulation of VEGFR2 in stalk cells, suppressing their tip cell potential. Angiogenic sprouting is also regulated through increased levels of sVEGFR1 that serves as a ligand trap for VEGF-A to dampen pro-angiogenic guidance cues.

Interplay between the Notch and VEGF Signaling Pathways is Critical for Endothelial Cell Morphogenesis

Angiogenesis is a complex process involving the coordination of cell specification, cell division, collective cell migration as well as dynamic cell shape change. Over the last decade, by using advanced confocal microscopy and gene modification techniques in animal models such as mouse and zebrafish, researchers have started to decipher the temporal-spatial coordination of these dynamic cell behaviors. At the molecular level, Notch-VEGF signaling feedback seems to be in the driver's seat, controlling differential endothelial cell morphogenesis.

Selective Specification of Endothelial Cells During Angiogenesis

Several distinct endothelial cell types have been classified based on their morphology, activation states and gene expression patterns during the angiogenic sprouting process (Wacker and Gerhardt, 2011). Angiogenesis is initiated with tip cell selection. In response to stimulatory factors, particularly VEGF-A, specialized endothelial cells are reactivated from their quiescent state and begin to extend abundant dynamic processes, called filopodia, to migrate away from the stabilized vessel wall (Gerhardt et al., 2003). During sprouting, the endothelial cell sitting in the leading position, termed a "tip cell," is thought to be phenotypically distinct from the endothelial cells lagging behind, called "stalk cells" (Wacker and Gerhardt, 2011) (Figure 24.4A). Once sprouting is complete and vessels are completely reconnected and perfused with blood flow, the endothelial cells return to their previous quiescent state, with cell-to-cell junctions and a stable state of homeostasis being re-established (Herbert and Stainier, 2011).

In response to microenvironmental stimuli, differentially regulated genetic programs are essential for fine-tuning selection and specialization of tip versus stalk cells. Accumulating evidence from studies both *in vitro* and *in vivo* using loss-of-function (LOF) and gain-of-function (GOF) tools suggest that paracrine secreted attractant and repellant factors such as VEGF and semaphorins play an essential role in initiating vascular sprouting (Herbert and Stainier, 2011). VEGF-A secreted from surrounding tissues is thought to induce tip cell specification, possibly though induction of high levels of Dll4 in these cells specifically. Dll4 can then trans-activate Notch receptor signaling in neighboring cells. High Notch activity in stalk cells transcriptionally downregulates VEGFR2 signaling, thereby efficiently suppressing their tip cell potential (Figure 24.4B). It is worth noting that recent experiments using time-lapse imaging techniques suggest that endothelial cells can compete for the tip cell position during angiogenic sprouting (Jakobsson et al., 2010; Tammela et al., 2011), suggesting that these cell "fates" are indeed plastic decisions.

24.4 EMERGENCE OF THE LYMPHATIC SYSTEM

In addition to the blood vascular system, vertebrates possess a completely separate and parallel network of endothelial vessels called the lymphatic vascular system. Unlike the blood circulatory system, the lymphatic system is a blind-ended system of vessels that protect and maintain the fluid environment of the body by filtering and draining away interstitial fluid. This fluid, which ranges from colorless to milky white depending on the fat content, is referred to as lymph, and commonly contains water, dissolved molecules, triglycerides and white blood cells. The lymphatic system is not closed and has no single, central pump. The mammalian and avian lymphatic systems begin with innumerable blind-ended, thin walled capillaries and larger vessels that drain lymphatic fluid from the extracellular spaces of all organs and tissues into larger collecting tubes (Oliver, 2004). These vessels are lined with a continuous, single layer of overlapping endothelial cells that form loose intercellular junctions, making them highly permeable to large macromolecules, pathogens and migrating cells. Larger lymph-collecting vessels ensure unidirectional transport of fluid with one-way, semilunar valves through which lymph moves slowly under low pressure due to the action of the surrounding skeletal muscles, which help to squeeze fluid through these vessels. This fluid is transported to progressively larger lymphatic vessels culminating in the right lymphatic duct (for lymph originating from the upper right body) and the thoracic duct (for the rest of the body). These ducts finally return this fluid to the blood circulatory system by draining into the right and left subclavian veins.

Under normal conditions, the lymphatic vascular system is necessary for the return of extravasated interstitial fluid and macromolecules to the blood circulation for immune defense and for the uptake of dietary fats. In addition to its important role during embryonic development, lymphatic vessels grow and proliferate as an essential component of tissue repair and inflammation in most organs (Leu et al., 2000). Impaired functioning of lymphatic vessels can result in the formation of lymphedema (Witte et al., 2001), while tumor-associated lymphangiogenesis may contribute to the spread of cancer cells from solid tumors. Thus far, it is unclear how tumor cells enter the lymphatic system; however, using lymphatic-specific molecular markers, many groups have shown that tumor cells activate peritumoral and intratumoral lymphangiogenesis (Mandriota et al., 2001; Skobe et al., 2001; Stacker et al., 2001).

Although the origins of the lymphatic system remained controversial for nearly a century, the accepted view of early lymphatic development was originally proposed by Florence Sabin in 1902 (Sabin, 1902). By performing ink injection experiments, Sabin postulated that the first lymphatic structures, the primitive jugular lymph sacs, originated from endothelial cells that bud from large veins. In the years since, two main experiments have provided conclusive evidence supporting this hypothesis. In 2006, Yaniv et al., traced the origin of lymphatic endothelial cells (LECs) that form the thoracic duct back to the cardinal vein by analyzing time-lapse movies of developing zebrafish (Yaniv et al., 2006) and in 2007, Srinivasan et al. performed detailed lineage tracing experiments of LECs in the murine model system (Srinivasan et al., 2007).

Tissue transplants in the avian system and expression analysis of developing *Xenopus* larvae have suggested mesenchymal cells are another potential source of LECs (Wilting et al., 2000; 2006; Ny et al., 2005). While currently there is insufficient evidence supporting the idea that mesenchymal cells give rise to LECs in mice or zebrafish, there is also no conclusive evidence excluding lymphangioblasts of the mesenchyme as a source. It remains to be seen whether the mesenchyme represents either a secondary or later source of lymphatic endothelial cells in vertebrates as originally postulated by Huntington and McClure in 1910 (Huntington and McClure, 1910).

With the recent advent of tools for molecular dissection of lymphatic development, we have begun to uncover some of the mechanisms responsible for the early specification of the lymphatic system (Figure 24.5). Some of these mechanisms are reviewed below.

Lymphatic Endothelial Cell Specification

In the mouse, the earliest known stages of LEC specification begin with the expression of the transcription factor Sex determining Region Y Box 18 (SOX18) in dorsolateral cells of the cardinal vein at embryonic day 9.0 (Francois et al., 2008). SOX18 induces expression of the transcription factor prospero homeobox 1 (PROX1) by E9.75 in these future LECs by binding and activating the proximal promoter of *Prox1* (Francois et al., 2008). SOX18 and PROX1 act as "master-regulators" of lymphatic fate, as evidenced by studies that showed that expression of either of these factors was sufficient for blood endothelial cells (BEC) to acquire an LEC fate *in vitro* (Hong et al., 2002; Petrova et al., 2002; Francois et al., 2008). PROX1 was additionally shown to be sufficient to induce BECs to acquire LEC fate *in vivo* when ectopically driven by the vascular endothelial-specific *Tie1* promoter (Kim et al., 2010). In the absence of SOX18 or PROX1, mice have normal blood vasculature but fail to specify LECs and therefore lack any lymphatic vasculature and subsequently die at E14.5 (Wigle and Oliver 1999; Wigle et al., 2002; Francois et al., 2008).

In humans, dominant mutations in *SOX18* are known to cause the lymphatic disorder hypotrichosis-lymphedema-telangiectasia (Irrthum et al., 2003). Although the function of SOX18 seems to be necessary only for the specification of LECs, PROX1 is also required for the maintenance of this identity, as eliminating *Prox1* expression in LECs results in a reversion to a BEC identity (Johnson et al., 2008). Expression of *Prox1* is also regulated by the orphan nuclear receptor NR2F2 (previously referred to as COUPTFII). In addition to its earlier role in establishing arterial/venous identity, NR2F2 has a role in the differentiation of LECs by regulating *Prox1* transcription and by forming heterodimers with PROX1 to coregulate downstream target genes (Lee et al., 2009; Yamazaki et al., 2009; Srinivasan et al., 2010; Aranguren et al., 2013).

Once specified, PROX1-positive LECs emerge from veins at E10.0 by actively budding dorsolaterally from the cardinal vein and venous intersegmental vessels (vISV) (Srinivasan et al., 2007; Yang et al., 2012). These LECs emerge and migrate

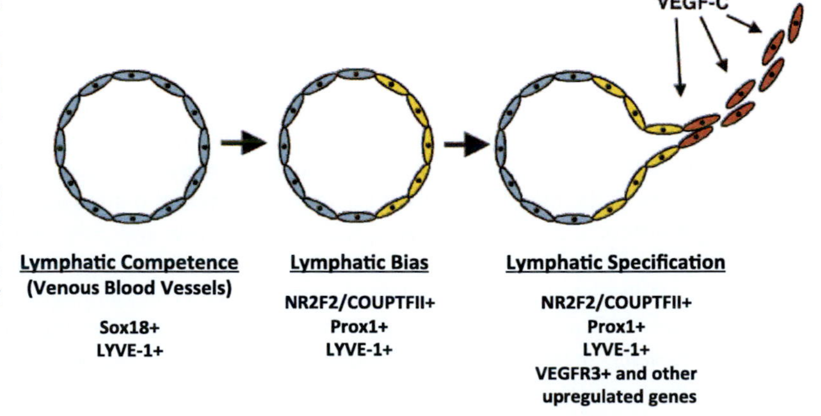

FIGURE 24.5 Proposed Model of Lymphatic Emergence from Venous Endothelium. Lymphatic endothelium emerges from competent venous endothelium expressing the markers LYVE-1 and Sox18. Polarized expression of Prox-1 on a subset of venous endothelial cells biases these cells toward producing lymphatic endothelium and induces or allows for the continued maintenance of the expression of a number of different lymphatic endothelial cell genes, including vascular endothelial growth factor receptor 3, the receptor for the lymphangiogenic factor vascular endothelial growth factor C. See Yang, et al., 2012 for a more comprehensive discussion of this model.

as streams of cells and reorganize by aggregating into larger, lumenized structures termed lymph sacs, which have been reported to soon after give rise to superficial lymphatic vessels, the peripheral longitudinal lymphatic vessel (pllv) and the primordial thoracic duct (pTD) (Wigle and Oliver 1999; Francois et al., 2012; Yang et al., 2012; Hagerling et al., 2013) (Figure 24.5).

In addition to the study of lymphatics in the mouse, the zebrafish model system also has a well-defined lymphatic system which shares characteristics of lymphatic vessels found in higher vertebrates (Yaniv et al., 2006). In the zebrafish, future LECs also sprout from the cardinal vein, and generate rostral lymph sacs in a manner similar to that found in mice (Isogai et al., unpublished findings). In the more posterior trunk proper, lymphangioblasts first migrate dorsally to form a lateral vessel known as the parachordal vessel. Cells located along the axis of the parachordal vessel then migrate ventrally to a position just between the dorsal aorta and cardinal vein where they next migrate laterally and establish the main trunk lymphatic vessel of the developing zebrafish, the thoracic duct. As in mice, zebrafish that lack *prox1* or *vegf-c* expression fail to form lymphatic vessels as evidenced by the lack of the thoracic duct (Kuchler et al., 2006; Yaniv et al., 2006) (Figure 24.5).

Lymphangiogenesis

A number of different molecules are critical to the process of LECs exiting veins and migrating in the direction where lymph sacs will form, but the best characterized to date are the growth factor VEGF-C, which is expressed by the surrounding mesenchyme (Kukk et al., 1996; Karkkainen et al., 2004) and the matrix interacting protein CCBE1 (Bos et al., 2011). In mice deficient for CCBE1 or VEGF-C signaling, PROX1-positive LECs are formed, but fail to exit the veins. These mice lack all lymphatic vasculature and as a result develop massive lymphedema and die by stage E17 (Karkkainen et al., 2004). VEGF-C is a secreted, paracrine-acting ligand that specifically binds to its high affinity receptor, VEGFR-3. VEGFR-3 is initially expressed in a subset of blood vascular EC and subsequently becomes restricted to LECs (Kaipainen et al., 1995; Oh et al., 1997). Exiting the veins seems to coincide with LECs adopting a more differentiated state and upregulating several LEC-specific genes including lymphatic endothelial hyaluronan receptor (*Lyve1*), podoplanin (*Pdpn*), Integrin α9 (*Itga9*), fibroblast growth factor receptor 3 (*Fgfr3*) and *Vegfr3* (Wigle et al., 2002; Shin et al., 2006; Mishima et al., 2007). Consistent with the earlier expression of *Vegfr3* in the blood vasculature, VEGFR3-deficient mice show defective blood vessel development at early embryonic stages and die at E9.5 (Dumont et al., 1998). Thus, VEGFR3 has an essential function in the remodeling of the primary capillary vasculature prior to formation of the lymphatic vessels. Expression of a dominant-negative *Vegfr3* in the skin of transgenic mice blocks lymphangiogenesis and induces the regression of already formed lymphatic vessels, demonstrating that VEGFR3 signaling is required not only for the initial formation but also for the maintenance of the lymphatic vasculature (Makinen et al., 2001). Mutations in *VEGFR3* result in the hereditary congenital lymphedema, Milroy disease, in humans (Ferrell et al., 1998).

The VEGFR3 ligand VEGF-C also binds to the neuropilin 2 (Nrp-2) receptor, which is specifically expressed by veins, and subsequently becomes restricted to lymphatic vessels (Karkkainen et al., 2001; Yuan et al., 2002). However, mice deficient for the Nrp-2 receptor specifically exhibit defects in the formation of smaller lymphatic vessels, suggesting that while Nrp-2 may assist in VEGF-C signaling, it is not as necessary for the initial budding of LECs from veins (Yuan et al., 2002). Less is known about the mechanism of action of the secreted CCBE1 protein, but mutations in the *CCBE1* gene are known to cause the Hennekam lymphangiectasia-lymphedema syndrome in humans (Alders et al., 2009).

Despite minor differences in lymphatic development between organisms, the amenability of the zebrafish to large-scale screens and *in vivo* imaging of organogenesis provides a promising venue for future insights in the development of the lymphatic system. To date, many of the molecular players involved in lymphatic development and lymphangiogenesis have been found to be evolutionarily conserved in the zebrafish, including the roles of PROX1, VEGF-C, VEGFR3, ANG2, SOX18, and Nrp-2 (Kuchler et al., 2006; Yaniv et al., 2006; Hermans et al., 2010; Cermenati et al., 2013). The zebrafish has also recently been particularly beneficial as a source to identify new factors involved in lymphangiogenesis, including the previously unknown role of the chemokine SDF-1 as a guidance cue in this process (Butler et al., 2009; Cha et al., 2012).

Blood and Lymphatic Vessel Separation

As previously mentioned, the blood vasculature and lymphatic vessels are connected at the thoracic and right ducts and the subclavian veins to allow drainage of lymph back into the blood circulation. Normally, valves exist at these junctions to ensure that there is no reflux into the lymphatic vasculature from the veins (Oliver, 2004). However during the early development of lymphatic vessels and prior to the establishment of these valves, it is believed that hematopoietic cells accomplish this task (Sebzda et al., 2006). Specifically, a number of mutations that affect development and signaling of platelets result in abnormal persistent connections between the blood and lymphatic vasculature. This in turn results in blood-filled lymphatics.

Platelets express a C-type lectin receptor named CLEC2 which recognizes podoplanin (PDPN), a glycosylated, transmembrane protein expressed by LECs (Schacht et al., 2003; Suzuki-Inoue et al., 2007). Activation of CLEC2 by PDPN signals the tyrosine kinase, SYK, through its adaptor protein LCP2 (formerly known as SLP76), to activate phospholipase C gamma 2 (PLCγ2) resulting in the aggregation of platelets at the sites of lymphatic/venous vessel connections (Suzuki-Inoue et al., 2006). Despite the fact that platelets aggregate at the site of connections between lymph sacs and the cardinal vein, it is believed that material secreted by activated platelets is responsible for the intended separation of these two vessel types by inhibiting LEC migration, proliferation and tube forming at these locations (Osada et al., 2012). Mice that fail to properly develop platelets or that are deficient in PDPN, CLEC2, SYK, PLCG2 or LCP2 signaling all have blood-filled lymphatic vessels, which often coincides with these vessels taking on a hyperproliferating, tortuous and dilated character (Abtahian et al., 2003; Ichise et al., 2009; Bertozzi et al., 2010; Bohmer et al., 2010; Carramolino et al., 2010; Uhrin et al., 2010; Finney et al., 2012).

The presence of blood-filled lymph sacs in many of these platelet mutants suggests that the lymphatic cells bud from the cardinal vein in a way that shares a lumen. Whether this is indicative of how LECs normally sprout remains to be determined, but a recent study has suggested an alternative hypothesis as to how LECs bud from the vein using a "ballooning" mechanism that would be consistent with a temporarily shared lumen (Francois et al., 2012). However, in transplants of irradiated mice with CLEC2 signaling-deficient platelets, the subjects are still seen to acquire blood-filled lymph sacs long after the initial stages of development where the final connections are made between the lymphatic system and the subclavian veins, suggesting that regardless of early connections, these mutants also acquire additional abnormal shunts between the lymphatic and venous systems (Abtahian et al., 2003; Ichise et al., 2009).

24.5 PATTERNING OF THE DEVELOPING VASCULATURE

Like any other organ system, the gross anatomy of the vascular system is characterized by a highly reproducible pattern of the major blood vessels. At least for the largest vessels (e.g., the aorta), features such as lumen size, branching angles and curvature along the vascular tree are quite reproducible and stereotypic; even the designated sites for the initial level of secondary sprouts (e.g., intersomitic vessels, main vessels penetrating different organs) are highly regulated. Conversely, microvessels and capillaries are mostly non-stereotyped, to ensure the maximal ability of vessels to reach growing and expanding tissue beds. The control of vessel branch patterning includes both attractive and repulsive guidance signals, and is regulated by both positive and negative cues.

Significant advances have been made in understanding the cellular and molecular mechanisms controlling blood vessel assembly at appropriate sites in the organism, yet much remains that is relatively poorly understood. In the next few sections, we will review some of the current understanding of the way that vascular patterning is established during embryonic development.

Assembly of the Primary Axial Vessels

The formation of the large axial vasculature in the embryonic trunk (e.g., the dorsal aorta and the cardinal vein) occurs though the process of vasculogenesis, during which angioblast progenitors arising from the mesoderm locally aggregate along the trunk midline, form vascular cords and subsequently undergo morphogenesis into lumenized blood vessels. Evidence suggests that there are midline cues that are important for the assembly of the initial vasculogenic vessels.

Studies in the zebrafish suggest a role the notochord in dorsal aorta formation. Zebrafish embryos with mutations in genes leading to a lack of formation of a differentiated notochord, such as in the floating head (encodes Xnot, a homeobox gene) or no tail (encodes Brachyury) mutants, also specifically lack a dorsal aorta, although they still form the cardinal vein (Schulte-Merker et al., 1994; Talbot et al., 1995; Fouquet et al., 1997; Sumoy et al., 1997). Wild-type notochord cells transplanted back into floating head mutants can locally rescue the assembly of aortic primordia (Fouquet et al., 1997), suggesting that notochord-derived signals are required for aortic specification. The floating head and no tail mutants also fail to form a hypochord, an endodermally derived thin strip of cells that lies just above the aorta and ventral to the notochord in fish and amphibian embryos. Studies in *Xenopus* and zebrafish have shown that the hypochord expresses a soluble short isoform of VEGF that could potentially act as a medium- to long-range graded signal for the medial migration and assembly of the angioblasts that contribute to dorsal aorta formation (Cleaver and Krieg, 1998; Liang et al., 2001; Lawson and Weinstein, 2002). However, experiments designed to directly test this idea have not yet been performed, and studies in the zebrafish suggest that VEGF expressed by the adjacent somites in response to notochord-derived Hedgehog signals may be the more critical signal driving dorsal aorta assembly (Lawson et al., 2002). In mice, loss of VEGF activity leads to the inability of the dorsal aorta to form, although it is unclear what tissues are specifically providing the midline vascular patterning cues in this system. Unlike fish and amphibians, mice and avian species lack a hypochord, although embryonic structures such as the somites or the primitive endoderm ventral to the aorta or the neural tube, which also express VEGF in these species, could be mediating the directional cues regulating aorta assembly (Ambler et al., 2001; Ambler et al., 2003; Hogan et al., 2004).

The role of the notochord in dorsal aorta formation in avian and mice species is also not totally clear, though increasing evidence suggests the notochord is a site of production of repulsive factors which help to maintain restrictive barriers at the midline. In avians and mice, the dorsal aortae are initially present as a pair of vessels on either side of the midline, rather than the single vessel that occurs in fish and amphibians. Studies using quail/chick chimeras in which axial structures were removed showed that the notochord acts as a midline barrier to impede avian angioblasts from crossing the midline (Klessinger and Christ, 1996), and subsequent reports showed that this was at least in part the result of notochord-expressed bone morphogenetic protein antagonists (Reese et al., 2004; Bressan et al., 2009).

In addition to VEGF, Hedgehog signaling also seems to be important for the assembly of the dorsal aorta in zebrafish and mice (Lawson et al., 2002; Vokes et al., 2004). As noted previously, notochord-derived Hedgehog signaling has been shown to be important in the pathway leading to arterial differentiation in zebrafish through induction of VEGF expression in the somites. However, unlike deficiencies in Notch signaling, the deficits in Hedgehog signaling lead to a complete loss of the dorsal aorta and not simply to its misspecification as vein, suggesting that Hedgehog signaling could play a dual role in regulating dorsal aorta assembly and arterial differentiation (Lawson et al., 2002; Williams et al., 2010). Recent studies in *Xenopus* and mice demonstrate similar results (Vokes et al., 2004), suggesting that Hedgehog signaling is critical for morphogenesis of vascular tubes.

The Role of Guidance Factors in Developmental Angiogenesis

After the assembly of the early embryonic vasculogenic plexus, including the dorsal aorta and the cardinal vein, further remodeling of the vasculature is achieved through angiogenic sprouting events, as described previously. Postnatally, angiogenesis is largely directed and guided by local tissue requirements for oxygen and nutrients, but during early development, angiogenesis is a highly stereotypic and evolutionarily well-conserved process, reminiscent of the initial highly patterned assembly of the nervous system and axon tracts. Indeed, the two systems display remarkable anatomic parallels, as vessels and nerves frequently course adjacent to one another. These similarities have led to the idea that the assembly of the vasculature might depend on similar (or even the same) attractive and repulsive guidance cues that axons use from surrounding tissues to help direct growing vessels along specific pathways (Larrivee et al., 2009; Adams and Eichmann, 2010; Melani and Weinstein, 2010) (Figure 24.6). A variety of studies have begun to show data to support this hypothesis.

The formation of the trunk intersegmental vessels (ISV) provides a good example of a stereotypically conserved process of vascular network assembly in all vertebrates (Figure 24.7A), forming bilaterally along each intersomitic boundary (vertical myoseptum). Studies in zebrafish have shown that these vessels grow and elongate in a choreographed and reproducible fashion. As a developmental model this has proven to be indispensable for the ability to image *in vivo* blood vessel formation in real time (Isogai et al., 2003). Initially, primary vascular sprouts emerge from the dorsal aorta bilaterally at intersomitic boundaries, and grow dorsally around the notochord and neural tube. Following the formation of the primary vascular lattice, a secondary wave of vascular sprouts emerge from the posterior cardinal vein. About half of these connect with the base of primary segments, becoming intersegmental veins while the rest of the primary segments, which remain connected only to the dorsal aorta, become intersegmental arteries. As previous mentioned, recent studies have shown that well-known neuronal guidance factors also play important roles in the guidance and patterning of vessels, and the current understand of these key points of these data will be discussed below.

Semaphorin-Plexin-Neuropilin Signaling

Some of the most definitive evidence for cross talk between neuronal and vascular guidance factors comes from the semaphorin field in both zebrafish and mice. Semaphorins are a family of both cell-associated and secreted proteins that signal through multimeric receptor associations (Bagri and Tessier-Lavigne, 2002). Membrane-associated semaphorins bind directly to plexin receptors, whereas secreted semaphorins bind to neuropilin-plexin receptor complexes. Recent work has shown that semaphorin-plexin signaling regulates the guidance and patterning of the vasculature in a manner similar to that of the repulsive guidance roles of semaphorins in the nervous system. ECs express various neuropilin and plexin receptors, including the endothelial-specific receptor plexinD1 (Soker et al., 1998; Miao et al., 1999; Basile et al., 2004; Gitler et al., 2004; Torres-Vazquez et al., 2004; Raimondi and Ruhrberg, 2013).

In zebrafish, type 3 semaphorins are expressed in the center of the somites, and promote semaphorin-plexinD1 signaling as a repulsive cue to the developing intersegmental vessels, restricting their growth to semaphorin-free paths along the

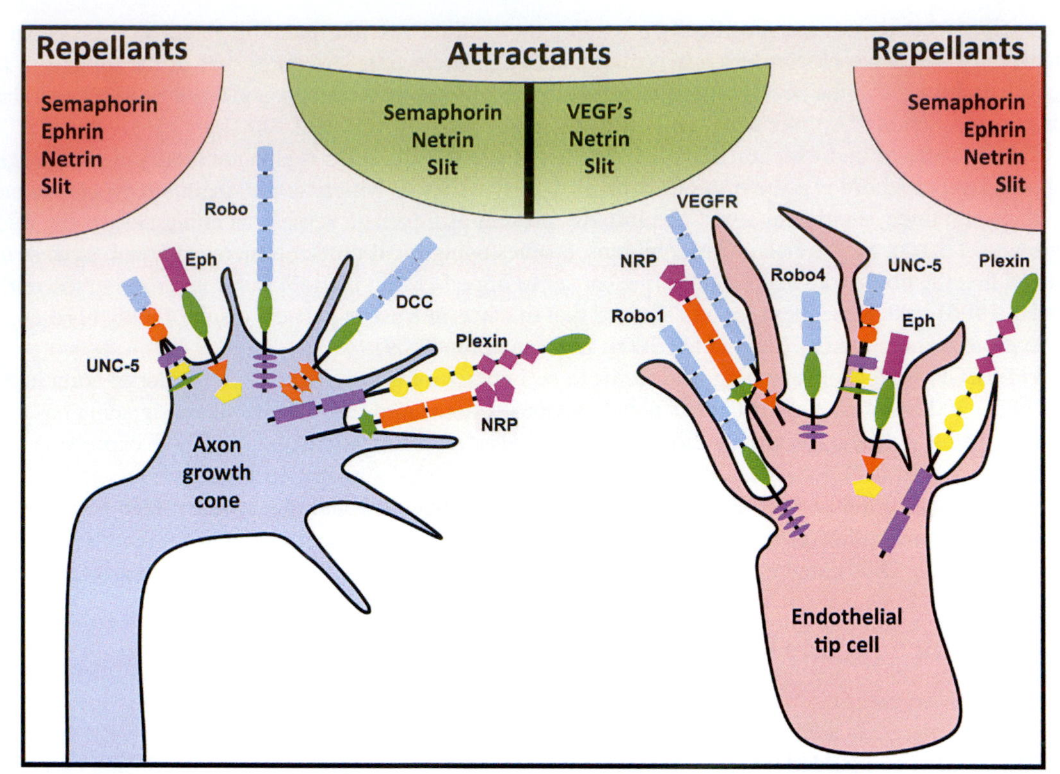

FIGURE 24.6 Similar ligand-receptor pairs help guide axonal growth cones and endothelial tip cells. Eph-Ephrin signaling is mostly repulsive for both growth cone and endothelial tip cell path finding. Semaphorins act mostly as repellent cues in growth cones and also in endothelial tip cells. Netrins can be both attractants and repellents depending on the cellular context and on the receptor to which they bind. DCC, Deleted in Colorectal Cancer; NRP, Neuropilin; Robo, Roundabout; UNC-5, Uncoordinated-5; VEGF, Vascular endothelial growth factor; VEGFR, Vascular endothelial growth factor receptor.

intersomitic boundaries (Torres-Vazquez et al., 2004; Melani and Weinstein, 2010; Zygmunt et al., 2011). The loss of function of either plexinD1 or trunk/somite expressed semaphorins in the zebrafish causes mispatterning of the intersegmental vessels, leading to aberrant vessel sprouting and growth (Figure 24.7 B–D). In mice, the targeted inactivation of plexinD1 also causes mispatterning of intersegmental vessels as well as increased vascularization of the somites, a normally avascular zone (Gitler et al., 2004; Gu et al., 2005; Zhang et al., 2009). In this case, the primary ligand appears to be Sema3E, which is normally expressed in a region of the somites adjacent to the intersegmental vessels (Gu et al., 2005). Sema3E, which binds plexinD1 directly, is not the only semaphorin to bind this receptor; others, such as Sema3A, also have plexinD1 binding capability in a neuropilin-dependent manner (Gitler et al., 2004; Torres-Vazquez et al., 2004). Interestingly, double homozygous mice for Nrp1^{Sema-} and Nrp2 null (retaining VEGF-Nrp1 signaling but presumably lacking Sema-Nrp signaling) form outwardly normal ISVs, suggesting potentially complex and redundant roles of semaphoring-plexin-nuropilin complexes in the vasculature.

Neuropilin receptors do seem to have an important role in vascular development, however. Knockout studies have confirmed that Nrp-1 null mice exhibit vascular defects, including impairments in neural vascularization, transposition of great vessels, disorganized and insufficient development of vascular networks in the yolk sac, and abnormal sprouting and growth of hindbrain vessels, among a number of other phenotypes (Kawasaki et al., 1999; Gerhardt et al., 2004). Mice with endothelium-specific targeted disruption of Nrp-1 also show defects in arterial differentiation (Gu et al., 2003). Nrp-2 knockout mice were initially reported not to have defects in the vascular system (Chen et al., 2000; Giger et al., 2000), however subsequent reports have shown that these mice do have defects in lymphatic capillary formation and reduced lymphatic EC proliferation (Yuan et al., 2002). The specific requirement for Nrp-2 function during development appears to be confined to lymphatic capillaries because arteries, veins, and larger, collecting lymphatic vessels develop normally. Interestingly, mice homozygous null for both Nrp-1 and -2 die *in utero* at E8.5 and have completely avascular yolk sacs (Takashima et al., 2002), suggesting that these to receptors might have some functional redundancy. The role of neuropilins in vascular development has also been extensively studied in the zebrafish, which have four Nrps: Nrp-1a, Nrp-1b, Nrp-2a, and Nrp-2b (Lee et al., 2002; Bovenkamp et al., 2004; Martyn and Schulte-Merker, 2004; Yu et al., 2004). Knocking down individual Nrp genes in the zebrafish through the use of morpholino antisense oligonucleotides does demonstrate vascular defects, however they are less severe than those noted in knockout mouse models and it remains unclear how many of the effects noted are endothelial-autonomous.

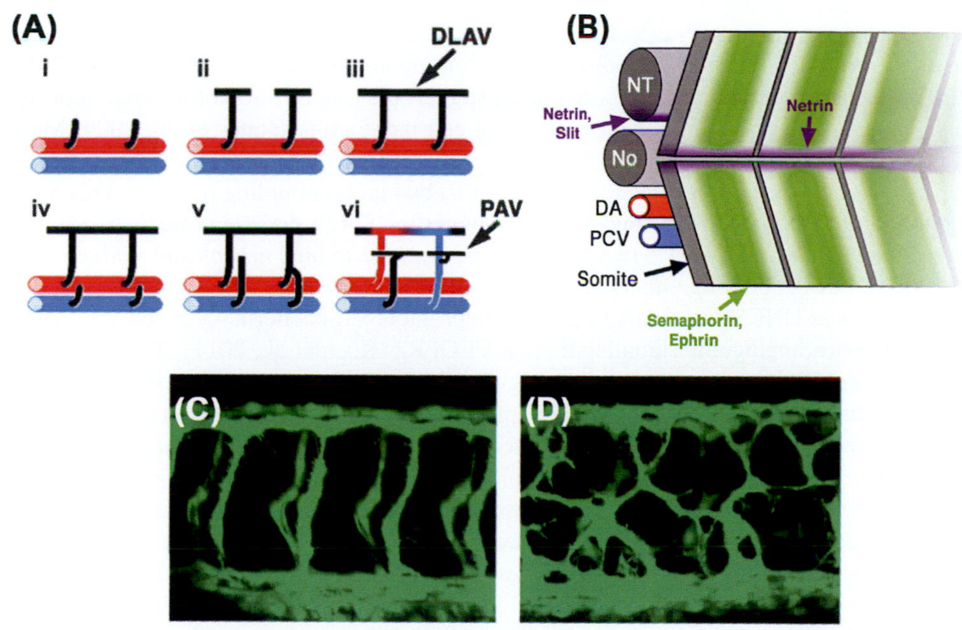

FIGURE 24.7 Zebrafish Trunk Vascular Network Assembly and Patterning Cues. (A) i. Primary sprouts emerge bilaterally exclusively from the dorsal aorta (red). ii. Primary sprouts grow dorsally, branching cranially and caudally at the level of the dorsal-lateral roof of the neural tube. iii. Branches interconnect on either side of the trunk to form two dorsal longitudinal anastomotic vessels (DLAV). iv. Secondary sprouts begin to emerge exclusively from the posterior cardinal vein (blue). v. Some secondary sprouts connect to the base of primary segments, whereas others do not. vi. Primary segments with patent connections to secondary segments become intersegmental veins (blue), whereas primary segments that remain connected only to the dorsal aorta become intersegmental arteries (red). Most of the secondary sprouts that do not connect to primary segments serve instead as ventral roots for the parachordal vessels (PAV). (B) Netrins and Slits are expressed in the ventral neural tube. Netrins are also expressed along the horizontal myoseptum, and semaphorins and ephrins are expressed within the somites. The localization of these signaling molecules to boundary regions between tissues leaves them well positioned to provide repulsive or attractive guidance cues for growing blood vessels growing throughout these tissue beds. NT, Neural tube; No, notochord; DA, dorsal aorta; PCV, posterior cardinal vein. Image from (Weinstein 2005). (C and D) Mid-trunk 48 hour postfertilization intersegmental vessels in control morpholino (C) or plexinD1 morpolino (D) injected fli1-eGFP transgenic zebrafish embryos. In plexinD1 deficient embryos (D), intersegmental vessels aberrantly sprout into semaphorin-rich regions of the somites. These regions are typically not vascularized (C) due to the repulsive nature of semaphoring-plexin signaling. Anterior to the left, dorsal up. Panels A/B are modified from (Isogai, Lawson et al. 2003). Panels C/D are modified from Torres-Vazquez, Gitler et al. 2004. *See references for further details. Isogai, S., N. D. Lawson, et al. (2003). Angiogenic network formation in the developing vertebrate trunk. Development 130(21): 5281–5290.*

Lawson, N. D., N. Scheer, et al. (2001). Notch signaling is required for arterial-venous differentiation during embryonic vascular development. Development 128(19): 3675–3683.

Lawson, N. D., A. M. Vogel, et al. (2002). sonic hedgehog and vascular endothelial growth factor act upstream of the Notch pathway during arterial endothelial differentiation. Dev Cell 3(1): 127–136.

Torres-Vazquez, J., A. D. Gitler, et al. (2004). Semaphorin-plexin signaling guides patterning of the developing vasculature. Dev Cell 7(1): 117–123.

Weinstein, B. M. (2005). Vessels and nerves: marching to the same tune. Cell 120(3): 299–302.

Yang, Y., J. M. Garcia-Verdugo, et al. (2012). Lymphatic endothelial progenitors bud from the cardinal vein and intersomitic vessels in mammalian embryos. Blood 120(11): 2340–2348.

A number of other semaphorins have recently been identified as being involved in both vascular and lymphatic development and patterning (Lamont et al., 2009; Bouvree et al., 2012; Meadows et al., 2012; Segarra et al., 2012) suggesting that the complexity within this field is only beginning to be understood, and that future studies will need to be done to work out the exact semaphorin-plexin-neuropilin binding partners during various stages of vascular development.

Slit-Robo-UNC5 Signaling

Additionally, several studies have implicated Slits and their receptors, Robos, in the regulation of angiogenesis. There are four Slit receptors ("Roundabout" receptors or Robos) in mammals. In the vasculature, the focus has been on Robo4, a structurally divergent member of the family that shows high endothelial cell specificity (Park et al., 2003). In adults, Robo4 is present at sites of both normal and pathologic angiogenesis, including in tumor vascular beds (Huminiecki et al., 2002; Jones et al., 2008). The role of Slit-Robo signaling in vascular guidance remains fairly controversial. One early study showed that Robo4 binds Slit2 and that it is able to inhibit the migration of Robo4-expressing cells (Park et al., 2003), but other studies have either failed to detect this particular Slit-Robo binding combination (Suchting et al., 2005; Koch et al.,

2011) or have demonstrated a promigratory effect of Slit2/Robo signaling on ECs (Wang et al., 2003), such as the defects in intersegmental vessel formation noted in zebrafish after the morpholino-mediated knockdown of Robo4 (Bedell et al., 2005). Complicating issues even further, more recent studies suggest the role of Slit-Robo function in regulating vascular barrier function and vascular integrity and not on endothelial cell pathfinding (Jones et al., 2009; Marlow et al., 2010; Koch et al., 2011).

Most recently, Koch et al. carried out a large-scale screen of Robo4 protein binding partners. This work clearly demonstrated the unique binding of Robo4 to UNC5b and not Slit2, suggesting Slit2's complex role in the vasculature is mediated through a different receptor. Classically, UNC5 receptors have been shown to bind netrins and mediate axon repulsion. This is the first demonstration of direct binding effects of Robo4 to UNC5 receptors. The authors then went on to demonstrate the ability of Robo4 to activate UNC5b associated signaling to promote VE-cadherin mediated vascular stability/integrity and decrease Src mediated pro-angiogenic signaling though VEGFR2 (Koch et al., 2011). These data provide substantial support to the hypothesis that Robo4 does not work as a guidance factor during vascular development, rather more as a promoter of long-term vascular stability. Substantial developmental vascular defects have not been reported in either Robo4 or Slit ligand knockout mice, although this could be explained by the extremely high levels of redundancy in the expression of both ligands and receptors (Long et al., 2004) or their role later in the developmental process (Jones et al., 2009; Koch et al., 2011). Mice with inactivation of UNC5 do have a vascular phenotype showing aberrant extension of endothelial tip cell filopodia, excessive vessel branching and abnormal navigation, consistent with the potential role of UNC5 mediated vascular stabilization (Lu et al., 2004). More recent studies of endothelial-specific knockdown of UNC5b, showed less robust vascular phenotypes, largely restricted to placental angiogenesis and formation of the parachordal in zebrafish, than those noted with the global knockdown, suggesting the role of UNC5b in the vasculature may not be completely cell autonomous (Navankasattusas et al., 2008). More work will need to be done, however, to discriminate the role of UNC5 binding partners, such as Robo4 versus netrins (discussed below), on promoting the different vascular phenotypes observed in UNC5 knockout conditions.

Additionally, a new microRNA, miR-218, has been shown to regulate Robo1 and 2 expression, leading to reduced vascular outgrowth and angiogenic sprouting in the murine retina (Small et al., 2010; Fish et al., 2011), adding a new level of complexity to this field. Again, continued studies will be required to work out the complex role that various Slit-Robo-UNC5 protein pairs play during vascular development.

Netrin Signaling

The final class of guidance cues that we will be considering here are the netrins, a family of highly conserved laminin-related secreted proteins (Hedgecock et al., 1990; Ishii et al., 1992; Serafini et al., 1994). Netrins have been shown to mediate the attraction of neurons by activating Deleted in colorectal cancer (DCC)/Neogenin receptor family members expressed on axons (Keino-Masu et al., 1996; Serafini et al., 1996; Fazeli et al., 1997; Leonardo et al., 1997). Conversely, netrin binding to members of the Uncoordinated-5 (UNC5) receptor family results in axon repulsion. Like Slits, the role of netrins in the guidance of the vasculature is still somewhat unclear. Several different studies both in mice and zebrafish have shown that netrin-1 is a pro-angiogenic factor for vascular ECs, likely though binding of DCC receptors (Park et al., 2004; Wilson et al., 2006), but other studies have shown that netrin-1 can act as a repellent in vessel guidance via the UNC5B receptor (Lu et al., 2004; Larrivee et al., 2007; Bouvree et al., 2008; Lejmi et al., 2008).

Data in the zebrafish most recently suggests that netrin-1a, which is expressed along the horizontal myoseptum dividing the dorsal and ventral halves of the somites, has little effect on primary vascular sprouts but does however prevent the formation of the parachordal vessels, a population of cells that eventually participate in the formation of the thoracic duct and lymphatic vessel formation. Loss of netrin-1a was shown to regulate secondary vascular sprout turning at the horizontal myoseptum, thus preventing parachordal formation and ultimately thoracic duct development. These data go on to further show evidence that netrins are likely acting indirectly to guide parachordal formation along the horizontal myoseptum though binding of *dcc* receptors expressed in motor neurons to regulate lymphatic vascular pathfinding (Lim et al., 2011). Although these data strongly support an indirect role for endothelial netrin-1-motor neuron dcc binding in lymphatic development, it does not exclude the possibility of other vascular phenotypes in loss of netrin situation being regulated though receptors other than dcc, such as UNC5. Differing conclusions of studies within the netrin field may at least partially reflect the biology of netrins, which are capable of both repulsive and attractive signaling depending on the cellular and environmental context in which they are presented. Further studies will be required to clarify the differences between these and other results regarding the activity of netrins and their receptors in the vasculature.

Clearly, there is much to be done before our understanding of the way that classical neuronal guidance factors regulate vascular patterning reaches the same level of understanding that it has in the neural field. The expression patterns of many of these factors in locations adjacent to or in opposition with developing vessels suggests that they play important roles restricting vessel boundaries and promoting stereotypically regulated vascular networks (Figures 24.6 and 24.7B), but conclusive *in vivo* evidence is still in many cases controversial and contradictory. As such, it is likely that the vascular system will be similar to the nervous system, with extensive redundant guidance factors coordinating their activities to spatially and temporally restrict blood vessel outgrowth. The strong parallels uncovered between these two systems already imply that studies carried out in the nervous system that are aimed at dissecting this interaction will continue to have relevance to the vascular system and, perhaps, over time vice versa.

24.6 CONCLUDING REMARKS

Significant research efforts have been carried out over the past several years aimed at shedding light on the multiple steps involved in the formation of a functional vascular network. The full stepwise process, from the emergence of endothelial cell progenitors to their coalescence into a directed and stereotypic vascular plexus with defined arteries, veins and lymphatics, has been extensively dissected and studied using a number of different developmental model systems. Genetic evidence from various knockout studies has highlighted the roles of a wide variety of molecules that affect vasculogenesis, angiogenesis and lymphangiogenesis in a complex and tightly regulated manner. In conjunction, significant amounts of new data are beginning to implicate that these same signaling pathways, that so tightly regulate vascular development, are likely the same pathways reactivated in vascular-based disease states. Because of these new implications regarding the mechanisms underlying vascular pathologies, it has become all the more critical to understand the normal baseline signaling regulation that drives vascular development to fully aid the identification of new therapeutic targets for diseases such as severe tissue ischemia, coronary heart disease, stroke and tumor-promoted angiogenesis.

A particularly interesting dichotomy that has arisen from recent research is aimed at trying to understand the heterogeneity displayed by ECs. Although it seems clear that all vessels share the same endothelial cell basis, within varying tissue beds throughout the organism ECs have acquired unique characteristics and functions that are vital to proper function of the cardiovascular system. This raised a number of interesting questions, such as: To what extent is the formation of different types of vessels defined solely by intrinsic programs or influenced by local cues? How does the intimate association of vessels with their cognate organs influence ECs to adopt functional specialties – such as at the blood-brain barrier and the fenestrated endothelium in the kidney glomeruli? The answers to these and a whole host of other questions are of vital importance when aiming to refine therapeutics targeted to specific subsets of the vasculature.

For much of the past decade, research in the field has focused on the pivotal role that VEGF and its receptors play in directing and promoting vascular development. Although there is still much more to uncover about VEGF signaling, over the last several years a large number of new regulators of vessel morphogenesis have been identified. One of the exciting new challenges for the field will be to work towards a full understanding of how these signaling pathways and molecules interact and fit together in the larger framework of vascular development. Genetic, molecular, and cell biologic tools are now available for the study of vessel formation in a wide variety of model systems, creating a broad and useful array of tools for vascular research moving forward.

24.7 CLINICAL RELEVANCE

- Antiangiogenic therapies are currently being pursued for the treatment of cancer. Over a number of years, research on vascular development has provided useful insights into potential new therapeutic targets.
- Understanding the molecular basis of blood vessel development and vascular stabilization has provided a platform to study more complex vascular dysfunctions, such as stroke and diabetes, and continues to provide new insights into detection, diagnosis and treatment of these disease states.
- There are also a number of genetic developmental vascular disorders, including cerebral cavernous malformation (CCM) and hereditary hemorrhagic telangiectasia (HHT), which lead to misspecification or separation of arterial and venous networks. These disorders have their basis in genes that also play important roles during vascular development, and insights gleaned from studying the developmental roles of these genes can lead to more effective diagnosis and/or treatment of these disorders.

ACKNOWLEDGMENTS

This work was supported by the intramural program of the NICHD.

REFERENCES

Abengozar, M.A., de Frutos, S., et al., 2012. Blocking ephrinB2 with highly specific antibodies inhibits angiogenesis, lymphangiogenesis, and tumor growth. Blood 119 (19), 4565–4576.

Abtahian, F., Guerriero, A., et al., 2003. Regulation of blood and lymphatic vascular separation by signaling proteins SLP-76 and Syk. Science 299 (5604), 247–251.

Adams, R.H., Diella, F., et al., 2001. The cytoplasmic domain of the ligand ephrinB2 is required for vascular morphogenesis but not cranial neural crest migration. Cell 104 (1), 57–69.

Adams, R.H., Eichmann, A., 2010. Axon guidance molecules in vascular patterning. Cold Spring Harb. Perspect. Biol. 2 (5), a001875.

Adams, R.H., Wilkinson, G.A., et al., 1999. Roles of ephrinB ligands and EphB receptors in cardiovascular development: demarcation of arterial/venous domains, vascular morphogenesis, and sprouting angiogenesis. Genes Dev. 13 (3), 295–306.

Alders, M., Hogan, B.M., et al., 2009. Mutations in CCBE1 cause generalized lymph vessel dysplasia in humans. Nat. Genet. 41 (12), 1272–1274.

Ambati, B.K., Nozaki, M., et al., 2006. Corneal avascularity is due to soluble VEGF receptor-1. Nature 443 (7114), 993–997.

Ambler, C.A., Nowicki, J.L., et al., 2001. Assembly of trunk and limb blood vessels involves extensive migration and vasculogenesis of somite-derived angioblasts. Dev. Biol. 234 (2), 352–364.

Ambler, C.A., Schmunk, G.M., et al., 2003. Stem cell-derived endothelial cells/progenitors migrate and pattern in the embryo using the VEGF signaling pathway. Dev. Biol. 257 (1), 205–219.

Aranguren, X.L., Beerens, M., et al., 2013. COUP-TFII orchestrates venous and lymphatic endothelial identity by homo- or hetero-dimerisation with PROX1. J. Cell. Sci. 126 (Pt 5), 1164–1175.

Artavanis-Tsakonas, S., Rand, M.D., et al., 1999. Notch signaling: cell fate control and signal integration in development. Science 284 (5415), 770–776.

Avraham-Davidi, I., Ely, Y., et al., 2012. ApoB-containing lipoproteins regulate angiogenesis by modulating expression of VEGF receptor 1. Nat. Med. 18 (6), 967–973.

Bagri, A., Tessier-Lavigne, M., 2002. Neuropilins as Semaphorin receptors: *in vivo* functions in neuronal cell migration and axon guidance. Adv. Exp. Med. Biol. 515, 13–31.

Barrantes, I.B., Elia, A.J., et al., 1999. Interaction between Notch signaling and Lunatic fringe during somite boundary formation in the mouse. Curr. Biol. 9 (9), 470–480.

Basile, J.R., Barac, A., et al., 2004. Class IV semaphorins promote angiogenesis by stimulating Rho-initiated pathways through plexin-B. Cancer Res. 64 (15), 5212–5224.

Bedell, V.M., Yeo, S.Y., et al., 2005. roundabout4 is essential for angiogenesis *in vivo*. Proc. Natl. Acad. Sci. USA 102 (18), 6373–6378.

Benedito, R., Rocha, S.F., et al., 2012. Notch-dependent VEGFR3 up-regulation allows angiogenesis without VEGF-VEGFR2 signaling. Nature 484 (7392), 110–114.

Bertozzi, C.C., Schmaier, A.A., et al., 2010. Platelets regulate lymphatic vascular development through CLEC-2-SLP-76 signaling. Blood 116 (4), 661–670.

Bertrand, J.Y., Traver, D., 2009. Hematopoietic cell development in the zebrafish embryo. Curr. Opin. Hematol. 16 (4), 243–248.

Bohmer, R., Neuhaus, B., et al., 2010. Regulation of developmental lymphangiogenesis by Syk^{+} leukocytes. Dev. Cell. 18 (3), 437–449.

Bos, F.L., Caunt, M., et al., 2011. CCBE1 is essential for mammalian lymphatic vascular development and enhances the lymphangiogenic effect of vascular endothelial growth factor-C *in vivo*. Circ. Res. 109 (5), 486–491.

Bouvree, K., Brunet, I., et al., 2012. Semaphorin3A, Neuropilin-1, and PlexinA1 are required for lymphatic valve formation. Circ. Res. 111 (4), 437–445.

Bouvree, K., Larrivee, B., et al., 2008. Netrin-1 inhibits sprouting angiogenesis in developing avian embryos. Dev. Biol. 318 (1), 172–183.

Bovenkamp, D.E., Goishi, K., et al., 2004. Expression and mapping of duplicate neuropilin-1 and neuropilin-2 genes in developing zebrafish. Gene. Expr. Patterns 4 (4), 361–370.

Bressan, M., Davis, P., et al., 2009. Notochord-derived BMP antagonists inhibit endothelial cell generation and network formation. Dev. Biol. 326 (1), 101–111.

Butler, M.G., Isogai, S., et al., 2009. Lymphatic development. Birth Defects Res C Embryo. Today 87 (3), 222–231.

Carmeliet, P., Collen, D., 2000. Molecular basis of angiogenesis. Role of VEGF and VE-cadherin. Ann. N. Y. Acad. Sci. 902, 249–262. discussion 262–244.

Carmeliet, P., Conway, E.M., 2001. Growing better blood vessels. Nat. Biotechnol. 19 (11), 1019–1020.

Carmeliet, P., Ferreira, V., et al., 1996. Abnormal blood vessel development and lethality in embryos lacking a single VEGF allele. Nature 380 (6573), 435–439.

Carramolino, L., Fuentes, J., et al., 2010. Platelets play an essential role in separating the blood and lymphatic vasculatures during embryonic angiogenesis. Circ. Res. 106 (7), 1197–1201.

Cermenati, S., Moleri, S., et al., 2013. Sox18 genetically interacts with VegfC to regulate lymphangiogenesis in zebrafish. Arterioscler. Thromb. Vasc. Biol 33 (6), 1238–1247.

Cha, Y.R., Fujita, M., et al., 2012. Chemokine signaling directs trunk lymphatic network formation along the pre-existing blood vasculature. Dev. Cell. 22 (4), 824–836.

Chappell, J.C., Taylor, S.M., et al., 2009. Local guidance of emerging vessel sprouts requires soluble Flt-1. Dev. Cell. 17 (3), 377–386.

Chen, H., Bagri, A., et al., 2000. Neuropilin-2 regulates the development of selective cranial and sensory nerves and hippocampal mossy fiber projections. Neuron 25 (1), 43–56.

Childs, S., Chen, J.N., et al., 2002. Patterning of angiogenesis in the zebrafish embryo. Development 129 (4), 973–982.

Chitnis, A.B., 1995. The role of Notch in lateral inhibition and cell fate specification. Mol. Cell. Neurosci. 6 (6), 311–321.

Choi, K., 1998. Hemangioblast development and regulation. Biochem. Cell. Biol. 76 (6), 947–956.

Chung, Y.S., Zhang, W.J., et al., 2002. Lineage analysis of the hemangioblast as defined by FLK1 and SCL expression. Development 129 (23), 5511–5520.

Cleaver, O., Krieg, P.A., 1998. VEGF mediates angioblast migration during development of the dorsal aorta in Xenopus. Development 125 (19), 3905–3914.

Coffin, J.D., Poole, T.J., 1988. Embryonic vascular development: immunohistochemical identification of the origin and subsequent morphogenesis of the major vessel primordia in quail embryos. Development 102 (4), 735–748.

Coultas, L., Chawengsaksophak, K., et al., 2005. Endothelial cells and VEGF in vascular development. Nature 438 (7070), 937–945.

Davis, S., Aldrich, T.H., et al., 1996. Isolation of angiopoietin-1, a ligand for the TIE2 receptor, by secretion-trap expression cloning. Cell 87 (7), 1161–1169.

Davis-Dusenbery, B.N., Wu, C., et al., 2011. Micromanaging vascular smooth muscle cell differentiation and phenotypic modulation. Arterioscler. Thromb. Vasc. Biol. 31 (11), 2370–2377.

Doetschman, T.C., Eistetter, H., et al., 1985. The *in vitro* development of blastocyst-derived embryonic stem cell lines: formation of visceral yolk sac, blood islands and myocardium. J. Embryol. Exp. Morphol. 87, 27–45.

Donovan, J., Kordylewska, A., et al., 2002. Tetralogy of fallot and other congenital heart defects in Hey2 mutant mice. Curr. Biol. 12 (18), 1605–1610.

Drake, C.J., Fleming, P.A., 2000. Vasculogenesis in the day 6.5 to 9.5 mouse embryo. Blood 95 (5), 1671–1679.

Duarte, A., Hirashima, M., et al., 2004. Dosage-sensitive requirement for mouse Dll4 in artery development. Genes Dev. 18 (20), 2474–2478.

Dumont, D.J., Fong, G.H., et al., 1995. Vascularization of the mouse embryo: a study of *flk-1, tek, tie*, and vascular endothelial growth factor expression during development. Dev. Dyn. 203 (1), 80–92.

Dumont, D.J., Jussila, L., et al., 1998. Cardiovascular failure in mouse embryos deficient in VEGF receptor-3. Science 282 (5390), 946–949.

Eichmann, A., Corbel, C., et al., 1997. Ligand-dependent development of the endothelial and hemopoietic lineages from embryonic mesodermal cells expressing vascular endothelial growth factor receptor 2. Proc. Natl. Acad. Sci. USA 94 (10), 5141–5146.

Eichmann, A., Yuan, L., et al., 2005. Vascular development: from precursor cells to branched arterial and venous networks. Int. J. Dev. Biol. 49 (2–3), 259–267.

Emerick, K.M., Krantz, I.D., et al., 2005. Intracranial vascular abnormalities in patients with Alagille syndrome. J. Pediatr. Gastroenterol. Nutr. 41 (1), 99–107.

Fazeli, A., Dickinson, S.L., et al., 1997. Phenotype of mice lacking functional Deleted in colorectal cancer (Dcc) gene. Nature 386 (6627), 796–804.

Felcht, M., Luck, R., et al., 2012. Angiopoietin-2 differentially regulates angiogenesis through TIE2 and integrin signaling. J. Clin. Invest. 122 (6), 1991–2005.

Ferkowicz, M.J., Starr, M., et al., 2003. CD41 expression defines the onset of primitive and definitive hematopoiesis in the murine embryo. Development 130 (18), 4393–4403.

Ferrara, N., 1996. Vascular endothelial growth factor. Eur. J. Cancer 32A (14), 2413–2422.

Ferrara, N., 1999. Vascular endothelial growth factor: molecular and biological aspects. Curr. Top. Microbiol. Immunol. 237, 1–30.

Ferrara, N., Gerber, H.P., et al., 2003. The biology of VEGF and its receptors. Nat. Med. 9 (6), 669–676.

Ferrell, R.E., Levinson, K.L., et al., 1998. Hereditary lymphedema: evidence for linkage and genetic heterogeneity. Hum. Mol. Genet. 7 (13), 2073–2078.

Fina, L., Molgaard, H.V., et al., 1990. Expression of the CD34 gene in vascular endothelial cells. Blood 75 (12), 2417–2426.

Finney, B.A., Schweighoffer, E., et al., 2012. CLEC-2 and Syk in the megakaryocytic/platelet lineage are essential for development. Blood 119 (7), 1747–1756.

Fischer, A., Leimeister, C., et al., 2002. Hey bHLH factors in cardiovascular development. Cold. Spring. Harb. Symp. Quant. Biol. 67, 63–70.

Fish, J.E., Wythe, J.D., et al., 2011. A Slit/miR-218/Robo regulatory loop is required during heart tube formation in zebrafish. Development 138 (7), 1409–1419.

Fisher, A., Caudy, M., 1998. The function of hairy-related bHLH repressor proteins in cell fate decisions. Bioessays 20 (4), 298–306.

Flamme, I., Frolich, T., et al., 1997. Molecular mechanisms of vasculogenesis and embryonic angiogenesis. J. Cell. Physiol. 173 (2), 206–210.

Folkman, J., D'Amore, P.A., 1996. Blood vessel formation: what is its molecular basis? Cell 87 (7), 1153–1155.

Fong, G.H., Rossant, J., et al., 1995. Role of the Flt-1 receptor tyrosine kinase in regulating the assembly of vascular endothelium. Nature 376 (6535), 66–70.

Foo, S.S., Turner, C.J., et al., 2006. Ephrin-B2 controls cell motility and adhesion during blood-vessel-wall assembly. Cell 124 (1), 161–173.

Fouquet, B., Weinstein, B.M., et al., 1997. Vessel patterning in the embryo of the zebrafish: guidance by notochord. Dev. Biol. 183 (1), 37–48.

Francois, M., Caprini, A., et al., 2008. Sox18 induces development of the lymphatic vasculature in mice. Nature 456 (7222), 643–647.

Francois, M., Short, K., et al., 2012. Segmental territories along the cardinal veins generate lymph sacs via a ballooning mechanism during embryonic lymphangiogenesis in mice. Dev. Biol. 364 (2), 89–98.

Fryxell, K.J., Soderlund, M., et al., 2001. An animal model for the molecular genetics of CADASIL. (Cerebral autosomal dominant arteriopathy with subcortical infarcts and leukoencephalopathy). Stroke 32 (1), 6–11.

Gale, N.W., Baluk, P., et al., 2001. Ephrin-B2 selectively marks arterial vessels and neovascularization sites in the adult, with expression in both endothelial and smooth-muscle cells. Dev. Biol. 230 (2), 151–160.

Gale, N.W., Dominguez, M.G., et al., 2004. Haploinsufficiency of delta-like 4 ligand results in embryonic lethality due to major defects in arterial and vascular development. Proc. Natl. Acad. Sci. USA 101 (45), 15949–15954.

Gale, N.W., Thurston, G., et al., 2002. Angiopoietin-2 is required for postnatal angiogenesis and lymphatic patterning, and only the latter role is rescued by Angiopoietin-1. Dev. Cell. 3 (3), 411–423.

Gale, N.W., Yancopoulos, G.D., 1999. Growth factors acting via endothelial cell-specific receptor tyrosine kinases: VEGFs, angiopoietins, and ephrins in vascular development. Genes. Dev. 13 (9), 1055–1066.

Gavard, J., Patel, V., et al., 2008. Angiopoietin-1 prevents VEGF-induced endothelial permeability by sequestering Src through mDia. Dev. Cell. 14 (1), 25–36.

Gerber, H.P., Hillan, K.J., et al., 1999. VEGF is required for growth and survival in neonatal mice. Development 126 (6), 1149–1159.

Gerety, S.S., Anderson, D.J., 2002. Cardiovascular ephrinB2 function is essential for embryonic angiogenesis. Development 129 (6), 1397–1410.

Gerety, S.S., Wang, H.U., et al., 1999. Symmetrical mutant phenotypes of the receptor EphB4 and its specific transmembrane ligand ephrin-B2 in cardiovascular development. Mol. Cell. 4 (3), 403–414.

Gerhardt, H., Golding, M., et al., 2003. VEGF guides angiogenic sprouting utilizing endothelial tip cell filopodia. J. Cell. Biol. 161 (6), 1163–1177.

Gerhardt, H., Ruhrberg, C., et al., 2004. Neuropilin-1 is required for endothelial tip cell guidance in the developing central nervous system. Dev. Dyn. 231 (3), 503–509.

Giger, R.J., Cloutier, J.F., et al., 2000. Neuropilin-2 is required *in vivo* for selective axon guidance responses to secreted semaphorins. Neuron 25 (1), 29–41.

Gitler, A.D., Lu, M.M., et al., 2004. PlexinD1 and semaphorin signaling are required in endothelial cells for cardiovascular development. Dev. Cell. 7 (1), 107–116.

Goishi, K., Klagsbrun, M., 2004. Vascular endothelial growth factor and its receptors in embryonic zebrafish blood vessel development. Curr. Top. Dev. Biol. 62, 127–152.

Gore, A.V., Swift, M.R., et al., 2011. Rspo1/Wnt signaling promotes angiogenesis via Vegfc/Vegfr3. Development 138 (22), 4875–4886.

Gridley, T., 2003. Notch signaling and inherited disease syndromes. Hum. Mol. Genet 1 (12 Spec No 1), R9–13 .

Grunewald, M., Avraham, I., et al., 2006. VEGF-induced adult neovascularization: recruitment, retention, and role of accessory cells. Cell 124 (1), 175–189.

Gu, C., Rodriguez, E.R., et al., 2003. Neuropilin-1 conveys semaphorin and VEGF signaling during neural and cardiovascular development. Dev. Cell. 5 (1), 45–57.

Gu, C., Yoshida, Y., et al., 2005. Semaphorin 3E and plexin-D1 control vascular pattern independently of neuropilins. Science 307 (5707), 265–268.

Hagerling, R., Pollmann, C., et al., 2013. A novel multistep mechanism for initial lymphangiogenesis in mouse embryos based on ultramicroscopy. EMBO. J. 32 (5), 629–644.

Hedgecock, E.M., Culotti, J.G., et al., 1990. The *unc-5, unc-6*, and *unc-40* genes guide circumferential migrations of pioneer axons and mesodermal cells on the epidermis in *C. elegans*. Neuron 4 (1), 61–85.

Herbert, S.P., Huisken, J., et al., 2009. Arterial-venous segregation by selective cell sprouting: an alternative mode of blood vessel formation. Science 326 (5950), 294–298.

Herbert, S.P., Stainier, D.Y., 2011. Molecular control of endothelial cell behavior during blood vessel morphogenesis. Nat. Rev. Mol. Cell. Biol. 12 (9), 551–564.

Hermans, K., Claes, F., et al., 2010. Role of synectin in lymphatic development in zebrafish and frogs. Blood 116 (17), 3356–3366.

Herzog, Y., Kalcheim, C., et al., 2001. Differential expression of neuropilin-1 and neuropilin-2 in arteries and veins. Mech. Dev. 109 (1), 115–119.

Hirschi, K.K., 2012. Hemogenic endothelium during development and beyond. Blood 119 (21), 4823–4827.

His, W. 1900. Lecithoblast und Angioblast der Wirbeltiere. Abh. Math-Phys. Kl. Saechs. Ges. 26, 171–328.

Hogan, K.A., Ambler, C.A., et al., 2004. The neural tube patterns vessels developmentally using the VEGF signaling pathway. Development 131 (7), 1503–1513.

Hong, Y.K., Harvey, N., et al., 2002. Prox1 is a master control gene in the program specifying lymphatic endothelial cell fate. Dev. Dyn. 225 (3), 351–357.

Huber, T.L., Kouskoff, V., et al., 2004. Haemangioblast commitment is initiated in the primitive streak of the mouse embryo. Nature 432 (7017), 625–630.

Huminiecki, L., Gorn, M., et al., 2002. Magic roundabout is a new member of the roundabout receptor family that is endothelial specific and expressed at sites of active angiogenesis. Genomics 79 (4), 547–552.

Huntington, G., McClure, C., 1910. The anatomy and development of the jugular lymph sac in the domestic cat (*Felis domestica*). Am. J. Anat. 10, 177–311.

Ichise, H., Ichise, T., et al., 2009. Phospholipase Cgamma2 is necessary for separation of blood and lymphatic vasculature in mice. Development 136 (2), 191–195.

Ilan, N., Mahooti, S., et al., 1998. Distinct signal transduction pathways are utilized during the tube formation and survival phases of *in vitro* angiogenesis. J. Cell. Sci. 111 (Pt 24), 3621–3631.

Irrthum, A., Devriendt, K., et al., 2003. Mutations in the transcription factor gene SOX18 underlie recessive and dominant forms of hypotrichosis-lymphedema-telangiectasia. Am. J. Hum. Genet. 72 (6), 1470–1478.

Ishii, N., Wadsworth, W.G., et al., 1992. UNC-6, a laminin-related protein, guides cell and pioneer axon migrations in C. elegans. Neuron 9 (5), 873–881.

Isogai, S., Horiguchi, M., et al., 2001. The vascular anatomy of the developing zebrafish: an atlas of embryonic and early larval development. Dev. Biol. 230 (2), 278–301.

Isogai, S., Lawson, N.D., et al., 2003. Angiogenic network formation in the developing vertebrate trunk. Development 130 (21), 5281–5290.

Jaffredo, T., Bollerot, K., et al., 2005. Tracing the hemangioblast during embryogenesis: developmental relationships between endothelial and hematopoietic cells. Int. J. Dev. Biol. 49 (2–3), 269–277.

Jaffredo, T., Nottingham, W., et al., 2005. From hemangioblast to hematopoietic stem cell: an endothelial connection? Exp. Hematol. 33 (9), 1029–1040.

Jakeman, L.B., Armanini, M., et al., 1993. Developmental expression of binding sites and messenger ribonucleic acid for vascular endothelial growth factor suggests a role for this protein in vasculogenesis and angiogenesis. Endocrinology 133 (2), 848–859.

Jakobsson, L., Franco, C.A., et al., 2010. Endothelial cells dynamically compete for the tip cell position during angiogenic sprouting. Nat. Cell. Biol. 12 (10), 943–953.

Jiang, Y.J., Brand, M., et al., 1996. Mutations affecting neurogenesis and brain morphology in the zebrafish, Danio rerio. Development 123, 205–216.

Johnson, N.C., Dillard, M.E., et al., 2008. Lymphatic endothelial cell identity is reversible and its maintenance requires Prox1 activity. Genes. Dev. 22 (23), 3282–3291.

Jones, C.A., London, N.R., et al., 2008. Robo4 stabilizes the vascular network by inhibiting pathologic angiogenesis and endothelial hyperpermeability. Nat. Med. 14 (4), 448–453.

Jones, C.A., Nishiya, N., et al., 2009. Slit2-Robo4 signaling promotes vascular stability by blocking Arf6 activity. Nat. Cell. Biol. 11 (11), 1325–1331.

Kaipainen, A., Korhonen, J., et al., 1995. Expression of the fms-like tyrosine kinase 4 gene becomes restricted to lymphatic endothelium during development. Proc. Natl. Acad. Sci. USA 92 (8), 3566–3570.

Kalimo, H., Ruchoux, M.M., et al., 2002. CADASIL: a common form of hereditary arteriopathy causing brain infarcts and dementia. Brain. Pathol. 12 (3), 371–384.

Kallianpur, A.R., Jordan, J.E., et al., 1994. The SCL/TAL-1 gene is expressed in progenitors of both the hematopoietic and vascular systems during embryogenesis. Blood 83 (5), 1200–1208.

Kamath, B.M., Spinner, N.B., et al., 2004. Vascular anomalies in Alagille syndrome: a significant cause of morbidity and mortality. Circulation 109 (11), 1354–1358.

Kamei, M., Saunders, W.B., et al., 2006. Endothelial tubes assemble from intracellular vacuoles *in vivo*. Nature 442 (7101), 453–456.

Kappas, N.C., Zeng, G., et al., 2008. The VEGF receptor Flt-1 spatially modulates Flk-1 signaling and blood vessel branching. J. Cell. Biol. 181 (5), 847–858.

Karkkainen, M.J., Haiko, P., et al., 2004. Vascular endothelial growth factor C is required for sprouting of the first lymphatic vessels from embryonic veins. Nat. Immunol. 5 (1), 74–80.

Karkkainen, M.J., Saaristo, A., et al., 2001. A model for gene therapy of human hereditary lymphedema. Proc. Natl. Acad. Sci. USA 98 (22), 12677–12682.

Kawai-Kowase, K., Owens, G.K., 2007. Multiple repressor pathways contribute to phenotypic switching of vascular smooth muscle cells. Am. J. Physiol. Cell. Physiol. 292 (1), C59–C69.

Kawasaki, T., Kitsukawa, T., et al., 1999. A requirement for neuropilin-1 in embryonic vessel formation. Development 126 (21), 4895–4902.

Kearney, J.B., Ambler, C.A., et al., 2002. Vascular endothelial growth factor receptor Flt-1 negatively regulates developmental blood vessel formation by modulating endothelial cell division. Blood 99 (7), 2397–2407.

Keino-Masu, K., Masu, M., et al., 1996. Deleted in Colorectal Cancer (DCC) encodes a netrin receptor. Cell 87 (2), 175–185.

Kim, H., Nguyen, V.P., et al., 2010. Embryonic vascular endothelial cells are malleable to reprogramming via Prox1 to a lymphatic gene signature. BMC. Dev. Biol. 10, 72.

Kinder, S.J., Loebel, D.A., et al., 2001. Allocation and early differentiation of cardiovascular progenitors in the mouse embryo. Trends. Cardiovasc. Med. 11 (5), 177–184.

Klagsbrun, M., D'Amore, P.A., 1996. Vascular endothelial growth factor and its receptors. Cytokine. Growth. Factor. Rev. 7 (3), 259–270.

Klein, R., 2012. Eph/ephrin signaling during development. Development 139 (22), 4105–4109.

Klessinger, S., Christ, B., 1996. Axial structures control laterality in the distribution pattern of endothelial cells. Anat. Embryol. (Berl.) 193 (4), 319–330.

Koch, A.W., Mathivet, T., et al., 2011. Robo4 maintains vessel integrity and inhibits angiogenesis by interacting with UNC5B. Dev. Cell. 20 (1), 33–46.

Koshiba-Takeuchi, K., Mori, A.D., et al., 2009. Reptilian heart development and the molecular basis of cardiac chamber evolution. Nature 461 (7260), 95–98.

Krebs, L.T., Xue, Y., et al., 2000. Notch signaling is essential for vascular morphogenesis in mice. Genes Dev. 14 (11), 1343–1352.

Kuchler, A.M., Gjini, E., et al., 2006. Development of the zebrafish lymphatic system requires VEGFC signaling. Curr. Biol. 16 (12), 1244–1248.

Kukk, E., Lymboussaki, A., et al., 1996. VEGF-C receptor binding and pattern of expression with VEGFR-3 suggests a role in lymphatic vascular development. Development 122 (12), 3829–3837.

Kullander, K., Klein, R., 2002. Mechanisms and functions of Eph and ephrin signaling. Nat. Rev. Mol. Cell. Biol. 3 (7), 475–486.

Lamont, R.E., Lamont, E.J., et al., 2009. Antagonistic interactions among Plexins regulate the timing of intersegmental vessel formation. Dev. Biol. 331 (2), 199–209.

Larrivee, B., Freitas, C., et al., 2009. Guidance of vascular development: lessons from the nervous system. Circ. Res. 104 (4), 428–441.

Larrivee, B., Freitas, C., et al., 2007. Activation of the UNC5B receptor by Netrin-1 inhibits sprouting angiogenesis. Genes. Dev. 21 (19), 2433–2447.

Lawson, N.D., Scheer, N., et al., 2001. Notch signaling is required for arterial-venous differentiation during embryonic vascular development. Development 128 (19), 3675–3683.

Lawson, N.D., Vogel, A.M., et al., 2002. Sonic hedgehog and vascular endothelial growth factor act upstream of the Notch pathway during arterial endothelial differentiation. Dev. Cell. 3 (1), 127–136.

Lawson, N.D., Weinstein, B.M., 2002. Arteries and veins: making a difference with zebrafish. Nat. Rev. Genet. 3 (9), 674–682.

le Noble, F., Moyon, D., et al., 2004. Flow regulates arterial-venous differentiation in the chick embryo yolk sac. Development 131 (2), 361–375.

Lee, P., Goishi, K., et al., 2002. Neuropilin-1 is required for vascular development and is a mediator of VEGF-dependent angiogenesis in zebrafish. Proc. Natl. Acad. Sci. USA 99 (16), 10470–10475.

Lee, S., Kang, J., et al., 2009. Prox1 physically and functionally interacts with COUP-TFII to specify lymphatic endothelial cell fate. Blood 113 (8), 1856–1859.

Lejmi, E., Leconte, L., et al., 2008. Netrin-4 inhibits angiogenesis via binding to neogenin and recruitment of Unc5B. Proc. Natl. Acad. Sci. USA 105 (34), 12491–12496.

Leonardo, E.D., Hinck, L., et al., 1997. Guidance of developing axons by netrin-1 and its receptors. Cold. Spring. Harb. Symp. Quant. Biol. 62, 467–478.

Leu, A.J., Berk, D.A., et al., 2000. Absence of functional lymphatics within a murine sarcoma: a molecular and functional evaluation. Cancer. Res. 60 (16), 4324–4327.

Liang, D., Chang, J.R., et al., 2001. The role of vascular endothelial growth factor (VEGF) in vasculogenesis, angiogenesis, and hematopoiesis in zebrafish development. Mech. Dev. 108 (1–2), 29–43.

Liang, D., Xu, X., et al., 1998. Cloning and characterization of vascular endothelial growth factor (VEGF) from zebrafish, *Danio rerio*. Biochim. Biophys. Acta. 1397 (1), 14–20.

Liao, W., Bisgrove, B.W., et al., 1997. The zebrafish gene *cloche* acts upstream of a *flk*-1 homologue to regulate endothelial cell differentiation. Development 124 (2), 381–389.

Lim, A.H., Suli, A., et al., 2011. Motoneurons are essential for vascular pathfinding. Development 138 (17), 3847–3857.

Limaye, N., Wouters, V., et al., 2009. Somatic mutations in angiopoietin receptor gene TEK cause solitary and multiple sporadic venous malformations. Nat. Genet. 41 (1), 118–124.

Long, H., Sabatier, C., et al., 2004. Conserved roles for Slit and Robo proteins in midline commissural axon guidance. Neuron 42 (2), 213–223.

Lu, X., Le Noble, F., et al., 2004. The netrin receptor UNC5B mediates guidance events controlling morphogenesis of the vascular system. Nature 432 (7014), 179–186.

Maisonpierre, P.C., Suri, C., et al., 1997. Angiopoietin-2, a natural antagonist for Tie2 that disrupts *in vivo* angiogenesis. Science 277 (5322), 55–60.

Makinen, T., Adams, R.H., et al., 2005. PDZ interaction site in ephrinB2 is required for the remodeling of lymphatic vasculature. Genes. Dev. 19 (3), 397–410.

Makinen, T., Jussila, L., et al., 2001. Inhibition of lymphangiogenesis with resulting lymphedema in transgenic mice expressing soluble VEGF receptor-3. Nat. Med. 7 (2), 199–205.

Mandriota, S.J., Jussila, L., et al., 2001. Vascular endothelial growth factor-C-mediated lymphangiogenesis promotes tumor metastasis. EMBO. J. 20 (4), 672–682.

Marlow, R., Binnewies, M., et al., 2010. Vascular Robo4 restricts proangiogenic VEGF signaling in breast. Proc. Natl. Acad. Sci. USA 107 (23), 10520–10525.

Marron, M.B., Hughes, D.P., et al., 2000. Evidence for heterotypic interaction between the receptor tyrosine kinases TIE-1 and TIE-2. J. Biol. Chem. 275 (50), 39741–39746.

Martyn, U., Schulte-Merker, S., 2004. Zebrafish neuropilins are differentially expressed and interact with vascular endothelial growth factor during embryonic vascular development. Dev. Dyn. 231 (1), 33–42.

Meadows, S.M., Fletcher, P.J., et al., 2012. Integration of repulsive guidance cues generates avascular zones that shape mammalian blood vessels. Circ. Res. 110 (1), 34–46.

Melani, M., Weinstein, B.M., 2010. Common factors regulating patterning of the nervous and vascular systems. Annu. Rev. Cell. Dev. Biol. 26, 639–665.

Miao, H.Q., Soker, S., et al., 1999. Neuropilin-1 mediates collapsin-1/semaphorin III inhibition of endothelial cell motility: functional competition of collapsin-1 and vascular endothelial growth factor-165. J. Cell. Biol. 146 (1), 233–242.

Mishima, K., Watabe, T., et al., 2007. Prox1 induces lymphatic endothelial differentiation via integrin alpha9 and other signaling cascades. Mol. Biol. Cell. 18 (4), 1421–1429.

Moyon, D., Pardanaud, L., et al., 2001. Plasticity of endothelial cells during arterial-venous differentiation in the avian embryo. Development 128 (17), 3359–3370.

Moyon, D., Pardanaud, L., et al., 2001. Selective expression of angiopoietin 1 and 2 in mesenchymal cells surrounding veins and arteries of the avian embryo. Mech. Dev. 106 (1–2), 133–136.

Murray, P.D.F. 1932. The development in vitro of the blood of the early chick embryo. Proc. R. Soc. B London, Ser. B 111, 497–521.

Nakagawa, O., McFadden, D.G., et al., 2000. Members of the HRT family of basic helix-loop-helix proteins act as transcriptional repressors downstream of Notch signaling. Proc. Natl. Acad. Sci. USA 97 (25), 13655–13660.

Nakagawa, O., Nakagawa, M., et al., 1999. HRT1, HRT2, and HRT3: a new subclass of bHLH transcription factors marking specific cardiac, somitic, and pharyngeal arch segments. Dev. Biol. 216 (1), 72–84.

Nakayama, M., Berger, P., 2013. Coordination of VEGF receptor trafficking and signaling by coreceptors. Exp. Cell. Res. 319 (9), 1340–1347.

Navankasattusas, S., Whitehead, K.J., et al., 2008. The netrin receptor UNC5B promotes angiogenesis in specific vascular beds. Development 135 (4), 659–667.

Nikolova, G., Lammert, E., 2003. Interdependent development of blood vessels and organs. Cell. Tissue. Res. 314 (1), 33–42.

Ny, A., Koch, M., et al., 2005. A genetic *Xenopus laevis* tadpole model to study lymphangiogenesis. Nat. Med. 11 (9), 998–1004.

Oh, S.J., Jeltsch, M.M., et al., 1997. VEGF and VEGF-C: specific induction of angiogenesis and lymphangiogenesis in the differentiated avian chorioallantoic membrane. Dev. Biol. 188 (1), 96–109.

Oliver, G., 2004. Lymphatic vasculature development. Nat. Rev. Immunol. 4 (1), 35–45.

Orlandi, A., Bennett, M., 2010. Progenitor cell-derived smooth muscle cells in vascular disease. Biochem. Pharmacol. 79 (12), 1706–1713.

Osada, M., Inoue, O., et al., 2012. Platelet activation receptor CLEC-2 regulates blood/lymphatic vessel separation by inhibiting proliferation, migration, and tube formation of lymphatic endothelial cells. J. Biol. Chem. 287 (26), 22241–22252.

Othman-Hassan, K., Patel, K., et al., 2001. Arterial identity of endothelial cells is controlled by local cues. Dev. Biol. 237 (2), 398–409.

Owens, G.K., Kumar, M.S., et al., 2004. Molecular regulation of vascular smooth muscle cell differentiation in development and disease. Physiol. Rev. 84 (3), 767–801.

Pan, W., Pham, V.N., et al., 2012. CDP-diacylglycerol synthetase-controlled phosphoinositide availability limits VEGFA signaling and vascular morphogenesis. Blood 120 (2), 489–498.

Pardanaud, L., Dieterlen-Lievre, F., 1999. Manipulation of the angiopoietic/hemangiopoietic commitment in the avian embryo. Development 126 (4), 617–627.

Park, K.W., Crouse, D., et al., 2004. The axonal attractant Netrin-1 is an angiogenic factor. Proc. Natl. Acad. Sci. USA 101 (46), 16210–16215.

Park, K.W., Morrison, C.M., et al., 2003. Robo4 is a vascular-specific receptor that inhibits endothelial migration. Dev. Biol. 261 (1), 251–267.

Partanen, T.A., Vuola, P., et al., 2013. Neuropilin-2 and vascular endothelial growth factor receptor-3 are up-regulated in human vascular malformations. Angiogenesis 16 (1), 137–146.

Petrova, T.V., Makinen, T., et al., 2002. Lymphatic endothelial reprogramming of vascular endothelial cells by the Prox-1 homeobox transcription factor. EMBO. J. 21 (17), 4593–4599.

Pitulescu, M.E., Adams, R.H., 2010. Eph/ephrin molecules – a hub for signaling and endocytosis. Genes. Dev. 24 (22), 2480–2492.

Pola, R., Ling, L.E., et al., 2001. The morphogen Sonic hedgehog is an indirect angiogenic agent upregulating two families of angiogenic growth factors. Nat. Med. 7 (6), 706–711.

Puri, M.C., Rossant, J., et al., 1995. The receptor tyrosine kinase TIE is required for integrity and survival of vascular endothelial cells. EMBO. J. 14 (23), 5884–5891.

Raimondi, C., Ruhrberg, C., 2013. Neuropilin signaling in vessels, neurons and tumors. Semin. Cell. Dev. Biol. 24 (3), 172–178.

Reese, D.E., Hall, C.E., et al., 2004. Negative regulation of midline vascular development by the notochord. Dev. Cell. 6 (5), 699–708.

Risau, W., 1997. Mechanisms of angiogenesis. Nature 386 (6626), 671–674.

Risau, W., Flamme, I., 1995. Vasculogenesis. Annu. Rev. Cell. Dev. Biol. 11, 73–91.

Rossant, J., Hirashima, M., 2003. Vascular development and patterning: making the right choices. Curr. Opin. Genet. Dev. 13 (4), 408–412.

Rossant, J., Howard, L., 2002. Signaling pathways in vascular development. Annu. Rev. Cell. Dev. Biol. 18, 541–573.

Ruchoux, M.M., Chabriat, H., et al., 1994. Presence of ultrastructural arterial lesions in muscle and skin vessels of patients with CADASIL. Stroke 25 (11), 2291–2292.

Ruchoux, M.M., Domenga, V., et al., 2003. Transgenic mice expression mutant Notch3 develop vascular alterations characteristic of cerebral autosomal dominant arteriopathy with subcortical infarcts and leukoencephalopathy. Am. J. Pathol 162 (1), 329–342.

Sabin, F.R., 1902. On the origin of the lymphatic system from the veins and the development of the lymph hearts and thoracic duct in the pig. Am. J. Anat. 1, 367–389.

Saharinen, P., Kerkela, K., et al., 2005. Multiple angiopoietin recombinant proteins activate the Tie1 receptor tyrosine kinase and promote its interaction with Tie2. J. Cell. Biol. 169 (2), 239–243.

Sakata, Y., Kamei, C.N., et al., 2002. Ventricular septal defect and cardiomyopathy in mice lacking the transcription factor CHF1/Hey2. Proc. Natl. Acad. Sci. USA 99 (25), 16197–16202.

Sato, T.N., Qin, Y., et al., 1993. Tie-1 and tie-2 define another class of putative receptor tyrosine kinase genes expressed in early embryonic vascular system. Proc. Natl. Acad. Sci. USA 90 (20), 9355–9358.

Sato, T.N., Tozawa, Y., et al., 1995. Distinct roles of the receptor tyrosine kinases Tie-1 and Tie-2 in blood vessel formation. Nature 376 (6535), 70–74.

Scappaticci, F.A., 2002. Mechanisms and future directions for angiogenesis-based cancer therapies. J. Clin. Oncol. 20 (18), 3906–3927.

Schacht, V., Ramirez, M.I., et al., 2003. T1alpha/podoplanin deficiency disrupts normal lymphatic vasculature formation and causes lymphedema. EMBO. J. 22 (14), 3546–3556.

Schröder, J.M., Sellhaus, B., et al., 1995. Identification of the characteristic vascular changes in a sural nerve biopsy of a case with cerebral autosomal dominant arteriopathy with subcortical infarcts and leukoencephalopathy (CADASIL). Acta Neuropathol 89 (2), 116–121.

Schulte-Merker, S., van Eeden, F.J., et al., 1994. no tail (ntl) is the zebrafish homologue of the mouse T (Brachyury) gene. Development 120 (4), 1009–1015.

Sebzda, E., Hibbard, C., et al., 2006. Syk and Slp-76 mutant mice reveal a cell-autonomous hematopoietic cell contribution to vascular development. Dev. Cell. 11 (3), 349–361.

Segarra, M., Ohnuki, H., et al., 2012. Semaphorin 6A regulates angiogenesis by modulating VEGF signaling. Blood 120 (19), 4104–4115.

Serafini, T., Colamarino, S.A., et al., 1996. Netrin-1 is required for commissural axon guidance in the developing vertebrate nervous system. Cell 87 (6), 1001–1014.

Serafini, T., Kennedy, T.E., et al., 1994. The netrins define a family of axon outgrowth-promoting proteins homologous to C. elegans UNC-6. Cell 78 (3), 409–424.

Shalaby, F., Rossant, J., et al., 1995. Failure of blood-island formation and vasculogenesis in Flk-1-deficient mice. Nature 376 (6535), 62–66.

Shin, J.W., Min, M., et al., 2006. Prox1 promotes lineage-specific expression of fibroblast growth factor (FGF) receptor-3 in lymphatic endothelium: a role for FGF signaling in lymphangiogenesis. Mol. Biol. Cell. 17 (2), 576–584.

Shirayoshi, Y., Yuasa, Y., et al., 1997. Proto-oncogene of int-3, a mouse Notch homologue, is expressed in endothelial cells during early embryogenesis. Genes. Cells. 2 (3), 213–224.

Shivdasani, R.A., Mayer, E.L., et al., 1995. Absence of blood formation in mice lacking the T-cell leukaemia oncoprotein tal-1/SCL. Nature 373 (6513), 432–434.

Shutter, J.R., Scully, S., et al., 2000. Dll4, a novel Notch ligand expressed in arterial endothelium. Genes Dev. 14 (11), 1313–1318.

Shweiki, D., Itin, A., et al., 1993. Patterns of expression of vascular endothelial growth factor (VEGF) and VEGF receptors in mice suggest a role in hormonally regulated angiogenesis. J. Clin. Invest. 91 (5), 2235–2243.

Simons, M., 2012. An inside view: VEGF receptor trafficking and signaling. Physiol. (Bethesda) 27 (4), 213–222.

Skobe, M., Hawighorst, T., et al., 2001. Induction of tumor lymphangiogenesis by VEGF-C promotes breast cancer metastasis. Nat. Med. 7 (2), 192–198.

Small, E.M., Sutherland, L.B., et al., 2010. MicroRNA-218 regulates vascular patterning by modulation of Slit-Robo signaling. Circ. Res. 107 (11), 1336–1344.

Soker, S., Takashima, S., et al., 1998. Neuropilin-1 is expressed by endothelial and tumor cells as an isoform-specific receptor for vascular endothelial growth factor. Cell 92 (6), 735–745.

Srinivasan, R.S., Dillard, M.E., et al., 2007. Lineage tracing demonstrates the venous origin of the mammalian lymphatic vasculature. Genes. Dev. 21 (19), 2422–2432.

Srinivasan, R.S., Geng, X., et al., 2010. The nuclear hormone receptor Coup-TFII is required for the initiation and early maintenance of Prox1 expression in lymphatic endothelial cells. Genes Dev. 24 (7), 696–707.

Stacker, S.A., Caesar, C., et al., 2001. VEGF-D promotes the metastatic spread of tumor cells via the lymphatics. Nat. Med. 7 (2), 186–191.

Stainier, D.Y., Weinstein, B.M., et al., 1995. Cloche, an early acting zebrafish gene, is required by both the endothelial and hematopoietic lineages. Development 121 (10), 3141–3150.

Stratman, A.N., Davis, G.E., 2012. Endothelial cell-pericyte interactions stimulate basement membrane matrix assembly: influence on vascular tube remodeling, maturation, and stabilization. Microsc. Microanal. 18 (1), 68–80.

Stratman, A.N., Davis, M.J., et al., 2011. VEGF and FGF prime vascular tube morphogenesis and sprouting directed by hematopoietic stem cell cytokines. Blood 117 (14), 3709–3719.

Suchting, S., Heal, P., et al., 2005. Soluble Robo4 receptor inhibits *in vivo* angiogenesis and endothelial cell migration. FASEB. J. 19 (1), 121–123.

Sumoy, L., Keasey, J.B., et al., 1997. A role for notochord in axial vascular development revealed by analysis of phenotype and the expression of VEGR-2 in zebrafish flh and ntl mutant embryos. Mech. Dev. 63 (1), 15–27.

Suri, C., Jones, P.F., et al., 1996. Requisite role of angiopoietin-1, a ligand for the TIE2 receptor, during embryonic angiogenesis. Cell 87 (7), 1171–1180.

Suri, C., McClain, J., et al., 1998. Increased vascularization in mice overexpressing angiopoietin-1. Science 282 (5388), 468–471.

Suzuki-Inoue, K., Fuller, G.L., et al., 2006. A novel Syk-dependent mechanism of platelet activation by the C-type lectin receptor CLEC-2. Blood 107 (2), 542–549.

Suzuki-Inoue, K., Kato, Y., et al., 2007. Involvement of the snake toxin receptor CLEC-2, in podoplanin-mediated platelet activation, by cancer cells. J. Biol. Chem. 282 (36), 25993–26001.

Swiatek, P.J., Lindsell, C.E., et al., 1994. Notch1 is essential for postimplantation development in mice. Genes. Dev. 8 (6), 707–719.

Takakura, N., Huang, X.L., et al., 1998. Critical role of the TIE2 endothelial cell receptor in the development of definitive hematopoiesis. Immunity 9 (5), 677–686.

Takashima, S., Kitakaze, M., et al., 2002. Targeting of both mouse neuropilin-1 and neuropilin-2 genes severely impairs developmental yolk sac and embryonic angiogenesis. Proc. Natl. Acad. Sci. USA 99 (6), 3657–3662.

Talbot, W.S., Trevarrow, B., et al., 1995. A homeobox gene essential for zebrafish notochord development. Nature 378 (6553), 150–157.

Tammela, T., Zarkada, G., et al., 2011. VEGFR-3 controls tip to stalk conversion at vessel fusion sites by reinforcing Notch signaling. Nat. Cell. Biol. 13 (10), 1202–1213.

Thompson, M.A., Ransom, D.G., et al., 1998. The cloche and spadetail genes differentially affect hematopoiesis and vasculogenesis. Dev. Biol. 197 (2), 248–269.

Thurston, G., Wang, Q., et al., 2005. Angiopoietin 1 causes vessel enlargement, without angiogenic sprouting, during a critical developmental period. Development 132 (14), 3317–3326.

Torres-Vazquez, J., Gitler, A.D., et al., 2004. Semaphorin-plexin signaling guides patterning of the developing vasculature. Dev. Cell. 7 (1), 117–123.

Torres-Vazquez, J., Kamei, M., et al., 2003. Molecular distinction between arteries and veins. Cell. Tissue. Res. 314 (1), 43–59.

Tsiamis, A.C., Morris, P.N., et al., 2002. Vascular endothelial growth factor modulates the Tie-2:Tie-1 receptor complex. Microvasc. Res. 63 (2), 149–158.

Uhrin, P., Zaujec, J., et al., 2010. Novel function for blood platelets and podoplanin in developmental separation of blood and lymphatic circulation. Blood 115 (19), 3997–4005.

Uyttendaele, H., Ho, J., et al., 2001. Vascular patterning defects associated with expression of activated Notch4 in embryonic endothelium. Proc. Natl. Acad. Sci. USA 98 (10), 5643–5648.

Vajkoczy, P., Menger, M.D., et al., 1999. Inhibition of tumor growth, angiogenesis, and microcirculation by the novel Flk-1 inhibitor SU5416 as assessed by intravital multi-fluorescence videomicroscopy. Neoplasia 1 (1), 31–41.

Vargesson, N., Patel, K., et al., 1998. Expression patterns of Notch1, Serrate1, Serrate2 and Delta1 in tissues of the developing chick limb. Mech. Dev. 77 (2), 197–199.

Vikkula, M., Boon, L.M., et al., 1996. Vascular dysmorphogenesis caused by an activating mutation in the receptor tyrosine kinase TIE2. Cell 87 (7), 1181–1190.

Villa, N., Walker, L., et al., 2001. Vascular expression of Notch pathway receptors and ligands is restricted to arterial vessels. Mech. Dev. 108 (1–2), 161–164.

Visvader, J.E., Fujiwara, Y., et al., 1998. Unsuspected role for the T-cell leukemia protein SCL/tal-1 in vascular development. Genes. Dev. 12 (4), 473–479.

Vittet, D., Prandini, M.H., et al., 1996. Embryonic stem cells differentiate *in vitro* to endothelial cells through successive maturation steps. Blood 88 (9), 3424–3431.

Vokes, S.A., Yatskievych, T.A., et al., 2004. Hedgehog signaling is essential for endothelial tube formation during vasculogenesis. Development 131 (17), 4371–4380.

Wacker, A., Gerhardt, H., 2011. Endothelial development taking shape. Curr. Opin. Cell. Biol. 23 (6), 676–685.

Wang, B., Xiao, Y., et al., 2003. Induction of tumor angiogenesis by Slit-Robo signaling and inhibition of cancer growth by blocking Robo activity. Cancer Cell. 4 (1), 19–29.

Wang, H.U., Chen, Z.F., et al., 1998. Molecular distinction and angiogenic interaction between embryonic arteries and veins revealed by ephrin-B2 and its receptor Eph-B4. Cell 93 (5), 741–753.

Wang, R., Clark, R., et al., 1992. Embryonic stem cell-derived cystic embryoid bodies form vascular channels: an *in vitro* model of blood vessel development. Development 114 (2), 303–316.

Wang, Y., Nakayama, M., et al., 2010. Ephrin-B2 controls VEGF-induced angiogenesis and lymphangiogenesis. Nature 465 (7297), 483–486.

Weinstein, B.M., 2002. Plumbing the mysteries of vascular development using the zebrafish. Semin. Cell. Dev. Biol. 13 (6), 515–522.

Wettstein, D.A., Turner, D.L., et al., 1997. The *Xenopus* homolog of *Drosophila* Suppressor of Hairless mediates Notch signaling during primary neurogenesis. Development 124 (3), 693–702.

Wigle, J.T., Harvey, N., et al., 2002. An essential role for Prox1 in the induction of the lymphatic endothelial cell phenotype. EMBO. J. 21 (7), 1505–1513.

Wigle, J.T., Oliver, G., 1999. Prox1 function is required for the development of the murine lymphatic system. Cell 98 (6), 769–778.

Wiles, M.V., Keller, G., 1991. Multiple hematopoietic lineages develop from embryonic stem (ES) cells in culture. Development 111 (2), 259–267.

Willett, C.G., Boucher, Y., et al., 2004. Direct evidence that the VEGF-specific antibody bevacizumab has antivascular effects in human rectal cancer. Nat. Med. 10 (2), 145–147.

Williams, C., Kim, S.H., et al., 2010. Hedgehog signaling induces arterial endothelial cell formation by repressing venous cell fate. Dev. Biol. 341 (1), 196–204.

Williams, N., Burgess, A.W., 1980. The effect of mouse lung granulocyte-macrophage colony-stimulating factor and other colony-stimulating activities on the proliferation and differentiation of murine bone marrow cells in long-term cultures. J. Cell. Physiol. 102 (3), 287–295.

Wilson, B.D., Ii, M., et al., 2006. Netrins promote developmental and therapeutic angiogenesis. Science 313 (5787), 640–644.

Wilting, J., Aref, Y., et al., 2006. Dual origin of avian lymphatics. Dev. Biol. 292 (1), 165–173.

Wilting, J., Papoutsi, M., et al., 2000. The lymphatic endothelium of the avian wing is of somitic origin. Dev. Dyn. 217 (3), 271–278.

Witte, M.H., Bernas, M.J., et al., 2001. Lymphangiogenesis and lymphangiodysplasia: from molecular to clinical lymphology. Microsc. Res. Tech. 55 (2), 122–145.

Wouters, V., Limaye, N., et al., 2010. Hereditary cutaneomucosal venous malformations are caused by TIE2 mutations with widely variable hyperphosphorylating effects. Eur. J. Hum. Genet. 18 (4), 414–420.

Xue, Y., Gao, X., et al., 1999. Embryonic lethality and vascular defects in mice lacking the Notch ligand Jagged1. Hum. Mol. Genet. 8 (5), 723–730.

Yamazaki, T., Yoshimatsu, Y., et al., 2009. COUP-TFII regulates the functions of Prox1 in lymphatic endothelial cells through direct interaction. Genes Cells 14 (3), 425–434.

Yancopoulos, G.D., Davis, S., et al., 2000. Vascular-specific growth factors and blood vessel formation. Nature 407 (6801), 242–248.

Yancopoulos, G.D., Klagsbrun, M., et al., 1998. Vasculogenesis, angiogenesis, and growth factors: ephrins enter the fray at the border. Cell 93 (5), 661–664.

Yang, Y., Garcia-Verdugo, J.M., et al., 2012. Lymphatic endothelial progenitors bud from the cardinal vein and intersomitic vessels in mammalian embryos. Blood 120 (11), 2340–2348.

Yaniv, K., Isogai, S., et al., 2006. Live imaging of lymphatic development in the zebrafish. Nat. Med. 12 (6), 711–716.

Yu, H.H., Houart, C., et al., 2004. Cloning and embryonic expression of zebrafish neuropilin genes. Gene. Expr. Patterns 4 (4), 371–378.

Yuan, L., Moyon, D., et al., 2002. Abnormal lymphatic vessel development in neuropilin 2 mutant mice. Development 129 (20), 4797–4806.

Zape, J.P., Zovein, A.C., 2011. Hemogenic endothelium: origins, regulation, and implications for vascular biology. Semin. Cell. Dev. Biol. 22 (9), 1036–1047.

Zhang, Y., Singh, M.K., et al., 2009. Tie2Cre-mediated inactivation of plexinD1 results in congenital heart, vascular and skeletal defects. Dev. Biol. 325 (1), 82–93.

Zhong, T.P., Childs, S., et al., 2001. Gridlock signaling pathway fashions the first embryonic artery. Nature 414 (6860), 216–220.

Zhong, T.P., Rosenberg, M., et al., 2000. Gridlock, an HLH gene required for assembly of the aorta in zebrafish. Science 287 (5459), 1820–1824.

Zovein, A.C., Hofmann, J.J., et al., 2008. Fate tracing reveals the endothelial origin of hematopoietic stem cells. Cell. Stem. Cell. 3 (6), 625–636.

Zygmunt, T., Gay, C.M., et al., 2011. Semaphorin-PlexinD1 signaling limits angiogenic potential via the VEGF decoy receptor sFlt1. Dev. Cell. 21 (2), 301–314.

Chapter 25

Blood Induction and Embryonic Formation

Xiaoying Bai* and Leonard I. Zon†
*University of Texas Southwestern Medical Center, Dallas, TX; †Howard Hughes Medical Institute, Children's Hospital of Boston, Boston, MA

Chapter Outline

GLOSSARY

Hematopoiesis The developmental process by which various types of blood cells are formed.

Hematopoietic stem cell (HSC) The precursor cells that give rise to all types of blood cells. As stem cells, they are defined by their ability to maintain themselves (self-renewal) and to differentiate to form multiple cells types. In mammals, HSCs are located in the adult bone marrow.

Hemangioblast A hypothesized common precursor for both blood cells and blood vessel endothelium cells.

Hemogenic endothelium A rare population of vascular endothelial cells that are able to differentiate into hematopoietic stem cells during embryogenesis, usually located in the lumen side of artery wall.

Primitive hematopoiesis A transient wave of hematopoiesis that generates the first blood cells in embryos. Most primitive blood cells are red cells. Macrophages and megakaryocytes are also found in this wave.

Definitive hematopoiesis Occurs shortly after the primitive hematopoiesis and contains two waves. The first definitive wave generates only red cell and myeloid lineages. The second wave generates HSCs that differentiate into all adult blood lineages.

SUMMARY

- Vertebrate blood derives from mesoderm.
- Blood formation (hematopoiesis) occurs in successive waves. The primitive wave predominantly generates red cells to transiently provide oxygen for the rapidly growing embryo. The definitive wave produces hematopoietic stem cells (HSCs) that ultimately differentiate into all the blood lineages in the adult.
- Hematopoiesis is tightly regulated by complex interactions between extrinsic and intrinsic factors. Transcription factors play intrinsic roles to define HSC generation and lineage differentiation.
- The process of hematopoiesis is highly conserved throughout vertebrate evolution. The genetic manipulation of animal models, such as the mouse and zebrafish, has identified many essential genes controlling normal and malignant hematopoiesis.

Principles of Developmental Genetics.

25.1 INTRODUCTION

Each day our body produces billions of new white blood cells, red blood cells, and platelets to replace the blood cells lost during the process of cell turnover. These mature blood cells are generated by a small population of stem cells, termed hematopoietic stem cells (HSCs) that reside in the bone marrow. HSCs are defined by their capacity for self-renewal and multi-lineage differentiation which can give rise to all the blood cells in our body, including erythrocytes, granulocytes (basophils, eosinophils and neutrophils), lymphocytes (B cells, T cells and natural killer (NK) cells), monocytes/macrophages and platelets. When transplanted into a host that has been lethally irradiated to remove endogenous HSCs, the donor-derived HSCs can reconstitute all blood lineages throughout life (Fleischman et al., 1982; Harrison et al., 1988; Spangrude et al., 1988; Jordan and Lemischka, 1990; Chaddah et al., 1996; Osawa et al., 1996).

The emergence of HSCs and the blood system can be traced back to the early stages of embryogenesis. Pioneering studies addressing the formation of the blood system were performed in chick embryos due to their easy accessibility for observation and manipulation. Later, this work was extended into mammalian models, such as the mouse. More recently, *Xenopus* and zebrafish models have been extensively used to decipher the molecular pathways involved in hematopoiesis. These studies have revealed a remarkably conserved developmental program of hematopoiesis. In addition, *in vitro* studies using hematopoietic progenitor cell culture and the embryonic stem (ES) cell-derived embryoid body (EB) have greatly contributed to our understanding of the hematopoietic hierarchy. In this chapter, we will review the general process of hematopoiesis in vertebrate model organisms, and will highlight the important molecules that regulate the development of blood.

25.2 ORIGIN OF BLOOD CELLS DURING EMBRYOGENESIS

Induction of Hematopoiesis in the Early Embryo

In vertebrates, blood cells derive from mesoderm. During gastrulation, the mesoderm is induced and subsequently patterned to adopt a dorsal or ventral fate. Studies in *Xenopus* and zebrafish have discovered mesoderm-inducing factors, including members of the transforming growth factor β (TGFβ) family and the fibroblast growth factors (FGF) (Munoz-Sanjuan, 2001) (see also Chapter 12). In addition, the T-box transcription factor *VegT* has also been found to be important for the mesoderm formation (Zhang et al., 1998). The zebrafish mutant *spadetail*, caused by the loss of function of *tbx16*, the homolog of *VegT*, displays a severe defect in the mesodermal and endodermal derivatives in the trunk, including an absence of blood (Thompson et al., 1998; Kimmel et al., 1989; Ho et al., 1990).

Patterning of the mesoderm is regulated by antagonistic interactions between the ventralizing signals and dorsalizing signals (Thomsen, 1997). Ventralizing signals are mediated by members of the bone morphogenetic proteins (BMP), a subgroup in the TGFβ superfamily. Overexpression of *bmp2, bmp4* or *bmp7* results in loss of dorsal derivatives such as the muscle and the notochord and expands ventral mesoderm fates such as the blood and the kidney (Clement et al., 1995; Fainsod et al., 1994; Jones et al., 1992; Mead et al., 1996; Maeno et al., 1996; Wang S et al., 1997). Conversely, several zebrafish mutants with disruptions in the BMP pathway, such as *swirl* (*bmp2* mutant), *snailhouse* (*bmp7* mutant), and *somitobun* (mutant of *smad5*, the downstream signal transducer of the BMP pathway), lack the ventral tissues (i.e., are dorsalized mutants) and have defects in the blood formation (Kishimoto et al., 1997; Schmid et al., 2000; Dick et al., 2000; Mullins et al., 1996; Nguyen et al., 1998; Hild et al., 1999). Similarly, targeted knockout of *Bmp4, Bmp2* or *Alk3* (a BMP receptor) genes in mice results in severe mesoderm deficiency and in certain genetic backgrounds, primitive hematopoiesis in these mutant embryos is severely disrupted (Lawson et al., 1999; Winnier et al., 1995). In addition to its general effect on hematopoiesis through dorsal/ventral patterning, a study in zebrafish showed that *Alk8*, a BMP receptor, regulates the specification of the myeloid lineage during early embryogenesis, suggesting that the BMP/TGFβ signaling may also regulate hematopoiesis in a lineage specific pattern (Hogan et al., 2006).

The BMP pathways are possibly mediated by the *Mix* family of transcription factors, which belongs to the paired class of homeobox genes (Mead et al., 1998). Overexpression of *Mix.1* in *Xenopus* induces excessive blood formation and ventralizes the embryos, similarly to the BMP overexpression phenotype (Jones et al., 1992; Dale et al., 1992). Using *Mixl1*-null ES cells, Elefanty and colleagues demonstrated that *Mixl1* is required for efficient hematopoiesis and BMP4-induced ventral mesoderm patterning (Ng et al., 2005). Conversely, induction of *Mixl* in ES-cell-derived EBs results in acceleration of mesoderm development and expansion of hematopoietic progenitors (Willey et al., 2006). Taken together, these studies suggest that the *Mix* family may participate in the BMP signaling pathways in patterning the mesoderm toward a ventral and hematopoietic fate.

From Hemangioblast to Hematopoietic Stem Cells

During embryonic development, hematopoietic and endothelial cells emerge simultaneously in close association with each other from the mesoderm. This developmental proximity has led to the hypothesis that they arise from a common progenitor, termed the hemangioblast. Numerous studies have suggested such a common origin for hematopoietic and endothelial lineages. For example, mice lacking *Flk-1*, an early endothelial marker, fail to generate either endothelial or hematopoietic cells (Shalaby et al., 1995). Conversely, ES cells deficient in *Scl*, an early hematopoietic marker, failed to contribute to the formation of the vitelline vessels in the mouse yolk sac (Visvader et al., 1998). *In vitro* studies on single cell-derived colonies that can form both hematopoietic and endothelial cells also support this hypothesis (Eichmann et al., 1997; Choi et al., 1998; Huber et al., 2004). Furthermore, a fate mapping study in zebrafish using laser activation of caged fluorescein dextran showed that a single cell labeled during gastrulation could give rise to both hematopoietic and endothelial lineages, providing direct evidence for the presence of such bipotential progenitors *in vivo* (Vogeli et al., 2006). In support of this, a zebrafish mutant *cloche* lacks both endothelial and hematopoietic cells (Stainier et al., 1995; Liao et al., 1997). Although the mutant gene has not been cloned, overexpression of *scl* was found to partially rescue both blood and vascular defects, suggesting that the mutant gene functions upstream of *scl*, perhaps at the hemangioblast level (Liao et al., 1998).

Primitive Hematopoiesis

Hematopoiesis in vertebrate embryos is characterized by two sequential waves occurring at anatomically distinct sites. The first wave is the primitive wave, generating mainly red cells (erythrocytes) with some macrophages and megakaryocytes. In mouse and avian embryos, this wave appears extra-embryonically within the mesoderm-derived yolk sac (YS) blood island (Moore and Metcalf, 1970; Palis et al., 1995; 1999) (Figure 25.1). In mouse embryos, expression of the hematopoietic genes *Scl* and *Lmo2* marks the initiation of YS hematopoiesis at 7 days postcoitum (dpc) (Palis et al., 2001). Subsequently, *Gata1*, an erythroid-specific transcription factor, is detected in the YS (Palis et al., 2001; Pevny et al., 1995). Between 8.5 and 9 dpc, the primitive erythrocytes enter the circulation, and by 9 dpc the primitive erythropoietic potential of the YS has disappeared (Palis et al., 1999). In birds, differentiated primitive red blood cells are detected at day 1.5, and embryonic circulation starts by day 2 (Evans, 1997).

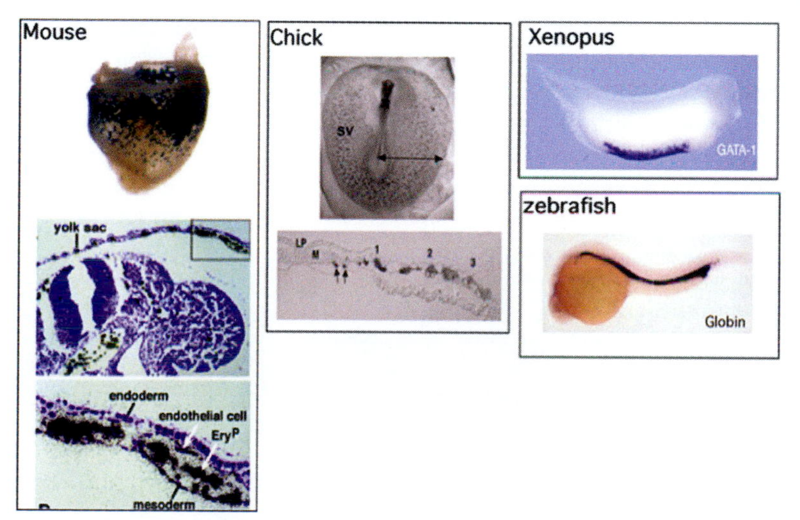

FIGURE 25.1 **The Site of Primitive Hematopoiesis in Different Species Mouse:** (Upper panel) a mouse embryo (~8.5 dpc) from a ε-globin lacZ transgenic mouse line stained for expression of β-galactosidase (dark blue) in primitive erythroid cells in the yolk sac. (Mid-panel) *in situ* hybridization of an embryonic globin probe to a 10.5 dpc mouse embryo section showing the cellular composition of the yolk sac (YS). The YS is seen as a membrane surrounding the embryo and contains a series of "blood islands." Two of these (rectangle) are shown at higher magnification in the lower panel. (Lower panel) Visceral endoderm and mesoderm components are seen enveloping clusters of primitive erythroblasts (EryP) surrounded by endothelial cells. *Figures are adapted from Baron 2001, with permission.* **Chick:** (Upper panel) two-day-old (10 somite pairs) chick embryo showing the expression of the *Lmo2* gene. The dotted pattern reveals the distribution of the blood islands that will give rise to the first (primitive) generation of erythrocytes. SV: sinus venosus. (Lower panel) Cross section at the level indicated above showing the maturation steps of the blood islands (BI). Arrows indicate the hemangioblasts that will give rise to the endothelial cell (EC) and hematopoietic cell (HC). 1: This immature BI is full of HC and EC are already differentiated. 2: The BI has matured; several cells have been freed creating a hole. 3: The BI is fully mature. HC are free and detached from one another. LP: lateral plate. M: mesoderm. *Figures are adapted from Jaffredo et al. 2003, with permission.* ***Xenopus*:** *In situ* hybridization of *gata1* probe to a swimming tadpole stage embryo reveals high expression of *gata1* in the VBI (ventral blood island) region. *Figures are adapted from Mead et al. 2001, with permission.* **Zebrafish:** *In situ* hybridization of embryonic globin probe to a 24 hpf embryo reveals the ICM (intermediate cell mass) region.

In *Xenopus*, the ventral blood island (VBI) is functionally equivalent to the mammalian and avian YS, and also develops from the ventral mesoderm (Mangia et al., 1970). By 24 hours post-fertilization (hpf), *Scl, c-myb* and *Gata1* can be detected in the developing VBI (Turpen et al., 1997) (Figure 25.1). Primitive erythrocytes start to express hemoglobin at 40 hpf, and by 50 hpf the circulation is established (Mangia et al., 1970).

Zebrafish primitive blood cells are generated in the embryo proper at the ventral lateral mesoderm. The anterior lateral mesoderm (ALM) is a major site of primitive myelopoiesis, whereas the posterior lateral mesoderm (PLM) predominantly gives rise to erythrocytes and some myeloid cells. The expression of the hematopoietic marker *Scl* is first detected around the 2-somite stage (11 hpf), in bilateral stripes of cells flanking the paraxial mesoderm and a subset of these *scl*$^{+}$ cells commits to an erythorid fate by expressing *gata1* (Davidson et al., 2003). These erythroid precursors then migrate towards the trunk midline to form the intermediate cell mass (ICM) and begin to express embryonic globins at 15 hpf (Willett et al., 1999) (Figure 25.1). Between 24 and 26 hpf, the heart starts beating and the erythroblasts enter the circulation, where they subsequently mature into primitive erythrocytes (Willett et al., 1999).

Primitive erythroid cells express embryonic globin proteins. For many years, murine primitive red cells were thought to retain their nuclei and therefore be more similar to the red cells in zebrafish, amphibians, and birds, which remain nucleated throughout their life span. Using antibodies specific to embryonic globins, Kingsley et al. have found that primitive erythroblasts in mouse embryos also undergo enucleation between 12.5 and 16.5 dpc, suggesting that primitive erythrocytes in mice are more related to their definitive counterparts (Kingsley et al., 2004).

Definitive Hematopoiesis and HSC

Primitive hematopoiesis is transient and is subsequently replaced by the definitive wave of hematopoiesis. In both mammals and zebrafish, two definitive waves have been characterized. The first wave begins with committed erythromyeloid progenitors (EMP) than can only differentiate into erythroid and myeloid lineages. They lack lymphoid potential and their differentiation is independent of the Notch signaling (Bertrand et al., 2005; 2007; 2010; Yokota et al., 2006). In contrast, the second definitive wave generates HSCs that self-renew and give rise to all adult blood lineages, including lymphocytes. In mouse embryos, definitive HSCs can be detected at 9 dpc in the intra-embryonic aorta-gonad-mesonephros (AGM) region (Figure 25.1). The long-term repopulating potential has been demonstrated by transplantation into lethally irradiated adult recipients lacking endogenous HSCs (Muller et al., 1994; Medvinsky et al., 1996). The AGM-like origin of the definitive HSCs has also been demonstrated in other vertebrates, including zebrafish (Thompson et al., 1998) (Figure 25.2). Most recently, human HSCs were demonstrated to emerge first in the AGM region (Ivanovs et al., 2011) (Figure 25.1).

In addition to the AGM region, significant numbers of definitive HSCs have been detected in the mouse placenta at a similar stage of AGM hematopoiesis (Alvarez-Silva et al., 2003; Gekas et al., 2005; Ottersbach et al., 2005). Between 11.5–12.5 dpc, the placental HSC pool expands, resulting in more than 15-fold more HSCs than in the AGM region (Gekas

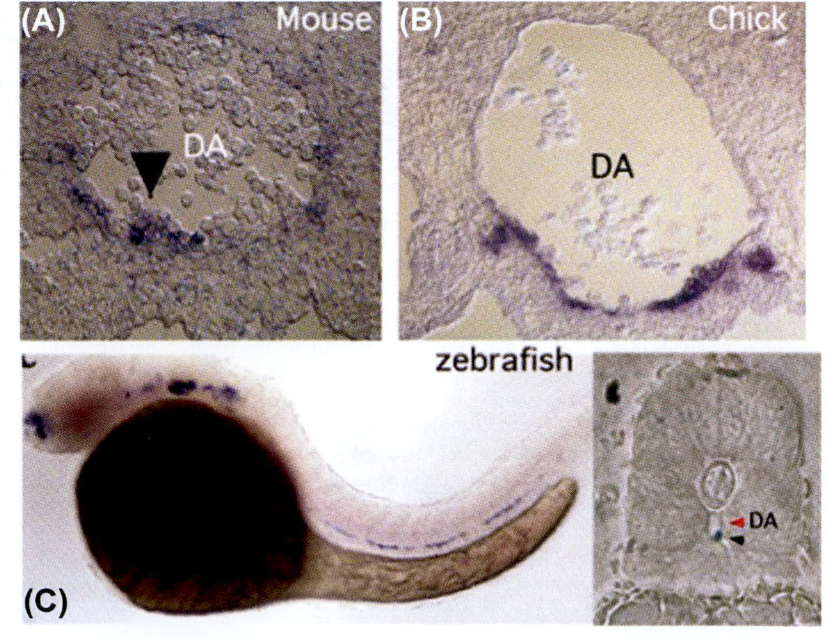

FIGURE 25.2 The Site of Definitive Hematopoiesis in Different Species. *In situ* hybridization of *runx1* probe in mouse (A), chick (B) and zebrafish (C) embryos reveals *runx1* expression in the ventral endothelium of the dorsal aorta (DA). In C, the cross section of DA is shown on right, with DA (red arrow head) and *runx1* in situ signal (blue arrow head) indicated. *Figures are adapted from Jeffredo et al. 2005, with permission.*

et al., 2005). This expansion in placental HSCs may result from continuous generation of HSCs in the placenta, or homing of HSCs to the placenta from other sites. Blood precursors with multi-lineage activity have also been found in the mouse YS, the umbilical arteries and the allantois (Samokhvalov et al., 2007; Inman et al., 2007), although further studies are required to test the capacity of *de novo* HSC generation in these sites.

At the cellular level, HSCs have long been hypothesized to originate from the "hemogenic endothelium." Early studies depicted the appearance of hematopoietic cells as clusters budding from the ventral wall of the dorsal aorta as well as the endothelium of vitelline/umbilical arteries, and expressing HSC markers such as *Flk-1, Scl* and *Runx1* (Garcia-Porrero et al., 1995; Tavian et al., 1999; Marshall et al., 1999; North et al., 1999; Jaffredo et al., 1998) (Figure 25.2). Functional analyses of these clusters have indicated the presence of HSC activity in these cells (Nishikawa et al., 1998; de Bruijn et al., 2002; North et al., 2002). However, the direct visualization of the birth of HSCs was only achieved recently by live imaging studies. By culturing mesodermal precursors derived from murine embryonic stem cells (mESCs), Eilken et al. were able to follow the fate of a single cell over several days using microscopic time-lapse imaging, and they detected the endothelial-to-hemogenic transition *in vitro* (Eilken et al., 2009). Boisset et al. used time-lapse confocal imaging to visualize the deeply located aorta in mouse embryos, and they detected the *de novo* emergence of HSCs directly from the ventral wall of the dorsal aorta starting at embryonic day 10.5 dpc (Boisset et al., 2010). A similar endothelial-to-blood transition was also visualized in zebrafish embryos and shown to be dependent on RUNX1 (Bertrand et al., 2010; Kissa and Herbomel, 2010). Using a permanent lineage tracing strategy, Bertrand et al. also demonstrated that the HSCs generated from AGM hemogenic endothelium eventually colonize the adult hematopoietic system in zebrafish (Bertrand et al., 2010). Although these studies clearly demonstrate the origin of HSCs in the endothelium, alternative models for HSC origin are not excluded. For example, a previous study identified hematopoietic cells with HSC activity within the mesenchyme underneath the aortic floor (Bertrand et al., 2005), suggesting a possibility in which HSCs emerge from mesenchymal cells and poke through the aorta endothelium to form HSC clusters along the ventral wall.

Colonization of Hematopoietic Organs by Embryonic Precursors

After their birth, HSCs migrate to the newly forming hematopoietic organs, where they reside in specific locations termed the stem cell niches, which provide a microenvironment for HSC self-renewal and differentiation. In these niches, HSCs undergo extensive proliferation and differentiation, giving rise to all blood lineages to support the growth of the organism.

In mouse embryos, the migration of HSCs is illustrated by a decrease in HSC activity in the AGM after 11 dpc and an exponential increase in HSC activity in the fetal liver (FL) from 12 dpc to 15 dpc (Muller et al., 1994; Sanchez et al., 1996; Morrison et al., 1995). The fetal liver predominates hematopoiesis from 12 dpc through birth, giving rise to all blood lineages that support the growing fetus (Delassus and Cumano, 1996; Mebius and Akashi, 2000; Mebius et al., 2001). At the end of fetal life, the spleen becomes a predominant erythropoietic organ and aids in the transition from the fetal liver to the bone marrow hematopoiesis (Sasaki and Matsumura, 1988; Godin et al., 1999). The bone marrow is the last hematopoietic organ to develop in the fetus, but it is the primary niche for HSCs in the adult. The bone marrow HSC reservoir is largely established during postnatal life when liver hematopoiesis ceases. Other vertebrates use different hematopoietic organs in fetal and adult stages. For example, in *Xenopus*, the liver is the main hematopoietic organ in both larval and adult stages (Chen and Turpen, 1995); whereas in the avian system, the site of hematopoiesis shifts directly from the AGM to the bone marrow (Dieterlen-Lievre and Martin, 1981). In zebrafish, fate mapping studies have shown that AGM-derived HSCs appear in the caudal hematopoietic tissue (CHT) from 40 hpf where they undergo extensive proliferation and further differentiation, suggesting a fetal liver-like function for the CHT (Jin et al., 2007; Murayama et al., 2006). Later on these cells were found in kidney, which is the mammalian bone marrow equivalent which maintains larval and adult hematopoiesis (Willett et al., 1999). In addition, AGM-derived cells also colonize the thymus by direct migration or through circulation to initiate lymphopoiesis (Murayama et al., 2006).

While the pathways regulating HSC migration and colonization are largely unknown, several studies have suggested that integrin-mediated adhesion is important for HSC movement in embryogenesis. HSCs lacking β1 integrin cannot colonize mouse FL (Hirsch et al., 1996), whereas loss of GPIIb integrin CD41 results in impaired retention of HSCs in their niche and increased numbers of HSCs in various embryonic sites, partially due to the loss of binding to fibronectin (Emambokus and Frampton, 2003). Chemokines and their receptors have also been implicated in the migration of HSCs. Mice deficient in the chemokine stromal cell-derived factor-1α (*SDF-1α*), or its receptor, *CXCR4*, fail to establish bone marrow hematopoiesis, although FL hematopoiesis is normal (Ara et al., 2003; Zou et al., 1998), suggesting a critical role of *SDF-1α-CXCR4* interaction in HSC migration to the bone marrow (BM). Consistent with these results, Weissman and colleagues recently demonstrated that FL HSCs migrate in response to *SDF-1α in vitro* (Wright et al., 2002). Moreover, the migratory response of FL HSCs is greatly enhanced in the presence of another chemokine, stem cell factor (*SCF*) (Christensen et al., 2004), which has well-established roles in HSC maintenance, survival and proliferation (Russell, 1979).

25.3 TRANSCRIPTIONAL REGULATION OF BLOOD DEVELOPMENT

Hematopoiesis is regulated by complex networks that integrate signaling pathways with transcription factors, chromatin modifiers, and regulatory RNAs. Here we discuss the transcriptional control of HSC generation and lineage determination during blood development, focusing on the central players which have been discovered through genetic studies in the mouse and zebrafish (Figure 25.3).

Transcription Factors Acting at the Hemagioblast Level

FLK-1 is a tyrosine kinase receptor for the vascular endothelial growth factor (VEGF). It is first expressed in the extra-embryonic mesoderm destined to give rise to both the vascular and hematopoietic components of the yolk sac blood island of the mouse embryo (Yamaguchi et al., 1993). Mice homozygous for disruption of *Flk-1* die between 8–9 dpc, lacking both the vascular and the primitive blood progenitors (Shalaby et al., 1995). Furthermore, *Flk-1* null ES cells fail to contribute to either the vessels or the blood cells in chimeric mouse embryos (Shalaby et al., 1995). These early findings suggested that *Flk-1* plays essential functions in both blood and endothelial development, perhaps at the hemangioblast stage. More recently, Lee et al. found that loss of function of Etv2 (also known as Er71), a member of the ETS (E-twenty-six) family of transcription factors, greatly reduced FLK1 expression and displayed severe blood and vessel defects that are highly reminiscent of the Flk1 null mouse phenotype (Lee et al., 2008). Further studies showed that mesodermal cells expressing both Flk-1 and Etv2 have significantly higher hemato-endothelial potential than cells expressing Flk-1 only, and enforced Etv2 expression in embryonic stem cells positively regulates hematopoietic and endothelial gene expression, suggesting that Etv2 specifies a hemangioblast fate in Flk1-expressing cells (Liu et al., 2012; Kataoka et al., 2011).

The existence of the hemangioblast is also supported by a zebrafish mutant *cloche (clo)*, which completely lacks blood and vascular cells, as well as the heart endocardium but not other mesodermal organs (Stainier et al., 1995). Gene expression analyses have revealed a near complete absence of HSC and angioblast markers including *scl, lmo2, gata2, runx1, fli1* and *flk1* (Liao et al., 1998). Although the mutant gene in *clo* has not been identified due to its telomeric location, uncovering the mutation responsible for *clo* mutant is expected to provide insight into the molecular events that direct commitment of mesoderm towards blood and endothelial fates.

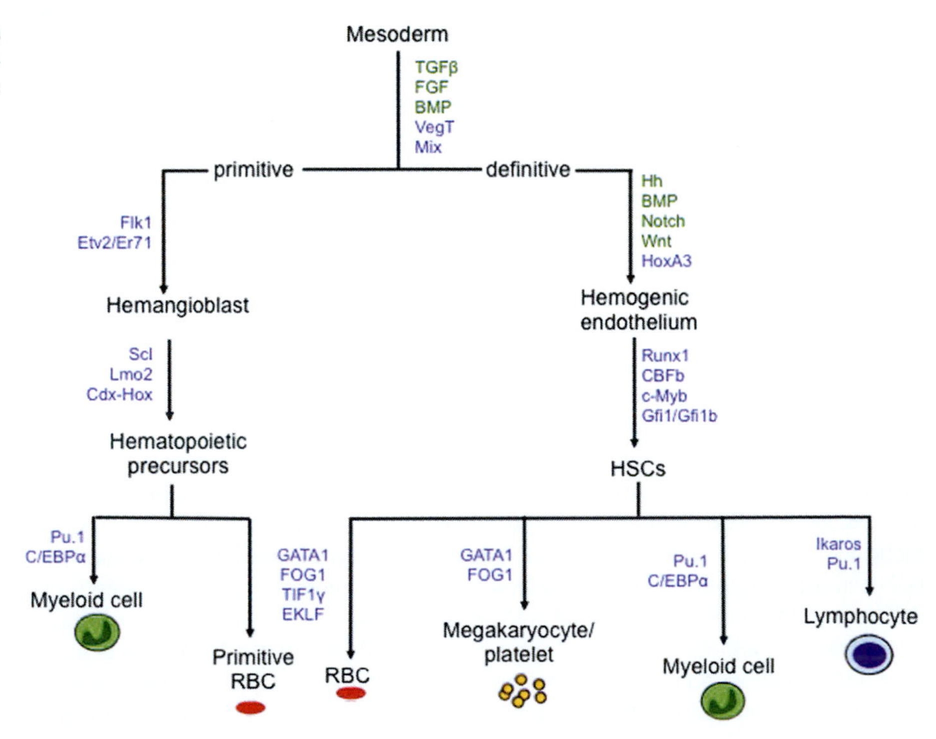

FIGURE 25.3 Regulation of Hematopoiesis in Vertebrate Development. Important signaling pathways (green) and transcription factors (blue) are listed for each developmental stage.

Master Regulators of Hematopoiesis in Early Embryos

SCL (also known as Tal-1*)* and LMO2 are essential transcription factors for both primitive and definitive hematopoiesis. SCL is a basic helix-loop-helix (bHLH) transcription factor that can directly bind to DNA at the E-box motifs (CANNTG) (Shivdasani et al., 1995), whereas LMO2 is a Lim-domain protein that acts as a bridge between DNA binding factors such as SCL and GATA-1 (Wadman et al., 1997). Disruption of either SCL or *Lmo2* in mice causes early lethality with complete lack of yolk sac hematopoiesis, and *Scl-* or *Lmo2*-null ES cells fail to contribute to any definitive hematopoietic lineage in chimeric mouse embryos (Porcher et al., 1996; Shivdasani et al., 1997; Warren et al., 1994; Yamada et al., 1998). They are also required for the late events in endothelial development, as null ES cells of either factor fail to contribute to the formation of the vitelline vessels in chimeric mouse embryos (Yamada et al., 1998; Visvader et al., 1998). Conditional deletion of *Scl* in adult mice does not disturb the HSC function, suggesting that *Scl* is not continually required for the maintenance of HSCs (Mikkola et al., 2003).

In zebrafish, Scl and Lmo2 are expressed in both hematopoietic and endothelial progenitors, and have been shown to act together to specify the hemangioblasts (Patterson et al., 2007). Mutants of *scl* and *lmo2* have been identified in zebrafish and both have severe defects in blood and vasculature with complete loss of primitive and definitive hematopoiesis (Bussmann et al., 2007; Weiss et al., 2012), demonstrating the evolutionarily conserved functions of these factors in fish and mammals.

The specification of hematopoietic cell fate also relies on the Cdx-Hox pathway, as elucidated by the zebrafish mutant *kugelig (kgg)*, which is characterized by severe anemia, shortened tail and reduced yolk tube extension (Hammerschmidt et al., 1996). The defective gene *cdx4* belongs to the caudally-related homeobox transcription factor family that is implicated in anterior-posterior axis patterning through regulation of the *hox* genes. In *kgg* mutants, the expression pattern of several *hox* genes is altered and the number of *scl*-expressing blood precursors is severely reduced, whereas the number of *flk+* angioblasts remains normal and the adjacent pronephric tissue is specified correctly (Davidson et al., 2003). Simultaneous knocking down of another Cdx member, *cdx1a*, further interrupts Hox expression patterns and results in an almost complete loss of *scl* expression (Davidson and Zon, 2006). Overexpression of *hoxb7* and *hoxa9* rescues the blood defect without correcting the tail morphology in *kgg* (Davidson et al., 2003), suggesting an integral role of the Cdx-Hox pathway in hematopoiesis. Inspired by the study in zebrafish, Wang et al. found that in mouse ES cells, ectopic *Cdx4* expression promotes hematopoietic mesoderm specification, increases hematopoietic progenitor formation, and together with *HoxB4*, enhances multi-lineage hematopoietic engraftment of lethally irradiated adult mice (Wang et al., 2005). Taken together, these studies demonstrate the specific function of the Cdx-Hox pathway in vertebrate blood development.

Regulation of Definitive HSC Formation

CBF (core binding factor) is a heterodimeric transcriptional factor consisting of a DNA-binding subunit RUNX1 (also known as AML1/CBFA2/PEBP2αB), and a non-DNA-binding subunit CBFb (Ogawa et al., 1993; Wang et al., 1993). RUNX1 belongs to the *Runx* family of transcription factors that bind to DNA through an evolutionarily conserved runt domain. CBFb associates with RUNX1 and enhances its DNA binding affinity. Mice deficient in either subunit lack all definitive blood lineages, but primitive hematopoiesis is not affected, indicating their specific roles in definitive HSCs (Okuda et al., 1996; Wang et al., 1996a,b; Sasaki et al., 1996). *Runx1* expression is found in the endothelial cells lining the ventral aspect of the dorsal aorta in the AGM region as well as other intra-aortic sites where hematopoietic clusters are thought to emerge (Figure 25.2). In *Runx1*-deficient mice, no hematopoietic clusters are generated, suggesting that *Runx1* is required for the "budding" of definitive HSCs from the intra-aortic endothelium (North et al., 1999). This was demonstrated recently by studies showing direct endothelial-to-hematopoietic transition (EHT) (Kissa and Herbomel, 2010; Lancrin et al., 2009). *Runx1*-deficiency in mouse embryonic stem cells or zebrafish embryos resulted in very few cells budding from the endothelium, and those few cells that did bud out died immediately, suggesting that Runx1 is critical for generation of HSCs from hemogenic endothelium. More recently, Gfi1 and Gfi1b have been identified as direct targets of Runx1 and critical regulators of EHT. Overexpression of Gfi1 and Gfi1b in cultured *Runx1−/−* cells can trigger the downregulation of endothelial markers and the morphological changes characteristic of the EHT. Conversely, blood progenitors in Gfi1- and Gfi1b-deficient embryos maintain the expression of endothelial genes (Lancrin et al., 2012). Additionally, HoxA3 has recently been proposed as a negative regulator of EHT to restrains hematopoietic differentiation and induces reversion of the earliest hematopoietic progenitors into endothelial cells (Iacovino et al., 2011).

C-*myb* is the cellular homolog of the *v-myb* oncogene. It is highly expressed in immature hematopoietic progenitors, but decreases as they differentiate (Shivdasani and Orkin, 1996). Mice lacking *c-Myb* have normal primitive hematopoiesis but marked loss of definitive progenitors in the fetal liver, resulting in death at 15 dpc (Mucenski et al., 1991). AGM cells from *c-Myb*$^{-/-}$ embryos do not generate hematopoietic cells *in vitro* (Mukouyama et al., 1999), indicating an essential function of *c-Myb* in early definitive hematopoiesis.

Similar to that in mouse embryos, zebrafish definitive HSCs in the AGM region are also marked with *runx1* and *c-myb* expression. Recently, both *runx1* and *c-myb* mutants were identified in zebrafish. Neither mutation affects primitive hematopoiesis but both severely impair AGM HSCs. Surprisingly, while *c-myb* mutants completely fail in definitive hematopoiesis, and thus lose all adult blood cells (Soza-Ried et al., 2010), ~ 20% of *runx1* mutant embryos are able to recover from the larval "bloodless" phase and generate CD41+ precursors that contribute to multi-lineage hematopoiesis in adulthood (Sood et al., 2010). The mutation of the *runx1* gene generates a premature truncation in the runt domain, and thus removes most of the residues important in Runx1 activity, suggesting a loss of Runx1 function in these mutants. One possible explanation for the recovery of hematopoiesis in the absence of functional Runx1 is that other Runx family members (Runx2a, Runx2b, and Runx3) may compensate. As noted, Runx3 was found to regulate both primitive and definitive hematopoiesis in zebrafish (Kalev-Zylinska et al., 2003). Knocking down Runx3 decreases AGM HSCs, whereas overexpression of Runx3 leads to an increase. Whether or not there are any Runx1-independent pathways existing in mammals is not clear but Runx1 knockout embryonic stem cells have been reported to be capable of contributing to adult hematopoiesis in chimeric mice at a very low level (Kundu et al., 2005).

The ontogeny of HSCs requires a coordinated interaction between the signal transduction pathways and transcription factors. In zebrafish embryos, inhibition of the Hedgehog (Hh) pathway by genetic mutants or chemical inhibition was found to greatly reduce *runx1*+ HSCs in the AGM while leaving the primitive erythrocytes intact, suggesting a specific requirement of Hh signaling in definitive hematopoiesis (Gering and Patient, 2005). Impaired migration of dorsal aorta angioblasts in these embryos indicates that the loss of HSCs results from improper patterning of the aorta. The positive role of Hh in AGM HSC induction has also been found in murine organ culture (Peeters et al., 2009). More recently, BMP signaling has been found to work together with the Hh pathway to polarize the dorsal aorta for HSC emergence and is required for *runx1* expression in the ventral arterial endothelium (Wilkinson et al., 2009). Other signaling pathways required for artery identity and HSC induction are VEGF and Notch (Gering and Patient, 2005; Burns et al., 2005). In the zebrafish *mindbomb* mutant, which lacks Notch signaling, the expression of both artery markers and HSC markers are greatly reduced, but this can be rescued by Runx1 overexpression, placing *runx1* downstream or in parallel with the Notch pathway (Burns et al., 2005). More recently, Notch signaling was found to be required even earlier for the establishment of the hemogenic endothelium during somitogenesis in zebrafish. These early Notch signals are controlled by non-canonical Wnt signaling and are mediated by somatic expression of the Notch ligands *deltaC (dlc)* and *deltaD (dld)*, therefore linking Notch signaling with non-canonical Wnt pathways (Clements et al., 2011).

Regulation of Lineage Specification

GATA1 and its cofactor FOG-1 are required for the differentiation of erythrocytes and megakaryocytes. As the founding member of the GATA family of zinc finger transcription factors, GATA1 has been found to bind to virtually all characterized erythroid-specific genes. Disrupting GATA1 function in mice results in embryonic lethality at 11.5 dpc from fatal anemia due to a block in erythroid differentiation accompanied by apoptosis (Pevny et al., 1995; Fujiwara et al., 1996). In addition, selective knockout of *Gata1* expression in megakaryocytes blocks megakaryocyte differentiation (Shivdasani et al., 1997). The regulation of most, but not all, GATA-1-dependent genes requires binding to FOG-1, a multi-type zinc finger protein that is co-expressed with *Gata1* during hematopoietic development. Mutant GATA1, which is unable to interact with FOG-1, fails to support erythroid and megakaryocytic differentiation (Crispino et al., 1999; Nichols et al., 2000), and targeted disruption of *Fog-1* in mice leads to phenotypes similar to *Gata1* knockout mice (Tsang et al., 1998), suggesting a crucial role for GATA1-FOG1 interaction in erythropoiesis and megakaryopoiesis. In zebrafish, the *gata1* gene is mutated in the *vlad tepes (vlt)* mutant. These mutant embryos have normal expression of early hematopoietic markers such as *scl* and *lmo2*, but a great reduction or absence of many erythroid markers throughout development, leading to a complete loss of red cells (Lyons et al., 2002). Furthermore, a cross-inhibitory action was found between Gata1 and the myeloid transcription factor Pu.1. Loss of *gata1* function transforms erythroid precursors into myeloid cells, and conversely, *pu.1* knockdown switches myeloid cells to red cell fate (Galloway et al., 2005; Rhodes et al., 2005).

Studies of transcription regulators for lineage differentiation have been mainly focused on factors that affect transcription initiation of polymerase II (Pol II). However, recent genome-wide studies have revealed a critical role for transcription elongation in gene regulation, suggested by the observation that Pol II is commonly "paused" on a large number of developmentally regulated genes (Guenther et al., 2007; Muse et al., 2007; Zeitlinger et al., 2007). A study of the zebrafish mutant *moonshine (mon)* directly supports this view (Bai et al., 2010). The mutant gene in *mon* is *tif1γ* (transcriptional intermediary factor 1 gamma), which encodes a ubiquitously expressed transcription cofactor that is highly enriched in the ICM blood island (Ransom et al., 2004). Loss of *tif1γ* results in a profound anemia in zebrafish embryos due to the blockage of erythroid differentiation. Through a genetic suppressor screen approach, Bai et al. found that the erythroid gene expression in *mon* can be restored by removing the factors that induce Pol II pausing. This study suggests that by physically interacting with both Scl-Gata1 transcription complexes and Pol II elongation factors, TIF1γ recruits Pol II elongation machinery to erythroid genes and releases Pol II from pausing to promote their expression. This function is conserved in murine erythropoiesis, as revealed by defective erythropoiesis in bone marrow when *tif1γ* is deleted (Bai et al., 2013). TIF1γ also plays an inhibitory role in myelopoiesis. The *mon* embryos have increased definitive myeloid cells and conditional knockout mice develop a MPD (myeloid proliferative disease)-like phenotype (Bai et al., 2013; Monteiro et al., 2011; Aucagne et al., 2011; Kusy et al., 2011). Moreover, the *tif1γ* gene is found silenced in a subset of chronic myeloid leukemia patients (Aucagne et al., 2011), suggesting a tumor suppressor function of TIF1γ in myeloid leukemia. The role of TIF1γ in gene regulation may be broader than regulating Pol II elongation as a recent study has found that it can also function as a chromatin factor to mediate TGFβ signaling in mouse ES cells by recognizing specific histone modification markers (Xi et al., 2011).

Epigenetic Regulators of Blood Development

Increasing evidence suggests that epigenetic regulators act in concert with transcription factors and signaling pathways to regulate the development and homeostasis of the hematopoietic system. The GATA-1/FOG1 complex has been shown to physically interact with subunits of the nucleosome remodeling and histone deacetylase (NuRD) complex through the N-terminus motif in FOG1 (Hong et al., 2005). Disruption of the FOG-1/NuRD interaction resulted in defects similar to *Gata1*- and *Fog1*-deficient mice, including anemia and macrothrombocytopaenia (Hong et al., 2005; Miccio et al., 2010; Gao et al., 2010). Gene expression analysis has revealed an essential function for NuRD during both the repression and activation of select GATA-1/FOG-1 target genes, demonstrating cooperation between lineage specific transcription factors and chromatin remodeling factors in the re-enforcement of lineage commitment. The NuRD complex has also been shown to interact with Ikaros (Kim et al., 1999), a zinc finger DNA-binding protein that serves as a key regulators of lymphocyte development. A recent study by Zhang et al. found that Ikaros regulates both the genomic distribution and the functional activity of the NuRD complex (Zhang et al., 2012), suggesting that site-specific transcription factors not only recruit chromatin factors to the target genes but can also modify their remodeling activity.

Brg1, the *ATPase* subunit of the SWI/SNF chromatin remodeling complex, has also been shown to participate in hematopoiesis to specifically regulate erythroid differentiation (Bultman et al., 2005). Mouse embryos carrying a hypomorphic Brg1 mutation that still retains ATPase activity fail to activate β-globin gene expression, leading to a differentiation block in the development of the erythroid lineage and embryonic death at 11.5 dpc. Further studies showed that the Brg1 mutation abrogates a cell type-specific loop between the β-globin locus control region and the downstream βmajor promoter (Kim et al., 2009), suggesting that BRG1 functions early in chromatin domain activation to mediate looping in a way that is independent of its ATPase activity.

The Mixed-Lineage Leukemia (MLL) gene encodes a Trithorax-related chromatin factor that is involved in histone H3 lysine 4 methylation. The yolk sac and fetal livers of $Mll^{-/-}$ mouse embryos are reduced in cellularity and have reduced colony-forming capacity (Hess et al., 1997; Yagi et al., 1998). Cells from the AGM region of Mll-deficient embryos are unable to repopulate the bone marrow of irradiated mice and $Mll^{-/-}$ ES cells show no contribution to any of the hematopoietic lineages seen in adult chimeras, suggesting a loss of definitive HSC activity (Ernst et al., 2004). Later studies using a conditional knockout model revealed the requirement for MLL in adult hematopoiesis by regulating HSC self-renewal (McMahon et al., 2007).

In zebrafish, the morpholino knockdown approach and available genetic mutants have facilitated the identification of novel chromatin players in hematopoiesis. Li et al. have found that the Mta3-NuRD complex is essential for the initiation of primitive hematopoiesis in zebrafish embryos (Li et al., 2009). Inhibition of NuRD activity through depletion of Mta3 or HDAC inhibitors abolishes primitive hematopoietic lineages and causes abnormal angiogenesis, whereas overexpression of NuRD components enhances the expression of *scl* and *lmo2* in zebrafish embryos. In addition, using an insertional mutagenesis screen for factors affecting definitive HSCs in the AGM region, Burns et al. have shown that *hdac1* acts downstream of Notch signaling but upstream or in parallel to *runx1* to promote AGM HSC formation (Burns et al., 2009).

To systematically identify epigenetic regulators of zebrafish hematopoiesis, Huang et al. performed the first genetic screen for chromatin factors using a large-scale morpholino knockdown approach (Huang, 2012). By injecting morpholinos targeting over 400 chromatin factors into the zebrafish embryos, they were able to identify more than 70 factors that affect primitive erythropoiesis, or definitive HSCs, or both. Their hits included some of the chromatin factors known from previous studies to regulate hematopoiesis, such as components of SWI/SNF complexes and HDACs, but most are novel regulators that have not been studied in this process. One of them is CHD7, a chromodomain chromatin remodeler that is mutated in patients with CHARGE syndrome (Vissers et al., 2004). Knocking down *chd7* significantly increased both primitive and definitive blood production, and this effect was cell autonomous as determined by blastula transplantation assays. While the mechanism by which CHD7 represses hematopoiesis is still the subject of the ongoing research, their study has provided great insights into the role of chromatin factors in regulating gene transcription in hematopoietic cells.

25.4 THERAPEUTIC USE OF HSCS

Understanding the basic biology of blood development and HSC formation has paved the way for the clinical usage of HSCs. The capacity of HSCs to self-renew and ultimately give rise to all blood lineages make them uniquely situated as a powerful tool for the treatment of a variety of blood diseases that are untreatable by traditional approaches. The best example is HSC transplantation, which has been used for the treatment of cancer-related hematopoietic deficiency and BM failure states.

Transplantation of HSCs first of all requires their purification. This is achieved by fluorescence activated cell sorting (FACS) of HSCs with monoclonal antibodies that recognize their unique surface markers. The original isolation strategy for mouse HSCs selects a population marked by the composite phenotype of c-kit$^+$, lineage markers $^{-/lo}$, and Sca-1$^+$; termed the LKS population (Okada et al., 1992). However, only ~10% of LKS cells contain HSCs that are capable of long-term hematopoietic reconstitution. To obtain HSCs of higher purity, several additional selection markers have been developed recently, such as SLAM markers, CD34, Flk2 and the fluorescent vital dye Hoechst (Osawa et al., 1996; Christensen et al., 2001; Goodell et al., 1997; Adolfsson et al, 2001; Kiel et al., 2005). Different combinations of these new markers have been demonstrated to further enrich HSCs within the LKS population and shown to increase the frequency of long-term hematopoietic reconstitution in lethally irradiated recipients. There are significant differences in surface marker expression between mouse and human HSCs. In humans, the combination of CD34$^+$ CD38$^-$ Thy1$^+$ CD45RA$^-$ results in a highly purified HSC population (Lansdorp et al., 1990; Bhatia et al., 1997; Baum et al., 1992; Negrin et al., 2000). To further enrich HSCs for transplantation, injection of growth factors such as granulocyte/macrophage colony-stimulating factor (GM-CSF) or granulocyte colony-stimulating factor (G-CSF) is commonly used to stimulate the proliferation of HSCs and mobilize HSCs out of the BM into the peripheral circulation. This is followed by collection and sorting of HSCs from the peripheral blood (Murray et al., 1995).

A major obstacle for the clinical use of HSC is that the fact that the cells are extremely rare. Only 1 in 10^6 cells in the human BM is a transplantable HSC (Wang et al., 1997). Due to the intrinsic self-renewal ability of HSCs observed *in vivo*, great efforts have been made to identify factors that can induce HSC expansion *in vitro*. Several factors, including insulin-like growth factor 2 (IGF-2), IGF binding protein 2 (IGFBP-2) and angiopoietin-like proteins (Angptls), have been shown to greatly induce the expansion of both mouse and human HSCs when combined with certain cytokines in culture (Zhang et al., 2004; 2006; 2008; Huynh et al., 2008). In addition, a high-throughput chemical screen in zebrafish has identified prostaglandin E_2 (PGE_2) as a positive regulator of HSC self-renewal (North et al., 2007). Preclinical analyses have demonstrated the therapeutic potential of PGE2 treatment using human and nonhuman primate HSCs (Goessling et al., 2011), and a clinical trial has been initiated to test if PGE2 can enhance human HSC engraftment in patients treated with cord blood transplants.

HSC transplantation is also valuable for gene therapy. For example, genetic hematopoietic disorders that are caused by mutation at a single locus can be treated by introducing a functional copy of the gene into the isolated HSCs, followed by the transplantation of these "corrected" HSCs back into the patient. Such gene therapy approaches have been performed in clinical trials on patients with severe combined immunodeficiency (SCID) disease and immune restoration has been observed after the transplantation (Aiuti et al., 2009; Baum, 2011). Due to the difficulty of HSC expansion *in vitro* as discussed above, an alternative source is induced pluripotent stem cells (iPSCs) (Takahashi et al., 2007; Yu et al., 2007). These somatic cells can be induced to acquire pluripotent stem cell characteristics by the ectopic expression of stem cell factors and then they can differentiate into HSCs *in vitro*. As a proof of principle, a study was performed in a humanized mouse model of sickle cell anemia (Hanna et al., 2007). Sickle iPSCs were corrected by gene-specific targeting and differentiated into hematopoietic progenitors *in vitro*. After transplantation into lethally irradiated mice, enhanced α-globin levels and the absence of sickle cells in the peripheral blood were observed. In another study, iPSCs were derived from a β-thalassemia patient. Transplantation of the genetically corrected iPSCs-derived hematopoietic progenitors into sub-lethally irradiated immune deficient mice showed the presence of human β-globin (Wang et al., 2012). These studies demonstrated the potential clinical use of iPSCs in the treatment of hematological disorders.

25.5 CLINICAL RELEVANCE

In this chapter, we have reviewed the development of vertebrate blood system. Using different model organisms, many factors controlling hematopoiesis have been identified and a highly conserved genetic program is beginning to emerge. Studies in different vertebrate models in the past few decades have significantly contributed to our understanding of the nature of HSCs. We believe that future studies using these model organisms will continually help us to understand the mechanisms by which HSCs differentiate into mature, functional cells. This will ultimately improve the treatment of human blood diseases.

- Many factors regulating normal hematopoiesis are implicated in hematopoietic malignancies through chromosomal translocations or somatic mutations. Elucidating their function in normal blood development is therefore crucial for us to understand the molecular mechanisms behind leukemiagenesis.
- Genetic manipulations in mice and zebrafish to model human blood diseases have greatly advanced our understanding of disease pathogenesis and contributed to the discovery of novel therapeutic targets.
- Novel pharmaceutical compounds can be discovered using zebrafish as a drug screening platform due to the easy manipulation of embryos, *in vivo* physiological conditions and conserved biology.
- Understanding the mechanisms of HSC formation and self-renewal will ultimately improve the treatment of hematological disorders, as well as shed light on the use of other stem cells in clinical settings.

RECOMMENDED RESOURCES

Davidson, A.J., Zon, L.I., 2004. The 'definitive' (and 'primitive') guide to zebrafish hematopoiesis. Oncogene 23 (43), 7233–7246.

Orkin, S., Zon, L.I., 2008. Hematopoiesis: an evolving paradigm for stem cell biology. Cell 132 (4), 631–644.

Doulatov, S., Notta, F., Laurenti, E., Dick, J.E., 2012. Hematopoiesis: a human perspective. Cell Stem Cell 10 (2), 120–136.

REFERENCES

Adolfsson, J., Borge, O.J., Bryder, D., Theilgaard-Monch, K., Astrand-Grundstrom, I., Sitnicka, E., et al., 2001. Upregulation of Flt3 expression within the bone marrow $Lin^{-}Sca1^{+}c\text{-}kit^{+}$ stem cell compartment is accompanied by loss of self-renewal capacity. Immunity 15 (4), 659–669.

Aiuti, A., Brigida, I., Ferrua, F., Cappelli, B., Chiesa, R., Marktel, S., et al., 2009. Hematopoietic stem cell gene therapy for adenosine deaminase deficient-SCID. Immunol. Res. 44 (1–3), 150–159.

Alvarez-Silva, M., Belo-Diabangouaya, P., Salaun, J., Dieterlen-Lievre, F., 2003. Mouse placenta is a major hematopoietic organ. Development 130 (22), 5437–5444.

Ara, T., Itoi, M., Kawabata, K., Egawa, T., Tokoyoda, K., Sugiyama, T., et al., 2003. A role of CXC chemokine ligand 12/stromal cell-derived factor-1/pre-B cell growth stimulating factor and its receptor CXCR4 in fetal and adult T cell development *in vivo*. J. Immunol. 170 (9), 4649–4655.

Aucagne, R., Droin, N., Paggetti, J., Lagrange, B., Largeot, A., Hammann, A., et al., 2011. Transcription intermediary factor 1gamma is a tumor suppressor in mouse and human chronic myelomonocytic leukemia. J. Clin. Invest. 121 (6), 2361–2370.

Bai, X., Kim, J., Yang, Z., Jurynec, M.J., Akie, T.E., Lee, J., et al., 2010. TIF1gamma controls erythroid cell fate by regulating transcription elongation. Cell 142 (1), 133–143.

Bai, X., Trowbridge, J.J., Riley, E., Lee, J.A., DiBiase, A., Kaartinen, V.M., et al., 2013. TiF1-gamma plays an essential role in murine hematopoiesis and regulates transcriptional elongation of erythroid genes. Dev. Biol. 373 (2), 422–430.

Baum, C., 2011. Gene therapy for SCID-X1: focus on clinical data. Mol. Ther. 19 (12), 2103–2104.

Baum, C.M., Weissman, I.L., Tsukamoto, A.S., Buckle, A.M., Peault, B., 1992. Isolation of a candidate human hematopoietic stem-cell population. Proc. Natl. Acad. Sci. USA 89 (7), 2804–2808.

Bertrand, J.Y., Chi, N.C., Santoso, B., Teng, S., Stainier, D.Y., Traver, D., 2010. Haematopoietic stem cells derive directly from aortic endothelium during development. Nature 464 (7285), 108–111.

Bertrand, J.Y., Giroux, S., Golub, R., Klaine, M., Jalil, A., Boucontet, L., et al., 2005. Characterization of purified intraembryonic hematopoietic stem cells as a tool to define their site of origin. Proc. Natl. Acad. Sci. USA 102 (1), 134–139.

Bertrand, J.Y., Kim, A.D., Violette, E.P., Stachura, D.L., Cisson, J.L., Traver, D., 2007. Definitive hematopoiesis initiates through a committed erythromyeloid progenitor in the zebrafish embryo. Development 134 (23), 4147–4156.

Bhatia, M., Wang, J.C., Kapp, U., Bonnet, D., Dick, J.E., 1997. Purification of primitive human hematopoietic cells capable of repopulating immune-deficient mice. Proc. Natl. Acad. Sci. USA 94 (10), 5320–5325.

Boisset, J.C., van Cappellen, W., Andrieu-Soler, C., Galjart, N., Dzierzak, E., Robin, C., 2010. *In vivo* imaging of haematopoietic cells emerging from the mouse aortic endothelium. Nature 464 (7285), 116–120.

Bultman, S.J., Gebuhr, T.C., Magnuson, T., 2005. A Brg1 mutation that uncouples ATPase activity from chromatin remodeling reveals an essential role for SWI/SNF-related complexes in beta-globin expression and erythroid development. Genes Dev. 19 (23), 2849–2861.

Burns, C.E., Galloway, J.L., Smith, A.C., Keefe, M.D., Cashman, T.J., Paik, E.J., et al., 2009. A genetic screen in zebrafish defines a hierarchical network of pathways required for hematopoietic stem cell emergence. Blood 113 (23), 5776–5782.

Burns, C.E., Traver, D., Mayhall, E., Shepard, J.L., Zon, L.I., 2005. Hematopoietic stem cell fate is established by the Notch-Runx pathway. Genes Dev. 19 (19), 2331–2342.

Bussmann, J., Bakkers, J., Schulte-Merker, S., 2007. Early endocardial morphogenesis requires Scl/Tal1. PLoS Genet. 3 (8), e140.

Chaddah, M.R., Wu, D.D., Phillips, R.A., 1996. Variable self-renewal of reconstituting stem cells in long-term bone marrow cultures. Exp. Hematol. 24 (4), 497–508.

Chen, X.D., Turpen, J.B., 1995. Intraembryonic origin of hepatic hematopoiesis in *Xenopus laevis*. J. Immunol. 154 (6), 2557–2567.

Choi, K., Kennedy, M., Kazarov, A., Papadimitriou, J.C., Keller, G., 1998. A common precursor for hematopoietic and endothelial cells. Development 125 (4), 725–732.

Christensen, J.L., Weissman, I.L., 2001. Flk-2 is a marker in hematopoietic stem cell differentiation: a simple method to isolate long-term stem cells. Proc. Natl. Acad. Sci. USA 98 (25), 14541–14546.

Christensen, J.L., Wright, D.E., Wagers, A.J., Weissman, I.L., 2004. Circulation and chemotaxis of fetal hematopoietic stem cells. PLoS Biol. 2 (3), E75.

Clement, J.H., Fettes, P., Knochel, S., Lef, J., Knochel, W., 1995. Bone morphogenetic protein 2 in the early development of *Xenopus laevis*. Mech. Dev. 52 (2–3), 357–370.

Clements, W.K., Kim, A.D., Ong, K.G., Moore, J.C., Lawson, N.D., Traver, D., 2011. A somitic Wnt16/Notch pathway specifies haematopoietic stem cells. Nature 474 (7350), 220–224.

Crispino, J.D., Lodish, M.B., MacKay, J.P., Orkin, S.H., 1999. Use of altered specificity mutants to probe a specific protein-protein interaction in differentiation: the GATA-1:FOG complex. Mol. Cell. 3 (2), 219–228.

Dale, L., Howes, G., Price, B.M., Smith, J.C., 1992. Bone morphogenetic protein 4: a ventralizing factor in early *Xenopus* development. Development 115 (2), 573–585.

Davidson, A.J., Ernst, P., Wang, Y., Dekens, M.P., Kingsley, P.D., Palis, J., et al., 2003. cdx4 mutants fail to specify blood progenitors and can be rescued by multiple hox genes. Nature 425 (6955), 300–306.

Davidson, A.J., Zon, L.I., 2006. The caudal-related homeobox genes cdx1a and cdx4 act redundantly to regulate hox gene expression and the formation of putative hematopoietic stem cells during zebrafish embryogenesis. Dev. Biol. 292 (2), 506–518.

de Bruijn, M.F., Ma, X., Robin, C., Ottersbach, K., Sanchez, M.J., Dzierzak, E., 2002. Hematopoietic stem cells localize to the endothelial cell layer in the midgestation mouse aorta. Immunity 16 (5), 673–683.

Delassus, S., Cumano, A., 1996. Circulation of hematopoietic progenitors in the mouse embryo. Immunity 4 (1), 97–106.

Dick, A., Hild, M., Bauer, H., Imai, Y., Maifeld, H., Schier, A.F., et al., 2000. Essential role of Bmp7 (snailhouse) and its prodomain in dorsoventral patterning of the zebrafish embryo. Development 127 (2), 343–354.

Dieterlen-Lievre, F., Martin, C., 1981. Diffuse intraembryonic hemopoiesis in normal and chimeric avian development. Dev. Biol. 88 (1), 180–191.

Eichmann, A., Corbel, C., Nataf, V., Vaigot, P., Breant, C., Le Douarin, N.M., 1997. Ligand-dependent development of the endothelial and hemopoietic lineages from embryonic mesodermal cells expressing vascular endothelial growth factor receptor 2. Proc. Natl. Acad. Sci. USA 94 (10), 5141–5146.

Eilken, H.M., Nishikawa, S., Schroeder, T., 2009. Continuous single-cell imaging of blood generation from haemogenic endothelium. Nature 457 (7231), 896–900.

Emambokus, N.R., Frampton, J., 2003. The glycoprotein IIb molecule is expressed on early murine hematopoietic progenitors and regulates their numbers in sites of hematopoiesis. Immunity 19 (1), 33–45.

Ernst, P., Fisher, J.K., Avery, W., Wade, S., Foy, D., Korsmeyer, S.J., 2004. Definitive hematopoiesis requires the mixed-lineage leukemia gene. Dev. Cell. 6 (3), 437–443.

Evans, T., 1997. Developmental biology of hematopoiesis. Hematol. Oncol. Clin. North Am. 11 (6), 1115–1147.

Fainsod, A., Steinbeisser, H., De Robertis, E.M., 1994. On the function of BMP-4 in patterning the marginal zone of the *Xenopus* embryo. EMBO. J. 13 (21), 5015–5025.

Fleischman, R.A., Custer, R.P., Mintz, B., 1982. Totipotent hematopoietic stem cells: normal self-renewal and differentiation after transplantation between mouse fetuses. Cell 30 (2), 351–359.

Fujiwara, Y., Browne, C.P., Cunniff, K., Goff, S.C., Orkin, S.H., 1996. Arrested development of embryonic red cell precursors in mouse embryos lacking transcription factor GATA-1. Proc. Natl. Acad. Sci. USA 93 (22), 12355–12358.

Galloway, J.L., Wingert, R.A., Thisse, C., Thisse, B., Zon, L.I., 2005. Loss of gata1 but not gata2 converts erythropoiesis to myelopoiesis in zebrafish embryos. Dev. Cell. 8 (1), 109–116.

Gao, Z., Huang, Z., Olivey, H.E., Gurbuxani, S., Crispino, J.D., Svensson, E.C., 2010. FOG-1-mediated recruitment of NuRD is required for cell lineage re-enforcement during haematopoiesis. EMBO. J. 29 (2), 457–468.

Garcia-Porrero, J.A., Godin, I.E., Dieterlen-Lievre, F., 1995. Potential intraembryonic hemogenic sites at pre-liver stages in the mouse. Anat. Embryol. 192 (5), 425–435.

Gekas, C., Dieterlen-Lievre, F., Orkin, S.H., Mikkola, H.K., 2005. The placenta is a niche for hematopoietic stem cells. Dev. Cell. 8 (3), 365–375.

Gering, M., Patient, R., 2005. Hedgehog signaling is required for adult blood stem cell formation in zebrafish embryos. Dev. Cell. 8 (3), 389–400.

Godin, I., Garcia-Porrero, J.A., Dieterlen-Lievre, F., Cumano, A., 1999. Stem cell emergence and hemopoietic activity are incompatible in mouse intraembryonic sites. J. Exp. Med. 190 (1), 43–52.

Goessling, W., Allen, R.S., Guan, X., Jin, P., Uchida, N., Dovey, M., et al., 2011. Prostaglandin E2 enhances human cord blood stem cell xenotransplants and shows long-term safety in preclinical nonhuman primate transplant models. Cell Stem Cell 8 (4), 445–458.

Goodell, M.A., Rosenzweig, M., Kim, H., Marks, D.F., DeMaria, M., Paradis, G., et al., 1997. Dye efflux studies suggest that hematopoietic stem cells expressing low or undetectable levels of CD34 antigen exist in multiple species. Nat. Med. 3 (12), 1337–1345.

Guenther, M.G., Levine, S.S., Boyer, L.A., Jaenisch, R., Young, R.A., 2007. A chromatin landmark and transcription initiation at most promoters in human cells. Cell 130 (1), 77–88.

Hammerschmidt, M., Pelegri, F., Mullins, M.C., Kane, D.A., Brand, M., van Eeden, F.J., et al., 1996. Mutations affecting morphogenesis during gastrulation and tail formation in the zebrafish, *Danio rerio*. Development 123, 143–151.

Hanna, J., Wernig, M., Markoulaki, S., Sun, C.W., Meissner, A., Cassady, J.P., et al., 2007. Treatment of sickle cell anemia mouse model with iPS cells generated from autologous skin. Science 318 (5858), 1920–1923.

Harrison, D.E., Astle, C.M., Lerner, C., 1988. Number and continuous proliferative pattern of transplanted primitive immunohematopoietic stem cells. Proc. Natl. Acad. Sci. USA 85 (3), 822–826.

Hess, J.L., Yu, B.D., Li, B., Hanson, R., Korsmeyer, S.J., 1997. Defects in yolk sac hematopoiesis in Mll-null embryos. Blood 90 (5), 1799–1806.

Hild, M., Dick, A., Rauch, G.J., Meier, A., Bouwmeester, T., Haffter, P., et al., 1999. The smad5 mutation somitabun blocks Bmp2b signaling during early dorsoventral patterning of the zebrafish embryo. Development 126 (10), 2149–2159.

Hirsch, E., Iglesias, A., Potocnik, A.J., Hartmann, U., Fassler, R., 1996. Impaired migration but not differentiation of haematopoietic stem cells in the absence of beta1 integrins. Nature 380 (6570), 171–175.

Ho, R.K., Kane, D.A., 1990. Cell-autonomous action of zebrafish spt-1 mutation in specific mesodermal precursors. Nature 348 (6303), 728–730.

Hogan, B.M., Layton, J.E., Pyati, U.J., Nutt, S.L., Hayman, J.W., Varma, S., et al., 2006. Specification of the primitive myeloid precursor pool requires signaling through Alk8 in zebrafish. Curr. Biol. 16 (5), 506–511.

Hong, W., Nakazawa, M., Chen, Y.Y., Kori, R., Vakoc, C.R., Rakowski, C., et al., 2005. FOG-1 recruits the NuRD repressor complex to mediate transcriptional repression by GATA-1. EMBO. J. 24 (13), 2367–2378.

Huang, H.-T., 2012. Epigenetic Regulation of Hematopoiesis in Zebrafish. PhD Thesis.

Huber, T.L., Kouskoff, V., Fehling, H.J., Palis, J., Keller, G., 2004. Haemangioblast commitment is initiated in the primitive streak of the mouse embryo. Nature 432 (7017), 625–630.

Huynh, H., Iizuka, S., Kaba, M., Kirak, O., Zheng, J., Lodish, H.F., et al., 2008. Insulin-like growth factor-binding protein 2 secreted by a tumorigenic cell line supports *ex vivo* expansion of mouse hematopoietic stem cells. Stem Cells 26 (6), 1628–1635.

Iacovino, M., Chong, D., Szatmari, I., Hartweck, L., Rux, D., Caprioli, A., et al., 2011. HoxA3 is an apical regulator of haemogenic endothelium. Nat. Cell. Biol. 13 (1), 72–78.

Inman, K.E., Downs, K.M., 2007. The murine allantois: emerging paradigms in development of the mammalian umbilical cord and its relation to the fetus. Genesis 45 (5), 237–258.

Ivanovs, A., Rybtsov, S., Welch, L., Anderson, R.A., Turner, M.L., Medvinsky, A., 2011. Highly potent human hematopoietic stem cells first emerge in the intraembryonic aorta-gonad-mesonephros region. J. Exp. Med. 208 (12), 2417–2427.

Jaffredo, T., Gautier, R., Eichmann, A., Dieterlen-Lievre, F., 1998. Intraaortic hemopoietic cells are derived from endothelial cells during ontogeny. Development 125 (22), 4575–4583.

Jin, H., Xu, J., Wen, Z., 2007. Migratory path of definitive hematopoietic stem/progenitor cells during zebrafish development. Blood 109 (12), 5208–5214.

Jones, C.M., Lyons, K.M., Lapan, P.M., Wright, C.V., Hogan, B.L., 1992. DVR-4 (bone morphogenetic protein-4) as a posterior-ventralizing factor in *Xenopus* mesoderm induction. Development 115 (2), 639–647.

Jordan, C.T., Lemischka, I.R., 1990. Clonal and systemic analysis of long-term hematopoiesis in the mouse. Genes Dev. 4 (2), 220–232.

Kalev-Zylinska, M.L., Horsfield, J.A., Flores, M.V., Postlethwait, J.H., Chau, J.Y., Cattin, P.M., et al., 2003. Runx3 is required for hematopoietic development in zebrafish. Dev. Dyn. 228 (3), 323–336.

Kataoka, H., Hayashi, M., Nakagawa, R., Tanaka, Y., Izumi, N., Nishikawa, S., et al., 2011. Etv2/ER71 induces vascular mesoderm from Flk1$^+$PDGFRalpha$^+$ primitive mesoderm. Blood 118 (26), 6975–6986.

Kiel, M.J., Yilmaz, O.H., Iwashita, T., Yilmaz, O.H., Terhorst, C., Morrison, S.J., 2005. SLAM family receptors distinguish hematopoietic stem and progenitor cells and reveal endothelial niches for stem cells. Cell 121 (7), 1109–1121.

Kim, J., Sif, S., Jones, B., Jackson, A., Koipally, J., Heller, E., et al., 1999. Ikaros DNA-binding proteins direct formation of chromatin remodeling complexes in lymphocytes. Immunity 10 (3), 345–355.

Kim, S.I., Bultman, S.J., Kiefer, C.M., Dean, A., Bresnick, E.H., 2009. BRG1 requirement for long-range interaction of a locus control region with a downstream promoter. Proc. Natl. Acad. Sci. USA 106 (7), 2259–2264.

Kimmel, C.B., Kane, D.A., Walker, C., Warga, R.M., Rothman, M.B., 1989. A mutation that changes cell movement and cell fate in the zebrafish embryo. Nature 337 (6205), 358–362.

Kingsley, P.D., Malik, J., Fantauzzo, K.A., Palis, J., 2004. Yolk sac-derived primitive erythroblasts enucleate during mammalian embryogenesis. Blood 104 (1), 19–25.

Kishimoto, Y., Lee, K.H., Zon, L., Hammerschmidt, M., Schulte-Merker, S., 1997. The molecular nature of zebrafish swirl: BMP2 function is essential during early dorsoventral patterning. Development 124 (22), 4457–4466.

Kissa, K., Herbomel, P., 2010. Blood stem cells emerge from aortic endothelium by a novel type of cell transition. Nature 464 (7285), 112–115.

Kundu, M., Compton, S., Garrett-Beal, L., Stacy, T., Starost, M.F., Eckhaus, M., et al., 2005. Runx1 deficiency predisposes mice to T-lymphoblastic lymphoma. Blood 106 (10), 3621–3624.

Kusy, S., Gault, N., Ferri, F., Lewandowski, D., Barroca, V., Jaracz-Ros, A., et al., 2011. Adult hematopoiesis is regulated by TIF1gamma, a repressor of TAL1 and PU.1 transcriptional activity. Cell Stem Cell 8 (4), 412–425.

Lancrin, C., Mazan, M., Stefanska, M., Patel, R., Lichtinger, M., Costa, G., et al., 2012. GFI1 and GFI1B control the loss of endothelial identity of hemogenic endothelium during hematopoietic commitment. Blood 120 (2), 314–322.

Lancrin, C., Sroczynska, P., Stephenson, C., Allen, T., Kouskoff, V., Lacaud, G., 2009. The haemangioblast generates haematopoietic cells through a haemogenic endothelium stage. Nature 457 (7231), 892–895.

Lansdorp, P.M., Sutherland, H.J., Eaves, C.J., 1990. Selective expression of CD45 isoforms on functional subpopulations of CD34+ hemopoietic cells from human bone marrow. J. Exp. Med. 172 (1), 363–366.

Lawson, K.A., Dunn, N.R., Roelen, B.A., Zeinstra, L.M., Davis, A.M., Wright, C.V., et al., 1999. Bmp4 is required for the generation of primordial germ cells in the mouse embryo. Genes. Dev. 13 (4), 424–436.

Lee, D., Park, C., Lee, H., Lugus, J.J., Kim, S.H., Arentson, E., et al., 2008. ER71 acts downstream of BMP, Notch, and Wnt signaling in blood and vessel progenitor specification. Cell Stem Cell 2 (5), 497–507.

Li, X., Jia, S., Wang, S., Wang, Y., Meng, A., 2009. Mta3-NuRD complex is a master regulator for initiation of primitive hematopoiesis in vertebrate embryos. Blood 114 (27), 5464–5472.

Liao, E.C., Paw, B.H., Oates, A.C., Pratt, S.J., Postlethwait, J.H., Zon, L.I., 1998. SCL/Tal-1 transcription factor acts downstream of cloche to specify hematopoietic and vascular progenitors in zebrafish. Genes Dev. 12 (5), 621–626.

Liao, W., Bisgrove, B.W., Sawyer, H., Hug, B., Bell, B., Peters, K., et al., 1997. The zebrafish gene cloche acts upstream of a flk-1 homologue to regulate endothelial cell differentiation. Development 124 (2), 381–389.

Liu, F., Kang, I., Park, C., Chang, L.W., Wang, W., Lee, D., et al., 2012. ER71 specifies Flk-1+ hemangiogenic mesoderm by inhibiting cardiac mesoderm and Wnt signaling. Blood 119 (14), 3295–3305.

Lyons, S.E., Lawson, N.D., Lei, L., Bennett, P.E., Weinstein, B.M., Liu, P.P., 2002. A nonsense mutation in zebrafish gata1 causes the bloodless phenotype in vlad tepes. Proc. Natl. Acad. Sci. USA 99 (8), 5454–5459.

Maeno, M., Mead, P.E., Kelley, C., Xu, R.H., Kung, H.F., Suzuki, A., et al., 1996. The role of BMP-4 and GATA-2 in the induction and differentiation of hematopoietic mesoderm in *Xenopus laevis*. Blood 88 (6), 1965–1972.

Mangia, F., Procicchiami, G., Manelli, H., 1970. On the development of the blood island in *Xenopus laevis* embryos: light and electron microscope study. Acta. Embryol. Exp. 2, 163–184.

Marshall, C.J., Moore, R.L., Thorogood, P., Brickell, P.M., Kinnon, C., Thrasher, A.J., 1999. Detailed characterization of the human aorta-gonad-mesonephros region reveals morphological polarity resembling a hematopoietic stromal layer. Dev. Dyn. 215 (2), 139–147.

McMahon, K.A., Hiew, S.Y., Hadjur, S., Veiga-Fernandes, H., Menzel, U., Price, A.J., et al., 2007. Mll has a critical role in fetal and adult hematopoietic stem cell self-renewal. Cell Stem Cell 1 (3), 338–345.

Mead, P.E., Brivanlou, I.H., Kelley, C.M., Zon, L.I., 1996. BMP-4-responsive regulation of dorsal-ventral patterning by the homeobox protein Mix.1. Nature 382 (6589), 357–360.

Mead, P.E., Zhou, Y., Lustig, K.D., Huber, T.L., Kirschner, M.W., Zon, L.I., 1998. Cloning of Mix-related homeodomain proteins using fast retrieval of gel shift activities, (FROGS), a technique for the isolation of DNA-binding proteins. Proc. Natl. Acad. Sci. USA 95 (19), 11251–11256.

Mebius, R., Akashi, K., 2000. Precursors to neonatal lymph nodes: LT beta+CD45+CD4+CD3− cells are found in fetal liver. Curr. Top. Microbiol. Immunol. 251, 197–201.

Mebius, R.E., Miyamoto, T., Christensen, J., Domen, J., Cupedo, T., Weissman, I.L., et al., 2001. The fetal liver counterpart of adult common lymphoid progenitors gives rise to all lymphoid lineages, CD45+CD4+CD3− cells, as well as macrophages. J. Immunol. 166 (11), 6593–6601.

Medvinsky, A., Dzierzak, E., 1996. Definitive hematopoiesis is autonomously initiated by the AGM region. Cell 86 (6), 897–906.

Miccio, A., Blobel, G.A., 2010. Role of the GATA-1/FOG-1/NuRD pathway in the expression of human beta-like globin genes. Mol. Cell. Biol. 30 (14), 3460–3470.

Mikkola, H.K., Klintman, J., Yang, H., Hock, H., Schlaeger, T.M., Fujiwara, Y., et al., 2003. Haematopoietic stem cells retain long-term repopulating activity and multipotency in the absence of stem-cell leukaemia SCL/tal-1 gene. Nature 421 (6922), 547–551.

Monteiro, R., Pouget, C., Patient, R., 2011. The gata1/pu.1 lineage fate paradigm varies between blood populations and is modulated by tif1gamma. EMBO. J. 30 (6), 1093–1103.

Moore, M.A., Metcalf, D., 1970. Ontogeny of the haemopoietic system: yolk sac origin of *in vivo* and *in vitro* colony forming cells in the developing mouse embryo. Br. J. Haematol. 18 (3), 279–296.

Morrison, S.J., Hemmati, H.D., Wandycz, A.M., Weissman, I.L., 1995. The purification and characterization of fetal liver hematopoietic stem cells. Proc. Natl. Acad. Sci. USA 92 (22), 10302–10306.

Mucenski, M.L., McLain, K., Kier, A.B., Swerdlow, S.H., Schreiner, C.M., Miller, T.A., et al., 1991. A functional c-myb gene is required for normal murine fetal hepatic hematopoiesis. Cell 65 (4), 677–689.

Mukouyama, Y., Chiba, N., Mucenski, M.L., Satake, M., Miyajima, A., Hara, T., et al., 1999. Hematopoietic cells in cultures of the murine embryonic aorta-gonad-mesonephros region are induced by c-Myb. Curr. Biol. 9 (15), 833–836.

Muller, A.M., Medvinsky, A., Strouboulis, J., Grosveld, F., Dzierzak, E., 1994. Development of hematopoietic stem cell activity in the mouse embryo. Immunity 1 (4), 291–301.

Mullins, M.C., Hammerschmidt, M., Kane, D.A., Odenthal, J., Brand, M., van Eeden, F.J., et al., 1996. Genes establishing dorsoventral pattern formation in the zebrafish embryo: the ventral specifying genes. Development 123, 81–93.

Munoz-Sanjuan, I., 2001. A HB. Early posterior/ventral fate specification in the vertebrate embryo. Dev. Biol. 237 (1), 1–17.

Murayama, E., Kissa, K., Zapata, A., Mordelet, E., Briolat, V., Lin, H.F., et al., 2006. Tracing hematopoietic precursor migration to successive hematopoietic organs during zebrafish development. Immunity 25 (6), 963–975.

Murray, L., Chen, B., Galy, A., Chen, S., Tushinski, R., Uchida, N., et al., 1995. Enrichment of human hematopoietic stem cell activity in the CD34+Thy-1+Lin- subpopulation from mobilized peripheral blood. Blood 85 (2), 368–378.

Muse, G.W., Gilchrist, D.A., Nechaev, S., Shah, R., Parker, J.S., Grissom, S.F., et al., 2007. RNA polymerase is poised for activation across the genome. Nat. Genet. 39 (12), 1507–1511.

Negrin, R.S., Atkinson, K., Leemhuis, T., Hanania, E., Juttner, C., Tierney, K., et al., 2000. Transplantation of highly purified CD34+Thy-1+ hematopoietic stem cells in patients with metastatic breast cancer. Biol. Blood Marrow Transplant. 6 (3), 262–271.

Ng, E.S., Azzola, L., Sourris, K., Robb, L., Stanley, E.G., Elefanty, A.G., 2005. The primitive streak gene Mixl1 is required for efficient haematopoiesis and BMP4-induced ventral mesoderm patterning in differentiating ES cells. Development 132 (5), 873–884.

Nguyen, V.H., Schmid, B., Trout, J., Connors, S.A., Ekker, M., Mullins, M.C., 1998. Ventral and lateral regions of the zebrafish gastrula, including the neural crest progenitors, are established by a bmp2b/swirl pathway of genes. Dev. Biol. 199 (1), 93–110.

Nichols, K.E., Crispino, J.D., Poncz, M., White, J.G., Orkin, S.H., Maris, J.M., et al., 2000. Familial dyserythropoietic anaemia and thrombocytopenia due to an inherited mutation in GATA1. Nat. Genet. 24 (3), 266–270.

Nishikawa, S.I., Nishikawa, S., Hirashima, M., Matsuyoshi, N., Kodama, H., 1998. Progressive lineage analysis by cell sorting and culture identifies FLK1+VE-cadherin+ cells at a diverging point of endothelial and hemopoietic lineages. Development 125 (9), 1747–1757.

North, T., Gu, T.L., Stacy, T., Wang, Q., Howard, L., Binder, M., et al., 1999. Cbfa2 is required for the formation of intra-aortic hematopoietic clusters. Development 126 (11), 2563–2575.

North, T.E., de Bruijn, M.F., Stacy, T., Talebian, L., Lind, E., Robin, C., et al., 2002. Runx1 expression marks long-term repopulating hematopoietic stem cells in the midgestation mouse embryo. Immunity 16 (5), 661–672.

North, T.E., Goessling, W., Walkley, C.R., Lengerke, C., Kopani, K.R., Lord, A.M., et al., 2007. Prostaglandin E2 regulates vertebrate haematopoietic stem cell homeostasis. Nature 447 (7147), 1007–1011.

Ogawa, E., Maruyama, M., Kagoshima, H., Inuzuka, M., Lu, J., Satake, M., et al., 1993. PEBP2/PEA2 represents a family of transcription factors homologous to the products of the Drosophila runt gene and the human AML1 gene. Proc. Natl. Acad. Sci. USA 90 (14), 6859–6863.

Okada, S., Nakauchi, H., Nagayoshi, K., Nishikawa, S., Miura, Y., Suda, T., 1992. *In vivo* and *in vitro* stem cell function of c-kit- and Sca-1-positive murine hematopoietic cells. Blood 80 (12), 3044–3050.

Okuda, T., van Deursen, J., Hiebert, S.W., Grosveld, G., Downing, J.R., 1996. AML1, the target of multiple chromosomal translocations in human leukemia, is essential for normal fetal liver hematopoiesis. Cell 84 (2), 321–330.

Osawa, M., Hanada, K., Hamada, H., Nakauchi, H., 1996. Long-term lymphohematopoietic reconstitution by a single CD34-low/negative hematopoietic stem cell. Science 273 (5272), 242–245.

Ottersbach, K., Dzierzak, E., 2005. The murine placenta contains hematopoietic stem cells within the vascular labyrinth region. Dev. Cell. 8 (3), 377–387.

Palis, J., Chan, R.J., Koniski, A., Patel, R., Starr, M., Yoder, M.C., 2001. Spatial and temporal emergence of high proliferative potential hematopoietic precursors during murine embryogenesis. Proc. Natl. Acad. Sci. USA 98 (8), 4528–4533.

Palis, J., McGrath, K.E., Kingsley, P.D., 1995. Initiation of hematopoiesis and vasculogenesis in murine yolk sac explants. Blood 86 (1), 156–163.

Palis, J., Robertson, S., Kennedy, M., Wall, C., Keller, G., 1999. Development of erythroid and myeloid progenitors in the yolk sac and embryo proper of the mouse. Development 126 (22), 5073–5084.

Patterson, L.J., Gering, M., Eckfeldt, C.E., Green, A.R., Verfaillie, C.M., Ekker, S.C., et al., 2007. The transcription factors Scl and Lmo2 act together during development of the hemangioblast in zebrafish. Blood 109 (6), 2389–2398.

Peeters, M., Ottersbach, K., Bollerot, K., Orelio, C., de Bruijn, M., Wijgerde, M., et al., 2009. Ventral embryonic tissues and Hedgehog proteins induce early AGM hematopoietic stem cell development. Development 136 (15), 2613–2621.

Pevny, L., Lin, C.S., D'Agati, V., Simon, M.C., Orkin, S.H., Costantini, F., 1995. Development of hematopoietic cells lacking transcription factor GATA-1. Development 121 (1), 163–172.

Porcher, C., Swat, W., Rockwell, K., Fujiwara, Y., Alt, F.W., Orkin, S.H., 1996. The T cell leukemia oncoprotein SCL/tal-1 is essential for development of all hematopoietic lineages. Cell 86 (1), 47–57.

Ransom, D.G., Bahary, N., Niss, K., Traver, D., Burns, C., Trede, N.S., et al., 2004. The zebrafish moonshine gene encodes transcriptional intermediary factor 1gamma, an essential regulator of hematopoiesis. PLoS Biol. 2 (8), E237.

Rhodes, J., Hagen, A., Hsu, K., Deng, M., Liu, T.X., Look, A.T., et al., 2005. Interplay of pu.1 and gata1 determines myelo-erythroid progenitor cell fate in zebrafish. Dev. Cell. 8 (1), 97–108.

Russell, E.S., 1979. Hereditary anemias of the mouse: a review for geneticists. Adv. Genet. 20, 357–459.

Samokhvalov, I.M., Samokhvalova, N.I., Nishikawa, S., 2007. Cell tracing shows the contribution of the yolk sac to adult haematopoiesis. Nature 446 (7139), 1056–1061.

Sanchez, M.J., Holmes, A., Miles, C., Dzierzak, E., 1996. Characterization of the first definitive hematopoietic stem cells in the AGM and liver of the mouse embryo. Immunity 5 (6), 513–525.

Sasaki, K., Matsumura, G., 1988. Spleen lymphocytes and haemopoiesis in the mouse embryo. J. Anat. 160, 27–37.

Sasaki, K., Yagi, H., Bronson, R.T., Tominaga, K., Matsunashi, T., Deguchi, K., et al., 1996. Absence of fetal liver hematopoiesis in mice deficient in transcriptional coactivator core binding factor beta. Proc. Natl. Acad. Sci. USA 93 (22), 12359–12363.

Schmid, B., Furthauer, M., Connors, S.A., Trout, J., Thisse, B., Thisse, C., et al., 2000. Equivalent genetic roles for bmp7/snailhouse and bmp2b/swirl in dorsoventral pattern formation. Development 127 (5), 957–967.

Shalaby, F., Rossant, J., Yamaguchi, T.P., Gertsenstein, M., Wu, X.F., Breitman, M.L., et al., 1995. Failure of blood-island formation and vasculogenesis in Flk-1-deficient mice. Nature 376 (6535), 62–66.

Shivdasani, R.A., Fujiwara, Y., McDevitt, M.A., Orkin, S.H., 1997. A lineage-selective knockout establishes the critical role of transcription factor GATA-1 in megakaryocyte growth and platelet development. EMBO. J. 16 (13), 3965–3973.

Shivdasani, R.A., Mayer, E.L., Orkin, S.H., 1995. Absence of blood formation in mice lacking the T-cell leukaemia oncoprotein tal-1/SCL. Nature 373 (6513), 432–434.

Shivdasani, R.A., Orkin, S.H., 1996. The transcriptional control of hematopoiesis. Blood 87 (10), 4025–4039.

Sood, R., English, M.A., Belele, C.L., Jin, H., Bishop, K., Haskins, R., et al., 2010. Development of multilineage adult hematopoiesis in the zebrafish with a runx1 truncation mutation. Blood 115 (14), 2806–2809.

Soza-Ried, C., Hess, I., Netuschil, N., Schorpp, M., Boehm, T., 2010. Essential role of c-myb in definitive hematopoiesis is evolutionarily conserved. Proc. Natl. Acad. Sci. USA 107 (40), 17304–17308.

Spangrude, G.J., Heimfeld, S., Weissman, I.L., 1988. Purification and characterization of mouse hematopoietic stem cells. Science 241 (4861), 58–62.

Stainier, D.Y., Weinstein, B.M., Detrich 3rd, H.W., Zon, L.I., Fishman, M.C., 1995. Cloche, an early acting zebrafish gene, is required by both the endothelial and hematopoietic lineages. Development 121 (10), 3141–3150.

Takahashi, K., Tanabe, K., Ohnuki, M., Narita, M., Ichisaka, T., Tomoda, K., et al., 2007. Induction of pluripotent stem cells from adult human fibroblasts by defined factors. Cell 131 (5), 861–872.

Tavian, M., Hallais, M.F., Peault, B., 1999. Emergence of intraembryonic hematopoietic precursors in the pre-liver human embryo. Development 126 (4), 793–803.

Thompson, M.A., Ransom, D.G., Pratt, S.J., MacLennan, H., Kieran, M.W., Detrich 3rd, H.W., et al., 1998. The cloche and spadetail genes differentially affect hematopoiesis and vasculogenesis. Dev. Biol. 197 (2), 248–269.

Thomsen, G.H., 1997. Antagonism within and around the organizer: BMP inhibitors in vertebrate body patterning. Trends Genet. 13 (6), 209–211.

Tsang, A.P., Fujiwara, Y., Hom, D.B., Orkin, S.H., 1998. Failure of megakaryopoiesis and arrested erythropoiesis in mice lacking the GATA-1 transcriptional cofactor FOG. Genes Dev 12 (8), 1176–1188.

Turpen, J.B., Kelley, C.M., Mead, P.E., Zon, L.I., 1997. Bipotential primitive – definitive hematopoietic progenitors in the vertebrate embryo. Immunity 7 (3), 325–334.

Vissers, L.E., van Ravenswaaij, C.M., Admiraal, R., Hurst, J.A., de Vries, B.B., Janssen, I.M., et al., 2004. Mutations in a new member of the chromodomain gene family cause CHARGE syndrome. Nat. Genet. 36 (9), 955–957.

Visvader, J.E., Fujiwara, Y., Orkin, S.H., 1998. Unsuspected role for the T cell leukemia protein SCL/tal-1 in vascular development. Genes Dev. 12 (4), 473–479.

Vogeli, K.M., Jin, S.W., Martin, G.R., Stainier, D.Y., 2006. A common progenitor for haematopoietic and endothelial lineages in the zebrafish gastrula. Nature 443 (7109), 337–339.

Wadman, I.A., Osada, H., Grutz, G.G., Agulnick, A.D., Westphal, H., Forster, A., et al., 1997. The LIM-only protein Lmo2 is a bridging molecule assembling an erythroid, DNA-binding complex which includes the TAL1, E47, GATA-1 and Ldb1/NLI proteins. EMBO. J. 16 (11), 3145–3157.

Wang, J.C., Doedens, M., Dick, J.E., 1997. Primitive human hematopoietic cells are enriched in cord blood compared with adult bone marrow or mobilized peripheral blood as measured by the quantitative *in vivo* SCID-repopulating cell assay. Blood 89 (11), 3919–3924.

Wang, Q., Stacy, T., Binder, M., Marin-Padilla, M., Sharpe, A.H., Speck, N.A., 1996a. Disruption of the Cbfa2 gene causes necrosis and hemorrhaging in the central nervous system and blocks definitive hematopoiesis. Proc. Natl. Acad. Sci. USA 93 (8), 3444–3449.

Wang, Q., Stacy, T., Miller, J.D., Lewis, A.F., Gu, T.L., Huang, X., et al., 1996b. The CBFbeta subunit is essential for CBFalpha2 (AML1) function *in vivo*. Cell 87 (4), 697–708.

Wang, S., Krinks, M., Kleinwaks, L., Moos Jr., M., 1997. A novel *Xenopus* homologue of bone morphogenetic protein-7 (BMP-7). Genes Funct. 1 (4), 259–271.

Wang, S., Wang, Q., Crute, B.E., Melnikova, I.N., Keller, S.R., Speck, N.A., 1993. Cloning and characterization of subunits of the T-cell receptor and murine leukemia virus enhancer core-binding factor. Mol. Cell Biol. 13 (6), 3324–3339.

Wang, Y., Yates, F., Naveiras, O., Ernst, P., Daley, G.Q., 2005. Embryonic stem cell-derived hematopoietic stem cells. Proc. Natl. Acad. Sci. USA 102 (52), 19081–19086.

Wang, Y., Zheng, C.G., Jiang, Y., Zhang, J., Chen, J., Yao, C., et al., 2012. Genetic correction of beta-thalassemia patient-specific iPS cells and its use in improving hemoglobin production in irradiated SCID mice. Cell Res. 22 (4), 637–648.

Warren, A.J., Colledge, W.H., Carlton, M.B., Evans, M.J., Smith, A.J., Rabbitts, T.H., 1994. The oncogenic cysteine-rich LIM domain protein rbtn2 is essential for erythroid development. Cell 78 (1), 45–57.

Weiss, O., Kaufman, R., Michaeli, N., Inbal, A., 2012. Abnormal vasculature interferes with optic fissure closure in lmo2 mutant zebrafish embryos. Dev. Biol. 369 (2), 191–198.

Wilkinson, R.N., Pouget, C., Gering, M., Russell, A.J., Davies, S.G., Kimelman, D., et al., 2009. Hedgehog and Bmp polarize hematopoietic stem cell emergence in the zebrafish dorsal aorta. Dev. Cell 16 (6), 909–916.

Willett, C.E., Cortes, A., Zuasti, A., Zapata, A.G., 1999. Early hematopoiesis and developing lymphoid organs in the zebrafish. Dev. Dyn. 214 (4), 323–336.

Willey, S., Ayuso-Sacido, A., Zhang, H., Fraser, S.T., Sahr, K.E., Adlam, M.J., et al., 2006. Acceleration of mesoderm development and expansion of hematopoietic progenitors in differentiating ES cells by the mouse Mix-like homeodomain transcription factor. Blood 107 (8), 3122–3130.

Winnier, G., Blessing, M., Labosky, P.A., Hogan, B.L., 1995. Bone morphogenetic protein-4 is required for mesoderm formation and patterning in the mouse. Genes Dev. 9 (17), 2105–2116.

Wright, D.E., Bowman, E.P., Wagers, A.J., Butcher, E.C., Weissman, I.L., 2002. Hematopoietic stem cells are uniquely selective in their migratory response to chemokines. J. Exp. Med. 195 (9), 1145–1154.

Xi, Q., Wang, Z., Zaromytidou, A.I., Zhang, X.H., Chow-Tsang, L.F., Liu, J.X., et al., 2011. A poised chromatin platform for TGF-beta access to master regulators. Cell 147 (7), 1511–1524.

Yagi, H., Deguchi, K., Aono, A., Tani, Y., Kishimoto, T., Komori, T., 1998. Growth disturbance in fetal liver hematopoiesis of Mll-mutant mice. Blood 92 (1), 108–117.

Yamada, Y., Warren, A.J., Dobson, C., Forster, A., Pannell, R., Rabbitts, T.H., 1998. The T cell leukemia LIM protein Lmo2 is necessary for adult mouse hematopoiesis. Proc. Natl. Acad. Sci. USA 95 (7), 3890–3895.

Yamaguchi, T.P., Dumont, D.J., Conlon, R.A., Breitman, M.L., Rossant, J., 1993. flk-1, an flt-related receptor tyrosine kinase is an early marker for endothelial cell precursors. Development 118 (2), 489–498.

Yokota, T., Huang, J., Tavian, M., Nagai, Y., Hirose, J., Zuniga-Pflucker, J.C., et al., 2006. Tracing the first waves of lymphopoiesis in mice. Development 133 (10), 2041–2051.

Yu, J., Vodyanik, M.A., Smuga-Otto, K., Antosiewicz-Bourget, J., Frane, J.L., Tian, S., et al., 2007. Induced pluripotent stem cell lines derived from human somatic cells. Science 318 (5858), 1917–1920.

Zeitlinger, J., Stark, A., Kellis, M., Hong, J.W., Nechaev, S., Adelman, K., et al., 2007. RNA polymerase stalling at developmental control genes in the *Drosophila melanogaster* embryo. Nat. Genet. 39 (12), 1512–1516.

Zhang, C.C., Kaba, M., Ge, G., Xie, K., Tong, W., Hug, C., et al., 2006. Angiopoietin-like proteins stimulate *ex vivo* expansion of hematopoietic stem cells. Nat. Med. 12 (2), 240–245.

Zhang, C.C., Kaba, M., Iizuka, S., Huynh, H., Lodish, H.F., 2008. Angiopoietin-like 5 and IGFBP2 stimulate *ex vivo* expansion of human cord blood hematopoietic stem cells as assayed by NOD/SCID transplantation. Blood 111 (7), 3415–3423.

Zhang, C.C., Lodish, H.F., 2004. Insulin-like growth factor 2 expressed in a novel fetal liver cell population is a growth factor for hematopoietic stem cells. Blood 103 (7), 2513–2521.

Zhang, J., Houston, D.W., King, M.L., Payne, C., Wylie, C., Heasman, J., 1998. The role of maternal VegT in establishing the primary germ layers in *Xenopus* embryos. Cell 94 (4), 515–524.

Zhang, J., Jackson, A.F., Naito, T., Dose, M., Seavitt, J., Liu, F., et al., 2012. Harnessing of the nucleosome-remodeling-deacetylase complex controls lymphocyte development and prevents leukemogenesis. Nat. Immunol. 13 (1), 86–94.

Zou, Y.R., Kottmann, A.H., Kuroda, M., Taniuchi, I., Littman, D.R., 1998. Function of the chemokine receptor CXCR4 in haematopoiesis and in cerebellar development. Nature 393 (6685), 595–599.

Chapter 26

How to Build a Kidney

Mor Grinstein and Thomas M. Schultheiss
Department of Anatomy and Cell Biology, Rappaport Faculty of Medicine, Technion-Israel Institute of Technology, Haifa, Israel

Chapter Outline

GLOSSARY

Collecting duct The segment of the nephron that conveys the urine to the outside.
Glomerulus The filtration component of the nephron.
Intermediate mesoderm (IM) A strip of mesenchyme adjacent to the somites that is the embryonic source of all kidney tissue.
Nephric duct An epithelial tube that is the source of the collecting ducts and that plays an essential role in inducing formation of the tubule in most kidney types.
Nephron The functional unit of the kidney. Kidneys are comprised of as few as one or as many as one million or more nephrons. Nephrons are divided into segments, including the glomerulus, tubule, and collecting duct.
Podocyte Specialized cell type of the glomerulus that plays an essential role in filtration and lies at the interface between glomerular capillaries and the entrance to the kidney tubule.
Tubule The segment of the nephron that processes the glomerular filtrate and produces urine.

SUMMARY

- Kidneys are built from individual functional units called nephrons.
- Nephrons are segmented into several distinct parts, including the glomerulus, tubule, and collecting duct.
- There is an underlying unity between the nephrons of all vertebrates, but the details of the nephron segments are variable in accordance with the physiological needs of the organism.
- Tubules develop from undifferentiated nephrogenic mesenchyme via a mesenchymal-to-epithelial transition (MET).
- There is increasing evidence that glomerular and tubule precursors are initially distinct and subsequently become incorporated into a unified nephron.

26.1 INTRODUCTION

Kidneys are major organs that function to control the internal milieu of the body. They play essential roles in maintaining water and metabolite balance and in excreting toxic substances to the outside. Because of the vastly differing environments in which animals live, it is natural that there should be a great variation in kidney physiology and anatomy amongst different animal species, and also between different types of kidney tissue that are generated during distinct stages of the life cycle within a single species. In light of these variations, it is not surprising that there are significant differences in the developmental programs that generate the kidneys in different vertebrate species. Yet there are also many aspects of kidney development that are shared across different kidney tissues and different species.

In the current discussion, we shall review vertebrate kidney development from a comparative point of view. The focus in this review will be on development of the nephron, the basic building block of the kidney, an issue that has rarely been the central focus of comparative reviews. A thorough understanding of nephron development is essential in order to provide basic information that will be essential for attempts to regenerate or repair diseased kidney tissue using the tools of regenerative medicine. As will be seen in the discussion, certain aspects of nephron formation, such as initial nephron induction and branching of the collecting system, are much better understood than others, such as the specification of the different nephron subunits or they way in which nephrons are assembled into functional kidneys. Excellent recent reviews exist for some of these topics, and the reader is referred to them for a detailed treatment of those issues (Dressler, 2006, 2009; Costantini and Kopan, 2010; Little and McMahon, 2012). In the current discussion, while covering vertebrate kidney development in general, we will focus on aspects of nephron formation that have received less attention in recent reviews.

Nephrons

The functional unit of vertebrate kidneys is the nephron (Figure 26.1). Each nephron comprises three basic subdivisions: the **glomerulus**, the **tubule**, and the **collecting system**. The glomerulus (from the Latin, meaning ball of thread, because of its histological appearance) is the interface between the internal body environment and the outside world (Saxen, 1987; Quaggin and Kreidberg, 2008). In essence, the glomerulus consists of a tuft of capillaries; specialized epithelial cells called podocytes; and a glomerular basement membrane that lies between them. Together, these three glomerular components comprise the glomerular filtration barrier. Fenestrations between glomerular endothelial cells, physical properties of the glomerular basement membrane, and specialized podocyte intercellular junctions called slit diaphragms, together regulate the passage of substances from the capillary and into the tubular lumen. A properly functioning glomerular basement membrane will retain cells and large molecules in the blood system, and will allow water, metabolites, and other small molecules to pass out of the capillaries and into the tubular luminal space. It should be noted that in addition to podocytes and

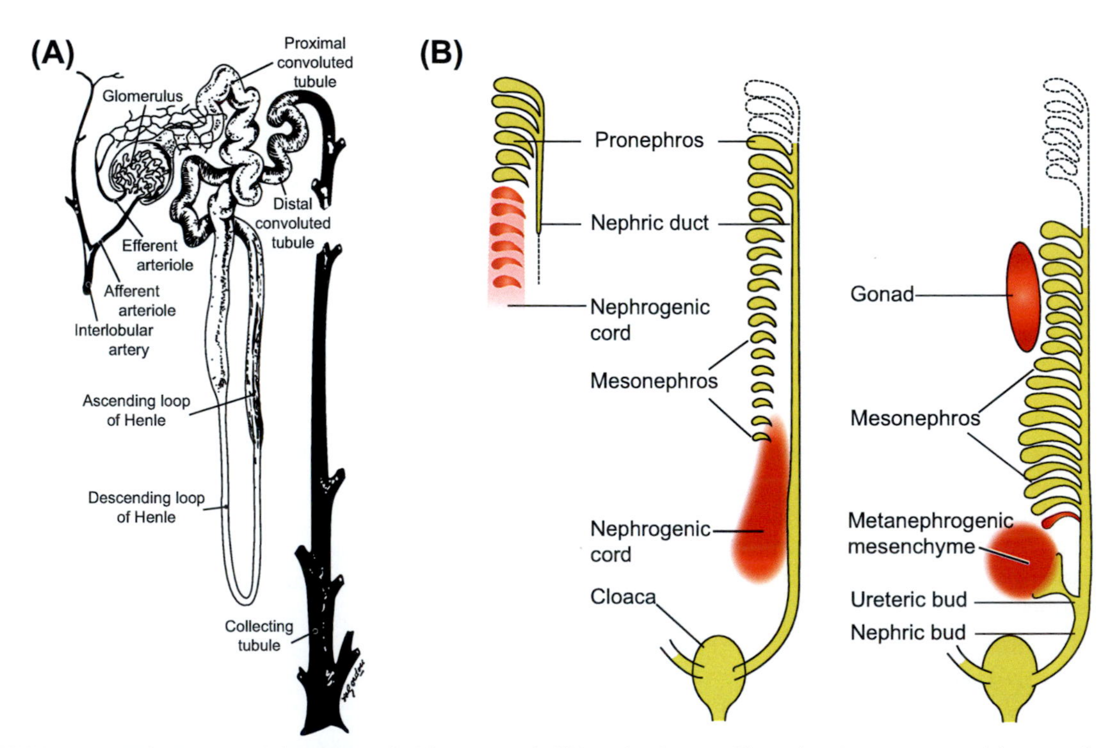

FIGURE 26.1 (A) Human metanephric nephron. (B) Three stages in kidney development illustrating the appearance of the pronephros, mesonephros and metanephros. *(A) Adapted from Ham and Cormack, Histology, Philadelphia:Lippincott, 1979. (B) Adapted from Gilbert, Developmental Biology Sunderland:Sinauer, 2006.*

endothelial cells, the glomerulus also includes mesangial cells whose functions are not entirely clear but which are thought to play a role in maintaining glomerular morphology.

The podocyte is the guardian of the entrance to the second component of the nephron, the **tubule**. The primary role of the tubule is to process the glomerular filtrate. These tasks are accomplished by selective exchange of substances between the tubular lumen and the epithelial cells that line the tubule. Among the types of processing that occur in the tubule are: water balance (accomplished by absorption of sodium ions and consequent passive absorption of water), mineral homeostasis, blood pH regulation (through hydrogen and bicarbonate ion exchange), and excretion of toxic metabolites. The tubule is typically divided into distinct segments, each of which carries out one or more of these specialized functions (Dressler, 2002, 2006). Specialized cell types within the tubular epithelium carry out specific functions (see below for more detailed discussion of the specialized tubular domains and their development). Once the filtrate has passed through the tubule, it enters the final part of the nephron, the collecting duct, which acts to convey the filtrate, now called urine, to the outside.

Types of Kidney

Vertebrate kidneys can consist of as few as one or as many as one million or more nephrons. Traditionally, three types of kidneys are described: **pronephros, mesonephros**, and **metanephros** (Figure 26.1) (Fraser, 1950). Pronephroi typically consist of one or a few nephrons (Vize et al., 1995; 1997; 2003). They develop early during development in the anterior, neck region of the embryo. The **pronephros** typically serves as the functional embryonic kidney in anamniotes (fish and amphibians). In amniotes, it is rudimentary and in many species it is doubtful as to whether it is ever functional. Indeed, it is likely that the main reason that the pronephros persists in amniotes is because it is the embryonic source of the nephric duct (see below). The mesonephros develops somewhat later and more posterior to the pronephros, and generally contains more functional nephrons (Sainio, 2003). It serves as the functional adult kidney in anamniotes, and as the fetal kidney in amniote embryos. In males, some of the mesonephric tubules become converted into the rete testis, which convey spermatozoa from the testis to the spermatic duct (which develops from the nephric duct). Finally, the metanephros develops last and in the most posterior location. The metanephros is the most complex of the kidney types, and can contain thousands and even millions of nephrons. It is found only in amniotes, where it serves as the functional adult kidney.

The existence of the different types of kidneys is likely connected to the different environments to which vertebrate animals were exposed during the course of evolution (Smith, 1953). The demands for water balance are very different in animals that live in salt water, fresh water, or on land, and it is not surprising that different kidney types are needed to survive and function in these differing environments (Goodrich, 1895; Romer, 1955). Thus, the metanephros, with its adaptations for retention of water, is found only in animals that live on land. The mesonephros is likely adequate for function in the aqueous environment of adult fish and amphibians, or the fetal environment of amniotes.

The simplest known vertebrate kidney is that of the hagfish embryo. The hagfish embryo kidney is called a **holonephros**, because it contains a single nephron in each body segment (Figure 26.1) (Price, 1897, 1904–5; Dean, 1899). Because of the primitive organization of the hagfish kidney, and because the hagfish is considered to retain features representative of the most basal vertebrates (Takezaki et al., 2003), the holonephros has been proposed to be representative of the most primitive type of vertebrate kidney (Goodrich, 1895; Romer, 1955). According to this line of thinking, the pro-, meso-, and metanephros evolved as specializations of the kidney-forming region at various positions along the anterior-posterior axis of the embryo.

Invertebrate animals also have excretory organs, including the nephridia of *C. elegans* and many other species, and the Malpigian tubules of *Drosophila*. Until recently it was not at all clear whether these represented homologs of vertebrate kidneys. Recent molecular and ultra-structural investigations, particularly of the glomerulus, have uncovered increasing support for the hypothesis that vertebrate and invertebrate kidneys are indeed homologous structures (Weavers et al., 2009).

The Intermediate Mesoderm

All vertebrate kidney tissue is derived from the intermediate mesoderm (IM), a strip of mesenchyme located lateral to the somites in the developing embryo (Figure 26.2). Investigations by a number of laboratories have begun to uncover some of the mechanisms of IM formation (Seufert et al., 1999; Mauch et al., 2000; James and Schultheiss, 2003; Wilm et al., 2004; James and Schultheiss, 2005; Preger-Ben Noon et al., 2009; Callery et al., 2010; Fleming et al., 2013). The earliest specific markers of the IM are the transcription factors Pax2 and the closely related Pax8 (Dressler et al., 1990; Plachov et al., 1990). Pax2 is expressed in the IM before it has begun to morphologically differentiate into kidney tissue (Carroll and Vize, 1999; Carroll et al., 1999; James and Schultheiss, 2003). No kidney tissue is formed in Pax2/8 double mutant mice (Bouchard et al., 2002). The genes Osr1, Eya1, Six2, and Wt1 are also expressed in the early IM, although some of these genes are also expressed in adjacent non-IM tissues (Osr1, Wt1), or are required only for formation of metanephric

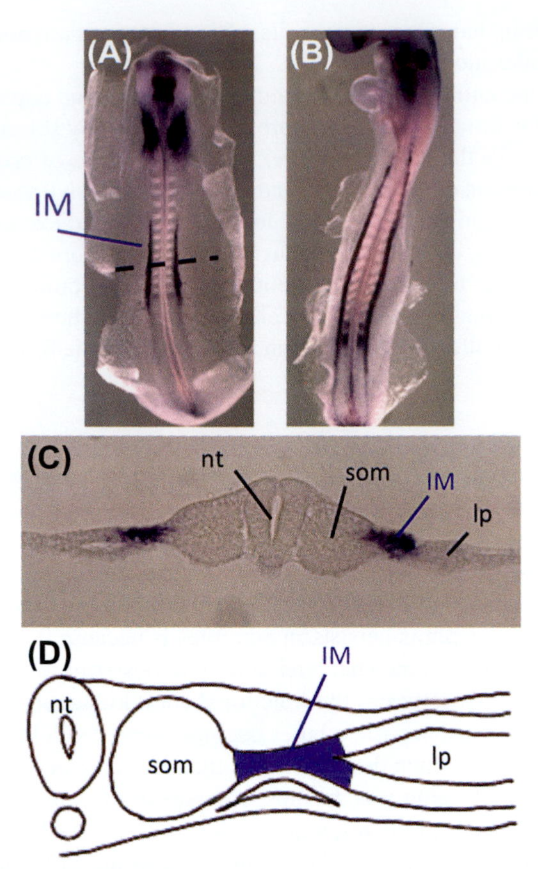

FIGURE 26.2 **The Intermediate Mesoderm.** (A-C) Expression of *Pax2* by whole mount *in situ* hybridization in Stage 10 (A) and Stage 13 (B) chick embryos. (C) is a section through a Stage 10 embryo at the axial level indicated by the dashed line in A. *Pax2* is expressed in the intermediate mesoderm (IM) adjacent to the somites. (D) Diagram of a section as shown in (C), illustrating the IM. lp, lateral plate; nt, neural tube; som, somite. *(A,B) From Soueid-Baumgarten et al.,* Dev. Biol. *285:122–35, 2014.*

kidney tissue and not for pro- or mesonephros formation (Dressler et al., 1990; Kalatzis et al., 1998; Ohto et al., 1998; So and Danielian, 1999; Xu et al., 1999, 2003; James and Schultheiss, 2005; Sajithlal et al., 2005). Intermediate levels of bone morphogenetic protein (BMP) signaling (James and Schultheiss, 2005), as well as Nodal-like and Retinoic Acid signaling (Preger-Ben Noon et al., 2009; Fleming et al., 2013) are required for the initiation of IM gene expression. Consistent with these findings, ectopic administration of a combination of Activin (which acts through the same receptors as Nodal-like signals) and Retinoic Acid has been found to induce robust kidney differentiation in *Xenopus* animal caps that would not normally form kidney tissue (Asashima et al., 2000). The results of these studies have begun to be applied to develop strategies to generate IM-like tissue from ES cells (Kim and Dressler, 2005; Takasato et al., 2014).

26.2 THE NEPHRIC DUCT

The Nephric Duct: Formation and Migration

We begin the discussion of segments of the nephron by considering the nephric duct (ND), because it is the first part of the nephric system to differentiate in most vertebrate kidneys and because it plays a critical role in inducing the differentiation of the other nephron segments.

The nephric duct is the embryonic source of the collecting system of the kidney (Schultheiss et al., 2003). The duct forms as part of the differentiation of the pronephros. In chicken embryos, which are generally representative of the process of duct formation in amniotes, the first evidence of duct formation occurs when a portion of the IM bulges dorsally (towards the ectoderm) and subsequently buds off from the IM to form the **nephric duct rudiment**, which lies between the ectoderm and the underlying IM (Figure 26.3) (James and Schultheiss, 2003; Attia et al., 2012). The process of nephric duct rudiment formation occurs only at the axial level of approximately somites 6–10. Subsequently, the duct rudiment extends posteriorly. The posterior extension of the nephric duct continues until its posterior tip reaches the region of the cloaca, where it fuses with the cloacal epithelium that will subsequently form the urinary bladder. Because of this process of extension, the

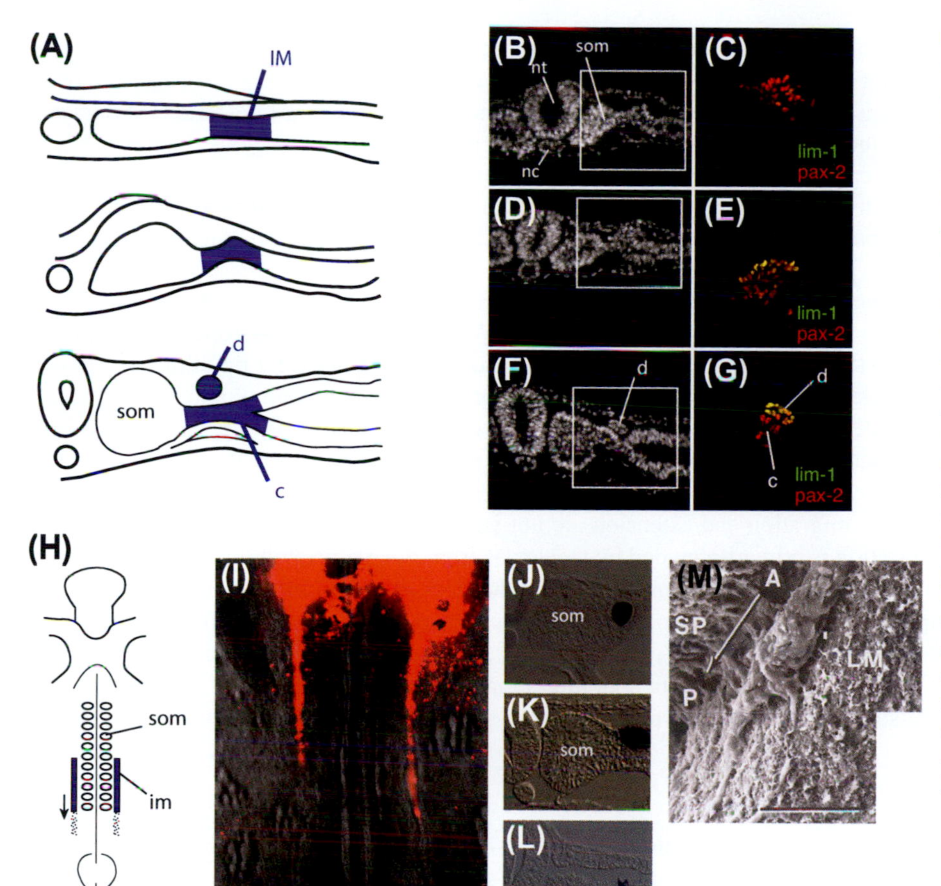

FIGURE 26.3 Formation and Migration of the Nephric Duct. (A–G) Stages in nephric duct formation. The nephric duct rudiment (d) bulges dorsally from the IM (blue in A), leaving behind the nephrogenic cord (c). (B–G) depict three stages in nephric duct formation, stained for the duct marker *lim1* and the general IM marker *Pax2*. (H–M) Migration of the nephric duct. (H) The chick nephric duct is formed in the IM at the axial levels of somites 6–10, and then migrates posteriorly. (I) The IM between the axial levels of somites 6–10 was labeled with DiI and the embryos were grown for 18 hours. Note that the nephric duct has extended posteriorly from the area of labeling. (J–L) Sections through the nephric duct at anterior, middle, and posterior axial levels. At the posterior tip (L), the duct is composed of loosely organized cells. In the mid-region (K), duct cells form a compact cord. More anteriorly (J), the duct cells have undergone Epithelial-to-mesenchymal transition (EMT) and a small lumen can be observed. (M) Scanning electron microscopy (EM) of the migrating duct, showing the mesenchymal-like nature of the duct tip cells. *(B–G) from James and Schultheiss* Dev. Biol. *253:109–24, 2003. (M) from Bellairs et al.,* Dev. Dynam. *202:333–342, 1995.*

nephric duct serves as the source of the collecting system not only for the pronephros, but also for the mesonephros and metanephros (in a process to be described below, in the section "Functions of the nephron").

The existing evidence indicates that the whole length of the nephric duct derives from IM of the pronephric region. If a barrier is placed posterior to the pronephros before the duct rudiment begins to extend, no nephric duct is found posterior to the barrier (Gruenwald, 1937; Waddington, 1938; Soueid-Baumgarten et al., 2014). In quail-chick chimeras in which the nephric duct rudiment is derived from quail cells and the posterior IM is derived from chick cells, the whole length of the nephric duct is derived solely from quail cells (Attia et al., 2012). Experiments in *Xenopus* also indicate that the nephric duct is derived entirely from cells that migrate posteriorly from the pronephros, with no contribution from more posterior cells (Lynch and Fraser, 1990). In zebrafish, there is a one report of posterior cells contributing to posterior regions of the duct (Serluca and Fishman, 2001). However, in that study it is not clear whether the labeling process could have labeled the migrating duct cells themselves. The situation in zebrafish and other species bears re-examination in order to establish whether the exclusive origin of the nephric duct from the pronephros is a universal phenomenon within the vertebrates.

Given that the ability to form a nephric duct is a unique property of the pronephric IM, the question arises as to how the pronephric IM acquires this property. Two general types of explanation can be considered: either the region of the pronephros (axial levels 6–10 in the chick embryo) constitutes a special duct inducing environment, or the pronephric IM is somehow predetermined to generate duct tissue. In order to distinguish between these possibilities, Attia and colleagues constructed chick-quail chimeras in which prospective non-pronephric IM was transplanted into the pronephric area or, prospective pronephric IM was transplanted into more posterior regions (Attia et al., 2012). These experiments found that prospective pronephric IM would form duct tissue even if transplanted into posterior regions that would not normally generate duct tissue, whereas prospective IM from more posterior regions would not form duct tissue even if transplanted into the pronephric region. The clear implications from these studies was that the ability to form duct is a property that is fixed in prospective IM very early in its development (indeed it is already fixed when prospective IM cells are gastrulating through the primitive streak) and that duct specification does not require a special duct inducing environment that is located in the region of the pronephros.

Attia et al., proceeded to show that the ability to form nephric duct was influenced by Hox genes (Attia et al., 2012). HoxA6, whose anterior border of expression is approximately at the axial level of somite 12 (just posterior to the duct-forming region), inhibits duct formation if misexpressed in the pronephric region. In contrast, HoxB4, which is normally expressed in the pronephros, does not inhibit nephric duct formation if overexpressed in that region. This result suggests that the ability to form nephric duct is under negative control, and that one reason that more posterior IM does not form duct is the presence of inhibitors of duct formation, including Hox genes, in more posterior regions. Interestingly, in the hagfish holonephros, discussed above, each body segment is thought to form a small portion of the nephric duct, with these segments being subsequently connected to form a single nephric duct. Thus, originally the ability to form duct may have been a general property of the IM. Subsequently, through the evolution of Hox regulatory control, duct formation may have become restricted to the most anterior, pronephric region of the IM.

Nephric duct extension is a striking example of the phenomenon of collective cell migration, in which groups of cells migrate while remaining connected to each other (Friedl and Gilmour, 2009; Weijer, 2009). Several zones can be distinguished in the extending duct (Figure 26.3) (Schultheiss et al., 2003). Cells of the advancing posterior tip of the duct have a mesenchymal morphology, with cellular extensions that are typical for cells migrating on a substrate (Poole and Steinberg, 1981; Bellairs et al., 1995). Cells immediately anterior to the mesenchymal tip cells have a cord-like appearance, without features of migrating cells but also without a classical epithelial histology. Anteriorly to the cord-like zone, duct cells undergo a MET to generate an epithelial duct with a central lumen.

The molecular factors that regulate duct migration and extension are not yet well understood. Studies by Steinberg's group in amphibians argue that duct migration is primarily guided by local cues from the extracellular matrix in the region of the duct tip and not by long-range signals emitted from the posterior end of the embryo (Poole and Steinberg, 1982). Duct migration has been reported to be impeded by blocking antibodies to fibronectin and by chemical interference with glial cell-derived neurotrophic factor (GDNF) signaling (Jacob et al., 1991; Drawbridge et al., 2000; Morris et al., 2003). Mouse mutations in the transcription factor Gata3 also show defects in duct migration (Grote et al., 2006). But a comprehensive understanding of the molecular mechanism that regulates duct extension awaits future studies.

Functions of the Nephric Duct

The nephric duct has two major functions during kidney development: it is a critical source of signals that induce tubule formation from the IM, and it is the exclusive source of cells for the collective system. These two interconnected functions have been most extensively studied in the context of metanephros formation.

In the region of the metanephros, signals from the metanephric mesenchyme (MM) induce the formation of an epithelial bud (ureteric bud, UB) that branches off from the ND and invades the MM (Figure 26.4) (Yu et al., 2004). GDNF signals in the MM and cRet receptors in the ND are critical for UB formation (Moore et al., 1996; Pichel et al., 1996; Sanchez et al., 1996; Shakya et al., 2005). Subsequently, the UB and the MM undergo a series of characteristic changes. The UB undertakes a series of sequential branches, resulting in a tree-like structure of tubes, all connected to the nephric duct. Each terminal branch of the UB becomes the collecting duct of an individual nephron. Thus, in the human metanephros, which contains approximately 1 million nephrons, the UB must branch sequentially at least 20 times ($2^{20} = \sim 1{,}000{,}000$) to create 1 million collecting ducts. UB branching is controlled by intrinsic features of the UB epithelium (Qiao et al., 1999) as well as general and local signals within the MM (Michos et al., 2010). GDNF and Fibroblast Growth Factor (FGF) signaling have been found to play crucial roles in regulating UB branching (Michos et al., 2010). UB branching is an excellent example of the more general phenomenon of branching morphogenesis. The reader is referred to recent reviews on duct branching for a comprehensive treatment of the subject (Costantini and Kopan, 2010; see also Chapter 14).

In order to obtain properly-integrated nephrons, the 1,000,000 collecting ducts must be properly linked to 1,000,000 tubules. This is accomplished through the role of the branching UB in initiating tubule formation. Metanephric kidney tubule formation initiates when localized areas of the MM aggregate to form pretubular aggregates (PTAs) and subsequently undergo mesenchymal-to-epithelial transition (MET) to form renal vesicles (RVs) (Figure 26.4) (Iino et al., 2001). Signals from the branching UB are critical for the initiation of tubulogenesis; in the absence of the UB, the MM does not undergo tubulogenesis and quickly degenerates. Wnt signals from the branching UB are an important component of the tubule-inducing activity of the UB. A source of canonical Wnt signaling can rescue tubule formation in isolated MM (Herzlinger et al., 1994), and Wnt9b, which is expressed in the branching UB, is required for formation of RV's and tubules (Carroll et al., 2005).

Differentiation of the MM into tubules must be carefully controlled. A too-general initial wave of tubule formation could run the risk of depleting MM cells, thus not leaving sufficient cells for additional generations of tubules (Karner et al., 2011). The fact that Wnts generally have a very short range of action renders them suitable for the task of initiating tubule formation. PTAs and RVs are typically localized on the underside of the tip of the branching UB (Figure 26.4). According to current models, Wnt signaling from the tips of branching UB initiates localized aggregation and formation of RVs, while

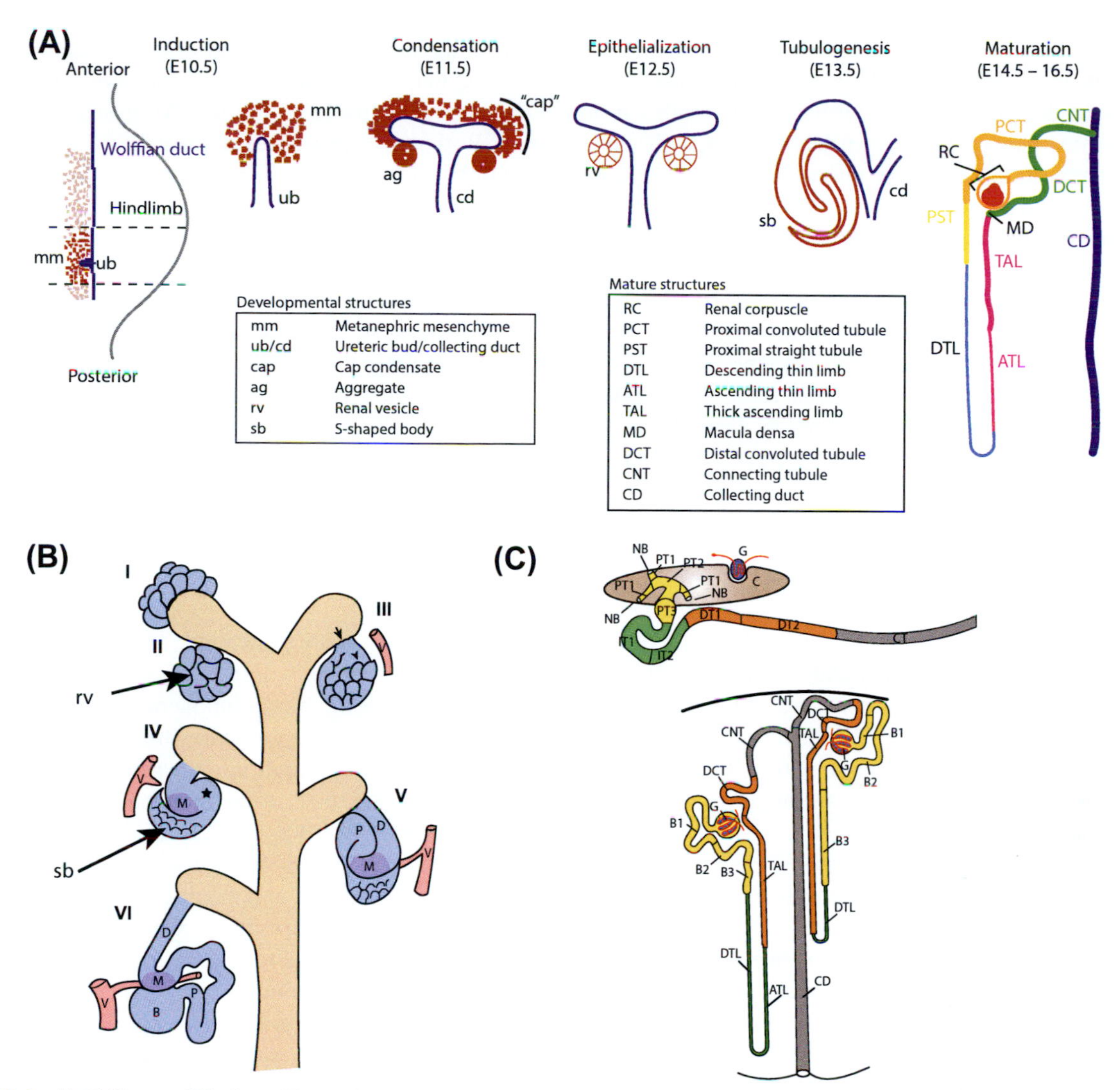

FIGURE 26.4 **(A-B) Stages of Nephron Formation in the Mouse Mesonephros.** (A) illustrates stages in the generation of an individual nephron. (B) illustrates the arrangement of a set of nephrons in various stages of formation (with the most mature on the bottom) and their relationship to the branched ureteric bud. P, m, and d indicate proximal, middle, and distal segments of the tubule. rv, renal vesicle; sb, S-shaped body. See text for explanation. *(A) From Yu et al., 2004. (B) From Iino et al., 2001.* (C) Comparison between the *Xenopus* pronephros (top) and the mammalian metanephros (bottom). The color coding of analogous nephron segments is based on the comparison of marker gene expression as shown in Raciti et al., 2008. On the basis of molecular markers, four distinct tubular compartments can be recognized. In the pronephros (top) each tubule may be further subdivided into distinct segments: proximal tubule (PT, yellow; PT1, PT2, and PT3), intermediate tubule (IT, green; IT1 and IT2), distal tubule (DT, orange; DT1 and DT2), and connecting tubule (CT, gray). The nephrostomes (NS) are ciliated peritoneal funnels that connect the coelomic cavity (C) to the nephron. Abbreviations used for the mammalian nephron segments are as follows: ATL, ascending thin limb; CD, collecting duct; CNT, connecting tubule; DCT, distal convoluted tubule; DTL, descending thin limb; S1, S2, and S3, segments of the proximal tubule; TAL, thick ascending limb. *Figure and legend adapted from Raciti et al.,* Genome Biol. *9:R84, 2008.*

leaving surrounding MM cells in an undifferentiated state (Costantini and Kopan, 2010). Significantly, MM differentiation appears to be under general negative control. The transcription factor Six2 is expressed throughout the undifferentiated MM and is downregulated upon PTA and RV formation in response to Wnt signals (Park et al., 2012). Animals with mutations in Six2 exhibit precocious and ectopic initiation of tubule formation (Self et al., 2006). One important pathway through which Wnt signaling initiates tubule formation is through localized repression of Six2 expression, thus allowing tubulogenesis to proceed (Park et al., 2012).

After their formation, RV's proceed to expand, generating morphological structures known as comma-shaped bodies (CBs), and S-shaped bodies (SBs) (Iino et al., 2001; Georgas et al., 2009). At the SB stage, characteristic markers of differentiated tubules and glomeruli can be detected in the distal (closest to the UB) and proximal (farthest from the UB) parts of the SB (Georgas et al., 2009). It is generally held that RVs transform into CBs and SBs through growth and morphogenesis, and thus that the RV contains precursors to the glomerulus as well as the tubule. However, as discussed below, it cannot be ruled out that additional cells join the SB during its formation, and thus that some of these elements, particularly the podocytes of the glomerulus, may have their origins in mesenchymal cells outside of the RV. At around the SB stage, the distal-most part of the nascent tubule fuses with the adjacent UB, thus merging the distal tubule with the collecting tubule to form an integrated nephron (Georgas et al., 2009). In parallel, signals, including Vascular Endothelial Growth Factor (VEGF), from the proximal end of the SB attract endothelial cells from the surrounding mesenchyme (Quaggin and Kreidberg, 2008). The endothelial cells integrate with differentiating podocytes at the proximal end of the SB to initiate formation of the glomerulus.

While these stages in nephron induction and differentiation have been best described in the metanephros, a similar process occurs in the mesonephros (Soueid-Baumgarten et al., 2014). In the case of the mesonephros, no UB is formed, and tubule formation appears to be induced by the ND itself. Tubule formation passes through the same morphological stages in the mesonephros as in the metanephros. As in the metanephros, removal of the ND completely blocks metanephric tubule formation (Gruenwald, 1937; Waddington, 1938; Soueid-Baumgarten et al., 2014). The ND in the area of the mesonephros expresses Wnt9b (Soueid-Baumgarten et al., 2014), and Wnt9b mutants do not initiate mesonephric tubule formation (Carroll et al., 2005). In the pronephros, duct and tubule are generated at roughly the same time, and it is not clear whether the duct or a duct precursor population plays an inductive role in pronephric tubule formation (Brennan et al., 1998).

26.3 NEPHRON SEGMENTATION

Nephron segmentation is the process by which the different parts of the nephron are generated. This includes the crude division into glomerulus, tubule, and collecting duct, as well as further subdivisions within these sections. As described above, the nephric duct and its derivative, the ureteric bud, give rise to the collecting duct component of each nephron. The other segments of the nephron – the tubule and the glomerulus – are derived from the nephrogenic mesenchyme of the pro-, meso- or metanephric regions (Figure 26.1). In the following sections, we review research into the processes by which the tubule and glomerular segments of the nephron are determined. In order to better understand the nephron segmentation in different species we should refer the reader to the evolutionary processes of nephrons as part of the kidney types (Fraser, 1950). Several kidney types have arisen during vertebrate evolution to meet the different physiological demands of aquatic and terrestrial life. Some current theories claim that tetrapods (mammals, birds, reptiles and amphibians) descended from a common ancestor that was a fresh water fish and had an ancestral kidney, the archinephros. As a result, the epithelial filtering units (tubules) of the vertebrate kidneys share common elements: a proximal tubule to absorb and secrete solutes, and distal segments to adjust the filtrate composition (Davidson, 2011). Nephrons of the mammalian metanephric kidney have been modified for life out of freshwater and therefore they differ morphologically from the mesonephric nephrons by the existence of a loop called the loop of Henle. This structure allows mammals to conserve water more easily by concentrating their urine. The most basic function of the tubule is to process and drain the filtrate arriving from the filtration unit – the glomerulus. In general, tubulogenesis starts from signals that originate from the nephric duct or the uretic bud that induce mesenchymal cells condensation and transition of these mesenchymal cells into epithelial tubule. However, the tubulogenesis process is different in the three sequential excretory organs. The molecular mechanisms of nephron segmentation are only beginning to be elucidated. Many questions remain, and we will highlight some of the most important of these.

Kidney Tubules

The Pronephric Tubule

The pronephric nephrotomes are located between the somatic mesenchyme and the lateral plate, and in the amphibian *Xenopus laevis*, in which they have been most extensively studied, they are found at an axial level of approximately the third somite. The precise axial level and the number of the pronephric tubules vary in different species. The tubules open through the nephrostome into the coelomic cavity, and their distal ends may fuse to form the most anterior portion of the pronephric duct (Saxen, 1987; Vize et al., 1995).

The pronephric tubules are subdivided into proximal, intermediate, and distal segments. Each segment has its own distinctive function as mirrored by the different cell morphologies and gene expression profiles along the length of the tubule. Raciti (2008) performed cross species gene expression comparisons to identify similarities between the *Xenopus* pronephros and the mammalian metanephros. These workers found 23 marker genes with highly regionalized expression in the *Xenopus* pronephroi that are similar to the mammalian metanephros. The genes include 18 *slc* genes, *calb1*, *cldn 8*, *cldn16*, *clcnk,* and *kcnj1*. Their study revealed that the pronephric nephron is composed of four basic domains: the proximal, the intermediate tubule, the distal tubule and the connecting tubule. These domains share signatures at the molecular level that are typical of the mammalian nephron. The striking structural and functional similarities between the pronephric and the metanephric nephron indicate that the mesonephros and the metanephros are based on the same gene and molecular mechanisms as the pronephros.

The proximal tubules are divided into three main domains, PT1, PT2, and PT3. Similar to the metanephros tubules (see Figure 26.4C), the main functions of the proximal tubule are reabsorbing ions, amino acids, glucose, and water (Raciti, 2008; Wessely, 2011). Beyenbach (2004) postulated that fish lacking a glomerulus as their filtration unit used the proximal tubule as a filtration segment for the NaCl and water excretion processes in order to increase their renal excretory capacity. The intermediate tubules, IT1 and IT2, express the same genes as the thin limb of the metanephric loop of Henle, as well as the distal tubule. The loop of Henle serves to help concentrate the filtrate, a feature not found in amphibians due to the wet environments in which they live; in amphibians the IT1 and IT2 are formed in order to reabsorb salt and ions. The pronephric distal tubules, DT1 and DT2, are equivalent to the thick ascending limb of the loop of Henle and the distal convoluted tubule; both are required for reabsorption processes.

The signaling and gene regulation of the pronephric tubules has been mainly investigated in two animal models: the zebrafish and the frog *Xenopus*. In *Xenopus*, pronephros development begins around stage 12.5. The pronephric tubules are induced from the intermediate mesoderm. MET is a crucial part of tubulogenesis and the terminal differentiation of these processes will end two days post fertilization. Studies of the mesonephros and metanephros of chick and mice provide a partial understanding of the signaling pathways involved in forming these organs. Interestingly, these studies showed that the signaling molecules that participate in the building of the mesonephros and metanephros also play a part in the pronephros development, indicating that some of these genes are evolutionarily conserved and have similar functions in the three kidney types. Wnt4 and FGF have some analogous functions in pronephros and metanephros tubulogenesis (Urban et al., 2006). According to several studies, Wnt 11/11r signals from the duct induce Wnt4 expression and promote tubule formation, particularly at the proximal end of the tubule. FGF8 is expressed in the pronephric primordium and is required for pronephric tubule and duct formation. BMP also has a central role in the pronephros. Bracken (2008) found that BMP signaling affects the developing duct and tubules, but not the glomus. They speculated that BMP functions to mediate morphogenesis of the specified renal field during vertebrate embryogenesis. The role of Notch signaling was also investigated in the *Xenopus* pronephros. Taelman et al., (2006) provided evidence indicating that early-phase pronephric Notch signaling promotes glomus formation, whereas late phase pronephric notch signaling promotes proximal tubule formation. Naylor (2009) showed that Notch signaling regulates the medial-lateral aspect of tubule formation and that the Notch-dependent transcription factor Hey1 regulates the proximal-distal patterning of the tubule. These activities determine whether cells are fated towards podocytes and proximal tubule or distal tubule. In zebrafish, Retinoic Acid and the transcription factor Irx3b promote rostral (i.e., proximal) tubule formation, whereas the transcription factor Caudal promotes the generation of distal tubules (Reggiani et al., 2007).

Mesonephric Tubules

The second segmental excretory organ is the mesonephros. In the avian embryo this kidney appears as a linear organ that stretches between the axial levels of the 15th to the 30th somites. Mesonephros differentiation proceeds from the anterior to the posterior and is dependent on signals from the nephric duct, which migrates from anterior to posterior through the mesonephric region (Soueid-Baumgarten et al., 2014). However, there are great differences in size, distribution, and functional maturity of the mesonephric nephrons between avian and mammalian species. The total number of nephrons in avians is around 100, whereas in mammals, the number of mesonephric tubules vary from approximately 30 (human) to more than 50 (sheep, pig) (Saxen, 1987). Morphogenesis of the mesonephric tubules starts with the formation of mesenchymal condensates, which soon develop into renal vesicles and S-shaped bodies – much like similar developmental events in the metanephros. The early condensation, the renal vesicle and the S-shaped body development, and further maturation of the mesonephric tubule closely resemble that of the metanephric tubule, but some specific, mainly structural, differences between the two have been observed in a mature nephron (Soueid-Baumgarten et al., 2014). The juxtaglomerular apparatus is missing from the mesonephric tissue, and the loop of Henle is shorter than the metanephros (Sainio, 2003).

The mesonephros has some experimental advantages for studying kidney induction. In the avian embryo it is possible to completely block interaction of the nephric duct with the mesonephric mesenchyme that will later become the tubule. Moreover, it is relatively simple to dissect and manipulate several sections of the mesonephric tubules due to their simple linear appearance that consist of several tubules per embryonic segment. Soueid-Baumgarten et al. (2014) investigated the maturation and induction of the mesenchyme into tubules. Because of these experimental advantages, they were able to more carefully determine which kidney genes are dependent on signals from the nephric duct than had been previously reported. They found that the *Pax2*, *Eya1*, *Wnt4*, and *Lhx1* genes are all activated in the mesonephric area in parallel with the nephric duct invasion to this region. They found that *Wnt4* and *Lhx1*, which are associated with the initiation of tubule formation, are dependent on the nephric duct, whereas expression of *Eya1* and *Pax2*, expressed in the undifferentiated mesenchyme, are duct independent. In agreement with their results (Grote et al., 2006) showed that Gata3–/– mice embryos, which exhibit defects in nephric duct migration, still activate *Pax2* in the mesonephric region. As in the mammalian metanephros, Wnt signaling also plays an important role in inducing mesonephric tubules (Soueid-Baumgarten et al., 2014). Another interesting observation was made by Diep et al., who demonstrated *in vivo* tubule regeneration in the adult zebrafish mesonephros (Diep, 2011). They traced the source of new nephrons in the adult zebrafish to small cellular aggregates containing nephron progenitors. They transplanted single aggregates comprising 10–30 cells and succeeded to generate multiple nephrons. Serial transplantation experiments to test self-renewal in the zebrafish revealed that nephron progenitor cells are long lived and have significant replicative potential. They speculated that the mesonephros kidney of the mature zebrafish contains self-renewing nephron stem/progenitor cells.

Metanephric Tubules

As discussed above, metanephric tubules proceed through the stages of PTA, RV, and S-shaped body (SB). Cells in the proximal domain of the S-shaped body differentiate into specialized epithelial cells types of the mature renal corpuscle. The more distal domain of the S-shaped body differentiates into different tubule segments including the proximal tubule, the loop of Henle, and distal tubular domains. The final stages of nephron maturation are characterized by extensive elongation of the tubular segments and the expression of specialized and specific transport proteins (Nakai, 2003; Grieshammer et al., 2005). The early stages of this process have been extensively characterized. Wnt9B, which is released from the developing collecting duct, induces nephrogenesis. Wnt4, produced by nephron progenitors within the PTAs is required for the MET that leads to the formation of renal vesicle (Grieshammer et al., 2005; Stark and Kulesa, 2005). FGF8 is also required for the pretubular aggregation, and in fact is upstream to Wnt4 in the MET process (Grieshammer et al., 2005). Wnt4 signaling is specifically required for the activation of Lhx1, an important transcription factor of the nephron in its developing stages, and in the distal nephron in later stages (Kobayashi et al., 2005).

By the S-shaped body stage, distinct domains can be molecularly identified that pre-shadow the future segments of the nephron (Georgas et al., 2009). Distinct expression domains can be seen for distal markers such as Lhx1 and proximal podocyte markers such as Wt1, Foxc2, and Nphs2. However, nephron segmentation prior to the S-shaped body stage is much less clear. It has been argued that proximal-distal patterning of the nephron already occurs at the RV stage. Thus, it has been reported that genes such as Lhx1 and Pou3f3 are expressed in the distal side of the RV whereas genes like the future podocyte gene Wt1 are restricted to the proximal side of the RV (Georgas et al., 2009). However, it is important to note that gene expression is not equivalent to lineage tracing. In addition, gene expression in the RV through SB stages is quite dynamic, and thus it can be difficult to identify nephron compartments solely on gene expression patterns at these early stages. For example, Lhx1 is first expressed in the complete RV, but later on is restricted to the distal tubule.

The Glomerulus

The glomerulus is the most proximal end of the nephron. It contains vascular capillaries and specialized epithelial cells, the podocytes. The combination of fenestrated slits between the capillary endothelial cells, the basement membrane between endothelial cells and podocytes, and specialized foot processes of the podocytes together form the glomerular filtration barrier. Small molecules can pass through the barrier into the interior of the nephron, while larger molecules and cells remain in the vascular system. Therefore the main function of the glomerulus is to filter the blood, and to allow the filtrate to enter the tubule. After passing the glomerular filtration barrier, the filtrate enters the tubule, where it is processed by absorption and secretion carried out by tubular epithelial cells (Grinstein et al., 2013). Glomerular filtration is an essential function of the kidney, to the extent that out of the 45,000 species of vertebrates existing today, only a minority – no more than 30 species – have kidneys without glomeruli. The aglomerular species (mainly fish) filter their fluids with the help of tubular secretion (Beyenbach, 2004). Although the aglomerular group is a very interesting topic to study in kidney organogenesis, we will focus only on the main glomerular kidney group.

The Pronephric Glomerulus

In the pronephros, the early vertebrate nephron, the glomerulus or glomus is responsible for blood filtration. The glomerular blood supply is derived from capillaries branching from the dorsal aorta. These capillary tufts either hang in the body cavity as a multi-segmental vascular structure, or exist in a closed nephron, being surrounded by a nephrocoele (opening in the coelomic lining), or Bowman's space. The pronephric glomerulus is composed of cell types typical of kidneys of higher vertebrates, including fenestrated endothelial cells in capillary tufts, and podocytes with extensive foot processes (Ellis and Youson, 1989). Pronephric glomerulus development starts with the partitioning of cells from the intermediate mesoderm that will serve as podocyte progenitors. Subsequently progenitors generate foot processes and begin to form the glomerular basement membrane. In the next stage, the podocytes interact with nearby aortic endothelial cells to form the glomerular tuft, and finally the supporting mesangial cells integrate into the glomerular architecture (Majumdar and Drummond, 1999).

The signals and the transcription factors that are responsible for the development of the pronephric glomerulus are only incompletely understood. Some evidence suggests that the pronephric tubule and the glomus are specified at the same time and are regulated by the same signals (Brennan et al., 1998), whereas other studies find that the glomus can develop after removal of the tubule and duct primordia, thus suggesting independent origins and signals for these structures (Vize et al., 1995). Lyons et al., (2009) showed that Wnt/b-catenin signaling is required for the formation of the glomus in the pronephros. Hyperactivation or inhibition of these signals caused inhibition of glomus creation as well as tubulogenesis in *Xenopus*. These results are in contrast to the role of Wnt in the mesonephros or metanephros glomerulus (see below). Wesseley and colleagues have found that a network of Wt1, FoxC2, and Notch signaling regulates podocyte differentiation in the *Xenopus* pronephros. Majumdar and Drummond (1999) studied an avascular zebrafish mutant, and showed that podocytes are able to differentiate properly in the absence of endothelium or endothelial derived signals.

The Mesonephric Glomerulus

The mesonephric nephrons are similar in morphology to their metanephric counterparts and pass through similar stages in their development, including PTAs and renal vesicle stages (Grinstein et al., 2013; Soueid-Baumgarten et al., 2014). However, the mesonephros is much simpler in structure than the metanephros, being a linear organ with only a few nephrons per body segment. The mesonephric glomerulus contains vascular capillaries and specialized podocyte epithelial cells. As opposed to some pronephric glomeruli, the mesonephros glomerulus is connected directly to its tubule and the filtrate that passes the three barriers enters the tubule directly. A recent study by Grinstein et al., (2013) provides several lines of evidence indicating that the developmental pathways of the podocytes and tubular components of the mesonephric nephron become distinct earlier than was previously thought. In this study, the dorsal and ventral regions of the intermediate mesoderm were separated before PTAs or renal vesicle formation, and were cultured separately. The dorsal region generated tubules without podocytes, and the ventral region generated cyst-like structures expressing the podocyte marker Nphs2 but did not express tubular genes. In addition, blocking migration of the nephric duct and thus the formation of tubules did not completely prevent the appearance of podocyte-like cells. This study did not determine precisely when the tubule and glomerular lineages become specified. It is possible that the lineages are already specified at the PTA stage. That Pod1 is expressed specifically in the ventral region and later is expressed specifically in podocytes suggests that the podocytes and tubule lineages may already be distinct from each other in this early stage. However, it is also possible that one or both of the IM regions continues to receive signals during nephron development from other neighboring tissues (Figure 26.5). Nevertheless, this study shows that the tubule and the podocyte lineages can develop independently to some extent from a stage before the first morphological signs of nephron formation.

The signaling pathways that govern mesonephric glomerulus specification and its distinction from the tubule are still incompletely understood. A recent study found an inhibitory role for Wnt signaling in regulating glomerular podocyte differentiation. Wnt signaling had been known as an inducer of PTAs and tubules, and in the absence of Wnt9b, mesonephric

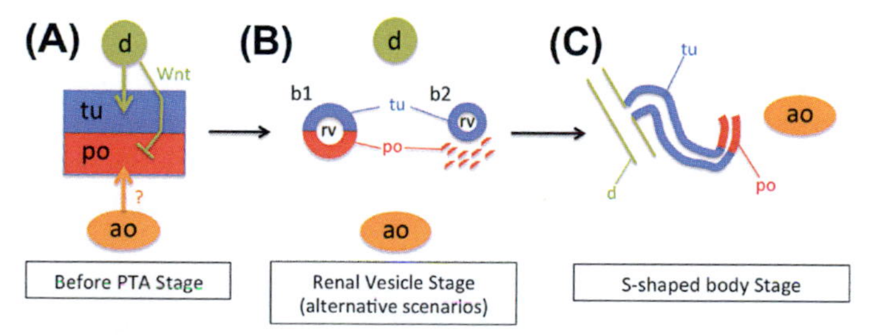

FIGURE 26.5 Models of Intermediate Mesoderm Regionalization and Nephron Patterning. (A) Wnt signaling from the nephric duct prior to the initiation of pretubular aggregate (PTA) formation promotes dorsalization of the mesonephric mesenchyme and specification of tubules, and inhibits specification of podocytes. There may be also podocyte-promoting signals from ventral tissues such as the aorta, although this is currently not clear. (B–C) Alternative models for the contribution of the renal vesicle to the S-shaped body. In b1, the RV is a hybrid structure comprised of dorsal tubule precursors and ventral podocyte precursors. In b2, the RV is a precursor structure to the tubule, and the pre-podocytes reside in the mesenchyme ventral to the RV. Regardless of which model from part B is correct, the S-shaped body (C) is comprised of contributions from both the ventral and the dorsal mesenchyme compartments. ao, aorta; d, nephric duct; po, podocytes; rv, renal vesicle; tu, tubule. *Adapted from Grinstein et al.,* Development *140:4565–73, 2013.*

tubules do not develop (see above). Grinstein et al. (2013) investigated the role of Wnt signaling in podocyte formation in the chick mesonephros, and found that canonical Wnt signaling is a strong repressor of podocyte creation. Integrating this result with the earlier finding of a role for Wnt signaling in inducing tubules suggests a model in which Wnt signaling plays a role in distinguishing between tubule and glomerular compartments of the nephron, with regions that receive high Wnt signaling levels becoming tubules and those receiving less Wnt signaling developing into podocytes (Figure 26.5). Consistent with this model, the glomerulus develops from the proximal side of the S-shaped body, which is farthest from the nephric duct, a source of Wnt signals. Inhibition of Notch signaling has been reported to repress podocyte differentiation in the metanephros (see below) (Cheng et al., 2003; Cheng and Kopan, 2005), and this finding was confirmed in the chick mesonephros (Grinstein et al., 2013). The relationship between Wnt and Notch signaling in specifying podocytes has not yet been investigated and is an important area for future studies.

The Metanephric Glomerulus

Although glomerulus-specific genes do not begin to be expressed until the S-shaped body stage, in most models of metanephric nephron formation the RV is postulated to be the source for both the tubule and the glomerulus. This is partly based on the expression of a few segment-specific genes in portions of the renal vesicle (Georgas et al., 2009). However, most of these genes are later expressed in different segments of the tubule and not in the glomerulus. Thus, the source of glomerular podocytes at RV stages and earlier is still not clear. A number of recent studies have indicated that Notch signaling is critical for differentiation of the metanephric glomerulus as well as the proximal-most parts of the tubule. The loss of Notch2 and the ligand Jag1 in mutant mice results in the loss of proximal nephron segments, including the capillaries and the mesangial cells of the glomerulus (Cheng et al., 2007). It is interesting to note that in these mutant mice the renal vesicle are created normally, and the absence of Notch in these mice is apparent only at the S-shaped body stage.

26.4 KIDNEY STEM CELLS?

Stem cells play an important part in the self-renewal of adult organs throughout life and after injury. The rate of this process varies in different tissues. Epithelial cells of the intestine and skin have high cell turnover rates and can almost completely self renew within days. Other organs with lower turnover rates, like the kidney and the lung, can still proliferate and repair after injury. However, it remains unclear whether the kidney has stem cells that contribute to the self-renewal of its nephrons (Humphreys and Bonventre, 2007). The basal tubule cell turnover in kidney is exceptionally low, and the turnover that can be detected appears to happen by division of terminally differentiated tubular epithelial cells. Humphreys et al. (2008) used genetic fate mapping techniques to identify the cells that contribute to tubule self-renewal after kidney injury. Transgenic mice were generated in which most of the tubular epithelial cells were labeled with red fluorescent protein (RFP). Two days after ischemia reperfusion injury, 50% of the tubular epithelial cells co-expressed Ki67 and RFP, indicating that differentiated epithelial cells that survived injury underwent proliferative expansion. Moreover, after the repair was complete, 67% of the cells had incorporated BrdU, compared to 3.5% of cells in the uninjured kidney. These results indicate that regeneration by surviving tubular epithelial cells is the predominant mechanism of repair after ischemic tubular injury. Details of the dedifferentiation processes are still not clear, but the current model holds that tubular cells themselves are the source of nephron repair (Humphreys, 2013).

Some studies have argued for the presence of a persistent pool of progenitor cells in the kidney. nFatC1, whose expression is associated with skin precursor cells, is activated in kidney cells recovering from acute kidney injury, suggesting that these cells might be an intratubular progenitor population (Langworthy et al., 2009). CD24 and CD133 have been identified in the parietal epithelium and in the proximal tubule and also might constitute a progenitor tubular population. These cells can form spheres in an *ex vivo* model and there is some evidence that they can ameliorate recovery after acute kidney injury (Lindgren et al., 2011; Angelotti et al., 2012). LGR5, which is expressed in stem cells of the gut, was recently shown to mark distal tubule cells during development (Barker et al., 2012). However, none of these candidates is well characterized, and further research is needed to determine whether progenitor cells indeed exist in the mature kidney. Other studies have investigated bone marrow derived mesenchymal stem cells (MSCs) as potential contributors to kidney regeneration after injury. However, there is currently no direct evidence that bone marrow derived MSCs are linked to kidney injury and repair processes (Duffield et al., 2005).

While tubule repair processes have been explored in the past several years, the glomerulus and especially the podocytes, which are the most important part of the filtration unit, have been less explored. Pavenstadt et al. (2003) hypothesized that the podocytes, in analogy to neurons, are unable to proliferate. An inability to repopulate a damaged glomerulus with functional podocytes was in good agreement with the progressive ultra-structural lesions seen in podocytes during filtration barrier failure. Some analyses of the cell cycle regulatory molecules in podocytes have revealed tight control of cell cycle quiescence, in sharp contrast to the proliferative capacity of the neighboring mesangial cells (Pavenstadt et al., 2003). However, podocytes are able to proliferate in a defined set of glomerular diseases, such as HIV nephropathy (Barisoni et al., 1999). To date, there has not been well-developed research dealing with podocyte turnover rates in health and disease, or their repair mechanisms after glomerulus injury. Future research aimed at characterizing podocyte turnover rates will be very important for understanding basic podocyte injury and repair mechanisms.

26.5 SUMMARY AND FUTURE DIRECTIONS

As a summary, we highlight several aspects of vertebrate nephron formation and kidney development that emerge from taking an approach that emphasizes the comparison between different kidney types and different organisms.

One issue involves formation of podocytes and the glomerular compartment of the nephron. In the amniote metanephros, the glomerulus is tightly integrated with the tubule, a situation that promotes efficient filtration and processing in animals with high rates of metabolism. Because of this tight integration, it has been natural to assume that the development of the glomerulus and tubule are tightly linked. Thus, most current models of metanephric nephron formation postulate that glomerular and tubular epithelial components derive from a common pool of unsegmented nephric mesenchyme which begins to be patterned at the RV stage. However, if one examines lower anamniote vertebrate and invertebrate kidneys, one learns that there is great variety between different animals in terms of the relationship between the glomerulus and tubule. In many species, the glomerulus is located at significant distance from the tubule (Romer, 1955; Ruppert, 1994; Weavers et al., 2009). Our recent studies, discussed above, found that in the chick mesonephros, which has an integrated nephron very similar to that of the mammalian metanephros, the glomerulus and tubule developmental pathways can be separated much earlier than was previously thought. This raises the possibility that the distinctness of the glomerular and tubular differentiation pathways is a fundamental feature of all kidneys, one which is obscured by the close apposition of the two compartments as seen in amniote kidneys. In future work, it will be important to test this hypothesis by doing single-cell fate mapping of kidney precursor cells in various kidney types, including metanephric kidneys. Testing this hypothesis will also be aided by the identification of additional markers for nephron progenitors, particularly for podocytes for which there is currently a lack of early, specific markers.

A second issue relates to the development and evolution of the nephric duct. In the holonephros of hagfish embryos, which is thought to represent something close to the primitive vertebrate kidney phenotype, the duct does not migrate from the anterior region to the posterior. Rather, each nephron is thought to form its own segment of duct tissue, and the individual duct segments subsequently fuse to generate the nephric duct (the caudal-most segment of the duct may indeed migrate to reach the caudal end of the embryo) (Goodrich, 1895, 1930; Dean, 1899). All embryos that generate a true pronephros in the head region appear to have evolved a migrating nephric duct in order solve the problem of transferring the urine to the outside. Thus, evolution of specialized nephric tissue along the rostral-caudal axis has been accompanied by a need to generate a migrating duct. Recent work, reviewed above, has found that in the chick embryo, *Hox* genes regulate the ability of IM tissue to generate a duct, with posterior *Hox* genes repressing the ability to form duct from more posterior regions (Attia et al., 2012). It is interesting to speculate that evolution of Hox control of IM patterning may be involved in the loss of duct-forming ability in more posterior regions as has occurred between the hagfish holonephros and the more specialized vertebrate kidneys. Similar changes in genetic regulation of IM along the A-P axis may be involved in generating the great variety of kidney morphologies that are seen within a single species and across species (Wellik et al., 2002; Mugford et al., 2008). An interesting future challenge will be to uncover the molecular basis of these differences in kidney morphology.

CLINICAL RELEVANCE

- Chronic kidney disease (CKD) is a major source of morbidity and mortality.
- Embryonic nephron number has been linked to future development of hypertension and kidney disease.
- Congenital anomalies of the kidney and urinary tract are among the most common birth defects, occurring 1 in 500 births.
- The kidney has only a limited capacity for repair and regeneration.
- Stem cell and regenerative medicine approaches have great potential for the treatment of kidney disease through the regeneration and repair of damaged kidney tissue.
- A thorough understanding of normal kidney development is the essential basis for the formulation of novel therapeutic approaches to repairing and regenerating kidney tissue.

ACKNOWLEDGMENTS

The authors would like to thank Jenny Schneider for help in preparing the manuscript. This work was supported by grants to T.M.S. from the March of Dimes, the Israel Science Foundation, the Rappaport Family Foundation, the Charles Krown Research Fund, and Technion Medical School Research Fund.

RECOMMENDED RESOURCES

GenitoUrinary Development Molecular Anatomy Project (GUDMAP): a web resource containing a wealth of information relating to kidney development, including gene expression patterns and mouse lines. http://www.gudmap.org (accessed 10.03.14).

Costantini, F., Kopan, R., 2010. Patterning a complex organ: branching morphogenesis and nephron segmentation in kidney development. Dev. Cell 18 (5), 698–712. A thorough review of several topics that are not covered in depth in the current chapter, including branching morphogenesis.

Little, M.H., McMahon, A.P., 2012. Mammalian kidney development: principles, progress, and projections. Cold Spring Harb. Perspect. Biol. 4 (5). pii: a008300. An up-to-date review of kidney development, with an emphasis on current research challenges.

REFERENCES

Angelotti, M.L., Ronconi, E., Ballerini, L., Peired, A., Mazzinghi, B., Sagrinati, C., Parente, E., Gacci, M., Carini, M., Rotondi, M., Fogo, A.B., Lazzeri, E., Lasagni, L., Romagnani, P., 2012. Characterization of renal progenitors committed toward tubular lineage and their regenerative potential in renal tubular injury. Stem Cells 30, 1714–1725.

Asashima, M., Ariizumi, T., Malacinski, G.M., 2000. In vitro control of organogenesis and body patterning by activin during early amphibian development. Comp. Biochem. Physiol. - B Biochem. Mol. Biol. 126, 169–178.

Attia, L., Yelin, R., Schultheiss, T.M., 2012. Analysis of nephric duct specification in the avian embryo. Development 139, 4143–4151.

Barisoni, L., Kriz, W., Mundel, P., D'Agati, V., 1999. The dysregulated podocyte phenotype: a novel concept in the pathogenesis of collapsing idiopathic focal segmental glomerulosclerosis and HIV-associated nephropathy. J. Am. Soc. Nephrol. 10, 51–61.

Barker, N., Rookmaaker, M.B., Kujala, P., Ng, A., Leushacke, M., Snippert, H., van de Wetering, M., Tan, S., Van Es, J.H., Huch, M., Poulsom, R., Verhaar, M.C., Peters, P.J., Clevers, H., 2012. Lgr5+ve Stem/Progenitor Cells Contribute to Nephron Formation during Kidney Development. Cell Rep. 2, 540–552.

Bellairs, R., Lear, P., Yamada, K.M., Rutishauser, U., Lash, J.W., 1995. Posterior extension of the chick nephric (Wolffian) duct: the role of fibronectin and NCAM polysialic acid. Dev. Dyn. 202, 333–342.

Beyenbach, K.W., 2004. Kidneys sans glomeruli. Am. J. Physiol. Renal Physiol. 286, F811–F827.

Bouchard, M., Souabni, A., Mandler, M., Neubüser, A., Busslinger, M., 2002. Nephric lineage specification by Pax2 and Pax8. Genes Dev. 16, 2958–2970.

Bracken, C.M., Mizeracka, K., McLaughlin, K.A., 2008. Patterning the embryonic kidney: BMP signaling mediates the differentiation of the pronephric tubules and duct in Xenopus laevis. Dev. Dyn. 237, 132–144.

Brennan, H.C., Nijjar, S., Jones, E.A., 1998. The specification of the pronephric tubules and duct in Xenopus laevis. Mech. Dev. 75, 127–137.

Callery, E.M., Thomsen, G.H., Smith, J.C., 2010. A divergent Tbx6-related gene and Tbx6 are both required for neural crest and intermediate mesoderm development in Xenopus. Dev. Biol. 340, 75–87.

Carroll, T.J., Park, J.S., Hayashi, S., Majumdar, A., McMahon, A.P., 2005. Wnt9b plays a central role in the regulation of mesenchymal to epithelial transitions underlying organogenesis of the mammalian urogenital system. Dev. Cell 9, 283–292.

Carroll, T.J., Vize, P.D., 1999. Synergism between Pax-8 and lim-1 in embryonic kidney development. Dev. Biol. 214, 46–59.

Carroll, T.J., Wallingford, J.B., Vize, P.D., 1999. Dynamic patterns of gene expression in the developing pronephros of Xenopus laevis. Dev. Genet. 24, 199–207.

Cheng, H.-T., Kim, M., Valerius, M.T., Surendran, K., Schuster-Gossler, K., Gossler, A., McMahon, A.P., Kopan, R., 2007. Notch2, but not Notch1, is required for proximal fate acquisition in the mammalian nephron. Development 134, 801–811.

Cheng, H.-T., Miner, J.H., Lin, M., Tansey, M.G., Roth, K., Kopan, R., 2003. Gamma-secretase activity is dispensable for mesenchyme-to-epithelium transition but required for podocyte and proximal tubule formation in developing mouse kidney. Development 130, 5031–5042.

Cheng, H.T., Kopan, R., 2005. The role of Notch signaling in specification of podocyte and proximal tubules within the developing mouse kidney. Kidney Int. 68, 1951–1952.

Costantini, F., Kopan, R., 2010. Patterning a complex organ: Branching morphogenesis and nephron segmentation in kidney development. Dev. Cell 18, 698–712.

Davidson, A.J., 2011. Uncharted waters: Nephrogenesis and renal regeneration in fish and mammals. Pediatr. Nephrol. 26, 1435–1443.

Dean, B., 1899. On the Embryology of Bdellostoma stouti. Verlag von Gustav Fish, Jena 1. 221–276.

Diep, C.Q., Ma, D., Deo, R.C., Holm, T.M., Naylor, R.W., Arora, N., Wingert, R.A., Bollig, F., Djordjevic, G., Lichman, B., Zhu, H., Ikenaga, T., Ono, F., Englert, C., Cowan, C.A., Hukriede, N.A., Handin, R.I., Davidson, A.J., 2011. Identification of adult nephron progenitors capable of kidney regeneration in zebrafish. Nature 470, 95–100.

Drawbridge, J., Meighan, C.M., Mitchell, E.A., 2000. GDNF and GFRalpha-1 are components of the axolotl pronephric duct guidance system. Dev. Biol. 228, 116–124.

Dressler, G.R., 2002. Tubulogenesis in the developing mammalian kidney. Trends Cell Biol. 12, 390–395.

Dressler, G.R., 2006. The cellular basis of kidney development. Annu. Rev. Cell Dev. Biol. 22, 509–529.

Dressler, G.R., 2009. Advances in early kidney specification, development and patterning. Development 136, 3863–3874.

Dressler, G.R., Deutsch, U., Chowdhury, K., Nornes, H.O., Gruss, P., 1990. Pax2, a new murine paired-box-containing gene and its expression in the developing excretory system. Development 109, 787–795.

Duffield, J.S., Park, K.M., Hsiao, L.-L., Kelley, V.R., Scadden, D.T., Ichimura, T., Bonventre, J.V., 2005. Restoration of tubular epithelial cells during repair of the postischemic kidney occurs independently of bone marrow-derived stem cells. J. Clin. Invest. 115, 1743–1755.

Ellis, L.C., Youson, J.H., 1989. Ultrastructure of the pronephric kidney in upstream migrant sea lamprey, Petromyzon marinus L. Am. J. Anat. 185, 429–443.

Fleming, B.M., Yelin, R., James, R.G., Schultheiss, T.M., 2013. A role for Vg1/Nodal signaling in specification of the intermediate mesoderm. Development 140, 1819–1829.

Fraser, E.A., 1950. The development of the vertebrate excretory system. Biol. Rev. 25, 159–187.

Friedl, P., Gilmour, D., 2009. Collective cell migration in morphogenesis, regeneration and cancer. Nat. Rev. Mol. Cell Biol. 10, 445–457.

Georgas, K., Rumballe, B., Valerius, M.T., Chiu, H.S., Thiagarajan, R.D., Lesieur, E., Aronow, B.J., Brunskill, E.W., Combes, A.N., Tang, D., Taylor, D., Grimmond, S.M., Potter, S.S., McMahon, A.P., Little, M.H., 2009. Analysis of early nephron patterning reveals a role for distal RV proliferation in fusion to the ureteric tip via a cap mesenchyme-derived connecting segment. Dev. Biol. 332, 273–286.

Gilbert, S.F., 2006. Developmental Biology, 8th ed. Sinauer, Sunderland.

Goodrich, E.A., 1930. Studies on the structure and development of vertebrates. Macmillan, London.

Goodrich, E.S., 1895. Coelom, genital ducts, and nephridia. Quart. J. Micro. Sci. 37, 56–123.

Grinstein, M., Yelin, R., Herzlinger, D., Schultheiss, T.M., 2013. Generation of the podocyte and tubular components of an amniote kidney: timing of specification and a role for Wnt signaling. Development 140, 4565–4573.

Grote, D., Souabni, A., Busslinger, M., Bouchard, M., 2006. Pax 2/8-regulated Gata 3 expression is necessary for morphogenesis and guidance of the nephric duct in the developing kidney. Development 133, 53–61.

Gruenwald, P., 1937. Zur entwicklungsmechanik der Urogenital-systems beim Huhn. Roux Arch. Entw. Mech. 136, 786–813.

Ham, A.W., Cormack, D.H., 1979. Histology, 8th ed. Lippincott, Philadelphia.

Herzlinger, D., Qiao, J., Cohen, D., Ramakrishna, N., Brown, A.M., 1994. Induction of kidney epithelial morphogenesis by cells expressing Wnt-1. Dev. Biol. 166, 815–818.

Humphreys, B.D., Bonventre, J.V., 2007. The contribution of adult stem cells to renal repair. Nephrol. Ther 3, 3–10.

Humphreys, B.D., Valerius, M.T., Kobayashi, A., Mugford, J.W., Soeung, S., Duffield, J.S., McMahon, A.P., Bonventre, J.V., 2008. Intrinsic Epithelial Cells Repair the Kidney after Injury. Cell Stem Cell 2, 284–291.

Humphreys, B.D., Xu, F., Sabbisetti, V., Grgic, I., Naini, S.M., Wang, N., Chen, G., Xiao, S., Patel, D., Henderson, J.M., Ichimura, T., Mou, S., Soeung, S., McMahon, A.P., Kuchroo, V.K., Bonventre, J.V., 2013. Chronic epithelial kidney injury molecule-1 expression causes murine kidney fibrosis. J. Clin. Invest. 123, 4023–4035.

Iino, N., Gejyo, F., Arakawa, M., Ushiki, T., 2001. Three-dimensional analysis of nephrogenesis in the neonatal rat kidney: light and scanning electron microscopic studies. Arch. Histol. Cytol. 64, 179–190.

Jacob, M., Christ, B., Jacob, H.J., Poelmann, R.E., 1991. The role of fibronectin and laminin in development and migration of the avian Wolffian duct with reference to somitogenesis. Anat. Embryol. (Berl). 183, 385–395.

James, R.G., Schultheiss, T.M., 2003. Patterning of the avian intermediate mesoderm by lateral plate and axial tissues. Dev. Biol. 253, 109–124.

James, R.G., Schultheiss, T.M., 2005. Bmp signaling promotes intermediate mesoderm gene expression in a dose-dependent, cell-autonomous and translation-dependent manner. Dev. Biol. 288, 113–125.

Kalatzis, V., Sahly, I., El-Amraoui, A., Petit, C., 1998. Eya1 expression in the developing ear and kidney: Towards the understanding of the pathogenesis of Branchio-Oto-Renal (BOR) syndrome. Dev. Dyn. 213, 486–499.

Karner, C.M., Das, A., Ma, Z., Self, M., Chen, C., Lum, L., Oliver, G., Carroll, T.J., 2011. Canonical Wnt9b signaling balances progenitor cell expansion and differentiation during kidney development. Development 138, 1247–1257.

Kim, D., Dressler, G.R., 2005. Nephrogenic factors promote differentiation of mouse embryonic stem cells into renal epithelia. J. Am. Soc. Nephrol. 16, 3527–3534.

Kobayashi, A., Kwan, K.-M., Carroll, T.J., McMahon, A.P., Mendelsohn, C.L., Behringer, R.R., 2005. Distinct and sequential tissue-specific activities of the LIM-class homeobox gene Lim1 for tubular morphogenesis during kidney development. Development 132, 2809–2823.

Langworthy, M., Zhou, B., de Caestecker, M., Moeckel, G., Baldwin, H.S., 2009. NFATc1 identifies a population of proximal tubule cell progenitors. J. Am. Soc. Nephrol. 20, 311–321.

Lindgren, D., Bostrom, A.K., Nilsson, K., Hansson, J., Sjolund, J., Moller, C., Jirstrom, K., Nilsson, E., Landberg, G., Axelson, H., Johansson, M.E., 2011. Isolation and characterization of progenitor-like cells from human renal proximal tubules. Am. J. Pathol. 178, 828–837.

Little, M.H., McMahon, A.P., 2012. Mammalian Kidney Development: Principles, Progress, and Projections. Cold Spring Harb. Perspect. Biol. 4. a008300–a008300.

Lynch, K., Fraser, S.E., 1990. Cell migration in the formation of the pronephric duct in Xenopus laevis. Dev. Biol. 142, 283–292.

Lyons, J.P., Miller, R.K., Zhou, X., Weidinger, G., Deroo, T., Denayer, T., Park, J., Il, Ji, H., Hong, J.Y., Li, A., Moon, R.T., Jones, E.A., Vleminckx, K., Vize, P.D., McCrea, P.D., 2009. Requirement of Wnt/beta-catenin signaling in pronephric kidney development. Mech. Dev. 126, 142–159.

Majumdar, A., Drummond, I.A., 1999. Podocyte differentiation in the absence of endothelial cells as revealed in the zebrafish avascular mutant, cloche. Dev. Genet. 24, 220–229.

Mauch, T.J., Yang, G., Wright, M., Smith, D., Schoenwolf, G.C., 2000. Signals from trunk paraxial mesoderm induce pronephros formation in chick intermediate mesoderm. Dev. Biol. 220, 62–75.

Michos, O., Cebrian, C., Hyink, D., Grieshammer, U., Williams, L., D'Agati, V., Licht, J.D., Martin, G.R., Costantini, F., 2010. Kidney development in the absence of Gdnf and Spry1 requires Fgf10. PLoS Genet. 6. e1000809.

Moore, M.W., Klein, R.D., Fariñas, I., Sauer, H., Armanini, M., Phillips, H., Reichardt, L.F., Ryan, A.M., Carver-Moore, K., Rosenthal, A., 1996. Renal and neuronal abnormalities in mice lacking GDNF. Nature 382, 76–79.

Morris, A.R., Drawbridge, J., Steinberg, M.S., 2003. Axolotl pronephric duct migration requires an epidermally derived, laminin 1-containing extracellular matrix and the integrin receptor alpha6beta1. Development 130, 5601–5608.

Mugford, J.W., Sipil, P., Kobayashi, A., Behringer, R.R., McMahon, A.P., 2008. Hoxd11 specifies a program of metanephric kidney development within the intermediate mesoderm of the mouse embryo. Dev. Biol. 319, 396–405.

Nakai, S., Sugitani, Y., Sato, H., Ito, S., Miura, Y., Ogawa, M., Nishi, M., Jishage, K., Minowa, O., Noda, T., 2003. Crucial roles of Brn1 in distal tubule formation and function in mouse kidney. Development 130, 4751–4759.

Naylor, R.W., Jones, E.A., 2009. Notch activates Wnt-4 signalling to control medio-lateral patterning of the pronephros. Development 136, 3585–3595.

Ohto, H., Takizawa, T., Saito, T., Kobayashi, M., Ikeda, K., Kawakami, K., 1998. Tissue and developmental distribution of Six family gene products. Int. J. Dev. Biol. 42, 141–148.

Park, J.S., Ma, W., O'Brien, L.L., Chung, E., Guo, J.J., Cheng, J.G., Valerius, M.T., McMahon, J.A., Wong, W.H., McMahon, A.P., 2012. Six2 and Wnt Regulate Self-Renewal and Commitment of Nephron Progenitors through Shared Gene Regulatory Networks. Dev. Cell 23, 637–651.

Pavenstädt, H., Kriz, W., Kretzler, M., 2003. Cell biology of the glomerular podocyte. Physiol. Rev. 83, 253–307.

Pichel, J.G., Shen, L., Sheng, H.Z., Granholm, A.C., Drago, J., Grinberg, A., Lee, E.J., Huang, S.P., Saarma, M., Hoffer, B.J., Sariola, H., Westphal, H., 1996. Defects in enteric innervation and kidney development in mice lacking GDNF. Nature 382, 73–76.

Plachov, D., Chowdhury, K., Walther, C., Simon, D., Guenet, J.L., Gruss, P., 1990. Pax8, a murine paired box gene expressed in the developing excretory system and thyroid gland. Development 110, 643–651.

Poole, T.J., Steinberg, M.S., 1981. Amphibian pronephric duct morphogenesis: segregation, cell rearrangement and directed migration of the Ambystoma duct rudiment. J. Embryol. Exp. Morphol. 63, 1–16.

Poole, T.J., Steinberg, M.S., 1982. Evidence for the guidance of pronephric duct migration by a craniocaudally traveling adhesion gradient. Dev. Biol. 92, 144–158.

Preger-Ben Noon, E., Barak, H., Guttmann-Raviv, N., Reshef, R., 2009. Interplay between activin and Hox genes determines the formation of the kidney morphogenetic field. Development 136, 1995–2004.

Price, G.C., 1897. Development of the excretory organs of a myxinoid Bdellostoma stouti Lockington. Zool. Jahrb. 10, 205–226.

Price, G.W., 1904. NoA further study of the development of the excretory organs of Bdellostoma stouti. Amer. J. Anat. 4, 117–138.

Qiao, J., Uzzo, R., Obara-Ishihara, T., Degenstein, L., Fuchs, E., Herzlinger, D., 1999. FGF-7 modulates ureteric bud growth and nephron number in the developing kidney. Development 126, 547–554.

Quaggin, S.E., Kreidberg, J.A., 2008. Development of the renal glomerulus: good neighbors and good fences. Development 135, 609–620.

Raciti, D., Reggiani, L., Geffers, L., Jiang, Q., Bacchion, F., Subrizi, A.E., Clements, D., Tindal, C., Davidson, D.R., Kaissling, B., Brändli, A.W., 2008. Organization of the pronephric kidney revealed by large-scale gene expression mapping. Genome Biol. 9, R84.

Reggiani, L., Raciti, D., Airik, R., Kispert, A., Brändli, A.W., 2007. The prepattern transcription factor Irx3 directs nephron segment identity. Genes Dev. 21, 2358–2370.

Romer, A.S., 1955. The vertebrtae body. Saunders, Philadelphia.

Ruppert, E.E., 1994. Evolutionary origin of the vertebrate nephron. Integr. Comp. Biol. 34, 542–553.

Sainio, K., 2003. Development of the mesonephric kidney. In: Vize, P., Woolf, A.S., Bard, J.B.L. (Eds.), The Kidney: From Normal Development to Congenital Disease. Academic Press, San Diego, pp. 75–86.

Sajithlal, G., Zou, D., Silvius, D., Xu, P.X., 2005. Eya1 acts as a critical regulator for specifying the metanephric mesenchyme. Dev. Biol. 284, 323–336.

Sánchez, M.P., Silos-Santiago, I., Frisén, J., He, B., Lira, S.A., Barbacid, M., 1996. Renal agenesis and the absence of enteric neurons in mice lacking GDNF. Nature 382, 70–73.

Saxen, L., 1987. Organogenesis of the Kidney. Cambridge University Press, London.

Schultheiss, T.M., James, R.G., Listopadova, A., Herzlinger, D., 2003. No Title. In: Vize, P.D., Woolf, A.S., Bard, J.B.L. (Eds.), The Kidney: From Normal Development to Congenital Disease. Academic Press, San Diego, pp. 51–60.

Self, M., Lagutin, O.V., Bowling, B., Hendrix, J., Cai, Y., Dressler, G.R., Oliver, G., 2006. Six2 is required for suppression of nephrogenesis and progenitor renewal in the developing kidney. EMBO J. 25, 5214–5228.

Serluca, F.C., Fishman, M.C., 2001. Pre-pattern in the pronephric kidney field of zebrafish. Development 128, 2233–2241.

Seufert, D.W., Brennan, H.C., DeGuire, J., Jones, E.A., Vize, P.D., 1999. Developmental basis of pronephric defects in Xenopus body plan phenotypes. Dev. Biol. 215, 233–242.

Shakya, R., Watanabe, T., Costantini, F., 2005. The role of GDNF/Ret signaling in ureteric bud cell fate and branching morphogenesis. Dev. Cell 8, 65–74.

Smith, H.W., 1953. From Fish to Philosopher. Little, Brown, Boston.

So, P.L., Danielian, P.S., 1999. Cloning and expression analysis of a mouse gene related to Drosophila odd-skipped. Mech. Dev. 84, 157–160.

Soueid-Baumgarten, S., Yelin, R., Davila, E.K., Schultheiss, T.M., 2014. Parallel waves of inductive signaling and mesenchyme maturation regulate differentiation of the chick mesonephros. Dev. Biol. 385, 122–135.

Stark, D.A., Kulesa, P.M., 2005. In vivo marking of single cells in chick embryos using photoactivation of GFP. Curr. Protoc. Cell Biol. Chapter 12, Unit 12.8.

Taelman, V., Van Campenhout, C., Sölter, M., Pieler, T., Bellefroid, E.J., 2006. The Notch-effector HRT1 gene plays a role in glomerular development and patterning of the Xenopus pronephros anlagen. Development 133, 2961–2971.

Takasato, M., Er, P.X., Becroft, M., Vanslambrouck, J.M., Stanley, E.G., Elefanty, a G., Little, M.H., 2014. Directing human embryonic stem cell differentiation towards a renal lineage generates a self-organizing kidney. Nat. Cell Biol. 16, 118–126.

Takezaki, N., Figueroa, F., Zaleska-Rutczynska, Z., Klein, J., 2003. Molecular phylogeny of early vertebrates: Monophyly of the agnathans as revealed by sequences of 35 genes. Mol. Biol. Evol. 20, 287–292.

Urban, A.E., Zhou, X., Ungos, J.M., Raible, D.W., Altmann, C.R., Vize, P.D., 2006. FGF is essential for both condensation and mesenchymal-epithelial transition stages of pronephric kidney tubule development. Dev. Biol. 297, 103–117.

Vize, P.D., Carroll, T.J., Wallingford, J.B., 2003. Induction, development, and physiology of pronephric tubules. In: The Kidney: From Normal Development to Congenital Disease. Academic Press, San Diego, pp. 19–50.

Vize, P.D., Jones, E.A., Pfister, R., 1995. Development of the Xenopus pronephric system. Dev. Biol. 171, 531–540.

Vize, P.D., Seufert, D.W., Carroll, T.J., Wallingford, J.B., 1997. Model systems for the study of kidney development: use of the pronephros in the analysis of organ induction and patterning. Dev. Biol. 188, 189–204.

Waddington, C.H., 1938. The Morphogenetic Function of a Vestigial Organ in the Chick. J. Exp. Biol. 15, 371–376.

Weavers, H., Prieto-Sánchez, S., Grawe, F., Garcia-López, A., Artero, R., Wilsch-Bräuninger, M., Ruiz-Gómez, M., Skaer, H., Denholm, B., 2009. The insect nephrocyte is a podocyte-like cell with a filtration slit diaphragm. Nature 457, 322–326.

Weijer, C.J., 2009. Collective cell migration in development. J. Cell Sci. 122, 3215–3223.

Wellik, D.M., Hawkes, P.J., Capecchi, M.R., 2002. Hox11 paralogous genes are essential for metanephric kidney induction. Genes Dev. 16, 1423–1432.

Wessely, O., Tran, U., 2011. Xenopus pronephros development-past, present, and future. Pediatr. Nephrol. 26, 1545–1551.

Wilm, B., James, R.G., Schultheiss, T.M., Hogan, B.L.M., 2004. The forkhead genes, Foxc1 and Foxc2, regulate paraxial versus intermediate mesoderm cell fate. Dev. Biol. 271, 176–189.

Xu, P.-X., Zheng, W., Huang, L., Maire, P., Laclef, C., Silvius, D., 2003. Six1 is required for the early organogenesis of mammalian kidney. Development 130, 3085–3094.

Xu, P.X., Adams, J., Peters, H., Brown, M.C., Heaney, S., Maas, R., 1999. Eya1-deficient mice lack ears and kidneys and show abnormal apoptosis of organ primordia. Nat. Genet. 23, 113–117.

Yu, J., McMahon, A.P., Valerius, M.T., 2004. Recent genetic studies of mouse kidney development. Curr. Opin. Genet. Dev. 14, 550–557.

Chapter 27

Development of the Genital System

Hongling Du* and Hugh S. Taylor†

**Division of Reproductive Endocrinology and Infertility, Department of Obstetrics, Gynecology and Reproductive Sciences, Yale University School of Medicine, New Haven, CT; †Department of Obstetrics, Gynecology and Reproductive Sciences Yale University School of Medicine, New Haven, CT*

Chapter Outline

GLOSSARY

Müllerian ducts Two embryonic tubes that extend along the mesonephros that become the uterine tubes, the uterus, and part of the vagina in the female and that form the prostatic utricle in the male. Also known as paramesonephric ducts.

Sex chromosome Either of a pair of chromosomes (usually designated as X or Y) in the germ cells of most animals and some plants that combine to determine the sex and sex-linked characteristics of an individual. In mammals, XX results in a female and XY results in a male.

Sex determination The process by which the sex of an organism is determined. In many species, the sex of an individual is dictated by the two sex chromosomes (X and Y) that it receives from its parents. In mammals, some plants, and a few insects, males are XY and females are XX; in birds, reptiles, some amphibians, and butterflies, the reverse is true. In bees and wasps, males are produced from unfertilized eggs, and females are produced from fertilized eggs. In 1991, *SRY* was identified as the sex determination gene of the Y chromosome. Environmental factors can also affect sex determination in some fish and reptiles. In turtles, for example, sex is influenced by the temperature at which the eggs develop.

Primordial germ cell An embryonic cell that gives rise to a germ cell from which a gamete (i.e., an egg or a sperm) develops.

Wolffian duct The duct in the embryo which drains the mesonephric tubules. It becomes the vas deferens in the male, and it forms vestigial structures in the female. It is also known as the mesonephric duct.

SUMMARY

- In mammals, sex determination is accomplished by a chromosomal mechanism. It begins at the time of fertilization through the coupling of two gametes: either two X chromosomes (XX in females) or an X and a Y chromosome (XY in males).
- The *SRY* gene is the sex determination gene of the Y chromosome.
- Gonadal differentiation begins after the migration of the primordial germ cells (PGCs) into the indifferent gonad. Testis determination is normally initiated in males by the expression of the *SRY* gene. In the absence of *SRY* expression, the bipotential gonad develops as an ovary. Ovarian differentiation is dependent on the presence of germ cells. However, germ cells are not necessary for testicular differentiation.
- The indifferent genital ducts consist of the mesonephric (Wolffian) ducts and the paramesonephric (Müllerian) ducts. In mammalian embryos, the testis secretes several hormones, such Anti-Müllerian hormone (AMH), testosterone, and insulin-like factor 3 (INSL3), which promote Wolffian duct differentiation into the male reproductive tract, whereas the Müllerian ducts degenerate in males. In the absence of the male hormones, the Wolffian ducts degenerate, whereas the Müllerian ducts persist and differentiate into the female internal reproductive tract. The fate of the indifferent genital ducts depends on the gender of the gonad.
- Genital duct development is a hormonally and genetically controlled process. *Hox* gene family members are involved in the development of genital ducts in both males and females. Along the anterior–posterior axis of the genital duct, *Hox* genes are expressed according to their 3'-5' order in the *Hox* clusters. *Wnt* gene family members are also involved in both female gonadal differentiation and female genital duct development.
- The external genitalia pass through an undifferentiated state before the distinguishing sexual characteristics appear. The development of the genital tubercle is initially regulated by *Hox* gene expression. Because the genital tubercle is located at the terminal part of the urogenital system, it expresses the 5'-most genes from the *Hox* gene clusters, specifically *Hoxa-13* and *Hoxd-13.*
- Malformations of the genital system are intrinsic defects in the developing human that result in localized abnormalities during the development of the reproductive duct system. The genetic abnormalities, such as chromosomal anomalies and gene mutations, influence the development of the genital system and cause malformations.

27.1 INTRODUCTION

In mammals, sex development is a genetically and hormonally controlled process, which involves three main sequential processes. It begins with the establishment of chromosomal or genetic sex at fertilization, when a sperm (with a Y or an X chromosome) fertilizes an ovum (with an X chromosome). This initial phase of genital development represents genetic sex determination, without any morphological indication of sex. The early embryo is phenotypically identical in both sexes; therefore this is referred to as the indifferent stage of sexual development. Next, gonadal differentiation is initiated in accordance with the expression of sex differentiation genes. During this phase, the gonads begin to acquire sexual characteristics and differentiate into either testes or ovaries. Subsequently, the development of internal or external sexual duct systems takes place. The morphological differentiation of sex is considered to begin with gonadal differentiation and to progress with sexual duct system development, which is influenced by the gonads. In this chapter, we will list the genetic factors that lead to sexual and gonadal differentiation and compare them among different species. Genetic and hormonal factors, which play a role in the formation of the internal and external genitals of both sexes, will be discussed. Malformations of the genital system and its mechanisms will also be described.

27.2 GENETIC SEX DETERMINATION

Sex Chromosomes

In mammals, sex determination is accomplished by a chromosomal mechanism. Whether a mammalian embryo develops into a male or a female is determined by its chromosome complement: XX embryos become females, and XY embryos become males. In humans, as in other mammals, females have two copies of a large X chromosome, and males have a single X and a much smaller and heterochromatic Y chromosome. The Y chromosome has a strong testis-determining effect on the indifferent stage, thus determining male gonad development. The hypothetical factor on the Y chromosome required for male differentiation is called the testis-determining factor (TDF). The male-specific Y chromosome in mammals, in addition to playing a vital role in sex determination, also harbors the genes that are required for spermatogenesis (Tiepolo et al., 1976; Welshons et al., 1959). The number of X chromosomes appears to be unimportant in sex determination, as indicated by the loss of an X chromosome in patients with Turner's syndrome (45, X or 45, XO). These patients present gonadal dysgenesis, but they are phenotypically female, with the ovaries represented by gonadal streaks. To summarize, the presence of a Y chromosome results in the differentiation of the embryonic somatic cells of the gonad into testes rather than ovaries. The absence of a Y chromosome results in female gonad development and the formation of the ovaries. The gonads then determine the type of sexual differentiation in the internal genital ducts and the external genitalia.

Non-mammalian vertebrate species have a variety of sex chromosomal systems, such as ZZ–WZ, WX–XX–WY, and XY (Nanda et al., 2002; Schartl, 2004). For example, birds have differentiated Z and W chromosomes in which the W is usually small and heterochromatic. As opposed to mammals, male birds are homogametic, which means that this sex produces sperm and has two copies of the same sex chromosome. Those with two copies of the Z chromosome (ZZ) are male, whereas those with a one each of the Z and W chromosomes (WZ) are female. In some reptiles, such as crocodilians and marine turtles, the development of the embryonic gonad into testis or ovary is dependent on temperature. By contrast, snakes have a fixed genetic sex determination system. The chromosomal system is the WZ type, and the incubation temperature of the eggs cannot influence the development of the embryonic gonads.

In amphibians, most species have homomorphic sex chromosomes, which are morphologically identical members of a homologous pair of chromosomes. However, there are also some species with heteromorphic sex chromosomes, which are a chromosome pair with some homology but that differ with regard to size, shape, and staining properties. In the Japanese frog *Rana rugosa*, populations with heteromorphic XY, homomorphic XY, and heteromorphic WZ sex chromosomes have been identified, even within the same species.

In fish, the chromosomal mechanisms show an enormous variety. XY and WZ systems are the most common. In addition, XO, ZO, X1X2Y, XY1Y2, and Y autosome fusion have been described. Systems exist in which multiple sex chromosomes are present in a population. For example, three types of sex chromosomes (X, W, and Y) coexist in a population of platyfish (*Xiphophorus maculatus*). Individuals with WX, XX, and WY become females, whereas XY and YY fish become males. However, in fish and amphibians, sex can be reverted by age, social factors, and temperature or hormone treatment, or can even change spontaneously.

Thus, in non-mammalian vertebrates, the variety of the chromosomal system is the consequence of the dynamic process of the evolution of sex determination mechanisms.

The Sex Determination Gene

Evolution of Sex Chromosomes

In mammals, male sex determination is controlled by genes on the Y chromosome. Evolutionary comparisons show that the X and Y chromosomes were originally homologous. Although the X chromosome is consistent in size and gene content, the Y chromosome is much more variable among mammalian species. A comparison of the gene content of sex chromosomes from the three major groups of extant mammals (placentals, marsupials, and monotremes) shows that part of the X chromosome and a corresponding region of the Y chromosome is shared by all mammals and thus must be very ancient. In humans, the X and Y chromosomes share two short regions at either end (the pseudoautosomal regions [PARs]) throughout which they are homologous and thus may recombine (Burgoyne, 1998). The PAR on the short arms of the X and Y chromosomes (PAR1) undergoes pairing and recombination at meiosis. By contrast, the PARs on the long arms of the X and Y chromosomes (PAR2) pair only infrequently, and the homology observed in this region is probably maintained by gene conversion. In addition, many genes and pseudogenes on the Y chromosomes have homologs on the X chromosomes. The evolution of the mammalian Y took place in several cycles of addition and attrition as autosomal regions were added to the pseudoautosomal region of one sex chromosome, recombined onto the other, and degraded on the Y.

This indicates that, no matter how different they are in size and gene content today, the X and Y chromosomes were once homologs. The Y chromosome seems to have degraded progressively over the last 200 million years, perhaps as a consequence of keeping the sex-determining gene together with allied male-specific genes. Similar degradation occurs in single chromosomes that do not undergo recombination, irrespective of sex. Snakes have a ZZ male and a ZW female system. Like the mammalian X chromosome, the Z chromosome of snakes is large, containing about 6% of the genome in all snake families. Birds that are as distantly related as the chicken and the emu have Z chromosomes that are nearly identical genetically (Shetty et al., 1999). The W chromosome is considered more variable, and it has different sizes in various families of birds and snakes. The W chromosome is largely homologous to the Z chromosome in the emu, but it has become small and heterochromatic in the chicken. The process of W chromosome degradation has therefore taken place to different extents and independently in different bird and snake lineages.

The differentiation of the X and Y chromosomes is thought to have been initiated in an ancestral mammal when an allele at a single locus on the proto-Y took over a male-determining function from an ancestral genetic or environmental sex-determining system. The comparison of sequences of X-Y shared genes across species shows that the Y copy changes far more rapidly than the X copy. Comparative mapping studies show no homology between bird Z and mammalian X sex chromosomes, which indicates that the two sex chromosomes systems evolved independently. Other vertebrate classes show a wide variety of genetic and environmental sex determination mechanisms, so it is not possible to infer the sex-determining system of the common reptilian ancestor.

Sex-Determining Genes on the Y Chromosome in Mammals

In the mammalian XY chromosomal sex-determining system, sex differentiation depends on Y chromosome-specific genes that trigger male development. Several testis-determining candidate genes have been identified: a minor male-specific antigen (HYA); the zinc finger Y chromosome (ZFY); and the sex-determining region of the Y chromosome (SRY).

HYA

HYA is a minor male-specific antigen. It was originally discovered during the mid-1950s as a transplantation or histocompatibility antigen that caused female mice of a certain inbred strain to reject male skin of their own strain. The *HYA/Hya* gene encoding the HYA has been mapped to the long arm of the Y chromosome in humans and the short arm of the Y chromosome in mice. During the 1970s, HYA was proposed to be the testis-determining factor as a result of two traits: HYA is uniquely expressed on male cells, and the presence of HYA is associated with testis determination as indicated by XX sex-reversed male mice. These mice have two X chromosomes, but they are clearly phenotypically male. As a result of a chromosome rearrangement, the *Hya* gene is expressed in these XX mice, which provides evidence that is consistent with HYA being the testis-determining Y-encoded gene (Wachtel et al., 1975). However, during the 1980s, it was found that HYA is absent from certain mice that develop testes and that are of indisputably male phenotype; this finding disputed the function of the HYA as the TDF (McLaren et al., 1984).

ZFY

Another candidate gene, *ZFY*, which is located on the Y chromosome, encodes a zinc finger protein. It lies close to the pseudoautosomal boundary on the short arm of the human Y chromosome. In the mouse, *Zfy* was found to consist of two duplicated genes, *Zfy-1* and *Zfy-2*, which are both present on the normal human Y chromosome. ZFY was initially considered as a TDF candidate (Page et al., 1987), but its lack of conservation on the Y chromosome in marsupials made it an unlikely candidate for a universal mammalian TDF. More refined mapping of the human Y chromosome defined a new minimum sex-determining region that lacked ZFY, and this ultimately excluded ZFY from a role in sex determination (Palmer et al., 1989).

SRY

In 1991, the *SRY* gene was cloned from a region closely linked to ZFY, and it has been confirmed as the TDF needed from the Y chromosome to establish male development (Figure 27.1; Koopman et al., 1990; Sinclair et al., 1990). *SRY/Sry* is a small intronless gene that encodes a protein with a conserved DNA-binding high mobility group (HMG) box. *SRY* is a member of a large family of *SRY*-like HMG box containing genes. The presence of an *SRY* mutation in about 15% of human XY females supported the proposition that this gene represented the TDF. The identity of *Sry* as the TDF was determined in the mouse. The *Sry* gene is absent in a strain of XY mice that are phenotypically female. In the transgenic mouse, *SRY* can cause XX mice to undergo sex reversal and develop as males. This occurs despite the fact that they lack all other genes of the Y chromosome. *Sry* is transcribed in the genital ridges of embryos just before testis differentiation, but it is not expressed in the gonads of female mice embryos. In addition, *sry* was cloned from marsupials and shown to map to the Y chromosome, which indicates that it represents the common ancestral mammalian TDF. The *SRY* gene encodes a transcription factor that regulates the genes that are responsible for testicular development.

Sex-Determining Genes on the Z or Y Chromosome in Non-Mammalian Vertebrates

Sex determination systems are diverse in vertebrates. As described previously, in mammals, sex differentiation depends on sex-determining genes. However, in non-mammalian vertebrates, sex is also determined by heredity, the environment, or both. Interestingly, SRY-like sex-determining genes have also been identified in non-mammalian vertebrates, and they are considered to be one of the sex-determining factors.

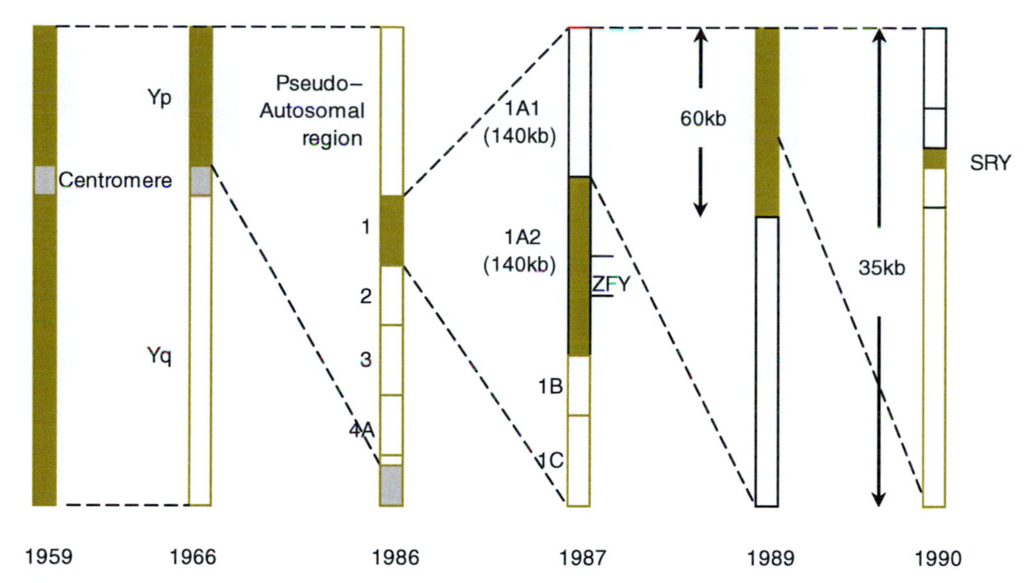

FIGURE 27.1 **A History of the Sex-Determination Gene (Sry) Localized on the Y Chromosome.** The central role of the Y chromosome in male sex determination has been recognized for many years. In 1987, ZFY was isolated and initially equated with the TDF. In 1989, another area located near ZFY was found to play role in male sex determination. In 1991, SRY gene was isolated in this area and identified as the TDF. *Reproduced with permission from Sultan et al. (1991).*

DMRTI

Doublesex and mab-3-related transcription factor 1 (Dmrt1) is a Z-linked candidate for the male sex-determining gene. In birds, there is no copy of *dmrt1* on the W chromosome, and males have two copies. *dmrt1* has been implicated in male sexual development in many vertebrate species, and it is considered to be the "*Sry* gene" of non-mammalian vertebrates (Zarkower, 2001). Recent advances in the characterization of the human *DMRT1* gene show that multiple transcripts are expressed in the human testis. Thus, *dmrt1* is likely to have an important role in the evolution of the sexual development mechanisms of many species (Cheng et al., 2006).

DMY

The DM domain gene on the Y chromosome (DMY) has been found in the sex-determining region of the Y chromosome of the teleost medaka fish *Oryzias latipes*. Mutations of the DMY cause a simple sex reversal in medaka, and this has been confirmed by the observation of naturally occurring mutant females in several wild populations. DMY appears to be closely related to Dmrt1 with regard to both nucleotide sequence (93% identity) and function (Matsuda, 2003). However, compared to the *Sry* gene, which is distributed widely in mammals, DMY is absent in most *Oryzias latipes* and closely related species (Takehana et al., 2007). So far, the evolutionary comparison of sex determination genes indicates that molecular similarities among phyla are only present in the fly doublesex, the worm mab-3, and the vertebrate *dmrt1* (dsx- and mab3-related transcription factor 1)/*dmy* genes (see also Chapter 9).

27.3 GONADAL DIFFERENTIATION

Primordial Germ Cell Migration

Primordial germ cells (PGCs) are the embryonic precursors of the gametes. In all systems, PGCs form far from the site of the developing gonads and migrate to the sites of developing ovaries or testes. This isolation may be important in the maintenance of their unique characters. It is accepted that PGC migration occurs in three phases: separation, migration, and colonization. Several mechanisms have been hypothesized to explain PGC migration, including self-movement, attraction by chemotactic factors, PGC–PGC interactions, substrate guidance, and interaction with extracellular matrix molecules.

In human embryos, PGCs are visible early in the fourth week of gestation, among the endodermal cells from the posterior wall of the yolk sac, near the origin of the allantois. During the folding of the embryo, part of the yolk sac is incorporated into the embryo, and PGCs migrate along the wall of the hindgut and through the dorsal mesentery, until they approach the newly appearing genital ridges late in the fifth week. During the sixth week, PGCs migrate into the underlying mesenchyme, and they become incorporated into the primary sex cords. The migration of PGCs is considered to be accomplished by the active amoeboid movement of the cells in response to a permissive extracellular matrix substrate.

In the mouse, PGCs are induced to form in the proximal epiblast, where their formation is dependent on the expression of bone morphogenetic proteins 4 and 8b in extra-embryonic tissue (Lawson et al., 1999; Ying et al. 2001). During gastrulation, they move through the primitive streak and invade the definitive endoderm, the parietal endoderm, and the allantois. In both XX and XY mice, PGCs can first be detected at 7.5 days post coitum (dpc) at the base of the allantois. By 9.0 dpc, the PGCs in the definitive endoderm become incorporated into the hindgut. Between 9.0 dpc to 9.5 dpc, PGCs migrate through the dorsal side of the hindgut to colonize the developing genital ridge. At 10.5 dpc, PGCs begin to cluster, forming a network of migrating cells. By 11.5 dpc, most PGCs have colonized the genital ridge. The entire migration takes approximately 4 days (Figure 27.2).

In mouse embryos, two genes required for the migration of PGCs have been identified: *C-kit* and *Steel*. *C-kit* encodes a protein receptor that is located on the surface of the migrating PGC; *Steel* encodes a growth factor that is expressed in the somatic cells that are placed in the pathway of the migrating PGC and that functions as the ligand of C-KIT. The STEEL/C-KIT interaction forms a ligand-receptor pathway that is required for PGC colonization, survival, and migration (Motro et al. 1991). *Fragillis/Mil-1* was recently identified as the gene that initiates PGC motility. It is a member of an interferon-inducible family of genes that is implicated in homotypic cell-cell adhesion and cell-cycle control. At 7.25 dpc, *Fragilis* is expressed in the posterior epiblast, with the highest level of expression overlapping the region where PGCs are formed. Twenty-four hours later, its expression is downregulated as PGCs scatter and move into the endoderm (Saitou et al., 2002). The expression of two other members of the fragilis gene family (*fragilis 2* and *fragilis 3*) was found in nascent PGCs that play a role in maintaining PGC migration.

PGC migration in flies can be divided into four stages:

(1) Internalization of the pole cells;
(2) Emigration of PGCs from the gut;

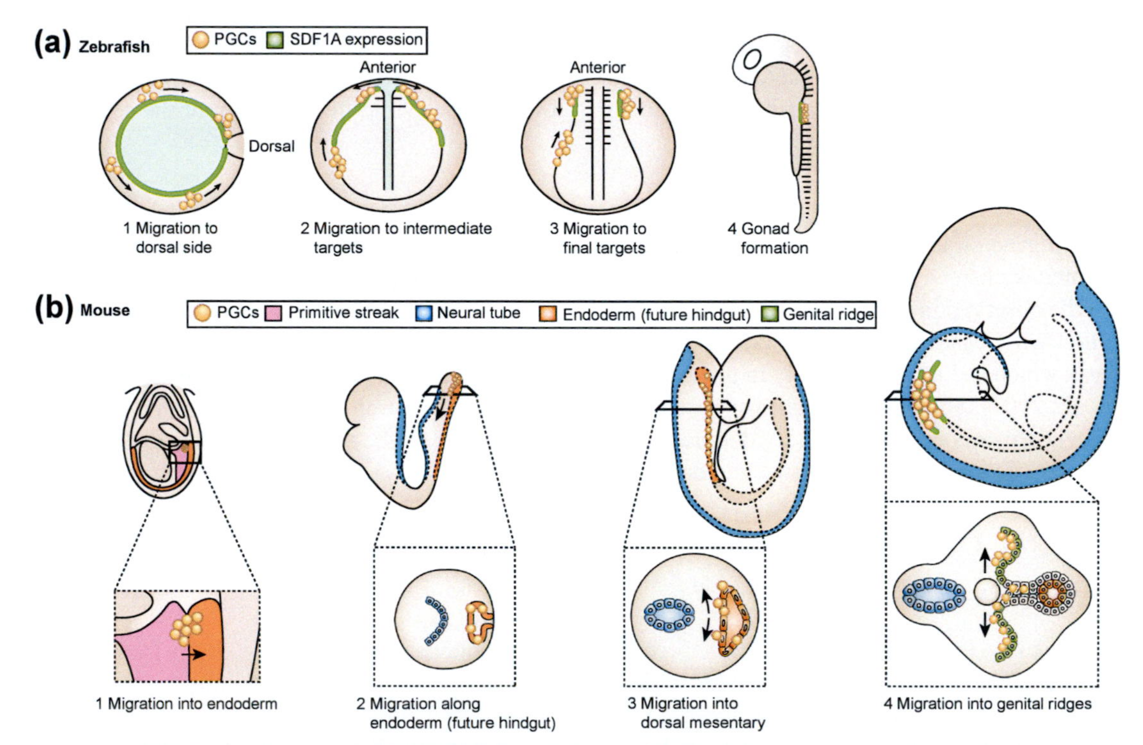

FIGURE 27.2 Stages of Primordial Germ Cell Migration (A) In zebrafish, PGCs migrate dorsally, following specification at four random locations (step 1; animal pole view – the animal pole refers to the portion of the blastula embryo that differentiates into mesoderm and ectoderm). At gastrulation, 4.5 hours post-fertilization (4.5 hpf), PGCs follow the expression of stromal-derived factor 1a (SDF1A) and at 10.5 hpf somites 1–3 act as intermediate targets (step 2; lateral view, left side). PGCs migrate towards the final target tissue at somites 8–10 at 13 hpf (step 3; frontal view). At 24 hpf, PGCs coalesce with the somatic cells of the gonad (step 4; lateral view, left side).(B) In mice, PGCs, specified in the proximal epiblast, migrate from the primitive streak to the endoderm (future hindgut) at embryonic day 7.5 (E7.5; step 1). A close-up is shown in the bottom panel. At E8, PGCs migrate along the endoderm (step 2). At E9.5, PGCs migrate bilaterally towards the dorsal body wall (step 3). At E10.5, PGCs reach the genital ridges to form the embryonic gonad (step 4). Lateral views (top panels) and transverse sections (bottom panels) are shown. *Reproduced with permission from Richardson and Lehmann (2010).*

(3) The lateral migration of PGCs; and
(4) Gonad coalescence.

PGCs arise from the posterior pole of the developing embryo, where localized maternal components become segregated into pole cells. Next, PGCs migrate out of the ventral side of the gut along the basal surface and into the lateral mesoderm, where they coalesce with the somatic cells of the gonad. Torso is a maternally inherited transcript encoding a tyrosine kinase. Its activity is required for the efficient incorporation of pole cells into the hindgut pocket. Torso signaling in flies initiates pole cell multiplication, and it may be analogous to the role of c-kit in PGC development in mice. The loosening of cell-cell contacts between the cells of midgut epithelium allows PGCs to emigrate from the gut, forcing them away from the gut and toward the overlying mesoderm by a repulsive signal mediated by *wunen/wunen2*. Thus, *wunen* gene expression appears to be responsible for directing the migration of the PGC in flies (Santos et al., 2004).

Similarly, in zebrafish, PGCs are specified by maternal components that become segregated into four clusters within the cleaving embryo. During gastrulation, these PGCs cluster more dorsally and align at the border between the head and trunk mesoderm, or they align within the lateral mesoderm. Next, PGCs migrate posteriorly to colonize the developing gonad (Figure 27.2). *Dead end* is a gene that has been identified in zebrafish that plays a specific role in the initiation of PGC motility. *Dead end* homologs have been identified in PGCs in *Xenopus*, the chicken, and the mouse. However, in the mouse, *Dead end* is expressed in PGCs after the migratory stages; hence, its function in PGC development may not be entirely conserved. In zebrafish, migration of PGCs is regulated by SDF-1/CXCR4 chemokine signaling. This pathway also has been shown to have an important role in PGC migration in the mouse, the chicken and *Xenopus* (Takeuchi et al., 2010).

Because there are enormous ethical, technical, and logistical problems with regard to *in vivo* studies of the movement of PGCs in humans, many experiments have been carried out in other animal models. The initiation of PGC motility is currently poorly understood, and it may be controlled by species-specific mechanisms. Despite their different origins, the early development of PGCs in flies and mice is quite similar, and the survival and early migration of PGCs in these systems require signaling via tyrosine kinase receptors (*torso* and *c-kit*, respectively). A tyrosine kinase receptor with a role similar to that of *torso/c-kit* has yet to be identified in zebrafish. The initiation of PGC motility in zebrafish is controlled by the mRNA binding protein Dead end. PGC guidance mechanisms have been well studied in all three species, and they require chemoattractants that signal via G-protein-coupled receptors, cell-cell adhesion, and, probably, specific interactions between PGCs and the extracellular matrix.

Despite all this, evidence now suggests that there is no active migration of PGCs in the human embryo and that the displacement of germ cells can be explained by the global growth and movement of the embryo (Freeman, 2003). The analysis of recent data suggests that human PGCs do not actively migrate at any stage (either up or down) but rather that they are embedded in the caudal tissues of the embryo (e.g., allantois, hindgut, coelomic serosa) and carried along passively during the curvilinear unrolling of this caudal region. The similarities between mouse and human embryos are so impressive that it has been doubted whether PGCs are moving independently in the mouse embryo, either. It has been proposed that, in the mammalian embryo, there is a passive carriage of multiplying cells in a caudal direction rather than an active ascent of PGCs. Such carriage of cells accompanies the normal development of the caudal part of the embryo (the so-called embryonic unrolling). This new hypothesis for a passive migration of PGCs in the human embryo encourages a re-examination of the evidence for the previously widely accepted active migration of PGCs in other species.

Origin of the Gonads

The gonads are derived from three sources: the coelomic epithelium, the underlying mesenchyme, and the PGCs. Gonadal organogenesis begins with the appearance of the genital ridge. Initially, a thickened area of coelomic epithelium develops on the medial aspect of the mesonephros. After PGCs begin to colonize the ventral region of the urogenital ridge, cells of the coelomic epithelium and the underlying mesenchymal cells undergo active proliferation. Branches of blood vessels from the dorsal aorta and the cardinal veins send endothelial cells into the genital ridges. Proliferating coelomic and mesenchymal cells, together with arriving PGCs and endothelial cells, form a cluster of condensed cells that gradually become the undifferentiated gonad.

In the mouse, the urogenital ridge develops beginning around 9.5 dpc. The morphologic establishment of the undifferentiated gonad takes approximately 48 hours (10 to 12 dpc). During this stage, the initially amorphous cluster of condensed cells becomes segregated into two compartments:

(1) An epithelial compartment formed by PGCs and epithelial-like cells surrounded by a basal membrane; and
(2) A stromal compartment formed by mesenchymal cells, fibroblasts, and blood vessels.

In the human, gonadal development is first identified during the fifth week of gestation. The gonads arise from an elongated region of mesoderm along the ventromedial border of the mesonephros. The indifferent gonad develops in close association with the mesonephros, an embryonic kidney that contributes to both the male and female reproductive tracts. Germ cells induce cells of the mesonephros to form the genital ridge. Cells in the cranial part of this region condense to form the adrenocortical primordia, and those of the caudal part become the genital ridges. Soon, finger-like epithelial cords called primary sex cords grow into the underlying mesenchyme.

The Bipotential Gonad and the Sex-Determining Switch

In mammals, both testis and ovary share a common origin – the bipotential gonad – that possesses neither distinctly male nor female characteristics. The indifferent gonad can differentiate into either testes or ovaries, depending on the presence or absence of a Y chromosome, respectively. The fate of the gonad is specified by the SRY gene product. In all other reproductive organs (discussed later), sexually dimorphic development does not depend directly on chromosomal complement but rather on the presence of either the male or female gonad.

There are essentially three different cell lineages present in the gonads in addition to the germ cells. Each lineage has a bipotential fate and is capable of differentiating along either the male or female pathway. The supporting cell lineage will give rise to Sertoli cells in the testis and follicle cells in the ovary. These cells surround the germ cells and provide an appropriate growth environment. The steroidogenic cell lineage produces the sex steroid hormones that will contribute to the development of the secondary sexual characteristics of the embryo. In the male, these correspond with the Leydig cells; in the female, these correspond with the theca cells. Finally, a connective cell lineage contributes to the formation of the organ as a whole in both the testes and the ovaries. Early testis development is characterized by the formation of testicular cords that contain Sertoli and germ cells, with the Leydig cells excluded to the interstitium. The connective cell lineage is a major contributor to cord formation as the peritubular myoid cells surround the Sertoli cells; together, they lay down the basal lamina. The testis is also characterized by rapid and prominent vascularization. Organization of the ovary takes place later in development and is less structured, with the connective tissue lineage giving rise to stromal cells and with no myoid cell equivalent.

There are three genes which have been identified as being necessary for the development of the undifferentiated or bipotential gonad in the mouse: *Wt-1*, *Sf1*, and *Lim-1* (Luo et al., 1994; Pritchard-Jones et al., 1990; Shawlot et al., 1995). The *WT-1* gene was isolated from humans who developed the Wilms' kidney tumor, and it is specifically expressed within the developing genital ridge and the kidney in mouse embryos. *Wt-1* knockout mouse embryos lack kidneys and a genital ridge. The orphan nuclear receptor steroidogenic factor 1 gene (*Sf1*) is a regulator of the tissue-specific expression of cytochrome P-450 steroid hydroxylases, and it is a member of the nuclear hormone receptor family. It is present in the mouse urogenital ridge at the earliest stage of organogenesis. The targeted disruption of *Sf-1* in mice prevents the establishment of undifferentiated gonads, and all *Sf-1* null mutants develop a female phenotype. *Lim-1*, which is part of the LIM-homeodomain subclass of LIM proteins, is also important for urogenital development. *Lim-1* null mutant mice typically die approximately 10 dpc, and the few that develop to term lack kidneys and gonads.

Differentiation of Testes

In mammals, testes differentiation begins at an earlier stage of development than does the differentiation of ovaries. Testis determination is normally initiated in males by the expression of the *SRY* gene on the Y chromosome in the bipotential gonad that is common to both males and females (Figure 27.3). In the human, SRY expression can first be detected at 41 days. In the mouse, gonads, SRY expression can be detected by 10.5 dpc, and testis differentiation occurs approximately 36 hours later, between 12.0 dpc and 12.5 dpc (Hacker et al., 1995). SRY expression in gonadal somatic cells initiates the differentiation of Sertoli cells, which are known as the supporting cell lineage of the testis that is essential for its subsequent differentiation. Sertoli cells polarize and aggregate around the germ cells to support their growth and maturation. After SRY expression, sry-related HMG box-9 (sox9), which is another definitive testis differentiation gene, is activated (Morais et al., 1996). *Sox9* upregulation is the earliest marker of pre-Sertoli cells. *Sry* and *Sox9* expression overlap in Sertoli cells, co-localizing to the nucleus of pre-Sertoli cells as early as 11.5 dpc. As the sox9-expressing population expands between 11.5 dpc and 12.5 dpc, the number of cells that co-express SRY decreases until 12.5 dpc, when SRY expression is extinguished and SOX9 expression is confined to Sertoli cells inside of the cords. Several male-specific cellular events (e.g., glycogenesis, coelomic epithelium proliferation, mesonephric migration, vasculogenesis) are induced in XY gonads after the onset of SRY and SOX9 expression. Although the *SRY* gene was discovered more than ten years ago, the mechanism of how SRY functions as a TDF remains unknown. Heterozygous human *SOX9* mutations cause campomelic dysplasia,

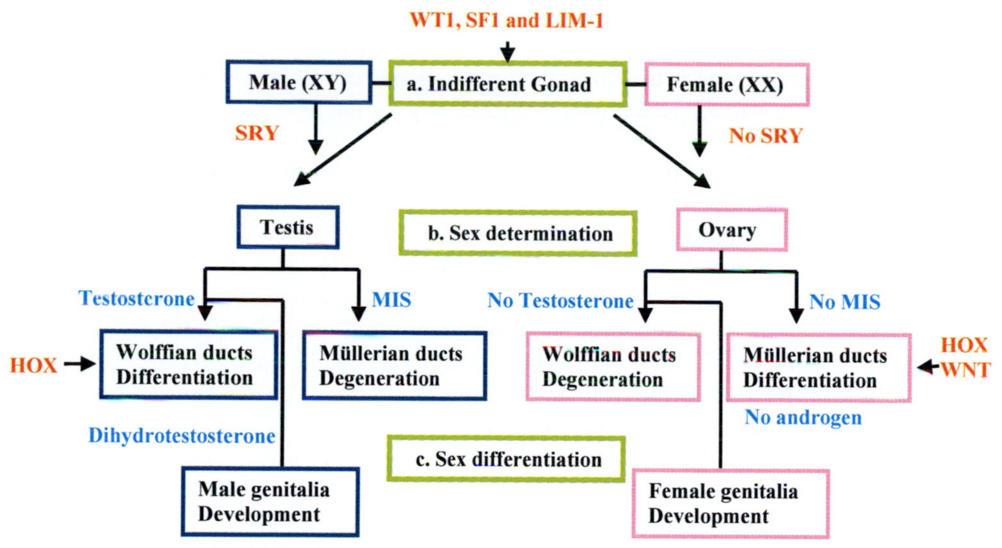

FIGURE 27.3 Genetic and Hormonal Factors in the Mammalian Sex Determination and Differentiation. (A) There are three genes identified as necessary for development of the undifferentiated or bipotential gonad in the mouse: wt1, sf1, and lim-1. (B) Testis determination is initiated in males by expression of the sry gene on the Y chromosome; in the absence of SRY expression, the bipotential gonad develops as an ovary. (C) Male hormones promote both Wolffian duct and male genitalia development, whereas Müllerian ducts degenerate in male. In the absence of male hormones, both Müllerian ducts and female genitalia differentiate, whereas the Wolffian ducts degenerate in female. *Hox* genes are involved in the development genital ducts both in male and female. *Wnt* genes are involved in female genital duct development.

a severe skeletal disorder that involves defective cartilage development. Many of these male patients also have gonadal dysgenesis. Heterozygous mice that are haplo-insufficient for *Sox9* die perinatally as a result of skeletal malformation. DAX1 is an X-linked orphan nuclear receptor that is expressed in the ventromedial hypothalamus, the pituitary gonadotropes, the adrenal cortex, the testis, and the ovary. The duplication of the X-chromosomal-region-spanning *Dax1* results in dosage-sensitive, male-to-female sex reversal. Alternatively, *Dax1* deficiency influences testis cord formation. It appears that appropriate DAX1 levels are critical for normal testis development: too little or too much of the factor can have an anti-testis effect. Recently, DAX1 was reported to function as an early mediator of testis development downstream of SRY. The analysis of $Dax1^{-/-}$ mice implies that DAX1 functions at an early stage downstream of SRY, or even possibly in a parallel pathway of SRY to establish Sertoli cell differentiation. The data also show that DAX1 may play an important functional role in Leydig cells (Meeks et al., 2003).

The lifespan of fetal Leydig cells can be divided into three stages: differentiation, fetal maturity, and regression. Leydig cells differentiate after Sertoli cells during fetal development under their paracrine action. There are two signaling systems that are essential for fetal Leydig cell differentiation: the Desert hedgehog (DHH)-Patched system and the platelet-derived growth factor (PDGF)-receptor A system (Brennan et al., 2003; Clark et al., 2000). The ligands DHH and PDGF-A are produced by fetal Sertoli cells, whereas fetal Leydig cells express their cognate receptors (Patched for Dhh and PDGF-Ra for PDGF-A). The recognition of the involvement of these signaling systems in the process of fetal Leydig cell differentiation came from gene inactivation experiments in mice. Genetic analysis has placed PDGF-Ra upstream of DHH. Mutants of both *Dhh* and *Pdgf-ra* have early defects in the partitioning of the testis cord and the interstitial compartment and severely impaired differentiation of fetal Leydig cells. A similar human phenotype was described for a 46, XY patient with a *DHH* mutation. In both the mouse knockout and the human mutation, female external genitalia with a blind vagina were observed. Moreover, the Wolffian duct derivatives and prostate were decreased in size, indicating insufficient androgen production by fetal Leydig cells. Finally, the X-linked aristaless-related homeobox gene (*Arx*) gene is a transcription factor that is barely detectable in fetal Leydig cells; however, in the $Arx^{-/-}$ knockout mice, the Leydig cell population was severely diminished suggesting an important role for this gene as well (Kitamura et al. 2002).

Anti-Müllerian hormone (AMH), which is also known as Müllerian inhibitor substance, is a key peptide hormone produced by Sertoli cells, and it belongs to the Transforming Growth Factor β family. It is the earliest marker of testis formation. In the human, AMH is secreted by fetal Sertoli cells from around 8 weeks. In the mouse, *Amh* transcripts are first present in the pre-Sertoli cells at 12.5 dpc, when these cells are forming cords. This factor acts as an important molecular switch to turn on a network of downstream factors that are involved in testicular development as well as male sex differentiation (Munsterberg et al., 1991).

Differentiation of Ovaries

In mammalian embryos, the testis pathway is the active pathway in gonad development. In mice lacking a Y chromosome (more specifically the *Sry* gene), gonadal development occurs slowly. In the absence of SRY expression, the bipotential gonad develops as an ovary.

In the female mouse, the gonads retain an undifferentiated state for longer than in the male. Strings of oocytes form indistinct cords that are observed at 14 dpc. By 15 dpc, these are separated by stromal cell partitions and blood vessels. In humans, the ovary is not positively identifiable until about the tenth week of gestation. Ovarian differentiation depends on the presence of germ cells, so if primordial germ cells (PGCs) fail to reach the genital ridges, streak ovaries are formed. In the absence of oocytes, somatic cells transdifferentiate toward testicular tissue and this includes the appearance of XX Sertoli cells (Nilsson et al., 2002; 2004). Similarly to the testis, the ovary contains primitive sex cords in the medullary region, but these are not as well developed as they are in the testis. In the female, the initial sex cords degenerate, but they are replaced by new epithelial proliferation that results in a new set of sex cords. These cortical sex cells superficially penetrate the mesenchyme, remaining near the outer cortex of the ovary, which is the location of the female germ cells. Rather than forming an interconnected network, these cords form distinct clusters surrounding germ cells. This is the initial origin of the ovarian follicles. The epithelial sex cords differentiate into granulosa cells, whereas the mesenchymal cells form the thecal cells.

Relatively few genes have been shown to be specific to early female gonadal development. *DAX1* was cloned from an X-chromosomal region in humans that was responsible for dosage-sensitive sex reversal. It is specifically expressed in XX gonads after 11.5 dpc, and it was initially suggested as a pro-ovarian or antitestis candidate gene. However, DAX1 loss of function on the XX background does not prevent ovary development. Subsequent studies have shown an expected role for *Dax1* in testis development, which indicates that its actions are highly dependent on the timing and level of expression (Bardoni et al., 1994).

The existence of sex reversal XX individuals that develop as males in the absence of the *SRY* gene led to the proposal that sry normally represses a factor (Z) that functions at the top of a genetic cascade as a repressor of male development. Thus, according to this theory, the Z factor would be repressed by SRY in the male, and it would be independent of *SRY* in an XX genetic background. Loss of a Z factor would be sex-reversing on XX (female-to-male). One candidate for this Z factor is Wingless-related integration site 4 (*Wnt4*), which acts as a partial antitestis gene by repressing aspects of male development in the female gonad. In the mouse, *Wnt* genes are expressed in the mesonephric mesenchyme between 9.5 and 10.5 dpc and in the gonadal mesenchyme of both sexes at 11 dpc. By 11.5 dpc, gonadal *Wnt4* expression is downregulated in the male but maintained in the female. *Wnt4* signaling is required to maintain the female germ line and to suppress the differentiation of Leydig cell precursors (McElreavey et al., 1993). In humans, a loss-of-function mutation in *WNT4* causes Mayer-Rokitansky-Kuster-Hauser syndrome, which is characterized by the defective development of Müllerian derivatives and the duplication of a chromosomal region containing *WNT4*; this condition was associated with a case of human XY sex reversal. Recently R-spondin1 (Rspo1), a novel, soluble activator for *Wnt/β-catenin* signaling, has been shown to be involved in early ovarian differentiation in both humans and mice (Chassot et al., 2008; Tomaselli et al., 2008; 2011).

Testicular differentiation can occur in the absence of germ cells, but these cells are essential for the formation and maintenance of follicles in the ovary. In their absence, follicles degenerate into cord-like structures, and XX cells express male markers such as SOX9 and AMH (Brennan et al., 2004).

27.4 DEVELOPMENT OF THE GENITAL DUCTS

The Indifferent Stage

Like the gonads, the sexual ducts pass through an early indifferent stage. Unlike the gonads, in which a single tissue is bipotential, the indifferent genital ducts involve two options: the mesonephric (Wolffian) ducts and the paramesonephric (Müllerian) ducts. Both male and female embryos have two pairs of genital or sexual ducts. These ducts can differentiate into male or female reproductive organs according to the hormonal status of the fetus. In mammalian embryos, the testis secretes several hormones that promote Wolffian duct differentiation into the male reproductive tract. They form the epididymides, the ductus, and the ejaculatory system when the Müllerian ducts degenerate. In the absence of male hormones, the Wolffian ducts degenerate, but the Müllerian ducts persist and differentiate into the female internal reproductive tract, which is made up of the fallopian tubes, the uterus, and the superior portion of the vagina. The fate of the indifferent genital ducts depends on the gender of the gonad (Figure 27.4).

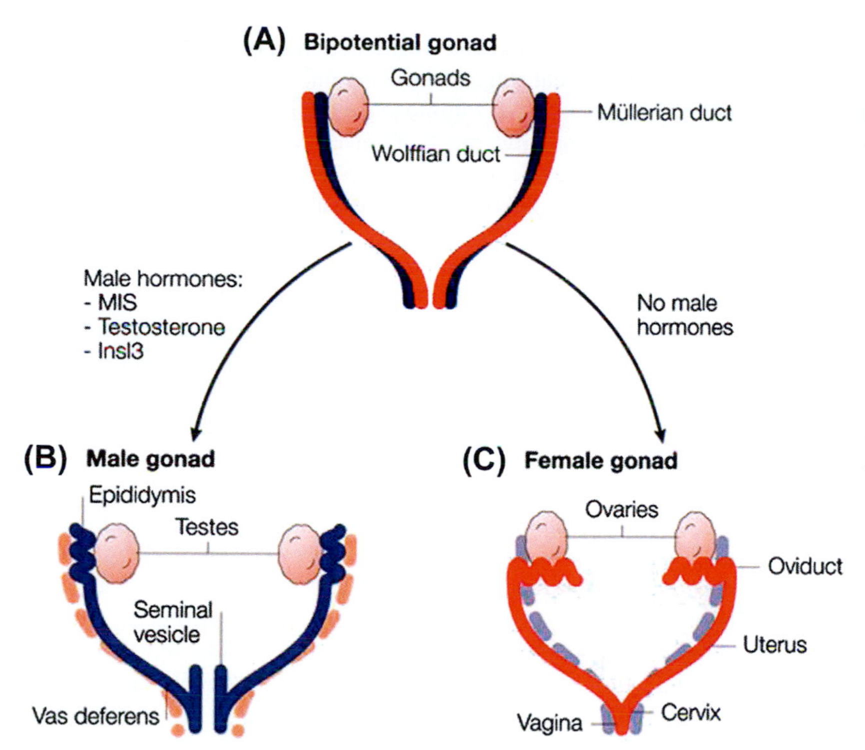

FIGURE 27.4 Sexual Differentiation of the Mammalian Reproductive System. Before sexual differentiation, both male and female embryos have bipotential gonads, as they possess both Wolffian and Müllerian ducts (A). These ducts can differentiate into male or female reproductive organs according to the hormonal status of the fetus. Owing to the expression of Sry, the bipotential gonad of males becomes the testis, which secretes several hormones including testosterone, MIS (aka AMH) and Insl3 (B). Testosterone promotes Wolffian duct differentiation into the male reproductive tract, and MIS eliminates the Müllerian ducts. In females, the bipotential gonad becomes the ovary (C). In the absence of male hormones, the Wolffian ducts regenerate, whereas the Müllerian ducts persist and differentiate into the female reproductive tract. *From Matzuk and Lamb (2003).*

Development of the Male Genital Duct

Owing to the expression of SRY, the bipotential gonad of males becomes the testis. Hormones play an essential role in regulating male sexual development after the testis has formed. In mammals, this regulation depends on three key hormones produced by the fetal testis: AMH, testosterone, and insulin-like factor 3 (INSL3). In the absence of these critical testicular hormones, female sex differentiation occurs.

In males, the Müllerian duct system forms early on but subsequently regresses. The elimination of the Müllerian ducts in the male fetus is driven by AMH (Viger et al., 2005), which is secreted by Sertoli cells. The expression of *AMH* starts around 12.5 dpc in the mouse and at around eight weeks in the human. It is maintained throughout fetal development, and it then declines markedly after birth. In XY embryos, AMH regulates male sex differentiation by triggering the regression of the paramesonephric (Müllerian) duct. In the human, the primary role of AMH in sex development is to cause a gradient of cranial-to-caudal regression of the Müllerian ducts during a short period (from 8 to 10 weeks of gestation). This is achieved by the protein binding to a similarly expressed gradient of AMH type II receptor in mesenchymal cells, which induces apoptosis of the epithelial cells of the Müllerian ducts. In the human, the absence of *AMH* expression or an inactivating mutation of the AMH type II receptor gene causes persistent Müllerian duct syndrome in males. In mice, the elimination of the Müllerian duct system in male fetuses is essentially complete by 16.5 dpc. Several transcription factors are involved in the regulation of *Amh* expression. The first factor shown to be crucial for *Amh* expression is the orphan nuclear receptor SF-1, which was identified as an essential regulator of the sex steroid synthesis in the adrenal glands and gonads (Shen et al., 1994). In the mouse, the targeted disruption of the *Sf-1* gene prevents Müllerian duct regression (Luo et al., 1994). In transgenic mice, an intact SF-1 binding site is required for the sex-specific expression of *Amh* (Giuili et al., 1997). The ability of SF-1 to activate *Amh* transcription has been demonstrated. This regulation is modulated through direct protein interactions with the factors WNT1, SOX9, and DAX1.

Androgens are also essential for normal male sex differentiation. Testosterone is secreted by the Leydig cells of the testes. During fetal life, testosterone promotes virilization of the urogenital tract in two ways. First, it stimulates the mesonephric (Wolffian) ducts to develop and differentiate into the epididymides, the seminal vesicles, and the vasa deferens. Second, in the early urogenital sinus and external genitalia, testosterone is rapidly transformed into dihydrotestosterone by the enzyme steroid 5α-reductase, where it induces the development of the male urethra, prostate, penis, and scrotum. In humans, 5α-reductase II deficiency is a cause of pseudohermaphroditism. Affected individuals are 46, XY males who have an autosomal-recessive disorder that is characterized by an external female phenotype at birth, bilateral testes, and normally virilized Wolffian structures that terminate in the vagina. Testosterone is also essential for testis descent into the scrotum during fetal development. Testis descent is an essential step in the male sex differentiation process. In mammals, this process follows two distinct and sequential stages: the intra-abdominal stage and the inguinoscrotal stage.

The INSL3 or relaxin-like factor, which is a member of the insulin-like hormone superfamily, seems to be required to regulate the intra-abdominal stage of testicular descent, although the mechanism remains poorly understood. INSL3 is expressed early in fetal mouse Leydig cells, and *Insl3* knockout male mice are bilaterally cryptorchid; the gubernacular bulbs fail to develop, and they resemble normal female gubernacular structures. Evidence shows that the steroid hormones estradiol and diethylstilbestrol could downregulate INSL3 expression in the fetal Leydig cells (Nef et al., 2000).

Hox genes encode homeodomain proteins that act as transcriptional regulators. *Hox* genes have a well-characterized role in embryonic development, during which they determine identity along the anterior–posterior body axis. In mice and humans, *Hox* genes are clustered in four unlinked genomic loci (designated *Hoxa-d* or *HOXA-D*), which contain a subset of nine to 13 genes each. The role of mammalian *Hox* genes in regulating segmental patterning of axial structures and the limb is well established. *Hox* genes also play a similar role in the specification of developmental fate in the individual regions of the male reproductive tract. *Hoxa-10* is expressed along the mesonephric duct from the caudal epididymis to the point at which the ductus deferens inserts into the urethra. Mutants of both *Hoxa-10* and *Hoxa-11* exhibit a homeotic transformation that results in the partial transformation of the ductus deferens to the epididymis (Bomgardner et al., 2003; Podlasek et al., 1999). *Hoxa-13* and *Hoxd-13* are the 5'-most members of the *Hox* A and D clusters. *HOXA*-13 is expressed in the terminal part of the digestive and urogenital tracts during embryogenesis, specifically in the genital tubercle from which the male accessory sex organs derive (Dolle et al., 1991; Warot et al., 1997). A *Hoxd-13* loss-of-function mutant demonstrated that *Hoxd-13* is essential to external genital development. *Hoxd-13* is the most caudally expressed *Hox* gene in the genitourinary tract. It is expressed in both the mesenchyme and the epithelium of the Wolffian duct and the urogenital sinus (Fromental-Ramain et al., 1996; Oefelein et al., 1996). Another homeobox gene, *Emx2*, which is a mammalian homolog of the *Drosophila empty spiracles*, is also expressed in the epithelial component of the intermediate mesoderm. *Emx2*$^{-/-}$ mutants completely lack reproductive tracts and gonads. In the mutant embryos, the Wolffian duct forms on embryonic day 10.5, but it subsequently degenerates on embryonic day 11.5. *Emx2* expression is only detected during the early stages of reproductive duct formation, suggesting that this gene is only required for a specific time period during the development of the intermediate mesoderm, possibly providing a survival signal.

Development of the Female Genital Duct

In contrast with the Wolffian duct, the Müllerian duct persists in the absence of external signals, and it must be directed to degenerate in males. In females, the absence of *SRY* results in ovarian organization. In the absence of male hormones, the Wolffian ducts degenerate, whereas the Müllerian ducts persist and differentiate into the female reproductive tract, including the oviduct (fallopian tube), the uterus, and the upper portion of the vagina. Several genes are involved in the development of the female genital duct. Although few genes have been identified in humans, mouse knockout studies have shed light on a set of genes that are essential for the regulation of Müllerian duct formation.

The *Wnt* gene family, which is homologous to the *Drosophila* segment polarity gene *wingless*, encodes secreted glycoproteins. They are involved in sex determination and the development of several female reproductive organs. In mice, a subset of *Wnt* gene family members has been found to regulate Müllerian duct development. For example, *Wnt4* expression is crucial for the formation of the Müllerian ducts, because it is required for tubule formation. Both *Wnt4*-deficient male and female mice completely lack Müllerian ducts, and *Wnt4*-mutant females even differentiate a normal male reproductive tract. Although *Wnt4* initiates Müllerian duct formation, *Wnt7a* may regulate its further outcome. It is expressed in the Müllerian duct epithelium in both sexes from 12.5 to 14.5 dpc. After Müllerian duct regression, expression is lost in males. In females, it persists in the epithelium of the Müllerian duct derivatives throughout life (Heikkila et al., 2001). From *Wnt7a* mutant female mice studies, it has been learned that *Wnt7a* is required for the proper differentiation of the oviduct and the uterus. In *Wnt7a* mutant male mice, the Müllerian ducts do not regress. This suggests that *Wnt* genes have a function not only in Müllerian duct formation but also in the regression of these ducts in males. Another member of this family, *Wnt5a*, is also critical for female reproductive tract development; it is needed for the formation of the genital tubercle. *Wnt5a*-deficient females have a coiled and shortened uterus and a poorly defined cervix and vagina.

The *Hox* gene family has also been found to regulate female reproductive tract development. As described previously, *Hox* genes play a role in male genital duct development. In the developing Müllerian duct, a number of posterior Abdominal B (Abd B) homeobox genes were found to be expressed in partially overlapping patterns along the anterior–posterior axis. *Abd B* genes are expressed according to their 3'-5' order in the *Hox* clusters: *Hoxa-9* is expressed in what will become

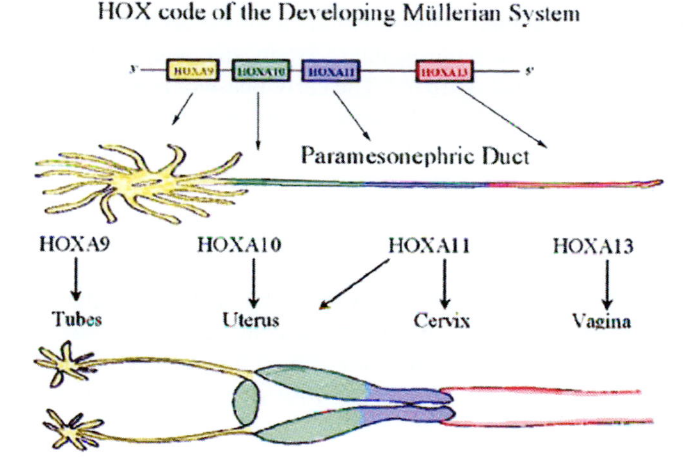

FIGURE 27.5 Expression of *Hox* Genes is Arranged a Linear Fashion along the Paramesonephric Duct (*Hum Reprod Update* 2000; 6:75–9). *Hoxa9* is expressed in areas destined to become fallopian tubes. *Hoxa10* is expressed in the developing uterus. *Hoxa11* is expressed in the primordial of the lower uterine segment and cervix and upper vagina in the developing uterus and cervix. *Hoxa13* is expressed in the upper vagina.

the oviduct, *Hoxa-10* is expressed in the developing uterus, *Hoxa-11* is found in the primordial lower uterus and cervix, and *Hoxa-13* is seen in the upper vagina (Figure 27.5; Taylor et al., 1997). The targeted mutagenesis of these genes results in region-specific defects along the female reproductive tract. *Hoxa-10* deficiency causes the homeotic transformation of the anterior part of the uterus into an oviduct-like structure, and it also causes reduced fertility in females. *Hoxa-13* null embryos show agenesis of the posterior portion of the Müllerian duct. When the *Hoxa-11* gene is replaced by the *Hoxa-13* gene, posterior homeotic transformation occurs in the female reproductive tract: the uterus, in which *Hoxa-11* but not *Hoxa-13* is normally expressed, becomes similar to the more posterior cervix and vagina, in which *Hoxa-13* is normally expressed.

It has also been revealed that *Wnt7a* is required for the maintenance of *Hoxa-10* and *Hoxa-11* expression in the uterus in mice. *Wnt5a* resides in the same genetic pathway as *Wnt7a* and *Hoxa* genes during female reproductive tract development, and *Emx2* also plays a role in female reproductive tract development. In contrast with the Wolffian duct, the Müllerian duct never forms in *Emx2* mutants.

Genes such as *Lim1* and Paired-box gene 2 (*Pax2*) are also indispensable for the early steps of Müllerian duct development. *Lim1* plays an essential role in mouse head and urogenital system development. It has a dynamic expression pattern in the Müllerian duct as early as embryonic day 11.5, which suggests that lim1 function is crucial for the initial formation of the Müllerian duct. The analysis of *Lim1* null mice revealed that Müllerian duct derivatives were completely absent. Furthermore, no Wolffian duct derivatives were ever observed in *Lim1* null neonatal males, thus demonstrating that *Lim1* is required for the formation of both sexual ducts. *Pax2*, which is a member of the Pax gene family, encodes a homeodomain transcription factor that is homologous to the *Drosophila pair-rule* genes. *Pax2* null mice also lack a reproductive tract. Unlike the *Lim1* null embryos, *Pax2* null mutants initially show both Wolffian and Müllerian duct formation, but these subsequently degenerate. This phenotype correlates with *Pax2* expression at this stage in both reproductive tracts on embryonic day 13.5, thus indicating a cell-autonomous role for *Pax2* in the developing Wolffian and Müllerian ducts.

The development of the female reproductive tract also depends on estrogenic hormones produced by the fetal ovaries. Estrogen action is mediated by estrogen receptors (ERs), which belong to the nuclear receptor superfamily of ligand-inducible transcription factors. In fetal mice, an ERα signal was detected in the nuclei of the surrounding cells of the Müllerian ducts as early as 11.5 dpc. No expression of ERβ was detected in the mouse Müllerian duct. However, ERβs were detected in rat Müllerian ducts from 15.5 to 21.5 dpc. The expression levels of ERβs are much lower than those of ERαs. These studies show that ERα is likely a dominant ER subtype that is used in Müllerian duct development (Greco et al., 1991). ERα knockout mice have a small but normally patterned reproductive tract.

27.5 DEVELOPMENT OF THE EXTERNAL GENITALIA

The Indifferent Stage

The external genitalia also pass through an undifferentiated state before distinguishing sexual characteristics appear. The external genitalia are derived from a complex of mesodermal tissue located around the cloaca. In the human, a very early midline elevation called the genital eminence is situated just cephalic to the proctodeal depression. This structure soon develops into a prominent genital tubercle at the cranial end of the cloacal membrane early during the fourth week. Genitalia swellings and urogenital folds soon develop on each side of the cloacal membrane. The development of the genital tubercle is initially regulated by *Hox* gene expression. Located at the terminal part of the urogenital system, the genital tubercle expresses the 5'-most genes of the *Hox* gene clusters, specifically *Hoxa-13* and *Hoxd-13*. The early phase of outgrowth of the genital tubercle also depends on *Wnt* signaling and the interacting signals of Sonic hedgehog (Carlson, 2004; Lin et al., 2008). Interestingly, a recent study has shown that fibroblast growth factors 8 and 10, which were believed to be involved in genital development, have no role in external female reproductive tract development (Seifert et al., 2009).

External Genitalia

The masculinization of the indifferent external genitalia is caused by androgens produced by the testis. Unlike the internal genitalia in the male, the external genitalia do not respond directly to testosterone. 5α-reductase converts testosterone to dihydrotestosterone in the external genitalia. Under the influence of dihydrotestosterone, the genital tubercle in the male undergoes a second phase of elongation to form the penis, and the genital swellings enlarge to form the scrotal pouches. 5α-reductase plays an important role in external genitalia development. As described previously, 5α-reductase II deficiency will cause male patients to have an external female phenotype at birth and to exhibit normally virilized Wolffian structures that terminate in the vagina.

In the absence of androgen signaling, the feminization of the indifferent external genitalia occurs. The genital tubercle in the female becomes the clitoris, the genital folds become the labia minora, and the genital swellings develop into the labia majora.

27.6 MALFORMATIONS OF THE GENITAL SYSTEM

Malformations of the genital system are intrinsic defects in the developing human embryo that result in localized abnormalities during the development of the reproductive duct system. Genetic events can result in congenital genital malformations during very early development stages. In the human, numerous factors can also affect the development of the reproductive tract, such as infectious agents, drugs or pharmaceutical products, environmental chemicals, physical agents, and maternal diseases. Despite their different origins, all of these factors cause some type of genetic or gene expression abnormalities that ultimately induce genital malformation. As described previously, genetic abnormalities occurring in animal models, such as chromosomal anomalies and gene mutations, influence the development of the genital system, thereby causing congenital malformations.

Abnormalities of Sexual Differentiation

Turner's Syndrome (Gonadal Dysgenesis)

Turner's syndrome is characterized by defective gonadal development in women with a karyotypic sex chromosome abnormality (45, X or 45, XO). These individuals are phenotypically females. Individuals with this syndrome possess PGCs that degenerate shortly after they reach the gonads. Affected individuals are generally of short stature, and they present with undifferentiated (streak) gonads. As expected, the internal and external reproductive structure develops as female as a result of the absence of AMH and testosterone. The Müllerian structures are small, and breast development is typically absent due to the lack of a functional ovary capable of producing estrogen.

Swyer Syndrome

Swyer syndrome, which is also known as XY gonadal dysgenesis, is a heterogeneous condition with variant forms that are caused, in most cases, by a structural abnormality on the Y chromosome that leads to *SRY* loss of function. Swyer syndrome has also been associated with autosomal mutations such as chromosome 9p deletions. Patients with Swyer syndrome are born without functional gonads; instead, they present simply with gonadal streaks. Affected individuals are phenotypically females at birth. However, because the streak gonads are incapable of producing the sex hormones that are essential for puberty, these patients do not develop most secondary sex characteristics without hormone replacement.

True Hermaphroditism

Hermaphroditism is a rare condition in which ovarian and testicular tissues exist in the same person. The testicular tissue contains seminiferous tubules and spermatozoa, and the ovarian tissue contains follicles or corpora albicantia. Hermaphrodite patients typically show a chromosomal male-female mosaicism in which both the male XY and the female XX chromosome pairs are present. External genitalia may show traits from both sexes.

Female Pseudohermaphroditism

Female pseudohermaphroditism is characterized by male or ambiguous genitalia coupled with a female karyotype (46, XX). The XX male syndrome is a heterogeneous disorder. The presence of the *sry* gene transposed with the X chromosome leads to male differentiation. About 80% of XX males express SRY. Some cases of SRY-negative XX males have been reported. This may be the result of an unrecognized XX/XY chimerism or an XX/XXY mosaicism, although an autosomal-recessive disorder has also been proposed as the intrinsic cause of these less common cases.

Male Pseudohermaphroditism

Male pseudohermaphroditism refers to a condition that affects 46, XY individuals with differentiated testes who exhibit varying degrees of feminization. In cases of male pseudohermaphroditism, there is a spectrum of external genitalia; some individuals are completely phenotypically female, whereas others appear to be normal males with varying spermatogenesis and/or pubertal virilization. Between these two extremes is a wide area of ambiguity. Deficiency of the enzyme 5α-reductase can result in external genitalia that appear to be female in an individual with an XY karyotype. High testosterone production at puberty can drive male external differentiation and an apparent "sex change" during adolescence.

Testicular Feminization (Androgen Insensitivity) Syndrome

Individuals with testicular feminization syndrome have a normal XY chromosomal complement; however, they are resistant to androgens (testosterone). This usually results from a mutation in the androgen receptor, which consequently leads to some extent of – or even total – external genitalia feminization. Complete testicular feminization results in an individual who looks outwardly female. Because these individuals produce AMH, the Müllerian ducts degenerate; however, the Wolffian ducts lack the ability to respond to testosterone, and they therefore regress. Patients with androgen insensitivity have no internal genitalia. The vagina in these individuals is a short structure that lacks communication with any internal organ. In these patients, breast development proceeds normally as a result of the lack of androgen action and the ability to convert testosterone to estrogen by the enzyme aromatase.

Vestigial Structures from the Embryonic Genital Ducts

Mesonephric Duct Remnants

Mesonephric duct remnants are also known as Gartner's ducts. In females, the remains of the cranial parts of the mesonephros may persist as the epoophoron or the paroophoron. The caudal parts of the mesonephric ducts are often seen in histological sections along the uterus or the upper vagina as Gartner's ducts. Portions of these duct remnants sometimes enlarge to form cysts.

Paramesonephric Duct Remnants

Remnants of the paramesonephric (Müllerian) ducts can be found in the male as a uterus-like structure. Historically, this has been called a masculine uterus. The remnants appear as one or two thin, uterus-like tubes that are medial to the ducti deferentes, with or without a medial corpus lying between the ampullae. The paramesonephric remnants resemble a normal female uterus, but the endometrium consists primarily of amorphous extracellular matrix.

Other Abnormalities of the Genital Duct System

Failed Müllerian Duct Fusion or Cannulation with Clinical Correlation

Normal uterine development depends on both the fusion of the two paramesonephric ducts and the absorption of the fused walls to create a single cavity. Failed Müllerian duct fusion occurs in females when the ducts do not meet or fuse. Fusion abnormalities can occur at any point along the Müllerian ducts; they may involve an isolated junction or the entire duct. The condition can have a range of results, from a small branch of the apex of the duct being affected to a complete duplication of the uterus into separate structures, each with a single fallopian tube. Failure to absorb the intervening wall between the two fused paramesonephric ducts results in a septum that separates the canals of a fused uterus. Both fusion defects and a septum can result in pregnancy complications such as miscarriage and premature delivery.

Congenital Absence of the Vas Deferens

The congenital bilateral absence of the vas deferens occurs in males when the tubes that carry sperm from the testes (the vas deferens) fail to develop normally. This condition can occur alone or in association with cystic fibrosis. The testes usually develop and function normally, but sperm cannot be transported out of the epididymis. This condition occurs in men with a cystic fibrosis gene mutation; however, they often do not have any of the other health problems associated with that disease (e.g., progressive lung damage, chronic digestive system problems).

27.7 CONCLUSION

Reproductive tract development follows from an ordered set of divergent signals that begins with chromosomal complement. It is one of the few instances of such a clear bimodal heterogeneity in development. The adaptation of conservation of these mechanisms in multiple organisms attests to their evolutionary value. The entire developmental process includes three stages:

(1) Sex determination;
(2) The differentiation of the internal genital ducts; and
(3) The differentiation of the external genitalia.

Interestingly, there are indifferent stages before the differentiation of distinguishing sexual characteristics in which numerous genetic factors are involved. The reasons that development requires this indifferent stage and the ways in which the required genes are initiated are still unclear. There are some genes, such as *Hox*, *lim1*, and *pax2*, which are involved in the development of the genital ducts in both males and females. How do they direct these two different structures? How is the expression of the *sry* gene initiated in the XY embryo? Finally, although the traditional dogma considers that the migration of PGCs occurs via self-movement, recent new theories question this mode of cell movement. Many questions in this field remain to be answered, and the exploitation of new genetic and genomic information will be critical in the answering of these important questions.

RECOMMENDED RESOURCES

Human Embryology, Genital System: http://www.embryology.ch/anglais/ugenital/planmodgenital.html (Last accessed on 01/01/2014)

Developmental Biology: 6th ed.: http://www.ncbi.nlm.nih.gov/books/bv.fcgi?rid=dbio (Last accessed on 01/01/2014)

Recent Progress in Hormone Research, http://rphr.endojournals.org/cgi/reprint/57/1/1 (Last accessed on 01/01/2014)

Molecular Regulation of Müllerian Development by Hox Genes: http://onlinelibrary.wiley.com/doi/10.1196/annals.1335.018/pdf (Last accessed on 01/01/2014)

Development, Differentiation and Derivatives of the Wolffian and Müllerian Ducts: http://www.intechopen.com/books/the-human-embryo/development-differentiation-and-derivatives-of-the-wolffian-and-m-llerian-ducts (Last accessed on 01/01/2014)

Development of the External Genitalia: Conserved and Divergent Mechanisms of Appendage Patterning: http://onlinelibrary.wiley.com/doi/10.1002/dvdy.22631/pdf (Last accessed on 01/01/2014)

REFERENCES

Bardoni, B., Zanaria, E., Guioli, S., Floridia, G., Worley, K.C., Tonini, G., Ferrante, E., Chiumello, G., McCabe, E.R., Fraccaro, M., Zuffardi, O., Camerino, G., 1994. A dosage sensitive locus at chromosome Xp21 is involved in male to female sex reversal. Nat. Genet. 7, 497–501.

Bomgardner, D., Hinton, B.T., Turner, T.T., 2003. 5' hox genes and meis 1, a hox-DNA binding cofactor, are expressed in the adult mouse epididymis. Biol. Reprod. 68, 644–650.

Brennan, J., Capel, B., 2004. One tissue, two fates: molecular genetic events that underlie testis versus ovary development. Nat. Rev. Genet. 5, 509–521.

Brennan, J., Tilmann, C., Capel, B., 2003. Pdgfr-alpha mediates testis cord organization and fetal Leydig cell development in the XY gonad. Genes. Dev. 17, 800–810.

Burgoyne, P.S., 1998. The mammalian Y chromosome: a new perspective. Bioessays 2, 363–366.

Carlson, B.M., 2004. Urogenital system. In: Ozols, I., Milnes, J. (Eds.), Human embryology and developmental biology. Elsevier, Philadelphia, pp. 393–419.

Chassot, A.A., Ranc, F., Gregoire, E.P., Roepers-Gajadien, H.L., Taketo, M.M., Camerino, G., de Rooij, D.G., Schedl, A., Chaboissier, M.C., 2008. Activation of beta-catenin signaling by Rspo1 controls differentiation of the mammalian ovary. Hum. Mol. Genet. 9, 1264–1277.

Cheng, H.H., Ying, M., Tian, Y.H., et al., 2006. Transcriptional diversity of DMRT1 (dsx- and mab3-related transcription factor 1) in human testis. Cell. Res. 16, 389–393.

Clark, A.M., Garland, K.K., Russell, L.D., 2000. Desert hedgehog (Dhh) gene is required in the mouse testis for formation of adult-type Leydig cells and normal development of peritubular cells and seminiferous tubules. Biol. Reprod. 63, 1825–1838.

Dolle, P., Izpisua-Belmonte, J.C., Brown, J.M., et al., 1991. HOX-4 genes and the morphogenesis of mammalian genitalia. Genes. Dev. 5, 1767–1776.

Freeman, B., 2003. The active migration of germ cells in the embryos of mice and men is a myth. Reproduction 125, 635–643.

Fromental-Ramain, C., Warot, X., Messadecq, N., et al., 1996. Hoxa-13 and Hoxd-13 play a crucial role in the patterning of the limb autopod. Development 122, 2997–3011.

Giuili, G., Shen, W.H., Ingraham, H.A., 1997. The nuclear receptor SF-1 mediates sexually dimorphic expression of Müllerian inhibiting substance, *in vivo*. Development 124, 1799–1807.

Greco, T.L., Furlow, J.D., Duello, T.M., et al., 1991. Immunodetection of estrogen receptors in fetal and neonatal female mouse reproductive tracts. Endocrinology 129, 1326–1332.

Hacker, A., Capel, B., Goodfellow, P., et al., 1995. Expression of Sry, the mouse sex determining gene. Development 121, 1603–1614.

Heikkila, M., Peltoketo, H., Vainio, S., 2001. Wnts and the female reproductive system. J. Exp. Zool. 290, 616–623.

Kitamura, K., Yanazawa, M., Sugiyama, N., Miura, H., Iizuka-Kogo, A., Kusaka, M., Omichi, K., Suzuki, R., Kato-Fukui, Y., Kamiirisa, K., Matsuo, M., Kamijo, S., Kasahara, M., Yoshioka, H., Ogata, T., Fukuda, T., Kondo, I., Kato, M., Dobyns, W.B., Yokoyama, M., Morohashi, K., 2002. Mutation of ARX causes abnormal development of forebrain and testes in mice and X-linked lissencephaly with abnormal genitalia in humans. Nat. Genet. 3, 359–369.

Koopman, P., Munsterberg, A., Capel, B., et al., 1990. Expression of a candidate sex-determining gene during mouse testis differentiation. Nature 348, 450–452.

Lawson, K.A., Dunn, N.R., Roelen, B.A., et al., 1999. Bmp4 is required for the generation of primordial germ cells in the mouse embryo. Genes. Dev. 13, 424–436.

Lin, C., Yin, Y., Long, F., Ma, L., 2008. Tissuespecific requirements of beta-catenin in external genitalia development. Development 135, 2815–2825.

Luo, X., Ikeda, Y., Parker, K.L., 1994. A cell-specific nuclear receptor is essential for adrenal and gonadal development and sexual differentiation. Cell 77, 481–490.

Matsuda, M., 2003. Sex determination in fish: lessons from the sex-determining gene of the teleost medaka, *Oryzias latipes*. Dev. Growth. Differ. 45, 397–403.

McElreavey, K., Vilain, E., Abbas, N., et al., 1993. A regulatory cascade hypothesis for mammalian sex determination: SRY represses a negative regulator of male development. Proc. Natl. Acad. Sci. USA 90, 3368–3372.

McLaren, A., Simpson, E., Tomonari, K., et al., 1984. Male sexual differentiation in mice lacking H-Y antigen. Nature 312, 552–555.

Meeks, J.J., Russell, T.A., Jeffs, B., et al., 2003. Leydig cell-specific expression of DAX1 improves fertility of the Dax1-deficient mouse. Biol. Reprod. 69, 154–160.

Morais da Silva, S., Hacker, A., Harley, V., et al., 1996. Sox9 expression during gonadal development implies a conserved role for the gene in testis differentiation in mammals and birds. Nat. Genet. 14, 62–68.

Motro, B., van der Kooy, D., Rossant, J., et al., 1991. Contiguous patterns of c-kit and steel expression: analysis of mutations at the Wand Sl loci. Development 113, 1207–1221.

Munsterberg, A., Lovell-Badge, R., 1991. Expression of the mouse anti-Müllerian hormone gene suggests a role in both male and female sexual differentiation. Development 113, 613–624.

Nanda, I., Haaf, T., Schartl, M., et al., 2002. Comparative mapping of Z-orthologous genes in vertebrates: implications for the evolution of avian sex chromosomes. Cytogenet. Genome. Res. 99, 178–184.

Nef, S., Shipman, T., Parada, L.F., 2000. A molecular basis for estrogen-induced cryptorchidism. Dev. Biol. 224, 354–361.

Nilsson, E.E., Kezele, P., Skinner, M.K., 2002. Leukemia inhibitory factor (LIF) promotes the primordial to primary follicle transition in rat ovaries. Mol. Cell. Endocrinol. 188, 65–73.

Nilsson, E.E., Skinner, M.K., 2004. Kit ligand and basic fibroblast growth factor interactions in the induction of ovarian primordial to primary follicle transition. Mol. Cell. Endocrinol. 214, 19–25.

Oefelein, M., Chin-Chance, C., Bushman, W., 1996. Expression of the homeotic gene Hox-d13 in the developing and adult mouse prostate. J. Urol. 155, 342–346.

Page, D.C., Mosher, R., Simpson, E.M., et al., 1987. The sex-determining region of the human Y chromosome encodes a finger protein. Cell 51, 1091–1104.

Palmer, M.S., Sinclair, A.H., Berta, P., et al., 1989. Genetic evidence that ZFY is not the testis-determining factor. Nature 342, 937–939.

Podlasek, C.A., Seo, R.M., Clemens, J.Q., et al., 1999. Hoxa-10 deficient male mice exhibit abnormal development of the accessory sex organs. Dev. Dyn. 214, 1–12.

Pritchard-Jones, K., Fleming, S., Davidson, D., et al., 1990. The candidate Wilms0 tumor gene is involved in genitourinary development. Nature 346, 194–197.

Saito, T., Kuang, J.Q., Bittira, B., et al., 2002. Xenotransplant cardiac chimera: immune tolerance of adult stem cells. Ann. Thorac. Surg. 74, 19–24.

Santos, A.C., Lehmann, R., 2004. Germ cell specification and migration in *Drosophila* and beyond. Curr. Biol. 14, R578–R589.

Schartl, M., 2004. Sex chromosome evolution in non-mammalian vertebrates. Curr. Opin. Genet. Dev. 14, 634–641.

Seifert, A.W., Yamaguchi, T., Cohn, M.J., 2009. Functional and phylogenetic analysis shows that Fgf8 is a marker of genital induction in mammals but is not required for external genital development. Development 136. 2643–265.

Shawlot, W., Behringer, R.R., 1995. Requirement for Lim1 in head-organizer function. Nature 374, 425–430.

Shen, W.H., Moore, C.C., Ikeda, Y., et al., 1994. Nuclear receptor steroidogenic factor 1 regulates the Müllerian inhibiting substance gene: a link to the sex determination cascade. Cell 77, 651–661.

Shetty, S., Griffin, D.K., Graves, J.A., 1999. Comparative painting reveals strong chromosome homology over 80 million years of bird evolution. Chromosome. Res. 7, 289–295.

Sinclair, A.H., Berta, P., Palmer, M.S., et al., 1990. A gene from the human sex-determining region encodes a protein with homology to a conserved DNA-binding motif. Nature 346, 240–244.

Sultan, C., Lobaccaro, JM., Medlej, R., Poulat, F., Berta, P., 1991. SRY and male sex determination. Horm. Res. 36, 1–3.

Takehana, Y., Demiyah, D., Naruse, K., Hamaguchi, S., Sakaizumi, M., 2007. Evolution of different Y chromosomes in two medaka species, *Oryzias dancena* and *O. latipes*. Genetics 175, 1335–1340.

Takeuchi, T., Tanigawa, Y., Minamide, R., Ikenishi, K., Komiya, T., 2010. Analysis of SDF-1/CXCR4 signaling in primordial germ cell migration and survival or differentiation in *Xenopus laevis*. Mech. Dev. 127 (1–2), 146–158.

Taylor, H.S., Vandenheuvel, G.B., Igarashi, P., 1997. A conserved Hox axis in the mouse and human female reproductive system – late establishment and persistent adult expression of the Hoxa cluster genes. Biol. Reprod. 57, 1338–1345.

Tiepolo, L., Zuffardi, O., 1976. Localization of factors controlling spermatogenesis in the nonfluorescent portion of the human Y chromosome long arm. Hum. Genet. 34, 119–124.

Tomaselli, S., Megiorni, F., De Bernardo, C., Felici, A., Marrocco, G., et al., 2008. Syndromic true hermaphroditism due to an R-spondin1 (RSPO1) homozygous mutation. Hum. Mutat. 29, 220–226.

Tomaselli, S., Megiorni, F., Lin, L., Mazzilli, M.C., Gerrelli, D., Majore, S., Grammatico, P., 2011. Achermann JC.Human RSPO1/R-spondin1 is expressed during early ovary development and augments β-catenin signaling. PLoS One 1, e16366.

Viger, R.S., Silversides, D.W., Tremblay, J.J., 2005. New insights into the regulation of mammalian sex determination and male sex differentiation. Vitam. Horm. 70, 387–413.

Wachtel, S.S., Ono, S., Koo, G.C., et al., 1975. Possible role for H–Y antigen in the primary determination of sex. Nature 257, 235–236.

Warot, X., Fromental-Ramain, C., Fraulob, V., et al., 1997. Gene dosage-dependent effects of the Hoxa-13 and Hoxd-13 mutations on morphogenesis of the terminal parts of the digestive and urogenital tracts. Development 124, 4781–4791.

Welshons, W.J., Russell, L.B., 1959. The Y-chromosome as the bearer of male determining factors in the mouse. Proc. Natl. Acad. Sci. USA 45, 560–566.

Ying, Y., Qi, X., Zhao, G.Q., 2001. Induction of primordial germ cells from murine epiblasts by synergistic action of BMP4 and BMP8B signaling pathways. Proc. Natl. Acad. Sci. USA 98, 7858–7862.

Zarkower, D., 2001. Establishing sexual dimorphism: conservation amidst diversity? Nat. Rev. Genet. 2, 175–185.

Chapter 28

Skeletal Development

Mark T. Langhans, Peter G. Alexander and Rocky S. Tuan
Center for Cellular and Molecular Engineering, Department of Orthopedic Surgery, University of Pittsburgh School of Medicine, Pittsburgh, PA

Chapter Outline

GLOSSARY

Endochondral ossification A mechanism of bone formation in which a cartilage model is replaced with ossified tissue produced by invading osteoblasts.

Growth plate The area of developing tissue between the diaphysis and epiphysis responsible for the longitudinal growth of bones.

Hypertrophy An increase in bulk without concomitant multiplication of parts.

Interzone A region of non-chondrogenic cells that forms at the site of a future joint.

Perichondrium The border region of primordial cartilage anlagen that eventually differentiate into osteoblasts.

Somite An embryonic tissue that patterns and maintains the metamism of axial tissues.

Resegmentation The process by which the sclerotomes of neighboring somites reorganize to form vertebral bodies that span the space between the dermomyotomal derivatives of the same somites.

SUMMARY

- The bones and connective tissues of the body are derived from two embryonic origins, the mesoderm and neural crest. The appendicular skeleton arises from the lateral plate mesoderm, the axial skeleton from the paraxial mesoderm, and the craniofacial skeleton from a combination of cephalic paraxial mesoderm and neural crest cells.
- Bones are formed through one of two processes:
  - Endochondral ossification involves the formation of primordial cartilaginous anlagen that are subsequently replaced by bone.
  - Intramembranous ossification is characterized by the direct differentiation of mesenchymal cells to osteogenic tissue.

- The growth plate can be divided into a resting zone, a proliferating zone, a pre-hypertrophic zone, and a hypertrophic zone, reflecting a gradient of chondrocyte differentiation that is regulated by antagonism between bone morphogenetic protein (BMP) and fibroblast growth factor (FGF) signaling pathways and gradients of Indian hedgehog (IHH) and parathyroid hormone related protein (PTHrP).
- Replacement of cartilage by osteoblasts is dependent upon prior mineralization of the cartilage matrix and vascular invasion. The activities of the hypertrophic chondrocytes and of the osteoblasts are regulated by RUNX2.
- Osteoclasts are of hematopoietic origin (monocyte), and with osteoblasts, they remodel bone throughout an individual's lifetime to adapt to mechanical demands.
- Joints and intra-articular structures arise from a cohort of progenitor cells distinct from those of the bones and are labeled by GDF5 and TGFBRII expression.
- Somites are formed from the segmental plate through the interaction of a cell-autonomous oscillator and a morphogenetic wave that coordinates somite boundary formation via the Notch-Delta juxtacrine system and epithelialization via wingless-type MMTV integration site (WNT) paracrine system and cadherin activity. Dysregulation of this developmental process leads to congenital scoliosis.
- The axial skeleton is derived from the ventro-medial sclerotome of somites. Differentiation of the somite is regulated by paracrine factors from surrounding tissues and the sclerotome in particular is induced by sonic hedgehog (SHH).
- Tendons and ligaments develop from the interaction of mesodermal mesenchyme with myoblasts.
- The nucleus pulposus of the intervertebral disc is initially derived from the notochord. The gradual loss of "embryonic cells" from the disc may be related to disc degeneration and back pain.

28.1 INTRODUCTION

This chapter will focus on the development of the axial and appendicular skeleton, with particular emphasis on the cartilaginous and bony skeleton that develops via endochondral ossification. Components of the skeleton derived from intramembranous ossification include the bones of the craniofacial skeleton, which is covered in another chapter (see Chapter 35). The chapter on limb development discusses the first two stages of limb development (limb initiation and limb patterning; see Chapter 29), and this chapter will focus on the final stage of limb development (late limb morphogenesis). Late limb morphogenesis includes the formation of the primordial anlagen via mesenchymal condensation, perichondrium and cartilage differentiation, formation of the growth plate, differentiation of osteoblasts and osteoclasts, and formation of articular joints and related structures. Additionally, this chapter will discuss the formation of the axial skeleton, with emphasis on the molecular regulation of somitogenesis, differentiation of the sclerotome, neural arch and rib development, resegmentation, and development of the intervertebral joints.

Because the skeleton is present only in vertebrates, the use of alternative model systems to study skeletal development is limited. Although useful paradigms and conserved cellular and molecular pathways in morphogenesis and development have been noted in studies of invertebrates, this chapter will focus on the genetics of skeletal development in humans, mice, and chickens. The generally non-lethal nature of mutations that affect the skeleton results in a multitude of human syndromes, as well as mouse knockout models that include skeletal dysmorphogenesis vital to our understanding of skeletogenesis. The development of the Cre-lox conditional knockout system in mice has overcome lethal null and dominant negative mutations in genes that affect multiple organ systems to reveal their function in skeletogenesis (Gu et al., 1993). The mouse genome can be manipulated to overexpress proteins involved in skeletogenesis in a general or tissue-specific manner as well. Experimentally, the chicken embryo model is of great utility, as it is easily manipulated *in ovo*, particularly at very early stages of limb bud outgrowth. Finally, the zebrafish is also a convenient experimental model, as morpholino antisense technology conveniently allows researchers to alter gene expression (Nasevicius and Ekker, 2000).

28.2 THE APPENDICULAR SKELETON

Origins of the Appendicular Skeleton

A brief summary of the three stages of limb development (initiation, outgrowth/patterning, and late morphogenesis) is provided here. During limb initiation, at a location specified by expression of Hox genes and T-box transcription factors along the axis of the body, cells from the lateral plate mesoderm migrate to the ectodermal surface of the embryo and begin proliferating, forming the limb buds. The hind limbs are marked by Tbx4 expression, while the forelimbs are marked by Tbx5 expression (Rodriguez-Esteban et al., 1999). Limb outgrowth and patterning relies on the formation of the zone of polarizing activity (ZPA) and the apical ectodermal ridge (AER), which determine the anterior-posterior (A-P) and proximal-distal (P-D) polarities of the developing skeletal elements, respectively, as well as influence cell proliferation and migration (Zeller et al., 2009).

Following the initial patterning of the limb into zeugopod, stylopod, and autopod, the final stage of late limb morphogenesis begins. As the limb bud grows larger, the mesenchymal cells most distant from the AER stop proliferating and condense, thus beginning the process of endochondral ossification that ultimately results in adult bone. Endochondral ossification can be summed up as follows: mesenchymal cells condense, undergo chondrogenesis, hypertrophy, calcify, and apoptose. Blood vessels invade the space left behind by chondrocyte apoptosis, and osteoblasts and osteoclasts take up residence. Additionally, joints are created as single cartilage anlage segments and form where two skeletal elements articulate.

Chondrogenesis

Condensation of Limb Bud Mesenchyme

As the size of the limb bud increases, the AER will eventually be positioned such that the proximal mesenchymal cells will no longer be under the influence of the FGF secreted by the AER. This change in FGF signaling, together with the WNT and BMP gradients established during limb patterning, will signal the mesenchymal cells of the limb bud to begin the process of condensation.

Members of the transforming growth factor-β (TGFβ) superfamily, including BMPs, are critical for the initiation of late limb morphogenesis. For example, injection of retrovirus expressing the BMP inhibitor Noggin into chick embryos *in ovo* prevents mesenchymal condensation, thus demonstrating the necessity of BMP signaling in early skeletogenesis (Pizette and Niswander, 2000). Similarly, mice overexpressing Noggin in their limbs fail to produce most of their skeletal elements (Tsumaki et al., 2002). Bmp2 present throughout the mesenchyme, turns on the expression of the cell-cell adhesion protein, neural cadherin (N-cadherin). N-Cadherin on one cell binds with N-cadherin on adjacent cells, initiating a signaling cascade that is one of several that initiates condensation. In micromass cultures of developing limb buds, treatment with anti-N-cadherin antibodies (Oberlender and Tuan, 1994) or infection with mutated N-cadherin prevents condensation, which ultimately inhibits chondrogenesis, demonstrating the importance of the condensation event (DeLise and Tuan, 2002).

Mesenchymal cells are surrounded by an extracellular matrix (ECM) rich in hyaluronic acid (HA), collagen type I, and an alternatively spliced form of collagen type II (IIA) (Dessau et al., 1980; Maleski and Knudson, 1996). Prior to condensation, cells begin to express hyaluronidase, an enzyme that digests the HA in the ECM, denuding the mesenchymal cells and allowing them to communicate with each other via cell surface proteins such as N-cadherin. Other cell-cell interaction proteins, such as neural cell adhesion molecule (N-CAM), as well as multiple ECM proteins and their cell surface receptors, are implicated in the condensation process. TGFβ initiates the expression of fibronectin (FN), an extracellular protein vital to condensation. There are multiple splice variants of FN, but two are particularly interesting in chondrogenesis (White et al., 2003). Prior to condensation, FN expresses exon IIIA. Cells spread less in the presence of the IIIA exon, allowing the cells to round up, a process that allows better packing, positively influencing condensation. Once condensation is complete, exon IIIA is spliced out. Versican is another ECM protein expressed in mesenchyme that, like FN IIIA, prevents cell spreading (Williams et al., 2005). In addition to fibronectin and versican, the thrombospondins, including cartilage oligomeric matrix protein (Comp) (Kipnes et al., 2003), tenascins, and many other ECM proteins are necessary for proper condensation.

At this time of development another member of the TGFβ superfamily, growth/differentiation factor-5 (Gdf5), is expressed (Chang et al., 1994; Storm et al., 1994). In micromass models, Gdf5 regulates condensation by increasing the ability of cells to communicate via gap junctions. Connexin 43 is expressed in developing limbs in an overlapping pattern with Gdf5, suggesting it may be the dominant gap junction protein (Coleman et al., 2003; Coleman and Tuan, 2003). Overexpression of Gdf5 by infecting the chick limb *in ovo* with Gdf5 constructs leads to larger cartilage elements with increased cell number. It is proposed that the increased number of cells is due to increased cell adhesiveness, as opposed to an increase in proliferation (Francis-West et al., 1999). In mice null for Gdf5, chondrocyte condensation size is significantly reduced. Because the size of the initial condensations regulates the size of the future skeletal elements, these mice are severely dwarfed. Human mutations in the GDF5 gene lead to a spectrum of dwarfisms, the most severe of which leads to a near absence of fingers.

The size of the condensation is critical in skeletogenesis. A condensation that is too small may fail to undergo chondrogenesis. If it is too large, the final bone will also be too large. The early cartilaginous skeleton determines the size and shape of the future bones, and thus its development must be very well controlled. Growth factors and proteins responsible for cell-cell and cell-matrix interactions are critical in determining the size of the condensations by altering cell adhesion and migration. However, the cells in the condensations also proliferate, and the proteins that guide proliferation are thus noteworthy. Homeobox genes have been implicated in the control of proliferation of the early mesenchyme (Boulet and Capecchi, 2004; Goff and Tabin, 1997). HoxA11/HoxD11 double mutant mice have shortened forelimbs due to decreased FGF expression in the AER (Boulet and Capecchi, 2004). Because FGFs in the AER direct outward growth, any changes in FGF expression will alter limb bud growth. Interestingly, Hox genes are activated by BMPs, further demonstrating their importance in skeletogenesis (Duprez et al., 1996).

At the stage of mesenchymal chondensation, the growing limb bud contains mesenchymal cells that have begun to adhere to one another and communicate, readying themselves to differentiate into chondrocytes. Before that happens, the edges of the condensation must be defined (Hall and Miyake, 2000). Syndecan is an integral membrane proteoglycan found in the mesenchymal cells surrounding the initial condensation (Koyama et al., 1995). Syndecan binds to FN, which is found in the ECM surrounding cells in the early condensation. The interaction of syndecan and FN leads to an intracellular signaling cascade that instructs the mesenchymal cell to downregulate N-CAM (Hall and Miyake, 2000). The point at which cells expressing N-CAM meet cells not expressing it becomes the boundary between the future cartilage anlage and the surrounding mesenchyme. Along this boundary, mesenchymal cells will flatten and become the perichondrium, the thin layer of cells surrounding the cartilage skeletal elements that will later become the periosteum.

Cartilage Differentiation

The Sry-related HMG box containing transcription factor Sox9 is critical for every step of chondrogenesis, and it is widely used as the marker to detect chondrocytes. Using the Cre-lox system, Sox9 can be eliminated specifically in early mouse limbs, prior to condensation (Akiyama et al., 2002). In these animals, no condensations form in the limb bud, and ultimately they are born with a complete absence of limbs. In micromass cultures created using Sox9–/– chimeric limb buds, Sox9–/– cells cannot be found in condensing regions. In Sox9–/–: wild-type chimeric animals, Sox9–/– cells are excluded from the cartilage primordium (Bi et al., 1999). Mice that overexpress Sox9 in mesenchymal precursors exhibit delayed cartilage mineralization (Akiyama et al., 2004), and ectopic expression of Sox9 in hypertrophic chondrocytes prevents vascularization and formation of the marrow cavity (Hattori et al., 2010). The mechanism of Sox9 induction is not well understood, but has been suggested to involve the control and sensing of ECM stiffness (Allen et al., 2012). Additionally, introduction of a bead soaked in Bmp2 into the developing chick limb leads to an upregulation of Sox9 expression, suggesting that the BMPs are important in initiating chondrogenesis (Healy et al., 1999).

The chondrogenic potential of Sox9 is mediated by the induction of Sox5 and Sox6 (Akiyama et al., 2002). In animals lacking Sox9, Sox5 and Sox6 are never expressed; Sox5 and Sox6 are known to have Sox9 binding sites in their enhancers, explaining this finding. Sox5 and Sox6 are redundant, and thus mutations in either gene lead to no overt phenotype. However, animals lacking both Sox5 and Sox6 have severely reduced long bones (Smits et al., 2001). Sox9 also binds the enhancer regions of collagen type II and type XI and turns on their expression early in chondrogenesis (Bi et al., 1999). Sox5 and Sox6 are thought to initiate the formation of the growth plate by inducing proliferation between the epiphysis and diaphysis, in the areas in which the columns of proliferating cells will appear (Smits et al., 2004). In animals null for Sox5 and Sox6, there are no proliferating columnar cells present, and growth plates fail to form. Together with Sox5 and Sox6, Sox9 increases the expression of collagen type IX, aggrecan, and link protein (Lefebvre et al., 1998). The expression of these ECM proteins is an indicator of mature cartilage. Collagen type II is the dominant fibril form of collagen in cartilage, and it is crosslinked by collagen type IX. Mutations in either collagen type II or type IX lead to disorganized growth plates and early onset osteoarthritis (Cremer et al., 1998). Aggrecan is the core protein of sulfated proteoglycans important in maintaining the high water content of cartilage (Dudhia, 2005). These structural components are absolutely vital in maintaining the mechanical properties of cartilage. This is particularly important in the articular cartilage at the ends of bones in joints. In addition, growth factors can be sequestered in the ECM, and cell surface receptors have important interactions with ECM proteins.

Growth Plate Regulation and Long Bone Development

The cartilage growth plate is an orderly arrangement of chondrocytes as they proliferate, differentiate, and apoptose, leaving behind a mineralized matrix for osteoblasts to invade (Figure 28.1). The growth plate is separated into resting, proliferating, pre-hypertrophic, and hypertrophic zones. The formation of the growth plate begins with the continued differentiation of chondrocytes in the center of the anlagen into hypertrophic chondrocytes to form the primary center of ossification at what will be the center of the diaphysis. A gradient of hypertrophic differentiation is formed from resting chondrocytes in the epiphysis to stacked proliferating chondrocytes to pre-hypertrophic chondrocytes to hypertrophic chondrocytes at the center of the diaphysis. The hypertrophic chondrocytes eventually apoptose and are replaced by invading chondroclasts, osteoblasts, and blood vessels. Later in development a secondary center of ossification forms at the center of the epiphysis,

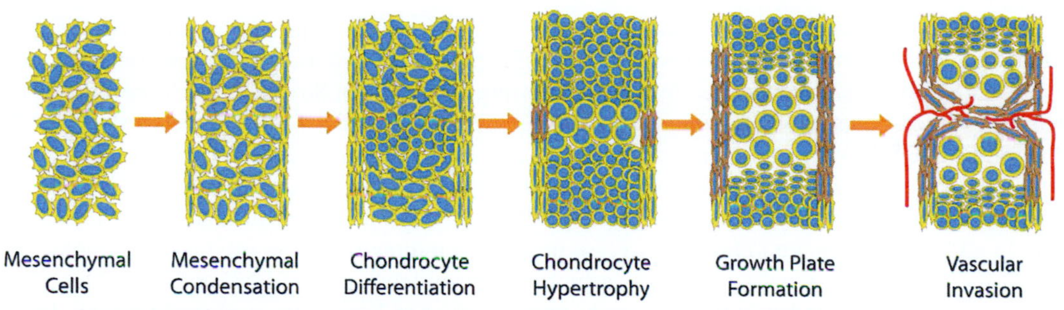

FIGURE 28.1 **Schematic of Endochondral Ossification.** The process of endochondral ossification forms the appendicular skeleton and the majority of the axial skeleton. Primordial cartilage anlagen are formed from condensing mesenchymal cells. Chondrocytes at the center of the anlagen within the perichondrial border undergo differentiation through hypertrophy to initiate the formation of the growth plate. Osteoblasts differentiate within the perichondrium (brown cells) and invade mineralized hypertrophic cartilage with vascular blood vessels and associated chondroclasts to establish the primary and secondary ossification centers and eventually the marrow cavity.

setting up a similar gradient of differentiation in the opposite direction. This leads to the formation of growth plates containing the immature resting chondrocytes between the two centers of ossification at the metaphysis. As the animal matures, the resting chondrocytes are depleted and progress to hypertrophic differentiation and apoptosis, resulting in the final closure of the growth plate at the end of puberty. A well-functioning growth plate is necessary for lengthening of the long bones, and any perturbations of the system will lead to gross deformities (Karsenty et al., 2009).

Upon the initiation of hypertrophy, the cells are committed and will ultimately undergo apoptosis. Once the first pre-hypertrophic cells are established, they begin secreting signals that will interact with proliferating cells, setting up the cross-talk that balances proliferation and hypertrophy. Several important studies have revealed some of the genes that regulate this orderly arrangement of chondrocyte differentiation. Two systems important for the maintenance of the growth plate and the carefully balanced act of proliferation versus hypertrophy are the Indian hedgehog (Ihh)/parathyroid hormone related protein (PTHrP) negative feedback loop and BMP/FGF antagonism.

Ihh is a secreted morphogen important in both limb bud patterning and growth plate regulation. Post-mitotic, pre-hypertrophic cells express Ihh. Ihh null animals show decreased proliferation of the growth plate chondrocytes, as well as an increase in hypertrophy. Ectopic expression of Ihh leads to a decrease in hypertrophy (St-Jacques et al., 1999). The Ihh receptor, patched (Ptch1), is found in the perichondrium surrounding the growth plate, as well as in the proliferating zone. Ptch1 is localized to a cellular organelle known as the primary cilium, and loss of any gene necessary for formation of the primary cilia abrogates the ability of growth plate chondrocytes to transduce the Ihh signal, resulting in skeletal dysplasias observed in ciliopathies such as Jeune asphyxiating thoracic dystrophy (Haycraft and Serra, 2008). Ihh activation of Ptch1 in the perichondrium turns on the expression of TGFβ, which in turn activates PTHrP expression, also in the perichondrium (Alvarez et al., 2001; 2002). PTHrP is a secreted protein that shares homology and a receptor with parathyroid hormone. Like the Ihh knockout, the PTHrP null animal exhibits increased hypertrophy (Amizuka et al., 1994). Similarly, ectopic expression of PTHrP gives rise to a completely cartilaginous skeleton, indicating a complete absence of hypertrophy. The PTHrP receptor (PTHR) is found only in proliferating cells on the edge of pre-hypertrophy. PTHrP acts on the proliferating cells to prevent them from entering into hypertrophy. Interestingly, Ihh null animals never express PTHrP, indicating that Ihh is necessary for PTHrP induction and demonstrating that the phenotype of increased hypertrophy seen in Ihh null mice is due in part to the absence of PTHrP (St-Jacques et al., 1999). In addition, these findings suggest Ihh acts directly on the cells of the proliferative zone to stimulate proliferation. Taken together, Ihh, in concert with PTHrP, acts in a negative feedback loop: cells committed to hypertrophy secrete a protein that prevents hypertrophy in proliferating cells.

The story becomes more complicated with the addition of BMP/FGF antagonism and its control over the Ihh/PTHrP feedback loop (Minina et al., 2002; 2001). Within the growth plate, FGFs encourage differentiation, decrease proliferation and suppress Ihh expression. As an example, activating mutations in FGFR3 result in accelerated hypertrophic differentiation and early closure of the growth plate resulting in achondroplasia, the most common genetic cause of dwarfism in humans (L'Hote and Knowles, 2005). In contrast, knockout of Fgfr3 in mice results in increased chondrocyte proliferation, delayed hypertrophic differentiation, and abnormally long bones. Fgfr3 is found in proliferating and pre-hypertrophic cells, whereas Fgfr1 is found only in hypertrophic cells. Fgf2, -8, -9, -17, and -18 are all found in cartilage, although their actions are dependent on which receptors they bind and where those receptors are. Of the FGF knockout animals, only the Fgf18 null animal has an abnormal skeletal phenotype, and it closely resembles the Fgfr3 knockout. Both have expanded proliferative and hypertrophic zones, while the Fgf18 null also has delayed bone formation, indicating it may also signal through Fgfr1. FGFs have been shown to decrease proliferation by direct action, as well as by decreasing Ihh expression. Decreased

Ihh expression increases differentiation. FGFs not only increase commitment to hypertrophy, they also accelerate hypertrophy by directly acting on hypertrophic cells. Presumably the effects on Ihh expression and on proliferation are mediated by Fgfr33, while the latter effect is also mediated by Fgfr1 (Karsenty et al., 2009). Sox5 and Sox6 prevent pre-hypertophy in resting chondrocytes at least in part by inhibiting the expression of FGF receptor 3 (Fgfr3) (Smits et al., 2004) (Figure 28.2).

In contrast to FGFs, BMPs suppress differentiation, increase proliferation, and encourage IHH expression. The BMPs are present throughout the cartilage and perichondrium. BMPs and Ihh form a positive feedback loop: BMPs upregulate Ihh, Ihh in turn upregulates them. By upregulating Ihh, BMPs indirectly increase proliferation and prevent hypertrophic differentiation. In addition, BMPs are known to increase proliferation directly, and to inhibit hypertrophic differentiation independently of the Ihh/PTHrP signaling cascade. In limb culture systems, adding Fgf2 and Bmp2 together induces no change to the bones, while independently Fgf2 increases hypertrophy and Bmp2 increases proliferation. Interestingly, the addition of Bmp2 to limb cultures expressing constitutively active Fgfr3 rescues the phenotype (Minina et al., 2002; 2001). These findings suggest that BMPs and FGFs not only oppose one another, but also form a feedback loop to influence expression of the other.

Another protein critical to differentiation of chondrocytes to hypertrophy is core-binding factor 1 (CBFA1), also known as Runx2 (Fujita et al., 2004; Otto et al., 1997). Runx2 is found in pre-hypertrophic and hypertrophic chondrocytes, as well as in osteoblasts. The primary phenotype of mice null for Runx2 is a failure to form bone, due to an absence of osteoblasts. However, infecting chick or mouse chondrocytes with antisense constructs for Runx2 prevents chondrocyte hypertrophy, confirming its role in endochondral ossification. In addition, infection of immature chondrocytes with Runx2 accelerates the rate of hypertrophy. This acceleration of hypertrophy is due in part to the upregulation of collagen type X and matrix metalloproteinase-13 (MMP13), which will be covered below (Enomoto et al., 2000; Zheng et al., 2003). In addition, Runx2 upregulates vascular endothelial growth factor (VEGF), which is necessary for vascular invasion of the hypertrophic zone.

Growth hormone (GH) and insulin-like growth factors (IGFs) are important in controlling the length of long bones, although their targets, molecular effects, and expression patterns are as yet unclear (Robson et al., 2002). GH is responsible for gigantism and dwarfism, due to over- or underexpression, respectively. It is known that GH is released from the pituitary gland and stimulates the liver to release circulating Igf1. Eliminating liver-derived circulating Igf1 stunts growth, suggesting the liver is the source of chondrocyte-activating IGF. On the other hand, directly injecting GH into the tibia increases length, both by directly influencing resting chondrocytes to proliferate, as well as by increasing local expression of Igf1. Interestingly, the growth plates of Igf1 null mice have no alteration in cell proliferation, but show a decrease in hypertrophic cell height. This finding suggests that perhaps Igf2 is important in stimulating proliferation. Complicating matters further is the fact that multiple IGF binding proteins (IGFBPs), which inhibit the activities of the IGFs, as well as the two IGF receptors are found in overlapping patterns throughout the growth plate. While it is clear that the GH/IGF system is important in long bone length, more work needs to be done to clarify the process.

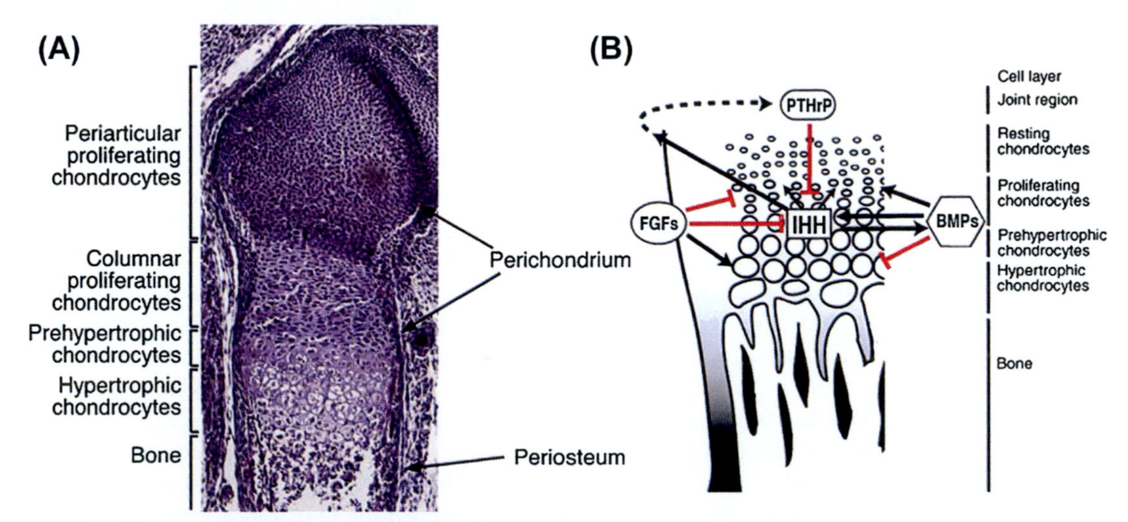

FIGURE 28.2 **Regulation of Cartilage Growth Plate Maturation.** (A) The growth plate is an orderly arrangement of chondrocytes as they proliferate, differentiate, and apoptose, leaving behind a mineralized matrix for osteoblasts to invade. (B) The Indian hedgehog (IHH)/parathyroid hormone related protein (PTHrP) negative feedback loop and BMP/FGF antagonism: IHH and BMPs increase proliferation and prevent hypertrophy, while FGFs prevent proliferation and promote hypertrophy. *(Adapted from Goldring et al., 2006).*

The term hypertrophy has been mentioned multiple times, but has not been defined. Hypertrophic cells are easily identified histologically, as they are the largest cells of the growth plate, due to an increase in intracellular organelles. In fact, the increased height of hypertrophic cells is responsible for upwards of 50% of bone lengthening. In molecular terms, gene expression of collagen type X and alkaline phosphatase defines hypertrophy. Both of these proteins are important in mineralization. Collagen type X is thought to influence mineralization by stimulating calcium accumulation in matrix vesicles and by acting as a docking site for the matrix vesicles (Kwan et al., 1997). Alkaline phosphatase is found in matrix vesicles and leads to an accumulation of pyrophosphate, a mineral important in the formation of calcium hydroxyapatite crystals. Together, collagen type X and alkaline phosphatase act to initiate mineralization of cartilage, which prevents any further growth in the hypertrophic zone of cartilage and sets up a scaffold for osteoblast invasion.

Once cells undergo hypertrophy, their matrix is remodeled, they undergo apoptosis, and capillaries invade the cartilage model, although not necessarily in that order. It is believed that matrix remodeling by MMPs is the initiator of apoptosis and vascular invasion. Mouse knockout models for both Mmp9 and Mmp13 show a reduction in angiogenesis, ECM remodeling, and apoptosis (Stickens et al., 2004; Vu et al., 1998). MMP13, turned on by Runx2, cleaves collagen type II, which is still highly expressed in the hypertrophic zone. Upon degradation of collagen type II, vascular endothelial growth factor (Vegf) previously expressed under the influence of Runx2 and sequestered in the ECM, is released. Vegf acts as an activator of vasculogenesis, as well as a recruiter of osteoclasts (Zelzer et al., 2002). Osteoclasts will be covered in detail later, but for now it is important to know that their primary function is ECM degradation. Osteoclasts express Mmp9, which further processes the collagen type II initially cleaved by Mmp13, and together with Mmp13, acts to degrade aggrecan. There are two requirements for capillary invasion: pro-angiogenic growth factors and space. The actions of MMPs provide both. In addition, MMPs are responsible for promoting apoptosis. Either directly or indirectly, MMP action either releases a pro-apoptotic factor from the ECM, or the ECM protein fragments stimulate the cells to apoptose (Dirckx et al., 2013).

Osteogenesis

Osteoblasts

Once a hypertrophic zone is established in the growth plate, cells in the perichondrium surrounding the newly hypertrophic cells begin to differentiate into osteoblasts, or bone forming cells, and the bone collar is formed. Lineage tracing experiments have revealed that the perichondrium is a source of invading osteoblasts, but have not ruled out the transdifferentiation of some hypertrophic chondrocytes into osteoblasts. Osteoblasts undergo a well-defined program of differentiation from mesenchymal progenitor cells to osteo-chondroprogenitors to mature osteoblasts and, finally, matrix embedded osteocytes. Stage specific Cre recombinase mouse lines have enabled specific interrogation of the steps and signaling involved in the osteoblast differentiation program (Figure 28.3). Mature osteoblasts are characterized by expression of osteocalcin, alkaline phosphatase, and collagen type I. The extracellular matrix produced by these cells, which forms the osteoid structure, later accumulates calcium phosphate to form hydroxyapatite, characteristic of the mineralized bone matrix. Additionally, osteoblasts and osteocytes are responsible for some control of osteoclastogenesis via their production of Rankl. The differentiation of osteoblasts is tightly controlled by a number of signaling pathways and transcription factors (Dirckx et al., 2013; Long, 2012).

Runx2 is the master switch that differentiates perichondrial precursor cells into pre-osteoblasts (Fujita et al., 2004; Otto et al., 1997). Pre-osteoblasts do not form bone, but they do express osteoblast specific proteins. They also express chondrocyte-specific genes, suggesting that they are not yet irreversibly committed to the osteoblast lineage. Runx2 acts upstream of osterix (Osx), another transcription factor involved in osteoblast differentiation. While Runx2 null mice do not express Osx, the effect is not thought to be direct. The effect may be mediated by Bmp2, which is known to directly activate Osx expression. Osx is responsible for converting pre-osteoblasts into bone forming osteoblasts. Mice lacking Osx have pre-osteoblasts that express Runx2 and have completely normal growth plates that vascularize appropriately. However, they fail to form bone (Nakashima et al., 2002). Atf4 is an important transcription factor in mature osteoblasts that directly regulates expression of osteocalcin and Rankl. In fact, misregulation of ATF4 underlies skeletal abnormalities seen in human patients with Coffin-Lowry syndrome (Long, 2012).

Ihh and BMP signaling are critical to bone formation. Ihh is secreted by pre-hypertrophic cells, but the Ihh receptor, Ptch1, is expressed throughout the perichondrium. In Ihh null mice, there are no osteoblasts, highlighting how vital this protein is in bone production (St-Jacques et al., 1999). Ihh affects differentiation by working upstream of the transcription factor Runx2, as well as by increasing expression of BMPs in a positive feedback loop. Animals null for most BMPs are early embryonic lethal, although Bmp6 and Bmp7 knockouts have mild skeletal phenotypes (Jena et al., 1997; Solloway et al., 1998). Ectopic expression of the BMP inhibitor, Noggin, or dominant negative BMP receptors leads to decreased

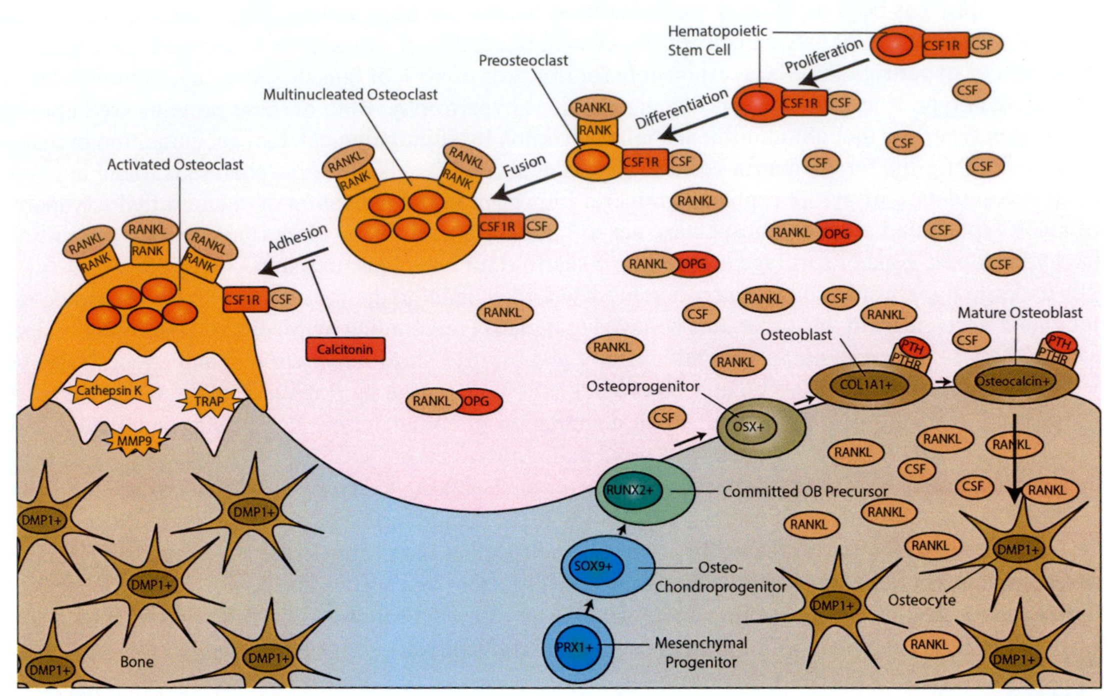

FIGURE 28.3 **Osteoblastic and Osteoclastic Differentiation.** All osteoblasts are descended from mesenchymal progenitor cells and proceed through an orderly process of differentiation. RUNX2 and OSX2 are two of the key transcription factors necessary for osteoblast differentiation. Mature osteoblasts eventually embed themselves within their own matrix formed from collagen type IA (COL1A1) and osteocalcin (OCN). OCN acts to initiate the formation of hydroxyapatite, and the embedded cells begin to express dentin matrix acidic phosphoprotein 1 (DMP1) and are known as osteocytes. Multinucleated osteoclasts differentiate from hematopoietic stem cells under the influence of receptor antagonist of nuclear factor kappa beta (Rankl) and colony stimulating factor (CSF) secreted from osteoblasts. Rankl and CSF are secreted by osteoblasts and osteocytes upon stimulation by parathyroid hormone (PTH). CSF primarily stimulates proliferation and Rankl differentiation and fusion. Osteoprotegerin (OPN) is secreted by many cells in the body and serves to bind and inhibit Rankl. The circulating hormone, calcitonin, inhibits osteoclast adhesion to bone. Osteoclasts adhere to bone via integrins, forming a cavity between themselves and bone into which they secreted matrix metalloproteinases (MMP9), tartrate-resistant acid phosphatase (TRAP), and proteases such as cathepsin K to break down the bony matrix. The proteins expressed at each stage of differentiation have been used to generate Cre recombinase lines for specific stages in osteoblast and osteoclast differentiation, allowing precise interrogation of the signals driving the differentiation process.

osteoblast function and differentiation. Conversely, ectopic expression of Bmp2 and Bmp4 leads to heterotopic ossification. In fact, fibrodysplasia ossifican progressiva (FOP), a human disease caused by ectopic activation of the BMP pathway via mutations in an Activin receptor (ACVR1), leads to horrible disfigurations as ossification occurs throughout the soft tissues of patients' bodies. BMPs are also capable of upregulating Runx2, possibly acting as the effector of Ihh upregulation of Runx2 (Long, 2012).

An important pathway for negatively regulating bone formation is the Notch pathway. The Notch pathway consists of the jagged (Jag1–2 in mammals) and delta (Dll1, -3, -4 in mammals) ligands on one cell surface signaling to adjacent cells via binding of these proteins to notch receptors (Notch1–4 in mammals), resulting in cleavage of Notch by presenilin (PS1–2 in mammals) to allow interaction with RBPJκ transcription factor (CBF1 in mammals) to induce expression of target gene Hey1. Mouse knockout of presenilin and Notch both enhanced osteoblast formation. Additionally, gain of function mutations in Notch2 are responsible for the severe and progressive bone loss disorder, Hadju-Cheney syndrome (Long, 2012).

Finally, communications between the perichondrial cells and invading capillaries are required for differentiation of osteoblasts. In one study, investigators placed cartilaginous long bones, complete with perichondrium, into kidney capsules of host animals (Colnot et al., 2004). When a filter was placed around the anlage such that the vasculature of the kidney could not invade, maturation proceeded normally until osteoblasts were expected to differentiate and form bone. While endothelial cells exist in the perichondrium, they can only form vessels and invade the hypertrophic zone upon contact with pre-exisiting blood vessels. In other words, the cells of the perichondrium are not sufficient for osteogenesis. Communications between preexisting capillaries and the perichondrium are essential for osteoblast differentiation. A number of signaling molecules mediate the migration of osteoblasts in both development and bone repair including TGFβ1, Pgdf (platelet derived growth factor), and Vegf. Vegf is produced by hypertrophic chondrocytes in response to hypoxia via Hif1α signaling and is essential for vascular and osteoblast invasion (Dirckx et al., 2013).

Osteoclasts and Bone Remodeling

As mentioned above, osteoclasts are the cells in bone responsible for ECM digestion. A similar cell type, the chondroclast, is important early in bone development when the hypertrophic zone of the cartilage growth plate is remodeled for capillary and osteoblast invasion. For the rest of an organism's life, osteoclasts will work hand-in-hand with osteoblasts to digest and rebuild the skeleton. In fact, every year ten percent of the skeleton is digested and remodeled, resulting in a new skeleton every ten years. This skeletal digestion must be carefully controlled, as too much can lead to osteopenia, osteoporosis, or Paget's disease, and too little can lead to osteopetrosis. Bone remodeling is not just meant to repair and refresh the skeletal structure; it also regulates serum calcium and phosphorus homeostasis, as bone is the body's storage depot for both elements. This additional level of complexity relies on communication with the gut, kidney, thyroid, and parathyroid (Boyle et al., 2003).

Osteoclasts derive from circulating monocytes that commit to the macrophage lineage. A small percentage of these circulating cells find their way to the bone and adhere. Interaction with bone stromal cells (pre-osteoblasts) is necessary for them to differentiate into osteoclasts. Two cell surface proteins produced by the osteoblasts, colony stimulating factor-1 (Csf1) and RANK ligand (Rankl) known to be required for osteoclastogenesis. Csf1 alone promotes proliferation, whereas Csf1 and Rankl together lead to differentiation, survival, and fusion. Osteoclasts are multinucleated, and fusion is a prerequisite for function. Continued exposure to Rankl leads to osteoclast activation. Mice deficient in Rankl or RANK (Dougall et al., 1999), the cell surface protein on osteoclasts that binds Rankl, have a complete absence of osteoclasts, while other monocyte-derived cells are normal (Boyle et al., 2003) While pre-osteoblasts, osteoblasts, and osteocytes all produce Rankl, experiments with different inducible knockout strains revealed that osteocytes are the essential producers of Rankl (Xiong et al., 2011).

Activated osteoclasts express a host of proteins involved in the degradation of bone and mineralized cartilage. To degrade bone, the osteclast adheres to the bone surface, creating a space into which it will secrete an acidic cocktail of digestive juices. In order to create this tight seal, the cell must express $\alpha v\beta 3$ integrin, which attaches to the collagen in bone. A vacuolar ATPase then pumps protons into the space between the osteoclast and bone, decreasing the pH to around 4.5, and allowing the dissolution of the inorganic mineral in bone. Tartrate-resistant acid phosphatase (TRAP) and cathepsin K (CATK) are also secreted into the space and are responsible for digesting the organic components of bone (Boyle et al., 2003).

While RANK and Rankl are essential for osteoclast differentiation and activation, other signaling molecules influence osteoclastogenesis. Osteoprotegrin (OPG) is a decoy receptor for Rankl, functionally blocking the interaction of RANK and Rankl. Osteoblasts can be stimulated to produce OPG by estrogen and BMPs. Estrogen also downregulates the expression of Rankl in osteoblasts, as well as the expression of the pro-inflammatory cytokines interleukin-1 (IL-1), interleukin-6 (IL-6), and tumor necrosis factor-α (TNFα), all of which positively influence osteoclast activity. Calcitonin, a protein secreted by the thyroid when serum calcium is elevated, prevents osteoclasts from attaching to bone. On the other hand, when serum calcium is low, the parathyroid gland releases parathyroid hormone (PTH), which stimulates osteoclast activity by inducing osteoblast expression of Rankl and Csf1. Many additional factors, including vitamin D, corticosteroids, and PTHrP, positively regulate osteoclasts as well (Boyle et al., 2003; O'Brien et al., 2013).

Although fewer signaling pathways are known to negatively regulate osteoclast activity, one prominent example is estrogen, as mentioned above. The impact of estogen is particularly prominent in post-menopausal women, who frequently suffer from osteoporosis, characterized by loss of bone mass and caused by increased osteoclast activity without compensatory bone growth. While osteoporosis is much more common in women, men are also susceptible. As men age, there is a reduction of testosterone. Testosterone is the precursor of estrogen, and it directly promotes osteoblast activity, thus both an osteoanabolic factor and an inhibitor of osteocatabolism are lost. Drug companies have developed pharmacological agents that target osteoclast activity at multiple steps. Bisphosphonate, which stimulates osteoclast death, and estrogen treatments are most commonly prescribed, but drugs targeting osteoclast adhesion and digestion are also in use or development (Boyle et al., 2003; O'Brien et al., 2013).

Joint Development

Interzone Specification and Cavitation

The initiation of the synovial joint formation is still an incompletely understood developmental process. The morphological changes are well known, but the molecular controls are still being elucidated. It is known that at specific points where cartilage anlagen meet, cells begin to flatten and form what is known as the interzone. These interzone cells do not produce chondrocyte-specific proteins collagen type IIB and aggrecan, but express a distinct ECM that contains collagen type I and HA.

Downregulation of Sox9 is responsible for the phenotypic switch of cells from chondrocytes to interzone cells. The cells at the center of the interzone begin to secrete an ECM that separates the cells, ultimately leading to cavitation. The remaining flanking interzone cells are thought to become articular chondrocytes. A mature joint consists of two opposing articular surfaces wrapped in a joint capsule, stabilized by ligament and tendon. Synovial cells line the inside of the joint and are responsible for producing joint fluid, which nourishes the avascular articular cartilage and lubricates the structure (Iwamoto et al., 2013; Koyama et al., 2010). Any perturbations in the joint structure, particularly as a function of age, lead to painful and often irreversible joint diseases, such as osteoarthritis (Khan et al., 2007).

Cells destined for incorporation into the joint have a distinct lineage from those destined for the epiphyseal growth plate. Lineage tracing experiments in mice reveal that matrilin-1 is a specific marker of epiphyseal chondrocytes. Articular chondrocytes, cruciate ligaments, menisci, synovium and their precursors are all derived from matrilin-1 negative precursors (Hyde 2007). Conversely, lineage tracing experiments utilizing Gdf5-Cre reveal that Gdf5 expression labels the interzone precursors, articular chondrocytes, synovial lining, and other joint tissues, but fails to label growth plate chondrocytes. Additional joint structures in the knee include the medial and lateral menisci, which serve as shock absorbers, and the cruciate ligaments, which add stability (Khan et al., 2007). Collagen type IIA lineage tracing experiments reveal that the cruciate ligaments and inner medial meniscus are derived from collagen type IIA labeled precursors that arise from the interzone while the lateral meniscus and outer medial meniscus are derived from collagen type IIA negative precursors that invade the joint (Hyde, 2008).

The location of the interzone is established quite early. If the presumptive elbow region is removed prior to any morphological changes, the joint fails to form. Specification of the location of the joint interzone is similar to specification of the location of limb bud outgrowth in the early embryo. Not surprisingly, mutations in homeobox genes, known to be important in embryonic patterning, lead to fusions in the joints of the wrists, implicating them in this specification (Khan et al., 2007). Two novel homeobox proteins, Cux1 (Lizarraga et al., 2002) and Barx1 (Meech et al., 2005), have recently been found in early joints. Interestingly, exogenous expression of Cux1 in the limb bud micromass system leads to downregulation of chondrogenic markers, one of the first steps in interzone formation. Similarly, triple knockout of Hox11A, Hox11C, and Hox11D in a mouse results in disrupted knee and elbow joints with alteration in the geometry of the articulating surface of long bones and decrease in expression of joint markers (Koyama et al., 2010).

The Wnt proteins are candidate master genes that regulate gene expression in the interzone. Wnt14 (Wnt9A) is of particular interest (Guo et al., 2004; Hartmann and Tabin, 2001). While Wnt4, -5A, -14, and -16 are all found in joints, only Wnt14 is confirmed to have a direct effect on the presumptive joint. How Wnt14 signals is controversial, but its effects are agreed upon: it upregulates the expression of joint markers (Gdf5, Chordin, and CD44) and downregulates epiphyseal chondrocyte markers (Sox9 and Noggin) – essential steps for the specification of interzone cells. Wnt14 also induces expression of genes necessary for HA synthesis (Onyekwelu et al., 2009). Additionally, if exogenous Wnt14 is retrovirally expressed by virus, the region adjacent to the infection fails to form a joint where a joint would normally be. This implies that Wnt14-stimulated proteins act to prevent new joint initiation within a certain distance of an established joint. In other words, Wnt signaling and downstream BMP signaling may be responsible for both joint specification and joint spacing.

TGFβ signaling plays an important role in joint development downstream of Wnt14. TGFβ type II receptor (Tgfbr2) is specifically expressed in progenitor cells of the joint interzone in synchrony with joint specific markers (Noggin, Gdf5, Notch1 and Jag1). Tgfbr2 labels the meniscal surface, ligaments, and the synovial lining of the infrapatellar fat pad, and limb specific (Prrx1-Cre) knockout of Tgfbr2 results in failure of interzone formation along with disruption of joint ligaments and meniscus development (Li, 2012). TGFβ family proteins Gdf5, Gdf6, Bmp2, and the inhibitor, Chordin, are all expressed in the interzone, while the inhibitor Noggin is specifically excluded. Upon cavitation, expression shifts to Bmp2, -4, -7, and Chordin. BMPs are thought to cause apoptosis in the interzone. Curiously, Chordin null mice have no joint phenotype, whereas Noggin null mice fail to form joints and never express Gdf5 (Brunet et al., 1998). This failure to form joints is thought to be due to either a failure of Gdf5 to be produced in joints or to an increase in BMP bioavailability, which could drive recruitment of cells into the cartilage model. That the interzone expresses agonists and their antagonists in the same spatiotemporal pattern is curious. Since cells rarely produce substances in a wasteful manner, there are likely additional controls on BMP/GDF or Noggin/Chordin in action.

One of the ways Gdf5 facilitates joint formation is via induction of Erg/C-1-1. Erg/C-1-1 is an Ets transcription factor expressed in developing limb joints. Erg/C-1-1 expression is induced by ectopic Gdf5, and overexpression of Erg/C-1-1 in mice results in a completely cartilaginous skeleton with chondrocytes actively expressing collagen type IIA and aggrecan with low expression of maturation and hypertrophy markers (Ihh, collagen type X, MMP13). Erg/C-1-1 also stimulates expression of PTHrP, providing a possible mechanism by which Erg/C-1-1 may counteract hypertrophy. Erg/C-1-1 is believed to be important for articular cartilage differentiation (Pitsillides and Ashhurst, 2008).

Cavitation is mediated by HA, which can act to condense cells or separate them, depending on the quantity of HA produced. HA binds to CD44, a receptor on the surface of interzone cells. Interzone cells produce large quantities of HA, leading to cell separation, which sets the joint up for cavitation. Overexpression of the joint marker Gdf5 fails to induce joint formation and leads to joint fusion, which is thought to be the result of greatly decreased HA production and concomitant failure to cavitate. Recent studies suggest activation of MEK-ERK signaling at sites of cavitation is responsible for changes in HA synthesis and binding necessary for cavitation. Candidate genes that function upstream in MEK-ERK signaling include Fgf2, -4, and -10. HA and CD44 up-regulation is also dependent on movement (Pitsillides and Ashhurst, 2008). Mechanical stimulation of cells increases CD44 and HA expression, and paralyzed animals form interzones but fail to cavitate (Iwamoto et al., 2013).

Articular cartilage differs in significant ways from the epiphyseal cartilage found in growth plates. Recent microarray studies comparing expression between the two populations show that articular chondrocytes have significantly decreased expression of hypertrophy markers (ALP and collagen type X). Articular chondrocytes also express unique markers including Tenascin C, Six1 (sine-oculis-related homeobox transcription factor), Grem1 (Gremlin 1), Dkk1 (Dickkopf 1), Frzb3 (Frizzled-related protein 3), Prg4 (proteoglycan 4), Thbs4 (thrombospond-4), Abi3bp (Abl interactor family member) (Iwamoto et al., 2013), and Scrg1 (Ochi et al., 2006). The significance and role of these factors in specifying and maintaining the articular chondrocyte phenotype remain to be elucidated. The mechanical properties of articular cartilage as are also significantly different, with articular cartilage ECM being stiffer and more resistant to compressive force. Greater understanding of the molecular genetics underlying the formation and maintenance of articular cartilage will be important for the development of regenerative therapies for joint cartilage repair (Iwamoto et al., 2013).

28.3 AXIAL SKELETON

Origin of the Axial Vertebrate Skeleton

The vertebrate skeleton is the product of cells from three distinct embryonic lineages. The craniofacial skeleton is derived from the cephalic paraxial mesoderm and invading cranial neural crest cells, the axial skeleton is derived from paraxial mesoderm in the vicinity of the notochord and neural tube, and the appendicular skeleton is the product of lateral plate mesodermal cells. The primordia of these tissues are specified early in ontogeny. In the chick embryo, three germ layers are formed 12 hours after fertilization with the onset of gastrulation, during which cells migrate towards and ingress through the primitive streak. This movement of cells has been shown to be controlled by the activities of the FGF regulated transcription factor Snail and SRY-box containing 3 gene or Sox3 (Acloque et al., 2011). In the Hamburger Hamilton stage 5 head process embryo (20 hours) (Hamburger and Hamilton, 1951) (Figure 28.4A), several different mesodermal compartments can already be identified: the axial chordamesoderm that will give rise to the notochord, and the paraxial, intermediate and lateral plate mesoderms. Within the paraxial mesoderm, two major regions are recognized: the segmental plate, which forms along the length of the primitive streak on either side of the notochord, and cephalic mesoderm on either side of the chordamesoderm, or head process (Figure 28.4A). The paraxial mesoderm of the trunk and the cephalic mesoderm, in combination with invading cranial neural crest, give rise to many tissues including bone, muscle, tendon, and dermis.

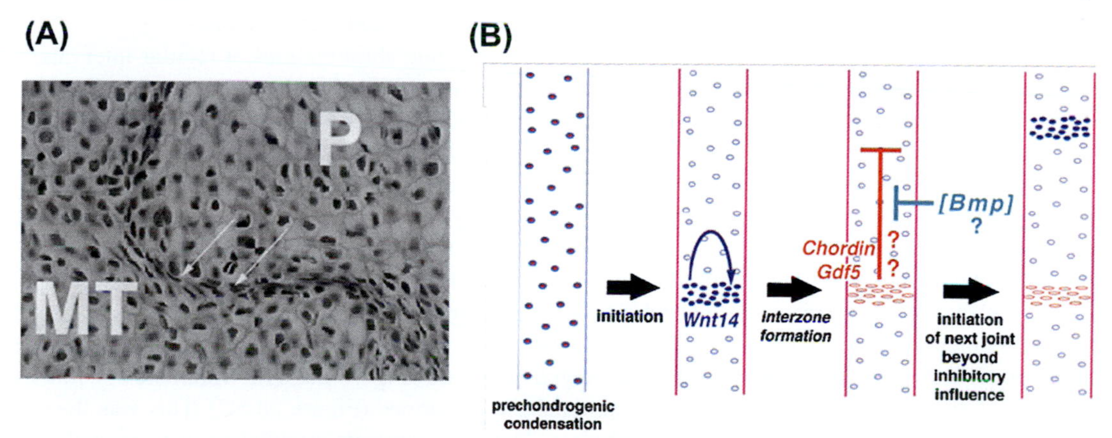

FIGURE 28.4 **Joint Formation.** (A) Morphological description: Cells of the interzone begin to flatten and stop producing chondrocyte-specific proteins (Archer 2003). (B) Molecular regulation: Wnt14 (aka Wnt9A) appears before the interzone is established and is required for interzone formation. Proteins expressed by the interzone, such as Gdf5 and Chordin, prevent new joint initiation within a certain distance of an established joint. *(Adapted from Archer et al., 2003, Hartmann et al., 2001, and Alexander, PG., 2001, "The role of Paraxis in somitogenesis and carbon monoxide-induced axial skeletal teratogenesis." Ph.D. Thesis, Thomas Jefferson University, Philadelphia, PA.)*

One of the earliest markers of presomitic mesoderm, Tbx6, is a T-box transcription factor that labels all of these structures, and a transgenic mouse line expressing all Cre recombinase under control of the Tbx6 promoter has aided in the elucidation of the processes underlying somitogenesis (White et al., 2005). While the axial skeleton is completely derived from cells of the paraxial mesoderm, cranial neural crest cells contribute heavily to the craniofacial skeleton. These cells are formed during the specification of the cephalic neural plate from ectoderm and begin de-epithelializing from the dorsal lip of the cephalic neural folds as they fuse to form the neural tube (see also Chapter 18). This section will concentrate on the axial skeleton.

Somitogenesis: Axial Skeletal Patterning

Somites are transient embryonic structures that are critical in realizing the metameric pattern characteristic of the vertebral column, and associated tissues. They are segmental units that bud from the segmental plate as the node passes rostrocaudally down the primitive streak (Figure 28.4B). In the node's wake, the notochord forms from the chordamesoderm, and the neural plate differentiates from the ectoderm, thickening, folding and fusing to form the neural tube. Following these events, somites form from the rostral end of the paraxial mesoderm through a mesenchymal-epithelial transition in a species-specific, time-dependent fashion. In the chick, a new somite is formed every 90 minutes. Since development of the embryo proceeds rostrocaudally, several stages of somitogenesis can be observed in one embryo (Figure 28.4C). The mesenchymal-epithelial transition involves an initially loosely associated ball of cells (the condensed somite), that reorganizes into a spherical epithelial structure (the epithelial somite) and ultimately differentiates (Figure 28.4D). There is a small population of cells inside the presomitic mesoderm that retain their mesenchymal organization to form the cells of the somitic core, the somitocoele.

Somite Patterning

The process of somitogenesis is divided into several distinct phases: patterning, morphogenesis, differentiation and maturation of somite-derived tissues. Although the morphogenesis of the somite was described early in the history of embryology, the exact mechanism of the patterning process remains elusive. The regularity of somite architecture and formation during embryogenesis suggests that it must be controlled by a clock or somitic oscillator within the embryo. The earliest experiments attempted to alter somite formation through surgical manipulations. These included removing individual tissues from the embryo or observing individual tissues in isolation. Among the most important observations was that the segmental plate could undergo somitogenesis when separated from both the last previously formed somite and the axial structures (Packard, 1976). This finding demonstrated that segmentation does not require a continuous flow of information rostrocaudally, and that the positional information or pattern for somitogenesis was established at gastrulation, possibly as the cells pass through the node to populate the paraxial mesoderm. Another enlightening observation was that implanted quail nodes could induce a secondary axis with a somite pattern dependent on the mediolateral position of the implant, suggesting a possible morphogen emanating from Hensen's node (Hornbruch et al., 1979). With the application of scanning electron microscopy, investigators identified segementation of the entire paraxial mesoderm in a number and regularity that pre-patterned the somite (Packard and Meier, 1983). These segments were coined somitomeres.

Another important observation was that heat-shock could induce somite abnormalities at regular intervals, suggesting a timing mechanism-based regulation of somitogenesis (Primmett et al., 1989). Similar segmentation defects were also observed when cell cycle disruptors were applied to the embryo, which indicated a link between the cell cycle and the somite oscillator (Primmett et al., 1989). This led to the *clock and wavefront* model of somitogenesis, which invoked a molecular clock within the cells of the paraxial mesoderm that coordinated events in response to a signal wavefront moving rostrocaudally through it (Cooke and Zeeman, 1976). This quickly became the favored model of somitogenesis, and subsequent genetic studies have proven that central components of this theory are true.

A major breakthough in understanding the somitic oscillator was made with the cloning of the chick hairy gene, c-Hairy1 (Palmeirim et al., 1997). Whole mount *in situ* hybridization for cHairy1 mRNA revealed a dynamic caudorostral expression pattern in the segmental plate (Figure 28.5). During the 90-minute cycle of chick somitogenesis, c-Hairy1 expression begins within a broad domain in the caudal third of the segmental plate (Figure 28.5A), which slowly moves rostrally, becoming restricted to the posterior boundary of the next somite to be formed (Figure 28.5C). This was the first molecular evidence of a somitic oscillator within the segmental plate. These experiments established that the cycles of c-Hairy1 expression are autonomous to the paraxial mesoderm, occurring independently of surrounding tissues. These cycles are also independent of protein expression, which indicates that this particular gene responds to the putative somitic oscillator. Although its sequence suggests it is a transcription factor, the function of c-Hairy1 remains elusive as does its connection to the somitic oscillator.

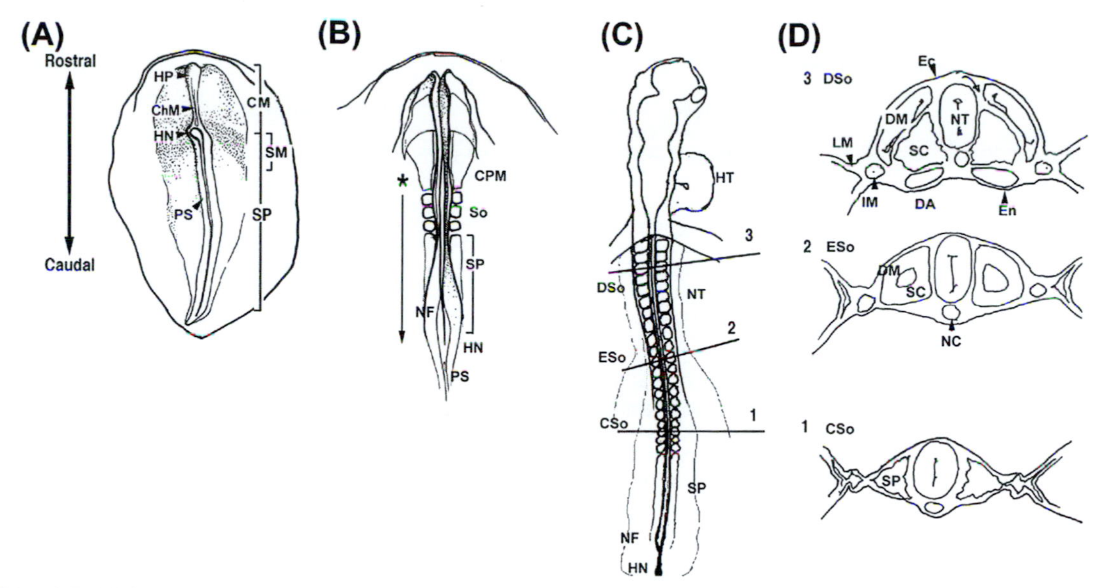

FIGURE 28.5 **A Brief Summary of Somitogenesis in the Chick Embryo.** (A) A head process stage embryo after 20 hours of development. Two regions of paraxial mesoderm are identified: the segmental plate (SP) lateral to the primitive streak (PS), and the cephalic mesoderm (CM) lateral to the head process (HP), a ridge formed by the underlying chordamesoderm (ChM) that also runs along the length of the embryo under the primitive streak. The chick node, Hensen's node (HN), is currently located at the rostral end of the primitive streak defining the border of head and trunk structures. The thickening of the rostral segmental plate emanating posteriolaterally from Hensen's node (HN) indicates cellular condensation characteristic of somitic mesoderm (SM) that here will give rise to the first somite. (B) A stage 6, four-somite embryo in which morphogenesis of the neural folds (NF) has narrowed and elongated the embryo. The arrow indicates the rostral-to-caudal migration of Hensen's node (HN) down the primitive streak (PS) from its rostral extreme (*). In the node's wake, neural folds (NF) have formed and the segmental plate (SP) has lengthened while somites (So) have formed from its rostral end at regular intervals. The cephalic mesoderm (CPM) has thickened with the invasion of neural crest cells but remains overtly unsegemented. (C) A stage 13 embryo with defined brain vesicles, optic and otic cups, a folded heart tube (HT) and approximately 20 somites. Hensen's node (HN) has almost reached the caudal end of the primitive streak (PS). Because development of the embryo proceeds rostrocaudally, somites at different levels of development can be visualized: condensed somites (CSo, 1), epithelial somites (ESo, 2) and differentiated somites (DSo, 3). (D) Tranverse sections through the stage 13 embryo in (C) at different rostrocaudal levels as indicated in (C) revealing (D1) the condensed somite (CSo), (D2) the epithelial somite (ESo) with specified dorsolateral dermomyotome (DM) and ventromedial sclerotome (SC) induced by the notochord (NC), and (D3) the differentiated somite (DSo) in which the sclerotome has de-epithelialized leaving the epithelial dermomyotome (DM) behind in the context of surrounding inductive tissues: notochord (NC), neural tube (NT), ectoderm (Ec), endoderm (En), intermediate (IM) and lateral plate mesoderm (LM) and the dorsal aortae (DA). *(Adapted from Alexander, PG., 2001, "The role of Paraxis in somitogenesis and carbon monoxide-induced axial skeletal teratogenesis." Ph.D. Thesis, Thomas Jefferson University, Philadelphia, PA.)*

Somite Border Formation

In the formation of the somite that follows, we recognize an initial condensed somite composed of unorganized mesodermal cells and its subsequent transformation into a spherical epithelial structure. These two processes are linked; however, different adhesion systems may be involved in each event. The boundaries of the somite are established through the Notch-Delta juxtacrine signaling system. Many components of this system exhibit cyclic expression autonomous to the segmental plate in phase with somite formation (Kuan et al., 2004; Saga and Takeda, 2001). The importance of somite border formation in axial skeletal development is reflected in mutations of the Notch-Delta signaling system that cause somite dysmorphogenesis and dramatic forms of scoliosis. For example, the *pudgy* (*pu*) mouse is characterized by a highly disorganized axial skeleton with many hemi- and fused vertebrae and fused rib elements (Kalter, 1980; Kusumi et al., 1998). These abnormalities are prefigured by highly irregular somites in 9.5–10.0 p.c. *pu* embryos. The mutation was shown to be in Dll3, a heterotypic binding partner of Notch (Dunwoodie et al., 2002). Mutations in the human homolog, DLL3, are known to cause spondylocostal dysplasia, a severe congenital scoliosis with a disorganized vertebral column characterized by multiple hemi-vertebrae, fused vertebrae and fused rubs.

Disturbances in the formation and maintenance of the axial skeleton lead to abnormal spinal curvatures in the neonate or adult. Abnormal axial skeletal curvature may be in the mediolateral or coronal plane, i.e., scoliosis, or in the dorsolateral or sagittal plane, i.e., kyphosis. However, most abnormal spinal curvature is mixed with one direction of curvature more dominant than another. Scoliosis is the more common of these two abnormal curvatures and will be used as a model in this discussion. There are two major classes of scoliosis. Idiopathic scoliosis is a lateral curvature of the spine with no known cause. Depending on the definition of curvature, the incidence of idiopathic scoliosis may be as high as 2–3 in 1,000 live

TABLE 28.1 Selected Human Syndromes with Scoliosis

Syndrome	Gene	Vertebral Abnormalities	Other Anomalies	OMIM #
Spondyloscostal dystosis and Jarcho-Levin syndrome	Dll3 Mesp2 Lnfg	Multiple hemi-vertebrae, fused vertebrae along the length of the spine nd fused ribs	Cranial neural tube defects, genitourinary defects	227300
Alagille syndrome	Jgd1	Butterfly vertebrae and shortened interpedicular space	Craniofacial, cardiac, and ocular defects and liver disease	118450
Klippel-Feil syndrome		Fused cervical vertebrae	Deafness, ear malformation, asymmetric facies	118110
MURCS		Shortened and fused vertebrae of the cervical and thoracic spine	Muellerian duct aplasia, unilateral renal aplasia	601076
VATER		Hemi-vertebrae and fused vertebrae, usually thoracic and lumbar and includes cases of lumbar-sacral agenesis	Anal atresia, tracheo-esophageal fistula with esophageal atresia, and renal dysplasia)	192350
VACTERL		Hemi-vertebrae and fused vertebrae, usually thoracic and lumbar and includes cases of lumbar-sacral agenesis	Anal atresia, cardiac malformations, tracheo-esophageal fistula, renal dysplasia and limb deformities	192350
Goldenhar syndrome/OAV		Fused vertebrae, usually cervical or thoracic	1st and 2nd branchial arch derivative hypoplasia, including facial clefts, esophageal atresia, cardiac and central nervous system (CNS), eye and ear defects	164210

births. It becomes evident postnatally, frequently during adolescence and mostly in females. Although the molecular genetics of idiopathic scoliosis remains unknown, there is a strong inheritable component. Many cases occur in the context of other muscuoloskeletal syndromes, including osteogenesis imperfecta, Marfan syndrome, Ehlers-Danlos, musculodystrophies, cerebral palsy, and spinal trauma in which variations in the muscles and connective tissues that surround the axial skeleton cause imbalances in tension that lead to abnormal curvatures (Giampietro et al., 2003).

Congenital scoliosis is evident at birth due to an underlying axial skeletal dysmorphogenesis. The incidence of congenital scoliosis is 0.5–1 per 1,000 live births, with a slightly greater incidence in males. Abnormalities of the vertebrae include hemi-vertebrae and wedge vertebrae, block vertebrae and vertebral bars and butterfly vertebrae, among others. Hemi-vertebrae may be due to ectopic or additional segments, while block vertebrae and vertebral bars are likely caused by a failure to produce or maintain segmentation (Erol et al., 2004). While congenital scoliosis can be associated with genetic syndromes such as spondylocostal dysplasia, spondylocostal dystosis, Alagille syndrome, Klippel-Feil syndrome, and Jarcho-Levin syndrome, up to 60% of congenital scoliosis cases are of an unknown etiology (Giampietro et al., 2003). Involvement of other organ systems is common in these cases, which include spinal/neural tissues, urogenital, gastrointestinal, and cardiovascular tissues, and craniofacial or appendicular skeleton. These syndromes include Klippel-Feil syndrome, Goldenhar's syndrome, VATER, VACTERL, and the MURCS association (Table 28.1). In contrast to idiopathic scoliosis, these cases usually arise spontaneously without a strong genetic component.

It has been hypothesized that the production and degradation of inhibitors or activators to Notch may reflect or constitute the somitogenic oscillator itself. Lunatic fringe is an excellent candidate since its expression oscillates at the posterior border of the next somite to be formed (Evrard et al., 1998) (Figure 28.6). Cells expressing lunatic fringe can be transplanted into the caudal segmental plate and induce ectopic border formation. Transgenic overexpression of lunatic fringe in the mouse embryo overwhelms endogenous lunatic fringe oscillations in the segmental plate and inhibits all segmentation, suggesting that the degradation of this component may constitute the somitic oscillator itself (Dale et al., 2003).

The concept of a wavefront in the *clock and wavefront* model implies an anterior-posterior gradient of a signal through the paraxial mesoderm. Two molecules with gradients of protein concentration in the paraxial mesoderm have been shown to be especially important in somitogenesis (Maroto et al., 2012), FGF8 (Sawada et al., 2001) and Wnt3a (Liu et al., 1999). The message and protein of the first candidate wavefront molecule, FGF8, exhibits a caudorostral gradient in the segmental plate (Dubrulle et al., 2001) (Figure 28.6) that may be produced through an RNA decay mechanism

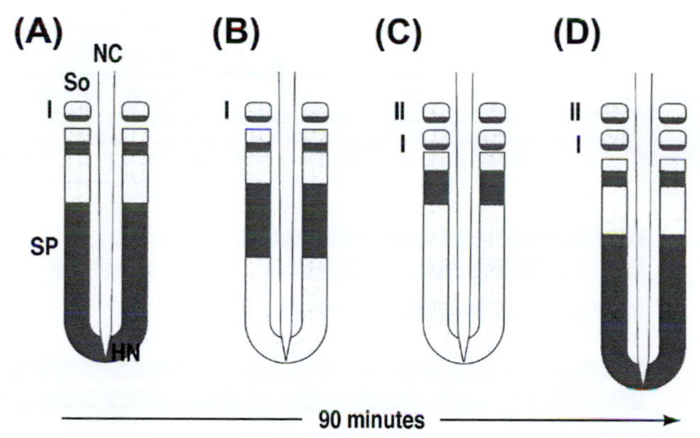

FIGURE 28.6 Dynamic c-hairy1 Expression in the Chick Segmental Plate. (A) The expression of c-hairy1 in the caudal half of the segmental plate (SP), the posterior border of the last formed somite (So) and in a region approximating the posterior half of the next somite at the beginning of the proposed somite oscillator cycle, time 0. Chick somites bud from the rostral segmental plate every 90 minutes – the temporal definition of the somite oscillator. (B) Within 30 minutes, the caudal expression zone of c-hairy1 expression is moving rostrally within the segemental plate while the anterior zone is consolidating into a band prefiguring the posterior border of the next somite (somitomere). (C) At 60 minutes the posterior border of the next somite has formed, while the caudal c-hairy1 expression zone has moved further rostral. (D) The somite oscillator cycle is complete when c-hairy1 expression returns to the caudal half of the segmental plate and an anterior zone prefiguring the posterior half of the next somite as in A. NC, notocord; So, somite; HN, Hensen's node. Somite numbering system adopted from Christ et al., 1998 in which the last formed somite is always numbered Roman numeral I. *(Adapted from Palmieram et al., 1997)*

(Dubrulle and Pourquie, 2004). Increasing the local concentration of FGF8 results in smaller somites, while inhibition of FGF8 signaling produces larger somites, suggesting that this signal determines the position of the somite boundary in response to a spatially progressive signal. The second candidate wavefront molecule, Wnt3a, is only expressed in the mesoderm as it migrates from the node (Aulehla et al., 2003) (Figure 28.6). A caudorostral gradient is possibly created through extracellular degradation of this protein. In the spontaneous Wnt3a mutant mouse *vestigial tail* (*vt*), FGF8 is downregulated in the tailbuds and segmental plate, suggesting that FGF8 acts downstream of Wnt3a (Greco et al., 1996). In either case, as the concentration of the gradient molecule is reduced in more rostral positions, a threshold is reached in which border formation and epithelialization are permitted.

Somite Epithelialization

The epithelialization of the somite is dependent upon a different signaling and adhesion system. Although the cyclic expression of somitogenic genes is observed in isolated segmental plates, overt somite morphogenesis is absent because signals emanating from the overlying ectoderm are required (Figure 28.7). Removal of the ectoderm is known to inhibit somite formation (but not alter the cycling expression of the Notch-Delta system). The signal responsible for this activity is Wnt6, produced by the ectoderm overlying the segmental plate (Rodriguez-Niedenfuhr et al., 2003; Schmidt et al., 2004). An important transducer of Wnt signaling is β-catenin, which acts both as a transcriptional regulator in the canonical Wnt pathway and a component of focal adhesion complexes. During somitogenesis, β-catenin translocates to the plasma membrane and associates with N-cadherin as somitic cells increase their affinity for each each other and form a spherical epithelium. Disruption of N-cadherin function inhibits somite epithelialization (Linask et al., 1998). This disruption may reflect inhibition of β-catenin's transcriptional function and/or inhibition of cadherin-based cell-cell adhesion via focal adhesion complexes. The interplay between the two functions of β-catenin is not clearly understood.

Somite epithelialization requires increased cell-cell and cell-matrix interactions in the segmental plate. Cadherins responsible for homotypic cell-cell adhesion play a vital role in this and include several members such as N-cadherin (Hatta et al., 1987), cadherin 11 (Kimura et al., 1995) and proto-cadherin (Kim et al., 2000). ECM molecules important in epithelial structures, such as FN and laminin, are also upregulated at this time (Solursh et al., 1979). These molecules are expressed in the rostral half of the segmental plate and coincide with the expression of several transcription factors such as paraxis (Figure 28.6), whose expression prefigures somite formation and epithelialization (Barnes et al., 1997; Burgess et al., 1995) and regulates β-catenin activity during somitogenesis (Linker et al., 2005). Knock-down and knockout of paraxis in various animal models have revealed both the importance of paraxis in epithelialization, but also the importance of the somite in organizing the axial tissues (Barnes et al., 1997; Burgess et al., 1996; Sosic et al., 1997). Although evidence of segmentation is still observed, somite differentiation in paraxis knockout mice is delayed and less precise, resulting in disorganized tissues. Homozygotes lacking paraxis die at birth due to an inability to breathe caused by disorganization of the intercostal musculature (Burgess et al., 1996).

The expression pattern and function of axin2 suggests a link between the somitic gradients of Wnt3a and FGF8 and the somitic oscillator acting upon the Notch-Delta signaling pathway (Maroto et al., 2012). Axin2, a cytoplasmic component of

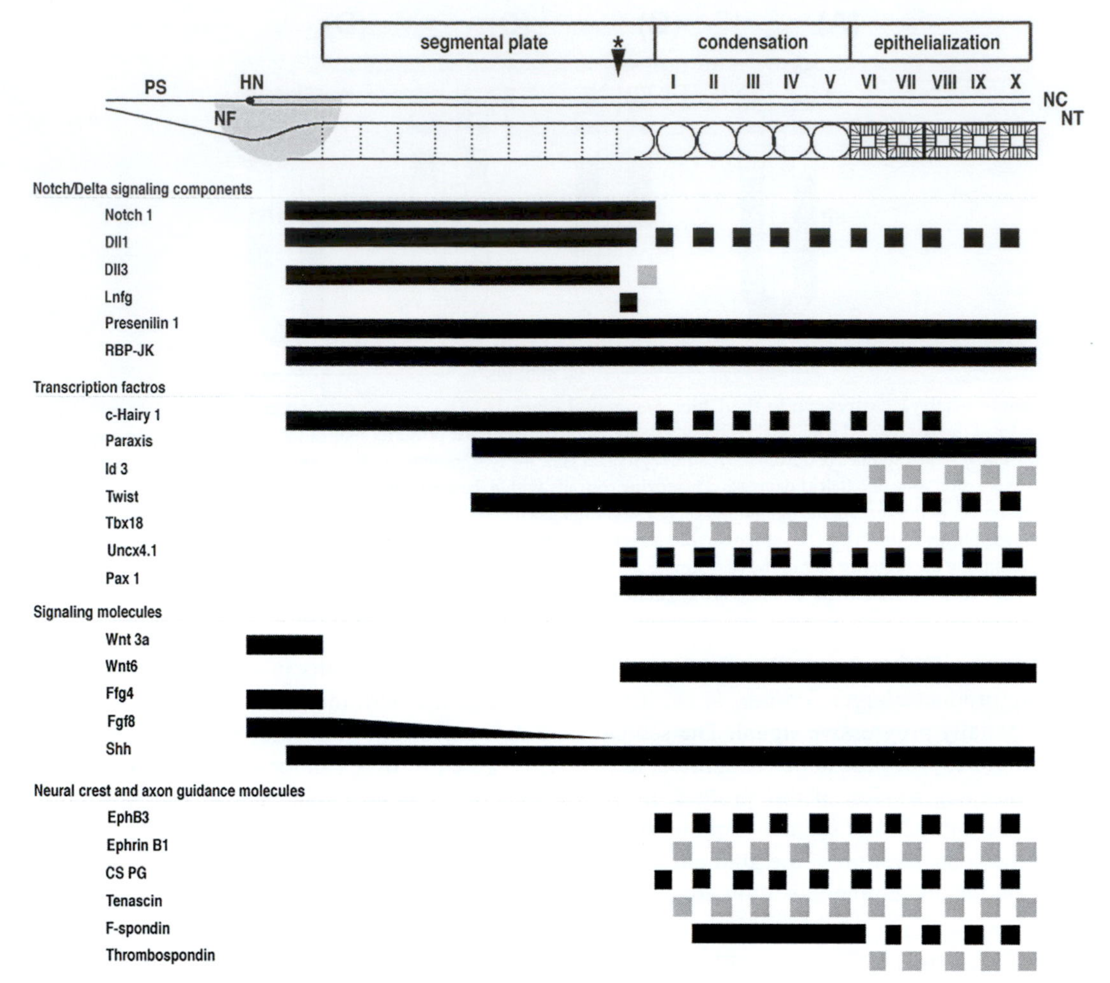

FIGURE 28.7 **Expression Patterns of Several Important Genes in Somitogenesis.** A schematic representation of a chick embryo with various stages of somitogenesis including the segmental plate with somitomeres, condensed somites (I–V), and epithelial somites (VI-X) notochord (NC) and neural tube (NT), neural folds (NF), Hensen's node (HN) and primitive streak (PS). The asterisk indicates the posterior border of the next somite to be formed. Spatial gene expression pattern is indicated by shaded areas. *(Adapted from Kuan et al, 2004)*

canonical Wnt pathways, is expressed in a caudorostral gradient under the control of Wnt3a. It is also expressed in a cyclic manner in the segmental plate at the posterior margin of the next somite to be formed, like lunatic fringe. However, axin2 expression alternates with lunatic fringe (Aulehla et al., 2003). In mice null for the Notch ligand Dll1, Notch signaling is impaired but axin2 oscillatory expression is still detectable. In contrast, the lack of axin2 in *vt* mice stops the oscillatory expression of lunatic fringe, but not its spatial expression pattern, which suggests a mechanistic link between the gradient and the clock. Axin2 downregulation during somite border formation may permit the influence of canonical Wnt6 pathway upon the segmental plate, suggesting a mechanistic link with border formation and epithelialization as well. Results of microsurgical and gene knock-down techniques support this model; however, the axin2 knockout mice showed no axial skeletal defect. On the otherhand, mutations in axin1 in the *fused* (*fu*) mouse produces profound axial skeletal dysmorphogenesis (Zeng et al., 1997). This discrepancy may reflect fundamental problems with previous experimental methods, misinterpretation of phenomena, or a species-specific difference between mice, and other developmental models

Differentiation of the Sclerotome

Once the patterning of the somite is accomplished, overt differentiation begins. The sclerotome that gives rise to the axial skeleton is derived from the ventromedial compartment of the somite (Figure 28.4). During the differentiation of the sclerotome, the dorsolateral component of the epithelial somite, the dermomyotome, retains its epithelial structure while the ventromedial sclerotome undergoes an epithelio-mesenchymal transition (EMT). This de-epithelialization is preceded by the expression of Pax1 (Barnes et al., 1996) (Figure 28.6), a paired-box transcription factor that is disrupted in the *undulated* mouse (Balling et al., 1988), whose phenotype includes scoliosis and missing vertebral bodies. The induction of Pax1expression is mediated by Shh which is produced by the notochord and floor plate of the neural tube (Johnson et al., 1994) (Figure 28.7). Ablation of the notochord will prevent expression of Pax1 in the somite and differentiation of the sclerotome. When a notochord or a Shh-soaked bead is implanted dorsal to the somite, Pax1 is induced in the dorsal somite at the expense of myotomal markers (Brand-Saberi et al., 1993; Ebensperger et al., 1995). BMP4 produced by the dorsal neural tube and intermediate mesoderm limits or modifies the effect of Shh to constrain sclerotome differentiation in the somite dorsally and laterally (McMahon et al., 1998; Tonegawa and Takahashi, 1998) (Figure 28.4).

After the sclerotome delaminates, it migrates or relocates to the region around the notochord. The function of Pax1 and other early markers of sclerotome, such as Pax9 and MFH1, is to maintain the sclerotome by promoting proliferation and preventing cell death (Christ et al., 2004). Functionally, this maintains the sclerotome in a mesenchymal state and presumably prevents differentiation. As the sclerotome develops, the posterior half becomes increasingly dense, while the anterior half remains loosely populated reflecting an important anterior-posterior polarity. Differentiation of the sclerotome begins in the posterior half in close association with the notochord. This is coincident with a down-regulation of Pax1 and Pax9 and an upregulation in Sox9, a master regulator of chondrogenic differentiation that controls expression of major cartilage constituents such as collagen type II and type X and aggrecan, in a manner similar to that of long bone development in the limb. The wave of differentiation proceeds radially from the notochord dorsally to surround the neural tube and ventrally towards the ventral body wall. The sclerotomal cells that inhabit the area around the notochord will give rise to the centrum of the vertebral body. Other structures of the axial skeleton are also derived from the sclerotome, but require additional regulatory mechanisms.

Neural Arch and Rib Development

An example of this alternative sclerotomal development is that of the vertebral neural arches that surround the spinal cord laterally and dorsally (Watanabe et al., 1998) (Figure 28.8). The dorsal-medial sclerotome remains in a mesenchymal state, proliferating and migrating towards the dorsal ectoderm to surround the neural tube. These cells express Msx2, which maintains the mesenchymal phenotype of the sclerotome and is required for subsequent chondrogenic differentiation (Monsoro-Burq et al., 1996; Monsoro-Burq and Le Douarin, 2000). This differentiation is dependent on BMP4 signaling from the dorsal neural tube and overlying ectoderm. Experimental dorsoventral rotation of the neural tube that removes the dorsal influence of BMP4 upon the dorsal sclerotome results in a failure of neural arch development (Nifuji et al., 1997).

The transverse processes and ribs are also derived from the Pax1 expressing sclerotome. The body wall in which the ribs reside is derived from the lateral plate that grows ventrally to encase the trunk and abdominal organs. During the time of rib mesenchyme outgrowth, Pax3 expressing myoblasts and sclerotomal cells both migrate towards the expanding lateral plate mesoderm (Henderson et al., 1999). Quail-chick chimeras, in which the sclerotome or dermomyotome of a chick were replaced by homotypic quail compartments, showed that all parts of the transverse processes and ribs are derived from the sclerotome (Evans, 2003; Olivera-Martinez et al., 2000). When the overlying ectoderm was removed from the body wall, rib development was severely impaired, suggesting an additional, uncharacterized inductive mechanism that regulates muscle and cartilage development (Sudo et al., 2001).

Much less is known about pectoral or pelvic girdle development, errors in which cause disorders such as congenital hip dysplasia. Recent work looking at scapular development has revealed a dual origin: the blade of the scapula is derived from somitic mesoderm, whereas the head and neck are derived from the lateral plate mesoderm (Ehehalt et al., 2004; Huang et al., 2000b). The somitic cells forming the scapular blade come from the chondrogenic differentiation of dermomyotomal

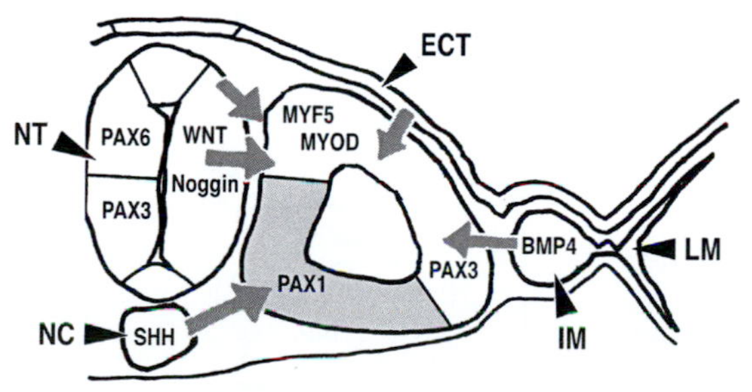

FIGURE 28.8 Tissue Interactions Involved in Specification of the Sclerotome in the Ventromedial Portion of the Epithelial Somite. WNT6 produced by the overlying ectoderm (ECT) induces epithelialization of the somite. SHH emitted from the notochord (NC) and floor plate of the neural tube (NT) induce PAX1 in the sclerotome (shaded region). PAX1 expression is inhibited dorsally and ventrally by BMP4. MYF5, MYOD and PAX3 are maintained in the dorsolateral portion of the epithelial somite. IM, intermediate mesoderm; LM, lateral plate mesoderm. *(Adapted from Alexander, PG., 201, "The role of Paraxis in somitogenesis and carbon monoxide-induced axial skeletal teratogenesis." Ph.D. Thesis, Thomas Jefferson University, Philadelphia, PA.)*

cells that are induced to express Paxl. In contrast to the vertebral body, Pax1 is induced, not inhibited, by BMP signals from the somatopleure (Wang et al., 2005). The identity of other factors important in the coordination of appendicular and axial skeletal articulation remains unknown.

Tendon Development

The bony elements of the axial skeleton develop in close association with the overlying musculature. The interaction is mutually dependent and critical for the development of other connective tissues: the tendon and ligaments. Characterization of a marker of ligament and tendon matrix, Tenascin C, has suggested that interactions between muscle and bone are important in the development of these connective tissues (Edom-Vovard and Duprez, 2004). Recent studies have reported the discovery of a cellular marker for tendons and ligaments, scleraxis, a basic helix-loop-helix (bHLH) protein closely related to MyoD and paraxis, expressed throughout their development (Brent and Tabin, 2004; Schweitzer et al., 2001). In the axial skeleton, scleraxis is first expressed at the cranial and caudal borders of the dorsal-most sclerotome in close proximity to the overlying myotome. This compartment has been coined the syndetome (Brent et al., 2003), and it is formed subsequent to sclerotomal differentiation in a manner dependent on FGF8 emitted by the myotome. Removal of the myotome prevents scleraxis induction and tendon/ligament development, while an implanted FGF8 soaked bead can rescue this deficit *in vivo* (Brent and Tabin, 2004). An in-depth presentation of the regulation of tendon and ligament development is presented in a recent review (Yang et al., 2013).

Resegmentation

A rostral-caudal polarity of the pre-vertebral sclerotome was first shown in 1850. Visually, one can see that the posterior half of the mesenchymal sclerotome is more dense than the anterior half. In addition, early embryologists identified a group of elongated and transversely oriented sclerotomal cells separating the rostral and caudal halves of the somite (von Ebner's fissures). The existence of these features agreed with the observation that the two halves of the epithelial somite have different adhesive qualities and influence on migrating neural crest cell and axons (Norris et al., 1989; Stern et al., 1986; Tosney, 1991). These cells migrate exclusively over the cranial portion (Oakley and Tosney, 1991). Somite anterior-posterior polarity is acquired from the very beginning with the differential localization of the Notch ligands, Dll1 and Dll3, and is reiterated by other molecules (Figure 28.6). The polarity of the somites may serve to direct the migration of neural crest and neuronal axons and to organize other tissues associated with the somites in preparation for resegmentation of the sclerotomes. Resegmentation is the process by which the bony elements are realigned out-of-phase with respect to other somite-derived tissues (muscle, tendon, ligament, blood vessels) and the neural ganglia (Figure 28.9) to form a functional vertebral joint. Quail-chick chimeras have shown that, to form a new segment, the posterior half of the mesenchymal sclerotome derived from one somite recombines with the anterior half of the mesenchymal sclerotome derived from the anterior somite (Huang et al., 2000a). In this way, the soft-tissue elements, including the muscles, now span the space between two

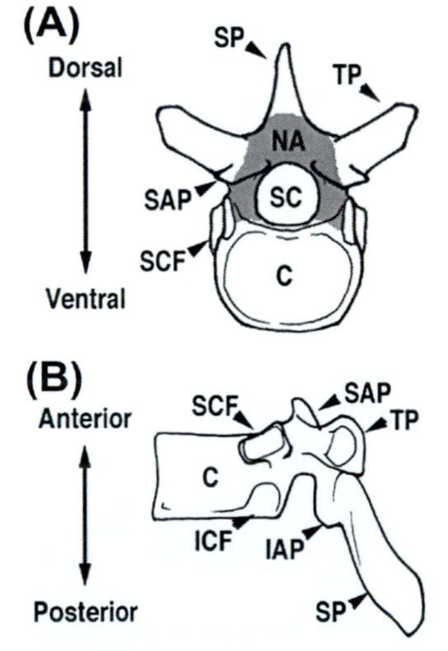

FIGURE 28.9 **Anatomy of the Vertebral Body.** (A) Anterior view of a thoracic vertebral body. Important anatomical features include the centrum (C), neural arch (NA), transverse process (TP), spinous process (SP) spinal canal (SC) the superior articular process or pedicle (SAP), and superior costal fovea (SCF). (B) Lateral view of a thoracic vertebral body. The gray area in A indicates a PAX1-expressing sclerotomal origin. Additional anatomical features of a vertebral body include the inferior articular process or pedicle (IAP), the costal fovea of the transverse process (CFTP) and the inferior costal fovea (ICF).

vertebral bodies to produce an articulating joint (Aoyama and Asamoto, 2000; Bagnall, 1992; Bagnall et al., 1988; Huang et al., 2000a). The loosely populated half of the new sclerotomal segment gives rise to the vertebral body (Figure 28.9). The contribution of the denser, anterior half is not as clear. Part of this half does contribute to the bony vertebral body, protecting nerves of the dorsal and ventral ganglia and blood vessels that supply a particular axial segment passing over and through it. The other portion contributes to the soft tissues and annulus fibrosus of the intervertebral disc, although how this is accomplished remains unknown.

Intervertebral Joint

During the resegmentation process, the mesenchymal somitocoele cells once located within the epithelial somite are relocated to intervertebral space (Figure 28.10). Vital dye staining has revealed that these cells or their progeny are involved in joint development. Removal of the original somitocoelic cells and replacement with an acellular spacer resulted in fused vertebral centra and pedicles, providing more evidence that these cells either give rise to the tissues of the vertebral joint, or direct its development (Mittapalli et al., 2005).

The origin of the intervertebral disc, the specialized joint between the centra of two vertebrae (Figure 28.9), is still incompletely understood (Hunter et al., 2003). There are three components to the intervertebral disc:

(1) the articular surfaces of the anterior and posterior vertebral bodies,
(2) the annulus fibrosus a specialized ligament that consists of concentric layers of collagen fibers that surround the third tissue,
(3) the nucleus pulposus, which is a more highly hydrated ECM-rich center of the intervertebral disc.

As mentioned above, the annulus fibrosis is derived from the densely packed sclerotomes after resegmentation. The origin of the nucleus pulposus which absorbs the compressive forces along the axis is elusive. In some vertebrates such as the chick,

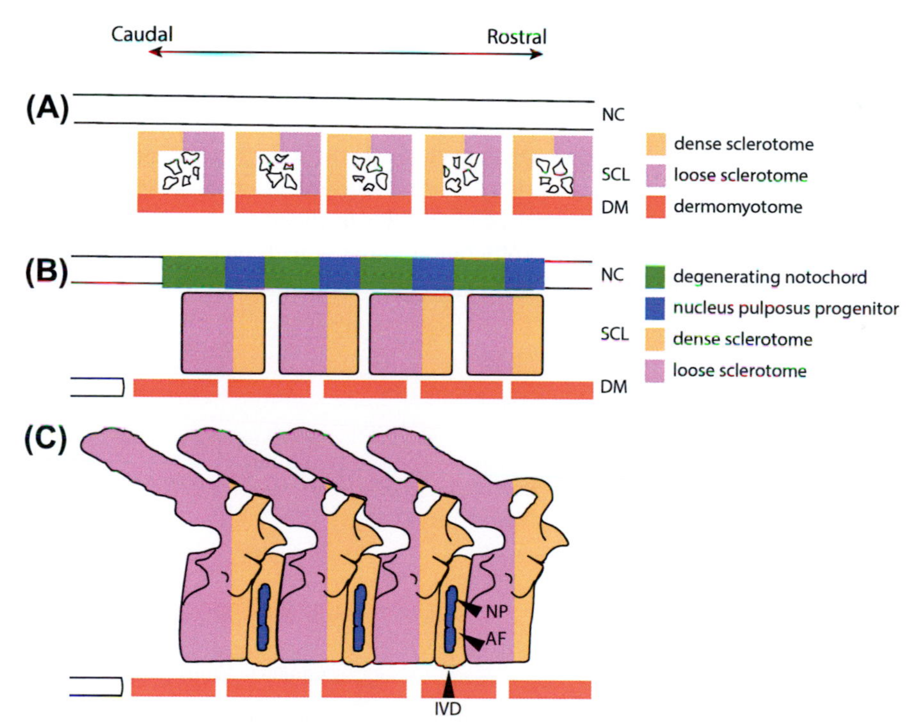

FIGURE 28.10 **Resegmentation of the Somites in Relation to the Adult Axial Skeleton.** (A) schematic of epithelial somites composed of specified sclerotome (SCL) and dermomyotome (DM) and the caudal more densely packed portion of the sclerotome. (B) Schematic of resegmented, de-epithelialized sclerotomes with the densely packed sclerotome (orange) which contributes to the annulus fibrosus of the intervertebral disc and possibly anterior portions of the vertebral body. The loosely packed regions of the sclerotomes (purple) undergo chondrogenesis and endochondral ossification to form the majority of the vertebral body, including the centrum, processes and ribs. Regional fates of the notochord (NC) are specified with green represnting the degenerating notochord neighboring the cartilaginous regions of the sclerotome, and blue representing the regions that will give rise to the nucleus pulposus of the intervertebral disc. (C) Four thoracic vertebrae with the contribution of the posterior somites in (A) and corresponding densely packed, anterior sclerotome to these vertebrae and annulus fibrosus (AF) (orange). The contribution of the notochord to the nucleus pulposus (NP) within the intervertebral disc (IVD) is highlighted in blue. Relation of the vertebral bodies to their musculature derived from the dermomytome is indicated in red.

the notochordal cells persist throughout life, while these cells disappear via apoptosis, terminal chondrocytic differentiation, or another phenotypic change in larger species such as mice and humans. The original notochord-derived cells may be replaced by mesenchymal cells from the annulus fibrosis or other sources. Clinically, it has been conjectured that the loss of the notochordal cells is closely associated with the development of intervertebral disc degeneration and associated back pain.

Rostro-Caudal Vertebral Identity

The axial skeleton is characterized by its metamerism; a pattern of similar, but not identical, repeating units. Comparison of a cervical, thoracic, lumbar and sacral vertebra demonstrates the rostrocaudal variances. The regionalization of the somites and constituent sclerotomes is determined by specific combinations of Hox genes in the paraxial mesoderm. In humans there are 13 paralogous groups aligned in four clusters. These clusters are expressed in a co-linear fashion rostrocaudally (3' to 5' along the genome), suggesting an epigenetic mechanism of gene regulation (Christ et al., 2000; Gruss and Kessel, 1991; see also Chapter 9). Refinement of the anterior-posterior expression of these genes may occur through the action of retinoic acid, a physiological metabolite of vitamin A expressed throughout development in regions undergoing patterning and morphogenesis known to affect skeletal development. In particular, exogenous administration of RA can alter Hox gene expression concomitant with homeotic transformations of axial segments (Kessel, 1992; Kessel and Gruss, 1991). Such transformations are manifest as mandibular hypoplasia, cervical ribs, or 13th ribs. In extreme cases, RA can produce caudal agenesis (Padmanabhan, 1998). Genetic knockout of particular RA receptors or dehydrogenases can lead to regional skeletal abnormalities (Cui et al., 2003; Niederreither et al., 2002), while knockout of RA metabolizing enzymes such as Cyp26A1 cause more global skeletal dysmorphogenesis (Sakai et al., 2001). Sclerotomal regional identity is determined very early in development (Fomenou et al., 2005). Heterotopic segmental plate grafts between chick and quail have shown that when cervical regions are transplanted into thoracic regions, ribs and scapula do not develop (Kieny, 1972; Christ, 1978). Retinoic acid is produced by cells of Hensen's node, and recent work has begun to link RA with the somitic oscillator and segment identity (Diez del Corral et al., 2003).

28.4 CONCLUSION

The process of endochondral ossification gives rise to the appendicular and many components of the axial skeleton. Recent advances utilizing genetic knockout approaches have begun to elucidate the role of some specific signaling pathways, transcription factors, and important regulatory molecules. Global and tissue-specific knockouts and knockins have elegantly elucidated the functional roles of: BMPs, cadherins, and ECM proteins in chondrocyte condensation; Ihh/PTHrP and FGF/BMP antagonism in chondrocyte differentiation and proliferation in the growth plate; Osx and Runx in osteoblast differentiation; Rankl and Csf1 in osteoclast differentiation; Hairy1 and Notch signaling in somitogenesis; and RA signaling

TABLE 28.2 Selected Mutations and Associated Human Syndromes

Gene	Mutation Type	Syndrome	Phenotype
FGFR3	Aberrant Activation	Achondroplasia	Dwarfism, shortened limbs with premature closure of growth plate
IFT80 (ciliopathy)	Null allele	Jeune Asphyxiating Throacic Dystrophy	Shortened long bones, constrictively small rib cage
RANKL, RANK	Null allele	Osteopetrosis (Albers-Schonberg)	Bones harden and eventually growth ablates bone marrow cavity and impinges nerves; deafness, blindness, stunted growth, anemia
PTHR1	Aberrant Activation	Jansen's metaphyseal chondrodysplasia	Dwarfism, hypercalcemia, irregular formation of long bones
PTHR1	Null allele	Chondrodysplasia Blomstrand	Ossification of the endocrine system and advanced skeletal maturation
NOTCH2	Aberrant Activation	Hadju-Cheney syndrome	Severe and progressive bone loss
ACVR	Aberrant Activation	Fibrodysplasia ossifican progessiva	Disfigurations as ossification occurs throughout soft tissues of body

in rostrocaudal vertebral identity among many others. As experimental feasibility to manipulate the expression of genes important to skeletogenesis with increasingly sophisticated techniques is integrated with human studies to identify associations between genes and specific skeletal malformation, our understanding of the processes controlling and shaping skeletogenesis will evolve. Eventually, these studies may be integrated into sophisticated models that can be utilized to diagnose and treat pathologies associated with diseases of the musculoskeletal system such as osteoarthritis, osteoporosis, and many others (Table 28.2).

28.5 CLINICAL RELEVANCE

- Skeletal dysplasias and malformation syndromes have been and continue to be associated with mutations affecting pathways critical for skeletal development (see Tables 28.1 and 28.2).
- The underlying pathology of diseases of the skeletal tissues such as osteoporosis and osteoarthritis can be directly linked to misregulation of the genetic pathways important to development of these tissues.
- Tissue engineering and clinical approaches are just beginning to utilize understanding gleaned from studies of the genetics of skeletal development to help rebuild and reinforce bones, tendons, and cartilage. As understanding of skeletal development grows, increasingly sophisticated approaches will become available.

ACKNOWLEDGMENTS

The authors would like to acknowledge research support from the US Department of Defense, NIH Cardiovascular Bioengineering Training Program (5T32HL076124–08), the Pittsburgh Foundation, and Pennsylvania Commonwealth Department of Health (SAP 4100050913).

RECOMMENDED RESOURCES

Online Mendelian Inheritance in Man: www.ncbi.nlm.nih.gov/omim/ (accessed 26.02.14).

Chondrogenesis/endochondral ossification

Dirckx, N., Van Hul, M., Maes, C., 2013. Osteoblast recruitment to sites of bone formation in skeletal development, homeostasis, and regeneration. Birth Defects Res. C Embryo Today 99, 170–191.

Hall, B.K., Miyake, T., 2000. All for one and one for all: condensations and the initiation of skeletal development. Bioessays 22, 138–147.

Iwamoto, M., Ohta, Y., Larmour, C., Enomoto-Iwamoto, M., 2013. Toward regeneration of articular cartilage. Birth Defects Res. C Embryo Today 99, 192–202.

Karsenty, G., Kronenberg, H.M., Settembre, C., 2009. Genetic control of bone formation. Annu. Rev. Cell Devel. Biol. 25, 629–648.

Shum, L., Coleman, C.M., Hatakeyama, Y., Tuan, R.S., 2003. Morphogenesis and dysmorphogenesis of the appendicular skeleton. Birth Defects Res. C Embryo Today 69, 102–122.

Somitogenesis

Alexander, P.G., Tuan, R.S., 2009. Environmental factors and axial skeletal dysmorphogenesis. In: Kusumi, K., Dunwoodie, S.L. (Eds.), The Genetics and Development of Scoliosis. Chapter 3. Springer, New York.

Aulehla, A., Herrmann, B.G., 2004. Segmentation in vertebrates: clock and gradient finally joined. Genes Dev. 18, 2060–2067.

Christ, B., Huang, R., Scaal, M., 2004. Formation and differentiation of the avian sclerotome. Anat. Embryol. 208, 333–350.

Giampietro, P.F., Blank, R.D., Raggio, C.L., Merchant, S., Jacobsen, F.S., Faciszewski, T., Shukla, S.K., Greenlee, A.R., Reynolds, C., Schowalter, D.B., 2003. Congenital and idiopathic scoliosis: clinical and genetic aspects. Clin. Med. Res. 1, 125–136.

Kalter, H., 1980. A compendium of the genetically induced congenital malformations of the house mouse. Teratol. 21, 397–429.

Maroto, M., Bone, R.A., Dale, J.K., 2012. Somitogenesis. Development 139, 2453–2456.

Saga, Y., Takeda, H., 2001. The making of the somite: Molecular events in vertebrate segmentation. Nat. Rev. Genet. 2, 835–845.

REFERENCES

Acloque, H., Ocana, O.H., Matheu, A., Rizzoti, K., Wise, C., Lovell-Badge, R., Nieto, M.A., 2011. Reciprocal repression between Sox3 and snail transcription factors defines embryonic territories at gastrulation. Dev. Cell 21, 546–558.

Akiyama, H., Chaboissier, M.C., Martin, J.F., Schedl, A., de Crombrugghe, B., 2002. The transcription factor Sox9 has essential roles in successive steps of the chondrocyte differentiation pathway and is required for expression of Sox5 and Sox6. Genes Dev. 16, 2813–2828.

Akiyama, H., Lyons, J.P., Mori-Akiyama, Y., Yang, X., Zhang, R., Zhang, Z., Deng, J.M., Taketo, M.M., Nakamura, T., Behringer, R.R., et al., 2004. Interactions between Sox9 and beta-catenin control chondrocyte differentiation. Genes Dev. 18, 1072–1087.

Alexander, P.G., 2001. The role of Paraxis in somitogenesis and carbon monoxide-induced axial skeletal teratogenesis. Ph.D. Thesis. Thomas Jefferson University, Philadelphia, PA.

Allen, J.L., Cooke, M.E., Alliston, T., 2012. ECM stiffness primes the TGFbeta pathway to promote chondrocyte differentiation. Mol. Biol. Cell 23, 3731–3742.

Alvarez, J., Horton, J., Sohn, P., Serra, R., 2001. The perichondrium plays an important role in mediating the effects of TGF-beta1 on endochondral bone formation. Dev. Dyn. 221, 311–321.

Alvarez, J., Sohn, P., Zeng, X., Doetschman, T., Robbins, D.J., Serra, R., 2002. TGFbeta2 mediates the effects of hedgehog on hypertrophic differentiation and PTHrP expression. Development 129, 1913–1924.

Amizuka, N., Warshawsky, H., Henderson, J.E., Goltzman, D., Karaplis, A.C., 1994. Parathyroid hormone-related peptide-depleted mice show abnormal epiphyseal cartilage development and altered endochondral bone formation. J. Cell Biol. 126, 1611–1623.

Archer, C.W., Dowthwaite, P.G., Francis-West, P., 2003. Development of synovial joints. Birth Defects Research Part C: Embryo Today: Reviews 69, 144–155.

Aoyama, H., Asamoto, K., 2000. The developmental fate of the rostral/caudal half of a somite for vertebra and rib formation: experimental confirmation of the resegmentation theory using chick-quail chimeras. Mech. Dev. 99, 71–82.

Aulehla, A., Wehrle, C., Brand-Saberi, B., Kemler, R., Gossler, A., Kanzler, B., Herrmann, B.G., 2003. Wnt3a plays a major role in the segmentation clock controlling somitogenesis. Dev. Cell 4, 395–406.

Bagnall, K.M., 1992. The migration and distribution of somite cells after labelling with the carbocyanine dye, Dil: the relationship of this distribution to segmentation in the vertebrate body. Anat. Embryol. 185, 317–324.

Bagnall, K.M., Higgins, S.J., Sanders, E.J., 1988. The contribution made by a single somite to the vertebral column: experimental evidence in support of resegmentation using the chick-quail chimaera model. Development 103, 69–85.

Balling, R., Deutsch, U., Gruss, P., 1988. Undulated, a mutation affecting the development of the mouse skeleton, has a point mutation in the paired box of PAX1. Cell 55, 531–535.

Barnes, G.L., Alexander, P.G., Hsu, C.W., Mariani, B.D., Tuan, R.S., 1997. Cloning and characterization of chicken Paraxis: a regulator of paraxial mesoderm development and somite formation. Dev. Biol. 189, 95–111.

Barnes, G.L., Hsu, C.W., Mariani, B.D., Tuan, R.S., 1996. Chicken Pax-1 gene: structure and expression during embryonic somite development. Differentiation. 61, 13–23.

Bi, W., Deng, J.M., Zhang, Z., Behringer, R.R., de Crombrugghe, B., 1999. Sox9 is required for cartilage formation. Nat. Genet. 22, 85–89.

Boulet, A.M., Capecchi, M.R., 2004. Multiple roles of Hoxa11 and Hoxd11 in the formation of the mammalian forelimb zeugopod. Development 131, 299–309.

Boyle, W.J., Simonet, W.S., Lacey, D.L., 2003. Osteoclast differentiation and activation. Nature 423, 337–342.

Brand-Saberi, B., Ebensperger, C., Wilting, J., Balling, R., Christ, B., 1993. The ventralizing effect of the notochord on somite differentiation in chick embryos. Anat. Embryol. 188, 239–245.

Brent, A.E., Schweitzer, R., Tabin, C.J., 2003. A somitic compartment of tendon progenitors. Cell 113, 235–248.

Brent, A.E., Tabin, C.J., 2004. FGF acts directly on the somitic tendon progenitors through the Ets transcription factors Pea3 and Erm to regulate scleraxis expression. Development 131, 3885–3896.

Brunet, L.J., McMahon, J.A., McMahon, A.P., Harland, R.M., 1998. Noggin, cartilage morphogenesis, and joint formation in the mammalian skeleton. Science 280, 1455–1457.

Burgess, R., Cserjesi, P., Ligon, K.L., Olson, E.N., 1995. Paraxis: a basic helix-loop-helix protein expressed in paraxial mesoderm and developing somites. Dev. Biol. 168, 296–306.

Burgess, R., Rawls, A., Brown, D., Bradley, A., Olson, E.N., 1996. Requirement of the paraxis gene for somite formation and musculoskeletal patterning. Nature 384, 570–573.

Chang, S.C., Hoang, B., Thomas, J.T., Vukicevic, S., Luyten, F.P., Ryba, N.J., Kozak, C.A., Reddi, A.H., Moos Jr., M., 1994. Cartilage-derived morphogenetic proteins. New members of the transforming growth factor-beta superfamily predominantly expressed in long bones during human embryonic development. J. Biol. Chem. 269, 28227–28234.

Christ, B., Jacobs, H.J., Jacob, M., 1978. Regional determination of early embryonic muscle primordium. Experimental studies on quail and chick embryos (demonstration). Verh. Anat. Ges. 72, 353–357.

Christ, B., Huang, R., Scaal, M., 2004. Formation and differentiation of the avian sclerotome. Anat. Embryol. 208, 333–350.

Christ, B., Huang, R., Wilting, J., 2000. The development of the avian vertebral column. Anat. Embryol. 202, 179–194.

Coleman, C.M., Loredo, G.A., Lo, C.W., Tuan, R.S., 2003. Correlation of Gdf5 and connexin 43 mRNA expression during embryonic development. Anat. Rec. A Discov. Mol. Cell. Evol. Biol. 275, 1117–1121.

Coleman, C.M., Tuan, R.S., 2003. Functional role of growth/differentiation factor 5 in chondrogenesis of limb mesenchymal cells. Mech. Dev. 120, 823–836.

Colnot, C., Lu, C., Hu, D., Helms, J.A., 2004. Distinguishing the contributions of the perichondrium, cartilage, and vascular endothelium to skeletal development. Dev. Biol. 269, 55–69.

Cooke, J., Zeeman, E.C., 1976. A clock and wavefront model for control of the number of repeated structures during animal morphogenesis. J. Theor. Biol. 58, 455–476.

Cremer, M.A., Rosloniec, E.F., Kang, A.H., 1998. The cartilage collagens: a review of their structure, organization, and role in the pathogenesis of experimental arthritis in animals and in human rheumatic disease. J. Mol. Med. 76, 275–288.

Cui, J., Michaille, J.J., Jiang, W., Zile, M.H., 2003. Retinoid receptors and vitamin A deficiency: differential patterns of transcription during early avian development and the rapid induction of RARs by retinoic acid. Dev. Biol. 260, 496–511.

Dale, J.K., Maroto, M., Dequeant, M.L., Malapert, P., McGrew, M., Pourquie, O., 2003. Periodic notch inhibition by lunatic fringe underlies the chick segmentation clock. Nature 421, 275–278.

DeLise, A.M., Tuan, R.S., 2002. Alterations in the spatiotemporal expression pattern and function of N-cadherin inhibit cellular condensation and chondrogenesis of limb mesenchymal cells *in vitro*. J. Cell. Biochem. 87, 342–359.

Dessau, W., von der Mark, H., von der Mark, K., Fischer, S., 1980. Changes in the patterns of collagens and fibronectin during limb-bud chondrogenesis. J. Embryol. Exp. Morphol. 57, 51–60.

Diez del Corral, R., Olivera-Martinez, I., Goriely, A., Gale, E., Maden, M., Storey, K., 2003. Opposing FGF and retinoid pathways control ventral neural pattern, neuronal differentiation, and segmentation during body axis extension. Neuron 40, 65–79.

Dirckx, N., Van Hul, M., Maes, C., 2013. Osteoblast recruitment to sites of bone formation in skeletal development, homeostasis, and regeneration. Birth Defects Res. C Embryo Today 99, 170–191.

Dougall, W.C., Glaccum, M., Charrier, K., Rohrbach, K., Brasel, K., De Smedt, T., Daro, E., Smith, J., Tometsko, M.E., Maliszewski, C.R., et al., 1999. RANK is essential for osteoclast and lymph node development. Genes Dev. 13, 2412–2424.

Dubrulle, J., McGrew, M.J., Pourquie, O., 2001. FGF signaling controls somite boundary position and regulates segmentation clock control of spatiotemporal Hox gene activation. Cell 106, 219–232.

Dubrulle, J., Pourquie, O., 2004. Fgf8 mRNA decay establishes a gradient that couples axial elongation to patterning in the vertebrate embryo. Nature 427, 419–422.

Dudhia, J., 2005. Aggrecan, aging and assembly in articular cartilage. Cellular and molecular life sciences. CMLS 62, 2241–2256.

Dunwoodie, S.L., Clements, M., Sparrow, D.B., Sa, X., Conlon, R.A., Beddington, R.S., 2002. Axial skeletal defects caused by mutation in the spondylocostal dysplasia/pudgy gene Dll3 are associated with disruption of the segmentation clock within the presomitic mesoderm. Development 129, 1795–1806.

Duprez, D.M., Kostakopoulou, K., Francis-West, P.H., Tickle, C., Brickell, P.M., 1996. Activation of Fgf-4 and HoxD gene expression by BMP-2 expressing cells in the developing chick limb. Development 122, 1821–1828.

Ebensperger, C., Wilting, J., Brand-Saberi, B., Mizutani, Y., Christ, B., Balling, R., Koseki, H., 1995. Pax-1, a regulator of sclerotome development is induced by notochord and floor plate signals in avian embryos. Anat. Embryol. 191, 297–310.

Edom-Vovard, F., Duprez, D., 2004. Signals regulating tendon formation during chick embryonic development. Dev. Dyn. 229, 449–457.

Ehehalt, F., Wang, B., Christ, B., Patel, K., Huang, R., 2004. Intrinsic cartilage-forming potential of dermomyotomal cells requires ectodermal signals for the development of the scapula blade. Anat. Embryol. 208, 431–437.

Enomoto, H., Enomoto-Iwamoto, M., Iwamoto, M., Nomura, S., Himeno, M., Kitamura, Y., Kishimoto, T., Komori, T., 2000. Cbfa1 is a positive regulatory factor in chondrocyte maturation. J. Biol. Chem. 275, 8695–8702.

Erol, B., Tracy, M.R., Dormans, J.P., Zackai, E.H., Maisenbacher, M.K., O'Brien, M.L., Turnpenny, P.D., Kusumi, K., 2004. Congenital scoliosis and vertebral malformations: characterization of segmental defects for genetic analysis. J. Pediatr. Orthop. 24, 674–682.

Evans, D.J., 2003. Contribution of somitic cells to the avian ribs. Dev. Biol. 256, 114–126.

Evrard, Y.A., Lun, Y., Aulehla, A., Gan, L., Johnson, R.L., 1998. lunatic fringe is an essential mediator of somite segmentation and patterning. Nature 394, 377–381.

Fomenou, M.D., Scaal, M., Stockdale, F.E., Christ, B., Huang, R., 2005. Cells of all somitic compartments are determined with respect to segmental identity. Dev. Dyn. 233, 1386–1393.

Francis-West, P.H., Abdelfattah, A., Chen, P., Allen, C., Parish, J., Ladher, R., Allen, S., MacPherson, S., Luyten, F.P., Archer, C.W., 1999. Mechanisms of GDF-5 action during skeletal development. Development 126, 1305–1315.

Fujita, T., Azuma, Y., Fukuyama, R., Hattori, Y., Yoshida, C., Koida, M., Ogita, K., Komori, T., 2004. Runx2 induces osteoblast and chondrocyte differentiation and enhances their migration by coupling with PI3K-Akt signaling. J. Cell Biol. 166, 85–95.

Giampietro, P.F., Blank, R.D., Raggio, C.L., Merchant, S., Jacobsen, F.S., Faciszewski, T., Shukla, S.K., Greenlee, A.R., Reynolds, C., Schowalter, D.B., 2003. Congenital and idiopathic scoliosis: clinical and genetic aspects. Clin. Med. Res. 1, 125–136.

Goff, D.J., Tabin, C.J., 1997. Analysis of Hoxd-13 and Hoxd-11 misexpression in chick limb buds reveals that Hox genes affect both bone condensation and growth. Development 124, 627–636.

Goldring, M.B., Tsuchimochi, K., Ijiri, K., 2006. The control of chondrogenesis. J. Cell. Biochem. 97, 33–44.

Greco, T.L., Takada, S., Newhouse, M.M., McMahon, J.A., McMahon, A.P., Camper, S.A., 1996. Analysis of the vestigial tail mutation demonstrates that Wnt-3a gene dosage regulates mouse axial development. Genes Dev. 10, 313–324.

Gruss, P., Kessel, M., 1991. Axial specification in higher vertebrates. Curr. Opin. Genet. Dev. 1, 204–210.

Gu, H., Zou, Y.R., Rajewsky, K., 1993. Independent control of immunoglobulin switch recombination at individual switch regions evidenced through Cre-loxP-mediated gene targeting. Cell 73, 1155–1164.

Guo, X., Day, T.F., Jiang, X., Garrett-Beal, L., Topol, L., Yang, Y., 2004. Wnt/beta-catenin signaling is sufficient and necessary for synovial joint formation. Genes dev. 18, 2404–2417.

Hall, B.K., Miyake, T., 2000. All for one and one for all: condensations and the initiation of skeletal development. BioEssays. 22, 138–147.

Hartmann, C., Tabin, C.J., 2001. Wnt-14 plays a pivotal role in inducing synovial joint formation in the developing appendicular skeleton. Cell 104, 341–351.

Hatta, K., Takagi, S., Fujisawa, H., Takeichi, M., 1987. Spatial and temporal expression pattern of N-cadherin cell adhesion molecules correlated with morphogenetic processes of chicken embryos. Dev. Biol. 120, 215–227.

Hattori, T., Muller, C., Gebhard, S., Bauer, E., Pausch, F., Schlund, B., Bosl, M.R., Hess, A., Surmann-Schmitt, C., von der Mark, H., et al., 2010. SOX9 is a major negative regulator of cartilage vascularization, bone marrow formation and endochondral ossification. Development 137, 901–911.

Hamburger, V., Hamilton, H.L., 1951. A series of normal stages in the development of the chick embryo. Journal of Morphology 88, 49–92.

Haycraft, C.J., Serra, R., 2008. Cilia involvement in patterning and maintenance of the skeleton. Curr. Top. Dev. Biol. 85, 303–332.

Healy, C., Uwanogho, D., Sharpe, P.T., 1999. Regulation and role of Sox9 in cartilage formation. Dev. Dyn. 215, 69–78.

Henderson, D.J., Conway, S.J., Copp, A.J., 1999. Rib truncations and fusions in the Sp2H mouse reveal a role for Pax3 in specification of the ventro-lateral and posterior parts of the somite. Dev. Biol. 209, 143–158.

Hornbruch, A., Summerbell, D., Wolpert, L., 1979. Somite formation in the early chick embryo following grafts of Hensen's node. J. Embryol. Exp. Morphol. 51, 51–62.

Huang, R., Zhi, Q., Brand-Saberi, B., Christ, B., 2000a. New experimental evidence for somite resegmentation. Anat. Embryol. 202, 195–200.

Huang, R., Zhi, Q., Patel, K., Wilting, J., Christ, B., 2000b. Dual origin and segmental organisation of the avian scapula. Development 127, 3789–3794.

Hunter, C.J., Matyas, J.R., Duncan, N.A., 2003. The notochordal cell in the nucleus pulposus: a review in the context of tissue engineering. Tissue Eng. 9, 667–677.

Hyde, G., Dover, S., Aszodi, A., Wallis, G.A., Boot-Handford, R.P., 2007. Lineage tracing using matrilin-1 gene expression reveals that articular chondrocytes exist as the joint interzone forms. Dev. Biol. 304, 825–833.

Hyde, G., Boot-Handford, R.P., Wallis, G.A., 2008. Col2a1 lineage tracing reveals that the meniscus of the knee joint has a complex cellular origin. J. Anat. 213, 531–538.

Iwamoto, M., Ohta, Y., Larmour, C., Enomoto-Iwamoto, M., 2013. Toward regeneration of articular cartilage. Birth Defects Res.C Embryo Today 99, 192–202.

Jena, N., Martin-Seisdedos, C., McCue, P., Croce, C.M., 1997. BMP7 null mutation in mice: developmental defects in skeleton, kidney, and eye. Exp. Cell Res. 230, 28–37.

Johnson, R.L., Laufer, E., Riddle, R.D., Tabin, C., 1994. Ectopic expression of Sonic hedgehog alters dorsal-ventral patterning of somites. Cell 79, 1165–1173.

Kalter, H., 1980. A compendium of the genetically induced congenital malformations of the house mouse. Teratology 21, 397–429.

Karsenty, G., Kronenberg, H.M., Settembre, C., 2009. Genetic control of bone formation. Annu. Rev. Cell and Dev. Biol. 25, 629–648.

Kessel, M., 1992. Respecification of vertebral identities by retinoic acid. Development 115, 487–501.

Kessel, M., Gruss, P., 1991. Homeotic transformations of murine vertebrae and concomitant alteration of Hox codes induced by retinoic acid. Cell 67, 89–104.

Khan, I.M., Redman, S.N., Williams, R., Dowthwaite, G.P., Oldfield, S.F., Archer, C.W., 2007. The development of synovial joints. Curr. Top. Dev. Biol. 79, 1–36.

Kieny, M., Mauger, A., Sengel, P., 1972. Early regionalization of somatic mesoderm as studied by the development of axial skeleton of the chick embryo. Dev. Biol. 28, 162–175.

Kim, S.H., Jen, W.C., De Robertis, E.M., Kintner, C., 2000. The protocadherin PAPC establishes segmental boundaries during somitogenesis in *Xenopus* embryos. Curr. Biol. 10, 821–830.

Kimura, Y., Matsunami, H., Inoue, T., Shimamura, K., Uchida, N., Ueno, T., Miyazaki, T., Takeichi, M., 1995. Cadherin-11 expressed in association with mesenchymal morphogenesis in the head, somite, and limb bud of early mouse embryos. Dev. Biol. 169, 347–358.

Kipnes, J., Carlberg, A.L., Loredo, G.A., Lawler, J., Tuan, R.S., Hall, D.J., 2003. Effect of cartilage oligomeric matrix protein on mesenchymal chondrogenesis *in vitro*. Osteoarthritis cartilage 11, 442–454.

Koyama, E., Leatherman, J.L., Shimazu, A., Nah, H.-D., Pacifici, M., 1995. Syndecan-3, tenascin-C, and the development of cartilaginous skeletal elements and joints in chick limbs. Dev. Dyn. 203, 152–162.

Koyama, E., Yasuda, T., Minugh-Purvis, N., Kinumatsu, T., Yallowitz, A.R., Wellik, D.M., Pacifici, M., 2010. Hox11 genes establish synovial joint organization and phylogenetic characteristics in developing mouse zeugopod skeletal elements. Development 137, 3795–3800.

Kuan, C.Y., Tannahill, D., Cook, G.M., Keynes, R.J., 2004. Somite polarity and segmental patterning of the peripheral nervous system. Mech. Dev. 121, 1055–1068.

Kusumi, K., Sun, E.S., Kerrebrock, A.W., Bronson, R.T., Chi, D.C., Bulotsky, M.S., Spencer, J.B., Birren, B.W., Frankel, W.N., Lander, E.S., 1998. The mouse pudgy mutation disrupts Delta homologue Dll3 and initiation of early somite boundaries. Nat. Genet. 19, 274–278.

Kwan, K.M., Pang, M.K., Zhou, S., Cowan, S.K., Kong, R.Y., Pfordte, T., Olsen, B.R., Sillence, D.O., Tam, P.P., Cheah, K.S., 1997. Abnormal compartmentalization of cartilage matrix components in mice lacking collagen X: implications for function. J. Cell Biol. 136, 459–471.

L'Hote, C.G., Knowles, M.A., 2005. Cell responses to FGFR3 signaling: growth, differentiation and apoptosis. Exp. Cell Res. 304, 417–431.

Lefebvre, V., Li, P., de Crombrugghe, B., 1998. A new long form of Sox5 (L-Sox5), Sox6 and Sox9 are coexpressed in chondrogenesis and cooperatively activate the type II collagen gene. EMBO J. 17, 5718–5733.

Linask, K.K., Ludwig, C., Han, M.D., Liu, X., Radice, G.L., Knudsen, K.A., 1998. N-cadherin/catenin-mediated morphoregulation of somite formation. Dev. Biol. 202, 85–102.

Linker, C., Lesbros, C., Gros, J., Burrus, L.W., Rawls, A., Marcelle, C., 2005. beta-Catenin-dependent Wnt signaling controls the epithelial organisation of somites through the activation of paraxis. Development 132, 3895–3905.

Liu, P., Wakamiya, M., Shea, M.J., Albrecht, U., Behringer, R.R., Bradley, A., 1999. Requirement for Wnt3 in vertebrate axis formation. Nat. Genet. 22, 361–365.

Lizarraga, G., Lichtler, A., Upholt, W.B., Kosher, R.A., 2002. Studies on the role of Cux1 in regulation of the onset of joint formation in the developing limb. Dev. Biol. 243, 44–54.

Li, T., Longobardi, L., Myers, T.J., Temple, J.D., Chandler, R.L., Ozkan, H., Contaldo, C., Spagnoli, A., 2013. Joint TGF- type II receptor-expressing cells: ontogeny and characterization as joint progenitors. Stem Cells Dev. 22, 1342–1359.

Long, F.X., 2012. Building strong bones: molecular regulation of the osteoblast lineage. Nat. Rev. Mol. Cell Bio. 13, 27–38.

Maleski, M.P., Knudson, C.B., 1996. Hyaluronan-mediated aggregation of limb bud mesenchyme and mesenchymal condensation during chondrogenesis. Exp. Cell Res. 225, 55–66.

Maroto, M., Bone, R.A., Dale, J.K., 2012. Somitogenesis. Development 139, 2453–2456.
McMahon, J.A., Takada, S., Zimmerman, L.B., Fan, C.M., Harland, R.M., McMahon, A.P., 1998. Noggin-mediated antagonism of BMP signaling is required for growth and patterning of the neural tube and somite. Genes Dev. 12, 1438–1452.
Meech, R., Edelman, D.B., Jones, F.S., Makarenkova, H.P., 2005. The homeobox transcription factor Barx2 regulates chondrogenesis during limb development. Development 132, 2135–2146.
Minina, E., Kreschel, C., Naski, M.C., Ornitz, D.M., Vortkamp, A., 2002. Interaction of FGF, Ihh/Pthlh, and BMP signaling integrates chondrocyte proliferation and hypertrophic differentiation. Dev. Cell 3, 439–449.
Minina, E., Wenzel, H.M., Kreschel, C., Karp, S., Gaffield, W., McMahon, A.P., Vortkamp, A., 2001. BMP and Ihh/PTHrP signaling interact to coordinate chondrocyte proliferation and differentiation. Development 128, 4523–4534.
Mittapalli, V.R., Huang, R., Patel, K., Christ, B., Scaal, M., 2005. Arthrotome: a specific joint forming compartment in the avian somite. Dev. Dyn. 234, 48–53.
Monsoro-Burq, A.H., Duprez, D., Watanabe, Y., Bontoux, M., Vincent, C., Brickell, P., Le Douarin, N., 1996. The role of bone morphogenetic proteins in vertebral development. Development 122, 3607–3616.
Monsoro-Burq, A.H., Le Douarin, N., 2000. Duality of molecular signaling involved in vertebral chondrogenesis. Curr. Top. Dev. Biol. 48, 43–75.
Nakashima, K., Zhou, X., Kunkel, G., Zhang, Z., Deng, J.M., Behringer, R.R., de Crombrugghe, B., 2002. The novel zinc finger-containing transcription factor osterix is required for osteoblast differentiation and bone formation. Cell 108, 17–29.
Nasevicius, A., Ekker, S.C., 2000. Effective targeted gene 'knockdown' in zebrafish. Nat. Genet. 26, 216–220.
Niederreither, K., Fraulob, V., Garnier, J.M., Chambon, P., Dolle, P., 2002. Differential expression of retinoic acid-synthesizing (RALDH) enzymes during fetal development and organ differentiation in the mouse. Mech. Dev. 110, 165–171.
Nifuji, A., Kellermann, O., Kuboki, Y., Wozney, J.M., Noda, M., 1997. Perturbation of BMP signaling in somitogenesis resulted in vertebral and rib malformations in the axial skeletal formation. J. Bone Miner. Res. 12, 332–342.
Norris, W.E., Stern, C.D., Keynes, R.J., 1989. Molecular differences between the rostral and caudal halves of the sclerotome in the chick embryo. Development 105, 541–548.
O'Brien, C.A., Nakashima, T., Takayanagi, H., 2013. Osteocyte control of osteoclastogenesis. Bone 54, 258–263.
Oakley, R.A., Tosney, K.W., 1991. Peanut agglutinin and chondroitin-6-sulfate are molecular markers for tissues that act as barriers to axon advance in the avian embryo. Dev. Biol. 147, 187–206.
Oberlender, S.A., Tuan, R.S., 1994. Expression and functional involvement of N-cadherin in embryonic limb chondrogenesis. Development 120, 177–187.
Ochi, K., Derfoul, A., Tuan, R.S., 2006. A predominantly articular cartilage-associated gene, SCRG1, is induced by glucocorticoid and stimulates chondrogenesis *in vitro*. Osteoarthritis Cartilage 14, 30–38.
Olivera-Martinez, I., Coltey, M., Dhouailly, D., Pourquie, O., 2000. Mediolateral somitic origin of ribs and dermis determined by quail-chick chimeras. Development 127, 4611–4617.
Onyekwelu, I., Goldring, M.B., Hidaka, C., 2009. Chondrogenesis, joint formation, and articular cartilage regeneration. J. Cell. Biochem. 107, 383–392.
Otto, F., Thornell, A.P., Crompton, T., Denzel, A., Gilmour, K.C., Rosewell, I.R., Stamp, G.W., Beddington, R.S., Mundlos, S., Olsen, B.R., et al., 1997. Cbfa1, a candidate gene for cleidocranial dysplasia syndrome, is essential for osteoblast differentiation and bone development. Cell 89, 765–771.
Packard Jr., D.S., 1976. The influence of axial structures on chick somite formation. Dev. Biol. 53, 36–48.
Packard Jr., D.S., Meier, S., 1983. An experimental study of the somitomeric organization of the avian segmental plate. Dev. Biol. 97, 191–202.
Padmanabhan, R., 1998. Retinoic acid-induced caudal regression syndrome in the mouse fetus. Reprod. Toxicol. 12, 139–151.
Palmeirim, I., Henrique, D., Ish-Horowicz, D., Pourquie, O., 1997. Avian hairy gene expression identifies a molecular clock linked to vertebrate segmentation and somitogenesis. Cell 91, 639–648.
Pitsillides, A.A., Ashhurst, D.E., 2008. A critical evaluation of specific aspects of joint development. Dev. Dyn. 237, 2284–2294.
Pizette, S., Niswander, L., 2000. BMPs are required at two steps of limb chondrogenesis: formation of prechondrogenic condensations and their differentiation into chondrocytes. Dev. Biol. 219, 237–249.
Primmett, D.R., Norris, W.E., Carlson, G.J., Keynes, R.J., Stern, C.D., 1989. Periodic segmental anomalies induced by heat shock in the chick embryo are associated with the cell cycle. Development 105, 119–130.
Robson, H., Siebler, T., Shalet, S.M., Williams, G.R., 2002. Interactions between GH, IGF-I, glucocorticoids, and thyroid hormones during skeletal growth. Pediatr. Res. 52, 137–147.
Rodriguez-Esteban, C., Tsuki, T., Yonei, S., Magallon, J., Tamura, K., Izpisua Belmonte, J.C., 1999. The T-box genes Tbx4 and Tbx5 regulate limb outgrowth and identity. Nature 398, 814–818.
Rodriguez-Niedenfuhr, M., Dathe, V., Jacob, H.J., Prols, F., Christ, B., 2003. Spatial and temporal pattern of Wnt-6 expression during chick development. Anat. Embryol. 206, 447–451.
Saga, Y., Takeda, H., 2001. The making of the somite: molecular events in vertebrate segmentation. Nat. Rev. Genet. 2, 835–845.
Sakai, Y., Meno, C., Fujii, H., Nishino, J., Shiratori, H., Saijoh, Y., Rossant, J., Hamada, H., 2001. The retinoic acid-inactivating enzyme CYP26 is essential for establishing an uneven distribution of retinoic acid along the anterio-posterior axis within the mouse embryo. Genes Dev. 15, 213–225.
Sawada, A., Shinya, M., Jiang, Y.J., Kawakami, A., Kuroiwa, A., Takeda, H., 2001. Fgf/MAPK signaling is a crucial positional cue in somite boundary formation. Development 128, 4873–4880.
Schmidt, C., Stoeckelhuber, M., McKinnell, I., Putz, R., Christ, B., Patel, K., 2004. Wnt 6 regulates the epithelialisation process of the segmental plate mesoderm leading to somite formation. Dev. Biol. 271, 198–209.

Schweitzer, R., Chyung, J.H., Murtaugh, L.C., Brent, A.E., Rosen, V., Olson, E.N., Lassar, A., Tabin, C.J., 2001. Analysis of the tendon cell fate using Scleraxis, a specific marker for tendons and ligaments. Development 128, 3855–3866.

Smits, P., Dy, P., Mitra, S., Lefebvre, V., 2004. Sox5 and Sox6 are needed to develop and maintain source, columnar, and hypertrophic chondrocytes in the cartilage growth plate. J. Cell Biol. 164, 747–758.

Smits, P., Li, P., Mandel, J., Zhang, Z., Deng, J.M., Behringer, R.R., de Crombrugghe, B., Lefebvre, V., 2001. The transcription factors L-Sox5 and Sox6 are essential for cartilage formation. Dev. Cell 1, 277–290.

Solloway, M.J., Dudley, A.T., Bikoff, E.K., Lyons, K.M., Hogan, B.L., Robertson, E.J., 1998. Mice lacking Bmp6 function. Dev. Genet. 22, 321–339.

Solursh, M., Fisher, M., Meier, S., Singley, C.T., 1979. The role of extracellular matrix in the formation of the sclerotome. J. Embryol. Exp. Morphol. 54, 75–98.

Sosic, D., Brand-Saberi, B., Schmidt, C., Christ, B., Olson, E.N., 1997. Regulation of paraxis expression and somite formation by ectoderm- and neural tube-derived signals. Dev. Biol. 185, 229–243.

St-Jacques, B., Hammerschmidt, M., McMahon, A.P., 1999. Indian hedgehog signaling regulates proliferation and differentiation of chondrocytes and is essential for bone formation. Genes Dev. 13, 2072–2086.

Stern, C.D., Sisodiya, S.M., Keynes, R.J., 1986. Interactions between neurites and somite cells: inhibition and stimulation of nerve growth in the chick embryo. J. Embryol. Exp. Morphol. 91, 209–226.

Stickens, D., Behonick, D.J., Ortega, N., Heyer, B., Hartenstein, B., Yu, Y., Fosang, A.J., Schorpp-Kistner, M., Angel, P., Werb, Z., 2004. Altered endochondral bone development in matrix metalloproteinase 13-deficient mice. Development 131, 5883–5895.

Storm, E.E., Huynh, T.V., Copeland, N.G., Jenkins, N.A., Kingsley, D.M., Lee, S.J., 1994. Limb alterations in brachypodism mice due to mutations in a new member of the TGF beta-superfamily. Nature 368, 639–643.

Sudo, H., Takahashi, Y., Tonegawa, A., Arase, Y., Aoyama, H., Mizutani-Koseki, Y., Moriya, H., Wilting, J., Christ, B., Koseki, H., 2001. Inductive signals from the somatopleure mediated by bone morphogenetic proteins are essential for the formation of the sternal component of avian ribs. Dev. Biol. 232, 284–300.

Tonegawa, A., Takahashi, Y., 1998. Somitogenesis controlled by Noggin. Dev. Biol. 202, 172–182.

Tosney, K.W., 1991. Cells and cell-interactions that guide motor axons in the developing chick embryo. BioEssays. 13, 17–23.

Tsumaki, N., Nakase, T., Miyaji, T., Kakiuchi, M., Kimura, T., Ochi, T., Yoshikawa, H., 2002. Bone morphogenetic protein signals are required for cartilage formation and differently regulate joint development during skeletogenesis. J. Bone Miner. Res. 17, 898–906.

Vu, T.H., Shipley, J.M., Bergers, G., Berger, J.E., Helms, J.A., Hanahan, D., Shapiro, S.D., Senior, R.M., Werb, Z., 1998. MMP-9/gelatinase B is a key regulator of growth plate angiogenesis and apoptosis of hypertrophic chondrocytes. Cell 93, 411–422.

Wang, B., He, L., Ehehalt, F., Geetha-Loganathan, P., Nimmagadda, S., Christ, B., Scaal, M., Huang, R., 2005. The formation of the avian scapula blade takes place in the hypaxial domain of the somites and requires somatopleure-derived BMP signals. Dev. Biol. 287, 11–18.

Watanabe, Y., Duprez, D., Monsoro-Burq, A.H., Vincent, C., Le Douarin, N.M., 1998. Two domains in vertebral development: antagonistic regulation by SHH and BMP4 proteins. Development 125, 2631–2639.

White, D.G., Hershey, H.P., Moss, J.J., Daniels, H., Tuan, R.S., Bennett, V.D., 2003. Functional analysis of fibronectin isoforms in chondrogenesis: Full-length recombinant mesenchymal fibronectin reduces spreading and promotes condensation and chondrogenesis of limb mesenchymal cells. Differentiation 71, 251–261.

White, P.H., Farkas, D.R., Chapman, D.L., 2005. Regulation of Tbx6 expression by Notch signaling. Genesis 42, 61–70.

Williams Jr., D.R., Presar, A.R., Richmond, A.T., Mjaatvedt, C.H., Hoffman, S., Capehart, A.A., 2005. Limb chondrogenesis is compromised in the versican deficient hdf mouse. Biochem. Biophys. Res. Commun. 334, 960–966.

Xiong, J., Onal, M., Jilka, R.L., Weinstein, R.S., Manolagas, S.C., O'Brien, C.A., 2011. Matrix-embedded cells control osteoclast formation. Nat. Med. 17, 1235–1241.

Yang, G., Rothrauff, B.B., Tuan, R.S., 2013. Tendon and ligament regeneration and repair: Clinical relevance and developmental paradigm. Birth Defects Res. C Embryo Today 99, 203–222.

Zeller, R., Lopez-Rios, J., Zuniga, A., 2009. Vertebrate limb bud development: moving towards integrative analysis of organogenesis. Nat. Rev. Genet. 10, 845–858.

Zelzer, E., McLean, W., Ng, Y.S., Fukai, N., Reginato, A.M., Lovejoy, S., D'Amore, P.A., Olsen, B.R., 2002. Skeletal defects in VEGF(120/120) mice reveal multiple roles for VEGF in skeletogenesis. Development 129, 1893–1904.

Zeng, L., Fagotto, F., Zhang, T., Hsu, W., Vasicek, T.J., Perry 3rd, W.L., Lee, J.J., Tilghman, S.M., Gumbiner, B.M., Costantini, F., 1997. The mouse Fused locus encodes Axin, an inhibitor of the Wnt signaling pathway that regulates embryonic axis formation. Cell 90, 181–192.

Zheng, Q., Zhou, G., Morello, R., Chen, Y., Garcia-Rojas, X., Lee, B., 2003. Type X collagen gene regulation by Runx2 contributes directly to its hypertrophic chondrocyte-specific expression *in vivo*. J. Cell Biol. 162, 833–842.

Chapter 29

Formation of Vertebrate Limbs

Yingzi Yang
Genetic Disease Research Branch, National Human Genome Research Institute, NIH, Bethesda, MD

Chapter Outline

GLOSSARY

Polydactyly Also known as hyperdactyly, this is the anatomical variant consisting of more than the usual number of digits on the hands and/or feet. When each hand or foot has six digits, it is sometimes called hexadactyly.

Brachydactyly This is a term which literally means "shortness of the fingers and toes" (digits). The shortness is relative to the length of other long bones and other parts of the body.

Syndactyly This is a condition in which the fingers fail to separate into individual appendages. This separation occurs during embryologic development.

Gorlin syndrome This is an autosomal dominant cancer syndrome. Patients with this rare syndrome often have anomalies in multiple organs, many of which are subtle. Gorlin syndrome patients have the propensity to develop multiple neoplasms, including basal cell carcinomas and medulloblastomas, and the patients' are extremely sensitive to ionizing radiation, including sunlight. Extra digits on the hands or feet also occur in this syndrome. Mutations in the PATCHED gene have been found to cause the Gorlin syndrome.

Bardet-Biedl syndrome This is mainly characterized by obesity, pigmentary retinopathy, polydactyly, mental retardation, hypogonadism, and renal failure in fatal cases.

Robinow syndrome This is a severe skeletal dysplasia with generalized shortening of all endochondral bones, segmental defects of the spine, brachydactyly, orodental abnormalities, hypoplastic genitalia and facial dysmorphism.

SUMMARY

Vertebrate limb development has been a fertile field for understanding the functional mechanisms of cell-cell signaling in controlling embryonic development. The rapid advancement of molecular genetic studies of vertebrate limb development has benefited tremendously from classical embryological experiments in the chick. These studies have revealed the molecular networks that control limb development in the following aspects:

- Limb bud formation at a specific position along the anterior-posterior body axis may be determined by the expression of *Hox* genes.
- Limb initiation is controlled by transcription factors Tbx4 (hindlimb) and Tbx5 (forelimb) and Wnt/Fgf signaling pathways.
- Limb outgrowth and patterning along the three axes (proximal-distal [PD], anterior-posterior [AP] and dorsal-ventral [DV]) are controlled by three signaling centers (apical ectoderm ridge [AER], zone of polarizing activity [ZPA] and non-AER limb ectoderm) respectively. Coordination between the three signaling centers occurs through interactions among the signaling pathways that mediate the function of the respective signaling centers.
- Signaling pathways controlling limb development often also play critical roles in regulating other important developmental processes.
- Genetic variations in humans which disrupt limb development lead to congenital limb malformation and other defects.

2015 Published by Elsevier Inc.

29.1 INTRODUCTION

There are considerable morphological and functional differences among the limbs of different vertebrate species. Most species use their limbs to support their body weight, to walk, and to run. However, birds and bats use their forelimbs to fly and whales use their limbs to swim. Human beings use their arms and legs, hands and feet to perform more complicated skills and artistic tasks. However, closer examination of all vertebrate limbs reveals that their structures are, in fact, remarkably similar. The origin of these similarities is believed to stem from the possession of a common ancestor. Most vertebrates have four limbs. No matter what the purposes to which the limbs are adapted, the skeletal elements constructing them, the muscles operating them and the nerves controlling the muscles always bear basic similarities. For instance, both the mouse forelimb and chick wing have a shoulder, girdle, humerus, radius, ulna and digits, although the number of digits differs. More importantly, the structural and functional similarities of vertebrate limbs are determined by similar developmental processes when the limbs form in the embryo.

Vertebrate limbs develop from an embryonic structure called the limb bud. The limb bud forms by localized proliferation of the lateral plate mesoderm at certain axial levels at the dorsal-ventral boundary. In the developing mouse embryo, a visible forelimb bud appears at 9.5 days post coitum (dpc). The hindlimb bud forms slightly later. The limb mesenchyme gives rise to patterned limb structures that form later in development and include cartilage, bone, tendon, and ligaments. Muscles, nerves and blood vessels are derived from cells that migrate into the limb bud during development. According to the molecular regulation and morphogenetic events, limb development can be divided into three stages: limb initiation, limb patterning and late limb morphogenesis. This chapter will focus on the first two stages.

Investigation of limb development is a very active field of developmental biology. Before the tools of molecular developmental biology were developed, classical embryological studies using mainly chick limb buds contributed significantly to our current understanding of limb development by identifying the signaling centers that control limb pattern formation. The chick limb bud has been one of the main systems of choice for embryological studies of limb development because it is accessible, large in size and can be easily manipulated *in ovo* without affecting other developmental processes. For instance, signaling centers that control three-dimensional limb patterning were identified through tissue recombination experiments in the chick limb bud, which could not have been performed in mouse embryos, which develop *in utero*. More recently, the availability of genetic tools in the mouse, combined with embryological studies in the chick, has led to the identification of the signaling molecules and pathways that mediate the function of the previously identified signaling centers, as will be elaborated in this chapter. The fundamental mechanisms uncovered in limb development are readily applicable to other developmental processes. Importantly, because of the pleiotrophic roles of the signaling pathways identified in limb development, molecular human and mouse genetic studies have identified genetic variants in the limb signaling pathways which lead to human birth defects and diseases such as polydactyly, brachydactyly, the Greig cephalopolysyndactyly syndrome, the Gorlin syndrome and the Bardet-Biedle Syndrome.

29.2 LIMB INITIATION

The positions of limb bud formation along the anterior/posterior axis are determined genetically in embryonic development. The area in the lateral plate mesoderm competent to form a limb is called the limb field. This was identified by tissue graft experiments in classical embryonic studies (Kieny, 1960; 1968). In the chick embryo, when grafted to an ectopic location, grafted limb field tissue directs the development of an entire limb. It is interesting to note that the size of limb field is bigger than the actual size of the lateral plate that forms a limb bud during normal development. It has been suggested that *Hox* genes expressed at different levels along the anterior/posterior axis determine where a limb bud will form (Cohn et al., 1997; Davidson et al., 1991; Izpisua-Belmonte et al., 1991). There is a narrow time window in which the grafted lateral plate mesoderm can induce the formation of an ectopic limb. Younger mesoderm requires associated somite tissue to induce limb formation, whereas older mesoderm from the already formed endogenous limb bud loses such inductive ability (Dhouailly and Kieny, 1972).

As the early limb bud is simply a mesoderm core covered by the surface ectoderm, limb initiation requires extensive epithelial-mesenchymal interactions. However, classical embryological work has demonstrated that it is the lateral mesenchymal cells that provide the signals to initiate the process of limb development, and that determine the specific identities of the limb type (arm/leg or wing/leg) and the axial levels of limb bud formation (Detwiler, 1933; Hamburger, 1938). In the pre-limb bud mesoderm, the -box transcription factors Tbx5 and Tbx4 are expressed in the future forelimb and hindlimb area, respectively (Agarwal et al., 2003; Naiche and Papaioannou, 2003; Takeuchi et al., 2003) (see also Chapter 34). However, despite the fact that the DNA sequences and functions of Tbx5 and Tbx4 are highly conserved, Tbx5 but not Tbx4 is required for limb bud initiation. No forelimb bud forms in $Tbx5^{-/-}$ mutant embryos, but in $Tbx4^{-/-}$ embryos, a hindlimb bud forms and then degenerates. The function of Tbx5 is cell autonomous as a transcription factor. However, limb initiation requires signaling from the lateral mesoderm to the overlying ectoderm. Fibroblast growth factor 10 (Fgf10), which has been found to play a critical role in limb initiation, appears to mediate the role of Tbx5 in limb bud initiation by signaling to the overlying ectoderm. First, it was found that beads soaked in recombinant Fgf1, Fgf2 and Fgf4 can induce the formation of a complete and morphologically normal limb when implanted to the lateral plate mesoderm (Cohn et al., 1995). Second, it was found that Fgf10 is expressed in the limb field and Fgf10$^{-/-}$ mice develop without limbs. Similarly to what had been observed in the Tbx4$^{-/-}$ mouse embryos, a tiny limb bud does form, but quickly degenerates (Axelrod et al., 1998; Sekine et al., 1999). Third, in the Tbx5$^{-/-}$ and Tbx4$^{-/-}$ mouse embryos, Fgf10 expression is either never induced or is weakly expressed and then quickly lost after initial expression (Naiche and Papaioannou, 2003). Fourth, Fgf receptor 2 (Fgfr2) appears to mediate Fgf10 signaling in controlling limb initiation. Fgfr2$^{-/-}$ mutant embryos also form very small limb buds that degenerate quickly. The expression of Tbx5 is intact in the Fgfr2$^{-/-}$ embryos (Agarwal et al., 2003). Therefore, Tbx5 and Tbx4 act upstream of Fgf10 during limb initiation.

Another signaling pathway that plays a critical role in limb initiation is the Wnt/β-catenin signaling pathway. Central to this pathway is the stabilization of β-catenin, which is phosphorylated and degraded in the absence of Wnt signaling. When Wnt signaling is active, β-catenin phosphorylation is inhibited and β-catenin then translocates to the nucleus where it binds LEF/TCF factors to activate downstream gene expression. *Lef1* and *Tcf1* are coexpressed with *Tbx5* or *Tbx4* in the prospective and early limb mesoderm. Expression of *Lef1/Tcf1* is also lost in the prospective limb field of the *Tbx5*$^{-/-}$ embryos. However, the expression of *Tbx5* is intact in *Lef1*$^{-/-}$;*Tcf1*$^{-/-}$ double mutant embryos (Agarwal et al., 2003). The *Lef1*$^{-/-}$;*Tcf1*$^{-/-}$ double mutant embryos also form much smaller limb buds that degenerate soon after formation (Galceran et al., 1999).

The epithelial-mesenchymal interaction during limb initiation is likely to be mediated by the interaction between Fgf and the Wnt/β-catenin signaling. After *Tbx5* is expressed, Wnt signaling acts in concert with *Tbx5* to activate *Fgf10* expression fully in the developing limb bud. *Tbx5* alone can activate *Fgf10* expression, whereas the canonical Wnt signaling sustains high levels of *Fgf10* expression during limb initiation (Agarwal et al., 2003) (Figure 29.1). *Tbx5, Fgf10* and *Wnt3* are initially expressed in a broader area than just the presumptive limb bud. It is likely that Tbx5 establishes a limb field by activating the expression of *Fgf10* in the mesenchyme of the limb field. Since *Wnt3* is expressed in the ectoderm overlying the presumptive limb bud mesoderm, *Wnt3* and *Fgf10* may mediate the extensive ectoderm-mesenchymal interaction during early limb initiation and maintain the expression of each other by forming a positive feedback loop (Figure 29.1). This is supported by the finding that *Wnt3* is required for limb initiation (Barrow et al., 2003). It appears that Wnt3 signals to both the limb mesenchyme and ectoderm to regulate limb initiation and elongation. It has been demonstrated that Wnt3 signals through β-catenin within the surface ectoderm to control the formation of the apical ectoderm ridge (AER) (Figure 29.1). The AER is a thickened epithelial structure that forms at the dorsal-ventral boundary of the early limb bud and is required for limb bud outgrowth. AER formation is a critical developmental event during limb initiation. Removal of either *Wnt3* or *β-catenin* genetically in the early limb ectoderm blocks AER formation (Barrow et al., 2003) and the limb development in these mutant embryos phenocopies that in the *Fgf10*$^{-/-}$ and the *Lef1*$^{-/-}$;*Tcf1*$^{-/-}$ mouse embryos. In the *Lef1*$^{-/-}$;*Tcf1*$^{-/-}$ mouse embryos, all Wnt/β-catenin activity in the presumptive limb region may be blocked. Since the early limb mesoderm as well as Fgf10 induces AER formation that can also be induced by Wnt/β-catenin signaling within the ectoderm without *Fgf10*, it appears that *Wnt3* expression is induced by the early presumptive limb mesoderm, possibly through Fgf10.

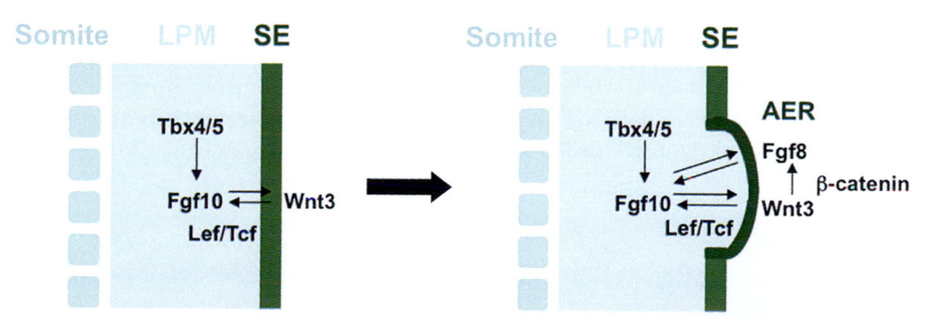

FIGURE 29.1 Signaling Pathways Regulating Limb Initiation and AER Formation. *Tbx4/5* is expressed in the lateral plate mesoderm (LPM) prior to the expression of *Fgf10, Lef1* and *Tcf1*. *Tbx4/5* directly activates *Fgf10* expression. *Fgf10* and *Wnt3* expressed in the surface ectoderm (SE) may form a positive feedback loop to achieve a high expression level required for limb bud out growth and AER induction. At this stage, Wnt3 signaling in regulating *Fgf10* expression is mediated by *Lef1* and *Tcf1*. Wnt3 signals in SE itself through β-catenin, which leads to AER induction and *Fgf8* expression. Once *Fgf8* is induced, it joins the *Wnt3 /Fgf* positive feedback loop by forming a positive feedback loop with *Fgf10*. This new *Wnt3 /Fgf* positive feedback loop is required for AER maintenance and proximal-distal limb outgrowth.

More recently, live imaging and lineage tracing in different vertebrate models has shown that the lateral plate contributes mesoderm to the early limb bud through directional cell movement. Wnt5a provides positional cue in the emerging limb bud to establish cell polarity that is likely to underlie the oriented cell behaviors (Wyngaarden et al., 2010).

29.3 LIMB BUD OUTGROWTH AND PATTERNING

The limb is a three-dimensional structure with three axes: anterior-posterior (AP; thumb to little finger), proximal-distal (PD; shoulder to finger tip) and dorsal-ventral (DV; back of hand to palm). Proper limb bud development requires patterning along these three axes, which is controlled by three signaling centers. The three signaling centers are established after limb initiation and then the limb develops autonomously by the coordination among these three signaling centers.

The AER is the signaling center that directs PD limb outgrowth. Its function in PD limb development was identified by the classical experiments performed by John Saunders in 1948 (Saunders, 1948; Summerbell, 1974). In these series of experiments, it was found that AER removal in the chick limb bud led to limb truncation along the PD axis. Earlier AER removal leads to limb truncation at more proximal limb levels.

Apart from the substantial growth of the limb bud along the PD axis, the limb bud is also patterned along the PD axis. The humerus forms at the most proximal part of the limb bud (also called styropod), whereas the radius and ulna form from its middle segment (also called zeugopod). The distal limb bud (autopod) forms the hand plate which includes the metatarsal/tarsal and digits (Figure 29.2). For a long time, patterning along the PD axis was thought to be regulated by the progress zone (progress zone model). In this model (Summerbell and Lewis, 1975), cells in the progress zone are progenitors of cartilage and connective tissues (see also Chapter 28). Parts of the limb are specified in proximo-distal sequence by an autonomous timing mechanism operating in a "progress zone" of undifferentiated growing mesenchyme under the influence of the AER, which serves to keep the cells in the progress zone actively dividing and relatively undifferentiated (Globus and Vethamany-Globus, 1976). If the AER is surgically removed, the limb bud will stop growing and truncation occurs along the PD axis, with earlier AER removal causing more proximal limb truncation. This progress model is consistent with the order of skeletal formation in the limb, which occurs first in the more proximal limb region. A central tenet of the progress zone model is that cells in the progress zone undergo a progressive change in positional information such that their specification depends on when they leave the progress zone and undergo differentiation. It is thought that cells which leave the progress zone earlier will adopt more proximal limb fates.

However, recent advances in molecular and genetic studies of limb development have brought new concepts and led to a significant modification of the progress zone model. The progress zone model was first challenged by a fate mapping study (Dudley et al., 2002). It was proposed in this study that, instead of being specified progressively as the limb bud grows out, PD specification of the limb occurs early in limb development and patterning along the PD axis is determined at the same time. Cell labeling experiments performed in early chick limb bud suggest that cells in the limb bud are specified as "proximal" or "distal" early on because early labeled limb cells were rarely found in two adjacent territories. Then, controlled by the AER, the regions of different cells expand at different times before differentiation to form the complete limb. This model was then challenged and modified by another study that showed only the most proximal limb (the stylopod that gives rise to the humerus) is regionalized at the early stage of limb development (Sato et al., 2007). A mixable unregionalized cell population is maintained in the distal area of the limb bud and regionalization progresses from proximal to distal limb bud during development. More recently, two studies have shown that PD limb patterning is dynamically determined by diffusible signals, not cell autonomous mechanisms. It was demonstrated that early limb mesenchyme in the chick is exposed to a combination of both proximal and distal signals and thus is maintained in a state capable of generating all limb segments (Cooper et al., 2011; Roselló-Díez A et al., 2011). As the limb bud grows, the proximal limb is regionalized by continued exposure to flank-derived proximal signal(s), whereas the medial and distal segments are initiated in domains that grow beyond proximal influence. Although retinoic acid (RA) is sufficient to act as a proximal signal for PD patterning, there is also evidence that RA signaling is not required for PD limb patterning (Cunningham et al., 2013).

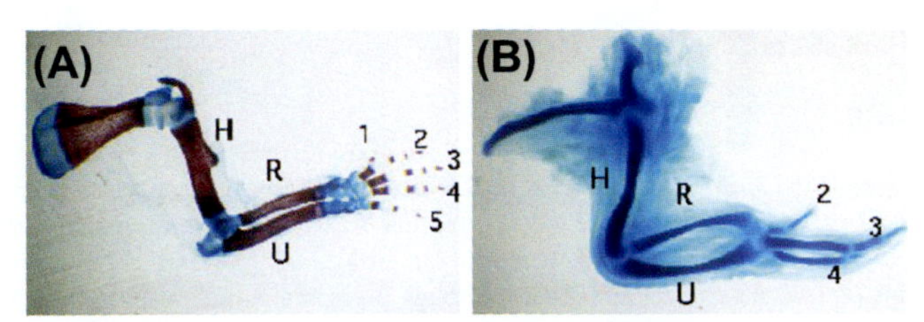

FIGURE 29.2 Skeletal Preparation of a Developing Forelimb in a Mouse Embryo (Arm, A) and a Chick Embryo (Wing, B). Mouse and chick limbs have similar structures. The forelimbs of the mouse and chick both contain styropod (humerus, H), zeugopod (radius (R) and ulna (U)) and autopod (digits). The mouse limbs have five digits (1–5) whereas the chick wing has three digits (2–4).

Fgf family members that are expressed in the AER, *Fgf4, Fgf8, Fgf9,* and *Fgf17* in the mouse have been identified as mediating the function of AER by molecular genetic studies in both the mouse and the chick. In the chick limb bud, heparin beads coated with either Fgf4 or Fgf8 can rescue limb truncation caused by AER removal when implanted to the distal edge of the AER-stripped limb bud (Crossley et al., 1996; Niswander et al., 1993). In the mouse, limbs lacking *Fgf4, Fgf9* and *Fgf17* have a normal skeletal pattern, indicating that *Fgf8* is sufficient among AER-Fgfs to sustain normal limb formation. Inactivation of *Fgf8* alone causes a mild skeletal phenotype (Lewandoski et al., 2000). When different combinations of the other AER-*Fgf* genes are removed, skeletal phenotypes of increasing severity are observed, reflecting the contribution that each Fgf can make to the total AER-Fgf signal. In addition to sustaining cell survival, AER-Fgfs regulate PD patterning gene expression during early limb bud development, providing genetic evidence that AER-Fgfs function to specify a distal domain (Mariani et al., 2008). Therefore, Fgf signaling from the AER has both permissive and instructive roles in PD patterning of the limb. First, it is required for limb bud cells to survive. Second, it is required to distalize the limb bud cells by antagonizing the proximalizing signal(s). It also appears that both Fgf and Wnt signaling, which are strongest in the distal limb, distalize limb bud cells by keeping their fate plastic (Ten Berge et al., 2008; Cooper et al., 2011; Tzchori et al., 2009).

The combination of oriented cell divisions and movements further drives the proximal-distal elongation of the early limb bud that is necessary to set the stage for subsequent morphogenesis. The Fgf signaling pathway, emanating from the AER, does not regulate cell orientation in the early limb bud, but instead establishes a gradient of cell velocity enabling continuous rearrangement of the cells at the distal tip of the limb (Gros et al., 2010). In contrast, Wnt5a signaling is necessary for the proper orientation of cell movements and cell division in the early limb bud. Furthermore, Wnt5a appears to act as a global cue to control PD limb elongation through establishing planar cell polarity (PCP) (Gao et al., 2011). As limb elongation is mainly driven by cartilage elongation, it is not surprising that Wnt5a controlled PCP is most prominent in the newly formed chondrocytes (Gao et al., 2011).

Patterning along the PD axis of the limb is determined by the interaction of regional specific transcription factors expressed in the limb mesoderm and the signaling pathways in the AER and the limb mesoderm. Restriction of expression of the evolutionally conserved homeobox genes *Meis 1* and *Meis2* to proximal regions of the limb bud is essential for limb development (Capdevila et al., 1999; Mercader et al., 2000). Ectopic *Meis2* expression in the distal limb bud severely disrupts limb outgrowth by repressing distal genes, whereas bone morphogenetic proteins (Bmps) and *Hoxd* genes restrict *Meis2* expression to the proximal limb bud. Combinations of Bmps and AER factors are sufficient to distalize proximal limb cells. *Hox* genes are major determinants of the animal body plan. In the limb, they regulate patterning along the PD limb axis. During mouse limb development, *Hoxd* genes are transcribed in two waves: early on, when the arm and forearm are specified, and later, when digits form. The transition between early and late regulations involves a functional switch between two opposite topological chromosomal domains. This transition between independent regulatory landscapes underlies the modular structures of the limbs along the PD axis and the distinct evolutionary histories of its various pieces. It also allows the formation of an intermediate area of low Hox proteins, which forms the wrist, during the transition between arms and hands. This regulatory strategy results in collinear Hox gene regulation in land vertebrate appendages (Andrey G et al., 2013).

The role of RA in limb patterning is still quite controversial. Genetic elimination of RA production has not been found to produce a discernible limb patterning defect. However, several pieces of evidence support a role of RA as an upstream activator of the proximal determinant genes *Meis1* and *Meis2*. RA can promote the proximalization of limb cells, and endogenous RA signaling appears to be required to maintain the proximal *Meis* domain in the limb. RA synthesis and signaling range is restricted to proximal limb domains after limb initiation by the AER activity that is mainly mediated by Fgfs. Fgfs may have a specific function in promoting distalization through the inhibition of RA production and signaling.

Although the AER serves to provide the growth signal along the PD axis, the limb bud type is controlled by the mesoderm. In this regard, the AER signal is permissive, but not instructive for limb type specification. When the chick wing bud mesoderm is recombined with the leg bud ectoderm, a wing develops (Zwilling, 1955). It was also found that development of cross-species recombination of limbs such as chick/rat, was typical of the species contributing the mesoderm (Jorquera and Pugin, 1971). Moreover, if the AP axis of the AER is reversed with reference to the mesoderm, the pattern of mesoderm differentiation is unchanged (Zwilling, 1956). Indeed, it has been found that *Tbx5* and *Tbx4* are expressed in the forelimb and hindlimb mesoderm, respectively, not in the ectoderm. In addition, the hindlimb identity is determined by a mesoderm-specific transcription factor Pitx1 (Lanctot et al., 1999; Logan and Tabin, 1999; Szeto et al., 1999).

The interaction between the AER and limb mesoderm is not one way. Equally important is the maintenance of AER structure and function by the underlying limb mesoderm. The presumed mesoderm-derived maintenance factor is called the apical ectoderm maintenance factor (AEMF). In the tissue recombination experiments, it was found that an older AER would change its morphology to resemble a young AER if it is recombined with a young mesoderm. The normal configuration of AER is also controlled by the limb mesoderm (Zwilling, 1956). If limb ectoderm is recombined with non-limb mesoderm (flank lateral plate or posterior somites), the AER degenerates within two days of the operation. But if a small piece of limb mesoderm is added beneath a part of the AER, that part of the AER survives (Zwilling, 1972). It has been suggested that the AEMF is not distributed evenly through the mesoderm. It is more concentrated in the posterior half of the limb bud, which is covered by a thicker and longer AER. As *Fgf10* forms a positive feedback loop with *Fgf8* during limb elongation stage, and this feedback loop is required for limb outgrowth and AER maintenance (Figure 29.1), Fgf10 is obviously qualified to be an AEMF. However, Fgf10 may not be the only AEMF. A classic example of AEMF deficiency in limb development is found in the *limb deformity* (*ld*) mutant mice, in which AER degenerates prematurely due to a mesodermal defect (Kuhlman and Niswander, 1997). The molecular nature of the defective AEMF in the *ld* limb has been identified to be a secreted signaling molecule, Gremlin (Khokha et al., 2003; Zuniga et al., 1999). Gremlin is an antagonist of Bmp signaling that is expressed in the distal limb mesoderm. As Bmp signaling promotes AER degeneration, Gremlin is required for the maintenance of AER integrity by antagonizing Bmp signaling.

The second signaling center is the zone of polarizing activity (ZPA), which is a group of mesenchymal cells located at the posterior limb margin, immediately adjacent to the AER. This polarizing activity was also discovered by John Saunders, who found that ZPA tissue grafted to the anterior limb bud leads to digit duplication in mirror image of the endogenous ones (Saunders and Gasseling, 1968). When limb mesoderm cells are dissociated and packed into the ectoderm jacket, the resulting limb has digits, but their specific identities cannot be determined. However, when a polarizing posterior mesenchyme graft is added, this results in a much more normal skeleton with recognizable digits, the most posterior digit forming closest to the polarizing posterior mesenchyme graft (MacCabe et al., 1973). MacCabe has also demonstrated that there is a gradient of the polarizing posterior mesenchymal activity in the chick limb bud and found that such activity appears to be mediated by a diffusible factor (Calandra and MacCabe, 1978; MacCabe and Parker, 1976). It was proposed by Summerbell (Summerbell, 1979) that the polarizing posterior mesenchyme emits a diffusible signal which forms a gradient and specifies the AP axis in a dose dependent manner.

Molecular genetic studies have identified that a vertebrate Hedgehog family member, Sonic hedgehog (Shh), mediates the activity of ZPA, as ectopically expressed Shh in the anterior mouse limb bud via *Shh* expressing cells and Shh protein coated beads implanted in the anterior chick limb bud leads to mirror image digit duplication (Chan et al., 1995; Riddle et al., 1993) (Figure 29.3). The active Shh signal corresponds to a 19 kD N-terminal peptide generated by autoproteolytic cleavage that is modified by the covalent addition of cholesterol and palmitate (Chamoun et al., 2001; Porter et al., 1995). Many studies have underscored the general long-range signaling capacity of Shh, and loss-of-function genetics has established essential Shh functions during embryogenesis, maintenance of stem cells, and disease in vertebrates. Shh has been demonstrated to act as a morphogen that patterns the dorsal-ventral axis of the developing neural tube by forming a morphogen gradient (Ericson et al., 1996). However, in the limb, although there is an absolute requirement for Shh signaling for AP patterning of the distal limb skeleton, as lack of Shh in various vertebrate species results in loss of the posterior zeugopod (ulna) and digits 5 to 2 (Chiang et al., 2001), it is still not a settled issue whether Shh produced by the ZPA also patterns the limb bud by acting as a diffusible morphogen. Shh protein can diffuse from the ZPA to elicit a response at a distance in the limb bud mesenchyme (Lewis et al., 2001; Yang et al., 1997). Cells responding to Shh signaling activate the Gli1 transcription factor, but genetic analysis in the mouse shows that Gli1 is not essential for limb bud development (Park et al., 2000). The related Gli2 protein also functions as a positive downstream mediator of Shh signaling, but is again not essential for limb bud development (Mo et al., 1997). Rather, Shh signaling seems to enable distal progression of limb bud morphogenesis and formation of the digit arch by inhibiting proteolytic production of the repressor form of another Gli family member, the Gli3 protein (te Welscher et al., 2002b; Wang et al., 2000), which is expressed primarily in cells that do not express Shh. However, when the direct contribution of ZPA cells to digit primordial was analyzed by genetic fate mapping (Harfe et al., 2004), it was found that ZPA cells give rise to descendants that either remain ZPA cells or join a distal-anteriorly expanding population of descendants that no longer express Shh. Together, these two populations of Shh descendant cells give rise to the most posterior digits (5 and 4), parts of digit 3, and the ulna. Therefore, digit 2 is the only skeletal element that is not derived from cells that have expressed Shh at some stage. These results challenge the relevance of a spatial morphogen gradient, as digit 2 may be the only one that depends on long-range Shh signaling. Indeed, reduction of Shh mobility across the limb bud affects digit 2, whereas digits 3 to 5 form normally. In addition, when kinetic studies

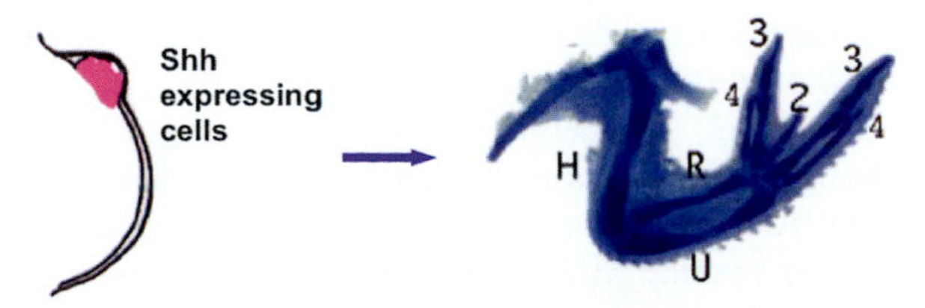

FIGURE 29.3 **Implantation of *Shh*-Expressing Cells in the Anterior Limb Bud Leads to Mirror Image Digit Duplication.** Shh expressing cell pellets are implanted in the anterior chick limb bud under the AER. The ectopic Shh activity causes mirror image digit duplication (from 2-3-4 to 4-3-2-3-4).

were performed to address how the identities of digits 3 to 5 are specified, it was revealed that the fates of Shh descendants are progressively restricted posteriorly. Descendant cells that do not express Shh and are born early contribute to all three digits, whereas the ZPA cells expressing Shh longest contribute exclusively to digit 5. It appears that limb bud cells somehow acquire a kinetic memory of the Shh signal received (Yang et al., 1997). Thus, the length of time for which the Shh expressing cells and their non-expressing descendants are exposed to Shh signaling determines the identities of the three posterior digits. This seems to be inconsistent with the spatial morphogen gradient model, according to which long-range Shh signaling specifies digits 4, 3 and 2, not just digit 2. However, during the dynamic process of AP limb patterning, the cells remaining in the ZPA or the non-Shh expressing descendants closer to ZPA obviously receive higher doses of Shh. Therefore, their more posterior identity can also be explained by a spatial Shh gradient model.

It is very likely that digit identity is patterned by both the spatial gradient and temporal duration of Shh morphogen (Yang et al., 1997). Indeed, when limb bud cells responding to Shh signaling were marked by analyzing transcriptional activation of Gli1, a direct transcriptional target of Shh signaling, at specific time points of limb development, it was found that the mesenchymal cells giving rise to digits 5 to 2 and the ulna respond to Shh signaling (Ahn and Joyner, 2004). However, although the cumulative Shh response was highest in the posterior mesenchyme (digit 5) and progressively lower toward the anterior, no specific thresholds of response were found as predicted by the morphogen gradient model. In contrast, the Shh responsiveness of the most posterior cells (fated to digit 5), which are exposed to Shh the longest, was reduced with time. In addition, the number of marked Gli1 expressing cells is reduced and their distribution is altered in limb buds of mouse embryos lacking the Gli2 gene, despite the fact that Gli2-deficient limbs develop completely normally. These results, together with the analysis of Gli3-deficient limb buds, indicate that it is not the positive response to Shh as mediated by Gli1 and Gli2, but its inhibitory effects on Gli3 repressor (Gli3R) formation that determines digit identities. In particular, the most anterior digit 1 is specified in the absence of Shh by high levels of Gli3R, whereas cumulative high levels of Shh response effectively repress Gli3R formation and thereby specify the most posterior digit 5. Taken together, these studies indicate that the vertebrate limb bud is patterned by a kinetic memory integrating the cumulative length and strength of Shh signaling that cells receive. These temporal and spatial gradients pattern digits 2 to 5 and the ulna. These processes are regulated by controlling the inhibition of Gli3R formation at different locations along the limb AP axis and over time.

Surprisingly, a rather different conclusion was reached in a study of limb AP patterning regulated by Shh (Zhu et al., 2008). This study argues that AP patterning is a *fait accompli* prior to expansion of mesenchymal progenitors. When Shh activity was removed at different stages of mouse limb development, it was found that earlier removal led to the formation of fewer digits. What was truly surprising is that the order of digit loss did not reflect their position along the AP axis, but rather, the time of appearance of a given digit condensation (from first to last for Shh-dependent digits: 4, 2, 5, 3). Therefore, a digit 3 condensation appears last and digit 3 is the first to be lost upon Shh removal. Zhu et al. argue that if Shh acts over a long period to progressively pattern more posterior digits, as suggested by earlier studies, this should be reflected in the order of digit reduction rather than the order in which digit condensations are lost. The principle of "last in first out" only applies to the location of the digits that are lost. However, unless one can find definitive molecular markers that allow one to distinguish different digit identities in early limb development, the following critical question cannot be answered: Can one infer digit identity from digit position alone? For example, a short exposure of Shh leads to a digit forming in the fourth digit position, does this have the character of a normal digit 4?

Although Shh plays a critical role in AP patterning of the limb, the AP axis of the limb appears to be established prior to Shh signaling. Expression of *Shh* in the posterior limb bud is controlled by a basic helix-loop-helix (bHLH) transcription factor dhand. Gli3 restricts *dhand* expression to the posterior mesenchyme prior to activation of Shh signaling in mouse limb buds. dhand, in turn, excludes anterior genes such as *Gli3* and *Alx4* from posterior mesenchyme. These interactions polarize the AP axis of the newly formed limb bud mesenchyme prior to Shh signaling (te Welscher et al., 2002a). Twist1 is also a bHLH transcription factor expressed in the developing limb bud and is required for maintaining the AER. Autosomal dominant mutations in the *twist1* gene are associated with limb and craniofacial defects in humans with Saethre-Chotzen syndrome (Jabs, 2004). Ectopic expression of *dhand* phenocopies *Twist1* loss-of-function in the limb, and the two factors have a gene dosage-dependent antagonistic interaction. Dimerization of Twist1 and dhand can be modulated by protein kinase A- and protein phosphatase 2A-regulated phosphorylation.

The third signaling center is the non-AER limb ectoderm that covers the limb bud. It sets up the DV polarity of not only the ectoderm but also the underlying mesoderm of the limb (review by Niswander, 2002; Tickle, 2003). In the early embryo before the limb bud forms, the DV polarity of the future limb is determined by the mesoderm, as shown by ectoderm-mesoderm recombination experiments (Geduspan and MacCabe, 1989). However, prior to limb bud initiation, the ability to determine the DV polarity is transferred from the mesoderm to the ectoderm. If the DV polarity of the limb ectoderm is reversed relative to the mesenchyme in the early limb bud, the DV polarity of the mesenchyme changes in accordance with that of the ectoderm. At much later stages, when skeletal morphogenesis starts, the capacity of the mesoderm to respond to ectodermal control is lost.

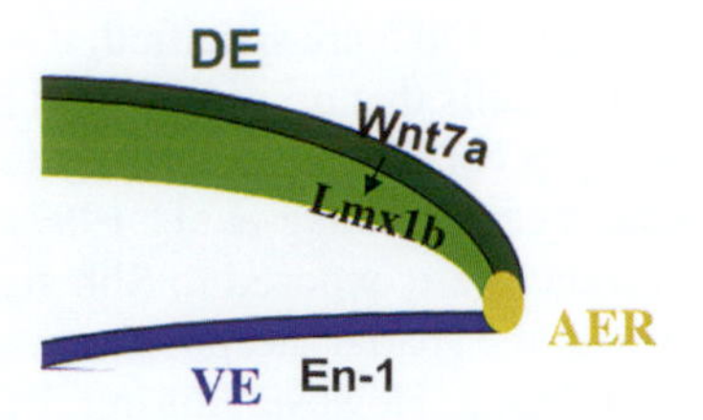

FIGURE 29.4 A schematic Section of the Limb Bud Along the Dorsal-Ventral Axis. *Wnt7a* expressed in the dorsal ectoderm signals to the distal limb mesoderm to determine the dorsal limb identity by activating the expression of *Lmx1b*. *En-1* is expressed in the ventral ectoderm and it is required for ventral limb development by inhibiting *Wnt7a* expression. DE: dorsal ectoderm; VE: ventral ectoderm.

Wnt and Bmp signaling are required to control DV limb polarity in both the limb ectoderm and mesoderm. *Wnt7a* is expressed specifically in the dorsal limb ectoderm and it activates the expression of *Lmx1b*, which encodes a dorsal-specific transcription factor that determines the dorsal identity (Parr et al., 1993; Riddle et al., 1995) (Figure 29.4). Mutant mice with a targeted *Wnt7a* null mutation develop ventralized limbs and the expression of *Lmx1b* is lost (Cygan et al., 1997; Parr and McMahon, 1995). On the other hand, ectopic expression of Wnt7a in the ventral limb dorsalizes the limb and activates *Lmx1b* expression. During normal limb development, *Wnt7a* expression in the ventral ectoderm is suppressed by *En-1*, which encodes a transcription factor that is expressed specifically in the ventral ectoderm (Loomis et al., 1996) (Figure 29.4). In the *En-1* loss-of-function mutant limb, *Wnt7a* is ectopically expressed in the ventral ectoderm and the limb is dorsalized. In the *En-1* and *Wnt7a* double mutant mouse, the limb is similarly ventralized like the *Wnt7a* single mutant (Cygan et al., 1997; Loomis et al., 1998). These studies indicate that either ventral cell fates are default (independent of active gene regulation by *Wnt7a* and *En-1*) or the function of *Wnt7a* is to actively suppress another ventralizing regulatory pathway.

It appears that this ventralizing regulatory pathway is mediated by Bmp signaling. During early limb development in the chick, Bmp2, Bmp4 and Bmp7 are expressed in the mesenchyme in an unrestricted manner along the DV axis, and in the AER (Francis et al., 1994; Francis-West et al., 1995). But in the ectoderm, these Bmps are preferentially expressed in the ventral ectoderm, coincident with En-1 and in a complementary pattern to that of Wnt7a in the dorsal ectoderm at the time the ectoderm provides DV information to the underlying mesenchyme. In the mouse limb bud, Bmp2 is also expressed in the early ventral limb ectoderm (Lyons et al., 1990). Therefore, the ventral limb mesoderm and ectoderm receive more Bmp signaling. These Bmps signal through Bmp receptor IA (BMPR-IA) in the limb ectoderm to establish normal DV limb pattern by activating En-1 expression in the ventral ectoderm (Ahn et al., 2001; Pizette et al., 2001). In both mouse and chick embryonic limb bud, the loss of Bmp signaling results in loss of En-1 expression and dorsalized limbs, whereas activated Bmp signaling leads to ventralized limbs with ectopic En-1 expression in the dorsal ectoderm.

It appears that the effects of Bmp signaling are mediated by Msx1 and Msx2, two transcription factors that are also themselves transcriptionally regulated by Bmp signaling. Loss of both *Msx1* and *Msx2* function in the mouse embryo also leads to the loss of *En-1* expression in the ventral ectoderm, resulting in the expansion of *Wnt7a* expression, which in turn caused an expansion of *Lmx1b* expression into the ventral mesenchyme, thus leading to limb dorsalization (Lallemand et al., 2005). Therefore, the function of Bmp signaling in the early limb ectoderm is upstream of *En-1* in controlling DV limb polarity.

It is interesting to note that like Bmp signaling, the Wnt/β–catenin signaling pathway also acts directly in the limb ectoderm to control DV pattern by controlling the expression of *En-1* (Barrow et al., 2003). In contrast to *Wnt7a*, which is a dorsalizing factor, it appears that *Wnt3*, which is expressed throughout the early limb ectoderm, signals through β–catenin to ventralize the limb. Wnt3/β–catenin signaling is required in the ventral ectoderm for the expression of *En-1*. Loss of *Wnt3* or *β–catenin* in the ventral limb ectoderm both resulted in a dorsalized limb, in a manner that that can be attributed to loss of *En-1* expression. However, it appears that Bmp and Wnt/β–catenin also signal to the limb mesoderm directly to control the DV pattern. When BmpRIA is specifically inactivated only in the mouse limb bud mesoderm, the distal limb is also dorsalized without altering the expression of *Wnt7a* and *En-1* in the limb ectoderm (Ovchinnikov et al., 2006). Likewise, in the mouse limb bud mesoderm, the loss of *β–catenin*, like the loss of *Wnt7a* in the dorsal ectoderm, leads to limb ventralization, whereas activated β–catenin, like *Wnt7a* ectopic expression, causes limb dorsalization. Since Wnt7a signals to the limb mesoderm directly to control *Lmx1b* expression and DV limb patterning, it is possible that Wnt7a signals through β–catenin in the limb mesoderm and the Wnt/β–catenin and Bmp signaling pathways antagonize each other in controlling the expression of *Lmx1b* directly in the mesoderm.

The three signaling centers that control growth and patterning along the three axes are not functionally independent. The interaction between these three signaling centers ensures that three-dimensional growth and patterning are coordinated during vertebrate limb development (Figure 29.5). It has previously been shown that polarizing posterior mesenchyme must

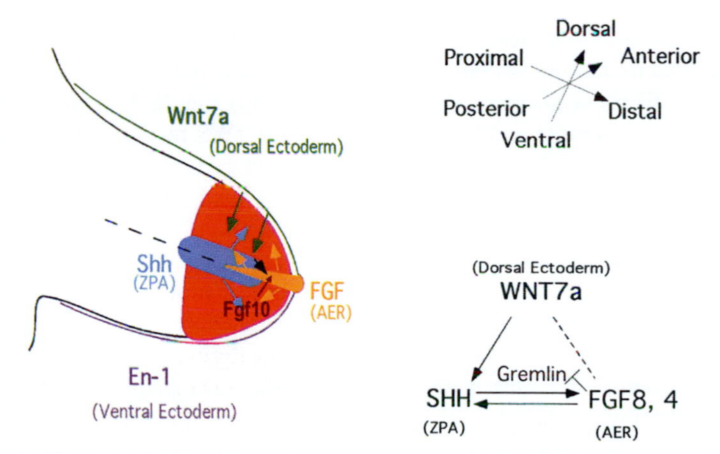

FIGURE 29.5 **Patterning Along the Three Limb Axes is Coordinately Regulated by Interactions among the Signaling Pathways.** Both DV signals (Wnt7a) and PD signals (Fgfs) regulate AP limb patterning by controlling *Shh* expression. The AP signal Shh is also required for PD limb outgrowth by regulating the expression of *Fgfs* in the AER through *Gremlin*. Limb bud outgrowth is self-terminated by Fgf inhibition of *Gremlin* expression.

be grafted under the AER in the anterior limb in order to cause mirror image digit duplication (Tickle et al., 1975). This highlights an interaction between the AP and PD signaling centers. It has also been demonstrated that disruption of the dorsal ectoderm, which controls DV patterning, results in shortening of limb skeletal elements along the PD axis (Martin and Lewis, 1986). This suggests that the DV signaling center also interacts with the PD signaling center. At the molecular level, it is now clear that the three signaling centers indeed interact with each other through interactions of the mediating signaling molecules (Figure 29.5). First, there is a positive feedback loop between *Shh* and *Fgfs* expressed in the AER, which connects AP limb patterning with PD limb outgrowth. Fgf signaling from the AER is required for *Shh* expression and Shh signals through *Gremlin* to maintain the normal expression of *Fgfs* in the AER (Khokha et al., 2003; Laufer et al., 1994; Niswander et al., 1994). Limb bud outgrowth is driven by signals in a positive feedback loop involving *Fgfs*, *Shh* and *Gremlin*. Precise termination of these signals is essential to restrict limb bud size (Scherz et al., 2004; Verheyden and Sun, 2008). *Shh* expressing cells and their descendants cannot express *Gremlin*. The proliferation of these descendants forms a barrier separating the Shh signal from *Gremlin* -expressing cells, which breaks down the Shh-Fgf4 loop and thereby affects limb size and provides a mechanism explaining regulative properties of the limb bud (Scherz et al., 2004). In addition, genetic evidence has shown that FGF signaling can repress *Gremlin* expression, revealing another Fgf/*Gremlin* inhibitory loop. This repression occurs both in mouse and chick limb buds, and is dependent on high FGF activity. These studies show that Grem1/Fgf feedback signaling is required for regulation of the temporal kinetics of the mesenchymal response to Shh signaling (Panman et al., 2006). They support a mechanism in which the positive Fgf/*Shh* loop drives outgrowth, and an increase in Fgf signaling triggers the Fgf/*Gremlin* inhibitory loop that terminates outgrowth signals (Verheyden et al., 2008). Also the dorsalizing signal Wnt7a is required for maintaining the expression of *Shh* that patterns the AP axis (Parr and McMahon, 1995; Yang and Niswander, 1995).

29.4 LIMB DEVELOPMENT AND DISEASES

The three-dimensional development of the limb is orchestrated by proper regulation of cell signaling and transcriptional networks. Molecular genetic studies of limb development in model organisms, mainly the mouse and the chick, have provided significant insights into both mutation identification and pathological mechanisms of human congenital limb deformities. In many cases, since limb deformities are not associated with other more detrimental defects, a wealth of clinical descriptions of different limb abnormalities has accumulated. Since most of the signaling pathways controlling limb development also play roles in other developing processes, identifying the molecular nature of limb deformities has also enhanced our understanding of congenital diseases affecting other systems, such as the kidney and the central nervous system.

Similarly to what has been shown in mouse genetic studies, mutations in the Wnt, Fgf and Shh signaling pathways in humans also affect limb formation, outgrowth and patterning along the three axes. *Wnt3* is required for limb initiation in the mouse. In humans, a nonsense mutation that truncates Wnt3 protein at its amino terminus has been found to be a very likely cause of a rare human genetic disorder, Tetra-amelia (OMIM #273395), in four affected fetuses of a consanguineous Turkish family (Niemann et al., 2004). Tetra-amelia is characterized by complete absence of all four limbs and other anomalies in craniofacial and urogenital development.

Fgf signaling plays multiple roles in limb development. The significance of FGF signaling in the developing human limb is highlighted by the findings that mutations in FGF receptors lead to limb and skeletal deformities in the Apert syndrome (AS, OMIM# 101200), Jackson-Weiss syndrome (JWS, OMIM# 123150) and the Pfeiffer syndrome (PS, OMIM#101600) (reviewed in Naski and Ornitz, 1998; Ornitz and Marie, 2002). AS and JWS result from mutations in FGFR2. These syndromes are characterized by syndactyly of the hand and feet and premature fusion of the cranial sutures. PS results from a single mutation in FGFR1 or one of several mutations in FGFR2. This syndrome is characterized by broad great toes and thumbs in addition to premature fusion of cranial sutures. Mechanistic studies of the FGF receptor mutations in these syndromes further indicate that some, and perhaps all, of the mutations causing JWS and PS are the result of activating mutations in an FGF receptor. In addition, it has been found that the S252W mutation of the AS alters the ligand binding affinity of the FGFR2 (Yu et al., 2000). The increased affinity for FGF ligands caused by this mutation may result in enhanced activation of the receptor in the mesenchymal cells destined to form the digits.

The Shh signaling pathway controls AP limb patterning. One of the most frequently observed human limb malformations is preaxial polydactyly (PPD) which has disrupted AP limb patterning. Patients with PPD (particularly triphalangeal thumb-polysyndactyly [OMIM#190605] and preaxial polydactyly type II [OMIM#174500]) have extra digits on the sides of their thumb or great toe, similarly to what has been observed in mouse mutants with ectopic *Shh* expression in the anterior limb bud. From the insight gained from the molecular genetic studies of the Sasquatch (Ssq) (Sharpe et al., 1999) and Hemimelic extratoe (Hx) (Clark et al., 2000) mouse mutants with preaxial supernumerary digits, it is found that point mutations in the long-range, limb-specific regulatory element of the SHH gene are responsible for the human limb abnormality of PPD (Lettice et al., 2002).

GLI3 transduces the Hedgehog signal, and mutations in *GLI3* result in Greig cephalopolysyndactyly syndrome (GCPS; OMIM# 175700) or Pallister-Hall syndrome (PHS; OMIM# 146510). Mutations such as deletions or translocations resulting in haplo-insufficiency of GLI3 are found in association with GCPS (Brueton et al., 1988; Pettigrew et al., 1991; Vortkamp et al., 1991), whereas mutations resulting in dominant negative GLI3 are found in Pallister-Hall syndrome (Johnston et al., 2005; Kang et al., 1997). In addition, a mutation at codon 764 of the *GLI3* gene, which is 3-prime to the conserved domain termed Pzf1 (post zinc finger-1), is found to cause postaxial polydactyly type A1 (PAPA1; OMIM# 174200) (Radhakrishna et al., 1997). PAPA1 is an autosomal trait characterized by an extra digit in the ulnar and/or fibular side of the upper and/or lower extremities. The extra digit is usually well formed and articulates with the fifth or extrametacarpal/metatarsal. Thus, different truncated GLI3 proteins are associated with different clinical syndromes, highlighting the different functions of specific GLI3 variant proteins during limb development. As GLI2 is also required to transduce the Hedgehog signaling, mutations in the human *GLI2* gene have also been found to cause postaxial polydactyly along with developmental defects that affect the pituitary gland and the brain (Roessler et al., 2003).

Dorsal-ventral limb patterning is controlled by Wnt7a/Lmx1b signaling in mice. In humans, there is evidence that the Fuhrmann syndrome (OMIM# 228930) and the Al-Awadi/Raas-Rothschild/Schinzel phocomelia syndrome (OMIM# 276820) are caused by mutations in the *WNT7A* gene (Woods et al., 2006). These two syndromes have been considered to have distinct limb malformations characterized by various degrees of limb aplasia/hypoplasia and joint dysplasia. It has been suggested that mutations causing a partial loss of WNT7A function lead to Fuhrmann syndrome, whereas null mutations lead to the more severe limb truncation phenotypes observed in Al-Awadi/Raas-Rothschild/Schinzel phocomelia syndrome. Again, the findings in human limb malformations provide insight into the role of WNT7A in multiple aspects of vertebrate limb development. As LMX1B is required for determining dorsal limb identity, mutations resulting in haplo-insufficiency of LMX1B have been found to cause the nail-patella syndrome (NPS, OMIM# 161200) that affects dorsal-ventral limb development and is characterized by dysplasia of the nails and absent or hypoplastic patellae. Other limb features of NPS include elbow stiffness with limitations in pronation and supination. Consistent with the findings in *Lmx1b*$^{-/-}$ mutant mice (Chen et al., 1998; Pressman et al., 2000), mutations in the *LMX1B* gene also cause renal abnormalities and open angle glaucoma (OAG, OMIM# 137760).

29.5 CONCLUSIONS AND PERSPECTIVES

Studies of classical embryology and molecular genetics in model organisms have led to the identification of molecular pathways that control coordinated, three-dimensional, limb development. These studies have also provided insight in understanding the molecular and pathological mechanisms of congenital human limb defects. The general principles will be relevant to devising new approaches for tissue repair. In addition, because of the pleiotrophic effects of the limb development pathways in other development processes, as has been shown in mouse and human genetic studies, the developing limb has provided an excellent model system for understanding how different signaling pathways operate and interact with each other in embryonic development. Although many genes and pathways with roles in limb development have been identified, we are still far from a full picture of how limb development is controlled at the molecular level. There are considerable gaps in our current understanding. To gain a deeper understanding of the cellular basis of limb development, we also need to find creative new ways of visualizing the behavior of extracellular signaling molecules and measuring their concentrations and the cellular responses they trigger. Furthermore, it is not yet clear how the three-dimensional positional information, controlled by distinct signaling pathways, is integrated in each cell of the developing limb. Now that both the mouse and human genomes have been sequenced, recent developments in functional genomic studies in mice, such as chemical and insertional mutagenesis and large scale gene targeting, will provide a tremendous amount of new information for understanding limb development. These studies, combined with the ever improved, powerful techniques in genetic mapping of human diseases, have allowed the advancement of our understanding of limb development at an unprecedented speed.

29.6 CLINICAL RELEVANCE

- Understanding the control of limb development is essential to understanding the pathological mechanisms underlying limb malformations in human development; such as polydactyly, brachydactyly and syndactyly.
- The vertebrate limb is a great system to dissect the molecular pathways that also control critical processes in other developmental or physiological processes, abnormalities of which lead to genetic diseases such as Golin syndrome, Bardet-Biedl syndrome and Robinow syndrome.
- Understanding the molecular mechanisms of signal transduction that control limb development have led to an in-depth understanding of critical signaling pathways such as those mediated by Wnts and Hedgehogs. Misregulation of these signaling pathways leads to common diseases, including cancer.
- Skeletal development is an integral part of limb development. Knowledge gained in the study of limb development is essential to the understanding of cartilage and bone diseases such as osteoarthritis and osteoporosis.

RECOMMENDED RESOURCES

Andrey, G., Montavon, T., Mascrez, B., Gonzalez, F., Noordermeer, D., Leleu, M., Trono, D., Spitz, F., Duboule, D., 2013. A switch between topological domains underlies HoxD genes collinearity in mouse limbs. Science 7, 340.

Tickle, C., 2006. Making digit patterns in the vertebrate limb. Nat. Rev. Mol. Cell. Biol. 7, 45–53.

Mariani, F.V., Martin, G.R., 2003. Deciphering skeletal patterning: clues from the limb. Nature 423, 319–325.

Niswander, L., 2002. Interplay between the molecular signals that control vertebrate limb development. Int. J. Dev. Biol. 46, 877–881.

REFERENCES

Agarwal, P., Wylie, J.N., Galceran, J., Arkhitko, O., Li, C., Deng, C., Grosschedl, R., Bruneau, B.G., 2003. Tbx5 is essential for forelimb bud initiation following patterning of the limb field in the mouse embryo. Development 130, 623–633.

Ahn, K., Mishina, Y., Hanks, M.C., Behringer, R.R., Crenshaw 3rd, E.B., 2001. BMPR-IA signaling is required for the formation of the apical ectodermal ridge and dorsal-ventral patterning of the limb. Development 128, 4449–4461.

Ahn, S., Joyner, A.L., 2004. Dynamic changes in the response of cells to positive hedgehog signaling during mouse limb patterning. Cell 118, 505–516.

Andrey, G., Montavon, T., Mascrez, B., Gonzalez, F., Noordermeer, D., Leleu, M., Trono, D., Spitz, F., Duboule, D., 2013. A switch between topological domains underlies HoxD genes collinearity in mouse limbs. Science 7 (340), 1234167-1–1234167-10.

Axelrod, J.D., Miller, J.R., Shulman, J.M., Moon, R.T., Perrimon, N., 1998. Differential recruitment of Dishevelled provides signaling specificity in the planar cell polarity and Wingless signaling pathways. Genes. Dev. 12, 2610–2622.

Barrow, J.R., Thomas, K.R., Boussadia-Zahui, O., Moore, R., Kemler, R., Capecchi, M.R., McMahon, A.P., 2003. Ectodermal Wnt3/beta-catenin signaling is required for the establishment and maintenance of the apical ectodermal ridge. Genes. Dev. 17, 394–409.

Brueton, L., Huson, S.M., Winter, R.M., Williamson, R., 1988. Chromosomal localisation of a developmental gene in man: direct DNA analysis demonstrates that Greig cephalopolysyndactyly maps to 7p13. Am. J. Med. Genet. 31, 799–804.

Calandra, A.J., MacCabe, J.A., 1978. The in vitro maintenance of the limb-bud apical ridge by cell-free preparations. Dev. Biol. 62, 258–269.

Capdevila, J., Tsukui, T., Rodriquez Esteban, C., Zappavigna, V., Izpisua Belmonte, J.C., 1999. Control of vertebrate limb outgrowth by the proximal factor Meis2 and distal antagonism of BMPs by Gremlin. Mol. Cell. 4, 839–849.

Chamoun, Z., Mann, R.K., Nellen, D., von Kessler, D.P., Bellotto, M., Beachy, P.A., Basler, K., 2001. Skinny hedgehog, an acyltransferase required for palmitoylation and activity of the hedgehog signal. Science 293, 2080–2084.

Chan, D.C., Laufer, E., Tabin, C., Leder, P., 1995. Polydactylous limbs in Strong's Luxoid mice result from ectopic polarizing activity. Development 121, 1971–1978.

Chen, H., Lun, Y., Ovchinnikov, D., Kokubo, H., Oberg, K.C., Pepicelli, C.V., Gan, L., Lee, B., Johnson, R.L., 1998. Limb and kidney defects in Lmx1b mutant mice suggest an involvement of LMX1B in human nail patella syndrome. Nat. Genet. 19, 51–55.

Chiang, C., Litingtung, Y., Harris, M.P., Simandl, B.K., Li, Y., Beachy, P.A., Fallon, J.F., 2001. Manifestation of the limb prepattern: limb development in the absence of sonic hedgehog function. Dev. Biol. 236, 421–435.

Clark, R.M., Marker, P.C., Kingsley, D.M., 2000. A novel candidate gene for mouse and human preaxial polydactyly with altered expression in limbs of Hemimelic extra-toes mutant mice. Genomics 67, 19–27.

Cohn, M.J., Izpisua-Belmonte, J.C., Abud, H., Heath, J.K., Tickle, C., 1995. Fibroblast growth factors induce additional limb development from the flank of chick embryos. Cell 80, 739–746.

Cohn, M.J., Patel, K., Krumlauf, R., Wilkinson, D.G., Clarke, J.D., Tickle, C., 1997. Hox9 genes and vertebrate limb specification. Nature 387, 97–101.

Cooper, K.L., Hu, J.K., ten Berge, D., Fernandez-Teran, M., Ros, M.A., Tabin, C.J., 2011. Initiation of proximal-distal patterning in the vertebrate limb by signals and growth. Science 332, 1083–1086.

Crossley, P.H., Minowada, G., MacArthur, C.A., Martin, G.R., 1996. Roles for FGF8 in the induction, initiation, and maintenance of chick limb development. Cell 84, 127–136.

Cunningham, T.J., Zhao, X., Sandell, L.L., Evans, S.M., Trainor, P.A., Duester, G., 2013. Antagonism between retinoic acid and fibroblast growth factor signaling during limb development. Cell. Rep. 3, 1503–1511.

Cygan, J.A., Johnson, R.L., McMahon, A.P., 1997. Novel regulatory interactions revealed by studies of murine limb pattern in Wnt-7a and En-1 mutants. Development 124, 5021–5032.

Davidson, D.R., Crawley, A., Hill, R.E., Tickle, C., 1991. Position-dependent expression of two related homeobox genes in developing vertebrate limbs. Nature 352, 429–431.

Detwiler, S.R., 1933. On the time of determination of the antero-posterior axis of the forelimb in Amblystoma. J. Exp. Zool. 64, 405–414.

Dhouailly, D., Kieny, M., 1972. The capacity of the flank somatic mesoderm of early bird embryos to participate in limb development. Dev. Biol. 28, 162–175.

Dudley, A.T., Ros, M.A., Tabin, C.J., 2002. A re-examination of proximodistal patterning during vertebrate limb development. Nature 418, 539–544.

Ericson, J., Morton, S., Kawakami, A., Roelink, H., Jessell, T.M., 1996. Two critical periods of Sonic Hedgehog signaling required for the specification of motor neuron identity. Cell 87, 661–673.

Francis, P.H., Richardson, M.K., Brickell, P.M., Tickle, C., 1994. Bone morphogenetic proteins and a signaling pathway that controls patterning in the developing chick limb. Development 120, 209–218.

Francis-West, P.H., Robertson, K.E., Ede, D.A., Rodriguez, C., Izpisua-Belmonte, J.C., Houston, B., Burt, D.W., Gribbin, C., Brickell, P.M., Tickle, C., 1995. Expression of genes encoding bone morphogenetic proteins and sonic hedgehog in talpid (ta3) limb buds: their relationships in the signaling cascade involved in limb patterning. Dev. Dyn. 203, 187–197.

Galceran, J., Farinas, I., Depew, M.J., Clevers, H., Grosschedl, R., 1999. Wnt3a$^{-/-}$-like phenotype and limb deficiency in Lef1$^{-/-}$cf1$^{-/-}$ mice. Genes. Dev. 13, 709–717.

Gao, B., Song, H., Bishop, K., Elliot, G., Garrett, L., English, M.A., Andre, P., Robinson, J., Sood, R., Minami, Y., Economides, A.N., Yang, Y., 2011. Wnt signaling gradients establish planar cell polarity by inducing Vangl2 phosphorylation through Ror2. Dev Cell 20 (2), 163–176.

Geduspan, J.S., MacCabe, J.A., 1989. Transfer of dorsoventral information from mesoderm to ectoderm at the onset of limb development. Anat. Rec. 224, 79–87.

Globus, M., Vethamany-Globus, S., 1976. An in vitro analogue of early chick limb bud outgrowth. Differentiation 6, 91–96.

Gros, J., Hu J.K., Vinegoni C., Feruglio, P.F., Weissleder, R., Tabin, C.J., 2010. WNT5A/JNK and FGF/MAPK pathways regulate the cellular events shaping the vertebrate limb bud. Curr. Biol. 20, 1993–2002.

Hamburger, V., 1938. Morphogenetic and axial self-differentiation of transplanted limb primordia of 2-day chick embryos. J. Exp. Zool. 77, 337–489.

Harfe, B.D., Scherz, P.J., Nissim, S., Tian, H., McMahon, A.P., Tabin, C.J., 2004. Evidence for an expansion-based temporal Shh gradient in specifying vertebrate digit identities. Cell 118, 517–528.

Izpisua-Belmonte, J.C., Tickle, C., Dolle, P., Wolpert, L., Duboule, D., 1991. Expression of the homeobox Hox-4 genes and the specification of position in chick wing development. Nature 350, 585–589.

Jabs, E.W., 2004. TWIST and Saethre-Chotzen Syndrome. Oxford, Oxford University Press. pp. 401–409.

Johnston, J.J., Olivos-Glander, I., Killoran, C., Elson, E., Turner, J.T., Peters, K.F., Abbott, M.H., Aughton, D.J., Aylsworth, A.S., Bamshad, M.J., et al., 2005. Molecular and clinical analyses of Greig cephalopolysyndactyly and Pallister-Hall syndromes: robust phenotype prediction from the type and position of GLI3 mutations. Am. J. Hum. Genet. 76, 609–622.

Jorquera, B., Pugin, E., 1971. Behavior of the mesoderm and ectoderm of the limb bud in the exchanges between chicken and rat. C. R. Acad. Sci. Hebd. Seances. Acad. Sci. D. 272, 1522–1525.

Kang, S., Graham Jr., J.M., Olney, A.H., Biesecker, L.G., 1997. GLI3 frameshift mutations cause autosomal dominant Pallister-Hall syndrome. Nat. Genet. 15, 266–268.

Khokha, M.K., Hsu, D., Brunet, L.J., Dionne, M.S., Harland, R.M., 2003. Gremlin is the BMP antagonist required for maintenance of Shh and Fgf signals during limb patterning. Nat. Genet. 34, 303–307.

Kieny, M., 1960. Inductive role of the mesoderm in the early differentiation of the limb bud in the chick embryo. J. Embryol. Exp. Morphol. 8, 457–467.

Kieny, M., 1968. Variation in the inductive capacity of mesoderm and the competence of ectoderm during primary induction in the chick embryo limb bud. Arch. Anat. Microsc. Morphol. Exp. 57, 401–418.

Kuhlman, J., Niswander, L., 1997. Limb deformity proteins: role in mesodermal induction of the apical ectodermal ridge. Development 124, 133–139.

Lallemand, Y., Nicola, M.A., Ramos, C., Bach, A., Cloment, C.S., Robert, B., 2005. Analysis of Msx1; Msx2 double mutants reveals multiple roles for Msx genes in limb development. Development 132, 3003–3014.

Lanctot, C., Moreau, A., Chamberland, M., Tremblay, M.L., Drouin, J., 1999. Hindlimb patterning and mandible development require the Ptx1 gene. Development 126, 1805–1810.

Laufer, E., Nelson, C.E., Johnson, R.L., Morgan, B.A., Tabin, C., 1994. Sonic hedgehog and Fgf-4 act through a signaling cascade and feedback loop to integrate growth and patterning of the developing limb bud. Cell 79, 993–1003.

Lettice, L.A., Horikoshi, T., Heaney, S.J., van Baren, M.J., van der Linde, H.C., Breedveld, G.J., Joosse, M., Akarsu, N., Oostra, B.A., Endo, N., et al., 2002. Disruption of a long-range cis-acting regulator for Shh causes preaxial polydactyly. Proc. Natl. Acad. Sci. USA 99, 7548–7553.

Lewandoski, M., Sun, X., Martin, G.R., 2000. Fgf8 signaling from the AER is essential for normal limb development. Nat. Genet. 26, 460–463.

Lewis, P.M., Dunn, M.P., McMahon, J.A., Logan, M., Martin, J.F., St-Jacques, B., McMahon, A.P., 2001. Cholesterol modification of sonic hedgehog is required for long-range signaling activity and effective modulation of signaling by Ptc1. Cell 105, 599–612.

Logan, M., Tabin, C.J., 1999. Role of Pitx1 upstream of Tbx4 in specification of hindlimb identity. Science 283, 1736–1739.

Loomis, C.A., Harris, E., Michaud, J., Wurst, W., Hanks, M., Joyner, A.L., 1996. The mouse Engrailed-1 gene and ventral limb patterning. Nature 382, 360–363.

Loomis, C.A., Kimmel, R.A., Tong, C.X., Michaud, J., Joyner, A.L., 1998. Analysis of the genetic pathway leading to formation of ectopic apical ectodermal ridges in mouse Engrailed-1 mutant limbs. Development 125, 1137–1148.

Lyons, K.M., Pelton, R.W., Hogan, B.L., 1990. Organogenesis and pattern formation in the mouse: RNA distribution patterns suggest a role for bone morphogenetic protein-2A (BMP-2A). Development 109, 833–844.

MacCabe, J.A., Parker, B.W., 1976. Polarizing activity in the developing limb of the Syrian hamster. J. Exp. Zool. 195, 311–317.

MacCabe, J.A., Saunders Jr., J.W., Pickett, M., 1973. The control of the anteroposterior and dorsoventral axes in embryonic chick limbs constructed of dissociated and reaggregated limb-bud mesoderm. Dev. Biol. 31, 323–335.

Mariani, F.V., Ahn, C.P., Martin, G.R., 2008. Genetic evidence that FGFs have an instructive role in limb proximal-distal patterning. Nature 453, 401–405.

Martin, P., Lewis, J., 1986. Normal development of the skeleton in chick limb buds devoid of dorsal ectoderm. Dev. Biol. 118, 233–246.

Mercader, N., Leonardo, E., Piedra, M.E., Martinez, A.C., Ros, M.A., Torres, M., 2000. Opposing RA and FGF signals control proximodistal vertebrate limb development through regulation of Meis genes. Development 127, 3961–3970.

Mo, R., Freer, A.M., Zinyk, D.L., Crackower, M.A., Michaud, J., Heng, H.H., Chik, K.W., Shi, X.M., Tsui, L.C., Cheng, S.H., et al., 1997. Specific and redundant functions of Gli2 and Gli3 zinc finger genes in skeletal patterning and development. Development 124, 113–123.

Naiche, L.A., Papaioannou, V.E., 2003. Loss of Tbx4 blocks hindlimb development and affects vascularization and fusion of the allantois. Development 130, 2681–2693.

Naski, M.C., Ornitz, D.M., 1998. FGF signaling in skeletal development. Front. Biosci. 3, d781–d794.

Niemann, S., Zhao, C., Pascu, F., Stahl, U., Aulepp, U., Niswander, L., Weber, J.L., Muller, U., 2004. Homozygous WNT3 mutation causes tetra-amelia in a large consanguineous family. Am. J. Hum. Genet. 74, 558–563.

Niswander, L., 2002. Interplay between the molecular signals that control vertebrate limb development. Int. J. Dev. Biol. 46, 877–881.

Niswander, L., Jeffrey, S., Martin, G.R., Tickle, C., 1994. A positive feedback loop coordinates growth and patterning in the vertebrate limb. Nature 371, 609–612.

Niswander, L., Tickle, C., Vogel, A., Booth, I., Martin, G.R., 1993. FGF-4 replaces the apical ectodermal ridge and directs outgrowth and patterning of the limb. Cell 75, 579–587.

Ornitz, D.M., Marie, P.J., 2002. FGF signaling pathways in endochondral and intramembranous bone development and human genetic disease. Genes. Dev. 16, 1446–1465.

Ovchinnikov, D.A., Selever, J., Wang, Y., Chen, Y.T., Mishina, Y., Martin, J.F., Behringer, R.R., 2006. BMP receptor type IA in limb bud mesenchyme regulates distal outgrowth and patterning. Dev. Biol. 295, 103–115.

Panman, L., Galli, A., Lagarde, N., Michos, O., Soete, G., Zuniga, A., Zeller, R., 2006. Differential regulation of gene expression in the digit forming area of the mouse limb bud by SHH and gremlin 1/FGF-mediated epithelial-mesenchymal signaling. Development 133, 3419–3428.

Park, H.L., Bai, C., Platt, K.A., Matise, M.P., Beeghly, A., Hui, C.C., Nakashima, M., Joyner, A.L., 2000. Mouse Gli1 mutants are viable but have defects in SHH signaling in combination with a Gli2 mutation. Development 127, 1593–1605.

Parr, B.A., McMahon, A.P., 1995. Dorsalizing signal Wnt-7a required for normal polarity of DV and AP axes of mouse limb. Nature 374, 350–353.

Parr, B.A., Shea, M.J., Vassileva, G., McMahon, A.P., 1993. Mouse Wnt genes exhibit discrete domains of expression in the early embryonic CNS and limb buds. Development 119, 247–261.

Pettigrew, A.L., Greenberg, F., Caskey, C.T., Ledbetter, D.H., 1991. Greig syndrome associated with an interstitial deletion of 7p: confirmation of the localization of Greig syndrome to 7p13. Hum. Genet. 87, 452–456.

Pizette, S., Abate-Shen, C., Niswander, L., 2001. BMP controls proximodistal outgrowth, via induction of the apical ectodermal ridge, and dorsoventral patterning in the vertebrate limb. Development 128, 4463–4474.

Porter, J.A., von Kessler, D.P., Ekker, S.C., Young, K.E., Lee, J.J., Moses, K., Beachy, P.A., 1995. The product of hedgehog autoproteolytic cleavage active in local and long-range signaling. Nature 374, 363–366.

Pressman, C.L., Chen, H., Johnson, R.L., 2000. LMX1B, a LIM homeodomain class transcription factor, is necessary for normal development of multiple tissues in the anterior segment of the murine eye. Genesis 26, 15–25.

Radhakrishna, U., Wild, A., Grzeschik, K.H., Antonarakis, S.E., 1997. Mutation in GLI3 in postaxial polydactyly type A. Nat. Genet. 17, 269–271.

Riddle, R.D., Ensini, M., Nelson, C., Tsuchida, T., Jessell, T.M., Tabin, C., 1995. Induction of the LIM homeobox gene Lmx1 by WNT7a establishes dorsoventral pattern in the vertebrate limb. Cell 83, 631–640.

Riddle, R.D., Johnson, R.L., Laufer, E., Tabin, C., 1993. Sonic hedgehog mediates the polarizing activity of the ZPA. Cell 75, 1401–1416.

Roessler, E., Du, Y.Z., Mullor, J.L., Casas, E., Allen, W.P., Gillessen-Kaesbach, G., Roeder, E.R., Ming, J.E., Ruiz i Altaba, A., Muenke, M., 2003. Loss-of-function mutations in the human GLI2 gene are associated with pituitary anomalies and holoprosencephaly-like features. Proc. Natl. Acad. Sci. USA 100, 13424–13429.

Roselló, Díez, Ros, M.A., Torres, M., 2011. Diffusible Signals, Not Autonomous Mechanisms, Determine the Main Proximodistal Limb Subdivision. Science 27, 1086–1088.

Sato, K., Koizumi, Y., Takahashi, M., Kuroiwa, A., Tamura, K., 2007. Specification of cell fate along the proximal-distal axis in the developing chick limb bud. Development 134, 1397–1406.

Saunders, J.W.J., 1948. The proximo-distal sequence of origin of the parts of the chick wing and the role of the ectoderm. J. Exp. Zool. 108, 363–403.

Saunders, J.W.J., Gasseling, M.T., 1968. Ectoderm-mesenchymal interaction in the origin of wing symmetry. In: Fleischmajer, R., Billingham, R.E. (Eds.), Epithelia-Mesenchymal Interactions. Williams and Wilkins, Baltimore, pp. 78–97.

Scherz, P.J., Harfe, B.D., McMahon, A.P., Tabin, C.J., 2004. The limb bud Shh-Fgf feedback loop is terminated by expansion of former ZPA cells. Science 305, 396–399.

Sekine, K., Ohuchi, H., Fujiwara, M., Yamasaki, M., Yoshizawa, T., Sato, T., Yagishita, N., Matsui, D., Koga, Y., Itoh, N., et al., 1999. Fgf10 is essential for limb and lung formation. Nat. Genet. 21, 138–141.

Sharpe, J., Lettice, L., Hecksher-Sorensen, J., Fox, M., Hill, R., Krumlauf, R., 1999. Identification of sonic hedgehog as a candidate gene responsible for the polydactylous mouse mutant Sasquatch. Curr. Biol. 9, 97–100.

Summerbell, D., 1974. A quantitative analysis of the effect of excision of the AER from the chick limb-bud. J. Embryol. Exp. Morphol. 32, 651–660.

Summerbell, D., 1979. The zone of polarizing activity: evidence for a role in normal chick limb morphogenesis. J. Embryol. Exp. Morphol. 50, 217–233.

Summerbell, D., Lewis, J.H., 1975. Time, place and positional value in the chick limb-bud. J. Embryol. Exp. Morphol. 33, 621–643.

Szeto, D.P., Rodriguez-Esteban, C., Ryan, A.K., O'Connell, S.M., Liu, F., Kioussi, C., Gleiberman, A.S., Izpisua-Belmonte, J.C., Rosenfeld, M.G., 1999. Role of the Bicoid-related homeodomain factor Pitx1 in specifying hindlimb morphogenesis and pituitary development. Genes. Dev. 13, 484–494.

Takeuchi, J.K., Koshiba-Takeuchi, K., Suzuki, T., Kamimura, M., Ogura, K., Ogura, T., 2003. Tbx5 and Tbx4 trigger limb initiation through activation of the Wnt/Fgf signaling cascade. Development 130, 2729–2739.

te Welscher, P., Fernandez-Teran, M., Ros, M.A., Zeller, R., 2002a. Mutual genetic antagonism involving GLI3 and dHAND prepatterns the vertebrate limb bud mesenchyme prior to SHH signaling. Genes. Dev. 16, 421–426.

te Welscher, P., Zuniga, A., Kuijper, S., Drenth, T., Goedemans, H.J., Meijlink, F., Zeller, R., 2002b. Progression of vertebrate limb development through SHH-mediated counteraction of GLI3. Science 298, 827–830.

Ten Berge, D., Brugmann, S.A., Helms, J.A., Nusse, R., 2008. Wnt and FGF signals interact to coordinate growth with cell fate specification during limb development. Development 135, 3247–3257.

Tickle, C., 2003. Patterning systems – from one end of the limb to the other. Dev. Cell. 4, 449–458.

Tickle, C., Summerbell, D., Wolpert, L., 1975. Positional signaling and specification of digits in chick limb morphogenesis. Nature 254, 199–202.

Tzchori, I., Day, T.F., Carolan, P.J., Zhao, Y., Wassif, C.A., Li, L., Lewandoski, M., Gorivodsky, M., Love, P.E., Porter, F.D., et al., 2009. LIM homeobox transcription factors integrate signaling events that control three-dimensional limb patterning and growth. Development 136, 1375–1385.

Verheyden, J.M., Sun, X., 2008. An Fgf/Gremlin inhibitory feedback loop triggers termination of limb bud outgrowth. Nature 454, 638–641.

Vortkamp, A., Gessler, M., Grzeschik, K.H., 1991. GLI3 zinc-finger gene interrupted by translocations in Greig syndrome families. Nature 352, 539–540.

Wang, B., Fallon, J.F., Beachy, P.A., 2000. Hedgehog-regulated processing of Gli3 produces an anterior/posterior repressor gradient in the developing vertebrate limb. Cell 100, 423–434.

Woods, C.G., Stricker, S., Seemann, P., Stern, R., Cox, J., Sherridan, E., Roberts, E., Springell, K., Scott, S., Karbani, G., et al., 2006. Mutations in WNT7A cause a range of limb malformations, including Fuhrmann syndrome and Al-Awadi/Raas-Rothschild/Schinzel phocomelia syndrome. Am. J. Hum. Genet. 79, 402–408.

Wyngaarden, L.A., Vogeli, K.M., Ciruna, B.G., Wells, M., Hadjantonakis, A.K., Hopyan, S., 2010. Oriented cell motility and division underlie early limb bud morphogenesis. Development 137, 2551–2558.

Yang, Y., Drossopoulou, G., Chuang, P.T., Duprez, D., Marti, E., Bumcrot, D., Vargesson, N., Clarke, J., Niswander, L., McMahon, A., et al., 1997. Relationship between dose, distance and time in Sonic Hedgehog-mediated regulation of anteroposterior polarity in the chick limb. Development 124, 4393–4404.

Yang, Y., Niswander, L., 1995. Interaction between the signaling molecules WNT7a and SHH during vertebrate limb development: dorsal signals regulate anteroposterior patterning. Cell 80, 939–947.

Yu, K., Herr, A.B., Waksman, G., Ornitz, D.M., 2000. Loss of fibroblast growth factor receptor 2 ligand-binding specificity in Apert syndrome. Proc. Natl. Acad. Sci. USA 97, 14536–14541.

Zhu, J., Nakamura, E., Nguyen, M.T., Bao, X., Akiyama, H., Mackem, S., 2008. Uncoupling Sonic hedgehog control of pattern and expansion of the developing limb bud. Dev. Cell. 14, 624–632.

Zuniga, A., Haramis, A.P., McMahon, A.P., Zeller, R., 1999. Signal relay by BMP antagonism controls the SHH/FGF4 feedback loop in vertebrate limb buds. Nature 401, 598–602.

Zwilling, E., 1955. Ectoderm – mesoderm relationship in the development of the chick embryo limb bud. J. Exp. Zool. 128, 423–441.

Zwilling, E., 1956. Interaction between limb bud ectoderm and mesoderm in the chick embryo. I. Axis establishment. J. Exp. Zool. 132, 157–171.

Zwilling, E., 1972. Limb morphogenesis. Dev. Biol. 28, 12–17.

Chapter 30

Patterning the Embryonic Endoderm into Presumptive Organ Domains

Anna M. Method and James M. Wells

Division of Developmental Biology, Cincinnati Children's Hospital Research Foundation and University of Cincinnati College of Medicine, Cincinnati, OH

Chapter Outline

GLOSSARY

Cloaca A transient embryonic cavity in mammals, lined with endoderm, which contributes to the epithelial lining of the urogenital and anorectal systems.

Endoderm The inner layer of cells in embryonic development that contributes to digestive and respiratory organs. Note that early in the development of birds and mammals, the definitive endoderm is initially part of the outermost layer of cells.

Extrahepatic biliary system A ductal network for the movement of bile that aids in digestion. It includes the common duct, extrahepatic bile ducts, and the gallbladder.

Foregut Anterior-most of three divisions of the digestive tract (foregut, midgut, hindgut).

Hindgut Posterior-most region of the digestive tract.

Liver The largest gland in the vertebrate body which has a role in digestion, glucose regulation and storage, blood clotting and removal of wastes from the blood.

Lung Organ containing sac-like structures where blood and air exchange oxygen and carbon dioxide.

Organogenesis Development of organs during embryonic development.

Pancreas A gland in the abdominal cavity with both exocrine and endocrine function that secretes digestive enzymes into the duodenum and also secretes the hormones insulin and glucagon into the blood.

Parathyroid Calcium homeostatic organ that secretes parathyroid hormone.

Patterning The act of subdividing embryos or tissues into distinct domains along the embryonic axes.

Primordia Cells, tissues or organs at the earliest stage of development.

Principles of Developmental Genetics.

Thyroid Endocrine organ with the function of regulating metabolism, growth and development.

Ultimobranchial bodies Small glands that develop separately from the thyroid that will fuse with the thyroid, and form the parafollicular cells of the thyroid.

SUMMARY

- The definitive endoderm germ layer in the late gastrula embryo (E7.5 in the mouse, HH4–5 in the chick) is a single layer of cells that are unspecified. Within two days of development (E9.5 mouse, HH18 chick), the definitive endoderm has formed a primitive gut tube with budding organ primordia.
- The late gastrula definitive endoderm is regionalized into axial and lateral domains along the anterior-posterior axis that give rise to the foregut, midgut and hindgut.
- By the early somite stage (E8.5 mouse, HH10 chick) the developing gut tube has become highly patterned at the molecular level. Genes are expressed in overlapping and distinct domains that indicate where organ buds will form.
- Several signaling pathways are involved in these early patterning mechanisms, including the fibroblast growth factors (FGFs), hedgehog, bone morphogenetic proteins (BMPs), retinoic acid (RA) and Wnt pathways.
- Disruption of these early patterning mechanisms can directly lead to defects in the development of endodermic organs including the thymus, parathyroid, lungs, liver, pancreas, and intestines.
- Genomics strategies are currently being used to identify molecular pathways involved in endoderm organogenesis.
- Information from endoderm organ development is being used to differentiate human embryonic stem cells into endoderm organ cell types, ultimately to treat degenerative diseases such as diabetes.

30.1 INTRODUCTION

The process of gastrulation subdivides the embryonic epiblast into the three primary germ layers: the ectoderm, mesoderm and endoderm. During organogenesis, the endoderm germ layer contributes to many vital organs including the liver, pancreas, thymus, and thyroid, and also to the epithelial lining of the gastrointestinal, genitourinary, and respiratory tracts (for review see Wells and Melton, 1999; Zorn and Wells, 2007; 2009, and chapters in this book). The endoderm cells of the post-gastrulation vertebrate embryo are largely unspecified and become regionalized along the anterior-posterior (A-P) and dorsal-ventral (D-V) axes in the process of forming a primitive gut tube. The process through which endoderm cells obtain positional identity along the A-P and D-V axes is called patterning, and is fundamental for directing endoderm cells into specific organ lineages. In addition, the endoderm itself acts as a source of signals that regulate the proper development of mesoderm and ectoderm organs, such as the anterior central nervous system (CNS) and heart. Therefore, failure to appropriately pattern the endoderm can considerably impact the developmental outcome of the entire embryo.

Several animal model systems have contributed greatly to our understanding of early endoderm organogenesis, including *Xenopus*, zebrafish, chicken, quail, and mouse (Zorn and Wells, 2009). More recently, human pluripotent stem cell derived endoderm cultures have been included in this list of models of endoderm development (Howell and Wells, 2011; Mayhew and Wells, 2009; Spence and Wells, 2007). It is remarkable that many aspects of early endoderm organogenesis are conserved across vertebrate species. For example, within a few days after gastrulation in amphibians, fish, birds and mammals, unspecified endoderm cells form a primitive gut tube with distinct A-P and D-V domains that predict the formation of organ primordia. Furthermore, there is increasing evidence that signals and the corresponding target genes that direct the formation of a patterned gut tube are also largely conserved across vertebrate species (Zorn and Wells, 2009). An example of this is the fact that embryonic pathways that are involved in endoderm formation and patterning have been used to direct the differentiation of human pluripotent stem cells into intestinal tissue *in vitro* (Spence et al., 2011b). In this chapter, we will discuss the morphogenetic and molecular processes that are involved in subdividing the endoderm into functional presumptive organ domains along the A-P axis, with a focus on studies in chick and mouse and the emerging use of human pluripotent stem cells (both embryonic and induced pluripotent cells) to model endoderm organ development in humans. These early stages of endoderm organogenesis are a key step for the proper development of the gastrointestinal, genitourinary, and respiratory tracts and the ontogeny of its associated vital organs (Figure 30.1).

Formation and patterning of the primitive gut tube is a dynamic morphogenetic process involving cell migration and tissue remodeling (Spence et al., 2011a). For simplicity we have subdivided the gut tube into the foregut, the midgut and the hindgut (Figure 30.1). The anterior segment of the gut tube, the foregut, forms first and comes from the endoderm extending from the most anterior portion of the embryo (headfold in the post-gastrula embryo, E7.5–8 in mouse) to just below the cardiac mesoderm. The foregut endoderm contributes to a remarkable number and diversity of organs, including the taste buds, inner ear, thyroid, thymus, esophagus, trachea, lungs, liver, extrahepatic biliary system, and pancreas. The hindgut

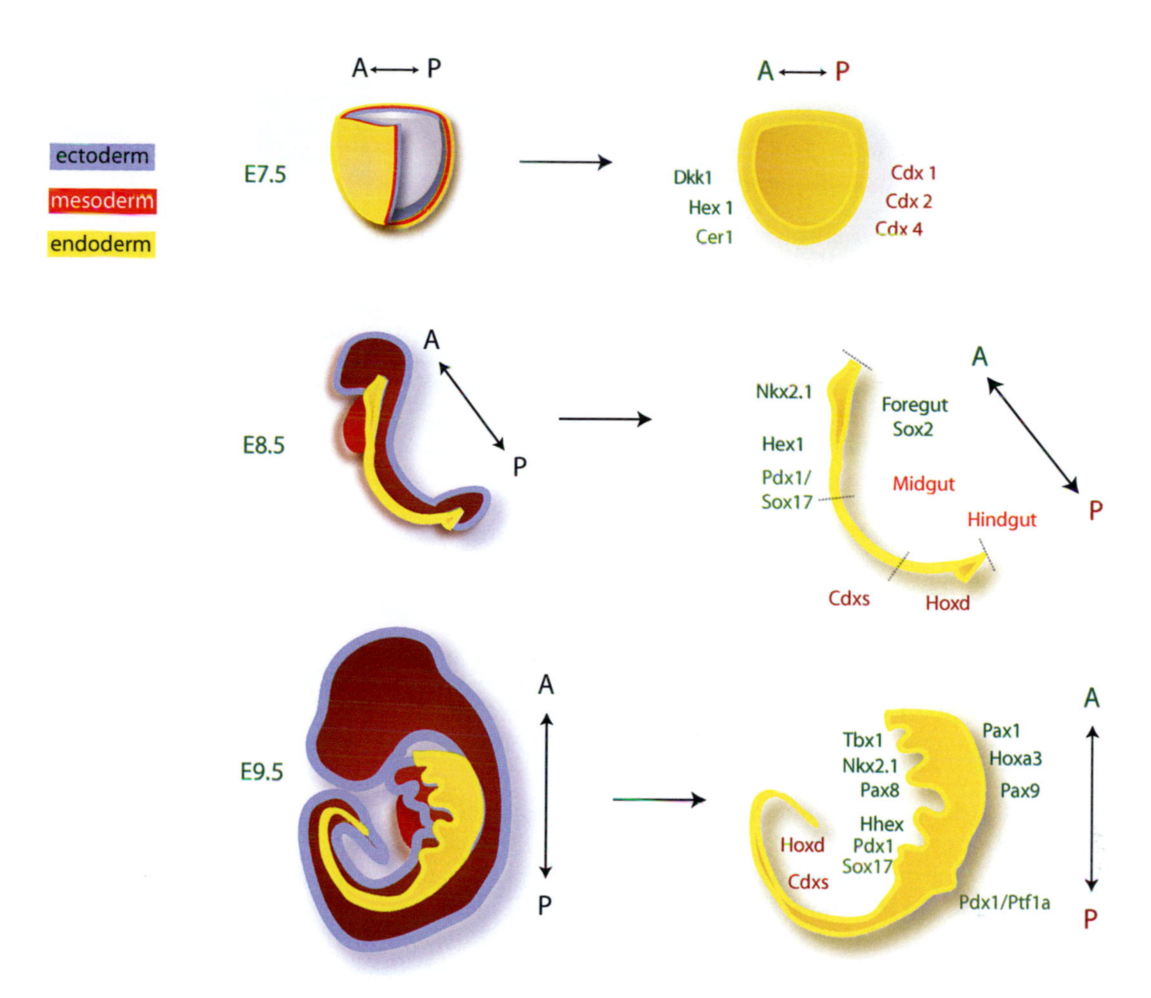

FIGURE 30.1 Early Stages of Endoderm Organogenesis in the Mouse. The left panels show a schematic of mouse embryos and the right panels highlight the endoderm and genes that are regionally expressed along the anterior to posterior (A-P) and dorsal to ventral (D-V) axes. The right panels highlight the endoderm layer and highlights several genes that are expressed in restricted domains along the A-P and D-V axes. After gastrulation, which is complete 7.5 days after fertilization (E7.5) in the mouse, the endoderm is the outermost layer of cells (yellow) surrounding the embryo. The middle layer is the mesoderm (red) and inner layer is the ectoderm (blue). After one day of embryonic development (E8.5), morphogenetic movements at the anterior and posterior initiate tube formation and start the process of internalizing the endoderm. These processes result in the formation of a primitive gut tube with a foregut and hindgut tube. The middle section of the primitive gut tube (midgut) remains open at this stage. Between E9.5 and E10.5 formation and internalization of the primitive gut tube is complete and organ primordia become morphologically apparent and express specific genes; lungs (Nkx2.1), liver (Hhex), thyroid (Hhex, Pax8), gallbladder (Sox17), and pancreas (Pdx1, Ptf1a).

forms shortly after the foregut and derives from the most posterior region of the gastrula stage endoderm overlying the primitive streak. The hindgut contributes to the small and large intestine as well as the bladder and urogenital tract. For the purposes of this chapter we will refer to the open portion of the gut tube at these early stages of gut tube morphogenesis as the midgut. This is the last region of the endoderm to form a tube.

30.2 FATE MAP OF THE EMBRYONIC ENDODERM

The endoderm organ primordia that evaginate from the primitive gut tube become morphologically apparent within two days of gut tube formation (E10.5 in the mouse). To better understand the early stages of endoderm organogenesis, we will discuss in this section the embryonic origins of the different A-P and D-V domains of the gut tube. Numerous studies have begun to identify the cellular origins of the endoderm organ domains by utilizing a method called lineage tracing. Also known as fate mapping, lineage tracing is a method in which cells are labeled and followed from an early progenitor stage to a more mature stage of development (Figure 30.2). Thus, an endoderm cell labeled at the late gastrula stage can be traced to specific organ domains as embryonic development progresses. Cell lineage analyses performed in the mouse, chick and frog have mapped the movement of gastrulation stage endoderm cells to broad domains of the developing gut tube and

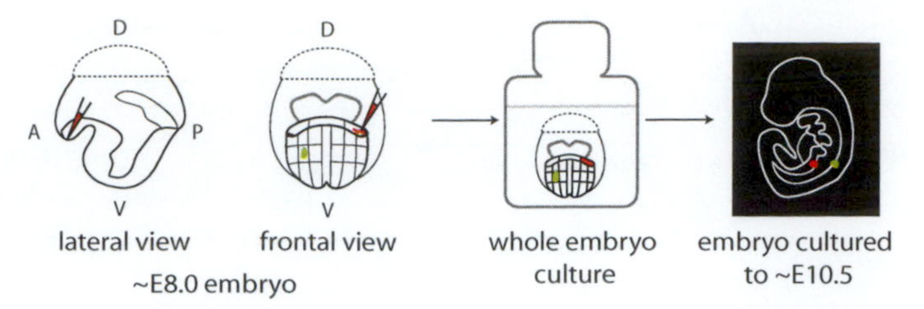

FIGURE 30.2 **Labeling Endoderm Cells for Lineage Analysis and Generating Fate Maps.** Cells can be labeled by injecting vital dyes (red – DiI) or by expressing fluorescent proteins (green). Labeled endoderm cells can be tracked through subsequent stages of development. The left two panels show a lateral and frontal view of an early somite stage embryo (~3 somites, E8 mouse). Embryos can be cultured *in vitro* and the positions of the labeled cells can be determined after 1–3 days of development. This approach has been used to identify which regions of the early endoderm give rise to specific gut tube and organ domains and generate fate maps.

early organ primordia (Chalmers and Slack, 2000; Kimura et al., 2006; Lawson and Schoenwolf, 2003; Lawson et al., 1986; Rosenquist, 1971; Tam et al., 2004). These studies suggest that many aspects of early endoderm patterning and gut tube formation are remarkably conserved across vertebrate species. Below we will compare the fate mapping studies performed in two model organisms, the chick and the mouse (Figure 30.2).

Mapping Definitive Endoderm Cell Lineage During Gut Tube Morphogenesis: From Gastrula Stages to the Early Somite Stages in Chick and Mouse

Definitive Endoderm Fate Maps at the Gastrula Stages

In the mouse and chick, definitive endoderm cells form during gastrulation and migrate out to integrate into the outer layer of cells, thus forming the endoderm germ layer (see also Chapter 13). Gastrulation in the mouse occurs between embryonic days 6 and 7.5 (E6.5–7.5, also known as days post coitum or dpc), and in the chick this occurs between 7 and 19 hours of development (Hamburger and Hamilton [HH] stages 2–5). Cell lineage analyses of early to mid-gastrula embryos of both species suggest that the timing of the exit of the definitive endoderm cells from the primitive streak influences where they end up along the A-P axis at the end of gastrulation. For example, in the early mouse gastrula (E6–6.5), the first endoderm cells to leave the primitive streak preferentially end up in the anterior endoderm overlying the anterior neural plate and trunk at the end of gastrulation (E7.5) (Lawson and Pedersen, 1987). At the mid-late gastrula stages (E7–7.5), definitive endoderm cells exiting the primitive streak often colonize more posterior domains and remain adjacent to the node and primitive streak. These fate mapping studies suggest that definitive endoderm cells migrate through the primitive streak, however genetic cell lineage experiments, combined with live imaging, suggest that definitive endoderm cells can delaminate from the epiblast and directly intercalate into the overlying visceral endoderm (Kwon et al., 2008).

Studies in the chick have added to our understanding about when different domains of definitive endoderm are formed, and have raised some interesting questions (Kimura et al., 2006; Lawson and Schoenwolf, 2003). One experiment that traced the lineage of axial/midline definitive endoderm cells reached a similar conclusion to parallel studies in the mouse; namely that the anterior endoderm forms first, followed by more posterior trunk endoderm (Lawson and Schoenwolf, 2003). However another study suggested that posterior/lateral endoderm is the first to incorporate into the hypoblast layer at the early gastrula stage (Kimura et al., 2006). These studies suggest that the midline and lateral endoderm arise from different populations of endoderm precursors, possibly consistent with recent studies in mice (Kwon et al., 2008). Another interesting finding from this study was that a population of endoderm cells exit the epiblast during gastrulation and migrate for several hours before inserting into the endoderm germ layer at the late gastrula stage (Kimura et al., 2006).

Mapping Late Gastrula Stage Definitive Endoderm to the Developing Gut Tube

The studies described above indicate that the late gastrula definitive endoderm (E7.5 mouse, HH 4–5 chick) might have distinct A-P and lateral domains. This section will describe how these domains within a two-dimensional sheet of endoderm will give rise to the A-P and D-V domains of a three-dimensional gut tube (Wells and Melton, 1999). Figure 30.3 compares the chick and mouse endoderm fate map and illustrates the relationship between domains of the gastrula stage definitive endoderm (E7.5 mouse, HH4–5 chick), the developing gut tube at early somite stages (E8.5 mouse, HH10 chick), and the gut tube at a stage when organ primordia start to become morphologically apparent (E9.5 mouse, HH 17 chick). Figure 30.3 is a composite of the results from several cell lineage studies that have begun to reveal the complex

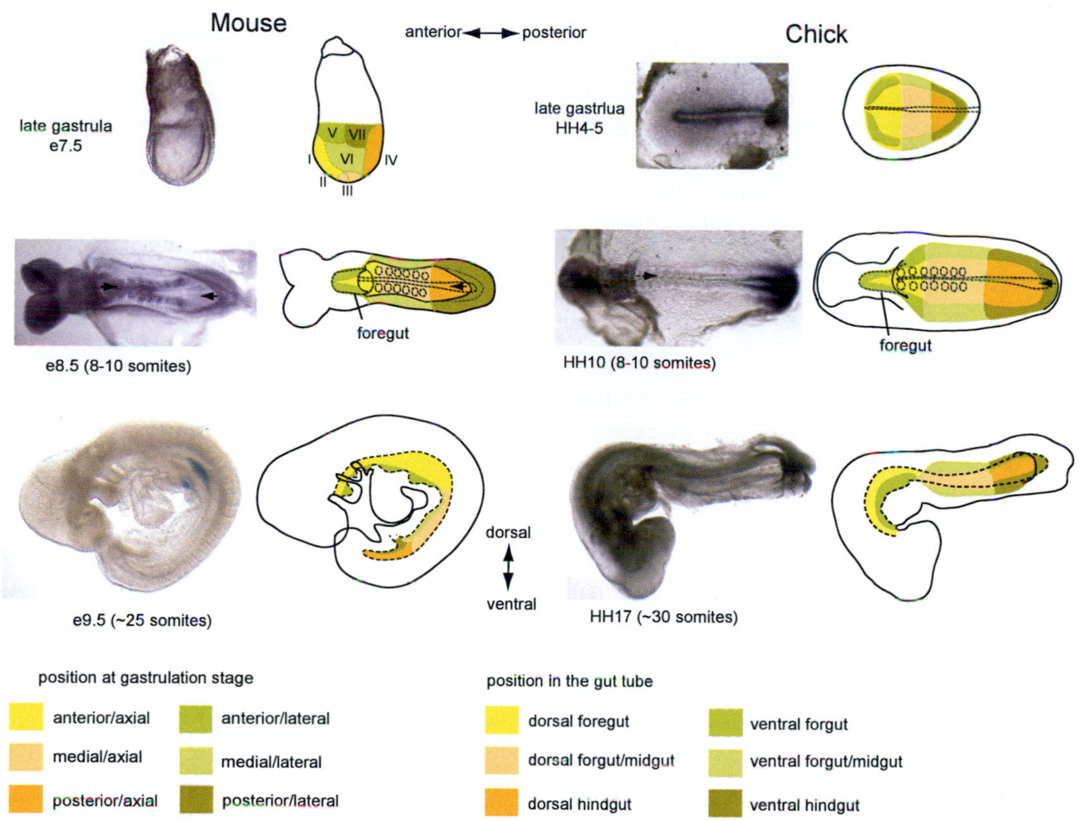

FIGURE 30.3 Fate Map of the Endoderm and Gut Tube in Mouse and Chick. The top panels: late gastrula stage mouse (lateral view) and chick (ventral view, endoderm facing up) embryos. The dotted line in the chick shows the developing notochord. The roman numerals (I through VII) indicate the domains identified in the mouse fate map studies (Lawson et al., 1986; Tam et al., 2004). The middle panels: ventral view of 8–10 somite stage embryos (E8.5 mouse, HH 10 chick) in which the foregut has started to form and be internalized. The arrows indicate the folding/migration direction of the foregut and hindgut. The dotted circles in the middle panels are somites, the dotted line along the midline is the developing notochord. The dotted line in the anterior outlines the developing foregut. The bottom panels: lateral view of mouse (E9.5, blue staining cells are *Pdx1-LacZ* expression) and chick (HH17) embryos at ~ 30 somite stage. At this stage the endoderm has been internalized and folded into a gut tube with the exception of the hindgut in chick, which has not fully formed at this stage. The colored domains of the gastrula stage embryos roughly correspond with the equivalent colored domains of the later stage embryos. Axial endoderm of the gastrula embryo (domains I, II, III and IV) contributes to both the dorsal and ventral gut tube, although only domains I and IV contribute ventrally. The lateral domains (V, VI, and VII) fold over to principally contribute to the ventral gut tube. For all images, anterior is left and posterior is right. In mouse, the fate map of the foregut between E8.5–E10 has been well studied (Tremblay and Zaret, 2005). The posterior domains at these stages have not been well studied and the domains in this figure are extrapolated from the earlier stage fate map. The *Pdx1-LacZ* mice were from Chris Wright, Vanderbilt University.

morphogenetic processes involved in forming the gut tube (Kimura et al., 2006; Lawson et al., 1986; Rosenquist, 1971; Tam et al., 2004; Tremblay and Zaret, 2005).

In mouse, cell lineage experiments performed over 25 years ago, together with more recent studies have resulted in a detailed fate map of the late gastrula stage (E7.5) mouse endoderm (Lawson et al., 1986; Tam et al., 2004). These studies have identified regions of endoderm at E7.5 that will contribute to distinct A-P and D-V domains of the developing gut tube (Figure 30.3). One recent study distinguishes between endoderm cells that are located along the midline versus those that are located more laterally. Midline endoderm cells tend to remain along the midline in both ventral and dorsal regions of the gut tube. In contrast, endoderm cells that are found laterally in the E7.5 endoderm preferentially end up in the ventral portions of the gut. These fate mapping studies concluded that the most anterior midline endoderm cells overlying the neural plate (Figure 30.3, region I), as well as anterior lateral endoderm (region V) both contribute to the ventral foregut which gives rise to the liver, ventral pancreas, and lungs (Lawson et al., 1986). Axial endoderm cells just anterior to the node (region II) contribute to the dorsal foregut, which forms the dorsal component of the stomach, pancreas, and duodenum, whereas lateral endoderm from region VI contributes to ventral midgut. Axial endoderm overlying the node and anterior primitive streak (region III) contributes to the dorsal midgut and hindgut eventually contributing to the small intestines. The most posterior endoderm overlying the primitive streak (Figure 30.3, regions IV and VII) contributes to both dorsal and ventral portions of the hindgut, which contribute to posterior intestinal derivatives and the urogenital tract.

Again, studies suggest that the chick fate map roughly corresponds with that of the mouse (Kimura et al., 2006; Lawson and Schoenwolf, 2003; Rosenquist, 1971): the anterior axial and lateral domains of stage HH5 endoderm contribute to the ventral foregut at HH10, the endoderm overlying the middle primitive streak contributes to dorsal mid and hindgut, and the most posterior endoderm over the primitive streak contributes to the ventral portions of the mid and hindgut. The combined evidence suggests that presumptive A-P and D-V regions of the gut tube are established as early as the gastrula stage, prior gut tube morphogenesis (Figure 30.3). Furthermore, the lineage tracing experiments demonstrate that gut tube morphogenesis involves the folding and migration of anterior/lateral endoderm to make the foregut and posterior/lateral endoderm to make the hindgut. The folding/migration of the foregut continues toward the posterior, and similar expansion of the hindgut continues toward the anterior, resulting in the closure of the primitive gut tube.

Regions of the Foregut that give Rise to Developing Organ Primordia

Less is known about domains of the early gut tube that give rise to developing the mid-gestation stage organ primordia. However, a study describing lineage tracing experiments performed on early somite stage mouse embryos identified domains in the developing foregut that contribute to specific foregut derivatives, including the liver and pancreas (Angelo et al., 2012; Tremblay and Zaret, 2005). In these studies, foregut endoderm cells were labeled at E8–8.5 (1–7 somites), and their descendants were analyzed at the early liver bud stage (E9.5). In the liver, one medial and two lateral domains of foregut endoderm were found to contribute to the liver bud. In addition, cells of the medial domain at the lip of the developing foregut gave rise to descendants all along the ventral midline, whereas the lateral domains contribute to specific ventral regions such as the liver. This study supports the idea that medial and lateral endoderm cells continue to migrate independently, then later converge during gut tube morphogenesis during formation of the liver and other ventral foregut derivatives such as the pancreas. The dorsal pancreas derives from dorsal endoderm adjacent to somites 2–4.

These experiments, as well as studies in other organisms including *Xenopus* and zebrafish, highlight the complex morphogenetic processes that occur during the formation and patterning of the primitive gut tube. The sections below discuss specific signaling pathways and target genes that functionally regulate these early stages of endoderm organogenesis.

30.3 GENE EXPRESSION DOMAINS PREDICT AND DETERMINE ENDODERM ORGAN PRIMORDIA

Genes have been identified that are expressed in discrete A-P and D-V domains during the development of the gut tube. These markers are invaluable in understanding how the endoderm is patterned at the molecular level. Many of these marker genes both predict where the endoderm organ primordia will form, and play a functional role in establishing organ boundaries (Figure 30.4). In this section we will discuss molecular studies that describe how patterns are established in the early endoderm and refined in the primitive gut tube, as well as the role of several of these genes during endoderm organogenesis.

At the end of gastrulation in *Xenopus*, the chick and the mouse, most reported gene expression patterns exist in broad overlapping domains along the A-P axis. For example, in all three species *Hhex (Hematopoietically expressed homeobox1)*, *Otx2* (*orthodenticle homolog 2*), *Dkk1* (*dickkopf-1*) and *Cer1 (Cerberus-1)* are expressed along the anterior half of the embryo, whereas the *Caudal type homeobox* genes *Cdx1*, *2*, and *4* (*CdxA* and *C* in chick) are broadly expressed in the posterior (Figure 30.1) (Ang et al., 1993; Chapman et al., 2002; Frumkin et al., 1993; Gamer and Wright, 1993; Jones et al., 1999; Northrop and Kimelman, 1994). The expression domains of these factors, as well as new markers identified through expression screens (Moore-Scott et al., 2007), suggest that the endoderm is broadly patterned along the A-P axis at this stage. Moreover, gene targeting experiments have demonstrated that *Hhex* is vital for anterior patterning as well as subsequent thyroid and liver formation (Martinez Barbera et al., 2000), and loss of *Cdx2* function results in posterior truncations around the time of gut tube formation (Chawengsaksophak et al., 2004). *Foxa2* is functionally important for the formation of the anterior primitive streak and foregut endoderm (Dufort et al., 1998). An endoderm-specific knockout of *Foxa2* and a closely related factor *Foxa1* results in failure to initiate liver development (Lee et al., 2005). In addition, *FoxA2* has been shown to directly activate *Gata4* to establish a *Gata4* expressing pool of progenitors that then contribute to the entire definitive endoderm (Rojas et al., 2010). Although *FoxA2* is also expressed throughout the midgut and hindgut, it is not required for posterior gut development (McKnight et al., 2010).

By the time of gut tube formation (E8.5 – mouse; HH10 – chick), a complex pattern of gene expression has emerged in the foregut (*Nkx2.1, Hhex, Sox2*), the posterior foregut (albumin) and midgut (*Pdx1*), and in the hindgut (*Cdx2, Cdx4; Cdx A* and *B* in chick) (Figures 30.1 and 30.6) (Chalmers et al., 2000; Gamer and Wright, 1993; Gittes and Rutter, 1992; Ohlsson et al., 1993; Zeng et al., 1998). The extensive patterning of the primitive gut tube suggests a high degree of gene regulation. Two well-studied examples of highly regulated genes are *Pdx1* and *Cdx*, which are part of the parahox cluster that arise

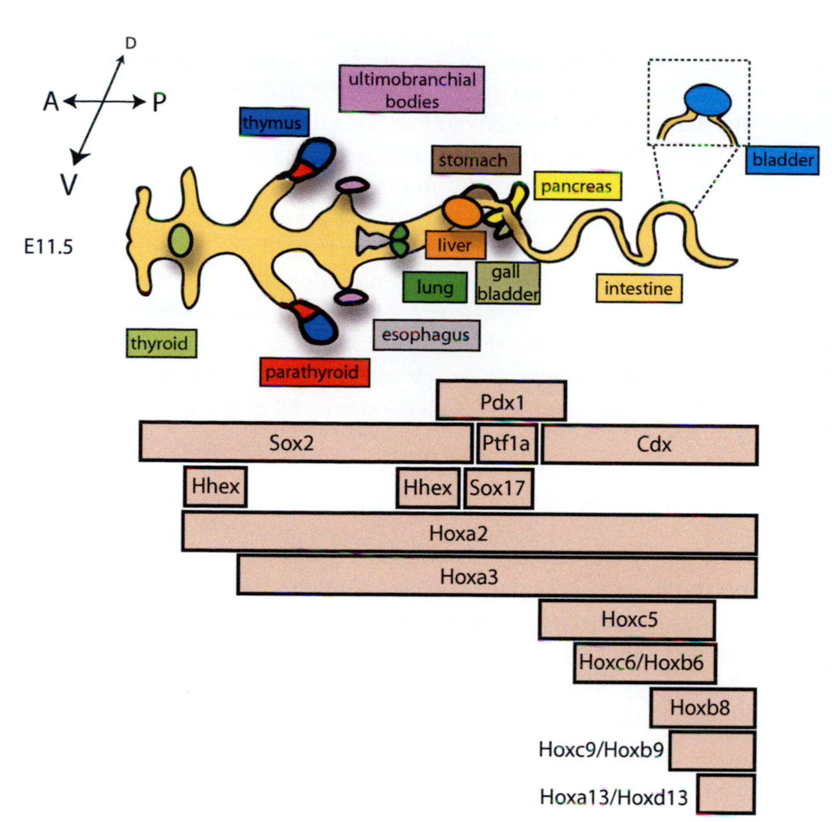

FIGURE 30.4 Overlapping expression domains of transcription factors predict the emergence of organ primordia along the A-P axis of the fetal gut tube. The upper schematic shows a ventral view of a fetal gut tube around the time of organ bud formation (E10.5–11 mouse). Anterior is left and posterior is right. The lower panel indicates the relative A-P expression boundaries of several transcription factors at E10.5 in mouse. *Hhex* and *Cdx* factors (*Cdx1,2,4* in mouse, *CdxA* and *B* in chick) are expressed in the gastrula stage endoderm (E7.5 in mouse) where as other genes such as *Pdx1* are first expressed later in the gut tube (E8.5 in mouse). The anterior and posterior expression limits of some of these factors is important in establishing organ domains. *The lower panel was adapted from (Grapin-Botton, 2005).*

from an ancestral protohox cluster. The parahox cluster exhibits conserved spatial expression along the A-P axis in divergent animal phyla. Here, *Cdx* expression defines the posterior (small and large intestine) and *Pdx1* expression defines the domain anterior to *Cdx* (duodenum, pancreas and caudal stomach). These and other transcription factors, including *Sox2* and *Hox* genes, may play a role in establishing each other's expression domains (Figure 30.4). For example in the chick, misexpression of *Pdx1* in the posterior results in repression of the *Cdx* genes (Grapin-Botton et al., 2001).

By E9.5 the gut tube has expanded anteriorly and posteriorly. At this stage the pattern in the anterior foregut becomes further refined, as indicated by the restricted expression of several genes (Figure 30.1). *Nkx2.1* is expressed in two domains marking the presumptive thyroid and lung. *Pax1, Pax9, Tbx1, Hoxa3* and *Pax8* are expressed in presumptive pharyngeal domains. By embryonic day 10.5 in the mouse, organ primordia for the thyroid, lung, pancreas (ventral and dorsal), gallbladder and the liver have begun to bud from these patterned domains, and the cloacal cavity, which contributes to anorectal and urogenital epithelium, is defined. Subsequent stages of endoderm organ development will be discussed in more detail in other chapters of this book.

A-P Patterning by Hox Genes

One mechanism by which parahox transcription factors are believed to control patterning is through regulation of the Hox genes. Hox genes are key transcriptional regulators of embryonic patterning in vertebrates, and are well known for their role in patterning the mesoderm and ectoderm. Hox genes are also expressed in defined domains along the A-P axis of the gut (Figure 30.4), and emerging evidence suggests their involvement in the development of the primitive gut and its derivatives (Grapin-Botton, 2005). In the mouse, the Hox cluster consists of 39 genes in four linkage groups, designated A, B, C and D, with 13 paralogous families. In both the mouse and the chick, the expression domains of Hox genes have been identified along the A-P axis of the developing fetal gut (Grapin-Botton, 2005). In the pharyngeal region, *Hoxa2, 3* and *4* have overlapping patterns of expression with different anterior expression limits, defining the identity of the pharyngeal arches (Graham and Smith, 2001). *Hoxc5* is expressed in a broad domain between the caudal stomach and anterior intestine, whereas *Hoxc9, Hoxb8* and *Hoxb9* are expressed in the most posterior portion of the fetal gut. *Hoxb6* and *c6* are expressed in restricted domains of the small and large intestines. In several cases, the A-P limits of Hox gene expression precisely correlate with endoderm organ boundaries; for example between the presumptive stomach and duodenum, and between the small and large intestine (Figure 30.4).

It is clear that the Hox genes have the ability to regulate the A-P patterning of the gut. For example, transgenic misexpression of the *Hox3.1* (*Hoxc8*) gene more anteriorly results in profound gastrointestinal tissue malformations and loss of positional identity (Grapin-Botton, 2005). In the pharyngeal region, the deletion of *Hoxa3* results in loss of the thymus and parathyroid, and hypoplasia of the thyroid (Manley and Capecchi, 1995; 1998). Similar gene targeting studies have demonstrated that *Hoxc4* mutants have esophageal defects, *Hoxa5* is necessary for development of the stomach, and *Hoxa13/Hoxd13* compound mutants have hindgut defects and anorectal malformations (reviewed in Grapin-Botton, 2005). Of these gut-related defects linked to Hox genes, only *Hoxa13* has been shown to specifically function in the endoderm. Novel *HOXA13* mutations have been associated with hand-foot-genital syndrome (HFGS), a rare, dominantly inherited condition in humans. Consistent with this finding, expression of this mutant form of *Hoxa13* in the endoderm of chick embryos results in hindgut and genitourinary patterning defects (de Santa Barbara and Roberts, 2002). Other Hox mutations that affect gut development (*Hoxa3, Hoxc4, Hoxa5*) are due to primary defects in gut mesenchyme or neural crest cells that contribute to the developing gut. This highlights the importance of signaling between endoderm and neighboring mesoderm and ectoderm for proper patterning and development of the gut.

Signaling Processes Involved in the Patterning and Formation of the Fetal Gut

Despite the cell lineage and molecular evidence which suggests that endoderm cells have positional identity by the end of gastrulation, these cells are not yet committed to specific lineages along the A-P axis. In fact, there is increasing evidence that a continuum of signals after the gastrulation stage progressively restricts endoderm cell fate into specific organ lineages. Signaling pathways that regulate endoderm patterning include growth factors such as Fibroblast Growth Factors (FGFs), Wnts, Bone Morphogenetic Proteins (BMPs), hedgehogs (hh), and retinoic acid (RA). These signaling pathways regulate the expression of transcription factors including the parahox and Hox genes, which as discussed above are important mediators of positional identity.

The gastrula stage endoderm is in close proximity to both the mesoderm and ectoderm, both of which can influence the A-P fate of endoderm (Horb and Slack, 2001; Kenny et al., 2012; Le Douarin, 1968; Le Douarin, 1988; Wells and Melton, 2000). Experiments in the mouse and frog have demonstrated that endoderm co-cultured with different A-P populations of mesoderm will adopt the A-P character of the co-cultured mesoderm (Figure 30.5) (Horb and Slack, 2001; Wells and

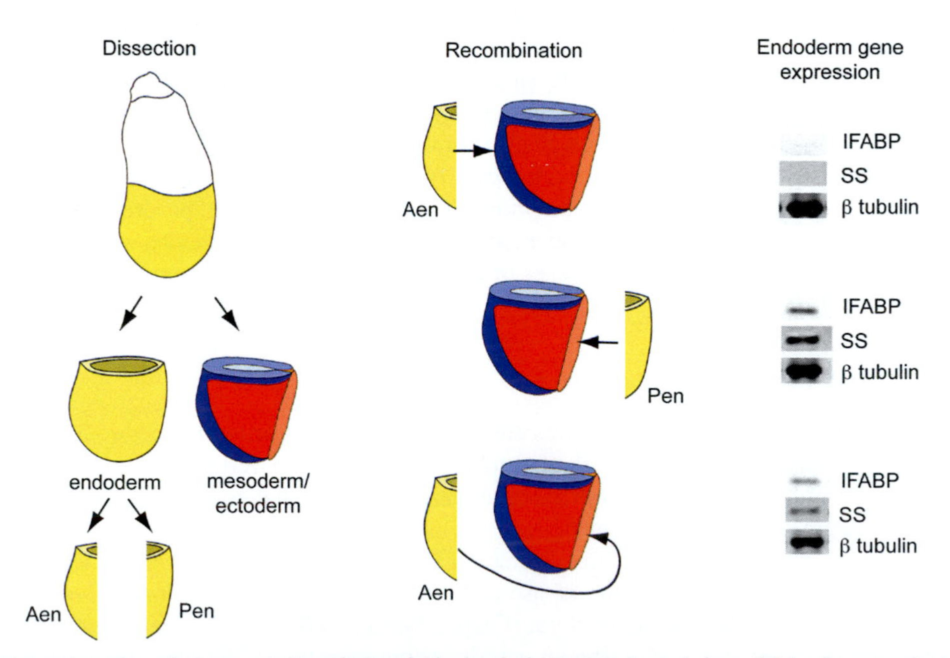

FIGURE 30.5 By using embryonic explant assays to investigate soluble signals that pattern the endoderm, E7.5 embryos can be isolated and the germ layers dissected and cultured. The left panel schematically shows the dissection embryos and the separation of endoderm, mesoderm and ectoderm germ layers. These layers can be further dissected into anterior or posterior endoderm. The middle panel illustrates the recombination of anterior endoderm (Aen) or posterior (Pen) endoderm with mesoderm/ectoderm. The right panel shows an RT-PCR analysis of several posterior endoderm markers IFABP and SS. Anterior endoderm does not normally express posterior markers when cultured with anterior mesoderm/ectoderm. However, when cultured with posterior mesoderm/ectoderm adopts a posterior fate as indicated by *de novo* expression of IFABP (intestinal fatty acid binding protein) and SS (somatostatin). Pen will also adopt an anterior fate when cultured with anterior mesoderm/ectoderm (not shown). *Data adapted from (Wells and Melton, 2000).*

Melton, 2000). Consistent with this, grafting posterior mouse definitive endoderm, which would normally become hindgut, into the presumptive pancreatic domain of a late gastrula chick embryo causes that endoderm to be re-specified to express *Pdx1* (J.M. Wells, unpublished data). These data suggest that endoderm cells at the gastrula stage have not acquired an A-P identity, but rather retain a high degree of plasticity (Figure 30.5).

Signals that Pattern the Gastrula Stage Endoderm

The temporal and spatial expression patterns of several peptide growth factors are consistent with a role in A-P patterning. In particular, the primitive streak, a structure that defines the posterior of the chick and mouse embryo, expresses several FGF, BMP and Wnt genes including *Fgf2, 3, 4, 5* and *8*; *Bmp2, 4, 5,* and *Wnt3, 5* and *11*. During gastrulation, all are expressed in the posterior and thus could function in the A-P patterning of the endoderm. As discussed below, all three of these pathways play a role in endoderm patterning as well as maintaining A-P domains in the developing gut tube (Dessimoz et al., 2006; Kenny et al., 2012; McLin et al., 2007; Spence et al., 2011b; Wells and Melton, 2000).

The FGF family of growth factors comprises 23 genes in mammals and is known to regulate cellular growth and differentiation processes, both throughout embryonic development and in adult tissues (Dailey et al., 2005). Of the FGF ligands that are expressed in the late gastrula embryo, only FGF4 (eFGF in frog) has been implicated in endoderm patterning (Dessimoz et al., 2006; Wells and Melton, 2000). In the mouse, embryos lacking either the *Fgf4* or the *Fgfr1* gene arrest at gastrulation (Ciruna et al., 1997; Yamaguchi et al., 1994). This suggests that *Fgf4* has a unique activity not compensated for by other FGF ligands that are expressed in this region of the embryo. *Fgf4* is expressed in the posterior mesoderm and ectoderm adjacent to the presumptive midgut and hindgut endoderm. At this stage, receptors for FGF4 and the FGF target genes *Sprouty 1* and *2* are broadly expressed in the endoderm along the A-P axis, suggesting that it is receiving an FGF signal (Dessimoz et al., 2006). Using mouse embryonic explants and chick cultures, FGF4 protein, but not other FGF ligands, has been shown to repress anterior cell fate (*Hhex*) and promote posterior foregut, midgut and hindgut cell fates (*Pdx1, Cdx*). FGF4-mediated inhibition of anterior cell identity corresponds with disrupted foregut morphogenesis, suggesting that a direct link exists between these two processes (Dessimoz et al., 2006; Wells and Melton, 2000). This suggests that FGF signaling is necessary for promoting posterior and inhibiting anterior cell fate (Figure 30.6). Consistent with this, a *Cdx* homolog in *Xenopus*, *xCad3*, is a direct target of the FGF signaling pathway (Haremaki et al., 2003).

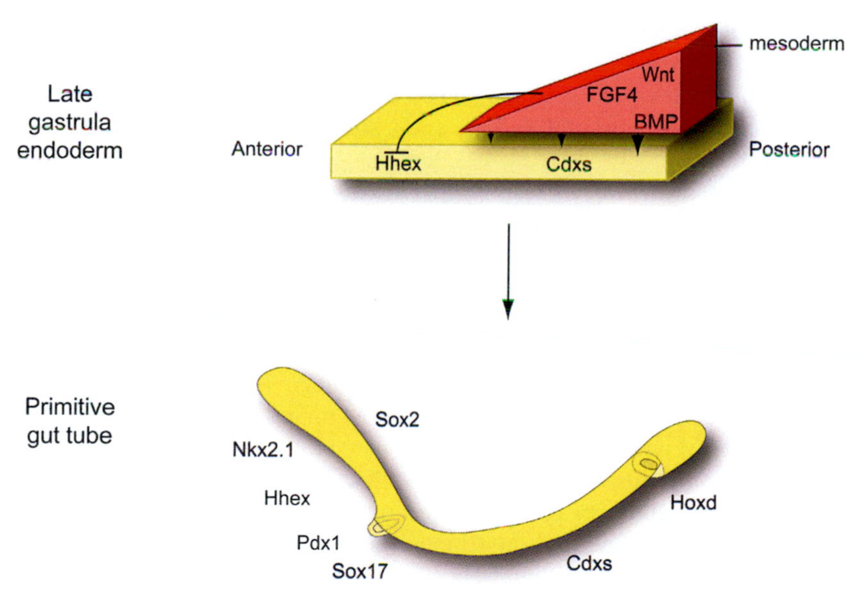

FIGURE 30.6 Signals from mesoderm at the gastrula stage are necessary for establishing A-P patterning during early gut tube development. The upper panel schematically shows the endoderm (yellow) and adjacent mesoderm (red). Several FGF, BMP and Wnt ligands are expressed in the posterior (primitive streak) in the late gastrula embryo. At this stage, FGF and Wnt have been shown to promote posterior gene expression and represses anterior gene (*Hhex*) expression. Although not shown in the context of endoderm patterning, it is known in other contexts that FGF4 is a direct target of the canonical Wnt signaling pathway (Kratochwil et al., 2002). FGF and Wnt signaling synergize to promote posterior patterning of human endoderm (Spence et al., 2011b). The lower panel shows several of the genes that are expressed along the A-P axis of the primitive gut tube. Many of these genes including *Nkx2.1, Hhex, Pdx1* and *Cdx,* are expressed in restricted domains that predict where organ primordia will form. Alterations in FGF and Wnt signaling disrupts both early endoderm patterning and causes morphological and gene expression changes in the developing foregut, midgut and hindgut domains (Dessimoz et al., 2006; McLin et al., 2007).

Although it is not known what regulates the posterior expression of *Fgf4* in the gastrula embryo, *Fgf4* was shown to be a direct target of the Wnt signaling pathway in the developing tooth (Kratochwil et al., 2002) and was shown to synergize with RA signaling to repress *Hhex* expression in anterior endoderm (Bayha et al., 2009). Several Wnt ligands are expressed in the primitive streak and at later stages in the anterior, suggesting a role in regulating A-P patterning of endoderm (Figure 30.6). Consistent with this, studies in *Xenopus* have demonstrated that high levels of canonical Wnt signaling promote a posterior endoderm fate while at the same time repressing anterior (McLin et al., 2007). However, low levels of Wnt are required for anterior endoderm and foregut development (Zhang et al., 2013). This is in part mediated by direct regulation of anteriorizing factors like *Hhex* (Rankin et al., 2011). In addition, BMP signaling is necessary for maintaining foregut progenitors, as demonstrated by the role of *Sizzled* and *Tolloid-like* 1, two modulators of BMP signaling, in regulating the size of the foregut domain and maintenance of foregut progenitors (Kenny et al., 2012).

Establishing Presumptive Organ Domains in the Primitive Gut Tube

Foregut Patterning

The foregut endoderm contributes to several organs including the thyroid, lungs, liver, extrahepatic biliary tree, gallbladder, and pancreas. The dynamic nature of foregut morphogenesis brings the endoderm into proximity with several mesodermally-derived tissues that are now known to pattern the foregut. At the early stages of foregut morphogenesis (E8 mouse, HH 6 chick) the ventral foregut endoderm is adjacent to the cardiac mesoderm. At later stages (E8.5 in mouse HH 9 in chick), the ventral foregut endoderm is in contact with septum transversum mesenchyme and the endothelial cells of the ventral veins. Dorsal foregut endoderm is brought into contact with the notochord and the dorsal aorta. Using a combination of tissue recombination, embryology and genetics, specific signaling molecules emanating from these tissues have been shown to pattern the foregut.

The liver derives exclusively from the ventral foregut and is the first foregut derivative to form. Studies in the chick and mouse suggest that initial specification of liver depends on signals from adjacent cardiac mesoderm (reviewed in McLin and Zorn, 2006). In mouse ventral foregut explant cultures, it has been shown that the cardiac mesoderm signals to the ventral foregut to induce liver between the 3 and 8 somite stage. Interestingly, in more posterior regions of the gut, negative signals were necessary to prevent precocious liver albumin gene expression in non-hepatic trunk endoderm.

A combination of studies in mice, chicks, frogs and fish have identified a role for FGF, Wnt, BMP and RA signaling in ventral foregut organ development. Through the use of mouse foregut explant cultures and genetically modified embryos, FGF signals from the cardiac mesoderm were shown to pattern the ventral foregut into liver, pancreas and lung (Calmont et al., 2006; Deutsch et al., 2001; Gualdi et al., 1996; Serls et al., 2005), with high levels promoting liver, moderate levels promoting lung, and low levels promoting ventral pancreas. It is also possible that the length of time that an endoderm cell is exposed to the cardiac mesoderm influences its fate. This possibility was explored in studies using *Xenopus*, which also showed that FGF signaling via the mesoderm is required for liver and lung specification. This study demonstrated the temporal effects of mesodermal contact, in which it was discovered that the pancreas requires little to no exposure to mesoderm or FGF, the liver requires an intermediate amount, and the lung requires the highest FGF levels and longest contact with the mesoderm (Shifley et al., 2012).

Similarly to the temporal and spatial requirements of FGF, BMP ligands promote the specification of hepatic progenitors starting at the 3–4 somite stage in the mouse and continuing through 7–9 somites (Wandzioch and Zaret, 2009). In contrast, the ventral pancreas progenitors only require BMP signaling for a short duration, from the 5–6 somite stage. This suggests that different spatiotemporal amounts of BMP signaling are required for the development of foregut organs like the liver and pancreas. Consistent with this, studies in zebrafish that used single cell lineage tracing in combination with modulating levels of BMP signaling demonstrated that bi-potential pancreatic-liver progenitors became liver in response to high BMP and pancreas in a low BMP environment (Chung et al., 2008). Lastly, RA signaling has been shown to positively regulate liver specification by inducing *wnt2bb* (Negishi et al., 2010), which is required for hepatoblast specification and proliferation (Ober et al., 2006; Poulain and Ober, 2011). These signaling pathways act in part through regulating key transcription factors, like *Hhex*, *Gata* factors, *Foxa1* and *Foxa2,* all of which are known to play a role in liver development (Reviewed in McLin and Zorn, 2006).

The foregut endoderm that gives rise to the liver and pancreas also becomes the extrahepatic biliary system (McCracken and Wells, 2012). This system, which gives rise to the common duct, extrahepatic bile ducts, and the gallbladder, was shown to share a common origin with the ventral pancreas and not the liver as previously believed (Spence et al., 2009). These pancreatobiliary progenitor cells co-express the transcription factors *Pdx1* and *Sox17* and are located at the ventral lip of the developing foregut at the 6 somite stage. Lineage tracing of pancretobiliary precursors demonstrated that they eventually segregate into a Pdx1+/Sox17− ventral pancreatic bud and a Sox17+/Pdx1− gallbladder primordium by E10.5. *Sox17* was shown to be critical for lineage segregation since maintained expression of *Sox17* suppressed pancreas formation and loss of *Sox17* resulted in gallbladder agenesis and formation of ectopic tissue in the extrahepatic biliary ducts (EHBDs). Ectopic pancreas in EHBDs was also observed in animals lacking the Notch effector *Hes1* (Fukuda et al., 2006; Sumazaki et al., 2004). Interestingly, *Sox17* expression domains were disrupted in *Hes1* mutant embryos (Spence et al., 2009) suggesting that *Sox17* and *Hes1* may coordinately function in pancreatobiliary lineage segregation. These data support a model in which *Sox17* co-expression with other factors defines progenitor domains. For example, early co-expression of *Sox17* with *Hhex1* defines foregut progenitors competent to form liver, pancreas and the EHBD. Down-regulation of *Sox17* in the *Hhex* domain coincides with liver specification. Similarly downregulation of *Sox17* in the pancreatobiliary domain coincides with segregation of the pancreatic and extrahepatic biliary lineages.

In addition to its role in early endoderm patterning, RA signaling has been shown to play a role in global A-P patterning of the foregut endoderm. For example, it has been shown that RA treatment was sufficient to cause an anterior shift of *Pdx1* expression in embryonic chick explants of lateral endoderm plus mesoderm (Kumar et al., 2003). In this example however, the presence of mesenchyme was required, suggesting the effect on endoderm was indirect. Other growth factors had similar posteriorizing activity, including BMPs and Activin; thus, it is possible that RA regulates the mesenchymal expression of these factors, which then signal to the endoderm. A global foregut patterning role of RA was also suggested in the mouse, in which embryos deficient in *retinaldehyde dehydrogenase 2* (*Raldh2*) and lacking active RA signaling in the foregut region fail to develop lungs, stomach, and dorsal pancreas, and also have impaired liver growth (Wang et al., 2006).

Additional studies indicate that RA signaling is active in both the dorsal pancreatic epithelium and mesenchyme, suggesting that other mesenchymal factors might play a role in the loss of the dorsal pancreas in *Raldh2−/−* embryos (Martin et al., 2005). Consistent with this, RA signaling in the foregut has been shown to act upstream of *Fgf10* and *Hox* genes, suggesting that RA signaling regulates the early patterning of the foregut mesenchyme. In zebrafish however, disruption of RA receptor function specifically in the endoderm perturbs normal pancreas development, suggesting a cell-autonomous role for RA signaling in the endoderm (Stafford et al., 2006). Studies in *Xenopus* suggest that RA does not regulate expression of *Pdx1* as it does in chicks and zebrafish (Bayha et al., 2009; Zeynali and Dixon, 1998). These apparent discrepancies over the role of RA signaling in the development of various foregut derivatives might be due to species differences, or it could suggest that RA signaling has multiple roles during the early stages of foregut patterning and organ development. A recent study of lung specification in *Xenopus* suggests RA and FGF signaling are required for the expression of two transcriptional repressors, *Odd-skipped-related* (*Osr*) *1* and *2* in the foregut endoderm and surrounding mesoderm prior to lung specification. The role of *Osr1/2* is to repress *Bmp4* in the lateral plate mesoderm, which allows the expression of *Wnt2b* and the subsequent induction of lung gene expression in the adjacent foregut endoderm (Rankin et al., 2012).

The role of hedgehog signaling in pancreas development is more extensively discussed in Chapter 33. However, there is evidence to suggest that a primary role of hedgehog signaling is patterning the gut tube. In transgenic mice in which *Sonic hedgehog* (*Shh)* was mis-expressed in the *Pdx1* domain, initiation of pancreas development and formation of the endocrine and exocrine lineages was relatively normal. However, the adjacent mesenchyme underwent a homeotic transformation toward an intestinal fate, forming intestinal smooth muscle and interstitial cells of Cajal. Gene targeting studies in mice also show that *Shh* mutants display intestinal transformation of the stomach epithelium, as evidenced by the expression of duodenal markers in the stomach (Ramalho-Santos et al., 2000). Studies in the chick have also suggested that *Shh* is involved in patterning the foregut and establishing the gizzard/stomach organ boundaries. This study suggests that Shh from the endoderm regulates the expression of *Nkx2.5* in the gizzard, and that ectopic Shh expression causes inappropriate *Nkx2.5* expression in the pancreas. Shh did not directly induce *Nkx2.5* but rather induced expression of *Bmp4* in adjacent mesenchyme. Cell-autonomous activation of BMP signaling in the mesenchyme through expression of activated BMP receptors was sufficient to cause upregulation of *Nkx2.5* in the gizzard, and resulted in perturbed smooth muscle differentiation.

Patterning the Anterior Foregut/Pharyngeal Domain

The anterior portion of the foregut is commonly referred to as the branchial or pharyngeal region. This region of the vertebrate embryo is anatomically hallmarked by a series of bilateral bulges called arches, which decrease in size from anterior to posterior (Figure 30.7). The pharyngeal endoderm lines the interior of the pharyngeal arches and will transiently fuse with the surface ectoderm to form pouches (interior endoderm) and clefts (exterior ectoderm) that mark the anterior and posterior boundaries of the arches. Pharyngeal pouches form from anterior to posterior and exhibit regionally restricted identity as marked by the differential expression of *Fgf8* and *Pax1* (*paired box 1*). During normal development, the pharyngeal endoderm will contribute to the trachea, esophagus, thymus, parathyroid, thyroid, ultimobranchial bodies (which form the calcitonin producing parafollicular cells of the thyroid), and the taste buds. However, multiple human disorders are associated with perturbations in pharyngeal development, including esophageal atresia, tracheoesophageal fistula,

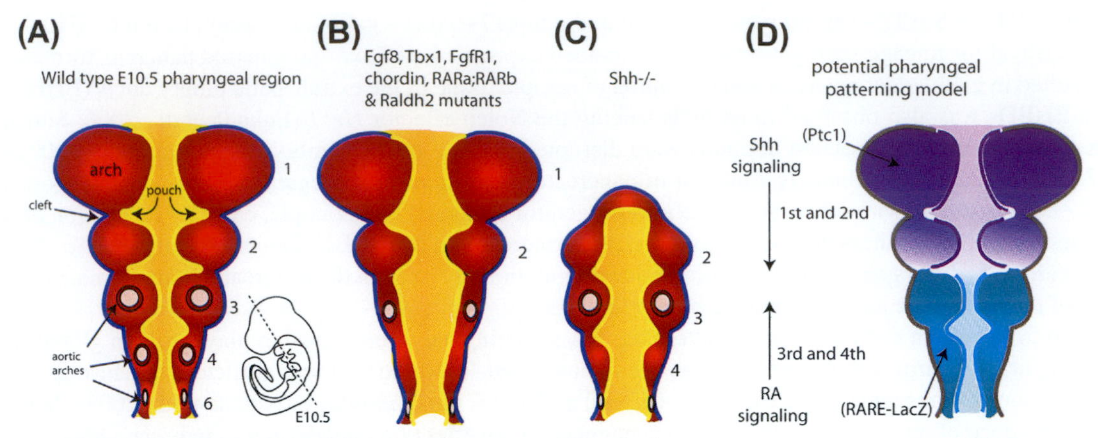

FIGURE 30.7 Patterning the Anterior Foregut/Pharyngeal Domain. (A) Ventral/coronal view of normal pharyngeal arches. The outline of an E10.5 embryo shows the corresponding plane of section (dotted line) of the larger image. The endoderm is in yellow, the surface ectoderm is blue, and the mesenchyme of the arches is red. Arches are numbered 1–4 then 6 from anterior to posterior. The pouches and clefts are numbered by that of the preceding arch. (B) The E10.5 pharyngeal phenotype of the *Fgf8*, *Tbx1*, *Fgfr1*, chordin, *RARalpha/RARbeta*, and *RALDH2* mutant mice where the caudal third and fourth arches are missing. (C) The E10.5 pharyngeal phenotype of the *Shh* null mutant mice where the first arch is missing and the second is severely hypomorphic. Note: The *Shh* mutant embryos initiate development of the first and second arches at E9.5 but lose the first arch by E10.5 and the second continues to atrophy. (D) A putative model for A-P patterning of the pharyngeal region with Shh and RA mediated signals regulate anterior and posterior regions of the pharyngeal region.

VACTERL, DiGeorge, and Velocardiofacial syndromes, emphasizing the clinical importance of understanding anterior foregut patterning. Furthermore from a teratological standpoint, pharyngeal region morphology appears to be particularly sensitive to alcohol exposure, possibly due to competitive interactions of alcohol with RA metabolizing enzymes present in the pharyngeal tissues (Wang, 2005).

One key role of the pharyngeal endoderm is as a regional source of patterning signals for the adjacent mesoderm and ectoderm (reviewed in Graham and Smith, 2001). Signaling molecules expressed by endoderm include Sonic hedgehog (Shh), multiple members of the FGF family, BMPs, the BMP antagonists Chordin and Noggin, and several retinoic acid receptors. Although it is not known exactly how these pathways intersect to pattern the pharyngeal region, we will discuss a potential patterning model based on an analysis of mouse mutants and gene expression studies below (Figure 30.7).

Both *Fgf3* and *Fgf8* are expressed in the pharyngeal pouch endoderm and are required for proper pouch formation (Crump et al., 2004). Mutants lacking *Fgf8* and *FgfR1* have severe phenotypes exhibiting deletions of both the third and fourth arches resulting in the loss of the thymus and parathyroid (Abu-Issa et al., 2002; Trokovic et al., 2003). *Fgf8* expression in the pharyngeal endoderm is lost in embryos lacking *Tbx1* (*T-box 1*; discussed in more detail in Chapter 36) suggesting *Tbx1* is upstream of *Fgf8*. Furthermore *Tbx1* mutations have been shown to contribute to DiGeorge syndrome in humans (Jerome and Papaioannou, 2001; see also Chapter 38). Patients with DiGeorge syndrome often have numerous defects in craniofacial and cardiovascular development and defects in organs that are derived from pharyngeal pouch endoderm including the thymus and parathyroid.

Due to the importance of *Tbx1* in human disease, there is much interest in the signaling pathways that regulate *Tbx1* expression. Several reports indicate that *Tbx1* may be regulated by the morphogen SHH, as *Shh* mutant mice show reduced expression of *Tbx1* in the pharyngeal region (Riccomagno et al., 2002; Yamagishi et al., 2003). Interestingly, several phenotypic differences between *Tbx1* and *Shh* mutants suggest that the hedgehog pathway is more critical for the development of the anterior pharyngeal region, whereas the *Tbx1* phenotypes are predominantly in the caudal pharyngeal arches. For example, *Shh* mutants form a thymus, unlike the *Tbx1*-null embryos, whereas parathyroid development is perturbed in both mutants (Ivins et al., 2005; Moore-Scott and Manley, 2005). This phenotypic overlap suggests that *Shh* may act as a modifier of *Tbx1* expression in the pharyngeal region.

In the anterior foregut, RA signaling, mediated by the retinoid receptors, RARα, RARβ and RARγ, is essential for patterning of the pharyngeal endoderm. Evidence for the dynamic nature of RA signaling in the foregut came from the identification of a retinoic acid inducible enhancer ($RAIDR_5$) in the *Hoxb1* gene (Graham and Smith, 2001; Grapin-Botton, 2005). Retinoic acid dependent expression of *Hoxb1* occurs broadly throughout the foregut at E8.5, but becomes restricted to the pharyngeal endoderm of the anterior foregut at E9.5. Exposing embryos to exogenous RA causes an anterior expansion of *Hoxb1*. Moreover RA regulates the expression of *Shh* and *Tbx1* (reviewed in Graham and Smith, 2001). For example, in *Tbx1* mutants, *Raldh2* expression expands anteriorly (Ivins et al., 2005). Embryos exposed to a pan RAR antagonist, a pan RAR agonist, vitamin A or retinoic acid lose the caudal third and fourth pharyngeal arches. These treatments phenocopy those observed in the compound *RARα* and *RARβ* knockouts and the hypomorphic *Raldh2* mutant and illustrate the importance of proper RA signaling in pharyngeal region patterning (Niederreither et al., 2003; Vermot et al., 2003). In addition to patterning, RA is important for pharyngeal pouch morphogenesis and segmentation (Kopinke et al., 2006).

BMPs are members of the TGFβ superfamily of growth factors and have also been implicated in patterning and pharyngeal organogenesis. *Bmp2*, *4* and *7* as well as the BMP antagonists *noggin* and *chordin* are expressed in the pharyngeal region in all three germ layers, including the endoderm. The *chordin* null, like the *Fgf8* and *Tbx1* null and the compound *RARα/RARβ* mutants, fail to form the posterior arches, implying that inhibition of BMP signaling is part of a complex pathway necessary for posterior patterning of the pharyngeal endoderm (Bachiller et al., 2003).

These data establish that *Tbx1, Fgf8*, RA and *chordin* are all required for the formation of the posterior pharyngeal region, whereas *Shh* appears to be required for establishing the anterior pharyngeal endoderm (Figure 30.7B and C) (Graham and Smith, 2001). Together, these data support a model in which hedgehog signaling in the anterior, and RA/BMP signaling in the posterior, are required for patterning and formation of the pharyngeal segments (Figure 30.7D).

Midgut and Hindgut Patterning

Early in the development of the gut tube, the posterior endoderm folds ventrally, so generating the hindgut. The midgut and hindgut form the small and large intestines. The hindgut also contributes to the cloaca, which forms the urogenital tract. Studies have begun to identify the molecular mechanisms that regulate posterior identity in the hindgut. As discussed above, FGF signaling prior to gut tube morphogenesis is involved in establishing posterior endoderm identity (Dessimoz et al., 2006). These studies also found that FGF signaling is required to maintain the A-P expression boundaries of *Pdx1* and *Cdx* genes at later stages of gut tube development. Shifting FGF4-mediated signaling to the anterior caused an anterior shift in expression of *Pdx1* and *Cdx* genes in the primitive gut tube, suggesting that anterior regions of the gut tube were being transformed to a posterior fate. Conversely, inhibiting FGF signaling caused loss of posterior identity. Moreover these studies demonstrated that FGF signaling was acting directly on the endoderm.

Several other signaling molecules, including BMPs, Wnts, Activin and RA also have the ability to posteriorize lateral endoderm that contributes to the ventral gut, but only in the presence of lateral plate mesoderm, suggesting that these factors might act indirectly (Kumar et al., 2003). The Wnt signaling pathway plays distinct temporal roles during the development of the gastrointestinal tract. As mentioned earlier, canonical Wnt signaling promotes posterior fate between the gastrula and gut tube stages of development. At later stages, both canonical and non-canonical Wnt signaling have been implicated in both the patterning and elongation of the developing gut tube. The canonical Wnt signaling pathway is mediated by its transcriptional effectors, the TCF/LEF family of transcription factors. In the chick, *TCF4* and *Lef1* were expressed in distinct A-P domains at different stages of development (Theodosiou and Tabin, 2003). In the mouse, E9.5 *Tcf1/Tcf4*$^{-/-}$ embryos have posterior truncations and loss of posterior hindgut, and at later stages there is an anterior transformation of the duodenum (Gregorieff et al., 2004). *Tcf1/Tcf4*$^{-/-}$ embryos have a similar phenotype to mouse embryos lacking *Caudal Homeobox* genes (*Cdxs*), which are involved in posterior development, across species which range from flies to mammals. Consistent with this, TCF4 directly regulates the transcription of *Cdx1* in the developing mouse intestine (Lickert et al., 2000), *Cdx2* is required for early posterior development in the mouse (Chawengsaksophak et al., 2004), and endoderm-specific deletion of *Cdx2* results in colon dysgenesis (Gao et al., 2009). At later stages, loss of *Cdx2* results in a loss of intestinal identity and expression of gastric markers (Stringer et al., 2008). In addition to Wnt/TCF regulating expression of Cdx factors, *Cdx2* was shown to directly regulate expression of *Wnt3a* (Savory et al., 2009) suggesting a complex regulatory loop.

These findings illustrate the fact that gut patterning involves reciprocal signaling between endodermally-derived epithelium and mesodermally-derived tissues, including the notochord and gut mesenchyme. Additional reciprocal interactions have been identified between the hindgut endoderm and the mesoderm that result in spatially restricted *Hox* gene expression in the hindgut, and establishment of posterior identity in the endoderm (Roberts et al., 1995; 1998). Specifically, hindgut endoderm expresses *Shh*, which is sufficient to induce *Bmp4* and *Hoxd13* expression in adjacent posterior mesoderm, but not in more anterior mesoderm adjacent to the stomach. If *Hoxd13* is mis-expressed in more anterior mesoderm, the adjacent stomach endoderm is transformed into an intestinal type of endoderm, as assayed by morphology and marker expression. It is possible that the signals that emanate from *Hoxd13* expressing mesenchyme and signal back to endoderm include factors such as FGF10, since *Fgf10*$^{-/-}$ embryos have colonic atresia (Sala et al., 2006).

The most posterior portion of the hindgut gives rise to a transient embryonic structure in mammals known as the cloaca. During early stages of development, the cloaca is a cavity with a simple endodermally-derived epithelium, surrounded by splanchnic mesenchyme. In later stages, the endoderm of the cloaca gives rise to the epithelial lining of the rectum, the anterior anus, and the urogenital system, including the bladder, urethra, and prostate in males (Seifert et al., 2008). This cavity is specified into these lineages through a combination of molecular and morphological processes that subdivide it into distinct A-P and D-V regions.

One of the better understood aspects of cloaca development is the process of septation. Initially, the cloacal is a simple cavity that forms around E10.5 at the most posterior end of the developing gut tube. A thickening of epithelium and mesenchyme marks the formation of the urorectal septum in the anterior of the cloaca. The urorectal septum pushes posteriorly into the cloacal cavity, such that after three days it meets with the posterior of the embryo, thus subdividing the cloaca into the dorsally located digestive tract and the ventrally located urogenital tracts. Defects in this process lead to clinically significant malformations, including imperforate anus, hypospadias, various fistulas, and most severely, persistent cloaca. Many anorectal and urogenital malformations are of unknown cause, but are likely linked to defects in a small number of signaling pathways.

To date there has been relatively little work done on early stages of cloaca patterning. There are however a few reports that have identified signaling pathways involved in septation. Work using multiple mouse lines has shown that there is a temporal role for Shh signaling during the process of anorectal and urogenital, or anogenital development. From E10.5–E13 Shh is required for proper septation of the cloaca, as well as outgrowth of the genital tubercle that forms the external genitalia. The degree of septation is determined by the duration of active Shh signaling, such that the longer Shh is active the further septation has progressed, and the more posterior the connection of the gut tube to the cloaca. In contrast, an absence of Shh signaling results in arrested septation and a persistent cloaca (Seifert et al., 2009). Shh is also required at later stages, from E13.5–E16.5 for proper closure of the urethral tube. BMP signaling has also been shown to be important for proper septation and formation of the anogenital lineages. Studies using *Bmp7* null mice show that *Bmp7* plays a role in cloacal development by promoting cell proliferation and survival of both endoderm and the surrounding mesoderm of the urogenital sinus (Wu et al., 2009). In addition, *Bmp7* also acts to promote directional outgrowth of the septum transversum by regulating cell polarity and the apical basal orientation of cell division during septation (Xu et al., 2012). Interestingly, *Wnt5a*, a noncanoical Wnt ligand that is well known for its role in regulating planar cell polarity, also has been shown to be critical for proper septation, as knockouts have imperforate anus and shortened intestinal tracts, which is believed to be in part due to their defects in cell proliferation (Cervantes et al., 2009; Tai et al., 2009). Another study suggests that *Wnt5a* may be involved in D-V patterning based on the localized expression of *Wnt5a* on the dorsal side of the cloaca in human embryo (Li et al., 2011). While both *Wnt5a* and *BMP7* regulate cell polarity during cloaca septation, it is not known if they act epistatically or separately.

Two transcription factors, *Six1* and *Six2*, have been shown to be asymmetrically expressed in the peri-cloacal mesenchymal progenitors and that they are required for proper development of the anogenital system. It is hypothesized that this asymmetrical expression leads to the asymmetrical growth of the mesenchyme and then acts as the driving force to septate the cloaca (Wang et al., 2013). Interestingly, these transcription factors also influence both BMP4 and BMP7 levels, as well as Wnt signaling. Proper expression of Six factors is required for proper development of the anogenital system, likely due to their ability to disrupt very important signaling pathways required for development.

Impact of Genomics on Our Understanding of Early Endoderm Organogenesis In Vivo

Over the past decade, there has been an explosion of gene expression data shedding light on endoderm organ development, from the specification of embryonic endoderm to formation of functioning adult cell types such as insulin producing β cells (see also Chapter 33). These databases allow researchers to analyze cell type specific expression of a particular gene, or to look for entire functional classes of genes, such as "DNA binding" genes expressed in developing or adult cell types. More importantly these types of analyses have provided a molecular foundation for identifying disease-causing alleles, as well as for studying the directed differentiation of human pluripotent stem cells into pancreatic endocrine cells, liver hepatocytes and intestinal tissue.

Given the technical difficulty of isolating enough material from early stage embryos, there are relatively few genomic studies of endoderm patterning and organ specification. One of the first examples of a genomics-based approach to studying endocrine pancreas organogenesis used a microarray approach, where the expression of > 12,000 genes was measured simultaneously to identify the transcripts that are both up- and downregulated as cells transition from one developmental stage to the next (Gu et al., 2004). The stages and cells analyzed in this experiment were:

1) E7.5 pre-pancreatic endoderm;
2) E10.5 pancreatic progenitor cells as defined by *Pdx1* expression;
3) E12.5 endocrine progenitor cells as defined by *Ngn3* expression; and
4) Adult islets of Langerhan's.

This approach was highly successful in generating a gene expression database of the endoderm and developing endocrine pancreas. More importantly, this study identified many novel genes expressed at each stage of endocrine pancreas development, and functional analysis of one of these genes demonstrated its involvement in endocrine cell development.

The molecular pathways that regulate later stages of endocrine pancreas development will be discussed in Chapter 33. However, the genes that showed a >3 fold expression change during this endoderm patterning stage were grouped into three functional classes; transcription factors, growth factors/receptors, and cell adhesion/matrix proteins to focus on regulatory molecules that control cell fate. As expected from previous studies, the FGF, Notch and Activin signaling pathways were all implicated in specification of the Pdx1 lineages (Kim and Hebrok, 2001) since *Pdx1* cells expressed *Fgf receptors 1, 2* and *4*, *activin receptor IIb* (data not shown), *notch 3* and *delta-like 1*. However, these studies also implicated the Wnt signaling pathway in early pancreas development due to the expression of Wnt receptors, *frizzled 2* and *4*, as well as secreted forms of Frizzled that antagonize Wnt signaling. The identification of Wnt signaling components in the early pancreas by functional studies demonstrated the importance of this pathway for development of the exocrine pancreas (Dessimoz et al., 2005; Murtaugh et al., 2005; Wells et al., 2007).

These signals culminate in expression of several transcription factors in the pancreatic endoderm including *Pdx1*, *Hlxb1* and *Ptf1a/p48*, all of which are necessary for proper pancreatic development (reviewed in Grapin-Botton, 2005). These studies also identified many new transcription factors of the basic helix-loop-helix type (*Id, Foxa2* and *3*), homeobox type (*Hoxa5*), zinc finger type (*Myt1*) as well as other classes. Several of these factors have been shown to play a role in the development and/or postnatal function of the liver (Lee et al., 2005) and pancreas (Wang et al., 2007).

With the advent of new technologies, more recent studies have analyzed the impact of specific mutations and epigenetic modifications on early endoderm development. One example analyzed mouse embryos lacking *Foxa2* (Tamplin et al., 2008) and identified *Foxa2*-dependent transcriptional programs in the axial mesendoderm and definitive endoderm. Moreover, *Foxa2* plays a role in the subsequent specification of the pancreas and liver by acting as a pioneer factor, which binds to target sites on liver and pancreatic genes promoting a more open chromatin configuration (Cirillo et al., 2002). In addition to genetic effects on development, epigenetic modifications of chromatin have been implicated in regulating foregut endoderm differentiation into pancreatic and hepatic lineages (Xu et al., 2011). In this study, foregut endoderm cells were shown to have an epigenetic "prepattern," where *Pdx1*, which is transcriptionally silent, was epigenetically marked by both a silencing and activating histone methylation pattern that allowed its subsequent activation in pancreas, but maintained a silent state in differentiated liver.

30.4 TRANSLATIONAL EMBRYOLOGY: THE IMPACT OF EMBRYONIC STUDIES ON HUMAN HEALTH

Activation of Embryonic Endoderm Pathways During Disease

A significant number of diseases affect endodermally-derived organs, particularly the lungs, liver and pancreas. Moreover childhood diseases such as asthma and diabetes which affect the respiratory and GI tracts are increasing at an alarming rate. The study of the molecular mechanisms underlying endoderm organogenesis have already led to the identification of genes involved in diseases of the endoderm. Moreover the fields of regenerative medicine and tissue engineering have greatly benefited from research in developmental biology. For example, it has become increasingly apparent that adult organs reactivate embryonic pathways in the process of regeneration. Injury to the lung and tracheal epithelium causes a regenerative response that is accompanied by a dramatic increase in the expression of embryonic endoderm regulatory genes including *Sox17*, *Foxa1*, *Foxa2* and others, and some of these genes might aid in regeneration (Park et al., 2006a,b; Whitsett et al., 2011). Progenitor cells expressing *FoxA2*, *Sox17*, *Sox9* and other embryonic transcription factors have been isolated from the adult extrahepatic biliary tree and these are able to differentiate into multiple organ lineages including liver and pancreas (Cardinale et al., 2011). These data suggest that transcriptional networks involved in early endoderm development might have a role in adult progenitor cell plasticity.

Human Pluripotent Stem Cells (hPSCs): New Models of Human Development and Disease

Another means to translate embryonic studies into clinical applications is through the directed differentiation of embryonic stem cells into endoderm derivatives. In this area, the decades of studies of early endoderm organogenesis have been invaluable, as they have provided a molecular framework that has been used to direct the differentiation of human PSCs into a variety of endoderm organ cells and tissues, including pancreatic endocrine cells, hepatocytes, and intestinal tissue. While the long term goal of this approach is to generate therapeutic tissues for transplantation, in the short term, hPSC cultures have provided developmental biologists with their first opportunity to study and manipulate development of human endoderm organogenesis. The TGFβ-growth factor Activin was demonstrated to be capable of mimicking the activity of Nodal and direct the differentiation of human and mouse embryonic stem cells into definitive endoderm in culture (D'Amour et al., 2005; Kubo et al., 2004). This exciting advance has allowed researchers to study how hPSC-derived endoderm is patterned and directed to differentiate into specific organ lineages including lungs, liver, pancreas and intestine (Cai et al., 2007; D'Amour et al., 2006; Mou et al., 2012; Spence et al., 2011b). The further differentiation into specific lineages will be covered in other chapters. However in the area of endoderm patterning, several reports have identified Activin, FGF and Wnt signaling as key regulators of A-P patterning of human PSC-derived endoderm (Ameri et al., 2010; Nostro et al., 2011; Sherwood et al., 2011; Spence et al., 2011b). The conclusions were similar to findings in embryonic studies, in that Wnt and FGF signaling promoted a posterior endoderm fate whereas an increase in Activin signaling favored an anterior definitive endoderm fate. Importantly, human PSCs are proving to be a powerful *in vitro* model to study early stages of endoderm organ development, allowing for easy manipulation, rapid analysis, and providing large amounts of material for biochemical and molecular studies.

CLINICAL RELEVANCE

- Understanding early endoderm organ development is vital to identifying underlying causes of congenital defects like DiGeorge and Velocariofacial syndromes, VACTERL, as well as hindgut defects like persistent cloaca.
- Studies of early endoderm organ development provide the knowledge to direct differentiation of human pluripotent stem cells into organ cell types and tissues for future transplantation-based therapies.

ACKNOWLEDGMENTS

JMW and AMM receive support from NIH grants 1U18NS080815, R01DK080823, and R01DK092456.

REFERENCES

Abu-Issa, R., Smyth, G., Smoak, I., Yamamura, K., Meyers, E.N., 2002. Fgf8 is required for pharyngeal arch and cardiovascular development in the mouse. Development 129, 4613–4625.

Ameri, J., Stahlberg, A., Pedersen, J., Johansson, J.K., Johannesson, M.M., Artner, I., Semb, H., 2010. FGF2 specifies hESC-derived definitive endoderm into foregut/midgut cell lineages in a concentration-dependent manner. Stem Cells 28, 45–56.

Ang, S.L., Wierda, A., Wong, D., Stevens, K.A., Cascio, S., Rossant, J., Zaret, K.S., 1993. The formation and maintenance of the definitive endoderm lineage in the mouse: involvement of HNF3/forkhead proteins. Development 119, 1301–1315.

Angelo, J.R., Guerrero-Zayas, M.I., Tremblay, K.D., 2012. A fate map of the murine pancreas buds reveals a multipotent ventral foregut organ progenitor. PLoS One 7, e40707.

Bachiller, D., Klingensmith, J., Shneyder, N., Tran, U., Anderson, R., Rossant, J., De Robertis, E.M., 2003. The role of chordin/BMP signals in mammalian pharyngeal development and DiGeorge syndrome. Development 130, 3567–3578.

Bayha, E., Jorgensen, M.C., Serup, P., Grapin-Botton, A., 2009. Retinoic acid signaling organizes endodermal organ specification along the entire antero-posterior axis. PLoS One 4, e5845.

Cai, J., Zhao, Y., Liu, Y., Ye, F., Song, Z., Qin, H., Meng, S., Chen, Y., Zhou, R., Song, X., Guo, Y., Ding, M., Deng, H., 2007. Directed differentiation of human embryonic stem cells into functional hepatic cells. Hepatology 45, 1229–1239.

Calmont, A., Wandzioch, E., Tremblay, K.D., Minowada, G., Kaestner, K.H., Martin, G.R., Zaret, K.S., 2006. An FGF response pathway that mediates hepatic gene induction in embryonic endoderm cells. Dev. Cell 11, 339–348.

Cardinale, V., Wang, Y., Carpino, G., Cui, C.B., Gatto, M., Rossi, M., Berloco, P.B., Cantafora, A., Wauthier, E., Furth, M.E., Inverardi, L., Dominguez-Bendala, J., Ricordi, C., Gerber, D., Gaudio, E., Alvaro, D., Reid, L., 2011. Multipotent stem/progenitor cells in human biliary tree give rise to hepatocytes, cholangiocytes, and pancreatic islets. Hepatology 54, 2159–2172.

Cervantes, S., Yamaguchi, T.P., Hebrok, M., 2009. Wnt5a is essential for intestinal elongation in mice. Dev. Biol. 326, 285–294.

Chalmers, A.D., Slack, J.M., 2000. The *Xenopus* tadpole gut: fate maps and morphogenetic movements. Development 127, 381–392.

Chalmers, A.D., Slack, J.M., Beck, C.W., 2000. Regional gene expression in the epithelia of the *Xenopus* tadpole gut. Mech. Dev. 96, 125–128.

Chapman, S.C., Schubert, F.R., Schoenwolf, G.C., Lumsden, A., 2002. Analysis of spatial and temporal gene expression patterns in blastula and gastrula stage chick embryos. Dev. Biol. 245, 187–199.

Chawengsaksophak, K., de Graaff, W., Rossant, J., Deschamps, J., Beck, F., 2004. Cdx2 is essential for axial elongation in mouse development. Proc. Nat. Acad. Sci. USA 101, 7641–7645.

Chung, W.S., Shin, C.H., Stainier, D.Y., 2008. Bmp2 signaling regulates the hepatic versus pancreatic fate decision. Dev. Cell 15, 738–748.

Cirillo, L.A., Lin, F.R., Cuesta, I., Friedman, D., Jarnik, M., Zaret, K.S., 2002. Opening of compacted chromatin by early developmental transcription factors HNF3 (FoxA) and GATA-4. Mol. Cell 9, 279–289.

Ciruna, B.G., Schwartz, L., Harpal, K., Yamaguchi, T.P., Rossant, J., 1997. Chimeric analysis of fibroblast growth factor receptor-1 (Fgfr1) function: a role for FGFR1 in morphogenetic movement through the primitive streak. Development 124, 2829–2841.

Crump, J.G., Maves, L., Lawson, N.D., Weinstein, B.M., Kimmel, C.B., 2004. An essential role for Fgfs in endodermal pouch formation influences later craniofacial skeletal patterning. Development 131, 5703–5716.

D'Amour, K.A., Agulnick, A.D., Eliazer, S., Kelly, O.G., Kroon, E., Baetge, E.E., 2005. Efficient differentiation of human embryonic stem cells to definitive endoderm. Nat. Biotechnol. 23, 1534–1541.

D'Amour, K.A., Bang, A.G., Eliazer, S., Kelly, O.G., Agulnick, A.D., Smart, N.G., Moorman, M.A., Kroon, E., Carpenter, M.K., Baetge, E.E., 2006. Production of pancreatic hormone-expressing endocrine cells from human embryonic stem cells. Nat. Biotechnol. 24, 1392–1401.

Dailey, L., Ambrosetti, D., Mansukhani, A., Basilico, C., 2005. Mechanisms underlying differential responses to FGF signaling. Cytokine Growth Factor Rev. 16, 233–247.

de Santa Barbara, P., Roberts, D.J., 2002. Tail gut endoderm and gut/genitourinary/tail development: a new tissue-specific role for Hoxa13. Development 129, 551–561.

Dessimoz, J., Bonnard, C., Huelsken, J., Grapin-Botton, A., 2005. Pancreas-specific deletion of beta-catenin reveals Wnt-dependent and Wnt-independent functions during development. Curr. Biol. 15, 1677–1683.

Dessimoz, J., Opoka, R., Kordich, J.J., Grapin-Botton, A., Wells, J.M., 2006. FGF signaling is necessary for establishing gut tube domains along the anterior-posterior axis *in vivo*. Mech. Dev. 123, 42–55.

Deutsch, G., Jung, J., Zheng, M., Lora, J., Zaret, K.S., 2001. A bipotential precursor population for pancreas and liver within the embryonic endoderm. Development 128, 871–881.

Dufort, D., Schwartz, L., Harpal, K., Rossant, J., 1998. The transcription factor HNF3beta is required in visceral endoderm for normal primitive streak morphogenesis. Development 125, 3015–3025.

Frumkin, A., Haffner, R., Shapira, E., Tarcic, N., Gruenbaum, Y., Fainsod, A., 1993. The chicken CdxA homeobox gene and axial positioning during gastrulation. Development 118, 553–562.

Fukuda, A., Kawaguchi, Y., Furuyama, K., Kodama, S., Horiguchi, M., Kuhara, T., Koizumi, M., Boyer, D.F., Fujimoto, K., Doi, R., Kageyama, R., Wright, C.V., Chiba, T., 2006. Ectopic pancreas formation in Hes1-knockout mice reveals plasticity of endodermal progenitors of the gut, bile duct, and pancreas. J. Clin. Invest. 116, 1484–1493.

Gamer, L.W., Wright, C.V., 1993. Murine Cdx-4 bears striking similarities to the *Drosophila* caudal gene in its homeodomain sequence and early expression pattern. Mech. Dev. 43, 71–81.

Gao, N., White, P., Kaestner, K.H., 2009. Establishment of intestinal identity and epithelial-mesenchymal signaling by Cdx2. Dev. Cell 16, 588–599.

Gittes, G.K., Rutter, W.J., 1992. Onset of cell-specific gene expression in the developing mouse pancreas. Proc. Natl. Acad. Sci. USA 89, 1128–1132.

Graham, A., Smith, A., 2001. Patterning the pharyngeal arches. Bioessays 23, 54–61.

Grapin-Botton, A., 2005. Antero-posterior patterning of the vertebrate digestive tract: 40 years after Nicole Le Douarin's PhD thesis. Int. J. Dev. Biol. 49, 335–347.

Grapin-Botton, A., Majithia, A.R., Melton, D.A., 2001. Key events of pancreas formation are triggered in gut endoderm by ectopic expression of pancreatic regulatory genes. Genes Dev. 15, 444–454.

Gregorieff, A., Grosschedl, R., Clevers, H., 2004. Hindgut defects and transformation of the gastro-intestinal tract in $Tcf4^{-/-}/Tcf1^{-/-}$ embryos. EMBO J. 23, 1825–1833.

Gu, G., Wells, J.M., Dombkowski, D., Preffer, F., Aronow, B., Melton, D.A., 2004. Global expression analysis of gene regulatory pathways during endocrine pancreatic development. Development 131, 165–179.

Gualdi, R., Bossard, P., Zheng, M., Hamada, Y., Coleman, J.R., Zaret, K.S., 1996. Hepatic specification of the gut endoderm *in vitro*: cell signaling and transcriptional control. Genes Dev. 10, 1670–1682.

Haremaki, T., Tanaka, Y., Hongo, I., Okamoto, H., 2003. Integration of multiple signal transducing pathways on FGF response elements of the *Xenopus* caudal homologue Xcad3. Development 130, 4907–4917.

Horb, M.E., Slack, J.M., 2001. Endoderm specification and differentiation in *Xenopus* embryos. Dev. Biol. 236, 330–343.

Howell, J.C., Wells, J.M., 2011. Generating intestinal tissue from stem cells: potential for research and therapy. Regenerative Med. 6, 743–755.

Ivins, S., Lammerts van Beuren, K., Roberts, C., James, C., Lindsay, E., Baldini, A., Ataliotis, P., Scambler, P.J., 2005. Microarray analysis detects differentially expressed genes in the pharyngeal region of mice lacking Tbx1. Dev. Biol. 285, 554–569.

Jones, C.M., Broadbent, J., Thomas, P.Q., Smith, J.C., Beddington, R.S., 1999. An anterior signaling center in *Xenopus* revealed by the homeobox gene XHex. Curr. Biol. 9, 946–954.

Jerome, L.A., Papaioannou, V.E., 2001. DiGeorge syndrome phenotype in mice mutant for the T-box gene, Tbx1. Nat. Genet. 27, 286–291.

Kenny, A.P., Rankin, S.A., Allbee, A.W., Prewitt, A.R., Zhang, Z., Tabangin, M.E., Shifley, E.T., Louza, M.P., Zorn, A.M., 2012. Sizzled-tolloid interactions maintain foregut progenitors by regulating fibronectin-dependent BMP signaling. Dev. Cell 23, 292–304.

Kim, S.K., Hebrok, M., 2001. Intercellular signals regulating pancreas development and function. Genes Dev. 15, 111–127.

Kimura, W., Yasugi, S., Stern, C.D., Fukuda, K., 2006. Fate and plasticity of the endoderm in the early chick embryo. Dev. Biol. 289, 283–295.

Kopinke, D., Sasine, J., Swift, J., Stephens, W.Z., Piotrowski, T., 2006. Retinoic acid is required for endodermal pouch morphogenesis and not for pharyngeal endoderm specification. Dev. Dyn. 235, 2695–2709.

Kratochwil, K., Galceran, J., Tontsch, S., Roth, W., Grosschedl, R., 2002. FGF4, a direct target of LEF1 and Wnt signaling, can rescue the arrest of tooth organogenesis in $Lef1^{-/-}$ mice. Genes Dev. 16, 3173–3185.

Kubo, A., Shinozaki, K., Shannon, J., Kouskoff, V., Kennedy, M., Woo, S., Fehling, H., Keller, G., 2004. Development of definitive endoderm from embryonic stem cells in culture. Development 131, 1651–1662.

Kumar, M., Jordan, N., Melton, D., Grapin-Botton, A., 2003. Signals from lateral plate mesoderm instruct endoderm toward a pancreatic fate. Dev. Biol. 259, 109–122.

Kwon, G.S., Viotti, M., Hadjantonakis, A.K., 2008. The endoderm of the mouse embryo arises by dynamic widespread intercalation of embryonic and extraembryonic lineages. Dev. Cell 15, 509–520.

Lawson, A., Schoenwolf, G.C., 2003. Epiblast and primitive-streak origins of the endoderm in the gastrulating chick embryo. Development 130, 3491–3501.

Lawson, K.A., Meneses, J.J., Pedersen, R.A., 1986. Cell fate and cell lineage in the endoderm of the presomite mouse embryo, studied with an intracellular tracer. Dev. Biol. 115, 325–339.

Lawson, K.A., Pedersen, R.A., 1987. Cell fate, morphogenetic movement and population kinetics of embryonic endoderm at the time of germ layer formation in the mouse. Development 101, 627–652.

Le Douarin, N., 1968. [Synthesis of glycogen in hepatocytes undergoing differentiation: role of homologous and heterologous mesenchyma]. Dev. Biol. 17, 101–114.

Le Douarin, N.M., 1988. On the origin of pancreatic endocrine cells. Cell 53, 169–171.

Lee, C.S., Friedman, J.R., Fulmer, J.T., Kaestner, K.H., 2005. The initiation of liver development is dependent on Foxa transcription factors. Nature 435, 944–947.

Li, F.F., Zhang, T., Bai, Y.Z., Yuan, Z.W., Wang, W.L., 2011. Spatiotemporal expression of Wnt5a during the development of the hindgut and anorectum in human embryos. Int. J. Colorectal Dis. 26, 983–988.

Lickert, H., Domon, C., Huls, G., Wehrle, C., Duluc, I., Clevers, H., Meyer, B.I., Freund, J.N., Kemler, R., 2000. Wnt/(beta)-catenin signaling regulates the expression of the homeobox gene Cdx1 in embryonic intestine. Development 127, 3805–3813.

Manley, N.R., Capecchi, M.R., 1995. The role of Hoxa-3 in mouse thymus and thyroid development. Development 121, 1989–2003.

Manley, N.R., Capecchi, M.R., 1998. Hox group 3 paralogs regulate the development and migration of the thymus, thyroid, and parathyroid glands. Dev. Biol. 195, 1–15.

Martin, M., Gallego-Llamas, J., Ribes, V., Kedinger, M., Niederreither, K., Chambon, P., Dolle, P., Gradwohl, G., 2005. Dorsal pancreas agenesis in retinoic acid-deficient Raldh2 mutant mice. Dev. Biol. 284, 399–411.

Martinez Barbera, J.P., Clements, M., Thomas, P., Rodriguez, T., Meloy, D., Kioussis, D., Beddington, R.S., 2000. The homeobox gene Hex is required in definitive endodermal tissues for normal forebrain, liver and thyroid formation. Development 127, 2433–2445.

Mayhew, C.N., Wells, J.M., 2009. Converting human pluripotent stem cells into beta-cells: recent advances and future challenges. Curr. Opin. Organ Trans. 15, 54–60.

McCracken, K.W., Wells, J.M., 2012. Molecular pathways controlling pancreas induction. Semin. Cell Dev. Biol. 23, 656–662.

McKnight, K.D., Hou, J., Hoodless, P.A., 2010. Foxh1 and Foxa2 are not required for formation of the midgut and hindgut definitive endoderm. Dev. Biol. 337, 471–481.

McLin, V.A., Rankin, S.A., Zorn, A.M., 2007. Repression of Wnt/{beta}-catenin signaling in the anterior endoderm is essential for liver and pancreas development. Development 134, 2207–2217.

McLin, V.A., Zorn, A.M., 2006. Molecular control of liver development. Clin. Liver Dis. 10, 1–25. v.

Moore-Scott, B.A., Manley, N.R., 2005. Differential expression of Sonic hedgehog along the anterior-posterior axis regulates patterning of pharyngeal pouch endoderm and pharyngeal endoderm-derived organs. Dev. Biol. 278, 323–335.

Moore-Scott, B.A., Opoka, R., Lin, S.C., Kordich, J.J., Wells, J.M., 2007. Identification of molecular markers that are expressed in discrete anterior-posterior domains of the endoderm from the gastrula stage to mid-gestation. Dev. Dyn. 236, 1997–2003.

Mou, H., Zhao, R., Sherwood, R., Ahfeldt, T., Lapey, A., Wain, J., Sicilian, L., Izvolsky, K., Musunuru, K., Cowan, C., Rajagopal, J., 2012. Generation of multipotent lung and airway progenitors from mouse ESCs and patient-specific cystic fibrosis iPSCs. Cell Stem. Cell 10, 385–397.

Murtaugh, L.C., Law, A.C., Dor, Y., Melton, D.A., 2005. Beta-catenin is essential for pancreatic acinar but not islet development. Development 132, 4663–4674.

Negishi, T., Nagai, Y., Asaoka, Y., Ohno, M., Namae, M., Mitani, H., Sasaki, T., Shimizu, N., Terai, S., Sakaida, I., Kondoh, H., Katada, T., Furutani-Seiki, M., Nishina, H., 2010. Retinoic acid signaling positively regulates liver specification by inducing wnt2bb gene expression in medaka. Hepatology 51, 1037–1045.

Niederreither, K., Vermot, J., Le Roux, I., Schuhbaur, B., Chambon, P., Dolle, P., 2003. The regional pattern of retinoic acid synthesis by RALDH2 is essential for the development of posterior pharyngeal arches and the enteric nervous system. Development 130, 2525–2534.

Northrop, J.L., Kimelman, D., 1994. Dorsal-ventral differences in *Xcad-3* expression in response to FGF-mediated induction in *Xenopus*. Dev. Biol. 161, 490–503.

Nostro, M.C., Sarangi, F., Ogawa, S., Holtzinger, A., Corneo, B., Li, X., Micallef, S.J., Park, I.H., Basford, C., Wheeler, M.B., Daley, G.Q., Elefanty, A.G., Stanley, E.G., Keller, G., 2011. Stage-specific signaling through TGFbeta family members and WNT regulates patterning and pancreatic specification of human pluripotent stem cells. Development 138, 861–871.

Ober, E.A., Verkade, H., Field, H.A., Stainier, D.Y., 2006. Mesodermal Wnt2b signaling positively regulates liver specification.[see comment]. Nature 442, 688–691.

Ohlsson, H., Karlsson, K., Edlund, T., 1993. IPF1, a homeodomain-containing transactivator of the insulin gene. Embo J. 12, 4251–4259.

Park, K.S., Wells, J.M., Zorn, A.M., Wert, S.E., Laubach, V.E., Fernandez, L.G., Whitsett, J.A., 2006a. Transdifferentiation of ciliated cells during repair of the respiratory epithelium. Am. J. Res. Cell Mol. Biol. 34, 151–157.

Park, K.S., Wells, J.M., Zorn, A.M., Wert, S.E., Whitsett, J.A., 2006b. Sox17 influences the differentiation of respiratory epithelial cells. Dev. Biol. 294, 192–202.

Poulain, M., Ober, E.A., 2011. Interplay between Wnt2 and Wnt2bb controls multiple steps of early foregut-derived organ development. Development 138, 3557–3568.

Ramalho-Santos, M., Melton, D.A., McMahon, A.P., 2000. Hedgehog signals regulate multiple aspects of gastrointestinal development. Development 127, 2763–2772.

Rankin, S.A., Gallas, A.L., Neto, A., Gomez-Skarmeta, J.L., Zorn, A.M., 2012. Suppression of Bmp4 signaling by the zinc-finger repressors Osr1 and Osr2 is required for Wnt/beta-catenin-mediated lung specification in *Xenopus*. Development 139, 3010–3020.

Rankin, S.A., Kormish, J., Kofron, M., Jegga, A., Zorn, A.M., 2011. A gene regulatory network controlling hhex transcription in the anterior endoderm of the organizer. Dev. Biol. 351, 297–310.

Riccomagno, M.M., Martinu, L., Mulheisen, M., Wu, D.K., Epstein, D.J., 2002. Specification of the mammalian cochlea is dependent on Sonic hedgehog. Genes Dev. 16, 2365–2378.

Roberts, D.J., Johnson, R.L., Burke, A.C., Nelson, C.E., Morgan, B.A., Tabin, C., 1995. Sonic hedgehog is an endodermal signal inducing Bmp-4 and Hox genes during induction and regionalization of the chick hindgut. Development 121, 3163–3174.

Roberts, D.J., Smith, D.M., Goff, D.J., Tabin, C.J., 1998. Epithelial-mesenchymal signaling during the regionalization of the chick gut. Development 125, 2791–2801.

Rojas, A., Schachterle, W., Xu, S.M., Martin, F., Black, B.L., 2010. Direct transcriptional regulation of Gata4 during early endoderm specification is controlled by FoxA2 binding to an intronic enhancer. Dev. Biol. 346, 346–355.

Rosenquist, G.C., 1971. The location of the pregut endoderm in the chick embryo at the primitive streak stage as determined by radioautographic mapping. Dev. Biol. 26, 323–335.

Sala, F.G., Curtis, J.L., Veltmaat, J.M., Del Moral, P.M., Le, L.T., Fairbanks, T.J., Warburton, D., Ford, H., Wang, K., Burns, R.C., Bellusci, S., 2006. Fibroblast growth factor 10 is required for survival and proliferation but not differentiation of intestinal epithelial progenitor cells during murine colon development. Dev. Biol. 299, 373–385.

Savory, J.G., Bouchard, N., Pierre, V., Rijli, F.M., De Repentigny, Y., Kothary, R., Lohnes, D., 2009. Cdx2 regulation of posterior development through non-Hox targets. Development 136, 4099–4110.

Seifert, A.W., Bouldin, C.M., Choi, K.S., Harfe, B.D., Cohn, M.J., 2009. Multiphasic and tissue-specific roles of sonic hedgehog in cloacal septation and external genitalia development. Development 136, 3949–3957.

Seifert, A.W., Harfe, B.D., Cohn, M.J., 2008. Cell lineage analysis demonstrates an endodermal origin of the distal urethra and perineum. Dev. Biol. 318, 143–152.

Serls, A.E., Doherty, S., Parvatiyar, P., Wells, J.M., Deutsch, G.H., 2005. Different thresholds of fibroblast growth factors pattern the ventral foregut into liver and lung. Development 132, 35–47.

Sherwood, R.I., Maehr, R., Mazzoni, E.O., Melton, D.A., 2011. Wnt signaling specifies and patterns intestinal endoderm. Mech. Dev. 128, 387–400.

Shifley, E.T., Kenny, A.P., Rankin, S.A., Zorn, A.M., 2012. Prolonged FGF signaling is necessary for lung and liver induction in *Xenopus*. BMC Dev. Biol. 12, 27.

Spence, J.R., Lange, A.W., Lin, S.C., Kaestner, K.H., Lowy, A.M., Kim, I., Whitsett, J.A., Wells, J.M., 2009. Sox17 regulates organ lineage segregation of ventral foregut progenitor cells. Dev. Cell 17, 62–74.

Spence, J.R., Lauf, R., Shroyer, N.F., 2011a. Vertebrate intestinal endoderm development. Dev. Dyn. 240, 501–520.

Spence, J.R., Mayhew, C.N., Rankin, S.A., Kuhar, M.F., Vallance, J.E., Tolle, K., Hoskins, E.E., Kalinichenko, V.V., Wells, S.I., Zorn, A.M., Shroyer, N.F., Wells, J.M., 2011b. Directed differentiation of human pluripotent stem cells into intestinal tissue *in vitro*. Nature 470, 105–109.

Spence, J.R., Wells, J.M., 2007. Translational embryology: using embryonic principles to generate pancreatic endocrine cells from embryonic stem cells. Dev. Dyn. 236, 3218–3227.

Stafford, D., White, R.J., Kinkel, M.D., Linville, A., Schilling, T.F., Prince, V.E., 2006. Retinoids signal directly to zebrafish endoderm to specify insulin-expressing {beta}-cells. Development 133, 949–956.

Stringer, E.J., Pritchard, C.A., Beck, F., 2008. Cdx2 initiates histodifferentiation of the midgut endoderm. FEBS Lett. 582, 2555–2560.

Sumazaki, R., Shiojiri, N., Isoyama, S., Masu, M., Keino-Masu, K., Osawa, M., Nakauchi, H., Kageyama, R., Matsui, A., 2004. Conversion of biliary system to pancreatic tissue in Hes1-deficient mice. Nat. Genet. 36, 83–87.

Tai, C.C., Sala, F.G., Ford, H.R., Wang, K.S., Li, C., Minoo, P., Grikscheit, T.C., Bellusci, S., 2009. Wnt5a knock-out mouse as a new model of anorectal malformation. J. Surgical Res. 156, 278–282.

Tam, P.P., Khoo, P.L., Wong, N., Tsang, T.E., Behringer, R.R., 2004. Regionalization of cell fates and cell movement in the endoderm of the mouse gastrula and the impact of loss of Lhx1(Lim1) function. Dev. Biol. 274, 171–187.

Tamplin, O.J., Kinzel, D., Cox, B.J., Bell, C.E., Rossant, J., Lickert, H., 2008. Microarray analysis of Foxa2 mutant mouse embryos reveals novel gene expression and inductive roles for the gastrula organizer and its derivatives. BMC Gen. 9, 511.

Theodosiou, N.A., Tabin, C.J., 2003. Wnt signaling during development of the gastrointestinal tract. Dev. Biol. 259, 258–271.

Tremblay, K.D., Zaret, K.S., 2005. Distinct populations of endoderm cells converge to generate the embryonic liver bud and ventral foregut tissues. Dev. Biol. 280, 87–99.

Trokovic, N., Trokovic, R., Mai, P., Partanen, J., 2003. Fgfr1 regulates patterning of the pharyngeal region. Genes Dev. 17, 141–153.

Vermot, J., Niederreither, K., Garnier, J.M., Chambon, P., Dolle, P., 2003. Decreased embryonic retinoic acid synthesis results in a DiGeorge syndrome phenotype in newborn mice. Pro. Nat. Acad. Sci. USA 100, 1763–1768.

Wandzioch, E., Zaret, K.S., 2009. Dynamic signaling network for the specification of embryonic pancreas and liver progenitors. Science 324, 1707–1710.

Wang, C., Wang, J., Borer, J.G., Li, X., 2013. Embryonic origin and remodeling of the urinary and digestive outlets. PLoS One 8, e55587.

Wang, S., Zhang, J., Zhao, A., Hipkens, S., Magnuson, M.A., Gu, G., 2007. Loss of Myt1 function partially compromises endocrine islet cell differentiation and pancreatic physiological function in the mouse. Mech. Dev. 124, 898–910.

Wang, X.D., 2005. Alcohol, vitamin A, and cancer. Alcohol 35, 251–258.

Wang, Z., Dolle, P., Cardoso, W., Niederreither, K., 2006. Retinoic acid regulates morphogenesis and patterning of posterior foregut derivatives. Dev. Biol. 297, 433–445.

Wells, J.M., Esni, F., Boivin, G.P., Aronow, B.J., Stuart, W., Combs, C., Sklenka, A., Leach, S.D., Lowy, A.M., 2007. Wnt/beta-catenin signaling is required for development of the exocrine pancreas. BMC Dev. Biol. 7, 4.

Wells, J.M., Melton, D.A., 1999. Vertebrate endoderm development. Annu. Rev. Cell Dev. Biol. 15, 393–410.

Wells, J.M., Melton, D.A., 2000. Early mouse endoderm is patterned by soluble factors from adjacent germ layers. Development 127, 1563–1572.

Whitsett, J.A., Haitchi, H.M., Maeda, Y., 2011. Intersections between pulmonary development and disease. Am. J. Res. Crit. Care Med. 184, 401–406.

Wu, X., Ferrara, C., Shapiro, E., Grishina, I., 2009. Bmp7 expression and null phenotype in the urogenital system suggest a role in re-organization of the urethral epithelium. Gene Exp. Patterns 9, 224–230.

Xu, C.R., Cole, P.A., Meyers, D.J., Kormish, J., Dent, S., Zaret, K.S., 2011. Chromatin "prepattern" and histone modifiers in a fate choice for liver and pancreas. Science 332, 963–966.

Xu, K., Wu, X., Shapiro, E., Huang, H., Zhang, L., Hickling, D., Deng, Y., Lee, P., Li, J., Lepor, H., Grishina, I., 2012. Bmp7 functions via a polarity mechanism to promote cloacal septation. PLoS One 7, e29372.

Yamagishi, H., Maeda, J., Hu, T., McAnally, J., Conway, S.J., Kume, T., Meyers, E.N., Yamagishi, C., Srivastava, D., 2003. Tbx1 is regulated by tissue-specific forkhead proteins through a common Sonic hedgehog-responsive enhancer. Genes Dev. 17, 269–281.

Yamaguchi, T.P., Harpal, K., Henkemeyer, M., Rossant, J., 1994. fgfr-1 is required for embryonic growth and mesodermal patterning during mouse gastrulation. Genes Dev. 8, 3032–3044.

Zeng, X., Yutzey, K.E., Whitsett, J.A., 1998. Thyroid transcription factor-1, hepatocyte nuclear factor-3beta and surfactant protein A and B in the developing chick lung. J. Anat. 193, 399–408.

Zeynali, B., Dixon, K.E., 1998. Effects of retinoic acid on the endoderm in *Xenopus* embryos. Dev. Genes Evol. 208, 318–326.

Zhang, Z., Rankin, S.A., Zorn, A.M., 2013. Different thresholds of Wnt-Frizzled 7 signaling coordinate proliferation, morphogenesis and fate of endoderm progenitor cells. Dev. Biol. 378, 1–12.

Zorn, A.M., Wells, J.M., 2007. Molecular basis of vertebrate endoderm development. Int. Rev. Cytol. 259, 49–111.

Zorn, A.M., Wells, J.M., 2009. Vertebrate endoderm development and organ formation. Annu. Rev. Cell Dev. Biol. 25, 1–31.

Chapter 31

Pancreas Development and Regeneration

Kimberly G. Riley* **and Maureen Gannon*,†**

**Department of Cell and Developmental Biology, Vanderbilt University Medical Center, Nashville, TN; †VA Medical Center, Nashville, TN and Departments of Medicine and Molecular Physiology and Biophysics, Vanderbilt University Medical Center, Nashville, TN*

Chapter Outline

GLOSSARY

β-cell The predominant endocrine cell type in the pancreatic islets; it is the only cell in the body that is capable of producing and secreting the hormone insulin.

Diabetes A heterogeneous group of diseases in which the pancreas fails to produce insulin in sufficient amounts to maintain euglycemia, thus causing a dramatic rise in blood glucose levels.

Directed differentiation The process by which embryonic stem cells or induced pluripotent stem cells are progressively differentiated down a particular lineage by using signals that attempt to recapitulate embryonic development.

Endocrine cells Within the pancreas, the cells that produce hormones such as insulin and glucagon that are secreted directly into the bloodstream, where they travel to their target tissues.

Euglycemia The state of maintaining blood glucose levels within the normal range (70 to 120 mg/dL).

Principles of Developmental Genetics.

Exocrine cells Within the pancreas, the cells that produce the digestive enzymes that are released into the pancreatic duct and eventually into the duodenum.

Glucose intolerance An impaired ability to clear glucose efficiently from the bloodstream after a meal; a prediabetic state.

Glucose-stimulated insulin secretion The increase in insulin released by pancreatic islets is proportional to the elevation in blood glucose; glucose must be metabolized by the β-cell for insulin to be released.

Insulin resistance The physiologic state in which insulin target tissues in the periphery (mainly liver and muscle) become insensitive to insulin, thereby requiring elevated plasma insulin levels to elicit the same biologic effect.

Islets Spherical clusters of pancreatic endocrine cells that are comprised mainly of insulin-producing β-cells.

Islet transplantation The experimental surgical procedure that is used to treat some patients with type 1 diabetes; cadaver donor islets are transplanted into the liver of patients via the portal vein, where they then begin to secrete insulin and reverse diabetes.

Mature onset diabetes of the young (MODY) A dominant monogenic form of type 2 diabetes.

Neogenesis The process of generating new β-cells from stem or progenitor cells.

Partial pancreatectomy A surgical procedure used in experimental animal models to stimulate pancreas and β-cell regeneration; a portion of the pancreas is removed, and new pancreas tissue is generated from the remnant organ.

Type 1 diabetes An autoimmune disease in which the insulin-producing β-cells are specifically destroyed.

Type 2 diabetes A disease that is usually associated with obesity in which the β-cells fail to produce enough insulin to overcome peripheral insulin resistance.

SUMMARY

- The pancreas is composed of two main cell types: exocrine cells and endocrine cells, both of which are derived from the endodermal germ layer.
- It is currently unclear if there exists a multipotent progenitor cell within the embryonic pancreas that is capable of giving rise to both exocrine and endocrine cell lineages.
- Progenitor cells for both the exocrine and endocrine cells reside within the embryonic pancreatic ductal epithelium.
- Many transcription factors have been identified that are critical for pancreas formation and the differentiation of the endocrine cell populations. However, less is known about the genes that control exocrine cell fate and differentiation.
- The β-cell mass has the capacity to expand and contract throughout the life of the organism in response to changing metabolic demands.
- In adults, the majority of new insulin-producing β-cells come from the proliferation of pre-existing β-cells, although, under some conditions (e.g., pancreatic injury, obesity), β-cell neogenesis may occur from cells that are located within the ductal epithelium.
- There are currently no good protocols for generating mature, glucose-responsive insulin-producing β-cells *in vitro* from embryonic stem (ES) or induced pluripotent stem cells (IPS) cells.
- Pancreatic progenitor cells generated *in vitro* from ES or iPS cells can differentiate into mature, glucose-responsice β-cells when transplanted *in vivo*. Thus, as yet unidentified *in vivo* factors promote terminal differentiation of β-cells from stem cell sources.

31.1 INTRODUCTION

As of 2010 approximately 26 million Americans (8.3% of the population) have diabetes, a heterogeneous group of disorders characterized by a decrease in functional insulin-producing β-cells resulting in increased blood glucose. Diabetes results from absolute (type 1) or relative (type 2) loss of functional β-cell mass. Whereas type 1 diabetes is characterized by the selective autoimmune destruction of β-cells (Roep and Peakman, 2012), type 2 diabetes occurs when the β-cell population fails to compensate for increased peripheral insulin resistance that is typically associated with obesity (Fox, 2010). Both forms of the disease would greatly benefit from treatment strategies that enhance β-cell regeneration, proliferation, and function. Although there were initial encouraging results from islet transplantation in achieving remission of type 1 diabetes (Ryan et al., 2001; Shapiro et al., 2000), this was limited by the low amount of donor tissue obtainable. In addition, islet transplantation is short-lived and inefficient, requiring multiple islet donors per patient. Nearly all islet transplants fail with the first three years due to a combination of factors, including ongoing autoimmunity and a less than ideal microenvironment in the liver (McCall and Shapiro, 2012).

The ability to induce β-cells or whole islets from resident adult pancreatic non-β-cells or other cell types *in vivo* or from embryonic stem (ES) cells or induced pluripotent stem (iPS) cells *in vitro* would provide an alternative source for replenishing β-cell mass in individuals with diabetes (Daley, 2012). Studies addressing the potential of adult pancreatic cell types to undergo proliferation, regeneration, transdifferentiation, and neogenesis to generate β-cells could lead to the restoration of β-cell mass in individuals with type 1 diabetes and enhanced β-cell compensation in patients with type 2 diabetes. Any approach to generating islet endocrine cells *in vivo* or *in vitro* will benefit greatly from a thorough understanding of the normal developmental processes that occur during pancreatic organogenesis (e.g., transcription factors, cell-signaling molecules, and cell-cell interactions that regulate endocrine differentiation and proliferation).

Much progress has been made in identifying the factors involved in normal development and differentiation of the various pancreatic cell types. Interestingly, mutations in several developmentally important factors have been identified in individuals with diabetes. In this chapter, we summarize what is known about the regulation of pancreas development, and the factors that control β-cell maturation. Knowledge of the pancreas development program is being used to generate insulin-producing cells or islets from stem cell sources (ES or iPS cells) *in vitro* or via transdifferentiation of embryonically related cells *in vivo*. Here we discuss the potential and challenges of different cell sources and strategies for generating glucose-responsive insulin-producing cells.

31.2 THE INITIAL STAGES OF PANCREATIC BUD FORMATION

The mature pancreas is composed of two distinct functional units: exocrine and endocrine. The exocrine component of the pancreas consists of clusters of acinar cells that produce and secrete digestive enzymes such as amylase and elastase, and the ductal network, which transports the acinar secretions into the rostral duodenum. The exocrine portion makes up ~98% of the adult organ. The endocrine compartment is composed of spherical clusters of five hormone-producing cell types: insulin (β-cells), glucagon (α-cells), somatostatin (δ-cells), ghrelin (ε-cells), and pancreatic polypeptide (PP cells; Figures 31.1 and 31.2). These endocrine clusters comprise micro-organs known as *islets of Langerhans*. Together, these islet hormones regulate glucose homeostasis by facilitating the uptake of ingested glucose into cells and stimulating glucose production by the liver during times of fasting. Acinar, ductal, and endocrine cells are all derived from the endoderm during embryonic development (Percival and Slack, 1999, see also Chapters 13 and 30 in this book).

The epithelial component of the pancreas arises from dorsal and ventral outgrowths of the posterior foregut endoderm just caudal to the liver diverticulum in all vertebrates examined, including humans (Figures 31.1 and 31.2; Field et al., 2003; Kelly and Melton, 2000; Kim et al., 1997; Ober et al., 2003; Wessells and Cohen, 1966). The dorsal and ventral pancreatic buds later fuse to form a single organ (this occurs on embryonic day (e) 12.5 in the mouse). Pancreatic bud outgrowth can be observed as early as day 25 of gestation in humans (Ashraf et al., 2005), e9.5 in the mouse (Slack, 1995), and 24 hours postfertilization in zebrafish (Figure 31.2; Field et al., 2003). In mammals, frogs, and chickens, both pancreatic buds consist of primary multipotent progenitor cells (MPCs) that generate exocrine and endocrine cells. (It is currently not clear whether these MPCs represent a homogeneous or heterogeneous population.) MPCs of the early pancreatic buds are demarked by

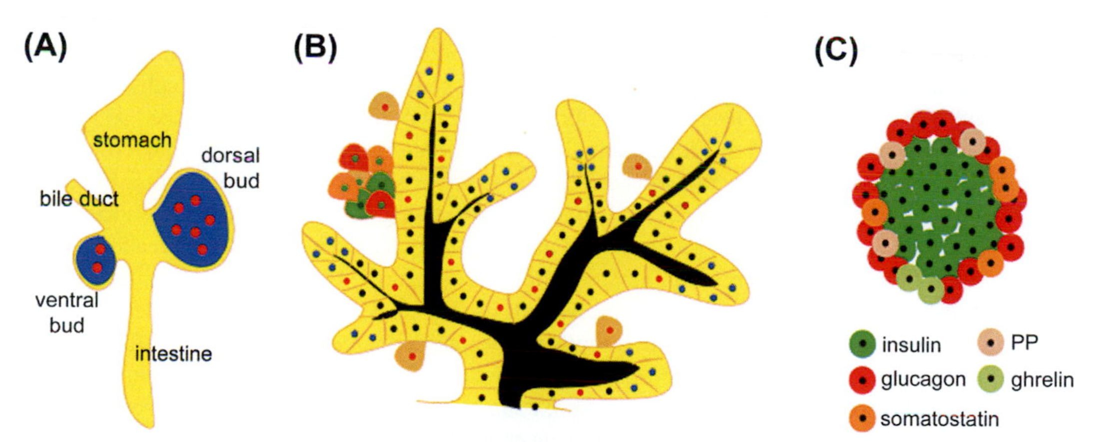

FIGURE 31.1 Schematic of Pancreas and Islet Development. (A) The pancreas arises as dorsal and ventral evaginations (buds) from the posterior foregut endoderm on embryonic day 9.5, which is marked by the expression of the *Pdx1* transcription factor *(yellow)*. Specific markers of the early multipotent pancreatic progenitor cells (MPCs) include the transcription factors *Ptf1a* and *Hb9 (blue)*. Within the developing buds, a subset of cells expresses markers of the endocrine lineage, including *Ngn3, Isl-1,* and *Pax6 (red)*. (B) As development proceeds, the pancreatic epithelium *(yellow)* becomes a highly branched ductal network. Secondary MPCs are located at the tips of branches (*blue nuclei*). Endocrine cells *(green nuclei)* and duct cells (*black nuclei*) arise from the trunk portion of the epithelium, while exocrine cells arise from cells at the tips of the branches. Endocrine progenitors are scattered throughout the epithelium, and they express *Ngn3 (red nuclei)*. These cells maintain *Ngn3* expression as they delaminate from the epithelium *(tan cells)*. *Ngn3* is downregulated as hormone expression begins *(green, red, and orange cells)*, and more general endocrine markers such as *Isl-1* and *Pax6* are expressed *(green nuclei)*. (C) Mature islets begin to form during late gestation. In the mouse, insulin-producing β-cells *(green)* are found at the islet core, and all other hormone-producing cells are located at the periphery.

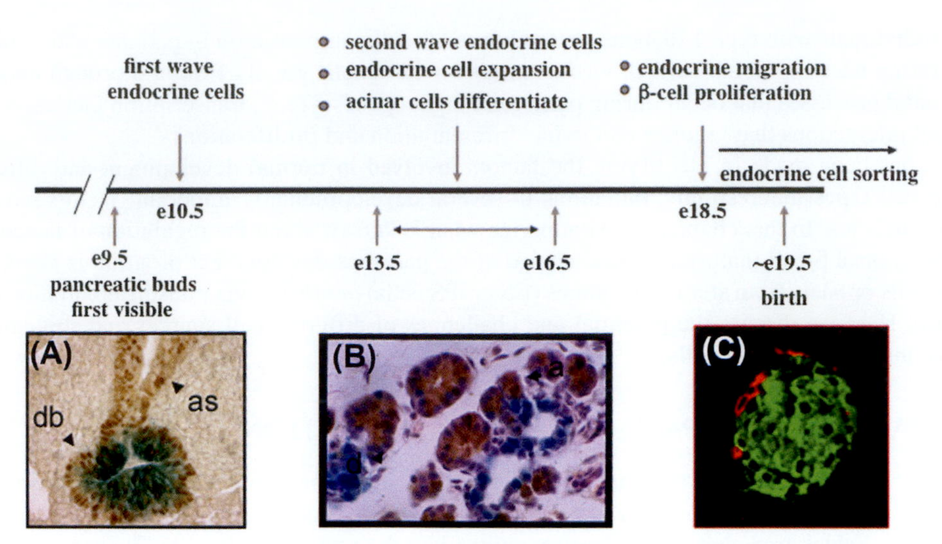

FIGURE 31.2 **Timeline of Pancreas Development in the Mouse.** Key events in mouse pancreas development are shown. (A) Pancreatic bud evagination can first be detected on embryonic day 9.5. In this image, the *Pdx1* expression domain is marked by the brown nuclei, and it includes the antral stomach *(as)* and the dorsal pancreatic bud *(db)*, which is also marked by a pancreas-specific lacZ transgene *(blue)*. Endocrine differentiation occurs in two waves. The first begins on embryonic day 10.5 and extends to embryonic day 13.5. The second wave begins on embryonic day 13.5 and continues until neonatal stages. (B) Acinar gene expression *(a; amylase in brown)* begins on embryonic day 14.5. Both acinar cells and endocrine cells bud off of the ductal epithelium *(d; blue)*. (C) During late gestation, β-cell proliferation increases, endocrine cells cluster, migrate away from the ductal epithelium, and organize into islets with β-cells at the core and other hormone cell types at the periphery. *Green,* Insulin, *red,* glucagon.

expression of *Pdx1, Ptf1a, Nkx6.1, Sox9, Hnf6, and Hnf1β*, each of which will be discussed later. By contrast, in zebrafish, the posterior (ventral) bud generates the endocrine tissue, which usually consists of a single islet, whereas the anterior (dorsal) bud gives rise primarily to the pancreatic duct and the acinar cells, although endocrine cells do arise from dorsal bud derivatives later during development (Field et al., 2003).

Morphogenesis of the pancreatic epithelium yields a highly branched ductal network within which multipotent progenitors of both exocrine and endocrine cells reside (Figure 31.1; Zhou et al., 2007). At approximately e12.5 in the mouse, the branching pancreatic epithelium becomes divided into "tip" and "trunk" domains via morphological and gene expression pattern changes. "Tip" portions of the developing pancreas retain secondary MPCs that express a unique combination of molecular markers compared with those found in the emerging buds earlier in development. The "tip" MPCs express *Ptf1a, c-myc,* and *Cpa1,* and continue to generate all pancreatic cell types (endocrine, duct or acinar). In addition, this secondary MPC population expresses *Hnf6* and *Pdx1*. *Hnf6* and *Pdx1* are expressed at lower levels in tip cells than observed in the primary pancreatic bud MPCs. "Tip" cells lose their multipotency by e14.5, a period known as the "secondary transition" which is discussed in more detail below. At this time point, the "tip" cell fate becomes restricted solely to the acinar lineage. The "trunk" domain cells originate from the main duct and extend to the border of the "tip" progenitor population. "Trunk" cells express *Hnf6*, *Hnf1β*, *Sox9*, *Muc1*, and *Nkx6.1* and are capable of giving rise to either duct or endocrine cells.

Expression of endocrine hormones such as glucagon and insulin is detected even at early pancreatic bud stages (e10.5); exocrine-specific gene transcription does not commence until e14.5 (Figure 31.2; Gittes and Rutter, 1992). Pancreatic endocrine differentiation occurs in two waves during embryogenesis (Figure 31.2; reviewed in Guney and Gannon, 2009). The first wave occurs between e9.5 and e13.5. Unlike second wave endocrine cells, these early differentiating hormone-producing cells can develop in the absence of the pancreatic duodenal homeobox 1 *(Pdx1)* gene, a critical pancreatic transcription factor (Ahlgren et al., 1996; Offield et al., 1996; Pang et al., 1994), and lack other genetic markers of mature islet endocrine cells (Lee et al., 1999; Pang et al., 1994; Wilson et al., 2002). These early hormone-positive cells have been shown by lineage tracing analyses to not contribute to mature islets (Herrera, 2000; Herrera et al., 1994). During the second wave, or the "secondary transition," of endocrine differentiation, which commences on e13.5 in the mouse, the numbers of endocrine cells greatly increase. These endocrine cells go on to populate the mature islets. The mechanism for the increase in endocrine differentiation at this stage is unknown. The formation of mature, optimally functional islets requires the generation of appropriate numbers of each endocrine cell type, and this process is likely regulated by positive and negative factors that influence cell proliferation and differentiation. As an example, recent studies indicate a role for two planar cell polarity (PCP) genes, Celsr2 and Celsr3, in β-cell differentiation during development; loss of these gene results in a decrease in β-cell number, while other endocrine cell types are not affected (Cortijo et al., 2012).

In a process reminiscent of neurogenesis in *Drosophila* and other organisms, islet organogenesis involves the delamination of specified endocrine cells from the ductal epithelium, migration away from the ducts, and the formation of adherent islet clusters (Figure 31.1; reviewed in Pan and Wright, 2011). As in *Drosophila* neurogenesis, the specification of endocrine progenitors within the ductal epithelium is dependent on cell-cell communication using Notch-Delta signaling. Cells experiencing higher levels of Notch signaling remain within the epithelium and actively repress *neurogenin 3* (*Ngn3*; a transcription factor expressed in endocrine progenitors), whereas those cells in which Delta levels become elevated activate *Ngn3*, exit from the epithelium, and ultimately give rise to the endocrine population (Figure 31.1) (Afelik et al., 2012; Apelqvist et al., 1999; Golson et al., 2009; Gu et al., 2002; Jensen et al., 2000a,b). Recent studies have shown that the Ptf1a transcription factor activates the Notch ligand, Dll1 in MPCs, thus providing an alternative, Ngn3-independent pathway to regulate Notch signaling and subsequent pancreas cell fate decsions (Ahnfelt-Ronne et al., 2012).

After delamination, the endocrine cells begin to organize into clusters that are initially still located close to ducts (Figure 31.1). Beginning around e18.5, these clusters lose their proximity to the ductal epithelium, and begin to form mature islets (Figures 31.1 and 31.2). As islets form, the endocrine cells segregate such that, in mice, the insulin-producing β-cells are located at the core, and glucagon-, somatostatin-, ghrelin- and PP-producing cells are located at the periphery or mantle (Figures 31.1 and 31.2). Little is known about how the different endocrine cell types and their precursors interact with one another to form functional islets. The processes of endocrine delamination and islet formation likely include changes in the expression of lineage-specific transcription factors, cell adhesion molecules, and extracellular matrix components. For example, in the developing human pancreas, cell adhesion molecules and certain integrins mark endocrine progenitor cells within and delaminating from the ductal epithelium (Cirulli et al., 2000; Yebra et al., 2011).

31.3 INDUCTIVE INTERACTIONS DURING PANCREAS DEVELOPMENT

Pancreas development is dependent on an interaction between epithelial (endodermal) and mesodermal components (reviewed in Guney and Gannon, 2009). For example, signals from the notochord have been implicated in pancreas specification and early bud formation (Hebrok et al., 1998; 2000; Kim et al., 1997; Kim and Melton, 1998), whereas pancreatic mesenchyme stimulates the overall growth of the endodermal epithelium (Ahlgren et al., 1997; Wessells and Cohen, 1966). In turn, the endoderm influences the character of the overlying mesoderm (Apelqvist et al., 1997; Slack, 1995).

Early Inductive Events in Pancreatic Endocrine Differentiation: The Role of the Notochord

Wessells and Cohen (1966) suggested that signals derived from dorsal axial tissue such as the notochord might be involved in inducing the outgrowth of the dorsal pancreatic bud. Experimental manipulations in chick embryos revealed that, in the absence of the notochord, the dorsal pancreatic bud undergoes only limited outgrowth and branching, and fails to activate the expression of pancreatic transcription factors (e.g., *Pdx1, Isl-1, Pax6*) and markers of differentiated endocrine or exocrine cells (Kim et al., 1997). By contrast, the outgrowth and differentiation of the ventral pancreatic bud occurs normally in the absence of the notochord. It is currently unclear what tissue interactions promote ventral pancreas development, although genes involved in ventral bud development have been identified (discussed later). Activin βB and fibroblast growth factor (FGF)-2 are likely to be the endogenous notochord-derived signals that induce dorsal pancreas bud outgrowth and differentiation (Hebrok et al., 1998). One of the main functions of notochord-derived factors is repression of endodermal Sonic hedgehog (Shh) expression in the region that is destined to give rise to the pancreas (Hebrok et al., 1998). Transplantation of an ectopic notochord to nonpancreatic regions of the developing gut tube results in decreased Shh expression in the region that is adjacent to the transplant (Hebrok et al., 1998). Maintenance of Shh in presumptive pancreatic endoderm using a transgenic approach results in impaired development of the pancreatic epithelium and altered character of the overlying mesoderm such that it expresses markers consistent with small intestine smooth muscle (Apelqvist et al., 1997).

Early Inductive Events in Pancreatic Endocrine Differentiation: The Role of Endothelial Cells

The vasculature of the pancreas is derived from the mesodermal germ layer. Although islets represent only approximately 2% of the total mass of the pancreas in an adult, they receive up to 15% of the blood flow (Brissova and Powers, 2008), facilitating secretion of hormones directly into the bloodstream. The morphology and architecture of endothelial cells differs among the different capillary beds (LeCouter et al., 2002). Capillaries in endocrine glands such as the islets are fenestrated (Cleaver and Melton, 2003; LeCouter et al., 2002). Early differentiating pancreatic endocrine cells (both first and second wave cells) produce angiogenic factors including vascular endothelial growth factor (VEGF) and angiopoietin 1 (Brissova et al., 2006); the expression of these factors is maintained in adult islets, suggesting that maintenance of a fenestrated endothelium is critical for mature islet function (Christofori et al., 1995; Lammert et al., 2001; Rooman et al., 1997). In addition to islet endocrine cells communicating with vascular endothelial cells via secreted growth factors, endothelial cells also signal to the pancreatic epithelium, influencing the differentiation of endocrine cells (Guney et al., 2011; Lammert et al., 2003b; Yoshitomi and Zaret, 2004).

Pancreatic bud outgrowth is initiated at sites in the posterior foregut endoderm, where it contacts the endothelium of major blood vessels. Endocrine differentiation initially occurs in cells that have direct contact with endothelial cells (Lammert et al., 2001; Matsumoto et al., 2001). Endothelial cells also produce basement membrane components for developing islets (Nikolova et al., 2006). The importance of vascular endothelial cells in pancreatic endocrine differentiation has been demonstrated both in tissue recombination experiments and in genetically modified mice. For example, e8.5 endoderm cultured in the absence of endothelial cells failed to activate either Pdx1or insulin protein expression; undifferentiated endoderm cultured in combination with dorsal aorta, both Pdx1 and insulin were induced (Lammert et al., 2001). Likewise, coculture of mouse ES cells with human endothelial cells enhances the formation of insulin-producing cells (Talavera-Adame et al., 2011). In VEGF receptor type 2 (VEGFR-2/flk-1) null mutant mice, which die before the second wave of endocrine differentiation, early insulin- and glucagon positive cells fail to develop (Lammert et al., 2001; Yoshitomi and Zaret, 2004). These mice express most pancreatic/endocrine transcription factors *(Pdx1, Hnf6, Ngn3, NeuroD, Prox1,* and *Hb9)*, with the exception of the early pancreatic bud marker, *Ptf1a* (Yoshitomi and Zaret, 2004). Transgenic embryos expressing $VEGF_{164}$ throughout the entire pancreatic bud early during development show greatly increased vasculature in the pancreas and a corresponding three-fold increase in islet number and islet area (Lammert et al., 2003b). The increase in endocrine area was at the expense of acinar tissue, suggesting endothelial factors promote the endocrine lineage (Brissova et al., 2006). However, forced continuous over-expression of VEGF in embryonic β-cells results in impaired islet morphology and decreased β-cell proliferation and mass (Cai et al., 2012; Magenheim et al., 2011), indicating the need for tight regulation of angiogenic factors.

These studies highlight the reciprocal communication between pancreatic endocrine cells and endothelial cells. Vascular endothelial cells produce several different secreted signaling molecules, including spingosine-1-phosphate, FGF, transforming growth factor (TGF)-β, Wnt, and hepatocyte growth factor (HGF; (Edsbagge et al., 2005; Lammert et al., 2003a). Pancreatic endothelial cells express HGF and HGF is mitogenic to β-cells (Araujo et al., 2012; Demirci et al., 2012; Garcia-Ocana et al., 2000). Thus, in the pancreas, endocrine-produced VEGF signaling through VEGF receptors on the endothelium may induce the expression of HGF, which in turn promotes endocrine proliferation.

31.4 GENES THAT AFFECT PANCREATIC BUD DEVELOPMENT

Pancreatic development requires factors that act autonomously within the endodermally derived epithelium as well as factors that function within adjacent mesodermally-derived tissues. Gene inactivation in mice has identified transcription factors that affect the differentiation of all or of a subset of the pancreatic cell types. Factors that regulate the differentiation and function of islet β-cells are candidates for susceptibility genes in type 2 diabetes. Indeed, most of the genetic lesions that result in a dominant form of the disease called maturity onset diabetes of the young (MODY) are associated with mutations in the transcription factors that are expressed in adult β-cells: *HNF4α* (MODY1), *HNF1α* (MODY3), *Pdx1* (MODY4), *HNF1β* (MODY5), *Beta2/NeuroD* (MODY6), *KLF11* (MODY7), *CEL* (MODY8), *Pax4* (MODY9), *ins* (MODY10), and *blk* (MODY11) (Bonnefond et al., 2011; Hattersley, 2000; Teo et al., 2013). Some of these factors play critical roles in distinct stages of pancreatic/endocrine development, including pancreatic bud outgrowth and branching, endocrine differentiation, and/or mature islet function (Guney and Gannon, 2009; Pan and Wright, 2011). Promoter analysis of islet-specific genes such as insulin has also helped to identify *trans*-acting factors that are critical for normal pancreas and/or endocrine development (Bonnefond et al., 2012; Edlund, 1998; Madsen et al., 1997; Matsuoka et al., 2003; Sander and German, 1997). An understanding of how these transcription factors function during normal islet differentiation should facilitate the manipulation of this pathway to induce the differentiation of functional β-cells and/or islets from *in vivo* or *in vitro* cell sources. This section will summarize some of the results from gene inactivation and/or over-expression studies aimed at determining the role of these factors in pancreas development and endocrine differentiation.

Endodermally Expressed Genes that Affect Pancreatic Bud Formation

Pdx1

The homeodomain transcription factor *Pdx1* (*Ipf1* in the human) is one of the earliest known markers of the developing pancreas (Figures 31.1 and 31.2; Gannon, 1999), and lineage tracing analysis reveals that all cells within the endodermal

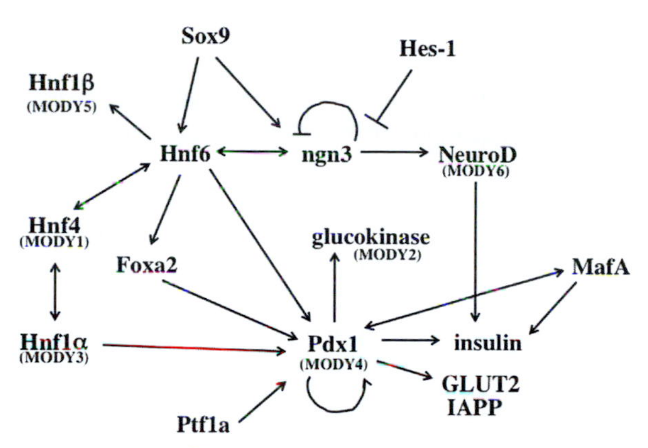

FIGURE 31.3 Simplified Pancreas Transcription Factor Network. Some of the factors that are important in the specification of the endocrine lineage and subsequent differentiated β-cells are shown. In particular, the interactions between different Mature Onset Diabetes of the Young (MODY) genes, transcriptional targets of *Hnf6* and *Pdx1,* and transcription factors that transactivate the insulin promoter are highlighted. Arrows indicate direct transcriptional targets. Bars indicate direct repression.

component of the pancreas are derived from a *Pdx1*$^{+}$ cell (Gu et al., 2002). *Pdx1* is expressed early (e9.5) throughout the endoderm of the pancreatic buds as well as in the antral stomach, the rostral duodenum, and the common bile duct. It becomes highly enriched in insulin-producing β-cells during late gestation, with lower levels of expression found in acinar cells (Guz et al., 1995; Offield et al., 1996; Wu et al., 1997). Pdx1 binds to and activates the insulin promoter and other β-cell-specific genes such as GLUT2, glucokinase, and islet amyloid polypeptide (IAPP) (Figure 31.3). In both mice and humans, homozygous inactivation of *Pdx1* results in pancreatic agenesis, whereas heterozygotes have impaired β-cell function and are prone to diabetes (Dutta et al., 1998; Jonsson et al., 1994; Nicolino et al., 2010; Offield et al., 1996; Stoffers et al., 1998; 1997). Mutations in *Pdx1* have been identified in some patients with type 2 diabetes (Hani et al., 1999; Stoffers et al., 1997). In *Pdx1* null mouse mutants, the dorsal bud undergoes limited proliferation and outgrowth to form a small, irregularly branched ductule (Offield et al., 1996). Transient insulin-positive cells and longer-lived glucagon positive clusters are found in the mutant epithelium (Ahlgren et al., 1996; Offield et al., 1996); these represent first wave (immature) endocrine cells. Thus, *Pdx1* is not absolutely essential for insulin or glucagon expression, although it is required to form mature, functional endocrine cells. *Pdx1* inactivation specifically in adult β-cells results in a loss of the β-cell phenotype, reduced β-cell proliferation and type 2 diabetes, thus demonstrating a role for *Pdx1* in the maintenance of β-cell function (Ahlgren et al., 1998; Gannon et al., 2008).

Ptf1a

Ptf1a/p48 is the pancreas-specific component of the heterotrimeric basic helix-loop-helix (bHLH) complex pancreas transcription factor 1 (PTF1; (Cockell et al., 1989; Krapp et al., 1996). The two other components of the PTF1 complex are the constitutively expressed HeLa E-box binding factor (HEB) (Cockell et al., 1989), and the mammalian Suppressor of Hairless (RBP-J; (Obata et al., 2001) or its paralog RBP-L (Beres et al., 2006). Although it was first identified as a regulator of exocrine-specific genes (Krapp et al., 1996; Rose et al., 2001), PTF1 was subsequently shown to be essential early during development for normal endocrine and exocrine pancreas formation in both mice and humans (Kawaguchi et al., 2002; Krapp et al., 1998; Obata et al., 2001; Sellick et al., 2004). PTF1 binds to and activates the *Pdx1* promoter in the early stages of development, but is not required for *Pdx1* expression (Burlison et al., 2008; Wiebe et al., 2007).

Ptf1a expression is detected as early as e9.5 throughout the developing pancreatic buds (Figure 31.1; Kawaguchi et al., 2002). Mice lacking *Ptf1a* have no detectable ventral bud outgrowth; the dorsal bud initiates but then arrests as a duct-like structure that lacks differentiated acinar cells (Kawaguchi et al., 2002; Krapp et al., 1998). Small clusters of first wave insulin and glucagon expressing cells are present within this structure, and a few isolated hormone-positive cells can be detected within the spleen. Lineage tracing studies revealed that acinar, ductal, and endocrine cells are all derived from a *Ptf1a*-expressing cell (Kawaguchi et al., 2002). In the absence of *Ptf1a,* presumptive ventral pancreatic cells instead become duodenal cells (Kawaguchi et al., 2002). These data along with studies showing that Ptf1a, in combination with Pdx1, can convert other regions of endoderm to a pancreatic fate suggests a high degree of plasticity in the endoderm in early development (Afelik et al., 2006). In zebrafish, *Ptf1a* is required for the endocrine and exocrine tissue that arises from the anterior (dorsal) bud, but it is not required for posterior (ventral) bud formation and outgrowth, which differentiates into endocrine tissue (Lin et al., 2004).

Hlxb9/Mnx1

Hlxb9/Mnx1 is the gene that encodes the Mnx1 homeodomain protein, which is expressed transiently throughout the pre-pancreatic epithelium during the early stages of mouse pancreatic development (e8.0 to e10.5) in both the dorsal and the ventral anlagen coincident with *Pdx1* and *Ptf1a* expression (Figure 31.1). *Mnx1* is re-expressed later (at e17.5) in differentiated β-cells and is prominently expressed in adult islet β-cells (Li et al., 1999). In the absence of Mnx1, the dorsal pancreatic bud fails to develop (Harrison et al., 1999; Li et al., 1999). By contrast, the ventral pancreatic bud develops, forming both endocrine and exocrine tissues; however, the endocrine cells are disorganized, and there are reduced numbers of insulin-producing cells. In zebrafish, loss of *Mnx1* results in β-cell precursors adopting an α-cell fate, suggesting that *Mnx1* may suppress the α-cell fate in developing β-cells (Dalgin et al., 2011).

Hex

The *Hex* homeobox gene is expressed in the ventral posterior foregut in the region that will give rise to both the liver bud and the ventral (but not the dorsal) pancreatic bud. The examination of *Hex* null embryos reveals a requirement for Hex in the proliferation and outgrowth of the liver bud and in the specification of the ventral pancreas (Bort et al., 2004). In the absence of Hex, the ventral pancreatic bud does not form, and it fails to express *Pdx1, Ptf1a,* and *Mnx1.* Dorsal pancreas specification appears to be normal in these mutants. Interestingly, explants of presumptive ventral pancreas endoderm differentiate normally in the absence of *Hex,* expressing *Pdx1* and endocrine lineage markers such as *Isl-1*, *Ngn3*, and *Beta2*. These results suggest that *Hex* null endoderm is fully competent to differentiate according to its normal fate, but that it is susceptible to influences from surrounding tissues within the intact embryonic environment that prevent ventral pancreas specification and differentiation. Indeed, when cocultured with cardiogenic mesoderm, *Hex*$^{-/-}$ ventral endoderm no longer expresses any pancreatic markers (Bort et al., 2004).

GATA4 and GATA6

GATA4 and GATA6 are zinc finger transcription factors expressed early in development in the foregut endoderm and later throughout the pancreatic epithelium (Decker et al., 2006). However, *Gata4* and *Gata6* expression diverges during pancreas progenitor differentiation, with *Gata4* being expressed in acinar cells, and *Gata6* expression limited to endocrine cell types (Ketola et al., 2004). Global null or pancreas-specific deletions of either *Gata4* or *Gata6* result in a failure to initate pancreatic development or near complete pancreatic agenesis, respectively (Carrasco et al., 2012; Decker et al., 2006; Watt et al., 2007; Xuan et al., 2012). In addition, *Gata6* haplo-insufficiency and functional mutations cause pancreatic agenesis or overt diabetes, respectively, in humans (De Franco et al., 2013; Lango et al., 2008).

Mesodermally Expressed Genes that Affect Pancreatic Bud Formation

Isl-1

The LIM homeodomain transcription factor *Islet-1 (Isl-1)* was first identified as a factor that binds to and transactivates the insulin gene promoter (Karlsson et al., 1990). It is expressed in all islet endocrine cell types in the embryo (Figure 31.1) and in the adult and in the mesodermally-derived mesenchyme surrounding the dorsal (but not the ventral) pancreatic bud (Ahlgren et al., 1997). Similar to the phenotype found in *Mnx1* mutant pancreata, the dorsal pancreatic bud does not develop and differentiate in *Isl-1* null embryos (Ahlgren et al., 1997). The ventral bud grows normally, but fails to produce any endocrine hormone-positive cell types; acinar cell differentiation appears normal. Unlike *Mnx1* mutant pancreata, in which the dorsal mesenchyme remains intact, the absence of dorsal bud derivatives in *Isl-1* null embryos is specifically the result of the loss of dorsal mesenchyme, because mesenchyme from wild type embryos can rescue the differentiation of exocrine cells from *Isl-1* mutant dorsal pancreatic endoderm in culture. Importantly, hormone-positive cells are not rescued by wild type mesenchyme, which suggests that *Isl-1* has a non-cell-autonomous role in the mesenchyme for dorsal bud outgrowth and a cell-autonomous role in the endoderm for endocrine differentiation. These results underscore the requirement for the pancreatic mesenchyme in pancreas development. In studies in which *Isl-1* was removed in the pancreatic epithelium at e13.5, endocrine precursors failed to mature into functional islet cells and postnatal endocrine cell mass expansion was impaired, resulting in diabetes (Du et al., 2009). Over-expression of *Isl-1* in the developing pancreas resulted in improved β-cell function postnatally, with no changes in β-cell mass (Liu et al., 2011).

N-Cadherin

During the early stages of pancreas development (e9.0 to e12.5), the cell adhesion molecule N-cadherin is expressed in both the pancreatic epithelium and the surrounding mesenchyme. After e12.5, expression becomes restricted to the formation of clusters of endocrine cells. In embryos lacking N-cadherin, the dorsal pancreatic bud fails to form, although genes such as *Mnx1* and *Isl-1* are expressed normally, and Shh expression is repressed in the dorsal endoderm (Esni et al., 2001). N-cadherin function is not required within the pancreatic endoderm, because coculture with wild type mesenchyme can rescue branching morphogenesis, exocrine, and endocrine differentiation (Johansson et al., 2010). In addition, restoring cardiac and circulatory function in N-cadherin null mice by the cardiac-specific transgenic expression of N-cadherin rescues the formation of the dorsal pancreas (Edsbagge et al., 2005). On the basis of this observation, it was proposed that soluble factors present in plasma are critical for the formation and/or maintenance of the dorsal pancreatic mesenchyme. Sphingosine 1-phosphate present in plasma was found to promote the budding of pancreatic endoderm by stimulating pancreatic mesenchymal cell proliferation; sphingosine 1-phosphate receptors are located within the mesenchyme (Edsbagge et al., 2005).

Retinoic Acid

Retinoic acid (RA) is produced by the foregut mesoderm and regulates the anterior and posterior boundaries of the posterior foregut. In zebrafish and *Xenopus* embryos, exogenous RA expression induced anterior expansion of pancreatic endoderm specification (Stafford et al., 2004). Later in development, RA signaling promotes both dorsal and ventral bud formation. When an enzyme required for RA synthesis, retinaldehyde dehydrogenase 2 (Raldh2), is removed in the mouse, the dorsal bud of the pancreas is lost (Martin et al., 2005; Molotkov et al., 2005). In addition, RA signaling induces the generation of endocrine progenitor cells and promotes their differentiation into β-cells later in development (Ostrom et al., 2008).

Fibroblast Growth Factors (FGFs)

The pancreatic mesenchyme and cardiac mesoderm at e8.5 highly express several FGFs in early pancreas development. FGF2 produced by the notochord, along with Activinβ-B, inhibits Shh signaling in the underlying endoderm to allow pancreas differentiation and to inhibit liver development (Deutsch et al., 2001; Shin et al., 2007; Zhang et al., 2004). Later in development at e9.5, the dorsal aorta induces the expression of FGF10, which is required for the proliferation of pancreatic progenitors and proper branching of the epithelium. Removal of *Fgf10* or its receptor *Fgfr2b* results in blunted pancreatic outgrowth and branching (Bhushan et al., 2001; Jacquemin et al., 2006).

31.5 GENES THAT AFFECT THE DIFFERENTIATION OF PARTICULAR PANCREATIC CELL TYPES

Markers of Multipotent Pancreatic Progenitors

Multipotent pancreatic progenitor cells (MPCs) have the potential to give rise to all pancreatic cell types. Several lineage tracing studies have analyzed the cell fate potentials of these MPCs. Since these studies were performed on populations of cells and not clones derived from single cells, it remains unclear whether a single MPC has the potential to give rise to all the different pancreatic cell types. The MPC pool maintains itself through *Sox9*-mediated proliferation, and inactivation of *Sox9* in the developing pancreas results in severe pancreatic hypoplasia (Seymour et al., 2007). Lineage tracing experiments have revealed that *Sox9*+ MPCs give rise to exocrine, endocrine and ductal cells (Kopp et al., 2011). *Pdx1* is also a marker of MPCs, as cells that at one time expressed *Pdx1* are able to give rise to all types of pancreatic cells (Gu et al., 2002; Herrera, 2000). Similarly, lineage tracing experiments showed that *Ptf1a* expressing cells have the capability of maturing into all pancreatic cell types, although not all endocrine cells derive from a *Ptf1a*-positive cell (Kawaguchi et al., 2002). Another marker of MPCs in the early pancreatic bud is hepatic nuclear factor 6 (Hnf6). Loss of *Hnf6* results in delayed *Pdx1* expression, pancreatic hypoplasia, and impaired endocrine and duct cell differentiation (Jacquemin et al., 2000; Pierreux et al., 2006; Zhang et al., 2009). The roles that these and other transcription factors play in pancreas differentiation will be discussed in more detail in the following sections.

Genes Involved in Exocrine Differentiation

There is actually very little known about the factors that specify the exocrine cell population from progenitors within the developing pancreatic trunk epithelium. Notch-Delta signaling appears to regulate differentiation of trunk progenitor cells into endocrine and duct cells (Apelqvist et al., 1999; Jensen et al., 2000a,b), in a manner somewhat similar to *Drosophilia* neurogenesis (Fisher and Caudy, 1998). Therefore, it was previously thought that cells that fail to activate Notch became endocrine cells, whereas those that activate the Notch pathway became duct cells. It is now clear that Notch activation maintains cells in a proliferative, progenitor state, and that other signals are required to generate duct cells from this pool. Broad activation of Notch in the developing pancreas inhibits exocrine differentiation in both mouse and zebrafish (Esni et al., 2004; Hald et al., 2003; Murtaugh et al., 2003). However, recent studies have indicated that Notch does not function in an "all or none" mode, but that a gradient of Notch activity is present in pancreatic progenitors, which induces specific cell fate decisions (Shih et al., 2012).

Pdx1 and Ptf1a

Despite the severe defects in pancreas development that occur in the absence of either *Pdx1* or *Ptf1a*, a small number of differentiated endocrine cells can be detected in both mutants. By contrast, there is a complete absence of exocrine tissue in both the *Pdx1* and *Ptf1a* null mutant animals, which highlights the fact that these genes are essential for exocrine development. As mentioned previously, *Ptf1a* is highly enriched in acinar cells after e13.5, and the PTF1 heterotrimeric transcription factor complex binds several exocrine-specific gene promoters, such as *Elastase* and *Trypsin* (Krapp et al., 1996; Rose et al., 2001). The loss of RBP-J, another component of the PTF1 complex, results in accelerated differentiation of endocrine cells at the expense of acinar cells (Apelqvist et al., 1999), thus, supporting the idea that a functional PTF1 is essential for acinar differentiation, but not endocrine formation (Krapp et al., 1998). Impaired acinar cell differentiation is also observed in mice heterozygous for a hypomorphic allele of *Ptf1a* (Fukuda et al., 2008). In zebrafish, *Ptf1a* expression levels play a role in lineage allocation of pancreatic progenitors to either an exocrine or endocrine cell fate. Low levels of *Ptf1a* promote endocrine fates, while high levels repress endocrine cell fates and promote exocrine cell fates (Dong et al., 2008). *Pdx1,* although expressed at much lower levels in mature acinar cells than in β-cells, regulates acinar gene expression in cooperation with two other transcription factors, *Pre-B-cell leukemia homeobox 1* (*Pbx1*) and *Homeobox protein Meis 2* (*Meis2*) (Swift et al., 1998). The maintenance of *Pdx1* expression in the exocrine lineage during embryonic development is required for acinar differentiation (Hale et al., 2005).

Mist1

The *Mist1* bHLH transcription factor is expressed in acinar cells, and activates acinar genes involved in gap junction communication and coordinated exocytosis (Pin et al., 2001; Rukstalis et al., 2003). *Mist1* null mutant mice exhibit extensive exocrine tissue disorganization and defects in regulated exocytosis, resulting in inappropriate intracellular zymogen activation (Kowalik et al., 2007). This occurs in addition to defects in cell proliferation, and a lack of appropriate cell-cell junctions (Jia et al., 2008). The *Mist1* null phenotype is reminiscent of some forms of pancreatic injury, such as chronic pancreatitis. Thus, *Mist1* is a key regulator of acinar cell function, stability, and identity.

Hnf6 and Prox1

Hepatic nuclear factor 6 *(HNF6/Onecut-1/OC-1)*, as the name implies, was first identified in the liver, but is more broadly expressed in the developing endoderm. Despite its early broad expression pattern in the early MPCs, *Hnf6* function is not required to generate a pancreas. In the absence of *Hnf6, Pdx1* gene activation and pancreas bud outgrowth are delayed, resulting in a slightly hypoplastic pancreas at birth (Jacquemin et al., 2003a,b). Hnf6 expression becomes restricted to ducts and acinar cells as endocrine cells begin to differentiate and is maintained at high levels in mature ducts and at lower levels in acinar cells throughout the life of the organism (Zhang et al., 2009). Hnf6 is required for proper duct differentiation. Loss of *Hnf6* results in dilated pancreatic ducts and cysts due to a lack of primary cilia on ductal epithelia cells (Pierreux et al., 2006; Zhang et al., 2009). *Hnf6* expression is also required for endocrine pancreas development, as discussed below. Although loss of *Hnf6* does not seem to adversely affect acinar differentiation, possibly due to functional redundancy of the closely related factors OC-2 and OC-3, pancreas-wide *Hnf6* inactivation does result in acinar-to-ductal metaplasia after birth (Zhang et al., 2009). Thus, like Mist1, Hnf6 may be required to maintain the acinar fate.

The homeobox gene *Prospero homeobox 1* (*Prox1*) is expressed in the posterior foregut endoderm in the presumptive pancreas region before bud outgrowth (Burke and Oliver, 2002), and is essential for normal liver bud outgrowth (Sosa-Pineda et al., 2000). Prox1 is normally expressed in pancreatic duct cells (Wang et al., 2005), but its expression is lost in *Hnf6* mutant pancreatic ducts (Zhang et al., 2009). Prox1 has recently been shown to promote proper balance between duct and acinar differentiation. Pancreas-wide deletion of *Prox1* results in premature differentiation of acinar cells and impaired outgrowth of the pancreatic epithelium. These findings indicate that *Prox1* may be critical for duct cell proliferation and morphogenesis (Westmoreland et al., 2012).

Transcription Factors Involved in Endocrine Specification and Lineage Determination

Many lineage-restricted or lineage-specific transcription factors are expressed more broadly in the pancreatic epithelium or in the endocrine population early during development, gradually becoming restricted as development proceeds to refine the pattern of gene expression to what is observed in the adult islet. Gene expression and mutational analyses in mice strongly correlate with gene function in humans; mutations in several genes involved in β-cell differentiation, including *Pdx1, Pax6,* and *NeuroD*, have been identified in individuals with type 2 diabetes.

Prox1

Prox1 is expressed in the posterior foregut endoderm in the presumptive pancreas region before bud outgrowth (Burke and Oliver, 2002), and is essential for normal liver bud outgrowth (Sosa-Pineda et al., 2000). On e13.5, *Prox1* is expressed in most cells throughout the pancreatic epithelium. As the second wave of endocrine differentiation commences after e13.5, *Prox1* becomes more highly expressed in Ngn3-positive (an endocrine progenitor marker) and Isl-1-positive (a marker of all islet endocrine) cells, and is downregulated in differentiating acinar cells (Wang et al., 2005). After birth, *Prox1* expression is maintained at high levels in the ductal epithelium and in peripheral islet cell types, with lower levels found in β-cells. At e15.5 (the time at which *Prox1*-deficient embryos die as a result of complications in other organ systems), the *Prox1* mutant pancreatic epithelium is less branched than that of wild type, and contains many fewer endocrine cells (Wang et al., 2005). By contrast, the number of differentiated acinar cells is relatively high, and the pancreas has increased levels of *Ptf1a* with decreased levels of endocrine lineage markers (e.g., *Ngn3*). Thus, *Prox1* may be required within the bipotential endocrine/duct progenitors within the "trunk" population to repress the acinar fate and promote differentiation down the endocrine lineage.

Hnf6

Similar to *Pdx1* and *Ptf1a*, *Hnf6* is initially expressed throughout the early MPCs. Target genes for Hnf6 in the developing pancreatic region include *Foxa2, Pdx1,* and *Hnf4*, which are critical endodermal regulators: *Foxa2* is involved in the β-cell specific expression of *Pdx1,* and Hnf4 activates *Hnf1α*. In turn, Hnf1α activates *Pdx1,* and Hnf4 regulates *Hnf6* (Figure 31.3; Jensen, 2004). Thus, alterations in the expression of a single Hnf can affect the expression of multiple genes in this hierarchy.

As development proceeds, *Hnf6* is maintained in the ductal epithelium and in acinar cells, but becomes downregulated in endocrine cells as they begin to express hormones (Landry et al., 1997; Rausa et al., 1997; Zhang et al., 2009). This downregulation is critical for normal islet ontogeny and function: continued *Hnf6* expression in islets impairs the migration of endocrine cells from the ductal epithelium, disrupts the organization of endocrine cell types within the islet (core vs. mantle), and severely compromises β-cell maturation and function, leading to overt diabetes (Gannon et al., 2000; Tweedie et al., 2006).

Hnf6 is an upstream activator of *Ngn3* (Figure 31.3; Jacquemin et al., 2000; Oliver-Krasinski et al., 2009), a transcription factor expressed in endocrine precursors. *Ngn3*$^{-/-}$ mice lack endocrine cells in both the pancreas and the small intestine (Gradwohl et al., 2000). *Hnf6*$^{-/-}$ mice have smaller numbers of *Ngn3*-positive cells during embryogenesis, and lack islets at birth (Jacquemin et al., 2000). Islets do develop later, but are abnormal, and the mice are diabetic. The presence of islets in the absence of *Hnf6* suggests that other factors can partially compensate for *Hnf6,* and, indeed, other closely related factors are also expressed in the developing pancreas (Jacquemin et al., 2003a). Conditional inactivation of *Hnf6* specifically in endocrine progenitors (cells that have activated *Ngn3* expression) results in fewer bipotential duct/endocrine progenitors committing to the endocrine lineage (Zhang et al., 2009). *Hnf6* function is therefore required to generate endocrine progenitors in the appropriate numbers and its sustained function is required to irreversibly commit cells to the endocrine lineage.

Ngn3 *and* NeuroD

Neurogenin 3 (*Ngn3*) and *Beta2/NeuroD* (MODY6) are two closely related bHLH transcription factors involved in the early stages of pancreatic endocrine development. Early trunk progenitors express low levels of Notch and its ligand, Delta. Through as yet undetermined mechanisms, cells stochastically begin to express elevated levels of either Notch or Delta. Higher levels of Notch result in transcriptional repression of the pro-endocrine transcription factor, *Ngn3* (Jensen et al., 2000b). *Ngn3*-expressing cells are first detected scattered throughout the pancreatic epithelium at e9.5. Their numbers reach a peak at e15.5 before decreasing to nearly undetectable levels by birth (Gradwohl et al., 2000). Ngn3-positive cells are found within or adjacent to the ductal epithelium, and do not co-express islet endocrine hormones (Figure 31.1). In the absence of *Ngn3,* endocrine markers within the pancreas (including both broad and lineage-restricted transcription factors and all hormones) are missing (Gradwohl et al., 2000). Thus, *Ngn3* represents the only known gene that specifically marks the endocrine progenitor population. Unfortunately, transgenic over-expression of Ngn3 in the pancreatic epithelium does not result in increased islet mass. Rather, expression of *Ngn3* throughout the pancreatic epithelium results in premature differentiation of glucagon-producing cells, with little to no β-cells being formed (Apelqvist et al., 1999; Grapin-Botton et al., 2001; Schwitzgebel et al., 2000). Interestingly, the endocrine phenotype observed after *Ngn3* over-expression is identical to that seen in *Hes1*-deficient animals. In the absence of *Hes-1,* which represses *Ngn3* transcription in response to Notch signaling, *Ngn3* is over-expressed, thereby leading to an excess of glucagon-producing cells and the depletion of endocrine progenitors (Jensen et al., 2000a,b).

Ngn3 expression is biphasic, correlating with the first and second waves of endocrine cell differentiation (Villasenor et al., 2008). Competence of *Ngn3*-expressing progenitor cells changes during the secondary transition, with early differentiating endocrine cells favoring an α-cell fate and later cells preferentially differentiating as β-cells or other endocrine cell types (Johansson et al., 2007). In addition, through highly sensitive lineage tracing experiments, it was determined that *Ngn3*-expressing progenitor cells are unipotent and thus give rise to only one endocrine cell type (Desgraz and Herrera, 2009). This suggests that the Ngn3-positive progenitor population is heterogeneous, although molecularly indistinguishable at this time.

NeuroD was isolated as a transactivator of the insulin gene (Figure 31.3), but is expressed in all islet endocrine cell types during development and in the adult. It is a direct transcriptional target of Ngn3 (Figure 31.3). Loss of *NeuroD* results in a dramatic decrease in all islet endocrine cell types, suggesting that *NeuroD* functions in the expansion of the endocrine population or in endocrine cell survival (Naya et al., 1997). The remaining endocrine cells fail to organize into normal spherical islet structures, indicating that *NeuroD* also functions in islet morphogenesis (Naya et al., 1997). Removal of *NeuroD* in developing β-cells results in an immature β-cell phenotype and greatly impaired β-cell function (Gu et al., 2010). This agrees with the defects observed in humans with NeuroD mutations (MODY6) (Malecki et al., 1999).

Nkx2.2 and Nkx6.1

Members of the NKX class of homeodomain proteins also have roles in the pancreatic endocrine lineage. Both Nkx2.2 and Nkx6.1 are expressed in most pancreatic epithelial cells during early stages of development; however, by e15.5, Nkx2.2 becomes restricted to the endocrine cell population, and Nkx6.1 is found exclusively in insulin-producing cells and scattered cells within the ductal epithelium (Nelson et al., 2005; Sander et al., 2000; Sussel et al., 1998). During late gestation, Nkx2.2 can be detected in nearly all hormone-positive cells, except for the somatostatin-producing cells. After birth, both genes are expressed in the β-cell population.

Nkx2.2 primarily functions as a transcriptional repressor in regulating endocrine differentiation (Doyle et al., 2007). Mice lacking *Nkx2.2* have no detectable insulin$^+$ cells at any stage examined, and they also have a dramatic reduction in the number of glucagon expressing cells and a more modest reduction in the number of PP$^+$ cells (Sussel et al., 1998). Expression of Isl-1 and synaptophysin, general markers of islet endocrine cells, is normal in *Nkx2.2* mutants, suggesting that loss of *Nkx2.2* does not result in a dramatic loss of endocrine cells in general. Subsequent analysis revealed that these "extra" endocrine cells in the *Nkx2.2* mutant pancreas were increased numbers of the ghrelin-producing ε-cell population (Prado et al., 2004). Thus, *Nkx2.2* is required to generate β-cells, maintain and expand α- and PP cells, and repress ε-cell fate. Experiments in mice show that *Nkx2.2* and another transcription factor, *Arx*, work in concert to regulate the population of δ- and ε-cells in the developing pancreas (Mastracci et al., 2011).

Nkx6.1 gene inactivation results in a highly specific, profound loss of second wave insulin-positive cells (after e13.5), with no alteration in the numbers of other islet endocrine cell types (Sander et al., 2000). Thus, in the absence of *Nkx6.1,* putative β-cells do not adopt an alternate islet endocrine cell fate. Genetic epistasis experiments have determined that *Nkx6.1* functions downstream of *Nkx2.2* in the expansion and terminal differentiation of the β-cell lineage (Sander et al., 2000). However, transgenic over-expression of *Nkx6.1* does not enhance β-cell mass or function (Schaffer et al., 2011). Upon deletion of both *Nkx6.1* and *Nkx2.2*, both α- and β-cells were decreased in number, indicating a role for both transcription factors in endocrine cell development (Henseleit et al., 2005; Nelson et al., 2007). Nkx6.1 actively represses the exocrine cell fate while promoting the endocrine cell fate. This is accomplished by Nkx6.1-mediated repression of *Ptf1a* expression (Schaffer et al., 2010).

Pax4 *and* Pax6

Pax4 and Pax6, two members of the paired class of homeodomain-containing transcription factors, function in pancreatic endocrine differentiation. In the pancreas, Pax4 is specifically expressed in first and second wave insulin-producing cells, and is maintained at very low levels in adult β-cells (Sosa-Pineda et al., 1997). Pax4 is thought to act mainly as a transcriptional repressor to promote the β-cell fate (Collombat et al., 2003; Ritz-Laser et al., 2002; Wang et al., 2008). In the absence of Pax4, first wave insulin-producing cells are observed, but mature β- and δ-cells fail to form, and there is an increase in glucagon- and ghrelin-producing cells (Prado et al., 2004; Sosa-Pineda et al., 1997; Wang et al., 2004). These data suggest that either Pax4 is separately required in the β- and δ-cell lineages or that these two islet cell types arise from a common progenitor that is dependent on Pax4. In addition, the increased numbers of α- and ε-cells suggests that cells that would have become β-or δ-cells have instead adopted one of these two cell fates or, alternatively, that β- and δ-cells normally produce something that inhibits the expansion of the α- and ε-cell populations. Despite Pax4 being dispensable for the formation of α- and PP-cell types, lineage tracing experiments have shown that *Pax4*-expressing cells give rise to α-, β-, and ε-cells, suggesting that Pax4 is expressed in a pluripotent endocrine progenitor (Wang et al., 2008). In addition, Pax4 promotes proper β-cell mass expansion in humans, with mutations in Pax4 resulting in diabetes (Brun et al., 2007).

In contrast with Pax4, Pax6 is expressed in all endocrine cell types within the pancreas, both during embryonic development and in adults; however, global loss of Pax6 has a specific effect on the α-cell lineage. In the absence of Pax6, there is a complete loss of glucagon-producing cells; the other endocrine cell types are present in reduced numbers, and they fail to organize into normal islet structures (Heller et al., 2004; Sander and German, 1997; St-Onge et al., 1997). Mice that lack both Pax4 and Pax6 completely lack all pancreatic endocrine cell types (St-Onge et al., 1997). After birth, Pax6 functions to maintain islet function; gene inactivation in mature endocrine cells results in decreased expression of β-cell-specific genes and diabetes, with an increase in ghrelin-expressing cells (Ashery-Padan et al., 2004; Hart et al., 2013).

Arx

The Aristaless related homeobox (Arx) homeodomain-containing protein is expressed in scattered cells throughout the pancreatic buds between e10.5 and e12.5 and is later co-expressed with glucagon at e14.5 and is also found in PP cells. Inactivation of *Arx* causes a complete loss of second wave glucagon-producing α-cells, resulting in severe postnatal hypoglycemia and death (Collombat et al., 2005; 2003); glucagon stimulates the liver to release glucose into the bloodstream during times of fasting. The decrease in α-cell number is accompanied by an increase in both δ-cells and β-cells. Subsequent analyses strongly suggest that *Arx*$^{-/-}$ cells are diverted from an α-cell fate toward a β- or δ-cell fate instead. Indeed, the β-cell transcription factor Pax4 is upregulated in *Arx* mutants, whereas *Arx* is upregulated in Pax4 mutants (Collombat et al., 2005; 2007). Thus, these two genes have opposing actions within the endocrine lineage to establish β or δ-cells (*Pax4*) and α-cells (*Arx*).

MafA *and* MafB

The large Musculoaponeurotic fibrosarcoma (Maf) proteins are basic leucine zipper transcription factors that were first identified in an avian retrovirus. MafA was identified by several independent groups as an activator of insulin gene transcription (Figure 31.3; Kajihara et al., 2003; Kataoka et al., 2004; Matsuoka et al., 2004; Olbrot et al., 2002; Zhao et al., 2005) and is specifically expressed in second wave insulin-positive cells beginning on e13.5 and continuing into adulthood, thus making it a marker of more mature β-cells (Artner et al., 2010; Nishimura et al., 2006). Despite its indication as a critical β-cell maturation factor, global inactivation of *MafA* had no effect on the number of insulin-producing cells generated during embryonic development. Instead, loss of MafA causes defects in β-cell gene expression and postnatal β-cell function, thus leading to diabetes (Zhang et al., 2005). The lack of a developmental islet phenotype in *MafA* knockout animals may be the result of compensation by another closely related Maf family member that is also expressed in developing endocrine cells, MafB (Artner et al., 2010; 2006).

MafB is also capable of activating insulin reporter gene transcription in tissue culture cells, although, in adult islets, it is expressed only in α-cells, where it regulates the expression of the glucagon gene (Artner et al., 2006). During embryonic development, MafB is expressed in both first and second wave insulin- and glucagon-producing cells, becoming restricted to α-cells soon after birth. MafB promotes the activation of β-cell differentiation genes including *Pdx1, Nkx6.1, Glut2,* and *MafA* (Artner et al., 2010). Loss of *MafB* results in a reduction of insulin and glucagon positive cells (Hang and Stein, 2011).

Pancreas and β-Cell Regeneration: Neogenesis vs. Replication

The mammalian pancreas (including that of humans) has the ability to partially regenerate after insult or injury, although not to the same extent as the liver (Risbud and Bhonde, 2002; Tsiotos et al., 1999). Regeneration of duct and acinar tissue is undisputed, but controversy exists as to whether new β-cells can arise from a stem cell or facultative progenitor cell within the pancreas during regeneration. While β-cell neogenesis does not seem to occur under normal conditions in the adult pancreas (Dor et al., 2004), neogenesis has been reported in several models of pancreas regeneration, including after β-cell destruction using chemical toxins such as alloxan or streptozotocin (Guz et al., 2001; McEvoy and Hegre, 1977, Risbud and Bhonde, 2002), induction of pancreatitis (Gress et al., 1994; Risbud and Bhonde 2002; Taguchi et al., 2002), cellophane wrapping (Wang et al., 1995), partial pancreatectomy (Ackermann and Gannon, 2007; Liu et al., 2000), and the targeting of inflammatory cytokines to the β-cells (Gu and Sarvetnick, 1993). However, recent lineage tracing studies have challenged the idea that β-cell neogenesis can occur after embryogenesis, even in the setting of injury or β-cell destruction (Kopp et al., 2011; Solar et al., 2009; Xiao et al., 2013).

β-cell mass expansion in the adult pancreas appears to occur through replication of existing β-cells and not from a resident stem or progenitor cell. The genes and pathways involved in maintaining or altering β-cell mass are candidates for being affected in diabetic individuals. Functional analysis of these genes and pathways may lead to new *in vivo* therapeutic strategies for increasing existing β-cell mass in diabetic patients, and may facilitate the production of β-cells *in vitro* from embryonic or induced pluripotent stem cells for cell-based therapies.

β-Cell Regeneration in Adult Pancreas

In several models of pancreas regeneration and ductal metaplasia, upregulation of the Pdx1 transcription factor is associated with increased ductal proliferation and reported islet neogenesis (Sharma et al., 1999; Song et al., 1999). It was thought that after insult, neogenesis may occur from ductal cells in a manner recapitulating embryonic pancreas development (Bonner-Weir et al., 1993). Indeed, after partial pancreatectomy, there is a proliferation of ductal epithelium and formation of new ductal complexes (Bonner-Weir et al., 1993). However, lineage tracing studies have generated contradictory results (Bonner-Weir et al., 2010; Criscimanna et al., 2011; Xu et al., 2007). One study suggested that β-cell neogenesis from a ductal origin still occurs in early postnatal mice, but not in adult mice (Kopp et al., 2011). However, other lineage tracing studies show little to no contribution of ductal cells to neonatal islets (Furuyama et al., 2011; Solar et al., 2009).

There is very limited information regarding pancreas regeneration in humans (Bonner-Weir et al., 2010). Regeneration is thought to occur via neogenesis from ductal cells in the neonatal period, but new pancreas tissue formation after injury occurs to a limited degree in adults (Tsiotos et al., 1999). Evidence suggests that β-cell mass does not regenerate after partial pancreatectomy in humans (Menge et al., 2008).

β-Cell Replication Overview

The endocrine pancreas undergoes substantial remodeling during the neonatal period, including replication, apoptosis, and ongoing neogenesis (Gunasekaran et al., 2012; Kulkarni et al., 2012; Lysy et al., 2012). Between months 1 and 3 in the rat, the β-cell number doubles each month. After 3 months, the number of dying β-cells approaches the number of replicating cells (Teta et al., 2005). β-cell replication declines with age, and the β-cell mass plateaus once the organism reaches its mature body mass (Ackermann and Gannon, 2007).

The majority of experimental evidence in mice indicates that most if not all new adult β-cells arise via proliferation of pre-existing β-cells, with little to no contribution from resident pancreatic stem cells or undifferentiated progenitors within the ductal epithelium (Desai et al., 2007; Dor et al., 2004; Teta et al., 2007). However, endocrine neogenesis may be ongoing in the adult human pancreas, based on biopsy and autopsy studies in which endocrine cells were observed within the pancreatic ducts of humans of different ages (Martin-Pagola et al., 2008; Meier et al., 2006; Reers et al., 2009).

The β-cell mass is dynamic, adapting to changing physiological needs and increased functional demands. For example, β-cell mass increases during pregnancy and with the insulin resistance associated with obesity, whereas it decreases after parturition and after insulinoma implantation (Bernard-Kargar and Ktorza, 2001; Butler et al., 2010a,b). In addition to increased β-cell proliferation, β-cells can also compensate for increased insulin demand via improved function of individual cells. Functional adaptations include a reduced threshold for glucose-stimulated insulin secretion and increased glucokinase activity, both of which lead to enhanced insulin secretion (Porat et al., 2011; Weinhaus et al., 2007). Changes in β-cell mass are achieved by both hyperplasia (an increased number of cells) and hypertrophy (an increased individual cell size).

Glucose is one of the best stimuli for β-cell replication both *in vitro* and *in vivo* (Alonso et al., 2007; Path et al., 2004; Topp et al., 2004). Thus, in a normally functioning pancreas, sustained elevations in blood glucose levels should lead to increased β-cell mass, thereby providing compensation for the increased glucose load. Autopsies of human patients reveal a 40% increase in β-cell mass in obese non-diabetic individuals compared with lean subjects (Butler et al., 2003), suggesting that β-cell compensation does indeed occur with increasing insulin resistance. However, it is also possible that individuals who fail to progress to diabetes after becoming obese had a greater β-cell mass to begin with and that this is the reason they remain euglycemic. Currently there is no reliable method of measuring β-cell mass longitudinally or in living humans.

Overall, β-cells appear to be very long lived with decreased replicative potential as they age (Cnop et al., 2010; 2011). Gene products affecting the renewal, proliferation, or turnover of β-cells are therefore candidate factors in diabetes susceptibility and therapeutics. In general, human β-cells are more recalcitrant to proliferative stimuli than rodent islets (Kulkarni et al., 2012). However, introduction of specific cell cycle regulators does increase human β-cell proliferation in culture (Davis et al., 2010; Fiaschi-Taesch et al., 2010).

Secreted Factors that Affect β-Cell Replication

Several secreted factors and their receptors play a role in β-cell mass dynamics. For example, the epidermal-growth-factor-related ligand, Betacellulin, promotes differentiation and proliferation of β-cells *in vivo* and in pancreatic bud culture (Demeterco et al., 2000; Huotari et al., 1998; Li et al., 2001; Oh et al., 2011; Yechoor et al., 2009). The TGFβ family member, Activin, plays a role in β-cell specification and/or proliferation *in vivo* as well; defects in Activin-receptor signaling during embryogenesis result in islet hypoplasia (Shiozaki et al., 1999; Szabat et al., 2010). HGF is a mesenchyme-derived growth factor that stimulates the proliferation of both fetal and adult islets in culture (Hayek et al., 1995; Otonkoski et al., 1994), acting through the c-Met receptor, which is highly expressed in β-cells (Watanabe et al., 2003). In transgenic mice, overexpression of HGF specifically in insulin-producing cells leads to increased β-cell proliferation and β-cell hypertrophy, and protects β-cells from apoptosis (Araujo et al., 2012; Garcia-Ocana et al., 2000). Loss of HGF/c-Met signaling during pregnancy impairs β-cell mass expansion resulting in gestational diabetes (Demirci et al., 2012). Placental Lactogen is a circulating factor that directly stimulates β-cell replication and is required for maternal β-cell mass expansion during pregnancy (Sorenson and Brelje, 1997; Vasavada et al., 2000). More recently, Betatrophin, produced by the liver in response to insulin resistance, was identified as a factor that can promote β-cell proliferation and mass expansion in mice (Yi et al., 2013).

Secreted gut hormones such as Glucagon-like peptide-1 (GLP-1) and Gastrin also promote β-cell neogenesis and regeneration after injury. In animal models, the long-acting GLP-1 analog Exendin 4 stimulates both β-cell neogenesis from ductal progenitors and the proliferation of existing β-cells (De Leon et al., 2003; Suarez-Pinzon et al., 2008). Compounds that mimic GLP-1 action or prevent its degradation in the circulation are currently being used for the treatment of type 2 diabetes. However, there remains controversy as to whether this treatment also stimulates proliferation of other epithelial cells in the pancreas leading to increased susceptibility to pancreatitis and pancreatic cancer (Kahn, 2013).

Cell Cycle Regulators and β-Cell Proliferation

Progression through the cell cycle requires the activity of the heterodimeric Cyclin/Cyclin-dependent kinase (CDK) complexes. Progression from the G1 phase to the S phase is mediated by the D class of Cyclins and their partners, CDK4 and CDK6. Murine β-cells are particularly sensitive to loss of CDK4 and Cyclin D2, resulting in decreased postnatal β-cell proliferation (Georgia and Bhushan, 2004; Kushner et al., 2004; Mettus and Rane, 2003; Rane et al., 1999). CDK4 null mutant mice have a 50% reduction in body and organ size, are infertile, and develop diabetes at ~two months old (Mettus and Rane, 2003; Rane et al., 1999). Although CDK4 does not seem to be required for embryonic β-cell proliferation, it is essential for establishing the pool of ngn3-positive endocrine progenitor cells from which β-cells arise (Kim and Rane, 2011). Likewise, Cyclin D2 inactivation has no effect on embryonic β-cell proliferation since mutants are born with normal β-cell mass; however postnatal β-cell expansion is impaired as a result of decreased proliferation, and mutants become diabetic by three months (Georgia and Bhushan, 2004; Kushner et al., 2004).

Transcription Factors that Affect β-Cell Proliferation

The FoxM1 winged helix transcription factor regulates cell cycle progression genes including several *Cyclins*, *Cdc25B*, and *Survivin* (Costa et al., 2003; Leung et al., 2001; Wang et al., 2002). FoxM1 is dispensable for embryonic β-cell proliferation and islet development, but is essential for postnatal β-cell replication and maintenance of β-cell mass (Ackermann and Gannon, 2007; Zhang et al., 2006). Pancreas-specific inactivation of *Foxm1* results in diabetes by nine weeks of age as a result of greatly impaired post-weaning β-cell replication, whereas the absence of *Foxm1* in maternal islets during pregnancy results in gestational diabetes due to impaired β-cell mass expansion (Zhang et al., 2010). Human islets transduced with a FoxM1-containing adenovirus show increased β-cell replication (Davis et al., 2010).

The Pdx1 transcription factor is required for embryonic β-cell replication and postnatal β-cell expansion in response to insulin resistance (Gannon et al., 2008; Kulkarni et al., 2004). The winged helix transcription factor FoxO1 inhibits *Pdx1* expression and β-cell proliferation (Kikuchi et al., 2012; Kitamura et al., 2002). Thus, inhibition of FoxO1 activity in β-cells is necessary for proliferation.

The transcription factor, Nuclear factor of activated T cells (NFAT), links increased glucose metabolism and a rise in intracellular calcium with β-cell replication. Inactivation of the NFAT pathway specifically in β-cells results in reduced β-cell proliferation and subsequent decreases in β-cell mass in mice, while constitutive activation promotes β-cell proliferation, in part through FoxM1 induction (Goodyer et al., 2012; Heit et al., 2006).

As discussed above, differential requirements exist for β-cell proliferative factors during embryonic development versus postnatal stages. Thus, redundant or parallel pathways are likely present during embryogenesis to ensure the generation of appropriate numbers of β-cells, whereas mature β-cells are highly susceptible to perturbations in cell cycle gene expression. For example, while many cell types express both CDK4 and CDK6, pancreatic β-cells express only one (CDK4 in mouse, CDK6 in human) (Kulkarni et al., 2012). The ability to activate these cell cycle genes or to prevent the age-dependent decline in their expression may facilitate the expansion of β-cell mass *in vivo* or *in vitro*. Indeed, the expression of an activated form of CDK4 in islet β-cells using the rat insulin promoter results in β-cell hyperplasia and improved insulin secretion without hypoglycemia and without the formation of insulinomas (Hino et al., 2004; Marzo et al., 2004). Similarly, human islets transduced with a lentivirus expressing activated CDK4 show increased β-cell proliferation (Marzo et al., 2004).

31.6 GENERATING ISLETS/β-CELLS FROM STEM OR PROGENITOR CELLS

There are several potential sources for the large number of insulin-producing cells that would be needed to make a cell-based therapy an option for individuals with diabetes (Figure 31.4), including:

(1) Proliferation and expansion of existing β-cells *in vivo* or *in vitro*;
(2) Induction of β-cell differentiation from endogenous progenitors;
(3) Induction of β-cell differentiation from embryonic stem cells (ESCs) or induced pluripotent stem cells (iPSCs); and
(4) Transdifferentiation of closely related cell types such as acinar, liver, and intestinal enteroendocrine cells (this concept is not discussed further in this chapter).

All of these avenues are experimentally feasible, and are currently being examined in animal models. It may be that more than one of these methods will ultimately be used to derive a steady supply of insulin-producing cells. Regardless of the source, experimentally derived insulin-producing cells must be rigorously tested for their survival, engraftment, physiological function *in vivo*, their ability to reverse diabetes and maintain euglycemia, and their lack of tumorigenicity in a relevant animal model. The ability of *in vitro* derived insulin-producing cells to function as mature β-cells in an animal model *in vivo* is critical to translating this eventually to human patients.

Generation of Insulin-Producing Cells from Adult Cell Sources

Several models of pancreatic or β-cell injury suggest that the adult ductal epithelium retains some capacity for the production of new β-cells (neogenesis). The isolation and culture of ductal epithelium yields islet-like clusters that contain

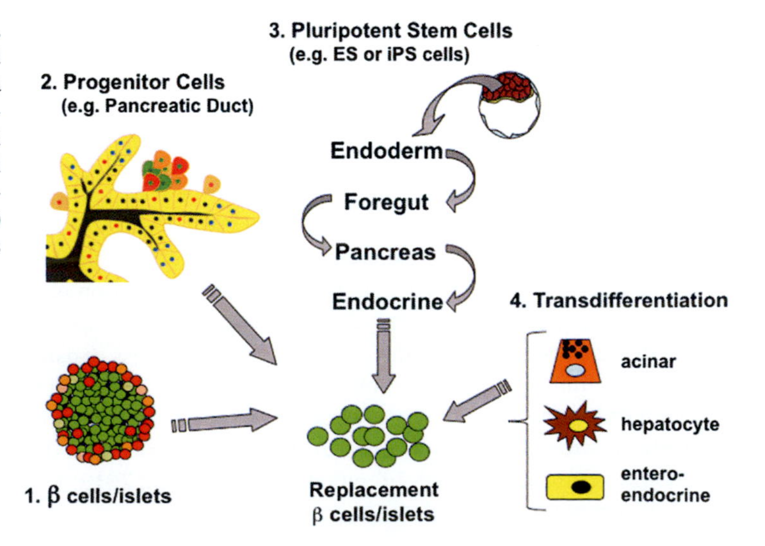

FIGURE 31.4 **Potential Sources of Replacement β-Cells/Islets.** There are several potential avenues being explored to generate and expand mature β-cells or functional islets *in vitro* and *in vivo* as a cell-based therapy to replace β-cells lost in diabetes: (1) proliferation and expansion of existing β-cells *in vivo*; (2) induction of β-cell differentiation from endogenous progenitors (embryonic ductal cells) or adult ductal epithelium; (3) induction of β-cell differentiation from embryonic or induced pluripotent stem cells; and (4) transdifferentiation of embryonically related cells. These strategies are more fully discussed in the text.

functional insulin-producing cells (Bonner-Weir et al., 2000; Ogata et al., 2004; Ramiya et al., 2000; Trivedi et al., 2001). These studies suggest that a facultative (if not a genuine) endocrine stem cell exists in the adult pancreatic ducts that, when properly activated, is capable of giving rise to new, functional β-cells. However, the presence of an adult pancreatic "stem cell" is highly controversial, as discussed prior in the regeneration section. Some reports suggest that the adult mouse pancreas does contain multipotent precursor cells that represent only a very small fraction of the total cell population in the adult (Seaberg et al., 2004; Smukler et al., 2011) and that these may represent centroacinar cells (Rovira et al., 2010).

Generation of Insulin-Producing Cells from ES and iPS Cells

Several methods for the *in vitro* directed differentiation of insulin-expressing cells from murine and human ESCs have been described. These protocols are mainly based on information gained from processes occurring during *in vivo* pancreas development (D'Amour et al., 2006; Nostro et al., 2011). In addition, high-throughput screening of small molecules that promote β-cell differentiation has shown promise (Borowiak et al., 2009).

Although much is known about the factors involved in pancreatic endocrine differentiation during murine embryonic development, there are still serious limitations to differentiating human ES or iPS cells into β-cells. Current protocols generate insulin-producing cells with a low efficiency of ~10–15% at the final *in vitro* differentiation step (D'Amour et al., 2006; Kroon et al., 2008). In addition, the functionality of *in vitro* generated insulin-producing cells is limited; they lack the ability to secrete insulin in response to glucose (D'Amour et al., 2006). Gene expression analyses reveal that these insulin-producing cells are immature, lacking MafA and Nkx6.1 transcription factors (Kroon et al., 2008; Nostro et al., 2011). However, transplantation of cells at an earlier stage of differentiation, specifically the pancreatic progenitor stage, generates mature, fully functional β-cells after about five months (Kroon et al., 2008). These results suggest that unidentified factors or micro-environmental features exist *in vivo* that are still lacking in the current directed differentiation protocols.

31.7 CLINICAL RELEVANCE

- The factors that control the specification and differentiation of the different pancreatic lineages continue to be elaborated. A complete understanding of normal pancreas development and endocrine differentiation will undoubtedly facilitate the generation of functional islets *in vitro* for use in the treatment of diabetes.
- Clearly, it is not sufficient to merely observe insulin reactivity in cultured cells derived from ESCs or any other source and conclude that these cells are β-cells. The potential application of these cells to human disease demands a stringency whereby these cells are shown to endogenously produce insulin and to secrete it in a regulatable manner in response to alterations in extracellular glucose concentrations.
- Although these types of studies can initially be performed *in vitro*, the ultimate test is to determine whether derived insulin-producing cells can fully rescue (i.e., reverse diabetes and maintain glucose homeostasis) an animal whose endogenous β-cell population has been destroyed.

ACKNOWLEDGMENTS

We would like to thank members of the Gannon lab for helpful discussions. Space restrictions prevented us from including all areas of pancreas development and pancreas stem cell research. There are many laboratories around the world that are contributing greatly to these areas, and we wish we could refer to them all. MG was supported by grants from the Veterans' Health Administration (1I01 BX000990–01A1) the Juvenile Diabetes Research Foundation (1–2011–592) and NIH/NIDDK (1U01 DK089540–01). K.G.R. was supported by the Molecular Endocrinology Training Program (5T32 DK756325).

REFERENCES

Ackermann, A.M., Gannon, M., 2007. Molecular regulation of pancreatic β-cell mass development, maintenance, and expansion. J. Mol. Endocrinol. 38, 193–206.

Afelik, S., Chen, Y., Pieler, T., 2006. Combined ectopic expression of Pdx1 and Ptf1a/p48 results in the stable conversion of posterior endoderm into endocrine and exocrine pancreatic tissue. Genes Dev. 20, 1441–1446.

Afelik, S., Qu, X., Hasrouni, E., Bukys, M.A., Deering, T., et al., 2012. Notch-mediated patterning and cell fate allocation of pancreatic progenitor cells. Development 139, 1744–1753.

Ahlgren, U., Jonsson, J., Edlund, H., 1996. The morphogenesis of the pancreatic mesenchyme is uncoupled from that of the pancreatic epithelium in IPF1/PDX1-deficient mice. Development 122, 1409–1416.

Ahlgren, U., Jonsson, J., Jonsson, L., Simu, K., Edlund, H., 1998. β-cell-specific inactivation of the mouse Ipf1/Pdx1 gene results in loss of the β-cell phenotype and maturity onset diabetes. Genes Dev. 12, 1763–1768.

Ahlgren, U., Pfaff, S.L., Jessell, T.M., Edlund, T., Edlund, H., 1997. Independent requirement for ISL1 in formation of pancreatic mesenchyme and islet cells. Nature 385, 257–260.

Ahnfelt-Ronne, J., Jorgensen, M.C., Klinck, R., Jensen, J.N., Fuchtbauer, E.M., et al., 2012. Ptf1a-mediated control of Dll1 reveals an alternative to the lateral inhibition mechanism. Development 139, 33–45.

Alonso, L.C., Yokoe, T., Zhang, P., Scott, D.K., Kim, S.K., et al., 2007. Glucose infusion in mice: a new model to induce β-cell replication. Diabetes 56, 1792–1801.

Apelqvist, A., Ahlgren, U., Edlund, H., 1997. Sonic hedgehog directs specialized mesoderm differentiation in the intestine and pancreas. Curr. Biol. 7, 801–804.

Apelqvist, A., Li, H., Sommer, L., Beatus, P., Anderson, D.J., et al., 1999. Notch signaling controls pancreatic cell differentiation. Nature 400, 877–881.

Araujo, T.G., Oliveira, A.G., Carvalho, B.M., Guadagnini, D., Protzek, A.O., et al., 2012. Hepatocyte growth factor plays a key role in insulin resistance-associated compensatory mechanisms. Endocrinology 153, 5760–5769.

Artner, I., Hang, Y., Mazur, M., Yamamoto, T., Guo, M., et al., 2010. MafA and MafB regulate genes critical to β-cells in a unique temporal manner. Diabetes 59, 2530–2539.

Artner, I., Le Lay, J., Hang, Y., Elghazi, L., Schisler, J.C., et al., 2006. MafB: an activator of the glucagon gene expressed in developing islet α- and β-cells. Diabetes 55, 297–304.

Ashery-Padan, R., Zhou, X., Marquardt, T., Herrera, P., Toube, L., et al., 2004. Conditional inactivation of Pax6 in the pancreas causes early onset of diabetes. Dev. Biol. 269, 479–488.

Ashraf, A., Abdullatif, H., Hardin, W., Moates, J.M., 2005. Unusual case of neonatal diabetes mellitus due to congenital pancreas agenesis. Pediatr. Diabetes 6, 239–243.

Beres, T.M., Masui, T., Swift, G.H., Shi, L., Henke, R.M., MacDonald, R.J., 2006. PTF1 is an organ-specific and Notch-independent basic helix-loop-helix complex containing the mammalian Suppressor of Hairless (RBP-J) or its paralogue, RBP-L. Mol. Cell. Biol. 26, 117–130.

Bernard-Kargar, C., Ktorza, A., 2001. Endocrine pancreas plasticity under physiological and pathological conditions. Diabetes 50 (Suppl. 1), S30–S35.

Bhushan, A., Itoh, N., Kato, S., Thiery, J.P., Czernichow, P., et al., 2001. Fgf10 is essential for maintaining the proliferative capacity of epithelial progenitor cells during early pancreatic organogenesis. Development 128, 5109–5117.

Bonnefond, A., Lomberk, G., Buttar, N., Busiah, K., Vaillant, E., et al., 2011. Disruption of a novel Kruppel-like transcription factor p300-regulated pathway for insulin biosynthesis revealed by studies of the c.-331 INS mutation found in neonatal diabetes mellitus. J. Biol. Chem. 286, 28414–28424.

Bonnefond, A., Sand, O., Guerin, B., Durand, E., De Graeve, F., et al., 2012. GATA6 inactivating mutations are associated with heart defects and, inconsistently, with pancreatic agenesis and diabetes. Diabetologia 55, 2845–2847.

Bonner-Weir, S., Baxter, L.A., Schuppin, G.T., Smith, F.E., 1993. A second pathway for regeneration of adult exocrine and endocrine pancreas. A possible recapitulation of embryonic development. Diabetes 42, 1715–1720.

Bonner-Weir, S., Li, W.C., Ouziel-Yahalom, L., Guo, L., Weir, G.C., Sharma, A., 2010. β-cell growth and regeneration: replication is only part of the story. Diabetes 59, 2340–2348.

Bonner-Weir, S., Taneja, M., Weir, G.C., Tatarkiewicz, K., Song, K.H., et al., 2000. *In vitro* cultivation of human islets from expanded ductal tissue. Proc. Natl. Acad. Sci. USA 97, 7999–8004.

Borowiak, M., Maehr, R., Chen, S., Chen, A.E., Tang, W., et al., 2009. Small molecules efficiently direct endodermal differentiation of mouse and human embryonic stem cells. Cell Stem Cell 4, 348–358.

Bort, R., Martinez-Barbera, J.P., Beddington, R.S., Zaret, K.S., 2004. Hex homeobox gene-dependent tissue positioning is required for organogenesis of the ventral pancreas. Development 131, 797–806.

Brissova, M., Powers, A.C., 2008. Revascularization of transplanted islets: can it be improved? Diabetes 57, 2269–2271.

Brissova, M., Shostak, A., Shiota, M., Wiebe, P.O., Poffenberger, G., et al., 2006. Pancreatic islet production of vascular endothelial growth factor-a is essential for islet vascularization, revascularization, and function. Diabetes 55, 2974–2985.

Brun, T., Duhamel, D.L., Hu He, K.H., Wollheim, C.B., Gauthier, B.R., 2007. The transcription factor PAX4 acts as a survival gene in INS-1E insulinoma cells. Oncogene 26, 4261–4271.

Burke, Z., Oliver, G., 2002. Prox1 is an early specific marker for the developing liver and pancreas in the mammalian foregut endoderm. Mech. Dev. 118, 147–155.

Burlison, J.S., Long, Q., Fujitani, Y., Wright, C.V., Magnuson, M.A., 2008. Pdx-1 and Ptf1a concurrently determine fate specification of pancreatic multipotent progenitor cells. Dev. Biol. 316, 74–86.

Butler, A.E., Cao-Minh, L., Galasso, R., Rizza, R.A., Corradin, A., et al., 2010a. Adaptive changes in pancreatic β-cell fractional area and β-cell turnover in human pregnancy. Diabetologia 53, 2167–2176.

Butler, A.E., Galasso, R., Matveyenko, A., Rizza, R.A., Dry, S., Butler, P.C., 2010b. Pancreatic duct replication is increased with obesity and type 2 diabetes in humans. Diabetologia 53, 21–26.

Butler, A.E., Janson, J., Bonner-Weir, S., Ritzel, R., Rizza, R.A., Butler, P.C., 2003. β-cell deficit and increased β-cell apoptosis in humans with type 2 diabetes. Diabetes 52, 102–110.

Cai, Q., Brissova, M., Reinert, R.B., Pan, F.C., Brahmachary, P., et al., 2012. Enhanced expression of VEGF-A in β-cells increases endothelial cell number but impairs islet morphogenesis and β-cell proliferation. Dev. Biol. 367, 40–54.

Carrasco, M., Delgado, I., Soria, B., Martin, F., Rojas, A., 2012. GATA4 and GATA6 control mouse pancreas organogenesis. J. Clin. Invest. 122, 3504–3515.

Christofori, G., Naik, P., Hanahan, D., 1995. Vascular endothelial growth factor and its receptors, flt-1 and flk-1, are expressed in normal pancreatic islets and throughout islet cell tumorigenesis. Mol. Endocrinol. 9, 1760–1770.

Cirulli, V., Beattie, G.M., Klier, G., Ellisman, M., Ricordi, C., et al., 2000. Expression and function of $\alpha_V\beta_3$ and $\alpha_V\beta_5$ integrins in the developing pancreas: roles in the adhesion and migration of putative endocrine progenitor cells. J. Cell. Biol. 150, 1445–1460.

Cleaver, O., Melton, D.A., 2003. Endothelial signaling during development. Nat. Med. 9, 661–668.

Cnop, M., Hughes, S.J., Igoillo-Esteve, M., Hoppa, M.B., Sayyed, F., et al., 2010. The long lifespan and low turnover of human islet β-cells estimated by mathematical modelling of lipofuscin accumulation. Diabetologia 53, 321–330.

Cnop, M., Igoillo-Esteve, M., Hughes, S.J., Walker, J.N., Cnop, I., Clark, A., 2011. Longevity of human islet α- and β-cells. Diabetes Obes. Metab. 13 (Suppl. 1), 39–46.

Cockell, M., Stevenson, B.J., Strubin, M., Hagenbuchle, O., Wellauer, P.K., 1989. Identification of a cell-specific DNA-binding activity that interacts with a transcriptional activator of genes expressed in the acinar pancreas. Mol. Cell. Biol. 9, 2464–2476.

Collombat, P., Hecksher-Sorensen, J., Broccoli, V., Krull, J., Ponte, I., et al., 2005. The simultaneous loss of Arx and Pax4 genes promotes a somatostatin-producing cell fate specification at the expense of the α- and β-cell lineages in the mouse endocrine pancreas. Development 132, 2969–2980.

Collombat, P., Hecksher-Sorensen, J., Krull, J., Berger, J., Riedel, D., et al., 2007. Embryonic endocrine pancreas and mature beta cells acquire alpha and PP cell phenotypes upon Arx misexpression. J. Clin. Invest. 117, 961–970.

Collombat, P., Mansouri, A., Hecksher-Sorensen, J., Serup, P., Krull, J., et al., 2003. Opposing actions of Arx and Pax4 in endocrine pancreas development. Genes Dev. 17, 2591–2603.

Cortijo, C., Gouzi, M., Tissir, F., Grapin-Botton, A., 2012. Planar cell polarity controls pancreatic β-cell differentiation and glucose homeostasis. Cell Reports 2, 1593–1606.

Costa, R.H., Kalinichenko, V.V., Holterman, A.X., Wang, X., 2003. Transcription factors in liver development, differentiation, and regeneration. Hepatology 38, 1331–1347.

Criscimanna, A., Speicher, J.A., Houshmand, G., Shiota, C., Prasadan, K., et al., 2011. Duct cells contribute to regeneration of endocrine and acinar cells following pancreatic damage in adult mice. Gastroenterology 141, 1451–1462. 62 e1–6.

D'Amour, K.A., Bang, A.G., Eliazer, S., Kelly, O.G., Agulnick, A.D., et al., 2006. Production of pancreatic hormone-expressing endocrine cells from human embryonic stem cells. Nat. Biotechnol. 24, 1392–1401.

Daley, G.Q., 2012. The promise and perils of stem cell therapeutics. Cell Stem Cell 10, 740–749.

Dalgin, G., Ward, A.B., Hao le, T., Beattie, C.E., Nechiporuk, A., Prince, V.E., 2011. Zebrafish mnx1 controls cell fate choice in the developing endocrine pancreas. Development 138, 4597–4608.

Davis, D.B., Lavine, J.A., Suhonen, J.I., Krautkramer, K.A., Rabaglia, M.E., et al., 2010. FoxM1 is up-regulated by obesity and stimulates β-cell proliferation. Mol. Endocrinol. 24, 1822–1834.

De Franco, E., Shaw-Smith, C., Flanagan, S.E., Shepherd, M.H., Hattersley, A.T., Ellard, S., 2013. GATA6 mutations cause a broad phenotypic spectrum of diabetes from pancreatic agenesis to adult-onset diabetes without exocrine insufficiency. Diabetes 62, 993–997.

De Leon, D.D., Deng, S., Madani, R., Ahima, R.S., Drucker, D.J., Stoffers, D.A., 2003. Role of endogenous glucagon-like peptide-1 in islet regeneration after partial pancreatectomy. Diabetes 52, 365–371.

Decker, K., Goldman, D.C., Grasch, C.L., Sussel, L., 2006. Gata6 is an important regulator of mouse pancreas development. Dev. Biol. 298, 415–429.

Demeterco, C., Beattie, G.M., Dib, S.A., Lopez, A.D., Hayek, A., 2000. A role for activin A and β-cellulin in human fetal pancreatic cell differentiation and growth. J. Clin. Endocrinol. Metab. 85, 3892–3897.

Demirci, C., Ernst, S., Alvarez-Perez, J.C., Rosa, T., Valle, S., et al., 2012. Loss of HGF/c-Met signaling in pancreatic β-cells leads to incomplete maternal β-cell adaptation and gestational diabetes mellitus. Diabetes 61, 1143–1152.

Desai, B.M., Oliver-Krasinski, J., De Leon, D.D., Farzad, C., Hong, N., et al., 2007. Preexisting pancreatic acinar cells contribute to acinar cell, but not islet beta cell, regeneration. J. Clin. Invest. 117, 971–977.

Desgraz, R., Herrera, P.L., 2009. Pancreatic neurogenin 3-expressing cells are unipotent islet precursors. Development 136, 3567–3574.

Deutsch, G., Jung, J., Zheng, M., Lora, J., Zaret, K.S., 2001. A bipotential precursor population for pancreas and liver within the embryonic endoderm. Development 128, 871–881.

Dong, P.D., Provost, E., Leach, S.D., Stainier, D.Y., 2008. Graded levels of Ptf1a differentially regulate endocrine and exocrine fates in the developing pancreas. Genes Dev. 22, 1445–1450.

Dor, Y., Brown, J., Martinez, O.I., Melton, D.A., 2004. Adult pancreatic β-cells are formed by self-duplication rather than stem-cell differentiation. Nature 429, 41–46.

Doyle, M.J., Loomis, Z.L., Sussel, L., 2007. Nkx2.2-repressor activity is sufficient to specify α-cells and a small number of β-cells in the pancreatic islet. Development 134, 515–523.

Du, A., Hunter, C.S., Murray, J., Noble, D., Cai, C.L., et al., 2009. Islet-1 is required for the maturation, proliferation, and survival of the endocrine pancreas. Diabetes 58, 2059–2069.

Dutta, S., Bonner-Weir, S., Montminy, M., Wright, C., 1998. Regulatory factor linked to late-onset diabetes? Nature 392, 560.

Edlund, H., 1998. Transcribing pancreas. Diabetes 47, 1817–1823.

Edsbagge, J., Johansson, J.K., Esni, F., Luo, Y., Radice, G.L., Semb, H., 2005. Vascular function and sphingosine-1-phosphate regulate development of the dorsal pancreatic mesenchyme. Development 132, 1085–1092.

Esni, F., Ghosh, B., Biankin, A.V., Lin, J.W., Albert, M.A., et al., 2004. Notch inhibits Ptf1 function and acinar cell differentiation in developing mouse and zebrafish pancreas. Development 131, 4213–4224.

Esni, F., Johansson, B.R., Radice, G.L., Semb, H., 2001. Dorsal pancreas agenesis in N-cadherin- deficient mice. Dev. Biol. 238, 202–212.

Fiaschi-Taesch, N.M., Salim, F., Kleinberger, J., Troxell, R., Cozar-Castellano, I., et al., 2010. Induction of human β-cell proliferation and engraftment using a single G1/S regulatory molecule, cdk6. Diabetes 59, 1926–1936.

Field, H.A., Ober, E.A., Roeser, T., Stainier, D.Y., 2003. Formation of the digestive system in zebrafish. I. Liver morphogenesis. Dev. Biol. 253, 279–290.

Fisher, A., Caudy, M., 1998. The function of hairy-related bHLH repressor proteins in cell fate decisions. BioEssays 20, 298–306.

Fox, C.S., 2010. Cardiovascular disease risk factors, type 2 diabetes mellitus, and the Framingham Heart Study. Trends Cardiovasc. Med. 20, 90–95.

Fukuda, A., Kawaguchi, Y., Furuyama, K., Kodama, S., Horiguchi, M., et al., 2008. Reduction of Ptf1a gene dosage causes pancreatic hypoplasia and diabetes in mice. Diabetes 57, 2421–2431.

Furuyama, K., Kawaguchi, Y., Akiyama, H., Horiguchi, M., Kodama, S., et al., 2011. Continuous cell supply from a Sox9-expressing progenitor zone in adult liver, exocrine pancreas and intestine. Nat. Genet. 43, 34–41.

Gannon, M., Ables, E.T., Crawford, L., Lowe, D., Offield, M.F., et al., 2008. pdx-1 function is specifically required in embryonic beta cells to generate appropriate numbers of endocrine cell types and maintain glucose homeostasis. Dev. Biol. 314, 406–417.

Gannon, M., Ray, M.K., Van Zee, K., Rausa, F., Costa, R.H., Wright, C.V., 2000. Persistent expression of HNF6 in islet endocrine cells causes disrupted islet architecture and loss of beta cell function. Development 127, 2883–2895.

Gannon MAW, C., 1999. Endodermal Patterning and Organogenesis. In: Moody, SA. (Ed.), Cell Lineage and Fate Determination. Academic Press, San Diego, California, pp. 583–615.

Garcia-Ocana, A., Takane, K.K., Syed, M.A., Philbrick, W.M., Vasavada, R.C., Stewart, A.F., 2000. Hepatocyte growth factor overexpression in the islet of transgenic mice increases beta cell proliferation, enhances islet mass, and induces mild hypoglycemia. J. Biol. Chem. 275, 1226–1232.

Georgia, S., Bhushan, A., 2004. β-cell replication is the primary mechanism for maintaining postnatal β-cell mass. J. Clin. Invest. 114, 963–968.

Gittes, G.K., Rutter, W.J., 1992. Onset of cell-specific gene expression in the developing mouse pancreas. Proc. Natl. Acad. Sci. USA 89, 1128–1132.

Golson, M.L., Le Lay, J., Gao, N., Bramswig, N., Loomes, K.M., et al., 2009. Jagged1 is a competitive inhibitor of Notch signaling in the embryonic pancreas. Mech. Dev. 126, 687–699.

Goodyer, W.R., Gu, X., Liu, Y., Bottino, R., Crabtree, G.R., Kim, S.K., 2012. Neonatal β-cell development in mice and humans is regulated by calcineurin/NFAT. Dev. Cell. 23, 21–34.

Gradwohl, G., Dierich, A., LeMeur, M., Guillemot, F., 2000. neurogenin3 is required for the development of the four endocrine cell lineages of the pancreas. Proc. Natl. Acad. Sci. USA 97, 1607–1611.

Grapin-Botton, A., Majithia, A.R., Melton, D.A., 2001. Key events of pancreas formation are triggered in gut endoderm by ectopic expression of pancreatic regulatory genes. Genes Dev. 15, 444–454.

Gress, T., Muller-Pillasch, F., Elsasser, H.P., Bachem, M., Ferrara, C., et al., 1994. Enhancement of transforming growth factor beta 1 expression in the rat pancreas during regeneration from caerulein-induced pancreatitis. Eur. J. Clin. Invest. 24, 679–685.

Gu, C., Stein, G.H., Pan, N., Goebbels, S., Hornberg, H., et al., 2010. Pancreatic beta cells require NeuroD to achieve and maintain functional maturity. Cell. Metab. 11, 298–310.

Gu, D., Sarvetnick, N., 1993. Epithelial cell proliferation and islet neogenesis in IFN-g transgenic mice. Development 118, 33–46.

Gu, G., Dubauskaite, J., Melton, D.A., 2002. Direct evidence for the pancreatic lineage: NGN3+ cells are islet progenitors and are distinct from duct progenitors. Development 129, 2447–2457.

Gunasekaran, U., Hudgens, C.W., Wright, B.T., Maulis, M.F., Gannon, M., 2012. Differential regulation of embryonic and adult beta cell replication. Cell Cycle 11, 2431–2442.

Guney, M.A., Gannon, M., 2009. Pancreas cell fate. Birth Defects Res. Part C Embryo. Today Rev. 87, 232–248.

Guney, M.A., Petersen, C.P., Boustani, A., Duncan, M.R., Gunasekaran, U., et al., 2011. Connective tissue growth factor acts within both endothelial cells and beta cells to promote proliferation of developing beta cells. Proc. Natl. Acad. Sci. USA 108, 15242–15247.

Guz, Y., Montminy, M.R., Stein, R., Leonard, J., Gamer, L.W., et al., 1995. Expression of murine STF-1, a putative insulin gene transcription factor, in beta cells of pancreas, duodenal epithelium and pancreatic exocrine and endocrine progenitors during ontogeny. Development 121, 11–18.

Guz, Y., Nasir, I., Teitelman, G., 2001. Regeneration of pancreatic beta cells from intra-islet precursor cells in an experimental model of diabetes. Endocrinology 142, 4956–4968.

Hald, J., Hjorth, J.P., German, M.S., Madsen, O.D., Serup, P., Jensen, J., 2003. Activated Notch1 prevents differentiation of pancreatic acinar cells and attenuate endocrine development. Dev. Biol. 260, 426–437.

Hale, M.A., Kagami, H., Shi, L., Holland, A.M., Elsasser, H.P., et al., 2005. The homeodomain protein PDX1 is required at mid-pancreatic development for the formation of the exocrine pancreas. Dev. Biol. 286, 225–237.

Hang, Y., Stein, R., 2011. MafA and MafB activity in pancreatic beta cells. Trends Endocrinol. Metab. 22, 364–373.

Hani, E.H., Stoffers, D.A., Chevre, J.C., Durand, E., Stanojevic, V., et al., 1999. Defective mutations in the insulin promoter factor-1 (IPF-1) gene in late-onset type 2 diabetes mellitus. J. Clin. Invest. 104, R41–R48.

Harrison, K.A., Thaler, J., Pfaff, S.L., Gu, H., Kehrl, J.H., 1999. Pancreas dorsal lobe agenesis and abnormal islets of Langerhans in Hlxb9-deficient mice. Nat. Genet. 23, 71–75.

Hart, A.W., Mella, S., Mendrychowski, J., van Heyningen, V., Kleinjan, D.A., 2013. The developmental regulator Pax6 is essential for maintenance of islet cell function in the adult mouse pancreas. PloS One 8, e54173.

Hattersley, A.T., 2000. Diagnosis of maturity-onset diabetes of the young in the pediatric diabetes clinic. Journal of pediatric endocrinology and metabolism. J. Pediatr. Endocrinol. Metab. 13 (Suppl. 6), 1411–1417.

Hayek, A., Beattie, G.M., Cirulli, V., Lopez, A.D., Ricordi, C., Rubin, J.S., 1995. Growth factor/matrix-induced proliferation of human adult β-cells. Diabetes 44, 1458–1460.

Hebrok, M., Kim, S.K., Melton, D.A., 1998. Notochord repression of endodermal Sonic hedgehog permits pancreas development. Genes Dev. 12, 1705–1713.

Hebrok, M., Kim, S.K., St Jacques, B., McMahon, A.P., Melton, D.A., 2000. Regulation of pancreas development by hedgehog signaling. Development 127, 4905–4913.

Heit, J.J., Apelqvist, A.A., Gu, X., Winslow, M.M., Neilson, J.R., et al., 2006. Calcineurin/NFAT signaling regulates pancreatic β-cell growth and function. Nature 443, 345–349.

Heller, R.S., Stoffers, D.A., Liu, A., Schedl, A., Crenshaw 3rd, E.B., et al., 2004. The role of Brn4/Pou3f4 and Pax6 in forming the pancreatic glucagon cell identity. Dev. Biol. 268, 123–134.

Henseleit, K.D., Nelson, S.B., Kuhlbrodt, K., Hennings, J.C., Ericson, J., Sander, M., 2005. NKX6 transcription factor activity is required for α- and β-cell development in the pancreas. Development 132, 3139–3149.

Herrera, P.L., 2000. Adult insulin- and glucagon-producing cells differentiate from two independent cell lineages. Development 127, 2317–2322.

Herrera, P.L., Huarte, J., Zufferey, R., Nichols, A., Mermillod, B., et al., 1994. Ablation of islet endocrine cells by targeted expression of hormone-promoter-driven toxigenes. Proc. Natl. Acad. Sci. USA 91, 12999–13003.

Hino, S., Yamaoka, T., Yamashita, Y., Yamada, T., Hata, J., Itakura, M., 2004. *In vivo* proliferation of differentiated pancreatic islet beta cells in transgenic mice expressing mutated cyclin-dependent kinase 4. Diabetologia 47, 1819–1830.

Huotari, M.A., Palgi, J., Otonkoski, T., 1998. Growth factor-mediated proliferation and differentiation of insulin-producing INS-1 and RINm5F cells: identification of β-cellulin as a novel β-cell mitogen. Endocrinology 139, 1494–1499.

Jacquemin, P., Durviaux, S.M., Jensen, J., Godfraind, C., Gradwohl, G., et al., 2000. Transcription factor hepatocyte nuclear factor 6 regulates pancreatic endocrine cell differentiation and controls expression of the proendocrine gene ngn3. Mol. Cell. Biol. 20, 4445–4454.

Jacquemin, P., Lemaigre, F.P., Rousseau, G.G., 2003a. The Onecut transcription factor HNF-6 (OC-1) is required for timely specification of the pancreas and acts upstream of Pdx-1 in the specification cascade. Dev. Biol. 258, 105–116.

Jacquemin, P., Pierreux, C.E., Fierens, S., van Eyll, J.M., Lemaigre, F.P., Rousseau, G.G., 2003b. Cloning and embryonic expression pattern of the mouse Onecut transcription factor OC-2. Gene Expr. Patterns 3, 639–644.

Jacquemin, P., Yoshitomi, H., Kashima, Y., Rousseau, G.G., Lemaigre, F.P., Zaret, K.S., 2006. An endothelial-mesenchymal relay pathway regulates early phases of pancreas development. Dev. Biol. Dev. Biol. 290, 189–199.

Jensen, J., 2004. Gene regulatory factors in pancreatic development. Dev. Dyn. 229, 176–200.

Jensen, J., Heller, R.S., Funder-Nielsen, T., Pedersen, E.E., Lindsell, C., et al., 2000a. Independent development of pancreatic α- and β-cells from neurogenin3-expressing precursors: a role for the notch pathway in repression of premature differentiation. Diabetes 49, 163–176.

Jensen, J., Pedersen, E.E., Galante, P., Hald, J., Heller, R.S., et al., 2000b. Control of endodermal endocrine development by Hes-1. Nat. Genet. 24, 36–44.

Jia, D., Sun, Y., Konieczny, S.F., 2008. Mist1 regulates pancreatic acinar cell proliferation through p21 CIP1/WAF1. Gastroenterology 135, 1687–1697.

Johansson, J.K., Voss, U., Kesavan, G., Kostetskii, I., Wierup, N., et al., 2010. N-cadherin is dispensable for pancreas development but required for β-cell granule turnover. Genesis 48, 374–381.

Johansson, K.A., Dursun, U., Jordan, N., Gu, G., Beermann, F., et al., 2007. Temporal control of neurogenin3 activity in pancreas progenitors reveals competence windows for the generation of different endocrine cell types. Dev. Cell. 12, 457–465.

Jonsson, J., Carlsson, L., Edlund, T., Edlund, H., 1994. Insulin-promoter-factor 1 is required for pancreas development in mice. Nature 371, 606–609.

Kahn, S.E., 2013. Incretin Therapy and Islet Pathology – A Time for Caution. Diabetes 62(7), 2178–2180.

Kajihara, M., Sone, H., Amemiya, M., Katoh, Y., Isogai, M., et al., 2003. Mouse MafA, homologue of zebrafish somite Maf 1, contributes to the specific transcriptional activity through the insulin promoter. Biochem. Biophys. Res. Commun. 312, 831–842.

Karlsson, O., Thor, S., Norberg, T., Ohlsson, H., Edlund, T., 1990. Insulin gene enhancer binding protein Isl-1 is a member of a novel class of proteins containing both a homeo- and a Cys-His domain. Nature 344, 879–882.

Kataoka, K., Shioda, S., Ando, K., Sakagami, K., Handa, H., Yasuda, K., 2004. Differentially expressed Maf family transcription factors, c-Maf and MafA, activate glucagon and insulin gene expression in pancreatic islet α- and β-cells. J. Mol. Endocrinol. 32, 9–20.

Kawaguchi, Y., Cooper, B., Gannon, M., Ray, M., MacDonald, R.J., Wright, C.V., 2002. The role of the transcriptional regulator Ptf1a in converting intestinal to pancreatic progenitors. Nat. Genet. 32, 128–134.

Kelly, O.G., Melton, D.A., 2000. Development of the pancreas in *Xenopus laevis*. Dev. Dyn. 218, 615–627.

Ketola, I., Otonkoski, T., Pulkkinen, M.A., Niemi, H., Palgi, J., et al., 2004. Transcription factor GATA-6 is expressed in the endocrine and GATA-4 in the exocrine pancreas. J. Mol. Endocrinol. 226, 51–57.

Kikuchi, O., Kobayashi, M., Amano, K., Sasaki, T., Kitazumi, T., et al., 2012. FoxO1 gain of function in the pancreas causes glucose intolerance, polycystic pancreas, and islet hypervascularization. PloS One 7, e32249.

Kim, S.K., Hebrok, M., Melton, D.A., 1997. Pancreas development in the chick embryo. Cold Spring Harbor. Symp. Q. Biol. 62, 377–383.

Kim, S.K., Melton, D.A., 1998. Pancreas development is promoted by cyclopamine, a hedgehog signaling inhibitor. Proc. Natl. Acad. Sci. USA 95, 13036–13041.

Kim, S.Y., Rane, S.G., 2011. The Cdk4-E2f1 pathway regulates early pancreas development by targeting Pdx1+ progenitors and Ngn3+ endocrine precursors. Development 138, 1903–1912.

Kitamura, T., Nakae, J., Kitamura, Y., Kido, Y., Biggs 3rd, W.H., et al., 2002. The forkhead transcription factor Foxo1 links insulin signaling to Pdx1 regulation of pancreatic beta cell growth. J. Clin. Invest. 110, 1839–1847.

Kopp, J.L., Dubois, C.L., Schaffer, A.E., Hao, E., Shih, H.P., et al., 2011. Sox9+ ductal cells are multipotent progenitors throughout development but do not produce new endocrine cells in the normal or injured adult pancreas. Development 138, 653–665.

Kowalik, A.S., Johnson, C.L., Chadi, S.A., Weston, J.Y., Fazio, E.N., Pin, C.L., 2007. Mice lacking the transcription factor Mist1 exhibit an altered stress response and increased sensitivity to caerulein-induced pancreatitis. Am. J. Physiol. Gastrointest. Liver Physiol. 292, G1123–G1132.

Krapp, A., Knofler, M., Frutiger, S., Hughes, G.J., Hagenbuchle, O., Wellauer, P.K., 1996. The p48 DNA-binding subunit of transcription factor PTF1 is a new exocrine pancreas-specific basic helix-loop-helix protein. EMBO. J. 15, 4317–4329.

Krapp, A., Knofler, M., Ledermann, B., Burki, K., Berney, C., et al., 1998. The bHLH protein PTF1-p48 is essential for the formation of the exocrine and the correct spatial organization of the endocrine pancreas. Genes Dev. 12, 3752–3763.

Kroon, E., Martinson, L.A., Kadoya, K., Bang, A.G., Kelly, O.G., et al., 2008. Pancreatic endoderm derived from human embryonic stem cells generates glucose-responsive insulin-secreting cells *in vivo*. Nat. Biotechnol. 26, 443–452.

Kulkarni, R.N., Jhala, U.S., Winnay, J.N., Krajewski, S., Montminy, M., Kahn, C.R., 2004. PDX-1 haploinsufficiency limits the compensatory islet hyperplasia that occurs in response to insulin resistance. J. Clin. Invest. 114, 828–836.

Kulkarni, R.N., Mizrachi, E.B., Ocana, A.G., Stewart, A.F., 2012. Human β-cell proliferation and intracellular signaling: driving in the dark without a road map. Diabetes 61, 2205–2213.

Kushner, J.A., Haj, F.G., Klaman, L.D., Dow, M.A., Kahn, B.B., et al., 2004. Islet-sparing effects of protein tyrosine phosphatase-1b deficiency delays onset of diabetes in IRS2 knockout mice. Diabetes 53, 61–66.

Lammert, E., Cleaver, O., Melton, D., 2001. Induction of pancreatic differentiation by signals from blood vessels. Science 294, 564–567.

Lammert, E., Cleaver, O., Melton, D., 2003a. Role of endothelial cells in early pancreas and liver development. Mech. Dev. 120, 59–64.

Lammert, E., Gu, G., McLaughlin, M., Brown, D., Brekken, R., et al., 2003b. Role of VEGF-A in vascularization of pancreatic islets. Curr. Biol. 13, 1070–1074.

Landry, C., Clotman, F., Hioki, T., Oda, H., Picard, J.J., et al., 1997. HNF-6 is expressed in endoderm derivatives and nervous system of the mouse embryo and participates to the cross-regulatory network of liver-enriched transcription factors. Dev. Biol. 192, 247–257.

Lango, H., Palmer, C.N., Morris, A.D., Zeggini, E., Hattersley, A.T., et al., 2008. Assessing the combined impact of 18 common genetic variants of modest effect sizes on type 2 diabetes risk. Diabetes 57, 3129–3135.

LeCouter, J., Lin, R., Ferrara, N., 2002. Endocrine gland-derived VEGF and the emerging hypothesis of organ-specific regulation of angiogenesis. Nat. Med. 8, 913–917.

Lee, Y.C., Damholt, A.B., Billestrup, N., Kisbye, T., Galante, P., et al., 1999. Developmental expression of proprotein convertase 1/3 in the rat. Mol. Cell. Endocrinol. 155, 27–35.

Leung, T.W., Lin, S.S., Tsang, A.C., Tong, C.S., Ching, J.C., et al., 2001. Over-expression of FoxM1 stimulates cyclin B1 expression. FEBS Lett. 507, 59–66.

Li, H., Arber, S., Jessell, T.M., Edlund, H., 1999. Selective agenesis of the dorsal pancreas in mice lacking homeobox gene Hlxb9. Nat. Genet. 23, 67–70.

Li, L., Seno, M., Yamada, H., Kojima, I., 2001. Promotion of β-cell regeneration by β-cellulin in ninety percent-pancreatectomized rats. Endocrinology 142, 5379–5385.

Lin, J.W., Biankin, A.V., Horb, M.E., Ghosh, B., Prasad, N.B., et al., 2004. Differential requirement for ptf1a in endocrine and exocrine lineages of developing zebrafish pancreas. Dev. Biol. 274, 491–503.

Liu, J., Hunter, C.S., Du, A., Ediger, B., Walp, E., et al., 2011. Islet-1 regulates Arx transcription during pancreatic islet α-cell development. J. Biol. Chem. 286, 15352–15360.

Liu, Y.Q., Nevin, P.W., Leahy, J.L., 2000. β-cell adaptation in 60% pancreatectomy rats that preserves normoinsulinemia and normoglycemia. Am. J. Physiol. Endocrinol. Metab. 279, E68–E73.

Lysy, P.A., Weir, G.C., Bonner-Weir, S., 2012. Concise review: pancreas regeneration: recent advances and perspectives. Stem Cell Translational Med. 1, 150–159.

Madsen, O.D., Jensen, J., Petersen, H.V., Pedersen, E.E., Oster, A., et al., 1997. Transcription factors contributing to the pancreatic β-cell phenotype. Horm. Metab. Res. 29, 265–270.

Magenheim, J., Ilovich, O., Lazarus, A., Klochendler, A., Ziv, O., et al., 2011. Blood vessels restrain pancreas branching, differentiation and growth. Development 138, 4743–4752.

Malecki, M.T., Jhala, U.S., Antonellis, A., Fields, L., Doria, A., et al., 1999. Mutations in NEUROD1 are associated with the development of type 2 diabetes mellitus. Nat. Genet. 23, 323–328.

Martin-Pagola, A., Sisino, G., Allende, G., Dominguez-Bendala, J., Gianani, R., et al., 2008. Insulin protein and proliferation in ductal cells in the transplanted pancreas of patients with type 1 diabetes and recurrence of autoimmunity. Diabetologia 51, 1803–1813.

Martin, M., Gallego-Llamas, J., Ribes, V., Kedinger, M., Niederreither, K., et al., 2005. Dorsal pancreas agenesis in retinoic acid-deficient Raldh2 mutant mice. Dev. Biol. 284, 399–411.

Marzo, N., Mora, C., Fabregat, M.E., Martin, J., Usac, E.F., et al., 2004. Pancreatic islets from cyclin-dependent kinase 4/R24C (Cdk4) knockin mice have significantly increased beta cell mass and are physiologically functional, indicating that Cdk4 is a potential target for pancreatic beta cell mass regeneration in Type 1 diabetes. Diabetologia 47, 686–694.

Mastracci, T.L., Wilcox, C.L., Arnes, L., Panea, C., Golden, J.A., et al., 2011. Nkx2.2 and Arx genetically interact to regulate pancreatic endocrine cell development and endocrine hormone expression. Dev. Biol. 359, 1–11.

Matsumoto, K., Yoshitomi, H., Rossant, J., Zaret, K.S., 2001. Liver organogenesis promoted by endothelial cells prior to vascular function. Science 294, 559–563.

Matsuoka, T.A., Artner, I., Henderson, E., Means, A., Sander, M., Stein, R., 2004. The MafA transcription factor appears to be responsible for tissue-specific expression of insulin. Proc. Natl. Acad. Sci. USA 101, 2930–2933.

Matsuoka, T.A., Zhao, L., Artner, I., Jarrett, H.W., Friedman, D., et al., 2003. Members of the large Maf transcription family regulate insulin gene transcription in islet β-cells. Mol. Cell. Biol. 23, 6049–6062.

McCall, M., Shapiro, A.M., 2012. Update on islet transplantation. Cold Spring Harbor. Perspect. Med. 2, a007823.

McEvoy, R.C., Hegre, O.D., 1977. Morphometric quantitation of the pancreatic insulin-, glucagon-, and somatostatin-positive cell populations in normal and alloxan-diabetic rats. Diabetes 26, 1140–1146.

Meier, J.J., Lin, J.C., Butler, A.E., Galasso, R., Martinez, D.S., Butler, P.C., 2006. Direct evidence of attempted β-cell regeneration in an 89-year-old patient with recent-onset type 1 diabetes. Diabetologia 49, 1838–1844.

Menge, B.A., Tannapfel, A., Belyaev, O., Drescher, R., Muller, C., et al., 2008. Partial pancreatectomy in adult humans does not provoke β-cell regeneration. Diabetes 57, 142–149.

Mettus, R.V., Rane, S.G., 2003. Characterization of the abnormal pancreatic development, reduced growth and infertility in Cdk4 mutant mice. Oncogene 22, 8413–8421.

Molotkov, A., Molotkova, N., Duester, G., 2005. Retinoic acid generated by Raldh2 in mesoderm is required for mouse dorsal endodermal pancreas development. Dev. Dyn. 232, 950–957.

Murtaugh, L.C., Stanger, B.Z., Kwan, K.M., Melton, D.A., 2003. Notch signaling controls multiple steps of pancreatic differentiation. Proc. Natl. Acad. Sci. USA 100, 14920–14925.

Naya, F.J., Huang, H.P., Qiu, Y., Mutoh, H., DeMayo, F.J., et al., 1997. Diabetes, defective pancreatic morphogenesis, and abnormal enteroendocrine differentiation in BETA2/neuroD-deficient mice. Genes Dev. 11, 2323–2334.

Nelson, S.B., Janiesch, C., Sander, M., 2005. Expression of Nkx6 genes in the hindbrain and gut of the developing mouse. J. Histochem. Cytochem. 53, 787–790.

Nelson, S.B., Schaffer, A.E., Sander, M., 2007. The transcription factors Nkx6.1 and Nkx6.2 possess equivalent activities in promoting β-cell fate specification in Pdx1+ pancreatic progenitor cells. Development 134, 2491–2500.

Nicolino, M., Claiborn, K.C., Senee, V., Boland, A., Stoffers, D.A., Julier, C., 2010. A novel hypomorphic PDX1 mutation responsible for permanent neonatal diabetes with subclinical exocrine deficiency. Diabetes 59, 733–740.

Nikolova, G., Jabs, N., Konstantinova, I., Domogatskaya, A., Tryggvason, K., et al., 2006. The vascular basement membrane: a niche for insulin gene expression and β-cell proliferation. Dev. Cell. 10, 397–405.

Nishimura, W., Kondo, T., Salameh, T., El Khattabi, I., Dodge, R., et al., 2006. A switch from MafB to MafA expression accompanies differentiation to pancreatic β-cells. Dev. Biol. 293, 526–539.

Nostro, M.C., Sarangi, F., Ogawa, S., Holtzinger, A., Corneo, B., et al., 2011. Stage-specific signaling through TGFβ family members and WNT regulates patterning and pancreatic specification of human pluripotent stem cells. Development 138, 861–871.

Obata, J., Yano, M., Mimura, H., Goto, T., Nakayama, R., et al., 2001. p48 subunit of mouse PTF1 binds to RBP-Jkappa/CBF-1, the intracellular mediator of Notch signaling, and is expressed in the neural tube of early stage embryos. Genes Cells 6, 345–360.

Ober, E.A., Field, H.A., Stainier, D.Y., 2003. From endoderm formation to liver and pancreas development in zebrafish. Mech. Dev. 120, 5–18.

Offield, M.F., Jetton, T.L., Labosky, P.A., Ray, M., Stein, R.W., et al., 1996. PDX-1 is required for pancreatic outgrowth and differentiation of the rostral duodenum. Development 122, 983–995.

Ogata, T., Park, K.Y., Seno, M., Kojima, I., 2004. Reversal of streptozotocin-induced hyperglycemia by transplantation of pseudoislets consisting of β-cells derived from ductal cells. Endocrinol. Jpn. 51, 381–386.

Oh, Y.S., Shin, S., Lee, Y.J., Kim, E.H., Jun, H.S., 2011. β-cellulin-induced β-cell proliferation and regeneration is mediated by activation of ErbB-1 and ErbB-2 receptors. PloS One 6, e23894.

Olbrot, M., Rud, J., Moss, L.G., Sharma, A., 2002. Identification of β-cell-specific insulin gene transcription factor RIPE3b1 as mammalian MafA. Proc. Natl. Acad. Sci. USA 99, 6737–6742.

Oliver-Krasinski, J.M., Kasner, M.T., Yang, J., Crutchlow, M.F., Rustgi, A.K., et al., 2009. The diabetes gene Pdx1 regulates the transcriptional network of pancreatic endocrine progenitor cells in mice. J. Clin. Invest. 119, 1888–1898.

Ostrom, M., Loffler, K.A., Edfalk, S., Selander, L., Dahl, U., et al., 2008. Retinoic acid promotes the generation of pancreatic endocrine progenitor cells and their further differentiation into β-cells. PloS One 3, e2841.

Otonkoski, T., Beattie, G.M., Lopez, A.D., Hayek, A., 1994. Use of hepatocyte growth factor/scatter factor to increase transplantable human fetal islet cell mass. Transplant. Proc. 26, 3334.

Pan, F.C., Wright, C., 2011. Pancreas organogenesis: from bud to plexus to gland. Dev. Dyn. 240, 530–565.

Pang, K., Mukonoweshuro, C., Wong, G.G., 1994. β-cells arise from glucose transporter type 2 (Glut2)-expressing epithelial cells of the developing rat pancreas. Proc. Natl. Acad. Sci. USA 91, 9559–9563.

Path, G., Opel, A., Knoll, A., Seufert, J., 2004. Nuclear protein p8 is associated with glucose-induced pancreatic β-cell growth. Diabetes 53 (Suppl. 1), S82–S85.

Percival, A.C., Slack, J.M., 1999. Analysis of pancreatic development using a cell lineage label. Exp. Cell. Res. 247, 123–132.

Pierreux, C.E., Poll, A.V., Kemp, C.R., Clotman, F., Maestro, M.A., et al., 2006. The transcription factor hepatocyte nuclear factor-6 controls the development of pancreatic ducts in the mouse. Gastroenterology 130, 532–541.

Pin, C.L., Rukstalis, J.M., Johnson, C., Konieczny, S.F., 2001. The bHLH transcription factor Mist1 is required to maintain exocrine pancreas cell organization and acinar cell identity. J. Cell. Biol. 155, 519–530.

Porat, S., Weinberg-Corem, N., Tornovsky-Babaey, S., Schyr-Ben-Haroush, R., Hija, A., et al., 2011. Control of pancreatic β-cell regeneration by glucose metabolism. Cell. Metab. 13, 440–449.

Prado, C.L., Pugh-Bernard, A.E., Elghazi, L., Sosa-Pineda, B., Sussel, L., 2004. Ghrelin cells replace insulin-producing β-cells in two mouse models of pancreas development. Proc. Natl. Acad. Sci. USA 101, 2924–2929.

Ramiya, V.K., Maraist, M., Arfors, K.E., Schatz, D.A., Peck, A.B., Cornelius, J.G., 2000. Reversal of insulin-dependent diabetes using islets generated *in vitro* from pancreatic stem cells. Nat. Med. 6, 278–282.

Rane, S.G., Dubus, P., Mettus, R.V., Galbreath, E.J., Boden, G., et al., 1999. Loss of Cdk4 expression causes insulin-deficient diabetes and Cdk4 activation results in β-islet cell hyperplasia. Nat. Genet. 22, 44–52.

Rausa, F., Samadani, U., Ye, H., Lim, L., Fletcher, C.F., et al., 1997. The cut-homeodomain transcriptional activator HNF-6 is coexpressed with its target gene HNF-3 β in the developing murine liver and pancreas. Dev. Biol. 192, 228–246.

Reers, C., Erbel, S., Esposito, I., Schmied, B., Buchler, M.W., et al., 2009. Impaired islet turnover in human donor pancreata with aging. Eur. J. Endocrinol. 160, 185–191.

Risbud, M.V., Bhonde, R.R., 2002. Models of pancreatic regeneration in diabetes. Diabetes Res. Clin. Pract. 58, 155–165.

Ritz-Laser, B., Estreicher, A., Gauthier, B.R., Mamin, A., Edlund, H., Philippe, J., 2002. The pancreatic beta-cell-specific transcription factor Pax-4 inhibits glucagon gene expression through Pax-6. Diabetologia 45, 97–107.

Roep, B.O., Peakman, M., 2012. Antigen targets of type 1 diabetes autoimmunity. Cold Spring Harbor. Perspect. Med. 2, a007781.

Rooman, I., Schuit, F., Bouwens, L., 1997. Effect of vascular endothelial growth factor on growth and differentiation of pancreatic ductal epithelium. Lab. Invest. 76, 225–232.

Rose, S.D., Swift, G.H., Peyton, M.J., Hammer, R.E., MacDonald, R.J., 2001. The role of PTF1-P48 in pancreatic acinar gene expression. J. Biol. Chem. 276, 44018–44026.

Rovira, M., Scott, S.G., Liss, A.S., Jensen, J., Thayer, S.P., Leach, S.D., 2010. Isolation and characterization of centroacinar/terminal ductal progenitor cells in adult mouse pancreas. Proc. Natl. Acad. Sci. USA 107, 75–80.

Rukstalis, J.M., Kowalik, A., Zhu, L., Lidington, D., Pin, C.L., Konieczny, S.F., 2003. Exocrine specific expression of Connexin32 is dependent on the basic helix-loop-helix transcription factor Mist1. J. Cell. Sci. 116, 3315–3325.

Ryan, E.A., Lakey, J.R., Rajotte, R.V., Korbutt, G.S., Kin, T., et al., 2001. Clinical outcomes and insulin secretion after islet transplantation with the Edmonton protocol. Diabetes 50, 710–719.

Sander, M., German, M.S., 1997. The β-cell transcription factors and development of the pancreas. J. Mol. Med. (Berl.) 75, 327–340.

Sander, M., Sussel, L., Conners, J., Scheel, D., Kalamaras, J., et al., 2000. Homeobox gene Nkx6.1 lies downstream of Nkx2.2 in the major pathway of β-cell formation in the pancreas. Development 127, 5533–5540.

Schaffer, A.E., Freude, K.K., Nelson, S.B., Sander, M., 2010. Nkx6 transcription factors and Ptf1a function as antagonistic lineage determinants in multipotent pancreatic progenitors. Dev. Cell 18, 1022–1029.

Schaffer, A.E., Yang, A.J., Thorel, F., Herrera, P.L., Sander, M., 2011. Transgenic overexpression of the transcription factor Nkx6.1 in β-cells of mice does not increase β-cell proliferation, β-cell mass, or improve glucose clearance. Mol. Endocrinol. 25, 1904–1914.

Schwitzgebel, V.M., Scheel, D.W., Conners, J.R., Kalamaras, J., Lee, J.E., et al., 2000. Expression of neurogenin3 reveals an islet cell precursor population in the pancreas. Development 127, 3533–3542.

Seaberg, R.M., Smukler, S.R., Kieffer, T.J., Enikolopov, G., Asghar, Z., et al., 2004. Clonal identification of multipotent precursors from adult mouse pancreas that generate neural and pancreatic lineages. Nat. Biotechnol. 22, 1115–1124.

Sellick, G.S., Barker, K.T., Stolte-Dijkstra, I., Fleischmann, C., Coleman, R.J., et al., 2004. Mutations in PTF1A cause pancreatic and cerebellar agenesis. Nat. Genet. 36, 1301–1305.

Seymour, P.A., Freude, K.K., Tran, M.N., Mayes, E.E., Jensen, J., et al., 2007. SOX9 is required for maintenance of the pancreatic progenitor cell pool. Proc. Natl. Acad. Sci. USA 104, 1865–1870.

Shapiro, A.M., Lakey, J.R., Ryan, E.A., Korbutt, G.S., Toth, E., et al., 2000. Islet transplantation in seven patients with type 1 diabetes mellitus using a glucocorticoid-free immunosuppressive regimen. N. Engl. J. Med. 343, 230–238.

Sharma, A., Zangen, D.H., Reitz, P., Taneja, M., Lissauer, M.E., et al., 1999. The homeodomain protein IDX-1 increases after an early burst of proliferation during pancreatic regeneration. Diabetes 48, 507–513.

Shih, H.P., Kopp, J.L., Sandhu, M., Dubois, C.L., Seymour, P.A., et al., 2012. A Notch-dependent molecular circuitry initiates pancreatic endocrine and ductal cell differentiation. Development 139, 2488–2499.

Shin, D., Shin, C.H., Tucker, J., Ober, E.A., Rentzsch, F., et al., 2007. Bmp and Fgf signaling are essential for liver specification in zebrafish. Development 134, 2041–2050.

Shiozaki, S., Tajima, T., Zhang, Y.Q., Furukawa, M., Nakazato, Y., Kojima, I., 1999. Impaired differentiation of endocrine and exocrine cells of the pancreas in transgenic mouse expressing the truncated type II activin receptor. Biochim. Biophys. Acta 1450, 1–11.

Slack, J.M., 1995. Developmental biology of the pancreas. Development 121, 1569–1580.

Smukler, S.R., Arntfield, M.E., Razavi, R., Bikopoulos, G., Karpowicz, P., et al., 2011. The adult mouse and human pancreas contain rare multipotent stem cells that express insulin. Cell Stem Cell 8, 281–293.

Solar, M., Cardalda, C., Houbracken, I., Martin, M., Maestro, M.A., et al., 2009. Pancreatic exocrine duct cells give rise to insulin-producing β-cells during embryogenesis but not after birth. Dev. Cell 17, 849–860.

Song, S.Y., Gannon, M., Washington, M.K., Scoggins, C.R., Meszoely, I.M., et al., 1999. Expansion of Pdx1-expressing pancreatic epithelium and islet neogenesis in transgenic mice overexpressing transforming growth factor α. Gastroenterology 117, 1416–1426.

Sorenson, R.L., Brelje, T.C., 1997. Adaptation of islets of Langerhans to pregnancy: beta-cell growth, enhanced insulin secretion and the role of lactogenic hormones. Horm. Metab. Res. 29, 301–307.

Sosa-Pineda, B., Chowdhury, K., Torres, M., Oliver, G., Gruss, P., 1997. The Pax4 gene is essential for differentiation of insulin-producing β-cells in the mammalian pancreas. Nature 386, 399–402.

Sosa-Pineda, B., Wigle, J.T., Oliver, G., 2000. Hepatocyte migration during liver development requires Prox1. Nat. Genet. 25, 254–255.

St-Onge, L., Sosa-Pineda, B., Chowdhury, K., Mansouri, A., Gruss, P., 1997. Pax6 is required for differentiation of glucagon-producing α-cells in mouse pancreas. Nature 387, 406–409.

Stafford, D., Hornbruch, A., Mueller, P.R., Prince, V.E., 2004. A conserved role for retinoid signaling in vertebrate pancreas development. Dev. Genes Evol. 214, 432–441.

Stoffers, D.A., Stanojevic, V., Habener, J.F., 1998. Insulin promoter factor-1 gene mutation linked to early-onset type 2 diabetes mellitus directs expression of a dominant negative isoprotein. J. Clin. Invest. 102, 232–241.

Stoffers, D.A., Zinkin, N.T., Stanojevic, V., Clarke, W.L., Habener, J.F., 1997. Pancreatic agenesis attributable to a single nucleotide deletion in the human IPF1 gene coding sequence. Nat. Genet. 15, 106–110.

Suarez-Pinzon, W.L., Power, R.F., Yan, Y., Wasserfall, C., Atkinson, M., Rabinovitch, A., 2008. Combination therapy with glucagon-like peptide-1 and gastrin restores normoglycemia in diabetic NOD mice. Diabetes 57, 3281–3288.

Sussel, L., Kalamaras, J., Hartigan-O'Connor, D.J., Meneses, J.J., Pedersen, R.A., et al., 1998. Mice lacking the homeodomain transcription factor Nkx2.2 have diabetes due to arrested differentiation of pancreatic beta cells. Development 125, 2213–2221.

Swift, G.H., Liu, Y., Rose, S.D., Bischof, L.J., Steelman, S., et al., 1998. An endocrine-exocrine switch in the activity of the pancreatic homeodomain protein PDX1 through formation of a trimeric complex with PBX1b and MRG1 (MEIS2). Mol. Cell. Biol. 18, 5109–5120.

Szabat, M., Johnson, J.D., Piret, J.M., 2010. Reciprocal modulation of adult β-cell maturity by activin A and follistatin. Diabetologia 53, 1680–1689.

Taguchi, M., Yamaguchi, T., Otsuki, M., 2002. Induction of PDX-1-positive cells in the main duct during regeneration after acute necrotizing pancreatitis in rats. J. Pathol. 197, 638–646.

Talavera-Adame, D., Wu, G., He, Y., Ng, T.T., Gupta, A., et al., 2011. Endothelial cells in co-culture enhance embryonic stem cell differentiation to pancreatic progenitors and insulin-producing cells through BMP signaling. Stem Cell Rev. 7, 532–543.

Teo, A.K., Windmueller, R., Johansson, B.B., Dirice, E., Njolstad, P.R., et al., 2013. Derivation of human induced pluripotent stem cells from patients with maturity onset diabetes of the young. J. Biol. Chem. 288, 5353–5356.

Teta, M., Long, S.Y., Wartschow, L.M., Rankin, M.M., Kushner, J.A., 2005. Very slow turnover of β-cells in aged adult mice. Diabetes 54, 2557–2567.

Teta, M., Rankin, M.M., Long, S.Y., Stein, G.M., Kushner, J.A., 2007. Growth and regeneration of adult β-cells does not involve specialized progenitors. Dev. Cell 12, 817–826.

Topp, B.G., McArthur, M.D., Finegood, D.T., 2004. Metabolic adaptations to chronic glucose infusion in rats. Diabetologia 47, 1602–1610.

Trivedi, N., Hollister-Lock, J., Lopez-Avalos, M.D., O'Neil, J.J., Keegan, M., et al., 2001. Increase in β-cell mass in transplanted porcine neonatal pancreatic cell clusters is due to proliferation of β-cells and differentiation of duct cells. Endocrinology 142, 2115–2122.

Tsiotos, G.G., Barry, M.K., Johnson, C.D., Sarr, M.G., 1999. Pancreas regeneration after resection: does it occur in humans? Pancreas 19, 310–313.

Tweedie, E., Artner, I., Crawford, L., Poffenberger, G., Thorens, B., et al., 2006. Maintenance of hepatic nuclear factor 6 in postnatal islets impairs terminal differentiation and function of beta-cells. Diabetes 55, 3264–3270.

Vasavada, R.C., Garcia-Ocana, A., Zawalich, W.S., Sorenson, R.L., Dann, P., et al., 2000. Targeted expression of placental lactogen in the beta cells of transgenic mice results in beta cell proliferation, islet mass augmentation, and hypoglycemia. J. Biol. Chem. 275, 15399–15406.

Villasenor, A., Chong, D.C., Cleaver, O., 2008. Biphasic Ngn3 expression in the developing pancreas. Dev. Dyn. 237, 3270–3279.

Wang, J., Elghazi, L., Parker, S.E., Kizilocak, H., Asano, M., et al., 2004. The concerted activities of Pax4 and Nkx2.2 are essential to initiate pancreatic β-cell differentiation. Dev. Biol. 266, 178–189.

Wang, J., Kilic, G., Aydin, M., Burke, Z., Oliver, G., Sosa-Pineda, B., 2005. Prox1 activity controls pancreas morphogenesis and participates in the production of "secondary transition" pancreatic endocrine cells. Dev. Biol. 286, 182–194.

Wang, Q., Elghazi, L., Martin, S., Martins, I., Srinivasan, R.S., et al., 2008. Ghrelin is a novel target of Pax4 in endocrine progenitors of the pancreas and duodenum. Dev. Dyn. 237, 51–61.

Wang, R.N., Kloppel, G., Bouwens, L., 1995. Duct- to islet-cell differentiation and islet growth in the pancreas of duct-ligated adult rats. Diabetologia 38, 1405–1411.

Wang, X., Krupczak-Hollis, K., Tan, Y., Dennewitz, M.B., Adami, G.R., Costa, R.H., 2002. Increased hepatic Forkhead Box M1B (FoxM1B) levels in old-aged mice stimulated liver regeneration through diminished p27Kip1 protein levels and increased Cdc25B expression. J. Biol. Chem. 277, 44310–44316.

Watanabe, H., Sumi, S., Kitamura, Y., Nio, Y., Higami, T., 2003. Immunohistochemical analysis of vascular endothelial growth factor and hepatocyte growth factor, and their receptors, in transplanted islets in rats. Surg. Today 33, 854–860.

Watt, A.J., Zhao, R., Li, J., Duncan, S.A., 2007. Development of the mammalian liver and ventral pancreas is dependent on GATA4. BMC. Dev. Biol. 7, 37.

Weinhaus, A.J., Stout, L.E., Bhagroo, N.V., Brelje, T.C., Sorenson, R.L., 2007. Regulation of glucokinase in pancreatic islets by prolactin: a mechanism for increasing glucose-stimulated insulin secretion during pregnancy. J. Endocrinol. 193, 367–381.

Wessells, N.K., Cohen, J.H., 1966. The influence of collagen and embryo extract on the development of pancreatic epithelium. Exp. Cell. Res 43, 680–684.

Westmoreland, J.J., Kilic, G., Sartain, C., Sirma, S., Blain, J., et al., 2012. Pancreas-specific deletion of Prox1 affects development and disrupts homeostasis of the exocrine pancreas. Gastroenterology 142, 999–1009. e6.

Wiebe, P.O., Kormish, J.D., Roper, V.T., Fujitani, Y., Alston, N.I., et al., 2007. Ptf1a binds to and activates area III, a highly conserved region of the Pdx1 promoter that mediates early pancreas-wide Pdx1 expression. Mol. Cell. Biol. 27, 4093–4104.

Wilson, M.E., Kalamaras, J.A., German, M.S., 2002. Expression pattern of IAPP and prohormone convertase 1/3 reveals a distinctive set of endocrine cells in the embryonic pancreas. Mech. Dev. 115, 171–176.

Wu, K.L., Gannon, M., Peshavaria, M., Offield, M.F., Henderson, E., et al., 1997. Hepatocyte nuclear factor 3β is involved in pancreatic beta-cell-specific transcription of the pdx-1 gene. Mol. Cell. Biol. 17, 6002–6013.

Xiao, X., Chen, Z., Shiota, C., Prasadan, K., Guo, P., et al., 2013. No evidence for β-cell neogenesis in murine adult pancreas. J. Clin. Invest. 123, 2207–2217.

Xu, Y., Xu, G., Liu, B., Gu, G., 2007. Cre reconstitution allows for DNA recombination selectively in dual-marker-expressing cells in transgenic mice. Nucleic Acids Res. 35, e126.

Xuan, S., Borok, M.J., Decker, K.J., Battle, M.A., Duncan, S.A., et al., 2012. Pancreas-specific deletion of mouse Gata4 and Gata6 causes pancreatic agenesis. J. Clin. Invest. 122, 3516–3528.

Yebra, M., Diaferia, G.R., Montgomery, A.M., Kaido, T., Brunken, W.J., et al., 2011. Endothelium-derived Netrin-4 supports pancreatic epithelial cell adhesion and differentiation through integrins $\alpha_2\beta_1$ and $\alpha_3\beta_1$. PloS One 6, e22750.

Yechoor, V., Liu, V., Paul, A., Lee, J., Buras, E., et al., 2009. Gene therapy with neurogenin 3 and β-cellulin reverses major metabolic problems in insulin-deficient diabetic mice. Endocrinology 150, 4863–4873.

Yi, P., Park, J.S., Melton, D.A., 2013. β-trophin: A hormone that controls pancreatic β-cell proliferation. Cell 153, 747–758.

Yoshitomi, H., Zaret, K.S., 2004. Endothelial cell interactions initiate dorsal pancreas development by selectively inducing the transcription factor Ptf1a. Development 131, 807–817.

Zhang, C., Moriguchi, T., Kajihara, M., Esaki, R., Harada, A., et al., 2005. MafA is a key regulator of glucose-stimulated insulin secretion. Mol. Cell. Biol. 25, 4969–4976.

Zhang, H., Ables, E.T., Pope, C.F., Washington, M.K., Hipkens, S., et al., 2009. Multiple, temporal-specific roles for HNF6 in pancreatic endocrine and ductal differentiation. Mech. Dev. 126, 958–973.

Zhang, H., Ackermann, A.M., Gusarova, G.A., Lowe, D., Feng, X., et al., 2006. The FoxM1 transcription factor is required to maintain pancreatic β-cell mass. Mol. Endocrinol. 20, 1853–1866.

Zhang, H., Zhang, J., Pope, C.F., Crawford, L.A., Vasavada, R.C., et al., 2010. Gestational diabetes mellitus resulting from impaired β-cell compensation in the absence of FoxM1, a novel downstream effector of placental lactogen. Diabetes 59, 143–152.

Zhang, Y.Q., Cleary, M.M., Si, Y., Liu, G., Eto, Y., et al., 2004. Inhibition of activin signaling induces pancreatic epithelial cell expansion and diminishes terminal differentiation of pancreatic β-cells. Diabetes 53, 2024–2033.

Zhao, L., Guo, M., Matsuoka, T.A., Hagman, D.K., Parazzoli, S.D., et al., 2005. The islet β-cell-enriched MafA activator is a key regulator of insulin gene transcription. J. Biol. Chem. 280, 11887–11894.

Zhou, Q., Law, A.C., Rajagopal, J., Anderson, W.J., Gray, P.A., Melton, D.A., 2007. A multipotent progenitor domain guides pancreatic organogenesis. Dev. Cell 13, 103–114.

Section IV

Selected Clinical Problems

Chapter 32

Diaphragmatic Embryogenesis and Human Congenital Diaphragmatic Defects

Kate G. Ackerman
Department of Pediatrics and Biomedical Genetics, University of Rochester Medical Center, Rochester, NY

Chapter Outline

GLOSSARY

Central tendon (of the diaphragm) The central tendon is the unmuscularized region of the diaphragm that has differentiated into a mature tendon. It makes up the center of the diaphragm, and it is attached to the liver by the falciform ligament. Distinct myotendonous junctions form the boundary between the central tendon and the peripherally muscularized regions.

Congenital diaphragmatic hernia (CDH) This term is used inconsistently to describe a variety of diaphragmatic defects. Among physicians, it is usually used to describe the severe hernias that present at birth. These hernias almost always involve the posterolateral, lateral, or posteromedial diaphragm. They may also present as aplasia of one side of the diaphragm. The term "CDH" is sometimes used interchangeably with "Bochdalek hernia," a hernia that should only arise from a weak spot along the posterolateral wall of the thorax.

Nitrofen (2,4-dichlorophenyl 4-nitrophenyl ether) An herbicide (and banned carcinogen) used to create a rodent model of congenital diaphragmatic hernia. Embryos from pregnant dams exposed to nitrofen develop diaphragm, lung, and other structural malformations. The chemical affects the retinoic acid signaling pathway.

Non-communicating defect or muscularization defect A non-communicating diaphragmatic defect is one in which there is no direct communication of the abdominal contents with the thoracic cavity. These include muscularization defects of the diaphragm. These defects include massive herniations covered by a membrane contiguous with the diaphragm (a sac hernia) and mild diffuse muscularization defects causing a slight elevation of the diaphragm (sometimes called eventration).

Pleuroperitoneal fold (PPF) and post-hepatic mesenchymal plate (PHMP) These terms describe primordial diaphragmatic mesenchymal tissue that is most easily visualized at mouse embryonic days 11.5–12.5 in the thoracoabdominal region. There is disagreement in the literature about which tissues grow to make up the majority of the posterior diaphragm. This is partially complicated by inconsistent application of the terms to the embryonic anatomy, as plane of sectioning artifacts and other factors can make it very difficult to delineate these structures.

SUMMARY

- The diaphragm is an essential organ for the transition to postnatal life. Serious developmental defects of the diaphragm cause pulmonary hypoplasia, respiratory distress and cardiopulmonary failure at birth. Pulmonary hypoplasia may result both from abnormal mechanical forces and from primary defects in development due to abnormal gene function. Although only a minority of diaphragmatic hernia patients have an obvious cardiac malformation, evidence from mouse models suggests that more occult developmental issues may be present in the heart, and these issues may contribute to long term disease.
- The most serious diaphragmatic defects occur in the posterolateral (most common), posteromedial, or lateral diaphragm. These are often called "congenital diaphragmatic hernia" or CDH. The nomenclature for these defects is not yet standardized. Defects of the anterior diaphragm are usually not associated with severe illness. Defects that are limited to only the central tendon are extremely rare in humans.
- The posterior diaphragm develops from primordial mesenchymal tissue best visualized in the mouse thoracoabdominal region at embryonic day 11.5–12.5. This tissue includes the post-hepatic mesenchymal plate and the pleuroperitoneal folds. The embryonic anatomy is difficult to interpret, and the use of these terms is inconsistent.
- The anterior diaphragm forms from the transverse septum. Migrating muscle progenitor cells reach the diaphragm after the development of a contiguous, connective tissue sheet. Tendon progenitor cells differentiate and mature to form the central tendon, a structure separated from muscle by myotendonous junctions. Central tendon patterning defects and posterior hernias may occur together in the same genetic mouse model or the same human patient.
- The non-muscle mesenchyme of the diaphragm is most often implicated in hernia models. The diaphragm is covered by mesothelium, and the mesothelial cells express genes important for diaphragm development. The etiology of non-muscle cells and the contribution that they make to different portions of the diaphragm is an important area of investigation, which is currently limited by the availability of specific markers and mouse tools.

32.1 INTRODUCTION

The respiratory diaphragm is a structure that is critical to mammalian survival, as it provides the majority of the workforce for inflating the lungs. It also provides an essential barrier between the thoracic and peritoneal cavities. The establishment of this barrier during embryogenesis is essential for the maintenance of proper organ position. Defects in diaphragm formation that occur during human embryogenesis cause devastating neonatal disease. The most common and most medically burdensome problem related to the diaphragm is an absence of a portion of the diaphragm, commonly called "congenital diaphragmatic hernia" or CDH. This heterogeneous group of disorders can arise from a variety of mechanisms (Sluiter et al., 2012). As a group, population-based mortality rates are high, and these patients usually require both intensive care resources and long term specialized care (van den Hout et al. 2010). The major cause of morbidity and mortality in these patients is severe pulmonary hypoplasia, pulmonary hypertension, and cardiorespiratory failure (Pennaforte, et al., 2013; Stege et al., 2003).

The diaphragm develops from embryonic mesenchymal structures, and many genes and genetic pathways are now recognized to be neccesary for proper development (Russell et al., 2012). It is both interesting and highly relevant to consider that these genes and pathways are also important for other aspects of organogenesis, thus, we now have evidence that patients with the "CDH" spectrum of disease also have developmental anomalies in other organ systems. These issues may be obvious or occult and perhaps not evident until later in childhood or in adulthood.

In addition to the widespread influence that mutations in important genes may have on development, abnormal mechanical forces also contribute to disease. For example, lungs that are compressed by herniated abdominal organs will be hypoplastic given a long enough and severe enough exposure (van Loenhout et al., 2009). It is likely that the influence of mechanics persists postnatally, as children with large diaphragmatic defects have impaired diaphragmatic function. This loss of function prevents optimal lung stretch, a major stimulus for compensatory growth (Ysasi et al., 2013). Since comorbidities may arise from both primary (altered program of developmental signaling) and secondary (mechanical) mechanisms, it is important to develop models of disease that can be used to dissect out these relationships.

This chapter will review our current understanding of normal diaphragmatic development and of mechanisms known to be involved in abnormal development. The recent availability of advanced genetic research tools has improved the identification of CDH models, expanded our knowledge of genes required for diaphragm development, and has allowed for gene-disease association in both mice and humans. Going forward, it is expected that improved understanding of gene function and genetic regulation will improve our ability to predict and modify the course of disease.

32.2 DIAPHRAGMATIC ANATOMY

Anatomy of the Diaphragm

The mature diaphragm is essentially a large tendon surrounded by skeletal muscle with attachments to the body wall (Figure 32.1). The central tendon is easily identifiable, as it is present in a distinct and characteristic pattern. The majority of the diaphragm contains costal muscle, the skeletal muscle that joins the central tendon to the chest wall. The posterior medial diaphragm is referred to as the crural diaphragm, and consists of the muscle that surrounds the esophageal orifice. Since the crural region appears to have a distinct anatomy and structure, it is sometimes referred to as the second diaphragm (Pickering and Jones, 2002). It is responsible for maintaining appropriate esophageal closure, that is, to prevent the stomach contents from refluxing up, and to allow for vomiting when necessary.

Both muscular regions receive motor nerve innervation from the phrenic nerve, classically considered to have roots arising from the cervical spinal cord at C3 to C5 (Bordoni and Zanier, 2013). Recent evidence suggests that the phrenic nerve and its numerous branches have a more complex structure than previously appreciated, and the diaphragm is divided into different neuromuscular compartments. These are delineated by the distribution of innervation from the primary phrenic nerve branches. The anterior diaphragm receives innervation from an anterior branch whereas the lateral and posterior regions are innervated primarily by the posterior branches (An et al., 2012). The crural diaphragm and its phrenoesophageal ligament also receive innervation from the vagus nerve (Young et al., 2010).

The motor nerves are closely associated with arteriolar blood supply in the diaphragmatic muscle (Correa and Segal, 2012). The diaphragmatic vascular supply includes arterial blood supply from aortic collaterals as well as arteries originating from the internal mammary and intercostal vessels. Venous drainage occurs via the musculophrenic and inferior diaphragmatic veins (Bordoni and Zanier, 2013). The diaphragm also contains a network of lymphatic vasculature that is necessary to aid and presumably regulate the movement of pleural fluid (Negrini and Moriondo, 2013).

Multiple ligamentous structures anchor the diaphragm to the thorax and to closely related organs such as the liver. For the most part, the ligaments are named intuitively, that is, the pulmonary ligaments anchor the lung to the diaphragm whereas the phereno-pericardial ligaments anchor the diaphragm to the pericardium. Since the pericardial anchor is central, it is thought to be the most critical for normal distribution of contractile tension. On the abdominal side, the falciform, coronary, and triangular ligaments anchor diaphragm to liver and other ligaments anchor diaphragm to colon and duodenojejunum (Bordoni and Zanier, 2013).

On a microscopic level, the tissue and cell types present in the diaphragm include skeletal myocytes and myofibrils, tendon, non-tendon connective tissue, nerves, blood vessels, mesothelium, and lymphatics. All of these cell types or structures have been found to be abnormal in disease model systems.

Developmental Diaphragmatic Defects

Since patterns of abnormal development may inform mechanisms of normal embryogenesis, it is important to consider the range of phenotypic defects observed in humans or in mammalian animal models. A comprehensive and universal method for describing diaphragmatic defects does not currently exist, but adequate phenotyping should include the following categories: exact location, size, presence or absence of a membrane that is contiguous with the diaphragm, evidence of muscle or tendon patterning defects including integrity of the myotendonous junctions, presence of bilateral defects, evidence that the defect arises from a specific structure or location and presence of normal vascularity and innervation (Ackerman et al., 2012). The locations of typical diaphragmatic defects are shown in Figure 32.1.

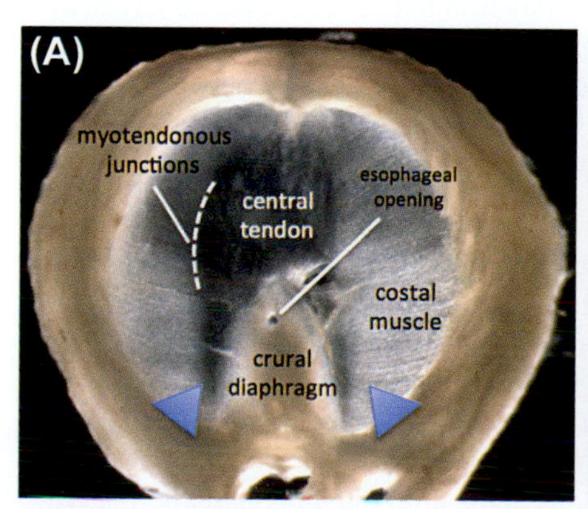

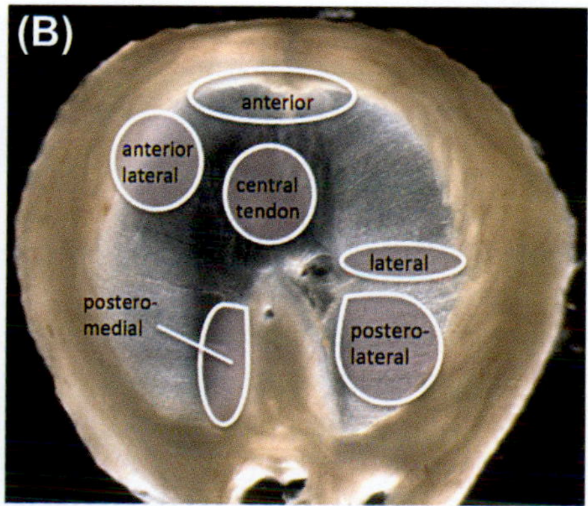

FIGURE 32.1 Mature mouse diaphragms showing anatomy (A) and common regions of herniation (B). The blue triangles show approximate areas of natural weakness along the posterior chest wall.

Location: Posterior Diaphragm

Defects in the posterior diaphragm are considered to be the most serious, as they are most often associated with critical illness in the newborn. In humans, defects involving the posterior or lateral diaphragm are often called Bochdalek hernias, a term adopted from Dr. Victor Bochdalek's description of hernias in 1848 (Kachlik and Cech, 2011). While the Bochdalek hernia was described to specifically arise from the lumbocostal triangle (Kachlik and Cech, 2011), a natural point of weakness in the diaphragm found adjacent to the posterolateral chest wall, the term has been adopted to describe a wide range of phenotypes including those with no involvement in that region (Ackerman et al., 2012). This is problematic for translational scientists hoping to understand mechanisms of development in human patients and in mouse models. For example, imperfect phenotyping prevents us from investigating whether patients with defects that arise from the lumbocostal triangle behave differently from defects not involving that region. Ideally, future phenotyping efforts will include more specific information about involvement of specific anatomic regions (anterolateral, lateral, posterolateral, posteromedial, or hemidiaphragmatic aplasia).

Consideration of defect size is important for prognosis since large lesions are associated with worse outcomes (Congenital Diaphragmatic Hernia Study, 2013). It is unknown whether specific genetic mechanisms determine the size or location of a defect, and this is an active area of investigation in animal models. Mouse genetic models show some specificity between the gene involved and the defect size and location, however there is phenotypic variability observed with changes of gene dose or background strain.

Location: Anterior Diaphragm

Defects in the anterior diaphragm may occur completely within the muscularized portion of the anterior diaphragm (Morgagni and Larrey hernias) or they may involve the tendon posterior to that region (Figure 32.1). This distinction is important because the former are not associated with cardiopulmonary failure and are often discovered incidentally on imaging done for reasons unrelated to lung disease. Although the terms Morgagni and Larrey hernia are used inconsistently, *Morgagni hernia,* described in the 1700s by Giovanni Battista Morgagni, occurs to the right of the xyphoid process, whereas Larrey hernias occur to the left. Hernias that cross the midline are best described as being sub-xyphoid (Salman et al., 1999; Loong and Kocher, 2005).

Anterior lesions that involve only central tendon or extend into the central tendon are different, in that they are more likely to cause severe illness. Like the posterior defects, central tendon defects arise through a variety of disease mechanisms, and they are heterogeneous with respect to morbidity, mortality, and associated malformations or diseases. Central tendon defects may be observed in patients with diseases associated with weak connective tissue such as Marfan's Disease (Slavotinek, 2007) and with multiple malformation syndromes such as Pentalogy of Cantrell (Mallula et al., 2013). They are also observed in mouse models with mutations in important developmental genes (Liu et al., 2003; Yuan et al., 2003).

Location: Esophageal Hiatus

Herniation of the gastroesophageal junction and the stomach through the esophageal hiatus is referred to as paraesophageal or hiatal hernia. These hernias are not traditionally considered to be a result of failed fetal diaphragmatic development, since presentation is typically in the adult, not the child. It is unknown whether this is a correct assumption as it is possible that occult abnormalities in development create a predisposition to adult hiatal hernia. Genetic factors paired with environmental risk factors such as pregnancy, obesity, and excessive coughing or vomiting may lead to late onset disease (Hyun and Bak, 2011). Childhood congenital abnormalities of the esophageal hiatus are rare, although they have been reported (Jetley et al., 2009). Perhaps the most severe abnormality in this spectrum includes a developmentally shortened esophagus with an intrathoracic stomach (Hendrickson et al., 2003; Leung et al., 2008).

Type of Defect: Defining the Continuity of the Diaphragm

Diaphragm defects are associated with variable severity of abdominal organ herniation into the thoracic cavity. The degree of herniation is a dynamic process, complicating its use for prediction of disease severity or disease mechanism. The term *eventration* is a synonym for herniation and is most often used to describe abnormal elevation of the diaphragm in a situation where the diaphragm remains intact. This results from a number of mechanisms including a lack of diaphragmatic function secondary to nerve or muscle dysfunction, or from abnormal muscular development of the diaphragm. The terminology becomes more confusing when a severe herniation is identified arising from an intact but undermuscularized portion of the diaphragm. In these cases, clinical descriptions include Bochdalek or congenital diaphragmatic hernia with a sac membrane, sac hernia, or eventration hernia (Ackerman et al., 2012). From a developmental biology perspective, it is important to differentiate these lesions since they likely occur via different mechanisms.

To improve phenotypic accuracy, it is recommended that the type of defect be described as either having no membrane contiguous with the diaphragm, thus allowing communication between the abdominal and thoracic cavities (communicating type) or as having a contiguous membrane that prohibits communication (non-communicating type).

Defect type is associated with defect location in humans. Defects in the anterior and anterolateral diaphragm are almost always of the non-communicating type. Defects in the posterior or lateral diaphragm are usually communicating, although it is not uncommon to see either one. Both types may be seen with variable degrees of organ herniation, although this is most limited in the lesions that occur within the muscularized region of the anterior diaphragm (Ackerman et al., 2012).

Bilateral Defects

Both mouse models and human patients may have bilateral defects. Since "mild" contralateral defects are not likely to be ascertained during typical clinical management, the true incidence is unknown. The limited field of view during surgical repair of congenital diaphragmatic hernia usually prohibits a complete bilateral inspection. Autopsy studies show that occult contralateral defects do occur, and while the locations may mirror each other, the types are variable. For example, a patient with a communicating posteromedial defect with severe herniation on the left might have a posteromedial defect on the right with minimal herniation. On pathology examination, the occult lesion might appear to be a thin, membranous diaphragm with absent muscularization. Since the presence of a contralateral defect is not systematically ascertained, the clinical significance of a potentially "underperforming" contralateral diaphragm is yet to be determined. The presence of a variety of types of defects within the same patient, the same syndrome, or within patients from the same family suggests that some phenotypes are developmentally related. This is supported by phenotypic variability observed in mouse models (Ackerman et al., 2012; Jay et al., 2007; Wat et al., 2012).

32.3 DIAPHRAGMATIC DEVELOPMENT

Development of the Non-Muscularized Diaphragm

The diaphragm is derived from multiple embryonic sources, however not all of these sources have been definitively identified using gold standard techniques such as cell fate mapping in the mouse. Perhaps the most elusive component of the diaphragm is the mesenchymal substrate that contributes to a non-muscle mesenchyme scaffold. This tissue is required for proper patterning and development of tendon and muscle, and it is a focus in congenital diaphragmatic hernia models, as this tissue is considered to be essential for maintaining diaphragmatic integrity (Merrell and Kardon, 2013).

Available evidence suggests that the non-muscle mesenchyme is derived from at least two populations of precursor cells. The anterior region of the diaphragm is derived from the septum transversum, and this tissue later develops into the mature central tendon and other essential non-muscle mesenchyme (Mayer et al., 2011). The posterior regions are derived from populations of cells found in the embryonic thorax and upper posterior abdomen during mid-embryogenesis. The definition of these cell populations as either pleuroperitoneal fold tissue, post-hepatic mesenchymal plate tissue, or primordial diaphragmatic mesenchyme is dependent on the orientation of tissue sections examined, the embryonic age studied, and investigator preference (Russell et al., 2012; Ackerman and Greer, 2007; You et al., 2005). The lack of availability of specific markers for diaphragmatic precursor mesenchyme has prohibited the exact delineation of these cells or structures, and interpretation of size, shape, and location is largely impacted by research methodology. Despite this confusion, it is evident that both nerve and muscle precursors migrate to the diaphragm within these tissues and structural variations in the tissue size or shape are associated with diaphragmatic defects (Clugston and Greer, 2007). On tissue section, it appears that structures described as pleuroperitoneal fold and post-hepatic mesenchymal plate (PHMP) make up a contiguous mesenchyme (Figure 32.2). Examination of scanning electron microscopy data supports the view that the pleuroperitoneal canal is closed by growth and extension of the PHMP tissue to meet up with the lateral pleuroperitoneal folds. The failure of growth of the PHMP tissue in a chemically induced animal model is one proposed mechanism for formation of a diaphragmatic hernia (Mayer et al., 2011; Iritani, 1984). As in the pathogenesis of other structural birth defects, there are likely a variety of mechanisms at play that can contribute to a variety of overlapping phenotypes and severity of disease.

Development of the Central Tendon and Non-Muscle Mesenchyme

Fortunately, genetic tools and markers used for delineating tendon differentiation and maturation in the mammalian limb are also useful for understanding the development of the diaphragm. Although limb tendons connect muscle to bone and

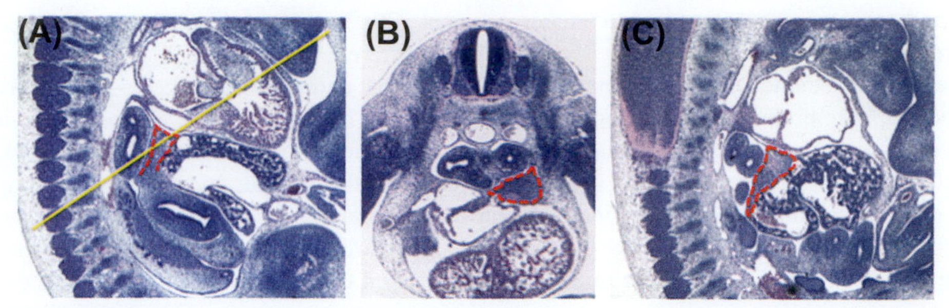

FIGURE 32.2 Primordial Diaphragm Tissue (Dotted Red Lines) Visualized in the Mouse Embryo at e11.5. Sagittal sections shown in A (left side) and C (right side). When an embryo is cut along the plane shown by the yellow line (A), the primordial diaphragm tissue on that transverse section appears to be wedge shaped tissue found between the lungs and heart (B, dotted line). When the plane of section is changed even slightly, the size and shape of the tissue changes. The tissue closest to the liver is called "post-hepatic mesenchymal plate" but is contiguous with both the pleuroperitoneal fold and the mesenchyme of the stomach.

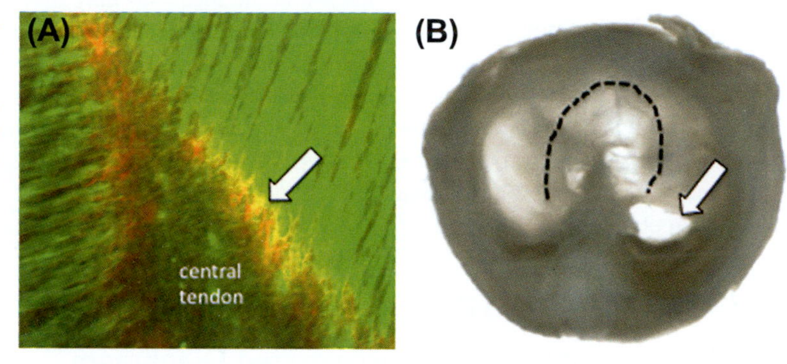

FIGURE 32.3 Magnified View of Normal Myotendonous Junctions (A, Arrow) Stained Red with Tenascin C. Muscle fibers (actin filaments) are stained green with fluorescent phalloidin. The myotendonous junctions form a distinct line between the central tendon and the muscle. (B) Diaphragm with lack of normal central tendon and posterior hernia (arrow). The dotted lines show where the myotendonous junctions should be found.

diaphragmatic central tendon joins muscle to muscle, both have characteristic myotendonous junctions (MTJ), in which muscle fibers are anchored to and delineated from tendon cells (Figure 32.3). Early in tendon development, undifferentiated mesenchymal cells expressing *SRY Box9 (Sox9)* become specified to a tendon cell fate after induction of *Scleraxis (Scx)*, a transcription factor required for tenocyte maturation (Schweitzer et al., 2010). In the mouse diaphragm, this occurs by day 12.5 of gestation and is followed by the establishment of myotendonous junctions along the tendon-muscle boundary. The myotendonous junctions express both *Scx* and an early tenocyte marker, Tenascin C (Tnc). When the central tendon is mature, it expresses the differentiated tenocyte marker, Tenomodulin (Tnmd). A normal mouse or human diaphragm that is double labeled with a skeletal muscle marker, such as Myosin heavy chain (MyHC), and Tenomodulin shows a sharp, distinct boundary between regions (Coles and Ackerman, 2013), and a disruption of this boundary may be observed in both mouse models and human patients (Coles and Ackerman, 2013; Ackerman et al., 2012).

A relatively new mouse genetic model has allowed for detailed analysis of diaphragmatic tendon development. Mice with loss of expression of the gene, *Kinesin family member 7 (Kif7)* have a central tendon that appears to be haphazardly overgrown by skeletal muscle (Coles and Ackerman, 2013). On closer inspection with molecular markers, the muscle overgrowth occurs when the tendon precursor cells do not differentiate properly. These mutant cells lack normal induction of *Scx,* and they have impaired tenocyte differentiation. The failed differentiation of the tendon is striking in the mature mutant diaphragm where there are no myotendonous junctions, no mature tenocytes, and no expression of *Collagen type 1, alpha-1 (Colla1)*, a marker of collagen expression in the diaphragmatic tendon. This model demonstrates that muscle has not simply overgrown the central tendon. Instead, the tendon is not developed and muscle overgrows the region meant to become a tendon, perhaps using undifferentiated connective tissue as a scaffold.

The muscle requires differentiated tendon to develop a normal and functional pattern. The potential consequences of mis-patterned muscle on diaphragmatic function are unknown, but could be quite pronounced. The association between poorly differentiated tendon and classic "congenital diaphragmatic hernia" is unknown, however *Kif7* mutant mice have both. Although the mechanisms responsible for the posterior hernias are not understood, the presence of tendon differentiation defects and posterior lateral hernia in the same animal suggests that tendon and other connective tissue maturation processes should be closely inspected in diaphragmatic hernia models. Since an overgrown central tendon region, or perhaps lack of central tendon, is occasionally observed in humans with posterolateral communicating hernias, it will be important to consider the consequences of "two hits" to the diaphragm: a hernia for which a patch repair renders a large portion non-functional, and an anterior diaphragm with poor function secondary to a tendon differentiation defect.

The scaffold tissue in the region of the central tendon must thicken and separate from the liver during development. After this occurs, the diaphragm has more mobility and remains attached to the central liver by the falciform ligament. Mouse models with deficiency in *Slit Homolog 3 (Slit3),* a secreted protein required for cell migration, fail to complete this process resulting in a thin central tendon that is subject to herniation (Liu et al., 2003; Yuan et al., 2003). Mice with mutation in the *Slit* receptors, called *Robos* (*Roundabouts*), also develop diaphragm defects, although these are reported to occur at the esophageal meatus and do not appear similar to typical human neonatal defects (Domyan et al., 2013).

The scaffold tissue in the posterior diaphragm may be disrupted to cause a hernia without influence of muscle cell populations. This was demonstrated when embryos genetically lacking diaphragmatic muscle were exposed to nitrofen (2,4-dichloro-phenyl-p-notrophenylether), a chemical frequently used to induce diaphragmatic defects in rodents. The amuscular diaphragm was reported to have hernias similar to those observed in muscularized diaphragms (Babiuk and Greer, 2002). This has led investigators to continue focusing on the non-muscular mesenchymal scaffold as the origin of serious diaphragmatic disease.

Muscularization and Innervation of the Diaphragm

The diaphragm contains skeletal muscle, and this muscle forms a distinct and functional pattern required for efficient function. In the mature diaphragm, the muscles must contract together to function as a bellows, causing the diaphragmatic dome to descend and the pleural pressure to fall. This facilitates air entry into the lungs (De Troyer and Boriek, 2011). Mouse genetic tools have allowed investigators to determine that the costal muscle comprising most of the diaphragm is derived from migratory precursor cell populations. These cells are derived from the cervical ventral lateral dermomyotome, a region that also gives rise to limb and tongue muscles. The diaphragmatic muscle shares many of the regulatory signals required to develop and direct other hypaxial muscle groups (see also Chapter 28), yet it also maintains its own identity, distinct from other muscle groups found in close proximity, such as the intercostal muscles. Manipulation of the genes required for the proliferation, maturation, and migration of muscle in mice and in the chick has provided some insight into the incredibly complex control of these processes (Birchmeier and Brohmann, 2000; Shih et al., 2008).

The diaphragm is able to form as an intact connective tissue membrane without receiving normal skeletal muscle cell populations. Mice with deletion of *Pax3 (Paired box gene 3)* have impaired establishment of muscle progenitor cells in the ventral dermomyotome. These mice do not form hypaxial muscles and have amuscular diaphragms (Li et al., 1999). Mice with deficiency of *Met*, the membrane receptor tyrosine kinase for HGF (Hepatocyte Growth Factor), also have amuscular diaphragms (Babiuk and Greer, 2002). This makes sense, as *HGF* and *Met* are required to control the delamination and migration of muscle precursor cells, and *Met* expression is at least partially transcriptionally controlled by *Pax3* (Epstein et al., 1996). HGF, a secreted factor required for motility and morphogenesis, is expressed in the mesenchymal cells along the path of muscle cell migration to the diaphragm. Altered expression of *HGF* is associated with abnormal muscle patterning in the diaphragm. This has been shown in mice with mutations of the *Zfpm2* gene (*Zinc Finger Protein, FOG family member 2*). Primordial diaphragm tissue from *Zfpm2* mutant embryos has reduced HGF expression, and their mature diaphragms have severe muscle patterning defects. These defects include a complete lack of muscle in the posterolateral regions, and a loss of normal costal muscle fiber orientation (Ackerman et al., 2005).

The characterization of muscle precursors that will contribute to the diaphragm is easier than defining the non-muscle mesenchyme because of the extensive availability of identified markers. Expression analysis of cells in the primordial diaphragmatic tissue, such as the pleuroperitoneal fold, demonstrates the presence of both proliferating and differentiating muscle precursor cells. The earlier stage tissue expresses Pax3, marking a myogenic population, while differentiated proliferating myoblasts express MyoD (Babiuk et al., 2003). Myoblasts form multi-nucleated myotubes that fuse to become mature muscle fibers. MyHC is an excellent marker for mature skeletal muscle fibers in the mouse diaphragm after e16.5 (Coles and Ackerman, 2013). While it is tempting to consider the muscle development and the associated markers and transcription factors in isolation from the rest of the diaphragm, recent evidence suggests that there is overlap of diaphragmatic gene expression in muscle and connective tissue lineages. For example, in cell culture, *Pax3* and *Tcf21* (*Transcription Factor 21*) may be expressed by cells in non-myogenic cells (Stuelsatz et al., 2012). Diaphragms during and after the muscularization process are shown in Figure 32.4.

Mouse Modeling of Diaphragmatic Defects

Mouse models have provided much valuable information to increase our understanding of abnormal diaphragm development, however in the large world of international mouse genome collaboratives, the diaphragm has been slow to appear as an organ of interest. Searches of databases or reviews of mouse mutant literature reveals many models with absent or

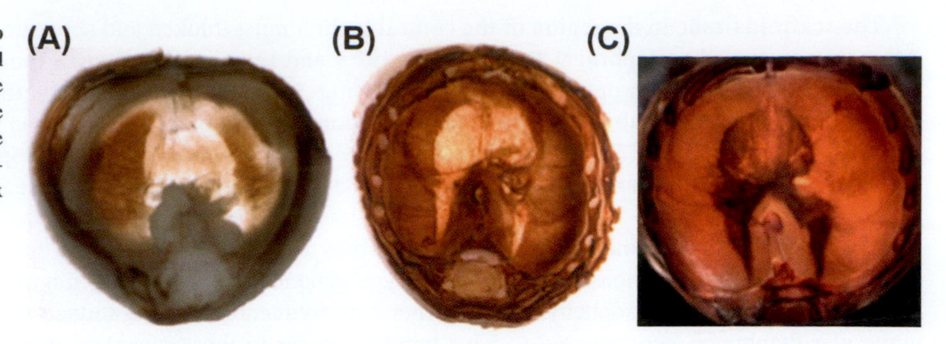

FIGURE 32.4 Diaphragms Stained to Show Muscularization (A, B) and Central Tendon (C). Migration of muscle to the diaphragm is incomplete in (A) and mature in (B). The diaphragm in (C) shows whole mount *in situ* hybridization for *fibromodulin*, a gene encoding an extracellular matrix protein.

significantly decreased muscularization, whereas relatively few models have human-like communicating diaphragmatic defects or tendon patterning anomalies, which are of most interest to translational investigators. Currently, it is difficult to find publically available precise embryonic expression data or information about lines with *Cre recombinase* expression in sites appropriate for diaphragmatic specific gene deletion.

The lack of characterized models, when compared to other organs such as the heart or brain, may result from a lack of awareness of the potential significance for human health, the difficult technical barriers to correct ascertainment of an extremely delicate organ, and the high degree of early to mid-embryonic lethality. Forward genetic screening in mice, for example, with the use of *N*-ethyl-*N*-nitrosourea (ENU), has produced models not identified during the characterization of null alleles. The development of successful human disease models in forward genetic mouse screens is dependent on careful and expert phenotypic screening for diaphragmatic defects in mid to late gestation embryos. Discoveries from these screens have capitalized on the fortunate ability of ENU to induce point mutations that often alter gene dose or function without complete gene deletion. Models discovered by these methods include those with mutations in *Zfpm2*, *Kif7*, and *Frem-1* (Coles and Ackerman, 2013; Ackerman et al., 2005; Beck et al., 2013).

As anticipated from progress made in other models of organogenesis, many different genes are required to function for proper diaphragm development. Two things are striking when looking at the list of genes so far implicated. First, most or perhaps all of the candidate genes function in the retinoic acid pathway (Beurskens et al., 2007; Clugston et al., 2010). A normal balance of retinoic acid function is required for proper generalized embryonic development, as well as for specific organ development. The diaphragm, like the heart, lungs, and other organs requires intact retinoic acid signaling (Rhinn and Dolle, 2012). This first became evident when it was discovered that a rodent fed a diet deficient in vitamin A was predisposed to have offspring with diaphragmatic defects (Andersen, 1941). In the chemical model of CDH, the mechanism of nitrofen (and similar chemicals) includes a downregulation of a major retinoic acid synthesizing enzyme. Phenotypes in this model can be at least partially rescued by administering vitamin A or retinoic acid (Babiuk et al., 2004). Most of the mouse genetic models identified to date have mutations in genes that require, utilize, or promote retinoic acid signaling, including *Gata4*, *Zfpm2*, *Nr2f2*, *Kif7*, *Wt1* (*Wilms Tumor 1*), and *Pdgfra*. A selected list of mouse genetic models, many associated with human disease, is shown in Table 32.1. The identification of a diverse group of genetic models allows for careful dissection of the developmental pathways required for different aspects of diaphragm maturation and function. This is important, as genetic findings in mice may be translated to human genetic studies. Consideration of all of the models together is ideal. For example, retinoic acid signaling pathway abnormalities in animal models have informed studies in humans in which retinoic acid pathway abnormalities have been associated with CDH (Beurskens et al., 2010).

Since mouse models with 100% gene deletion tend to be embryonic lethal prior to diaphragm formation, it is helpful to study gene alleles on different backgrounds at different gene dosages, and it is optimal to utilize time and site-specific gene manipulation (see also Chapter 6). Two conditional gene models have been used successfully to create diaphragmatic phenotypes. First, *Nkx3.2* (*NK3 Homeobox 2*) was used to drive *Cre recombinase* in a cross with *Nr2f2* floxed mice. The *Cre* expression, driven by the *Nkx3.2* abdominal mesenchyme transcription factor, was left dominant, and embryos had left sided hernias (You et al., 2005). It is important when evaluating conditional models to consider that the phenotype is constrained to the region of deletion in the model. That is, the predominance of left sided lesions in the *Nr2f2* model does not necessarily occur for the same reason that human lesions are predominantly left sided. A second model utilizes *Wt1* driving a CreERT2 fusion protein, thus recombination occurs after exposure to tamoxifen. The introduction of a time-sensitive response element allows for careful dissection of developmental timing (Paris et al., 2010). There are other promising genes that may serve to drive site-specific recombination. These include non-muscle mesenchyme connective tissues such as the collagens and other matrix coding genes.

TABLE 32.1 Major Mouse Models of Congenital Diaphragmatic Hernia

Gene Name	Function / Encodes	Major Developmental Functions	Human Significance
Frem-1 (Beck et al., 2013)	Extracellular matrix protein	Craniofacial eyes, kidneys, lung digits	Unclear (1 patient, not yet replicated, new discovery)
Gata4* (Jay et al., 2007; Beck et al., 2013;)	Transcription factor	Cardiovascular, lung, gonads, liver	Yes (non-isolated)
Kif7* (Coles and Ackerman, 2013)	Kinesin motor protein, primary cilia function	Brain, heart, lung, limbs/digits	Unknown (new discovery)
Lox (Hornstra et al., 2003)	Enzyme to crosslink collagen and elastin	Connective tissue integrity	Connective tissue disease (not neonatal CDH)
Nr2f2* (couptf2) (You et al., 2005)	Nuclear hormone receptor, transcription factor	Cardiovascular	Unclear (Hotspot, no validated sequence changes)
Pdgfrα (Bleyl et al., 2007)	Tyrosine-protein kinase, growth factor receptor	Cardiovascular, lung, skeleton, kidneys	Unclear (1 patient, not yet replicated, unclear sig)
Slit3 (Slit-Robos) (Liu et al., 2003; Yuan et al., 2003; Domyan et al., 2013)	Ligand for Robo receptor, multiple functions	Cardiovascular, lung, kidneys	Not identified
Sox7 (Wat et al., 2012)	Transcription factor	Cardiovascular	Unclear (Hotspot, no sequence changes)
Wt1* (Kreidberg et al., 1993; Clugston et al., 2006)	Transcription factor	Cardiovascular, gonads, kidneys	Yes (non-isolated)
Zfpm2* (was Fog2) (Ackerman et al., 2005)	Transcription cofactor	Cardiovascular, lung, gonads,	Yes (replicated)

**Mouse models demonstrated to have loss of posterior diaphragmatic integrity (most like severe neonatal congenital diaphragmatic hernia)*

32.4 GENETICS OF HUMAN CONGENITAL DIAPHRAGMATIC DEFECTS

Human Spectrum of Defects

Most patients with the classic posterior region congenital diaphragmatic hernia do not have other obvious structural birth defects, that is, the defects are considered to be isolated. Hernias associated with other defects may be classified as non-isolated or complex, and when findings match a recognized pattern, they are classified as being syndromic (Pober et al., 2010). This classification is a bit misleading as virtually all non-isolated patients have pulmonary hypoplasia, and many have cardiac dysfunction. From a developmental perspective, both the heart and lungs are very likely to harbor primary developmental problems, and these comorbidities are discussed later in this chapter.

Cardiovascular malformations are the most common structural birth defect identified in patients with congenital diaphragmatic hernia, occurring in 10–25% of patients, depending on the cohort examined. Although left sided outflow tract issues are frequently diagnosed, most of these are false calls due to the distorted anatomy and blood flow associated with herniation and mediastinal shifting. When a cardiovascular malformation is present, septal defects are most common. Conotruncal defects such as tetralogy of Fallot are also commonly observed (Lin et al., 2007; Colvin et al., 2005).

Other malformations associated with diaphragmatic defects include neural tube defects, hydrocephalus, limb defects, genitourinary abnormalities, and small or absent eyes. Patients with one of the most characteristic syndromes, Fryns Syndrome, have hypoplasia of the nails and distal phalanges as well as abnormal facies, genitourinary anomalies, cardiovascular defects, and more (Slavotinek, 2004). For a comprehensive review of syndromes associated with diaphragmatic defects, see the Congenital Diaphragmatic Hernia Overview at http://www.genereviews.org (Pober et al., 2010).

Genetic Abnormalities

The approach to genetic research and clinical testing for congenital diaphragmatic hernia patients is in a rapid state of change. These complex defects are usually non-Mendelian with low rates of precurrence and recurrence (Pober et al., 2005).

Although heritability is low, families with multiple affected members in large pedigrees have been identified. In these families, it has been possible to perform high density genotyping and to identify shared associated regions using genetic tools such as homozygosity exclusion rare allele mapping (HERAM) and phased haplotype sharing (HapShare) (Arrington et al., 2012). Both large unrelated cohorts and families have been studied using whole genome array genomic hybridization methods to identify abnormal copy number variants. From these studies, cytogenetic hotspots have been identified that are thought to contain diaphragmatic hernia candidate genes (Yu et al., 2012).

While cytogenetic studies were once performed only in syndromic type CDH patients, the availability of higher density arrays has allowed for discovery of small genetic aberrations in patients who have CDH without other malformations. These include loci on chromosome 15 (microdeletions of 15q26.2, OMIM DIH1), the short arm of chromosome 8 (8p23.1, OMIM DIH2), and the long arm of chromosome 8 (8q22.3–23.1, OMIM DIH3). The identified cytogenetic imbalances have served as targets for investigation. Multiple independent groups have replicated experiments to identify and narrow these hotspot regions in different cohorts. Genes within these regions have been sequenced with hope of identifying causal genes. In the 15q26.2 locus, the region was narrowed to a handful of candidate genes including NR2F2, an important developmental gene for which the mouse model forms a diaphragmatic hernia (You et al., 2005; Klaassens et al., 2005; Scott et al., 2007). Unfortunately, despite multiple independent efforts, patients have not been identified with mutations in the coding region of this gene (Scott et al., 2007). Expression analysis of genes in the region has been used to further identify a top candidate gene, however all of the genes were found to have expression in the developing diaphragm (Clugston et al., 2008). The causal gene, if there is one, remains undiscovered. It is interesting to note that the *Kif7* candidate gene identified in a mouse model sits very close to the 15q26.2 region. It is possible that this gene or others close by are influenced by transcriptional regulation sequences in the deleted region.

It is probably not coincidental that the 8p23.1 and 8q22.3–23.1 CDH cytogenetic hotspots contain genes that are known partners in regulating transcription. The candidate gene at 8q, ZFPM2, has been validated in multiple independent reports, as damaging sequence variants have been identified in CDH patients. This includes mutations that are *de novo* (Ackerman et al., 2005) or inherited (Wat et al., 2011), however the utility of genetic screening is unclear since there is variable penetrance and coding mutations or cytogenetic changes are fairly rare. Cytogenetic variants at 8p are more common than at 8q, however the role of sequence mutations in 8p candidate genes in potentially causing human CDH is unclear (Arrington et al., 2012; Longoni et al., 2012). In this region, GATA4 is a major candidate gene (Yu et al. 2013) as it forms a transcriptional complex with ZFPM2 to promote normal organogenesis (Ackerman et al., 2007). There is evidence that multiple genes in the 8p region also interact and are required to form a normal diaphragm (Wat et al., 2012). This has, however, not been validated in humans, since sequencing data do not support an association of compound heterozygous mutations in 8p candidate genes with congenital diaphragmatic hernia (Wat et al., 2012).

Whole exome and whole genome approaches to understanding the genetic contribution to a non-Mendelian disease are difficult without informed decision-making about probable candidate genes. The best information will come through compilation of the data from mouse models, cytogenetic studies, sequence analysis, and other assays. The luxury of access to sequence for any gene in a large cohort of samples will advance our understanding of mechanisms of disease when a team approach is used to understand gene function.

32.5 CARDIOPULMONARY DEVELOPMENT AND THE DIAPHRAGM

Pulmonary Development

Although development of the diaphragm is a critically important focus, investigation of cardiopulmonary disease mechanisms associated with congenital diaphragmatic hernia provides an opportunity to modulate outcome in the postnatal period. Diaphragmatic hernia involving the posterolateral (most common), posteromedial, or lateral diaphragm is associated with severe pulmonary hypoplasia (Ackerman et al., 2012). This arises from at least two major mechanisms. First, herniation of abdominal organs through a diaphragmatic defect compresses the lung during a time of critical development. The impact from a non-working or less than optimally working diaphragm (Abolmaali et al., 2010) also likely plays a role in that there is a suboptimal stimulus for stretch. The effect of these mechanical forces has been validated in a number of studies where diaphragmatic hernias have been surgically created in fetal lambs (reviewed in van Loenhout et al., 2009). It is now widely recognized that the lungs may also have an independent developmental defect, and this likely varies from model to model or patient to patient. This independent developmental lung defect was demonstrated in the nitrofen model, as pulmonary hypoplasia could be induced independently of a diaphragmatic defect (Keijzer et al., 2000). Genetic evidence for this was first shown in the *Zfpm2* model, in which the mutated gene is required for patterning and growth of the lung even when isolated and cultured outside of the environment of the diaphragm (Ackerman et al., 2005). Since diaphragmatic hernia is a genetically heterogeneous disease, it is expected that a subset of patients have both a mechanical developmental insult and a developmental genetic insult to the lung and pulmonary vascular systems. More interestingly, many of the identified candidate genes have postnatal functions in the lung and may inform a number of therapeutic targets or perhaps help to predict long term disease.

Cardiovascular Development

Although the most common birth defect associated with the diaphragm is a heart defect, most patients do not have one and are not considered to have congenital heart disease. This is an ongoing area of investigation, as most of the genes identified to be important for diaphragm development in the mouse are also critically important for cardiac development. For example, mice with *Gata4* or *Zfpm2* mutations have both cardiac and diaphragmatic defects. Humans however, seem to have one or the other, and coding sequence changes are fairly rare. There are many possible reasons for the phenotypic discrepancies. First, damaging coding mutations are likely to cause high rates of embryonic lethality, thus, it is possible that patients with multiple affected organs die *in utero*. Second, these and other candidate genes have some degree of organ-specific regulation (He et al., 2011; Burch et al., 2005). Patients may have gene regulatory mutations that cause organ-specific disease. The future discovery of regulatory "treasures" within so called "junk DNA" and the advancement of whole genome sequencing will help to address these questions in the future.

Developmentally, it is interesting that genes required for morphogenesis of the heart may also be required for long term ventricular function and stress adaptation. Morphological anomalies might also be subtle and clinically irrelevant until an environmental stress occurs. This is best described in *Zfpm2* and *Gata4* models; gene expression is required for the normal regulation of cardiac function, the adaptation of the heart to a pressure stress, and for the regulation of coronary angiogenesis (Zhou et al., 2009; Bisping et al., 2006). These findings are interesting, as cardiac failure is the usual cause of death in diaphragmatic hernia patients. This has been thought to occur in a "normal heart" responding to the severe stress of pulmonary hypoplasia associated pulmonary hypertension.

Thinking developmentally, we can now piece together hypotheses that could help explain the heterogeneity of disease. Based on mouse models and candidate genes that mediate development in the diaphragm, heart, and lung, some patients may have genetically determined dysfunction of the pulmonary vascular system. It is also probable that the morphologically normal heart of some CDH patients has a reduced ability to tolerate hypoxia, pulmonary hypertension, or other stressors.

CONCLUSIONS

The development of the diaphragm and the role that the genes required for diaphragm development play in other organs is an important area of basic and translational research. Children with serious isolated structural birth defects, like congenital diaphragmatic hernia, have acute dysfunction in the newborn period as well as long term potential for dysfunction. It is possible that the heterogeneity in outcomes can be partially attributed to mutations in many different genes and pathways, and it is probable that understanding this biology will enhance long term health. There are rodent models which can be used to understand diaphragm development, but investigations are limited by early embryonic lethality. Going forward, investigations using conditional gene deletion will advance our understanding of the contribution of specific genes to development in the diaphragm as well as in the lungs, the heart, and other organs that cause comorbid disease.

CLINICAL RELEVANCE

- Congenital diaphragmatic defects are a heterogeneous group of birth defects. The most common defects involve the posterior diaphragm and present with respiratory distress at birth.
- Diaphragmatic hernias are associated with pulmonary hypoplasia. The etiology of pulmonary hypoplasia is multifactorial and may include: compression from herniated organs, decreased mechanical stretch stimuli from abnormal diaphragmatic function, and primary developmental issues due to gene mutations. Many of the genes required for diaphragm development are also necessary for lung and heart development.
- Congenital heart defects are the most common malformations associated with congenital diaphragmatic hernias. Animal models suggest that CDH patients without structural heart malformations may have occult issues with cardiac function.
- The diaphragm muscle is anchored to a central tendon. The tendon must be patterned normally for maximal efficiency. Some human patients with diaphragmatic hernias also have tendon patterning defects.

REFERENCES

Abolmaali, N., et al., 2010. Lung volumes, ventricular function and pulmonary arterial flow in children operated on for left-sided congenital diaphragmatic hernia: long-term results. Euro. Radio. 20, 1580–1589.

Ackerman, K.G., Greer, J.J., 2007. Development of the diaphragm and genetic mouse models of diaphragmatic defects. Am. J. Med. Genet. C. Semin. Med. Genet. 145C, 109–116.

Ackerman, K.G., et al., 2012. Congenital diaphragmatic defects: proposal for a new classification based on observations in 234 patients. Pediatr. Dev. Pathol. 15, 265–274.

Ackerman, K.G., et al., 2005. Fog2 is required for normal diaphragm and lung development in mice and humans. PLoS Genet. 1, 58–65.

Ackerman, K.G., et al., 2007. Gata4 is necessary for normal pulmonary lobar development. Am. J. Respir. Cell Mol. Biol. 36, 391–397.

An, X., et al., 2012. Intramuscular distribution of the phrenic nerve in human diaphragm as shown by Sihler staining. Muscle Nerve 45, 522–526.

Andersen, D.H., 1941. Incidence of congenital diaphragmatic hernia in the young of rats bred on a diet deficient in vitamin A. Am. J. Dis. Child. 62, 888–889.

Arrington, C.B., et al., 2012. A family-based paradigm to identify candidate chromosomal regions for isolated congenital diaphragmatic hernia. Am. J. Med. Genet. A. 158A, 3137–3147.

Babiuk, R.P., Greer, J.J., 2002. Diaphragm defects occur in a CDH hernia model independently of myogenesis and lung formation. Am. J. Physiol. Lung. Cell Mol. Physiol. 283, L1310–1314.

Babiuk, R.P., Thebaud, B., Greer, J.J., 2004. Reductions in the incidence of nitrofen-induced diaphragmatic hernia by vitamin A and retinoic acid. Am. J. Physiol. Lung. Cell Mol. Physiol. 286, L970–973.

Babiuk, R.P., Zhang, W., Clugston, R., Allan, D.W., Greer, J.J., 2003. Embryological origins and development of the rat diaphragm. J. Comp. Neurol. 455, 477–487.

Beck, T.F., et al., 2013. Novel frem1-related mouse phenotypes and evidence of genetic interactions with gata4 and slit3. PLoS One 8, e58830.

Beurskens, L.W., et al., 2010. Retinol status of newborn infants is associated with congenital diaphragmatic hernia. Pediatrics 126, 712–720.

Beurskens, N., Klaassens, M., Rottier, R., de Klein, A., Tibboel, D., 2007. Linking animal models to human congenital diaphragmatic hernia. Birth Defects Res. A. Clin. Mol. Teratol. 79, 565–572.

Birchmeier, C., Brohmann, H., 2000. Genes that control the development of migrating muscle precursor cells. Curr. Opin. Cell Biol. 12, 725–730.

Bisping, E., et al., 2006. Gata4 is required for maintenance of postnatal cardiac function and protection from pressure overload-induced heart failure. Pro. Nat. Acad. Sci. USA 103, 14471–14476.

Bleyl, S.B., et al., 2007. Candidate genes for congenital diaphragmatic hernia from animal models: sequencing of FOG2 and PDGFRalpha reveals rare variants in diaphragmatic hernia patients. Euro. J. Human. Gen. EJHG 15, 950–958.

Bordoni, B., Zanier, E., 2013. Anatomic connections of the diaphragm: influence of respiration on the body system. J. Multidiscip. Healthc. 6, 281–291.

Burch, J.B., 2005. Regulation of GATA gene expression during vertebrate development. Semin. Cell Dev. Bio. 16, 71–81.

Clugston, R.D., Greer, J.J., 2007. Diaphragm development and congenital diaphragmatic hernia. Semin. Pediatr. Surg. 16, 94–100.

Clugston, R.D., et al., 2006. Teratogen-induced, dietary and genetic models of congenital diaphragmatic hernia share a common mechanism of pathogenesis. Am. J. Pathol. 169, 1541–1549.

Clugston, R.D., Zhang, W., Greer, J.J., 2008. Gene expression in the developing diaphragm: significance for congenital diaphragmatic hernia. Am. J. Physiol. Lung. Cell Mol. Physiol. 294, L665–675.

Clugston, R.D., Zhang, W., Alvarez, S., de Lera, A.R., Greer, J.J., 2010. Understanding abnormal retinoid signaling as a causative mechanism in congenital diaphragmatic hernia. Am. J. Respir. Cell Mol. Biol. 42, 276–285.

Coles, G.L., Ackerman, K.G., 2013. Kif7 is required for the patterning and differentiation of the diaphragm in a model of syndromic congenital diaphragmatic hernia. Pro. Nat. Acad. Sci. USA 110, E1898–1905.

Colvin, J., Bower, C., Dickinson, J.E., Sokol, J., 2005. Outcomes of congenital diaphragmatic hernia: a population-based study in Western Australia. Pediatrics 116, e356–363.

Congenital Diaphragmatic Hernia Study, G., et al., 2013. Congenital diaphragmatic hernia: defect size correlates with developmental defect. J. Pediatr. Surg. 48, 1177–1182.

Correa, D., Segal, S.S., 2012. Neurovascular proximity in the diaphragm muscle of adult mice. Microcirculation 19, 306–315.

De Troyer, A., Boriek, A.M., 2011. Mechanics of the respiratory muscles. Compr. Physiol. 1, 1273–1300.

Domyan, E.T., et al., 2013. Roundabout receptors are critical for foregut separation from the body wall. Dev. Cell 24, 52–63.

Epstein, J.A., Shapiro, D.N., Cheng, J., Lam, P.Y., Maas, R.L., 1996. Pax3 modulates expression of the c-Met receptor during limb muscle development. Pro. Nat. Acad. Sci. USA 93, 4213–4218.

He, A., Kong, S.W., Ma, Q., Pu, W.T., 2011. Co-occupancy by multiple cardiac transcription factors identifies transcriptional enhancers active in heart. Pro. Nat. Acad. Sci. USA 108, 5632–5637.

Hendrickson, R.J., Fenton, L., Hall, D., Karrer, F.M., 2003. Congenital paraesophageal hiatal hernia. J. Am. Coll. Surg. 196, 483.

Hornstra, I.K., et al., 2003. Lysyl oxidase is required for vascular and diaphragmatic development in mice. J. Biol. Chem. 278, 14387–14393.

Hyun, J.J., Bak, Y.T., 2011. Clinical significance of hiatal hernia. Gut. liver 5, 267–277.

Iritani, I., 1984. Experimental study on embryogenesis of congenital diaphragmatic hernia. Anat. Embryol. (Berl.) 169, 133–139.

Jay, P.Y., et al., 2007. Impaired mesenchymal cell function in Gata4 mutant mice leads to diaphragmatic hernias and primary lung defects. Dev. Biol. 301, 602–614.

Jetley, N.K., Al-Assiri, A.H., Al Awadi, D., 2009. Congenital para esophageal hernia: a 10 year experience from Saudi Arabia. Ind. J. Ped. 76, 489–493.

Kachlik, D., Cech, P., 2011. Vincenz Alexander Bochdalek (1801–83). J. Med. Biog. 19, 38–43.

Keijzer, R., Liu, J., Deimling, J., Tibboel, D., Post, M., 2000. Dual-hit hypothesis explains pulmonary hypoplasia in the nitrofen model of congenital diaphragmatic hernia. Am. J. Pathol. 156, 1299–1306.

Klaassens, M., et al., 2005. Congenital diaphragmatic hernia and chromosome 15q26: determination of a candidate region by use of fluorescent in situ hybridization and array-based comparative genomic hybridization. Am. J. Hum. Genet. 76, 877–882.

Kreidberg, J.A., et al., 1993. WT-1 is required for early kidney development. Cell 74, 679–691.

Leung, A.W., et al., 2008. Congenital intrathoracic stomach: short esophagus or hiatal hernia? Neonatology 93, 178–181.

Li, J., Liu, K.C., Jin, F., Lu, M.M., Epstein, J.A., 1999. Transgenic rescue of congenital heart disease and spina bifida in Splotch mice. Development 126, 2495–2503.

Lin, A.E., Pober, B.R., Adatia, I., 2007. Congenital diaphragmatic hernia and associated cardiovascular malformations: type, frequency, and impact on management. Am. J. Med. Genet. C. Semin. Med. Genet. 145C, 201–216.

Liu, J., et al., 2003. Congenital diaphragmatic hernia, kidney agenesis and cardiac defects associated with Slit3-deficiency in mice. Mech. Dev. 120, 1059–1070.

Longoni, M., et al., 2012. Congenital diaphragmatic hernia interval on chromosome 8p23.1 characterized by genetics and protein interaction networks. Am. J. Med. Genet. A. 158A, 3148–3158.

Loong, T.P., Kocher, H.M., 2005. Clinical presentation and operative repair of hernia of Morgagni. Postgrad. Ned. J. 81, 41–44.

Mallula, K.K., Sosnowski, C., Awad, S., 2013. Spectrum of Cantrell's Pentalogy: Case series from a single tertiary care center and review of the literature. Pediatric. Cardiol. 34, 1703–1710.

Mayer, S., Metzger, R., Kluth, D., 2011. The embryology of the diaphragm. Semin. Pediatr. Surg. 20, 161–169.

Merrell, A.J., Kardon, G., 2013. Development of the diaphragm – a skeletal muscle essential for mammalian respiration. FEBS J. 280, 4026–4035.

Negrini, D., Moriondo, A., 2013. Pleural function and lymphatics. Acta physiol. 207, 244–259.

Paris, N., Baglia, L., Zhang, X., Pu, W.T., Ackerman, K.G., 2010. Wilms Tumor 1 contributes to the mesothelium of the anterior and posterior diaphragm and is associated with a variety of diaphragmatic hernia phenotypes. E-PAS2010:3701.25.

Pennaforte, T., et al., 2013. The long-term follow-up of patients with a congenital diaphragmatic hernia: review of the litterature. Arch. de Pediat. 20 (Suppl. 1), S11–18.

Pickering, M., Jones, J.F., 2002. The diaphragm: two physiological muscles in one. J. Anat. 201, 305–312.

Pober, B.R., et al., 2005. Infants with Bochdalek diaphragmatic hernia: sibling precurrence and monozygotic twin discordance in a hospital-based malformation surveillance program. Am. J. Med. Genet. A. 138A, 81–88.

Pober, B.R., Russell, M.K., Ackerman, K.G., et al., 2010. In: Pagon, R.A. (Ed.), GeneReviews.

Rhinn, M., Dolle, P., 2012. Retinoic acid signaling during development. Development 139, 843–858.

Russell, M.K., et al., 2012. Congenital diaphragmatic hernia candidate genes derived from embryonic transcriptomes. Pro. Nat. Acad. Sci. USA 109, 2978–2983.

Salman, A.B., Tanyel, F.C., Senocak, M.E., Buyukpamukcu, N., 1999. Four different hernias are encountered in the anterior part of the diaphragm. Turkish. J. Ped. 41, 483–488.

Schweitzer, R., Zelzer, E., Volk, T., 2010. Connecting muscles to tendons: tendons and musculoskeletal development in flies and vertebrates. Development 137, 2807–2817.

Scott, D.A., et al., 2007. Genome-wide oligonucleotide-based array comparative genome hybridization analysis of non-isolated congenital diaphragmatic hernia. Hum. Mol. Genet. 16, 424–430.

Shih, H.P., Gross, M.K., Kioussi, C., 2008. Muscle development: forming the head and trunk muscles. Acta. Histochem. 110, 97–108.

Slavotinek, A.M., 2004. Fryns syndrome: a review of the phenotype and diagnostic guidelines. Am. J. Med. Genet. A. 124A, 427–433.

Slavotinek, A.M., 2007. Single gene disorders associated with congenital diaphragmatic hernia. Am. J. Med. Genet. C. Semin. Med. Genet. 145C, 172–183.

Sluiter, I., et al., 2012. Etiological and pathogenic factors in congenital diaphragmatic hernia.. Eur. J. Pediatr. Surg. 22(5), 345–354.

Stege, G., Fenton, A., Jaffray, B., 2003. Nihilism in the 1990s: the true mortality of congenital diaphragmatic hernia. Pediatrics 112, 532–535.

Stuelsatz, P., Keire, P., Almuly, R., Yablonka-Reuveni, Z., 2012. A contemporary atlas of the mouse diaphragm: myogenicity, vascularity, and the Pax3 connection. J. Histochem. Cytochem. 60, 638–657.

van den Hout, L., et al., 2010. Risk factors for chronic lung disease and mortality in newborns with congenital diaphragmatic hernia. Neonatology 98, 370–380.

van Loenhout, R.B., Tibboel, D., Post, M., Keijzer, R., 2009. Congenital diaphragmatic hernia: comparison of animal models and relevance to the human situation. Neonatology 96, 137–149.

Wat, M.J., et al., 2011. Genomic alterations that contribute to the development of isolated and non-isolated congenital diaphragmatic hernia. J. Med. Genet. 48, 299–307.

Wat, M.J., et al., 2012. Mouse model reveals the role of SOX7 in the development of congenital diaphragmatic hernia associated with recurrent deletions of 8p23.1. Hum. Mol. Genet. 21, 4115–4125.

You, L.R., et al., 2005. Mouse lacking COUP-TFII as an animal model of Bochdalek-type congenital diaphragmatic hernia. Pro. Nat. Acad. Sci. USA 102, 16351–16356.

Young, R.L., Page, A.J., Cooper, N.J., Frisby, C.L., Blackshaw, L.A., 2010. Sensory and motor innervation of the crural diaphragm by the vagus nerves. Gastroenterology 138, 1091–1101, e1091–1095.

Ysasi, A.B., Belle, J.M., Gibney, B.C., Fedulov, A.V., Wagner, W., AkiraTsuda, et al., 2013. Effect of unilateral diaphragmatic paralysis on post-pneumonectomy lung growth. Am. J. Physiol. Lung. Cell. Mol. Physiol. 305(6), L439–L445.

Yu, L., et al., 2012. De novo copy number variants are associated with congenital diaphragmatic hernia. J. Med. Genet. 49, 650–659.

Yu, L., et al., 2013. Variants in GATA4 are a rare cause of familial and sporadic congenital diaphragmatic hernia. Hum. Genet. 132, 285–292.

Yuan, W., et al., 2003. A genetic model for a central (septum transversum) congenital diaphragmatic hernia in mice lacking Slit3. Pro. Nat. Acad. Sci. USA 100, 5217–5222.

Zhou, B., et al., 2009. Fog2 is critical for cardiac function and maintenance of coronary vasculature in the adult mouse heart. J. Clin. Invest. 119, 1462–1476.

Chapter 33

Genetic and Developmental Basis of Congenital Cardiovascular Malformations

John W. Belmont

Departments of Pathology and Immunology and Pediatrics, Baylor College of Medicine, Houston, TX

Chapter Outline

GLOSSARY

Conotruncus Regions of the bulbus cordis that forms the conus cordis and the truncus arteriosus.

Heterotaxy Discordance of sidedness of the organ systems.

Secondary heart field A more medial population of mesoderm cells just anterior to the anterior lateral splanchnic mesoderm of the cardiac crescent. These cells give rise to the common outflow tract and most of the right ventricle in the more mature heart.

Tetralogy of Fallot A heart defect characterized by: stenosis of the pulmonary trunk; ventricular septal defect; overriding aorta; and enlarged right ventricle. A patent ductus arteriosus is also present.

ABBREVIATIONS

AI Aortic Insufficiency
AS Aortic Valve Stenosis
ASD Atrial Septal Defect
AVSD Atrioventricular Septal Defect
BAV Bicuspid Aortic Valve
CoA Coarctation of the Aorta
DORV Double Outlet Right Ventricle
HLHS Hypoplastic Left Heart Syndrome
IAA Interrupted Aortic Arch
IAA-B Interrupted Aortic Arch Type B
LVH Left Ventricular Hypertrophy

Principles of Developmental Genetics.

MR Mitral Regurgitation
MS Mitral Stenosis
MVP Mitral Valve Prolapse
PAIVS Pulmonary Artery Stenosis with Intact Ventricular Septum
PAS Pulmonary Artery Stenosis
PDA Patent Ductus Arteriosus
PFO Patent Foramen Ovale
PS Pulmonary Valve Stenosis
SVC Superior Vena Cava
SVAS Supravalvar Aortic Stenosis
TAPVR Total Anomalous Pulmonary Venous Return
TGA Transposition of the Great Arteries
TOF Tetralogy of Fallot
TS Tricuspid Stenosis
VSD Ventricular Septal Defect
WPW Wolf-Parkinson-White Conduction Defect

SUMMARY

- Heart malformations are the most common class of birth defects.
- There are numerous known causes of congenital heart defects including genetic abnormalities and teratogens.
- A large number of chromosomal, genomic and single gene (Mendelian) disorders that cause heart malformations have been identified.
- Patterns of inheritance include *de novo* dominant, transmitted autosomal dominant, autosomal recessive and X-linked recessive; mutations in more than 150 different human genes have been found to cause heart defects.
- There is an association of classes of heart defects with classes of specific genes, for example, genes encoding cilia components may cause left-right patterning abnormalities.
- Mutations in some genes lead to a wide spectrum of heart defects encompassing multiple disparate classes.

33.1 UNDERSTANDING CONGENITAL HEART DEFECTS IN THE CONTEXT OF NORMAL CARDIAC DEVELOPMENT

Cardiac development is one of the most studied processes in classical embryology, and over the last two decades our concepts of heart development have been transformed to include many specific molecular mechanisms (see Chapter 23). Understanding the genetic control of heart development is extremely important to clinical medicine, not only because congenital cardiovascular malformations (CVM) are exceptionally common but also the fact that the same networks are required for maintenance of the normal adult heart. Congenital CVMs are exceptionally common (Hoffman and Kaplan, 2002, 2004), affecting 0.5–0.7% of all live born infants. Even in the era of modern surgery, some cardiac defects continue to have very poor prognosis (Hoffman and Kaplan, 2004) and constitute the largest fraction of infant mortality attributable to birth defects (Boneva et al., 2001). Nevertheless, it is estimated that there are now more than 1 million adults with a history of significant CVM (Marelli et al., 2007; van der Bom et al., 2011).

Despite important advances over the last 15 years, the etiology of the vast majority of CVM cases is unknown. There is a distinct lack of data concerning molecular mechanisms that are required for normal human cardiac development and very little direct observation of abnormal human embryos and fetuses. There is additional difficulty in making the connection between normal cardiac development and CVMs because there are no direct methods for determining in any single affected individual which developmental step(s) were disturbed. Many CVMs have an ambiguous embryological origin and they may be interpreted as arising from multiple alternative early events or later and more specific processes. Given these difficulties, genetic analysis both for research into the etiology of CVM and for diagnosis of individual patients is likely to play a central role in the future.

In trying to define the origin of heart defects one might consider two very broad and non-exclusive models. In the **"embryonic insult"** model, a single inciting event in a specific developmental field or process is followed by a cascade of disturbed anatomic relationships, abnormal flow-, oxygen- and pressure-dependent remodeling, and abnormal maintenance of the cardiac muscle, valve, and vessel tissues. Embryonic insults could involve genetic and/or environmental agents. The second model invokes **"developmental pleiotropy"**; i.e., the inciting factor(s) affect multiple independent processes in heart development. In this model, the anatomic outcome reflects the specificity of the disturbed developmental process, e.g., truncus arteriosus resulting from direct impairment of aortico-pulmonary septation rather than some earlier abnormality in cardiac precursor differentiation or growth. Genetic factors, either causal mutations or risk-increasing variants, could easily operate through either of these mechanisms.

Genetic Epidemiology of Congenital Cardiovascular Malformations

Congenital heart defects (CHD) are among the most common of all medically significant birth defects, and are a leading contributor to infant mortality in the US (Hoffman and Kaplan, 2002). The etiologies of CHD are complex and heterogeneous (Hoffman and Kaplan, 2004). Numerous environmental and genetic factors have been implicated in heart defects; these include congenital rubella infection and maternal retinoid exposure. Cytogenetic abnormalities, exemplified by Trisomy 21, often cause CVM. Single gene disorders have also been identified, e.g., Noonan syndrome and Holt-Oram syndrome. However, together these conditions account for only a small fraction of congenital cardiovascular malformations (CCVMs) and the causative factors contributing to most cases are obscure. The high birth incidence (0.5–0.7%) together with the substantial sibling recurrence risk (1–3%) has suggested the hypothesis that CVMs have a multifactorial etiology (Boneva et al., 2001). CVMs can also be classified based on the likely developmental mechanisms disturbed during embryonic heart formation (Marelli et al., 2007). Therefore, a plausible strategy for investigating the etiology of CVMs is to classify patients with regard to the likely embryological defect and to then search for genes contributing to susceptibility in that subgroup. Clinically important cardiovascular malformations occur in about 0.7% of live births. A much larger number of infants are found to have minor anomalies of the heart at birth, including atrial and ventricular septal defects if imaging studies are performed without regard to symptoms. Bicuspid aortic valve (BAV) is a very common anomaly that does not cause symptoms in early life. BAV may be the most common congenital heart defect, with studies in healthy adults indicating a prevalence of 0.9% (Insley, 1987). This anatomical variant constitutes an important risk factor for subacute bacterial endocarditis and late onset aortic valve calcification/stenosis in adults. As such it may account for more excess mortality than all other CVMs combined. Patent Foramen Ovale has been considered a normal variant, in that it may occur in up to 30% of the population, but PFO also has a strong association with migraine and stroke.

About 20% of infants born with a CVM have non-cardiac malformations or neurodevelopmental delays. These children are considered to have either 'multiple congenital anomalies' or 'syndromic CVM' to contrast them with those who have only 'isolated' CVM. Epidemiological studies usually distinguish between these groups, but the published literature is inconsistent in the criteria that are applied. The high birth incidence together with the substantial sibling recurrence risk (1–4%) has suggested the hypothesis that CVMs have a multifactorial etiology (Gill et al., 2003; Burn et al., 1998). Supporting this supposition is the fact that a much larger number of infants have minor anomalies of the heart at birth, such as small atrial and ventricular septal defects, if imaging studies are performed without regard to symptoms.

There are generally very similar rates of CVM in all major geographical regions. For some CVMs there are measurable differences in rates between racial groups (Canfield et al., 2006; Grech, 1998; 1999; Ho, 1991; Pradat et al., 2003; McBride et al., 2005a; Fixler et al., 1990; Muir, 1960; Schrire, 1963; Shann, 1969). The most likely explanation for these differences lies in the distinctive genetic history of these groups. There may be some increase in rate of CVM over the last decade (Oyen et al., 2009a). This will require verification, as previous studies are consistent, showing stable birth prevalence rates of CVM over the last 60 years (Hoffman JI, Kaplan, 2002). Changes in rates over relatively short time frames have typically been interpreted as reflecting environmental factors. However, the rapid change in mean parental age must also be accounted for, and this could still involve increased rates of genomic and single gene mutation mechanisms.

The high heritability of CHDs means that genetic factors play a very large role in the overall occurrence of heart defects in the population, and that the aggregate impact of genetic factors is apparently far larger than all environmental factors combined. Studies of some types of CVM have been consistent with a heritability of 50–90% (Burn et al., 1998; Hinton et al., 2007; McBride et al., 2005b; Cripe et al., 2004; Insley, 1987). Several factors cause heritability to be underestimated in CVM:

1) Heart defects are approximately 10-fold more common in miscarried pregnancies and, as a consequence, many affected offspring may not be counted in population surveys;
2) Affected individuals have fewer children than people without heart defects;
3) Families whose firstborn is affected may decide against having further offspring.

A family history of CVM is one of the most consistently identified risk factors in CVM (Wollins et al., 2001; Zavala et al., 1992; Loffredo et al., 2000; 2001; Oyen et al., 2009a); the rate of occurrence in close relatives of affected individuals being substantially (five to 40-fold) higher than the general population rate. Across all CVM sibling and or offspring, the recurrence risk is estimated to be 1–4% (Gill et al., 2003; Burn et al., 1998; Oyen et al., 2009b; Meijer et al., 2005; Siu et al., 1998; Hoess et al., 2002; Hoffman, 1990; Whittemore et al., 1994; Lewin et al., 2004; Piacentini et al., 2005; Digilio et al., 2001; Oyen et al., 2011). Several studies have also demonstrated increased rates of cardiovascular malformations in populations with increased inbreeding and consanguineous parentage (Ramegowda S, Ramachandra, 2006; Chehab et al., 2007; Nabulsi et al., 2003; Badaruddoza et al., 1994; Becker S, Al Halees, 1999; Becker et al., 2001). This is most likely to result from autosomal recessive inheritance of CVM-causing mutations. Inbreeding and the consequent reduced effective population size also makes it more likely that CVM risk-increasing genetic variants could be present in one or both parents, thus increasing the occurrence of oligogenic traits.

Patterns of Inheritance of CVM

Chromosomal abnormalities are an important cause of CVM, accounting for 12–14% of all cases. Trisomy 21 is the most common chromosomal abnormality in live born infants, occurring in about 1 per 1000. Approximately 50% of children with Trisomy 21 (Down syndrome) have cardiac malformations (especially of the atrioventricular canal). Trisomy 13 (Patau syndrome) Trisomy 18 (Edwards syndrome), and Monosomy X (Turner syndrome) are common causes of severe CVM. Submicroscopic cytogenetic disorders are also important contributors (see Table 33.1). These disorders most likely affect several contiguous genes, but often a single gene within the region, e.g., *TBX1* in the 22q11del syndrome, is thought to be the major factor in causing the cardiac developmental defect.

More than 200 syndromes and single gene disorders have been associated with cardiac malformations. Mutations in more than 100 genes that are involved in these disorders have also been identified (Table 33.1). Most of these conditions cause complex phenotypes or syndromes such that the cardiac component is more or less common. Examples include Noonan (PTPN11), Holt-Oram (TBX5), CHARGE (CHD7) and CHAR (TFA2B) syndromes. A less common scenario, so far, is the identification of Mendelian genetic disorders in which the cardiac phenotype is either isolated or defining. Examples include Familial Atrial Septal Defect caused by NKX2.5, GATA4, and MYHC6, as well as Familial Calcific Bicuspid Aortic Valve caused by NOTCH1 mutation. A third category includes genes that were initially identified as causing syndromic CVM but have also been shown to play a role in patients with isolated heart defects. Examples include JAG1, originally identified in Alagille syndrome but also found in subjects with apparently isolated TOF, and ZIC3. This causes X-linked heterotaxy but has also been observed in familial isolated TGA. Tables 33.1 and 33.2 provides a list of human disorders and their associated genes. The list is meant to convey the range of defects seen rather than being a quantification of which heart defects are most common or characteristic. Most of these conditions are inherited as autosomal dominant with variable expression and incomplete penetrance. Recent data using exome sequencing suggest that around 10% of CHD cases could be caused by *de novo* dominant mutations. Confirmation of this possibility will require larger case sample series and functional analysis of rare gene mutations. Because many of these disorders are very rare, it may not be possible to know their true phenotype spectrum. Mechanisms that lead to cardiac phenotype heterogeneity in these disorders include:

1) Allelic heterogeneity;
2) Locus heterogeneity;
3) Epistasis-modifier loci;
4) Gene environment interactions;
5) Pleiotropy – the gene products play more than one role in discrete embryological processes; and
6) Early disturbance in gastrula-stage mesoderm leads to disturbed early heart patterning with a large stochastic component.

These same factors contribute to the apparent complex inheritance of CVM. There are few estimates of heritability (the relative contribution of genetic factors) to the overall liability to develop a CVM. Heritability is probably underestimated in CVM due to several additional considerations:

1) CVMs are approximately 10-fold more common in miscarried pregnancies, indicating that many affected offspring may be unobserved;
2) Due to the burden of illness, families with an affected child may choose to not to have more children; and
3) Affected individuals have reduced reproductive fitness.

But familial aggregation, higher offspring recurrence risk compared to sibling recurrence risk, and relatively stable rates of CVM across time and in various populations is strongly suggestive that genetic factors are important. Evidence for recent slight increases in the rate of heart defects could be consistent with either changes in environmental factors or genetic mechanisms such as increased paternal age. The reduced reproductive fitness of individuals with CVM is predicted to reduce the frequency of causative alleles across all involved genes. The most extreme scenario is that most CVM arise as the result of

TABLE 33.1 Genomic Disorders Associated with CVM

OMIM	Deletion Syndrome	CVM Reported in the Literature
123450	Cri-Du-Chat	VSD, ASD, PS, TOF, AS, PDA
136570	16p12.1 deletion	VSD, PS, DORV, BAV, HLHS, MA, APVR, HTX
141750	alpha-thalassemia/mental retardation	PDA
146390	18p deletion	VSD, AS, HLHS, PDA, HTX
147791	Jacobsen	VSD, ASD, TA, DORV, BAV, AS, HLHS, MS, CoA
154230	9p24.3 deletion	VSD, ASD, CoA, PDA
179613	Recombinant 8	VSD, ASD, PS, TOF, DORV
182290	Smith-Magenis	VSD, ASD, PS, PA, TOF, AS, MS, APVR
192430	Velocardiofacial	VSD, PA, TOF, DORV
194050	Williams-Beuren	SVAS, PBS
194190	Wolf-Hirschhorn	VSD, ASD
247200	Miller-Dieker	ASD, PDA
274000	1q21.1 deletion	VSD, ASD, TOF, CoA
600383	mesomelia-synostoses	VSD, ASD, CoA, PDA
601808	18q deletion	VSD, ASD, PS, PA, AS, PDA, Aortopathy
607872	1p36 deletion	VSD, ASD, PDA, Aortopathy
609425	3q29 deletion	PDA, PH
609625	10q26 deletion	ASD, PDA
610253	Kleefstra	VSD, ASD, PS, TOF, BAV, CoA, PDA
610443	17q21.31 deletion	VSD, ASD, PS, BAV, Aortopathy
610543	16p13.3 deletion	HLHS
611867	Distal 22q11.2 deletion	VSD, PS, TA, BAV, PDA, hypoplastic aortic arch
611913	16p11.2 deletion	BAV, AS
612001	15q13.3 deletion	TOF
612242	10q23 deletion	VSD, ASD
612474	1q21.1 deletion	VSD, TA, TGA, BAV, CoA, PDA
612513	2p16.1-p15 deletion	BAV, MVP
612530	1q41-q42 deletion	TOF, PDA
612582	6pter-p24 deletion	ASD, TOF, PDA
612626	15q26-qter deletion	VSD, ASD, CoA
612863	6q24-q25 deletion	VSD, ASD, PS
613355	17q23.1-q23.2 deletion	ASD, BAV, PDA
613406	15q24 deletion	Aortopathy
613457	14q11-q22 deletion	VSD, PDA
613604	16p12.2-p11.2 deletion	PA, TOF
NA	Microduplication of 8q12	VSD, ASD

Continued

TABLE 33.1 Genomic Disorders Associated with CVM—cont'd

OMIM	Deletion Syndrome	CVM Reported in the Literature
613458	16p13.3 duplication	ASD, TOF
610883	Potocki-Lupski	VSD, ASD, BAV, AS, PBS
609757	7q11.23 microduplication	APVR
608363	22q11.2 duplication	TGA, DORV, CoA

TABLE 33.2 Human Genes Associated with Cardiovascular Malformations

GENE	Syndrome	CVM
ABCC6, ENPP1	Generalized Arterial Calcification of Infancy	
ABCC9	Cantu Syndrome	BAV
ACTB	Baraitser-Winter Syndrome	BAV,AS, PDA
ADAM17	Neonatal Inflammatory Skin and Bowel Disease	LV dilation
ADAMTS10	Weill-Marchesani Syndrome	AS, PS, VSD, MR, PDA
ADAMTSL2	Geleophysic Dysplasia	AS, AI, MS, MR
AKT3	Megalencephaly-Polymicrogyria-Polydactyly-Hydrocephalus Syndrome	ASD, VSD, Vascular Ring
ALX3	Frontonasal Dysplasia	TOF
AMER1	Osteopathia Striata with Cranial Sclerosis	ASD, VSD
ANKS6, INVS, NPHP4	Nephronophthisis	AS, PS, PDA, Heterotaxy
ARHGAP31, DOCK6, EOGT, RBPJ	Adams-Oliver Syndrome	ASD, VSD, PS, TOF
ARL6, BBS1,-2,-4,-5,-7,-9,-10,-12, CCDC28B, CEP290, LZTFL1, MKKS, MKS1, TMEM67, TRIM32, TTC21B, TTC8, WDPCP	Bardet-Biedl Syndrome	complex
ASXL1	Bohring-Opitz Syndrome	ASD, VSD
ATRX	Alpha-Thalassemia/Mental Retardation Syndrome	VSD
B3GALT6	Spondyloepimetaphyseal Dysplasia with Joint Laxity	ASD,VSD, MR, BAV
B3GALTL	Peters-Plus Syndrome	ASD,VSD, PS
B3GAT3	Multiple Joint Dislocations, Short Stature, Craniofacial Dysmorphism,	LVH, BAV, MVP, ASD, PFO
BCOR	Microphthalmia, Syndromic	ASD, VSD, MVP
BMP2	Wolf-Parkinson-White	WPW
BRAF	Cardiofaciocutaneous Syndrome	ASD, PS
CCBE1	Hennekam Lymphangiectasia-Lymphedema Syndrome	ASD, VSD, Pericardial lymphangiectasia

TABLE 33.2 Human Genes Associated with Cardiovascular Malformations—cont'd

GENE	Syndrome	CVM
AK7, CCDC103, CCDC114, CCDC39, CCDC40, DNAAF3, DNAH11, DNAH5, DNAI1, DNAI2, DNAL1, DYNC2H1, HEATR2, HYDIN, LRRC50, LRRC6, NME8, RSPH4A, RSPH6A, RSPH9	Primary Ciliary Dyskinesia	Heterotaxy
ACVR2B, CCDC11, CFC1, CRELD1, FOXH1, GALNT11, GDF1, MED13L, MEGF8, NEK2, NKX2–5, NODAL, INVS, NPHP3, NPHP4, NUP188, PKD2, ROCK2, SHROOM3, TGFBR2, LZTFL1, BCL9L	Heterotaxy Syndrome	heterotaxy
CD96	C Syndrome	VSD, PDA
CEP57	Mosaic Variegated Aneuploidy Syndrome	ASD, VSD, AI, CoA, subaortic stenosis
CHD7	CHARGE Syndrome	TOF, ASD, VSD, DORV, PDA, PS
CHRM3	Absence of Abdominal Muscles with Urinary Tract Abnormality	PDA
CHRNA1, CHRND, CHRNG	Multiple Pterygium Syndrome	Cardiac hypoplasia
CHST3	Spondyloepiphyseal Dysplasia With Congenital Joint Dislocations	VSD, MS, MR, TS, AS, AI, PS
ACTC1, CITED2, GATA4, GATA6, NKX2–5, TBX20, TLL1	Atrial Septal Defect	ASD, VSD
COL1A1	Sillence Type I Osteogenesis Imperfecta	MR, AS
COL1A2	Ehlers-Danlos Syndrome, Cardiac Valvular Form	MVP, MR, AI
COL2A1	Stickler Syndrome, type i; stl1	MVP
COL3A1	Ehlers-Danlos Syndrome, Vascular	MVP
COX7B	Aplasia Cutis Congenita, Reticulolinear, with Microcephaly	ASD, VSD, TOF
CREBBP	Rubinstein-Taybi Syndrome	ASD, VSD, PDA
CRELD1	Atrioventricular Septal Defect AVSD2	AVSD
DDX11	Warsaw Breakage Syndrome	VSD, TOF
DHCR24	Desmosterolosis	TAPVR
DHCR7	Smith-Lemli-Opitz syndrome	ASD, VSD, CoA, PDA, HLHS
DIS3L2	Perlman Syndrome	IAA
ACTC1, DTNA, LDB3, MIB1, MYBPC3, MYH7, PRDM16, TNNT2, TPM1	Left Ventricular Noncompaction	VSD, MR, HLHS
EFTUD2	Mandibulofacial Dysostosis, Guion-Almeida Type	ASD, VSD
EHMT1	Kleefstra Syndrome	VSD, TOF, DORV
ELN	Cutis Laxa, Autosomal Dominant	MR, AI, SVAS
ESCO2	Roberts Syndrome	AS
EVC, EVC2	Ellis-van Creveld Syndrome	Common Atrium, ASD

Continued

TABLE 33.2 Human Genes Associated with Cardiovascular Malformations—cont'd

GENE	Syndrome	CVM
FBLN5	Cutis Laxa, Autosomal Recessive	SVAS, vascular tortuosity
FBN2	Distal Arthrogryposis, type 9	ASD, VSD, BAV
FGFR2	Antley-Bixler Syndrome, Apert Syndrome	VSD
FIG4	Yunis-Varon Syndrome	TOF, VSD
FLNA	Periventricular Heterotopia	TOF, VSD
FLNB	Larsen Syndrome	ASD, VSD
FOXC1	Axenfeld-Rieger Syndrome	PDA, VSD
FOXC2	Lymphedema-Distichiasis Syndrome	TOF
G6PC3	Severe Congenital Neutropenia	ASD
GLI3	Pallister-Hall Syndrome; phs	VSD, CoA, PDA
GPC3	Simpson-Golabi-Behmel Syndrome	VSD, PS
GPC6	Omodysplasia	ASD, VSD
HDAC4	Brachydactyly-Mental Retardation Syndrome	AS
DLL3, HES7, LFNG, MESP2, TBX6	Spondylocostal Dysostosis	VSD, PDA, Dextrocardia, Complex
HOXD13	VATER Association	VSD, TOF, TGA
HYLS1, KIF7	Hydrolethalus Syndrome	AVSD, VSD
IFT122, IFT43	Cranioectodermal Dysplasia	BAV, PS
IGFBP7	Retinal Arterial Macroaneurysm with Supravalvular Pulmonic Stenosis;	PS
IRX5	Hamamy Syndrome	AVSD, ASD, PDA
JAG1	Alagille Syndrome 1; algs1	PS, TOF
KAT6B	Genitopatellar Syndrome; Ohdo Syndrome	ASD, VSD
KCNJ2	Andersen Cardiodysrhythmic Periodic Paralysis	BAV, AS
KDM6A, KMT2D	Kabuki Syndrome	ASD, VSD, CoA
KIF7	Acrocallosal Syndrome	ASD, VSD, PS
LMNA	*Heart*-Hand Syndrome, Slovenian type	Conduction
LRP2	Donnai-Barrow Syndrome	VSD
MEGF8	Carpenter Syndrome	heterotaxy
MGP	Keutel Syndrome	VSD
MID1	Opitz GBBB Syndrome	VSD
MKKS	McKusick-Kaufman Syndrome	VSD
MYCN	Feingold Syndrome	PDA
MYH11	Familial Thoracic Aortic Aneurysm	PDA
NAA10	Ogden Syndrome	VSD, ASD, Conduction
NEK1	Short Rib-Polydactyly Syndrome	VSD, ASD

TABLE 33.2 Human Genes Associated with Cardiovascular Malformations—cont'd

GENE	Syndrome	CVM
NF1	Neurofibromatosis-Noonan Syndrome	PS
NFIX	Marshall-Smith Syndrome	ASD, PDA
NIPBL	Cornelia de Lange Syndrome	ASD, VSD
NOTCH2	Hajdu-Cheney Syndrome	ASD, VSD, PDA
NPHP3	Renal-Hepatic-Pancreatic Dysplasia	ASD, AS, PDA, heterotaxy
NSD1	Sotos Syndrome	PAIVS, ASD, VSD, PDA
OFD1	Orofaciodigital Syndrome	VSD
PAFAH1B1	Miller-Dieker Lissencephaly Syndrome	VSD
PEX1	Peroxisome Biogenesis Disorder	HLHS
PIGA, PIGN, PIGT	Multiple Congenital Anomalies-Hypotonia-Seizures Syndrome	ASD, PDA
PIK3CA	Megalencephaly-Capillary Malformation-Polymicrogyria Syndrome	VSD
PIK3R2	Megalencephaly-Polymicrogyria-Polydactyly-Hydrocephalus Syndrome	ASD, VSD, Vascular Ring
PKD1	Polycystic Kidney Disease, Autosomal Dominant	AS, MVP
PQBP1	Renpenning Syndrome 1; rens1	ASD, VSD, TOF, heterotaxy
PTCH1	Basal Cell Nevus Syndrome; bcns	Cardiac fibroma
PTPN11, KRAS, SOS1, RAF1, NRAS, BRAF, MEK1 MEK2, SHOC2, CBL, SPRED1	Noonan Syndrome	PS, ASD, CoA
RAB23	Carpenter Syndrome 1; crpt1	ASD, VSD, PS, TOF
FANCC, FANCA, FANCD2, RAD51C	Fanconi Anemia	AD, VSD
RAI1	Smith-Magenis Syndrome	VSD
RBM10	TARP Syndrome	ASD
ROR2	Robinow Syndrome, autosomal recessive	PS, PAS
RPS19	Diamond-Blackfan Anemia	VSD
RPS6KA3	Coffin-Lowry Syndrome	MR
SALL1	Townes-Brocks Syndrome	TOF, VSD
SALL4	Duane-Radial Ray Syndrome	ASD, VSD
SETBP1	Schinzel-Giedion Syndrome	ASD
SF3B4	Acrofacial Dysostosis, Nager Type	TOF
SKIV2L, TTC37	Trichohepatoenteric Syndrome	AI, PS, VSD, TOF, PS
SLC2A10	Arterial Tortuosity Syndrome; ats	AS, PAS
SMAD3, TGFB2, TGFBR1, TGFBR2	Loeys-Dietz Syndrome	ASD
SMAD4	Myhre Syndrome	ASD, VSD, AS, PDA, CoA, Pericardial effusion and fibrosis

Continued

TABLE 33.2 Human Genes Associated with Cardiovascular Malformations—cont'd

GENE	Syndrome	CVM
NOTCH1, SMAD6, TAB2	Aortic Valve Disease	AS, BAV
SMOC1	Microphthalmia with Limb Anomalies	Inferior vena cava interruption with azygos continuation
SNIP1	Psychomotor Retardation, Epilepsy, and Craniofacial Dysmorphism	AS, BAV
SOX9	Campomelic Dysplasia	ASD, VSD
SRCAP	Floating-Harbor Syndrome	ASD, Mesocardia, CoA, Persistent left SVC
STAMBP	Microcephaly-Capillary Malformation Syndrome	ASD, VSD
TAZ	Barth Syndrome	VSD
TBX1	DiGeorge Syndrome	VSD, TOF, DORV, IAA-B, TGA
TBX3	Ulnar-Mammary Syndrome	VSD, WPW
TBX5	Holt-Oram Syndrome; hos	ASD, VSD, HLHS, TOF, TAPVR
TCTN3, TMEM237	Joubert Syndrome	VSD
TFAP2B	Char Syndrome	PDA
TP63	Ankyloblepharon-Ectodermal Defects-Cleft Lip/Palate	VSD, PDA
TRIM37	Mulibrey Nanism	Myocardial fibrosis
TSC1, TSC2	Tuberous Sclerosis	Cardiac Rhabdomyoma
TTC7A	Multiple Intestinal Atresia	VSD
TWIST1	Saethre-Chotzen Syndrome; scs	ASD, VSD
UBE3B	Blepharophimosis-Ptosis-Intellectual Disability Syndrome	ASD, VSD, CoA
UBR1	Johanson-Blizzard Syndrome	ASD, VSD, heterotaxy
VPS13B	Cohen Syndrome	MVP
WNT3	Tetraamelia, Autosomal Recessive	Peripheral pulmonary vessel aplasia
WNT5A	Robinow Syndrome, Autosomal Dominant	PS
ZEB2	Mowat-Wilson Syndrome	VSD, ASD, PS, PDA, PAS, pulmonary artery sling
ZMPSTE24	Lethal Restrictive Dermopathy	ASD, PDA
ZNF469	Brittle Cornea Syndrome	MVP

private mutations, their relative frequency being mainly due to the large number of genes that play a role in normal cardiac development. This could be called the 'large mutational target' hypothesis. However, even a weakly penetrant gene variant might contribute a substantial part of the overall population risk if it were common. At this time a model involving both rare and common variants possibly with gene and environmental interactions is the most likely.

Genetic disorders make up the most complex and numerically significant category of known causes of CVM. A broad range of genetic mechanisms are either known to participate or are strongly suspected in causing cardiovascular malformations. Like most traits that exhibit complex inheritance, there are still many unknowns and the relative importance of various genetic factors (common variants, rare variants, epistasis, epigenetics, etc.) remains to be defined.

33.2 HEART TUBE AND CARDIAC LOOPING

Patterning and Growth of the Heart Tube

Commitment of precardiac mesoderm is covered in detail elsewhere (see Chapter 23) and will not be reviewed here. Human heart development begins at day 18 as the cardiogenic mesoderm coalesces toward the midline. Splanchnic mesenchyme aggregates to form a pair of elongated strands that canalize forming paired endothelial tubes. These fuse in the midline proceeding anterior to posterior to form a single heart tube. As the heart tubes fuse splanchnic mesenchyme proliferates and forms a myoepicardial mantle (Kaufman MH, Navaratnam, 1981). The future cardiac myocytes are separated from the endothelial lining of the heart tube by cardiac jelly (DeRuiter et al., 1992), a gelatinous connective tissue matrix which forms a subendocardial zone. The Wnt inhibitory molecules, Crescent and Dickkoph, are expressed by the cardiac mesoderm and ectopic expression of these in posterior mesoderm is sufficient to induce heart formation (Kawano, Kypta, 2003; Lickert et al., 2002; Zorn, 2001; Olson, 2001; Marvin et al., 2001). Cardiac induction is mediated in part by BMP4 (Zhao and Rivkees, 2000), BMP7 (Solloway and Robertson, 1999; Kim et al., 2001), and BMP2 (Schlange et al., 2000; Yamagishi et al., 1999) produced by adjacent neural ectoderm and foregut endoderm. In mice, the cardiogenic mesoderm expresses the transcription factor Nkx2.5, an NK homeobox family member with specific functions in both early and late cardiac development (Patterson et al., 1998; Prall et al., 2002). Nkx2.5 is the vertebrate homolog of *Drosophila* Tinman, which is required for the formation of the heart-like vessel in flies (Tanaka et al., 1998; 1999). The early cardiomyocytes express heart-specific differentiation markers, including Mef-2C, cardiac Actin, Desmin, eHand, and Cardiac-specific Ankyrin Protein (CARP) which probably lie downstream of Nkx2.5 (Schwartz RJ, Olson, 1999; Zou et al., 1997). Another important family of transcription factors, Gata4, Gata5, and Gata6, are all expressed in the precardiac mesoderm (Patient and McGhee, 2002).

Heart tube. By day 22, the single heart tube elongates and segmentations marked by slight constrictions that define the truncus arteriosus, bulbus cordis, primitive ventricles, atrium, and sinus venosus. These indicate the anterior to posterior patterning in the otherwise symmetrical heart tube (Christoffels et al., 2000; Lo and Frasch, 2003). At the anterior end, the truncus arteriosus is continuous with the aortic sac and aortic arch arteries. At the posterior end of the heart tube, the sinus venosus receives respectively the umbilical, vitelline, and common cardinal veins from the chorion (primitive placenta), yolk sac, and embryo proper (Mavrides et al., 2001). Mutants in Gata4 exhibit failure to fuse the bilateral heart tubes (Molkentin et al., 1997; Narita et al., 1997; Kuo et al., 1997). A similar cardia bifida phenotype is seen in Zebrafish faust/Gata5 (Reiter et al., 1999), OEP/Cryptic (Reiter et al., 2001), and casanova (Dickmeis et al., 2001) mutants, which all affect endoderm development and reinforce the role of endodermal-mesoderm interactions in shaping the early formation of the heart.

Shortly after coalescence into this heart tube, rhythmic contractions commence, resulting from the intrinsic contractile and rhythmic properties of the myocardial cells (Hirota et al., 1985; Yada et al., 1985). In the absence of valves, the flow of blood is directed from the inflow to the outflow by a kind of peristaltic action beginning in the sinus venosus. The contraction of the outflow portion is longer and more sustained, thus preventing backflow. The heart tube is sensitive to flow, in that normal flow is required for its proper growth and remodeling. Experimental obstruction to flow leads to morphological abnormalities that are suggestive of several human malformations (Hove et al., 2003).

Concomitant processes are involved in specifying chamber identity. The transcription factors Mesp1 and Mesp2, for example, are required for primitive streak mesoderm migration into the cardiac crescent (Saga et al., 1999; 2000; Kitajima et al., 2000), but also play a cell-autonomous role in ventricular but not atrial chamber formation. Irx4, a homeodomain transcription factor, is also involved in ventricular specification and is highly expressed on the outer curvature of the heart tube where there is rapid growth and remodeling (Christoffels et al., 2000; Bao et al., 1999; Bruneau et al., 2000). Irx4 seems to be regulated by Nkx2.5 and another transcription factor dHand (Yamagishi et al., 2001). dHand and eHand exhibit generally complementary patterns of expression. eHand is predominantly expressed in the left ventricle and outflow tract. Both are required for normal ventricle growth. RV growth is specifically impaired in dHand mutant embryos, leading to defective looping stage development (Thomas et al., 1998; Srivastava et al., 1995; 1997). Another transcription factor, Mef2c, is also required for both LV and RV growth (Lin et al., 1997; Bi et al., 1999; Iida et al., 1999), while COUP-TFII is required for atrial and sinus venosus precursor development (Pereira et al., 1999).

Secondary Heart Field. A second heart forming field has been defined over the last decade (Yutzey et al., 2002; Kelly and Buckingham, 2002). It arises from a more medial population of mesoderm cells just anterior to the anterior lateral splanchnic mesoderm of the cardiac crescent (Kelly, 2012). Descendents of the precursors in the anterior heart forming field give rise to the common outflow tract and most of the RV in the more mature heart. Due to their medial position, second heart field (SHF) cells form the dorsal wall of the pericardial cavity where they continue to proliferate. Contiguity with the heart tube is maintained at the arterial and venous poles following breakdown of the dorsal mesocardium. During the period of rightward looping and heart tube elongation, SHF cells contribute to these poles. SHF cells interact with cardiac neural crest cells in the pharyngeal region, as both cell populations add to the heart tube. Expression of *Fgf8*, *Fgf10*, *Isl1*, and

Tbx1 define a core genetic signature for the SHF pharyngeal mesoderm. The transcription factor *Six1* is expressed in pharyngeal mesoderm (Guo et al., 2011). Lineage marking indicates that progenitor cells that have expressed *Six1* contribute to the outflow tract and atria. Embryos lacking both SIX1 and EYA1 exhibit conotruncal defects, such as double outlet right ventricle and truncus arteriosus. These findings are partially explained by the discovery of a special role for FGF receptors and ligands in SHF cells. *Fgf8* expression is downregulated in *Six1/Eya1* double mutant embryos. *Fgf8* expression is also reduced or lost in *Tbx1* conditional KO embryos (Vitelli et al., 2010) and those with reduced Notch signaling (High et al., 2009). Together these results suggest that FGF8 operates downstream of the most important SHF transcriptional regulators. An additional role for FGF3 and FGF10 has also been demonstrated. Compound and double null *Fgf3/Fgf10* mutants embryos have heart defects reflecting disturbed SHF and neural crest development (Urness et al., 2011).

Looping and Laterality Defects

The arterial and venous ends of the heart tube are fixed by the branchial arches and the septum transversum. By day 23 a slight asymmetric growth of the left sino-atrial areas produces a "leftward jog" that is the first manifestation of the future left-right (LR) asymmetry of the mature heart (Manner, 2000). Because the bulbus cordis and the outer curvature of the right ventricle grow faster than the other segments, the heart tube bends ventrally, forming the bulboventricular loop. This proceeds to a rightward folding of the ventricular segment, forming the C-looped heart (Figure 33.1). The direction of looping is controlled by underlying LR patterning in the cardiac mesoderm, the position of the heart tube relative to the midline, and the intrinsic differential growth of the heart tube segments. As growth continues, an S-shaped heart forms as the atrium and sinus venous are pushed dorsal (i.e., behind) to the bulbus cordis, truncus arterious, and future ventricles. The relative positions of the future left and right ventricles are also established at this stage. The heart is the first organ to demonstrate the future LR asymmetry of the thoracic and abdominal organs. There is now an extensive literature on the establishment of the LR axis in the various animal model systems (Yost, 2001; Mercola, 1999; Boorman and Shimeld, 2002). There are several linked processes that mediate this cascade:

1) Symmetry breaking;
2) Formation of the node which will relay LR positional information from the organizer to the lateral plate mesoderm (LPM);
3) Stabilization of lateralized patterning in the LPM;
4) Expression of the Nodal-dependent signal transduction pathway in the left LPM;
5) Transfer of positional information to organ primordia.

The node is a critical organizer that forms at the anterior most portion of the primitive streak (Tabin and Vogan, 2003; Brennan and Norris, 2002; Vincent et al., 2003). Signals that pattern the embryo to establish the LR axis emanate from the node and mutations that affect node formation alter LR patterning. The process of symmetry breaking is poorly understood

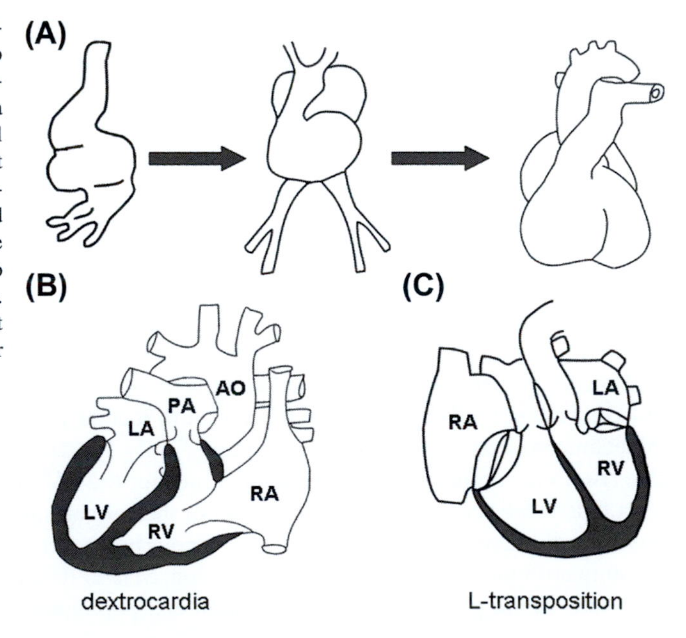

FIGURE 33.1 Heart Tube and Looping Stage Defects. (A) The primitive heart tube is initially symmetric but undergoes rightward looping to form the C-loop stage heart. Later, dorsal movement of the atria and differential growth of the ventricle leads to the final anatomic position of the atria rostral to the ventricles. (B) Dextrocardia results from complete reversal of the LR axis so that all of the heart structures are concordant. LA, Left atrium; LV, left ventricle; RA, right atrium; RV, right ventricle; PA, pulmonary artery; Ao, aorta. C, L-Transposition (sometimes called corrected transposition) also results from leftward looping, but there is discordance of the ventricles and atria. Note that the anatomic right atrium connects to the anatomic left ventricle, which pumps blood to the pulmonary artery. The corresponding atrioventricular valve has two leaflets. The anatomic left atrium connects to the anatomic right ventricle, which acts as the pump for the systemic circulation through the aorta.

in mammalian embryogenesis and may occur in the node itself. Earlier establishment of LR expression domain in *Xenopus* raises the possibility that LR orientation has already been established before the node is formed. In *Xenopus,* LR axis formation involves H+/K+-ATPase (Levin et al., 2002), PKCγ, and Syndecan-2 before gastrulation (Kramer et al., 2002). The node is composed of specialized epithelial cells with ventral monocilia. The monocilia exhibit a gyrorotatory motion that is required for the organizer functions of the node. This has led to the "nodal flow hypothesis," in which the motion of the monocilia generates a morphogen gradient necessary to induce left side identity (Watanabe et al., 2003; Nonaka et al., 2002). The requirement for directional fluid flow at the node has been elegantly demonstrated in embryo culture experiments (Nonaka et al., 2002). Likewise, mutants that affect monocilia development or movement also result in LR patterning defects (Brody et al., 2000; Takeda et al., 1999; Nonaka et al., 1998). A competing hypothesis has been put forward, however, in which some of the monocilia act as mechanosensors for the fluid flow (Tabin and Vogan, 2003). In that scenario it is not the actual establishment of a morphogen gradient, but rather the transduction of the flow directional information that triggers the left side program. Nevertheless, each of these mechanisms seems to converge on the induction of Nodal expression. Nodal is a TGFβ family signaling molecule that plays a variety of roles in the early embryo (Juan and Hamada, 2001; Hamada et al., 2002), and whose specific receptors have intrinsic serine threonine kinase activity. Activation of Nodal receptors depends on membrane-bound coreceptors that then trigger phosphorylaton of Smad2/3, releasing it from the inhibitory Smad and allowing interaction with Smad4 (Miyazawa et al., 2002). Phospho-Smad2/3 translocates to the nucleus, where it interacts with both inhibitory and activating cotranscription factors TGIF and FAST1 (Wotton and Massague, 2001; Weisberg et al., 1998; Chen et al., 1998). These in turn activate Nodal, Lefty-2, and PITX2 expression. Nodal expression in the node is required for proper LR patterning. Establishment of the midline acts as a barrier to left signals crossing into the right LPM. The barrier involves the prechordal plate, notochord, and spinal cord floor plate. These structures are dependent on dorsal-ventral patterning of Shh, and mutations in Shh lead to left isomerisms due to failure of the midline (Zhang et al., 2001; Tsukui et al., 1999). Lefty-1 is a Nodal antagonist that is expressed in medial left LPM and is required to delimit the area of Nodal activity; mutations in Lefty-1 lead to left isomerism (Meno et al., 1997; 1998). Transfer of positional information to organ primordia is mediated in part by PITX2, a homeobox transcription factor that plays a central role in cardiac growth and development of the common outflow tract (Logan et al., 1998; Piedra et al., 1998; Ryan et al., 1998). PITX2 is directly regulated by Nodal to mediate the LR patterning inherent in the looping process. PITX2 is also directly regulated by the Wnt/β-catenin pathway to control development of the outflow tract via interaction with neural crest cells (Kioussi et al., 2002).

Improvements in our understanding of the embryological events that control the formation of the major embryonic axes are leading to new insights into the causes of heterotaxy. Situs inversus indicates complete left-right reversal of organ position and is not usually associated with structural anomalies. Situs ambiguus means discordance in the relationship between the normally asymmetric organs of the thorax and abdomen. The term heterotaxy derives from the Greek roots "other arrangement." Heterotaxy arises from abnormal LR embryonic patterning with consequent abnormal segmental arrangements of cardiac chambers, vessels, lungs, and/or abdominal organs. Incomplete or failed LR patterning may lead to anatomical discordances (e.g., transposition of the great arteries), loss of structures (e.g., asplenia), improper symmetry or lateralization (e.g., right atrial isomerism in which left atrial development is concomitantly lost), or failure to regress symmetrical embryonic structures (e.g., persistent left superior vena cava).

Heterotaxy syndromes are characterized by a wide variety of congenital anomalies of both lateralized internal organs and midline structures. CCVMs in heterotaxy are often complex. These typically combine malposition and TGA, atrial septal defects, ventricular septal defects, bilateral superior vena cava, partial anomalous pulmonary venous return, intrahepatic interruption of the inferior vena cava with connection to the azygous or hemiazygous vein, double outlet right ventricle, common atrium, atrioventricular (AV) canal defects, pulmonary atresia and stenosis, single ventricle, and left ventricular outflow tract obstruction. Some of the CCVMs seen in heterotaxy are readily interpreted as arising from abnormal LR specification, such as L-loop with dextrocardia, in which a mirror-image heart resides in the right thoracic compartment but is otherwise completely normal. Ventricular inversion is a rare defect in which there is discordance of the atrioventricular connection alone. Interestingly, it is not typically associated with the other anatomic abnormalities of heterotaxy. Associated cardiac defects, like partial anomalous pulmonary venous return (PAPVR), are presumably secondary, but this is not certain.

33.3 FORMATION OF THE ATRIOVENTRICULAR CANAL

Growth and Differentiation of the AV Canal

Blood flows from the sinus venosus to the common atrium and passes through the atrioventricular canal into the future ventricles (Figures 33.2, 33.3). Partition of the atrioventicular canal at the common atrium and the ventricles begins about the middle of the fourth week in the development of the human embryo (Figure 33.4). Thickening of the subendocardial

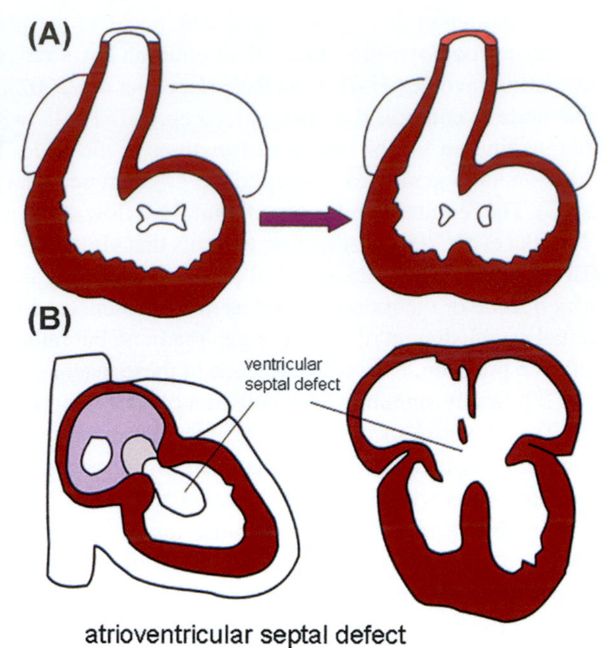

FIGURE 33.2 Formation of the Atrioventricular Canal. (A) Before the partitioning of the primitive ventricle in the right ventricle and the left ventricle, blood flows from the common atrium through the atrioventricular canal. Growth from the endocardial cushion and differential growth of the ventricular myocardium lead to the formation of the ventricular septum. See Figure 33.3 for a sagittal view of the endocardial cushion. (B) Failure of the complete closure of the atrioventricular canal leads to atrioventricular septal defects.

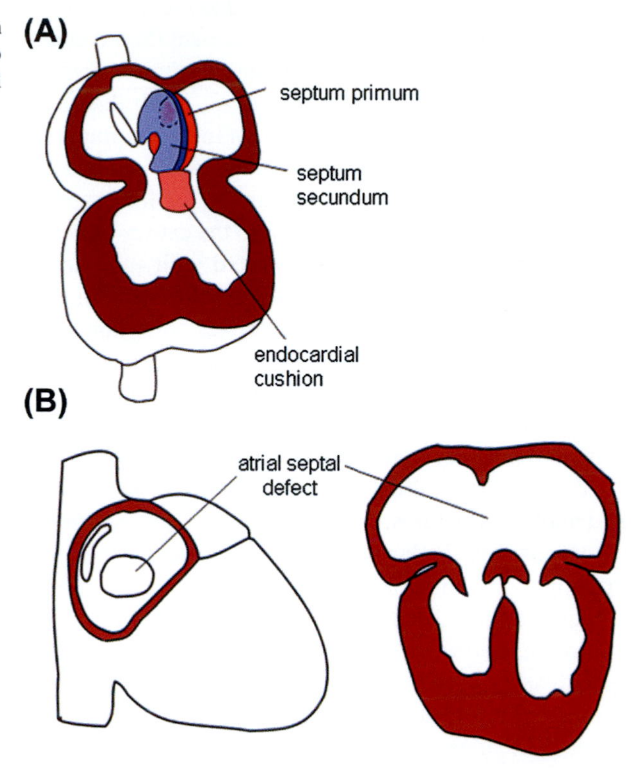

FIGURE 33.3 Atrial Septation. (A) Sequential growth of the septum primum and the septum secundum leads to the partitioning of the common atrium into the left and right atria. (B) The failure of the closure of the septum leads to atrial septal defects.

tissue, i.e., the endocardial cushions, emerge from the dorsal and ventral walls of the atrioventricular canal. By the fifth week, the endocardial cushions fuse in the dorsal-ventral midline to divide the AV canal into the right and left canals that precede the formation of the mature AV valves. Two transcription factors, PITX2 and FOG-2 (Tevosian et al., 2000; Svensson et al., 2000), also are involved in the formation of the atrioventricular canal and thus lead to abnormal atrial septation. FOG2 specifically interacts with GATA4 and its deficiency leads to defects that resemble tetralogy of Fallot (Crispino et al., 2001). Atrioventricular canal defects are also observed in Tolloid-1 mutants (Clark et al., 1999). Recently, new insight has been gained into the developmental origins of the AV canal and a role for SHF progenitors has been identified. Mesocardial tissue attaches the heart tube at its venous and arterial poles to the embryo proper. Venous pole dorsal mesocardium is the

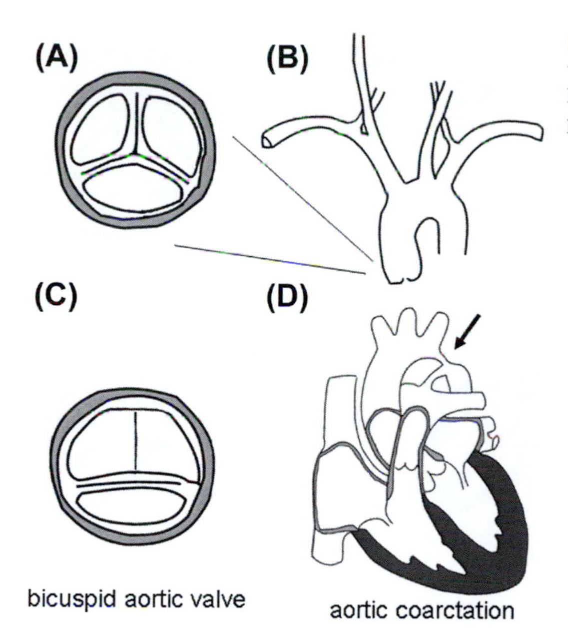

FIGURE 33.4 The Left Ventricular Outflow Tract. (A) A normal, three-leaflet aortic valve. (B), A normal aortic arch. (C) The bicuspid aortic valve. Note the raphe in the upper leaflet, which indicates a fusion of the leaflets. (D) Coarctation of the aorta (arrow). The position of the narrowing is often at the ductus.

entry point for protrusion into the common atrium of SHF-derived mesenchymal cells. This structure, called the dorsal mesenchymal protrusion (DMP), together with two other tissues, the endocardial cushions and cap of the septum primum, form the AV mesenchymal complex. DMP growth and differentiation is dependent on Shh signaling and requires expression of *Tbx5* (Hoffmann et al., 2009; Xie et al., 2012).

Atrioventricular Septal Defects

Atrioventricular septal defects include a family of malformations that involve the inferior atrial septum and the superior ventricular septum. These have also been called endocardial cushion defects or atrioventricular canal defects. This class of anomalies is characteristic of Down syndrome and approximately 50% of children with complete AVSD have Down syndrome (Barlow et al., 2001). Mutations in CRELD1 have been associated with non-syndromic ASD, but the mechanistic basis of this and confirmatory animal models are not yet available. AVSD is also strongly associated with 3p deletion syndrome, which subsequently led to the identification of *CRELD1* as a specific contributor (Maslen, 2004; Robinson et al., 2003). CRELD1 variants have also been implicated as modifiers of the risk of AVSD in Down syndrome (Ackerman et al., 2012).

33.4 TARGETED GROWTH OF THE PULMONARY VEINS

Pulmonary Veins and Dorsal Mesocardium

The primitive pulmonary venous plexus coalesces into the pulmonary veins which in turn merge into the common pulmonary vein (Figure 33.5). This vein grows toward the primitive left atrium in a process of targeted growth and the venous wall is gradually incorporated into the wall of the left atrium. Later, the proximal branches of the pulmonary veins are incorporated into the dorsal wall of the chamber, resulting in four pulmonary veins with separate openings into the atrium. The left auricle, which is a remnant derived from the primitive atrium, develops trabeculations reflecting its distinct embryonic origin compared the remainder of the smooth-walled left atrium. At the present there are almost no molecular data from any of the animal model systems that reflect on these interesting processes, despite their importance in human CVM.

Anomalous Pulmonary Veins

Although families have been found to segregate an autosomal dominant TAPVR linked chromosome 4, to date no gene identification has been reported. TAPVR and PAPVR are frequently found in conjunction with left and right isomerism sequences, respectively. Interestingly, TAPVR has been observed in a variety of disorders which do not have an obvious connection to either targeted growth of the pulmonary veins or LR patterning. This suggests that it could result as a

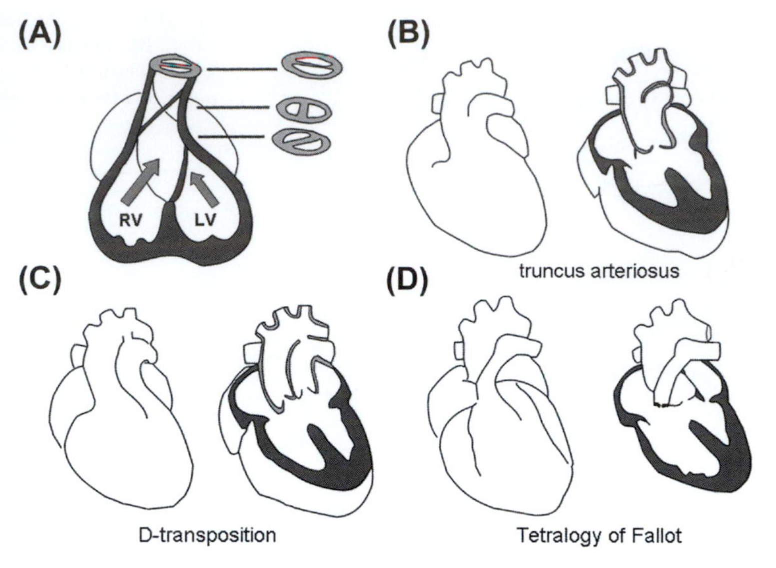

FIGURE 33.5 The Formation of the Common Outflow Tract and Aorticopulmonary Septation. (A) Spiral growth of the conal ridges leads to the partitioning of the outflow tract to form the aorta and the pulmonary artery. (B) The complete failure of the partition leads to truncus arterious. (C) Anterior malpositioning of the aorta leads to the connection of the aorta to the right ventricle and of the pulmonary artery with the left ventricle. The resulting lesion can be either d-TGA or DORV (not depicted). (D) Deficient growth of the proximal outflow tract leads to stenotic pulmonary valve (arrow), VSD, and overriding aorta. This lesion is considered TOF.

secondary abnormality in cardiac looping or other early atrial growth defects. This could explain the association with AVSD in some cases. Anomalous pulmonary veins (APVR) indicate malformations in which there is complete or partial failure of the establishment of the pulmonary vein connections to left atrium. There is a specific association of APVR with Cat Eye syndrome (tetrasomy 22q usually resulting from dicentric and bisatellited inv dup (22) supernumerary marker chromosome), but it is difficult to attribute this to a single gene. APVR has been described in rare genomic disorders involving ring 12p, deletion 11q24-ter, Williams syndrome and Smith-Magenis syndrome. Families have been found to segregate an autosomal dominant total anomalous pulmonary veins (TAPVR) linked to chromosome 4 (Bleyl et al., 1995). Using a very novel mapping procedure based on the likely single origin of the mutation, Bleyl et al. (2010) were able to identify the underlying mechanism as disrupted regulation of the *PDGFRA* gene. Disrupted regulation of the *ANKRD1* gene has also been demonstrated in an individual with APVR and bearing a *de novo* 10;21 balanced translocation (Cinquetti et al., 2008). There is insufficient data from model systems to determine whether these genetic disorders directly disturb pulmonary venous development, i.e., represent pathway specificity. TAPVR has been reported in rare cases with *SEMA3D* mutation (Degenhardt et al., 2013). Total anomalous pulmonary venous return (TAPVR) and partial anomalous pulmonary veins (PAPVR) are frequently found in conjunction with left and right isomerism sequences, respectively, as a feature of heterotaxy. In those cases the APVR is clearly secondary to aberrant LR patterning.

33.5 ATRIAL AND VENTRICULAR SEPTATION

Differential Growth and Remodeling in Cardiac Septation

Atrial Septation. At the level of the common atrium, the septum primum grows from the dorsal anterior wall of the atrial chamber (Dalgleish, 1976; Hendrix and Morse, 1977; Morse and Hendrix, 1980; Allwork, 1982; Morse et al., 1984; Rogers and Morse, 1986a,b). The foramen primum is formed by the gap between the septum primum and the endocardial cushions. Before this gap is completely closed, perforations appear in the anterior central septum primum and coalesce to produce the foramen secundum. Late in the fifth week, the septum secundum emerges from the ventral anterior wall of the common atrium to the right of the septum primum. The septum secundum grows toward the endocardial cushions and thereby covers the foramen secundum. The septum secundum is incomplete, leaving an opening between left and right sides called the foramen ovale. The upper part of the septum primum regresses from the anterior wall of the left atrium so that the remaining part of the septum primum forms a flap valve for the foramen ovale. Before birth, the foramen ovale lets most of the blood entering the right atrium from the inferior vena cava across to the left atrium. After birth the foramen ovale fuses and the atrial septum is complete.

Ventricular Septation. Partition of right and left ventricles begins with a muscular fold in the constriction (interventricular groove) between the primitive ventricles. The fold increases in prominence as the result of the differential growth of the ventricles on each side. Active upward growth of myocardium forms the muscular interventricular septum. The wall between the ventricles remains incomplete through the seventh week. The final closure of the interventricular septum is coupled with the partitioning of the common outflow tract. Ridges from both the right and left side of bulbus cordis emerge and these in turn fuse with the ridge produced by the endocardial cushions of the AV canal. The membranous interventricular septum derives from the right side of the endocardial cushions joining the aorticopulmonary septum and the muscular part of the interventricular septum. At closure of the interventricular septum, the pulmonary trunk connects with the right ventricle and the aorta with the left ventricle. Several transcription factors have been shown to play a role in ventricular septation. These include retinoic acid coreceptor (RXRα; Sucov et al., 1994), TEF1 (Chen et al., 1994), N-myc (Moens et al., 1993), and Sox4 (Ya et al., 1998).

Atrial Septal Defects

There are two broad classes of atrial septal defects; isolated and atrioventricular septal defects. About 30% of normal individuals retain a patent but valve-competent foramen ovale and this has recently been recognized as a risk factor for migraine and stroke. Defects in the fossa ovalis or secundum ASD are the most common form of ASD, but abnormal valve-incompetent foramen ovale is also important.

Common atrium is an early defect that presumably results from the failure of the growth of the septum primum. This anomaly is characteristic of the Ellis-van Creveld syndrome. Atrioventricular canal and common atrium may also be observed in Smith-Lemli-Opitz syndrome. This condition results in defective cholesterol biosynthesis and the developmental defects resulting from a defect in cholesterol modification of sonic hedgehog (SHH), a secreted protein that is involved in establishing cell fates in diverse tissues and organs. The very diverse defects associated with this condition are striking and strongly suggest very early abnormalities in heart development.

Holt-Oram syndrome is the prototype genetic disorder causing ASD. This condition is caused by mutations in the transcription factor TBX5, and is almost always associated with thumb hand and radial ray malformations (see also Chapter 34). In the past few years, additional ASD families have been identified with mutations in NKX2.5. These individuals often have associated cardiac conduction defects. Mutations in yet another key cardiac transcription factor, GATA4, cause isolated ASD in a few families (Hirayama-Yamada et al., 2005) and, recently, mutation in MYHC6 encoding a structural protein (Ching et al., 2005; Granados-Riveron et al., 2010) that is a transcriptional regulatory target of all three of the above transcription factors has been identified.

Ventricular Septal Defects

Ventricular septal defects are anatomically heterogeneous and although the cardiology literature is clear on this, genetic studies most often fail to make important distinctions about critical anatomic details. Perimembranous VSDs occur within and adjacent to the membranous septum, which is formed in the final stages of formation of the septum by the fusion of the endocardial cushion tissue of the conus septum with the superior portion of the muscular septum. Perimembranous VSDs can be divided into outlet and inlet, which are AV canal type and trabecular. Defects in the outlet septum are thought to be caused by failure of fusion of the conus septum, whereas outlet defects may be caused by the failure of complete fusion of the right superior endocardial cushion with the muscular septum. Muscular defects in the trabecular septum are probably due to excessive remodeling of the interventricular wall or inadequate merger of the medial walls.

VSDs are observed in almost all genetic disorders affecting heart development. Some represent a continuum of defects of the common outflow tract as in Velocardiofacial syndrome, while others presumably are due to defects in cardiomyocyte growth as in Holt-Oram syndrome. VSDs also commonly accompany more complex defects and so only represent a secondary anatomical defect. The origin of some associations such as VSD with coarctation of the aorta is not well understood.

33.6 VALVULOGENESIS AND OUTFLOW TRACT DEVELOPMENT

Semilunar and Atrioventricular Valves

The semilunar (aortic and pulmonary) valves develop from three ridges of the endocardial tissue at the orifices of the aorta and pulmonary trunk. These swellings become hollowed out and reshaped to form the three thin walled cusps. The atrioventricular valves (tricuspid and mitral) develop similarly from localized proliferation of subendocardial tissue around the atrioventricular canals. Mutation of mouse NFATc leads to defective semilunar valve formation (Ranger et al., 1998; de la Pompa et al., 1998). A double *Egfr/Ptpn11* mutation in mice also leads to defective semilunar valve growth (Chen et al., 2000). Heparin-binding epidermal growth factor (HB-EGF) and betacellulin (BTC) are activating ligands for the EGF receptor. Mutation of these ligands leads to defective endocardial valve precursor growth (Jackson et al., 2003). Defective valvulogenesis occurs in tumor necrosis factor-alpha converting enzyme (TACE) mutants. All these results highlight the importance of epidermal growth factor receptor, EGFR, activation in valvulogenesis.

Aorticopulmonary Septation

During the fifth week, ridges of the subendocardial tissue form in the bulbus cordis. Similar ridges also form in the truncus arteriosus and are continuous with those in the bulbus cordis. The spiral orientation of the ridges results in a spiral aorticopulmonary septum when these ridges fuse. This septum divides the bulbus cordis and the truncus arteriosus into two channels, the aorta and the pulmonary trunk. Blood from the aorta now passes into the third and fourth pairs of aortic arch arteries (future aortic arch) and blood from the pulmonary trunk flows into the sixth pair of aortic arch arteries (future pulmonary arteries). Several mouse mutants including disheveled-2, semaphorin3C, and c-Jun exhibit failure of aorticopulmonary septation (Hamblet et al., 2002; Feiner et al., 2001; Eferl et al., 1999). Various combinations of the RAR alpha 1, RAR beta, and RXR alpha gene mutations also result in muscular ventricular septal defects, double outlet right ventricle, transposition, and truncus arteriosus (Lee et al., 1997). In addition, mutations in *Fgf8* and *Tbx1* (Frank et al., 2002; Vitelli et al., 2002; Jerome and Papaioannou, 2001) demonstrate common outflow tract developmental abnormalities that are suggestive of the type seen in DiGeorge syndrome. Moreover, mutation of *Pitx2c* isoform provides functional evidence that the anterior heart field contributes to the common outflow tract and aortic arch remodeling (Liu et al., 2002).

Neural Crest Contribution to the Outflow Tract

At the time of formation of the neural tube, neural ectodermal cells at the most dorsal ridge migrate from the tube. Neural crest cell migrate to their ultimate developmental field and there differentiate into a wide variety of neural and mesenchymal cell types (see Chapter 18). Two broad domains of neural crest cells may be defined – cephalic and truncal. In addition, related cells emerge from the cephalic placodes and give rise to craniofacial structures such as the inner ear (see Chapter 19). Neural crest cells provide autonomic innervation to the heart, with the sympathetic system arising from the truncal neural crest (Baptista and Kirby, 1997). Cardiac neural crest extends from the otic placode to the third somite. Cardiac neural crest provides mesenchymal cells to the heart and outflow tract. This same population of cells plays a critical role in the development of the thymus and parathyroid glands. These cells transit through the third, fourth, and sixth pharyngeal arches. They participate in the secondary heart field related structures and are required for formation of the aorticopulmonary cushions. These cells also contribute to the walls of the aorta and pulmonary arteries distal to the outflow tract. Various mutant models (Creazzo et al., 1998; Hutson and Kirby, 2003; Jiang et al., 2000; van den Hoff et al., 2000; Waldo et al., 1998) including Pax3 (Conway et al., 1997) have been inferred to affect cardiac neural crest migrations or differentiation. Pathways such as that requiring semaphorin3C (Feiner et al., 2001; Maschhoff and Baldwin, 2000) are directly involved in controlling neural crest cell migration. Other pathways, exemplified by the endothelin system, Tbx1, and Pitx2, may be involved in differentiation and growth or indirect mechanisms (Vitelli et al., 2002; Liu et al., 2002; Yanagisawa et al., 2000; Sinning, 1998).

Conotruncal Heart Defect

Several common defects have their origins in failure of development of the common outflow tract. These include truncus arteriosus (TA), transposition of the great arteries (TGA), double outlet right ventricle (DORV), tetralogy of Fallot (TOF – consisting of VSD, overriding aorta, pulmonic stenosis, and right ventricular hypertrophy), and interrupted aortic arch type B (IAA-B). Truncus arteriosus represents a failure of aorticopulmonary septation and consequently there is a single semilunar valve. As noted above, d-TGA may be associated with LR patterning defects, but may also arise from a much later failure of the conotruncal septum to properly orient to the ventricles. Such failure leaves the aorta anterior and rightward of the pulmonary artery, creating a connection of aorta to RV and pulmonary artery to LV. DORV is similar, in that there is malposition of the aorta causing it to receive flow from the RV. IAA-B refers to an interruption of the aorta between the take-off positions of the carotid and subclavian arteries. Velocardiofacial/DiGeorge syndrome (22q11 deletion) is the prototypical conotruncal disorder, occurring in 1 in 4000 live births (Burn and Goodship, 1996). About 75% of individuals with 22q11.2 deletion have CVM. The typical ~3 Mb deletion which encompasses ~60 known and predicted genes, is mediated by meiotic non-allelic recombination events. These events occur due to the flanking segmental duplications termed LCR22, leading to aberrant interchromosomal exchanges. The deletion is also characterized by several extra-cardiac abnormalities, including palatal abnormalities, hypocalcemia, immune deficiency, renal anomalies, and learning difficulties (see also Chapter 36). It is estimated that one of every eight cases of TOF, one of every five cases of truncus arteriosus, and one of every two cases of interrupted aortic arch type B in the population are attributable to the 22q11.2 deletion (Botto et al., 2003). Although most deletions occur *de novo*, approximately 7% are inherited from an affected parent. Rare mutations in *TBX1* indicate that haplo-insufficiency of this transcription factor plays a major role in the occurrence of heart defects in 22q11 deletion (Yagi et al., 2003).

Recurrent duplications of 1q21.1 are often reported in conjunction with TOF (Soemedi et al., 2012a). The *GJA5* gene, which encodes a gap junction protein, Connexin40, has been proposed as a candidate gene within this ~1.5 Mb region. Conotruncal malformations have been demonstrated in Cx40-deficient mice (Gu et al., 2003). The reciprocal 1q21.1 deletions are associated with a more heterogeneous phenotype characterized by incomplete penetrance, with fewer reports of conotruncal abnormalities (Soemedi et al., 2012b). It is estimated that in all cases of sporadic nonsyndromic TOF, *de novo* copy number variants CNVs can be identified in about 10% of cases (Greenway et al., 2009).

Conotruncal heart defects are characteristic of CHARGE (*CHD7*) (Vissers et al., 2004), and Alagille (*JAG1*, *NOTCH2* (McDaniell et al., 2006)) syndromes. JAG1 is a ligand for NOTCH family receptors, and the finding of mutations in this gene in patients with Alagille syndrome and isolated TOF (Bauer et al., 2010) demonstrates an important role for Notch signaling in outflow tract development. FOG2 specifically interacts with GATA4 and mutations in it lead to TOF (De Luca et al., 2010). Mutations in *GATA6* have been reported in truncus arterious and TOF (Lin et al., 2010). Rare mutations in *NKX2.5*, *TBX5*, *FOXH1* causing isolated TOF emphasize the variety of defects that may result from alterations in these key cardiac transcription factors. Other transcription factors have also been implicated as rare causes of syndromic conotruncal defects. These include mutation in *PROSIT240* (Muncke et al., 2003), *HOXA1,* and *RAI1*. Bosley-Salih-Alorainy syndrome (BSAS) is an autosomal recessive disorder caused by *HOXA1* mutation (Bosley et al., 2008). Mutation of *RAI1* (Slager et al., 2003) overlaps significantly with Smith-Magenis syndrome (deletion 17q) consistent with the concept that this gene is the keystone for that disorder.

Right Ventricular Outflow Tract Obstruction

Abnormal development of the pulmonary valve often leads to obstruction of flow from the right ventricle. The characteristic lesions are called pulmonary stenosis or atresia. These lesions often occur in combination with other defects and are a component, for example, of tetralogy of Fallot (VSD, overriding aorta, pulmonic stenosis, and right ventricular hypertrophy). These lesions are very frequent in syndromic cardiovascular malformations, suggesting that they have diverse origins. The pulmonary valve dysplasia of Noonan syndrome is a well known example in which mutations in PTPN11 lead to constitutive activation of mitogen activated protein kinase, MAPK, signal transduction pathways. The Velocardiofacial/DiGeorge syndromes associated with deletion of 22q11 give another good example of aberrant developmental mechanisms leading to pulmonary valve stenosis or atresia. The TBX1 gene is probably largely determinative for these disorders and rare patients have been found with point mutations in that gene.

Left Ventricular Outflow Tract Obstruction

Left ventricular outflow tract obstruction (LVOTO) type CVM includes aortic valve stenosis (AS), coarctation of the aorta (CoA), hypoplastic left heart syndrome (HLHS), complicated mitral valve stenosis with HLHS and CoA (Shone complex), and bicuspid aortic valve (BAV). Severe outflow obstruction caused by aortic valve or aortic abnormality is thought to lead to poor growth of the left ventricle in HLHS. CoA, AS, and HLHS are the most common CVMs seen in Turner syndrome (45,X), apparent in about 30% of cases of this syndrome (Korpal-Szczyrska et al., 2005; van Egmond et al., 1988; Volkl et al., 2005). Aortopathy including dilatation of the ascending aorta, aortic aneurysms, and aortic dissection has been exemplified in several studies, supporting the need for vigilant cardiac follow-up into adulthood for several individuals with Turner syndrome (Volkl et al., 2005; Bondy, 2008). Several single gene disorders are associated with LVOTO defects, including Smith-Lemli-Opitz (*DHCR7*), X-linked heterotaxy (Ware et al., 2004) (*ZIC3*) and Holt-Oram syndrome (*TBX5*). Multiplex families have been reported with CoA (MIM21000). Families with multiple occurrences of HLHS, AS, CoA, and BAV strongly suggest the existence of one or more discrete susceptibility genes common to all these defects (Ferencz et al., 1997). Recent heritability and linkage studies however indicate that the inheritance is most often oligogenic and there is significant locus heterogeneity. Rare occurrences of familial AS and sporadic LVOTO defects have been associated with mutation in *NOTCH1* (Garg et al., 2005). Notch signaling is apparently critical for normal valve leaflet development, tying congenital aortic valve dysplasia with both BAV and late onset aortic valve calcification. Deletions in 16q24 that affect a FOX gene cluster (*FOXF1*, *FOXL1*, and *FOXC1*) give a characteristic syndrome of alveolar capillary dysplasia and HLHS (Stankiewicz et al., 2009). The gene *FOXF1* is apparently required for lung development, but patients with point mutations in that gene do not exhibit cardiac defects. A report characterizing patients with deletion in 6q24 has identified the MAPK signaling cofactor *TAB2* as being responsible for the associated AS and BAV with aortic dilation (Thienpont et al., 2010). The 11q terminal deletions between 5–20 Mb (Jacobsen syndrome) occur in conjunction with CVM in about 56% of cases (Grossfeld et al., 2004). Left sided obstructive lesions are frequently described in Jacobsen syndrome. A transcription factor, *ETS-1* within this region has been implicated in cardiac defects (Ye et al.,

2010). Cytogenetically visible terminal deletions of 15q26 are frequently associated with LVOTO, including CoA, AS, and HLHS (Davidsson et al., 2008; Tumer et al., 2004). At least two genes within this region, *COUP-TFII* (Pereira et al., 1999) and *MCTP2* (Lalani et al., 2013), have been linked to heart development. Kabuki syndrome is one of the most interesting syndromic associations with LVOTO defects (frequent occurrence of CoA and VSD), caused by mutations in the chromatin modifier *MLL2* and the lysine-specific demethylase 6A, *KDM6A* genes (Ng et al., 2010; Lederer et al., 2012). Supravalvar aortic stenosis is a rare form of LVOTO commonly observed in 7q11.23 deletion (Williams-Beuren syndrome) (Zalzstein et al., 1991). Duplication 17p11.2 syndrome (Potocki-Lupski syndrome) is the homologous recombination reciprocal of the Smith-Magenis syndrome, often presenting with BAV and dilated aortic root (Sanchez-Valle et al., 2011; Potocki et al., 2000; Jefferies et al., 2012). HLHS has also been reported in a very small number of patients (Sanchez-Valle et al., 2011).

Persistent Ductus Arteriosus

The ductus arteriosus is a normal structure that allows flow of oxygenated blood from the venous circulation to enter the systemic circulation *in utero*. After birth and the inflation of the lungs, the ductus closes, allowing for establishment of the separate venous and arterial circulations. Persistent patent ductus arteriosus (PDA) results when the ductus fails to undergo its normal physiological closure and involution. PDA is seen in numerous genetic disorders and the causal mechanisms are very poorly understood. Char syndrome (*TFAP2B*) (Mani et al., 2005) is an example of a relatively specific association in that other heart defects are not typically observed in that condition. Mowat-Wilson syndrome is caused by mutation in the transcription factor *ZEB2* (Zweier et al., 2005). Patients often have PDA, and it is the most specific CVM associated with that disorder. Rare families with thoracic aortic aneurysm plus PDA have been found to segregate mutations in the vascular smooth muscle contractile protein *MYH11* (Zhu et al., 2006).

33.7 INTRACARDIAC CONDUCTION SYSTEM

Differentiation and Patterning of the Sino-Atrial Node and Conduction System

At the heart tube and looping stages the primitive common atrium acts as the pacemaker. As the chambers are formed, the atrioventricular constriction forms fibers that are specialized for conduction. The Purkinje fibers form the sino-atrial node, the atrioventricular node, and the atrioventricular bundle. They are innervated by making connections with neural crest derived autonomic ganglia that invade the subendocardial tissue. The sino-atrial node originates in the sinus venosus later incorporating into the wall of the right atrium at the entrance of the superior vena cava. The AV node develops in the lower part of the interatrial septum. The atrioventricular bundle in the interventricular septum consists of Purkinje fibers which extend from the AV node to the ventricles. A specific role for NKX2.5 in the development of the conduction system is demonstrated by both mouse and human mutations (Ikeda et al., 2002; Schott et al., 1998; Benson et al., 1999; Kasahara et al., 2001; Biben et al., 2000). Connexin40 is likely to be a key downstream target in that effect (VanderBrink et al., 2000; van Rijen et al., 2001; Verheule et al., 1999; Hagendorff et al., 1999; Simon et al., 1998). In addition, mutation in the transcription factor HF-1b also leads to disrupted conduction system differentiation (Nguyen-Tran et al., 2000).

Developmental Anomalies of Conduction

So far, at least three human disease genes have been associated with abnormal development of the cardiac conduction system. The first was mentioned in relation to ASD and is caused by mutations in NKX2.5. Recently, a gene for familial Wolf-Parkinson White syndrome was identified. The fact that some individuals with this disorder also have cardiomyopathy suggests more complex abnormalities in the maintenance of cardiomyocytes. Tachyarrhythmias are also observed in some patients with Costello syndrome. This condition has been shown to be due to mutations in *HRAS*. Arrhythmias in each of these conditions are likely due to either abnormal development or subsequent loss of specialized conducting tissue within the heart.

33.8 CLINICAL RELEVANCE

- Chromosomal abnormalities cause about 10% of cardiovascular malformations.
- Newborns with heart defects accompanied by non-cardiac anatomic defects should have microarray testing for genomic disorders.

- Most Mendelian forms of cardiac defects found to date show an autosomal dominant pattern of inheritance.
- *De novo* dominant mutations may account for up to 10% of heart defects.
- Study of model organisms has uncovered critical pathways required for normal heart development; mutations in genes encoding components of those pathways have been shown to cause human heart defects.

RECOMMENDED RESOURCES

CHDwiki http://homes.esat.kuleuven.be/~bioiuser/chdwiki/index.php/Main_Page (accessed 11/04/14)

American Heart Association http://www.heart.org/HEARTORG/Conditions/CongenitalHeartDefects/Congenital-Heart-Defects_UCM_001090_SubHomePage.jsp (accessed 11/04/14)

Centers for Disease Control http://www.cdc.gov/ncbddd/heartdefects/facts.html (accessed 11/04/14)

Congenital Heart Public Health Consortium http://www.chphc.org/hostedsites/Pages/default.aspx (accessed 11/04/14)

REFERENCES

Ackerman, C., Locke, A.E., Feingold, E., Reshey, B., Espana, K., Thusberg, J., et al., 2012. An excess of deleterious variants in VEGF-A pathway genes in Down-syndrome-associated atrioventricular septal defects. Am. J. Hum. Genet. 91 (4), 646–659.

Allwork, S.P., 1982. Anatomical-embryological correlates in atrioventricular septal defect. Br. Heart J. 47 (5), 419–429.

Badaruddoza, Afzal M., 1994. Akhtaruzzaman. Inbreeding and congenital heart diseases in a north Indian population. Clin. Genet. 45 (6), 288–291.

Bao, Z.Z., Bruneau, B.G., Seidman, J.G., Seidman, C.E., Cepko, C.L., 1999. Regulation of chamber-specific gene expression in the developing heart by Irx4. Science 283 (5405), 1161–1164.

Baptista, C.A., Kirby, M.L., 1997. The cardiac ganglia: cellular and molecular aspects. Kaohsiung. J. Med. Sci. 13 (1), 42–54.

Barlow, G.M., Chen, X.N., Shi, Z.Y., Lyons, G.E., Kurnit, D.M., Celle, L., et al., 2001. Down syndrome congenital heart disease: a narrowed region and a candidate gene. Genet. Med. 3 (2), 91–101.

Bauer, R.C., Laney, A.O., Smith, R., Gerfen, J., Morrissette, J.J., Woyciechowski, S., et al., 2010. Jagged1 (JAG1) mutations in patients with tetralogy of Fallot or pulmonic stenosis. Hum. Mutat. 31 (5), 594–601.

Becker, S., Al Halees, Z., 1999. First-cousin matings and congenital heart disease in Saudi Arabia. Community Genet. 2 (2–3), 69–73.

Becker, S.M., Al Halees, Z., Molina, C., Paterson, R.M., 2001. Consanguinity and congenital heart disease in Saudi Arabia. Am. J. Med. Genet. 99 (1), 8–13.

Benson, D.W., Silberbach, G.M., Kavanaugh-McHugh, A., Cottrill, C., Zhang, Y., Riggs, S., et al., 1999. Mutations in the cardiac transcription factor NKX2.5 affect diverse cardiac developmental pathways. J. Clin. Invest. 104 (11), 1567–1573.

Bi, W., Drake, C.J., Schwarz, J.J., 1999. The transcription factor MEF2C-null mouse exhibits complex vascular malformations and reduced cardiac expression of angiopoietin 1 and VEGF. Dev. Biol. 211 (2), 255–267.

Biben, C., Weber, R., Kesteven, S., Stanley, E., McDonald, L., Elliott, D.A., et al., 2000. Cardiac septal and valvular dysmorphogenesis in mice heterozygous for mutations in the homeobox gene Nkx2–5. Circ. Res. 87 (10), 888–895.

Bleyl, S., Nelson, L., Odelberg, S.J., Ruttenberg, H.D., Otterud, B., Leppert, M., et al., 1995. A gene for familial total anomalous pulmonary venous return maps to chromosome 4p13-q12. Am. J. Hum. Genet. 56 (2), 408–415.

Bleyl, S.B., Saijoh, Y., Bax, N.A., Gittenberger-de Groot, A.C., Wisse, L.J., Chapman, S.C., et al., 2010. Dysregulation of the PDGFRA gene causes inflow tract anomalies including TAPVR: integrating evidence from human genetics and model organisms. Hum. Mol. Genet. 19 (7), 1286–1301.

Bondy, C.A., 2008. Aortic dissection in Turner syndrome. Curr. Opin. Cardiol. 23 (6), 519–526.

Boneva, R.S., Botto, L.D., Moore, C.A., Yang, Q., Correa, A., Erickson, J.D., 2001. Mortality associated with congenital heart defects in the United States: trends and racial disparities, 1979–1997. Circulation 103 (19), 2376–2381.

Boorman, C.J., Shimeld, S.M., 2002. The evolution of left-right asymmetry in chordates. Bioessays 24 (11), 1004–1011.

Bosley, T.M., Alorainy, I.A., Salih, M.A., Aldhalaan, H.M., Abu-Amero, K.K., Oystreck, D.T., et al., 2008. The clinical spectrum of homozygous HOXA1 mutations. Am. J. Med. Genet. A. 146A (10), 1235–1240.

Botto, L.D., May, K., Fernhoff, P.M., Correa, A., Coleman, K., Rasmussen, S.A., et al., 2003. A population-based study of the 22q11.2 deletion: phenotype, incidence, and contribution to major birth defects in the population. Pediatrics 112 (1 Pt 1), 101–107.

Brennan, J., Norris, D.P., Robertson, E.J., 2002. Nodal activity in the node governs left-right asymmetry. Genes. Dev. 16 (18), 2339–2344.

Brody, S.L., Yan, X.H., Wuerffel, M.K., Song, S.K., Shapiro, S.D., 2000. Ciliogenesis and left-right axis defects in forkhead factor HFH-4-null mice. Am. J. Respir. Cell. Mol. Biol. 23 (1), 45–51.

Bruneau, B.G., Bao, Z.Z., Tanaka, M., Schott, J.J., Izumo, S., Cepko, C.L., et al., 2000. Cardiac expression of the ventricle-specific homeobox gene Irx4 is modulated by Nkx2–5 and dHand. Dev. Biol. 217 (2), 266–277.

Burn, J., Brennan, P., Little, J., Holloway, S., Coffey, R., Somerville, J., et al., 1998. Recurrence risks in offspring of adults with major heart defects: results from first cohort of British collaborative study. Lancet 351 (9099), 311–316.

Burn, J., Goodship, J., 1996. Developmental genetics of the heart. Curr. Opin. Genet. Dev. 6 (3), 322–325.

Canfield, M.A., Honein, M.A., Yuskiv, N., Xing, J., Mai, C.T., Collins, J.S., et al., 2006. National estimates and race/ethnic-specific variation of selected birth defects in the United States, 1999–2001. Birth Defects Res. A. Clin. Mol. Teratol. 76 (11), 747–756.

Chehab, G., Chedid, P., Saliba, Z., Bouvagnet, P., 2007. Congenital cardiac disease and inbreeding: specific defects escape higher risk due to parental consanguinity. Cardiol. Young. 17 (4), 414–422.

Chen, B., Bronson, R.T., Klaman, L.D., Hampton, T.G., Wang, J.F., Green, P.J., et al., 2000. Mice mutant for Egfr and Shp2 have defective cardiac semilunar valvulogenesis. Nat. Genet. 24 (3), 296–299.

Chen, Y.G., Hata, A., Lo, R.S., Wotton, D., Shi, Y., Pavletich, N., et al., 1998. Determinants of specificity in TGF-beta signal transduction. Genes. Dev. 12 (14), 2144–2152.

Chen, Z., Friedrich, G.A., Soriano, P., 1994. Transcriptional enhancer factor 1 disruption by a retroviral gene trap leads to heart defects and embryonic lethality in mice. Genes. Dev. 8 (19), 2293–2301.

Ching, Y.H., Ghosh, T.K., Cross, S.J., Packham, E.A., Honeyman, L., Loughna, S., et al., 2005. Mutation in myosin heavy chain 6 causes atrial septal defect. Nature genetics 37 (4), 423–428.

Christoffels, V.M., Habets, P.E., Franco, D., Campione, M., de Jong, F., Lamers, W.H., et al., 2000. Chamber formation and morphogenesis in the developing mammalian heart. Dev. Biol. 223 (2), 266–278.

Cinquetti, R., Badi, I., Campione, M., Bortoletto, E., Chiesa, G., Parolini, C., et al., 2008. Transcriptional deregulation and a missense mutation define ANKRD1 as a candidate gene for total anomalous pulmonary venous return. Hum. Mutat. 29 (4), 468–474.

Clark, T.G., Conway, S.J., Scott, I.C., Labosky, P.A., Winnier, G., Bundy, J., et al., 1999. The mammalian Tolloid-like 1 gene, Tll1, is necessary for normal septation and positioning of the heart. Development 126 (12), 2631–2642.

Conway, S.J., Henderson, D.J., Kirby, M.L., Anderson, R.H., Copp, A.J., 1997. Development of a lethal congenital heart defect in the splotch (Pax3) mutant mouse. Cardiovasc. Res. 36 (2), 163–173.

Creazzo, T.L., Godt, R.E., Leatherbury, L., Conway, S.J., Kirby, M.L., 1998. Role of cardiac neural crest cells in cardiovascular development. Annu. Rev. Physiol. 60, 267–286.

Cripe, L., Andelfinger, G., Martin, L.J., Shooner, K., Benson, D.W., 2004. Bicuspid aortic valve is heritable. J. Am. Coll. Cardiol. 44 (1), 138–143.

Crispino, J.D., Lodish, M.B., Thurberg, B.L., Litovsky, S.H., Collins, T., Molkentin, J.D., et al., 2001. Proper coronary vascular development and heart morphogenesis depend on interaction of GATA-4 with FOG cofactors. Genes. Dev. 15 (7), 839–844.

Dalgleish, A.E., 1976. The development of the septum primum relative to atrial septation in the mouse heart. J. Morphol. 149 (3), 369–382.

Davidsson, J., Collin, A., Bjorkhem, G., Soller, M., 2008. Array based characterization of a terminal deletion involving chromosome subband 15q26.2: an emerging syndrome associated with growth retardation, cardiac defects and developmental delay. BMC. Med. Genet. 9, 2.

de la Pompa, J.L., Timmerman, L.A., Takimoto, H., Yoshida, H., Elia, A.J., Samper, E., et al., 1998. Role of the NF-ATc transcription factor in morphogenesis of cardiac valves and septum. Nature 392 (6672), 182–186.

De Luca, A., Sarkozy, A., Ferese, R., Consoli, F., Lepri, F., Dentici, M.L., et al., 2010. New mutations in ZFPM2/FOG2 gene in tetralogy of Fallot and double outlet right ventricle. Clin. Genet.

Degenhardt, K., Singh, M.K., Aghajanian, H., Massera, D., Wang, Q., Li, J., et al., 2013. Semaphorin 3d signaling defects are associated with anomalous pulmonary venous connections. Nat. Med. 19 (6), 760–765.

DeRuiter, M.C., Poelmann, R.E., VanderPlas-de Vries, I., Mentink, M.M., Gittenberger-de Groot, A.C., 1992. The development of the myocardium and endocardium in mouse embryos. Fusion of two heart tubes? Anat. Embryol. (Berl). 185 (5), 461–473.

Dickmeis, T., Mourrain, P., Saint-Etienne, L., Fischer, N., Aanstad, P., Clark, M., et al., 2001. A crucial component of the endoderm formation pathway, CASANOVA, is encoded by a novel sox-related gene. Genes. Dev. 15 (12), 1487–1492.

Digilio, M.C., Casey, B., Toscano, A., Calabro, R., Pacileo, G., Marasini, M., et al., 2001. Complete transposition of the great arteries: patterns of congenital heart disease in familial precurrence. Circulation 104 (23), 2809–2814.

Eferl, R., Sibilia, M., Hilberg, F., Fuchsbichler, A., Kufferath, I., Guertl, B., et al., 1999. Functions of c-Jun in liver and heart development. J. Cell. Biol. 145 (5), 1049–1061.

Feiner, L., Webber, A.L., Brown, C.B., Lu, M.M., Jia, L., Feinstein, P., et al., 2001. Targeted disruption of semaphorin 3C leads to persistent truncus arteriosus and aortic arch interruption. Development 128 (16), 3061–3070.

Ferencz, C., Loffredo, C.A., Corea-Villasenor, A., Wilson, P.D., 1997. Left-sided obstructive lesions, In: Genetic and environmental risk factors for major cardiovasuclar malformations: The Baltimore-Washington Infant Study 1981–1989. Futura Publishing Co., Inc. p. 165–225.

Fixler, D.E., Pastor, P., Chamberlin, M., Sigman, E., Eifler, C.W., 1990. Trends in congenital heart disease in Dallas County births. 1971–1984. Circulation 81 (1), 137–142.

Frank, D.U., Fotheringham, L.K., Brewer, J.A., Muglia, L.J., Tristani-Firouzi, M., Capecchi, M.R., et al., 2002. An Fgf8 mouse mutant phenocopies human 22q11 deletion syndrome. Development 129 (19), 4591–4603.

Garg, V., Muth, A.N., Ransom, J.F., Schluterman, M.K., Barnes, R., King, I.N., et al., 2005. Mutations in NOTCH1 cause aortic valve disease. Nature 437 (7056), 270–274.

Gill, H.K., Splitt, M., Sharland, G.K., Simpson, J.M., 2003. Patterns of recurrence of congenital heart disease: an analysis of 6,640 consecutive pregnancies evaluated by detailed fetal echocardiography. J. Am. Coll. Cardiol. 42 (5), 923–929.

Granados-Riveron, J.T., Ghosh, T.K., Pope, M., Bu'Lock, F., Thornborough, C., Eason, J., et al., 2010. Alpha-cardiac myosin heavy chain (MYH6) mutations affecting myofibril formation are associated with congenital heart defects. Human molecular genetics 19 (20), 4007–4016.

Grech, V., 1998. An excess of tetralogy of Fallot in Malta. J. Epidemiol. Community Health 52 (5), 280–282.

Grech, V., 1999. Decreased prevalence of hypoplastic left heart syndrome in Malta. Pediatr. Cardiol. 20 (5), 355–357.

Greenway, S.C., Pereira, A.C., Lin, J.C., DePalma, S.R., Israel, S.J., Mesquita, S.M., et al., 2009. De novo copy number variants identify new genes and loci in isolated sporadic tetralogy of Fallot. Nat. Genet. 41 (8), 931–935.

Grossfeld, P.D., Mattina, T., Lai, Z., Favier, R., Jones, K.L., Cotter, F., et al., 2004. The 11q terminal deletion disorder: a prospective study of 110 cases. Am. J. Med. Genet. A. 129A (1), 51–61.

Gu, H., Smith, F.C., Taffet, S.M., Delmar, M., 2003. High incidence of cardiac malformations in connexin40-deficient mice. Circ. Res. 93 (3), 201–206.

Guo, C., Sun, Y., Zhou, B., Adam, R.M., Li, X., Pu, W.T., et al., 2011. A Tbx1-Six1/Eya1-Fgf8 genetic pathway controls mammalian cardiovascular and craniofacial morphogenesis. J. Clin. Invest. 121 (4), 1585–1595.

Hagendorff, A., Schumacher, B., Kirchhoff, S., Luderitz, B., Willecke, K., 1999. Conduction disturbances and increased atrial vulnerability in Connexin40-deficient mice analyzed by transesophageal stimulation. Circulation 99 (11), 1508–1515.

Hamada, H., Meno, C., Watanabe, D., Saijoh, Y., 2002. Establishment of vertebrate left-right asymmetry. Nat. Rev. Genet. 3 (2), 103–113.

Hamblet, N.S., Lijam, N., Ruiz-Lozano, P., Wang, J., Yang, Y., Luo, Z., et al., 2002. Dishevelled 2 is essential for cardiac outflow tract development, somite segmentation and neural tube closure. Development 129 (24), 5827–5838.

Hendrix, M.J., Morse, D.E., 1977. Atrial septation. I. Scanning electron microscopy in the chick. Dev. Biol. 57 (2), 345–363.

High, F.A., Jain, R., Stoller, J.Z., Antonucci, N.B., Lu, M.M., Loomes, K.M., et al., 2009. Murine Jagged1/Notch signaling in the second heart field orchestrates Fgf8 expression and tissue-tissue interactions during outflow tract development. J. Clin. Invest. 119 (7), 1986–1996.

Hinton Jr., R.B., Martin, L.J., Tabangin, M.E., Mazwi, M.L., Cripe, L.H., Benson, D.W., 2007. Hypoplastic left heart syndrome is heritable. J. Am. Coll. Cardiol. 50 (16), 1590–1595.

Hirayama-Yamada, K., Kamisago, M., Akimoto, K., Aotsuka, H., Nakamura, Y., Tomita, H., et al., 2005. Phenotypes with GATA4 or NKX2.5 mutations in familial atrial septal defect. Am. J.Med. genetics Part A 135 (1), 47–52.

Hirota, A., Kamino, K., Komuro, H., Sakai, T., Yada, T., 1985. Early events in development of electrical activity and contraction in embryonic rat heart assessed by optical recording. J. Physiol. 369, 209–227.

Ho, N.K., 1991. Congenital malformations in Toa Payoh hospital–a 18 year experience (1972–1989) Ann. Acad. Med. Singap. 20 (2), 183–189.

Hoess, K., Goldmuntz, E., Pyeritz, R.E., 2002. Genetic counseling for congenital heart disease: new approaches for a new decade. Curr. Cardiol. Rep. 4 (1), 68–75.

Hoffman, J.I., Kaplan, S., Liberthson, R.R., 2004. Prevalence of congenital heart disease. Am. Heart J. 147 (3), 425–439.

Hoffman, J.I., Kaplan, S., 2002. The incidence of congenital heart disease. J. Am. Coll. Cardiol. 39 (12), 1890–1900.

Hoffman, J.I., 1990. Congenital heart disease: incidence and inheritance. Pediatr. Clin. North. Am. 37 (1), 25–43.

Hoffmann, A.D., Peterson, M.A., Friedland-Little, J.M., Anderson, S.A., Moskowitz, I.P., 2009. sonic hedgehog is required in pulmonary endoderm for atrial septation. Development 136 (10), 1761–1770.

Hove, J.R., Koster, R.W., Forouhar, A.S., Acevedo-Bolton, G., Fraser, S.E., Gharib, M., 2003. Intracardiac fluid forces are an essential epigenetic factor for embryonic cardiogenesis. Nature 421 (6919), 172–177.

Hutson, M.R., Kirby, M.L., 2003. Neural crest and cardiovascular development: a 20-year perspective. Birth Defects Res. Part C Embryo. Today 69 (1), 2–13.

Iida, K., Hidaka, K., Takeuchi, M., Nakayama, M., Yutani, C., Mukai, T., et al., 1999. Expression of MEF2 genes during human cardiac development. Tohoku. J. Exp. Med. 187 (1), 15–23.

Ikeda, Y., Hiroi, Y., Hosoda, T., Utsunomiya, T., Matsuo, S., Ito, T., et al., 2002. Novel point mutation in the cardiac transcription factor CSX/NKX2.5 associated with congenital heart disease. Circ. J. 66 (6), 561–563.

Insley, J., 1987. The heritability of congenital heart disease. Br. Med. J. (Clin. Res. Ed.) 294 (6573), 662–663.

Jackson, L.F., Qiu, T.H., Sunnarborg, S.W., Chang, A., Zhang, C., Patterson, C., et al., 2003. Defective valvulogenesis in HB-EGF and TACE-null mice is associated with aberrant BMP signaling. Embo. J. 22 (11), 2704–2716.

Jefferies, J.L., Pignatelli, R.H., Martinez, H.R., Robbins-Furman, P.J., Liu, P., Gu, W., et al., 2012. Cardiovascular findings in duplication 17p11.2 syndrome. Genet. Med. 14 (1), 90–94.

Jerome, L.A., Papaioannou, V.E., 2001. DiGeorge syndrome phenotype in mice mutant for the T-box gene, Tbx1. Nat. Genet. 27 (3), 286–291.

Jiang, X., Rowitch, D.H., Soriano, P., McMahon, A.P., Sucov, H.M., 2000. Fate of the mammalian cardiac neural crest. Development 127 (8), 1607–1616.

Juan, H., Hamada, H., 2001. Roles of nodal-lefty regulatory loops in embryonic patterning of vertebrates. Genes. Cells. 6 (11), 923–930.

Kasahara, H., Wakimoto, H., Liu, M., Maguire, C.T., Converso, K.L., Shioi, T., et al., 2001. Progressive atrioventricular conduction defects and heart failure in mice expressing a mutant Csx/Nkx2.5 homeoprotein. J. Clin. Invest. 108 (2), 189–201.

Kaufman, M.H., Navaratnam, V., 1981. Early differentiation of the heart in mouse embryos. J. Anat. 133 (Pt 2), 235–246.

Kawano, Y., Kypta, R., 2003. Secreted antagonists of the Wnt signaling pathway. J. Cell. Sci. 116 (Pt 13), 2627–2634.

Kelly, R.G., Buckingham, M.E., 2002. The anterior heart-forming field: voyage to the arterial pole of the heart. Trends. Genet. 18 (4), 210–216.

Kelly, R.G., 2012. The second heart field. Curr. Top. Dev. Biol. 100, 33–65.

Kim, R.Y., Robertson, E.J., Solloway, M.J., 2001. Bmp6 and Bmp7 are required for cushion formation and septation in the developing mouse heart. Dev. Biol. 235 (2), 449–466.

Kioussi, C., Briata, P., Baek, S.H., Rose, D.W., Hamblet, N.S., Herman, T., et al., 2002. Identification of a Wnt/Dvl/beta-Catenin → Pitx2 pathway mediating cell-type-specific proliferation during development. Cell. 111 (5), 673–685.

Kitajima, S., Takagi, A., Inoue, T., Saga, Y., 2000. MesP1 and MesP2 are essential for the development of cardiac mesoderm. Development 127 (15), 3215–3226.

Korpal-Szczyrska, M., Aleszewicz-Baranowska, J., Dorant, B., Potaz, P., Birkholz, D., Kaminska, H., 2005. Cardiovascular malformations in Turner syndrome. Endokrynol. Diabetol. Chor. Przemiany. Materii. Wieku. Rozw. 11 (4), 211–214.

Kramer, K.L., Barnette, J.E., Yost, H.J., 2002. PKCgamma regulates syndecan-2 inside-out signaling during xenopus left-right development. Cell. 111 (7), 981–990.

Kuo, C.T., Morrisey, E.E., Anandappa, R., Sigrist, K., Lu, M.M., Parmacek, M.S., et al., 1997. GATA4 transcription factor is required for ventral morphogenesis and heart tube formation. Genes. Dev. 11 (8), 1048–1060.

Lalani, S.R., Ware, S.M., Wang, X., Zapata, G., Tian, Q., Franco, L.M., et al., 2013. MCTP2 is a dosage-sensitive gene required for cardiac outflow tract development. Hum. Mol. Genet. 22 (21), 4339–4348.

Lederer, D., Grisart, B., Digilio, M.C., Benoit, V., Crespin, M., Ghariani, S.C., et al., 2012. Deletion of KDM6A, a histone demethylase interacting with MLL2, in three patients with Kabuki syndrome. Am. J. Hum. Genet. 90 (1), 119–124.

Lee, R.Y., Luo, J., Evans, R.M., Giguere, V., Sucov, H.M., 1997. Compartment-selective sensitivity of cardiovascular morphogenesis to combinations of retinoic acid receptor gene mutations. Circ. Res. 80 (6), 757–764.

Levin, M., Thorlin, T., Robinson, K.R., Nogi, T., Mercola, M., 2002. Asymmetries in H^+/K^+-ATPase and cell membrane potentials comprise a very early step in left-right patterning. Cell. 111 (1), 77–89.

Lewin, M.B., McBride, K.L., Pignatelli, R., Fernbach, S., Combes, A., Menesses, A., et al., 2004. Echocardiographic evaluation of asymptomatic parental and sibling cardiovascular anomalies associated with congenital left ventricular outflow tract lesions. Pediatrics 114 (3), 691–696.

Lickert, H., Kutsch, S., Kanzler, B., Tamai, Y., Taketo, M.M., Kemler, R., 2002. Formation of multiple hearts in mice following deletion of beta-catenin in the embryonic endoderm. Dev. Cell. 3 (2), 171–181.

Lin, Q., Schwarz, J., Bucana, C., Olson, E.N., 1997. Control of mouse cardiac morphogenesis and myogenesis by transcription factor MEF2C. Science 276 (5317), 1404–1407.

Lin, X., Huo, Z., Liu, X., Zhang, Y., Li, L., Zhao, H., et al., 2010. A novel GATA6 mutation in patients with tetralogy of Fallot or atrial septal defect. J. Hum. Genet. 55 (10), 662–667.

Liu, C., Liu, W., Palie, J., Lu, M.F., Brown, N.A., Martin, J.F., 2002. Pitx2c patterns anterior myocardium and aortic arch vessels and is required for local cell movement into atrioventricular cushions. Development 129 (21), 5081–5091.

Lo, P.C., Frasch, M., 2003. Establishing a-p polarity in the embryonic heart tube. A conserved function of hox genes in *Drosophila* and vertebrates? Trends Cardiovasc. Med. 13 (5), 182–187.

Loffredo, C.A., Ferencz, C., Wilson, P.D., Lurie, I.W., 2000. Interrupted aortic arch: an epidemiologic study. Teratology 61 (5), 368–375.

Loffredo, C.A., Hirata, J., Wilson, P.D., Ferencz, C., Lurie, I.W., 2001. Atrioventricular septal defects: possible etiologic differences between complete and partial defects. Teratology 63 (2), 87–93.

Logan, M., Pagan-Westphal, S.M., Smith, D.M., Paganessi, L., Tabin, C.J., 1998. The transcription factor Pitx2 mediates situs-specific morphogenesis in response to left-right asymmetric signals. Cell. 94 (3), 307–317.

Mani, A., Radhakrishnan, J., Farhi, A., Carew, K.S., Warnes, C.A., Nelson-Williams, C., et al., 2005. Syndromic patent ductus arteriosus: evidence for haploinsufficient TFAP2B mutations and identification of a linked sleep disorder. Proc. Natl. Acad. Sci. U S A 102 (8), 2975–2979.

Manner, J., 2000. Cardiac looping in the chick embryo: a morphological review with special reference to terminological and biomechanical aspects of the looping process. Anat. Rec. 259 (3), 248–262.

Marelli, A.J., Mackie, A.S., Ionescu-Ittu, R., Rahme, E., Pilote, L., 2007. Congenital heart disease in the general population: changing prevalence and age distribution. Circulation 115 (2), 163–172.

Marvin, M.J., Di Rocco, G., Gardiner, A., Bush, S.M., Lassar, A.B., 2001. Inhibition of Wnt activity induces heart formation from posterior mesoderm. Genes. Dev. 15 (3), 316–327.

Maschhoff, K.L., Baldwin, H.S., 2000. Molecular determinants of neural crest migration. Winter Am. J. Med. Genet. 97 (4), 280–288.

Maslen, C.L., 2004. Molecular genetics of atrioventricular septal defects. Curr. Opin. Cardiol. 19 (3), 205–210.

Mavrides, E., Moscoso, G., Carvalho, J.S., Campbell, S., Thilaganathan, B., 2001. The anatomy of the umbilical, portal and hepatic venous systems in the human fetus at 14–19 weeks of gestation. Ultrasound. Obstet. Gynecol. 18 (6), 598–604.

McBride, K.L., Marengo, L., Canfield, M., Langlois, P., Fixler, D., Belmont, J.W., 2005a. Epidemiology of noncomplex left ventricular outflow tract obstruction malformations (aortic valve stenosis, coarctation of the aorta, hypoplastic left heart syndrome) in Texas, 1999–2001. Birth Defects Res. A. Clin. Mol. Teratol. 73 (8), 555–561.

McBride, K.L., Pignatelli, R., Lewin, M., Ho, T., Fernbach, S., Menesses, A., et al., 2005b. Inheritance analysis of congenital left ventricular outflow tract obstruction malformations: Segregation, multiplex relative risk, and heritability. Am. J. Med. Genet. A. 134 (2), 180–186.

McDaniell, R., Warthen, D.M., Sanchez-Lara, P.A., Pai, A., Krantz, I.D., Piccoli, D.A., et al., 2006. NOTCH2 mutations cause Alagille syndrome, a heterogeneous disorder of the notch signaling pathway. Am. J. Hum. Genet. 79 (1), 169–173.

Meijer, J.M., Pieper, P.G., Drenthen, W., Voors, A.A., Roos-Hesselink, J.W., van Dijk, A.P., et al., 2005. Pregnancy, fertility, and recurrence risk in corrected tetralogy of Fallot. Heart 91 (6), 801–805.

Meno, C., Ito, Y., Saijoh, Y., Matsuda, Y., Tashiro, K., Kuhara, S., et al., 1997. Two closely-related left-right asymmetrically expressed genes, lefty-1 and lefty-2: their distinct expression domains, chromosomal linkage and direct neuralizing activity in *Xenopus* embryos. Genes. Cells. 2 (8), 513–524.

Meno, C., Shimono, A., Saijoh, Y., Yashiro, K., Mochida, K., Ohishi, S., et al., 1998. lefty-1 is required for left-right determination as a regulator of lefty-2 and nodal. Cell. 94 (3), 287–297.

Mercola, M., 1999. Embryological basis for cardiac left-right asymmetry. Semin. Cell. Dev. Biol. 10 (1), 109–116.

Miyazawa, K., Shinozaki, M., Hara, T., Furuya, T., Miyazono, K., 2002. Two major Smad pathways in TGF-beta superfamily signalling. Genes. Cells. 7 (12), 1191–1204.

Moens, C.B., Stanton, B.R., Parada, L.F., Rossant, J., 1993. Defects in heart and lung development in compound heterozygotes for two different targeted mutations at the N-myc locus. Development 119 (2), 485–499.

Molkentin, J.D., Lin, Q., Duncan, S.A., Olson, E.N., 1997. Requirement of the transcription factor GATA4 for heart tube formation and ventral morphogenesis. Genes. Dev. 11 (8), 1061–1072.

Morse, D.E., Hendrix, M.J., 1980. Atrial septation. II. Formation of the foramina secunda in the chick. Dev. Biol. 78 (1), 25–35.

Morse, D.E., Rogers, C.S., McCann, P.S., 1984. Atrial septation in the chick and rat: a review. J. Submicrosc. Cytol. 16 (2), 259–272.

Muir, C.S., 1960. Incidence of congenital heart disease in Singapore. Br. Heart J. 22, 243–254.

Muncke, N., Jung, C., Rudiger, H., Ulmer, H., Roeth, R., Hubert, A., et al., 2003. Missense mutations and gene interruption in PROSIT240, a novel TRAP240-like gene, in patients with congenital heart defect (transposition of the great arteries). Circulation 108 (23), 2843–2850.

Nabulsi, M.M., Tamim, H., Sabbagh, M., Obeid, M.Y., Yunis, K.A., Bitar, F.F., 2003. Parental consanguinity and congenital heart malformations in a developing country. Am. J. Med. Genet. A. 116 (4), 342–347.

Narita, N., Bielinska, M., Wilson, D.B., 1997. Wild-type endoderm abrogates the ventral developmental defects associated with GATA-4 deficiency in the mouse. Dev. Biol. 189 (2), 270–274.

Ng, S.B., Bigham, A.W., Buckingham, K.J., Hannibal, M.C., McMillin, M.J., Gildersleeve, H.I., et al., 2010. Exome sequencing identifies MLL2 mutations as a cause of Kabuki syndrome. Nat. Genet. 42 (9), 790–793.

Nguyen-Tran, V.T., Kubalak, S.W., Minamisawa, S., Fiset, C., Wollert, K.C., Brown, A.B., et al., 2000. A novel genetic pathway for sudden cardiac death via defects in the transition between ventricular and conduction system cell lineages. Cell. 102 (5), 671–682.

Nonaka, S., Shiratori, H., Saijoh, Y., Hamada, H., 2002. Determination of left-right patterning of the mouse embryo by artificial nodal flow. Nature 418 (6893), 96–99.

Nonaka, S., Tanaka, Y., Okada, Y., Takeda, S., Harada, A., Kanai, Y., et al., 1998. Randomization of left-right asymmetry due to loss of nodal cilia generating leftward flow of extraembryonic fluid in mice lacking KIF3B motor protein. Cell. 95 (6), 829–837.

Olson, E.N., 2001. Development. The path to the heart and the road not taken. Science 291 (5512), 2327–2328.

Oyen, N., Poulsen, G., Boyd, H.A., Wohlfahrt, J., Jensen, P.K., Melbye, M., 2009a. National time trends in congenital heart defects, Denmark, 1977–2005. Am. Heart J. 157 (3), 467–473. e1.

Oyen, N., Poulsen, G., Boyd, H.A., Wohlfahrt, J., Jensen, P.K., Melbye, M., 2009b. Recurrence of congenital heart defects in families. Circulation 120 (4), 295–301.

Oyen, N., Poulsen, G., Wohlfahrt, J., Boyd, H.A., Jensen, P.K., Melbye, M., 2011. Recurrence of discordant congenital heart defects in families. Circ. Cardiovasc. Genet. 3 (2), 122–128.

Patient, R.K., McGhee, J.D., 2002. The GATA family (vertebrates and invertebrates). Curr. Opin. Genet. Dev. 12 (4), 416–422.

Patterson, K.D., Cleaver, O., Gerber, W.V., Grow, M.W., Newman, C.S., Krieg, P.A., 1998. Homeobox genes in cardiovascular development. Curr. Top Dev. Biol. 40, 1–44.

Pereira, F.A., Qiu, Y., Zhou, G., Tsai, M.J., Tsai, S.Y., 1999. The orphan nuclear receptor COUP-TFII is required for angiogenesis and heart development. Genes. Dev. 13 (8), 1037–1049.

Piacentini, G., Digilio, M.C., Capolino, R., Zorzi, A.D., Toscano, A., Sarkozy, A., et al., 2005. Familial recurrence of heart defects in subjects with congenitally corrected transposition of the great arteries. Am. J. Med. Genet. A. 137 (2), 176–180.

Piedra, M.E., Icardo, J.M., Albajar, M., Rodriguez-Rey, J.C., Ros, M.A., 1998. Pitx2 participates in the late phase of the pathway controlling left-right asymmetry. Cell. 94 (3), 319–324.

Potocki, L., Chen, K.S., Park, S.S., Osterholm, D.E., Withers, M.A., Kimonis, V., et al., 2000. Molecular mechanism for duplication 17p11.2- the homologous recombination reciprocal of the Smith-Magenis microdeletion. Nat. Genet. 24 (1), 84–87.

Pradat, P., Francannet, C., Harris, J.A., Robert, E., 2003. The epidemiology of cardiovascular defects, part I: a study based on data from three large registries of congenital malformations. Pediatr. Cardiol. 24 (3), 195–221.

Prall, O.W., Elliott, D.A., Harvey, R.P., 2002. Developmental paradigms in heart disease: insights from tinman. Ann. Med. 34 (3), 148–156.

Ramegowda, S., Ramachandra, N.B., 2006. Parental consanguinity increases congenital heart diseases in South India. Ann. Hum. Biol. 33 (5–6), 519–528.

Ranger, A.M., Grusby, M.J., Hodge, M.R., Gravallese, E.M., de la Brousse, F.C., Hoey, T., et al., 1998. The transcription factor NF-ATc is essential for cardiac valve formation. Nature 392 (6672), 186–190.

Reiter, J.F., Alexander, J., Rodaway, A., Yelon, D., Patient, R., Holder, N., et al., 1999. Gata5 is required for the development of the heart and endoderm in zebrafish. Genes. Dev. 13 (22), 2983–2995.

Reiter, J.F., Verkade, H., Stainier, D.Y., 2001. Bmp2b and Oep promote early myocardial differentiation through their regulation of gata5. Dev. Biol. 234 (2), 330–338.

Robinson, S.W., Morris, C.D., Goldmuntz, E., Reller, M.D., Jones, M.A., Steiner, R.D., et al., 2003. Missense mutations in CRELD1 are associated with cardiac atrioventricular septal defects. Am. J. Hum. Genet. 72 (4), 1047–1052.

Rogers, C.S., Morse, D.E., 1986a. Atrial septation in the rat. I. A light microscopic and histochemical study. J. Submicrosc. Cytol. 18 (2), 313–324.

Rogers, C.S., Morse, D.E., 1986b. Atrial septation in the rat. II. An electron microscopic study. J. Submicrosc. Cytol. 18 (2), 325–334.

Ryan, A.K., Blumberg, B., Rodriguez-Esteban, C., Yonei-Tamura, S., Tamura, K., Tsukui, T., et al., 1998. Pitx2 determines left-right asymmetry of internal organs in vertebrates. Nature 394 (6693), 545–551.

Saga, Y., Kitajima, S., Miyagawa-Tomita, S., 2000. Mesp1 expression is the earliest sign of cardiovascular development. Trends. Cardiovasc. Med. 10 (8), 345–352.

Saga, Y., Miyagawa-Tomita, S., Takagi, A., Kitajima, S., Miyazaki, J., Inoue, T., 1999. MesP1 is expressed in the heart precursor cells and required for the formation of a single heart tube. Development 126 (15), 3437–3447.

Sanchez-Valle, A., Pierpont, M.E., Potocki, L., 2011. The severe end of the spectrum: Hypoplastic left heart in Potocki-Lupski syndrome. Am. J. Med. Genet. A. 155A (2), 363–366.

Schlange, T., Andree, B., Arnold, H.H., Brand, T., 2000. BMP2 is required for early heart development during a distinct time period. Mech. Dev. 91 (1–2), 259–270.

Schott, J.J., Benson, D.W., Basson, C.T., Pease, W., Silberbach, G.M., Moak, J.P., et al., 1998. Congenital heart disease caused by mutations in the transcription factor NKX2–5. Science 281 (5373), 108–111.

Schrire, V., 1963. Experience with congenital heart disease at Groote Schuur Hospital, Cape Town. an analysis of 1,439 patients over an eleven year period. S. Afr. Med. J. 37, 1175–1180.

Schwartz, R.J., Olson, E.N., 1999. Building the heart piece by piece: modularity of cis-elements regulating Nkx2–5 transcription. Development 126 (19), 4187–4192.

Shann, M.K., 1969. Congenital heart disease in Taiwan, Republic of China. Circulation 39 (2), 251–258.

Simon, A.M., Goodenough, D.A., Paul, D.L., 1998. Mice lacking connexin40 have cardiac conduction abnormalities characteristic of atrioventricular block and bundle branch block. Curr. Biol. 8 (5), 295–298.

Sinning, A.R., 1998. Role of vitamin A in the formation of congenital heart defects. Anat. Rec. 253 (5), 147–153.

Siu, S., 1998. The risk of congenital heart disease recurrence in the offspring of adults with major heart defects. Evid. Based Cardiovasc. Med. 2 (3), 70.

Slager, R.E., Newton, T.L., Vlangos, C.N., Finucane, B., Elsea, S.H., 2003. Mutations in RAI1 associated with Smith-Magenis syndrome. Nat. Genet. 33 (4), 466–468.

Soemedi, R., Topf, A., Wilson, I.J., Darlay, R., Rahman, T., Glen, E., et al., 2012a. Phenotype-specific effect of chromosome 1q21.1 rearrangements and GJA5 duplications in 2436 congenital heart disease patients and 6760 controls. Hum. Mol. Genet. 21 (7), 1513–1520.

Soemedi, R., Wilson, I.J., Bentham, J., Darlay, R., Topf, A., Zelenika, D., et al., 2012b. Contribution of global rare copy-number variants to the risk of sporadic congenital heart disease. Am. J. Hum. Genet. 91 (3), 489–501.

Solloway, M.J., Robertson, E.J., 1999. Early embryonic lethality in Bmp5;Bmp7 double mutant mice suggests functional redundancy within the 60A subgroup. Development 126 (8), 1753–1768.

Srivastava, D., Cserjesi, P., Olson, E.N., 1995. A subclass of bHLH proteins required for cardiac morphogenesis. Science 270 (5244), 1995–1999.

Srivastava, D., Thomas, T., Lin, Q., Kirby, M.L., Brown, D., Olson, E.N., 1997. Regulation of cardiac mesodermal and neural crest development by the bHLH transcription factor, dHAND. Nat. Genet. 16 (2), 154–160.

Stankiewicz, P., Sen, P., Bhatt, S.S., Storer, M., Xia, Z., Bejjani, B.A., et al., 2009. Genomic and genic deletions of the FOX gene cluster on 16q24.1 and inactivating mutations of FOXF1 cause alveolar capillary dysplasia and other malformations. Am. J. Hum. Genet. 84 (6), 780–791.

Sucov, H.M., Dyson, E., Gumeringer, C.L., Price, J., Chien, K.R., Evans, R.M., 1994. RXR alpha mutant mice establish a genetic basis for vitamin A signaling in heart morphogenesis. Genes. Dev. 8 (9), 1007–1018.

Svensson, E.C., Huggins, G.S., Lin, H., Clendenin, C., Jiang, F., Tufts, R., et al., 2000. A syndrome of tricuspid atresia in mice with a targeted mutation of the gene encoding Fog-2. Nature genetics 25 (3), 353–356.

Tabin, C.J., Vogan, K.J., 2003. A two-cilia model for vertebrate left-right axis specification. Genes. Dev. 17 (1), 1–6.

Takeda, S., Yonekawa, Y., Tanaka, Y., Okada, Y., Nonaka, S., Hirokawa, N., 1999. Left-right asymmetry and kinesin superfamily protein KIF3A: new insights in determination of laterality and mesoderm induction by kif3A-/- mice analysis. J. Cell. Biol. 145 (4), 825–836.

Tanaka, M., Chen, Z., Bartunkova, S., Yamasaki, N., Izumo, S., 1999. The cardiac homeobox gene Csx/Nkx2.5 lies genetically upstream of multiple genes essential for heart development. Development 126 (6), 1269–1280.

Tanaka, M., Kasahara, H., Bartunkova, S., Schinke, M., Komuro, I., Inagaki, H., et al., 1998. Vertebrate homologs of tinman and bagpipe: roles of the homeobox genes in cardiovascular development. Dev. Genet. 22 (3), 239–249.

Tevosian, S.G., Deconinck, A.E., Tanaka, M., Schinke, M., Litovsky, S.H., Izumo, S., et al., 2000. FOG-2, a cofactor for GATA transcription factors, is essential for heart morphogenesis and development of coronary vessels from epicardium. Cell. 101 (7), 729–739.

Thienpont, B., Zhang, L., Postma, A.V., Breckpot, J., Tranchevent, L.C., Van Loo, P., et al., 2010. Haplo-insufficiency of TAB2 causes congenital heart defects in humans. Am. J. Hum. Genet. 86 (6), 839–849.

Thomas, T., Yamagishi, H., Overbeek, P.A., Olson, E.N., Srivastava, D., 1998. The bHLH factors, dHAND and eHAND, specify pulmonary and systemic cardiac ventricles independent of left-right sidedness. Dev. Biol. 196 (2), 228–236.

Tsukui, T., Capdevila, J., Tamura, K., Ruiz-Lozano, P., Rodriguez-Esteban, C., Yonei-Tamura, S., et al., 1999. Multiple left-right asymmetry defects in Shh(-/-) mutant mice unveil a convergence of the shh and retinoic acid pathways in the control of Lefty-1. Proc. Natl. Acad. Sci. U. S. A. 96 (20), 11376–11381.

Tumer, Z., Harboe, T.L., Blennow, E., Kalscheuer, V.M., Tommerup, N., Brondum-Nielsen, K., 2004. Molecular cytogenetic characterization of ring chromosome 15 in three unrelated patients. Am. J. Med. Genet. A. 130A (4), 340–344.

Urness, L.D., Bleyl, S.B., Wright, T.J., Moon, A.M., Mansour, S.L., 2011. Redundant and dosage sensitive requirements for Fgf3 and Fgf10 in cardiovascular development. Dev. Biol. 356 (2), 383–397.

van den Hoff, M.J., Moorman, A.F., 2000. Cardiac neural crest: the holy grail of cardiac abnormalities? Cardiovasc. Res. 47 (2), 212–216.

van der Bom, T., Zomer, A.C., Zwinderman, A.H., Meijboom, F.J., Bouma, B.J., Mulder, B.J., 2011. The changing epidemiology of congenital heart disease. Nat. Rev. Cardiol. 8 (1), 50–60.

van Egmond, H., Orye, E., Praet, M., Coppens, M., Devloo-Blancquaert, A., 1988. Hypoplastic left heart syndrome and 45X karyotype. Br. Heart J. 60 (1), 69–71.

van Rijen, H.V., van Veen, T.A., van Kempen, M.J., Wilms-Schopman, F.J., Potse, M., Krueger, O., et al., 2001. Impaired conduction in the bundle branches of mouse hearts lacking the gap junction protein connexin40. Circulation 103 (11), 1591–1598.

VanderBrink, B.A., Sellitto, C., Saba, S., Link, M.S., Zhu, W., Homoud, M.K., et al., 2000. Connexin40-deficient mice exhibit atrioventricular nodal and infra-Hisian conduction abnormalities. J. Cardiovasc. Electrophysiol. 11 (11), 1270–1276.

Verheule, S., van Batenburg, C.A., Coenjaerts, F.E., Kirchhoff, S., Willecke, K., Jongsma, H.J., 1999. Cardiac conduction abnormalities in mice lacking the gap junction protein connexin40. J. Cardiovasc. Electrophysiol. 10 (10), 1380–1389.

Vincent, S.D., Dunn, N.R., Hayashi, S., Norris, D.P., Robertson, E.J., 2003. Cell fate decisions within the mouse organizer are governed by graded Nodal signals. Genes. Dev. 17 (13), 1646–1662.

Vissers, L.E., van Ravenswaaij, C.M., Admiraal, R., Hurst, J.A., de Vries, B.B., Janssen, I.M., et al., 2004. Mutations in a new member of the chromodomain gene family cause CHARGE syndrome. Nat. Genet. 36 (9), 955–957.

Vitelli, F., Lania, G., Huynh, T., Baldini, A., 2010. Partial rescue of the Tbx1 mutant heart phenotype by Fgf8: genetic evidence of impaired tissue response to Fgf8. J. Mol. Cell. Cardiol. 49 (5), 836–840.

Vitelli, F., Morishima, M., Taddei, I., Lindsay, E.A., Baldini, A., 2002. Tbx1 mutation causes multiple cardiovascular defects and disrupts neural crest and cranial nerve migratory pathways. Hum. Mol. Genet. 11 (8), 915–922.

Volkl, T.M., Degenhardt, K., Koch, A., Simm, D., Dorr, H.G., Singer, H., 2005. Cardiovascular anomalies in children and young adults with Ullrich-Turner syndrome the Erlangen experience. Clin. Cardiol. 28 (2), 88–92.

Waldo, K., Miyagawa-Tomita, S., Kumiski, D., Kirby, M.L., 1998. Cardiac neural crest cells provide new insight into septation of the cardiac outflow tract: aortic sac to ventricular septal closure. Dev. Biol. 196 (2), 129–144.

Ware, S.M., Peng, J., Zhu, L., Fernbach, S., Colicos, S., Casey, B., et al., 2004. Identification and functional analysis of ZIC3 mutations in heterotaxy and related congenital heart defects. Am. J. Hum. Genet. 74 (1), 93–105.

Watanabe, D., Saijoh, Y., Nonaka, S., Sasaki, G., Ikawa, Y., Yokoyama, T., et al., 2003. The left-right determinant Inversin is a component of node monocilia and other 9+0 cilia. Development 130 (9), 1725–1734.

Weisberg, E., Winnier, G.E., Chen, X., Farnsworth, C.L., Hogan, B.L., Whitman, M., 1998. A mouse homologue of FAST-1 transduces TGF beta superfamily signals and is expressed during early embryogenesis. Mech. Dev. 79 (1–2), 17–27.

Whittemore, R., Wells, J.A., Castellsague, X., 1994. A second-generation study of 427 probands with congenital heart defects and their 837 children. J. Am. Coll. Cardiol. 23 (6), 1459–1467.

Wollins, D.S., Ferencz, C., Boughman, J.A., Loffredo, C.A., 2001. A population-based study of coarctation of the aorta: comparisons of infants with and without associated ventricular septal defect. Teratology 64 (5), 229–236.

Wotton, D., Massague, J., 2001. Smad transcriptional corepressors in TGF beta family signaling. Curr. Top. Microbiol. Immunol. 254, 145–164.

Xie, L., Hoffmann, A.D., Burnicka-Turek, O., Friedland-Little, J.M., Zhang, K., Moskowitz, I.P., 2012. Tbx5-hedgehog molecular networks are essential in the second heart field for atrial septation. Dev. Cell. 23 (2), 280–291.

Ya, J., Schilham, M.W., de Boer, P.A., Moorman, A.F., Clevers, H., Lamers, W.H., 1998. Sox4-deficiency syndrome in mice is an animal model for common trunk. Circ. Res. 83 (10), 986–994.

Yada, T., Sakai, T., Komuro, H., Hirota, A., Kamino, K., 1985. Development of electrical rhythmic activity in early embryonic cultured chick double-heart monitored optically with a voltage-sensitive dye. Dev. Biol. 110 (2), 455–466.

Yagi, H., Furutani, Y., Hamada, H., Sasaki, T., Asakawa, S., Minoshima, S., et al., 2003. Role of TBX1 in human del22q11.2 syndrome. Lancet 362 (9393), 1366–1373.

Yamagishi, H., Yamagishi, C., Nakagawa, O., Harvey, R.P., Olson, E.N., Srivastava, D., 2001. The combinatorial activities of Nkx2.5 and dHAND are essential for cardiac ventricle formation. Dev. Biol. 239 (2), 190–203.

Yamagishi, T., Nakajima, Y., Miyazono, K., Nakamura, H., 1999. Bone morphogenetic protein-2 acts synergistically with transforming growth factor-beta3 during endothelial-mesenchymal transformation in the developing chick heart. J. Cell. Physiol. 180 (1), 35–45.

Yanagisawa, H., Hammer, R.E., Richardson, J.A., Emoto, N., Williams, S.C., Takeda, S., et al., 2000. Disruption of ECE-1 and ECE-2 reveals a role for endothelin-converting enzyme-2 in murine cardiac development. J. Clin. Invest. 105 (10), 1373–1382.

Ye, M., Coldren, C., Liang, X., Mattina, T., Goldmuntz, E., Benson, D.W., et al., 2010. Deletion of ETS-1, a gene in the Jacobsen syndrome critical region, causes ventricular septal defects and abnormal ventricular morphology in mice. Hum. Mol. Genet. 19 (4), 648–656.

Yost, H.J., 2001. Establishment of left-right asymmetry. Int. Rev. Cytol. 203, 357–381.

Yutzey, K.E., Kirby, M.L., 2002. Wherefore heart thou? Embryonic origins of cardiogenic mesoderm. Dev. Dyn. 223 (3), 307–320.

Zalzstein, E., Moes, C.A., Musewe, N.N., Freedom, R.M., 1991. Spectrum of cardiovascular anomalies in Williams-Beuren syndrome. Pediatr. Cardiol. 12 (4), 219–223.

Zavala, C., Jimenez, D., Rubio, R., Castillo-Sosa, M.L., Diaz-Arauzo, A., Salamanca, F., 1992. Isolated congenital heart defects in first degree relatives of 185 affected children. Prospective study in Mexico City. Arch. Med. Res. 23(4):177–182.

Zhang, X.M., Ramalho-Santos, M., McMahon, A.P., 2001. Smoothened mutants reveal redundant roles for Shh and Ihh signaling including regulation of L/R asymmetry by the mouse node. Cell. 105 (6), 781–792.

Zhao, Z., Rivkees, S.A., 2000. Programmed cell death in the developing heart: regulation by BMP4 and FGF2. Dev. Dyn. 217 (4), 388–400.

Zhu, L., Vranckx, R., Khau Van Kien, P., Lalande, A., Boisset, N., Mathieu, F., et al., 2006. Mutations in myosin heavy chain 11 cause a syndrome associating thoracic aortic aneurysm/aortic dissection and patent ductus arteriosus. Nat. Genet. 38 (3), 343–349.

Zorn, A.M., 2001. Wnt signaling: antagonistic Dickkopfs. Curr. Biol. 11 (15), R592–R595.

Zou, Y., Evans, S., Chen, J., Kuo, H.C., Harvey, R.P., Chien, K.R., 1997. CARP, a cardiac ankyrin repeat protein, is downstream in the Nkx2–5 homeobox gene pathway. Development 124 (4), 793–804.

Zweier, C., Thiel, C.T., Dufke, A., Crow, Y.J., Meinecke, P., Suri, M., et al., 2005. Clinical and mutational spectrum of Mowat-Wilson syndrome. Eur. J. Med. Genet. 48 (2), 97–111.

Chapter 34

T-Box Genes and Developmental Anomalies

Nataki C. Douglas*, Andrew J. Washkowitz*, L.A. Naiche† and Virginia E. Papaioannou*

*Department of Genetics and Development, Columbia University Medical Center, New York, NY; †National Cancer Institute, Cancer and Developmental Biology Lab, Frederick, MD

Chapter Outline

GLOSSARY

Conditional allele An allele that is engineered to function normally until it is recombined by the introduction of a recombinase to produce a mutant allele in a tissue- or temporally-specific manner.

Dominant-negative A mutated protein that interferes with the function of normal copies of the same protein.

Half site One half of the palindromic T-box binding element.

Haplo-insufficiency A situation in which a mutation in one copy of a gene reduces the level of a gene product below threshold levels and a phenotypic effect results.

Hypomorphic allele A mutant allele that results in reduced activity or level of the protein.

Online Mendelian Inheritance in Man (OMIM) A catalog of human genes and genetic disorders developed by the National Center for Biotechnology Information.

T-box binding element (TBE) The palindromic DNA consensus sequence with a high affinity for binding the T domain.

T-box domain Also called the T domain, it is the defining feature of the T-box gene family, and it consists of a conserved sequence encoding a polypeptide domain that extends across a region of 180 to 200 amino acid residues.

SUMMARY

- The T-box gene family consists of 17 related genes that encode DNA binding proteins that function as transcription factors important in many developmental processes.
- *TBX1* has been identified as a major factor in the DiGeorge deletion syndrome and mutations in other T-box genes result in developmental syndromes (*TBX3*, ulnar mammary; *TBX4*, small patella; *TBX5*, Holt-Oram; *TBX15*, Cousin; *TBX22*, cleft palate with ankyloglossia) or developmental anomalies (*TBX6*, spondylocostal dysostosis; *TBX19*, isolated ACTH deficiency; *TBX20*, congenital heart disease).
- Most congenital abnormalities resulting from mutations in T-box genes have effects in heterozygotes with more severe effects in homozygotes, indicating dosage sensitivity.

- Mouse models are being used to explore the mechanism of action of these and other T-box genes during development and to point the way to recognizing the full range of possible phenotypes. Sophisticated genetic engineering experiments in mice and the study of naturally occurring mutations in humans will continue to reveal the multiple developmental roles of T-box genes.

34.1 T-BOX GENES: TRANSCRIPTION FACTOR GENES WITH MANY DEVELOPMENTAL ROLES

In 1927, a mouse gene discovered in a mutagenesis screen captured the interest of geneticists and developmental biologists. During the succeeding decades, it gathered many claims to fame. This gene, named *Brachyury* or *T* for tail, was recognized by its heterozygous mutant phenotype, a short tail, and was found to be lethal when homozygous, making it one of the earliest known examples of a mammalian developmental gene causing embryonic lethality (Dobrovolskaïa-Zavadskaïa, 1927). *T* was the first mammalian gene to be positionally cloned and when the sequence of the gene and the structure of the protein were studied, it was found to code for a protein with a novel DNA-binding motif called the T domain (Herrmann, 1995). Shortly after its cloning, *T* was recognized as the proband of a previously unknown family of transcription factor genes, now called the T-box gene family, which is present in all metazoans. Thus, although *T* has been the subject of intense study for more than half a century, other members of the T-box gene family are relative newcomers but have garnered attention due to their involvement in human congenital defects. The T-box genes are essential for all manner of developmental processes, with multiple sites of expression and multiple roles in a wide variety of tissues (Naiche, 2005; Papaioannou, 2001; Wardle and Papaioannou, 2008; Hariri et al., 2012; Naiche and Papaioannou, 2007a).

The T-box genes can be grouped phylogenetically on the basis of sequence comparisons of the region coding for the T domain, thus providing a family tree that reflects the evolution of the genes (Papaioannou and Goldin, 2004). In mammals, 17 genes are organized into five subfamilies (Figure 34.1). Although their phylogenetic relationships and their relationship to orthologs in other species is fairly clear, we still know relatively little about how the functions of T-box genes have evolved, because even very closely related family members have divergent functions. For example, *Tbx5* has a critical role in heart development, whereas the closely related *Tbx4* is not expressed there and has no role in the heart. Conversely, *Tbx4* and *Tbx5* cooperate during lung development. What is clear is that T-box proteins function as transcription factors in a wide variety of developmental settings. Most T-box gene mutations display heterozygous phenotypes and thus have been readily recognized as being responsible for a variety of human developmental syndromes and anomalies.

In this chapter, we explore the multiple roles of T-box genes in developmental processes through an examination of human developmental syndromes and anomalies caused by their mutations (Table 34.1) and the corresponding mouse mutations that model these syndromes. We consider how the study of the mouse models has led to our understanding of the

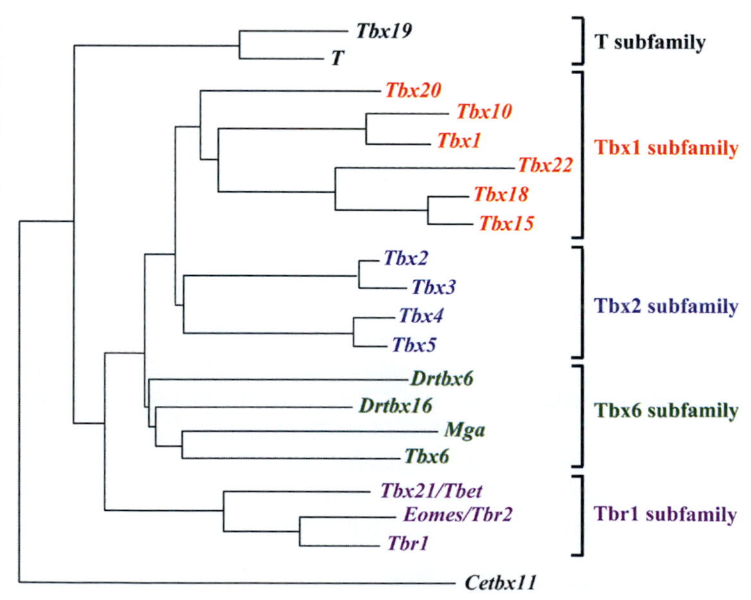

FIGURE 34.1 Schematic phylogenetic tree of the T-box gene family of vertebrates based on the phylogenetic analysis described by Papaioannou and Goldin (2004) showing the relationship of genes in the five subfamilies, which are indicated by brackets on the right. All genes are present in mammals except orthologs of the zebrafish genes *Drtbx6* and *Drtbx16*, which do not exist in mammals. *(Reprinted with permission from the* Annual Review of Genetics, *Volume 39, © 2005 by Annual Reviews [www.annualreviews.org:Naiche et al., 2005].)*

TABLE 34.1 Human T-Box Genes and Known or Suspected Associations with Human Developmental Abnormalities

Human Gene	Chromosomal Location	Mutant Syndrome/ Anomaly	Gene/locus MIM #	Phenotype MIM #
TBX1	22q11.21	DiGeorge syndrome (DGS)	602054	188400
TBX3	12q24.21	Ulnar mammary syndrome (UMS)	601621	181450
TBX4	17q23.2	Small patella syndrome (SPS)	601719	147891
TBX5	12q24.21	Holt-Oram syndrome (HOS)	601620	142900
TBX6	16p11.2	Dominant spondylocostal dysostosis	602427	122600
TBX15	1p12	Cousin syndrome	604127	260660
TBX19	1q24.2	Isolated adrenocorticotrophic hormone deficiency (IAD)	604614	201400
TBX20	7p14.2	Atrial septal defect4 (ASD4)	606061	611363
TBX22	Xq21.1	Cleft palate with or without ankyloglossia (CPX)	300307	303400
T	6q27	Susceptibility to neural tube defects	601397	182940
TBX21	17q21.32	Asthma susceptibility	604895	208550
TBX2	17q23.2	unknown	600747	na
TBX10	11q13.2	unknown	604648	na
TBX18	6q14–q15	unknown	604613	na
EOMES	3p24.1	unknown	604615	na
TBR1	2q24.2	unknown	604616	na
MGA	15q14	unknown	na	na

basis of the human syndromes, and how the variety of mutations in both human and mouse have led to a better understanding of gene function. The current state of the field emphasizes the need for conditional alleles and multiple allelic series to fully explore the function of members of this important gene family in all sites of expression.

34.2 DNA BINDING AND TRANSCRIPTIONAL REGULATION BY T-BOX PROTEINS

The T-box DNA binding motif that characterizes all T-box proteins binds DNA in a sequence-specific manner. A palindromic DNA consensus sequence with high affinity for T protein was first defined and called the T-box binding element (TBE). The T protein binds this sequence as a dimer, with each monomer binding a half site called a T-half site (5'-AGGTGTGAAATT-3') (Kispert and Herrmann, 1993). The crystal structures of both T and TBX3 T-domain homodimers bound to the canonical TBE show that despite differences in ternary structure, both T-box proteins make the same DNA contacts with the same amino acids, which indicates strong conservation of the underlying DNA binding functions (Coll et al., 2002; Müller and Herrmann, 1997). Work with other T-box proteins indicates they are all capable of binding the T-half site, although some have different optimal target sequences (e.g., Ghosh et al., 2001; Lingbeek et al., 2002; Sinha et al., 2000). Moreover, different proteins have preferences for different combinations of T-half sites varying with regard to orientation, number, and spacing (Sinha et al., 2000; Conlon et al., 2001; El Omari et al., 2011). In the context of the promoters of downstream target genes, these differences between T-box proteins in structure, binding preferences, and affinity may constitute the basis for target gene specificity, and they may be the basis for interactions among different T-box proteins in areas of expression overlap.

There is evidence that T-box proteins function in combination with a variety of other transcription factors – including other T-box proteins – to regulate their activity. The homeobox protein Nkx2–5, for example, has been shown to heterodimerize with Tbx2, Tbx5, and Tbx20 in the developing heart (Habets et al., 2002; Hiroi et al., 2001; Stennard et al., 2003). In this system, competition between Tbx2 and Tbx5 for Nkx2–5 heterodimerization is thought to regulate positional specificity in the developing heart (Habets et al., 2002). These competitive interactions may be the basis for another level of regulation of the activity of the T-box proteins based on the kinetics of their physical interactions.

The heterodimerization of T-box proteins also leads to regional specificity of their function. For example, in the heart, Tbx2 binds to Nkx2–5, while in melanoma cells, Tbx2 was found to interact with the tumor suppressor Rb1 (Vance et al., 2010). These differences likely play a role in the differential regulation of target genes in diverse contexts and allow the T-box proteins to have broad functions throughout development.

T-box proteins can act as both activators and repressors of the transcription of target genes. Activation domains have been mapped to the C-terminal domains of several T-box proteins (Stennard et al., 2003; Kispert, 1995; Zaragoza et al., 2004), and Tbx2 and Tbx3 have been shown to repress transcription (Lingbeek et al., 2002; Habets et al., 2002; Carreira et al., 1998). Some T-box genes contain both activation and repression domains in their C-terminal domains, and may act in either fashion, depending on promoter context (Stennard et al., 2003; Kispert, 1995; Ouimette et al., 2010; Paxton et al., 2002).

In addition to directly regulating transcription, there may be a role for T-box proteins in the epigenetic regulation of target genes. Tbx2, Tbx4, Tbx5, and Tbx6 have all been shown to be able to specifically recognize the N-terminal tail of histone H3 (Demay et al., 2007). Tbx3 was shown to interact with the polycomb repressive complex ECS/E(Z) (PRC2), the catalytic enzyme for mediating H3K27 methylation. Moreover, Tbx3 was shown to regulate the H3K27 methylation of target genes (Lu et al., 2011). Together, these results indicate that T-box proteins may regulate transcription through epigenetic modification in addition to recruitment of the transcriptional machinery.

34.3 HUMAN SYNDROMES AND MOUSE MODELS

TBX1 and the DiGeorge Syndrome

(See also Chapter 36.) A convergence of research on *Tbx1* in mouse and a number of microdeletion syndromes in humans collectively known as 22q11.2 deletion syndrome (22q11.2DS) resulted in the recognition of *TBX1* as a major factor in the etiology of this syndrome. Variously described as DiGeorge syndrome, velocardiofacial syndrome, Shprintzen velocardiofacial syndrome and conotruncal face syndrome, 22q11.2DS is one of the more common birth defects (approximately 1 in 4,000 live births). It is highly variable and is characterized by cardiovascular anomalies including outflow tract and aortic arch defects, craniofacial anomalies including cleft palate, ear defects, thymic and parathyroid hypoplasia or aplasia, skeletal anomalies, kidney abnormalities, and cognitive and neuropsychological problems. Most commonly, the syndrome results from a 3 megabase pair (Mbp) deletion that includes more than 30 genes or, less commonly, from a 1.5 Mbp deletion that includes about 20 genes; both deletions include *TBX1*. Although there is the potential for the involvement of a large number of genes, compelling evidence from mouse models and the discovery of *TBX1* mutations in human patients with the characteristic syndrome but without microdeletions (Yagi et al., 2003; Paylor et al., 2006; Torres-Juan et al., 2007; Zweier et al., 2007) implicates haplo-insufficiency for *TBX1* as one of the major genetic causes of the syndrome.

Tbx1 has an expression pattern during embryogenesis that places the gene product in a position to affect pharyngeal, heart, ear, skeletal and brain development. It is expressed in the core mesoderm of the pharyngeal arches, pharyngeal endoderm, otic placodes, head mesenchyme, tooth buds, lung epithelium, developing vertebrae, and brain, which are areas that encompass the precursors of virtually all of the tissues affected in 22q11.2DS patients (Figure 34.2) (Chapman et al., 1996; Hiramoto et al., 2011). Chromosomal engineering to produce a deletion of the 22q11.2 syntenic region in the mouse provided a haplo-insufficiency model that recapitulated some of the features of 22q11.2DS (Lindsay et al., 1999), including learning and memory impairments (Paylor et al., 2001; 2006). Loss-of-function mutations in mouse *Tbx1* further support the contention that *TBX1* is a major causative factor of 22q11.2DS, because heterozygotes demonstrate haplo-insufficiency defects in thymus and aortic arch artery development and homozygotes show a full spectrum of phenotypic abnormalities typical of 22q11.2DS, albeit with more severe phenotypes than in human patients (Jerome and Papaioannou, 2001; Lindsay et al., 2001; Merscher et al., 2001; Kelly et al., 2004).

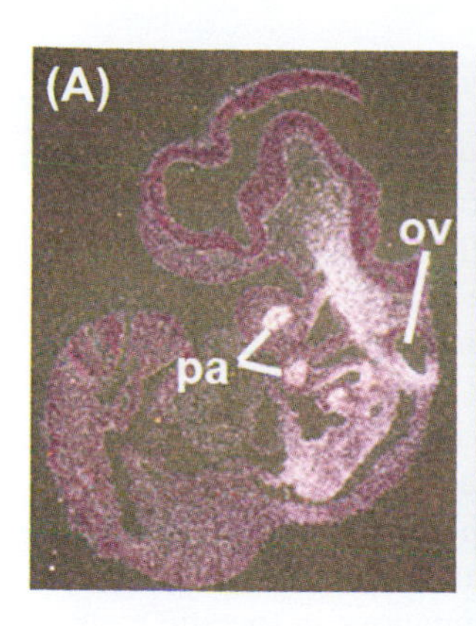

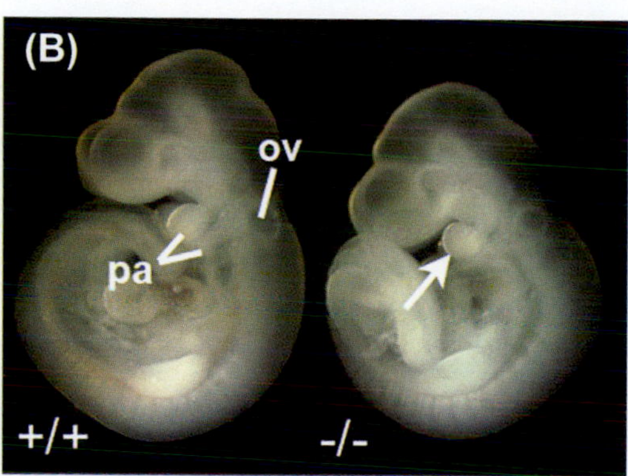

FIGURE 34.2 (A) Expression of *Tbx1* in a parasagittal section showing expression in the otic vesicle (ov), the core of the pharyngeal arches (pa), the pharyngeal endoderm and the head mesenchyme (adapted and reprinted with permission from (Chapman et al., 1996). (B) Whole mount wild type (+/+) and *Tbx1* mutant (–/–) midgestation embryos. The mutant embryo has only one pharyngeal arch (arrow) and a small otic vesicle. *(A: Adapted and reprinted with permission from Wiley-Liss, Inc., a subsidiary of John Wiley & Sons, Inc., from Chapman et al.,* Developmental Dynamics *206:379 © 1996 Wiley-Liss, Inc. B: adapted and reprinted with permission from Macmillan Publishers Ltd. From Jerome and Papaioannou,* Nature Genetic *27:286 © 2001.*

Despite this evidence, it is still the case that the vast majority of patients also have a large number of other genes deleted, many with developmental roles (Tan et al., 2010), that could contribute to the phenotype either independently as a contiguous gene syndrome or by regulating or interacting with *TBX1*. Exploration of other 22q11.2DS candidate genes led to the discovery of an interaction between *Tbx1* and *Crkl*, the ortholog of which is located within the human 3 Mbp deletion. Mice homozygous for a null mutation in *Crkl* phenocopy multiple aspects of 22q11.2DS, whereas heterozygous mutants show only rare craniofacial and thymic anomalies. Double heterozygotes for *Crkl* and *Tbx1* show a greatly increased penetrance of aortic arch, thymic, and parathyroid defects compared with either type of single heterozygotes (Jerome and Papaioannou, 2001; Guris et al., 2001; 2006), which indicates that a genetic interaction between these two genes affects the development of the pharyngeal apparatus. Defects in these mice are associated with functional aberrations in two major signaling pathways, retinoic acid (RA) and fibroblast growth factor (FGF) signaling. Triple heterozygous embryos with mutations in *Crkl*, *Tbx1*, and *Aldh1a2* (*Raldh2*) (a gene that encodes a key enzyme in the synthesis of RA) show an amelioration of the phenotype, which indicates that the gain of function of RA seen in double heterozygotes is at least partially responsible for the phenotype. Similarly, in triple heterozygotes with *Crkl*, *Tbx1*, and *Fgf8* mutations, the reduction of Fgf8 exacerbates the severity of the double heterozygous phenotype (Guris et al., 2006), thereby further confirming a link between *Tbx1* and FGF signaling and its importance in 22q11.2DS (Baldini, 2005).

Studies using conditional alleles to determine the requirement for *Tbx1* in different tissues and at different times in development have highlighted the multiple roles of *Tbx1* (Scambler, 2010) and have led to the discovery of novel phenotypes. For example, early loss of *Tbx1* has a severe effect on the second through sixth pharyngeal arches, while loss one day later results in normal first through third arches (Xu et al., 2005); loss of expression in the pharyngeal endoderm alone accounts for most of the pharyngeal abnormalities, although the gene is expressed in other pharyngeal tissues (Xu et al., 2004; 2005; Arnold et al., 2006); and finally, loss of *Tbx1* specifically from endothelial cells results in failure of maintenance of gastrointestinal lymphatic vasculature, which is not a phenotype that has been as yet associated with 22q11.2DS (Chen et al., 2010).

As a transcription factor, Tbx1 has the potential to affect many different downstream target genes, and because of its expression in multiple tissues it could have different targets in different tissues. Microarray studies have indicated that a large number of genes are affected by alterations in *Tbx1* gene dosage (Liao et al., 2008; Pane et al., 2012; van Bueren et al., 2010) and follow-up studies have identified a number of Tbx1 targets that contribute to the mutant phenotype. Tbx1 regulates a number of genes which are important in pharyngeal arch artery (PAA) development, such as the indirect targets *Hes1*, an effector of Notch signaling necessary for vascular smooth muscle development (van Bueren et al., 2010), and *Gbx2*, which affects fourth PAA development through effects on neural crest cell migration (Calmont et al., 2009), and the direct target, *Smad7*, which is required for the remodeling that occurs in the hypoplastic fourth PAA in Tbx1 heterozygotes (Papangeli and Scambler, 2013). In the cardiac progenitors of the second heart field (SHF), Tbx1 directly controls transcription of *Fgf8* (Hu et al., 2004), *Fgf10* (Xu et al., 2004; Watanabe et al., 2012) and *Wnt5a* (Chen et al., 2012). The asymmetric expression of *Pitx2* in the left SHF is also directly controlled by Tbx1, which is asymmetrically expressed in the posterior SHF (Nowotschin et al., 2006). In contrast to the other Tbx1 targets, which are activated by Tbx1, *Mef2c*, a cardiogenic transcription factor involved in cardiomyocyte differentiation, is repressed by Tbx1 by a combination of direct and indirect regulation (Posch et al., 2010).

Thus Tbx1 has multiple targets and roles that affect the development of many tissues and organs. Its transcriptional activity in any given tissue will be highly context dependent, depending on the availability of co-factors and the expression of modifier genes. The myriad interactions and tissues in which Tbx1 functions may explain the large variation in phenotypes associated with 22q11.2DS.

TBX3 and the Ulnar Mammary Syndrome

Ulnar mammary syndrome (UMS) is an autosomal dominant disorder originally mapped to 12q23–q24.1, a region that contains both *TBX3* and *TBX5*. The syndrome was named for two of the more common abnormalities that make up this disorder: defects in the posterior part of the hand and arm (particularly the ulna) and a deficiency of mammary gland development. Other features are variably present in UMS, including a lack of axillary apocrine glands, a lack of axial hair, delayed puberty in males, a variety of anomalies of the internal and external urogenital system, obesity, abnormalities of teeth and palate, laryngeal stenosis, ventricular septal defects, and foot defects (Washkowitz et al., 2012; Linden et al., 2009; Meneghini et al., 2006; Joss et al., 2011). The availability of large multigenerational kindreds with UMS allowed for the fine mapping of the disorder and the discovery of causative mutations in *TBX3* (Bamshad et al., 1997). Within each kindred, family members showed different abnormalities, thus illustrating the highly variable phenotype even in individuals with the same mutation. Conversely, in a study of UMS kindreds with ten different *TBX3* mutations scattered throughout the gene, no correlation was found between the frequency or nature of different defects and mutations in specific parts of the gene. All kindreds had the same wide range of developmental abnormalities whether the mutations were predicted to be loss-of-function or dominant-negative alleles (Bamshad et al., 1999). This suggests that all of these mutations result in haplo-insufficiency and that interacting genes and/or epigenetic factors are responsible for the variability in phenotype. No homozygous mutant *TBX3* individuals have been identified.

Concurrent studies of the mouse *Tbx3* gene have provided an explanation for the wide range of structures affected in cases of UMS. Like many T-box genes, *Tbx3* is expressed in a highly dynamic pattern throughout development. Notably, it is present in the developing limbs, mammary gland, craniofacial region, heart, kidney, Wolffian duct epithelium, and mesenchyme around both Wolffian and Müllerian ducts, pituitary gland, dorsal root ganglia, yolk sac, and lungs (Chapman et al., 1996; Douglas et al., 2012). Expression is found in the embryonic precursors of many – if not all – of the tissues affected in UMS, as well as in a number of other tissues not known to be affected in patients with UMS.

A mutation in *Tbx3* that results in a loss of function provides a mouse model for UMS (Davenport et al., 2003). Some homozygous mutants die early as a result of abnormal cell death in the yolk sac and deficient vitelline circulation, but others have normal circulation and survive to late gestation. In addition to the life-threatening and sometimes lethal yolk sac deficiency, these mutant embryos display defects that are characteristic of UMS, such as ulnar and fibular ray deficiencies, a lack of mammary gland induction, and heart defects (Washkowitz et al., 2012; Davenport et al., 2003; Mesbah et al., 2008). Although the mutant phenotype is generally more severe in the homozygous mice than in heterozygous humans, the defects are of a similar type to those seen in UMS. However, the homozygous mice reveal a possible human-mouse species difference in the relative severity of the effect in the forelimbs as compared with the hind limbs. In UMS, the arms or hands are always affected, but the feet are only rarely involved; in *Tbx3* homozygous mutant mice this is reversed: forelimbs have a variable loss of posterior digits and shortening or loss of the ulna, whereas the hind limbs are severely truncated with loss of all but the first digit. Molecular studies of limb development in *Tbx3* mutant mice reveal that the posterior defects are coincident with a downregulation of *sonic hedgehog* (*Shh*) in the posterior of the limbs, suggesting that *Tbx3* functions in the Shh signaling pathway to establish posterior patterning and proliferation in the limb. Alternatively, *Tbx3* may be required to maintain cell proliferation in the *Shh*-expressing posterior limb tissue (Davenport et al., 2003).

Tbx3 heterozygous mice, which are genetically analogous to patients with UMS, are viable and fertile. They show no signs of either forelimb or hind limb abnormalities, although there are mild structural abnormalities in the external genitalia of adult females. They do, however, provide a good model for the mammary gland hypoplasia that is characteristic of UMS. Heterozygous females show a variable lack of development of the first and second pairs of mammary glands, and those glands that do form are hypoplastic, thereby mirroring the nipple and mammary defects seen in patients with UMS (Figure 34.3) (Jerome-Majewska et al., 2005). Furthermore, the availability of a mouse mutation in *Tbx2* (Harrelson et al., 2004), a closely related T-box gene that is co-expressed in the mammary mesenchyme, led to the discovery of an interaction between *Tbx3* and *Tbx2* in mammary gland induction in which double heterozygotes have a more severe mammary gland reduction than single *Tbx3* heterozygotes (Jerome-Majewska et al., 2005).

The observation of an interaction between *Tbx2* and *Tbx3* may provide a link between the mammary hypoplasia in UMS and the hyperplasia of mammary tumors. *TBX2* is amplified and overexpressed in a large fraction of *BRCA1* and *BRCA2* related breast tumors (Sinclair et al., 2002). *TBX3* is overexpressed in breast cancer cell lines, and mutations have been characterized in a subset of primary breast tumors (Douglas and Papaioannou, 2013). While both *TBX2* and *TBX3* have been molecularly linked to the cell cycle machinery via the ARF-MDM2-p53 pathway in cell lines, although not in fetal or prepubertal mouse development (Jerome-Majewska et al., 2005; Brummelkamp et al., 2002; Jacobs et al., 2000), TBX3 overexpression causes mammary gland hyperplasia and an increase in mammary stem-like cells through inhibition of the NF-kappaB pathway (Liu et al., 2011). Further work, including the production of over- and misexpression alleles in mice,

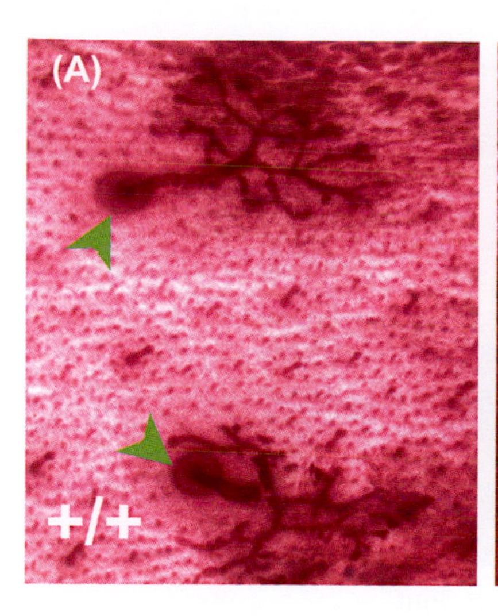

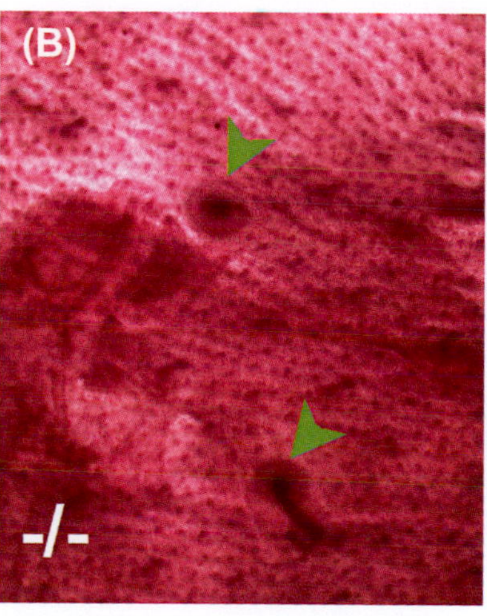

FIGURE 34.3 Mammary Gland Hypoplasia in *Tbx3* Heterozygous Mice. Carmine alum stain of the isolated skin showing the second and third mammary glands of late gestation female mouse embryos either wild type (A) or heterozygous (B) for a null mutation in *Tbx3*. Wild type embryos have arborized ductal trees attached to a nipple (arrowheads), whereas a high proportion of heterozygous mutants have only a small duct with one or no branches and sometimes lack the nipple. Double heterozygosity for *Tbx2* and *Tbx3* exacerbates this defect resulting in a higher proportion of animals with nipple and duct hypoplasia. *(Adapted and reprinted with permission from Wiley-Liss, Inc, a subsidiary of John Wiley & Sons, Inc., from Jerome-Majewska et al.,* Developmental Dynamics *234:922, © 2005 Wiley-Liss, Inc.)*

will help to determine how these genes contribute to normal mammary development and how misregulation can contribute to mammary tumorigenesis.

The mouse model of UMS, despite revealing species differences in dose sensitivity and penetrance of the phenotype in the limbs, provides valuable material for the study of UMS. By virtue of the greater severity of defects in the homozygous mutants, the mouse model for UMS has revealed additional sites of *Tbx3* activity not known or observed in UMS patients, such as the yolk sac. The mouse model also demonstrates that different structures differ in their sensitivity to haplo-insufficiency: mammary gland and genital defects are observed in heterozygous mice, but limb and yolk sac problems are seen only in homozygotes. For further exploration of these issues, the mouse model system can be used to manipulate *Tbx3* levels or to allow for tissue-specific ablation of gene function at specific times in development. One such study using a series of *Tbx3* hypomorphic and conditional mutants revealed a role for *Tbx3* in development, maturation and homeostasis of the cardiac conduction system in the fetus and the adult (Frank et al., 2012).

TBX4 and the Small Patella Syndrome

Mutations in human *TBX4* cause small patella syndrome (SPS) (Bongers et al., 2004). In this rare syndrome, skeletal defects are variably present at all axial levels of the leg, including abnormal pelvic ossification, small or missing patella, and a large gap between first and second toes, suggesting that *TBX4* is required for the correct development of all of the hind limb elements. Some patients with Okihiro syndrome (caused by mutations in *SALL4*) also present with the enlarged gap between the first toes that is characteristic of small patella syndrome, which suggests that *SALL4* may function downstream of *TBX4* in hind limb development (Kohlhase et al., 2003) (Figure 34.4).

During development, *Tbx4* is expressed primarily in the hind limb, genital papilla, mesenchyme surrounding the lung bud and trachea, and the allantois, which will form the umbilical cord and fetal component of the placenta (Chapman et al., 1996). *Tbx4* heterozygous mutants are normal, whereas homozygous mutants are embryonic lethal at midgestation. The hind limbs are initiated but fail to outgrow. However, the critical defect leading to lethality is the failure of the allantois to connect with the chorion to form the chorio-allantoic placenta (Naiche and Papaioannou, 2003). A requirement for *Tbx4* for the growth and development of proximal skeletal elements of the hind limb is revealed when the gene is conditionally mutated after limb bud initiation or specifically ablated only in the hind limbs (Naiche and Papaioannou, 2007b). Lowering the level of *Tbx4* expression in the hind limbs through mutation of a limb-specific enhancer has the effect of decreasing the size of hind limb skeletal elements (Menke et al., 2008).

In spite of the considerable contribution of *Tbx4*-expressing tissue to lung mesenchyme (Naiche et al., 2011), no lung phenotype is observed in *Tbx4* mutant embryos. However, studies that mutated both *Tbx4* and *Tbx5*, a closely related gene that is also expressed in the lung mesenchyme, show severely reduced lung growth and branching in embryos deficient for both genes (Arora et al., 2012). Recently, a serious lung phenotype, childhood-onset pulmonary arterial hypertension, has been associated with deletions or mutations of *TBX4* in small patella syndrome patients (Kerstjens-Frederikse et al., 2013), emphasizing the value of using the animal model as an indicator of potential human phenotypes that might not otherwise be recognized as being part of a syndrome. The mechanistic link between the childhood-onset pulmonary arterial hypertension and deficiency for *TBX4* has not yet been determined, but the animal model should prove valuable for exploring this connection. As with other T-box mutations, the severity of the mouse homozygous null phenotype indicates that human

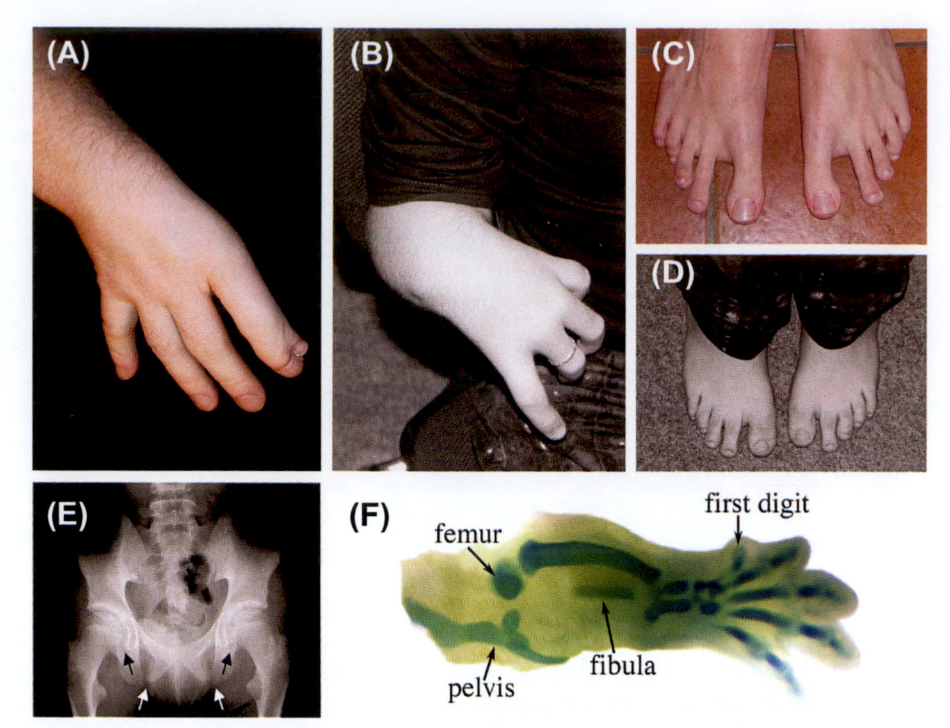

FIGURE 34.4 **Limb Malformations Caused by Mutations in T-Box Gene Pathways.** (A) Holt-Oram patient with a mutation in *TBX5* exhibits both the loss and partial duplication of the anterior digits (He et al., 2004). (B) Mutation of *SALL4*, a putative *TBX5* target, causes similar limb malformations in Okihiro syndrome (Kohlhase et al., 2003). (C) Patients with small patella syndrome, caused by *TBX4* mutations, show a characteristic gap between the first two toes and short fourth and fifth toes (Bongersv et al., 2004). (D) *SALL4* may also be a *TBX4* target, as this patient with Okihiro syndrome has similar foot abnormalities (Kohlhase et al., 2003). (E) Patients with small patella syndrome show abnormal ossifications of the pelvis (arrows) as well as foot defects (C) and the knee defects for which the syndrome is named (Bongersv et al., 2004). (F) Mice in which *Tbx4* has been ablated immediately after limb development starts also show defects at all axial levels of the hindlimb (Naiche and Papaioannou, 2007). *(A: reprinted with permission from Wiley-Liss, Inc, a subsidiary of John Wiley & Sons, Inc., from He et al.,* American Journal of Medical Genetics *126:93, © 2004 Wiley-Liss, Inc. B and D reprinted with permission from the BMJ Publishing Group from Kohlhase et al.,* J Med Genet *40:473 © 2003. C and E reprinted with permission from the University of Chicago Press from Bongers et al.,* Am J Hum Genet *174:1239 © 2004 by the American Society of Human* Genetics.*)*

homozygotes would not survive, which may be an important consideration for genetic counseling of small patella syndrome patients.

TBX5 and Holt-Oram Syndrome

Holt-Oram syndrome (HOS) is a developmental syndrome that affects heart and arm development and occurs in approximately 1 in 100,000 births. Individuals with HOS show a range of cardiovascular defects, including both atrial and ventricular septal defects and problems with cardiac conduction, as well as forelimb defects ranging from minor thumb elongation to severe arm truncation. Physical mapping in affected families linked this disorder to chromosome 12q2, after which a candidate gene approach identified *TBX5* as the causative gene (Basson et al., 1997; Li et al., 1997). Many reports of HOS cases comprising several dozen *TBX5* mutations have been published (Heinritz et al., 2005). The profound effect of *TBX5* function on heart development is also illustrated by a study of patients with non-syndromic septal defects in which mutations in *TBX5* were found in 9 out of 68 hearts, but not in peripheral blood (Reamon-Buettner et al., 2004), suggesting that somatic mutation of *TBX5* is a relatively common cause of heart malformations.

Expression studies in the mouse reveal that *Tbx5* is expressed in both the heart and the forelimb from their inception through late development. In the heart, Tbx5 was shown to interact with a number of heart-specific transcription factors including Gata4, Gata6, Mef2c, and Nkx2-5 to activate transcription of target genes (Habets et al., 2002; Ghosh et al., 2009; Maitra et al., 2009). The necessity of these interactions is supported by the fact that point mutations in *Gata4*, *Tbx5* or *Nkx2-5* that affect only their mutual interaction but not their DNA binding capacity are sufficient to recapitulate heart malformations caused by full deletion (Hiroi et al., 2001; Ding et al., 2006; Fan et al., 2003; Garg et al., 2003; Linhares et al., 2004). These factors have all been found to co-occupy the enhancers of genes specific to the cardiac gene program (He et al., 2011). Moreover, transfection of *Tbx5*, *Gata4*, and *Mef2c* was shown to be able to induce cardiomyocyte-like cells in infarct hearts, indicating the importance of *Tbx5* in differentiation of cardiac tissue (Inagawa et al., 2012). Together, these transcription factors are thought to form a core module for the cardiac differentiation transcriptional program to become active (Zhou et al., 2012).

Given this role, it is not surprising that mice lacking *Tbx5* were found to have severe malformations of the heart (Bruneau et al., 2001). Heterozygous mouse mutants closely recapitulate the HOS heart phenotype, with frequent atrial septal defects, abnormalities of the conduction system, and occasional ventricular septal defects. Several genes involved in heart development, such as *Gja5* (*Cx40*) and *Nppa* (*Anf*), are downregulated in the *Tbx5* heterozygotes, thus demonstrating that these genes are highly sensitive to *Tbx5* dosage. *Tbx5* homozygous mutant embryos have an even more dramatic phenotype, in which the heart is severely hypoplastic, does not loop, fails to express *Nppa*, and has dramatically lower levels of an array of cardiac differentiation markers. The *Tbx5* homozygous mutant heart does nonetheless express cardiac-specific transcription factors and contractile proteins, which indicates that *Tbx5* is required for cardiac differentiation, but not specification (Bruneau et al., 2001).

The mouse *Tbx5* mutation also revealed the importance of *Tbx5* in forelimb development. Mice that are heterozygous for *Tbx5* have hypoplastic wrist bones and elongated phalanges of the first forelimb digit, which is similar to what has been reported in some HOS patients. In *Tbx5* homozygous mutant embryos, the forelimb bud fails to form due to the absence of *Fgf10*, a key gene in limb initiation and outgrowth, in the forelimb field (Ahn et al., 2002). This is probably a direct effect, as *Tbx5* can directly regulate the *Fgf10* promoter (e.g., Ahn et al., 2002). The forelimb abnormalities seen in HOS patients may be the result of a partial loss of *Fgf10* function, although this remains to be shown.

In general, the mouse model of a human disorder provides opportunities to investigate the molecular function of a gene with a variety of techniques that are unavailable for use in human patients (e.g., *in situ* hybridization, tissue or organ explants). In an unusual twist, the large number of *TBX5* mutations found in HOS patients has provided a wealth of data regarding the molecular function of *TBX5* that is comparable to that provided by the mouse model. Comparison of the severity of limb defects with the severity of heart defects in human pedigrees with mutations in different parts of *TBX5* has provided suggestions regarding which protein domains are key to the phenotypes in each organ (Basson et al., 1999), although these relationships have not held up in all analyses (Brassington et al., 2003). Several point mutations in *TBX5* have been found to cause HOS, which has allowed investigators to compare normal and disease-causing variants in biochemical assays to ascertain the mechanistic nature of the causative defect. Not surprisingly, many disease-causing point mutations that ablate TBX5 DNA binding or activation domains have been linked to HOS. However, other mutations causing elongated versions of TBX5 or with mutations in the *TBX5* regulatory region have also been linked to HOS or other heart defects (Muru et al., 2011; Shan et al., 2012; Smemo et al., 2012). More remarkably, mutations in which TBX5 protein binds DNA successfully but is not capable of interaction with the homeobox transcription factor NKX2–5 or the co-activator protein WWTR1 (TAZ) are also disease-causing, which shows that transcriptional co-factors are key components of T-box transcriptional function (Fan et al., 2003; Murakami et al., 2005), as has been shown in the mouse.

Mutations in *TBX5* have not been found in all individuals diagnosed with HOS. In some cases, this may be due to causative mutations in regulatory regions of *TBX5*, which were not sequenced. In others cases, individuals have mutations in other genes that produce the same phenotype, which suggests that they may be in the same molecular pathway as *TBX5*. Mutations in *SALL4*, which is normally associated with Okihiro syndrome, are sometimes misdiagnosed as HOS (Kohlhase et al., 2003). Mouse *Sall4* has been shown to be a downstream target of *Tbx5* in both the forelimb and the heart (Harvey and Logan, 2006; Koshiba-Takeuchi et al., 2006).

An interesting family has been identified in which a hemizygous deletion encompasses both *TBX3* and *TBX5* (Borozdin et al., 2006). The affected members of this family show symptoms of both HOS and UMS, characterized by the posterior limb defects and hypoplastic nipples typical of UMS and the anterior limb defects and heart abnormalities typical of HOS. Despite the fact that *TBX3* and *TBX5* are co-expressed in regions of the heart and limb, the loss of both genes did not appear to exacerbate the phenotype of either syndrome. Although it is impossible to draw firm conclusions from a single family, the combined UMS/HOS phenotype does indicate that there is little or no redundancy between *TBX3* and *TBX5* in functions that are sensitive to haplo-insufficiency.

TBX6 and Spondylocostal Dysostosis

The effects of mutation for mouse *Tbx6* have been known for some time but mutations in the human gene, *TBX6*, have only recently been associated, in a single family, with autosomal dominant spondylocostal dysostosis (SCD) (Sparrow et al., 2013), a defect characterized by segmental defects of the vertebrae. *TBX6* was a candidate gene for this and other vertebral patterning defects due to the phenotype of mouse mutants. A hypomorphic allele, $Tbx6^{rv}$, shows disrupted somite patterning resulting in fused ribs and vertebrae, with defects mediated through regulation of the Notch/Delta signaling pathway (Watabe-Rudolph et al., 2002). Complete loss of *Tbx6* in the mouse is embryonic lethal as there are additional effects on presomitic mesoderm cell fate and left/right axis determination that are not seen in mice with the hypomorphic allele (Chapman and Papaioannou, 1998; Hadjantonakis et al., 2008). As with other T-box genes, multiple developmental processes are affected through multiple pathways, each with different dosage sensitivity. Thus the degree of correspondence of a mouse model with a human syndrome will depend on zygosity as well as the nature of the mutation and how it affects gene dosage. The human mutation which causes SCD lowers the transcriptional activation activity of the protein, presumably due to haplo-insufficiency, such that this syndrome resembles the hypomorphic $Tbx6^{rv}$ model rather than the more severe homozygous null model.

TBX15 and Cousin Syndrome

Cousin syndrome, an autosomal recessive disorder characterized by dysplasia of the pelvis and scapula with epiphyseal anomalies (Figure 34.5), dwarfism, and craniofacial malformations, was initially described in siblings born from parents who were first cousins (Cousin et al., 1982). In 2008, *TBX15* was sequenced in two unrelated girls with skeletal anomalies resembling those of the spontaneous mouse mutant, *droopy ear* ($Tbx15^{de}$) that results from *Tbx15* deficiency (Candille et al., 2004; Lausch et al., 2008; Stein, 2009). Sequencing of *TBX15* in these families revealed that Cousin syndrome results from single nucleotide deletions causing loss of the C-terminal of TBX15, while leaving the T-box intact. Although *TBX15* expression in humans has not as yet been described, the mouse expression pattern correlates well with the phenotype observed in Cousin syndrome.

Tbx15 is a member of the Tbx1 subfamily, which includes *Tbx18* and *Tbx22* (Agulnik et al., 1998; Kraus et al., 2001). *Tbx15* and *Tbx18* form homodimers, heterodimerize with one another, and interact with Groucho proteins, a family of highly conserved corepressors, to repress transcription (Farin et al., 2007). *Tbx15* is expressed in the medial regions of the developing fore and hind limbs, the craniofacial region, branchial arches, and flank (Candille et al., 2004; Agulnik et al., 1998; Singh et al., 2005b). When formation of cartilage begins at mid-gestation, *Tbx15* expression is detectable in the maxillary and mandibular processes and nasal cartilage, and is highly expressed in the joints and mesenchymal cells surrounding cartilage of the developing limbs and vertebral column. Analyses of embryonic flank regions reveal *Tbx15* expression in both dorsal and ventral skin, in the dermis, superficial mesenchyme, subcutaneous muscles, and abdominal muscles (Candille et al., 2004; Agulnik et al., 1998; Singh et al., 2005b).

While mice heterozygous for a targeted null allele of *Tbx15* appear normal and are fertile, homozygous mice die a few days after birth, likely due to impaired suckling. If allowed to suckle without wild type littermates, *Tbx15* homozygotes survive and reach adulthood but females have reduced fertility and poor nursing behavior (Singh et al.,

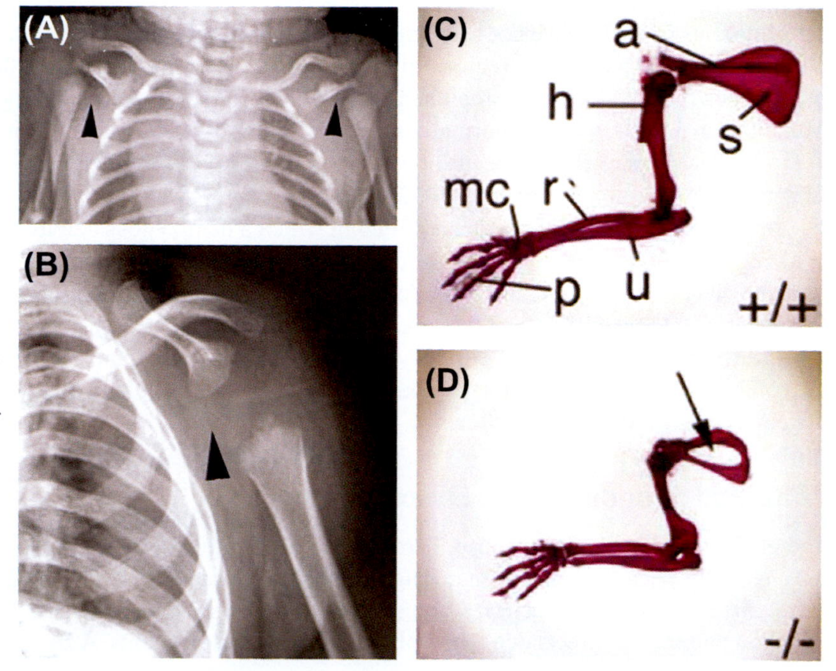

FIGURE 34.5 Similar Skeletal Features in Humans and Mice with *Tbx15* Deficiency. (A) Radiographic image of a five year old with Cousin syndrome showing aplasia of both scapular blades (arrowheads). (B) Radiographic image of a three year old with Cousin syndrome showing aplasia of the scapular blade (arrowhead). (C) Alizarin red and alcian blue staining showing the forelimb of a six week old wild type (+/+) mouse with a normal acromion (a), blade of the scapula (s), humerus (h), radius (r), ulna (u), metacarpals (mc), and phalanges (p). (D) All forelimb bones are smaller in mice homozygous for targeted disruption in *Tbx15* (–/–) as compared to wild type mice. The arrow highlights a hole in the scapular blade. *(A and B adapted and reprinted with permission from Elsevier from Lausch et al.,* Am J Hum Genet *83:649 © 2008 by the American Society of Human* Genetics. *C and D adapted and reprinted with permission from Elsevier from Singh et al.,* Mech Dev *122:131 © 2004 by Elsevier Ireland Ltd.)*

2005b). The most striking features of mice homozygous for the *Tbx15*^{de} mutation and for targeted disruption of *Tbx15* are holes in the posterior half of the scapulae (Figure 34.5) and craniofacial malformations similar to those seen in Cousin syndrome (Candille et al., 2004; Singh et al., 2005b; Kuijper et al., 2005). Additional skeletal defects in *Tbx15* mutant mice include overall reduction in bone size and changes in bone shape; abnormal spinous processes of cervical and thoracic vertebrae; and abnormally shaped bones in the head. Most defects are apparent by mid-gestation in the cartilaginous pre-skeleton of *Tbx15* homozygous null embryos. *Tbx15* is required for proliferation of mesenchymal precursor cells and chondrocytes, but not for differentiation of chondrocytes or formation of bone (Singh et al., 2005b). Thus, *Tbx15* likely serves as an important regulator of the number of mesenchymal precursor cells and chondrocytes during development of the skeleton.

Compound mutants of *Tbx15* and *Gli3*, a zinc finger transcription factor, or *Tbx15* and *Pax3*, a member of the paired box family of transcription factors, reveal genetic interactions, in the formation of the shoulder girdle. *Tbx15; Gli3* and *Tbx15; Pax3* double homozygous null embryos have such severe disruption of the shoulder girdle that only a remnant of the scapular blade remains. Compound heterozygous and homozygous mutants demonstrate that the scapular phenotype depends on gene dosage of both *Tbx15* and *Gli3* (Kuijper et al., 2005). Loss of *Pax3* in a *Tbx15* mutant background also results in a dose-dependent increase in the scapula defects (Singh et al., 2005a). While these findings highlight important genetic interactions in shoulder girdle formation, upstream mediators and downstream targets of *Tbx15* in skeletal development remain to be elucidated.

In addition to a role in skeletal development, *Tbx15* is involved in formation of the dorsal-ventral boundary of skin and coat pigmentation through control of the expression domain of *agouti signaling protein* (*a*) (Candille et al., 2004). Although there is no known human counterpart to this particular coat pigmentation phenotype, human *AGOUTI SIGNALING PROTEIN* (*ASIP*) is expressed in adipose tissue, as well as skin, and is associated with risk for melanoma and type 2 diabetes (Smith et al., 2003; Maccioni et al., 2013). This could point to as yet undiscovered aspects of Cousin syndrome if *TBX15* regulates *ASIP* in humans, similar to the mouse.

TBX19 and Adrenocorticotrophic Hormone Deficiency

One of the predominant causes of early-onset isolated adrenocorticotrophic hormone (ACTH) deficiency, which leads to adrenal insufficiency in infants, is mutation of the *TBX19* gene, also known as *TPIT*. The absence of pituitary pro-opiomelanocortin (POMC), the biochemical precursor to ACTH, leads to adrenal insufficiency in these patients. Unlike other known human T-box related disorders, loss-of-function alleles of *TBX19* that cause disease are recessive (Pulichino et al., 2003a; Couture et al., 2012). However, an association of genetic variation in *TBX19* with angry/hostility personality trait among suicide attempters highlights the potential behavioral effects of more subtle variations in *TBX19* activity (Wasserman et al., 2007).

During embryonic development in the mouse, *Tbx19* has a very restricted expression pattern in only two lineages of the pituitary gland, the corticotrophs of the anterior lobe and the melanotrophs of the intermediate lobe. Like humans, mice that are heterozygous for loss-of-function *Tbx19* mutations have no ACTH deficiency, and they have normal numbers of POMC-expressing cells in their pituitaries with no evidence of any haplo-insufficiency effects (Pulichino et al., 2003a,b). The phenotype of homozygous mutants of *Tbx19*, on the other hand, indicates that *Tbx19* is required for the terminal differentiation of corticotrophs and melanotrophs and for the upregulation of POMC. *Tbx19* activates POMC transcription in cooperation with another transcription factor, *Pitx1*, through the recruitment of co-activators to its cognate DNA target, the Tpit/Pitx regulatory element in the POMC promoter (Maira et al., 2003; Lamolet et al., 2001; Liu et al., 2001). *Tbx19* also represses gonadotroph differentiation, thus leading to the hypothesis that it controls alternative cell fates during pituitary development (Pulichino et al., 2003b; Lamolet et al., 2001; Liu et al., 2001).

TBX20 and Cardiovascular Anomalies

Congenital heart defects (CHD) are common in the human population and encompass a large variety of structural abnormalities. *TBX20* has been associated with CHD and has been shown to be mutated in cases of atrial septal defects, ventral septal defects, dilated myopathy and tetralogy of Fallot (OMIM 611363, 108800, 187500, 115200) (Posch et al., 2010; Kirk et al., 2007; Qian et al., 2008; Liu et al., 2008; Qiao et al., 2012). Study of the murine ortholog of *TBX20*, *Tbx20*, has shed light on the role that it plays in the development of CHD. *Tbx20* is expressed in the early developing heart in both the myocardial and endocardial cells, with highest expression on the ventral side of the heart tube and graded expression along the rostral-caudal axis. By mid-gestation, *Tbx20* expression is highest in the atrioventricular canal and outflow tract, specifically in the endothelial cells that have migrated into the endocardial cushion matrix (Stennard et al., 2003). Tbx20 has been shown to physically interact with other cardiac transcription factors including Nkx2–5, Gata4, and Gata5, as well as with SMAD proteins, indicating a critical role for *Tbx20* in the regulation of heart growth and differentiation, though not in specification (Stennard et al., 2003; Singh et al., 2009).

Targeted mutations have been generated that show embryonic lethality during mid-gestation due to structural heart defects (Singh et al., 2005a; Stennard et al., 2005). Embryonic hearts are linear and show no signs of looping or chamber ballooning. Additionally, there is a failure to recruit SHF cells that results in shortened heart tubes that lack a proper outflow tract. Despite the failure of heart development, mutant hearts retain markers of chamber specification and differentiation such as Tbx5, Nppa, and Gata4 in the correct locations, though at reduced levels. Interestingly, mutants show ectopic expression of *Tbx2*, a gene that can repress the formation of cardiac chambers. However, this ectopic *Tbx2* is not responsible for the lack of chamber formation in *Tbx20* mutants as double *Tbx2; Tbx20* homozygotes mutants still have chamber formation defects (Singh et al., 2009).

The importance of *Tbx20* in the development of the heart underlies the association of *TBX20* and CHD. In the human population, a number of missense and nonsense mutations in *TBX20* have been described that are associated with CHD (Kirk et al., 2007; Qian et al., 2008). While some of these mutations alter the T-box directly, others are in more distant parts of the protein raising the possibility that interactions with other factors are also critical for the proper function of *TBX20*. Other gain-of-function mutations have also been described, leading to the hypothesis that *TBX20* dosage is critical for proper function (Posch et al., 2010). Supporting this hypothesis, mutations in the *TBX20* promoter that alter the levels of *TBX20* have also been associated with CHD (Qiao et al., 2012).

TBX22 and X-Linked Cleft Palate with Ankyloglossia

X-linked cleft palate with ankyloglossia (CPX), an X-linked semi-dominant disorder affecting males and approximately one third of carrier females (Marçano et al., 2004; Pauws et al., 2009) is caused by mutations in *TBX22*. *TBX22* was initially discovered during transcriptional mapping of the X chromosome to identify genes involved in X-linked mental retardation (Laugier-Anfossi et al., 2000). The phenotypic spectrum, noted both within individual CPX families and in unrelated kindreds, includes complete or partial cleft of the secondary palate, submucous cleft, and bifid uvula with or without ankyloglossia. In addition, ankyloglossia may be present alone, especially in carrier females (Marçano et al., 2004; Braybrook et al., 2001, 2002). Mutations in *TBX22* have also been described in individuals with non-syndromic, isolated cleft palate (Suphapeetiporn et al., 2007). CPX and non-syndromic cleft palate are caused by a number of different mutations in *TBX22*, all of which interfere with DNA binding or impair sumoylation, resulting in loss of TBX22 function as a transcriptional repressor (Marçano et al., 2004; Braybrook et al., 2001; Suphapeetiporn et al., 2007; Andreou et al., 2007).

Isolation and characterization of the murine *Tbx22* gene contributed to defining the role of *TBX22* in mammalian palatogenesis. *Tbx22* is located on the mouse X chromosome and the protein shares greater than 70% amino acid sequence identity with the human TBX22 protein (Braybrook et al., 2002; Bush et al., 2002). Murine and human *TBX22* are both highly expressed in the mesenchyme of the embryonic palatal shelves. The expression domain also includes the first branchial arch, developing nose, and base of the tongue in embryos of both species. At the beginning of the fetal period in humans and in late gestation in mice, expression in the palate disappears as elevation and midline fusion of bilateral palatal shelves occurs. *TBX22* expression is maintained in the nasal cartilage, odontogenic mesenchyme, and tooth buds in both mice and humans, but expression at the base of the nasal septum persists only in mice (Braybrook et al., 2002). The expression domain of *TBX22* correlates well with the CPX phenotype. Unlike in the human, female heterozygous mutant mice have no craniofacial defects and both male and female *Tbx22* homozygous mutant mice are affected. Most *Tbx22* homozygous mutants have a submucous cleft palate instead of an overt cleft, ankyloglossia, and additional craniofacial defects, such as choanal atresia which leads to impaired suckling and lethality in about 50% of newborn mice (Pauws et al., 2009). Although the palate of *Tbx22* mutant mice lacks mineralized bone due to abnormal osteoblast maturation, the expression of genes involved in palate development, including *Msx1* and *Bmp4*, was not altered in *Tbx22* mutants (Pauws et al., 2009). Thus, the downstream targets of *Tbx22* critical for proper palatogenesis remain to be discovered.

Investigation of the molecular mechanisms involving *Tbx22* in craniofacial development revealed that *Tbx22* is regulated by genes, such as *Meningioma 1* (*Mn1*), *Fgf8* and *Bmp4*, that are known to be important for normal palate development (Liu W et al., 2008; Fuchs et al., 2010). *Mn1* and *Tbx22* have a similar expression pattern during normal palate development and *Mn1* directly activates *Tbx22* expression *in vitro*. *Tbx22* expression is reduced in palatal shelves in *Mn1* mutant embryos (Liu W et al., 2008) and like *Tbx22, Mn1* has an essential role in the maturation and function of osteoblasts (Zhang et al., 2009). Taken together, these data indicate that *Tbx22* and *Mn1* function in the same molecular pathway during palatogenesis. Impaired FGF and bone morphogenetic protein (BMP) signaling can result in cleft palate in humans and mice (Fuchs et al., 2010; Liu et al., 2005; Riley et al., 2007). *Tbx22* is induced by Fgf8 and inhibited by Bmp4 signaling (Fuchs et al., 2010). The similar phenotype of humans and mice lacking *Tbx22* provides insight about active signaling pathways in human palatogenesis.

34.4 FUTURE DIRECTIONS

Mutations in different T-box genes are a significant contributor to human malformation and disease, and they result in both common and rare developmental syndromes. Further understanding of this gene family will not only greatly enhance our understanding of the mechanisms of embryonic development, but it may also allow us to treat these disorders, particularly those that have postnatal outcomes, such as asthma or hormonal deficiencies. Advances in sequencing and elucidation of HOS etiology have already allowed for genetic screening of *in vitro* fertilized human oocytes and resulted in the birth of two healthy infants to parents with HOS (He et al., 2004).

However, much remains unknown about T-box gene function and regulation. Some target genes controlled by T-box protein transcriptional regulatory activity have been identified, but there are likely to be many more to be discovered. Biochemical research has suggested that individual T-box proteins have different targets as a result of differential preferences for half site spacing and binding partners, but little is known about how these factors play out *in vivo*. Mutations in individual T-box genes have shown defects in tissues where other T-box genes are expressed, thus indicating that different members of this family can play different roles even in the same cell. Gene swaps have shown that there can be functional redundancy between T-box genes, complicating analysis of single gene mutations. Isolated reports have indicated that T-box genes can regulate each other's expression in some tissues, but most aspects of the regulation of T-box gene expression remain uncharacterized.

Model systems are being used to address these issues. Conditional alleles are being developed to precisely ablate gene function in order to isolate the source of defects when multiple interacting affected tissues express the same T-box gene. Conditional alleles are also being used to bypass early defects and to examine the results of loss of gene function later in embryonic development or in the adult. Over- and underexpression constructs are being used to create allelic series involving varying gene dosages in different tissues. Mutations in individual T-box genes are being combined through breeding to clarify specific and redundant effects of the loss of multiple genes from the same tissue. Mutant tissues can be used in comparative transcriptome analyses and chromosome immunoprecipitation assays will isolate targets of T-box transcriptional regulation. All of these experiments will expand our knowledge of T-box gene function and of the developmental processes that they control.

CLINICAL RELEVANCE

- Most T-box genes are associated with human syndromes. These syndromes have diverse features and affect a wide variety of structures.
- Expression studies of T-box genes in the developing mouse embryo have given insights into the role that these genes may play in congenital disease.
- Mouse models that contain mutations in T-box genes largely phenocopy human mutations and have been valuable tools in understanding the development of these syndromes.
- The wide variety of interactions that T-box proteins have with other transcription factors as well as the variable levels of T-box protein expression may play a role in their function and their contribution to disease.

ACKNOWLEDGMENTS

This work was supported by grant R37HD033082 (V.E.P.) from the National Institutes of Health Eunice Kennedy Shriver Institute for Child Health and Human Development, grant K12HD000849 (N.C.D.) from the Reproductive Scientist Development Program and a grant from the Amos Medical Faculty Development Program/Robert Wood Johnson Foundation (N.C.D.). L.A.N. was supported by a fellowship from the Mandl Foundation. A. J. W was supported by a National Institute of Health Ruth L. Kirschstein predoctoral training grant GM008798.

RECOMMENDED RESOURCES

Naiche, L.A., Harrelson, Z., Kelly, R.G., Papaioannou, V.E., 2005. T-box genes in vertebrate development. Ann. Rev. Genet. 39, 219–239.

Packham, E.A., Brook, J.D., 2003. T-box genes in human disorders. Human Mol. Genet. 12, R37–R44.

Online Mendelian Inheritance in Man (OMIM): http://www.ncbi.nlm.nih.gov/entrez/query.fcgi?db=OMIM (accessed 31 Dec 2013).

Mouse Genome Informatics: http://www.informatics.jax.org/ (accessed 31 Dec 2013).

The e-Mouse Atlas Project: http://www.emouseatlas.org/emap/home.html (accessed 31 Dec 2013).

REFERENCES

Agulnik, S.I., Papaioannou, V.E., Silver, L.M., 1998. Cloning, mapping and expression analysis of TBX15, a new member of the T-box gene family. Genomics 51, 68–75.

Ahn, D., Kourakis, M.J., Rohde, L.A., Silver, L.M., Ho, R.K., 2002. T-box gene tbx5 is essential for formation of the pectoral limb bud. Nature 417, 754–758.

Andreou, A.M., Pauws, E., Jones, M.C., Singh, M.K., Bussen, M., Doudney, K., et al., 2007. TBX22 missense mutations found in patients with X-linked cleft palate affect DNA binding, sumoylation, and transcriptional repression. Am. J. Hum. Genet. 81 (4), 700–712.

Arnold, J.S., Werling, U., Braunstein, E.M., Liao, J., Nowotschin, S., Edelmann, W., et al., 2006. Inactivation of Tbx1 in the pharyngeal endoderm results in 22q11DS malformations. Development 133, 977–987.

Arora, R., Metzger, R., Papaioannou, V.E., 2012. Multiple roles of Tbx4 and Tbx5 in the developing lung and trachea. PLos Genet. 8 (8). e1002866.

Baldini, A., 2005. Dissecting contiguous gene defects: TBX1. Curr. Opin. Genet. Dev. 15, 279–284.

Bamshad, M., Le, T., Watkins, W.S., Dixon, M.E., Kramer, B.E., Roeder, A.D., et al., 1999. The spectrum of mutations in TBX3: genotype/phenotype relationship in ulnar-mammary syndrome. Am. J. Hum. Genet. 64, 1550–1562.

Bamshad, M., Lin, R.C., Law, D.J., Watkins, W.S., Krakowiak, P.A., Moore, M.E., et al., 1997. Mutations in human TBX3 alter limb, apocrine and genital development in ulnar-mammary syndrome. Nat. Genet. 16, 311–315.

Basson, C.T., Bachinsky, D.R., Lin, R.C., Levi, T., Elkins, J.A., Soults, J., et al., 1997. Mutations in human TBX5 [corrected] cause limb and cardiac malformation in Holt-Oram syndrome. Nat. Genet. 15 (1), 30–35.

Basson, C.T., Huang, T., Lin, R.C., Bachinsky, D.R., Weremowicz, S., Vaglio, A., et al., 1999. Different TBX5 interactions in heart and limb defined by Holt-Oram syndrome mutations. Pro. Nat. Acad. Sci. USA 96, 2919–2924.

Bongers, E.M.H.F., Duijf, P.H.G., van Beersum, S.E.M., Schoots, J., van Kampen, A., Burckhardt, A., et al., 2004. Mutations in the human TBX4 gene cause small patella syndrome. Am. J. Hum. Genet. 74, 1239–1248.

Borozdin, W., Bravo-Ferrer Acosta, A.M., Seemanova, E., Leipoldt, M., Bamshad, M.J., Unger, S., et al., 2006 Sep 1. Contiguous hemizygous deletion of TBX5, TBX3, and RBM19 resulting in a combined phenotype of Holt-Oram and ulnar-mammary syndromes. Am. J. Med. Genet. A. 140A (17), 1880–1886.

Brassington, A.M.E., Sung, S.S., Toydemir, R.M., Le, T., Roeder, A.D., Rutherford, A.E., et al., 2003. Expressivity of Holt-Oram syndrome is not predicted by TBX5 genotype. Am. J. Hum. Genet. 73, 74–85.

Braybrook, C., Doudney, K., Marcano, A.C.B., Arnason, A., Bjornsson, A., Patton, M.A., et al., 2001. The T-box transcription factor gene TBX22 is mutated in X-linked cleft palate and ankyloglossia. Nat. Genet. 29, 179–183.

Braybrook, C., Lisgo, S., Doudney, K., Henderson, D., Marcano, A.C.B., Strachan, T., et al., 2002 Oct 15. Craniofacial expression of human and murine TBX22 correlates with the cleft palate and ankyloglossia phenotype observed in CPX patients. Hum. Mol. Genet. 11 (22), 2793–2804.

Brummelkamp, T.R., Kortlever, R.M., Lingbeek, M., Trettel, F., MacDonald, M.E., van Lohuizen, M., et al., 2002. TBX-3, the gene mutated in ulnar-mammary syndrome, is a negative regulator of p19ARF and inhibits senescence. J. Biol. Chem. 277, 6567–6572.

Bruneau, B.G., Nemer, G., Schmitt, J.P., Charron, F., Robitaille, L., Caron, S., et al., 2001. A murine model of Holt-Oram syndrome defines roles of the T-box transcription factor Tbx5 in cardiogenesis and disease. Cell 106, 709–721.

Bush, J.O., Lan, Y., Maltby, K.M., Jiang, R., 2002. Isolation and developmental expression analysis of Tbx22, the mouse homolog of the human X-linked cleft palate gene. Dev. Dyn. 225 (3), 322–326.

Calmont, A., Ivins, S., Van Bueren, K.L., Papangeli, I., Kyriakopoulou, V., Andrews, W.D., et al., 2009. Tbx1 controls cardiac neural crest cell migration during arch artery development by regulating Gbx2 expression in the pharyngeal ectoderm. Development 136 (18), 3173–3183.

Candille, S.I., Van Raamsdonk, C.D., Chen, C., Kuijper, S., Chen-Tsai, Y., Russ, A., et al., 2004. Dorsoventral patterning of the mouse coat by Tbx15. PLos Biol. 2 (1), 0030–0042.

Carreira, S., Dexter, T.J., Yavuzer, U., Easty, D.J., Goding, C.R., 1998. Brachyury-related transcription factor Tbx2 and repression of the melanocyte-specific TRP-1 promoter. Mol. Cell. Biol. 18, 5099–5108.

Chapman, D.L., Garvey, N., Hancock, S., Alexiou, M., Agulnik, S., Gibson Brown, J.J., et al., 1996. Expression of the T-box family genes, Tbx1-Tbx5, during early mouse development. Dev. Dyn. 206, 379–390.

Chapman, D.L., Papaioannou, V.E., 1998. Three neural tubes in mouse embryos with mutations in the T-box gene, Tbx6. Nature 391, 695–697.

Chen, L., Fulcoli, F.G., Ferrentino, R., Martucciello, S., Illingworth, E.A., Baldini, A., 2012. Transcriptional control in cardiac progenitors: Tbx1 interacts with the BAF chromatin remodeling complex and regulates Wnt5a. PLos Genet. 8 (3), E1002571.

Chen, L., Mupo, A., Huynh, T., Cioffi, S., Woods, M., Jin, C., et al., 2010 May 3. Tbx1 regulates Vegfr3 and is required for lymphatic vessel development. J. Cell. Biol. 189 (3), 417–424.

Coll, M., Seidman, J.G., Muller, C.W., 2002. Structure of the DNA-bound T-box domain of human TBX3, a transcription factor responsible for ulnar-mammary syndrome. Structure 10, 343–356.

Conlon, F.L., Fairclough, L., Price, B.M.J., Casey, E.S., Smith, J.C., 2001. Determinants of T-box protein specificity. Development 128, 3749–3758.

Cousin, J., Walbaum, R., Cegarra, P., Huguet, J., Louis, J., Pauli, A., et al., 1982. [Familial pelvi-scapulary dysplasia with anomalies of the epiphyses, dwarfism and dysmorphy: a new syndrome? (author's transl)]. Arch. Fr. Pediatr. 39 (3), 173–175.

Couture, C., Saveanu, A., Barlier, A., Carel, J.C., Fassnacht, M., Fluck, C.E., et al., 2012. Phenotypic homogeneity and genotypic variability in a large series of congenital isolated ACTH-deficiency patients with TPIT gene mutations. J. Clin. Endocrinol. Metab. 97 (3). E486–EE95.

Davenport, T.G., Jerome-Majewska, L.A., Papaioannou, V.E., 2003. Mammary gland, limb, and yolk sac defects in mice lacking Tbx3, the gene mutated in human ulnar mammary syndrome. Development 130, 2263–2273.

Demay, F., Bilican, B., Rodriguez, M., Carreira, S., Pontecorvi, M., Ling, Y., et al., 2007. T-box factors: targeting to chromatin and interaction with the histone H3 N-terminal tail. Pigment Cell Res. 20 (4), 279–287.

Ding, B., Liu, C., Huang, Y., Hickey, R.P., Yu, J., Kong, W., et al., 2006. p204 is required for the differentiation of p19 murine embryonal carcinoma cells to beating cardiac myocytes. Its expression is activated by the cardiac Gata4, Nkx2, 5, and Tbx5 proteins. J. Biol. Chem. 281 (21), 14882–14892.

Dobrovolskaïa-Zavadskaïa, N., 1927. Sur la mortification spontanée de la queue che la souris nouveau-née et sur l'existence d'un caractère (facteur) héréditaire "non viable". C R Seanc. Soc. Biol. 97, 114–116.

Douglas, N.C., Heng, K., Sauer, M.V., Papaioannou, V.E., 2012. Dynamic expression of Tbx2 subfamily gnees in development of the mouse reproductive system. Dev. Dyn. 241, 365–375.

Douglas, N.C., Papaioannou, V.E., 2013. The T-box transcription factors TBX2 and TBX3 in mammary gland development and breast cancer. J. Mammary Gland Biol. Neoplasia 18, 143–147.

El Omari, K., De Mesmaeker, J., Karia, D., Ginn, H., Bhattacharya, S., Mancini, E.J., 2011. Structure of the DNA-bound T-box domain of human TBX1, a transcription factor associated with the DiGeorge syndrome. Proteins 80 (2), 655–660.

Fan, C., Liu, M., Wang, Q., 2003. Functional analysis of TBX5 missense mutations associated with Holt-Oram syndrome. J. Biol. Chem. 278 (10), 8780–8785.

Farin, H.F., Bussen, M., Schmidt, M.K., Singh, M.K., Schuster-Gossler, K., Kispert, A., 2007. Transcriptional repression by the T-box proteins Tbx18 and Tbx15 depends on groucho corepressors. J. Biol. Chem. 282 (35), 25748–25759.

Frank, D.U., Carter, K.L., Thomas, K.R., Burr, R.M., Bakker, M.L., Coetzee, W.A., et al., 2012. Lethal arrhythmias in Tbx3-deficient mice reveal extreme dosage sensitivity of cardiac conduction system function and homeostasis. Proc. Nat. Acad. Sci. 109 (3), E154–E163.

Fuchs, A., Inthal, A., Herrmann, D., Cheng, S., Nakatomi, M., Peters, H., et al., 2010. Regulation of Tbx22 during facial and palatal development. Dev. Dyn. 239 (11), 2860–2874.

Garg, V., Kathiriya, I.S., Barnes, R., Schluterman, M.K., King, I.N., Butler, C.A., et al., 2003. GATA4 mutations cause human congenital heart defects and reveal an interaction with TBX5. Nature 424 (6947), 443–447.

Ghosh, T.K., Packham, E.A., Bonser, A.J., Robinson, T.E., Cross, S.J., Brook, J.D., 2001. Characterization of the TBX5 binding site and analysis of mutations that cause Holt-Oram syndrome. Hum. Mol. Genet. 10 (18), 1983–1994.

Ghosh, T.K., Song, F.F., Packham, E.A., Buxton, S., Robinson, T.E., Ronksley, J., et al., 2009. Physical interaction between TBX5 and MEF2C is required for early heart development. Mol. Cell. Biol. 29 (8), 2205–2218.

Guris, D.L., Duester, G., Papaioannou, V.E., Imamoto, A., 2006. Dose-dependent interaction of Tbx1 and Crkl and locally aberrant RA signaling in a model of del22q11 syndrome. Dev. Cell 10, 81–92.

Guris, D.L., Fantes, J., Tara, D., Druker, B.J., Imamoto, A., 2001. Mice lacking the homologue of the human 22q11.2 gene CRKL phenocopy neurocristopathies of DiGeorge syndrome. Nat. Genet., 27.

Habets, P.E.M.H., Moorman, A.F.M., Clout, D.E.W., van Roon, M.A., Lingbeek, M., van Lohuizen, M., et al., 2002. Cooperative action of Tbx2 and Nkx2.5 inhibits ANF expression in the atrioventricular canal: implications for cardiac chamber formation. Genes Dev. 16, 1234–1246.

Hadjantonakis, A.-K., Pisano, E., Papaioannou, V.E., 2008. Tbx6 regulates left/right patterning in mouse embryos through effects on nodal cilia and perinodal signaling. Plos ONE 3 (6). e2511.

Hariri, F., Nemer, M., Nemer, G., 2012. T-box factors: Insights into the evolutionary emergence of the complex heart. Ann. Med. 44 (7), 680–693.

Harrelson, Z., Kelly, R.G., Goldin, S.N., Gibson-Brown, J.J., Bollag, R.J., Silver, L.M., et al., 2004. Tbx2 is essential for patterning the atrioventricular canal and for morphogenesis of the outflow tract during heart development. Development 131, 5041–5052.

Harvey, S.A., Logan, M.P., 2006. sall4 acts downstream of tbx5 and is required for pectoral fin outgrowth. Development 133, 1165–1173.

He, A., Kong, S.W., Ma, Q., Pu, W.T., 2011. Co-occupancy by multiple cardiac transcription factors identifies transcriptional enhancers active in heart. Pro. Nat. Acad. Sci. 108 (14), 5632–5637.

He, J., McDermott, D.A., Song, Y., Gilbert, F., Kligman, I., Basson, C.T., 2004. Preimplantation genetic diagnosis of human congenital heart malformation and Holt-Oram syndrome. Am. J. Med. Genet. 126, 93–98.

Heinritz, W., Shou, L., Moschik, A., Froster, U.G., 2005. The human TBX5 gene mutation database. Hum. Mutat. 26, 397–401.

Herrmann, B.G., 1995. The mouse Brachyury (T) gene. Semi. Dev. Biol. 6, 385–394.

Hiramoto, T., Kang, G., Suzuki, G., Satoh, Y., Kucherlapati, R., Watanabe, Y., et al., 2011 Dec 15. Tbx1: identification of a 22q11.2 gene as a risk factor for autism spectrum disorder in a mouse model. Hum. Mol. Genet. 20 (24), 4775–4785.

Hiroi, Y., Kudoh, S., Monzen, K., Ikeda, Y., Yazaki, Y., Nagai, R., et al., 2001. Tbx5 associates with Nkx2–5 and synergistically promotes cardiomyocyte differentiation. Nat. Genet. 28, 276–280.

Hu, T., Yamagishi, H., Maeda, J., McAnally, J., Yamagishi, C., Srivastava, D., 2004. Tbx1 regulates fibroblast growth factors in the anterior heart field through a reinforcing autoregulatory loop involving forkhead transcription factors. Development 131, 5491–5502.

Inagawa, K., Miyamoto, K., Yamakawa, H., Muraoka, N., Sadahiro, T., Umei, T., et al., 2012. Induction of cardiomyocyte-like cells in infarct hearts by gene transfer of Gata4, Mef2c, and Tbx5. Circ. Res. 111 (9), 1147–1156.

Jacobs, J.J.L., Keblusek, P., Robanus-Maandag, E., Kristel, P., Lingbeek, M., Nederlof, P.M., et al., 2000. Senescence bypass screen identifies TBX2, which represses Cdkn2a (p19ARF) and is amplified in a subset of human breast cancers. Nat. Genet. 26, 291–299.

Jerome, L.A., Papaioannou, V.E., 2001. DiGeorge syndrome phenotype in mice mutant for the T-box gene, Tbx1. Nat. Genet. 27, 286–291.

Jerome-Majewska, L.A., Jenkins, G.P., Ernstoff, E., Zindy, F., Sherr, C.J., Papaioannou, V.E., 2005. Tbx3, the ulnar-mammary syndrome gene, and Tbx2 interact in mammary gland development through a p19Arf/p53-independent pathway. Dev. Dyn. 234, 922–933.

Joss, S., Kini, U., Fisher, R., Mundlos, S., Prescott, K., Newbury-Ecob, R., et al., 2011. The face of Ulnar Mammary syndrome? Eur. J. Med. Genet. May-Jun; 54 (3), 301–305.

Kelly, R.G., Jerome-Majewska, L.A., Papaioannou, V.E., 2004. The del 22q11.2 candidate gene Tbx1 regulates branchiomeric myogenesis. Hum. Mol. Genet. 13 (22), 2829–2840.

Kerstjens-Frederikse, W.S., Bongers, E.M.H.F., Roofthooft, M.T.R., Leter, E.M., Douwes, J.M., Van Dijk, A., et al., 2013. TBX4 mutations (small patella syndrome) are associated with childhood-onset pulmonary arterial hypertension. J. Med. Genet. O, 1–7.

Kirk, E.P., Sunde, M., Costa, M.W., Rankin, S.A., Wolstein, O., Castro, M.L., et al., 2007. Mutations in cardiac T-box factor gene TBX20 are associated with diverse cardiac pathologies, including defects of septation and valvulogenesis and cardiomyopathy. Am. J. Med. Genet. 81 (2), 280–291.

Kispert, A., 1995. The Brachyury protein: a T-domain transcription factor. Semi. Dev. Biol. 6, 395–403.

Kispert, A., Herrmann, B.G., 1993. The Brachyury gene encodes a novel DNA binding protein. Eur. Molec. Biol. Organ J. 12, 3211–3220.

Kohlhase, J., Schubert, L., Liebers, M., Rauch, A., Becker, K., Mohammed, S.N., et al., 2003. Mutations at the SALL4 locus on chromosome 20 result in a range of clinically overlapping phenotypes, including Okihiro syndrome, Holt-Oram syndrome, acro-renal-ocular syndrome, and patients previously reported to represent thalidomide embryopathy. J. Med. Genet. 40, 473–478.

Koshiba-Takeuchi, K., Takeuchi, J.K., Arruda, E.P., Kathiriya, I.S., Mo, R., Hui, C., et al., 2006. Cooperative and antagonistic interactions between Sall4 and Tbx5 pattern the mouse limb and heart. Nat. Genet. 38, 175–183.

Kraus, F., Haenig, B., Kispert, A., 2001. Cloning and expression analysis of the mouse T-box gene Tbx18. Mech. Dev. 100, 83–86.

Kuijper, S., Beverdam, A., Kroon, C., Brouwer, A., Candille, S., Barsh, G., et al., 2005. Genetics of shoulder girdle formation: roles of Tbx15 and aristaless-like genes. Development 132, 1601–1610.

Lamolet, B., Pulichino, A.-M., Lamonerie, T., Gauthier, Y., Brue, T., Enjalbert, A., et al., 2001. A pituitary cell-restricted T box factor, Tpit, activates POMC transcription in cooperation with Pitx homeoproteins. Cell 104, 849–859.

Laugier-Anfossi, F., Villard, L., 2000. Molecular characterization of a new human T-box gene (TBX22) located in Xq21.1 encoding a protein containing a truncated T-domain. Gene 255 (2), 289–296.

Lausch, E., Hermanns, P., Farin, H.F., Alanay, Y., Unger, S., Nikkel, S., et al., 2008. TBX15 mutations cause craniofacial dysmorphism, hypoplasia of scapula and pelvis, and short stature in Cousin syndrome. Am. J. Hum. Genet. 83, 649–655.

Li, Q.Y., Newbury-Ecob, R.A., Terrett, J.A., Wilson, D.I., Curtis, A.R.J., Yi, C.H., et al., 1997. Holt-Oram syndrome is caused by mutations in TBX5, a member of the Brachyury (T) gene family. Nat. Genet. 15, 21–29.

Liao, J., Aggarwal, V.S., Nowotschin, S., Bondarev, A., Lipner, S., Morrow, B.E., 2008 Apr 15. Identification of downstream genetic pathways of Tbx1 in the second heart field. Dev. Biol. 316 (2), 524–537.

Linden, H., Williams, R., King, J., Blair, E., Kini, U., 2009. Ulnar Mammary syndrome and TBX3: expanding the phenotype. Am. J. Med. Genet. A. 149A (12), 2809–2812.

Lindsay, E.A., Botta, A., Jurecic, V., Carattini-Rivera, S., Cheah, Y.-C., Rosenblatt, H.M., et al., 1999. Congenital heart disease in mice deficient for the DiGeorge syndrome region. Nature 401, 379–383.

Lindsay, E.A., Vitelli, F., Su, H., Morishima, M., Huynh, T., Pramparo, T., et al., 2001. Tbx1 haplo-insufficiency in the DiGeroge syndrome region causes aortic arch defects in mice. Nature 410, 97–101.

Lingbeek, M.E., Jacobs, J.J.L., van Lohuizen, M., 2002. The T-box repressors TBX2 and TBX3 specifically regulate the tumor-suppressor p14ARF via a variant T-site in the initiator. J. Biol. Chem. 277, 26120–26127.

Linhares, V.L.F., Almeida, N.A.S., Menezes, D.C., Elliot, D.A., Lai, D., Beyer, E.C., et al., 2004. Transcriptional regulation of the murine Connexin40 promoter by cardiac factors Nkx2–5, GATA4, and Tbx5. Cardiovascular Res. 64, 402–411.

Liu, C., Shen, A., Li, X., Jiao, W., Zhang, X., Li, Z., 2008. T-box transcription factor TBX20 mutations in Chinese patients with congenital heart disease. Eur. J. Med. Genet. 51 (6), 580–587.

Liu, J., Esmailpour, T., Shang, X., Gulsen, G., Liu, A., Huang, T., 2011. TBX3 over-expression causes mammary gland hyperplasia and increases mammary stem-like cells in an inducible transgenic mouse model. BMC Dev. Biol. 11, 65.

Liu, J., Lin, C., Gleiberman, A., Ohgi, K.A., Herman, T., Huang, H.-P., et al., 2001. Tbx19, a tissue-selective regulator of POMC gene expression. Proc. Nat. Acad. Sci. 98 (15), 8674–8679.

Liu, W., Lan, Y., Pauws, E., Meester-Smoor, M.A., Stanier, P., Zwarthoff, E.C., et al., 2008. The Mn1 transcription factor acts upstream of Tbx22 and preferentially regulates posterior palate growth in mice. Development 135 (23), 3959–3968.

Liu, W., Sun, X., Braut, A., Mishina, Y., Behringer, R.R., Mina, M., et al., 2005. Distinct functions for Bmp signaling in lip and palate fusion in mice. Development 132 (6), 1453–1461.

Lu, R., Yang, A., Jin, Y., 2011 Mar 11. Dual functions of T-box 3 (Tbx3) in the control of self-renewal and extraembryonic endoderm differentiation in mouse embryonic stem cells. J. Biol. Chem. 286 (10), 8425–8436.

Maccioni, L., Rachakonda, P.S., Scherer, D., Bermejo, J.L., Planelles, D., Requena, C., et al., 2013. Variants at chromosome 20 (ASIP locus) and melanoma risk. Int. J. Cancer 132 (1), 42–54.

Maira, M., Couture, C., Le Martelot, G., Pulichino, A.-M., Bilodeau, S., Drouin, J., 2003. The T-box factor Tpit recruits SRC/p160 co-activators and mediates hormone action. J. Biol. Chem. 278, 46523–46532.

Maitra, M., Schluterman, M.K., Nichols, H.A., Richardson, J.A., Lo, C.W., Srivastava, D., et al., 2009 2/15/. Interaction of Gata4 and Gata6 with Tbx5 is critical for normal cardiac development. Dev. Biol. 326 (2), 368–377.

Marçano, A.C.B., Doudney, K., Braybrook, C., Squires, R., Patton, M.A., Lees, M.M., et al., 2004. TBX22 mutations are a frequent cause of cleft palate. J. Med. Genet. 41, 68–74.

Meneghini, V., Odent, S., Platonova, N., Egeo, A., Merlo, G.R., 2006. Novel TBX3 mutation data in families with Ulnar-Mammary syndrome indicate a genotype-phenotype relationship: mutations that do not disrupt the T-domain are associated with less severe limb defects. Eur. J. Med. Genet. 49 (2), 151–158.

Menke, D.B., Guenther, C., Kingsley, D.M., 2008. Dual hindlimb control elements in the Tbx4 gene and region-specific control of bone size in vertebrate limbs. Development 135, 2543–2553.

Merscher, S., Funke, B., Epstein, J.A., Heyer, J., Puech, A., Lu, M.M., et al., 2001. TBX1 is responsible for cardiovascular defects in velo-cardio-facial/DiGeroge syndrome. Cell 104, 619–629.

Mesbah, K., Harrelson, Z., Theveniau-Ruissy, M., Papaioannou, V.E., Kelly, R.G., 2008. Tbx3 is required for outflow tract development. Circ. Res. 103, 743–750.

Müller, C.W., Herrmann, B.G., 1997. Crystallographic structure of the T domain-DNA complex of the Brachyury transcription factor. Nature 389, 884–888.

Murakami, M., Nakagawa, M., Olson, E.N., Nakagawa, O., 2005. A WW domain protein TAZ is a critical coactivator for TBX5, a transcription factor implicated in Holt-Oram syndrome. Pro. Nat. Acad. Sci. USA 102 (50), 18034–18039.

Muru, K., Kalev, I., Teek, R., Sõnajalg, M., Kuuse, K., Reimand, T., et al., 2011. A boy with Holt-Oram syndrome caused by novel mutation c.1304delT in the TBX5 gene. Mol. Syndromol. 1 (6), 307–310.

Naiche, L.A., Arora, R., Kania, A., Lewandoski, M., Papaioannou, V.E., 2011. Identity and fate of Tbx4-expressing cells reveal developmental cell fate decisions in the allantois, limb, and external genitalia. Dev. Dyn. 240 (10), 2290–2300.

Naiche, L.A., Harrelson, Z., Kelly, R.G., Papaioannou, V.E., 2005. T-box genes in vertebrate development. Ann. Rev. Genet. 39, 219–239.

Naiche, L.A., Papaioannou, V.E., 2007. Multiple roles of T-box genes. In: Moody, S.A. (Ed.), Principles of Developmental Genetics. Academic Press.

Naiche, L.A., Papaioannou, V.E., 2003. Loss of Tbx4 blocks hindlimb development and affects vascularization and fusion of the allantois. Development 130, 2681–2693.

Naiche, L.A., Papaioannou, V.E., 2007. Tbx4 is not required for hindlimb identity or post-bud hindlimb outgrowth. Development 134, 93–103.

Nowotschin, S., Liao, J., Gage, P.J., Epstein, J.A., Campione, M., Morrow, B.E., 2006. Tbx1 affects asymmetric cardiac morphogenesis by regulating Pitx2 in the secondary heart field. Development 133 (8), 1565–1573.

Ouimette, J.-F., Jolin, M.L., L'honore, A., Gifuni, A., Drouin, J., 2010. Divergent transcriptional activities determine limb identity. Nat. Comm. 1, 35.

Pane, L.S., Zhang, Z., Ferrentino, R., Huynh, T., Cutillo, L., Baldini, A., 2012 Jun 1. Tbx1 is a negative modulator of Mef2c. Hum. Mol. Genet. 21 (11), 2485–2496.

Papaioannou, V.E., 2001. T-box genes in development: From hydra to humans. Int. Rev. Cytol. 207, 1–70.

Papaioannou, V.E., Goldin, S.N., 2004. Introduction to the T-box genes and their roles in developmental signaling pathways. In: Epstein, C.J., Erickson, R.P., Wynshaw-Boris, A. (Eds.), Inborn Errors of Development: The Molecular Basis of Clinical Disorders of Morphogenesis. Oxford Monographs on Medical Genetics No 49. Oxford University Press, Oxford.

Papangeli, I., Scambler, P.J., 2013 Jan 4. Tbx1 genetically interacts with the transforming growth factor-b/bone morphogenetic protein inhibitor Smad7 during great vessel remodeling. Circ. Res. 112 (1), 90–102.

Pauws, E., Hoshino, A., Bentley, L., Prajapai, S., Keller, C., Hammond, P., et al., 2009. Tbx22null mice have a submucous cleft palate due to reduced palatal bone formation and also display ankyloglossia and choanal atresia phenotypes. Hum. Mol. Genet. 18 (21), 4171–4179.

Paxton, C., Zhao, H., Chin, Y., Langner, K., Reecy, J., 2002. Murine Tbx2 contains domains that activate and repress gene transcription. Gene 283, 117–124.

Paylor, R., Glaser, B., Mupo, A., Ataliotis, P., Spencer, C., Sobotka, A., et al., 2006 May 16. Tbx1 haplo-insufficiency is linked to behavioral disorders in mice and humans: implications for 22q11 deletion syndrome. Proc. Natl. Acad. Sci. USA 103 (20), 7729–7734.

Paylor, R., McIlwain, K.L., McAninch, R., Nellis, A., Yuva-Paylor, L.A., Baldini, A., et al., 2001. Mice deleted for the DiGeroge/velocardiofacial syndrome region show abnormal sensorimotor gating and learning and memory impairments. Hum. Mol. Genet. 10, 2645–2650.

Posch, M.G., Gramlich, M., Sunde, M., Schmitt, K.R., Lee, S.H.Y., Richter, S., et al., 2010. A gain-of-function TBX20 mutation causes congenital atrial septal defects, patent foramen ovale and cardiac valve defects. J. Med. Genet. 47, 230–235.

Pulichino, A.M., Vallette-Kasic, S., Couture, C., Gauthier, Y., Brue, T., David, M., et al., 2003a. Human and mouse TPIT gene mutations cause early onset pituitary ACTH deficiency. Genes Dev. 17, 711–716.

Pulichino, A.M., Vallette-Kasic, S., Tsai, J.P.-Y., Couture, C., Gauthier, Y., Drouin, J., 2003b. Tpit determines alternate fates during pituitary cell differentiation. Gene Dev. 17, 738–747.

Qian, L., Mohapatra, B., Akasaka, T., Liu, J., Ocorr, K., Towbin, J.A., et al., 2008 Dec 16. Transcription factor neuromancer/TBX20 is required for cardiac function in *Drosophila* with implications for human heart disease. Proc. Natl. Acad. Sci. USA 105 (50), 19833–19838.

Qiao, Y., Wanyan, H., Xing, Q., Xie, W., Pang, S., Shan, J., et al., 2012 May 25. Genetic analysis of the TBX20 gene promoter region in patients with ventricular septal defects. Gene 500 (1), 28–31.

Reamon-Buettner, S.M., Borlak, J., 2004. TBX5 mutations in non-Holt-Oram Syndrome (HOS) malformed hearts. Hum. Mutat. 24 (1), 1–7.

Riley, B.M., Mansilla, M.A., Ma, J., Daack-Hirsch, S., Maher, B.S., Raffensperger, L.M., et al., 2007 Mar 13. Impaired FGF signaling contributes to cleft lip and palate. Proc. Natl. Acad. Sci. USA 104 (11), 4512–4517.

Scambler, P.J., 2010. 22q11 deletion syndrome: A role for TBX1 in pharyngeal and cardiovascular development. Pediatr. Cardiol 31, 378–390.

Shan, J., Pang, S., Qiao, Y., Ma, L., Wang, H., Xing, Q., et al., 2012 9//. Functional analysis of the novel sequence variants within TBX5 gene promoter in patients with ventricular septal defects. Trans. Res. 160 (3), 237–238.

Sinclair, C.S., Adem, C., Naderi, A., Soderberg, C.L., Johnson, M., Wu, K., et al., 2002. TBX2 is preferentially amplified in BRCA1- and BRCA2-related breast tumors. Cancer Res. 62, 3587–3591.

Singh, M.K., Christoffels, V.M., Dias, J.M., Trowe, M.-O., Petry, M., Schuster-Gossler, K., et al., 2005a. Tbx20 is essential for cardiac chamber differentiation and repression of Tbx2. Development 132 (12), 2697–2707.

Singh, M.K., Petry, M., Haenig, B., Lescher, B., Leitges, M., Kispert, A., 2005b. The T-box transcription factor Tbx15 is required for skeletal development. Mech. Dev. 122, 131–144.

Singh, R., Horsthuis, T., Farin, H.F., Grieskamp, T., Norden, J., Petry, M., et al., 2009 Aug 28. Tbx20 interacts with smads to confine Tbx2 expression to the atrioventricular canal. Circ. Res. 105 (5), 442–452.

Sinha, S., Abraham, S., Gronostajski, R.M., Campbell, C.E., 2000. Differential DNA binding and transcription modulation by three T-box proteins, T, TBX1 and TBX2. Gene 258, 15–29.

Smemo, S., Campos, L.C., Moskowitz, I.P., Krieger, J.E., Pereira, A.C., Nobrega, M.A., 2012. Regulatory variation in a TBX5 enhancer leads to isolated congenital heart disease. Hum. Mol. Genet. 21 (14), 3255–3263.

Smith, S.R., Gawronska-Kozak, B., Janderova, L., Nguyen, T., Murrell, A., Stephens, J.M., et al., 2003. Agouti expression in human adipose tissue: functional consequences and increased expression in type 2 diabetes. Diabetes 52, 2914–2922.

Sparrow, D.B., McInerney-Leo, A., Gucev, Z.S., Gardiner, B., Marshall, M., Leo, P.J., et al., 2013. Autosomal dominant spondylocostal dysostosis is caused by mutation in TBX6. Hum. Mol. Genet. 22 (8), 1625–1631.

Stein, R.A., 2009. A new TBX gene linked to human disease. Clin. Genet. 76 (1), 23–24.

Stennard, F.A., Costa, M.W., Elliott, D.A., Rankin, S., Haast, S.J.P., Lai, D., et al., 2003. Cardiac T-box factor Tbx20 directly interacts with Nkx2–5, GATA4, and GATA5 in regulation of gene expression in the developing heart. Dev. Biol. 262, 206–224.

Stennard, F.A., Costa, M.W., Lai, D., Biben, C., Furtado, M.B., Solloway, M.J., et al., 2005. Murine T-box transcription factor Tbx20 acts as a repressor during heart development and is essential for adult heart integrity, function and adaption. Development 132, 2451–2462.

Suphapeetiporn, K., Tongkobpetch, S., Siriwan, P., Shotelersuk, V., 2007. TBX22 mutations are a frequent cause of non-syndromic cleft palate in the Thai population. Clin. Genet. 72 (5), 478–483.

Tan, T.Y., Gordon, C.T., Amor, D.J., Farlie, P.G., 2010. Developmental perspectives on copy number abnormalities of the 22q11.2 region. Clin. Genet. 78 (3), 201–218.

Torres-Juan, L., Rosell, J., Morla, M., Vidal-Pou, C., Garcia-Algas, F., de la Fuente, M.A., et al., 2007. Mutations in TBX1 genocopy the 22q11.2 deletion and duplication syndromes: a new susceptibility factor for mental retardation. Eur. J. Hum. Genet. 15 (6), 658–663.

van Bueren, K.L., Papangeli, I., Rochais, F., Pearce, K., Roberts, C., Calmont, A., et al., 2010 Apr 15. Hes1 expression is reduced in Tbx1 null cells and is required for the development of structures affected in 22q11 deletion syndrome. Dev. Biol. 340 (2), 369–380.

Vance, K.W., Shaw, H.M., Rodriguez, M., Ott, S., Goding, C.R., 2010. The retinoblastoma protein modulates Tbx2 functional specificity. Mol. Biol. Cell 21, 2770–2779.

Wardle, F., Papaioannou, V.E., 2008. Teasing out T-box targets in early mesoderm. Curr. Opin. Genet. Dev. 18, 1–8.

Washkowitz, A.J., Gavrilov, S., Begum, S., Papaioannou, V.E., 2012. Diverse functional networks of Tbx3 in development and disease. WIREs Syst. Biol. Med. 4 (3), 273–283.

Wasserman, D., Geijer, T., Sokolowski, M., Rozanov, V., Wasserman, J., 2007. Genetic variation in the hypothalamic-pituitary-adrenocortical axis regulatory factor, T-box 19, and the angry/hostility personality trait. Genes. Brain Behav. 6 (4), 321–328.

Watabe-Rudolph, M., Schlautmann, N., Papaioannou, V.E., Gossler, A., 2002. The mouse rib-vertebrae mutation is a hypomorphic Tbx6 allele. Mech. Dev. 119, 251–256.

Watanabe, Y., Zaffran, S., Kuroiwa, A., Higuchi, H., Ogura, T., Harvey, R.P., et al., 2012 Nov 6. Fibroblast growth factor 10 gene regulation in the second heart field by Tbx1, Nkx2–5, and Islet1 reveals a genetic switch for down-regulation in the myocardium. Proc. Natl. Acad. Sci. USA 109 (45), 18273–18280.

Xu, H., Cerrato, F., Baldini, A., 2005. Timed mutation and cell-fate mapping reveal reiterated roles of Tbx1 during embryogenesis, and a crucial function during segmentation of the pharyngeal system via regulation of endoderm expansion. Development 132, 4387–4395.

Xu, H., Morishima, M., Wylie, J.N., Schwartz, R.J., Bruneau, B.G., Lindsay, E.A., et al., 2004. Tbx1 has a dual role in the morphogenesis of the cardiac outflow tract. Development 131, 3217–3227.

Yagi, H., Furutani, Y., Hamada, H., Sasaki, T., Asakawa, S., Minoshima, S., et al., 2003. Role of TBX1 in human del22q11.2 syndrome. Lancet 362, 1366–1373.

Zaragoza, M.V., Lewis, L.E., Sun, G., Wang, E., Li, L., Said-Salman, I., et al., 2004. Identification of the TBX5 transactivating domain and the nuclear localization signal. Gene 330, 9–18.

Zhang, X., Down, D.R., Moore MCK, T.A., Meester-Smoor, M.A., Zwarthoff, E.C., MacDonald, P.N., 2009. TMeningioa 1 is required for appropriate osteoblast proliferation, motility, differentiation, and function. J. Biol. Chem. 284, 18174–18183.

Zhou, L., Liu, Y., Lu, L., Lu, X., Dixon, R.A.F., 2012. Cardiac gene activation analysis in mammalian non-myoblasic cells by Nkx2–5, Tbx5, Gata4 and Myocd. Plos ONE 7 (10). e48028.

Zweier, C., Sticht, H., Aydin-Yaylagul, I., Campbell, C.E., Rauch, A., 2007. Human TBX1 missense mutations cause gain of function resulting in the same phenotype as 22q11.2 deletions. Am. J. Hum. Genet. 80 (3), 510–517.

Chapter 35

Craniofacial Syndromes: Etiology, Impact and Treatment

Ching-Fang Chang*,†, Elizabeth N. Schock*,†, David A. Billmire* and Samantha A. Brugmann*,†

**Division of Plastic Surgery, Department of Surgery, Cincinnati Childrens Hospital Medical Center, Cincinnati, OH; †Division of Developmental Biology, Department of Pediatrics, Cincinnati Childrens Hospital Medical Center, Cincinnati, OH*

Chapter Outline

GLOSSARY

Coloboma A condition where an individual is missing part of his or her eye structure, including, but not limited to, the iris, retina, and eyelid. Treacher Collins syndrome patients usually have coloboma of the eyelid and have a notch in the upper or lower eyelid.

Fontanelles Embryonic and early childhood structures in the skull commonly referred to as soft spots. These allow the skull plates to shift during delivery. Once the skull becomes ossified, around the age of two, the fontanelles close and become the skull sutures.

Le Fort I–III Named after the French surgeon, Le Fort fractures are common patterns of facial fractures that occur based on naturally occurring fusion patterns of facial bones.

Le Fort I results from an impact to the lower mandible in a downward direction. Fractures extend from the nasal septum, travel horizontally above the teeth, and cross below the zygomaticomaxillary junction.

Le Fort II fractures occur from a blow to the lower or mid-maxilla and usually involve the inferior orbital rim. A Le Fort II fracture has a pyramidal shape and extends from the nasal bridge at or below the nasofrontal suture through the frontal processes of the maxilla.

Le Fort III fractures result from impact to the nasal bridge or upper maxilla. These fractures start at the nasofrontal and frontomaxillary sutures and extend posteriorly along the medial wall of the orbit through the nasolacrimal groove and ethmoid bones, and are otherwise known as craniofacial dissociation.

Surgeons use the Le Fort fractures to assist in resetting bones after surgical procedures like distraction osteogenesis or craniosynostosis repair.

Micrognathia and retrognathia Conditions in which the jaw is underdeveloped. Micrognathia is specific to the lower jaw (mandible), whereas those with retrognathia have a smaller, often receding, upper (maxilla) and lower jaw.

Microtia A condition in which the ear is underdeveloped, misshapen, and lacking the external auditory meatus. There are varying degrees of microtia, from mild, in which the ears are smaller, but have all the main features of the ear, to severe, in which no ear is formed (anotia).

Tessier clefts Craniofacial clefts that are classified on the basis of their position at either the soft tissue or bone level. This is in contrast to the Van der Meulen classification system for clefts, which bases the classification upon embryogenesis. There are different types of Tessier cleft classifications that use a 0–14 numbering system, with zero designating the midline.

SUMMARY

- The craniofacial complex is derived from seven separate embryonic swellings (paired maxillary, mandibular and lateral nasal prominences and a singular frontonasal prominence) that grow and fuse together to form a seamless face.
- The craniofacial complex is derived from tissues of all embryonic origins including ectoderm, neural crest, mesoderm, and endoderm.
- The most common craniofacial defects affect the growth and development of first arch structures (maxillary and mandibular prominences) and the development of the oral cavity.
- Midline defects are caused by aberrant growth and development of the frontonasal prominence. Midline defects are among the most difficult to repair because they often involve not only cranial bones and soft tissue, but also the brain.

35.1 INTRODUCTION

"It is the common wonder of all men, how among so many million faces, there should be none alike."

Sir Thomas Browne

Impact of Craniofacial Syndromes on Society

Often our initial perception of a person is shaped by a superficial survey of his or her face. With one look, we examine; we classify; we judge. To a new face, we assign personality traits, levels of intelligence, and a place in the social/economic stratum. We see examples in literature and film where individuals with facial defects are characterized as villainous, estranged, or plebeian. These fictional archetypes captivate our society and result in preconceived notions about individuals based on the way they look. But perception is often far from the reality.

While many other types of defects or disease are internal and can be concealed from the rest of the world, a craniofacial defect cannot be hidden. Other syndromes can be dealt with through the use of medication or one-time surgery; however, individuals with craniofacial defects undergo multiple surgeries across a lifetime, even multiple surgeries a year. Within the first few years of life, the home health expenditure for children with even minor orofacial clefts is nearly 45 times higher than unaffected children (Cassell et al., 2008). Families of children with craniofacial syndromes experience financial burdens, sometimes lasting a lifetime. According to the Center for Disease Control, for cleft lip and cleft palate alone the lifetime cost for treating children born each year is estimated to be $697 million in the United States. In addition to surgery, patients may need speech therapy, extensive dental care, and counseling (Wehby et al., 2010), thus increasing the biomedical burden. As a general rule, the severity of the craniofacial defect is not an indicator of the psychological state of the individual (Pruzinsky, 1992). Those with the most severe craniofacial defects are at higher risk of developing social and psychological issues, but many have positive social and psychological interactions and attitudes (Lansdown, 1991). Those individuals confronted with negative social interactions, discrimination, and prejudice commonly experience loneliness, stress, and even shame (Pruzinsky, 1992). Thus, not only are craniofacial syndromes a medical burden, they are equally financially, socially, and psychologically burdensome.

The impact of craniofacial syndromes can extend beyond childhood and last well into adulthood. Many innovative surgical procedures are able to reduce the severity of the craniofacial defect and improve quality of life for afflicted individuals. While many focus on treatment options once the child is born, real value toward understanding these syndromes can be found by studying the embryonic formation of the face. It is in the earliest stages of life that the facial features form – slight molecular nuances giving rise to an individual and unique face. The objectives of this chapter are:

(a) to give a basic understanding to how the human face is formed under normal conditions;
(b) to describe the genetic and molecular etiology of various craniofacial syndromes;
(c) to outline animal models that assist in our study of craniofacial syndromes;
(d) to describe surgical intervention and treatment options for patients with craniofacial defects.

Overview of Craniofacial Development

The craniofacial complex begins forming as early as 21 days of development. It is then that the first progenitor cells of the facial skeleton are born and start on their path to give rise to what we recognize as a face. The major events in facial development occur throughout the 4th–11th week of gestation. It is during this time that the embryonic structures that make up the face must move and grow in concert and eventually fuse to form what we recognize as the human face (Figure 35.1).

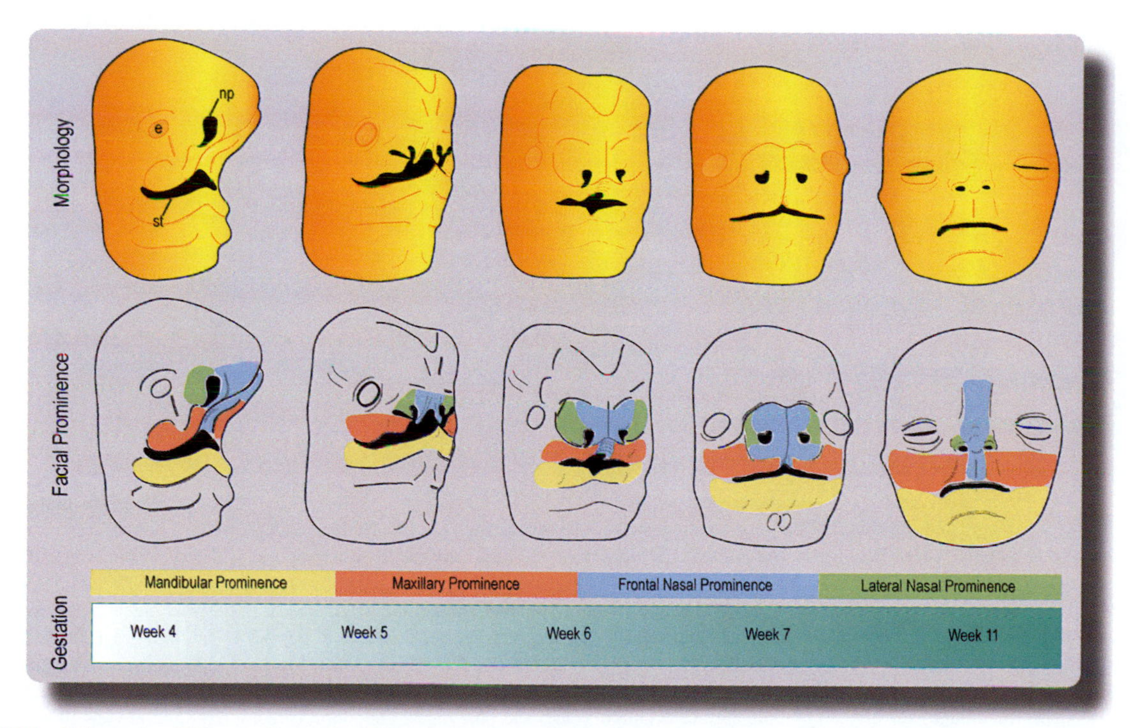

FIGURE 35.1 **Schematic of Human Craniofacial Development from 4–11 Weeks of Gestation.** Changes in morphology and growth of the facial prominences are illustrated. np, nasal pit; e, eye; st, stomodeum.

Craniofacial development is perhaps one of the most complex embryological processes, as it requires integration and interactions between multiple organ systems and tissue types. For example, development of the craniofacial complex depends on the presence of, and signals from, the developing brain. Forebrain signaling centers that sit adjacent to the developing face determine growth zones and shape the facial midline. Signals from the prechordal mesoderm induce visual and auditory development and contribute to the development of the upper third of the face. Thus, craniofacial development is highly dependent upon proper development of adjacent tissues.

Another complexity associated with craniofacial development is its embryonic origin. Many of the structures of the craniofacial complex are of a dual embryonic origin. For example, the skeletal scaffold and most of the connective tissues including blood vessels originate from a group of cells called the cranial neural crest (see also Chapter 18), whereas the musculature and some posterior parts of the skull originate from mesoderm (see also Chapter 12). Furthermore, many ectodermal specializations, including teeth and hair, require molecular interactions between the overlying ectoderm and the underlying mesenchyme (see also Chapter 22). Thus, proper orientation of tissues and expression of signaling molecules in adjacent tissues are essential for the formation of many hallmarks of the craniofacial complex.

A third specific feature of craniofacial development is the rapid and dramatic morphogenic changes that occur during development. A staggering amount of cell movement and reorganization, both passive cell displacement and active cell migration, occur in a very short period of time (Figure 35.1). These morphogenic changes take the face from being a compilation of disconnected parts to an integrated, fully recognizable human face. For example, the cells that form the facial skeleton (cranial neural crest) must migrate from the dorsal neural tube into the facial prominences. Their subsequent proliferation then guides the growth and movement of the facial prominences from a lateral to a medial location. Formation of the palate also requires highly choreographed tissue movements. Early in development, the palatal shelves extend vertically on either side of the tongue, rotate to a horizontal plane dorsal to the tongue and fuse. Any event, genetic or environmental, that disrupts the rate, timing, or the extent of these complex cellular behaviors can result in a craniofacial anomaly. In this chapter we will introduce the craniofacial complex and the basic aspects of its development.

35.2 FIRST ARCH DERIVED STRUCTURES

Formation of the First Arch

The pharyngeal arches are bilaterally orientated, paired embryonic structures that give rise to various structures of the head and neck (Figure 35.2). Proper growth and development of the first pharyngeal arch has a tremendous impact on

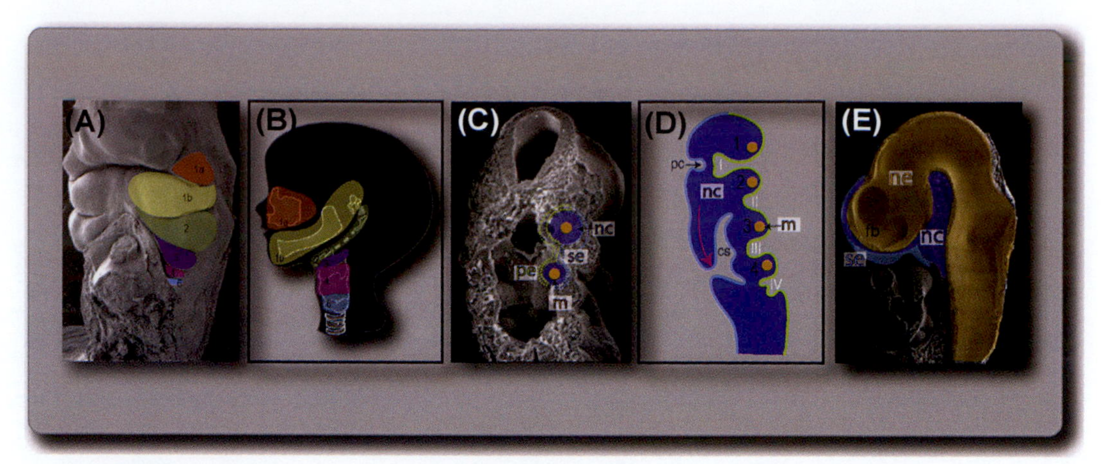

FIGURE 35.2 **The Pharyngeal Arches.** (A) Scanning electron image of the pharyngeal arches, indicated by different colors. The first arch has split into the maxillary (1a, orange) and mandibular (1b, yellow) components. The second arch (2, green); third arch (3, purple); fourth arch (4, pink) and the residual sixth arch (6, blue). (B) The skeletal derivatives of the pharyngeal arches. The arches and their corresponding derivatives are marked by matching colors. (C) The tissue composition of the pharyngeal arches. All the pharyngeal arches are surrounded by an epithelial layer. The outer epithelial layer is derived from the surface ectoderm (se, dotted blue line) whereas the inner (lumenal) epithelium originates from the pharyngeal endoderm (pe, dotted green line). Neural crest cells (nc, purple) migrate into the epithelial lined pouches and surround a core of mesoderm (m, orange). (D) The outgrowth of the pharyngeal clefts (pc) and pouches (white Roman numerals). The second arch grows downward (pink arrow) overlapping the second, third and fourth pharyngeal clefts forming the lateral cervical sinus of Hiss (cs). As previously depicted, the arches (1, 2, 3, 4) are composed of a core of neural crest (nc, purple) and mesoderm (m, orange) surrounded by both surface ectoderm (blue) and pharyngeal endoderm (green). (E) Sagittal view illustrating the tissue components of the frontonasal prominence. The frontonasal prominence consists of an outer layer of surface ectoderm (se, blue), and an inner layer of neural crest (nc, purple). The lumenal lining of the frontonasal prominence is the neural ectoderm (ne, orange) of the forebrain (fb) as opposed to the pharyngeal endoderm in the pharyngeal arches.

craniofacial development, as the majority of craniofacial anomalies involve aberrant development of this structure (Gorlin et al., 1990). As development continues, the first arch subdivides into a maxillary and mandibular component, and grows medially. These components then fuse to one another, as well as to other facial prominences.

The first arch is composed of an organized group of tissues of various embryonic origins. The central core of the first arch is mesodermally derived mesenchyme. First arch mesenchyme contributes to head musculature, connective tissues and dermis of the skin. Surrounding the mesodermally derived mesenchyme is a second, neural crest-derived mesenchyme. First arch neural crest mesenchyme gives rise to the skeletal elements of the first arch including the mandible and the maxilla. The mesenchyme is encased outwardly by the surface ectoderm and inwardly by the pharyngeal endoderm. Surface ectoderm gives rise to the skin and is required for the induction of teeth. The pharyngeal endoderm gives rise to the oro- and laryngopharyngeal mucosa, pharyngeal taste buds, and minor salivary glands.

The Maxillary Prominence

The maxillary prominences give rise to the upper jaw (maxilla), lateral aspects of the upper lip, and the secondary palate (Figures 35.1 and 35.2). Most developmental texts support the view that the maxillary prominence arises from only a small, cranio-ventral portion of the first arch. However, more recent fate mapping studies hypothesize that maxillary components arise from a separate maxillary condensation that forms between the eye and the maxillo-mandibular cleft (Lee et al., 2004). The maxillary prominences must grow medially and fuse to other facial prominences to establish seamless continuity of the upper jaw and lip. Medially, the maxillary prominence will fuse to the lateral nasal and frontonasal prominence to give structure to the nostril and philtrum of the upper lip. Within the developing oral cavity, the bilateral maxillary prominences (now referred to as the palatal shelves) will fuse medially with one another to give rise to the secondary palate. The secondary palate separates the nasal passage from the oral cavity and pharynx. Initially, the palatal shelves are lined by the ectoderm that encases the maxillary prominences. Prior to fusion, this epithelium is partially sloughed off, exposing the more basal medial edge epithelium (MEE). As the maxillary derived palatal shelves grow towards the midline, the MEE of each shelf approximates and forms the midline epithelial seam (MES). The MES is then removed through the process of mesenchymal to epithelial transition, which establishes confluence between the mesenchyme of the palatal shelves. A myriad of genetic, mechanical, and/or environmental factors can affect the growth and fusion of the maxillary prominences. Disruptions in these processes result in palatal insufficiency (inadequate outgrowth of the maxillary prominences) and subsequent clefts in the lip and/or palate.

The Mandibular Prominence

The mandibular prominence is also a first arch derivative. The major derivatives of the mandibular prominence are the lower lip, lower jaw and the anterior two-thirds of the tongue. Similar to the maxillary prominences, the mandibular prominences begin as bilaterally paired structures that must grow medially and fuse together to form a seamless lower jaw (Figures 35.1 and 35.2). The mandibular prominences fuse medially with one another and cranio-laterally they fuse to the maxillary prominences to form the corners of the mouth. Unlike the maxillary prominences, disruptions in the fusion process of the mandibular prominences are extremely rare.

The scaffolding of the lower jaw begins with a unique cartilaginous structure that runs along the medial aspect of the mandible from the middle ear region to its distal tip, which is called Meckel's cartilage. Meckel's cartilage is formed prior to formation of the ossified mandible and almost completely degenerates prior to birth. Therefore, Meckel's has been recognized as a temporal structure of the lower jaw before the formation of the mandible in mammals. Upon the fusion of the mandibular prominences, the distal ends of Meckel's cartilage fuse to form the mandibular symphysis. The distal to intermediate portion of Meckel's is invaded by the membranous ossification of the mandible, which forms the permanent supporting structure of the lower jaw. After resorption of the cartilaginous matrix, the proximal part of Meckel's transforms into fibroblasts forming a dense connective tissue of the sphenomandibular ligament (Ishizeki et al., 1996; 2001).

Development of the tongue also relies on proper formation of the mandibular prominence. At the site of fusion of the mandibular prominences, the median tongue bud, (tuberculum impar) and two lateral lingual swellings form. The lingual swellings expand and give rise to the mucosa and anterior two-thirds of the tongue. The taste buds form as a result of interactions between the overlying oral ectoderm and underlying mesenchyme, and the tongue musculature forms via interactions between neural crest and mesodermally derived mesenchyme.

Syndromes of the First Arch

Treacher Collins Syndromes

Treacher Collins syndrome (TCS, also known as Mandibulofacial Dysotosis or Franceschetti-Klein syndrome) affects the first and second pharyngeal arches. It presents clinically with bilateral Tessier Clefts 6, 7 and 8 and a hypoplastic mandible. Penetrance ranges from very mild to severe with a compromised airway, colobomas of the eyelids, downward slanting eyelids (palpebral fissures), microtia or atresia of the ear, middle ear abnormalities, bony abnormalities of the zygoma, maxilla, mandible, and pterygoids. There is also 30% to 40% incidence of clefting of the lip and/or palate (Gorlin et al., 1990; Dixon et al., 2006).

Consistent with the phenotype, the etiology of TCS is rooted in aberrant development of cranial neural crest cells, the progenitors of the facial skeleton. The survival of neural crest cells is compromised in TCS patients. In both zebrafish and murine models of TCS there is increased apoptosis specifically in the neural crest progenitor cells and neuroectoderm (Dixon et al., 2006; Weiner et al., 2012). Additionally, there is a decrease in proliferation of neuroectoderm, neural crest derived mesenchyme, and paraxial mesenchyme cells (Dixon et al., 2006). Consistent with the specificity of the affected cell populations, cranial neural crest derivatives such as the facial bones and the middle/external ear are missing or severely hypoplastic.

Approximately 1:50,000 newborns are diagnosed with TCS. Of these individuals, 40% have inherited the disease while the other 60% have *de novo* mutations (Passos-Bueno et al., 2009; Sakai and Trainor, 2009). The majority of TCS cases are caused by mutations in *TCOF1*, a gene that encodes the nucleolar protein, Treacle. Small deletions have been detected throughout the gene, thus any mutation causing truncation of the Treacle protein results in TCS (Schlump et al., 2012). Haplo-insufficiency of *TCOF1* results in full penetrance of TCS (Dixon et al., 2006).

Treacle regulates ribosome biosynthesis and plays an important role in ribosome maturation, specifically in the neuroectoderm and neural crest cells. Treacle mutations leading to TCS can disrupt several of the protein's functions. Truncations of the Treacle protein result in the loss of its nuclear import signal and the subsequent inability of the protein to localize to the nucleolus (Winokur et al., 1998). Reduced levels of nucleolar Treacle then cause a decrease in the expression of rRNA genes essential for ribosome biosynthesis. Low levels of mature ribosomes impede the cells ability to generate enough protein to meet the metabolic demands of the highly migratory, highly proliferative neural crest (Dixon et al., 2006). In addition to *TCOF1* mutations, mutations in *POLR1D* and *POLR1C*, genes that encode for subunits of RNA polymerase I and III, have been linked to TCS. *POLR1D* and *POLR1C* TCS mutations exhibit an autosomal recessive pattern of inheritance (Dauwerse et al., 2011). Phenotypes for these patients are indistinguishable from those with *TCOF1* mutation, suggesting involvement in a common pathway.

It is suggested that disruption in the ribosome biosynthesis pathway can arrest the cell cycle during epithelial to mesenchymal transition (EMT), an essential step in neural crest development (Dixon et al., 2006) (see also Chapter 18). Aberrant

EMT could result in increased apoptosis and decreased neural crest cell proliferation. This hypothesized mechanism for TCS is supported in zebrafish models, in which expression changes are detected in genes involved in migration, apoptosis, cell proliferation, and craniofacial development (Weiner et al., 2012). The mechanism through which Treacle is involved in the regulation of these genes is not yet clear. Mutations in *POLR1D* and *POLR1C* prevent the dimerization of RNA polymerase I and III subunits (Dauwerse et al., 2011). Failure to dimerize can have drastic consequences on the enzymes ability to properly translate rRNA into a ribosomal subunit. Thus, although POLR1D and POLR1C have distinct roles from Treacle, it is likely that defects in these proteins affect the same biological pathway and mechanism relative to the onset of TCS. Further studies regarding the molecular mechanism of TCS are now possible, as a neural crest cell line has been created from induced pluripotent stem cells (iPSCs) taken from TCS patients (Menedez et al., 2013).

Surgical intervention/treatment: Treatment for TCS varies with the presentation of the disorder and can range from virtually none to multiple stages of reconstruction. The first concern in the surgical treatment of TCS is always the airway. The hypoplastic mandible can cause obstruction of the airway and require intervention at birth with immediate intubation followed typically by either tracheostomy and/or moving the mandible forward (distraction osteogenesis). Depending on the extent of the defects, or the absence of eyelids, urgent eyelid reconstruction to protect the globe may be necessary. Once the child is stabilized with a safe airway and eye protection, a staged reconstruction can be planned based on the childs presentation, development, and needs. Further surgery may include reconstruction of the small cheekbones (hypoplastic zygomas), orbital walls, eyelids, and further reconstruction of the hypoplastic mandible with advancement, often in conjunction with repositioning, rotation and advancement of the maxilla. Surgeries are tailored for the individual; however, most undergo orthodontic treatment, bone grafts, ear reconstruction and have bone conduction hearing aids. Although there are no pharmacological compounds currently used in the treatment of TCS, ongoing research suggests that p53 is a potential target for therapeutic options, as an increase in apoptosis is responsible for the phenotype (Jones et al., 2008). As with all pharmacological treatment options, specific targeting is of the utmost importance. Specific targeting of the cranial neural crest cells may prove difficult. Currently there is no prenatal diagnosis available for TCS.

Miller/Nager

Individuals with Miller and Nager syndromes (postaxial acrofacial dysostosis syndrome, POADS) have craniofacial defects and limb anomalies. Craniofacial defects primarily affect derivatives of the first and second pharyngeal arch. In the mandible, the most significant and severe defects are characterized by micrognathia or retrognathia with a posteriorly rotated mandible. Individuals also have a hypoplastic malar area and dental defects including agenesis of permanent teeth. Other significant craniofacial defects include: anklosoed tongue (abnormal anchoring of tongue to the floor of the mouth), down-slanted palpebral fissures, hypertelorism, protruding and small ears, and absence or clefting of the soft palate (Halonen et al., 2006; Bernier et al., 2012) (Figure 35.3). Unlike Treacher Collins in which the defects are restricted to the face, individuals with POADS also have limb defects. Nager syndrome patients may have shortened forearms, vertebral anomalies and decreased movement of the elbow. Miller syndrome patients tend to have more severe limb anomalies with postaxial (ulnar) deficiencies of the limbs and small, distorted lower extremities that may require amputation and use of prosthesis (Miller et al., 1979). While both Nager and Miller syndromes are both classified as POADS, they have a unique etiology and the defective genes responsible for the syndrome differ.

Nager syndrome (NS) is an extremely rare disorder. The most common mutation associated with NS is within the *SF3B4* gene. *SF3B4* is a highly conserved gene that encodes the SAP49 protein (Bernier et al., 2012). The actual mutation in the gene of NS patients varies throughout the sample population; frameshift, missense, nonsense, and splicing mutations have all been detected (Bernier et al., 2012). It is hypothesized that haplo-insufficiency of *SF3B4* results in NS, but there have been reports supporting both an autosomal dominant and an autosomal recessive mode of inheritance with varying degrees of penetrance and expressivity (Halonen et al., 2006). While the exact mechanism related to SAP49 mode of action in NS individuals has not been determined, SAP49 has been shown to function through two distinct mechanisms *in vivo*. First, SAP49 functions in a splicesome complex required for intron removal from pre-mRNA (Bernier et al., 2012). In a second hypothesized mechanism, SAP49 is suggested to have a role outside of the nucleus. It binds directly to the intercellular domain of BMPR-1A, the receptor for BMP-2/4 ligands (Watanabe et al., 2007). The BMPs have been shown to be important for osteogenesis. *In vitro* studies suggest that over-expression of *SF3B4* delays or inhibits osteogenic and chondrocytic differentiation (Watanabe et al., 2007). As the underlying bone structure in the face of NS patients is disrupted, this provides insights into the molecular mechanism responsible for the observed craniofacial defects.

Although the phenotypes of NS and Miller syndrome (MS) are very similar, the causative gene for each disease differs. MS is caused by compound heterozygous mutation in the *DHODH* gene. Inheritance of MS is autosomal recessive and missense mutations in the gene are most commonly observed (Rainger et al., 2012). *DHODH* encodes for an enzyme called

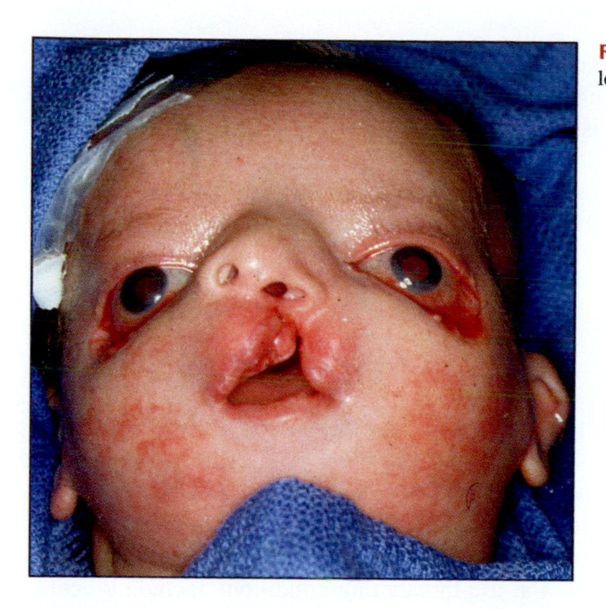

FIGURE 35.3 **Child with Miller Syndrome.** Note down-slanting palpebral fissures, low set ears and clefting.

dihydroorotate dehydrogenease (DHODH) that localizes to the inner mitochondrial membrane (Kinoshita et al., 2011). There is an upper and lower threshold for DHODH activity, as greater than 50% activity results in a normal phenotype, but too little activity leads to death *in utero* (Rainger et al., 2012). DHODH, and other members of the *de novo* pyrimidine synthesis pathway are expressed in the limbs and pharyngeal arches of mice (Rainger et al., 2012). This specific expression pattern is consistent with the MS phenotype. In addition, these areas are highly proliferative and may have a greater requirement for DNA replication reagents. In the absence of sufficient amounts of functional DHODH, it is likely that the cells are unable to keep up with the metabolic requirements for increased proliferation. If this is the case, then it follows that an increase in apoptosis would be observed in these areas, similar to Treacher Collins. It has been noted that dysfunction in DHODH causes disrupted cell migration, decreased proliferation, and increased apoptosis through inhibition of NF-κB activity, which provides preliminary support for the above hypothesis (Kinoshita et al., 2011).

Surgical intervention/treatment: From a technical standpoint, Miller and Nager syndromes are a severe manifestations of Treacher Collins syndrome with the additional concerns of associated limb abnormalities. As with TCS, the management of both Miller and Nager syndromes almost invariably requires tracheostomy and eyelid reconstruction immediately after birth, followed by staged reconstruction as was described for Treacher Collins.

35.3 MIDFACE

Formation of the Midface

The frontonasal prominence (also known as medial nasal prominence) establishes the facial midline. The frontonasal prominence (FNP), which is composed of the forehead, the bridge and tip of the nose, the philtrum, and the primary palate, is comprised of a cranial neural crest-derived mesenchyme (which gives rise to the majority of the skeletal elements of the face), surface (facial) ectoderm, and neuroectoderm from the prosencephalic region. Early in development, the FNP is separated into the medial nasal prominence and two lateral nasal prominences by the emergence and invagination of the nasal pits. The medial nasal prominence, also known as intermaxillary segment, acts as the facial "keystone" as it is the site of fusion between itself and the bilateral maxillary prominences. Fusion between these three structures forms recognizable facial features such as the philtrum of the upper lip and primary palate (Figure 35.1) (Sperber et al., 2010).

The development of the FNP is largely affected by the interactions between forebrain neuroectoderm, neural crest cells and facial ectoderm. In the early stage of facial formation, the facial prominences sit adjacent to the anterior neural tube. The growth and development of the neural tube into the regionalized central nervous system physically impacts the growth of the facial prominences in early stage human embryos (Diewert et al.,1993; Diewert and Lozanoff, 1993a,b). In addition to the physical effects from the expanding brain, the signaling interactions between neuroectoderm, neural crest cells, and surface ectoderm are also important for the development of the FNP. The frontonasal ectodermal zone (FEZ), defined by a Sonic Hedgehog (Shh) expression domain in the facial ectoderm that is flanked by a fibroblast growth factor 8 (FGF8) expression domain, greatly influences the growth of the FNP (Hu et al., 2003). Most of the information we have regarding the molecular signaling that guides FNP growth and development was obtained through the use of avian model systems. In avians, the FEZ is established by a series of inductive events that culminate with a Shh-dependent signaling center in the developing telencephalon. It is this Shh activity in telencephalon, along with yet to be determined signaling in migrating neural crest cells, that initiate Shh expression in the FEZ. The FEZ is the initial site for FNP outgrowth. Proliferation of frontonasal neural crest cells is promoted by Shh activity in the FEZ. The FEZ also controls the dorsoventral patterning within the face at later development stages. The FEZ has been identified in mammals (Hu and Marcucio, 2009), and Shh is expressed in a similar domain during human midfacial growth. Furthermore, mutations in human *SHH* are tightly linked to midline defects such as holoprosencephaly (Odent et al., 1999).

Midline Syndromes

Holoprosencephaly (HPE)

Holoprosencephaly (HPE) represents a spectrum of malformations characterized by specific but variable defects of the brain and midface (Roessler et al., 1996; Belloni et al., 1996). It is one of the most common congenital malformations, occurring in 1:250 fetuses and approximately 1:10,000 to 1:16,000 live births (Matsunaga and Shiota, 1977; Edison and Muenke, 2003; Leoncini et al., 2008). Eighty percent of HPE cases present with facial malformations (Solomon et al., 1993; Geng and Oliver, 2009). Facial malformations of HPE range from severe conditions including cyclopia, ethmocephaly, cebocephaly, and premaxillary agenesis with median cleft lip to mild, non-life-threatening conditions including ocular hypotelorism, midfacial hypoplasia with a flat nasal bridge, cleft lip and/or palate and a single central maxillary incisor (Muenke, 2001) (Figures 35.4 and 35.5). Neurologically, HPE is classified by aberrant cleavage of the embryonic forebrain. The plane of disrupted cleavage, as well as timing of the impairment, determines the type and severity of the HPE phenotype. Impaired sagittal cleavage manifests as failure or incomplete separation of the cerebral hemispheres. Impaired transverse cleavage manifests as a failure to separate the telencephalon from diencephalon. Finally, impaired horizontal cleavage manifests as a failure to form the olfactory bulbs and optic vesicles.

Associated brain malformations also occur with different degrees of severity, ranging from total lack of vesicle cleavage to microforms of HPE (e.g., alobar, semilobar, lobar, microvariant). Clinicians usually adhere to the dogma that "the face predicts the brain." That is, the severity of the facial malformation is indicative of the severity of the brain defects. However, with the improved understanding of phenotypes and genetics of HPE, there is a growing appreciation that HPE is a multi-factorial disease spectrum that requires not only a genetic mutation within a certain locus, but also an interaction of this mutation with factors within the patients genetic background and perhaps environmental insult (Roessler and Muenke, 2010).

The phenotypes associated with all forms of HPE are tightly restricted to the growth and development of forebrain neuroectoderm and the facial midline. As mentioned earlier in this chapter, the midface develops from the FNP. The forebrain serves as a scaffold on which the FNP develops. The FNP is unique among the facial prominences as it is the only singular, midline prominence and its growth and development is tightly coupled to that of the developing forebrain. The proximity of the FNP and the developing forebrain allow for molecular signals emanating from the forebrain neuroectoderm to affect FNP growth and development. Specifically, signals from the forebrain are hypothesized to affect the proliferation/survival of neural crest-derived cranial mesenchyme within the FNP. A significant reduction in the amount of neural crest-derived mesenchyme in the midfacial region results in hypoplasia of the FNP and subsequent loss of midline structures (Shiota and Yamada, 2010). In contrast to the FNP, the other facial prominences (maxillary and mandibular prominences) are typically

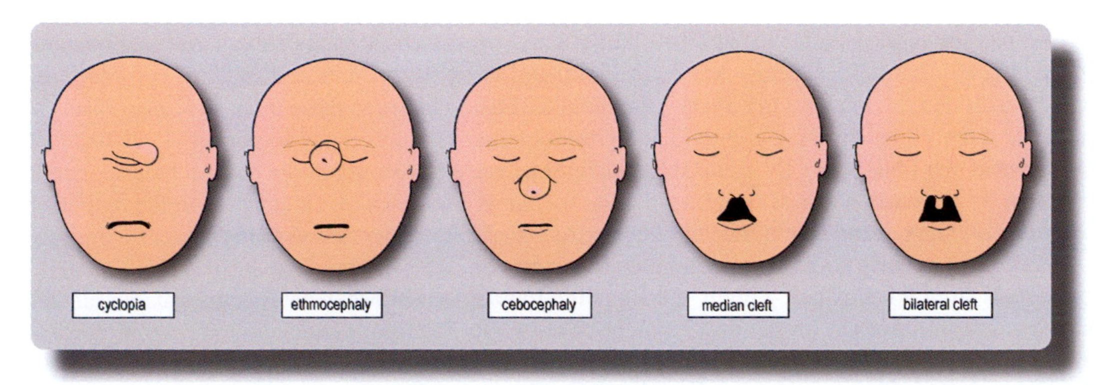

FIGURE 35.4 **Schematic Representation of the Holoprosencephaly Spectrum.** HPE forms are arranged from most severe (cyclopia) to least severe (bilateral cleft).

unaffected in HPE. The physical location and cellular makeup of these prominences prevent the forebrain signaling center from exerting a molecular influence on their growth.

The majority of HPE cases are sporadic in nature, resulting from exposure to teratogens or *de novo* genetic mutations; however, familial cases of HPE are also well documented. Approximately 13 genetic loci have been linked to HPE (Roessler and Muenke, 2010) and mutations in nine genes have been identified in patients with HPE (Muenke and Cohen, 2000; Muenke and Beachy, 2000; Krauss, 2007; Roessler et al., 2009a). These loci and genes represent signaling pathways that play an essential role in forebrain and facial development.

Although the majority of HPE cases are of an unknown etiology, some genetic progress has been made in the past decade. Mutations in many genes within and associated with the Sonic Hedgehog (Shh) pathway are among the most commonly associated with HPE. Variants in the *SHH* gene itself are the most commonly detected mutations in a sample of liveborn HPE patients (Roessler et al., 1996; 1997); however, a significant number of HPE cases are associated with receptors and modifiers of the pathway. Mutations in the Shh receptor Patched (Ptch) (Ming et al., 2002; Ribeiro et al., 2006), the downstream transcription factor Gli2 (Wannasilp et al., 2011; Roessler et al., 2005; 2003), the ligand transporter Dispatched (Disp1) (Roessler et al., 2009b; Ma et al., 2002) and the transmembrane protein Cdo (Zhang et al., 2006) have all been identified as causal genetic elements in HPE cases. Disruptions within the *Shh* locus or in these receptors and co-factors disrupt the formation of the prechordal plate, which subsequently causes defects in axial patterning and forebrain defects (Warr et al., 2008). In mice, defects in the Shh co-receptor Smoothened (Smo), which are associated with the response component of the Shh signaling pathway, result in neuroectoderm that is not competent to respond to the Shh signal and lead to HPE (Fuccillo et al., 2004). Taken together these examples support a hypothesis that any disruption in the generation or transduction of a Shh signal within the developing forebrain can result in a form of HPE.

Additional mutations within genes that are hypothesized to interact with the Shh pathway have also been linked to human HPE cases. After Shh itself, the second most common detectable genetic mutations in HPE patients occur within the transcription factor ZIC2 (Roessler et al., 2009a). The molecular mechanism of Zic2 mediated HPE remains unclear. One hypothesis is that Zic2 functions within the Shh pathway. Thus, defects in Zic2 cause HPE via disruption of Shh transduction. However, recent genetic studies have argued against this hypothesis. The alternative molecular mechanism for Zic2-mediated HPE hypothesizes that Zic2 functions within the embryonic organizer at mid-gastrulation (Warr et al., 2008). This hypothesis places Zic2 function, relative to onset of HPE, upstream of Shh mediated prechordal plate development. Heterozygous mutations in the homeodomain containing transcription factor Six3 is the third most common defect in HPE patients (Wallis et al., 1999). Six3 is essential for the development of the anterior neural plate and eye field in humans. Additionally, it is hypothesized that Six3 functions to regulate Shh expression in the ventral forebrain (Geng et al., 2008; Geng and Oliver, 2009) via physical interactions with the Shh forebrain enhancer. Failure of this interaction results in compromised Shh expression in the ventral forebrain (Jeong et al., 2008).

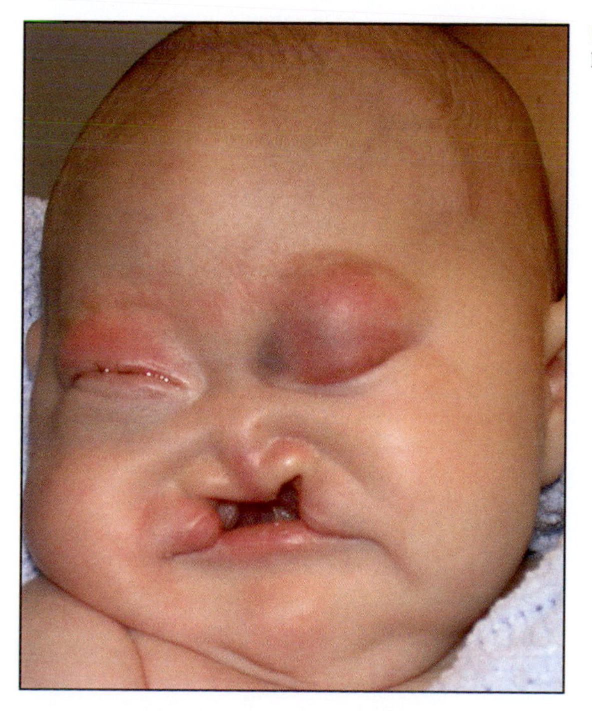

FIGURE 35.5 **Child with Holoprosencephaly.** Note median clefting and ocular hypotelorism.

There is an additional set of HPE-associated mutations that either indirectly regulate the Shh pathway or disrupt other pathways that regulate forebrain development. *TGIF1* (Thymine/Guanine-Interacting Factor) maps to the HPE4 locus, and heterozygous loss-of-function TGIF1 mutations are associated with HPE (Gripp et al., 2000). TGIF1 interacts with Transforming Growth Factor (TGF) β-activated Smad proteins (Bertolino et al., 1995; Wotton et al., 1999). Recent mouse models suggest that Tgif regulation of the Nodal-Smad2 pathway is required for the correct balance between Gli3 and Shh activity in the Shh pathway. Mutations in TGIF limit its regulation of Nodal and thus result in inappropriate levels of Shh activity in the forebrain (Taniguchi et al., 2012). The GPI-anchored protein TDFG1 and the transcription factor FoxH1 (forkhead box H1) function in the Nodal pathway and have also been linked to HPE (Roessler et al., 2008; McKean and Niswander, 2012). Disruption of the GPI-anchored protein Cripto (mouse) and TDGF1 (human ortholog) result in HPE. Both factors are hypothesized to act as obligate Nodal co-factors involved in TGFβ signaling.

Phenotypic and genetic complexities associated with the HPE spectrum have made a genotype/phenotype correlation difficult. Recent studies that expanded the diagnostic approach to include more in-depth analysis of the proband, the probands family and family history have hypothesized that the most severe cases of HPE are more likely associated with mutations in either *SHH* or *ZIC2* (Solomon et al., 2010).

A plethora of animal models have been used to study HPE (Geng et al., 2009) (Table 35.1). Most models are transgenic mouse or fish lines that represent various subclasses of HPE. These models disrupt the function of one of the nine linked genes or the major pathways associated with the condition. In addition there are drug-inducible models that utilize small

TABLE 35.1 Animal models of HPE

Species	Gene/Pathway		References
Mouse	Nodal	Nodal Signaling	Lowe *et al.*, 2001
	Cripto3		Chu *et al.*, 2005
	Smad2		Nomura and Li, 1998
	ActR2A		Song *et al.*, 1999
	Gdf1		Andersson *et al.*, 2006
	Bmp4	Bmp Signaling	Spoelgen *et al.*, 2005
	Bmpr1a/b		Fernandes *et al.*, 2007
	Chd		Bachiller *et al.*, 2000
	Nog		Anderson *et al.*, 2002
	Tsg		Petryk *et al.*, 2004; Zakin and De Robertis, 2004
	Shh	Shh Signaling	Chiang *et al.*, 1996
	Disp 1		Ma *et al.*, 2002
	Smo		Zhang *et al.*, 2001
	Gli2		Hardcastle *et al.*, 1998
	Megalin		Willnow *et al.*, 1996
	Gas 1		Seppala *et al.*, 2007
	Cdo		Zhang *et al.*, 2006; Cole and Krauss, 2003
	Talpid3		Bangs *et al.*, 2011
	Fgf8	Fgf Signaling	Storm *et al.*, 2006
	Fgfr1		Gutin *et al.*, 2006
	Otx2		Jin *et al.*, 2001
	Hnf3b		Jin *et al.*, 2001
	Six3		Geng *et al.*, 2008
	Zic2		Warr *et al.*, 2008
	ACTB		Cheng *et al.*, 2006
	Tgif		Kuang *et al.*, 2006
	Cdc42		Chen *et al.*, 2006
Chicken	Talpid3		Buxton *et al.*, 2004
Zebrafish	cyp26a1		Gongal and Waskiewicz, 2008

molecules which inhibit the key pathways linked to HPE (Table 35.2) (Mercier et al., 2013). Although these models can be useful for studying the molecular mechanism of the condition, it is important to recognize that they do not reliably reconstruct the genetic architecture of human HPE due to the genetic complexities associated with the human genome.

Surgical intervention/treatment: Clinically, holoprosencephaly occurs in a wide spectrum. Those that present to a clinician typically have hypotelorism (narrowed midface with close set eyes), a midline cleft lip and/or palate with a small to nonexistent central strut of the nose (columella). HPE infants tend to have low muscle tone, poor feeding, and failure to thrive. They may also have an underlying endocrine deficiency, requiring assessment and treatment. Most HPE patients succumb during the first or second year of life. Surgical correction of the lip to decrease the oral and nasal opening is usually delayed until the child is strong enough to tolerate the surgery and its consequences. Depending upon whether the child can manage the smaller opening and associated airway problems, some cases may necessitate a tracheostomy to reestablish the airway. A small subset of the children are of normal intelligence, and those cases are managed as any cleft lip and palate child.

Opitz BBB

Opitz BBB syndrome (OS) is characterized by craniofacial, cardiac, laryngo-tracheo-esphogeal, and urogenital defects. While phenotypic variability occurs frequently among OS patients, nearly all have midline defects, specifically hypertelorism (expansion of the facial midline). Other common syndromic features include: cleft lip/palate, cardiac abnormalities, imperforate or ectopic anus, and some instances of developmental delay/mental retardation (Cordero and Holmes, 1978; Cappa et al., 1987; Robin, 1995). OS is genetically heterogeneous, as both mutations in the *MID1* gene on the X chromosome and mutations at the 22q11.2 locus cause the same disease phenotype. Nearly 50% of all OS cases are the X-linked form of the disease. Female carriers have hypertelorism, but few other abnormalities beyond that (De Falco et al., 2003). In contrast, their male offspring have the much more severe phenotype.

Many different tissue types and organ systems are affected in OS patients. Expression studies in avian and murine models have shown that Mid1 is ubiquitously expressed (Richman et al., 2002; Dal Zotto et al., 1998). Robust *Mid1* expression is observed in the head mesenchyme, especially in the FNP and brachial aches (Richman et al., 2002; Dal Zotto et al., 1998). Lower levels of *Mid1* are also detectable in the central nervous system, eyes, and midgut (Richman et al., 2002; Dal Zotto et al., 1998). These regions of expression correspond to the locations where the most severe and common phenotypes occur. Avian studies have shown that there is partial, but not complete overlap of *Mid1's* expression domain with migrating neural crest cells. This partial co-expression may serve to explain the severe craniofacial phenotype. Interestingly, species variations exist regarding *Mid1's* expression. Avians have very robust expression of *Mid1* in the limbs and heart, whereas mice lack any cardiac expression (Richman et al., 2002; Dal Zotto et al., 1998; Quaderi et al., 1997). Further variation in expression is likely in humans, as very few OS patients have limb defects. Although not all OS patients present with cardiac

TABLE 35.2 Drugs that induce HPE like phenotypes

Drug	Signaling Pathway	Molecular Mechanism	References
Cyclopamine	Hedgehog	Inhibitor of Shh via Smo binding	Mercier *et al.,* 2013
SB-505124	Nodal	Inhibitor of TGF-β type I receptors	Mercier *et al.,* 2013
Ectopic Bmp 4/5	Bmp	Stimulates Bmp signaling	Golden *et al.,*1999
Retinoic Acid	Retinoic Acid	Stimulates retinoic acid signalling	Sulik *et al.,*1995
Ochratoxin A	Unknown	Multiple and varied	Wei and Sulik, 1993
BM15.766	Cholesterol synthesis	Inhibits 7-dehydrocholesterol conversion to cholesterol	Lanoue *et al.,* 1997
Ethanol	Multiple	Apoptosis; Inhibition of retinoic acid synthesis; Disruption of Shh signaling; Free radial damage; Intracellular calcium changes	Féré,1899; LePlat,1913; Stockard,1910; Sulik & Johnston,1982/1983; Webster *et al.,*1983; Chen & Sulk,1996 Schambra *et al.,*1990; Godin *et al.,*2010; Loucks & Ahlgren,2009; Deltour *et al.,*1996;Debelak-Kragtorp *et al.,*2003; Sulik, 1984

defects, nearly 35% of cases do have cardiac defects making increased expression of *Mid1* in the heart biomedically relevant (De Falco et al., 2003).

Defects in the *MID1* gene are the most commonly observed among OS patients. *MID1* is located at locus Xp22.3 and codes the MID1 protein (Quaderi et al., 1997). Mutations in this gene in OS patients are not specific for a domain; however, most patients have a truncated form of the protein which causes a loss of function. Genotype/phenotype correlations have not indicated a direct link between the specific mutation and individual phenotype (Fontanella et al., 2008). Monozygous OS twins harboring the same mutation presented with variations in severity of syndromic features (De Falco et al., 2003). Although the nature of the mutation does not seem to be responsible for the properties of the phenotype, some of the functions and protein-protein interactions of each domain of MID1 have been characterized. MID1 is an E3 ubiquitin ligase that facilitates ubiquitin-mediated degradation. MID1 also specifically binds microtubule complexes via a B-Box1 domain (Trockenbacher et al., 2001). Mutations within this domain prevent proper transport along the microtubules. It has been suggested that MID1 transport along microtubules is important for establishing cell polarity and/or directing migration (Aranda-Orgilles et al., 2008a,b). Mutations in the C-terminal region of MID1 impede binding to microtubules. Failure to bind to microtubules leads to the formation of cytoplasmic aggregates. Interestingly, the C-terminal domain of MID1 also associates with RNA, suggesting that MID1 may help to direct and coordinate translation along the cytoskeleton (Aranda-Orgilles et al., 2008a,b). It may be that this function of MID1 is important for localizing mRNAs to a specific portion of the cell, contributing to the emergence of cell polarity or directional migration. While the function of normal and mutant MID1 has been under study for over 20 years, the direct contribution of the MID1 complex to the pathogenesis of OS remains unknown. It is hypothesized that Sonic Hedgehog (Shh) is a direct or indirect target of MID1 (Granata and Quaderi, 2003) however, the relevance and important of this interaction in the pathogenesis of OS is only theoretical at this time.

Surgical intervention/treatment: Opitz BBB is a genetic disorder involving midline structures and presents with hypertelorism, cleft lip and palate, cleft of the larynx and cleft of the penis (hypospadias) in males. The patients are managed like any other cleft lip and palate patients, with special attention being paid to correcting the laryngeal cleft, if necessary. The lip is repaired in early infancy, the palate at approximately one year of age and the hypospadias at approximately one to two years of age. It is rare that the hypertelorism is of such a degree that it necessitates correction in these children. Severity of hypertelorism is frequently determined by measuring the distance between the medial corners of the eyes (intercanthal distance). The average intercanthal distance is between 28–34 mm. Distances up to the high 30s are usually well tolerated; however, when distances are over 40 mm, surgery is usually considered. Timing of the surgical correction for hypertelorism can vary depending upon the type of surgery required.

Ciliopathies

Ciliopathies are a group of human genetic diseases associated with primary cilia, a microtubule-based organelle that extends from the cell surface and transduces molecular signals from the extracellular environment (Goetz and Anderson, 2010). Defects in primary cilia often result in pleiotropic effects, and thus patients with ciliopathies exhibit a broad spectrum of symptoms. Baker and Beales reviewed 102 conditions involving known or possible ciliopathies and identified nine common clinical features, including retinitis pigmentosa, renal cystic disease, polydactyly, *situs inversus*, mental retardation, hypoplasia of corpus callosum, dandy-walker malformation, posterior encephalocoele, and hepatic disease. They hypothesized that the likelihood of a disease being a ciliopathy is increased with the number of core clinical features present within an individual. With further evaluation of the 102 conditions Baker and Beales put forth, it was found that 30% of ciliopathies are defined primarily by craniofacial phenotypes (Zaghloul and Brugmann, 2011). Of these "craniofacial ciliopathies," 70% display defects in the facial midline, with a high frequency in hypertelorism.

Although primary cilia extend from almost all cells in the body, the clinical features observed in patients with ciliopathies highlight the importance of primary cilia in the development of certain types of tissue. The phenotypes associated with craniofacial ciliopathies include midline, cephalic, oral and dental anomalies (Zaghloul and Brugmann, 2011). Transgenic mice whose primary cilia were specifically deleted in neural crest cells showed midline defects including widened FNP and cleft secondary palate (Kolpakova-Hart et al., 2007), suggesting that neural crest cells require primary cilia for proper facial development. Primary cilia may be utilized by neural crest cells during early migration, as shown by a study in zebrafish, in which knockdown of Bardet-Biedel syndrome (*bbs*) by morpholinos resulted in neural crest migration defects and craniofacial phenotypes (Tobin et al., 2008). Brain malformations are a common feature of craniofacial ciliopathies, suggesting a role of primary cilia in neurodevelopment. Mice with mutations in genes that encode ciliary components essential for ciliogenesis exhibit malformed brains (Gorivodsky et al., 2009; Stottmann et al., 2009; Willaredt et al., 2008; Tran et al., 2008). It has been suggested that these ciliary components are required for cilia-dependent Hedgehog signaling in the regulation of dorsal/ventral and anterior/posterior patterning during brain development. As for dental tissue, it has been shown that some ciliary genes are expressed in odontoblasts, and primary cilia are extended from odontoblasts, parallel to the dentin walls (Thivichon-Prince et al., 2009). The importance of primary cilia in tooth development was supported by the fact that *Evc*$^{-/-}$ mutant mice have defects in molar development (Nakatomi et al., 2013).

Primary cilia play an important role in the context-dependent regulation of multiple signaling pathways and are considered to be a cellular antenna. Signaling pathways such as Hedgehog, Wnt, and PDGF have all been linked to primary cilia. Hedgehog signaling plays an important role in craniofacial development, in that Shh is expressed in the ventral neuroectoderm, facial ectoderm and foregut endoderm, and is required for neural crest cell proliferation, survival, migration, and patterning (reviewed in Zaghloul and Brugmann, 2011). Aberrant Shh expression often leads to a spectrum of patterning defects within the facial midline. The role of primary cilia in regulating Shh signaling has been extensively studied in the past decade. Multiple components of the Shh signaling pathway, including receptors and co-receptors, localize to primary cilia. Primary cilia are also required for converting Gli proteins to a full-length activator and truncated repressor (Hui and Angers, 2011). Depending on tissue context, loss of primary cilia could have gain-of-function (e.g., hypertelorism, polydactyly) or loss-of-function phenotypes (e.g., brain malformations). The Wnt pathway is also linked to the primary cilia and is important in craniofacial development. Wnt activity is important for the fusion of facial prominences, and disruptions in this pathway result in facial clefting, micrognathia, and hypertelorism (Song et al., 2009; Lan et al., 2006; Brugmann et al., 2007). The linkage between primary cilia and Wnt signaling has been reported in multiple model systems (reviewed in Zaghloul and Brugmann, 2011). However, it is less clear how primary cilia regulate Wnt compared to Hedgehog signaling. There is some evidence suggesting the involvement of ciliary proteins in switching between canonical and non-canonical Wnt signaling (Gerdes et al., 2007; Simons et al., 2005) and regulation of downstream components (Corbit et al., 2008).

Most of the genetic mutations identified in patients with ciliopathies encode ciliary components (Waters and Beales, 2011). Oro-facial-digital symptom type I (OFDS1), an X-linked dominant disorder characterized by facial, oral cavity and digit abnormalities, is caused by mutations in the *OFD1* gene on chromosome Xp22 (Ferrante et al., 2001; Thauvin-Robinet et al., 2006). *OFD1* encodes a centrosomal protein localized to the basal bodies of primary cilia (Romio et al., 2004). Most mutations identified in OFDS1 patients often result in truncated, non-functional gene products (Thauvin-Robinet et al., 2006). In a murine model system, knockout of *Ofd1* results in abnormal or absent primary cilia and reduced Shh activity in the ventral neural tube (Ferrante et al., 2006). In zebrafish, knockdown of *Ofd1* affects primary cilia formation and convergent extension, consistent with the role of primary cilia in regulation of non-canonical Wnt signaling (Ferrante et al., 2009).

Meckel-Gruber syndrome (MKS) is an autosomal recessive perinatal lethal disorder characterized by severe craniofacial defects, polycystic kidneys, polydactyly, and heart defects. Mutations in *MKS1, MKS2/TMEM216, MKS3/TMEM67, RPGRIP1L, CEP290,* and *CC2D2A* have been identified as genetic causes for the six MKS subtypes (Logan et al., 2011). MKS1 is localized to basal bodies and is required for ciliogenesis. *Mks1* loss-of-function mice (krc mutants) phenocopy human MKS type 1. Data from *Mks1* mutant mice indicate that the Shh pathway is the molecular mechanism underlying most phenotypic defects in MKS type 1 (Weatherbee et al., 2009). Mutations of *TMEM216* are found in both MKS type 2 and Joubert syndrome (JS) type 2 (Valente et al., 2010). TMEM216 is a transmembrane protein localized to basal bodies and is required for ciliogenesis and centrosomal docking (Valente et al., 2010). MKS type 3 is associated with mutations in *TMEM67* (or *Meckelin*), which encodes a transmembrane protein localized to the ciliary membrane. Mutations in *CEP290* have been linked to a variety of ciliopathies including MSK type 4, JS, Senior-Loken syndrome, nephronophthisis and Bardet-Biedel syndrome. *CEP290* encodes a centrosomal protein localized to basal bodies, but the exact cellular function of CEP290 is not clear. Mutations in *RPGRIP1L* are associated with MKS type 5 and Joubert syndrome. *RPGRIP1L* encodes a RPGR-interacting protein 1-like protein, which localizes to basal bodies and may be involved in stabilization of Disheveled for non-canonical Wnt/PCP signaling (Mahuzier et al., 2012). *CC2D2A* was identified as the genetic causes of MKS type 6 and Joubert syndrome type 9. The biological function of CC2D2A is not characterized. However, the calcium-binding domain (C2) predicts it has a function in sensing calcium signals.

Ellis-van Creveld (EVC) syndrome is an autosomal recessive disorder characterized by craniofacial defects, short limbs, short stature, polydactyly, congenital heart defects, and nail and teeth dysplasia. The genetic mutations in EVC syndrome are mapped to two non-homologous genes *EVC1* and *EVC2* (Ruiz-Perez et al., 2000; 2003). Evc1 and Evc2 are ciliary proteins mutually required for their co-localization to basal bodies and ciliary membrane, and are both positive regulators for Shh signaling (Ruiz-Perez et al., 2007; Blair et al., 2011). Evc1/Evc2 are required for dissociation of the Sufu/Gli3FL complex upon Hedgehog activation (Caparros-Martin et al., 2013). It is hypothesized that Evc1/Evc2 are involved in regulating Hedgehog signaling during endochondral bone formation (Caparros-Martin et al., 2013).

Surgical intervention/treatment: Ciliopathies encompass a broad range of disorders, a number of which are fatal or associated with multiple congenital anomalies including facial clefting, polydactyly (extra digits), renal cysts, and neurological disorders. Treatment, correlated with developmental progress, is directed towards the specific deformity (i.e., repair of the cleft, correction of limb deformity, treatment of the renal, hepatic, and cardiac disease). Hypertelorism associated with ciliopathies is frequently to such a degree (intercanthal distance over 40 mm) to recommend correction. Oro-facial-digital syndrome, as the name implies, involves structures of the oral cavity, face and hands. They often present with a cleft palate or cleft lip/palate. Intraorally they often have oral frenulas; thick bands of tissue connecting the cheeks to the gums. The lip and palate are treated as in any child with clefts of this nature. The oral frenulas are divided to allow for greater oral mobility. The hand deformities include polydactyly, syndactyly, and brachydactyly (short fingers). Treatment again would be removal of the extra digits and separation of fused digits. Brachydactyly is not treated. Of special interest in these children is the polycystic renal disease that requires evaluation and treatment. Ellis-van Creveld syndrome may be associated with cleft lip and palate, polydactyly, skeletal dysplasia, dental abnormalities, epispadias (dorsal clefting of the penis), and undescended testicles. Treatment is centered on repair of the clefts, both oral and penile. Frontonasal dysplasia is associated with major midline deformities of the face resulting in significant hypertelorism frequently associated with agenesis of the corpus callosum. Treatment usually requires correction of the hypertelorism and nose usually around the age of eight to ten years.

35.4 CRANIAL VAULT

Formation of Cranium

The skull, or cranium, can be subdivided into two parts: the neurocranium and the viscerocranium. The neurocranium, which encases the brain, is composed of the cranial vault (or calvaria) and cranial base. The viscerocranium contributes to the skeleton of lower face and lower jaw (mandible). Neurocranial development initiates from a three-layered membranous structure surrounding the brain called the meninges. The meninges are derived from mesoderm and neural crest-derived mesenchyme, and are subdivided into outer ectomeninx and inner endomeninx. Several bones of the cranial vault, including the frontal and parietal bones, form via intramembranous ossification centers initiated at the outer ectomeninx. Sutures form on the edges of the opposed bones, whereas fontanelles are the fibrous tissue at the intersection between sutures. These membranous junctions are still present at birth and provide the flexibility for postnatal bone growth. The closure of posterior fontanelles (the junction of sagittal and lambdoid suture) occurs by three months postnatally and the closure of the anterior fontanelles (the junction of sagittal, coronal and frontal sutures) by 20 months postnatally.

The initial development of the skull bones is tightly regulated by signals from the underlying brain. One of these signals is the tensile strain generated by the growing brain. In order to accommodate this growth and to compensate for the tensile strain, the cranial bone has to grow and remodel to fit the shape of the brain. The functional matrix hypothesis was first proposed by Moss et al. in 1960, describing bone growth as a result of external forces (strain) rather than intrinsic genetic factors. Although the mechanisms are still unknown, it is suggested that signaling pathways regulated by extracellular matrix molecules and the cytoskeleton may be involved. The functional matrix hypothesis is supported by clinical observations that the absence of brain (anencephaly) results in a roofless skull (acalvaria), fissured cranium (cranioschisis), or absence of skull (acrania).

Syndromes of the Cranial Vault

Cleidocranial Dysotosis

Cleidocranial dysplasia (CCD) is a rare, autosomal dominant genetic disease characterized by skeletal abnormalities. The common clinical features include open or delayed closure of fontanelles and sutures; midface hypoplasia; hypoplastic or aplastic clavicles, which result in hypermobile shoulders; dental abnormalities including supernumerary teeth, unerupted primary teeth and malocclusion. Brachydactyly and moderate short stature are also features of CCD patients (Gorlin et al., 1990). Other medical problems include hearing loss, increased susceptibility to sinus and ear infections and osteoporosis. Intelligence is not affected in CCD patients.

To date, mutations in *RUNX2* are the only know genetic cause of CCD. *RUNX2* encodes for a transcription factor containing a DNA-binding domain in the N-terminus and a nuclear matrix-associated regulatory domain in the C-terminus. RUNX2 is required for several key steps during skeletal development and maintenance, including osteoblast differentiation, chondrocyte hypertrophy, and vascular invasion of cartilage. Therefore, aberrant function of RUNX2 affects both endochondral and intramembranous bone formation (Zheng et al., 2005; Schroeder et al., 2005). Deletion, missense, nonsense, and frameshift mutations of *RUNX2* have been identified in CCD patients. The majority of these mutations are clustered in the N-terminal DNA-binding domain (Otto et al., 2002). A murine model system of CCD suggested a requirement of RUNX2 dosage for normal bone development. Homozygous mutant mice with RUNX2 activity at 70–79% showed CCD phenotypes, whereas heterozygous mutant mice with 79–84% of RUNX2 activity appeared normal (Lou et al., 2009). This finding implies that the level of RUNX2 activity may correlate with the severity of CCD phenotypes in human.

In CCD patients, mutations in RUNX2 mostly affect the cranial vault and teeth, which are derived from interactions between ectoderm, mesoderm and neural crest cells during development. RUNX2 is a master regulator for preosteoblast differentiation from mesenchymal progenitors. Transgenic mice over-expressing *Runx2* showed osteopenia (low bone mineral density) (Liu et al., 2001), whereas *Runx2* null mice had no bone but a normal cartilage pattern with perturbed chondrocyte maturation (Komori et al., 1997; Otto et al., 1997). Despite the well-known cellular function of Runx2, our understanding of Runx2 regulation remains nebulous, since multiple signaling pathways involved in skeletal development including BMP/TGF, Wnt, FGF, and parathyroid hormone (PTH) can regulate Runx2 activity (Schroeder et al., 2005).

Surgical intervention/treatment: Cleidocranial dysotosis is characterized by a prominent forehead (frontal bossing), a rather impressive anterior frontal soft spot (fontanelle), mild hypertelorism and absent clavicles. Despite the fact that the fontanelle is large and remains patent well into adolescence, it does not require treatment, nor does the mild hypertelorism. Associated loose joints require monitoring. Likewise, the patients have delayed eruption of the permanent adult teeth as well as numerous supernumerary teeth, which require dental management.

Craniosynostosis

Craniosynostosis results from premature ossification and fusion of one or multiple cranial sutures. It is a common malformation with prevalence of 1 in 2,500 births (Gorlin et al., 1990). Premature fusion of cranial sutures results in deformation of the skull (Figure 35.6). The skull normally grows perpendicularly to the sutures, whereas the skull of craniosynostotic patients shows a compensatory growth parallel to the closed suture. The sagittal suture is the most commonly affected suture in craniosynostosis (in approximately 40–58% of cases). The premature closure of the sagittal suture results in an increase in anterior-posterior diameter of skull (dolichocephaly or scaphocephaly) (Figure 35.6). Coronal synostosis accounts for 20–29% of cases, and is the second most common form of craniosynostosis. Unilateral fusion of the coronal suture results in an asymmetric skull (plagiocephaly) (Figures 35.6 and 35.7), whereas bilateral fusion of the coronal suture results in brachycephaly and severe cases result in cloverleaf skull. Metopic synostosis is due to the fusion of two frontal bones and is characterized by a triangularly shaped forehead and a prominent ridge along the forehead (trigonocephaly) (Figure 35.6). Lambdoidal synostosis is the least common type of craniosynostosis. Most cases are unilateral and result in asymmetric posterior plagiocephaly. Fusion occurring in multiple sutures accounts for approximately 5% of

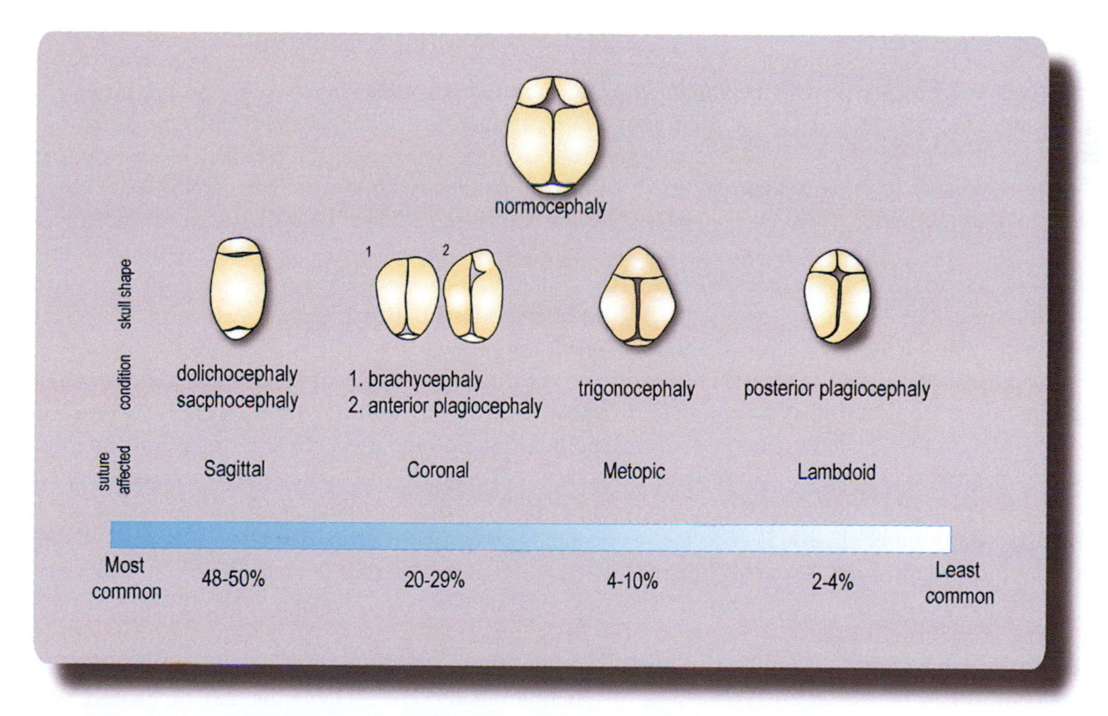

FIGURE 35.6 **Schematic Representation of Various Types of Craniosynostosis.** Resulting skull shape, affected sutures and prevalence are noted.

FIGURE 35.7 **Child with Unilateral Fusion of the Coronal Suture.** Note plagiocephaly.

craniosynostoses. Pansynostosis involves the fusion of all sutures. In severe cases, protrusion of the brain through the open anterior and parietal fontanelles results in a cloverleaf skull (Kleeblattschädel) phenotype.

Advances in our understanding of the human genetics of craniosynostosis, along with studies of transgenic mouse models over the past two decades have illuminated the biology and mechanisms of normal cranial suture development (Morriss-Kay and Wilkie, 2005). Sutures are spaces between opposed calvarial bones, where osteogenic mesenchymal cells reside. The sutural edges are lined by osteoprogenitors that proliferate and undergo osteoblast differentiation. Therefore, the intramembranous bone growth of skull occurs at suture margins. Differentiation of osteoprogenitors/osteoblasts and ossification at the sutural margins are regulated by signaling from the mesenchyme, the osteogenic bone front, and underlying dura mater (Levi et al., 2012; Gagan et al., 2007). Genetic mutations identified in craniosynostosis patients are associated with osteogenic processes and result in hyper-ossification along sutural margins and eventual premature closure of the suture (see below).

Craniosynostosis can be classified as isolated/non-syndromic (without other anomalies) or syndromic (concomitant with other developmental defects). There are more than 180 syndromes associated with craniosynostosis. Genetic mutations have been identified in both syndromic and non-syndromic craniosynostoses. A single-nucleotide mutation in *MSX2* was the first genetic linkage to craniosynostosis. It was found in patients with Boston-type craniosynostosis, a rare syndrome confined to a single family (Jabs et al., 1993). *MSX2* is expressed in osteoblasts adjacent to the calvarial sutures. *Msx2*$^{-/-}$ mice show defects in skull ossification (Satokata et al., 2000), whereas over-expression of *Msx2* in mice results in premature fusion of calvarial sutures (Liu et al., 1995) The mutation identified in Boston-type craniosynostosis results in a Pro7His substitution in the N-terminal end of the MSX2 homeodomain. This mutation results in an increased binding affinity between MSX2 and its target genes (Ma et al., 1996).

FGF signaling plays a central role in the genetics of craniosynostosis in that gain-of-function mutations in all three receptors in the FGF family were found in syndromic craniosynostosis. FGFRs (FGFR1, FGFR2 and FGFR3) are transmembrane receptor tyrosine kinases, each containing an extracellular ligand-binding region, a single transmembrane domain and two intracellular tyrosine kinase domains. Most of the mutations of FGFR2 found in Aperts syndrome are missense mutations localized to the extracellular domain of the receptor, which result in increased affinity and reduced specificity of FGF-ligand binding (Ibrahimi et al., 2005). The missense mutations of FGFR2 found in Pfeiffer and Crouzon syndromes are in the extracellular domain, which leads to constitutive activation of the receptor (Kan et al., 2002). Mutations in FGFR3 are associated with Muenke syndrome and a rare variant of Crouzon syndrome (Muenke et al., 1997). Mutation of FGFR1 is associated with a mild form of Pfeiffer syndrome (Rossi et al., 2003).

The majority of patients with Saethre-Chotzen syndrome, a syndromic craniosynostosis characterized by unilateral or bilateral coronal synostosis, have a heterozygous loss-of-function mutation in *TWIST1* that can be caused by a number of genetic insults including deletion, nonsense, frameshift and missense mutations (el Ghouzzi et al., 1997; Howard et al.,

1997). *TWIST1* encodes a basic helix-loop-helix (bHLH) transcription factor, and is expressed in coronal suture mesenchyme. It is suggested that Twist1 plays a role in maintaining the suture boundary between neural crest derived (frontal bone) and mesoderm-derived (parietal bone) osteogenic mesenchyme (Merrill et al., 2006). Haplo-insufficiency of Twist1 in *Twist1*$^{+/-}$ mice results in expansion of *Msx2* expression and reduction of *Ephrin-A4* distribution in the coronal suture mesenchyme (Merrill et al., 2006). Reduction of *Msx2* expression in the *Twist1*$^{+/-}$ background rescues suture boundaries and *Ephrin-A4* expression, suggesting that *Twist1, Msx2*, and *Ephrin-A4* coordinate in a common pathway to regulate coronal suture formation and maintenance (Merrill et al., 2006).

Surgical intervention/treatment: Treatment for craniosynostosis is centered around three major concerns; the resultant elevated intracranial pressure and associated sequelae from restricted growth of the cranium, corneal exposure secondary to shallow orbits, and airway obstruction from the hypoplastic midface. Surgical correction requires the release of the fused suture and correction of the associated skull deformity. Timing of surgery can vary from early (birth–4 mo.) to late (9–12 mo.) depending on surgical technique, the suture(s) involved, and urgent concerns about elevated intracranial pressure. Correction of craniosynostosis resulting in a symmetrical deformity (i.e., sagittal and metopic) will tend to give better results than correction of craniosynostosis resulting in an asymmetrical deformity (i.e., unilateral coronal or lambdoidal). In cases involving multiple sutures and those with extension in the cranial base there will be a tendency for refusion and inability to correct the deformity. The probability of refusion is frequently correlated with whether the child is syndromic or not and the age of the child at the time of surgery. Early surgical dogma suggested intervention at four months of age. Surgeries at this stage of development have a refusion rate of approximately 20%. In the event of refusion, a second surgery is performed at approximately 12 months of age. High rates of refusions have prompted surgeons to now wait to approximately nine months of age. This delay in surgical intervention reduced the refusion rate to 2%. Involvement of skull base and midfacial deformities necessitate additional jaw surgery to move the maxilla and orbits forward to correct the airway and provide globe protection. Clinicians typically perform these surgeries utilizing the Le Fort system of fracture patterns (Le Fort I–III), which collectively involve the separation of all or a portion of the facial bones from the skull base.

Foramina Parietalia Permagna (Cranium Occultum)

In stark contrast to the excessive ossification of craniosynostosis, Foramina Parietalia Permagna (FPP) is an autosomal dominant condition occurring when there is an insufficient amount of ossification. FPP patients typically present with persistent, wide fontanelles (cranium bifidum) as well as aplasia of the scalp and other abnormalities including cleft lip/palate. FPP is a rare congenital skull defect typically identified on physical examination and confirmed radiographically. Persistent parietal fontanelles are typically bilateral oval defects situated on each parietal bone. Although small parietal foramina are common variants in up to 60% to 70% of normal skulls, large parietal foramina ranging from 5 mm to multiple centimeters are less common, with a prevalence of 1:15,000 to 1:25,000 (Kortesis et al., 2003).

Two transcription factors have been associated with the onset of FPP. First, the homeobox gene *MSX2* has been identified as a protein important for craniofacial development due to its expression in several domains of the developing skull (Davidson et al., 1995). Mutations in *MSX2* have been identified in numerous FPP families. Analysis of five affected families identified missense and nonsense mutations in *MSX2* that result in a protein that lacks the functional DNA-binding homeodomain (Wuyts et al., 2000b). Interestingly, mutations that cause stabilizing effects between MSX2 and DNA lead to a gain-of-function in MSX2 (Ma et al., 1996). Mutations such as these have been identified in Boston-type craniosynostosis patients and produce excessive bone growth (Jabs et al., 1993). Thus, MSX2 appears to be a critical protein during calvarial bone growth and development. Notwithstanding, FPP patients without any mutation in MSX2 have been identified, suggesting genetic heterogeneity for this disorder.

Mutations in ALX4, a homeobox containing transcription factor associated with skeletal development, have also been detected in three unrelated FPP families without the MSX2 mutation (Wuyts et al., 2000a). Similar to MSX2, mutations in ALX4 that affect DNA-binding and result in a loss-of-function or haplo-insufficiency of the protein have been associated with FPP. Although expression studies of ALX4 suggest an important role for the protein in differentiation of craniofacial and limb mesenchyme, some data also suggest that a negative feedback loop between ALX4 and SHH exists (Takahashi et al., 1998). Thus, although two causal genetic elements have been linked to FPP, the molecular mechanism is not fully understood.

Surgical intervention/treatment: FPP usually manifests as bony deficits in the parietal region. While these defects will often become smaller over time, they usually require surgery to close the remaining openings and protect the brain. A number of techniques have been employed to rectify persistent skull openings in these patients, but the most common is the split cranial graft. Split cranial graft consists of separating the outer table of the skull from the inner table, essentially splitting the thickness of the skull. The inner table is left *in situ* while the outer table is used as a graft to close the open area.

35.5 CONCLUSIONS

HISTORY, FUTURE AND OPTIONS FOR PHARMACOLOGICAL TREATMENT OF CRANIOFACIAL SYNDROMES

Craniofacial surgery was originally performed in adults to treat the deformities of childhood. Success in adults allowed surgeons to start treating individuals in childhood, often with better outcomes and using less complicated procedures. Surgical intervention in early childhood obviates many of the problems that we subsequently see in untreated adults. The greater plasticity of the tissue of children allows for significant remodeling and subsequent growth not observed in adult tissues. Furthermore, procedures are frequently better tolerated in children and often have better results when done on children compared to adults (i.e., amount of bone growth in children vs. adults and scarring in children vs. adults). Treatment in adults is planned based on the symptomatology and functional concerns of the patient, whereas treatment in children can be based on expected concerns.

While diagnostic ultrasound and *in utero* sampling have greatly increased our prognostic abilities, our ability to perform *in utero* preemptive intervention has not, to date, been as successful as hoped. There have been some *in utero* interventions on cleft lip patients in the pursuit of scarless healing, but these have not been clinically successful and endanger both the mother and fetus. Clinicians have been successful in predicting severe micrognathia leading to airway obstruction and have been able to plan safe delivery of these children with exit procedures. This technique allows the child to remain on placental circulation while a safe airway is established either by intubation or tracheostomy.

There are currently no pharmacological treatments for craniofacial syndromes. Virtually all treatment is of a surgical nature. However, advancing research has garnered some hope that pharmacological treatments may be an option in the future. For drug treatments to be viable, the genetic basis of the disorder or a the mechanisms of the defective processes must be understood. Currently fledgling treatments are being proposed for a lack of, as well as excessive, bone growth.

Crouzon syndrome is a craniosynostotic syndrome due to a mutation that causes a constitutive activation of fibroblast growth factor receptor 2 (*FGFR2*) in the absence of FGF ligand (Mooney et al., 2002; Bodo et al., 1999; Mansukhani et al., 2000; Mangasarian et al., 1997). Although surgical remodeling of the cranial vault is currently the only treatment for patients, animal models of the syndrome have been designed to further understand the molecular etiology of the syndrome (Perlyn et al., 2006a) and current studies are exploring the strategy of using tyrosine kinase inhibition as a treatment for constitutive FGFR activation (Perlyn et al., 2006b). Using the generated craniosynostotic animal models and tyrosine kinase inhibition, preliminary data showed that the coronal sutures in calvaria treated with a tyrosine kinase inhibitor remained patent and were without evidence of synostosis (Perlyn et al., 2006b). Further studies are required to develop viable *in vivo* treatment protocols in animal models that do not negatively affect other organ systems or biological processes.

Another possible avenue for pharmacological treatment is the use of Bone Morphogenic Protein to induce bone growth in conditions that lack bone (clefting, Treacher Collins, Cranium Occultum, Miller/Nager Syndrome). Use of Bone Morphogenetic Protein-2 (BMP-2) has been widely reported in the osseous reconstruction of large calvarial and craniofacial defects (Smith et al., 2008; Wehrhan et al., 2012; Levi et al., 2010; Docherty-Skogh et al., 2010; Koh et al., 2008; Farhadieh et al., 2004; Chang et al., 2003: Alden et al., 2000; Sailer et al., 1994). Although these studies have proven very promising, extremely high doses of BMP-2 are required, and bone growth cannot be guided for the detail-orientated repair required for intricate craniofacial structures (Smith et al., 2012). With growing evidence that BMP-2 is capable of enhancing bone growth, evaluating the safety of *in vivo* BMP-2 application becomes the most pertinent issue. Treating craniofacial defects with BMP-2 poses significant safety concerns because of the effects BMP-2 has on the developing brain (Ebendal et al., 1998). Furthermore, as is the case with a number of mitogens, the potential carcinogenic effects of BMP-2 must be evaluated prior to *in vivo* application. In summary, BMP application does facilitate bone formation and graft success, but it is currently an adjunct treatment. There are a number of current initiatives in tissue engineering for creating missing structures (such as ears) and tissues (primarily bone), but most of these remain investigational. The safety and efficacy of these compounds in the context of pediatric craniofacial surgery remain the major concerns.

Craniofacial anomalies are some of the most common and most debilitating disorders in our society. Before we establish effect treatments for these disorders, we must have a firm understanding of craniofacial biology on multiple levels. We must understand the gene function and molecular mechanism within the cells and tissues of the craniofacial complex. Furthermore, we must understand how the various tissues of the complex (neuroectoderm, neural crest and surface ectoderm) respond to genetic insult and how this affects the tissue-tissue interactions that are essential for proper craniofacial development.

Although we have made great strides in both understanding the etiology of craniofacial disorders and in surgical treatments, our eyes are still keen to pick up the slightest deviation from the accepted norm. Children with highly successful surgical interventions can often still be identified as "different." While progress continues to be made, *in utero* treatments are far from being either standard or practical at this point. A more complete understanding of the molecular pathways involved in craniofacial development will open doors toward therapeutic potential. The field of tissue engineering holds great promise towards viable and "passable" treatment options for children with severe disorders that require both hard and soft tissue grafts.

35.6 CLINICAL RELEVANCE

- Craniofacial defects are among the most common congenital defects affecting humans, with cleft lip with or without cleft palate (CL ± P) being the most common. Prevalence varies within different ethnic groups. The highest prevalence rates for (CL ± P) are reported for Native Americans and Asians. Africans have the lowest rates.
- Surgical repair of craniofacial defects is among the most difficult to perform, as the craniofacial complex is made up of various tissue types from several different embryonic origins. Whereas craniofacial surgeries most often involve manipulation of bone, surgical procedures will most likely also involve skin, muscle, and epidermal specializations such as teeth.
- Advances in tissue engineering and knowledge of molecular mechanisms underlying craniofacial syndromes will assist in better surgical repair and *in utero* treatment of this devastating class of developmental anomalies.

RECOMMENDED RESOURCES

http://embryology.med.unsw.edu.au (accessed Aug 2013)
http://www.craniofacial.net/resources-craniofacial-glossary (accessed Dec 2013)
http://www.youtube.com/watch?v=DgZ_tqucdI4 (accessed Jul 2012)
http://www.youtube.com/watch?v=SG3do_BeB0M (accessed Jan 2013)

REFERENCES

Alden, T.D., Beres, E.J., Laurent, J.S., Engh, J.A., Das, S., London, S.D., Jane Jr., J.A., Hudson, S.B., Helm, G.A., 2000. The use of bone morphogenetic protein gene therapy in craniofacial bone repair. J. Craniofac. Surg. 11 (1), 24–30.

Aranda-Orgilles, B., Aigner, J., Kunath, M., Lurz, R., Schneider, R., Schweiger, S., 2008a. Active transport of the ubiquitin ligase MID1 along the microtubules is regulated by protein phosphatase 2A. PLoS One 3 (10), e3507.

Aranda-Orgilles, B., Trockenbacher, A., Winter, J., Aigner, J., Kohler, A., Jastrzebska, E., Stahl, J., Muller, E.C., Otto, A., Wanker, E.E., et al., 2008b. The Opitz syndrome gene product MID1 assembles a microtubule-associated ribonucleoprotein complex. Hum. Genet. 123 (2), 163–176.

Belloni, E., Muenke, M., Roessler, E., Traverso, G., Siegel-Bartelt, J., Frumkin, A., Mitchell, H.F., Donis-Keller, H., Helms, C., Hing, A.V., et al., 1996. Identification of Sonic hedgehog as a candidate gene responsible for holoprosencephaly. Nat. Genet. 14 (3), 353–356.

Bernier, F.P., Caluseriu, O., Ng, S., Schwartzentruber, J., Buckingham, K.J., Innes, A.M., Jabs, E.W., Innis, J.W., Schuette, J.L., Gorski, J.L., et al., 2012. Haploinsufficiency of SF3B4, a component of the pre-mRNA spliceosomal complex, causes Nager syndrome. Am. J. Hum. Genet. 90 (5), 925–933.

Bertolino, E., Reimund, B., Wildt-Perinic, D., Clerc, R.G., 1995. A novel homeobox protein which recognizes a TGT core and functionally interferes with a retinoid-responsive motif. J. Biol. Chem. 270 (52), 31178–31188.

Blair, H.J., Tompson, S., Liu, Y.N., Campbell, J., MacArthur, K., Ponting, C.P., Ruiz-Perez, V.L., Goodship, J.A., 2011. Evc2 is a positive modulator of Hedgehog signaling that interacts with Evc at the cilia membrane and is also found in the nucleus. BMC Biol. 9, 14.

Bodo, M., Baroni, T., Carinci, F., Becchetti, E., Bellucci, C., Conte, C., Pezzetti, F., Evangelisti, R., Tognon, M., Carinci, P., 1999. A regulatory role of fibroblast growth factor in the expression of decorin, biglycan, betaglycan and syndecan in osteoblasts from patients with Crouzons syndrome. Eur. J. Cell. Biol. 78 (5), 323–330.

Brugmann, S.A., Goodnough, L.H., Gregorieff, A., Leucht, P., ten Berge, D., Fuerer, C., Clevers, H., Nusse, R., Helms, J.A., 2007. Wnt signaling mediates regional specification in the vertebrate face. Development 134 (18), 3283–3295.

Caparros-Martin, J.A., Valencia, M., Reytor, E., Pacheco, M., Fernandez, M., Perez-Aytes, A., Gean, E., Lapunzina, P., Peters, H., Goodship, J.A., et al., 2013. The ciliary Evc/Evc2 complex interacts with Smo and controls Hedgehog pathway activity in chondrocytes by regulating Sufu/Gli3 dissociation and Gli3 trafficking in primary cilia. Hum. Mol. Genet. 22 (1), 124–139.

Cappa, M., Borrelli, P., Marini, R., Neri, G., 1987. The Opitz syndrome: a new designation for the clinically indistinguishable BBB and G syndromes. Am. J. Med. Genet. 28 (2), 303–309.

Cassell, C.H., Meyer, R., Daniels, J., 2008. Health care expenditures among Medicaid enrolled children with and without orofacial clefts in North Carolina, 1995–2002. Birth Defects Res. A. Clin. Mol. Teratol. 82 (11), 785–794.

Chang, S.C., Wei, F.C., Chuang, H., Chen, Y.R., Chen, J.K., Lee, K.C., Chen, P.K., Tai, C.L., Lou, J., 2003. *Ex vivo* gene therapy in autologous critical-size craniofacial bone regeneration. Plast Reconstr. Surg. 112 (7), 1841–1850.

Cole, F., Krauss, R.S., 2003. Microform Holoprosencephaly in Mice that Lack the Ig Superfamily Member Cdon. Curr. Biol 13, 411–415.

Corbit, K.C., Shyer, A.E., Dowdle, W.E., Gaulden, J., Singla, V., Chen, M.H., Chuang, P.T., Reiter, J.F., 2008. Kif3a constrains beta-catenin-dependent Wnt signaling through dual ciliary and non-ciliary mechanisms. Nat. Cell. Biol. 10 (1), 70–76.

Cordero, J.F., Holmes, L.B., 1978. Phenotypic overlap of the BBB and G syndromes. Am. J. Med. Genet. 2 (2), 145–152.

Dal Zotto, L., Quaderi, N.A., Elliott, R., Lingerfelter, P.A., Carrel, L., Valsecchi, V., Montini, E., Yen, C.H., Chapman, V., Kalcheva, I., et al., 1998. The mouse Mid1 gene: implications for the pathogenesis of Opitz syndrome and the evolution of the mammalian pseudoautosomal region. Hum. Mol. Genet. 7 (3), 489–499.

Dauwerse, J.G., Dixon, J., Seland, S., Ruivenkamp, C.A., van Haeringen, A., Hoefsloot, L.H., Peters, D.J., Boers, A.C., Daumer-Haas, C., Maiwald, R., et al., 2011. Mutations in genes encoding subunits of RNA polymerases I and III cause Treacher Collins syndrome. Nat. Genet. 43 (1), 20–22.

Davidson, D., 1995. The function and evolution of Msx genes: pointers and paradoxes. Trends. Genet. 11 (10), 405–411.

De Falco, F., Cainarca, S., Andolfi, G., Ferrentino, R., Berti, C., Rodriguez Criado, G., Rittinger, O., Dennis, N., Odent, S., Rastogi, A., et al., 2003. X-linked Opitz syndrome: novel mutations in the MID1 gene and redefinition of the clinical spectrum. Am. J. Med. Genet. A. 120A (2), 222–228.

Diewert, V.M., Lozanoff, S., 1993a. Growth and morphogenesis of the human embryonic midface during primary palate formation analyzed in frontal sections. J. Craniofac. Genet. Dev. Biol. 13 (3), 162–183.

Diewert, V.M., Lozanoff, S., 1993b. A morphometric analysis of human embryonic craniofacial growth in the median plane during primary palate formation. J. Craniofac. Genet. Dev. Biol. 13 (3), 147–161.

Diewert, V.M., Lozanoff, S., Choy, V., 1993. Computer reconstructions of human embryonic craniofacial morphology showing changes in relations between the face and brain during primary palate formation. J. Craniofac. Genet. Dev. Biol. 13 (3), 193–201.

Dixon, J., Jones, N.C., Sandell, L.L., Jayasinghe, S.M., Crane, J., Rey, J.P., Dixon, M.J., Trainor, P.A., 2006. Tcof1/Treacle is required for neural crest cell formation and proliferation deficiencies that cause craniofacial abnormalities. Proc. Natl. Acad. Sci. USA 103 (36), 13403–13408.

Docherty-Skogh, A.C., Bergman, K., Waern, M.J., Ekman, S., Hultenby, K., Ossipov, D., Hilborn, J., Bowden, T., Engstrand, T., 2010. Bone morphogenetic protein-2 delivered by hyaluronan-based hydrogel induces massive bone formation and healing of cranial defects in minipigs. Plast. Reconstr. Surg. 125 (5), 1383–1392.

Ebendal, T., Bengtsson, H., Soderstrom, S., 1998. Bone morphogenetic proteins and their receptors: potential functions in the brain. J. Neurosci. Res. 51 (2), 139–146.

Edison, R., Muenke, M., 2003. The interplay of genetic and environmental factors in craniofacial morphogenesis: holoprosencephaly and the role of cholesterol. Congenit. Anom. Kyoto. 43 (1), 1–21.

el Ghouzzi, V., Le Merrer, M., Perrin-Schmitt, F., Lajeunie, E., Benit, P., Renier, D., Bourgeois, P., Bolcato-Bellemin, A.L., Munnich, A., Bonaventure, J., 1997. Mutations of the TWIST gene in the Saethre-Chotzen syndrome. Nat. Genet. 15 (1), 42–46.

Farhadieh, R.D., Gianoutsos, M.P., Yu, Y., Walsh, W.R., 2004. The role of bone morphogenetic proteins BMP-2 and BMP-4 and their related postreceptor signaling system (Smads) in distraction osteogenesis of the mandible. J. Craniofac. Surg. 15 (5), 714–718.

Ferrante, M.I., Giorgio, G., Feather, S.A., Bulfone, A., Wright, V., Ghiani, M., Selicorni, A., Gammaro, L., Scolari, F., Woolf, A.S., et al., 2001. Identification of the gene for oral-facial-digital type I syndrome. Am. J. Hum. Genet. 68 (3), 569–576.

Ferrante, M.I., Romio, L., Castro, S., Collins, J.E., Goulding, D.A., Stemple, D.L., Woolf, A.S., Wilson, S.W., 2009. Convergent extension movements and ciliary function are mediated by ofd1, a zebrafish orthologue of the human oral-facial-digital type 1 syndrome gene. Hum. Mol. Genet. 18 (2), 289–303.

Ferrante, M.I., Zullo, A., Barra, A., Bimonte, S., Messaddeq, N., Studer, M., Dolle, P., Franco, B., 2006. Oral-facial-digital type I protein is required for primary cilia formation and left-right axis specification. Nat. Genet. 38 (1), 112–117.

Fontanella, B., Russolillo, G., Meroni, G., 2008. MID1 mutations in patients with X-linked Opitz G/BBB syndrome. Hum. Mutat. 29 (5), 584–594.

Fuccillo, M., Rallu, M., McMahon, A.P., Fishell, G., 2004. Temporal requirement for hedgehog signaling in ventral telencephalic patterning. Development 131 (20), 5031–5040.

Gagan, J.R., Tholpady, S.S., Ogle, R.C., 2007. Cellular dynamics and tissue interactions of the dura mater during head development. Birth Defects Res. C. Embryo Today 81 (4), 297–304.

Geng, X., Oliver, G., 2009. Pathogenesis of holoprosencephaly. J. Clin. Invest. 119 (6), 1403–1413.

Geng, X., Speirs, C., Lagutin, O., Inbal, A., Liu, W., Solnica-Krezel, L., Jeong, Y., Epstein, D.J., Oliver, G., 2008. Haplo-insufficiency of Six3 fails to activate Sonic hedgehog expression in the ventral forebrain and causes holoprosencephaly. Dev. Cell 15 (2), 236–247.

Gerdes, J.M., Liu, Y., Zaghloul, N.A., Leitch, C.C., Lawson, S.S., Kato, M., Beachy, P.A., Beales, P.L., DeMartino, G.N., Fisher, S., et al., 2007. Disruption of the basal body compromises proteasomal function and perturbs intracellular Wnt response. Nat. Genet. 39 (11), 1350–1360.

Goetz, S.C., Anderson, K.V., 2010. The primary cilium: a signaling center during vertebrate development. Nat. Rev. Genet. 11 (5), 331–344.

Gongal, P.A., Waskiewicz, A.J., 2008. Zebrafish model of holoprosencephaly demonstrates a key role for TGIF in regulating retinoic acid metabolism. Hum. Mol. Genet 17, 525–538.

Gorivodsky, M., Mukhopadhyay, M., Wilsch-Braeuninger, M., Phillips, M., Teufel, A., Kim, C., Malik, N., Huttner, W., Westphal, H., 2009. Intraflagellar transport protein 172 is essential for primary cilia formation and plays a vital role in patterning the mammalian brain. Dev. Biol. 325 (1), 24–32.

Gorlin, R.J., Cohen, M.M., Levin, L.S., 1990. Syndromes of the Head and Neck. Oxford University Press, New York.

Granata, A., Quaderi, N.A., 2003. The Opitz syndrome gene MID1 is essential for establishing asymmetric gene expression in Hensens node. Dev. Biol. 258 (2), 397–405.

Gripp, K.W., Wotton, D., Edwards, M.C., Roessler, E., Ades, L., Meinecke, P., Richieri-Costa, A., Zackai, E.H., Massague, J., Muenke, M., et al., 2000. Mutations in TGIF cause holoprosencephaly and link NODAL signaling to human neural axis determination. Nat. Genet. 25 (2), 205–208.

Halonen, K., Hukki, J., Arte, S., Hurmerinta, K., 2006. Craniofacial structures and dental development in three patients with Nager syndrome. J. Craniofac. Surg. 17 (6), 1180–1187.

Howard, T.D., Paznekas, W.A., Green, E.D., Chiang, L.C., Ma, N., Ortiz de Luna, R.I., Garcia Delgado, C., Gonzalez-Ramos, M., Kline, A.D., Jabs, E.W., 1997. Mutations in TWIST, a basic helix-loop-helix transcription factor, in Saethre-Chotzen syndrome. Nat. Genet. 15 (1), 36–41.

Hu, D., Marcucio, R.S., 2009. Unique organization of the frontonasal ectodermal zone in birds and mammals. Dev. Biol. 325 (1), 200–210.

Hu, D., Marcucio, R.S., Helms, J.A., 2003. A zone of frontonasal ectoderm regulates patterning and growth in the face. Development 130 (9), 1749–1758.

Hui, C.C., Angers, S., 2011. Gli proteins in development and disease. Annu. Rev. Cell Dev. Biol. 27, 513–537.

Ibrahimi, O.A., Chiu, E.S., McCarthy, J.G., Mohammadi, M., 2005. Understanding the molecular basis of Apert syndrome. Plast Reconstr. Surg. 115 (1), 264–270.

Ishizeki, K., Takahashi, N., Nawa, T., 2001. Formation of the sphenomandibular ligament by Meckels cartilage in the mouse: possible involvement of epidermal growth factor as revealed by studies in vivo and in vitro. Cell Tissue Res. 304 (1), 67–80.

Ishizeki, K., Takigawa, M., Nawa, T., Suzuki, F., 1996. Mouse Meckels cartilage chondrocytes evoke bone-like matrix and further transform into osteocyte-like cells in culture. Anat. Rec. 245 (1), 25–35.

Jabs, E.W., Muller, U., Li, X., Ma, L., Luo, W., Haworth, I.S., Klisak, I., Sparkes, R., Warman, M.L., Mulliken, J.B., 1993. A mutation in the homeodomain of the human MSX2 gene in a family affected with autosomal dominant craniosynostosis. Cell 75 (3), 443–450.

Jeong, Y., Leskow, F.C., El-Jaick, K., Roessler, E., Muenke, M., Yocum, A., Dubourg, C., Li, X., Geng, X., Oliver, G., et al., 2008. Regulation of a remote Shh forebrain enhancer by the Six3 homeoprotein. Nat. Genet. 40 (11), 1348–1353.

Jones, N.C., Lynn, M.L., Gaudenz, K., Sakai, D., Aoto, K., Rey, J.P., Glynn, E.F., Ellington, L., Du, C., Dixon, J., et al., 2008. Prevention of the neurocristopathy Treacher Collins syndrome through inhibition of p53 function. Nat. Med. 14 (2), 125–133.

Kan, S.H., Elanko, N., Johnson, D., Cornejo-Roldan, L., Cook, J., Reich, E.W., Tomkins, S., Verloes, A., Twigg, S.R., Rannan-Eliya, S., et al., 2002. Genomic screening of fibroblast growth-factor receptor 2 reveals a wide spectrum of mutations in patients with syndromic craniosynostosis. Am. J. Hum. Genet. 70 (2), 472–486.

Kinoshita, F., Kondoh, T., Komori, K., Matsui, T., Harada, N., Yanai, A., Fukuda, M., Morifuji, K., Matsumoto, T., 2011. Miller syndrome with novel dihydroorotate dehydrogenase gene mutations. Pediatr. Int. 53 (4), 587–591.

Koh, J.T., Zhao, Z., Wang, Z., Lewis, I.S., Krebsbach, P.H., Franceschi, R.T., 2008. Combinatorial gene therapy with BMP2/7 enhances cranial bone regeneration. J. Dent. Res. 87 (9), 845–849.

Kolpakova-Hart, E., Jinnin, M., Hou, B., Fukai, N., Olsen, B.R., 2007. Kinesin-2 controls development and patterning of the vertebrate skeleton by Hedgehog- and Gli3-dependent mechanisms. Dev. Biol. 309 (2), 273–284.

Komori, T., Yagi, H., Nomura, S., Yamaguchi, A., Sasaki, K., Deguchi, K., Shimizu, Y., Bronson, R.T., Gao, Y.H., Inada, M., et al., 1997. Targeted disruption of Cbfa1 results in a complete lack of bone formation owing to maturational arrest of osteoblasts. Cell 89 (5), 755–764.

Kortesis, B., Richards, T., David, L., Glazier, S., Argenta, L., 2003. Surgical management of foramina parietalia permagna. J. Craniofac. Surg. 14 (4), 538–544.

Krauss, R.S., 2007. Holoprosencephaly: new models, new insights. Expert Rev. Mol. Med. 9 (26), 1–17.

Lan, Y., Ryan, R.C., Zhang, Z., Bullard, S.A., Bush, J.O., Maltby, K.M., Lidral, A.C., Jiang, R., 2006. Expression of Wnt9b and activation of canonical Wnt signaling during midfacial morphogenesis in mice. Dev. Dyn. 235 (5), 1448–1454.

Lansdown, R., 1991. This sporting life: mental and physical health in schoolchildren today. J. R. Soc. Med. 84 (5), 318–319.

Lee, S.H., Bedard, O., Buchtova, M., Fu, K., Richman, J.M., 2004. A new origin for the maxillary jaw. Dev. Biol. 276 (1), 207–224.

Leoncini, E., Baranello, G., Orioli, I.M., Anneren, G., Bakker, M., Bianchi, F., Bower, C., Canfield, M.A., Castilla, E.E., Cocchi, G., et al., 2008. Frequency of holoprosencephaly in the International Clearinghouse Birth Defects Surveillance Systems: searching for population variations. Birth Defects Res. A. Clin. Mol. Teratol. 82 (8), 585–591.

Levi, B., James, A.W., Nelson, E.R., Vistnes, D., Wu, B., Lee, M., Gupta, A., Longaker, M.T., 2010. Human adipose derived stromal cells heal critical size mouse calvarial defects. PLoS One 5 (6), e11177.

Levi, B., Wan, D.C., Wong, V.W., Nelson, E., Hyun, J., Longaker, M.T., 2012. Cranial suture biology: from pathways to patient care. J. Craniofac. Surg. 23 (1), 13–19.

Liu, W., Toyosawa, S., Furuichi, T., Kanatani, N., Yoshida, C., Liu, Y., Himeno, M., Narai, S., Yamaguchi, A., Komori, T., 2001. Overexpression of Cbfa1 in osteoblasts inhibits osteoblast maturation and causes osteopenia with multiple fractures. J. Cell. Biol. 155 (1), 157–166.

Liu, Y.H., Kundu, R., Wu, L., Luo, W., Ignelzi Jr., M.A., Snead, M.L., Maxson Jr, R.E., 1995. Premature suture closure and ectopic cranial bone in mice expressing Msx2 transgenes in the developing skull. Proc. Natl. Acad. Sci. USA 92 (13), 6137–6141.

Logan, C.V., Abdel-Hamed, Z., Johnson, C.A., 2011. Molecular genetics and pathogenic mechanisms for the severe ciliopathies: insights into neurodevelopment and pathogenesis of neural tube defects. Mol. Neurobiol. 43 (1), 12–26.

Lou, Y., Javed, A., Hussain, S., Colby, J., Frederick, D., Pratap, J., Xie, R., Gaur, T., van Wijnen, A.J., Jones, S.N., et al., 2009. A Runx2 threshold for the cleidocranial dysplasia phenotype. Hum. Mol. Genet. 18 (3), 556–568.

Ma, L., Golden, S., Wu, L., Maxson, R., 1996. The molecular basis of Boston-type craniosynostosis: the Pro148–>His mutation in the N-terminal arm of the MSX2 homeodomain stabilizes DNA binding without altering nucleotide sequence preferences. Hum. Mol. Genet. 5 (12), 1915–1920.

Ma, Y., Erkner, A., Gong, R., Yao, S., Taipale, J., Basler, K., Beachy, P.A., 2002. Hedgehog-mediated patterning of the mammalian embryo requires transporter-like function of dispatched. Cell 111 (1), 63–75.

Mahuzier, A., Gaude, H.M., Grampa, V., Anselme, I., Silbermann, F., Leroux-Berger, M., Delacour, D., Ezan, J., Montcouquiol, M., Saunier, S., et al., 2012. Dishevelled stabilization by the ciliopathy protein Rpgrip1l is essential for planar cell polarity. J. Cell. Biol. 198 (5), 927–940.

Mangasarian, K., Li, Y., Mansukhani, A., Basilico, C., 1997. Mutation associated with Crouzon syndrome causes ligand-independent dimerization and activation of FGF receptor-2. J. Cell. Physiol. 172 (1), 117–125.

Mansukhani, A., Bellosta, P., Sahni, M., Basilico, C., 2000. Signaling by fibroblast growth factors (FGF) and fibroblast growth factor receptor 2 (FGFR2)-activating mutations blocks mineralization and induces apoptosis in osteoblasts. J. Cell. Biol. 149 (6), 1297–1308.

Matsunaga, E., Shiota, K., 1977. Holoprosencephaly in human embryos: epidemiologic studies of 150 cases. Teratology 16 (3), 261–272.

McKean, D.M., Niswander, L., 2012. Defects in GPI biosynthesis perturb Cripto signaling during forebrain development in two new mouse models of holoprosencephaly. Biol. Open. 1 (9), 874–883.

Menendez, L., Kulik, M.J., Page, A.T., Park, S.S., Lauderdale, J.D., Cunningham, M.L., Dalton, S., 2013. Directed differentiation of human pluripotent cells to neural crest stem cells. Nat. Protoc. 8 (1), 203–212.

Mercier, S., David, V., Ratie, L., Gicquel, I., Odent, S., Dupe, V., 2013. NODAL and SHH dose-dependent double inhibition promotes an HPE-like phenotype in chick embryos. Dis. Model. Mech. 6 (2), 537–543.

Merrill, A.E., Bochukova, E.G., Brugger, S.M., Ishii, M., Pilz, D.T., Wall, S.A., Lyons, K.M., Wilkie, A.O., Maxson Jr., R.E., 2006. Cell mixing at a neural crest-mesoderm boundary and deficient ephrin-Eph signaling in the pathogenesis of craniosynostosis. Hum. Mol. Genet. 15 (8), 1319–1328.

Miller, M., Fineman, R., Smith, D.W., 1979. Postaxial acrofacial dysostosis syndrome. J. Pediatr. 95 (6), 970–975.

Ming, J.E., Kaupas, M.E., Roessler, E., Brunner, H.G., Golabi, M., Tekin, M., Stratton, R.F., Sujansky, E., Bale, S.J., Muenke, M., 2002. Mutations in PATCHED-1, the receptor for SONIC HEDGEHOG, are associated with holoprosencephaly. Hum. Genet. 110 (4), 297–301.

Mooney, M.D., Siegel, M.I., 2002. Understanding Craniofacial Anomalies; The etiopathogenesis of craniosynotoses and facial clefting. Wiley-Liss, Inc, New York.

Morriss-Kay, G.M., Wilkie, A.O., 2005. Growth of the normal skull vault and its alteration in craniosynostosis: insights from human genetics and experimental studies. J. Anat. 207 (5), 637–653.

Muenke, M., Beachy, P.A., 2000. Genetics of ventral forebrain development and holoprosencephaly. Curr. Opin. Genet. Dev. 10 (3), 262–269.

Muenke M, B.P., 2001. Holoprosencephaly. In: Scriver CR, B.A., Sly, W.S., Valle, D. (Eds.), The metabolic and molecular bases of inherited disease. McGraw-Hill, New York.

Muenke, M., Cohen Jr, M.M., 2000. Genetic approaches to understanding brain development: holoprosencephaly as a model. Ment. Retard. Dev. Disabil. Res. Rev. 6 (1), 15–21.

Muenke, M., Gripp, K.W., McDonald-McGinn, D.M., Gaudenz, K., Whitaker, L.A., Bartlett, S.P., Markowitz, R.I., Robin, N.H., Nwokoro, N., Mulvihill, J.J., et al., 1997. A unique point mutation in the fibroblast growth factor receptor 3 gene (FGFR3) defines a new craniosynostosis syndrome. Am. J. Hum. Genet. 60 (3), 555–564.

Nakatomi, M., Hovorakova, M., Gritli-Linde, A., Blair, H.J., MacArthur, K., Peterka, M., Lesot, H., Peterkova, R., Ruiz-Perez, V.L., Goodship, J.A., et al., 2013. Evc regulates a symmetrical response to Shh signaling in molar development. J. Dent. Res. 92 (3), 222–228.

Nomura, M., Li, E., 1998. Smad2 role in mesoderm formation, left-right patterning and craniofacial development. Nature 393, 786–790.

Odent, S., Atti-Bitach, T., Blayau, M., Mathieu, M., Aug, J., Delezo de, A.L., Gall, J.Y., Le Marec, B., Munnich, A., David, V., et al., 1999. Expression of the Sonic hedgehog (SHH) gene during early human development and phenotypic expression of new mutations causing holoprosencephaly. Hum. Mol. Genet. 8 (9), 1683–1689.

Otto, F., Kanegane, H., Mundlos, S., 2002. Mutations in the RUNX2 gene in patients with cleidocranial dysplasia. Hum. Mutat. 19 (3), 209–216.

Otto, F., Thornell, A.P., Crompton, T., Denzel, A., Gilmour, K.C., Rosewell, I.R., Stamp, G.W., Beddington, R.S., Mundlos, S., Olsen, B.R., et al., 1997. Cbfa1, a candidate gene for cleidocranial dysplasia syndrome, is essential for osteoblast differentiation and bone development. Cell 89 (5), 765–771.

Passos-Bueno, M.R., Ornelas, C.C., Fanganiello, R.D., 2009. Syndromes of the first and second pharyngeal arches: A review. Am. J. Med. Genet. A. 149A (8), 1853–1859.

Perlyn, C.A., DeLeon, V.B., Babbs, C., Govier, D., Burell, L., Darvann, T., Kreiborg, S., Morriss-Kay, G., 2006a. The craniofacial phenotype of the Crouzon mouse: analysis of a model for syndromic craniosynostosis using three-dimensional MicroCT. Cleft. Palate. Craniofac. J. 43 (6), 740–748.

Perlyn, C.A., Morriss-Kay, G., Darvann, T., Tenenbaum, M., Ornitz, D.M., 2006b. A model for the pharmacological treatment of Crouzon syndrome. Neurosurgery 59 (1), 210–215. discussion 210–5.

Pruzinsky, T., 1992. Social and psychological effects of major craniofacial deformity. Cleft. Palate. Craniofac. J. 29 (6), 578–584. discussion 570.

Quaderi, N.A., Schweiger, S., Gaudenz, K., Franco, B., Rugarli, E.I., Berger, W., Feldman, G.J., Volta, M., Andolfi, G., Gilgenkrantz, S., et al., 1997. Opitz G/BBB syndrome, a defect of midline development, is due to mutations in a new RING finger gene on Xp22. Nat. Genet. 17 (3), 285–291.

Rainger, J., Bengani, H., Campbell, L., Anderson, E., Sokhi, K., Lam, W., Riess, A., Ansari, M., Smithson, S., Lees, M., et al., 2012. Miller (Genee-Wiedemann) syndrome represents a clinically and biochemically distinct subgroup of postaxial acrofacial dysostosis associated with partial deficiency of DHODH. Hum. Mol. Genet. 21 (18), 3969–3983.

Ribeiro, L.A., Murray, J.C., Richieri-Costa, A., 2006. PTCH mutations in four Brazilian patients with holoprosencephaly and in one with holoprosencephaly-like features and normal MRI. Am. J. Med. Genet. A. 140 (23), 2584–2586.

Richman, J.M., Fu, K.K., Cox, L.L., Sibbons, J.P., Cox, T.C., 2002. Isolation and characterization of the chick ortholog of the Opitz syndrome gene, Mid1, supports a conserved role in vertebrate development. Int. J. Dev. Biol. 46 (4), 441–448.

Robin, N.H., 1995. Opitz syndrome is genetically heterogeneous, with one locus on Xp22, and a second locus on 22q11.2. Nat. Genet. 11, 459–461.

Roessler, E., Belloni, E., Gaudenz, K., Jay, P., Berta, P., Scherer, S.W., Tsui, L.C., Muenke, M., 1996. Mutations in the human Sonic Hedgehog gene cause holoprosencephaly. Nat. Genet. 14 (3), 357–360.

Roessler, E., Belloni, E., Gaudenz, K., Vargas, F., Scherer, S.W., Tsui, L.C., Muenke, M., 1997. Mutations in the C-terminal domain of Sonic Hedgehog cause holoprosencephaly. Hum. Mol. Genet. 6 (11), 1847–1853.

Roessler, E., Du, Y.Z., Mullor, J.L., Casas, E., Allen, W.P., Gillessen-Kaesbach, G., Roeder, E.R., Ming, J.E., Ruiz, i., Altaba, A., Muenke, M., 2003. Loss-of-function mutations in the human GLI2 gene are associated with pituitary anomalies and holoprosencephaly-like features. Proc. Natl. Acad. Sci. USA 100 (23), 13424–13429.

Roessler, E., Ermilov, A.N., Grange, D.K., Wang, A., Grachtchouk, M., Dlugosz, A.A., Muenke, M., 2005. A previously unidentified amino-terminal domain regulates transcriptional activity of wild-type and disease-associated human GLI2,. Hum. Mol. Genet. 14 (15), 2181–2188.

Roessler, E., Lacbawan, F., Dubourg, C., Paulussen, A., Herbergs, J., Hehr, U., Bendavid, C., Zhou, N., Ouspenskaia, M., Bale, S., et al., 2009a. The full spectrum of holoprosencephaly-associated mutations within the ZIC2 gene in humans predicts loss-of-function as the predominant disease mechanism. Hum. Mutat. 30 (4), E541–E554.

Roessler, E., Ma, Y., Ouspenskaia, M.V., Lacbawan, F., Bendavid, C., Dubourg, C., Beachy, P.A., Muenke, M., 2009b. Truncating loss-of-function mutations of DISP1 contribute to holoprosencephaly-like microform features in humans. Hum. Genet. 125 (4), 393–400.

Roessler, E., Muenke, M., 2010. The molecular genetics of holoprosencephaly,. Am. J. Med. Genet. C. Semin. Med. Genet. 154C (1), 52–61.

Roessler, E., Ouspenskaia, M.V., Karkera, J.D., Velez, J.I., Kantipong, A., Lacbawan, F., Bowers, P., Belmont, J.W., Towbin, J.A., Goldmuntz, E., et al., 2008. Reduced NODAL signaling strength via mutation of several pathway members including FOXH1 is linked to human heart defects and holoprosencephaly. Am. J. Hum. Genet. 83 (1), 18–29.

Romio, L., Fry, A.M., Winyard, P.J., Malcolm, S., Woolf, A.S., Feather, S.A., 2004. OFD1 is a centrosomal/basal body protein expressed during mesenchymal-epithelial transition in human nephrogenesis. J. Am. Soc. Nephrol. 15 (10), 2556–2568.

Rossi, M., Jones, R.L., Norbury, G., Bloch-Zupan, A., Winter, R.M., 2003. The appearance of the feet in Pfeiffer syndrome caused by FGFR1 P252R mutation. Clin. Dysmorphol. 12 (4), 269–274.

Ruiz-Perez, V.L., Blair, H.J., Rodriguez-Andres, M.E., Blanco, M.J., Wilson, A., Liu, Y.N., Miles, C., Peters, H., Goodship, J.A., 2007. Evc is a positive mediator of Ihh-regulated bone growth that localises at the base of chondrocyte cilia. Development 134 (16), 2903–2912.

Ruiz-Perez, V.L., Ide, S.E., Strom, T.M., Lorenz, B., Wilson, D., Woods, K., King, L., Francomano, C., Freisinger, P., Spranger, S., et al., 2000. Mutations in a new gene in Ellis-van Creveld syndrome and Weyers acrodental dysostosis. Nat. Genet. 24 (3), 283–286.

Ruiz-Perez, V.L., Tompson, S.W., Blair, H.J., Espinoza-Valdez, C., Lapunzina, P., Silva, E.O., Hamel, B., Gibbs, J.L., Young, I.D., Wright, M.J., et al., 2003. Mutations in two nonhomologous genes in a head-to-head configuration cause Ellis-van Creveld syndrome. Am. J. Hum. Genet. 72 (3), 728–732.

Sailer, H.F., Kolb, E., 1994. Application of purified bone morphogenetic protein (BMP) in cranio-maxillo-facial surgery. BMP in compromised surgical reconstructions using titanium implants. J. Craniomaxillofac. Surg. 22 (1), 2–11.

Sakai, D., Trainor, P.A., 2009. Treacher Collins syndrome: unmasking the role of Tcof1/treacle. Int. J. Biochem. Cell. Biol. 41 (6), 1229–1232.

Satokata, I., Ma, L., Ohshima, H., Bei, M., Woo, I., Nishizawa, K., Maeda, T., Takano, Y., Uchiyama, M., Heaney, S., et al., 2000. Msx2 deficiency in mice causes pleiotropic defects in bone growth and ectodermal organ formation. Nat. Genet. 24 (4), 391–395.

Schlump, J.U., Stein, A., Hehr, U., Karen, T., Moller-Hartmann, C., Elcioglu, N.H., Bogdanova, N., Woike, H.F., Lohmann, D.R., Felderhoff-Mueser, U., et al., 2012. Treacher Collins syndrome: clinical implications for the paediatrician–a new mutation in a severely affected newborn and comparison with three further patients with the same mutation, and review of the literature. Eur. J. Pediatr. 171 (11), 1611–1618.

Schroeder, T.M., Jensen, E.D., Westendorf, J.J., 2005. Runx2: a master organizer of gene transcription in developing and maturing osteoblasts. Birth Defects. Res. C. Embryo. Today 75 (3), 213–225.

Shiota, K., Yamada, S., 2010. Early pathogenesis of holoprosencephaly. Am. J. Med. Genet. C. Semin. Med. Genet. 154C (1), 22–28.

Simons, M., Gloy, J., Ganner, A., Bullerkotte, A., Bashkurov, M., Kronig, C., Schermer, B., Benzing, T., Cabello, O.A., Jenny, A., et al., 2005. Inversin, the gene product mutated in nephronophthisis type II, functions as a molecular switch between Wnt signaling pathways. Nat. Genet. 37 (5), 537–543.

Smith, D.M., Cooper, G.M., Mooney, M.P., Marra, K.G., Losee, J.E., 2008. Bone morphogenetic protein 2 therapy for craniofacial surgery. J. Craniofac. Surg. 19 (5), 1244–1259.

Smith, D.M., Cray Jr., J.J., Weiss, L.E., Dai Fei, E.K., Shakir, S., Rottgers, S.A., Losee, J.E., Campbell, P.G., Cooper, G.M., 2012. Precise control of osteogenesis for craniofacial defect repair: the role of direct osteoprogenitor contact in BMP-2-based bioprinting. Ann. Plast. Surg. 69 (4), 485–488.

Solomon, B.D., Gropman, A., Muenke, M., 1993. Holoprosencephaly Overview. In: Pagon, R.A., Bird, T.D., Dolan, C.R., Stephens, K., Adam, M.P. (Eds.). GeneReviews, Seattle (WA).

Solomon, B.D., Mercier, S., Velez, J.I., Pineda-Alvarez, D.E., Wyllie, A., Zhou, N., Dubourg, C., David, V., Odent, S., Roessler, E., et al., 2010. Analysis of genotype-phenotype correlations in human holoprosencephaly. Am. J. Med. Genet. C. Semin. Med. Genet. 154C (1), 133–141.

Song, L., Li, Y., Wang, K., Wang, Y.Z., Molotkov, A., Gao, L., Zhao, T., Yamagami, T., Wang, Y., Gan, Q., et al., 2009. Lrp6-mediated canonical Wnt signaling is required for lip formation and fusion. Development 136 (18), 3161–3171.

Sperber GH, S.S., Guttmann, G.D., Shelton, C.T., 2010. Craniofacial Embryogenetics And Development Craniofacial Embryogenetics and Development. People's Medical Publishing House, USA.

Stottmann, R.W., Tran, P.V., Turbe-Doan, A., Beier, D.R., 2009. Ttc21b is required to restrict sonic hedgehog activity in the developing mouse forebrain. Dev. Biol. 335 (1), 166–178.

Sulik, K.K., 1984. Craniofacial defects from genetic and teratogen-induced deficiencies in presomite embryos. Birth defects original article series 20, 79–98.

Takahashi, M., Tamura, K., Buscher, D., Masuya, H., Yonei-Tamura, S., Matsumoto, K., Naitoh-Matsuo, M., Takeuchi, J., Ogura, K., Shiroishi, T., et al., 1998. The role of Alx-4 in the establishment of anteroposterior polarity during vertebrate limb development. Development 125 (22), 4417–4425.

Taniguchi, K., Anderson, A.E., Sutherland, A.E., Wotton, D., 2012. Loss of Tgif function causes holoprosencephaly by disrupting the SHH signaling pathway. PLoS Genet. 8 (2), e1002524.

Thauvin-Robinet, C., Cossee, M., Cormier-Daire, V., Van Maldergem, L., Toutain, A., Alembik, Y., Bieth, E., Layet, V., Parent, P., David, A., et al., 2006. Clinical, molecular, and genotype-phenotype correlation studies from 25 cases of oral-facial-digital syndrome type 1: a French and Belgian collaborative study. J. Med. Genet. 43 (1), 54–61.

Thivichon-Prince, B., Couble, M.L., Giamarchi, A., Delmas, P., Franco, B., Romio, L., Struys, T., Lambrichts, I., Ressnikoff, D., Magloire, H., et al., 2009. Primary cilia of odontoblasts: possible role in molar morphogenesis. J. Dent. Res. 88 (10), 910–915.

Tobin, J.L., Di Franco, M., Eichers, E., May-Simera, H., Garcia, M., Yan, J., Quinlan, R., Justice, M.J., Hennekam, R.C., Briscoe, J., et al., 2008. Inhibition of neural crest migration underlies craniofacial dysmorphology and Hirschsprungs disease in Bardet-Biedl syndrome. Proc. Natl. Acad. Sci. USA 105 (18), 6714–6719.

Tran, P.V., Haycraft, C.J., Besschetnova, T.Y., Turbe-Doan, A., Stottmann, R.W., Herron, B.J., Chesebro, A.L., Qiu, H., Scherz, P.J., Shah, J.V., et al., 2008. THM1 negatively modulates mouse sonic hedgehog signal transduction and affects retrograde intraflagellar transport in cilia. Nat. Genet. 40 (4), 403–410.

Trockenbacher, A., Suckow, V., Foerster, J., Winter, J., Krauss, S., Ropers, H.H., Schneider, R., Schweiger, S., 2001. MID1, mutated in Opitz syndrome, encodes an ubiquitin ligase that targets phosphatase 2A for degradation. Nat. Genet. 29 (3), 287–294.

Valente, E.M., Logan, C.V., Mougou-Zerelli, S., Lee, J.H., Silhavy, J.L., Brancati, F., Iannicelli, M., Travaglini, L., Romani, S., Illi, B., et al., 2010. Mutations in TMEM216 perturb ciliogenesis and cause Joubert, Meckel and related syndromes. Nat. Genet. 42 (7), 619–625.

Wallis, D.E., Roessler, E., Hehr, U., Nanni, L., Wiltshire, T., Richieri-Costa, A., Gillessen-Kaesbach, G., Zackai, E.H., Rommens, J., Muenke, M., 1999. Mutations in the homeodomain of the human SIX3 gene cause holoprosencephaly. Nat. Genet. 22 (2), 196–198.

Wannasilp, N., Solomon, B.D., Warren-Mora, N., Clegg, N.J., Delgado, M.R., Lacbawan, F., Hu, P., Winder, T.L., Roessler, E., Muenke, M., 2011. Holoprosencephaly in a family segregating novel variants in ZIC2 and GLI2. Am. J. Med. Genet. A. 155A (4), 860–864.

Warr, N., Powles-Glover, N., Chappell, A., Robson, J., Norris, D., Arkell, R.M., 2008. Zic2-associated holoprosencephaly is caused by a transient defect in the organizer region during gastrulation. Hum. Mol. Genet. 17 (19), 2986–2996.

Watanabe, H., Shionyu, M., Kimura, T., Kimata, K., Watanabe, H., 2007. Splicing factor 3b subunit 4 binds BMPR-IA and inhibits osteochondral cell differentiation. J. Biol. Chem. 282 (28), 20728–20738.

Waters, A.M., Beales, P.L., 2011. Ciliopathies: an expanding disease spectrum. Pediatr. Nephrol. 26 (7), 1039–1056.

Weatherbee, S.D., Niswander, L.A., Anderson, K.V., 2009. A mouse model for Meckel syndrome reveals Mks1 is required for ciliogenesis and Hedgehog signaling. Hum. Mol. Genet. 18 (23), 4565–4575.

Wehby, G.L., Cassell, C.H., 2010. The impact of orofacial clefts on quality of life and healthcare use and costs. Oral. Dis. 16 (1), 3–10.

Wehrhan, F., Amann, K., Molenberg, A., Lutz, R., Neukam, F.W., Schlegel, K.A., 2012. PEG matrix enables cell-mediated local BMP-2 gene delivery and increased bone formation in a porcine critical size defect model of craniofacial bone regeneration. Clin. Oral. Implants. Res. 23 (7), 805–813.

Wei, X., Sulik, K.K., 1993. Pathogenesis of craniofacial and body wall malformations induced by ochratoxin A in mice. Am. j Med. Genet./ 47, 862–871.

Weiner, A.M., Scampoli, N.L., Calcaterra, N.B., 2012. Fishing the molecular bases of Treacher Collins syndrome. PLoS One 7 (1), e29574.

Willaredt, M.A., Hasenpusch-Theil, K., Gardner, H.A., Kitanovic, I., Hirschfeld-Warneken, V.C., Gojak, C.P., Gorgas, K., Bradford, C.L., Spatz, J., Wolfl, S., et al., 2008. A crucial role for primary cilia in cortical morphogenesis,. J. Neurosci. 28 (48), 12887–12900.

Winokur, S.T., Shiang, R., 1998. The Treacher Collins syndrome (TCOF1) gene product, treacle, is targeted to the nucleolus by signals in its C-terminus. Hum. Mol. Genet. 7 (12), 1947–1952.

Wotton, D., Lo, R.S., Lee, S., Massague, J., 1999. A Smad transcriptional corepressor. Cell 97 (1), 29–39.

Wuyts, W., Cleiren, E., Homfray, T., Rasore-Quartino, A., Vanhoenacker, F., Van Hul, W., 2000a. The ALX4 homeobox gene is mutated in patients with ossification defects of the skull (foramina parietalia permagna, OMIM 168500). J. Med. Genet. 37 (12), 916–920.

Wuyts, W., Reardon, W., Preis, S., Homfray, T., Rasore-Quartino, A., Christians, H., Willems, P.J., Van Hul, W., 2000b. Identification of mutations in the MSX2 homeobox gene in families affected with foramina parietalia permagna. Hum. Mol. Genet. 9 (8), 1251–1255.

Zaghloul, N.A., Brugmann, S.A., 2011. The emerging face of primary cilia. Genesis 49 (4), 231–246.

Zakin, L., De Robertis, E.M., 2004. Inactivation of mouse Twisted gastrulation reveals its role in promoting Bmp4 activity during forebrain development. Development 131, 413–424.

Zhang, W., Kang, J.S., Cole, F., Yi, M.J., Krauss, R.S., 2006. Cdo functions at multiple points in the Sonic Hedgehog pathway, and Cdo-deficient mice accurately model human holoprosencephaly. Dev. Cell. 10 (5), 657–665.

Zheng, Q., Sebald, E., Zhou, G., Chen, Y., Wilcox, W., Lee, B., Krakow, D., 2005. Dysregulation of chondrogenesis in human cleidocranial dysplasia. Am. J. Hum. Genet. 77 (2), 305–312.

Chapter 36

22q11 Deletion Syndrome: Copy Number Variations and Development

Alejandra Fernandez*,†,§, Daniel Meechan*,†, Jennifer L. Baker**, Beverly A. Karpinski*,‡, Anthony-Samuel LaMantia*,† and Thomas M. Maynard*,†

*GW Institute for Neuroscience; †Department of Pharmacology and Physiology; **Center for the Advanced Study of Hominid Paleobiology and Department of Anthropology; ‡Department of Anatomy and Regenerative Biology; §GW Institute for Biomedical Sciences, The George Washington University School of Medicine and Health Sciences, Washington DC

Chapter Outline

GLOSSARY

Copy number variation (CNV) disorder A genetic disorder caused by the heterozygosity of one or more genes (such as in a microdeletion). These disorders result from the reduced expression of the deleted genes, rather than complete absence or a mutation.

Hypoplasia A clinical term for the reduced size of an organ or structure, often caused by disruptions in embryonic development.

Low copy repeats (LCR) A term for the highly homologous segments of DNA, which are the result of segmental duplications and other recombination events, that mediate microdeletion events. There are four primary LCRs in the 22q11.2 region, which are named LCR A–D.

Microdeletion A chromosomal deletion of between hundreds to the low millions of base pairs. Unlike larger chromosomal deletions, microdeletions are not readily detectable in microscopic analysis of chromosomal banding patterns (karyotype)

Non-allelic homologous recombination (NAHR) A mechanism responsible for causing microdeletions and other chromosomal rearrangements. NAHR occurs during meiosis if inappropriate recombination events occur between non-homologous segments of chromosome (either on the same chromosome, or different chromosomes).

Polymicrogyria A disorder of the cerebral cortex, distinguished by the presence of smaller than normal folds (gyri) in the cortical surface. This condition can be very local, affecting only a small segment of the cortex, or it can encompass entire cortical lobes.

Segmental duplication (SD) A genomic process that results in the duplication of portions of a chromosome. This can occur due to errors in meiotic recombination.

SUMMARY

- 22q11.2 deletion syndrome (22q11DS), also known as DiGeorge syndrome, is a genetic disorder caused by the deletion of 1.5–3.0 MB of chromosome 22q11.2.

Principles of Developmental Genetics.

- The relatively high incidence of 22q11DS (1 in 2,000–4,000 live births) is due to the genomic instability of the 22q11.2 region. Several segments of highly repetitive DNA, called segmental duplications or low copy repeats, are found in the region, and non-homologous recombination between these sites leads to chromosomal deletions.
- 22q11.2 deletion leads to the disruption of numerous developmental processes, and is associated with cardiovascular defects, craniofacial anomalies, and cognitive and psychiatric disorders, including autism and schizophrenia.
- Craniofacial and cardiac defects associated with 22q11DS are largely the result of disrupted development of the embryonic branchial arches.
- The cognitive and psychiatric disorders associated with 22q11DS appear to result from disruptions in the proliferation and migration of neural progenitor cells during brain development.
- 22q11DS is associated with numerous anomalies and deficiencies in immune, endocrine, and metabolic functions.
- One key 22q11.2 gene, *Tbx1*, appears to be central for some (but not all) craniofacial and cardiac anomalies, but *Tbx1* is not by itself sufficient to account for many of the other 22q11DS phenotypes.
- The outcome of 22q11DS is highly variable, and this variability likely involves the interaction of multiple 22q11.2 genes with multiple developmental signaling pathways, and the severity of individual phenotypes may be ameliorated or exacerbated by environmental factors.

36.1 22Q11DS IS A GENOMIC DISORDER WITH WIDESPREAD CONSEQUENCES FOR DEVELOPMENT

The Initial Identification of 22q11DS Provides Clues to its Variability

22q11 deletion syndrome (22q11DS) is a genomic disorder that results from the deletion of a 1.5–3 megabase (MB) deletion of chromosome 22q11.2. 22q11DS is the most common survivable genetic deletion disorder known, with an estimated incidence of 1/2,000–4,000 live births (reviewed by Shprintzen, 2008; McDonald-McGinn and Sullivan, 2011; Robin and Shprintzen, 2005). The heterozygous loss of this chromosomal region is associated with a spectrum of congenital defects and psychiatric disorders that suggests that the syndrome has its root in the disruption of numerous developmental processes.

The identification of 22q11DS long predates the discovery of its genetic origins, and the convoluted history of the discovery of the syndrome belies its diverse constellation of anomalies and broad developmental origins (Robin and Shprintzen, 2005). Descriptions of individuals with what appear to be 22q11DS appear in the clinical literature at least as early as the 1950s, although it was not at the time recognized as a defined syndrome. However, beginning in the late 1960s, multiple clinical groups around the world began to define the syndrome from different perspectives. The pediatric endocrinologist Angelo DiGeorge described three children with immunological and endocrine deficiencies resulting from absent or severely hypoplastic thymus and parathyroid glands, which he suggested were due to disrupted development of the 3rd and 4th branchial arches. Although it was defined strictly by the loss of the thymus, as additional case reports of "DiGeorge syndrome" were reported in the literature it became very clear that DiGeorge syndrome was associated with other congenital anomalies, including cardiovascular defects and craniofacial anomalies (Lischner, 1972).

While DiGeorge syndrome was becoming established in the literature, other clinicians were describing other syndromes with features that overlapped with DiGeorge syndrome from different clinical perspectives (Robin and Shprintzen, 2005). Primary among these was a report in the late 1970s from the speech-language pathologist Robert Shprintzen (Shprintzen et al., 1978) that reported on the cases of several patients with submucosal or overt cleft palate, hypernasal speech, and learning disabilities; most of these individuals had cardiac malformations (ventricular-septal defects), other craniofacial anomalies (large nose, long ears, short mandible/overbite, ear malformations), and poor fine motor control. This constellation of anomalies was named Velocardiofacial syndrome (VCFS), and despite the similarities with DiGeorge syndrome, it was not clear that VCFS, DiGeorge, and other reported cases and syndromes were actually different perspectives on the same syndrome. In the early 1990s, Scambler (Scambler et al., 1992) and Driscoll (Driscoll et al., 1993), found microdeletions in chromosome 22q11.2 in a majority of cases identified as DiGeorge syndrome and VCFS.

After specific genetic tests became available for 22q11DS, it became possible to distinguish the array of phenotypes directly resulting from 22q11 deletion, and exclude non-22q11 deleted individuals with overlapping anomalies, in order to more clearly define the range of phenotypes associated with the syndrome. With a precise genetic lesion identified, it is now commonplace to refer to the syndrome as 22q11.2 deletion syndrome (22q11DS) – a clearly defined neurodevelopmental disorder that results in a distinct set of congenital anomalies, cognitive impairments, and risks of psychiatric disorders.

The molecular identification of 22q11DS has been crucial to understanding the syndrome, as the phenotypes of individuals with 22q11DS are widespread and highly variable. In the clinical literature, more than 180 clinical anomalies and features have been associated with the deletion (Robin and, Shprintzen, 2005), including phenotypes that disrupt numerous organ systems and physiological functions (Table 36.1). A number of these phenotypes are clearly hallmarks of this

TABLE 36.1 Spectrum of Phenotypes Observed in 22q11DS, and Their Likely Developmental Origins

Craniofacial anomalies
- cleft palate (overt or submucosal)
- Retrognathia (short lower jaw)
- Platybasia (flat skull base)
- Asymmetric face (structural and functional)
- Long face
- Dental anomalies (small or missing teeth, thin enamel)
- Skull anomalies (microcephaly, small posterior cranial fossa)
- Ear malformations (Protuberant, cup-shaped ears, overfolded helix, small ears)
- Prominent nasal bridge

Musculoskeletal anomalies
- Short stature
- Scoliosis
- Spina bifida occulta (incomplete spinal fusion)
- Vertebral anomalies (fused vertebrae, butterfly vertebrae)
- Hyperextensible joints
- Tapered digits with short fingernails
- Low muscle tone
- Contractures of hands and feet

Ear and Eye anomalies
- Mild conductive hearing loss
- Sensorineural hearing loss (often unilateral)
- Tortuous retinal vessels
- Strabismus ("crossed eyes")
- Narrow palpebral fissures ("eye slits")

Cardiovascular anomalies
- Defects in atrial or ventricular septum (ASD/VSD)
- Anomalies of cardiac vasculature (absent or constricted pulmonary artery, abnormal position of vasculature including right sided aorta, or stenosis
- Severe malformations of cardiac outflow tract and cervical arteries, including truncus arteriosus, interrupted aortic arch, aortic valve anomalies, displaced or malformed carotid arteries, circle of Willis anomalies)
- Tetralogy of Fallot (a combination of cardiac malformations highly associated with 22q11 DS)

Oral and speech dysfunction
- Velopharyngeal incompetence (weakness or lack of coordination in movment of pharynx and soft palate during speaking and swallowing)
- Feeding difficulty as infants (slow weight gain, frequent vomiting and gastroesophageal reflux)
- Frequent ear, sinus, and lung infections due to aspiration (particularly in infants)
- Severely hypernasal speech

Immunological and Endocrine dysfunction
- Absent or small thymus and parathyroid
- Reduced thymic and parathyroid function, including reduced T-cell populations, reduced thymic hormone, hypocalcemia

Neurological, behavioral, and psychiatric consequences
- Cysts and calcifications along brain ventricles
- Small cerebellar vermis
- Seizures
- Schizophrenia
- Autism Spectrum Disorders
- Attention deficit hyperactivity disorder
- Mood disorders s
- Learning disabilities

This list (adapted and condensed from Robin and Shprintzen et al., 2005) contains features commonly associated with 22q11DS in multiple case reports. Each phenotype occurs with varying penetrance, and each individual will display an independent subset of phenotypes. While extensive, this list is not exhaustive: many other findings have been reported in 22q11 DS cases, but are not yet conclusively identified as being characteristic of the disorder.

deletion disorder, such as the cardiovascular and craniofacial defects described above; however, even these defects appear with variable (and apparently independent) frequency (Gerdes et al., 2001), such that one deleted individual may present with a cardiac defect, and only mild or undetectable craniofacial anomalies, while another may be identified on the basis of craniofacial anomalies, but lack cardiac defects. In addition, there is a wide range of phenotypes identified sporadically in individuals with 22q11DS. While it is possible that these infrequent phenotypes are either completely independent, or secondary to other complications of 22q11DS, experimental evidence in animals models has supported a linkage for a subset of these phenotypes, including neural tube defects (Nickel and Magenis, 1996) and lymphatic anomalies (Mansir et al., 1999), which have been observed in experimental analysis of animal models of 22q11DS (Maynard et al., 2013; Chen et al., 2010) (see also Chapter 20). In addition, as described below, the wide extent, and high degree of variability of 22q11 phenotypes is present in mouse models of 22q11DS, despite their uniform genetic lesions and homogeneous genetic backgrounds, which further confirms that variability is a central characteristic of the syndrome, and not a function of clinical misdiagnosis or complication by additional genetic lesions.

The Genomic Landscape of Chromosome 22q11.2 Elucidates a Mechanism for Chromosomal Deletions

About 70% of 22q11.2 deletions occur *de novo*, with most probably arising during gametogenesis; and of these roughly 2/3 occur during spermatogenesis (Rommel et al., 2008). While the number of live births with 22q11DS is quite high, the embryonic incidence of 22q11.2 deletions is likely far higher. When single spermatids were analyzed for the deletion by fluorescent hybridization analysis, the incidence was far higher (1 in 370), suggesting that when maternal deletions are included the incidence of deletions in the fertilized egg may be closer to 1 in 250. It is not clear why the majority of deleted gametes do not produce viable offspring; it is possible that there are differences in the extent or nature of the deletion in the non-viable gametes, and/or that there is substantial early embryonic lethality associated with 22q11DS that has not yet been characterized. Nevertheless, it is clear that the 22q11.2 region possesses a degree of genomic instability that leaves it prone to deletions.

Studies of the genomic landscape of 22q11.2 have revealed that this genomic region is particularly unstable due to the presence of multiple fragments of repeated DNA known as segmental duplications (SDs), which are generally referred to as low copy repeats (LCRs) in the 22q11DS literature (Shaikh et al., 2000, 2001, 2007; Gotter et al., 2004; Torres-Juan et al., 2007). SDs are segments of DNA with almost identical sequences that map to two or more highly dynamic regions in the genome (Babcock et al., 2007). Because of the high degree of similarity between these regions, SDs can facilitate non-allelic homologous recombination (NAHR) events between two lengths of DNA with high sequence similarity (Hurles et al., 2006; Beckmann et al., 2007). NAHR events occur during meiosis as an error in the normal process of recombination. Instead of crossover events occurring between homologous regions of the two chromosomes, non-homologous regions line up (generally along short stretches of repeated DNA) and cross over, leading to an inappropriate splicing of chromosomal DNA. These recombination events cause rearrangements in the genome, leading to the deletion or duplication of segments of genes, such as those that occur in 22q11DS (Guo et al., 2011). In the case of 22q11, the typically deleted region contains four major SDs, referred to as LCR A–D (Figure 36.1A). The most frequent deletion (in approximately 85% of individuals) is a recombination between LCR A on one chromosome with LCR D on the other (Shaikh et al., 2001). This recombination leads to the loss of a 3 MB region, although other smaller deletions are commonly observed.

The genomic sequence within the LCRs on 22q11.2 are enriched with multiple types of transposable elements, including *Alu* repeat elements and *SINE* (short interspersed nuclear element) elements), as well as AT-rich repeats. It is likely that these transposable elements and repetitive sequences may have been involved in creating the complex genomic architecture of the 22q11.2 LCRs (Babcock et al., 2007) by mediating NAHR events that duplicated sequences in the LCRs themselves. In addition to the wholesale deletion/duplication events mediated by the LCRs, 22q11.2 also displays a great number of small-scale copy number variations as well. These variations are likely generated by recombination between the same types of transposable elements (e.g., *SINE* and *Alu* repeats) that are found in the LCRs. These genomic rearrangements are not thought to occur arbitrarily, however, as *Alu* sequences (and particularly the *AluY* subtype) are thought to enhance the frequency of recombination in adjacent sequences (Guo et al., 2011).

Interestingly, at least some 22q11DS cases appear to be caused by a different, non-meiotic mechanism, probably occurring during early embryogenesis. These non-meiotic recombinations result in mosaicism, with only a fraction of the individual's cells or tissues containing the deletion (Scheet et al., 2012; Maiti et al., 2011; Notini et al., 2008). While 22q11DS mosaicism is probably somewhat rare, it has been identified in several individuals that display 22q11DS phenotypes (Halder et al., 2008; Consevage et al., 1996).

The Instability of 22q11.2 Likely Results from Recent Rearrangement Events

While it is clear that SDs create a huge risk of disease, recent studies have suggested that genomic regions including segmental duplications may be "hotspots" for genomic evolution (Bailey and Eichler, 2006; Lorente-Galdos et al., 2013). Although LCRs are particularly unstable, this genomic instability can potentially be a mechanism for creating new genes

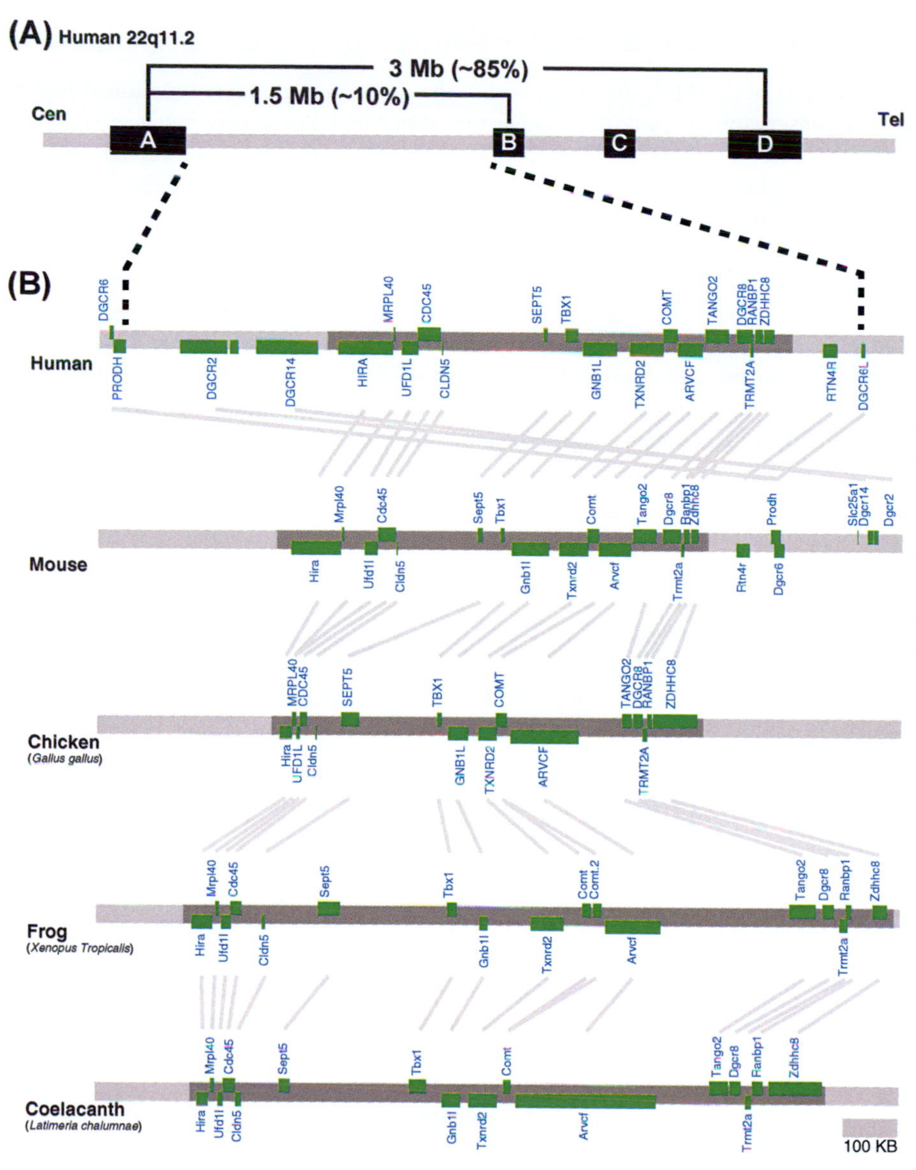

FIGURE 36.1 (A) Schematic diagram, illustrating relative locations of the four low copy repeats (LCRs) in human chromosome 22q11.2. These LCRs (referred to as A–D) are a genomic feature often referred to as segmental duplications, where fragments of DNA are duplicated as a result of evolutionary events and viral integration events. Because there are elements within each of these LCRs that have a high degree of homology with each other, recombination events can inappropriately cause recombination between disparate LCRs, leading to deletion or duplication of chromosomal segments. The recombination between LCR A and D accounts for the ~3 MB deletion found in most cases of 22q11DS; however, the less frequent recombination between LCR A and B is sufficient to cause all the key phenotypes of the syndrome. (B) The genomic region in the LCR A to B deletion is highly conserved between highly divergent vertebrate species. *Figure 1A is adapted from Shaikh et al., 2000.*

and modulating genomic function; and in some cases positive selection may act on such rearrangement events. Phylogenetic comparisons demonstrate that SDs have in fact contributed significantly to primate evolution in general, by allowing the generation of new genes as well as altering genetic variation that is thought to contribute to human disease susceptibility. As a whole, about 5% of the human genome contains SDs (She et al., 2004), whereas at 11.9%, human chromosome 22 contains more than twice that amount (Bailey and Eichler, 2006). As a result of multiple LCRs within a 9 MB region, this region of chromosome 22q11 chromosomal region is susceptible to microduplications and microdeletions (Beddow, 2012), including deletions and duplications proximal and distal to the ones involved in 22q11DS.

Investigating the 22q11 LCRs as they appear in the genomes of different hominoid species provides clues to their complex evolutionary history. There is a dramatic amplification of the complexity of the LCRs in the orthologous region of 22q11 in gibbons, and multiple orthologous LCRs appear in the telomeric regions of many chromosomes in gorillas. Analyses of the human 22q11.2 region indicate that several of its duplicated sequences appeared recently by both small and large scale duplications in either the human or the African great ape lineage and contribute to the meiotic instability and genome dynamics of human 22q11.2 (Guo et al., 2011). By tracing both the substrates and products of recombination events, Babcock et al. (2007) were able to ascertain the mechanisms responsible for shaping the LCRs in chromosome 22q11.2 during primate evolution, and determined that the complex genomic architecture of 22q11.2 occurred as a result of "genome shuffling," largely mediated by *Alu*-type transposable element sequences. Thus, the high incidence of 22q11DS appears to be in large part a consequence of recent evolutionary changes in the structure of 22q11.2.

The Genes within the "Minimal Deleted Region" are Highly Conserved in Model Organisms

Despite the complex changes in the LCR regions of 22q11.2, the genes within the deletion region have remained highly conserved across all vertebrate species. This conservation became apparent as the human and mouse genomes were mapped and sequenced (Puech et al., 1997), and this orthology is particularly robust for the so-called "minimal deleted region" between LCR A and B. This 1.5 MB deletion region is observed in approximately 10% of deleted individuals, but this segment is included within the larger 3 MB deletion observed in the vast majority (85%) of individuals with 22q11DS. This smaller 1.5 MB deletion has been the focus of most genetic studies, as no significant phenotypic differences have been identified that distinguish it from deletions of the larger (often called the "typical") 3 MB deletion observed in most 22q11DS individuals (Michaelovsky et al., 2012; Carlson et al., 1997). This segment is highly conserved, with only modest rearrangements between humans and rodents. The "core" of this minimal region remains syntenic and highly conserved in the phylogenically-divergent mouse, chicken, frog and coelacanth (primitive jawless fish) genomes (Figure 36.1B). Furthermore, the protein sequences of the individual genes within this region are highly conserved across vertebrates, a feature which allows the study of the function of individual 22q11 genes, and in fact, the function of the genomic region itself, in a variety of model organisms.

Because of the high degree of conservation of the individual genes within the "minimal deletion region," studies using model organisms have been particularly instructive in understanding the role of individual 22q11.2 genes in causing individual disease phenotypes. A major advance in the study of 22q11DS developmental biology occurred in the late 1990s and early 2000s, when two groups adapted the experimental Cre-Lox method, previously used for manipulating the expression of single genes, to generate mouse models of multiple gene deletions. Two transgenic mouse lines, the *Df1* model (Lindsay et al., 1999) and the *LgDel* model (Merscher et al., 2001), which model deletions orthologous to the minimal 1.5 MB region of 22q11.2, resulted from these initial exercises, and have been particularly useful in determining the developmental consequences of 22q11.2 deletion.

36.2 22Q11DS AS VIEWED FROM A DEVELOPMENTAL PERSPECTIVE

From the time of its initial discovery, it has been clear that 22q11DS is a developmental disorder that impacts the formation and function of multiple embryonic structures, across numerous tissue compartments and organ systems. As illustrated in Table 36.1, deleted individuals have a high incidence of congenital anomalies that reflect the disrupted development of numerous structures, including the face, mouth, teeth, ears, eyes, cardiovascular system, immune system, skeletal structures, and the brain and other neural structures. Although these disparate organs, tissues, and structures have little in common in an adult, they share many similarities in their developmental origins when investigated at the cellular and molecular level.

In the remainder of this review, we will discuss how 22q11DS disrupts individual developmental processes, and how these disruptions lead to the functional impairments associated with 22q11DS. Rather than covering the entirety of the 22q11DS literature in an exhaustive or superficial fashion, we have selected several developmental processes that are relatively well characterized and provide insight into key aspects and features of 22q11DS from a developmental viewpoint.

Cardiac Malformations Provide Insight into *Tbx1* Function

The best characterized and most frequently identified anomalies in 22q11DS are cardiovascular anomalies. Structural defects in the heart and cardiopulmonary vasculature are observed in approximately 75–80% of 22q11-deleted individuals (Botto et al., 2003; Ryan et al., 1997), and include ventriculoseptal and atrioseptal defects (failure of the cardiac septum that separates left and right heart chambers to close), defects in the cardiac valves, and malformations of the great vessels (the aorta, vena cava, and their branches). One particular variant that is closely identified with 22q11DS is tetralogy of Fallot, a combination of cardiac malformations, including a ventriculoseptal defect, enlarged right ventricle, stenosis (constriction) of the pulmonary artery, and an "overriding aorta" (misplacement of the aortic valve). These cardiac malformations often require significant neonatal intervention, and are a significant challenge for many 22q11DS patients.

Because cardiovascular anomalies are of such critical clinical importance in 22q11DS, significant attention was directed towards understanding their etiology. Thus, it is not very surprising that three independent groups, using independently generated mouse models, identified *Tbx1* as a key gene in the 22q11.2 region for causing cardiovascular anomalies in 2001 (Merscher et al., 2001; Lindsay et al., 2001; Jerome and Papaioannou, 2001). Two of these studies began by creating and characterizing mouse models that deleted the mouse ortholog of the minimal deletion region, the *Df1* (Lindsay et al., 1999; 2001) and the *LgDel* (Merscher et al., 2001) mouse models. These mice carrying deletions orthologous to the complete 1.5 MB deletion have cardiac anomalies that mirror those seen in human 22q11DS.

As illustrated in Figure 36.2, branchial (or pharyngeal) arch arteries initially appear along with the arches themselves, which occurs in the mouse embryo between E9.5–10.5. Each arch has its own associated artery, which is surrounded by mesenchymal cells derived both from the mesoderm and the neural crest. Over the next several days, these arch arteries are remodeled to form the complex network of cardiovascular arteries, including the aortic arch itself as well as its primary branches (including the carotid arteries), as well as the pulmonary artery. During this remodeling process, segments of the arch arteries expand or regress to form the mature cardiovascular structures of the aortic arch and the pulmonary artery. A failure of any segment to properly remodel results in cardiovascular malformations, with potentially lethal consequences.

In both the Df1 and *LgDel* models of 22q11DS, the initial formation of the branchial arch arteries is disrupted. In particular, the 4th aortic arch artery is hypoplastic in both the *Df1* (Lindsay et al., 2001;) and *LgDel* (Maynard et al., 2013) models of 22q11DS. This early cardiovascular anomaly prefigures severe aortic arch malformations observed in *LgDel* embryos just prior to birth; 15% of deleted embryos have vascular defects that would be critical or lethal if uncorrected in 22q11DS infants. In particular, 10% of embryos had a defect referred to as a type-B interrupted aortic arch, where the segment of the aortic arch normally derived from the 4th pharyngeal arch was absent, which severely interrupts blood flow to the descending aorta. While virtually all 22q11DS embryos show branchial arch artery malformations, the majority of them appear to recover and produce phenotypically normal cardiac vasculature during later embryonic stages; it appears that the early growth delay that results from 22q11 deletion can be overcome at later stages of embryonic growth (Lindsay and Baldini, 2001). These studies clearly showed that both the *Df1* and *LgDel* models, which recapitulate the full "minimal" 22q11 deletion, accurately model the cardiac anomalies associated with 22q11DS. However, a significant question remained: could a singular genetic cause for cardiovascular defects in 22q11DS? The answer seems to be a somewhat qualified "yes."

Using different genetic approaches, these two studies narrowed the genetic region responsible for these cardiovascular defects (Merscher et al., 2001; Lindsay et al., 2001). Both of these studies identified the same gene, *Tbx1*, as a single gene in the region that could replicate the cardiovascular phenotypes observed in the models of the full 1.5 MB deletion. When *Tbx1*$^{+/-}$ (heterozygote) embryos are assessed, the cardiovascular phenotypes identified in the *Df1* and *LgDel* models are recapitulated at similar frequencies. Thus, *Tbx1* is certainly a central gene that is primarily responsible for the incidence of cardiovascular defects in 22q11DS. (see also Chapter 20).

The mechanism by which reduced *Tbx1* function leads to cardiac malformations appears to be somewhat complex and indirect. The primary deficit in 22q11DS (and *Tbx1* deleted) embryos is believed to arise from a failure of cardiac neural crest cells to appropriately pattern the branchial arch arteries (and, in particular, the 4th arch artery) (Vitelli et al., 2002). While the initial migration of the neural crest is apparently not disrupted in either *Df1* or *Tbx1* embryos (Lindsay and Baldini, 2001; Vitelli et al., 2002), *Tbx1* disruption appears to prevent neural crest cells from appropriately distributing

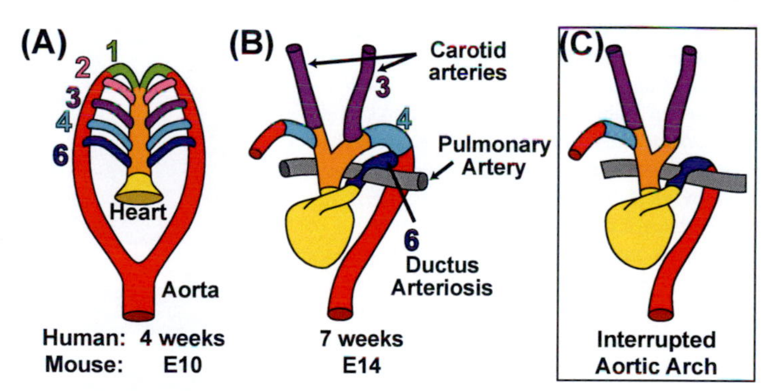

FIGURE 36.2 **Simplified Schematic of Cardiac Development, Illustrating the Remodeling of the Branchial Arch Arteries to Form the Arteries of the Aortic Arch.** (A) The cardiac vasculature arises from the arteries of the branchial arch. Each arch, except the fifth (which is only present as a transient structure) has its own artery that connects the outflow of the heart tube (truncus arteriosis, yellow, and aortic sac, orange) to the aorta (red). (B) As heart development proceeds, the branchial arch arteries are remodeled to produce the aortic arch arteries. The third arch artery is remodeled to form the carotid arteries, the left fourth arch artery becomes part of the aortic arch (also known as the ascending aorta), and the left sixth arch artery connects the right ventricle of the heart to the pulmonary arteries and to the aorta. This latter segment connecting the pulmonary artery to the aorta is an important transient structure called the ductus arteriosis. The ductus arteriosis allows blood flow from the right ventricle to bypass the non-functioning lungs during development, and it normally closes at birth. (C) Illustrated here is a simplified schematic of one of the many different types of cardiac malformations caused by 22q11, an interrupted aortic arch. In this malformation, the left fourth arch artery recedes and does not form the aortic arch. At birth, the lower half of the body shares its blood flow with the lungs, a condition that requires immediate surgical repair. *Figures 36.2A and B are adapted from Larsen's Human Embryology, 4th edition (2009), Figure 13–16. (Schoenwolf, G.C., Larsen, 2009)*

themselves around the arch arteries. *Tbx1* is primarily expressed in the endoderm of the branchial arches (Vitelli et al., 2002), and not in the neural crest cells themselves, and it appears that a loss of *Tbx1* disrupts the final stages of cardiac neural crest migration by altering the expression of some as yet unidentified migratory cues or growth factors that permit proper distribution of crest cells in the pouches of the branchial arches (Vitelli et al., 2002; Calmont et al., 2009). These branchial arch neural crest cells act as a signaling center that patterns the arch arteries by secreting retinoic acid (RA). Cardiac neural crest populations generate higher than normal levels of RA, and genetically reducing the level of RA in *Tbx1* heterozygotes (by generating a $Tbx1^{+/-}$; $Raldh2^{+/-}$ compound heterozygous embryo) ameliorates the hypoplasia of the 4th arch artery (Ryckebusch et al., 2010; Maynard et al., 2012).

While the evidence is convincing that *Tbx1* is central to mediating cardiovascular malformations in 22q11DS, recent evidence suggests that it may not act alone. While reducing RA levels can rescue 4th arch defects in *Tbx1* heterozygotes (Ryckebusch et al., 2010; Maynard et al., 2012), a similar reduction of RA levels leads to a worsening of 4th arch phenotypes in the *LgDel* model of 22q11DS (Ryckebusch et al., 2010). In fact, cardiovascular anomalies in the *LgDel* model of the complete 22q11.2 deletion are significantly worsened by modest decreases or increases in RA levels (Ryckebusch et al., 2010), even though these disruptions are subtle enough to leave their wild-type littermates largely unaffected (Figure 36.3). This increased susceptibility to RA manipulations suggests that 22q11DS leads to a dysregulation of RA signaling, possibly by altering the normal homeostatic mechanisms that allow a normal embryo to compensate for modest fluctuations in RA levels that likely occur during normal development.

While it is obvious that *Tbx1* is a central gene for cardiovascular defects in 22q11DS, it is clear that other 22q11.2 genes cannot be ignored. *Tbx1* deficiency may interact with disruptions in other genes, including other 22q11 genes, to mediate cardiovascular defects. Studies in *Xenopus* embryos show that depletion of *Tbx1* leads to downregulation of *Arvcf* (Tran et al., 2011), a 22q11.2 gene in the β-catenin family that has been implicated as a modulator of multiple developmental signaling pathways, including Wnt (Hong et al., 2010; Hatzfeld, 2005; Fang et al., 2004). Depletion of *Arvcf* levels can cause the same type of arch artery defect in *Xenopus* as does *Tbx1* depletion (Tran et al., 2011), suggesting that the heterozygosity of both genes in 22q11DS may lead to a concatenation of genetic effects. Furthermore (as described below), *Tbx1* does not appear to be a central gene in many of the other anomalies associated with 22q11DS. For example, cardiac dysfunction obviously has wide-ranging consequences, but there is no significant correlation between cardiac dysfunction in 22q11DS and the impairments in cognitive/behavioral function that often accompany 22q11DS (Yi et al., 2013; Butcher et al., 2012). Thus, *Tbx1* is a central gene for the etiology of 22q11DS, but it is not uniquely responsible for the widespread effects of the syndrome.

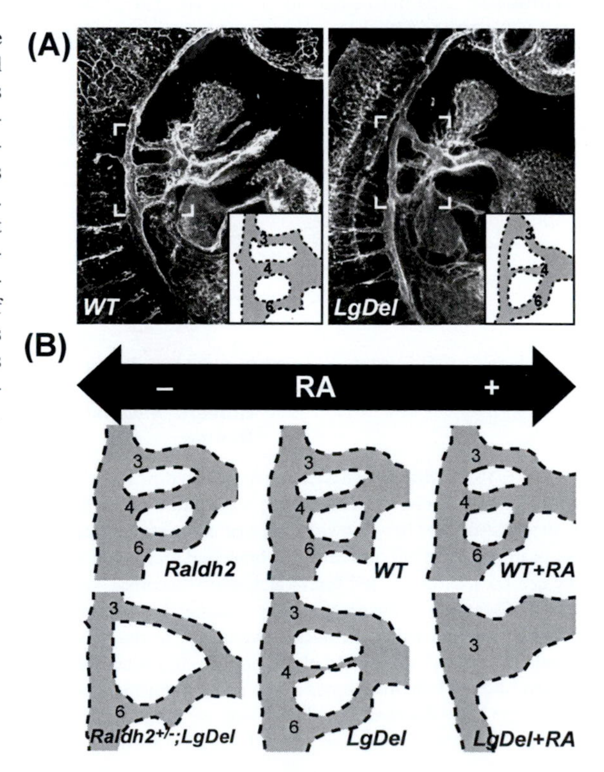

FIGURE 36.3 Aortic Arch Malformations are Observed in the *LgDel* Mouse Model of 22q11DS. (A) Developing vasculature in wild-type control (WT) and LgDel embryos at E10.5, visualized by immunofluorescent staining for PECAM. Brackets outline the region containing the 3rd, 4th, and 6th Most LgDel embryos have a constricted 4th arch artery (stenosis). A schematic version of the arch vasculature is provided as an inset for additional clarity. (B) Cardiac development in *LgDel* embyros is highly sensitive to even modest changes to embryonic retinoic acid (RA) levels. Center panels show WT and LgDel vasculature under normal conditions. A modest reduction in RA levels in the *Raldh2* heterozygote (left panels) does not cause detectable changes in aortic arch morphology by itself, but it severely disrupts arch development in a compound *LgDel;Raldh2* embyro. Likewise, injection of low-levels of RA (right panels) does not significantly impair arch development by itself, but causes catastrophic anomalies in a *LgDel* embryo. These results suggest that RA signaling is dysregulated in the LgDel embyro, and normal mechanisms that compensate for modest fluctuations in RA levels are impaired. *Figure adapted from Maynard et al., 2013.*

Disrupted Craniofacial Development in 22q11DS Leads to Feeding and Swallowing Complications

Craniofacial defects and anomalies of several different forms are strongly associated with 22q11DS. It has long been noted that a particular set of facial features is observed in individuals with 22q11DS, including a long face, retruded lower jaw, high nasal bridge, small ears, jaw, and mouth, and wide-set almond-shaped eyes (Shprintzen, 2008). Many deleted individuals have asymmetric facial features, including "asymmetric crying facies" in infants, which is a readily observable sag in the features on one side of the face when an infant is crying; this feature is observed in around 15% of 22q11DS infants (Pasick et al., 2013). Of particular interest are defects and anomalies of the palate, which can strongly impact feeding and swallowing, as well as speech. Most deleted individuals (50–75%; Ryan et al., 1997; Lay-Son et al., 2012) have some structural or functional orofacial anomaly, although these anomalies can be somewhat subtle in their presentation. An overt cleft palate is observed in approximately 10–15% of deleted individuals (Ryan et al., 1997; Lay-Son et al., 2012), and is found along with other structural anomalies, such as a failure to correctly form palatal bone and cartilage (submucosal cleft palate), or the formation of a high arched palate in the majority of deleted individuals (Ryan et al., 1997; Lay-Son et al., 2012). Tooth formation is also disrupted in 22q11DS, with the majority of individuals having dental issues, including thin enamel, missing or asymmetric teeth, or aberrantly shaped teeth (Klingberg et al., 2002). Further complicating oral function is that 1 in 3 deleted individuals have velopharyngeal insufficiency (VPI), an inability of the muscles of the soft palate to properly elevate to close off the nose during speech and swallowing. Further studies have shown that many deleted individuals have significant difficulty in coordinating orofacial muscles (including the palate and tongue) during swallowing, often leading to dysphagia – gagging, aspiration of food, and other complications (Eicher et al., 2000). For swallowing to properly occur, the muscles of the face, tongue, jaw, and throat need to move in a coordinated fashion in order to close off the nasopharynx and trachea and push food from the mouth and into the esophagus. Studies of children with 22q11DS have revealed that their swallowing is uncoordinated, and this reduced level of motor control often results in gagging or regurgitation (Eicher et al., 2000). These feeding and swallowing complications occur in deleted infants and children, even in the absence of any detectable cardiac or palate defects (Eicher et al., 2000), and are likely a key reason why 22q11DS infants fail to gain weight normally (Tarquinio et al., 2012), despite their normal weight at birth, and may contribute to the increased incidence of ear, nasal sinus, and lung infections in deleted infants and children (Oskarsdottir et al., 2005).

Feeding and swallowing complications are a fundamental issue in 22q11DS. Therefore, it is perhaps not surprising that feeding disruptions are also apparent in a mouse model of 22q11DS. *LgDel* mice gain weight slowly, frequently aspirate milk into the nose and lungs, and frequently have evidence of nasal sinus, ear, and lung infections (Karpinski et al., 2013). At least part of the cause of this dysphagia is apparently due to craniofacial dysmorphology: when *LgDel* embryos are examined at E14.5, the palatal shelves fail to elevate in around 30% of embryos (Karpinski et al., 2013). The palatal shelves, which begin as ventrally-oriented flaps located on either side of the tongue, normally elevate and extend by E14.5, and fuse together by E15 to form the palatal structures that separate the mouth from the nasopharynx. A failure of the palatal shelves to elevate and fuse leads to an overt cleft palate, but even in cases where the palatal shelves eventually fuse after significant delay, these early palatal abnormalities likely prefigure more subtle anomalies in palatal structure referred to as submucosal clefts. These submucosal clefts are common in 22q11DS patients (McDonald-McGinn et al., 1993), and impact the normal functions of the mouth and palate during swallowing and speech.

In addition to structural defects, there is also evidence that there are defects in the nervous system's control of swallowing. Observation of the mechanics of swallowing in infants and children with 22q11DS reveals a lack of coordination in the muscles controlling swallowing, including hypercontractility of the pharyngeal muscles (Eicher et al., 2000). One insight into the proximal source of this lack of coordination comes from studies of the *LgDel* mouse embryo; in these embryos, there are significant anomalies in the initial formation of key cranial nerves for chewing and swallowing: the trigeminal (V), glossopharyngeal (IX), and vagus (X) nerves (Karpinski et al., 2013). In *LgDel* embryos, the trigeminal nerve is dysmorphic, with shortened, sparse and defasciculated branches that fail to properly enter the tissues of the maxilla and mandible (Figure 36.4A). The embryonic glossopharyngeal and vagus nerves are also dysmorphic; in *LgDel* embryos these two nerves frequently form adherent connections, or actually fuse together to form a single dysmorphic ganglion. As each of these nerves normally has both sensory and motor functions for key structures necessary for swallowing – including the jaw, tongue, pharynx, and larynx – any disruption in the organization or projections of these nerves would be detrimental to the coordinated muscle movements required for swallowing.

While this constellation of complications includes many different structures throughout the head and neck, it appears that the genesis of both the craniofacial malformations and feeding disorders begins early in development, before the appearance of faces and jaws, with subtle disruptions in the patterning of the embryonic hindbrain (Karpinski et al., 2013). When the hindbrain is examined in *LgDel* embryos at E9.5, the anterior segments are abnormally shortened, and a molecular

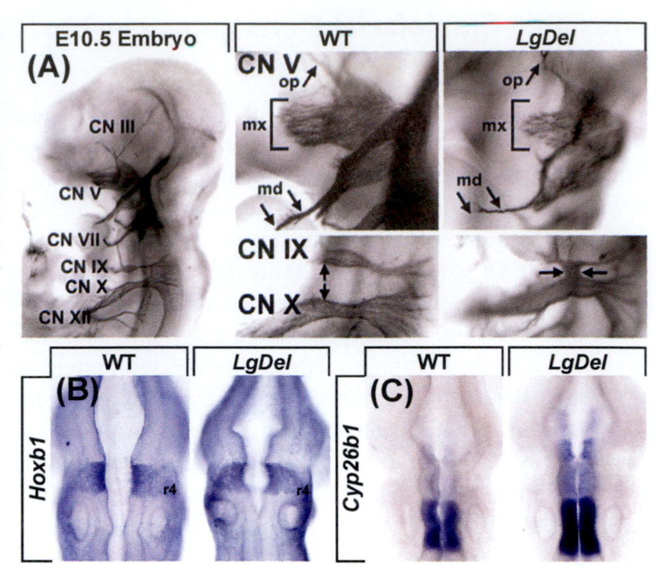

FIGURE 36.4 Feeding and Swallowing Problems in the LgDel Mouse Appear to Have a Developmental Origin. (A) To investigate the early development of the cranial nerves (CN), E10.5 embryos can be immoncytochemically stained for neurofilament to identify cranial nerve axons. Left: a representative E10.5 WT embryo, showing the normal appearance of the cranial nerves (CN) Right top panels: A close-up view of CN V and IX/X from control (WT) and LgDel embryos. In the *LgDel* embryo, CN V appears sparse and defasciculated, and the ganglia of CN IX and X are fused. (B) It is likely that some of the craniofacial anomalies in the *LgDel* mouse originate during hindbrain patterning. Examination of E9.5 WT and *LgDel* embyros for expression of *Hoxb1*, a homeobox gene that marks hindbrain rhombomere (r) 4, shows that the structures anterior to r4 are shortened in the *LgDel* embryo. (C) The expression of *Cyp26b1*, which is normally restricted to the posterior of the E9.5 hindbrain, is expanded into anterior regions in the *LgDel* hindbrain. *Figure adapted from Karpinski et al., 2013.*

marker that is normally associated with posterior hindbrain segments (*Cyp26b1*) extends more anteriorly than in wild type littermates (Figure 36.4B, C). The embryonic hindbrain not only gives rise to important central nervous system (CNS) structures of the medulla and pons that coordinate swallowing, but also to neural crest cells that are crucial for forming the cranial ganglia and craniofacial structures including the palate.

Because *Tbx1* has been shown to be crucial for cardiovascular development, numerous studies have explored the potential role for this gene in craniofacial development as well. While it is apparent that *Tbx1* is involved in aspects of craniofacial development that impact feeding and swallowing, heterozygosity of *Tbx1* alone cannot account for all the impacts of 22q11 deletion. $Tbx1^{+/-}$ (heterozygous) embryos do not show either the hindbrain dysmorphology or anterior expansion of posterior hindbrain markers (Karpinski et al., 2013). *Tbx1* heterozygosity partially recapitulates the cranial nerve dysmorphology observed in *LgDel* embryos – they have an identical incidence of fusions of the glossopharyngeal (IX) and vagus (X) nerves; however, they show no anomalies in the more anterior trigeminal (V) nerve or ganglion (Karpinski et al., 2013). Thus, 22q11 genes other than *Tbx1* appear to be essential for the proper patterning and organization of the hindbrain and cranial nerves.

In a similar fashion, it appears that *Tbx1* heterozygosity may be important, but not sufficient for causing palatal defects in 22q11DS. *Tbx1* is an important gene for palatal development, as *Tbx1* null embryos have significant defects in palatal development. *Tbx1* null embryos show decreased palatal shelf elevation (Jerome and Papaioannou, 2001; Goudy et al., 2010; Funato et al., 2012), and have inappropriate fusions between the epithelia surrounding the palatal shelf and adjacent oral epithelia (Funato et al., 2012). Although there is some evidence that *Tbx1* polymorphisms may be associated with non-syndromic cleft palate (Paranaiba et al., 2013), it does not appear likely that heterozygous loss of *Tbx1* is, by itself, sufficient to cause cleft palate. Heterozygous deletion of *Tbx1* does not lead to cleft palate in mouse models (Jerome and Papaioannou, 2001), although it is clearly sufficient to cause cardiovascular anomalies. Furthermore, there is no significant evidence that a weak or non-functional allele of the remaining copy of *Tbx1* in individuals with 22q11DS may underlie cleft palate in the syndrome (Herman et al., 2012), and there is no evidence for a correlation between cleft palate and cardiovascular defects in 22q11DS (Friedman et al., 2011), which might be expected if loss of *Tbx1* function were central to both phenotypes. Thus, it appears likely that *Tbx1* may be important for specific aspects of palate development, but its role in 22q11DS is dependent on other 22q11 genes.

22q11DS Disrupts Cortical Neurogenesis, with Behavioral Consequences

22q11DS is associated with a wide range of psychiatric conditions, including a high incidence of autism spectrum disorders and schizophrenia. As is the case with congenital anomalies of the heart and face, it has generally been assumed that 22q11DS disrupts aspects of brain development, and particularly of the cerebral cortex. Numerous studies have been undertaken to identify specific brain alterations in 22q11DS that may predispose patients to cognitive and behavioral disorders. Among these, a number of human magnetic resonance imaging (MRI) studies have revealed subtle yet complex anatomical perturbations in many deleted individuals, including thinning of the cerebral cortex, polymicrogyria, a reduction in the size

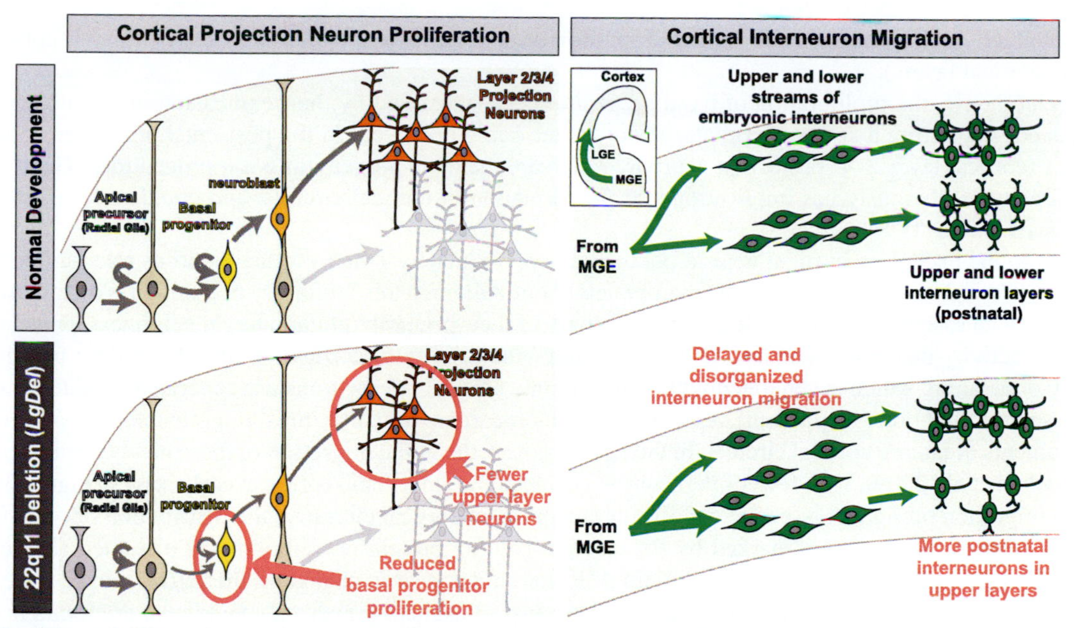

FIGURE 36.5 Schematic Illustration of Disrupted Development in the *LgDel* Cerebral Cortex. Left: The majority of neurons in the cerebral cortex are excitatory neurons that are generated within the cortex itself. The primary neuronal stem cells are self-renewing apical progenitors (radial glia) that generate both neurons that primarily populate the lower layers of the cortex, as well as a "transit amplifying" stem cell, the basal progenitors, that divide once to produce two neurons that primarily populate upper cortical layers. In the *LgDel* mouse, the basal progenitors do not proliferate as rapidly; this prefigures a reduction in upper layer excitatory neurons in the adult *LgDel* brain. Right: A second key population of neurons in the cortex are inhibitory neurons, which are primarily generated in a region of the ventral forebrain called the medial ganglionic eminence (MGE), and migrate into the cortex. In control embryos, as the embryonic interneurons migrate into the cortex, they form upper and lower streams, leading to their normal distribution in both upper and lower layers of the cortex. In the *LgDel* mouse, however, the migration is disorganized, and in the post-natal cortex interneurons are unevenly distributed, with more interneurons found in upper layers than in controls. These changes in cellular distribution suggest that the connectivity between cortical neurons and interneurons is likely disrupted, which may underlie cognitive and behavioral deficits in *LgDel* mice. *Figure adapted from Meechan et al., 2009; 2012.*

of the corpus callosum (the key nerve bundle linking the left and right sides of the brain), and other anomalies in structures along the midline of the brain (Bearden et al., 2007; Kraynack et al., 1999; Lee et al., 2003; van Amelsvoort et al., 2004). Analysis of postmortem brain tissue from a small sample of 22q11 schizophrenics have shown other anomalies in cortical structures (Kiehl et al., 2009), including the presence of neuronal cell bodies with cortical white matter (axonal tracts that normally lack neuronal cell bodies), as well as the presence of periventricular nodular heterotopias (PH). PH is a rare condition where newly born cortical neurons fail to migrate properly into the outer layers of the cortex, and instead form clumps (nodules) along the ventricle of the brain. These heterotopias may be a common feature of 22q11DS, as upon careful analysis others have identified similar heterotopias on the MRI scans of two 22q11DS children suffering from developmental delay (van Kogelenberg et al., 2010). These cortical alterations appear to be more frequent in 22q11DS children who have autistic spectrum disorders (ASD), ADHD (attention deficit hyperactivity disorder), and childhood-onset schizophrenia (Sporn et al., 2004). The appearance of cortical anomalies suggests that the high rates of psychiatric illness in 22q11DS are probably the consequence of disruptions in the formation of brain circuitry during embryonic development.

Because it is impossible to retrospectively identify what developmental process failed in a child or adult with psychiatric problems, our understanding of the role of 22q11DS in neural development comes from animal models. In several studies, Meechan et al. used the *LgDel* mouse model to identify disruptions in the development of the cerebral cortex, and test their functional consequences. They observed altered proliferation of a particular class of cortical stem cells in the embryonic *LgDel* cortex, basal progenitors (Figure 36.5, left). During cortical development, the majority of cortical cells (approximately 80%) are excitatory neurons that are generated from stem cells residing within the embryonic cortex, referred to as the pallium. The primary stem cell population resides along the apical surface (next to the ventricle), and cell divisions in this apical progenitor pool occur exclusively in cell nuclei that are directly positioned along the ventricle. While this apical progenitor stem cell pool is the only source of excitatory neurons in non-mammalian vertebrates such as chicken and turtles (Fietz and Huttner, 2011), the apical stem cells of mammals give rise to a secondary pool of "transit amplifying" cortical stem cells that reside on the basal side of the apical stem cell pool (Fietz and Huttner, 2011). While some of the excitatory neurons in the cortex are generated directly from apical cells (and particularly those in lower cortical layers), the basal progenitor stem cell pool produces a substantial proportion of cortical neurons (and particularly, though not exclusively, cells in upper cortical layers).

In the *LgDel* mouse, the proliferation of basal progenitor cells (identified by their expression of the molecular marker *Tbr2*) is reduced, suggesting that fewer excitatory neurons are being produced. In the post-natal brain there is a reduction in upper layer (cortical layer 2–4) projection neurons, which are the main progeny of basal progenitors. The loss of upper layer cortical neurons obviously has implications for the formation of cortical circuits, and could underlie cognitive and behavioral issues in 22q11DS.

In addition to the excitatory cortical neuron population, approximately 1 in 5 cortical neurons release inhibitory neurotransmitters (primarily GABA) onto their excitatory neighbors. Although this inhibitory population is quantitatively much smaller, they play an essential role in modulating the activity of their primarily glutamatergic neighbors, preventing waves of spontaneous activity that cause seizures, and "fine tuning" the activity of cortical circuits for proper function. Unlike the excitatory population, which is generated within the pallium, cortical interneurons are generated outside the embryonic cortex in the ventral forebrain (subpallium). As newly born interneurons form, they must migrate into the embryonic cortex to become elements of mature cortical circuits. In the *LgDel* mouse, the initial migration of these newly born interneurons is delayed (Meechan et al., 2009), with fewer interneurons visible in the embryonic cortex at early stages (Figure 36.5, right). This delay in migration prefigures a second significant disruption in cortical circuitry: the distribution of one of the major sub-classes of cortical interneurons (marked by the calcium-binding protein parvalbumin) is disrupted (Meechan et al., 2009). Although the absolute number of parvalbumin cells does not appear to change in the *LgDel* mouse cortex, more of them are found in upper cortical layers (where there is also a loss of excitatory neurons), and fewer are found in lower layers. These changes in the mouse model suggest that cellular phenotypes that arise embryonically can lead to alterations in post-natal cortical circuitry that may underlie disorders associated with 22q11DS, such as schizophrenia and autism.

The movement and integration of interneurons into the developing cortex is a complex endeavor that recruits sophisticated molecular control mechanisms to ensure that these processes are carried out successfully. To understand in more detail the cellular and molecular processes that explain aberrant migration and placement of *LgDel* cortical interneurons, Meechan et al. (2013) analyzed *LgDel* embryos crossed with a mouse line that genetically labels embryonic interneurons (Meechan et al., 2012). This approach allowed for the small fraction of embryonic interneurons to be isolated from the embryonic cortex by fluorescence-activated cell sorting (FACS), and the subsequent identification of transcriptional changes in this small population of migrating cells by microarray analysis. Reduced transcription of several genes known to regulate cellular migration was observed in the migrating interneurons; in particular the G protein-coupled cell surface receptor, *Cxcr4* was reduced in expression. *Cxcr4* is specifically expressed in interneurons migrating through the developing cortex and elicits a chemotactic response to the endogenously expressed ligand *Cxcl12*. *Cxcr4* was initially described as a key receptor involved in regulating directed migration of immune cells, and it has been shown to be central for mediating the proper migration of interneurons into the embryonic cortex (Wang et al., 2011). To see whether the reduction in Cxcr4 levels was responsible for the defects in interneuron migration in the *LgDel* mouse, Meechan et al. (2012) tested whether the *LgDel* interneuron migration phenotype could be intensified by genetically suppressing *Cxcr4* expression in interneurons (Meechan et al., 2012). Reducing *Cxcr4* expression in *LgDel* interneurons further disrupted interneuron migration, demonstrating that there is a genetic interaction between *Cxcr4* and 22q11DS within cortical interneurons. Thus, the disruption of interneuron migration in the *LgDel* mouse is likely a direct consequence of the disruption of normal Cxcr4 signaling. Together, these findings show that relatively subtle disruptions in cortical development – a modest reduction in the proliferation of the basal progenitor stem cell pool, and reduced Cxcr4 signaling in migrating interneurons – may underlie the anomalies in cortical structure that have been described in 22q11DS.

While determining cellular processes that could account for brain abnormalities in 22q11DS patients is valuable, it is also important to consider whether mouse models have traits that correlate to behavioral abnormalities in 22q11DS patients and whether these can be attributed to particular cellular defects in the model system. While there are obviously great differences between mouse and human behaviors, mouse models of 22q11DS do display several behavioral impairments that mirror those observed in their human counterparts. The *LgDel* mouse recapitulates "sensory gating" behavioral defects that are frequently observed in schizophrenics, whereby the brain adjusts its response to a startling stimulus based on an earlier weaker pre-pulse and which is referred to as pre-pulse inhibition (Long et al., 2006). Additionally, Sigurdsson et al. (2010) report altered neural synchrony between the hippocampus and the cortex in another 22q11 murine model with a similar deletion. Altered neural synchrony between different brain regions is thought to be disrupted in psychiatric illnesses such as schizophrenia (Spencer et al., 2003, 2004).

More recently, a state-of-the-art visual discrimination cognitive test was utilized to show that the *LgDel* model has cognitive deficits that can be associated with particular cortical neurons and the circuits they form (Meechan et al., 2013). This test measures the ability of the *LgDel* animal to learn a set of simple rules – to visually discriminate between patterns presented to it on an LCD screen, and select the pattern that provides a reward. While *LgDel* mice do not show dramatic impairments in the initial learning of this task, they are dramatically impaired at adapting when the rules change (Figure 36.6). This ability to use both long- and short-term memory, as well as other cognitive abilities, to solve problems (and, in particular, adapt to changes in the task at hand) is referred to as "executive function," and this cognitive process is often perturbed in psychiatric disorders. A subsequent analysis of cortices from the behaviorally tested *LgDel* animals revealed that their performance correlated with the degree of layer 2/3 projection neuron density in a specific area of the cortex considered critical for performing executive functions, the medial frontal cortex. Thus, in the *LgDel* mouse model, there is substantial evidence that specific anomalies that occur during the embryonic development of cortical neurons leads to a disruption of cortical circuitry, leading to identifiable behavioral impairments.

These identified disruptions in cortical circuitry, and their behavioral consequences, are likely only a subset of the anomalies that are present in the 22q11DS brain. For example, some defects in post-natal synapses have also been reported and attributed to some of these behavioral deficits in 22q11DS mouse models (Stark et al., 2008). It is likely that further research will identify other anomalies in cortical circuitry and cortical function. Further work is required to assess the relative contribution of synaptic defects, interneuron mispositioning, loss of neuronal cell types, and other cellular anomalies in causing the cognitive and behavioral disruptions associated with 22q11DS.

A key challenge for 22q11DS, as with all copy number variation (CNV) disorders, is determining which gene or genes are responsible for specific aspects of the syndrome. During embryonic development of the brain, several 22q11 genes are exquisitely regulated both temporally and cell-specifically. Several 22q11 genes with cell-cycle roles are expressed in the embryonic cortex, while others are selectively expressed in newly born neurons (Meechan et al., 2009; Maynard et al., 2003). As discussed before, the T-box transcription factor *Tbx1* has been implicated in other clinical symptoms of 22q11DS such as heart defects (Merscher et al., 2001), and as such has attracted attention as a candidate for causing disruptions of brain development and resulting cognitive and psychiatric illness. However, the role *Tbx1* may play in psychiatric illness susceptibility is less clear. There is one report that mice with heterozygous deletions of *Tbx1* have sensory gating deficits (Paylor et al., 2006), although this finding was not confirmed by others (Long et al., 2006). Furthermore, heterozygous deletion of *Tbx1* does not lead to the same neurodevelopmental proliferation and migration phenotypes that are observed in the *LgDel* line (Meechan et al., 2009). Distinguishing which 22q11.2 genes can cause the cellular and behavioral phenotypes observed in mouse models of 22q11DS, when deleted heterozygously, and contribute to schizophrenia susceptibility in 22q11DS patients remains a long-term goal.

22q11DS Leads to Immune, Endocrine and Metabolic Disruptions

Many individuals with 22q11DS have disruptions in endocrine, immune, and metabolic functions that may impact many aspects of their development, growth, cognitive function, and overall health. As noted previously, one of the initial observations that led to the identification of 22q11DS is a failure of the thymus to form correctly. Thymus abnormalities likely arise as a consequence of the disrupted development of the branchial arches, apparently by mechanisms parallel to those that cause the disrupted development of the aortic arch arteries (described above). 22q11DS often leads to a severe reduction in

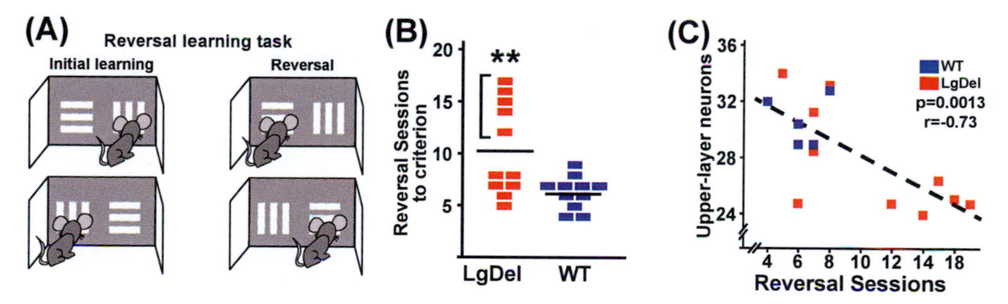

FIGURE 36.6 The Performance of the *LgDel* Mouse on a Behavioral Test of Executive Function is Correlated with the Density of Upper Layer Cortical Neurons. (A) *LgDel* and WT control mice were tested on a touch-screen based behavioral task. Mice were taught over multiple sessions to associate a specific stimulus (in this case, vertical bars) with a food reward. The mice learn this task relatively efficiently and over several sessions learn to efficiently associate this stimulus with the reward. Following initial learning, mice are challenged by reversing the rule, such that choosing the opposite stimulus (e.g., horizontal bars) is required to obtain the reward. (B) *LgDel* mice take significantly longer (more sessions) to learn the new rules for the task, as measured by the sessions required for each mouse to reach critierion. (C) After testing, the density of upper layer cortical neurons was measured in the same animals that were behaviorally tested. Low performance on the behavioral task was specifically correlated with low densities of upper layer neurons. Thus, it appears that the cortical circuitry anomalies can directly lead to behavioral impairments in the *LgDel* mouse model. *Figure adapted from (Meechan et al., 2013).*

the size of the thymus (hypoplasia), or even its complete absence; these thymus anomalies often lead to immune dysfunction (Davies, 2013). The thymus gland secretes thymosins, which stimulates the production of antibodies in the bone marrow. In addition, the thymus plays a key role in T cell development, as immature T cells migrate from the bone marrow into the thymus. To correct these deficiencies, newborns with 22q11DS often receive thymus transplants.

The same mechanisms that disrupt the aortic arch and thymus can also lead to parallel disruptions of the thyroid and parathyroid glands and, to a lesser extent, the anterior pituitary gland. Thyroid hypoplasia is the most common thyroid abnormality in 22q11DS; although other morphological anomalies (such as a partial loss of the thyroid, or the failure to form the thyroid isthmus) also occur in some cases (Stagi et al., 2010b). Abnormalities in the thyroid gland disrupts the regulation of many metabolic processes. The loss of thyroid hormone can cause a significant increase in blood lipids (hyperlipidemia) and cholesterol (Sharma et al., 2014), as thyroid hormones promote reverse cholesterol transport, and stimulate conversion of cholesterol to bile acids and to enhanced metabolic rate. Failure to properly form the parathyroid glands can also lead to endocrine dysfunction. Parathyroid hormone is essential for the regulation of calcium levels; about 60% of individuals with 22q11DS have very low calcium levels (hypocalcemia) (Stagi et al., 2010a). Parathyroid hormone is involved in vitamin D-dependent calcium reabsorption, as it promotes renal calcium reabsorption while promoting renal phosphate excretion (Al-Azem and Khan, 2012). The clinical consequences of parathyroid hormone deficiency include nervous system and muscular disorders (hyperexcitability, paresthesias, cramps, tetany, hyperreflexia, convulsions, and basal ganglia calcifications), as well as cataracts, weakened tooth enamel and brittle nails. On the other hand, some individuals with 22q11DS show few signs of hypocalcemia, even if their calcium levels are low. It is possible that the abnormal formation of teeth that is often observed in 22q11DS is at least partially caused by hypocalcemia in infancy (Stagi et al., 2010a).

Abnormal pituitary development is also observed in some individuals with 22q11DS. The development of the pituitary can be impacted by the same mechanisms that cause cleft palate and other midline abnormalities, as the pituitary forms in close proximity and at the same time as the palate. Abnormalities in the anterior pituitary, and resulting deficiencies in growth hormone production have been observed in many 22q11DS patients (Weinzimer et al., 1998; Bizzocchi et al., 2011; Brauner et al., 2003). As pituitary produced growth hormone is a key regulator of growth; it is notable that short stature is associated with 22q11.2 microdeletion syndrome (Weinzimer et al., 1998; Kitsiou-Tzeli et al., 2005). The effects of pituitary hormone are widespread; growth hormone can act both directly and indirectly through Insulin-like Growth Factor-1 (IGF-1) on many tissues, and it also influences regulation of cellular metabolism. Hence, pituitary abnormalities may be responsible for numerous and widespread complications associated with 22q11DS.

In addition to endocrine dysfunction, at least two genes in the 22q11-deleted region are thought to be involved in the metabolism of neurotransmitters, *Comt* and *Prodh* (Boot et al., 2011). *Comt* is responsible for degrading catecholamine neurotransmitters, such as dopamine, epinephrine, and norepinephrine. When a nerve is activated and releases its catecholamine neurotransmitters into the synapse, it is crucial that the neurotransmitter is quickly degraded and removed, in order to prevent continued firing of the neuron, and to prevent global elevation of neurotransmitter levels throughout the brain. The *Comt* gene in the 22q11.2 region encodes for catechol-*O*-methyl-transferase (COMT), one of the two major enzymes involved in dopamine degradation in humans. Individuals with 22q11DS have only one functional copy of the *Comt* gene, and as a result they have reduced expression of the COMT enzyme. In mouse models of *Comt* deletion, reduced COMT activity leads to significantly slower clearance of dopamine from the brain, and *Comt*-deleted mice display some behavioral changes in tests of anxiety and aggression (Gogos et al., 1998).

In addition to the deletion of one copy of Comt, 22q11DS individuals may have even further reductions in COMT activity due to genetic variability. The COMT gene contains a common single nucleotide polymorphism (Tunbridge et al., 2006), a valine-to-methionine substitution ($Val^{108/158}$ Met). The Val allele of COMT is stable at body temperature, but the Met allele is unstable and less enzymatically active (Lotta et al., 1995). The Val allele is therefore less able to regulate synaptic clearance. Individuals with 22q11DS that carrying the low-activity Val isoform on their remaining allele have an even greater reduction of COMT activity than their Met allele counterparts. Thus, it is not surprising that these 22q11DS/Val-isoform individuals are at a significantly higher risk for schizophrenia (Dunham et al., 1992) and other psychiatric conditions including ADHD and OCD.

Although the Val/Met polymorphism appears to be specific to humans, there is some evidence for high- and low-activity forms in mice. A genome-wide analysis showed that naturally occurring genomic variants of *Comt* are associated with differences in several behaviors that are observed in different inbred mice strains (Segall et al., 2010). Unlike human *Comt* polymorphisms, the genomic variants observed in mice do not involve amino acid changes in the COMT protein; instead, alterations in the 3'-untranslated region (3'-UTR) of the gene appear to alter levels of protein expression. The alteration in the 3'-UTR is due to the insertion of a genomic fragment called a short interspersed nuclear element (SINE), which is a fragment of DNA that can move about in the genome. Mice with the SINE element have higher COMT activity, which correlates with decreased pain sensitivity and anxiety in behavioral tests. This variation mirrors the behavioral changes observed in humans due to the Val/Met polymorphism. Thus, for both mice and men, variations in *Comt* activity may be an important modulator of certain behaviors.

Another 22q11.2 gene, *Prodh*, is also implicated in metabolic functions that may influence brain function and behavior. *Prodh* encodes the mitochondrially-localized proline dehydrogenase protein (also known as proline oxidase), which catalyzes the first step in the degradation of the amino acid proline. When *Prodh* is not expressed at proper levels or functionally mutated, the result is a harmful increase in the amino acid proline, called hyperprolinemia. The clinical symptoms of hyperprolinemia can vary significantly, depending largely on exactly how *Prodh* is mutated, how severely its expression is reduced, or whether other factors further hinder proline metabolism (Woody et al., 1969). In most cases, *Prodh* mutation only results in minor increases in proline levels, and the resulting hyperprolinemia is considered to be a benign condition under most circumstances (Phang et al., 2001). The most severe cases of hyperprolinemia, which have very high accumulations of of proline in the blood plasma, have severe cognitive and motor impairments, and are prone to epileptic seizures (Jacquet et al., 2002). In addition to hyperprolinemia, *Prodh* may also influence other aspects of neuronal function, as a secondary function of *Prodh* appears to be the conversion of proline to glutamate (Boot et al., 2011; Thompson et al., 1985), which is the primary neurotransmitter for the most abundant neurons in the cerebral cortex. A mouse model of Prodh deficiency shows some impairments in cognitive function, particularly in tests of sensory gating (Gogos et al., 1999). Thus, reduced expression of *Prodh* (along with *Comt*) may underlie many aspects of the cognitive and behavioral impairments associated with 22q11DS.

36.3 GENES AND PHENOTYPES: 22Q11DS AS A PROTOTYPE GENOMIC DISEASE

22q11DS is the most thoroughly studied syndrome that results from a microdeletion, but it is far from alone. Genetic evidence has been accumulating for a fundamental role for these small deletions, as well as duplications, of segments of chromosomes, which are generally referred to as "copy number variations" (CNVs). Because these CNVs (both deletions and duplications) are below the threshold of detection for traditional chromosomal analysis (such as karyotyping analysis), the true understanding of their incidence and significance has only begun to become clear following the introduction of whole-genome analysis methods (including "next-generation sequencing" and high-density SNP [single nucleotide polymorphism] arrays).

CNVs, which range in size from thousands to the low millions of base pairs, are apparently a frequent feature in the human genome, and are not only associated with genetic diseases (Iafrate et al., 2004; Sebat et al., 2004; McCarroll et al., 2006). There have been over 500 CNVs identified in the population, with roughly half appearing to be normal genomic variants, as they occur at relatively high frequency in the population (McCarroll et al., 2006). It is clear, however, that many of these CNVs are not benign, as a high frequency of CNVs are observed in individuals with cognitive and psychiatric disorders (Morrow, E.M. (2010). In addition to 22q11.2DS, there are multiple neurodevelopmental syndromes that have been closely associated with defined deletion or duplication CNVs (Maynard et al., 2002), such as Williams syndrome (OMIM 194050, caused by a deletion at 7q11.23), Smith-Magenis syndrome (OMIM 182290, caused by a deletion at 17q11.2), or Prader-Willi syndrome (OMIM 176270; caused by a deletion at 15q11.2). Other deletion risk loci have been more recently identified for neurodevelopmental disorders such as autism (Girirajan et al., 2013) and schizophrenia (Purcell et al., 2014; Malhotra et al., 2011), for psychiatric disorders such as bipolar disorder (Malhotra et al., 2011) and attention deficit-hyperactivity disorder (Ramos-Quiroga et al., 2014), and associated with general structural birth defects (Southard et al., 2012). Both deletion and duplication CNVs have been described, and for at least one case (the linkage of 16p11.2 CNVs to autism spectrum disorders) both deletions and duplications of the same locus have both been associated with the same disorder (Weiss et al., 2008; Kumar et al., 2008). As is the case for 22q11DS, other CNVs have been associated with a wide range of phenotypes and neurodevelopmental disorders. 22q11DS, therefore, can be considered a prototype CNV disorder, and the most well characterized of the CNV disorders.

One aspect that appears to be shared by many of the CNV disorders is a high degree of phenotypic variability. Although it is tempting to dismiss or ignore variability, variability is a key feature that likely reflects the etiology of these disorders. For LCR mediated CNVs like 22q11DS, it has been hypothesized that phenotypic variability may result from heterogeneity in the size of the genomic deletion. Several groups have attempted to associate specific phenotypes with deletions of specific regions in the 22q11 region, or with specific candidate genes within the region. In general, these attempts at establishing genotype-phenotype correlations have only been successful when they were based on small samples of patients with uncommon deletions, and the analysis limited to a narrow subset of 22q11DS phenotypes (Weksberg et al., 2007). It is possible that some of the phenotypic variability may come from non-genetic sources, such as environmental factors (including placental location or maternal diet), or stochastic factors; however, such factors are by their very nature very difficult to characterize.

While several key genes have been identified as central to individual 22q11DS phenotypes, it is unlikely that the syndrome will ever be truly characterized by interpreting the consequences of single gene deletions. For example, *Comt* and *Prodh* have been identified as key genes for 22q11DS psychiatric phenotypes, and *Tbx1* is obviously a central gene for the thymic and cardiovascular phenotypes; however, the spectrum of anomalies of 22q11DS far exceeds the sum of these individual mutations. It appears that 22q11DS instead reflects combinatorial effects of reduced dosage of multiple genes acting in concert at all phenotypically compromised sites (Meechan et al., 2007). Rather than a simple concatenation of single, independent mutations, 22q11DS is best described as the disruption of a set of interacting genes. Finally, understanding and studying 22q11DS as a polymorphic, multigenic syndrome may lead to a better characterization of other CNVs.

36.4 CLINICAL RELEVANCE

- 22q11.2 Deletion Syndrome (22q11DS) / DiGeorge syndrome is the most common microdeletion disorder, affecting 1 in 2,000–4,000 live births.
- 22q11.2DS is a significant cause of severe cardiovascular defects, including aortic arch malformations, ventricular-septal defects, and the complex combination of cardiovascular defects known as tetralogy of Fallot.
- Most infants with 22q11DS have feeding and swallowing difficulties (dysphagia), which result from both physical craniofacial defects (overt or submucosal cleft palate) and neuromuscular control issues that prevent proper coordination of the tongue, pharyngeal, and throat muscles.
- 22q11DS is strongly associated with mild cognitive impairment and psychiatric issues, including autism spectrum disorders, schizophrenia, and obsessive-compulsive disorder. Recent evidence suggests that a key source of these brain disorders is the disruption of cortical circuitry due to the disrupted proliferation and migration of neuronal precursors.
- 22q11DS is the best described copy number variation (CNV) disorder, which result from the heterozygous deletion of hundreds to the low millions of base pairs of chromosomal DNA. 22q11DS, like most other CNV disorders that have been described, has a highly variable presentation, and disrupts the development of multiple systems.

ACKNOWLEDGEMENTS

This work was supported by the National Institutes of Health (HD042182).

REFERENCES

Al-Azem, H., Khan, A.A., 2012. Hypoparathyroidism. Best Pract. Res. Clin. Endocrinol. Metab. 26, 517–522.

Babcock, M., Yatsenko, S., Hopkins, J., Brenton, M., Cao, Q., de Jong, P., Stankiewicz, P., Lupski, J.R., Sikela, J.M., Morrow, B.E., 2007. Hominoid lineage specific amplification of low-copy repeats on 22q11.2 (LCR22s) associated with velo-cardio-facial/DiGeorge syndrome. Hum. Mol. Genet. 16, 2560–2571.

Bailey, J.A., Eichler, E.E., 2006. Primate segmental duplications: crucibles of evolution, diversity and disease. Nat. Rev. Genet. 7, 552–564.

Bearden, C.E., van Erp, T.G., Dutton, R.A., Tran, H., Zimmermann, L., Sun, D., Geaga, J.A., Simon, T.J., Glahn, D.C., Cannon, T.D., Emanuel, B.S., Toga, A.W., Thompson, P.M., 2007. Mapping cortical thickness in children with 22q11.2 deletions. Cereb. Cortex. 17, 1889–1898.

Beckmann, J.S., Estivill, X., Antonarakis, S.E., 2007. Copy number variants and genetic traits: closer to the resolution of phenotypic to genotypic variability. Nat. Rev. Genet. 8, 639–646.

Beddow, R., 2012. Paediatric Teratoid/Rhabdoid Tumours: Germline Deletions of Chromosome 22q11.2 In: Pediatric Cancer, Volume 3: Diagnosis, Therapy, and Prognosis, Springer Science+Business Media: Dordrecht

Bizzocchi, A., Genoni, G., Petri, A., Prodam, F., Negro, M., Bellone, S., Bona, G., 2011. Growth hormone deficiency associated with 22q11.2 deletion: a case report. Minerva. Pediatr. 63, 527–531.

Boot, E., Booij, J., Abeling, N., Meijer, J., da Silva Alves, F., Zinkstok, J., Baas, F., Linszen, D., van Amelsvoort, T., 2011. Dopamine metabolism in adults with 22q11 deletion syndrome, with and without schizophrenia – relationship with COMT Val(1)(0)(8)/(1)(5)(8)Met polymorphism, gender and symptomatology. J. Psychopharmacol. 25, 888–895.

Botto, L.D., May, K., Fernhoff, P.M., Correa, A., Coleman, K., Rasmussen, S.A., Merritt, R.K., O'Leary, L.A., Wong, L.Y., Elixson, E.M., Mahle, W.T., Campbell, R.M., 2003. A population-based study of the 22q11.2 deletion: phenotype, incidence, and contribution to major birth defects in the population,. Pediatrics 112, 101–107.

Brauner, R., Le Harivel de Gonneville, A., Kindermans, C., Le Bidois, J., Prieur, M., Lyonnet, S., Souberbielle, J.C., 2003. Parathyroid function and growth in 22q11.2 deletion syndrome. J. Pediatr. 142, 504–508.

Butcher, N.J., Chow, E.W., Costain, G., Karas, D., Ho, A., Bassett, A.S., 2012. Functional outcomes of adults with 22q11.2 deletion syndrome. Genet. Med. 14, 836–843.

Calmont, A., Ivins, S., Van Bueren, K.L., Papangeli, I., Kyriakopoulou, V., Andrews, W.D., Martin, J.F., Moon, A.M., Illingworth, E.A., Basson, M.A., Scambler, P.J., 2009. Tbx1 controls cardiac neural crest cell migration during arch artery development by regulating Gbx2 expression in the pharyngeal ectoderm. Development 136, 3173–3183.

Carlson, C., Sirotkin, H., Pandita, R., Goldberg, R., McKie, J., Wadey, R., Patanjali, S.R., Weissman, S.M., Anyane-Yeboa, K., Warburton, D., Scambler, P., Shprintzen, R., Kucherlapati, R., Morrow, B.E., 1997. Molecular definition of 22q11 deletions in 151 velo-cardio-facial syndrome patients. Am. J. Hum. Genet. 61, 620–629.

Chen, L., Mupo, A., Huynh, T., Cioffi, S., Woods, M., Jin, C., McKeehan, W., Thompson-Snipes, L., Baldini, A., Illingworth, E., 2010. Tbx1 regulates Vegfr3 and is required for lymphatic vessel development. J. Cell. Biol. 189, 417–424.

Consevage, M.W., Seip, J.R., Belchis, D.A., Davis, A.T., Baylen, B.G., Rogan, P.K., 1996. Association of a mosaic chromosomal 22q11 deletion with hypoplastic left heart syndrome. Am. J. Cardiol. 77, 1023–1025.

Davies, E.G., 2013. Immunodeficiency in DiGeorge syndrome and options for treating cases with complete athymia. Front. Immunol. 4, 322.

Driscoll, D.A., Salvin, J., Sellinger, B., Budarf, M.L., McDonald-McGinn, D.M., Zackai, E.H., Emanuel, B.S., 1993. Prevalence of 22q11 microdeletions in DiGeorge and velocardiofacial syndromes: implications for genetic counselling and prenatal diagnosis. J. Med. Genet. 30, 813–817.

Dunham, I., Collins, J., Wadey, R., Scambler, P., 1992. Possible role for COMT in psychosis associated with velo-cardio-facial syndrome. Lancet 340, 1361–1362.

Eicher, P.S., McDonald-Mcginn, D.M., Fox, C.A., Driscoll, D.A., Emanuel, B.S., Zackai, E.H., 2000. Dysphagia in children with a 22q11.2 deletion: unusual pattern found on modified barium swallow. J. Pediatr. 137, 158–164.

Fang, X., Ji, H., Kim, S.W., Park, J.I., Vaught, T.G., Anastasiadis, P.Z., Ciesiolka, M., McCrea, P.D., 2004. Vertebrate development requires ARVCF and p120 catenins and their interplay with RhoA and Rac. J. Cell. Biol. 165, 87–98.

Fietz, S.A., Huttner, W.B., 2011. Cortical progenitor expansion, self-renewal and neurogenesis-a polarized perspective. Curr. Opin. Neurobiol. 21, 23–35.

Friedman, M.A., Miletta, N., Roe, C., Wang, D., Morrow, B.E., Kates, W.R., Higgins, A.M., Shprintzen, R.J., 2011. Cleft palate, retrognathia and congenital heart disease in velo-cardio-facial syndrome: a phenotype correlation study. Int. J. Pediatr. Otorhinolaryngol. 75, 1167–1172.

Funato, N., Nakamura, M., Richardson, J.A., Srivastava, D., Yanagisawa, H., 2012. Tbx1 regulates oral epithelial adhesion and palatal development. Hum. Mol. Genet. 21, 2524–2537.

Gerdes, M., Solot, C., Wang, P.P., McDonald-McGinn, D.M., Zackai, E.H., 2001. Taking advantage of early diagnosis: preschool children with the 22q11.2 deletion. Genet. Med. 3, 40–44.

Girirajan, S., Johnson, R.L., Tassone, F., Balciuniene, J., Katiyar, N., Fox, K., Baker, C., Srikanth, A., Yeoh, K.H., Khoo, S.J., Nauth, T.B., Hansen, R., Ritchie, M., Hertz-Picciotto, I., Eichler, E.E., Pessah, I.N., Selleck, S.B., 2013. Global increases in both common and rare copy number load associated with autism. Hum. Mol. Genet. 22, 2870–2880.

Gogos, J.A., Morgan, M., Luine, V., Santha, M., Ogawa, S., Pfaff, D., Karayiorgou, M., 1998. Catechol-O-methyltransferase-deficient mice exhibit sexually dimorphic changes in catecholamine levels and behavior. Proc. Natl. Acad. Sci. U S A 95, 9991–9996.

Gogos, J.A., Santha, M., Takacs, Z., Beck, K.D., Luine, V., Lucas, L.R., Nadler, J.V., Karayiorgou, M., 1999. The gene encoding proline dehydrogenase modulates sensorimotor gating in mice. Nat. Genet. 21, 434–439.

Gotter, A.L., Shaikh, T.H., Budarf, M.L., Rhodes, C.H., Emanuel, B.S., 2004. A palindrome-mediated mechanism distinguishes translocations involving LCR-B of chromosome 22q11.2. Hum. Mol. Genet. 13, 103–115.

Goudy, S., Law, A., Sanchez, G., Baldwin, H.S., Brown, C., 2010. Tbx1 is necessary for palatal elongation and elevation. Mech. Dev. 127, 292–300.

Guo, X., Freyer, L., Morrow, B., Zheng, D., 2011. Characterization of the past and current duplication activities in the human 22q11.2 region. BMC. Genomics. 12, 71.

Halder, A., Jain, M., Kabra, M., Gupta, N., 2008. Mosaic 22q11.2 microdeletion syndrome: diagnosis and clinical manifestations of two cases. Mol. Cytogenet. 1, 18.

Hatzfeld, M., 2005. The p120 family of cell adhesion molecules. Eur. J. Cell. Biol. 84, 205–214.

Herman, S.B., Guo, T., McGinn, D.M., Blonska, A., Shanske, A.L., Bassett, A.S., International Chromosome 22q, C, et al., 2012. Overt cleft palate phenotype and TBX1 genotype correlations in velo-cardio-facial/DiGeorge/22q11.2 deletion syndrome patients. Am. J. Med. Genet. A. 158A, 2781–2787.

Hong, J.Y., Park, J.I., Cho, K., Gu, D., Ji, H., Artandi, S.E., McCrea, P.D., 2010. Shared molecular mechanisms regulate multiple catenin proteins: canonical Wnt signals and components modulate p120-catenin isoform-1 and additional p120 subfamily members. J. Cell. Sci. 123, 4351–4365.

Hurles, M.E., Lupiski, J.R., 2006. In: Lupiski, J.R., Stankiewicz, P.T. (Eds.), Genomic disorders: the genomic basis of disease. Humana Press, Totowa, N.J., pp. 341–355.

Iafrate, A.J., Feuk, L., Rivera, M.N., Listewnik, M.L., Donahoe, P.K., Qi, Y., Scherer, S.W., Lee, C., 2004. Detection of large-scale variation in the human genome. Nat. Genet. 36, 949–951.

Jacquet, H., Raux, G., Thibaut, F., Hecketsweiler, B., Houy, E., Demilly, C., Haouzir, S., Allio, G., Fouldrin, G., Drouin, V., Bou, J., Petit, M., Campion, D., Frebourg, T., 2002. PRODH mutations and hyperprolinemia in a subset of schizophrenic patients. Hum. Mol. Genet. 11, 2243–2249.

Jerome, L.A., Papaioannou, V.E., 2001. DiGeorge syndrome phenotype in mice mutant for the T-box gene, Tbx1. Nat. Genet. 27, 286–291.

Karpinski, B.A., Maynard, T.M., Fralish, M.S., Nuwayhid, S., Zohn, I., Moody, S.A., Lamantia, A.S., 2013. Dysphagia and disrupted cranial nerve development in a mouse model of DiGeorge/22q11 Deletion Syndrome. Dis. Model. Mech 7(2), 245–257.

Kiehl, T.R., Chow, E.W., Mikulis, D.J., George, S.R., Bassett, A.S., 2009. Neuropathologic features in adults with 22q11.2 deletion syndrome. Cereb. Cortex. 19, 153–164.

Kitsiou-Tzeli, S., Kolialexi, A., Mavrou, A., 2005. Endocrine manifestations in DiGeorge and other microdeletion syndromes related to 22q11.2. Hormones (Athens) 4, 200–209.

Klingberg, G., Oskarsdottir, S., Johannesson, E.L., Noren, J.G., 2002. Oral manifestations in 22q11 deletion syndrome. Int. J. Paediatr. Dent. 12, 14–23.

Kraynack, N.C., Hostoffer, R.W., Robin, N.H., 1999. Agenesis of the corpus callosum associated with DiGeorge-velocardiofacial syndrome: a case report and review of the literature. J. Child Neurol. 14, 754–756.

Kumar, R.A., KaraMohamed, S., Sudi, J., Conrad, D.F., Brune, C., Badner, J.A., Gilliam, T.C., Nowak, N.J., Cook Jr., E.H., Dobyns, W.B., Christian, S.L., 2008. Recurrent 16p11.2 microdeletions in autism. Hum. Mol. Genet. 17, 628–638.

Lay-Son, G., Palomares, M., Guzman, M.L., Vasquez, M., Puga, A., Repetto, G.M., 2012. Palate abnormalities in Chilean patients with chromosome 22q11 microdeletion syndrome. Int. J. Pediatr. Otorhinolaryngol. 76, 1726–1728.

Lee, S.I., Jung, H.Y., Kim, S.H., Kim, D.K., Mok, J., 2003. Midline brain anomalies in a young schizophrenic patient with 22q11 deletion syndrome. Schizophr. Res. 60, 323–325.

Lindsay, E.A., Botta, A., Jurecic, V., Carattini-Rivera, S., Cheah, Y.C., Rosenblatt, H.M., Bradley, A., Baldini, A., 1999. Congenital heart disease in mice deficient for the DiGeorge syndrome region. Nature 401, 379–383.

Lindsay, E.A., Baldini, A., 2001. Recovery from arterial growth delay reduces penetrance of cardiovascular defects in mice deleted for the DiGeorge syndrome region. Hum. Mol. Genet. 10, 997–1002.

Lindsay, E.A., Vitelli, F., Su, H., Morishima, M., Huynh, T., Pramparo, T., Jurecic, V., Ogunrinu, G., Sutherland, H.F., Scambler, P.J., Bradley, A., Baldini, A., 2001. Tbx1 haploinsufficieny in the DiGeorge syndrome region causes aortic arch defects in mice. Nature 410, 97–101.

Lischner, H.W., 1972. DiGeorge syndrome(s). J. Pediatr. 81, 1042–1044.

Long, J.M., Laporte, P., Merscher, S., Funke, B., Saint-Jore, B., Puech, A., Kucherlapati, R., Morrow, B.E., Skoultchi, A.I., Wynshaw-Boris, A., 2006. Behavior of mice with mutations in the conserved region deleted in velocardiofacial/DiGeorge syndrome. Neurogenetics. 7(4), 247–257.

Lorente-Galdos, B., Bleyhl, J., Santpere, G., Vives, L., Ramirez, O., Hernandez, J., Anglada, R., Cooper, G.M., Navarro, A., Eichler, E.E., Marques-Bonet, T., 2013. Accelerated exon evolution within primate segmental duplications. Genome. Biol. 14, R9.

Lotta, T., Vidgren, J., Tilgmann, C., Ulmanen, I., Melen, K., Julkunen, I., Taskinen, J., 1995. Kinetics of human soluble and membrane-bound catechol O-methyltransferase: a revised mechanism and description of the thermolabile variant of the enzyme. Biochemistry (Mosc) 34, 4202–4210.

Maiti, S., Kumar, K.H., Castellani, C.A., O'Reilly, R., Singh, S.M., 2011. Ontogenetic *de novo* copy number variations (CNVs) as a source of genetic individuality: studies on two families with MZD twins for schizophrenia. PLoS One. 6, e17125.

Malhotra, D., McCarthy, S., Michaelson, J.J., Vacic, V., Burdick, K.E., Yoon, S., et al., 2011. High frequencies of *de novo* CNVs in bipolar disorder and schizophrenia. Neuron 72, 951–963.

Mansir, T., Lacombe, D., Lamireau, T., Taine, L., Chateil, J.F., Le Bail, B., Demarquez, J.L., Fayon, M., 1999. Abdominal lymphatic dysplasia and 22q11 microdeletion. Genet. Couns. 10, 67–70.

Maynard, T.M., Gopalakrishna, D., Meechan, D.W., Paronett, E.M., Newbern, J., LaMantia, A.S., 2013. 22q11 gene dosage establishes an adaptive range for sonic hedgehog and retinoic acid signaling during early development. Hum. Mol. Genet. 22(2), 300–312.

Maynard, T.M., Gopalakrishna, D., Meechan, D.W., Paronett, E.M., Newbern, J.M., LaMantia, A.S., 2013. 22q11 Gene dosage establishes an adaptive range for sonic hedgehog and retinoic acid signaling during early development. Hum. Mol. Genet. 22, 300–312.

Maynard, T.M., Haskell, G.T., Lieberman, J.A., LaMantia, A.S., 2002. 22q11 DS: genomic mechanisms and gene function in DiGeorge/velocardiofacial syndrome. Int. J. Dev. Neurosci. 20, 407–419.

Maynard, T.M., Haskell, G.T., Peters, A.Z., Sikich, L., Lieberman, J.A., LaMantia, A.S., 2003. A comprehensive analysis of 22q11 gene expression in the developing and adult brain. Proc. Natl. Acad. Sci. U S A 100, 14433–14438.

McCarroll, S.A., Hadnott, T.N., Perry, G.H., Sabeti, P.C., Zody, M.C., Barrett, J.C., Dallaire, S., Gabriel, S.B., Lee, C., Daly, M.J., Altshuler, D.M., International HapMap, C., 2006. Common deletion polymorphisms in the human genome. Nat. Genet. 38, 86–92.

McDonald-McGinn, D.M., Emanuel, B.S., Zackai, E.H., 1993. In: Pagon, R.A., Adam, M.P., Bird, T.D., Dolan, C.R., Fong, C.T., Stephens, K. (Eds.), GeneReviews. Seattle (WA).

McDonald-McGinn, D.M., Sullivan, K.E., 2011. Chromosome 22q11.2 deletion syndrome (DiGeorge syndrome/velocardiofacial syndrome). Medicine (Baltimore) 90, 1–18.

Meechan, D.W., Maynard, T.M., Gopalakrishna, D., Wu, Y., LaMantia, A.S., 2007. When half is not enough: gene expression and dosage in the 22q11 deletion syndrome. Gene. Expr. 13, 299–310.

Meechan, D.W., Rutz, H.L., Fralish, M.S., Maynard, T.M., Rothblat, L.A., Lamantia, A.S., 2013. Cognitive ability is associated with altered medial frontal cortical circuits in the LgDel mouse model of 22q11.2DS. Cereb. Cortex. In Press.

Meechan, D.W., Tucker, E.S., Maynard, T.M., LaMantia, A.S., 2009. Diminished dosage of 22q11 genes disrupts neurogenesis and cortical development in a mouse model of 22q11 deletion/DiGeorge syndrome. Proc. Natl. Acad. Sci. U S A 106, 16434–16445.

Meechan, D.W., Tucker, E.S., Maynard, T.M., Lamantia, A.S., 2012. Cxcr4 regulation of interneuron migration is disrupted in 22q11.2 deletion syndrome. Proc. Natl. Acad. Sci. U S A 109, 18601–18606.

Merscher, S., Funke, B., Epstein, J.A., Heyer, J., Puech, A., Lu, M.M., et al., 2001. TBX1 is responsible for cardiovascular defects in velo-cardio-facial/DiGeorge syndrome. Cell 104, 619–629.

Michaelovsky, E., Frisch, A., Carmel, M., Patya, M., Zarchi, O., Green, T., Basel-Vanagaite, L., Weizman, A., Gothelf, D., 2012. Genotype-phenotype correlation in 22q11.2 deletion syndrome. BMC. Med. Genet. 13, 122.

Morrow, E.M., 2010. Genomic copy number variation in disorders of cognitive development. J. Am. Acad. Child Adolesc. Psychiatry. 49, 1091–1104.

Nickel, R.E., Magenis, R.E., 1996. Neural tube defects and deletions of 22q11. Am. J. Med. Genet. 66, 25–27.

Notini, A.J., Craig, J.M., White, S.J., 2008. Copy number variation and mosaicism. Cytogenet. Genome. Res. 123, 270–277.

Oskarsdottir, S., Persson, C., Eriksson, B.O., Fasth, A., 2005. Presenting phenotype in 100 children with the 22q11 deletion syndrome. Eur. J. Pediatr. 164, 146–153.

Paranaiba, L.M., de Aquino, S.N., Bufalino, A., Martelli-Junior, H., Graner, E., Brito, L.A., e Passos-Bueno, M.R., Coletta, R.D., Swerts, M.S., 2013. Contribution of polymorphisms in genes associated with craniofacial development to the risk of nonsyndromic cleft lip and/or palate in the Brazilian population. Med. Oral. Patol. Oral. Cir. Bucal. 18, e414–e420.

Pasick, C., McDonald-McGinn, D.M., Simbolon, C., Low, D., Zackai, E., Jackson, O., 2013. Asymmetric crying facies in the 22q11.2 deletion syndrome: implications for future screening. Clin. Pediatr. (Phila) 52, 1144–1148.

Paylor, R., Glaser, B., Mupo, A., Ataliotis, P., Spencer, C., Sobotka, A., et al., 2006. Tbx1 haploinsufficiency is linked to behavioral disorders in mice and humans: implications for 22q11 deletion syndrome. Proc. Natl. Acad. Sci. U S A 103, 7729–7734.

Phang, J.M., Hu, C.-A., Valle, D., 2001. In: Scriver, C.R., Beaudet, A.L., Sly, W.S., Valle, D. (Eds.), The metabolic and molecular bases of inherited disease. McGraw Hill, New York, pp. 1821–1838.

Puech, A., Saint-Jore, B., Funke, B., Gilbert, D.J., Sirotkin, H., Copeland, N.G., 1997. Comparative mapping of the human 22q11 chromosomal region and the orthologous region in mice reveals complex changes in gene organization. Proc. Natl. Acad. Sci. U S A 94, 14608–14613.

Purcell, S.M., Moran, J.L., Fromer, M., Ruderfer, D., Solovieff, N., Roussos, P., et al., 2014. A polygenic burden of rare disruptive mutations in schizophrenia. Nature 506, 185–190.

Ramos-Quiroga, J.A., Sanchez-Mora, C., Casas, M., Garcia-Martinez, I., Bosch, R., Nogueira, M., et al., 2014. Genome-wide copy number variation analysis in adult attention-deficit and hyperactivity disorder. J. Psychiatr. Res. 49, 60–67.

Robin, N.H., Shprintzen, R.J., 2005. Defining the clinical spectrum of deletion 22q11.2. J. Pediatr. 147, 90–96.

Rommel, N., Davidson, G., Cain, T., Hebbard, G., Omari, T., 2008. Videomanometric evaluation of pharyngo-oesophageal dysmotility in children with velocardiofacial syndrome. J. Pediatr. Gastroenterol. Nutr. 46, 87–91.

Ryan, A.K., Goodship, J.A., Wilson, D.I., Philip, N., Levy, A., Seidel, H., et al., 1997. Spectrum of clinical features associated with interstitial chromosome 22q11 deletions: a European collaborative study. J. Med. Genet. 34, 798–804.

Ryckebusch, L., Bertrand, N., Mesbah, K., Bajolle, F., Niederreither, K., Kelly, R.G., Zaffran, S., 2010. Decreased levels of embryonic retinoic acid synthesis accelerate recovery from arterial growth delay in a mouse model of DiGeorge syndrome. Circ. Res. 106, 686–694.

Scambler, P.J., Kelly, D., Lindsay, E., Williamson, R., Goldberg, R., Shprintzen, R., Wilson, D.I., Goodship, J.A., Cross, I.E., Burn, J., 1992. Velo-cardio-facial syndrome associated with chromosome 22 deletions encompassing the DiGeorge locus. Lancet 339, 1138–1139.

Scheet, P., Ehli, E.A., Xiao, X., van Beijsterveldt, C.E., Abdellaoui, A., Althoff, R.R., et al., 2012. Twins, tissue, and time: an assessment of SNPs and CNVs. Twin. Res. Hum. Genet, 1–9.

Schoenwolf, G.C., Larsen, W.J., 2009. Larsen's human embryology, fourth ed. Elsevier/Churchill Livingstone, Philadelphia, PA. pp. xix, 687 p.

Sebat, J., Lakshmi, B., Troge, J., Alexander, J., Young, J., Lundin, P., Maner, S., Massa, H., Walker, M., Chi, M., Navin, N., Lucito, R., Healy, J., Hicks, J., Ye, K., Reiner, A., Gilliam, T.C., Trask, B., Patterson, N., Zetterberg, A., Wigler, M., 2004. Large-scale copy number polymorphism in the human genome. Science 305, 525–528.

Segall, S.K., Nackley, A.G., Diatchenko, L., Lariviere, W.R., Lu, X., Marron, J.S., et al., 2010. Comt1 genotype and expression predicts anxiety and nociceptive sensitivity in inbred strains of mice. Genes. Brain Behav. 9, 933–946.

Shaikh, T.H., Kurahashi, H., Emanuel, B.S., 2001. Evolutionarily conserved low copy repeats (LCRs) in 22q11 mediate deletions, duplications, translocations, and genomic instability: an update and literature review. Genet. Med. 3, 6–13.

Shaikh, T.H., Kurahashi, H., Saitta, S.C., O'Hare, A.M., Hu, P., Roe, B.A., et al., 2000. Chromosome 22-specific low copy repeats and the 22q11.2 deletion syndrome: genomic organization and deletion endpoint analysis. Hum. Mol. Genet. 9, 489–501.

Shaikh, T.H., O'Connor, R.J., Pierpont, M.E., McGrath, J., Hacker, A.M., Nimmakayalu, M., et al., 2007. Low copy repeats mediate distal chromosome 22q11.2 deletions: sequence analysis predicts breakpoint mechanisms. Genome. Res. 17, 482–491.

Sharma, P., Levesque, T., Boilard, E., Park, E.A., 2014. Thyroid hormone status regulates the expression of secretory phospholipases. Biochem. Biophys. Res. Commun. 444(1), 56–62.

She, X., Jiang, Z., Clark, R.A., Liu, G., Cheng, Z., Tuzun, E., Church, D.M., Sutton, G., Halpern, A.L., Eichler, E.E., 2004. Shotgun sequence assembly and recent segmental duplications within the human genome. Nature 431, 927–930.

Shprintzen, R.J., 2008. Velo-cardio-facial syndrome: 30 Years of study. Dev. Disabil. Res. Rev. 14, 3–10.

Shprintzen, R.J., Goldberg, R.B., Lewin, M.L., Sidoti, E.J., Berkman, M.D., Argamaso, R.V., Young, D., 1978. A new syndrome involving cleft palate, cardiac anomalies, typical facies, and learning disabilities: velo-cardio-facial syndrome. Cleft. Palate. J. 15, 56–62.

Sigurdsson, T., Stark, K.L., Karayiorgou, M., Gogos, J.A., Gordon, J.A., 2010. Impaired hippocampal-prefrontal synchrony in a genetic mouse model of schizophrenia. Nature 464, 763–767.

Southard, A.E., Edelmann, L.J., Gelb, B.D., 2012. Role of copy number variants in structural birth defects. Pediatrics 129, 755–763.

Spencer, K.M., Nestor, P.G., Niznikiewicz, M.A., Salisbury, D.F., Shenton, M.E., McCarley, R.W., 2003. Abnormal neural synchrony in schizophrenia. J. Neurosci. 23, 7407–7411.

Spencer, K.M., Nestor, P.G., Perlmutter, R., Niznikiewicz, M.A., Klump, M.C., Frumin, M., Shenton, M.E., McCarley, R.W., 2004. Neural synchrony indexes disordered perception and cognition in schizophrenia. Proceedings of the National Academy of Sciences of the United States of America 101, 17288–17293.

Sporn, A., Addington, A., Reiss, A.L., Dean, M., Gogtay, N., Potocnik, U., et al., 2004. 22q11 deletion syndrome in childhood onset schizophrenia: an update. Molecular psychiatry 9, 225–226.

Stagi, S., Lapi, E., Gambineri, E., Manoni, C., Genuardi, M., Colarusso, G., et al., 2010a. Bone density and metabolism in subjects with microdeletion of chromosome 22q11 (del22q11). Eur. J. Endocrinol. 163, 329–337.

Stagi, S., Lapi, E., Gambineri, E., Salti, R., Genuardi, M., et al., 2010b. Thyroid function and morphology in subjects with microdeletion of chromosome 22q11 (del(22)(q11)). Clin. Endocrinol. (Oxf) 72, 839–844.

Stark, K.L., Xu, B., Bagchi, A., Lai, W.S., Liu, H., Hsu, R., et al., 2008. Altered brain microRNA biogenesis contributes to phenotypic deficits in a 22q11-deletion mouse model. Nat. Genet. 40, 751–760.

Tarquinio, D.C., Jones, M.C., Jones, K.L., Bird, L.M., 2012. Growth charts for 22q11 deletion syndrome. Am. J. Med. Genet. A. 158A, 2672–2681.

Thompson, S.G., Wong, P.T., Leong, S.F., McGeer, E.G., 1985. Regional distribution in rat brain of 1-pyrroline-5-carboxylate dehydrogenase and its localization to specific glial cells. J. Neurochem. 45, 1791–1796.

Torres-Juan, L., Rosell, J., Sanchez-de-la-Torre, M., Fibla, J., Heine-Suner, D., 2007. Analysis of meiotic recombination in 22q11.2, a region that frequently undergoes deletions and duplications. BMC. Med. Genet. 8, 14.

Tran, H.T., Delvaeye, M., Verschuere, V., Descamps, E., Crabbe, E., Van Hoorebeke, L., et al., 2011. ARVCF depletion cooperates with Tbx1 deficiency in the development of 22q11.2DS-like phenotypes in. *Xenopus*. Dev. Dyn. 240, 2680–2687.

Tunbridge, E.M., Harrison, P.J., Weinberger, D.R., 2006. Catechol-o-methyltransferase, cognition, and psychosis: Val158Met and beyond. Biol. Psychiatry. 60, 141–151.

van Amelsvoort, T., Daly, E., Henry, J., Robertson, D., Ng, V., Owen, M., et al., 2004. Brain anatomy in adults with velocardiofacial syndrome with and without schizophrenia: preliminary results of a structural magnetic resonance imaging study. Arch. Gen. Psychiatry. 61, 1085–1096.

van Kogelenberg, M., Ghedia, S., McGillivray, G., Bruno, D., Leventer, R., Macdermot, K., et al., 2010. Periventricular heterotopia in common microdeletion syndromes. Molecular syndromology 1, 35–41.

Vitelli, F., Morishima, M., Taddei, I., Lindsay, E.A., Baldini, A., 2002. Tbx1 mutation causes multiple cardiovascular defects and disrupts neural crest and cranial nerve migratory pathways. Hum. Mol. Genet. 11, 915–922.

Wang, Y., Li, G., Stanco, A., Long, J.E., Crawford, D., Potter, G.B., Pleasure, S.J., Behrens, T., Rubenstein, J.L., 2011. CXCR4 and CXCR7 have distinct functions in regulating interneuron migration. Neuron 69, 61–76.

Weinzimer, S.A., McDonald-McGinn, D.M., Driscoll, D.A., Emanuel, B.S., Zackai, E.H., Moshang Jr, T., 1998. Growth hormone deficiency in patients with 22q11.2 deletion: expanding the phenotype. Pediatrics 101, 929–932.

Weiss, L.A., Shen, Y., Korn, J.M., Arking, D.E., Miller, D.T., Fossdal, R., et al., 2008. Association between microdeletion and microduplication at 16p11.2 and autism. N. Engl. J. Med. 358, 667–675.

Weksberg, R., Stachon, A.C., Squire, J.A., Moldovan, L., Bayani, J., Meyn, S., Chow, E., Bassett, A.S., 2007. Molecular characterization of deletion breakpoints in adults with 22q11 deletion syndrome. Hum. Genet. 120, 837–845.

Woody, N.C., Snyder, C.H., Harris, J.A., 1969. Hyperprolinemia: clinical and biochemical family study. Pediatrics 44, 554–563.

Yi, J.J., Tang, S.X., McDonald-McGinn, D.M., Calkins, M.E., Whinna, D.A., Souders, M.C., et al., 2013. Contribution of congenital heart disease to neuropsychiatric outcome in school-age children with 22q11.2 deletion syndrome. Am. J. Med. Genet. B. Neuropsychiatr. Genet.

Chapter 37

Neural Tube Defects

Irene E. Zohn
Center for Neuroscience Research, Children's Research Institute, Children's National Medical Center, Washington DC

Chapter Outline

GLOSSARY

Anencephaly A neural tube defect (NTD) that results from failure of neurulation in the cephalic region, and characterized by the absence of a large part of the brain and overlying skull.

Apical constriction A process by which contraction of the apical surface of neural epithelial cells results in the cell taking on a wedge shape. Apical constriction is essential for hinge point formation during neural fold elevation.

Convergent extension (CE) The process in which the cells of a tissue rearrange to converge towards the midline and extend the tissue along the anterior-posterior axis of the embryo.

Craniorachischisis A neural tube defect (NTD) that results from failure of neural tube closure along the entire neural axis.

Exencephaly A neural tube defect (NTD) that results from failure of neural tube closure in the cephalic region. As fetal development continues, the exposed brain tissue will degenerate and exencephaly will progress to anencephaly.

Neurulation The process by which the neural tissue is induced and transformed from a flat neural plate into the neural tube that will develop into the central nervous system. Neurulation can be primary or secondary, where the flat neural plate rolls into a tube, or mesenchyme cells coalesce into a medullary cord that cavitates to form a tube, respectively. Failures in neurulation result in neural tube defects (NTDs).

Planar cell polarity (PCP) This establishes the polarity of cells within an epithelium, or for migrating cells controls the localization of polarized protrusions. PCP signaling is essential for the convergent extension movement that drives neural tube closure.

Spina bifida A neural tube defect (NTD) that results in the failure of neural tube closure in the posterior region of the embryo. Can be classified into a number of forms including: meningocele where the meninges protrude from an opening in the vertebrae; myelomeningocele where the meninges and spinal cord protrude from an opening in the vertebrae, and spina bifida occulta where the dorsal part of the vertebrae does not form but neither the meninges nor the spinal cord protrude.

SUMMARY

- Neural tube defects arise from failures of neurulation, and represent some of the most common congenital malformations in humans.

- Neural tube defects have a complex etiology, having both genetic and environmental contributing factors.
- Neurulation involves a complex series of precisely choreographed, morphogenetic movements that transform the neural plate into the neural tube.
- Mouse, chicken, *Xenopus* and zebrafish model systems have been instrumental in elucidating the molecular and cellular mechanisms underlying neurulation.
- The study of these model systems has led to the discovery of a large number of candidate genes for sequencing in human patients.

37.1 INTRODUCTION

Neural tube defects (NTDs) are a group of congenital malformations of the central nervous system that occur when the neural tube fails to close properly during embryogenesis. They represent some of the most common structural birth defects in humans. The frequency of NTDs varies greatly in different regions of the world from 3.4 in 10,000 live births in the United States to 1 in 100 live births in Northern China (Gu et al., 2007; Parker et al., 2010). NTDs arise from failures in neurulation, a complex morphogenetic process that results in transformation of the flat neural plate into the neural tube. The consequence of NTDs vary with the size and placement of the lesion and can result in a spectrum of outcomes from death to severe or moderate disability, and can even be asymptomatic. The causes of NTDs are multifactorial, having both genetic and environmental components. This chapter reviews the molecular and cellular basis of neurulation and how the study of this process using animal models has lead to identification of the causes and improved prevention and treatments for NTDs in humans.

37.2 DISCUSSION

NTDs

NTDs can be classified into a number of different malformations (Figure 37.1). The most common in humans is spina bifida, in which the neural tube fails to close in the posterior region of the neural axis or spine. Spina bifida is compatible with life and the consequence varies greatly, depending on the size and placement of the defect and the involvement of the spinal nerves and meninges. Spina bifida can be categorized into multiple types (Figure 37.2). Myelomeningocele is the most severe form of spina bifida and involves protrusion of the meninges and spinal cord through an opening in the vertebrae. Meningocele involves protrusion of the meninges but not the spinal cord. Spina bifida occulta is a mild form of spina bifida where the dorsal part of vertebrae does not form. Neither the meninges nor the spinal cord protrude and spina bifida

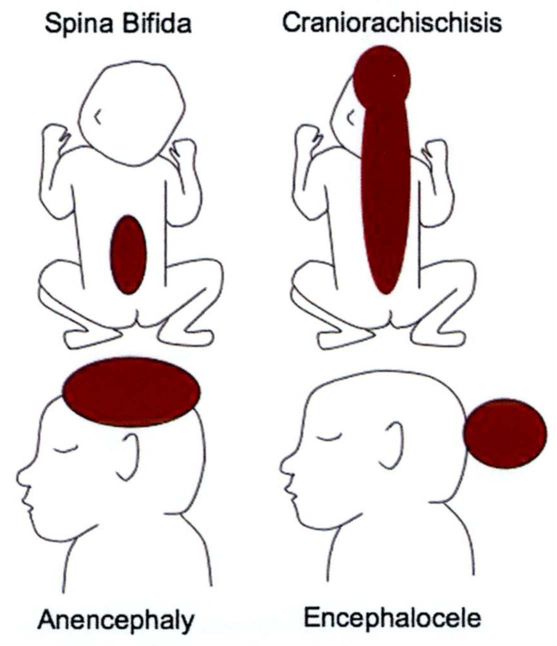

FIGURE 37.1 **Neural Tube Defects are Classified into a Number of Different Malformations.** Spina bifida occurs when the neural tube fails to close in the posterior region. Anencephaly occurs when neurulation fails in the cranial region. When the neural tube fails to close along the entire neural axis, craniorachischisis results. Encephalocele occurs when the brain and meninges protrude from an opening in the skull.

occulta is often asymptomatic. Craniorachischisis is the most severe NTD, resulting from failure of neural tube closure along the entire neural plate. Exencephaly (the embryonic precursor to anencephaly) occurs when closure fails in the anterior neural plate or brain. Anencephaly and craniorachischisis are fatal, resulting in prenatal death of the fetus or demise of the newborn shortly after birth. Encephalocele occurs when the cranial vault fails to form properly around the closed neural tube and the brain and meninges protrude through an opening in the skull.

Diagnosis and Treatment of Spina Bifida

Most commonly, NTDs are diagnosed before birth by standard prenatal screening tests. As part of the triple screen for multiple birth defects, levels of alpha fetal protein (AFP) in maternal serum are measured between the 15th and 20th week of pregnancy. High levels of AFP in this test are correlated with an NTD, and signal the need for further testing. Additionally, high levels of AFP in amniotic fluid obtained by amniocentesis are indicative of the presence of a NTD. Most NTDs can be identified by ultrasound during the routine anatomy scan performed between 18 and 22 weeks.

Babies with spina bifida are delivered by cesarean section to minimize further damage to the lesion (reviewed in Thompson, 2009). Secondary defects of the central nervous system also frequently develop with spina bifida including Arnold-Chiari malformations with hindbrain herniation, hydrocephalus requiring placement of a shunt and tethered spinal cord where the spinal cord becomes attached to scar tissue at the site of the repaired lesion (Adzick and Walsh, 2003; Dias and McLone, 1993; Oakeshott and Hunt, 2003). A tethered spinal cord results in progressive pain, incontinence, and weakness of the lower extremities, foot and spinal deformities as the patient grows during childhood, and requires additional surgeries to correct (Hertzler et al., 2010; Hudgins and Gilreath, 2004; Yamada et al., 2004). Nerve damage results in neurogenic bladder and/or bowel and paralysis of the lower extremities requiring many spina bifida patients be confined to a wheelchair or to use braces or crutches to assist with walking (reviewed in Thompson, 2009). Because of reduced sensation in the lower extremities, patients are susceptible to unrealized bacteria and fungal infections of the legs that can lead to the need to perform amputations. Other complications include learning disabilities, social issues and latex allergies (reviewed in Thompson, 2009). Improvements in management of the complications of spina bifida have resulted in the majority of patients surviving well into adulthood (reviewed in Adzick and Walsh, 2003; Thompson, 2009).

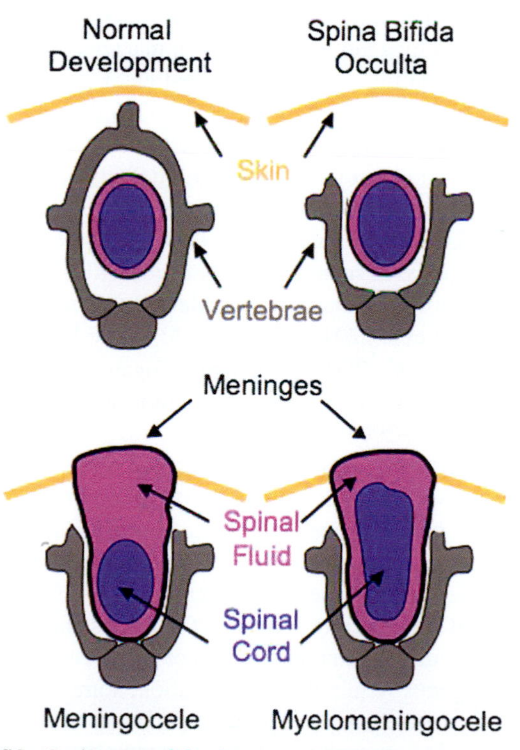

FIGURE 37.2 **Common Types of Spina Bifida.** During normal development of the spinal column, vertebrae form around the closed neural tube. In spina bifida occulta, the vertebral arches do not form correctly. Skin and the neural tissue cover the defect and meninges do not protrude. Meningocele occurs when the meninges but not spinal cord protrudes from the open vertebrae. Myelomeningocele is the most common and severe form of spina bifida. Both the spinal cord and meninges protrude from the open vertebrae.

Myelomeningocele is corrected with surgery either *in utero* or shortly after birth. A "two hit" model has been proposed, in which some deficits arise from the initial malformation and deficits worsen during later gestation, where exposed spinal cord tissue and nerves progressively deteriorate. This has been supported by studies in mouse models, which demonstrate further deterioration of tissue during late gestation (Stiefel et al., 2007). These experiments suggested that repair of the lesion *in utero* should improve outcomes. Numerous studies have compared outcomes with prenatal versus postnatal surgery (Bruner et al., 1999a,b; Farmer et al., 2003; Sutton et al., 1999; Tulipan et al., 1998; 1999a,b); however, in most of these, the postnatal group consisted of historical controls. A recent randomized clinical trial, the MOMs study (Management of Myelomeningocele Study), compared outcomes of prenatal (before 26 weeks) and postnatal surgery in a large patient sample (Adzick et al., 2011). This study revealed that prenatal closure of the NTD results in significant reductions in deficits compared to postnatal surgery including the need to place a shunt for hydrocephalus, better mental and motor development and improvements in other outcomes such as hindbrain herniation. However, prenatal surgery increased the risk of preterm birth, uterine rupture, and other pregnancy complications. Overall the benefits of prenatal surgery outweigh the risks and *in utero* surgery is becoming the standard of care.

Causes of NTDs

The causes of NTDs are complex, with both environmental and genetic contributions. a multifactorial threshold model has been proposed to account for the pattern of NTD inheritance observed in humans (Harris and Juriloff, 2007; Zohn and Sarkar, 2008). This model postulates that neural tube closure is a threshold event that occurs either successfully – resulting in normal neural tube closure – or not – resulting in an NTD. One threshold may be the timing of the completion of neural tube closure. Live imaging studies in the mouse and chicken suggest that disruption of the processes required for neural tube closure may result in a slowing down of the neurulation process (Massarwa and Niswander, 2013; Yamaguchi et al., 2011). Following this deadline, fusion may no longer be able to occur or expansion of the brain ventricle counteracts closure and can cause reopening of an incompletely closed neural tube.

A number of environmental factors contribute to NTDs, with one of the most important being maternal nutrition during the periconception period (reviewed in Zohn and Sarkar, 2010). Numerous nutrients have been implicated as potentially important for the prevention of NTDs, of which folic acid is the most noteworthy (Schorah, 2009). Others include inositol, choline, retinoic acid, and iron (reviewed in Greene and Copp, 2005; Zeisel, 2009; Zohn and Sarkar, 2010). Exposure to certain medications such as valproic acid is associated with increased NTD incidence (Lammer et al., 1987). Ingestion of fumonisins, a group of mycotoxins produced by a mold that grows on maize, also causes NTDs (reviewed in Gelineau-van Waes et al., 2009). Auto-antibodies against the folic acid receptor that block the interaction of folate with the receptor are found at higher levels in maternal serum from NTD affected pregnancies (Rothenberg et al., 2004). Hyperthermia as a result of hot tub usage or maternal fever during neural tube closure can result in an NTD (Cabrera et al., 2004). Maternal diabetes and/or obesity increase the likelihood of a number of structural birth defects including NTDs (reviewed in Reece, 2012). Thus the current global epidemic of obesity could potentially increase the incidence of NTDs in the near future.

Multiple lines of evidence suggest a genetic component of NTDs. Chromosomal abnormalities are often present in NTD-affected fetuses, and NTDs are often noted in spontaneous abortions with abnormal karyotypes (Chen, 2007a,b; 2009; Detrait et al., 2005). NTDs are associated with syndromes such as Meckel-Gruber, Waardenburg and DiGeorge/22q11 deletion syndromes (Baldwin et al., 1994; Canda et al., 2012; Chen, 2007c; 2008; Forrester et al., 2002; Kinoshita et al., 2010; Kousseff, 1984; Logan et al., 2011; Maclean et al., 2004; Nickel and Magenis, 1996; Nickel et al., 1994; Seller et al., 2002; Toriello et al., 1985). Finally, twin studies indicate a 5% concordance rate and NTD risk is increased in patients with a NTD or a previous affected pregnancy (Detrait et al., 2005; Risch, 1990a,b).

The identification of causative mutations for NTDs has been hampered by the paucity of large families with multiple affected individuals. Linkage analyses of some smaller multiplex families implicates specific chromosomal regions as susceptibility loci, but causative mutations have not been identified (Rampersaud et al., 2005; Stamm et al., 2006; 2008). Most studies have taken a candidate gene approach to identifying mutations in samples from NTD patients (reviewed in Au et al., 2010). Thus far, studies of mouse models have identified over 200 genes that when mutated result in NTDs, providing a long list of candidate genes to test (reviewed in Harris, 2009; Harris and Juriloff, 2007; 2010; Juriloff and Harris, 2012). This list is far from complete as loss-of-function phenotypes of many genes in the mouse genome have not yet been characterized. With the large number of genes required for neurulation, some known and many not, the candidate gene approach has only identified a handful of putative causative mutations. The genes involved in humans likely represent those required for neurulation and include null or hypomorphic variants. The majority of variants would not cause an NTD unless combined with other genetic or environmental factors, and would either occur commonly in the population or represent rare mutations. Whole genome and exome sequencing is now feasible, as is microarray analysis to detect copy number variants, and many groups are beginning to use these strategies to find mutations in an unbiased fashion (Au et al., 2010; Bassuk et al., 2013). These approaches promise to identify many more potential causes of NTDs, and may also identify multiple variants in a single individual that could contribute to the NTD in a multifactorial fashion.

Once variants are identified, the challenge remains to determine whether they disrupt gene function. Multiple variants in the same gene are typically not found and consistent with the multifactorial nature of NTDs, variants are also present in unaffected parents and siblings. Thus, multiple innovative approaches are needed to determine whether the mutation contributes to the NTD. The use of model systems such as the zebrafish, along with cell and biochemical assays based upon prior characterization of gene function, are proving to be essential for validating the deleterious nature of variants. Thus a coherent understanding of the genes and pathways required for neural tube closure provides candidate genes for this birth defect both *a priori* when selecting which genes to sequence, or after the fact when analyzing whole genome sequencing data. Most importantly, information gained from basic research on the mechanisms of neurulation provides a framework for determining the deleterious nature of the variants identified.

NTDs Result from Failure of Neurulation

Primary neurulation is a complex morphogenetic process that results in the transformation of the flat neural plate into the neural tube. Neural tube formation involves the coordinated growth and morphogenesis of multiple tissues. Forces that drive neural tube closure arise from the neural tissue itself (intrinsic forces) as well as from the surrounding surface ectoderm and mesoderm (extrinsic forces; reviewed in Colas and Schoenwolf, 2001). Primary neurulation in mouse and chicken embryos begins after gastrulation, as the neuroepithelium is induced from the embryonic ectoderm (Figure 37.3). Individual

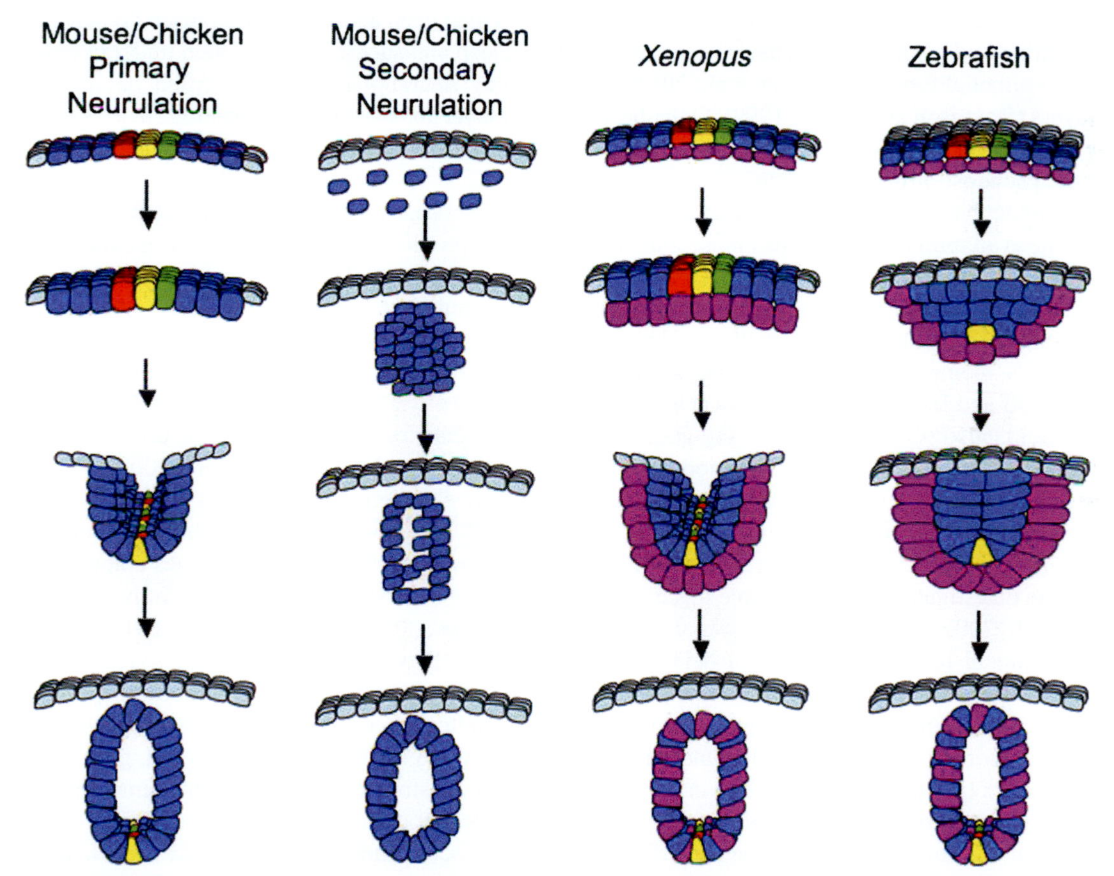

FIGURE 37.3 Neurulation in Mouse, Chicken, *Xenopus* and Zebrafish. Primary neurulation in mouse and chicken begins with the induction of neural ectoderm (blue) separating it from non-neural ectoderm (light blue). Neural epithelial cells are arranged in a monolayer and then elongate along the apical-basal axis. Convergent extension and hinge point formation occur simultaneously driving elevation of the neural folds and lengthening and thinning of the neural plate. Medial-lateral intercalation of cells is illustrated by rearrangement of red, yellow and green labeled cells in the midline. As the neural folds rise and meet in the dorsal midline they fuse and the neural and surface epidermis remodel to from two separate epithelial sheets. Secondary neurulation occurs in the most caudal regions of the spinal cord and involves coalescence of mesenchyme cells to form a rod that undergoes epithelialization to form the neural tube. In *Xenopus*, deep (pink) and superficial (blue) cells remodel as the neural tube forms. Only cells in the superficial layer undergo apical constriction. As closure progresses, medial migration and radial intercalation of the deep and superficial cells result in establishment of the neural tube. In zebrafish, deep and superficial cells elongate and converge to form the neural keel. Deep and superficial cells interdigitate and epithelize to form the neural tube.

neuroepithelial cells elongate and appear as a thickening of the ectoderm on the dorsal side of the embryo. Following neural induction, two coordinated morphogenetic movements intrinsic to the neural plate drive elevation of the neural folds to aid rolling the plate into a tube: 1) lengthening and narrowing of the neural plate due to convergent extension (CE) movements; and 2) formation of hinge points around which the neural plate bends. A single hinge point forms in the midline of the neural plate (medial hinge point, MHP) followed by formation of paired dorsal lateral hinge points (DLHPs) in lateral regions of the neural plate. Extrinsic forces from the surface epithelium and surrounding mesenchyme also promote elevation of the neural folds. As the paired neural folds meet in the dorsal midline they fuse, and the neural and surface epithelium remodel to form two separate epithelial sheets. Finally, the skull and vertebrae form over the closed neural tube. Disruptions in any of these processes can result in an NTD. In the following sections, the molecular and cellular basis of each of these important events will be discussed in detail, and relationships with the causes and prevention of NTDs in humans will be pointed out.

Animal Models Used to Study Neurulation

A number of animal models are used to study the genetic, cellular and molecular basis of neural tube morphogenesis, including the mouse, chicken, frog (African clawed frog; *Xenopus laevis*) and fish (zebrafish; *Danio rerio*). Neurulation in the chicken and mouse is most similar to that of humans, whereas *Xenopus* and zebrafish utilize many of the same cellular processes but show important differences in overall morphogenesis (Figure 37.3). In the mouse, primary neurulation results in the formation of the majority of the central nervous system, while secondary neurulation forms of a portion of the spine including and caudal to the sacral vertebrae (Morriss-Kay et al., 1994; Schoenwolf, 1984). The majority of the neural tube in chicken is also formed by primary neurulation (Schoenwolf and Delongo, 1980). In contrast, zebrafish utilize a modified secondary neurulation process along the entire neural axis, in which deep and superficial mesenchymal cells converge towards the midline and coalesce into a neural keel intermediate. Deep and superficial cells then undergo radial intercalation to form an epithelial tube (Harrington et al., 2010). The *Xenopus'* neural plate is also stratified, having both deep and superficial layers (Davidson and Keller, 1999). Only the superficial layer shows apical constriction to drive elevation of the neural folds (Schroeder, 1970). Once the folds fuse in the dorsal midline, deep and superficial cells undergo radial intercalation to form the pseudostratified epithelium.

The mouse model has the advantage of being amenable to genetic approaches for studying the genes required for neural tube closure. The availability of numerous mouse mutants with NTDs provides a rich source of diverse models to elucidate the genes and pathways required for neural tube closure (reviewed in Harris, 2009; Harris and Juriloff, 2007; 2010; Juriloff and Harris, 2012). Furthermore, mouse models can simulate the multifactorial genetics of NTDs observed in humans (reviewed in Zohn, 2012; Zohn and Sarkar, 2008). Numerous examples exist of compound mutants or modifier variants that do not cause NTDs themselves but increase the penetrance and/or severity of NTDs in combination (Harris and Juriloff, 2007; 2010; Juriloff and Harris, 2012; Kooistra et al., 2012; Korstanje et al., 2008; Letts et al., 1995; Neumann et al., 1994). The mouse model is also very useful for testing gene-environment interactions. For example, the effectiveness of nutrient supplementation to suppress NTDs in mouse models has been tested (reviewed in Gray and Ross, 2009; Greene and Copp, 2005). Mouse models of diabetes have been developed and provide novel insight as to the genes and pathways that interact with hyperglycemia to cause NTDs (reviewed in Zabihi and Loeken, 2010). However, because the mouse embryo develops *in utero,* examination of the cell behaviors that underlie neurulation presents significant challenges compared to the other models. Yet some investigators have overcome this hurdle by combining embryo culture with innovative live imaging techniques (Jones et al., 2002; Massarwa and Niswander, 2013; Pyrgaki et al., 2010; Yamaguchi et al., 2011).

The chicken model has the advantage of being more amenable to live imaging experiments. Gene function can be manipulated by electroporation of DNA constructs, antisense morpholino oligonucleotides (morpholinos) or small interfering RNAs (siRNAs) or culturing embryos with pharmacological inhibitors. *Xenopus* and zebrafish models are very useful for studying the conserved cell behaviors that underlie neurulation through live imaging. In addition, gene function can be perturbed by genetics (in the zebrafish system) or expression of DNA constructs, morpholinos and small interfering RNAs (siRNAs). Thus, much of our knowledge of the morphogenetic events that drive neurulation comes from combined information gained from the study of all of these models. This chapter will focus primarily on the mouse model, and will discuss findings from the chicken, *Xenopus* and zebrafish where these systems have contributed mechanistic insight.

Elongation of the Neuroepithelium and the Microtubule Cytoskeletion

The thickening of the neuroepithelium occurs as cells transform from a cuboidal to columnar epithelial organization and elongate along the apical-basal axis (reviewed in Colas and Schoenwolf, 2001; Suzuki et al., 2012). Elongation involves increased polymerization and reorientation of microtubules along the apical-basal axis (Burnside, 1973; Handel and Roth,

1971; Karfunkel, 1971; Lee et al., 2007a). Non-centrosomal gamma-tubulin becomes concentrated apically during elongation, acting as nucleating centers for microtubule polymerization (Burnside, 1973; Lee et al., 2007a; Schroeder, 1970; Suzuki et al., 2012). Microtubule polymerization inhibitors prevent thickening of the neural epithelium and neural fold elevation (Handel and Roth, 1971; Karfunkel, 1971). Additionally, gamma-tubulin localization and microtubule organization are disrupted with inhibition of Shroom3, a microtubule binding protein, in the *Xenopus* neural tube (Lee et al., 2007a).

Convergent Extension Movements

Following apical-basal elongation of the neuroepithelium, the neural plate undergoes convergent extension (CE) movements, which result in lengthening and narrowing of the neural plate along the anterior-posterior and medial-lateral axes, respectively. CE movements have been best characterized in *Xenopus* in which polarized cell movements drive the intercalation of cells to transform the neural plate into a longer, thinner structure and elongate the body axis (reviewed in Wallingford, 2012). Polarized cell behaviors that mediate CE movements are controlled by the planar cell polarity (PCP) pathway (Heisenberg et al., 2000; Tada and Smith, 2000; Wallingford et al., 2000). PCP was first described in *Drosophila,* where it controls the polarity of cells within an epithelial plane and the proper positioning of asymmetrically localized structures such as the hairs on the wing (reviewed in Vladar et al., 2009). PCP depends on the asymmetric distribution of membrane associated protein complexes comprised of both core and effector PCP proteins (reviewed in Wallingford, 2012). Core PCP proteins include: frizzled (fz), Disheveled (Dvl), prickle (pk), Van Gogh (Vang), flamingo (fmi), and diego (dgo); and effector PCP proteins include: Disheveled-associated activator of morphogenesis 1 (Daam1), Ras homolog gene family, member A (RhoA), Rho-associated protein kinase (Rock). and c-Jun N-terminal kinase (Jnk).

The PCP pathway is conserved in vertebrates and is controlled by the noncanonical Wnt pathway. The Wnt ligands Wnt5a and Wnt11 along with coreceptors Ptk7 and Ror2 are required for PCP (Gao et al., 2011; Heisenberg et al., 2000; Hikasa et al., 2002; Lu et al., 2004; Qian et al., 2007; Tada and Smith, 2000; Wallingford and Harland, 2001). Upon ligand stimulation, the Fz receptor recruits Dvl and Daam1 to activate the small GTPase RhoA resulting in Rock and Jnk pathway activation to regulate the actin cytoskeleton (Habas et al., 2001; Nishimura et al., 2012). The asymmetrically localized Vangl2/Celsr1/Pk complex located on the other side of the cell inhibits the Fz/Dvl complex resulting in asymmetric activation of the pathway and polarized protrusions (reviewed in Tissir and Goffinet, 2010).

During neurulation, PCP regulates the polarization of mediolateral protrusions that direct CE movements in the neural plate (Wallingford and Harland, 2002). When CE is defective in the mouse or *Xenopus*, the embryos are shorter with a wider midline and floorplate resulting in neural folds that are too far apart to fuse in the dorsal midline (Wallingford and Harland, 2002; Wang et al., 2006a; Ybot-Gonzalez et al., 2007b). In embryos with defects in CE movements, the neural tube fails to close along the entire anterior-posterior neural axis, resulting in a condition known as craniorachischisis. This is best studied in the *Looptail* (*Lp*) mouse line that carries a missense mutation in *Van Gogh like-2* (*Vangl2*) (Kibar et al., 2001; Murdoch et al., 2001). *Vangl2*^{Lp} mutants have a shorter body axis and a broader notochord and floorplate (Gerrelli and Copp, 1997; Greene et al., 1998; Smith and Stein, 1962; Wilson and Wyatt, 1994). Further studies have demonstrated that *Vangl2* is necessary for CE movements in the notochord and neural plate (Ybot-Gonzalez et al., 2007b). Mutations in other PCP genes that result in NTDs have since been identified in the mouse. Embryos homozygous for mutations in *Celsr1,* the vertebrate homolog of *flamingo/starry night,* exhibit craniorachischisis (Curtin et al., 2003). Similarly, compound mutants for the vertebrate homologs of *Disheveled* (*Dvl*), *Dvl1*$^{-/-}$;*Dvl2*$^{-/-}$, *Dvl2*$^{-/-}$;*Dvl3*$^{+/-}$, *Dvl2*$^{+/-}$;*Dvl3*$^{-/-}$, or the frizzled receptor *Fzd3*$^{-/-}$;*Fzd6*$^{-/-}$ show craniorachischisis (Etheridge et al., 2008; Hamblet et al., 2002; Wang et al., 2006b) as do targeted knockouts of other PCP pathway components such as *Wnt5a and Ptk7* (Lu et al., 2004; Paudyal et al., 2010; Qian et al., 2007). Mutation of PCP genes can also result in milder forms of NTDs, such as spina bifida and exencephaly. For example, *Dvl2*, *Vangl2*$^{Lp/+}$, *Vangl1*, *Fuzzy*, and *Inturned* mutants exhibit variably penetrant spina bifida and/or exencephaly (Copp et al., 1994; Gray et al., 2009; Hamblet et al., 2002; Heydeck and Liu, 2011; Heydeck et al., 2009; Zeng et al., 2010). Interestingly, human embryos with craniorachischisis are also shorter and have a broader floorplate (Kirillova et al., 2000), indicating that similar mechanisms may underlie craniorachischisis in humans.

Consistent with the multifactorial threshold model for NTDs, a number of loci can increase penetrance and severity of NTDs in *Vangl2*^{Lp} heterozygotes. For example, *Vangl2*^{Lp} can genetically interact with other PCP genes, including *Wnt5a*; *Vangl1, Dvl2, Dvl3, Celsr1, Fz1, Fz2,* and *Protein tyrosine kinase-7* (*Ptk7)* in compound mutants to cause NTDs (Curtin et al., 2003; Etheridge et al., 2008; Lu et al., 2004; Paudyal et al., 2010; Qian et al., 2007; Torban et al., 2008; Wang et al., 2006a,b; Yu et al., 2010). Additionally, *Vangl2*^{Lp} can genetically interact with mutations in genes not previously identified as regulating PCP pathways to give NTDs. These include: *Grhl3*, *Bardet-Biedl syndrome-1 (BBS1)*, *BBS4*, *BBS6*, *cordon blue* (*cobl*) and *Scribble (Scrbl*), *Syndecan 4* (*Sdc4*) and *Sec24b* (Caddy et al., 2010; Carroll et al., 2003; Escobedo et al., 2013; Merte et al., 2010; Murdoch et al., 2003; Ross et al., 2005). Interestingly, heterozygous *Vangl2*$^{Lp/+}$ embryos show a slightly wider and shorter midline and delayed neural tube closure (Ybot-Gonzalez et al., 2007b), providing the basis for the development of NTDs in heterozygous embryos and a likely mechanism for NTDs in genetic interaction experiments.

PCP Controls Ciliogenesis in Vertebrates

In addition to controlling polarized protrusions, PCP also controls the positioning of cilia on the cell (reviewed in Wallingford and Mitchell, 2011). Many of the genes that interact with *Vangl2*^{Lp} regulate cilia formation, for example the BBS (Bardet-Biedl syndrome) proteins. BBS is a rare pleiotropic disorder caused by abnormal cilia formation. NTDs are not a described component of BBS, yet mouse mutants for some of the genes that cause BBS show a low penetrance of NTDs or interact with other genes to cause NTDs (Eichers et al., 2006; Ross et al., 2005). Other ciliopathies, such as Meckel-Gruber (MKS) and Joubert syndromes are also associated with NTDs in mouse models but not in the human syndrome (Abdelhamed et al., 2013; Caspary et al., 2007; Weatherbee et al., 2009). Furthermore, mutation of the PCP effector proteins *Fuzzy* and *Inturned* results in defects in cilia, Sonic hedgehog (Shh) signaling, and neural tube closure (Heydeck and Liu, 2011; Heydeck et al., 2009; Zeng et al., 2010). In vertebrates, cilia play an essential role in the transduction of Shh signaling. The formation of cilia and movement of Shh signaling components in the cilia require intraflagellar transport (IFT), and mutation of IFT proteins disrupts Shh signaling and causes NTDs (reviewed in Bay and Caspary, 2012; Briscoe and Therond, 2013; see also Chapter 1). The Shh pathway relies on multiple inhibitory interactions that ultimately result in the conversion of Gli proteins into activator and repressor forms to effect the transcription of Shh target genes (Haycraft et al., 2005; Huangfu and Anderson, 2005; Liu et al., 2005; May et al., 2005). Thus both activation and inhibition of Shh is affected in IFT mutants.

Mutation of PCP Pathway Components Cause NTDs in Human Patients

The realization that mutation of the genes involved in PCP pathways can cause NTDs in mouse models in a multifactoral fashion prompted the sequencing of PCP genes in human patients with NTDs. Thus far, mutations in numerous PCP-related genes have been found in these patients. Predicted and/or proven deleterious mutations in *CELSR1, FZD6, PRICKLE1, VANGL1, VANGL2, FUZ, SCRIB,* and *DACT1* occur in patients with NTDs (Allache et al., 2012; Bosoi et al., 2011; De Marco et al., 2012; Doudney et al., 2005; Kibar et al., 2007; 2011; Lei et al., 2010; 2013; Reynolds et al., 2010; Robinson et al., 2012; Seo et al., 2011; Shi et al., 2012). The deleterious nature of some of these mutations has been determined by failure to rescue PCP phenotypes in zebrafish, disrupted binding to known binding partners or mislocalization in polarized epithelium (Bosoi et al., 2011; Kibar et al., 2007; Reynolds et al., 2010; Robinson et al., 2012; Seo et al., 2011).

Formation of Hinge Points

The MHP and paired DLHPs are formed as cells of the neural epithelium become wedge-shaped. The cellular mechanisms contribute to transformation of neural epithelial cells into this wedge shape include: 1) retention of the nucleus in the basal region of the cell, 2) localized disruption of apical-basal polarity and 3) apical constriction (reviewed in Colas and Schoenwolf, 2001; Eom et al., 2013). The pseudostratified neuroepithelium in the chicken and mouse consists of bipolar neural progenitor cells with attachments at the apical surface (future ventricle) and the basal basement membrane. The cell body containing the nucleus moves between apical and basal positions dependent upon the phase of the cell cycle in a process called interkinetic nuclear migration. During mitosis, the nucleus is localized at the apical surface, and during other phases of the cell cycle it is positioned more basally. Typically, adjacent cells in the neuroepithelium are at different stages in the cell cycle and the nuclei of adjacent cells are located at different positions. However, in the MHP of the chicken embryo, greater numbers of nuclei are localized at basal positions (Eom et al., 2011; 2012; Smith and Schoenwolf, 1987). This is accomplished at least in part by slowing down the cell cycle and arresting cells in non-mitotic phases of the cell cycle (Eom et al., 2011; 2012). Since the majority of cells in the MHP have basally positioned nuclei, rather than randomly distributed between apical and basal positions, the net effect is multiple wedge shaped cells that contribute to bending the epithelium at the MHP.

Disruption of apical-basal polarity at hinge points also helps to bend the neural plate. The apical-basal polarity of the neuroepithelium is established by the asymmetrical distribution of polarity protein complexes (reviewed in Eom et al., 2013). Tight and adherens junctions are multi-protein complexes localized in the apical region of the cell and exclude proteins normally localized to the basal-lateral regions. Tight junctions include the PAR and Crumbs complexes which link adjacent cells, forming a barrier to the paracellular diffusion of solutes. Adherens junctions are localized slightly basal to the tight junction and link adjacent cells via membrane-localized adhesion molecules (e.g., N-cadherin) to the actin and microtubule cytoskeletons. In the basal-lateral region, proteins such as Scribble, Disks large and Lethal giant larva (LGL) localize. To form the MHP of the chicken neural plate apical-basal polarity is modified as adherens and tight junctional complexes reorganize (Eom et al., 2011; 2012). PAR3 dissociates from the adherens junction and increased levels are found in the cytoplasm. N-cadherin is endocytosed at the tight junction and the basal-lateral LGL becomes localized to the apical region of the cell. This reorganization causes a localized destabilization of apical-basal polarity and may contribute to the ability of the rigid neural plate to bend around the hinge point.

In addition to their role in establishing the apical-basal polarity of the neuroepithelium, apical junctions play essential roles in apical constriction during hinge point formation and bending of the neural plate (reviewed in Eom et al., 2013). The adherens junction nucleates filamentous actin (F-actin) polymerization at the apical end of the cell where actin interacts with the non-muscle myosin motor protein to form circumferential actomyosin cables. The contractile force needed for apical constriction is generated as myosin moves along the actin filaments, pulling them closer together. Contraction of these cables results in constriction of the apical end of the cell. During neurulation, these actomyosin cables become thickened as apical constriction occurs.

The signaling pathways that control activation of myosin to promote the apical constriction that takes place during neural fold elevation involve the small GTPase RhoA, which initiates a kinase cascade that ultimately results in phosphorylation and activation of myosin. RhoA activates Rock, which phosphorylates and activates myosin light chain kinase (P-MLC) (reviewed in Suzuki et al., 2012). P-MLC in turn phosphorylates and activates myosin, stimulating its movement along actin filaments and generating contractile force. Many of these proteins (F-actin, RhoA, phosphorylated MLC and myosin) are concentrated in overlapping patterns in the apical region of neuroepithelial cells during neural fold elevation (Haigo et al., 2003; Hildebrand and Soriano, 1999; Lee et al., 2007b; Nishimura and Takeichi, 2008).

The requirement of this pathway in apical constriction of the neural epithelium is illustrated by inhibitor studies. Treatment of chicken embryos with inhibitors of myosin or RhoA disrupts neural tube closure (Kinoshita et al., 2008). Similarly, morpholino-mediated inhibition of myosin in *Xenopus* prevents the apical accumulation of F-actin and apical constriction resulting in NTDs (Rolo et al., 2009). Treatment of neurulation-stage chicken or mouse embryos in culture with Rock inhibitors causes NTDs (Gray and Ross, 2011; Gray et al., 2013; Kinoshita et al., 2008) and inhibition of actin polymerization with cytochalasin D results in failure of apical constriction (Kinoshita et al., 2008). Genetic studies in mice also illustrate the essential role of actin in neurulation. Compound mutants in actin-binding proteins including *Mena*, *VASP* and *profilin* exhibit NTDs (Lanier et al., 1999; Menzies et al., 2004). Xena, the *Xenopus* ortholog of Mena is required for neural tube closure, apical constriction, and apical accumulation of F-actin (Roffers-Agarwal et al., 2008), suggesting that NTDs in the mouse may also be due to defects in apical constriction.

Two independent upstream activators are linked to the regulation of apical constriction: Shroom3 and Lulu1/2. Shroom3 was identified as a regulator of apical constriction, hinge point formation and neural tube closure in a genetrap mouse model (Hildebrand and Soriano, 1999) and has since been shown to regulate apical constriction and neural tube closure in *Xenopus* and chicken embryos (Haigo et al., 2003; Nishimura and Takeichi, 2008; Plageman et al., 2010; 2011). In the mouse, Shroom3 is localized at the apical adherens junction, where it recruits Rock and myosin to locally regulate actomyosin contraction (Hildebrand, 2005; Hildebrand and Soriano, 1999; Nishimura and Takeichi, 2008). Inhibition of Shroom3 in chicken also prevents apical accumulation of F-actin and phosphorylated MLC (Nishimura and Takeichi, 2008). Furthermore, Shroom3 is sufficient to cause apical constriction when overexpressed in the *Xenopus* neural tube (Haigo et al., 2003). Interestingly, live imaging experiments in mice show that bending of the neural plate in *Shroom3* mutants does occur but is slowed (Massarwa and Niswander, 2013).

Additional regulators of apical constriction in the neuroepithelium are the FREM-domain containing proteins Lulu1 and Lulu2. *Lulu1* was originally identified in an ENU mutagenesis screen in mice as a regulator of gastrulation and neural fold morphogenesis (Lee et al., 2007b). Mutation of *Lulu1* results in neural folds that fail to elevate with abnormal apical accumulation of myosin and F-actin and abnormal localization of mitotic nuclei away from apical positions (Lee et al., 2007b). Lulu2 activates 114RhoGEF, a guanine nucleotide exchange factor for RhoA at the apical junction (Nakajima and Tanoue, 2011), thus initiating the cascade of events resulting in actomyosin contractility. In epithelial cells, overexpression of Lulu2 is sufficient to cause apical constriction, whereas, siRNA mediated knockdown results in disorganized actomyosin cables (Nakajima and Tanoue, 2009).

Apical Constriction is Coordinated with PCP Activation in the Neural Plate

Dynamic integration of PCP and apical constriction pathways cause apical constriction preferentially along the mediolateral axis, allowing the neural plate to bend rather than to pucker as it would if constriction occurred radially (Nishimura et al., 2012). Typically, apical constriction is visualized in cross sections of the epithelial sheet; however, viewing the epithelial surface of the neural plate in the chicken embryo reveals that actomyosin cables are polarized along the mediolateral facing edge of neuroepithelial cells (Nishimura et al., 2012; Nishimura and Takeichi, 2008). Here components of PCP and apical constriction pathways such as Celsr1 and Shroom3 co-localize to locally activate RhoA (Nishimura et al., 2012). This results in initiation of the protein kinase cascade and myosin activation preferentially along the mediolateral axis. This coordination of PCP with apical constriction drives simultaneous CE and bending of the neural plate. The finding that PCP and apical-basal polarity pathways are coordinated provides an explanation for the genetic interaction of *Vangl2*^{Lp} with the basal-lateral Scribble and why *Scrbl*^{Crc} mutants show craniorachischisis (Murdoch et al., 2003).

Formation of Hinge Points is Regulated by Shh and Bone Morphogenic Protein (BMP)

The relative contribution of the MHP and DLHPs to neurulation differs along the anterior-posterior axis of the spinal cord (Figure 37.4; Shum and Copp, 1996). In the anterior spinal cord, MHPs are most prominent and DLHPs fail to form; this pattern of neurulation is referred to as "Mode 1." In this mode of neurulation the neural plate folds over the MHP and fuses in the dorsal midline. In more caudal regions the MHP and paired DLHPs are prominent and the neural plate rolls around these hinge points. This is referred to as "Mode 2" neurulation. In the posterior spinal cord a prominent MHP does not form and the neural folds roll around the DLHP as they rise. This is referred to as "Mode 3" neurulation. In the cranial region, both MHP and DLHPs form; however, here DLHP formation is a dynamic process. Live imaging experiments reveal that DLHPs form, disappear and then reform as the neural folds elevate (Massarwa and Niswander, 2013).

The dynamic expression of secreted factors along the anterior-posterior axis of the spinal cord influences the mode of neurulation (Figure 37.4; Ybot-Gonzalez et al., 2002). Shh is expressed at highest levels in the anterior regions of the spinal cord and is almost nonexistent in the most caudal regions (Ybot-Gonzalez et al., 2002). Moreover, Shh and BMP inhibit formation of the DLHPs (Ybot-Gonzalez et al., 2002; 2007a). BMP is secreted from the surface ectoderm and its expression

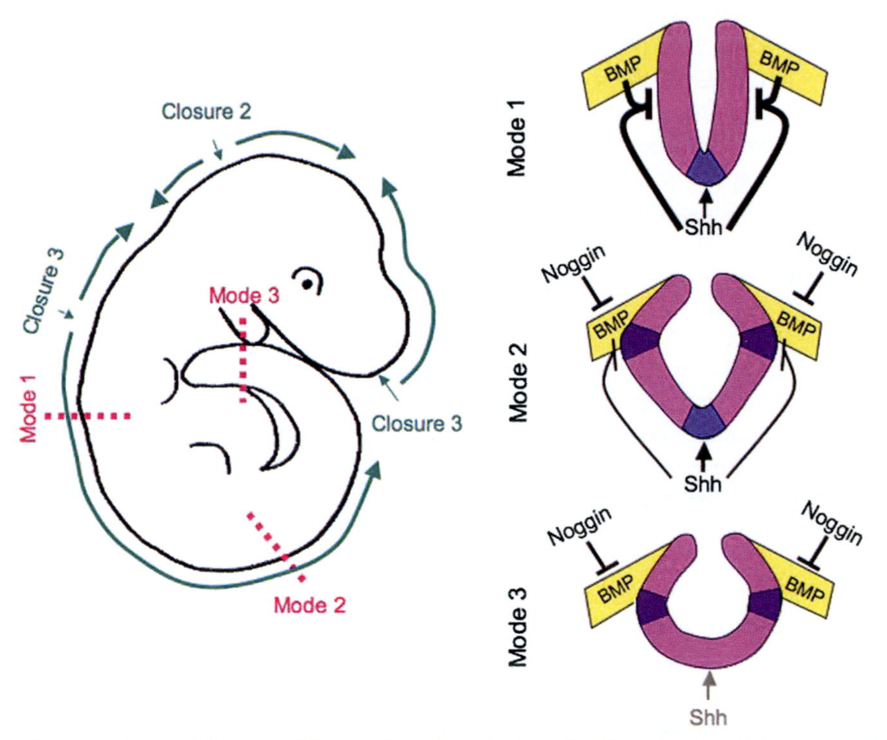

FIGURE 37.4 **Different Closure Points and Modes of Neurulation Along the Anterior-Posterior Axis.** Primary neurulation initiates at multiple closure points from which the neural tube zips closed in both anterior and posterior directions (represented by teal arrows). Closure point 1 forms at the hindbrain/spinal cord boundary, closure 2 forms in the midbrain region and closure 3 at the most anterior aspect of the neural tube. Different modes of neurulation occur along the anterior-posterior axis. In the anterior spinal cord (Mode 1) only the medial hinge point (MHP; blue) forms. In the mid-spinal cord region (Mode 2) both a MHP and paired dorsal lateral hinge points (DLHP; purple) form. In the posterior spinal region (Mode 3) the MHP does not form but exaggerated DLHPs are responsible for bringing the neural folds together. The formation of the MHP and DLHPs are regulated by Shh from the notochord and BMP/Noggin from the non-neural ectoderm. In the anterior spinal cord, BMP and Shh are expressed and inhibit DLHP but promote MHP formation. In the mid-spinal cord region, Noggin is also expressed and antagonizes the DLHP inhibiting activity of BMP and both MHP and DLHPs form. In the posterior spinal region, Shh expression is weak or nonexistent and no MHP forms. The absence of Shh and the presence of Noggin promotes the formation of prominent DLHPs.

remains essentially constant along the embryonic axis. However, the BMP antagonist Noggin is expressed in middle and posterior regions where it promotes DLHP formation by inhibiting BMP.

Disruption of either BMP or Shh signaling during neurulation can lead to NTDs. Since loss of Shh results in exaggerated hinge point formation, mutation of Shh itself does not lead to NTDs. On the other hand, mutation of negative regulators of Shh such as IFT or Gli proteins results in failure of DLHP formation and neural tube closure where DLHPs are critical (Murdoch and Copp, 2010). Similarly mutation of BMP pathway components results in NTDs including null mutants for BMP2, Noggin, Bmp5/7 and Smad 5 conditional BMPR1A mutants or transgenics expressing *caBmpr-1b* (Castranio and Mishina, 2009; Chang et al., 1999; McMahon et al., 1998; Panchision et al., 2001; Solloway and Robertson, 1999; Stottmann et al., 2006; Stottmann and Klingensmith, 2011). While the cellular basis of NTDs in these mouse models has not been addressed, experiments in chicken have shed light on how BMP regulates hinge point formation.

Attenuation of BMP signaling promotes hinge point formation by allowing bending of the neural plate as well as promoting basal nuclei accumulation. This is accomplished in the MHP by destabilization of apical-basal polarity and regulation of the cell cycle with reduced BMP (Eom et al., 2011; 2012). Examination of BMP activation by immunostaining with an anti-Phosphorylated-Smad1/5/9 (P-Smad) antibody in the neural epithelium reveals a dorsal-ventral gradient of P-Smad in the neural plate. Additionally, higher levels of P-Smad are found in apical and lower in basal regions of the neural epithelium, correlating with interkinetic nuclear migration and the cell cycle. Hence, in the MHP region where BMP inhibition is greatest, a greater percentage of cells are in non-mitotic phases of the cell cycle, and they remain there longer contributing to the majority of these cells exhibiting a wedge shape. In addition, decreased BMP activity results in destabilization of apical-basal polarity by removal of the PAR complex and N-cadherin from junctional complexes via endocytosis and movement of basal-lateral proteins such as LGL to the apical compartment. The combination of an increase in the proportion of wedge-shaped cells and destabilization of the polarity of the epithelium allows the neural plate to bend.

Role of the Cranial Mesenchyme in Neural Fold Elevation

In the cranial region of the neural tube, the cranial mesenchyme plays an essential role in elevation of the neural folds (reviewed in Zohn and Sarkar, 2012). These cells underlie the neural plate and undergo stereotypical rearrangements that are hypothesized to drive elevation of the neural folds. As the neural folds rise, the hyaluronate-rich extracellular matrix expands, resulting in increased space between individual cranial mesenchyme cells (Morris-Wiman and Brinkley, 1990a; Morriss and Solursh, 1978a,b). Based on inhibitor studies, expansion of the extracellular matrix has been implicated in driving neural fold elevation (Morris-Wiman and Brinkley, 1990a,b,c; Morriss-Kay et al., 1986; Schoenwolf and Fisher, 1983). In addition, mutation of genes expressed exclusively in the cranial mesenchyme results in exencephaly associated with abnormal organization of the cranial mesenchyme. Abnormal cranial mesenchyme morphogenesis has been suggested to underlie NTDs in a number of mouse mutants including *Ski*, *Cart1/Alx1*, *Alx3*, *Inka1* and *Twist* (Berk et al., 1997; Chen and Behringer, 1995; Lakhwani et al., 2010; Reid et al., 2010; Zhao et al., 1993; Zhao et al., 1996). Of these, the *Twist* mutant has been the best characterized. Embryos with a homozygous mutation in the basic helix-loop-helix transcription factor *Twist* exhibit exencephaly associated with abnormal morphology of the cranial mesenchyme and expanded extracellular space (Chen and Behringer, 1995). Another mouse mutant implicating the cranial mesenchyme in neural fold elevation is *openmind,* which harbors a mutation in the novel ubiquitin ligase *Hectd1*. The neural folds fail to elevate in *Hectd1*^{opm} mutants and cranial mesenchyme cells fail to undergo their characteristic expansion (Sarkar and Zohn, 2012; Zohn et al., 2007).

Role of the Non-Neural Ectoderm in Neural Fold Elevation

The non-neural ectoderm is required for neural tube closure in the chicken and in *Xenopus* (Alvarez and Schoenwolf, 1992; Hackett et al., 1997; Jacobson and Moury, 1995; Morita et al., 2012; Moury and Schoenwolf, 1995). It potentially contributes to neural tube closure by 1) providing an inductive signal for DLHP formation; 2) providing a driving force for elevation of the neural folds; and/or 3) fusion of the neural folds. In the chicken, removal of the surface epithelium results in failure of DLHP formation and neural fold elevation (Moury and Schoenwolf, 1995). This could reflect either an inductive or mechanical role in DLHP formation and elevation of the neural folds. In support of an inductive role, removal of all but a small strip of surface epithelium is sufficient to induce DLHPs (Moury and Schoenwolf, 1995). BMP and Noggin are expressed in the surface ectoderm, and culture with a Noggin-coated bead will induce DLHPs (Ybot-Gonzalez et al., 2007a). On the other hand, orientated cell divisions in surface epithelium of the chicken embryo cause medial-lateral expansion of the epidermis (Sausedo et al., 1997) and the surface epithelium in *Xenopus* migrates medially during neural tube closure (Morita et al., 2012), potentially provide a mechanical force for neural fold elevation.

Grhl2 and *Grhl3* are expressed almost exclusively in the surface ectoderm during neurulation and are required for proper development of the epidermis and neural tube closure (Auden et al., 2006; Brouns et al., 2011; Gustavsson et al., 2007; Pyrgaki et al., 2011; Rifat et al., 2010; Ting et al., 2003; Werth et al., 2010; Yu et al., 2006). *Grhl3* is also expressed in the hindgut epithelium where a role in regulation of proliferation is thought to contribute to spina bifida in the hypomorphic *Grhl3*^{ct} mouse line. This leads to a NTD by creating an imbalance in proliferation between the posterior neural tube and the underlying hindgut epithelium (Copp et al., 1988), *Grhl3* null mutants have defects in more anterior regions of the spinal cord, where it is likely that additional mechanisms are responsible. *Grhl2*$^{-/-}$ mutants do not form DLHPs in the spinal neural tube (Rifat et al., 2010); however, factors known to induce DLHPs (*BMP2* and *Noggin*) are expressed normally in the surface epidermis (Rifat et al., 2010). On the other hand, live imaging experiments with the ENU-induced *Grhl2*^{m1Nisw} mutant demonstrate that elevation and apposition of the neural folds can occur but fusion fails (Pyrgaki et al., 2011). This finding suggests that *Grhl2* and the non-neural ectoderm also contribute to fusion of the neural folds.

Fusion of the Neural Folds

The neural plate does not roll into a tube all at once; rather closure is initiated at discrete points followed by "zipping" in both anterior and posterior directions (Figure 37.4; Copp et al., 2003). Closure 1 initiates at the hindbrain/spinal cord boundary and extends from there. This is followed by formation of closure points in the cranial region: Closure 2 occurs at the midbrain/forebrain boundary; and Closure 3 at the anterior aspect of the forebrain. The position of Closure 2 is variable between mouse strains and its position is correlated with susceptibility to development of exencephaly (Fleming and Copp, 2000; Juriloff et al., 1991). Closure 2 may also be variable during human neurulation, as it has been identified in some but not other human embryo samples (reviewed in Greene and Copp, 2009).

During fusion of the neural folds, cells extend finger-like projections that find protrusions from cells on the opposing neural fold. These intercalate, draw the folds towards one another and fasten them together (Pyrgaki et al., 2010). Live imaging experiments in the mouse suggest that closure in the midbrain region occurs not by "zipping" but by formation of multiple intermediate closure points that "button up" the folds (Pyrgaki et al., 2010). The neural folds consist of both neural and non-neural ectoderm. The tissue layer that makes initial contact differs with the anterior-posterior level. Between closure points 1 and 2, fusion is initiated by cells of the non-neural ectoderm followed by cells of the neural ectoderm (Geelen and Langman, 1979; Pyrgaki et al., 2010). Between closure points 2 and 3, both layers contact at the same time while initiation at Closure 3 is performed by neural ectoderm (Geelen and Langman, 1979).

Defects in neural fold fusion result in NTDs in mouse models. For example, *Ephrin-A5* and *EphA7* are expressed in the tips of the neural folds, and mutants show failure of fusion (Holmberg et al., 2000). Similarly, inhibition of EphA2 activity results in delay and/or failure of neural fold fusion (Abdul-Aziz et al., 2009). The essential role of the non-neural ectoderm in mouse is illustrated by mutation of genes with restricted expression. As mentioned earlier, *Grhl2* expression is restricted to non-neural ectoderm and *Grhl2*^{m1Nisw} mutants show defective fusion of the neural folds (Auden et al., 2006; Pyrgaki et al., 2011). *Protease activated receptor 2* (*Par2*) is expressed specifically in the non-neural ectoderm of the neural folds and *Par1/2* compound mutants shows failure of neural fold fusion and NTDs (Camerer et al., 2010).

High levels of apoptosis occur specifically in the neural folds beginning slightly before and continue after fusion (Copp et al., 2003; Geelen and Langman, 1979; Massa et al., 2009). Despite this intriguing pattern, the role of apoptosis in neural fold fusion has been controversial. Apoptosis is clearly required for neural tube closure, as mutations of *Caspase 3, Caspase 9* and *Apaf1* in mouse models result in NTDs (Cecconi et al., 1998; Honarpour et al., 2001; Houde et al., 2004; Leonard et al., 2002). Similarly, culture of chicken embryos with apoptosis inhibitors resulted in failure of neural tube closure (Weil et al., 1997). However, the NTD phenotype in these examples has not been analyzed in sufficient detail to determine which step in neurulation is affected. More recent experiments, in which mouse embryos were cultured with lower dosages of the same inhibitors specifically during fusion of the neural folds suggest that apoptosis is not required for fusion, as these embryos did not exhibit NTDs (Massa et al., 2009). Furthermore, the use of an apoptosis indicator mouse in embryo culture experiments with caspase inhibitors demonstrates inhibition of apoptosis; yet, NTDs were not found (Yamaguchi et al., 2011). Importantly, live imaging of these embryos shows delayed closure and reduced bending of the neural plate (Yamaguchi et al., 2011), suggesting that apoptosis is required for shaping of the neural plate and neural fold elevation but not fusion.

Formation of the Skull Over the Closed Neural Tube

Encephalocele is characterized by protrusion of the brain and meninges through an opening in the skull. Encephaloceles always occur in the dorsal midline where fusion occurs and is covered with skin. The developmental pathways underlying encephalocele remain unknown but this defect has been proposed to arise from incomplete fusion and/or abnormal separation of neural and non-neural epidermis following fusion (Hoving, 1990). Hints of the mechanism come from the study of mouse models of Meckel Syndrome (MKS). MKS is a rare ciliopathy and accounts for the largest group of syndromic encephalocele in humans (reviewed in Logan et al., 2011). Consistent with the role in ciliogenesis, mouse models of MKS show defects in Shh signaling (reviewed in Logan et al., 2011) and exencephaly (Abdelhamed et al., 2013; Weatherbee et al., 2009) supporting the idea that encephalocele may result from failures of neurulation.

Formation of the Vertebrae Over the Closed Neural Tube

Once neurulation is complete, vertebrae form around the closed spinal neural tube (reviewed in Zohn and Sarkar, 2008). If the neural tube fails to close properly, the bones cannot form around the lesion. Additionally, spina bifida occulta is caused by a failure of the vertebrae to form around the closed neural tube. Vertebrae develop from the sclerotome portion of the somite, transient embryonic mesodermal structures located on either side of the neural tube (reviewed in Christ et al., 2004). Cells delaminate from the sclerotome and migrate to positions around the notochord and neural tube to form the vertebral body and arch. Errors in patterning or migration of the sclerotome cells can result in defects in vertebrae formation and spina bifida occulta.

Both *Pax1* and *Pdgfra* transcripts are expressed in the sclerotome portion of the somite and are required for the development of the vertebrae (Balling et al., 1988; Deutsch et al., 1988; Orr-Urtreger et al., 1992; Payne et al., 1997; Pickett et al., 2008). Mutation of *Pax1* the *undulated* (*un*) mouse results in malformations of the vertebrae, but not spina bifida (Balling et al., 1988). The *Patch* (*Ph*) mutant mouse line carries a deletion that includes *Platelet derived growth factor receptor-alpha* (*Pdgfra*; (Smith et al., 1991; Stephenson et al., 1991). Homozygous *Ph/Ph* mutants show spina bifida depending on the genetic background and $Pax1^{un/un}$;*Ph/+* compound mutant embryos exhibit highly penetrant spina bifida (Helwig et al., 1995; Morrison-Graham et al., 1992; Orr-Urtreger et al., 1992; Payne et al., 1997). Histological analysis of $Pax1^{un/un}$;*Ph/+* compound mutants or *Ph/Ph* homozygotes demonstrates that neural tube closure occurs normally, but the vertebrae do not form properly causing spina bifida occulta (Helwig et al., 1995; Payne et al., 1997).

Pax1 acts as a transcriptional activator at the *Pdgfra* promoter (Joosten et al., 1998). Importantly, a point mutation in *PAX1* was found in a single fetus that presented with spina bifida (Hol et al., 1996) and the human and $Pax1^{un}$ mutation reduced the transcriptional activity of Pax1 (Joosten et al., 1998). Since unaffected relatives were also heterozygous for mutant *PAX1*, other environmental and/or genetic insults may contribute to the NTD. From the analysis of $Pax1^{un}$ and *Ph* mutant mouse lines, one genetic risk factor may be at the *PDGFRA* locus. Certain promoter haplotypes of *PDGFRA* are associated with decreased activity of the pathway and increased risk of NTDs in some but not other human populations (Au et al., 2005; Joosten et al., 2001; Toepoel et al., 2009; Zhu et al., 2004). Environmental factors are also likely candidates and one study investigating the effect of environmental factors on penetrance of NTDs with the susceptible *PDGFRA* haplotype found an interaction with obesity, glucose, and inositol (Toepoel et al., 2009).

Prevention of NTDs: Folic Acid and Inositol

Folic acid deficiency was first implicated in contributing to NTD risk with the finding that women with NTD-affected pregnancies had lower serum folate levels and consumed less folic acid than women with unaffected pregnancies (Laurence et al., 1968; Smithells et al., 1976). This led to a series of small clinical trials to test whether folic acid supplementation could prevent NTDs in women with previous affected pregnancies (Holmes-Siedle et al., 1992; Laurence et al., 1981; Seller and Nevin, 1984; Smithells et al., 1980; 1981; Vergel et al., 1990). In 1991, the Medical Research Council (MRC) published results from a double blind randomized trial demonstrating a 72% reduction of NTDs in a large number of women with previously affected pregnancies (Wald et al., 1991). Further trials to determine whether folic acid supplementation could prevent NTDs in women of average risk demonstrated that improvement is greater depending upon the initial NTD rate of the population (reviewed in Heseker et al., 2009). For example, in Northern China where the NTD rate is very high (48/10,000) the rate was reduced to 7/10,000 with supplementation (Berry et al., 1999). In contrast, the reduction was smaller in Southern China where the NTD rate is 10/10,000 and was reduced to 6/10,000 (Berry et al., 1999). Following the MRC trial many countries now fortify grains and cereals with folic acid. In the US, fortification has resulted in increased folate status of the population and significant reductions in the incidence of NTDs (Honein et al., 2001).

Common polymorphisms in the metabolic pathways that utilize folic acid are associated with increased risk of NTDs. The most striking example of this is polymorphisms in the MTHFR gene (reviewed in Blom et al., 2006). 40% of the population is heterozygous and 10% homozygous for the MTHFR 667C>T polymorphism. Mothers who are homozygous for the MTHFR 667C>T allele have a 60% increased risk of an NTD-affected pregnancy and their homozygous offspring have a 90% chance of an NTD. Heterozygote mothers also have a slightly increased risk of a NTD-affected pregnancy. In addition, folic acid receptor auto-antibodies that inhibit folic acid uptake have been found in mothers with NTD affected pregnancies (Rothenberg et al., 2004). The risk associated with MTHFR 667C>T polymorphism or the presence of folic acid receptor auto-antibodies can potentially be overcome with folic acid supplementation.

In spite of the clear benefit of folic acid supplementation in preventing NTDs, it is not clear how folic acid works (reviewed in Blom et al., 2006). Folates are not synthesized by the body and thus needs to be included in the diet. Folates are important cofactors in numerous metabolic reactions, including DNA and amino acid synthesis and production of methyl donor S-adenosyl methionine (SAM) used for methylation of histones, proteins, lipids and DNA. Thus folic acid supplementation may have an effect on proliferation or transcriptional and epigenetic regulation.

Mouse models with mutations in genes required for folic acid metabolism show NTDs including *Folr1* and *Slc19a1/RFC* (Gelineau-van Waes et al., 2008; Piedrahita et al., 1999). Folic acid supplementation can rescue NTDs in some models such as *Folbp1*, *RFC1*, *Cart1*, *Lrp6*^{Cd}, *Gcn5* and *Pax3*^{2H} (Barbera et al., 2002; Carter et al., 2005; Fleming and Copp, 1998; Gelineau-van Waes et al., 2008; Lin et al., 2008; Piedrahita et al., 1999; Spiegelstein et al., 2004; Wlodarczyk et al., 2006; Zhao et al., 1996; 2001). Just as not all NTDs in humans can be prevented by folic acid supplementation, some mouse lines are not rescued by folic acid supplementation (Chi et al., 2005; Essien and Wannberg, 1993; Seller, 1994; Stottmann et al., 2006; Ting et al., 2003; Wong et al., 2008); in others, folic acid supplementation results in early loss of mutant embryos (Gray et al., 2010; Marean et al., 2011). On the other hand, folic acid deficiency is not sufficient to induce NTDs in humans or mouse models (Beaudin et al., 2012; Burgoon et al., 2002; Burren et al., 2008, 2010; De Castro et al., 2010; Heid et al., 1992; Kirke et al., 1993). Rather gene-environment interactions (e.g., suboptimal folate status plus a genetic predisposition) likely combine to result in a NTD. For example, folate deficiency increased the NTD frequency in *Pax3*^{Sp} mutants and in a NTD susceptible mouse background strain (Burren et al., 2008, 2010). Similarly, mutation of a gene required for folate metabolism (*Shmt1*) does not result in NTDs with a normal folate diet but a folate deficiency causes NTDs (Beaudin et al., 2011; 2012). Altered folate metabolism has been documented in cell lines derived from NTD affected fetuses and the *Pax3*^{Sp} and *Lrp6*^{Cd} mouse models of NTDs (Dunlevy et al., 2007; Ernest et al., 2006; Fleming and Copp, 1998). Finally, *Pax3*$^{Sp/+}$;*Shmt1*$^{-/+}$ compound mutants exhibit increased penetrance and severity of NTDs (Beaudin et al., 2011).

Homozygous *Pax3*^{Sp} mutant embryos show NTDs such as spina bifida and exencephaly with variable penetrance and expressivity depending upon the background strain (Greene et al., 2009). NTDs including spina bifida and exencephaly have been associated with *PAX3* mutations in the autosomal dominant Waardenburg Syndrome (Baldwin et al., 1994; Begleiter and Harris, 1992; Carezani-Gavin et al., 1992; Chatkupt et al., 1993; da-Silva, 1991; de Saxe et al., 1984; Hol et al., 1995; Hoth et al., 1993; Kujat et al., 2007; Moline and Sandlin, 1993; Nye et al., 1998; Pantke and Cohen, 1971; Shim et al., 2004). While the majority of individuals are heterozygous for mutant *PAX3*, homozygous individuals have been identified. Yet, not all fetuses homozygous for mutant *PAX3* present with NTDs (Ayme and Philip, 1995; Wollnik et al., 2003; Zlotogora et al., 1995), suggesting that modifiers likely influence the penetrance and expressivity of NTDs with *PAX3* mutations in humans as has been found in mice (Estibeiro et al., 1993; Lakkis et al., 1999; Pani et al., 2002a,b).

Pax3 is also an important gene that is regulated in NTDs and also associated with diabetes. Pregnancies in diabetic women have a five-fold greater incidence of structural birth defects including NTDs (reviewed in Chappell et al., 2009). Maternal diabetes results in dramatic alterations in gene expression in the embryo (Pavlinkova et al., 2009; Reece et al., 2006; Salbaum and Kappen, 2010) with reduced expression of *Pax3* the best characterized (Fine et al., 1999; Phelan et al., 1997). Similar to the *Pax3* mutants, supplementation of diabetic mice with folic acid can reduced frequency of NTDs (Oyama et al., 2009).

Another supplement with promise to affect NTD rates is inositol. Hyperglycemia also results in inositol depletion and inositol supplementation suppresses diabetes-induced NTDs (Khandelwal et al., 1998; Reece et al., 1997). Mouse embryos grown in culture and *Grhl3*^{ct} mutants in particular develop NTDs with reduced inositol concentrations in the media (Cockroft, 1988; Cockroft et al., 1992). NTDs in *Grhl3*^{ct} mutants are suppressed by supplementation with inositol but not folic acid (Greene and Copp, 1997; van Straaten and Copp, 2001) Additionally, disruption of genes involved in inositol metabolism, including *Inositol 1,3,4-trisphosphate 5/6-kinase* (*Itpk1*) or *Phosphatidylinositol-4-phosphate 5-kinase, type 1 gamma* (*Pip5k1c*) results in NTDs (Wang et al., 2007; Wilson et al., 2009). Studies in humans also suggest that inositol may prevent NTDs. Low serum concentrations of inositol are associated with increased NTD risk and low serum levels are found in children with spina bifida (Groenen et al., 2003). Clinical trials are currently underway to determine whether inositol supplementation can prevent NTD recurrence. At this point no NTDs have occurred in these trials; however, the sample size is too low to discern whether or not this is a real effect (Cavalli and Copp, 2002; Cavalli et al., 2008; 2011).

37.3 FUTURE DIRECTIONS

In spite of improvements in prevention and treatment of NTDs, they still remain one of the most common structural birth defects. For example, even when periconception folic acid supplementation is administered, approximately 30% of NTDs are not prevented. Thus, future work to identify and test other prevention strategies is needed. With the advent of rapid DNA sequencing technologies and microarrays to identify copy number variants, the identification of the genetic causes of NTDs in patients will be greatly accelerated. In addition, plans to characterize the deletion phenotype of all of the genes in the mouse genome will identify the genes required for neurulation. Furthermore, testing the effect and mechanism of action of folic acid and other environmental factors on these models will elucidate how they prevent NTDs. The development of live imaging techniques to analyze how specific gene disruptions affect neural tube closure in real time will greatly improve our understanding of the cellular mechanisms underlying neurulation. Together these approaches will provide a better understanding of the causes of NTDs, which should significantly improve the way that the prevention of this devastating birth defect is approached in the future.

37.4 CLINICAL RELEVANCE

- NTDs are common structural birth defects in humans.
- NTDs are caused by interaction of genetic and environmental factors.
- NTDs are prevented by periconception folic acid supplementation.
- Animal models of NTDs have been essential for the identification and characterization of genetic and environmental factors that cause NTDs in humans.

RECOMMENDED RESOURCES

Blom, H.J., Shaw, G.M., den Heijer, M., Finnell, R.H., 2006. Neural tube defects and folate: case far from closed. Nat. Rev. Neurosci. 7, 724–731.

Wallingford, J.B., 2012. Planar cell polarity and the developmental control of cell behavior in vertebrate embryos. Annu. Rev. Cell. Dev. Biol. 28, 627–653.

Wallingford, J.B., Niswander, L.A., Shaw, G.M., Finnell, R.H., 2013. The continuing challenge of understanding, preventing, and treating neural tube defects. Science 339, 1222002.

Wyszynski, D.F., 2005. Neural Tube Defects: From Origin to Treatment. Oxford University Press.

REFERENCES

Abdelhamed, Z.A., Wheway, G., Szymanska, K., Natarajan, S., Toomes, C., Inglehearn, C., Johnson, C.A., 2013. Variable expressivity of ciliopathy neurological phenotypes that encompass Meckel-Gruber syndrome and Joubert syndrome is caused by complex de-regulated ciliogenesis, Shh and Wnt signaling defects. Hum. Mol. Genet. 22, 1358–1372.

Abdul-Aziz, N.M., Turmaine, M., Greene, N.D., Copp, A.J., 2009. EphrinA-EphA receptor interactions in mouse spinal neurulation: implications for neural fold fusion. Int. J. Dev. Biol. 53, 559–568.

Adzick, N.S., Thom, E.A., Spong, C.Y., Brock 3rd, J.W., Burrows, P.K., Johnson, M.P., Howell, L.J., Farrell, J.A., Dabrowiak, M.E., Sutton, L.N., et al., 2011. A randomized trial of prenatal versus postnatal repair of myelomeningocele. N. Engl. J. Med. 364, 993–1004.

Adzick, N.S., Walsh, D.S., 2003. Myelomeningocele: prenatal diagnosis, pathophysiology and management. Semin. Pediatr. Surg. 12, 168–174.

Allache, R., De Marco, P., Merello, E., Capra, V., Kibar, Z., 2012. Role of the planar cell polarity gene CELSR1 in neural tube defects and caudal agenesis. Birth Defects Res. A. Clin. Mol. Teratol. 94, 176–181.

Alvarez, I.S., Schoenwolf, G.C., 1992. Expansion of surface epithelium provides the major extrinsic force for bending of the neural plate. J. Exp. Zool. 261, 340–348.

Au, K.S., Ashley-Koch, A., Northrup, H., 2010. Epidemiologic and genetic aspects of spina bifida and other neural tube defects. Dev. Disabil. Res. Rev. 16, 6–15.

Au, K.S., Northrup, H., Kirkpatrick, T.J., Volcik, K.A., Fletcher, J.M., Townsend, I.T., Blanton, S.H., Tyerman, G.H., Villarreal, G., King, T.M., 2005. Promotor genotype of the platelet-derived growth factor receptor-alpha gene shows population stratification but not association with spina bifida meningomyelocele. Am. J. Med. Genet. A. 139, 194–198.

Auden, A., Caddy, J., Wilanowski, T., Ting, S.B., Cunningham, J.M., Jane, S.M., 2006. Spatial and temporal expression of the Grainyhead-like transcription factor family during murine development. Gene. Expr. Patterns 6, 964–970.

Ayme, S., Philip, N., 1995. Possible homozygous Waardenburg syndrome in a fetus with exencephaly. Am. J. Med. Genet. 59, 263–265.

Baldwin, C.T., Lipsky, N.R., Hoth, C.F., Cohen, T., Mamuya, W., Milunsky, A., 1994. Mutations in PAX3 associated with Waardenburg syndrome type I. Hum. Mutat. 3, 205–211.

Balling, R., Deutsch, U., Gruss, P., 1988. undulated, a mutation affecting the development of the mouse skeleton, has a point mutation in the paired box of Pax 1. Cell 55, 531–535.

Barbera, J.P., Rodriguez, T.A., Greene, N.D., Weninger, W.J., Simeone, A., Copp, A.J., Beddington, R.S., Dunwoodie, S., 2002. Folic acid prevents exencephaly in Cited2 deficient mice. Hum. Mol. Genet. 11, 283–293.

Bassuk, A.G., Muthuswamy, L.B., Boland, R., Smith, T.L., Hulstrand, A.M., Northrup, H., Hakeman, M., Dierdorff, J.M., Yung, C.K., Long, A., et al., 2013. Copy number variation analysis implicates the cell polarity gene glypican 5 as a human spina bifida candidate gene. Hum. Mol. Genet. 22, 1097–1111.

Bay, S.N., Caspary, T., 2012. What are those cilia doing in the neural tube? Cilia 1, 19.

Beaudin, A.E., Abarinov, E.V., Malysheva, O., Perry, C.A., Caudill, M., Stover, P.J., 2012. Dietary folate, but not choline, modifies neural tube defect risk in Shmt1 knockout mice. Am. J. Clin. Nutr. 95, 109–114.

Beaudin, A.E., Abarinov, E.V., Noden, D.M., Perry, C.A., Chu, S., Stabler, S.P., Allen, R.H., Stover, P.J., 2011. Shmt1 and *de novo* thymidylate biosynthesis underlie folate-responsive neural tube defects in mice. Am. J. Clin. Nutr. 93, 789–798.

Begleiter, M.L., Harris, D.J., 1992. Waardenburg syndrome and meningocele. Am. J. Med. Genet. 44, 541.

Berk, M., Desai, S.Y., Heyman, H.C., Colmenares, C., 1997. Mice lacking the ski proto-oncogene have defects in neurulation, craniofacial, patterning, and skeletal muscle development. Genes. Dev. 11, 2029–2039.

Berry, R.J., Li, Z., Erickson, J.D., Li, S., Moore, C.A., Wang, H., Mulinare, J., Zhao, P., Wong, L.Y., Gindler, J., et al., 1999. Prevention of neural-tube defects with folic acid in China. China-US Collaborative Project for Neural Tube Defect Prevention. N. Engl. J. Med. 341, 1485–1490.

Blom, H.J., Shaw, G.M., den Heijer, M., Finnell, R.H., 2006. Neural tube defects and folate: case far from closed. Nat. Rev. Neurosci. 7, 724–731.

Bosoi, C.M., Capra, V., Allache, R., Trinh, V.Q., De Marco, P., Merello, E., Drapeau, P., Bassuk, A.G., Kibar, Z., 2011. Identification and characterization of novel rare mutations in the planar cell polarity gene PRICKLE1 in human neural tube defects. Hum. Mutat. 32, 1371–1375.

Briscoe, J., Therond, P.P., 2013. The mechanisms of Hedgehog signaling and its roles in development and disease. Nat. Rev. Mol. Cell. Biol. 14 (7), 416–429.

Brouns, M.R., De Castro, S.C., Terwindt-Rouwenhorst, E.A., Massa, V., Hekking, J.W., Hirst, C.S., Savery, D., Munts, C., Partridge, D., Lamers, W., et al., 2011. Over-expression of Grhl2 causes spina bifida in the Axial defects mutant mouse. Hum. Mol. Genet. 20, 1536–1546.

Bruner, J.P., Richards, W.O., Tulipan, N.B., Arney, T.L., 1999. Endoscopic coverage of fetal myelomeningocele *in utero*. Am. J. Obstet. Gynecol. 180, 153–158.

Bruner, J.P., Tulipan, N., Paschall, R.L., Boehm, F.H., Walsh, W.F., Silva, S.R., Hernanz-Schulman, M., Lowe, L.H., Reed, G.W., 1999. Fetal surgery for myelomeningocele and the incidence of shunt-dependent hydrocephalus. J. Am. Med. Assoc. 282, 1819–1825.

Burgoon, J.M., Selhub, J., Nadeau, M., Sadler, T.W., 2002. Investigation of the effects of folate deficiency on embryonic development through the establishment of a folate deficient mouse model. Teratology 65, 219–227.

Burnside, B., 1973. Microtubules and microfilaments in amphibian neurulation. Am. Zool. 13, 989–1006.

Burren, K.A., Savery, D., Massa, V., Kok, R.M., Scott, J.M., Blom, H.J., Copp, A.J., Greene, N.D., 2008. Gene-environment interactions in the causation of neural tube defects: folate deficiency increases susceptibility conferred by loss of Pax3 function. Hum. Mol. Genet. 17, 3675–3685.

Burren, K.A., Scott, J.M., Copp, A.J., Greene, N.D., 2010. The genetic background of the curly tail strain confers susceptibility to folate-deficiency-induced exencephaly. Birth Defects Res. A. Clin. Mol. Teratol. 88, 76–83.

Cabrera, R.M., Hill, D.S., Etheredge, A.J., Finnell, R.H., 2004. Investigations into the etiology of neural tube defects. Birth Defects Res. C. Embryo. Today 72, 330–344.

Caddy, J., Wilanowski, T., Darido, C., Dworkin, S., Ting, S.B., Zhao, Q., Rank, G., Auden, A., Srivastava, S., Papenfuss, T.A., et al., 2010. Epidermal wound repair is regulated by the planar cell polarity signaling pathway. Dev. Cell. 19, 138–147.

Camerer, E., Barker, A., Duong, D.N., Ganesan, R., Kataoka, H., Cornelissen, I., Darragh, M.R., Hussain, A., Zheng, Y.W., Srinivasan, Y., et al., 2010. Local protease signaling contributes to neural tube closure in the mouse embryo. Dev. Cell. 18, 25–38.

Canda, M.T., Demir, N., Bal, F.U., Doganay, L., Sezer, O., 2012. Prenatal diagnosis of a 22q11 deletion in a second-trimester fetus with conotruncal anomaly, absent thymus and meningomyelocele: Kousseff syndrome. J. Obstet. Gynaecol. Res. 38, 737–740.

Carezani-Gavin, M., Clarren, S.K., Steege, T., 1992. Waardenburg syndrome associated with meningomyelocele. Am. J. Med. Genet. 42, 135–136.

Carroll, E.A., Gerrelli, D., Gasca, S., Berg, E., Beier, D.R., Copp, A.J., Klingensmith, J., 2003. Cordon-bleu is a conserved gene involved in neural tube formation. Dev. Biol. 262, 16–31.

Carter, M., Chen, X., Slowinska, B., Minnerath, S., Glickstein, S., Shi, L., Campagne, F., Weinstein, H., Ross, M.E., 2005. Crooked tail (Cd) model of human folate-responsive neural tube defects is mutated in Wnt coreceptor lipoprotein receptor-related protein 6. Proc. Natl. Acad. Sci. USA 102, 12843–12848.

Caspary, T., Larkins, C.E., Anderson, K.V., 2007. The graded response to Sonic Hedgehog depends on cilia architecture. Dev. Cell. 12, 767–778.

Castranio, T., Mishina, Y., 2009. Bmp2 is required for cephalic neural tube closure in the mouse. Dev. Dyn. 238, 110–122.

Cavalli, P., Copp, A.J., 2002. Inositol and folate resistant neural tube defects. J. Med. Genet. 39, E5.

Cavalli, P., Tedoldi, S., Riboli, B., 2008. Inositol supplementation in pregnancies at risk of apparently folate-resistant NTDs. Birth Defects Res. A. Clin. Mol. Teratol. 82, 540–542.

Cavalli, P., Tonni, G., Grosso, E., Poggiani, C., 2011. Effects of inositol supplementation in a cohort of mothers at risk of producing an NTD pregnancy. Birth Defects Res. A. Clin. Mol. Teratol. 91, 962–965.

Cecconi, F., Alvarez-Bolado, G., Meyer, B.I., Roth, K.A., Gruss, P., 1998. Apaf1 (CED-4 homolog) regulates programd cell death in mammalian development. Cell 94, 727–737.

Chang, H., Huylebroeck, D., Verschueren, K., Guo, Q., Matzuk, M.M., Zwijsen, A., 1999. Smad5 knockout mice die at mid-gestation due to multiple embryonic and extra-embryonic defects. Development 126, 1631–1642.

Chappell Jr., J.H., Wang, X.D., Loeken, M.R., 2009. Diabetes and apoptosis: neural crest cells and neural tube. Apoptosis 14, 1472–1483.

Chatkupt, S., Chatkupt, S., Johnson, W.G., 1993. Waardenburg syndrome and myelomeningocele in a family. J. Med. Genet. 30, 83–84.

Chen, C.P., 2007a. Chromosomal abnormalities associated with neural tube defects (I): full aneuploidy. Taiwan J. Obstet. Gynecol. 46, 325–335.

Chen, C.P., 2007b. Chromosomal abnormalities associated with neural tube defects (II): partial aneuploidy. Taiwan J. Obstet. Gynecol. 46, 336–351.

Chen, C.P., 2007c. Meckel syndrome: genetics, perinatal findings, and differential diagnosis. Taiwan J. Obstet. Gynecol. 46, 9–14.

Chen, C.P., 2008. Syndromes, disorders and maternal risk factors associated with neural tube defects (III). Taiwan J. Obstet. Gynecol. 47, 131–140.

Chen, C.P., 2009. Prenatal sonographic features of fetuses in trisomy 13 pregnancies (II). Taiwan J. Obstet. Gynecol. 48, 218–224.

Chen, Z.F., Behringer, R.R., 1995. Twist is required in head mesenchyme for cranial neural tube morphogenesis. Genes. Dev. 9, 686–699.

Chi, H., Sarkisian, M.R., Rakic, P., Flavell, R.A., 2005. Loss of mitogen-activated protein kinase kinase kinase 4 (MEKK4) results in enhanced apoptosis and defective neural tube development. Proc. Natl. Acad. Sci. USA 102, 3846–3851.

Christ, B., Huang, R., Scaal, M., 2004. Formation and differentiation of the avian sclerotome. Anat. Embryol. (Berl.) 208, 333–350.

Cockroft, D.L., 1988. Changes with gestational age in the nutritional requirements of postimplantation rat embryos in culture. Teratology 38, 281–290.

Cockroft, D.L., Brook, F.A., Copp, A.J., 1992. Inositol deficiency increases the susceptibility to neural tube defects of genetically predisposed (curly tail) mouse embryos *in vitro*. Teratology 45, 223–232.

Colas, J.F., Schoenwolf, G.C., 2001. Towards a cellular and molecular understanding of neurulation. Dev. Dyn. 221, 117–145.

Copp, A.J., Brook, F.A., Roberts, H.J., 1988. A cell-type-specific abnormality of cell proliferation in mutant (curly tail) mouse embryos developing spinal neural tube defects. Development 104, 285–295.

Copp, A.J., Checiu, I., Henson, J.N., 1994. Developmental basis of severe neural tube defects in the loop-tail (Lp) mutant mouse: use of microsatellite DNA markers to identify embryonic genotype. Dev. Biol. 165, 20–29.

Copp, A.J., Greene, N.D., Murdoch, J.N., 2003. The genetic basis of mammalian neurulation. Nat. Rev. Genet. 4, 784–793.

Curtin, J.A., Quint, E., Tsipouri, V., Arkell, R.M., Cattanach, B., Copp, A.J., Henderson, D.J., Spurr, N., Stanier, P., Fisher, E.M., et al., 2003. Mutation of Celsr1 disrupts planar polarity of inner ear hair cells and causes severe neural tube defects in the mouse. Curr. Biol. 13, 1129–1133.

da-Silva, E.O., 1991. Waardenburg I syndrome: a clinical and genetic study of two large Brazilian kindreds, and literature review. Am. J. Med. Genet. 40, 65–74.

Davidson, L.A., Keller, R.E., 1999. Neural tube closure in Xenopus laevis involves medial migration, directed protrusive activity, cell intercalation and convergent extension. Development 126, 4547–4556.

De Castro, S.C., Leung, K.Y., Savery, D., Burren, K., Rozen, R., Copp, A.J., Greene, N.D., 2010. Neural tube defects induced by folate deficiency in mutant curly tail (Grhl3) embryos are associated with alteration in folate one-carbon metabolism but are unlikely to result from diminished methylation. Birth Defects Res. A. Clin. Mol. Teratol. 88, 612–618.

De Marco, P., Merello, E., Rossi, A., Piatelli, G., Cama, A., Kibar, Z., Capra, V., 2012. FZD6 is a novel gene for human neural tube defects. Hum. Mutat. 33, 384–390.

de Saxe, M., Kromberg, J.G., Jenkins, T., 1984. Waardenburg syndrome in South Africa. Part I. An evaluation of the clinical findings in 11 families. S. Afr. Med. J. 66, 256–261.

Detrait, E.R., George, T.M., Etchevers, H.C., Gilbert, J.R., Vekemans, M., Speer, M.C., 2005. Human neural tube defects: developmental biology, epidemiology, and genetics. Neurotoxicol. Teratol. 27, 515–524.

Deutsch, U., Dressler, G.R., Gruss, P., 1988. Pax 1, a member of a paired box homologous murine gene family, is expressed in segmented structures during development. Cell 53, 617–625.

Dias, M.S., McLone, D.G., 1993. Hydrocephalus in the child with dysraphism. Neurosurg. Clin. N. Am. 4, 715–726.

Doudney, K., Ybot-Gonzalez, P., Paternotte, C., Stevenson, R.E., Greene, N.D., Moore, G.E., Copp, A.J., Stanier, P., 2005. Analysis of the planar cell polarity gene Vangl2 and its co-expressed paralogue Vangl1 in neural tube defect patients. Am. J. Med. Genet. A. 136, 90–92.

Dunlevy, L.P., Chitty, L.S., Burren, K.A., Doudney, K., Stojilkovic-Mikic, T., Stanier, P., Scott, R., Copp, A.J., Greene, N.D., 2007. Abnormal folate metabolism in foetuses affected by neural tube defects. Brain 130, 1043–1049.

Eichers, E.R., Abd-El-Barr, M.M., Paylor, R., Lewis, R.A., Bi, W., Lin, X., Meehan, T.P., Stockton, D.W., Wu, S.M., Lindsay, E., et al., 2006. Phenotypic characterization of Bbs4 null mice reveals age-dependent penetrance and variable expressivity. Hum. Genet. 120, 211–226.

Eom, D.S., Amarnath, S., Agarwala, S., 2013. Apicobasal polarity and neural tube closure. Dev. Growth. Differ. 55, 164–172.

Eom, D.S., Amarnath, S., Fogel, J.L., Agarwala, S., 2011. Bone morphogenetic proteins regulate neural tube closure by interacting with the apicobasal polarity pathway. Development 138, 3179–3188.

Eom, D.S., Amarnath, S., Fogel, J.L., Agarwala, S., 2012. Bone morphogenetic proteins regulate hinge point formation during neural tube closure by dynamic modulation of apicobasal polarity. Birth Defects Res. A. Clin. Mol. Teratol. 94, 804–816.

Ernest, S., Carter, M., Shao, H., Hosack, A., Lerner, N., Colmenares, C., Rosenblatt, D.S., Pao, Y.H., Ross, M.E., Nadeau, J.H., 2006. Parallel changes in metabolite and expression profiles in crooked-tail mutant and folate-reduced wild-type mice. Hum. Mol. Genet. 15, 3387–3393.

Escobedo, N., Contreras, O., Munoz, R., Farias, M., Carrasco, H., Hill, C., Tran, U., Pryor, S.E., Wessely, O., Copp, A.J., et al., 2013. Syndecan 4 interacts genetically with Vangl2 to regulate neural tube closure and planar cell polarity. Development 140, 3008–3017.

Essien, F.B., Wannberg, S.L., 1993. Methionine but not folinic acid or vitamin B-12 alters the frequency of neural tube defects in Axd mutant mice. J. Nutr. 123, 27–34.

Estibeiro, J.P., Brook, F.A., Copp, A.J., 1993. Interaction between splotch (Sp) and curly tail (ct) mouse mutants in the embryonic development of neural tube defects. Development 119, 113–121.

Etheridge, S.L., Ray, S., Li, S., Hamblet, N.S., Lijam, N., Tsang, M., Greer, J., Kardos, N., Wang, J., Sussman, D.J., et al., 2008. Murine dishevelled 3 functions in redundant pathways with dishevelled 1 and 2 in normal cardiac outflow tract, cochlea, and neural tube development. PLoS Genet. 4, e1000259.

Farmer, D.L., von Koch, C.S., Peacock, W.J., Danielpour, M., Gupta, N., Lee, H., Harrison, M.R., 2003. In utero repair of myelomeningocele: experimental pathophysiology, initial clinical experience, and outcomes. Arch. Surg. 138, 872–878.

Fine, E.L., Horal, M., Chang, T.I., Fortin, G., Loeken, M.R., 1999. Evidence that elevated glucose causes altered gene expression, apoptosis, and neural tube defects in a mouse model of diabetic pregnancy. Diabetes 48, 2454–2462.

Fleming, A., Copp, A.J., 1998. Embryonic folate metabolism and mouse neural tube defects. Science 280, 2107–2109.

Fleming, A., Copp, A.J., 2000. A genetic risk factor for mouse neural tube defects: defining the embryonic basis. Hum. Mol. Genet. 9, 575–581.

Forrester, S., Kovach, M.J., Smith, R.E., Rimer, L., Wesson, M., Kimonis, V.E., 2002. Kousseff syndrome caused by deletion of chromosome 22q11–13. Am. J. Med. Genet. 112, 338–342.

Gao, B., Song, H., Bishop, K., Elliot, G., Garrett, L., English, M.A., Andre, P., Robinson, J., Sood, R., Minami, Y., et al., 2011. Wnt signaling gradients establish planar cell polarity by inducing Vangl2 phosphorylation through Ror2. Dev. Cell. 20, 163–176.

Geelen, J.A., Langman, J., 1979. Ultrastructural observations on closure of the neural tube in the mouse. Anat. Embryol. (Berl.) 156, 73–88.

Gelineau-van Waes, J., Heller, S., Bauer, L.K., Wilberding, J., Maddox, J.R., Aleman, F., Rosenquist, T.H., Finnell, R.H., 2008. Embryonic development in the reduced folate carrier knockout mouse is modulated by maternal folate supplementation. Birth Defects Res. A. Clin. Mol. Teratol. 82, 494–507.

Gelineau-van Waes, J., Voss, K.A., Stevens, V.L., Speer, M.C., Riley, R.T., 2009. Maternal fumonisin exposure as a risk factor for neural tube defects. Adv. Food. Nutr. Res. 56, 145–181.

Gerrelli, D., Copp, A.J., 1997. Failure of neural tube closure in the loop-tail (Lp) mutant mouse: analysis of the embryonic mechanism. Brain. Res. Dev. Brain. Res. 102, 217–224.

Gray, J., Ross, M.E., 2011. Neural tube closure in mouse whole embryo culture. J. Vis. Exp.

Gray, J.D., Kholmanskikh, S., Castaldo, B.S., Hansler, A., Chung, H., Klotz, B., Singh, S., Brown, A.M., Ross, M.E., 2013. LRP6 exerts non-canonical effects on Wnt signaling during neural tube closure. Hum. Mol. Genet.

Gray, J.D., Nakouzi, G., Slowinska-Castaldo, B., Dazard, J.E., Rao, J.S., Nadeau, J.H., Ross, M.E., 2010. Functional interactions between the LRP6 WNT co-receptor and folate supplementation. Hum. Mol. Genet. 19, 4560–4572.

Gray, J.D., Ross, M.E., 2009. Mechanistic insights into folate supplementation from Crooked tail and other NTD-prone mutant mice. Birth Defects Res. A. Clin. Mol. Teratol. 85, 314–321.

Gray, R.S., Abitua, P.B., Wlodarczyk, B.J., Szabo-Rogers, H.L., Blanchard, O., Lee, I., Weiss, G.S., Liu, K.J., Marcotte, E.M., Wallingford, J.B., et al., 2009. The planar cell polarity effector Fuz is essential for targeted membrane trafficking, ciliogenesis and mouse embryonic development. Nat. Cell. Biol. 11, 1225–1232.

Greene, N.D., Copp, A.J., 1997. Inositol prevents folate-resistant neural tube defects in the mouse. Nat. Med. 3, 60–66.

Greene, N.D., Copp, A.J., 2005. Mouse models of neural tube defects: investigating preventive mechanisms. Am. J. Med. Genet. C. Semin. Med. Genet. 135C, 31–41.

Greene, N.D., Copp, A.J., 2009. Development of the vertebrate central nervous system: formation of the neural tube. Prenat. Diagn. 29, 303–311.

Greene, N.D., Gerrelli, D., Van Straaten, H.W., Copp, A.J., 1998. Abnormalities of floor plate, notochord and somite differentiation in the loop-tail (Lp) mouse: a model of severe neural tube defects. Mech. Dev. 73, 59–72.

Greene, N.D., Massa, V., Copp, A.J., 2009. Understanding the causes and prevention of neural tube defects: Insights from the splotch mouse model. Birth Defects Res. A. Clin. Mol. Teratol. 85, 322–330.

Groenen, P.M., Peer, P.G., Wevers, R.A., Swinkels, D.W., Franke, B., Mariman, E.C., Steegers-Theunissen, R.P., 2003. Maternal myo-inositol, glucose, and zinc status is associated with the risk of offspring with spina bifida. Am. J. Obstet. Gynecol. 189, 1713–1719.

Gu, X., Lin, L., Zheng, X., Zhang, T., Song, X., Wang, J., Li, X., Li, P., Chen, G., Wu, J., et al., 2007. High prevalence of NTDs in Shanxi Province: a combined epidemiological approach. Birth Defects Res. A. Clin. Mol. Teratol. 79, 702–707.

Gustavsson, P., Greene, N.D., Lad, D., Pauws, E., de Castro, S.C., Stanier, P., Copp, A.J., 2007. Increased expression of Grainyhead-like-3 rescues spina bifida in a folate-resistant mouse model. Hum. Mol. Genet. 16, 2640–2646.

Habas, R., Kato, Y., He, X., 2001. Wnt/Frizzled activation of Rho regulates vertebrate gastrulation and requires a novel Formin homology protein Daam1. Cell 107, 843–854.

Hackett, D.A., Smith, J.L., Schoenwolf, G.C., 1997. Epidermal ectoderm is required for full elevation and for convergence during bending of the avian neural plate. Dev. Dyn. 210, 397–406.

Haigo, S.L., Hildebrand, J.D., Harland, R.M., Wallingford, J.B., 2003. Shroom induces apical constriction and is required for hingepoint formation during neural tube closure. Curr. Biol. 13, 2125–2137.

Hamblet, N.S., Lijam, N., Ruiz-Lozano, P., Wang, J., Yang, Y., Luo, Z., Mei, L., Chien, K.R., Sussman, D.J., Wynshaw-Boris, A., 2002. Dishevelled 2 is essential for cardiac outflow tract development, somite segmentation and neural tube closure. Development 129, 5827–5838.

Handel, M.A., Roth, L.E., 1971. Cell shape and morphology of the neural tube: implications for microtubule function. Dev. Biol. 25, 78–95.

Harrington, M.J., Chalasani, K., Brewster, R., 2010. Cellular mechanisms of posterior neural tube morphogenesis in the zebrafish. Dev. Dyn. 239, 747–762.

Harris, M.J., 2009. Insights into prevention of human neural tube defects by folic acid arising from consideration of mouse mutants. Birth Defects Res. A. Clin. Mol. Teratol. 85, 331–339.

Harris, M.J., Juriloff, D.M., 2007. Mouse mutants with neural tube closure defects and their role in understanding human neural tube defects. Birth Defects Res. A. Clin. Mol. Teratol. 79, 187–210.

Harris, M.J., Juriloff, D.M., 2010. An update to the list of mouse mutants with neural tube closure defects and advances toward a complete genetic perspective of neural tube closure. Birth Defects Res. A. Clin. Mol. Teratol. 88, 653–669.

Haycraft, C.J., Banizs, B., Aydin-Son, Y., Zhang, Q., Michaud, E.J., Yoder, B.K., 2005. Gli2 and Gli3 localize to cilia and require the intraflagellar transport protein polaris for processing and function. PLoS Genet. 1, e53.

Heid, M.K., Bills, N.D., Hinrichs, S.H., Clifford, A.J., 1992. Folate deficiency alone does not produce neural tube defects in mice. J. Nutr. 122, 888–894.

Heisenberg, C.P., Tada, M., Rauch, G.J., Saude, L., Concha, M.L., Geisler, R., Stemple, D.L., Smith, J.C., Wilson, S.W., 2000. Silberblick/Wnt11 mediates convergent extension movements during zebrafish gastrulation. Nature 405, 76–81.

Helwig, U., Imai, K., Schmahl, W., Thomas, B.E., Varnum, D.S., Nadeau, J.H., Balling, R., 1995. Interaction between undulated and Patch leads to an extreme form of spina bifida in double-mutant mice. Nat. Genet. 11, 60–63.

Hertzler 2nd, D.A., DePowell, J.J., Stevenson, C.B., Mangano, F.T., 2010. Tethered cord syndrome: a review of the literature from embryology to adult presentation. Neurosurg. Focus 29, E1.

Heseker, H.B., Mason, J.B., Selhub, J., Rosenberg, I.H., Jacques, P.F., 2009. Not all cases of neural-tube defect can be prevented by increasing the intake of folic acid. Br. J. Nutr. 102, 173–180.

Heydeck, W., Liu, A., 2011. PCP effector proteins inturned and fuzzy play nonredundant roles in the patterning but not convergent extension of mammalian neural tube. Dev. Dyn. 240, 1938–1948.

Heydeck, W., Zeng, H., Liu, A., 2009. Planar cell polarity effector gene Fuzzy regulates cilia formation and Hedgehog signal transduction in mouse. Dev. Dyn. 238, 3035–3042.

Hikasa, H., Shibata, M., Hiratani, I., Taira, M., 2002. The *Xenopus* receptor tyrosine kinase Xror2 modulates morphogenetic movements of the axial mesoderm and neuroectoderm via Wnt signaling. Development 129, 5227–5239.

Hildebrand, J.D., 2005. Shroom regulates epithelial cell shape via the apical positioning of an actomyosin network. J. Cell. Sci. 118, 5191–5203.

Hildebrand, J.D., Soriano, P., 1999. Shroom, a PDZ domain-containing actin-binding protein, is required for neural tube morphogenesis in mice. Cell 99, 485–497.

Hol, F.A., Geurds, M.P., Chatkupt, S., Shugart, Y.Y., Balling, R., Schrander-Stumpel, C.T., Johnson, W.G., Hamel, B.C., Mariman, E.C., 1996. PAX genes and human neural tube defects: an amino acid substitution in PAX1 in a patient with spina bifida. J. Med. Genet. 33, 655–660.

Hol, F.A., Hamel, B.C., Geurds, M.P., Mullaart, R.A., Barr, F.G., Macina, R.A., Mariman, E.C., 1995. A frameshift mutation in the gene for PAX3 in a girl with spina bifida and mild signs of Waardenburg syndrome. J. Med. Genet. 32, 52–56.

Holmberg, J., Clarke, D.L., Frisen, J., 2000. Regulation of repulsion versus adhesion by different splice forms of an Eph receptor. Nature 408, 203–206.

Holmes-Siedle, M., Lindenbaum, R.H., Galliard, A., 1992. Recurrence of neural tube defect in a group of at risk women: a ten year study of Pregnavite Forte F. J. Med. Genet. 29, 134–135.

Honarpour, N., Gilbert, S.L., Lahn, B.T., Wang, X., Herz, J., 2001. Apaf-1 deficiency and neural tube closure defects are found in fog mice. Proc. Natl. Acad. Sci. USA 98, 9683–9687.

Honein, M.A., Paulozzi, L.J., Mathews, T.J., Erickson, J.D., Wong, L.Y., 2001. Impact of folic acid fortification of the US food supply on the occurrence of neural tube defects. J. Am. Med. Assoc. 285, 2981–2986.

Hoth, C.F., Milunsky, A., Lipsky, N., Sheffer, R., Clarren, S.K., Baldwin, C.T., 1993. Mutations in the paired domain of the human PAX3 gene cause Klein-Waardenburg syndrome (WS-III) as well as Waardenburg syndrome type I (WS-I). Am. J. Hum. Genet. 52, 455–462.

Houde, C., Banks, K.G., Coulombe, N., Rasper, D., Grimm, E., Roy, S., Simpson, E.M., Nicholson, D.W., 2004. Caspase-7 expanded function and intrinsic expression level underlies strain-specific brain phenotype of caspase-3-null mice. J. Neurosci. 24, 9977–9984.

Hoving, E.W., 1990. Separation of neural and surface ectoderm in relation to the pathogenesis of encephaloceles. Z. Kinderchir. 45 (Suppl. 1), 40.

Huangfu, D., Anderson, K.V., 2005. Cilia and Hedgehog responsiveness in the mouse. Proc. Natl. Acad. Sci. USA 102, 11325–11330.

Hudgins, R.J., Gilreath, C.L., 2004. Tethered spinal cord following repair of myelomeningocele. Neurosurg. Focus 16, E7.

Jacobson, A.G., Moury, J.D., 1995. Tissue boundaries and cell behavior during neurulation. Dev. Biol. 171, 98–110.

Jones, E.A., Crotty, D., Kulesa, P.M., Waters, C.W., Baron, M.H., Fraser, S.E., Dickinson, M.E., 2002. Dynamic *in vivo* imaging of postimplantation mammalian embryos using whole embryo culture. Genesis 34, 228–235.

Joosten, P.H., Hol, F.A., van Beersum, S.E., Peters, H., Hamel, B.C., Afink, G.B., van Zoelen, E.J., Mariman, E.C., 1998. Altered regulation of platelet-derived growth factor receptor-alpha gene-transcription *in vitro* by spina bifida-associated mutant Pax1 proteins. Proc. Natl. Acad. Sci. USA 95, 14459–14463.

Joosten, P.H., Toepoel, M., Mariman, E.C., Van Zoelen, E.J., 2001. Promoter haplotype combinations of the platelet-derived growth factor alpha-receptor gene predispose to human neural tube defects. Nat. Genet. 27, 215–217.

Juriloff, D.M., Harris, M.J., 2012. A consideration of the evidence that genetic defects in planar cell polarity contribute to the etiology of human neural tube defects. Birth Defects Res. A. Clin. Mol. Teratol. 94, 824–840.

Juriloff, D.M., Harris, M.J., Tom, C., MacDonald, K.B., 1991. Normal mouse strains differ in the site of initiation of closure of the cranial neural tube. Teratology 44, 225–233.

Karfunkel, P., 1971. The role of microtubules and microfilaments in neurulation in *Xenopus*. Dev. Biol. 25, 30–56.

Khandelwal, M., Reece, E.A., Wu, Y.K., Borenstein, M., 1998. Dietary myo-inositol therapy in hyperglycemia-induced embryopathy. Teratology 57, 79–84.

Kibar, Z., Salem, S., Bosoi, C.M., Pauwels, E., De Marco, P., Merello, E., Bassuk, A.G., Capra, V., Gros, P., 2011. Contribution of VANGL2 mutations to isolated neural tube defects. Clin. Genet. 80, 76–82.

Kibar, Z., Torban, E., McDearmid, J.R., Reynolds, A., Berghout, J., Mathieu, M., Kirillova, I., De Marco, P., Merello, E., Hayes, J.M., et al., 2007. Mutations in VANGL1 associated with neural-tube defects. N. Engl. J. Med. 356, 1432–1437.

Kibar, Z., Vogan, K.J., Groulx, N., Justice, M.J., Underhill, D.A., Gros, P., 2001. Ltap, a mammalian homolog of Drosophila Strabismus/Van Gogh, is altered in the mouse neural tube mutant Loop-tail. Nat. Genet. 28, 251–255.

Kinoshita, H., Kokudo, T., Ide, T., Kondo, Y., Mori, T., Homma, Y., Yasuda, M., Tomiyama, J., Yakushiji, F., 2010. A patient with DiGeorge syndrome with spina bifida and sacral myelomeningocele, who developed both hypocalcemia-induced seizure and epilepsy. Seizure 19, 303–305.

Kinoshita, N., Sasai, N., Misaki, K., Yonemura, S., 2008. Apical accumulation of Rho in the neural plate is important for neural plate cell shape change and neural tube formation. Mol. Biol. Cell. 19, 2289–2299.

Kirillova, I., Novikova, I., Auge, J., Audollent, S., Esnault, D., Encha-Razavi, F., Lazjuk, G., Attie-Bitach, T., Vekemans, M., 2000. Expression of the sonic hedgehog gene in human embryos with neural tube defects. Teratology 61, 347–354.

Kirke, P.N., Molloy, A.M., Daly, L.E., Burke, H., Weir, D.G., Scott, J.M., 1993. Maternal plasma folate and vitamin B12 are independent risk factors for neural tube defects. QJM 86, 703–708.

Kooistra, M.K., Leduc, R.Y., Dawe, C.E., Fairbridge, N.A., Rasmussen, J., Man, J.H., Bujold, M., Juriloff, D., King-Jones, K., McDermid, H.E., 2012. Strain-specific modifier genes of Cecr2-associated exencephaly in mice: genetic analysis and identification of differentially expressed candidate genes. Physiol. Genomics. 44, 35–46.

Korstanje, R., Desai, J., Lazar, G., King, B., Rollins, J., Spurr, M., Joseph, J., Kadambi, S., Li, Y., Cherry, A., et al., 2008. Quantitative trait loci affecting phenotypic variation in the vacuolated lens mouse mutant, a multigenic mouse model of neural tube defects. Physiol. Genomics. 35, 296–304.

Kousseff, B.G., 1984. Sacral meningocele with conotruncal heart defects: a possible autosomal recessive trait. Pediatrics 74, 395–398.

Kujat, A., Veith, V.P., Faber, R., Froster, U.G., 2007. Prenatal diagnosis and genetic counseling in a case of spina bifida in a family with Waardenburg syndrome type I. Fetal. Diagn. Ther. 22, 155–158.

Lakhwani, S., Garcia-Sanz, P., Vallejo, M., 2010. Alx3-deficient mice exhibit folic acid-resistant craniofacial midline and neural tube closure defects. Dev. Biol. 344, 869–880.

Lakkis, M.M., Golden, J.A., O'Shea, K.S., Epstein, J.A., 1999. Neurofibromin deficiency in mice causes exencephaly and is a modifier for Splotch neural tube defects. Dev. Biol. 212, 80–92.

Lammer, E.J., Sever, L.E., Oakley Jr., G.P., 1987. Teratogen update: valproic acid. Teratology 35, 465–473.

Lanier, L.M., Gates, M.A., Witke, W., Menzies, A.S., Wehman, A.M., Macklis, J.D., Kwiatkowski, D., Soriano, P., Gertler, F.B., 1999. Mena is required for neurulation and commissure formation. Neuron 22, 313–325.

Laurence, K.M., Carter, C.O., David, P.A., 1968. Major central nervous system malformations in South Wales. II. Pregnancy factors, seasonal variation, and social class effects. Br. J. Prev. Soc. Med. 22, 212–222.

Laurence, K.M., James, N., Miller, M.H., Tennant, G.B., Campbell, H., 1981. Double-blind randomized controlled trial of folate treatment before conception to prevent recurrence of neural-tube defects. Br. Med. J. (Clin. Res. Ed.) 282, 1509–1511.

Lee, C., Scherr, H.M., Wallingford, J.B., 2007a. Shroom family proteins regulate gamma-tubulin distribution and microtubule architecture during epithelial cell shape change. Development 134, 1431–1441.

Lee, J.D., Silva-Gagliardi, N.F., Tepass, U., McGlade, C.J., Anderson, K.V., 2007b. The FERM protein Epb4.115 is required for organization of the neural plate and for the epithelial-mesenchymal transition at the primitive streak of the mouse embryo. Development 134, 2007–2016.

Lei, Y., Zhu, H., Duhon, C., Yang, W., Ross, M.E., Shaw, G.M., Finnell, R.H., 2013. Mutations in planar cell polarity gene SCRIB are associated with spina bifida. PLoS One 8, e69262.

Lei, Y.P., Zhang, T., Li, H., Wu, B.L., Jin, L., Wang, H.Y., 2010. VANGL2 mutations in human cranial neural-tube defects. N. Engl. J. Med. 362, 2232–2235.

Leonard, J.R., Klocke, B.J., D'Sa, C., Flavell, R.A., Roth, K.A., 2002. Strain-dependent neurodevelopmental abnormalities in caspase-3-deficient mice. J. Neuropathol. Exp. Neurol. 61, 673–677.

Letts, V.A., Schork, N.J., Copp, A.J., Bernfield, M., Frankel, W.N., 1995. A curly-tail modifier locus, mct1, on mouse chromosome 17. Genomics 29, 719–724.

Lin, W., Zhang, Z., Srajer, G., Chen, Y.C., Huang, M., Phan, H.M., Dent, S.Y., 2008. Proper expression of the Gcn5 histone acetyltransferase is required for neural tube closure in mouse embryos. Dev. Dyn. 237, 928–940.

Liu, A., Wang, B., Niswander, L.A., 2005. Mouse intraflagellar transport proteins regulate both the activator and repressor functions of Gli transcription factors. Development 132, 3103–3111.

Logan, C.V., Abdel-Hamed, Z., Johnson, C.A., 2011. Molecular genetics and pathogenic mechanisms for the severe ciliopathies: insights into neurodevelopment and pathogenesis of neural tube defects. Mol. Neurobiol. 43, 12–26.

Lu, X., Borchers, A.G., Jolicoeur, C., Rayburn, H., Baker, J.C., Tessier-Lavigne, M., 2004. PTK7/CCK-4 is a novel regulator of planar cell polarity in vertebrates. Nature 430, 93–98.

Maclean, K., Field, M.J., Colley, A.S., Mowat, D.R., Sparrow, D.B., Dunwoodie, S.L., Kirk, E.P., 2004. Kousseff syndrome: a causally heterogeneous disorder. Am. J. Med. Genet. A. 124A, 307–312.

Marean, A., Graf, A., Zhang, Y., Niswander, L., 2011. Folic acid supplementation can adversely affect murine neural tube closure and embryonic survival. Hum. Mol. Genet. 20, 3678–3683.

Massa, V., Savery, D., Ybot-Gonzalez, P., Ferraro, E., Rongvaux, A., Cecconi, F., Flavell, R., Greene, N.D., Copp, A.J., 2009. Apoptosis is not required for mammalian neural tube closure. Proc. Natl. Acad. Sci. USA 106, 8233–8238.

Massarwa, R., Niswander, L., 2013. In toto live imaging of mouse morphogenesis and new insights into neural tube closure. Development 140, 226–236.

May, S.R., Ashique, A.M., Karlen, M., Wang, B., Shen, Y., Zarbalis, K., Reiter, J., Ericson, J., Peterson, A.S., 2005. Loss of the retrograde motor for IFT disrupts localization of Smo to cilia and prevents the expression of both activator and repressor functions of Gli. Dev. Biol. 287, 378–389.

McMahon, J.A., Takada, S., Zimmerman, L.B., Fan, C.M., Harland, R.M., McMahon, A.P., 1998. Noggin-mediated antagonism of BMP signaling is required for growth and patterning of the neural tube and somite. Genes. Dev. 12, 1438–1452.

Menzies, A.S., Aszodi, A., Williams, S.E., Pfeifer, A., Wehman, A.M., Goh, K.L., Mason, C.A., Fassler, R., Gertler, F.B., 2004. Mena and vasodilator-stimulated phosphoprotein are required for multiple actin-dependent processes that shape the vertebrate nervous system. J. Neurosci. 24, 8029–8038.

Merte, J., Jensen, D., Wright, K., Sarsfield, S., Wang, Y., Schekman, R., Ginty, D.D., 2010. Sec24b selectively sorts Vangl2 to regulate planar cell polarity during neural tube closure. Nat. Cell. Biol. 12, 41–46. ; sup pp 1–8.

Moline, M.L., Sandlin, C., 1993. Waardenburg syndrome and meningomyelocele. Am. J. Med. Genet. 47, 126.

Morita, H., Kajiura-Kobayashi, H., Takagi, C., Yamamoto, T.S., Nonaka, S., Ueno, N., 2012. Cell movements of the deep layer of non-neural ectoderm underlie complete neural tube closure in *Xenopus*. Development 139, 1417–1426.

Morris-Wiman, J., Brinkley, L.L., 1990a. Changes in mesenchymal cell and hyaluronate distribution correlate with *in vivo* elevation of the mouse mesencephalic neural folds. Anat. Rec. 226, 383–395.

Morris-Wiman, J., Brinkley, L.L., 1990b. The role of the mesenchyme in mouse neural fold elevation. I. Patterns of mesenchymal cell distribution and proliferation in embryos developing *in vitro*. Am. J. Anat. 188, 121–132.

Morris-Wiman, J., Brinkley, L.L., 1990c. The role of the mesenchyme in mouse neural fold elevation. II. Patterns of hyaluronate synthesis and distribution in embryos developing *in vitro*. Am. J. Anat. 188, 133–147.

Morrison-Graham, K., Schatteman, G.C., Bork, T., Bowen-Pope, D.F., Weston, J.A., 1992. A PDGF receptor mutation in the mouse (Patch) perturbs the development of a non-neuronal subset of neural crest-derived cells. Development 115, 133–142.

Morriss, G.M., Solursh, M., 1978a. Regional differences in mesenchymal cell morphology and glycosaminoglycans in early neural-fold stage rat embryos. J. Embryol. Exp. Morphol. 46, 37–52.

Morriss, G.M., Solursh, M., 1978b. The role of primary mesenchyme in normal and abnormal morphogenesis of mammalian neural folds. Zoon 6, 33–38.

Morriss-Kay, G., Wood, H., Chen, W.H., 1994. Normal neurulation in mammals. Ciba. Found. Symp. 181, 51–63. discussion 63–9.

Morriss-Kay, G.M., Tuckett, F., Solursh, M., 1986. The effects of Streptomyces hyaluronidase on tissue organization and cell cycle time in rat embryos. J. Embryol. Exp. Morphol. 98, 59–70.

Moury, J.D., Schoenwolf, G.C., 1995. Cooperative model of epithelial shaping and bending during avian neurulation: autonomous movements of the neural plate, autonomous movements of the epidermis, and interactions in the neural plate/epidermis transition zone. Dev. Dyn. 204, 323–337.

Murdoch, J.N., Copp, A.J., 2010. The relationship between sonic Hedgehog signaling, cilia, and neural tube defects. Birth Defects Res. A. Clin. Mol. Teratol. 88, 633–652.

Murdoch, J.N., Doudney, K., Paternotte, C., Copp, A.J., Stanier, P., 2001. Severe neural tube defects in the loop-tail mouse result from mutation of Lpp1, a novel gene involved in floor plate specification. Hum. Mol. Genet. 10, 2593–2601.

Murdoch, J.N., Henderson, D.J., Doudney, K., Gaston-Massuet, C., Phillips, H.M., Paternotte, C., Arkell, R., Stanier, P., Copp, A.J., 2003. Disruption of scribble (Scrb1) causes severe neural tube defects in the circletail mouse. Hum. Mol. Genet. 12, 87–98.

Nakajima, H., Tanoue, T., 2009. Epithelial cell shape is regulated by Lulu proteins via myosin-II. J. Cell. Sci. 123, 555–566.

Nakajima, H., Tanoue, T., 2011. Lulu2 regulates the circumferential actomyosin tensile system in epithelial cells through p114RhoGEF. J. Cell. Biol. 195, 245–261.

Neumann, P.E., Frankel, W.N., Letts, V.A., Coffin, J.M., Copp, A.J., Bernfield, M., 1994. Multifactorial inheritance of neural tube defects: localization of the major gene and recognition of modifiers in ct mutant mice. Nat. Genet. 6, 357–362.

Nickel, R.E., Magenis, R.E., 1996. Neural tube defects and deletions of 22q11. Am. J. Med. Genet. 66, 25–27.

Nickel, R.E., Pillers, D.A., Merkens, M., Magenis, R.E., Driscoll, D.A., Emanuel, B.S., Zonana, J., 1994. Velo-cardio-facial syndrome and DiGeorge sequence with meningomyelocele and deletions of the 22q11 region. Am. J. Med. Genet. 52, 445–449.

Nishimura, T., Honda, H., Takeichi, M., 2012. Planar cell polarity links axes of spatial dynamics in neural-tube closure. Cell 149, 1084–1097.

Nishimura, T., Takeichi, M., 2008. Shroom3-mediated recruitment of Rho kinases to the apical cell junctions regulates epithelial and neuroepithelial planar remodeling. Development 135, 1493–1502.

Nye, J.S., Balkin, N., Lucas, H., Knepper, P.A., McLone, D.G., Charrow, J., 1998. Myelomeningocele and Waardenburg syndrome (type 3) in patients with interstitial deletions of 2q35 and the PAX3 gene: possible digenic inheritance of a neural tube defect. Am. J. Med. Genet. 75, 401–408.

Oakeshott, P., Hunt, G.M., 2003. Long-term outcome in open spina bifida. Br. J. Gen. Pract. 53, 632–636.

Orr-Urtreger, A., Bedford, M.T., Do, M.S., Eisenbach, L., Lonai, P., 1992. Developmental expression of the alpha receptor for platelet-derived growth factor, which is deleted in the embryonic lethal Patch mutation. Development 115, 289–303.

Oyama, K., Sugimura, Y., Murase, T., Uchida, A., Hayasaka, S., Oiso, Y., Murata, Y., 2009. Folic acid prevents congenital malformations in the offspring of diabetic mice. Endocrinol. Jpn. 56, 29–37.

Panchision, D.M., Pickel, J.M., Studer, L., Lee, S.H., Turner, P.A., Hazel, T.G., McKay, R.D., 2001. Sequential actions of BMP receptors control neural precursor cell production and fate. Genes. Dev. 15, 2094–2110.

Pani, L., Horal, M., Loeken, M.R., 2002a. Polymorphic susceptibility to the molecular causes of neural tube defects during diabetic embryopathy. Diabetes 51, 2871–2874.

Pani, L., Horal, M., Loeken, M.R., 2002b. Rescue of neural tube defects in Pax-3-deficient embryos by p53 loss of function: implications for Pax-3- dependent development and tumorigenesis. Genes. Dev. 16, 676–680.

Pantke, O.A., Cohen Jr., M.M., 1971. The Waardenburg syndrome. Birth Defects Orig. Artic. Ser. 7, 147–152.

Parker, S.E., Mai, C.T., Canfield, M.A., Rickard, R., Wang, Y., Meyer, R.E., Anderson, P., Mason, C.A., Collins, J.S., Kirby, R.S., et al., 2010. Updated National Birth Prevalence estimates for selected birth defects in the United States, 2004–2006. Birth Defects Res. A. Clin. Mol. Teratol. 88, 1008–1016.

Paudyal, A., Damrau, C., Patterson, V.L., Ermakov, A., Formstone, C., Lalanne, Z., Wells, S., Lu, X., Norris, D.P., Dean, C.H., et al., 2010. The novel mouse mutant, chuzhoi, has disruption of Ptk7 protein and exhibits defects in neural tube, heart and lung development and abnormal planar cell polarity in the ear. BMC. Dev. Biol. 10, 87.

Pavlinkova, G., Salbaum, J.M., Kappen, C., 2009. Maternal diabetes alters transcriptional programs in the developing embryo. BMC. Genomics. 10, 274.

Payne, J., Shibasaki, F., Mercola, M., 1997. Spina bifida occulta in homozygous Patch mouse embryos. Dev. Dyn. 209, 105–116.

Phelan, S.A., Ito, M., Loeken, M.R., 1997. Neural tube defects in embryos of diabetic mice: role of the Pax-3 gene and apoptosis. Diabetes 46, 1189–1197.

Pickett, E.A., Olsen, G.S., Tallquist, M.D., 2008. Disruption of PDGFRalpha-initiated PI3K activation and migration of somite derivatives leads to spina bifida. Development 135, 589–598.

Piedrahita, J.A., Oetama, B., Bennett, G.D., van Waes, J., Kamen, B.A., Richardson, J., Lacey, S.W., Anderson, R.G., Finnell, R.H., 1999. Mice lacking the folic acid-binding protein Folbp1 are defective in early embryonic development. Nat. Genet. 23, 228–232.

Plageman Jr., T.F., Chauhan, B.K., Yang, C., Jaudon, F., Shang, X., Zheng, Y., Lou, M., Debant, A., Hildebrand, J.D., Lang, R.A., 2011. A Trio-RhoA-Shroom3 pathway is required for apical constriction and epithelial invagination. Development 138, 5177–5188.

Plageman Jr., T.F., Chung, M.I., Lou, M., Smith, A.N., Hildebrand, J.D., Wallingford, J.B., Lang, R.A., 2010. Pax6-dependent Shroom3 expression regulates apical constriction during lens placode invagination. Development 137, 405–415.

Pyrgaki, C., Liu, A., Niswander, L., 2011. Grainyhead-like 2 regulates neural tube closure and adhesion molecule expression during neural fold fusion. Dev. Biol. 353, 38–49.

Pyrgaki, C., Trainor, P., Hadjantonakis, A.K., Niswander, L., 2010. Dynamic imaging of mammalian neural tube closure. Dev. Biol. 344, 941–947.

Qian, D., Jones, C., Rzadzinska, A., Mark, S., Zhang, X., Steel, K.P., Dai, X., Chen, P., 2007. Wnt5a functions in planar cell polarity regulation in mice. Dev. Biol. 306, 121–133.

Rampersaud, E., Bassuk, A.G., Enterline, D.S., George, T.M., Siegel, D.G., Melvin, E.C., Aben, J., Allen, J., Aylsworth, A., Brei, T., et al., 2005. Whole genome-wide linkage screen for neural tube defects reveals regions of interest on chromosomes 7 and 10. J. Med. Genet. 42, 940–946.

Reece, E.A., 2012. Diabetes-induced birth defects: what do we know? What can we do? Curr. Diab. Rep. 12, 24–32.

Reece, E.A., Ji, I., Wu, Y.K., Zhao, Z., 2006. Characterization of differential gene expression profiles in diabetic embryopathy using DNA microarray analysis. Am. J. Obstet. Gynecol. 195, 1075–1080.

Reece, E.A., Khandelwal, M., Wu, Y.K., Borenstein, M., 1997. Dietary intake of myo-inositol and neural tube defects in offspring of diabetic rats. Am. J. Obstet. Gynecol. 176, 536–539.

Reid, B.S., Sargent, T.D., Williams, T., 2010. Generation and characterization of a novel neural crest marker allele, Inka1-LacZ, reveals a role for Inka1 in mouse neural tube closure. Dev. Dyn. 239, 1188–1196.

Reynolds, A., McDearmid, J.R., Lachance, S., De Marco, P., Merello, E., Capra, V., Gros, P., Drapeau, P., Kibar, Z., 2010. VANGL1 rare variants associated with neural tube defects affect convergent extension in zebrafish. Mech. Dev. 127, 385–392.

Rifat, Y., Parekh, V., Wilanowski, T., Hislop, N.R., Auden, A., Ting, S.B., Cunningham, J.M., Jane, S.M., 2010. Regional neural tube closure defined by the Grainy head-like transcription factors. Dev. Biol. 345, 237–245.

Risch, N., 1990a. Linkage strategies for genetically complex traits. II. The power of affected relative pairs. Am. J. Hum. Genet. 46, 229–241.

Risch, N., 1990b. Linkage strategies for genetically complex traits. III. The effect of marker polymorphism on analysis of affected relative pairs. Am. J. Hum. Genet. 46, 242–253.

Robinson, A., Escuin, S., Doudney, K., Vekemans, M., Stevenson, R.E., Greene, N.D., Copp, A.J., Stanier, P., 2012. Mutations in the planar cell polarity genes CELSR1 and SCRIB are associated with the severe neural tube defect craniorachischisis. Hum. Mutat. 33, 440–447.

Roffers-Agarwal, J., Xanthos, J.B., Kragtorp, K.A., Miller, J.R., 2008. Enabled (Xena) regulates neural plate morphogenesis, apical constriction, and cellular adhesion required for neural tube closure in *Xenopus*. Dev. Biol. 314, 393–403.

Rolo, A., Skoglund, P., Keller, R., 2009. Morphogenetic movements driving neural tube closure in *Xenopus* require myosin IIB. Dev. Biol. 327, 327–338.

Ross, A.J., May-Simera, H., Eichers, E.R., Kai, M., Hill, J., Jagger, D.J., Leitch, C.C., Chapple, J.P., Munro, P.M., Fisher, S., et al., 2005. Disruption of Bardet-Biedl syndrome ciliary proteins perturbs planar cell polarity in vertebrates. Nat. Genet. 37, 1135–1140.

Rothenberg, S.P., da Costa, M.P., Sequeira, J.M., Cracco, J., Roberts, J.L., Weedon, J., Quadros, E.V., 2004. Auto-antibodies against folate receptors in women with a pregnancy complicated by a neural-tube defect. N. Engl. J. Med. 350, 134–142.

Salbaum, J.M., Kappen, C., 2010. Neural tube defect genes and maternal diabetes during pregnancy. Birth Defects Res. A. Clin. Mol. Teratol. 88, 601–611.

Sarkar, A.A., Zohn, I.E., 2012. Hectd1 regulates intracellular localization and secretion of Hsp90 to control cellular behavior of the cranial mesenchyme. J. Cell. Biol. 196, 789–800.

Sausedo, R.A., Smith, J.L., Schoenwolf, G.C., 1997. Role of nonrandomly oriented cell division in shaping and bending of the neural plate. J. Comp. Neurol. 381, 473–488.

Schoenwolf, G.C., 1984. Histological and ultrastructural studies of secondary neurulation in mouse embryos. Am. J. Anat. 169, 361–376.

Schoenwolf, G.C., Delongo, J., 1980. Ultrastructure of secondary neurulation in the chick embryo. Am. J. Anat. 158, 43–63.

Schoenwolf, G.C., Fisher, M., 1983. Analysis of the effects of Streptomyces hyaluronidase on formation of the neural tube. J. Embryol. Exp. Morphol. 73, 1–15.

Schorah, C., 2009. Dick Smithells, folic acid, and the prevention of neural tube defects. Birth Defects Res. A. Clin. Mol. Teratol. 85, 254–259.

Schroeder, T.E., 1970. Neurulation in *Xenopus laevis*. An analysis and model based upon light and electron microscopy. J. Embryol. Exp. Morphol. 23, 427–462.

Seller, M.J., 1994. Vitamins, folic acid and the cause and prevention of neural tube defects. Ciba. Found. Symp. 181, 161–173. discussion 173–9.

Seller, M.J., Mohammed, S., Russell, J., Ogilvie, C., 2002. Microdeletion 22q11.2, Kousseff syndrome and spina bifida. Clin. Dysmorphol. 11, 113–115.

Seller, M.J., Nevin, N.C., 1984. Periconceptional vitamin supplementation and the prevention of neural tube defects in south-east England and Northern Ireland. J. Med. Genet. 21, 325–330.

Seo, J.H., Zilber, Y., Babayeva, S., Liu, J., Kyriakopoulos, P., De Marco, P., Merello, E., Capra, V., Gros, P., Torban, E., 2011. Mutations in the planar cell polarity gene, Fuzzy, are associated with neural tube defects in humans. Hum. Mol. Genet. 20, 4324–4333.

Shi, Y., Ding, Y., Lei, Y.P., Yang, X.Y., Xie, G.M., Wen, J., Cai, C.Q., Li, H., Chen, Y., Zhang, T., et al., 2012. Identification of novel rare mutations of DACT1 in human neural tube defects. Hum. Mutat. 33, 1450–1455.

Shim, S.H., Wyandt, H.E., McDonald-McGinn, D.M., Zackai, E.Z., Milunsky, A., 2004. Molecular cytogenetic characterization of multiple intrachromosomal rearrangements of chromosome 2q in a patient with Waardenburg's syndrome and other congenital defects. Clin. Genet. 66, 46–52.

Shum, A.S., Copp, A.J., 1996. Regional differences in morphogenesis of the neuroepithelium suggest multiple mechanisms of spinal neurulation in the mouse. Anat. Embryol. (Berl.) 194, 65–73.

Smith, E.A., Seldin, M.F., Martinez, L., Watson, M.L., Choudhury, G.G., Lalley, P.A., Pierce, J., Aaronson, S., Barker, J., Naylor, S.L., et al., 1991. Mouse platelet-derived growth factor receptor alpha gene is deleted in W19H and patch mutations on chromosome 5. Proc. Natl. Acad. Sci. USA 88, 4811–4815.

Smith, J.L., Schoenwolf, G.C., 1987. Cell cycle and neuroepithelial cell shape during bending of the chick neural plate. Anat. Rec. 218, 196–206.

Smith, L.J., Stein, K.F., 1962. Axial elongation in the mouse and its retardation in homozygous looptail mice. J. Embryol. Exp. Morphol. 10, 73–87.

Smithells, R.W., Sheppard, S., Schorah, C.J., 1976. Vitamin dificiencies and neural tube defects. Arch. Dis. Child. 51, 944–950.

Smithells, R.W., Sheppard, S., Schorah, C.J., Seller, M.J., Nevin, N.C., Harris, R., Read, A.P., Fielding, D.W., 1980. Possible prevention of neural-tube defects by periconceptional vitamin supplementation. Lancet 1, 339–340.

Smithells, R.W., Sheppard, S., Schorah, C.J., Seller, M.J., Nevin, N.C., Harris, R., Read, A.P., Fielding, D.W., 1981. Apparent prevention of neural tube defects by periconceptional vitamin supplementation. Arch. Dis. Child. 56, 911–918.

Solloway, M.J., Robertson, E.J., 1999. Early embryonic lethality in Bmp5;Bmp7 double mutant mice suggests functional redundancy within the 60A subgroup. Development 126, 1753–1768.

Spiegelstein, O., Mitchell, L.E., Merriweather, M.Y., Wicker, N.J., Zhang, Q., Lammer, E.J., Finnell, R.H., 2004. Embryonic development of folate binding protein-1 (Folbp1) knockout mice: Effects of the chemical form, dose, and timing of maternal folate supplementation. Dev. Dyn. 231, 221–231.

Stamm, D.S., Rampersaud, E., Slifer, S.H., Mehltretter, L., Siegel, D.G., Xie, J., Hu-Lince, D., Craig, D.W., Stephan, D.A., George, T.M., et al., 2006. High-density single nucleotide polymorphism screen in a large multiplex neural tube defect family refines linkage to loci at 7p21.1-pter and 2q33.1-q35. Birth Defects Res. A. Clin. Mol. Teratol. 76, 499–505.

Stamm, D.S., Siegel, D.G., Mehltretter, L., Connelly, J.J., Trott, A., Ellis, N., Zismann, V., Stephan, D.A., George, T.M., Vekemans, M., et al., 2008. Refinement of 2q and 7p loci in a large multiplex NTD family. Birth Defects Res. A. Clin. Mol. Teratol. 82, 441–452.

Stephenson, D.A., Mercola, M., Anderson, E., Wang, C.Y., Stiles, C.D., Bowen-Pope, D.F., Chapman, V.M., 1991. Platelet-derived growth factor receptor alpha-subunit gene (Pdgfra) is deleted in the mouse patch (Ph) mutation. Proc. Natl. Acad. Sci. USA 88, 6–10.

Stiefel, D., Copp, A.J., Meuli, M., 2007. Fetal spina bifida in a mouse model: loss of neural function in utero. J. Neurosurg. 106, 213–221.

Stottmann, R.W., Berrong, M., Matta, K., Choi, M., Klingensmith, J., 2006. The BMP antagonist Noggin promotes cranial and spinal neurulation by distinct mechanisms. Dev. Biol. 295, 647–663.

Stottmann, R.W., Klingensmith, J., 2011. Bone morphogenetic protein signaling is required in the dorsal neural folds before neurulation for the induction of spinal neural crest cells and dorsal neurons. Dev. Dyn. 240, 755–765.

Sutton, L.N., Adzick, N.S., Bilaniuk, L.T., Johnson, M.P., Crombleholme, T.M., Flake, A.W., 1999. Improvement in hindbrain herniation demonstrated by serial fetal magnetic resonance imaging following fetal surgery for myelomeningocele. J. Am. Med. Assoc. 282, 1826–1831.

Suzuki, M., Morita, H., Ueno, N., 2012. Molecular mechanisms of cell shape changes that contribute to vertebrate neural tube closure. Dev. Growth. Differ. 54, 266–276.

Tada, M., Smith, J.C., 2000. Xwnt11 is a target of *Xenopus Brachyury*: regulation of gastrulation movements via Dishevelled, but not through the canonical Wnt pathway. Development 127, 2227–2238.

Thompson, D.N., 2009. Postnatal management and outcome for neural tube defects including spina bifida and encephalocoeles. Prenat. Diagn. 29, 412–419.

Ting, S.B., Wilanowski, T., Auden, A., Hall, M., Voss, A.K., Thomas, T., Parekh, V., Cunningham, J.M., Jane, S.M., 2003. Inositol- and folate-resistant neural tube defects in mice lacking the epithelial-specific factor Grhl-3. Nat. Med. 9, 1513–1519.

Tissir, F., Goffinet, A.M., 2010. Planar cell polarity signaling in neural development. Curr. Opin. Neurobiol. 20, 572–577.

Toepoel, M., Steegers-Theunissen, R.P., Ouborg, N.J., Franke, B., Gonzalez-Zuloeta Ladd, A.M., Joosten, P.H., van Zoelen, E.J., 2009. Interaction of PDGFRA promoter haplotypes and maternal environmental exposures in the risk of spina bifida. Birth Defects Res. A. Clin. Mol. Teratol. 85, 629–636.

Torban, E., Patenaude, A.M., Leclerc, S., Rakowiecki, S., Gauthier, S., Andelfinger, G., Epstein, D.J., Gros, P., 2008. Genetic interaction between members of the Vangl family causes neural tube defects in mice. Proc. Natl. Acad. Sci. USA 105, 3449–3454.

Toriello, H.V., Sharda, J.K., Beaumont, E.J., 1985. Autosomal recessive syndrome of sacral and conotruncal developmental field defects (Kousseff syndrome). Am. J. Med. Genet. 22, 357–360.

Tulipan, N., Hernanz-Schulman, M., Bruner, J.P., 1998. Reduced hindbrain herniation after intrauterine myelomeningocele repair: A report of four cases. Pediatr. Neurosurg. 29, 274–278.

Tulipan, N., Bruner, J.P., Hernanz-Schulman, M., Lowe, L.H., Walsh, W.F., Nickolaus, D., Oakes, W.J., 1999a. Effect of intrauterine myelomeningocele repair on central nervous system structure and function. Pediatr. Neurosurg. 31, 183–188.

Tulipan, N., Hernanz-Schulman, M., Lowe, L.H., Bruner, J.P., 1999b. Intrauterine myelomeningocele repair reverses preexisting hindbrain herniation. Pediatr. Neurosurg. 31, 137–142.

van Straaten, H.W., Copp, A.J., 2001. Curly tail: a 50-year history of the mouse spina bifida model. Anat. Embryol. (Berl.) 203, 225–237.

Vergel, R.G., Sanchez, L.R., Heredero, B.L., Rodriguez, P.L., Martinez, A.J., 1990. Primary prevention of neural tube defects with folic acid supplementation: Cuban experience. Prenat. Diagn. 10, 149–152.

Vladar, E.K., Antic, D., Axelrod, J.D., 2009. Planar cell polarity signaling: the developing cell's compass. Cold Spring Harb. Perspect. Biol. 1, a002964.

Wald, N., Sneddon, J., Densem, J., Frost, C., Stone, R., Group, M.V.S.R., 1991. Prevention of neural tube defects: results of the Medical Research Council Vitamin Study. MRC Vitamin Study Research Group. Lancet 338, 131–137.

Wallingford, J.B., 2012. Planar cell polarity and the developmental control of cell behavior in vertebrate embryos. Annu. Rev. Cell. Dev. Biol. 28, 627–653.

Wallingford, J.B., Harland, R.M., 2001. *Xenopus* Dishevelled signaling regulates both neural and mesodermal convergent extension: parallel forces elongating the body axis. Development 128, 2581–2592.

Wallingford, J.B., Harland, R.M., 2002. Neural tube closure requires Dishevelled-dependent convergent extension of the midline. Development 129, 5815–5825.

Wallingford, J.B., Mitchell, B., 2011. Strange as it may seem: the many links between Wnt signaling, planar cell polarity, and cilia. Genes. Dev. 25, 201–213.

Wallingford, J.B., Rowning, B.A., Vogeli, K.M., Rothbacher, U., Fraser, S.E., Harland, R.M., 2000. Dishevelled controls cell polarity during *Xenopus* gastrulation. Nature 405, 81–85.

Wang, J., Hamblet, N.S., Mark, S., Dickinson, M.E., Brinkman, B.C., Segil, N., Fraser, S.E., Chen, P., Wallingford, J.B., Wynshaw-Boris, A., 2006a. Dishevelled genes mediate a conserved mammalian PCP pathway to regulate convergent extension during neurulation. Development 133, 1767–1778.

Wang, Y., Guo, N., Nathans, J., 2006b. The role of Frizzled3 and Frizzled6 in neural tube closure and in the planar polarity of inner-ear sensory hair cells. J. Neurosci. 26, 2147–2156.

Wang, Y., Lian, L., Golden, J.A., Morrisey, E.E., Abrams, C.S., 2007. PIP5KI gamma is required for cardiovascular and neuronal development. Proc. Natl. Acad. Sci. USA 104, 11748–11753.

Weatherbee, S.D., Niswander, L.A., Anderson, K.V., 2009. A mouse model for Meckel syndrome reveals Mks1 is required for ciliogenesis and Hedgehog signaling. Hum. Mol. Genet. 18, 4565–4575.

Weil, M., Jacobson, M.D., Raff, M.C., 1997. Is programd cell death required for neural tube closure? Curr. Biol. 7, 281–284.

Werth, M., Walentin, K., Aue, A., Schonheit, J., Wuebken, A., Pode-Shakked, N., Vilianovitch, L., Erdmann, B., Dekel, B., Bader, M., et al., 2010. The transcription factor grainyhead-like 2 regulates the molecular composition of the epithelial apical junctional complex. Development 137, 3835–3845.

Wilson, D.B., Wyatt, D.P., 1994. Analysis of neurulation in a mouse model for neural dysraphism. Exp. Neurol. 127, 154–158.

Wilson, M.P., Hugge, C., Bielinska, M., Nicholas, P., Majerus, P.W., Wilson, D.B., 2009. Neural tube defects in mice with reduced levels of inositol 1,3,4-trisphosphate 5/6-kinase. Proc. Natl. Acad. Sci. USA 106, 9831–9835.

Wlodarczyk, B.J., Tang, L.S., Triplett, A., Aleman, F., Finnell, R.H., 2006. Spontaneous neural tube defects in splotch mice supplemented with selected micronutrients. Toxicol. Appl. Pharmacol. 213, 55–63.

Wollnik, B., Tukel, T., Uyguner, O., Ghanbari, A., Kayserili, H., Emiroglu, M., Yuksel-Apak, M., 2003. Homozygous and heterozygous inheritance of PAX3 mutations causes different types of Waardenburg syndrome. Am. J. Med. Genet. A. 122, 42–45.

Wong, R.L., Wlodarczyk, B.J., Min, K.S., Scott, M.L., Kartiko, S., Yu, W., Merriweather, M.Y., Vogel, P., Zambrowicz, B.P., Finnell, R.H., 2008. Mouse Fkbp8 activity is required to inhibit cell death and establish dorso-ventral patterning in the posterior neural tube. Hum. Mol. Genet. 17, 587–601.

Yamada, S., Won, D.J., Siddiqi, J., Yamada, S.M., 2004. Tethered cord syndrome: overview of diagnosis and treatment. Neurol. Res. 26, 719–721.

Yamaguchi, Y., Shinotsuka, N., Nonomura, K., Takemoto, K., Kuida, K., Yosida, H., Miura, M., 2011. Live imaging of apoptosis in a novel transgenic mouse highlights its role in neural tube closure. J. Cell. Biol. 195, 1047–1060.

Ybot-Gonzalez, P., Cogram, P., Gerrelli, D., Copp, A.J., 2002. Sonic hedgehog and the molecular regulation of mouse neural tube closure. Development 129, 2507–2517.

Ybot-Gonzalez, P., Gaston-Massuet, C., Girdler, G., Klingensmith, J., Arkell, R., Greene, N.D., Copp, A.J., 2007a. Neural plate morphogenesis during mouse neurulation is regulated by antagonism of Bmp signaling. Development 134, 3203–3211.

Ybot-Gonzalez, P., Savery, D., Gerrelli, D., Signore, M., Mitchell, C.E., Faux, C.H., Greene, N.D., Copp, A.J., 2007b. Convergent extension, planar-cell-polarity signaling and initiation of mouse neural tube closure. Development 134, 789–799.

Yu, H., Smallwood, P.M., Wang, Y., Vidaltamayo, R., Reed, R., Nathans, J., 2010. Frizzled 1 and frizzled 2 genes function in palate, ventricular septum and neural tube closure: general implications for tissue fusion processes. Development 137, 3707–3717.

Yu, Z., Lin, K.K., Bhandari, A., Spencer, J.A., Xu, X., Wang, N., Lu, Z., Gill, G.N., Roop, D.R., Wertz, P., et al., 2006. The Grainyhead-like epithelial transactivator Get-1/Grhl3 regulates epidermal terminal differentiation and interacts functionally with LMO4. Dev. Biol. 299, 122–136.

Zabihi, S., Loeken, M.R., 2010. Understanding diabetic teratogenesis: where are we now and where are we going? Birth Defects Res. A. Clin. Mol. Teratol. 88, 779–790.

Zeisel, S.H., 2009. Importance of methyl donors during reproduction. Am. J. Clin. Nutr. 89, 673S–677S.

Zeng, H., Hoover, A.N., Liu, A., 2010. PCP effector gene Inturned is an important regulator of cilia formation and embryonic development in mammals. Dev. Biol. 339, 418–428.

Zhao, G.Q., Zhou, X., Eberspaecher, H., Solursh, M., de Crombrugghe, B., 1993. Cartilage homeoprotein 1, a homeoprotein selectively expressed in chondrocytes. Proc. Natl. Acad. Sci. USA 90, 8633–8637.

Zhao, Q., Behringer, R.R., de Crombrugghe, B., 1996. Prenatal folic acid treatment suppresses acrania and meroanencephaly in mice mutant for the Cart1 homeobox gene. Nat. Genet. 13, 275–283.

Zhao, R., Russell, R.G., Wang, Y., Liu, L., Gao, F., Kneitz, B., Edelmann, W., Goldman, I.D., 2001. Rescue of embryonic lethality in reduced folate carrier-deficient mice by maternal folic acid supplementation reveals early neonatal failure of hematopoietic organs. J. Biol. Chem. 276, 10224–10228.

Zhu, H., Wicker, N.J., Volcik, K., Zhang, J., Shaw, G.M., Lammer, E.J., Suarez, L., Canfield, M., Finnell, R.H., 2004. Promoter haplotype combinations for the human PDGFRA gene are associated with risk of neural tube defects. Mol. Genet. Metab. 81, 127–132.

Zlotogora, J., Lerer, I., Bar-David, S., Ergaz, Z., Abeliovich, D., 1995. Homozygosity for Waardenburg syndrome. Am. J. Hum. Genet. 56, 1173–1178.

Zohn, I.E., 2012. Mouse as a model for multifactorial inheritance of neural tube defects. Birth Defects Res. C. Embryo. Today 96, 193–205.

Zohn, I.E., Anderson, K.V., Niswander, L., 2007. The Hectd1 ubiquitin ligase is required for development of the head mesenchyme and neural tube closure. Dev. Biol. 306, 208–221.

Zohn, I.E., Sarkar, A.A., 2008. Modeling neural tube defects in the mouse. Curr. Top. Dev. Biol. 84, 1–35.

Zohn, I.E., Sarkar, A.A., 2010. The visceral yolk sac endoderm provides for absorption of nutrients to the embryo during neurulation. Birth Defects Res. A. Clin. Mol. Teratol. 88, 593–600.

Zohn, I.E., Sarkar, A.A., 2012. Does the cranial mesenchyme contribute to neural fold elevation during neurulation? Birth Defects Res. A. Clin. Mol. Teratol. 94, 841–848.

Glossary

β-cell The predominant endocrine cell type in the pancreatic islets; it is the only cell in the body that is capable of producing and secreting the hormone insulin.

Adenohypophysis The anterior pituitary gland, derived from the adenohypophyseal placode, which contains cells that secrete a number of peptide hormones.

Alternative splicing The differential processing of pre-messenger (nuclear) RNA transcripts involving exon and intron sequences.

Ameloblasts Epithelial cells producing dental enamel.

Ampulla The swollen end of a branch, that will give rise to new branches.

Anencephaly A neural tube defect (NTD) that results from failure of neurulation in the cephalic region, and characterized by the absence of a large part of the brain and overlying skull.

Angioblast Mesoderm-derived endothelial precursors that have certain characteristics of endothelial cells but that have not yet assembled into functional vessels.

Angiogenesis The formation of new blood vessels from pre-existing vessels by endothelial sprouting and splitting. This process of remodeling the primary capillary network leads to the formation of mature arteries and veins.

Anterior visceral endoderm (AVE) A population of visceral endoderm cells found in the anterior region of the conceptus, opposite the primitive streak at mouse E6–E6.5, and characterized by the expression of a specific gene repertoire including *Hhex*, *Hesx1*, *Cer1* and *Lefty1*. The AVE is formed after cells of the distal visceral endoderm (DVE) have moved, thus breaking the radial symmetry of the egg cylinder. This population is heterogeneous as shown by distinct gene expression in medial and lateral domains, but also dynamic in nature, as its cellular composition changes during the general movement of the Visceral Endoderm (VE) which occurs between mouse E5.5 and E6.5.

Apical constriction A process by which contraction of the apical surface of neural epithelial cells results in the cell taking on a wedge shape. Apical constriction is essential for hinge point formation during neural fold elevation.

Artiodactyls A group of herbivorous mammals with even-toed hooves, including pigs, hippopotamuses, and ruminants.

Atoh1 A basic helix-loop-helix transcription factor that is necessary and sufficient for hair cell formation. Atonal Homolog 1a (Atoh1a) and Atonal Homolog 1b (Atoh1b) are the zebrafish homologs of *Drosophila* Atonal, a bHLH, transcription factor that also determines the specification of mechanoreceptors in *Drosophila*. Atoh1a expression gives cells in protoneuromasts or neuromasts the potential to become sensory hair cell progenitors.

Bardet-Biedl syndrome This human congenital disease is mainly characterized by obesity, pigmentary retinopathy, polydactyly, mental retardation, hypogonadism, and renal failure in fatal cases.

Bifurcation Division into two (as in one branch bifurcating to become two).

Blastoporal-ablastoporal axis An idealized pan-vertebrate axis that runs orthogonally to the zone of internalizing mesendodermal cells. The zone of internalizing mesendodermal cells defines the blastoporal end of this axis.

Brachydactyly This is a term which literally means "shortness of the fingers and toes" (digits). The shortness is relative to the length of other long bones and other parts of the body.

Branchio-otic renal syndrome An autosomal dominant syndrome in humans that presents with craniofacial defects, hearing loss, and kidney or urinary tract defects.

Branchio-otic syndrome An autosomal dominant syndrome in humans that presents with branchial cleft fistulas and hearing loss. Some cases are caused by mutations in the *Eya1* gene and some are caused by mutations in *Six* genes.

Cardia bifida The formation of two beating and looping heart tubes as a result of the improper fusion of the heart primordia.

Cell fate The neuronal subtype that a progenitor will become. This term does not imply commitment or differentiation, only that the cell will eventually become a certain type.

Cell migration A process during which cells translocate their cell body to a new location according to various signals present in their local environment. Such signals include extracellular matrix, guidance cues and contacts with other cells.

Central tendon (of the diaphragm) The central tendon is the unmuscularized region of the diaphragm that has differentiated into a mature tendon. It makes up the center of the diaphragm, and it is attached to the liver by the falciform ligament. Distinct myotendonous junctions form the boundary between the central tendon and the peripherally muscularized regions.

Chemokines A class of chemotactic cytokine signaling molecules secreted by cells that direct the chemotactic response of other cells. They are divided into CXC, CC, CX3C and XC subfamilies.

ChIA-PET An adaptation of 3C technologies that employs high throughput sequencing. This method utilizes ChIP to find regions of DNA looping associated with a specific transcription factor.

ChIP (chromatin immunoprecipitation) A biochemical technique in which an antibody is bound to chromatin (DNA + protein), genomic DNA is cleaved into small fragments, and DNA bound by the antibody is pulled down and sequenced (ChIP-seq) or hybridized to a microarray (ChIP-chip).

ChIP-seq A high throughput sequencing method employing immunoprecipitation of fragmented chromatin to identify sites bound by a protein, genome-wide.

Chromosome conformation capture (3C or CCC) An experimental method that employs quantitative polymerase chain reaction (PCR) to identify transcriptional targets of enhancer elements on the basis of physical interactions between enhancers and promoters.

Ciliary margin The most peripheral region of the retina. In the frog and fish, this region contains dividing retinal progenitor cells that contribute to retinal growth and/or repair.

***Cis*-regulatory elements (CREs)** A distinct region of DNA of a gene that influences its transcription. CREs can be located upstream of the transcription start site or may be located downstream. These sequences can be modular in nature.

Clefting Division of an ampulla by ingrowth of a sharp cleft.

Cloaca A transient embryonic cavity in mammals, lined with endoderm, which contributes to the epithelial lining of the urogenital and anorectal systems.

Collecting duct The segment of the nephron that conveys the urine to the outside.

Coloboma A condition where an individual is missing part of his or her eye structure, including, but not limited to, the iris, retina, and eyelid. Treacher Collins syndrome patients usually have coloboma of the eyelid and have a notch in the upper or lower eyelid.

Commitment An irreversible decision to produce or become a particular cell type. This is defined operationally as the inability of a cell to change its fate when exposed to different signaling cues. Commitment is always discussed in the context of a particular differentiation step.

Competence The ability of a cell to respond to a cue or set of cues to produce a defined outcome.

Conceptus Contains all the derivatives of the zygote, embryonic (epiblast) and extra-embryonic tissues.

Conditional allele An allele that is engineered to function normally until it is recombined by the introduction of a recombinase to produce a mutant allele in a tissue- or temporally-specific manner.

Congenital diaphragmatic hernia (CDH) This term is used inconsistently to describe a variety of diaphragmatic defects. Among physicians, it is usually used to describe the severe hernias that present at birth. These hernias almost always involve the posterolateral, lateral, or posteromedial diaphragm. They may also present as aplasia of one side of the diaphragm. The term "CDH" is sometimes used interchangeably with "Bochdalek hernia," a hernia that should only arise from a weak spot along the posterolateral wall of the thorax.

Conotruncus Regions of the bulbus cordis that forms the conus cordis and the truncus arteriosus.

Convergent extension (CE) The process in which the cells of a tissue rearrange to converge towards the midline and extend the tissue along the anterior-posterior axis of the embryo.

Copy number variation (CNV) disorder A genetic disorder caused by the heterozygosity of one or more genes (such as in a microdeletion). These disorders result from the reduced expression of the deleted genes, rather than complete absence or a mutation.

Craniorachischisis A neural tube defect (NTD) that results from failure of neural tube closure along the entire neural axis.

Cytoneme Thin, actin-based cellular protrusion several cell diameters long, extending from the apical surface of a cell toward a source of signaling protein (ligand); first observed in *Drosophila* wing disc cells extending toward Dpp and Wg.

Cytoplasmic determinant A cytoplasmic protein/mRNA acting to promote a specific cell fate.

Data mining The use of statistical and graphical tools to compare expression patterns between data sets.

Definitive endoderm Endoderm tissue that contributes to the embryo proper, in contrast to the extra-embryonic endoderm tissues (visceral and primitive endoderm) which may be required for correct development or embryo survival but do not contribute to the overall body plan.

Definitive hematopoiesis Occurs shortly after the primitive hematopoiesis and contains two waves. The first definitive wave generates only red cell and myeloid lineages. The second wave generates hematopoietic stem cells (HSC) that differentiate into all adult blood lineages.

Determination An irreversible decision made by an undifferentiated cell that restricts its differentiation.

Determined Cells are considered to be determined if they are able to assume their natural fate when placed in an antagonistic environment. Therefore, determination represents a higher level of commitment.

Deuterostome Deuterostome literally means "second mouth" (deutero – two; stome – mouth). The blastopore is formed first during gastrulation and the mouth is formed secondarily. This mode of development applies to all deuterostomes except Tunicata. Echinodermata, Hemichordata, Xenoturbellida and Chordata are considered deuterostome phyla.

Diabetes A heterogeneous group of diseases in which the pancreas fails to produce insulin in sufficient amounts to maintain euglycemia, thus causing a dramatic rise in blood glucose levels.

Differentiation In developmental biology, differentiation is the process through which a stem cell or progenitor cell becomes more specialized. The process is the elaboration of particular characteristics expressed by an end-stage cell type, or by a cell en route to becoming an end-stage cell. This term is not synonymous with commitment, but differentiation features are used to determine when a cell is committed.

Dimorphism The expression of two different forms within a population or species; in contrast to monomorphism (one form).

Dipodial Branching in which all branches are equal, with no central trunk.

Directed differentiation The process by which embryonic stem cells or induced pluripotent stem cells are progressively differentiated down a particular lineage by using signals that attempt to recapitulate embryonic development.

Distal visceral endoderm (DVE) A group of visceral endoderm cells found at the distal tip of the mouse egg cylinder at E5.5, characterized by its columnar morphology and the expression of a specific gene repertoire including *Hhex*, *Cer1* and *Lefty1*. The DVE is formed in two waves. The first wave consists of descendants of precursors already expressing *Lefty1* and *Cer1* in the primitive endoderm and migrate first anteriorly. The second wave is induced at the distal pole of the mouse embryo after E5.5. As the DVE migrates towards one side, the anterior pole becomes defined, and the group of cells is referred to as the anterior visceral endoderm (AVE).

DNAse-seq A high throughput sequencing method to identify sites of open chromatin based on their relative accessibility to cleavage by DNAse I.

Dominant-negative A mutated protein that interferes with the function of normal copies of the same protein.

Dosage compensation The control of X chromosome gene expression chromosomes dependent upon genotype (XX vs. XY).

Ectoderm One of the three germ layers of the embryo. It gives rise to the epidermis, the neural plate, cranial placodes and the neural crest.

Ectodermal dysplasias Syndromes that affect two or more ectodermal organs; the most common syndromes involving dental defects.

Egg cylinder The mouse conceptus after implantation, from E5 to gastrulation. The conceptus at these stages has the shape of a cylinder and comprises a proximal and a distal region.

Embryonic induction The interaction between an inducing tissue and a responding tissue. As a result, the responding tissue undergoes a change in fate.

Enamel knots Signaling centers in dental epithelium that regulate tooth morphogenesis.

Endochondral ossification A mechanism of bone formation in which a cartilage model is replaced with ossified tissue produced by invading osteoblasts.

Endocrine cells Within the pancreas, the cells that produce hormones such as insulin and glucagon that are secreted directly into the bloodstream, where they travel to their target tissues.

Endocytosis A process by which cell membranes envelope and internalize transmembrane and/or extracellular signaling proteins into cytoplasmic vesicles. Cytoplasmic proteins including dynamin mediate the inward budding of vesicles from the plasma membrane; dynamin is often targeted to block endocytosis in genetic studies. Once internalized, endocytic vesicles are sorted into different groups which undergo recycling to the plasma membrane or lysosomal degradation.

Endoderm The innermost germ layer of vertebrate embryos that gives rise to the epithelial cells lining the digestive and respiratory system as well as visceral organs including the liver, thyroid and pancreas. Note that early in the development of birds and mammals, the definitive endoderm is initially part of the outermost layer of cells.

Endostyle This is an endodermal structure found in invertebrate chordates in the pharyngeal area. The endostyle secretes mucus to capture small particles and so increase the efficiency of filter feeding. In lancelets and tunicates, the endostyle also accumulates iodine and is considered homologous to the vertebrate thyroid gland.

Enhancer element A region of DNA that impacts gene transcription in *cis* through the recruitment of transcription factors or other DNA binding/modifying proteins; also called cis-regulatory modules.

Epiblast A tissue formed of pluripotent precursors that will generate all the fetal organs as well as the extra-embryonic mesoderm.

Epistasis and genetic interactions Epistatic and genetic interactions are functional interactions between mutations in non-allelic genes suggestive of the gene products acting together in a given process or pathway. A *genetic interaction* is manifested through a compound mutant phenotype that is more pronounced than the sum of the two single mutant phenotypes, e.g., disrupted development in an animal heterozygous for two mutations for which either heterozygous mutation (in isolation) does not result in a developmental phenotype. An *epistatic relationship/interaction* is manifest through the ability of one allele to suppress the phenotypic consequences of a second mutation, typically indicating dominance of the epistatic mutation or the result of its being downstream in a common genetic pathway.

Epithelial polarity Regional differences within cells forming an epithelial tissue.

Epithelial-mesenchymal transition (EMT) A series of changes at the molecular level that converts an epithelial cell into a mesenchymal cell. EMT involves changes in cell-cell and cell-matrix adhesion properties as well as changes in cell polarity.

Euglycemia The state of maintaining blood glucose levels within the normal range (70 to 120 mg/dL).

Exencephaly A neural tube defect (NTD) that results from failure of neural tube closure in the cephalic region. As fetal development continues, the exposed brain tissue will degenerate and exencephaly will progress to anencephaly.

Exocrine cells Within the pancreas, the cells that produce the digestive enzymes that are released into the pancreatic duct and eventually into the duodenum.

Extra-embryonic ectoderm (ExE) An extra-embryonic tissue derived from the (polar) trophectoderm, in contact with the inner cell mass in the mammalian blastocyst. Unlike cells of the ectoplacental cone (another trophectoderm derivative), which differentiate into giant cells, cells of the ExE retain stem cell potential. The ExE contributes to the placenta.

Extrahepatic biliary system A ductal network for the movement of bile that aids in digestion. It includes the common duct, extrahepatic bile ducts, and the gallbladder.

Extrinsic factors Extracellular molecules such as secreted signaling proteins and their transmembrane receptors that provide a cell with information about its environment.

Eye field or eye primordia A domain in the anterior neural plate that is specified to give rise to eye tissue. This domain expresses a combination of eye field transcription factors that are essential for eye formation.

Fate domain An embryonic region whose cellular constituents undertake reproducible pathways of differentiation, *under normal circumstances*. These cells may or may not be committed and changes in their environment can reveal this.

Feedback regulation A mechanism by which a signaling pathway regulates its own activity, e.g., by activating a regulatory factor that alters signal transduction, by altering sensitivity of the pathway to upstream signals, and/or by modifying the activity or interactions of proteins in the pathway. In some cases, multiple signaling pathways generate a feedback regulatory circuit, in which activation of one pathway results in regulation of another, and vice versa.

FGFs Fibroblast Growth Factors are a family of growth factors, whose members are involved in angiogenesis, wound healing, embryonic development and various endocrine signaling pathways.

Filopodia Thin, short cellular protrusions containing both actin and microtubules that are not polarized toward a source of signaling protein.

First heart field (FHF) Early differentiating myocardial cells that contribute to the linear heart tube and ultimately to the left ventricle.

Fontanelles Embryonic and early childhood structures in the skull commonly referred to as soft spots. These allow the skull plates to shift during delivery. Once the skull becomes ossified, around the age of two, the fontanelles close and become the skull sutures.

Foregut Anterior-most of three divisions of the digestive tract (foregut, midgut, hindgut).

Fractal Self-similar (looks the same at all scales).

Gene regulatory network (GRN) The interactions between transcription factors and their targets that govern developmental processes. A network of transcription factors with cross-regulatory (e.g., repression, activation) relationships, e.g., that translate morphogen signals into discrete gene expression boundaries.

Genome-wide association study (GWAS) A genetic approach that uses high throughput sequencing to compare control and affected populations and identify genetic variants (e.g., SNPs, mutated alleles) that are associated with a given phenotype (disease).

Glomeruli A specialized "globe" of neuropil that consists of dendrites from olfactory bulb mitral cells, dendrites from olfactory bulb granule cells, axon endings and synaptic boutons from olfactory receptor neurons, and axonal and dendritic processes from periglomerular cells, whose cell bodies define the limits of each glomerulus.

Glucose intolerance An impaired ability to clear glucose efficiently from the bloodstream after a meal; a prediabetic state.

Glucose-stimulated insulin secretion The increase in insulin released by pancreatic islets is proportional to the elevation in blood glucose; glucose must be metabolized by the β-cell for insulin to be released.

Gonadotropin Releasing Hormone (GnRH) Also know as luteinizing releasing hormone. This peptide hormone is synthesized by a small number of cells in the hypothalamus that are generated initially in the olfactory epithelium. They migrate subsequently to the hypothalamus. GnRH is released in pulses from the hypothalamic neurons that produce it, in both males and females. It regulates the secretion of follicle stimulating hormone (FSH) and luteinizing hormone (LH), which are essential for generation of gametes in both females and males.

Gorlin syndrome This is an autosomal dominant cancer syndrome. Patients with this rare syndrome often have anomalies in multiple organs, many of which are subtle. Gorlin syndrome patients have the propensity to develop multiple neoplasms, including basal cell carcinomas and medulloblastomas, and the patients' are extremely sensitive to ionizing radiation, including sunlight. Extra digits on the hands or feet also occur in this syndrome. Mutations in the PATCHED gene have been found to cause the Gorlin syndrome.

Graptolites These abundant Cambrian fossils have been shown to be colonial hemichordates, or members of the hemichordate class Pterobranchia.

Growth plate The area of developing tissue between the diaphysis and epiphysis responsible for the longitudinal growth of bones.

Haplo-insufficiency The condition in which having only single functional allele in a diploid organism results in pathology. A situation in which a mutation reduces the level of a gene product below threshold levels and a phenotypic effect results.

Hemangioblast A hypothesized common precursor for both blood cells and blood vessel endothelium cells.

Hematopoiesis The developmental process by which various types of blood cells are formed.

Hematopoietic stem cells (HSCs) The precursor cells that give rise to all types of blood cells. As stem cells, they are defined by their ability to maintain themselves (self-renewal) and to differentiate to form multiple cells types. In mammals, HSCs are located in the adult bone marrow.

Hemichordates This phylum includes enteropneust worms and colonial pterobranchs. Hemichordates are tripartite as adults, having three body regions. The most anterior is the proboscis (protostome), then the collar (mesosome) and the posterior trunk (metasome). These three regions are reduced and modified in colonial Pterobranchia.

Hemogenic endothelium A rare population of vascular endothelial cells that are able to differentiate into hematopoietic stem cells during embryogenesis, usually located in the lumen side of artery wall.

Heparin sulfate proteoglycan (HSPG) A macromolecule comprised of a core protein and glycosaminoglycan side chains of the heparin sulfate (polysaccharide) family, abundant in the extracellular matrix and sometimes associated with plasma membranes via lipid moieties. HSPGs are important for many signaling events as revealed by the effects of mutations in HSPG core proteins and synthesis enzymes (e.g., involved in appending the HS side groups).

Heterotaxy Discordance of sidedness of the organ systems.

Hindgut Posterior-most region of the digestive tract.

Hypertrophy An increase in bulk without concomitant multiplication of parts.

Hypodontia Tooth agenesis (the congenital absence of one or more teeth).

Hypomorphic allele A mutant allele that results in reduced activity or level of the protein.

Hypoplasia A clinical term for the reduced size of an organ or structure, often caused by disruptions in embryonic development.

Induced Pluripotent Stem (iPS) Cells A type of stem cell that is derived from somatic or non-pluripotent cell sources by misexpressing transcription factors present in pluripotent stem cells.

Induction The process by which one group of cells signals to a second group in a way that influences their development. This is achieved through signaling by diffusible growth factors.

Insulin resistance The physiologic state in which insulin target tissues in the periphery (mainly liver and muscle) become insensitive to insulin, thereby requiring elevated plasma insulin levels to elicit the same biologic effect.

Intermediate mesoderm (IM) A strip of mesenchyme adjacent to the somites that is the embryonic source of all kidney tissue.

Interzone A region of non-chondrogenic cells that forms at the site of a future joint.

Intussusceptive branching Branching by longitudinal division of a tube (by local ingrowth and meeting of the walls).

Islet transplantation The experimental surgical procedure that is used to treat some patients with type 1 diabetes; cadaver donor islets are transplanted into the liver of patients via the portal vein, where they then begin to secrete insulin and reverse diabetes.

Islets Spherical clusters of pancreatic endocrine cells that are comprised mainly of insulin-producing β-cells.

Lancelets The common name for cephalochordates. These animals are frequently referred to by the incorrect term amphioxus.

Large scale genomic analysis Methods or techniques to analyze a significant fraction of the information coded in RNA or DNA present in cells.

Lateral inhibition A conserved developmental process in which a population of equipotential cells become sorted into one of more unique phenotypes through a process in which some cells acquire a particular fate and then express membrane bound ligands that bind to membrane receptors on neighboring cells to prevent those cells from developing with the same fate.

Le Fort I–III Named after the French surgeon, Le Fort fractures are common patterns of facial fractures that occur based on naturally occurring fusion patterns of facial bones. Le Fort I results from an impact to the lower mandible in a downward direction. Fractures extend from the nasal septum, travel horizontally above the teeth, and cross below the zygomaticomaxillary junction. Le Fort II fractures occur from a blow to the lower or mid-maxilla and usually involve the inferior orbital rim. A Le Fort II fracture has a pyramidal shape and extends from the nasal bridge at or below the nasofrontal suture through the frontal processes of the maxilla. Le Fort III fractures result from impact to the nasal bridge or upper maxilla. These fractures start at the nasofrontal and frontomaxillary sutures and extend posteriorly along the medial wall of the orbit through the nasolacrimal groove and ethmoid bones, and are otherwise known as craniofacial dissociation. Surgeons use the Le Fort fractures to assist in resetting bones after surgical procedures like distraction osteogenesis or craniosynostosis repair.

Lef1 Lymphoid enhancer-binding factor 1, a transcription factor that binds Lef/TCF sites in the DNA to mediate ß catenin Wnt signaling.

Lineage restricted precursor cells Cells that are still dividing but have a more restricted developmental potential than stem cells.

Liver The largest gland in the vertebrate body which has a role in digestion, glucose regulation and storage, blood clotting and removal of wastes from the blood.

Low copy repeats (LCR) A term for the highly homologous segments of DNA, which are the result of segmental duplications and other recombination events, that mediate microdeletion events. There are four primary LCRs in the 22q11.2 region, which are named LCR A–D.

Lung Organ containing sac-like structures where blood and air exchange oxygen and carbon dioxide.

Lymphatic system A blind-ended system of vessels that protects and maintains the fluid environment of the body by filtering and draining away lymphatic fluid. Under normal conditions, the lymphatic vascular system is necessary for the return of extravasated interstitial fluid and macromolecules to the blood circulation, for immune defense, and for the uptake of dietary fats.

Mantle cells (outer support cells) These are cells that surround the inner support cells and serve as a proliferative population that can replenish the population of inner support cells, which are progressively lost as they are specified as sensory hair cell progenitors.

Mature onset diabetes of the young A dominant monogenic form of type 2 diabetes.

Mechanosensory hair cell Specialized sensory cells found in the inner ear of birds, fish, amphibians, reptiles and mammals. Modified microvilli located at the lumenal surface of each cell detect small movements or pressure waves.

Mesendoderm A transient vertebrate tissue that forms during gastrulation and is equally capable of differentiating into mesoderm and endoderm.

Mesoderm One of the three vertebrate germ layers formed during gastrulation that gives rise to connective tissue, muscle, and the skeletal and circulatory systems.

Metastable epiallele Genes that are variably expressed in genetically identical individuals due to epigenetic modifications established during early development.

Microdeletion A chromosomal deletion of between hundreds to the low millions of base pairs. Unlike larger chromosomal deletions, microdeletions are not readily detectable in microscopic analysis of chromosomal banding patterns (karyotype).

Micrognathia and retrognathia Conditions in which the jaw is underdeveloped. Micrognathia is specific to the lower jaw (mandible), whereas those with retrognathia have a smaller, often receding, upper (maxilla) and lower jaw.

Microtia A condition in which the ear is underdeveloped, misshapen, and lacking the external auditory meatus. There are varying degrees of microtia, from mild, in which the ears are smaller, but have all the main features of the ear, to severe, in which no ear is formed (anotia).

Misexpression The addition of RNA, DNA or protein into a developing embryo that naturally receives less or none of that same product.

MoA The mechanism of action (MoA) of a small molecule refers to the mechanism by which the small molecule produces its effect.

Monopodial Branching in which side-branches form from a main trunk.

Morphogen A protein that acts on target cells at a distance from its cell of origin, that forms an expression or activity gradient over a field of responsive cells, and that drives different cellular responses at different distances from the signal source. Morphogens can have non-synonymous or synonymous activity gradients, based on whether the signaling downstream of ligand-receptor activation exhibits a gradient that correlates with the extracellular ligand concentration.

Müllerian ducts Two embryonic tubes that extend along the mesonephros that become the uterine tubes, the uterus, and part of the vagina in the female and that form the prostatic utricle in the male. Also known as paramesonephric ducts.

Multipotent The ability of a progenitor to generate more than one class of terminally differentiated cells.

Neogenesis The process of generating new β-cells from stem or progenitor cells.

Nephric duct An epithelial tube that is the source of the collecting ducts and that plays an essential role in inducing formation of the tubule in most kidney types.

Nephron The functional unit of the kidney. Kidneys are comprised of as few as one or as many as one million or more nephrons. Nephrons are divided into segments, including the glomerulus, tubule, and collecting duct.

Neural crest The transient tissue found in vertebrate embryos that is formed at the border of the neural plate and that undergoes a transformation into migrating cells, which differentiate into a huge variety of cells. It gives rise to most of the cartilage and bones of the head, the peripheral nervous system, and the pigments of the skin, among other things.

Neural induction Intercellular communication that converts ectodermal cells into neural tissue.

Neural retina The sensory neuroepithelium of the eye that contains multiple neuronal cell types organized into three cellular layers. This tissue is part of the central nervous system, and it is specialized for transducing light information and transmitting it to the brain.

Neural stem cells Self-renewing, multipotent cells that can differentiate into neurons, oligodendrocytes and astrocytes.

Neurogenesis Generation of neurons from uncommitted ectodermal cells.

Neuromasts Sensory organs deposited by the migrating PLLp to initiate establishment of the Posterior Lateral Line (PLL) system. They have paired sensory hair cells at the center. Support cells (also called inner support cells) surround sensory hair cells and mantle cells (also called outer support cells).

Neuropil A collection of axons, dendrites, synaptic specialization, and supporting glial processes. Cell bodies are not included in the neuropil. Most synapses in invertebrate as well as vertebrate brains are made in the neuropil.

Neurosensory progenitor cells Multipotent progenitor cells found in the ventral anterior region of the otocyst. These cells have the ability to develop as both the neurons comprising the VIIIth ganglion or the hair cells and supporting cells that constitute the sensory epithelia of the inner ear.

Neurosphere An *in vitro*, usually clonally derived, accumulation of proliferative neuronal precursors grown from primary stem or progenitor cells taken from the central or peripheral nervous system. Primary neurosphere cultures rely upon selection by beta fibroblast growth factor (bFGF) and epidermal growth factor (EGF) to stabilize and propagate dividing cells from the primary neural tissue. Once established, primary neurospheres can be dissociated and these dissociated cells will go onto generate additional secondary neurospheres.

Neurulation The developmental process in vertebrate embryos through which the flat, disc-shaped neural plate ectoderm folds into the elongated neural tube, which will develop into the central nervous system. Neurulation can be primary or secondary, where the flat neural plate rolls into a tube, or mesenchyme cells coalesce into a medullary cord that cavitates to form a tube, respectively. Failures in neurulation result in neural tube defects (NTDs).

Nitrofen (2,4-dichlorophenyl 4-nitrophenyl ether) An herbicide (and banned carcinogen) used to create a rodent model of congenital diaphragmatic hernia. Embryos from pregnant dams exposed to nitrofen develop diaphragm, lung, and other structural malformations. The chemical affects the retinoic acid signaling pathway.

Nodal A TGF-beta signaling ligand that is a key regulator of the vertebrate endoderm GRN controlling first the induction of mesendoderm and then its segregation into distinct endoderm and mesoderm germ layers.

Non-allelic homologous recombination (NAHR) A mechanism responsible for causing microdeletions and other chromosomal rearrangements. NAHR occurs during meiosis if inappropriate recombination events occur between non-homologous segments of chromosome (either on the same chromosome, or different chromosomes).

Non-communicating defect or muscularization defect A non-communicating diaphragmatic defect is one in which there is no direct communication of the abdominal contents with the thoracic cavity. These include muscularization defects of the diaphragm. These defects include massive herniations covered by a membrane contiguous with the diaphragm (a sac hernia) and mild diffuse muscularization defects causing a slight elevation of the diaphragm (sometimes called eventration).

Notochord The key morphological character of chordates is the notochord. The notochord forms a stiff rod running from anterior to posterior in chordates beneath the dorsal neural tube, usually surrounded by a sheath of extracellular matrix. The gut is found just under the notochord in vertebrates and lancelets, but there is only an endodermal strand in the nonfeeding ascidian tadpole larvae. In lancelets and appendicularians, the notochord persists in the adult, while in ascidians the notochord undergoes apoptosis before metamorphosis and tail resorption. In vertebrates, the notochord persists as the intervertebral discs (nucleus pulposa) as the vertebrae develop from somites.

Odontoblasts Mesenchymal cells producing dentin.

Odontogenesis Tooth development.

Oligodontia Severe tooth agenesis that affects more than six teeth (wisdom teeth not included).

Online Mendelian Inheritance in Man (OMIM) A catalog of human genes and genetic disorders developed by the National Center for Biotechnology Information.

Optic vesicle An evagination of the eye domain from the lateral walls of the embryonic forebrain that will give rise to eye tissues, including the neural retina and retinal pigment epithelium.

Organ of Corti The specialized sensory epithelium of the mammalian cochlea. It contains a highly regular mosaic of hair cells and supporting cells. Hair cells are arranged in four distinct rows while supporting cells separate each hair cell from each of its neighbors.

Organizer The mesodermal region of the vertebrate embryo that is the source of signaling molecules that dorsalize and pattern both the mesoderm and the ectoderm.

Organogenesis Development of organs during embryonic development.

Otic placode An ectodermal thickening that develops in the dorsal anterior region of the embryo adjacent to the developing hindbrain.

Otocyst A hollow sphere derived from the otic placode. As development proceeds, it will undergo a series of morphological changes to give rise to all of the structures within the inner ear.

Oto-facial-cervical syndrome Patients present with hearing loss, a long narrow face and various facial and cervical structural abnormalities. Some cases are caused by mutations in the *Eya1* gene.

Pancreas A gland in the abdominal cavity with both exocrine and endocrine function that secretes digestive enzymes into the duodenum and also secretes the hormones insulin and glucagon into the blood.

Parathyroid Calcium homeostatic organ that secretes parathyroid hormone.

Partial pancreatectomy A surgical procedure used in experimental animal models to stimulate pancreas and β-cell regeneration; a portion of the pancreas is removed, and new pancreas tissue is generated from the remnant organ.

Patterning The act of subdividing embryos or tissues into distinct domains along the embryonic axes.

Perichondrium The border region of primordial cartilage anlagen that eventually differentiate into osteoblasts.

Pharyngeal gill bars Cartilaginous elements made of extracellular matrix and located between the pharyngeal endoderm, giving structure to the pharynx of hemichordates, lancelets and vertebrates. Pharyngeal gill bars are secreted from endoderm in hemichordates and lancelets, but develop from neural crest cells in vertebrates.

Pharyngeal The area of the digestive system that serves as a respiratory and feeding organ in hemichordates, tunicates and lancelets. The vertebrate homolog is the pharynx, which develops into gills in early vertebrates, but is the area of the throat, including the thyroid gland and thymus in amniotes (birds and mammals).

Placodes An area of an ectodermal thickening from which cells delaminate and eventually achieve a cell fate that is not epidermal. There are both neurogenic and non-neurogenic cranial placodes, which are associated with the peripheral nervous system in vertebrates. Placodes were thought to be found only in vertebrates, but have recently been described in tunicates, using both molecular markers and careful morphological analyses.

Planar cell polarity (PCP) This establishes the polarity of cells within an epithelium, or for migrating cells controls the localization of polarized protrusions. PCP signaling is essential for the convergent extension movement that drives neural tube closure.

Pleoitropy The phenomenon observed when one gene influences multiple traits and phenotypes.

Pleuroperitoneal fold (PPF) and post-hepatic mesenchymal plate (PHMP) These terms describe primordial diaphragmatic mesenchymal tissue that is most easily visualized at mouse embryonic days 11.5–12.5 in the thoracoabdominal region. There is disagreement in the literature about which tissues grow to make up the majority of the posterior diaphragm. This is partially complicated by inconsistent application of the terms to the embryonic anatomy, as plane of sectioning artifacts and other factors can make it very difficult to delineate these structures.

PLLp Posterior Lateral Line Primordium, a cluster of approximately 100 cells that migrates collectively from the ear to the tip of the tail periodically depositing neuromasts.

Pluripotent stem cells Pluripotent stem cells, including the embryonic stem cells, epiblast stem cells, and induced pluripotent stem cells, possess the abilities of infinite self-renewal and differentiation into all the cell types of three germ layers.

Pluripotent, multipotent Refers to cells or tissues that retain the potential to differentiate into more than one cell or tissue type.

Podocyte Specialized cell type of the glomerulus that plays an essential role in filtration and lies at the interface between glomerular capillaries and the entrance to the kidney tubule.

Polydactyly Also known as hyperdactyly, this is the anatomical variant consisting of more than the usual number of digits on the hands and/or feet. When each hand or foot has six digits, it is sometimes called hexadactyly.

Polymicrogyria A disorder of the cerebral cortex, distinguished by the presence of smaller than normal folds (gyri) in the cortical surface. This condition can be very local, affecting only a small segment of the cortex, or it can encompass entire cortical lobes.

Pre-placodal ectoderm The region of the embryonic ectoderm that surrounds the anterior neural plate and that will give rise to all of the cranial sensory placodes. It is characterized by the expression of *Six* and *Eya* genes.

Primary cilium Microtubule-based cytoplasmic extension comprised of a membrane ensheathed axoneme extending from the basal body, which is a modified centriole. In primary cilia, the axoneme is comprised of nine microtubule (MT) pairs in a 9+0 arrangement (in contrast to motile cilia which have 9 MT pairs surrounding two central fibrils, i.e., 9+2 arrangement). Proteins are moved into and out of the primary cilium via intraflagellar transport (IFT) along the axoneme. Plasma membranes around primary cilia are distinct/segregated from the cellular plasma membrane. Disruption of ciliary or IFT proteins results in a broad range of human ciliopathies.

Primary sex differentiation Refers to the specification and determination of the gonads (testes, ovaries) and associated production of sexually distinct gametes (spermatogenesis vs. oogenesis).

Primitive endoderm A polarized epithelium covering the epiblast precursors and facing the blastocoel cavity in the implanting mouse blastocyst (E3.75 to E4.5). The primitive endoderm is an extra-embryonic tissue derived from inner cell mass precursors. It will give rise to the visceral endoderm.

Primitive hematopoiesis A transient wave of hematopoiesis that generates the first blood cells in embryos. Most primitive blood cells are red cells. Macrophages and megakaryocytes are also found in this wave.

Primitive streak The primitive streak forms opposite to the anterior visceral endoderm. It is a region of the epiblast along which precursor cells of the mesoderm and the definitive endoderm ingress during gastrulation when they undergo an epithelial to mesenchymal transition. Cells are progressively recruited into the primitive streak as it elongates distally, to give rise to regionally distinct precursors in an orderly fashion, both along the medio-lateral and anterior–posterior axes. Epiblast cells located at the anterior pole, do not ingress trough the primitive streak, remain epithelial and contribute to the anterior surface ectoderm and neurectoderm.

Primordia Cells, tissues or organs at the earliest stage of development.

Primordial germ cell An embryonic cell that gives rise to a germ cell from which a gamete (i.e., an egg or a sperm) develops.

Progenitor cell A mitotic cell that is not capable of indefinite self-renewal and which will produce a limited repertoire of cell types.

Proneural gene A basic helix-loop-helix transcription factor that is both necessary and sufficient to initiate the development of neuronal lineages and to promote the generation of progenitor cells which are committed to neuronal differentiation.

Protoneuromasts These are nascent neuromasts forming within the migrating PLLp. The migrating PLLp contains 2–3 protoneuromasts at progressive stages of maturation. It consists of a cluster of about 15–30 cells that form epithelial rosettes in which a sensory hair cell progenitor gets specified at the center.

Pterobranch Class Pterobranchia refers to a group of colonial hemichordates, or pterobranchs. Colonial hemichordates reproduce both sexually and asexually and have feeding tentacles to capture small particles for feeding. There are many fossil pterobranchs, called graptolites, but only two extant families, Rhabdopleuridae and Cephalodiscidae.

qPCR, or quantitative polymerase chain reaction (PCR) An assay based on detection of increasing fluorescence with cycle number as amplified DNA fragments increase in concentration in the reaction.

Ramogen A signaling molecule that promotes branching.

Reference standard A reference sample that allows comparison of data across laboratories and different platforms.

Regulatory state The status of a cell's nucleus as defined by the sum total of trans-acting factors engaging it and the epigenetic marks it carries.

Resegmentation The process by which the sclerotomes of neighboring somites reorganize to form vertebral bodies that span the space between the dermomyotomal derivatives of the same somites.

Retinal pigment epithelium (RPE) A single layer of pigmented epithelial cells that lie behind the neural retina. These cells are necessary for the maintenance of visual function and form a part of the blood/brain barrier in the eye.

Retinal progenitor cell A dividing cell that will give rise to multiple cell types in the neural retina but that is limited in potential and not capable of prolonged self-renewal.

Retinoic acid Abbreviated as RA, it is the acidic form of vitamin A (retinol), usually synthesized from dietary intake of vitamin A. RA is a member of the steroid/thyroid family of lipid-soluble ligands that activate receptor transcription factors to directly control gene expression. RA is a major morphogenetic signal – one of the first to be discovered – as well as a regulator of adult regeneration and tissue repair.

RNA-seq A high throughput sequencing method that sequences cDNA fragments derived from RNA populations. This method allows for determination of gene structure and quantitation of RNA abundances. An alternative to microarray analysis for expression profiling, in which RNA from a tissue of interest is isolated and subjected to high throughput sequencing, and the number of transcripts for each gene is quantified.

Robinow syndrome This is a severe skeletal dysplasia with generalized shortening of all endochondral bones, segmental defects of the spine, brachydactyly, orodental abnormalities, hypoplastic genitalia and facial dysmorphism.

SAR A structure-and-activity-relationship (SAR) usually refers to the relationship between the structure of a series of small molecules and their biological activity.

Secondary heart field (SHF) A more medial population of mesoderm cells just anterior to the anterior lateral splanchnic mesoderm of the cardiac crescent. These cells give rise to the common outflow tract, the atria, and most of the right ventricle in the more mature heart.

Secondary sex differentiation The development of external (genitalia) and internal characteristics that distinguish male and female forms.

Segmental duplication (SD) A genomic process that results in the duplication of portions of a chromosome. This can occur due to errors in meiotic recombination.

Self-renewal Self-renewal of stem cells describes the process of a stem cell undergoing mitotic division and giving rise to daughter cells that are identical to the mother cell.

Semaphorin3a A cell surface adhesion molecule, one of a family of semaphorins. The semaphorins are considered to be primarily anti-adhesion or repulsive molecules that cause growing axons to stop and turn in an opposite direction to the location of the semaphorin. Semaphorin receptors include two families of molecules, the plexins, and the neuropilins.

Sensory hair cell progenitor A cell expressing relatively high levels of Atoh1a that has the potential to undergo a final division to produce a pair of sensory hair cells. The first sensory hair cell progenitor is specified at the center of a protoneuromast. However, additional sensory hair cell progenitors continue to be specified from support cells that surround the central sensory hair cells during growth and regeneration of the neuromasts.

Sex chromosome Either of a pair of chromosomes (usually designated as X or Y) in the germ cells of most animals and some plants that combine to determine the sex and sex-linked characteristics of an individual. In mammals, XX results in a female and XY results in a male.

Sex determination The process by which the sex of an organism is determined. In many species, the sex of an individual is dictated by the two sex chromosomes (X and Y) that it receives from its parents. In mammals, some plants, and a few insects, males are XY and females are XX; in birds, reptiles, some amphibians, and butterflies, the reverse is true. In bees and wasps, males are produced from unfertilized eggs, and females are produced from fertilized eggs. In 1991, *SRY* was identified as the sex determination gene of the Y chromosome. Environmental factors can also affect sex determination in some fish and reptiles. In turtles, for example, sex is influenced by the temperature at which the eggs develop.

SILAC Stable isotope labeling by amino acids in cell culture (SILAC) is a mass spectrometry based quantitative proteomic approach that is widely used for whole cellular proteome comparison. It relies on metabolic incorporation of normal amino acids and amino acids labeled with stable heavy isotopes into proteins.

Spemann-Mangold Organizer (SMO) proximodistal axis An idealized pan-vertebrate axis that runs parallel to the zone of internalizing mesendodermal cells. The Spemann-Mangold Organizer (SMO) defines the proximal end of this axis.

SNP Single nucleotide polymorphism.

Somite An embryonic tissue that patterns and maintains the metamism of axial tissues.

Specification The cell intrinsic integration of both extra and intracellular patterning cues to determine a specific transcriptional identity. Cells are considered to be specified if they are able to assume their natural fate when placed in a neutral environment (e.g., tissue culture media).

Spina bifida A neural tube defect (NTD) that results in the failure of neural tube closure in the posterior region of the embryo. Can be classified into a number of forms including: meningocele where the meninges protrude from an opening in the vertebrae; myelomeningocele where the meninges and spinal cord protrude from an opening in the vertebrae, and spina bifida occulta where the dorsal part of the vertebrae does not form but neither the meninges nor the spinal cord protrude.

Spiral ganglion A cluster of afferent neurons that develop from neurosensory cell progenitors and act as the primary relay between hair cells and the central nervous system.

Stomochord The stomochord is a projection of the endoderm that juts forward into the hemichordate proboscis, against which the hemichordate heart beats. The hemichordate stomochord cells are vacuolated and make an extracellular sheath, but there is no molecular evidence that it is a notochord homolog.

Support cells (inner support cells) A population of cells that surround the central sensory hair cell progenitors or central sensory hair cells. While providing support to inner sensory hair cells, they also serve as a population of undifferentiated progenitors from which sensory hair cell progenitors are specified during the course of growth and regeneration of neuromasts. In contrast to sensory hair cell progenitors, these cells do not express high levels *atoh1a* and have relatively high levels of Notch activity.

Syndactyly This is a condition in which the fingers fail to separate into individual appendages. This separation occurs during embryologic development.

T-box binding element (TBE) The palindromic DNA consensus sequence with a high affinity for binding the T domain.

T-box domain Also called the T domain, it is the defining feature of the T-box gene family, and it consists of a conserved sequence encoding a polypeptide domain that extends across a region of 180 to 200 amino acid residues.

Tessier clefts Craniofacial clefts that are classified on the basis of their position at either the soft tissue or bone level. This is in contrast to the Van der Meulen classification system for clefts, which bases the classification upon embryogenesis. There are different types of Tessier cleft classifications that use a 0–14 numbering system, with zero designating the midline.

Tetralogy of Fallot A heart defect characterized by: stenosis of the pulmonary trunk; ventricular septal defect; overriding aorta; and enlarged right ventricle. A patent ductus arteriosus is also present.

Thyroid Endocrine organ with the function of regulating metabolism, growth and development.

Tissue-specific stem cells Tissue-specific stem cells are multipotent stem cells. They reside in specific tissues and possess limited abilities of self-renewal and differentiation into specialized cell types of the lineage they originate. Tissue-specific stem cells maintain tissue homeostasis and allow tissue regeneration in response to tissue injuries.

Transcriptional profiling Analysis of mRNA expression, e.g., using microarrays or RNA-seq; in cell signaling studies, comparative transcriptional profiling is often used to assess the transcriptional targets of a signaling pathway.

Transdifferentiation When differentiated cells transform directly from one cell type to another by the addition of extrinsic or intrinsic factors.

Triploblasts A classification encompassing animals that form all three embryonic germ layers (mesoderm, ectoderm, and endoderm) during embryogenesis. Most, if not all, triploblasts display bilateral symmetry during embryogenesis. As such, *Bilaterians* refers to the same subset of animals as *triploblasts*.

Tubule The segment of the nephron that processes the glomerular filtrate and produces urine.

Tunicates A monophyletic group of animals that includes ascidians, appendicularians, and thaliaceans. This group of animals is also sometimes called "urochordates," but Tunicata is the preferred term. There are over 2,800 described species of tunicates.

Type 1 diabetes An autoimmune disease in which the insulin-producing β-cells are specifically destroyed.

Type 2 diabetes A disease that is usually associated with obesity in which the β-cells fail to produce enough insulin to overcome peripheral insulin resistance.

Ultimobranchial bodies Small glands that develop separately from the thyroid that will fuse with the thyroid, and form the parafollicular cells of the thyroid.

Vascular endothelial growth factor The first growth factor described as being a specific mitogen for endothelial cells. It is important for the migration, proliferation, maintenance, and survival of endothelial cells, and it is critical during both vasculogenesis and angiogenesis. It also plays an important role in arterial differentiation.

Vasculogenesis The process of the formation of primitive vessels in vertebrate embryos during which mesodermal cells differentiate into endothelial precursor cells called angioblasts. These angioblasts then differentiate *in situ* into endothelial cells and coalesce to form the earliest vessels.

Visceral Endoderm (VE) A polarized absorptive epithelium of the mouse egg cylinder derived from the primitive endoderm. It is specified in two regions covering the epiblast and the extra-embryonic ectoderm, referred to as EPI-VE and ExE-VE respectively. The ExE-VE and the proximal third of the EPI-VE contribute to the formation of the yolk sac. Cells in the EPI-VE are dispersed during gastrulation by primitive streak derived definitive endoderm cells and are incorporated into the embryonic gut.

Whole exome sequencing Selective sequencing of the entire known coding sequence complement of a genome, usually after targeted capture by hybridization.

Wnts Signaling molecules in an evolutionarily conserved signaling pathway that get their name from the homolog, Wingless, in *Drosophila*, and mammalian Int1, discovered in mice.

Wolffian duct The duct in the embryo which drains the mesonephric tubules. It becomes the vas deferens in the male, and it forms vestigial structures in the female. It is also known as the mesonephric duct.

Yeast two-hybrid (Y2H) An experimental method used to assay for direct protein-protein interactions; in this method, a "bait" protein is fused to the DNA binding domain of a transcription factor (TF), and a series of "fish" proteins are fused to the activation domain of the TF. When the bait and fish proteins physically interact, the proximity of the two TF domains render the complex capable of driving expression of a reporter construct.

Index

Note: Page numbers followed by "*f*" indicate figures; "*t*", tables.

A

E

F

N

T

U

V

W

X

Y

Z